T0191469

NEW ZEALAND INVENTORY OF BIODIVERSITY

NEW ZEALAND INVENTORY OF Biodiversity

VOLUME TWO

KINGDOM ANIMALIA

Chaetognatha, Ecdysozoa, Ichnofossils

Edited by

DENNIS P. GORDON

CANTERBURY UNIVERSITY PRESS

First published in 2010 by
CANTERBURY UNIVERSITY PRESS
University of Canterbury
Private Bag 4800
Christchurch
NEW ZEALAND

www.cup.canterbury.ac.nz

ISBN 978-1-877257-93-3

A catalogue record for this book is available from the National Library of New Zealand.

Pre-production by Rachel Scott

Printed in China through Bookbuilders

Canterbury University Press gratefully acknowledges grants in aid of publication from the National Institute of Water & Atmospheric Research (NIWA), the Ministry of Foreign Affairs & Trade, the Ministry of Fisheries, the Department of Conservation, and MAF Biosecurity New Zealand.

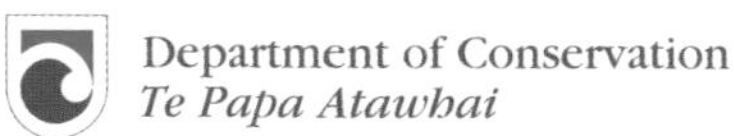

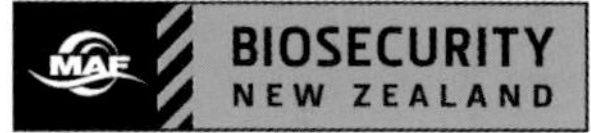

Cover: Velvet worm *Peripatoides novaezealandiae* (phylum Onychophora), an endemic species whose relationships evoke the ancient supercontinent of Gondwana.
Maria Minor and Alastair Robertson, Massey University

Endpapers: Creatures of fresh water – *Paracalliope fluviatilis*, an amphipod crustacean (upper left) and larvae of *Austrolestes* sp., a damselfly (lower left) and *Atelophlebioides cromwelli*, a mayfly.
Stephen Moore, Landcare Research

Half-title page: Average sea-surface temperature derived from AVHRR satellite. Warmest temperatures (around New Guinea) are ~30ºC; coolest temperatures (around Antarctica) are ~-1.5ºC.

Title page: The introduced damselfly *Ischnura aurora*.
Stephen Moore, Landcare Research

CONTENTS

FOREWORD

Two years ago in the Preface to the first volume of this series, I celebrated the enormous achievement that the project behind the book represented – a stocktake of the known biodiversity of New Zealand at the beginning of the 21st century. This volume brings that achievement closer to completion, listing the species known in the largest group of animals, the Ecdysozoa (animals that regularly shed their exoskeleton during growth) as well as the Chaetognatha (arrow worms) and trace fossils (burrows, footprints, etc.) of extinct animals – 'whispers of ancient life' that inform us about the history of our home.

Biologist J. S. B. Haldane liked to say that if biology had taught him anything about the nature of the Creator, it was that he had 'an inordinate fondness for beetles'. Globally, beetles account for more than 40% of known insect species, and about 30% of all animals. This volume proves that that statement holds true within New Zealand, but not quite as much as it does globally. Out of a total recorded animal diversity of at least 34,636 species in New Zealand, 21,418 are Arthropods, of those the majority (14,438) are insects, and nearly half of the insects (5062 named, about 420 known unnamed, and perhaps 3000 undiscovered species) are beetles.

The superphylum Ecdysozoa includes the numerically dominant arthropods (insects, crustaceans and their relatives), as well as lesser-known and dramatically named groups such as the Tardigrada (water bears), Onychophora (velvet worms), Kinorhyncha (mud dragons), Loricifera (corset worms), Priapulida (penis worms), Nematoda (roundworms, eelworms) and Nematomorpha (horsehair worms, hairworms).

Some of these are animals of considerable direct economic significance, such as rock lobsters and honeybees, but this importance is completely dwarfed by the ecosystem services they provide to allow life on earth, including human life, to continue: they are food for other animals, controllers of pest species, and recyclers of waste.

Minimising losses to New Zealand's overall biodiversity requires that we know about the smaller, less spectacular, and often less well-known groups as well as the kiwi, kakapo and other iconic species. To conserve biodiversity is one of my department's primary functions. To conserve species we must know they exist, and to conserve functioning ecosystems we need to know what the components of those ecosystems are and what their roles are. Faced with incomplete knowledge, the wise course is a precautionary approach.

This publication, like the ones preceding and following it, significantly advances our understanding of New Zealand's biodiversity. It will help us to make better-informed decisions in choosing places where ecosystem management has the greatest benefit, and ensure that species do not slip through the cracks to extinction. The information in this volume provides the starting point for listing species according to threat of extinction; for mapping the distribution of biodiversity across the landscape and thereby identifying ecologically important sites; and for prioritising species recovery work so that the security of threatened species unique to New Zealand is improved.

Again, Dennis Gordon deserves our gratitude for managing contributions from more than 60 authors and bringing this volume to publication.

Alastair Morrison
Director-General of Conservation

Life on this planet can stand no more plundering. Quite apart from obedience to the universal and moral imperative of saving the Creation, based upon religion and science alike, conserving biodiversity is the best economic deal humanity has ever had placed before it since the invention of agriculture. The time to act, my respected friend, is now. The science is sound, and improving. Those living today will either win the race against extinction or lose it, the latter for all time. They will earn either everlasting honor or everlasting contempt …

So: we are but one of many species on a little-known planet. Nearly 250 years ago Carolus Linnaeus introduced the practice of giving each species a two-part Latinized name, thus *Homo sapiens* for humanity. He advocated the complete exploration of life on Earth. For the adventure of exploring a little-known planet, and for our own security, it will be wise to press on to finish the great enterprise Linnaeus began. The effort to accomplish a full accounting would be a scientific moon shot, the equivalent of the Human Genome Project.

E. O. Wilson, *The Creation: An appeal to save life on Earth*

INTRODUCTION

With the publication of this volume, some of the authors may justly exclaim, Whew! Peter Maddison, the compiler of the beetle species list would be one such person. The last time we had a complete inventory of Coleoptera was in the early 1920s, when 4323 species were recognised. Today we have 5062 named species, with about 420 additional unnamed forms in museum collections. But there could be upwards of 3000 more undiscovered beetle species in New Zealand. Such is the wonder and mystery of life's variety; we understand only a fraction of the roles that even the known species play in New Zealand's diverse ecosystems.

This volume completes the review and inventory of Kingdom Animalia in New Zealand. All but one of the phyla belong to the major branch of the animal kingdom known as Ecdysozoa ('moulting animals'). Ecdysis is the periodic shedding of the cuticle that takes place in not only insects, crustaceans, and arachnids – the best-known moulting animals – but also roundworms and their kin. The one phylum not belonging to this group is the Chaetognatha (arrow worms), which does not fit comfortably into any of the major sub-branches of bilaterally symmetrical Animalia. Arrow worms are important predators of larval fish and small zooplankters in ocean ecosystems. They are mostly small, and generally quite transparent, and few people other than planktologists are familiar with them.

Each of the phyla reviewed in this volume has stories to tell. Some of these pertain to the economic or ecological importance of the group. Insect pests, for example, cost New Zealand multiple millions of dollars a year in damage to agricultural and horticultural crops, pasture, and stored foods, while others provide beneficial ecosystem services like pollination or are used in biological control. Some of the stories raise interesting questions; we may wonder, for example, why Arthur's Pass National Park is *the* southern-hemisphere hotspot for water bears (phylum Tardigrada). Is it merely an artefact of sampling, or is there some underlying biological reason? Questions of distribution also feature significantly in the distributions of groups of lesser taxonomic rank. For example, the most archaic living centipedes (order Craterostigmomorpha) comprise just two living species – one in Tasmania and one in New Zealand. This fact is no mere piece of academic trivia, for it bears on a major controversy concerning New Zealand's geological history. The genetic pattern of New Zealand's *Craterostigmus crabilli* supports a Gondwanan origin, not long-distance dispersal, and hence does not support the hypothesis of complete submergence of the New Zealand landmass in the late Oligocene some 23 million years ago.

The archaic creatures known as velvet worms, with stumpy limbs like those of tardigrades, were once thought to be a kind of intermediate between annelid worms and the jointed-limbed arthropods. They have not changed much in overall appearance since the Paleozoic, when their ancestors lived in the sea. All modern New Zealand species belong to a family whose distribution also evokes the ancient supercontinent Gondwana. Certainly there is a close affinity between New Zealand and Tasmanian forms.

Some spider families suggest an ancient Gondwanan origin too, but there are too few data to be absolutely sure. Either way, they are a special group in New Zealand, with a huge diversity in relation to the land area, and 95% of them are found nowhere else in the world. Arachnophobes among us may not like spiders, but their ecological role is of critical importance in helping to control insects. For this reason, some have potential as biological-control agents. Their

Diversity of New Zealand biota

Kingdom	Total species*
Bacteria	699
Protozoa	2,598
Chromista	1,868
Plantae	7,071
Fungi	7,065
Animalia	34,636–35,001
Total	**53,937–54,302**

* Approximate

cousins the mites and ticks are perhaps even more important – certainly they are more costly to humans than are spiders, and what they lack in size they make up for in numbers. They can swarm in their countless millions and are major pests of crops, foodstuffs, and horticultural plants. Their small size means they are easily carried about, like the bee mite *Varroa destructor*. Since its discovery in New Zealand, it has cost the New Zealand bee industry an estimated $400–900 million.

Hexapods – which have six legs as their name implies – mostly comprise insects, and they are far and away the most speciose group of organisms in the world and in New Zealand. Between 720,000 and one million of the world's 1.8 million named species are insects, and there may be 4–6 million more species still undiscovered, mostly in tropical rainforests. No one really knows. Twenty-five insect orders are found in New Zealand, varying in diversity from just one species of scorpionfly (Mecoptera) to the hyper-diverse beetles, the most diverse of the 'big-five' orders, the others being the flies (Diptera – 2483 named), the moths and butterflies (Lepidoptera – 1686 named), the bugs, cicadas, scale insects, aphids, and relatives that comprise Hemiptera (1056 named), and the bees, wasps, and ants (Hymenoptera – 717 named). New Zealand entomologists estimate that about 9000 additional species coould be found in collections and discovered in the wild. For anyone interested in *Guinness Book of Records*-type facts, one of New Zealand's insect species has been documented as the heaviest in the world. This is the Little Barrier giant weta *Deinacrida heteracantha*, weighing in at just over 70 grams. This is the limit. Not only are insects constrained by their tracheal respiratory system (tubular invaginations of the external surface), but gravity imposes another limitation. When a very large insect moults, its new skeleton is too soft to hold it up; an insect the size of a large mammal would collapse under its own weight (conveniently overlooked by Hollywood movies).

Academic interest aside, insects play a huge role in the New Zealand economy, as the Minister for Science and Innovation, Dr Wayne Mapp, mentioned in his address to the 59th Annual Meeting of the New Zealand Entomological Society in April 2010, citing the sensitivity of our primary industries to alien pests, among them the clover root weevil, which causes hundreds of millions of dollars of damage to pasture. We have pests of grasslands and pastures; of fruit, field, and cereal crops; and of stored products; also pests of forests and humans, all of which underscore the need for vigilance in border biosecurity. On the other hand, some insects are beneficial, especially honey bees and other pollinators and those used in biological control (like the parasitoid wasp used against clover root weevil). The checklists in this volume are annotated to indicate alien species and insect biocontrol agents.

The most disparate group of organisms covered in this volume is the Crustacea. Whereas insects are far more numerous than crustaceans in terms of species, they depart less from their generalised body plan. Crustaceans have an amazing diversity of form. It is partly for this reason that there has never been a single comprehensive treatment of all New Zealand Crustacea until now. Crustaceans are the dominant arthropods in the sea, whereas insects are the dominant arthropods on land. Interestingly, the historical question among zoologists as to why crustaceans were not especially successful in 'invading' the terrestrial realm (except as small inhabitants of forest litter and other moist habitats) appears to have been answered by molecular-phylogenetic studies in an unexpected way. They did so as insects, for the evidence points towards a crustacean ancestor of Hexapoda.

The basic body plan of crustaceans has been so highly modified in some groups as to be unrecognisable (except, frequently, in the earliest larval stages). Among others, the group includes the familiar crabs, lobsters, and shrimps, the bivalved seed shrimps (which look like miniature clams), sessile forms like acorn and stalked barnacles, and a range of strange parasites. These include the tongue worms of vertebrate respiratory tracts and the weird root-like ascothoracicans and rhizocephalans that ramify deeply into the bodies of their hosts.

The marine environment also lends itself to a greater range of body size among crustaceans than is possible among terrestrial hexapods, which are limited by physiological constraints as mentioned above. Crustaceans range in size from a tenth of a millimetre to half a metre in length and breadth and weighing up to 20 kilograms. This large size means that many are prized for their edibility. Crustaceans are especially important in the economy of the sea, being among the most numerous organisms in the open-water environment, especially copepods and krill, which are critical components of oceanic food webs. On the seafloor, benthic crustaceans play important roles as scavengers, predators, and prey.

It appears likely, however, that as numerous as arthropods are, they are eclipsed in numbers of individuals by nematodes – the eelworms of plants and roundworms of vertebrates. One nematologist famously wrote, in 1905: '[If] all the matter in the universe except the nematodes were swept away, our world would still be dimly recognisable, and if, as disembodied spirits, we could then investigate it, we should find its mountains, hills, vales, rivers, lakes, and oceans represented by a film of nematodes. The location of towns would be decipherable, since for every massing of human beings there would be a corresponding massing of certain nematodes. Trees would stand in ghostly rows representing our streets and highways. The location of various plants and animals would still be decipherable, and, had we sufficient knowledge, in many cases even their species could be determined by an examination of their erstwhile nematode parasites.'

Nematodes feature abundantly in the meiofauna – the microscopic life between sand grains, and there, the numbers of species can only be guessed at. Three other marine-meiofaunal groups feature in the present volume – these are the Kinorhyncha (mud dragons), Loricifera (corset animals), and penis worms (Priapulida). These groups are unknown to most New Zealanders, and even many professional zoologists have never laid eyes on them. This is not surprising; most are small and are buried in sediments, and there are not many species. Mud dragons have a higher diversity than the other two groups. Although apparently very common, they are minute, and only specialists of the meiofauna are likely to encounter them. What they lack in size, they make up for in scientific interest in terms of their evolutionary relationships with penis worms and corset animals. The curiously named latter creatures are, on average, smaller than mud dragons and rarer. They were not discovered until the 1970s and the phylum was named only in 1983. The 'corset' refers to the plated exoskeleton of the trunk. Like mud dragons and penis worms, corset worms also have a spine-bearing head. The chapter devoted to them in this volume is the first report on New Zealand Loricifera. Penis worms are the least speciose of the three groups but achieve the largest size, ranging from tiny meiofaunal individuals to giants many centimetres in length. They are also the most ancient, and studies of the living forms are helpful in shedding light on their Cambrian relatives.

The last of the group of moulting animals to be considered in this volume is the curious horsehair worms, which, as their name suggests, are long, thin, and mostly brown. As with the tardigrades, the New Zealand fauna is special – not because there are many species (only four mainland and one marine species), but because some are extremely abundant, and the South Island is one of the best places in the world to study them. Large numbers can be obtained throughout the year, which is ideal for conducting studies on their behaviour, life histories, host–parasite relations, and within-species variation. As the author of the chapter on horsehair worms points out, we are concerned enough to accord protection to our unique weta species, but what about their most important parasites the hairworms? These parasitic associations have existed for many millions of years.

In this now-completed review and inventory of New Zealand's animal diversity we have not overlooked their prehistory. Many of the ancestors of our modern species are to be sought in the fossil record. None of the shelly or bony groups features here, however – they were covered in volume 1. But many arthropodan taxa have stout exoskeletons and these preserve well, not only in the case of crabs with their thick carapaces, limbs, and claws, but also the tiny valves of the seed shrimps, which are important in stratigraphy. Our oldest animal fossils are trilobites, found in Cambrian rocks of the northwest Nelson district, and a short chapter is devoted to this group. Fossil organisms have also left traces of their past activities, and these 'whispers' of ancient life, known as ichnofossils, which have their own specialist taxonomy, are covered here too, lest the record be incomplete.

The larger context

As mentioned in the introduction to volume 1, the New Zealand Inventory of Biodiversity is a part of a regional contribution to the global Catalogue of Life, a one-stop shop for all named species. So far, 1.25 million species names have been validated, and the catalogue constitutes the authoritative names list for the Global Biodiversity Information Facility. The New Zealand names list that is published in the three volumes of this inventory underpins the New Zealand Organisms Register (NZOR) that is being developed. It will be linked with digital taxonomic data from databases maintained by Crown Research Institutes, universities, museums, and other agencies.

These are exciting and challenging times for understanding biodiversity. On the one hand, more and more knowledge is being gained about the extraordinary variety of living things with which we share this planet. On the other, biodiversity loss must be ranked among the most serious issues to confront human beings, yet most of the world's governments pay little more than lip service to the goals of the Convention on Biological Diversity; or if there is a response, it is inadequate in the face of the crisis that confronts us. A recent report[1] in the prestigious journal *Science* lamented that none of the 2010 goals to halt the rate of decline of global biodiversity had been met. Using over 30 indicators, a comprehensive assessment found not only no evidence of significant reduction in biodiversity loss but rather increasing pressures. One of the authors of the report, from the United Nations Environment Programme World Conservation Monitoring Centre, observed: 'Our data show that 2010 will not be the year that biodiversity loss was halted, but it needs to be the year in which we start taking the issue seriously and substantially increase our efforts to take care of what is left of our planet.' One positive outcome of this recognition is the formation of IPBES – the Intergovernmental Platform on Biodiversity and Ecosystem Services, to be modelled after the IPCC (Intergovernmental Panel on Climate Change). In this International Year of Biodiversity, this is an appropriate and much-needed step forward.

Dennis P. Gordon
NIWA, Wellington

1. Walpole, M.; Almond, R. E. A.; Besançon, C. et al. 2010: Tracking progress toward the 2010 biodiversity target and beyond. *Science 325*: 1503–1504.

Relief map of the New Zealand region

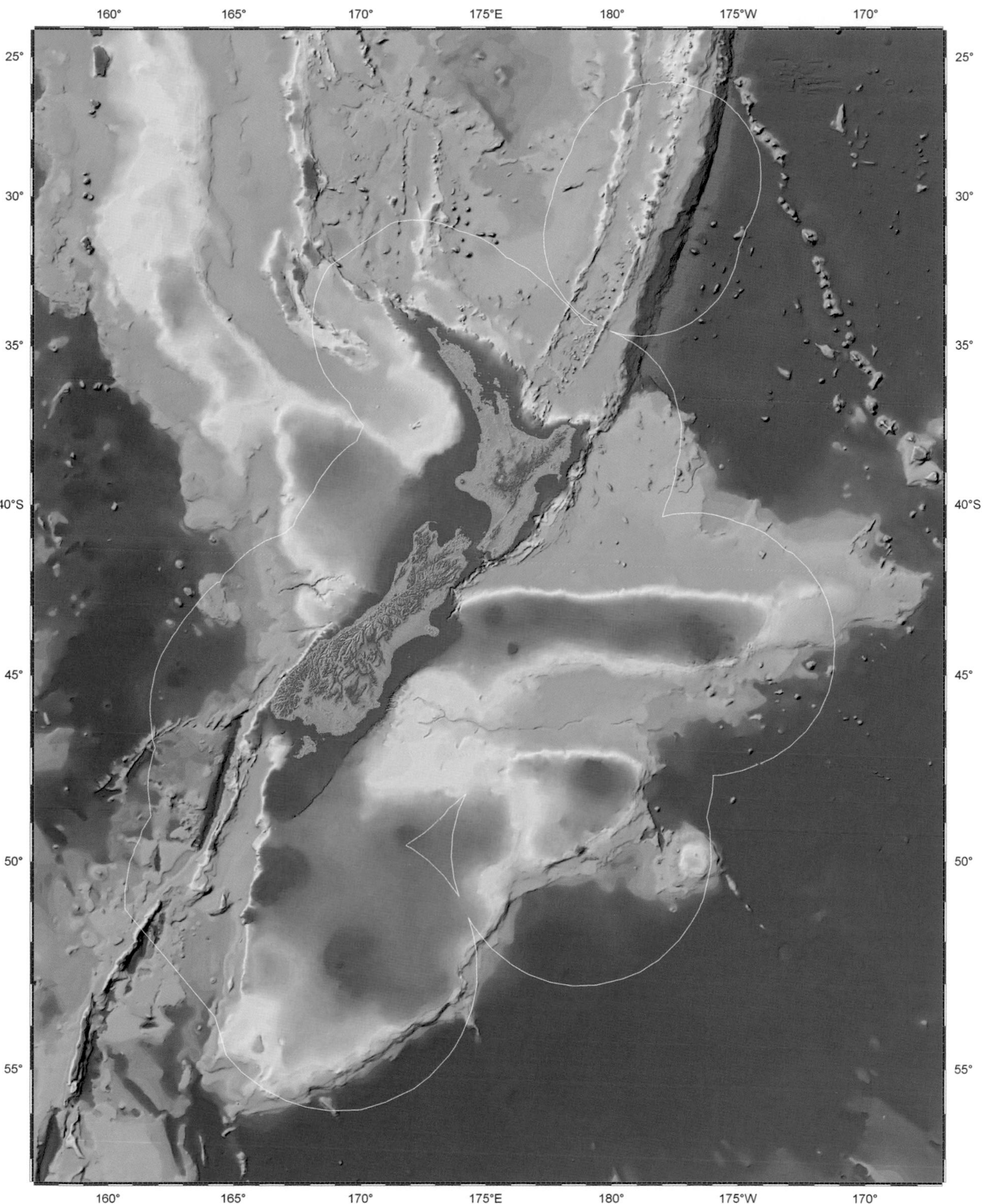

All species inventoried in this volume are found within the marine, terrestrial, freshwater, and fossil habitats bounded by the 200 nautical mile Exclusive Economic Zone (the thin white line). Red indicates the upper kilometre of the ocean, with orange and yellow pertaining to deeper areas of the New Zealand continental mass; blue indicates abyssal depths of 4–6 kilometres, and the purple-magenta of the Kermadec Trench depths of 6–10 kilometres. NIWA

The New Zealand geological timescale

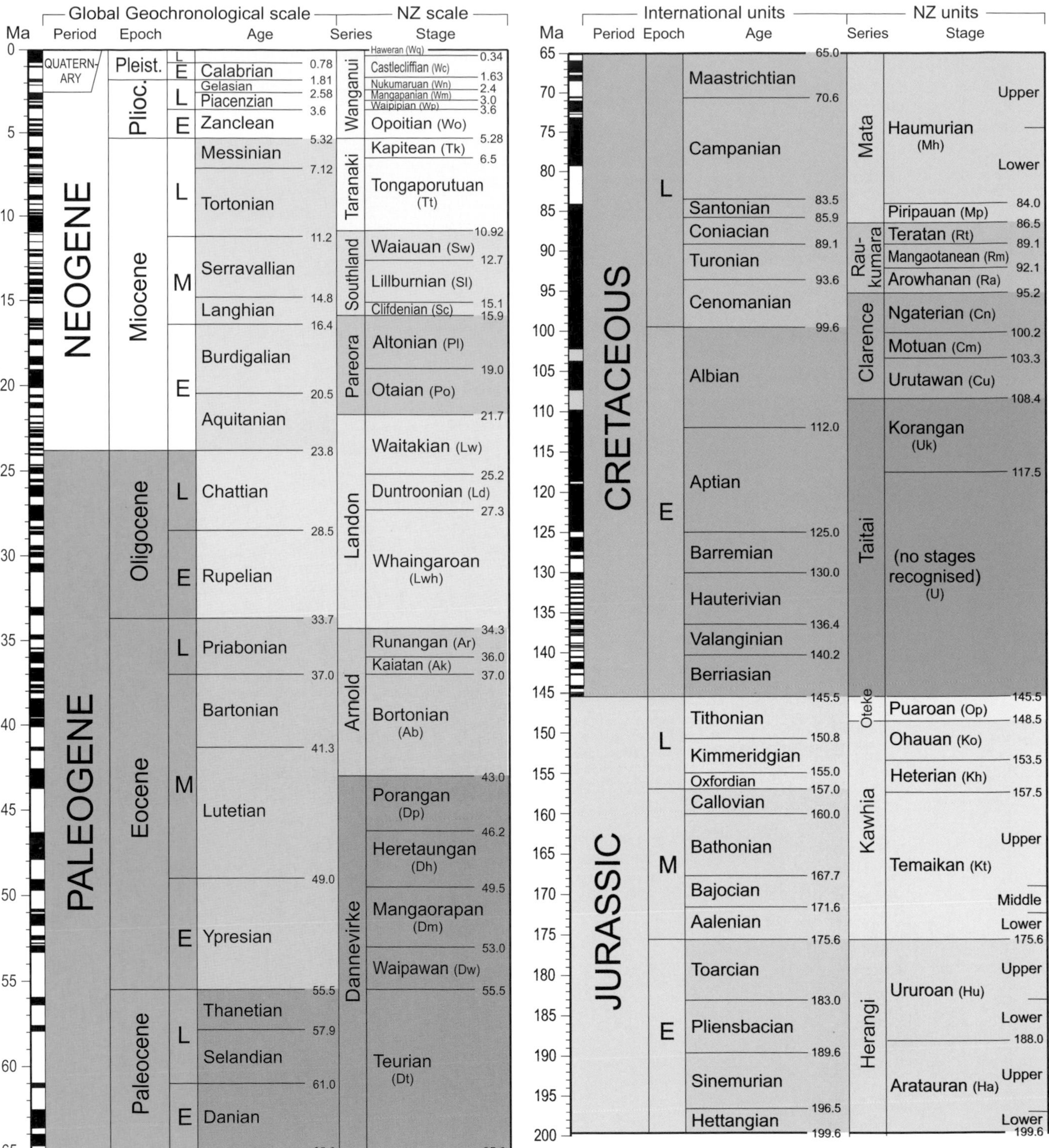

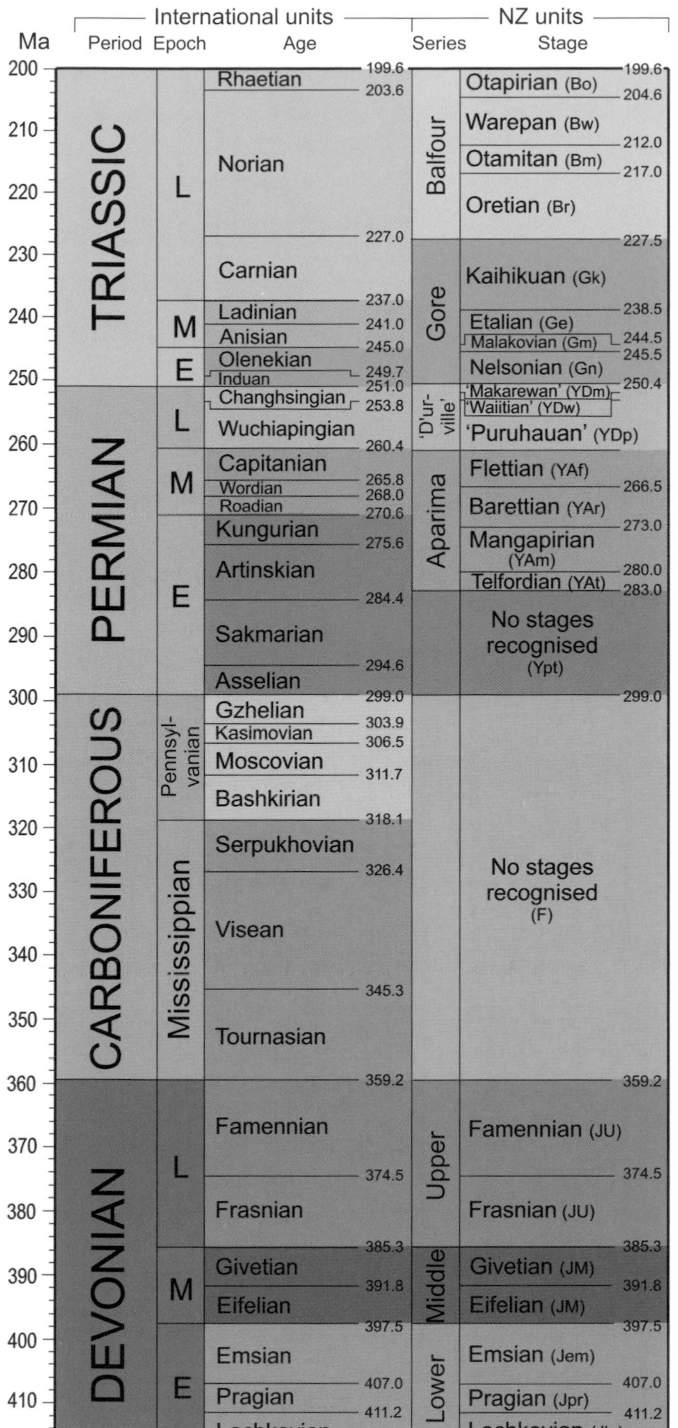

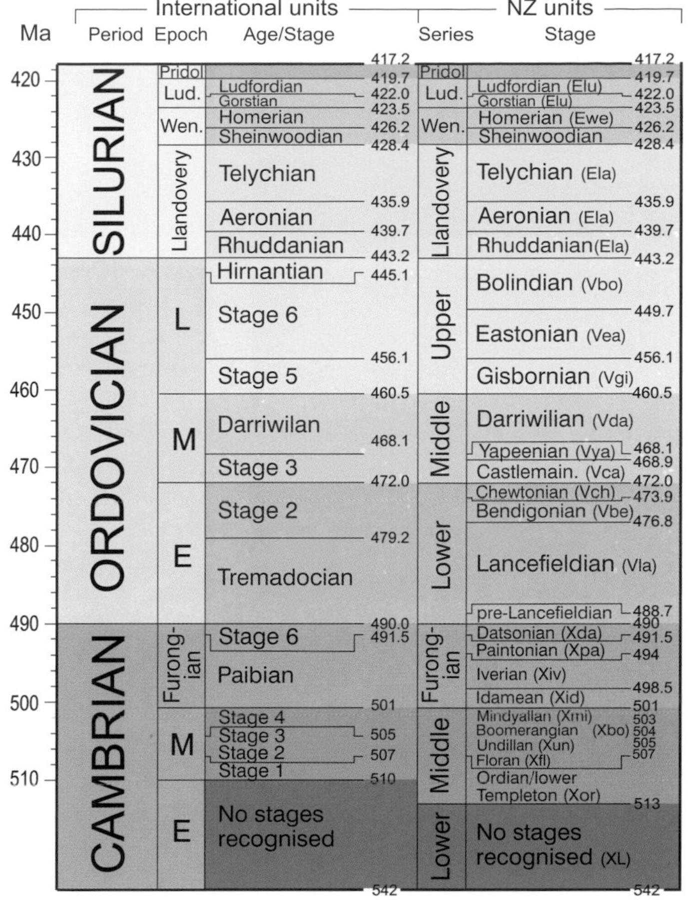

New Zealand divisions, ages of unit boundaries, and error ranges on ages are from Cooper (ed., *The New Zealand Geological Timescale*, Institute of Geological and Nuclear Sciences Monograph 22, 2004), who gives a full description of the stratigraphic basis for the New Zealand scale and its calibration.

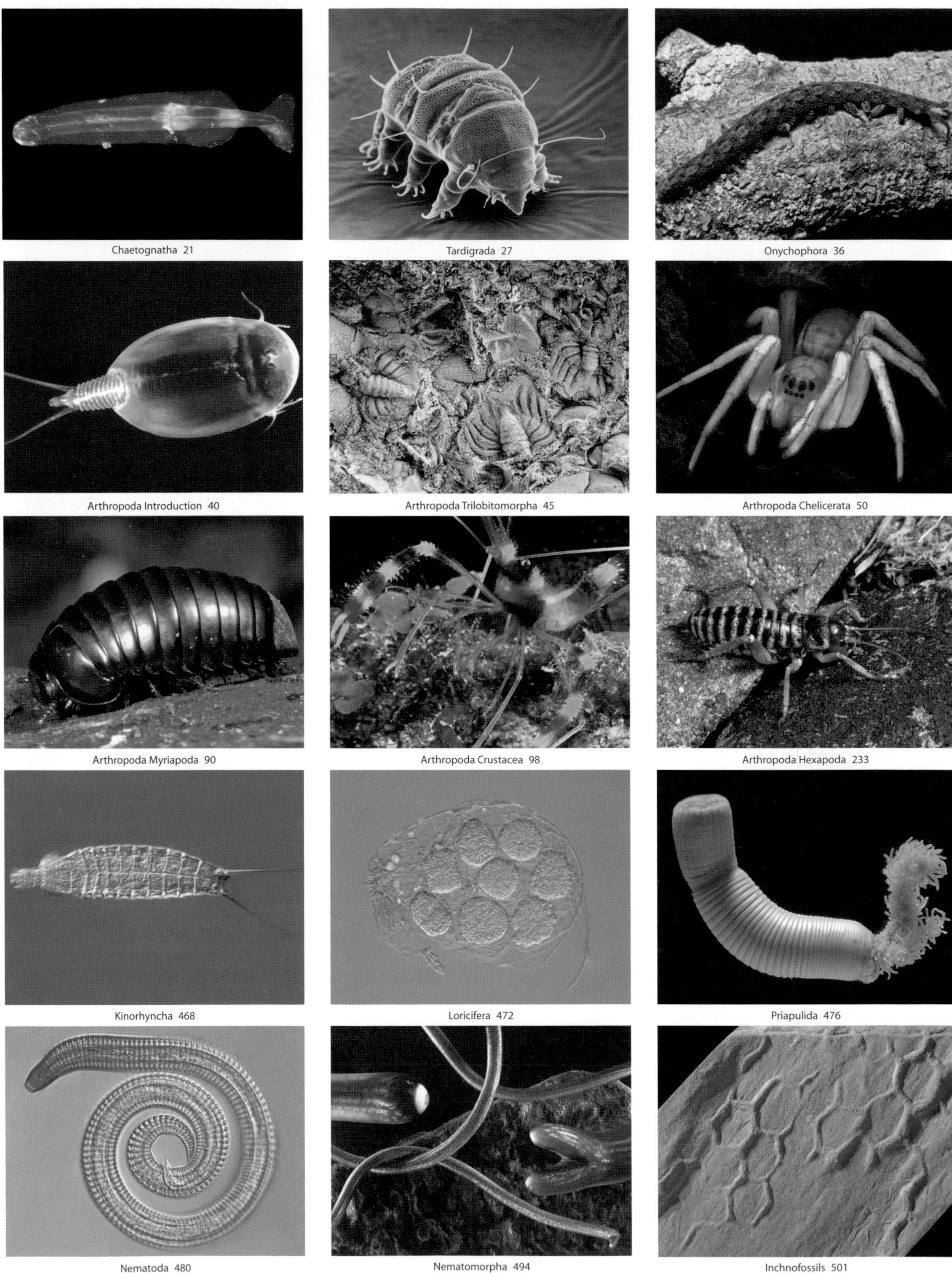

Chaetognatha 21

Tardigrada 27

Onychophora 36

Arthropoda Introduction 40

Arthropoda Trilobitomorpha 45

Arthropoda Chelicerata 50

Arthropoda Myriapoda 90

Arthropoda Crustacea 98

Arthropoda Hexapoda 233

Kinorhyncha 468

Loricifera 472

Priapulida 476

Nematoda 480

Nematomorpha 494

Inchnofossils 501

Kingdom Animalia: A 'family tree'

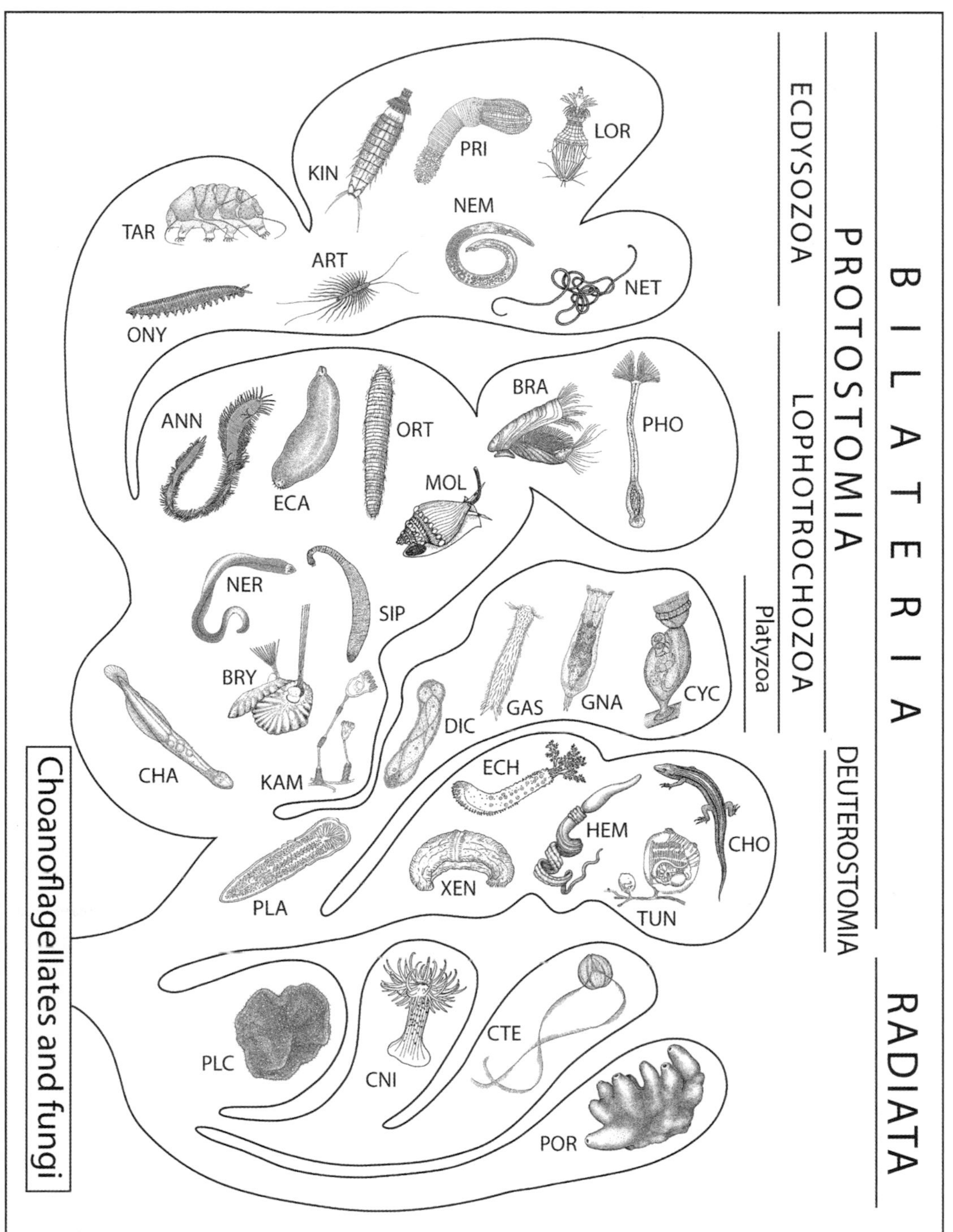

This diagram is a kind of family tree of the animal kingdom. Each creature illustrated represents one of the 33 phyla in the kingdom, showing its general relationship to other phyla by its inclusion in a particular grouping (or branch of the kingdom). We cannot be more precise in depicting connections within the branches. The tree would need to be shown three-dimensionally, and there is in any case some debate concerning the relationships of many of the phyla (explained in the chapters that follow).

Radiate animals include Porifera (POR), Placozoa (PLC), Cnidaria (CNI), and Ctenophora (CTE). Deuterostomes comprise Xenoturbellida (XEN), Echinodermata (ECH), Hemichordata (HEM), Tunicata (TUN), and Chordata (CHO). Platyzoans are Platyhelminthes (PLA), Dicyemida (DIC), Gastrotricha (GAS), Gnathifera (GNA), and Cycliophora (CYC). The balance of Lophotrochozoa includes the Kamptozoa (KAM), Bryozoa (BRY), Sipuncula (SIP), Mollusca (MOL), Brachiopoda (BRA), Phoronida (PHO), Annelida (ANN), Echiura (ECA), Orthonectida (ORT), and Nemertea (NER). Ecdysozoans comprise Onychophora (ONY), Tardigrada (TAR), Arthropoda (ART), Kinorhyncha (KIN), Priapulida (PRI), Loricifera (LOR), Nematoda (NEM) and Nematomorpha (NET). Near the base of the Protostomia is the Chaetognatha (CHA), of uncertain affinities.

Species diversity of New Zealand living Animalia

Taxon	Described species	Known undescribed/ undetermined species	Totals	Estimated undiscovered species
Porifera*	472	257	729	530
Placozoa*	0	0	0	1
Ctenophora*	15	4	19	12
Cnidaria*	796	330	1,126	1,230
Platyhelminthes*	397	142	539	832
Dicyemida*	5	1	6	50
Gastrotricha*	0	5	5	25
Gnathifera*	486	44	530	83
Cycliophora*	0	0	0	1
Mollusca*	2,891	1,697	4,588	432
Brachiopoda*	35	3	38	17
Phoronida*	3	0	3	3
Bryozoa*	630	331	961	300
Kamptozoa*	6	6	12	25
Sipuncula*	26	0	26	5
Echiura*	5	2	7	15
Annelida*	786	268	1,054	761
Orthonectida*	0	1	1	3
Nemertea*	37	2	39	140
Xenoturbellida*	0	0	0	1
Echinodermata*	556	67	623	45
Hemichordata*	5	2	7	4
Tunicata*	189	3	192	196
Chordata*	1,571	212	1,783	835
Chaetognatha†	14	1	14	25
Tardigrada†	85	3	88	100
Onychophora†	9	25	34	10
Arthropoda†	18,067	3,351–3,717	21,418–21,784	25,125–28,065
Kinorhyncha†	6	39	45	30
Loricifera†	0	4	4	10
Priapulida†	3	1	4	4
Nematoda†	708	52	760	1,800
Nematomorpha†	5	0	5	2
Totals	**27,808**	**6,828–7,194**	**34,635–35,001**	**32,650–35,600**

* Data from *New Zealand Inventory of Biodiversity* Vol. 1, 2009.
† Data from this volume.

Species diversity of New Zealand Animalia species by major environment and as fossils

Taxon	Marine	Terrestrial	Freshwater	Fossil
Porifera*	724	0	5	~83
Placozoa*	0	0	0	0
Ctenophora*	19	0	0	0
Cnidaria*	1,112	0	14	204
Platyhelminthes*	324	137	78	0
Dicyemida*	6	0	0	0
Gastrotricha*	4	0	1	0
Gnathifera*	44	2	484	0
Cycliophora*	0	0	0	0
Mollusca*	3,593	906	89	6,945
Brachiopoda*	38	0	0	525
Phoronida*	3	0	0	0
Bryozoa*	953	0	8	~471
Kamptozoa*	12	0	0	0
Sipuncula*	26	0	0	0
Echiura*	7	0	0	0
Annelida*	792	207	55	33
Orthonectida*	1	0	0	0
Nemertea*	29	6	4	0
Xenoturbellida*	0	0	0	0
Echinodermata*	623	0	0	180
Hemichordata*	7	0	0	309
Tunicata*	192	0	0	>1
Chordata*	1,427	264	96	637
Chaetognatha†	14	0	0	0
Tardigrada†	5	–	83††	0
Onychophora†	0	34	0	0
Arthropoda†	2,926	17,446–17,609	1,242	803
Kinorhyncha†	45	0	0	0
Loricifera†	4	0	0	0
Priapulida†	4	0	0	0
Nematoda†	173–211	540	9	0
Nematomorpha†	1	0	4	0
Totals	**13,146**	**19,508–19,671**	**2,172**	**~10,192**

Note that some species live in more than one major environment during their life-cycle, hence the totals for these across all environments will exceed the total species in the table opposite.

* Data from *New Zealand Inventory of Biodiversity* Vol. 1, 2009.
† Data from this volume.
†† Including terrestrial species as these occur in interstitial water.

Diversity of New Zealand Animalia and percentage endemism

Taxon	Total species	Adventive species	Endemic species	Percentage endemic
Porifera*	729	0	336	46.1
Placozoa*	0	0	0	0
Ctenophora*	19	0	5	20.8
Cnidaria*	1,126	24	234	21.3
Platyhelminthes*	539	35	199	36.9
Dicyemida*	6	0	6	100
Gastrotricha*	5	0	0	0
Gnathifera*	530	0**	19	3.6
Cycliophora*	0	0	0	0
Mollusca*	4,588	50	3,880	84.6
Brachiopoda*	38	0	15	39.5
Phoronida*	3	0	0	0
Bryozoa*	961	24	581	60.5
Kamptozoa*	12	1	2	16.7
Sipuncula*	26	0	2	7.7
Echiura*	7	0	0	0
Annelida*	1,054	80	434	41.2
Orthonectida*	1	0	1	100
Nemertea*	39	0	28	71.8
Xenoturbellida*	0	0	0	0
Echinodermata*	623	0	237	38.0
Hemichordata*	7	0	1	14.3
Tunicata*	192	11	125	65.1
Chordata*	1,783	96	457	25.6
Chaetognatha†	14	0	0	0
Tardigrada†	88	0**	25	29.8
Onychophora†	34	0	34	100
Arthropoda†	21,418–21,784	1,773	16,147–16,161	74.1–75.4
Kinorhyncha†	45	0	6	13.3
Loricifera†	4	0	2	50.0
Priapulida†	4	0	0	0
Nematoda†	760	152	90	11.8
Nematomorpha†	5	0	4	80.0
Totals	**34,635–35,001**	**2,246**	**22,884–22,898**	**65.3–66.1**

* Data from *New Zealand Inventory of Biodiversity* Vol. 1, 2009.
** Not easily determinable.
† Data from this volume.

ONE

Phylum CHAETOGNATHA

arrow worms

JANET BRADFORD-GRIEVE, JEAN-PAUL CASANOVA

Arrow worms are well named. They are streamlined predators that dart about in the sea, stabilised by fins that resemble the vanes on an arrow. They are striking creatures. The phylum name means 'bristle-jawed', referring to the ferocious-looking, moveable hooks with which these animals catch their food. Appropriately, a group of arrow worms is called a 'quiver'. Shallow-water forms are generally transparent, making them well placed to capture prey and avoid being eaten, although they can in turn be important food for some fish. Some deep-water forms are highly coloured, probably from the prey they eat. Arrow worms are among the most important predators of small zooplankton in the open ocean. Prey includes copepod crustaceans and fish larvae.

In both body design and anatomy arrow worms are relatively simply constructed. There is a short, rounded head, a long trunk, and medium-length tail region. The mouth is located on the underside of the head, which usually carries a pair of eyes on its upper side. A hood-like fold of the body wall behind the head is capable of being pulled forward to enclose the entire head. This hood probably protects the hooks when they are not in use and reduces water resistance when the animal is swimming. Along the trunk there are one or two pairs of lateral fins and a tail fin. The fins are very delicate and are often damaged in animals caught in plankton nets. Arrow worms range in length from a third of a centimetre to 15 centimetres and are hermaphrodite. Despite fossil evidence of the existence of arrow worms as far back as the Carboniferous and Cambrian (e.g. Chen & Huang 2002), they are altogether puzzling from an evolutionary point of view. Based on their embryonic development, they were thought to be related to deuterostome animals (Margulis & Schwartz 1998). On the other hand, recent molecular work suggests either a relationship with moulting animals (Ecdysozoa) like nematodes or priapulids (Helmkampf et al. 2008) or that they are a sister group to protostomes and that deuterostome features, like the mouth being formed from a secondary opening in the embryo, are ancestral for bilateral animals and retained by chaetognaths and deuterostomes (Casanova & Duvert 2002; Papillon et al. 2004; Helfenbein et al. 2004). Most species are permanently pelagic, i.e. living in the open ocean, although some live at or above the sea floor as part of the epibenthos (*Spadella, Paraspadella*), or are benthopelagic, rising into the water just above the seafloor to feed and mate (*Heterokrohnia*), or live in isolated marine-cave (anchialine) situations like *Paraspadella anops* (Shin 1997).

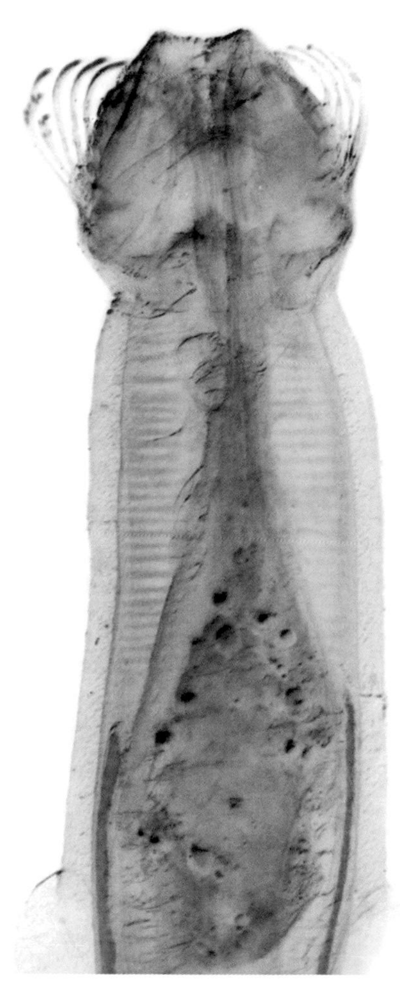

Anterior end of a species of *Sagitta* with a captured copepod crustacean in its gut.
Erik Thuesen

Many arrow worms are completely transparent and their internal organs can be seen in the live animal. The body cavity is divided into three compartments in the head, trunk, and tail. Longitudinal muscles are arranged in two dorsal and ventral bands and there are special muscles in the head that operate the hood, teeth, hooks, and other structures. Although arrow worms cannot swim

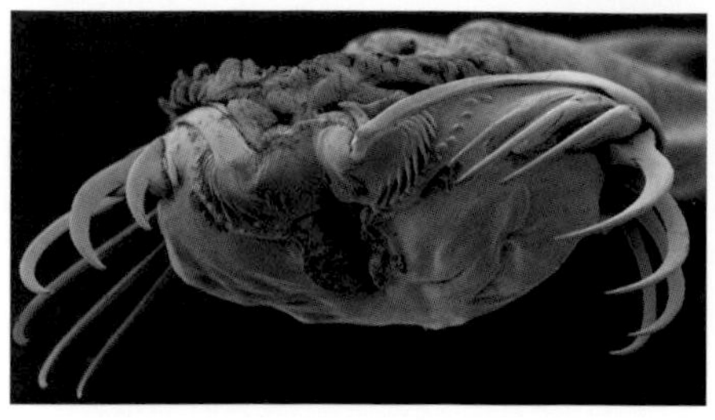

Head of a species of *Sagitta* with jaw-like hooks in prey-capture mode.

Erik Thuesen

Summary of New Zealand chaetognath diversity

Taxon	Described species	Known undescribed/ undetermined species	Estimated unknown species	Endemic species	Endemic genera
Syngonata	0	0	2	0	0
Chorismogonata	14	0	23	0	0
Totals	**14**	**0**	**25**	**0**	**0**

against currents they are capable of darting movements by flexing their tails and contracting their longitudinal muscles, with the fins acting as stabilisers and flotation devices. Species in the sea-floor genera *Spadella* and *Paraspadella* adhere to bottom objects by means of special adhesive papillae, but can swim short distances.

All arrow worms are carnivores (Bone et al. 1991). Planktonic species feed on other planktonic organisms, particularly copepods (Terazaki & Marumo 1982), and some are even cannibalistic (Pearre 1981). They find food by detecting the vibrations arising from the movements of their prey. It appears that they paralyse their prey using tetrodoxin venom that is produced by bacteria at an unknown location in the arrow-worm's body (Thuesen 1991). Chaetognaths are capable of eating animals as big as themselves and may consume up to approximately one third of their body weight in a day (Nagasawa & Marumo 1972; Canino & Grant 1985). Prey is swallowed whole and solid waste is eliminated from the anus located at the septum between the trunk and tail. Arrow worms consume 3–10% of annual planktonic secondary production (Falkenhaug & Sakshaug 1991; Terazaki 1998).

Arrow worms do not have gas exchange or excretory organs; the coelomic fluid acts as a circulatory medium. The nervous system of chaetognaths is within the epidermis. It consists of a large ventral ganglion linked to a smaller cerebral ganglion. Nerves lead to trunk muscles, tail, gut, hooks, and eyes (Margulis & Schwartz 1998). Sensitive external cilia do not sense touch but instead enable the worm to detect vibrations, chemicals, or water flow. Each eye consists of five inverted pigment-cup ocelli that sense motion and differences in light intensity; they are probably not able to form visual images. In some species of *Eukrohnia* (e.g. *E. hamata*), they have the appearance of ommatidia, i.e. like the individual units of a compound eye. Eyes are absent in most of the deep-sea species.

The key predators of chaetognaths are larval and juvenile fish and other chaetognaths (Feigenbaum 1991). There is also evidence of species of the pelagic copepod *Oncaea* feeding on chaetognaths (Go et al. 1998).

Reproduction in arrow worms is sexual. Ovaries lie on each side of the intestine in the trunk coelom and testes lie in the tail coelom. Sperm mature before the eggs and are formed into single sperm packets (spermatophores) in the seminal vesicles. Both self- and cross-fertilisation may take place. Mating takes place when two arrow worms lie side by side, facing in opposite directions. In the case of *Spadella*, each individual attaches a spermatophore to the neck of the other. Individual sperm stream along each arrow worm's back and through the openings of the oviduct on either side (Barnes 1980). The eggs are fertilised when they pass into the oviduct. In some species the eggs are planktonic whereas in other species eggs may be attached to the body surface or on other objects. *Paraspadella gotoi* has an elaborate mating behaviour (courtship), correlated with the presence of limb-like appendages (Casanova et al. 1995).

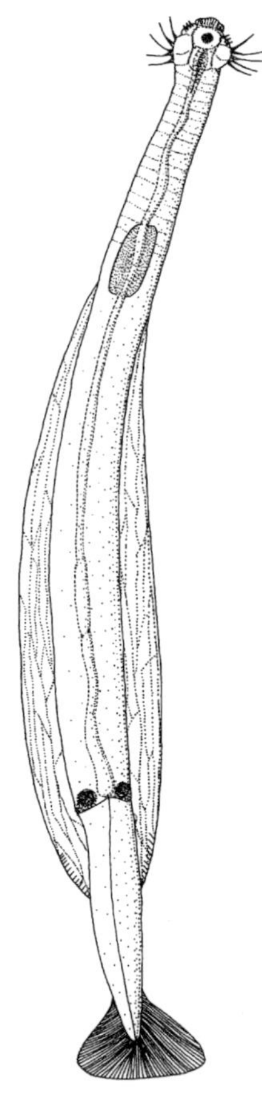

Eukrohnia hamata.

From Lutschinger 1993

Chaetognatha were thought to be incapable of bioluminescence. Nevertheless, *Sagitta macrocephala* (a cosmopolitan species generally found at depths greater than 700 m) was recently found to luminesce (Haddock & Case 1994).

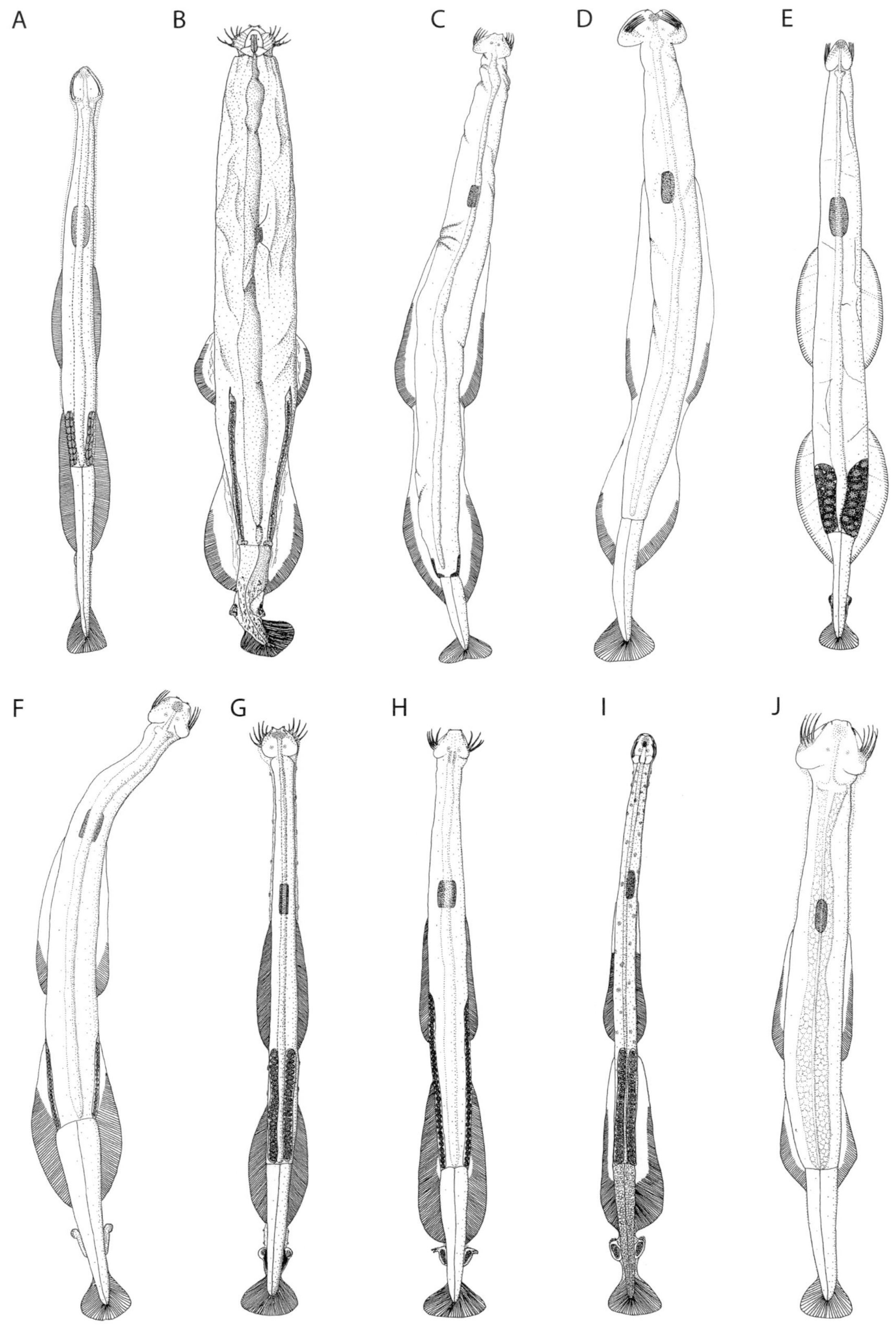

New Zealand arrow worms of the genus *Sagitta*.
A: *Sagitta regularis*. B: *S. hexaptera*. C: *S. gazellae*. D: *S. maxima*. E: *S. minima*. F: *S. decipiens*. G: *S. bipunctata*. H: *S. serrata*. I *S. tasmanica*. J: *S. zetesios*.
From Lutschinger 1993

The discoverers of this phenomenon in these animals determined that light is produced by a coelenterazine-luciferase reaction, making it the seventh animal phylum known to employ this type of luminescence.

Arrow worms live over a wide depth range in the pelagic environment. They live in the surface 200 metres (epipelagic), at depths greater than 1000 metres (bathypelagic), and at intermediate depths (mesopelagic) (Cheney 1985; Pierrot-Bults & Nair 1991). They may perform daily vertical migrations, swimming to the surface at night and, in some cases, annual vertical migrations. Pierrot-Bults (1982) showed that chaetognaths near Bermuda were most abundant from 50 to 100 metres depth by day and from 0 to 10 metres depth at night, although the limits of distribution and levels of maximum density were greatly influenced by the stage of maturity, as most species show ontogenetic vertical distribution. On the other hand, *Sagitta gazellae*, a large antarctic/subantarctic species, is found in maximum abundance between 50 and 100 metres. It does not undergo daily migration, but migrates to deep water in winter (David 1955).

Arrow-worm classification

The phylum comprises a single class, Sagittoidea, with two subclasses and three orders. Subclass Syngonata contains the order Biphragmophora; subclass Chorismogonata contains orders Monophragmophora and Aphragmophora (Casanova 1985). Ordinal classification is based on the presence or absence of transverse lateroventral musculature, and whether these muscles are found in the trunk and the tail. The phylum contains 6–9 families, 11–24 genera, and about 125 species (Bieri 1991; Thuesen 2009). Casanova (1999) pointed out that some of the proposed genera of Sagittidae comprise unrelated species (i.e. are no monophyletic), so he prefers to use the genus name *Sagitta* in its broadest sense (cf. Lutschinger 1993).

Subclass Syngonata – species in this class have ducts between the genitals in the trunk and tail regions of the body.

Order Biphragmophora – arrow worms in this order are characterised by transverse lateroventral musculature (interupted ventrally in the space between the two longitudinal lateroventral muscles) in the trunk and the tail segment (Casanova 1985). There is only one family, Heterokrohniidae, with three genera – *Heterokrohnia* (16 species), *Archeterokrohnia* (3 species), and *Xenokrohnia* (1 species).

Subclass Chorismogonata – ducts are lacking between genitals in the trunk and tail.

Order Monophragmophora – arrow worms in this order have transverse lateroventral musculature limited to the trunk (Casanova 1985). There are two certain families. The Spadellidae contains three genera – *Spadella* (11 species), *Paraspadella* (9 species), and *Hemispadella* (1 species). Monospecific *Bathyspadella* may also belong in the Spadellidae but is sometimes included in its own family. The Eukrohniidae contains one genus, *Eukrohnia*, with 10 species.

Order Aphragomophora – transverse musculature is absent and there are weakly developed glandular structures on the body (Casanova 1985). Suborder Ctenodontina contains two families – the large family Sagittidae with 1–11 genera (see Bieri 1991; Casanova 1999) and about 60 species, and monospecific Pterosagittidae (*Pterosagitta draco*). Suborder Flabellodontina contains the Krohnittidae with sole genus *Krohnitta* (2 species).

The genera *Bathybelos, Krohnitella*, and *Pterokrohnia* need further description in order to be properly classified (Casanova 1999).

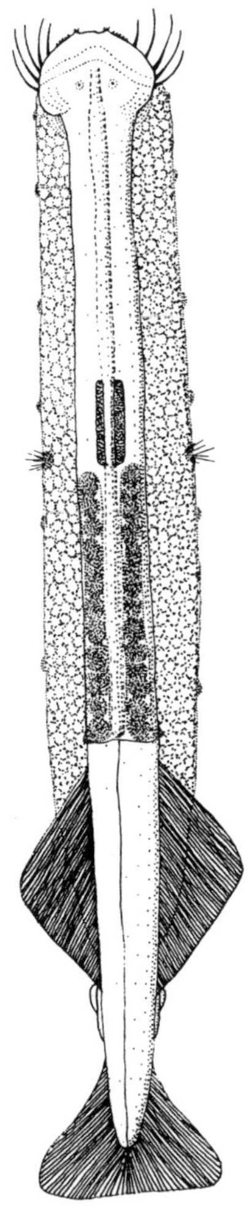

Pterosagitta draco.
From Lutschinger 1993

New Zealand arrow-worm studies

Existing knowledge of the chaetognath fauna of New Zealand is based primary on the monograph of Lutschinger (1993). In this work she summarised the historical data from expedition reports, studies on particular species, or plankton studies in general. The new records in her monograph came entirely from plankton samples, therefore epibenthic and benthopelagic species are still unknown from the New Zealand region.

The New Zealand pelagic fauna has 14 species of arrow worm, *Sagitta tasmanica* and *S. minima* being the species most commonly encountered. There are four types of distribution encountered in the region. *Sagitta regularis* appears to be a stray from a tropical distribution. The majority of species have a subtropical distribution with their southern limit at or just north of the Subtropical Front (STF), viz *Krohnitta subtilis, Pterosagitta draco, Sagitta bipunctata, S. decipiens, S. hexaptera, S. lyra, S. minima*, and *S. zetesios* (Lutschinger 1993). *Sagitta tasmanica* has been found in both subtropical and subantarctic waters, with the majority found north of the STF. *Eukrohnia hamata, S. maxima,* and *S. gazellae* have been found in subtropical, subantarctic, and antarctic waters. *Eukrohnia hamata* and *S. maxima* live at great depths in the tropics, occurring progressively nearer the surface at higher latitudes.

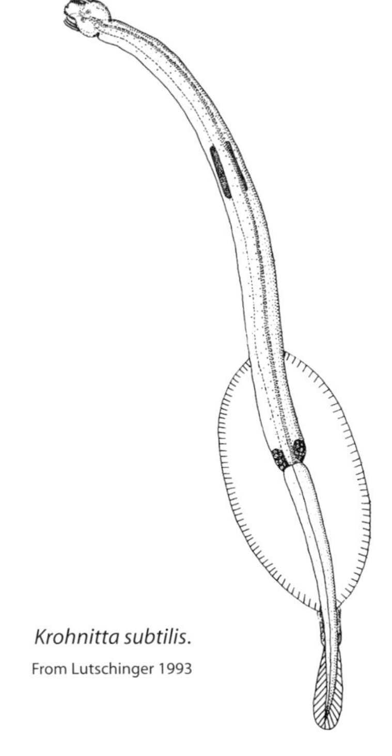

Krohnitta subtilis.
From Lutschinger 1993

Authors

Dr Janet M. Bradford-Grieve National Institute of Water & Atmospheric Research, Private Bag 14901, Kilbirnie, Wellington, New Zealand [j.grieve@niwa.co.nz]

Dr Jean-Paul Casanova Université de Provence, Biologie Animale (Plancton), case 18, 3 Place Victor Hugo, 1331 Marseille, Cedex 3 France [bioplankton@up.univ-mrs.fr]

References

BARNES, R. D. 1980: *Invertebrate Zoology*. Holt-Saunders International Editions, Tokyo. 1089 p.

BIERI, R. 1991: Systematics of Chaetognatha. Pp. 122–136 *in*: Bone, Q.; Kapp, H.; Pierrot-Bults, A.C. (eds), *The Biology of Chaetognaths*. Oxford University Press, Oxford. 173 p.

BONE, Q.; KAPP, H.; PIERROT-BULTS, A. C. (Eds) 1991: *The Biology of Chaetognaths*. Oxford University Press, Oxford. 173 p.

CANINO, M. F.; GRANT, C. C. 1985: The feeding and diet of *Sagitta tenuis* (Chaetognatha) in the lower Chesapeake Bay. *Journal of Plankton Research 7*: 175–188.

CASANOVA, J.-P. 1985. Description de l'appareil génital primitif du genre *Heterokrohnia* et nouvelle classification des Chaetognathes. *Compte Rendu des Séances de l'Académie des Sciences, Paris, sér. 3, 301*: 397–402.

CASANOVA, J.-P. 1999: Chaetognatha. Pp. 1353–1389 *in*: Boltovsky, D. (ed.), *Zooplankton of the South Atlantic*. Backhuys Publishers, Leiden. xvi + 1706 p.

CASANOVA, J.-P.; DUVERT, M. 2002: Comparative studies and evolution of muscles in chaetognaths. *Marine Biology 141*: 925–938.

CASANOVA, J.-P. ; DUVERT, M.; GOTO, T. 1995: Emergence of limb-like appendages from fins in chaetognaths. *Compte Rendu des Séances de l'Academie des Sciences, Paris, sér. 3, 318*: 1167–1172.

CHEN, J.-Y.; HUANG, D.-Y. 2002: A possible lower Cambrian chaetognath (arrow worm). *Science 298*: 187.

CHENEY, J. 1985: Spatial and temporal abundance patterns of oceanic chaetognaths in the western North Atlantic. 2. Vertical distribution and migrations. *Deep-Sea Research 32(9A)*: 1061–1075.

DAVID, P. M. 1955: The distribution of *Sagitta gazellae* Ritter-Zahony. *Discovery Reports 27*: 235–278.

FALKENHAUG, I.; SAKSHAUG, E. 1991: Prey composition and feeding rate of *Sagitta elegans* var. *arctica* (Chaetognatha) in the Barents Sea in early summer. *Polar Research 10*: 487–506.

FEIGENBAUM, D. 1991: Food and feeding behaviour. Pp. 45–54 *in*: Bone, Q.; Kapp, H.; Pierrot-Bults, A.C. (eds), *The Biology of Chaetognaths*. Oxford University Press, Oxford. 173 p.

GO, Y.-B.; OH, B.-C.; TERAZAKI, M. 1998: Feeding behaviour of the poecilostomatoid copepods *Oncaea* spp. on chaetognaths. *Journal of Marine Systems 15*: 475–482.

HADDOCK, S. H. D.; CASE, J. F. 1994: A bioluminescent chaetognath. *Nature 367*: 225–226.

HELFENBEIN, K. G.; FOURCADE, H. M.; VANJANI, R. G.; BOORE, J. L. 2004: The mitochondrial genome of *Paraspadella gotoi* is highly reduced and reveals that chaetognaths are a sister group to protostomes. *Proceedings of the National Academy of Sciences of the USA 101*: 10639–10643.

HELMKAMPF, M.; BRUCHHAUS, I.; HAUSDORF, B. 2008: Multigene analysis of lophophorate and chaetognath phylogenetic relationships. *Molecular Phylogenetics and Evolution 46*: 206–214.

LUTSCHINGER, S. 1993: The marine fauna of New Zealand: Chaetognatha (arrow worms). *New Zealand Oceanographic Institute Memoir 101*: 1–61.

MARGULIS, L.; SCHWARTZ, K.V. 1998: *Five Kingdoms: An Illustrated Guide to the Phyla of Life on Earth*. W.H. Freeman and Company, New York. 520 p.

NAGASAWA, S.; MARUMO, R. 1972: Feeding of a pelagic chaetognath, *Sagitta nagae* Alvariño, in Suruga Bay, central Japan. *Journal of the Oceanographic Society, Japan 28*: 181–186.

PAPILLON, D.; PEREZ, Y.; CAUBIT, X.; Le PARCO, Y. 2004: Identification of chaetognaths as protostomes is supported by the analysis of their mitochondrial genome. *Molecular Biology and Evolution 21*: 2122–2129.

PEARRE, S. 1981: Feeding by Chaetognatha: Energy balance and importance of various components of the diet of *Sagitta elegans*. *Marine Ecology Progress Series 5*: 45–54.

PIERROT-BULTS, A. C.; NAIR, V. R. 1991: Distribution patterns in Chaetognatha. Pp. 86–116 *in*: Bone, Q.; Kapp, H.; Pierrot-Bults, A.C. (eds), *The Biology of Chaetognaths*. Oxford University Press, Oxford. 173 p.

PIERROT-BULTS, A. C. 1982: Vertical distribution of Chaetognatha in the central northwest Atlantic near Bermuda. *Biological Oceanography 2*: 31–60.

SHIN, G. L. 1997: Chaetognatha. Pp. 103–220 *in*: Harrison, F. W.; Ruppert, E. E. (eds), *Microscopic Anatomy of Invertebrates: Hemichordata, Chaetognatha, and the Invertebrate Chordates*. Wiley-Liss, New York.

TERAZAKI, M. 1998: Life history, distribution, seasonal variability and feeding of the pelagic chaetognath *Sagitta elegans* in the subarctic Pacific: A review. *Plankton Biology and Ecology 45*: 1–17.

TERAZAKI, M.; MARUMO, R. 1982: Feeding habits of meso- and bathypelagic Chaetognatha, *Sagitta zetesios* Fowler. *Oceanologia Acta 5*: 461–464.

THUESEN, E. V. 1991: The tetrodotoxin venom of chaetognaths. Pp. 55–60 *in*: Bone, Q.; Kapp, H.; Pierrot-Bults, A. C. (eds), *The Biology of Chaetognaths*. Oxford University Press, Oxford. 173 p.

THUESEN, E. V. 2009: Chaetognatha. http://192.211.16.13/t/thuesene/chaetognaths/chaetognaths.htm

Checklist of New Zealand Chaetognatha

The list below contains records only from the New Zealand Exclusive Economic Zone.

PHYLUM CHAETOGNATHA
Class SAGITTOIDEA
Subclass CHORISMOGONATA
Order MONOPHRAGMOPHORA
EUKRONIIDAE
Eukrohnia hamata (Möbius, 1875)

Order APHRAGMOPHORA
Suborder CTENODONTINA
PTEROSAGITTIDAE
Pterosagitta draco (Krohn, 1853)
SAGITTIDAE
Sagitta bipunctata (Quoy & Gaimard, 1828)
Sagitta decipiens (Fowler, 1905)
Sagitta gazellae (Ritter-Zahony, 1909)
Sagitta hexaptera (d'Orbigny, 1836)
Sagitta lyra (Krohn, 1853)
Sagitta maxima (Conant, 1896)
Sagitta minima (Grassi, 1881)
Sagitta regularis (Aida, 1897)
Sagitta serratodentata (Krohn, 1853)
Sagitta tasmanica (Thomson, 1947)
Sagitta zetesios (Fowler, 1905)

Suborder FLABELLODONTINA
KROHNITTIDAE
Krohnitta subtilis (Grassi, 1881)

TWO

Phylum TARDIGRADA

water bears

DONALD S. HORNING, RICARDO L. PALMA, WILLIAM R. MILLER

The animal phylum Tardigrada (literally 'slow walkers') is little known outside the zoological world. There are more than 750 described species worldwide (Kinchin 1994), but these creatures are tiny, mostly ranging in length from one tenth to one half of a millimetre; giants reach 1.2 millimetres long. Their bear-like appearance, legs with claws, and slow side-to-side lumbering gait gave them the common name of water bears. Following their discovery in a drop of fresh water more than 220 years ago, these ubiquitous invertebrates have been found almost everywhere there is free water. The three basic groups are based on the type of aquatic environment in which they are found – in marine and brackish water, in fresh water, and even in terrestrial situations if there is at least a film of water to surround the body. Without water they are inactive. They range from the Arctic to Antarctica, from beach sands to abyssal depths, in pools, streams, rivers and lakes, in soil and leaf litter, and on mosses, liverworts, lichens, algae, and vascular plants. Algal scrapings from stone walls or debris from roof spoutings will almost certainly yield tardigrades for the inquisitive student and amateur microscopist.

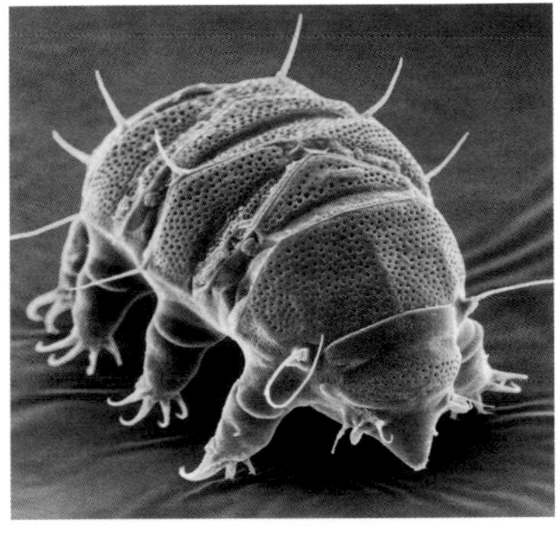

Echiniscus elaeinae from Kowhai Bush, near Kaikoura.

Diane Nelson

Some marine species are interstitial, living between grains of sand; other species are epiphytic, living on the surfaces of algae and plants, and others are epizoic or commensal, living on or in other marine invertebrates such as the mantle of mussels. Deep-sea tardigrades are known from 3000 metres depth off the coasts of Antarctica and Spain and in the Caribbean Sea, but they are rarely collected. In fresh water, tardigrades prefer to live among decaying vegetation, where they feed on a range of items including organic debris, cell contents of plants (especially mosses), microalgae, protozoans, and other tiny invertebrates like rotifers. More than half of the terrestrial species share their microhabitats with a variety of other organisms, primarily rotifers and nematodes but also bacteria, protozoans, mites, and small insect larvae. Shortage of food is almost never a problem – water is the critical factor. Many species have the ability to withstand environmental extremes, however, whether freezing or drying. In such conditions they dehydrate and form a cyst or tun (so-called because it is shaped like a large wine cask). The process is known as anhydrobiosis and enables tardigrades (and other invertebrates such as some rotifers and nematodes) to withstand the periodic drying that occurs naturally in their environment. Tuns are also a means for their dispersal, by wind or in dried mud clinging to waterfowl and other creatures.

Through an alternation of active and anhydrobiotic periods, a tardigrade may have a life-span of less than a year to more than 100 years. For instance, a museum specimen of a dried moss that was in a herbarium for 120 years yielded active tardigrades when soaked in water (Kinchin 1994). Not only can they resist

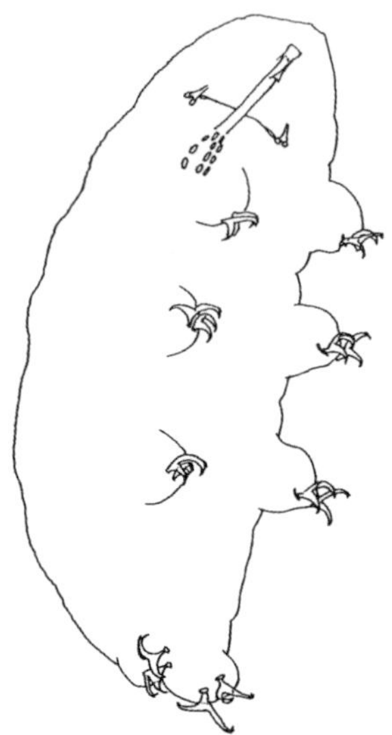

Isohypsibius palmai.
From Pilato 1996

desiccation, but tuns may show tolerances to abnormally high concentrations of carbon dioxide, hydrogen sulphide, and potassium cyanide. In addition, anhydrobiotic tardigrades have been experimentally subjected to temperatures as low as –272° C and as high as +150° C (Kinchin 1994). They have been stored at –200° C for 20 months, exposed to high vacuum, and survived 6000 atmospheres of pressure.

Anhydrobiosis normally occurs in response to daily or seasonal weather conditions and can be induced in the laboratory. Exposed to drying conditions, a tardigrade begins to lose the water that makes up more than 85 per cent of its bodyweight, until only two per cent or less body-weight water is left (Crowe 1975). As the water is lost, the tun shape is assumed, which reduces the surface area of the specimen. Very little is known of what happens during anhydrobiosis. Metabolism is almost imperceptible and appears to cease altogether in oxygen-deficient environments. Although oxygen uptake occurs even under very low levels of relative humidity, evidence now indicates that oxygen may not be essential to the anhydrobiotic tuns and indeed may be harmful. Some anhydrobiotic tuns maintained in an oxygen-free environment may survive longer than those kept in oxygen (Crowe 1975). Crowe et al. (1984) postulated that the production of highly reactive molecular fragments – free radicals – may be involved. Free radicals produced by the cell's own oxidative reactions may cause cross-linkage of large, important radicals or even loss or change in genetic information of cells. Perhaps ageing and 'natural' death in tardigrades is partly a consequence of the accumulation of free radicals and their products.

Despite these notable capabilities, tardigrades are still little understood. No specific medical, commercial, or environmental effects have been found for tardigrades (Miller 1997). They co-exist with other micro-invertebrates and form part of the food chain, but generally they have low population numbers.

The phylogenetic relationships of the phylum, named by Doyère (1840), remain debatable. Because of their tiny size, tardigrades were originally classed as Infusoria, a miscellaneous assemblage of microscopic organisms including protozoans, but they have also been classified among annelids and arthropods, the latter attribution reflecting their similarities to the onychophoran–arthropod and aschelminth complexes (Nelson 1991; Roeding et al. 2007). The tardigrade body is bilaterally symmetrical, generally with a rounded back and flattened belly, with four pairs of legs terminating in claws or paddles. Segmentation is indistinct, but the head is followed by three trunk segments, each with a pair of legs, and a caudal segment with the fourth pair of legs projecting backward.

The feeding apparatus consists of a mouth tube, a muscular bulbous pharynx, and a pair of calcareous piercing stylets for puncturing the walls of plant cells, or the bodies of other tiny animals such as nematodes and rotifers, and sucking out the contents. Many tardigrade eggs have characteristic surface pores and conical projections. In some cases a species is reliably identified only by these egg patterns, especially in *Macrobiotus* and *Minibiotus*. Important characteristics in separating species of *Echiniscus* include the size and shape of pores in the dorsal armoured plates and the development of cirri and clavas. The form

Diversity by environment

Taxon	Marine	Freshwater/ Terrestrial
Heterotardigrada	5	21+1*
Eutardigrada	0	62
Totals	**5**	**83+1***

* subspecies

Summary of New Zealand tardigrade diversity

Taxon	Described living species + subspecies	Known undescr. species	Estimated unknown species	Adventive species	Endemic species	Endemic genera
Heterotardigrada	24+1	2	–	–	11	0
Eutardigrada	61	1	–	–	14	1
Totals	**85+1**	**3**	**>100?**	**0?**	**25**	**1**

of the claws or paddles is important in all genera. Nelson's (1991) account of the phylum is an excellent source of details of anatomy, physiology (latent states), reproduction, development, and ecology.

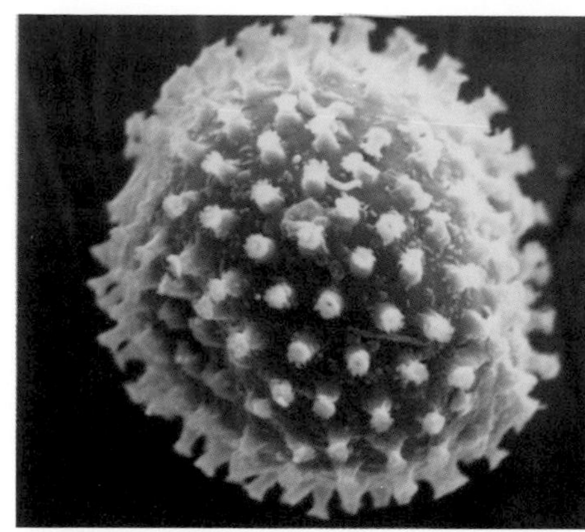

Tardigrade eggs: *Minibiotus* sp. (upper) and *Macrobiotus* sp. (lower).

Don Horning

Distribution and ecology of New Zealand tardigrades

It is very difficult to give an informed discussion of tardigrade geographic distribution because of the paucity of collecting in many areas of the world, especially in the Southern Hemisphere. Distributions are really a reflection of where tardigradologists have collected. Some species appear to be cosmopolitan, e.g. *Milnesium tardigradum*, but may in fact be a series of closely related or sibling species. This is especially true in the genus *Macrobiotus*. Other species are seemingly endemic and are rarely encountered. But one species, *Hypsibius allisoni*, described from a lawn at Rangiora, South Island, has been found to be common in the Vestfold Hills of Antarctica (Miller et al. 1988). On the other hand, *Isohypsibius cameruni* was known only from Cameroon (Africa) before it was recorded in New Zealand by Horning et al. (1978) (see McInnes 1994). Some of the truly remarkable disjunct distributions may possibly be explained by lack of collections in critical geographical areas. Of the 89 terrestrial, freshwater, and marine species and subspecies known from New Zealand, only 25 have been found nowhere else. Two of these, *Echiniscus velaminis* and *Macrobiotus montanus*, were described by Murray in 1910. Considering the dearth of records from southern Africa, Tasmania, the subantarctic islands, Chile, and even Argentina, we cannot be sure that these species are endemic to New Zealand.

Very little is known about the ecology of New Zealand tardigrades, despite the efforts of Horning et al. (1978) to limit ecological parameters. They dealt with tardigrade-positive samples only, determining the numbers of tardigrade species per sample, their co-occurrence, the moisture content of the sample collected, its aspect and exposure in its natural habitat, and tardigrade–plant species relationships. They concluded: 'Our inability to more precisely predict where water bears will be found or to better understand why they are present emphasises how little is known of the habitat requirements of the minute animals.' They found that foliose lichens provided the most suitable habitat for tardigrades, followed by mosses, liverworts, and hornworts, and fruticose and crustose lichens. Among the mosses, *Racomitrium* species yielded 24 species of tardigrades, and 18 species were recovered from *Hypnum cupressiforme*. Only three species of tardigrade were isolated from 10 collections of *Sphagnum*, namely *Diphascon* (*Adropion*) *scotium*, *Macrobiotus occidentalis*, and *Milnesium tardigradum*. Possibly the peat-moss habitat is too acidic.

History of discovery of New Zealand tardigrades

Richters (1908) first reported tardigrades from New Zealand, describing a new species, *Echiniscus novaezeelandiae* (now in the genus *Pseudechiniscus*), with type locality 'Nordinsel von Neuseeland', and reported *Macrobiotus hufelandi* and *Echiniscus gladiator* (now *Hypechiniscus exarmatus*) from North and South Islands. All of these species have been subsequently collected in New Zealand. Richters (1908) also mentioned two unidentifiable species of *Macrobiotus* and one of *Echiniscus*. An extensive search was made for this unidentified material in the 1970s but it could not be found. Murray (1910) listed 21 species of water bears, which included two new species, *Echiniscus velaminis* and *Macrobiotus montanus*, from Nuns Veil Mountain, South Island. The former species has been collected only in New Zealand but the latter species is now known from many other localities in Europe, West Africa, the Himalayas, Canada, California, Galápagos Islands, and Antarctica (Grigarick et al. 1973a; McInnes 1994; Ramazzotti & Maucci 1995). Murray (1910) also listed four species that could not be identified – three of *Echiniscus* and one of *Macrobiotus*. Three of the species he reported were not collected by Horning et al. (1978). Murray (1910) stated that *Macrobiotus arcticus* (now *Hypsibius arcticus*) was considered to be in some doubt as to its identification

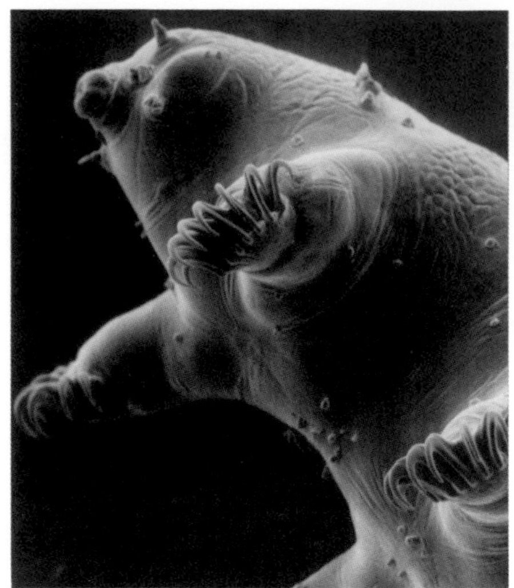

Echiniscoides sp.
William Miller

because the egg was unavailable, and *Macrobiotus echinogenitus* was too vaguely defined to be recognised even if it was re-collected. *Macrobiotus hufelandi* sensu Murray is possibly the species that Horning et al. (1978) identified as *Macrobiotus hibiscus*. *Hypsibius arcticus* and *Macrobiotus echinogenitus* have been deleted from the list of New Zealand Tardigrada.

Grigarick et al. (1973b) published scanning electron micrographs of the eggs of three *Macrobiotus* species from New Zealand – *M. recens*, *M. anderssoni*, and a species close to *M. hibiscus*. Martin and Yeates (1975) reported the effects of four insecticides on nematode, rotifer, and tardigrade populations in a New Zealand pasture ecosystem. Putative *Macrobiotus liviae*, possibly one of three species – *M. diffusus*, *M. hieronimi*, or *M. montanus* – was recorded from Nelson, South Island. Interestingly, the four insecticides tested had no direct effects on the rotifers and tardigrades but two of them considerably reduced nematode populations. Nelson (1977) reported three types of fungal parasites that infected several *Macrobiotus* and *Hypsibius* species from Kowhai Bush, near Kaikoura, South Island. Stout (1978) obtained eight cultures of unidentified tardigrades – six from water-irrigated soil cores and two from effluent-irrigated cores. Toftner et al. (1975) analysed eggs of different species in the *hufelandi* group of *Macrobiotus*. Two eggs from Rangatira Island, Chatham Islands, were cited as having 'a MS name ...'. These eggs belonged to *Macrobiotus rawsoni*. When Horning et al. (1978) described *M. rawsoni*, they stated that the active stages of this species were indistinguishable from specimens of *M. anderssoni* but that the eggs were distinctive. Bertolani and Rebecchi (1993) assigned *M. rawsoni* to the *M. hufelandi* group and stated that *M. rawsoni* may be readily distinguished by egg morphology alone.

Horning et al. (1978) provided a comprehensive survey of New Zealand tardigrades. Areas covered during the survey included the Three Kings Islands, D'Urville Island (and small islands in Fiordland), Chatham Islands, Open Bay Islands, Stewart Island, the Snares, and Campbell Island. They reported 55 species (12 new) and one new genus (*Limmenius*). Nelson and Horning (1979) conducted an intensive survey in Kowhai Bush, Kaikoura. They found 21 species of which three were new to New Zealand – *Pseudechiniscus suillus facettalis* (now *P. facettalis*), *Hypsibius* (*Hypsibius*) *baumanni* (now *Ramazzottius baumanni*), and *Hypsibius* (*Diphascon*) *higginsi* (now *Diphascon* (*Diphascon*) *higginsi*). Altogether, 58 species of New Zealand tardigrades were recognised, but Nelson and Horning (1979) stated: 'Undoubtedly additional species of *Macrobiotus* in the *hufelandi* group are present, however these are not easily identified by adult characters alone. A further search for eggs is necessary in order to differentiate these species.'

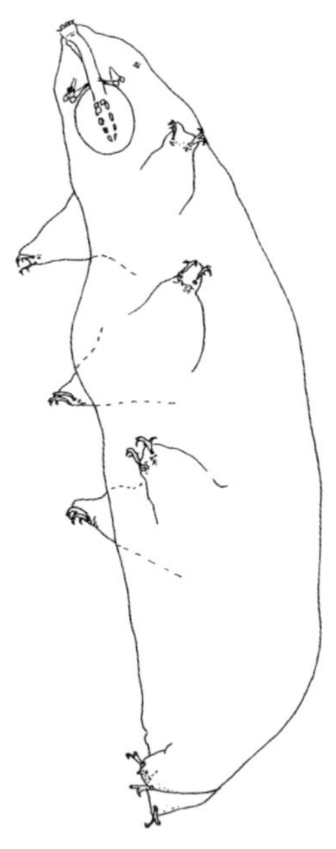

Macrobiotus armatus.
From Pilato & Binda 1996

Horning and Schuster (1983) re-examined one of the species of Horning et al. (1978), namely *Pseudechiniscus lateromamillatus*, and subdivided it into three new species – *Pseudechiniscus conversus*, *P. parvisentus*, and *P. perplexus*. Kristensen (1987) described four new genera and placed these three species into the new genus *Antechiniscus*. He also stated that *Pseudechiniscus lateromamillatus* of Horning et al. (1978) was not that species but consisted of these three species. Hence *Antechiniscus lateromamillatus* has been deleted from the list of New Zealand Tardigrada. Palma et al. (1989) listed the holotypes and gave label information on the slides of the 15 species described by Horning et al. (1978) and Horning and Schuster (1983). Suren (1991, 1992) recorded *Macrobiotus dispar* (now *Dactylobiotus dispar*) in bryophytes from unshaded alpine streams in Arthurs Pass National Park, South Island. This species was not collected by Horning et al. (1978). Dastych (1991) examined a few of Richter's specimens that had been found at the Senckenberg Museum in Frankfurt. The slides contained several species, including the very controversial and misunderstood *Isohypsibius sattleri*. Dastych redescribed this species and stated: 'The presence of *sattleri* in New Zealand should be confirmed as well because the identification of this species was based on a general and potentially inadequate definition of the taxon given by Ramazzottii ...' We have retained this species until material can be re-examined. By 1994, 65 species of Tardigrada had been listed from New Zealand (McInnes 1994).

A number of other authors have re-examined specimens in the Horning et al. (1978) collection, resulting in additional new species. For example, Pilato (1996) redescribed *Isohypsibius wilsoni* from the type series at Open Bay Islands and described two new species – *Isohypsibius campbellensis* from subantarctic Campbell Island and *I. palmai* from Piriaka, North Island from specimens previously identified as *Hypsibius* (*Isohypsibius*) *wilsoni* by Horning et al. (1978). Pilato and Binda (1996) described two new species of *Macrobiotus* from material identified as *Macrobiotus liviae* by Horning et al. (1978) – *Macrobiotus pseudoliviae* from Taumaka Island, Open Bay Islands, South Island (type locality) and *M. armatus* from Gillespies Beach, South Island (type locality). They also recorded three additional species from the *M. liviae* material as new to New Zealand – *M. diffusus*, *M. hieronimi*, and *M. montanus*. Pilato and Binda (1997) examined another small collection of Horning et al. (1978) and described *Hypsibius novaezeelandiae* from D'Urville Island, Catherine Cove, South Island. They also mentioned five additional species, reporting one of them, *Hypsibius marcelli* from the Snares, as newly recorded from New Zealand. The species is actually new for the Australasian region – the type series (five specimens) of *H. marcelli* has previously been collected from Tierra del Fuego (Pilato & Binda 1997).

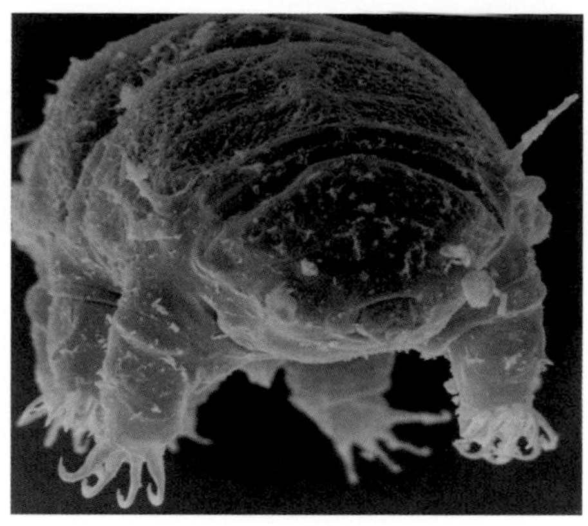

Echiniscus sp., anterior view.
William Miller

Dastych (1997) described a new species, *Echiniscus palmai* (type locality Pegleg Flat, Arthurs Pass National Park, South Island), from a series labelled *Echiniscus bigranulatus* by Horning et al. (1978). Claxton (1998) recorded eight species of *Minibiotus* from New Zealand. Five species (*Minibiotus taiti*, *M. hispidus*, *M. asteris*, *M. scopulus*, and *M. maculatus*) were new records for New Zealand. She also synonymised *Macrobiotus intermedius subjulietae* Horning et al. 1978 with *Minibiotus aculeatus* Murray, 1910. Dastych et al. (1998) questionably synonymised *Oreella minor* with *O. mollis*. Two specimens from North and South Islands were used in this study. This synonymy will require confirmation after new specimens, particularly eggs, are examined from South America (Dastych et al. 1998).

The first decade of the 21st century has seen much taxonomic work on New Zealand tardigrades, with a notable increase in the number of newly recognised species. Tumanov (2004) described *Macrobiotus kovalevi* moss on trees in the vicinity of Karamea, South Island. From the same sample, he also recorded *Macrobiotus* cf. *coronatus* and *Calcarobiotus* (*Discrepunguis*) sp. This latter name is a new record for New Zealand. Binda et al. (2005) described the new species *Macrobiotus divergens* (type locality Turangi, North Island) from some material identified as *M. furciger* by Horning et al. (1978). This new species was also recorded from New South Wales.

Pilato et al. (2005) described another new species, *Echiniscus arthuri* (type locality Upper Otira Valley, Arthur's Pass National Park) from a series of slides labelled *E. velaminis* by Horning et al. 1978. They also described a second new species, *Echiniscus elaeinae* (type locality Taumaka Island, Open Bay Islands, South Island West Coast), from some specimens identified as *E. spiniger*, and recorded *Echiniscus curiosus* from the Featherston Summit, North Island, from specimens identified by as *E. quadrispinosus brachyspinosus* by Horning et al. (1978). Additional descriptions and remarks were given for *Pseudechiniscus novaezelandiae* (Richters, 1908) and several species described by Horning et al. (1978) and Horning & Schuster (1983).

Pilato and Lisi (2006) described *Macrobiotus rigidus* (type locality The Plateau, Egmont National Park) from some specimens identified as *Macrobiotus harmsworthi coronatus* by Horning et al. (1978). Pilato et al. (2006) described *Macrobiotus hapukuensis* (type locality Hapuku Scenic Reserve, north of Kaikoura) from specimens identified as *M. areolatus* by Horning et al. (1978). They also described *Macrobiotus semmelweisi* (type locality northwest of Te Karaka, North

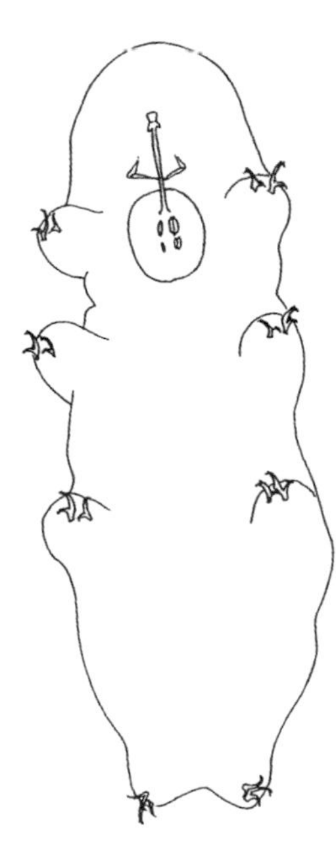

Hypsibius novaezeelandiae.
From Pilato & Binda 1997

Minibiotus sp., anterior view.
Don Horning

Island) from specimens identified as *M. recens* and suggested integrations or corrections of some descriptions of some other species recorded from New Zealand. They examined two specimens and one egg from material identified as *M. harmsworthi coronatus* and referred them to *M.* cf. *coronatus*. They gave a re-description of *M. rawsoni* and examined some specimens of *M. anderssoni*, concluding that they were not that species. They gave a short description and three photographs of *Doryphoribius zyxiglobus* (Horning et al. 1978). They examined two specimens of putative *Diphascon alpinum* and concluded they belong to *D. langhovdense*. Pilato et al. (2006) examined a specimen previously identified as *D. chilenensis* and concluded it belonged to *Diphascon (Diphascon) pingue* – a new record for New Zealand. They concluded their paper with a short description of *D. (D.) bullatum* and noted that the specimens that they examined from New Zealand were perfectly similar, for many characters, to specimens from North Africa with eye spots.

Several of the above revisers suggested that some of the species attributed to the New Zealand fauna by Horning et al. (1978) and Nelson and Horning (1979) be removed from the list of New Zealand Tardigrada, but not all of the specimens in the collections of these authors have been re-examined. Accordingly, we retain the following species in the end-chapter checklist – *Macrobiotus anderssoni, M. coronatus, M. liviae, Echiniscus spiniger*, and *E. quadrispinosus brachyspinosus*.

Among marine species, Horning et al. (1978) listed *Halechiniscus* sp. as occurring in New Zealand, based on a personal communication from L. E. Pollock. It must still be listed as such because the specimen has not been identified further (L. E. Pollock, pers. comm. 2000). Kristensen and Higgins (1984) recorded a juvenile of *Styraconyx kristenseni* from Leigh, North Island. There are two further species that are very common in the intertidal zone in New Zealand – *Archechiniscus marci* and *Echiniscoides sigismundi*. Specimens of *Tanarctus* sp. were collected off Dunedin in 1991 from shell gravel at 100 metres depth (R. Kristensen, pers. comm. 2000). The latter three species are newly recorded from New Zealand (see end-chapter checklist).

Tardigrade collections

Major repositories/holders of New Zealand Tardigrada are: R. M. Bohart Museum, University of California, Davis; Museum of New Zealand Te Papa Tongarewa, Wellington; Dr Reinhardt Kristensen, University of Copenhagen, Denmark; Dr Diane Nelson, East Tennessee State University, Johnson City, Tennessee; and Sandra Claxton, South Camden, New South Wales, Australia.

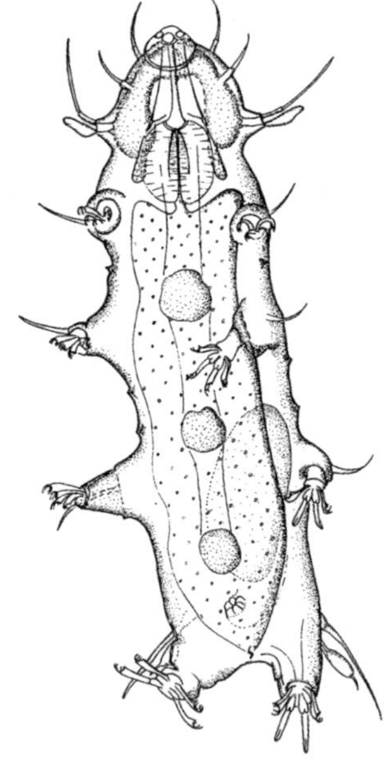

Marine tardigrade *Styraconyx kristenseni*.
From Renaud-Mornant 1981

Gaps in knowledge and scope for future research

What can be done to further the knowledge of New Zealand Tardigrada? Species distribution is poorly known. It would be interesting to re-collect one or more areas extensively, using more refined techniques. This would provide knowledge of possible changes in species composition within limited areas in different seasons. One such area could be the Snares Islands. A detailed study here would contribute to a better understanding of the distribution of tardigrades in the subantarctic zone. The tardigrade 'hotspot' of New Zealand and, at present, of the Southern Hemisphere, is Arthur's Pass National Park. Here, the two most tardigrade-diverse areas are Halpins Creek on the east side and Kellys Creek on the west side of the divide. These two places have yielded more species of tardigrades than any other locality in New Zealand. In fact, there were six more species collected there than in the entire North Island. It would also be worthwhile to collect in areas missed in the 1978 survey, such as Fiordland and offshore islands around Stewart Island. In other parts of the world, it has been found that there is an abundance of tardigrades in limestone country. No limestone areas were sampled during the study made by Horning et al. (1978).

Other habitats that need more sampling to find unrecorded marine species are ocean sand beaches, the mantle of mussels such as *Perna*, and red and green algae. The freshwater habitats of ponds, lakes, streams, and rivers have scarcely been sampled.

There are no resident New Zealanders who have worked specifically on the systematics of tardigrades. Most of the work has been carried out by scientists from Germany, Great Britain, France, Italy, and North and South America. It would be most worthwhile to have at least one New Zealander involved with the taxonomy, ecology, and biogeography of Tardigrada. Now is the time to set up a position while there are overseas workers who are willing to help mentor a new recruit. Zhang (2006) reported that 35 papers dealing with the systematic of tardigrades were published in *Zootaxa* during 2002–05, two of them (by Italians and a Russian) pertaining to New Zealand species. Collecting and taxonomic techniques have improved considerably over the past 20 years, which should make it easier for a new student to become established in this specialist field. One of the potential values of the work done by Horning et al. (1978) was to provide large collections of tardigrades from many areas of New Zealand for others to study later using more refined techniques than were available to the authors at the time. Subsequent studies by Pilato, Dastych, and Claxton, among others, are tangible examples of that value. All of the remaining Horning et al. (1978) material should be re-examined and re-identified where needed. After this is accomplished, another look at the ecology of the species may better delimit habitats. Perhaps this would be a good starting point for a budding tardigradologist.

Echiniscus sp., posterior view.

William Miller

Acknowledgements

The authors wish to thank the following colleagues for providing references and information for the improvement of this paper: Drs Hieronymus Dastych (Hamburg Museum), Reinhardt Kristensen (University of Copenhagen), Leland Pollard (Drew University), and Brian Mackness (Beerwah, Queensland, Australia).

Authors

Dr Donald S. Horning Tumblegum Research Laboratory, 2133 Duri-Dungowan Road, Loomberah, NSW 2340, Australia [tumblegu@optusnet.com.au]

Dr Ricardo L. Palma Museum of New Zealand Te Papa Tongarewa, P.O. Box 467, Wellington, New Zealand [ricardop@tepapa.govt.nz]

Dr William R. Miller Baker University, P.O. Box 65, Baldwin City, Kansas 66006-0065, USA [william.miller@bakeru.edu]

References

BERTOLANI, R.; REBECCHI, L. 1993: A revision of the *Macrobiotus hufelandi* group (Tardigrada, Macrobiotidae), with some observations on the taxonomic characters of eutardigrades. *Zoologica Scripta 22*: 127–152.

BINDA, M. G.; PILATO, G.; LISI, O. 2005: Remarks on *Macrobiotus furciger* Murray, 1906 and description of three new species of the *furciger* group (Eutardigrada, Macrobiotidae). *Zootaxa 1075*: 55–68.

CLAXTON, S. K. 1998: A revision of the genus *Minibiotus* (Tardigrada: Macrobiotidae) with descriptions of eleven new species from Australia. *Records of the Australian Museum 50*: 125–160.

CROWE, J. H. 1975: The physiology of cryptobiosis in tardigrades. Pp. 37–59 *in*: Higgins, R. P. (ed.), *International Symposium on Tardigrades. Memorie dell'Istituto Italiano di Idrobiologia 32 (Suppl.)*: 1–469.

CROWE, J. H.; CROWE, L.; CHAPMAN, D. 1984. Preservation of membranes in anhydrobiotic organisms: the role of trehalose. *Science 223*: 701–703.

DASTYCH, H. 1991: *Isohypsibius sattleri* (Richters, 1902), a valid species (Tardigrada). *Senckenbergiana Biologica 71*: 181–189.

DASTYCH, H. 1997: A new species of the genus *Echiniscus* (Tardigrada) from New Zealand. *Entomologische Mitteilungen aus dem Zoologischen Museum Hamburg 12*: 209–215.

DASTYCH, H.; McINNES, S. J.; CLAXTON, S. K. 1998: *Oreella mollis* Murray, 1910 (Tardigrada): a redescription and revision of *Oreella*. *Mitteilungen aus dem Hamburgischen Zoologischen Museum und Institut 95*: 89–113.

DEGMA, P.; BERTOLANI, R.; GUIDETTI, R. 2009: Actual checklist of Tardigrada species (Version 6: 18-04-2009). www.tardigrada. modena.unimo. it/miscellanea/Actual%20checklist%20of%20 Tardigrada.pdf.

DEGMA, P.; GUIDETTI, R. 2007: Notes to the current checklist of Tardigrada. *Zootaxa 1579*: 41–53.

DOYÈRE, L. 1840: Mémoire sur les Tardigrades. *Annales des Sciences Naturelles, Zoologique, sér. 2, 14*: 269–362.

GRIGARICK, A. A.; SCHUSTER, R. O.; TOFTNER, E. C. 1973a: *Macrobiotus montanus* from California. *Pan-Pacific Entomologist 49*: 229–231.

GRIGARICK, A. A.; SCHUSTER, R. O.; TOFTNER, E. C. 1973b: Descriptive morphology of eggs of some species in the *Macrobiotus hufelandi* group. *Pan-Pacific Entomologist 49*: 258–263.

GUIDETTI, R.; BERTOLANI, R. 2005: Tardigrada taxonomy: an updated check list of the taxa and a list of characters for their identification. *Zootaxa 845*: 1–46.

HORNING, D. S.; SCHUSTER, R. O. 1983: Three new species of New Zealand tardigrades (Tardigrada: Echiniscidae). *Pan-Pacific Entomologist 59*: 108–112.

HORNING, D. S.; SCHUSTER, R. O.; GRIGARICK, A. A. 1978: Tardigrada of New Zealand. *New Zealand Journal of Zoology 5*: 185–280.

KINCHIN, I. M. 1994: *The Biology of Tardigrades*. Portland Press, London. 186 p.

KRISTENSEN, R. M. 1987: Generic revision of the Echiniscidae (Heterotardigrada), with a discussion of the origin of the family. Pp. 261–335 *in*: Bertolani, R. (ed.), *Biology of Tardigrades. Selected Symposia and Monographs U.Z.I. 1*. Mucchi, Modena.

KRISTENSEN, R. M.; HIGGINS, R. P. 1984: Revision of *Styraconyx* (Tardigrada: Halechiniscidae), with descriptions of two new species from Disko Bay, West Greenland. *Smithsonian Contributions to Zoology 391*: 1–40.

McINNES, S. J. 1994: Zoogeographic distribution of terrestrial/ freshwater tardigrades from current literature. *Journal of Natural History 28*: 257–352.

MARTIN, N. A.; YEATES, G. W. 1975: Effect of four insecticides on the pasture ecosystem. III. Nematodes, rotifers and tardigrades. *New Zealand Journal of Agricultural Research 18*: 307–312.

MILLER, J. D.; HORNE, P.; HEATWOLE, H.; MILLER, W. R.; BRIDGES, L. 1988: A survey of the terrestrial Tardigrada of the Vestfold Hills, Antarctica. *Hydrobiologia 165*: 197–208.

MILLER, W. R. 1997: Tardigrades: bears of the moss. *Kansas School Naturalist 43*: 1–16.

MURRAY, J. 1910: Tardigrada. *British Antarctic Expedition 1907–09. Reports on the Scientific Investigations. Volume 1, Biology 5*: 1–185.

NELSON, D. R. 1977: Fungal parasites of New Zealand tardigrades (phylum: Tardigrada). *Bulletin of the Association of Southeastern Biologists 24*: 1–173.

NELSON, D. R. 1991: Tardigrada. Pp. 501–521 *in*: Thorp, J. H.; Covich, A. P. (eds), *Ecology and Classification of North American Freshwater Invertebrates*. Academic Press, New York. 911 p.

NELSON, D. R.; HORNING, D. S. 1979: Tardigrada of the Kowai Bush, Kaikoura, New Zealand. *Zeszyty Naukowe Uniwersytetu Jagiellonskiego 529, Prace Zoologiczne Zeszyt 25*: 125–142.

PALMA, R. L.; LOVIS, P. M.; TITHER, C. 1989: An annotated list of primary types of the phyla Arthropoda (except Crustacea) and Tardigrada held in the National Museum of New Zealand. *National Museum of New Zealand Miscellaneous Series 20*: 1–49.

PILATO, G. 1996: Redescription of *Isohypsibius wilsoni* and description of two new species of *Isohypsibius* (Eutardigrada) from New Zealand. *New Zealand Journal of Zoology 23*: 67–71.

PILATO, G.; BINDA, M. G. 1996: Two new species and new records of *Macrobiotus* (Eutardigrada) from New Zealand. *New Zealand Journal of Zoology 23*: 375–379.

PILATO, G.; BINDA, M. G. 1997: Remarks on some tardigrades of New Zealand with the description of *Hypsibius novaezeelandiae* n. sp. (Eutardigrada). *New Zealand Journal of Zoology 24*: 65–67.

PILATO, G.; LISI, O. 2006: *Macrobiotus rigidus* sp. nov., new species of eutardigrade from New Zealand. *Zootaxa 1109*: 49–55.

PILATO, G.; BINDA, M. G.; LISI, O. 2005: Remarks on some Echiniscidae (Heterotardigrada) from New Zealand with the description of two new species. *Zootaxa 1027*: 27–45.

PILATO, G.; BINDA, M. G.; LISI, O. 2006: Eutardigrada from New Zealand, with descriptions of two new species. *New Zealand Journal of Zoology 33*: 49–63.

RAMAZZOTTI, G.; MAUCCI, W. 1995: *The Phylum Tardigrada*. Beasley, C. W. (transl.). McMurray University, Abilene, Texas. [English translation of: *Il Phylum Tardigrada*: III edizione riveduta e aggiomata, 1983. *Memorie dell'Istituto Italiano di Idrobiologia 41*: 1–1011.]

RICHTERS, F. 1908: Beitrag zur Kenntnis der Moosfauna Australiens und der Inseln des Pazifischen Ozeans. *Zoologische Jahrbuecher, Jena 26*: 196–213.

ROEDING, F.; HAGNER-HOLLER, S.; RUHBERG, H.; EBERSBERGER, I.; von HAESELER, A.; KUBE, M.; REINHARDT, R.; BURMESTER, T. 2007: EST sequencing of Onychophora and phylogenetic analysis of Metazoa. *Molecular Phylogenetics and Evolution 45*: 942–951.

STOUT, J. D. 1978: Effect of irrigation with municipal water or sewage effluent on the biology of soil cores. II. Protozoan fauna. *New Zealand Journal of Agricultural Research 21*: 11–20.

SUREN, A. M. 1991: Bryophytes as invertebrate habitat in two New Zealand alpine streams. *Freshwater Biology 26*: 399–418.

SUREN, A. M. 1992: Meiofaunal communities associated with bryophytes and gravels in shaded and unshaded alpine streams in New Zealand. *New Zealand Journal of Marine and Freshwater Research 26*: 115–125.

TOFTNER, E. C.; GRIGARICK, A. A.; SCHUSTER, R. O. 1975: Analysis of scanning electron microscope images of *Macrobiotus* eggs. Pp. 395–411 *in*: Higgins, R. P. (ed.), *International Symposium on Tardigrades. Memorie dell'Istituto Italiano di Idrobiologia 32 (Supplement)*: 1–469.

TUMANOV, D.V. 2004: *Macrobiotus kovalevi* , a new species of Tardigrada from New Zealand (Eutardigrada, Macrobiotidae). *Zootaxa 406*: 1–8.

ZHANG, Z.-Q. 2006: Tardigrada nova: a bibliographic analysis and catalogue of new species described in *Zootaxa* before 2006. *Zootaxa 1369*: 63–68.

Checklist of New Zealand Tardigrada

This checklist follows the format of Degma and Guidetti (2007) and Degma et al. (2009). Marine species in the list pertain only to the New Zealand Exclusive Economic Zone. An underlined name indicates an endemic genus; E indicates an endemic species, and M, marine; all others are found in freshwater bodies or interstitial water in terrestrial settings.

PHYLUM TARDIGRADA
Class HETEROTARDIGRADA
Order ARTHROTARDIGRADA
HALECHINISCIDAE
ARCHECHINISCINAE
Archechiniscus marci Schultz, 1953* M
HALECHINISCINAE
Halechiniscus sp. L. E. Pollock M
STYRACONYXINAE
Styraconyx kristenseni kristenseni Renaud-Mornant, 1981 M
TANARCTINAE
Tanarctus sp.* R. Kristensen M

Order ECHINISCOIDEA
OREELLIDAE
Oreella mollis Murray, 1910
ECHINISCIDAE
Antechiniscus conversus (Horning & Schuster, 1983) E
Antechiniscus parvisentus (Horning & Schuster, 1983) E
Antechiniscus perplexus (Horning & Schuster, 1983 E
Echiniscus arctomys Ehrenberg, 1853
Echiniscus arthuri Pilato, Binda & Lisi, 2005 E
Echiniscus bigranulatus Richters, 1907
Echiniscus blumi blumi Richters, 1903
Echiniscus curiosus Claxton, 1996
Echiniscus elaeinae Pilato, Binda & Lisi, 2005 E
Echiniscus nigripustulus Horning, Schuster & Grigarick, 1978 E
Echiniscus palmai Dastych, 1997 E
Echiniscus porabrus Horning, Schuster & Grigarick, 1978 E
Echiniscus quadrispinosus brachyspinosus Bartoš, 1930
Echiniscus q. quadrispinosus Richters, 1902
Echiniscus spiniger Richters, 1904
Echiniscus velaminis Murray, 1910 E
Echiniscus vinculus Horning, Schuster & Grigarick, 1978 E
Echiniscus zetotrymus Horning, Schuster & Grigarick, 1978 E
Hypechiniscus exarmatus (Murray, 1907)
Pseudechiniscus facettalis Petersen, 1951
Pseudechiniscus novaezeelandiae (Richters, 1908
Pseudechiniscus suillus (Ehrenberg, 1853)
ECHINISCOIDIDAE
Echiniscoides sigismundi Plate, 1989* M

Class EUTARDIGRADA
Order APOCHELA
MILNESIIDAE
<u>*Limmenius*</u> *porcellus* Horning, Schuster & Grigarick, 1978 E
Milnesium tardigradum tardigradum (Doyère, 1840)

Order PARACHELA
CALOHYPSIBIIDAE
Calohypsibius ornatus (Richters, 1900)
HYPSIBIIDAE
DIPHASCONINAE
Diphascon (Adropion) prosirostre Thulin, 1928
Diphascon (A.) scotium Murray, 1905
Diphascon (Diphascon) alpinum Murray, 1906
Diphascon (D.) bullatum Murray, 1905
Diphascon (D.) chilenense chilenense Plate, 1888
Diphascon (D.) higginsi Binda, 1971
Diphascon (D.) langhordense (Sudzuki, 1964)
Diphascon (D.) pingue Marcus, 1936
Diphascon (D.) nodulosum (Ramazzottii, 1957)
Hebesuncus conjungens (Thulin, 1911)
HYPSIBIINAE
Doryphoribius zyxiglobus (Horning, Schuster & Grigarick, 1978 E
Hypsibius allisoni Horning, Schuster & Grigarick, 1978
Hypsibius convergens (Urbanowicz, 1925)
Hypsibius dujardini (Doyère, 1840)
Hypsibius marcello Pilato, 1990
Hypsibius novaezeelandiae Pilato & Binda, 1997 E
Isohypsibius annulatus (Murray, 1905)
Isohypsibius cameruni (Iharos, 1969)
Isohypsibius campbellensis Pilato, 1996 E
Isohypsibius nodosus (Murray, 1907)
Isohypsibius palmai Pilato, 1996 E
Isohypsibius papillifer papillifer (Murray, 1905)
Isohypsibius sattleri (Richters, 1902)
Isohypsibius wilsoni (Horning, Schuster & Grigarick, 1978) E
Ramazzottius baumanni (Ramazzotti, 1962)
Ramazzottius oberhaeuseri (Doyère, 1840)
MACROBIOTIDAE
Calcarobiotus (Discrepunguis) sp. Tumanov 2004
Macrobiotus andersoni Richters, 1907
Macrobiotus areolatus Murray, 1907
Macrobiotus armatus Pilato & Binda, 1996 E
Macrobiotus diffusus Binda & Pilato, 1987
Macrobiotus divergens Binda, Pilato & Lisi, 2005 E
Macrobiotus furciger Murray, 1907
Macrobiotus cornatus Barros, 1942
Macrobiotus hapukuensis Pilato, Binda & Lisi, 2006 E
Macrobiotus hibiscus Barros, 1942
Macrobiotus hieronimi Pilato & Claxton, 1988
Macrobiotus hufelandi hufelandi Schultze, 1833
Macrobiotus kovalevi Tumanov, 2004 E
Macrobiotus liviae Ramazzottii, 1962
Macrobiotus montanus Murray, 1910
Macrobiotus occidentalis occidentalis Murray, 1910
Macrobiotus orcadensis Murray, 1902
Macrobiotus pseudoliviae Pilato & Binda, 1996 E
Macrobiotus rawsoni Horning, Schuster & Grigarick, 1978 E
Macrobiotus recens Cuénot, 1932
Macrobiotus rictersi Murray, 1911
Macrobiotus rigidus Pilato & Lisi, 2006 E
Macrobiotus semelweisi Pilato, Binda & Lisi, 2006 E
Macrobiotus snaresensis Horning, Schuster & Grigarick, 1978
Minibiotus aculeatus (Murray, 1910)
Minibiotus asteris Claxton, 1998
Minibiotus bisoctus (Horning, Schuster & Grigarick, 1978)
Minibiotus hispidus Claxton, 1998
Minibiotus intermedius (Plate, 1889)
Minibiotus maculartus Pilato & Claxton, 1988
Minibiotus scopulus Claxton, 1998
Minibiotus taiti Claxton, 1998
MURRAYIDAE
Dactylobiotus dispar (Murray, 1907)

THREE

Phylum ONYCHOPHORA

velvet worms, peripatus

DIANNE M. GLEESON, HILKE RUHBERG

Peripatoides novaezealandiae.
Hilke Ruhberg

Onychophorans (Greek *onyx*, claw; *phoreus*, bearer) are a small group of terrestrial animals, somewhat caterpillar-like in general appearance but commonly known as velvet worms from their velvety skin. They are readily identifiable by their 'oncopods' – unjointed stumpy walking limbs similar to those in tardigrades – and the prominent antennae that adorn the head.

This group has long been regarded as important in evolutionary biology owing to their phylogenetic position, ancient history, and Gondwanan distribution. Labelled 'living fossils' and 'missing links' in some popular accounts, onychophorans appear transitional between annelid worms and jointed-legged arthropods. Internally, onychophorans are segmented like annelids (earthworms and relatives). The excretory and reproductive organs, the nervous system, and the muscle arrangement are also annelid-like, but the circulatory and respiratory systems are similar to arthropods. As such, they have historically been allied with annelids or arthropods or treated as a separate phylum. The latest phylogenies, which are a synthesis of both morphological and molecular data, clearly show the Onychophora to be a sister group to the arthropods and a member of the Ecdysozoa, a group of phyla known collectively as the moulting animals (Adoutte et al. 2000; Roeding et al. 2007).

They are truly archaic creatures, having scarcely changed in appearance since 550 million years ago, judging from compressed specimens in the Cambrian Burgess Shale sediments in present-day British Columbia. These fossils indicate that primitive onychophorans lived in shallow marine environments, had longer legs than present-day forms, and possessed various spines and body plates. The most famous onychophoran-like fossils include *Xenusion auerswaldae* from the Early Cambrian Baltic (Krumbiegel et al. 1980) and *Aysheaia pedunculata* from the Burgess Shale (Whittington 1978). Poinar (2000) proposed a classification of the Onychophora that includes all known living and fossil forms. Today, the phylum consists entirely of two families of terrestrial species – the Peripatidae (found in the Antilles, Mexico, Central America, northern South America, equatorial West Africa, Assam, and Southeast Asia) and the Peripatopsidae (found in Chile, South Africa, West Irian, New Guinea, Australia, and New Zealand). The presence of both families in Africa suggests that they diverged prior to the break-up of Gondwana.

Living onychophorans are restricted to moist dark microhabitats such as can be found in rotten logs, under leaf litter and rocks, and in crevices in rocks and soil. They are nocturnal predators that trap their prey using a sticky substance expelled from a pair of modified limbs, the oral papillae, located on either side of the head. Reproductive strategies within this group are surprisingly diverse. These include oviparity, ovoviviparity, viviparity with yolk-free eggs, and placental viviparity (Dendy 1902; Anderson & Manton 1972; Walker & Campiglia 1990).

All placental forms are found within the neotropical Peripatidae, which produce minute yolkless eggs. Oviparity and ovoviviparity are the common reproductive strategies of the Peripatopsidae, but there are also viviparous, non-placental representatives. Many aspects of onychophoran reproductive behaviour are still poorly known, although recent observations in the field and in cultures are providing further clues (Ruhberg unpubl. data).

A worldwide estimate of species numbers within the Onychophora is difficult to ascertain as some taxonomic revisions are in progress. In particular, the application of scanning electron microscopy has proven a powerful tool in uncovering an extensive radiation of species within Australia (Ruhberg et al. 1988; Reid 1996). These recent discoveries have led to an estimate of over 200 species worldwide, with a disproportionate number existing in Australia owing to the monograph by Reid (1996). Furthermore, the application of molecular methods in identifying further species has also proven extremely useful (Tait et al. 1990; Rowell et al. 1995; Tait & Briscoe 1995).

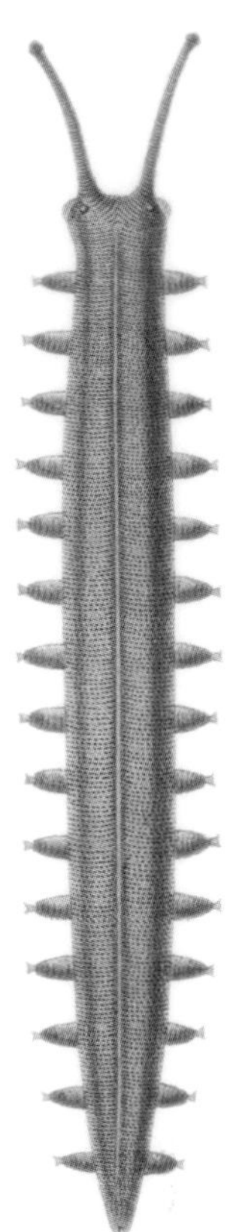

Reproduction of a colour painting of a rare green peripatus, representing a new genus and species, from a mountainous region of the South Island.

A. Balfour in E. L. Bouvier 1907

New Zealand onychophorans

In New Zealand, five species within two genera are currently recognised using morphological characters (Hutton 1876; Dendy 1894; Ruhberg 1985; Gleeson 1996). A further four 'taxa' have been suggested, based on differences in electromorph patterns revealed using isozyme electrophoresis (Tait & Briscoe 1995; Trewick 1998). The validity and utility of these proposed taxa is questionable and they are currently being revised as part of a larger treatment of the fauna by the authors.

The two described genera are separated mainly on the basis of alternative reproductive strategies, *Peripatoides* being ovoviviparous, and *Ooperipatellus* oviparous. *Peripatoides* is endemic to New Zealand, whereas representatives of *Ooperipatellus* occur in both Australia and New Zealand. This is not surprising, as DNA sequence data in determining the phylogenetic relationships among Australasian onychophorans provides convincing support for the close affinity between New Zealand and Tasmanian taxa (Gleeson et al. 1998). Onychophora are widely distributed in New Zealand. Our data from more than 1000 specimens have revealed interesting patterns, which may reflect past geological events. Species within *Ooperipatellus* are predominantly found in the alpine regions of the South Island, with only a few records from the North Island. *Peripatoides* species are found throughout the North Island and in northwest Nelson, Otago, and Southland districts (excluding Fiordland). Gaps in distribution include the Kaikoura Ranges and Stewart Island. Of the described species, the most widespread is *Peripatoides novaezealandiae,* which is likely to be a complex of species. Other described members of *Peripatoides* have relatively restricted distributions, with *P. indigo* found only in a small area of northwest Nelson (including inside limestone caves) and *P. suteri* limited to a small number of sites in the vicinity of Mt Taranaki.

Where the two reproductive forms co-occur, they do not occupy exactly the same microhabitat or niche. *Peripatoides* is found usually within rotting logs, while *Ooperipatellus* is found mainly under logs and among leaf litter. Native Onychophora are not restricted to indigenous forest. Specimens of *Peripatoides*

Summary of New Zealand onychophoran diversity

Taxon	Described living species	Known undescribed species	Estimated unknown species	Adventive species	Endemic species	Endemic genera
Onychophora	9	25	10	0	9	1

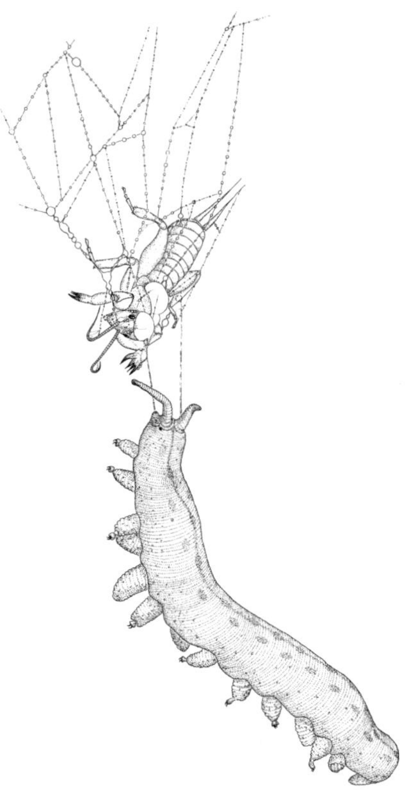

An onychophoran catching its prey, using an adhesive glue discharged from slime glands via the oral papillae.
Claudia Brockmann

can be found in tussock land in Otago and Southland (Gleeson unpubl. data) and in suburban Wellington and Dunedin (Harris 1991).

Taxonomy has proven difficult within the Onychophora, owing in part to the lack of non-overlapping morphological characters and the uniform anatomy. Another factor is the apparent rarity, in collections, of males, which possess the most informative morphological characters. This is a consequence of their biology and behaviour. Sex ratios tend to skew towards females (Tutt 1997), females are larger and more conspicuous, and males are often more likely found in leaf litter as they tend to disperse while females remain in logs (Scott & Rowell 1991). The application of molecular data is useful in identifying monophyletic groups that may be reflective of genera and/or species. Once these groupings are defined, morphological characters consistent with phylogeny can be more easily identified. This is currently the approach we are implementing in the revision of New Zealand's Onychophora. It is estimated that a further 25 species will be described.

The Peripatopsidae thus represents a distinctive element of New Zealand's fauna. Their antiquity is of particular interest as, having survived New Zealand's turbulent geological history, they should provide clues to relationships among Gondwanan species. In addition, the distribution and phylogenetic relationship of species will provide significant insight into the processes that have been involved in speciation within this group. Onychophorans are classified in the IUCN Invertebrate Red Data Book as 'vulnerable' (Wells et al. 1983). A useable and sound taxonomy is therefore critical in order to determine the conservation status of these unique and intriguing organisms in different parts of New Zealand.

Authors

Dr Dianne M. Gleeson Landcare Research, Private Bag 92170, Auckland Mail Centre, Auckland 1142, New Zealand [gleesond@landcareresearch.co.nz]

Prof. Dr Hilke Ruhberg Biozentrum Grindel und Zoologisches Museum, Universität Hamburg, Martin-Luther-King-Platz 3, D-20146 Hamburg, Bundesrepublik Deutschland [ruhberg@zoologie.uni-hamburg.de]

References

ANDERSON, D. T.; MANTON, S. M. 1972: Studies on the Onychophora. VIII. The relationship between the embryos and the oviduct in the viviparous placental onychophorans *Epiperipatus trinidadensis* Bouvier and *Macroperipatus torquatus* Kennel from Trinidad. *Philosophical Transactions of the Royal Society, B, 264*: 161–189.

ADOUTTE, A.; GUILLAUME, B.; LARTILLOT, N.; LESPINET, O.; PRUD'HOMME, B.; de ROSA, R. 2000: The new animal phylogeny: Reliability and implications. *Proceedings of the National Academy of Science 97*: 4453–4456.

DENDY, A. 1902: On the oviparous species of Onychophora. *Quarterly Journal of Microscopical Science, n.s., 45*: 363–415, pls 19–22.

GLEESON, D. M. 1996: Onychophora of New Zealand; past, present and future. *New Zealand Entomologist 19*: 51–55.

GLEESON, D. M.; ROWELL, D. M.; TAIT, N. N.; BRISCOE, D. A.; HIGGINS, A. V. 1998: Phylogenetic relationships among Onychophora from Australasia inferred from the mitochondrial cytochrome oxidase subunit I gene. *Molecular Phylogenetics and Evolution 10*: 237–248.

HARRIS, A. C. 1991: A large aggregation of *Peripatoides novaezealandiae* (Hutton, 1876) (Onychophora: Peripatopsidae). *Journal of the Royal Society of New Zealand 21*: 405–406.

HUTTON, F. W. 1876: On *Peripatus novae-zealandiae*. *Annals and Magazine of Natural History, ser. 4, 18*: 361–369.

KRUMBIEGEL, G.; DEICHFUSS, H.; DEICHFUSS, H. 1980: Ein neuer Fund von *Xenusion*. *Hallesches Jahrbuch für Geowissenschaften 5*: 97–99.

POINAR, G. Jr 2000: Fossil onychophorans from Dominican and Baltic amber: *Tertiapatus dominicanus* n.g., n.sp. (Tertiapatidae) n.fam.) and *Succinipatopsis balticus* n.g., n.sp. (Succinipatopsidae n.fam.) with a proposed classification of the subphylum Onychophora. *Invertebrate Biology 119*: 104–109.

REID, A. L. 1996: Review of the Peripatopsidae (Onychophora) in Australia, with comments on peripatopsid relationships. *Invertebrate Taxonomy 10*: 663–936.

ROEDING, F.; HAGNER-HOLLER, S.; RUHBERG, H.; EBERSBERGER, I.; von HAESELER, A. ; KUBE, M.; REINHARDT, R.; BURMESTER, T. 2007: EST sequencing of Onychophora and phylogenetic analysis of Metazoa. *Molecular Phylogenetics and Evolution 45*: 942–951.

RUHBERG, H. 1985: Die Peripatopsidae (Onychophora). Systematik, Ökologie, Chorologie und phylogenetische Aspekte. *Zoologica 137*: 1–184.

RUHBERG, H.; TAIT, N. N.; BRISCOE, D. A.; STORCH, V. 1988: *Cephalofovea tomahmontis* n. gen., n. sp., an Australian peripatopsid (Onychophora) with a unique cephalic pit. *Zoologischer Anzeiger 221*: 117–133.

ROWELL, D. M.; HIGGINS, A. V.; BRISCOE, D. A.; TAIT, N. N. 1995: The use of chromosomal data in the systematics of viviparous onychophorans from Australia (Onychophora: Peripatopsidae). *Zoological Journal of the Linnean Society 114*: 139–153.

SCOTT, I. A. W.; ROWELL, D. M. 1991: Population biology of *Euperipatoides leuckartii* (Onychophora: Peripatopsidae). *Australian Journal of Zoology 39*: 499–508.

TAIT, N. N.; STUTCHBURY, R. J.; BRISCOE, D. A. 1990: Review of the discovery and identification of Onychophora in Australia. *Proceedings of the Linnean Society of New South Wales 112*: 153–171.

TAIT, N. N.; BRISCOE, D. A. 1995: Genetic differentiation within New Zealand Onychophora and their relationships to the Australian fauna. *Zoological Journal of the Linnean Society 114*: 91–102.

TREWICK, S. A. 1998: Sympatric cryptic species in New Zealand Onychophora. *Biological Journal of the Linnean Society 63*: 307–329.

TUTT, K. 1997: The life history and reproductive cycle of *Peripatoides novaezealandiae* (Onychophora: Peripatopsidae). Unpublished M.Sc. thesis, Victoria University of Wellington. 93 p.

WALKER, M.; CAMPIGLIA, S. 1990: Some observations on the placenta and embryonic cuticle during development in *Peripatus acacioi* Marcus and Marcus (Onychophora, Peripatidae). Pp. 449–459 *in*: Minelli, A. (ed.), *Proceedings of the 7th International Congress of Myriapodology*. E.J. Brill, Leiden. 495 p.

WELLS, S. M.; PYLE, R. M.; COLLINS, N. M. 1983: *The IUCN Invertebrate Red Data Book*. IUCN, Cambridge. 682 p.

WHITTINGTON, H. B. 1978: The lobopod animal *Aysheaia pedunculata* Walcott, Middle Cambrian, Burgess Shale, British Columbia. *Philosophical Transactions of the Royal Society of London, B, 284*: 165–197.

Checklist of New Zealand Onychophora

Endemic genera are underlined (first entry only); E indicates endemic species.

PHYLUM ONYCHOPHORA
PERIPATOPSIDAE
<u>*Peripatoides*</u> *aurorbis* Trewick, 1998 E
Peripatoides indigo Ruhberg, 1985 E
Peripatoides kawekaensis Trewick, 1998 E
Peripatoides morgani Trewick, 1998 E
Peripatoides novaezealandiae (Hutton, 1876) E
Peripatoides suteri (Dendy, 1894) E
Peripatoides sympatrica Trewick, 1998 E
Peripatoides n. sp. Kapiti Island, H. Ruhberg E
Peripatoides n. sp. Omeru Scenic Reserve, Auckland H. Ruhberg E
Ooperipatellus nanus Ruhberg, 1985 E
Ooperipatellus viridimaculatus (Dendy, 1900) E
Gen. nov. et n. sp. = Bouvier's 'green' *P. novaezealandiae* H. Ruhberg E
Gen. nov. et n. spp. Dunedin H. Ruhberg

FOUR

Phylum ARTHROPODA

INTRODUCTION

jointed-limbed animals

SHANE T. AHYONG, DENNIS P. GORDON

New Zealand scampi *Metanephrops challengeri*.
Shane Ahyong

Arthropods are the most speciose and abundant animals on earth, occupying almost every habitat capable of supporting multicellular life. They include insects, crustaceans, centipedes, spiders, and the extinct trilobites. They are distributed from the highest mountains to the deepest ocean trenches and from warm springs to the cold wastes of Antarctica. They can even be found throughout the atmosphere, where silk-borne spiderlings and flying insects can be captured as a kind of aerial plankton and nekton. There is also a remarkable diversity of form, including giant two-metre-long sea-scorpions (Eurypterida) from the Paleozoic and the modern Japanese spider crab with a span of 3.7 metres from claw-tip to claw-tip, multi-legged centipedes 30 centimetres long, large four-winged dragonflies, worm-like parasitic crustaceans found in the nasal passages of birds and snakes, and microscopic mites that live in the hair shafts of human eyebrows. In 1995, the United Nations Environment Programme estimated that there were nearly 1.1 million described species of arthropod (Hammond 1995), representing about 63% of the 1.75 million species of all life then known. Current estimates of living diversity stand at 1.8 million described species, of which 823,500 to 1,189,700 are arthropods and 720,000 to more than 1,000,000 are insects (Chapman 2009). Judging from insect diversity in tropical rainforests, their species may number in the millions – one definitive study estimated that the total (described and undiscovered) may be 4–6 million (Novotny et al. 2002), encouragingly fewer than an earlier extrapolation of 30 million (Erwin 1982)! Arthropod biomass and numbers are also large. Including ocean waters, in which tiny copepod crustaceans swarm in vast numbers, one estimate puts the total population of arthropods on earth at a billion billion individuals, but even this figure may be conservative.

As implied by their name (Greek *arthron*, joint, *podos*, foot), arthropods are characterised by jointed limbs, along with jointed bodies. Body segmentation in arthropods is much more strongly regionalised than in annelids. Some segments are fused to form special body regions called tagmata, like head, thorax, and abdomen – quite unlike the simple repetitive segments of an earthworm, for example – and the process and condition of fusion (sometimes with loss of segments) is called tagmosis. The arthropod body is covered by a jointed external skeleton (exoskeleton) made up primarily of the protein chitin, often impregnated with lipids, other proteins, and calcium carbonate. Primitively, each segment has paired appendages (e.g. antennae, mouthparts, legs) on each segment, but in the course of evolution many of these appendages have become highly modified or even lost. Some arthropods, like centipedes, millipedes, and insects, have legs with a single branch (uniramous). The rest had (or retain, in the case of most Crustacea) legs with two branches (biramous). The outer branch is often a flattened gill, while the inner branch is used for walking or modified for grasping, chewing, or reproduction. Arthropods such as spiders and scorpions, with uniramous limbs, are descended from marine ancestors that had biramous appendages; insects, too, with uniramous limbs, may have an ancestry within the biramous Crustacea. Growth is achieved by shedding the exoskeleton, a process known as moulting or ecdysis.

The arthropod nervous system is annelid-like, with a brain (cerebral ganglion) and a nerve ring surrounding the pharynx that connects the brain with a pair of ventral nerve cords, nominally with a ganglion in each segment. As in onychophorans, the arthropod circulatory system is open, with a dorsal tubular heart perforated by pores (ostia) that open into the general body cavity or haemocoel, i.e. an open body cavity filled with tissue, sinuses, and blood. Unlike annelids, but as in nematodes and other Ecdysozoa, ciliated tissues and ciliated larvae are absent. Respiration takes place through the body surface and/or by means of gills, tracheae, or book-lungs. Most arthropods have separate sexes with internal reproduction. Most lay eggs, and development often involves some kind of metamorphosis.

Arthropoda is generally recognised to include five or six major groups or clades: Hexapoda (e.g. insects); Crustacea (e.g. crabs, shrimps, lobsters, slaters, and barnacles); Myriapoda (e.g. centipedes, millipedes); Chelicerata (e.g. horseshoe crabs, spiders, mites, scorpions, and sea spiders, though the latter (Pycnogonida) are often treated as a separate group; and the extinct Trilobitomorpha (trilobites). Not surprisingly, in a group as large and diverse as the arthropods, scientific controversies abound. Phylogenetic controversies past and present can be summarised in three main questions. First, are the arthropods monophyletic, that is, do they form a natural group that originated from a single common ancestor, or have the major groups of arthropods evolved independently from different and unrelated ancestors? Second, what are the closest relatives of the arthropods? Third, how are the major groups of arthropods related to each other – are the insects more closely related to the crustaceans, or to other hexapods, or to something else?

Arthropod monophyly (i.e. from a single ancestor) is not currently controversial, unlike the situation in the late nineteenth and twentieth centuries. Until the late 19th century, arthropod monophyly was unquestioned by most zoologists. The segmented bodies of arthropods and annelid worms had been thought to link the two groups, reflected in the collective name Articulata (see Cuvier 1812), but increasing difficulties in explaining arthropod interrelationships via prevailing recapitulationary models, along with growing interest in convergence as an explanatory tool, led to increasing speculation about arthropod polyphyly – perhaps the major arthropod groups each evolved separately out of different annelid-like ancestors. The polyphyly debate even reached as far afield as New

Summary of New Zealand arthropod diversity

Taxon	Described living species + subspecies	Known undescr. species	Estimated unknown species	Adventive species	Endemic species	Endemic genera named + unnamed
Chelicerata	2,842+71	704	15,130–16,280	114	2,693	242+3
Myriapoda	164+2	131	200	28	266	113+9
Crustacea	2,488+3	485	5,060	46	1,097	98+10
Hexapoda	12,573+110	2,231–2,397	4,735–6,525	1,585+4	12,105–12,091+91	1,188+48
Totals	**18,067+186**	**3,351–3,717**	**25,125–28,065**	**1,773+4**	**16,161–16,147+91**	**1,641+70**

Diversity by environment

Taxon	Marine	Terrestrial	Freshwater	Fossil[††]
Trilobitomorpha*	–	–	–	80
Chelicerata	115+2	3,273+69	158	0
Myriapoda	0	295+2	0	0
Crustacea	2,615	123	236	602
Hexapoda[†]	197+10	13,755–13,918+68	848+2	121
Totals[†]	**2,926+12**	**17,446–17,609+139**	**1,242+2**	**803**

* wholly extinct
† Some species live in more than one major environment during their life-cycles, hence the totals for these across all environments will exceed the total species in the preceding table.
†† pre-Holocene

Zealand, where in 1897, F. W. Hutton prompted the British journal *Natural Science* to publish a series of papers themed on the question 'Are the Arthropoda a Natural Group?' (Bowler 1994). The early twentieth century saw little further progress in the field, and during this time American entomologist R. E. Snodgrass cogently defended arthropod monophyly, with particular emphasis on the homologies of hexapod and crustacean mandibles (e.g. Snodgrass 1938). In response to Snodgrass, arthropod polyphyly was returned to centre-stage in the 1950s–70s by Sidnie Manton and her students, essentially reprising arguments of a previous generation of invertebrate zoologists (e.g. Manton 1977). Further research through the 1980s and 90s, however, using considerably more data, and most importantly, appropriate methodology, has shown convincingly that arthropods are monophyletic (Edgecombe 1998).

But what about the closest relatives of the arthropods? Previous thought linked the arthropods with annelids via the onychophorans in Articulata, and this has found continued support among a number of morphological analyses (see Jenner & Scholtz 2005 for review). Almost all recent studies, however, primarily using molecular data, show that arthropods belong within a wider clade of moulting animals – Ecdysozoa. These include Nematoda (round worms), Nematomorpha (horsehair worms), Kinorhyncha (mud dragons), Onychophora (velvet worms), Priapulida (penis worms), and Loricifera (corset animals) (Alguinado et al. 1997; Colgan et al. 2008; Podsiadlowski et al. 2008; Hejnol et al. 2009). Among these, the closest relatives to the arthropods are the velvet worms (Onychophora) and water bears (Tardigrada). Segmentation has been lost in most ecdysozoan clades, although aspects of segmentation remain in the kinorhynchs. The annelids are more closely related to molluscs, nemerteans and sipunculans.

Whether or not Arthropoda may be treated as a phylum or a superphylum is a moot point. Here, Arthropoda is treated as a phylum, with all New Zealand species, living and fossil, distributed among five subphyla – Trilobitomorpha, Chelicerata, Myriapoda, Crustacea, and Hexapoda. The interrelationships of the subphyla are controversial, but as observed by Giribet et al. (2005), little dispute exists over the validity of the major arthropod clades, the subphyla of this work. Most disagreements about how these major groups are related essentially amount to uncertainty over the position of the root of the arthropod tree.

Although consensus is yet to be reached on the interrelationships of the major arthropod clades, some patterns are emerging. A close relationship between hexapods and crustaceans is widely recognised in phylogenetic analyses (e.g. Field et al. 1988; Schultz & Regier 2000; Giribet et al. 2005; Colgan et al. 2008; Hejnol et al. 2009), although this is not universally accepted (e.g. Bitsch & Bitsch 2004, 2005). Whether or not Crustacea and Hexapoda are themselves strictly monophyletic is subject to debate, with the strong possibility that all or most hexapods are derived from within the crustaceans (Carapelli et al. 2007; Timmermans et al. 2008). Developmental studies using *Hox* (homeobox) genes (involved in morphogenesis) and an analysis of arthropod haemoglobins suggest that primitive crustaceans were actually the ancestors of insects and other hexapods (Akam 2000; Burmester 2001; Cook et al. 2001; Mallat & Giribet 2006). Collectively, the insect + crustacean grouping is known as Pancrustacea or, more formally, Tetraconata (Dohle 2001). The Pancrustacea concept is very attractive as it appears to solve a long-standing puzzle – that is, why insects, which have been the most successful colonisers of land, were not able to diversify successfully in the marine environment. The inverse is true with respect to the Crustacea – why were they not able to diversify on land more than they have, given the antiquity of the group? The Pancrustacea hypothesis neatly proposes that crustaceans successfully invaded land *as insects* (Regier et al. 2005; Glenner et al. 2006).

The positions of the Myriapoda and Chelicerata remain in question. The increasing support for Pancrustacea has removed Myriapoda from a position alongside Hexapoda. Currently, the most viable hypotheses on the position of the myriapods are as either the sister to Chelicerata or the sister to Pancrustacea (which would make chelicerates sister to the remaining extant arthropods). A myriapod + pancrustacean clade, termed Mandibulata, is morphologically parsimonious, suggesting a single derivation of mandibles. Conversely, strong molecular support has been found for a chelicerate + myriapod grouping (e.g. Friedrich & Tautz 1995; Nardi et al. 2003; Mallatt & Giribet 2006; Podsiadlowski et al. 2008), which has been termed Paradoxopoda, reflecting the apparent lack of synapomorphies (shared derived characters) for the clade. Recent studies, however, have discovered possible embryological synapomorphies for Paradoxopoda (Mayer & Whitington 2009). No doubt much remains to be discovered.

Unlike other arthropod subphyla, Trilobitomorpha has the distinction of being extinct and is therefore not amenable to molecular study. Suffice it to say, the trilobites and their relatives have a varied history of interpretation. The discovery of antennae and biramous limbs in trilobites in the late nineteenth century led trilobites to be allied with the crustaceans. In contrast, Heider (1913) saw a close relationship between trilobites and chelicerates and this perspective has been maintained by most workers ever since (Scholtz & Edgecombe 2005). Bergstrom (1979) emphasised appendage similarities with chelicerates; in particular the imbricated (overlapping) lamellar setae on trilobite exopods is similar to the book gills of some chelicerates. An alternative hypothesis, proposed by Boudreaux (1979) and extended by Scholtz and Edgecombe (2005), follows detailed study of cephalic homologies and tagmosis patterns. The trilobites are posited as the stem lineage of Mandibulata; that is, in the 'ancestry' of the clade containing Myriapoda, Hexapoda and Crustacea.

The phylogeny of the arthropods is far from settled, but whatever their interrelationships, they are remarkable creatures that delight, amaze, or scare us. They have been responsible for some of the most devastating plagues and famines that humankind has known. On the other hand, many species are essential for our existence, directly or indirectly providing us with food, clothing, medicines, and protection from harmful organisms. Arthropods simply have to be reckoned with.

Authors

Dr Shane T. Ahyong National Institute of Water & Atmospheric Research, Private Bag 14901, Kilbirnie, Wellington, New Zealand [s.ahyong@niwa.co.nz]

Dr Dennis P. Gordon National Institute of Water & Atmospheric Research, Private Bag 14901, Kilbirnie, Wellington, New Zealand [d.gordon@niwa.co.nz]

References

AKAM, M. 2000: Arthropods: developmental diversity within a (super) phylum. *Proceedings of the National Academy of Sciences 97*: 4438–4441.

ALGUINADO, A. M.; TURBEVILLE, J. M.; LINFORD, L. S.; RIVERA, M. C.; GAREY, J. R.; RAFF, R. A.; LAKE, J. A. 1997: Evidence for a clade of nematodes, arthropods and other moulting animals. *Nature 387*: 498–493.

BERGSTRÖM, J. 1979: Morphology of fossil arthropods as a guide to phylogenetic relationships. Pp. 3–56 in: Gupta, A. (ed) *Arthropod Phylogeny*. Van Nostrand Reinhold Co., New York.

BITSCH, C.; BITSCH, J. 2004: Phylogenetic relationships of basal hexapods among the mandibulate arthropods: a cladistic analysis based on comparative morphological characters. *Zoologica Scripta 33*: 511–550.

BITSCH, C.; BITSCH, J. 2005: Evolution of eye structure and arthropod phylogeny. *in:* Koenemann, S.; Jenner, R.A. (eds), *Crustacea and Arthropod Relationships. Crustacean Issues*, 16: 185–214.

BOUDREAUX, H. B. 1979: *Arthropod phylogeny – with special reference to insects*. Wiley, New York.

BOWLER, P. J. 1994. *Life's splendid drama: evolutionary biology and the reconstruction of life's ancestry, 1860–1940*. University of Chicago Press, Chicago & London.

BURMESTER, T. 2001: Molecular evolution of the arthropod hemocyanin superfamily. *Molecular Biology and Evolution 18*: 184–195.

CARAPELLI, A.; LIÒ, P.; NARDI, F.; van der WATH, E.; FRATI, F. 2007: Phylogenetic analysis of mitochondrial protein coding genes confirms the reciprocal paraphyly of Hexapoda and Crustacea. *BMC Evolutionary Biology 7 (Suppl. 2):S8*: [1–13]. doi:1-1186/1471-2148-7-S2-S8.

CHAPMAN, A. D. 2009: *Numbers of Living Species in Australia and the World*. 2nd edn. Department of the Environment, Water, Heritage and the Arts, Canberra. 80 p.

COLGAN, D. J.; HUTCHINGS, P. A.; BEACHAM, E. 2008: Multi-gene analyses of the phylogenetic relationships among the Mollusca, Annelida, and Arthropoda. *Zoological Studies 47*: 338–351.

COOK, C. E.; SMITH, M. L.; TELFORD, M. J.; BATIONELLO, A.; AKAM, M. 2001: *Hox* genes and the phylogeny of the arthropods. *Current Biology 11*: 759–763.

CUVIER, G. 1812: Sur un nouveau rapprochement à établir entre les classes qui composant le Règne Animal. *Annales du Museum d'Histoire Naturelle de Paris* 19: 73–84.

DOHLE, W. 2001: Are the insects terrestrial crustaceans? A discussion of some new facts and arguments and the proposal of the proper name Tetraconata for the monophyletic unit Crustacea + Hexapoda. *Annales de la Société Entomologique de France 37*: 85–103.

EDGECOMBE, G. D. 1998: Introduction: the role of extinct taxa in arthropod phylogeny. Pp. 1–7 in: Edgecombe, G. D. (eds), *Arthropod Fossils and Phylogeny*. Columbia University Press, New York.

ERWIN, T. L. 1982: Tropical forests: their richness in Coleoptera and other species. *Coleopterist's Bulletin 36*: 74–75.

FIELD, K. G.; OLSEN, G. J.; LANE, D. J.; GIOVANNONI, S. J.; GHISELIN, M. T.; RAFF, E. C.; PACE, N. R.; RAFF, R. A. 1988: Molecular phylogeny of the animal kingdom. *Science 239*: 748–753.

FRIEDRICH, M. & TAUTZ, D. 1995: Ribosomal DNA phylogeny of the major extant arthropod classes and the evolution of myriapods. *Nature* 376: 165–167.

GIRIBET, G.; EDGECOMBE, G. D.; WHEELER, W.C. 2001: Arthropod phylogeny based on eight molecular loci and morphology. *Nature 413*: 157–161.

GLENNER, H.; THOMSEN, P. F.; HEBSGAARD, M. B.; SØRENSEN, M.V.; WILLERSLEV, E. 2006: The origin of insects. *Science 314*: 1883–1884.

HAMMOND, P. M. 1995: The current magnitude of biodiversity. Pp. 113–138 *in*: Heywood, V. H.; Watson, R. T. (eds), *Global Diversity Assessment*. Cambridge University Press for United Nations Environment Programme, Cambridge. xi + 1140 p.

HEIDER, K. 1913: Entwicklungsgeschichte und Morphologie der Wirbellosen. Pp. 176–332 *in*: Hinneberg, P. (ed) *Die Kultur der Gegenwart*, teil 3, Abt. 4, Bd. 2. Teubner, Leipzig

HEJNOL, A.; OBST, M.; STAMATAKIS, A.; OTT, M.; ROUSE, G. W.; EDGECOMBE, G. D.; MARTINEZ, P.; BAGUNA, J.; BAILLY, X.; JONDELIUS, U.; WIENS, M.; MÜLLER, W.; SEAVER, E.; WHEELER, W. C.; MARTINDALE, M. Q.; GIRIBET, G.; DUNN, C. 2009: Assessing the root of bilaterian animals with scalable phylogenetic methods. *Proceedings of the Royal Society ser. B*. doi:10.1098/rspb.2009.0896.

JENNER, R. A.; SCHOLTZ, G. 2005: Playing another round of metazoan phylogenetics: historical epistemology, sensitivity analysis, and the position of Arthropoda within the Metazoa on the basis of morphology. *in:* Koenemann, S.; Jenner, R.A. (eds), *Crustacea and Arthropod Relationships. Crustacean Issues*, 16: 307–352.

MALLATT. J.; GIRIBET, G. 2006: Further use of nearly complete, 28S and 18S rRNA genes to classify Ecdysozoa: 37 more arthropods and a kinorhynch. *Molecular Phylogenetics and Evolution. 40*: 772–794.

MANTON, S. M. 1977. *The Arthropoda: Habits, Functional Morphology, and Evolution*. Clarendon, Oxford.

MAYER, G.; WHITINGTON, P. M. 2009: Velvet worm development links myriapods with chelicerates. *Proceedings of the Royal Society, ser. B* 276: 3579–3579.

NARDI, F.; SPINSANTI, G.; BOORE, J.L.; CARAPELLI, A.; DALLAI, R.; FRATI, F. 2003: Hexapod origins: monophyletic or paraphyletic? *Science 299*: 1887–1889.

NOVOTNY, V.; BASSET, Y.; MILLER, S. E.; WEIBLENS, G. D.; BREMER, B.; CIZEK, L.; DROZD, P. 2002: Low host specificity of herbivorous insects in a tropical forest. *Nature 416*: 841–844.

PODSIADLKOWSKI, L.; BRABAND, A.; MAYER, G. 2008: The complete mitochondrial genome of the onychophoran *Euperipatus biollei* reveals a unique transfer RNA set and provides further support for the Ecdysozoa hypothesis. *Molecular Biology and Evolution* 25: 42–51.

REGIER, J.C.; SHULTZ, J.W.; KAMBIC, R.E. 2005: Pancrustacean phylogeny: hexapods are terrestrial crustaceans and maxillopods are not monophyletic. *Proceedings of the Royal Society, ser. B, 272*: 395–401.

SCHOLTZ, G.; EDGECOMBE, G. D. 2005: Heads, Hox and the phylogenetic position of trilobites. *in:* Koenemann, S.; Jenner, R. A. (eds), *Crustacea and Arthropod Relationships. Crustacean Issues*, 16: 139–166.

SCHULTZ, J. W.; REGIER, J. C. 2000: Phylogenetic analysis of arthropods using two nuclear protein-encoding genes supports a crustacean + hexapod clade. *Proceedings of the Royal Society of London, ser. B, 267*: 1011–1019.

SNODGRASS, R. E. 1938: The evolution of the Annelida, Onychophora, and Arthropoda. *Smithsonian Miscellaneous Collections* 97: 1–59.

TIMMERMANS, M. J. T. N.; ROELOFS, D.; MARIËN, J.; van STRAALEN, N. M. 2008: Revealing pancrustacean relationships: phylogenetic analysis of ribosomal protein genes places Collembola (springtails) in a monophyletic Hexapoda and reinforces the discrepancy between mitochondrial and nuclear DNA markers. *BMC Evolutionary Biology 8:83*: [1–10]. doi:10.1186/1471-2148-8-83.

FIVE

Phylum

ARTHROPODA

TRILOBITOMORPHA

trilobites

ANTHONY J. WRIGHT, ROGER A. COOPER

Trilobites are an extinct group of marine arthropods with calcareous exoskeletons. They were extremely abundant in the Cambrian and Ordovician periods, becoming extinct at the end of the Paleozoic during the devastating Permo-Triassic extinction event. They had a flattened oval body divided into a head-shield (cephalon) with a pair of antennae, compound eyes, and biramous appendages; a trunk (thorax) bearing paired biramous limbs, and a terminal segment (pygidium). Most of the known species ranged in length from one millimetre to 10 centimetres, but some giants achieved 76 centimetres. The name trilobite alludes to the three longitudinal lobes of the body, namely a central axial lobe flanked to left and right by a pleural (side) lobe.

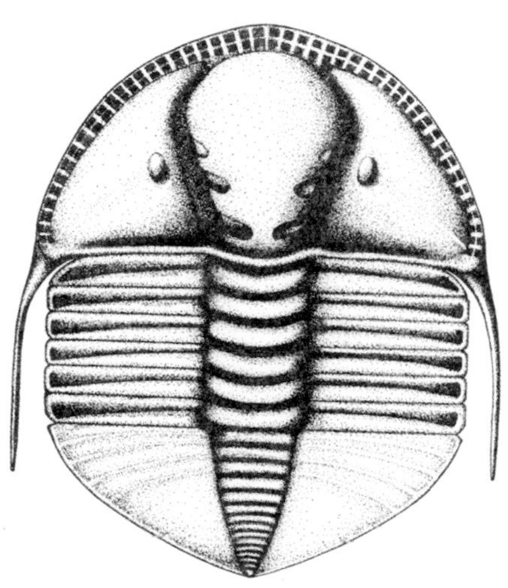

The Late Ordovician trilobite *Incaia bishopi* (order Asaphida), from the Paturau River area of northwest Nelson.

Anthony Wright

The group diversified extensively during its more than 300-million-year duration in those ancient seas, both in body form and numbers of species, such that specialists recognise 10 orders, more than 150 families, about 5000 genera, and over 17,000 described species. While most were wide-bodied, some were narrow-bodied and streamlined. Trilobite fossils are often found enrolled like modern pillbugs or woodlice for protection. Some species were extremely spiny, while others were smooth. Some had huge compound eyes, while others were eyeless. While uncertainty remains concerning the function and ecological role of some of the body shapes, it is clear that trilobites were extremely successful, found in a very wide variety of habitats, and probably occupied many, if not all, of the ecological niches that marine crustaceans do today, including planktonic, free-swimming, benthic, burrowing, and reef-dwelling, but possibly not parasitic. An excellent website for information about all aspects of trilobites is http://www.trilobites.info/.

There were two main groups. Agnostids were characterised by small skeletons and two thoracic segments; many authorities have interpreted them as planktonic, and they have great significance for intercontinental correlation of Cambrian strata, in which they are best developed. The larger, mostly seafloor-dwelling, trilobites had more than two thoracic segments. Known as'polymerids', they are mostly useful for intrabasinal correlation.

The New Zealand trilobite fossil record is very limited, largely owing to the small amount and highly deformed nature of Early Paleozoic strata. Elsewhere in the world, rich and diverse trilobite faunas are found in platform strata or shallow shelf strata. Paleozoic developments of such environments were widespread in Australia but not in New Zealand.

Some 80 species-level taxa have been recorded from New Zealand, of which the most important trilobite faunas are from the Cambrian, Ordovician, and Devonian. Silurian trilobites (with the exception of an indeterminate specimen

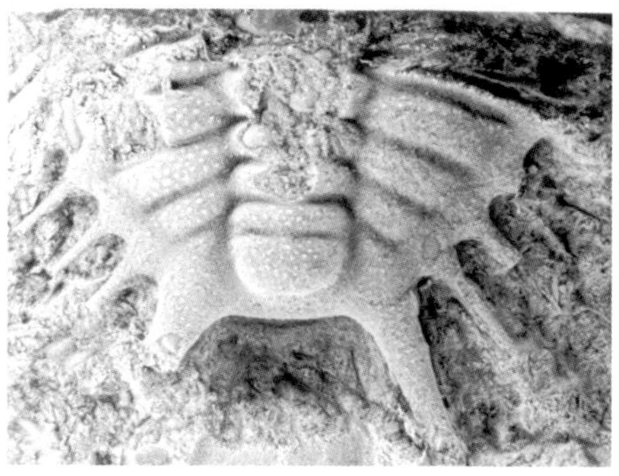

The cephalon (upper photo) and pygidium (lower photo) of an undetermined genus and species of dolichometopid trilobite (order Corynexochida) from the Middle Cambrian of the Cobb Valley area of northwest Nelson.
Roger Cooper

from the possibly latest Silurian *Notoconchidium* fauna from the Wangapeka River) and Carboniferous trilobites are as yet unknown. There are two reported Permian trilobite occurrences. Occurrences of trilobites in the New Zealand Paleozoic sedimentary record are patchy in the sense that there are no sequences of trilobite-bearing strata. All records are from the northwest Nelson district of the South Island, with the exception of rare Permian occurrences from Southland.

Stratigraphic distribution

Cambrian

All identified Cambrian trilobites are from the Takaka Terrane of northwest Nelson. The Kootenia-Peronopsis fauna in the Heath Creek Beds represents the oldest New Zealand fossils, of middle Middle Cambrian (probably Floran) age, about 508 million years ago (Münker & Cooper 1999). In the Middle Cambrian Tasman Formation, allochthonous limestone lenses, derived from a carbonate shelf or platform of which no other trace remains, contain a varied fauna of large polymerid trilobites including *Dorypyge, Koptura, Pianaspis, 'Solenoparia'*, and thick-shelled agnostids (Henderson & MacKinnon 1981). The best-known lens is 'Trilobite Rock' in Cobb Valley, first reported by Benson (1956) and from which 15 species have been recorded (Cooper 1979). The enclosing shales represent an off-shelf environment and contain agnostids and small polymerids such as nepeids preserved in in situ concretionary bands. The in situ trilobites in the lower part of the Tasman Formation are of probable Undillan (Drumduan) age and those from the top represent the following Boomerangian (Guzhangian) age (Münker & Cooper 1999), both in Epoch 3 of the Global Geochronological Scale (equivalent to the late Middle Cambian in the traditional scale). The Cambrian trilobites, which have not yet been formally described, are most closely related to those of Australia and Antarctica (Cooper 1979; Cooper & Shergold 1991).

Cambrian–Ordovician

The next youngest important occurrence of trilobites is that in the Mount Patriarch area, also in northwest Nelson. Trilobites were first collected here in the 1920s by members of the Geological Survey of New Zealand; on the basis of the initial poor material, Reed (1926) described one species from near Mount Patriarch as *Dionide hectori* (now placed in *Hysterolenus*).

A large fauna (some 40 species including *H. hectori*) was described by Wright et al. (1994) from the sequence at Mount Patriarch. This succession spans the Cambrian–Ordovician boundary, with the bulk of the fauna from the Patriarch Formation considered latest Cambrian (late Furongian) to earliest Ordovician (Tremadocian). A smaller, post-Tremadoc fauna occurs in the overlying Summit Limestone. A few taxa are endemic but the fauna is largely cosmopolitan at the genus level. Several genera and species (e.g. *Kainella meridionalis, Onychopyge* cf. *riojana*) highlight a close relationship with South American (Argentina, Bolivia) faunas.

Ordovician

One of the two common species of Gisbornian (early Late Ordovician) age from the Paturau Formation on the Paturau River in northwest Nelson was described by Reed (1926) as *Ogygites collingwoodensis*, an asaphid now assigned to *Basiliella* following Wright (2009). A second species was described from this locality by Hughes and Wright (1970) as *Incaia bishopi*, the only trinucleid yet known from New Zealand. The finding of this genus is of some paleogeographic interest, as its only other occurrences are in Peru, Bolivia, and China.

Cocks and Cooper (2004) reported several trilobite taxa characteristic of the cosmopolitan *Hirnantia* (latest Ordovician, Hirnantian) faunal assemblage, based on a poorly preserved shelly fauna from the Wangapeka Formation collected from the Wangapeka River area. The fauna includes *Eoleonaspis*? sp., *Mucronaspis mucronata*, a possible aulacopleurid, and a possible panderid.

Silurian

One generically indeterminate trilobite specimen has been collected from *Notoconchidium*-bearing sandstones (Wright & Garratt 1991) that outcrop near the junction of the Rolling and Wangapeka Rivers in Nelson Province. The most recent assessment of the age of this distinctive faunal assemblage from New Zealand, Tasmania, and Victoria is latest Silurian (Wright & Garratt 1991).

Devonian

Low-diversity Early Devonian trilobite faunas have been described from the Reefton region and the older Baton Formation in northwest Nelson. The fauna from the latter area, of Pragian (early Early Devonian) age, was described by Shirley (1938) and Wright (1990). This fauna strongly resembles coeval eastern Australian faunas. Typical trilobite genera include *Acaste*, *Acastella* and *Calymene*. Trilobites from the Emsian (late Early Devonian) Reefton sequence were described by Hector (1876) and Allan (1935). These belong mostly to cosmopolitan genera, but the Reefton fauna is of much greater biogeographic interest than the 'Old World' Baton River fauna, as the brachiopod fauna at Reefton includes coldwater 'Malvinokaffric' elements. The Reefton fauna includes the somewhat unexpected trilobite *Dechenella* (*Eudechenella*) *mackayi*, which is clearly anomalous, as no dechenellids are known from the Devonian of eastern Australia. Other genera described by Allan (1935) were assigned to *Digonus*. Sandford (2005) clarified the status of *Wenndorfia expansa* (and its synonyms) from Reefton.

Permian

Rare, undescribed Late Permian trilobites are known from two localities in Southland, one from near Mossburn within the Countess Formation of the Dun Mountain–Maitai Terrane (Hyden et al. 1982; H. J. Campbell pers. comm.), and the other from Pleasant Creek, a tributary of Aparima River, eastern Takitimu Mountains, within a faulted outlier of the Productus Creek Group of Brook Street Terrane (Begg 1981; H. J. Campbell pers. comm.). Permian trilobites from southeastern Australia are similarly very rare.

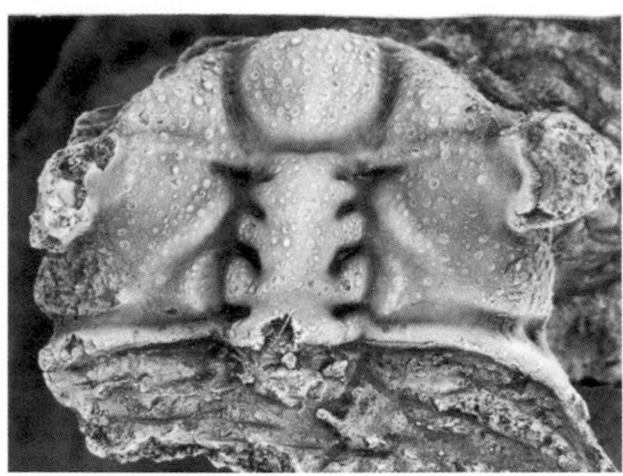

The cephalon of an undescribed species of *Nepea* (order Ptychopariida) from the Middle Cambrian of the Cobb Valley area of northwest Nelson.

Roger Cooper

Future work

The only major undescribed faunas are the above-mentioned Cambrian ones; it is likely that further species will be found in Cambrian rocks, which so far have been investigated in detail only in the Cobb Valley area. In younger strata, apart from any further revision of the Devonian faunas and the description of the Permian faunas, there appears to be little prospect of major finds. Acid leaching of Cambrian and Ordovician limestones could produce further important phosphatised or silicified material as described by Wright et al. (1994) and Percival et al. (2009). As far as detailed correlation value is concerned, the Cambrian agnostids are globally important such fossils and offer the greatest potential for improving the correlation of New Zealand rocks with those of adjacent continents.

Authors

Dr Anthony J. Wright School of Earth and Environmental Sciences, University of Wollongong, Wollongong NSW 2522, Australia [awright@uow.edu.au]

Dr Roger A. Cooper Institute of Geological and Nuclear Sciences, P.O. Box 30368, Lower Hutt, New Zealand [r.cooper@gns.cri.nz]

References

ALLAN, R. S. 1935: The fauna of the Reefton Beds, (Devonian) New Zealand. *New Zealand Geological Survey, Palaeontological Bulletin 14*: 1–72.

BEGG, J. G. 1981: The basement geology and paleontology of the Wairaki Hills, Southland. Unpublished PhD thesis, University of Otago, Dunedin. 400 p.

BENSON, W. N. 1956: Cambrian rocks and fossils in New Zealand (preliminary note). Pp. 285–288 *in*: Rodgers, J. (ed.) *Symposium sobre el Systema Cámbrico, su Paleogeografía y el Problema de sur Base. Vol. 2(2) (Australia, America)*. [*Proceedings of the 20th International Geological Congress*.] International Geological Congress, Mexico City.

COCKS, L. R. M.; COOPER, R. A. 2004: Late Ordovician (Hirnantian) shelly fossils from New Zealand and their significance. *New Zealand Journal of Geology and Geophysics 47*: 71–80.

COOPER, R. A. 1979: Lower Paleozoic rocks of New Zealand. *Journal of the Royal Society of New Zealand 9*: 29–84.

COOPER, R. A. 1985: The Cambrian System in Australia, Antarctica and New Zealand. *In*: Shergold, J. H.; Jago, J. B.; Cooper, R. A.; Laurie, J. R. (eds), *International Union of Geological Sciences Publication 19*: 59–60.

COOPER, R. A.; BRADSHAW, M. A. 1986: Lower Paleozoic of Nelson-Westland. Excursion guide, Hornibrook Symposium, 1985. *Geological Society of New Zealand Miscellaneous Publication 33C*: 1–42.

COOPER, R. A.; SHERGOLD, J. H. 1991: Palaeozoic invertebrates of Antarctica. Pp. 455–481 *in*: Tingey, R.J. (ed.), *Geology of Antarctica*. Oxford University Press, Oxford.

FORTEY, R. 1997: Classification. Pp. 289–302 *in*: Whittington, H.B. *et al*. (eds), *Treatise on Invertebrate Paleontology, Part O. Volume 1, Trilobita, Revised*. Geological Society of America and University of Kansas, Boulder & Lawrence. 530 p.

HECTOR, J. 1876: On a new trilobite (*Homalonotus expansus*). *Transactions of the New Zealand Institute 9*: 602.

HENDERSON, R. A.; MacKINNON, D. I., 1981: New Cambrian inarticulate Brachiopoda from Australasia and the age of the Tasman Formation. *Alcheringa 5*: 289–309.

HUGHES, C. P.; WRIGHT, A. J. 1970: The trilobites *Incaia* Whittard, 1955 and *Anebolithus* gen. nov. *Palaeontology 13*: 677–690.

HYDEN, G.; BEGG, J. G.; CAMPBELL, H. J.; CAMPBELL, J. D. 1982: Permian fossils from the Countess Formation, Mossburn, Southland. *New Zealand Journal of Geology and Geophysics 25*: 101–108.

MÜNKER, C.; COOPER, R. A. 1999: The Cambrian arc complex of the Takaka Terrane, New Zealand: an integrated stratigraphical, palaeontological and geochemical approach. *New Zealand Journal of Geology and Geophysics 42*: 415–455.

PERCIVAL, I. G.; WRIGHT, A. J.; SIMES, J. E.; COOPER, R. A.; ZHEN, Y. Y. 2009: Middle Ordovician (Dariwillian) brachiopods and trilobites from Thompson Creek, northwest Nelson, New Zealand. *Memoirs of the Association of Australasian Paleontologists 37*: 611–639.

REED, F. R. C. 1926: New trilobites from the Ordovician Beds of New Zealand. *Transactions of the New Zealand Institute 57*: 310–314.

SANDFORD, A. C. 2005: Homalonotid trilobites from the Silurian and lower Devonian of south-eastern Australia and New Zealand (Arthropoda: Trilobita: Homalonotidae). *Memoirs of Museum Victoria 62*: 1–66.

SHIRLEY, J. 1938: The fauna of the Baton River Beds, New Zealand. *Quarterly Journal of the Geological Society of London 94*: 459–506.

WAISFELD, B. G.; VACCARI, N. E.; CHATTERTON, B. D. E.; EDGECOMBE, G. E. 2001: Systematics of Shumardiidae (Trilobita), with new species from the Ordovician of Argentina. *Journal of Paleontology 75*: 827–859.

WRIGHT, A. J. 1979: Evaluation of a New Zealand Tremadocian trilobite. *Geological Magazine 116*: 353–365.

WRIGHT, A. J. 1990: Acastid trilobites from the Baton Formation (Early Devonian), New Zealand. *New Zealand Journal of Geology and Geophysics 33*: 49–53.

WRIGHT, A. J.; GARRATT, M. J. 1991: *Notoconchidium*: species, age, distribution and affinities. Pp. 23–30 *in*: McKinnon, D. I.; Lee, D. E.; Campbell, J. D. (eds), *Brachiopods through Time*. [Proceedings of the 2nd International Brachiopod Congress.] A.A. Balkema, Rotterdam.

WRIGHT, A. J.; COOPER, R. A.; SIME, J. 1994: Cambrian and Ordovician faunas and stratigraphy, Mount Patriarch, New Zealand. *New Zealand Journal of Geology and Geophysics 37*: 437–476.

Checklist of New Zealand Trilobita

In the following list, taxonomic order of suprageneric taxa is as given by Fortey (1997: *Treatise on Invertebrate Paleontology, Part O, revised*). Age assignments, abbreviations, and localities for the relatively few trilobite-bearing levels are as follows: CB, Undillan to Boomerangian (Drumduan to Guzhangian), Epoch 3 of the Global Geochronological Scale (equivalent to late Middle Cambrian of the traditional scale), Cobb Valley; CP, latest Cambrian (Payntonian) and OT, earliest Ordovician (Tremadoc), Mount Patriarch; OD, Middle Ordovician, Thompson Creek; OG, early Late Ordovician (Gisbornian), Paturau River; OH, Hirnantian (latest Ordovician), Wangapeka River; DP, Early Devonian (Pragian), Baton River; DE, late Early Devonian (early Emsian), Reefton; and PL, Late Permian, Southland. E indicates an endemic species or genus.

PHYLUM ARTHROPODA
SUBPHYLUM TRILOBITOMORPHA
Class TRILOBITA
Order PTYCHOPARIIDA
Suborder PTYCHOPARIINA
ACROCEPHALITIDAE
Ketyna? sp. indet. Wright et al. 1994 CP-OT
CATILLICEPHALIDAE
Onchonotellus sp. indet. Wright et al. 1994 CP?
CONOKEPHALINIDAE
Suludella? sp. Cooper 1979 C
EULOMIDAE
Amzasskiella kupenga Wright, 1994 CP E
Proteuloma ahu Wright, 1994 CP E
NEPEIDAE
Nepea cf. *avara* Opik, 1970 Münker & Cooper 1999 CB
Nepea sp. Cooper & Bradshaw 1986 CB
PAPYRIASPIDIDAE
Pianaspis sp. Cooper 1979 CB
PROASAPHISCIDAE
Koptura sp. Cooper & Bradshaw 1986 CB (= kopturids of Cooper *in* Shergold et al. 1985)
Sudanomocarina sp. Münker & Cooper 1999 CB
SHUMARDIIDAE
Shumardia (Conophrys) tauwhena Wright, 1994 CP E
Shumardia (Conophrys) wrighti Waisfeld et al., 2001 OT E (= *S.* (*C.*) sp. indet. of Wright et al. 1994)
SOLENOPLEURIDAE
Hystricurus? sp. indet. Wright et al. 1994 OT
'Solenoparia' sp. Cooper 1979 CB

Suborder OLENINA
OLENIDAE
Leptoplastides grindleyi Wright, 1994 OT E
Parabolinella sp. indet. A Wright et al. 1994 OT
Parabolinella sp. indet. B Wright et al. 1994 OT
Plicatolina sp. indet. Wright et al. 1994 OT

Order AGNOSTIDA
AGNOSTIDAE
Lotagnostus (Lotagnostus) cf. *asiaticus* (Troedsson, 1937) Wright et al. 1994 CP
DIPLAGNOSTIDAE
Diplagnostus sp. Münker & Cooper 1999 CB
Neoagnostus sp. Wright et al. 1994 OT
Oidalagnostus n. sp. Cooper 1979 CB
Pseudagnostus (Pseudagnostus) sp. indet. Wright et al. 1994 CP
Tasagnostus sp. Cooper *in* Shergold et al. 1985 CB
METAGNOSTIDAE
Geragnostus? sp. indet. A Wright et al. 1994 CP
Geragnostus? sp. indet. B Wright et al. 1994 OT
PERONOPSIDAE
Hypagnostus of *clipeus* type Münker & Cooper 1999 CB
Hypagnostus of *parvifrons* type Münker & Cooper 1999 CB
Peronopsis of *elkedraensis-longiqua* type Münker & Cooper 1999 CB
PTYCHAGNOSTIDAE
Goniagnostus aculeatus Angelin, 1851 Cooper *in* Shergold et al. 1985 CB
Goniagnostus cf. *nathorsti* (Brögger, 1878) Cooper 1979 CB
INCERTAE SEDIS
Grandagnostus sp. Cooper in Shergold et al. 1985 CB
Valenagnostus sp. Münker & Cooper 1999 CB

Order ASAPHIDA
ALSATASPIDIDAE
Hupalopleura? sp. indet. Wright et al. 1994 CP-OT
Skjarella sp. indet. Wright et al. 1994 OT
ASAPHIDAE
cf. *Niobella* sp. indet. Wright et al. 1994 CP
Basiliella collingwoodensis (Reed, 1926) OG E
CERATOPYGIDAE
Hedinaspis regalis Troedsson, 1937 Wright et al. 1994 CP
Hysterolenus hectori (Reed, 1926) Wright et al. 1994 CP
Onychopyge aff. *riojana* Harrington, 1938 Wright et al. 1994 OT
cf. *Onychopyge* sp. indet. A Wright et al. 1994 CP
Onychopyge? sp. indet. B Wright et al. 1994 OT
Pianaspis sp. Cooper *in* Shergold et al. 1985 CB
Gen. et sp. indet. A Wright et al. 1994 OT
Gen. et sp. Indet. B Wright et al. 1994 CP
HUNGAIIDAE
Asaphopsoides sp. indet. Wright et al. 1994 CP
NILEIDAE
Ordosaspis? sp. indet. Wright et al. 1994 CP
REMOPLEURIDIDAE
Apatokephalops sp. Wright et al. 1994 OT
Apatokephalus cf. *tibicen* Pribyl & Vanék, 1980 Wright et al. 1994 OT
Apatokephalus? sp. indet. A Wright et al. 1994 OT
Apatokephalus? sp. indet. B Wright et al. 1994 OT
Eorobergia? sp. indet. Wright et al. 1994 OT
Kainella meridionalis Kobayashi, 1935 Wright et al. 1994 OT
Kainella cf. *conica* Kobayashi, 1935 Wright et al. 1994 OT
Pseudokainella lata (Kobayashi, 1935) Wright et al. 1994 CP
TRINUCLEIDAE
Incaia bishopi Hughes & Wright, 1970 OG E

Order CORYNEXOCHIDA
Suborder CORYNEXOCHINA
DOKIMOCEPHALIDAE
Wuhuia sp. indet. Wright et al. 1994 OT
DOLICHOMETOPIDAE
Gen. et sp. indet. Cooper 1979 CB
DORYPYGIDAE
Dorypyge sp. Cooper 1979 CB
Kootenia sp. Münker & Cooper 1999 CB
Olenoides sp. Cooper 1979 CB

Suborder ILLAENINA
PANDERIIDAE
Panderiidae? gen. et sp. indet. Cocks & Cooper 2004 OH
STYGINIDAE
Scutellum sp. (as *Goldius* sp.) Shirley 1938 DP

Order PHACOPIDA
Suborder CALYMENINA
CALYMENIDAE
Calymene (Gravicalymene) ?*angustior* Chapman, 1915 Shirley 1938 DP
HOMALONOTIDAE
Homalonotus sp. Shirley 1938 DP
Wenndorfia expanda (Hector, 1876) DE E

Suborder CHEIRURINA
PILEKIIDAE
Gogoella sp. indet. OD
cf. *Metapilekia bilirata* Harrington, 1938 Wright et al. 1994 OT
Gen. et sp. indet. A Wright et al. 1994 CP
Gen. et sp. indet. B Wright et al. 1994 CP
?Pilekiidae gen. et sp. indet. Wright et al. 1994 CP

Suborder PHACOPINA
ACASTIDAE
Acaste sp. indet. Wright 1990 DP
Acastella sp. indet. Wright 1990 DP
DALMANITIDAE
Mucronaspis mucronata (Brongniart, 1822) Cocks & Cooper 2004 OH
PHACOPIDAE
?*Phacops* sp. Willis 1965 DP

Order PROETIDA
AULACOPLEURIDAE?
Gen. et sp. indet. Cocks & Cooper 2004 OH
Dechenella (Eudechenella) mackayi Allan, 1935 DE E
PHILLIPSIIDAE
Gen. et sp. indet. A Begg 1981 PL
Gen. et sp. indet. B Hyden et al. 1982 PL

Order ODONTOPLEURIDA
ODONTOPLEURIDAE
Eoleonaspis? sp. Cocks & Cooper 2004 OH

SIX

Phylum ARTHROPODA

CHELICERATA

horseshoe crabs, arachnids, sea spiders

PHILIP J. SIRVID, ZHI-QIANG ZHANG, MARK S. HARVEY, BIRGIT E. RHODE, DAVID R. COOK, ILSE BARTSCH, DAVID A. STAPLES

Uliodon sp., an unidentified zoropsid spider from Palmerston North.

Alastair Robertson and Maria Minor, Massey University

Chelicerates comprise three distinctive groups of creatures – horseshoe crabs (class Xiphosura), spiders, mites, and kin (class Arachnida), and the sea spiders (class Pycnogonida). In contrast with hexapods, which have three body sections, chelicerates have only two. The head and thorax are fused into a single unit, the cephalothorax or prosoma (Latin *pro*, forward; Greek *soma*, body), and there is a hind-body or abdomen, which is usually referred to as the opisthosoma (Greek *opisthen*, behind). Chelicerates are the only arthropods to lack antennae and they have fangs or pincers (the first pair of appendages) instead of mandibles. They also differ from crustaceans and insects in having no compound eyes, only simple ones. Members of this ancient arthropod group also possess pedipalps, a second pair of appendages near the mouth. These vary in different groups but are usually used to manipulate food. The prosoma has six or more fused segments, with at least six pairs of appendages including four pairs of legs. The abdomen lacks legs but may contain such specialised features as book gills or book lungs (pages of respiratory flaps), a telson (a long spike for balance), spinnerets (in spiders and some other arachnids for spinning silk), or a sting (in scorpions), apparently a modified telson).

Chelicerates include in their number both the tiniest (mites) and largest (extinct eurypterids) of all Arthropoda. Eurypterid (sea scorpion) fossils have not yet been found in New Zealand. Nor are there fossil or living representatives of the Xiphosura.

Xiphosurans (the so-called horseshoe crabs) are the most primitive living chelicerates. They are often referred to as 'living fossils' because they have changed so little in appearance from the earliest forms in the Cambrian period. No one can say with certainty why they have been able, with no 'modern improvements', to survive in competition with more highly developed arthropods. Perhaps their success results in part from a combination of unobtrusive habits and a heavy hoodlike carapace that forms a protective roof over the body and limbs. As archaic forms they give evidence of what the ancestral aquatic arachnid may have been like, with five pairs of walking legs, book gills, and a long spike-like telson. Living horseshoe crabs (family Limulidae) live in shallow sandy coastal areas of the eastern seaboard of the Atlantic from Nova Scotia to the Gulf of Mexico (genus *Limulus*) or along Asian coasts from Japan and Korea south through the Philippines and Southeast Asia to India (*Tachypleus* and *Carcinoscorpius*).

Although New Zealand lacks living horseshoe crabs in its fauna, there have nevertheless been two records of accidental live introductions. In 1908, a biologist found on the stone facing of Calliope Dock, Auckland Harbour, a specimen of

Carcinoscorpius rotundicauda that had clearly been in the dock for some time (Chilton 1910). And a 31.5-centimetre male of *Limulus polyphemus* was found on low-tidal rocks at Great Barrier Island, Hauraki Gulf, in 1940 (Powell 1949). Astonishingly, in spite of their size, both of these specimens must have travelled to New Zealand attached to the hulls or lodged in some recesses of vessels (Cranfield et al. 1998).

Horseshoe 'crab' *Limulus polyphemus.*
After Tudge 2000

Class Arachnida

The class Arachnida includes the spiders, mites and ticks, harvestmen, pseudoscorpions and other lesser-known groups, and usually the scorpions although these may be more closely related to the extinct eurypterids. Nearly 100,000 living arachnid species have so far been discovered worldwide (Harvey 2002). The class takes its name from Arachne of Greek mythology, a girl turned into a spider for challenging the goddess Athena to a weaving competition. Fossils of most living arachnid orders are known from at least the Carboniferous, and scorpions are known even earlier, in the Silurian, while the earliest mites, spiders, and pseudoscorpions are Devonian in age. Apart from some which are secondarily adapted to aquatic life (especially among the mites), arachnids are terrestrial, and aspects of their body design reflect this – like the book lungs and tracheae, particular sense organs, and unique innovations like silk glands, found in spiders, pseudoscorpions, and some mites. Given the arachnophobia that many people experience, arachnids are probably the most objectionable group of arthropods as far as the layperson is concerned but this bias is largely unwarranted.

Summary of New Zealand chelicerate diversity

Taxon	Described species + subspecies	Known undescr./ undeterm. species	Estimated unknown species	Adventive species	Endemic species	Endemic genera and subgenera
Araneae*	1,153	547	940–2,040	38	1,628	137
Palpigradi	0	3?	3?	0?	0?	0
Acari	1,375+2	73	14,000	71	706	70
Opiliones	157+63	80	100	2	257+63	23+3
Pseudoscorpiones	67+5	0	50–100	3	59+5	11
Pycnogonida	90+1	1	35	0	43	1
Totals	**2,842+71**	**704**	**15,130–16,280**	**>114**	**~2,693**	**242+3**

* including 31 *nomina dubia* and 5 *incertae sedis*

Chelicerate diversity by environment†

Taxon	Marine	Freshwater	Terrestrial
Araneae	1	0	1,699
Palpigradi	0	0	3?
Acari	23+1	158	1,267+1*
Opiliones	0	0	237+63
Pseudoscorpiones	0	0	67+5
Pycnogonida	91+1	0	0
Totals	**115+2**	**158**	**3,273+69**

† species + subspecies
* includes parasitic forms

There is some controversy over the relationship of the various arachnid groups to one another, especially the scorpions and mites. Representatives of five of the traditional orders are found in New Zealand – the Acari (mites and ticks), Opiliones (harvestmen), Pseudoscorpiones (pseudoscorpions), Palpigradi (micro whip scorpions), and Araneae (spiders). Their classification in this chapter reflects schemes developed by Shultz (2007), Weygoldt (1998), and Wheeler and Hayashi (1998), based on a combination of morphological and molecular evidence. Nevertheless, it must be recognised that any classification scheme is imperfect, trying as it does to provide utility and practicality while reflecting evolutionary relationships. The scheme used here is somewhat of a compromise. For example, owing to their antiquity and diversity, the taxon Acari is sometimes treated as a subclass of Arachnida, especially by mite specialists. On the other hand, in the official terminology of the Arachnology Nomenclature Committee, Acari is an order. Here it is treated as an infraclass. In some schemes, mites and harvestmen (with tubular respiratory tracheae) and pseudoscorpions (with sieve tracheae), are linked with micro whip scorpions as Apulmonata, i.e. lacking book lungs. Another character that links mites and pseudoscorpions in the Apulmonata is the similarity in mouthparts, arranged in a kind of beak or rostrum. In some schemes, micro whip scorpions are closer to spiders. The spiders are linked with other orders in having book lungs, of which the first pair is large (Megoperculata) and the second pair (sometimes even the first pair) may in some groups be converted to tracheae. The jury is still out on arachnid relationships.

Araneae: Spiders

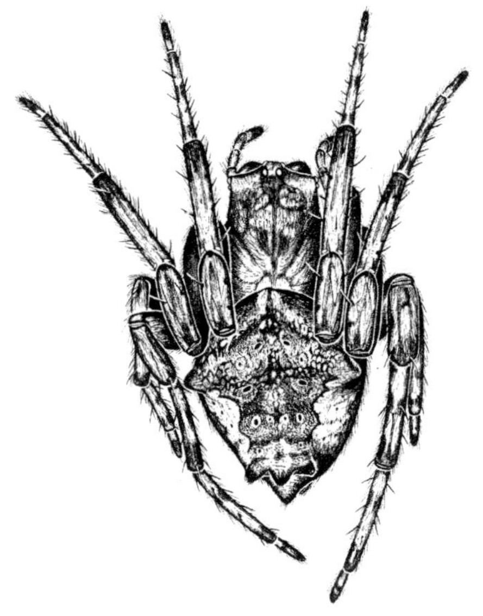

Orbweb spider *Eriophora pustulosa*.

From Forster & Forster 1973

'The spider is the dominant arachnid; it surpasses all others in the number and variety of its species, in the complexity of habits, and in the breadth of its range across the world. So well is it advertised by the beauty of the orb-web that all men know the spider, and for many it represents the whole class of Arachnida. The spider may be encountered in mythology, in history, in literature; its reputation is not unspotted and its merits are seldom recognized' (Savory 1977).

Spiders are predatory arachnids that occur naturally in all land areas with the exception of Antarctica. They can be found from the intertidal zone to the summits of the highest mountains. It would be a gross understatement to say that spiders are not held in the same esteem by the general public as whales, dolphins, and butterflies. Along with snakes, spiders are one of the most feared and loathed animal groups on the planet. This view is unfair, however, when one considers the role of spiders as beneficial organisms. They are frequently the dominant non-vertebrate predators in fields and forests (Platnick 1995) and thus have an important role in the regulation of insect numbers. For example, W. S. Bristowe (1958) estimated that the spider population of Great Britain annually consumes a weight of insects that exceeds that of the human population. Given a statistic like this, it is no surprise to learn that spiders are regarded as potential biological-control agents (Reichert & Lockley 1984).

Humans also have a history of benefiting from spiders more directly. Hillyard (1994) provides numerous examples, including accounts of New Guinea tribesmen making nets from *Nephila* webs and people using spiders medicinally, including as a remedy for constipation if 'eaten in *handfuls* on bread and butter'! Modern technology has allowed humans to begin making use of spiders in new ways. Spider silk is stronger than steel yet more elastic than nylon, and so it is being investigated as a superior replacement for Kevlar in bulletproof vests. Spider venoms are also being investigated for a number of purposes, including insecticides (Quicke 1988) and in neurobiological research (Jackson & Parks 1989). Finally, spiders are also used as biological indicators of pollutants (Clausen 1986; Hodge & Vink 2000) and other signs of environmental change (Yen 1995).

The behaviour and functional anatomy of these ubiquitous animals has received wide coverage. Interested readers are referred to Forster and Forster (1999), who gave an account in their comprehensive book on New Zealand spiders. More detailed information is given by Foelix (1996).

With the exception of members of the family Uloboridae (represented in New Zealand by *Waitkera waitakerensis*), spiders use venom to subdue their prey. While the venom of a small number of species is dangerous to humans, that of the vast majority is harmless. Many species are simply too small to bite humans while others either do not possess venom capable of harming a human, or if they do, simply cannot inject enough to produce any serious effects. In New Zealand the species with the most dangerous bites are almost certainly representatives of the genus *Latrodectus*, commonly known as the widow spiders. There are two *Latrodectus* species in New Zealand – the endemic *L. katipo* (katipo) and the introduced *L. hasseltii* (Australian redback). The latter is commonly intercepted by quarantine authorities, yet surprisingly is known to have established only in central Otago (Forster & Forster 1999). White-tail spiders (*Lampona cylindrata* and *L. murina*) are two long-established Australian species that have received much media attention because of the supposedly necrotic effects of their venom (Banks et al. 2004; Derraik et al. 2008). However, a study of 130 verified bite cases in Australia indicates that while a white-tail spider's bite is typically very painful, the resulting symptoms are otherwise minor (Isbister & Gray 2003). It appears other explanations should be considered before blaming white-tail spiders in cases of unexplained skin lesions or necrosis (White 2003). Alternative explanations for the effects attributed to their bites include inflammatory dermatoses, infections, inflammatory immune responses, and insect bites (Rademaker & Derraik 2009). Note that there is no evidence to show that white-tail bites are routinely accompanied by infection (Isbister & grey 2003) although any skin breakage is still a potential avenue for subsequent infection.

Waitkera waitakerensis.
From Forster & Forster 1973

If the use of venom amongst spiders is not quite universal, the use of silk certainly is. Spiders and their webs are inextricably linked in the minds of most of us. From single strands (*Phoroncidia*) (Forster & Forster 1999) to sheetwebs (*Cambridgea*), orbwebs (Araneidae and other families), and many others, the range of web forms is extensive. But a snare is not the only use for silk. Indeed, many groups of spiders (e.g. Salticidae, jumping spiders) construct no snare at all, yet silk still has an integral part to play in their lives. A spider starts its life as one of many eggs wrapped in a silken sac. As a juvenile, a spider might disperse to a new area carried by air currents on gossamer strands, although this is uncommon in New Zealand's predominantly forest-dwelling spider fauna (Forster 1975). Regardless of age, a spider trails a silken dragline as it moves about, providing a safety line should it fall, and some species, including many that do not construct prey-catching webs, will shelter for part of each day in a silk-lined lair.

History of New Zealand studies

New Zealand spiders were first described by Walckenaer (1837) in his *Histoire Naturelle des Insectes* (Forster 1967). He was followed by White (1849), who described several species collected during the 1840–43 *Erebus* and *Terror* expedition. In the 1870s, the New Zealand fauna received more substantial attention when L. Koch's monumental work *Die Arachniden Australiens* (Koch 1871–83) described some 35 species from New Zealand (Forster 1967). The most significant early New Zealand-based worker was undoubtedly A. T. Urquhart. He published about 200 descriptions of New Zealand spiders in the *Transactions and Proceedings of the New Zealand Institute* during the 1880s and 1890s (e.g. Urquhart 1885, 1888, 1892). Many of his names were, however, synonymised in revisions by later workers such as Bryant (1933, 1935). Other important contributions to the knowledge of New Zealand's spider diversity include those made by Pickard-Cambridge (1879), Goyen (1887, 1888), Hogg (1909, 1911), Dalmas (1917), Berland (1931), Chamberlain (1944, 1946), Parrott

Hexathele huttoni, Palmerston North.
Alastair Robertson and Maria Minor, Massey University

(1946), B. J. Marples (1956, 1962), and R. R. Marples (1959). This list is by no means comprehensive. Without doubt, New Zealand's foremost arachnologist was the late Ray Forster, who authored or coauthored the descriptions of approximately 700 new species from New Zealand alone. As well as numerous scientific papers, Forster and his wife Lyn published a number of popular articles and books, culminating in their 1999 volume, *Spiders of New Zealand and their Worldwide Kin.* This is likely to be the standard general reference work on New Zealand spiders for some years to come. More recently, an illustrated key to spider families has been published (Paquin et al. 2010).

New Zealand collections

All of New Zealand's major city museums, along with Landcare Research in Auckland and the Entomology Research Museum at Lincoln University, hold significant numbers of spider specimens, including numerous primary types. The largest collection is held at Otago Museum, the institutional base of the late Ray Forster. Much of the historic A. T. Urquhart collection survives in the Canterbury Museum and several of L. Koch's types are still available at the Zoologisches Museum, Hamburg.

Composition of New Zealand's spider fauna

Worldwide, some 41,000 species of spider have so far been described (Platnick 2009). Some 1700 species are known from New Zealand, but the total fauna is estimated to be somewhere in the range of 2500 (Forster & Forster 1999) to 3600 (Platnick 1992). The latter figure gives New Zealand a fauna similar in size to that of the continental United States (Platnick 1992). Regardless of which estimate one chooses to believe, the undescribed portion of the New Zealand spider fauna is clearly very substantial, particularly when one considers that many families – including very large families such as the Salticidae and Theridiidae – have never been formally revised. By way of comparison, the United Kingdom, an island nation of similar size and with a far better documented fauna, has about 700 described species. While 1700 species may seem to represent a substantial effort thus far, it should be noted that many species are known from only a handful of specimens from one or two localities – or worse, from individuals of one sex. While many species still await description in museum collections, sampling cannot be regarded as comprehensive and it is still easy to collect undescribed species in the field. Clearly, much work remains to be done. Not only is the basic task of naming and classifying New Zealand's fauna nowhere near complete, but we also remain ignorant of the true distributions of many species that have previously been described. Consequently, all work that is underpinned by basic taxonomy, such as bioprospecting and ecological surveys, is impeded by the lack of such basic knowledge.

Cave spider *Spelungula cavernicola*, Kahurangi National Park.
Andrew McLachlan

The Nelson cave spider *Spelungula cavernicola* is a noteworthy species from many standpoints. With a legspan of some 15 centimetres it is New Zealand's largest spider and one of only two protected native species (the other being the katipo). It is also a member of the family Gradungulidae, a group that has a number of characters that link the two infraorders of opisthotheline spiders, the Mygalomorphae and the Araneomorphae. Mygalomorphs are spiders with primitive characters such as paraxial chelicerae (the chelicerae are in line with the body and the spider must rear up and expose the fangs in order to bite) and four book lungs for respiration. Mygalomorphs include tarantulas (Theraphosidae) and their relatives (such as the New Zealand tunnelweb genera *Porrhothele* and *Hexathele*).

Araneomorph spiders are considered to be more highly evolved and make up the great majority of spider species. They typically have two book lungs and/or spiracles and have diaxial chelicerae (the fangs are orientated sideways and the spider does not have to rear up to bite). The Gradungulidae, with four book lungs and diaxial chelicerae are considered to be intermediate in characters. Gradungulids are known only from New Zealand and eastern Australia.

So far, 52 of the world's 109 spider families are known from New Zealand. Only one family, Huttoniidae is endemic. About 7% of the world's approximately 3800 spider genera are known from New Zealand, with 142 out of 252 recorded genera considered to be endemic. Of New Zealand's known species, 1628 are considered endemic, 27 are considered native but not endemic (with many naturally occurring in both Australia and New Zealand), and 37 are regarded as exotic. The New Zealand fauna is characterised by a high degree of endemism at the species level. As will be seen in the end-chapter checklist, which includes recent updates on some taxa (e.g. Vink 2002; Zabka & Pollard 2002; Fitzgerald & Sirvid, 2009), the endemic component of New Zealand's spider fauna is approximately 95%. A frequently updated list of all spider species can be found online at the World Spider Catalog (Platnick 2009).

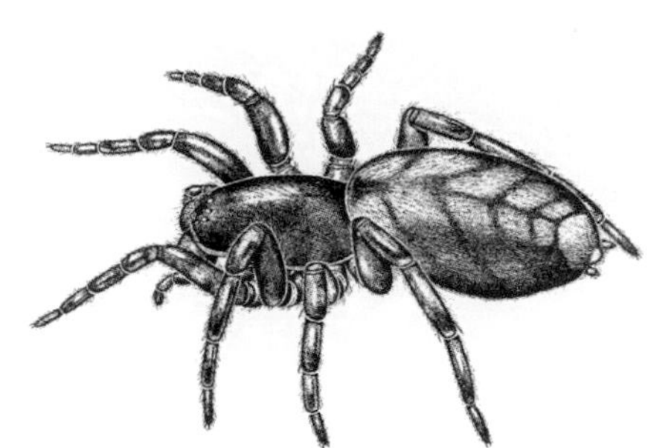

Huttonia palpimanoides.
From Forster & Forster 1973

While some introduced species are cosmopolitan (e.g. *Pholcus phalangioides*), the majority of them are Australian (e.g. *Supunna picta, Lampona cylindrata*). Some may have established here after ballooning over the Tasman Sea e.g. *Oxyopes gracilipes* (Vink & Sirvid 1998, 2000), but the majority of exotic species have probably arrived in New Zealand by stowing away with humans or their cargo, because most of them appear to be associated with modified environments.

New Zealand's faunal connections with other parts of the world are by no means confined to recent invasions by exotic species. The relationships may be far older and the presence of a number of families confined to the Southern Hemisphere suggests Gondwanan linkages. For example, Orsolobidae are confined to Chile, Australia, and New Zealand (Forster & Platnick 1985) as is the Synotaxidae (Forster et al. 1990). The Gondwanan model has not gone unchallenged, however. Paleontological evidence indicates that the modern austral distributions of some groups are relictual, with northern hemisphere members of these groups going extinct (Eskov 1990). Molecular analysis may eventually reveal much about the evolutionary history of New Zealand spiders but is still at an early stage. Work on Lycosidae (wolf spiders) indicates the genus *Anoteropsis* probably arrived in New Zealand after the breakup of Gondwana and subsequently radiated during the past five million years (Vink & Paterson 2003), suggesting that dispersal may be influential in shaping our spider fauna. However, similar studies on families postulated as Gondwanan (e.g. Orsolobidae) are yet to be conducted.

Palpigradi: Micro whip scorpions

These minute eyeless arachnids with whip-like antenniform tails were not known from New Zealand until Forster and Forster (1999) recorded their presence. They alluded to several species and noted that one is undescribed but were not certain whether the other species are native or exotic (possibly arriving with other soil animals from Europe many years ago). Unfortunately, the authors did not specify the number of genera or species they had found and both are now deceased.

Palpigrade *Eukoenenia austriaca* (non-New Zealand).
Rollin Verlinde

The 80-odd palpigrade species known worldwide belong to two families – the Eukoeneniidae and Prokoeneniidae. None exceeds three millimetres in length. They live beneath stones, in soil and litter, and in caves where humidity is high. Their biology is very little known. The carapace of the cephalothorax is divided into two plates. Curiously, the undifferentiated pedipalps are used as walking legs (hence the name Palpigradi – palp walkers) whereas the first pair of walking legs is held above the ground during walking.

Acari: Mites and ticks

Acari are everywhere – they live in all major terrestrial and aquatic habitats, including the deep ocean where their rivals, the insects, have failed to invade (Krantz 1978; Lindquist 1994; Walter & Proctor 1999). They have invaded our houses – the dust mite *Dermatophagoides pteronyssinus* – and you have them in

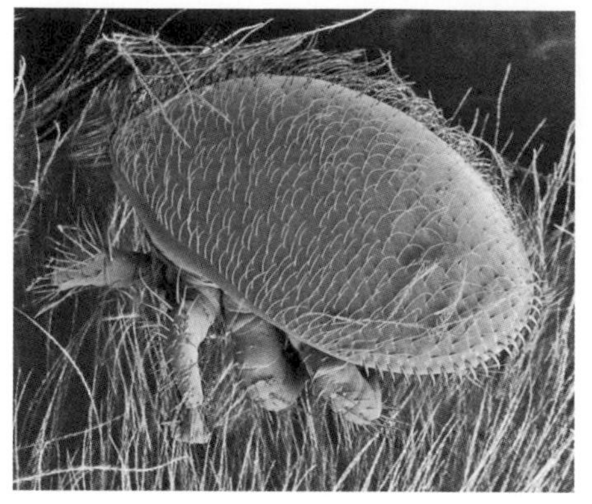

Varroa destructor on a honey bee.

Eric Erbe and Chris Pooley, Electron Microscope Laboratory Agricultural Research Service, USDA

your face, in the hair follicles of your eyebrows (*Demodex folliculorum*). Although mites are small, they are often visibly encountered – like the tiny bright red creatures (family Anystidae) commonly seen moving rapidly on concrete paths (Ramsay 1975). Although nearly 50,000 species have been described worldwide, this hyper-diverse group of minute creatures is unfortunately far less known than the majority of arthropods. The known species of Acari probably represent only a small fraction of the total – estimates range as high as half a million to one million species worldwide (Halliday *et al.* 1999; Walter & Proctor 1999).

The minute size of mites is conducive to their diversity, but unfortunately is also a deterrent to the advancement of their study. Most species are less than a millimetre long. Males of a mite parasitic on bees (*Acarapis woodi*), for example, are less than a tenth of a millimetre long. At the other end of the scale, the males of the tick *Ornithodoros acinus* may grow to three centimetres in length after a full blood meal. An average mite, however, is about half a millimetre in length.

Mites have a prodigious range of food preferences and habits, as predators, ectoparasites and animal hitchhikers, plant-eaters and fungal-feeders, and those species attracted to decaying plant material, dung, and dead animal bodies. Not surprisingly, in spite of their small size, mites are important ecologically and economically as pests of crops, food, and horticultural plants, natural enemies of pests, parasites of domestic animals and humans, nutrient recyclers in the food chain, and vectors and agents of diseases. Their economic importance can be seen from a single species, the honey bee parasitic mite *Varroa destructor*, relatively recently described by Anderson and Trueman (2000). Since its discovery in New Zealand it has cost the New Zealand bee industry an estimated $400–900 million (Zhang 2000).

Mites have four pairs of legs in the adult stage and a body in which the segments have been more or less completely fused so that there is no separation between head and abdomen. Body parts are variously modified, however, depending on a mite's life-style. For example, legs and claws are modified for gripping fur and feathers in some parasitic species, the claws may become suckers, and suckers may even be developed on the body. The jaws may be changed from a biting, chewing type into stylets able to penetrate plant or animal tissues. The body may become wormlike and the legs reduced to two. Most mites lay eggs that hatch into a six-legged larva but the eggs of some ticks may develop into adults within a greatly swollen mother.

Ectoparasites and epizoites

Ticks are among the better-known Acari because of their relatively large size and because they attack humans and domestic animals (Ramsay 1975). The cattle tick *Haemaphysalis bispinosa*, which also attacks other animals, is the only economically significant tick pest in New Zealand and is present over much of the North Island. A few species of ticks occur on seabirds such as gulls, shags, penguins, and petrels and are found mainly in nesting areas. Another tick is found on native land birds including ducks, kiwis, and wekas, one occurs on native bats, and one is restricted to the tuatara. The elongated mites that are specialised for living in hair follicles (see Desch 1989) can be quite harmless but others can cause mange and similar skin disorders in domestic pets. Some, like the scabies mite *Sarcoptes scabiei* that infects humans (and is also found on hedgehogs – Gorton et al. 1999), burrow into the skin. Other parasitic mites live in the ears of cats and dogs. These and other such species are cosmopolitan. Birds have an interesting fauna of parasitic mites, including at least one species of feather mite for each species of bird. Others are blood feeders, like the poultry mite *Dermanyssus gallinae*, which sometimes develops vast populations during nesting. After the nests are abandoned these mites will travel widely for food and can invade houses where there are birds nesting in the walls or roof. Chiggers (family Trombiculidae) are relatively unknown in New Zealand but have been found on several reptiles, kiwi, and a rat species. Other mites live beneath the

Scabies mite, *Sarcoptes scabiei*.

Natural History Museum, London

wing covers of native beetles and on millipedes and wetas and their mouthparts may be permanently embedded in the host. Some are internal parasites, living in the nasal cavities of marine mammals and air sacs of birds, and even in the respiratory tracheae of insects (Ramsay 1975). Not all mites attached to other arthropods are parasitic – some merely hitchhike to a new location or food source.

Commercial pests

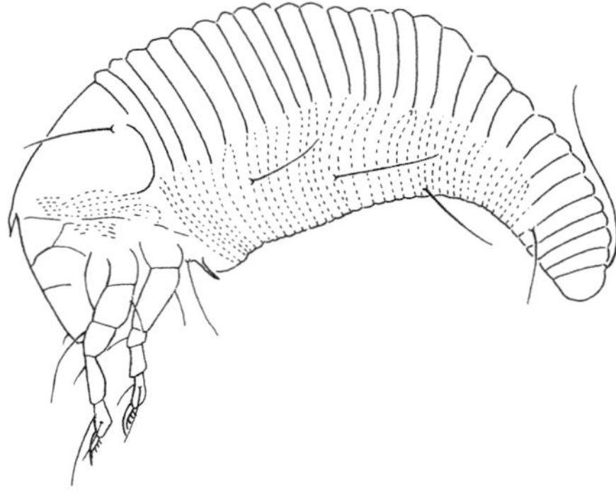

Phyllocoptes metrosideri, an eriophyid mite of pohutukawa.
From Manson 1989

Plant-feeding mites have considerable economic importance, particularly as horticultural pests. Ramsay (1975) briefly reviewed the New Zealand species, which include a number of eriophyoid gall mites. Eriophyoids are the creatures that cause a variety of changes to plants (gall, erineum, blister, rust, etc.), and economic pests include the tomato russet mite *Aculops lycopersici*), pear leaf blister mite *Eriophyes pyri*, blackcurrant bud mite *Cecidophyopsis ribis*, walnut erineum mite *Aceria erinea*, and grape erineum mite *Colomerus vitis*, among others. Another gall mite, blackberry mite *Acalitus essigi*, causes raspberry disease in blackberries, resulting in small, hard, red fruits. Spider mites are also very important plant pests. The European red mite *Panonychus ulmi* is a major pest of apple and other fruit trees, affecting leaves and sometimes fruit. Clover mite *Bryobia cristata* likewise damages leaves, and two-spotted spider mite *Tetranychus urticae* affects strawberries and other plants. A few mite species of cosmopolitan distribution cause considerable damage to stored grains and cereal products (Fan & Zhang 2004, 2007). They are spread through a non-feeding nymph stage (hypopus) specially adapted for dispersal, aided by long body hairs. Their numbers can build up tremendously in damp cupboards, furniture padding, animal bedding, and birds' nests. Two pasture species of oribatid mite are vectors of the sheep tapeworm *Moniezia expansa*, owing to their ability to ingest tapeworm eggs (Ramsay 1975).

Aquatic mites

One successful group is the water mites and a discussion of these will serve to highlight other aspects of the New Zealand mite fauna, such as biogeography, endemism, and scientifically interesting features. New Zealand has 158 named freshwater species in 59 genera and 21 families, and unnamed species of several other genera have been collected. The groundwater and substream interstitial faunas are particularly diverse and include 34 endemic genera. Many freshwater groups are very ancient, having originated in the Mesozoic, but other groups have probably recently dispersed into New Zealand on insect hosts. Some members of the orders Astigmata, Mesostigmata, and Oribatida occur in fresh water but the most speciose group is the Hydrachnidia (sometimes referred to as the Hydrachnellae and Hydrachnida) with more than 5000 species worldwide (Harvey 1998). Water mites occur in virtually all freshwater environments, and even cursory collections from New Zealand lakes and streams yield specimens of more than one species.

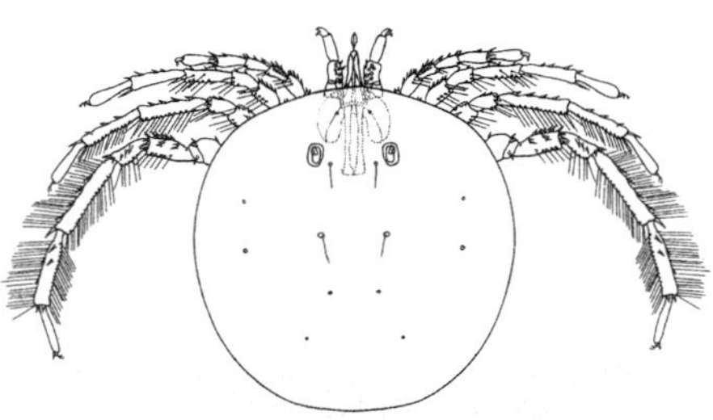

Water mite *Hydrarachna maramauensis* (female).
After Stout 1976

Water mites make up a subgroup of the Parasitengona, a large assemblage of mites that, as their name suggests, typically possess a parasitic stage during their life-cycle, in this case a parasitic larva that attaches to another animal for a blood meal prior to the nymphal stage. The water-mite life-cycle is limited in comparison with other arachnids, with only three active stages – the larva, deutonymph, and adult. The six-legged larva of most species attaches to an adult insect, engorging a meal of blood and other body fluids before detaching itself back into the water. In a few species the females deposit eggs with such adequacy of food reserves that the parasitic stage may be bypassed. In these cases the larva may metamorphose while still in the egg and emerge as a deutonymph or as a larva that metamorphoses without feeding. Water mites rarely attain a body length greater than five millimetres, and the adults of numerous species are less than a millimetre in length. Although most

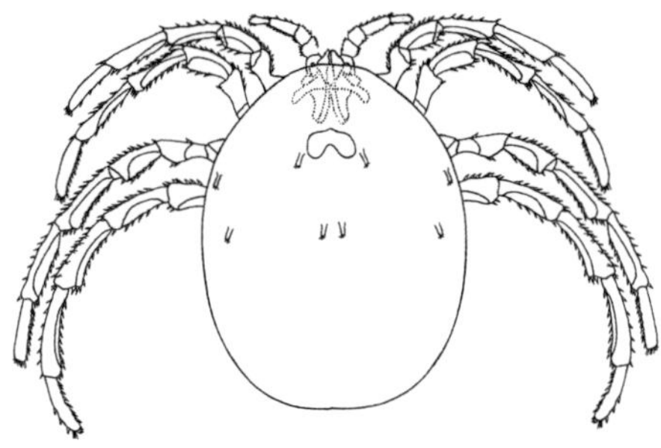

Water mite *Eylais schauinslandi* (female).
After Stout 1976

water mites are either red or yellow, many are green, blue, purple, or brown. Interstitial species are usually devoid of colour and their eyes are often reduced in size or entirely absent.

The early discovery and description of water mites from New Zealand was based upon stream species belonging to cosmopolitan genera. The first was the eylaid *Eylais schauinslandi* from D'Urville Island (Koenike 1900), followed nearly half a century later by *Piona uncata exigua* from an unspecified locality (Viets 1949). Shortly after, Stout (1953a, b) described three additional species, giving significant information about their life history and anatomy, and Womersley (1953) described a species of *Panisopsis* (later transferred to *Wandesia*) from several larvae attached to the apterous stonefly *Apteryoperla monticola*. Hopkins (1966a,b, 1967, 1969, 1975) and Hopkins and Schminke (1970) began the description of the rich mite fauna of New Zealand lakes, recording 18 new species in a variety of families, and Imamura (1977, 1978, 1979) added four species from cave waters. Cook (1983, 1991, 1992) and Cook and Hopkins (1998) described 97 new species, mostly from lake and interstitial habitats, redescribed many of the older taxa, and consolidated our understanding of the relationships of the New Zealand fauna with that of other Gondwanan regions. Schwoerbel (1984), Smit (1996a, b, 2002), and Pesic et al. (2010) have described additional species and Crowell (1990) discovered the only New Zealand species of Unionicolidae taken in association with a freshwater sponge.

New Zealand's freshwater mites are highly endemic at the species level, the sole non-endemic, *Limnochares australica*, also occurring in Australia. Of the 59 genera currently recorded from New Zealand, 35 genera (or 59%) are endemic. This total includes 20 endemic genera of Aturidae and six genera of Hygrobatidae. A further nine genera occur in Australia, South America, or southern Africa, thus highlighting the strong Gondwanan links of the New Zealand fauna. The remaining 12 genera are generally widespread, seven being cosmopolitan and five having distinctive bipolar distributions, i.e. in extreme portions of both hemispheres but lacking in the tropics. These genera are from ancient Pangaean groups that have changed little since the Mesozoic and which are mostly incapable of any form of long-distance dispersal. Although the New Zealand fauna shares many strong links with that of Australia, there are several large families in which there is no overlap in genera (Harvey 1998). Of particular interest is the aturid subfamily Notoaturinae, which is abundant in many Gondwanan regions, where it has diversified into a plethora of genera including 12 in New Zealand, eight in Australia, three in South America, and three in southern and eastern Africa. Several family-rank taxa are currently thought to be endemic to New Zealand, including the families Zelandothyadidae and Stygotoniidae, the aturid subfamily Zelandopsinae, and the pionid subfamily Schminkeinae.

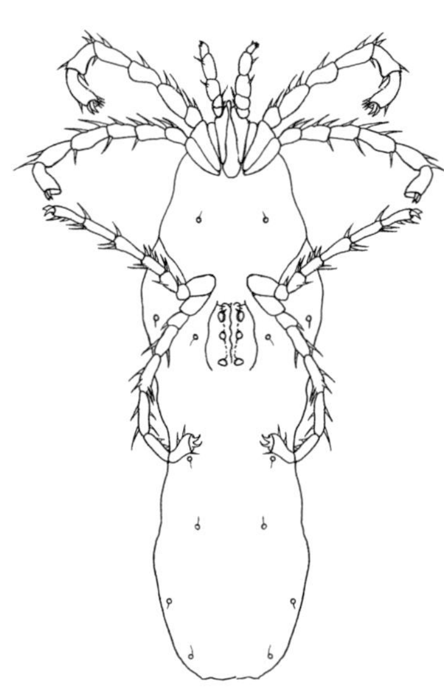

Euwandesia tenebrio from running water (male, ventral).
After Hopkins & Schminke 1970

Some other peculiar distributions emphasise the importance of the New Zealand fauna to an understanding of the world biogeography of the group, especially as each group has not yet been found in Australia, although this may simply reflect a lack of collecting in suitable habitats (Harvey 1998). The family Ctenothyadidae, previously based on a single species from Java, has recently been found in New Zealand, with two described species in the endemic genus *Stellulathyas* (Cook & Hopkins 1998). The sole New Zealand member of the Sperchontidae, *Apeltosperchon zelandicus* (Cook 1983), belongs to a subfamily and genus otherwise known only from South America (Besch 1964; Cook 1980). The presence of New Zealand species of Nudomideopsidae (e.g. Cook 1992; Cook & Hopkins 1998) is highly unusual, as the family is elsewhere restricted to the Holarctic region (Smith 1990).

Overall, the high degree of endemism at both genus (60%) and species levels is of extreme biogeographic interest, emphasising the poor dispersal powers of these small, often overlooked, arachnids. The total size of the New Zealand fauna is difficult to estimate without additional collecting and taxonomic work, but several new species remain undescribed and many more species and genera await discovery. The diverse and distinct hyporheic fauna (i.e. that found in the interface region of soil or sediment where stream water mixes with ground water) is only barely sampled, with new discoveries awaiting anybody with a shovel and a net. Only Stout (1953a,b) has provided details on the life history of some New Zealand species. Apart from a consideration of their intercontinental relationships (e.g. Cook 1984), the biogeography of the New Zealand water mites has been largely unanalysed and no distribution maps have been published for any species. Water mites are virtually unknown from the various New Zealand offshore islands, and the only published record is that of *Arrenurus rotoensis* from Chatham Island (Smit 1996a).

The marine-mite family Halacaridae in New Zealand is slowly becoming documented (Bartsch 1979, 1986a, b, 2007). So far, 23 species and one subspecies have been found in the sea, and four other species have been discovered in subterranean fresh water. These tiny (0.2–1.0 millimetre-long) creatures occur as free-living predators or as commensals on other marine organisms (Bartsch 1996). More than 250 species are likely in New Zealand seas.

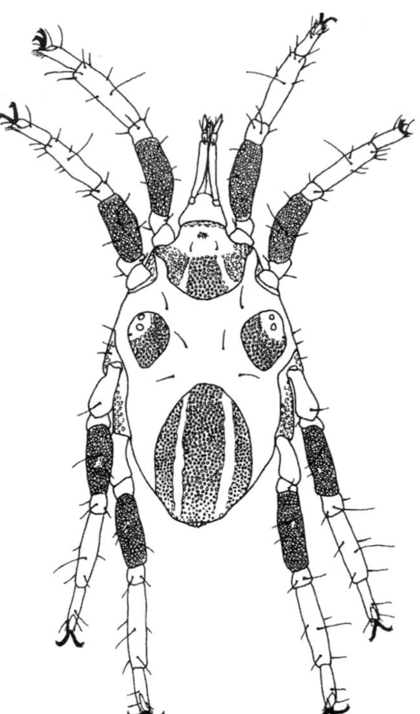

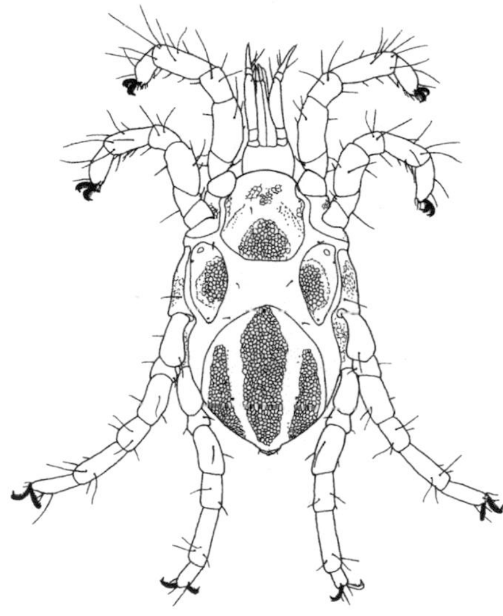

Marine (halacarid) mites *Agaue parva* (upper) and *Halacarellus lubricus* (lower).

Ilse Bartsch

Classification

The ordinal-level classification of Acari is not settled (Krantz 1978; Johnston 1982; Lindquist 1984; Evans 1992; Walter & Proctor 1999). Evans (1992) recognised seven orders arranged in three superorders – Opilioacariformes (Opilioacarida), Parasitiformes (Holothyrida, Mesostigmata, Ixodida), and Acariformes (Prostigmata, Oribatida, Astigmata). Walter and Proctor (1999) treated the superorders of Evans (1992) as orders and his orders as suborders. The Astigmata was shown to be a group within the Oribatida (Norton 1998); this view has gained support (Walter & Proctor 1999) and was adopted in a new classification in Krantz and Walter's (2009) third edition of *A Manual of Acarology* in which the Opilioacarida is considered as an order of Parasitiformes. We follow Krantz and Walter in recognising Astigmata as a part of Oribatida but we discuss this diverse group in a separate section for ease of comparison with previous works, e.g. Spain and Luxton (1971). Five orders of Acari are now recognised as present in New Zealand. The Opilioacarida is a small order of about 20 species distributed in Africa, Asia, Australia, Central America, and Europe, but it has not yet been recorded from New Zealand.

The end-chapter checklist of Acari updates the last one given by Spain and Luxton (1971) and the order-level classification follows that in Krantz and Walter (2009). The *Zoological Record* has been scanned for mites recorded from New Zealand since 1971. Many original papers have been consulted and The Acarine Reprint Collection (more than 10,000 papers/books) in the New Zealand Arthropod Collection (NZAC) at Landcare Research was used extensively for compiling the list. It should be noted that only names in published literature up to 2009 are included. Recognised species with specimens in NZAC or any other collection that have no published record are not included. Many species in the old literature are now placed in different genera or families. In the new checklist, all species are placed in the currently appropriate genera and families based on recent monographs, revisions, reviews, or catalogues of various taxa. Efforts were made to check the placement and spelling of names using original papers and current monographs/revisions, but some original papers were not available to the authors.

Diversity of the New Zealand mite fauna

A total of 1447 mite species in 614 genera is now reported from New Zealand, belonging to 197 families in five orders. A few have been recorded as Holocene subfossils (Ramsay 1960). The distribution of species richness of the New Zealand mite fauna is quite uneven among the orders, however – Trombidiformes (exclusively Prostigmata) 660 species, Sarcoptiformes 598 (Endeostigmata 5, Oribatida 593 including 184 species of Astigmata), Holothyrida 1, Ixodida 11, and Mesostigmata 178 species. In comparison, Spain and Luxton (1971) listed 47 Astigmata, 380 Oribatida (Cryptostigmata), 271 Prostigmata, 1 Holothyrida (Notostigmata), 12 Ixodida (Metastigmata), and 121 Mesostigmata. The last four decades saw relatively more growth in knowledge of the diversity of New Zealand Trombidiformes and Sarcoptiformes.

The distribution of taxon richness among families is also very patchy, with few large families but many small ones. On average, there are about seven species per family. Of the 188 families, 141 or three-quarters of them have seven or fewer species. The 10 largest families are Eriophyidae (126 species), Oppiidae (89), Aturidae (68), Stigmaeidae (65), Phytoseiidae (35), Tetranychidae (35), Ologamasidae (29), Acaridae (28), Halacaridae (28) and Tenuipalpidae (27). These 10 families alone account for 38% of all the species. The sizes of the Eriophyidae, Tetranychidae, Tenuipalpidae, Phytoseiidae, and Stigmaeidae are relatively large. This may not be simply because they are more diverse than other families, but could result from more studies on these mites owing to their economic importance as pests or natural enemies of pests.

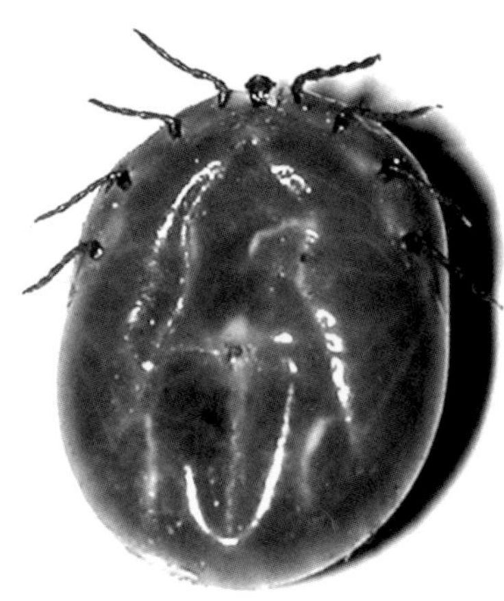

Engorged female cattle tick *Haemaphysalis longicornis* (ventral).
Wikimedia Commons

Superorder Parasitiformes

Order Holothyrida

Holothyrids are large (2–7 millimetres) predators inhabiting litter and soil in the Southern Hemisphere. Some 30 described species are placed in nine genera and three families, but some 160–320 species are estimated globally (Walter & Proctor 1999). Only a single species, *Allothyrus australasiae* (Womersley 1935), has been formally recorded from New Zealand. Clark (1997), however, claimed that this species 'is most certainly absent from New Zealand' but nevertheless recognised many other unnamed species in this country. Numerous unnamed species are also recognised by Zhi-Qiang Zhang in NZAC.

Order Ixodida

Known commonly as ticks, these are large parasites of vertebrates. Some 880 species are known in the world, distributed in 12 genera and three families (Walter & Proctor 1999). The New Zealand tick fauna includes 11 species in five genera and two families. The relatively low number of species in New Zealand is probably related to the low diversity of natural mammalian hosts.

Order Mesostigmata

These medium-sized (0.2–1.2 millimetres) free-living predators occur in a variety of habitats including on parasites (or associates) of vertebrates and invertebrates. Many species are well sclerotised, with a brownish dorsal shields. Some 11,615 described species are placed in about 558 genera and 72 families (Walter & Proctor 1999). The New Zealand fauna includes 178 species in 91 genera and 24 families. Since Spain and Luxton (1971), studies have been published on the Digamasellidae (Luxton 1984, 1989), Macrochelidae (Emberson 1973), Macronyssidae (Domrow et al. 1980), Ologamasidae (Lee & Hunter 1974), and Phytoseiidae (Collyer 1982). Many families of New Zealand Mesostigmata remain to be systematically examined.

Superorder Acariformes

Order Trombidiformes

This order consists of two suborders – Sphaerolichida (not yet recorded from New Zealand) and Prostigmata. This is the most diverse mite suborder, exhibiting great variation in body length (0.1–16 millimetres). Species are found across the wide range of habitats occupied by all the above orders as predators, parasites, phytophages, saprophytes, and fungivores. Some 17,170 described species globally are placed in 1348 genera and 131 families (Walter & Proctor 1999). The New Zealand fauna as reviewed by Qin and Henderson (2001) included 539 species up to 1996; currently, 659 species in 235 genera and 64 families are recognised. Recent major contributions to the New Zealand fauna include those on water mites (Cook 1983, 1991, 1992; Hopkins 1975; Imamura 1977, 1978; Schwoerbel 1984; Smit 1996a, b; 2002; Pesic et al. 2010) Eriophyoidea (Manson 1984a, 1989; Manson & Gerson 1986; Xue & Zhang 2008), Demodecidae (Desch 1989), Cunaxidae (Smiley 1992), Erythraeidae (Southcott 1988), Allotanaupodidae (Zhang & Fan 2007), Halacaridae (Bartsch 1979, 1986a; Newell 1984), Siteroptidae (Martin 1978), Tenuipalpidae (Collyer 1973a, b, c), Eupodoidea (Qin 1988a, b), and Raphignathoidea (Fan & Zhang 2005).

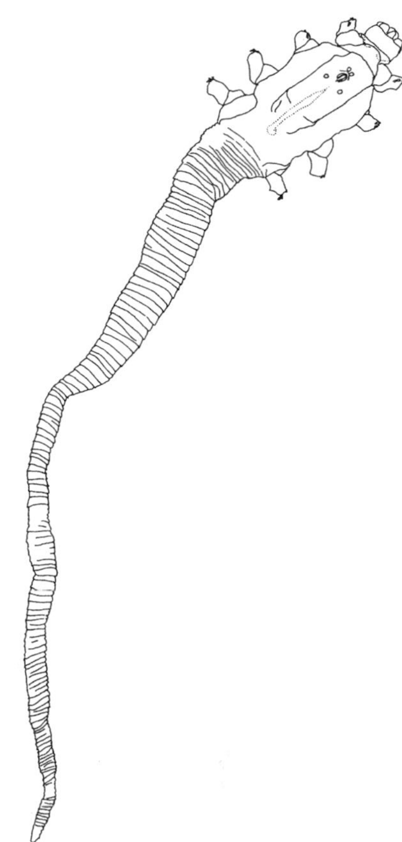
Hair-follicle mite *Demodex mystacina* from short-tailed bat.
From Desch 1989

Order Sarcoptiformes
This order consists of two suborders – Endeostigmata and Oribatida. The Endeostigmata was previously included in Prostigmata. New Zealand has five species in four genera placed in three families. Many more species remain undescribed. Suborder Oribatida now includes both Oribatida in its classical sense (beetle mites) and Astigmata, which in the past was a separate order/suborder.

Oribatida except Astigmata. 'Beetle mites' are medium-sized (0.2–1.2 millimetres), often well sclerotised, and very common in soil and litter. They feed mostly on fungi and decaying plants and are important for litter decomposition and soil formation. Some 11,000 described species globally are placed in about 1100 genera and 150 families (Walter & Proctor 1999). The New Zealand fauna includes 409 species in 172 genera and 67 families, most of which (366 species) were summarised in *Fauna of New Zealand* volume seven by Luxton (1984). Major contributions were made by Hammer (1966, 1967, 1968). Recent studies include those by Niedbała (1993, 2006), Niedbała and Colloff (1997), and Colloff and Cameron (2009).

Astigmata. These are weakly (or sometimes well-) sclerotised, medium-sized (0.2–1.2 millimetre-long) mites found in a diverse range of habitats. Members of the 'Acaridia' are free-living or parasites, associated with insects or crustaceans, and only rarely parasites of mammals, whereas those of the 'Psoroptidia' are parasites of birds (e.g. feather mites) and mammals, rarely of insects or free-living. Some 4500 described species are placed in 627 genera and 70 families (Walter & Proctor 1999). The New Zealand fauna comprises 184 recorded species in 107 genera and 36 families. Sporadic contributions (with descriptions of a few species) since the last checklist (Spain & Luxton 1971) have been made on *Rhizoglyphus* (Acaridae) (Manson 1972), the Atopomelidae (Fain 1972; Fain & Domrow 1974; Domrow 1992), Avenzoariidae (Mironov 1990, 1994; Mironov & Dabert 1997), Echimyopodidae, Glycyphagidae, Histiosomatidae, and Winterschmidtiidae (Fain & Galloway 1993), and Hyadesiidae (Luxton 1984, 1989). Two recent major contributions are those of Fan & Zhang (2004) on *Rhizoglyphus* and Fan & Zhang (2007) on *Tyrophagus*.

Endemism
It is difficult to estimate the degree of endemism for the New Zealand mite fauna as a whole because most groups have not yet been studied systematically here, in Australia, or elsewhere. On the other hand, for most of the groups that have been well studied, endemism is very high. One-third of New Zealand oribatid genera and 82% of the species are endemic (Hammer & Wallwork 1979; Hammer 1968). Fifty-nine percent of the 59 New Zealand water-mite genera are found nowhere else, and all species save one are endemic. Some 89% of the 65 New Zealand stigmaeid species are endemic. The degree of endemism for New Zealand spider mites, however, is relatively low (29%), possibly because of the relatively high rate of dispersal of these phytophagous mites with economic plants around the world and the lack of systematic collection on native plants.

Gaps in knowledge
The ixodid (tick) fauna of New Zealand and the rest of the world is well known,

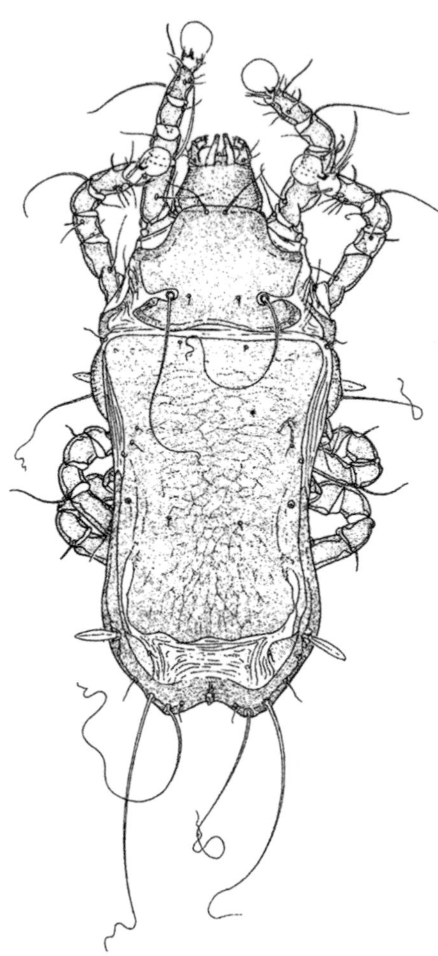

Feather mite *Coraciacarus muellermotzfeldi* from extinct huia *Heteralocha acutirostris*.
After Dabert & Alberti 2008

with most species already described. The oribatid fauna of New Zealand is relatively well studied thanks to Hammer (1966, 1967, 1968) and others (e.g. Luxton 1985) but many new species are anticipated. The New Zealand fauna of Holothyrida is rich but virtually unstudied. Most families of New Zealand Mesostigmata and Astigmata await systematic study. Most families of the most diverse suborder, Prostigmata, have many more species to be described, including water mites and eriophyoid mites, which have recently been revised (Cook 1983; Manson 1984a, b).

It is difficult to evaluate the total size of the New Zealand mite fauna without further systematic collecting and taxonomic research. Halliday et al. (1999) calculated the size of the mite fauna of Australia, North America, and Great Britain (plus Ireland) using a multiplying factor estimated from the average ratio between the number of species after and before revisions of selected taxa. Using the same approach, the multiplying factor for New Zealand mite fauna is 11.2 (ranging from 2.9 to 44.6). It may be expected, therefore, that the New Zealand mite fauna could be ca. 11.2 x 1429 = 16,005 species. Halliday et al. (1999) assessed the multiplying factor for the mite fauna of Australia as 2.9, of Great Britain and Ireland as 1.5, and of North America as 3.7, with the estimated sizes of the total fauna to be 20,000 for Australia, 2590 for Great Britain and Ireland, and 29,800 for North America. The very high multiplying factor for the New Zealand mite fauna owes most to the excellent work of Hammer (1966, 1967, 1968) who increased the oribatid fauna of New Zealand 44-fold.

Another way of looking at the size of the New Zealand mite fauna is by comparing it with that of the world. Halliday et al. (1999) calculated that 48,200 species were described in the world by 1997 and some 788 new species have been described on average every year since 1978. Thus there are now almost 50,000 described species worldwide, or about 5–10 % of the half to one million species estimated to be present (Halliday 1999; Walter & Proctor 1999). The described New Zealand fauna is about 2.5% of the global total. If this rate holds true for undescribed species as well, then we could expect some 12,500–25,000 species present in New Zealand. The estimate of 14,000 adopted in this review is within this range and may be considered conservative.

Opiliones: Harvestmen

Described by T. H. Savory (1977) as 'ludicrous arachnids with two eyes perched on a small body, often bizarre in form, and supported by legs too long for convenience and somewhat insecurely attached', harvestmen are wonderfully diverse and exotic-looking creatures. Important morphological characters include a compact, undivided body (usually with some segmentation), not more than two eyes (typically on an eyemound), chelate chelicerae, and, while lacking the means to produce poison or silk, they do possess repugnatorial glands (Forster & Forster 1999). While primarily scavengers, they are known to catch and eat small insects. The most up-to-date published source of information about harvestmen for the interested reader is that of Pinto da Rocha et al. (2008).

While Opiliones are found all over the world, the New Zealand fauna has its strongest relationships with other Gondwanan regions, particularly southern Africa and Australia. For example, the family Triaenonychidae can be described as Gondwanan even though it has extensions into North America, Japan, and Korea (Hunt 1996). The southern hemisphere fauna is also characterised by sexual polymorphism – marked sexual dimorphism is commonplace in the Eupnoi (Forster & Forster 1999) and the presence of different male morphs within the same species seems to occur frequently, particularly in the New Zealand members of the suborder Laniatores (Forster 1954).

The most frequently encountered species is the so-called 'daddy longlegs', *Phalangium opilio* (not to be confused with the daddy longlegs spider *Pholcus phalangioides*). *Phalangium opilio* is a common European species that has been established in New Zealand for some time and is a resident of fields and gardens throughout the country. It is often met roaming about on bright sunny days. In contrast, New Zealand's endemic species are typically nocturnal forest dwellers, with a number of species also recorded from caves and other habitats.

The late Ray Forster was the leading New Zealand authority on these animals, publishing several revisions of different elements of the fauna (e.g. Forster 1944, 1948, 1952). These fascinating animals are yet another group that had suffered from taxonomic neglect, with Forster's last description of new species published more than four decades ago (Forster 1965). However, the appearance of papers since 2000 describing new species of Cyphophthalmi (e.g. Boyer & Giribet 2003) and revisory work on the genus *Pantopsalis* (Taylor 2004) is encouraging.

Harvestmen are well represented in New Zealand's main entomological collections. The majority of primary types can be found in the Canterbury Museum and the Museum of New Zealand Te Papa Tongarewa.

New Zealand Opiliones have traditionally been classified in three suborders – Cyphophthalmi, Palpatores, and Laniatores – but this arrangement has been called into question. For example, several workers, most notably Martens (1980), have proposed uniting Cyphophthalmi and Palpatores into a single suborder, Cyphopalpatores. Other studies (e.g. Giribet et al. 2002) have suggested that Palpatores is not an evolutionary cohesive (monophyletic) group. Currently, instead of Palpatores, the subordinal names Eupnoi and Dyspnoi are used in recognition of two phylogenetically well-supported palpatorean lineages (Giribet & Kury 2007). Despite common usage of the term Palpatores in the New Zealand literature (e.g. Forster & Forster 1999), all New Zealand species previously placed in this group are now considered part of Eupnoi (Taylor 2004). The suborder Dyspnoi is not known to occur in New Zealand.

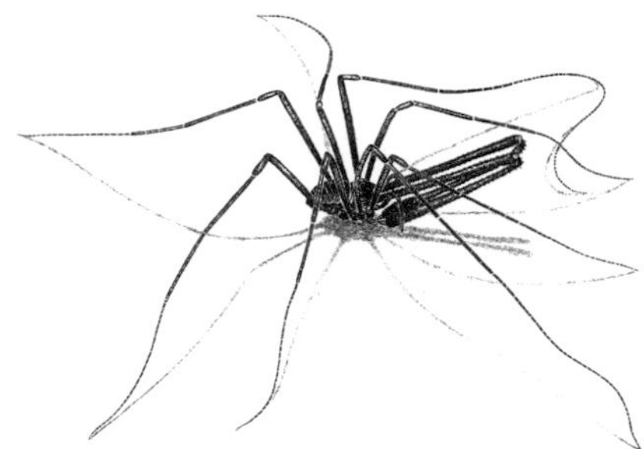

Male (upper) and female (lower) *Megalopsalis* sp. (Eupnoi).
From Forster & Forster 1973

Suborder Eupnoi

Known as long-legged harvestmen, these are perhaps the most outlandish of all New Zealand's arachnids. Upon seeing one of these creatures with legs like long lengths of thread supporting a tiny body that may be armed with massive chelicerae one may be forgiven for wondering how such an animal can move, let alone thrive. They can move very quickly when alarmed, however, if in a somewhat ungainly fashion. Forster and Forster (1999) observed that these creatures can autotomise legs at the end of the first joint if attacked, leaving them to make their escape while the would-be predator is left with a limb. Sexual dimorphism is particularly marked in New Zealand Palpatores. Males are commonly black with a hard carapace and massively developed chelicerae while females have soft bodies patterned with shades of green or brown and chelicerae that are tiny by comparison (Forster 1962). The sexes are sometimes so different that, for many years, males and females were treated as separate species (Forster & Forster 1999).

Currently, 28 species and one subspecies have been described from New Zealand, with all but two of them regarded as endemic. It is clear, however, that many more species are yet to be described, perhaps as many as 100 (Forster pers. comm.). Taylor (2004, 2009) added one new species and a new genus.

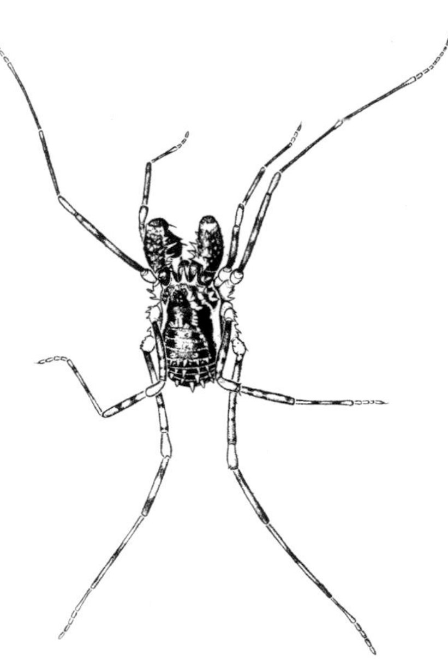

Algidia chiltoni (Laniatores).
From Forster & Forster 1973

Suborder Laniatores

Commonly known as short-legged harvestmen, they can further be distinguished from other harvestmen by their well-developed spinous pedipalps, with other spines and structures adorning the body in ways that border on the extreme. Indeed, Forster and Forster (1973) were moved to remark, 'these are the harvestmen whose profusion of seemingly useless spines, knobs and pustules could raise some doubts as to the advantages of evolution within this group'. The Forsters later speculated that these structures assist in camouflaging the animals (Forster & Forster 1999). Male polymorphism is a common feature in this suborder, although why this should be so is not certain. Regardless of the explanation, it appears that the sex ratio of males to females is still about 1 : 1 (Forster 1962) and females do not appear to exhibit any mating preference for one male morph over another (Forster & Forster 1999).

Laniatores are New Zealand's most diverse opilionids with more than 160 species and subspecies known so far. All are considered endemic. It is likely that many more species still await description (Forster & Forster 1999). The most recent formal descriptions of New Zealand Laniatores are those of Forster (1965).

Suborder Cyphophthalmi

These are the mite-like harvestmen, typically found in leaf litter. Their resemblance to mites clearly differentiates them from the other opilionid suborders. They are also tiny (no more than 5 millimetres in length), blind (except for one family that does not occur in New Zealand), and have very obvious repugnatorial glands on the flanks of the carapace (Forster & Forster 1999). Segmentation on the abdomen immediately distinguishes them from mites. New Zealand Cyphophthalmi are all members of the family Pettalidae, which includes three endemic genera containing 29 species and subspecies. Petallids have also been cited as an example of a model Gondwanan group (Boyer & Giribet 2007). Only one new description of a New Zealand member of this suborder has appeared (Boyer & Giribet 2003) since the work of Forster (1952).

Rakaia pauli (Cyphophthalmi).
From Forster & Forster 1973

Pseudoscorpiones: False scorpions

The name 'pseudoscorpion' would suggest that these creatures are closely related to true scorpions but the similarity extends mainly to the pincer-like pedipalps that both groups of creatures possess. In fact, pseudoscorpions are regarded as a sister group to the Solifugae (sun spiders) (Weygoldt & Paulus 1979; Shultz 2007; Harvey 1992), an order of arachnids not found in New Zealand or in nearby Australia. Not only are pseudoscorpions much smaller (the largest is just over 10 millimetres long) than scorpions (the smallest is 13 millimetres long), but they lack a differentiated postabdomen (tail), sting, and pectines (comb-like sensory structures under the preabdomen of scorpions). On the other hand, pseudoscorpions have a number of distinctive features, such as a silk-spinning apparatus in the moveable finger of each chelicera and an enlarged patella on the walking legs. Nymphs and adult females of most species use silk to construct small igloo-shaped chambers in which moulting and egg-laying frequently occur.

Adults of most pseudoscorpions are 2–7 millimetres long; the largest include species of the seashore genus *Garypus* (family Garypidae). Pseudoscorpions occur in a wide range of terrestrial habitats in most parts of the world including oceanic islands, but are absent from extreme northern areas and from Antarctica. They are most commonly found in leaf litter and upper soil layers and can easily be collected in large numbers by the use of extraction techniques such as Tullgren funnels. They often cling to the underside of rocks and logs embedded in the ground and some species are restricted to the upper littoral zone at the seashore. Their powers of dispersal are generally limited as they lack any obvious means of travel other than walking. But some species in a number of families use close associations with other arthropods, in particular flying insects, to disperse between favourable habitats (Beier 1948a; Muchmore 1971; Poinar et al. 1998).

The world fauna contains more than 3300 named species in 439 genera (Harvey 2009). Harvey (1992) provided a synopsis and re-evaluation of the classification of the order, recognising 24 families distributed among seven superfamilies and two suborders, the Epiocheirata and the Iocheirata, displacing earlier higher classifications proposed by Chamberlin (1929a, 1930, 1931) and Beier (1932a,b). Although the composition of the order is small in comparison with other orders of terrestrial invertebrates, pseudoscorpions are common enough to be encountered in many biotic surveys using suitable techniques. Their small size, however, ensures that identification requires specialised techniques, such as slide-mounting of dissected appendages, for detailed examination.

Historical overview

The early discovery and description of New Zealand pseudoscorpions was based almost entirely upon collections belonging to the British Museum (Natural History) (now Natural History Museum), London. The first was a rudimentary description of *Chelifer pallipes* from an unspecified New Zealand locality by White (1849) based upon a single specimen amongst a consignment of arthropods sent to the British Museum. Despite some confusion over the identity of this species, it is now recognised as belonging to the genus *Philomaoria* Chamberlin, 1931. The holotype is still preserved (Judson 1997). The next species were described nearly 60 years later, when Carl With (1907) published a paper based upon further British Museum material – *Olpium pacificum* (now *Xenolpium pacificum*) from Stewart Island, and *Chelifer vigil* (now *Maorichernes vigil*) and *C. taierensis* (now *Thalassochernes taierensis*) from Taieri.

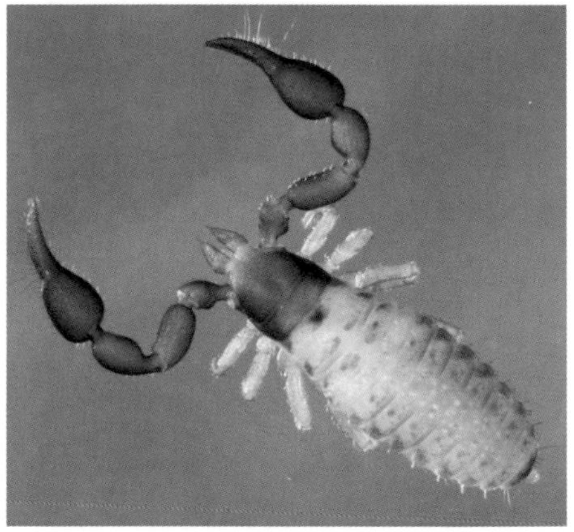

Unidentified chernetid pseudoscorpion.
Alastair Robertson and Maria Minor, Massey University

The pre-eminent pseudoscorpion expert Joseph Chamberlin erected the first genus based upon a New Zealand type species when he described *Maorichthonius mortenseni* from Auckland (Chamberlin 1925). He later described *Tyrannochthonius grimmeti* and *Ideobisium peregrinum* from Wellington and *Maorigarypus melanochelatus* (now placed in the genus *Synsphyronus*) from Okahune (Chamberlin 1929b, 1930). These collections were also partly based upon material loaned from the British Museum.

Apart from two species of *Tyrannochthonius* described by Moyle (1989a,b), the rest of the indigenous New Zealand pseudoscorpions were described by Max Beier of the Naturhistorisches Museum in Vienna. As part of a life-long career in which he described hundreds of pseudoscorpions from all over the world, he also described numerous species from New Zealand (Beier 1932b, 1948b, 1962, 1964a,b, 1966, 1967, 1969, 1976), 53 of which are currently recognised as valid. All of the species described prior to the 1960s were based upon sporadic collections that were lodged in European museums, whereas greater collaboration with American and New Zealand entomologists during the 1960s and 1970s placed large amounts of material at his disposal, culminating in his final paper on New Zealand pseudoscorpions that provided a synopsis and keys to the fauna (Beier 1976). This paper is still the starting point of any work on the New Zealand pseudoscorpions. As part of a revision of the garypid genus *Synsphyronus*, Harvey (1987a) redescribed *S. melanochelatus* and *S. lineatus* and later (Harvey 1989) transferred several species from *Morikawia* or *Paraliochthonius* to the genus *Tyrannochthonius*.

Endemism and biogeography

The New Zealand pseudoscorpion fauna is highly endemic at the species level, but less so at genus level and there are no endemic families or subfamilies. Of the 64 named native species (and five subspecies), 60 species (94% of the fauna) are endemic. Only four (*Philomaoria pallipes*, *Tyrannochthonius kermadecensis*, *T. norfolkensis*, and *Xenolpium pacificum*) have been reported outside of New Zealand – *T. norfolkensis* from Norfolk Island and the remaining three from Lord Howe Island.

Several genera – the chthoniids *Maorichthonius*, *Sathrochthoniella*, and *Tyrannochthoniella*, the garypinid *Nelsoninus*, and the chernetids *Heterochernes*, *Maorichernes*, *Nesiotochernes*, *Opsochernes*, *Phaulochernes*, *Systellochernes*, and *Thalassochernes* – are currently thought to be endemic to New Zealand. Two chernetid genera, *Apatochernes* and *Nesochernes*, are found only in New Zealand and Norfolk Island, while the curious genus *Philomaoria*, currently assigned to the Cheliferidae (Harvey 1992), is found in New Zealand and Lord Howe Island with one or more undescribed species from New Caledonia (M. Harvey, unpubl.). The distinctive chernetid genus *Reischekia* is elsewhere found only in West Papua. There is a distinctive Gondwanan element in the New Zealand fauna, with several genera on one or more of the southern continents. The garypid genus *Synsphyronus* and the cheliferid *Protochelifer* are also found in Australia, while the former has recently been found in New Caledonia (M. Harvey unpubl.). The chthoniid *Austrochthonius* is found in Australia, South America, and South Africa, while *Sathrochthonius* is found in Australia, Lord Howe Island, South America and New Caledonia. The remaining genera are generally widespread, either in the Indo-Pacific region (the chernetids *Nesidiochernes* and *Smeringochernes* and the olpiid *Xenolpium*) or are more widespread (the chthoniid *Tyrannochthonius*, the syarinid *Ideobisium*, and the cheiridiid *Apocheiridium*).

The various offshore islands of New Zealand harbour a wide variety of pseudoscorpions, but very few appear to be endemic. The most distant islands from the New Zealand mainland, in the Kermadec and Chatham groups, possess two and nine species respectively, but few species appear to be endemic to any particular island.

Ecology and distribution

Pseudoscorpions are found in most terrestrial environments and occur most abundantly in forest litter and soil. There have been no ecological studies on any New Zealand species. Although taxonomic papers list numerous collecting localities, there has been no summary of the distributional patterns of pseudoscorpions and distribution maps have been published for only two species (Harvey 1987a).

Three cosmopolitan species – *Chelifer cancroides*, *Lamprochernes savignyi*, and *Withius piger* – have been recorded from human settlements and undoubtedly represent accidental introductions (Beier 1976; Harvey 1987b). Like many mites, some pseudoscorpions can hitchhike (a phenomenon called phoresy); *L. savignyi* is known to use houseflies for transport, grasping hold of them with its pedipalps.

Gaps in knowledge and scope for future research

Despite the small number of described species, the New Zealand pseudoscorpion fauna is unique and inherently fascinating. The high percentage of endemic species and, to a lesser extent, genera, highlights the relictual nature of these small, relatively immobile creatures. The biogeographic origins of the majority of genera lie with other parts of Gondwana, particularly Australia and New Caledonia.

The true size of the New Zealand fauna is difficult to estimate without additional taxonomic work. Several new undescribed species of *Apatochernes* and *Tyrannochthonius* have been recognised in collections but many pseudoscorpion specimens are difficult to place in existing species or even genera. This may reflect the cursory way in which many species have been described in the past. The currently named fauna of 64 native and three introduced species (see end-chapter checklist) will probably be doubled with additional revisionary studies, particularly when diverse, frequently collected genera such as *Apatochernes*, *Austrochthonius*, and *Tyrannochthonius* are studied in detail.

There is also a paucity of detail in numerous morphological features of New Zealand pseudoscorpions, which hinders a full assessment of the fauna in relation to ongoing phylogenetic work on the group. A critical example is the lack of any illustrations of the spermathecae of New Zealand chernetids.

Class Pycnogonida

Pycnogonids – sea spiders – are curious-looking creatures that appear to be all legs and no body. In fact their alternative name Pantopoda ('all legs') reflects that fact. They only superficially resemble spiders and, although there are about 1500 species so far known worldwide, it is mostly only snorkellers, divers, and marine biologists that have encountered them. Most scientific studies on pycnogonids have been taxonomic, i.e. naming and classifying them, and the natural history and biology of most species is unknown. They are exceedingly rare as fossils, there being only about eight records from the Jurassic, Early Devonian and Silurian (Siveter et al. 2004). There has been some debate about the relationships of pycnogonids to other chelicerates. Studies of the fossils along with morphological and molecular analysis of living forms (Wheeler & Hayashi 1998; Edgecombe et al. 1999; Giribet & Ribera 1999; Giribet et al. 2001; Dunlop & Arango 2004; Maxmen et al. 2005) are either in support of pycnogonids as chelicerates or that they are a sister group to the rest of the Arthropoda. It is acknowledged that they do have unusual features, including a proboscis, highly reduced abdomen, ovigers, multiple gonopores, and gut branches into the legs.

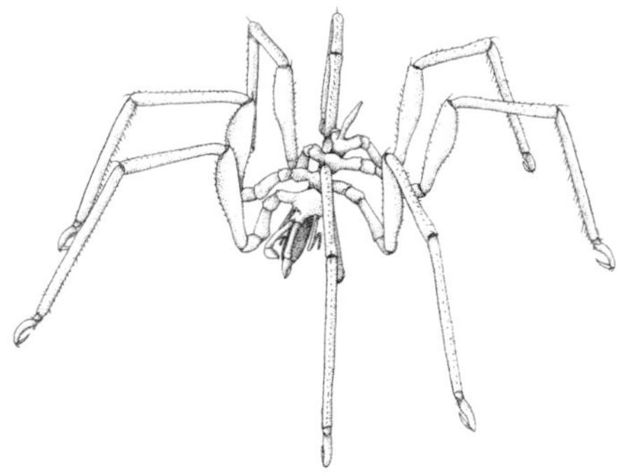

Stylised drawing showing the general form of a sea spider.

Danielle Archer

They are harmless, fragile, and very slow-moving. They are found in all oceans, from the seashore to more than 7000 metres deep, and are generally commoner in cooler waters. Mostly they are small, with leg spans ranging from a few millimetres to about six centimetres, but in very deep water and in the Antarctic there are giants with total leg spans exceeding half a metre. Intertidally, they are found among seaweeds and beneath boulders, frequently associated with sea anemones and tubeworms. Subtidally, they are most often found clinging to hydroids, bushy bryozoans, algae, and sponges. Some live surrounded by sand.

The body of sea spiders may be smooth, spiny, or tubercular and some appear to secrete a sticky substance that allows sand grains and detritus to adhere. Such species can be particularly difficult to find when outstretched on a sponge surface or on sand. Many are semitransparent and tend to display the colour of their gut contents, again providing camouflage. Like other arthropods, sea spiders moult at regular intervals. They can also repair damaged limbs.

It is difficult to generalise about sea spiders, as invariably there are exceptions to every character. The size and shape of the body vary enormously, as do the limbs and other appendages, either by their presence, absence, shape, or number of segments. The body is barely thicker than the legs. It consists of a cephalothorax of usually four segments, a proboscis with a mouth at the tip, and a reduced one-segment abdomen. Some deep-sea species have 10 legs (one found in New Zealand waters – *Pentanymphon antarcticum*) or 12 but all shallow-water species have eight. Each leg has eight moveable segments and a strong terminal claw that may be flanked by a pair of smaller claws. Eyes are placed on the ocular tubercle, a raised process on the mid-dorsal surface of the most anterior segment. Usually there are four simple eyes, but some species have eight and in some deep-sea species both the ocular tubercle and eyes are lacking altogether.

Classification is primarily dependent on the presence, absence, or degree of development of the three pairs of appendages attached to the cephalothorax. These are the chelifores, originating above the proboscis; palps, carried either side of the proboscis; and the ovigers, situated ventrally, posterior to the proboscis. The chelifores comprise one or two basal segments to which are attached the chelae. The inner margins of the chela fingers can be smooth or armed with sharp denticles or teeth suitable for grasping and restraining prey. Chelae are present in all larval pycnogonids but may atrophy to non-functional knobs or be lost completely with the last juvenile moult.

When present, the palps are typically 5–10-segmented and are sensory (touch and taste). They are also used to manipulate prey to facilitate feeding. Wyer and King (1973) recorded the use of palps by *Achelia echinata* to prevent closure of the operculum of bryozoan zooids so as to insert the proboscis to suck out the contents. Similarly, the palps of a species of *Ammothella* that is known to feed on the tubeworm *Galeolaria* are placed behind the operculum of the worm, preventing withdrawal into its tube (Staples pers. obs.).

Ovigers are used by male sea spiders for carrying the eggs and usually by both sexes for grooming. Typically they are 10-segmented but they may be present in the male only or reduced to such an extent that their function is problematic. In some species of *Pycnogonum* they are absent in both sexes. In these species eggs are cemented directly onto the ventral surface of the male. The ovigers also assist in the transfer of eggs from the female to the male, where they are typically gathered into balls and carried until hatching. The relationship between egg numbers and mortality has not been studied.

A newly hatched sea-spider larva is called a protonymphon. The protonymphon may develop to a post-larval stage on the male, at which time it can live independently, usually feeding on the same host as the adult. Alternatively, they may leave the adult to parasitise a diverse range of soft-bodied invertebrates including opisthobranch and bivalve molluscs, sea cucumbers, and polychaete worms. Observations of tightly folded protonymphons occupying gall-like swellings on hydroids are common (Staples & Watson 1987). Similar gall-like bodies have also been recorded on a gorgonian (Stock 1953).

The proboscis with the mouth at its tip, articulates with the cepahalothorax by means of a flexible membrane. The proboscis is highly variable in size, in some cases, as with the deep-sea genus *Colossendeis*, reaching a length considerably greater than the trunk itself. The proboscis shape may also demonstrate a high level of specialisation, none more so than that found in the Austrodecidae. Species in this family are characterised by possession of a pipette-shaped proboscis with a biradiate ventrodistal mouth (triradiate and distal in all other families). This shape allows access to the withdrawn hydranths of hydroids possessing long, curved hydrothecae, such as in species of *Dictyocladium* found in New Zealand waters.

Thick-set robust forms like species of *Pycnogonum* are limited in their dispersal and live a relatively sedentary life compared to their slender long-legged relatives. The latter can generate enough lift by hurriedly treading water (or swimming) to take advantage of tides and currents. By employing this swimming action periodically, they can remain in the water column and be carried over considerable distances with little effort. Descent is regulated by folding all legs dorsally thus reducing the surface area and causing the animal to sink. Bathypelagic species of *Pallenopsis* are recorded at great depths parasitising jellyfish. Long setae on the legs of these deepwater species are thought to assist in providing lift and maintaining their suspension in the water column. There are also several shallow-water records of protonymphon larvae attached to hydroid anthomedusae, each with the proboscis piercing the host tissue. Whilst the initial attachment has not been documented, it is not difficult to speculate that drifting pycnogonids would readily come into contact with a swarm of recently liberated medusae or that the larvae attach when a medusa touches bottom.

Pycnogonum planum.
Danielle Archer, after Stock 1954

The significance of sea spiders in the marine food chain has not been assessed, but there is no doubt that the impact on individual hydroid colonies is significant, and bryozoan authority John Ryland (1976) went so far as to write that sea spiders are the most important predators of that group, although that assertion was based more on physical association than actual observations of predation (Lidgard 2008).

The morphological extremes found in New Zealand sea spiders are represented by the genera *Nymphon* and *Pycnogonum*. *Nymphon* species all have long slender legs attached to an equally slender body, 5-segmented palps and 10-segmented ovigers in both sexes. Well-developed chelifores over-reach the proboscis. The chelae are often delicate, the fingers possessing rows of fine, sharp teeth well suited to grasping and restraining prey. *Nymphon* is known to feed on diverse creatures such as small spionid bristleworms and hydroids. Wyer and King (1974) described the feeding of *Nymphon* using the palps to locate a hydranth which is then held in front of the mouth whilst being partially macerated by the scissor action of the chela. The genus *Pycnogonum* is one of the more distinctive forms, having a broad body and short robust legs.

Its species lack palps, chelifores, and, in the two known New Zealand species, ovigers. Individuals are typically found clinging to sea anemones, often in the intertidal zone and/or shallow water. Species feed on anemones almost exclusively – the adults ectoparasitically on the outer body wall of the host and juveniles endoparasitically in the gastrovascular cavity. Strong claws are used both to anchor the sea spider and to tear the body wall of the anemone, allowing insertion of the proboscis to feed on the soft internal tissue.

Child (1998) provided an excellent review of the pycnogonid fauna of the New Zealand region, a key to nine families, and a list of 94 species belonging to 27 genera but he neglected to mention the deep-water records of Fage (1956a,b) from the Kermadec Trench and Tasman Sea. Based on all published and unpublished (NIWA Invertebrate Collection, Wellington) records, 92 pycnogonid species are here recognized as living within the 200 nautical-mile New Zealand EEZ. Some 21 species have a depth range of 1000 metres. Specimens from recent surveys, often associated with the biota of wharf pilings, indicates that many more species await description. About 40 of the New Zealand species and one genus, *Oorhynchus*, appear endemic to the EEZ but this number could reduce following more collecting in the wider region and Southwest Pacific generally. Most of the New Zealand species that are routinely encountered are relatively small and delicate, with long spindly legs. The largest New Zealand species is the aptly named *Colossendeis colossea.* It is cosmopolitan in deep water and has been recorded in the New Zealand EEZ between 1100 and 2640 metres depth. One specimen held in the Smithsonian Institute has a leg span exceeding 70 centimetres and is the largest recorded pycnogonid.

The classification used in this chapter (Bamber 2007) is based on molecular and morphological characters. There is much variation in pycnogonids, and the existence of forms apparently intermediate between major morphologies has challenged the specialist. Classifications recognise eight to eleven families in a single living order Pantopoda (there are also three extinct orders) in a class Pycnogonida.

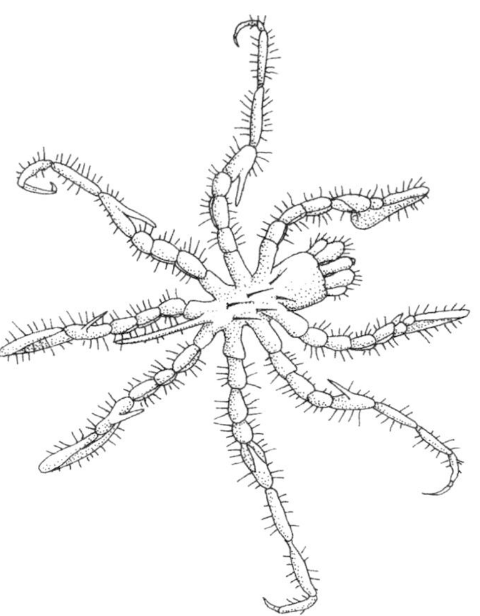

Oorhynchus aucklandiae.
Danielle Archer, after Hoek 1881

Acknowledgements

Thanks are due to the following people for helpful comments on aspects of the review of the mite fauna: Dr Trevor Crosby (Landcare Research, Auckland, New Zealand), Dr Bruce Halliday (CSIRO, Canberra, Australia), Dr Nick Martin (Crop & Food Research, Auckland, New Zealand), Dr Roy Norton (SUNY College of Environmental Science and Forestry, Syracuse, NY, USA), Dr Barry O'Connor (University of Michigan, Ann Arbor, Michigan, USA), Dr Ting-Kui Qin (Biosecurity Australia, Canberra, Australia), Dr Richard Robbins (US Department of Defence Armed Forces Pest Management Board, Washington, DC), and Lynn Royce (formerly L. A. Richards), Oregon, USA. The Acari review was funded in part by the Foundation for Research, Science, and Technology, New Zealand, under contract number C09617. Thanks are due to Cor Vink (Agresearch, Lincoln, New Zealand) for highlighting taxonomic problems affecting the systematics of the interesting New Zealand pseudoscorpion fauna. PJS would like to thank Cor Vink, Pierre Paquin (SWCA Environmental Consultants Austin Texas, USA), Mike Fitzgerald (Museum of New Zealand Te Papa Tongarewa, Wellington), Robert Raven (Queensland Museum, Brisbane, Australia), and the late Ray Forster (formerly Otago Museum, Dunedin, New Zealand) for their input into the list of spiders. The overall review of New Zealand arachnids is dedicated to the memory of Ray Forster, whose work on the native fauna inspired an entire generation. Doug Forster is thanked for permission to reproduce some of the Forster illustrations of arachnids.

Authors

Dr Ilse Bartsch Forschungsinstitut Senckenberg, Notkestrasse 85, D-22607 Frankfurt, Germany [bartsch@meeresforschung.de]
marine mites

Dr David R. Cook 7836 North Invergordon Place, Paradise Valley, Arizona 85253, USA [watermites2@live.com]
water mites

Dr Mark S. Harvey Department of Terrestrial Zoology, Western Australian Museum, Locked Bag 49, Welshpool DC, WA 6986, Australia [mark.harvey@museum.wa.gov.au]
water mites, pseudoscorpions

Dr Birgit E. Rhode Landcare Research, Private Bag 92-170, Auckland, New Zealand [rhodeb@landcareresearch.co.nz]
mites and ticks

Philip J. Sirvid Museum of New Zealand Te Papa Tongarewa, PO Box 467, Wellington, New Zealand [phils@tepapa.govt.nz]
spiders, opiliones, microwhip scorpions

David A. Staples Museum Victoria, GPO Box 666, Melbourne 3001, Victoria Australia [dstaples@museum.vic.gov.au]
Pycnogonida

Dr Zhi-Qiang Zhang Landcare Research, Private Bag 92-170, Auckland, New Zealand [zhangz@landcareresearch.co.nz]
mites and ticks

References

ANDERSON, D.; TRUEMAN, J. W. H. 2000: *Varroa jacobsoni* (Acari: Varroidae) is more than one species. *Experimental and Applied Acarology 24*: 165–189.

ATYEO, W. T. 1963: The Bdellidae (Acarina) of the Australian Realm part I. New Zealand, Campbell Island, and the Auckland Islands. *Bulletin of the University of Nebraska Museum 4*: 113–166.

BALOGH, J.; BALOGH, P. 1985: Studies on the Anderemaeidae J. Balogh, 1972 (Acari, Oribatei). *Opuscula Zoologica (Budapest), 19–20*: 41–48.

BALOGH, J.; BALOGH, P. 1986: Some oribatid mites collected in the western Pacific area. *Acta Zoologica Academiae Scientiarum Hungaricae, 32(3–4)*: 263–280.

BAMBER, R. N. 2007: A holistic re-interpretation of the phylogeny of the Pycnogonida Latreille, 1810 (Arthropoda). *Zootaxa 1668*: 295–312.

BANKS. J.; SIRVID, P. J.; VINK, C. J. 2004: White-tailed spider bites – arachnophobic fallout? *New Zealand Medical Journal 117(1188)*: [1–7].

BARTSCH, I. 1979: Five new species of Halacaridae (Acari) from New Zealand. *New Zealand Journal of Marine and Freshwater Research 13*: 175–185.

BARTSCH, I. 1986a: New species of Halacaridae (Acari) from New Zealand. *New Zealand Journal of Zoology 12*: 547–560.

BARTSCH, I. 1986b: A new species of *Halixodes* and a review of the New Zealand species. *Journal of the Royal Society of New Zealand 16*: 51–56.

BARTSCH, I. 1996: Halacarids (Halacaroidea, Acari) in freshwater. Multiple invasions from the Paleozoic onwards? *Journal of Natural History 30*: 67–99.

BARTSCH, I. 2007: Freshwater Halacaridae (Acari) from New Zealand rivers and lakes, with notes on character variability. *Mitteilungen aus dem hamburgischen Zoologischen Museum und Institut 104*: 73–87.

BEIER, M. 1932a: Pseudoscorpionidea I. Subord. Chthoniinea et Neobisiinea. *Tierreich 57*: xx, 1–258.

BEIER, M. 1932b: Pseudoscorpionidea II. Subord. C. Cheliferinea. *Tierreich 58*: xxi, 1–294.

BEIER, M. 1948a: Phoresie und Phagophilie bei Pseudoscorpionen. *Österreichische Zoologische Zeitschrift 1*: 441–497.

BEIER, M. 1948b: Über Pseudoscorpione der australischen Region. *Eos, Madrid 24*: 525–562.

BEIER, M. 1962: On some Pseudoscorpionidea from New Zealand. *Records of the Canterbury Museum 7*: 399–402.

BEIER, M. 1964a: False scorpions (Pseudoscorpionidea) from the Auckland Islands. *Pacific Insects Monograph, Supplement 7*: 628–629.

BEIER, M. 1964b: Insects of Campbell Island. Pseudoscorpionidea. *Pacific Insects Monograph 7*: 116–120.

BEIER, M. 1966: Zur Kenntnis der Pseudoscorpioniden-Fauna Neu-Seelands. *Pacific Insects 8*: 363–379.

BEIER, M. 1967: Contributions to the knowledge of the Pseudoscorpionidea from New Zealand. *Records of the Dominion Museum* 5: 277–303.

BEIER, M. 1969: Additional remarks to the New Zealand Pseudoscorpionidea. *Records of the Auckland Institute and Museum 6*: 413–418.

BEIER, M. 1976: The pseudoscorpions of New Zealand, Norfolk and Lord Howe. *New Zealand Journal of Zoology 3*: 199–246.

BERLAND, L. 1931: Araignées des îles Auckland et Campbell. *Records of the Canterbury Museum 3*: 357–365.

BESCH, W. 1964: Systematik und Verbreitung der südamerikanischen rheophilen Hydrachnellen. *Beiträge zur Neotropischen Fauna 3*: 77–194.

BOYER, S. L.; GIRIBET, G. 2003: A new *Rakaia* species (Opiliones, Cyphopthalmi, Pettalidae) from New Zealand. *Zootaxa 133*: 1–14.

BOYER, S. L.; GIRIBET, G. 2007: A new model Gondwanan taxon: systematics and biogeography of the harvestman family Pettalidae (Arachnida, Opiliones, Cyphophthalmi), with a taxonomic revision of genera from Australia and New Zealand. *Cladistics 23*: 337–361.

BRISTOWE, W. S. 1958: *The World of Spiders*. Collins, London & Glasgow. 304 p.

BRYANT, E .B. 1933: Notes on the types of Urquhart's spiders. *Records of the Canterbury Museum 4*: 1–27.

BRYANT, E. B. 1935: Notes on some of Urquhart's species of spiders. *Records of the Canterbury Museum 4*: 53–70.

CHAMBERLAIN, G. 1944: Revision of the Araneae of New Zealand. Part 1. *Records of the Auckland Institute and Museum 3*: 69–78.

CHAMBERLAIN, G. 1946: Revision of the Araneae of New Zealand. Part 2. *Records of the Auckland Institute and Museum 3*: 85–97.

CHAMBERLIN, J. C. 1925: Notes on the status of genera in the Chelonethid family Chthoniidae together with a description of a new genus and species from New Zealand. *Videnskabelige Meddelelser fra Dansk Naturhistorisk Forening i Kjøbenhavn 81*: 333–338.

CHAMBERLIN, J. C. 1929a: On some false scorpions of the suborder Heterosphyronida (Arachnida – Chelonethida). *Canadian Entomologist 61*: 152–155.

CHAMBERLIN, J. C. 1929b: A synoptic classification of the false scorpions or chela-spinners, with a report on a cosmopolitan collection of the same. Part 1. The Heterosphyronida (Chthoniidae) (Arachnida–Chelonethida). *Annals and Magazine of Natural History, ser. 10*, 4: 50–80.

CHAMBERLIN, J. C. 1930: A synoptic classification

of the false scorpions or chela-spinners, with a report on a cosmopolitan collection of the same. Part II. The Diplosphyronida (Arachnida-Chelonethida). *Annals and Magazine of Natural History, ser. 10, 5*: 1–48, 585–620.

CHAMBERLIN, J. C. 1931: The arachnid order Chelonethida. *Stanford University Publications, Biological Sciences 7*: 1–284.

CHILD, C. A. 1998: The marine fauna of New Zealand: Pycnogonida (sea spiders). *NIWA Biodiversity Memoir 109*: 1–71.

CHILTON, C. 1910: Note on the dispersal of marine Crustacea by means of ships. *Transactions and Proceedings of the New Zealand Institute 43*: 131–133.

CLARK, J. M. 1997: Holothyrid mites: occasional allergenic parasites of humans? *New Zealand Journal of Zoology 24*: 299.

CLAUSEN, I. H. S. 1986: The use of spiders as biological indicators. *Bulletin of the British Arachnological Society 7*: 83–86.

COLLOFF, M. J.: CAMERON, S. L. 2009: Revision of the oribatid mite genus *Austronothrus* Hammer (Acari : Oribatida): sexual dimorphism and a re-evaluation of the phylogenetic relationships of the family Crotoniidae. *Invertebrate Systematics 23*: 87–110.

COLLOFF, M. J.; HALLIDAY, R. B. 1998: *Oribatid Mites: A Catalogue of the Australian Genera and Species*. CSIRO Publications, Melbourne. 223 p.

COLLYER, E. 1973a: Two new species of the genus *Colopalpus* (Acari: Tenuipalpidae). *New Zealand Journal of Science 16*: 529–532.

COLLYER, E. 1973b: New species of the genus *Tenuipalpus* (Acari: Tenuipalpidae) from New Zealand, with a key to the world fauna. *New Zealand Journal of Science 16*: 915–955.

COLLYER, E. 1973c: New species of the genus *Tenuipalpus* (Acari: Tenuipalpidae) from New Zealand, with a key to the world fauna. *New Zealand Journal of Science* 16: 915–955.

COLLYER, E. 1982: The Phytosciidae of New Zealand (Acarina). 1. The genera *Typhlodromus* and *Amblyseius*—keys and new species. *New Zealand Journal of Zoology 9*: 185–206.

COOK, D. R. 1980: Studies on Neotropical water mites. *Memoirs of the American Entomological Institute 31*: 1–645.

COOK, D. R. 1983. Rheophilic and hyporheic water mites of New Zealand. *Contributions of the Entomological Institute 21(2)*: 1–224.

COOK, D. R. 1984: Preliminary review of the relationships of the water mite faunas of South America, Australia, and New Zealand. Pp. 959–964 *in*: Griffiths, D. A.; Bowman, C. E. (eds), *Acarology VI*. Vol. 2. Ellis Norwood, New York.

COOK, D. R. 1986: Water mites from Australia. *Memoirs of the American Entomological Institute 40*: 1–568.

COOK, D. R. 1991: Water mites from driven wells in New Zealand, the subfamily Notoaturinae Besch (Acarina, Aturidae). *Stygologia 6*: 235–253.

COOK, D. R. 1992: Water mites (Hydracarina), mostly from driven wells in New Zealand: taxa other than the Notoaturinae Besch. *Stygologia 7*: 43–62.

COOK, D. R.; HOPKINS, C. L. 1998: New water mite species (Acari: Hydracarina) from New Zealand. *Acarologia 39*: 257–263.

CRANFIELD, H. J.; GORDON, D. P.; WILLAN, R. C.; MARSHALL, B. A.; BATTERSHILL, C. N.; FRANCIS, M. P.; NELSON, W. A.; GLASBY, C. J.; READ, G. B. 1998: Adventive marine species in New Zealand. *NIWA Technical Report 34*: 1–48.

CROWELL, R. M. 1990: *Unionicola* (*Pentatax*) *billieaehonore* n. sp. a sponge-associated Hydracarina (Acari: Unionicolidae) from New Zealand. *New Zealand Journal of Zoology 17*: 265–269.

DABERT, J.; ALBERTI, G. 2008: A new species of the genus *Coraciacarus* (Gabuciniidae, Pterolichoidea) from the huia *Heteralocha acutirostris* (Callaeatidae, Passeriformes), an extinct bird species from New Zealand. *Journal of Natural History 42*: 2763–2776.

DALMAS, Comte de 1917: Araignées de Nouvelle-Zélande. *Annales de la Société Entomologique du France 86*: 317–436.

DERRAIK, J. G. B.; SIRVID, P. J.; VINK, C. J.; HALL, G. 2008: White-tail tails.*New Zealand Medical Journal 121(1269)*: 85–86.

DESCH, C. E. Jr 1989: Two new species of *Demodex* (Acari: Demodicidae) from the New Zealand short-tailed bat, *Mystacina tuberculata* Gray, 1843 (Chiroptera, Mystacinidae). *New Zealand Journal of Zoology 16*: 221–230.

DOMROW, R. 1988: Acari Mesostigmata parasitic on Australian vertebrates: an annotated checklist, keys and bibliography. *Invertebrate Taxonomy 1*: 817–948.

DOMROW, R. 1992: Acari Astigmata (excluding feather mites) parasitic on Australian vertebrates: an annotated checklist, keys and bibliography. *Invertebrate Taxonomy 6*: 1459–1606.

DOMROW, R.; HEATH, A. C. G.; KENNEDY, C. 1980: Two new species of *Ophionyssus* (Acari: Dermanyssidae) from New Zealand lizards. *New Zealand Journal of Zoology 7*: 291–297.

DOMROW, R.; LESTER, L. N. 1985: Chiggers of Australia (Acari: Trombiculidae): an annotated checklist, keys and bibliography. *Australian Journal of Zoology Supplementary Series 114*: 1–111.

DUNLOP, J. A.; ARANGO, C. P. 2004: Pycnogonid affinities: a review. *Journal of Zoological Systematics and Evolutionary Research 43*: 8–21.

EDGECOMBE, G. D.; WILSON, G. D. F.; COLGAN, D. J.; GRAY, M. R.; CASSIS, G. 1999: Arthropod cladistics: combined analysis of histone H3 and U2 snRNA sequences and morphology. *Cladistics 16*: 155–203.

EMBERSON, R. M. 1973: Additions to the macrochelid mites in New Zealand (Acarina: Mesostigmata: Macrochelidae). *New Zealand Entomologist 5*: 294–302.

ESKOV, K. Y. 1990: Spider palaeontology: present trends and future expectations. *Acta Zoologica Fennica 190*: 123–128.

EVANS, G. O. 1992: *Principles of Acarology*. CAB International, Wallingford. 563 p.

FAGE, L. 1956a: Les pycnogonides du genre *Nymphon*. *Galathea Report 2*: 159–166.

FAGE, L. 1956b: Les pycnogondes [*sic*] (excl. le genre *Nymphon*). *Galathea Report 2*: 167–182.

FAIN, A. 1972: Les Listrophorides d'Australie et de Nouvelle-Guinée (Acarina: Sarcoptiformes). *Bulletin de l'Institut Royal des Sciences Naturelles de Belgique 48(5)*: 1–196.

FAIN, A. 2003 : Two new species of Cheyletidae (Acari Prostigmata) of the genera *Neoeucheyla* Radford, 1950 and *Granulocheyletus* Fain & Bochkov, 2002. *Bulletin de la Societe Royale Belge d'Entomologie* 139: 97-101.

FAIN, A.; BARKER, G. M. 2003: A new genus and species of mite of the family Ereynetidae (Acari Prostigmata) from the pallial cavity of a New Zealand terrestrial gastropod (Athoracophoridae). *Bulletin de la Société Royale Belge d'Entomologie 139*: 233–238.

FAIN, A.; DOMROW, R. 1974: Acariens parasités de marsupiaux et de rongeurs (Listrophoroidea). *In*: Fain, A. (ed.), Mission zoologique du fonds Léopold III en Irian (Nouvelle-Guinée occidentale). *Bulletin de l'Institut Royal des Sciences Naturelles de Belgique 50(7)*: 1–22.

FAIN, A.; GALLOWAY, T. D. 1993: Mites (Acari) from nests of seabirds in New Zealand. 1. Description and developmental stages of *Psylloglyphus parapsyllus* n. sp. (Winterschmidtiidae). *Acarologia 34*: 159–166.

FAN, Q.-H.; ZHANG, Z.-Q. 2002: *Primagistemus* gen. nov. (Acari: Prostigmata: Stigmaeidae). *Zootaxa 29*: 1–8.

FAN, Q.-H.; ZHANG, Z.-Q. 2004: *Revision of Rhizoglyphus Claparède (Acari: Acaridae) of Australasia and Oceania*. Systematic & Applied Acarology Society, London. 374 p.

FAN, Q.-H.; ZHANG, Z.-Q. 2005: Raphignathoidea (Acari: Prostigmata). *Fauna of New Zealand 52*: 1–400.

FAN, Q.-H.; ZHANG, Z.-Q. 2007: *Tyrophagus* (Acari: Astigmata: Acaridae). *Fauna of New Zealand 56*: 1–291.

FITZGERALD, B. M.; SIRVID, P. J. 2009: A revision of *Nomaua* (Araneae: Synotaxidae) and description of a new synotaxid genus from New Zealand. *Tuhinga 20*: 137–158.

FOELIX, R. F. 1996: *Biology of Spiders* 2nd edn. Oxford University Press, New York and Oxford. 330 p.

FORSTER, R. R. 1944: The genus *Megalopsalis* Roewer in New Zealand, with keys to the New Zealand genera of Opiliones. *Records of the Dominion Museum 1*: 183–192.

FORSTER, R. R. 1948: The sub-order Cyphophthalmi in New Zealand. *Dominion Museum Records in Entomology 1*: 79–119.

FORSTER, R. R. 1952: Supplement to the sub-order Cyphophthalmi. *Dominion Museum Records in Entomology 1*: 179–211.

FORSTER, R. R. 1954: The New Zealand Harvestmen (suborder Laniatores). *Canterbury Museum Bulletin 2*: 1–329.

FORSTER, R. R. 1962: A key to the New Zealand harvestmen. Part 1. *Tuatara 10*: 129–137.

FORSTER, R. R. 1964: The Araneae and Opiliones of the subantarctic islands of New Zealand. *Pacific Insects Monographs 7*: 58–115.

FORSTER, R. R. 1965: Harvestmen of the sub-order Laniatores from New Zealand caves. *Records of the Otago Museum 2*: 1–18.

FORSTER, R. R. 1967: The spiders of New Zealand. Part 1. *Otago Museum Bulletin 1*: 1–124.

FORSTER, R. R. 1975: The spiders and harvestmen. Pp. 493–505 *in*: Kuschel, G. (ed.), *Biogeography and Ecology in New Zealand*. W. Junk, The Hague.

FORSTER, R. R.; FORSTER, L. M. 1973: *New Zealand Spiders. An Introduction.* Collins, Auckland. 254 p.

FORSTER, R. R.; FORSTER, L. M. 1999: *Spiders of New Zealand and their Worldwide Kin.* University of Otago Press, Dunedin. 270 p.

FORSTER, R. R.; PLATNICK, N. I. 1985: A review of the austral spider family Orsolobidae (Arachnida, Araneae), with notes on the superfamily Dysderoidea. *Bulletin of the American Museum of Natural History 181*: 1–230.

FORSTER, R. R.; PLATNICK, N. I.; CODDINGTON, J.A. 1990: A proposal and review of the spider family Synotaxidae (Araneae, Araneoidea), with notes on theridiid interrelationships. *Bulletin of the American Museum of Natural History 193*: 1–116.

GIRIBET, G.; EDGECOMBE, G. D.; WHEELER, W. C. 2001: Arthropod phylogeny based on eight molecular loci and morphology. *Nature 413*: 157–161.

GIRIBET, G.; EDGECOMBE, G. D.; WHEELER, W. C.; BABBITT, C. 2002: Phylogeny and systematic position of Opiliones: a combined analysis of chelicerate relationships using morphological and molecular data. *Cladistics 18*: 5–70.

GIRIBET, G.; KURY, A. B. 2007: Phylogeny and biogeography. Pp. 61–87 *in*: Pinto-da-Rocha, R.; Machado, G.; Giribet, G. (eds), *Harvestmen – The Biology of Opiliones.* Harvard University Press, Cambridge. 597 p.

GIRIBET, G.; RIBERA, C. 1999: A review of arthropod phylogeny: new data based on ribosomal DNA sequences and direct character optimization. *Cladistics 16*: 204–231.

GORTON, R. J.; CHARLESTON, W. A. G.; MORRIS, R. S. 1999: [Abstract] *Sarcoptes scabiei* infestation of New Zealand hedgehogs. *New Zealand Journal of Zoology 26*: 80.

GOYEN, P. 1887: Descriptions of New Spiders. *Transactions and Proceedings of the New Zealand Institute 19*: 201–212.

GOYEN, P. 1888: Descriptions of new species of New Zealand Araneae. *Transactions and Proceedings of the New Zealand Institute 20*: 133–139.

HALLIDAY, R. B. 1998: Mites of Australia. A checklist and bibliography. *Monographs on Invertebrate Taxonomy 5*: 1–317.

HALLIDAY, R. B.; O'CONNOR, B. M.; BAKER, A. S. 1999: Global diversity of mites. Pp. 192–203 *in*: Raven, P. H.; Williams, T. (eds), *Nature and Human Society*. National Academy Press, Washington, D.C.

HAMMER, M. 1966: Investigations on the oribatid fauna of New Zealand. Part I. *Biologiske Skrifter Kongelige Danske Videnskabernes Selskab 15(2)*: 1–108.

HAMMER, M. 1967: Investigations on the oribatid fauna of New Zealand. Part II. *Biologiske Skrifter Kongelige Danske Videnskabernes Selskab 15(4)*: 1–64.

HAMMER, M. 1968: Investigations on the oribatid fauna of New Zealand, with a comparison between the oribatid fauna of New Zealand and that of the Andes Mountains, South America. Part III. *Biologiske Skrifter Kongelige Danske Videnskabernes Selskab 16(2)*: 1–96.

HAMMER, M.; WALLWORK, J. A. 1979: A review of the world distribution of oribatid mites in relation to continental drift. *Biologisker Skrifter 22(4)*: 1–31.

HARVEY, M. S. 1987a: A revision of the genus *Synsphyronus* Chamberlin (Garypidae: Pseudoscorpionida: Arachnida). *Australian Journal of Zoology, Supplementary Series 126*: 1–99.

HARVEY, M. S. 1987b: Redescription and new synonyms of the cosmopolitan species *Lamprochernes savignyi* (Simon) (Chernetidae: Pseudoscorpionida). *Bulletin of the British Arachnological Society 7*: 111–116.

HARVEY, M. S. 1989: Two new cavernicolous chthoniids from Australia, with notes on the generic placement of the south-western Pacific species attributed to the genera *Paraliochthonius* Beier and *Morikawia* Chamberlin (Pseudoscorpionida: Chthoniidae). *Bulletin of the British Arachnological Society 8*: 21–29.

HARVEY, M. S. 1992: The phylogeny and systematics of the Pseudoscorpionida (Chelicerata: Arachnida). *Invertebrate Taxonomy 6*: 1373–1435.

HARVEY, M. S. 1998: *The Australian Water Mites: A Guide to the Families and Genera*. CSIRO Publishing, Melbourne.

HARVEY, M. S. 2002: The neglected cousins: what do we know about the smaller arachnid orders? *Journal of Arachnology 30*: 357–372.

HARVEY, M. S. 2009: *Pseudoscorpions of the World, Version 1.2.* Western Australian Museum, Perth. Available at http://museum.wa.gov.au/arachnids/pseudoscorpions/.

HEATH, A. C. G. 2002: Recently introduced exotic animals and their parasites: what risk to New Zealand's biosecurity? *Surveillance 29(4)*: 15–17.

HILLYARD, P. 1994: *The Book of the Spider.* Hutchinson, London. 196 p.

HODGE, S.; VINK, C. J. 2000: An evaluation of *Lycosa hilaris* as a bioindicator of organophosphate insecticide contamination. *New Zealand Plant Protection 53*: 226–229.

HOGG, H. R. 1909: Spiders and Opiliones. Pp. 155–181 *in*: Chilton, C. (ed.), *The Subantarctic Islands of New Zealand*. Philosophical Institute of Canterbury, Christchurch. 848 p.

HOGG, H. R. 1911: On some New Zealand spiders. *Proceedings of the Zoological Society of London 1911*: 297–313.

HOPKINS, C. L. 1966a: A new species of *Limnesia* (Acari, Hydrachnellae) from New Zealand. *Transactions of the Royal Society of New Zealand 8 (Zoology)*: 1–4.

HOPKINS, C. L. 1966b: Two new species of water-mite (Acari, Hydrachnellae) from New Zealand. *Transactions of the Royal Society of New Zealand 8 (Zoology)*: 111–117.

HOPKINS, C. L. 1967: New genera and species of water mites (Acari, Hydrachnellae) from New Zealand. *Transactions of the Royal Society of New Zealand (Zoology) 10*: 33–44.

HOPKINS, C. L. 1969: New species of *Limnesia* and *Tryssaturus* (Acari, Hydrachnellae) from New Zealand. *Transactions of the Royal Society of New Zealand 11 (Biological Sciences)*: 89–92.

HOPKINS, C. L. 1975: New species of Hygrobatidae and Lebertiidae (Acari: Hydrachnella) from New Zealand. *Journal of the Royal Society of New Zealand 5*: 5–11.

HOPKINS, C. L.; SCHMINKE, H. K. 1970: A species of *Euwandesia* (Acari: Hydrachnellae) from New Zealand. *Acarologia 12*: 357–359.

HUNT, G. S. 1996: A preliminary phylogenetic analysis of Australian Triaenonychidae (Arachnida: Opiliones). *Revue Suisse de Zoologie, Hors Série*: 295–308.

HUSBAND, R. W.; ZHANG, Z.-Q. 2002: *Wetapolipus jamiesoni* gen. nov., spec. nov. (Acari: Podapolipidae), an ectoparasite of the mountain stone weta, Hemideina maori (Orthoptera: Anostostomatidae) from New Zealand. *Zootaxa* 125: 1–12.

IMAMURA, T. 1977: Two new water-mites (Acari, Hydrachnellae) from cave waters in New Zealand. *Journal of the Speleological Society of Japan 2*: 9–12.

IMAMURA, T. 1978: A new subgenus and species of troglobiontic water-mite from New Zealand. *Journal of the Speleological Society of Japan 3*: 41–43.

IMAMURA, T. 1979: One more new subgenus and a new species of troglobiontic water-mite from New Zealand. *Journal of the Speleological Society of Japan 4*: 27–30.

ISBISTER, G. K.; GRAY, M. R. 2003: White-tail spider bite: a prospective study of 130 definite bites by *Lampona* species. *Medical Journal of Australia 179*: 199–202..

JACKSON, H.; PARKS, T. N. 1989: Spider toxins: Recent applications in Neurobiology. *Annual Review of Neuroscience 12*: 405–414.

JOHNSTON, D. E. 1982: Acari. Pp. 111–117 *in*: Parker, S.P. (ed.), *Synopsis and Classification of Living Organisms*, Vol. 2. McGraw-Hill, New York. 1232 p.

JUDSON, M. L. I. 1997: Catalogue of the pseudoscorpion types (Arachnida: Chelonethi) in the Natural History Museum, London. *Occasional Papers on Systematic Entomology 11*: 1–54.

KOCH, L. 1871–83: *Die Arachniden Australiens, nach der Natur Beschreiben und Abgebildet.* Bauer & Raspe, Nürnburg. 1,489 p.

KOENIKE, F. 1900: Eine unbekannte *Eylais*-Form nebst einer Notiz zur Synonymie einer verwandten Art. *Zoologische Jahrbücher, Abtheilung für Systematik, Geographie und Biologie der Thiere 13*: 125–132.

KRANTZ, G. W. 1978: *A Manual of Acarology*. 2nd edn. Oregon State University, Corvallis. 509 p.

KRANTZ, G. W.; WALTER, D. E. (Eds) 2009: *A Manual of Acarology*. 3rd edn. Texas Tech University Press, Lubbock. 807 p.

LEE, D. C.; HUNTER, P. E. 1974: Arthropoda of the subantarctic islands of New Zealand. 6. Rhodacaridae (Acari: Mesostigmata). *New Zealand Journal of Zoology 1*: 295–328.

LIDGARD, S. 2008: Predation on marine bryozoan colonies: taxa, traits and tropic groups. *Marine Ecology Progress Series 359*: 117–131.

LIN, J. Z.; ZHANG, Z.-Q. 2005a: New Zealand species of *Fungitarsonemus* Cromroy (Acari: Tarsonemidae). *Systematic and Applied Acarology 10*: 83–110.

LIN, J.Z.; ZHANG, Z.-Q. 2005b. New Zealand species of *Steneotarsonemus* Beer (Acari: Tarsonemidae). *Zootaxa 1028*: 1–22.

LINDQUIST, E. E. 1984: Current theories on the evolution of major groups of Acari and on their relationships with other groups of Arachnida, with consequent implications for their classification. Pp. 28–62 *in*: Griffiths, D. A.; Bowman, C. E. (eds), *Acarology VI*. Vol. 1. John Wiley & Sons, New York.

LINDQUIST, E. E. 1994: Foreword. Pp. ix–xii *in*: Houck, M. A. (ed.), *Mites–Ecological and Evolutionary Analysis of Life-history Patterns*. Chapman & Hall, New York & London.

LUXTON, M. 1984: More marine littoral mites from New Zealand. *New Zealand Journal of Marine and Freshwater Research 18*: 291–302.

LUXTON, M. 1985: Cryptostigmata (Arachnida: Acari)–a concise review. *Fauna of New Zealand 7*: 1–106.

LUXTON, M. 1989: New taxa of intertidal mites (Acari). *Journal of Natural History 23*: 407–428.

MACFARLANE, R. P. 2005: Mites associated with bumble bees (*Bombus*: Apidae) in New Zealand. *Records of the Canterbury Museum 19*: 29–34.

MANSON, D. C. M. 1972: New species and new records of eriophyid mites (Acarina: Eriophyidae) from New Zealand and the Pacific area. *Acarologia 13*: 351–360.

MANSON, D. C. M. 1984a: Eriophyoidea except Eriophyinae (Arachnida: Acari). *Fauna of New Zealand 4*: 1–142.

MANSON, D. C. M. 1984b: Eriophyinae (Arachnida: Acari: Eriophyoidea). *Fauna of New Zealand 5*: 1–123.

MANSON, D. C. M. 1989: New species and records of eriophyid mites from New Zealand. *New Zealand Journal of Zoology 16*: 37–49.

MANSON, D. C. M.; GERSON, U. 1986: Eriophyid mites associated with New Zealand ferns. *New Zealand Journal of Zoology 13*: 117–129.

MARPLES, B. J. 1944: A new harvestman of the genus *Megalopsalis. Transactions of the Royal Society of New Zealand* 73: 313–314.

MARPLES, B. J. 1956: Spiders from the Three Kings Islands. *Records of the Auckland Institute and*

Museum 4: 329–342.
MARPLES, B. J. 1962: The Matachiinae, a group of cribellate spiders. *Journal of Zoology 44*: 701–720.
MARPLES, R. R. 1959: The dictynid spiders of New Zealand. *Transactions of the Royal Society of New Zealand 87*: 333–361.
MARTENS, J. 1980: Versuch eines phylogenetischen systems der Opiliones. *Verhandlungen 8 Internationaler Arachnologen-Kongress abgehalten ander Universität für Bodenkultur Wien*: 355–360.
MARTIN, N. A. 1978: *Siteroptes* (*Siteroptoides*) species with *Pediculaster*-like phoretomorphs (Acari: Tarsonemida: Pygemphoridae) from New Zealand and Polynesia. *New Zealand Journal of Zoology 5*: 121–155.
MAXMEN, A.; BROWNE, W. E.; MARTINDALE, M. Q.; GIRIBET, G. 2005: Neuroanatomy of sea spiders implies an appendicular origin of the protocerebral segment. *Nature 437*: 1144–1148.
MIRONOV, S.V. 1990: A review of feather mites of the genus *Scutomegninia* (Analgoidea: Avenzoariidae) living on cormorants. *Parasitologiya 24*: 43–55.
MIRONOV, S.V. 1994: Three new species of the feather mite genus *Bychovskiata* (Analgoidea: Avenzoariidae) from exotic plovers (Charadriiformes). *International Journal of Acarology 20*: 45–51.
MIRONOV, S.V.; DABERT, J. 1997: A systematic review of the feather mite genus *Bychovskiata* Dubinin (Analgoidea: Avenzoariidae) with description of 11 new species. *Mitteilungen aus dem Hamburger Zoologischen Museum und Institut 94*: 91–123.
MIRONOV S.V.; DABERT, J. 2007: Three new feather mite genera of the *Protolichus* generic group (Astigmata, Pterolichidae) from parrots (Aves, Psittaciformes) of the Old World. *Acta Parasitologica 52*: 386–402
MIRONOV, S.V.; GALLOWAY, T. D. 2001: New feather mite taxa (Acari: Angloidea) and mites collected from native and introduced birds of New Zealand. *Acarologia* 42: 85–101.
MIRONOV, S.V.; GALLOWAY, T. D. 2002: *Nymphicilichus perezae* gen. nov., sp. nov., a new feather mite (Astigmata: Pterolichidae) from the cockatiel, *Nymphicus hollandicus* (Psittaciformes: Cacatuidae). *Journal of the Royal Society of New Zealand 32*: 1–6.
MOYLE, B. 1989a: A description of *Tyrannochthonius noaensis* (Arachnida: Pseudoscorpionida: Chthoniidae). *New Zealand Entomologist 12*: 58–59.
MOYLE, B. 1989b: A description of *Tyrannochthonius tekauriensis* (Pseudoscorpionida: Chthoniidae). *New Zealand Entomologist 12*: 60–62.
MUCHMORE, W. B. 1971: Phoresy by North and Central American pseudoscorpions. *Proceedings of the Rochester Academy of Science 12*: 79–97.
NEWELL, I. M. 1984: Biology of the Antarctic Seas. 15. Antarctic Halacaroidea. *Antarctic Research Series, Washington 40*: 1–284.
NIEDBAŁA, W. 1993. New species of Euptyctima (Acari, Oribatida) from New Zealand. *New Zealand Journal of Zoology 20*: 137–159.
NIEDBAŁA, W. 2006: Supplement to the knowledge of ptyctimous mites (Acari: Oribatida) from Australian region. *Annales Zoologici (Warszawa) 56 (Suppl. 1)*: 99–156.
NIEDBAŁA, W.; COLLOFF, M. J. 1997: Euptyctime oribatid mites from Tasmanian rainforest (Acari: Oribatida). *Journal of Natural History 31*: 489–538.
NORTON, R. A. 1998: Morphological evidence for the evolutionary origin of Astigmata (Acari: Acariformes). *Experimental and Applied Acarology 22*: 559–594.
PAQUIN, P.; VINK, C. J.; DUPÉRRÉ, N. 2010: *Spiders of New Zealand: Annotated family key and species checklist.* Manaaki Whenua Press, Lincoln. 118 p
PARROTT, A. W. 1946: A systematic catalogue of New Zealand spiders. *Records of the Canterbury Museum 5*: 51–93.
PESIC, V.; SMIT, H.; DATRY, T. 2010: Water mites (Acari: Hydrachnidia) from the hyporheic waters of the Selwyn River (New Zealand), with descriptions of nine new species. *Zootaxa 2355*: 1–34.
PICKARD-CAMBRIDGE, O. 1879: On some new and rare spiders from New Zealand with characteristics of four new genera. *Proceedings of the Zoological Society of London 1879*: 279–293.
PINTO-DA-ROCHA, R.; MACHADO, G., GIRIBET, G. (Eds) 2008: *Harvestmen – The Biology of Opiliones.* Harvard University Press, Cambridge. 597 p.
PLATNICK, N. I. 1992: Patterns of biodiversity. Pp. 15–24 *in*: Eldredge, N. (ed.), *Systematics, Ecology and the Biodiversity Crisis.* Columbia University Press, New York. 220 p.
PLATNICK, N. I. 1995: An abundance of spiders! *Natural History 104(3)*: 50–53.
PLATNICK, N. I. 2009: The World Spider Catalog, Version 10.0. American Museum of Natural History. http://research.amnh.org/ entomology/spiders/catalog/index.html.
POINAR, G. O. Jr; CURCIC, B. P. M.; COKENDOLPHER, J. C. 1998: Arthropod phoresy involving pseudoscorpions in the past and present. *Acta Arachnologica 47*: 79–96.
POWELL, A. W. B. 1949: A second record of a king-crab from New Zealand waters. *Records of the Auckland Institute and Museum 3*: 372.
QIN, T.-K. 1998a: *Callipenthalodes*, a new genus of Penthalodidae (Acariformes: Eupodoidea) from New Zealand. *International Journal of Acarology 24*: 221–225.
QIN, T.-K. 1998b: A checklist and key to species of Eupodoidea (Acari: Prostigmata) from Australia and New Zealand, and their subantarctic islands. *Journal of the Royal Society of New Zealand 28*: 295–307.
QIN, T.-K.; HENDERSON, R. 2001: Taxonomy of New Zealand Prostigmata: Past, present, and future. Pp. 35–39 *in* Halliday, R.B.; Walter, D.E.; Proctor, H.C.; Norton, R.A.; Colloff, M.J. 2001: *Acarology 2000. Proceedings of the 10th International Congress of Acarology.* CSIRO Publishing, Melbourne.
QUICKE, D. 1988: Spiders bite their way towards safer insecticides. *New Scientist 1640*: 38–41.
RADEMACHER, M.; DERRAIK, J. G. B. 2009: White-tail spider bites. *ACC Review 42*: 1–2. [http://www.acc.co.nz/PRD_EXT_CSMP/groups/external_providers/documents/reports_results/prd_ctrb109760.pdf]
RAMSAY, G. W. 1960: Subfossil mites from the Hutt Valley. *Transactions of the Royal Society of New Zealand 88*: 575–576
RAMSAY, G. 1975: Mites and ticks. *New Zealand's Nature Heritage 5(64)*: 1790–1796.
RAVEN, R. J.; STUMKAT, K. 2003: Problem solving in the spider families Miturgidae, Ctenidae and Psechridae (Araneae) in Australia and New Zealand. *Journal of Arachnology 31*: 105–121.
RIECHERT, S. E.; LOCKLEY, T. C. 1984: Spiders as biological control agents. *Annual Review of Entomology 29*: 299–320.
ROEWER, C. F. 1923: *Die Weberknechte der Erde.* Gustav Fischer, Jena. v +1,116 p.
ROSENZWEIG, M. L. 1995: *Species Diversity in Space and Time.* Cambridge University Press, Cambridge. 436 p.
RYLAND, J. S. 1976: Physiology and ecology of marine bryozoans. *Advances in Marine Biology 14*: 417–421.
SAVORY, T. H. 1977: *Arachnida.* 2nd edn. Academic Press, London. 399 p.
SCHON, N. L.; MACKAY, A. D.; MINOR, M. A.; YEATES, G. W.; HEDLEY, M. J. 2008: Soil fauna in grazed New Zealand hill country pastures at two management intensities. *Applied Soil Ecology 40*: 218–228.
SCHWOERBEL, J. 1984: Subterrane Wassermilben aus Fliessgewässern Neuseelands (Acari: Actinedida). *Archiv für Hydrobiologie, Suppl. 66*: 293–306.
SCHULTZ, J. W. 1998: Phylogeny of Opiliones (Arachnida): an assessment of the the 'Cyphopalpatores' concept. *The Journal of Arachnology* 26: 257–272.
SCHULTZ, J. W. 2007: A phylogenetic analysis of the arachnid orders based on morphological characters. *Zoological Journal of the Linnean Society 150*: 221–265.
SIVETER, D. J.; SUTTON, M. D.; BRIGGS, D. E. G.; SIVETER, D. J. 2004: A Silurian sea spider. *Nature 431*: 978–980.
SMILEY, R. L. 1992: *The Predatory Mite Family Cunaxidae (Acari) of the World with a New Classification.* Indira Publishing, West Bloomfield, Michigan. 356 p.
SMIT, H. 1996a: The first record of water mites from the Chatham Islands, New Zealand (Acari, Hydrachnellae). *New Zealand Entomologist 19*: 57–59.
SMIT, H. 1996b: New species of water mites from New Zealand, with remarks on the water mites from ponds and lakes (Acari, Hydrachnellae). *Acarologia 37*: 45–53.
SMIT, H. 2002: Two new water mite species from New Zealand (Acari: Hydrachnidia). *Zootaxa 61*: 1–10.
SMITH, I. M. 1990: Proposal of Nudomideopsidae fam. nov. (Acari: Arrenuroidea) with a review of North American taxa and description of a new subgenus and species of *Nudomideopsis* Szalay, 1945. *Canadian Entomologist 122*: 229–252.
SOUTHCOTT, R.V. 1986: Australian larvae of the genus *Trombella* (Acarina: Trombidioidea). *Australian Journal of Zoology 34*: 611–646.
SOUTHCOTT, R.V. 1988: Two new larval Erythraeinae (Acarina: Erythraeidae) from New Zealand, and the larval Erythraeinae revised. *New Zealand Journal of Zoology 15*: 223–233.
SOUTHCOTT, R.V. 1991a: A further revision of *Charletonia* (Acarina: Erythraeidae) based on larvae, protonymphs and deutonymphs. *Invertebrate Taxonomy 5*: 61–131.
SOUTHCOTT, R.V. 1991b: Descriptions of larval *Leptus* (Acarina: Erythraeidae) ectoparasitic on Australian Diptera, and two earlier described Australian larvae. *Invertebrate Taxonomy 5*: 717–763.
SOUTHCOTT, R.V. 1993: Revision of the taxonomy of the larvae of the subfamily Eutrombidiinae (Acarina: Microtrombidiidae). *Invertebrate Taxonomy 7*: 885–959.
SOUTHCOTT, R.V. 1996: On some Australian and other larval Callidosomatinae (Acari: Erythraeidae). *International Journal of Acarology 22*: 253–278.
SOUTHCOTT, R.V. 1997: Revision of the larvae of *Paratrombium* (Acarina: Trombidiidae) of

Australia and Papua New Guinea, with notes on life histories. *Records of the South Australian Museum 29*: 95–120.
SPAIN, A.V.; LUXTON, M. 1971: Catalog and bibliography of the Acari of the New Zealand subregion. *Pacific Insects Monograph 25*: 179–226.
STAPLES, D. A.; WATSON, J. E. 1987: Associations between pycnogonids and hydroids. Pp. 215–226 *in*: Bouillon, J. (ed.), *Modern Trends in the Systematics, Ecology and Evolution of Hydroids and Hydroidomedusae*. Oxford University Press, Oxford.
STOCK, J. H. 1953: Biological results of the Snellius Expedition. XVII. Contribution to the knowledge of the pycnogonid fauna of the East Indian Archipelago. *Temminckia 9*: 276–313.
STOUT, V. M. 1953a: *Eylais waikawae* n.sp. (Hydracarina) and some features of its life history and anatomy. *Transactions of the Royal Society of New Zealand 81*: 389–416.
STOUT, V. M. 1953b: New species of Hydracarina, with a description of the life history of two. *Transactions of the Royal Society of New Zealand 81*: 417–466.
TAYLOR, C. K. 2004: New Zealand harvestmen of the subfamily Megalopsalidinae (Opiliones: Monoscutidae) the genus *Pantopsalis*. *Tuhinga 15*: 53–76.
TAYLOR, C. 2008: A new species of Monoscutinae (Arachnida, Opiliones, Monoscutidae) from New Zealand, with a redescription of *Monoscutum titirangiense*. *Journal of Arachnology 36*: 176–179.
URQUHART, A.T. 1885: On the spiders of New Zealand. *Transactions and Proceedings of the New Zealand Institute 17*: 31–53.
URQUHART, A.T. 1888: On new species of Araneidea. *Transactions and Proceedings of the New Zealand Institute 20*: 109–125.
URQUHART, A.T. 1892: Descriptions of new species of Araneae. *Transactions and Proceedings of the New Zealand Institute 24*: 230–253.
VIETS, K. 1949: Nomenklatorische und taxonomische Bemerkungen zur Kenntnis der Wassermilben (Hydrachnellae, Acari), 1–10. *Abhandlungen hrsg. vom Naturwissenschaften Verein zu Bremen 23*: 292–327.
VINK, C. J. 2002: Lycosidae (Arachnida: Araneae). *Fauna of New Zealand 44*: 1–94.
VINK, C. J.; PATERSON, A. M. 2003: Combined molecular and morphological phylogenetic analyses of the New Zealand wolf spider genus Anoteropsis (Araneae: Lycosidae). *Molecular Phylogenetics and Evolution. 28*: 576–587.
VINK, C. J.; SIRVID, P. J. 1998: The Oxyopidae (lynx spiders) of New Zealand. *New Zealand Entomologist 21*: 1–9.
VINK, C. J.; SIRVID, P. J. 2000: New synonymy between *Oxyopes gracilipes* (White) and *Oxyopes mundulus* L. Koch (Oxyopidae: Araneae). *Memoirs of the Queensland Museum 45*: 637–640.
WALCKANAER, C. A. 1837: *Histoire naturelle des Insectes. Aptères*. Vol. 1. Paris. 682 p.
WALOSZEK, D.; DUNLOP, J. A. 2002: A larval sea spider (Arthropoda: Pycnogonida) from the Upper Cambrian 'Orsten' of Sweden, and the phylogenetic position of pycnogonids. *Palaeontology 45*: 421–446.
WALTER, D. E.; PROCTOR, H. C. 1999: *Mites–Ecology, Evolution and Behaviour*. University of New South Wales Press and CABI Publishing, Sydney and Wallingford. 322 p.
WEBBER, J.; CHAPMAN, R. B. 2008: Ttiming of sulphur spray application for control of hazelnut big bud mites (*Phytoptus avellanae* and *Cecidophyopsis vermiformis*). *New Zealand Plant Protection 61*: 191–196.
WEYGOLDT, P. 1998: Review: Evolution and systematics of the Chelicerata. *Experimental and Applied Acarology 22*: 63–79.
WEYGOLDT, P.; PAULUS, H. F. 1979: Untersuchungen zur Morphologie, Taxonomie und Phylogenie der Chelicerata. II. Cladogramme und die Entfaltung der Chelicerata. *Zeitschrift für die Zoologische Systematik und Evolutionforschung 17*: 177–200.
WHEELER, W. C.; HAYASHI, C.Y. 1998: The phylogeny of the extant chelicerate orders. *Cladistics 14*: 173–192.
WHITE, A. 1849: Descriptions of apparently new species of Aptera from New Zealand. *Proceedings of the Zoological Society of London 17*: 3–6.
WHITE, J. 2003: Debunking spider bite myths. *Medical Journal of Australia 179*: 180–181.
WITH, C. J. 1907: On some new species of the Cheliferidae, Hans., and Garypidae, Hans., in the British Museum. *Journal of the Linnean Society of London, Zoology 30*: 49–85.
WOMERSLEY, H. 1935: A species of Acarina of the genus *Holothyrus* from Australia and New Zealand. *Annals and Magazine of Natural History, ser. 10, 16*: 154–157.
WOMERSLEY, H. 1953: An interesting new larval species of *Panisopsis* (Thyasidae, Acarina) from New Zealand. *Records of the Canterbury Museum, n.s., 6*: 233–235.
WYER, D.; KING, P. E. 1973: Relationships between British littoral and sublittoral bryozoans and pycnogonids. Pp. 199–207 *in*: Larwood, G.P. (ed.), *Living and Fossil Bryozoa*. Academic Press, London & New York. xvii + 634 p.
WYER, D.; KING, P. E. 1974: Feeding in British littoral pycnogonids. *Estuarine and Coastal Marine Studies 2*: 177–184.
XUE, X.-F.; ZHANG, Z.-Q. 2008: New Zealand Eriophyoidea (Acari: Prostigmata): an update with descriptions of one new genus and six new species. *Zootaxa 1962*: 1–32.
YEN, A. L. 1995: Australian spiders: an opportunity for conservation. *Records of the Western Australian Museum, Suppl. 52*: 39–47.
ZABKA, M.; POLLARD, S. D. 2002: A checklist of Salticidae (Arachnida: Araneae) of New Zealand. *Records of the Canterbury Museum 16*: 73–82.
ZHANG, Z.-Q. 2000: Notes on *Varroa destructor* (Acari Varroidae) parasitic on honeybees in New Zealand. *Systematic and Applied Acarology Special Publication 5*: 9–14.
ZHANG, Z.-Q.; BENNETT, S. 2009: Filling the gap in distribution of the genus *Sonotetranychus* (Acari: Tetranychidae) with a description of a new species from New Zealand. *Systematic and Applied Acarology 14*: 161–170.
ZHANG, Z.-Q.; FAN, Q.-H. 2005: A new genus of Neothrombiidae (Acari: Trombidioidea) from New Zealand. *Systematic and Applied Acarology 10*: 155–162.
ZHANG, Z.-Q.; FAN, Q.-H. 2007. Allotanaupodidae, a new family of early derivative Parasitengona (Acari: Prostigmata). *Zootaxa 1517*: 1–52.
ZHANG, Z.-Q.; LIANG, L. R. 1997: *An Illustrated Guide to Mites of Agricultural Importance*. Tongji University Press, Shanghai. 228 p.
ZHANG, Z.-Q.; MARTIN, N.A. 2001: A review of *Schizotetranychus*-like mites from New Zealand. *Journal of the Royal Society of New Zealand 31*: 307–325.
ZHANG, Z.-Q. & RHODE, B. E. 2003: A faunistic summary of acarine diversity in New Zealand. *Systematic and Applied Acarology 8*: 75–84.

Checklist of New Zealand Chelicerata

Genera endemic to New Zealand are underlined (first entry only). Single-letter codes indicate endemic species (E), adventive species (A), and new records (*). All species are taken to be terrestrial unless otherwise indicated, i.e. F, freshwater; M, marine. The mite (Acari) checklist is correct to 2009. For hosts of ectoparasites and phoretic (hitchhiking) species: Ma, mammal; B, bird; R, reptile; A, arthropod. † = Holocene extinction; *n.d.* = *nomen dubium*.

SUBPHYLUM CHELICERATA
Class ARACHNIDA
Subclass MEGOPERCULATA
Infraclass TETRAPULMONATA
Superorder LABELLATA
Order ARANEAE
Suborder OPISTHOTHELAE
Infraorder MYGALOMORPHAE
HEXATHELIDAE
Hexathele cantuaria Forster, 1968 E
Hexathele cavernicola Forster, 1968 E
Hexathele exemplar Parrott, 1960 E
Hexathele hochstetteri Ausserer, 1871 E
Hexathele huka Forster, 1968 E
Hexathele huttoni Hogg, 1908 E
Hexathele kohua Forster, 1968 E
Hexathele maitaia Forster, 1968 E
Hexathele nigra Forster, 1968 E
Hexathele otira Forster, 1968 E
Hexathele para Forster, 1968 E
Hexathele petriei Goyen, 1887 E
Hexathele pukea Forster, 1968 E

Hexathele putuna Forster, 1968 E
Hexathele ramsayi Forster, 1968 E
Hexathele rupicola Forster, 1968 E
Hexathele taumara Forster, 1968 E
Hexathele waipa Forster, 1968 E
Hexathele waita Forster, 1968 E
Hexathele wiltoni Forster, 1968 E
Porrhothele antipodiana (Walckenaer, 1837) E
Porrhothele blanda Forster, 1968 E
Porrhothele moana Forster, 1968 E
Porrhothele modesta Forster, 1968 E
Porrhothele quadrigyna Forster, 1968 E
IDIOPIDAE
Cantuaria abdita Forster, 1968 E
Cantuaria allani Forster, 1968 E
Cantuaria aperta Forster, 1968 E
Cantuaria apica Forster, 1968 E
Cantuaria assimilis Forster, 1968 E
Cantuaria borealis Forster, 1968 E
Cantuaria catlinensis Forster, 1968 E
Cantuaria cognata Forster, 1968 E
Cantuaria collensis (Todd, 1945) E
Cantuaria delli Forster, 1968 E
Cantuaria dendyi (Hogg, 1901) E
Cantuaria depressa Forster, 1968 E
Cantuaria dunedinensis Forster, 1968 E
Cantuaria gilliesi (O. Pickard-Cambridge, 1877) E
Cantuaria grandis Forster, 1968 E
Cantuaria huttoni (O. Pickard-Cambridge, 1879) E
Cantuaria insulana Forster, 1968 E
Cantuaria isolata Forster, 1968 E
Cantuaria johnsi Forster, 1968 E
Cantuaria kakahuensis Forster, 1968 E
Cantuaria kakanuiensis Forster, 1968 E
Cantuaria lomasi Forster, 1968 E
Cantuaria magna Forster, 1968 E
Cantuaria marplesi (Todd, 1945) E
Cantuaria maxima Forster, 1968 E
Cantuaria medialis Forster, 1968 E
Cantuaria minor Forster, 1968 E
Cantuaria myersi Forster, 1968 E
Cantuaria napua Forster, 1968 E
Cantuaria orepukiensis Forster, 1968 E
Cantuaria parrotti Forster, 1968 E
Cantuaria pilama Forster, 1968 E
Cantuaria prina Forster, 1968 E
Cantuaria reducta Forster, 1968 E
Cantuaria secunda Forster, 1968 E
Cantuaria sinclairi Forster, 1968 E
Cantuaria stephenensis Forster, 1968 E
Cantuaria stewarti (Todd, 1945) E
Cantuaria sylvatica Forster, 1968 E
Cantuaria toddae Forster, 1968 E
Cantuaria vellosa Forster, 1968 E
Cantuaria wanganuiensis (Todd, 1945) E
Gen. et n. spp. (20) 20E
MIGIDAE
Migas australis Wilton, 1968 E
Migas borealis Wilton, 1968 E
Migas cambridgei Wilton, 1968 E
Migas cantuarius Wilton, 1968 E
Migas centralis Wilton, 1968 E
Migas cumberi Wilton, 1968 E
Migas distinctus O. Pickard-Cambridge, 1879 E
Migas gatenbyi Wilton, 1968 E
Migas giveni Wilton, 1968 E
Migas goyeni Wilton, 1968 E
Migas hesperus Wilton, 1968 E
Migas hollowayi Wilton, 1968 E
Migas insularis Wilton, 1968 E
Migas kirki Wilton, 1968 E
Migas kochi Wilton, 1968 E
Migas linburnensis Wilton, 1968 E
Migas lomasi Wilton, 1968 E
Migas marplesi Wilton, 1968 E
Migas minor Wilton, 1968 E
Migas otari Wilton, 1968 E
Migas paradoxus L. Koch, 1873 E
Migas quintus Wilton, 1968 E
Migas sandageri Goyen, 1891 E
Migas saxatilis Wilton, 1968 E
Migas secundus Wilton, 1968 E
Migas solitarius Wilton, 1968 E
Migas taierii Todd, 1945 E
Migas tasmani Wilton, 1968 E
Migas toddae Wilton, 1968 E
Migas tuhoe Wilton, 1968 E
NEMESIIDAE
Stanwellia bipectinata (Todd, 1945) E
Stanwellia hapua (Forster, 1968) E
Stanwellia hollowayi (Forster, 1968) E
Stanwellia houhora (Forster, 1968) E
Stanwellia kaituna (Forster, 1968) E
Stanwellia media (Forster, 1968) E
Stanwellia puna (Forster, 1968) E
Stanwellia regia (Forster, 1968) E
Stanwellia taranga (Forster, 1968) E
Stanwellia tuna (Forster, 1968) E

Infraorder ARANEOMORPHAE
AGELENIDAE
Ahua dentata Forster & Wilton, 1973 E
Ahua insula Forster & Wilton, 1973 E
Ahua kaituna Forster & Wilton, 1973 E
Ahua vulgaris Forster & Wilton, 1973 E
Huka alba Forster & Wilton, 1973 E
Huka lobata Forster & Wilton, 1973 E
Huka minima Forster & Wilton, 1973 E
Huka minuta Forster & Wilton, 1973 E
Huka pallida Forster & Wilton, 1973 E
Mahura accola Forster & Wilton, 1973 E
Mahura bainhamensis Forster & Wilton, 1973 E
Mahura boara Forster & Wilton, 1973 E
Mahura crypta Forster & Wilton, 1973 E
Mahura detrita Forster & Wilton, 1973 E
Mahura hinua Forster & Wilton, 1973 E
Mahura musca Forster & Wilton, 1973 E
Mahura rubella Forster & Wilton, 1973 E
Mahura rufula Forster & Wilton, 1973 E
Mahura scuta Forster & Wilton, 1973 E
Mahura sorenseni Forster & Wilton, 1973 E
Mahura southgatei Forster & Wilton, 1973 E
Mahura spinosa Forster & Wilton, 1973 E
Mahura spinosoides Forster & Wilton, 1973 E
Mahura takahea Forster & Wilton, 1973 E
Mahura tarsa Forster & Wilton, 1973 E
Mahura turris Forster & Wilton, 1973 E
Mahura vella Forster & Wilton, 1973 E
Neoramia allanae Forster & Wilton, 1973 E
Neoramia alta Forster & Wilton, 1973 E
Neoramia charybdis (Hogg, 1910) E
Neoramia childi Forster & Wilton, 1973 E
Neoramia crucifera (Hogg, 1909) E
Neoramia finschi (L. Koch, 1872) E
Neoramia fiordensis Forster & Wilton, 1973 E
Neoramia hoggi (Forster, 1964) E
Neoramia hokina Forster & Wilton, 1973 E
Neoramia janus (Bryant, 1935) E
Neoramia koha Forster & Wilton, 1973 E
Neoramia komata Forster & Wilton, 1973 E
Neoramia mamoea Forster & Wilton, 1973 E
Neoramia marama Forster & Wilton, 1973 E
Neoramia margaretae Forster & Wilton, 1973 E
Neoramia matua Forster & Wilton, 1973 E
Neoramia minuta Forster & Wilton, 1973 E
Neoramia nana Forster & Wilton, 1973 E
Neoramia oroua Forster & Wilton, 1973 E
Neoramia otagoa Forster & Wilton, 1973 E
Neoramia raua Forster & Wilton, 1973 E
Neoramia setosa (Bryant, 1935) E
Neorepukia hama Forster & Wilton, 1973 E
Neorepukia pilama Forster & Wilton, 1973 E
Oramia chathamensis (Simon, 1899) E
Oramia littoralis Forster & Wilton, 1973 E
Oramia mackerrowi (Marples, 1959) E
Oramia marplesi Forster, 1964 E
Oramia occidentalis (Marples, 1959) E
Oramia rubrioides (Hogg, 1909) E
Oramia solanderensis Forster & Wilton, 1973 E
Oramiella wisei Forster & Wilton, 1973 E
Orepukia alta Forster & Wilton, 1973 E
Orepukia catlinsensis Forster & Wilton, 1973 E
Orepukia dugdalei Forster & Wilton, 1973 E
Orepukia egmontensis Forster & Wilton, 1973 E
Orepukia florae Forster & Wilton, 1973 E
Orepukia geophila Forster & Wilton, 1973 E
Orepukia grisea Forster & Wilton, 1973 E
Orepukia insula Forster & Wilton, 1973 E
Orepukia nota Forster & Wilton, 1973 E
Orepukia nummosa (Hogg, 1909) E
Orepukia orophila Forster & Wilton, 1973 E
Orepukia pallida Forster & Wilton, 1973 E
Orepukia poppelwelli Forster & Wilton, 1973 E
Orepukia prina Forster & Wilton, 1973 E
Orepukia rakiura Forster & Wilton, 1973 E
Orepukia redacta Forster & Wilton, 1973 E
Orepukia riparia Forster & Wilton, 1973 E
Orepukia sabua Forster & Wilton, 1973 E
Orepukia similis Forster & Wilton, 1973 E
Orepukia simplex Forster & Wilton, 1973 E
Orepukia sorenseni Forster & Wilton, 1973 E
Orepukia tanea Forster & Wilton, 1973 E
Orepukia tonga Forster & Wilton, 1973 E
Orepukia virtuta Forster & Wilton, 1973 E
Paramyro apicus Forster & Wilton, 1973 E
Paramyro parapicus Forster & Wilton, 1973 E
Porotaka detrita Forster & Wilton, 1973 E
Porotaka florae Forster & Wilton, 1973 E
Tararua celeripes (Urquhart, 1891) E
Tararua clara Forster & Wilton, 1973 E
Tararua diversa Forster & Wilton, 1973 E
Tararua foordi Forster & Wilton, 1973 E
Tararua puna Forster & Wilton, 1973 E
Tararua rutuma Forster & Wilton, 1973 E
Tararua versuta Forster & Wilton, 1973 E
Tegenaria domestica (Clerck, 1757) A
Tuapoka cavata Forster & Wilton, 1973 E
Tuapoka ovalis Forster & Wilton, 1973 E
Gen. et n. spp. (6) 6E
AMAUROBIIDAE
Auhunga pectinata Forster & Wilton, 1973 E
Maloides cavernicola (Forster & Wilton, 1973) E
Muritaia kaituna Forster & Wilton, 1973 E
Muritaia longispinata Forster & Wilton, 1973 E
Muritaia orientalis Forster & Wilton, 1973 E
Muritaia parabusa Forster & Wilton, 1973 E
Muritaia suba Forster & Wilton, 1973 E
Otira canasta Forster & Wilton, 1973 E
Otira indura Forster & Wilton, 1973 E
Otira liana Forster & Wilton, 1973 E
Otira parva Forster & Wilton, 1973 E
Otira satura Forster & Wilton, 1973 E
Otira terricola Forster & Wilton, 1973 E
Pakeha buechlerae Forster & Wilton, 1973 E
Pakeha duplex Forster & Wilton, 1973 E
Pakeha hiloa Forster & Wilton, 1973 E
Pakeha inornata Forster & Wilton, 1973 E
Pakeha insignita Forster & Wilton, 1973 E
Pakeha kirki (Hogg, 1909) E
Pakeha lobata Forster & Wilton, 1973 E
Pakeha manapouri Forster & Wilton, 1973 E
Pakeha maxima Forster & Wilton, 1973 E
Pakeha media Forster & Wilton, 1973 E
Pakeha minima Forster & Wilton, 1973 E
Pakeha paratecta Forster & Wilton, 1973 E

Pakeha parrotti Forster & Wilton, 1973 E
Pakeha protecta Forster & Wilton, 1973 E
Pakeha pula Forster & Wilton, 1973 E
Pakeha stewartia Forster & Wilton, 1973 E
Pakeha subtecta Forster & Wilton, 1973 E
Pakeha tecta Forster & Wilton, 1973 E
Paravoca opaca Forster & Wilton, 1973 E
Paravoca otagoensis Forster & Wilton, 1973 E
Poaka graminicola Forster & Wilton, 1973 E
Waitetola huttoni Forster & Wilton, 1973 E
AMPHINECTIDAE
Akatorea gracilis (Marples, 1959) E
Akatorea otagoensis Forster & Wilton, 1973 E
Amphinecta decemmaculata Simon, 1898 E
Amphinecta dejecta Forster & Wilton, 1973 E
Amphinecta luta Forster & Wilton, 1973 E
Amphinecta mara Forster & Wilton, 1973 E
Amphinecta milina Forster & Wilton, 1973 E
Amphinecta mula Forster & Wilton, 1973 E
Amphinecta pika Forster & Wilton, 1973 E
Amphinecta pila Forster & Wilton, 1973 E
Amphinecta puka Forster & Wilton, 1973 E
Amphinecta tama Forster & Wilton, 1973 E
Amphinecta tula Forster & Wilton, 1973 E
Aorangia agama Forster & Wilton, 1973 E
Aorangia ansa Forster & Wilton, 1973 E
Aorangia fiordensis Forster & Wilton, 1973 E
Aorangia isolata Forster & Wilton, 1973 E
Aorangia kapitiensis Forster & Wilton, 1973 E
Aorangia mauii Forster & Wilton, 1973 E
Aorangia muscicola Forster & Wilton, 1973 E
Aorangia obscura Forster & Wilton, 1973 E
Aorangia otira Forster & Wilton, 1973 E
Aorangia pilgrimi Forster & Wilton, 1973 E
Aorangia poppelwelli Forster & Wilton, 1973 E
Aorangia pudica Forster & Wilton, 1973 E
Aorangia semita Forster & Wilton, 1973 E
Aorangia silvestris Forster & Wilton, 1973 E
Aorangia singularis Forster & Wilton, 1973 E
Aorangia tumida Forster & Wilton, 1973 E
Dunstanoides angustiae (Marples, 1959) E
Dunstanoides hesperis (Forster & Wilton, 1973) E
Dunstanoides hinawa (Forster & Wilton, 1973) E
Dunstanoides hova (Forster & Wilton, 1973) E
Dunstanoides kochi (Forster & Wilton, 1973) E
Dunstanoides mira (Forster & Wilton, 1973) E
Dunstanoides montana (Forster & Wilton, 1973) E
Dunstanoides nuntia (Marples, 1959) E
Dunstanoides salmoni (Forster & Wilton, 1973) E
Holomamoea foveata Forster & Wilton, 1973 E
Huara antarctica (Berland, 1931) E
Huara chapmanae Forster & Wilton, 1973 E
Huara decorata Forster & Wilton, 1973 E
Huara dolosa Forster & Wilton, 1973 E
Huara grossa Forster, 1964 E
Huara hastata Forster & Wilton, 1973 E
Huara inflata Forster & Wilton, 1973 E
Huara kikkawa Forster & Wilton, 1973 E
Huara marplesi Forster & Wilton, 1973 E
Huara mura Forster & Wilton, 1973 E
Huara ovalis (Hogg, 1909) E
Huara pudica Forster & Wilton, 1973 E
Makora calypso (Marples, 1959) E
Makora detrita Forster & Wilton, 1973 E
Makora diversa Forster & Wilton, 1973 E
Makora figurata Forster & Wilton, 1973 E
Makora mimica Forster & Wilton, 1973 E
Mamoea assimilis Forster & Wilton, 1973 E
Mamoea bicolor (Bryant, 1935) E
Mamoea cantuaria Forster & Wilton, 1973 E
Mamoea cooki Forster & Wilton, 1973 E
Mamoea florae Forster & Wilton, 1973 E
Mamoea grandiosa Forster & Wilton, 1973 E
Mamoea hesperis Forster & Wilton, 1973 E
Mamoea hughsoni Forster & Wilton, 1973 E
Mamoea inornata Forster & Wilton, 1973 E
Mamoea mandibularis (Bryant, 1935) E
Mamoea maorica Forster & Wilton, 1973 E
Mamoea montana Forster & Wilton, 1973 E
Mamoea monticola Forster & Wilton, 1973 E
Mamoea otira Forster & Wilton, 1973 E
Mamoea pilosa (Bryant, 1935) E
Mamoea rakiura Forster & Wilton, 1973 E
Mamoea rufa (Berland, 1931) E
Mamoea unica Forster & Wilton, 1973 E
Mamoea westlandica Forster & Wilton, 1973 E
Maniho australis Forster & Wilton, 1973 E
Maniho cantuarius Forster & Wilton, 1973 E
Maniho centralis Forster & Wilton, 1973 E
Maniho insulanus Forster & Wilton, 1973 E
Maniho meridionalis Forster & Wilton, 1973 E
Maniho ngaitahu Forster & Wilton, 1973 E
Maniho otagoensis Forster & Wilton, 1973 E
Maniho pumilio Forster & Wilton, 1973 E
Maniho tigris Marples, 1959 E
Maniho vulgaris Forster & Wilton, 1973 E
Marplesia dugdalei Forster & Wilton, 1973 E
Marplesia pohara Forster & Wilton, 1973 E
Neolana dalmasi (Marples, 1959) E
Neolana pallida Forster & Wilton, 1973 E
Neolana septentrionalis Forster & Wilton, 1973 E
Neororea homerica Forster & Wilton, 1973 E
Neororea sorenseni (Forster, 1955) E
Oparara karamea Forster & Wilton, 1973 E
Oparara vallus (Marples, 1959) E
Paramamoea aquilonalis Forster & Wilton, 1973 E
Paramamoea arawa Forster & Wilton, 1973 E
Paramamoea incerta Forster & Wilton, 1973 E
Paramamoea incertoides Forster & Wilton, 1973 E
Paramamoea insulana Forster & Wilton, 1973 E
Paramamoea pandora Forster & Wilton, 1973 E
Paramamoea paradisica Forster & Wilton, 1973 E
Paramamoea parva Forster & Wilton, 1973 E
Paramamoea urewera Forster & Wilton, 1973 E
Paramamoea waipoua Forster & Wilton, 1973 E
Rangitata peelensis Forster & Wilton, 1973 E
Reinga apica Forster & Wilton, 1973 E
Reinga aucklandensis (Marples, 1959) E
Reinga grossa Forster & Wilton, 1973 E
Reinga media Forster & Wilton, 1973 E
Reinga waipoua Forster & Wilton, 1973 E
Rorea aucklandensis Forster & Wilton, 1973 E
Rorea otagoensis Forster & Wilton, 1973 E
Waterea cornigera Forster & Wilton, 1973 E
Gen et. n. sp. E
ANAPIDAE
Novanapis spinipes (Forster, 1951) E
Paranapis insula (Forster, 1951) E
Paranapis isolata Platnick & Forster, 1989 E
Zealanapis armata (Forster, 1951) E
Zealanapis australis (Forster, 1951) E
Zealanapis conica (Forster, 1951) E
Zealanapis insula Platnick & Forster, 1989 E
Zealanapis kuscheli Platnick & Forster, 1989 E
Zealanapis matua Platnick & Forster, 1989 E
Zealanapis montana Platnick & Forster, 1989 E
Zealanapis otago Platnick & Forster, 1989 E
Zealanapis punta Platnick & Forster, 1989 E
Zealanapis waipoua Platnick & Forster, 1989 E
ANYPHAENIDAE
Amaurobioides major Forster, 1970 E
Amaurobioides maritima O. Pickard-Cambridge, 1883 E
Amaurobioides minor Forster, 1970 E
Amaurobioides pallida Forster, 1970 E
Amaurobioides picuna Forster, 1970 E
Amaurobioides piscator Hogg, 1909 E
Amaurobioides pleta Forster, 1970 E
Amaurobioides pohara Forster, 1970 E
ARANEIDAE
Acroaspis decorosa (Urquhart, 1894)
Arachnura feredayi (L. Koch, 1871)
Argiope protensa L. Koch, 1872
Backobourkia brounii (Urquhart, 1885)
Celaenia atkinsoni (O. Pickard-Cambridge, 1879)
Celaenia excavata (L. Koch, 1867) A
Celaenia hectori (O. Pickard-Cambridge, 1879) E
Celaenia olivacea (Urquhart, 1885) E
Celaenia penna (Urquhart, 1887) E
Celaenia tuberosa (Urquhart, 1889) E
Colaranea brunnea Court & Forster, 1988 E
Colaranea melanoviridis Court & Forster, 1988 E
Colaranea verutum (Urquhart, 1887) E
Colaranea viriditas (Urquhart, 1887) E
Cryptaranea albolineata (Urquhart, 1893) E
Cryptaranea atrihastula (Urquhart, 1891) E
Cryptaranea invisibilis (Urquhart, 1892) E
Cryptaranea stewartensis Court & Forster, 1988 E
Cryptaranea subalpina Court & Forster, 1988 E
Cryptaranea subcompta (Urquhart, 1887) E
Cryptaranea venustula (Urquhart, 1891) E
Cyclosa trilobata (Urquhart, 1885)
Eriophora pustulosa (Walckenaer, 1841)
Neoscona orientalis (Urquhart, 1887) E
Novakiella trituberculosa (Roewer, 1942)
Novaranea queribunda (Keyserling, 1887) E
Poecilopachys australasia (Griffith & Pidgeon, 1833) A
Prasonica plagiata (Dalmas, 1917) E
Zealaranea crassa (Walckenaer, 1841) E
Zealaranea prina Court & Forster, 1988 E
Zealaranea saxatilis (Urquhart, 1887) E
Zealaranea trinotata (Urquhart, 1890) E
Zygiella x-notata (Clerck, 1757) A
Gen et. n. sp. E
CLUBIONIDAE
Clubiona blesti Forster, 1979 E
Clubiona cada Forster, 1979 E
Clubiona cambridgei L. Koch, 1873 E
Clubiona chathamensis Simon, 1905 E
Clubiona chevronia Urquhart, 1892 E *n.d.*
Clubiona clima Forster, 1979 E
Clubiona consensa Forster, 1979 E
Clubiona contrita Forster, 1979 E
Clubiona convoluta Forster, 1979 E
Clubiona delicata Forster, 1979 E
Clubiona huttoni Forster, 1979 E
Clubiona nitida Urquhart, 1893 E *n.d.*
Clubiona peculiaris L. Koch, 1873 E
Clubiona producta Forster, 1979 E
Clubiona scatula Forster, 1979 E
Clubiona torta Forster, 1979 E
CORINNIDAE
Supunna picta (L. Koch), 1873) A
CYATHOLIPIDAE
Hanea paturau Forster, 1988 E
Tekella absidata Urquhart, 1894 E
Tekella bisetosa Forster, 1988 E
Tekella lineata Forster, 1988 E
Tekella nemoralis (Urquhart, 1889) E
Tekella unisetosa Forster, 1988 E
Tekelloides australis Forster, 1988 E
Tekelloides flavonotatus (Urquhart, 1891) E
CYCLOCTENIDAE
Cycloctenus agilis Forster, 1979 E
Cycloctenus centralis Forster, 1979 E
Cycloctenus duplex Forster, 1979 E
Cycloctenus fiordensis Forster, 1979 E
Cycloctenus fugax Goyen, 1890 E
Cycloctenus lepidus Urquhart, 1890 E
Cycloctenus nelsonensis Forster, 1979 E
Cycloctenus paturau Forster, 1979 E
Cycloctenus pulcher Urquhart, 1891 E
Cycloctenus westlandicus Forster, 1964 E
Plectophanes altus Forster, 1964 E
Plectophanes archeyi Forster, 1964 E

Plectophanes frontalis Bryant, 1935 E
Plectophanes hollowayae Forster, 1964 E
Plectophanes pilgrimi Forster, 1964 E
Toxopsiella alpina Forster, 1964 E
Toxopsiella australis Forster, 1964 E
Toxopsiella centralis Forster, 1964 E
Toxopsiella dugdalei Forster, 1964 E
Toxopsiella horningi Forster, 1979 E
Toxopsiella lawrencei Forster, 1964 E
Toxopsiella medialis Forster, 1964 E
Toxopsiella minuta Forster, 1964 E
Toxopsiella nelsonensis Forster, 1979 E
Toxopsiella orientalis Forster, 1964 E
Toxopsiella parrotti Forster, 1964 E
Toxopsiella perplexa Forster, 1964 E
Uzakia unica (Forster, 1970) E
Gen et. n. spp. (2) 2E
DESIDAE
Badumna insignis (L. Koch, 1872) A
Badumna longinqua (L. Koch, 1867) A
Desis marina (Hector, 1877) M
Gasparia busa Forster, 1970 E
Gasparia coriacea Forster, 1970 E
Gasparia delli (Forster, 1955) E
Gasparia dentata Forster, 1970 E
Gasparia edwardsi Forster, 1970 E
Gasparia kaiangaroa Forster, 1970 E
Gasparia littoralis Forster, 1970 E
Gasparia lomasi Forster, 1970 E
Gasparia mangamuka Forster, 1970 E
Gasparia manneringi (Forster, 1964) E
Gasparia montana Forster, 1970 E
Gasparia nava Forster, 1970 E
Gasparia nebulosa Marples, 1956 E
Gasparia nelsonensis Forster, 1970 E
Gasparia nuntia Forster, 1970 E
Gasparia oparara Forster, 1970 E
Gasparia parva Forster, 1970 E
Gasparia pluta Forster, 1970 E
Gasparia rupicola Forster, 1970 E
Gasparia rustica Forster, 1970 E
Gasparia tepakia Forster, 1970 E
Gasparia tuaiensis Forster, 1970 E
Gohia clarki Forster, 1964 E
Gohia falxiata (Hogg, 1909) E
Gohia isolata Forster, 1970 E
Gohia parisolata Forster, 1970 E
Goyenia electa Forster, 1970 E
Goyenia fresa Forster, 1970 E
Goyenia gratiosa Forster, 1970 E
Goyenia lucrosa Forster, 1970 E
Goyenia marplesi Forster, 1970 E
Goyenia multidentata Forster, 1970 E
Goyenia ornata Forster, 1970 E
Goyenia sana Forster, 1970 E
Goyenia scitula Forster, 1970 E
Goyenia sylvatica Forster, 1970 E
Hapona amira Forster, 1970 E
Hapona aucklandensis (Forster, 1964) E
Hapona crypta (Forster, 1964) E
Hapona insula (Forster, 1964) E
Hapona marplesi (Forster, 1964) E
Hapona moana Forster, 1970 E
Hapona momona Forster, 1970 E
Hapona muscicola (Forster, 1964) E
Hapona otagoa (Forster, 1964) E
Hapona paihia Forster, 1970 E
Hapona reinga Forster, 1970 E
Hapona salmoni (Forster, 1964) E
Hapona tararua Forster, 1970 E
Helsonia plata Forster, 1970 E
Hulua convoluta Forster & Wilton, 1973 E
Hulua manga Forster & Wilton, 1973 E
Hulua minima Forster & Wilton, 1973 E
Hulua pana Forster & Wilton, 1973 E
Laestrygones albiceres Urquhart, 1894 E
Laestrygones chathamensis Forster, 1970 E
Laestrygones minutissimus (Hogg, 1909) E
Laestrygones otagoensis Forster, 1970 E
Laestrygones westlandicus Forster, 1970 E
Lamina minor Forster, 1970 E
Lamina montana Forster, 1970 E
Lamina parana Forster, 1970 E
Lamina ulva Forster, 1970 E
Mesudus frondosa (Forster, 1970) E
Mesudus setosa (Forster, 1970) E
Mesudus solitaria (Forster, 1970) E
Mangareia maculata Forster, 1970 E
Mangareia motu Forster, 1970 E
Matachia australis Forster, 1970 E
Matachia livor (Urquhart, 1893) E
Matachia marplesi Forster, 1970 E
Matachia ramulicola Dalmas, 1917 E
Matachia similis Forster, 1970 E
Myro marinus (Goyen, 1890) E
Neomyro amplius Forster & Wilton, 1973 E
Neomyro circe Forster & Wilton, 1973 E
Neomyro scitulus (Urquhart, 1891) E
Notomatachia cantuaria Forster, 1970 E
Notomatachia hirsuta (Marples, 1962) E
Notomatachia wiltoni Forster, 1970 E
Nuisiana arboris (Marples, 1959) E
Otagoa chathamensis Forster, 1970 E
Otagoa nova Forster, 1970 E
Otagoa wiltoni Forster, 1970 E
Panoa contorta Forster, 1970 E
Panoa fiordensis Forster, 1970 E
Panoa mora Forster, 1970 E
Panoa tapanuiensis Forster, 1970 E
Rapua australis Forster, 1970 E
Toxopsoides huttoni Forster & Wilton, 1973 E
Tuakana mirada Forster, 1970 E
Tuakana wiltoni Forster, 1970 E
DICTYNIDAE
Arangina cornigera (Dalmas, 1917) E
Arangina pluva Forster, 1970 E
'Dictyna' urquharti Roewer, 1951 E *n.d.*
Paradictyna ilamia Forster, 1970 E
Paradictyna rufoflava (Chamberlain, 1946) E
Viridictyna australis Forster, 1970 E
Viridictyna kikkawai Forster, 1970 E
Viridictyna nelsonensis Forster, 1970 E
Viridictyna parva Forster, 1970 E
Viridictyna picata Forster, 1970 E
DYSDERIDAE
Dysdera crocata C.L. Koch, 1838 A
GNAPHOSIDAE
Anzacia gemmea (Dalmas, 1917) A
Hemicloea rogenhoferi L. Koch, 1875 A
Hemicloea sundevalli Thorell, 1870 A
Hypodrassodes apicus Forster, 1979 E
Hypodrassodes courti Forster, 1979 E
Hypodrassodes crassus Forster, 1979 E
Hypodrassodes dalmasi Forster, 1979 E
Hypodrassodes insulanus Forster, 1979 E
Hypodrassodes isopus Forster, 1979 E
Hypodrassodes maoricus (Dalmas, 1917) E
Intruda signata (Hogg, 1900) A
Kaitawa insulare (Marples, 1956) E
Matua festiva Forster, 1979 E
Matua valida Forster, 1979 E
Nauhea tapa Forster, 1979 E
Notiodrassus distinctus Bryant, 1935 E
Notiodrassus fiordensis Forster, 1979 E
Scotophaeus pretiosus (L. Koch, 1873) E
Zelanda elongata (Forster, 1979) E
Zelanda erebus (L. Koch, 1873) E
Zelanda kaituna (Forster, 1979) E
Zelanda miranda(Forster, 1979) E
Zelanda obtusa (Forster, 1979) E
Zelanda titirangia (Ovtsharenko, Fedoryak & Zakharov, 2006 E
GRADUNGULIDAE
Gradungula sorenseni Forster, 1955 E
Pianoa isolata Forster, 1987 E
Spelungula cavernicola Forster, 1987 E
HAHNIIDAE
Alistra centralis (Forster, 1970) E
Alistra inanga (Forster, 1970) E
Alistra mangareia (Forster, 1970) E
Alistra napua (Forster, 1970) E
Alistra opina (Forster, 1970) E
Alistra reinga (Forster, 1970) E
Alistra tuna (Forster, 1970) E
Kapanga alta Forster, 1970 E
Kapanga festiva Forster, 1970 E
Kapanga grana Forster, 1970 E
Kapanga hickmani (Forster, 1964) E
Kapanga isulata (Forster, 1970) E
Kapanga luana Forster, 1970 E
Kapanga mana Forster, 1970 E
Kapanga manga Forster, 1970 E
Kapanga solitaria (Bryant, 1935) E
Kapanga wiltoni Forster, 1970 E
Porioides rima (Forster, 1970) E
Porioides tasmani (Forster, 1970) E
Rinawa bola Forster, 1970 E
Rinawa cantuaria Forster, 1970 E
Rinawa otagoensis Forster, 1970 E
Rinawa pula Forster, 1970 E
Scotospilus divisus (Forster, 1970) E
Scotospilus nelsonensis (Forster, 1970) E
Scotospilus plenus (Forster, 1970) E
Scotospilus westlandicus (Forster, 1970) E
Gen et. n. spp. (5) 5E
HOLARCHAEIDAE
Holarchaea novaeseelandiae (Forster, 1949) E
Holarchaea n. spp. (5) 5E
HUTTONIIDAE E
Huttonia palpimanoides O. Pickard-Cambridge, 1879 E
Huttonia n. spp. (21) 21E
LAMPONIDAE
Lampona cylindrata (L. Koch, 1866) A
Lampona murina L. Koch, 1873 A
LINYPHIIDAE
Araeoncus humilis (Blackwall, 1841) A
Cassafroneta forsteri Blest, 1979 E
Diplocephalus cristatus (Blackwall, 1833) A
Diploplecta adjacens Millidge, 1988 E
Diploplecta communis Millidge, 1988 E
Diploplecta duplex Millidge, 1988 E
Diploplecta nuda Millidge, 1988 E
Diploplecta opaca Millidge, 1988 E
Diploplecta proxima Millidge, 1988 E
Diploplecta simplex Millidge, 1988 E
Drapetisca australis Forster, 1955 E
Dunedinia decolor Millidge, 1988 E
Dunedinia denticulata Millidge, 1988 E
Dunedinia pullata Millidge, 1988 E
Erigone prominens Bösenberg & Strand, 1906 A
Erigone wiltoni Locket, 1973 A
Haplinis abbreviata (Blest, 1979) E
Haplinis alticola Blest & Vink, 2002
Haplinis anomala Blest & Vink, 2003 E
Haplinis antipodiana Blest & Vink, 2002 E
Haplinis attenuata Blest & Vink, 2002 E
Haplinis australis Blest & Vink, 2003 E
Haplinis banksi (Blest, 1979) E
Haplinis brevipes (Blest, 1979) E
Haplinis chiltoni (Hogg, 1911) E
Haplinis contorta (Blest, 1979) E
Haplinis diloris (Urquhart, 1886) E
Haplinis dunstani (Blest, 1979) E
Haplinis exigua Blest & Vink, 2002 E

Haplinis fluviatilis (Blest, 1979) E
Haplinis fucatinia (Urquhart, 1894) E
Haplinis fulvolineata Blest & Vink, 2002 E
Haplinis horningi (Blest, 1979) E
Haplinis inexacta (Blest, 1979) E
Haplinis innotabilis (Blest, 1979) E
Haplinis insignis (Blest, 1979) E
Haplinis major (Blest, 1979) E
Haplinis marplesi Blest & Vink, 2003 E
Haplinis minutissima (Blest, 1979) E
Haplinis morainicola Blest & Vink, 2002 E
Haplinis mundenia (Urquhart, 1894) E
Haplinis paradoxa (Blest, 1979) E
Haplinis redacta (Blest, 1979) E
Haplinis rufocephala (Urquhart, 1888) E
Haplinis rupicola (Blest, 1979) E
Haplinis silvicola (Blest, 1979) E
Haplinis similis (Blest, 1979) E
Haplinis subclathrata Simon, 1894 E
Haplinis subdola (O. Pickard-Cambridge, 1879) E
Haplinis subtilis Blest & Vink, 2002 E
Haplinis taranakii (Blest, 1979) E
Haplinis tegulata (Blest, 1979) E
Haplinis titan (Blest, 1979) E
Haplinis tokaanuae Blest & Vink, 2002 E
Haplinis wairarapa Blest & Vink, 2002 E
Hyperafroneta obscura Blest, 1979 E
Laetesia amoena Millidge, 1988 E
Laetesia aucklandensis (Forster, 1964) E
Laetesia bellissima Millidge, 1988 E
Laetesia chathami Millidge, 1988 E
Laetesia distincta Millidge, 1988 E
Laetesia germana Millidge, 1988 E
Laetesia intermedia Blest & Vink, 2003 E
Laetesia minor Millidge, 1988 E
Laetesia olvidada Blest & Vink, 2003 E
Laetesia paragermana Blest & Vink, 2003 E
Laetesia peramoena (O. Pickard-Cambridge, 1879) E
Laetesia prominens Millidge, 1988 E
Laetesia pseudamoena Blest & Vink, 2003 E
Laetesia pulcherrima Blest & Vink, 2003 E
Laetesia trispathulata (Urquhart, 1886) E
Laperousea blattifera (Urquhart, 1887)
Lessertia dentichelis (Simon, 1884) A
'*Linyphia*' *albiapiata* Urquhart, 1891 E *n.d.*
'*Linyphia*' *cruenta* Urquhart, 1891 E *n.d.*
'*Linyphia*' *multicolor* Urquhart, 1891 E *n.d.*
'*Linyphia*' *pellos* Urquhart, 1891 E *n.d.*
Maorineta acerba Millidge, 1988 E
Maorineta gentilis Millidge, 1988 E
Maorineta minor Millidge, 1988 E
Maorineta mollis Millidge, 1988 E
Maorineta sulcata Millidge, 1988 E
Maorineta tibialis Millidge, 1988 E
Maorineta tumida Millidge, 1988 E
Megafroneta dugdaleae Blest & Vink, 2002 E
Megafroneta elongata Blest, 1979 E
Megafroneta gigas Blest, 1979 E
Mermessus fradeorum (Berland, 1932) A
Metafroneta minima Blest, 1979 E
Metafroneta sinuosa Blest, 1979 E
Metafroneta subversa Blest & Vink, 2002 E
Metamynoglenes absurda Blest & Vink, 2002 E
Metamynoglenes attenuata Blest, 1979 E
Metamynoglenes flagellata Blest, 1979 E
Metamynoglenes gracilis Blest, 1979 E
Metamynoglenes helicoides Blest, 1979 E
Metamynoglenes incurvata Blest, 1979 E
Metamynoglenes magna Blest, 1979 E
Metamynoglenes ngongotaha Blest & Vink, 2002 E
Microctenonyx subitaneus (O. Pickard-Cambridge, 1875) A
Novafroneta annulipes Blest, 1979 E
Novafroneta gladiatrix Blest, 1979 E
Novafroneta nova Blest & Vink, 2003 E
Novafroneta parmulata Blest, 1979 E
Novafroneta truncata Blest & Vink, 2003 E
Novafroneta vulgaris Blest, 1979 E
Novalaetesia anceps Millidge, 1988 E
Novalaetesia atra Blest & Vink, 2003 E
Ostearius melanopygius (O. Pickard-Cambridge, 1879) A
Parafroneta ambigua Blest, 1979 E
Parafroneta confusa Blest, 1979 E
Parafroneta demota Blest & Vink, 2002 E
Parafroneta haurokoae Blest & Vink, 2002 E
Parafroneta hirsuta Blest & Vink, 2003 E
Parafroneta insula Blest, 1979 E
Parafroneta marrineri (Hogg, 1909) E
Parafroneta minuta Blest, 1979 E
Parafroneta monticola Blest, 1979 E
Parafroneta persimilis Blest, 1979 E
Parafroneta pilosa Blest & Vink, 2003 E
Parafroneta subalpina Blest & Vink, 2002 E
Parafroneta subantarctica Blest, 1979 E
Parafroneta westlandica Blest & Vink, 2002 E
Poecilafroneta caudata Blest, 1979 E
Promynoglenes grandis Blest, 1979 E
Promynoglenes minuscula Blest & Vink, 2002 E
Promynoglenes minuta Blest & Vink, 2002 E
Promynoglenes nobilis Blest, 1979 E
Promynoglenes parvula Blest, 1979 E
Promynoglenes silvestris Blest, 1979 E
Protoerigone obtusa Blest, 1979 E
Protoerigone otagoa Blest, 1979 E
Pseudafroneta frigida Blest, 1979 E
Pseudafroneta incerta (Bryant, 1935) E
Pseudafroneta lineata Blest, 1979 E
Pseudafroneta maxima Blest, 1979 E
Pseudafroneta pallida Blest, 1979 E
Pseudafroneta perplexa Blest, 1979 E
Pseudafroneta prominula Blest, 1979 E
Tenuiphantes tenuis (Blackwall, 1852) A
LYCOSIDAE
Allotrochosina schauinslandi (Simon, 1889) E
Anoteropsis adumbrata (Urquhart, 1887) E
Anoteropsis aerescens (Goyen, 1887) E
Anoteropsis alpina Vink, 2002 E
Anoteropsis arenivaga (Dalmas, 1917) E
Anoteropsis blesti Vink, 2002 E
Anoteropsis canescens (Goyen, 1887) E
Anoteropsis cantuaria Vink, 2002 E
Anoteropsis flavescens L. Koch, 1878 E
Anoteropsis forsteri Vink, 2002 E
Anoteropsis hallae Vink, 2002 E
Anoteropsis hilaris (L. Koch, 1877) E
Anoteropsis insularis Vink, 2002 E
Anoteropsis lacustris Vink, 2002 E
Anoteropsis litoralis Vink, 2002 E
Anoteropsis montana Vink, 2002 E
Anoteropsis okatainae Vink, 2002 E
Anoteropsis ralphi (Simon, 1905) E
Anoteropsis senica (L. Koch, 1877) E
Anoteropsis urquharti (Simon, 1898) E
Anoteropsis westlandica Vink, 2002 E
Artoria hospita Vink, 2002 E
Artoria segrega Vink, 2002 E
Artoria separata Vink, 2002 E
Hogna crispipes (L. Koch, 1877)
Notocosa bellicosa (Goyen, 1888) E
Venatrix konei (Berland, 1924)
MALKARIDAE
Gen. et n. spp. (12) 12E
MECYSMAUCHENIIDAE
Aotearoa magna (Forster, 1949) E
Zearchaea clypeata Wilton, 1946 E
Zearchaea fiordensis Forster, 1955 E
Gen. et n. spp. (13) 13E
MICROPHOLCOMMATIDAE
Algidiella aucklandica (Forster, 1955) E
Pua novaezealandiae (Forster, 1959) E
Rayforstia antipoda (Forster, 1959) E
Rayforstia insula (Forster, 1959) E
Rayforstia mcfarlanei (Forster, 1959) E
Rayforstia plebeia (Forster, 1959) E
Rayforstia propinqua (Forster, 1959)
Rayforstia salmoni (Forster, 1959) E
Rayforstia scuta (Forster, 1959) E
Rayforstia signata (Forster, 1959) E
Rayforstia vulgaris (Forster, 1959) E
Rayforstia wisei (Forster, 1964) E
Taliniella nigra (Forster, 1959) E
Taliniella vinki Rix & Harvey, 2010 E
Taphiassa punctata (Forster, 1959) E
Tinytrella pusilla (Forster, 1959) E
Gen. et spp. (10) 10E
MIMETIDAE
Australomimetus mendicus (O. Pickard-Cambridge, 1879) E
Australomimetus sennio (Urquhart, 1891) E
Gen. et n. spp. (7) 6E
MITURGIDAE
Cheiracanthium stratioticum L. Koch, 1873
Pacificana cockayni Hogg, 1904 E
Zealoctenus cardronaensis Forster & Wilton, 1973 E
Gen. et n. spp. (5) 5E
MYSMENIDAE
Gen. et n. spp. (16) 16E
NEPHILIDAE
Nephila edulis (Labillardière, 1799)
NICODAMIDAE
Forstertyna marplesi (Forster, 1970) E
Megadictyna thilenii Dahl, 1906 E
OECOBIIDAE
Oecobius navus Blackwall, 1859 A
OONOPIDAE
Kapitia obscura Forster, 1956 E
ORSOLOBIDAE
Anopsolobus subterraneus Forster & Platnick, 1985 E
Ascuta australis Forster, 1956 E
Ascuta cantuaria Forster & Platnick, 1985 E
Ascuta inopinata Forster, 1956 E
Ascuta insula Forster & Platnick, 1985 E
Ascuta leith Forster & Platnick, 1985 E
Ascuta media Forster, 1956 E
Ascuta monowai Forster & Platnick, 1985 E
Ascuta montana Forster & Platnick, 1985 E
Ascuta musca Forster & Platnick, 1985 E
Ascuta ornata Forster, 1956 E
Ascuta parornata Forster & Platnick, 1985 E
Ascuta taupo Forster & Platnick, 1985 E
Ascuta tongariro Forster & Platnick, 1985 E
Ascuta univa Forster & Platnick, 1985 E
Bealeyia unicolor Forster & Platnick, 1985 E
Dugdalea oculata Forster & Platnick, 1985 E
Duripelta alta Forster & Platnick, 1985 E
Duripelta australis Forster, 1956 E
Duripelta borealis Forster, 1956 E
Duripelta egmont Forster & Platnick, 1985 E
Duripelta hunua Forster & Platnick, 1985 E
Duripelta koomaa Forster & Platnick, 1985 E
Duripelta mawhero Forster & Platnick, 1985 E
Duripelta minuta Forster, 1956 E
Duripelta monowai Forster & Platnick, 1985 E
Duripelta otara Forster & Platnick, 1985 E
Duripelta pallida (Forster, 1956) E
Duripelta paringa Forster & Platnick, 1985 E
Duripelta peha Forster & Platnick, 1985 E
Duripelta scuta Forster & Platnick, 1985 E
Duripelta totara Forster & Platnick, 1985 E
Duripelta townsendi Forster & Platnick, 1985 E
Duripelta watti Forster & Platnick, 1985 E
Maoriata magna (Forster, 1956) E
Maoriata montana Forster & Platnick, 1985 E
Maoriata vulgaris Forster & Platnick, 1985 E

Orongia medialis Forster & Platnick, 1985 E
Orongia motueka Forster & Platnick, 1985 E
Orongia whangamoa Forster & Platnick, 1985 E
Paralobus salmoni (Forster, 1956) E
Pounamuella australis (Forster, 1964) E
Pounamuella complexa (Forster, 1956) E
Pounamuella hauroko Forster & Platnick, 1985 E
Pounamuella hollowayae (Forster, 1956) E
Pounamuella insula Forster & Platnick, 1985 E
Pounamuella kuscheli Forster & Platnick, 1985 E
Pounamuella ramsayi (Forster, 1956) E
Pounamuella vulgaris (Forster, 1956) E
Subantarctia centralis Forster & Platnick, 1985 E
Subantarctia dugdalei Forster, 1956 E
Subantarctia fiordensis Forster, 1956 E
Subantarctia florae Forster, 1956 E
Subantarctia muka Forster & Platnick, 1985 E
Subantarctia penara Forster & Platnick, 1985 E
Subantarctia stewartensis Forster, 1956 E
Subantarctia trina Forster & Platnick, 1985 E
Subantarctia turbotti Forster, 1955 E
Tangata alpina (Forster, 1956) E
Tangata furcata Forster & Platnick, 1985 E
Tangata horningi Forster & Platnick, 1985 E
Tangata kohuka Forster & Platnick, 1985 E
Tangata murihiku Forster & Platnick, 1985 E
Tangata nigra Forster & Platnick, 1985 E
Tangata orepukiensis (Forster, 1956) E
Tangata otago Forster & Platnick, 1985 E
Tangata parafurcata Forster & Platnick, 1985 E
Tangata plena (Forster, 1956) E
Tangata pouaka Forster & Platnick, 1985 E
Tangata rakiura (Forster, 1956) E
Tangata stewartensis (Forster, 1956) E
Tangata sylvester Forster & Platnick, 1985 E
Tangata tautuku Forster & Platnick, 1985 E
Tangata townsendi Forster & Platnick, 1985 E
Tangata waipoua Forster & Platnick, 1985 E
Tautukua isolata Forster & Platnick, 1985 E
Turretia dugdalei Forster & Platnick, 1985 E
Waiporia algida (Forster, 1956) E
Waiporia chathamensis Forster & Platnick, 1985 E
Waiporia egmont Forster & Platnick, 1985 E
Waiporia extensa (Forster, 1956) E
Waiporia hawea Forster & Platnick, 1985 E
Waiporia hornabrooki (Forster, 1956) E
Waiporia mensa (Forster, 1956) E
Waiporia modica (Forster, 1956) E
Waiporia owaka Forster & Platnick, 1985 E
Waiporia ruahine Forster & Platnick, 1985 E
Waiporia tuata Forster & Platnick, 1985 E
Waiporia wiltoni Forster & Platnick, 1985 E
Waipoua gressitti (Forster, 1964) E
Waipoua hila Forster & Platnick, 1985 E
Waipoua insula Forster & Platnick, 1985 E
Waipoua montana Forster & Platnick, 1985 E
Waipoua otiana Forster & Platnick, 1985 E
Waipoua ponanga Forster & Platnick, 1985 E
Waipoua toronui Forster & Platnick, 1985 E
Waipoua totara (Forster, 1956) E
Wiltonia elongata Forster & Platnick, 1985 E
Wiltonia eylesi Forster & Platnick, 1985 E
Wiltonia fiordensis Forster & Platnick, 1985 E
Wiltonia graminicola Forster & Platnick, 1985 E
Wiltonia lima Forster & Platnick, 1985 E
Wiltonia nelsonensis Forster & Platnick, 1985 E
Wiltonia pecki Forster & Platnick, 1985 E
Wiltonia porina Forster & Platnick, 1985 E
Wiltonia rotoiti Forster & Platnick, 1985 E
OXYOPIDAE
Oxyopes gracilipes (White, 1849)
PARARCHAEIDAE
Forstrarchaea rubra (Forster, 1949) E
Ozarchaea forsteri (Rix, 2006)
Pararchaea alba Forster, 1955 E
PERIEGOPIDAE
Periegops suterii (Urquhart, 1892) E
Periegops sp. E
PHOLCIDAE
Pholcus phalangioides (Fuesslin, 1775) A
Psilochorus simoni (Berland, 1911) A
PISAURIDAE
Dolomedes aquaticus Goyen, 1888 E
Dolomedes dondalei Vink & Duperre, 2010 E
Dolomedes minor L. Koch, 1876 E
Dolomedes schauinslandi Simon, 1899 E
SALTICIDAE
Adoxotoma forsteri Zabka, 2004 E
'Attus' abbreviatus Walckenaer, 1837 E *n.d.*
'Attus' aquilis Urquhart, 1886 E *n.d.*
'Attus' cooki Walckenaer, 1837 E *n.d.*
'Attus' monticolus Urquhart, 1891 E *n.d.*
'Attus' montinus Urquhart, 1891 E *n.d.*
'Attus' pullus Urquhart, 1890 E *n.d.*
'Attus' scindus Urquhart, 1890 E *n.d.*
'Attus' subfuscus Urquhart, 1887 E *n.d.*
'Attus' tenebrosus Urquhart, 1893 E *n.d.*
'Attus' valentulus Urquhart, 1891 E *n.d.*
Bianor compactus (Urquhart, 1885) E
Bianor maculatus (Keyserling, 1883) A
Clynotis barresi Hogg, 1909 E
Clynotis knoxi Forster, 1964 E
Clynotis saxatilis (Urquhart, 1886) E
Cosmophasis archeyi Berland, 1931 E
Hasarius adansoni (Audouin, 1826) A
Helpis minitabunda (L. Koch, 1880) A
Hinewaia embolica Zabka & Pollard, 2002 E
Holoplatys appressus (Powell, 1873) E
Holoplatys planissima (L. Koch, 1879) A
Hypoblemum albovitattum (Keyserling, 1882) A
Jotus ravus (Urquhart, 1893) E
Laufeia aerihirta (Urquhart, 1888) E
Marpissa arenaria Urquhart, 1888 E
Marpissa armifera Urquhart, 1892 E
Marpissa cineracea Urquhart, 1891 E
Marpissa leucophaina Urquhart, 1888 E
Marpissa marina (Goyen, 1892) E
Marpissa nemoralis Urquhart, 1892 E
Ocrisiona cinerea (L. Koch, 1879) E
Ocrisiona leucocomis (L. Koch, 1879)
Plexippus sylvarus Urquhart, 1892
'Salticus' albobarbatus Powell, 1873 E *n.d.*
'Salticus' albopalpis Urquhart, 1885 E *n.d.*
'Salticus' alpinus Urquhart, 1885 E *n.d.*
'Salticus' atratus Powell, 1873 E *n.d.*
'Salticus' curvus Urquhart, 1885 E *n.d.*
'Salticus' fumosus Powell, 1873 E *n.d.*
'Salticus' furvus Urquhart, 1885 E *n.d.*
'Salticus' tabinus Urquhart, 1885 E *n.d.*
'Salticus' tenebricus Urquhart, 1885 E *n.d.*
'Salticus' v-notatus Powell, 1873 E *n.d.*
'Salticus' zanthofrontalis Urquhart, 1885 E *n.d.*
Trite auricoma (Urquhart, 1886) E
Trite herbigrada (Urquhart, 1889) E
Trite mustilina (Powell, 1873) E
Trite parvula (Bryant, 1935) E
Trite planiceps Simon, 1899 E
Trite urvillei (Dalmas, 1917) E
Gen. et n. spp. (>160) 160E
SEGESTRIIDAE
Ariadna barbigera Simon, 1905 E
Ariadna bellatoria Dalmas, 1917 E
Ariadna septemcincta (Urquhart, 1891) E
Segestria saeva Walckenaer, 1837 E
Gen. et n. spp. (11) 11E
SPARASSIDAE
Delena cancerides Walckenaer, 1837 A
Isopeda villosa L. Koch, 1875 A
STIPHIDIIDAE
Cambridgea agrestis Forster & Wilton, 1973 E
Cambridgea ambigua Blest & Vink, 2000 E
Cambridgea annulata Dalmas, 1917 E
Cambridgea antipodiana (White, 1849) E
Cambridgea arboricola (Urquhart, 1891) E
Cambridgea australis Blest & Vink, 2000 E
Cambridgea decorata Blest & Vink, 2000 E
Cambridgea elegans Blest & Vink, 2000 E
Cambridgea elongata Blest & Vink, 2000 E
Cambridgea fasciata (L. Koch, 1872) E
Cambridgea foliata (L. Koch, 1872) E
Cambridgea inaequalis Blest & Vink, 2000 E
Cambridgea insulana Blest & Vink, 2000 E
Cambridgea longipes Blest & Vink, 2000 E
Cambridgea mercurialis Blest & Vinck, 2000 E
Cambridgea obscura Blest & Vink, 2000 E
Cambridgea occidentalis Forster & Wilton, 1973 E
Cambridgea ordishi Blest & Vink, 2000 E
Cambridgea pallidula Blest & Vink, 2000 E
Cambridgea peculiaris Forster & Wilton, 1973 E
Cambridgea peelensis Blest & Vink, 2000 E
Cambridgea plagiata Forster & Wilton, 1973 E
Cambridgea quadromaculata Blest & Taylor, 1995 E
Cambridgea ramsayi Forster & Wilton, 1973 E
Cambridgea reinga Forster & Wilton, 1973 E
Cambridgea secunda Forster & Wilton, 1973 E
Cambridgea solanderensis Blest & Vink, 2000 E
Cambridgea sylvatica Forster & Wilton, 1973 E
Cambridgea tuiae Blest & Vink, 2000 E
Cambridgea turbotti Forster & Wilton, 1973 E
Ischalea spinipes L. Koch, 1872 E
Nanocambridgea gracilipes Forster & Wilton, 1973 E
Procambridgea grayi Davies *in* Davies & Lambkin, 2001 A
Stiphidion facetum Simon, 1902 A
SYNOTAXIDAE
Mangua caswell Forster, 1990 E
Mangua convoluta Forster, 1990 E
Mangua flora Forster, 1990 E
Mangua forsteri (Brignoli, 1983) E
Mangua gunni Forster, 1990 E
Mangua hughsoni Forster, 1990 E
Mangua kapiti Forster, 1990 E
Mangua makarora Forster, 1990 E
Mangua medialis Forster, 1990 E
Mangua oparara Forster, 1990 E
Mangua otira Forster, 1990 E
Mangua paringa Forster, 1990 E
Mangua sana Forster, 1990 E
Mangua secunda Forster, 1990 E
Meringa australis Forster, 1990 E
Meringa borealis Forster, 1990 E
Meringa centralis Forster, 1990 E
Meringa conway Forster, 1990 E
Meringa hinaka Forster, 1990 E
Meringa leith Forster, 1990 E
Meringa nelson Forster, 1990 E
Meringa otago Forster, 1990 E
Meringa tetragyna Forster, 1990 E
Nomaua arborea Forster, 1990 E
Nomaua cauda Forster, 1990 E
Nomaua crinifrons (Urquhart, 1891) E
Nomaua nelson Forster, 1990 E
Nomaua perdita Forster, 1990 E
Nomaua rakiura Fitzgerald & Sirvid, 2009 E
Nomaua repanga Fitzgerald & Sirvid, 2009 E
Nomaua rimutaka Fitzgerald & Sirvid, 2009 E
Nomaua taranga Fitzgerald & Sirvid, 2009 E
Nomaua urquharti Fitzgerald & Sirvid, 2009 E
Nomaua waikaremoana Forster, 1990 E
Pahora cantuaria Forster, 1990 E
Pahora graminicola Forster, 1990 E
Pahora kaituna Forster, 1990 E
Pahora media Forster, 1990 E
Pahora montana Forster, 1990 E
Pahora murihiku Forster, 1990 E

Pahora rakiura Forster, 1990 E
Pahora taranaki Forster, 1990 E
Pahora wiltoni Forster, 1990 E
Pahoroides courti Forster, 1990 E
Pahoroides whangarei Forster, 1990 E
Pahoroides n. ssp. (6) 6 E
Runga akaroa Forster, 1990 E
Runga flora Forster, 1990 E
Runga moana Forster, 1990 E
Runga nina Forster, 1990 E
Runga raroa Forster, 1990 E
Zeatupua forsteri (Fitzgerald & Sirvid, 2009) E
Gen. et n. spp. (3) 3E
TENGELLIDAE
Wiltona filicicola (Forster & Wilton, 1973) E
TETRAGNATHIDAE
Eryciniolia purpurapunctata (Urquhart, 1889) E
Leucauge dromedaria (Thorell, 1881)
Meta rufolineata (Urquhart, 1889) E
Nanometa gentilis Simon, 1908
Nanometa n. spp. [8] 8 E
Orsinome lagenifera (Urquhart, 1888) E
Tetragnatha flavida Urquhart, 1891 E
Tetragnatha multipunctata Urquhart, 1891 E
Tetragnatha nigricans Dalmas, 1917 E
Tetragnatha nitens (Audouin, 1826) A
Gen. et n. spp. (9) 9E
THERIDIIDAE
Argyrodes antipodianus O. Pickard-Cambridge, 1880
Argyrodes lepidus O. Pickard-Cambridge, 1879 E
Arianme triangulatus Urquhart, 1887 E
Cryptachaea blattea (Urquhart, 1886) A
Cryptachaea veruculata (Urquhart, 1886) A
Episinus antipodianus O. Pickard-Cambridge, 1879 E
Episinus similanus Urquhart, 1893 E
Episinus similitudus Urquhart, 1893 E
Euryopis oecobioides (O. Pickard-Cambridge, 1879) E
Icona alba Forster, 1955 E
Icona drama Forster, 1964 E
Latrodectus hasseltii Thorell, 1870 A
Latrodectus katipo Powell, 1871 E
Moneta conifera (Urquhart, 1887) E
Nesticodes rufipes (Lucas, 1846) A
Parasteatoda tepidariorum (C. L. Koch, 1841) A
Pholcomma antipodianum (Forster, 1955) E
Pholcomma hickmani Forster, 1964 E
Pholcomma turbotti (Marples, 1956) E
Phoroncidia pukeiwa (Marples, 1955) E
Phoroncidia puketoru (Marples, 1955) E
Phoroncidia quadrata (O. Pickard-Cambridge, 1879) E
Phycosoma oecobioides O. Pickard-Cambridge, 1879 E
Rhomphaea urquharti (Bryant, 1933) E
Steatoda capensis Hann, 1990 A
Steatoda grossa (C.L. Koch, 1838) A
Steatoda lepida (O. Pickard-Cambridge, 1879) E
Steatoda nubilosa Urquhart, 1893 E *n.d.*
Steatoda truncata (Urquhart, 1888) E
Theridion albocinctum Urquhart, 1892 E
Theridion ampliatum Urquhart, 1892 E
Theridion argentatulum (Roewer, 1942) E
Theridion cruciferum Urquhart, 1886 E
Theridion flabelliferum Urquhart, 1887 E
Theridion gibbosum Urquhart, 1894 E *n.d.*
Theridion gracilipes Urquhart, 1889 E
Theridion longicrure Marples, 1956 E
Theridion porphyreticum Urquhart, 1889 E
Theridion pumilio Urquhart, 1886 E
Theridion punicapunctatum Urquhart, 1891 E
Theridion squalidum Urquhart, 1886 E
Theridion viridanum Urquhart, 1887 E
Theridion zantholabio Urquhart, 1886 E
Gen. et. spp. nov. (>160) 160E
THERIDIOSOMATIDAE
Gen. et n. spp. (9) 9E
THOMISIDAE
Cymbachina albobrunnea (Urquhart, 1893) E
Diaea albolimbata L. Koch 1875 E
Diaea ambara (Urquhart, 1885) E
Diaea sphaeroides (Urquhart, 1885) E
'*Philodromus*' *rubrofrontus* (Urquhart, 1891) E *n.d.*
Sidymella angularis (Urquhart, 1885) E
Sidymella angulata (Urquhart, 1885) E
Sidymella benhami (Hogg, 1911) E
Sidymella longipes (L. Koch, 1874)
Synema suteri Dahl, 1907 E
Gen. et n. spp. (15) 14E
ULOBORIDAE
Waitkera waitakerensis (Chamberlain, 1946) E
ZODARIIDAE
Forsterella faceta Jocqué, 1991 E
ZORIDAE
Argoctenus aureus (Hogg, 1911) E
Gen. et n. spp. (5) 5E
ZOROPSIDAE
Uliodon albopunctatus (L. Koch, 1873) E
Uliodon cervinus L. Koch, 1873 E
Uliodon frenatus L. Koch, 1873 E
Gen. et n. spp. (40) 40E
INCERTAE SEDIS
Anzacia scitula (Urquhart, 1893) E
Anzacia viridicoma (Urquhart, 1892) E
Epeira viridana Urquhart, 1893 E
Haplinis chiltoni Hogg, 1911 E
Nomaua waikanae (Forster, 1990) E

Infraclass MICROTHELYPHONIDA
Superorder PALPIGRADIDA
Order PALPIGRADI
EUKOENENIIDAE
Gen. et spp. indet.

Subclass ACAROMORPHA
Infraclass ACARI
Superorder ACARIFORMES
Order SARCOPTIFORMES
Suborder ENDEOSTIGMATA
ALICORHAGIIDAE
Alicorhagia usitata Theron, Meyer & Ryke, 1971
NANORCHESTIDAE
Nanorchestes antarcticus Strandtmann, 1963
Nanorchestes dicrosetus Luxton, 1984
Speleorchestes sp.
PACHYGNATHIDAE
Bimichaelia novazealandica Womersley, 1944

Suborder ORIBATIDA
ORIBATIDA [(excluding ASTIGMATA])
ACARONYCHIDAE
Acaronychus sp.
Stomacarus campbellensis (Wallwork, 1966) E
Stomacarus ciliosus Luxton, 1982 E
Stomacarus ligamentifer (Hammer, 1967) E
Stomacarus watsoni (Travé, 1964)
ADHAESOZETIDAE
Adhaesozetes barbarae Hammer, 1966 E
AMERIDAE
Andesamerus novaezealandiae Balogh & Balogh, 1986 E
Hymenobelba diversisetosa Luxton, 1988 E
Hymenobelba flexisetosa Luxton, 1988 E
APHELACARIDAE
Aphelacarus sp.
ACHIPTERIIDAE
Achipteria sp.
ASTEGISTIDAE
Cultroribula lata Aoki, 1961
AUSTRACHIPTERIIDAE
Austrachipteria furcatus (Hammer, 1967) E
Austrachipteria giganteus (Hammer, 1967) E
Austrachipteria grandis (Hammer, 1967) E
Austrachipteria lobatus (Hammer, 1967) E
Austrachipteria macrodentatus (Hammer, 1967) E
Austrachipteria maximus (Hammer, 1967) E
Austrachipteria quadridentatus (Hammer, 1967) E
Parahypozetes bidentatus (Hammer, 1967) E
AUTOGNETIDAE
Austrogneta multipilosa Balogh & Csiszár, 1963
Austrogneta quadridentata Hammer, 1966 E
BRACHYCHTHONIIDAE
Brachychthonius jugatus Jacot, 1938
Brachychthonius novazealandicus (Hammer, 1966) E
Liochthonius altimonticola (Hammer, 1958) E
Liochthonius altus (Hammer, 1958) E
Liochthonius fimbriatissimus Hammer, 1962 E
Liochthonius idem Hammer, 1966 E
Liochthonius saltaensis (Hammer, 1958) E
CALEREMAEIDAE
Anderemaeus forsteri Balogh & Balogh, 1985 E
CAMISIIDAE
Camisia segnis (Hermann, 1804)
Heminothrus microclava Hammer, 1966 E
Platynothrus major Hammer, 1966 E
Platynothrus peltifer (Koch, 1840)
Platynothrus skottsbergii Trägårdh, 1931
Platynothrus tenuiclava Hammer, 1966 E
Platynothrus traversus (Hammer, 1966) E
CARABODIDAE
Austrocarabodes elegans Hammer, 1966 E
Austrocarabodes maculatus Hammer, 1966 E
Austrocarabodes nodosus Hammer, 1966 E
Carabodes ornatissimus Hammer, 1966 E
Carabodes variabilis Hammer, 1966 E
CERATOZETIDAE
Ceratozetes bicornis Hammer, 1967 E
Ceratozetes gracilis (Michael, 1884)
Ceratozetes hamobatoides Hammer, 1967 E
Ceratozetes mediocris Berlese, 1908
Edwardzetes novazealandicus Hammer, 1967 E
Macrogena crassa Hammer, 1967 E
Macrogena monodactyla Wallwork, 1966 E
Macrogena rudentiger Hammer, 1967 E
Magellozetes clathratus Hammer, 1967 E
Parafurcobates cuspidatus Hammer, 1967 E
Tutorozetes termophilus Hammer, 1967 E
CHAMOBATIDAE
Pedunculozetes andinus Hammer, 1962 E
Pedunculozetes minutus Hammer, 1967 E
CHARASSOBATIDAE
Topalia clavata Hammer, 1966 E
Topalia granulata Hammer, 1966 E
Topalia velata Hammer, 1966 E
COMPACTOZETIDAE
Bornebuschia binodosa Luxton, 1988 E
Bornebuschia peculiaris Hammer, 1966 E
Compactozetes niger Hammer, 1966 E
Compactozetes rotoruensis Hammer, 1966 E
Compactozetes zeugus Luxton, 1988 E
Dicrotegaeus mirabilis Luxton, 1988 E
Sadocepheus foveolatus Luxton, 1988 E
COSMOCHTHONIIDAE
Cosmochthonius semiareolatus Hammer, 1966 E
CROTONIIDAE
Austronothrus clarki Colloff & Cameron, 2009 E
Austronothrus curviseta Hammer, 1966 E
Austronothrus flagellatus Colloff & Cameron, 2009 E
Crotonia brevicornuta (Wallwork, 1966) E
Crotonia caudalis (Hammer, 1966) E
Crotonia cervicorna Luxton, 1982 E
Crotonia cophinaria (Michael, 1908)
Crotonia cupulata Luxton, 1982 E
Crotonia longibulbula Luxton, 1982 E
Crotonia obtecta (Pickard-Cambridge, 1875)
Crotonia reticulata Luxton, 1982 E
Crotonia tuberculata Luxton, 1982 E
Crotonia unguifera (Michael, 1908)

Holonothrus concavus Wallwork, 1966 E
Holonothrus foliatus Wallwork, 1963 E
Holonothrus gracilis Olszanowski, 1997
Holonothrus naskreckii Olszanowski, 1997
Holonothrus pulcher Hammer, 1966 E
CYMBAEREMAEIDAE
Bulleremaeus reticulatus Hammer, 1966 E
Bulleremaeus tuberculatus Hammer, 1966 E
Capillibates stagaardi Hammer, 1966 E
Scapheremaeus emarginatus Hammer, 1966 E
Scapheremaeus insularis Hammer, 1966 E
Scapheremaeus patella (Berlese, 1910)
DAMAEIDAE
Metabelba obtusus Hammer, 1966 E
DAMAEOLIDAE
Fosseremus laciniatus (Berlese, 1905)
ENIOCHTHONIIDAE
Hypochthoniella minutissima (Berlese, 1904)
EREMULIDAE
Eremulus flagellifer Berlese, 1908
Eremulus serratus Hammer, 1966 E
EUPHTHIRACARIDAE
Microtritia contraria Niedbała, 1993
Microtritia novazealandiensis Niedbała, 1996
Rhysotritia bifurcata Niedbała, 1993
EUTEGAEIDAE
Compactozetes zeugus Luxton, 1989 E
Eutegaeus bostocki (Michael, 1908)
Eutegaeus curviseta Hammer, 1966
(= *E. bostocki* Michael, 1908) E
Eutegaeus membraniger Hammer, 1966 E
Eutegaeus pinnatus Hammer, 1966 E
Eutegaeus radiatus Hammer, 1966 E
Eutegaeus stylesi Hammer, 1966 E
Neseutegaeus angustus Hammer, 1966 E
Neseutegaeus consimilis Hammer, 1966 E
Neseutegaeus distentus Hammer, 1966 E
Neseutegaeus latus Hammer, 1966 E
Neseutegaeus spinatus Wolley, 1965 E
FORTUYNIIDAE
Fortuynia elamellata Luxton, 1967 E
GALUMNIDAE
Acrogalumna longipluma (Berlese, 1904)
Allogalumna novazealandica Hammer, 1968 E
Galumna microfissum Hammer, 1968 E
Galumna rugosa Hammer, 1968 E
Galumna scaber Hammer, 1968 E
Pergalumna remota (Hammer, 1968) E
Pergalumna reniformis Hammer, 1968 E
Pergalumna silvestris Hammer, 1968 E
HAPLOZETIDAE
Angullozetes rostratus Hammer, 1967 E
Incabates angustus Hammer, 1967 E
Lauritzenia acutirostrum Hammer, 1968 E
Lauritzenia rotundirostrum Hammer, 1968 E
Magnobates flagellifer Hammer, 1967 E
Peloribates fragilis Hammer, 1967 E
Peloribates magnisetosus Hammer, 1967 E
Protoribates capucinus Berlese, 1908 E
Rostrozetes foveolatus Sellnick, 1925
Sicaxylobates sicafer (Hammer, 1968) E
Totobates acutissimus (Hammer, 1967) E
Totobates anareensis (Dalenius & Wilson, 1958)
Totobates antarcticus Wallwork, 1966 E
Totobates capita Hammer, 1968 E
Totobates communis Hammer, 1967 E
Totobates elegans (Hammer, 1958) E
Totobates latus Hammer, 1967 E
Totobates macroonyx Hammer, 1967 E
Totobates microseta Hammer, 1968 E
Totobates minimus Hammer, 1967 E
Totobates ovalis Hammer, 1967 E
HERMANNIIDAE
Phyllhermannia foliata Hammer, 1966 E
Phyllhermannia forsteri Balogh, 1985
Phyllhermannia mollis Hammer, 1966 E
Phyllhermannia phyllophora (Michael, 1908)
Phyllhermannia rubra Hammer, 1966 E
HERMANNIELLIDAE
Hermanniella clavasetosa Hammer, 1966 E
Hermanniella diversisetosa Hammer, 1966 E
Hermanniella longisetosa Hammer, 1966 E
Hermanniella microsetosa Hammer, 1966 E
HYDROZETIDAE
Hydrozetes lemnae (de Coggi, 1899)
HYPOCHTHONIIDAE
Hypochthonius luteus Oudemans, 1913
LIODIDAE
Liodes nigricans Hammer, 1966 E
MACHUELLIDAE
Machuella pyriformis Hammer, 1968 E
Machuella ventrisetosa Hammer, 1961 E
MALACONOTHRIDAE
Fossonothrus novaezelandiae Hammer, 1966 E
Malaconothrus indifferens Hammer, 1966 E
Malaconothrus zealandicus Hammer, 1966 E
Trimalaconothrus angustirostrum Hammer, 1966 E
Trimalaconothrus crispus Hammer, 1962 E
Trimalaconothrus longirostrum Hammer, 1966 E
Trimalaconothrus novus (Sellnick, 1921)
Trimalaconothrus opisthoseta Hammer, 1966 E
Trimalaconothrus oxyrhinus Hammer, 1962 E
Trimalaconothrus plathyrhinus Hammer, 1962 E
Trimalaconothrus sacculus Hammer, 1966 E
Trimalaconothrus tonkini Luxton, 1982 E
Zeanothrus elegans Hammer, 1966 E
MAORIZETIDAE
Maorizetes ferox Hammer, 1966 E
MICROZETIDAE
Cuspitegula stellifer Hammer, 1966 E
MYCOBATIDAE
Antarctozetes intermedius (Hammer, 1967) E
Antarctozetes longicaulis (Hammer, 1967) E
Antarctozetes luteus (Hammer, 1967) E
Baloghobates nudus Hammer, 1967 E
Baloghobates parvoglobosus Hammer, 1967 E
Cryptobothria monodactyla Wallwork, 1963 E
Minunthozetes semirufus (Koch, 1841)
Mycozetes oleariae Spain, 1968
Neomycobates tridentatus Wallwork, 1963 E
Punctoribates manzanoensis Hammer, 1958 E
Punctoribates punctum (Koch, 1839)
NANHERMANNIIDAE
Nanhermannia acutisetosa Berlese, 1913 E
Nanhermannia nana (Nicolet, 1855)
Nanhermannia tenuicoma Hammer, 1966 E
NEOTRICHOZETIDAE
Neotrichozetes spinulosa (Michael, 1908)
NOTHRIDAE
Nothrus biciliatus Koch, 1841
Nothrus pupuensis (Hammer, 1966) E
Nothrus silvestris Nicolet, 1855
Novonothrus flagellatus Hammer, 1966 E
NOTOCEPHEIDAE
Nodocepheus dentatus Hammer, 1958 E
ONYCHOBATIDAE
Onychobates nidicola Hammer, 1967 E
OPPIIDAE
Acutoppia crassiseta (Hammer, 1968) E
Acutoppia jelevae (Hammer, 1968) E
Amerioppia longiclava Hammer, 1962 E
Amerioppia woolleyi Hammer, 1968 E
Arcoppia arcualis (Berlese, 1913)
Arcoppia winkleri (Hammer, 1968) E
Austroppia crozetensis (Richters, 1908)
Baioppia moritzi (Hammer, 1968) E
Belloppia beemanensis (Wallwork, 1964) E
Belloppia evansi Hammer, 1968 E
Belloppia shealsi Hammer, 1968 E
Belloppia wallworki Hammer, 1968 E
Brachiopiella hartensteini (Hammer, 1968) E
Brachiopiella higginsi (Hammer, 1968) E
Brachiopiella rajskii Hammer, 1968 E
Brachiopiella walkeri (Hammer, 1968) E
Campbelloppia diaphora (Wallwork, 1964) E
Convergoppia pletzeni (Hammer, 1968) E
Globoppia campbellensis Wallwork, 1964 E
Globoppia gressitti Wallwork, 1964 E
Globoppia nidicola Hammer, 1968 E
Gressittoppia baderi (Hammer, 1968) E
Hamoppia lionis Hammer, 1968 E
Hamoppia thamdrupi Hammer, 1968 E
Insculptoppia suciui (Hammer, 1968) E
Laminoppia blocki Hammer, 1968 E
Lanceoppia banksi Hammer, 1968 E
Lanceoppia becki Hammer, 1968 E
Lanceoppia berlesi Hammer, 1968 E
Lanceoppia bertheti Hammer, 1968 E
Lanceoppia csiszarae Hammer, 1968 E
Lanceoppia ewingi Hammer, 1968 E
Lanceoppia feideri (Hammer, 1968) E
Lanceoppia haarlovi (Hammer, 1968) E
Lanceoppia jacoti Hammer, 1968 E
Lanceoppia knullei Hammer, 1968 E
Lanceoppia luxtoni Hammer, 1968 E
Lanceoppia maerkeli Hammer, 1968 E
Lanceoppia mahunkai Hammer, 1968 E
Lanceoppia menkei Hammer, 1968 E
Lanceoppia perezinigoi (Hammer, 1968) E
Lanceoppia piffli Hammer, 1968 E
Lanceoppia poppi Hammer, 1968 E
Lanceoppia ramsayi Hammer, 1968 E
Lanceoppia rigidiseta Hammer, 1968 E
Lanceoppia schusteri Hammer, 1968 E
Lanceoppia schweizeri Hammer, 1968 E
Lanceoppia sellnicki Hammer, 1968 E
Lanceoppia seydi Hammer, 1968 E
Lanceoppia strenzkei Hammer, 1968 E
Lanceoppia thori Hammer, 1968 E
Lanceoppia turki (Hammer, 1968) E
Lanceoppia vanderhammeni Hammer, 1968 E
Lanceoppia vaneki Hammer, 1968 E
Lanceoppia willmanni Hammer, 1968 E
Lanceoppia woodringi Hammer, 1968 E
Loboppia covarrubiasi (Hammer, 1968) E
Membranoppia karppineni Hammer, 1968 E
Membranoppia krivoluzkyi Hammer, 1968 E
Membranoppia sitnilovae Hammer, 1968 E
Microppia minus (Paoli, 1908) A
Microppia minutissima (Sellnick, 1950)
Microppia zealandica Hammer, 1968 E
Multioppia sp.
Nesoppia tuxeni (Hammer, 1968) E
Operculoppia kunsti Hammer, 1968 E
Oppiella dubia Hammer, 1962 E
Oppiella nova (Oudemans, 1902)
Oppiella obsoleta (Paoli, 1908)
Oxyoppia suramericana (Hammer, 1958) E
Paroppia lebruni Hammer, 1968 E
Pletzenoppia rafalskii (Hammer, 1968) E
Polyoppia baloghi Hammer, 1968 E
Pravoppia disjuncta (Wallwork, 1964) E
Processoppia oudemansi (Hammer, 1968) E
Ptiloppia bulanovae (Hammer, 1968) E
Quadroppia circumita (Hammer, 1961) E
Quadroppia quadricarinata (Michael, 1885)
Ramusella sengbuschi Hammer, 1968 E
Rhaphoppia mihelcici (Hammer, 1968) E
Setuloppia newelli (Hammer, 1968) E
Solenoppia grandjeani Hammer, 1968 E
Solenoppia taberlyi Hammer, 1968 E
Solenoppia travei Hammer, 1968 E
Tripiloppia aokii Hammer, 1968 E
Tripiloppia dalenii Hammer, 1968 E
Tripiloppia forsslundi Hammer, 1968 E

Tripiloppia tarraswahlbergi Hammer, 1968 E
Tripiloppia traegardhi Hammer, 1968 E
ORIBATELLIDAE
Lamellobates palustris Hammer, 1958 E
ORIBATULIDAE
Crassoribatula maculosa Hammer, 1967 E
Ingella bullager Hammer, 1967 E
Liebstadia similis (Michael, 1888)
Maculobates longipilosus Hammer, 1967 E
Maculobates longus Hammer, 1967 E
Maculobates luteomarginatus Hammer, 1967 E
Maculobates luteus Hammer, 1967 E
Maculobates magnus Hammer, 1967 E
Maculobates minor Hammer, 1967 E
Maculobates vulgaris Hammer, 1967 E
Paraphauloppia novazealandica Hammer, 1967 E
Pontiobates denigratus Luxton, 1989 E
Subphauloppia dentonyx Hammer, 1967 E
Zygoribatula connexa (Berlese, 1904)
Zygoribatula magna Ramsay, 1966 E
Zygoribatula novazealandica Hammer, 1967 E
Zygoribatula terricola von der Hammen, 1952
ORIBOTRITIIDAE
Indotritia aotearoana Ramsay, 1966 E
Oribotritia brevis Niedbała et Colloff, 1997
Oribotritia contortula Niedbała, 1993
Oribotritia contraria Niedbała, 1993
Oribotritia paraincognita Niedbała, 2006
Oribotritia teretis Niedbała, 1993
OTOCEPHEIDAE
Clavazetes decorus Hammer, 1966 E
Neotocepheus colliger Hammer, 1966 E
Plenotocepheus curtiseta Hammer, 1966 E
Plenotocepheus mollicoma Hammer, 1966 E
Pseudotocepheus curtiseta Hammer, 1966 E
Pseudotocepheus foveolatus Hammer, 1966 E
Pseudotocepheus punctatus Hammer, 1966 E
Pseudotocepheus tenuiseta Hammer, 1966 E
PARAKALUMMIDAE
Neoribates barbatus Hammer, 1968 E
Porokalumma rotunda (Wallwork, 1963) E
PELOPPIIDAE
Macquarioppia striata (Wallwork, 1963) E
Pseudoceratoppia asetosa Hammer, 1967 E
Pseudoceratoppia clavasetosa Hammer, 1967 E
Pseudoceratoppia diversa Hammer, 1967 E
Pseudoceratoppia microsetosa Hammer, 1967 E
Pseudoceratoppia sexsetosa Hammer, 1967 E
PHENOPELOPIDAE
Nesopelops monodactylus (Hammer, 1966) E
Nesopelops punctatus (Hammer, 1966) E
Peloptulus sp.
PLATEREMAEIDAE
Pedrocortesella cryptonotus Hammer, 1966 E
Pedrocortesella gymnonotus Hammer, 1966 E
Pedrocortesella latoclava Hammer, 1966 E
Pedrocortesella nigroclava Hammer, 1966 E
Pedrocortesella sexpilosus Hammer, 1966 E
Pedrocortesia australis Hammer, 1962 E
Pedrocortesia luteomarginata Hammer, 1966 E
Pedrocortesia rotoruensis Hammer, 1966 E
PHTHIRACARIDAE
Austrophthiracarus pulchellus Niedbała, 1993
Hoplophorella sp.
Hoplophthiracarus bisulcus Niedbała, 1993
Notophthiracarus australis Ramsay, 1966 E
Notophthiracarus brachys Niedbała, 2006
Notophthiracarus claviger Niedbała, 1993
Notophthiracarus conspiquus Niedbała, 1989
Notophthiracarus paracapillatus Niedbała, 2006
Notophthiracarus perlucundus Niedbała, 2000
Notophthiracarus rotoitiensis Niedbała, 2006
Notophthiracarus paraunicarinatus Niedbała & Penttinen, 2007
Notophthiracarus quietus Niedbała, 1989
Notophthiracarus repostus Niedbała, 1989
Phthiracarus pellucidus Ramsay, 1966 E
Protophthiracarus neotrichus (Wallwork, 1966) E
PLATEREMAEIDAE
Pedrocortesella cryptonotus Hammer, 1966 E
Pedrocortesella gymnonotus Hammer, 1966 E
Pedrocortesella latoclava Hammer, 1966 E
Pedrocortesella nigroclava Hammer, 1966 E
Pedrocortesella sexpilosus Hammer, 1966 E
Pedrocortesia australis Hammer, 1962 E
Pedrocortesia luteomarginata Hammer, 1966 E
Pedrocortesia rotoruensis Hammer, 1966 E
PODACARIDAE
Alaskozetes antarcticus (Michael, 1903)
Halozetes bathamae Luxton, 1984 E
Halozetes belgicae (Michael, 1903)
Halozetes crozetensis (Richters, 1908)
Halozetes intermedius Wallwork, 1963 E
Halozetes macquariensis (Dalenius & Wilson, 1958)
Halozetes marinus (Lohmann, 1908)
Halozetes otagoensis Hammer, 1966 E
Halozetes plumosus Wallwork, 1966 E
Podacarus auberti Grandjean, 1955
PTEROZETIDAE
Pterozetes novazealandicus Hammer, 1966 E
RAMSAYELLIDAE
Ramsayellus grandis (Hammer, 1967) E
SCHELORIBATIDAE
Campbellobates acanthus Wallwork, 1964 E
Campbellobates aureus Hammer, 1967 E
Campbellobates latohumeralis Hammer, 1967 E
Campbellobates occultus Hammer, 1967 E
Grandjeanobates novazealandicus Hammer, 1967 E
Incabates angustus Hammer, 1967 E
Scheloribates aequalis Hammer, 1967 E
Scheloribates anzacensis Hammer, 1967 E
Scheloribates conjuges Hammer, 1967 E
Scheloribates crassus Hammer, 1967 E
Scheloribates flagellatus Wallwork, 1966 E
Scheloribates keriensis Hammer, 1967 E
Scheloribates maoriensis Hammer, 1968 E
Scheloribates pacificus Hammer, 1967 E
Scheloribates zealandicus Hammer, 1967 E
Setobates medius Hammer, 1967 E
Setobates scheloribatoides (Ramsay, 1966) E
Zeascheloribates palustris Luxton, 1982 E
SCUTOVERTICIDAE
Scutovertex sculptus Michael, 1879
SELLNICKIIDAE
Sellnickia caudata (Michael, 1908)
SUCTOBELBIDAE
Suctobelba falcata Forsslund, 1941
Suctobelba longicurva Hammer, 1966 E
Suctobelba nasalis Forsslund, 1941
Suctobelba nondivisa Hammer, 1966 E
Suctobelba plumata Hammer, 1966 E
Suctobelba subcornigera Forsslund, 1941
Suctobelbila dentata (Hammer, 1961) E
Zeasuctobelba arcuata Hammer, 1968 E
Zeasuctobelba nodosa Hammer, 1966 E
Zeasuctobelba quinquenodosa Hammer, 1966 E
Zeasuctobelba trinodosa Hammer, 1966 E
TECTOCEPHEIDAE
Tectocepheus velatus (Michael, 1880)
TEGORIBATIDAE
Paraphysobates monodactylus (Hammer, 1966) E
TIKIZETIDAE
Tikizetes spinipes Hammer, 1967 E
TRHYPOCHTHONIIDAE
Allonothrus sp.
Mucronothrus nasalis Willmann, 1929
Trhyphochthoniellus crassus (Wharburton & Pearce, 1905)
TUMEROZETIDAE E
Tumerozetes bifurcatus Hammer, 1966 E
Tumerozetes circularis Hammer, 1966 E
Tumerozetes indistinctus Hammer, 1966 E
Tumerozetes parallelus Hammer, 1966 E
Tumerozetes pumilis Hammer, 1966 E
TUPAREZETIDAE E
Tuparezetes christineae Spain, 1969 E
Tuparezetes philodendrus Spain, 1969 E

ORIBATIDA – ASTIGMATA
ACARIDAE
Acarus farris (Oudemans, 1905) A
Acarus immobilis Griffiths, 1964 A
Acarus siro (Linnaeus, 1758) A
Cosmoglyphus laarmani Samsinak, 1966
Forcellinia galleriella (Womersley, 1963)
Kuzinia levis (Dujardin, 1849) A
Mycetoglyphus fungivorus (Oudemans, 1932)
Neotropacarus bakeri (Collyer, 1967)
Rhizoglyphus echinopus (Fumouze & Robin, 1868)
Rhizoglyphus minutus Manson, 1972 A
Rhizoglyphus ogdeni Fan & Zhang, 2004 E
Rhizoglyphus ranunculi Manson, 1972
Rhizoglyphus robini Claparède, 1869
Sancassania berlesi (Michael, 1903)
Schwiebea talpa Oudemans, 1916
Thyreophagus australis Clark, 2009
Thyreophagus entomophagus (Laboulbène, 1852)
Thryeophagus australis Clark, 2009
Tyrophagus casei (Oudemans, 1910)
Tyrophagus communis Fan & Zhang, 2007
Tyrophagus curvipenis Fain & Fauvel, 1993
Tyrophagus lini (Oudemans, 1924)
Tyrophagus longior (Gervais, 1844)
Tyrophagus macfarlanei Fan & Zhang, 2007
Tyrophagus neiswanderi Johnston & Brice, 1965
Tyrophagus putrescentiae (Schrank, 1781) A
Tyrophagus robertsonae Lynch, 1989
Tyrophagus savasi Lynch, 1989
Tyrophagus similis Volgin, 1949
Tyrophagus vanheurni Oudemans, 1924
ALLOPTIDAE
Alloptes (Sternalloptes) bisetatus (Haller, 1882)
Alloptes mucronatus Trouessart, 1899
Alloptes oxylobus Dubinin, 1951
Alloptes mucronatus Trouessart, 1899
Alloptes oxylobus Dubinin, 1951
Alloptes stercorarii Dubinin, 1952
Brephosceles constrictus Peterson, 1971
Brephosceles disjunctus Peterson, 1971
Brephosceles lunatus Peterson, 1971
Brephosceles puffini Peterson 1971
Echinacarus rubidus (Trouessart, 1886)
Laminalloptes minor (Trouessart, 1885)
Laminalloptes simplex (Trouessart, 1885)
Microspalax brevipes (Mégnin & Trouessart, 1884)
ANALGIDAE
Analges anthi Mironov, 1985 A
Analges nitzschi Haller, 1878 A
Analges passerinus (Linnaeus, 1758) A
Analges turdinus Mironov, 1985 A
Anhemialges gracillimus (Bonnet, 1924)
Anhimomegninia longipes (Trouessart, 1899)
Hemialges pilgrimi Mironov & Galloway, 2001
Kiwialges haastii Bishop, 1985 E
Kiwialges palametrichus Gaud & Atyeo, 1970 E
Kiwialges phalagotrichus Gaud & Atyeo, 1970 E
Megninia androgyna Trouessart 1886
Megninia californica Mironov & Galloway, 2001
Megninia cubitalis (Mégnin *in* Robin & Mégnin, 1877)
Megninia elongata Trouessart, 1886
Megninia gallinulae (Buchholz 1869)
Megninia ginglymura (Megnin, 1877)
Megninia scapularis Trouessart, 1899
Megniniella sp.
Metanalges sp.

Protalges australis Trouessart, 1885
Protalges psittacinus Trouessart, 1885
Strelkoviacarus sp. aff. *quadratus* (Haller, 1882)
ATOPOMELIDAE A
Atellana papilio Domrow, 1958 A
Chirodiscoides caviae Hirst, 1917 A
Murichirus anabiotus Domrow, 1992 A
Petrogalochirus dycei (Domrow, 1960) A
Petrogalochirus macropus Fain & Domrow, 1974 A
Petrogalochirus tasmaniensis Fain, 1972 A
AVENZOARIIDAE
Bychovskiata obscura Mironov & Dabert, 1997
Bychovskiata subcharadrii Dubinin, 1951
Bychovskiata thinorni Mironov, 1994
Scutomegninia chathamensis Mironov, 1990
Scutomegninia phalacrocoracis (W. Dubinin & M. Dubinin, 1940)
Zachvatkinia puffini (Buchholz, 1869)
CARPOGLYPHIDAE A
Carpoglyphus lactis (Linnaeus, 1767) A
CHORTOGLYPHIDAE A
Chortoglyphus arcuatus (Troupeau, 1879) A
CYTODITIDAE
Cytodites nudus (Vizioli, 1868)
ECHIMYOPODIDAE
Blomia thori novaezealandiae Fain & Galloway, 1993
Marsupiopus trichosuri Fain, 1968 A
FALCULIFERIDAE
Falculifer rostratus (Buchholz, 1869)
Hemiphagacarus spinosus (Trouessart, 1898) E
Spilolichus sp.
Pterophagus strictus Mégnin *in* Robin & Mégnin, 1877
FREYANIIDAE
Freyana anatina (Koch, 1844)
Michaelia sp.
Morinyssus simplex Gaud & Atyeo, 1982
Sulanyssus caputmedusae (Trouessart, 1887)
GABUCINIDAE
Coraciacarus muellermotzfeldi Dabert & Alberti, 2008 E †
Gabucinia sculpturata (Hirst, 1920)
GLYCYPHAGIDAE
Glycyphagoides spheniscicola Fain & Galloway, 1993
Glycyphagus destructor (Schrank, 1781)
Glycyphagus domesticus (de Geer, 1778)
Glycyphagus sp.
Gohieria fusca (Oudemans, 1902)
HEMISARCOPTIDAE A
Hemisarcoptes coccophagus Meyer, 1962 A
Hemisarcoptes cooremani (Thomas, 1961) A
Hemisarcoptes malus (Shimer, 1868) A
HISTIOSTOMATIDAE
Histiostoma feroniarum (Dufour, 1839)
Histiostoma montanum Clark, 2010 E
Myianoetus antipodus Fain & Galloway, 1993
HYADESIIDAE
Amhyadesia punctulata Luxton, 1989
Hyadesia mollis Luxton, 1989
Hyadesia microseta Luxton, 1989
Hyadesia plicata Luxton, 1989
Hyadesia reticulata Luxton, 1989
Hyadesia tessellata Luxton, 1989
Hyadesia zelandica Luxton, 1984
HYPODERATIDAE
Neottialges hughesae antipodus (Fain & Clark, 1994)
Neottialges h. hughesae (Fain, 1969)
KIWILICHIDAE E
Kiwilichus cryptosikyus Gaud & Atyeo, 1970 E
Kiwilichus delosikyus Gaud & Atyeo, 1970 E
KNEMIDOKOPTIDAE
Knemidokoptes mutans Robin & Lanquentin, 1859
Knemidokoptes pilae Lavoipierre & Griffiths, 1951
Neocnemidocoptes gallinae (Railliet, 1887)
Picicnemidocoptes laevis (Railliet, 1885)
KYTODITIDAE
Kytodites nudus (Vizioli, 1868)
LAMINOSIOPTIDAE A
Laminosioptes cisticola (Vizioli, 1868) A
LISTROPHORIDAE A
Leporacarus gibbus (Pagenstecher, 1861) A
Listrophorus mustelae Mégnin, 1885 A
Lynxacarus radovskyi Tenorio, 1974 A
MYOCOPTIDAE A
Myocoptes musculinus (Koch, 1836) A
PROCTOPHYLLODIDAE
Exochojoubertia lobulata (Trouessart, 1885)
Monojoubertia microphylla (Robin, 1877)
Neodectes sp.
Proctophyllodes anthi (Vitzthum, 1922)
Proctophyllodes ceratophyllus Atyeo & Braasch, 1966
Proctophyllodes ciae Bauer, 1939
Proctophyllodes gerygonae Mironov & Galloway, 2001 E
Proctophyllodes pinnatus (Nitzsch, 1818)
Proctophyllodes troncatus Robin, 1877
Proctophyllodes sp.
Pterodectes sp.
PSOROPTIDAE A
Caparinia tripilis (Michael, 1889) A
Chorioptes bovis (Hering, 1845) A
Otodectes cynotis (Hering, 1838) A
Psoroptes cuniculi (Delafond, 1859) A
Psoroptes equi (Hering, 1838) A
Psoroptes natalensis Hirst, 1919 A
Psoroptes ovis (Hering, 1838) A
PSOROPTOIDIDAE
Mesalges diaphanoxus Bonnet, 1924
Mesalges lyrurus Trouessart & Neumann, 1888
Mesalges oscinum (Koch, 1841)
Pandalura strigisoti Buchholz, 1869
Temnalges mesalgoides Gaud & Atyeo, 1967
PTEROLICHIDAE
Apexolichus affinis (Mégnin & Trouessart, 1884) A
Calyptolichus favettei (Trouessart, 1899)
Grallobia sp.
Grallolichus sp.
Kakapolichus strigopis Mironov & Perez, 2003 E
Nestorilichus atyeoi Mironov & Dabert, 2007 E
Nymphicilichus perezae Mironov & Galloway, 2002 A
Pseudalloptinus sp.
Scopusacarus sp.
Titanolichus chiragricus (Mégnin & Trouessart, 1884)
PTERONYSSIDAE
Mouchetia novaezealandica Mironov & Galloway, 2001 E
Pteronyssoides striatus (Robin, 1877)
Pteronyssus sp.
Pteroherpus zosteropis Mironov, 1992
Scutulanyssus truncatus (Trouessart, 1885)
PTILOXENIDAE
Sokoloviana gracilis (Mégnin & Trouessart, 1884)
PYROGLYPHIDAE A
Dermatophagoides farinae Hughes, 1961 A
Dermatophagoides pteronyssinus (Trouessart, 1898) A
SARCOPTIDAE
Chirophagoides mystacopis Fain, 1963
Notoedres cati (Hering, 1838) A
Notoedres muris (Mégnin, 1877) A
Notoedres sp.
Sarcoptes scabiei (Linnaeus, 1758)
SUIDASIIDAE
Suidasia nesbitti Hughes, 1948
Suidasia reticulata Manson, 1973
TROUESSARTIIDAE
Calcealges yunkeri Gaud, 1962
Trouessartia megadisca Gaud, 1962
Trouessartia microcaudata Mironov, 1983
Trouessartia rhipidurae Mironov & Galloway, 2001 E
Trouessartia rosteri (Berlese, 1886)
Trouessartia sp.
TURBINOPTIDAE
Turbinoptes strandtmanni Boyd, 1949
XOLALGIDAE
Dubininia sp.
Ingrassiella sp.
Pteralloptes sp.
WINTERSCHMIDTIIDAE
Psylloglyphus parapsyllus Fain & Galloway, 1993
Czenspinskia sp.

Order TROMBIDIFORMES
Order Suborder PROSTIGMATA
ALLOTANAUPODIDAE
Allotanaupodus williamsi Zhang & Fan, 2007 E
Allotanaupodus orete Zhang & Fan, 2007 E
Allotanaupodus winksi Zhang & Fan, 2007 E
Nanotanaupodus andrei Zhang & Fan, 2007 E
Nanotanaupodus gracehallae Zhang & Fan, 2007 E
ANISITSIELLIDAE
Anisitsiellides arraphus Cook, 1983 F E
Anisitsiellides partitus Cook, 1983 F E
Anisitsiellides zealandicus Hopkins, 1967 F E
Mamersella anomaola Hopkins, 1967 F E
Zelandatonia orion Cook, 1992 F E
ANYSTIDAE
Anystis baccarum Linnaeus, 1758
ARRENURIDAE
Arrenurus (*Arrenurus*) *lacus* Stout, 1953 F E
Arrenurus (*A.*) *longipetiolatus* Smit, 1996 F E
Arrenurus (*A.*) *rotoensis* Stout, 1953 F E
Arrenurus (*A.*) *schuckardi* Smit, 1996 F E
Arrenurus (*A.*) *stouti* Cook, 1983 F E
Arrenurus (*A.*) *zelandicus* Cook, 1983 F E
ATHIENEMANNIIDAE
Anamundamella zelandica Cook, 1992 F E
ATURIDAE
Abelaturus cornophorus Cook, 1983 F E
Abelaturus ogalus Cook, 1991 F E
Abelaturus schwoerbeli Pesic & Smit *in* Pesic, Smit & Datry, 2010 F E
Bleptaturus magnipalpis Cook, 1991 F E
Canterburaturus cooki Pesic & Smit *in* Pesic, Smit & Datry, 2010 F E
Canterburaturus minutus Pesic & Smit *in* Pesic, Smit & Datry, 2010 F E
Canterburaturus novaeseelandicus Pesic & Smit *in* Pesic, Smit & Datry, 2010 F E
Colobaturus daimphida Cook, 1991 F E
Colobaturus selwynus Pesic & Smit *in* Pesic, Smit & Datry, 2010 F E
Evidaturus exilis Cook, 1983 F E
Evidaturus scopticus Cook, 1983 F E
Hestaturus ovalis Cook, 1991 F E
Kritaturus (*Caudaturus*) *jacundus* Cook, 1983 F E
Kritaturus (*C.*) *rucabus* Cook, 1983 F E
Kritaturus (*C.*) *tenonus* Cook, 1983 F E
Kritaturus (*C.*) *uncipalpis* Cook, 1983 F E
Kritaturus (*Kritaturus*) *dornarus* Cook, 1983 F E
Kritaturus (*K.*) *gennadus* Cook, 1983 F E
Kritaturus (*K.*) *ianthus* Cook, 1983 F E
Kritaturus (*K.*) *sornus* Cook, 1983 F E
Kritaturus (*K.*) *vinnulus* Cook, 1983 F E
Neotryssaturus inusitatus (Hopkins, 1967) F E
Neotryssaturus pallidus Cook, 1983 F E
Omegaturus longipalpis Cook, 1983 F E
Paratryssaturus cantermus Cook, 1983 F E
Paratryssaturus minutus (Hopkins, 1969) F E
Paratryssaturus morimotoi Imamura, 1979 F E
Paratryssaturus zodelus Cook, 1983 F E
Pilosaturus villosus (Hopkins, 1967) F E
Piotaturus alvecaudatus Cook, 1983 F E
Piotaturus bovalus Cook, 1983 F E
Planaturus lundbladi Cook, 1983 F E
Planaturus rugosus Cook, 1991 F E
Planaturus setipalpis Cook, 1983 F E

Pseudotryssaturus acutus Cook, 1983 F E
Pseudotryssaturus anchistus Cook, 1983 F E
Pseudotryssaturus dapsilus Cook, 1983 F E
Pseudotryssaturus dictydermis Cook, 1983 F E
Pseudotryssaturus indentatus (Hopkins, 1967) F E
Pseudotryssaturus papillidermis Cook, 1983 F E
Pseudotryssaturus planus Cook, 1983 F E
Taintaturus abditus Cook, 1983 F E
Taintaturus accidens Cook, 1983 F E
Taintaturus brevipalpis Cook, 1991 F E
Taintaturus hopkinsi Cook, 1983 F E
Taintaturus lembus Cook, 1991 F E
Taintaturus livingstoni Cook, 1991 F E
Taintaturus projectus Cook, 1983 F E
Taintaturus rostratus Pesic & Smit *in* Pesic, Smit & Datry, 2010 F E
Taintaturus selwynus Pesic & Smit *in* Pesic, Smit & Datry, 2010 F E
Taintaturus stoutae Cook, 1991 F E
Taintaturus waikirikiri Pesic & Smit *in* Pesic, Smit & Datry, 2010 F E
Taintaturus zelandicus Cook, 1991 F E
Tryssaturopsis asopos Cook, 1983 F E
Tryssaturopsis novus (Hopkins, 1967) F E
Tryssaturopsis parvicaudatus Cook, 1983 F E
Tryssaturopsis solivagus Cook, 1983 F E
Tryssaturus spinipes Hopkins, 1967 F E
Uralbia filipalpis Schwoerbel, 1984 F E
Uralbia gracilipes Cook, 1983 F E
Uralbia parva Cook, 1983 F E
Uralbia projecta Hopkins, 1967 F E
Zelandalbia acuta Cook, 1991 F E
Zelandalbia hopkinsi (Imamura, 1978) F E
Zelandalbia imamurai Cook, 1983 F E
Zelandalbia longipalpis Pesic & Smit *in* Pesic, Smit & Datry, 2010 F E
Zelandopsis aturoides Schwoerbel, 1984 F E
Zelandopsis morimotoi Imamura, 1977 F E
BDELLIDAE
Bdella iconica Berlese, 1923
Bdellodes (Bdellodes) harpax Atyeo, 1963
Bdellodes (B.) oraria Atyeo, 1963
Bdellodes (B.) tanta Atyeo, 1963
Bdellodes (B.) vireti Atyeo, 1963
Bdellodes (Hoploscirus) agrestis Atyeo, 1963
Bdellodes (H.) ancalae Atyeo, 1963
Bdellodes (H.) bryi Atyeo, 1963
Bdellodes (H.) camellae Atyeo, 1963
Bdellodes (H.) conformis Atyeo, 1963
Bdellodes (H.) copiosa Atyeo, 1963
Bdellodes (H.) curvus Atyeo, 1963
Bdellodes (H.) flexuosa Atyeo, 1963
Bdellodes (H.) gressitti Atyeo, 1963
Bdellodes (H.) intricata Atyeo, 1963
Bdellodes (H.) intricatoria Atyeo, 1963
Bdellodes (H.) lapidaria (Kramer, 1881)
Bdellodes (H.) multicia Atyeo, 1963
Bdellodes (H.) petila Atyeo, 1963
Bdellodes (H.) procincta Atyeo, 1963
Bdellodes (H.) reticulata (Atyeo, 1960)
Bdellodes (H.) serpentinus Atye, 1963
Bdellodes (H.) tellustris Atyeo, 1963
Cyta latirostris (Hermann, 1804)
Thoribdella reticulata Atyeo, 1960
CAECULIDAE
Neocaeculus luxtoni Coineau, 1967
CALYPTOSTOMATIDAE
Calyptostoma sp.
CAMEROBIIDAE
Neophyllobius sturmerwoodi Bolland, 1991
Tycherobius aotearoa Fan & Zhang, 2005
CHEYLETIDAE
Cheletomorpha lepidopterorum (Shaw, 1794)
Cheyletus eruditus (Schrank, 1781)
Cheyletiella blakei Smiley, 1970
Cheyletiella parasitivorax Megnin, 1878
Cheyletiella yasguri Smiley, 1965
Granulocheyletus gallowayi Fain, 2003
Oudemansicheyla coprosmae Thewke & Enns, 1976
Psorergates ovis Womersley, 1941
CHYZERIIDAE
Chyzeria novaezealandiae Hirst, 1924 E
Nothotrombicula deinacridae Dumbleton, 1947 E
CRYPTOGNATHIDAE
Cryptognathus leopardus Luxton, 1973 E
Cryptognathus striatus Luxton, 1973 E
Cryptognathus vulgaris Luxton, 1973 E
Favognathus leopardus Luxton, 1973 E
CTENOTHYADIDAE
Stellulathyas lundbladi Cook & Hopkins, 1998 F E
Stellulathyas magnifica Cook & Hopkins, 1998 F E
CUNAXIDAE
Bonzia woodi Smiley, 1992 E
Cunaxa parvirostris (Berlese, 1910)
Cunaxa reevesi Smiley, 1992 E
Cunaxoides sp.
Dactyloscirus inermis (Tragardh, 1905)
Denheyernaxoides martini Smiley, 1992
Neoscirula luxtoni Smiley, 1992 E
Neoscirula proctorae Smiley, 1992 E
Neoscirula sevidi Den Heyer, 1980
Paracunaxoides newzealandicus Smiley, 1992 E
Pseudobonzia breviscuta Luxton, 1982 E
Pseudobonzia newzealandicus Smiley, 1992 E
DEMODECIDAE
Demodex aries Desch, 1986 A
Demodex bovis Stiles, 1892 A
Demodex brevis Akbulatova, 1963 A
Demodex caballi Railliet, 1895 A
Demodex canis Leydig, 1859 A
Demodex caprae Railliet, 1893 A
Demodex cati Hirst, 1919 A
Demodex equi Railliet, 1893 A
Demodex erminae Hirst, 1919 A
Demodex folliculorum Simon, 1842 A
Demodex mystacina Desch, 1989 E
Demodex novazelandica Desch, 1989 E
Demodex ovis Railliet, 1895 A
Demodex phylloides Csokor, 1879
Marsupiopus trichosuri Fain, 1968 A
DIPTILOMIOPIDAE
Abacoptes ulmivagrans (Keifer, 1939)
Brevulacus reticulatus Manson, 1984 E
Dacundiopus stylosus Manson, 1984 E
Diptacus gigantorhynchus (Nalepa, 1892)
Lambella cerina (Lamb, 1953) E
Levonga papaitongensis Manson, 1984 E
Rhyncaphytoptus fagacis Boczek, 1964
EREYNETIDAE
Austreynetes maudensis Fain & Barker, 2003
Ereynetes macquariensis Fain, 1962
Riccardoella limacum (Schrank, 1776)
Riccardoella novaezealandiae Fain& Barker, 2004
ERIOPHYIDAE
Abacarus hystrix (Nalepa, 1896)
Acalitus australis (Lamb, 1952) E
Acalitus avicenniae (Lamb, 1952) E
Acalitus carpatus Manson, 1984 E
Acalitus cottieri (Lamb, 1952) E
Acalitus dissimus Manson, 1984 E
Acalitus essigi (Hassan, 1928)
Acalitus excelsus Manson, 1984 E
Acalitus intertextus Manson, 1984 E
Acalitus kohus Manson, 1984 E
Acalitus lowei Manson, 1972 E
Acalitus morrisoni Manson, 1970 E
Acalitus orthomerus (Keifer, 1951)
Acalitus rubensis Manson, 1970 E
Acalitus spinus Manson, 1984 E
Acalitus taurangensis (Manson, 1965) E
Aceria bipedis Manson, 1984 E
Aceria calystegiae (Lamb, 1952) E
Aceria capreae Manson, 1984 E
Aceria carmichaeliae Lamb, 1952 E
Aceria clianthi Lamb, 1952 E
Aceria depressae Manson, 1984 E
Aceria diospyri Keifer, 1944
Aceria erinea (Nalepa, 1891)
Aceria flynni Xue & Zhang, 2008 E
Aceria genistae (Nalepa, 1891)
Aceria gersoni Manson, 1984 E
Aceria gleicheniae Manson, 1984 E
Aceria hagleyensis Manson, 1984 E
Aceria healyi Manson, 1970 E
Aceria korelli Manson, 1984 E
Aceria manukae Lamb, 1952 E
Aceria mayae Manson, 1984 E
Aceria melicopis Manson, 1984 E
Aceria melicyti Lamb, 1953 E
Aceria microphyllae Manson, 1984 E
Aceria parvensis Manson, 1972 E
Aceria pimeliae Manson, 1984 E
Aceria plagianthi Manson, 1984 E
Aceria rubifaciens Lamb, 1953 E
Aceria sheldoni (Ewing, 1937) A
Aceria simonensis Manson, 1984 E
Aceria strictae Manson, 1984 E
Aceria tenuifolii Manson, 1984 E
Aceria titirangiensis Lamb, 1953 E
Aceria tulipae (Keifer, 1938)
Aceria zoysima Manson, 1989 E
Aceria victoriae Ramsay, 1958 E
Aceria waltheri (Keifer, 1939)
Acerimina pyrrosiae Manson, 1984 E
Aculodes mckenziei (Keifer, 1944)
Aculops albensis Manson, 1984 E
Aculops beeveri Manson & Gerson, 1986 E
Aculops gaultheriae (Lamb, 1953) E
Aculops hilli Manson, 1989 E
Aculops lycopersici (Massee, 1937) E
Aculops pittospori Manson, 1984 E
Aculops propinquae (Manson, 1984) E
Aculops serrati Manson, 1984 E
Aculops wahlenbergiae Manson, 1984 E
Aculus cornutus (Banks, 1905)
Aculus corynocarpi (Manson, 1984) E
Aculus fockeui (Nalepa & Trouessart, 1891)
Aculus haloragi (Lamb, 1953) E
Aculus heatherae (Manson, 1984) E
Aculus lalithi Xue & Zhang, 2008 E
Aculus mansoni Amrine & Stasny, 1994
Aculus schlechtendali (Nalepa, 1890) A
Arectus bidwillius Manson, 1984 E
Asetilobus hodgkinsi (Manson, 1965) E
Calacarus carinatus (Green, 1890)
Calepitrimerus baileyi Keifer, 1938
Calepitrimerus vitis (Nalepa, 1905)
Cecidophyes violae (Nalepa, 1902)
Cecidophyopsis hendersoni (Keifer, 1954)
Cecidophyopsis ribis (Westwood, 1869)
Cecidophyopsis vermiformis (Nalepa, 1889)
Chrecidus quercipodus Manson, 1984 E
Colomerus coplus Manson, 1984 E
Colomerus nudi Manson, 1984 E
Colomerus vitis (Pagenstecher, 1857)
Cosella simplicis Manson 1984 E
Cosetacus camelliae (Keifer, 1945)
Cymeda zealandica Manson & Gerson, 1986 E
Cymoptus waltheri (Keifer, 1939)
Disella rebeeveri Xue & Zhang, 2008 E
Epitrimerus pyri (Nalepa, 1892)
Eriophyes bennetti Xue & Zhang, 2008 E
Eriophyes dracophylli Lamb, 1953 E
Eriophyes duguidae Manson, 1972 E
Eriophyes georgeae Xue & Zhang, 2008 E

Eriophyes hoheriae Lamb, 1952 E
Eriophyes lambi Manson, 1965 E
Eriophyes leptophyllae Manson, 1984 E
Eriophyes mali Burts, 1970
Eriophyes paratrophis Lamb, 1953 E
Eriophyes parsonsiae Manson, 1984 E
Eriophyes plaginus Manson, 1984 E
Eriophyes planchonellus Manson, 1984 E
Eriophyes pyri (Pagenstecher, 1857)
Eriophyes sexstylosae Manson, 1984 E
Eriophyes totarae Manson, 1984 E
Litaculus acutus Manson & Gerson, 1986 E
Litaculus antapicus Manson & Gerson, 1986 E
Litaculus gillianae Manson & Gerson, 1986 E
Litaculus khandus Manson, 1984 E
Litaculus pennigerus Manson & Gerson, 1986 E
Litaculus squarrosus Manson & Gerson, 1986 E
Nameriophyes sapidae Xue & Zhang, 2008 E
Nothacus tuberculatus Manson, 1984 E
Pangacarus grisalis Manson 1984 E
Phyllocoptes abaenus Keifer, 1940
Phyllocoptes coprosmae Lamb, 1952 E
Phyllocoptes hazelae Manson, 1984 E
Phyllocoptes metrosideri Manson, 1989 E
Phyllocoptruta oleivora (Ashmead, 1879) E
Ramaculus mahoe Manson, 1984 E
Rectalox falita Manson, 1984 E
Rhombacus chatelaini Manson, 1984 E
Tegolophus alicis Manson, 1984 E
Tegolophus australis Keifer, 1964
Tegolophus meliflorus Manson, 1984 E
Tegolophus poriruensis Manson, 1984 E
Tetra martini Manson, 1984 E
Tetra cookiana Manson, 1989 E
Vittacus mansoni Keifer, 1969 E
ERYTHRAEIDAE
Abrolophus zelandicus Luxton, 1989 E
Erythrites jacksoni Southcott, 1988 E
Microsmaris mirandus Hirst, 1926 E
Neosmaris novaezelandiae Hirst, 1926 E
Ramsayella rangitata Zhang, 2000 E
Taranakia lambi Southcott, 1988 E
EUPODIDAE
Claveupodes delicatus Strandtmann & Prasse, 1977
Cocceupodes sp.
Eupodes longisetatus Strandtmann, 1964
Eupodes minutus (Strandtmann, 1967)
Linopodes sp.
EYLAIDAE
Eylais schauinslandi Koenike, 1900 F
HALACARIDAE
Agaue insignata Bartsch, 1979 M
Agaue parva (Chilton, 1883) M
Agauopsis luxtoni Bartsch, 1986 M
Agauopsis novaezelandiae Bartsch, 1986 M
Agauopsis similis Bartsch, 1979 M
Bathyhalacarus angustops (Newell, 1984) M
Copidognathus lohmanni Trouessart, 1889 M
Copidognathus lubricus Bartsch, 1986 M
Halacarellus antipodianus (Newell, 1984) M
Halacarellus decipiens (Newell, 1984) M
Halacarellus lubricus Bartsch, 1986 M
Halacarus elegans Newell, 1984 M
Halacarus nitidus Bartsch, 1979 M
Halacarus zealandicus Newell, 1984 M
Halixodes chitonis chitonis (Brucker, 1897) M
Halixodes c. stoutae Viets, 1959 M
Halixodes novaezelandiae Bartsch, 1986 M
Halixodes truncipes (Chilton, 1883) M
Lobohalacarus subterraneus Bartsch, 1995 F E
Lobohalacarus weberi (Romijn & Viets, 1924) F
Porohalacarus alpinus (Thor, 1910) F
Rhombognathus fractus Bartsch, 1979 M
Rhombognathus lacunosus Bartsch, 1979 M
Rhombognathus novaezelandicus Bartsch, 1986 M
Simognathus glaber Bartsch, 1986 M
Simognathus glareus Bartsch, 1986 M
Soldanellonyx monardi Walter, 1919 F
Werthella parvirostris (Trouessart, 1889) M
HYDRACHNIDAE
Hydrachna evansensis Stout, 1953 F E
Hydrachna maramauensis Stout, 1953 F E
Hydrachna wainuiensis Stout, 1953 F E
HYDRYPHANTIDAE
Euwandesia tenebrio Hopkins & Schminke, 1970 F E
Notopanisus shewell Smit, 1996 F E
Pseudohydryphantes bebelus Cook, 1983 F E
Wandesia wiselyi (Womersley, 1953) F E
HYGROBATIDAE
Aciculacarus amilis Cook, 1983 F E
Aciculacarus papillosus Hopkins, 1975 F E
Aspidiobates orbiculatus Hopkins, 1975 F E
Australiobates lacustris Smit, 1996 F E
Australiobates savanus Cook, 1983 F E
Australiobates setipalpis Cook, 1983 F E
Australiobates solomis Cook, 1983 F E
Australiobates vietsi Cook, 1983 F E
Hopkinsobates suzannae Cook, 1983 F E
Notohygrobates kathrynae Cook, 1983 F E
Zelandobatella naias Hopkins, 1975 F E
Zelandobates clevatus Cook, 1983 F E
Zelandobates crinitus Hopkins, 1966 F E
Zelandobates imamurai Schwoerbel, 1984 F E
Zelandobates tornus Cook, 1983 F E
LABIDOSTOMMIDAE
Labidostomma circinus Atyeo & Crossley, 1961
Labidostomma fictiluteum Atyeo & Crossley, 1961
Labidostomma glandula Atyeo & Crossley, 1961
Labidostomma luteum Kramer, 1879
Labidostomma malleolus Atyeo & Crossley, 1961
Labidostomma multifarum Atyeo & Crossley, 1961
Labidostomma ocellatum Atyeo & Crossley, 1961
Labidostomma striatum Atyeo & Crossley, 1961
LEEUWENHOEKIIDAE
Odontacarus lygosomae (Dumbleton, 1947)
LIMNESIIDAE
Limnesia auspexa Cook, 1983 F E
Limnesia birgelda Cook, 1983 F E
Limnesia conroyi Cook, 1983 F E
Limnesia crowelli Cook, 1983 F E
Limnesia foldoma Cook, 1983 F E
Limnesia halcarda Cook, 1983 F E
Limnesia reptans Hopkins, 1966 F E
Limnesia rotoruaensis Smit, 2002 F E
Limnesia testacea Hopkins, 1969 F E
Limnesia zelandica Cook, 1983 F E
LIMNOCHARIDAE
Limnochares (Cyclothrix) australica Lundblad, 1941 F
MECOGNATHIDAE
Mecognatha hirsuta Wood, 1967
Mecognatha parilis Fan & Zhang, 2005 E
Mecognatha rara Fan & Zhang, 2005 E
MICRODISPIDAE
Brennadania sp. b.
Brennadania sp. m.
Brennadania sp. r.
MICROTROMBIDIIDAE
Ettmulleria townsendi Dumbleton, 1962 E
Holcotrombidium scalaris (Womersley, 1936)
Microtrombidium aucklandicum Luxton, 1989 E
Microtrombidium karriensis Womersley, 1934
Microtrombidium otagoensis Luxton, 1989 E
Microtrombidium zelandicum Womersley, 1936
Platytrombidium pritchardi (Womersley, 1936)
MIDEOPSIDAE
Guineaxonopsis confusus (Schwoerbel, 1984) F E
Guineaxonopsis ramsayi (Cook, 1983) F E
Guineaxonopsis serratipalpis (Cook, 1983) F E
Kuschelacarus ovalis Cook, 1992 F E
MOMONIIDAE
Momonia (Kondia) hopkinsi Schwoerbel, 1984 F E
Neomomonia benova Cook, 1983 F E
Neomomonia hopkinsi Cook, 1983 F E
Neomomonia paramecia Cook, 1983 F E
Neomomonia rotunda Schwoerbel, 1984 F E
Neomomonia torquipes (Hopkins, 1966) F E
Partidomomonia polyplacophora Cook, 1983 F E
Partidomomonia ramsayi Cook, 1992 F E
MYOBIIDAE
Myobia musculi (Schrank, 1781)
Radfordia affinis (Poppe, 1896)
Radfordia davisi (Radford, 1938)
Radfordia ensifera (Poppe, 1896)
NEOTHROMBIIDAE
Opiliotrombium akatarawa Zhang & Fan, 2005 E
NUDOMIDEOPSIDAE
Nudomideopsis forkensis (Imamura, 1977) F E
Nudomideopsis kuscheli Cook, 1992 F E
Nudomideopsis nobilis Cook, 1992 F E
Nudomideopsis parva Cook, 1992 F E
Paramideopsis kyphus Cook, 1992 F E
Paramideopsis regalis Cook & Hopkins, 1998 F E
OXIDAE
Oxus (Flabellifrontipoda) bravana Cook, 1983 F E
Oxus (Flabellifrontipoda) crameri Cook, 1983 F E
Oxus (Flabellifrontipoda) hadinoma Cook, 1983 F E
Oxus (Flabellifrontipoda) lacustris Cook, 1983 F E
Oxus (Flabellifrontipoda) ladilofa Cook, 1983 F E
Oxus (Flabellifrontipoda) mastigophora Cook, 1983 F E
Oxus (Flabellifrontipoda) minutipalpis (Smit, 2002) F E
Oxus (Flabellifrontipoda) reductipalpa Cook, 1983 F E
Oxus (Flabellifrontipoda) smithi Cook, 1983 F E
Oxus (Flabellifrontipoda) zelandica Hopkins, 1975 F E
PENTHALEIDAE
Halotydeus destructor (Tucker, 1925)
Linopenthaloides novazealandicus Strandtmann, 1981
Penthaleus major (Duges, 1834)
PENTHALODIDAE
Callipenthalodes pennyae Qin, 1998 E
Stereotydeus nudisetus Strandtmann, 1964
Stereotydeus pulcher Strandtmann, 1964
Stereotydeus undulatus Strandtmann, 1964
PHYTOPTIDAE
Phytoptus avellanae Nalepa, 1889
Phytoptus rufensis Manson, 1970 E
PIONIDAE
Piona exigua Viets, 1949 F
Piona pseudouncata (Piersig, 1906) F
Schminkea pacifica Schwoerbel, 1984 F E
Twinforksella tura Cook, 1992 F E
PODAPOLIPIDAE
Locustacarus buchneri Stammer, 1951
Locustacarus masoni Husband, 1974
Locustacarus trachealis Ewing, 1924
Podapolipoides ramsayi Husband, 1990
Wetapolipus jamiesoni Husband & Zhang, 2002 E
PSORERGATIDAE A
Psorobia ovis (Womersley, 1941) A
Psorergates (Psorobia) mustelae Lukoschus, 1969 A
PTERYGOSOMATIDAE
Geckobia haplodactyli haplodactyli Womersley, 1941
Geckobia naultini Womersley, 1941
PYEMOTIDAE
Pyemotes tritici (La Greze-Fossat & Montagne, 1851)
Pyemotes ventricosus (Newport, 1850)
PYGMEPHORIDAE
Bakerdania arvorum (Jacot, 1936)
Bakerdania luxtoni Mahunka, 1970
Bakerdania mirabilis (Mahunka, 1969)
Bakerdania novaezelandicus Mahunka, 1970
Bakerdania quadratus (Ewing, 1917)
Bakerdania sellnicki (Krczal, 1958)
Bakerdania tarsalis (Hirst, 1921)
Bakerdania togatus (Willmann, 1942)
Bakerdania sp. a

Bakerdania sp. e
Bakerdania sp. f
Bakerdania sp. p
Bakerdania sp. q
Pygmephorus lambi Krczal, 1964
RAPHIGNATHIDAE
Raphignathus atomatus Fan & Zhang, 2005
Raphignathus collegiatus Atyeo, Baker & Crossley, 1961
Raphignathus crustus Fan & Zhang, 2005
Raphignathus gracilis (Rack, 1962)
RHAGIDIIDAE
Coccorhagidia clavifrons (Canestrini, 1886)
Rhagidia campbellensis Zacharda, 1980
Rhagidia mildredae Strandtmann, 1964
SCUTACARIDAE
Imparipes (Imparipes) insignis Mahunka, 1970
Scutacarus extrovertus Mahunka, 1970
Scutacarus (Variatipes) quadrangularis (Paoli, 1911)
SITEROPTIDAE
Siteroptes (Pediculaster) kneeboni (Wicht, 1970)
Siteroptes (P.) mesembrinae (Canestrini, 1881)
Siteroptes (P.) microsaniae Martin, 1978
Siteroptes (P.) morellae (Rack, 1975)
Siteroptes (P.) muscarius Martin, 1978
Siteroptes (P.) portatus Martin, 1978
Siteroptes (Siteroptes) avenae Mueller, 1905
Siteroptes (S.) crossi Mahunka, 1969
Siteroptes (S.) longisetosus Mahunka, 1970
Siteroptes (S.) sp. a
Siteroptes (S.) sp. c
Siteroptes (S.) sp. e
Siteroptes (S.) sp. g
SMARIDIDAE
Hirstiosoma novaehollandiae Womersley, 1936
SPERCHONTIDAE
Apeltosperchon zelandicus Cook, 1983 F E
STIGMAEIDAE
Agistemus collyerae González-Rodríguez, 1963
Agistemus longisetus González-Rodríguez, 1963
Agistemus mecotrichus Fan & Zhang, 2005
Agistemus novazelandicus González-Rodríguez, 1963
Agistemus subreticulata (Wood, 1967) E
Cheylostigmaeus luxtoni Wood, 1968 E
Eryngiopus arboreus Wood, 1967 E
Eryngiopus bifidus Wood, 1967
Eryngiopus nelsonensis Wood, 1971 E
Eryngiopus similis Wood, 1967 E
Eustigmaeus brevisetosus (Wood, 1966) E
Eustigmaeus clavigerus (Wood, 1966) E
Eustigmaeus corticolus (Wood, 1966) E
Eustigmaeus distinctus (Wood, 1966) E
Eustigmaeus dumosus (Wood, 1966) E
Eustigmaeus eburneus Fan & Zhang, 2005 E
Eustigmaeus edentatus Fan & Zhang, 2005 E
Eustigmaeus granulosus (Wood, 1966) E
Eustigmaeus kermesinus (Koch, 1841)
Eustigmaeus manapouriensis (Wood, 1966) E
Eustigmaeus mixta mixtus (Wood, 1966)
Eustigmaeus ptilosetus Fan & Zhang, 2005 E
Eustigmaeus simplex (Wood, 1966) E
Ledermuelleriopsis incisa Wood, 1967 E
Ledermuelleria spinosa Wood, 1967 E
Mediolata brevisetis Wood, 1967 E
Mediolata delicata Fan & Zhang, 2005 E
Mediolata favulosa Wood, 1967 E
Mediolata mollis Wood, 1971 E
Mediolata oleariae Wood, 1971 E
Mediolata polylocularis Fan & Zhang, 2005 E
Mediolata robusta González-Rodríguez, 1965 E
Mediolata simplex Wood, 1967 E
Mediolata whenua Fan & Zhang, 2005 E
Mediolata woodi Fan & Zhang, 2005 E
Mediolata xerxes Fan & Zhang, 2005 E
Mediolata zonaria Fan & Zhang, 2005 E
Mullederia arborea Wood, 1964 E
Mullederia procurrens Fan & Zhang, 2005 E
Mullederia scutellaris Fan & Zhang, 2005 E
Primagistemus loadmani (Wood), 1967 E
Pseudostigmaeus collyerae Wood, 1967 E
Pseudostigmaeus longisetis Wood, 1971 E
Pseudostigmaeus schizopeltatus Fan & Zhang, 2005 E
Pseudostigmaeus striatus Wood, 1967 E
<u>*Scutastigmaeus*</u> *confusus* (Wood, 1967) E
Scutastigmaeus longisetis (Wood, 1967) E
Stigmaeus arboricola Wood, 1981 E
Stigmaeus brevisetis Wood, 1967 E
Stigmaeus campbellensis Wood, 1971 E
Stigmaeus luxtoni Wood, 1981 E
Stigmaeus montanus (Wood, 1981) E
Stigmaeus novazealandicus Wood, 1981 E
Stigmaeus rotundus Wood, 1967 E
Stigmaeus rupicola Wood, 1967 E
Stigmaeus summersi Wood, 1967 E
Storchia robustus (Berlese, 1885)
Storchia hendersonae Fan & Zhang, 2005 E
Summersiella coprosmae (Wood, 1967) E
Zetzellia antipoda Wood, 1967) E
Zetzellia biscutata Fan & Zhang, 2005 E
Zetzellia gonzalezi Wood, 1967 E
Zetzellia maori González-Rodríguez, 1965
Zetzellia oudemansi Wood, 1967 E
Zetzellia spiculosa Fan & Zhang, 2005 E
STYGOTONIIDAE E
<u>*Stygotonia*</u> *ambigua* Cook, 1992 F E
TARSONEMIDAE
Acarapis dorsalis Morgenthaler, 1934 A
Acarapis externus Morgenthaler *in* Morison, 1931 A
Acarapis vagans Schneider, 1941 A
Fungitarsonemus kawakawa Lin & Zhang, 2005 E
Fungitarsonemus kohia Lin & Zhang, 2005 E
Fungitarsonemus kerikeri Lin & Zhang, 2005 E
Fungitarsonemus tawa Lin & Zhang, 2005 E
Phytonemus pallidus (Banks, 1899)
Polyphagotarsonemus latus (Banks 1904)
Steneotarsonemus (Mahunkacarus) mayae Lin & Zhang, 2005 E
Steneotarsonemus (Neosteneotarsonemus) ramus Lin & Zhang, 2005 E
Steneotarsonemus (Steneotarsonemus) spirifex (Marchal, 1902)
Steneotarsonemus sp. B.
Suskia mansoni Lindquist, 1986 E
Tarsonemus fusarii Cooreman, 1941 A
Tarsonemus parawaitei Kim, Qin & Lindquist, 1998
Tarsonemus rakowiensis (Kropczynska, 1965)
Tarsonemus talpae Schaarschmidt, 1959
Tarsonemus waitei Banks, 1912
Tarsonemus sp. c
Tarsonemus sp. d
Tarsonemus sp. f
Tarsonemus sp. nr *ellipticus* Schaarschmidt, 1959
Tarsonemus sp. nr *setifer* Baker & Wharton, 1952
Xenotarsonemus sp.
TENUIPALPIDAE
Aegyptobia pomaderrisae Collyer, 1969
Brevipalpus californicus (Banks, 1904)
Brevipalpus essigi (Baker, 1949)
Brevipalpus obovatus Donnadieu, 1875
Brevipalpus phoenicis (Geijskes, 1939)
Brevipalpus russulus (Boisduval, 1867)
Dolichotetranychus alpinus Collyer, 1973
Dolichotetranychus ancistrus Baker & Pritchard, 1956 E
Tenuipalpus antipodus Collyer, 1973
Tenuipalpus elegans (Collyer, 1973)
Tenuipalpus mahoensis Collyer, 1964
Tenuipalpus mansoniculus Ghai & Shenhmar, 1985
Tenuipalpus montanus Collyer, 1973
Tenuipalpus rangiorae Collyer, 1964
Tenuipalpus senecionis Collyer, 1973
Tenuipalpus venustus Collyer, 1973
Tenuipalpus womersleyi Pritchard & Baker, 1958
Ultratenuipalpus aberrans (Collyer, 1973)
Ultratenuipalpus arboreus (Collyer, 1973)
Ultratenuipalpus asteliae (Collyer, 1973)
Ultratenuipalpus asteliicola (Collyer, 1973)
Ultratenuipalpus carpodeti (Collyer, 1973)
Ultratenuipalpus coprosmae (Collyer, 1964)
Ultratenuipalpus coprosmicus (Collyer, 1973)
Ultratenuipalpus cyatheae (Gerson & Collyer, 1984)
Ultratenuipalpus nothofagi (Collyer, 1973)
Ultratenuipalpus rubi (Collyer, 1964)
TETRANYCHIDAE
Bryobia annatensis Manson, 1967 E
Bryobia gramium (Schrank, 1781)
Bryobia kissophila van Eyndhoven 1955
Bryobia lagodechiana Reck, 1953
Bryobia praetiosa Koch, 1836
Bryobia rubrioculus (Scheuten, 1857)
Bryobia variabilis Manson, 1967 E
Bryobia vasiljevi Reck, 1953
Bryobia watersi Manson, 1967
Eotetranychus sexmaculatus (Riley, 1890)
Oligonychus brevipodus Targioni Tozzetti, 1878 A
Oligonychus hondoensis (Ehara, 1954)
Oligonychus ununguis (Jacobi, 1905)
Panonychus citri (McGregor, 1916)
Panonychus ulmi (Koch, 1836)
Petrobia (Petrobia) latens (Mueller, 1776)
Petrobia (Tetranychina) harti (Ewing, 1909)
Schizotetranychus kaspari Manson, 1967 E
Schizotetranychus levinensis Manson, 1967 E
Sonotetranychus tawhairauriki Zhang & Bennett, 2009 E
Tetranychus cinnabarinus (Boisduval, 1867)
Tetranychus collyerae Manson, 1967 E
Tetranychus elsae Manson, 1967 E
Tetranychus eyrewellensis Manson, 1967 E
Tetranychus lambi Pritchard & Baker, 1955
Tetranychus lintearius Dufour, 1832 A
Tetranychus ludeni Zacher, 1913
Tetranychus moutensis Manson, 1970 E
Tetranychus neocaledonicus Andre, 1933
Tetranychus turkestani (Ugarov & Nikolskii, 1937)
Tetranychus urticae Koch, 1836 A
<u>*Tribolonychus*</u> *collyerae* Zhang & Martin, 2001 E
Yezonychus brevipilus Zhang & Martin, 2001 E
Yezonychus cornus (Pritchard & Baker, 1955) E
Yezonychus falsicornus Zhang & Martin, 2001
TROMBICULIDAE
Guntheria (Derrickiella) apteryxi Loomis & Goff, 1983
Neotrombicula hoplodactyla Goff, Loomis & Ainsworth, 1987
Neotrombicula naultini (Dumbleton, 1947)
Neotrombicula sphenodonti Goff, Loomis & Ainsworth, 1987
Whartonacarus sp.
TUCKERELLIDAE
Tuckerella flabellifera Miller, 1964
Tuckerella litoralis Collyer, 1969
TYDEIDAE
Australotydeus kirstenae Spain, 1969
Lorryia collyerae Baker, 1968
Microtydeus beltrani Baker, 1944
Novzelorryia deserta Kazmierski, 1996 E
Pseudolorryia Paralorryia mansoni (Baker, 1968)
Tydaeolus sp. nr *krantzi* Naegele, 1965
Tydaeolus sp. nr *loadmani* Wood, 1965
Tydaeolus sp. nr *tenuiclaviger* Thor, 1931
Tydeus californicus (Banks, 1904)
Tydeus caudatus (Duges, 1834)
Tydeus interruptus Thor, 1932
Tydeus lambi Baker, 1970
UNIONICOLIDAE
Unionicola billieaehonore Crowell, 1990 F E

Unionicola longiseta Walter, 1915 F
ZELANDOTHYADIDAE E
Zelandothyas diamphida Cook, 1983 F E
Zelandothyas hyporheica Smit, 1996 F E

Superorder PARASITIFORMES
Order HOLOTHYRIDA
ALLOTHYRIDAE
Allothyrus australasiae (Womersley, 1935)

Order IXODIDA
ARGASIDAE
Ornithodoros (Alectorobius) capensis Neumann, 1901
IXODIDAE
Amblyomma (Aponomma) sphenodonti Dumbleton, 1943 E
Haemaphysalis (Kaiseriana) longicornis Neumann, 1901 A
Ixodes (Ceratixodes) jacksoni Hoogstraal, 1967 E
Ixodes (C.) uriae White, 1852
Ixodes (Multidentatus) auritulus zealandicus Dumbleton, 1961 E
Ixodes (M.) eudyptidis Maskell, 1885
Ixodes (M.) kerguelenensis André & Colas Belcour, 1942
Ixodes (M.) kohlsi Arthur, 1955
Ixodes (Scaphixodes) unicavatus Neumann, 1908
Ixodes (Sternalixodes) anatis Chilton, 1904 E

Order MESOSTIGMATA
AMEROSEIIDAE
Ameroseius sp.
Hattena tongana (Manson, 1974)
Neocypholaelaps novahollandiae Evans, 1961
ASCIDAE
Asca aphidioides (Linnaeus, 1758)
Asca arboriensis Wood, 1966
Asca brevisetosa Wood, 1965
Asca duosetosa Fox, 1946
Asca foliata Womersley, 1956
Asca novazelandica Wood, 1965
Asca plumosa Wood, 1966
Asca porosa Wood, 1966
Asca tuberculata Wood, 1965
Iphidozercon sp.
Leioseius australis Luxton, 1984
Laelaptoseius novaezelandiae Womersley, 1960
Proctolaelaps pygmaeus (Müller, 1859)
BLATTISOCIIDAE
Blattisocius dentriticus (Berlese, 1918)
Blattisocius tarsalis (Berlese, 1918)
Lasioseius penicilliger Berlese, 1916
Platyseius mackerrasae Womersley, 1956
DERMANYSSIDAE
Ayersacarus gelidus Hunter, 1965
Ayersacarus gressitti Hunter, 1964
Ayersacarus plumapilus Hunter, 1964
Ayersacarus strandtmanni Hunter, 1964
Dermanyssus gallinae (de Geer, 1778)
Hirstionyssus talpae Zemskaya, 1955
Liponyssoides eudyptulae Fain & Galloway, 1993
DIGAMASELLIDAE
Digamasellus kargi (Hirschmann, 1966)
Digamasellus schusteri (Hirschmann, 1966)
Digamasellus watsoni (Hirschmann, 1966)
Digamasellus sp.
Pontiolaelaps salinus Luxton, 1989
Pontiolaelaps terebratus (Luxton, 1984)
EVIPHIDIDAE
Alliphis siculus (Oudemans, 1905)
Thinoseius ramsayi Evans, 1969
EURYPARASITIDAE
Acugamasus watsoni (Hirschmann, 1966)
Acugamasus sp.
GAMASELLIDAE
Gamasellus sp.
LAELAPIDAE
Androlaelaps casalis (Berlese, 1887)
Androlaelaps fahrenholzi (Berlese, 1911)
Androlaelaps pachyptilae (Zumpt & Till, 1956)
Chirolaelaps mystacinae Heath, Bishop & Daniel, 1987
Dicrocheles eothenes Treat, 1970
Dicrocheles scedastes Treat, 1969
Euelaelaps oudemansi Turk, 1945
Eulaelaps stabularis (Koch, 1836)
Gaeolaelaps queenslandicus (Womersley, 1956)
Gymnolaelaps annectans Womersley, 1955
Haemogamasus pontiger (Berlese, 1904)
Hypoaspis evansi Hunter, 1964
Hypoaspis (Cosmolaelaps) sp.
Hypoaspis sp.
Laelaps nuttalli Hirst, 1915
Leptolaelaps macquariensis (Womersley, 1937)
Mesolaelaps australiensis (Hirst, 1926)
Mellitiphis alvearius (Berlese, 1896)
Ololaelaps paratasmanicus Ryke, 1963
Pneumolaelaps bombicolens G. Canestrini, 1885
Pneumolaelaps breviseta (Evans & Till, 1966)
Trichosurolaelaps crassipes Womersley, 1956
LEPTOLAELAPIDAE
Leptolaelaps reticulatus Evans, 1957
MACROCHELIDAE
Geholaspis longispinosus (Kramer, 1876)
Glyptholapis americana (Berlese, 1888)
Glyptholapis confusa (Foà, 1900)
Macrocheles caelatus (Berlese, 1918)
Macrocheles glaber (Müller, 1860)
Macrocheles hyatti Krantz & Filipponi, 1964
Macrocheles matrius (Hull, 1925)
Macrocheles merdarius (Berlese, 1889)
Macrocheles muscaedomesticae (Scopoli, 1772)
Macrocheles novaezelandiae Emberson, 1973
Macrocheles penicilliger (Berlese, 1904)
Macrocheles perglaber Filipponi & Pegazzano, 1962
Macrocheles robustulus (Berlese, 1904)
Macrocheles scutatus (Berlese, 1904)
Macrocheles subbadius (Berlese, 1904)
MACRONYSSIDAE A
Ophionyssus galeotes Domrow, Heath & Kennedy, 1980 A
Ophionyssus scincorum Domrow, Heath & Kennedy, 1980 A
Ornithonyssus bacoti (Hirst, 1913) A
Ornithonyssus bursa (Berlese, 1888) A
Ornithonyssus spinosa Manson, 1973 A
Ornithonysus sylviarum (Canestrini & Fanzago, 1877) A
OLOGAMASIIDAE
Antennolaelaps sp.
Athiasella longiseta Lee & Hunter, 1974
Athiasella pecten Lee & Hunter, 1974
Athiasella scaphosternum Lee & Hunter, 1974
Athiasella viripileus Lee & Hunter, 1974
Caliphis novaezelandiae (Womersley, 1956)
Caliphis schusteri (Hirschmann, 1966)
Cymiphis cymosus (Lee, 1966)
Cymiphis dumosus (Lee, 1966)
Cymiphis leptosceles (Lee, 1966)
Cymiphis mansoni (Lee, 1966)
Cymiphis nucilis (Lee, 1966)
Cymiphis validus (Lee, 1966)
Cymiphis watsoni (Hirschmann, 1966)
Euepicrius caesariatus Lee & Hunter, 1974
Evanssellus foliatus Ryke, 1961
Gamasiphis sp.
Gamasiphoides costai Lee & Hunter, 1974
Gamasiphoides macquariensis (Hirschmann, 1966)
Heydeniella markmitchelli (Lee, 1970)
Heydeniella sherrae Lee & Hunter, 1974
Heydeniella womersleyi Lee & Hunter, 1974
Hydrogamasellus antarcticus (Trägårdh, 1907)
Hydrogamasus kensleri Luxton, 1967
Litogamasus setosus (Kramer, 1898)
Litogamasus falcipes Lee & Hunter, 1974
Parasitiphis aurora Lee, 1970
Parasitiphis jeanneli (André, 1947)
Pilellus rugipellis Lee & Hunter, 1974
PARANTENNULIDAE
Micromegistus gourlayi Womersley, 1958
PARASITIDAE
Eugamasus cornutus (Canestrini & Canestrini, 1882)
Eugamasus sp.
Parasitus intermedius (Berlese, 1882)
Pergamasus crassipes (Linnaeus, 1758)
Pergamasus longicornis (Berlese, 1906)
Pergamasus runcatellus (Berlese, 1906)
Pergamasus sp.
Phorytocarpais americanus (Berlese, 1906)
PHYSALOZERCONIDAE
Physalozercon raffray (Wasmann, 1902)
PHYTOSEIIDAE
Amblyseius largoensis (Muma, 1955)
Amblyseius martini (Collyer, 1982) E
Amblyseius obtusus (Koch, 1839)
Amblyseius perlongisetus Berlese, 1916
Euseius ovalis (Evans, 1953)
Galendromus occidentalis (Nesbitt, 1951) A
Iphiseius bidibidi Collyer, 1964 E
Macmurtryseius christinae (Schicha, 1981)
Neoseiulella ashleyae Chant & Yoshida-Shaul, 1988 E
Neoseiulella cassiniae (Collyer, 1982) E
Neoseiulella cottieri (Collyer, 1964)
Neoseiulella dachanti (Collyer, 1964)
Neoseiulellea manukae (Collyer, 1964)
Neoseiulella myopori (Collyer, 1982) E
Neoseiulella nesbitti Womersley, 1954
Neoseiulella novaezealandiae (Collyer, 1964)
Neoseiulella oleariae (Collyer, 1982) E
Neoseiulella spaini (Collyer, 1982) E
Neoseiulus barkeri Hughes, 1948
Neoseiulus cucumeris (Oudemans, 1930)
Neoseiulus fallacis (Garman, 1948) A
Neoseiulus harrowi (Collyer, 1964)
Neoseiulus longispinosus (Evans, 1952)
Phytoscutus acaridophagus (Collyer, 1964)
Phytoseiulus persimilis Athias-Henriot, 1957 A
Phytoseius fotheringhamiae Denmark & Schicha, 1975
Phytoseius leaki Schicha, 1977
Proprioseiopsis expodalis (Kennett, 1958)
Proprioseiopsis mexicanus (Garman, 1958)
Typhlodromalus limonicus (Garman & McGregor, 1956)
Typhlodromina eharai Muma & Denmark, 1969
Typhlodromina tropica (Chant, 1959)
Typhlodromus (Anthoseius) bakeri (Garman, 1948)
Typhlodromus (A.) caudiglans (Schuster, 1959)
Typhlodromus (Typhlodromus) pyri Scheuten, 1857
POLYASPIDAE
Calotrachytes fimbriatipes (Michael, 1908)
Calotrachytes sclerophyllus (Michael, 1908)
RHINONYSSIDAE
Ptilonyssus cractici Domrow, 1964
Ptilonyssus emberizae Fain, 1956
Ptilonyssus euroturdi Fain & Hyland, 1963
Rallinyssus gallinulae Fain, 1960
Rhinonyssus rhinolethrum (Trouessart, 1895)
Sternostoma tracheacolum Lawrence, 1948
Tinaminyssus halcyonus (Domrow, 1965)
Tinaminyssus melloi (de Castro, 1948)
RHODACARIDAE
Rhodacarellus sp.
Rhodacarus sp.
Tangaroellus porosus Luxton, 1968
SEJIDAE

Sejus novaezealandiae Fain & Galloway, 1993
UROPODIDAE
Oodinychus sp.
Urodinychus sp.
Uroobovella sp.
Uropoda vegetans (De Geer, 1768)
Uropoda sp.
VARROIDAE A
Varroa destructor Anderson & Trueman, 2000 A
VEIGAIIDAE
Cyrthydrolaelaps watsoni Hirschmann, 1966
Veigaia sp.

Subclass DROMOPODA
Infraclass PHALANGIDA
Superorder OPILIONIDA
Order OPILIONES
Suborder EUPNOI
CADDIDAE
Zeopsopilio neozealandiae E
GAGRELLIDAE
Nelima doriae (Canestrini, 1871) A
MONOSCUTIDAE
Acihasta salebrosa Forster, 1948 E
Megalopsalis chiltoni chiltoni Hogg, 1909 E
Megalopsalis c. nigra Forster, 1944 E
Megalopsalis distincta (Forster, 1964) E
Megalopsalis fabulosa (Grimmett & Phillipps, 1932) E
Megalopsalis grayi (Hogg, 1920) E
Megalopsalis grimmetti Forster, 1944 E
Megalopsalis inconstans Forster, 1944 E
Megalopsalis marplesi Forster, 1944 E
Megalopsalis triascuta Forster, 1944 E
Megalopsalis tumida Forster, 1944 E
Megalopsalis turneri Marples, 1944 E
Megalopsalis wattsi (Hogg, 1920) E
Monoscutum titirangiense Forster, 1948 E
Pantopsalis albipalpis Pocock, 1903 E
Pantopsalis cheliferoides (Colenso, 1893) E
Pantopsalis coronata Pocock, 1903 E
Pantopsalis halli Hogg, 1920 E
Pantopsalis johnsi Forster, 1964 E
Pantopsalis listeri (White, 1849) E
Pantopsalis luna (Forster, 1944) E
Pantopsalis phocator Taylor, 2004 E
Pantopsalis pococki Hogg, 1920 E
Pantopsalis renelli Forster, 1964 E
Pantopsalis snaresensis Forster, 1964 E
Phalangium opilio Linneus, 1761 A
Templar incongruens Taylor, 2008 E
Gen. et spp. indet. (~100) R. R. Forster pers. comm. ~100E

Suborder LANIATORES
TRIAENONYCHIDAE
Algidia chiltoni chiltoni Roewer, 1931 E
Algidia c. longispinosa Forster, 1954 E
Algidia c. oconnori Forster, 1954 E
Algidia cuspidata cuspidata Hogg, 1920 E
Algidia c. multispinosa Forster, 1954 E
Algidia homerica Forster, 1954 E
Algidia interrupta interrupta Forster, 1954 E
Algidia i. solatia Forster, 1954 E
Algidia marplesi Forster, 1954 E
Algidia nigriflava (Loman, 1902) E
Algidia viridata bicolor Forster, 1954 E
Algidia v. viridata Forster, 1954 E
Cenefia aediformis (Roewer, 1931) E
Cenefia delli Forster, 1954 E
Cenefia sorenseni hawea Forster, 1954 E
Cenefia s. sorenseni Forster, 1954 E
Cenefia westlandica Forster, 1954 E
Hedwigia manubriata Roewer, 1931 E
Hendea aurora Forster, 1968 E
Hendea bucculenta Forster, 1954 E
Hendea coatesi Forster, 1968 E
Hendea fiordensis Forster, 1954 E
Hendea hendei (Hogg, 1920) E
Hendea maini Forster, 1968 E
Hendea maitaia Forster, 1954 E
Hendea myersi assimilis Forster, 1954 E
Hendea m. cavernicola Forster, 1954 E
Hendea m. myersi (Phillipps & Grimmett, 1932) E
Hendea m. ochrea Forster, 1954 E
Hendea m. roeweri Forster, 1954 E
Hendea oconnori Forster, 1954 E
Hendea nelsonensis Forster, 1954 E
Hendea phillippsi phillippsi Forster, 1954 E
Hendea p. stiphra Forster, 1954 E
Hendea spina Forster, 1968 E
Hendea takaka Forster, 1968 E
Hendea townsendi Forster, 1968 E
Hendeola bullata bullata Forster, 1954 E
Hendeola b. pterna Forster, 1954 E
Hendeola woodwardi Forster, 1954 E
Karamea lobata aurea Forster, 1954 E
Karamea l. australis Forster, 1954 E
Karamea l. lobata Forster, 1954 E
Karamea trailli (Hogg, 1902) E
Karamea tricerata Forster, 1954 E
Karamea tuthilli Forster, 1954 E
Muscicola picta Forster, 1954 E
Neonuncia blacki Forster, 1954 E
Neonuncia campbelli Forster, 1954 E
Neonuncia eastoni Forster, 1954 E
Neonuncia enderbyi (Hogg, 1909) E
Neonuncia opaca Forster, 1954 E
Nuncia (Corinuncia) coriacea cockayni Forster, 1954 E
Nuncia (C.) c. coriacea (Pocock, 1903) E
Nuncia (C.) elongata Forster, 1954 E
Nuncia (C.) frustrata Forster, 1954 E
Nuncia (C.) levis Forster, 1954 E
Nuncia (C.) nigriflava nigriflava Forster, 1954 E
Nuncia (C.) n. parva Forster, 1954 E
Nuncia (C.) n. parvocula Forster, 1954 E
Nuncia (C.) pallida Forster, 1954 E
Nuncia (C.) planocula Forster, 1954 E
Nuncia (C.) smithi Forster, 1954 E
Nuncia (C.) sublaevis Forster, 1954 E
Nuncia (C.) stewartia stewartia (Hogg, 1909) E
Nuncia (C.) s. tumosa Forster, 1954 E
Nuncia (C.) tumidarta Forster, 1954 E
Nuncia (C.) variegata australis Forster, 1954 E
Nuncia (C.) v. delli Forster, 1954 E
Nuncia (C.) v. granulata Forster, 1954 E
Nuncia (C.) v. variegata (Hogg, 1954) E
Nuncia (Micronuncia) alpha Forster, 1954 E
Nuncia (M.) contrita Forster, 1954 E
Nuncia (M.) roeweri callida Forster, 1954 E
Nuncia (M.) r. demissa Forster, 1954 E
Nuncia (M.) r. gelida Forster, 1954 E
Nuncia (M.) r. humilis Forster, 1954 E
Nuncia (M.) r. moderata Forster, 1954 E
Nuncia (M.) r. pilgrimi Forster, 1954 E
Nuncia (M.) r. roeweri Forster, 1954 E
Nuncia (M.) r. seditosa Forster, 1954 E
Nuncia (M.) r. unica Forster, 1954 E
Nuncia (Nuncia) arcuata aorangiensis Forster, 1954 E
Nuncia (N.) a. arcuata Forster, 1954 E
Nuncia (N.) conjuncta conjuncta Forster, 1954 E
Nuncia (N.) c. fiordensis Forster, 1954 E
Nuncia (N.) c. magnopercula Forster, 1954 E
Nuncia (N.) constantia Forster, 1954 E
Nuncia (N.) dentifera Forster, 1954 E
Nuncia (N.) fatula Forster, 1954 E
Nuncia (N.) grandis Forster, 1954 E
Nuncia (N.) heteromorpha heteromorpha Forster, 1954 E
Nuncia (N.) h. prolobula Forster, 1954 E
Nuncia (N.) inopinata Forster, 1954 E
Nuncia (N.) kershawi Forster, 1962 E
Nuncia (N.) marchanti Forster, 1962 E
Nuncia (N.) obesa grimmetti Forster, 1954 E
Nuncia (N.) o. magna Forster, 1954 E
Nuncia (N.) o. obesa (Simon, 1899) E
Nuncia (N.) o. rotunda Forster, 1954 E
Nuncia (N.) oconneri connocula Forster, 1954 E
Nuncia (N.) o. kopua Forster, 1954 E
Nuncia (N.) o. oconnori Forster, 1954 E
Nuncia (N.) paucispinosa Forster, 1954 E
Nuncia (N.) stabilis Forster, 1954 E
Nuncia (N.) sulcata Forster, 1954 E
Nuncia (N.) tapanuiensis Forster, 1954 E
Nuncia (N.) townsendi Forster, 1962 E
Nuncia (N.) tumula Forster, 1954 E
Nuncia (N.) vidua Forster, 1954 E
Prasma tuberculata intermedia Forster, 1954 E
Prasma t. mulsa Forster, 1954 E
Prasma t. mearosa Forster, 1954 E
Prasma t. tuberculata (Hogg, 1920) E
Prasma sorenseni regalia Forster, 1954 E
Prasma s. sorenseni Forster, 1954 E
Prasmiola unica Forster, 1954 E
Pristobunus acentrus acentrus Forster, 1954 E
Pristobunus a. hilus Forster, 1954 E
Pristobunus a. insulanus Forster, 1954 E
Pristobunus a. nodosus Forster, 1954 E
Pristobunus acuminatus acantheis Forster, 1954 E
Pristobunus a. acuminatus (Hogg, 1920) E
Pristobunus a. hamiltoni Forster, 1954 E
Pristobunus a. tragulus Forster, 1954 E
Pristobunus barnardi Forster, 1954 E
Pristobunus ceratias Forster, 1954 E
Pristobunus hadrus Forster, 1954 E
Pristobunus henopeus gorensis Forster, 1954 E
Pristobunus h. henopeus Forster, 1954 E
Pristobunus h. ileticus Forster, 1954 E
Pristobunus h. pelorus Forster, 1954 E
Pristobunus heterus Forster, 1954 E
Pristobunus ignavus Forster, 1954 E
Pristobunus laminus Forster, 1954 E
Pristobunus synaptus Forster, 1954 E
Psalenoba nunciaeformes Roewer, 1931 E
Soerensenella bicornis bicornis Forster, 1954 E
Soerensenella b. parva Forster, 1954 E
Soerensenella b. waikanae Forster, 1954 E
Soerensenella prehensor nitida Forster, 1954 E
Soerensenella p. obesa Forster, 1954 E
Soerensenella p. prehensor Pocock, 1903 E
Soerensenella rotara Phillips and Grimmett, 1932
Triregia bilineata (Forster, 1943) E
Triregia fairburni fairburni (Forster, 1943) E
Triregia f. grata Forster, 1954 E
Triregia monstrosa Forster, 1948 E
SYNTHETONYCHIDAE E
Synthetonychia acuta Forster, 1954 E
Synthetonychia cornua Forster, 1954 E
Synthetonychia fiordensis Forster, 1954 E
Synthetonychia florae Forster, 1954 E
Synthetonychia glacialis Forster, 1954 E
Synthetonychia hughsoni Forster, 1954 E
Synthetonychia minuta Forster, 1954 E
Synthetonychia obtusa Forster, 1954 E
Synthetonychia oliveae Forster, 1954 E
Synthetonychia oparara Forster, 1954 E
Synthetonychia proxima Forster, 1954 E
Synthetonychia ramosa Forster, 1954 E
Synthetonychia sinuosa Forster, 1954 E
Synthetonychia wairarapae Forster, 1954 E

Suborder CYPHOPHTHALMI
PETALLIDAE
Aoraki calcarobtusa calcarobtusa (Forster, 1952) E
Aoraki c. westlandica (Forster, 1952) E
Aoraki crypta (Forster, 1948) E
Aoraki denticulata denticulata (Forster, 1948) E

Aoraki d. major (Forster, 1948) E
Aoraki granulosa (Forster, 1952) E
Aoraki healyi (Forster, 1948) E
Aoraki inerma inerma (Forster, 1948) E
Aoraki i. stephenensis (Forster, 1952) E
Aoraki longitarsa (Forster, 1952) E
Aoraki tumidarta (Forster, 1948) E
Neopurcellia salmoni Forster, 1948 E
Rakaia antipodiana Hirst, 1925 E
Rakaia dorothea (Phiilipps & Grimmett, 1932) E
Rakaia florensis (Forster, 1948) E
Rakaia isolata Forster, 1952 E
Rakaia lindsayi Forster, 1952 E
Rakaia macra Boyer & Giribet, 2003 E
Rakaia magna australis Forster, 1952 E
Rakaia m. magna Forster, 1948 E
Rakaia media insula Forster, 1952 E
Rakaia m. media Forster, 1948 E
Rakaia minutissima (Forster, 1948) E
Rakaia pauli Forster, 1952 E
Rakaia solitaria Forster, 1948 E
Rakaia stewartensis Forster, 1948 E
Rakaia sorenseni digitata Forster, 1952 E
Rakaia s. sorenseni Forster, 1952 E
Rakaia uniloca Forster, 1952 E

Infraclass NONOGENUATA
Superorder HAPLOCNEMATA
Order PSEUDOSCORPIONES
Suborder EPIOCHEIRATA
CHTHONIIDAE
Austrochthonius mordax Beier, 1967 E
Austrochthonius rapax Beier, 1976 E
Austrochthonius zealandicus obscurus Beier, 1966 E
Austrochthonius z. zealandicus Beier, 1966 E
Maorichthonius mortenseni Chamberlin, 1925 E
Sathrochthoniella zealandica Beier, 1967 E
Sathrochthonius maoricus Beier, 1976 E
Tyrannochthoniella zealandica foveauxana Beier, 1966 E
Tyrannochthoniella z. zealandica Beier, 1967 E
Tyrannochthonius caecatus (Beier, 1976) E
Tyrannochthonius densedentatus (Beier, 1967) E
Tyrannochthonius grimmeti Chamberlin, 1929 E
Tyrannochthonius horridus (Beier, 1976) E
Tyrannochthonius kermadecensis (Beier, 1976)
Tyrannochthonius luxtoni (Beier, 1967) E
Tyrannochthonius noaensis Moyle, 1989
Tyrannochthonius norfolkensis (Beier, 1976)
Tyrannochthonius tekauriensis Moyle, 1989 E

Suborder IOCHEIRATA
CHEIRIDIIDAE
Apocheiridium validissimum Beier, 1976 E
Apocheiridium validum Beier, 1967 E
Apocheiridium zealandicum Beier, 1976 E
CHELIFERIDAE
Chelifer cancroides (Linnaeus, 1758) A
Philomaoria hispida Beier, 1976 E
Philomaoria pallipes (White, 1849)
Protochelifer exiguus Beier, 1976 E
Protochelifer novaezealandiae Beier, 1948 E
CHERNETIDAE
Apatochernes antarcticus antarcticus Beier, 1964 E
Apatochernes a. knoxi Beier, 1976 E
Apatochernes a. pterodromae Beier, 1964 E
Apatochernes chathamensis Beier, 1976 E
Apatochernes cheliferoides Beier, 1948 E
Apatochernes cruciatus Beier, 1976 E
Apatochernes curtulus Beier, 1948 E
Apatochernes gallinaceus Beier, 1967 E
Apatochernes insolitus Beier, 1976 E
Apatochernes kuscheli Beier, 1976 E
Apatochernes maoricus Beier, 1966 E
Apatochernes nestoris Beier, 1962 E
Apatochernes obrieni Beier, 1966 E
Apatochernes proximus Beier, 1948 E
Apatochernes solitarius Beier, 1976 E
Apatochernes turbotti Beier, 1969 E
Apatochernes vastus Beier, 1976 E
Apatochernes wisei Beier, 1976 E
Heterochernes novaezealandiae (Beier, 1932) E
Lamprochernes savignyi (Simon, 1881) A
Maorichernes vigil (With, 1907) E
Nesidiochernes kuscheli Beier, 1976 E
Nesidiochernes scutulatus Beier, 1969 E
Nesidiochernes zealandicus Beier, 1966 E
Nesiotochernes stewartensis Beier, 1976 E
Nesochernes gracilis Beier, 1932 E
Opsochernes carbophilus Beier, 1966 E
Phaulochernes howdenensis Beier, 1976 E
Phaulochernes jenkinsi Beier, 1976 E
Phaulochernes kuscheli Beier, 1976 E
Phaulochernes maoricus Beier, 1976 E
Phaulochernes townsendi Beier, 1976 E
Reischekia coracoides Beier, 1948 E
Reischekia exigua exigua Beier, 1976 E
Reischekia e. sentiens Beier, 1976 E
Smeringochernes zealandicus Beier, 1976 E
Systellochernes alacki Beier, 1976 E
Systellochernes zonatus Beier, 1964 E
Thalassochernes kermadecensis Beier, 1976 E
Thalassochernes taierensis (With, 1907) E
GARYPIDAE
Synsphyronus lineatus Beier, 1966 E
Synsphyronus melanochelatus (Chamberlin, 1930) E
GARYPINIDAE
Nelsoninus maoricus Beier, 1967 E
OLPIIDAE
Xenolpium pacificum (With, 1907)
SYARINIDAE
Ideobisium peregrinum Chamberlin, 1930 E
WITHIIDAE
Withius piger (Simon, 1878) A

Class PYCNOGONIDA
Order PANTOPODA
Suborder STIRIPASTERIDA
AUSTRODECIDAE
Austrodecus (Austrodecus) breviceps Gordon, 1938
Austrodecus (A.) cestum Child, 1994 E
Austrodecus (A.) enzoi Clark, 1971 E
Austrodecus (A.) frigorifugum Stock, 1954 E
Austrodecus (A.) glaciale Hodgson, 1907
Austrodecus (A.) gordonae Stock, 1954 E
Austrodecus (A.) sinuatum Stock, 1957 E
Austrodecus (Microdecus) confusum Stock, 1957 E
Austrodecus (M.) fryi Child, 1994
Austrodecus (M.) minutum Clark, 1971 E
Pantopipetta australis (Hodgson, 1914)

Suborder EUPANTOPODIDA
ACHELIIDAE
Achelia assimilis (Haswell, 1885)
Achelia australiensis Miers, 1884
Achelia dohrni (Thomson, 1884) E
Achelia transfuga Stock, 1954 E
AMMOTHEIDAE
Ammothea adunca Child, 1994
Ammothea antipodensis Clark, 1972 E
Ammothea longispina Gordon, 1932
Ammothea magniceps Thomson, 1884
Ammothea makara Clark, 1977 E
Ammothea uru Clark, 1977 E
Oorhynchus aucklandiae Hoek, 1881 E
Pycnosomia asterophila Stock, 1981
AMMOTHELLIDAE
Cilunculus cactoides Fry & Hedgpeth, 1969
Cilunculus sewelli Calman, 1938
Cilunculus spinicristus Child, 1987 E
ANOPLODACTYLIDAE
Anoplodactylus laciniosus Child, 1995 E
Anoplodactylus pycnosoma (Helfer, 1938)
Anoplodactylus speculus Child, 1995
Anoplodactylus typhlops Sars, 1888
Anoplodactylus xenus Stock, 1980 E
ASCORHYNCHIDAE
Ascorhynchus antipodus Child, 1987 E
Ascorhynchus cooki Child, 1987
Ascorhynchus insularum Clark, 1971 E
Ascorhynchus orthostomum Child, 1998 E
CALLIPALLENIDAE
Austropallene cristata (Bouvier, 1911)
Austropallene tibicina Calman, 1915
Callipallene emaciata Stock, 1954 E
Callipallene novaezealandiae (Thomson, 1884)
Cheilopallene trappa Clark, 1971 E
Neopallene antipoda Stock, 1954 E
Oropallene dolichodera Child, 1995
Oropallene metacaula Child, 1995 E
Parapallene exigua Stock, 1954 E
COLOSSENDEIDAE
Colossendeis angusta G.O. Sars, 1877
Colossendeis arcuata H. Milne-Edwards, 1885
Colossendeis australis Hodgson, 1907
Colossendeis bruuni Fage, 1956 E
Colossendeis colossea Wilson, 1881
Colossendeis cucurbita Cole, 1909
Colossendeis hoeki Gordon, 1944
Colossendeis japonica Hoek, 1898
Colossendeis longirostris Gordon, 1938
Colossendeis macerrima Wilson, 1881
Colossendeis megalonyx Hoek, 1881
Colossendeis cf. *mycterismos* Bamber, 2004
Colossendeis stramenti Hoek, 1881*
Colossendeis tortipalpis Gordon, 1932
NYMPHONIDAE
Heteronymphon exiguum (Hodgson, 1927)
Nymphon australe Hodgson, 1902
Nymphon a. caecum Gordon, 1944
Nmphon bicuspidum Child, 1995 E
Nymphon compactum Hoek, 1881
Nymphon galatheae Fage, 1956 E
Nymphon immane Stock, 1954
Nymphon inerme Fage, 1956 E
Nymphon longicollum Hoek, 1881
Nymphon longicoxa Hoek, 1881
Nymphon maoriana Clark, 1958 E
Nymphon punctum Child, 1995 E
Nymphon trispinum Child, 1998 E
Nymphon typhlops (Hodgson, 1915)
Nymphon uncatum Child, 1998 E
Pentanymphon antarcticum Hodgson, 1904
PALLENOPSIDAE
Bathypallenopsis antipoda Clark, 1971 E
Bathypellenopsis californica Schimkewitsch, 1893
Pallenopsis kupei Clark, 1971
Pallenopsis latus Child, 1998 E
Pallenopsis mauii Clark, 1958 E
Pallenopsis obliqua (Thomson, 1884)
Pallenopsis pilosa (Hoek, 1881)
Pallenopsis triregia Clark, 1971 E
PYCNOGONIDAE
Pycnogonum anovigerum Clark, 1956 E
Pycnogonum magellanicum Hoek, 1881*
Pycnogonum planum Stock, 1954 E
RHOPALORHYNCHIDAE
Hedgpethia eleommata Child, 1998 E
RHYNCHOTHORACIDAE
Rhynchothorax articulatus Stock, 1968 E
Rhynchothorax australis Hodgson, 1907
Rhynchothorax percivali Clark, 1976 E
TANYSTYLIDAE
Tanystylum antipodum Clark, 1977 E
Tanystylum excuriatum Stock, 1954 E
Tanystylum neorhetum Marcus, 1940

SEVEN

Phylum

ARTHROPODA

MYRIAPODA

centipedes, millipedes, pauropods, and symphylans

PETER M. JOHNS

Zelanion sp., a geophilid centipede, Rimu Valley, Riwaka, South Island.

Alastair Robertson and Maria Minor, Massey University

The myriapods ('many feet') comprise four closely related groups – Chilopoda (centipedes), Diplopoda (millipedes), and the lesser-known Pauropoda and Symphyla, together comprising nearly 11,000 species worldwide. All have jointed bodies, with a head and an abdomen or trunk made up of very similar units or segments. Most of the abdominal units have legs, totalling 12 to 177 pairs, more or less. Although a few centipedes have around 100 legs (the range is 15–171 pairs), no millipede has 1000 legs, so these creatures are really misnamed. A surprising feature, however, is that there is nearly always an odd number of leg pairs.

All myriapods are litter or soil dwellers and are among the pushers and shovers of the animal kingdom, burrowing and pushing aside soil and litter particles. Few are fast-running surface dwellers. The multiplicity of legs enables them to exert a small force with each, yet the sum of those forces pushes the body through the soil and litter. The smoothness of the body surface and cylindrical shape of the burrowing myriapods particularly helps them in this process. Others are flat, especially the centipedes, and, depending on their size, can push their way between particles of soil, litter, and leaves and twist around much sharper corners than millipedes.

Because most myriapods have a thin, light cuticle and live in environments where fossilisation is unlikely, the paleontological record of myriapods is quite sparse. What there is tells us, however, that the group is very ancient. The oldest myriapod-like fossils have been found in marine sediments from the Cambrian. Some fossil burrows from the Ordovician have been claimed as myriapod in origin; this speculation is hard to test, but if it is correct, then myriapods might have been living on land as early as 400 million years ago. The oldest definite body fossils of myriapods come from the Late Silurian and the oldest definite centipedes come from the Devonian of New York State (Shear & Bonamo 1990).

Although myriapods are more closely related to each other (Gai et al. 2006) than to chelicerates or hexapods, the major groups differ significantly enough to be considered as separate classes. Like insects, myriapods have a tracheal respiratory system, a dorsal heart, and malpighian tubules for excretion. Myriapods lack a waxy cuticle on the outside of their bodies, however, and so need a relatively humid environment. The head bears a pair of antennae and sometimes simple eyes (ocelli). True compound eyes are found only in some centipedes.

There are very few predators of myriapods. The flatworm *Coenoplana coerulea* is the single known predator of millipedes in New Zealand. The ant *Amblypone australis*, which forms very small colonies (12–20 workers), requires centipedes or ground-beetle larvae as food for its young and shifts its brood to the prey each time one is caught. Fungi, however, are known to attack myriapods during their long moulting period. Parasites are also known (Clark 1978). The paucity of millipede predators may be related to the presence of very strong-smelling phenolic defence compounds at the openings of the repugnatory glands (ozopores) of most trunk segments. These stain the hands and irritate the skin of collectors and no doubt have stronger effects on predators. Nevertheless, the New Zealand robin has been seen using a species of *Eumastigonus* in anting behaviour (Sherley pers. comm.) – it wiped its feathers with it and presumably the phenols affected its feather lice and mites.

Centipedes are known to nurse their eggs and young larvae, gently cleaning them with their mouthparts. It is quite common to find *Tuoba xylophaga* nursing its young under logs lying above the high tide on sandy beaches (Jones 1998; P. M. Johns pers. obs.). Dalodesmid millipedes carefully construct small, dome-shaped chambers into which they lay 30–50 eggs. Presumably the hardened soil-and-saliva mixture prevents fungal attack as well as preventing predators getting the eggs.

The total known New Zealand myriapod fauna comprises around 295 species but many more are expected to be found. New Zealand has relatively few centipedes (41 species including adventives) but an extremely rich millipede fauna – around 600 species are likely, although only 214 have been found so far. Symphylans and pauropods together comprise 40 species.

Eumastigonus sp., a spirostreptid millipede, Rimu Valley, South Island.
Alastair Robertson and Maria Minor, Massey University

Class Chilopoda

Centipedes are probably the best-known myriapods. Worldwide, there are about 3000 species. The New Zealand fauna of 41 species is small but reasonably diverse, being distributed among all five centipede orders and including one species of the most archaic living genus, *Craterostigmus*.

Four main groups of centipedes have long been recognised and are easily distinguished by their body form. The orders Scolopendrida and Lithobiida contain flattened heavy-bodied forms. In the former, the tergal plates along the dorsal side are similar-sized and there are 21–23 pairs of legs; in the latter, large and small tergal plates alternate and there are only 15 pairs of legs in the adult. Members of the Scutigerida also have 15 pairs of legs, but these are very long, as are the antennae. They are the only myriapods with compound eyes, which, intriguingly, appear homologous with those of crustaceans and insects (Müller et al. 2003). In the eyeless Geophilida the body is relatively long and slender and there are 31–171 pairs of legs. *Craterostigmus*, in its own order, was formerly classified among the Lithobiida. Brooding of eggs is also typical in the Scolopendrida and Geophilida.

Cryptops sp., a scolopendrid centipede, Craigieburn Forest Park.
Alastair Robertson and Maria Minor, Massey University

Summary of New Zealand myriapod diversity

All are terrestrial and there are no known fossils.

Taxon	Described species + subspecies	Known undescr./ undeterm. species	Estimated unknown species	Adventive species	Endemic species	Endemic genera named + unnamed
Diplopoda	87+2	127	200	13	203	12+>9
Chilopoda	39	2	0	8	33?	1
Pauropoda	19	1	0	3?	17	0
Symphyla	19	1	0	4	16	0
Totals	164+2	131	200	~28	~266	113+>9

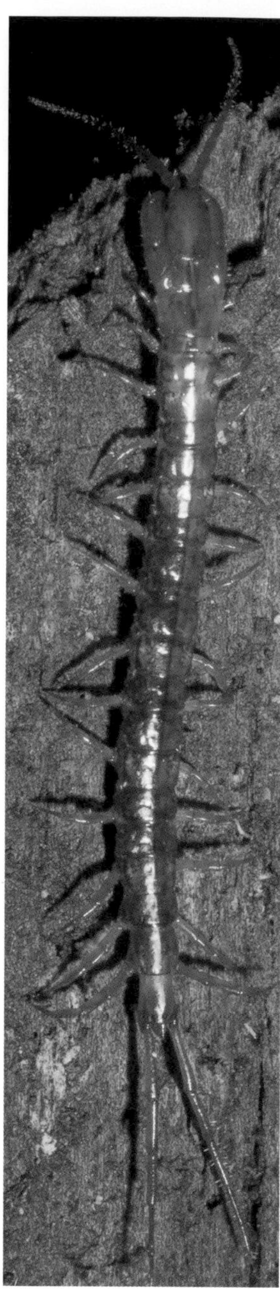

Craterostigmus crabilli, one of only two craterostigmid centipedes in the world, Flora Saddle, Kahurangi National Park.

Alastair Robertson and Maria Minor, Massey University

Centipedes, like spiders, tend to have a negative image, but only one New Zealand species needs handling with care. In fact, because of their predatory habits, centipedes must be regarded as beneficial, especially when they eat insect pests and their eggs; the very long and flexible species push their way through soil searching for these prey as well as for small earthworms. Shorter species are faster and can capture moving prey. Centipedes are easily recognised by their protruding pincers or jaws, which are a modified pair of legs. Prey is caught with the pincers, often injected with a poison, and then sucked dry or slowly masticated by the very small jaws that lie under the head. Small centipedes catch other arthropods or find the eggs and small larvae of slugs, snails, and insects. The very large, mainly tropical, centipedes (ca. 25 centimetres long) are able to catch the young of birds, rats, and mice. One such species is known to produce poison droplets from each of its legs, and the poison is said to be such that human skin dies and the person is left with suppurating sores at each leg-fall spot; these gradually increase in size and coalesce, and healing takes months. Although the two relatively large New Zealand centipedes are close relatives of those tropical species (both also occur in Australia), no one is recorded as having suffered ill effects after a bite. Nevertheless, the large size of *Cormocephalus rubriceps* (up to 20 centimetres in Northland and nearby offshore islands) certainly frightens some people. It is only six centimetres shorter than the largest species in the world, from Tropical America.

One of the smallest New Zealand centipedes, *Lamyctes emarginatus*, has affected centipede terminology in its travels around the world. It was first found and described from New Zealand (Newport 1844). Soon after, it was discovered in England, described again, and given another name. Under that name (and several other later names) it was recorded from various sites eastwards in Europe and then in New York and westwards through North America. It has no relatives at all in the Northern Hemisphere. Such was the influence of the European workers that several New Zealanders gave the European name to the New Zealand population while others used the New Zealand name without knowing that the species was also overseas. Finally, in a review of the fauna of Kamchatka, eastern Russia, more than 160 years after it was first discovered, Eason (1996) synonymised all other names and the species is now rightly recognised as a New Zealand species that was introduced to the rest of the world. It is a soil dweller and tolerant of what humans do to the environment, but the important feature that enabled it to become a global traveller is its ability to become parthenogenetic – all but a few populations overseas have no males, as well as some in the modified habitats of New Zealand.

Many Tasmanian centipedes have been identified as New Zealand species (Mesibov 1986) but a thorough review of both faunas may well separate some of them. Certainly *Henicops maculatus* and *Lamyctes emarginatus* are shared. *Craterostigmus* in New Zealand was recently segregated from the Tasmanian form as *C. crabilli* (Edgecombe & Giribet 2008). These two archaic centipedes are the only representatives of the order Craterostigmomorpha and are thus known only from Tasmania and New Zealand. The New Zealand species has additional significance in that its genetic pattern supports a Gondwanan origin, not long-distance dispersal, and hence does not give support to the hypothesis of complete submergence of the New Zealand landmass in the late Oligocene (Landis et al. 2008). These forms are the key to understanding the evolution of modern centipedes (see Shear & Bonamo 1990). The family/subfamily placement of New Zealand geophilids needs reviewing, and *Ballophilus hounselli* needs re-evaluating with respect to New Caledonian forms.

Class Diplopoda

Millipedes are the most diverse and speciose myriapods, with over 7500 species worldwide. A distinguishing feature of the class is a doubling of trunk segments into 'diplosegments', each of which bears two pairs of legs, from which the class name is derived. The diplosegmented condition is also evident internally, for there are two pairs of nerve ganglia and two pairs of heart openings within each segment. The largest species are tropical forms that attain 28 centimetres in length but some North Island species of *Eumastigonus* reach 10 centimetres. On the other hand, the smallest millipedes anywhere, including New Zealand, are only a few millimetres long.

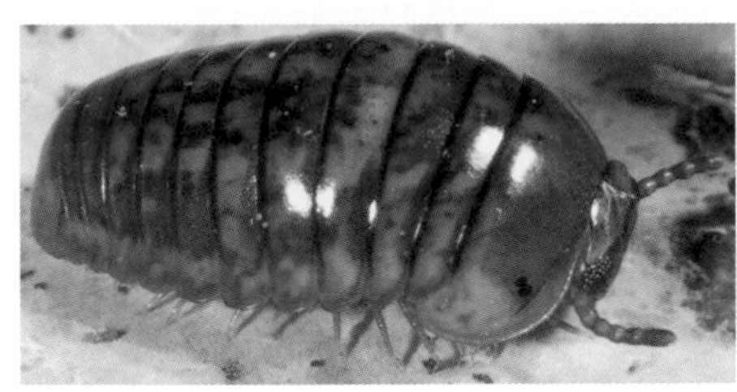

Procyliosoma striolatum, a sphaerotheriid millipede, Kaituna Track, Nelson.
Alastair Robertson and Maria Minor, Massey University

At the family level, New Zealand's millipede fauna is typical of the Southern Hemisphere, more so than for the Australian fauna. Compared with most other parts of the world, New Zealand's millipedes probably are more important in the breakdown of litter. Giant pill millipedes (Sphaerotheriidae) live in forests and their frass often forms a deep layer under logs. The long cylindrical millipedes of the families Spirobolellidae and Iulomorphidae burrow into rotting wood. The dominant and most diverse family is the Dalodesmidae. Millipedes often chew whole pieces of rotting leaf or wood, leaving piles of finely divided frass behind. The resultant increase in surface area of the organic particles allows bacterial action to be efficacious in the recycling of the soil nutrients. The millipedes, with their many legs, smooth body surfaces, and rounded heads, are able to barge their way through the soil creating passageways for other small organisms and allowing oxygen into the deeper layers. In this respect they are almost as important as earthworms, perhaps more so in New Zealand, where the majority of earthworms are litter dwellers rather than deep burrowers.

There is great disparity in the numbers of species in each of the four classes of Myriapoda, globally and in New Zealand, with millipedes by far the most numerous. In New Zealand they show gradations in body form from one site to another. Johns (1979) demonstrated such clinal variation in Canterbury, and more examples have been recognised since. Although reported only for the Dalodesmidae, they are probably present in other families. No genetic studies of this variation have yet been made. Clinal variation, which has led to the many species in some genera that are now present, may be related to the changes that have occurred in the New Zealand landmass and climate over the past 80 million years. The great number of species and clines in *Icosidesmus* (> 40 species) is a very good example of this rampant process. *Icosidesmus* is also represented in Tasmania by at least one species (pers. obs.) but other Dalodesmidae do not seem closely related to those in mainland Australia or New Caledonia. On the other hand, *Spirobolellus antipodarus* (see synonyms below) is very similar to a New Caledonian species and the genus is also represented in coastal Queensland, New South Wales, and Lord Howe Island. The estimate of undescribed species in the tabular summary is a minimum, less than that given by Johns (1992), which was probably closer to reality.

Chamberlin (1920) described several New Zealand millipedes. His types, and those of some others before him, are distributed in many museums. The material on which Hutton's (1877, 1878) 11 species were based was stated to be in the Canterbury Museum (Hutton 1904) but little resides there. His material identified as *Polydesmus gervaisii* and *Iulus antipodarus* is present, the former being a mixture of *Icosidesmus* species and the latter a new species of *Eumastigonus*. Some of his material was sent to the Natural History Museum, London, but it is still uncertain whether that institution contains the types. His species are presented below as new combinations that are based only on the best interpretations of features noted in the descriptions. Mauries (1978, 1983, 1987) has redescribed several of the older species and added a new species, while also changing their family placement. Mauries (1983) considered *Eumastigonus insulanus* (Attems), *E. fasciatus* Chamberlin, and *Dimerogonus kaorinus* Chamberlin as varieties of the same species. Korsós and Johns (2009) have re-examined all of Chamberlin's types and have added two more species; more descriptions are in preparation.

A symphylan, Kawhatau Base, Manawatu.
Alastair Robertson and Maria Minor, Massey University

Class Symphyla

The Symphyla is a small class of about 160 known species that at one time evoked interest among zoologists as being myriapods that display some characteristics reminiscent of insects, especially in the mouth parts. The modern consensus is that any resemblance is only superficial. The symphylan body has 12 leg-bearing segments, which are covered by 15–22 tergal plates along the dorsal side.

Symphylans are small – adults are usually between five and ten millimetres long. Their body is very flexible and they are able to force their way around hard particles, bending the body beyond a right angle, thanks to the extra tergal plates. They can also squeeze through cracks smaller than their body depth. Their habitat is litter through to deep soil, and they are one of the first groups of animals to enter a coffin in the ground (worms come later), remaining there for years feeding on the rich bacterial and fungal meal.

At 20 species (16 native), the New Zealand symphylan fauna is relatively diverse, comprising around 13% of the world fauna. Very little is known about the biology of the native species.

Class Pauropoda

Pauropods are even smaller than symphylans, about the same size as the many interstitial spaces in which they live between soil particles, i.e. about one half to two millimetres. They live on bacterial and fungal films within the soil, humus, and the bodies of dead animals. On the whole, they are seldom seen and are very difficult to identify. There are about 380 species worldwide. All but three of the 20 New Zealand species are thought to be endemic, but there are still doubts as to whether the older records have been correctly identified. For example, a New Zealand record of European *Pauropus furcifer* was based on only a single juvenile (Scheller 1976). Hilton's (1943) species are yet to be recognised and may be synonyms of others. Also, some species, notably *Allopauropus maoriorum* and *Scleropauropus dugdalei*, are widely distributed around the world (Scheller 1968, 1976, 1977). Although they were first described from New Zealand, they may already have been introduced.

The grub-like pauropods more closely resemble millipedes than any other myriapod group – nine of the 11 segments bear pairs of legs. The antennae, on the other hand, are forked. As in millipedes, young hatch with only three pairs of legs.

Pauropods, Whanganui National Park.
Alastair Robertson and Maria Minor, Massey University

Special habitats – caves and alpine screes

A number of species of millipedes and centipedes live in caves and there has been considerable speciation associated with the subterranean habitat. Johns (1991) listed species present in South Island caves and they are certainly known also from northern caves. Whereas *Icosidesmus* millipedes are almost ubiquitous in New Zealand forests and shrublands, they are uncommon in caves. *Tongodesmus*, however, has few species in forests but several in lowland caves, as has *Schedotrigona*. Centipedes are represented by species of *Cryptops* and *Haasiella*. Cave centipedes are often pale, having lost much of their integumental pigment. Eyes too, are reduced or absent. The antennae and legs are elongate and some setae are extremely long. These setae are highly sensitive sense organs, sensing any air vibration from possible live food, mates, or predators. Surprisingly, the most modified myriapods are in lowland caves close to the sea. The one exception is a new genus and species of millipede from high-altitude caves on The Twins–Mt Arthur massif in Kahurangi National Park, Northwest Nelson District.

Adventive species

About 28 myriapod species are naturalised in New Zealand. Among the eight alien centipede species is the only house centipede, *Scutigera coleoptrata*. It is not huge but has very long legs and can give the unwary person quite a start when encountered. Among the 13 alien millipede species the Australian paradoxosomatids *Aulacoporus pruvoti* (new record) and *Akamptogonus novarae* must have arrived in New Zealand relatively early in the period of European settlement. This may have been during the spar-cutting period in the kauri forests (*Agathis australis*) of Northland as *A. pruvoti* is still confined to the Kerikeri area and *A. novarae* has spread from Auckland southwards over a century and only reached Christchurch in the 1960s. A southeast Asian paradoxosomatid, *Oxidus gracilis*, has also been present in Northland for at least 80 years. Three European blaniulids, *Blaniulus guttulatus*, *Choneiulus palmatus* (new record), and *Nopiulus kochii* are easily recognised – the very long body is relatively pale and their red or brownish repugnatory glands are easily seen as large spots on each side of the segments. The first is widespread in gardens and orchards, and the last two are more restricted to gardens in the North Island. Two European polydesmids, *Polydesmus kochii* and *Brachydesmus superus*, are also widespread in both main islands of New Zealand. *Brachydesmus superus* has hardly moved away from much-modified environments, but *P. kochii* has at least entered modified native shrublands in Nelson.

Akamptogonus novarae, an introduced polydesmid millipede, Ohinetonga Scenic Reserve, Tongariro National Park.
Alastair Robertson and Maria Minor, Massey University

Scope for future research

Large numbers of undescribed species are already known among the New Zealand Myriapoda, especially among the millipedes, and further collecting is expected to increase this group to around 600 species. Other new taxa are known and/or expected for the other groups. But this is only the beginning. Little is known about the biology of the native species and one can speak of their roles in forest and other ecosystems in only very general terms. There is no-one in New Zealand currently specialising in the systematics and biology of any myriapodous group.

Author

Peter M. Johns Canterbury Museum, Rolleston Avenue, Christchurch, New Zealand [pjohns@canterburymuseum.com]

References

ATTEMS, C. 1953: Myriopoden von Indochina. *Mémoires du Muséum National d'Histoire Naturelle, Zoologie, n. sér. A, 5*: 133–230.

CHAMBERLIN, R.V. 1920: The Myriapoda of the Australasian Region. *Bulletin of the Museum of Comparative Zoology, Harvard 64*: 1–269.

CLARK, W. C. 1978: New species of rhigonematid and thelostomatid nematodes from the pill millipede *Procyliosoma tuberculata* (Diplopoda: Oniscomorpha). *New Zealand Journal of Zoology 5*: 1–6.

EASON, E. H. 1996: Lithobiomorpha from Sakhalin Island, Kamchatka Peninsula and the Kurile Islands (Chilopoda). *Arthropoda Selecta 5*: 117–123.

EDGECOMBE, G. D.; GIRIBET, G. 2008: A New Zealand species of the trans-Tasman centipede order Craterostigmomorpha (Arthropoda: Chilopoda) corroborated by molecular evidence. *Invertebrate Systematics 22*: 1–15.

GAI,Y.-H.; SONG, D.-X.; SUN, H.-Y.; ZHOU, K.-Y. 2006: Myriapod monophyly and relationships among the myriapod classes based on nearly complete 28S and 18S rDNA sequences. *Zoological science 23*: 1101–1108.

HILTON, W. A. 1943: Some Pauropoda from New Zealand. *Journal of Entomology and Zoology 35*: 33–37.

HUTTON, F. W. 1877: Descriptions of new species of New Zealand Myriapoda. *Annals and Magazine of Natural History, ser. 4, 20*: 114–117.

HUTTON, F. W. 1878: Notes on the New Zealand Myriapoda in the Otago Museum. *Transactions of the New Zealand Institute 10*: 288–293.

HUTTON, F. W. [1904]. Types in the Hutton Collection, Canterbury Museum. Unpublished MS in Hutton's handwriting. Canterbury Museum, Christchurch.

JOHNS, P. M. 1964: The Sphaerotrichopidae (Diplopoda) of New Zealand. I. Introduction, revision of some known species and description of new species. *Records of the Canterbury Museum 8*: 1–49.

JOHNS, P. M. 1970: New genera of New Zealand Dalodesmidae (Diplopoda). *Transactions of the Royal Society of New Zealand (Biological Sciences) 12*: 217–237.

JOHNS, P. M. 1979: Speciation in New Zealand Diplopoda. Pp. 49–57 *in*: Camatini, M. (ed.), *Myriapod Biology*. Academic Press, London.

JOHNS, P. M. 1991: Distribution of cave species in northwest Nelson, Westland and Canterbury. *The Weta 14*: 11–21.

JOHNS, P. M. 1992: A review of the New Zealand Diplopoda and Chilopoda. Unnumbered pp. 45–48 *in* Heath, A. (comp.), *Entomological Society of New Zealand, 41st Annual Conference Proceedings*. Entomological Society of New Zealand, Lower Hutt.

JONES, R. E. 1998: On the species of *Tuoba* (Chilopoda: Geophilomorpha) in Australia, New Zealand, New Caledonia, Solomon Islands and New Britain. *Records of the Western Australian Museum 18*: 333–346.

KORSÓS, Z.; JOHNS, P. M. 2009: Introduction to the taxonomy of Iulomorphidae of New Zealand, with descriptions of two new species of *Eumastigonus* Chamberlin, 1920 (Diplopoda: Spirostreptida: Epinannolenidea). *Zootaxa 2065*: 1–24.

LANDIS, C. A.; CAMPBELL, H. J.; BEGG, J. G.; MILDENHALL, D. C.; PATERSON, A. M.; TREWICK, S. A. 2008: The Waipounamu Erosion Surface: questioning the antiquity of the New Zealand land surface and terrestrial fauna and flora. *Geological Magazine 145*: 173–197.

MAURIES, J.-P. 1978: Le genre neozelandais *Schedotrigona* Silvestri, 1903: révision et place dans une nouveau classification des Craspedosomides (Myriapoda, Diplopoda, Craspedosomida). *Bulletin du Muséum Nationale d'Histoire Naturelle, sér. 3, Zoologie 351*: 43–66.

MAURIES, J.-P. 1983: Cambalides nouveaux et peu connus d'Asie, d'Amérique et d'Océanie. I. Cambalidae et Cambalopsidae (Myriapoda: Diplopoda). *Bulletin du Muséum Nationale d'Histoire Naturelle, sér. 4, 5A*: 247–276.

MAURIES, J.-P. 1987: Cambalides nouveaux et peu connus d'Asie, d'Amérique et d'Océanie II. Pseudonannolenidae, Choctellidae (Myriapoda, Diplopoda). *Bulletin du Muséum Nationale d'Histoire Naturelle, sér. 4, 9A*: 169–199.

MESIBOV, R. 1986: *A Guide to the Tasmanian Centipedes*. Hobart, R. Mesibov.

MÜLLER, C. H. G.; ROSENBERG, J.; RICHTER, S.; MEYER-ROCHOW, V. B. 2003: The compound eye of *Scutigera coleoptrata* (Linnaeus, 1758) (Chilopoda: Notostigmophora): an ultrastructural reinvestigation that adds support to the Mandibulata concept. *Zoomorphology 122* : 191–209.

SCHELLER, U. 1968: Chilean and Argentinian Pauropoda. *Biologie de 'lAmérique Australe 4*: 275–306.

SCHELLER, U. 1976: The Pauropoda and Symphyla of the Geneva Museum II. A review of the Swiss Pauropoda (Myriapoda). *Revue Suisse de Zoologie 83*: 3–37.

SCHELLER, U. 1977: The Pauropoda and Symphyla of the Geneva Museum IV. A basic list of the Pauropoda of Greece (Myriapoda). *Revue Suisse de Zoologie 84*: 361–408.

SHEAR, W. A., BONAMO, P. M. 1990: Fossil centipeds from the Devonian of New York State, U.S.A. Pp. 89–96 *in*: Minelli, A. (ed.), *Proceedings of the 7th International Congress of Myriapodology*. E. J. Brill, Leiden.

Checklist of New Zealand Myriapoda

In the following list, new combinations and one new name are indicated by the symbol ¶. Endemic genera are underlined. Single-letter codes **E** and **A** are used to indicate endemic and adventive species, respectively; * = new record.

SUBPHYLUM MYRIAPODA
Class CHILOPODA
Subclass ANAMORPHA
Order SCUTIGERIDA
SCUTIGERIDAE
Scutigera coleoptrata (Lamarck, 1801) A
Scutigera smithi (Newport, 1844) E?

Order CRATEROSTIGMIDA
CRATEROSTIGMIDAE
Craterostigmus crabilli Edgecombe & Giribet, 2008 E

Order LITHOBIIDA
HENICOPIDAE
Anopsobius neozelandicus Silvestri, 1909 E
Haasiella halli (Archey, 1917) E
Haasiella insularis (Haase, 1887) E
Haasiella trailli (Archey, 1917) E
Haasiella n. sp. E
Henicops maculatus Newport, 1844 E
Lamyctes emarginatus (Newport, 1844) E
Paralamyctes harrisi Archey, 1922 E
Paralamyctes validus Archey, 1917 E

LITHOBIIDAE
Lithobius ?transmarinus Latzel, 1880 A*
Lithobius forficatus (Linnaeus, 1758) A
Lithobius melanops Newport, 1845 A
Lithobius microps Meinert, 1868 A
Lithobius obscurus Meinert, 1872 A

Subclass EPIMORPHA
Order SCOLOPENDRIDA
CRYPTOPIDAE
Cryptops arapuni Archey, 1922 E
Cryptops australis Newport, 1845 E
Cryptops dilagus Archey, 1921 E
Cryptops hortensis Leach, 1815 A*
Cryptops lamprethus Chamberlin, 1920 E
Cryptops megaloporus Haase, 1887 E
Cryptops polyodontus Attems, 1903 E
Cryptops spinipes Pocock, 1891 E
SCOLOPENDRIDAE
Cormocephalus rubriceps Newport, 1844 E
Cormocephalus westwoodi huttoni Pocock, 1893 E

Order GEOPHILIDA
GEOPHILIDAE [Incl. Chilenophilidae]
Australiophilus ferrugineus (Hutton, 1877) E
Maoriella aucklandica Attems, 1903 E
Maoriella ecdema Crabill, 1964 E
Maoriella macrostigma Attems, 1903 E
Maoriella zelanica (Chamberlin, 1920) E
Pachymerium ferrugineum (C.L. Koch, 1835) A
Schizoribautia brittoni Archey, 1922 E
Tasmanophilus spenceri (Pocock, 1901) E
Tuoba xylophaga (Attems, 1903) E
Zelanion antipodus (Pocock, 1891) E
Zelanion dux Chamberlin, 1920 E
Zelanion morbosus (Hutton, 1877) E
Zelanophilus provocator (Pocock, 1891) E
SCHENDYLIDAE
Ballophilus hounselli Archey, 1936 E

Class DIPLOPODA
Subclass PENICILLATA
Order POLYXENIDA
POLYXENIDAE
Propolyxenus forsteri Conde, 1951 E
Propolyxenus n. sp. E

Subclass PENTAZONIA
Order SPHAEROTHERIIDA
SPHAEROTHERIIDAE
Procyliosoma delacyi delacyi (White, 1859) E
Procyliosoma d. striolatum (Pocock, 1895) E
Procyliosoma leiosomum (Hutton, 1877) E
Procyliosoma tuberculatum tuberculatum Silvestri, 1917 E
Procyliosoma t. westlandicum Holloway, 1956 E

Subclass HELMINTHOMORPHA
Superorder OMMATOPHORA
Order POLYZONIIDA
POLYZONIIDAE
Siphonethus bellus Chamberlin, 1920 E
Siphonethus enotatus Chamberlin, 1920 E
Siphonethus n. spp. (5+) 6E

Superorder ANOCHETA
Order SPIROBOLIDA
SPIROBOLELLIDAE
Spirobolellus antipodarus (Newport, 1844) E
Spirobolellus n. sp. E

Superorder DIPLOCHETA
Order SPIROSTREPTIDA
IULOMORPHIDAE
Eumastigonus ater (Chamberlin, 1920) E
Eumastigonus distinctior Chamberlin, 1920 E
Eumastigonus hallelujah Korsós & Johns, 2009 E
Eumastigonus hemmingseni Mauries, 1983 E
Eumastigonus insulanis (Attems, 1903) E
Eumastigonus kaorinus Chamberlin, 1920 E
Eumastigonus maior Chamberlin, 1920 E
Eumastigonus otekauri Korsós & Johns, 2009 E
Eumastigonus parvus Chamberlin, 1920 E
Eumastigonus striatus (Hutton, 1877) E¶
Eumatigonus waitahae Korsós & Johns, 2009 E
Eumastigonus n. spp. (20+) 21E

Order JULIDA
BLANIULIDAE
Blaniulus guttulatus (Fabricius, 1798) A
Choneiulus palmatus (Nemec, 1895) A*
Nopoiulus kochii (Gervais, 1847) A
JULIDAE
Brachiulus pusillus (Leach, 1815) A
Cylindroiulus britannicus (Verhoeff, 1891) A
Cylindroiulus londinensis Leach, 1814 A
Ophyiulus pilosus (Newport, 1842) A
Ophyiulus verruculiger Verhoeff, 1910 A

Superorder TYPHLOGENA
Order SIPHONOPHORIDA
SIPHONOPHORIDAE
Siphonophora zelandica Chamberlin, 1920 E
Siphonophora n. spp. (5+) 6E

Superorder COELOCHETA
Order CHORDEUMATIDA
SCHEDOTRIGONIDAE
Schedotrigona crucifer Mauries, 1978 E
Schedotrigona johnsi Mauries, 1978 E
Schedotrigona smithi Silvestri, 1903 E
Schedotrigona tremblayi Mauries, 1978 E
Schedotrigona trisetosum (Hutton, 1877) E
Schedotrigona n. spp. (10+) 11E

Superorder MEROCHETA
Order POLYDESMIDA
DALODESMIDAE
Blysmopeltis figurata Johns, 1970 E
Blysmopeltis systropha Johns, 1970 E
Blysmopeltis n. sp. E
Dityloura brevipes Johns, 1970 E
Dityloura cothonognatha Johns, 1970 E
Dityloura dasygnatha Johns, 1970 E
Dityloura dealbata Johns, 1970 E
Dityloura ditylognatha Johns, 1970 E
Dityloura edaphica Johns, 1970 E
Dityloura ignava Johns, 1970 E
Dityloura lissognatha Johns, 1970 E
Dityloura notohyla Johns, 1970 E
Dityloura spicativentris Johns, 1970 E
Dityloura trachypyga Johns, 1970 E
Dityloura unicostata Johns, 1970 E
Dityloura macrocephala (Hutton, 1877) E¶
Erythrodemus echinopogon Johns, 1970 E
Icosidesmus barathrodes Johns, 1964 E
Icosidesmus cismontanus Johns, 1964 E
Icosidesmus collinus Johns, 1964 E
Icosidesmus eratochlorus Johns, 1964 E
Icosidesmus falcatus Johns, 1964 E
Icosidesmus hochstetteri Humbert & Saussure, 1869 E
Icosidesmus holcus Johns, 1964 E
Icosidesmus latidens Johns, 1964 E
Icosidesmus longisetosus Johns, 1964 E
Icosidesmus montanus Johns, 1964 E
Icosidesmus nanus Carl, 1902 E
Icosidesmus olivaceus Carl, 1902 E
Icosidesmus orescius Johns, 1964 E
Icosidesmus saxatilis Johns, 1964 E
Icosidesmus schenkeli Carl, 1902 E
Icosidesmus suteri Carl, 1902 E
Icosidesmus tumidus Johns, 1964 E
Icosidesmus variegatus Carl, 1902 E
Icosidesmus wheeleri Chamberlin, 1920 E
Icosidesmus worthingtoni (Hutton, 1877) E¶
Icosidesmus n. spp. (20+) 21E
Notonaia aucklandica Johns, 1970 E
Notonaia campbellensis Johns, 1970 E
Notonesiotes aucklandensis Johns, 1970 E
Pacificosoma yaldwyni Schubart, 1963 E
Pacificosoma n. sp. E
Pseudoprionopeltis caesius (Karsch, 1881) E
Pseudoprionopeltis cinereus Carl, 1902 E
Pseudoprionopeltis elaphrus Johns, 1964 E
Pseudoprionopeltis grassator Johns, 1964 E
Pseudoprionopeltis haastii (Humbert & Saussure, 1869) E
Pseudoprionopeltis macrocephalus (Hutton, 1877) E
Pseudoprionopeltis ravidus Johns, 1964 E
Pseudoprionopeltis serratus (Hutton, 1877) E
Pseudoprionopeltis n. spp. (8) 8E
Serangodes strongylosomoides Attems, 1898 E
Serangodes n. spp. 2E
Tongodesmus stilifer Schubart, 1963 E
Tongodesmus n. spp. (9) 9E
Gen. nov. A et n. spp. (9) 9E
Gen. nov. B et spp. (2) 2E
Gen. nov. C et n. spp. (2) 2E
Gen. nov. D et n. spp. (2) 2E
Gen. nov. E et n. sp. E
Gen. nov. F et n. sp. E
Gen. nov. G et n. sp. E
Gen. nov. H et n. sp. E
HAPLODESMIDAE
Genera nov. et n. spp. (20+) 21E
PARADOXOSOMATIDAE
Akamptogonus novarae (Humbert & Saussure, 1869) A
Aulacoporus pruvoti (Brolemann, 1931) A*
Oxidus gracilis (C.L. Koch, 1847) A
POLYDESMIDAE
Brachydesmus superus Latzel, 1884 A
Polydesmus inconstans Latzel, 1884 A

Class PAUROPODA
Order PAUROPODA
BRACHYPAUROPODIDAE
Brachypauropoides pistillifer Rémy, 1954 E
Brachypauropoides praestans Rémy 1956 E
EURYPAUROPODIDAE
Eurypauropus maurius Hilton, 1943 E
Hansenauropus gratus Rémy, 1954 E
PAUROPIDAE
Allopauropus maoriorum Rémy, 1956 E?
Allopauropus muscicolus Rémy, 1956 E
Pauropus furcifer Silvestri, 1902 A
Pauropus huxleyi Lubbock A
Pauropus confinus Rémy, 1956 E
Pauropus cf. *confinus* Rémy, 1956 E
Pauropus dolosus Rémy, 1956 E
Pauropus hirtus Rémy, 1952 E
Pauropus forsteri Rémy, 1952 E
Pauropus furcillatus Rémy, 1952 E
Pauropus zelandus Hilton, 1943 E
Stylopauropus tiegsi Rémy, 1952 A?
Stylopauropus duplex Rémy, 1956 E
Stylopauropus infidus Rémy, 1956 E
Stylopauropus zelandus Hilton, 1943 E
SCLEROPAUROPIDAE
Scleropauropus dugdalei Rémy, 1956 E?

Class SYMPHYLA
Order SYMPHYLA
SCOLOPENDRELLIDAE
Symphylella essigi Michelbacher, 1939 A
Symphylella maorica Adam & Burtel, 1956 E
Symphylella vulgaris (Hansen, 1903) A
SCUTIGERELLIDAE
Hanseniella brachycerca Adam & Burtel, 1956 E
Hanseniella caldaria (Hansen, 1903) A
Hanseniella campbelli Juperthie-Jupeau, 1964 E
Hanseniella confusa Adam & Burtel, 1956 E
Hanseniella crassisetosa Adam & Burtel, 1956 E
Hanseniella dugdalei Adam & Burtel, 1956 E
Hanseniella echinata Adam & Burtel, 1956 E
Hanseniella forsteri Adam & Burtel, 1956 E
Hanseniella glabra Adam & Burtel, 1956 E
Hanseniella indecisa (Attems, 1911) A
Hanseniella mutila Adam & Burtel, 1956 E
Hanseniella neozelandica Chamberlin, 1920 E
Hanseniella proxima Adam & Burtel, 1956 E
Hanseniella cf. *proxima* Adam & Burtel, 1956 E
Hanseniella setigera Adam & Burtel, 1956 E
Hanseniella southgatei Adam & Burtel, 1956 E
Hanseniella vulgata Adam & Burtel, 1956 E

New combinations and synonyms

Two new combinations are presented here (indicated by ¶ in checklist). *Spirobolellus antipodarus* (Newport, 1844) has two new synonyms: *Spirobolellus dryomophilus* Chamberlin, 1920 and *Desmocricellus reischeki* Attems, 1953.

Icosidesmus worthingtoni (Hutton, 1877) n. comb.
Dityloura macrocephala (Hutton, 1877) n. comb.

EIGHT

Phylum

ARTHROPODA

SUBPHYLUM CRUSTACEA

shrimps, crabs, lobsters, barnacles, slaters, and kin

W. RICHARD WEBBER, GRAHAM D. FENWICK, JANET M. BRADFORD-GRIEVE, STEPHEN H. EAGAR, JOHN S. BUCKERIDGE, GARY C. B. POORE, ELLIOT W. DAWSON, LES WATLING, J. BRIAN JONES, JOHN B. J. WELLS, NIEL L. BRUCE, SHANE T. AHYONG, KIM LARSEN, M. ANNE CHAPMAN, JØRGEN OLESEN, JU-SHEY HO, JOHN D. GREEN, RUSSELL J. SHIEL, CARLOS E. F. ROCHA, ANNE-NINA LÖRZ, GRAHAM J. BIRD, W. A. CHARLESTON

Scyphax ornatus, an endemic coastal slater.
Shane Ahyong

'No group of plants or animals on the planet exhibits the range of morphological diversity seen among the extant Crustacea.' This provocative quote from Martin and Davis (2001) highlights at least one attribute of the group. Nevertheless, the body plan of the Crustacea has a number of unifying characteristics, including a five-segmented head with two pairs of antennae and an elongate body that may be divided into two more-or-less distinct sections – generally the thorax or 'body' and the pleon or 'abdomen'. Each of these sections bears multisegmented appendages (mostly limbs) that are primitively biramous (forked) but some are uniramous in many groups. Brusca and Brusca (2002) gave a succinct summary of the characteristics of the subphylum. In addition to enormous diversity of form, crustaceans exhibit a great range of sizes (exceeded only by molluscs, which can claim the largest individual invertebrate in the form of the colossal squid), from minute interstitial and parasitic forms (e.g. Tantulocarida) measuring as little as a tenth of a millimetre to giant crabs, lobsters, and isopods with a body size of up to half a metre in length or breadth and weighing up to 20 kilograms. By virtue of their edibility, many crustaceans are prized items on restaurant menus around the world.

They are an ancient group, dating from at least the Early Cambrian (Chen et al. 2001), and have diversified abundantly since then. Calculations of the number of named living species of Crustacea range from approximately 50,000 to 67,000. Estimates of the potential number of species range from 10 to 100 times that number. The smaller species, such as those of the Peracarida and Copepoda may eventually be found in numbers comparable to those of the insects on land. By way of an example, the Isopoda currently number approximately 11,000 species, but estimates suggest that as many as 50,000 species of Isopoda could exist on coral-reef habitats alone (Kensley 1988), a figure close to the current total for all Crustacea, while Wilson (2003) estimated a total of 400,000 deep-sea species! Clearly, with thorough documentation, crustacean diversity will be found to be huge.

Five (Brusca & Brusca 2002) or six (Martin & Davis 2001) classes of Crustacea are recognised. Whichever classification is used, only the cave-dwelling

Remipedia have not yet been found in New Zealand waters. As one moves down the taxonomic hierarchy from class to species, the level of endemism increases. The New Zealand fauna currently stands at 2974 known species, of which at least 485 have not yet been named or described. This number is very conservative, and more than a thousand additional species will surely be discovered. Most major groups of Crustacea (orders) are to be found in New Zealand waters, though many families and genera will be found to be absent, particularly among those groups with strong warm-water representation, such as the commercially and gastronomically desirable 'prawns'. Prawns of the family Penaeidae (notably *Penaeus* and *Metapenaeus*) and portunid crabs of the genera *Portunus* and *Scylla* are rare or absent.

Class Branchiopoda: Fairy shrimps, water fleas, and kin

The approximately 1000 species of branchiopods ('gill feet') mostly inhabit fresh water (Dumont & Negrea 2002). They cover a wide range of body form from many-segmented, ancient-looking taxa – generally the larger-bodied forms such as Anostraca (fairy shrimps), Notostraca (tadpole shrimps), and 'Conchostraca' (clam shrimps) – to more-modified short-bodied taxa like the Cladocera (water fleas). The larger Branchiopoda do not collectively form a natural, evolutionary group but have a general similarity (many segments and same structure of trunk limbs) and are almost all adapted to a short life-span in temporary pools.

Tadpole shrimp
Lepidurus apus viridis (Notostraca).
Stephen Moore

There are more than 250 species of Anostraca (fairy shrimps) worldwide (Dumont & Negrea 2002), none of which is naturally represented in New Zealand (Chapman & Lewis 1976) although the brine shrimp *Artemia franciscana* has apparently been introduced into saline Lake Grassmere near Blenheim. They are all relatively slow and graceful forms that swim with the back facing the bottom (opposite to most other Crustacea) while they use their 11 pairs of trunk limbs, beating in metachronal (wave-like) fashion, for both swimming and filtration.

The Notostraca (tadpole shrimps) comprises about 10 species worldwide, one of which (*Lepidurus apus viridis*) is found in New Zealand. One of the most striking features of notostracans is the large, flattened dorsal carapace that originates immediately behind the head and overhangs a part of the body. Behind the carapace is a relatively long (sometimes *very* long), flexible and limbless abdomen that ends in a pair of superficially segmented tail-like processes. At the front end, the carapace has a conspicuous so-called 'dorsal organ' (used for osmoregulation). The first and second antennae – which often have sensory functions in the Crustacea – are much reduced in size in the adult, and the sensory function has been taken over by the very long endites (innermost branches) of the first pair of biramous trunk limbs. All notostracans have basically the same lifestyle. In contrast to most other branchiopods, notostracans are not filter-feeders, but remain near the bottom, where they use the heavily chitinised parts of the anterior trunk limbs to handle detritus and small organisms (Fryer 1988).

It has recently been shown that the former order 'Conchostraca' is most likely to be paraphyletic, having given rise to descendant evolutionary lineages (Braband et al. 2002; Olesen 1998, 2000; Spears & Abele 2000; Richter et al. 2007). The taxonomic rearrangement of Martin and Davis (2001) recognises the order Diplostraca, with four suborders – Laevicaudata, Spinicaudata, Cyclestherida, and Cladocera – of which only the Cladocera and Spinicaudata are represented in New Zealand, the latter by a species of *Eulimnadia*. All diplostracans have the body and legs enclosed between a large, sometimes bivalved carapace. The biramous second antennae are used for swimming, while the phyllopodous (leaf-like), often serially similar, trunk limbs are used for filtration. The most speciose group in New Zealand is the Cladocera, discussed below.

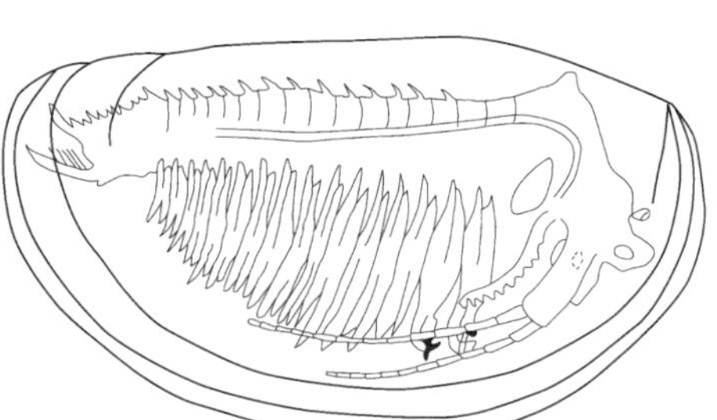

Eulimnadia marplesi (Diplostraca).
After Timms & McLay 2005

Summary of New Zealand crustacean diversity

A query (?) following an entry in the column for alien species indicates that alien status is suspected for some but not confirmed.

Taxon	Described living species + subspecies	Known undescribed/ undetermined species	Estimated unknown species	Adventive species named + unnamed	Endemic species	Endemic genera
Branchiopoda	44	5	7	3?	5	0
Anostraca	1	0	0	1?	0	0
Notostraca	1	0	0	0	0	0
Diplostraca	42	5	7	2?	5	0
Cephalocarida	1	0	1	0	1	1
Maxillopoda	661+2	139	2,067	16?	153	5
Ascothoracida	2	1	7	0	1	0
Acrothoracica	1	0	2	0	1	0
Rhizocephala	8	3	30	0	4	0
Thoracica	77	6	20	3	34	2*
Tantulocarida	3	0	8	0	2	0
Branchiura	1	0	0	1	0	0
Pentastomida	1	0	0	1	0	0
Copepoda	568	129	2,000	11?	111	3
Calanoida	252+1	9	290	6?	10	0
Cyclopoida	100	4	500	5?	8	0
Mormonilloida	1	0	3	0	0	0
Harpacticoida	130	99	850	0	63	3***
Siphonostomatoida	85+1	16	330	0	30	0
Monstrilloida	0	1	27	0	0	0
Ostracoda	356	86	320	3	89	7
Palaeocopida	3	0	0	0	3	0
Podocopida	275	82	200	3	61	6
Myodocopida	78	4	120	0	24	1
Malacostraca	1,425+1	255	2,665	23	850	85+10
Leptostraca	3	2	2	0	0	0
Stomatopoda	8	0	20	1	2	0
Anaspidacea	2	4	5	0	5	1
Bathynellacea	5	3	5	0	8	0
Lophogastrida	5	1	3	0	0	0
Mysida	17	1	50	0	11	0
Thermosbaenacea	0	0	5	0	0	0
Amphipoda	439	64	800	11	268	48+10
Isopoda	358	67	1,000	7	331	19**
Tanaidacea	40	77	300	0	12	0
Cumacea	51	24	110	1?	66	7*
Euphausiacea	19+1	0	15	0	0	0
Decapoda	480	12	150	4	147	10
Totals	**2,488+3**	**485**	**~5,060**	**46?**	**1,097**	**98+10**

* including one new undescribed genus
** including two new undescribed genera
*** including three new undescribed genera

Order Diplostraca: Suborder Cladocera – water fleas

The Cladocera is generally believed to be a monophyletic group within the Branchiopoda (Martin & Cash-Clark 1995; Olesen 1998; Taylor et al. 1999; Spears & Abele 2000; Martin & Davis 2001), a notion that was called into question by Fryer (1987) when providing detailed diagnoses for all branchiopod 'orders' (the rank was changed by Martin & Davis 2001). The Cladocera is by far the most diverse and speciose group within the Branchiopoda, with approximately 640 species worldwide (Korovchinsky 2000), which is more than half of all branchiopod species described.

Historically, Sars (1865) had recognised four tribes within the Cladocera – the Haplopoda, Ctenopoda, Anomopoda, and Onychopoda – which are basically still accepted as monophyletic groups; these groups are now treated as infraorders (Martin & Davis 2001). The Anomopoda is the most species-rich, with at least five families (the number varies depending on the author), 75 genera (Dumont & Negrea 2002), and approximately 560 species (Korovchinsky 2000); the Ctenopoda has eight genera and 47 species (Korovchinsky 2000), the Onychopoda 10 genera with 34 species (Rivier 1998), and the Haplopoda is monotypic with only one species (*Leptodora kindtii* – not represented in New Zealand).

The four infraorders are rather different in their general morphology, which means that cladocerans are difficult to characterise overall. They are in general small, free-living crustaceans ranging from about 0.2–5.0 millimetres in length (with the exception of *Leptodora kindtii,* which is a giant at one centimetre long). Most are somewhat compact in appearance (except for *L. kindtii* and some Cercopagididae, an onychopod family not represented in New Zealand). They have a bivalved carapace (sometimes modified) with one compound eye, small tubular unsegmented antennules (*Ilyocryptus* excepted), large branching antennae, and a distinctive pair of so-called 'postabdominal setae' (similar setae are seen in other branchiopods). They swim using their antennae. The Ctenopoda and Anomopoda are somewhat alike and both have a bivalved carapace that covers the body (but not the head), a pair of curved caudal claws, and five to six (Anomopoda) or always six (Ctenopoda) flattened leaf-like trunk limbs that are used to filter food particles from the water. In the Ctenopoda the six trunk limbs show serial similarity (as in the 'large' branchiopods), while the trunk limbs of the Anomopoda have undergone remarkable evolutionary modifications in relation to food selection, with each limb in many cases being different from its neighbour limb (Fryer 1963, 1968, 1974, 1991). The remaining two groups, the Haplopoda and Onychopoda, are also somewhat alike, having, in contrast to all other branchiopods, narrow-footed segmented trunk limbs – four pairs in the Onychopoda and six pairs in the Haplopoda, used for predation or at least for selective feeding. Olesen et al. (2001) have shown how the segmented trunk limbs of the Haplopoda (*Leptodora kindtii*) have been derived secondarily from the typical phyllopodous limbs of other branchiopods. Both the Haplopoda and the Onychopoda have a relatively small carapace that does not cover the trunk limbs.

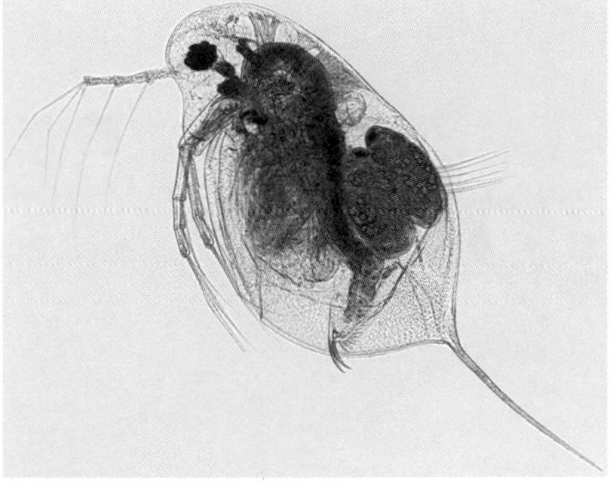

Water flea *Daphnia dentifera* (Cladocera).

Barry O'Brien

In New Zealand, as elsewhere, freshwater cladocerans (water fleas) can often be found in great abundance in open water or at the weedy edges and bottom deposits of lakes, ponds, and stream backwaters (Chapman & Lewis 1976). A child with a scoop-net can easily capture a good supply for a home aquarium. A few species are known from brackish and nearshore ocean environments (Rivier 1998). Among the freshwater species, some are strictly planktonic, others are bottom-dwelling, and *Scapholeberis* (Daphniidae) lives against the surface film. *Simocephalus* (Daphniidae) has the distinctive habit of interrupting its swimming and hanging down from algal filaments by a hooked bristle on one of the swimming antennae (e.g. Fryer 1991). Daphniids are specialist filter-feeders, while chydorids and many macrothricids feed by scraping particles off substrata

Summary of New Zealand crustacean diversity by environment

Taxon	Terrestrial species	Fully freshwater species	Marine/ estuarine species
Branchiopoda	0	41	8
Anostraca	0	0	1
Notostraca	0	1	0
Diplostraca	0	40	7
Cephalocarida	0	0	1
Maxillopoda	2	68	730
Ascothoracida	0	0	3
Acrothoracica	0	0	1
Rhizocephala	0	0	11
Thoracica	0	0	83
Tantulocarida	0	0	3
Branchiura	0	1	0
Pentastomida	1*	0	0
Copepoda	1	67	629
Calanoida	0	11	250
Cyclopoida	0	21	83
Mormonilloida	0	0	1
Harpacticoida	1**	35	193
Siphonostomatoida	0	0	101
Monstrilloida	0	0	1
Ostracoda	1	37	404
Palaeocopida	0	0	3
Podocopida	1**	37	319
Myodocopida	0	0	82
Malacostraca	120	90	1,470
Leptostraca	0	0	5
Stomatopoda	0	0	8
Anaspidacea	0	6	0
Bathynellacea	0	8	0
Lophogastrida	0	0	6
Mysida	0	0	18
Amphipoda	47***	54	402
Isopoda	72	17	336
Tanaidacea	0	1	116
Cumacea	0	0	75
Euphausiacea	0	0	19
Decapoda	1	4	487
Totals	**123**	**236**	**2,614**

* internal parasite of mammal
** damp forest litter
*** including 11 supralittoral species

using their trunk limbs. Genera in the infraorders Onychopoda and Haplopoda are predaceous or at least raptorial feeders (Rivier 1998).

Cladocerans are able to produce non-fertilised (parthenogenetic) eggs that develop in a brood-pouch under the carapace and hatch as miniature adults. Females may continue to moult and grow after reaching sexual maturity, unlike copepods and ostracods. Cladocerans reproduce sexually as well as asexually and produce resting eggs after males have appeared in the population; these eggs undergo a period of dormancy before development begins. In the case of the Anomopoda, resting eggs are protected by a part of the mother's carapace, which is shed together with the eggs as an ephippium. The appearance of males is probably triggered by environmental conditions.

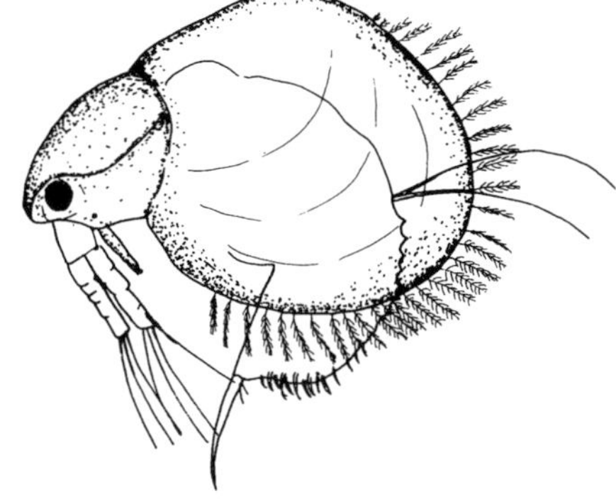

Water flea
Ilyocryptus sordidus (Cladocera).
From Chapman & Lewis 1976

Summary of New Zealand fossil crustacean diversity

Taxon	Described fossil species + subspecies	Known undescribed/ undetermined species	Endemic species	Endemic genera
Maxillopoda	61	19	60	2
Acrothoracica	0	4	1	0
Rhizocephala	0	1	0	0
Thoracica	61+3	14	59	2**
Ostracoda*	284	127	22	5
Archaeocopida	0	2	0	0
Palaeocopida	1	0	1	0
Podocopida	283	124	21	5
Myodocopida	0	1	0	0
Malacostraca	67	44	61	8
Phyllocarida	7+1	1	7	0
Eumalacostraca	60	43	54	8
Isopoda	4	0	4	1
Decapoda	56	43	50	7
Totals	412	190	143	15

* Several species range to the present day; these are also in the Recent checklist.
** undescribed new genera

The end-chapter list of New Zealand Cladocera is based on the work of Chapman and Lewis (1976) for freshwater species and the records of Krämer (1895) and Jillett (1971) for marine species. The marine forms particularly need revising, as most of Krämer's species are not well known. The zoogeography of freshwater zooplankton in Australasia (Bayly 1995 and references therein) suggests that the New Zealand cladoceran fauna reflects the fact that New Zealand split from Antarctica during the Late Cretaceous. New Zealand, Australia, and South America completely lack the predaceous-raptorial families Polyphemidae and Cercopagididae (Onychopoda), the Leptodoridae (Haplopoda), and the Holopedidae (Ctenopoda). It seems likely that these families evolved in Laurasia after splitting from Pangaea (Bayly 1995). On the other hand, the Anomopoda, well-represented in New Zealand, are a very ancient group (from at least 130 million years ago) that was probably distributed over Pangaea.

Class Cephalocarida

The Cephalocarida was introduced as a new crustacean subclass by Sanders (1955) for a tiny, primitive-looking species taken off the Atlantic coast of North America. Since then, very few additional species have been discovered, and the most recent treatments recognise only one family with five genera and 10 species worldwide (Hessler & Wakabara 2000; Martin & Davis 2001). All are very small, measuring only 2–4 millimetres in length. The swimming limbs barely differ from one another, with the endemic New Zealand genus *Chiltoniella* being the least modified. The class is generally regarded as one of the more primitive of the living Crustacea.

Most species have been recorded from silty seafloors. In general, their biology is poorly known. New Zealand's sole species, endemic *Chiltoniella elongata*, is known from the Hawke's Bay region (Knox & Fenwick 1977).

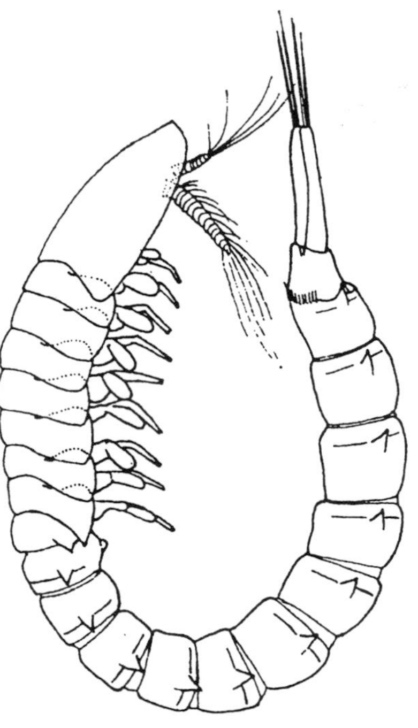

Chiltoniella elongata (Cephalocarida).
From Knox & Fenwick 1977

Class Maxillopoda

Barnacles, seed shrimps, oar-footed bugs (copepods), and related parasitic groups – these are all examples of maxillopod crustaceans. They are a disparate lot, and carcinologists (crustacean specialists) are still arguing over whether or not they are a single evolutionary lineage (monophyletic). Apart from some barnacles, most species are small or minute. Most feed by means of mouthparts called maxillae (instead of using trunk limbs as filtration devices), barnacles again being a notable exception. Other characteristics of maxillopods include a basic body plan of five head and 10 trunk segments followed by a terminal telson. Abdominal segments usually lack appendages; elsewhere on the body, appendages are usually branched (biramous). As a group, maxillopod crustaceans are very important – economically, as in the case of many marine-fouling barnacle species, and more especially ecologically because of their shear abundance. Copepods, for example, are the most numerous crustaceans in open-ocean waters.

Subclass Thecostraca

This subclass comprises representatives of two infraclasses in New Zealand – the Ascothoracica and Cirripedia ('curly footed'). The latter includes barnacles, sessile crustaceans that use their trunk limbs to catch food particles. Most New Zealanders will be familiar with the acorn barnacles that carpet the upper zones of rocky seashores or, annoyingly, boat hulls, and perhaps the stalked goose barnacles that attach to floats and other buoyant objects, but few will know of the tiny burrowing and parasitic thecostracans.

Minute borings in mollusc shells, attributed to barnacles, have been well documented since Darwin (1854a) collected and described specimens during his voyage on HMS *Beagle*. Originally a number of parasitic organisms were included within this group of 'burrowing barnacles', e.g. the Ascothoracica and Rhizocephala (Newman et al. 1969), but these latter two taxa have been subsequently shown to possess spermatozoa, nauplius larvae, and newly settled cypris stages that are very different from barnacles. Following the re-evaluation of the Cirripedia by Newman (1987, 1996), the Ascothoracica and Rhizocephala are no longer considered as barnacles by some specialists; on the other hand, Martin and Davis (2001), Buckeridge and Newman (2006), and Lützen et al. (2009) treat the Rhizocephala as a superorder of Cirripedia. Ascothoracicans are represented in New Zealand by two species of starfish parasites (Palmer 1997); living rhizocephalans, virtually unknown in New Zealand until very recently, comprise 11 species (Brockerhoff et al. 2006; Lörz et al. 2008; Lützen et al. 2009).

The burrowing acrothoracicans possess a soft carapace, with calcareous plates reduced or absent. There are about 40 known species worldwide, including one endemic New Zealand species. All live buried in calcareous shells of a wide range of marine invertebrates, including molluscs, echinoderms, corals, bryozoans, and other barnacles. The group has a fossil record extending back to the Devonian (Tomlinson 1987), although no pre-Mesozoic taxa are known from New Zealand. As the fossil record of acrothoracicans is based solely upon burrows, two distinct acrothoracican nomenclatures have developed, one ichnomorphic, the other biological. This may lead to some confusion, as trace-fossil names such as *Zapfella* have equivalents such as *Australophialus*. Both systems are used in this review of the New Zealand fauna because the relationship between fossils and living species is unclear.

The familiar thoracican barnacles are classified into four orders with 81 living species in New Zealand – the stalked (pedunculate) Ibliformes, Lepadiformes, and Scalpelliformes, and the generally squat, nonstalked Sessilia, comprising the acorn (balanomorph) barnacles, wart (verrucomorph) barnacles, and the

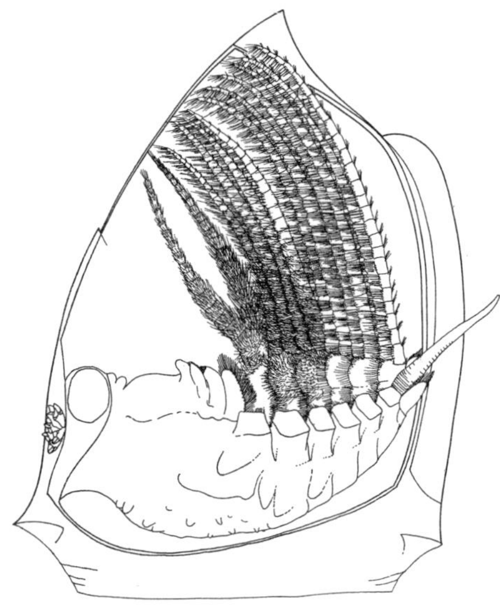

Cutaway view of *Calantica spinilatera* showing the long bristly feeding limbs (cirri) with smaller mouthparts to the lower left of the cirri.

From Foster 1979

Brachylepadomorpha (confined to deep-ocean hydrothermal vents and not yet known from New Zealand).

Most barnacles are hermaphrodites, although in some species the 'typical' hermaphrodite form may also carry minute or dwarf males within the capitulum (see below). These dwarf males possess either reduced or no appendages and capitular plates, being essentially packages of male gonads. Sexual differentiation does occur in some species, e.g. endemic *Idioibla idiotica*, (Ibliformes).

Idioibla idiotica.
John Buckeridge

The pedunculate forms are the most ancient of the barnacles. They are characterised by a stalk (peduncle), by which they attach themselves to the substratum. A series of calcareous plates, together forming a capitulum, are found on top of the peduncle of most species, enclosing most of the soft tissue of the animal. A careful examination of this area verifies the evolutionary placement of the barnacles within the crustaceans, as the animal is effectively arranged head down, with its jointed limbs (cirri) extending out through a slit (orifice) in the capitulum wall. When the barnacle is submerged, the cirri extend into the surrounding water, netting planktonic food.

As the number and arrangement of capitular plates varies considerably between taxa, they are of considerable value in classification. In the goose barnacle *Lepas* (Lepadiformes) there are five plates: paired terga and scuta with a single carina, arranged in a single whorl. However, in species like *Calantica spinosa* (Scalpelliformes) the number of capitular plates varies from 11 to more than 50, and these are arranged in two or more whorls. In taxa like *Calantica* and *Anguloscalpellum,* the peduncle is armoured with small overlapping plates or scales. In contrast, there are no plates or overlapping scales on the peduncle in Lepadiformes. The most primitive order of living thoracicans is the Ibliformes, with predominantly chitinous rather than calcareous plates. Of the five living genera, three of them are found in New Zealand, including the endemic genus *Chitinolepas* from Spirits Bay (Buckeridge & Newman 2006).

The Verrucomorpha are a group of barnacles that, because of their asymmetry, have intrigued cirripede workers since Darwin (1854b). Although they are amongst the most primitive Sessilia that are likely to be encountered as fossils, they are as yet unconfirmed from the New Zealand Mesozoic. They are, however, known from the Cretaceous of Australia (Buckeridge 1983). The Verrucidae are represented in New Zealand waters by species of *Altiverruca* and *Metaverruca,* both of which possess six calcareous plates. The lid (operculum) comprises just two articulating plates, the shell wall being made up of the remaining four: a fixed tergum and fixed scutum, plus rostrum and carina. Unlike other Sessilia, each wall plate in verrucids joins with its adjacent plate by interlocking ribs. The distribution of verrucid genera tends to conform to depth, with *Verruca* species characteristic of shallow coastal waters, *Metaverruca* to midshelf environments, and *Altiverruca* to the continental slope and deeper. Some verrucid species also have symbiotic or commensal relationships with other invertebrates, and these may be host-specific, e.g. *Brochiverruca* on cnidarians and *Rostratoverruca* on cidaroid urchins (Buckeridge 1997). This appears to be the situation with an as-yet-undescribed verrucid from northern New Zealand waters that inhabits the coral *Ellanopsammia rostrata*.

Chitinolepas spiritsensis.
From Buckeridge & Newman 2006

When one considers balanomorph or acorn barnacles, the image many people have is of a limpet-like creature commonly attached to vessel hulls. Although barnacle fouling on ships is well known, it represents only a small proportion of their distribution. They are best seen as ubiquitous opportunists of the marine environment attached to a great variety of living and inanimate objects. Barnacles include species specialised for attachment to whales, sea snakes, turtles, corals, sponges, and other crustaceans.

Many shallow-water acorn barnacles are known to have variable tolerances to both high temperatures and desiccation. Because of this, species in the intertidal zone may be found distributed in distinctive bands, e.g. on exposed

Coronula diadema, a barnacle that grows on whales.

John Buckeridge

rocky shores, where *Chamaesipho brunnea* forms bands in the uppermost intertidal and *Epopella plicata* at mid- to low tide.

The balanomorph shell is made up of two parts: a rigid calcareous wall comprising four or more parietal plates, and an operculum or lid generally made up of paired scuta and terga. The opercular plates articulate to permit extension of the cirri between them during feeding. They also enable the animal to seal itself off from the environment in times of stress (e.g. predation, desiccation). As with the stalked barnacles, the plates are very important in identifying species. Parietal plates may be solidly calcified (e.g. *Austrominius*), calcareous with internal chitinous laminae (e.g. *Epopella*), calcareous with one row of vertical tubes (e.g. *Balanus*), or calcareous with chitin, arranged in multiple rows of tubes as in *Tetraclitella* (Buckeridge 2008). The number of parietal plates is also significant, with four in *Austrominius, Epopella,* and *Tetraclitella* and six in *Austromegabalanus, Balanus, Chamaesipho, Coronula, Megabalanus,* and *Notobalanus*.

The elements of barnacle anatomy and morphology, forming the basis of our modern classification and understanding, were elucidated by none other than Charles Darwin. His outstanding work on these creatures had a very strong influence on the ideas that eventually led to his revolutionary book *On the Origin of Species*. Indeed, Darwin was so amazed by the profusion and ubiquity of barnacles in the Cenozoic that he described Tertiary seas as 'abounding with species of *Balanus* to an extent now quite unparalleled in any quarter of the world'. (In Darwin's time, although most sessile cirripedes were ascribed to the genus *Balanus*, he was able to demonstrate groupings of similar taxa through the use of 'varieties'.)

That Darwin was infatuated with barnacles is clear, and he put much else aside to work on them: 'I have for the present given up Geology, and am hard at work at pure Zoology and am dissecting various genera of Cirripedia, and am extremely interested in the subject.' [Letter to Dieffenbach, February 1847]. But it was not always an agreeable infatuation: 'I have now for a long time been at work on the fossil cirripedes, which take up more time than the recent: confound and exterminate the whole tribe; I can see no end to my work.' [Letter to Hooker, 1850]. Darwin did persist, both with his monographs on fossil and living cirripedes (Darwin 1851a,b, 1854a,b) and his *Origin of Species*. Darwin's second cirripede volume was dated 1851 but came out quite late in 1852. His works endure as a monument to scholarship, and remarkably, one and a half centuries later, still provide the intellectual platform from which we are able to develop our present-day understanding of Earth's biodiversity.

Infraclass Ascothoracica

Adult female of *Dendrogaster otagoensis*.

From Palmer 1997

These curious creatures are primitive among thecostracans, ectoparasitic on feather stars and sea urchins, and endoparasitic within some corals and sea stars. Females have a much-reduced thorax and abdomen and a simplification or loss of limbs. The carapace is enlarged and grossly distorted, being much-branched and unrecognisable as belonging to a crustacean. Males are tiny and more recognisably crustacean in form, resembling larvae. They have a well-segmented body enclosed in a carapace and greatly elongated testes and and are found within the mantle cavity of females.

Ascothoracicans were unknown in New Zealand until Palmer (1997) found two species inhabiting sea stars off the Otago coast. *Dendrogaster otagoensis* was described as a new species, infesting *Asterodon miliaris*. Of a collection of 159 sea stars taken from the coast over an 11-month period, 124 (78%) were infested with the parasite. Found inside the arms and disc of the sea star, there can be as many as 15 female parasites, with their convoluted carapaces over 20 millimetres across, causing some atrophy of the sea-star's digestive caecae and gonads. Up to 19 creamy-white males 2.9–3.5 millimetres long occur inside the female parasite.

A second species, *Dendrogaster argentinensis*, was also found off Otago, infesting 96% of 152 specimens of the sea star *Allostichaster insignis* quite severely.

This particular parasite, previously known from southern South America and the Falkland Islands, can fill much of the sea-star's body cavity, comprising up to 28% of the wet weight. Gonads in such specimens are absent, and digestive caecae are severely atrophied. Curiously, specimens of *A. insignis* in other parts of its range (Cook Strait to the Auckland Islands) have never been noted as having such parasites, so it would be interesting to know what conditions promote such infestations in Otago waters.

Dendrogaster belongs to one of three families in the ascothoracican order Dendrogastrida. Palmer (1997) also mentioned an unpublished Te Papa (Museum of New Zealand) record of an undescribed member of the Synagogidae, one of three families in the only other ascothoracican order, Laurida.

Infraclass Cirripedia: Barnacles

Superorder Acrothoracica

Apart from the study by Batham and Tomlinson (1965) on *Australophialus melampygos*, there has been little work done on New Zealand acrothoracicans. They are a very difficult group to work with, particularly as most occurrences are known only by their tiny borings. *Australophialus melampygos* is often found infesting paua (*Haliotis iris*) and mussel (*Perna canaliculus*) shells, commonly in very large numbers (up to 3350 borings noted in a single paua shell. The family Cryptophialidae was revised by Tomlinson (1969), who introduced *Australophialus* to incorporate the austral members (including *A. melampygos*) of *Cryptophialus* that possessed four rather than three pairs of terminal cirri (feeding appendages).

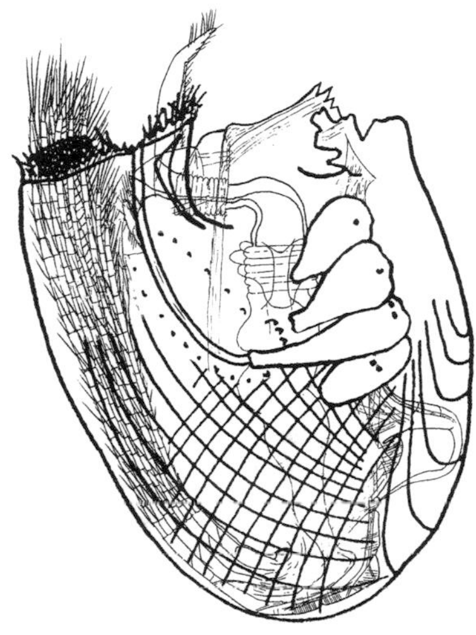

Australophialus melampygos removed from its excavation in a shell; five dwarf males attached middle right.

Modified from Batham & Tomlinson 1965

Existing literature infers that acrothoracicans have very low diversity in the New Zealand region. Further, they appear to be somewhat host-specific, and whilst this is not generally a problem where a host is a common marine invertebrate, there is cause for concern if the host is over-fished. Both *Haliotis iris* (paua) and *Perna canaliculus* (green-lipped mussel) are extensively harvested as a food source, and although they are now widely cultured in marine farms, the new aquacultural environment does not appear to provide the habitat so favoured by *Australophialus melampygos* in nature. The likelihood that the shell-infesting population represents more than one species should not be overlooked, especially in light of acrothoracicans' poorly mobile larval phase (which may account for its absence from the Chatham Islands). The distribution of these molluscs extends from Northland to Stewart Island; although both species range well into the subtidal, *A. melampygos* is not known much below low tide, its preferred habitat.

Australophialus melampygos falls within a group of southern acrothoracicans including *A. tomlinsoni* from the Antarctic and *A. turbonis* from South Africa. Newman and Ross (1971) considered the cirral arrangement of these taxa to be more generalised (and therefore phylogenetically older) than other Cryptophialidae, inferring a Southern Hemisphere origin for the family. However, rather than a South African centre of cryptophialid diversification, abundant cryptophialids in some turritellid gastropods within the Pakaurangi Formation (Early Miocene), Kaipara Harbour, should not rule out the New Zealand region as a potential centre of dispersal.

Briarosaccus callosus, a saccular rhizocephalan parasite under the abdomen of the king crab *Paralomis hirtella*.

Dianne Tracey

Superorder Rhizocephala

Rhizocephalans are wholly parasitic. They have little similarity with other cirripedes, or indeed other crustacean adults, as there are neither appendages nor segmentation (e.g. Høeg & Lützen 1995, 1996). A rhizocephalan consists of a sac-shaped body, the externa, which is mainly involved in reproduction and is attached to the outside of the host's abdomen. The host is always another crustacean, in most instances an anomuran or brachyuran crab. A mouth and a digestive tract are absent and nutrients are taken up from the host's interior by an internal trophic root system (or interna) which is distributed

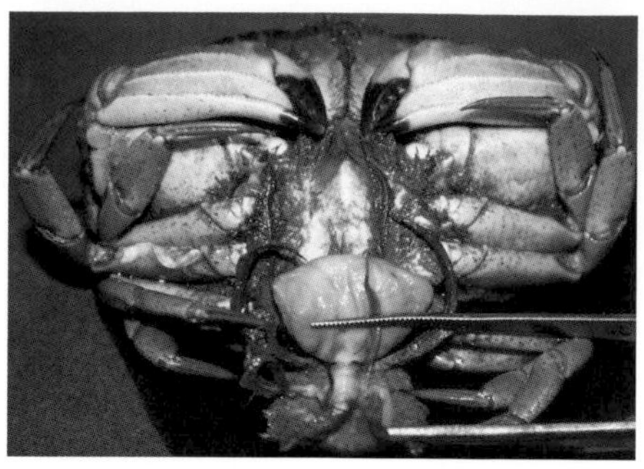

Sacculina sp., a saccular rhizocephalan parasite under the abdomen (folded back) of the crab *Metacarcinus novaezelandiae*.
Annette Brockerhoff

within the haemolymph of the host (Høeg & Lützen 1995). The externae are most often attached singly or a few together to the host's abdomen, but some rhizocephalans are colonial and in such species many small externae may attach to the abdomen, appendages, or other parts of the host body (Høeg & Lützen 1993, 1996). Despite their bizarre appearance, rhizocephalans are related to the non-parasitic barnacles, which they resemble in reproducing via short-lived planktonic nauplii and/or cypris larvae (Høeg & Lützen 1993).

Apart from sparse records in the literature, rhizocephalans were almost unknown in New Zealand until the 2000s; there are now at least 10 genera and 11 species (Brockerhoff et al. 2006; Lörz et al. 2008; Lützen et al. 2009). Decapod host species belong to the families Paguridae, Lithodidae, Galatheidae, Chirostylidae, and Callianassidae. *Parthenopea vulcanophila* (Lützen et al. 2009), is the first rhizocephalan recorded from the vicinity of active cold seeps.

The recently discovered New Zealand rhizocephalans are registered in the invertebrate collections of the National Institute of Water and Atmospheric Research (NIWA) and the National Museum of New Zealand Te Papa Tongarewa, Wellington (NMNZ). Some of the specimens could not be identified because they were in turn infected by species of Cryptoniscinae, a subfamily of hyperparasitic isopods. In the final stage of this relationship of a parasite on a parasite the rhizocephalan host is no longer recognisable (Øksnebjerg 2000).

Recent gene-sequencing studies on the Rhizocephala have indicated that the conventional grouping of its members is in need of rearrangement (Glenner et al. 2003; Glenner & Hebsgaard 2006). Since these findings have not yet resulted in a taxonomic revision, the traditional division of the Rhizocephala into the orders Kentrogonida and Akentrogonida is followed in the end-chapter checklist; as a consequence of the study by Glenner and Hebsgaard (2006), however, *Parthenopea* is included in the Akentrogonida.

Superorder Thoracica

On 3 October 1769, in calm seas some 300 kilometres off what is now known as Mahia Peninsula, HM Bark *Endeavour*, under the command of James Cook, retrieved 'one peice of wood coverd with Striated Barnacles *Lepas Anserina*?' (Banks 1962). This was not only the first record of barnacles from New Zealand seas, but also one of the first records of marine life from the region. In an editorial footnote to Banks's journal, J. C. Beaglehole stated that Daniel Solander (the naturalist who accompanied Banks) considered the species to be *Lepas anserifera*. The next major scientific expedition to New Zealand was in 1827, when the *Astrolabe* collected extensive natural history material, including barnacles. The barnacles were subsequently described by Quoy and Gaimard (1834) as *Anatifera spinosa*, *Anatifera elongata*, and *Anatifera tubulosa* (now respectively known as *Calantica spinosa* (Quoy & Gaimard), *Lepas testudinata* Aurivillius, and *Heteralepas quadrata* (Aurivillius)). The first endemic New Zealand barnacle to be described was, therefore, *C. spinosa*.

In 1839 the New Zealand Company appointed Ernst Dieffenbach as surgeon and naturalist on the *Tory*. Dieffenbach made extensive biological collections during his time in New Zealand, and included in these were barnacles. These were later compiled by J. E. Gray into a *Fauna of New Zealand* and listed as an appendix to Dieffenbach's *Travels in New Zealand* (Gray 1843). Gray recorded nine thoracicans, now known as *C. spinosa*, *L. testudinata*, *H. quadrata*, *Coronula diadema*, *Epopella plicata*, *Tetraclitella depressa*, *Tubinicella major*, and two unidentified species of *Balanus*.

Shortly after this, Darwin's four comprehensive monographs on living and fossil cirripedes were published. Darwin had collected New Zealand barnacles from the Bay of Islands during the voyage of HMS *Beagle*, which, along with British institutional material, resulted in 14 species being listed from the New Zealand region. Ten were new to science, of which *Austrominius modestus*,

Notobalanus vestitus, and *Notomegabalanus decorus* are endemic to New Zealand. Darwin included a complete description of the endemic species *Chamaesipho columna*, which had previously been described from material supposedly collected from Tahiti (Spengler 1790). Spengler's original description was, however, incomplete, as the shells he possessed were without opercula or soft tissue. In Foster and Anderson (1986), the status of *C. columna* was reviewed and it was concluded that Spengler's material came from New Zealand, where it is endemic. (They renamed the Australian species previously attributed to *C. columna* as *Chamaesipho tasmanica*.)

The last major systematic work of the 19th century that dealt with New Zealand barnacles was based upon specimens obtained during the 1873–76 HMS *Challenger* expedition. In an expedition report, Hoek (1883) described five new species, now known as *Amigdoscalpellum costellatum*, *Anguloscalpellum pedunculatum*, *Gymnoscalpellum intermedium*, *Smilium acutum*, and *Verum novaezelandiae*. During the early to mid-20th century, numerous descriptions of new records for the region, generally for single species, were published and a full list of these was given by Foster (1979). The latter work is the most comprehensive study ever written on living New Zealand Thoracica. In it, Foster listed a fauna of 61 species, nine (including a new subspecies) of which were new, one was a new name, and 15 species were recorded for the first time from New Zealand waters. Foster also made valuable observations on the geographic distribution, zonation, and ecology of barnacle species. In the 14 years following his 1979 monograph, Foster described a further two new species and add records of eight taxa not previously known from New Zealand waters (Foster & Willan 1979; Foster 1980, 1981; Foster & Anderson 1986). Brian Foster died suddenly in 1992, tragically cutting short what was, up to that time, a prolific and invaluable career in barnacle systematics and biology. Since then, J. S. Buckeridge, a student of Foster, has continued study of the New Zealand fauna, frequently in collaboration with W. Newman. The systematics of barnacles was reviewed by Buckeridge and Newman (2006), in which the Iblidae was identified as the most ancient family of Thoracica. Significantly, it was the discovery of an extraordinary but minute new species from New Zealand, *Chitinolepas spiritsensis*, that provided the impetus for this work, which demonstrated that the New Zealand region not only has a diverse living thoracican fauna but also one of the most primitive.

Although not specifically focussing on the New Zealand fauna, Newman's (1979) publication is an inspired revision of the phylogenetic and biogeographic relationships between barnacles of the Southern Ocean. His work led to a reappraisal of the entire fauna, with many of the proposed taxonomic concepts incorporated in Buckeridge (1983). The evolving nature of systematic biology results from an ongoing reappraisal of relationships between taxa. As our understanding of barnacle phylogeny becomes more sophisticated, this often creates the need to provide new names for species. The overview herein is based upon the comprehensive review of Cirripedia by Newman (1996), in which subgenera are elevated to full generic status. Consequently, species like *Elminius modestus* and *Austromegabalanus decorus* are now listed as *Austrominius modestus* and *Notomegabalanus decorus* respectively. A recent publication reviews the status of the Elminiinae and identifies *Austrominius* as a tetraclitoid, returning it closer to *Epopella*, where Darwin (1854) had originally perceived it to be (Buckeridge & Newman 2010).

There are 81 species of Recent thoracican cirripedes known from the New Zealand EEZ. Of these, six are currently undescribed. Four are stalked barnacles, comprising two species of *Scillaelepas* (Calanticidae) one of which conforms to a southern group of primitive scalpellids, and two species of Scalpellidae; an unusual undescribed verrucid is likely to represent a new genus; and a possible new species of *Acasta* (Archaeobalanidae) remains to be determined (J. Buckeridge is currently reviewing this genus of sponge-inhabiting barnacles). All

Chamaesipho columna.
Dennis Gordon

Smilium zancleanum, with plates on the right-hand side removed to show the cirri.
John Buckeridge

Metaverruca recta.
John Buckeridge

species referred to as new in the end-chapter checklist are held in the collections of the NIWA Invertebrate Collection, Wellington.

The vertical zonation of thoracican barnacles on New Zealand surf shores has been well documented (e.g. Morton & Miller 1968). The zonation is not always consistent, however, with ranges expanding/contracting in the absence/presence of other taxa (Foster 1979). Nevertheless, there are generalisations that can be made, and these provide useful ecological benchmarks: chthamalids are found higher on the shore than all other thoracicans; below them, and overlapping somewhat, are the tetraclitids; further down the shore the lower range of the tetraclitids overlaps the balanids. This chthamalid-tetraclitid-balanid arrangement appears to be fairly uniform on both temperate and tropical shores (Foster 1974, 1979). *Cantellius septimus*, a widespread Indo-Pacific species, has been found in *Montipora* coral off Raoul Island (Kermadec Ridge), representing the most southerly record of a coral-inhabiting barnacle (Achituv 2004).

Some species are epizoic on cetaceans. *Conchoderma auritum*, *C. virgatum*, and *Coronula* species attach to whales and three species of the latter genus are found in the New Zealand fossil record.

The isolation of New Zealand since the late Mesozoic has led to high regional endemism in taxa that evolved during the Late Cretaceous–Early Cenozoic. This is no more evident than in the thoracican barnacles (Buckeridge 1996a,b, 1999a). Although 40% of the Recent species listed are endemic, the figure is a little misleading, as the current distribution of New Zealand species such as *Austrominius modestus* to include Australia and Europe has almost certainly been achieved via shipping. What is particularly significant about the New Zealand region is the high proportion of endemics that are phylogenetically primitive. The percentage of balanomorph and verrucid taxa that have their earliest (fossil) records in New Zealand is impressive, with 73% of all primitive sessilians with a generic age earlier than the Miocene being first recorded here (Buckeridge 1996a).

There are several species of thoracican barnacles that may be termed 'living fossils', i.e. they have fossil records extending back at least to the Early Miocene. Two of these, *Chionelasmus darwini* and *Notobalanus vestitus* extend back to the Eocene and Oligocene, respectively; two others, *Metaverruca recta* and *Chamaesipho brunnea*, to the earliest Miocene. The order Ibliformes extends back to the Permian and the Neolepadinae to the Jurassic.

Sampling of deep-sea cirripedes from the New Zealand EEZ is far from comprehensive, but 13 species are known from depths greater than 1500 m, the deepest of which are *Gymnoscalpellum intermedium* (to 2505 m) *Amygdoscalpellum costellatum* (to 3120 m), and *Verum raccidium* (to 4405 m) according to NIWA database records. Specimens have often been made available as bycatch from the fishing industry or from research cruises. Recent discoveries include the neolepadine *Vulcanolepas osheai* from ca. 1500 metres depth in the volcanically active Brothers Caldera (in the Havre Trough northeast of the Bay of Plenty) and a related taxon, *Ashinkailepas kermadecensis* (Buckeridge 2009), from a cold-water seep at 1165 m on the western flank of the Kermadec Ridge. Both of these taxa have specialisations, like long filamentous cirri, that permit them to feed on bacteria, the most abundant food source in the area, living on the barnacle exteriors and around the vents and seeps (Suzuki et al. 2009). Bathylasmatids such as *Tetrachaelasma tasmanicum*, although not yet formally recorded from within the New Zealand EEZ, almost certainly occur here. This taxon was recently described from 3600 metres on the southeastern Tasman Rise (Buckeridge 1999b) where it is widely distributed as disassociated shells that are very similar to isolated plates collected from New Zealand waters; in the absence of living tissue the latter material has not been placed to species.

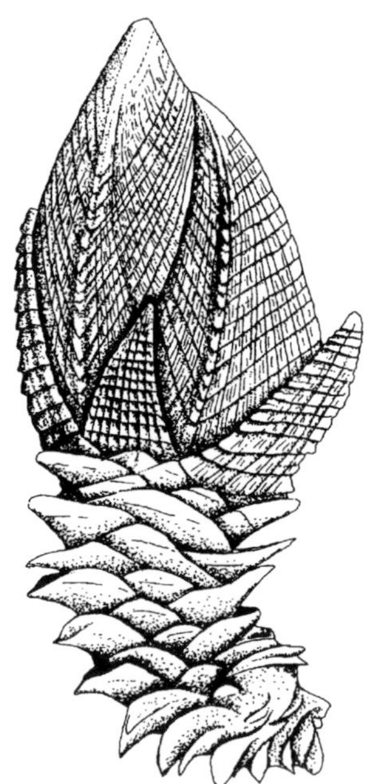

Ashinkailepas kermadecensis.
From Buckeridge 2009

Although the total number of thoracican barnacle species from New Zealand is not high compared with the numbers of species of taxa such as the Bryozoa and Mollusca, it is high compared with cirripede faunas from other regions. In particular there is a broader representation of known cirripede taxa (especially phylogenetically primitive taxa) than in any region of comparable size, and there is a disproportionately large number of species, both living and fossil, that have their earliest records in New Zealand (Buckeridge 1996a).

Palaeontology and paleoecology

Acrothoracica

Acrothoracican burrows are known to occur in thick-shelled bivalves (e.g. trigoniids) of Late Triassic age from Nelson and Southland (H. J. Campbell pers. comm.) and belemnite guards (e.g. *Belemnopsis alfurica*) of Late Jurassic age from Kawhia. These can be attributed to the ichnogenus *Zapfella*, to which the burrow shapes generally conform; however, their true biological relationships remain unclear and, as such, no move is made to classify them at ordinal level or below. The Triassic record extends the range of *Zapfella* from that provided in Häntzschel (1975) of 'Jurassic to Tertiary'. Burrows are also known in Early Miocene deposits from the Auckland region, e.g. Waiheke Island (J. A. Grant-Mackie pers. comm.), and in turritellid gastropods from the Pakaurangi Formation, Kaipara Harbour. The later burrows appear indistinguishable from modern *Australophialus* borings, to which genus they are tentatively assigned.

Rhizocephala

Perhaps surprisingly, given their parasitic lifestyle, rhizocephalans are detectable in the fossil record and are known from the New Zealand Miocene. Feldmann (1998) studied a large number of beautifully preserved specimens of the large xanthoid crab *Tumidocarcinus giganteus*. Several males had abnormally broad abdomens, which is normally attributable to the parasitic castration induced by the parasite.

Thoracica

Thoracican barnacles have a fossil record extending back to the Paleozoic, but not in New Zealand. The pedunculate order Cyprilepadiformes is known from the Silurian, attached to a eurypterid, and other thoracicans are known from the Early Devonian and the Pennsylvanian (upper Carboniferous) (Newman et al. 1969; Buckeridge 1983; Foster & Buckeridge 1987; Newman 1996; Buckeridge & Newman 2006). There is no record of Paleozoic cirripedes from the entire New Zealand–Australian–Antarctic region, the first such record being *Eolepas*? *novaezelandiae* from Middle Triassic strata of Southland (Buckeridge 1983).

Although there are rare scalpellomorphs of Jurassic age, it is not until the Cretaceous that significant records are known – locally abundant, as-yet-undescribed remains of *Cretiscalpellum*? are known from Middle Cretaceous rocks in the Coverham area. These scalpellomorphs are preserved in association with species of the large bivalve *Inoceramus*, upon which they appear to have been growing. Hence, apart from a new verrucid from the Cretaceous of the Waipara River in central Canterbury, the only barnacles known from the New Zealand Mesozoic are stalked ones. Surprisingly, even though there are barnacle-rich horizons in the Paleocene of the Chatham Islands, there are no barnacles of Mesozoic age known from there. This is not likely to have resulted from a paucity of appropriate facies, as there are some excellent Late Cretaceous fossiliferous horizons present on Pitt Island that could have been expected to have provided an appropriate environment for scalpellomorphs. At present, it must be concluded that the absence of a Cretaceous barnacle fauna reflects incomplete paleontological knowledge, and this provides an impetus for further fieldwork on the islands.

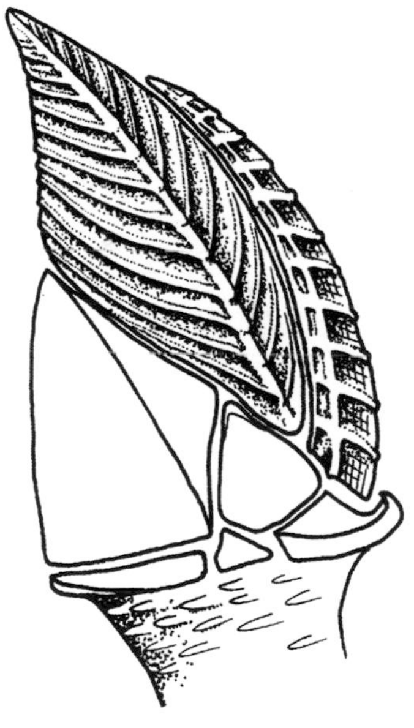

Reconstruction of the fossil barnacle *Anguloscalpellum euglyphum* (Oligocene).
John Buckeridge

Cenozoic barnacles

The New Zealand Cenozoic barnacle fauna is dominated by balanomorphs. The first fossil cirripede to be described from New Zealand strata was the giant balanomorph *Bathylasma aucklandicum*, from Early Miocene strata near Auckland. The locally abundant, but generally disarticulated plates of this sessile barnacle were however, initially described as a pedunculate (Hector 1888). A quarter of a century was to pass before the true nature of the remains was established, in a paper wherein the author also described two new endemic species now known as *Anguloscalpellum ungulatum* and *Smilium subplanum* (Withers 1913) (see Jones 1992). In the early 1920s, Withers, working from the British Museum, was commissioned by the then Geological Survey of New Zealand to produce a monograph of the fossil cirripedes of New Zealand (Withers 1924). This listed 18 species, of which only 15 were truly fossil, and seven of these were both new and endemic to New Zealand. In 1953, he published his last major work that dealt specifically with cirripedes from New Zealand (Withers 1953). This included a revised list of the New Zealand fossil fauna, arranged according to stratigraphic horizons. He listed 15 species, none of which was new. Interestingly, he omitted the record for '*Balanus amphitrite*' that he included in his 1924 monograph, but added the record for what is now *Pristinolepas harringtoni*. No reason is given for his omission of '*Balanus amphitrite*', which is now recognised in the New Zealand fossil record as *Amphibalanus variegatus*. In all, Withers described nine fossil cirripedes from the region, all of which are endemic.

Many limestones are so enriched with balanomorph remains that they may justifiably be termed 'barnacle coquinas'. The first horizons with locally abundant balanomorphs are of late Paleocene age, occurring as lenses in the Red Bluff Tuff of the Chatham Islands. In some of these lenses, the barnacle *Pachylasma veteranum* is also the dominant macrofossil, with the other macrofauna primarily being teeth of the elasmobranch fish *Isurus* sp. plus brachiopod and bivalve shells. Although barnacle-rich horizons are also recorded in the Early Oligocene (Cobden Limestone, West Coast), and Early Miocene (basal Cape Rodney Formation, Auckland), it is the Pliocene coquina limestones of the North Island East Coast that are singularly spectacular, e.g. the Pukenui and Castlepoint Limestones, which contain extensive horizons dominated by *Fosterella tubulatus* and *Notobalanus vestitus*. These coquinas outcrop at Rangitumau and Castlepoint respectively (both in the Wairarapa), and have extensive beds in which *F. tubulatus* comprises more than 50% of the total mass. There are no modern equivalents of these deposits, although lesser shell banks of *N. vestitus* and *Notomegabalanus decorus* are today accumulating in the outer Hauraki Gulf near the Mokohinau Islands. It is inferred by Beu et al. (1980) that these deposits originated in subtidal settings dominated by strong currents, in a Pliocene sea occupying the East Coast Inland Depression. These Pliocene 'barnacle coquinas' are not only impressive from a cirripedological perspective, they are also the greatest accumulation of fossil crustaceans known!

Because barnacle species tend to be distributed along clearly delineated depth, salinity, and temperature zones, their presence as fossils can be most useful in paleoecological reconstruction. There are, however, some trends in the 'preferred' environments of some taxa over time, e.g. species of the genus *Pachylasma* are currently restricted to deep water, with the shallowest living species of the group not known from less than 55 metres. In the Paleocene, however, *Pachylasma veteranum* is known to have lived in very shallow water, along with a diverse fauna of bryozoans, molluscs, and cnidarians, well within the photic zone (Buckeridge 1983, 1999a). A similar pattern can be observed with species of *Bathylasma*, which also occupied upper subtidal environments in the Paleogene, but are now exclusively mid- to outer-shelf species. Indeed, this change, which was interpreted by Buckeridge (1983) as 'migratory', is now viewed more as a result of having been excluded (or outcompeted) from the shallower-water environments by 'modern' balanomorphs. Modern taxa such as *Austrominius modestus* have a higher metabolism and an earlier onset of sexual maturity, which has permitted the species to aggressively exploit desirable shallow-water niches. This has left refugial chthamalids (such as *Chamaesipho columna* and *Chamaesipho brunnea*) occupying upper littoral niches, and pachylasmatines (such as *Pachylasma scutistriata* and *Bathylasma alearum*) mid- to outer-shelf environments (Buckeridge 1999a).

By the Late Miocene, it appears that thoracican barnacles occupied much the same habitats as their modern counterparts (including as epibionts on other crustaceans – Glaessner 1960, 1969). As a consequence, the zonation of modern balanomorphs is useful in the reconstruction of the fossil depositional environments that existed in the Late Cenozoic, e.g. in the barnacle-rich Titiokura Limestone of the eastern North Island Te Aute Limestone Complex. The Titiokura Limestone (Beu 1995), outcropping in the northwest of Hawke's Bay, is characterised by a mixed assemblage of barnacles, including *Pachylasma* sp., *Notomegabalanus miodecorus*, and the inferred intertidal taxon *Epopella* cf. *plicata*. The depositional environment at that time is, however, considered to have been at more than 100 metres depth. The geological processes operating at the time resulted in the build-up of shallow-water sediments on the upper shelf to a point at which the accumulation became unstable. Sediments and faunas were then mobilised, to be transported and deposited alongside deeper-water elements as a mixed thanatocoenosis (death assemblage).

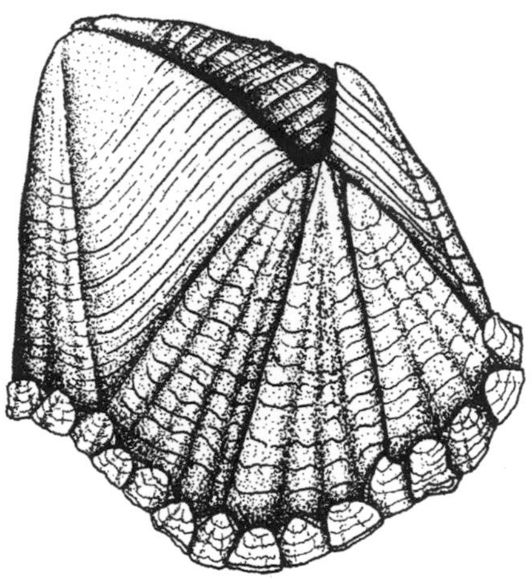

Waikalasma juneae (Miocene).
From Buckeridge 1983

The sessile Balanomorpha are not known from strata older than the Paleocene, with the first of these, *Bathylasma rangatira* and *Pachylasma veteranum*, being recorded from the Chatham Islands (Buckeridge 1983). There has been considerable conjecture concerning the origins of the balanomorphs, which diversified and spread very rapidly in the Early Cenozoic. Buckeridge (1996a, 1999a) proposed that the Chatham Islands was a centre of sessilian diversification during the Paleogene, with taxa evolving in the warm shallow seas that characterised the environmental conditions for strata like the Red Bluff Tuff. New Zealand has a remarkable fossil cirripede fauna, with the phylogenetically early taxa *Eolasma, Chionelasmus, Waikalasma, Pachylasma, Bathylasma, Tetraclitella, Palaeobalanus, Notobalanus, Chamaesipho,* and *Notomegabalanus* having their earliest records here.

As with the Recent fauna, there are a number of publications describing single new species of New Zealand fossil Thoracica. These are listed in the historical review provided in Buckeridge (1983), which also revised and improved current knowledge of the New Zealand and Australian fossil cirripede faunas. Buckeridge listed 69 fossil taxa from New Zealand, of which 36 were new. Of these, 94% (i.e. all but two) are endemic to New Zealand. Since 1983, Buckeridge has described a further six species of fossil cirripedes (Buckeridge 1984a,b, 1991, 1999a, 2008), and in addition has a further four new taxa awaiting formal description.

Economic aspects of barnacles

Marine fouling

The first 'close encounter' some New Zealanders may have with barnacles is when they need to remove fouling organisms from the hulls of their recreational or fishing vessels. Barnacles are opportunistic organisms that colonise almost any available surface in the marine environment. Boats and ships provide excellent surfaces for suspension-feeders – a platform within the upper subtidal zone that generally coincides with oxygenated, predator-poor, plankton-rich waters. In addition, the mobile substratum facilitates dispersal.

Exotic fouling species in the New Zealand environment are generally introduced through commercial shipping. It is in this way that the widespread species *Amphibalanus amphitrite, A. variegatus,* and *Lepas anatifera* were introduced many decades ago. *Lepas anserifera, Fistulobalanus albicostatus, Amphibalanus reticulatus, Megabalanus rosa, M. volcano,* and *Tetraclita squamosa japonica* were introduced on oil-drilling platforms (Foster & Willan 1979) but none appears to have become naturalised in New Zealand waters. Hosie and Ahyong (2008) reported the establishment of the Australian species *Austromegabalanus nigrescens* and its South American congener *A. psittacus* at Taharoa and Wellington respectively.

Research into the development of antifouling systems has intensified as a result of a greater understanding of the deleterious ecological impact of traditional antifouling paints such as tributyltin (Buckeridge 1998). Preliminary results indicate that low-level ultrasonic transmitters have the potential to restrict organic accumulation on certain hulls.

Barnacles as a food source

Although balanomorph barnacles such as the very large South American *Austromegabalanus psittacus* are considered a delicacy, they do not occupy a similar place in modern New Zealand cuisine. There is evidence, however, that barnacles were once eaten by Maori, as they are often found in middens (Foster 1986). In most cases, it appears that this was not through deliberate harvesting; rather it was incidental to the harvesting of other seafood such as *Perna canaliculus* (green-lipped mussel). This is no doubt a reflection of the small size of most shallow-water New Zealand barnacles – many hundreds of *Austrominius modestus* would need to be collected to make even a small meal. Nevertheless, somewhat larger species such as *Notomegabalanus decorus* and *Epopella plicata* may occasionally have been deliberately collected as a dietary supplement (Foster 1986).

Environmental monitoring

Thoracican barnacles have a number of properties that may prove to be invaluable to humans. One that is currently under development is their use as environmental indicators. Common shallow-water fouling species such as *Austrominius modestus* and *Epopella plicata* are invaluable in monitoring environmental changes to marine systems during urbanisation (e.g. at Auckland's Long Bay–Okura Marine Reserve). A high metabolic rate, rapid onset of maturity, and frequent spawning make *Austrominius modestus* an excellent species for gauging the impact of human activities.

Biotechnology

Another feature of thoracican barnacles that has intrigued scientists is the means by which they attach themselves to surfaces. Barnacles are known to grow on a very wide range of materials, both natural and synthetic. Their ability to successfully adhere to flexible and elastic materials like plastic sheeting and fibreglass is of specific interest, for if the nature of this 'organic adhesive' is determined and commercially manufactured, it will have obvious use in fields such as dentistry.

Barnacles that are commensal or symbiotic with other marine organisms may need to produce chemicals to prevent the host overgrowing them. This is particularly the case with sponge-inhabiting taxa like *Acasta* and coral-inhabiting taxa like *Brochiverruca*. Isolation of chemical deterrents may be invaluable in the design of new drugs for restricting or reducing cell growth in other species, including humans.

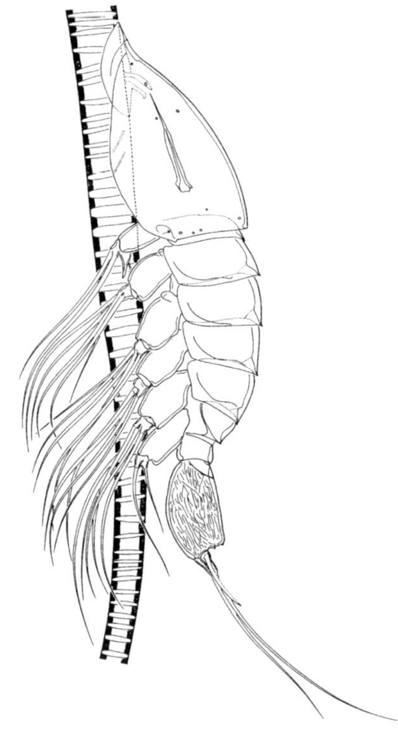

Tantulus larva of *Deoterthron dentatum* attached to an antenna seta of its ostracod host.
From Huys 1990

Subclass Tantulocarida: Tantulocarids

Nearly 30 years ago, a new maxillopodan subclass was created by Boxshall and Lincoln (1983) to accommodate, amongst others, three tiny parasitic crustaceans discovered in the New Zealand region (Bradford & Hewitt 1980; Boxshall & Lincoln 1983; Lincoln & Boxshall 1983). They infect benthic and hyperbenthic crustaceans such as amphipods. Tantulocarids are minute ectoparasites, not exceeding half a millimetre (0.04–0.40 millimetre) in length, with a unique dual life cycle that is completed, without moulting, on a crustacean host (Huys et al.

1993). There are now five recognised families with more than 20 genera and about 30 species worldwide (Ohtsuka & Boxshall 1998), notably with several taxa being recently documented from Japan (Huys et al. 1992; Huys et al. 1994; Ohtsuka & Boxshall 1998).

While there have been no further records of tantulocarids from New Zealand, it is very likely that more species of this subclass will be discovered as the benthic and benthopelagic fauna of the New Zealand region becomes better studied.

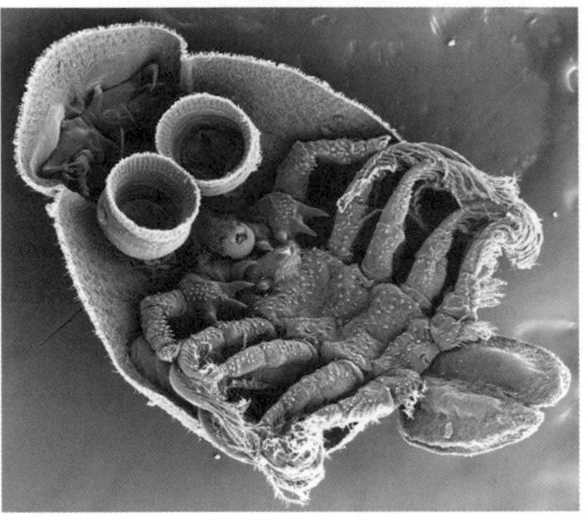

Argulus japonicus.
Note the paired suckers.
Kenneth M. Bart

Subclass Branchiura: Branchiurans

Branchiurans are parasitic on marine and freshwater fishes. They resemble copepods in many respects but differ in some important features. Unlike copepods, they have compound eyes and lateral head lobes, the opening of the genital ducts lies between the fourth pair of thoracic limbs, and they have a proximal extension to some of the exopodites (outer branch) of the thoracic limbs. They are good swimmers and females deposit their eggs on stones and other objects. The larvae differ little from the adult. *Argulus* has a pair of suckers on the maxillae and a poison spine in front of the proboscis. One introduced species has been recorded from goldfish in New Zealand (Hine et al. 2000). It is likely that more species will be discovered.

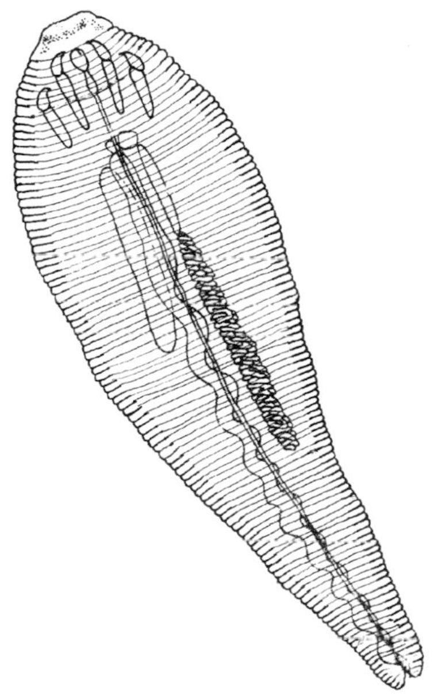

Tongue worm *Linguatula serrata.*
Composite from various sources

Subclass Pentastomida: Tongue worms

Tongue worms are obligatory parasites of reptiles, mammals, and birds, inhabiting their respiratory tracts (nasal passages and lungs). Particularly prevalent in the tropics, there are no native species in New Zealand, but one introduced species has been reported (Tenquist & Charleston 2001). This is *Linguatula serrata*, whose most regular host is the dog. It is rare in New Zealand, but developmental stages have also been reported from the brown hare, European rabbit, house cat, and sheep (Thomson 1922; Gurr 1953; Sweatman 1962).

Globally, there are about 130 species, ranging in length from about 3 to 150 millimetres or more and generally transparent or yellow to red-coloured. Like most parasites, their body form is simple and wormlike. Blood is their only food. The jawless mouth (sometimes protruding) and two pairs of lobe-like appendages with claws give the appearance of five orifices, hence, *penta-* (five) *stomida* (mouths). Long treated as a separate phylum of invertebrates, tongue worms are now regarded as highly modified crustaceans, based on sperm and larval morphology, the nervous system, and DNA studies. Some very convincing fossils of apparent larval pentastomids from the Late Cambrian give no evidence of a crustacean relationship, leading Maas and Waloszek (2001) to question it. On the other hand, recent mitochondrial DNA sequencing supports the evidence from sperm that pentastomids are most closely related to the Branchiura (Lavrov *et al.* 2004).

Subclass Copepoda: Copepods

Copepoda (oar-footed bugs) are small crustaceans that are common in aquatic and semi-aquatic environments, both marine and freshwater. Zoogeographical data indicate that copepods are ancient arthropods (Dussart & Defaye 1995) and fossils are known from the lower Cretaceous (Huys & Boxshall 1991). They have undergone extensive adaptive radiation and include a wide variety of open-water, bottom-dwelling, herbivorous, predatory, and parasitic forms. Copepods can often be extremely abundant and have been estimated to be among the most numerous animals on earth, mostly because of their dominance in the plankton of oceans and lakes. There are a number of excellent accounts that give general information on copepods. The comprehensive monograph by Huys and Boxshall (1991) deals especially with morphology and evolution, while Williamson

(1991) and Dussart and Defaye (1995) concentrate on the structure, function, and taxonomy of freshwater species. Coull and Hicks (1983) and Mauchline (1998) provide detailed information on the biology of harpacticoid and calanoid copepods, respectively, especially the marine species. These references are the main sources of the following notes.

The name 'Copepoda' is derived from two Greek words (*kope*, oar, and *podos*, foot), hence oar-footed. Copepods are typically small, mostly in the range 0.5–5.0 millimetres. Free-swimming forms may achieve a mimimum size of only 0.2 millimetres (some *Oncaea*) or a remarkable 18 millimetres (a *Valdiviella* species), but some parasites are even larger. The body is usually approximately cylindrical and segmented, and divided into three parts—cephalosome, metasome, and urosome (equivalent to head, trunk, and abdomen). There are 10 pairs of appendages on both the cephalosome and metasome, used for both feeding and locomotion (some of these appendages also have a sensory function), and the urosome ends in two bristle-bearing caudal rami. Uniquely among crustaceans, copepods have a flat plate that connects the basal segments of each pair of swimming legs. This plate is probably why copepods can have a rapid jumping mode of movement. In all copepods the first thoracic segment (bearing the maxillipeds) is incorporated in the cephalosome, unlike other maxillopodans.

The presence of a uniramous (unbranched) antennule is also a fairly reliable copepod characteristic. In male copepods the first antennae can be typically geniculate (with a prominent elbow), and are used to grasp the female during mating. The antennae, mandibles, maxillules, maxillae and maxillipeds are used in feeding. A wide variety of food types are utilised, including detritus, bacteria, algae, rotifers, nematodes, naidid oligochaete worms, crustaceans, and larval fish, and the structure of the feeding appendages varies in association with diet. The mechanics of feeding are complex, although copepods are probably fundamentally raptorial and use their mouthparts to grasp food particles. Many species, however, especially calanoids, are suspension-feeders and use the mouthparts to create water currents that bring food particles towards the copepod. Smaller particles are then captured passively and directed towards the mouth by bristles on the maxillipeds, maxillae, and maxillules, while larger particles are individually grasped by 'fling and clap' movements of the maxillae that grasp both the particle and a packet of water surrounding it and remove the water by an inward squeeze.

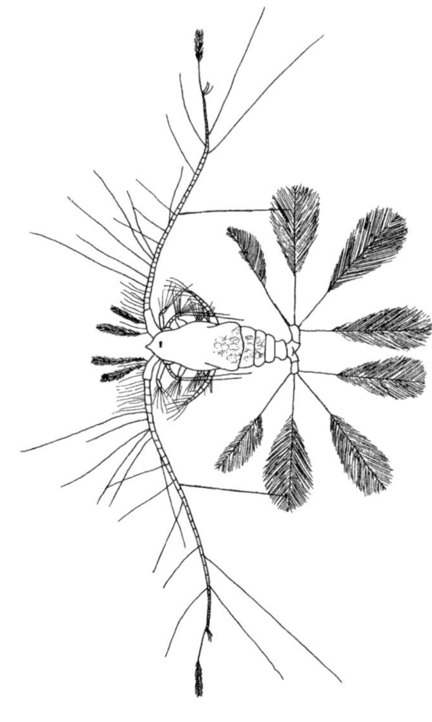

Calocalanus pavo.
After Giesbrecht 1893

Reproduction is usually sexual, and sperm are transferred from male to female in a sac-like spermatophore (a few harpacticoids can reproduce parthenogenetically). Egg sacs are probably not an ancestral condition of Copepoda as many groups lack true egg sacs. Nevertheless, in many copepods the eggs are carried in one or two egg masses, sacs, or strings until hatching. Under favourable conditions, multiple clutches of eggs can be produced, at intervals of a few days or weeks, so that each female may produce tens to hundreds of eggs in a lifetime. The egg hatches into a nauplius larva and the life-cycle typically includes six naupliar stages and six copepodite stages, the last of which is the adult stage. There is a marked metamorphosis between the last nauplius and the first copepodite stage. Development may sometimes be abbreviated, especially in parasites. Copepods are relatively long-lived compared to other microcrustaceans. Development times from egg to adult are typically in the order of 1–6 weeks, but may take several months, and the lifespan of adults may be from one to several months. Developmental times are markedly affected by temperature and food levels. Some copepods have resting stages that enable avoidance of detrimental environmental conditions and dispersal. Calanoids and harpacticoids produce resting eggs that have a thick shell and which can survive extended periods of dormancy and dryness. In cyclopoids and some harpacticoids, copepodites may enter diapause and encyst in bottom sediments.

There are 11 orders, approximately 213 families, 1763 genera, and 11,956 species worldwide (Humes 1994; Ho 2003). The Harpacticoida alone comprises

54 families, about 599 genera, and about 4400 species (J. Wells, unpublished data updating Wells 2007). The Calanoida has 42 families with about 2000 species (Boltovskoy et al. 1999); in the Poecilostomatoida there are 55 families, 359 genera, and about 1770 species (Ho 2003); and in the Siphonostomatoida there are 45 families, 377 genera, and about 1840 species (Ho 2003). The known New Zealand copepod fauna comprises 698 species, of which the Calanoida is the best known with 261 species, nine of which are undescribed. There are only 230 species of Harpacticoida, with about 99 of them undescribed; the remaining orders are also very poorly known.

Copepods live in a remarkable number of environments. These include not only marine and freshwater planktonic realms but in or on aquatic sediments, in association with plants, forest litter, and damp moss, in subterranean habitats or anchialine (isolated-marine) caves, and deep-sea hydrothermal-vent settings, but also in association with other animals as commensals or parasites.

In the marine plankton, calanoid copepods ('insects' of the sea) are extremely abundant. Some typical New Zealand examples are *Acartia ensifera, Calanus australis, Centropages aucklandicus,* and *Paracalanus indicus.* They are adapted to swimming in the water column and are fine-particle feeders in near-surface waters, eating mainly phytoplankton and protozoans. Carnivorous or detritivorous forms occupy deeper water-layers down to the deepest trenches. In the water column we also find forms that are not strictly free-living but live associated in some way with surfaces – the sea floor, the underside of sea ice, or on other planktonic animals.

The freshwater plankton in New Zealand is dominated by calanoid copepods of the family Centropagidae, which are widespread and very abundant in lakes, ponds, and the lower reaches of larger rivers. Many of the species also occur in Australia, although there are at least three endemic species. *Calamoecia lucasi* and *Boeckella dilatata* are typical lake dwellers while *B. triarticulata* is found in ponds. As in marine habitats, the freshwater calanoids are suspension-feeders on algae and protozoans, although at least some of the boeckellids are also predatory on small zooplankters such as rotifers and nauplii. A few cyclopoid copepods also live in fresh water, although they are usually sparser than the calanoids. They are probably mostly omnivores, consuming both animals and algae. Some are found mainly in the bottom waters and are probably strays from the benthic and littoral areas.

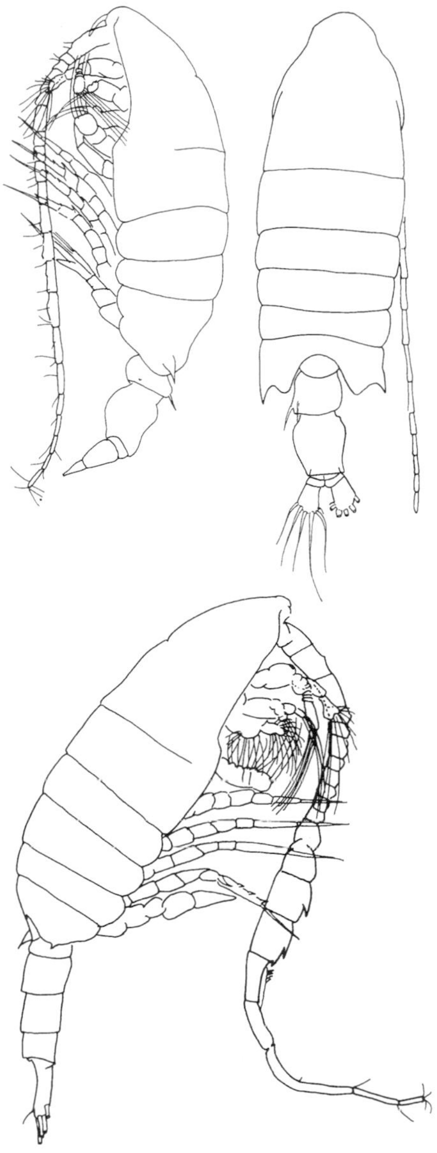

Centropages aucklandicus – female at top (left profile and dorsal views), male below, with modified antenna for copulation.

From Bradford-Grieve 1999

In aquatic sediments, copepods (mainly harpacticoids) live either permanently within the sediment or alternate between the sediment and its surface, browsing on the microflora associated with the sediment particles or with the accompanying detritus. In well-oxygenated coarse-grained sediments such as beach sand, specialised copepods (again, mainly harpacticoids) are part of the 'interstitial fauna' that lives within the interstices of this habitat. This habitat is commoner in marine sediments than in freshwater sediments, although it does exist in river systems and their ground waters where a strong intra-sediment water flow occurs. Most families of Harpacticoida have representatives in all of the above habitats, with specialisations for the interstitial habitat having evolved many times in different lineages. These trends exist among the New Zealand fauna to the same extent as they do elsewhere and are represented by numerous endemic and non-endemic species. An extremely important characteristic of this fauna is that, with very few exceptions, the entire life-cycle is benthic and the larvae are not dispersed large distances by water movements. This not only must affect their ecology but must also impact on population genetics and eventually on phylogeny. As a result we should expect a high level of endemism.

Many copepods are associates of plants. In the marine intertidal zone many harpacticoids live in association with seaweeds and sea grasses and are highly specialised for life on the surface of the fronds. Members of the Porcellidiidae, Peltidiidae, and Tegastidae, for example, are especially adapted to this environment; each family is well represented in New Zealand. In the littoral areas of freshwater lakes, ponds, and running waters, cyclopoids and harpacticoids are abundant on and amongst macrophytes. Damp terrestrial situations are exploited by cyclopoid and harpacticoid copepods. These include damp soil, forest litter, sphagnum bogs, liverwort and moss clumps, and the pools between the leaves of bromeliads. Only the harpacticoids from this cryptic fauna have been extensively studied in New Zealand, and in these the same trends exist as elsewhere in the world; most species belong to cosmopolitan genera in the predominantly freshwater family Canthocamptidae, and most are endemic.

Copepods live in groundwater and can be caught in springs, wells, and pools in caves. In New Zealand these habitats have not been extensively surveyed (Chapman & Lewis 1976) and nothing is known about the copepods except that parastenocaridids have not been found, despite extensive searching (Schminke 1981a). Overseas, the Parastenocarididae (Harpacticoida) is a large family of ca. 270 species (190 of them currently placed in the genus *Parastenocaris*) that mostly inhabit the interstices of groundwater. These habitats range from the water table beneath beaches and sand banks, including a few fully marine beaches, to brackish systems such as the Baltic Sea, and riverine and lacustrine inland systems, above and below ground.

Recently the study of deep-sea hydrothermal vents and marine caves has revealed many interesting copepods of great importance to the study of evolutionary relationships between the various groups of copepods, as they are amongst the most primitive forms. Because isolated marine caves are not yet known in New Zealand and the microscopic fauna of New Zealand hydrothermal vents has not yet been studied, these types of copepods have not been recorded here.

In thermal waters of the central North Island only one copepod, the endemic cyclopoid *Paracyclops waiariki*, is known. It is restricted to Lake Rotowhero, which has seasonal temperatures varying between 29.5° and 37.5° C and an average pH of 3.1.

Nearly half of all known copepod species live in symbiotic relationships with other organisms. It is evident that commensalism and parasitism have evolved independently several times in the class, even within an order. Copepods parasitise virtually every phylum of animals from sponges and cnidarians to vertebrates including mammals. They also have a range of associations from external and internal parasitism to varied forms of commensalism. For example, two species of endemic New Zealand harpacticoids are associated with macroinvertebrates – *Porcellidium tapui* on hermit crabs and *Alteuthoides kootare* on sponges. It is interesting to note that these genera are highly adapted for clinging to a substratum and are genuinely 'phytal' in this respect. This particular association with macroinvertebrates is almost certainly of the same type as with marine plants, i.e. using them as a substratum on which bacteria, fungi, and microalgae grow abundantly. Similarly, *Paramphiascopsis waihonu* is known only from a sample of spent elasmobranch embryo cases (taken at 1116 m), where many specimens occurred along with a gastropod mollusc; an association with the gastropod is unlikely and it is most probable that both are feeding on detritus and decay products within the case. *Paramphiascopsis* comprises several other species that have been taken in association with ascidians, polychaetes, gorgonians, and decapod crustaceans but many species are also known from algae and sediments.

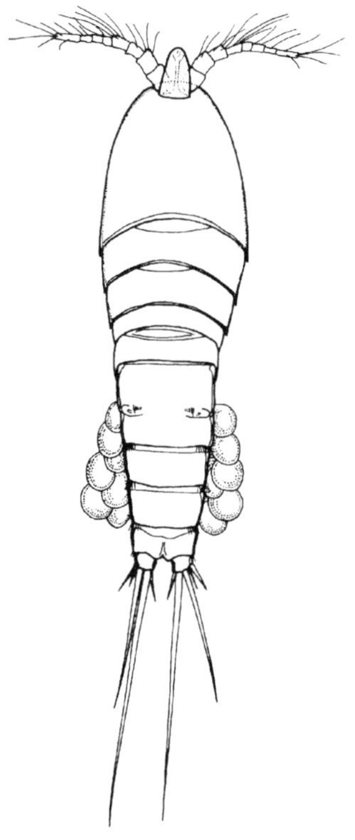

Paramphiascopsis waihonu.
From Hicks 1986

Harpacticoids are also found in burrows in wood inhabited by the gribble (*Limnoria* spp.), where the nature of the association is unclear (Hicks 1988a), with some authors arguing for an obligate commensal relationship and others believing the attraction for the copepod is the microhabitat created by the gribble. Evidence for the latter is the presence of copepods in decaying wood no longer occupied by *Limnoria*, but the fact remains that the copepod species have never been found in habitats that have not been associated with the gribble. Five species, of which four are endemic, occupy this habitat in New Zealand waters.

Importance of copepods

In both marine and fresh waters worldwide, abundant copepods form a vital link in the food web that leads from minute algal cells or phytoplankton and small protozoans (e.g. Chapman & Green 1987; Bradford-Grieve et al. 1998) to the largest fishes, and some whales in the oceans. Many commercial and non-commercial marine fish (and some crustaceans) are utterly dependent on copepods as a food source during a portion of their larval life. For example, in New Zealand it has been shown that the larvae of hoki (*Macruronus novaezelandiae*), which forms the basis of the largest New Zealand fishery, feed on copepod adults (e.g. *Calocalanus*) and copepodites almost exclusively (Murdoch 1990). With their large mouth size, hoki larvae actively select copepods such as *Calocalanus* and *Paracalanus* (Murdoch & Quigley 1994). For inshore benthos and for migratory fish, estuaries and lagoons are typically the critical location for this life-history phase. In a New Zealand estuary, *Parastenhelia megarostrum* is a principal prey item for young post-metamorphic flatfish during the first six months of their lives (Hicks 1984). The very smallest fish feed on the naupliar stages while larger specimens have an increasing proportion of older copepods in their guts. In lakes, copepods are an important part of the diet of smelt (e.g. Stephens 1984, Chapman & Green 1987), which in turn form a major part of the diet of rainbow trout. Copepods can be so abundant that their faecal pellets, produced at a rate of several per hour, are an important source of food for detritus feeders. Copepod grazing can significantly reduce the densities of at least some algal species (e.g. Edgar & Green 1994) and it has been suggested that they may have potential in the biomanipulation of the effects of eutrophication in lakes (Edgar 1993). Copepods are increasingly being used as test organisms in ecotoxicological testing. In New Zealand, the freshwater species *Calamoecia lucasi, Boeckella delicata,* and *Mesocyclops* sp. have been shown to be very sensitive to pentachlorophenol (Willis 1998) and the latter two species have been recommended as suitable candidates for the development of routine testing protocols involving acute and chronic endpoints (Willis 1999).

Copepods can be important economic pests when they parasitise commercial species. This is especially the case overseas, where ectoparasitic copepods of the families Ergasilidae and Caligidae ('sealice') infect salmonids reared in sea cages, causing damage and sometimes death of valuable aquacultured product reared in marine areas (Johnson et al. 1997). In New Zealand, copepod 'sealice' are not yet a problem in salmon culture (Hine & Jones 1994) but the causative copepod genera are present in the farms (Jones 1988a). Copepods of the family Sphyriidae are also of economic importance in that the anterior portion of the copepod is buried in the musculature of the host fish, while the posterior portion bearing egg strings trails from a hole in the skin. Skinning machines do not remove the 'head' from the fillet causing wastage and customer complaints.

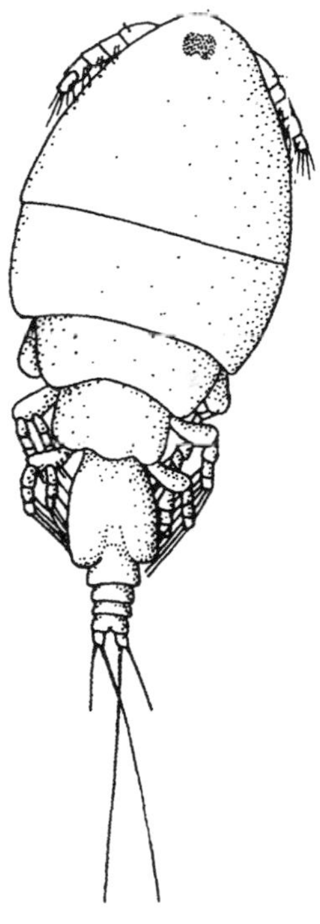

Abergasilus amplexus.
From Jones 1981.

In freshwaters, the ergasilid *Abergasilus amplexus* infests a wide variety of fish including longfinned and shortfinned eels, smelt, inanga, goldfish, and perch (e.g. Jones 1981). Two other parasitic copepods, *Thersitina inopinata* and *Paeonodes nemaformis*, are rather enigmatic (McDowall 1990). *Thersitina inopinata* is known only from its free-swimming males, while *P. nemaformis,* although endemic, is known to parasitise only introduced brown trout and salmon. The exotic copepod *Lernaea cyprinacea* has been recorded from introduced goldfish. Free-living copepods are also known to be intermediate hosts in the life-cycles of tapeworms of freshwater fish. The initial stages of *Amurotaenia decidua,* which parasitises bullies, occur in *Macrocyclops albidus* (Weekes 1986) and planktonic copepods are secondary hosts in the life-cycle of *Ligula intestinalis,* the pleurocercoid of which infests both rainbow trout and bullies (Weekes & Penlington 1986).

Copepods can be disease vectors for human parasites in tropical climates. But conversely they can also carry the fungi or sporozoans that parasitise malarial mosquitoes. Copepods have been implicated in the spread of viruses through fish populations (Mulcahy et al. 1990). Freshwater copepods of the genera *Mesocyclops* and *Macrocyclops* have been used for control of the container-breeding mosquito species of *Aedes, Anopheles,* and *Culex*. So far, no examples of these kinds of relationships have been noted in New Zealand.

Zoogeography of the New Zealand copepod fauna

Marine plankton

Very few marine planktonic copepods are endemic to the New Zealand region. The distribution of pelagic Copepoda (Bradford & Jillett 1980; Bradford *et al.* 1983; Bradford-Grieve 1994, 1999a) in the region appears to be maintained by a combination of factors probably related to their occurrence in water masses in some way or other. The physiological requirements of a species (temperature tolerances, ability to breed in differing temperature regimes, nutritional requirements for growth and breeding) and their behaviour (vertical migration in relation to particular water masses or physical-oceanographic phenomena) all contribute to the patterns we observe. An additional factor (plate tectonics) was probably important in the occurrence of some neritic plankton species in the New Zealand region.

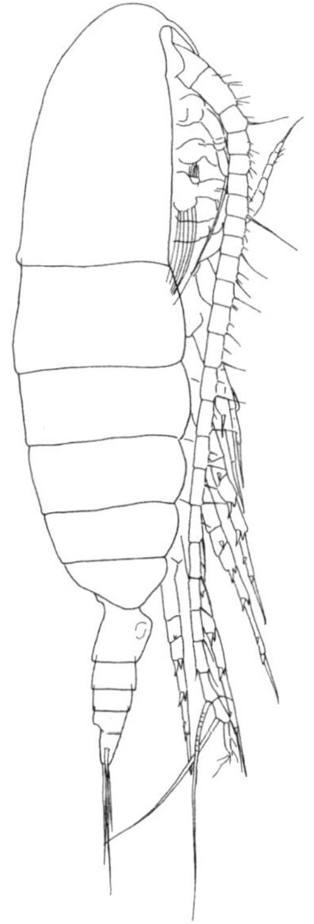

Calanus australis (female)
From Bradford-Grieve 1994

Some species have a clearly coastal distribution. Among the New Zealand epipelagic calanoids, only species of Acartiidae, Calanidae, Centropagidae, Clausocalanidae, Paracalanidae, Pontellidae, and Temoridae contain coastal forms that are rarely encountered in oceanic waters. Endemic coastal species such as the calanoids *Acartia ensifera, A. jilletti, A. simplex,* and *Centropages aucklandicus* and the poecilostomatoid *Corycaeus aucklandicus* are confined to New Zealand waters, whereas *Gladioferens pectinatus, Labodocera cervi,* and *Sulcanus conflictus* are confined to Australia and New Zealand. *Calanus australis* is found in at least New Zealand and southeastern Australian coastal waters, where it is essentially restricted to the mid-shelf (Bradford 1985). It seems possible that many of these species had common ancestors with close relatives in other temperate neritic parts of the world as far back as the Oligocene, when equatorial sea temperatures were low (Bradford 1979). *Paracalanus indicus* is restricted to coastal waters, with maximum concentrations occurring close to shore (Bradford 1985), although this species possibly has a broad tropical/subtropical distribution. *Clausocalanus jobei* and *Temora turbinata* also have a tropical/subtropical distribution whereas *Drepanopus pectinatus* has a coastal distribution around subantarctic islands.

Relationships to water masses are most clearly seen among oceanic epipelagic species. Nevertheless, in the New Zealand region some oceanic species are capable of responding rapidly to the heightened productivity of coastal waters and may attain maximum numbers close to the coast, obscuring their oceanic affinities. Examples of this type of distribution are seen in the calanoids *Nannocalanus minor* and *Clausocalanus ingens* and the cyclopoid *Oithona similis*.

Warm-water (tropical) oceanic epipelagic species usually have a cosmopolitan distribution if they are able to breed at a range of latitudes extending to 40° S, whereas those with breeding ranges restricted to lower latitudes (e.g. *Euchaeta rimana*) are not circumglobal in their distribution because of the geographical barriers (South America and Africa) presented to their distribution. In tropical or subtropical waters, epipelagic calanoid species with distributions extending to 40° S and sometimes as far as the Subtropical Front are *Aetideus giesbrechti*, many *Calocalanus* species, *Clausocalanus arcuicornis, C. lividus, C. parapergens, C. paululus, C. pergens, Eucalanus hyalinus, Mecynocera clausi, Nannocalanus minor, Neocalanus gracilis, Pareucalanus sewelli, Pareuchaeta acuta, P. media, Rhincalanus nasutus,* and *Subeucalanus crassus*. Species with a warm-temperature (transition zone) Southern Hemisphere distribution include *Aetideus pseudarmatus, Clausocalanus ingens, Pareucalanus langae,* and possibly *Neocalanus tonsus* and *Calanoides macrocarinatus*. Species with subantarctic distributions include *Cala-*

nus simillimus, Clausocalanus brevipes, Neocalanus tonsus, and *Subeucalanus longiceps*. Species with Antarctic–subantarctic distributions include *Aetideus australis, Clausocalanus laticeps*, and *Rhincalanus gigas*.

Marine sediments

Throughout the world the copepod fauna of marine sediments (predominantly harpacticoids) is well known only for the intertidal and shallow sea areas. Detailed data are available for only a few sites of more than a few metres in depth, mostly in Europe, although scattered information is known for all depths down to almost the bottom of the deepest trenches. Even for intertidal and sublittoral areas, most of the world outside Atlantic Europe, the western Mediterranean, and a few locations on the eastern coast of the Americas is poorly known or even totally unknown. A reasonably comprehensive survey of the North and South Islands of New Zealand has been carried out, but the results have yet to be fully published and many species remain unnamed. Furthermore, assessment of the zoogeographic relationships of the New Zealand fauna is made impossible by the almost complete absence of information from Australia and New Caledonia. All that can be said at this time is that it seems unlikely that New Zealand will harbour many endemic genera (though that will depend on the attitude of future taxonomists towards taxon definitions).

Freshwater plankton

In New Zealand, most freshwater calanoids (eight species of *Boeckella* and one of *Calamoecia*) belong to the family Centropagidae, the non-marine members of which are mainly confined to Australasia, the subantarctic, the Antarctic Peninsula, and parts of South America (Bayly 1992). Only three of these species are found only in New Zealand (Jamieson 1998); the others also occur in Australia. A further four species are considered to be resident natives (*Boeckella dilatata, B. propinqua, B. triarticulata*, and *Calamoecia lucasi*) whereas *B. minuta* and *B. symmetrica* may have invaded New Zealand since European colonisation (Banks & Duggan 2009). Recently, the diaptomid cross-hemisphere invaders *Skistodiaptomus pallidus* and *Sinodiaptomus valkanovi* have been recorded in constructed water bodies (Duggan et al. 2006; Banks & Duggan 2009; Makino et al. 2009).

Bayly (1995 and references therein) concluded that the present day distribution of freshwater and brackish Centropagidae can be interpreted as being a result of the colonisation of southern-hemisphere inland waters from marine and then brackish-water ancestors at a time when Australia, New Zealand, and South America were still linked to Antarctica, and Africa, Madagascar, and India had already drifted northwards. The absence of the Diaptomidae from New Zealand, most of Australia, and all of Antarctica also appears to be related to the timing of the separation of these landmasses from Pangaea in relation to the evolution of this family.

The distribution of calanoids in the major lakes is probably well known (Chapman & Green 1987; Jamieson 1988, 1998; Bayly 1992; Banks & Duggan 2009) but has yet to be fully examined in smaller habitats, especially ephemeral pools and the less-accessible high-country tarns. Most species show relatively clear habitat segregation. *Calamoecia lucasi* is widespread in northern, central, and western parts of the North Island, where it is found in streams, ponds, and large rivers. It also lives in a few small lakes in Northern Nelson. *Calamoecia ampulla*, a widespread species in Australia, is known only from one unverified South Island record (Bayly pers. comm.). Of the *Boeckella* species, *B. minuta, B. symmetrica*, and *B. tanea* have restricted distributions in the North Island. *Boeckella tanea* is found only in Northland, *B. symmetrica* in a pond near Auckland, and *B. minuta* in the Waikato River hydroelectric reservoirs and water-supply reservoirs in Wellington. It has been suggested that *B. symmetrica* and *B. minuta* may be

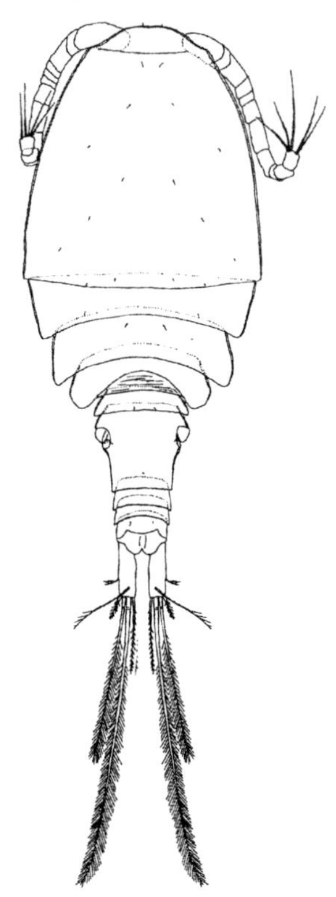

Abdiacyclops cirratus, an endemic cyclopoid genus andspecies from a subterranean well in Canterbury.

From Karanovic 2005

recent immigrants from Australia (Chapman & Green 1987) and this may apply to *C. ampulla* too. *Boeckella propinqua* occurs mainly in central and northern areas of the North Island but, like *C. lucasi*, its distribution also extends to the tip of the South Island. *Boeckella hamata* occurs throughout the southeastern part of the North Island, the eastern part of the South Island, and southern Westland, mainly in reservoirs and coastal lakes. *Boeckella triarticulata* has a similar distribution but apparently does not co-occur with *B. hamata*. It is found mainly in ponds and reservoirs in eastern parts of the South Island from Canterbury to Otago, with one record from Hawke's Bay in the North Island. *Boeckella delicata* has a disjunct distribution, occurring in Northland and the Waikato region of the North Island and also on the west coast of the South Island. *Boeckella dilatata* occurs only in the South Island, mainly in glacial lakes and in associated reservoirs. It also has a disjunct distribution and is found only in northern and southern areas of this island. Unlike the usual situation elsewhere in the world, co-occurrences of two or more species of calanoids in one lake are rare, and most lakes have only one calanoid. In the North Island, there are a few co-occurrences of *C. lucasi* and *B. delicata*, *C. lucasi* and *B. propinqua*, and *C. lucasi* and *B. minuta*, and in the South Island *B. triarticulata* and *B. dilatata*, *B. triarticulata* and *B. hamata*, and *C. lucasi* and *B. propinqua* in a few habitats (Chapman & Green 1987; Jamieson 1998; Banks & Duggan 2009).

Various attempts have been made to explain the distributional patterns of the New Zealand freshwater calanoids (summarised by Jamieson 1998) and, until recently, most of these used dispersalist biogeographical ideas. Banks and Duggan (2009) have highlighted the role of constructed lakes and ponds in facilitating inter-and intracontinental invasions of calanoid species. Maly (1984) suggested that distributions resulted from probabilities of immigration and extinction that were assessed from clutch sizes and the likelihood of predation by fish. Maly (1991) modified these ideas to include the number of existing populations and concluded that dispersal was probably not important over long distances but may be important at local scales. Jamieson (1988) explained the distribution of *Boeckella dilatata*, *B. hamata*, and *B. triarticulata* by relating differences in their ecological requirements and dispersal abilities to vicariant events. More recently, Jamieson (1998) has provided a convincing explanation for the distribution of these three species and *B. delicata* based on panbiogeographic methods. She showed that their distributions are correlated with the three principal pre-Late Cretaceous technostratigraphic terranes that, over the last 150–200 million years, have come together to make up New Zealand. *Boeckella dilatata* and *B. delicata* occur in lakes and ponds on the Tuhua and Caples Terranes and *B. hamata* and *B. triarticulata* on the Torlesse Terrane. The species overlap at the terrane margins. The present-day disjunct distributions of *B. dilatata* and *B. delicata* are thus thought to result from tracks arcing out to sea.

The species pairs on the different terrane groups are thought to differ in ecology; in particular *B. delicata* and *B. hamata* are suggested to have a higher salt tolerance than either *B. dilatata* or *B. triarticulata*, thus enabling sympatry. Localised dispersal presumably explains the overlap of species at the terrane margins. Jamieson's panbiogeographic approach would seem to have considerable potential for explaining distributions of the remaining calanoids. It is clear, however, that ecological information remains important for explaining distributions of sympatric species. Ecological studies of life-histories and food requirements have been made of some species (e.g. Green 1975; Forsyth & James 1984; Jamieson 1986; Chapman & Green 1987; Burns 1988; Jamieson & Burns 1988; Xu & Burns 1991; Burns & Xu 1990; Twombly et al. 1998; Couch et al. 1999), but much more remains to be done. The effects of post-European colonisation, with altered fish communities and changing trophic status of lakes, on distributional patterns are not known.

The cyclopoid copepod fauna is very poorly known taxonomically and ecologically. A few cyclopoids are found in the lake plankton, but their populations are usually either sparse or seasonal and little is known about them. There are no equivalents of the large-bodied *Cyclops* (in the strict sense) of many Northern Hemisphere lakes.

Mesocyclops leuckarti has been recorded from various North Island lakes (Green 1974, 1976; Jamieson 1977; Chapman & Green 1987; Greenwood et al. 1999), but it is likely that these records were not of the nominate species as *M. leuckarti* does not occur in the Southern Hemisphere (Kiefer 1981). Bayly (1995) has suggested that its correct identity is possibly *M. australiensis*. *Macrocyclops albidus* occurs in low numbers in the Rotorua and Taupo lakes (e.g. Chapman 1973; Forsyth & McCallum 1980), in the lakes of the Waitaki River system, and in other South Island lakes (Stout 1978; Burns & Mitchell 1980). *Eucyclops serrulatus* is found in the plankton of Lakes Hayes and Johnson (Burns & Mitchell 1980) and *Acanthocyclops robustus* in the plankton of Lake Mahinerangi (Mitchell 1975). It still can be concluded that, until a revision is made of the freshwater cyclopoids, no valid assessments of biogeographical relationships can be made. Nevertheless, Karanovic (2005) held it to be highly likely that the cosmopolitan cyclopoids *Acanthocyclops robustus, Diacyclops bisetosus, Eucyclops serrulatus,* and *Paracyclops fimbriatus* were accidentally introduced to New Zealand by early European settlers in barrels of fresh water. Jamieson (1980a, b) conducted experimental studies of predatory feeding and development rates of *Mesocyclops* sp.

Plant associates

In marine systems the term 'plant associates' means the fauna associated with macroalgae and sea grasses and is usually called the phytal habitat. In addition, a few species have been found associated only with decaying wood (from wharf piles to driftwood dredged from depths of 1100 metres). These perhaps should be included in the phytal fauna as it is most probable that the role of the living or dead plant is primarily as a substratum for the copepods' food supply, namely bacteria, fungi, and microalgae attached to the plant. However, in this regard the phytal fauna is little different from the true benthos, which relies on these food sources attached to particles of the sediment.

Most of the species do not show obvious morphological adaptations to the phytal habitat. In those that do, the adaptations are usually to enable the animal to attach itself more effectively to the plant. Very few species seem actually to damage the plant or to be directly feeding on its tissues. Many genera that contain species found among algae have other species living on or in the adjacent benthic sediment. Many species are found equally often among algae and in sediments without associated plant growth. Also, it is known that many of the species washed from samples of macroalgae and sea grasses are actually associated with the sediment and detritus that becomes trapped in the interstices of the plant and thus are really part of the sediment fauna. Even many of the truly phytal species that do show adaptations to that environment have been shown to leave the plant for mating; this may partially explain the relative rarity of males in collections of these species.

In the marine system, about 45% of the described phytal species are endemic. Only a few undescribed species currently exist in collections, which may partly be a consequence of inadequate collecting and cataloguing. Notwithstanding, the phytal fauna is quite well known ecologically (e.g. Hicks 1977, 1988b) and, while it is very probable that many species remain to be discovered, the main outlines of the fauna are well known. Unfortunately, the phytal fauna of adjacent marine regions is as poorly known as their sediment fauna and similar remarks about understanding zoogeographical relationships apply. The comments below on endemism in the sediment fauna apply equally to the phytal but the lack of regional collecting makes it futile to try to estimate the true level of endemism.

The situation in freshwater and terrestrial systems is much the same. Some copepods (cyclopoids and harpacticoids) probably use plants mainly as the substratum on which their food grows, but much less is known about

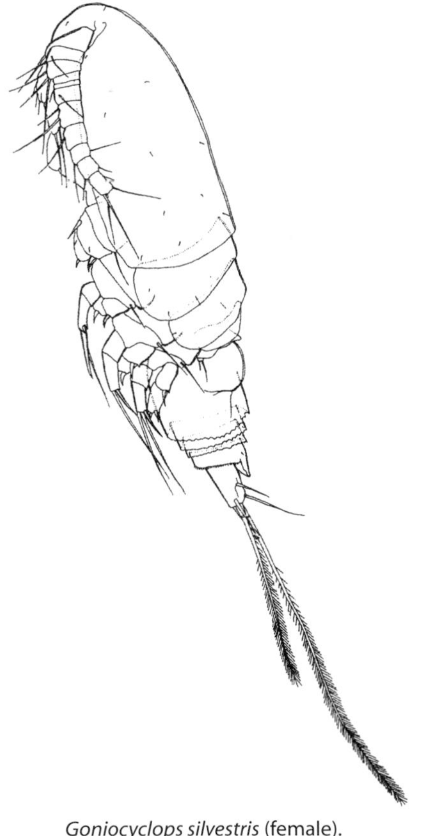

Goniocyclops silvestris (female).
From Karanovic 2005

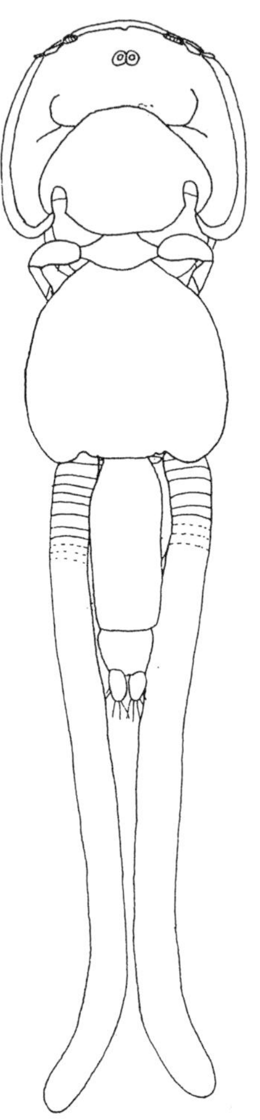

The fish parasite *Caligus pelamydis*, from barracouta.
From Hewitt 1963

their ecology. Certain copepods are found associated with aquatic vegetation in lakes and ponds, and with mosses (Harding 1958; Chapman & Lewis 1976). In semiterrestrial situations such as mossy banks and the edges of waterfalls or in damp forest litter and decaying wood, some copepods (such as *Goniocyclops silvestris* and a variety of harpacticoids) are found; most are apparently endemic but this fauna has still to be properly examined (Chapman & Lewis 1976).

Animal associates

It is difficult to make any definitive statement about the zoogeography of animal associates because the commensal and parasitic copepod fauna of marine invertebrates in New Zealand and neighbouring seas is very poorly known. For example, known New Zealand siphonostomatoid species diversity is only 29% of that in European seas, and even less for cyclopoids and harpacticoids, whereas, based on what is known for well-studied high-level Animalia taxa in both regions, New Zealand species diversity matches or exceeds that in European waters (Gordon et al. in press). The end-chapter checklist of New Zealand species in these copepod orders is annotated to indicate the type of relationship and host.

Species identifications of parasitic copepods from fishes of neighbouring seas are, in many cases, awaiting critical review. For example, *Trifur lotellae* in New Zealand would appear to be identical to *Trifur physiculi* from Australia. There are many other such examples. Also, the parasitic copepod fauna of marine invertebrates in New Zealand and neighbouring seas is almost totally unknown. Nevertheless, Jones (1988b) examined the then known parasitic copepod fauna and concluded that endemism on teleosts at the generic level was very low (2%) and there were no endemic genera on elasmobranchs (sharks).

The freshwater parasitic copepod fauna consists of only three species – *Abergasilus amplexus* and two very rare or extinct species, *Thersitina inopinata* and *Paeonodes nemaformis*. *Abergasilus* is an endemic estuarine genus common in, and known only from, Lake Ellesmere and the Chatham Islands lagoon. It has close affinities with South American genera. *Thersitina* has been found only once, in a plankton sample from Lake Poerua (Percival 1937). *Paeonodes nemaformis* has been found only twice, both times in South Westland on introduced salmonids (Hewitt 1969). The genus has also been found in Africa and is apparently closely related to *Mugilicola*, found in South Africa, India, and Australia (Boxshall 1986). The native hosts of *Thersitina* and *Paeonodes* are unknown, despite extensive searching. It is concluded that the parasitic copepod fauna of marine vertebrates is derived from the wandering of host fishes and reflects the strong links with Australia and the island chains to the north (Jones 1988a,b).

Endemism

One key element in the occurrence of endemism in New Zealand is the paleogeography of the region. The freshwater, brackish, and inshore copepod faunas illustrate the key elements of such reconstructions (Lewis 1984; Bayly 1995). The absence of the calanoid family Diaptomidae and presence of freshwater species of Centropagidae in Australia, New Zealand, South America, and Antarctica indicates that the period when these land masses were still linked but already separated from Africa, Madagascar, and India (120–80 million years ago) is crucial in reconstructing the evolution of *Boeckella*, *Calamoecia*, and *Gladioferens* in New Zealand and other southern hemisphere regions. These events, and the subsequent submergence of New Zealand in the Oligocene (35 million years ago) were probably responsible for speciation and the currently observed endemism (Bayly 1995).

The connection between New Zealand and Antarctica was broken during the Late Cretaceous. Three of eight New Zealand species of *Boeckella* are endemic to New Zealand (Maly & Bayly 1991) and it is likely that this genus inhabited the fresh waters of the ancestral landmass when it separated from Antarctica.

By the Late Oligocene, nearly all of the New Zealand landmass (possibly all of it according to Landis et al. 2008) was submerged. Significant extinctions will have occurred at this time, accounting for the relatively impoverished fauna of New Zealand compared with that of Tasmania. On the other hand, the multiple vicariant events associated with the production of a diminishing New Zealand archipelago in the Oligocene might have been expected to result in some speciation and the currently observed endemism if not all of the landmass was in fact submerged.

We predict that a higher degree of endemism than is currently recorded will be discovered amongst freshwater and benthic copepods when the less well-known groups are revised. But we need to introduce here a note of caution in this discussion of endemism. While the number of endemic species indeed reflects the evolutionary history of a particular fauna, in practice the number of such species recognised by past and present taxonomists depends on the interpretation of morphological variability within a species, especially where there is discontinuous distribution and not enough morphomolecular information for phylogenetic analysis.

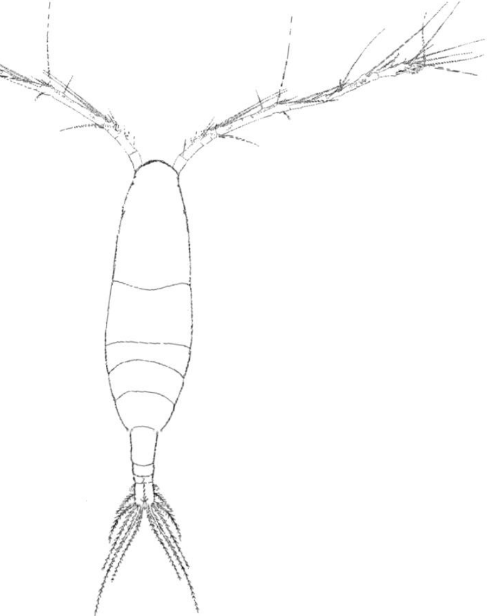
Acartia ensifera.
After Bradford-Grieve 1994

Marine plankton

Very few marine planktonic species are endemic to New Zealand. The main reason for this is that most species are oceanic and are relatively widespread in a global sense, ranging from circumglobal subantarctic and Indo-Pacific to distributions encompassing all the world's oceans. Only a few coastal calanoid or cyclopoid species are endemic to New Zealand waters (*Acartia ensifera, A. jilletti, A. simplex, Centropages aucklandicus,* and *Corycaeus aucklandicus*). The cyclopoid *Corycaeus aucklandicus* is endemic to coastal waters of northern New Zealand.

Freshwater plankton and benthos

Only three freshwater calanoid species are endemic – *Boeckella dilatata, B. hamata,* and *B. tanea*; the other seven species also occur in Australia. Only two (*Metacyclops monacanthus, Paracyclops waiariki*) of the 19 cyclopoid species are known to be endemic to New Zealand. All others are supposedly cosmopolitan or Australasian. Notably, several genera recorded from Australia, some with multiple species (*Apocyclops, Australocyclops, Ectocyclops, Mixocyclops, Neocyclops, Thermocyclops*), have not yet been recorded from New Zealand. Some studies (see Bayly 1995) have shown much greater degrees of differentiation and endemicity than previously recognised in microcrustaceans, and it is evident that more stringent resolution of morphotypic variation of the New Zealand freshwater cyclopoids is required before their status can be assessed. Presumed 'cosmopolitan' species may be so only because of widespread and indiscriminate misuse of authoritative (?northern hemisphere) taxonomic references. As noted earlier for *Mesocyclops leuckarti* (discovered to be a species complex by Kiefer (1981) and not represented by the nominate species in the Southern Hemisphere), comparable species groups may be found in other 'cosmopolitan' species. An on-going global revision of the Cyclopoida (e.g. Dussart & Defaye 1995; Einsle 1996) will help resolve some of the problems. This series should be consulted as a guide to the global literature on cyclopoid genera and families, and in particular for the accepted modern level of taxonomic discrimination.

Marine sediments

Approximately 50% of the described harpacticoid species are endemic, but at least three times as many species remain undescribed in collections, and it is reasonable to estimate that at least 75% of these will prove to be endemic new species. It would seem, therefore, that the rate of endemism in New Zealand is high compared, for example, to the British Isles (as an example of another island group of comparable size), where probably it is less than 10%. But this comparison is meaningless. The British fauna has been investigated for much longer and at much greater intensity. As a result, it is known to contain at least four times as many species. Further, and very importantly, the British Isles are close to the shores of northwestern Europe, where the fauna is also very well known and shares many species with Britain. New Zealand is distant from its nearest neighbours. This, and its geological history since separation from the rest of Gondwana, may well have increased the level of endemism, but the lack of data from Australia (where the fauna is very poorly known) undoubtedly inflates the current estimates.

The limited amount that is known about the benthopelagic calanoid fauna indicates that there may be some degree of endemism (e.g. Bradford 1969; Bradford-Grieve 1999b) in the New Zealand region. Nevertheless, in the deep sea the perception of endemicity may reflect the paucity of sampling of near-bottom faunas worldwide.

Cryptic habitats

Freshwater harpacticoids in New Zealand have been collected mainly from clumps of moss or liverworts or similar vegetation in streams, the littoral of ponds and lakes, or from wet banks close to water bodies and in damp forest in leaf litter. Of the 19 named species in the end-chapter checklist, 17 are endemic, but relatively little collecting has been carried out and large areas of the country remain unexplored. The total fauna is likely to be many times the recorded number of species, but it is probable that a very high level of endemism, and of localised distribution of species, may be found. It will be interesting to see if their distribution supports the panbiogeographic explanation for the distribution of freshwater planktonic Calanoida (Jamieson 1998). The presence of small cyclopoid species has also been noted, but only one has been identified to species and the true extent of this fauna cannot be estimated at this time (Chapman & Lewis 1976).

Gaps in taxonomic knowledge of copepods and scope for future research

Platycopioida

This order is not known in the New Zealand region. It is possible that platycopioids will be found when the benthopelagic realm is properly sampled, because they have been found in other temperate, shallow-water, near-bottom habitats. Other genera have been found in marine caves in Bermuda so their relatives might not be expected to occur in New Zealand.

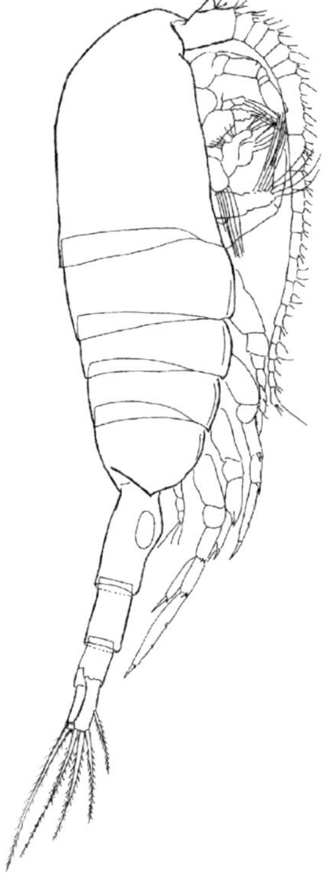

Metridia lucens (Calanoida).
From Bradford-Grieve 1999

Calanoida

The marine pelagic calanoid copepod fauna of New Zealand is fairly well known, mainly from the work of Janet Bradford-Grieve. The end-chapter crustacean species list incorporates results from Bradford and Jillett (1980), Bradford et al. (1983), and Bradford-Grieve (1994, 1999a,b). Their data are augmented by information in the revisions of the Aetideidae (Markhaseva 1996) and Euchaetidae (Park 1995). All these works incorporate other records of 19th- and 20th-century workers.

A number of calanoid families have not been recorded in the New Zealand region. This may partly reflect lack of extensive sampling. For example, the poor sampling of benthopelagic habitats at all depths is probably responsible for the absence of the Diaixidae, Discoidae, Hyperbionychidae, Mesaiokeratidae, Parkiidae, Pseudocyclopiidae, Ridgewayiidae, and Ryocalanidae, although it is likely that the New Zealand fauna does include some species from a number of these families. The apparent absence of isolated marine (anchialine) caves in New Zealand probably explains the absence of the Boholinidae, Epacteriscidae, and Fosshageniidae.

Species of Parapontellidae have been recorded only from the North Atlantic

Ocean and from deep waters of the Malay Archipelago, so this rare family may not occur in the New Zealand region.

Other families are absent from the New Zealand fauna for paleogeographic reasons. The Diaptomidae are known from fresh waters in most of the world apart from New Zealand, most of Australia, and all of Antarctica (Bayly 1995). Pseudodiaptomids are brackish to marine species, widespread in other parts of the world but present in the Australasian region only in northern Australia.

The taxonomy of the freshwater planktonic calanoids is reasonably well known (Chapman & Green 1987), although genetic studies using modern techniques are required to assess whether there has been cryptic speciation in any of the geographically widespread and disjunct species and in those shared with Australia (cf. Boileau 1991). Ecological studies are still in their infancy, and for all species much more needs to be known about autecology (e.g. growth and reproduction, feeding rates, behaviour, life-history strategies, population dynamics, etc.), and contributions to community and ecosystem dynamics (e.g. competitive interactions, predation effects, production rates, contribution to food chains, nutrient cycling, etc.).

Misophrioida

Members of this order have not been recorded from New Zealand. It is possible that they might be found when marine benthopelagic habitats are more extensively sampled.

Cyclopoida

This order now includes the Poecilostomatoida (Boxshall & Halsey 2004). Cyclopoids have been relatively little studied in New Zealand – knowledge of the marine, freshwater, and brackish non-parasitic Cyclopoida is very scattered and inadequate.

Oncaea media (Cyclopoida).
From Heron & Bradford-Grieve 1995

Early records of freshwater Cyclopoida were summarised by Hutton (1904) and amplified by Chapman and Lewis (1976). The synonymies and taxonomic arrangement given by Dussart and Defaye (1985) in their checklist of the world free-living Cyclopoida were taken into account in compiling the New Zealand list. In addition, the revision of the *Paracyclops fimbriatus* complex (Karaytug & Boxshall 1998) and the records of Roper et al. (1983) were noted. The commoner New Zealand taxa in ponds and lakes are known but both their generic and species status need re-examination in view of the recent taxonomic revisions of supposedly cosmopolitan genera (Morton 1985; Dussart & Defaye 1995). The underground and cryptic fauna is unknown taxonomically apart from *Goniocyclops silvestris* in forest litter (Harding 1958), and genera and species described by Karanovic (2005), but other undescribed species are known. Entries in the end-chapter checklist accompanied by a question mark are doubtful old records that need further investigation.

Checklists entries of the free-living marine planktonic families Oithonidae, Corycaeidae, and Sapphirinidae of the New Zealand region are based on the unpublished records of Janet Bradford-Grieve; the identities of the species need more detailed study. The species of Oncaeidae are known from the work of Heron and Bradford-Grieve (1995).

Another group of families comprises mainly marine parasites or associates of other animals. For example, *Hemicyclops* (a near relative has been discovered in New Zealand but is undescribed) has a typical cyclopoid body form and lives in loose associations with other marine organisms (e.g. polychaetes), sharing their burrows. There has been some work on fish parasites in New Zealand but the fauna is essentially unknown or undescribed – an extensive collection of *Sarcotaces* spp., made by Jones in the 1980s and 1990s from around New Zealand, remains in the Auckland Museum collection awaiting description.

The parasitic families Archinotodelphyidae, Chordeumiidae, Cucumari-colidae, Mantridae, Ozmanidae, and Thespesiopsyllidae and the marine benthic

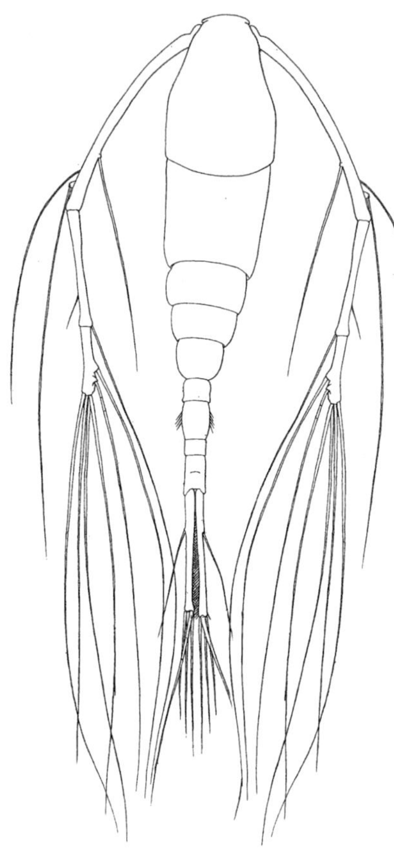

Mormonilla phasma (Mormonilloida).
After Giesbrecht 1893

family Cyclopinidae are not known from New Zealand. The freshwater parasitic family Lernaeidae is represented by only *Lernaea cyprinacea*, which was introduced with ornamental fish (Boustead 1982). The commensal Ascidicolidae and Notodelphyidae, living in association with tunicates, are known from only two collections (Schellenberg 1922a, b; Jones 1974, 1979). It is certain that many more cyclopoid associates of marine invertebrates remain to be found and described.

Data on the occurrence of commensal and parasitic forms have been collated here using the works of Thomson, Hewitt, Jones, Pilgrim, and Ho as described above. In general, we can say that the symbiotic copepods of New Zealand are very poorly known, particularly those occurring in association with marine invertebrates. Certainly, those parasitic on marine fishes are better known than those parasitic or commensal on/in other hosts, but we still cannot say that fish copepods are well known in New Zealand. There is currently nobody working on symbiotic copepods in New Zealand.

Gelyelloida

The two known species of this order are found in subterranean waters of France and the order is unlikely to be found in New Zealand.

Mormonilloida

This order contains only two species that are usually found at mesopelagic depths. *Mormonilla phasma* has been recorded off the east coast of northern New Zealand.

Harpacticoida

Early contributions to knowledge of New Zealand's fauna were made by Thomson (1878a,b, 1882), Brady (1899), Sars (1905), Brehm (1928, 1929), Farran (1929), Lang (1934), and Harding (1958). More recent additions to the fauna have been made by Barclay (1969), Hicks (1971, 1976, 1986, 1988a,c), Lewis (1972a,b; 1984), Wells et al. (1982), Hicks and Webber (1983), and a number of other authors. Hicks has also contributed a body of ecological and biological information on the phytal harpacticoid fauna. Included herein are unpublished records of freshwater species from Dr Maureen Lewis, and marine species from Drs John Wells and Geoff Hicks. When the presently undescribed species in existing collections are worked up, our knowledge of the sediment-dwelling harpacticoids of seashores will be reasonably good, but much work still needs to be done on the marine phytal fauna (mainly nationwide collecting to establish distributional patterns). As is common worldwide, there is very little knowledge of the sediment or phytal faunas of the sublittoral and deeper.

Lack of extensive exploration may be responsible for the absence of some families. It is highly probable that Argestidae, Cerviniinae (Aegisthidae), Cletopsyllidae, and Nannopodidae will be found in shelf and deep-water sediments and Longipediidae and Metidae associated with seashore plants and algae. On the other hand, the absence of the Parastenocarididae may be for geological reasons.

Only a fraction of New Zealand's freshwater and damp terrestrial locations has been surveyed. It is to be expected that the number of species in the fauna will be at least tripled, and New Zealand's geological history makes it likely that a number of intriguing questions of zoogeography and phylogeny will arise as a result. The harpacticoid fauna of New Zealand's ground waters is completely unknown, yet cave systems exist that are comparable to the species-rich karst formations of Europe.

Of particular note is the paucity of information on the fauna of the far offshore islands from the Kermadecs to the Chathams and subantarctic islands.

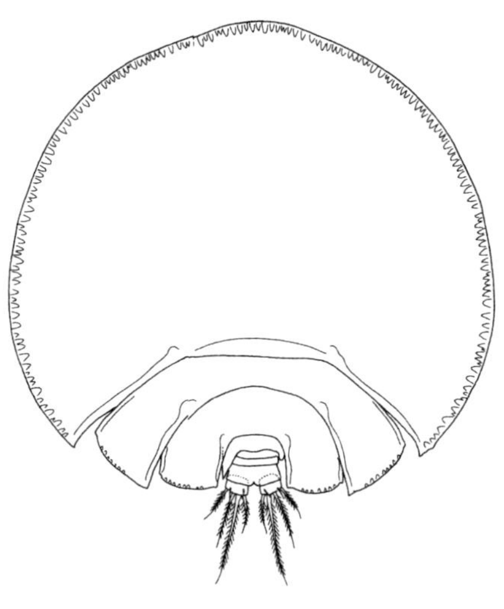

Artotrogus gordoni (Siphonostomatoida).
From Kim 2009

Siphonostomatoida

All Siphonostomatoida are parasites or associates of other animals and the order is mainly marine. Most work has been done in New Zealand on the parasites of fish, but this work is nowhere near complete. Almost nothing is known of the vast proportion of this order likely to live in association with marine invertebrates. We estimate that there are many species waiting to be discovered in the New Zealand siphonostomatoid fauna. There is currently nobody working on symbiotic copepods in New Zealand.

Commensal and parasitic forms have been collated here using the works of Thomson, whose major work was published in 1890 and whose collection is still housed in the Otago Museum (Thomson 1890). Gordon Hewitt also published extensively in the 1960s (Hewitt 1963, 1967, 1968, 1969) and, later, one of his students, Brian Jones, continued (1979, 1981, 1985, 1988b, 1991); his collection, including many undescribed species, is now in the Auckland Museum. A large collection was amassed at Kaikoura by students of the University of Canterbury under Bob Pilgrim (Pilgrim 1985) and some of that material was worked up by Ju-Shey Ho (Ho 1975, 1991; Ho & Dojiri 1987). The compilation given in the end-chapter crustacean species is based on the parasite list of Hewitt and Hine (1972), Pilgrim (1985), and the unpublished collection records of Jones.

An unidentified species of *Monstrilloida*.
Geoff Read

Monstrilloida

All Monstrilloida have internal parasitic naupliar and early postnaupliar stages and free-swimming, non-feeding adults. The known hosts are polychaete worms and prosobranch molluscs. Members of this order have been noted in the New Zealand fauna although there are no published records and descriptions.

Conclusions

There are few copepod taxonomists in New Zealand and none is able to work full-time on the subject. The greatest gaps in our knowledge copepod diversity are in the orders Cyclopoida, Harpacticoida, Siphonostomatoida, and Poecilostomatoida, especially concerning copepods as symbionts and parasites. These can be filled only by sampling little-studied environments, namely phytal, freshwater, deep-water, damp-terrestrial groundwater, and offshore islands. Sampling of benthopelagic and deep-sea habitats will yield records of hitherto undiscovered families and orders.

Because copepods are ecologically and economically so important, there is tremendous scope to understand the roles they play in the different ecosystems that they occupy, and to understand their impact on the other organisms with which they live in association, some of which are directly exploited by humans.

Class Ostracoda: Seed shrimps, mussel shrimps

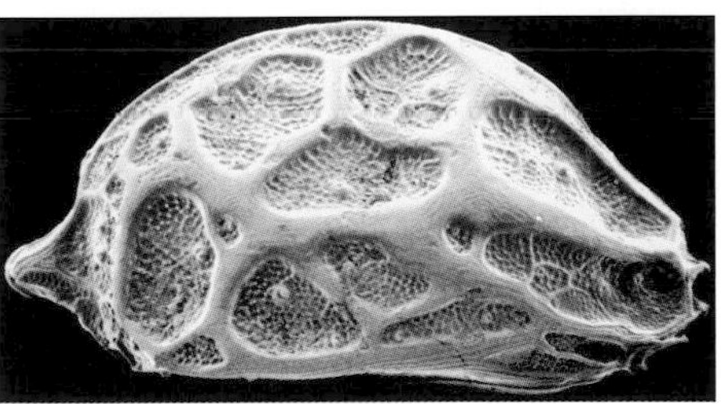

Hemicytherura pentagona (Pleistocene).
Stephen Eagar

Ostracods are tiny bivalved crustaceans that are widely distributed in the oceans, in fresh waters, and, rarely, in terrestrial situations. Food-mediated seasonal blooms in some freshwater habitats can result temporarily in vast numbers. Their shape confers on them the common name seed shrimps or mussel shrimps. Species subclass Podocopa range from 0.2 to 1.5 millimetres in length, while modocopids are often much longer, reaching an extreme of 30mm in *Gigantocypris*. Their shells, strengthened by deposition of calcium carbonate amongst the layers of cuticle, also fossilise well; in fact, ostracods are the most abundant arthropods in the fossil record, with a body plan that has been conserved at least since the Silurian. The shells can be brightly coloured and highly sculptured, making them attractive creatures to study, especially with a scanning electron microscope. They have an indistinctly segmented body like most arthropods, with paired appendages that are adapted for a variety of functions. Their identification is normally a specialist occupation.

They are very useful organisms, as knowledge of their taxonomy and distribution can be applied to studies of ecology and to environmental monitoring in relation to water quality, water depth, salinity levels, and temperature, as well as in stratigraphy. The number of specialists studying this group of animals is declining even though there is great potential for their usefulness. There are approximately 22,000 living and fossil species in the Catalog of Ostracoda published by the American Museum of Natural History and estimates of likely global diversity suggest more than 62,000 species in total. Of the described living species, 7000 belong to subclass Podocopa and 600 to subclass Myodocopa (Cohen 1998). There are many more species yet to be found in New Zealand, both living and fossil, in all environments.

Ostracods live in most aquatic environments and even, in the case of one New Zealand species – the bright yellow *Scottia audax* – in the damp leaf litter of the forest (Chapman 1961). Freshwater species live for between one season (as ponds dry) and three years. Marine species similarly live for one season to two years. Many marine planktonic ostracods constitute food for fish and species of one family (Entocytheridae, represented in New Zealand by a single species) are commensal on fish and other arthropods. Some myodocopids are bioluminescent but none have yet been found in New Zealand.

The first description of an ostracod, by Carl Linnaeus (1746), was very generalised. A figure was published in 1753, but the 'father' of the study of ostracods is regarded as O. F. Müller who, in a 1785 monograph on Entomostraca from Denmark and Norway, produced good descriptions and figures of freshwater ostracods.

History of study in New Zealand

Currently, the New Zealand living ostracod fauna stands at 442 species (including 86 undetermined), mostly marine but also comprising 37 freshwater and one terrestrial species. This tally is the product of many zoological studies since 1843; actual descriptive taxonomy has proceeded in pulses. The first species to be studied, by William Baird, was a relatively large (1.94 millimetres body length) freshwater species (*Candonocypris novaezelandiae*), often found in ponds and drinking troughs for farm animals (Baird *in* White & Doubleday 1843). It was collected by naturalist-explorer Ernst Dieffenbach. Baird (1850) was also responsible for describing the large (6.5 millimetres) marine species *Leuroleberis zealandica* sent to him by Rev. Richard Taylor of Waimate, one of the early settlers. George M. Thomson, teacher, Member of Parliament, and an amateur naturalist, produced the first locally published paper on ostracods from the Dunedin district in 1879. The first global oceanographic voyage of HMS *Challenger* (1873–1876) brought the ship into New Zealand waters and into Wellington Harbour for sampling. The results were published by Brady (1880). With the general establishment of the New Zealand colony, there was by the end of the 19th century an exchange of information between naturalists in New Zealand and Europe who were keen to document the fauna. So material was sent away for identification. Norwegian G. O. Sars (1894) published on freshwater species contained in dried mud and Brady (1898), living in Newcastle, England, received some marine specimens from New Zealand. Owing to the paucity of New Zealand ostracod taxonomists, this practice continued well into the 20th century with Brehm (1929) in Austria, Kornicker (1975) in the USA, and Hartmann (1982) in Germany providing identifications. One consequence is that many of the type specimens of New Zealand species reside in overseas institutions.

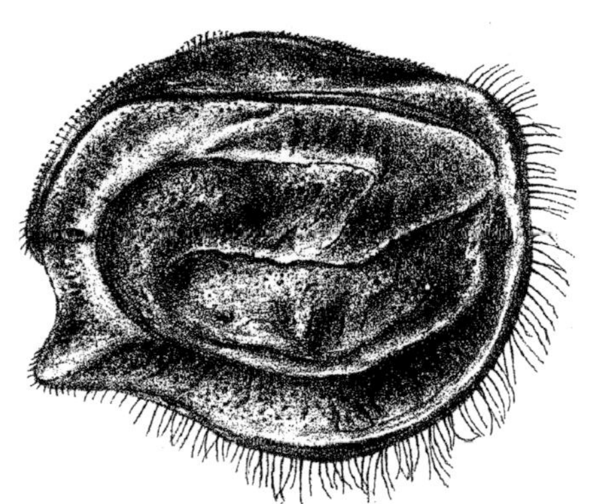

Cymbicopa hanseni.
From Brady 1898

The freshwater ostracod fauna was reviewed by Chapman (1963) and Chapman and Lewis (1976), and Scarsbrook et al. (2003) briefly summarised the ecology of New Zealand groundwaters in which ostracods occur but which are poorly known.

The podocopids and platycopids from the shallow intertidal to outer shelf have been the most intensively studied ostracods because they are also the most accessible (e.g. Morley & Hayward 2007). As mentioned above, ostracods are useful for environmental monitoring. They are sensitive to small changes in salinity and water quality and respond negatively to pollution. One study of a New Zealand waste outfall has shown the effects of sewage on a coastal ostracod fauna (Eagar 1999).

The planktonic myodocopids, which require specialist zoological knowledge, has been treated in monographs by Poulsen (1962, 1965) and Kornicker (1975, 1979) and in research studies by Deevey (1982). The first halocyprids were not recorded until Barney (1929). This group, together with the deep-sea podocopids, had received the least attention, but the recent study by Jellinek and Swanson (2003) has significantly increased knowledge of the latter.

Fossil species have followed a similar pattern of study. The earliest paper was by Jones (1860) on some tertiary species from Orakei. A bulletin by Chapman (1926) was issued by the New Zealand Geological Survey for Cretaceous and Tertiary species, but he used European names. His records are therefore not explicitly included in the following checklist, but the species are probably still represented there as synonyms of other workers' identifications. Benson (1956) recorded the occurrence of ostracods in late Middle Cambrian rocks from New Zealand, based on F. H. T. Rhodes's identification of their remains in a limestone. The preservation did not permit accurate identification. Simes (1977) recorded a phosphatic or phosphatised specimen from the limestone of the Upper Cambrian Anatoki Formation, and silicified ostracods were recorded by Marden et al. (1987) from the Triassic (Norian age). No other records whatsoever are available for any specimens from the Ordovician to the Jurassic.

Good fossil faunas are now known from sediments of Cretaceous age at several localities and these have been published recently (Dingle 2009). There have been a large number of papers on the systematics and paleoecology of New Zealand region Tertiary Ostracoda from the mid-1950s onwards (Swanson, 1969; Ayress 1990, 1991, 1993a,b,c, 1995, 1996; Ayress & Warne 1993; Ayress *et al.* 1994, 1995, 1997, 1999; Ayress & Drapala 1996). These faunas are rich, easily obtained, and interesting as they can be tied into other paleontological work. Most of the ostracod species in the end-chapter fossil checklist are therefore Tertiary species. The first publications to illustrate New Zealand ostracods using scanning electron microscopy came later (Swanson 1979a,b, 1980). The end-chapter checklist following builds on the one published by Eagar (1971).

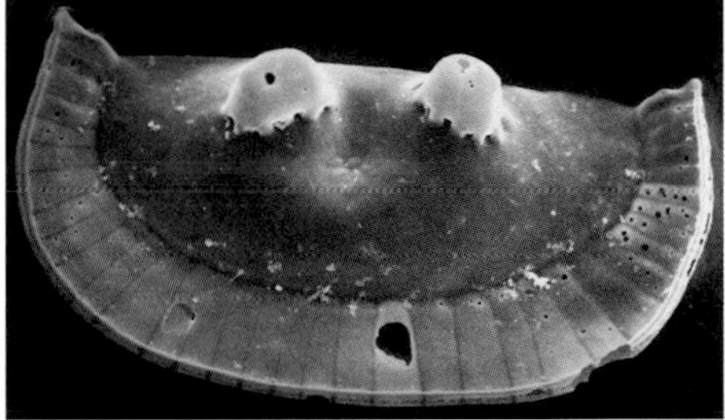

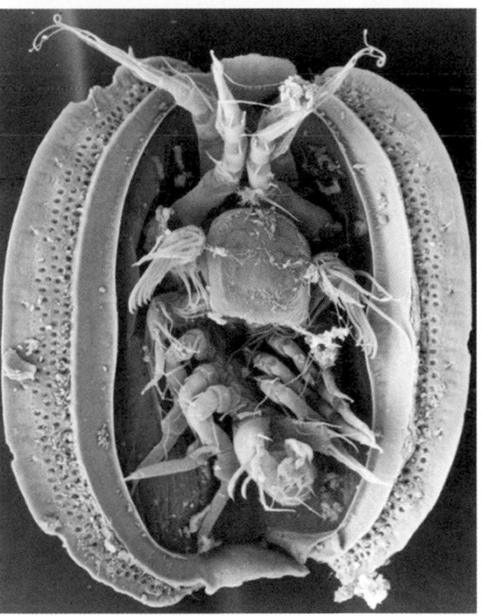

Lateral view of valve of *Puncia* sp. (upper) and ventral view of *Manawa staceyi*, both from Cavalli Islands.

Kerry Swanson

Features of the New Zealand ostracod fauna

Many Cenozoic marine species are endemic, long-ranging, and even still living. Presuming that they have not evolved a tolerance to changed ecological conditions, it can be assumed that the paleoeviromental conditions in which they lived were the same as now. Of particular interest are species of the endemic living-fossil genera *Manawa* and *Puncia* (Punciidae). Similar in shape and ornamentation to some Paleozoic genera, they are found living in shallow water off the north and east coasts of New Zealand. They provide insight into the soft-part anatomy of a group of ostracods (order Palaeocopida) that has otherwise been extinct for a long time (Hornibrook 1963; Swanson 1990; Horne et al. 2005).

Freshwater species are rare as fossils. Many species are swamp- or pond-dwellers and are not found on lake margins; inasmuch as ostracod shells are very soluble in the acid conditions of swamp deposits, their chances of preservation there are small. Further, most of New Zealand was submerged by the Late Oligocene and there were relatively few lakes, along with limited means of dispersal, available in the geological past (Hornibrook 1955; Eagar 1995a). Once colonisation from Europe was established, trout, salmon, and carp were introduced from Europe via Australia and it is likely that ostracod eggs travelled as hitchhikers to New Zealand on the damp media used to transport the fish (Eagar 1994). There is one non-marine saline species – *Diacypris thomsoni* (see Bayly & Williams 1973) – from Sutton, Otago, in salinity conditions of up to 15 parts per thousand. Guise (2001) discovered in the Avon-Heathcote Estuary, Christchurch, a new endemic genus of brackish-water ostracod (*Swansonella*) that tolerates higher salinities.

There are now more opportunities for introducing ostracods into New Zealand. Resting eggs that can withstand desiccation may even be transported by aircraft on footwear and camping gear. In addition to European freshwater species, several other species have an Australasian distribution. One marine species discovered close to shipping ports in the North and South Islands may have been brought in ballast water (Eagar 1999).

Few studies have been made of the anatomy of New Zealand ostracods. These were mostly on myodocopids (Poulsen 1962, 1965, Kornicker 1975, 1979) and to a lesser extent to the freshwater species (Podocopida: Cyprididae) (Chapman 1963; Eagar 1995b; Rosetti et al. 1998), with a few ventures into the marine podocopids (e.g. Brady 1902; Swanson & Ayress 1999).

Class Malacostraca

This class contains more than half of all known species of crustaceans, including the aristocrats – the giant spider crabs of Japan with their 3-metre leg span (vying with fossil eurypterids as the largest of all arthropods) and the New Zealand packhorse rock-lobster (*Sagmariasus verreauxi*) at 20 kilograms – and krill, one of the most ecologically critical malacostracans in marine food webs, slaters, and tiny sand-hoppers. Malacostracans are very unevenly divided into three subclasses – Phyllocarida, Hoplocarida, and Eumalacostraca.

Levinebalia fortunata.
From Wakabara 1976

Subclass Phyllocarida: Phyllocarids

Order Leptostraca

The Leptostraca is the sole living order of the Phyllocarida, a group of Crustacea with a long geological history (Rolfe 1969), possibly extending back as far as the Cambrian, some 600 million years ago (Briggs 1992). Despite new conclusions from DNA analyses as to their place in crustacean evolution (Spears & Abele 1999), the Leptostraca may still be regarded as 'living fossils' indicative of the times and conditions in which the so-called primitive arthropods lived (Hessler & Schram 1984; Dawson 2003b). They are known from the New Zealand Ordovician (Chapman 1934), and the presence of several living species of Leptostraca in the region is of considerable interest. Using the small-subunit 18S ribosomal-DNA gene of 10 representative foliaceous-limbed Crustacea, Spears and Abele (1999) concluded that the Phyllocarida are true malacostracans, which diverged fairly early from the main lineage. This result is consistent with the pioneer work of Claus (1888) and Calman (1909) and with Manton's (1934) study of embryology, and also corroborates the views of Dahl (1987, 1991) of the Leptostraca as an early offshoot.

The late British zoologist Sir Alistair Hardy (1956) vividly recalled the excitement of his first encounter with one of the little crustaceans, *Nebaliopsis typica*, found in great depths but rarely collected, and then usually dead and very damaged. It had only ever been seen alive on one occasion – on the Swedish Antarctic Expedition in 1904 – until a second specimen was collected from the *Discovery II* fifty years later. The Leptostraca, wherever they have been found subsequently, have continued to excite and interest zoologists and paleontologists alike.

A paleontological summary of the Phyllocarida was made by Rolfe (1969). Monographs on the Leptostraca as a whole have been made by Claus (1888) and Cannon (1960), and these still have their usefulness, but a new and compact text has been produced (Dahl & Wägele 1996). More recently, the relationships of the leptostracan genera were examined by Olesen (1999) and by Walker-Smith and Poore (2001), who revised the families and genera. The latter authors also provided a complete listing of all species of Leptostraca together with keys to the families and genera. Some 42 species of living Leptostraca are recognised at present, divided into three families – Nebaliopsidae (genera *Nebaliopsis, Pseudonebaliopsis*), Paranebaliidae (named only in 2001, containing *Paranebalia, Levinebalia,* and *Saronebalia*), and Nebaliidae (with five other genera). Many species of *Nebalia* and *Paranebalia* remain undescribed as yet (Dahl & Wägele 1996).

Leptostracans are small, usually 4–12 millimetres in length although one species, *Nebaliopsis typica,* can exceed 35 millimetres. They are characterised by the possession of a relatively large, bivalved carapace, hinged on the midline and held together by an adductor muscle. The carapace loosely covers the abdomen and part of the thorax, and is attached by a hinged rostral plate covering the head and closing the anterior gap of the carapace itself. Long anteriorly projecting antennae are used for swimming, the antennal flagellum in males being as long as the body. There are eight pairs of foliaceous, leaf-like thoracic limbs that also provide a feeding mechanism and may be modified in the female in the form of a fan of plumose setae forming a basket-like chamber for brooding eggs between the ventral regions of the valves of the carapace. The first four pairs of pleopods are well developed and biramous whereas the 5th and 6th pairs are small and uniramous. The abdomen ends in two characteristic long and articulated tail spines or furci. In contrast with all the six abdominal segments possessed by all other Malacostraca, the Leptostraca have a 7th segment and this lacks any appendages. The telson may be considered an 8th segment.

Relatively little is known of the life-history, growth rates, or physiology of most leptostracans. Useful observations have been made by Cannon (1927), Rowett (1943, 1946), Martin et al. (1996), Vetter (1996a), and Wägele (1983). Manton (1934) worked on the embryology of *Nebalia bipes,* helping to elucidate phylogenetic relationships of the Phyllocarida (Dahl 1987; Spears & Abele 1999). Linder (1943) described some larval stages, which could be useful for recognition in sorting plankton samples. Leptostracans play a significant role in benthic production (Rainer & Unsworth 1991; Vetter 1996a,b; MacLeod et al. 2007). The unusual marine rotifer *Seison* is often found epizoic on leptostracans. None has yet been discovered in New Zealand but it would be worth checking local *Nebalia* to ascertain their presence or absence.

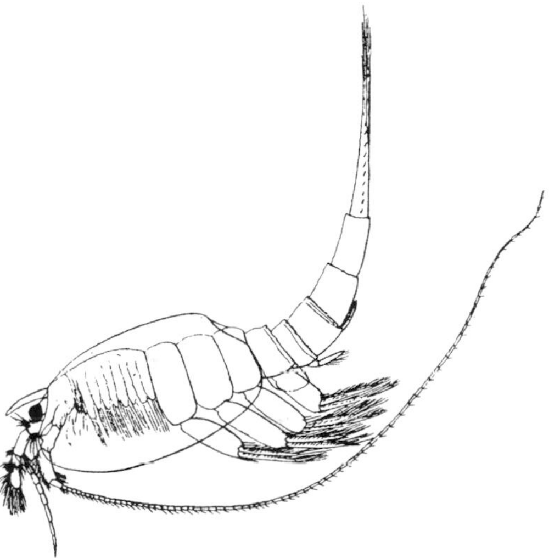

Nebalia longicornis.
From Thomson 1879

Leptostracans are widely distributed as a group. Individual species may be limited or widespread in depth range and geographically, but taxonomic caution needs to be observed in the case of the purportedly wide-ranging species. Dahl's (1990) analysis of the *Nebalia longicornis* complex showed that it comprised at least 10 different species. Walker-Smith (1998) reviewed the genus *Nebaliella,* describing the first known Australian species. In her unpublished Honours thesis, she recognised six new species and a new genus of Leptostraca from Australia (Walker-Smith pers. comm. 2000).

Present-day leptostracans live in a variety of habitats, including under intertidal stones, with decaying seaweed or dead shell, in crab pots, on mangrove shores and coral reefs, and in subtidal sandy plains or muddy sand. A non-New Zealand species, *Speonebalia cannoni,* is the only leptostracan to be recorded from a groundwater habitat. *Nebalia hessleri* lives in enriched sediments and detrital mats with low oxygen levels in submarine canyons off southern California. Here they form the highest density ever reported for a macrofaunal assemblage, namely 1.5 million per square metre. In northwestern Spain, Moreira et al. (2009) reported six species of leptostacans in subtidal sediments, the largest number of species recorded in a single area. *Dahlella caldariensis* occurs among mussels and vestimentiferan worm tubes, swimming above clumps of animals at hydrothermal vents.

The New Zealand leptostracan fauna

The New Zealand fauna currently consists of five species in four of the 10 known genera. Unfortunately, little is known of the true numbers of taxa represented in any one geographic area, but the indications are that New Zealand could well be shown to have a higher diversity.

The first to be recorded and named in New Zealand was *Nebalia longicornis*, based on a single mature male collected in Otago Harbour (Thomson 1879a). It was subsequently described in more detail, based on records from 8–10 metres depth in Dunedin Harbour and 20 metres at Stewart Island (Thomson 1881). This later paper by Thomson (with its slightly different figure) appears to have been overlooked by all subsequent authors. *Nebalia longicornis* was inadequately described and illustrated according to Dahl (1990), and great taxonomic confusion subsequently resulted from attempts to apply this name to later records of *Nebalia* from other parts of the world. Since Thomson's type specimen could no longer be found, Dahl redescribed the species based on a female collected from Otago Harbour in 1965, thereby fixing *Nebalia longicornis* Thomson, 1879a as a member of the New Zealand fauna. Thomson (1913) noted his *Nebalia longicornis* as found in Otago Harbour and frequently taken outside the Otago Heads in trawl-nets.

Thiele (1904) reported a specimen of what he considered to be *Nebalia longicornis* from Akaroa Harbour. Dahl (1990) examined this specimen and found it to be a species of *Nebalia* (then in his genus *Sarsinebalia*) but in too damaged a condition to be able to describe further. Thiele had also recorded juvenile *Nebaliella antarctica* from Akaroa Harbour but apparently this specimen has not been re-examined.

In 1907, W. Benham collected a juvenile *Nebalia* from Musgrave Harbour on the Auckland Islands that Chilton (1909) attributed to *N. longicornis* as then understood. Another specimen was taken at Port Ross, Auckland Island, in 1914 during the Mortensen Expedition (Stephensen 1927). Calman (1917) reported two immature specimens of Leptostraca collected in 1911 at *Terra Nova* Stations 130 and 135 off Three Kings Islands and in Spirits Bay [given incorrectly by Dahl (1990) as Stns 10 and 15]. Dahl (1990) has since examined these specimens, concluding that one is a *Nebalia* and the other a *Sarsinebalia*.

Morton and Miller (1968) described a *Nebalia* as a member of the protected sandy-beach fauna, one of the small filter-feeding Crustacea that live in the fine sands of the lower beach. They also illustrated it as the prey of the small shallow-water cephalopod *Sepioloidea pacifica*.

The only other work on New Zealand leptostracans has been the description of *Levinebalia fortunata* (Wakabara 1976, as *Paranebalia*) based on 16 females collected by trawl nets at 420–660 metres depth in canyons off Otago Peninsula, representing a marked extension to the known bathymetric range of the genus. Apart from Prof. John Jillett at Otago (see Dahl 1990) no-one has conscientiously searched New Zealand habitats for leptostracans. It is likely that deliberately intensive collecting will reveal not only great extensions of the range of the already listed forms but undescribed species as well. Morton (2004) suggested searching for leptostracans in black anaerobic sediments with decaying algae and carrion-baited traps may also be useful (Lee & Morton 2005), especially for assessing population densities.

Chapman (1934) described several species from Ordovician rocks in Fiordland, based on numerous specimens. They have never been studied since and are listed in the end-chapter checklist of fossil New Zealand Crustacea under the generic names recommended by Rolfe (1969).

Mantis shrimp *Heterosquilla tricarinata*.
Shane Ahyong

Subclass Hoplocarida

Order Stomatopoda: Mantis shrimps

Mantis shrimps are among the most aggressive and behaviourally complex crustaceans. All are active predators and mark one of the very few radiations of obligate carnivores within the Crustacea. The general morphology of mantis shrimps has been described by Holthuis and Manning (1969), and characteristic features are the triflagellate antennules, well-developed stalked eyes, and the greatly enlarged, raptorial second maxillipeds. The name mantis shrimp stems from these large and powerful raptorial claws. Prey is captured by 'spearing' or 'smashing', depending on whether the dactyl of the raptorial claw is extended or kept folded during the strike. (Think of the dactyl as a finger, opposing the thicker 'thumb' of the claw.) Hence the two modes of prey-capture define the 'smashers' and the 'spearers' among mantis shrimps (Caldwell & Dingle 1976). The strike of the raptorial claw is among the fastest known of animal movements, being completed in 3–5 milliseconds, and the strike of large species of 'smashers' may break aquarium glass.

Vision in mantis shrimps is strongly developed. In most species, the cornea is divided into two halves by a midband of ommatidia, enabling binocular vision with each eye. Additionally, the midband ommatidia in many families enable colour vision and detection of polarised light (Marshall 1988).

Pterygosquilla schizodontia.
Shane Ahyong

Most stomatopods live in temperate or tropical shallow marine habitats, but several species also range into subantarctic waters, and a few tropical species may occur in brackish water. Seven superfamilies are recognised: Bathysquilloidea, Erythrosquilloidea, Eurysquilloidea, Gonodactyloidea, Parasquilloidea, Lysiosquilloidea, and Squilloidea. Most members of the Gonodactyloidea occur on coral reefs where they shelter in or under boulders and coral. The bathysquilloids are known only from deep outer-shelf waters. Members of other superfamilies generally burrow in flat sandy and muddy harbour bottoms and sea-floors.

The Stomatopoda comprises the only living order of Hoplocarida, two other orders (Aeschronectida and Palaeostomatopoda) being known only as fossils. Compared with other major crustacean groups such as the Decapoda, the fossil record of the Hoplocarida is relatively poor but it appears that the hoplocarids originated in the Devonian and the Stomatopoda proper first appeared during the Carboniferous. Recognisably modern stomatopods, with well-developed raptorial claws, did not appear until the Mesozoic (Holthuis & Manning 1969; Hof 1998; Hof & Schram 1998).

Over the past three decades, the taxonomy of the Stomatopoda has been extensively revised, largely through the work of the late R. B. Manning, who recognised five living superfamilies (Manning 1995). Ahyong and Harling (2000) provided the most recent phylogenetic study. At present, more than 450 species in more than 100 genera, 19 families, and 7 superfamilies are recognised.

The stomatopods of the Atlantic have been monographed and are well known (Manning 1969, 1977), while those of the eastern Pacific were treated relatively comprehensively by Schmitt (1940) and Hendrickx and Salgado-Barragán (1991). Stomatopod diversity in the Indo-West Pacific region, however, is more poorly known. The most important major works for this region are those of Kemp (1913) on the Indian fauna, Manning (1995) on the Vietnamese fauna, and Ahyong (2001) on the Australian fauna. The Indo-West Pacific fauna has been extensively studied in the past decade (e.g. Ahyong 2002a,b,c; Ahyong & Naiyanetr 2002; Ahyong et al. 2008).

The New Zealand fauna

New Zealand's mantis shrimps are known from only a few studies, the most important of which are those of Miers (1876), Chilton (1891, 1911a) and Manning (1966). Manning (1966) recognised three species from New Zealand and its offshore islands: *Pterygosquilla schizodontia*, *Heterosquilla tricarinata*, and *Acaenosquilla brazieri* (as *Heterosquilla brazieri*). He also remarked that *Squilla tridentata* Thomson, 1882, synonymised with *H. tricarinata* by Chilton (1891), was probably a distinct species. Ahyong (2001) recognised Thomson's species as distinct under the combination *Heterosquilla tridentata*. Other additions to the

New Zealand stomatopod fauna are *Hemisquilla australiensis* (Stephenson 1967), *Odontodactylus brevirostris* (Manning 1991), and the striking 30-centimetre-long, scarlet deep-sea species *Bathysquilla microps* (O'Shea et al. 2000). Therefore, seven species are presently recorded from New Zealand.

The commonest species are *Heterosquilla tricarinata* (known around both main islands and Chatham, Stewart, Campbell, and Auckland Islands, generally in intertidal sand or mudflat burrows) and *Pterygosquilla schizodontia* (central New Zealand to the Auckland Islands, burrowing in subtidal sand and mud). Their biology has received little scientific study. Larval development of *Pterygosquilla schizodontia* was studied by Pyne (1972). Several studies have been conducted on *H. tricarinata* including those of Fussell (1979), Greenwood and Williams (1984), and Williams et al. (1985).

The New Zealand stomatopod fauna is relatively small, and this is consistent with the primarily tropical distribution of most species. Neverthless, low diversity may also reflect low collecting effort. Study of collections from northern island groups in New Zealand territorial waters should reveal numerous additional faunal records. The Japanese mantis shrimp *Oratosquilla oratoria* has become established in some North Island estuaries and is the first exotic species of Stomatopoda to be detected in New Zealand waters. New species and numerous additional distribution records will be reported in a forthcoming revision of the New Zealand Stomatopoda by Shane Ahyong.

Subclass Eumalacostraca

Superorder Syncarida

Orders Anaspidacea, Bathynellacea

The Syncarida constitutes a group of tiny crustaceans that may be regarded as living fossils, with a geological history extending as far back as the Carboniferous (Dover 1953; Drummond 1959; Brooks 1969; Schram & Hessler 1984; Uhl 1999, 2002; Jarman & Elliott 2000; Dawson 2003a). They are little known to most biologists, the exception being the large-sized *Anaspides*, found in Tasmania, which has attracted much interest and attention largely because of its accessibility in open waters rather than the subterranean habitat in which most syncarids live.

The Syncarida were first made known to science by the report of a fossil species, *Uronectes fimbriatus*, in Europe. Their relationships and place in the crustacean hierarchy remained a matter of contention until Packard (1885, 1886) gave them separate status as the Syncarida. Much later, Brooks (1962, 1969) finally settled the status of the fossil as one of three orders constituting the superorder Syncarida, and Schminke (1975) related them to the living orders. Schram (1984) subsequently reviewed and revised the fossil species, which range in time from the Early Carboniferous (Uhl 2002) to the Early Permian in Europe and North America, the Late Permian of Brazil, and the Triassic of Australia, corresponding to the former landmass of Laurentia prior to the formation of Pangaea.

New Zealander George Malcolm Thomson, a noted amateur scientist, teacher, and politician, is generally credited with the discovery and description of the first living syncarid – *Anaspides tasmaniae*, which he discovered when visiting Tasmania in January 1892. He was of the opinion that his discovery was a schizopod shrimp (Thomson 1894). However, Calman (1896) said this new crustacean was no schizopod and supplemented Thomson's description in some detail, comparing *Anaspides* with fossils from Illinois and Germany that Packard (1885) had already placed in his new group, Syncarida. Calman concluded that *Anaspides* was, in fact, a living representative of primitive malacostracans that had flourished widely in Paleozoic times

Ironically, however, living syncarids had in fact been discovered some years previously when Vejdovský (1882, 1889) published a description of the tiny *Bathynella* that he had found two years earlier in a well in Prague. Calman (1899)

subsequently recognised *Bathynella* as a syncarid, but little more was known until 1913 when Chappuis (1915) found more specimens in a well near Basle. He placed them in a new taxon, Bathynellacea. Syncarids were soon found to occur in many places throughout Europe, in wells, springs, or streams in caves (Chappuis 1939) as well as in Australia, New Zealand, Japan, North and South America, and elsewhere.

Although Thomson turned out not to be the first discoverer of a living syncarid, the finding of such an ancient form of crustacean living in Tasmania did excite many subsequent workers (up to the present day), resulting in a substantial number of publications on aspects of their morphology, development, ecology, and relationships – and even a poem in the style of Longfellow dedicated to *Anaspides* (Mesibov 2000). In essence, there have been two approaches to the study of the Syncarida, one concentrating on the relatively tiny subterranean and interstitial forms (basically the order Bathynellacea), and the larger, open-water taxa of Australia (order Anaspidacea, which also includes the subterranean Stygocarididae). General accounts of the Syncarida can be found in Siewing (1959), Noodt (1964), McLaughlin (1980), Schminke (1982), Schram (1986), and Coineau (1996, 1998).

Within the Eumalacostraca, the Syncarida are distinguished by the absence of a carapace, an elongate body form (more or less cylindrical in the subterranean forms), with a thorax consisting of seven or eight segments, the first segment being fused to the head in some groups. The abdomen consists of six segments and a telson, or five segments followed by a pleotelson formed from the fusion of the 6th segment with the telson.

The order Anaspidacea contains four families: Anaspididae, Kooningidae, Psammaspididae, and Stygocarididae. Only the last of these has been found in New Zealand. They include the largest of the syncarids, with a body length ranging from about 1 to 50 millimetres. The Bathynellacea contains two families, the Bathynellidae and the Parabathynellidae, which are both represented in the New Zealand fauna as it is presently known. They are very much smaller in size than the anaspidaceans, ranging from about 0.4 to 3.5 millimetres.

The body form of syncarids is reflected in the habitats in which they are found: the tiny forms, with slender, cylindrical bodies, devoid of pigment and eyes, are found in caves and underground waters, whereas the much larger forms, such as *Anaspides*, found in surface waters are shrimp-like.

Living syncarids comprise more than 200 species worldwide (Camacho & Valdecasas 2008), although fresh explorations and more refined collecting techniques are already increasing this number. There are many species of syncarids collected from eastern Australian caves and karst areas awaiting identification and description (Thurgate et al. 2001) and such may be the case for New Zealand.

Syncarids have the reputation of being rare animals, although the pioneer investigations by Chappuis (1943) on *Bathynella* in Hungary showed that numerically rich collections could be made at individual sites. Much of the alleged rarity is a consequence of their small size (which is why early investigators in New Zealand such as Chilton did not find them) and their largely subterranean habits. Schminke (1986) has said that those who know how to sample their habitats 'today have lost the impression of dealing with rare animals.' Syncarids are globally widespread; Schminke (1986) listed all the species then known, with their locations. New taxa continue to be described Camacho 2005a,b; Cho 2005; Cho et al. 2005, 2006; Camacho et al. 2006; Cho & Schminke 2006).

While some Syncarida inhabit open- and surface-water habitats (Camacho & Valdecasas 2008), it is acceptable to say that syncarids are characteristic of subterranean habits throughout the world, whether groundwater (as revealed by sampling wells, springs, and gravel river margins), or caves with streams and sandbanks providing living space in the interstitial spaces between sediment grains.

Syncarids have been recorded from springs in Australia (Knott & Lake

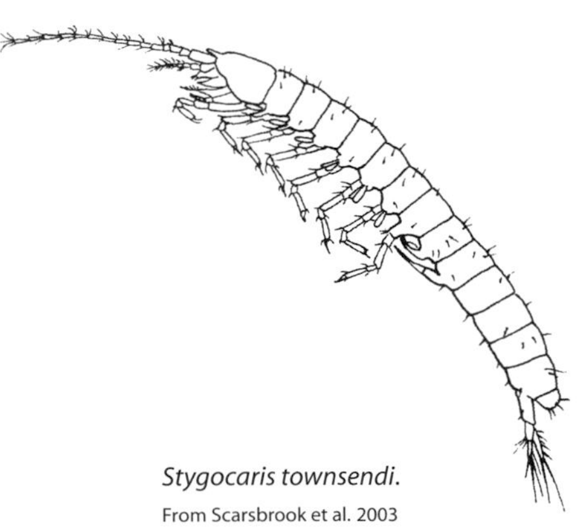

Stygocaris townsendi.
From Scarsbrook et al. 2003

(1980), and in New Zealand they occur in similar situations as well as from groundwater in wells (Scarsbrook et al. 2003), just as did the first-discovered European living syncarids. Many syncarids have been collected from caves, although in New Zealand only *Stygocaris townsendi* has been described from such a habitat (Morimoto 1977). Karst landscapes throughout the world provide habitats for syncarids.

Information on the development, life-history, and habits of syncarids is still quite limited. So far as the Anaspidacea are concerned, most of the developmental studies have been done on *Anaspides tasmaniae*, by Hickman (1937), with other aspects covered in other studies, for instance Dohle (2000). The biology of bathynellaceans is less well known, but what is known has been summarised by Coineau (1996). In feeding, *Anaspides* has a filtering mechanism, used in conjunction with collecting particles by scraping detritus with its limbs. Smith (1908) noted that *Anaspides* was an omnivorous feeder, eating dead insects as well as each other, but mainly feeding on algal slime and submerged mosses and liverworts. The habitat of Tasmanian anaspidaceans, notably *Allanaspides hickmani* and *A. helonomus*, is under continuing threat (Driessen et al. 2006).

Compared to the amount of information regarding the general biology and ecology of the anaspidacean syncarids, there is virtually nothing recorded about the lifestyle and habits of the Bathynellacea. What is known has been summarised by Coineau (1996), and Camacho (1992) has outlined the abiotic characters of the subterranean environment in which most of bathynellaceans live.

Camacho (2006) noted 256 species and subspecies of extant Syncarida, 95% of which are subterranean in habitat. In addition to the two living orders is the order Palaecaridacea, which is entirely fossil.

The order Anaspidacea comprises five families, of which three are confined to Australia. These include: Anaspididae, with five genera – *Allanaspides, Anaspides, Paranaspides, Anaspidites* (Triassic, Australia), *Koonaspides* (Lower Cretaceous, Australia); Koonungidae, with two genera – *Koonunga, Micraspides*; Psammaspididae, with two genera – *Eucrenonaspides, Psammaspides*; and Stygocarididae, with four genera – *Oncostygocaris* (Chile), *Parastygocaris* (Argentina), *Stygocarella* (New Zealand), and *Stygocaris* (Australia, New Zealand, Chile). The 21 living species of Anaspidacea are confined to the Southern Hemisphere. *Anaspides tasmaniae* is of particular interest in the context of mitochondrial DNA studies, in which it has been demonstrated that there may be at least three cryptic species (Jarman & Elliott 2000).

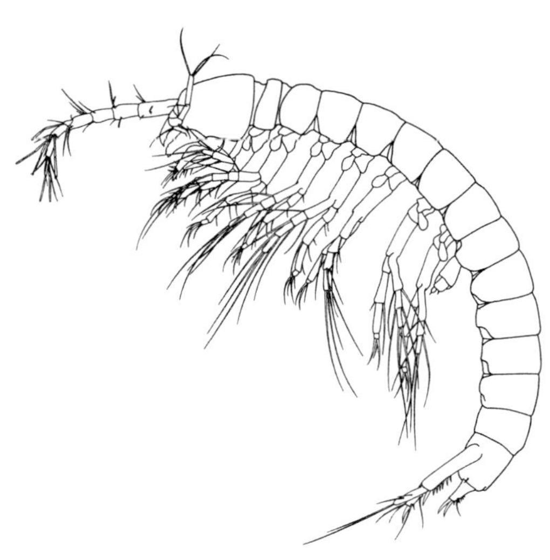

Notobathynella longipes.
From Schminke 1978

The order Bathynellacea comprises two families, both distributed widely throughout the world, totaling 66 genera an 219 species: Bathynellidae, with more than 20 genera (including *Bathynella*, of which there are New Zealand representatives) and more than 80 described species; and Parabathynellidae, with about 32 genera and more than 90 species (also recorded from New Zealand in the genera *Atopobathynella, Hexabathynella*, and *Notobathynella*). As discussed by Camacho et al. (2002), there have been two contrasting views as to the systematic position of the bathynellids as being either within the superorder Syncarida or as a separate suborder Podophallocarida in infraclass Eonomostraca. These Spanish researchers' molecular studies in Spain on a cave-dwelling bathynellid, *Iberobathynella* (*Espanobathynella*) *magna*, have now provided a nucleotide sequence that supports a basal position for the Bathynellacea with a clear distinction from the Syncarida, placing them in the Podophallocarida but retained in the Eumalacostraca.

Schminke (1986) postulated that the Syncarida originated in the marine environment from whence they invaded freshwater by two independent lines, living first in surface waters and then invading the groundwater habitat. He developed the 'zoea' theory (Schminke 1981b) in which it was suggested that the Syncarida originally passed through a series of larval stages and through neoteny reached sexual maturity at a stage corresponding to the zoea larva of the penaeid prawns (Decapoda). Schminke (1972) had previously demonstrated, by a study of all the then-known species of *Hexabathynella* (but which did not include the subsequently discovered *H. aotearoae* of New Zealand), all of which were known to occur close to the sea, that syncarids did not invade the freshwater interstitial habitat from sandy marine beaches. Presumably, some of the more recently discovered occurrences of *Hexabathynella aotearoae* indicate secondarily derived habitats. This species is closest evolutionarily to Australian *H. halophila* (Camacho 2003).

Biogeographically, the breakup of the ancient supercontinent Gondwana has been invoked to explain some of the distributions between northern and southern hemispheres and within the austral landmasses (Schminke 1973, 1974, 1975, 1980, 1981a; Williams 1986). Subsequent information about the distribution and phylogeny of the various syncarid groups can be found in Coineau (1996), Camacho and Coineau (1989), Camacho *et al.* (2000), and Guil and Comacho (2001).

The New Zealand fauna

In 1967 and 1968, visiting German scientist Kurt Schminke searched for syncarids quite widely throughout New Zealand, taking almost 200 samples from interstitial freshwaters at 11 different localities (Schminke & Noodt 1968; Schminke 1973). Of these, 36 yielded syncarids in the families Bathynellidae, Parabathynellidae, and Stygocarididae. In his unpublished thesis, Schminke (1971) included two new forms of *Bathynella*, a species and its subspecies (as yet not formally described), collected from the Tauherenikau River in the Wairarapa and from the Orari River in South Canterbury. In his major work on the evolution, taxonomy and biogeography of the world fauna of the Parabathynellidae, Schminke (1973) listed his collecting locations in New Zealand with descriptions and distribution maps of four new species from New Zealand: *Atopobathynella compagana, Hexabathynella aotearoae, Notobathynella chiltoni*, and *N. hineoneae*.

Schminke (1978) subsequently reported on a collection, made by by G. Kuschel of the former DSIR Entomology Division, which included a female bathynellid from a bore in Nelson, and two females of *Notobathynella*. He also noted two more specimens of *Atopobathynella compagana* and described *Notobathynella longipes* from wells at Motueka. In the Anaspidacea, Schminke (1973) mentioned at least three unidentified New Zealand species of Stygocarididae in one new genus, later describing *Stygocarella pleotelson* (Schminke 1980) and noting 16 localities from which other unidentified specimens had been collected. During a brief trip to New Zealand in 1975, Morimoto (1977) collected syncarids at four South Island locations, finding three species of *Stygocaris*, of which *S. townsendi* was described as new. More recently, in a NIWA study of the New Zealand groundwater fauna (Scarsbrook et al. 2000), syncarids appeared to be widespread in interstitial habitats in alluvial groundwaters in Hawkes Bay and Canterbury, both within the margins of gravel riverbeds and in the deeper (10–20 metres) ground water.

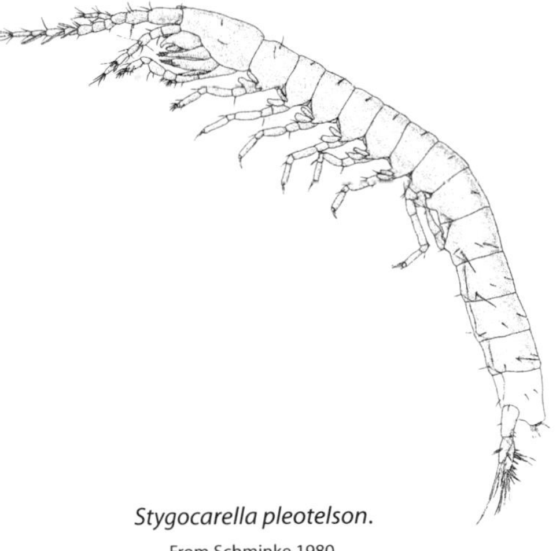

Stygocarella pleotelson.
From Schminke 1980

Thus, the New Zealand syncarid fauna, as presently known from limited sampling, consists of at least four species of Anaspidacea – *Stygocaris*, and one or possibly more species of *Stygocarella*. The Bathynellacea is represented by what appear to be quite abundant and widespread species of Bathynellidae (*Bathynella*), none formally described, and three genera of the Parabathynellidae – *Atopobathynella* and *Hexabathynella* (each with one described species), and *Notobathynella* (at least four species, three of them named). It is highly likely that the New Zealand syncarid fauna will be found to be much more extensive, if only in terms of the distribution of the already described species, all of which are endemic.

Gaps in knowledge of Syncarida

Not only is taxonomic knowledge of the New Zealand Syncarida incomplete; even less is known about their ecology and special adaptations to the several kinds of habitats in which they occur. It is apparent that a geographically widely distributed syncarid fauna exists in New Zealand's ground waters. The brief venture into cave collecting by Morimoto (1977), taken with what is known of the distribution of syncarids in Europe and Australia, suggests the prospect of further exciting discoveries locally in this particular habitat. Cave systems and karst-type landscapes with sink holes and sunken streams are common in many parts of New Zealand (Crossley et al. 1981), and there is a very strong fraternity of recreational cavers, some of whom have already contributed to scientific knowledge of cave life. There is a real challenge to use the technical expertise of such people to look for these fascinating'living fossils'; a preliminary guide to promote such work was issued by the New Zealand Department of Conservation (Hunt & Millar 2001). The results of a 15-year study of Spanish cave fauna by Camacho (2000) shows what could be achieved by a systematic approach towards elucidating New Zealand's subterranean syncarid fauna.

There is a growing appreciation worldwide of the importance of groundwater organisms as environmental indicators of water quality, not to mention the scientific interest of these organisms in their own right (Danielopol 1992; Marmonier et al. 1993; Danielopol et al. 2000; Gibert et al. 1994; Jones & Mulholland 2000; Scarsbrook et al. 2000, 2003) and the need to understand karst landscapes and their fauna from a conservation perspective (William & Wilde 1985) and cave life in general (Vandel 1964; Ford & Cullingford 1976; Sasowsky et al. 1997; Culver 1982; Camacho 1992; Juberthie & Decu 1994–2001). 'Living fossils' carry appealing overtones in the public imagination (Dawson 2003a), so the demonstration of the existence even of such tiny forms as the syncarids could be another highlight to make known.

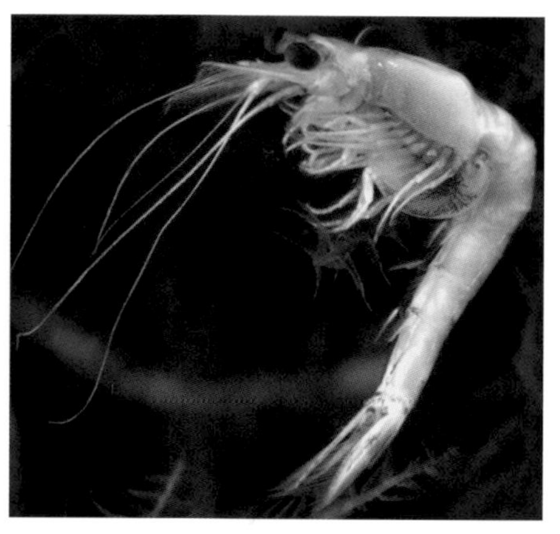

An estuarine species of *Tenagomysis*.
Stephen Moore

Orders Lophogastrida and Mysida ('Mysidacea'): Opossum shrimps

The Mysidacea are shrimp-like but have a number of characters, including a 'brood pouch', that distinguish them from other crustaceans of similar appearance. They are mainly marine, living in all oceans from great depths to brackish coastal waters, and there is a small number of freshwater species. They are of limited commercial importance and therefore not as familiar as the decapod shrimps and prawns. Mysidacea may, however, be very common, particularly in estuaries and coastal waters, where they often congregate in large swarms, and are of considerable importance as primary consumers and as food of fishes.

Historically, the Mysidacea comprised a single order with two suborders – Lophogastrida and Mysida. The two groups differ in important ways and there is now debate over whether they are mono- or polyphyletic (having one, or more than one, ancestor). Some workers question whether the Mysida, which contains the great majority of Mysidacea, even belongs in the large malacostracan superorder Peracarida, with the Lophogastrida; see Martin and Davis (2001) for a discussion of mysid classification. These authors discarded the Mysidacea, raising the two suborders to ordinal status, a decision followed here. Even so, the two groups have many characters in common and, since relatively few species (24) have been recorded from New Zealand waters, are discussed here collectively.

Historical studies

Mysidacea have been recognised since the late 18th century. The taxonomic literature is scattered and deals mostly with northern hemisphere faunas and least with that of the Indian Ocean and Australasia. Major contributors include Tattersall and Tattersall (1951), Gordan (1957), Mauchline and Murano (1977), Mauchline (1980), and Müller (1993).

The history of New Zealand mysidacean taxonomy is brief. The first published record is that of Thomson (1880), who described *Siriella denticulata*.

Kirk (1881) described *Mysis meinertzhagenii,* but the type and further evidence of its existence have not been found since. Thomson (1900) described *Tenagomysis novaezealandiae* from brackish water near Dunedin, and Calman (1908) attributed an immature mysidacean specimen to the genus *Pseudomma,* apparently not identified since. *Tenagomysis tenuipes* Tattersall, 1918, from Carnley Harbour, Auckland Islands, brought the early list of mysidaceans known with certainty to occur in New Zealand to three. Next, Walter Tattersall's (1923) report on the Mysidacea of the 1910 *Terra Nova* Expedition to Antarctica and the Southern Ocean added 12 species. Eight were new, seven of which belonged to *Tenagomysis,* and the remaining species was named *Gastrosaccus australis,* the first and so far only named species of the genus from New Zealand. New records for New Zealand of previously described species of Mysida included *Euchaetomera oculata, E. typica,* and *Siriella thompsoni.* Chilton (1926) presented an overview of New Zealand Mysidacea to that date. Later, Olive Tattersall (1955) identified *Boreomysis rostrata* and *Euchaetomera zurstrasseni* from New Zealand waters and Hodge (1964) redescribed *Tenagomysis chiltoni.* The most recent addition to the fauna was that of *Tenagomysis longosquama* (Fukuoka & Bruce 2005).

Walter Tattersall (1923) appears to be the first to have recorded a species of Lophogastrida, *Paralophogaster glaber,* in New Zealand. Apart from a record of *Lophogaster* sp. from Te Papa (Museum of New Zealand) collections, the remaining records are from Fage (1941) reporting on mysidaceans caught by the 1928–30 *Dana* Expedition, all in the family Gnathophausiidae: *Gnathophausia elegans, G. zoea, Neognathophausia ingens,* and *N. gigas.*

Clearly there are more mysidacean species to be described from New Zealand. Small numbers of specimens have been collected, with most material in New Zealand held at the University of Otago and Auckland University of Technology (Jocqué & Blom 2009).

Morphology, species, and endemism

The carapace is well developed in Mysidacea and covers the thorax, as it does in euphausiids and decapod shrimps, but is fused with the anterior three or four thoracic segments only; the back of the carapace can simply be lifted to expose the posterior four or five thoracic segments. They have a shrimp-like abdomen with fully developed or reduced pleopods, and the telson and paired uropods form a tail fan. Mysidacean eyes are compound and stalked although in a few deep-water species they are reduced to immovable plates. The antennules are always biramous and most have an antennal scale. Of the eight pairs of thoracic appendages, the first one or two are modified as maxillipeds. The remaining six or seven pairs form legs and generally bear swimming exopods. A feature of female mysidaceans is their large leaf-like oostegites, on the inner side of some or all of the legs, which overlap to form a brood chamber or marsupium (recalling the name opossum shrimp) beneath the thorax, in which eggs are laid and the young develop. Both orders have all these characters in common.

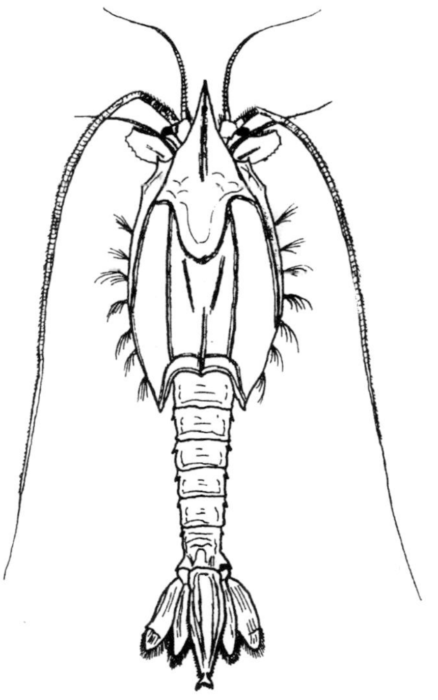

Female *Neognathophausia ingens* (Lophogastrida).
From Sars 1885

In the Lophogastrida, however, gills are present on some or all of the legs, pleura ('side plates') are present on the abdominal segments, and the pleopods are well developed and usually unmodified in both sexes. Lophogastrids also have seven pairs of oostegites but lack statocysts in the endopods of the uropods. All species of Lophogastrida live offshore in meso- and bathypelagic habitats, with many being hyperbenthic (living close to the bottom) in deep water. The largest mysidaceans belong to this order and most occur throughout the world's oceans but are less often seen than species of Mysida, because of their oceanic existence.

A characteristic of the Mysida (excluding all 33 species of the Petalophthalmidae) is the presence of a pair of balancing organs or statocysts, in the telson. Situated near the base of each uropodal endopod, statocysts are an obvious feature, distinguishing mysids from similar animals such as euphausiids (krill). Mysids also lack gills and the pleopods of females are reduced or rudimentary;

those of males are variously modified. Like the lophogastrids, many mysids have seven pairs of oostegites, but there are fewer pairs in some subfamilies of the Mysidae, including the Gastrosaccinae, Mysinae, and Siriellinae, which between them contain 16 of the 18 species of Mysida recorded from New Zealand. Mysids occur throughout the marine environment to deep oceanic trenches but are particularly concentrated in coastal regions, and 24 species have colonised fresh waters around the world.

Adult Mysidacea range considerably in size from 2–3 to 350 millimetres long. The largest are in the Lophogastrida but most species belong to the Mysida and are appreciably less than 100 millimetres long. The few species recorded from New Zealand almost cover this range, with mature females of *Tenagomysis macropsis* as small as 3.2 millimetres long (Greenwood *et al.* 1985) and the largest of all mysidaceans, *Neognathophausia gigas*, also being recorded in New Zealand waters (Fage 1941).

Around 1000 species of Mysidacea have been described worldwide, the great majority in the order Mysida, with some 51 in the Lophogastrida. Twenty-four species have been recorded in New Zealand waters, representing both orders (see end-chapter checklist). Of the three lophogastrid families, the Eucopiidae are not yet known here. Of the four families of Mysida, two are found in New Zealand – the Petalophthalmidae (one unnamed species) and Mysidae (all other species). Globally, this is a very large family, with ca. 870 species. Four of the six subfamilies occur in New Zealand.

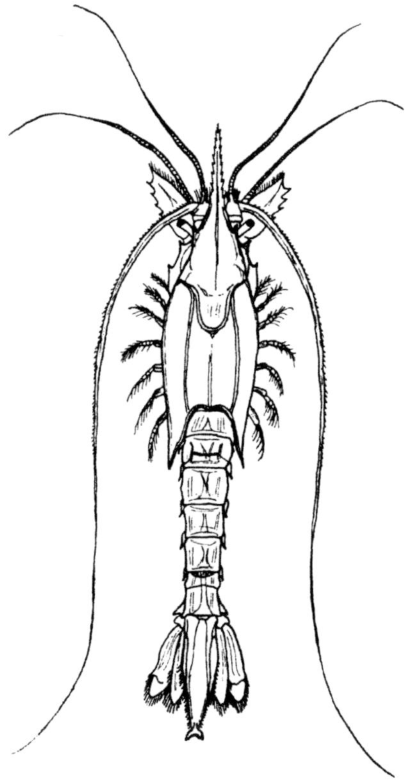

As might be expected in a worldwide group inhabiting a wide diversity of habitats, endemism reflects distribution; no species of the oceanic order Lophogastrida is confined to New Zealand whereas endemism is high in species occupying coastal and littoral waters. Twelve of the 18 species (~67%) of New Zealand Mysida are endemic, including all 10 species of *Tenagomysis* (Müller 1993), but although the genus was first described from Otago (Thomson 1900) it is no longer restricted to New Zealand; five species are known from either Australian or African shores. While *Siriella denticulata* is endemic, *S. thompsoni* is cosmopolitan in its distribution, as one of a minority of epipelagic Mysidacea. The five non-endemic New Zealand Mysida are offshore species, the shallowest among them being *Euchaetomera typica*, another pelagic species, found between the surface and 380 metres. The distributions of the two unnamed species of Mysidacea are not known. Neither *Petalophthalmus* sp. from deep offshore water nor *Lophogaster* sp. in a typically oceanic genus is likely to be endemic.

Male *Neognathophausia gigas* (Lophogastrida).
From Sars 1885

Ecology and distribution

Distributional records of named New Zealand Mysidacea are, for the most part, far from comprehensive, although there are probably records of littoral species in unpublished environmental reports from various parts of the country. Apparently the only records of *Paralophogaster glaber* are those of Tattersall (1923) offshore of Cape Maria van Diemen and the Three Kings Islands in the far north. Te Papa collections indicate that *Neognathophausia ingens* is common around central New Zealand at least as far south as Banks Peninsula, *N. gigas* is present off the east coast of the North Island, and *Gnathophausia zoea* in the Bay of Plenty and on the outer Challenger Plateau west of Cook Strait. The deepest record of any of the mysidacean species found in New Zealand waters is that of *G. zoea*, at 6050 metres (Müller 1993) at a non-New Zealand locality.

The majority of mysidacean species are found on the inner shelf and in coastal and littoral areas and form an abundant component of estuary zooplankton. Many have very localised distributions and can form dense concentrations among rocks and algal beds. Ingles (1973) encountered *Tenagomysis macropsis* in high numbers in association with red algae in Pauatahanui Inlet. All 10 *Tenagomysis* species in New Zealand are coastal pelagic or littoral, and in some cases freshwater dwellers. *Tenagomysis macropsis* is widespread around New

Zealand, from Spirits Bay eastwards almost to the Chatham Islands (Tattersall 1923) and south to Foveaux Strait although the maximum recorded depth of the species is only 24 metres (Bary 1956). New Zealand's southernmost species, *T. tenuipes*, is so far known only from Foveaux Strait and east of Stewart Island (Bary 1956), and from Carnley Harbour, Auckland Islands.

New Zealand has no strictly freshwater species but *Tenagomysis chiltoni* passes through its life-cycle in at least one completely freshwater locality – Lake Oturi, near Waverley, southwestern North Island (Hodge 1964). Thomson (1900) had originally collected *T. chiltoni* from fresh and saline waters in Otago. Jones et al. (1989) confirmed that this species also frequents saline waters in the Avon-Heathcote Estuary (Christchurch) but is an upper estuarine species and was seldom found in salinities greater than 20 parts per million (ppm). Chapman and Lewis (1976) reported *T. chiltoni* and *T. novaezealandiae* as living in brackish water below the *Paratya curvirostris* (Decapoda) zone in streams. Jones et al. (1989) indicated a salinity-correlated ecological separation between *T. chiltoni* and *T. novaezealandiae* in the Avon-Heathcote Estuary with the former in upper reaches and the latter mid- to upper estuarine. In this study and that of Greenwood et al. (1985), *T. macropsis* was found throughout the estuary and had no linear correlation with salinity range along a transect from 4.1 to 30.1 ppm. In his work in Pauatahanui Inlet, Ingles (1973) found distinct differences in distributions between three species in the Horokiwi Stream – *T. macropsis* occurred in the estuary proper, entering only the mouth of the stream, *T. novaezealandiae* centred around the mouth and lower part of the stream, while *Gastrosaccus australis* was highest upstream, not moving as far as the mouth.

Tenagomysis macropsis, the most abundant species in the Avon-Heathcote Estuary, occurs in greatest numbers at salinities between 16.9 and 19.2 ppm, but it is clearly euryhaline as Bary (1956) found it in great numbers in Foveaux Strait (ca. 60,000 individuals in one plankton tow). The results from overnight surface samples in a tideway, taken during his survey of mysidaceans and euphausiids east and south of the South Island, indicated daily vertical migrations by *T. macropsis* and *T. tenuipes*. The numbers of both species at the surface (including juvenile *T. macropsis*), peaked around 2 a.m. Bary's is the only study published to date on vertical distributions of New Zealand Mysidacea.

Swarming is characteristic of mysidaceans (though not as densely as euphausiids) and more complex than it may appear. Mauchline (1980) discussed possible reasons for this behaviour. Concentrations probably result when physical and chemical factors in the water make some areas more habitable than others. Salinity, food availability, light or dark, and age class are all components of swarming behaviour. Conditions change regularly in estuaries and dispersed populations can be forced to aggregate in restricted areas at low tide. Ingles's (1973) work on *T. macropsis* in Pauatahanui Inlet suggested a relationship between shoaling and the tidal cycle. Breeding aggregations also take place, probably more so in deep-water species because littoral mysidaceans regularly aggregate for other reasons but breed at the same time. Data gathered by Greenwood et al. (1985) suggested that *T. macropsis* may undertake seasonal migrations, in common with littoral mysids in other parts of the world (Mauchline 1980). Mature *T. macropsis* females move up-estuary with the rise of temperature in spring whereas Roper et al. (1983) found them closer to the estuary mouth in winter. Aggregation of females over the summer breeding season suggests that this is for breeding purposes. Swarming in currents can also lead to the segregation of age classes, which have differential swimming rates. Swarms are of all shapes from globular to elongated and can be very extensive horizontally in the water but only a few centimetres thick (Mauchline 1980).

Reproduction and development

Mysidacea do not have planktonic larvae as most euphausiids and decapods do. Instead, development of embryos and larvae takes place in the marsupium, from which they emerge as juveniles. Mating usually, if not always, involves the male using its fourth pleopods to deposit sperm in the female marsupium (Mauchline 1980) and eggs are fertilised as they are laid in the marsupium. The number and size of resulting embryos depends upon the size of the eggs and the female and on water temperature and season. Meso- and bathypelagic species tend to have larger eggs and produce somewhat fewer young than epipelagic and coastal species, relative to body size.

The embryo (developmental stage 1) grows to some extent in the egg membrane, moults into a stage 2 (eyeless) larva, and passes through a third (eyed) stage to moult into a juvenile ready to emerge from the marsupium. Juveniles grow to become adults without passing through further stages, although the abdomen increases in proportion to the cephalothorax, and the appendages and telson undergo gradual changes as well. Jones et al. (1989) found the sizes of embryos in Avon-Heathcote Estuary species to be in accord with the range generally found for coastal forms. A range of embryo numbers was also carried in the marsupia of the *Tenagomysis* species: 4–25 in *T. macropsis*, 6–19 in *T. novaezealandiae*, and 22–39 in *T. chiltoni*.

In *T. tenuipes* from Foveaux Strait, Bary (1956) found that females (up to 19.9 millimetres long) shed juveniles of 4.2 millimetres length. Those of *T. macropsis* from the same area were about 2.5 millimetres long, mature females of *T. macropsis* being less than half the length of *T. tenuipes*. Greenwood *et al.* (1985) found emerging larvae of *T. macropsis* to average only 1.47 millimetres in length in the Avon-Heathcote Estuary.

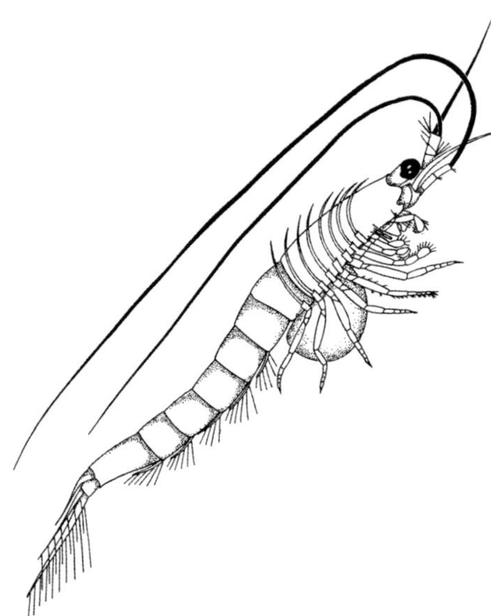
Tenagomysis chiltoni.
After Chapman & Lewis 1976

Food and predators

Dietary studies of Mysidacea are limited and none has been carried out on any New Zealand species, although Chapman and Lewis (1976) considered that *Tenagomysis chiltoni* and *T. novaezealandiae* might be detritus feeders. Chapman and Thomas (1998) subsequently reported predatory feeding in *T. chiltoni*. Mouthparts and thoracic appendages are variously modified in relation to diet. Some species are strict filter feeders, some specialise in grazing phytoplankton, and some are carnivores concentrating on certain substrata such as algae, but most are more opportunistic and eat a considerable range of the organic material in their environments. Mauchline (1980) tabulated the diets of 25 species of mainly northern hemisphere mysids. Though by no means comprehensive, his table showed the major importance of organic detritus, significantly supplemented by diatoms, other algae, copepods, and other crustaceans. Probably most New Zealand shallow-water species have similarly generalised diets, but a few specialised feeders are indicated. One of the most extreme modifications of feeding appendages is of the mandibular palp in *Petalophthalmus* species. It is greatly elongated, projecting well beyond the antennae. Carnivorous *Petalophthalmus armiger* pierces its prey and sucks out the internal contents (Mauchline 1980). *Lophogaster typicus* is incapable of filter-feeding, having mouthparts modified to feed on large lumps of food material on the surface of sediments, suggesting that New Zealand *Lophogaster* sp. could have a similar diet.

Filter-feeding is common in Mysidacea and is accomplished using setose mouthparts and thoracic appendages. The animals 'stand on their heads' above soft substrata, creating a current using the thoracic exopods and filtering particles from the stirred-up sediment. In common with euphausiids (see section on Euphausiacea), some Mysidacea employ a 'food basket', formed by the mouthparts and anterior thoracic appendages, in which food items collected using the mandibular palps are retained until they are chewed and swallowed.

Some species follow diel feeding rhythms, with certain species feeding by day, others only at night. *Gastrosaccus australis* individuals caught by Jones et al. (1989) were virtually all taken at night in the Avon-Heathcote Estuary, suggesting that they feed nocturnally instead of competing with the three *Tenagomysis* species by day.

Mysidacea are important links in the food web between primary producers (e.g. bacteria and microalgae) and secondary consumers, especially in coastal waters. They therefore play a critical role in the cycling of energy through the detrital pathway (Jones et al. 1989). Mysids especially are eaten by a very wide variety of fish and also by decapod crustaceans, seabirds, cetaceans, and other predators. Data on predation of lophogastrids is limited because they live offshore but the size and appearance of the largest species facilitates their recognition in stomach contents. Albacore tuna eat *Neognathophausia ingens*. Deep-sea hyperbenthic rattails eat mysidaceans including *N. gigas*, a species also reported in fin whale stomachs (Nemoto 1959 in Mauchline 1980) and *N. ingens* has been found in the stomachs of pigmy sperm whales stranded in New Zealand (Te Papa data). Weddell seals and gentoo penguins are known to eat the Antarctic mysid *Antarctomysis maxima* and, intriguingly, yellow-nosed albatross near Tristan da Cunha have been found with *N. ingens* and *N. zoea* in their guts (Mauchline 1980). It seems these otherwise extremely deep-living lophogastrids may undertake diel migrations near enough to the surface to be captured by albatrosses. Deep benthic and midwater prawns including *Aristaeopsis edwardsiana*, *Pasiphaea tarda*, and *Aristaeomorpha foliacea*, found in New Zealand waters, have also been found to eat mysidaceans.

Mauchline (1980) cited many studies of the diets of coastal fish that indicated the major significance of mysids as food items. He also noted that mysids tended to be underestimated as prey items because their remains were often mistaken for euphausiids. Little information on mysidaceans in the diets of New Zealand fish is apparent, although Griffiths (1976) reported that introduced European perch in the Selwyn River (Canterbury) eat high numbers of *T. novaezealandiae*. Estuaries such as Avon-Heathcote and Pauatahanui are important as fish nurseries and there is little doubt that the mysids that concentrate there are an important part of their diets. In lakes of the Waikato district, mysids are an important part of the diet of smelt (Northcote & Chapman 1999). Along the coast, seahorses (*Hippocampus abdominalis*) ingest *Tenagomysis similis* along with amphipods and the shrimp *Hippolyte bifidirostris* (Woods 2002), all found in the subtidal kelp beds in which seahorses live.

Mysidacea employ defensive strategies to avoid being eaten, including, as in shrimps and lobsters, tail flexing. While transparent and virtually invisible when swimming, mysids have chromatophores – pigment cells – that enable them to adopt camouflage colours and blend with algae, rocks, or sand. Lophogastrids are uniformly bright red, so can avoid detection by exploiting the lack of penetration of red light in sea water, as do many meso- and bathypelagic decapods. Swarming may also confer some protection on mysidaceans by reducing the number of targets apparent to their attackers.

A wide range of ecto- and endoparasites have been reported in Mysidacea. Very common endoparasites are ellobiopsid protozoans (phylum Myzozoa) such as *Thalassomyces fasciatus*, found in *N. gigas*, *N. ingens*, and *G. zoea*. Choniostomatid copepods parasitise mysidaceans, and epicaridean isopods, particularly of the family Dajidae, are common ectoparasites. Juvenile and small male dajids live in the host's marsupium among the developing larvae.

Economic aspects

Mauchline (1980) reported that thousands of tons of *Neomysis intermedia*, *N. japonica*, and *Acanthomysis mitsukurii* are harvested each year in Japan; *N. intermedia* from brackish lakes is the most important of these and is cooked, dried and eaten. There do not appear to be any other major fisheries for Mysidacea but several species are or have been fished in South-east Asia, China, and Korea by local fishers, who net them when they swarm. Some species have been reared successfully in laboratories, and freshwater species have been successfully transferred to other rivers or lakes as food for fish. It is also possible that some Mysidacea have colonised other habitats by transferring there on ships' hulls or in ballast water.

Future work

There is clearly a need for further taxonomy followed by biological research on the Mysidacea of New Zealand before we can gain a reasonable appreciation of their diversity. Historically, New Zealand has never had the services of a mysidacean specialist but the need for such work is surely increasing given the importance of mysids in the marine economy, particularly as a major food of fish. Once Mysidacea currently held in collections are analysed, further assessment of their diversity, numbers, and roles in the region will require sampling gear and strategies appropriate to the collecting of these generally small and easily damaged animals.

The freshwater amphipod *Paracalliope fluviatilis*.
Stephen Moore

Order Amphipoda: Beach fleas, sand hoppers, and kin

Amphipods are the among the most ubiquitous crustaceans, inhabiting diverse environments from the depths of the oceans' trenches to high altitudes on mountains, living in situations as varied as plankton in the open seas, burrowers in surf beaches, litter-dwellers on forest floors, epizoites on the skin of whales and dolphins, and cryptic inhabitants of subterranean aquifers more than 20 metres below ground level. Amphipods are likely to be found in almost all aquatic habitats, as well as on land wherever water is freely available or humidity is high. In many of these situations, species are numerous and numbers high, frequently overwhelmingly so. It is surprising, therefore, that they have received relatively little scientific attention.

The name of order is derived from two Greek words – *amphi*, both or of two kinds, alluding to the forward orientation of the anterior legs and the backward and/or lateral orientation of the posterior legs (Stebbing 1888), and *podos*, foot.

The relative neglect of amphipods as subjects for scientific study in New Zealand may be because of two related attributes – their biodiversity is bewildering and different species are often not easily distinguished by the untrained eye. The trained worker, on the other hand, finds the myriad variations on the basic morphology fascinating, continually generating questions about relationships between taxa and the selective value of the differences in morphological structures.

The basic amphipod body plan is difficult to define because of the group's diversity. Amphipods are distinguished from other peracarids (malacostracan crustaceans that brood their eggs and young) by the following combination of characteristics: body generally laterally compressed, carapace absent, eyes sessile and usually lacking cuticular facets, pereon (thorax) with seven pairs of unbranched limbs, pereopods (legs) 1–4 orientated anteriorly, pereopods 5–7 directed posteriorly, pereopods 1–2 usually modified as subchelate (grasping) gnathopods, coxal gills present on some pereopods, pleon (abdomen) segments 1–3 with multi-articulate swimming appendages (pleopods), usually biramous, pleon segments 4–6 (urosomites) with stouter, biramous uropods, the final urosomite with a distinct telson.

Some 6000 species in about 120 families are known worldwide (Barnard & Karaman 1991). Estimates suggest that several thousand species await discovery and scientific description, despite more than 100 new species being described annually, on average, during the mid-1980s. The order is divided into three suborders – Ingolfiellidea, Gammaridea, and Hyperiidea; caprellids (formerly Caprellidea) are now regarded as specialised gammarideans.

Paradexamine houtete.
From Barnard 1972a

Historical overview

Knowledge of the New Zealand amphipod fauna began with Dana's (1852, 1853–55) descriptions of a few species, but accelerated with G. M. Thomson's and Charles Chilton's work. Thomson's (1879b) first paper was followed by 14 more over the next 34 years; that of Chilton (1882a) was succeeded by 52 papers by 1926, although not all dealt with New Zealand species. Thomson and

Chilton's (1886) 'Critical list of the Crustacea' contained 71 amphipod species names: 63 gammarideans, four hyperiids, three caprellids, and one cyamid.

Chilton was the strongest influence on early New Zealand amphipod systematics. He himself was influenced by Della Valle's (1893) attempt to combine many of the world's Gammaridea into fewer species and he treated many New Zealand species as variants of extrinsic taxa (Barnard 1972a). This tendency was exacerbated in his later career by his acquaintance with research on phenotypic variation of amphipods at Plymouth (England). This led him to regard many New Zealand species as phenotypes of sub-cosmopolitan species (Barnard 1972a) or as variants of local species (Fenwick 2001a). Significant contributions were also made by Stebbing (1888, 1910) through his work on local collections made by the *Challenger* and *Thetis* Expeditions. Also notable are Walker's (1908) work on subantarctic material, K. H. Barnard's (1930) studies of *Terra Nova* Expedition collections from the far north of New Zealand, and Stephensen's (1927) and Nicholls's (1938) studies of subantarctic amphipods.

A new phase of New Zealand amphipod systematics began in the 1950s with D. E. Hurley's detailed papers (1954–75) on gammarideans, hyperiids, and cyamids. Several problems were resolved, new species described, and many previously described species clarified. Extensive collections from New Zealand's deep waters were made during the Danish Deep-Sea Expedition, 1950–52, on the *Galathea*. Dahl's (1959) and Barnard's (1961) reports on these collections added considerably to knowledge of our fauna. In none of the preceding investigations, however, was there any attempt to collect amphipods widely in New Zealand waters in order to gain understanding of species' distributions. This, however, was the approach followed by J. L. Barnard during his 1967–68 visit. The resulting monograph (Barnard 1972a) made a preliminary assessment of the biogeography of the New Zealand gammaridean fauna, described numerous new taxa, and provided the most comprehensive guide to date of the fauna (although its focus was algae-living amphipods). Barnard's visit and monograph stimulated much subsequent local interest in the gammaridean fauna (Cooper 1974; Cooper and Fincham 1974; Hurley and Cooper 1974; Fincham 1974, 1977; Lowry 1979, 1981; Fenwick 1976, 1977, 1983; Myers 1981; Lowry and Fenwick 1982, 1983; Moore 1983a,b, 1985; Lowry and Stoddart 1983a,b, 1984).

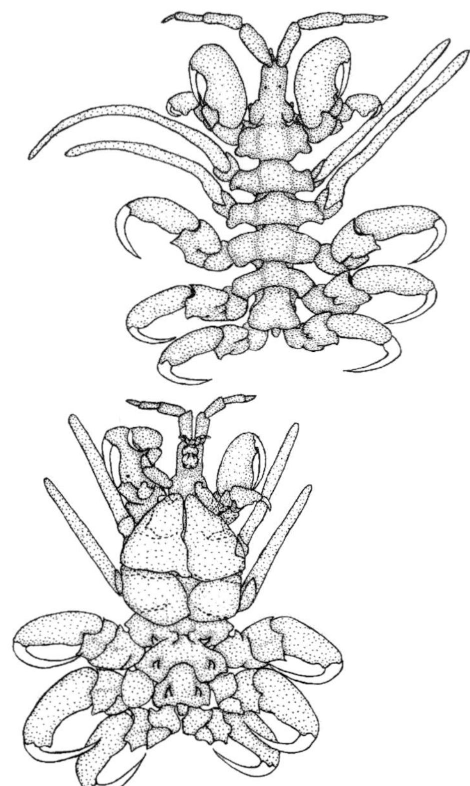

Cyamis boopis.

From Hurley 1952

New Zealand freshwater amphipods were studied by Hurley (1954a, c, f) over this period, as were terrestrial amphipods (Hurley 1955a, 1957a, c). Bousfield (1964) and Duncan (1968) also investigated the terrestrial amphipods. Subsequently, Duncan (1994) substantially reviewed this group, recognising several new genera and species.

Elements of the New Zealand hyperiid fauna were reported by Stebbing (1888) and K. H. Barnard (1930). After about 1950, hyperiids and caprellids were usually investigated and reported separately from gammarideans, with Fage (1960), Shih (1969), and Hurley (1955b) exploring the fauna more fully. Much of this information is brought together in Vinogradov's (Vinogradov et al. 1996) substantial review of the world hyperiids, with Zeidler (2003a, b, 2004a, b, 2006, 2009) refining the group's systematics and adding further new records. The New Zealand caprellids were reviewed by McCain (1969) and he described one new species subsequently (McCain 1979).

Amphipod diversity in New Zealand currently stands at 500 species, of which 64 are undetermined or undescribed.

Amphipods in the ecology of New Zealand

The general abundance of amphipods means that, despite their small individual size, collectively they are important in the ecology of many ecosystems, especially as food for larger animals. Huge densities of amphipods are found among New Zealand seaweeds, in which they often dominate the associated fauna (Fenwick 1976; Taylor 1998).

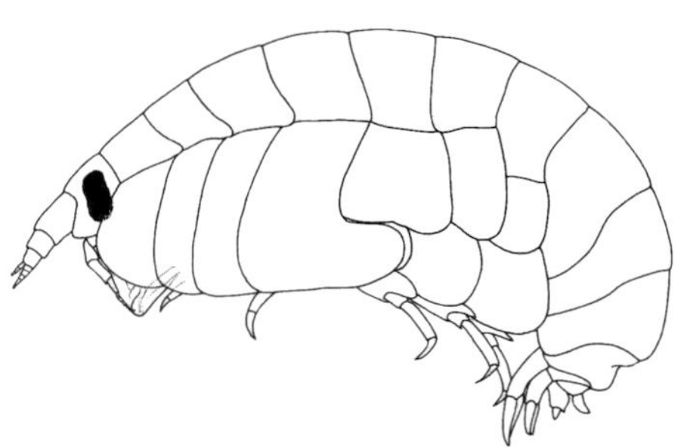

Parawaldeckia angusta.

From Lowry & Stoddart 1983a

Several studies have demonstrated the importance of gammaridean and

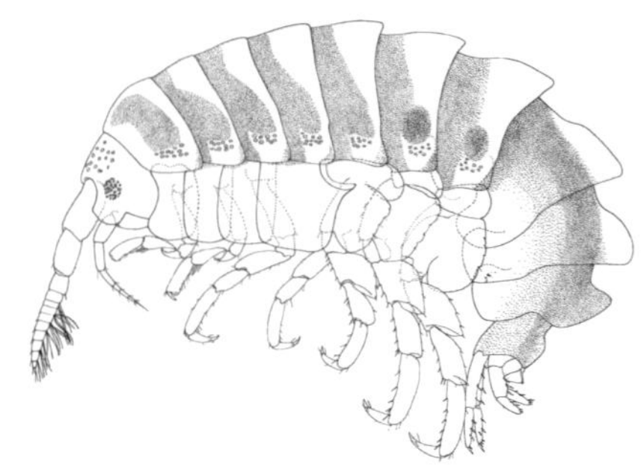
Waitomo manene.
From Barnard 1972a

hyperiid amphipods as food for fish and birds in New Zealand. Amphipods were the most frequently utilised food item among 26 species of common northern New Zealand reef fishes (Russell 1983). Indeed, Jones (1988, p. 454) considered 'the importance of gammaridean amphipods as a food source … startling' for juvenile fish. They were the principal food item for adults of several species and formed important secondary foods for the others (Russell 1983). These amphipods were mostly gammarideans and caprellids but some planktivorous fishes ate a few hyperiids. Small fish species were most dependent upon amphipods for food. Amphipods were eaten by 75–90% of specimens and comprised 40–60% of diet by volume in the various triplefin species (Russell 1983). A few large species also fed extensively on amphipods. Over half of all red moki, blue moki, trevally, goatfish, and juvenile snapper ate gammaridean amphipods, which made up 40%, 38%, 51%, 55%, and 62%, respectively, of their food by volume (Choat & Kingett 1982; Russell 1983). A similar study at Kaikoura (Duffy 1989) confirmed the importance of amphipods as food for fishes and showed their increased consumption by fishes inhabiting brown seaweeds of semi-sheltered, southern shores.

Amphipods are important food for some fishes inhabiting soft bottoms and estuaries also. Adults of nine species of fish in the Avon-Heathcote Estuary all ate amphipods, although they were a common (> 10%) food item for three species only – common sole (13%), cockabully (68%), and common bully (74%) (Webb 1973). Although amphipods were scarce in the diets of yellow-bellied and sand flounders in the estuary (Webb 1973), their juveniles fed almost exclusively (92–96% of food items) on the small tube-dwelling amphipod *Paracorophium excavatum* (Nairn 1998). Offshore, however, larger amphipods were common items (33%) in adult yellow-bellied flounders' diets (Knox & Fenwick 1978).

Fish also eat pelagic hyperiid amphipods. Warehou, banded rattails, javelin fish, black oreos, southern blue whiting, carinate rattails, small-scaled brown slickhead, and small-scaled nototheneids all include amphipods as substantial components of their diets. Many of these fishes fed extensively on amphipods when smaller (up to 37% of food weight and eaten by up to 75% of small fish), with individual fishes taking larger prey as their sizes increased (Gavrilov & Markina 1981; Clark 1985; Rosecchi et al. 1988; Clark et al. 1989). Amphipods were a minor element of the diets of several other deeper-water fishes, notably hoki, smooth oreos, smooth rattails, and orange roughy. Pelagic fishes are the usual predators of these amphipods, but benthic fishes may feed extensively on hyperiids when swarms are carried into shallow water. At The Snares, the demersal telescope fish, as well as spotties, banded wrasse, and benthic nototheneid cod, fed intensively on hyperiids (*Themisto gaudichaudi, Hyperietta luzoni*) and krill swarming close to the surface (Fenwick 1978).

The importance of amphipods in freshwater fishes' diets varies with species, amphipod abundance, abundance of other prey items, and fish size. Long-finned and short-finned eels, whitebait (*Galaxias maculatus*), mudfish, common smelt, and brown trout all eat small numbers of the common stream amphipod, *Paracalliope fluviatilis* (McDowall 1968; Eldon 1979; Ryan 1986; Jellyman 1989; Sagar & Glova 1995, 1998; Hicks 1997). Typically, amphipods comprise less than 5% of whitebait food, but more are eaten with increasing fish size (McDowall 1968). Amphipods are commoner in the diets of whitebait closer to estuaries than those further upstream and, in some rivers, amphipods comprise up to 45% of the diet (McDowall 1968). Similar variation in the consumption of amphipods occurs in eels. Amphipods (*Paracalliope fluviatilis* and the brackish *Paracorophium excavatum*) may be a major (70%) or minor (< 0.01%) food for short-finned eels, depending upon the specific habitat, season, and eel size, with amphipods being most important for small eels 100–190 millimetres long. Similarly, juvenile brown trout feed preferentially on amphipods, which make up 80% of food items of trout inhabiting tree-lined sections of some rivers.

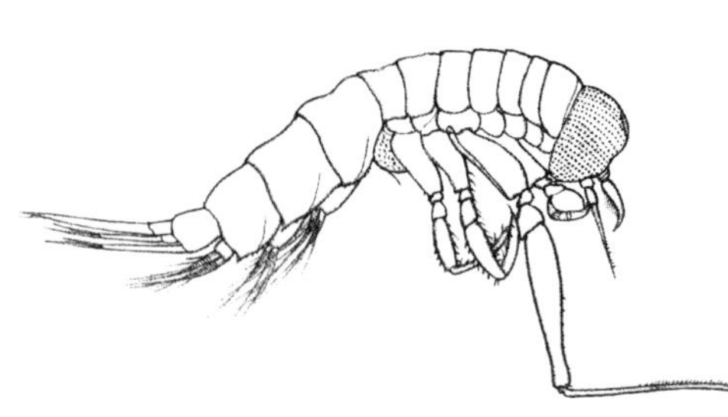
Themisto gaudichaudi.
From Stebbing 1888.

Birds also feed on marine and estuarine amphipods. A number of oceanic birds typically feed extensively on hyperiid amphipods. Red-billed gulls, cape pigeons, Buller's mollymawks, and sooty shearwaters fed on hyperiid swarms at The Snares, with the latter two diving below the surface to catch them at times (Fenwick 1978). Fairy prion chicks are fed a diet comprising 14% amphipods by weight, diving petrels consume 17% by weight of amphipods, and grey-faced storm petrels at the Chatham Islands include four species of amphipods in their diet (Prince & Morgan 1987).

Numerous other New Zealand birds eat amphipods as larger or smaller components of their diets. For example, most penguins are believed to include these crustaceans in their diets (Croxall & Lishman 1987). In North American estuaries, some migratory waders consume 10,000–22,000 corophiid amphipods per day (Wilson 1989). Related species (plovers, dotterels, and wrybills) in New Zealand probably eat appreciable quantities of amphipods. Ground-foraging, insectivorous birds (e.g. robins, fernbirds, tits, and wekas), as well as blackbirds and song thrushes, are almost certain to include land hoppers from among plant litter in their diets. In addition, gulls and other birds probably capture beach-fleas from amongst wrack at times.

Diversity of New Zealand amphipods

Ingolfiellidea

Ingolfiellids are highly specialised, mostly small (< 3 but up to 14 millimetres long), worm-like animals adapted to living interstitially in marine and freshwater sediments, as well as in groundwaters. Marine species occur from the intertidal to the deep sea. Widely regarded as very primitive amphipods, over 30 species are known from two families. They are reported from most continents, including Australia, and two species from New Zealand interstitial marine habitats (Schminke & Noodt 1968) remain undescribed.

Whale louse *Scutocyamus antipodensis*.
From Lincoln & Hurley 1980

Caprellidea

In a detailed cladistic analysis, Myers and Lowry (2003) demonstrated that caprellids and cyamids are specialised corophiidean amphipods. They are discussed separately here but the end-chapter checklist follows Myers and Lowry. The Caprellidea includes two distinct families, both found worldwide – the skeleton shrimps (Caprellidae) and whale lice (Cyamidae). Whale lice live ectoparasitically on whales and dolphins, whereas caprellids are benthic and often extremely abundant among algal fronds and on bryozoans, hydroids, and sea stars intertidally and on shallow marine bottoms. Each group's body form is very different, although both possess rudimentary abdomens and vestigial abdominal limbs. Whale lice have short, flattened bodies with powerful limbs adapted to grasp their hosts' skin firmly. Caprellids have long slender bodies and their last three pairs of legs, grouped posteriorly, are modified for grasping the substratum, leaving their anterior legs and antennae free for feeding.

Caprellids are quite diverse, with about 85 genera worldwide (McCain & Steinberg 1970; Laubitz 1993). The New Zealand skeleton-shrimp fauna comprises just eight species in six genera, belonging to two subfamilies (McCain 1969, 1979; Guerra-García 2003). Half (four) of these species are endemic. Eight species of whale lice in four genera are known from New Zealand (Hurley 1952; Lincoln & Hurley 1980), whereas the worldwide cyamid fauna comprises 27 described species in six genera (Martin & Heyning 1999). If, however, cyamids known to occur on whale and dolphin species reported from New Zealand waters are considered, the total cyamid fauna may number some 19 species in all six known genera.

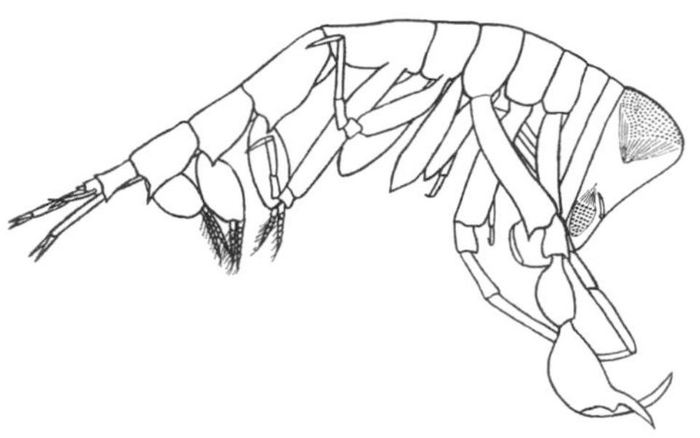

Phronima sedentaria.
From Hurley 1955b

Hyperiidea

Hyperiid amphipods are purely pelagic, living freely in the ocean or associated with other pelagic invertebrates, from the surface to abyssopelagic depths (> 7000 metres) (Vinogradov et al. 1996). Species living near the surface typically

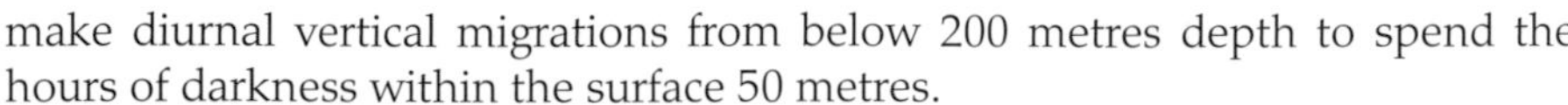

make diurnal vertical migrations from below 200 metres depth to spend the hours of darkness within the surface 50 metres.

A great variety of body shapes occurs within the hyperiids, making them extremely difficult to characterise. Large eyes and/or an inflated head and variously reduced first thoracic segments or pleon and urosome are common (e.g. Hyperiidae), although the opposite is true in others (e.g. Scinidae). The very compact forms of many surface dwellers (e.g. Platyscelidae) contrast with the needle-like shapes of *Rhabdosoma* species. Lengths also vary widely from 2.5 millimetres (e.g. *Hyperietta luzoni*) to over 150 millimetres for the extremely elongate *Rhabdosoma armatum*.

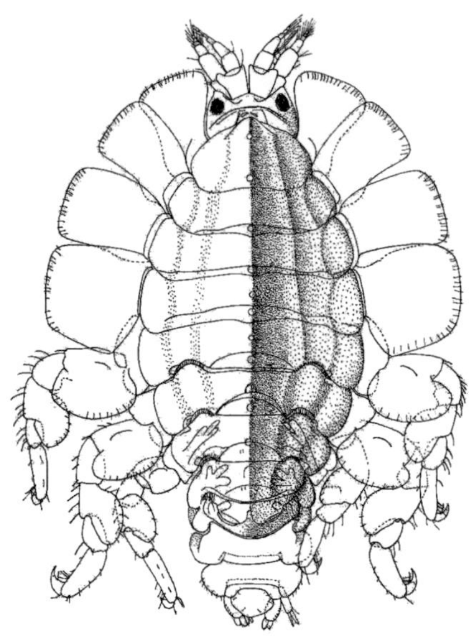

Iphinotus typicus.
From Barnard 1972a

Some hyperiids live on and within one or a few species of jellyfish, siphonophores, and ctenophores. The relationship between host and amphipod seems uncertain, but the consistent pairings of some species (e.g. *Hyperia macrocephala* is found only on the jellyfish *Desmonema gaudichaudi*) indicate commensalism. Host tissues and other prey items in the guts of these amphipods suggest that the amphipods behave opportunistically, with no obvious advantage to the host. Species of the family Phronimidae apparently eat the viscera of pelagic tunicates, siphonophores, and heteropods and use the prey's transparent covers as a refuge against predators and for rearing their eggs.

Over 240 species of hyperiid in more than 72 genera and 23 families are known from the world's oceans. It is difficult to characterise the New Zealand fauna because of the hyperiid pelagic habitat. Many hyperiids have very wide distributions (Vinogradov et al. 1996), so it seems inevitable that most widely distributed species will be found in local waters eventually (Zeidler 1992), depending upon movements of the specific water masses with which they tend to be associated (Young & Anderson 1987). Thus, New Zealand's hyperiid fauna probably exceeds the reported 94 species in 49 genera reported from our surrounding seas (Hurley 1955b; Kane 1962; Vinogradov et al. 1996; Zeidler (2003a, b, 2004a, b, 2006, 2009) and a total fauna in excess of 100 species seems probable.

Gammaridea

The Gammaridea is the most abundant, ubiquitous, and diverse of the amphipod suborders. More than 5800 species in about 1100 genera are known, some from hadal depths exceeding 10,000 metres (Dahl 1959) and others higher than 4000 m above sea level (Stebbing 1888). Gammarideans range in length from about 2–3 millimetres to a whopping 340 millimetres for the abyssal *Alicella gigantea* (Barnard & Ingram 1986). Large size appears to be associated with higher dissolved-oxygen concentrations in cold-water habitats, and warm-water faunas are dominated by very small species. These amphipods also seem most abundant and diverse in temperate to cool climates, with tropical faunas being relatively inconspicuous, although surprisingly diverse (Thomas 1993). Gammarideans are often referred to as the laterally compressed amphipods. Land-hoppers, beach-fleas, and many aquatic amphipods certainly have the typical shape. However, several tube-dwelling and nestling genera have elongated, more vermiform, shapes. Burrowers in surf beaches (urothoids and some phoxocephalids) are wide-bodied, presumably for stability in high-energy habitats. *Iphinotus typicus* is even more flattened. Its limpet-like shape adapts it for life on the fronds of smooth brown seaweeds on New Zealand's turbulent rocky shores.

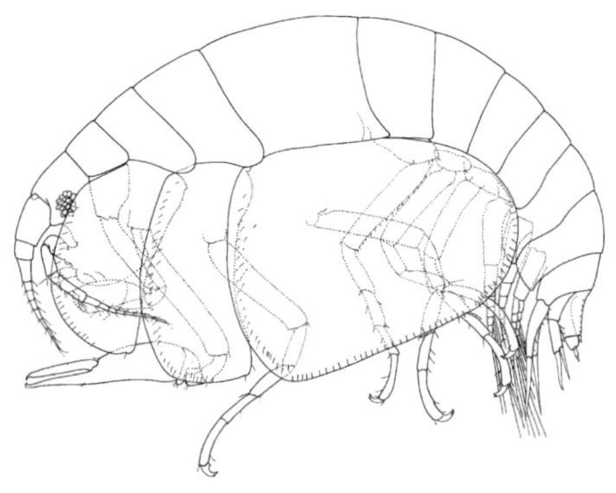

Raukumara rongo.
From Barnard 1972a

Marine and freshwater gammarideans are predominantly free-living and benthic. A few are planktonic and others form close associations with algae, hydroids, bryozoans, and a variety of other invertebrates. Members of some families build tubes, nests, or columns from strands of material secreted from glands in their anterior legs, variously incorporating mud, sand, shell, bryozoan fragments, and other particles from their habitats. Species of yet other families characteristically burrow in soft sediments, at times burrowing to more than 200 millimetres beneath the sediment surface. Scavenging, detritivory, and omnivory are the predominant feeding habits, but predation, ectoparasitism on fish, and herbivory also are known (Bousfield 1987; Enequist 1949; Lowry & Stoddart 1983b; Sainte-Marie & Lamarche 1985; Haggitt 1999).

The New Zealand gammaridean fauna (including caprellids and cyamids) comprises 401 species (62 undescribed) in 192 genera (10 unnamed), belonging to 55 families. [Figures below indicate that New Zealand's total gammaridean amphipod diversity is probably 3–4 times geater than the total reported here.] This equates to about 5.6% of the world's described species and 15.8% of world genera, representing over a third of all families. Some 74% of the species (296) are endemic, as are ~29% (55) of the genera. The fauna inhabiting each of three major habitats in New Zealand is discussed separately below.

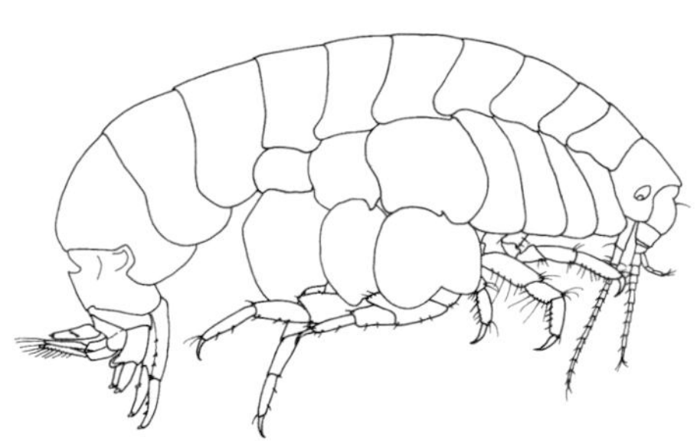

Paracentromedon? whero.
From Fenwick 1983

Terrestrial amphipods

All terrestrial species belong to the Talitridae, the only amphipod family to have successfully occupied terrestrial habitats worldwide. These amphipods inhabit gardens, forest floors, and grasslands, where they live in litter, under trees and rocks, or in burrows that they construct themselves. Some 36 species in 10 genera occur in New Zealand (Duncan 1994; Fenwick & Webber 2008). Beach fleas are usually considered with terrestrial species, and 11 species in three genera are known from shore environments, although their revision seems overdue. Most New Zealand talitrids are endemics, but there are at least three aliens. New Zealand species range in length from c. 5–6 to > 50 millimetres for the giant subantarctic *Notorchestia aucklandiae*. Land hoppers and beach fleas occur throughout New Zealand, including the subantarctic islands, from sea level to over 2000 metres.

Freshwater amphipods

Some 53 species (~30 undescribed) in nine named (and 10 additional unnamed new) genera belonging to eight families are reported from freshwater habitats in New Zealand (Fenwick 2001a,b). Several undescribed species from hypogean water (saturated sediments beneath or beside streams and rivers (hyporheic) and groundwater) are currently under investigation and others from epigean (surface) waters await description (Fenwick 2000). Within these habitats, amphipods are often surprisingly abundant, but have received little attention. This relative neglect probably reflects their small adult sizes (3–6 millimetres body length), although two hypogean species (*Phreatogammarus fragilis* and *Ringanui toonuiiti*) grow to over 20 millimetres long. All New Zealand freshwater species, five named genera, ca. 10 unnamed genera, and three families are endemic.

Marine and estuarine amphipods

The New Zealand marine and estuarine amphipod fauna comprises some 365 species. Amphipods inhabit every conceivable habitat in the sea, although few species live in estuaries. They are predominantly benthic, living in and on mud and sand and rocky bottoms, as well as among other invertebrates and algae. The total diversity of the New Zealand marine amphipod fauna is difficult to estimate, but is likely to comprise at least three times the presently known species. Of the known marine fauna, 194 species (~53%) and 35 genera (19%) are endemic.

Special features of the New Zealand gammaridean fauna

Biodiversity and abundance

Amphipods are frequently a major component of marine benthos, especially in cool-temperate to cold-water environments. New Zealand is no exception in this respect. A study of animals inhabiting the green alga *Caulerpa brownii* at

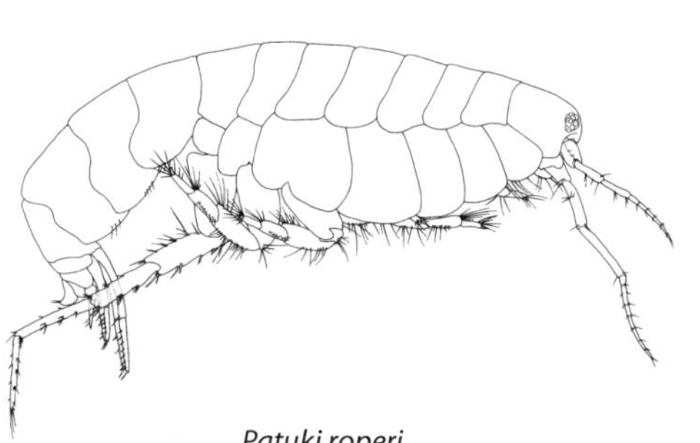

Patuki roperi.
From Fenwick 1983

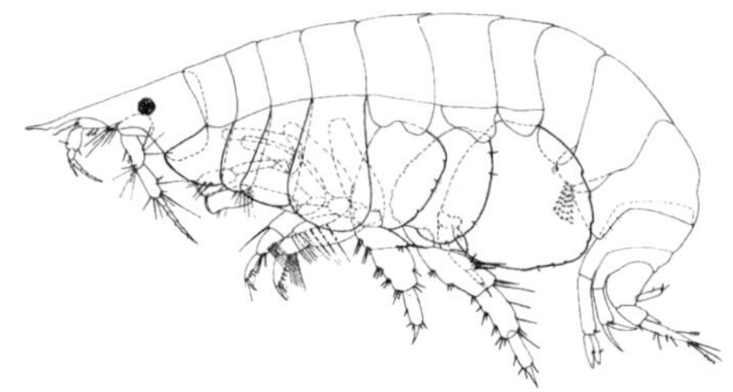

Ringaringa littoralis.
From Cooper & Fincham 1974

Kaikoura on the South Island east coast revealed a fauna dominated by huge numbers of amphipods – up to 12,000 per 200 grams wet weight (handful) of alga (Fenwick 1976). Some 61 species occurred in this specific habitat. Amphipod abundance increased dramatically with increased exposure to wave action, but fewer species predominated. Thus, the fauna at more sheltered sites comprised lower densities, with more species having more equal abundances.

Shallow sand bottoms at Kaikoura illustrate amphipod abundance in another near-shore habitat. Four species of amphipods and a large myodocopid ostracod comprise most of the fauna in this habitat. Amphipod densities average about 6000 per square metre, fluctuating from a winter low of 4000 to a summer high of more than 12,000 per square metre (Fenwick 1985). Crowding of these crustaceans is reduced by each species occupying a different depth in the sediment (Fenwick 1984) – cryptically coloured, surface-skipping *Patuki roperi* lives in the top 25 millimetres of sand, smaller white *Ringaringa littoralis* dwells at about 40 millimetres depth, bright red *Paracentromedon*? *whero* inhabits mid-depths (50–80 millimetres), and large *Protophoxus australis* overlaps at mostly 65–85 millimetres. *Leuroleberis zealandica*, a very large ostracod, is most abundant at 75–100 millimetres depth. Species' mean depths in the sediment change slightly between sand ripples (150–200 millimetres high) and troughs, as well as with season.

Amphipods are a significant component of surf-zone faunas on New Zealand's exposed beaches, such as in Pegasus Bay (Fenwick 1999). These small, frail-appearing crustaceans not only survive in these highly turbulent situations, but some species are found nowhere else. Amphipod densities peak just outside the zone of wave break, at about six metres depth in Pegasus Bay. Biodiversity of the amphipod fauna changes markedly with depth and, hence, changes in wave-induced turbulence, with most species abundant in only one depth zone. All but one of the abundant inshore (3–10 metres depth) species are free-living active burrowers of the family Phoxocephalidae.

These three studies demonstrate some key aspects of marine amphipod biodiversity. Perhaps most significantly, amphipods are a very important component of faunas inhabiting many of the shallow marine habitats around New Zealand. Not only are amphipods abundant in many of these habitats, but also their biodiversity is high. Individual species of amphipods are very sensitive to small changes or variations in their environments, resulting in marked changes in faunas within and between habitats. Species within some families exhibit very different tolerances of environmental factors, indicating that species or genus may be more useful levels of taxonomic resolution for amphipods in ecological investigations.

New Zealand Phoxocephalidae

Phoxocephalids are the typical amphipods of the surf beaches and sandy shores that make up so much of New Zealand's coastline. Fifteen (88%) of the 17 phoxocephalid species known from New Zealand are endemic. Eight (53%) of the 15 genera to which these species belong are endemic and monospecific. This generic diversity and endemism is remarkably high. Museum collections indicate that the fauna includes 15 or more undescribed species, indicating over 30 species of phoxocephalids in New Zealand.

The Australian shallow-water phoxocephalid fauna consists of 89 species in 26 genera (comprising 40% of the known phoxocephalid species worldwide), with 23 of these genera endemic (Barnard & Drummond 1978; Barnard & Karaman 1983). Despite the high biodiversities of both the Australian and the New Zealand phoxocephalid faunas, there is little overlap between the two. Only one shallow-water genus (*Booranus*?) seems to be shared between New Zealand and Australia, although three deep-water genera (*Cephaloxoides, Harpiniopsis, Protophoxus*) and two of their species are found on both sides of the Tasman Sea.

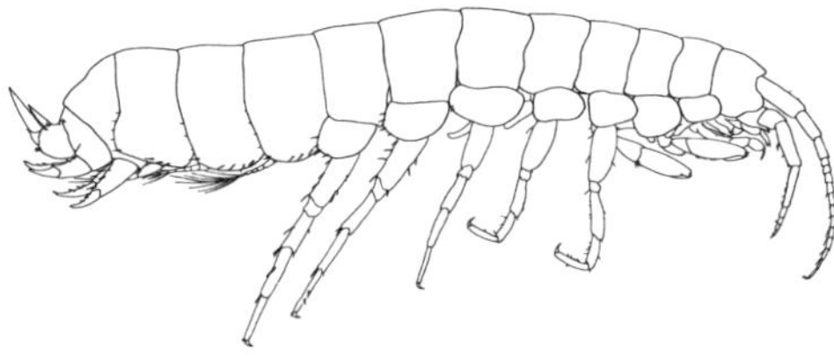

Paracrangonyx compactus.
From Fenwick 2001

Australia is regarded as the epicentre of phoxocephalid evolution because of high diversity of species and genera and high generic endemism (Barnard & Karaman 1983). The subantarctic islands of South America are the only other centre of phoxocephalid radiation, with distinctive attributes present among its species and genera. New Zealand's location between Australia and South America indicates that the New Zealand phoxocephalid fauna is likely to be both diverse and of special biogeographic interest.

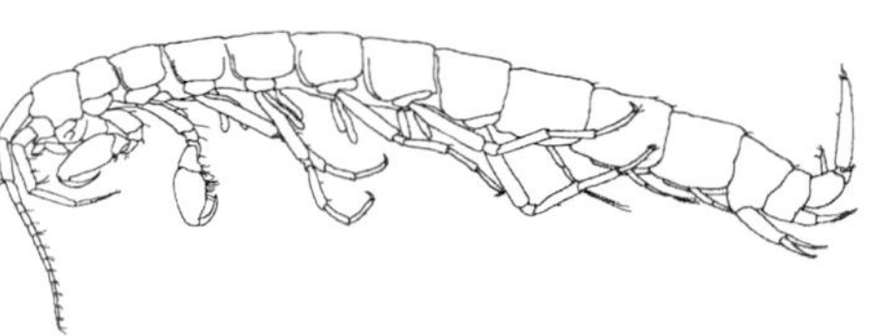

Paracrangonyx winterbourni.
From Fenwick 2001

Groundwater amphipods

Late in the 19th century the biological world was intrigued by Chilton's (1882a,b, 1884, 1894) reports of crustaceans living within aquifers of the Canterbury Plains. Following this initial work, the groundwater received scant attention. Subsequent workers, including Chilton himself (e.g. 1912, 1924), apparently assumed no additional species, assigning specimens to known taxa without critical examination.

During the 1970s Guillermo Kuschel of the former DSIR surveyed groundwater faunas by pumping wells throughout the country. Ten new gastropod mollusc, 71 mite, and two water-beetle species were described from these collections (Scarsbrook et al. 2003). The several amphipods from Kuschel's collections await full investigation, but preliminary work (Fenwick 2000) revealed several new taxa. Current collecting effort indicates the existence of a further 20–30 species of groundwater amphipods.

The described hypogean (groundwater) amphipod fauna of New Zealand comprises four species in three endemic genera (two of which have epigean representatives) each belonging to quite different families. Two of the hypogean families are endemic. Given the number of species, this fauna seems remarkably diverse at generic and familial levels. Preliminary work indicates that the New Zealand hypogean amphipod fauna appears dominated by paraleptamphopids and is very different to that of Australia, where hadzioids and crangonyctioids predominate (Bradbury & Williams 1997). Taxonomic work on these collections is required to determine the true diversity, to determine relationships with the Australian freshwater amphipod fauna, and to make the fauna accessible to ecologists.

Should we be surprised by a high diversity of groundwater amphipods in New Zealand? Groundwater volumes in New Zealand are huge and probably several times greater than volumes within surface waters (lakes and rivers). For example, the groundwater of the Golden Bay region is estimated to approximate the volume of water in Lake Taupo. There are extensive aquifers beneath most of the Canterbury Plains to depths of 350–550 metres. This is not simply all water, but variably porous gravels with water moving through interstices. Obviously, there is a huge volume of water beneath the plains. Other parts of the country also comprise large plains of porous alluvial gravels (e.g. Waimea Plains around Nelson, the Heretaunga Plains of Hawke's Bay) containing extensive aquifer systems. Given the very large habitable volumes available and the apparent barriers to dispersion between each groundwater system, a high amphipod biodiversity should not be unexpected.

Investigations at one site in Canterbury indicate that groundwater amphipods help to maintain the quality of Canterbury's groundwater (Fenwick et al. 2004). The three known amphipod species, as well as a large subterranean isopod (*Phreatoicus typicus*), congregate at sites of organic enrichment from sewage-oxidation-pond effluent. A series of field and laboratory experiments showed that these animals browse on non-living organic slime layers from sediment and stone surfaces (Fenwick 1987). Extrapolation of experimental results using conservative estimates of crustacean densities indicates that the two dominant amphipods remove large amounts of organic carbon annually in the vicinity of the disposal area.

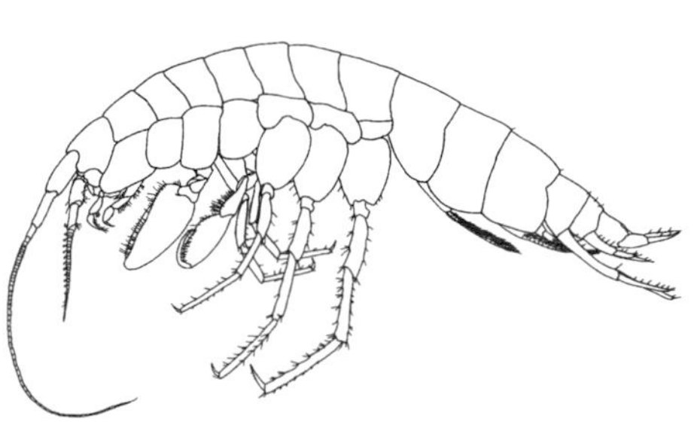

Ringanui toonuiiti.
From Fenwick 2006

Further understanding of the biology of these valuable groundwater systems depends on documenting and monitoring their biodiversity to facilitate

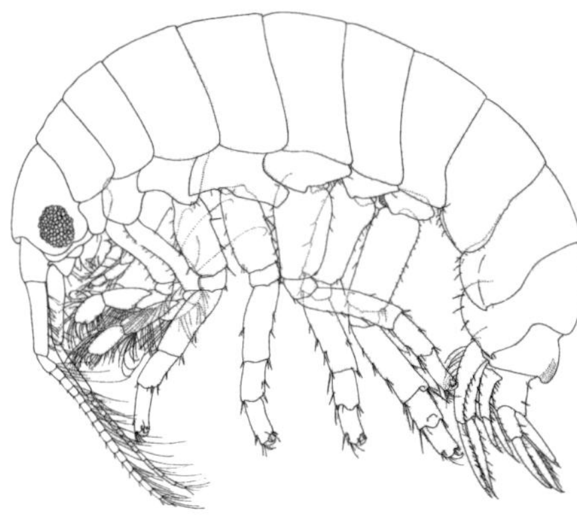

Polycheria obtusa.
From Barnard 1972a

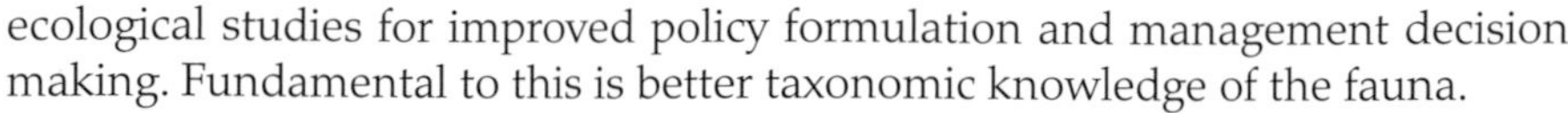

ecological studies for improved policy formulation and management decision-making. Fundamental to this is better taxonomic knowledge of the fauna.

Biogeography of the freshwater fauna

Some New Zealand freshwater amphipods have attracted considerable interest from workers seeking to untangle phylogenies and relationships between faunules of Gondwana and other landmasses. Two endemic genera are of special interest. *Phreatogammarus* was seen as 'an amazing antiboreal morphological counterpart of the Holarctic crangonyctids' (Barnard & Barnard 1983, p. 51), a group now largely confined to North America. This genus was considered to be 'perhaps the most primitive [living] gammarid' (*ibid.*, p. 420) that is 'now a perfect relict' (Barnard & Barnard 1982, p. 264). The absence of any significant amphipod fossils increases the significance of *Phreatogammarus* to evolutionary biologists. The morphologies of both *Phreatogammarus* and *Paraleptamphopus*, a modern derivative from a *Phreatogammarus*-like ancestor (Barnard & Barnard 1983), are incompletely known. Thus it is difficult to establish the relationships of these two genera with other genera.

Other New Zealand freshwater amphipod genera are also distinctive and have intriguing faunal relationships. *Paracalliope*, a genus with three New Zealand species and Australian, Philippine, New Caledonian, and Fijian representatives, is calliopiid-like, but sufficiently distinctive to justify placement in a separate family, the Paracalliopiidae, which has one other genus (Barnard & Karaman 1982, 1991). The endemic genus *Chiltonia*, together with the closely related *Afrochiltonia*, *Austrochiltonia*, and *Phreatochiltonia*, comprise the subfamily Chiltoniinae from New Zealand, Australia, and South Africa (Barnard 1972b). Yet another endemic genus poses biogeographic and phylogenetic problems. Bousfield (1977) moved the genus *Paracrangonyx* into his superfamily Bogidielloidea, re-assigned it to the Crangonyctoidea (Bousfield 1978), thence (Bousfield 1982, 1983), along with three other disparate genera, to the family Paracrangonyctidae within his superfamily Liljeborgioidea. Barnard & Barnard (1983, p. 52) placed *Paracrangonyx* among the bogidiellid gammaroids 'for the moment'. Following careful analysis, Koenemann and Holsinger (1999) found the genus to be most closely related to three genera from each of Western Australia, Madeira, and East Africa. After reviewing these placements and rediagnosing the genus, Fenwick (2001b) concluded that the relationship of *Paracrangonyx* to other genera remains uncertain, but that it belongs within the crangonycoid cluster and is close to the Paramelitidae, as well as showing relationships to other genera of Australian hypogean amphipods.

Many of these taxa have not been re-examined since their first collection. The original specimens of some species are in very poor condition and the illustrations and descriptions of some are inadequate. Consequently, many older taxa must be revised before descriptions of new taxa can take place.

Special associations

The ecology of New Zealand amphipods is generally poorly known and few associations with other invertebrates are reported. Gammaridean amphipod associations with other crustaceans, ascidians, sponges, hydroids, echinoids, molluscs, and other organisms elsewhere are well documented (e.g. Vader 1978, 1984, 1996) and some New Zealand amphipods probably live in similar associations.

The corophioid amphipod *Pagurisaea schembrii* occurs only on the hermit crab *Paguristes pilosus*, where up to 50 at a time live among the dense setae on the host's chelipeds, walking legs, and carapace (Moore 1983a). The amphipods apparently do not steal their host's food but use their specially modified antennae to capture food particles from the host's respiratory current whilst sheltering within the host's setae and shell.

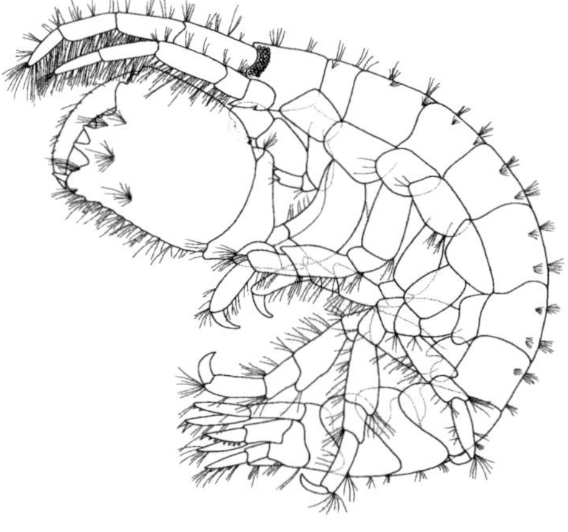

Rakiroa rima.
From Lowry & Fenwick 1982

Some amphipods are found almost exclusively on algae, but the nature of the relationships between amphipods and the algae is uncertain. Many species are found on more than one species of alga, as well as on foliose invertebrates (hydroids, bryozoans). This suggests that many amphipods use their hosts more as a substratum than as a partner in some interdependent association. Species of the tube-building genus *Notopoma* found at Kaikoura illustrate this apparently non-obligate relationship. *Notopoma fallohidea* lives only on the green alga *Caulerpa brownii* at relatively sheltered sites (Lowry 1981). One of its congeners, *N. harfoota*, is extremely abundant on the same alga in more severe wave action, but lives on other algae also. A third Kaikoura species, *N. stoora*, is most abundant on the foliose bryozoan *Costaticella solida*, although a few occur on *Caulerpa*.

Ocosingo fenwicki (anterior at left, head hidden by large lateral coxae).
from Lowry & Stoddart 1984

Another New Zealand amphipod, *Orchomenella aahu*, bores into stipes of the kelp *Ecklonia radiata* to eat up to 22 milligrams per day of the more palatable (low phenolic content) internal tissues (Haggitt 1999). These amphipods remain within the stipe, reproducing several times. Whole families of as many as 300 individuals, comprising several generations, live within most infected plants. This association seems opportunistic because *O. aahu* is also an active scavenger of animal tissue (Lowry & Stoddart 1983b).

The large subantarctic amphipod *Rakiroa rima* appears to live only within empty sponge-covered shells of a large barnacle (*Megabalanus campbelli*) (Lowry & Fenwick 1982). Similarly, some cryptic species such as *Acontiostoma tuberculata*, *Ocosingo fenwicki*, and *Stomacontion* spp. are known only from among collections of subtidal encrusting sponges (Lowry & Stoddart 1983b). It is uncertain whether these are commensal associations or whether the conditions sought by the amphipods are found coincidentally in close proximity to these other organisms. Some have, however, evolved specialised morphological and reproductive adaptations to their inquilinous life-styles. For example, species of *Ocosingo* and *Stomacontion* have specialised piercing mouthparts (Lowry & Stoddart 1984). *Acontiostoma* and some *Stomacontion* species undergo a sex change to ease the problems of finding a mate; small sexually mature males change into reproductive females as they grow larger (Lowry & Stoddart 1983b, 1984, 1986).

The place of some amphipods in various food-webs makes them ideal intermediate hosts for parasites. The common freshwater amphipod *Paracalliope fluviatilis* is the intermediate host for a parasitic nematode (*Hedruris spinigera*) commonly found in long-finned and short-finned eels, smelt, brown mudfish (Hine 1978, 1980; Jellyman 1989), and whitebait (McDowall 1968). Infection rates of the nematode in these fishes (up to 38% for short-finned and 70% for long-finned eels) are often directly related to abundances of the amphipod and the incidence of *Paracalliope fluviatilis* or smelt in the fishes' diets (McDowall 1968; Hine 1978). This amphipod is also the intermediate host for three additional parasites of freshwater fishes – *Acanthocephalus galaxii*, *Coitocaecum anaspides*, and at least one species of hymenolepid cystocercoid (Hine 1978). Similar amphipod–parasite relationships are almost certain to occur among marine species.

These observations show some of the diverse relationships between amphipods and other organisms. Other relationships, notably those between widely distributed hyperiid amphipods and various other planktonic invertebrates (salps, tunicates, medusae), plus those between cyamids and their cetacean hosts, are not considered. Numerous other relationships between New Zealand caprellid and gammaridean amphipods and various parasites, other invertebrates, and algae are likely to be described in the future.

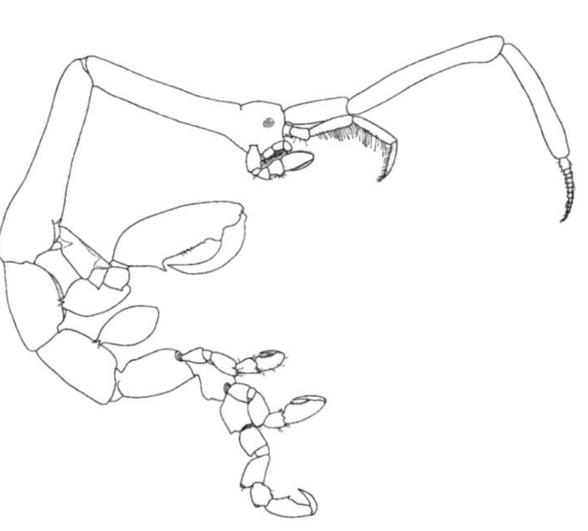

Caprella equilibra.
From McCain 1968

Alien species

Relatively few invasive amphipods (11 species) have been reported in New Zealand. Among the hyperiids, the potential for a species to invade seems extremely low; ships' ballast water seems the only feasible vector, but the likelihood of hyperiids surviving within ballast water for any appreciable time seems remote. Certainly, exotic species may arrive fortuitously as ephemeral

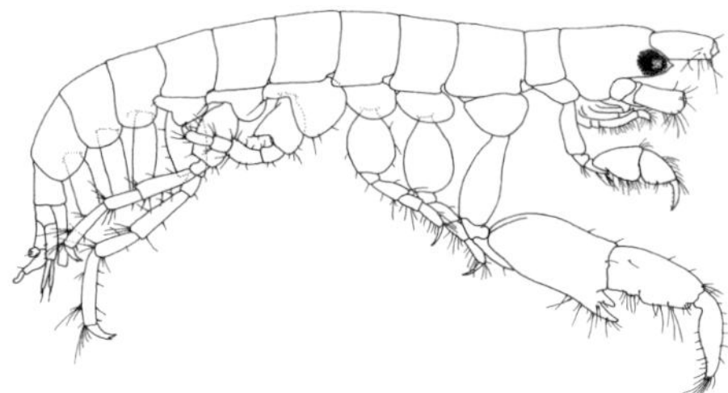

Ericthonius pugnax (antennae broken).
From Just 2009

transients within water masses not normally entering our region. Such arrivals seem destined to disappear when their water masses are displaced by the more usual regime.

One New Zealand caprellid, *Caprella mutica*, is a very recent invader (Willis et al. 2009), another species (*Caprella equilibra*) is cosmopolitan, and a third (*Caprellina longicollis*) is widespread in southern waters (McCain 1969, 1979). Caprellids' usual association with sessile fouling invertebrates at sites of high water movement suggests that the latter two caprellids could arrive on the fouled hulls of ships and, thus, may be invaders. Equally, several additional cyamids may be found in New Zealand in the future. Whale hosts of several more species are known from New Zealand waters, but these small, apparently rare, amphipods are collected infrequently.

One land hopper, *Arcitalitrus sylvaticus*, has been imported from Australia. It is established in urban and disturbed habitats of northern New Zealand, displacing native land hoppers to become the principal land hopper in domestic gardens in Wellington and Auckland (Duncan 1994). The species has failed to become established in Christchurch, despite at least two separate introductions via potted plants.

There is no evidence of any exotic amphipods invading New Zealand's fresh waters. A few gammarideans have been introduced to harbours, however, via ships. Two corophioids, *Monocorophium acherusicum* and *Apocorophium acutum*, are cosmopolitan and 'trace out some of the major shipping routes, particularly that from England through the Mediterranean and Suez Canal to South Africa' (Hurley 1954f), indicating that both are invaders. *Ericthonius pugnax*, another tube-building corophioid, is probably another invader because, although its distribution is less readily explained (New Zealand and Indonesia), the species was not discovered in New Zealand until 1923, some 70 years after its original description.

Two additional corophioids have been reported as invaders in New Zealand. *Paracorophium brisbanensis* and an unidentified species of *Corophium* were found in brackish waters of the upper reaches of Tauranga Harbour. Both were regarded as adventives because neither was reported from New Zealand previously, they were not found at any of 92 similar sites surveyed around the country, both Tauranga populations had 'remarkably limited genetic variability', and juveniles dominated their population structures (Stevens et al. 2002).

Another notable alien amphipod, distributed nearly globally, is the wood-boring *Chelura terebrans*. First found in New Zealand in Auckland Harbour (Chilton 1919), this small amphipod bores into most human-made wooden structures around the world (Barnard 1955). *Chelura*, along with *Limnoria* isopods and boring molluscs (*Teredo* species), wreaks havoc on wharf piles, rapidly boring into the timber and weakening any wooden structures. Apart from Chilton's (1919) original records, there appear to be no other reports of this species from New Zealand, although it is certain to be more widespread.

Three additional aliens were found in the sea chest (a large recess in a ship's hull for seawater intake pipes) of a vessel from the tropical Pacific that was slipped at Nelson in September 1999. These were *Stenothoe gallensis* and *Elasmopus rapax*, two known tropicopolitan species, and an unidentified species belonging to the cosmopolitan genus *Podocerus*. The first two species were abundant and included mature males, gravid females, and juveniles. There is no information on whether any of these species has become established in Nelson or elsewhere in New Zealand, despite repeat surveys.

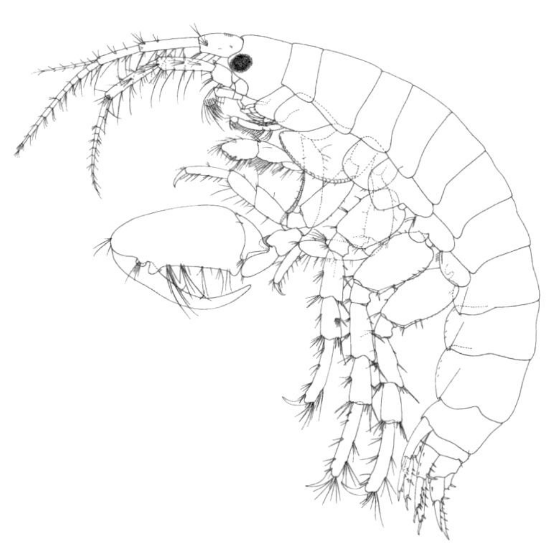

Gammaropsis typica.
From Barnard 1972a

In general, it seems extremely difficult to determine whether marine species with wide distributions are invaders (become established on new shores through dispersal by human activities) or simply arrived by natural dispersal. Several other New Zealand species have variably wide extrinsic distributions, but the ecologies of only a few seem likely to equip them for dispersal on the hulls of ships. Tube-builders and nestlers, especially corophioids, are the most likely candidates. For example, *Gammaropsis crassipes* was described from shallow bays and harbours in eastern Australia in 1881 but not reported from New Zealand until 1920, suggesting possible introduction. Recent invasions by algae, as well as long-term climatic changes, suggest that the potential for permanent establishment by amphipod invaders will increase in the future.

Monocorophium sextonae was considered to be a successful New Zealand invader of European shores (Hurley 1954f), although this has recently been questioned (Costello 1993; Bousfield & Hoover 1997). First described from Plymouth and Wembury in 1937, this amphipod was present, albeit unrecognised, in Chilton's (1921) material (Hurley 1954f). Crawford (1937) remarked that the 'abundance of this species is the more surprising since it is not present in the rich collections of *Corophium* made from the same dredging grounds in 1895–1911. It seems possible, therefore, that it is not indigenous at Plymouth … I cannot guess its original locality'. In revising these species of the family Corophiidae, Bousfield and Hoover (1997) considered that *M. sextonae* 'is almost certainly endemic to the eastern North Atlantic and Mediterranean regions, from whence it has been spread by commerce to world-wide temperate marine waters'.

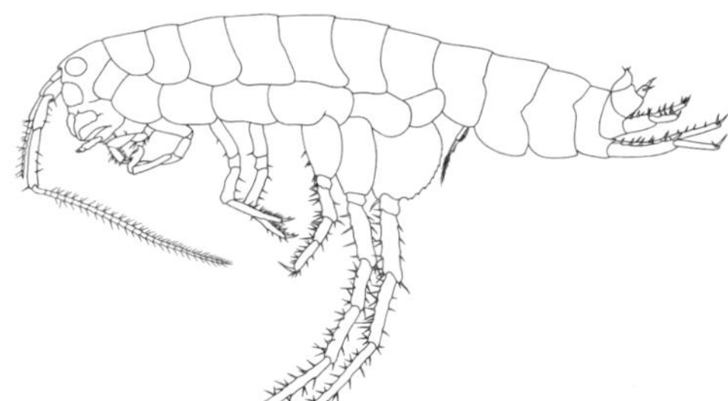

Puhuruhuru aotearoa.
From Fenwick & Webber 2008

Amphipods in environmental investigations

Diverse approaches are used to assess and manage human impacts on the aesthetic and life-sustaining qualities of natural environments. Use of plants and animals as bioindicators is increasingly common because of the sensitivities and broad-spectrum responses of some species. Amphipods are ideal bioindicators for shallow marine environments (Conradi et al. 1997) because they are ecologically (trophically) important, tend to be numerically dominant within many habitats, have specific niche requirements, have generally low mobility and dispersive capabilities, and are known to be sensitive to several pollutants and toxicants. Indeed, Thomas (1993) reported that '[a]mphipods are so useful as bioindicators that U.S. Government agencies now require their identification to species in permitting operations such as oil leases and outfalls.' In addition, individual species of amphipods may serve as very useful assays for pollutants (Lamberson et al. 1992). Several US agencies employ amphipods in bioassays to test toxicities and specific contaminant levels independent of chemical analyses and environmental surveys, particularly for marine environments.

Many of New Zealand's estuarine and marine amphipods fulfil all of Thomas's (1993) criteria for effective biomonitors (e.g. Fenwick 1976, 1985; Hickey & Martin 1995; Nipper & Roper 1995; Nipper et al. 1998). This is also true for some terrestrial (e.g. Rainbow et al. 1993) and freshwater species (Hunt 1974). Environmental survey research in New Zealand, however, continues to look at the total fauna and these investigations follow a trend of identifying and enumerating taxa to family level only (Somerfield & Clarke 1995) in attempts to reduce costs by minimising the taxonomic expertise and time required for identifications. Some workers (Thomas 1993; Conradi *et al.* 1997) advocated focusing on the amphipods alone in surveys of shallow marine environments and, certainly, their identification to species seems worthwhile in such surveys. There has been no specific examination of the merits of using amphipods alone for such surveys in New Zealand, and identification tools and knowledge of the group are inadequately developed for this to become a viable, standard approach in the short term.

New Zealand estuarine amphipods (*Paracorophium excavatum, P. lucasi*) have been used in bioassays of sediment contamination and toxicity (Nipper & Roper 1995). Additional studies (Nipper et al. 1998; De Witt et al. 1999) revealed the robustness of this assay approach, which is now used extensively. Only recently, however, has the taxonomy of these two species been resolved (Chapman et al. 2002), illustrating that taxonomic knowledge of New Zealand's amphipod fauna is often inadequate for reliable ecological applications.

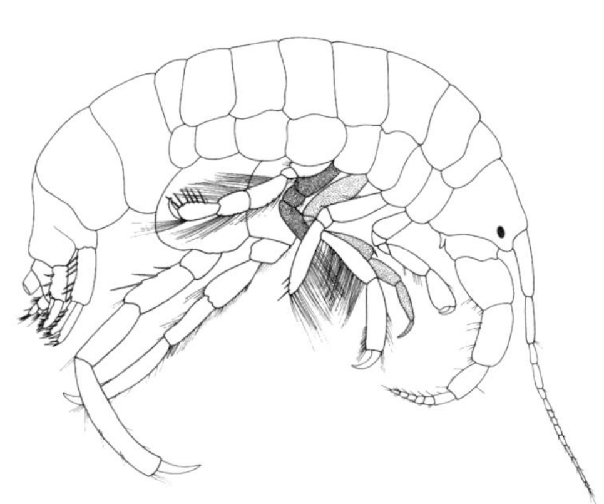

Paracorophium excavatum.
From Barnard 1969

Gaps in knowledge and future research

New Zealand's amphipod fauna is important ecologically on land, in fresh waters (especially groundwaters), and in marine habitats where species fill vital roles in food-webs and often provide appreciable direct or indirect economic benefits. Amphipods also offer considerable potential as bioindicators of environmental quality in many habitats. Obviously, the potential for ecological and environmental research using amphipods is huge, even when only the more urgent or applied issues are considered. Equally, the scope for academic investigation of amphipods is enormous.

Despite all this, their systematics is very incomplete, hindering attempts to work with the group. Certainly, the land-hoppers appear well known as a result of Duncan's (1994) work, but the beach fleas require equivalent treatment. Freshwater amphipods require urgent attention in view of our scant knowledge of this group and the huge environmental pressures on fresh waters. Known species require extensive redescription and revision to facilitate work on the >50 new taxa in collections. Several other new species exist in other freshwater habitats that await collecting.

The marine gammaridean amphipods of shallow and continental-shelf waters comprise another substantial gap. Collecting has been sparse and the fauna at no one location is well known. Even the distribution of the algal-dwelling species along New Zealand is poorly known, despite Barnard's (1972a) work. Amphipod faunas of shallow soft seafloors are very poorly known. A study in Pegasus Bay showed that 28% of species in the 4–10-metre depth zone are undescribed (Fenwick 1999). Similarly, less than 30% of the 98 species in a series of collections off Kaikoura are known and the unknown ones include several new genera. Also, just 24% (10 of 42 species identified by a leading taxonomist) of amphipods in another study of New Zealand shelf benthos were known to science (Probert & Grove 1998).

Amphipod research in New Zealand thus offers considerable scope for both economically important issues and questions of more academic interest. However, the present status of the group's taxonomy hinders the successful development of this work, as well as discouraging many ecologists from using amphipods as ideal subjects for environmental and ecological investigations. The future, therefore, requires not just more taxonomy, but also the development of interactive guides and keys to overcome these barriers and make the local fauna accessible to non-specialists. This is particularly true for hypogean and other freshwater amphipods, given their role in maintaining the quality of groundwaters and the urgent need for effective management of this economically important resource in the face of increasing demands and human-induced threats.

Order Isopoda: Slaters, fish lice, and kin

The most diverse range of body plans of all the nine peracaridan orders, if not of all crustacean orders, is shown by the Isopoda, named, however, for the relative sameness of limbs (Greek *isos*, equal, like; *podos*, foot).

Only one of the isopod suborders, Oniscidea, is familiar to most people. The oniscideans are commonly called woodlice, slaters, pillbugs, or roly-polies. However, the order is predominantly marine, being less well-represented in estuarine and fresh waters. There are fewer common names for the marine groups but sea-lice, fish doctors, tongue-biters, and sea-centipedes are applied to some families. No common name, except isopod, applies to all members of the order.

Life-styles vary. Free-living predators, marine filter-feeders, scavengers in forest leaf-litter and on the sea floor, and various parasitic forms are represented in the order. The isopods have succeeded in two unusual habitats besides the shallow marine environments where most crustaceans are typically found. One is the land, where woodlice, slaters, and phreatoicideans are most often the sole

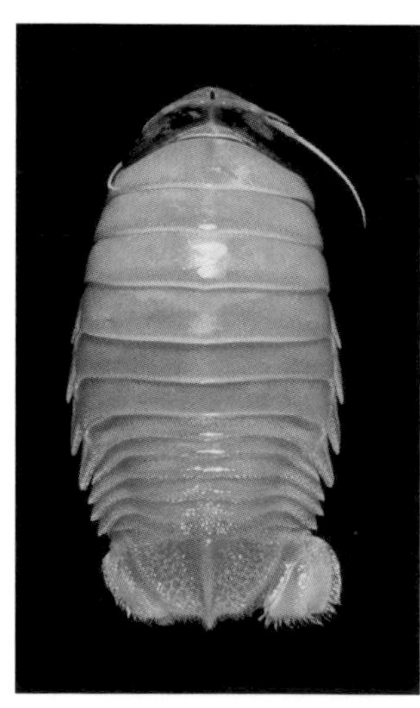

Fish micropredator
Aega monophthalma (Cymothoida).
From Bruce 2009

crustacean representatives in some habitats, and the other is the deep sea, where the suborder Asellota has radiated into a variety of bizarre forms.

Although they are often said to be 'dorsoventrally flattened' while their close relatives the amphipods are 'laterally flattened,' there are many exceptions; some are cylindrical, others laterally compressed, and others extraordinarily ornamented. The smallest isopod adults are c. 1 millimetre long, many are in the range 4–12 millimetres, and the largest are deep-sea scavengers of the genus *Bathynomus*, growing to an astonishing 400 millimetres!

The only sure way to tell an isopod from an amphipod is that isopods lack strongly chelate first legs and have only one pair of uropods (tail appendages) and a free second thoracic segment. Character interpretation can be difficult, however, because uropods vary considerably. They may be flat limbs that lie in the same plane as the pleotelson, or enclose the pleopodal gills, or have any of several other forms. Technically, Isopoda are defined as follows: eyes sessile (not stalked); carapace absent; one pair of maxillipeds; seven pairs of pereopods (legs), without exopods (an outer branch); abdomen clearly differentiated from thorax and divided into a pleon of five segments (sometimes some fused) and pleotelson (fused pleonite six and telson); pleopods 1–5 similar or anterior pair operculiform, forked; one pair of uropods.

Isopods are of interest to marine biologists because of the important roles they play in ecosystems, especially on the sea floor. Here, species of many families are important scavengers of decaying material. Isopods of the family Cirolanidae are critical in cleaning up decaying dead fish (Bruce 1986a; Brusca et al., 1995; Keable 1995). Fish-lice of the family Cymothoidae are flesh- and blood-feeders that attach to the skin of living fishes. Aegids and juvenile gnathiids are blood-sucking micropredators of fishes, and in the tropics gnathiids can be so abundant that fishes attend cleaning stations where wrasses remove and eat them. Sea-centipedes (Idoteidae) feed on algae. The diverse Sphaeromatidae feed on living and dead material of all sorts. Many isopods are ideal food for many bottom-living fishes such as flounders and skates.

One family of economic significance is the Limnoriidae (gribble). These are wood-borers, formerly of ships but now only of wooden piles and wharves. Like timber borers on land, gribble make galleries throughout the timber and weaken it considerably (Menzies 1957; Cookson 1991). Species of *Sphaeroma* (Sphaeromatidae) behave similarly. Another important group, at least to gardeners, is the terrestrial slaters or woodlice. While most feed innocuously on decaying leaves and wood they can become so abundant as to attack vegetables and other garden plants.

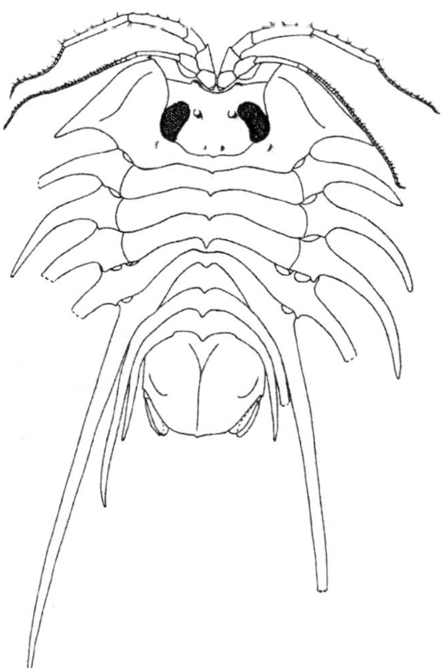

Brucerolis hurleyi (Sphaeromatidea).
From Storey & Poore 2009

Diversity of New Zealand Isopoda

The world's isopod fauna exceeds 10,000 described species but the actual number of species is certainly several times this. There are big gaps in knowledge of the deep sea, the tropics, and some families with small individuals. The New Zealand fauna totals only 426 living species (and four fossil species) but it appears that few shallow-water isopod groups are well covered taxonomically. It would not be surprising if many species of Sphaeromatidae, Cirolanidae, Gnathiidae, anthuroids, Asellota, and Valvifera remain to be discovered, especially from shelf depths. Even so, the number of already described species (353) somewhat exceeds that of South Africa (cf. 275 species in Kensley 1978) but, not surprisingly, is far fewer than in Australia (1,118 species; Poore 2002, 2005). South African and Australian isopods have attracted greater taxonomic attention than those in New Zealand. As is the case for many marine and terrestrial animals, New Zealand isopods are largely endemic.

The only habitat that is relatively well known is intertidal and subtidal rocky shores, but even here the Asellota have been largely ignored. Museum collections from The Snares (partly described by Poore 1981) contain several undescribed species of small asellotes and more such species could be expected throughout New Zealand. While the benthos of the New Zealand continental shelf has been thoroughly sampled, the gear used has not deliberately targeted small invertebrates, and collections available for study seem not particularly diverse for isopods. NIWA collections appear from superficial examination to be far less rich than, for example, those from comparable habitats in Bass Strait at similar latitudes in Australia. Museum Victoria, Melbourne, houses a benthic collection that includes c. 250 species of isopods from sediments (Poore unpubl.). There are even fewer species described from the continental slope. Poore et al. (1994) recognised 359 species, mostly undescribed, from this habitat off the southeastern coast of Australia and a similar number could be expected for the New Zealand slope. Several species from bathyal depths north of New Zealand were described from collections of the *Galathea* Expedition but the rest of the EEZ is virtually unsampled. Another habitat as yet largely unexplored is fresh water in limestone caves; sphaeromatids are known from this environment near Nelson, South Island (Sket & Bruce 2004).

Three species of isopod fossils have been recorded from New Zealand (Grant-Mackie et al. 1996; Hiller 1999; Feldmann & Rust 2006).

Numerous families, 120 at last count in the world fauna, are arranged in a complex hierarchy within suborders (Martin & Davis 2001). Most of the widely used suborders are monophyletic groups, but the one that has traditionally included the most familiar marine species, Flabellifera, is not (Wägele 1989; Brusca & Wilson 1991; Brandt & Poore 2003). Here, Brandt and Poore's (2003) classification is followed and the suborder Flabellifera is superseded by the three suborders Cymothoida, Limnoriidea, and Sphaeromatidea. Three other previously recognised suborders are subsumed within Cymothoida – Epicaridea as superfamilies Bopyroidea and Cryptoniscoidea, Anthuridea as superfamily Anthuroidea, and Gnathiidea as family Gnathiidae. Hurley and Jansen (1977) provided an effective key to identify some families but their classification is now out of date. Modern faunal treatments, also using the older classification, can be found in Kensley (1978) or Kensley and Schotte (1989). Only 49 families have so far been recorded from New Zealand.

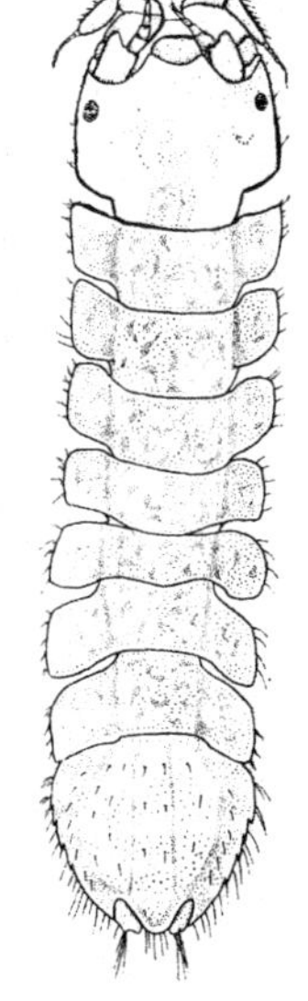

Joeropsis sp. (Asellota).
From Hurley & Jansen 1977

Suborder Asellota

Some 93 New Zealand species are known, of which 36 remain unnamed or not fully determined. They have diverse shapes. Diagnostic characters include: coxal plates usually minute; one (rarely two or three) pleonites free, others fused; uropods attached posteriorly. Asellotes are common but small, difficult to find, and even harder to identify. A microscopic examination of tufts of algae from sheltered marine environments will often reveal several species of asellotes, rarely more than two millimetres long. Others live in freshwater streams. Globally, almost 30 diverse families exhibit an exceptional range of form on the floor of the deep sea. Some species are quite bizarre, with extraordinary ornamentation. Several species from the deep sea near New Zealand were described from collections of the Danish research ship *Galathea* (Wolff 1956a, 1962) but only one family from this environment in New Zealand has been treated in detail (Lincoln 1985). The identity of many of the species recorded from subantarctic New Zealand may be in doubt until specimens are compared with those from other islands or continents. Globally, Wilson and Wägele (1994) listed all known asellote species and provided a key to the genera of Janiridae, an important shallow-water family, and Cohen (1998) did the same for Dendrotiidae. The diverse Munnopsididae has been treated in part by G. D. F. Wilson (1989), the Stenetriidae by Serov and Wilson (1995), Pseudojaniridae by Serov and Wilson (1999), Joeropsididae by Just (2001), and Paramunnidae by Just and Wilson (2004, 2006).

Suborder Phreatoicidea

Nine New Zealand species are known, all endemic, and in endemic genera. They are laterally flattened. Other diagnostic characters include: coxal plates extending ventrally; five pleonites free; uropods rod-like and attached posteriorly. Peculiar to southern continents and islands, phreatoicids comprise an unusual group of freshwater and terrestrial species. They superficially resemble amphipods but differ in having only one pair of uropods as well as other isopod features. Most of the New Zealand fauna was dealt with by Nicholls (1944), with one species described in detail by Wilson and Fenwick (1999). The suborder was reviewed by Wilson and Keable (2001).

Neophreatoicus assimilis (Phreatoicidea)

From Hurley & Jansen 1977

Suborder Cymothoida

Comprising sea-lice, fish-lice and other mobile scavengers, predators, and microparasites, 116 described and 16 undetermined New Zealand species are known. Diagnostic characters: usually dorsoventrally flattened but otherwise diverse; mandibular molar blade-like or reduced; coxal plates expanded and free or reduced; five pleonites free or variously fused; uropods usually forming tail fan with pleotelson, rotating in horizontal plane and in broad contact with pleopods. All are marine, but habits and shapes vary. Numerous authors have contributed to knowledge of cymothoidan families in New Zealand, notably the Cirolanidae (Jansen 1978; Bruce 1986a, 2003, 2004a; Svavarsson & Bruce 2000; Keable 2006), Cymothoidae (Bruce 1986b), Gnathiidae (Cohen & Poore 1994; Svarvasson 2006), Tridentellidae (Bruce 1988, 2002), and Aegidae (Bruce 1983, 2004b, 2009a). The suborder contains four superfamilies – Anthuroidea, Bopyroidea, Cryptoniscoidea, and Cymothooidea.

Some 21 described New Zealand species of Anthuroidea are known (in the families Anthuridae, Expanathuridae, Hyssuridae, Leptanthuridae, and Paranthuridae). Diagnostic characters include: shape elongate and cylindrical; coxal plates indistinguishable from pereon wall; pleonites fused or free; uropodal exopod attached proximally on peduncle and dorsally arched over pleotelson. Anthuroids live in sediment and on macroalgae, although the New Zealand species *Cruregens fontanus* is unusual in living in artesian and river waters (Wägele 1982). Very few species had been described until the work of Wägele (1985). The family arrangement follows Poore (2001a), who synthesised many papers and whose earlier work, principally on the Australian fauna, is relevant.

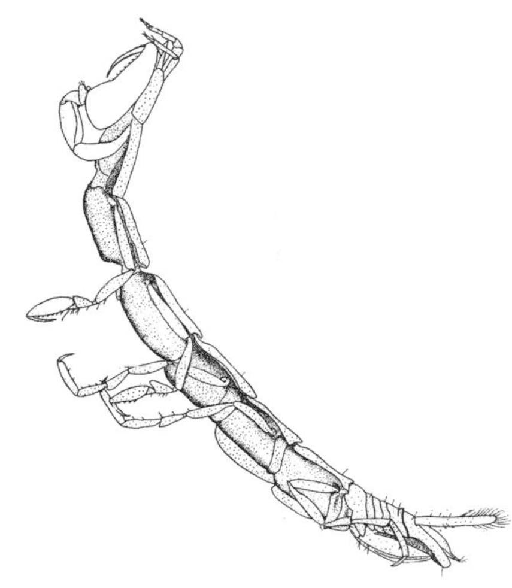

Cruregens fontanus (Cymothoida).

From Hurley & Jansen 1977

The superfamily Bopyroidea comprises parasitic isopods of crustaceans, with 13 described New Zealand species in the family Bopyridae. Diagnostic characters include: individuals sexually dimorphic, females usually asymmetrical, males minute; mouthparts reduced; branchial parasites of crabs, shrimps etc., but also of other crustaceans and some hyperparasites of other bopyroideans. Page (1985) studied New Zealand species. Few modern taxonomists have tackled this confusing group, but Markham (1985) and other papers by this author are a good introduction.

The largest superfamily in New Zealand is Cymothooidea, with 93 species (15 unnamed or not fully determined) in the families Aegidae, Anuropidae, Cirolanidae (with endemic genus *Pseudaega*), Cymothoidae, Gnathiidae, and Tridentellidae. The largest of these, with 37 species, is the recently monographed Aegidae (Bruce 2009a), a family of micropredators mostly associated with fishes. The Cryptoniscoidea has just five species in New Zealand, in the families Crinoniscidae and Hemioniscidae (Hosie 2008).

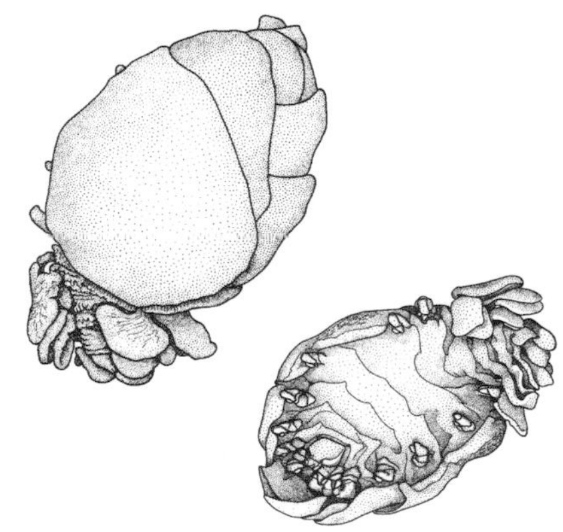

Dorsal (upper) and ventral (lower) views of *Athelges lacertosi* (Cymothoida), a parasite of the hermit crab *Lophopagurus lacertosus*.

From Pike 1961

Suborder Limnoriidea

These are wood-boring isopods, sometimes called gribble, with nine New Zealand species all in a single family, Limnoriidae, reviewed by Cookson (1991). Mandibles are specially modified, the body is cylindrical, and pleonites are free. Wood is not their only target in New Zealand. *Limnoria limnorum* caused the 1916 failure of the Cook Strait submarine cable when some individuals bored through the gutta-percha that was around the inner cable core.

Suborder Sphaeromatidea

These comprise marine pillbugs in general, with 81 described New Zealand species known, including 61 species of Sphaeromatidae. Diagnostic characters: usually dorsoventrally vaulted, occasionally flattened, sometimes able to enroll; coxal plates well developed; pleonites variously fused; uropods usually forming tail fan with pleotelson, rotating in vertical plane and excluded from branchial cavity. All are marine, but habits and shapes vary. Notable taxonomic contributions include those on the Sphaeromatidae (Hurley & Jansen 1977) and the enigmatic, sometime sphaeromatid, genus *Paravireia*, herein placed as *incertae sedis* (Jansen 1973; Brökeland et al. 2001). A sphaeromatid species is host to a fecampiid flatworm, *Kronborgia isopodicola*, described from Kaikoura, the adults of which live in the body cavity of *Exosphaeroma obtusum* (Blair & Williams 1987; Williams 1988).

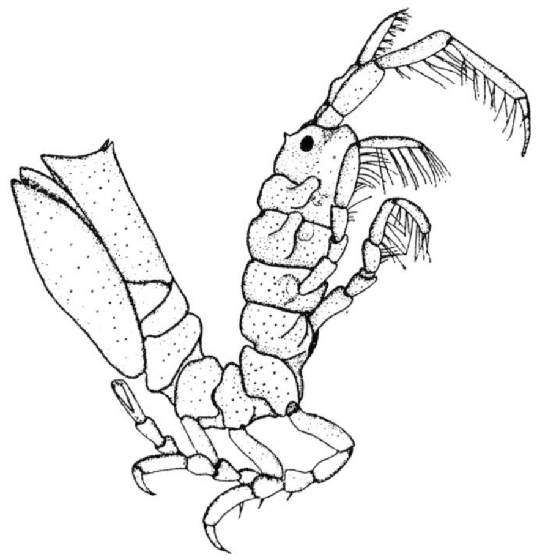

Pseudarcturella chiltoni (Valvifera)
From Hurley & Jansen 1977

Suborder Valvifera

These include the so-called sea-centipedes and other bizarre forms, comprising 25 described New Zealand species. The form of the uropods, as long plates attached to the side of the abdomen and tightly enclosing all the pleopods, defines the valviferans. Most are marine, but the three species of *Austridotea* are among the few freshwater members of the suborder (Chadderton et al. 2003). Some forms are ornately decorated. The only family-level reviews are by Poore and Lew Ton (1990, 1993) and Poore and Bardsley (1992). The family arrangement follows Poore (2001b).

Suborder Oniscidea

These are the land-dwelling woodlice, slaters, and pillbugs, with 72 described New Zealand species known. Four species are naturally occurring non-endemics and six others are introduced. Diagnostic characters: usually flattened but sometimes able to roll up; five pleonites usually free; pleopods highly modified for air-breathing. Oniscideans are exclusively terrestrial and are the only crustacean group to compete successfully with other arthropods on land. Seven pairs of legs immediately reveal that they are not insects or millipedes. There are examples high up on the seashore but none is truly marine. Although damp places, and under leaves and decaying logs, are favoured habitats, some overseas species are known from deserts. Like all isopods, oniscideans rely for respiration on their pleopods, which are kept damp with a variety of water-conservation measures. Most species are scavengers on dead plant litter but some can be pests in gardens. There are numerous families including five genera and many species endemic to New Zealand. But the most commonly seen species are introduced from Europe. The New Zealand fauna was reviewed by Hurley (1950) and one family revised by Green (1971). Some of the names listed by Hurley are now out of date and the present review follows the taxonomy of Green et al. (2002).

Historical overview of isopod studies

The first scientific collection of isopods in New Zealand was made by the French biologists J. R. C. Quoy and J. P. Gaimard when the *l'Astrolabe*, captained by Dumont d'Urville, visited in 1826. They discovered two shallow-water sphaeromatids from algae, described 13 years later as *Isocladus armatus* and *Cassidina typa* in a significant publication on isopods by H. Milne Edwards (1840). Earlier publication dates appear in the New Zealand checklist but these are of species either introduced to the country or of species described from elsewhere. Later, the United States Exploring Expedition visited New Zealand on its 1838–42 round-the-world voyage, and numerous species of marine animals were described by its chief scientist, James D. Dana. Among these are 19 species of isopods (Dana 1852b, 1853–55). The first review of the New Zealand crustacean fauna (Miers 1876) listed 28 isopod species in 16 genera. When a second review was completed 10 years later by Thomson and Chilton

(1886), 60 species of isopods had by then been described, many by these two authors. A third checklist and key (Hurley 1961) listed 168 species; the increase in the intervening years was contributed largely by results from foreign deep-sea expeditions like the British HMS *Challenger* (1873–76) and Danish *Galathea* (1952). By 2009 the number had grown again, largely as a result of the work of New Zealand-based taxonomists Desmond Hurley and Peter Jansen in the 1970s and Niel Bruce in the 2000s, as well as overseas workers with an interest in specific families (J. Just, R. Lincoln, G. C. B. Poore, and J.-W. Wägele).

Special features of the New Zealand isopod fauna

Some 38 isopod families have marine representatives in the New Zealand fauna. Gaps can be explained by inadequate collecting. For example, it is safe to say that most deep-water asellote families will be recorded once appropriate sampling is done. It is possible that the Ancinidae and Corallanidae might one day be found in New Zealand. Four small families from the southwestern Pacific (Bathynataliidae, Hadromastaciidae, Keuphyliidae, and Phoratopodidae) are so far not recorded from New Zealand. The Serolidae, rich in species in shelf environments in Australia (Harrison & Poore 1984; Poore 1985, 1987), the southwestern Pacific (Bruce 2009b), and Antarctica (Brandt 1988; Wägele 1994), is represented in New Zealand by only a relatively small number of deep-water species, several of which have been described (Bruce 2008; Storey & Poore 2009).

The Gondwanan affinities of the fauna are evident in the largest families, Sphaeromatidae and Cirolanidae, where genera found in other Gondwanan continents dominate. This is clear too in Plakarthriidae, a family known only from three species, one each in South America, New Zealand, and southern Australia (Poore & Brandt 2001). The same is true for the terrestrial families, with many Palaearctic oniscidean families absent and strong radiation of southern ones. The Phreatoicidea is a typical high-level Gondwanan taxon, being confined to New Zealand, Australia, and India.

New Zealand isopods are largely endemic – 100% of freshwater species, 86% of terrestrial species, and almost 77% of marine species. The endemicity of some taxa reflects the long isolation of the fauna from Australia, the continent from which it last separated 85 million years ago. Close relatives (perhaps sister species) of New Zealand species are found in Australia within several families, e.g. Austrarcturellidae, Idoteidae, Leptanthuridae, Phreatoicoidae, Plakarthriidae, and Sphaeromatidae. Much less is known about relationships among other apparent endemics. Many species from the shelf and deep sea are known only from type specimens from a single sample, so their true distribution is unknown. But even here evidence is emerging that endemism is truly high. For example, none of the anthurideans or haploniscid and dendrotiid asellotes described from New Zealand occurs in Australia (Cohen 1998; unpublished material and catalogues).

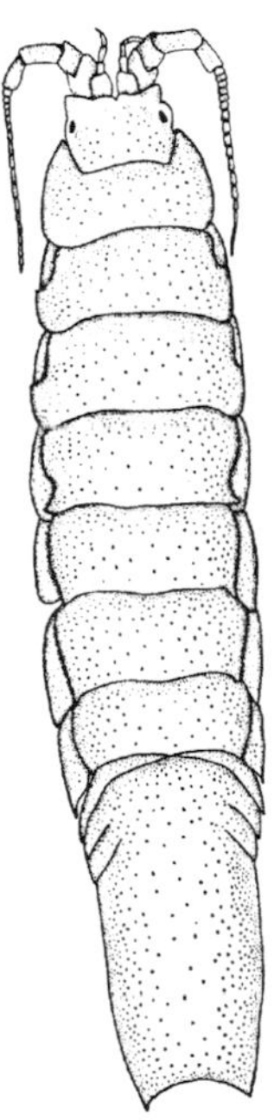

Paridotea ungulata (Valvifera).
From Hurley & Jansen 1977

Non-endemic species fall into two groups – those apparently naturally widespread, and those thought to be introduced. The idoteids *Batedotea elongata* and *Paridotea ungulata* have been identified from algal communities in New Zealand and Tasmania and another, *Idotea metallica*, is cosmopolitan on oceanic algal wrack (Poore & Lew Ton 1993). Several other species may occur naturally in New Zealand and Australia and sometimes also elsewhere, e.g. *Natatolana pellucida* (Cirolanidae), *Limnoria rugosissima*, *L. tripunctata* (Limnoriidae), and *Cymodoce convexa* (Sphaeromatidae). Several species of aegid micropredators of fishes and at least three species of cymothoid fish ectoparasites seem widespread in the Tasman Sea (and sometimes beyond), as are their host species. A deep-sea gnathiid, *Bathygnathia vollenhovia*, which occurs on both sides of the Tasman Sea (Cohen & Poore 1994), is certainly naturally distributed. For other seemingly widespread species, identifications are suspect until type material has been compared. Specimens of the New Zealand sphaeromatid *Pseudosphaeroma campbellense* identified from Australia (Harrison 1984) may be specifically different (Poore 1994; Bruce & Wetzer 2008). This suspicion is especially valid for some species recorded from the New Zealand subantarctic but whose type locality is elsewhere, e.g. the sphaeromatids *Exosphaeroma gigas* and *Cymodocella tubicauda* (Hurley & Jansen 1977; Brandt & Wägele 1989).

The most familiar isopods of gardens and farmland, the woodlice and pillbugs, are definite imports from Britain or continental Europe, namely *Armadillidium vulgare, Porcellionides pruinosus,* and *Porcellio scaber*. They arrived with garden plants or simply as stowaways with the first Europeans. An export of a slater has occurred, too – the styloniscid *Styloniscus otakensis* to Australia's Macquarie Island (van Klinken & Green 1992).

Alien marine isopods

For marine isopods the presence in New Zealand of exotics is ambivalent, although the ability to be transported to and from New Zealand with fouling on ships is certain. Cranfield et al. (1998) recorded three isopods as potentially introduced to New Zealand. The first, Australian species *Cymodoce tuberculata* (Sphaeromatidae), recorded by Chilton (1911b) from a plank of the ship *Terra Nova* in Lyttelton, seems not to have become established in New Zealand. The second, a species of wood-boring gribble, *Limnoria tripunctata* (Limnoriidae), has potentially been distributed by shipping between widespread localities around the world but its origin is unknown (Cookson 1991). The third, *Limnoria rugosissima,* is a borer of algal holdfasts, not of timber, so is more likely to be distributed between southern Australia and New Zealand by drifting kelp. On the other hand, *Limnoria quadripunctata* (not listed by Cranfield et al. 1998) was first described from Europe and now globally recognised; its origin is more probably Southern than Northern Hemisphere (Cookson 1989; Poore & Storey 1999). Likewise, *Sphaeroma quoianum* (Sphaeromatidae), another wood-borer and its commensal, *Iais californica* (Janiridae), could have been distributed similarly. *Eurylana arcuata* (Cirolanidae) is possibly a New Zealand species introduced to Australia (or vice versa) and to North America (Bowman et al. 1981).

The affinities of the New Zealand fauna can only be understood if the taxonomy is accurate. Two species of *Phalloniscus* (Oniscidae) erroneously recorded from Australia, *P. kenepurensis* and *P. punctatus,* were excluded by Bowley (1935) and Green (1961). *Deto marina* (Scyphacidae), recorded from New Zealand by Schultz (1972), is endemic to Australia.

Order Tanaidacea: Tanaids

Tanaids (there is no common name) are very small, shrimp-like creatures. They are mostly in the 2–5 millimetre range but adults of a few species can be as small as half a millimetre or as long as 75 millimetres (Gamo 1984). There are three living orders, the members of which exhibit characteristic morphologies and, to some extent, lifestyles. Species of Neotanaidomorpha are free-living surface dwellers, while those of Tanaidomorpha are largely tube dwellers and the Apseudomorpha are mostly burrowers or crawlers. The first two segments of the thorax are covered by a carapace forming, with the head, a cephalothorax. The first thoracic segment supports a small pair of maxillipeds, the second a distinctive pair of chelipeds, and each of the third to seventh segments bears a pair of pereopods. The first pereopod may be adapted for burrowing in the suborder Apseudomorpha, equipped with spinning glands for tube construction in the suborder Tanaidomorpha, or may be a simple 'walking leg' in the suborder Neotanaidomorpha. Sexual dimorphism is often evidenced in the chelipeds and the claw of the left cheliped can be greatly enlarged in the males of some species of Apseudomorpha. Each of the first five abdominal segments normally carries pleopods but these may be absent in many deep-sea species. The final pleonal segment is fused with the telson

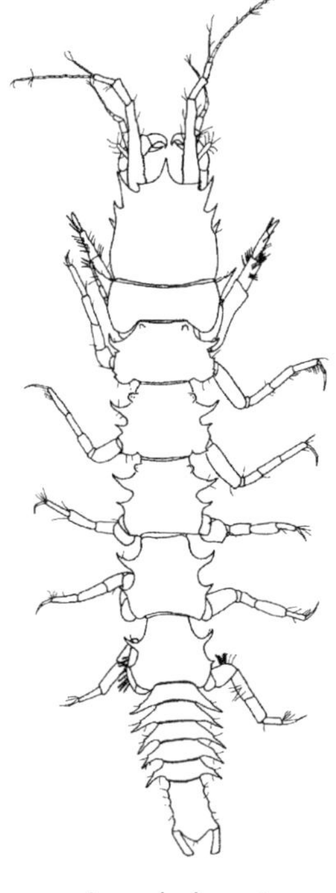

Apseudes larseni.
From Knight & Heard 2006

(forming a pleotelson) and carries a pair of uropods. Respiration takes place over the inner surface of the carapace.

As with other peracarid crustaceans such as isopods, amphipods, and cumaceans, tanaids carry their fertilised eggs and mancae (post-larval juveniles) within a ventral marsupium. In most groups this is formed out of four pairs of oostegites, attached to the first four pairs of pereopods. This is not the case in the Tanaidae, examples of which that are common in intertidal habitats; in this family the marsupium is seen as a ventral pair of elongate sacs (or sometimes just one sac). Similarly, species of Pseudotanaidae, common in the deep-sea, have only a single pair of oostegites arising from the fourth pair of pereopods. There is also some evidence to show that in some burrowing-tubicolous groups (such as the Typhlotanaidae) the female constructs a mucous brood pouch in which she and her young live (G. Bird unpubl.).

Tanaids are usually detritivores or grazers but some taxa are filter-feeders and opportunistic predation on smaller invertebrates (such as foraminiferans or juvenile echinoderms) may be common. Only a few species are considered to be parasitic but none are obligate parasites. Tanaids are preyed upon by a large number of other organisms including polychaetes, other crustaceans, migratory birds, and a large number of juvenile and adult fish such various rat-tails and grenadiers in the deep sea (Bird unpubl.)

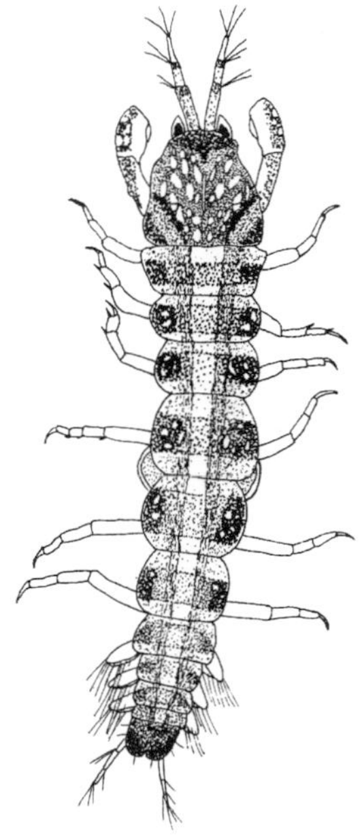

Sinelobus stanfordi.
From Chapman & Lewis 1976

Identification of tanaids is notoriously difficult, complicated by their small size and sexual and developmental variation (Larsen 2005) along with widespread and intense convergent evolution. So far, 25 families, more than 200 genera, and more than 1000 species have been described, but it is estimated that the order contains several thousand undescribed species, most of which are suspected to live in the deep sea. Tanaids live almost exclusively in marine or brackish habitats, with just a few species in fresh water. They occupy a wide range of depths. Marine species can be found intertidally among coralline algae, crevices, holdfasts, and in rock-pools. Shallow-water and shelf forms can be found in sand and mud, although tanaid sand-faunas are typically sparse. Tanaids are very common and species-rich in deep-sea oozes and some live in deep-ocean trenches to hadal depths exceeding 9000 metres (Kudinova-Pasternak 1972).

Apart from those species that are attached to floating objects, all tanaids are benthic, but some have short-lived males that can be found swimming above the seafloor in their search for females. Tanaids are free-living, tube-dwelling, burrowing, or live in association with other organisms in a variety of relationships. Some live as epifauna on solitary corals (Sieg & Zibrowius 1988), colonial corals and hydroids (Bacescu 1981), live scallops (Brown & Beckman 1992), oysters (Bamber 1990), barnacles (Reimer 1975), and even sea turtles (Caine 1986). Some species are true symbionts, living together with gastropods (Howard 1952), tube-dwelling sea cucumbers (Larsen 2005), in the canals of sponges (Hassack & Holdich 1987), and as cleaning commensals on mobile bryozoan colonies (Thurston et al. 1987). Tanaids may also have their own epifaunal associates such as stalked protozoans (Gardiner 1975) or bivalves (Warén & Carrozza 1994) and deep-sea species can carry foraminiferans embedded in the cuticle. They may be parasitised internally by nematodes and externally by copepod-like tantulocarids (Larsen 2005).

The New Zealand fauna is so poorly known that even an approximate assessment is difficult but, if comparison is made with a similar area and range of habitats, based on the Rockall-Biscay region of the Northeast Atlantic (G. Bird unpubl.), then 250–300 species are possible. The cryptic habits of the group and the small number of active specialists globally and in New Zealand suggest that this state of affairs may continue for some time although progress is now being made. Knowledge of the New Zealand fauna is still largely based on the older published records of Chilton (1882c, 1883), Thomson (1880, 1913), Stephensen (1927), Wolff (1956b), and Lang (1968). As a consequence, there are only about 20 authoritative records among the species in the end-chapter checklist. The remainder are unpublished records or undescribed species based on studies by Graham Bird, Elizabeth Hassack and the late Jürgen Sieg. Amongst these records are a number of undescribed species (indicated in the end-chapter checklist by bracketed numbers) and several new genera, the family affiliation of which is not currently available. This list is a snap-shot view and highly provisional. A few old records have been reappraised in the light of current tanaid taxonomy (Larsen & Wilson 1998, 2002; Knight & Heard 2006; Bird 2008). The New Zealand fauna also contains one of the few known freshwater tanaids – *Sinelobus stanfordi* from lakes in the Rotorua district.

Order Cumacea: Comma shrimps

The common name for cumaceans alludes to one of their distinctive features, i.e. resemblance to a comma when preserved. That is, they have an enlarged front section (head and part of the thorax) followed by a rather narrow posterior section (remainder of thorax and abdomen).

Comma shrimps live on the seafloor with their bodies generally slightly submerged in the sediment. They feed on diatoms, pieces of seaweed, foraminiferans, and detritus, which they collect from the sediment surface. For the most part, they will stay hidden in the sediment during the day, and some will make extended trips into the overlying water after sunset. The reasons for these excursions are not precisely known, but include moulting and searching for mates. In fact, in some cumacean families, the body morphology of the mature male is completely modified for swimming, suggesting that at that stage the animal rarely visits the sediment. Swimming cumaceans are vulnerable to fish predation, and mature males are commonly found in fish stomachs.

The cumacean body is one of the more modified of the higher crustaceans. Anteriorly, the head and three segments of the thorax are covered with a carapace. As a result, the normal feeding appendages of the head are augmented by three thoracic appendages (known as maxillipeds) that are also used for feeding. The first of these is also highly modified for respiration. That is, the epipod, which is not present in amphipods and is reduced in isopods, is greatly enlarged in cumaceans as a branchial lobe. Respiration occurs as the branchial lobe is moved back and forth underneath the sides of the carapace.

The remaining thoracic segments bear appendages that function as walking legs. In some cases, especially in mature males, these legs will also have an outer branch, the exopod, that is used to aid in swimming. The abdomen is generally long and thin. Abdominal appendages are either pleopods, if they occur on one or more of the first five segments, and uropods when present on the last segment. Pleopods are not present in the females of species that occur in New Zealand, and may or may not be present on some or all segments in the males. A final, post-abdominal segment, the telson, may be present as a separate structure, or it may be fused to the last abdominal segment.

Cumaceans are rare in the fossil record. There are two species known from the Jurassic, but they are more or less similar to a modern cumacean family, suggesting that the group as a whole is quite old. On the other hand, cumaceans are among the last of their line to have evolved, so it possible that all peracarids were present by the end of the Paleozoic.

As with other members of the superorder Peracarida, cumaceans carry their young in a brood pouch, with the young hatchling looking like a miniature version of the adult minus the last pair of thoracic legs. Because of this direct development, cumacean species are generally not very widespread, and some genera are restricted to individual continents or ocean basins. Some families, such as the Bodotriidae and Nannastacidae, are primarily warm-temperate to tropical, while others such as the Lampropidae and Diastylidae are most diverse in colder oceans. All families are represented in the deep sea, but lampropids show the greatest diversity in that environment.

New Zealand Cumacea

The first cumaceans known from New Zealand were described by George Thomson (1892), who had spent a couple of days dredging in the Bay of Islands in 1883. Not being able to sort the material for some time, his two species went undiscovered for several years. It would be another decade before Zimmer (1902) would describe an additional two species, collected by Prof. Dr Thilenius from the Bay of Plenty and deposited in the Berlin Museum. The biggest contribution, to this day, of our knowledge of New Zealand cumaceans was made by W. T. Calman, who, over a 10-year period (Calman 1907, 1908, 1911, 1917), described 17 species from material sent to him by G. M. Thomson and Henry Suter. Norman Jones, a prolific cumacean worker, described a new species and added a new record from the Chatham Islands area (Jones 1960). He added five new species and two new records to the New Zealand fauna in his now classic monograph covering material in the collections of the former New Zealand Oceanographic Institute (now part of NIWA), the Zoology Departments of Auckland and Canterbury Universities, and the then Dominion Museum, Wellington (Jones 1963). A further eight deep-water species were described by Jones (1969) from material collected in the Tasman Sea by the *Galathea* Expedition.

Over the intervening 31 years, many samples containing cumaceans have been taken in the waters of New Zealand's EEZ and stored in the NIWA Invertebrate Collection, Wellington. Until this present review, no one had taken the challenge of working up this material. Most of the new material examined was collected in the deep waters of the New Zealand microcontinent and contains much that is new, both at species and genus levels. From these collections, four new species of Gynodiastylidae were found and described in a recent monograph of the family by Gerken (2001). Several other new taxa have been sorted from the collections and will be described in future papers.

Of the eight currently recognised cumacean families, only six are represented in New Zealand waters. (The Ceratocumatidae is known only from abyssal depths in the Atlantic and Indian Oceans and the Pseudocumatidae are so far exclusively Eurasian–Atlantic in distribution.) The Gynodiastylidae is the smallest of the families represented in New Zealand, with only seven species, and the Diastylidae the largest, with 19 species formally known (and at least another six species remain to be characterised). Some remarks are now offered for each family, based on historical records as well as new findings from NIWA material.

Family Bodotriidae: Subfamily Bodotriinae. Members of this subfamily occur in all oceans, primarily in shallow water, but also in the deep sea. New Zealand is quite unusual in having only one (*Cyclaspis*) of the 13 genera represented in its fauna. This is most likely because the other genera are primarily warm-water and have invaded temperate waters only at the edges of their distributions. Because of the long isolation of the New Zealand microcontinent, temperate-water invasion would have been difficult. On the other hand, *Cyclaspis* is found in tropical to cool-temperate shelf waters as well as the cold waters of the deep sea, so its radiation in New Zealand waters might be expected. The level of endemicity is high in absolute numbers, but species in this genus are usually found in one, maybe two, zoogeographic provinces. Few new species are likely to be found in shelf waters, with most additions to the fauna coming from bathyal depths. If another genus is to be added, it will most likely be something completely new.

Family Bodotriidae: Subfamily Vaunthompsoniinae. This subfamily is largely austral in its distribution and is found from tropical-shelf habitats to cold bathyal waters. Only one New Zealand shelf species is known, and it is not endemic. One of the two bathyal species is endemic, as are both abyssal species. It is unlikely that more than one or two additional shelf species will be found, but the deep-water fauna could continue to contribute new genera and species.

Family Diastylidae. Of the seven genera represented, one (*Colurostylis*) is

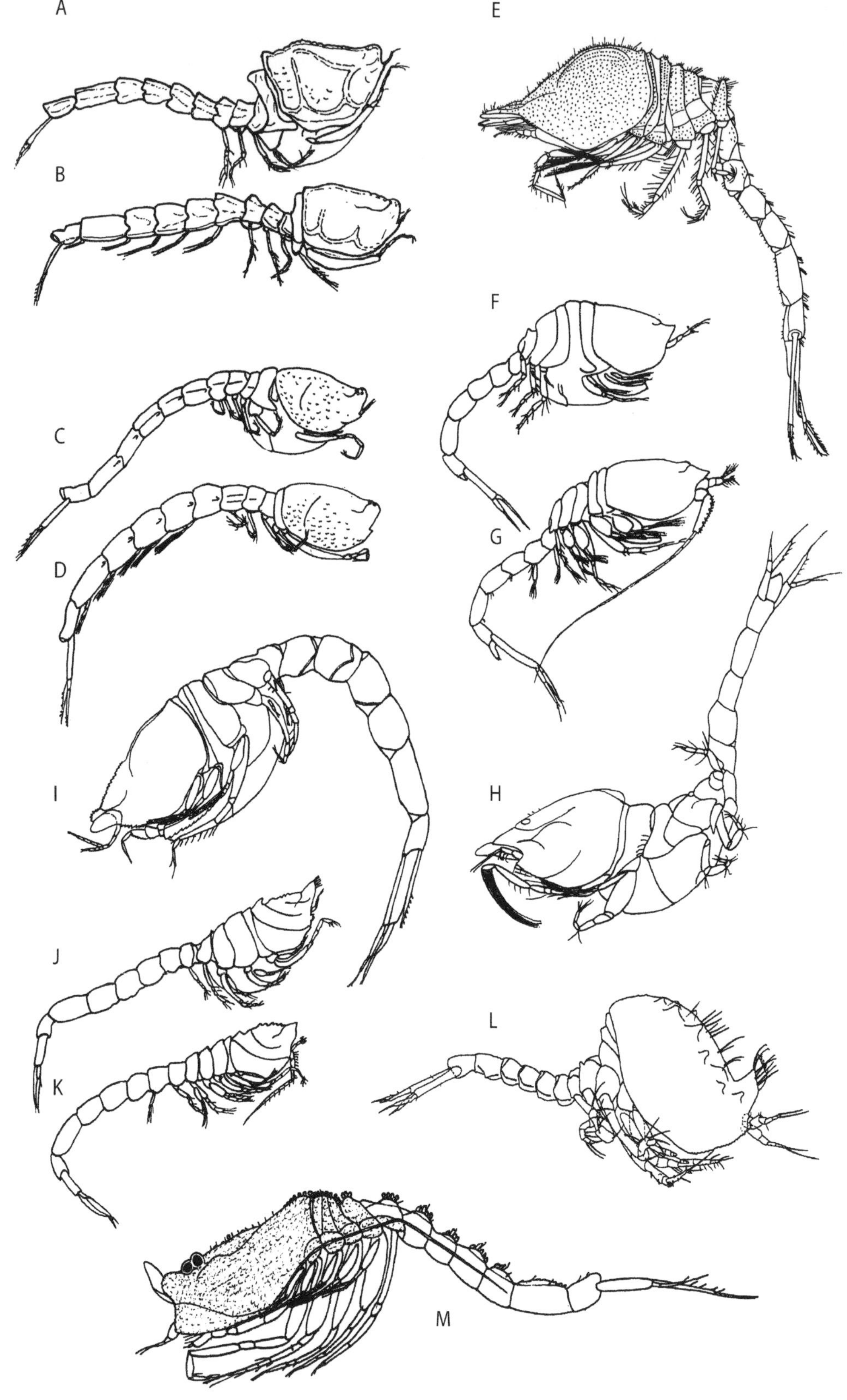

Some New Zealand representatives of cumacean families.
Bodotriidae: A (female), B (male), *Cyclaspis elegans*; C (female), D (male), *Cyclaspis thompsoni*.
Diastylidae: E (female), *Diastylis acuminata* (Diastylidae); F (female), G (male), *Colurostylis pseudocuma*.
Gynodiastylidae: H (female), *Gynodiastylis milleri*. Lampropidae: I (female), *Hemilamprops pellucida*.
Leuconidae: J (female), K (male), *Paraleucon suteri*. Nannastacidae: L (female), *Campylaspis rex*; M (male), *Nannastacus pilgrimi*.
A–K, M, from Jones 1960; L, from Gerken & Ryder 2002

endemic. The others are broadly distributed in the colder waters of the world ocean. The genera *Makrokylindrus* and *Vemakylindrus* are exclusively bathyal or deeper. Specific endemicity is very high (18 of 19 known species) for this family considering the widespread nature of the genera. In addition, diastylids are very abundant and at least one or two individuals can be found at any benthic sampling station.

Family Gynodiastylidae. This is a predominantly southern hemisphere family (but ranges as far west as the Persian Gulf and east to Japan) and exhibits its greatest radiation in southern Australia. There are seven endemic species in New Zealand shallow waters, of which three are in the widespread genus *Gynodiastylis*. One of the new species, in the genus *Allodiastylis*, was found at bathyal depths.

Family Lampropidae. The lampropids are a worldwide, cold-water, primarily deep-sea group. The taxonomy of the family is in need of serious revision, so some of the species found in the current study may be assigned to new endemic genera when revision is completed. Prior to this study only one lampropid, *Hemilamprops pellucidus*, was known from New Zealand. It is a widely distributed southern hemisphere species. Bathyal waters, however, have so far produced eight new species and one new genus (Gerken 2010), suggesting that the Chatham Rise and Campbell Plateau have much higher-than-average lampropid diversity.

Family Leuconidae. This family has very high generic endemicity (three of six genera) in New Zealand, especially in shelf waters. Further, the endemic genera are morphologically advanced within the family, anchoring a group (clade) where the male second antenna becomes reduced in length and modified so it can be used to grasp the female during mating. This trend continues in other eastern Pacific genera, with the second antenna possessing a more complete grasping structure in one Japanese genus and finally culminating in a western North American slope-dwelling genus where the grasping structure is all that is left of the appendage. All species of leuconids are endemic, with the single exception of *Eudorella truncatula*, which is surely an introduced species, broadly distributed in the North Atlantic and North Pacific. This family does not seem to be well represented in New Zealand bathyal samples, in contrast to what is seen in northern hemisphere waters.

Family Nannastacidae. There are two groups of genera in this family in New Zealand – deposit-feeding *Cumella* and its relatives and carnivorous *Campylaspis* and its relatives. Of the deposit-feeders, only one genus, *Scherocumella*, has been found in shallow waters, and two genera were found in the bathyal samples. This group seems to be under-represented in New Zealand. In contrast, there are at least six species of the carnivorous genus *Campylaspis* and two of *Procampylaspis*. The radiation within these genera is typical of that seen in other shelf and slope cold-water environments in both northern and southern hemispheres. All species in this family are endemic. The finding of a species of *Styloptocuma* extends the range of this genus into the Pacific.

In summary, there are two groups of cumaceans in the New Zealand fauna – the highly endemic species and genera of shallow water and the continental shelf, and the bathyal and abyssal species that belong to genera and families that are widespread throughout the cold deep waters of the world. Notably, within one family, the Leuconidae, there has developed a specialised morphology among the males that seems to have spread northwards in the eastern Pacific, culminating in advanced forms in Japan. Finally, New Zealand lacks representatives of many warm-temperate genera, even though it has a warm-temperate zoogeographic province and the Kermadec Islands within its EEZ. This may be a consequence of the geological history of the microcontinent, which, after it became isolated, went through a cooling period, thus eliminating resident warm-water species.

Gaps in knowledge of New Zealand Cumacea

The cumacean fauna of New Zealand's EEZ currently comprises 31 genera (two not yet named) and 74 species, not all formally named. Of these, about half, i.e.

15 genera and 37 species, are from shelf waters. In 1999, a brief collection by Les Watling in a few areas of the North and South Islands produced one new species of *Colurostylis*. Additional collecting is probably not likely to result in the addition of more than 10 new species from shelf depths, with the possible exception of Stewart Island and the subantarctic islands, which so far remain unexplored with respect to cumaceans. The relatively few samples (ca. 15) obtained by Watling have so far yielded 31 new species and two new genera, with the Diastylidae still to be studied in detail. None of the species in the new NIWA and Watling samples can be matched to the eight species Jones (1969) described from the Tasman Sea, suggesting either that there is a high level of endemicity between the east and west deep waters of New Zealand or that the deep-water fauna is very diverse. Neither of these hypotheses is unlikely. Because they brood their young, cumacean species are highly restricted to zoogeographic provinces in shallow water, and may well be restricted to individual tectonic plates in deep water. Since cumacean diversity is generally highest in the Southwestern Pacific, one might expect the overall diversity of bathyal waters to be much higher, at least by a factor of two, than that which has been observed to date. In addition, the lack of correspondence between the shallow New Zealand and southern Australian faunas lends credence to the fact that there is little natural water-borne transport of cumaceans. Most likely the shelf-dwelling cumaceans of New Zealand evolved in situ from whatever stock was present after Zealandia (the New Zealand continental mass) separated from Antarctica about 56 million years ago.

Order Euphausiacea: Krill

We've all heard of 'krill', shrimp-like crustaceans congregating in vast swarms in cooler latitudes of both hemispheres, and famous as whale food. The term krill was originally used by Norwegian whalers for the northern hemisphere cold-water euphausiids *Meganyctiphanes norvegica* and *Thysanoessa inermis* (Mauchline & Fisher 1969) but is now applied to all species of the order Euphausiacea. 'Euphausiids' is itself an unusual word because the ending '-ids' is commonly reserved for family names, not orders. But all except one species of Euphausiacea belong in just one family, the Euphausiidae and, based on long-term use, 'euphausiids' is here to stay. The Euphausiidae contains 85 species and the Bentheuphausiidae one species.

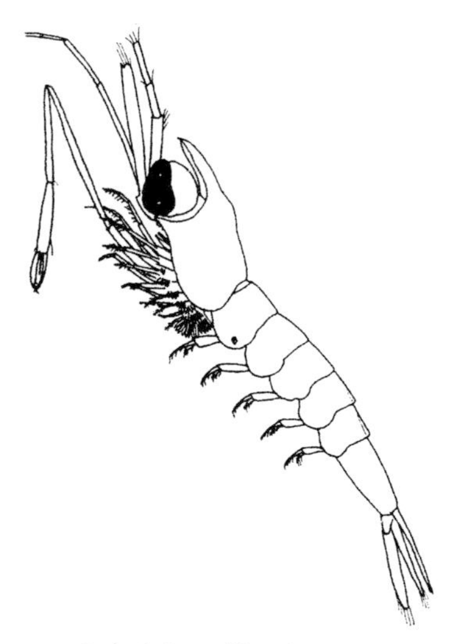

Stylocheiron abbreviatum.
After Sars 1885

The Euphausiacea is notable among the crustacean orders because all the species have conceivably been described. One or two new species may yet be discovered, but only eight have been added in the last 50 years, two in the last 30, with the very deep-water *Thysanopoda minyops* Brinton, 1987, the most recent. However, in some species, particularly in the genus *Stylocheiron*, up to six distinct 'forms' are recognised (Brinton et al. 1999). A few species such as *Euphausia similis* and *E. similis* var. *armata* are also extremely similar. In some cases these forms and species are geographically separate and in others overlapping. It is unclear what the taxonomic significance of the forms is, but new taxonomic techniques such as gene-sequence analysis may resolve this problem. If so, it seems likely that any future changes in the number of euphausiid species are more likely to result from redefinition of current taxa than from new discoveries. There is a further, informal subdivision of the family Euphausiidae, with Brinton et al. (1999) listing several 'species groups' within five of the larger genera based on morphological similarity. The 19 species found in New Zealand waters are named in one or another of these groups.

Krill are of great importance in the marine economy because of their vast numbers. They constitute a major proportion of oceanic biomass, are major grazers of phytoplankton and consumers of small zooplankton, and are themselves essential in the diets of whales, fish, seals, seabirds, and even people.

Morphology and distinguishing characters of krill

Krill are rather uniform in appearance and easily distinguished from other crustaceans. Their morphology is well illustrated and described in several publications, including Baker et al. (1990), who gave a particularly clear overview of their structure, and Brinton et al. (1999). Only the more distinctive characters are described here. Typical of shrimp-like crustaceans, krill are adapted to a natant (swimming) life-style, having an elongate body with the cephalothorax covered by a carapace, a six-segmented abdomen, and a telson with uropods that form a tail fan. They also have moveable eyes, biramous first and uniramous second antennae, and, behind the mandibles, two pairs of maxillae. There are eight pairs of thoracic limbs. Each has a two-segmented outer exopod and a five-segmented inner 'leg' but the posteriormost pair of limbs (eighth pair) is reduced to lobes in all but *Bentheuphausia amblyops*. The form of the seventh pair of limbs also varies between genera. While the first pair of limbs is used in the manner of maxillipeds they are similar in form to those behind. Abdominal segments 1–5 bear a pair of pleopods, the first pair in males being modified to form a handlike copulatory organ (petasma). This is used to transfer sperm packages to a midventral female structure (thelycum). The petasma and thelycum are diagnostic of species although they can be difficult to examine and other, more accessible, structures are generally used for this purpose if they are present and undamaged. Of particular use in this respect are the proximal three segments of the antennule (the antennular peduncle), which may bear a lappet having a characteristic shape or number of spines. The peduncle is usually present in collected specimens and used in combination with other characters.

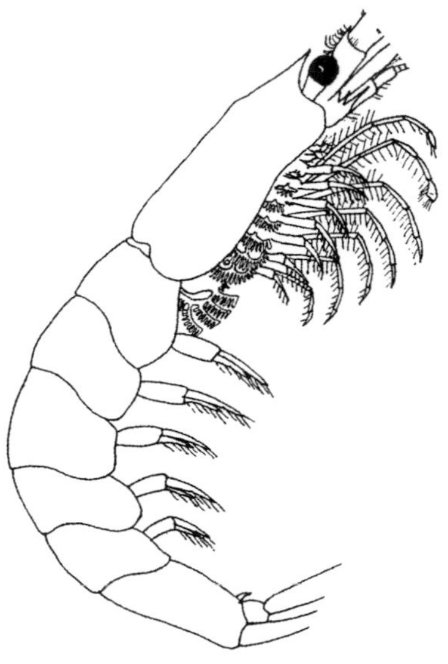

Thysanopoda acutifrons.
From Holt & Tatersall 1906

Krill are easily distinguished from other shrimp-like crustaceans in having the gills exposed below the edges of the carapace, rather than covered by it. Euphausiid gills stem laterally from the first (coxal) segment of the thoracic limbs and become larger, more branched, and more obvious posteriorly.

A second distinctive character is the presence of movable light organs called photophores (the name Euphausiidae indicates they emit 'true light'), which are distributed in the same pattern throughout the order. Only the two very deep-water species *Bentheuphausia amblyops* and *Thysanopoda minyops* lack photophores; all others have a photophore on the carapace beneath each eyestalk and two pairs ventrally on the thorax, adjacent to the second and seventh limbs. Most also have four single photophores ventrally on abdominal segments 1–4, but in species of *Stylocheiron* only one abdominal photophore is present, on the first segment.

The cuticle is thin, flexible, and mostly smooth, with a small spine behind the eye and one or two pairs of denticles (tiny spines) on the sides of the carapace in some species. The front is rounded or produced into a simple sharp rostrum that is small in comparison to many other shrimp-like crustaceans. Some species have a keel behind the rostrum, there may be low-profile dorsal spines and keels on the third to sixth abdominal segments, and, in a few species, some characteristic sculpturing of the abdominal pleura (side-plates). Krill otherwise lack the variety of rostra, spines, and keels found in many decapod shrimps and mysidaceans but they still have rather unusual, distinguishing characters.

Two groups of the Euphausiidae can be distinguished by the shape of the eyes, which are round or almost so in one group and divided by a constriction into upper and lower lobes in the other (Baker *et al.* 1990). The genera *Euphausia*, *Meganyctiphanes*, *Nyctiphanes*, *Pseudeuphausia*, and *Thysanopoda* have round eyes, while *Nematobrachion*, *Nematosceles*, and *Stylocheiron* have bilobed eyes. One genus, *Thysanoessa*, has a mixture of both eye types. There is also a consistent relationship between eye shape and the form of the thoracic limbs (Baker et al. 1990) – species with bilobed eyes have one or two pairs of thoracic limbs greatly elongated while round-eyed species do not. *Stylocheiron* eyes are the oddest of all – four New Zealand species have eyes with enlarged crystalline cones making them tube- or pear-shaped. While lacking obvious cones, the eyes of *S.*

abbreviatum are also pear-shaped and those of *S. maximum* dumbbell-shaped.

Fully grown krill range in length from < 10 millimetres (e.g. *Stylocheiron affine*) to the largest, *Thysanopoda spinicaudata*, which reaches 150 millimetres (Brinton et al. 1999). In New Zealand, the smallest is probably *S. suhmi* at 6–7 millimetres; the largest so far recorded is *Thysanopoda cornuta*, which can reach 120 millimetres.

Classification

Martin and Davis (2001) placed the order Euphausiacea, with the Decapoda and Amphionidacea, in the superorder Eucarida, well separated from the Mysidacea and other orders of the Peracarida. Brinton et al. (1999) recorded earlier recognition of the similarities between krill and the pelagic decapod shrimps of the Sergestidae (suborder Dendrobranchiata). Krill and sergestid shrimps have free-swimming nauplius larvae, metamorphose to the post-naupliar larval stage, have reduced posterior thoracic limbs, and have a petasma in the male and thelycum in the female. However, Brinton (1966) had suggested these similarities might reflect parallel evolution rather than a close relationship. A recent analysis of ribosomal DNA sequences in krill (Jarman et al. 2000) indicates that they may be more closely related to the Mysida than to the Sergestidae, which accords with Brinton's suggestion.

Discovery and diversity of New Zealand krill

Most krill are oceanic in distribution, with consequent low endemicity, and no species is confined to the New Zealand region, so the history of studies of species recorded in the region is mostly international. The first species recorded from the New Zealand region were those collected by the 1873–76 *Challenger* Expedition (Sars 1883, 1885). Sars's reports included 12 of the 21 species now known from New Zealand waters (see end-chapter checklist). H. J. Hansen (1905a,b–1911) described many species in several papers published in the early 20th century, including five species that occur in the New Zealand EEZ.

Tattersall (1924) provided the first list of seven New Zealand krill species gleaned from the reports of Sars (1883), Thomson (1900), and Hansen (1911) and added six more collected by the *Terra Nova* Expedition of 1910. Soon after, Chilton (1926) listed them again but included two species that Tattersall had reported, although rather unclearly, as occurring only in Australian waters (*Pseudeuphausia latifrons* and *Euphausia tenera*). Neither has been recorded from New Zealand since, meaning Chilton's (1926) list more accurately gives 13 New Zealand species. The remaining 12 recorded species have resulted from surveys of pelagic faunas and plankton off New Zealand's coasts (Roberts 1972; Bradford 1972; Bartle 1976; Robertson et al. 1978). The work of Bartle (1976) focused on krill in Cook Strait and is the most extensive study of the New Zealand fauna to date. Four new records are included in the current checklist from collections held at the Museum of New Zealand.

The only identification guide to krill that includes the New Zealand region was produced by Kirkwood (1982), but, apart from the early works listing New Zealand species and referred to above, no taxonomic works on krill in New Zealand waters have appeared. Sheard (1953) reported in detail on the taxonomy, distribution, and development of the Euphausiacea with particular emphasis on the Australasian species *Nyctiphanes australis*. A number of recent papers have reported on aspects of the biology of *N. australis* in southern New Zealand waters and/or included useful distributional and biological observations (e.g. Bary 1956, 1959; Jillett 1971; Bradford 1972; Dalley & McClatchie 1989; McClatchie et al. 1989, 1990, 1991a,b; Murdoch 1989; O'Driscoll 1998a,b; O'Driscoll & McClatchie 1998).

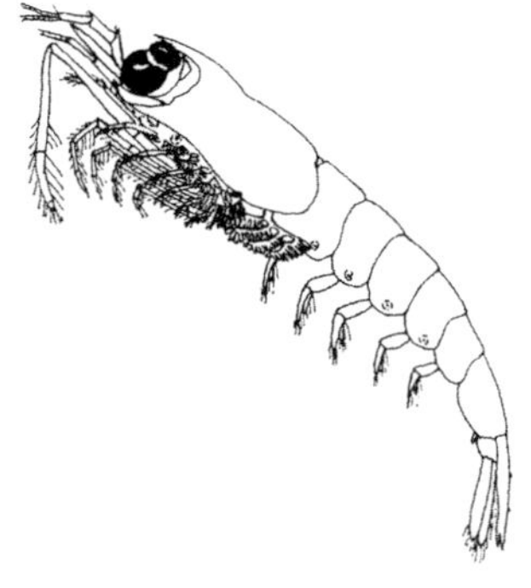

Thysanoessa gregaria.
After Sars 1885

Research on krill biology continues around the world, especially on species of economic importance such as *Euphausia superba*, but the review of Mauchline and Fisher (1969) remains the major source of information. These authors brought together a large and disparate literature on all aspects of euphausiid biology, and Mauchline (1980) updated this. Baker et al.'s (1990) guide to the world's krill species is indispensable. It includes a good brief description of euphausiid anatomy and well-illustrated keys to the species. A paper on krill fisheries of the world (Nicol & Endo 1997) was recently published by FAO, and an easy-to-use CD by Brinton et al. (1999), giving illustrated identification of species, synonymies, references and distribution maps, was published by UNESCO.

The genus best represented in the New Zealand region is *Stylocheiron*. Half of the 12 species known globally occur in New Zealand waters, whereas only two (20%) of 10 *Thysanoessa* species have been recorded here. *Nyctiphanes australis* is one of four and *Nematobrachion flexipes* one of three species in their genera. Two of seven species of *Nematosceles* (29%) and five of 14 *Thysanopoda* species (36%) are present. *Euphausia*, the largest euphausiid genus with 31 species, is represented in New Zealand waters by just six species and one subspecies (22%). Records from New Zealand include three 'round-eyed' genera (*Euphausia*, *Nyctiphanes*, *Thysanopoda*) and three genera with bilobed eyes and elongated legs (*Nematobrachion*, *Nematosceles*, *Stylocheiron*). Both *Thysanoessa* species found in New Zealand waters also have bilobed eyes.

Species recorded in the literature as present, and species believed to be correctly identified, are listed in the end-chapter checklist, but this probably does not give the full picture. Other species are very likely to occur in New Zealand waters. Brinton (1962a) and Brinton et al. (1999) have given Pacific-wide and worldwide distributions of krill. Because they are typically offshore and pelagic in habit, mostly with wide geographic distributions, these distributional data and maps are, of necessity, generalised. Records from outside New Zealand's EEZ suggest that some species may range within the EEZ boundary, and shading on some maps in both works (Brinton 1962a; Brinton et al. 1999) indicates that they do. It is possible, though unlikely, that one or two species have been recorded from New Zealand in food studies of their many predators (fish, birds and whales), not reviewed here. Unrecorded krill species likely to be present include some medium-to-large sized species that may escape capture; not all krill swarm, and swarming species are easier to catch. Some species also live at depths where fine mesh nets are seldom deployed. The deep-living species *Nematosceles tenella* and oceanic *N. atlantica* fit these criteria and have yet to be found in New Zealand waters.

More species of *Thysanopoda* are also likely to be present in New Zealand waters. Mesopelagic *T. astylata*, *T. cristata*, *T. orientalis*, and *T. pectinata* occur widely in the Pacific to about 35° S and a few *Thysanopoda* species are meso- or bathypelagic and seldom sampled, e.g. *T. spinicaudata*, found at 2000–3000 metres. Species such as *T. cristata* are sparsely distributed and not caught regularly. Distributional records in Brinton et al. (1999) suggest at least some of these species may occur in the deep offshore waters of New Zealand but have yet to be collected, which is also the case for *Bentheuphausia amblyops* (Bentheuphausiidae) found throughout the Pacific to 54° S.

The species considered above live either in tropical or subtropical waters or are bathypelagic. Several species present in colder, Antarctic circumpolar water lying south of the Subantarctic Convergence (*Euphausia superba*, *E. frigida*, *E. triacantha*, *Thysanoessa macrura*, and *T. vicina*) must also come close to encroaching on the southern areas of New Zealand's EEZ. However, Morris et al. (2001) have shown that the Subantarctic Front (Convergence) forms a boundary between the colder, fresher Antarctic water to the south and warmer saltier subantarctic water to the north of the front. This abrupt, hydrographic and biological barrier extends deeply into the water column and is apparently a permanent phenomenon. The front also skirts the southern edge of the Campbell Plateau, 200 kilometres south of Campbell Island. This suggests that these circumpolar species are unlikely to be found within the EEZ, except perhaps as stragglers.

Nyctiphanes australis, a small species with adults 10–17 mm long and first recorded in New Zealand more than a century ago (Thomson 1900), is probably the best-known euphausiid of New Zealand waters, being abundant around the main islands and south to The Snares. It has also been studied more than any other species occurring here or in Australian waters, where it is also plentiful from New South Wales to South Australia including Tasmania.

The New Zealand species of *Euphausia* are all small to medium-sized; as adults, *E. recurva* is smallest at 10–14 millimetres long; *E. longirostris*, the largest, can reach 34 millimetres. *Euphausia similis* and *E. similis armata*, both 22–26 millimetres long as adults, are difficult to distinguish but the latter is more often caught and is one of the commonest krill species encountered in New Zealand.

Three of the five species of *Thysanopoda* found in New Zealand waters are new records (*T. cornuta*, *T. egregia*, *T. monacantha*). The largest of these is *T. cornuta* at 50–120 millimetres adult length; purple-red *T. egregia* reaches 50–62 millimetres, and *T. obtusifrons* is the smallest at 18–23 millimetres (Brinton et al. 1999).

The identity of *Stylocheiron longicorne* is complicated by the existence of three 'forms' – a North Indian Ocean form, a short form, and a long form. The latter is present in New Zealand waters and throughout all three main oceans, while the short form is almost as widespread and may occur in northern New Zealand. *Stylocheiron longicorne* is also one of three species of the '*S. longicorne* species group' (Brinton *et al.* 1999) in New Zealand waters, the other two being *S. elongatum* and *S. suhmi*.

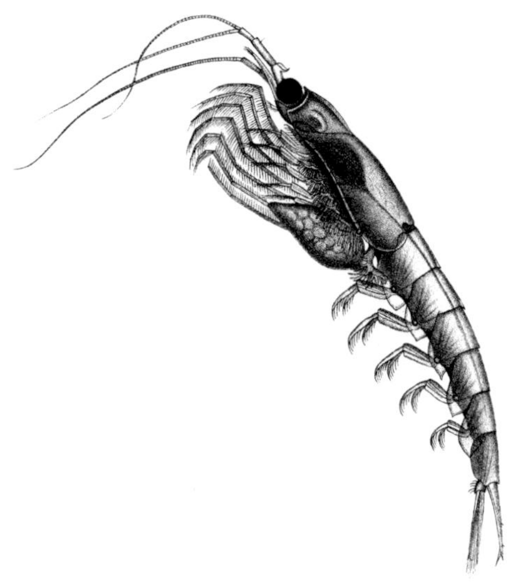

Nyctiphanes australis.
From Sars 1885

Ecology and distribution of New Zealand krill species

Most krill live in the upper layers of the oceans or in coastal areas. Because they are pelagic at all stages in their life cycles and strongly influenced by currents and environmental factors (light intensity, oxygen saturation, temperature, salinity, and food availability), they tend to be confined to certain water-masses. The majority of species undertake daily migrations, swimming upwards into shallower strata of the water column by night and back down before daylight. Most species are omnivores and feed day and night. Upward migration at night into shallower waters may enable consumption of phytoplankton, while retreat to deeper layers during daylight probably helps to avoid pelagic predators.

Krill are well known for swarming, which they do at regular seasonal intervals or irregularly (Mauchline 1984). Aggregations form at or below the surface for feeding or reproduction and swarming by *Nyctiphanes australis* during the breeding season is well developed. Swarms of *N. australis* have been found in harbour and coastal waters of Otago in summer and autumn and a very dense swarm of about four cubic metres was photographed off The Snares by Fenwick (1978). Such swarms tend to be patchy and ephemeral (O'Driscoll & McClatchie 1998) but can be huge and occasionally wash ashore. The largest of a series of strandings of *N. australis* on Otago Harbour beaches in January 1990 was estimated to be ca. 100 tonnes (McClatchie et al. 1991b). *Euphausia similis armata* also intermittently strands in large numbers. In March 1985 and February 2002, millions of live individuals were washed ashore at Waikanae Beach north of Wellington. Drifts were hundreds of metres long and 'ankle deep', as reported by locals, who also observed gulls gorging themselves on the windfall. The krill had apparently been brought ashore by unusual wind and current patterns in the Cook Strait area.

Although krill actively swim, they are classified as plankton because they are moved about by currents, but the larger-sized species may behave more as nekton. *Nyctiphanes australis* lives mainly over the continental shelf and further inshore than other species recorded in the New Zealand region (Bary 1956; Blackburn 1980; Brinton et al. 1999). Offshore transport of *N. australis* is limited by coastal currents running parallel to the coast and by behaviour generated by environmental factors, possibly including vertical movements that place the krill in currents that retain them near the coast (Bradford 1979). Murdoch (1989) and O'Driscoll and McClatchie (1998) found that *N. australis* off Otago became entrained in an anticlockwise gyre off Blueskin Bay and are most numerous in low-salinity coastal waters resulting from river runoff. Bary (1956) observed that the species tolerates a wide salinity range and also penetrates semi-enclosed waters such as Otago and Wellington Harbours and the Marlborough Sounds. *Nyctiphanes australis* undertakes diel vertical migrations from below 150 metres into the top 40 metres of the water column (Bartle 1976) and Bradford (1979) observed that *N. australis* off Kaikoura was able to exist in water temperatures from 8–10° to 23°C.

All species of *Euphausia* recorded in New Zealand waters are oceanic with a circumglobal distribution in the Southern Hemisphere. Only one subspecies, *Euphausia similis similis,* occurs in both hemispheres; the remaining New Zealand representatives of the genus are confined to the Southern Hemisphere, with each distributed in a circumglobal band. South of the Equator *E. similis similis* ranges from 25°S to 55°S (Brinton et al. 1999), which coincides with the northern and southern extremities of the EEZ and encompasses the distribution of its co-subspecies *E. similis armata*. Both subspecies inhabit depths of 0–300 metres but it is not clear if either migrates vertically. Baker (1965) observed what seems to be an inverse relationship between the numbers of the two subspecies and Bartle (1976) suggested this may reflect a difference in depth as he found *E. similis similis* mostly in the upper 100 metres of Cook Strait while *E. similis armata* was mainly deeper.

Euphausia longirostris, E. lucens, and *E. spinifera* also occur north and south of the Subtropical Convergence in New Zealand waters (Bary 1956; Bartle 1967; Robertson et al. 1978; James 1989). *Euphausia recurva* is a more tropical species found as far south as Cook Strait (Bartle 1976) and is bi-antitropical in the major oceans, meaning it is distributed both north and south of the Equator but not across it, although it can be found at lower latitudes than 20° S and 20° N. On the other hand, *E. vallentini* is a colder-water species, recorded by Brinton et al. (1999) from 50°–60° south of mainland New Zealand, but also found within or just to the north of the Subtropical Convergence Zone off Kaikoura (Bradford 1972).

Recognition of *Nematobrachion boopis* in New Zealand waters was only a matter of time since it is very widespread in the three main oceans from 42°N to 50°S. It is the deepest-living species in its genus, the adults being mesopelagic at 300 metres or more, but it also performs daily migrations. *Nematobrachion flexipes* is a deeper mesopelagic species (100–600 metres). It is very widespread though more patchily distributed than *N. boopis* (Brinton et al. 1999).

Two species of *Nematosceles* are found in New Zealand – *N. megalops* and *N. microps*. The former is a warm-temperate species found in all main ocean basins in the Southern Hemisphere and in the North Atlantic. *Nematosceles microps* is widespread in warm-temperate seas in all three main oceans between 40° N and 35° S (Brinton et al. 1999) but has been recorded only once off northern New Zealand (Tattersall 1924).

Stylocheiron elongatum is widespread in all oceans from 40° N to 35° S (Brinton et al. 1999) although Bartle (1976) collected two juvenile specimens from Cook Strait. He did not consider this unusual since waters of subtropical origin are known to penetrate southwards along the Hikurangi Trench into Cook Strait at 300–500 m, the appropriate depth for *S. elongatum*. *Stylocheiron carinatum, S. suhmi,* and *S. abbreviatum* have been recorded only in northern New Zealand waters (Tattersall 1924) but *S. maximum* is very widespread in the three main oceans. Its distribution encompasses New Zealand to 63° S in the Pacific Ocean (Brinton et al. 1999) although Robertson et al. (1978) found it only north of the Subtropical Convergence east of central New Zealand. *Stylocheiron maximum* is mesopelagic, being mostly caught at depths exceeding 400 metres, while *S. carinatum* occupies near-surface waters above 140 metres both day and night (Brinton et al. 1999).

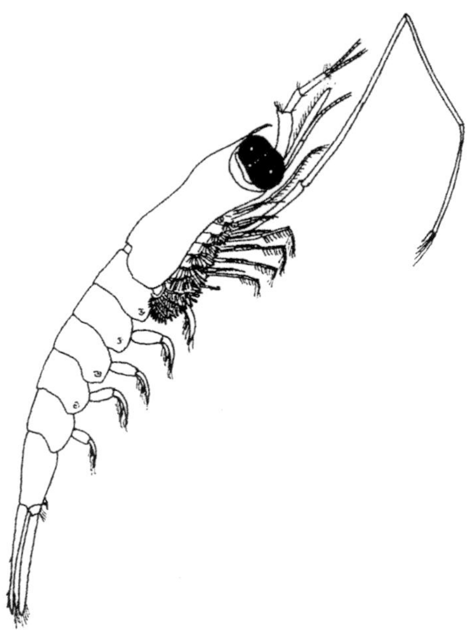

Nematoscelis megalops.
After Sars 1885

Thysanoessa gregaria is biantitropical in all three oceans, is found throughout New Zealand waters, and has been caught regularly in eastern and southern areas (Bartle 1976; Bary 1959; Bradford 1972; Murdoch 1989). While it is usually found above 150 metres depth, Bartle (1976) noted that it is deeper in subtropical than subantarctic waters and suggested it also undergoes extensive vertical migrations. Brinton et al. (1999) indicated that it occupies thermocline waters, rising and falling with them day and night, and that it has been found as deep as 1200 metres. Roberts (1972) identified *Thysanoessa macrura* at the Auckland Islands but Brinton et al. (1999) placed this species in circumpolar Antarctic waters south of 55° S. It seems likely that Roberts was dealing with *T. vicina* rather than *T. macrura* since the two species are difficult to distinguish and, according to Brinton et al. (1999), *T. vicina* overlaps and occurs north of *T. macrura* to 50° S.

Thysanopoda cornuta has been found at scattered locations in the three main oceans at 1200–2500 metres depth, while larvae and juveniles are present at 700 metres or deeper. *Thysanopoda egregia* occurs at 800–2000 metres, while *T. monacantha* is mesopelagic at 300–400 metres, rising into the upper layers at night. Like several other widespread krill found at these depths, *T. monacantha* requires water fully saturated with oxygen and is absent from oxygen-deficient areas of the northern Indian and eastern central Pacific Oceans (Brinton 1962b). *Thysanopoda obtusifrons* inhabits the low-nutrient central water masses of the main oceans and is found up to 140 metres deep at night, migrating below 300 metres during the day.

Breeding and development of krill

Krill sexes are separate. During mating, a sperm package is transferred to the female and sperm are stored in a reservoir until eggs are laid and fertilised externally. In the species of *Nematobrachion, Nematosceles, Nyctiphanes, Pseudeuphausia, Stylocheiron,* and *Tessarabrachion,* eggs are attached to the posterior three pairs of thoracic limbs until they hatch at the second nauplius (metanauplius) larval stage. As in other *Nyctiphanes* species, *N. australis* females not only retain their eggs until this stage, but also secrete a paired, membranous 'egg sac' to hold the eggs (Brinton et al. 1999). *Nematosceles megalops* lays 220–250 small eggs per brood and *Stylocheiron* species 2–50 larger eggs (Mauchline & Fisher 1969), both taxa being represented in New Zealand. In the remaining genera (58 species), the first nauplius hatches from eggs that are shed directly into the water. Thus krill have two nauplius stages, but in those with attached eggs the first stage is passed through in the egg.

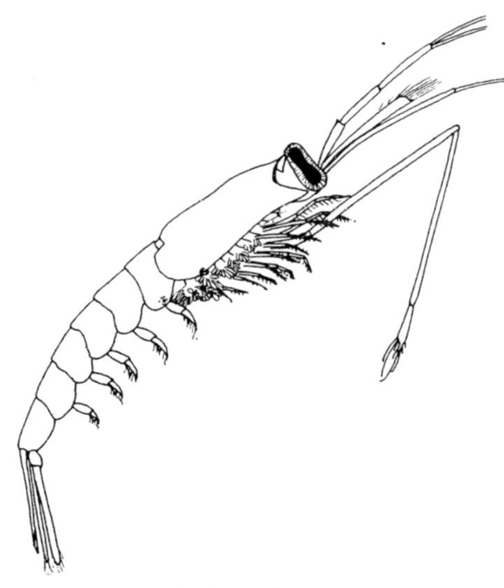

Stylocheiron longicorne.
After Sars 1885

Nauplius larvae swim using their antennae, and all subsequent developmental stages through to the adult are pelagic. The nauplius metamorphoses to the first of three calyptopus stages in which the abdomen develops its full complement of six segments, a telson and uropods. Throughout the calyptopus phase the eyes remain beneath the carapace, and locomotion continues to be provided by the antennae. The final calyptopus moults to the first of several furcilia stages in which the eyes become stalked and free of the carapace, the antennae are no longer natatory, the thoracic legs and gills appear, and, throughout a series of moults, the pleopods and photophores become fully developed. The furcilia passes through various numbers of moults both between and within species and the rate of addition of functional parts varies, depending on environmental conditions. *Euphausia superba* has the least number of furcilia stages of any euphausiid (six) while species of *Thysanoessa* may have as many as 11 stages (Mauchline & Fisher 1969).

Sheard (1953) described these complex larval phases of the life-cycle in several species that happen to occur in New Zealand waters, including a detailed description of those in *Nyctyphanes australis*. Typical of coastal species, the number of larval instars and the sequence of addition of morphological characters (the developmental pathways taken) in *N. australis* is variable, and more so than in oceanic species. The final furcilia moults to the first adolescent stage with little morphological change.

Food, predation, and parasitism

Krill are omnivorous, feeding on phytoplankton, zooplankton, and organic detritus from bottom sediments. Species with highly fringed feeding limbs use them to filter minute protozoans and algal plankton from the water. The bristles effectively form a fine net to strain food from currents created by the thoracic limbs and pleopods. Species with less setose appendages feed more on zooplankton.

The anterior thoracic limbs can be held in such a way as to form a 'food basket' between them and the mouthparts (Mauchline 1984). Bottom-feeding krill employ two methods of collecting food. In one, the animal approaches the bottom in a near-vertical position and, by beating the thoracic exopods, raises into suspension sediment that is filtered by the mouthparts. In the second method, the animal approaches the bottom at a flatter angle and ploughs up the soft sediments with its antennae to form a lump, which it 'sucks' into the food basket by a sudden lateral movement of the thoracic limbs. This method is also used repeatedly as the animals swim, to trap planktonic prey such as copepods or chaetognaths in the food basket.

Among New Zealand krill, 'round-eye' *Euphausia*, *Nyctiphanes*, and *Thysanopoda* species have more highly fringed feeding limbs than 'bilobed-eye' *Nematobrachion*, *Nematosceles*, *Stylocheiron*, and *Thysanoessa* species. In general, the former group is omnivorous, consuming bottom detritus as well as small plankton and non-living particles from the water column. The two large deep-sea species *Thysanopoda cornuta* and *T. egregia* are also known to eat live prey, having been found with copepods, arrow worms, and juvenile fish in their stomachs (Brinton et al. 1999). Carnivory had been suspected in the latter group of krill because bilobed eyes and elongated legs are thought to be adaptations for the capture of live prey (Mauchline & Fisher 1969). The two large deep-sea species *Thysanopoda cornuta* and *T. egregia* are also known to eat live prey, having been found with copepods, arrow worms, and juvenile fish in their stomachs (Brinton et al. 1999).

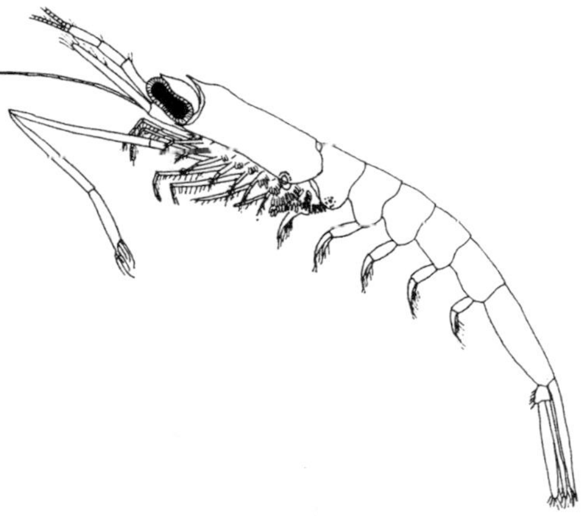

Stylocheiron elongatum.
After Sars 1885

Nyctiphanes australis is the only one among the above species whose feeding has been studied in New Zealand waters. Bradford (1972) found maximum numbers of this species in Kaikoura waters underneath concentrations of copepods, eating their faecal pellets. Blackburn (1980) listed diatoms, copepods, and copepod faecal pellets in its diet and McClatchie et al. (1991a) also confirmed omnivory in the species in Otago waters.

Dalley and McClatchie (1989) carried out a detailed study of the feeding morphology of *Nyctiphanes australis* in Otago, and McClatchie et al. (1991a) measured the spaces between setae of the food basket at 2–8 micrometres, the finest of any euphausiid measured to that time. This suggested *N. australis* is equipped to filter nanoplankton-sized particles. However, Dalley and McClatchie (1989) also concluded that the species is an 'opportunistic omnivore' since it has both a mandibular molar process typical of predators and a mandibular palp and stomach armature characteristic of herbivores. Gut contents, measured using a pigment fluorescence technique (McClatchie et al. 1991a), also revealed substantial amounts of chlorophyll pigments from phytoplankton much larger than nannoplankton, consumed directly, or secondarily in the gut contents of prey. The swarming of *N. australis* in Otago Harbour also coincides with the spring diatom bloom (McClatchie et al. 1991a).

Krill are eaten by a wide variety of cetaceans, fish, and birds. Mauchline (1980) listed the euphausiid species, their major predators, and whether they swarm or not, swarming being an important aspect of their consumption in large numbers. Little appears to be known about predators of *Euphausia longirostris* but five of the other six New Zealand *Euphausia* species that swarm are an important constituent in the diets of baleen whales. *Euphausia vallentini* was reported by Nemoto (1962b in Mauchline and Fisher 1969) to be eaten by fin and sei whales in waters south of New Zealand. Among the six species of *Stylocheiron*, only *S. abbreviatum* is reported as swarming, but all are known to be important food for planktivorous and micronektonic fish. Being mesopelagic, *S. maximum* is also found in the stomach contents of some demersal fish. Whales, planktivorous fish, and seabirds all eat *Thysanoessa gregaria* when it swarms at the sea surface but, while *T. macrura* has been found in whale stomachs, much less is known about it as a food item. *Nematosceles megalops* swarms but both it and *N. microps* apparently do not approach the surface and are preyed on by demersal and planktivorous fish. Pelagic and midwater fish feed on *Thysanopoda monacantha*, and whales and demersal fish on *T. acutifrons*.

Studies of feeding in New Zealand fish and seabirds have revealed that *Nyctiphanes australis* plays an important role in their diets. Kahawai (*Arripis trutta*) around Kaikoura depend on *N. australis* for much of their diet (Bradford 1972) and barracouta (*Thyrsites atun*) also eat this species (Bartle 1976). O'Driscoll and McClatchie (1998) used side-scan radar to study schooling behaviour in barracouta off Otago and came to the conclusion that 'schooling of barracouta seems to be a feeding strategy to exploit surface swarms of krill'. They also found that jack mackerel (*Trachurus murphyi*) and slender tuna (*Allothunnus fallai*) prey on *N. australis*. Blackburn (1980) reported that southern bluefin (*Thunnus thynnus maccoyii*) and skipjack tuna (*Katsuwonus pelamis*), common in New Zealand waters, eat *N. australis* off Australia. No doubt other pelagic fish prey on this species, and Fenwick (1978) saw six different species of bottom-dwelling fish attacking a swarm near The Snares.

With the exception of penguins, seabirds can exploit krill only at or near the sea surface. Rockhopper penguin (*Eudyptes chrysocome*) stomachs have been found with *N. australis* remains – mainly eyes, which seem to resist digestion longer than other body parts (Te Papa unpubl. data). Many flying birds also exploit this species, e.g. grey-faced petrels (*Pterodroma macroptera*), fairy prions (*Pachyptila turtur*) (Bartle 1976), and, importantly, black-billed gulls (*Larus bulleri*) (McClatchie et al. 1989). They are eaten at sea by red-billed gulls (*Larus novaehollandiae*) but not by black-backed gulls (*Larus dominicanus*), which prefer stranded krill (McClatchie et al. 1991b).

Krill are hosts to various parasites. Mauchline (1980) listed three types of ectoparasites – ellobiopsid and apostome protozoans and dajid isopods. The effects of ectoparasites on the host are not always obvious but it is thought that they impair swimming, increase the risk of predation, and damage the cuticle, allowing bacterial infections (McClatchie et al. 1990). Among krill species found in New Zealand, *Euphausia lucens*, *E. recurva*, *E. similis*, *E. vallentini*, *Nyctiphanes australis*, and *Thysanoessa gregaria* have been recorded as being infested with the ellobiopsid protozoan *Thalassomyces fagei* (phylum Myzozoa) (Mauchline 1980). Its precise life-history is not known, but *T. fagei* first appears under the upper carapace of the host, sends a root-like structure down among the organs to gain nourishment, then grows a 'neck', up through the carapace, that branches and produces spores. Dajid isopods attach themselves to the cephalothorax of the host. Among the krill recorded in New Zealand, dajids have been observed in *Nematosceles megalops*, *T. gregaria*, and *Stylocheiron longicorne*. McClatchie et al. (1990) discovered that a stalked pennate diatom also grows externally on *N. australis* caught in Otago Harbour, the first record of such an infestation; 50–70% of *N. australis* sampled in the Harbour were infested. The effects of the diatom on the host were unclear but diatom chlorophyll introduced error into their chlorophyll pigment fluorescence experiments on the krills' diet.

Commercial exploitation and resource potential of krill

The publication by Nicol and Endo (1997) on the world's krill fisheries is an accessible and essential reference for anyone interested in the subject. These authors listed six species of krill commercially harvested in various parts of the world – *Euphausia superba* in the Antarctic Ocean, *E. pacifica* off Japan and British Columbia, *E. nana* off southern Japan, *Thysanoessa inermis* off northern Japan and in the Gulf of St Lawrence (eastern Canada), and *T. raschi* and *Meganyctiphanes norvegica* also in the Gulf of St Lawrence. In 1997, the annual catch of krill for human use was estimated at 160,000 tonnes, with *E. superba* the most important species.

Japan is *the* major fishing nation of both Antarctic krill and northern species, but Ukraine and Poland also have an important stake in the Antarctic fishery. Russia, Korea, and Chile have also been involved at various times. Probably of more interest to New Zealand is research carried out in Tasmania on the potential for a fishery there for *Nyctiphanes australis*, since the species is abundant in New Zealand coastal waters as well.

Human uses of krill include food, bait for sport fishing, aquarium food, and aquaculture food, which is the major use. Krill are of high nutritional value and in Japan are also used to add colour to fish flesh for human consumption. Like the exploited species, *N. australis* has also been shown to have high nutritional value. Krill contain a wide variety of biochemicals, some of possible pharmaceutical value, and Nicol and Endo (1997) listed and discussed their properties and potential uses. They also outlined conservation needs for krill. Current catch rates are thought to be far below the potential for sustainable fishing but the importance of krill in marine food-webs is enormous. The probable effects of overfishing on the many bird, cetacean, and fish predators of krill was important in setting the regulatory Convention on the Conservation of Antarctic Marine Resources in 1980.

Scope for future work

New records of krill species found elsewhere can be expected in the New Zealand region and there is a need to clarify the status of species 'forms' and species groups. Compared to the northern Pacific and Atlantic Oceans there is a lack of data on krill in the SW Pacific. Should a fishery for *Nyctiphanes australis* prove commercially viable off Tasmania, investment in further research on this and other species in New Zealand waters will probably follow.

Order Decapoda: Shrimps, lobsters, crabs, and kin

Decapods ('10-footed') are the most familiar crustaceans, numbering more than 10,000 living species worldwide – almost half the named species of Crustacea. They occur in a great diversity of forms and habitats and some are highly specialised. Most decapods are marine, living from above high tide to depths of more than 5000 metres and at all levels of the ocean. Some live in fresh water and on land but all land dwellers, including the forest crabs of tropical latitudes, must have access to water to hatch their eggs and to drink. Decapods range in size from tiny shrimps about a millimetre long to the largest of all arthropods, the giant Japanese spider crab *Macrocheira kaempferi* with claws that can span up to four metres. There are tiny crabs that live out their lives within coral galls and the huge xanthid crab *Pseudocarcinus gigas* of southern Australia that reaches 15 kilograms in weight. While North American clawed lobsters are the heaviest of all crustaceans, the largest rock (spiny) lobster is the packhorse rock lobster *Sagmariasus verreauxi* of New Zealand and eastern Australia that can weigh 16 kilograms.

Behaviourally, some shrimps and prawns spend their whole lives swimming, while others associate with various bottom habitats. Lobsters and crabs inhabit all kinds of rocky or soft substrata, some bury themselves temporarily, and others live in permanent burrows in mud and sand. Certain genera of squat lobsters are found only on deep-sea branching corals, while small shrimps are often closely associated with algae, adjusting their colours to blend in. A small number

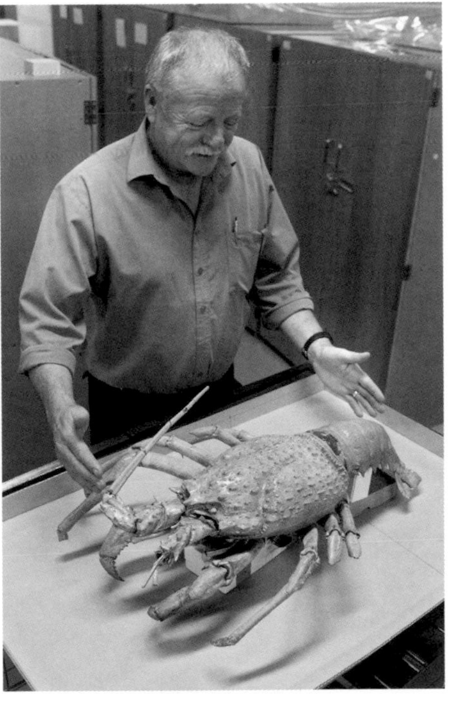

Carcinologist Rick Webber with a historic specimen of the large packhorse rock lobster *Sagmariasus verreauxi*.

Te Papa Tongarewa

of shrimp species have become specialised fish cleaners and a few decapods are confined to very circumscribed habitats such as coastal anchialine caves, underwater geothermal vents and cold-water or hydrocarbon seeps, or are specialised to live on decaying wood or whale bone.

The relationships of decapods with other orders of Malacostraca continue to be argued as do relationships among decapod groups (e.g. Martin et al. 2009). The classification followed here is that of De Grave et al. (2009). The traditional separation of decapods into natants and reptants has no formal status but is useful when discussing the 'swimming' and 'crawling' members of the order and is used here informally.

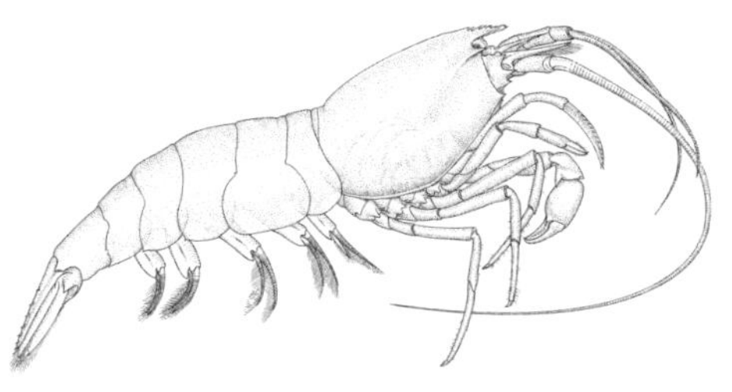

Alvinocaris niwa, a hot-vent shrimp.
From Webber 2004

The Decapoda is divided into two suborders, the Dendrobranchiata, which includes the penaeoid and sergestoid prawns with gill lamellae divided into many dendritic branches, and the Pleocyemata, including all remaining Decapoda, whose gill lamellae are not dendritic (gills are lamellar in the caridean shrimps and prawns, Brachyura and most Anomura; filamentous in crayfish, lobsters and some dromiid crabs – see McLaughlin 1980 for description of gill types). The Pleocyemata thus includes the majority of shrimp and prawn species as well as freshwater crayfish, clawed, slipper and rock lobsters, true crabs and king crabs, hermit crabs, and squat lobsters.

Along with all other members of the class Malacostraca, the decapod body consists of five cephalic (head) somites (six if the eyes are taken as representing a separate somite), eight thoracic, and six abdominal somites. Appendages of the anterior three thoracic somites are modified as food-handling maxillipeds, a principal diagnostic character of the Decapoda since other Crustacea have no more than two pairs of maxillipeds, while the legs articulate with the five posterior thoracic segments. In all decapods the cephalic and thoracic segments are fused, and protected by a carapace that extends down each side of the cephalothorax to enclose the gills and form branchial chambers. The carapace varies from more or less cylindrical in shrimps, prawns, and lobsters to rounded and flattened in crabs but it is the abdomen that has undergone the greatest modifications. In the natants, the decapod abdomen is at its largest, most muscular, and least flexible. It is substantial but proportionately smaller in the reptant lobsters and their relatives, and able to be curved under the cephalothorax, but is reduced to a flap normally held firmly beneath the cephalothorax, in crabs and crab-like Anomura. Despite this variation, all but males of a few hermit-crab species retain at least some abdominal pleopods. Pleopods provide propulsion in natant forms and penis-like organs in male decapods, and in female Pleocyemata remain large enough to carry eggs, even in the shell-inhabiting hermit crabs, whose abdomens are soft and pleopod numbers reduced.

The chitinous integument (exoskeleton) of crustaceans is variously hardened by the addition of calcium salts to increase its strength and rigidity. In crabs and lobsters the skeleton is generally hard and well calcified, except of course at the joints of appendages and abdominal segments, and most extreme in the huge claws of lobsters and mature male crabs. But calcium also adds weight and is therefore minimal in open-water shrimps and prawns. There is also little calcification in burrowing forms, particularly the callianassid 'ghost shrimps', which seldom if ever venture from their protective tunnels, and in hermit crabs the claws and front end of the body are well calcified while the abdomen remains membranous and flexible.

In decapods the sexes are usually separate, although protandry (in which males change to females as they grow) occurs in a number of species and protandric hermaphroditism (where male and female reproductive systems remain functional after the female system develops) has been observed in a shrimp genus. Mating involves the deposition of non-motile sperm, packaged in spermatophores, either externally on the cuticular surface of the female, or internally. Eggs are laid into the surrounding water by dendrobranchs but in the Pleocyemata are retained by the female's pleopods until hatching .

Historical overview of studies on New Zealand Decapoda

Sydney Parkinson, artist on Cook's first voyage to the South Pacific in 1769, illustrated the spider crab now known as *Notomithrax peronii* from material collected in New Zealand. Early settlers and explorers observed and collected intertidal and shallow-water Crustacea (Yaldwyn 1957a) and Cook and his crews traded 'crayfish' with Maori in the Bay of Plenty (Begg & Begg 1969), a hundred years before the species *Jasus edwardsii* (Hutton, 1875) was formally described.

In the last half-century, major reviews of some New Zealand decapod groups have appeared, summarising historical research on these taxa. Forest et al. (2000) monographed the hermit crabs (Diogenidae, Paguridae, Parapaguridae, and Pylochelidae). Their historical account documents an increasingly confused taxonomy of these families in New Zealand, a problem not confined to the hermits. Thirteen years earlier, McLay (1988) published his indispensable book on New Zealand crabs and listed previous contributors to the group. These included Melrose (1975) who reviewed the hitherto poorly known Hymenosomatidae, Griffin (1966) who reviewed the majid spider crabs and their research history, and Bennett (1964) who had himself monographed the Brachyura and provided a critical history of contributions to the group. In two unpublished theses, Yaldwyn (1954, 1959) detailed the history of contributions to New Zealand shrimp and prawn systematics. Wear and Fielder (1985) outlined the very brief history of local larval taxonomy in a monograph on New Zealand brachyuran larvae, a publication that probably advanced knowledge of New Zealand's crab larvae beyond that of any other region.

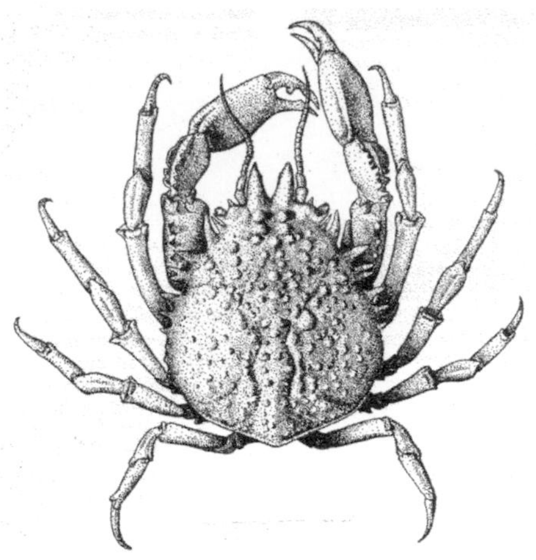

Spider crab *Notomithrax ursus*.
From Griffin 1966

The first decapod described from New Zealand is probably the shallow-water spider crab *Notomithrax ursus* (Herbst, 1788) collected on one of Cook's voyages. *Halicarcinus planatus* (Fabricius, 1775) may have been the first but McLay (1988) considered this unlikely. No further descriptions of New Zealand material appeared for 46 years (although 14 species now recorded in New Zealand were described from other localities prior to 1834). The mid-1830s saw an increase in taxonomic activity resulting from collections made during exploratory voyages by ships from Europe and North America visiting the New Zealand region.

Several explorations of the region provided early knowledge of decapod diversity. These included d'Urville's first voyage to New Zealand (1826–29) (decapods reported by H. Milne Edwards, e.g. 1834–1840); the U. S. Exploring Expedition (1838–42) (decapods reported by Dana, e.g. 1853–55); HMS *Erebus* and *Terror* (1839–43) (decapods reported by White, e.g. 1847); and the Austrian frigate *Novara* (1857–59) (some decapods reported by Heller, e.g. 1868). Decapoda from early exploratory work were first listed with the 'Annulose Animals' by White and Doubleday (1843) in Dieffenbach's *Travels in New Zealand*. The 1880s were the most significant decade of the 19th century in terms of additions to the fauna. The 1874 French Mission de l'Île Campbell made collections from Cook Strait, Stewart Island, and the subantarctic islands (decapods reported by Filhol, e.g. 1886). HMS *Challenger* visited New Zealand on its round-the-world journey (1873–76) and was the first to sample deep-water stations east and west of the country and off the Kermadec Islands (Yaldwyn 1957). Bate (1881, 1888) reported on the mostly meso- and bathypelagic natants, Henderson (1888) the Anomura, and Miers (1886) the Brachyura. Miers (1876) also compiled a *Catalog of the Stalk- and Sessile-eyed Crustacea of New Zealand* from the literature, museum collections, and a collection borrowed from the New Zealand Government.

New Zealanders began to contribute to local decapod taxonomy with the first publication of G. M. Thomson (1879b) describing two natant species. Thomson went on to make an important contribution to New Zealand crustacean studies, including revisions of the New Zealand hermit crabs (1898) and natants. With Charles Chilton he provided a list of New Zealand decapods for Hutton's (1904) *Index Faunae Novae Zealandiae*. Chilton made a valuable contribution to crustacean systematics in New Zealand in a career lasting more than 40 years. Beginning in 1882 he dealt with a variety of reptants and natants,

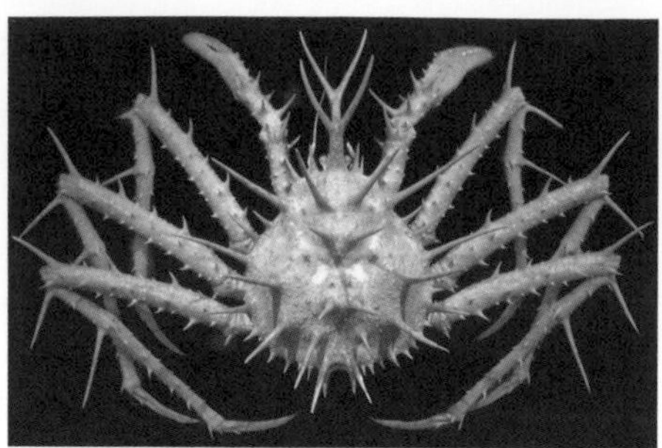

King crab *Lithodes aotearoa.*
From Ahyong 2010

from the Subantarctic to the Kermadec Islands and greatly increased knowledge of their distributions. Chilton (1911c) reported on the New Zealand Government *Nora Niven* Trawling Expeditions that covered most of New Zealand's coastlines. His 1910 paper on crustaceans from the Kermadec Islands, collected by Oliver in 1908, remained the major reference to the Decapoda of these islands until the 21st century. The British *Terra Nova* expedition of 1911 sampled a single but very valuable bottom station off Northland from which Borradaile (1916) described brachyurans, hermit crabs, chirostylids, and natants. Decapods collected from the Auckland and Campbell Islands by Mortensen's Pacific Expedition of 1914–16 were described by Stephensen (1927), and Balss (1929) reported on those collected by the 1924 German Expedition to the Subantarctic Islands led by Kohl-Larsen.

Foreign expeditions continue to visit New Zealand but the contribution of local surveys has greatly increased since World War II, such as those organised by university and museum researchers (e.g. Yaldwyn 1957) and the former New Zealand Oceanographic Institute of the DSIR (incorporated into NIWA since 1992). The Ministry of Fisheries' Observer Programme, in which onboard observers monitor commercial fish catches within the EEZ, has yielded a steady flow of interesting decapods from deep water. In addition, NIWA vessels are currently adding new and rare decapods taken in deep water, on and around seamounts and other locations not previously sampled.

In the postwar period, crab systematics was advanced by the work of Richardson (1949a,b) and Dell (e.g. 1960, 1963a,b, 1968a,b, 1971, 1972, 1974), sometimes in collaboration (e.g. Richardson & Dell 1964; Dell et al. 1970). The first recognition of lithodid king crabs in New Zealand waters came from the identification of *Paralomis zealandica* (as *Lithodes* sp.) from Cook Strait by King (1958), and, as deep-water investigations increased, five further species were added (Yaldwyn & Dawson 1970; Dawson & Yaldwyn 1970, 1971, 1985; Dawson 1989; O'Shea et al. 1999), with the total New Zealand fauna now numbering at least 13 species (Ahyong 2010). Schembri and McLay (1983) published an annotated key to hermit crabs of the Otago region that, in the absence of any similar publication, proved a particularly useful guide to identification until the comprehensive review by Forest et al. (2000).

John Yaldwyn of the Dominion (later National) Museum published on several decapod groups but his most extensive contribution concerned the New Zealand shrimp and prawn fauna. In 1957, he described the Sergestidae of Cook Strait, a significant contribution to this difficult group (Yaldwyn 1957b). He and L. R. Richardson published keys to New Zealand's natant decapods (Richardson & Yaldwyn 1958), now outdated but still the only comprehensive guide available. He added numerous new species to the fauna, notably those collected by the Chatham Islands 1954 Expedition (Yaldwyn 1960) and from the National Museum's collection (Yaldwyn 1971), and published or contributed to numerous other works (e.g. Yaldwyn 1954a,b, 1959, 1961, 1974; Yaldwyn & Dawson 1985).

Since 2000, the rate of publication on New Zealand decapod taxonomy has increased Papers on brachyuran crabs have predominated, with the emphasis on collections from the Kermadec Islands (e.g. Takeda & Webber 2006; McLay 2007; Ahyong 2008) and sea mounts and chemosynthetic habitats (Ahyong 2008). Reviews of the chirostylid squat lobsters (Schnabel 2009) and king crabs (Ahyong 2010) added many new species.

It appears the first systematically collected and recorded New Zealand collection of decapods (and other Crustacea) was that of Charles Chilton, who deposited his material in the Canterbury Museum. Another collection of note is that of A. W. B. Powell at the Auckland Institute and Museum, collected in the 1930s and '40s. After World War II, the collection of Decapoda at the then Dominion Museum increased steadily with the efforts of Moreland and Dell and was continued at greater pace by Yaldwyn between 1959 and 1969 and by Webber into the 1990s. This museum collection is particularly strong in offshore natants and decapod larvae but has a wide coverage of New Zealand decapods as well as some valuable material from Pacific Islands. A small collection made by Betty Batham in the 1940s and `50s is housed at the Portobello Marine Laboratory of Otago University. NIWA, Wellington, has a major collection of decapods, which has become the fastest growing in New Zealand.

The New Zealand decapod fauna

Some 591 decapod species (492 living, ~99 fossil) are known from New Zealand, not all of them formally named, and there are still more to be discovered. New Zealand's decapod fauna is generally considered depauperate compared to other regions (Dell 1968a), given the extent of the EEZ over 30 degrees of latitude, the exceptionally large area of continental shelf and slope, and the wide variety of seafloor relief and ecological niches available. It is difficult to find comparable areas but the numbers of New Zealand crabs have been compared with South Australia by Dell (1968a) and with the Galápagos, Chile, eastern USA, China, and Japan by Feldmann and McLay (1993). These comparisons certainly indicate the limited nature of New Zealand's crab fauna. This is more simply observed in the lack of variety and number of crabs found on seashores or the small number of locally caught crabs, shrimps, or lobsters in fish shops compared with neighbouring Australia and many places further afield. It is generally felt that this limited diversity of species has resulted from New Zealand's isolation from centres of diversity that might have acted as sources of recruitment. Dell (1968a) suggested that New Zealand's separation from Australia in the Early Tertiary occurred before evolutionary radiation gave Australia its diverse fauna but it is unclear why a similar process has not occurred in New Zealand. It is reasonable to view most of New Zealand's decapod taxa as depauperate but there are exceptions – New Zealand is well represented by southern hemisphere oceanic natants that live independently of shallow water and are less limited by constraints on dispersal, but there is also a high diversity of hermit crabs and some squat lobster genera and the two crab families Majidae and Hymenosomatidae are also well represented.

Taxonomic knowledge of New Zealand's present-day Decapoda is comprehensive for the hermit crabs and squat lobsters, and reasonably good for coastal and shelf natants and the Brachyura, but not so for the thalassinids and penaeoid and sergestoid shrimps and prawns. Present exploration of deep-sea rocky habitats, notably the many seamounts in the New Zealand region, is rapidly increasing our knowledge of decapods in these places. Geographically, the least well-known areas are the Kermadec Islands (although knowledge of the shallow-water crab fauna is rapidly increasing), and much of the west coast of New Zealand.

Decapods are an important component of the luxury food market worldwide. Despite New Zealand's limited variety of edible species, some nevertheless support very valuable fisheries, most notably the red rock lobster *Jasus edwardsii*. Interest in developing new crustacean fisheries is growing, and considerable research effort is now expended on ways of improving rock-lobster productivity and quality through habitat enhancement, ongrowing of juveniles, and the possibility of culturing.

The main collections of New Zealand decapods are held at the Museum of New Zealand and NIWA, but considerable historic collections and the majority of types are kept at the Natural History Museum (London) and the Muséum National d'Histoire Naturelle in Paris. Other significant collections are located in the Senckenburg Museum (Frankfurt) and the Australian Museum (Sydney), while further important material resides in a number of other institutions, notably Museum Victoria, the U. S. National Museum of Natural History, and the National Science Museum in Tokyo. The largest type collection in the country is housed at the Museum of New Zealand, where there are 202 lots, including just 42 primary types. A smaller collection of types is held by NIWA and type material is also kept at Auckland, Canterbury, and Otago Museums.

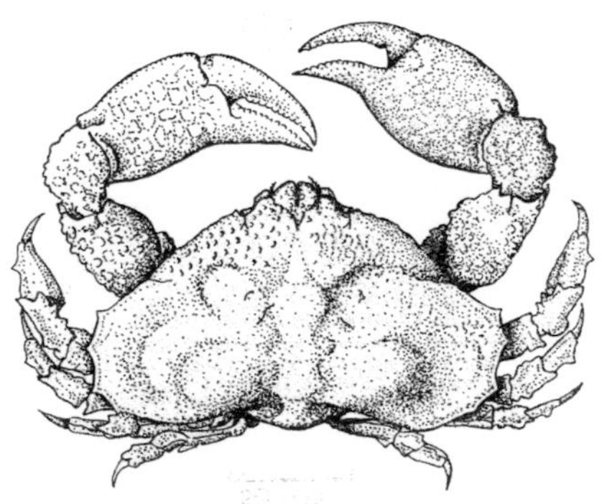

Endemic triangle crab *Eurynolambrus australis*.
From Griffin 1966

A total of 492 living decapod species have been recorded within New Zealand's EEZ (see end-chapter checklist). When the first Decapoda checklist was compiled for Species 2000 New Zealand in 2002 the classification used was that of Martin and Davis (2001). The greatest effect their revised classification had on the hierarchy of New Zealand decapods was to increase the number of families recognised locally, mainly by raising subfamilies to family status, especially in the Brachyura. Since then, there has been less change in the classification of shrimps and prawns and other non-brachyuran groups but changes continue to be made in brachyuran families (e.g. Ng et al. 2008). New Zealand has 84 of the 151 families of Martin and Davis (2001) although a large proportion of them (43%) contain only one or two species (20 with only one species, 15 with two). In contrast, the three most species-rich families contain 112 species, or almost a quarter of the decapod fauna. Of these three, the Galatheidae has the greatest number with 46 species, the Paguridae with 34 species and the Chirostylidae with 33. The Chirostylidae also includes the most speciose New Zealand genus, *Uroptychus*, with 27 named species. The largest natant family is the Oplophoridae with 18 species, all named. Among the subfamilies raised to family in Martin and Davis (2001) are those of the superfamily Majoidea (previously family Majidae), which contains 33 species. Despite this division, however, the previous subfamily Majinae (now the Majidae in the strict sense) contains 17 species, almost as many as the largest New Zealand brachyuran family, Xanthidae (18 species).

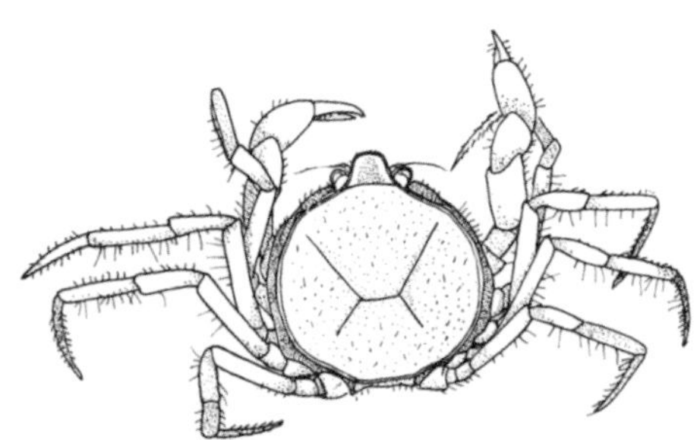

Freshwater hymenosomatid crab *Amarinus lacustris*.
From Melrose 1975

Endemism

Of the 492 living New Zealand decapods known, 12 are unnamed or not yet fully determined. The level of endemism is only ~30% (144 species). As might be expected, endemism is lowest in pelagic offshore species and highest among benthic and shallow-water forms. Thus all seven dendrobranch families (23 named species, two undetermined) contain no endemics at all and the four pelagic carid families Nematocarcinidae, Oplophoridae, Pandalidae, and Pasiphaeidae (44 species in total) include only one endemic species. New Zealand's dearth of nearshore pelagic natants in any of these groups is reflected in this low endemism and, although an estimated 35 additional penaeoid and sergestoid species may be anticipated for the fauna, few if any are likely to be restricted to New Zealand waters. Subtract offshore natant groups from the named decapods and the proportion of endemics rises. But lower endemism is not characteristic of all natants – of the 471 named living New Zealand Decapoda, 97 are carid shrimps of which 30 (~31%) are endemic, the same proportion as for the reptants alone, of which 106 (~31%) are confined to the New Zealand region. Ten of the 253 New Zealand decapod genera are endemic, viz the brachyurans *Eurynolambrus*, *Halimena*, *Heterozius*, *Jacquinotia*, *Neohymenicus*, *Neommatocarcinus*, *Nepinnotheres*, *Pteropeltarion*, and *Trichoplatus* and the slipper lobster genus *Antipodarctus* – all of which contain a single species. One family, Belliidae, is endemic.

Most New Zealand species of Crangonidae and Palaemonidae are endemic, as are both species of Spongicolidae, probably reflecting their close association with hexactinellid sponges. There is also higher-than-average endemism of Axiidea and Gebiidea (former Thalassinidea), Diogenidae, and Paguridae. This is in contrast to the deeper-water hermit crabs of the Pylochelidae and Parapaguridae, which each have only a single endemic species.

While the two freshwater parastacid crayfish *Paranephrops planifrons* and *P. zealandicus* and the only freshwater shrimp *Paratya curvirostris* are endemic, the freshwater hymenosomatid crab *Amarinus lacustris* is not, occurring also at Norfolk and Lord Howe Islands and in southeastern Australia and Tasmania.

Native paddle crab *Ovalipes catharus*.
Shane Ahyong

A number of rarely caught deep-sea species previously thought to be endemic to New Zealand have been found in greater numbers and further afield, particularly in southeast Australian waters (e.g. *Lipkius holthuisi*, *Teratomaia richardsoni*). The apparent endemism and rarity of some deep-sea species are probably the result of insufficient sampling. Endemism in New Zealand's second-largest crab family, Majidae, is rather low at 35% (six of 17 species) but includes intertidal (e.g. *Notomithrax peronii*) and shelf/slope (e.g. *Thacanophrys filholi*) taxa. Hymenosomatid crabs are well represented in New Zealand and 10 of the 14 species (71%) are also endemic. One of the non-endemics, *Halicarcinus innominatus*, is thought to be of New Zealand origin but accidentally introduced to Tasmania.

New Zealand's two species of Pinnotheridae (pea crabs) are both endemic, as might be expected of shallow-water associates of bivalve molluscs, but endemism in the crab families Portunidae (paddle crabs) and Xanthidae is quite low at less than 30%. Just three of 11 native portunids and three of 15 native xanthids (all found only at the Kermadec Islands) are endemic. Portunids and species of Varunidae tend to have long larval lives and some are able to travel great distances as adults so that most species are distributed widely. Even New Zealand's only terrestrial decapod, *Geograpsus grayi* of the Kermadec Islands, is widespread in the Indo-West Pacific.

Of New Zealand's 132 endemic decapods, 14 are recorded from the Kermadec Islands and nine are restricted there. Five are hermit crabs, all from moderately deep water except *Pagurixus kermadecensis*, which is found in rock pools. Like a number of other apparent endemics, the shrimp *Stylodactylus discissipes* is known from only a single station at 1100 m depth and is likely to be more widespread.

Ecological studies

Paddle crabs (*Ovalipes catharus*) are numerous enough to comprise a small fishery, encouraging investigation of marketing (Cameron 1984) and reproductive biology (Haddon 1994, 1995; Haddon & Wear 1993). University research has made a considerable contribution to decapod biology, particularly that carried out over the years by Malcolm Jones and Colin McLay (Canterbury) and Bob Wear (Wellington), with their students. The physiology of musculature, haemolymph, locomotion, and eye function in shore crabs have been addressed (e.g. Jones & Greenwood 1982; Bedford et al. 1991; Forster 1991; Meyer-Rochow & Reid 1994; Palmer & Williams 1993; Meyer-Rochow & Meha 1994; Depledge & Lundebye 1996) as have the effects of low oxygen and varying pH on freshwater shrimp (West et al. 1997; Dean & Richardson 1999). Feeding studies of shore crabs were carried out (e.g. Wear & Haddon 1987; Creswell & McLay 1990; Woods 1991; Woods & McLay 1994). Jones (1976, 1977, 1978, 1980, 1981), Jones and Winterbourn (1978), and Jones and Simons (1981, 1982, 1983) undertook significant work on intertidal crabs of the Avon-Heathcote Estuary and Kaikoura, and other ecological studies were made by McLay and McQueen (1995), Palmer (1995), and Morrisey et al. (1999). Several papers on the behaviour and associations of shore crabs have also appeared (e.g. Field 1990; Taylor 1991; Chatterton & Williams 1994; Woods & McLay 1994; Woods 1995; Woods & Page 1999) and Berkenbush and Rowden (1998, 1999) studied population dynamics and sediment turnover in the burrowing ghost shrimp *Callianassa filholi*.

Alien species

Interest in adventive species is growing rapidly in New Zealand (see Cranfield et al. 1998 for a list of adventive decapods and the Ministry of Fisheries for details of potential invaders (Marine Pest Identification Guide series)). Some decapods have been introduced intentionally but mostly without success; this is probably a good thing as some crab and lobster species are among the most destructive of invaders. The first such introduction appears to have been of the Australian penaeid prawn *Melicertus canaliculatus* (as *Penaeus canaliculatus*), released at Nelson in 1892 and at the entrance to Otago Harbour in 1894 (Thomson 1922). They were never seen again. Between 1906 and 1918, a more serious attempt

Alien paddle crab *Charybdis japonica.*
Shane Ahyong

was made to introduce the European lobster *Homarus gammarus* into New Zealand. A similar project was undertaken with the European edible crab *Cancer pagurus* between 1907 and 1914 (Thomson & Anderton 1921). Live crabs and lobsters were imported from the United Kingdom and kept at the Portobello Marine Fish-Hatchery in Otago Harbour. Several million crab larvae and more than 750,000 lobster larvae were hatched and liberated in the harbour during those years. Some young lobsters were reared for several years before release, and mature adults of both species were also liberated but no trace of free-living specimens of either species has been found in Otago or New Zealand waters since.

There was a short-term attempt in the early 1990s to farm a 'saltwater king prawn' from Hong Kong, probably the penaeid *Fenneropenaeus chinensis*, at South Kaipara Heads. Like the *H. gammarus* and *C. pagurus* experiments this also failed but in this case the stock was destroyed. So too was the entire stock at a pond farm of the Western Australian crayfish or marron, *Cherax tenuimanus*, at Warkworth, north of Auckland in the late 1980s and early 1990s (Hughes 1988; Lilly 1992). Fear of their escape into waterways led to this action but the same problem does not occur with large palaemonid prawns farmed at Wairakei, near Taupo. Here, *Macrobrachium rosenbergii* from South-east Asia and northern Australia is successfully farmed in artificially heated water. This is drawn from the Waikato River and warmed by a heat exchanger using hot-water runoff from a geothermal power station nearby. *Macrobrachium rosenbergii* cannot breed or survive in ambient New Zealand fresh waters.

Foreign decapods periodically appear accidentally in New Zealand, apparently introduced in ships' ballast water or on hulls. Some species disappear but others threaten to become established and compete with the local biota. The hymenosomatid crab *Halicarcinus ovatus*, normally found around western, southern, and eastern Australia, was recorded just once at Port Chalmers, Otago, by Filhol (1886) but has not been recorded in New Zealand since (Melrose 1975; McLay 1988). In 1978, the small inachoidid spidercrab *Pyromaia tuberculata*, originally from the Central American west coast but subsequently found in other localities in the Pacific and Atlantic Oceans, was discovered in the Firth of Thames (Webber & Wear 1981). It has become established but is not often found and does not seem to be a major threat to endemic species.

In the early 1990s live specimens of three species of crab were found in a ship's sea chest at a Nelson slipway – *Pilumnus minutus*, *Carupa tenuipes*, and *Charybdis hellerii* (Dodgshun & Coutts 1993). The significance of sea chests (recesses in ship hulls housing the intakes of ballast water) as a mode of introduction quickly became apparent. *Pilumnus minutus* is small and uncommon and *C. tenuipes* tropical, and neither is likely to become established, but the Asian and northern Indian Ocean portunid *C. hellerii* is a successful invader of the eastern Mediterranean and western Atlantic from Florida to Brazil. It is unlikely that *C. hellerii* could establish itself in New Zealand, except perhaps in the far north, but a close relative has. First reported from Waitemata Harbour in 2001, hundreds of *Charybdis japonica*, including egg-bearing females, have since been caught, and it is also present in the Firth of Thames (Webber 2001; Smith et al. 2003). Almost as large, and far more aggressive than the paddlecrab *Ovalipes catharus*, *C. japonica* is likely to exclude the local species from harbour and estuary mouths but is unlikely to spread to open sand beaches or much further south, as it is a warm-water species. Its behaviour in nets causes problems for flounder fishers but if it remains in large-enough numbers, it may at least become a new fishery.

Introductions have also occurred in the opposite direction. The small hymenosomatid crab *Halicarcinus innominatus* and the larger pie-crust crab *Metacarcinus novaezelandiae* were probably accidentally introduced to Tasmania when *Ostrea angasi* was imported from New Zealand to enhance the oyster fishery (Lucas 1980).

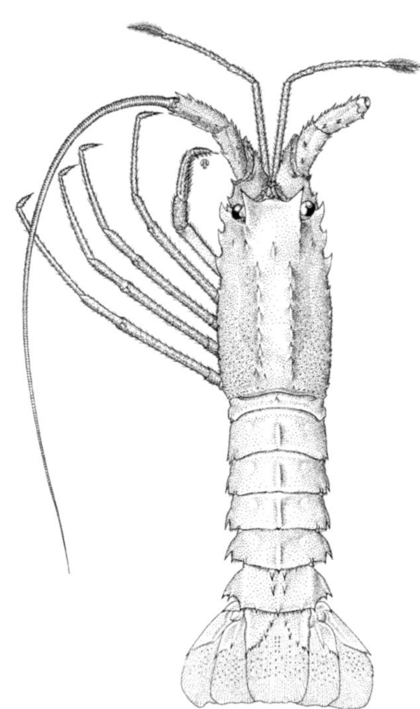

Projasus parkeri, a recent palinurid.
W. Richard Webber

New Zealand fossil Decapoda

The fossil decapod fauna comprises approximately 99 species, although only 56 of these are named unequivocally owing to the high proportion of small or unique specimens or their often incomplete or fragmentary state. There are 48 named genera in 27 families, and six of the seven Recent reptant infraorders (only Polychelida lacking), and only the Glypheidea (superorder Pleocyemata) among the natants. Nineteen of the 58 Recent reptant families include fossil species, with five families represented in New Zealand only by fossils. Some 22 fossil genera also occur in the present-day New Zealand fauna and four Recent species are represented in the New Zealand fossil record, possibly six, should fossil *Ctenocheles* cf. *maorianus* and *Ommatocarcinus* cf. *Neommatocarcinus huttoni* prove indistinguishable from their living namesakes.

Although the fossil decapod fauna of 99 species is small relative to the present-day fauna, recent research has revealed its significance to the origins of decapods in New Zealand and in the South Pacific (Feldmann 2003). The xanthid crab *Tumidocarcinus tumidus* was the first fossil decapod described from New Zealand, but 94 years were to elapse before additional records were published. Glaessner (1960) published his signal work on the New Zealand fossil Decapoda, recognising 29 species in eight genera, including a new genus and 16 new species. Most of these were brachyurans (22 crabs in seven families) but Glaessner also identified five callianassid ghost shrimps and three astacoidean lobsters of the families Glypheidae and Mecochiridae. In addition, he described the palinurid rock lobster *Sagmariasus flemingi* (as *Jasus flemingi*), the only fossil yet discovered among the nine Recent species of non-stridulating Palinuridae (*Jasus*, *Projasus*, and *Sagmariasus* species, all austral).

Glaessner's (1960) work remains the most important contribution in terms of numbers of taxa added to the fossil fauna, although subsequent work has trebled the known fauna. Only three more new species were added to the fauna during the 1960s and 1970s, but momentum and diversity then increased, with nine new species described in the 1980s and 16 in the 1990s. Crabs predominate among the new records, but several other new taxa have also been identified, leading to fresh interpretations of their origins and relationships to Recent forms. For example, New Zealand's first fossil nephropid lobster, *Metanephrops motunauensis*, was described from north Canterbury.

The first decapod added to the fauna by a New Zealand worker (*Trichopeltarion greggi*) was also the first fossil species of the extant family Atelecyclidae (Dell 1969). The tymoloid family Torynommidae was erected by Glaessner (1980) to contain several extinct Australasian crabs including two new New Zealand species, and in the same paper Glaessner named three new species of raninids for New Zealand. Hyden and Forest (1980) described the first, and so far the only named, fossil hermit crab from New Zealand (*Diacanthurus spinulimanus*), and the late Sir Charles Fleming (1981) described *Miograpsus papaka*, so far the only fossil grapsid recorded from New Zealand.

The description of the squat-lobster-like anomuran *Haumuriaegla glaessneri* was significant, both for the implications it had for the interpretation of New Zealand's fossil record and as the beginning of a major and continuing contribution to New Zealand decapod palaeontology by its author (Feldmann 1984). *Linuparus korura* was the second palinurid added to the New Zealand fossil fauna (Feldmann & Bearlin 1998) and Feldmann and Maxwell (1999) described five more decapods – two raninids, two majids, and a single goneplacid, the first New Zealand fossil of the genus *Carcinoplax*. At this point, a review of the fossil decapods of New Zealand by Feldmann and Keyes (1992) appeared, listing all previously published records, giving a detailed index of locality records and an updated checklist of taxa, and tabulating their stratigraphic ranges in the Mesozoic and Cenozoic. Some 81 decapods were recorded, although just 38 species were named. Forty genera were recorded in 21 or 22 families, a considerable increase from the eight genera in 11 families recognised by Glaessner (1960). Five more

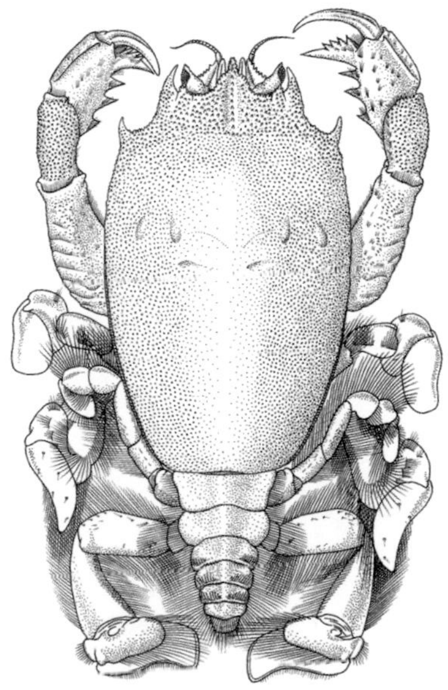

Native frog crab *Notosceles pepeke*.

From Yaldwyn & Dawson 2000

new species were soon added to the fauna by Feldmann (1993), including the first published record for New Zealand of the Calappidae (*Calappilia maxwelli*), the first record of the genus *Glyphea* (*G. stilwelli*), and one further species in each of the Holodromiidae, Torynommidae, and Majidae.

Feldmann and Keyes' (1992) review and McLay's (1988) survey of New Zealand's Recent crab fauna were closely followed by a substantial paper on the paleogeographic history of the New Zealand Brachyura (Feldmann & McLay 1993). In their analysis, these authors compared New Zealand's extant Brachyura with that of other, mostly Pacific, regions and went on to identify significant relationships not recognised previously between New Zealand's Recent and fossil faunas. A number of new taxa have come to light since these works, supporting their observations.

The first recognition of the family Parastacidae in the fossil record (*Paranephrops fordycei*) was published from a single specimen found in Miocene deposits of Central Otago (Feldmann & Pole 1994). Two further majids were added to the fauna by McLay et al. (1995) and a new cancrid by Feldmann and Fordyce (1996). The world's first fossil lithodid (king) crab (*Paralomis debodeorum*) was described only in the 1990s (Feldmann 1998), along with a glypheid lobster, *Glyphea christeyi* (Feldmann & Maxwell 1999), both from Canterbury.

The origins of New Zealand's decapod fauna are far from clear and continue to be debated, particularly because of fossil discoveries over the past 20 years in both New Zealand and Antarctica. Until the early 1980s it was believed that New Zealand's decapod fauna was primarily of Australian and Indo-Pacific origin. Glaessner's (1960) Tertiary material occurred no earlier than the middle Eocene (45–50 million years ago). He considered the presence of *Tumidocarcinus* in the middle Tertiary of Australia and in the Eocene and Miocene of New Zealand as indicative of a 'distinctive zoogeographical province' and that Australasian genera could be considered as originating in the ancient Tethys Sea. Fleming (1962, 1979) also concluded that New Zealand decapods were primarily of Tethyan origin and that typical New Zealand marine decapod faunas had appeared since the Oligocene. In his analysis of the distribution and composition of New Zealand's extant Brachyura, Dell (1968a) found that the strongest external elements in the present-day crab fauna are Australian and Malayo-Pacific in practically equal strength, which also implies a Tethyan origin.

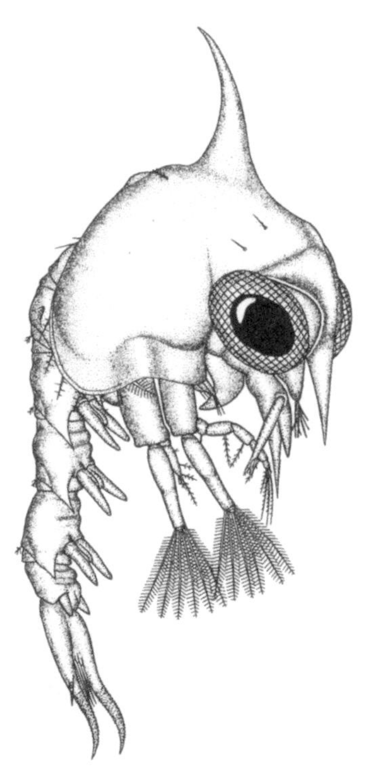

Planktonic zoea larva of the majid crab *Jacquinotia edwardsii*.
W. Richard Webber

The late Mesozoic *H. glaessneri* from North Canterbury was a shallow-water marine species and the earliest known representative of the extant freshwater anomuran family Aeglidae, which is confined to temperate latitudes of southern South America. This discovery, and analysis of fossil and recent species of *Lyreidus* (Raninidae), led Feldmann (1984, 1986, 1990) to believe that these and other decapod genera had evolved in high-latitude southern waters rather than originating in the Tethys. This occurred during the late Mesozoic prior to New Zealand's split from Australia and Australia's split from Antarctica, which also had a cool-temperate climate. Feldmann considered that species evolving along this coast would be dispersed eastwards by the southern Pacific gyre but that this would have discontinued with a cooling climate and the break of Australia from the Antarctic, allowing the circumpolar current to develop.

Newman (1991), however, questioned this view and suggested that taxa like the entirely austral *Jasus* species may have resulted by reliction (reduction in range) following an amphitropical (northern as well as southern hemisphere) distribution. He offered three hypotheses on how such southern hemisphere endemism could have come about – centres of origin, dispersal to the southern hemisphere, or vicariance (see Newman 1991).

This debate continues, with research on fossil decapods worldwide increasing in recent years. Schweitzer (2001) has summarized decapod paleobiogeography and the diverse literature on decapod fossils and their interpretation was reviewed by Feldmann (2003).

Decapod development

No discussion of decapod diversity would be complete without reference to their larvae. The morphology of decapod developmental stages is an important aspect of decapod systematics, and knowledge of larval biology and recruitment to adult populations is essential to managing decapod fisheries.

Development in the great majority of Decapoda, both natants and reptants, includes free-swimming planktonic larvae. In the penaeoid and sergestoid (dendrobranch) shrimps and prawns, eggs are laid into the surrounding water and tiny, motile nauplius larvae subsequently hatch into the plankton. All other decapod groups (the Pleocyemata) retain their eggs attached to the pleopods until larvae hatch. In the plankton, larvae grow through a series of instars until, at the final moult, they metamorphose into a post-larva, an intermediate form looking more or less like the adult but retaining the ability to swim. The role of the post-larva is to relocate itself to the milieu of the adult phase where it again moults to become a juvenile crab, lobster, shrimp, or prawn. Like their larvae, shrimps and prawns are pelagic. The transition from larva through post-larva to juvenile is less abrupt although the final larval moult is still marked in pelagic species by a fundamental change in locomotion from using appendages of the cephalothorax to propulsion by the abdominal appendages (pleopods).

Most decapod families have different though predictable numbers of larval growth stages and a single post-larva during development, but a few groups and species have either extended or abbreviated development. Some have even eliminated free-swimming larval or post-larval phases altogether, with juveniles hatching directly from the eggs. The number of larval stages relates to the duration of the larval phase, and those species with abbreviated or direct development usually occur in habitats where free-swimming larvae would be lost. Some of these different strategies are exemplified by New Zealand Decapoda.

Larval decapods are of taxonomic interest because they differ morphologically from adults. This is particularly so in benthic forms, which make up the majority of decapod species and occupy very different habitats from their offspring. Pelagic larvae have evolved their own adaptations to planktonic life, yet the medium they frequent is in many ways more uniform than the variety of substrata or depths occupied by the adult phase, which serves to emphasise the importance to taxonomy of differences in larval features.

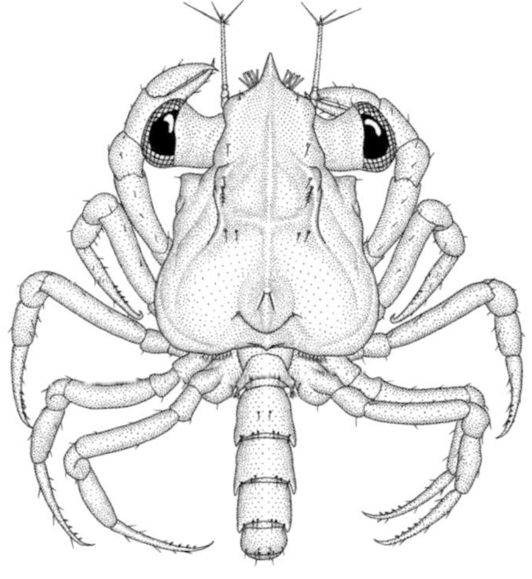

Megalopa larva of spider crab *Notomithrax minor.*
From Webber & Wear 1981

Limits to the use of larval features are more practical than theoretical, however; while larvae caught in plankton can usually be attributed confidently to higher taxa, incorrect identifications of genera and species based on morphology are often made (e.g. McWilliam et al. 1995). The only foolproof method of putting names to larvae caught in plankton is to hatch them from eggs of known parentage or rear planktonic larvae through to identifiable adults. Since Vaughan Thompson (1828) first put the provenance of decapod larvae beyond doubt by observing larvae hatching (see Gurney 1942), rearing techniques have improved, but maintaining ovigerous females and their delicate offspring in captivity, even when robust berried females can be caught, is always difficult and sometimes impossible. However, this impasse has begun to be resolved in the last few years as molecular analysis has enabled more precise matching of adult and larval forms. DNA analysis has even enabled the type species of some old larval genera and species to be matched to the adults they correctly belong with (Palero et al. 2008).

New Zealand's larval decapods, particularly the Brachyura, are comparatively well known, thanks largely to the work of Robert Wear and his students (1965–1985) at Victoria University in Wellington. Their efforts are summarised in two particularly useful publications. One (Wear & Fielder 1985) consists of a comprehensive illustrated atlas of all previously described New Zealand brachyuran larvae, with keys and some new descriptions; the other (Wear 1985), is an annotated list of all non-brachyuran New Zealand species whose larvae had been described to that time. Prior to 1985, numerous authors published descriptions of New Zealand decapod larvae but only the more significant are referred to here. Thomson and Anderson were the first New Zealanders to describe the larvae of brachyurans of the region, hatched at Portobello marine station. Prior to the 1960s, the most substantial contribution to New Zealand larval taxonomy was made by Gurney (1924, 1936, 1942), who described eight decapod species (in seven families) collected by the *Terra Nova* and *Discovery* Expeditions. Webber (1979) described the developmental stages of eight majid spider crabs, published later by Webber and Wear (1981). Larvae of 12 species of carid shrimps, in the families Crangonidae, Hippolytidae, and Palaemonidae, were described in detail by Packer (1983) who published a guide to these and six other shallow-water shrimp species in 1985. Since then, the output of larval taxonomy has slowed. Horn and Harms (1988) completed the larval description of *Halicarcinus varius*; Lemaitre and McLaughlin (1992) described the megalopa of the deep-water parapagurid *Sympagurus dimorphus*; the complete development of the packhorse rock lobster *Sagmariasus verreauxi* was described by Kittaka et al. (1997); and those of the red rock lobster *Jasus edwardsii* by Kittaka et al. (2005) from lobsters cultured in Japan; Cuesta et al. (2001) re-examined the zoeas of *Cyclograpsus lavauxi*, *Hemigrapsus sexdentatus*, and *H. crenulatus*; and detailed descriptions of the phyllosomas and nisto of a slipper lobster *Scyllarus* sp. Z (probably *S. aoteanus*) were published by Webber and Booth (2001).

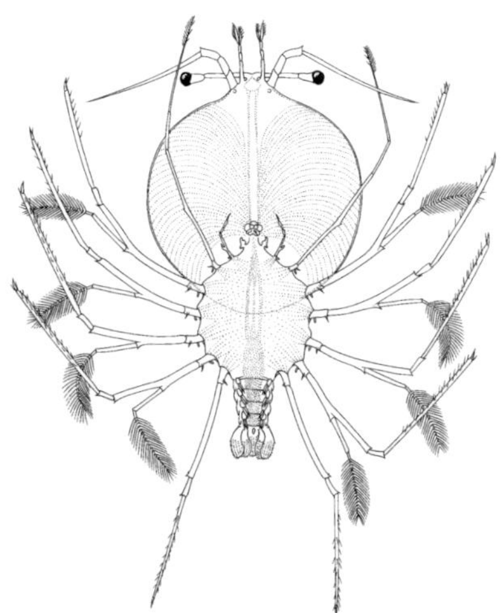

Final phyllosoma larval stage of the rock lobster *Jasus edwardsii*.
From Kittaka et al. 2005

Developmental stages of 94 species (21%) of New Zealand Decapoda have been described, but a much greater proportion of higher taxa is represented by this number. Descriptions of larvae, post-larvae, or both have been published from 45 (54%) of the 84 families recorded from New Zealand. These percentages reflect the high proportion of families containing only one species (larval descriptions of single species account for 27 families) but it also indicates the broad spectrum of decapods whose various larval forms are known to some degree. Best documented are the Brachyura, with 22 of New Zealand's 39 families represented by larval descriptions. The remaining 17 families contain 54 of the 167 brachyuran species, while, in the larger families, 11 of 14 hymenosomatid and five of 12 portunid species include larval descriptions.

Descriptions of all stages in the development of New Zealand's crayfish and lobsters were completed relatively recently, but commercial interest has now generated considerable investment in research into all aspects of their biology. The freshwater crayfish *Paranephrops planifrons* provides an example of direct development in which there are no larval stages and crayfish hatch from the eggs (Hopkins 1967). Young crayfish, with the cephalothorax packed with yolk, attach themselves to the female's pleopods and pass through three stages with the third having exhausted its supply of yolk. Development in scampi (*Metanephrops challengeri*) is not direct but apparently abbreviated. Wear (1976) found that while larvae hatch as prezoeas the prezoeal cuticle is quickly shed and the single-stage large zoea appears to last only two to three days or less before moulting to the post-larva. Scampi zoeas are not found in surface plankton and have a restricted ability to swim, which led Wear (1976) to suggest they are very short-lived and settle as a post-larva soon after hatching.

At the other end of the scale are the palinurid and scyllarid lobsters. New Zealand's rock lobsters *Jasus edwardsii* and *Sagmariasus verreauxi*, and slipper lobsters whose larval development is known (*Ibacus alticrenatus* and *Scyllarus* sp. Z), are typical of the Palinuroidea in having a long-lived larval phase. Longest of all is that of *J. edwardsii*, with 11 phyllosoma stages that can last more than a year, perhaps as long as 24 months, in the plankton (Booth & Phillips 1994). *Sagmariasus verreauxi* has a similar number of stages but of shorter duration (up to a year) (Booth & Phillips 1994), *I. alticranatus* still shorter (4–6 months) with seven stages (Atkinson & Boustead 1982), and *Scyllarus* sp. Z with 10 phyllosoma stages that probably have a duration as short as or shorter than *I. alticrenatus*. Planktonic larval sampling has concentrated on *J. edwardsii* because of its high economic value, but the incidental capture of phyllosomas of other species has enabled useful comparisons to be made. After hatching and shedding the naupliosoma cuticle, early-stage phyllosomas drift out to sea. Most sampled mid- to late-stage larvae of *J. edwardsii* appeared to become entrained in the Wairarapa Eddy southeast of the North Island, while those of *Scyllarus* sp. Z are found much closer to the North Island east and northeast coasts but also in oceanic waters to the north and northwest of New Zealand (Webber & Booth 2001). While mid- and late-stage *J. edwardsii* are rarely found inside the continental-shelf break, all stages of *Scyllarus* sp. Z are found there in good numbers, indicating that they go through larval development closer to shore. This accords with the much shorter larval duration in the scyllarid species and it is assumed that the widely scattered phyllosomas to the north and northwest are lost. The distribution of adult *Scyllarus* sp. Z is confined to the northeast coast of the North Island between Cape Maria van Diemen and Gisborne and is completely overlapped by *J. edwardsii*, yet the larvae they produce become distributed in different geographical areas. Phyllosomas have very limited ability to swim horizontally but they can move vertically through the water column. Coupled with changing phototactic responses during development, vertical mobility enables larvae to exploit currents flowing in different directions at different depths, a strategy that enables them to position themselves in water masses from which they can return to the coast as post-larvae (Webber & Booth 2001).

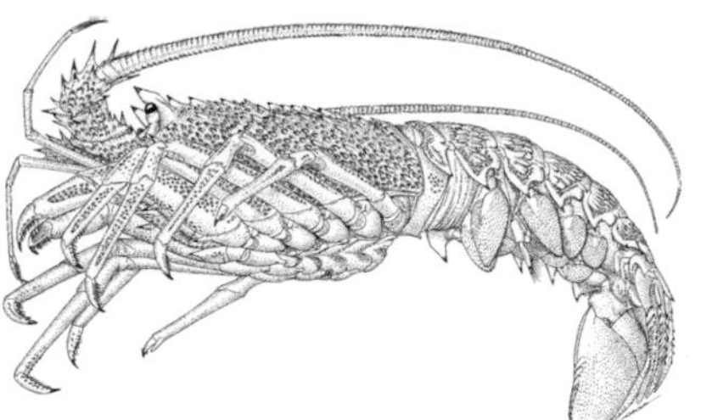

Rock lobster *Jasus edwardsii*.
W. Richard Webber

Commercial exploitation and resource potential of decapods

Studies of decapod biology and ecology have increased in the last half-century, especially of commercially important species. Early surveys of fishing potential included the southern spider crab *Jacquinotia edwardsii* (Ritchie 1970, 1971; Ryff & Voller 1976), prawns in the Bay of Plenty in the 1970s, and experiments aimed at culturing freshwater crayfish. As one of New Zealand's most valuable fisheries, *Jasus* rock lobsters are the subject of numerous and continuing studies. Their movements and migratory behaviour have been investigated for more than 30 years (e.g. Street 1969, 1971, 1973, 1994; Annala 1981; McCoy 1983; Booth 1984, 1997; MacDiarmid 1991, 1994; MacDiarmid *et al.* 1991; Andrew & MacDiarmid 1991; Annala & Bycroft 1993; Kelly 1995; Babcock et al. 1999; Butler et al. 1999; Kelly et al. 1999). Because rock lobsters have pelagic larvae and post-larvae, research has been carried out on the ecology and recruitment of developmental stages to adult populations (Booth 1979, 1986, 1994, 1995, 1997; Hayakawa et al. 1990; Booth & Grimes 1991; Booth et al. 1991; Booth & Stewart 1992; Booth & Phillips 1994; Booth & Kittaka 1994; Booth et al. 1998, 2000; Nishida et al. 1995; Chiswell & Booth 1999; Chiswell & Roemmich 1999). Rearing of New Zealand lobster larvae has advanced greatly (Kittaka 1994a,b; Kittaka et al. 1997; Tong et al. 1997, 2000a,b; Moss et al. 1999), while additional research on their biology and fisheries has also appeared (e.g. Booth & Breen 1994; James & Tong 1998; MacDiarmid & Butler 1999a,b). Genetic techniques have been employed to improve *Jasus* species stock identities (Ovenden et al. 1992; Ovenden & Brasher 1994; Booth & Ovenden 2000). Allozyme variation has also been identified in scampi populations around New Zealand.

Acknowledgements

Drs Paul Sagar (NIWA) and Wolfgang Zeidler (South Australian Museum) provided information on amphipods (literature on amphipods as prey for birds, and Hyperiidea, respectively). Dr Merlijn Jocqué (University of Leuven, Belgium) checked the section on Mysidacea and added a new endemic species. Thanks are due to Dr Bob McDowall (NIWA, Christchurch) for his constructive review of the Amphipoda section. Drs Michael Ayress (Ichron, UK) and Kerry Swanson (University of Canterbury, Christchurch) clarified aspects of ostracod taxonomy for the checklist; John Simes provided information on pre-Tertiary fossils.

Authors

Dr Shane T. Ahyong National Institute of Water & Atmospheric Research, Private Bag 14901, Kilbirnie, Wellington, New Zealand [s.ahyong@niwa.co.nz] Hoplocarida

Dr Graham J. Bird 8 Shotover Grove, Waikanae, Kapiti Coast 5036, New Zealand [zeuxo@clear.net.nz] Tanaidacea

Dr Janet M. Bradford-Grieve National Institute of Water & Atmospheric Research, Private Bag 14901, Kilbirnie, Wellington, New Zealand [j.grieve@niwa.co.nz] marine Copepoda, Branchiura, Tantulocarida

Dr Niel L. Bruce Museum of Tropical Queensland, 70–102 Flinders Street, Townsville, Queensland 4810, Australia [niel.bruce@qm.qld.gov.au] Isopoda

Professor John S. Buckeridge School of Civil, Environmental and Chemical Engineering, RMIT University, GPO Box 2476V, Melbourne, Victoria 3001, Australia [john.buckeridge@rmit.edu.au] Cirripedia

Dr M. Anne Chapman Deceased. Formerly Department of Biological Sciences, Waikato University, Private Bag 3105, Hamilton, New Zealand Freshwater crustacean ecology

Dr W. A. (Tony) Charleston 488 College Street, Palmerston North, New Zealand [charleston@inspire.net.nz] Pentastomida

Mr Elliot W. Dawson Museum of New Zealand Te Papa Tongarewa, P.O. Box 467, Wellington, New Zealand [edawson@xtra.co.nz] Leptostraca, Syncarida

Mr Stephen H. Eagar School of Earth Sciences, Victoria University of Wellington, P.O. Box 600, Wellington, New Zealand [stephen.eagar@paradise.net.nz] Ostracoda

Dr Graham D. Fenwick National Institute of Water & Atmospheric Research (NIWA), P.O. Box 8602, Christchurch, New Zealand [g.fenwick@niwa.co.nz] Amphipoda

Dr John D. Green 36 Paturoa Road, Titirangi, Waitakere, Auckland 0604, New Zealand [john.green@worldnet.co.nz] Freshwater copepod ecology

Dr Ju-Shey Ho Department of Biological Sciences, California State University, Long Beach, 1250 Bellflower Boulevard, Long Beach, California 90840-3702, USA [jsho@csulb.edu] Parasitic copepoda

Dr J. Brian Jones Fisheries WA, C/o Animal Health Lab., Agriculture WA, Locked Bag 4, Bentley Delivery Centre, WA 6983, Australia [bjones@agric.wa.gov.au] Branchiura, parasitic Copepoda

Dr Kim Larsen CIIMAR, University of Porto, Rua dos Bragas n. 289, 4050-123 Porto, Portugal [tanaids@hotmail.com] Tanaidacea

Dr Anne-Nina Lörz National Institute of Water & Atmospheric Research, Private Bag 14901, Kilbirnie, Wellington, New Zealand [a.lorez@niwa.co.nz] Rhizocephala

Dr Jørgen Olesen Zoological Museum, University of Copenhagen, Universitetsparken 15, DK-2100 Copenhagen, Denmark [J1Olesen@zmuc.ku.dk] Branchiopoda

Dr Gary C. B. Poore Museum Victoria, GPO Box 666E, Melbourne, Victoria 3001, Australia [gpoore@museum.vic.gov.au] Isopoda

Dr Carlos E. F. Rocha Universidade de São Paulo, Departamento de Zoologia, Caixa Postale 11461, CEP 05422 970, São Paulo, Brazil [cefrocha@usp.br] Copepoda: Oithonidae

Dr Russell J. Shiel Department of Environmental Biology, University of Adelaide, Adelaide, South Australia 5005, Australia [russell.shiel@adelaide.edu.au] Freshwater Copepoda

Dr Les Watling Department of Zoology, University of Hawaii at Manoa, Honolulu, HI 96822, USA [watling@hawaii.edu] Cumacea

Dr John B. J. Wells Department of Biological Sciences, Victoria University of Wellington, P.O. Box 600, Wellington, New Zealand [wellsjm@xtra.co.nz] Harpacticoida

Mr W. R. (Richard) Webber Museum of New Zealand Te Papa Tongarewa, P.O. Box 467, Wellington New Zealand [rickw@tepapa.govt.nz] Decapoda, Euphausiacea, Mysidacea

References

ACHITUV, Y. 2004: Coral-inhabiting barnacles (Cirripedia: Balanomorpha: Pyrgomatinae) from the Kermadec Islands and Niue Island, New Zealand. *New Zealand Journal of Marine and Freshwater Research 38*: 43–49.

AHYONG, S. T. 2001: Revision of the Australian Stomatopod Crustacea. *Records of the Australian Museum, Suppl. 26*: 1–326.

AHYONG, S. T. 2002a: Stomatopod Crustacea from the Marquesas Islands: results of MUSORSTOM 9. *Zoosystema 24*: 347–372.

AHYONG, S. T. 2002b: Stomatopod Crustacea of the Karubar Expedition in Indonesia. *Zoosystema 24*: 373–383.

AHYONG, S. T. 2002c: A new species and new records of Stomatopoda from Hawaii. *Crustaceana 75*: 827–840.

AHYONG, S. T. 2008: Deepwater crabs from seamounts and chemosynthetic habitats off eastern New Zealand (Crustacea: Decapoda: Brachyura). *Zootaxa 1708*: 1–72.

AHYONG, S. T. 2010: The marine fauna of New Zealand: King crabs of New Zealand, Australia and the Ross Sea (Crustacea: Decapoda: Lithodidae. *NIWA Biodiversity Memoir 123*: 1–194.

AHYONG, S. T.; HARLING, C. 2000: The phylogeny of the Stomatopod Crustacea. *Australian Journal of Zoology 48*: 607–642.

AHYONG, S. T.; NAIYANETR, P. 2002: Stomatopod Crustaceans from Phuket and the Andaman Sea. *Phuket Marine Biological Center Special Research Publication 23*: 281–312.

AHYONG, S. T.; CHAN, T.-Y.; LIAO, Y.-C. 2008: *A Catalog of the Mantis Shrimps (Stomatopoda) of Taiwan*. National Taiwan Ocean University, Keelung. iv + 190 p.

ANDREW, N. L.; MacDIARMID, A. B. 1991: Interrelations between sea urchins and spiny lobsters in northeastern New Zealand. *Marine Ecology Progress Series 70*: 211–222.

ANNALA, J. H. 1981: Movements of rock lobsters (*Jasus edwardsii*) tagged near Gisborne, New Zealand. *New Zealand Journal of Marine and Freshwater Research 15*: 437–443.

ANNALA, J. H.; BYCROFT, B. L. 1993: Movements of rock lobsters (*Jasus edwardsii*) tagged in Fiordland, New Zealand. *New Zealand Journal of Marine and Freshwater Research 27*: 183–190.

ATKINSON, J. M.; BOUSTEAD, N. C. 1982: The complete larval development of the scyllarid lobster *Ibacus alticrenatus* Bate, 1888 in New Zealand waters. *Crustaceana 42*: 275–287.

AYRESS, M. A. 1990: New cytheromatid Ostracoda from the Cenozoic of New Zealand. *New Zealand Natural Sciences 17*: 67–72.

AYRESS, M. A. 1991: Ostracod biostratigraphy and palaeoecology of the Kokoamu Greensand and Otekaike Limestone (Late Oligocene to Early Miocene), North Otago and South Canterbury, New Zealand. *Alcheringa 17*: 125–151.

AYRESS, M. A. 1993a: Ostracod biostratigraphy and paleontology of the Kokoamu Greensand and Otekaike Limestone (Late Oligocene to Early Miocene), North Otago and South Canterbury, New Zealand. *Alcheringa 17*: 125–151.

AYRESS, M. A. 1993b: *Crescenticythere*, a new enigmatic ostracode from the Tertiary of New Zealand. *Journal of Paleontology 67*: 905–906.

AYRESS, M. A. 1993c: Middle Eocene Ostracoda (Crustacea) from the coastal section, Bortonian Stage, at Hampden, South Island, New Zealand. *New Zealand Natural Sciences 20*: 15–21.

AYRESS, M. A. 1995: Late Eocene Ostracoda (Crustacea) from the Waihao District, South Canterbury. *New Zealand. Journal of Paleontology 69*: 897–921.

AYRESS, M. A. 1996: New species and biostratigraphy of late Eocene cytherurid Ostracoda from New Zealand. *Revista Española de Micropaleontología 28*: 11–36.

AYRESS, M. A.; DRAPALA, V. 1996: New Recent and fossil discoveries of *Cluthia* (Leptocytheridae) in the Southwest Pacific: implications on its origin and dispersal. Pp. 149–158 *in*: Keen, M. C. (ed.), *Proceedings of the 2nd European Ostracodologists Meeting, Glasgow 1993*. British Micropalaeontological Society, London.

AYRESS, M. A.; SWANSON, K. M. 1991: New fossil and Recent genera and species of cytheracean Ostracoda (Crustacea) from South Island, New Zealand. *New Zealand Natural Sciences 18*: 1–18.

AYRESS, M. A.; WARNE, M. T. 1993: *Vandiemencythere*, a new ostracod genus from the Cainozoic of New Zealand, Australia and the S.W. Pacific Ocean. *Revista Española de Micropaleontología 25*: 33–40.

AYRESS, M. A.; BARROWS, T.; PASSLOW, V.; WHATLEY, R. 1999: Neogene to Recent species of *Krithe* (Crustacea: Ostracoda) from the Tasman Sea and off Southern Australia with description of five new species. *Records of the Australian Museum 51*: 1–22.

AYRESS, M. A.; CORREGE, T.; PASSLOW, V.; WHATLEY, R. C. 1994: New bythocytherid and cytherurid ostracode species from the deep-sea, Australia, with enigmatic dorsal expansion. *Geobios 29*: 73–90.

AYRESS, M. A.; NEIL, H.; PASSLOW, V.; SWANSON, K. M. 1997: Benthonic ostracods and deep watermasses: a qualitative comparison of Southwest Pacific, Southern and Atlantic Oceans. *Palaeogeography, Palaeoclimatology, Palaeoecology 131*: 287–302.

AYRESS, M. A.; WHATLEY, R. C.; DOWNING, S. E.; MILLSON, K. J. 1995: Cainozoic and Recent deep sea cytherurid Ostracoda from the South Western Pacific and eastern Indian Ocean, Part 1: Cytherurinae. *Records of the Australian Museum 47*: 203–223.

BABCOCK, R. C.; KELLY, S.; SHEARS, N. T.; WALKER, J. W.; WILLIS, T. J. 1999: Changes in community structure in temperate marine reserves. *Marine Ecology Progress Series 189*: 125–134.

BACESCU, M. 1981: Contribution to the knowledge of the monokonophora (Crustacea, Tanaidacea) of the eastern Australian coral reefs. *Revue Roumaine de Biologie (Biologie Animale) 26*: 111–120.

BAIRD, W. 1850: Description of a new crustacean. *Proceedings of the Zoological Society of London 18*: 102, pl. 17.

BAKER, A. deC. 1965: The latitudinal distribution of *Euphausia* species in the surface waters of the Indian Ocean. *Discovery Reports 33*: 309–334.

BAKER, A. deC.; BODEN, B. P.; BRINTON, E. 1990: *A Practical Guide to the Euphausiids of the World*. British Museum (Natural History), London. 96 p.

BALSS, H. 1929: Die Decapoden (Crustaceen). Zoologische Ergebnisse der Reisen von Dr. L. Kohl-Larsen nach den subantarktischen Inseln bei Neuseeland und nach Südgeorgien. *Senckenbergiana 11*: 195–210.

BAMBER, R. N. 1990: A new species of *Zeuxo* (Crustacea: Tanaidacea) from the French Atlantic coast. *Journal of Natural History 24*: 1587–1596.

BANKS, C. M.; DUGGAN, I. C. 2009: Lake construction has facilitated calanoid copepod invasions in New Zealand. *Diversity and Distributions 15*: 80–87.

BANKS, Sir J. 1962: *The Endeavour Journal of Joseph Banks 1768–1771*. Beaglehole, J.C. (ed.), Public Library of New South Wales, Sydney. Vol. 1, 476 p.

BARCLAY, M. H. 1969: First records and a new species of *Phyllognathus* (Copepoda: Harpacticoida) in New Zealand. *New Zealand Journal of Marine and Freshwater Research 3*: 295–303.

BARNARD, J. L. 1955: Wood boring habits of *Chelura terebrans* Philippi in Los Angeles Harbor. Pp. 87–98 *in: Essays in Natural Science in Honor of Captain Allan Hancock on the occasion of his birthday July 26, 1955*. Allan Hancock Foundation, University of Southern California, Los Angeles. 345 p.

BARNARD, J. L. 1961: Gammaridean Amphipoda from depths of 400 to 6000 meters. *Galathea Report 5*: 23–128.

BARNARD, J. L. 1969: The families and genera of marine gammaridean Amphipoda. *United States National Museum Bulletin 271*: 1–535.

BARNARD, J. L. 1972a: The marine fauna of New Zealand: Algae-living littoral Gammaridea (Crustacea Amphipoda). *New Zealand Oceanographic Institute Memoir 62*: 1–215.

BARNARD, J. L. 1972b: Gammaridean Amphipoda of Australia, Part 1. *Smithsonian Contributions to Zoology 103*: 1–333.

BARNARD, J. L.; BARNARD, C. M. 1982: Biogeographical microcosms of world freshwater Amphipoda (Crustacea). *Polskie Archiwum Hydrobiologii 29*: 255–273.

BARNARD, J. L.; BARNARD, C. M. 1983: *Freshwater Amphipoda of the world. I. Evolutionary patterns. II. Handbook and bibliography.* Hayfield Associates, Mt Vernon, Virginia. 830 p.

BARNARD, J. L.; DRUMMOND, M. M. 1978: Gammaridean Amphipoda of Australia, part III: the Phoxocephalidae. *Smithsonian Contributions to Zoology 245*: 1–551.

BARNARD, J. L.; DRUMMOND, M. M. 1992: *Paracalliope*, a genus of Australian shorelines (Crustacea: Amphipoda: Paracalliopiidae). *Memoirs of the Museum of Victoria 53*: 1–29.

BARNARD, J. L.; INGRAM, C. L. 1986: The supergiant amphipod *Alicella gigantea* Chevreux from the North Pacific Gyre. *Journal of Crustacean Biology 6*: 825–839.

BARNARD J. L.; KARAMAN, G. S. 1982: Classificatory revisions in gammaridean Amphipoda (Crustacea), Part 2. *Proceedings of the Biological Society of Washington 95*: 167–187.

BARNARD, J. L.; KARAMAN, G. S. 1983: Australia as a major evolutionary center for Amphipoda. *Memoirs of the Australian Museum 18*: 45–61.

BARNARD, J. L.; KARAMAN, G. S. 1991: The families and genera of marine gammaridean Amphipoda (except gammaroids). *Records of the Australian Museum, Suppl. 13 (Parts 1–2)*: 1–866.

BARNARD, K. H. 1930: Amphipoda. *British Antarctic ('Terra Nova') Expedition, 1910. Natural History Reports. Zoology 8*: 307–454.

BARNEY, R. W. 1929: Crustacea, Ostracoda. *British Antarctic (Terra Nova) Expedition 1910 Natural History Report, Zoology 3(7), 5*: 175–189.

BARTLE, J. A. 1976: Euphausiids of Cook Strait: a transitional fauna? *New Zealand Journal of Marine and Freshwater Research 10*: 559–576.

BARY, B. 1956: Notes on ecology, systematics and development of some Mysidacea and Euphausiacea (Crustacea) from New Zealand. *Pacific Science 10*: 431–467.

BARY, B. 1959: Species of zooplankton as a means

of identifying different surface waters and demonstrating their movements and mixing. *Pacific Science 13*: 14–54.

BATE, C. S. 1881: On the Penaeidea. *Annals and Magazine of Natural History, ser. 5, 8*: 169–196.

BATE, C. S. 1888: Report on the Crustacea Macrura collected by H.M.S. Challenger during the years 1873–76. *Report on the Scientific Results of the Voyage of HMS* Challenger, *Zoology 24*: 1–942.

BATHAM, E. J. 1967: The first three larval stages and feeding behaviour of the New Zealand palinurid crayfish *Jasus edwardsii* (Hutton, 1875). *Transactions of the Royal Society of New Zealand, Zoology 9*: 53–64.

BATHAM, E. J.; TOMLINSON, J. T. 1965: On *Crytophialus melampygos* Berndt, a small boring barnacle of the order Acrothoracica abundant in some New Zealand molluscs. *Transactions of the Royal Society of New Zealand, Zoology 7*: 141–154.

BAYLY, I. A. E. 1963: A revision of the coastal water genus *Gladioferens* (Copepoda: Calanoida). *Australian Journal of Marine and Freshwater Research 14*: 194–217.

BAYLY, I. A. E. 1992: *The Non-marine Centropagidae (Copepoda: Calanoida) of the world*. SPB Academic Publishing, The Hague. 30 p.

BAYLY, I. A. E. 1995: Distinctive aspects of the zooplankton of large lakes in Australasia, Antarctica and South America. *Australian Journal of Marine and Freshwater Research 46*: 1109–2000.

BAYLY, I. A. E.; WILLIAMS, W. D. 1973: *Inland Waters and their Ecology*. Longman, Melbourne.

BEDFORD, J. J.; SMITH, R. A. J.; THOMAS, M.; LEADER, J. P. 1991: SUP 1 N-NMR and HPLC studies for the changes involved in volume regulation in the muscle fibres of the crab, *Hemigrapsus edwardsi*. *Comparative Biochemistry and Physiology, A, Comparative Physiology 100*: 145–149.

BEGG, A. C., BEGG, N. C. 1969: *James Cook and New Zealand*. Government Printer, Wellington. 169 p.

BENNETT, D. B. 1964: The marine fauna of New Zealand: Crustacea Brachyura. *New Zealand Oceanographic Institute Memoir 22*: 1–120.

BENSON, W. N. 1956: Cambrian rocks and fossils in New Zealand (preliminary note). Pp. 285–288 *in*: Rodgers, J. (ed.) *Symposium sobre el Systema Cámbrico, su Paleogeografía y el Problema de sur Base. Vol. 2(2) (Australia, America)*. [*Proceedings of the 20th Internnational Geological Congress*.] International Geological Congress, Mexico City.

BERKENBUSH, K.; ROWDEN, A. A. 1998: Population dynamics of the burrowing ghost shrimp *Callianassa filholi* on an intertidal sandflat in New Zealand. *Ophelia 49*: 55–69.

BERKENBUSH, K.; ROWDEN, A. A. 1999: Factors influencing sediment turnover by the burrowing ghost shrimp *Callianassa filholi* (Decapoda: Thalassinidea). *Journal of Experimental Marine Biology and Ecology 238*: 283–292.

BERNARD, F. 1953: Decapoda Eryonidae (*Eryoneicus* et *Willemoessia*). *Dana 37*: 1–93.

BEU, A. G.; GRANT-TAYLOR, T. L.; HORNIBROOK, N. deB. 1980: The Te Aute Limestone Facies, Poverty Bay to Northern Wairarapa, 1:250,000. *New Zealand Geological Survey Miscellaneous Series Map 13*: 1–36, 2 sheets.

BIRD, G. J. 2008: Untying the Gordian Knot: on *Tanais novaezealandiae* Thomson (Crustacea, Tanaidacea, Tanaidae) from New Zealand, with descriptions of two new *Zeuxoides* species. *Zootaxa 1877*: 1–36.

BLACKBURN, M. 1980: Observations on the distribution of *Nyctiphanes australis* Sars (Crustacea, Euphausiidae) in Australian waters. *CSIRO, Division of Fisheries and Oceanography 119*: 1–10.

BLAIR, D.; WILLIAMS, J. B. 1987: A new fecampiid of the genus *Kronborgia* (Platyhelminthes: Turbellaria: Neorhabdocoela) parasitic in the intertidal isopod *Exosphaeroma obtusum* (Dana) from New Zealand. *Journal of Natural History 21*: 115–1172.

BOILEAU, M. G. 1991: A genetic determination of cryptic species (Copepoda: Calanoida) and their postglacial biogeography in North America. *Zoological Journal of the Linnaean Society 102*: 375–396.

BOLTOVSKOY, D.; GIBBONS, M. J.; HUTCHINGS, L.; BINET, D. 1999: General biological features of the South Atlantic. Pp. 1–42 *in*: Boltovskoy, D. (ed.), *South Atlantic Zooplankton*. Backhuys Publishers, Leiden. 1705 p.

BOOTH, J. D. 1979: Settlement of the rock lobster *Jasus edwardsii* (Decapoda: Palinuridae), at Castlepoint, New Zealand. *New Zealand Journal of Marine and Freshwater Research 13*: 395–406.

BOOTH, J. D. 1984: Movements of packhorse rock lobsters (*Jasus verreauxi*) tagged along the eastern coast of the North Island, New Zealand. *New Zealand Journal of Marine and Freshwater Research 18*: 275–281

BOOTH, J. D. 1986: Recruitment of packhorse rock lobster *Jasus verreauxi* in New Zealand. *Canadian Joournal of fisheries and aquatic sciences 43*: 2212–2220.

BOOTH, J. D. 1994: *Jasus edwardsii* larval recruitment off the east coast of New Zealand. *Crustaceana 66*: 295–317.

BOOTH, J. D. 1995: Lobster phyllosomata from offshore NZ waters. *Lobster Newsletter 8(1)*: 8.

BOOTH, J. D. 1997: Long-distance movements of *Jasus* spp. and their role in larval recruitment. *Bulletin of Marine Science 61*: 111–128.

BOOTH, J. D.; BREEN, P. A. 1994: The New Zealand fishery for *Jasus edwardsii* and *J. verreauxi*. Pp. 64–75 *in*: Phillips, B. F.; Cobb, J. S.; Kittaka, J. (eds), *Spiny Lobster Management*. Fishing News Books, Oxford. xxiii +550 p.

BOOTH, J. D.; CARRUTHERS, A. D.; BOLT, C. D.; STEWART, R. A. 1991: Measuring depth of settlement in the red rock lobster *Jasus edwardsii*. *New Zealand Journal of Marine and Freshwater 25*: 123–132.

BOOTH, J. D.; FORMAN, J. S.; STOTTER, D. 1998: Abundance of early life history stages of the red rock lobster, *Jasus edwardsii* with management implications. *New Zealand Fisheries Assessment Research Document 98/10*: 1–36.

BOOTH, J. D.; FORMAN, J. S.; JAMES, P. 2000: Evaluations of experiments into collection and transport techniques for larval and newly-settled *Jasus edwardsii. NIWA Client Report WLG00/57.*

BOOTH, J. D.; GRIMES, P. 1991:Tangaroa's first research. *New Zealand Professional Fisherman 5(8)*: 61–62.

BOOTH, J. D.; KITTAKA, J. 1994: *Jasus edwardsii* larval recruitment off the east coast of New Zealand. *Crustaceana 66*: 295–317.

BOOTH, J. D.; OVENDEN, J. R. 2000: Distribution of *Jasus* spp. (Decapoda: Palinuridae) phyllosomas in southern waters: implications for larval recruitment. *Marine Ecology Progress Series 200*: 241–255.

BOOTH, J. D.; PHILLIPS, B. F. 1994: Early life history of spiny lobster. *Crustaceana 63*: 271–294.

BOOTH, J. D.; STEWART, R. A. 1992: Distribution of phyllosoma larvae of the red rock lobster *Jasus edwardsii* off the east coast of New Zealand in relation to oceanography. Pp. 138–148 *in*: Hancock, D. A. (ed), *Larval Biology. Proceedings No. 15, Australian Society for Fish Biology Workshop*. Australian Government Service, Canberra.

BORRADAILE, L. A. 1916: Crustacea. Part 1. Decapoda. *British Antarctic Terra Nova Expedition Zoology 3(2)*: 75–110.

BOUSFIELD, E. L. 1964: Insects of Campbell Island. Talitrid amphipod crustaceans. *Pacific Insects Monograph 7*: 45–57.

BOUSFIELD, E. L. 1977: A new look at the systematics of gammaroidean amphipods of the world. *Crustaceana, Suppl. 4*: 282–316.

BOUSFIELD, E. L. 1978: A revised classification and phylogeny of amphipod crustaceans. *Transactions of the Royal Society of Canada 4*: 343–390.

BOUSFIELD, E. L. 1982: An updated phyletic classification and palaeohistory of Amphipoda. Pp. 257–277 *in*: Schram, F. R. (ed.), *Crustacean Phylogeny*. A.A. Balkema, Rotterdam. 372 p.

BOUSFIELD, E. L. 1983: The amphipod superfamily Talitroidea in the northeastern Pacific region. I. Family Talitridae: systematics and distributional ecology. *National Museum of Natural Sciences Publications in Biological Oceanography 11*: 1–73.

BOUSFIELD, E. L. 1987: Amphipod parasites of fishes in Canada. *Canadian Bulletin of Fisheries and Aquatic Sciences 217*: 1–37.

BOUSFIELD, E. L.; HOOVER, P. M. 1997: The amphipod superfamily Corophioidea on the Pacific coast of North America. Part V. Family Corophiidae: Corophiinae, new subfamily. Systematics and distributional ecology. *Amphipacifica 2*: 67–139.

BOUSTEAD, N. 1982: Fish diseases recorded in New Zealand, with a discussion on potential sources and certification procedures. *Occasional Publication, Fisheries Research Division, New Zealand Ministry of Agriculture and Fisheries 43*: 1–19.

BOWLEY, E. A. 1935: A survey of the oniscoid genus *Phalloniscus* Budde-Lund, with a description of new species. *Journal of the Royal Society of Western Australia 21*: 45–73.

BOWMAN, T. E.; ABELE, L. G. 1982: Classification of the Recent Crustacea. Pp. 1–27 *in*: Abele, L. G. (ed.), *The Biology of Crustacea. Volume 1. Systematics, the Fossil record, and Biogeography*. Academic Press, London.

BOWMAN, T. E.; BRUCE, N. L.; STANDING, J. D. 1981: Recent introduction of the cirolanid isopod crustacean *Cirolana arcuata* into San Francisco Bay. *Journal of Crustacean Biology 1*: 545–557.

BOXSHALL, G. A. 1986: A new species of *Mugilicola* Tripathi (Copepoda: Poecilostomatoida) and a review of the family Therodamasidae. *Proceedings of the Linnean Society of New South Wales 108*: 183–186.

BOXSHALL, G. A.; HALSEY, S. H. 2004: *An Introduction to Copepod Diversity*. The Ray Society, London. 966 p.

BOXSHALL, G. A.; LINCOLN, R. J 1983: Tantulocarida, a new class of Crustacea ectoparasitic on other crustaceans. *Journal of Crustacean Biology 3*: 1–16.

BRABAND, A.; RICHTER S.; HIESEL, R; SCHOLTZ, G. 2002: Phylogenetic relationships within the Phyllopoda (Crustacea, Branchiopoda) based on mitochondrial and nuclear markers. *Molecular Phylogenetics and Evolution 25*: 229–244.

BRADBURY, J. H.; WILLIAMS, W. D. 1997: Amphipod (Crustacea) diversity in underground waters in Australia: an Aladdin's cave. *Memoirs of the Museum of Victoria 56*: 513–519.

BRADFORD, J. M. 1969: New genera and species of benthic calanoid copepods from the New Zealand slope. *New Zealand Journal of Marine and Freshwater Research 3*: 473–505.

BRADFORD, J. M. 1972: Systematics and ecology of New Zealand central east coast plankton sampled at Kaikoura. *New Zealand Oceanographic Institute Memoir 54*: 1–87.

BRADFORD, J. M. 1979: Zoogeography of some New Zealand neritic pelagic Crustacea and their close relatives. *In*: *Proceedings of the International Symposium on Marine Biogeography and Evolution in the Southern Hemisphere. New Zealand DSIR Information Series 137*: 593–612.

BRADFORD, J. M. 1985: Distribution of zooplankton off Westland, June 1979 and February 1982. *New Zealand Journal of Marine and Freshwater Research 19*: 311–326.

BRADFORD, J. M.; HAAKONSSEN, L.; JILLETT, J. B. 1983: The marine fauna of New Zealand: Pelagic Copepoda: Families Scolecithricidae, Phaennidae, Diaxidae, Tharybidae. *New Zealand Oceanographic Institute Memoir 90*: 1–146.

BRADFORD, J. M.; HEWITT, G. C. 1980: A new maxillopodan crustacean, parasitic on a myodocopid ostracod. *Crustaceana 38* : 67–72.

BRADFORD, J. M.; JILLETT, J. B. 1980: The marine fauna of New Zealand : Pelagic calanoid copepods : Family Aetideidae. *New Zealand Oceanographic Institute Memoir 86*: 1–101.

BRADFORD-GRIEVE, J. M. 1994: The marine fauna of New Zealand: Megacalanidae, Calanidae, Paracalanidae, Mecynoceridae, Eucalanidae, Spinocalanidae, Clausocalanidae. *New Zealand Oceanographic Institute Memoir 102*: 1–160.

BRADFORD-GRIEVE, J. M. 1999a: The marine fauna of New Zealand: Pelagic Calanoid Copepoda: Arietellidae, Augaptilidae, Heterorhabdidae, Lucicutiidae, Metridinidae, Phyllopodidae, Centropagidae, Pseudodiaptomidae, Temoridae, Candaciidae, Pontellidae, Sulcanidae, Acartiidae, Tortanidae. *NIWA Biodiversity Memoirs 111*: 1–268.

BRADFORD-GRIEVE, J. M. 1999b: New species of benthopelagic copepods of the genus *Stephos* (Calanoida: Stephidae) from Wellington Harbour, New Zealand. *New Zealand Journal of Marine and Freshwater Research 33*: 13–27.

BRADFORD-GRIEVE, J. M.; MURDOCH, R.; JAMES, M., OLIVER, M.; McLEOD, J. 1998: Mesozooplankton biomass, composition, and potential grazing pressure on phytoplankton during austral winter and spring 1993 in the Subtropical Convergence region near New Zealand. *Deep-Sea Research I, 45*: 1709–1737.

BRADY, G. S. 1880: Report on the Ostracoda dredged by HMS *Challenger* during the years 1873–1876. *Reports on the Scientific Results of the voyage of HMS* Challenger, *Zoology 1(3)*: 1–184, 44 pls.

BRADY, G. S. 1898: On new and imperfectly known species of Ostracoda, chiefly from New Zealand. *Transactions of the Zoological Society of London 14:* 429–452, pls 43–47.

BRADY, G. S. 1899: On the marine Copepoda of New Zealand. *Transactions of the Zoological Society of London 15*: 31–54, pls 9–13.

BRADY, G. S. 1902: On new or imperfectly known Ostracoda in the Zoological Museum, Copenhagen. *Transactions of the Zoological Society of London 16*: 179–206, pls219–23.

BRANDT, A. 1988: *Antarctic Serolidae and Cirolanidae (Crustacea: Isopoda): New Genera, New Species, and Redescriptions.* [Theses Zoologicae.] Koeltz Scientific Books, Koenigstein, Germany. 143 p.

BRANDT, A.; POORE, G. C. B. 2003: Higher classification of the flabelliferan and related Isopoda based on a reappraisal of relationships. *Invertebrate Systematics 17*: 893–923.

BRANDT, A.; WÄGELE, J. W. 1989: Redescriptions of *Cymodocella tubicauda* Pfeffer, 1887 and *Exosphaeroma gigas* (Leach, 1818) (Crustacea, Isopoda, Sphaeromatidae). *Antarctic Science 1*: 205–214.

BREHM, V. 1928: Vorfläufige Mitteilung über die Süsswasserfauna Neu–Seelands. *Zoologischer Anzeiger 75*: 223–225.

BREHM, V. 1929: Contribution to knowledge of freshwater fauna of New Zealand *Transactions of the New Zealand Institute 59*: 779–803.

BRIGGS, D.E.G. 1992: Phylogenetic significance of the Burgess Shale crustacean *Canadaspis. Acta Zoologica 73*: 293–300.

BRINTON, E. 1953: *Thysanopoda spinicaudata*, a new bathypelagic giant euphausiid crustacean, with comparative notes on *Thysanopoda cornuta* and *Thysanopoda egregia*. *Journal of the Washington Academy of Sciences 43*: 408–411.

BRINTON, E. 1962a: The distribution of Pacific euphausiids. *Bulletin of the Scripps Institution of Oceanography, University of California 8*: 51–270.

BRINTON, E. 1962b: Two new species of Euphausiacea, *Euphausia nana* and *Stylocheiron robustum* from the Pacific. *International Journal of Crustacean Research 4*: 167–179.

BRINTON, E. 1966: Remarks on euphausiacean phylogeny. *In*: Symposium on Crustacea. *Journal of the Marine Biological Association of India 1*: 255–259.

BRINTON, E. 1987: A new abyssal euphausiid, *Thysanopoda minyops*, with comparisons of eye size, photophores, and associated structures among deep-living species. *Journal of Crustacean Biology 7*: 636–666.

BRINTON, E.; OHMAN, M. D.; TOWNSEND, A. W.; KNIGHT, M. D.; BRIDGEMAN, A. L. 1999: Euphausiids of the world ocean. The Expert Centre for Taxonomic Identification CD-ROM. UNESCO.

BROCKERHOFF, A.; McLAY, C.; KLUZA, D. 2006: Defenders of the peace: New Zealand's marine parasites *versus* exotic crabs? *Biosecurity New Zealand 72*: 18–19.

BRÖKELAND, W.; WÄGELE, J.-W.; BRUCE, N. L. 2001: *Paravireia holdichi* n. sp., an enigmatic isopod crustacean from the Canary Islands with affinities to species from New Zealand. *Organisms, Diversity and Evolution 1*: 83–98.

BROOKS, H. K. 1962: On the fossil Anaspidacea, with a revision of the classification of the Syncarida. *Crustaceana 4*: 229–242.

BROOKS, H. K. 1969: Syncarida. Pp. R345–R359 *in*: Moore, R. C. (ed.), *Treatise on Invertebrate Paleontology, Part R, Arthropoda 4 (1)*. Geological Society of America and the University of Kansas Press, Lawrence. 398 p.

BROWN, D. W.; BECKMAN, P. A. 1992: Epizooic Foraminifera, tanaid, and polychaete species association on Antarctic scollop shell. *Antarctic Journal of the United States 27*: 134–135.

BRUCE, N. L. 1983: Aegidae (Isopoda: Crustacea) from Australia with descriptions of three new species. *Journal of Natural History 17*: 757–788.

BRUCE, N. L. 1986a: Cirolanidae (Crustacea: Isopoda) of Australia. *Records of the Australian Museum, Suppl. 6*: 1–239.

BRUCE, N. L. 1986b: Revision of the isopod crustacean genus *Mothocya* Costa, in Hope, 1851 (Cymothoidae: Flabellifera), parasitic on marine fishes. *Journal of Natural History 20*: 1089–1192.

BRUCE, N. L. 1988: Two new species of *Tridentella* (Crustacea, Isopoda, Tridentellidae) from New Zealand. *National Museum of New Zealand Records 3*: 71–79.

BRUCE, N. L. 2002: *Tridentella rosemariae* sp. nov. (Crustacea: Isopoda: Tridentellidae) from northern New Zealand waters. *Crustaceana 75*: 159–170.

BRUCE, N. L. 2003: A new deep-water species of *Natatolana* (Crustacea: Isopoda: Cirolanidae) from the Chatham Rise, eastern New Zealand. *Zootaxa 265*: 1–12.

BRUCE, N. L. 2004a: New species of the *Cirolana 'parva*-group' (Crustacea: Isopoda: Cirolanidae) from coastal habitats around New Zealand. *Species Diversity 9*: 47–66.

BRUCE, N. L. 2004b: Reassessment of the isopod crustacean *Aega deshaysiana* (Milne Edwards, 1840) (Cymothoida: Aegidae) – a world-wide complex of 21 species. *Zoological Journal of the Linnean Society 142*: 135–232.

BRUCE, N. L. 2008: Two new deep-water species of *Caecoserolis* Wägele, 1994 (Isopoda, Sphaeromatidea, Serolidae) from off North Island, New Zealand. *Zootaxa 1866*: 453–466.

BRUCE, N. L. 2009a: The marine fauna of New Zealand: Isopoda, Aegidae (Crustacea). *NIWA Biodiversity Memoir 122*: 1–252.

BRUCE, N. L. 2009b: New genera and species of the marine isopod family Serolidae (Crustacea, Sphaeromatidea) from the southwestern Pacific. *ZooKeys 18*: 17–76.

BRUCE, N. L.; WETZER, R. 2008: New Zealand exports: *Pseudosphaeroma* Chilton, 1909 (Isopoda: Sphaeromatidae), a Southern Hemisphere genus introduced to the Pacific coast of North America. *Zootaxa 1908*: 51–56.

BRUSCA, R. C.; BRUSCA, G. J. 2002: *Invertebrates.* 2nd edn. Sinauer Associates, Sunderland, Maryland. xx + 936 p.

BRUSCA, R. C.; WETZER, R.; FRANCE, S. C. 1995: Cirolanidae (Crustacea: Isopoda: Flabellifera) of the tropical eastern Pacific. *Proceedings of the San Diego Society of Natural History 30*: 1–96.

BRUSCA, R. C.; WILSON, G. D. F. 1991: A phylogenetic analysis of the Isopoda with some classificatory recommendations. *Memoirs of the Queensland Museum 31*: 143–204.

BUCKERIDGE, J. S. 1983: The fossil barnacles (Cirripedia: Thoracica) of New Zealand and Australia. *New Zealand Geological Survey Paleontological Bulletin 50*: 1–151, 14 pls.

BUCKERIDGE, J. S. 1984a: A new species of *Elminius* from Pomahaka River, Southland, New Zealand. *New Zealand Journal of Geology and Geophysics 27*: 217–219.

BUCKERIDGE, J. S. 1984b: Two new Tertiary scalpellid barnacles (Cirripedia: Thoracica) from the Chatham Islands, New Zealand. *Journal of the Royal Society of New Zealand 14*: 319–326.

BUCKERIDGE, J. S. 1991: *Pachyscalpellum cramptoni* a new genus and species of lepadomorph cirripede from the Cretaceous of northern Hawkes Bay, New Zealand. *Journal of the Royal Society of New Zealand 21*: 55–60.

BUCKERIDGE, J. S. [1995] 1996a: Phylogeny and biogeography of the primitive Sessilia and a consideration of a Tethyan origin for the group. *Crustacean Issues 10*: 255–267.

BUCKERIDGE, J.S. 1996b: A living fossil *Waikalasma boucheti* sp. nov. (Cirripedia: Balanomorpha) from Vanuatu (New Hebrides), Southwest Pacific. *Bulletin du Muséum national d'Histoire naturelle, sér. 4, 18*: 447–457.

BUCKERIDGE, J. S. 1997: Cirripedia: Thoracica. New ranges and species of Verrucomorpha from the Indian and Southwest Pacific Oceans. *Résultats des Campagnes* Musorstom 18. *Mémoires du Muséum national d'Histoire naturelle 176*: 125–149.

BUCKERIDGE, J. S. 1998: Monitoring and Management of Heavy Metals, Pesticides, PCBs, Dyes and Pigments. Regional Report: New Zealand. *Commonwealth Science Council (London) Workshop on Monitoring and Management of Heavy Metals, Pesticides, PCBs, Dyes and Pigments. Islamabad, Pakistan.* Unitec Publishing, Auckland. 14 p.

BUCKERIDGE, J. S. 1999a: Post Cretaceous biotic recovery: A case study on Crustacea : Cirripedia from the Chatham Islands, New Zealand. *Records of the Canterbury Museum 13*: 43–51.

BUCKERIDGE, J. S. 1999b: A new deep-sea barnacle, *Tetrachaelasma tasmanicum* sp. nov. (Cirripedia : Balanomorpha) from the South Tasman Rise, South Pacific Ocean. *New Zealand Journal of Marine and Freshwater Research 33*: 521–531.

BUCKERIDGE, J. S. 2008: Two new species and a new subspecies of *Tetraclitella* (Cirripedia: Thoracica) from the Cainozoic of Australia and New Zealand and a consideration of the significance of tubiferous walls. *Zootaxa 1897*: 43–52.

BUCKERIDGE, J. S. 2009: *Ashinkailepas kermadecensis*, a new species of deep-sea scalpelliform barnacle (Thoracica: Eolepadidae) from the Kermadec Islands, southwest Pacific. *Zootaxa 2021*: 57–65.

BUCKERIDGE, J. S.; NEWMAN, W. A. 2006: A revision of the Iblidae and the stalked barnacles (Crustacea: Cirripedia: Thoracica), including new ordinal, familial and generic taxa, and two new species from New Zealand and Tasmanian waters. *Zootaxa 1136*: 1–38.

BUCKERIDGE, J. S.; NEWMAN, W. A. 2010: A review of the subfamily Elminiinae (Cirripedia: Thoracica: Austrobalanidae), including a new genus, *Protelminius* nov., from the Oligocene of New Zealand. *Zootaxa 2349*: 39–54.

BURNS, C. W. 1988: Starvation resistance among copepod nauplii and adults. *Verhandlungen der Internationale Vereinigung für theoretische und angewandte Limnologie 23*: 2087–2091.

BURNS, C. W.; MITCHELL, S. F. 1980: Seasonal succession and vertical distribution of zooplankton in Lake Hayes and Lake Johnson. *New Zealand Journal of Marine and Freshwater Research 14*: 189–204.

BURNS, C. W.; XU, Z. 1990: Utilization of colonial cyanobacteria and algae by freshwater calanoid copepods: survivorship and reproduction of adult *Boeckella* species. *Archiv für Hydrobiologie 117*: 257–270.

BUTLER, M. J. IV; MacDIARMID, A. B.; BOOTH, J.D. 1999: The cause and consequences of ontogenetic changes in social aggregation in New Zealand spiny lobsters. *Marine Ecology Progress Series 188*: 179–191.

CAINE, E. A. 1986: Carapace epibionts of nesting loggerhead turtles: Atlantic coast of U.S.A. *Journal of Experimental Marine Biology and Ecology 95*: 15–26.

CALDWELL, R. L.; DINGLE, H. 1976: Stomatopods. *Scientific American 234*: 80–89.

CALMAN, W. T. 1896: On the genus *Anaspides* and its affinities with certain fossil Crustacea. *Transactions of the Royal Society of Edinburgh 38*: 787–802.

CALMAN, W.T. 1899: On the characters of the crustacean genus *Bathynella*. *Journal of the Linnean Society, London 27*: 338–345.

CALMAN, W. T. 1907: On new or rare Crustacea of the order Cumacea from the collection of the Copenhagen Museum. I. The families Bodotriidae, Vaunthompsoniidae and Leuconidae. *Transactions of the Zoological Society of London 18*: 1–58.

CALMAN, W. T. 1908: Notes on a small collection of plankton from New Zealand. Crustacea (excluding Copepoda). *Annals and Magazine of Natural History, ser. 8, 1*: 232–240.

CALMAN, W. T. 1908: Notes on a small collection of plankton from New Zealand. I. Crustacea. *Annals and Magazine of Natural History, ser. 8, 1*: 232–240.

CALMAN, W. T. 1909: Part V. Appendiculata. Third Fascicule. Crustacea. *In*: Lankester, E. R. (ed.) *A Treatise on Zoology*. A. & C. Black, London. 346 p.

CALMAN, W. T. 1911: On new or rare Crustacea of the Order Cumacea from the collection of the Copenhagen Museum. II. The families Nannastacidae and Diastylidae. *Transactions of the Zoological Society of London 18*: 341–398.

CALMAN, W. T. 1917: Crustacea. Part IV. – Stomatopoda, Cumacea, Phyllocarida and Cladocera. *British Antarctic ('Terra Nova') Expedition, 1910. Natural History Reports. Zoology III. Arthropoda 5*: 137–162.

CAMACHO, A. I. 1992: A classification of the aquatic and terrestrial subterranean environment and their associated fauna. Pp. 56–103 *in*: Camacho, A. I. (ed.), *The Natural History of Biospeleology. Monografías de Museo Nacional de Ciencias Naturales, Madrid 7*: 1–680.

CAMACHO, A. I. 1996: El mundo subterráneo, un reducto de biodiversidad. *Fronteras de la Ciencia y la Tecnología 13*: 49–53.

CAMACHO, A. I. 2000: La fauna subterránea de Lamasón y Peñamellera Baja: 15 años de investigaciones biospeleológicas. *Boletín Cántabro de Espeleología 14*: 153–164.

CAMACHO, A. I. 2003: Historical biogeography of *Hexabathynella*, a cosmopolitan genus of groundwater Syncarida (Crustacea, Bathynellacea, Parabathynellidae). *Biological Journal of the Linnean Society 78*: 457–466.

CAMACHO, A. I. 2004: An overview of *Hexabathynella* (Crustacea, Syncarida, Parabathynellidae) with the description of a new species. *Journal of Natural History 28*: 1249–1261.

CAMACHO, A. I. 2005a: One more piece in the genus puzzle: a new species of *Iberobathynella* Schminke, 1973 (Syncarida, Bathynellacea, Parabathynellidae) from the Iberian Peninsula. *Graellsia 61*: 123–133.

CAMACHO, A. I. 2005b: Expanding the taxonomic conundrum: three new species of groundwater crustacean (Syncarida, Bathynellacea, Parabathynellidae) endemic to the Iberian Peninsula. *Journal of Natural History 39*: 1819–1838.

CAMACHO, A. I. 2005c: Disentangling an Asian puzzle: two new bathynellid (Crustacea, Syncarida, Parabathynellidae) genera from Viet Nam. *Journal of Natural History 39*: 2861–2886.

CAMACHO, A. I. 2006: An annotated checklist of the Syncarida (Crustacea, Malacostraca) of the world. *Zootaxa 1374*: 1–54.

CAMACHO, A. I.; COINEAU, N. 1989: Les parabathynellacés (Crustacés syncarides) de la peninsule ibérique, repartition et paléobiogéographie. *Mémoires de Biospéologie 16*: 111–124.

CAMACHO, A. I.; REY, I., DORDA, B. A.; MACHORDOM, A.; VALDECASAS, A. G. 2002: A note on the systematic position of the Bathynellacea (Crustacea, Malacostraca) using molecular evidence. *Contributions to Zoology 71*: 123–129.

CAMACHO, A. I.; SERBAN, E.; GUIL, N. 2000: Phylogenetical review and biogeographic remarks on the interstitial and subterranean freshwater iberobathynells (Crustacea, Syncarida, Parabathynellidae). *Journal of Natural History 34*: 563–585.

CAMACHO, A. I.; TRONTELJ, P.; ZAGMAJSTER, M. 2006: First record of Bathynellacea (Crustacea, Syncarida, Parabathynellidae) in China: a new genus. *Journal of Natural History 40*: 1747–1760.

CAMACHO, A. I.; VALDECASAS, A. G. 2008: Global diversity of syncarids (Syncarida; Crustacea) in freshwater. *Hydrobiologia 595*: 257–266.

CAMERON, M.L. 1984: *The Paddle Crab Industry in New Zealand: Development of the US West Coast Market*. Winston Churchill Memorial Trust Fellowship Report, Department of Internal Affairs, Wellington. 38 p.

CANNON, H. G. 1927: On the feeding mechanism of *Nebalia bipes*. *Transactions of the Royal Society of Edinburgh 55*: 355–369.

CANNON, H. G. 1960: Leptostraca. *Bronn's Klassen und Ordnungen des Tierreiches 5, Abt. 1, Buch 4 (1)*: 1–81.

CHADDERTON, W. L.; RYAN, P. A.; WINTERBOURN, M. J. 2003: Distribution, ecology, and conservation status of freshwater Idoteidae (Isopoda) in southern New Zealand. *Journal of the Royal Society of New Zealand 33*: 529–548.

CHAPMAN, F. 1926: Cretaceous and Tertiary Foraminifera of New Zealand with an appendix on the Ostracoda. *New Zealand Geological Survey Palaeontological Bulletin 11*: 1–119.

CHAPMAN, F. 1934: On some phyllocarids from the Ordovician of Preservation Inlet and Cape Providence, New Zealand. *Transactions of the Royal Society of New Zealand 64*: 105–114.

CHAPMAN, M. A. 1961: The terrestrial ostracod of New Zealand, *Mesocypris audax* sp. nov. *Crustaceana 2*: 255–261.

CHAPMAN, M. A. 1963: A review of the freshwater ostracods of New Zealand. *Hydrobiologia 22*: 1–40.

CHAPMAN, M. A. 1973: *Calamoecia lucasi* (Copepoda: Calanoida) and other zooplankters in two Rotorua, New Zealand, lakes. *Internationale Revue der gesampten Hydrobiologie 58*: 79–104.

CHAPMAN, M. A. 2002: Australian species of *Paracorophium* (Crustacea: Amphipoda): the separate identities of *P. excavatum* (Thomson, 1884) and *P. brisbanensis* sp. nov. *Journal of the Royal Society of New Zealand 32*: 203–228.

CHAPMAN, M. A.; GREEN, J. D. 1987: Zooplankton ecology. *In*: Viner, A. B. (ed.), *Inland waters of New Zealand. Department of Scientific and Industrial Research Bulletin Bulletin 241*: 225–263.

CHAPMAN, M. A.; LEWIS, M. 1976: *An Introduction to the Freshwater Crustacea of New Zealand*. Collins, Auckland & London. 261 p.

CHAPMAN, M. A.; THOMAS, M. F. 1998: An experimental study of feeding in *Tenagomysis* chiltoni (Crustacea, Mysidacea). *Archiv für Hydrobiologie 143*: 197–209.

CHAPMAN, M. A.; HOGG, I. D.; SCHNABEL, K. E.; STEVENS, M. I. 2002: Synonymy of the New Zealand corophiid amphipod genus *Chaetocorophium* Karaman, 1979 with *Paracorophium* Stebbing, 1899: morphological and genetic evidence. *Journal of the Royal Society of New Zealand 32*: 229–241.

CHAPPUIS, P. A. 1915: *Bathynella natans* und ihre Stellung im System. *Zoologisches Jahrbuch, Abteilung für Systematik, Geographie, und Tierkunde 40*: 147–176.

CHAPPUIS, P. A. 1943: A talaj–és hasadék vizek állat világáról. *Allatani Közlemények 40*: 221–225.

CHATTERTON, T. D.; WILLIAMS, B. G. 1994: Activity patterns of the New Zealand cancrid crab *Cancer novaezelandiae* (Jacquinot) in the field and laboratory. *Journal of Experimental Marine Biology and Ecology 178*: 261–274.

CHEN, J.-Y.; VANNIER, J.; HUANG, D.-Y. 2001: The origin of crustaceans: new evidence from the Early Cambrian of China. *Proceedings of the Royal Society of London, ser. B, 268*: 2181–2187.

CHILTON, C. 1882a: On some subterranean Crustacea. *New Zealand Journal of Science 1*: 44.

CHILTON, C. 1882b: On some subterranean Crustacea. *Transactions and Proceedings of the New Zealand Institute 14*: 174–180, pls 9–10.

CHILTON, C. 1882c: Additions to the isopodan fauna of New Zealand. *Transactions of the New Zealand Institute 15*: 145–159.

CHILTON, C. 1883: Additions to the sessile-eyed Crustacea of New Zealand. *Transactions of the New Zealand Institute 16*: 249–265.

CHILTON, C. 1884: Subterranean Crustacea. *New Zealand Journal of Science 2*: 89.

CHILTON, C. 1891: Notes on the New Zealand Squillidae. *Transactions and Proceedings of the Royal Society of New Zealand Institute 23*: 58–68, pl. 10.

CHILTON, C. 1894: The subterranean Crustacea of New Zealand: with some general remarks on the fauna of caves and wells. *Transactions of the Linnean Society of London, ser. 2, Zoology 6*: 163–284, pls 16–23.

CHILTON, C. 1909: The Crustacea of the Subantarctic Islands of New Zealand. Pp. 601–671 *in*: Chilton, C. (ed.) *The Subantarctic Islands of New Zealand*. Philosophical Institute of Canterbury and Government Printer, Wellington. Vol. 2, pp. 389–848.

CHILTON, C. 1910: The Crustacea of the Kermadec Islands. *Transactions and Proceedings of the New Zealand Institute 43*: 544–573.

CHILTON, C. 1911a: Revision of the New Zealand Stomatopoda. *Transactions and Proceedings of the Royal Society of New Zealand Institute 43*: 134–139.

CHILTON, C. 1911b: Note on the dispersal of marine Crustacea by means of ships. *Transactions of the Linnean Society of London 43*: 131–133.

CHILTON, C. 1911c: Scientific results of the New Zealand Government Trawling Expedition 1907, Crustacea. *Records of the Canterbury Museum 1*: 285–312.

CHILTON, C. 1912: Miscellaneous notes on some New Zealand Crustacea. *Transactions of the New Zealand Institute 44*: 128–135.

CHILTON, C. 1919: Destructive boring Crustacea in New Zealand. *New Zealand Journal of Science and Technology 2*: 3–15.

CHILTON, C. 1921: Some New Zealand Amphipoda No. 2. *Transactions of the New Zealand Institute 53*: 220–234.

CHILTON, C. 1924: Some New Zealand Amphipoda: No. 4. *Transactions of the New Zealand Institute 55*: 269–280.

CHILTON, C. 1926: The New Zealand Crustacea Euphasiacea and Mysidacea. *Transactions of the New Zealand Institute 56*: 519–522.

CHISWELL, S. M.; BOOTH J. D. 1999: Rock lobster *Jasus edwardsii* larval retention by the Wairarapa Eddy off New Zealand. *Marine Ecology Progress Series 183*: 227–240.

CHISWELL, S. M.; ROEMMICH, D. 1999: The East Cape Current and two eddies: a mechanism for larval retention? *New Zealand Journal of Marine and Freshwater Research 32*: 385–397.

CH'NG, T. K. 1973: Aspects of the biology of the New Zealand freshwater shrimp *Paratya curvirostris* (Heller) in the Horokiwi Stream. Unpublished BSc project, Victoria University of Wellington.

CHO, J. L. 2005: A primitive representative of the Parabathynellidae (Bathynellacea, Syncarida) from Yilgarn Craton of Western Australia. *Journal of Natural History 39*: 3423–3433.

CHO, J. L.; HUMPHREYS, W. F.; LEE, S. D. 2006: Phylogenetic relationships within the genus *Atopobathynella* Schminke (Bathynellidae: Parabathynellidae). *Invertebrate Systematics 20*: 9–41.

CHO, J. L; PARK, J. G.; HUMPHREYS, W. F. 2005: A new genus and six new species of the Parabathyellidae (Bathynellacea, Syncarida) from the Kimberley Region, Western Australia. *Journal of Natural History 39*: 2225–2255.

CHO, J. L.; PARK, J. G.; REDDY, Y. R. 2006: *Brevisomabathynella* gen. nov. with two new species from Western Australia (Bathynellacea, Syncarida): the first definitive evidence of predation in Parabathynellidae. *Zootaxa 1247*: 25–42.

CHO, J. L.; SCHMINKE, H. K. 2006: A phylogenetic review of the genus *Hexabathynella* Schminke, 1972 (Crustacea, Malacostraca, Bathynellacea): with a description of four new species. *Zoological Journal of the Linnean Society 147*: 71–96.

CHOAT, J. H.; KINGETT, P. D. 1982: The influence of fish predation on the abundance cycles of an algal turf invertebrate fauna. *Oecologia 54*: 88–95.

CLARK, M. R. 1985: The food and feeding of seven fish species from the Campbell Plateau, New Zealand. *New Zealand Journal of Marine and Freshwater Research 19*: 339–363.

CLARK, M. R.; KING, K. J.; McMILLAN, P. J. 1989: The food and feeding relationships of black oreo, *Allocyttus niger*, smooth oreo, *Pseudocyttus maculatus*, and eight other fish species from the continental slope of the south-west Chatham Rise, New Zealand. *Journal of Fish Biology 35*: 465–484.

CLAUS, C. 1888: Über den Organismus der Nebaliden und die systematische Stellung der Leptostraken. *Arbeiten aus dem Zoologischen Institut der Universität Wien und der Zoologischen Station in Triest 8*: 1–148.

COHEN, A. C.; MARTIN, J. W.; KORNICKER, L. S. 1998: Homology of Holocene ostracode biramous appendages with those of other crustaceans: the protopod, epipod, exopod and endopod. *Lethaia 31*: 251–265.

COHEN, B. F. 1998: Dendrotiidae (Crustacea: Isopoda) of the southeastern Australian continental slope. *Memoirs of the Museum of Victoria 57*: 1–38.

COHEN, B. F.; POORE, G. C. B. 1994: Phylogeny and biogeography of the Gnathiidae (Crustacea: Isopoda) with descriptions of new genera and species, most from south-eastern Australia. *Memoirs of the Museum of Victoria 54*: 271–397.

COINEAU, N. 1996: Sous-classe des Eumalacostracés (Eumalacostraca Grobben, 1892) Superordre des Syncarides (Syncarida Packard, 1885). *Traité de Zoologie 7(2)*: 897–954.

COINEAU, N. 1998: Syncarida. Pp. 863–876 *in*: Juberthie, C.; Decu, V. (eds) *Encyclopaedia Biospeleologica*. Société de Biospéologie & Moulis, Bucharest. 1378 p.

CONRADI, M.; LOPEZ-GONZALEZ, P. J.; GARCÍA-GOMEZ, C. 1997: The amphipod community as a bioindicator in Algeciras Bay (Southern Iberian Peninsula) based on a spatio-temporal distribution. *Marine Ecology 18*: 97–111.

COOKSON, L. J. 1989: Taxonomy of the Limnoriidae (Crustacea: Isopoda), and its relevance to marine wood preservation in Australia. PhD Thesis, Monash University, Melbourne. 432 p., appended papers.

COOKSON, L. J. 1991: Australasian species of Limnoriidae (Crustacea: Isopoda). *Memoirs of the Museum of Victoria 52*: 137–262.

COOPER, R. A. (Comp.) 2004: New Zealand geological timescale 2004/2 wallchart. *Institute of Geological and Nuclear Sciences Information Series 64*.

COOPER, R. D. 1974: Preliminary diagnoses of three new amphipod species from Wellington Harbour (note). *New Zealand Journal of Marine and Freshwater Research 8*: 239–241.

COOPER, R. D.; FINCHAM, A. A. 1974: New species of Haustoriidae, Phoxocephalidae, and Oedicerotidae (Crustacea: Amphipoda) from northern and southern New Zealand. *Records of the Dominion Museum 8*: 159–179.

COSTELLO, M. J. 1993: Biogeography of alien amphipods occurring in Ireland, and interactions with native species. *Crustaceana 65*: 287–299.

COUCH, K. M.; BURNS, C. W.; GILBERT, J. J. 1999: Contribution of rotifers to the diet and fitness of *Boeckella* (Copepoda: Calanoida). *Freshwater Biology 41*: 107–118.

COULL, B. C.; HICKS, G. R. F. 1983: The ecology of marine meiobenthic copepods. *Oceanography and Marine Biology Annual Review 21*: 67–175.

CRANFIELD, H. J.; GORDON, D. P.; WILLAN, R. C.; MARSHALL, B. A.; BATTERSHILL, C. N.; FRANCIS, M. P.; NELSON, W. A.; GLASBY, C. J.; READ, G. B.1998: Adventive marine species in New Zealand. *NIWA Technical Report 34*: 1–48.

CRAWFORD, G. I. 1937: A review of the amphipod genus *Corophium*, with notes on the British species. *Journal of the Marine Biological Association of the United Kingdom 21*: 589–630.

CROSSLEY, P. C.; HURST, B. P.; WEST, R. G. 1981: *The New Zealand Cave Atlas*. New Zealand Speleological Society and Department of Geography, Auckland University, Auckland. 257 p.

CROXALL, J. P.; LISHMAN, G. S. 1987: The food and feeding ecology of penguins. Pp. 101–134 *in* Croxall, J.P. (ed.), *Seabirds: Feeding Ecology and Role in Marine Ecosystems*. Cambridge University Press, Cambridge. 408 p.

CUESTA, J. A.; DIESEL, R.; SCHUBART, C. D. 2001: Reexamination of the zoeal morphology of *Chasmagnathus granulatus*, *Cyclograpsus lavauxi*, *Hemigrapsus sexdentatus* and *H. crenulatus* confirms consistent chaetotaxy in the Varunidae (Decapoda, Brachyura). *Crustaceana 74*: 895–912.

CULVER, D. C. 1982: *Cave Life: Evolution and Ecology*. Harvard University Press, Cambridge. 189 p.

CULVER, D. C.; DEHARVANG, L.; GIBERT, J.; SASOWSKY, I. D. (Eds) 1980: Mapping subterranean biodiversity. *Karst Waters Institute Special Publication 6*: 1–82.

DAHL, E. 1959: Amphipoda from depths exceeding 6000 meters. *Galathea Report 1*: 211–241.

DAHL, E. 1987: Malacostraca maltreated – the case of the Phyllocarida. *Journal of Crustacean Biology 7*: 721–726.

DAHL, E. 1990: Records of *Nebalia* (Crustacea Leptostraca) from the Southern Hemisphere – a critical review. *Bulletin of the British Museum (Natural History), Zoology 56*: 73–91.

DAHL, E. 1991: Crustacea Phyllopoda and Malacostraca: a reappraisal of cephalic dorsal shield and fold systems. *Philosophical Transactions of the Royal Society of London, B, 334*: 1–26.

DAHL, E.; WÄGELE, J. W. 1996: Sous-classe des Phyllocarides (Phyllocarida Packard, 1879). *Traité de Zoologie 7(2)*: 865–896.

DALLEY, D. D.; McCLATCHIE, S. 1989: Functional feeding morphology of the euphausiid *Nyctiphanes australis*. *Marine Biology 101*: 195–203.

DANA, J. D. 1852a: Conspectus Crustaceorum quae in Orbis Terrarum circumnavigatione, Carolo Wilkes e Classe Reipublicae Foederatae Duce, lexit et descripsit Jacobus D. Dana. Pars III. Amphipoda. No. 1. *Proceedings of the American Academy of Arts and Science 2*: 6–28, 201–220.

DANA, J. D. 1852b: Conspectus Crustaceorum quae in Orbis Terrarum circumnavigatione, Carolo Wilkes e Classe Reipublicae Foederatae Duce, lexit et descripsit. *Proceedings of the Academy of Natural Science of Philadelphia 6*: 6–28.

DANA, J. D. 1853–55: Crustacea. Part II. *United States Exploring Expedition during the years 1838, 1839, 1840, 1841, 1842 under the command of Charles Wilkes, U.S.N. 13*: 691–1618, 96 pls. (Atlas).

DANIELOPOL, D. L. 1992: New perspectives in ecological research of groundwater organisms. Pp. 15–22 *in*: Stanford, J. A.; Simins, J. J. (eds), *Proceedings of the First International Conference on Ground Water Ecology*. American Water Resources Association, Bethesda. 419 p.

DANIELOPOL, D. L; POSPISIL, P.; ROUCH, R. 2000: Biodiversity in groundwater: a large-scale view. *Trends in Ecology and Evolution 15*: 223–224.

DARWIN, C. 1851: *A Monograph on the Fossil Lepadidae; or Pedunculated Cirripedes of Great Britain*. Palaeontological Society, London. 88 p.

DARWIN, C. 1851 [1852]: *A Monograph on the Subclass Cirripedia. The Lepadidae; or Pedunculated Cirripedes*. Ray Society, London. 400 p.

DARWIN, C. 1854a: *A Monograph on the Sub-class*

Cirripedia. The Balanidae and Verrucidae. Ray Society, London. 684 p.

DARWIN, C. 1854b: *A Monograph on the Fossil Balanidae and Verrucidae of Great Britain*. Palaeontological Society, London. 44 p.

DAWSON, E. W. 1989: King crabs of the world (Crustacea: Lithodidae) and their fisheries: a comprehensive bibliography. *NZOI Miscellaneous Publication 101*: 1–338.

DAWSON, E. W. 2003a: The Syncarida – 'bareback shrimps' – tiny crustacean survivors from ancient times. *Occasional Papers of the Hutton Foundation New Zealand 10*: ii, 1–36.

DAWSON. E. W. 2003b: 'Sea Fleas' – Crustacea/ Leptostraca – tiny living fossils of our seas and shores: a New Zealand perspective. *Occasional Papers of the Hutton Foundation New Zealand 11*: ii, 1–24.

DAWSON, E. W.; YALDWYN, J. C. 1970: Diagnosis of a new species of *Neolithodes* (Crustacea, Anomura, Lithodidae) from New Zealand (Note). *New Zealand Journal of Marine and Freshwater Research 4*: 227–228.

DAWSON, E. W.; YALDWYN, J. C. 1971: Diagnosis of a new species of *Paralomis* (Crustacea, Anomura, Lithodidae) from New Zealand. *Records of the Dominion Museum 7(7)*: 51–54.

DAWSON, E. W.; YALDWYN, J. C. 1985: King crabs of the world or the world of king crabs: An overview of identity and distribution, with illustrated diagnostic keys to the genera of the Lithodidae and to the species of *Lithodes*. *In*: Melteff, B. R. (co-ord.), Proceedings of the International King Crab Symposium, Anchorage, Alaska, USA, 22–24 January 1985. *University of Alaska, Alaska Sea Grant Report No. 85*: 69–106.

DEAN, T. L.; RICHARDSON, J. 1999: Responses of seven species of native freshwater fish and a shrimp to low levels of dissolved oxygen. *New Zealand Journal of Marine and Freshwater Research* 33: 99–106.

DEEVEY, G. B. 1982: A faunistic study of the planktonic ostracods (Myodocopa, Halocyprididae) collected on eleven cruises of the Eltanin between New Zealand, Australia, the Ross Sea and the South Indian Ocean. *Antarctic Research Series 22* [*Biology of the Antarctic Seas 10*]: 131–167.

DELL, R. K. 1960: Biological results of the Chatham Islands 1954 Expedition. Part 1. Crabs (Decapoda, Brachyura). *New Zealand Department of Scientific and Industrial Research Bulletin 139(1)* [*New Zealand Oceanographic Institute Memoir 4*]: 1–7, pls 1–2.

DELL, R. K. 1963a: *Nature in New Zealand: Native Crabs*. A.H. & A.W. Reed, Wellington. 64 p.

DELL, R. K. 1963b: *Pachygrapsus marinus* (Rathbun), a new crab for New Zealand waters. *Transactions of the Royal Society of New Zealand 3*: 179–180.

DELL, R. K. 1968a: Composition and distribution of the New Zealand brachyuran fauna. *Transactions of the Royal Society of New Zealand, Zoology 10*: 225–240.

DELL, R. K. 1968b: Notes on New Zealand crabs. *Records of the Dominion Museum 6*: 13–28.

DELL, R. K. 1969: A new Pliocene fossil crab of the genus *Trichopeltarion* from new Zealand. *Records of the Canterbury Museum 8*: 367–370.

DELL, R.K.1971: Two new species of crab of the genus *Cymonomus* from New Zealand (Crustacea: Brachyura). *Records of the Dominion Museum 7*(8): 55–64.

DELL, R. K.1972: A new genus and species of atelecyclid crab from New Zealand. *Journal of the Royal Society of New Zealand 2*: 55–59.

DELL, R. K. 1974: Crabs. *New Zealand's Nature Heritage 3(45)*: 1237–1244.

DELL, R. K.; GRIFFIN, D. J. G.; YALDWYN, J. C. 1970: A new swimming crab from the New Zealand subantarctic and a review of the genus *Nectocarcinus* A. Milne Edwards. *Transactions of the Royal Society of New Zealand 12*: 49–68.

DELLA VALLE, A. 1893: Gammarini del golfo di Napoli. Fauna und Flora des Golfes von Neapel und der angrenzenden Meers-Abschnitte. *Monographie 20*: 1–948.

DEPLEDGE, M. H.; LUNDEBYE, A. K. 1996: Physiological monitoring of contaminant effects in individual rock crabs, *Hemigrapsus edwardsi*: the ecotoxicological significance of variability in response. *Comparative Biochemistry and Physiology, C, Pharmacology, Toxicology and Endocrinology 113*: 277–282.

DE SIMÓN, M. 1979: Primeros estadios larvarios de *Pontocaris lacazei* (Gourret) (Decapoda, Macrura, Crangonidae) obtenidos en laboratorio. *Investigación Pesquera 43*: 565–580.

DE WITT, T. H.; HICKEY, C. W.; MORRISEY, D. J.; NIPPER, M. G.; ROPER, D. S.; WILLIAMSON, R. B.; VAN DAM, L.; WILLIAMS, E. K. 1999: Do amphipods have the same concentration response to contaminated sediment in situ as in vitro? *Environmental Toxicology and Chemistry 18*: 1026–1037.

DINGLE, R.V. 2009: Implications for high latitude gondwanide palaeozoogeographical studies of some new Upper Cretaceous marine ostracod faunas from New Zealand and the Antarctic Peninsula. *Revista Española de Micropaleontogia 41*: 145–196.

DOHLE, W. 2000: Hunting for *Anaspides* eggs. *Invertebrata* [Queen Victoria Museum & Art Gallery, Tasmania] *18*: 3.

DODGSHUN, T.; COUTTS, A. 2003: Opening the lid on sea chests. *Seafood New Zealand 11(2)*: 35.

DOJIRI, M.; SIEG, J. 1997: The Tanaidacea. Pp. 181–278 *in*: Blake, J. A.; Scott, P. H. (eds), *Taxonomic Atlas of the Benthic Fauna of the Santa Maria Basin and Western Santa Barbara Channel 11 – The Crustacea. Part 2. The Isopoda, Cumacea and Tanaidacea*. Santa Barbara Museum of Natural History, Santa Barbara. 278 p.

DOVER, C. 1953: The story of a 'living fossil': *Parabathynella malaya* Sars. *Nytt Magasin for Zoologi 1*: 87–97.

DRIESSEN, M. M.; MALLICK, S. A.; LEE, A.; THURSTANS, S. 2006: Loss of habitat through inundation and the conservation status of two endemic Tasmanian syncarid crustaceans: *Allanaspides hickmani* and *A. helonomus*. *Oryx 40*: 464–467.

DRUMMOND, F. H. 1959: The syncarid Crustacea, a living link with remote geological ages. *Australian Museum Magazine 13*: 63–64.

DUFFY, C. A. J. 1989: The fish fauna of subtidally fringing macroalgae sampled at Wairepo Flats, Kaikoura: species composition, distribution and abundance. Unpublished MSc thesis (Zoology), University of Canterbury. 137 p.

DUGGAN, I. C.; GREEN, J. D.; BURGER, D. F.; 2006: First New Zealand records of three non-indigenous zooplankton species: *Skistodiaptomus pallidus*, *Sinodiaptomus valkanovi*, and *Daphnia dentifera*. *New Zealand Journal of Marine and Freshwater Research 40*: 561–569.

DUMONT, H.; NEGREA, S.V. 2002. Introduction to the class Branchiopoda. *In*: Dumont, H. J. (coord. ed.), *Guides to the Identification of the Microinvertebrates of the Continental Waters of the World*. Backhuys Publishers, Leiden. 398 p.

DUNCAN, K.W. 1968: A description of a new species of terrestrial amphipod (Fam. Talitridae) from New Zealand. *Transactions of the Royal Society of New Zealand, Zoology 10*: 205–210.

DUNCAN, K. W. 1994: Terrestrial Talitridae (Crustacea: Amphipoda). *Fauna of New Zealand 31*: 1–128.

DUSSART, B.; DEFAYE, D. 1985: *Répertoire mondiale des Copépodes cyclopoïdes*. Editions du C.N.R.S., Paris. 236 p.

DUSSART, B. H.; DEFAYE, D. 1995: Copepoda. Introduction to the Copepoda. *Guides to the identification of the Microinvertebrates of the Continental Waters of the World 7*: 1–277.

EAGAR, S. H. 1971: A check list of the Ostracoda of New Zealand. *Journal of the Royal Society of New Zealand 1*: 53–64.

EAGAR, S. H. 1994: Freshwater Ostracoda from eastern North Island. New Zealand. *New Zealand Natural Sciences 21*: 71–86.

EAGAR, S. H. 1995a: Ostracoda from a Pleistocene lake deposit at Kourarau, Wairarapa. *New Zealand Natural Sciences 22*: 19–25.

EAGAR, S. H. 1995b. Myodocopid ostracods from New Zealand collected with a light trap. Pp. 399–406 *in*: Riha, J. (ed.), *Ostracoda and Biostratigraphy*. A.A. Balkema, Rotterdam. 453 p.

EAGAR, S. H. 1999: Distribution of Ostracoda around a coastal sewer outfall: a case study from Wellington, New Zealand. *Journal of the Royal Society of New Zealand 29*: 257–264.

EDGAR, N. B. 1993: Trophic manipulation in freshwater planktonic communities. Unpublished DPhil thesis, University of Waikato, Hamilton.

EDGAR, N. B.; GREEN, J. D. 1994: Calanoid copepod grazing on phytoplankton: seasonal experiments on natural communities. *Hydrobiologia 273*: 147–161.

EDWARDS, H. M. 1837: Crustacés. *In*: *Suites à Buffon, formant avec les oeuvres de cet auteur un cours complet d'Histoire Naturelle*. Vol. 2. Paris.

EDWARDS, H. M. 1840: *Histoire naturelle des Crustacés, Comprenant l'Anatomie, la Physiologie et la Classification de ces Animaux*. Librairie Encyclopédique de Roret, Paris. Vol. 3. 638 p.

EINSLE, U. 1996: Copepoda: Cyclopoida. Genera *Cyclops, Megacyclops, Acanthocyclops. Guides to the Identification of the Microinvertebrates of the Continental Waters of the World 10*: 1–82.

ELDON, G. A. 1979: Food of the Canterbury mudfish, *Neochanna burrowsius* (Salmoniformes: Galaxiidae). *New Zealand Journal of Marine and Freshwater Research 13*: 255–261.

ENEQUIST, P. 1949: Studies in the soft-bottom amphipods of the Skaggerak. *Zoologiska Bidrag från Uppsala 28*: 297–492.

FAGE, L. 1941: Mysidacea Lophogastrida – 1. *Dana Report 4 (19)*: 1–52.

FAGE, L. 1960: Oxycephalidae. *Dana Report 9(52)*: 1–145.

FARRAN, G. P. 1929: Crustacea. Part X. Copepoda. *Natural History Reports of the British Antarctic Terra Nova Expedition, Zoology 8(3)*: 203–306, pls 1–4.

FELDMANN, R. M. 1984: *Haumuriaegla glaessneri* n. gen. et sp. (Decapoda: Anomura: Aeglidae) from Haumurian (Late Cretaceous) rocks near Cheviot, New Zealand. *New Zealand Journal of Geology and Geophysics 27*: 379–385.

FELDMANN, R. M. 1986: Paleobiogeography of two decapod crustacean taxa in the Southern Hemisphere: Global conclusions with sparse data. *Crustacean Issues 4*: 5–20.

FELDMANN, R. M. 1990: Decapod crustacean paleobiogeography: resolving the problem of a small sample size. *In*: Mikulic, D. G. (ed.), *Arthropod Paleobiology. Paleontological Society Short Courses in Paleontology 3*: 303–315.

FELDMANN, R. M. 1993: Additions to the fossil decapod crustacean fauna of New Zealand. *New Zealand Journal of Geology and Geophysics 36*: 201–211.

FELDMANN, R. M. 1998: Parasitic castration of the crab, *Tumidocarcinus giganteus* Glaessner, from the Miocene of New Zealand: coevolution within the Crustacea. *Journal of Paleontology 72*: 493–498.

FELDMANN, R. M. 1998: *Paralomis debodeorum*, a new species of decapod crustacean from the Miocene of New Zealand: first notice of the Lithodidae in the fossil record. *New Zealand Journal of Geology and Geophysics 41*: 35–38.

FELDMANN, R. M. 2003: The Decapoda: New initiatives and novel approaches. *Journal of Paleontology 77*: 1021–1039.

FELDMANN, R. M.; BEARLIN, R. K. 1998: *Linuparus korura* n. sp. (Decapoda: Palinura) from the Bortonian (Eocene) of New Zealand. *Journal of Paleontology 62*: 245–250.

FELDMANN, R. M.; FORDYCE, R. E. 1996: A new cancrid crab from New Zealand. *New Zealand Journal of Geology and Geophysics 39*: 509–513.

FELDMANN, R. M.; KEYES, I. W. 1992: Systematic and stratigraphic review with catalogue and locality index of the Mesozoic and Cenozoic decapod Crustacea of New Zealand. *New Zealand Geological Survey Record 45*: 1–73.

FELDMANN, R. M.; MAXWELL, P. A. 1999: A new species of glypheid lobster, *Glyphea christeyi* (Decapoda: Palinura), from the Eocene (Bortonian) Waihao Greensand, South Canterbury, New Zealand. *New Zealand Journal of Geology and Geophysics 42*: 75–78.

FELDMANN, R. M.; McLAY, C. L. 1993: Geological history of brachyuran decapods from New Zealand. *Journal of Crustacean Biology 13*: 433–455.

FELDMANN, R. M.; POLE, M. S. 1994: A new species of *Paranephrops* White, 1842: a fossil freshwater crayfish (Decapoda: Parastacidae) from the Manuherikia Group (Miocene), Central Otago, New Zealand. *New Zealand Journal of Geology and Geophysics 37*: 163–167.

FENWICK, G. D. 1976: The effect of wave exposure on the amphipod fauna of the alga *Caulerpa brownii*. *Journal of Experimental Marine Biology and Ecology 25*: 1–18.

FENWICK, G. D. 1977: *Mesoproboloides excavata* n.sp. (Amphipoda: Gammaridea: Stenothoidae) from New Zealand. *New Zealand Journal of Marine and Freshwater Research 11*: 471–478.

FENWICK, G. D. 1978: Plankton swarms and their predators at the Snares Islands (Note). *New Zealand Journal of Marine and Freshwater Research 12*: 223–224.

FENWICK, G. D. 1983. Two new sand-dwelling amphipods from Kaikoura, New Zealand (Oedicerotidae and Lysianassidae). *New Zealand Journal of Zoology 10*: 133–145.

FENWICK, G. D. 1984: Partitioning of a rippled sand habitat by five infaunal crustaceans. *Journal of Experimental Marine Biology and Ecology 83*: 53–72.

FENWICK, G. D. 1985: Life-histories of five co-occurring amphipods from a shallow, sand bottom at Kaikoura, New Zealand. *New Zealand Journal of Zoology 12*: 71–105.

FENWICK, G. D. 1987: Organic carbon pathways in the Canterbury groundwater ecosystem and the role of phreatic crustaceans. Report to the National Water & Soil Conservation Organisation. 84 p.

FENWICK, G. D. 1999: The benthos off South Brighton, Pegasus Bay: a preliminary assessment. NIWA Client Report CHC99/53. 27 p.

FENWICK, G. D. 2000: Collections of New Zealand groundwater amphipods. *NIWA Technical Report 95*: 1–21.

FENWICK, G. D. 2001a: The freshwater Amphipoda (Crustacea) of New Zealand: a review. *Journal of the Royal Society of New Zealand 31*: 341–363.

FENWICK, G. D. 2001b: *Paracrangonyx* Stebbing, 1899, a genus of New Zealand amphipods (Crustacea: Amphipoda: Gammaridea). *Journal of the Royal Society of New Zealand 31*: 457–479.

FENWICK, G. D. 2006: *Ringanui*, a new genus of stygofaunal amphipods from New Zealand (Amphipoda: Gammaridea: Paraleptamphopidae). *Zootaxa 1148*: 1–25.

FENWICK, G. D.; THORPE, H. R.; WHITE, P. A. 2004: Groundwater systems. Pp. 29.1–29.18 *in*: Harding, J.; Mosely, P.; Pearson, C.; Sorrell, B. (eds), *Freshwaters of New Zealand*. New Zealand Hydrological Society & New Zealand Limnological Society, Christchurch. 700 p.

FENWICK, G. D.; WEBBER, R. 2008: Identification of New Zealand's terrestrial amphipods (Crustacea: Amphipoda: Talitridae). *Tuhinga 19*: 29–56.

FIELD, L. H. 1990: Aberrant defense displays of the big-handed crab, *Heterozius rotundifrons* (Brachyura: Belliidae). *New Zealand Journal of Marine and Freshwater Research 24*: 211–220.

FILHOL, H. 1886: Catalogue des Crustacés de la Nouvelle-Zélande, des Îles Auckland et Campbell. *Mission de l'Île Campbell 3*: 349–510.

FINCHAM, A. A. 1974: Intertidal sand-dwelling fauna of Stewart Island. *New Zealand Journal of Marine and Freshwater Research 8*: 1–14.

FINCHAM, A. A. 1977: Establishment of a new genus in the Phoxocephalidae (Crustacea: Amphipoda) and a description of a new species from North Island, New Zealand. *Bulletin of the British Museum (Natural History) 31*: 285–292.

FLEMING, C. A. 1962: New Zealand biogeography. A palaeontologist's view. *Tuatara 20*: 53–108.

FLEMING, C. A. 1979: *The Geological History of New Zealand and its Life*. Auckland University Press & Oxford University Press, Auckland. 141 p.

FLEMING, C.A. 1981: A new grapsid crab from the upper Miocene of New Zealand. *Journal of the Royal Society of New Zealand 11*: 103–108.

FORD, T. D.; CULLINGFORD, C. H. D. (Eds) 1976: *The Science of Speleology*. Academic Press, London & New York. xiv + 593 p.

FOREST, J.; de SAINT LAURENT, M.; McLAUGHLIN, P. A.; LEMAITRE, R. 2000: The marine fauna of New Zealand: Paguridea (Decapoda: Anomura) exclusive of the Lithodidae. *NIWA Biodiversity Memoir 114*: 1–250.

FORSTER, M.E. 1991: Haemolymph oxygenation and oxygen consumption in a high shore crab, *Leptograpsus variegatus*, breathing in air and water. *New Zealand Natural Sciences 18*: 19-23.

FORSYTH, D. J.; JAMES, M. R. 1984: Zooplankton grazing on lake bacterio-plankton and phytoplankton. *Journal of Plankton Research 6*: 803–810.

FORSYTH, D. J.; McCALLUM, I. D. 1980: Zooplankton of Lake Taupo. *New Zealand Journal of Marine and Freshwater Research 14*: 65–69.

FOSTER, B. A. 1974: The barnacles of Fiji with observations on the ecology of barnacles on tropical shores. *Pacific Science 28*: 35–56.

FOSTER, B.A. 1978[1979]: The marine fauna of New Zealand: Barnacles (Cirripedia: Thoracica). *New Zealand Oceanographic Institute Memoir 69*: 1–160.

FOSTER, B. A. 1980: Further records and classification of scalpellid barnacles (Cirripedia: Thoracica) from New Zealand. *New Zealand Journal of Zoology 7*: 523–531.

FOSTER, B. A. 1981: Cirripedes from ocean ridges north of New Zealand. *New Zealand Journal of Zoology 8*: 349–367.

FOSTER, B. A. 1986: Barnacles in Maori middens. *Journal of the Royal Society of New Zealand 16*: 43–49.

FOSTER, B. A.; ANDERSON, D. T. 1986: New names for two well-known shore barnacles (Cirripedia: Thoracica) from Australia and New Zealand. *Journal of the Royal Society of New Zealand 16*: 57–69.

FOSTER, B. A.; BUCKERIDGE, J. S. 1987: Barnacle palaeontology. Pp. 43–62 *in*: Southward, A. J. (ed.), *Crustacean Issues* 5: *Barnacle Biology*. A. A. Balkema Publishers, Rotterdam.

FOSTER, B. A.; WILLAN, R. C. 1979: Foreign barnacles transported to New Zealand on an oil platform. *New Zealand Journal of marine and Freshwater Research 13*: 143–149.

FRYER, G. 1963: The functional morphology and feeding mechanism of the chydorid cladoceran *Eurycercus lamellatus* (O. F. Müller). *Transactions of the Royal Society of Edinburgh 65*: 335–381.

FRYER, G. 1968: Evolution and adaptive radiation in the Chydoridae (Crustacea, Cladocera): a study in comparative functional morphology and ecology. *Philosophical Transactions of the Royal Society of London, B, 254*: 221–385.

FRYER, G. 1974: Evolution and adaptive radiation in the Macrothricidae (Crustacea: Cladocera): a study in comparative functional morphology and ecology. *Philosophical Transactions of the Royal Society of London, B, 269*: 137–274.

FRYER, G. 1987: A new classification of the branchiopod Crustacea. *Zoological Journal of the Linnean Society 91*: 357–383.

FRYER, G. 1991: Functional morphology and the adaptive radiation of the Daphniidae (Branchiopoda: Anomopoda). *Philosophical Transactions of the Royal Society of London, B, 331*: 1–99.

FUKUOKA, K.; BRUCE, N.L. 2005: A new species of *Tenagomysis* (Crustacea: Mysida: Mysidae) from New Zealand with notes on three *Tenagomysis* species. *Zootaxa 878*: 1–15.

FUSSELL, C. R. 1979: The biology of *Heterosquilla tricarinata* (Crustacea: Stomatopoda). Unpublished report, Portobello Marine Laboratory, University of Otago. 61 p.

GAMO, S. 1984: A new remarkably giant tanaid, *Gigantapseudes maximus* sp. nov. (Crustacea) from abyssal depths far off southeast of Mindanao, the Philippines. *Scientific Reports of Yokohama Natural University Series 11(31)*: 1–12.

GARDINER, L. F. 1975: The systematics, postmarsupial development, and ecology of the deep-sea family Neotanaidae (Crustacea: Tanaidacea). *Smithsonian Contributions to Zoology 170*: 1–265.

GAVRILOV, G. M.; MARKINA, N. P. 1981: The feeding ecology of fishes of the genus *Seriolella* (fam. Nomeidae) on the New Zealand plateau. *Journal of Ichthyology 19*: 128–135.

GERKEN, S. 2001: The Gynodiastylidae (Crustacea: Cumacea). *Memoirs of Museum Victoria 59*: 1–276.

GIBBONS, M. J.; BARANGE, M.; HUTCHINGS, L. 1995: Zoogeography and diversity of euphausiids around southern Africa. *Marine Biology 123*: 257–268.

GIBERT, J.; DANIELOPOL, D. L.; STANFORD, J.A. (eds) 1994: *Groundwater Ecology*. Academic Press, San Diego. 571 p.

GLAESSNER, M. F. 1960: The fossil decapod Crustacea of New Zealand and the evolution of the order Decapoda. *New Zealand Geological Survey Paleontological Bulletin 31*: 1–79.

GLAESSNER, M. F. 1960: New Cretaceous and Tertiary crabs (Crustacea: Brachyura) from Australia and New Zealand. *Transactions of the Royal Society of South Australia 104*: 171-192

GLAESSNER, M. F. 1969: Decapoda. Pp. R400-R533 *in*: Moore, R. C. (ed.), *Treatise on Invertebrate Paleontology, Pt. R, Arthropoda 4(2)*. University of Kansas and Geological Society of America, Lawrence, Kansas.

GLENNER, H.; LÜTZEN, J.; TAKAHASHI, T. 2003: Molecular evidence for a monophyletic clade of asexually reproducing Rhizocephala: *Polyascus*, a new genus (Cirripedia). *Journal of Crustacean*

Biology 23: 548–557.
GLENNER, H.; HEBSGAARD, M.B. 2006: Phylogeny and evolution of life history strategies of the parasitic barnacles (Crustacea, Cirripedia, Rhizocephala). *Molecular Phylogenetics and Evolution 41*: 528–538.
GORDAN, J. 1957: A bibliography of the order Mysidacea. *Bulletin of the American Museum of Natural History 112*: 281–393.
GORDON, D.P.; BEAUMONT, J.; MacDIARMID, A.; ROBERTSON, D.A.; ROWDEN, A.A.; CONSALVEY, M. In press: Marine biodiversity of Aotearoa New Zealand. *PLoS ONE*.
GRANT-MACKIE, J. A.; BUCKERIDGE, J. S.; JOHNS, P. M. 1996: Two new Upper Jurassic arthropods from New Zealand. *Alcheringa 20*: 31–39.
GRAY, J. E. 1843: List of the annulose animals hitherto recorded as found in New Zealand, with the descriptions of some new species by Messrs. Adam White and Edward Doubleday, Assistants in the Zoological Department of the British Museum. Pp. 265–291 *in*: Dieffenbach, E. *Travels in New Zealand*. John Murray, London. Vol. 2. 396 p.
GREEN, A. J. A. 1961: A study of Tasmanian Oniscoidea (Crustacea: Isopoda). *Australian Journal of Zoology 9*: 258–365.
GREEN, A. J. A. 1971: Styloniscidae (Isopoda, Oniscoidea) from Tasmania and New Zealand. *Papers and Proceedings of the Royal Society of Tasmania 105*: 59–74.
GREEN, A. J. A.; LEW TON, H. M.; POORE, G. C. B. 2002: Suborder: Oniscidea Latreille, 1802. *Zoological Catalogue of Australia 19.2A*: 279–344.
GREEN, J. D. 1974: The limnology of a New Zealand reservoir, with particular reference to the life histories of the copepods *Boeckella propinqua* Sars and *Mesocyclops leuckarti* Claus. *Internationale Revue der gesamten Hydrobiologie 59*: 441–487.
GREEN, J. D. 1975: Feeding and respiration in the New Zealand copepod *Calamoecia lucasi* Brady. *Oecologia 21*: 345–358.
GREEN, J. D. 1976: Plankton of Lake Ototoa, a sand dune lake in northern New Zealand. *New Zealand Journal of Marine and Freshwater Research 10*: 43–59.
GREENWOOD, J. G. 1965: The larval development of *Petrolisthes elongatus* (H. Milne Edwards) and *Petrolisthes novaezelandiae* Filhol (Anomura, Porcellanidae) with notes on breeding. *Crustaceana 8*: 285–307.
GREENWOOD, J. G. 1966: Some larval stages of *Pagurus novae-zelandiae* (Dana, 1852) (Decapoda, Anomura). *New Zealand Journal of Science 9*: 545–558.
GREENWOOD, J. G.; JONES, M. B.; GREENWOOD, J. 1985: Reproductive biology, seasonality and distribution of *Tenagomysis macropsis* W. Tattersall, 1923 (Crustacea, Mysidacea) in a New Zealand estuary. *Bulletin of Marine Science 37*: 538–555.
GREENWOOD, J. G.; WILLIAMS, B. G. 1984: Larval and early postlarval stages in the abbreviated development of *Heterosquilla tricarinata* (Claus, 1871) (Crustacea: Stomatopoda). *Journal of Plankton Research 6*: 615–635.
GREENWOOD, T. L.; GREEN, J. D.; HICKS, B. J.; CHAPMAN, M. A. 1999: Seasonal abundance of small cladocerans in Lake Mangakaware, Waikato, New Zealand. *New Zealand Journal of Marine and Freshwater Research 33*: 399–415.
GRIFFIN, D. J. G. 1966: The Marine fauna of New Zealand: spider crabs, family Majidae (Crustacea, Brachyura). *New Zealand Oceanographic Institute Memoir 35* [*New Zealand Department of Scientific and Industrial Research Bulletin 172*]: 1–112.
GRIFFITHS, W. E. 1976: Food and feeding habits of European perch in the Selwyn River, Canterbury, New Zealand. *New Zealand Journal of Marine and Freshwater Research 10*: 417–428.
GUIL, N.; CAMACHO, A. I. 2001: Historical biogeography of *Iberobathynella* (Crustacea, Syncarida, Bathynellacea), an aquatic subterranean genus of parabathynellids, endemic to the Iberian Peninsula. *Global Ecology and Biogeography 10*: 487–501.
GUISE, J. E. 2001: A new genus of brackish-water ostracod, *Swansonella*, from the Avon-Heathcote Estuary, Christchurch, New Zealand. *New Zealand NaturalSsciences 26*: 75–86.
GURNEY, R. 1924: Crustacea. Part 9. Decapod larvae. *British Antarctic Terra Nova Expedition, 1910, Natural History Reports, Zoology 8*: 37–202.
GURNEY, R. 1936: Larvae of decapod Crustacea: Part III. Phyllosoma. *Discovery Reports 12*: 337–440.
GURNEY, R. 1942: *Larvae of the Decapod Crustacea*. The Ray Society, London. 306 p.
GURNEY, R.; LEBOUR, M. V. 1940: Larvae of decapod Crustacea: Part VI. The genus *Sergestes*. *Discovery Reports 20*: 1–68.
GURR, L. 1953: A note on the occurrence of *Linguatula serrata* (Frohlich, 1789) in the wild rabbit, *Oryctolagus cuniculus*, in New Zealand. *New Zealand Journal of Science and Technology, B, 35*: 49–50.
HADDON, M. 1994: Size-fecundity relationships, mating behaviour and larval release in the New Zealand paddle crab, *Ovalipes catharus* (White, 1843) (Brachyura: Portunidae). *New Zealand Journal of Marine and Freshwater Research 28*. 329–334.
HADDON, M. 1995: Avoidence of post-coital cannibalism in the brachyuran paddle crab *Ovalipes catharus*. *Oceologia 104*: 256–258. [Abstr.]
HADDON, M.; WEAR, R. G. 1993: Seasonal incidence of egg-bearing in the New Zealand paddle crab *Ovalipes catharus* (Crustacea: Brachyura), and its production of multiple egg batches. *New Zealand Journalof Marine and Freshwater Research 27*: 287–293.
HAGGITT, T. 1999: Relationship between the kelp *Ecklonia radiata* and the stipe-boring amphipod *Orchomenella aahu*. New Zealand Marine Sciences Society Conference 1–3 September 1999. Abstracts.
HANSEN, H. J. 1905a: Preliminary report on the Schizopoda collected by H.S.H. Prince Albert of Monaco during the cruise of the 'Princess Alice' in the year 1904. *Bulletin du Musée Océanographique de Monaco 30*: 11–32.
HANSEN, H. J. 1905b: Further notes on Schizopoda. *Bulletin du Musée Océanographique de Monaco 42*: 1–32.
HANSEN, H. J. 1908: Sur quelques Crustacés pélagiques d'Amboine. *Revue Suisse de Zoologie 16*: 157–159.
HANSEN, H. J. 1910: The Schizopoda of the Siboga Expedition. *Siboga-Expedite 37*: 1–123.
HANSEN, H. J. 1911: The genera and species of the order Euphausiacea, with account of remarkable variation. *Bulletin de l'Institut Océanographique de Monaco 210*: 1–54.
HÄNTZSCHEL, W. 1975: Trace Fossils and Problematica. *In*: Teichert, C. (ed.), *Treatise on Invertebrate Paleontology, Part W, Miscellanea, Supplement 1*. Geological Society of America and University of Kansas, Lawrence. 269 p.
HARDING, J. P. 1958: *Bryocamptus stouti* and *Goniocyclops silvestris*; two new species of copepod crustacean from forest litter in New Zealand. *Annals and Magazine of Natural History, ser. 13, 1*: 309–330.
HARDY, A. C. 1956: *The Open Sea. Its Natural History: The World of Plankton*. [The New Naturalist 34.] Collins, London. xvi + 335 p.
HARRISON, K. 1984: Some sphaeromatid isopods (Crustacea) from southern and south-western Australia, with description of a new genus and two new species. *Records of the Western Australian Museum 11*: 259–286.
HARRISON, K.; POORE, G. C. B. 1984: *Serolis* (Crustacea, Isopoda, Serolidae) from Australia with a new species from Victoria. *Memoirs of the Museum of Victoria 45*: 13–31.
HART, D. G.; HART, C. W. 1974: The ostracod family Entocytheridae. *The Academy of Natural Sciences of Philadelphia, Monograph 18*: 1–239.
HARTMANN, G. 1982: Beitrag zur Ostracodenfauna Neuseelands (mit einem Nachtrag zur Ostracodenfauna der Westküste Australiens). *Mitteilungen Hamburgishen Museum Institut 79*: 119–150.
HASSACK, E.; HOLDICH, D. M. 1987: The tubicolous habit amongst the Tanaidacea (Crustacea, Peracarida) with particular reference to deep-sea species. *Zoologica Scripta 16*: 223–233.
HAYAKAWA, Y.; KITTAKA, J.; BOOTH, J. D.; NISHIDA, S.; SEKIGUCHI, H.; SAISHO, T. 1990: Daily settlement of the puerulus stage of the red rock lobster *Jasus edwardsii* at Castlepoint, New Zealand. *Nippon Suisan Gakkaishi 56*: 1703–1716.
HECTOR, J. 1888: Specimens of a large stalked cirripede. *Transactions of the New Zealand Institute 20*: 440.
HELLER, C. 1868: Crustaceen. *Reise der Osterreichischen Fregatte Novara um der Erde 1857–1859, Zoologischer Thiel 2*: 1–280.
HENDERSON, J. R. 1888: Report on the Anomura collected by H.M.S. Challenger during the years 1873–76. *Report on the Scientific Results of the Voyage of H.M.S. Challenger, Zoology 27*: 1–221.
HENDRICKX, M. E.; SALGADO-BARRAGÁN, J. 1991: Los estomatópos (Crustacea: Hoplocarida) del Pacifico Mexicano. *Instituto Cienias del Mar y Limnología, Universidad Nacional Autónoma de México, Publicaciones Especiales 10*: 1–200
HERON, G. A.; BRADFORD-GRIEVE, J. M. 1995: The marine fauna of New Zealand: Pelagic Copepoda: Poecilostomatoida: Oncaeidae. *New Zealand Oceanographic Institute Memoir 104*: 1–57.
HESSLER, R. R.; SCHRAM, F. R. 1984: Leptostraca as living fossils. Pp. 187–191 *in*: Eldredge, N.; Stanley, S. M. (eds), *Living Fossils*. Springer, New York. 291 p.
HESSLER, R. R.; WAKABARA, Y. 2000: *Hampsonellus brasiliensis* n. gen., n. sp., a cephalocarid from Brazil. *Journal of Crustacean Biology 20*: 550–558.
HEWITT, G. C. 1963: Some New Zealand parasitic Copepoda of the family Caligidae. *Transactions of the Royal Society of New Zealand 4*: 61–115.
HEWITT, G. C. 1967: Some New Zealand parasitic Copepoda of the family Pandaridae. *New Zealand Journal of Marine and Freshwater Research 1*: 180–264.
HEWITT, G. C. 1968: Some New Zealand parasitic Copepoda of the family Anthosomidae. *Zoology Publications of Victoria University, Wellington 47*: 1–31.
HEWITT, G. C. 1969: A new species of *Paeonodes* (Cyclopoida: Copepoda) parasitic on New Zealand freshwater fish with a re–examination of *Paeonodes exiguus* Wilson. *Zoology Publications of Victoria University, Wellington 50*: 32–39.
HEWITT, G. C.; HINE, P. M. 1972: Checklist of parasites of New Zealand fishes and of their hosts. *New Zealand Journal of Marine and Freshwater Research 6*: 69–114.
HICKEY, C. W.; MARTIN, L. 1995: Relative sensitivity of five benthic invertebrate species to reference toxicants and resin–acid contaminated

sediments. *Environmental Toxicology and Chemistry 14*: 1401–1409.
HICKMAN, V. V. 1937: The embryology of the syncarid crustacean *Anaspides tasmaniae*. *Papers and Proceedings of the Royal Society of Tasmania 1936*: 1–36.
HICKS, B. J. 1997: Food webs in forest and pasture streams in the Waikato region, New Zealand: a study based on analyses of stable isotopes of carbon and nitrogen and fish gut contents. *New Zealand Journal of Marine and Freshwater Research 31*: 651–664.
HICKS, G. R. F. 1971: Some littoral harpacticoid copepods, including five new species, from Wellingon, New Zealand. *New Zealand Journal of Marine and Freshwater Research 5*: 86–119.
HICKS, G. R. F. 1976: *Neopeltopsis pectinipes*, a new genus and species of seaweed-dwelling copepod (Harpacticoida: Peltidiidae) from Wellington, New Zealand. *New Zealand Journal of Marine and Freshwater Research 10*: 363–370.
HICKS, G. R. F. 1977: Species composition and zoogeography of marine phytal harpacticoid copepods from Cook Strait, and their contribution to the total phytal meiofauna. *New Zealand Journal of Marine and Freshwater Research 11*: 441–469.
HICKS, G. R. F. 1984: Spatio-temporal dynamics of a meiobenthic copepod and the impact of predation–disturbance. *Journal of Experimental Marine Biology and Ecology 81*: 47–72.
HICKS, G. R. F. 1986: Phylogenetic relationships within the harpacticoid copepod family Peltidiidae Sars, including the description of a new genus. *Zoological Journal of the Linnean Society 86*: 349–362.
HICKS, G. R. F. 1988a: Systematics of the Donsiellinae Lang (Copepoda, Harpacticoida). *Journal of Natural History 22*: 639–684.
HICKS, G. R. F. 1988b: Evolutionary implications of swimming behaviour in meiobenthic copepods. *Hydrobiologia 167/168*: 497–504.
HICKS, G. R. F. 1988c: Harpacticoid copepods from biogenic substrata in offshore waters of New Zealand. 1: New species of *Paradactylopodia*, *Stenhelia (St.)* and *Laophonte*. *Journal of the Royal Society of New Zealand 18*: 437–452.
HICKS, G. R. F.; WEBBER, R. 1983: *Porcellidium tapui*, new species (Copepoda Harpacticoida), associated with hermit crabs from New Zealand, with evidence of great morphological variability and a dimorphic male. *Journal of Crustacean Biology 3*: 438–453.
HILLER, N., 1999: A new genus and species of isopod from the late Cretaceous of Marlborough, New Zealand. *Records of the Canterbury Museum 13*: 53–56.
HINE, P. M. 1978: Distribution of some parasites of freshwater eels in New Zealand. *New Zealand Journal of Marine and Freshwater Research 12*: 179–187.
HINE, P. M. 1980: Distribution of helminths in the digestive tracts of New Zealand freshwater eels. 1. Distribution of digeneans. *New Zealand Journal of Marine and Freshwater Research 14*: 329–338.
HINE, P. M.; JONES, J. B. 1994: *Bonamia* and other aquatic parasites of importance to New Zealand. *New Zealand Journal of Zoology 21*: 49–56.
HINE, P. M.; JONES, J. B.; DIGGLES, B. K. 2000: A checklist of the parasites of New Zealand fishes, including previously unpublished records. *NIWA Technical Report 75*: 1–95.
HO, J. S. 1975: Cyclopoid copepods of the family Chondracanthidae parasitic on New Zealand marine fishes. *Publications of the Seto Marine Biological Laboratory 22*: 303–319.
HO, J. S. 1991: Two new species of chondracanthid copepods (Poecilostomatoida) parasitic on commercial fishes in the Pacific. *Publications of the Seto Marine Biological Laboratory 35*: 1–10.
HO, J. S. 2003: Why do symbiotic copepods matter? *Hydrobiologia 453/454*: 1–7.
HO, J. S., DOJIRI, M. 1987: *Mecaderochondria pilgrimi* gen. et sp. nov., a chondracanthid copepod parasitic on a New Zealand marine fish, *Kathetostoma giganteum* Haast (Teleostei: Uranoscopidae). *New Zealand Journal of Marine and Freshwater Research 21*: 615–620.
HODGE, D. 1964: A redescription of *Tenagomysis chiltoni* (Crustacea: Mysidacea) from a freshwater coastal lake in New Zealand. *New Zealand Journal of Science 7*: 387–395.
HØEG, J. T.; LÜTZEN, J. 1993: Comparative morphology and phylogeny of the family Thompsoniidae (Cirripedia, Rhizocephala, Akentrogonida), with descriptions of three new genera and seven new species. *Zoologica Scripta 22*: 363–386.
HØEG, J. T.; LÜTZEN, J. 1995: Life cycle and reproduction in the Cirripedia Rhizocephala. *Oceanography and Marine Biology 33*: 427–485.
HØEG, J. T.; LÜTZEN, J. 1996: Rhizocephala. *Traité de Zoologie 7(2) Crustacea*: 541–568.
HOEK, P. P. C. 1883: Report on the Cirripedia collected by H.M.S. Challenger during the years 1873–1876. *Report on the Scientific Results of the Voyage of H.M.S. Challenger during the years 1873–1876, Zoology 8 (25)*: 1–169.
HOF, C. H. J. 1998: Fossil stomatopods (Crustacea: Malacostraca) and their phylogenetic impact. *Journal of Natural History 32*: 1567–1576.
HOF, C. H. J.; SCHRAM, F. R. 1998: Stomatopods (Crustacea: Malacostraca) from the Miocene of California. *Journal of Palaeontology 72*: 317–331.
HOLDICH, D. M.; JONES, J. A. 1983: Tanaids. *Synopses of the British Fauna, n.s., 27*: 1–98.
HOLTHUIS, L. B.; MANNING, R. B. 1969: Stomatopoda. Pp. 535–552 *in*: Moore, R. C. (ed.), *Treatise on Invertebrate Palaeontology, Part R, Arthropoda 4*. Geological Society of America and University of Kansas, Lawrence.
HOPKINS, C. L. 1967: Breeding in the freshwater crayfish *Paranephrops planifrons* White. *New Zealand Journal of Marine and Freshwater Research 1*: 51–58.
HORN, R.; HARMS, J. 1988: Larval development of *Halicarcinus varius* (Decapoda: Hymenosomatidae). *New Zealand Journal of Marine and Freshwater Research 22*: 1–8.
HORNE, D. J.; SCHÖN, I.; SMITH, R. J.; MARTENS, K. 2005: What are Ostracoda? A cladistic analysis of the extant superfamilies of the subclasses Myodocopa and Podocopa (Crustacea: Ostracoda). Pp. 250–273 *in*: Koenemann, S.; Jenner, R. A. (eds), *Crustacean Issues 16. Crustacea and Arthropoda Relationships*. CRC Press, Taylor & Francis Group, Boca Raton. x + 423 p., 3 pls.
HORNIBROOK, N. deB. 1952a: *In:* Fleming, C. A.: A Foveaux Strait oyster bed. *New Zealand Journal of Science and Technology, B, 34*: 184–185.
HORNIBROOK, N. deB. 1952b: Tertiary and Recent marine Ostracoda of New Zealand. *New Zealand Geological Survey Palaeontological Bulletin 18*: 1–82.
HORNIBROOK, N. deB. 1953: Some New Zealand Tertiary Marine Ostracoda useful in stratigraphy. *Transactions of the Royal Society of New Zealand 81*: 303–311.
HORNIBROOK, N. deB. 1955: Ostracoda in the deposits of Pyramid Valley Swamp. *Records of the Canterbury Museum 6*: 267–278.
HORNIBROOK, N. deB. 1963: The New Zealand family Punciidae. *Micropaleontology 9*: 318–320.
HORWITZ, P. 1989: The faunal assemblage (or pholeteros) of some freshwater crayfish burrows in southwest Tasmania. *Bulletin of the Australian Society for Limnology 12*: 29–36.
HOSIE, A. M. 2008: Four new species and a new record of Cryptoniscoidea (Crustacea: Isopoda: Hemioniscidae and Crinoniscidae) parasitising stalked barnacles from New Zealand. *Zootaxa 1795*: 1–28.
HOSIE, A; AHYONG, S. T. 2008: First records of the giant barnacles *Austromegabalanus nigrescens* (Lamarck, 1818) and *A. psittacus* (Molina, 1782) (Cirripedia: Balanidae) from New Zealand with a key to New Zealand Balanidae. *Zootaxa 1674*: 59–64.
HOWARD, A. D. 1952. Molluscan shells occupied by tanaids. *Nautilus 65*: 74–75.
HUGHES, H. R. 1988: *Importation of Marron* (*Cherax tenuimanus*). Office of the Parliamentary Commissioner for the Environment, Wellington. 23 p.
HUMES, A. G 1994: How many copepods? *Hydrobiologia 292/293*: 1–7.
HUNT, D. 1974: The toxicity of paraquat to *Paracalliope fluviatilis* (Amphipoda). *Mauri Ora 2*: 67–72.
HUNT, M. R.; MILLAR, I. 2001: Cave invertebrate collecting guide. *New Zealand Department of Conservation Technical Series 26*: 1–29.
HURLEY, D.E. 1950: New Zealand terrestrial isopods. *Tuatara 3*: 115–127.
HURLEY, D. E. 1952: Studies on the New Zealand amphipodan fauna No. 1 – The family Cyamidae: the whale-louse *Paracyamus boopis*. *Transactions of the Royal Society of New Zealand 80*: 63–68.
HURLEY, D. E. 1954a: Studies on the New Zealand amphipodan fauna No. 2. The family Talitridae: the fresh-water genus *Chiltonia* Stebbing. *Transactions of the Royal Society of New Zealand 81*: 563–577.
HURLEY, D. E. 1954b: Studies on the New Zealand amphipodan fauna No. 3. The family Phoxocephalidae. *Transactions of the Royal Society of New Zealand 81*: 579–599.
HURLEY, D. E. 1954c: Studies on the New Zealand amphipodan fauna No. 4. The family Gammaridae, including a revision of the freshwater genus *Phreatogammarus* Stebbing. *Transactions of the Royal Society of New Zealand 81*: 601–618.
HURLEY, D. E. 1954d: Studies on the New Zealand amphipodan fauna No. 5. *Pleonexes lessoniae*, a new species of the family Amphithoidae. *Transactions of the Royal Society of New Zealand 81*: 619–626.
HURLEY, D. E. 1954e: Studies on the New Zealand amphipodan fauna No. 6. Family Colomastigidae, with descriptions of two new species of *Colomastix*. *Transactions of the Royal Society of New Zealand 82*: 419–429.
HURLEY, D. E. 1954f: Studies on the New Zealand amphipodan fauna. No. 7. The family Corophiidae, including a new species of *Paracorophium*. *Transactions of the Royal Society of New Zealand 82*: 431–460.
HURLEY, D. E. 1954g: Studies on the New Zealand amphipodan fauna No. 9. The families Acanthonotozomatidae, Pardaliscidae and Liljeborgiidae. *Transactions of the Royal Society of New Zealand 82*: 763–802.
HURLEY, D. E. 1954h: Studies on the New Zealand amphipodan fauna No. 10. A new species of *Cacao*. *Transactions of the Royal Society of New Zealand 82*: 803–811.
HURLEY, D. E. 1955a: Studies on the New Zealand amphipodan fauna No. 8. Terrestrial amphipods of the genus *Talitrus* Latr. *Pacific Science 9*:

144–157.

HURLEY, D. E. 1955b: Pelagic amphipods of the sub–order Hyperiidea in New Zealand waters I.–Systematics. *Transactions of the Royal Society of New Zealand 83*: 119–194.

HURLEY, D. E. 1955c: Studies on the New Zealand amphipodan fauna No. 12. The marine families Stegocephalidae and Amphilochidae. *Transactions of the Royal Society of New Zealand 83*: 195–221.

HURLEY, D. E. 1956: Studies on the New Zealand amphipodan fauna No. 13. Sandhoppers of the genus *Talorchestia*. *Transactions of the Royal Society of New Zealand 84*: 359–389.

HURLEY, D. E. 1957a: Studies on the New Zealand amphipodan fauna No. 14. – The genera *Hyale* and *Allorchestes* (family Talitridae). *Transactions of the Royal Society of New Zealand 84*: 903–933.

HURLEY, D. E. 1957b: Some Amphipoda, Isopoda and Tanaidacea from Cook Strait. *Zoological Publications, Victoria University College 21*: 1–20.

HURLEY, D. E. 1957c: Terrestrial and littoral amphipods of the genus *Orchestia*. Family Talitridae. *Transactions of the Royal Society of New Zealand 85*: 149–199.

HURLEY, D. E. 1961: A checklist and key to the Crustacea Isopoda of New Zealand and Subantartic Islands. *Transactions of the Royal Society of New Zealand (Zoology) 1*: 259–292.

HURLEY, D. E.; COOPER, R. D. 1974: Preliminary description of a new species of *Parawaldeckia* (Crustacea Amphipoda: Lysianassidae) from New Zealand (note). *New Zealand Journal of Marine and Freshwater Research 8*: 563–567.

HURLEY, D. E.; JANSEN, K. P. 1977: The marine fauna of New Zealand: family Sphaeromatidae (Crustacea: Isopoda: Flabellifera). *Memoirs of the New Zealand Oceanographic Institute 63*: 1–95.

HUTTON, F. W. 1875: Descriptions of two new species of Crustacea from New Zealand. *Annals and Magazine of Natural History, ser. 4, 15*: 41–42.

HUTTON, F. W. 1904: *Index Faunae Novae Zealandiae*. Dulau & Co., London for Philosophical Institute of Canterbury. viii + 372 p.

HUYS, R.; BOXSHALL, G. A. 1991: *Copepod Evolution*. The Ray Society, London. 468 p.

HUYS, R.; BOXSHALL, G.A.; LINCOLN, R.J. 1993: The tantulocaridan life cycle: the circle closed? *Journal of Crustacean Biology 13*: 432–442.

HUYS, R.; OHTSUKA, S.; BOXSHALL, G. A. 1994: A new tantulocaridan (Crustacea: Maxillopoda) parasitic on calanoid, harpacticoid and cyclopoid copepods. *Publications of the Seto Marine Biological Laboratory 36*: 197–209.

HUYS, R.; OHTSUKA, S.; BOXSHALL, G. A.; ITO, T. 1992: *Itoitantulus misophricola* gen. et sp. nov.: first record of Tantulocarida (Crustacea: Maxillopoda) in the North Pacific region. *Zoologica Scripta 9*: 875–886.

HYDEN, F. M.; FOREST, J. 1980. An in situ hermit crab from the early Miocene of southern New Zealand. *Palaeontology* 23: 471–474.

INGLES, R. J. 1973. Studies on the composition and distribution of the Mysidacea in Pauatahanui Inlet, Wellington. Unpublished B.Sc. Hon. Thesis, Victoria University of Wellington. 53 p.

JAMES, M. R. 1989: Role of zooplankton in the nitrogen cycle off the west coast of the South Island, New Zealand, winter 1987. *New Zealand Journal of Marine and Freshwater Research 23*: 507–518.

JAMES, P. J.; TONG, L. J. 1998: Feeding technique, critical size and size prefeence of *Jasus Edwardsii* fed cultured and wild mussels. *Marine and Freshwater Research 49*: 151–156.

JAMIESON, C. D. 1977: The feeding ecology of *Mesocyclops leuckarti* Claus. Unpublished M.Sc. thesis, University of Waikato, Hamilton. 126 p.

JAMIESON, C. D. 1980a: The predatory feeding of copepodid III to adult *Mesocyclops leuckarti*. Pp. 518–537 *in*: Kerfoot W.C. (ed.), *Evolution and Ecology of Zooplankton Communities*. [American Society for Limnology and Oceanography Special Symposium Volume 3.] University Press of New England, Hanover. 793 p.

JAMIESON, C. D. 1980b: Observations of the effect of diet and temperature on the rate of development of *Mesocyclops leuckarti* (Claus) (Copepoda: Cyclopoida). *Crustaceana 38*: 145–154.

JAMIESON, C. D. 1986: The effects of temperature and food on naupliar development, growth and metamorphosis in three species of *Boeckella* (Copepoda: Calanoida). *Hydrobiologia 139*: 277–286.

JAMIESON, C. D. 1988: The biogeography of three *Boeckella* species (Copepoda: Calanoida) in New Zealand. *Hydrobiologia 164*: 259–270.

JAMIESON, C. D. 1998: Calanoid copepod biogeography in New Zealand. *Hydrobiologia 367*: 189–197.

JAMIESON, C. D.; BURNS, C. W. 1988: The effects of temperature and food on copepodite development, growth and reproduction in three species of *Boeckella* (Copepoda: Calanoida). *Hydrobiologia 164*: 235–257.

JANSEN, K. P. 1973: Preliminary diagnosis of a new species of marine isopod from Stewart Island. *New Zealand Journal of Marine and Freshwater Research 7*: 261–262.

JANSEN, K. P. 1978: A revision of the genus *Pseudaega* Thomson (Isopoda: Flabellifera: Cirolanidae) with diagnoses of four new species. *Journal of the Royal Society of New Zealand 8*: 143–156.

JARMAN, S. N.; ELLIOTT, N. G. 2000: DNA evidence for morphological and cryptic Cenozoic speciations in the Anaspididae,'living fossils' from the Triassic. *Journal of Evolutionary Biology 13*: 624–633.

JARMAN, S. N.; NICOL, S.; E., N. G.; McMINN, A. 2000: 26S rDNA evolution in the Eumalacostraca and the phylogenetic position of krill. *Molecular Phylogenetics and Evolution 17*: 26–36.

JELLINEK, T.; SWANSON, K. M. 2003: Report on the taxonomy, biogeography and phylogeny of mostly living benthic Ostracoda (Crustacea) from deep-sea samples (Intermediate Water depths) from the Challenger Plateau (Tasman Sea) and Campbell Plateau (Southern Ocean), New Zealand. *Abhandlungen der Senckenbergischen Naturforschenden Gesellschaft Frankfurt am Main 558*: 1–329.

JELLYMAN, D. J. 1989: Occurrence of the nematode *Hedruris spinigera* in the stomachs of freshwater eels. *New Zealand Journal of Marine and Freshwater Research 16*: 185–189.

JILLETT, J. B. 1971: Zooplankton and hydrology of Hauraki Gulf New Zealand. *New Zealand Oceanographic Institute Memoir 53*: 1–103.

JOCQUÉ, M.; BLOM, W. 2009: Mysidae (Mysida) of New Zealand; a checklist, identification key to species and an overview of material in New Zealand collections. *Zootaxa 2304*: 1–20.

JOHNSON, S. C.; KENT, M. L.; MARGOLIS, L. 1997: Crustacean and helminth parasites of seawater-reared salmonids. *Aquaculture Magazine 23(2)*: 40–64.

JONES, G. P. 1988: Ecology of rocky reef fish of north-eastern New Zealand: a review. *New Zealand Journal of Marine and Freshwater Research 22*: 445–462.

JONES, J. B. 1974: New Notodelphyidae (Copepoda: Cyclopoida) from solitary ascidians. *New Zealand Journal of Marine and Freshwater Research 8*: 255–273.

JONES, J. B. 1979: New Notodelphyidae (Copepoda: Cyclopoida) from New Zealand solitary ascidians. *New Zealand Journal of Marine and Freshwater Research 13*: 533–544.

JONES, J. B. 1981: *Abergasilus amplexus* Hewitt 1978 (Ergasilidae: Copepoda) from New Zealand with a description of the male. *New Zealand Journal of Marine and Freshwater Research 15*: 275–278.

JONES, J. B. 1985: A revision of the genus *Hatschekia* (Copepoda: Hatschekiidae). *New Zealand Journal of Zoology 12*: 213–271.

JONES, J. B. 1988a: New Zealand Parasitic Copepoda; genus *Caligus* Müller, 1785 (Siphonostomatoida: Caligidae). *New Zealand Journal of Zoology 15*: 397–413.

JONES, J. B. 1988b: Zoogeography of parasitic Copepoda of the New Zealand region. *Hydrobiologia 167/168*: 623–627.

JONES, J. B. 1991: Parasitic copepods of albacore tuna (*Thunnus alalunga*) in the South Pacific. *Bulletin of the Plankton Society of Japan, Special Volume*: 419–428.

JONES, J. B.; MULHOLLAND, P. J. 2000: *Streams and Ground Waters*. Academic Press, New York. 425 p.

JONES, M. B. 1976: Limiting factors in the distribution of intertidal crabs (Crustacea: Decapoda) in the Avon-Heathcote estuary, Christchurch. *New Zealand Journal of Marine and Freshwater Research 10*: 577–587.

JONES, M. B. 1977: Breeding and seasonal population changes of *Petrolishtes elongatus* (Crustacea, Decapoda, Anomura) at Kaikoura, New Zealand. *Journal of the Royal Society of New Zealand 7*: 259–272.

JONES, M. B. 1978: Aspects of the biology of the big-handed crab, *Heterozius rotundifrons* (Decapoda: Brachyura) from Kaikoura, New Zealand.. *New Zealand Journal of Zoology 5*: 783–794.

JONES, M. B. 1980: Reproductive ecology of the estuarine burrowing mud crab *Helice crassa* (Grapsidae). *Estuarine and Coastal Marine Ecology 11*: 433–443.

JONES, M. B. 1981: Effect of temperature, season, and stage of life cycle on salinity tolereance of the estuarine crab *Helice crassa* Dana (Grapsidae). *Journal of Experimental Biology and Ecology 52*: 271–282.

JONES, M. B. GREENWOOD, J. G. 1982: Water loss of a porcelain crab, *Petrolisthes elongatus* (Milne Edwards, 1837) (Decapoda, Anomura) during atmospheric exposure. *Comparative Biochemisty and Physiology 72A*: 631–36.

JONES, M. B.; GREENWOOD, J. G.; GREENWOOD, J. 1989: Distribution, body size, and brood characteristics of four species of mysids (Crustacea: Peracarida) in the Avon–Heathcote Estuary, New Zealand. *New Zealand Journal of Marine and Freshwater Research 23*: 195–199.

JONES, M.; WINTERBOURN, M. 1978: Adaptation to environment in the mud crab. *NZ Science Teacher 19*: 4–11.

JONES, M. B.; SIMONS, M. J. 1981: Habitat preferences of two estuarine burrowing crabs *Helice crassa* Dana (Grapsidae) and *Macrophthalmus hirtipes* (Jacquinot) (Ocypodidae). *Journal of Experimental Marine Biology and Ecology 56*: 49–62.

JONES, M. B.; SIMONS, M. J. 1982: Water loss of a porcelain crab, *Petrolisthes elongatus* (Milne Edwards, 1837) (Decapoda, Anomura) during atmospheric exposure. *Comparative Biochemistry and Physiology 72A*: 631–636.

JONES, M. B.; SIMONS, M. J. 1983: Latitudinal variation in reproductive characteristics of a mud crab, *Helice crassa* (Grapsidae). *Bulletin of Marine Science 33*: 656–670.

JONES, N. S. 1960: The Cumacea of the Chatham Islands 1954 Expedition. *New Zealand Department of Scientific and Industrial Research Bulletin 139*: 9–11. [*New Zealand Oceanographic Institute Memoir 5*.]

JONES, N. S. 1963: The marine fauna of New Zealand: crustaceans of the Order Cumacea. *New Zealand Oceanographic Institute Memoir 23*: 9–81.

JONES, N. S. 1969: The systematics and distribution of Cumacea from depths exceeding 200 metres. *Galathea Reports 10*: 99–180.

JONES, T. R. 1860: Notes on fossils [from Orakei Creek, Auckland]. *In:* Heaphy, C. *Quarterly Journal of the Geological Society of London 17*: 242–251.

JORDAN, H. 1847: Entdeckung fossiler Crustaceen in Saarbrückenschen Steinkohlengebirge. *Verhandlungen des naturhistorischen Vereins der preussischen Rheinlande und Westfalens 4*: 89–92.

JUBERTHIE, C.; DECU, V. (Eds) 1994–2001: *Encyclopaedia Biospéologica.* Société de Biospéologie/CNRS, Moulis/Académie Roumaine, Bucharest. Vols I (1994) 834 p., II (1998) 1378 p., III (2001) 2294 p.

JUST, J. 2001: Bathyal Joeropsididae (Isopoda: Asellota) from south-eastern Australia, with descriptions of two new genera. *Memoirs of the Museum of Victoria 58*: 297–333.

JUST, J. 2009: Ischyroceridae. *In*: Lowry, J. K. & Myers, A. A. (eds), Benthic Amphipoda (Crustacea: Peracarida) of the Great Barrier Reef, Australia. *Zootaxa 2260*: 1–930.

JUST, J.; WILSON, G. D. F. 2004: Revision of the *Paramunna* complex (Isopoda: Asellota: Paramunnidae). *Invertebrate Systematics 18*: 377–466.

JUST, J.; WILSON, G. D. F. 2006: Revision of southern hemisphere *Austronanus* Hodgson, 1910, with two new genera and five new species of Paramunnidae (Crustacea: Isopoda: Asellota). *Zootaxa 1111*: 21–58.

KANE, J. E. 1962: Amphipoda from waters south of New Zealand. *New Zealand Journal of Science 5*: 295–315.

KARANOVIC, T. 2005: Two new genera and three new species of subterranean cyclopoids (Crustacea, Copepoda) from New Zealand, with redescription of *Goniocyclops sylvestris* Harding, 1958. *Contributions to Zoology 74*: 223–254.

KARAYTUG, S.; BOXSHALL, G. A. 1998: The *Paracyclops fimbriatus*-complex (Copepoda, Cyclopoida): a revision. *Zoosystema 20 (4)*: 563–602.

KEABLE, S. J. 1995: Structure of the marine invertebrate scavenging guild of a tropical reef ecosystem: field studies at Lizard Island, Queensland, Australia. *Journal of Natural History 29*: 27–45.

KEABLE, S. J. 2006: Taxonomic revision of *Natatolana* (Crustacea: Isopoda: Cirolanidae). *Records of the Australian Museum 58*: 133–244.

KELLY, S. 1995: Offshore movements of the spiny lobster *Jasus edwardsii*. *Lobster Newsletter 8(12)*: 11, 14.

KELLY, S.; MacDIARMID, A. B.; BABCOCK, R. C. 1999: Characteristics of spiny lobster, *Jasus edwardsii*, aggregations in exposed reef and sandy areas. *Marine and Freshwater Research 50*: 409–416.

KEMP, S. 1913: An account of the Crustacea Stomatopoda of the Indo-Pacific region, based on the collection in the Indian Museum. *Memoirs of the Indian Museum 4*: 1–217, pls 1–10.

KENSLEY, B. 1978: *Guide to the Marine Isopods of Southern Africa*. Trustees of the South African Museum, Cape Town. 173 p.

KENSLEY, B. 1988: Preliminary observation on the isopod crustacean fauna of Aldabra Atoll. *Bulletin of the Biological Society of Washington 8*: 40–44.

KENSLEY, B.; SCHOTTE, M. 1989: *Guide to the Marine Isopod Crustaceans of the Caribbean*. Smithsonian Institution Press, Washington, D.C., and London. 308 p.

KIEFER, F. 1981: Beitrag zur Kenntnis von Morphologie, Taxonomie und geographischer Verbreitung von *Mesocyclops leuckarti* auctorum. *Archive für Hydrobiologie, Suppl. 62*: 148–190.

KING, M. D. 1958: Close-up photography of small plants and animals. *Tuatara 7*: 63–70.

KIRK, T. W. 1881: Notice of new crustaceans. *Transactions of the Royal Society of New Zealand 13*: 236–237.

KIRKWOOD, J. M. 1982: A guide to the Euphausiacea of the Southern Ocean. *ANARE Research Notes 1*: 1–45.

KITTAKA, J. 1994a: Larval rearing. Pp. 402–423 *in*: Phillips, B. F.; Cobb, J. S.; Kittaka, J. (eds), *Spiny Lobster Management*. Blackwell, Oxford.

KITTAKA, J. 1994b: Culture of phyllosomas of spiny lobster and its application to studies of larval recruitment and acquaculture. *Crustaceana 66*: 258–270.

KITTAKA, J.; ONO, K.; BOOTH, J. D. 1997: Complete development of the green rock lobster, *Jasus verreauxi* from egg to juvenile. *Bulletin of Marine Science 61*: 57–71.

KITTAKA, J.; ONO, K.; BOOTH, J. D.; WEBBER, W. R. 2005: Development of the red rock lobster, *Jasus edwardsii*, from egg to juvenile. *New Zealand Journal of Marine and Freshwater Research 39*: 263–277.

KNIGHT, J. S.; HEARD, R. W. 2006: A new species, *Apseudes larseni* (Crustacea: Tanaidacea), from the marine waters of New Zealand. *Zootaxa 1306*: 57–67.

KNOTT, B.; LAKE, P. S. 1977: Of a wine cellar and psammaspids. *Australian Society for Limnology, Newsletter 15*: 49.

KNOTT, B.; LAKE, P.S . 1980: *Eucrenonaspides oinotheke* gen. et sp.n. (Psammaspidae) from Tasmania, and a new taxonomic scheme for Anaspidacea (Crustacea, Syncarida). *Zoological Scripta 9*: 25–33.

KNOX, G. A.; FENWICK, G. D. 1977: *Chiltoniella elongata* n.gen. et sp. (Crustacea: Cephalocarida) from New Zealand. *Journal of the Royal Society of New Zealand 7*: 425–432.

KNOX, G. A.; FENWICK, G. D. 1978: A quantitative study of the benthic fauna off Clive, Hawke Bay. *University of Canterbury Estuarine Research Unit Report 14*: 1–91.

KOENEMANN, S.; HOLSINGER, J. R. 1999: Phylogenetic analysis of the amphipod family Bogidiellidae s. lat., and revision of taxa above the species level. *Crustaceana 72*: 781–816.

KORNICKER, L. S. 1975: Antarctic Ostracoda (Myodocopina). *Smithsonian Contributions to Zoology 163*: 1–720 [in 2 vols].

KORNICKER, L. S. 1979: The marine fauna of New Zealand: Benthic Ostracoda (Suborder Myodocopina). *New Zealand Oceanographic Institute Memoir 82*: 1–58.

KORNICKER, L. S. 1981: A new bathyal myodocopine ostracode from New Zealand and a key to developmental stages of Sarsiellidae. *New Zealand Journal of Marine and Freshwater Research 15*: 385–390.

KOROVCHINSKY, N. M. 2000: Trends in Cladocera and Copepoda taxonomy. *Arthropoda Selecta 9*: 153–158.

KRÄMER, A. 1895: On the most frequent pelagic copepods and cladoceres of the Hauraki Gulf. *Transactions of the New Zealand Institute, Zoology 27*: 214–233, pls 15–23.

KUDINOVA-PASTERNAK, R. K. 1972: Notes about the tanaidacean fauna (Crustacea, Malacostraca) of the Keramadec Trench. *Complex Research of the Nature of the Ocean. Publications of Moscow University 3*: 257–258.

LAMBERSON, J. O.; DEWITT, T. H.; SWARTZ, R. C. 1992: Assessment of sediment toxicity to marine benthos. *U.S. Environmental Protection Agency Report 600/A-93/108*: 1–32.

LANDIS, C. A.; CAMPBELL, H. J.; BEGG, J.G.; MILDENHALL, D. C.; PATERSON, A. M.; TREWICK, S. A. 2008: The Waipounamu Erosion Surface: questioning the antiquity of the New Zealand land surface and terrestrial fauna and flora. *Geological Magazine 145*: 173–197.

LANG, K. 1934: Marine Harpacticiden von der Campbell-Insel und einigen anderen südlichen Inseln. *Acta Universitatis Lundensis, n.s., 30(14)*: 1–56.

LANG, K. 1968: *Deep-sea Tanaidacea*. Galathea Report 9: 23–209.

LARSEN, K. 2005: Deep-sea Tanaidacea (Peracarida) from the Gulf of Mexico. *Crustaceana Monographs 5*: x, 1–381.

LARSEN, K.; WILSON, G. D. F. 1998: Tanaidomorphan systematics.– Is it obsolete? *Journal of Crustacean Biology 18*: 346–362.

LARSEN, K.; WILSON, G. D. F. 2002: Tanaidacean phylogeny. The first step: The superfamily Paratanaidoidea. *Journal of Zoological Systematics and Evolutionary Research 40*: 205–222.

LAUBITZ, D. R. 1993: Caprellidea (Crustacea: Amphipoda): towards a new synthesis. *Journal of Natural History 27*: 965–976.

LAVROV, D.V.; BROWN, W. M.; BOORE, J. L. 2004: Phylogenetic position of the Pentastomida and (pan)crustacean relationships. *Proceedings of the Royal Society of London, ser. B, 271*: 537–544.

LEBOUR, M.V. 1955: First stage larvae hatched from New Zealand decapod Crustacea. *Annals and Magazine of Natural History, ser. 12, 8*: 43–48.

LEE, C. N.; MORTON, B. 2005: Demography of *Nebalia* sp. (Crustacea: Leptostraca) determined by carrion bait-trapping in Lobster Bay, Cape d'Aguilar Marine Reserve, Hong Kong. *Marine Biology 148*: 149–157.

LEMAITRE, R.; MCLAUGHLIN, P. A. 1992: Descriptions of megalopa and juveniles of *Sympagurus dimorphus* (Studer, 1883), with an account of the Parapaguridae (Crustacea: Anomura: Paguroidea) from Antarctic and Subantarctic waters. *Journal of Natural History 26*: 745–768.

LESSER, J. H. R. 1974: Identification of early larvae of New Zealand spiny and shovel-nosed lobsters (Decapoda, Palinuridae and Scyllaridae). *Crustaceana 27*: 259–277.

LEWIS, M. H. 1972a: Freshwater harpacticoid copepods of New Zealand 1. *Atthyella* and *Elaphoidella* (Canthocamptidae). *New Zealand Journal of Marine and Freshwater Research 6*: 23–47.

LEWIS, M. H. 1972b: Freshwater harpacticoid copepods of New Zealand 2. *Antarctobiotus* (Canthocamptidae). *New Zealand Journal of Marine and Freshwater Research 6*: 277–297.

LEWIS, M. H. 1984: The freshwater Harpactioida of New Zealand: A zoogeographical discussion. *Crustaceana, Suppl. 7*: 305–314.

LILLY, C. 1992: Massacre of the marron: the crushing of an innovative initiative. *North and South, May 1992*: 62–71.

LINCOLN, R. J. 1985: The marine fauna of New Zealand: Deep-sea Isopoda Asellota, family Haploniscidae. *New Zealand Oceanographic Institute Memoir 94*: 1–56.

LINCOLN, R. J.; BOXSHALL, G. A. 1983: A new species of *Deoterthron* (Crustacea: Tantulocarida) ectoparasitic on a deep-sea asellote from New Zealand. *Journal of Natural History 17*: 881–889.

LINCOLN, R. J.; HURLEY, D. E. 1980: *Scutocyamus*

antipodesis n. sp. (Amphipoda: Cyamidae) on Hector's dolphin (*Cephalorhynchus hectori*) from New Zealand. *New Zealand Journal of Marine and Freshwater Research 14*: 295–301.

LINDER, F. 1943: Über *Nebaliopsis typica* G.O. Sars nebst einigen allgemeinen Bemerkungen über die Leptostraken. *Dana Report 25*: 1–38.

LÖRZ, A.-N.; GLENNER, H.; LÜTZEN, J. 2008: First records of rhizocephalans from New Zealand, including first rhizocephalan records from hot vents and cold seeps. *Crustaceana 81*: 1013–1019.

LOWRY, J. K. 1979: New gammaridean Amphipoda from Port Pegasus, Stewart Island, New Zealand. *New Zealand Journal of Zoology 6*: 201–212.

LOWRY, J. K. 1981: The amphipod genus *Cerapus* in New Zealand and subantarctic waters (Corophioidea, Ischyroceridae). *Journal of Natural History 15*: 183–211.

LOWRY, J. K.; FENWICK, G. D. 1982: *Rakiroa*, a new amphipod genus from The Snares, New Zealand (Gammaridea, Corophiidae). *Journal of Natural History 16*: 119–125.

LOWRY, J. K.; FENWICK, G. D. 1983: The shallow-water gammaridean Amphipoda of the subantarctic islands of New Zealand and Australia: Melitidae, Hadziidae. *Journal of the Royal Society of New Zealand 13*: 201–260.

LOWRY, J. K.; STODDART, H. E. 1983a: The amphipod genus *Parawaldeckia* in New Zealand waters (Crustacea, Lysianassoidea). *Journal of the Royal Society of New Zealand 13*: 261–277.

LOWRY, J. K.; STODDART, H. E. 1983b: The shallow-water gammaridean Amphipoda of the subantarctic islands of New Zealand and Australia: Lysianassoidea. *Journal of the Royal Society of New Zealand 13*: 279–394.

LOWRY, J. K.; STODDART, H. E. 1984: Taxonomy of the lysianassoid genera *Phoxostoma* K.H. Barnard, *Conicostoma* Lowry & Stoddart, and *Ocosingo* J.L. Barnard (Amphipoda, Gammaridea). *Crustaceana 47*: 192–208.

LOWRY, J. K.; STODDART, H. E. 1986: Protandrous hermaphrodites among lysianassoid Amphipoda. *Journal of Crustacean Biology 6(4)*: 742–748.

LUCAS, J. S. 1980: Spider crabs of the family Hymenosomatidae (Crustacea; Brachyura) with particular reference to Australian species: systematics and biology. *Records of the Australian Museum 33*: 148–247.

LÜTZEN, J.; GLENNER, H.; LÖRZ, A.-N. 2008: Parasitic barnacles (Cirripedia: Rhizocephala) from New Zealand waters. *New Zealand Journal of Marine and Freshwater Research 43*: 613–621.

MAAS, A.; WALOSZEK, D. 2001: Cambrian derivatives of the early arthropod stem lineage, pentastomids, tardigrades and lobopodians – an 'Orsten' perspective. *Zoologischer Anzeiger 240*: 451–459.

MacDIARMID, A. B. 1991: Seasonal changes in depth distribution , sex ratio and size frequency of spiny lobster *Jasus edwadsii* on a coastal reef in northern New Zealand. *Marine Ecology Progress Series 70*: 129–141.

MacDIARMID, A. B. 1994: Cohabitation in the spiny lobster *Jasus edwardsii* (Hutton, 1875). *Crustaceana 66*: 341–355.

MacDIARMID, A. B.; BUTLER, M. J. IV 1999a: Sperm economy and limitation in spiny lobsters. *Behavioural Ecology and Sociobiology 46*: 14–24.

MacDIARMID, A.B.; BUTLER, M.J. IV 1999b: Sperm limitation in exploited spiny lobsters. *Lobster Newsletter 12(1)*: 2–3.

MacDIARMID, A. B.; HICKEY, B.; MALLER, R. A. 1991: Daily movement patterns of the spiny lobster *Jasus edwardsii* (Hutton) on a shallow reef in northern New Zealand. *Journalof Experimental Marine Biology and Ecology 147*: 185–205.

MacLEOD, C. K.; MOLTSCHANIWSKYJ, N. A.; CRAWFORD, C. M.; FORBES, S. E. 2007: Biological recovery from organic enrichment: some systems cope better than others. *Marine Ecology Progress Series 342*: 41–53.

MAKINO, W.; KNOX, M. A.; DUGGAN, I. C. 2009: Invasion genetic variation and species identity of the calanoid copepod *Sinodiaptomus valkanovi*. *Freshwater Biology*. doi:10.1111/j.1365-2427.2009.02287.x

MALY, E. J. 1984: Dispersal ability and relative abundance of *Boeckella* and *Calamoecia* (Copepoda: Calanoida) in Australian and New Zealand waters. *Oecologia 62*: 173–181.

MALY, E. J. 1991: Co-occurrence patterns among Australian centropagid copepods. *Hydrobiologia 222*: 213–221.

MALY, E. J.; BAYLY, I. A. E. 1991: Factors influencing biogeographic patterns of Australasian centropagid copepods. *Journal of Biogeography 18*: 455–461.

MANNING, R. B. 1966: Notes on some Australian and New Zealand stomatopod Crustacea, with an account of the species collected by the Fisheries Investigation Ship *Endeavour*. *Records of the Australian Museum 27*: 79–137.

MANNING, R. B. 1969: Stomatopod Crustacea of the western Atlantic. *Studies in Tropical Oceanography, Miami 8*: viii, 1–380.

MANNING, R. B. 1977: A monograph of the West African stomatopod Crustacea. *Atlantide Report 12*: 25–181.

MANNING, R. B. 1991: Stomatopod Crustacea collected by the *Galathea* Expedition, 1950–1952, with a list of Stomatopoda known from depths below 400 meters. *Smithsonian Contributions to Zoology 521*: 1–18.

MANNING, R. B. 1995: Stomatopod Crustacea of Vietnam: the legacy of Raoul Serène. *Crustacean Research, Special No. 4*: 1–339.

MANTON, S. M. 1934: On the embryology of the crustacean *Nebalia bipes*. *Philosophical Transactions of the Royal Society of London, B, 223*: 163–234.

MARDEN, M.; SIMES, J. E.; CAMPBELL, H. J. 1987: Two Mesozoic faunas from Torlesse melange terrane, (Ruahine Range), New Zealand, and new evidence for Oretian correlation. *New Zealand Journal of Geology and Geophysics 30*: 389–399.

MARKHAM, J. C. 1985: A review of the bopyrid isopods infesting caridean shrimps in the northwestern Atlantic Ocean, with special reference to those collected during the Hourglass Cruises in the Gulf of Mexico. *Memoirs of the Hourglass Cruises 7*: 1–156.

MARKHASEVA, E. L. 1996: *Calanoid copepods of the Family Aetideidae of the World Ocean*. Russian Academy of Sciences, Zoological Institute, St Petersburg. 331 p.

MARMONIER, P.; VERVIER, P.; GIBERT, J.; DOLE-OLIVIER, M.-J. 1993: Biodiversity in groundwater. *Trends in Ecology and Evolution 8*: 392–395.

MARSHALL, N. J. 1988: A unique colour and polarization system in mantis shrimps. *Nature 333*: 557–560.

MARTIN, J. W.; CASH-CLARK, C. 1995: The external morphology of the onychopod cladoceran' genus *Bythotrephes* (Crustacea, Branchiopoda, Onychopoda, Cercopagididae), with notes on the morphology and phylogeny of the order Onychopoda. *Zoologica Scripta 24*: 61–90.

MARTIN, J. W.; DAVIS, G. E. 2001. An updated classification of the Recent Crustacea. *Natural History Museum of Los Angeles County, Contributions in Science 39*: 1–124.

MARTIN, J. W.; HEYNING, J. E. 1999: First record of *Isocyamus kogiae* Sedlak-Weinstein, 1992 (Crustacea, Amphipoda, Cyamidae) from the eastern Pacific, with comments on morphological characters, a key to the genera of the Cyamidae, and a checklist of cyamids and their hosts. *Bulletin of the Southern California Academy of Sciences 98*: 26–38.

MARTIN, J.; VETTER, E. W.; CASH-CLARK, C. E. 1996: Description, external morphology, and natural history observations of *Nebalia hessleri*, new species (Phyllocarida: Leptostraca), from Southern California, with a key to the extant families and genera of the Leptostraca. *Journal of Crustacean Biology 16*: 347–372.

MAUCHLINE, J. 1980: The biology of mysids and euphausiids. *Advances in Marine Biology 18*: 1–681.

MAUCHLINE, J. 1984: Euphausiid, stomatopod and leptostracan crustaceans, keys and notes for the identification of the species. *Synopses of the British Fauna, n.s., 30*: vii, 1–91.

MAUCHLINE, J. 1998: *The Biology of Calanoid Copepods*. Academic Press, London. 710 p.

MAUCHLINE, J.; FISHER, L. R. 1969: The biology of euphausiids. *Advances in Marine Biology 7*: 1–454.

MAUCHLINE, J.; MURANO, M. 1977: World list of the Mysidacea, Crustacea. *Journal of the Tokyo University of Fisheries 64*: 39–88.

McCAIN, J. C. 1968: The Caprellidae (Crustacea: Amphipoda) of the western North Atlantic. *United States National Museum Bulletin 278*: 1–147.

McCAIN, J. C. 1969: New Zealand Caprellidae (Crustacea: Amphipoda). *New Zealand Journal of Marine and Freshwater Research 3*: 286–295.

McCAIN, J. C. 1979: A new caprellid (Crustacea: Amphipoda) associated with a starfish from Antipodes Island. *New Zealand Journal of Marine and Freshwater Research 13*: 471–473.

McCAIN, J.C.; STEINBERG, J. E. 1970: Amphipoda I. Caprellidea I. Fam. Caprellidae. Pp. 1–78 *in*: Gruner, H.-E.; Holthuis, L. B. (eds), *Crustaceorum Catalogus*. W. Junk, The Hague. 80 p.

McCLATCHIE, S.; HUTCHINSON, D.; NORDIN, K. 1989: Aggregation of avian predators and zooplankton prey in Otago shelf waters, New Zealand. *Journal of Plankton Research 11*: 361–374.

McCLATCHIE, S.; JAQUIERY, P.; KAWACHI, R.; PILDITCH, C. 1991: Grazing rates of *Nyctiphanes australis* (Euphausiacea) in the laboratory and Otago Harbour, New Zealand, measured using three independent methods. *Continental Shelf Research 11*: 1–2.

McCLATCHIE, S.; JILLETT, J.B.; GERRING, P. 1991: Observations of gulls foraging on beach-stranded plankton in Otago Harbour, New Zealand. *Limnology and Oceanography 36*: 1195–1200.

McCLATCHIE, S.; KAWACHI, R.; DALLEY, D.E. 1990: Epizoic diatoms on the euphausiid *Nyctiphanes australis*: consequences for gut pigment analysis of whole krill. *Marine Biology 104*: 227–232.

McCOY, J. L. 1983: Movements of rock lobsters, *Jasus edwardsii* (Decapoda: Palinuridae), tagged near Stewart Island, New Zealand. *New Zealand Journal of Marine and Freshwater Research 17*: 357–366.

McDOWALL, R. M. 1968: *Galaxias maculatus* (Jenyns), the New Zealand whitebait. *Fisheries Research Bulletin, n.s., 2*: 1–84.

McDOWALL, R. M. 1990: *New Zealand Freshwater Fishes*. Heinemann Reed, Wellington. 553 p.

McLAUGHLIN, P. A. 1980: *Comparative Morphology of Recent Crustacea*. W. H. Freeman and Co., San Fransisco.

McLAY, C. L. 1988: Crabs of New Zealand. *Leigh Laboratory Bulletin No. 22*: 1–463.

McLAY, C. 2007: New crabs from hydrothermal

vents of the Kermadec Ridge submarine volcanoes, New Zealand: *Galdalfus* gen. nov. (Bythograeidae) and *Xenograpsus* (Varunidae) (Decapoda: Brachyura). *Zootaxa 1524*: 1–22.

McLAY, C. L.; FELDMANN, R. M.; MacKINNON, D. I. 1995: New species of Miocene spider crabs from New Zealand, and a partial cladistic analysis of the genus *Leptomithrax* Miers, 1876 (Brachyura: Majidae). *New Zealand Journal of Geology and Geophysics 38*: 299–313.

McLAY, C. L.; McQUEEN, D. J. 1995: Intertidal zonation of *Cyclograpsus lavauxi* H. Milne Edwards, 1853 (Brachyura: Grapsidae) along the coast of the South Island of New Zealand. *Crustacean Research 24*: 49–64.

McWILLIAM, P. S.; PHILLIPS, B. F.; KELLY, S. 1995: Phyllosoma larvae of *Scyllarus* species (Decapoda, Scyllaridae) from the shelf waters of Australia. *Crustaceana 68*: 537–566.

MELROSE, M. J. 1975: The marine fauna of New Zealand: Family Hymenosomatidae (Crustacea, Decapoda, Brachyura). *New Zealand Oceanographic Institute Memoir 34*: 1–123.

MENZIES, R. J. 1957: The marine borer family Limnoriidae (Crustacea, Isopoda). Part I. Northern and Central America: systematics, distribution, and ecology. Part II: Additions to the systematics. *Bulletin of Marine Science of the Gulf and Caribbean 7*: 101–200.

MEYER-ROCHOW, V. B.; MEHA, W. P. 1994: Tidal rhythm and the role of vision in shelter-seeking behaviour of the half-crab *Petrolisthes elongatus* (Crustacea; Anomura; Porcellanidae). *Journalof the Royal Society of New Zealand 24*: 423–427.

MEYER-ROCHOW, V. B.; REID, W. A. 1994: The eye of the New Zealand freshwater crab *Halicarcinus lacustris,* and some eco-physiological predictions based on eye anatomy. *Journal of the Royal Society of New Zealand 24*: 133–142.

MESIBOV, B. 2000: Anaspides! *Invertebrata* (Queen Victory Museum & Art Gallery, Tasmania] *18*: 3.

MIERS, E. J. 1876: Catalogue of the stalk and sessile-eyed Crustacea of New Zealand. *Colonial Museum and Geological Department of New Zealand, National History Publication 10*: 1–133.

MIERS, E. J. 1886: Report on the Brachyura collected by HMS *Challenger* during the years 1873–76. *Report on the Scientific Results of the Voyage of HMS* Challenger, *Zoology 17*: 1–221.

MILNE EDWARDS, H. 1834–40: *Histoire Naturelle des Crustacés comprenant l'Anatomie, la Physiologie et la Classification de ces Animaux*. 3 vols + atlas. *Librairie Encyclopédique de Roret, Paris.*

MILNE EDWARDS, H. 1837: *Histoire Naturelle des Crustacés*. Librairie Encyclopedique de Roret, Paris. Vol. 2, 531 p.

MITCHELL, S. F. 1975: Some effects of agricultural development and fluctuations in water level on the phytoplankton productivity and zooplankton of a New Zealand reservoir. *Freshwater Biology 5*: 547–562.

MOORE, P. G. 1983a: *Pagurisaea schembrii* gen. et sp.n. (Crustacea, Amphipoda) associated with New Zealand hermit crabs, with notes on *Isaea elmhirsti* Patience. *Zoologica Scripta 12*: 47–56.

MOORE, P. G. 1983b: A revision of the *Haplocheira* group of genera (Amphipoda: Aoridae). *Zoological Journal of the Linnean Society 79*: 179–221.

MOORE, P. G. 1985: A new deep water species of Amphipoda (Crustacea) discovered off Otago, New Zealand and a note on another little known species. *Zoological Journal of the Linnean Society 83*: 229–240.

MOREIRA, J.; DÍAZ-AGRAS, G.; CANDÁS, M.; SEÑARÍS, M. P.; URGORRI, V. 2008: Leptostracans (Crustacea: Phyllocarida) from the Ría de Ferrol (Galicia, NW Iberian Peninsula), with description of a new species of *Nebalia* Leach, 1814. *Scientia Marina 73*: 269–285.

MORIMOTO, Y. 1977: A new *Stygocaris* (Syncarida, Stygocarididae) from New Zealand. *Bulletin of the National Science Museum, Tokyo (A), Zoology 3*: 19–24.

MORLEY, M. S.; HAYWARD, B. W. 2007: Intertidal and shallow-water Ostracoda of the Waitemata Harbour, New Zealand. *Records of the Auckland Museum 44*: 17–32.

MORRIS, M.; STANTON, B.; NEIL, H. 2001: Subantarctic oceanography around New Zealand: preliminary results from an ongoing survey. *New Zealand Journal of Marine and Freshwater Research 35*: 499–519.

MORRISEY, D. J.; DEWITT, T. H.; ROPER, D. S.; WILLIAMSON, R. B. 1999: Variation in the depth and morphology of the mud crab *Helice crassa* among different types of intertidal sediment in New Zealand. *Marine Ecology Progress Series 182*: 231–242.

MORTON, D. W. 1985: Revision of the Australian Cyclopidae (Copepoda: Cyclopoida). I. *Acanthocyclops* Kiefer, *Diacyclops* Kiefer and *Australocyclops* gen.nov. *Australian Journal of Marine and Freshwater Research 36*: 615–634.

MORTON, J. 2004: *Seashore Ecology of New Zealand and the Pacific*. David Bateman, Auckland. 504 p.

MORTON, J.; MILLER, J. 1968: *The New Zealand Sea Shore.* Collins, London and Auckland. 638 p.

MOSS, G. A.; TONG, L. J.; ILLINGWORTH, J. 1999: Effects of light levels and food density the growth and survival of early stage phyllosoma larvae of the rock lobster *Jasus edwardsii. Marine and Freshwater Research 50*: 129–134.

MULCAHY, D.; KLAYBOR, D.; BATTS, W. N. 1990: Isolation of infectious hematopoietic necrosis virus from a leech (*Piscicola salmositica*) and a copepod (*Salmincola* sp.) ectoparasites of sockeye salmon *Oncorhynchus nerka*. *Diseases of Aquatic Organisms 8*: 29–34.

MÜLLER, H.-G. 1993: *World Catalogue and Bibliography of the recent Mysidacea*. Laboratory for Tropical Ecosystems Research & Information Service, Wetzler. 491 p.

MURDOCH, R. C. 1989: The effects of a headland eddy on surface macro-zooplankton assemblages north of Otago Peninsula, New Zealand. *Estuarine, Coastal and Shelf Science 29*: 361–383.

MURDOCH, R. C. 1990: Diet of hoki (*Macruronus novaezeladiae*) off Westland, New Zealand. *New Zealand Journal of Marine and Freshwater Research 2*: 519–527.

MURDOCH, R. C.; QUIGLEY, B. 1994: A patch study of mortality, growth and feeding of the larvae of the southern gadoid *Macruronus novaezelandiae. Marine Biology 121*: 23–33.

MYERS, A. A. 1981: Studies on the genus *Lembos* Bate. X. Antiboreal species. *L. pertinax* sp. nov., *L. acherontis* sp. nov., *L. hippocrenes* sp. nov., *L. chiltoni* sp. nov. *Bollettino del Museo Civico di Storia Naturale Verona 8*: 85–111.

MYERS, A. A.; LOWRY, J. K. 2003: A phylogeny and a new classification of the Corophioidea Leach, 1814 (Amphipoda). *Journal of Crustacean Biology 23*: 443–485.

NAIRN, H. J. 1998. *Fish fauna of the Avon–Heathcote Estuary, Christchurch.* Unpublished MSc thesis, (Zoology), University of Canterbury, Christchurch. 73 p.

NEWMAN, W. A. 1979: On the biogeography of balanomorph barnacles of the southern ocean including new balanid taxa: a subfamily, two genera and three species. *Proceedings of the International Symposium on Marine Biogeography and Evolution in the Southern Hemisphere. New Zealand Department of Scientific and Industrial Research Information Series 137*: 279–306.

NEWMAN, W. A. 1987: Evolution of cirripedes and their major groups. Pp. 3–42 *in*: Southward, A. J. (ed.), *Crustacean Issues 5: Barnacle Biology*. A.A. Balkema Publishers, Rotterdam. 443 p.

NEWMAN, W. A. 1991: Origins of Southern Hemisphere endemism, especially among marine Crustacea. *Memoirs of the Queensland Museum 31*: 51–76.

NEWMAN, W. A. 1996: Sous-Classe des Cirripèdes (Cirripedia Burmeister, 1834). Super-Ordres des Thoraciques et des Acrothoraciques (Thoracica Darwin, 1854 – Acrothoracica Gruvel, 1905). *Traité de Zoologie, Anatomie, Systématique, Biologie 7(2)*: 453–540.

NEWMAN, W. A.; ROSS, A. 1971: Antarctic Cirripedia. *American Geophysical Union Antarctic Research Series 14*: 1–209 p.

NEWMAN, W. A.; ZULLO, V. A.; WITHERS, T. H. 1969: Cirripedia. Pp. 206–295 *in*: Moore, R. C. (ed.), *Treatise on Invertebrate Paleontology, Part R, Arthropoda 4, 1.* Geological Society of America and University of Kansas, Lawrence.

NG, P. K. L.; GUINOT, D.; DAVIE, J. F. 2008: Systema Brachyurorum: Part 1. An annotated checklist of extant brachyuran crabs of the world. *The Raffles Bulletin of Zoology 2008 17*: 1–286.

NICHOLLS, G. E. 1938: Amphipoda Gammaridea. *Australasian Antarctic Expedition 1911–14, Scientific Reports, Series C, Zoology and Botany 2(4)*: 1–145.

NICHOLLS, G. E. 1944: The Phreatoicoidea. Part II. The Phreatocoidae. *Papers and Proceedings of the Royal Society of Tasmania 1943*: 1–156.

NICOL, S.; ENDO, Y. 1997. Krill fisheries of the world. *FAO Fisheries Technical Paper 367*: ix, 1–100.

NIPPER, M. G.; ROPER, D. S. 1995: Growth of an amphipod and a bivalve in uncontaminated sediments: implications for chronic toxicity assessments. *Marine Pollution Bulletin 31*: 424–430.

NIPPER, M. G.; ROPER, D. S.; WILLIAMS, E. K.; MARTIN, M. L.; VAN DAM, L. F.; MILLS, G. N. 1998: Sediment toxicity and benthic communities in mildly contaminated mudflats. *Environmental Toxicology and Chemistry 17*: 502–510.

NISHIDA, S.; TAKAHASHI, Y.; KITTAKA, J. 1995: Structural changes in the hepatopancreas of the rock lobster *Jasus edwardsii* (Crustacea: Palinuridae) during development from the puerulus to post-puerulus. *Marine Biology 123*: 837–844.

NOODT, W. 1964: Natürliches System und Biogeographie der Syncarida (Crustacea Malacostraca). *Gewässer und Abwässer 37/38*: 77–186.

NORTHCOTE, T. G.; CHAPMAN, M. A. 1999: Dietary alterations in resident and migratory New Zealand common smelt (*Retropinna retropinna*) in lower Waikato lakes after two decades of habitat change. *New Zealand Journal of Marine and Freshwater Research 33*: 425–436.

O'DRISCOLL, R. L. 1998a: Feeding and schooling behaviour of barracouta (*Thyrsites atun*) off Otago, New Zealand. *Marine and Freshwater Research 49*: 19–24.

O'DRISCOLL, R. L. 1998b: Description of spatial pattern in seabird distributions along line transects using neighbour K statistics. *Marine Ecology Progress Series 165*: 81–94.

O'DRISCOLL, R. L.; McCLATCHIE, S. 1998: Spatial distribution of planktivorous fish schools in relation to krill abundance and local hydrography off Otago, New Zealand. *Deep-Sea Research II 45*: 1295–1325.

OHTSUKA, S.; BOXSHALL, G. A. 1998: Two new genera of Tantulocarida (Crustacea) infesting asellote isopods and siphonostomatoid copepods from western Japan. *Journal of Natural History 32*:

683–699.

ØKSNEBJERG, B. 2000: The Rhizocephala (Crustacea: Cirripedia) of the Mediterranean and Black seas: taxonomy, biogeography, and ecology. *Israel Journal of Zoology 46*: 1–102.

OLESEN, J. 1998: A phylogenetic analysis of the Conchostraca and Cladocera (Crustacea, Branchiopoda, Diplostraca). *Zoological Journal of the Linnean Society 122*: 491–536.

OLESEN, J. 1999: A new species of *Nebalia* (Crustacea, Leptostraca) from Unguja Island (Zanzibar), Tanzania, East Africa, with a phylogenetic analysis of the leptostracan genera. *Journal of Natural History 33* : 1789–1810.

OLESEN, J. 2000: An udated phylogeny of the Conchostraca-Cladocera clade (Branchiopoda, Diplostraca). *Crustaceana 73:* 869–886.

OLESEN, J.; RICHTER, S.; SCHOLTZ, G. 2001: The evolutionary transformation of phyllopodous to stenopodous limbs in the Branchiopoda (Crustacea)– Is there a common mechanism for early limb development in arthropods? *International Journal of Developmental Biology 45*: 869–876.

O'SHEA, S.; McKNIGHT, D.; CLARK, M. 1999: Bycatch – the common, unique and bizarre. *Seafood New Zealand 7(6)*: 45–51.

O'SHEA, S.; RAETHKE, N.; CLARK, M. 2000: *Bathysquilla microps* – a spectacular new deepsea crustacean from New Zealand. *Seafood New Zealand 8(9)*: 36.

OVENDEN, J. R.; BRASHER, D. J.; WHITE, R. W. G. 1992: Mitochondrial DNA analysis of the red rock lobster *Jasus edwardsii* supports an apparent absence of population subdivision throughout Australasia. *Marine Biology 112*: 319–326.

OVENDEN, J. R.; BRASHER, D. J. 1994: Stock identity of the red (*Jasus edwardsii*) and green (*J. verreauxi*) rock lobsters inferred from mitochondrial DNA analysis. Pp. 230–249 *in*: Phillips, B. F., Cobb, J. S., Kittaka, J. (eds), *Spiny Lobster Management*. Fishing News Books, Oxford, England. xxiii + 550 p.

PACKARD, A. S. 1885: The Syncarida, a group of Carboniferous Crustacea. *American Naturalist 19*: 700–703.

PACKARD, A. S. 1886: On the Syncarida, a hitherto undescribed synthetic group of extinct malacostracous Crustacea. *Memoirs of the National Academy of Sciences 3*: 123–128.

PACKER, H. A. 1983: Larval morphology of some New Zealand shallow water shrimps (Crustacea, Decapoda, Caridea) of the families Crangonidae, Hippolytidae and Palaemonidae. Unpublished MSc thesis, Victoria University of Wellington.

PACKER, H. A. 1985: A guide to the larvae of New Zealand's shallow water Caridea (Crustacea, Decapoda, Natantia). *Zoology publications from Victoria University of Wellington 78*: 1–16.

PAGE, R. D. M. 1985: Review of the New Zealand Bopyridae (Crustacea: Isopoda: Epicaridea). *New Zealand Journal of Zoology 12*: 185–212.

PALMER, J. D.; WILLIAMS, B. G. 1993: Comparative studies of tidal rhythms. XII: persistent photoaccumulation and locomotor rhythms in the crab *Cyclograpsus lavauxi*. *Marine Behaviour and Physiology 22*: 119–129.

PALMER, P. L. 1995: Occurrence of a New Zealand pea crab, *Pinnotheres novavezelandiae,* in five species of surf clams. *Marine and Freshwater Research 46*: 1071–1075.

PALMER, P. L. 1997: A new species of ascothoracid parasite (Maxillopoda) from the Otago Shelf, New Zealand, and a new host record. *Crustaceana 70*: 769–779.

PARK, T. 1995: Taxonomy and distribution of the marine calanoida copepod family Euchaetidae. *Bulletin of the Scripps Institution of Oceanography 29*: 1–203.

PERCIVAL, E. 1937: New species of Copepoda from New Zealand Lakes. *Records of the Canterbury Museum 4*: 169–175, pls 21–24.

PIKE, R. B.; WEAR, R. G. 1969: Newly hatched larvae of the genera *Gastroptychus* and *Uroptychus* (Crustacea, Decapoda, Galatheidea) from New Zealand waters. *Biological Sciences 11*: 189–195.

PIKE, R. B.; WILLIAMSON, D. I. 1966: The first zoeal stage of *Campylonotus rathbunae* Schmitt and its bearing on the systematic position of the Campylonotidae (Decapoda, Caridea). *Transactions of the Royal Society of New Zealand Zoology 7*: 209–213.

PILGRIM, R. L. C. 1985: Parasitic Copepoda from marine coastal fishes in the Kaikoura–Banks Peninsula region, South Island, New Zealand, with a key to their identification. *Mauri Ora 12*: 13–53.

POORE, G. C. B. 1981: Marine Isopoda of the Snares Islands, New Zealand – 1. Gnathiidea, Valvifera, Anthuridea, and Flabellifera. *New Zealand Journal of Zoology 8*: 331–348.

POORE, G. C. B. 1985: *Basserolis kimblae,* a new genus and species of isopod (Serolidae) from Australia. *Journal of Crustacean Biology 5*: 175–181.

POORE, G. C. B. 1987: *Serolina,* a new genus for *Serolis minuta* Beddard (Crustacea: Isopoda: Serolidae) with descriptions of eight new species from Australia. *Memoirs of the National Museum of Victoria 48*: 141–189.

POORE, G. C. B.: 1994. Marine biogeography of Australia. Pp. 189–213 *in*: Hammond, L. S.; Synnot, R. (eds), *Marine Biology.* Longman Cheshire, Melbourne.

POORE, G. C. B. 2001a: Families and genera of Isopoda Anthuridea. *Crustacean Issues 13*: 63–173.

POORE, G. C. B. 2001b: Isopoda Valvifera: diagnoses and relationships of the families. *Journal of Crustacean Biology 21*: 205–230.

POORE, G. C. B. 2002: Crustacea: Malacostraca: Syncarida, Peracarida: Isopoda, Tanaidacea, Mictacea, Thermosbaenacea, Spelaeogriphacea. *In*: Houston, W. W. K.; Beesley, P. L. (eds), *Zoological Catalogue of Australia*. CSIRO Publishing, Melbourne. Vol. 19.2A, xii + 434 p.

POORE, G. C. B. 2005: Supplement to the 2002 catalogue of Australian Crustacea: Malacostraca – Syncarida and Peracarida (Volume 19.2A): 2002–2004. *Museum Victoria Science Reports 7*: 1–15.

POORE, G. C. B.; BARDSLEY, T. M. 1992: Austrarcturellidae (Crustacea: Isopoda: Valvifera), a new family from Australasia. *Invertebrate Taxonomy 6*: 843–908.

POORE, G. C. B.; BRANDT, A. 2001: *Plakarthrium australiense*, a third species of Plakarthriidae (Crustacea: Isopoda). *Memoirs of Museum Victoria 58*: 373–382.

POORE, G. C. B.; JUST, J.; COHEN, B. F. 1994: Composition and diversity of Crustacea Isopoda of the southeastern Australian continental slope. *Deep-Sea Research 41*: 677–693.

POORE, G. C. B.; LEW TON, H. M. 1990: The Holognathidae (Crustacea: Isopoda: Valvifera) expanded and redefined on the basis of body-plan. *Invertebrate Taxonomy 4*: 55–80.

POORE, G.C.B.; LEW TON, H.M. 1993: Idoteidae of Australia and New Zealand (Crustacea: Isopoda: Valvifera). *Invertebrate Taxonomy 7*: 197–278.

POORE, G. C. B.; STOREY, M. 1999: Soft sediment Crustacea of Port Phillip Bay. *Centre for Research on Introduced Marine Pests, CSIRO Marine Research, Technical Report 20*: 150–170.

POULSEN, E. M. 1962: Ostracoda – Myodocopa, 1: Cypridiniformes – Cyprinidae. *Dana Report 57*: 1–414.

POULSEN, E. M. 1965: Ostracoda – Myodocopa, 2: Cypridiniformes – Rutidermatidae, Sarsiellidae and Asteropidae. *Dana Report 65*: 1–484.

PRINCE, P. A.; MORGAN, R. A. 1987: Diet and feeding ecology of the Procellariiformes. Pp. 135–172 *in* Croxall, J. P. (ed.), *Seabirds: Feeding Ecology and Role in Marine Ecosystems.* Cambridge University Press, Cambridge. 408 p.

PROBERT, P. K.; GROVE, S. L. 1998: Macrobenthic assemblages of the continental shelf and upper slope off the west coast of South Island, New Zealand. *Journal of the Royal Society of New Zealand 28*: 259–280.

PYNE, R. R. 1972: Larval development and behaviour of the mantis shrimp *Squilla armata* Milne Edwards (Crustacea: Stomatopoda). *Journal of the Royal Society of New Zealand 2*: 121–146.

QUOY, J. R. E.; GAIMARD, J. P. 1834: *Voyage d'Astrolabe.* Zoologie 3, Mollusques: 623–643.

RAINBOW, P. S.; EMSON, R. H.; SMITH, B. D.; MOORE, P. G.; MLADENOV, P.V. 1993: Talitrid amphipods as biomonitors of trace metals near Dunedin, New Zealand. *New Zealand Journal of Marine and Freshwater Research 27*: 201–207.

RAINER, S. F.; UNSWORTH, P. 1991: Ecology and production of *Nebalia* sp. (Crustacea: Leptostraca) in a shallow-water seagrass community. *Australian Journal of Marine and Freshwater Research 42*: 53–68.

REIMER, A. A. 1975: Description of a *Tetraclita stalactifera panamensis* community on a rock intertidal Pacific shore of Panama. *Marine Biology 35*: 225–238.

RICHARDSON, L. R. 1949a: A guide to the brachyrhynchous crabs. *Tuatara 2*: 29–36.

RICHARDSON, L.R. 1949b. A guide to the Oxyrhyncha, Oxystomata and lesser crabs. Tuatara *2*: 58–69.

RICHARDSON, L. R.; DELL, R. K. 1964: A new crab of the genus *Trichopeltarion* from New Zealand. *Transactions of the Royal Society of New Zealand 4*: 145–151.

RICHARDSON, L. R.; YALDWYN, J. C 1958: A guide to the natant Decapoda Crustacea (shrimps and prawns) of New Zealand *Tuatara 7*: 17–41.

RITCHIE, L. D. 1970: Southern spider crab (*Jacquinotia edwardsii* (Jacquinot, 1853)) survey –Auckland Islands and Campbell Island 30/1/70-23/2/70. *New Zealand Marine Department Fisheries Technical Report 52*: 1–111.

RITCHIE, L. D. 1971: Commercial fishing for southern spider crab (*Jacquinotia edwardsii*) at the Auckland Islands, October 1971. *New Zealand Marine Department Fisheries Technical Report 101*: 1–95.

RIVIER, I. K. 1998: *The predatory Cladocera (Onychopoda: Podonidae, Polyphemidae, Cercopagidae) and Leptodorida of the world.* Backhuys Publishing, Leiden. 213 p.

ROBERTS, P. E. 1971: Zoea larvae of *Pagurus campbelli* Filhol, 1885, from Perseverance Harbour, Campbell Island (Crustacea, Decapoda, Paguridae). *Journal of the Royal Society of New Zealand 1*: 187–196.

ROBERTS, P. E. 1972: The Plankton of Perseverence Harbour, Campbell Island, New Zealand. *Pacific Science 26*: 296–309.

ROBERTS, P. E. 1973: Larvae of *Munida subrugosa* White (1847) from Perseverance Harbour, Campbell Island. *Journal of the Royal Society of New Zealand 3*: 393–408.

ROBERTSON, D. A.; ROBERTS, P. E.; WILSON, J. B. 1978: Mesopelagic faunal transition across the Subtropical Convergence east of New Zealand. *New Zealand Journal of Marine and Freshwater Research 12*: 295–312.

ROLFE, W. D. I. 1969: Phyllocarida. Pp. R296–R331 *in*: Moore, R. C. (ed.), *Treatise on Invertebrate Paleontology. Part R. Arthropoda 4 (1).* University of Kansas and Geological Society of America, Lawrence.

ROLFE, W. D. I. 1981: Phyllocarida and the origin of the Malacostraca. *Geobios 14*: 17–24.

ROPER, D. S.; SIMONS, M. J.; JONES, M. B. 1983: Distribution of zooplankton in the Avon–Heathcote Estuary, Christchurch. *New Zealand Journal of Marine and Freshwater Research 17*: 267–278.

ROSECCHI, E.; TRACEY, D. M.; WEBBER, W. R. 1988: Diet of orange roughy, *Hoplostethus atlanticus* (Pisces: Trachichthyidae) on the Challenger Plateau, New Zealand. *Marine Biology 99*: 293–306.

ROSSETTI, G.; EAGAR, S. H.; MARTENS, K. 1998: On two new species of the genus *Darwinula* (Crustacea, Ostracoda) from New Zealand. *Italian Journal of Zoology 65*: 325–332.

ROWETT, H. C. Q. 1943: The gut of Nebaliacea. *Discovery Reports* 23: 1–17.

ROWETT, H. C. Q. 1946: A comparison of the feeding mechanisms of *Calma glaucoides* and *Nebaliopsis typica*. *Journal of the Marine Biological Association of the United Kingdom 26*: 352–357.

RUSSELL, B. C. 1983: The food and feeding habits of rocky reef fish of north-eastern New Zealand. *New Zealand Journal of Marine & Freshwater Research 17*: 121–145.

RYAN, P. A. 1986: Seasonal and size-related changes in the food of the short-finned eel, *Anguilla australis* in Lake Ellesmere, Canterbury, New Zealand. *Environmental Biology of Fishes 15*: 47–58.

RYFF, M. R.; VOLLER, R. W. 1976: Aspects of the southern spider crab (*Jacquinotia edwardsii*) fishery of southern New Zealand Islands and Pukaki Rise. *New Zealand Marine Department Fisheries Technical Report 143*: 1–65.

SAGAR, P. M.; GLOVA, G. J. 1995: Prey availability and diet of juvenile brown trout (*Salmo trutta*) in relation to riparian willows (*Salix* spp.) in three New Zealand streams. *New Zealand Journal of Marine and Freshwater Research 29*: 527–537.

SAGAR, P. M.; GLOVA, G. J. 1998: Diel feeding and prey selection of three size classes of short-finned eel (*Anguilla australis*) in New Zealand. *Marine and Freshwater Research 49*: 421–428.

SAINTE-MARIE, B.; LAMARCHE, G. 1985: The diets of six species of the carrion-feeding lysianassid amphipod genus *Anonyx* and their relation with morphology and swimming behaviour. *Sarsia 70*: 119–126.

SANDERS, H. L. 1955: The Cephalocarida, a new subclass of Crustacea from Long Island Sound. *Proceedings of the National Academy of Science 41*: 61–66.

SARS, G. O. 1865: Norges ferskvandskrebsdyr. Første afsnit. Branchiopoda I. Cladocera Ctenopoda (Fam. Sididae og Holopedidae). *Universitetsprogram Kristiania for 1ste halvår 1863*: 1–71, pls 1–4.

SARS, G. O. 1883: Preliminary notices on the Schizopoda of HMS *Challenger* Expedition. *Forhandlinger i Videnskabs-Selskabet i Kristiania 7*: 1–43.

SARS, G. O. 1885: Report on Schizopoda collected by HMS *Challenger* during the years 1873–1876. *Reports on the Scientific Results of the voyage of HMS* Challenger, *Zoology 13*: 1–228.

SARS, G. O. 1894: Contributions to the knowledge of the freshwater Entomostraca of New Zealand as shown by artificial hatching from dried mud. *Skrifter udg. af Videnskabsselskabets i Christiania 5*: 1–62, pls 1–8.

SARS, G. O. 1905: Pacifische Plankton – Crustacea. (Ergebnisse einer Reise nach dem Pacific. Schauinsland 1896–1897). II. Brackwasser-Crustaceen von dem Chatham-Inseln. *Zoologische Jahrbücher 21*: 371–414, pls 14–20.

SASOWSKY, I. D.; FONG, D. W.; WHITE, E. L. 1997: Conservation and protection of the biota of karst. *Karst Waters Institute Special Publication* 3: 1–125.

SCARSBROOK, M.; FENWICK, G. D.; RADFORD, J. 2000: Living groundwater: studying the fauna beneath our feet. *Water & Atmosphere 8(3)*: 15–16.

SCARSBROOK, M. R.; FENWICK, G. D.; DUGGAN, I. C.; HAASE, M. 2003: A guide to the groundwater invertebrates of New Zealand. *NIWA Science and Technology Series 51*: 1–59.

SCHELLENBERG, A. 1922a: Neue Notodelphyiden des Berliner und Hamburger Museums mit einer Ubersicht der ascidienbewohnenden Gattungen und Arten. Part I. *Mitteilungen aus dem Zoologisches Museum Berlin 10*: 217–274.

SCHELLENBERG, A. 1922b: Neue Notodelphyiden des Berliner und Hamburger Museums mit einer Ubersicht der ascidienbewohnenden Gattungen und Arten. Part II. *Mitteilungen aus dem Zoologisches Museum Berlin 10*: 275–298.

SCHEMBRI, P. J.; McLAY, C. L. 1983: An annotated key to the hermit crabs (Crustacea: Decapoda: Anomura) of the Otago region (south-eastern New Zealand). *New Zealand Journal of Marine and Freshwater Research 17*: 27–35.

SCHMINKE, H. K. 1971: Evolution, Natürliches System und Verbreitungsgeschichte der Bathynellacea (Crustacea, Malacostraca). PhD thesis, Zoologisches Institut der Universität Kiel, Kiel.

SCHMINKE, H. K. 1972: *Hexabathynella halophila* gen. n., sp. n. und die Frage nach der marinen Abkunft der Bathynellacea (Crustacea, Malacostraca). *Marine Biology 15*: 282–287.

SCHMINKE, H. K. 1973: Evolution, System und Verbreitungsgeschichte der Familie Parabathynellidae (Bathynellacea, Malacostraca). *Mikrofauna des Meeresboden 24*: 1–192.

SCHMINKE, H. K. 1974: Mesozoic intercontinental relationships as evidenced by bathynellid Crustacea (Syncarida: Malacostraca). *Systematic Zoology 23*: 157–164.

SCHMINKE, H. K. 1975: Phylogenie und Verbreitungsgeschichte der Syncarida. *Verhandlungen der Deutschen Zoologischen Gesellschaft 1974*: 384–388.

SCHMINKE, H. K. 1978: *Notobathynella longipes* sp. n and new records of other Bathynellacea (Crustacea, Syncarida) from New Zealand. *New Zealand Journal of Marine and Freshwater Research 12*: 457–462.

SCHMINKE, H. K. 1980: Zur Systematik der Stygocarididae (Crustacea, Syncarida) und Beschreibung zweier neuer Arten (*Stygocarella pleotelson* gen. n., sp. n. und *Stygocaris giselae* sp. n.) *Beaufortia 30*: 139–154.

SCHMINKE, H. K. 1981a: Perspectives in the study of the zoogeography of interstitial Crustacea: Bathynellacea (Syncarida) and Parastenocarididae (Copepoda). *International Journal of Speleology 11*: 83–89.

SCHMINKE, H. K. 1981b: Adaptation of Bathynellacea (Crustacea: Syncarida) to life in the interstitial ('Zoea Theory'). *Internationale Revue der gesamten Hydrobiologie* 66 (4): 575–837.

SCHMINKE, H. K. 1982: Syncarida. Pp 233–237 *in*: Parker, S. P. (ed.) *Synopsis and Classification of Living Organisms* 2. McGraw Hill, New York. 1232 p.

SCHMINKE, H. K. 1986: Syncarida. Pp 389–404 *in*: Botosaneanu, L. (ed.) *Stygofauno Mundi. A faunistic, distributional, and ecological synthesis of the world fauna inhabiting subterranean waters (including the marine interstitial)*. Brill/Backhuys, Leiden. 740 p.

SCHMINKE, H. K.; NOODT, W. 1968: Discovery of Bathynellacea, Stygocaridacea and other interstitial Crustacea in New Zealand. *Zeitschrift die Naturwissenschaften 54*: 184–185.

SCHMITT, W. L. 1940: The stomatopods of the west coast of America, based on collections made by the Allan Hancock Expedition, 1933–38. *Allan Hancock Pacific Expeditions 5(4)*: 129–225.

SCHNABEL, K. E. 2009: A review of the New Zealand Chirostylidae (Anomura: Galatheoidea) with description of six new species from the Kermadec Islands. *Zoological journal of the Linnean Society 155*: 542–582.

SCHRAM, F. R.; HESSLER, R. R. 1984: Anaspidid Syncarida. Pp. 192–195 *in*: Eldredge, N.; Stanley, S. M. *Living Fossils*. Springer, New York and Berlin. 291 p.

SCHRAM, F. R. 1984: Fossil Syncarida. *Transactions of the San Diego Society of Natural History 20*: 189–246.

SCHRAM, F. R. 1986: *Crustacea*. Oxford University Press. 606 p.

SCHULTZ, G. A. 1972: A review of the family *Scyphacidae* in the New World (Crustacea, Isopoda, Oniscoidea). *Proceedings of the Biological Society of Washington 84*: 477–488.

SCHWEITZER, C. E. 2001: Paleobiogeography of Cretaceous and Tertiary decapod crustaceans of the North Pacific Ocean. *Journal of Paleontology 75*: 808–826.

SEROV, P. A.; WILSON, G. D. F. 1995: A review of the Stenetriidae (Crustacea: Isopoda: Asellota). *Records of the Australian Museum 47*: 39–82.

SEROV, P. A.; WILSON, G. D. F. 1999: A revision of the Pseudojaniridae Wilson, with a description of a new genus of Stenetriidae Hansen (Crustacea: Isopoda: Asellota). *Invertebrate Taxonomy 13*: 67–116.

SHEARD, K. 1953: Taxonomy, distribution and development of the Euphausiacea (Crustacea). *Report of the British and New Zealand Antarctic Research Expedition, ser. B (Zoology and Botany) 8*: 1–72.

SHIH, C.-T. 1969: The systematics and biology of the family Phronomidae (Crustacea: Amphipoda). *Dana Report 74*: 1–100.

SIEG, J.; ZIBROWIUS, H. 1988: Association of a tube inhabiting tanaidacean, *Bifida scleractinicola* gen. nov., sp. nov., with bathyal scleractinians off New Caledonia (Crustacea Tanaidacea-Cnidaria Scleractininia). *Mésogéé 48*: 189–199.

SIEWING, R. 1959: Syncarida. *Bronn's Klassen und Ordnungen des Tierreichs (2 Auflage), 5, 1, 4(2)*: 1–121.

SIMES, J. E. 1977: The first record of Lower Paleozoic ostracods from New Zealand. *Geological Society of New Zealand Newsletter 44*: 9–10.

SKET, B.; BRUCE, N. L. 2004: Sphaeromatids (Isopoda, Sphaeromatidae) from New Zealand fresh and hypogean waters, with description of *Bilistra* n. gen. and three new species. *Crustaceana 76*: 1347–1370.

SMITH, G. W. 1908: Preliminary account of the habits and structure of the Anaspididae, with remarks on some other fresh-water Crustacea from Tasmania. *Proceedings of the Royal Society of London, B, 80*: 465–473.

SMITH, P. J.; WEBBER, W. R.; McVEAGH, S. M.; INGLID, G. S.; GUST, N. 2003: DNA and morphological identification of an invasive swimming crab, *Charybdis japonica*, in New Zealand waters. *New Zealand Journal of Marine and Freshwater Research 37*: 753–762.

SOMERFIELD, P. J.; CLARKE, K. R. 1995: Taxonomic levels, in marine community studies, revisited. *Marine Ecology Progress Series 127*: 113–119.

SPEARS, T.; ABELE, L. G. 1999: Phylogenetic relationships of crustaceans with foliaceous limbs:

an 18S rDNA study of Branchiopoda, Cephalocarida, and Phyllocarida. *Journal of Crustacean Biology 19*: 825–843.

SPEARS, T.; ABELE, L. G. 2000: Branchiopod monophyly and interordinal phylogeny inferred from 18S ribosomal DNA. *Journal of Crustacean Biology 20*: 1–24.

SPENGLER, L. 1790: Beskrivelse og Oplysning over den hidindtil lidet udarbeidede Sloegt af mangeskallende Konchylier som Linnaeus har kaldet *Lepas* med tilfoiede nye og ubeskrevne Arter. *Skrivter af Naturhistorie–Selskabet 1*: 158–212.

STEBBING, T. R. R. 1888: Report on the Amphipoda Collected by HMS *Challenger* during the years 1873–1876. *Report on the Scientific Results of the Voyage of HMS* Challenger *during the years 1873–1876, Zoology 29(1–2)*: 1–1737, 210 pls.

STEBBING, T. R. R. 1910: Crustacea. Part 5. Amphipoda. Scientific Results of the Trawling Expeditions of H.M.C.S. Thetis. *Memoirs of the Australian Museum 4*: 565–658, pls 47–60.

STEPHENS, R. T. T. 1984: Trout–smelt interactions in Lake Taupo. Unpublished DPhil thesis, University of Waikato, Hamilton.

STEPHENSEN, K. 1927: Crustacea from the Auckland and Campbell Islands. Papers from Dr. Th. Mortensen's Pacific Expedition 1914–1916. XI. *Videnskabelige Meddelelser fra Dansk Naturhistorisk Forening 83*: 289–390.

STEPHENSON, W. 1967: A comparison of Australasian and American specimens of *Hemisquilla ensigera* (Owen, 1832) (Crustacea: Stomatopoda). *Proceedings of the United States National Museum 120*: 1–18.

STEVENS, M. I.; HOGG, I. D.; CHAPMAN, M. A. 2002: The corophiid amphipods of Tauranga Harbour, New Zealand: evidence of an Australian crustacean invader. *Hydrobiologia 474*: 147–154.

STOUT, V. M. 1978: Effects of silt loads and of hydro-electric development on four large lakes. *Verhandlungen der Internationale Vereinigung für theoretische und angewandte Limnologie 20*: 1182–1185.

STREET, R. J. 1969: The New Zealand crayfish *Jasus edwardsii* (Hutton, 1875). *New Zealand Marine Department Fisheries Technical Report 30*: 1–53.

STREET, R.J. 1971: Rock lobster migration off Otago. *Commercial Fishing June 1971*: 16-17.

STREET, R.J. 1973: Trends in the rock lobster fishery in southern New Zealand 1970-71. *New Zealand Ministry of Agriculture and Fisheries Technical Report 116*. 1–13.

STREET, R. J. 1994: Rock lobster migrations in southern New Zealand. *Seafood New Zealand 2(2)*: 44–46.

SUZUKI, Y.; SUZUKI, M.; TSUCHIDA, S.; TAKAI, K.; SOUTHWARD, A. J.; NEWMAN, W. A.; YAMAGUCHI, T. 2009: Molecular investigations of the stalked barnacle *Vulcanolepas osheai* and the epibotic bacteria from the Brothers Caldera, Kermadec Arc, New Zealand. *Journal of the Marine Biological Association of the United Kingdom 89*: 727–733.

SVAVARSSON, J. 2006: New species of Gnathiidae (Crustacea, Isopoda, Cymothoida) from seamounts off northern New Zealand. *Zootaxa 1173*: 39–56.

SVAVARRSON, J.; BRUCE, N. L. 2000: Redescription of the cosmopolitan meso- and bathypelagic cirolanid *Metacirolana caeca* (Hansen, 1916), comb. nov. (Crustacea, Isopoda). *Steenstrupia 25*: 147–158.

SWANSON, K. M. 1969: Some Lower Miocene Ostracoda from the Middle Waipara District, New Zealand. *Transactions of the Royal Society of New Zealand (Geology) 7*: 33–48.

SWANSON, K. M. 1979a: Recent Ostracoda from Port Pegasus, Stewart Island, New Zealand. *New Zealand Journal of Marine and Freshwater Research 13*: 151–170.

SWANSON, K. M. 1979b: The marine fauna of New Zealand: Ostracods of the Otago Shelf. *New Zealand Oceanographic Institute Memoir 78*: 1–56.

SWANSON, K. M. 1980: Five new species of Ostracoda from Port Pegasus, Stewart Island. *New Zealand Journal of Marine and Freshwater Research 14*: 205–211.

SWANSON, K. M. 1990: The punciid ostracod – a new crustacean evolutionary window. *Courier Forschungsinstitut Senckenberg 123*: 11–18.

SWANSON, K. M.; AYRESS, M. A. 1999: *Cytheropteron testudo* and related species from the S W Pacific with analyses of their soft anatomies, relationships and distribution (Crustacea, Ostracoda, Cytheruridae). *Senckenbergiana Biologica 79*: 151–193.

SWEATMAN, G. K.; 1962: Parasitic mites of non-domesticated animals in New Zealand. *New Zealand Entomologist 3*: 15–23.

TAKEDA, M.; WEBBER, R. 2006: Crabs from the Kermadec Islands in the South Pacific. *National Science Museum Monographs 34*: 191–237.

TATTERSALL, O. S. 1955: Mysidacea. *Discovery Reports 28*: 1–190.

TATTERSALL, W. M. 1918: Euphausiacea and Mysidacea. *Scientific Reports of the Australasian Antarctic Expedition, ser. C, Zoology and Botany 5*: 1–15.

TATTERSALL, W. M. 1923: Crustacea. Pt.VII. Mysidacea. *British Antarctic 'Terra Nova' Expedition, 1910, Natural History Reports, Zoology 3*: 273–304.

TATTERSALL, W. M. 1924: Crustacea.VIII. Euphausiacea. *British Antarctic 'Terra Nova' Expedition, 1910, Natural History Reports, Zoology 8*: 1–36.

TATTERSALL, W. M.; TATTERSALL, O. S. 1951: *The British Mysidacea*. The Ray Society, London. viii + 460 p.

TAYLOR, D. J.; CREASE, T. J.; BROWN, W. M. 1999: Phylogenetic evidence for a single long-lived clade of crustacean cyclic parthenogens and its implications for the evolution of sex. *Proceedings of the Royal Society of London, B, 266*: 791–797.

TAYLOR, J. 2002: A review of the genus *Wildus* (Amphipoda: Phoxocephalidae) with a description of a new species from the Andaman Sea, Thailand. *Phuket Marine Biological Center Special Publication 23*: 253–263.

TAYLOR, P. D. 1991: Observations of symbiotic associations of bryozoans and hermit crabs from the Otago shelf of New Zealand. *In*: Bigey, F. P. (ed.), *Bryozoaires Actuels et Fossiles: Bryozoa Living and Fossil. Memoire de la Société des Sciences Naturelles de l'Ouest de la France, h.s., 1*: 487–495.

TAYLOR, R. B. 1998: Seasonal variation in assemblages of mobile epoifauna inhabiting three subtidal brown seaweeds in north-eastern New Zealand. *Hydrobiologica 361*: 25–35.

TENQUIST, J. D.; CHARLESTON, W. A. G. 2001: A revision of the annotated checklist of ectoparasites of terrestrial mammals in New Zealand. *Journal of the Royal Society of New Zealand 31*: 481–542.

THIELE, J. 1904: Die Leptostraken. *Wissenschaftliche Ergebnisse der Deutschen Tiefsee-Expedition auf dem Dampfer 'Valdivia' 1898–1899, 8(1)*: 1–26.

THOMAS, J. D. 1993: Biological monitoring and tropical diversity in marine environments: a critique with recommendations, and comments on the use of amphipods as bioindicators. *Journal of Natural History 27*: 795–806.

THOMSON, G. M. 1878a: New Zealand Crustacea, with descriptions of new species. *Transactions of the New Zealand Institute 11*: 230–248.

THOMSON, G. M. 1878b: On the New Zealand Entomostraca. *Transactions of the New Zealand Institute 11*: 251–263.

THOMSON, G. M. 1879a: On a new genus of *Nebalia* from New Zealand. *Annals and Magazine of Natural History, ser. 5, 4*: 418–419.

THOMSON, G. M. 1879b: New species of Crustacea from New Zealand. *Annals and Magazine of Natural History, ser. 5, 6*: 1–6.

THOMSON, G. M. 1880. New species of Crustacea from New Zealand. *Annals of Natural History 5*: 6.

THOMSON, G. M. 1881: Recent additions to and notes on New Zealand Crustacea. *Transactions and Proceedings of the New Zealand Institute 13*: 204–221.

THOMSON, G. M. 1882a: Additions to the crustacean fauna of New Zealand. *Transactions and Proceedings of the New Zealand Institute 14*: 230–238, pls 17, 18.

THOMSON, G. M. 1882b: On New Zealand Copepoda. *Transactions and Proceedings of the New Zealand Institute 15*: 93–116.

THOMSON, G. M. 1890: Parasitic Copepoda of New Zealand, with descriptions of new species. *Transactions and Proceedings of the New Zealand Institute 22*: 353–376.

THOMSON, G. M. 1892: On the occurrence of two species of Cumacea in New Zealand. *Journal of the Linnean Society (Zoology) 24*: 263–270.

THOMSON, G. M. 1893: Notes on Tasmanian Crustacea, with descriptions of new species. *Proceedings of the Royal Society of Tasmania for 1892*: 45–76.

THOMSON, G. M. 1894: On a freshwater schizopod from Tasmania. *Transactions of the Linnean Society, ser. 2, Zoology 6*: 285–303.

THOMSON, G. M. 1898: A revision of the Crustacea Anomura of New Zealand. *Transactions of the New Zealand Institute 31*: 169–197.

THOMSON, G. M. 1900: On some New Zealand Schizopoda. *Journal of the Linnean Society, Zoology 27*: 482–486.

THOMSON, G. M. 1913: The natural history of Otago Harbour and the adjacent sea, together with a record of the researches carried on at the Portobello Marine Fish-hatchery. *Transactions and Proceedings of the New Zealand Institute 45*: 225–251.

THOMSON, G. M. 1922: *The Naturalisation of Animals and Plants in New Zealand*. Cambridge University Press, Cambridge. 607 p.

THOMSON, G. M.; ANDERTON, T. 1921: History of the Portobello Marine Fish-hatchery and Biological Station. *Dominion of New Zealand, Board of Science and Art, Bulletin 2*: 1–131.

THOMSON, G. M.; CHILTON, C. 1886: Critical list of the Crustacea Malacostraca of New Zealand. *Transactions and Proceedings of the New Zealand Institute, Zoology 18*: 141–159.

THURGATE, M. E.; GOUGH, J. S.; CLARKE, A. K.; SEROV, P.; SPATE, A. 2001: Stygofauna diversity and distribution in Eastern Australian cave and karst areas. Pp. 49–62 *in*: Humphreys, W. F.; Harvey, M. S. (eds), *Subterranean Biology in Australia 2000. Records of the Western Australian Museum, Suppl. 64*: 1–242.

THURSTON, M. H.; BILLETT, D. S. M.; HASSACK, E. 1987: An association between *Exspina typica* Lang (Tanaidacea) and deep-sea holothurians. *Journal of the Marine Biological Association of the United Kingdom 67*: 11–15.

TOMLINSON, J. T. 1969: The burrowing barnacles (Cirripedia: Order Acrothoracica). *Bulletin of the United States National Museum 296*: 1–162.

TOMLINSON, J. T. 1987: The burrowing barnacles (Acrothoracica). Pp. 63–71 *in*: Southward, A. J.

(ed.), *Crustacean Issues 5: Barnacle Biology*. A.A. Balkema Publishers, Rotterdam.

TONG, L. J.; MOSS, G. A.; PAEWAI, M. M.; PICKERING, T. D. 1997: Effect of brine-shrimp numbers on the growth and survival of early-stage phyllosoma larvae of the rock lobster *Jasus edwardsii. Marine and Freshwater Research 48*: 935–940.

TONG, L. J.; MOSS, G. A.; PAEWAI, M. M.; PICKERING, T. D. 2000a: Effect of temperature and feeding rate on the growth and survival of early and mid- late stage phyllosomas of the spiny lobster *Jasus edwardsii. Marine and Freshwater Research 51*: 235–241.

TONG, L.J.; MOSS, G.A.; PICKERING, T.D.; PAEWAI, M.M. 2000b: Effect of temperature and feeding rate on the growth and survival of the early and late-stage phyllosomas of the spiny lobster *Jasus edwardsii. Marine and Freshwater Research 51*: 243–248.

TWOMBLY, S.; CLANCY, N.; BURNS, C. W. 1998: Life history consequences of food quality in the freshwater copepod *Boeckella triarticulata. Ecology 79*: 1711–1724.

UHL, D. 1999: Syncarids (Crustaceae, Malacostraca) from the Stephanian D (Upper Cretaceous) of the Saar–Nahe Basin (SW Germany). *Neues Jahrbuch für Geologie und Palaeontologie Monatsheft 111*: 679–697.

UHL, D. 2002: *Uronectes fimbriatus* Jordan (Syncarida, Malacostraca) aus dem Rotliegend (Ober-karbon und Unter-permian) des Saar-Nahe-Beckens (SW Deutschland). *Pollichia 89*: 43–56.

VADER, W. 1978: Associations between amphipods and echinoderms. *Astarte 11*: 123–134.

VADER, W. 1984: Notes on Norwegian marine Amphipoda. 8. Amphipods found in association with sponges and tunicates. *Fauna Norvegica, ser. A, 5*: 16–21.

VADER, W. 1996: *Liljeborgia* species (Amphipoda, Liljeborgiidae) as associates of hermit crabs. *Polskie Archiwum Hydrobiologii 42*: 517–525.

van KLINKEN, R. D.; GREEN, A. J. A. 1992: The first record of Oniscidea (terrestrial Isopoda) from Macquarie Island. *Polar Record 28*: 240–242.

VANDEL, A. 1964: *Biospéologie: La Biologie des Animaux cavernicoles. Géobiologie, Écologie, Aménagement. Collection internationale sous la direction de C. Delamare Deboutteville.* Gauthier–Villars, Paris. 619 p. [Also in English edition, 1965, Pergamon Press, Oxford.]

VEJDOVSKÝ, F.1882: *Thierische organismen der Brunnenwässer vom Prag*. Selbstverlag, Prague. 66, [4] p.

VEJDOVSKÝ, F. 1899: O systemickém umístení stud niccného korýše *Bathynella natans* [On the systematic position of the well shrimp *Bathynella natans*]. *Sitzungsberichte der Königlichen Böhmischen Gesellschaft der Wissenschaften 14*: 1–2.

VETTER, E. W. 1996a: Enrichment experiments and infaunal population cycles in a Southern California sand plain: response of the leptostracan *Nebalia daytoni* and other infauna. *Marine Ecology Progress Series 137*: 83–93.

VETTER, E. W. 1996b: Secondary production of a Southern California *Nebalia* (Crustacea: Leptostraca). *Marine Ecology Progress Series 137*: 95–101.

VETTER, E. W. 1996c: Life-history patterns of two Southern California *Nebalia* species (Crustacea: Leptostraca: the failure of form to predict function. *Marine Ecology Progress Series 127*: 131–141.

VINOGRADOV, M. E.; VOLKOV, A. F.; SEMENOVA, T. N. [1996] 1982: *Hyperiid Amphipods (Amphipoda, Hyperiidea) of the World Oceans*. Science Publishers, Lebanon, New Hampshire. 632 p.

WÄGELE, J.-W. 1982: The hypogean Paranthuridae *Cruregens* Chilton and *Curassanthura* Kensley (Crustacea, Isopoda), with remarks on their morphology and adaptations. *Bijdragen tot de Dierkunde 52*: 49–59.

WÄGELE, J.-W. 1983: *Nebalia marerubri* sp. nov., aus dem Roten Meer (Crustacea: Phyllocarida: Leptostraca). *Journal of Natural History 17*: 127–138.

WÄGELE, J.-W. 1985: Two new genera and twelve new species of Anthuridea (Crustacea: Isopoda) from off the west coast of New Zealand. *New Zealand Journal of Zoology 12*: 363–423.

WÄGELE, J.-W. 1989: Evolution und phylogenetisches System der Isopoda. Stand der Forschung und neue Erkenntnisse. *Zoologica 140*: 1–262.

WÄGELE, J.-W. 1994: Notes on Antarctic and South American Serolidae (Crustacea, Isopoda) with remarks on the phylogenetic biogeography and a description of new genera. *Zoologische Jahrbücher der Systematik 121*: 3–69.

WAKABARA, Y. 1976: *Paranebalia fortunata* n. sp. from New Zealand (Crustacea, Leptostraca, Nebaliacea). *Journal of the Royal Society of New Zealand 6*: 197–300.

WALKER, A. O. 1908: Amphipoda from the Auckland Islands. *Annals and Magazine of Natural History, ser. 8, 2*: 33–39.

WALKER-SMITH, G. K. 1998. A review of *Nebaliella* (Crustacea: Leptostraca) with description of a new species from the continental slope of southeastern Australia. *Memoirs of the Museum of Victoria 57*): 39–56.

WALKER-SMITH, G. K.; POORE, G. C. B. 2001: A phylogeny of the Leptostraca (Crustacea) with keys to families and genera. *Memoirs of Museum Victoria 58*: 383–410.

WARÉN, A.; CARROZZA, F. 1994: *Arculus sykesi* (Chaster), a leptonacean bivalve living on a tanaid crustacean in the Gulf of Genova. *Bolletino Malacologico 29*: 303–306.

WEAR, R. G. 1964a: Larvae of *Petrolisthes nova ezelandiae* Filhol, 1885 (Crustacea, Decapoda, Anomura). *Transactions of the Royal Society of New Zealand, Zoology 4*: 229–244.

WEAR, R. G. 1964b: Larvae of *Petrolisthes elongatus* (H. Milne Edwards, 1837) (Crustacea, Decapoda, Anomura). *Transactions of the Royal Society of New Zealand, Zoology 5*: 39–53.

WEAR, R. G. 1965a: Zooplankton of Wellington Harbour, New Zealand. *Zoology Publications from Victoria University of Wellington 38*: 1–31.

WEAR, R. G. 1965b: Larvae of *Petrocheles spinosus* Miers, 1876 (Crustacea, Decapoda, Anomura) with keys to New Zealand porcellanid larvae. *Transactions of the Royal Society of New Zealand, Zoology 5*: 147–168.

WEAR, R. G. 1965c: Pre–zoea larvae of *Petrolisthes novaezelandiae* Filhol, 1885 (Crustacea, Decapoda, Anomura) with keys to New Zealand porcellanid larvae. *Transactions of the Royal Society of New Zealand, Zoology 6*: 127–132.

WEAR, R. G. 1965d: Breeding cycles and pre-zoea larvae of *Petrolisthes elongatus* (H. Milne Edwards, 1837) (Crustacea, Decapoda). *Transactions of the Royal Society of New Zealand, Zoology 5*: 169–175.

WEAR, R. G. 1966: Pre-zoea larvae of *Petrocheles spinosus* Miers, 1876 (Crustacea, Decapoda, Anomura). *Transactions of the Royal Society of New Zealand, Zoology* 8: 119–124.

WEAR, R. G. 1976: Studies on the larval development of *Metanephrops challengeri* (Balss, 1914) (Decapoda, Nephropidae). *Crustaceana 30*: 113–122.

WEAR, R. G. 1985: Checklist and annotated bibliography of New Zealand decapod crustacean larvae (Natantia, Macrura Reptantia and Anomura). *Zoological Publications from Victoria University of Wellington 70*: 1–15.

WEAR, R. G.; FIELDER, D. R. 1985: The marine fauna of New Zealand: larvae of Brachyura (Crustacea: Decapoda). *New Zealand Oceanographic Institute Memoir 92*: 1–89.

WEAR, R. G.; HADDON, M. 1987: Natural diet of the crab *Ovalipes catharus* (Crustacea, Portunidae) around central and northern New Zealand. *Marine Ecology Progress Series 35*: 39–49.

WEAR, R. G.; YALDWYN, J. C. 1966: Studies on thalassinid Crustacea (Decapoda, Macrura Reptantia) with a description of a new *Jaxea* from New Zealand and an account of its larval development. *Zoology Publications from Victoria University of Wellington 41*: 1–27.

WEBB, B. F. 1973: Fish populations of the Avon–Heathcote Estuary 3. Gut contents. *New Zealand Journal of Marine and Freshwater Research 7*: 223–234.

WEBBER, W. R. 1979: Developmental stages of some New Zealand Majidae (Crustacea, Decapoda, Brachyura) with observations on the larval affinities of the Majidae. Unpublished MSc thesis, Victoria University of Wellington.

WEBBER, W. R. 2001: Space invaders; crabs that turn up in New Zealand unannounced. *Seafood New Zealand 9(10)*: 80–84.

WEBBER, W. R.; BOOTH, J. D. 2001: Larval stages, developmental ecology, and distribution of *Scyllarus* sp. Z (probably *Scyllrus aoteanus* Powell, 1949) (Decapoda: Scyllaridae). *New Zealand Journal of Marine and Freshwater Research 35*: 1025–1056.

WEBBER, W. R.; WEAR, R. G. 1981: Life history studies on New Zealand Brachyura 5. Larvae of the family Majidae. *New Zealand Journal of Marine and Freshwater Research 15*: 331–383.

WEEKES, P. J. 1986: Growth and development of *Amurotaenia decidua* Hine, 1977, in a copepod. P. 255 *in*: Howell, M. J. (ed.), *Parasitology – Quo vadit? Handbook, programmes and abstracts, Sixth International Congress of Parasitology, Brisbane, Australia, 25–29 August 1986.* Australian Academy of Sciences, Canberra.

WEEKES, P. J., PENLINGTON, B. P. 1986. First record of *Ligula intestinalis* (Cestoda) in rainbow trout, *Salmo gairdneri*, and common bully, *Gobiomorphus cotidianus*, in New Zealand. *Journal of Fish Biology 28*: 183–190.

WELLS, J. B. J. 2007: An annotated checklist and keys to the species of Copepoda Harpacticoida (Crustacea). *Zootaxa 1568*: 1–872..

WELLS, J.; HICKS, G. F.; COULL, B. 1982: Common harpacticoid copepods from New Zealand harbours and estuaries. *New Zealand of Zoology 9*: 151–184.

WEST, D. W.; BOUBEE, J. A. T.; BARRIER, R. F. G. 1997: Responses to pH of nine fishes and one shrimp native to New Zealand freshwaters. *New Zealand Journal of Marine and Freshwater Research 31*: 461–468.

WHATLEY, R. C.; MILLSON, K. J. 1992: *Marwickcythereis*, a new ostracod genus from the Tertiary of New Zealand. *New Zealand Natural Sciences 19*: 41–44.

WHITE, A. 1847: List of the specimens of decapod Crustacea in the collections of the British Museum. Trustees of the British Museum, London. viii + 143 p.

WHITE, A.; DOUBLEDAY, E. 1843: List of the annulose animals hitherto recorded as found in New Zealand, with the descriptions of some new species. Pp. 265–291 *in* Dieffenbach, E., *Travels in New Zealand; with contributions to the geography, geology, botany, and natural history of that country.* John Murray, London. Vol. 2, iv + 396 p.

WILLIAMS, B. G.; GREENWOOD, J. G.; JILLET, J. B. 1985: Seasonality and duration of the developmental stages of *Heterosquilla tricarinata* (Claus, 1871) (Crustacea: Stomatopoda) and the

replacement of the larval eye at metamorphosis. *Bulletin of Marine Science 36*: 104–114.
WILLIAMS, D. R.; WILDE, K. (Eds) 1985: *Cave Management in Australasia. Proceedings of the Sixth Australasian Conference on Cave Tourism and Management, Waitomo Caves, New Zealand, September 1985.* Tourist Hotel Corporation of New Zealand, Waitomo. 229 p.
WILLIAMS, J. B. 1988: Further observations on *Kronborgia isopodicola*, with notes on the systematics of the Fecampiidae (Turbellaria: Rhabdocoela). *New Zealand Journal of Zoology 15*: 211–21.
WILLIAMS, W. D. 1985: Subterranean occurrence of *Anaspides tasmaniae* (Thomson) (Crustacea, Syncarida). *International Journal of Speleology 1*: 333–337.
WILLIAMS, W. D. 1986: Amphipoda on landmasses derived from Gondwana. Pp. 553–559 *in*: Botosaneanu, L. (ed.), *Stygofauna Mundi. A Faunistic, Distributional, and Ecological Synthesis of the World Fauna Inhabiting Subterranean Waters (including the marine interstitial)*. Brill/Backhuys, Leiden. 740 p.
WILLIAMSON, C. E. 1991: *Copepoda*. Pp. 787–822 *in*: Thorp, J. H.; Covich, A. P. (eds), *Ecology and Classification of North American Freshwater Invertebrates.* Academic Press, San Diego. 911 p.
WILLIAMSON, D. I. 1965: Some larval stages of three Australian crabs belonging to the families Homolidae and Raninidae, and observations on the affinities of these families (Crustacea: Decapoda). *Australian Journal of Marine and Freshwater Research 16*: 369–98.
WILLIS, K. J. 1998: From single species to mesocosms: Responses of freshwater copepods and their community to PCP. Unpublished Ph.D. thesis, University of Waikato, Hamilton. 191 p.
WILLIS, K. J. 1999: Acute and chronic bioassays with New Zealand freshwater copepods using pentachlorophenol. *Environmental Toxicology and Chemistry 18*: 2580–2586.
WILLIS, K. J; WOODS, C. M. C.; ASHTON, G.V. 2009: *Caprella mutica* in the Southern Hemisphere: Atlantic origins, distribution, and reproduction of an alien marine amphipod in New Zealand. *Aquatic Biology 7*: 249–259.
WILSON. G. D. F. 1989: A systematic revision of the deep-sea subfamily of the isopod crustacean family Munnopsidae. *Bulletin of the Scripps Institution of Oceanography 27*: xiii, 1–138.
WILSON, G. D. F. 2003: Deep-sea biodiversity. Australian Museum http://www–personal.usyd.edu.au/~buz/deepsea.html.
WILSON, G. D. F.; FENWICK, G. D. 1999: Taxonomy and ecology of *Phreatoicus typicus* Chilton, 1883 (Crustacea, Isopoda, Phreatoicidae). *Journal of the Royal Society of New Zealand 29*: 41–64.
WILSON, G. D. F.; KEABLE, S. J. 2001: Systematics of the Phreatoicidea. *Crustacean Issues 13*: 175–194.
WILSON, G. D. F.; WÄGELE, J.-W. 1994: Review of the familiy Janiridae (Crustacea: Isopoda: Asellota). *Invertebrate Taxonomy 8*: 683–747.
WILSON, W. H. 1989: Predation and mediation of intraspecific competition in an infaunal community in the Bay of Fundy. *Journal of Experimental Marine Biology and Ecology 132*: 221–245.
WITHERS, T. H. 1913: Some Miocene cirripedes of the genera *Hexelasma* and *Scalpellum* from New Zealand. *Proceedings of the Zoological Society of London 56*: 840–854.
WITHERS, T. H. 1924: The fossil cirripedes of New Zealand. *New Zealand Geological Survey Paleontological Bulletin 10*: 1–47.
WITHERS, T. H. 1953: *Catalogue of Fossil Cirripedia in the Department of Geology v.3, Tertiary*. British Museum (Natural History), London. 369 p.
WOLFF, T. 1956a: Crustacea Tanaidacea from depths exceeding 6000 m. *Galathea Reports 2*: 85–157.
WOLFF, T. 1956b: Six new abyssal species of *Neotanais* (Crust. Tanaidacea). *Videnskabelige Meddelelser fra Danske Naturhistorisk Forening i København 118*: 41–52.
WOLFF, T. 1962: The systematics and biology of bathyal and abyssal Isopoda Asellota. *Galathea Reports 6*: 1–20.
WOODS, C. M. C. 1993: Natural diet of the crab *Notomithrax ursus* (Brachyura: Majidae) at Oaro, South Island, New Zealand. *New Zealand Journal of Marine and Freshwater Research 27*: 309–315.
WOODS, C. M. C. 1995: Masking in the spider crab *Trichoplatus huttoni* (Brachyura: Majidae). *New Zealand Natural Sciences 22*: 75–80.
WOODS, C. M. C. 2002: Natural diet of the seahorse *Hippocampus abdominalis. New Zealand Journal of Marine and Freshwater Research 36*: 655–660.
WOODS, C. M. C., McLAY, C. L. 1994: Masking and ingestion preferences of the spider crab *Notomithrax ursus* (Brachyura: Majidae). *New Zealand Journal of Marine and Freshwater Research*: *28*: 105–111.
WOODS, C. M. C.; PAGE, M. J. 1999: Sponge masking and related preferences in the spider crab *Thacanophrys filholi* (Brachyura: Majidae). *Marine and Freshwater Research 50*: 135–143.
XU, Z.; BURNS, C. W. 1991: Development, growth and survivorship of juvenile calanoid copepods on diets of cyanobacteria and algae. *Internationale Revue der gesamten Hydrobiologie 76*: 73–87.
YALDWYN, J. C. 1954a: A preliminary survey of the New Zealand Crustacea Decapoda Natantia Vol. 1, pp. 1–280, Vol. 2, pp. 281–544. Unpublished MSc thesis, Victoria University of Wellington.
YALDWYN, J. C. 1954b: *Nephrops challengeri* Balss, 1914, Crustacea, Decapoda, Reptantia) from New Zealand and Chatham Island waters. *Transactions of the Royal Society of New Zealand 82*: 721–732.
YALDWYN, J. C. 1957a: A review of deep-water biological investigation in the New Zealand area. *New Zealand Science Review 15*: 41–45.
YALDWYN, J. C. 1957b: Deep-water Crustacea of the genus *Sergestes* (Decapoda, Natantia) from Cook Strait, New Zealand. *Zoological Publications from Victoria University of Wellington 22*: 1–27.
YALDWYN, J. C. 1959: The New Zealand natant decapod Crustacea, systematics, distribution and relationships. Unpublished PhD thesis, Victoria University of Wellington. 435 p.
YALDWYN, J. C. 1960: Crustacea Decapoda Natantia from the Chatham Rise: a deep water bottom fauna from New Zealand. *New Zealand Oceanographic Institute Memoir 4*: 13–53.
YALDWYN, J. C. 1961: A scyllarid lobster, *Arctides antipodarum* Holthuis, new to New Zealand waters. *Records of the Dominion Museum 4*: 1–6.
YALDWYN, J. C. 1971: Preliminary descriptions of a new genus and twelve new species of natant decapod Crustacea from New Zealand. *Records of the Dominion Museum 7*: 85–94.
YALDWYN, J. C. 1974: Shrimps and prawns. *New Zealand's Nature Heritage 38*: 1041–1046.
YALDWYN, J. C., DAWSON, E. W. 1970: The stone crab *Lithodes murrayi* Henderson: the first New Zealand record. *Records of the Dominion Museum 6*: 275–284.
YALDWYN, J. C.; DAWSON, E. W. 1985: *Lithodes nintokuae* Sakai: a deep-water king crab (Crustacea, Anomura, Lithodidae) newly recorded from Hawaii. *Pacific Science 39*: 16–23.
YOUNG, J. W.; ANDERSON, D. T. 1987: Hyperiid amphipods (Crustacea: Peracarida) from a warm-core eddy in the Tasman Sea. *Australian Journal of Marine and Freshwater Research 38*: 711–725.
ZEIDLER, W. 1992: Hyperiid amphipods (Crustacea: Amphipoda: Hyperiidea) collected recently from eastern Australian waters. *Records of the Australian Museum 44*: 85–133.
ZEIDLER, W. 2003a: A review of the hyperiidean amphipod family Cystisomatidae Willemöes-Suhm, 1875 (Crustacea: Amphipoda: Hyperiidea). *Zootaxa 141*: 1–43.
ZEIDLER, W. 2003b: A review of the hyperiidean amphipod superfamily Vibilioidea Bowman and Gruner, 1973 (Crustacea: Amphipoda: Hyperiidea). *Zootaxa 280*: 1–104.
ZEIDLER, W. 2004a: A review of the hyperiidean amphipod superfamily Lycaeopsoidea Bowman & Gruner, 1973 (Crustacea: Amphipoda: Hyperiidea). *Zootaxa 520*: 1–18.
ZEIDLER, W. 2004b: A review of the families and genera of the hyperiidean amphipod superfamily Phronimoidea Bowman & Gruner, 1973 (Crustacea: Amphipoda: Hyperiidea). *Zootaxa 567*: 1–66.
ZEIDLER, W. 2006: A review of the hyperiidean amphipod superfamily Archaeoscinoidea Vinogradov, Volkov & Semenova, 1982 (Crustacea: Amphipoda: Hyperiidea). *Zootaxa 1125*: 1–37.
ZEIDLER, W. 2009: A review of the hyperiidean amphipod superfamily Lanceoloidea Bowman & Gruner, 1973 (Crustacea: Amphipoda: Hyperiidea). *Zootaxa 2000*: 1–117.
ZIMMER, C. 1902: Cumaceen. *Ergebnisse der Hamburger Magalhaensische Sammelreise 2*: 1–18.

Checklist of New Zealand living Crustacea

The following classification is based mostly on Martin and Davis (2001). All species are to be regarded as marine unless indicated otherwise by habitat codes.

All species: A, adventive; B, brackish/estuarine; C, commensal; E, endemic; F, freshwater; S, supralittoral; T, terrestrial; *, unpublished (new) record; ? after a genus name or before a species name indicates uncertainty or a possible misidentification. Endemic genera are underlined (first mention).

Notostraca: Hs, hypersaline environments.

Cirripedia: Letters in parentheses following new records indicate where material is held, i.e. AUT (Earth and Oceanic Sciences Research Centre, Auckland University of Technology); GNS (GNS Science, Lower Hutt); NIWA (National Institute of Water & Atmosphere, Wellington); UA (Geology Department, University of Auckland).

Other groups, especially Copepoda: Habitat codes – Be, benthic; L, littoral; Sl, sublittoral (to ca. 10 metres depth); Sh, shelf (ca. 10–200 metres depth); Ba, bathyal (> 200 metres depth); Bp, benthopelagic; Co, coastal; F, freshwater (including wells, as well as species found in terrestrial mosses as they comprise an essentially aquatic habitat); O, oceanic; P, parasitic; Pe, pelagic (planktonic); Ep, epipelagic; Me, mesopelagic; By, bathypelagic; Ph, phytal (if marine, usually in algal and seagrass communities in the littoral or sublittoral, but W indicates decaying or mollusc-bored wood, which may have been dredged from depths up to 2000 metres. If freshwater, usually in algal or flowering-plant communities but M indicates moss or liverwort and includes water courses and damp terrestrial situations. Zoogeography codes: Ant, Antarctic; Ca, Campbell Island; Ch, Chatham Islands; Sa, subantarctic; Sn, Snares Islands; Tr/St, tropical/subtropical; Tz, transition zone; W, widespread.

Amphipoda: Families of the section Gammaridea sensu Barnard and Barnard (1983) (Barnard's 1969 family Gammaridae), follow Barnard and Barnard (1983) and Barnard and Karaman (1991). Known unpublished amphipod taxa are not included in the list.

SUBPHYLUM CRUSTACEA
Class BRANCHIOPODA
Subclass PHYLLOPODA
Order ANOSTRACA
ARTEMIIDAE
Artemia franciscana Kellogg, 1906 Hs A?

Order NOTOSTRACA
TRIOPSIDAE
Lepidurus apus viridis Baird, 1850 F

Order DIPLOSTRACA
Suborder SPINICAUDATA
LIMNADIIDAE
Eulimnadia marplesi Timms & McLay, 2005 F E

Suborder CLADOCERA
Infraorder ANOMOPODA
BOSMINIDAE
Bosmina meridionalis Sars, 1904 F
CHYDORIDAE
Alona abbreviata Sars, 1896 F
Alona affinis s.l. (Leydig, 1860) F
Alona cambouei Guerne & Richard, 1893 F
Alona guttata s.l. Sars, 1862 F
Alona quadrangularis (Müller, 1785) F
Alona rectangula s.l. Sars, 1862 F
Armatolona macrocopa Sars, 1895 F
Camptocercus australis Sars, 1896 F
Camptocercus rectirostris Schödler, 1862 F
Chydorus sphaericus s.l. (Müller, 1785) F
Dunhevedia crassa King, 1853 F
Ephemeroporus barroisi s.l. (Richard, 1894) F
Graptoleberis testudinaria (Fischer, 1851) F
Leydigia ?australis Sars, 1885 F
Monospilus dispar Sars, 1861 F A?
Oxyurella tenuicaudis (Sars, 1862) F
Pleuroxus hastirostris Sars, 1904 F E
Pleuroxus helvenacus Frey, 1991 F E
Pleuroxus unispinus Henry, 1922 F

DAPHNIIDAE
Ceriodaphnia dubia Richard, 1895 F
Ceriodaphnia cf. *pulchella* Sars, 1862 F
Ceriodaphnia ?reticulata (Jurine, 1820) F
Daphnia carinata s.l. King, 1852 F
Daphnia dentifera Forbes, 1893 F A
Daphnia lumholtzi Sars, 1903 F
Daphnia obtusa Kurz, 1942 F
Scapholeberis kingi Sars, 1903 F
Simocephalus exspinosus (Koch, 1841) F
Simocephalus obtusatus (Thomson, 1894) F E
Simocephalus ?vetulus (Müller, 1776) F
ILYOCRYPTIDAE
Ilyocryptus sordidus s.l. (Lieven, 1848) F
MACROTHRICIDAE
Lathonura ?rectirostris (Müller, 1785) F
Macrothrix schauinslandi Sars, 1904 F
Pseudomoina lemnae (King, 1853) F
Streblocerus serricaudatus (Fischer, 1849) F
MOINIDAE
Moina australiensis Sars, 1896 F
Moina tenuicornis Sars, 1896 F
NEOTHRICIDAE
Neothrix armata Gurney, 1927
SAYCIIDAE
Saycia cooki novaezealandiae Frey, 1971 F E
SIDIDAE
Penilia avirostris Dana, 1852
Penilia pacifica Kraemer, 1895

Suborder ONYCHOPODA
PODONIDAE
Evadne nordmanni Loven, 1836
Evadne aspinosus Kraemer, 1895
Pleopis polyphaemoides (Leuckart, 1859)
Pleopis trisetosus Kraemer, 1895

Class CEPHALOCARIDA
Order BRACHYPODA
HUTCHINSONIELLIDAE
Chiltoniella elongata Knox & Fenwick, 1977 E

Class MAXILLOPODA
Subclass THECOSTRACA
Infraclass ASCOTHORACIDA
Order LAURIDA
SYNAGOGIDAE
Gen. et sp. indet. Te Papa Palmer 1997

Order DENDROGASTRIDA
DENDROGASTRIDAE
Dendrogaster argentinensis Grygier & Salvat, 1987
Dendrogaster otagoensis Palmer, 1997 E

Infraclass CIRRIPEDIA
Superorder ACROTHORACICA
Order PYGOPHORA
CRYPTOPHIALIDAE
Australophialus melampygos (Brandt, 1907) E

Superorder RHIZOCEPHALA
Order KENTROGONIDA
LERNAEODISCIDAE
Triangulus munidae Smith, 1906
PELTOGASTRIDAE
Boschmaia munidicola Reinhard, 1958
Briarosaccus callosus Boschma, 1930
Galatheascus babai Lützen, 1985
Peltogaster sp. Lörz et al. 2008 E
Tortugaster discoidalis Lützen, 1985 E
SACCULINIDAE
Sacculina sp. Brockerhoff, McLay & Kluza 2006

Order AKENTROGONIDA
THOMPSONIIDAE
?Thompsonia affinis Krüger, 1912
Thylacoplethus novaezealandiae Lützen, Glenner & Lörz, 2009 E
INCERTAE SEDIS
Parthenopea vulcanophila Lützen, Glenner & Lörz,

2009 E
Gen. et sp. indet. Lützen, Glenner & Lörz 2009

Superorder THORACICA
Order IBLIFORMES
IDIOIBLIDAE
Chaetolepas segmentata Studer, 1889 E
Chitinolepas spiritsensis Buckeridge & Newman, 2006 E
Idioibla idiotica (Batham, 1945) E

Order LEPADIFORMES
Suborder LEPADOMORPHA
LEPADIDAE
Alepas pacifica Pilsbry, 1907
Conchoderma auritum (Linné, 1767)
Conchoderma virgatum (Spengler, 1790)
Dosima fascicularis (Ellis & Solander, 1786)
Lepas anatifera Linné, 1758 A
Lepas australis Darwin, 1851
Lepas pectinata Spengler, 1793
Lepas testudinata Aurivillius, 1892
OXYNASPIDAE
Oxynaspis indica (Annandale, 1910)
Oxynaspis terranovae Totton, 1923 E
POECILASMATIDAE
Megalasma carinatum (Hoek, 1883)
Megalasma striatum (Hoek, 1883)
Poecilasma kaempferi (Darwin, 1851)
Trilasmis eburneum Hinds, 1883

Suborder HETERALEPADOMORPHA
ANELASMATIDAE
Anelasma squalicola Lovén, 1845*
HETERALEPADIDAE
Heteralepas japonica (Aurivillius, 1892)
Paralepas minuta (Philippi, 1836)
Paralepas quadrata (Aurivillius, 1894)

Order SCALPELLIFORMES
CALANTICIDAE
Calantica spinosa (Quoy & Gaimard, 1834) E
Calantica spinilatera Foster, 1979 E
Calantica villosa (Leach, 1824) E
Scillaelepas fosteri Newman, 1980 E
Scillaelepas studeri (Weltner, 1922)
Scillaelepas n. sp. 1* NIWA E
Scillaelepas n. sp. 2* NIWA E
Smilium acutum (Hoek, 1883)
Smilium zancleanum (Seguenza, 1876)
EOLEPADIDAE
Ashinkailepas kermadecensis Buckeridge, 2009 E
Vulcanolepas osheai (Buckeridge, 2000) E
SCALPELLIDAE
Alcockianum persona (Annandale, 1916)
Amigdoscalpellum costellatum (Withers, 1935)
Amigdoscalpellum vitreum (Hoek, 1883)
Anguloscalpellum pedunculatum (Hoek, 1883) E
Anguloscalpellum n. sp.* NIWA E
Arcoscalpellum trochelatum Foster, 1979 E
Arcoscalpellum affbricatum Foster, 1979 E
Arcoscalpellum pertosum Foster, 1979 E
Gymnoscalpellum intermedium (Hoek, 1883)
Verum novaezelandiae (Hoek, 1883)
Verum raccidium (Foster, 1979) E
Gen. indet. et n. spp. (2)* NIWA 2E

Order SESSILIA
Suborder VERRUCOMORPHA
VERRUCIDAE
Altiverruca galapagosa Zevina, 1978*
Altiverruca gibbosa (Hoek, 1883)
Altiverruca nitida (Hoek, 1883)*
Metaverruca recta (Aurivillius, 1898)
Metaverruca cf. *defayeae* Buckeridge, 1994*

Gen. nov. et n. sp.* J. Buckeridge E

Suborder BALANOMORPHA
ARCHAEOBALANIDAE
Acasta sp. *AUT
Notobalanus vestitus (Darwin, 1854) E
Solidobalanus auricoma (Hoek, 1913)
AUSTROBALANIDAE
Austrominius modestus (Darwin, 1854) E
Epopella kermadeca Foster, 1979 E
Epopella plicata (Gray, 1843) E
BALANIDAE
Amphibalanus amphitrite (Darwin, 1854) A
Amphibalanus variegatus (Darwin, 1854) A
Austromegabalanus nigrescens (Lamarck 1818)
Austromegabalanus psittacus (Molina, 1782)
Balanus trigonus Darwin, 1854
Notomegabalanus campbelli (Filhol, 1885) E
Notomegabalanus decorus (Darwin, 1854) E
Megabalanus tintinnabulum linzei (Foster, 1979)
BATHYLASMATIDAE
Bathylasma alearum (Foster, 1979)
Hexelasma gracilis Foster, 1981 E
Hexelasma nolearia (Foster, 1979) E
Mesolasma fosteri (Newman & Ross, 1971) E
Tetrachaelasma tasmanicum Buckeridge, 1999
CHIONELASMATIDAE
Chionelasmus crosnieri Buckeridge, 1998
CHTHAMALIDAE
Chamaesipho brunnea Moore, 1944 E
Chamaesipho columna (Spengler, 1790) E
CORONULIDAE
Coronula diadema (Linné, 1767)
Coronula reginae Darwin, 1854
Tubinicella major Lamarck, 1802
PACHYLASMATIDAE
Pachylasma auranticacum Darwin, 1854
Pachylasma scutistriata Darwin, 1854
PLATYLEPADIDAE
Platylepas hexastylos (Fabricus, 1798)
Stomatolepas elegans (Costa, 1838)
PYRGOMATIDAE
Cantellius septimus (Darwin, 1854)
Creusia spinulosa Leach, 1824
TETRACLITIDAE
Tesseropora rosea (Krauss, 1848)
Tetraclita aoranga Foster, 1979 E
Tetraclitella depressa Foster & Anderson, 1986 E

Subclass TANTULOCARIDA
DEOTERTHRIDAE
Deoterthron dentatum Bradford & Hewitt, 1980 P E (ostracod host)
Doryphallophora aselloticola (Boxshall & Lincoln, 1983) P (isopod host)
Doryphallophora megacephala (Lincoln & Boxshall, 1983) P (isopod host) E

Subclass BRANCHIURA
Order ARGULOIDA
ARGULIDAE
Argulus japonicus Thiele, 1900 F P (fish host) A

Subclass PENTASTOMIDA
Order POROCEPHALIDA
LINGUATULIDAE
Linguatula serrata (Leuckart, 1860) T P (mammal) A

Subclass COPEPODA
Order CALANOIDA
ACARTIIDAE
Acartia danae Giesbrecht, 1889 Pe O Ep Tr/St
Acartia negligens Dana, 1849 Pe O Ep Tr
Acartia ensifera Brady, 1899 Pe Co Ep St E
Acartia jilletti Bradford, 1976 Pe Co Ep St E
Acartia simplex Sars, 1905 Pe Co Ep St E

AETIDEIDAE
Aetideus acutus Farran, 1929 Pe Ep Tr
Aetideus australis (Vervoort, 1957) Pe Ep Sa
Aetideus giesbrechti Cleve, 1904 Pe Ep Tr/St
Aetideus pseudarmatus Bradford, 1971 Pe Ep Tr
Aetideopsis tumorosa Bradford, 1969 Pe/BP Me Sa
Bradyidius capax Bradford-Grieve, 2003 Ba Bp
Bradyidius spinifer Bradford, 1969 Ba Bp
Chiridius molestus Tanaka, 1957 Pe Ep/Me Tr/St
Chiridius pacificus Brodsky, 1950 Pe By Tr/St
Chiridius poppei Giesbrecht, 1892 Pe Me Tr
Chirundina streetsii Giesbrecht, 1895 Pe Me Tr/St
Comantenna crassa Bradford, 1969 Ba Bp
Crassantenna comosa Bradford, 1969 Ba Bp
Crassantenna mimorostrata Bradford, 1969 Ba Bp
Euchriella amoena Giesbrecht, 1888 Pe Me Tr
Euchirella bitumida With, 1915 Pe Me Tr
Euchirella curticauda Giesbrecht, 1888 Pe Me Tr/St
Euchirella formosa Vervoort, 1949 Pe Me Tr/St
Euchirella latirostris Farran, 1929 Pe Me Sa
Euchirella messinensis indica Vervoort, 1949 Pe Me Tr/St
Euchirella m. messinensis (Claus, 1863) Pe By Tr/St
Euchirella rostrata (Claus, 1866) Pe Me Tr/St/Sa
Euchirella rostromagna Wolfenden, 1911 Pe Me Sa/
Euchirella similis Wolfenden, 1911 Pe By Tr/St
Euchirella speciosa Grice & Hulsemann, 1968 Pe Me Tr/St
Euchirella truncata Esterly, 1911 Pe Me Tr/St
Euchirella venusta Giesbrecht, 1888 Pe Me Tr/St
Gaetanus brevicornis Esterly, 1906 Pe By Tr/St
Gaetanus brevispinus (Sars, 1900) Pe By Tr/St
Gaetanus kruppii Giesbrecht, 1903 Pe By Tr/St
Gaetanus latifrons Sars, 1905 Pe By Tr/St
Gaetanus minor Farran, 1905 Pe Me Tr/St
Gaetanus minutus (Sars, 1907) Pe Me Tr/St
Gaetanus pileatus Farran, 1903 Pe By Tr/St
Gaetanus secundus Esterly, 1911 Pe Me Tr/St
Gaetanus tenuispinus (Sars, 1900) Pe MR Tr/St/Sa
Lutamator hurleyi Bradford, 1969 Ba Bp
Pseudeuchaeta brevicauda Sars, 1905 Pe By W
Pseudeuchaeta flexuosa Bradford, 1969 Ba Bp
Pseudeuchaeta magna Bradford, 1969 Ba Bp
Pseudochirella dentata (A. Scott, 1909) Pe By Tr/St
Pseudochirella mawsoni Vervoort, 1957 Pe BySt/ Sa/ Ant
Pseudochirella notacantha (Sars, 1905) Pe By Tr/St
Pseudochirella obesa Sars, 1920 Pe By Tr/St
Pseudochirella obtusa (Sars, 1905) Pe By Tr/St
Pseudotharybis brevispinus (Bradford, 1969) Ba Bp
Pseudotharybis dentatus (Bradford, 1969) Ba Bp
Pseudotharybis robustus (Bradford, 1969) Ba Bp
Pseudotharybis spinibasis (Bradford, 1969) Ba Bp
Sursamucro spinatus Bradford, 1969 Ba Bp
Undeuchaeta incisa Esterly, 1911 Pe By Tr/St
Undeuchaeta major Giesbrecht, 1888 Pe Me Tr/St
Undeuchaeta plumosa (Lubbock, 1856) Pe Me Tr/St
Valdiviella insignis Farran, 1908 Pe By Tr/St
ARIETELLIDAE
Arietellus aculeatus (T. Scott, 1894b) Pe Me Tr
Arietellus setosus Giesbrecht, 1892 Pe Me/By Tr
Campaneria latipes Ohtsuka, Boxshall & Roe, 1994 Ba Bp St
Paramisophria n. sp.* Bp Sh
Paraugaptiloides magnus (Bradford, 1974) Ba Bp St
Paraugaptilus ?buchani Wolfenden, 1904 Pe Me Tr
Scutogerulus pelophilus Bradford, 1969 Ba Bp St
AUGAPTILIDAE
Augaptilus longicaudatus (Claus, 1863) Pe Me Tr/St
Centraugaptilus horridus (Farran, 1908) Pe By Tr/ St
Euaugaptilus bullifer (Giesbrecht, 1889) Pe By Tr/St/Sa
Euaugaptilus filigerus (Claus, 1963) Pe By T/St
Euaugaptilus hecticus (Giesbrecht, 1889) Pe Ep/

Me Tr
Euaugaptilus humilis Farran, 1926 Pe By Tr
Euaugaptilus laticeps (Sars, 1905) Pe By Tr/St
Euaugaptilus longimanus (Sars, 1905) Pe By Tr
Euaugaptilus nodifrons (Sars, 1905) Pe By Tr/St/Sa
Euaugaptilus oblongus (Sars, 1905) Pe By Tr/St
Euaugaptilus palumbii (Giesbrecht, 1889) Pe Me Tr
Haloptilus acutifrons (Giesbrecht, 1892) Pe Me Tr/St
Haloptilus fons Farran, 1908 Pe Me/By Tr/St/Sa
Haloptilus longicornis (Claus, 1893) Pe Ep/Me Tr/St/Sa
Haloptilus ornatus (Giesbrecht, 1892) Pe Ep/Me Tr/St
Haloptilus oxycephalus (Giesbrecht, 1889) Pe Ep/ Me Tr/St/Sa
Haloptilus spiniceps (Giesbrecht, 1892) Pe Ep/Me Tr
Pachyptilus eurygnathus (Sars, 1905) Pe By Tr/St
BATHYPONTIIDAE
Temorites elongata (Sars, 1905) Pe By W
CALANIDAE
Calanoides acutus (Giesbrecht, 1902) Pe Ep/Me Sa/Ant
Calanoides macrocarinatus Brodsky, 1972 Pe Ep/Me St
Calanus australis Brodsky, 1959 Pe Co Ep St/Sa
Calanus simillimus Giesbrecht, 1902 Pe Ep Sa
Canthocalanus pauper (Giesbrecht, 1888) Pe Ep Tr
Cosmocalanus darwinii (Lubbock, 1860) Pe Ep Tr
Mesocalanus tenuicornis (Dana, 1849) Pe Ep T/St/ Sa
Nannocalanus minor (Claus, 1863) Pe Ep Tr/St
Neocalanus gracilis Dana, 1849 Pe Ep Tr/St
Neocalanus tonsus (Brady, 1883) Pe Ep/Me St/Sa
CANDACIIDAE
Candacia bipinnata (Giesbrecht, 1888) Pe Ep/Me Tr/St
Candacia cheirura Cleve, 1904 Pe Ep/Me St/Sa
Candacia ethiopica (Dana, 1849) Pe Ep/Me Tr
Candacia longimana (Claus, 1863) Pe Ep/Me Tr/St
Candacia pachydactyla (Dana, 1849) Pe Ep/Me St
Candacia tenuimana (Giesbrecht, 1888) Pe Me Tr/St
Paracandacia simplex (Giesbrecht, 1889) Pe Ep T/St
Paracandacia worthingtoni Grice, 1981 Pe Ep Tr
CENTROPAGIDAE
Boeckella delicata Percival, 1937 F Pe
Boeckella dilatata Sars, 1904 F Pe E
Boeckella hamata Brehm, 1928 F Pe E
Boeckella minuta Sars, 1896 F Pe A
Boeckella propinqua Sars, 1904 F Pe
Boeckella symmetrica Sars, 1908 F Pe A
Boeckella tanea Chapman, 1973 F Pe E
Boeckella triarticulata (Thomson, 1883) F Pe
Calamoecia lucasi Brady, 1906 F Pe
Centropages aucklandicus Krämer, 1895 Pe Co Ep St E
Centropages bradyi Wheeler, 1900 Pe Me Tr/St
Centropages elegans Giesbrecht, 1895 Pe O Ep Tr
Centropages violaceus (Claus, 1863) Pe O Ep Tr
Gladioferens pectinatus (Brady, 1899) B Pe Ep St
Gladioferens spinosus Henry, 1919 B Pe Ep St
CLAUSOCALANIDAE
Clausocalanus arcuicornis (Dana, 1849) Pe Ep Tr/St
Clausocalanus brevipes Frost & Fleminger, 1968 Pe Ep Sa
Clausocalanus ingens Frost & Fleminger, 1968 Pe Ep Tr/St/Sa
Clausocalanus jobei Frost & Fleminger, 1968 Pe Ep St
Clausocalanus laticeps Farran, 1929 Pe Ep Sa
Clausocalanus lividus Frost & Fleminger, 1968 Pe Ep Tr/St
Clausocalanus parapergens Frost & Fleminger, 1968 Pe Ep Tr/St
Clausocalanus paululus Farran, 1926 Pe Ep Tr/St
Clausocalaus pergens Farran, 1926 Pe Ep St
Ctenocalanus vanus Giesbrecht, 1888 Pe Ep St
Drepanopus pectinatus Brady, 1883 Pe Ep Co Sa
DIAPTOMIDAE A
Sinodiaptomus valkanovi Kiefer, 1938 F Pe A
Skistodiaptomus pallidus (Herrick, 1879) F Pe A
EUCALANIDAE
Eucalanus hyalinus (Claus, 1866) Pe Ep/Me Tr/St
Pareucalanus langae (Fleminger, 1973) Pe Ep Tr
Pareucalanus sewelli (Fleminger, 1973) Pe Ep Tr/St
Rhincalanus gigas Brady, 1883 Pe Ep/Me Sa/Ant
Rhincalanus nasutus Giesbrecht, 1888 Pe Ep/Me St
Rhincalanus rostrifrons (Dana, 1852) Pe Ep Tr
Subeucalanus crassus (Giesbrecht, 1888) Pe Ep Tr/St
Subeucalanus longiceps (Matthews, 1925) Pe Ep Sa
Subeucalanus mucronatus (Giesbrecht, 1888) Pe Ep Tr
EUCHAETIDAE
Euchaeta acuta Giesbrecht, 1892 Pe Ep Tr/St
Euchaeta media Giesbrecht, 1888 Pe Ep Tr/St
Euchaeta longicornis Giesbrecht, 1888 Pe Ep T/St
Euchaeta rimana Bradford, 1974 Pe Ep T/St
Euchaeta pubera Sars, 1907 Pe Ep T/St
Euchaeta spinosa Giesbrecht, 1892 Pe Me Tr
Pareuchaeta biloba Farran, 1929 Pe Me Sa/Ant
Pareuchaeta bisinuata (Sars, 1907) Pe By Tr/St
Pareuchaeta comosa Tanaka, 1958 Pe By Tr/St
Pareuchaeta exigua (Wolfenden, 1911) Pe By Tr/St
Pareuchaeta hansenii (With, 1915) Pe Me Tr/St
Pareuchaeta pseudotonsa (Fontaine, 1967) Pe By Tr/St/Sa
Pareuchaeta sarsi (Farran, 1908) Pe By W
HETERORHABDIDAE
Disseta magna Bradford, 1971 Pe By St
Disseta palumbii Giesbrecht, 1889 Pe By Tr/St
Heterorhabdus abyssalis (Giesbrecht, 1889) Pe Me/By St
Heterorhabdus austrinus Giesbrecht, 1902 Pe Me/By Sa/Ant
Heterorhabdus caribbeanensis Park, 1970 Pe Me Tr
Heterorhabdus lobatus Bradford, 1971 Pe Me Tr
Heterorhabdus pacificus Brodsky, 1950 Pe By Tr/ St
Heterorhabdus papilliger (Claus, 1863) Pe Ep/me Tr
Heterorhabdus proximus Davis, 1949 Pe Me St
Heterorhabdus robustus Farran, 1908 Pe
Heterorhabdus spinifer Park, 1970 Pe Me Tr
Heterorhabdus spinifrons (Claus, 1863) Pe Me Tr/St
Heterohabdus spinosus Bradford 1971 Pe Me St
Heterostylites longicornis (Giesbrecht, 1889) Pe Me Tr/St
LUCICUTIIDAE
Lucicutia bicornuta Wolfenden, 1905 Pe Ep/Me Tr/St
Lucicutia clausi (Giesbrecht, 1889) Pe Me Tr/St
Lucicutia curta Farran, 1905 Pe Me W
Lucicutia flavicornis (Claus, 1863) Pe Ep/Me Tr/St
Lucicutia cf. *flavicornis*, Bradford-Grieve, 1999 Pe Ep/Me Tr/St
Lucicutia gemina Farran, 1926 Pe Ep/Me Tr
Lucicutia grandis (Giesbrecht, 1895) Pe By W
Lucicutia longiserrata (Giesbrecht, 1889) Pe By Tr
Lucicutia magna Wolfenden *in* Fowler, 1903 Pe By W
Lucicutia ovalis (Giesbrecht, 1889) Pe Ep/Me Tr
MECYNOCERIDAE
Mecynocera clausi Thompson, 1888 Pe Ep Tr/St
MEGACALANIDAE
Megacalanus longicornis Sars, 1925 Pe By W
METRIDINIDAE
Gaussia princeps T. Scott, 1894 Pe By Tr/St
Metridia brevicauda Giesbrecht, 1889 Pe Me/By Tr/St
Metridia curticauda Giesbrecht, 1889 Pe Me/By W
Metridia lucens Boeck, 1865 Pe Ep/Me Tr/St/Sa
Metridia princeps Giesbrecht, 1892 Pe By W
Metridia venusta Giesbrecht, 1892 Pe Me/By Tr/ St
Pleuromamma abdominalis (Lubbock, 1856) Pe Me Tr/St/Sa
Pleuromamma borealis (Dahl, 1893) Pe Me Tr/St/Sa
Pleuromamma gracilis (Claus, 1863) Pe Me Tr/St
Pleuromamma piseki Farran, 1929 Pe Me Tr/St
Pleuromamma quadrungulata (Dahl, 1893) Pe Me Tr/St/Sa
Pleuromamma robusta (Dahl, 1893) Pe Me Tr/St/Sa
Pleuromamma xiphias Giesbrecht, 1889 Pe Me Tr/St
NULLOSETIGERIDAE
Nullosetigera bidentatus (Brady, 1883) Pe Me W
Nullosetigera helgae (Farran, 1908) Pe Me/By W
PARACALANIDAE
Calocalanus longispinus Shmeleva, 1978 Pe Ep Tr/St
Calocalanus minutus Andronov, 1973 Pe Ep Tr/St
Calocalanus namibiensis Andronov, 1973 Pe Ep Tr/St
Calocalanus neptunus Schmeleva, 1965 Pe Ep Tr/St
Calocalanus pavo (Dana, 1849) Pe Ep Tr/St
Calocalanus plumulosus (Claus, 1863) Pe Ep T/St
Calocalanus styliremis Giesbrecht, 1888 Pe Ep Tr/St
Calocalanus tenuis Farran, 1926 Pe Ep Tr/St
Paracalanus aculeatus Giesbrecht, 1892 Pe Ep Tr/St
Paracalanus indicus Wolfenden, 1905 Pe Ep Tr/St
PHAENNIDAE
Cornucalanus chelifer (I.C. Thompson, 1903) Pe By Tr/St
Onchocalanus cristatus (Wolfenden, 1904) Pe By T/St
Onchocalanus trigoniceps Sars, 1905 Pe By Tr/St
Neoscolecithrix cf. *magna* (Grice, 1972) Bp
Neoscolecithrix ornata Bradford-Grieve, 2001 Bp
Phaenna spinifera Claus, 1863 Pe Me T/St
Xanthocalanus penicillatus Tanaka, 1960 Pe By Tr/St
PONTELLIDAE
Calanopia aurivilli Cleve, 1901 Pe O Ep Tr
Labidocera cervi Krämer, 1895 Pe Co Ep St
Labidocera detruncata (Dana, 1849) Pe O Ep Tr
Pontella novaezelandiae Farran, 1929 Pe Co Ep St E
Pontella valida Dana, 1852 Pe O Ep Tr
Pontella whiteleggei Krämer, 1896 Pe O Ep Tr
Pontellina plumata (Dana, 1849) Pe O Ep Tr
Pontellopsis grandis (Lubbock, 1853) Pe O Ep Tr
PSEUDOCYCLOPIDAE
Pseudocyclops n. sp.* Bp Sh
SCOLECITRICHIDAE
Amallothrix arcuata (Sars, 1920) Pe By Tr/St
Amallothrix dentipes (Vervoort, 1951) Pe Me Sa/Ant
Amallothirx emarginata (Farran, 1905) Pe By Tr/St
Amallothrix gracilis (Sars, 1905) Pe By Tr/St
Amallothrix parafalcifer (Park, 1980) Pe By St
Amallothrix pseudopropinqua (Park, 1980) Pe By St
Amallothrix valida (Farran, 1908) Pe By W
Lophothrix frontalis Giesbrecht, 1895 Pe By Tr/St
Lophothrix latipes (T. Scott, 1894) Pe Me Tr
Scaphocalanus affinis (Sars, 1905) Pe By W
Scaphocalanus brevicornis (Sars, 1900) Pe Me Tr/St
Scaphocalanus curtus (Farran, 1926) Pe Ep Tr
Scaphocalanus echinatus (Farran, 1905) Pe Ep Tr/St/Sa
Scaphocalanus longifurca (Giesbrecht, 1888) Pe Me Tr/St
Scaphocalanus magnus (T. Scott, 1894) Pe By W
Scaphocalanus major (T. Scott, 1894) Pe Me Tr/St
Scaphoclanaus subbrevicornis (Wolfenden, 1911) Pe Me W
Scolecithricella abyssalis (Giesbrecht, 1888) Pe Me Tr/St
Scolecithricella dentata (Giesbrecht, 1892) Pe Me Tr/St
'Scolecithricella' fowleri (Farran, 1926) Pe Me Tr
Scolecithricella minor (Brady, 1883) Pe Ep W
Scolecithricella ovata (Farran, 1905) Pe Me W
Scolecithricella schizosoma Park, 1980 Pe By Sa/Ant
Scolecithricella vittata (Giesbrecht, 1892) Pe Me Tr/St
Scolecithrix bradyi Giesbrecht, 1888 Pe Ep Tr
Scolecithrix danae (Lubbock, 1856) Pe Ep Tr
Scopalatum sp. Bradford *et al.* 1983 Pe Me St
Scottocalanus helenae (Lubbock, 1856) Pe Me Tr/St

Scottocalanus securifons (T. Scott, 1894) Pe By Tr/ St
Scottocalanus terranovae Farran, 1929 Pe By St
Scottocalanus thorii With, 1915 Pe By Tr/St
SPINOCALANIDAE
Spinocalanus longicornis Sars, 1900 Pe By W
Spinocalanus spinosus Farran, 1908 Pe By Tr
STEPHIDAE
Stephos angulatus Bradford-Grieve, 1999 Bp Sh E
Stephos hastatus Bradford-Grieve, 1999 BP Sh E
SULCANIDAE
Sulcanus conflictus Nicholls, 1945 B Pe Co Ep A?
TEMORIDAE
Temora turbinata (Dana, 1849) Pe Co Ep S/St
Temoropia minor Deevey, 1972 Pe By Tr
Gen. et sp. indet.* Bp Sh
THARYBIDAE
Tharybis inaequalis Bradford-Grieve, 2001 Ba Bp
Tharybis spp. (2)* Bp Sh
Undinella brevipes Farran, 1908 Pe Me Tr/St

Order CYCLOPOIDA
ASCIDICOLIDAE
Botryllophilus cf. *banyulensis* Brément, 1909*
Enteropsis onychophorus Schellenberg, 1922 P (tunicates)
Haplostoma gibberum (Shellenberg, 1922) P (tunicates)
Haplostomides otagoensis Ooishi, 2001 P (tunicates)
BOMOLOCHIDAE
Acanthocolax sp. Beresford 1991 P (fish)
Pseudoeucanthus australiensis Roubal, Armitage & Rohde, 1983* P (fish)
Pseudoeucanthus uniserratus Wilson, 1913 P (fish)
Unicolax chrysophryenus Roubal, Armitage & Rohde, 1983 P (fish)
CHITONOPHILIDAE
Cocculinika myzorama Jones & Marshall, 1986 P (molluscs)
CHONDRACANTHIDAE
Acanthochondria incisa Shiino, 1955 P (fish)
Chondracanthodes radiatus Müller, 1777 P (fish)
Chondracanthus australis Ho, 1991 P (fish)
Chondracanthus distortus Wilson, 1922 P (fish)
Chondracanthus genypteri Thomson, 1890 P (fish)
Chondracanthus lotellae Thomson, 1890 P (fish)
Chondracanthus yanezi Atria, 1980 P (fish)
Mecaderochondria pilgrimi Ho & Dojiri, 1987 P (fish)
Prochondracanthus platycephali Ho, 1975 P (fish)
Pseudochondracanthus chilomycteri (Thomson, 1890) P (fish)
CLAUSIDIIDAE
Hemicyclops? n. sp., n. gen.? * Be C
Teredicola typicus Wilson, 1942 P (boring molluscs)
CORYCAEIDAE
Corycaeus agilis Dana, 1849* Pe Ep Tr/St
Corycaeus aucklandicus Kramer, 1895 Pe Ep Co E
Corycaeus clausi F. Dahl, 1894* Pe Ep Tr/St
Corycaeus crassiusculus Dana, 1849* Pe Ep Tr/St
Corycaeus flaccus Giesbrecht, 1891* Pe Ep Tr/St
Corycaeus furcifer Claus, 1863* Pe Ep Tr/St
Corycaeus latus Dana, 1849* Pe Ep Tr/St
Corycaeus limbatus Brady, 1883* Pe Ep Tr/St
Corycaeus longistylis Dana, 1849* Pe Ep Tr
Corycaeus speciosus Dana, 1849* Pe Ep Tr/St
Corycaeus typicus Krøyer, 1849* Pe Ep Tr
Farranula rostata (Claus, 1863)* Pe Ep S/St
CYCLOPIDAE
Abdiacyclops cirratus Karanovic, 2005 F E
Acanthocyclops robustus (Sars, 1863) F Be A?
Acanthocyclops vernalis (Fischer, 1853) F Pe
Cyclops? strennus strennus Fischer, 1851 P
Diacyclops bicuspidatus (Claus, 1857) F Be
Diacyclops bisetosus (Rehberg, 1880) F Be A?
Eucyclops serrulatus (Fischer, 1851) F Pe A?
Euryte? longicauda Philippi, 1843 Be
Goniocyclops silvestris Harding, 1958 F Ph E
Halicyclops? magniceps (Lilljeborg, 1853) B Be
Halicyclops? neglectus Kiefer, 1935 F/B Be/Pe
Macrocyclops albidus (Jurine, 1820) F Be
Mesocyclops? australiensis (Sars, 1908) F Pe
Mesocyclops? leuckarti (Claus, 1857) F Pe
Metacyclops monacanthus (Kiefer, 1928) B Pe E
Microcyclops? varicans Sars, 1863 F Be/Pe
Paracyclops chiltoni (Thomson, 1883) F Be
Paracyclops fimbriatus (Fischer, 1853) F/B Be A?
Paracyclops waiariki Lewis, 1974 F Be E
Tropocyclops? prasinus (Fischer, 1860) F Be/Pe
Zealandcyclops fenwicki Karanovic, 2005 F E
Zealandcyclops haywardi Karanovic, 2005 F E
ERGASILIDAE
Abergasilus amplexus Hewitt, 1978 B P (fish)
Paeonodes nemaformis Hewitt, 1969 F P (fish, extinct?) E
Thersitina inopinata Percival, 1937 F Pe P (fish, extinct?)
LERNAEIDAE
Lernaea cyprinacea Linnaeus, 1758 F P (fish) A
LICHOMOLGIDAE
Lichomolgidium tupuhiae Jones, 1975 C (molluscs)
Lichomolgus uncus Jones, 1976 C (molluscs)
MYTILICOLIDAE
Pseudomyicola spinosus (Raffaele & Monticelli, 1885) C (molluscs)
NOTODELPHYIDAE
Pygodelphys novaeseelandius (Shellenberg, 1922) C (tunicates)
Doropygus globosus Jones, 1974 C (tunicates)
Doropygus louisae Jones, 1980 C (tunicates)
Doropygus platythorax Jones, 1974 C (tunicates)
Doropygus pulex Shellenberg, 1922 C (tunicates)
Doropygus spinosus Jones, 1980 C (tunicates)
Doropygus trisetosus Shellenberg, 1922 C (tunicates)
Ophioseides schellenbergi Jones, 1980 C (tunicates)
OITHONIDAE
Oithona atlantica Farran, 1908 Pe Ep St
Oithona nana Giesbrecht, 1892 Pe Ep Tr/St
Oithona plumifera Baird, 1843 Pe Ep Tr/St
Oithona similis Claus, 1866 Pe Ep W
ONCAEIDAE
Conaea rapax Giesbrecht, 1891 Pe Me W
Lubbockia aculeata Giesbrecht, 1891 Ep/Me Tr/St
Lubbockia squillimana Claus, 1863 Pe Ep/Me Tr/St
Oncaea antarctica Heron, 1977 Pe Ep/Me Sa/Ant
Oncaea conifera Giesbrecht, 1891 Pe Ep/Bap Tr/St
Oncaea derivata Heron & Bradford-Grieve, 1995 Pe Me Tr/St
Oncaea englishi Heron, 1977 Pe Ep/Bap W
Oncaea furcula Farran, 1936 Pe Me Tr/St
Oncaea inflexa Heron, 1977 Pe Ep/Me Sa
Oncaea media Giesbrecht, 1891 Pe Ep/Me Tr/St
Oncaea mediterranea (Claus, 1863) Pe Ep/Me W
Oncaea quadrata Heron & Bradford-Grieve, 1995 Pe Ep/Me St
Oncaea redacta Heron & Bradford-Grieve, 1995 Pe Ep/Me Tr
Oncaea scottodicarloi Heron & Bradford-Grieve, 1995 Pe Ep Tr/St
Oncaea similis Sars, 1918 Pe Ep/Me St
Oncaea venusta Philippe, 1843 Pe Ep/Me Tr/St
PHILICHTHYIDAE
Philichthys xiphiae Steenstrup, 1862 P (fish)
Sarcotaces sp. Avdeev & Avdeev 1975 P (fish)
SAPPHIRINIDAE
Copilia hendorffi Dahl, 1892* Pe Ep Tr/St
Copilia mirabilis Dana, 1849* Pe Ep/Me Tr/St
Copilia vitrea (Haeckel, 1864)* Pe Ep/Me Tr
Sapphirina angusta Dana, 1849* Pe Ep Tr/St
Sapphirina autonitens-sinuicauda Lehnhofer, 1929* Pe Ep Tr/St
Sapphirina ovatolanceolata-gemma Lehnhofer, 1929* Pe Ep Tr/St
Sapphirina intestinata Giesbrecht, 1891* Pe Ep T/St
Sapphirina iris Dana, 1849* Pe Ep Tr/St
Sapphirina opalina-darwini Lehnhofer, 1929* Pe Ep Tr/St
Sapphirina sali Farran, 1929* Pe Ep St
Sapphirina scarlata Giesbrecht, 1891* Pe Ep T/St
THAMNOMOLGIDAE
Thamnomolgus eurycephalus Humes & Kiss, 2004 P (black coral)

Order MORMONILLOIDA
Mormonilla phasma Giesbrecht, 1891* Pe

Order HARPACTICOIDA
AEGISTHIDAE
Aegisthus mucronatus Giesbrecht, 1891 Pe
AMEIRIDAE
Ameira minuta Boeck, 1864 Ph
Ameira parvula (Claus, 1866) Ph BeL
Ameira sp.* BeL
Ameiropsyllus (?) spp. (5)* BeL
Leptameira sp.* BeL
Nitocra fragilis Sars, 1905 Ch B Be
Nitocra sp. (2)* BeL
Parapseudoleptomesochra (?) sp.* BeL
Parevansula sp.* BeL
Psyllocamptus sp.* BeL
ANCORABOLIDAE
Laophontodes hamatus (Thomson, 1883) Ph E
Laophontodes whitsoni T. Scott, 1912 Ca Be
Paralaophontodes sp.* BeL
ARENOPONTIIDAE
Arenopontia sp.* BeL
CANTHOCAMPTIDAE
Antarctobiotus australis Lewis, 1972 F Ph(M) E
Antarctobiotus diversus Lewis, 1972 F Ph(M) E
Antarctobiotus elongatus Lewis, 1972 F Ph(M) E
Antarctobiotus exiguus Lewis, 1972 F Ph(M) E
Antarctobiotus ignobilis Lewis, 1972 F Ph(M) E
Antarctobiotus triplex Lewis, 1972 F Ph(M) E
Antarctobiotus n. sp.* F Ph(M)
Antipodiella chappuisi Brehm, 1928* F Ph(M)
Antipodiella n. spp. (3)* 3F Ph(M)
Attheyella (Chappuisiella) fluviatilis Lewis, 1972 F Ph(M) E
Attheyella (C.) maorica (Brehm, 1928) F Ph(M) E
Attheyella (C.) rotoruensis Lewis, 1972 F Pe E
Attheyella (Delachauxiella) bennetti Brehm, 1927 F Ph(M) E
Attheyella (D.) brehmi Kiefer, 1928 F Ph(M) E
Attheyella (D.) humidarum Lewis, 1972 F Ph(M) E
Attheyella (D.) stillicidarum Lewis, 1972 F Ph(M) E
Bryocamptus (Rheocamptus) pygmaeus (Sars, 1862)* F Ph(M)
Bryocamptus (Echinocamptus) stouti Harding, 1958 T (forest litter) E
Bryocamptus n. spp. (3)* 3F
Elaphoidella bidens coronata Sars, 1904 F BeL
Elaphoidella silvestris Lewis, 1972 F Ph(M) E
Elaphoidella sp.* F Be
Epactophanes richardi Mrázek, 1893 F Ph, Ph(M)
Loeflerella n. sp.* F Ph(M)
Mesochra flava Lang, 1933 Ph
Mesochra meridionalis Sars, 1905 B
Mesochra parva Thomson, 1946 B BeL BeSL
Mesochra pygmaea (Claus, 1863)* BeL
Mesochra spp. (2)* BeL
Gen. nov. (2) et n. spp. (7)* 7F
CANUELLIDAE
Brianola sp.* B BeL
CLETODIDAE
Enhydrosoma variabile Wells, Hicks & Coull, 1982 BeL BeSL E

Enhydrosoma spp. (2)* BeL
Enhydrosomella spp. (2)* BeL
Stylicletodes longicaudatus (Brady & Robertson, 1880) Ph
Stylicletodes sp.* BeL
DACTYLOPUSIIDAE
Dactylopusia frigida T. Scott, 1912 Ph
Dactylopusia tisboides (Claus, 1863) Ph BeL BeSL
Diarthrodes cystoecus Fahrenbach, 1954 Ph
Diarthrodes novaezealandiae Thomson, 1882 Ph E
Diarthrodes sp.* Ph
Paradactylopodia brevicornis (Claus, 1866) Ph
Paradactylopodia trioculata Hicks, 1988 Ph(W) E
DARCYTHOMPSONIIDAE
Gen. nov. et n. sp. Huys & Gee in press* BeL
ECTINOSOMATIDAE
Arenosetella sp. * BeL
Ectinosoma melaniceps Boeck, 1864 Ca Ch BeL
Ectinosoma sp.* BeL
Glabrotelson spp. (3)* BeL
Halectinosoma hydrofuge Wells, Hicks & Coull, 1982 BeL E
Halectinosoma otakoua Wells, Hicks & Coull, 1982 BeL E
Halectinosoma spp. (3)* BeL
Kliella (?) sp.* BeL
Microsetella norvegica (Boeck, 1864) Pe Ep W
Microsetella rosea (Dana, 1848) Pe Ep W
Noodtiella sp. * BeL
HARPACTICIDAE
Harpacticus furcatus Lang, 1936 Ph
Harpacticus glaber Brady, 1899 Pe SL E
Harpacticus pulvinatus Brady, 1910 Ph
Harpacticus spp. (2)* Ph
Perissocope litoralis Lang, 1934 Ph E
Tigriopus angulatus Lang, 1933 Ca Sn Ph
Tigriopus raki Bradford, 1967 Ph E
Zaus sp.* Ph
Zausopsis contractus (Thomson, 1883) Ph E
Zausopsis mirabilis Lang, 1934 Ph E
LAOPHONTIDAE
Afrolaophonte sp.* BeL
Apolethon sp.* BeL
Folioquinpes chathamensis (Sars, 1905) B E
Harrietella simulans (T. Scott, 1894) Ph(W)
Heterolaophonte campbelliensis (Lang, 1934) Ca Ph
Heterolaophonte tenuispina (Lang, 1934) Ca Ph
Klieonychocamptoides sp.* BeL
Laophonte australasica Thomson, 1883 E
Laophonte cornuta Philippi, 1840 Ca Ph
Laophonte elongata barbata Lang, 1934 Ph
Laophonte inornata A. Scott, 1902 Ph
Laophonte lignosa Hicks, 1988 Ph(W) E
Laophonte sima Gurney, 1927 Ph
Laophonte spp. (2)* BeL
Onychocamptus mohammed (Blanchard & Richard, 1891) B
Paeudonychocamptus sp.* BeL
Paralaophonte aenigmaticum Wells, Hicks & Coull, 1982 BeL E
Paronychocamptus exiguus (Sars, 1905) B E
Paralaophonte meinerti (Brady, 1899) Ca Ph
Paralaophonte spp. (4)* BeL
Pseudolaophonte spp. (2)* BeL
Quinquelaophonte candelabrum Wells, Hicks & Coull, 1982 BeL BeSL Ph E
Quinquelaophonte longifurcata (Lang, 1965) Ph
Quinquelaophonte sp.* BeL
Xanthilaophonte trispinosa (Sewell, 1940) BeL BeSL
LEPTASTACIDAE
Leptastacus sp.* BeL
LOURINIIDAE
Lourinia armata (Claus, 1866) Ph
MIRACIIDAE
Amonardia perturbata Lang, 1965 Ph
Amphiascoides nichollsi Lang, 1965 Ph
Amphiascoides sp.* BeL
Amphiascopsis cinctus (Claus, 1866) Ph
Amphiascopsis southgeorgiensis (Lang, 1936) Ph
Amphiascus waihonu (Hicks, 1986) Be (C?) E
Bulbamphiascus imus (Brady, 1872) Ph
Bulbamphiascus spp. (2)* BeL
Cladorostrata sp.* BeL
Delavalia spp. (3)* BeL
Helmutkunzia sp.* BeL
Macrosetella gracilis (Dana, 1847) Pe Ep Tr/St
Metamphiascopsis monardi (Lang, 1934) Ph E
Miscegenus heretaunga Wells, Hicks & Coull, 1982 BeL BeSL E
Miscegenus spp. (2)* BeL
Oculosetella gracilis (Dana, 1849) Pe Ep Tr/St
Pseudostenhelia sp.* BeL
Robertgurneya sp.* BeL
Robertsonia propinqua (T. Scott, 1893) Ph
Sarsamphiascus hirtus (Gurney, 1927) Ca Ph
Sarsamphiascus lobatus (Hicks, 1971)
Sarsamphiascus pacificus (Sars, 1905) Ch Ph
Sarsamphiascus tainui (Hicks, 1989) W E
Sarsamphiascus spp. (2)* BeL
Schizopera clandestina (Klie, 1924) B
Schizopera longicauda Sars, 1905 Ch B Be
Schizopera sp.* BeL
Stenhelia xylophila Hicks, 1988 Ph(W) E
Stenhelia sp. BeL
Teissierella (?) sp.* BeL
Typhlamphiascus unisetosus Lang, 1965 Ph
Typhlamphiascus sp.*BeL
NANNOPODIDAE
Gen. et sp. indet.* BeL
NORMANELLIDAE
Normanella incerta Lang, 1934 Ph E
ORTHOPSYLLIDAE
Orthopsyllus linearis (Claus, 1866) Ph
PARAMESOCHRIDAE
Apodopsyllus sp.* BeL
Diarthrodella sp.* BeL
Emertonia sp.* BeL
PARASTENHELIIDAE
Parastenhelia hornelli Thompson & A. Scott, 1903 BeL
Parastenhelia megarostrum Wells, Hicks & Coull, 1982 BeL BeSL E
Parastenhelia spinosa (Fischer, 1860) CaPh BeL BeSL
Parastenhelia sp.* BeL
PELTIDIIDAE
Alteutha depressa (Baird, 1837) Ph
Alteutha novaezealandiae (Brady, 1899) Ph E
Alteuthoides kootare Hicks, 1986 C (sponges) E
Clytemnestra rostrata (Brady, 1883) Pe Ep/Me Tr/St
Clytemnestra scutellata Dana, 1848 Pe Ep/Me Tr/St
Eupelte regalis Hicks, 1971 Ph E
Neopeltopsis pectinipes Hicks, 1976 Ph E
PHYLLOGNATHOPODIDAE
Phyllognathopus viguieri (Maupas, 1892) F Ph(M)
Phyllognathopus volcanicus Barclay, 1969 F BeL BeS Ph E
PORCELLIDIIDAE
Dilatatiocauda dilatatum (Hicks, 1971) Ph E
Porcellidium erythrum Hicks, 1971 Ph E
Porcellidium fulvum Thomson, 1883 Ph E
Porcellidium interruptum Thomson, 1883 Ph E
Porcellidium tapui Hicks & Webber, 1983 C E (hermit crabs)
PSAMMOPSYLLIDAE
Psammopsyllus sp.* BeL
PSEUDOTACHIDIIDAE
Dactylopodella flava (Claus, 1866) Ph(W)
Dactylopodella janetae Hicks, 1989 Ph(W) E
Dactylopodella sp.* Ph
Danielssenia sp.* Be L
Donsiella bisetosa Hicks, 1988 Ph(W) E
Paranannopus sp.* BeL
Pseudomesochra sp.* BeL
Pseudonsiella aotearoa Hicks, 1988 Ph(W) E
Xouthous intermedia (Lang, 1934) Ph E
Xouthous novaezealandiae (Thomson, 1882) Ph E
Xylora bathyalis Hicks, 1988 Ph(W) E
Xylora neritica Hicks, 1988 Ph(W)E
RHIZOTHRICIDAE
Rhizothricidae sp.* BeL
RHYNCHOTHALESTRIDAE
Rhynchothalestris campbelliensis Lang, 1934 Ph E
TACHIDIIDAE
Euterpina acutifrons (Dana, 1848) Pe Ep W
Geeopsis incisipes (Klie, 1913) B
Tachidius sp.* BeL
TEGASTIDAE
Syngastes clausii (Thomson, 1883) Ph E
TETRAGONICIPITIDAE
Phyllopodopsyllus minor (Thompson & A. Scott, 1903) Ph
Phyllopodopsyllus sp.* BeL
THALESTRIDAE
Flavia crassicornis Brady, 1899 E
Thalestris australis Brady, 1899 Ph? E
Thalestris ciliata Brady, 1899 Ph? E
TISBIDAE
Scutellidium armatum (Wiborg, 1964) Ph
Scutellidium idyoides (Brady, 1883) Ph?
Scutellidium macrosetum Branch, 1975 Ph
Scutellidium plumosum Brady, 1899 Ca Ph BeL
Scutellidium ringueleti Pallares, 1969 Ph
Tisbe furcata (Baird, 1837) Ch Ph
Tisbe gurneyi (Lang, 1934) Ph E
Tisbe holothuriae Humes, 1957 Ph
Tisbe sp.* Ph

Order SIPHONOSTOMATOIDA
ARTOTROGIDAE
Artotrogus gordoni Kim, 2009 E (bryozoan)
ASTEROCHERIDAE
Cecidomyzon conophorae Stock, 1981 P (coral) E
Cystomyzon dimerum Stock, 1981 P (coral) E
Oedomyzon tripodum Stock, 1981 P (coral) E
CANCERILLIDAE
Cancerilla neozelandica Stephensen, 1927 P (brittlestars) E
CALIGIDAE
Caligus aesopus Wilson, 1921 P (fish)
Caligus bonito Wilson, 1905 P (fish)
Caligus brevis Shiino, 1954 P (fish)
Caligus buechlerae Hewitt, 1964 P (fish) E
Caligus coryphaenae Steenstrup & Lütken, 1861 P (fish)
Caligus elongatus Nordmann, 1832 P (fish)
Caligus epidemicus Hewitt, 1971 P (fish)
Caligus kahawai Jones, 1988 P (fish) E
Caligus lalandei Barnard, 1948 P (fish)
Caligus longicaudatus Brady, 1899 P (fish) E
Caligus pelamydis Krøyer, 1863 P (fish)
Caligus productus Dana, 1852 P (fish) ?
Caligus sp. 1 Sharples & Evans 1995 P (fish)
Caligus sp. 2 Sharples & Evans 1995 P (fish)
Dentigryps sp.* P (fish)
Lepeophtheirus argentus Hewitt, 1963 P (fish) E
Lepeophtheirus crassus Wilson & Bere, 1936 P (fish)
Lepeophtheirus distinctus Hewitt, 1963 P (fish) E
Lepeophtheirus erecsoni Thomson, 1891 P (fish) E
Lepeophtheirus hastus Shiino, 1960 P (fish)
Lepeophtheirus heegaardi Hewitt, 1963 P (fish)
Lepeophtheirus histiopteridi Kazachenko, Korotaeva & Kurochkin, 1972 P (fish) E
Lepeophtheirus nordmanni (Edwards, 1840) P (fish)
Lepeophtheirus polyprioni Hewitt, 1963 P (fish) E

Lepeophtheirus scutiger Shiino, 1952 P (fish)
Lepeophtheirus sekii Yamaguti, 1936 P (fish)
Lepeophtheirus sp.* P (fish)
CECROPIDAE
Cecrops latreillei Leach, 1816 P (fish)
DICHELESTHIIDAE
Anthosoma crassum (Abildgaard, 1794) P (fish)
ENTOMOLEPIDAE
Entomolepis ovalis Brady, 1899 E
EUDACTALINIDAE
Eudactylina acanthii Scott, 1901 P (fish)
Jusheyus shogunus Deets & Benz, 1987 P (fish)
Nemesis lamma lamma Risso, 1826 P (fish)
Nemesis l. vermi Scott, 1929 P (fish)
Nemesis robusta (van Beneden, 1851) P (fish)
EURYPHORIDAE
Euryphorus brachypterus (Gerstaecker, 1853) P (fish)
Euryphorus nordmanni Milne-Edwards, 1840 P (fish)
Gloiopotes huttoni (Thomson, 1890) P (fish)
HATSCHEKIIDAE
Congericola kabatai Hewitt, 1975 P (fish) E
Hatschekia conifera Yamaguti, 1939 P (fish)
Hatschekia crenata Hewitt, 1969 P (fish) E
Hatschekia pagrosomi Yamaguti, 1939 P (fish)
Hatschekia quadrata Hewitt, 1969 P (fish) E
Hatschekia squamata Jones & Cabral, 1990 P E (fish)
HERPYLLOBIIDAE
Herpyllobius rotundus Lutzen & Jones, 1976 P (polychaete) E
KROYERIIDAE
Kroyeria carchariaeglauci Hesse, 1897* P (shark)
Kroyeria cf. *lineata* P (fish)
LERNAEOPODIDAE
Albionella sp.* P (fish)
Alella tarakihi Hewitt & Blackwell, 1987 P (fish) E
Brachiella thynni Cuvier, 1830 P (fish)
Brachiella sp.* P (fish)
Charopinus parkeri (Thomson, 1816) P (fish)
Clavella zini Kabata, 1979 P (fish) E
Clavella sp.* P (fish)
Clavellodes sp. Vooren & Tracey 1976 P (fish)
Clavellopsis sargi (Kurz, 1877) P (fish)
Dendrapta sp. Jones, 1988 P (fish)
Lernaeopoda musteli Thomson, 1890 P (fish) E
Lernaeopoda sp. *B. Jones unpubl. P (fish)
Naobranchia sp. Pilgrim 1985 P (fish)
Parabrachiella amphipacifica Ho, 1982 P (fish)
Parabrachiella insidiosa f. *lageniformes* (Heller, 1865) P (fish)
Parabrachiella sp. Pilgrim 1985 P (fish)
Pseudocharopinus bicaudatus (Kroyer, 1837) P (fish)
Schistobrachia pilgrimi Kabata, 1988 P (fish) E
Vanbenedenia sp. P (fish)
LERNANTHROPIDAE
Aethon garricki Hewitt, 1968 P (fish) E
Aethon morelandi Hewitt, 1968 P (fish)
Aethon percis (Thomson, 1890) P (fish) E
Lernanthropus microlamini Hewitt, 1968 P (fish) E
Lernanthropus sp.* P (fish)
Sagum foliaceus (Goggio, 1905) P (fish)
NICOTHOIDAE
Rhizorhina seriolis Green, 1959 P (isopod) E
Sphaeronella bradfordae Boxshall & Lincoln, 1983 P (isopod) E
Sphaeronella serolis Monod, 1930 P (isopod) E
Sphaeronellopsis littoralis Hansen, 1905 P (ostracod) E
PANDARIDAE
Demoleus latus Shiino, 1954 P (fish)
Dinemoura latifolia Steenstrup & Lütken, 1861 P (fish)
Dinemoura producta (Müller, 1785) P (fish)
Echthrogaleus denticulatus Smith, 1874 P (fish)
Echthrogaleus coleoptratus (Gúerin-Meneville, 1837) P (fish)
Nesippus orientalis Heller, 1865 P (fish)
Nogagus borealis (Steenstrup & Lütken, 1861) P (fish)
Pandarus bicolor Leach, 1816 P (fish)
Pandarus satyrus Dana, 1852 P (fish)
Perissopus dentatus Steenstrup & Lütken, 1861 P (fish)
Phyllothyreus cornutus (Edwards, 1840) P (fish)
PENNELLIDAE
Cardiodectes bellotti (Richiardi, 1882) P (fish)
Pennella histiophori Thomson, 1890 P (fish)
Trifur lotellae Thomson, 1890 P (fish)
PSEUDOCYCNIDAE
Pseudocycnus appendicualatus Heller, 1868 P (fish)
SPHYRIIDAE
Lophoura laticervix Hewitt, 1964 P (fish)
Lophoura spp. *B. Jones unpubl. P (fish)
Periplexis antarcticensis Hewitt, 1965 P (fish)
Sphyrion laevigatum (Quoy & Gaimard, 1824) P (fish)
Sphyrion lumpi (Kroyer, 1845) P (fish)?
Sphyrion quadricornis Gavevskaya & Kovaleva, 1984 P (fish)

Order MONSTRILLOIDA
MONSTRILLIDAE?
Monstrilla sp.* P

Class OSTRACODA
Order PALAEOCOPIDA
Suborder BEYRICHICOPIDA
PUNCIIDAE
Manawa staceyi Swanson, 1989 E
Manawa tryphena Hornibrook, 1949 E
Puncia novozealandica Hornibrook, 1949 E

Order PODOCOPIDA
Suborder PODOCOPINA
BAIRDIIDAE
Bairdoppilata kerryi Milau, 1993
Bairdoppilata villosa (Brady, 1880)
Bairdoppilata sp. Swanson 1979
Neonesidea amygdaloides (Brady, 1880)
Neonesidea crosskeiana (Brady, 1886)
Neonesidea fusca (Brady, 1880)
Neonesidea ovata (Bosquet, 1853)
Neonesidea sp. Ayress 1993
BYTHOCYPRIDIDAE
Orlovibairdia arcaforma (Swanson, 1979) E
Orlovibairdia aff. *angulata* (Brady, 1870)
Orlovibairdia aff. *fumata* (Brady, 1890)
Orlovibairdia sp. Swanson 1979
BYTHOCYTHERIDAE
Baltraella cf. *peterroyi* Yassini & Jones, 1995
Bythocythere arenacea Brady, 1880
Bythocythere bulba Swanson, 1979
Bythoceratina decepta Hornibrook, 1952
Bythoceratina edwardsoni Hornibrook, 1952
Bythoceratina fragilis Hornibrook, 1952
Bythoceratina hornibrooki Jellinek & Swanson, 2003
Bythoceratina maoria Hornibrook, 1952
Bythoceratina mestayerae Hornibrook, 1952
Bythoceratina powelli Hornibrook, 1952
Bythoceratina tuberculata Hornibrook, 1952
Bythoceratina utilazea Hornibrook, 1952
Microceratina quadrata Swanson, 1980
<u>Miracythere</u> novaspecta Hornibrook, 1952 E
Miracythere speciosa Jellinek & Swanson, 2003 E
CYPRIDIDAE
Candona aotearoa Chapman, 1963 F E
Candona inexpecta Chapman, 1963 F E
Candonocypris assimilis Sars, 1894 F
Candonocypris novaezelandiae (Baird *in* White & Doubleday, 1843) F E
Cypretta turgida (Sars, 1896) F E
Cypretta viridis (Thomson, 1879) F
Cyprinotus flavescens Brady, 1898 F E
Cyprinotus sarsi Brady, 1898 F E
Cypris kaiapoiensis Chapman, 1963 F E
Diacypris thomsoni (Chapman, 1963) F E
Eucypris lateraria (King, 1855) F
Eucypris sanguineus (Chapman, 1963) F E
Eucypris virens (Jurine, 1820) F A
Herpetocypris pascheri Brehm, 1929 F E
Heterocypris incongruens (Rhamdohr, 1808) F E
Ilyodromus stanleyanus (King, 1855) F
Ilyodromus obtusus Sars, 1894 F E
Ilyodromus smaragdinus Sars, 1894 F
Ilyodromus subsriatus Sars, 1894 F E
Ilyodromus varrovillius (King, 1855) F
Mesocypris insularis (Chapman, 1963) F E
Paracypria tenuis (Sars, 1905) F
Potamocypris sp. Hornibrook, 1955 F
Scottia audax (Chapman, 1961) T E
CYPRIDOPSIDAE
Cypridopsis obstinata Barclay, 1968 F E
Cypridopsis vidua (Müller, 1776) F A
Pleisiocypridopsis jolleae (Chapman, 1963) F E
Prionocypris marplesi Chapman, 1963 F E
CYTHERALISONIDAE
Cytheralison fava (Hornibrook, 1952) E
Cytheralison tehutui Jellinek & Swanson, 2003 E
Cytheralison sp. Jellinek & Swanson 2003
Debissonia fenestrata Jellinek & Swanson, 2003 E
Debissonia pravacauda (Hornibrook, 1952) E
Debissonia sp. Jellinek & Swanson 2003
CYTHERIDAE
Loxocythere crassa Hornibrook, 1952
Loxocythere hornibrooki McKenzie, 1967
Loxocythere kingi Hornibrook, 1952
Loxocythere sp. Hornibrook 1952
CYTHERIDEIDAE
Cytheridea aoteana Hornibrook, 1952 E
Hemicytheridea mosaica Hornibrook, 1952
Pseudeucythere sp. Jellinek & Swanson 2003
Pseudocythere (Pseudocythere) caudata Sars, 1866
Pseudocythere (Plenocythere) fragilis Swanson, 1979
Rotundracythere gravepuncta Hornibrook, 1952
Rotundracythere cf. *gravepunctata* Hornibrook, 1952
Rotundracythere inaequa Hornibrook, 1952
Rotundracythere mytila Hornibrook, 1952
Rotundracythere nux Jellinek & Swanson, 2003 E
Rotundracythere rotunda Hornibrook, 1952
Rotundracythere subovalis Hornibrook, 1952
Rotundracythere sp. A Jellinek & Swanson 2003
Rotundracythere sp. B Jellinek & Swanson 2003
Rotundracythere sp. C Jellinek & Swanson 2003
Rotundracythere sp. D Jellinek & Swanson 2003
Rotundracythere sp. E Jellinek & Swanson 2003
CYTHERURIDAE
Aversovalva aurea Hornibrook, 1952
Aversovalva sp. Ayress 1995
Cytheropteron anisovalva Ayress, Correge, Passlow & Whatley, 1996
Cytheropteron confusum (Hornibrook, 1952)
Cytheropteron curvicaudum Hornibrook, 1952
Cytheropteron dividentum (Hornibrook, 1952)
Cytheropteron dorsocorrugatum Ayress, Correge, Passlow & Whatley, 1996
Cytheropteron fornix (Hornibrook, 1952)
Cytheropteron hikurangiensis Swanson & Ayress, 1999 E
Cytheropteron latiscalpum Hornibrook, 1952
Cytheropteron obtusalum Hornibrook, 1952
Cytheropteron terecaudum Hornibrook,
Cytheropteron vertex Hornibrook, 1952
Cytheropteron wellingtoniense Brady, 1880
Cytheropteron wellmani Hornibrook, 1952
Cytheropteron willetti Hornibrook, 1952

Cytheropteron sp. Ayress 1993 ?Rec
Cytheropteron sp. Hartmann 1982
Cytherura clausi Brady, 1880
Eucytherura boomeri Ayress, Whatley, Downing & Millson, 1995
Eucytherura calabra (Colalongo & Pasini, 1980)
Eucytherura multituberculata Ayress, Whatley, Downing & Millson, 1995
Eucytherura? anoda Ayress, Whatley, Downing, & Millson, 1995
Hemicytherura (Hemicytherura) aucklandica Hornibrook, 1952
Hemicytherura (H.) delicatula Hornibrook, 1952
Hemicytherura (H.) fereplana Hornibrook, 1952
Hemicytherura (H.) gravis Hornibrook, 1952
Hemicytherura (H.) pandorae Hornibrook, 1952
Hemicytherura (H.) pentagona Hornibrook, 1952
Hemicytherura (H.) quadrazea Hornibrook, 1952
Hemicytherura (Kangarina) radiata (Hornibrook, 1952)
Microcytherura hornibrooki (McKenzie, 1967)*
Microcytherura (Elofsonia) sp. Hayward 1981
Oculocytheropteron acutangulum (Hornibrook, 1952)
Oculocytheropteron confusum (Hornibrook, 1952)
Oculocytheropteron improbum (Hornibrook, 1952)
Pterygocythere mucronalata (Brady, 1880)
Semicytherura arteria Swanson, 1979
Semicytherura cf. *costellata* (Brady, 1880)
Semicytherura hexagona (Hornibrook, 1952
Semicytherura sericava (Hornibrook, 1952
DARWINULIDAE
Penthesilenula aotearoa (Rossetti, Eagar & Martens, 1998) F E
Penthesilenula kohanga (Rossetti, Eagar & Martens, 1998) F E
Penthesilenula? repoa (Chapman, 1963) F E
Penthesilenula sphagna (Barclay, 1968) F E
ENTOCYTHERIDAE
Laccocythere aotearoa Hart & Hart, 1970 E
HEMICYTHERIDAE
Ambostracon pumila (Brady, 1880)
Aurila sp. Hartmann 1985
Bradleya arata (Brady, 1880)
Bradleya claudiae Jellinek & Swanson, 2003 E
Bradleya cupa Jellinek & Swanson, 2003
Bradleya deltoides Hornibrook, 1952
Bradleya dictyon (Brady, 1880)
Bradleya fenwicki Jellinek & Swanson, 2003
Bradleya glabra Jellinek & Swanson, 2003 E
Bradleya lordhowensis Whatley, Downing, Kesler & Harlow, 1984
Bradleya opima Swanson, 1979
Bradleya pelasgica Whatley, Downing, Kesler & Harlow, 1984
Bradleya cf. *pelasgica* Whatley, Downing, Kesler & Harlow, 1984
Bradleya perforata Jellinek & Swanson, 2003
Bradleya pygmaea Whatley, Downing, Kesler & Harlow, 1984
Bradleya reticlava Hornibrook, 1952
Bradleya silentium Jellinek & Swanson, 2003 E
Bradleya wyvillethomsoni (Brady, 1880)
Bradleya n. sp.'*dictyon*' Hornibrook 1952
Bradleya (Quasibradleya) cuneazea Hornibrook, 1952
Harleya ansoni (Whatley, Moguilevsky, Ramos & Coxill, 1998)
Harleya davidsoni Jellinek & Swanson, 2003 E
Harleya sp. Jellinek & Swanson 2003
Hemicythere brunnea (Brady, 1898)
Hemicythere foveolata (Brady, 1880)
Hemicythere fulvotincta (Brady, 1880)
Hemicythere kerguelensis (Brady, 1880)
Hemicythere munida Swanson, 1979
Hermanites andrewsi Swanson, 1979
Hermanites briggsi Swanson, 1979
Jacobella papanuiensis Swanson, 1979
Mutilus cf. *pumilus* (Brady, 1866)
Poseidonamicus major Benson, 1972
Poseidonamicus minor Benson, 1972
Poseidonamicus ocularis Whatley, Downing, Kesler & Harlow, 1986
Poseidonomicus sp. Jellinek & Swanson 2003
Poseidonamicus spp. Ayress, Neil, Passlow & Swanson 1997
Procythereis (Serratocythere) lytteltonensis Hartmann, 1982
Quadracythere biruga Hornibrook, 1952
Quadracythere mediaruga Hornibrook, 1952
Quadracythere radizea Hornibrook, 1952
Quadracythere truncula Hornibrook, 1952
Waiparacythereis joanae Swanson, 1969
ILYOCYPRIDIDAE
Ilyocypris fallax Brehm, 1929 F E
KRITHIDAE
Krithe antisawanensis Ishizaki, 1966
Krithe comma Ayress, Barrows, Passlow & Whatley, 1999
Krithe compressa (Seguenza, 1980)
Krithe dolichodeira Bold, 1946
Krithe marialusae Abate, Barra, Aiello & Bonaduce, 1993
Krithe minima Coles, Whatley & Moguilevsky, 1994
Krithe morkhoveni morkhoveni Bold, 1960
Krithe nitida Whatley & Downing, 1993 ?Rec
Krithe producta Brady, 1880
Krithe pseudocomma Ayress, Barrows, Passlow & Whatley, 1999
Krithe reversa Bold, 1958
Krithe swansoni Milau, 1993
Krithe trinidadensis Bold, 1958
Krithe sp. Ayress, Neil, Passlow & Swanson 1997
Krithe sp. 2 Ayress, Barrows, Passlow & Whatley 1999
Parakrithe sp. Swanson 1979
LEGUMINOCYTHERIDIDAE
Triginglymus? sp. Hornibrook 1952
LEPTOCYTHERIDAE
Callistocythere dorsotuberculata Hartmann, 1979
Callistocythere innominata (Brady, 1898)
Callistocythere mosleyi (Brady, 1880)
Callistocythere murrayana (Brady, 1880)
Callistocythere neoplana Swanson, 1979 E
Callistocythere obtusa Swanson, 1979 E
Callistocythere puri McKenzie, 1967
Callistocythere n. sp. cf. *crispata* Hornibrook, 1952
Callistocythere sp. Hornibrook 1952
Cluthia australis Ayress & Drapala, 1996
Kangarina unispinosa Swanson, 1980
Leptocythere hartmanni (McKenzie, 1967)
Leptocythere lacustris De Deckker, 1981
Leptocythere swansoni Hartmann, 1982 E
<u>Swansonella</u> novaezealandica (Hartmann, 1982) E
Swansonella newbrightonensis Guise, 2002 E
LIMNOCYTHERIDAE
Gomphocythere duffi (Hornibrook, 1955) F
Gomphocythere problematica (Brehm, 1932) F
<u>Kiwicythere</u> anneari Martens, 1992 F E
Kiwicythere vulgaris (McKenzie & Swanson, 1981) F E
Paralimnocythere vulgaris McKenzie & Swanson, 1981 F
LOXOCONCHIDAE
Loxoconcha anomala Brady, 1880
Loxoconcha parvifoveata Hartmann, 1980 A
Loxoconcha punctata Thomson, 1879
Loxoconcha suteri Hartmann, 1982
Loxoconcha tubmani Swanson, 1980
Loxoconcha sp. Swanson 1969
Loxoconcha sp. Hartmann 1982
MACROCYPRIDIDAE
Macrocyprina campbelli Jellinek & Swanson, 2003 E
Macrocyprina sp. Swanson 1979
Macrocyprina sp. A Jellinek & Swanson 2003
Macrocyprina sp. B Jellinek & Swanson 2003
Macrocyprina sp. C Jellinek & Swanson 2003
Macrocypris decora (Brady, 1866)
Macrocypris tumida Brady, 1880 (doubtful)
Macrocypris sp. Hornibrook 1952
Macrocypris sp. Swanson 1979
Macrocypris sp. Ayress 1993
Macromckensiea cf. *porcelica* Whatley & Downing, 1983
Macromckenziea swansoni Maddocks, 1990 E
Macropyxis andreseni Jellinek & Swanson, 2003
Macropyxis sonneae Jellinek & Swanson, 2003 E
'Macropyxis' thiedei Jellinek & Swanson, 2003 E
Macropyxis sp. Jellinek & Swanson 2003
Macrosarisa sp. Jellinek & Swanson 2003
Macroscapha procera Jellinek & Swanson, 2003 E
Gen et sp. indet. Jellinek & Swanson 2003
NEOCYTHERIDEIDAE
Copytus novaezealandiae (Brady, 1898) E
Neocytherideis muehlenhardtae Hartmann, 1982 E
Pontocythere hedleyi (Chapman, 1906)
NOTODROMADIDAE
Newnhamia fenestrata King, 1855
PARACYPRIDIDAE
Paracypris bradyi McKenzie, 1967
Phylctenophora zealandica Brady 1880
Tasmanocypris sp. Morley & Hayward 2007
PARADOXOSTOMATIDAE
Paradoxostoma spp. Hornibrook 1952
Sclerochillus littoralis (Thomson, 1879)
Sclerochillus sp. a Swanson 1979
Sclerochillus sp. b Swanson 1979
Sclerochillus sp. c Swanson 1979
PARVOCYTHERIDAE
Hemiparvocythere lagunicola Hartmann, 1982
PECTOCYTHERIDAE
Keijia demissa (Brady, 1968)
Kotoracythere formosa Swanson, 1979
Mckenzieartia sp. Morley & Hayward 2007
Munseyella aequa Swanson, 1979
Munseyella brevis Swanson, 1979
Munseyella dedeckeri (Swanson, 1980)
Munseyella modesta, Swanson, 1979
Munseyella punctata Whatley & Downing, 1983
Munseyella tumida Swanson, 1979
Munseyella sp. 10 Hartmann, 1982
Parakeijia aff. *thomi* (Yassini & Mikulandra, 1989)
<u>Swansonites</u> aequa (Swanson, 1979)
PONTOCYPRIDIDAE
Argilloecia clavata Brady, 1880 E
Argilloecia eburnea Brady, 1880
Argilloecia aff. *pusilla* (Brady, 1880)
Argilloecia sp. Swanson 1979
Propontocypris cf. *attenuata* Brady, 1868)
Propontocypris cf. *herdmani* (Scott, 1905)
Propontocypris (Ekpontocypris) epicyrta Maddocks, 1969
Propontocypris (Propontocypris) sp. Swanson 1979
Propontocypris (Schedopontocypris?) sp. 3 Maddocks 1969
TRACHYLEBERIDIDAE
<u>Abyssophilos</u> ktis Jellinek & Swanson, 2003
Actinocythereis thomsoni (Hornibrook, 1952)
Ambocythere christineae Jellinek & Swanson, 2003
Ambocythere recta Jellinek & Swanson, 2003
Apatihowella (Apatihowella) rustica Jellinek & Swanson, 2003 E
Apatihowella (A.) sp. Jellinek & Swanson 2003
Apatihowella (Fallacihowella) caligo Jellinek & Swanson, 2003
Apatihowella (F.) sol Jellinek & Swanson, 2003

Arculacythereis sp. Morley & Hayward 2007
Cletocythereis rastromarginata (Brady, 1880)
Clinocthereis australis Ayress & Swanson, 1991
Cythereis finlayi Hornibrook, 1952
Cythereis incerta Swanson, 1979
Dutoitella suhmi (Brady, 1880)
Henryhowella dasyderma (Brady, 1880)
Glencoeleberis armata Jellinek & Swanson, 2003
Glencoeleberis cf. *armata* Jellinek & Swanson, 2003
Glencoeleberis occultata Jellinek & Swanson, 2003 E
Glencoeleberis thomsoni (Hornibrook, 1952)
Legitimocythere acanthoderma (Brady, 1880)
Legitimocythere aculeata Jellinek & Swanson, 2003
Legitimocythere castanea Jellinek & Swanson, 2003
Legitimocythere sp. A Jellinek & Swanson 2003
Legitimocythere sp. B Jellinek & Swanson 2003
Philoneptunus gigas Jellinek & Swanson, 2003 E
Philoneptunus gravizea Hornibrook, 1952
Philoneptunus neesi Jelinek & Swanson, 2003
Philoneptunus paeminosus Whatley, Millson & Ayress, 1992
Philoneptunus paragravazea Whatley, Millson & Ayress, 1992
Philoneptunus planaltus (Hornibrook, 1952)
Philoneptunus provocator Jellinek & Swanson, 2003
Ponticocythereis decora Swanson, 1979
Ponticocythereis militaris (Brady, 1866)
Rugocythereis reticulata Ayress, 1993
Taracythere ayressi Jellinek & Swanson, 2003
Taracythere rhinoceros Jellinek & Swanson, 2003 E
Taracythere ulcus Jellinek & Swanson, 2003
Taracythere venusta Jellinek & Swanson, 2003 E
Taracythere sp. Jellinek & Swanson 2003
Trachyleberis cf. *clavigera* (Brady, 1880)
Trachyleberis lytteltonsis Harding & Sylvester-Bradley, 1953
Trachyleberis melobesoides (Brady, 1866)
Trachyleberis rugibrevis (Hornibrook, 1952)
Trachyleberis scabrocuneata (Brady, 1898)
Trachyleberis scutigera (Brady, 1880)
Trachyleberis tetrica (Brady, 1880)
Trachyleberis zeacristata Hornibrook, 1952
XESTOLEBERIDIDAE
Foveoleberis sp. Jellinek & Swanson 2003
Microxestoleberis triangulata Swanson, 1980
Semixestoleberis taiaroaensis Swanson, 1979
Xestoleberis africana Brady, 1880
Xestoleberis atra (Thomson, 1879) E
Xestoleberis aff. *chilensis austrocontinentalis* Hartmann, 1978
Xestoleberis compressa Brady, 1898
Xestoleberis cf. *curta* (Brady, 1865)
Xestoleberis foveolata Brady, 1880
Xestoleberis luxata Brady, 1898
Xestoleberis olivacea Brady, 1898
Xestoleberis margaretea Brady, 1865
Xestoleberis setigera Brady, 1880
Xestoleberis cf. *trimaculata* Hartmann, 1962
Xestoleberis sp. Hornibrook 1952
Xestoleberis sp. Swanson 1979
Xestoleberis sp. A Jellinek & Swanson 2003
Xestoleberis sp. B Jellinek & Swanson 2003
Xestoleberis sp. C Jellinek & Swanson 2003
INCERTAE SEDIS
Bisulcocythere novaezealandiae Ayress & Swanson, 1991 E
Saida torresi (Brady, 1880)*

Suborder PLATYCOPINA
CYTHERELLIDAE
Cytherella corpusculum Swanson, Jellinek, & Malz, 2003
Cytherella eburnea Brady, 1898 E
Cytherella hemipuncta Swanson, 1969
Cytherella hiatus Swanson, Jellinek & Malz, 2003
Cytherella intonsa Swanson, Jellinek & Malz, 2003
Cytherella lata Brady, 1880
Cytherella paranitida Whatley & Downing, 1983
Cytherella permutata Swanson, Jellinek & Malz, 2003
Cytherella plusminusve Swanson, Jellinek & Malz, 2003
Cytherella polita Brady, 1880
Cytherella pulchra Brady, 1880
Cytherella punctata Brady, 1880
Cytheretta sp. Morley & Hayward 2007
Cytherelloidea willetti Swanson, 1969* E
Cytherelloidea n. sp. van den Bold 1963
Grammcythella dyspnoea Swanson, Jellinek & Malz, 2003
Inversacytherella tanantia Swanson, Jellinek & Malz, 2003

Order MYODOCOPIDA
Suborder MYODOCOPINA
CYPRIDINIDAE
Bathyvargula walfordi Poulsen, 1963
Codonocera crueta Brady, 1902
Cypridina inermis (Müller, 1906)
Cypridinodes reticulata Poulsen, 1962 E
Cypridinodes concentrica Kornicker, 1979 E
Gigantocypris australis Poulsen, 1962 Pe
Gigantocypris danae Poulsen, 1962 Pe
Macrocypridina castanea (Brady, 1897) Pe
Metavargula iota Kornicker, 1975 E
Metavargula bradfordi Kornicker, 1979 E
Metavargula mazeri Kornickeri, 1979 E
Paracypridina aberrata Poulsen, 1962 E
Vargula ascensus Kornicker, 1979 E
Vargula stathme Kornicker, 1975 E
PHILOMEDIDAE
Euphilomedes agilis (Thomson, 1879)
Euphilomedes ferox Poulsen, 1962
Harbansus n. sp. Eagar 1995
Scleroconcha arcuata Poulsen, 1962 E
Scleroconcha sculpta (Brady, 1898) E
Scleroconcha flexilis (Brady, 1898) E
Scleroconcha wolffi Kornicker, 1975 E
CYLINDROLEBERIDIDAE
Bathyleberis oculata Kornicker, 1975 E
Cycloleberis bradyi Poulsen, 1965
Diasterope grisea (Brady, 1898) E
Dolasterope johansoni Poulsen, 1965 E
Leuroleberis zealandica (Baird, 1850) E
Parasterope pectinata Poulsen, 1965 E
Parasterope quadrata (Brady, 1898) E
Pasterope crinita Kornicker, 1975 E
Synasterope empoulseni Korniker, 1975 E
SARSIELLIDAE
Ancohenia n sp. Eagar 1995
Chelicopia tasmanensis Kornicker, 1981
Cymbicopia brevicostata Kornicker, 1975 E
Cymbicopia hanseni (Brady, 1898) E
Cymbicopia hispida (Brady, 1898) E
Cymbicopia zealandica (Poulsen, 1965) E
HALOCYPRIDIDAE
Archiconchoecia cuculata (Brady, 1802)
Archiconchoecia versicula (Deevey, 1978)
Conchoecia acuticostata Müller, 1906
Conchoecia amblypostha Müller, 1906
Conchoecia antipoda Müller, 1906
Conchoecia belgicae Müller, 1906
Conchoecia bispinosa Claus, 1890
Conchoecia brachyaskos Müller (1906)
Conchoecia chuni Müller 1906
Conchoecia ctenophora (Müller, 1906)
Conchoecia discorphora Müller, 1906
Conchoecia eltaninae Deevey, 1982
Conchoecia hyalophyllum Claus, 1890 Pe
Conchoecia loricata (Claus, 1894)
Conchoecia macrocheira Müller, 1906 Pe
Conchoecia magna Claus, 1874 Pe
Conchoecia major Müller, 1906
Conchoecia nasotuberculata Müller, 1906
Conchoecia parvidentata Müller, 1906 Pe
Conchoecia pusilla Müller, 1906
Conchoecia rhynchena Müller, 1906
Conchoecia serrulata laevis Brady, 1907
Conchoecia skogsbergi Iles, 1953
Conchoecia spinifera Clauss, 1890
Conchoecia subarcuata Claus, 1890 Pe
Conchoecia stigmata Müller, 1906
Conchoecia teretivalvata Iles, 1953
Conchoecia (*Alaca*) *hettacra* (Müller, 1906)
Conchoecia (*A.*) *valdiviae* (Müller, 1906)
Conchoecia (*Conchoecilla*) *chuni* (Müller, 1906)
Conchoecia (*C.*) *daphnoides* (Clauss, 1890)
Conchoecia (*Conchoecissa*) *ametra* (Müller, 1906)
Conchoecia (*C.*) *imbricata* (Brady, 1880)
Conchoecia (*C.*) *symmetrica* (Müller, 1906)
Conchoecia (*Discoconchoecia*) *elegans* Sars, 1865
Conchoecia (*Obtusoecia*) *antarctica* (Muller, 1906)
Conchoecia (*Orthoconchoecia*) *haddoni* Brady & Norman, 1896
Conchoecia (*Porroecia*) *spinirostris* Claus, 1874
Conchoecia (*P.*) *porrecta* Claus, 1890
Conchoecia (*Pseudoconchoecia*) *serrulata* Claus 1874
Fellia cornuta (Müller, 1906) Pe
Fellia dispar (Müller, 1906) Pe
Halocypris inflata (Dana, 1849) Pe
Halocypris globosa (Claus, 1874) Pe

Suborder CLADOCOPINA
POLYCOPIDAE
Polycope sp. Swanson 1979
Polycopsis cf. *loscobanosi* Hartmann, 1959

Class MALACOSTRACA
Subclass PHYLLOCARIDA
Order LEPTOSTRACA
NEBALIIDAE
Nebalia longicornis G.M. Thomson, 1879
Nebaliella antarctica Thiele, 1904
Sarsinebalia sp. 1 Dahl 1990
Sarsinebalia sp. 2 Dahl 1990
PARANEBALIIDAE
Levinebalia fortunata (Wakabara, 1976)

Subclass HOPLOCARIDA
Order STOMATOPODA
BATHYSQUILLIDAE
Bathysquilla microps (Manning, 1961)
HEMISQUILLIDAE
Hemisquilla australiensis Stephenson, 1967
ODONTODACTYLIDAE
Odontodactylus brevirostris (Miers, 1884)
SQUILLIDAE
Oratosquilla oratoria (de Haan, 1844) A
Pterygosquilla schizodontia (Richardson, 1953)
TETRASQUILLIDAE
Acaenosquilla brazieri (Miers, 1880)
Heterosquilla tricarinata (Claus, 1871) E
Heterosquilla tridentata (Thomson, 1882) E

Subclass EUMALOCOSTRACA
Superorder SYNCARIDA
Order ANASPIDACEA
STYGOCARIDIDAE
Stygocaris townsendi Morimoto, 1977 F E
Stygocaris sp. 1 Morimoto 1977 F E
Stygocaris sp. 2 Morimoto 1977 F E
Stygocaris sp. Schminke 1980 F
Stygocarella pleotelson Schminke, 1980 F E
Stygocarella sp. Schminke 1973 F E

Order BATHYNELLACEA
BATHYNELLIDAE
Bathynella sp. 1 Schminke 1971 F E
Bathynella sp. 2 Schminke 1971 F E
PARABATHYNELLIDAE
Atopobathynella compagana Schminke, 1973 F E
Hexabathynella aotearoae Schminke, 1973 F E
Notobathynella chiltoni Schminke, 1973 F E
Notobathynella hineoneae Schminke, 1973 F E
Notobathynella longipes Schminke, 1978 F E
Notobathynella sp. Schminke 1973 F E

Superorder PERACARIDA
Order LOPHOGASTRIDA
GNATHOPHAUSIIDAE
Gnathophausia elegans G.O. Sars, 1883
Gnathophausia zoea Willemoes-Suhm, 1875
Neognathophausia ingens (Dohrn, 1870)
Neognathophausia gigas (Willemoes-Suhm, 1875)
LOPHOGASTRIDAE
Lophogaster sp.* MNZ
Paralophogaster glaber Hansen, 1910

Order MYSIDA
MYSIDAE
Boreomysis rostrata Illig, 1906
Euchaetomera oculata Hansen, 1910
Euchaetomera typica G.O. Sars, 1884
Euchaetomera zurstrasseni (Illig, 1906)
Gastrosaccus australis W. Tattersall, 1923 E
Siriella denticulata (Thomson, 1880) E
Siriella thompsonii (H. Milne Edwards, 1837)
Tenagomysis chiltoni W. Tattersall, 1923 E
Tenagomysis longisquama Fukuoka & Bruce, 2005 E
Tenagomysis macropsis W. Tattersall, 1923 E
Tenagomysis novaezealandiae Thomson, 1900 E
Tenagomysis producta W. Tattersall, 1923 E
Tenagomysis robusta W. Tattersall, 1923 E
Tenagomysis scotti W. Tattersall, 1923 E
Tenagomysis similis W. Tattersall, 1923 E
Tenagomysis tenuipes W. Tattersall, 1918 E
Tenagomysis thomsoni W. Tattersall, 1923 E
PETALOPHTHALMIDAE
Petalophthalmus sp.* MNZ

Order AMPHIPODA
Suborder INGOLFIELLIDEA
INGOLFIELLIDAE
"Pseudoingolfiella" sp. a Schminke & Noodt 1968
"Pseudoingolfiella" sp. b Schminke & Noodt 1968

Suborder GAMMARIDEA
AMARYLLIDAE
Amaryllis macrophthalma Haswell, 1880
AMPELISCIDAE
Ampelisca albedo Barnard, 1961 E
Ampelisca chiltoni Stebbing, 1888 E
Byblisoides esferis Barnard, 1961 E
Haploops decansa Barnard, 1961 E
AMPHILOCHIDAE
Amphilochus filidactylus Hurley, 1955 E
Amphilochus marionis? Stebbing, 1888
Amphilochus opunake Barnard, 1972 E
Gitanopsis desmondi Barnard, 1972 E
Gitanopsis kupe Barnard, 1972 E
Gitanopsis squamosa (Thomson, 1880)
AMPITHOIDAE
Ampithoe hinatore Barnard, 1972 E
Ampithoe sp. Barnard 1972 E
Parampithoe aorangi (Barnard, 1972) E
<u>*Pseudopleonexes*</u> *lessoniae* (Hurley, 1954) E
AORIDAE
Aora maculata (Thomson, 1879)
Aora typica Kroyer, 1845
Aora sp. Barnard 1972

Camacho bathyplous Stebbing, 1888
Camacho nodderi Coleman & Lörz, 2010 E
Haplocheira barbimana (Thomson, 1879)
Haplocheira lendenfeldi Chilton, 1884 E
Lembos? sp. No. 1 Barnard 1972
Lembos? sp. No. 3 Barnard 1972
Lembos? sp. No. 4 Barnard 1972
<u>*Meridiolembos*</u> *acherontis* (Myers, 1981) E
Meridiolembos hippocrenes (Myers, 1981) E
Meridiolembos pertinax (Myers, 1981) E
Microdeutopus apopo Barnard, 1972 E
CAPRELLIDAE
Caprella equilibra Say, 1818
Caprella manneringi McCain, 1979 E
Caprella mutica Schurin, 1935 A
Caprellina longicollis (Nicolet, 1849)
Caprellaporema subantarctica Guerra-García, 2003 E
Caprellinoides mayeri (Pfeffer, 1888)
Pseudaeginella campbellensis Guerra-García, 2003 E
Pseudoprotomima hurleyi McCain, 1969 E
CEINIDAE
Ceina egregia (Chilton, 1883) E
<u>*Taihape*</u> *karori* Barnard, 1972 E
<u>*Waitomo*</u> *manene* Barnard, 1972 E
CHELURIDAE
Chelura terebrans Philippi, 1839 A
CHEVALIIDAE
Chevalia sp. Ahyong
CHILTONIIDAE
<u>*Chiltonia*</u> *enderbyensis* Hurley, 1954 F E
Chiltonia mihiwaka (Chilton, 1898) F E
Chiltonia minuta Bousfield, 1964 ?F E
Chiltonia rivertonensis Hurley, 1954 F E
COLOMASTIGIDAE
Colomastix magnirama Hurley, 1954 E
Colomastix subcastellata Hurley, 1954 E
COROPHIIDAE
Apocorophium acutum Chevreux, 1908 A
Monocorophium acherusicum (Costa, 1857) A
Monocorophium insidiosum (Crawford, 1937) A
Monocorophium sextonae (Crawford, 1937) A
Paracorophium brisbanensis Chapman, 2002 B A
Paracorophium excavatum (Thomson, 1884) F B E
Paracorophium lucasi Hurley, 1954 F B E
CYAMIDAE
Cyamus balaenopterae Barnard, 1931
Cyamus boopis Lutken, 1873
Cyamus erraticus Roussel de Vauzeme, 1834
Cyamus gracilis Roussel de Vauzeme, 1834
Cyamus ovalis Roussel de Vauzeme, 1834
Isocyamus delphini Guerin-Meneville, 1837
Neocyamus physeteris (Pouchet, 1888)
Scutocyamus antipodensis Lincoln & Hurley, 1980 E
CYPHOCARIDIDAE
Cyphocaris anonyx Boeck, 1871
Cyphocaris richardi Chevreux, 1905
CYPROIDEIDAE
<u>*Neocyproidea*</u> *otakensis* (Chilton, 1900) E
Neocyproidea pilgrimi Hurley, 1955 E
<u>*Peltopes*</u> *peninsulae* (Hurley, 1955) E
Peltopes productus K.H. Barnard, 1930 E
DEXAMINIDAE
Atylus reductus (K.H. Barnard, 1930) E
Atylus taupo Barnard, 1972 E
Guernea timaru Barnard, 1972 E
Lepechinella sucia Barnard, 1961
Lepechinella wolffi Dahl, 1959 E
Paradexamine barnardi Sheard, 1938 E
Paradexamine houtete Barnard, 1972 E
Paradexamine muriwai Barnard, 1972 E
Paradexamine pacifica (Thomson, 1879) E
Paradexamine sp. Barnard 1972 E
Polycheria obtusa Thomson, 1882 E
Syndexamine carinata Chilton, 1914 E
DOGIELINOTIDAE

Allorchestes compressa Dana, 1852
'Allorchestes compressus' Bousfield 1964 F? E
Allorchestes novizealandiae Dana, 1852 F E
ENDEVOURIDAE
Ensayara iara Lowry & Stoddart, 1983 E
Ensayara kermadecensis Kilgallen, 2009 E
Ensayara ursus Kilgallen, 2009 E
EOPHLIANTIDAE
Bircenna fulva Chilton, 1884 E
Bircenna macayai Lörz, Kilgallen & Thiel, 2009 E
Cylindryllioides kaikoura Barnard, 1972 E
Wandelia wairarapa Barnard, 1972 E
EPIMERIIDAE
Epimeria bruuni Barnard, 1961 E
Epimeria glaucosa Barnard, 1961 E
Epimeria horsti Lörz, 2008 E
Epimeria norfanzi Lörz, 2010
Epimeriella victoria Hurley, 1957 E
EUSIRIDAE
Atyloella moke Barnard, 1972 E
Bathyschraderia magnifica Dahl, 1959 E
Eusiroides monoculoides (Haswell, 1880)
Eusirus antarcticus Thomson, 1880
Gondogeneia bidentata (Stephensen, 1927)
Gondogeneia danai (Thomson, 1879) E
Gondogeneia rotorua Barnard, 1972 E
Gondogeneia subantarctica (Stephensen, 1938) E
Gondogeneia sp. Chilton 1909 E
Oradarea novaezealandiae (Thomson, 1879) E
Paramoera aucklandica (Walker, 1908) E
Paramoera chevreuxi (Stephensen, 1927) E
Paramoera fasciculata (Thomson, 1880) E
Paramoera fissicauda? (Dana, 1852)
Paramoera rangatira Barnard, 1972 E
Paramoera sp. Barnard 1972 E
Paramoera sp. Barnard 1972 F E
Prostebbingia? levis (Thomson, 1879) E
Regalia fascicularis Barnard, 1930 E
Rhachotropis chathamensis Lörz, 2010 E
Rhachotropis delicata Lörz, 2010 E
Rhachotropis levantis Barnard, 1961 E
Schraderia serraticauda (Stebbing, 1888)
<u>*Whangarusa*</u> *translucens* (Chilton, 1884) E
EXOEDICEROTIDAE
<u>*Patuki*</u> *breviuropodus* Cooper & Fincham, 1974 E
Patuki roperi Fenwick, 1983 E
HADZIIDAE
<u>*Zhadia*</u> *subantarctica* Lowry & Fenwick, 1983 E
HYALIDAE
Apohyale hirtipalma (Dana, 1852)
Apohyale media (Dana, 1853)
Apohyale novaezealandiae (Thomson, 1879) E
Protohyale (Protohyale) campbellica (Filhol, 1885) E
Protohyale (Boreohyale) grenfelli Chilton, 1916 E
Protohyale (B.) maroubrae Stebbing, 1899
Protohyale (B.) rubra (Thomson, 1879)
Hyale sp. Thomson 1899
IPHIMEDIIDAE
Amathillopsis grevei Barnard, 1961
<u>*Anisoiphimedia*</u> *haurakiensis* (Hurley, 1954) E
Curidia knoxi Lowry & Myers, 2003 E
Epimeria bruuni Barnard, 1961 E
Epimeria glaucosa Barnard, 1961 E
Epimeriella victoria Hurley, 1957 E
Iphimedia spinosa (Thomson, 1880) E
Labriphimedia hinemoa (Hurley, 1954) E
ISAEIDAE
Gammaropsis chiltoni (Thomson, 1897) E
Gammaropsis crassipes (Haswell, 1881)
Gammaropsis haswelli (Thomson, 1897)
Gammaropsis kermadeci (Stebbing, 1888) E
Gammaropsis longimana (Chilton, 1884) E
Gammaropsis tawahi Barnard, 1972 E
Gammaropsis thomsoni Stebbing, 1888
Gammaropsis typica (Chilton, 1884) E

Gammaropsis sp. Barnard 1972 E
Pagurisaea schembrii Moore, 1983 E
Photis brevicaudatus Norman, 1867
Photis nigrocula Lowry, 1979 E
Photis phaeocula Lowry, 1979 E
Photis sp. Barnard 1972 E
ISCHYROCERIDAE
Ericthonius pugnax (Dana, 1852) A
Ischyrocerus longimanus (Haswell, 1880)
Jassa alonsoae Conlan, 1990
Jassa fenwicki Conlan, 1990
Jassa hartmannae Conlan, 1990 E
Jassa justi Conlan, 1990
Jassa marmorata Conlan, 1990
Jassa slatteryi Conlan, 1990
Notopoma fallohidea (Lowry, 1981) E
Notopoma harfoota (Lowry, 1981) E
Notopoma stoora (Lowry, 1981) E
Parajassa andromedae Moore, 1985 E
Runanga coxalis Barnard, 1961 E
Runanga wairoa McCain, 1969 E
Ventojassa frequens (Chilton, 1883) E
KAMAKIDAE
Aorcho delgadus Barnard, 1961
LEUCOTHOIDAE
Leucothoe trailli Thomson, 1882 E
LILJEBORGIIDAE
Liljeborgia aequabilis Stebbing, 1888
Liljeborgia akaroica Hurley, 1954
Liljeborgia barhami Hurley, 1954 E
Liljeborgia dubia (Haswell, 1880)
Liljeborgia hansoni Hurley, 1954 E
LYSIANASSIDAE
Acheronia pegasus Lowry, 1984 E
Acontiostoma marionis Stebbing, 1888
Acontiostoma tuberculata Lowry & Stoddart, 1983 E
Acontiostoma sp.
Ambasiopsis robustus Barnard, 1961 E
Bruunosa bruuni (Dahl, 1959) E
Cheirimedon cansada (Barnard, 1961)
Eurythenes gryllus (Lichtenstein, 1822)
Hippomedon antitemplado Barnard, 1961 E
Hippomedon concolor Barnard, 1961 E
Hippomedon hake Lowry & Stoddart, 1983 E
Hippomedon hurleyi Kilgallen, 2009 E
Hippomedon incisus K.H. Barnard, 1930 E
Hippomedon iugum Kilgallen, 2009 E
Hippomedon kergueleni (Miers, 1875)
Hippomedon tasmanicus Barnard, 1961 E
Hirondella dubia Dahl, 1959 E
Kakanui punui Lowry & Stoddart, 1983 E
Lepidecreella bidens (Barnard, 1930) E
Lysianopsis tieke Lowry & Stoddart, 1983 E
Ocosingo fenwicki Lowry & Stoddart, 1983 E
Orchomene aahu Lowry & Stoddart, 1983 E
Orchomenella cavimanus (Stebbing, 1888)
Paracentromedon? manene (Lowry & Stoddart, 1983) E
Paracentromedon? matikuku (Lowry & Stoddart, 1983) E
Paracentromedon? whero (Fenwick, 1983) E
Paralicella similis Birnstein & Vinogradov, 1960
Parawaldeckia angusta Lowry & Stoddart, 1983 E
Parawaldeckia dabita Lowry & Stoddart, 1983 E
Parawaldeckia hirsuta Lowry & Stoddart, 1983 E
Parawaldeckia karaka Lowry & Stoddart, 1983 E
Parawaldeckia kidderi Lowry & Stoddart, 1983
Parawaldeckia parata Lowry & Stoddart, 1983 E
Parawaldeckia pulchra Lowry & Stoddart, 1983 E
Parawaldeckia stephenseni Hurley & Cooper, 1974 E
Parawaldeckia suzae Lowry & Stoddart, 1983 E
Parawaldeckia thomsoni (Stebbing, 1906) E
Parawaldeckia vesca Lowry & Stoddart, 1983 E
Pseudambasia rossii Stephensen, 1927 E
Schisturella abyssi tasmanensis (Barnard, 1961) E
Stomacontion hurleyi Lowry & Stoddart, 1983 E
Stomacontion pungapunga Lowry & Stoddart, 1983 E
Stomacontion sp.
Tryphosella moana Kilgallen, 2009 E
Tryphosella serans Lowry & Stoddart, 1983 E
Valettiopsis multidentata Barnard, 1961 E
MELITIDAE
Ceradocopsis macracantha Lowry & Fenwick, 1983 E
Ceradocopsis carnleyi (Stephensen, 1927) E
Ceradocopsis peke Barnard, 1972 E
Ceradocus chiltoni Sheard, 1939 E
Ceradocus rubromaculatus haumuri Barnard, 1972
Elasmopus bollonsi Chilton, 1915
Elasmopus neglectus Chilton, 1915 E
Elasmopus wahine Barnard, 1972 E
Gammarella hybophora Lowry & Fenwick, 1983 E
Hoho hirtipalma (Barnard, 1972) E
Linguimaera tias Krapp-Schickel, 2003
Maera incerta Chilton, 1883 E
Maera spp. Barnard 1972
Mallacoota nanaui Myers, 1985
Melita awa Barnard, 1972 B E
Melita festiva (Chilton, 1884)
Melita inaequistylis Dana, 1852 E
Melita? solada Barnard, 1961 E
Melita sp. Barnard 1972 E
Micramaera tepuni (Barnard, 1972) E
Parapherusa crassipes (Haswell, 1880)
Tagua aporema Lowry & Fenwick, 1983 E
MELPHIDIPPIDAE
Horniella whakatane (Barnard, 1972) E
NIHOTUNGIDAE
Nihotunga noa Barnard, 1972 E
OCHLESIDAE
Curidia knoxi Lowry & Myers, 2003 E
OEDICEROTIDAE
Bathymedon neozelanicus Barnard, 1930 E
Carolobatea novaezealandiae Chilton, 1909
Lopiceros forensia Barnard, 1961 E
Monoculodes abacus Barnard, 1961 E
Oediceroides apicalis Barnard, 1931
Oediceroides limpieza Barnard, 1961 E
Oediceroides microcarpa Barnard, 1930 E
Oediceroides wolffi Barnard, 1961
PARACALLIOPIIDAE
Paracalliope fluviatilis (Thomson, 1879) F E
Paracalliope karitane Barnard, 1972 F E
Paracalliope novizealandiae (Dana, 1853) E
PARACRANGONYCTIDAE E
Paracrangonyx compactus (Chilton, 1882) F E
Paracrangonyx winterbourni Fenwick, 2001 F E
Pseudoingolfiella Morimotoi Grosso, Peralta & Ruffo, 2006 F E
PARALEPTAMPHOPIDAE E
Paraleptamphopus caeruleus (Thomson, 1885) F E
Paraleptamphopus subterraneus (Chilton, 1882) F E
Paraleptamphopus spp. (10) 10E G. D. Fenwick
Ringanui koonuiroa Fenwick, 2006 F E
Ringanui toonuiiti Fenwick, 2006 F E
Gen. nov. (~10) et n. spp. (~20) ~ 20E G. D. Fenwick
PARDALISCIDAE
Arculfia trago Barnard, 1961 E
Halice macronyx (Stebbing, 1888)
Halice secunda (Stebbing, 1888)
Halice sublittoralis Lowry, 1979 E
Halicoides tambiella Barnard, 1961 E
Pardaliscoides longicaudatus Dahl, 1959 E
Princaxelia abyssalis Dahl, 1959
PHLIANTIDAE
Iphinotus typicus (Thomson, 1882) E
PHOXOCEPHALIDAE
Booranus? spinibasus (Cooper, 1974) E
Cephaloxoides keppeli (Barnard & Drummond, 1978) E
Cephalophoxus regium (Barnard, 1930) E
Harpiniopsis nadania (Barnard, 1961) E
Joubinella traditor Pirlot, 1932
Palabriaphoxus palabria Barnard, 1961 E
Parajoubinella concinna Gurjanova, 1977 E
Paraphoxus? pyripes Barnard, 1930 E
Protophoxus australis Barnard, 1930
Ringaringa littoralis (Cooper & Fincham, 1974) E
Synphoxus novaezelandicus Gurjanova, 1980 E
Torridoharpinia hurleyi (Barnard, 1958) E
Trichophoxus capillatus Barnard, 1930 E
Waitangi rakiura (Cooper & Fincham, 1974) E
Waitangi? brevirostris Fincham, 1977 E
Waitangi? chelatus (Cooper, 1974) E
Wildus waipiro (Barnard, 1972) E
PHREATOGAMMARIDAE E
Phreatogammarus fragilis (Chilton, 1882) F E
Phreatogammarus helmsi Chilton, 1918 F E
Phreatogammarus propinquus Chilton, 1907 F E
Phreatogammarus waipoua Chapman, 2003 F E
PLATYISCHNOPIDAE
Otagia neozelanicus (Chilton, 1987) E
PODOCERIDAE
Podocerus cristatus (Thomson, 1879) E
Podocerus karu Barnard, 1972 E
Podocerus manawatu Barnard, 1972 E
Podocerus sp. Chilton, 1926
Podocerus wanganui Barnard, 1972 E
RAKIROIDAE E
Rakiroa rima Lowry & Fenwick, 1982 E
SCOPELOCHEIRIDAE
Scopelocheirus? schellenbergi Bernstein & Vinogradov, 1958
SEBIDAE
Seba typica (Chilton, 1884)
STEGOCEPHALIDAE
Andaniotes corpulentus (Thomson, 1882)
Euandandania gigantea (Stebbing, 1888)
Phippsiella nipoma Barnard, 1961
Stegosoladidus simplex (Barnard, 1930) E
Tetradeion crassum (Chilton, 1883) E
STENOTHOIDAE
Mesoproboloides? excavata Fenwick, 1977 E
Parathaumatelson nasicum (Stephensen, 1927) E
Probolisca ovata (Stebbing, 1888)
Raukumara rongo (Barnard, 1972) E
Stenothoe aucklandicus Stephensen, 1927 E
Stenothoe gallensis Walker, 1904 A
Stenothoe moe Barnard, 1972 E
Stenothoe valida? Dana, 1853
STILIPEDIDAE
Alexandrella mixta (Nicholls, 1938)
Stilipes sanguineus (Hurley, 1954) E
SYNOPIIDAE
Syrrhoe affinis? Chevreux, 1908
TALITRIDAE
Arcitalitrus dorrieni (Hunt, 1925) T A
Arcitalitrus sylvaticus (Haswell, 1880) T A
Austroides sp. Fenwick & Webber 2008 T
Bellorchestia quoyana (Milne-Edwards, 1840) S E
Bellorchestia spadix Hurley, 1956 S E
Bellorchestia tumida Thomson, 1885 S E
Kanikania improvisa (Chilton, 1909) T E
Kanikania motuensis Duncan, 1994 T E
Kanikania rubroannulata (Hurley, 1957) T E
Makawe hurleyi (Duncan, 1968) T E
Makawe insularis (Chilton, 1909) T E
Makawe maynei (Chilton, 1909) T E
Makawe otamatuakeke Duncan, 1994 T E
Makawe parva (Chilton, 1909) T E
Makawe waihekensis Duncan, 1994 T E
Makawe sp. A Fenwick & Webber 2008 T E
Makawe sp. B Fenwick & Webber 2008 T E
Makawe sp. C Fenwick & Webber 2008 T E
Notorchestia aucklandiae (Bate, 1862) S E

Orchestia? recens (Thomson, 1884) F E
Orchestia? sp. A Hurley, 1975 F E
Orchestia? sp. B Hurley, 1975 F E
Parorchestia ihurawao Duncan, 1994 T E
Parorchestia lesliensis (Hurley, 1957) T E
Parorchestia longicornis (Stephensen, 1938) T E
Parorchestia tenuis (Dana, 1852) T E
Protorchestia campbelliana (Bousfield, 1964) T E
<u>*Puhuruhuru*</u> *aotearoa* Duncan, 1994 T E
Puhuruhuru patersoni (Stephensen, 1938) T E
Puhuruhuru sp. Fenwick & Webber 2008 T E
<u>*Talitroides*</u> *topitotum* (Burt, 1934) T A
<u>*Tara*</u> *hauturu* Duncan, 1994 T E
Tara simularis (Hurley, 1957) T E
Tara sinbadensis (Hurley, 1957) T E
Tara sylvicola (Dana, 1852) T E
Tara taranaki Duncan, 1994 T E
Tara sp. A Fenwick & Webber 2008 T E
Tara sp. B Fenwick & Webber 2008 T E
Transorchestia bollonsi (Chilton, 1909) S E
Transorchestia chathamensis (Hurley, 1956) S E
Transorchestia cookii Filhol, 1885 S E
Transorchestia dentata (Filhol, 1885) S E
Transorchestia kirki (Hurley, 1956) S E
Transorchestia miranda (Chilton, 1916) S E
Transorchestia serrulata (Dana, 1852) S E
Transorchestia telluris (Bate, 1862) S E
<u>*Waematau*</u> *kaitaia* Duncan, 1994 T E
Waematau manawatahi Duncan, 1994 T E
Waematau muriwhenua Duncan, 1994 T E
Waematau reinga Duncan, 1994 T E
Waematau unuwhao Duncan, 1994 T E
URISTIDAE
Abyssorchomene abyssorum (Stebbing, 1888)
<u>*Galathella*</u> *galatheae* (Dahl, 1959) E
Galathella solivagus Kilgallen, 2009 E
UROTHOIDAE
Carangolia puliciformis Barnard, 1961 E
Urothoe elizae Cooper & Fincham, 1974 E
Urothoe wellingtonensis Cooper, 1974 E
Urothoides lachneessa (Stebbing, 1888)

Suborder HYPERIIDEA
ARCHAEOSCINIDAE
Archaeoscina steenstrupi (Bovallius, 1885)
Paralanceola wolffi Zeidler, 2006
BRACHYSCELIDAE
Brachyscelus crusculum Bate, 1861
Brachyscelus rapacoides Stephensen, 1925
Brachyscelus rapax (Claus, 1871)
CHUNEOLIDAE
Chuneola paradoxa Woltereck, 1909
CYLLOPIDAE
Cyllopus magellanicus Dana, 1853
CYSTISOMATIDAE
Cystisoma fabricii Stebbing, 1888
Cystisoma magna (Woltereck, 1903)
Cystisoma pellucida (Willemoes-Suhm, 1873)
DAIRELLIDAE
Dairella californica (Bovallius 1887)
HYPERIIDAE
Hyperia gaudichaudii Milne-Edwards, 1840
Hyperia spinigera Bovallius, 1889
Hyperiella antarctica Bovallius, 1887
Hyperoche mediterranea Senna, 1908
Hyperoche medusarum (Kroyer, 1838)
Lestrigonus schizogeneios (Stebbing, 1888)
Themisto australis (Stebbing, 1888)
Themisto gaudichaudi Guerin, 1825
IULOPIDIDAE
Iulopis loveni Bovallius, 1887
LANCEOLIDAE
Lanceola clausi Bovallius, 1885
Lanceola grunneri Zeidler, 2009
Lanceola intermedia Vinogradov, 1960
Lanceola longidactyla Vinogradov, 1964
Lanceola loveni (Bovallius, 1885)
Lanceola pacifica Stebbing, 1888
Lanceola sayana Bovallius, 1885
Lanceola serrata Bovallius, 1885
Scypholanceola aestiva (Stebbing, 1888)
LESTRIGONIDAE
Hyperietta luzoni (Stebbing, 1888)
Hyperietta vosseleri (Stebbing, 1904)
Hyperioides longipes Chevreux, 1900
Hyperionyx macrodactylus (Stephensen, 1924)
LYCAEIDAE
Lycaea nasuta Claus, 1879
Lycaea pachypoda (Claus, 1879)
Lycaea pulex Marion, 1874
Simorhynchotus antennarius (Claus, 1871)
LYCAEOPSIDAE
Lycaeopsis themistoides Claus, 1879
Lycaeopsis zamboangae (Stebbing, 1888)
MEGALANCEOLIDAE
Megalanceola stephenseni (Chevreux, 1920)
MICROPHASMIDAE
Microphasma agassizi Woltereck, 1909
MIMONECTIDAE
Mimonectes gaussi (Woltereck, 1904)
OXYCEPHALIDAE
Calamorhynchus pellucidus Streets, 1878
Leptocotis tenuirostris (Claus, 1871)
Oxycephalus piscator Milne-Edwards, 1830
Streetsia challengeri Stebbing, 1888
Streetsia porcella (Claus, 1879)
PARAPHRONIMIDAE
Paraphronima crassipes Claus, 1879
Paraphronima gracilis Claus, 1879
PARASCELIDAE
Parascelus edwardsi Claus, 1879
PHRONIMIDAE
Phronima atlantica Guérin-Menéville, 1836
Phronima sedentaria (Forsskål, 1775)
Phronimella elongata (Claus, 1862)
PHROSINIDAE
Anchylomera blossevillei Milne-Edwards, 1830
Phrosina semilunata Risso, 1822
Primno macropa Guérin-Menéville, 1836
PROLANCEOLIDAE
Prolanceola vibiliformis Woltereck, 1907
PLATYSCELIDAE
Amphithyrus bispinosus Claus, 1879
Hemityphis tenuimanus Claus, 1879
Paratyphis parvus Claus, 1887
Paratyphis spinosus Spandl, 1924
Platyscelus armatus (Claus, 1879)
Platyscelus ovoides (Risso, 1816)
Platyscelus serratulus Stebbing, 1888
Tetrathyrus arafurae Stebbing, 1888
Tetrathyrus forcipatus Claus, 1879
PRONOIDAE
Eupronoe maculata Claus, 1879
Eupronoe minuta Claus, 1879
Paralycaea gracilis Claus, 1879
Parapronoe campbelli Stebbing, 1888
Parapronoe crustulum Claus, 1879
Parapronoe parva Claus, 1879
Pronoe capito Guérin-Menéville, 1836
SCINIDAE
Acanthoscina acanthodes (Stebbing, 1895)
Scina borealis (G.O. Sars, 1882)
Scina crassicornis (Fabricius, 1775)
Scina curvidactyla Chevreux, 1914
Scina pusilla Chevreux, 1919
Scina tullbergi (Bovallius, 1885)
Scina wagleri abyssalis Vinogradov, 1957
TRYPHANIDAE
Tryphana malmi Boeck, 1871
VIBILIIDAE
Vibilia antarctica Stebbing, 1888
Vibilia armata Bovallius, 1887
Vibilia borealis Bate & Westwood, 1868
Vibilia caeca Bulycheva, 1955
Vibilia chuni Behning & Woltereck, 1912
Vibilia cultripes Vosseler, 1901
Vibilia gibbosa Bovallius, 1887
Vibilia longicarpus Behning, 1913
Vibilia propinqua Stebbing, 1888
Vibilia pyripes Bovallius, 1887
Vibilia robusta Bovallius, 1887
Vibilia stebbingi Behning & Woltereck, 1912
Vibilia viatrix Bovallius, 1887

Order ISOPODA
Suborder ASELLOTA
ACANTHASPIDIDAE
Mexicope sushara Bruce, 2004 E
Acanthaspidia sp. E
DENDROTIIDAE
Acanthomunna proteus Beddard, 1886 E
Dendromunna mirabile Wolff, 1962 E
DESMOSOMATIDAE
Chelator spp. (3) N. Bruce 2008
Desmosoma sp. N. Bruce 2008
Eugerda sp. N. Bruce 2008
Eugerdella spp. (2) N. Bruce 2008
Mirabilicoxa sp. N. Bruce 2008
Prochelator tupuhi Brix & Bruce, 2008 E
HAPLONISCIDAE
Chauliodoniscus tasmanaeus Lincoln, 1985 E
Haploniscus kermadecensis Wolff, 1962 E
Haploniscus piestus Lincoln, 1985 E
Haploniscus miccus Lincoln, 1985 E
Haploniscus saphos Lincoln, 1985 E
Haploniscus silus Lincoln, 1985 E
Haploniscus tangaroae Lincoln, 1985 E
Hydroniscus lobocephalus Lincoln, 1985 E
Mastigoniscus pistus Lincoln, 1985 E
JANIRIDAE
Heterias n. sp. Scarsbrook et al. 2003 E
Iais californica (Richardson, 1904)
Iais pubescens (Dana, 1852)
Ianiropsis neglecta (Chilton, 1909) E
Iathrippa longicauda (Chilton, 1884) E
Iathrippa sp. NIWA N. Bruce
Mackinia sp. Scarsbrook et al. 2003
ISCHNOMESIDAE
Ischnomesus anacanthus Wolff, 1962 E
Ischnomesus birsteini Wolff, 1962 E
Ischnomesus bruuni Wolff, 1956 E
Ischnomesus spaercki Wolff, 1956 E
Mixomesus pellucidus Wolff, 1962 E
JOEROPSIDIDAE
Joeropsis neozealanica Chilton, 1892 E
Joeropsis palliseri Hurley, 1957 E
Joeropsis spp. (2) 2E
MUNNIDAE
Echinomunna sp. E
Munna neozelanica Chilton, 1892 E
Munna spp. (4) 4E
Uromunna schauinslandi (Sars, 1905) E
MUNNOPSIDIDAE
Bathybadistes andrewsi Merrin, Malyutina & Brandt, 2009
Disconectes madseni (Wolff, 1956) E
Echinozone n. sp. E
Epikopais mystax Merrin, 2009 E
Eurycope galatheae Wolff, 1956 E
Eurycope gibberifrons Wolff, 1962 E
Hapsidohedra aspidophora (Wolff, 1962) E
Ilyarachna kermadecensis Wolff, 1962 E
Ilyarachna n. spp. (7) 7E
Munneurycope harrietae Wolff, 1962 E
Munneurycope menziesi Wolff, 1962 E

Munnopsis gracilis Beddard, 1886 E
Notopais euaxos Merrin & Bruce, 2006 E
Notopais zealandica Merrin, 2004 E
Paropsurus giganteus Wolff, 1962
Pseudarachna nohinohi Merrin, 2006 E
Storthyngura benti Wolff, 1956 E
Vanhoeffenura abyssalis Wolff, 1962 E
Vanhoeffenura furcata Wolff, 1956 E
Vanhoeffenura kermadecensis Wolff, 1962 E
Vanhoeffenura novaezelandiae (Beddard, 1885) E
Sursumura affinis Malyutina, 2004
PARAMUNNIDAE
Allorostrata n. sp. NIWA N. Bruce E
Austronanus aucklandensis Just & Wilson, 2006
Austronanus sp. A Just & Wilson 2006
Omanana serraticoxa Just & Wilson, 2004 E
'*Paramunna serrata*' sensu Stephenson 1927 E
Paramunna snaresi Just & Wilson, 2004 E
Spiculonana petraea Just & Wilson, 2004 E
Spiculonana platysoma Just & Wilson, 2004 E
Sporonana concavirostra Just & Wilson, 2004 E
Sporonana litoralis Just & Wilson, 2004 E
Gen. nov. 1 N. Bruce 2008 E
Gen. nov. 2 N. Bruce 2008 E
PSEUDOJANIRIDAE
Schottea taupoensis Serov & Wilson, 1999 E
Schottea n. sp. E
SANTIIDAE
Halacarsantia uniramea (Menzies & Miller, 1955) E
Kuphomunna n. sp. NIWA N. Bruce E
Santia hispida (Vanhöffen, 1914)
Santia n. spp. (2) 2E
STENETRIIDAE
Protallocoxa abyssale (Wolff, 1962) E
Stenetrium fractum Chilton, 1884 E

Suborder PHREATOICIDEA
PHREATOICIDAE
Neophreatoicus assimilis (Chilton, 1894) F E
Notamphisopus benhami Nicholls, 1944 F E
Notamphisopus dunedinensis (Chilton, 1906) F E
Notamphisopus flavius Nicholls, 1944 F E
Notamphisopus kirkii (Chilton, 1906) F E
Notamphisopus littoralis Nicholls, 1944 F E
Notamphisopus percevali Nicholls, 1944 F E
Phreatoicus orarii Nicholls, 1944 F E
Phreatoicus typicus Chilton, 1883 F E

Suborder CYMOTHOIDA
AEGIDAE
Aega komai Bruce, 1996
Aega monophthalam Johnston, 1834
Aega semicarinata Miers, 1875
Aega stevelowei Bruce, 2009
Aega urotoma Barnard, 1914
Aegapheles alazon (Bruce, 2004)
Aegapheles birubi (Bruce, 2004)
Aegapheles copidis Bruce, 2009
Aegapheles hamiota (Bruce, 2004)
Aegapheles mahana Bruce, 2009 E
Aegapheles rickbruscai (Bruce, 2004)
Aegapheles umpara (Bruce, 2004)
Aegiochus coroo (Bruce, 1983)
Aegiochus gordoni Bruce, 2009 E
Aegiochus insomnis Bruce, 2009 E
Aegiochus kakai Bruce, 2009 E
Aegiochus kanohi Bruce, 2009
Aegiochus laevis (Studer, 1883)
Aegiochus nohinohi Bruce, 2009
Aegiochus piihuka Bruce, 2009
Aegiochus riwha Bruce, 2009
Aegiochus tara Bruce, 2009
Aegiochus vigilans (Haswell, 1881)
Aegiochus sp. Bruce 2009
Epulaega derkoma Bruce, 2009
Epulaega fracta (Hale, 1940)
Rocinela bonita Bruce, 2009 E
Rocinela garricki Hurley, 1957 E
Rocinela leptopus Bruce, 2009 E
Rocinela pakari Bruce, 2009 E
Rocinela resima Bruce, 2009 E
Rocinela runga Bruce, 2009 E
Rocinela satagia Bruce, 2009 E
Rocinela sp. Bruce 2009
Syscenus latus Richardson, 1909 Pe
Syscenus springthorpei Bruce, 1997 Pe
Syscenus sp. Bruce 2009
ANTHURIDAE
Haliophasma novaezelandiae Wägele, 1985 E
Haliophasma platytelson Wägele, 1985 E
Quantanthura pacifica Wägele, 1985 E
Quantanthura raoulia Poore & Lew Ton, 1986 E
Mesanthura affinis (Chilton, 1883) E
ANUROPIDAE
Anuropus novaezealandiae Jansen, 1981 Pe E
Anuropus sp. N. Bruce 2008
BOPYRIDAE
Athelges lacertosi Pike, 1961 E
Eophrixus shojii Shiino, 1941
Gigantione pikei Page, 1985 E
Gyge angularis Page, 1985 E
Hemiarthrus nematocarcini Stebbing, 1914
Pleurocryptella infecta Nierstrasz & Brender à Brandis, 1923
Pseudione affinis (Sars, 1882)
Pseudione hayi Nierstrasz & Brender à Brandis, 1931 E
Pseudione hyndmanni (Bate & Westwood, 1868)
Pseudione murawaiensis Page, 1985 E
Pseudione pontocari Page, 1985 E
Pseudostegias otagoensis Page, 1985 E
Rhopalione atrinicolae Page, 1985 E
CIROLANIDAE
Cirolana canaliculata Tattersall, 1921 E
Cirolana kokoru Bruce, 2004 E
Cirolana quechso Bruce, 2004 E
Cirolana quadripustulata Hurley, 19571 E
Cirolana n. spp. (5) 5E
Eurydice subtruncata Tattersall, 1921 E
Eurylana arcuata (Hale, 1925) E
Eurylana cooki (Filhol, 1885) E
Metacirolana caeca (Hansen, 1916) Pe
Metacirolana japonica (Hansen, 1890)
Natatolana amplocula Bruce, 1986
Natatolana aotearoa Keable, 2006 E
Natatolana honu Keable, 2006 E
Natatolana narica (Bowman, 1971) E
Natatolana paranarica Keable, 2006 E
Natatolana pellucida (Tattersall, 1921)
Natatolana rekohu Bruce, 2003 E
Natatolana rossi (Miers, 1876) E
Natatolana n. spp. (3) 3E
Pseudaega melanica Jansen, 1978 E
Pseudaega punctata Thomson, 1884 E
Pseudaega quarta Jansen, 1978 E
Pseudaega secunda Jansen, 1978 E
Pseudaega tertia Jansen, 1978 E
CRINONISCIDAE
Crinoniscus cephalatus Hosie, 2008 E
Crinoniscus politosummus Hosie, 2008 E
CYMOTHOIDAE
Ceratothoa imbricata (Fabricius, 1775)
Ceratothoa lineatus (Miers, 1876) E
Ceratothoa trillesi (Avdeev, 1979) E
Elthusa neocytta (Avdeev, 1975)
Elthusa propinqua (Richardson, 1904)
Elthusa raynaudii (Milne Edwards, 1840)
Mothocya ihi Bruce, 1986 E
Nerocila orbignyi (Guérin-Menéville, 1832)
EXPANATHURIDAE
Eisothistos adlateralis Knight-Jones & Knight-Jones, 2002 E
Heptanthura novaezealandiae Kensley, 1978 E
Rhiganthura spinosa Kensley, 1978 E
GNATHIIDAE
Bathygnathia tapinoma Cohen & Poore, 1994 E
Bathygnathia vollenhovia Cohen & Poore, 1994
Caecognathia akaroensis (Monod, 1926) E
Caecognathia nieli Svavarsson, 2005 E
Caecognathia pacifica (Monod, 1926) E
Caecognathia polythrix (Monod, 1926) E
Caecognathia regalis (Monod, 1926) E
Caecognathia sifae Svarvarsson, 2005 E
Caecognathia n. sp. E
Eunognathia n. sp. E
Gnathia brachyuropus Monod, 1926
Thaumastognathia diceros Monod, 1926 E
HEMIONISCIDAE
Scalpelloniscus nieli Hosie, 2008 E
Scalpelloniscus cf. *penicillatus* Grygier, 1981
Scalpelloniscus vomicus Hosie, 2008
HYSSURIDAE
Kupellonura proberti Wägele, 1985 E
LEPTANTHURIDAE
Albanthura rotunduropus Wägele, 1985 E
Albanthura stenodactyla Wägele, 1985 E
Bullowanthura crebrui Wägele, 1985 E
Cruregens fontanus Chilton, 1882 F E
Leptanthura chiltoni (Beddard, 1886) E
Leptanthura exilis Wägele, 1985 E
Leptanthura profundicola Wägele, 1985 E
Leptanthura truncatitelson Wägele, 1985 E
Psittanthura egregia Wägele, 1985 E
PARANTHURIDAE
Califanthura rima (Poore, 1981) E
Paranthura flagellata (Chilton, 1882) E
Paranthura longa Wägele, 1985 E
TRIDENTELLIDAE
Tridentella acheronae Bruce, 1988 E
Tridentella rosemariae Bruce, 2002 E
Tridentella tangaroae Bruce, 1988 E
Tridentella n. sp.

Suborder LIMNORIIDEA
LIMNORIIDAE
Limnoria convexa Cookson, 1991 E
Limnoria hicksi Schotte, 1989 E
Limnoria loricata Cookson, 1991 E
Limnoria quadripunctata Holthuis, 1949
Limnoria reniculus Schotte, 1989 E
Limnoria rugosissima Menzies, 1957
Limnoria segnis Chilton, 1883 E
Limnoria stephenseni Menzies, 1957 E
Limnoria tripunctata Menzies, 1951

Suborder SPHAEROMATIDEA
PLAKARTHRIIDAE
Plakarthrium typicum Chilton, 1883 E
SEROLIDAE
Acutiserolis sp. Poore & Storey 2009
Brucerolis brandtae Storey & Poore, 2009 E
Brucerolis howensis Storey & Poore, 2009 E
Brucerolis hurleyi Storey & Poore, 2009 E
Brucerolis osheai Storey & Poore, 2009 E
Myopiarolis bicolor (Bruce, 2008) E
Myopiarolis carinata (Bruce, 2008) E
Myopiarolis n. spp. (7) 7E
Spinoserolis latifrons (Miers, 1875) E
SPHAEROMATIDAE
Amphoroidea falcifer Thomson, 1879 E
Amphoroidea longipes Hurley & Jansen, 1977 E
Amphoroidea media Hurley & Jansen, 1971 E
Benthosphaera guaware Bruce, 1994
Bilistra cavernicola Sket & Bruce, 2004 F E
Bilistra millari Sket & Bruce, 2004 F E
Bilistra mollecopulans Sket & Bruce, 2004 F E

Cassidina typa Milne Edwards, 1840 E
Cassidinopsis admirabilis Hurley & Jansen, 1977 E
Cerceis trispinosa (Haswell, 1882)
Cilicaea angustispinata Hurley & Jansen, 1977 E
Cilicaea caniculata (Thomson, 1879) E
Cilicaea dolorosa Hurley & Jansen, 1977 E
Cilicaea tasmanensis Hurley & Jansen, 1977 E
Cilicaeopsis n. sp. N. Bruce 2008 E
Cymodoce allegra Hurley & Jansen, 1977 E
Cymodoce australis Hodgson, 1902 E
Cymodoce convexa Miers, 1876 E
Cymodoce hamata Stephensen, 1927 E
Cymodoce hodgsoni Tattersall, 1921 E
Cymodoce iocosa Hurley & Jansen, 1977 E
Cymodoce penserosa Hurley & Jansen, 1977 E
Cymodocella capra Hurley & Jansen, 1977 E
Cymodocella egregia (Chilton, 1892) E
Cymodocella tubicauda Pfeffer, 1887
Cymodopsis impudica Hurley & Jansen, 1977 E
Cymodopsis sphyracephalata Hurley & Jansen, 1977 E
Cymodopsis torminosa Hurley & Jansen, 1977 E
Dynamenoides decima Hurley & Jansen, 1977 E
Dynamenoides vulcanata Hurley & Jansen, 1977 E
Dynamenopsis varicolor Hurley & Jansen, 1971 E
Exosphaeroma chilense (Dana, 1853)
Exosphaeroma echinense Hurley & Jansen, 1977 E
Exosphaeroma falcatum Tattersall, 1921 E
Exosphaeroma gigas (Leach, 1818)
Exosphaeroma montis (Hurley & Jansen, 1977) E
Exosphaeroma obtusum (Dana, 1853) E
Exosphaeroma planulum Hurley & Jansen, 1971 E
Exosphaeroma waitemata Bruce, 2005 E
Exosphaeroma n. sp. N. Bruce E
Ischyromene condita (Hurley & Jansen, 1977) E
Ischyromene cordiforaminalis (Chilton, 1883) E
Ischyromene hirsuta (Hurley & Jansen, 1971) E
Ischyromene huttoni (Thomson, 1879) E
Ischyromene insulsa (Hurley & Jansen, 1977) E
Ischyromene kokotahi Bruce, 2006 E
Ischyromene mortenseni (Hurley & Jansen), 1977 E
Isocladus armatus (Milne Edwards, 1840) E
Isocladus calcareus (Dana, 1853) E
Isocladus dulciculus Hurley & Jansen, 1977 E
Isocladus inaccuratus Hurley & Jansen, 1977 E
Isocladus reconditus Hurley & Jansen, 1977 E
Isocladus spiculatus Hurley & Jansen, 1977 E
Makarasphaera amnicosa Bruce, 2005 F E
Pseudosphaeroma callidum Hurley & Jansen, 1977 E
Pseudosphaeroma campbellensis Chilton, 1909
Scutuloidea kutu Stephenson & Riley, 1996 E
Scutuloidea maculata Chilton, 1883 E
Sphaeroma laurensi Hurley & Jansen, 1977 E
Sphaeroma quoianum Milne Edwards, 1840
Syncassidina aestuaria Baker, 1929 A?
INCERTAE SEDIS
Paravireia typica Chilton, 1925 E
Paravireia pistus Jansen, 1973 E

Suborder VALVIFERA
ANTARCTURIDAE
Caecarcturus quadraspinosus Schultz, 1981 E
Chaetarcturus myops (Beddard, 1886) E
ARCTURIDAE
Neastacilla antipodea Poore, 1981 E
Neastacilla fusiformis (Hale, 1946) E
Neastacilla levis (Thomson & Anderton, 1921) E
Neastacilla tattersalli Lew Ton & Poore, 1986 E
Neastacilla tuberculata (Thomson, 1879) E
Neastacilla spp. (4) N. Bruce 2008
AUSTRARCTURELLIDAE
Dolichiscus opiliones (Schultz, 1981) E
Austrarcturella galathea Poore & Bardsley, 1992 E
Pseudarcturella chiltoni Tattersall, 1921 E
Pseudarcturella crenulata Poore & Bardsley, 1992 E
CHAETILIIDAE
Macrochiridothea uncinata Hurley & Murray, 1968 E
Maoridotea naylori Jones & Fenwick, 1978 E
Maoridotea n. sp. N. Bruce E
HOLOGNATHIDAE
Cleantis tubicola (Thomson, 1885) E
Holognathus karamea Poore & Lew Ton, 1990 E
Holognathus stewarti (Filhol, 1885) E
IDOTEIDAE
Austridotea annectens Nicholls, 1937 F E
Austridotea benhami Nicholls, 1937 F E
Austridotea lacustris (Thomson, 1879) F E
Batedotea elongata (Miers, 1876)
Euidotea durvillei Poore & Lew Ton, 1993 E
Idotea? festiva Chilton, 1881 E
Idotea metallica Bosc, 1802
Paridotea ungulata Pallas, 1772
PSEUDIDOTHEIDAE
Pseudidothea richardsoni Hurley, 1957 E

Suborder ONISCIDEA
Infraorder LIGIAMORPHA
ACTAECIIDAE
Actaecia euchroa Dana, 1853 T E
Actaecia opihensis Chilton, 1901 T E
ARMADILLIDAE
Acanthodillo spinosus (Dana, 1853) T E
Coronadillo hamiltoni (Chilton, 1901) T E
Coronadillo milleri (Chilton, 1917) T E
Coronadillo suteri (Chilton, 1915) T E
Cubaris ambitiosa (Budde-Lund, 1885) T E
Cubaris minima Vandel, 1977 T E
Cubaris murina Brandt, 1833 T A
Cubaris tarangensis (Budde-Lund, 1904) T E
Merulana chathamensis (Budde-Lund, 1904) T E
Sphaerilloides? antipodum Vandel, 1977 T E
Sphaerilloides? invisibilis Vandel, 1977 T E
Sphaerilloides? macmahoni (Chilton, 1901) T E
Sphaerilloides? minimus Vandel, 1977 T E
Sphaerilloides? rugulosus (Miers, 1876) T E
Sphaerilloides? tuberculatus Vandel, 1977 T E
Spherillo bipunctatus Budde-Lund 1904 T E
Spherillo brevis Budde-Lund, 1904 T E
Spherillo danae Heller, 1865 T E
Spherillo inconspicuus (Miers, 1876) T E
Spherillo marginatus Budde-Lund, 1904 T E
Spherillo monolinus Dana, 1853 T E
Spherillo rufomarginatus Budde-Lund, 1904 T E
Spherillo setaceus Budde-Lund, 1904 T E
Spherillo speciosus (Dana, 1853) T E
Spherillo squamatus Budde-Lund, 1904 T E
Reductoniscus watti Vandel, 1977 T E
ARMADILLIDIIDAE
Armadillidium vulgare (Latreille, 1804) T A
LIGIIDAE
Ligia exotica Roux, 1828 T
Ligia novizealandiae Dana, 1853 T E
ONISCIDAE
Phalloniscus armatus Bowley, 1935 T E
Phalloniscus bifidus Vandel, 1977 T E
Phalloniscus bowleyi Vandel, 1977 T E
Phalloniscus chiltoni Bowley, 1935 T E
Phalloniscus cooki (Filhol, 1885) T E
Phalloniscus forsteri Vandel, 1977 T E
Phalloniscus kenepurensis (Chilton, 1901) T E
Phalloniscus lamellatus Vandel, 1977 T E
Phalloniscus minimus Vandel, 1977 T E
Phalloniscus montanus Vandel, 1977 T E
Phalloniscus occidentalis Vandel, 1977 T E
Phalloniscus propinquus Vandel, 1977 T E
Phalloniscus punctatus (Thomson, 1879) T E
PHILOSCIIDAE
Adeloscia dawsoni Vandel, 1977 T E
Okeaninoscia oliveri (Chilton, 1911) T E
Papuaphiloscia hurleyi Vandel, 1977 T
Paraphiloscia brevicornis (Budde-Lund, 1912) T E
Paraphiloscia fragilis (Budde-Lund, 1904) T E
Philoscia novaezealandiae Filhol, 1885 T E
Philoscia pubescens (Dana, 1853) T E
Stephenoscia bifrons Vandel, 1977 T E
PORCELLIONIDAE
Porcellio scaber Latreille, 1804 T A
Porcellionides pruinosus (Brandt, 1833) T A
SCYPHACIDAE
Deto aucklandiae (Thomson, 1879) T E
Deto bucculenta (Nicolet, 1849) T
Scyphax ornatus Dana, 1853 T E
Scyphoniscus magnus Chilton, 1909 T E
Scyphoniscus waitatensis Chilton, 1901 T E
STYLONISCIDAE
Notoniscus australis (Chilton, 1909) T E
Notoniscus helmsii (Chilton, 1901) T E
Styloniscus commensalis (Chilton, 1910) T E
Styloniscus kermadecensis (Chilton, 1911) T E
Styloniscus magellanicus Dana, 1853 T
Styloniscus otakensis Chilton, 1901 T E
Styloniscus phormianus (Chilton, 1901) T E
Styloniscus thomsoni (Chilton, 1885) T E
Styloniscus phormianus (Chilton, 1901) T E
Styloniscus thomsoni (Chilton, 1885) T E
TRACHELIPODIDAE
Nagurus nanus (Budde-Lund, 1908) T A
TRICHONISCIDAE
Haplophthalmus danicus Budde-Lund, 1885 T A

Infraorder TYLOMORPHA
TYLIDAE
Tylos neozelanicus Chilton, 1901 T E

Order TANAIDACEA
Suborder APSEUDOMORPHA
APSEUDIDAE
Apseudes larseni Knight & Heard, 2006 E
Apseudes meridionalis Richardson, 1912*
Apseudes spectabilis Studer, 1883*
Apseudes spp. (9)
Gollumudes spp. (2?) NIWA G. Bird
Leviapseudes galatheae Wolff, 1956* E
Leviapseudes segonzaci Bacescu, 1981*
Spinosapseudes setosus (Lang, 1968) E
Taraxapseudes diversus (Lang, 1968)*
METAPSEUDIDAE
Apseudomorpha timaruvia (Chilton, 1882) E
Cyclopoapseudes latus (Chilton, 1883) E
Metapseudes aucklandiae Stephensen, 1927 E
Synapseudes n. spp. (2)*
PAGURAPSEUDIDAE
Pagurapseudes? sp.*
SPHYRAPIDAE
Kudinopasternakia dispar (Lang, 1968)*
INCERTAE SEDIS
Gen. et sp. indet. NIWA J. Sieg/G. Bird

Suborder NEOTANAIDOMORPHA
NEOTANAIDAE
Herpotanais kirkegaardi Wolff, 1956
Neotanais barfoedi Wolff, 1956
Neotanais hadalis Wolff, 1956
Neotanais mesostenoceps Gardiner, 1975*
Neotanais robustus Wolff, 1956
Neotanais vemae Gardiner, 1975*
Neotanais sp. NIWA G. Bird

Suborder TANAIDOMORPHA
AGATHOTANAIDAE
Agathotanais spinipoda Larsen, 1999*
Paragathotanais sp. NIWA G. Bird*
Paranarthrura fortispina Sieg, 1986*
Paranarthrura meridionalis Sieg, 1986*
Paranarthrura spp. (2)*

ANARTHRURIDAE
Siphonolabrum sp. NIWA G. Bird
Gen. et spp. indet. (2) NIWA G. Bird
COLLETTEIDAE
Collettea cylindratoides Larsen, 1999*
Leptognathiella spp. (2) NIWA G. Bird
Libanius sp. NIWA G. Bird
Macrinella spp. (2?) NIWA G. Bird
LEPTOCHELIIDAE
Konarus sp. G. Bird
Leptochelia mirabilis Stebbing, 1905
LEPTOGNATHIIDAE
Leptognathia spp. (>3)*
NOTOTANAIDAE
Nototanais sp. G. Bird Ca
PARATANAIDAE
Bathytanais spp. (2) NIWA G. Bird
Paratanais oculatus (Vanhoeffen, 1914) B
Paratanais tenuis (G.M.Thomson, 1880) E
Paratanais sp.* Auckland Is.
Paratanais spp. (3)*
PSEUDOTANAIDAE
Akanthinotanais sp. NIWA G. Bird
Cryptocopoides arcticus (Hansen, 1886)
Cryptocopoides sp. NIWA G. Bird
Mystriocentrus sp. NIWA G. Bird
Pseudotanais nordenskioldi (Sieg, 1977)
Pseudotanais spp. (3)*
TANAELLIDAE
Araphura spp. (2) NIWA G. Bird
Araphuroides sp. NIWA G. Bird
Arthrura monocanthus (Vanhoeffen, 1914) n. comb.*
Tanaella forcifera (Lang, 1968)*
Tanaella spp. (4) NIWA G. Bird
TANAIDAE
Pancoloides litoralis (Vanhöffen, 1914)*
Pancoloides sp.* NIWA G. Bird
Sinelobus stanfordi (Richardson, 1901) F B C (sponge)
Synaptotanais sp. NIWA G. Bird
Tanais sp.*
Zeuxo novaezealandiae (Thomson, 1879) E
Zeuxo phytalensis Sieg, 1980*
Zeuxoides aka Bird, 2008 E
Zeuxoides helleri Sieg, 1980*
Zeuxoides ohlini (Stebbing, 1914)*
Zeuxoides pseudolitoralis Sieg, 1980*
Zeuxoides rimuwhero Bird, 2008 E
Zeuxoides sp.*
TYPHLOTANAIDAE
Hamatipeda spp. (2) NIWA G. Bird
Larsenotanais sp. NIWA G. Bird
Meromonakanatha sp. NIWA G. Bird
Paratyphlotanais sp. NIWA G. Bird
Typhlotanais greenwichensis Shiino, 1970*
Typhlotanais spp. (10)*
INCERTAE SEDIS
Akanthophoreus spp. (2) NIWA G. Bird
Chauliopleona spp. (2) NIWA G. Bird
Exspina typica Lang, 1968 C (holothurian)
Mirandotanais vorax Kussakin & Tzareva, 1974*
Stenotanais sp. NIWA G. Bird
Tanaopsis spp. (2) NIWA G. Bird

Order CUMACEA
BODOTRIIDAE
Apocuma n. sp. 1 B E
Bathycuma longirostre Calman, 1905 B
Cyclaspis argus Zimmer, 1902 E
Cyclaspis coelebs Calman, 1907 E
Cyclaspis elegans Calman, 1907 E
Cyclaspis laevis Thomson, 1892
Cyclaspis similis Calman, 1907
Cyclaspis tasmanica Jones, 1969 B E
Cyclaspis thomsoni Calman, 1907
Cyclaspis triplicata Calman, 1907 E
Cyclaspis n. sp. 1 B E
Cyclaspis n. sp. 2 B E
Cyclaspis n. sp. ?3 E
Gaussicuma scabra Jones, 1969 B E
Gaussicuma n. sp. 1 B E
Pomacuma australiae (Zimmer, 1921)
DIASTYLIDAE
Colurostylis castlepointensis Gerken & Lörz, 2007 E
Colurostylis lemurum Calman, 1917 E
Colurostylis longicauda Jones, 1963 E
Colurostylis pseudocuma Calman, 1911 E
Colurostylis stenocuma Lomakina, 1968 E
Diastylis acuminata Jones, 1960 E
Diastylis delicata Jones, 1969 B E
Diastylis insularum (Calman, 1908) E
Diastylis neozelanica Thomson, 1892 E
Diastylopsis crassior Calman, 1911 E
Diastylopsis elongata Calman, 1911 E
Diastylopsis thileniusi (Zimmer, 1902) E
Leptostylis profunda Jones, 1969 E
Leptostylis recalvastrata Hale, 1945
Makrokylindrus? mersus Jones, 1969 B E
Makrokylindrus neptunius Jones, 1969 E (abyssal)
Makrokylindrus sp. 1 B E
Paradiastylis? bathyalis Jones, 1969 E
Vemakylindrus sp. 1 E
GYNODIASTYLIDAE
Allodiastylis acanthanasillos Gerken, 2001 E
Axiogynodiastylis fimbriata Gerken, 2001 B E
Axiogynodiastylis kopua Gerken, 2001 E
Gynodiastylis carinata Calman, 1911 E
Gynodiastylis koataata Gerken, 2001 E
Gynodiastylis milleri Jones, 1963 E
Litogynodiastylis laevis (Calman, 1911) E
LAMPROPIDAE
Hemilamprops pellucidus Zimmer, 1908 S B
Hemilamprops ?n. sp. 1 E
Hemilamprops n. sp. 2 B E
Mesolamprops sp. B E
Paralamprops sp. 1 B E
Paralamprops sp. 2 B E
Paralamprops? sp. 3 B E
Paralamprops? sp. 4 B E
Watlingia cassis Gerken, 2010 E
Watlingia chathamensis Gerken, 2010 E
LEUCONIDAE
Eudorella hurleyi Jones, 1963 E
Eudorella truncatula (Bate, 1856) ?A
Eudorellopsis resima Calman, 1907 E
Hemileucon comes Calman, 1907 E
Hemileucon uniplicatus Calman, 1907 E
Heteroleucon akaroensis Calman, 1907 E
Leucon (Alytoleucon) sp. B E
Leucon (Crymoleucon) heterostylis Calman, 1907 E
Leucon (C.) sp. B E
Leucon (Epileucon) latispina Jones, 1963 E
Leucon (?n. subgen.) sp. B E
Paraleucon suteri Calman, 1907 E
NANNASTACIDAE
Campylaspis inornata Jones, 1969 B E
Campylaspis rex Gerken & Ryder, 2002 B E
Campylaspis sp. 2 B E
Campylaspis sp. 3 B E
Campylaspis sp. 4 B E
Campylaspis sp. 5 B E
Procampylaspis sp. 1 B E
Procampylaspis sp. 2 B E
Scherocumella pilgrimi (Jones, 1963) E
Styloptocuma sp. 1 B E
Gen. nov. et n. sp. B

Order EUPHAUSIACEA
EUPHAUSIIDAE
Euphausia longirostris Hansen, 1908
Euphausia lucens Hansen, 1905
Euphausia recurva Hansen, 1905
Euphausia similis G.O. Sars, 1883
Euphausia s. armata Hansen, 1911
Euphausia spinifera G.O. Sars, 1883
Euphausia vallentini Stebbing, 1900.
Nematobrachion flexipes (Ortmann, 1893)
Nematosceles megalops G.O. Sars, 1883
Nematosceles microps G.O. Sars, 1883
Nyctiphanes australis G.O. Sars, 1883
Stylocheiron abbreviatum G.O.Sars, 1883
Stylocheiron elongatum G.O. Sars, 1883
Stylocheiron longicorne G.O. Sars, 1883
Stylocheiron maximum Hansen, 1908
Stylocheiron suhmi G.O. Sars, 1883
Thysanoessa gregaria G.O. Sars, 1883
Thysanoessa macrura G.O. Sars, 1883
Thysanopoda acutifrons Holt & Tattersall, 1905
Thysanopoda obtusifrons G.O. Sars, 1883

Order DECAPODA
Suborder DENDROBRANCHIATA
ARISTEIDAE
Aristaeomorpha foliacea (Risso, 1826)
Aristaeopsis edwardsiana (Johnson, 1867)
Aristeus semidentatus Bate, 1881
Austropenaeus cf. *nitidus* (Barnard, 1947)
BENTHESICYMIDAE
Benthesicymus cereus Burkenroad, 1936
Benthesicymus investigatoris Alcock & Anderson, 1899
Gennadas capensis Calman, 1925 Pe
Gennadas gilchristi Calman, 1925 Pe
Gennadas incertus (Balss, 1927)
Gennadas kempi Stebbing, 1914 Pe
Gennadas tinayrei Bouvier, 1906 Pe
LUCIFERIDAE
Lucifer typus H. Milne Edwards, 1837 Pe
PENAEIDAE
Funchalia villosa (Bouvier, 1905) Pe
Funchalia woodwardi Johnson, 1867 Pe
SERGESTIDAE
Sergestes arcticus Kröyer, 1855 Pe
Sergestes disjunctus Burkenroad, 1940 Pe
Sergestes index Burkenroad, 1940 Pe
Sergestes cf. *seminudus* Hansen, 1919 Pe
Sergia japonica (Bate, 1881) Pe
Sergia kroyeri (Bate, 1881) Pe
Sergia potens (Burkenroad, 1940) Pe
SICYONIIDAE
Sicyonia inflexa (Kubo, 1940)*
Sicyonia truncata (Kubo, 1949)
SOLENOCERIDAE
Haliporoides sibogae (de Man, 1907)
Hymenopenaeus obliquirostris (Bate, 1881)
Solenocera comata Stebbing 1915

Infraorder CARIDEA
ALPHEIDAE
Alpheopsis garricki Yaldwyn, 1971 E
Alpheus euphrosyne richardsoni Yaldwyn, 1971 E
Alpheus hailstonei Coutière, 1905
Alpheus novaezealandiae Miers, 1876
Alpheus socialis Heller, 1865
Athanas indicus Coutière, 1903
Betaeopsis aequimanus (Dana, 1852) E
ALVINOCARIDIDAE
Alvinocaris alexander Ahyong, 2009 E
Alvinocaris longirostris Kikuchi & Ohta, 1995
Alvinocaris niwa Webber, 2004 E
Nautilocaris saintlaurentae Komai & Segonzac, 2004
ATYIDAE
Paratya curvirostris (Heller, 1862) F E
CAMPYLONOTIDAE
Campylonotus rathbunae Schmitt, 1926
CRANGONIDAE

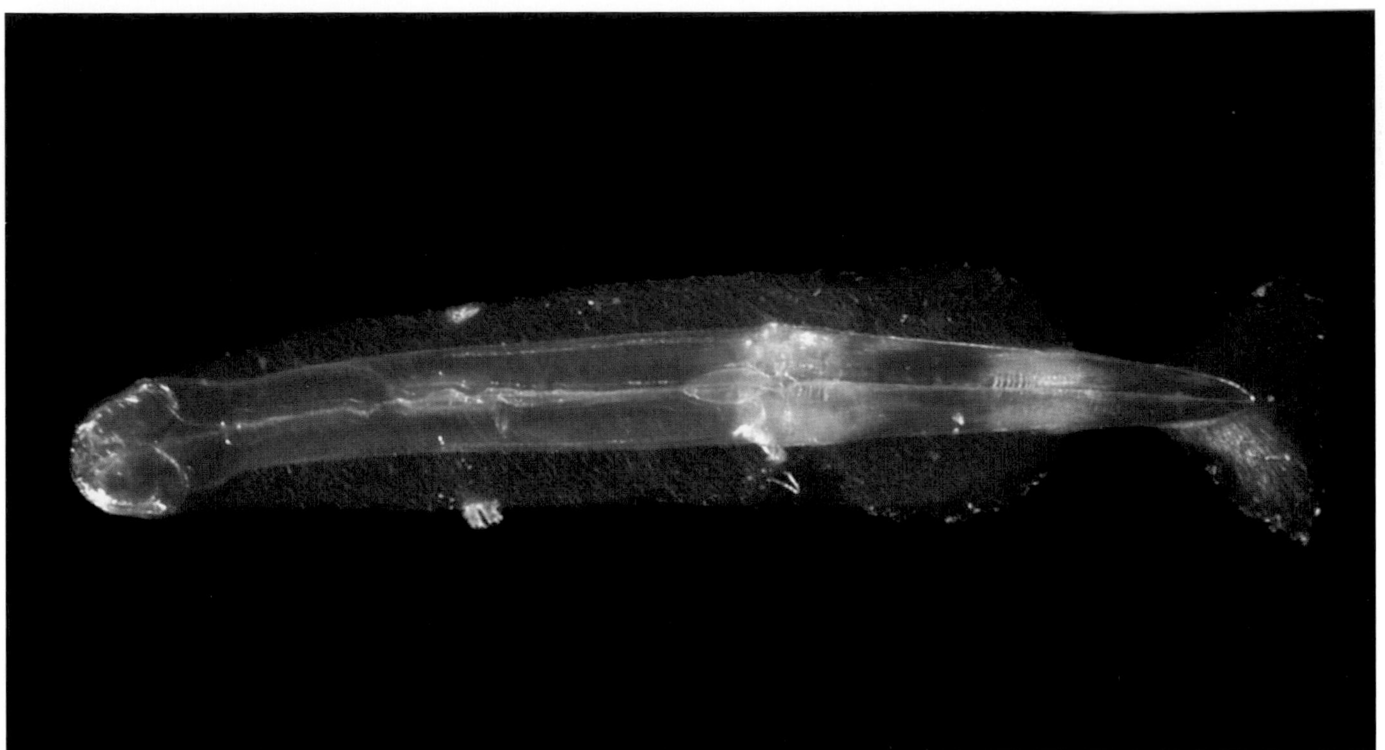

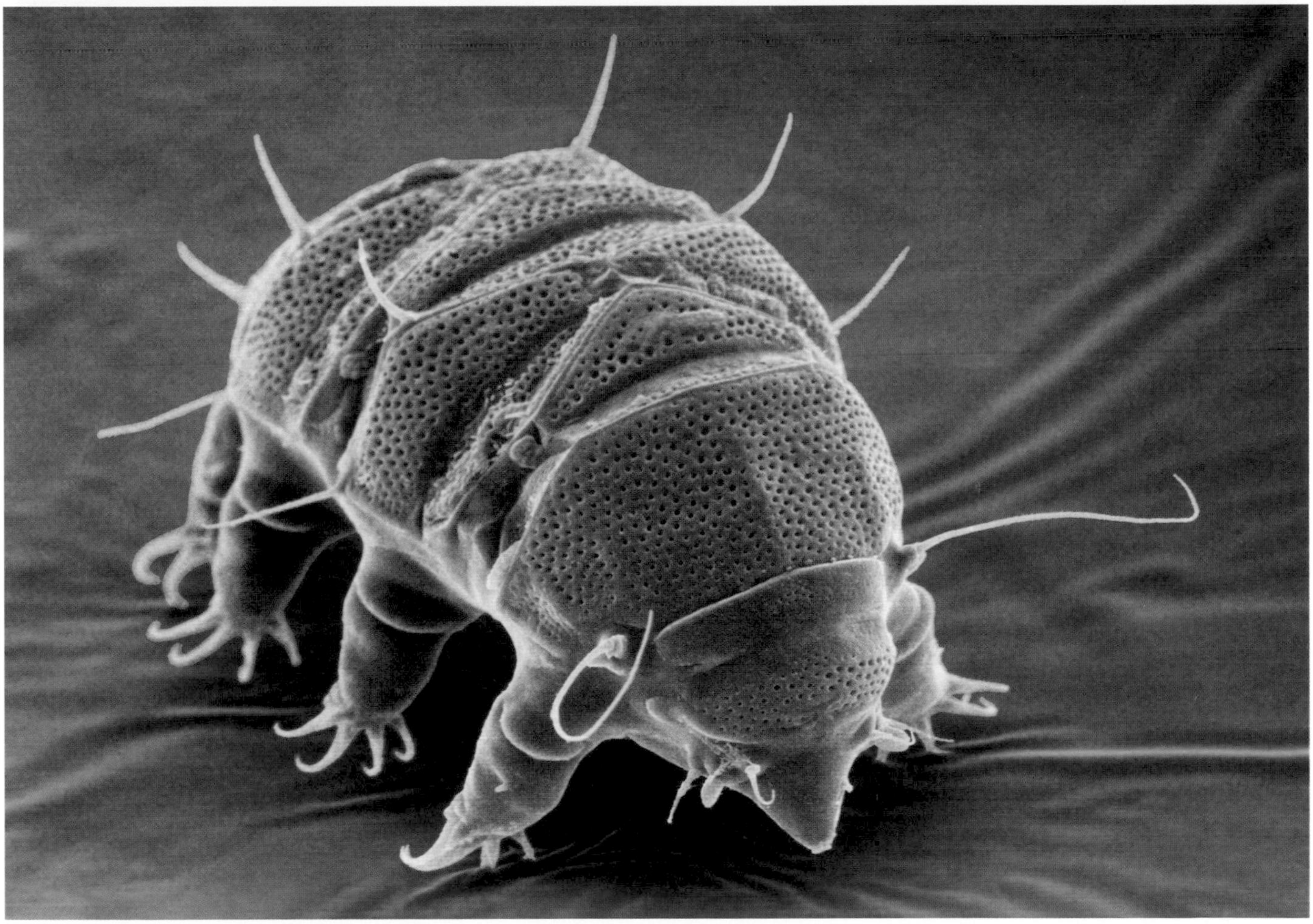

Plate 1a (top) Chaetognatha: arrow worm *Pterosagitta draco.* Cheryl Clarke, University of Alaska, Fairbanks
Plate 1b Tardigrada: water bear *Echiniscus elaeinae.* Diane Nelson, East Tennessee State University

Plate 2a (top) Onychophora: velvet worm *Ooperipatellus viridimaculatus*. Hilke Ruhberg and Hubert Bosch, Hamburg University

Plate 2b Arthropoda, Trilobitomorpha, Trilobita: fossil remains of the Cambrian trilobite *Koptura* sp. Marianna Terezow, GNS Science

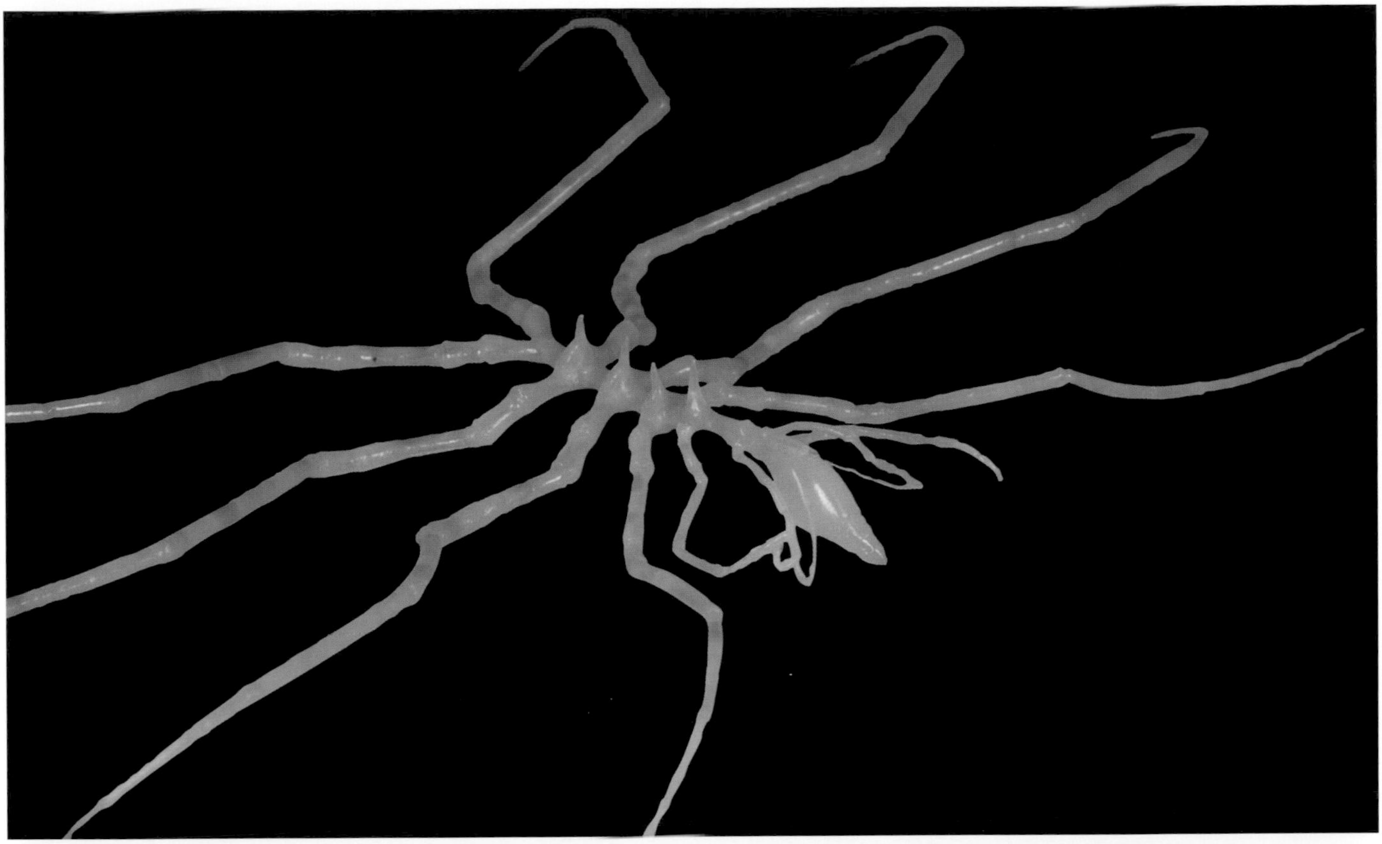

Plate 3a (top) Arthropoda, Chelicerata, Pycnogonida: subadult deep-sea sea spider *Ascorhynchus cooki*. Peter Marriott, NIWA
Plate 3b Arthropoda, Chelicerata, Arachnida: water spider *Dolomedes*. Steven Moore, Landcare Research

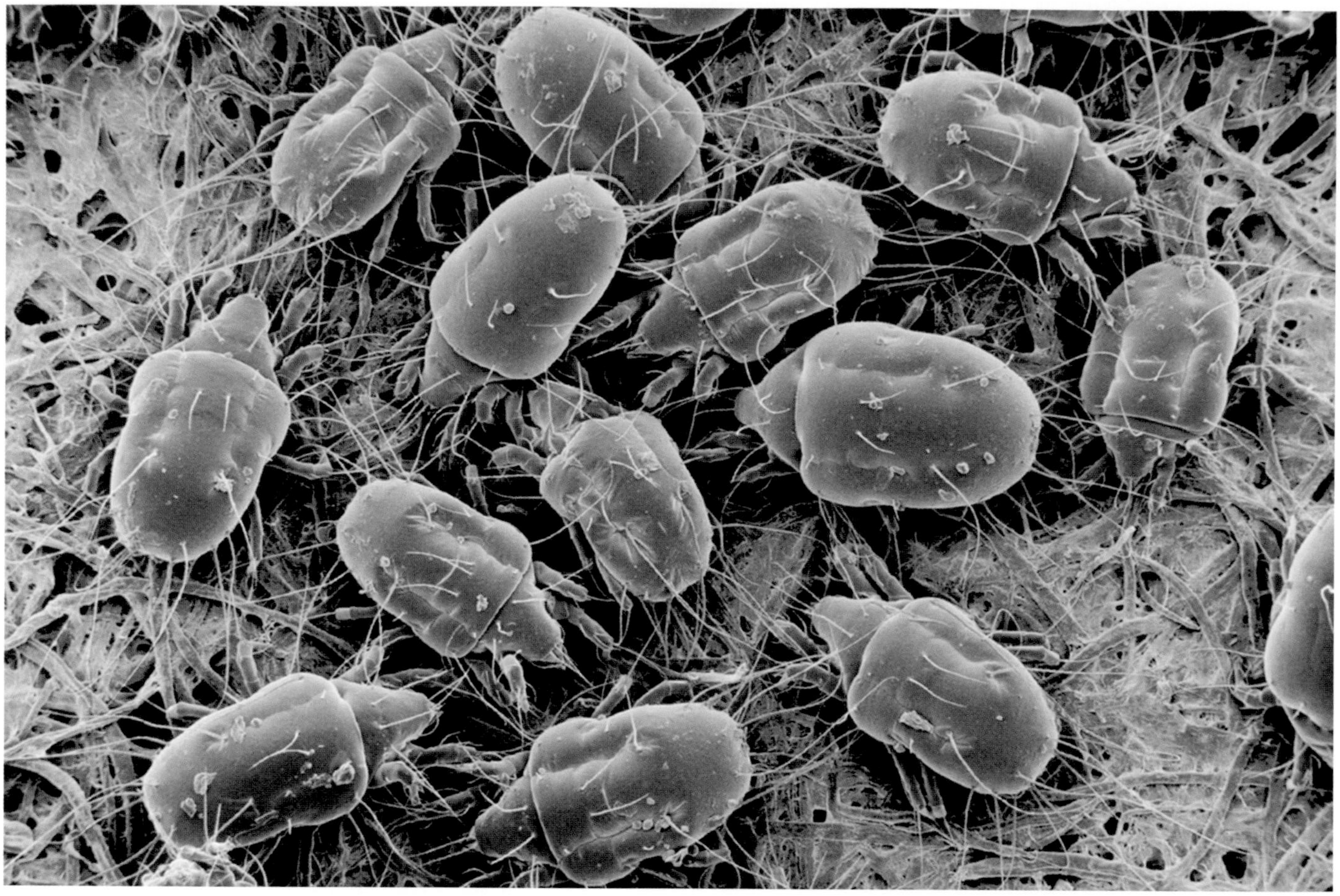

Plate 4a (top) Arthropoda, Chelicerata, Arachnida: cheese mite *Tyrophagus putrescentiae*. Eric Erbe and Chris Pooley, USDA by permission
Plate 4b Arthropoda, Chelicerata, Arachnida: an unidentified chernetid pseudoscorpion. Alastair Robertson and Maria Minor

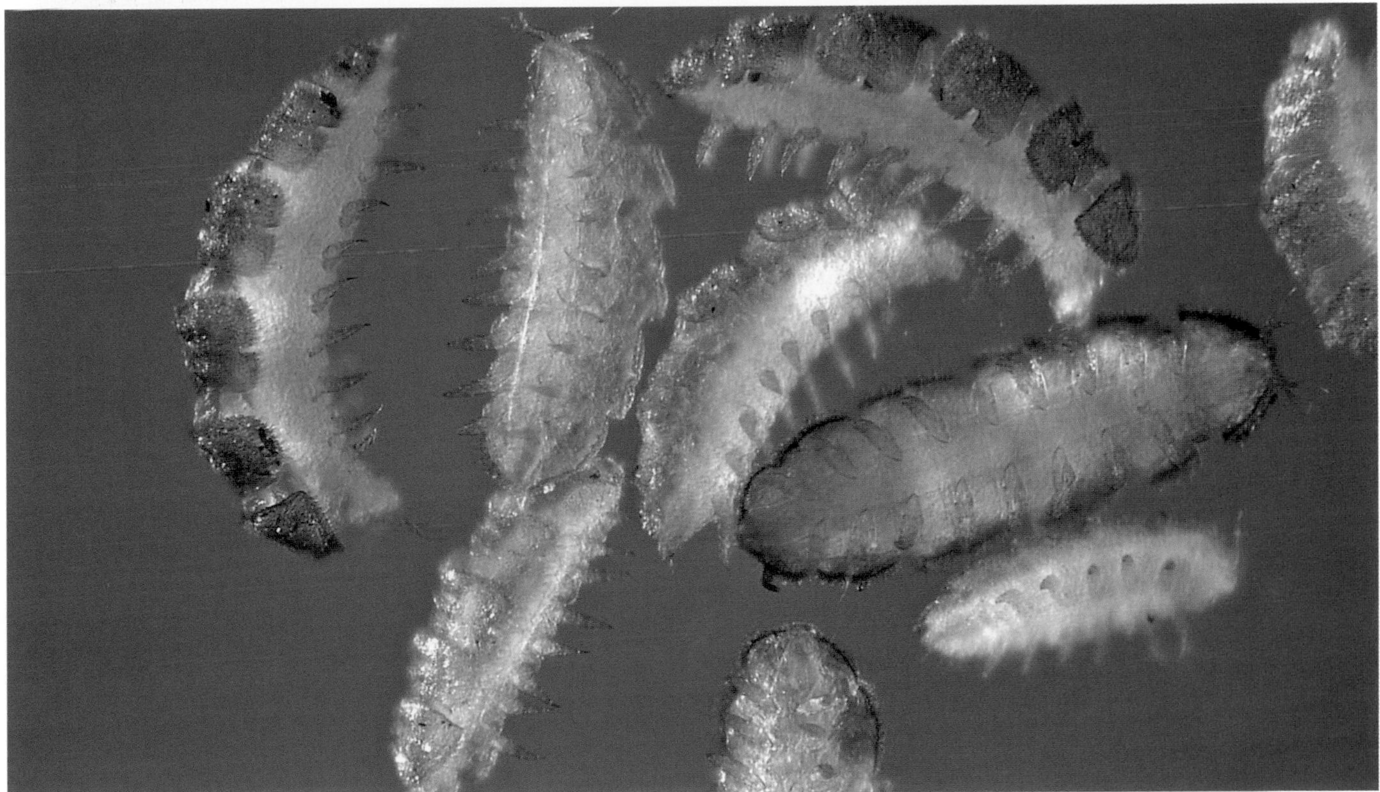

Plate 5a (top) Arthropoda, Myriapoda, Chilopoda: centipede *Craterostigmus crabilli*. Alastair Robertson and Maria Minor, Massey University

Plate 5b Arthropoda, Myriapoda, Pauropoda: an unidentified pauropod species. Alastair Robertson and Maria Minor, Massey University

Plate 6a (top) Arthropoda, Myriapoda, Symphyla: an unidentified symphylan. Alastair Robertson and Maria Minor, Massey University
Plate 6b Arthropoda, Myriapoda, Diplopoda: pill millipede *Procyliosoma striolatum*. Alastair Robertson and Maria Minor, Massey University

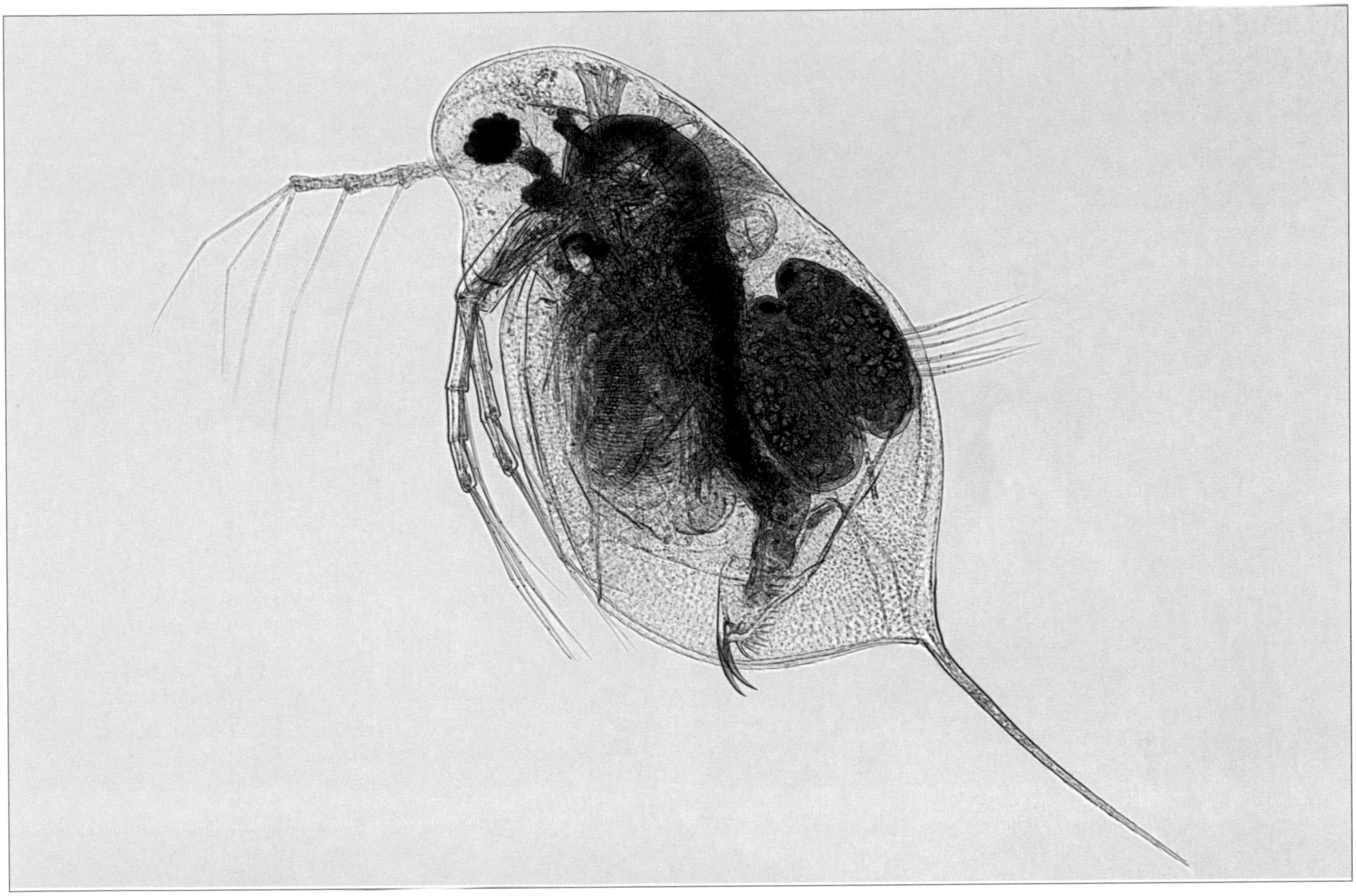

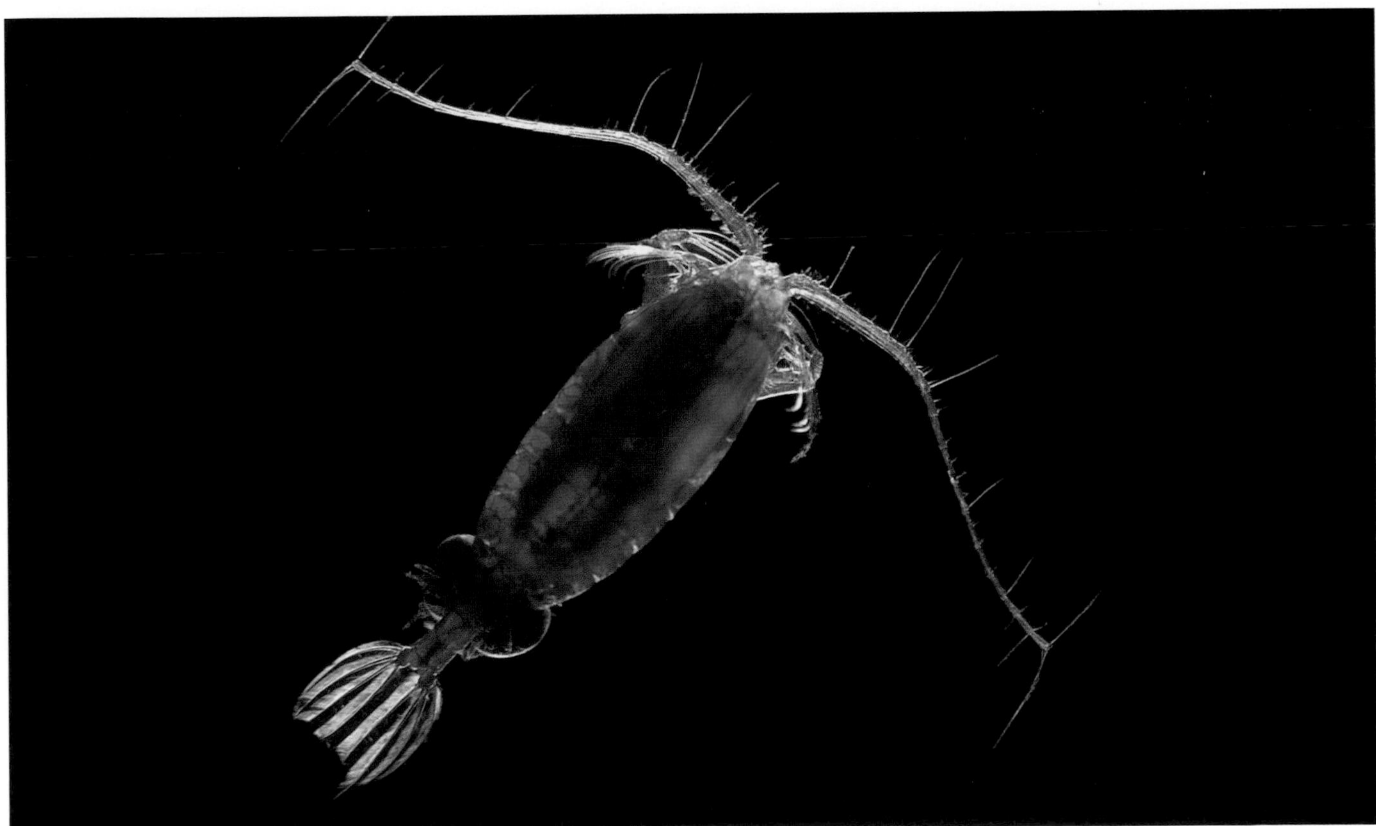

Plate 7a (top) Arthropoda, Crustacea, Branchiopoda: water flea, *Daphnia dentifera*. Barry O'Brien, University of Waikato
Plate 7b Arthropoda, Crustacea, Maxillopoda, Copepoda: *Valdiviella insignis*. Russell Hopcroft, University of Alaska, Fairbanks

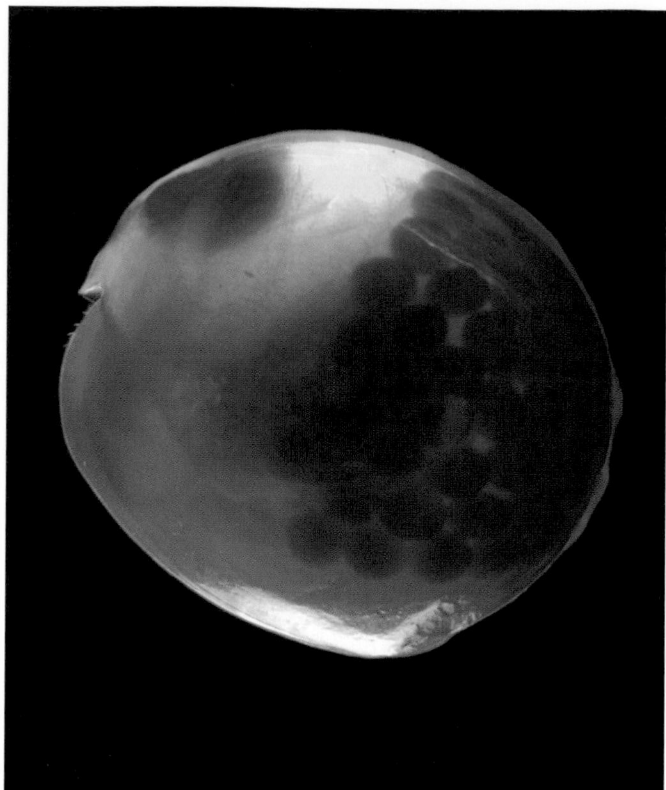

Plate 8a (top) Arthropoda, Crustacea, Ostracoda: *Conchoecissa imbricata* (left) and *Gigantocypris muelleri*. Russell Hopcroft, University of Alaska, Fairbanks

Plate 8b Arthropoda, Crustacea, Malacostraca: banded coral shrimp *Stenopus hispidus*. Roger Grace, Leigh

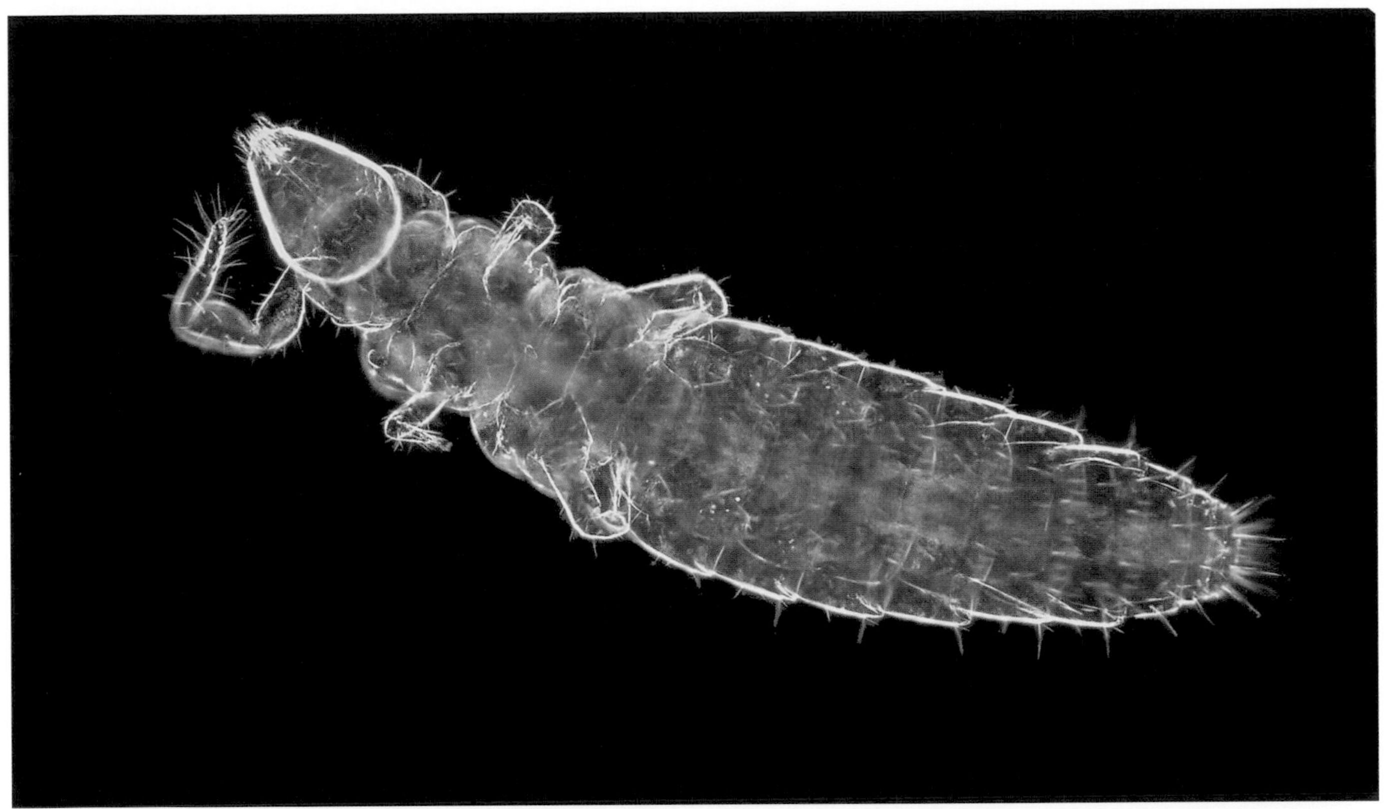

Plate 9a (top) Arthropoda, Hexapoda, Protura: proturan *Eosentomon australicum*. Alastair Robertson and Maria Minor, Massey University
Plate 9b Arthropoda, Hexapoda, Collembola: collembolon *Holacanthella brevispinosa*. Alastair Robertson and Maria Minor, Massey University

Plate 10a (top) Arthropoda, Hexapoda, Diplura: dipluran *Heterojapyx novaezealandiae*. Alastair Robertson and Maria Minor, Massey University
Plate 10b Arthropoda, Hexapoda, Insecta: damselfly *Ischnura aurora* (Odonata). Stephen Moore, Landcare Research

Plate 11a (top) Arthropoda, Hexapoda, Insecta: weta *Hemideina maori* (Orthoptera). Alastair Robertson and Maria Minor, Massey University
Plate 11b Arthropoda, Hexapoda, Insecta: alpine cicada *Maoricicada nigra frigida* (Hemiptera). Alastair Robertson, Massey University

Plate 12a (top) Arthropoda, Hexapoda, Insecta: water beetle *Hyphydrus elegans* (Coleoptera). Stephen Moore, Landcare Research
Plate 12b Arthropoda, Hexapoda, Insecta: water beetle *Onchohydrus hookeri*. Stephen Moore, Landcare Research

Plate 13a (top) Arthropoda, Hexapoda, Insecta: batwing fly *Exsul singularis* (Diptera). Brian Patrick, Central Stories Museum & Art Gallery, Alexandra
Plate 13b Arthropoda, Hexapoda, Insecta: moth *Asaphodes cinnabari* (Lepidoptera). Brian Patrick, Central Stories Museum & Art Gallery, Alexandra

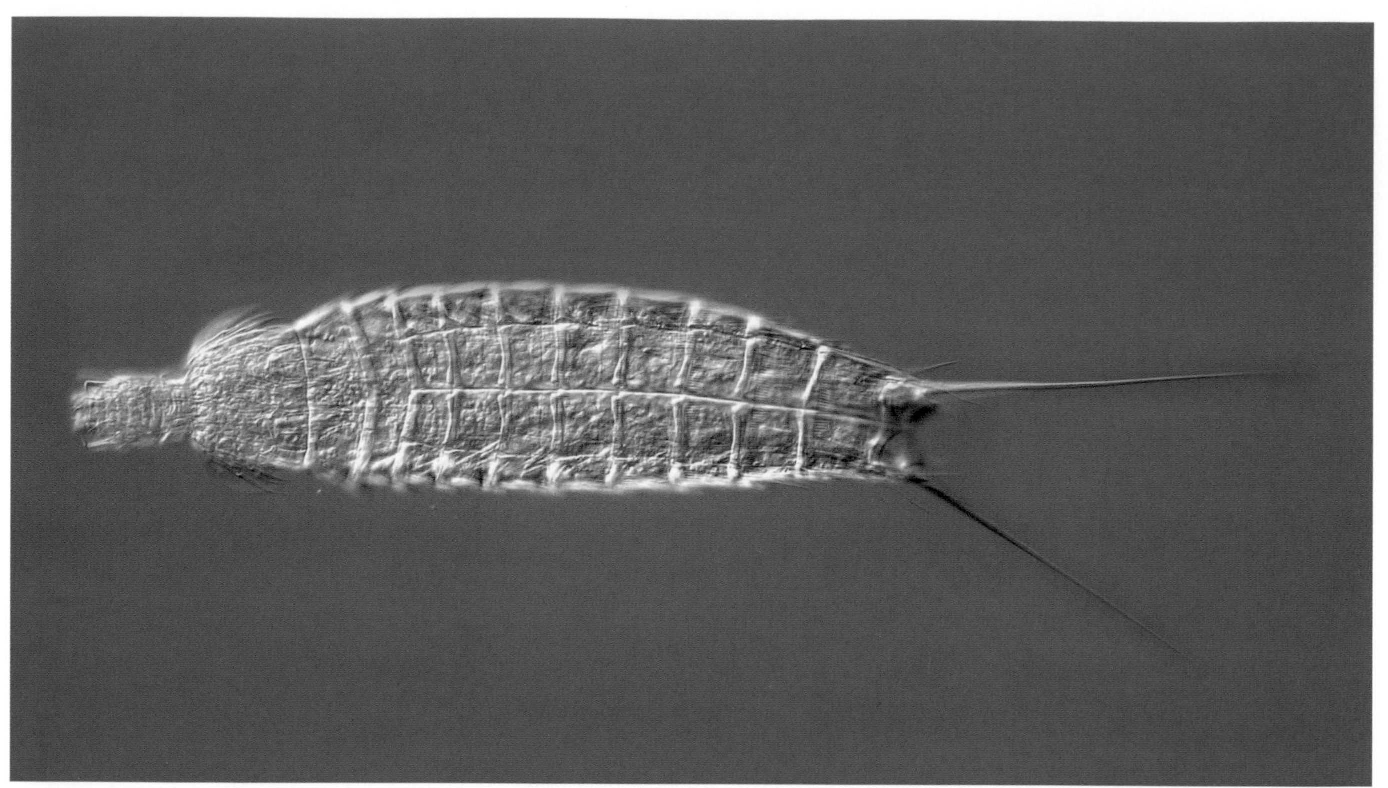

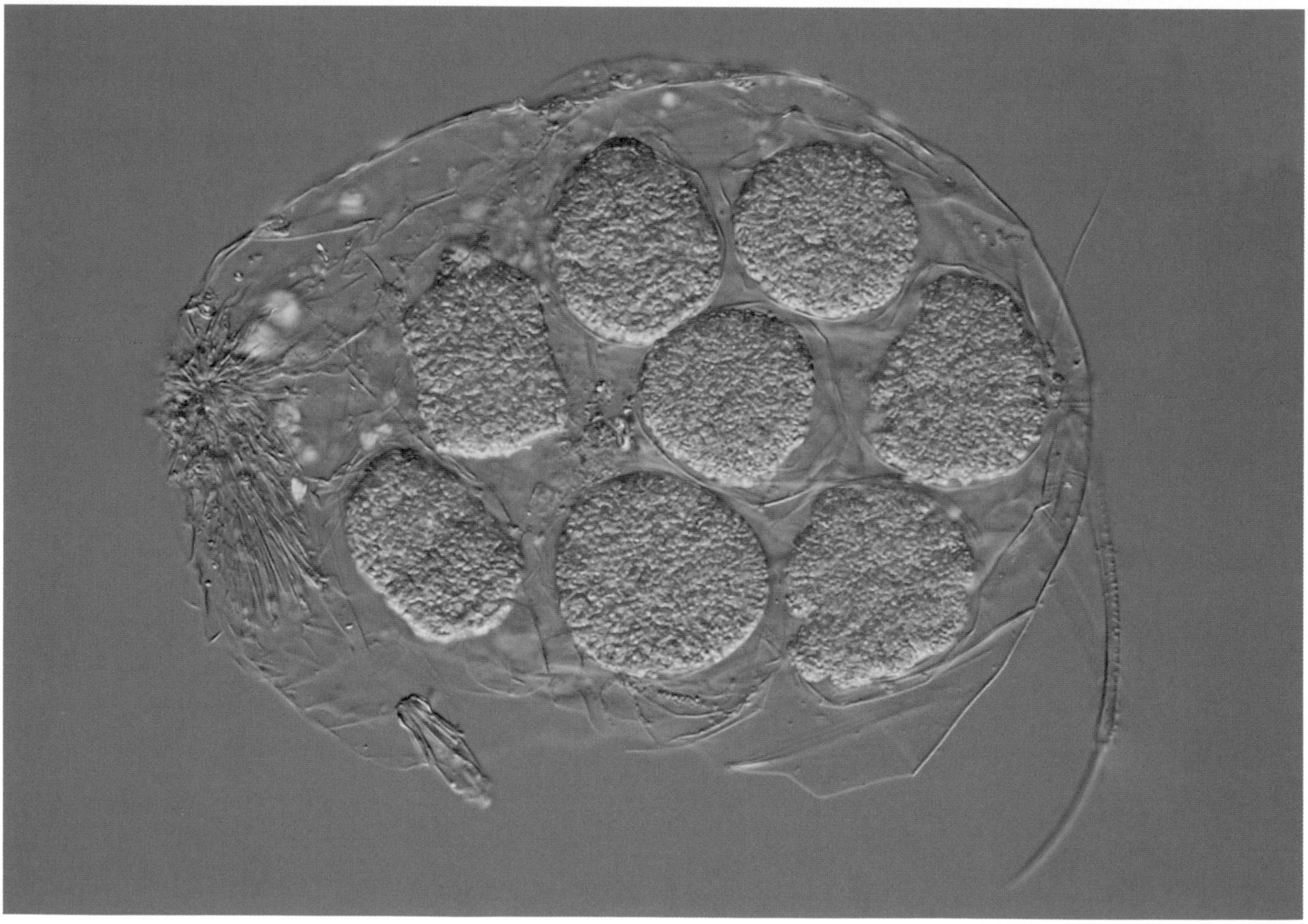

Plate 14a (top) Kinorhyncha: mud dragon *Echinoderes* sp. Birger Neuhaus, Museum für Naturkunde, Berlin
Plate 14b Loricifera: corset worm *Pliciloricus* sp. Iben Heimer, National Food Institute, Søborg, Denmark

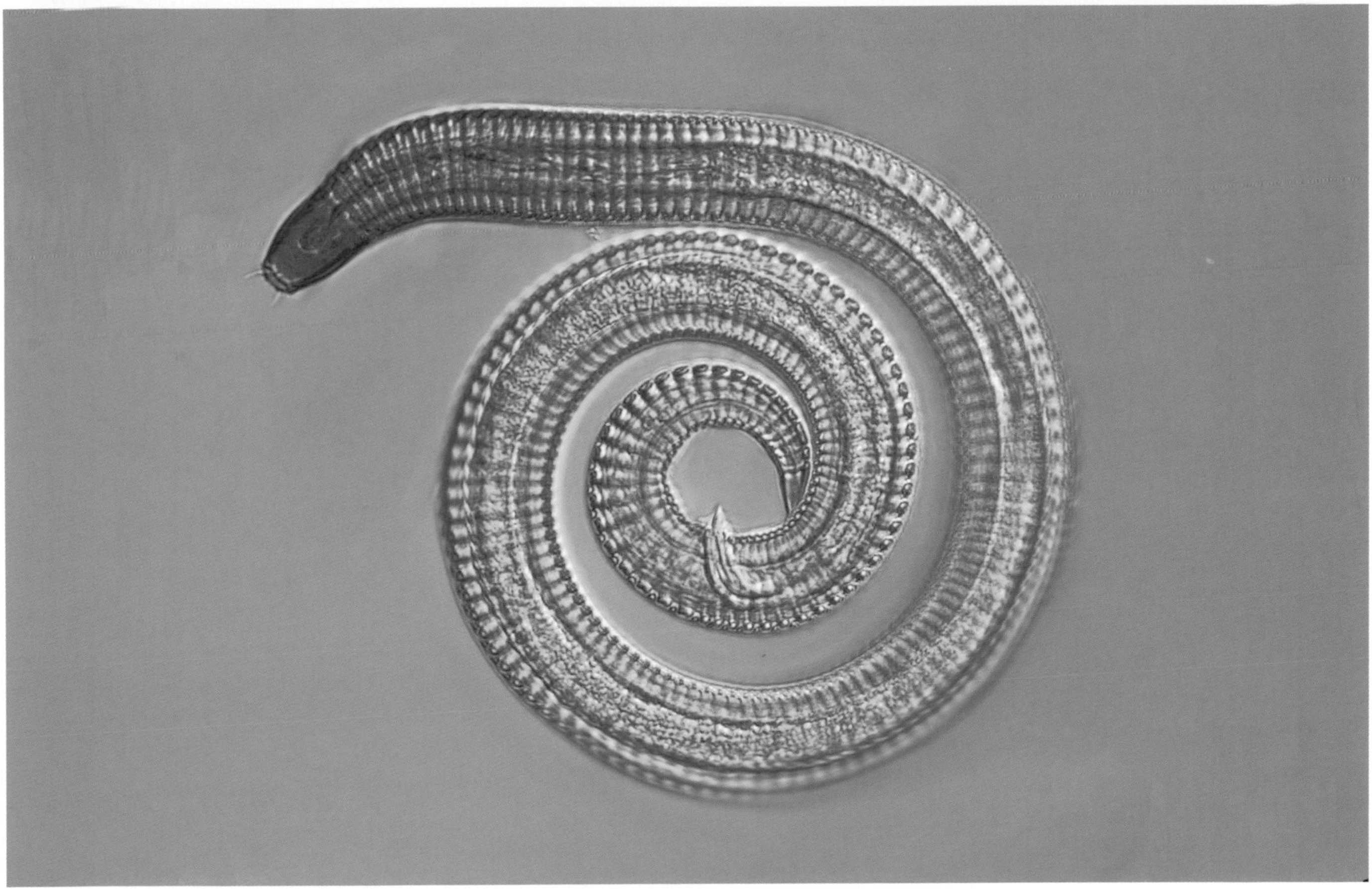

Plate 15a (top) Priapulida: penis worm *Priapulopsis australis.* Richard Taylor, University of Auckland
Plate 15b Nematoda: marine free-living nematode *Pselionema* sp. Daniel Leduc, NIWA, Wellington

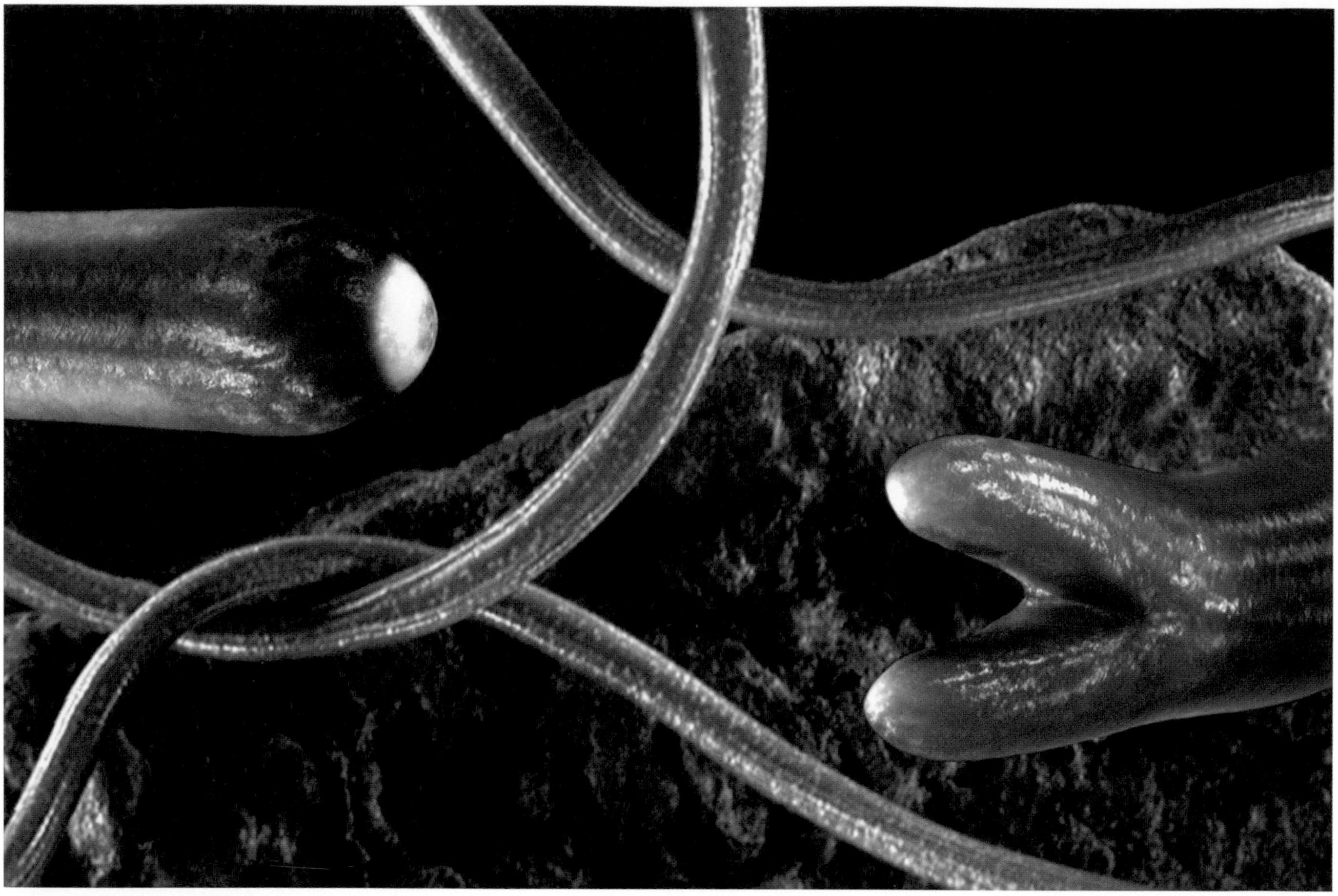

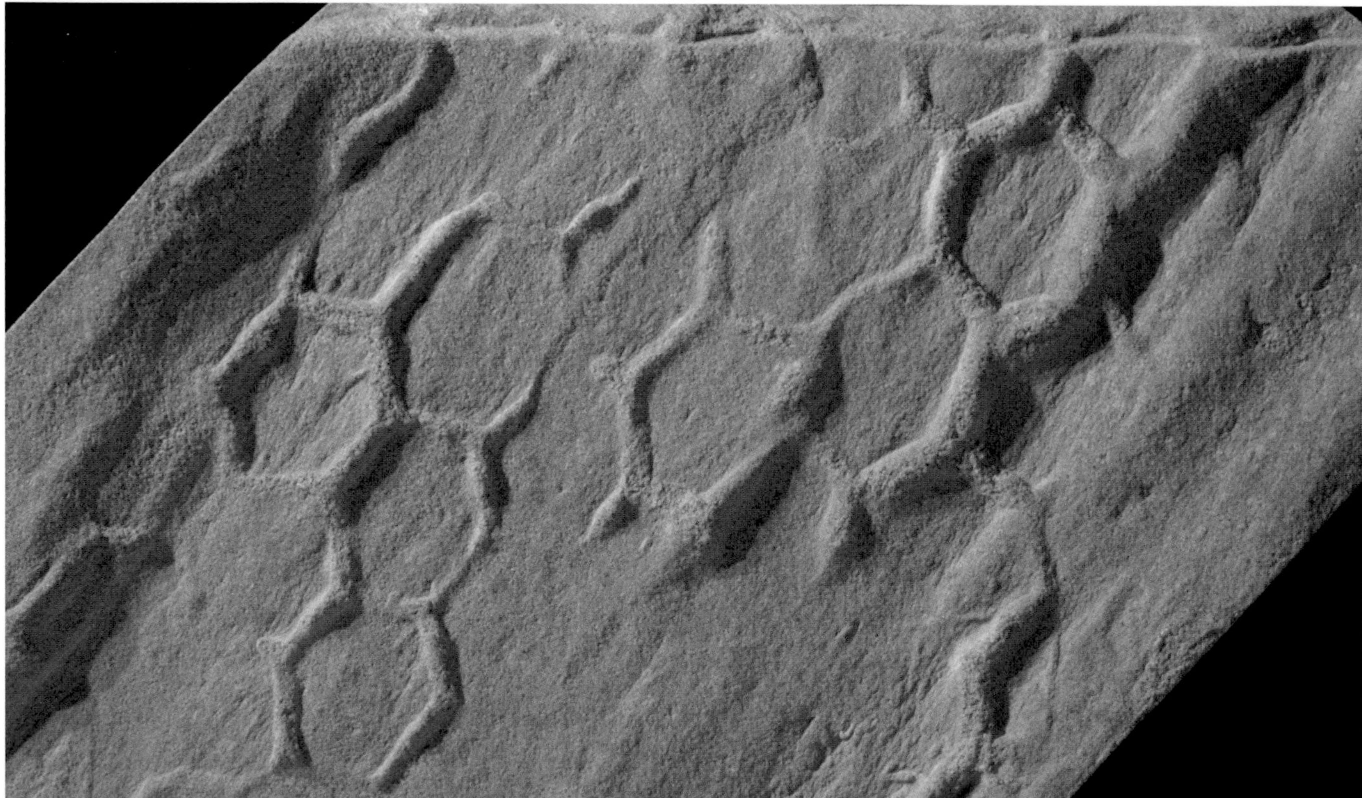

Plate 16a (top) Nematomorpha: male horsehair worm *Gordius dimorphus* (head, left inset; tail, right inset). Stephen Moore, Landcare Research
Plate 16b The trace fossil (ichnofossil) *Paleodictyon* (Tertiary, Mataikona). Marianna Terezow, GNS Science

Aegaeon lacazei (Gourret, 1888)
Metacrangon knoxi (Yaldwyn, 1960) E
Metacrangon richardsoni (Yaldwyn, 1960) E
Philocheras acutirostratus (Yaldwyn, 1960) E
Philocheras australis (Thomson, 1879) E
Philocheras chiltoni (Kemp, 1911) E
Philocheras hamiltoni (Yaldwyn, 1971) E
Philocheras pilosoides (Stephensen, 1927) E
Philocheras quadrispinosus (Yaldwyn, 1971) E
Philocheras yaldwyni (Zarenkov, 1968) E
Parapontophilus junceus Bate, 1888 E
Prionocrangon curvicaulis Yaldwyn, 1960
DISCIADIDAE
Discias cf. *exul* Kemp, 1920
HIPPOLYTIDAE
Alope spinifrons (H. Milne Edwards, 1837) E
Bathyhippolyte yaldwyni Hayashi & Miyake, 1970 E
Hippolyte bifidrostris (Miers, 1876) E
Hippolyte multicolorata Yaldwyn, 1971 E
Lebbeus cristatus Ahyong, 2009 E
Lebbeus wera Ahyong, 2009 E
Leontocaris alexander Poore, 2009
Leontocaris amplectipes Bruce, 1990
Leontocaris yarramundi Taylor & Poore, 1998
Lysmata morelandi (Yaldwyn, 1971)
Lysmata trisetacea (Heller, 1861)
Lysmata vittata (Stimpson, 1860)
Merhippolyte chacei Kensley, Tranter & Griffin, 1987
Nauticaris marionis Bate, 1888
Tozeuma novaezealandiae Borradaile, 1916 E
GLYPHOCRANGONIDAE
Glyphocrangon caeca Wood-Mason & Alcock, 1891
Glyphocrangon lowryi Kensley, Tranter & Griffin, 1987
Glyphocrangon regalis Bate, 1888
Glyphocrangon sculpta (Smith, 1882)
NEMATOCARCINIDAE
Lipkius holthuisi Yaldwyn, 1960
Nematocarcinus cf. *exilis* (Bate, 1888) ZMUC
Nematocarcinus gracilis Bate, 1888
Nematocarcinus hiatus Bate, 1888
Nematocarcinus longirostris Bate, 1888
Nematocarcinus novaezealandicus Burukovsky, 2006
Nematocarcinus serratus Bate, 1888
Nematocarcinus undulatipes Bate, 1888
Nematocarcinus webberi Burukovsky, 2006
Nematocarcinus yaldwyni Burukovsky, 2006
OGYRIDIDAE
Ogyrides delli Yaldwyn, 1971
OPLOPHORIDAE
Acanthephyra brevirostris Smith, 1885 Pe
Acanthephyra eximia Smith, 1884 Pe
Acanthephyra pelagica (Risso, 1816) Pe
Acanthephyra quadrispinosa Kemp, 1939 Pe
Acanthephyra smithi Kemp, 1939 Pe
Ephyrina figueirai Crosnier & Forest, 1973 Pe
Heterogenys microphthalma (Smith, 1885) Pe
Hymenodora glacialis (Buchholz, 1874) Pe
Janicella spinicauda (A. Milne Edwards, 1883) Pe
Kemphyra corallina (A. Milne Edwards, 1883) Pe
Meningadora mollis Smith, 1882 Pe
Meningadora vesca (Smith, 1886) Pe
Notostomus auriculatus Barnard, 1950 Pe
Notostomus japonicus Bate, 1888 Pe
Oplophorus novaezeelandiae de Man, 1931 Pe
Oplophorus spinosus (Brullé, 1839) Pe
Systellaspis debilis (A. Milne Edwards, 1881) Pe
Systellaspis pellucida (Filhol, 1885) Pe
PALAEMONIDAE
Hamiger novaezealandiae (Borradaile, 1916) E
Palaemon affinis H. Milne Edwards, 1937 E
Periclimenes fenneri Bruce, 2005
Periclimenes tangeroa Bruce, 2005
Periclimenes yaldwyni Holthuis, 1959 E
PANDALIDAE
Chlorotocus novaezealandiae (Borradaile, 1916)
Heterocarpus laevigatus Bate, 1888
Notopandalus magnoculus (Bate, 1888) E
Plesionika costelloi (Yaldwyn, 1971)
Plesionika martia (A.Milne Edwards, 1883)
Plesionika spinipes Bate, 1888
PASIPHAEIDAE
Alainopasiphaea australis (Hanamura, 1989)
Eupasiphae gilesii (Wood-Mason, 1892) Pe
Parapasiphae compta Smith, 1884 Pe
Parapasiphae sulcatifrons Smith, 1884 Pe
Pasiphaea barnardi Yaldwyn, 1971 Pe
Pasiphaea burukovskyi Wasmer, 1992 Pe
Pasiphaea grandicula Burukovsky, 1976 Pe
Pasiphaea notosivado Yaldwyn, 1971 Pe
Pasiphaea tarda Kröyer, 1845 Pe
Psathyrocaris infirma Alcock & Anderson, 1894 Pe
PROCESSIDAE
Processa moana Yaldwyn, 1971 E
RHYNCHOCINETIDAE
Rhynchocinetes balssi Gordon, 1936
Rhynchocinetes ikatere Yaldwyn, 1971 E
STYLODACTYLIDAE
Stylodactyloides crosnieri Cleva, 1990
Stylodactylus discissipes Bate, 1888 E

Suborder PLEOCYEMATA
Infraorder STENOPODIDEA
SPONGICOLIDAE
Spongicoloides novaezealandiae Baba, 1980 E
Spongiocaris yaldwyni Bruce & Baba, 1973 E
STENOPODIDAE
Stenopus hispidus (Olivier, 1811)

Infraorder ASTACIDEA
NEPHROPIDAE
Metanephrops challengeri (Balss, 1914) E
Nephropsis suhmi Bate, 1888
PARASTACIDAE
Paranephrops planifrons White, 1842 F E
Paranephrops zealandicus (White, 1847) F E

Infraorder AXIIDEA
AXIIDAE
Axius cf. *werribee* (Poore & Griffin, 1979) MNZ
Calocarides vigila Sakai, 1992 E
Calocaris isochela Zarenkov, 1898 E
Dorphinaxius kermadecensis (Chilton, 1911)
Eiconaxius kermadeci Bate, 1888 E
Eiconaxius parvus Bate, 1888
Eucalastacus torbeni Sakai, 1992 E
Spongiaxius novaezealandiae (Borradaile, 1916) E
CALLIANASSIDAE
Corallianassa articulata (Rathbun, 1906)
Corallianassa cf. *collaroy* (Poore & Griffin, 1979) MNZ
'Callianassa' filholi (A. Milne Edwards, 1879) E
Vulcanocalliax sp. E
CTENOCHELIDAE
Ctenocheles maorianus Powell, 1949 E

Infraorder GEBIIDEA
LAOMEDIIDAE
Jaxea novaezealandiae Wear & Yaldwyn, 1966 E
UPOGEBIIDAE
Acutigebia danai (Miers, 1876) E
Upogebia hirtifrons (White, 1847) E

Infraorder ACHELATA
PALINURIDAE
Jasus edwardsii (Hutton, 1875)
Sagmariasus verreauxi (H. Milne Edwards, 1851)
Projasus parkeri (Stebbing, 1902)

Infraorder POLYCHELIDA
POLYCHELIDAE
Pentacheles laevis Bate, 1878
Pentacheles validus A. Milne Edwards, 1880
Polycheles enthrix (Bate, 1878)
Polycheles kermadecensis (Sund, 1920)
Stereomastis nana (Smith, 1884)
Stereomastis sculpta (Smith, 1880)
Stereomastis suhmi Bate, 1878
Stereomastis surda (Galil, 2000)
Willemoesia pacifica Sund, 1920
SCYLLARIDAE
Antarctus mawsoni (Bage, 1938)
Antipodarctus aoteanus (Powell, 1949) E
Arctides antipodarum Holthuis, 1960
Ibacus alticrenatus Bate, 1888
Ibacus brucei Holthuis, 1977
Scyllarides haanii (de Haan, 1841)

Infraorder ANOMURA
ALBUNEIDAE
Albunea microps Miers, 1878
CHIROSTYLIDAE
Chirostylus novaecaledoniae Baba, 1991
Eumunida pacifica Gordon, 1930
Gastroptychus novaezelandiae (Baba, 1974)
Gastroptychus rogeri (Baba, 2000)
Uroptychodes epigaster Baba, 2004
Uroptychodes spinimarginatus (Henderson, 1885)
Uroptychus alcocki Ahyong & Poore, 2004
Uroptychus australis (Henderson, 1885)
Uroptychus bicavus Baba & de Saint Laurent, 1992
Uroptychus cardus Ahyong & Poore, 2004
Uroptychus empheres Ahyong & Poore, 2004
Uroptychus flindersi Ahyong & Poore, 2004
Uroptychus gracilimanus (Henderson, 1885)
Uroptychus kaitara Schnabel, 2009
Uroptychus latus Ahyong & Poore, 2004
Uroptychus longicheles Ahyong & Poore, 2004
Uroptychus longvae Ahyong & Poore, 2004
Uroptychus maori Borradaile, 1916 E
Uroptychus multispinosus Ahyong & Poore, 2004
Uroptychus novaezealandiae Borradaile, 1916 E
Uroptychus paku Schnabel, 2009
Uroptychus paracrassior Ahyong & Poore, 2004
Uroptychus pilosus Baba, 1981
Uroptychus politus (Henderson, 1885) E
Uroptychus raymondi Baba, 2000
Uroptychus rutua Schnabel, 2009
Uroptychus scambus Benedict, 1902
Uroptychus spinirostris (Ahyong & Poore, 2004)
Uroptychus thermalis Baba & de Saint Laurent, 1992
Uroptychus toka Schnabel, 2009
Uroptychus tomentosus Baba, 1975 E
Uroptychus webberi Schnabel, 2009
Uroptychus yaldwyni Schnabel, 2009
DIOGENIDAE
Calcinus imperialis Whitelegge, 1901
Cancellus frontalis Forest & McLaughlin, 2000 E
Cancellus laticoxa Forest & McLaughlin, 2000 E
Cancellus rhynchogonus Forest & McLaughlin, 2000 E
Cancellus sphraerogonus Forest & McLaughlin, 2000 E
Dardanus arroser (Herbst, 1796)
Dardanus hessii (Miers, 1884)
Paguristes barbatus (Heller, 1862) E
Paguristes pilosus (H. Milne Edwards, 1836) E
Paguristes setosus (H. Milne Edwards, 1848) E
Paguristes subpilosus Henderson, 1888 E
GALATHEIDAE
Agononida incerta (Henderson, 1888)
Agononida marini (Macpherson, 1994)
Agononida nielbrucei Vereshchaka, 2005 E
Agononida procera Ahyong & Poore, 2004
Agononida squamosa (Henderson, 1885)

Allogalathea elegans (Adams & White, 1848)
Galathea whiteleggii Grant & McCulloch, 1906
Galacantha quiquei Macpherson, 2007
Galacantha rostrata A. Milne Edwards, 1880
Leiogalathea laevirostris (Balss, 1913)
Munida acacia Ahyong, 2007
Munida chathamensis Baba, 1974 E
Munida collier Ahyong, 2007
Munida eclepsis Macpherson, 1994
Munida erato Macpherson, 1994
Munida endeavourae Ahyong & Poore, 2004
Munida exilis Ahyong, 2007
Munida gracilis Henderson, 1885 E
Munida gregaria (Fabricius, 1793)
Munida icela Ahyong, 2007
Munida isos Ahyong & Poore, 2004
Munida kapala Ahyong & Poore, 2004
Munida notata Macpherson, 1994
Munida psylla Macpherson, 1994
Munida notialis Baba, 2005
Munida rubrimana Ahyong, 2007
Munida spinicruris Ahyong & Poore, 2004
Munida zebra Macpherson, 1994
Munidopsis antonii (Filhol, 1884)
Munidopsis bractea Ahyong, 2007
Munidopsis comarge Taylor, Ahyong & Andreakis, 2010
Munidopsis kaiyoae Baba, 1974 E
Munidopsis marginata (Henderson, 1885)
Munidopsis maunga Schnabel & Bruce, 2006
Munidopsis papanui Schnabel & Bruce, 2006
Munidopsis proales Ahyong & Poore, 2004
Munidopsis cf. *serricornis* (Lovén, 1852)
Munidopsis tasmaniae Ahyong & Poore, 2004
Munidopsis treis Ahyong & Poore, 2004
Munidopsis valdiviae (Balss, 1913)
Munidopsis victoriae Baba & Poore, 2002
Onconida alaini Baba & de Saint Laurent, 1996
Paramunida antipodes Ahyong & Poore, 2004
Phylladiorhynchus integrirostris (Dana, 1852)
Phylladiorhynchus pusillus (Henderson, 1885)
Tasmanida norfolkae Ahyong, 2007
LITHODIDAE
Lithodes aotearoa Ahyong, 2010 E
Lithodes jessica Ahyong, 2010
Lithodes macquariae Ahyong, 2010
Lithodes robertsoni Ahyong, 2010 E
Neolithodes brodiei Dawson & Yaldwyn, 1970
Neolithodes bronwynae Ahyong, 2010
Paralomis dawsoni Macpherson, 2001
Paralomis echidna Ahyong, 2010
Paralomis hirtella Saint Laurent & Macpherson, 1997
Paralomis poorei Ahyong, 2010
Paralomis staplesi Ahyong, 2010
Paralomis webberi Ahyong, 2010 E
Paralomis zealandica Dawson & Yaldwyn, 1971 E
PAGURIDAE
Bathypaguropsis cruentus de Saint Laurent & McLaughlin, 2000 E
Bathypaguropsis yaldwyni McLaughlin, 1994
Catapagurus spinicarpus de Saint Laurent & McLaughlin, 2000 E
Diacanthurus ecphyma McLaughlin & Forest, 1997
Diacanthurus rubricatus (Henderson, 1888) E
Diacanthurus spinulimanus (Miers, 1876) E
Lophopagurus (Australeremus) cookii (Filhol, 1883) E
Lophopagurus (*A.*) *cristatus* (H. Milne Edwards, 1836) E
Lophopagurus (*A.*) *eltaninae* (McLaughlin & Gunn, 1992) E
Lophopagurus (*A.*) *kirkii* (Filhol, 1883 E
Lophopagurus (*A.*) *laurentae* (McLaughlin & Gunn, 1992) E
Lophopagurus (*A.*) *stewarti* (Filhol, 1883) E
Lophopagurus (*A.*) *triserratus* (Ortmann, 1892)
Lophopagurus (Lophopagurus) foresti McLaughlin & Gunn, 1992 E
Lophopagurus (*L.*) *lacertosus* (Henderson, 1888) E
Lophopagurus (*L.*) ?*nanus* (Henderson, 1888)
Lophopagurus (*L.*) *nodulosus* McLaughlin & Gunn, 1992 E
Lophopagurus (*L.*) *pumilis* de Saint Laurent & McLaughlin, 2000 E
Lophopagurus (*L.*) *thompsoni* (Filhol, 1885) E
Michelopagurus? sp. E
Pagurixus hectori (Filhol, 1883) E
Pagurixus kermadecensis de Saint Laurent & McLaughlin, 2000 E
Pagurodes inarmatus Henderson, 1888
Pagurojacquesia polymorpha (de Saint Laurent & McLaughlin, 1999)
Pagurus albidianthus de Saint Laurent & McLaughlin, 2000 E
Pagurus iridocarpus de Saint Laurent & Mclaughlin, 2000 E
Pagurus novizealandiae (Dana, 1852) E
Pagurus sinuatus (Stimpson, 1858)
Pagurus traversi (Filhol, 1885) E
Porcellanopagurus chiltoni de Saint Laurent & McLaughlin, 2000
Porcellanopagurus edwardsi Filhol, 1885 E
Porcellanopagurus filholi de Saint Laurent & McLaughlin, 2000
Porcellanopagurus tridentatus Whitelegge, 1900
Propagurus deprofundis (Stebbing, 1924)
PARAPAGURIDAE
Oncopagurus sp. E
Paragiopagurus diogenes (Whitelegge, 1900)
Paragiopagurus hirsutus (de Saint Laurent, 1972)
Parapagurus abyssorum (Filhol, 1885)
Parapagurus bouvieri Stebbing, 1910
Parapagurus latimanus Henderson, 1888
Parapagurus richeri Lemaitre, 1999
Sympagurus dimorphus (Studer, 1883)
Sympagurus papposus Lemaitre, 1996
PORCELLANIDAE
Pachycheles pisoides (Heller, 1865)
Petrocheles spinosus (Miers, 1876) E
Petrolisthes elongatus (H. Milne Edwards, 1837)
Petrolisthes lamarckii (Leach, 1820)
Petrolisthes novaezelandiae Filhol, 1885 E
PYLOCHELIDAE
Cheiroplatea pumicicola Forest, 1987
Pylocheles mortensenii Boas, 1926
Trizocheles brachyops Forest & de Saint Laurent, 1987
Trizocheles perplexus Forest, 1987 E
Trizocheles spinosus (Henderson, 1888)
Trizocheles pilgrimi Forest & McLaughlin, 2000

Infraorder BRACHYURA
AETHRIDAE
Actaeomorpha erosa Miers, 1877
ATELECYCLIDAE
<u>Pteropeltarion</u> novaezelandiae Dell, 1972 E
Trichopeltarion fantasticum Richardson & Dell, 1964 E
Trichopeltarion janetae Ahyong, 2008
BELLIIDAE E
<u>Heterozius</u> rotundifrons A. Milne Edwards, 1867 E
BYTHOGRAEIDAE
Gandalfus puia McLay, 2007
CALAPPIDAE
Mursia australiensis Campbell, 1971
Mursia microspina Davie & Short, 1989
CANCRIDAE
Glebocarcinus amphioetus (Rathbun, 1898) A
Metacarcinus novaezelandiae (Hombron & Jacquinot, 1846)
Romaleon gibbulosus (Rathbun, 1898) A
CRYPTOCHIRIDAE
Cryptochirus coralliodytes Heller, 1861
CYMONOMIDAE
Cymonomus aequilonius Dell, 1971 E
Cymonomus bathamae Dell, 1971 E
Cymonomas clarki Ahyong, 2008 E
DROMIIDAE
Cryptodromiopsis unidentata (Rüppell, 1830)
Metadromia wilsoni (Fulton & Grant, 1902)
Tumidodromia dormia (Linnaeus, 1763)
DYNOMENIDAE
Dynomene pilumnoides Alcock, 1900
Metadynomene tanensis (Yokoya, 1933)
EPIALTIDAE
Huenia heraldica (de Haan, 1839)
Leptomaia tuberculata Griffin & Tranter, 1986
Oxypleurodon wanganella Webber & Richer de Forges, 1995 E
Rochinia ahyongi McLay, 2009 E
Rochinia riversandersoni (Alcock, 1895)
ERIPHIIDAE
Bountiana norfolcensis (Grant & McCulloch, 1907)
ETHUSIDAE
Ethusina castro Ahyong, 2008 E
Ethusina rowdeni Ahyong, 2008 E
GERYONIDAE
Chaceon bicolor Manning & Holthuis, 1989
Chaceon yaldwyni Manning, Dawson & Webber, 1990 E
GONEPLACIDAE
Goneplax marivenae Komatsu & Takeda, 2004
<u>Neommatocarcinus</u> huttoni (Filhol, 1886) E
Pycnoplax meridionalis (Rathbun, 1923)
Pycnoplax victoriensis (Rathbun, 1923)
Thyroplax truncata Castro, 2007
GRAPSIDAE
Geograpsus grayi (H. Milne Edwards, 1853) T
Leptograpsus variegatus (Fabricius, 1793)
Pachygrapsus minutus A. Milne Edwards, 1873
Planes major (MacLeay, 1838)
Planes marinus Rathbun, 1914
HOMOLIDAE
Dagnaudus petterdi (Grant, 1905)
Homola orientalis Henderson, 1888
Homola ranunculus Guinot & Richer de Forges, 1995
Homolochunia kullar Griffin & Brown, 1976
Yaldwynopsis spinimanus (Griffin, 1965)
HOMOLODROMIIDAE
Dicranodromia delli Ahyong, 2008 E
Dicranodromia spinulata Guinot, 1995
Homolodromia kai Guinot, 1993
HYMENOSOMATIDAE
Amarinus lacustris (Chilton, 1882) F
Elamena longirostris Filhol, 1885 E
Elamena momona Melrose, 1975 E
Elamena producta Kirk, 1879 E
Halicarcinus cookii (Filhol, 1885) E
Halicarcinus innominatus Richardson, 1949
Halicarcinus ovatus Stimpson, 1858
Halicarcinus planatus (Fabricius, 1775)
Halicarcinus tongi Melrose, 1975 E
Halicarcinus varius (Dana, 1851) E
Halicarcinus whitei (Miers, 1876) E
<u>Halimena</u> aotearoa Melrose, 1975 E
Hymenosoma depressum Hombron & Jacquinot, 1846 E
<u>Neohymenicus</u> pubescens (Dana, 1851) E
INACHIDAE
Achaeus akanensis Sakai, 1938
Achaeus curvirostris (A. Milne Edwards, 1873)
Achaeus kermadecensis Webber & Takeda, 2005 E
Cyrtomaia cornuta Richer de Forges & Guinot, 1988
Cyrtomaia lamellata Rathbun, 1906

Dorhynchus ramusculus (Baker, 1906)
Platymaia maoria Dell, 1963
Platymaia wyvillethomsoni Miers, 1886
Trichoplatus huttoni A. Milne Edwards, 1876 E
Vitjazmaia latidactyla Zarenkov, 1994
INACHOIDIDAE
Pyromaia tuberculata (Lockington, 1877) A
LATREILLIIDAE
Eplumula australiensis (Henderson, 1888)
Latreillia metanesa Williams, 1982
LEUCOSIIDAE
Bellidilia cheesmani (Filhol, 1886) E
Ebalia humilis Takeda, 1977
Ebalia jordani Rathbun, 1906
Ebalia tuberculosa (A. Milne Edwards, 1873)
Ebalia webberi Komatsu & Takeda, 2007 E
Merocryptus lambriformis A. Milne Edwards, 1873
Tanaoa distinctus (Rathbun, 1893)
Tanaoa pustulosus (Wood-Mason *in* Wood-Mason & Alcock, 1891)
MACROPHTHALMIDAE
Macrophthalmus (Hemiplax) hirtipes (Jacquinot *in* Hombron & Jacquinot, 1846) E
MAJIDAE
Eurynolambrus australis H. Milne Edwards & Lucas, 1841 E
Eurynome bituberculata Griffin, 1964 E
Jacquinotia edwardsii (Jacquinot, 1853) E
Leptomithrax australis (Jacquinot, 1853) E
Leptomithrax garricki Griffin, 1966 E
Leptomithrax longimanus Miers, 1876
Leptomithrax longipes (Thomson, 1902)
Leptomithrax tuberculatus mortenseni Bennett, 1964
Naxia spinosa (Hess, 1865)
Notomithrax minor (Filhol, 1885)
Notomithrax peronii (H. Milne Edwards, 1834) E
Notomithrax spinosus (Miers, 1879)
Notomithrax ursus (Herbst, 1788)
Prismatopus filholi (A. Milne Edwards, 1876) E
Prismatopus goldsboroughi (Rathbun, 1906)
Schizophroida hilensis (Rathbun, 1906)
Teratomaia richardsoni (Dell, 1960)
MATHILDELLIDAE
Intesius richeri Crosnier & Ng, 2004
Mathildella mclayi Ahyong, 2008 E
Neopilumnoplax nieli Ahyong, 2008
OCYPODIDAE
Ocypode pallidula Jacquinot *in* Hombron & Jacquinot, 1846
OZIIDAE
Ozius truncatus H. Milne Edwards, 1834
PALICIDAE
Pseudopalicus declivis Castro, 2000
Pseudopalicus oahuensis (Rathbun, 1906)
Pseudopalicus undulatus Castro, 2000
PARTHENOPIDAE
Actaeomorpha erosa Miers, 1877
Garthambrus allisoni (Garth, 1992)
Garthambrus tani Ahyong, 2008
Platylambrus constrictus (Takeda & Webber, 2007)
PILUMNIDAE
Actumnus griffini Takeda & Webber, 2006 E
Pilumnopeus serratifrons (Kinahan, 1856)
Pilumnus fimbriatus H. Milne Edwards, 1834
Pilumnus lumpinus Bennett, 1964 E
Pilumnus novaezelandiae Filhol, 1886 E
PINNOTHERIDAE
Nepinnotheres atrinicola (Page, 1983) E
Nepinnotheres novaezelandiae (Filhol, 1885) E
PLAGUSIIDAE
Miersiograpsus australiensis Türkay, 1978
Percnon planissimum (Herbst, 1804)
Plagusia chabrus (Linnaeus, 1758)
Plagusia dentipes de Haan, 1835
Plagusia squamosa (Herbst, 1790)
PORTUNIDAE
Caphyra acheronae Takeda & Webber, 2006 E
Charybdis japonica (A. Milne Edwards, 1861) A
Liocarcinus corrugatus (Pennant, 1777)
Nectocarcinus antarcticus (Jacquinot, 1853) E
Nectocarcinus bennetti Takeda & Miyake, 1969 E
Ovalipes catharus (White, 1843)
Ovalipes elongatus Stephenson & Rees, 1968
Ovalipes molleri (Ward, 1933)
Portunus pelagicus (Linnaeus, 1766)
Scylla serrata (Forskål, 1775)
Thalamita danae Stimpson, 1858
Thalamita macrops Montgomery, 1931
RANINIDAE
Lyreidus tridentatus de Haan, 1841
Notosceles pepeke Yaldwyn & Dawson, 2000 E
TRAPEZIIDAE
Calocarcinus africanus Calman, 1909
Trapezia cymodoce (Herbst, 1801)
Trapezia guttata Rüppell, 1830
Trapezia septata Dana, 1852
VARUNIDAE
Austrohelice crassa (Dana, 1851) E
Cyclograpsus insularum Campbell & Griffin, 1966
Cyclograpsus lavauxi H. Milne Edwards, 1853 E
Hemigrapsus crenulatus (H. Milne Edwards, 1837)
Hemigrapsus sexdentatus (H. Milne Edwards, 1837) E
XANTHIDAE
Antrocarcinus petrosus Ng & Chia, 1994
Banareia armata A. Milne Edwards, 1869
Banareia banareias (Rathbun, 1911)
Euryxanthops chiltoni Ng & McLay, 2007 E
Gaillardiellus bathus Davie, 1997
Gaillardiellus rueppelli (Krauss, 1843)
Leptodius nudipes (Dana, 1852)
Liomera yaldwyni Takeda & Webber, 2006 E
Lybia leptochelis (Zehntner, 1894)
Medaeops serenei Ng & McLay, 2007 E
Miersiela haswelli (Miers, 1886)
Nanocassiope sp. Takeda & Webber 2006
Pilodius nigrochrinitus Dana, 1852
Platypodia delli Takeda & Webber, 2006 E
Pseudoliomera helleri (A. Milne Edwards, 1865)
Serenius actaeoides (A. Milne Edwards, 1834)
Xanthias dawsoni Takeda & Webber, 2006 E
Xanthias lamarckii (H. Milne Edwards, 1834)
XENOGRAPSIDAE
Xenograpsus ngatama McLay, 2007 E

Synonyms or possible synonyms in cyclopoid Copepoda
Diacyclops crassicaudoides (Kiefer, 1928) = *D. bisetosus* (Rehberg, 1880)
Eucyclops (Eucyclops) serrulatus (Fischer, 1851) (= *Cyclops novaezealandiae* Thomson, 1879)
?*Euryte longicauda* Philippi, 1843 (= *Thorellia brunnae* Boeck, 1864)
?*Cyclops strennus strennus* Fischer, 1851 (= *C. ewarti* Brady, 1888)
Diacyclops bicuspidatus (Claus, 1857) (= *Cyclops gigas*, Thomson, 1883)
?*Halicyclops magniceps* (Lilljeborg, 1853) (= ?*C. aequorus*, Thomson, 1883)
?*Macrocyclops distinctus* (Richard, 1887) = *M. albidus* (Jurine, 1820)
?*Mesocyclops australiensis* (Sars, 1908) (= ?*M. leuckarti*)

Checklist of New Zealand fossil Crustacea

Letters in parentheses following new records indicate where material is held, i.e. AUT (Earth and Oceanic Sciences Research Centre, Auckland University of Technology); GNS (Institute of Geological and Nuclear Sciences, Lower Hutt); NIWA (National Institute of Water and Atmospheric Sciences, Wellington); UA (Geology Department, University of Auckland). Stratigraphic ranges, using abbreviations for New Zealand stages (Cooper 2004), follow each fossil species listing.

SUBPHYLUM CRUSTACEA
Class MAXILLOPODA
Infraclass CIRRIPEDIA
Superorder ACROTHORACICA
Order PYGOPHORA
CRYPTOPHIALIDAE
Australophialus? sp. nov.* Po-Pl (AUT) E
Gen. et sp. indet..* Po-Pl (UoA)

INCERTAE SEDIS
Zapfella sp.* Bm (GNS)
Zapfella? sp.* Ko (UoA)

Superorder RHIZOCEPHALA
Order KENTROGONIDA
SACCULINIDAE?
Gen. et sp. indet. Feldmann 1998 Mio

Superorder THORACICA
Order LEPADIFORMES
LEPADIDAE
Lepas ?australis Darwin, 1851 Qu
Lepas clifdenica Buckeridge, 1983 Sl-Tt E
Lepas moturoaensis Maxwell, 1968 Po E
Pristinolepas harringtoni (Laws, 1948) Lw-Pl E
Pristinolepas haurakiensis (Buckeridge, 1983) Lw-Po E
Pristinolepas pakaurangiensis (Buckeridge, 1983) Po-Pl E
Pristinolepas waikatoica (Buckeridge, 1983) Ld-Lw E
Pristinolepas n. sp. Ar E

Order SCALPELLIFORMES
ARCOSCALPELLIDAE
Anguloscalpellum complanatum (Withers, 1924) Lwh-Ld E
Anguloscalpellum cf. *complanatum* (Withers, 1924) Po E
Anguloscalpellum crassiforme Buckeridge, 1983 Lwh

E
Anguloscalpellum euglyphum (Withers, 1924) Lwh-Ld E
Anguloscalpellum grantmackiei Buckeridge, 1983 Po-Sw E
Anguloscalpellum? striatulum (Withers, 1924) Lwh-Ld E
Anguloscalpellum ungulatum (Withers, 1913) Lwh-Sw E
CALANTICIDAE
Calantica spinilatera Foster, 1979 Ww-Rec E
Cretiscalpellum cf. *glabrum* (Roemer, 1841) Uk
Cretiscalpellum? sp. nov.* Cn (GNS) E
Cretiscalpellum? sp. Buckeridge 1983 Mp-Dt
Euscalpellum egmontense Buckeridge, 1983 Ww E
Pachyscalpellum cramptoni Buckeridge, 1991 Mp
Pachyscalpellum debodae Buckeridge, 1999 Mh E
Scillaelepas arguta (Withers, 1924) Lwh-Ld E
Scillaelepas? pittensis Buckeridge, 1984 Ab-Ar E
Scillaelepas cf. *studeri* (Weltner, 1922) Ab-Ar
Scillaelepas waitemata Buckeridge, 1983 Lw-Po E
Smilium calanticoideum Buckeridge, 1983 Dw-Dm
Smilium chathecum Buckeridge, 1984 Pl E
Smilium subplanum (Withers, 1913) Lw-Po E
Zeascalpellum crassum Buckeridge, 1983 Dm-Ab E
Gen. nov. et n. sp.* Mh-Dt (GNS) E
Gen. et sp. indet. Buckeridge 1983 Mp-Mh
EOLEPADIDAE
Eolepas? novaezelandiae Buckeridge 1983 Ce E
ZEUGMATOLEPADAE
Zeugmatolepas? sp. Buckeridge 1983 Kh

Order SESSILIA
Suborder VERRUCOMORPHA
VERRUCIDAE
Metaverruca recta (Aurivillius, 1898) Po-Rec
Verruca nuciformis Buckeridge, 1983 Dm-Po E
Verruca sauria Buckeridge, 2010 Mh E
Verruca tasmanica chatheca Buckeridge, 1983 Dw-Dm E
Verruca t. tasmanica Buckeridge, 1983 Lwh

Suborder BALANOMORPHA
ARCHAEOBALANIDAE
Armatobalanus motuketeketeensis Buckeridge, 1983 Po E
Armatobalanus? sp. Buckeridge 1983 Po E
Striatobalanus zelandicus (Withers, 1924) Sl-Tt E
Notobalanus vestitus (Darwin, 1854) Lw-Rec E
Palaeobalanus lornensis Buckeridge, 1983 Ab-Ak E
Palaeobalanus? waihaoensis Buckeridge, 1983 Ab E
Tasmanobalanus acutus acutus (Withers, 1924) Pl-Sw E
Tasmanobalanus a. clifdensensis Buckeridge, 1983 Sc E
Tasmanobalanus a. convexus Buckeridge, 1983 Pa E
Tasmanobalanus grantmackiei Buckeridge, 1983 Sw-Ww E
Zullobalanus everetti (Buckeridge, 1983) Lwh E
Zullobalanus novozelandicus (Buckeridge, 1983) Ld-Lw E
AUSTROBALANIDAE
Austrobalanus imperator aotea Buckeridge, 1983 Ld-Po E
Austrobalanus macdonaldensis Buckeridge, 1983 Lwh E
Epopella eoplicata Buckeridge, 1983 Po E
Epopella cf. *plicata* Gray, 1843* Wp (AUT) E
Protelminius pomahakensis (Buckeridge, 1984) Ld E
BATHYLASMATIDAE
Bathylasma aucklandicum (Hector, 1888) Lw-Ww E
Bathylasma rangatira Buckeridge, 1983 Dt-Dm E
BALANIDAE
Amphibalanus variegatus (Darwin, 1854) Ww-Rec
Fistulobalanus kondakovi (Tarasov & Zevina, 1957) ?Wn
Fosterella chathamensis Buckeridge, 1983 Wo-Wn E
Fosterella tubulatus (Withers, 1924) Wo-Wn E
Notomegabalanus decorus argyllensis (Buckeridge, 1983) Wn-Qu E
Notomegabalanus miodecorus (Buckeridge, 1983) Sw-Ww E
CHIONELASMATIDAE
Chionelasmus darwini (Pilsbry, 1907) Ak-Rec
CHTHAMALIDAE
Chamaesipho brunnea Moore, 1944 Po-Rec E
CORONULIDAE
Coronula aotea Fleming, 1959 Ww-Wm E
Coronula diadema (Linné, 1767) Wn-Rec
Coronula intermedia Buckeridge, 1983 Wn E
PACHYLASMATIDAE
Eolasma maxwelli Buckeridge, 1983 Dw-Dm E
Pachylasma distortum Buckeridge, 1983 Lwh E
Pachylasma? southlandicum Buckeridge, 1983 Ld-Po E
Pachylasma veteranum Buckeridge, 1983 Dt-Dm E
Pachylasma sp.* Wp (AUT)
Waikalasma juneae Buckeridge, 1983 Po-Pl E
TETRACLITIDAE
Tesseroplax? maorica Buckeridge, 1983 Lw-Po E
Tesseropora cf. *pacifica* (Pilsbry, 1928) Po
Tetraclitella nodicostata Buckeridge, 2008 Lw-Po

Class OSTRACODA
All the marine Tertiary species may be regarded as endemic.
Order ARCHAEOCOPIDA
Gen. et spp. indet. (2) Simes 1977 LPz

Order PALAEOCOPIDA
Suborder BEYRICHICOPIDA
PUNCIIDAE
Puncia goodwoodensis Hornibrook, 1963 Pl E

Order PODOCOPIDA
Suborder PODOCOPINA
BAIRDIIDAE
Bairdia canterburyensis Swanson, 1969 Pl E
Bairdoppilata kerryi Milau, 1993 Po-Rec
Bairdoppilata cf. *austracretacea* (Bate, 1972) Mh
Bairdoppilata sp. 5052 Dingle 2009 Mh
Neonesidea australis (Chapman, 1914) Ak-Lw
Neonesidea chapmani Whatley & Downing, 1983 Ak-Lw
Neonesidea waitematanensis Milau, 1993 Po E
Neonesidea sp. Ayress 1993 Ab-Rec
BYTHOCYPRIDIDAE
Bythocypris sudaustralis McKenzie, Reyment & Reyment, 1991 Ak
Bythocypris cf. *sudaustralis* McKenzie, Reyment & Reyment, 1991 Mh
Bythocypris cf. *chapmani* Neale, 1975 Mh
Bythocypris sp. Ayress, 1993 Lwh-Lw
BYTHOCYTHERIDAE
Abyssobythere inequivalva Ayress, Correge, Passlow & Whatley, 1996 Wc
Bythoceratina decepta Hornibrook, 1952 Wc-Rec
Bythoceratina cf. *dubia* (Müller, 1908) Ak
Bythoceratina edwardsoni Hornibrook, 1952 Wc-Rec
Bythoceratina maoria Hornibrook, 1952 Sc-Rec
Bythoceratina mestayerae Hornibrook, 1952 Pl-Rec
Bythoceratina powelli Hornibrook, 1952 Ar-Rec
Bythoceratina robusta Milau, 1993 Po
Bythoceratina utilazea Hornibrook, 1952 Pl-Rec
Bythoceratina sp. Ayress 1993 Ld-Lw
Miracythere novaspecta Hornibrook, 1952 Lw-Rec E
Neobuntonia oneroaensis Milau, 1993 Po
Pseudeucythere biplana Ayress, 1995 Ak-Wc
Vitjasiella duplicispina Avress, 1993 Lw-Pl
Vitjasiella ferox (Hornibrook, 1952) Ab-Wc
CYPRIDIDAE
Candona sp. Hornibrook 1955 Wc F
Candonocypris assimilis Sars, 1894 Wc-Rec F
Cypretta viridis (Thomson, 1879) Wc-Rec F
Cypris sp. Hornibrook 1955 Wc F
Heterocypris ciliata (Thomson, 1879) Wc-Rec F
Heterocypris incongruens (Rhamdohr, 1808) Wc-Rec F E
Ilyodromus stanleyanus (King, 1855) Wc-Rec F
CYTHERALISONIDAE
Cytheralison amiesi Hornibrook, 1953 Lwh-Ld
Cytheralison fava Hornibrook, 1952 Ab-Rec
Cytheralison parafava Ayress, 1993 Ld-Lw
Cytheralison spinosa Ayress, 1993 Ld-Lw
Cytheralison sp. Ayress 1995 Ak
Debissonia hornibrooki Ayress, 2003 Ld-Lw
Debissonia pravacauda (Hornibrook, 1952) Dm-Rec
CYTHERIDAE
Chejudocythere cf. *higashikawai* Ishizaki, 1981 Ak
Cythere allanthomsoni Chapman, 1926 Sw
Loxocythere crassa Hornibrook, 1952 Po-Rec
Loxocythere kingi Hornibrook, 1952 Pl-Rec
CYTHERIDEIDAE
Cytheridea aoteana Hornibrook, 1952 Wc-Rec E
Cytheridea symmetrica Swanson, 1969 Pl
Cytheridea (Clithrocytheridea) marwicki Hornibrook, 1953 Pl
Hemicytheridea mosaica Hornibrook, 1952 Dm-Rec
Eucythere sulcocostatula Ayress, 1995 Ak-Wc
Eucythere parapubera Whatley & Downing, 1983 Lwh-Ld
Eucythere cf. *parapubera* Whatley & Downing, 1983 Ak
Eucythere sp. Ayress 1995 Ak-Lw
Eucythere sp. 1 Ayress 1993 Lwh-Lw
Rostrocytheridea pukehouensis Dingle, 2009 E Mh
Rostrocytheridea aff. *allaruensis*? Krömmelbein, 1975 Cn
Rostrocytheridea? sp. 4992 Dingle 2009 Mh
Rotundracythere gravepuncta Hornibrook, 1952 Ar-Rec
Rotundracythre inaequa Hornibrook, 1952 Wc-Rec
Rotundracythere mytila Hornibrook, 1952 Ld-Rec
Rotundracythere rotunda Hornibrook, 1952 Ar-Rec
Rotundracythere subovalis Hornibrook, 1952 Ar-Rec
Pseudocythere (Pseudocythere) caudata Sars, 1866 Ld-Lw
Pseudocythere (P.) caudata Sars, 1866 Lw-Rec
CYTHEROMATIDAE
Malibaricythere oceanica Yassini & Jones, 1995 Lw
Paracytheroma stilwelli Ayress, 1990 Ld-Pl
Paracytheroma convexa Milau, 1993 Po
Pellucistoma coombsi Ayress, 1990 Ak-Pl
Pellucistoma fordycei Ayress, 1990 Ak-Pl
CYTHERURIDAE
Aversovalva aurea Hornibrook, 1952 Ab-Rec
Aversovalva pteroalata Ayress, 1993 Lwh-Ld n. nud.
Cytheropteron anisovalva Ayress, Correge, Passlow & Whatley, 1996 Ar-Rec
Cytheropteron cuneatum Ayress, 1996 Ak
Cytheropteron confusum (Hornibrook, 1952) Lwh-Rec
Cytheropteron crassicutum Ayress, 1998 Po-Wn
Cytheropteron curvicaudum Hornibrook, 1952 Lwh-Rec
Cytheropteron dividentum (Hornibrook, 1952) Lwh-Rec
Cytheropteron dorsocorrugatum Ayress, Correge, Passlow & Whatley, 1996 Wc
Cytheropteron fornix (Hornibrook, 1952) Ab-Rec
Cytheropteron obtusalum Hornibrook, 1952 Ar-Rec
Cytheropteron planalatum Guernet, 1985 Ak-Po
Cytheropteron terecaudum Hornibrook, 1952 Pl-Rec
Cytheropteron testudo Sars, 1869 Ak-Ar
Cytheropteron vertex Hornibrook, 1952 Wn-Rec

Cytheropteron wellmani Hornibrook, 1952 Mp-Rec
Cytheropteron willetti Hornibrook, 1952 Wo-Rec
Cytheropteron sp. Ayress 1993 Ab-?Rec
Cytheropteron sp. Ayress 1995 Ak
Cytheropteron sp. 1 Ayress 1993 Lwh-Lw
Cytheropteron sp. 1 Ayress 1996 Ar-Lw
Cytheropteron sp. 2 Ayress 1993 Lwh-Ld
Cytheropteron sp. 2 Ayress, 1996 Ak
Cytheropteron sp. 3 Ayress 1993 Lwh-Ld
Eocytheropteron? sp. Ayress 1993 Ld-Lw
Cytherura clausi Brady, 1880 Pl-Rec
Cytherura nonspinosa Ayress, 1996 Ak
Eucytherura boomeri Ayress, Whatley, Downing, & Millson, 1995 Wq
Eucytherura calabra (Colalongo & Pasini, 1980) Ak-Rec
Eucytherura downingae Ayress, Whatley, Downing, & Millson, 1995 Wc
Eucytherura elegantula Ayress, Whatley, Downing, & Millson, 1995 Ab
Eucytherura pacifica Ayress, Whatley, Downing, & Millson, 1995 Lw-Wc
Eucytherura tumida Ayress, Whatley, Downing, & Millson, 1995 Wo-Wc (homonym of *E. tumida* Bonnema, 1941)
Eucytherura bakeri Hornibrook, 1952 Po-Pl
Eucytherura batalaria Ayress, Whatley, Downing, & Millson, 1995 Lwh-Wc
Eucytherura multituberculata Ayress, Whatley, Downing, & Millson, 1995 Wo-Rec
Eucytherura sp. Ayress 1993 Ld
Eucytherura sp. 1 Ayress 1993 Ld-Lw
Eucytherura sp. 1 Ayress 1995 Ak
Eucytherura sp. 2 Ayress 1993 Ld
Eucytherura sp. 2 Ayress 1995 Ak
Eucytherura sp. 2 Ayress, Whatley, Downing, & Millson 1995 Wo
Eucytherura? *polydictyota* Ayress, Whatley, Downing, & Millson, 1995 Wc
Hemicytherura (Hemicytherura) aucklandica Hornibrook, 1952 Lw-Rec
Hemicytherura (H.) delicatula Hornibrook, 1952 Lwh-Rec
Hemicytherura (H.) fereplana Hornibrook, 1952 Ak-Rec
Hemicytherura (H.) gravis Hornibrook, 1952 Ak-Rec
Hemicytherura (H.) quadrazea Hornibrook, 1952 Lwh-Rec
Hemicytherura sp. Ayress 1993 Ld-Lw
Hemicytherura (Kangarina) radiata (Hornibrook, 1952) Ak-Rec
Hemiparacytheridae leopardina Ayress, Whatley, Downing & Millson, 1995 Wo
Hemiparacytheridea mediopunctata Ayress, Whatley, Downing & Millson, 1995 Wo-Wc
Hemiparacytheridae vanharteni Ayress, Whatley, Downing & Millson, 1995 Wc
Malabaricythere oceanica Yassini & Jones, 1995 Lw
Microcytherura alata Ayress, 1993 Lw n. nud.
Microcytherura sp. Ayress 1993 Lwh-Lw
Microcytherura haywardi Milau, 1993 Po
Microcytherura sp. Ayress 1993 Lwh-Lw
Microcytherura sp. 1 Ayress 1996 Ak-Ar
Microcytherura sp. 2 Ayress 1996 Ak-Ar
Oculocytheropteron aff. *abyssorum* (Brady, 1880) Ak
Oculocytheropteron acutangulum (Hornibrook, 1952) Lwh-Rec
Oculocytheropteron australopunctatarum McKenzie, Reyment & Reyment 1991 Ak
Oculocytheropteron confusum (Hornibrook, 1952) Lwh-Rec
Oculocytheropteron ferrieri Milau, 1993 Po
Oculocytheropteron grantmackei Milau, 1993 Lw-Po
Oculocytheropteron improbum (Hornibrook, 1952) Ak-Rec
Oculocytheropteron microfornix Whatley & Downing, 1983 Ak
Oculocytheropteron paratinctum Ayress, 1996 Ak
Oculocytheropteron waihoensis Ayress, 1996 Ak
Oculocytheropteron sp. Ayress 1993 Lwh-Lw
Paracytheridea sp. Ayres, 1993 Ld-Lw
Pedicythere ?australis Neale, 1975 Ak
Pelecocythere? sp. 5042 Dingle 2009 Mh
Semicytherura arteria Swanson, 1979 Ak-Rec
Semicytherura coeca Ciampo, 1980 Ak-Lw
Semicytherura cf. *costellata* (Brady, 1880) Ak-Rec
Semicytherura eocenica Ayress, 1996 Ak-Ar
Semicytherura hexagona (Hornibrook, 1952) Wn-Rec
Semicytherura okinawaensis Nohara, 1987 Ak
Semicytherura sericava (Hornibrook, 1952) Pl-Rec
Semicytherura sp. Ayress 1993 Ld-Lw
Semicytherura sp. 1 Ayress 1996 Ak
Semicytherura sp. 2 Ayress 1996 Ak
HEMICYTHERIDAE
Ambostracon sp. Ayress 1993 Lw
Ambostracon fredbrooki Milau, 1993 Po
Ambostracon (Patagonacythere) elongata Milau, 1993 Po
Bradleya arata (Brady, 1880) Wn-Rec
Bradleya clifdenensis Hornibrook, 1952 Ld-Pl
Bradleya dictyon (Brady, 1880) Dm-Rec
Bradleya kaiata Hornibrook, 1953 Ab-Ar
Bradleya opima Swanson, 1979 Ak-Rec
Bradleyla pakaurangia Hornibrook, 1952 Pl
Bradleya proarata Hornibrook, 1952 Ar-Lw
Bradleya pygmaea Whatley, Downing, Kesler & Harlow, 1984 Mio-Rec
Bradleya reticlava Hornibrook, 1952 Ld-Rec
Bradleya semiarata Hornibrook, 1952 Pl
Bradleya (Quasibradleya) cuneazea Hornibrook, 1952 Ar-Rec
Bradleya (Q.) dictyonites Benson, 1972 Ak-Lw
Bradleya sp. Ayress 1993 Ab-Lwh
Bradleya sp. Ayress, 1993 Ld-Lw
Caudites impostor Hornibrook, 1953 Dh-Ab
Caudites cf. *scopulicolus* Hartmann, 1981
Hemicythere hornibrooki Swanson, 1969 Pl
Hemicythere munida Swanson, 1979 Ak-Rec
Hermanites andrewsi Swanson, 1979 Ld-Rec
Hermanites ?briggsi Swanson, 1979 Ak
Hermanites rectidorsa Milau, 1993 Po
Hermanites spinosa Milau, 1993 Po
Jacobella sp. Ayress 1995 Ak
Jugosocythereis reticulospinosa Ayress, 1993 Lwh-Lw n. nud.
Limburgina quadrazea (Hornibrook, 1952) Dm-Ld
Patagonocythere tricostata Hartmann 1962 Ak
Patagonacythere waihaoensis Ayress, 1995 Ak
Patagonacythere parvitenuis (Hornibrook, 1953) Ak-Ar
Poseidonamicus spp. Ayress, Neil, Passlow & Swanson, 1997 Wc-Rec
Quadracythere alatazea Hornibrook, 1952 Pl-Sw
Quadracythere biruga Hornibrook, 1952 Ld-Rec
Quadracythere chattonensis Hornibrook, 1953 Ld-Lw
Quadracythere claremontensis Swanson, 1969 Pl
Quadracythere clavala Hornibrook, 1952 Lw-Sc
Quadracythere clifdenensis Hornibrook, 1952 Ak-Sl
Quadracythere longazea Hornibrook, 1952 Lwh-Sw
Quadracythere mediaplana Hornibrook, 1952 Po-Pl
Quadracythere mediaruga Hornibrook, 1952 Ak-Rec
Quadracythere planazea Hornibrook, 1952 Ld-Sl
Quadracythere radizea Hornibrook, 1952 Dm-Pl
Urocythereis opima Swanson, 1969 Lwh-Pl
Waiparacythereis caudata Swanson, 1969 Pl
Waiparacythereis decora Swanson, 1969 Pl
Waiparacythereis joanae Swanson, 1969 Pl-Rec
Waiparacythereis sp. Ayress 1993 Lwh
KRITHIDAE
Krithe antisawanensis Ishizaki, 1966 Sl-Rec
Krithe comma Ayress, Barrows, Passlow & Whatley, 1999 Sl-Rec
Krithe compressa (Seguenza, 1980) Sw-Rec
Krithe dolichodeira Bold, 1946 Sw-Rec
Krithe marialusae Abate, Barra, Aiello & Bonaduce, 1993 Tt-Rec
Krithe minima Coles, Whatley & Moguilevsky, 1994 Lw-Rec
Krithe morkhoveni morkhoveni Bold, 1960 Wo-Rec
Krithe nitida Whatley & Downing, 1993 Ak-?Rec
Krithe pseudocomma Ayress, Barrows, Passlow & Whatley, 1999 Lw-Rec
Krithe reversa Bold, 1958 Tk-Rec
Krithe swansoni Milau, 1993 Po-Rec
Krithe triangularis Ayress, Barrows, Passlow & Whatley, 1999 Wc
Krithe trinidadensis Bold, 1958 Ww-Rec
Krithe sp. Ayress 1993 Lwh-Lw
Krithe sp. Ayress 1995 Ak
Krithe sp. 1 Ayress, Barrows, Passlow & Whatley 1999 Wn
Krithe sp. 2 Ayress, Barrows, Passlow & Whatley 1999 Lw-Rec
Krithe sp. 5055 Dingle 2009 Mh
Krithe sp. 5056 Dingle 2009 Mh
Krithe sp. 5079 Dingle 2009 Mh
Parakrithe sp. Ayress 1993 Lwh-Lw
Parakrithella lethiersi Milau, 1993 Po
LEGUMINOCYTHERIDIDAE
Triginglymus? *hobsonensis* Milau, 1993 Po
LEPTOCYTHERIDAE
Bisulcocythere campbelli Ayress & Swanson, 1991 Sw
Bisulcocythere compressa Ayress & Swanson, 1991 Po-Sw
Bisulcocythere eocenica Ayress & Swanson, 1991 Ak
Bisulcocythere micropunctata Ayress & Swanson, 1991 Lwh-Pl
Bisulcocythere novaezealandiae Ayress & Swanson, 1991 Pl-Rec
Callistocythere hanai Swanson, 1969 Pl
Callistocythere kaiata (Hornibrook, 1953) Ar-Ar
Callistocythere mansari Milau, 1993 Po
Cluthia antiqua Ayress & Drapala, 1996 Ak-Ar
Cluthia australis Ayress & Drapala, 1996 Wn-Rec
Cluthia micra Ayress & Drapala, 1996 Pl
Cluthia novaezealandiae Ayress & Drapala, 1996 Wn
Cluthia sp. Ayress 1993 Ld-Lw
Leptocythere sp. Ayress 1993 Ld-Lw
Leptocythere sp. Ayress 2006 Lw-Po
Leptocythere sp. Milau 1993 Po
Vandiemencythere phleboides Ayress & Warne, 1993 Ak-Lw
LIMNOCYTHERIDAE
Gomphocythere duffi (Hornibrook, 1955) Wc-Rec F
Limnocythere mowbrayensis Chapman, 1914 Wc F
Paralimnocythere vulgaris McKenzie & Swanson, 1981 Qu-Rec F
LOXOCONCHIDAE
Kuiperiana juglandica Ayress, 1993 Pl
Kuiperiana cf. *lindsayi* McKenzie, Reyment & Reyment, 1991) Ak
Loxoconcha abrupta Hornibrook, 1952 Ld-Sw
Loxoconcha propunctata Hornibrook, 1952 Pl
Loxoconcha punctata Thomson, 1879 Ak-Rec
Loxoconcha sp. Milau 1969 Po
Microloxoconcha sp. Ayress 1995 Ak
Microloxoconcha sp. Ayress 1995 Ak
Palmoconcha juglandis Ayress, 1993 Lwh-Lw
Sagmatocythere carboneli Milau, 1993 Ak-Po
MACROCYPRIDIDAE
Macrocypris sp. Ayress 1993 Lwh-Lw
Macropyxis? sp. Ayress 2006 Lwh-Po

Macroscapha? sp. Ayress 1995 Ak
NEOCYTHERIDEIDIDAE
Copytus pseudoelongatus Ayress, 1995 Ak
Copytus sp. Ayress 1993 Ld-Lw
Neocytherideis mediata Swanson, 1969 Ld-Pl
Neocytherideis reticulata Ayress, 1995 Ak-Lw
Pontocythere hedleyi (Chapman, 1906) Ak-Rec
NOTODROMADIDAE
Newnhamia fenestrata King, 1855 Wc-Rec
PARACYPRIDIDAE
Aglaia? praecox Chapman, 1926 Ld.
Paracypris eocuneata (Hornibrook, 1953) Ab-Lwh
Paracypris sp. 5040 Dingle 2009 Mh
Paracypris? sp. 5080 Dingle 2009 Mh
Phylctenophora zealandica Brady 1880 Ld-Rec
PARADOXOSTOMATIDAE
Cytherois parallella Milau, 1993 Po
Paracytherois cf. *gracilis* (Chapman, 1915) Ak
Paracytherois sp. Ayress 1993 Ld
PECTOCYTHERIDAE
Ameghinocythere eagari Dingle, 2009 Mh
Ameghinocythere? sp. 5078 Dingle 2009 Mh
Keijia? hornibrooki Milau, 1993 Po
Keijia sp. Ayress 2006 Po
Munseyella brevis Swanson, 1979 Ld-Rec
Munseyella dunoona McKenzie, Reyment & Reyment, 1993 Ak
Munseyella modesta, Swanson, 1979 Ak-Rec
Munseyella pseudobrevis Ayress, 1995 Ak
Munseyella rectangulata Swanson, 1969 Pl
Munseyella cf. *splendida* Whatley & Downing, 1983 Ld-Lw
Swansonites aequa (Swanson, 1979) Ld-Rec E
Swansonites intermedia Milau, 1993 Po E
PONTOCYPRIDIDAE
Argilloecia acuticadata Whatley & Downing, 1983 Ak
Argilloecia australomiocenica Whatley & Downing, 1983 Ak
Argillaocia krithiformae Whatley & Downing, 1983 Ak
Argilloecia pusilla (Brady, 1880) Lwh-Lw
Australoecia sp. Ayress 1995 Ak-Lwh
Maddocksella argilloeciaformis (Whatley & Downing, 1883) Ak
Maddocksella tumefacta (Chapman, 1914) Lwh-Lw
Maddocksella sp. 5047 Dingle 2009 Mh
Pontocypria sp. Ayress 1993 Lw
Propontocypris cf. *herdmani* (Scott, 1905) Ab-Rec
PROGONOCYTHERIDAE
Majungaella waiparaensis Dingle, 2009 E Mh
Majungaella wilsoni Dingle, 2009 E Mh
Majungaella sp. 4978 Dingle 2009 Mh
Parahystricocythere ericea Dingle, 2009 E Mh
Parahystricocythere sp. 5070 Dingle 2009 Mp
ROCKALLIIDAE
Arcacythere chapmani Hornibrook, 1952 Mp-Sw
Arcacythere aff. *chapmani* Hornibrook, 1952 Lwh-Lw
Arcacythere eocenica (Whatley et al, 1980) Ak
SCHIZOCYTHERIDAE
Apateloschizocythere? colleni Dingle, 2009 Cn
TRACHYLEBERIDIDAE
Abyssocythere sp. Ayress 1993 Ld-Lw
Abyssophilos leptodictyotus (Ayress, 1995) Ar E
Actinocythereis microagrenon Ayress, 1995 Ak-Lw
Actinocythereis thomsoni (Hornibrook, 1952) Dw-Rec
Acanthocythereis? reticulospinosa Ayress, 1993 Ab
Actinocythereis sp. Ayress 1993 Ab
Alataleberis paranuda Milau, 1993 Po
Anebocythereis hostizea (Hornibrook, 1952) Dh-Ld
Cletocythereis cf. *bradyi* Holden, 1967 Pl
Cletocythereis rastromarginata (Brady, 1880) Ak-Rec
Clinocthereis australis Ayress & Swanson, 1991 Ak-Rec
Cythereis contigua Hornibrook, 1952 Dm-Pl
Cythereis inlayi Hornibrook, 1952 Pl-Rec
Cythereis planalta Hornibrook, 1952 Dh-Po
Cythereis cf. *brevicostata* Bate, 1972 Mh
Glencoeleberis? cf. *armata* Jellinek & Swanson, 2003 Lwh-Po
Glencoeleberis? cf. *brevicosta* (Hornibrook, 1952) Lwh-Po
Glencoeleberis? cf. *incerta* (McKenzie, Reyment & Reyment, 1991) Lwh-Po
Glencoeleberis? cf. *occultata* Jellinek & Swanson, 2003 Lwh-Po
Glencoeleberis thomsoni (Hornibrook, 1952) Pal-Rec
Limburgina postaurora Dingle, 2009 E Mh
Marwickcythereis marwicki (Hornibrook, 1952) Ab-Ar E
Marwickcythereis ordotormenta Whatley & Millson, 1992 Dw E
Oertliella semivera (Hornibrook, 1952) Dm-Ld
Oertliella echinata (McKenzie, Reyment & Reyment, 1993) Ak-Lw
Philoneptunus alagracilus Whatley, Millson & Ayress, 1992 Mh-Ab
Philoneptunus crassimurus Whatley, Millson & Ayress, 1992 Ld-Lw
Philoneptunus eagari Whatley, Millson & Ayress, 1992 Dh
Philoneptunus eocenicus Whatley, Millson & Ayress, 1992 Dw-Dh
Philoneptunus gravizea Hornibrook, 1952 Dm-Rec
Philoneptunus hornibrooki Whatley, Millson & Ayress, 1992 Ak-Ar
Philoneptunus paragravazea Whatley, Millson & Ayress, 1992 Lwh-Rec
Philoneptunus paeminosus Whatley, Millson & Ayress, 1992 Dh-Rec
Philoneptunus planaltus (Hornibrook, 1952) Lwh-Rec
Philoneptunus praeplanaltus Whatley, Millson & Ayress, 1992 Lwh
Philoneptunus reticulatus Whatley, Millson & Ayress, 1992 Ab-Ar
Philoneptunus swansoni Whatley, Ayress & Millson, 1992 Ab-Lwh
Philoneptunus tricostatus Whatley, Millson & Ayress, 1992 Dm-Dh
Philoneptunus sp. 1 Whatley, Millson & Ayress 1992 Lw
Philoneptunus sp. 2 Whatley, Millson & Ayress 1992 Pli-Ple
Philoneptunus sp. 3 Whatley, Millson & Ayress 1992 Ple
Philoneptunus sp. 5 Whatley, Millson & Ayress 1992 Lwh
Philoneptunus sp. 6 Whatley, Millson & Ayress 1992 Ak
Ponticocythereis praemilitaris Milau, 1993 Po
Protobuntonia hayi (Hornibrook, 1953) Ab-Ar
Rayneria? punctata Dingle, 2009 E Mh
Rugocythereis reticulata Ayress, 1993 Ab-Rec
Rugocythereis semicontigua (Hornibrook, 1953) Ab-Lwh
Scepticocythereis cf. *ornata* Bate, 1972 Mh
Scepticocythereis? sp. 5044 Dingle 2009 Mh
Taracythere conjunctispina Ayress, 1995 Ak-Po
Taracythere hampdenensis (Ayress, 1993) Ab-Ak
Taracythere proterva (Hornibrook, 1953) ?Dt-Lw
Taracythere sp. Ayress 1993 Ab
Trachleberis ayressi Milau, 1993 Po
Trachyleberis brevicostata Hornibrook, 1952 Ld-Sl
Trachleberis denticulata Milau, 1993 Po
Trachyleberis hornibrooki Dingle, 2009 E Mh
Trachyleberis jilletti Ayress, 1993 Lw
Trachyleberis lytteltonsis Harding & Sylvester-Bradley, 1953 Tt-Rec
Trachyleberis paucispinosa McKenzie, Reyment & Reyment, 1993 Ak
Trachyleberis probesiodes Hornibrook, 1952 Sc-Wp
Trachyleberis retizea Hornibrook, 1952 Po-Pl
Trachyleberis rugibrevis (Hornibrook, 1952) Ld-Rec
Trachyleberis tridens Hornibrook, 1952 Ar-Pl
Trachyleberis zeacristata Hornibrook, 1952 Lw-Rec
XESTOLEBERIDIDAE
Microxestoleberis sp. Ayress 1993 Ld-Lw
Uroleberis minutissima (Chapman, 1926) Ak-Lw
Xestoleberis basiplana McKenzie, Reyment & Reyment, 1993 Ak
Xestoleberis chilensis austrocontinentalis Hartmann, 1978 Ak
Xestoleberis cf. *curta* (Brady, 1865) Lwh-Rec
Xestoleberis paratruncata Whatley & Downing, 1983 Ak
Xestoleberis waihekeensis Milau, 1993 Po
Xestoleberis sp. 1 Ayress 1993 Lwh-Lw
Xestoleberis sp. 2 Ayress 1993 Lwh-Lw
Xestoleberis sp. Ayress 1995 Ak
INCERTAE SEDIS
Crescentocythere phoebe Ayress, 1993 Pl
Saidia limbata Colalongo & Passini, 1980 Ak
Saida torresi (Brady, 1880)*An-Rec
Saida sp. Ayress 1993 Lwh-Lw

Suborder PLATYCOPINA
CYTHERELLIDAE
Cytherella ballancei Milau, 1993 Po
Cytherella bisson Milau, 1993 Po-Pl
Cytherella chapmani Milau, 1993 Po
Cytherella elongata Swanson, 1969 Pl
Cytherella hemipunctata Swanson, 1969 Lw-Rec
Cytherella ?hemipunctata Swanson, 1969 Ak
Cytherella magna Ayress, 2006 Lw-Sc
Cytherella paranitida Whatley & Downing, 1983 Ab-Rec
Cytherella sp. Ayress, 1993 Ab-Lw
Cytherella sp. 5051 Dingle 2009 Mh
Cytherella sp. 5063 Dingle 2009 Cn
Cytherella sp. 5086 Dingle 2009 Mh
Cytherella sp. 1a Dingle 2009 Mh
Cytherelloidea paranitida Whatley & Downing, 1993 Lw
Cytherelloidea praeauricula (Chapman, 1926) Ak-Lw
Cytherelloidea willetti Swanson, 1969* Ak-Rec E
Cytherelloidea cf. *westaustraliensis* Bate, 1972 Mh
Cytherelloidea n. sp. van den Bold, 1963 Rec
Cytherelloidea sp. Ayress, 1993 Lwh-Lw
Cytherelloidea sp. 1 Ayress 2006 Ld-Lw
Healdia? sp. Milau, 1993 Po
Platella sp. 5048 Dingle 2009 Mh
Platella sp. 5071 Dingle 2009 Mh
Order MYODOCOPIDA
Suborder MYODOCOPINA
SARSIELLIDAE
Sarsiella sp. Milau, 1993 Po

Class MALACOSTRACA
Subclass PHYLLOCARIDA
Order HYMENOSTRACA
HYMENOCARIDIDAE
Hymenocaris bensoni Chapman, 1934 Ord
Hymenocaris lepadoides Chapman, 1934 Ord

Order ARCHAEOSTRACA
CERATIOCARIDIDAE
Caryocaris cf. *acuta* Bulman, 1931 Ord
Caryocaris bulmani (Chapman, 1934) Ord
Caryocaris maccoyi (Etheridge, 1892) Ord
Caryocaris m. tumida (Chapman, 1934) Ord
Caryocaris marrii Chapman, 1934 Ord

Caryocaris minima Chapman, 1934 Ord
Caryocaris wrightii Chapman, 1934 Ord

Subclass EUMALACOSTRACA
Superorder PERACARIDA
Order ISOPODA
Suborder VALVIFERA
HOLOGNATHIDAE
Debodea mellita Hiller, 1999 (not Cirolanidae) UCret E

Suborder CYMOTHOOIDA
CIROLANIDAE
Cirolana makikihi Feldmann, Schweitzer, Maxwell & Kelley, 2008 Wo E
Palaega kakatahi Feldmann & Rust, 2006 Wo-Wp E

INCERTAE SEDIS
URDIDAE
Urda zelandica Buckeridge & Johns, 1996 UJur E

Superorder EUCARIDA
Order DECAPODA
Suborder PLEOCYEMATA
Infraorder GLYPHEIDEA
ERYMIDAE
Gen. et sp. indet. Mp-Mh
GLYPHEIDAE
Glyphea christeyi Feldmann & Maxwell, 1999 Ab E
Glyphea stilwelli Feldmann, 1993 Dt E
Glypheopsis antipodum Glaessner 1960 Hu E
MECOCHIRIDAE
Mecochirus marwicki Glaessner, 1960 Kh
Mecochirus? sp. Bw, Kh-Op

Infraorder ASTACIDEA
NEPHROPIDAE
Hoploparia sp. Mp
Metanephrops motunauensis Jenkins, 1972 Sw-Tt E
PARASTACIDAE
Paranephrops fordycei Feldmann & Pole, 1994 Po-Sl E

Infraorder AXIIDEA
CALLIANASSIDAE
Callianassa awakina Glaessner, 1960 Po E
Callianassa waikurana Glaessner, 1960 Mh E
Callianassa sp. a Mh
Callianassa sp. b Tt
Callianassa sp. Cn, Mp-Mh
Callianassa sp. Ab, Lwh-Pl, Sw-Tt
Protocallianassa sp. Mp-Mh
CTENOCHELIDAE
Ctenocheles cf. *maorianus* Powell, 1949 Wc
Ctenocheles sp. Wc
INCERTAE SEDIS
Gen. et sp. indet. Feldmann, Schweitzer, Maxwell & Kelley, 2008 Wo E

Infraorder GEBIIDEA
UPOGEBIIDAE
Upogebia kowai Feldmann, Schweitzer, Maxwell & Kelley, 2008 Wo E
Upogebia sp. Ar-Lwh

Infraorder ACHELATA
PALINURIDAE
Jasus flemingi Glaessner, 1960 Pl
Linuparus korura Feldmann & Bearlin, 1988 Ab
Linuparus sp. Mp-Mh
Linuparus? sp. Mp-Mh

Infraorder ANOMURA
AEGLIDAE
Haumuriaegla glaessneri Feldmann, 1984 Mp-Mh E
GALATHEIDAE
Galathea sp. Wp-Wn
LITHODIDAE
Paralomis debodeorum Feldmann, 1998 MMio-LMio E
PAGURIDAE
Diacanthurus clifdenensis (Hyden & Forest, 1980) Pl E
Pagurus sp. Tt, Wp, Wn

Infraorder BRACHYURA
ATELECYCLIDAE
Trichopeltarion greggi Dell, 1969 Sw-Tt E
Trichopeltarion merrinae Schweitzer & Salva, 2000 L Mio E
CALAPPIDAE
Calappilia maxwelli Feldmann, 1993 Po E
CANCRIDAE
Lobocarcinus pustulosus Feldmann & Fordyce, 1996 Pl E
Metacarcinus novaezelandiae (Hombron & Jacquinot, 1846) Wo-Rec
Metacarcinus cf. *novaezelandiae* (Hombron & Jacquinot, 1846) Tk, Wp
Metacarcinus sp. Ak, Ld, Wp-Wn
GONEPLACIDAE
Carcinoplax temikoensis Feldmann & Maxwell, 1990 Ak-Ar E
Carcinoplax sp. Wp-Wn
Kowaicarcinus maxwellae Feldmann, Schweitzer, Maxwell & Kelley, 2008 Wo E
Ommatocarcinus arenicola Glaessner, 1960 Pl E
Ommatocarcinus cf. *arenicola* Glaessner, 1960 Pl
Ommatocarcinus cf. *Neommatocarcinus huttoni* (Filhol, 1886) Wp-Wn
Ommatocarcinus sp. Pl
HOMOLODROMIIDAE
Homolodromia novaezelandica Feldmann, 1993 Mp-Mh E
Homolodromia sp. Mp-Mh
MACROPHTHALMIDAE
Macrophthalmus (*Hemiplax*) *hirtipes* (Heller, 1862) Wq–Rec E
Hemiplax?major Glaessner, 1960 Wn E
Hemiplax cf. *major* Glaessner, 1960 Po, Wc
Hemiplax sp. Wn-Wc
MAJIDAE
Actinotocarcinus chidgeyorum Jenkins, 1974 Sc-Tt E
Actinotocarcinus maclauchlani Feldmann, 1993 Sw-Tt E
Jacquinotia edwardsii (Jacquinot, 1853) Wp-Rec E
Leptomithrax atavus Glaessner, 1960 Tk E
Leptomithrax elongatus McLay, Feldmann & MacKinnon, 1995 Sw E
Leptomithrax garthi McLay, Feldmann & MacKinnon, 1995 Sw-Tt E
Leptomithrax griffini Feldmann & Maxwell, 1990 Ab-Ar E
Leptomithrax irirangi Glaessner, 1960 Wo E
Leptomithrax aff. *irirangi* Glaessner, 1960 Sw
Leptomithrax uruti Glaessner, 1960 E Tt
Leptomithrax cf. *uruti* Glaessner, 1960 Tt
Leptomithrax sp. Tt
Micromithrax? minisculus Feldmann & Wilson, 1988 Dm-Dh
Notomithrax allani Feldmann & Maxwell, 1990 Ak-Ar E
Notomithrax minor (Filhol, 1885) Wc – Rec
Notomithrax sp. Wc
MENNIPIDAE
Galene proavita Glaessner, 1960 Pl-Sc E
Galene sp. Wp-Wn
Menippe sp. Pl
Pseudocarcinus sp. Tk
PORTUNIDAE
Ovalipes cf. *catharus* (White, 1843) Wn-Wc
Ovalipes sp. A Wp
Ovalipes sp. Wn-Wc
Pororaria eocenica Glaessner, 1980 Ak-Ar E
Portunus sp. Lwh, Lw
Rhachiosoma granuliferum (Glaessner, 1960) Dp-Ar E
Gen. et sp. indet. Dm-Dh, Ab-Ak
PSEUDOZIIDAE
Tongapapaka motunauensis Feldmann, Schweitzer, Maxwell & Kelley, 2008 Wo E
RANINIDAE
Hemioon novozelandicum Glaessner, 1980 Cn E
Laeviranina keyesi Feldmann & Maxwell, 1990 Ak-Ar E
Laeviranina perarmata Glaessner, 1960 Ab E
Laeviranina pororariensis (Glaessner, 1980) Ak-Ar E
Lyreidus bennetti Feldmann & Maxwell, 1990 Ak-Ar E
Lyreidus elegans Glaessner, 1960 Po-Pl E
Lyreidus waitakiensis Glaessner, 1980 Ab E
Lyreidus sp. Sw
Gen. et sp. indet. Ab
TORYNOMMIDAE
Eodorripe spedeni Glaessner, 1980 Mp-Mh E
Torynomma flemingi Glaessner, 1980 Mp-Mh E
Torynomma planata Feldmann, 1993 Mp-Mh E
TUMIDOCARCINIDAE
Tumidocarcinus dentatus Glaessner, 1960 Lwh-Ld E
Tumidocarcinus cf. *dentatus* (Glaessner, 1960) Lwh
Tumidocarcinus giganteus Glaessner, 1960 Pl-Tt E
Tumidocarcinus cf. *giganteus* Glaessner, 1960 Lw-Po, Sw-Tk
Tumidocarcinus tumidus (Woodward, 1876) Ab-Ld E
Tumidocarcinus cf. *tumidus* (Woodward, 1876) Lwh-Ld
Tumidocarcinus? sp. Ak-Ld, Po-Sc
VARUNIDAE
Austrohelice manneringi Feldmann, Schweitzer, Maxwell & Kelley, 2008 Wo E
Miograpsus papaka Fleming, 1981 Tt E

Developmental stages of New Zealand Decapoda

Compiled by W. R. Webber

Following are the larvae and/or pre- or post-larvae described to date, of species listed in the decapod species list above. Species named below are those with one or more developmental stages described in the literature. Names and dates in brackets indicate publications in which larvae are described, **not** species authorities. However, *Jaxea novaezealandiae* (Gebiidea) was described in the same paper as the adult and two polychelid species were described from the larvae, thus authors in brackets after these names are also the original authorities. Literature sources for the species list below are cited in the References section, above.

PHYLUM CRUSTACEA
Class MALACOSTRACA
Order DECAPODA
Suborder DENDROBRANCHIATA
SERGESTIDAE
Sergestes arcticus [Gurney & Lebour 1940; Wear 1985]
SOLENOCERIDAE
Solenocera comata [Gurney 1924; Wear 1985]

Suborder PLEOCYEMATA
Infraorder STENOPODIDEA
STENOPODIDAE
Stenopus hispidus [Gurney 1936, 1942]

Infraorder CARIDEA
ALPHEIDAE
Alpheus euphrosyne richardsoni [Packer 1983, 1985]
Alpheus socialis [Packer 1983, 1985]
Alpheopsis garricki [Packer 1983, 1985]
Betaeopsis aequimanus [Packer 1983, 1985]
ATYIDAE
Paratya curvirostris [Ch'ng 1973; Wear 1985]
CAMPYLONOTIDAE
Campylonotus rathbunae [Pike & Williamson 1966; Wear 1985]
CRANGONIDAE
Aegaeon lacazei [De Simón 1979; Packer 1983, 1985]
Philocheras australis [Thomson & Anderton 1921; Packer 1983, 1985]
Philocheras chiltoni [Packer 1983, 1985]
Philocheras hamiltoni [Packer 1983, 1985]
Philocheras pilosoides [Packer 1983, 1985]
HIPPOLYTIDAE
Alope spinifrons [Lebour 1955; Packer 1983, 1985]
Hippolyte bifidrostris [Packer 1983, 1985]
Hippolyte multicolorata [Packer 1983, 1985]
Nauticaris marionis [Packer 1983, 1985]
Tozeuma novaezealandiae [Packer 1983, 1985]
OGYRIDIDAE
Ogyrides delli [Packer 1983, 1985]
PALAEMONIDAE
Palaemon affinis [Lebour 1955; Packer 1983, 1985]
Periclimenes yaldwyni [Packer 1983, 1985]
Periclimenes (Periclimenes) sp. [Packer 1983, 1985]

Infraorder ASTACIDEA
NEPHROPIDAE
Metanephrops challengeri [Wear 1976]
PARASTACIDAE
Paranephrops planifrons [Hopkins 1967]

Infraorder AXIIDEA
CALLIANASSIDAE
Callianassa filholi [Gurney 1924; Lebour 1955; Wear 1965a]

Infraorder GEBIIDEA
LAOMEDIIDAE
Jaxea novaezealandiae [Wear & Yaldwyn 1966]
UPOGEBIIDAE
Acutigebia danai [Gurney 1924]

Infraorder PALINURA
PALINURIDAE
Jasus edwardsii [Batham 1967; Lesser 1974]
Sagmariasus verreauxi [Lesser 1974; Kittaka *et al.* 1997]
POLYCHELIDAE
Gen. et sp. indet. (as *Eryonicus fagei*) [Bernard 1953]
Gen. et sp. indet. (as *Eryonicus scharffi*) [Selbie 1914]
SCYLLARIDAE
Ibacus alticrenatus [Atkinson & Boustead 1982]
Scyllarus sp. Z [Webber & Booth 2001]

Infraorder ANOMURA
CHIROSTYLIDAE
Gastropyychus novaezelandiae [Pike & Wear 1969]
Uroptychus n. sp. [Pike & Wear 1969]
GALATHEIDAE
Munida gregaria [Roberts 1973]
PAGURIDAE
Pagurixus hectori [Roberts 1971; Wear 1985]
Pagurus novizealandiae [Greenwood 1966; Wear 1985]
Pagurus traversi [Thomson & Anderton 1921; Wear 1985]
Porcellanopagurus edwardsi [Roberts 1972; Wear 1985]
PARAPAGURIDAE
Sympagurus dimorphus [Lemaitre & McLaughlin 1992]
PORCELLANIDAE
Petrocheles spinosus [Wear 1965b, 1966]
Petrolisthes elongatus [Greenwood 1956; Wear 1964b, 1965c]
Petrolisthes novaezelandiae [Greenwood 1956; Wear 1964a, 1965d]

Infraorder BRACHYURA
ATELECYCLIDAE
Trichopeltarion fantasticum [Wear & Fielder 1985]
BELLIIDAE
<u>*Heterozius*</u> *rotundifrons* [Wear & Fielder 1985]
CANCRIDAE
Metacarcinus novaezelandiae [Wear & Fielder 1985]
CYMONOMIDAE
Cymonomus bathamae [Wear & Fielder 1985
DROMIIDAE
Metadromia wilsoni [Wear & Fielder 1985]
GONEPLACIDAE
<u>*Neommatocarcinus*</u> *huttoni* Wear & Fielder 1985
GRAPSIDAE
Leptograpsus variegatus [Wear & Fielder 1985]
Planes major [Wear & Fielder 1985]
Planes marinus [Wear & Fielder 1985]
HOMOLIDAE
Dagnaudus petterdi [Williamson 1965; Wear & Fielder 1985]
Homola orientalis [Wear & Fielder 1985]
HYMENOSOMATIDAE
Amarinus lacustris [Wear & Fielder 1985]
Elamena longirostris [Wear & Fielder 1985]
Elamena momona [Wear & Fielder 1985]
Elamena producta [Wear & Fielder 1985]
Halicarcinus cookii [Wear & Fielder 1985]
Halicarcinus innominatus [Wear & Fielder 1985]
Halicarcinus planatus [Wear & Fielder 1985]
Halicarcinus varius [Horn &Harms 1988]
Halicarcinus whitei [Wear & Fielder 1985]
Hymenosoma depressum [Wear & Fielder 1985]
<u>*Neohymenicus*</u> *pubescens* [Wear & Fielder 1985]
INACHIDAE
Achaeus curvirostris [Wear & Fielder 1985]
Cyrtomaia lamellata [Wear & Fielder 1985]
INACHOIDIDAE
Pyromaia tuberculata [Webber & Wear 1981; Wear & Fielder 1985]
LATREILLIIDAE
Eplumula australiensis (Wear &Fielder 1985)
LEUCOSIIDAE
Bellidilia cheesmani [Wear & Fielder 1985]
MACROPHTHALMIDAE
Macrophthalmus (Hemiplax) hirtipes [Wear & Fielder 1985]
MAJIDAE
Eurynolambrus australis [Webber & Wear 1981; Wear & Fielder 1985]
<u>*Jacquinotia*</u> *edwardsi* [Webber & Wear 1981; Wear & Fielder 1985]
Leptomithrax longimanus [Webber & Wear 1981; Wear & Fielder 1985]
Leptomithrax longipes [Webber & Wear 1981; Wear & Fielder 1985]
Leptomithrax tuberculatus mortenseni [Wear & Fielder 1985]
Notomithrax minor [Webber & Wear 1981; Wear & Fielder 1985]
Notomithrax peronii [Webber & Wear 1981; Wear & Fielder 1985]
Notomithrax ursus [Webber & Wear 1981; Wear & Fielder 1985]
OZIIDAE
Ozius truncatus (Wear & Fielder 1985)
PILUMNIDAE
Pilumnopeus serratifrons [Wear & Fielder 1985]
Pilumnus lumpinus [Wear & Fielder 1985]
Pilumnus novaezelandiae [Wear & Fielder 1985]
PINNOTHERIDAE
<u>*Nepinnotheres*</u> *novaezelandiae* [Wear & Fielder 1985]
PLAGUSIIDAE
Plagusia chabrus [Wear & Fielder 1985]
PORTUNIDAE
Liocarcinus corrugatus [Wear & Fielder 1985]
Nectocarcinus antarcticus [Wear & Fielder 1985]
Ovalipes catharus [Wear & Fielder 1985]
Portunus pelagicus [Wear & Fielder 1985]
Scylla serrata [Wear & Fielder 1985]
RANINIDAE
Lyreidus tridentatus [Wear & Fielder 1985]
VARUNIDAE
Austrohelice crassa [Wear & Fielder 1985]
Cyclograpsus insularum [Wear & Fielder 1985]
Cyclograpsus lavauxi [Wear & Fielder 1985]
Hemigrapsus crenulatus [Wear & Fielder 1985
Hemigrapsus sexdentatus [Wear & Fielder 1985]

NINE

Phylum
ARTHROPODA

SUBPHYLUM HEXAPODA

Protura, springtails, Diplura, and insects

ROD P. MACFARLANE, PETER A. MADDISON, IAN G. ANDREW, JOCELYN A. BERRY, PETER M. JOHNS, ROBERT J. B. HOARE, MARIE-CLAUDE LARIVIÈRE, PENELOPE GREENSLADE, ROSA C. HENDERSON, COURTENAY N. SMITHERS, RICARDO L. PALMA, JOHN B. WARD, ROBERT L. C. PILGRIM, DAVID R. TOWNS, IAN McLELLAN, DAVID A. J. TEULON, TERRY R. HITCHINGS, VICTOR F. EASTOP, NICHOLAS A. MARTIN, MURRAY J. FLETCHER, MARLON A. W. STUFKENS, PAMELA J. DALE, DANIEL BURCKHARDT, THOMAS R. BUCKLEY, STEVEN A. TREWICK

A defining feature of the Hexapoda, as the name suggests, is six legs. Also, the body comprises a head, thorax, and abdomen. The number of abdominal segments varies, however; there are only six in the Collembola (springtails), 9–12 in the Protura, and 10 in the Diplura, whereas in all other hexapods there are strictly 11. Insects are now regarded as comprising only those hexapods with 11 abdominal segments.

Whereas crustaceans are the dominant group of arthropods in the sea, hexapods prevail on land, in numbers and biomass. Altogether, the Hexapoda constitutes the most diverse group of animals – the estimated number of described species worldwide is just over 900,000, with the beetles (order Coleoptera) comprising more than a third of these. Today, the Hexapoda is considered to contain four classes – the Insecta, and the Protura, Collembola, and Diplura. The latter three classes were formerly allied with the insect orders Archaeognatha (jumping bristletails) and Thysanura (silverfish) as the insect subclass Apterygota ('wingless'). The Apterygota is now regarded as an artificial assemblage (Bitsch & Bitsch 2000). Though fewer in numbers of species than the beetles, Collembola (springtails) are perhaps the most abundant arthropods on earth, especially in soil litter. Found in the same environment, Protura are very small pale arthropods that are rarely encountered; Diplura include a few families of larger pale arthropods. The vast majority of hexapod species are insects, classified among 26–30 orders (some gene-sequencing studies suggest the amalgamation of some orders).

Celatoblatta vulgaris.
Alastair Robertson and Maria Minor, Massey University

The earliest known hexapods in the fossil record are a Collembolon and an insect, both terrestrial, from the Early Devonian Rhynie Chert in Scotland (Engel & Grimaldi 2004). Both have relatively advanced features, showing that insects must have originated earlier, in the Silurian. Traditionally, springtails, Protura, and Diplura were united in a group called Entognatha, so named because members of these classes all have the base of the mouthparts internalised, so that the

Summary of New Zealand hexapod taxonomic diversity

Taxon	Described living species+ subspecies	Known undescribed/ unrecorded species+ subspecies	Estimated unknown species	Adventive species+ subspecies	Endemic species+ subspecies	Endemic genera named +new
Protura	17	1	0–10	2	10	0
Collembola	346+34	0	650	44+3	266+29	20
Diplura	8+1	2	0–5	4?	6	0
Archaeognatha	2	0	0–2	0	2	0
Thysanura	3	8	>8	2	8	0+2
Ephemeroptera	48	3	10	0	51	20
Odonata	17	0	0	3	10	3
Plecoptera	99	21	20	0	120	19
Blattodea	35	9	2–5	13	31	5+1
Isoptera	9	0	0	5	3	0
Mantodea	2	0	0	1	1	0
Dermaptera	21	1	0–3	9?	13?	0
Orthoptera	116	69	<5	7?	173	27+1
Phasmatodea	22	7	7	1?	29	10
Hemiptera	1,079+15	78	235–440	280?	826?+14	119+1
Thysanoptera	122	1	>10	55	65?	11
Psocoptera	62	7	30	5	31	3
Phthiraptera	340+8	10	20*	90	29	1
Megaloptera	1	0	0	0	1	0
Neuroptera	14	0	0	7	7	0
Coleoptera	5,062+29	417–420+1	3,000	418+29	5,002–5,005+30	533+2
Mecoptera	1	0	0	0	1	0
Siphonaptera	28+6	0	0–5	11	17	0
Diptera	2,483+9	736–785	130–1,640	195	3,030–3,026+9	192+17
Strepsiptera	2	1	2?	0	4	0
Hymenoptera	721	820–934	>500	259?	740–742	55+21
Trichoptera	228	26	10–50	0	241	33+2
Lepidoptera	1,686+8	14	>100	139+1	1,389+9	140
Totals	**12,573+110**	**2,231–2,397**	**4,735–6,525**	**~1,585+33**	**12,105–12,091+91**	**1,188+48**

* Not including possible species on vagrant birds

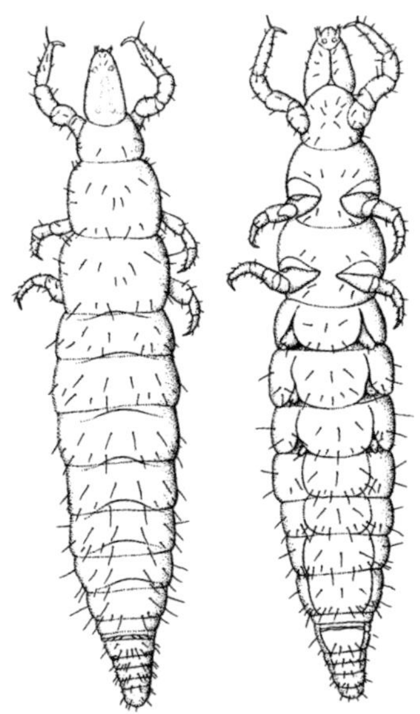

Dorsal and ventral views of *Amphientulus zelandicus*.

R. Nielsen, from Tuxen 1986

mandible and maxilla are partly contained within the head capsule. In addition to this similarity in mouth structure, these three classes share reduced Malpighian tubules. The precise relationships of these groups to one another and to Insecta are still uncertain, however; analyses of morphological and developmental characters and gene sequencing give conflicting results (Cook et al. 2001; Giribet et al. 2004; Regier et al. 2005; Carapelli et al. 2007; Dell'Ampio et al. 2008; Timmermans et al. 2008). The majority of studies support the monophyly of Hexapoda.

All hexapods undergo several moults as they develop and grow in size, necessitated by the fact that the exoskeleton, particularly in insects, is inelastic and cannot expand. The newly moulted insect has a soft, flexible, lightly pigmented cuticle; it swallows air or water and so increases its volume before the cuticle hardens and darkens and increases in thickness. Life-cycles vary. In some groups the young are called nymphs, which are similar in form to the adult except that the wings are not developed until the adult stage. This is called incomplete metamorphosis and insects showing this are termed hemimetabolous. Representative orders include Ephemeroptera (mayflies), Odonata (damselflies and dragonflies), Plecoptera (stoneflies), Orthoptera (crickets and kin), Dermaptera (earwigs), Hemiptera (true bugs, cicadas, and kin), and Thysanoptera (thrips) among others. Holometabolous insects pass through a complex metamorphosis always accompanied by a pupal stage. The wings develop internally and the larvae are usually specialised. Representative orders include the Coleoptera (beetles), Hymenoptera (ants, wasps, and kin), Diptera (true flies), Trichoptera (caddis), and Lepidoptera (moths and butterflies) among others.

Living hexapods range in size from tiny hymenopteran parasites less than a fifth of a millimetre to a slender 56 cm-long stick insect that lives on the island of Borneo. While the overall largest insects today are Goliath beetles, the heaviest documented species is the Little Barrier giant weta (*Deinacrida heteracantha*), which used to live on the northern North Island mainland; one specimen weighed more than 70 grams.

Although many hexapods are major pests, parasitising humans and livestock, causing damage to crops and stored products, or transmitting diseases, others are profoundly beneficial, as pollinators of economic crops or sources of products ranging from honey to silk. Scavenging hexapods help recycle dead animals and fallen trees; in fact insects of soil litter are responsible for much of the process by which topsoil is made. About 1200 species are used as human food.

Geographically, hexapods range from polar zones to the equator and can be found even in hot springs and deserts. Not only soils, but fresh waters support their own extensive insect faunas, and a number of species are adapted to life in the sea shore and, as larvae, in the shallow subtidal. There is not a single species of seed plant that does not provide food for one or more species of insect. Hexapods are so ubiquitous and important that our terrestrial planetary ecosystems could not survive without them.

Class Protura

Sometimes referred to as coneheads, Protura are possibly the simplest of all living hexapods. The name means 'first tail', first in this context implying original, alluding to the primitive form of the abdomen lacking specialised structures at the rear end. These small white to transparent creatures, only 0.6–1.5 millimetres long in New Zealand, are quite distinct because they lack antennae, eyes, and wings. In consequence of lacking antennae, the front legs, which are somewhat hairy, are raised in front of the body to act as sensory detectors. The abdomen is tubular with up to 12 segments and the legs are simple with five segments. Protura are unique among hexapods in that eggs hatch into larvae with only a few abdominal segments, with the number increasing with subsequent moults to the full adult complement. Proturan sperm is unique too, differing from any other hexapod sperm and also very different in the two proturan orders.

Protura live only in humid places, mainly in acid soils, sometimes in rotten wood, as part of the community of decomposers that help break down and recycle organic nutrients. They are frequently associated with leaf litter and moss under trees where they may feed on fungi.

The first Protura were not discovered until 1907, in New York State. About 500 species have been named so far, arranged in two orders – Eosentomoida and Acerentomoida – segregated not only on sperm characters but the presence or absence of spiracular openings to a tracheal respiratory system.

Tuxen (1986) described seven new species out of the 18 species currently known from New Zealand. *Eosentomon* is the most diverse genus, with seven species. Protura cannot disperse widely owing to their vulnerability to desiccation and saline water and their habitat restriction to soil or rotting wood. It is therefore not surprising that ~56% of the species are endemic and 30% are indigenous with Gondwanan distributions. Two species are accidental introductions from Europe. Within New Zealand, distributional records are very limited for 14 of the species, with only one to six locality records for each. Collection and identification involve extraction of specimens from soil or logs and careful preparation for mounting on microscope slides. The specimens in Tuxen's study were derived from three major collections – New Zealand Arthropod Collection (NZAC), Landcare Research; Museum of New Zealand Te Papa Tongarewa (MONZ); and Lincoln University (LUNZ), but geographic

Eosentomon australicum.
Alistair Robertson & Marie Minor, Massey University

Summary of New Zealand hexapod diversity by environment
(species plus subspecies)

Taxon	Marine*	Freshwater*	Terrestrial*	Fossil†	
				pre-Holocene	Holocene
Protura	0	0	18	0	0
Collembola	0	0	346+34	0	0
Diplura	0	0	10+1	0	0
Archaeognatha	0	0	2	0	0
Thysanura	0	0	11	0	0
Ephemeroptera	0	51	0	0	0
Odonata	0	15	2	0	0
Plecoptera	0	120	0	0	0
Blattodea	0	0	44	0	0
Isoptera	0	0	9	0	0
Mantodea	0	0	2	0	0
Dermaptera	0	0	22	0	0
Orthoptera	0	0	185	1	0
Phasmatodea	0	0	29	0	0
Hemiptera	1	10	1,146+15	2	0
Thysanoptera	0	0	123	0	0
Psocoptera	0	0	69	0	0
Phthiraptera	169+5	47+2	134+1	0	0
Megaloptera	0	1	0	0	0
Neuroptera	0	5	9	0	0
Coleoptera	5	83	5,391–5,394	113	106
Mecoptera	0	1	0	0	0
Siphonaptera	11+5	0	18+3	0	0
Diptera	12	265	2,942–2,991+9	2	0
Strepsiptera	0	0	3	0	0
Hymenoptera	0	0	1,541–1,655	0	0
Trichoptera	5	249	0	1	0
Lepidoptera	0	1	1,699+8	2	0
Totals	**197+10**	**848+2**	**13,755–13,918+68**	**121**	**106**

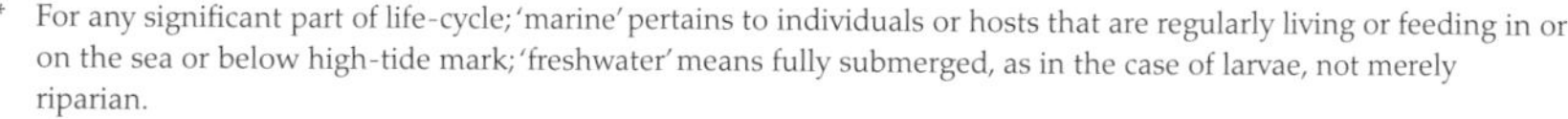
* For any significant part of life-cycle; 'marine' pertains to individuals or hosts that are regularly living or feeding in or on the sea or below high-tide mark; 'freshwater' means fully submerged, as in the case of larvae, not merely riparian.

† 219 species are known only from fossil remains. Eight of these are Pleistocene through Holocene in temporal distribution, the Holocene being taken as the latter part of the Quaternary that started approximately 12,000 years before the present day, i.e. around 10,000 BCE.

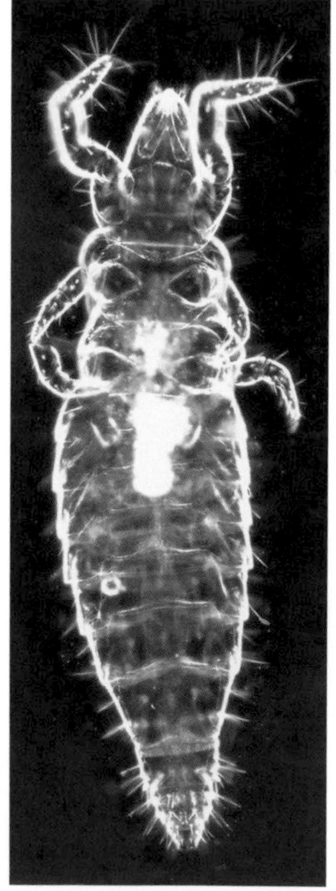
Gracilentulus gracilis.
Alastair Robertson and Maria Minor, Massey University

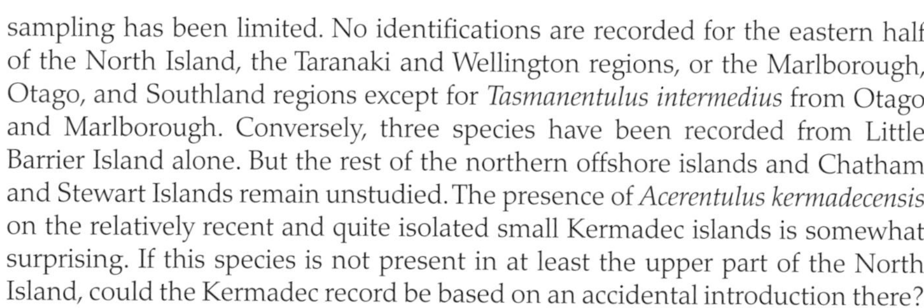
sampling has been limited. No identifications are recorded for the eastern half of the North Island, the Taranaki and Wellington regions, or the Marlborough, Otago, and Southland regions except for *Tasmanentulus intermedius* from Otago and Marlborough. Conversely, three species have been recorded from Little Barrier Island alone. But the rest of the northern offshore islands and Chatham and Stewart Islands remain unstudied. The presence of *Acerentulus kermadecensis* on the relatively recent and quite isolated small Kermadec islands is somewhat surprising. If this species is not present in at least the upper part of the North Island, could the Kermadec record be based on an accidental introduction there?

The most recent study, conducted in the Wanganui–Manawatu region (Minor 2008), compared proturan diversity in native and exotic (*Pinus radiata*) forests. The mean density of Protura was significantly higher in pine plantations than native broadleaf forests. Among native forests, Protura were most abundant under native beech (*Nothofagus*). These abundance patterns may reflect the association between Protura and mycorrhizal fungal communities in the soil, making Protura promising bioindicators of forest health, particular of exotic conifers in the Southern Hemisphere.

Class Collembola: Springtails

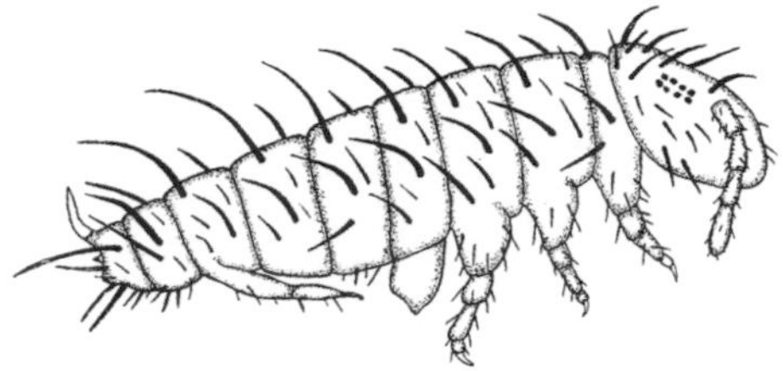

Triacanthella setacea (Arthropleona, Hypogastruridae).
From Salmon 1941

Collembola are minute soft-bodied arthropods that look superficially similar to the larvae of some insects. This is because they have a body divided into three parts – a head with antennae, a three-segmented thorax with a pair of jointed legs on each segment, and a segmented abdomen with paired appendages ventrally on some segments. They differ from insects in lacking a hard exoskeleton and wings, in the mouthparts being internal, and in possessing simple eyes, up to eight, on each side of the head. Their common name, springtail, refers to the ability of many species to leap considerable distances when disturbed. There are three orders of Collembola that can be easily distinguished by body shape. The Arthropleona, to which most species belong, are elongate with most body segments separate and equal in length. The Symphypleona are globular with anterior body segments fused, and the Neelipleona are a very small group of minute globular animals that live deep in the soil. There are about 25 families worldwide, mostly of Arthropleona.

Because of their small size and cryptic habits, springtails are not well known. The class as a whole is, however, extremely widespread, with a global distribution that includes polar and arid regions and is particularly diverse in tropical and temperate rainforests. Although springtails are predominantly soil and litter dwellers, they also occur in a wide range of other habitats, such as on vegetation including tree canopies, in caves, in the marine littoral zone, and in freshwater systems. As detritus feeders, springtails are important in nutrient cycling and thus can be generally considered to be beneficial as very few species feed on live plant material. The biology and widespread nature of springtails ought to warrant more attention from biologists.

Worldwide, about 7900 species of Collembola in more than 580 genera have been described. For information on the New Zealand fauna we are dependent on the pioneer work of J. T. Salmon who worked on the indigenous Collembola from about 1940 to 1970 (see, for example, Salmon 1964). Later, Wise (1977) meticulously documented the species known from New Zealand and provided a complete list of synonyms and new combinations together with all the relevant references. Only one species has been described from New Zealand since Wise's checklist, but there have been numerous new combinations and a number of synonyms published of genera and species (Deharveng & Wise 1987; Christiansen & Bellinger 2000; Greenslade 1982, 1984, 1986, 1989, 1994). Currently 380 species and subspecies in 103 genera are known from New Zealand, which is about five per cent of the world fauna. However, from more recent but unstudied collections that exist, it is clear that only a small proportion of the fauna has been described.

A new compilation of New Zealand Collembola has been made (see end-chapter checklist) that includes all name changes and the new species described since 1977. References to the changes can be found on the World List of Collembola (Bellinger et al. 2009). The taxonomic arrangement used in the checklist follows that of recent revisions, although recent morphological and molecular studies are finding that most of the tribes and subfamilies are not supported phylogenetically. They are used here for convenience and until an improved higher classification has been published and adopted.

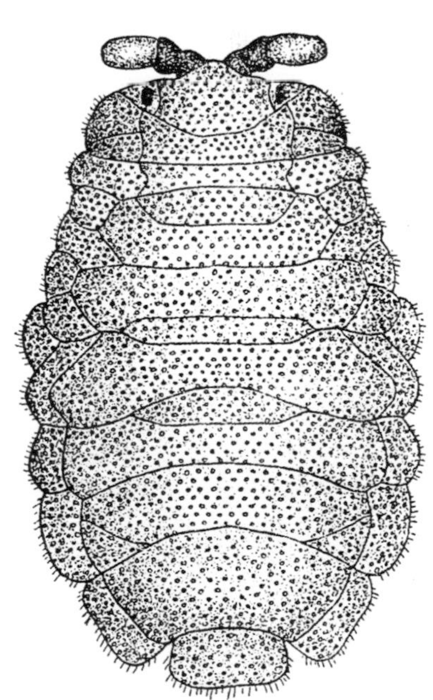

Platanurida marplesi (Arthropleona, Neanuridae).
From Salmon 1941

Diagnostic morphology of Collembola

Springtails are white or coloured, sometimes darkly pigmented and often patterned. Normally 1–2 millimetres long, they can range from 0.2 to 10 millimetres. Their bodies are furnished with chaetae (also called setae), which can be numerous or sparse, fine or thick, long or short, serrated, clublike, and with hair-like structures or smooth. Scales and bothriotricha (long chaetae inserted in pits) are found in some families. The mouth is anterior and the head is aligned with the main body axis, except in Symphypleona, and Neelipleona, where the head is at right angles to the body. Mouthparts are elongate and adapted either

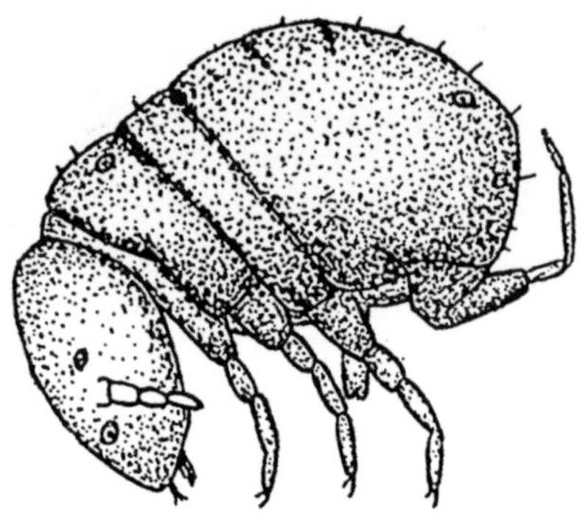

Zelandothorax novaezealandiae (Neelipleona, Neelidae).
From Salmon 1944

Percentages of Collembola species per family in New Zealand, Australia, the world

(Excluded are 14 small families with no Australasian species)

Family	New Zealand no. of species	% of total	Australia no. of species	% of total	World no. of species	% of total
Brachystomellidae	3	0.8	21	5.6	129	1.6
Entomobryidae	98	25.8	90	23.9	1661	21.1
Paronellidae	17	4.5	20	5.3	506	6.4
Hypogastruridae	22	5.8	28	7.4	666	8.4
Isotomidae	87	22.9	56	14.9	1,311	16.6
Neanuridae	41	10.8	46	12.2	1,410	17.9
Odontellidae	6	1.6	4	1.1	129	1.6
Onychiuridae	10	2.6	6	1.6	555	7.0
Tomoceridae	14	3.7	7	1.9	147	1.9
Tullbergiidae	7	1.8	17	4.5	214	2.7
Arrhopalitidae	1	0.3	2	0.5	131	1.7
Bourlettiellidae	5	1.3	24	6.4	245	3.1
Dicytomidae	4	1.1	4	1.1	199	2.5
Katiannidae	51	13.4	33	8.8	205	2.6
Sminthuridae	6	1.6	6	1.6	145	1.8
Sminthurididae	4	1.1	7	1.9	145	1.8
Spinothecidae	1	0.3	1	0.3	6	0.1
Neelidae	3	0.8	3	0.8	33	0.4
Onopoduridae	0	0	1	0.3	52	0.7
Totals	**380**	**100**	**376**	**100**	**7,889**	**100**
% of world fauna		**4.8**		**4.8**		

for biting and chewing/grinding, or for fluid feeding. They consist of a pair of maxillae, a median labium and labrum, and a pair of mandibles. The pleural folds, together with the labrum and the labium, completely enclose the mouthparts to form the buccal cone, which occasionally projects anteriorly. A postantennal organ is usually present, and up to eight, sometimes fewer, ocelli on each side of the head or these simple eyes may be totally absent. The antennae are four-segmented, sometimes subsegmented, with muscles within all segments.

The thorax has three segments; the first segment sometimes lacks chaetae and is shorter than the other segments. The thoracic segments are conspicuous in the Arthropleona but not in Symphypleona or Neelipleona, where they are fused and form a single mass with the abdomen. The legs and the abdomen are both six-segmented but two or three posterior segments in the abdomen are sometimes fused. Specialised appendages – ventral tube, tenaculum, and furca – are found ventrally on abdominal segments I, III, and IV respectively. The ventral tube consists of a column with a pair of enclosed tubes distally, which are normally retracted within the column, but which can be extruded. Their cuticle is permeable to water and the tubes can provide adhesion to the substratum after a leap. The jumping organ or furca is normally folded forward beneath the body and held in place by the teeth of the tenaculum. When suddenly released from the tenaculum, the furca springs backwards to the resting position and so hits the ground with considerable force, propelling the animal into the air. Jumping is both a reaction to disturbance and a means of dispersal, and some directional ability has been demonstrated. The genital opening is in a ventral position on abdominal segment V and the anal aperture is on abdominal segment VI. Cerci are absent, but anal spines are sometimes present. There is no metamorphosis.

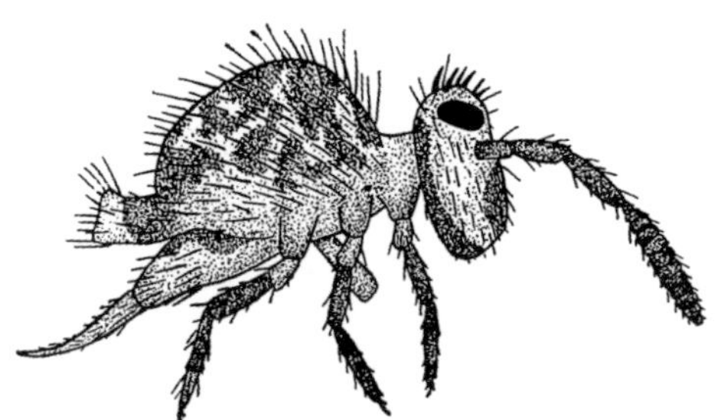

Katianna purpuravirida (Symphypleona, Katiannidae).
From Salmon 1941

Life-history

The life-histories of only a few springtails have been studied in detail anywhere in the world, but there is a general pattern. Most species reproduce sexually, although parthenogenesis is common in soil-inhabiting forms. Sperm transfer is indirect in most families, the males depositing a globular, stalked spermatophore on the ground in which sperm are encysted. In some Symphypleona (Bourletiellidae and Sminthurididae), there may be an elaborate dance in which the male grasps the female by means of the specialised spines and hairs on the antennae, head, or legs and directs her to the spermatophore. There is strong sexual dimorphism in these families and all truly aquatic species belong to this group. Springtails lay their eggs singly or in clusters in protected sites such as in soil, in leaf litter, under stones, or in crevices. Eggs are spherical and pale. In some Symphypleona they are covered with freshly eaten soil mixed with a rectal fluid voided through the anus after oviposition. This covering protects the eggs against dehydration and fungal attack. In other species, eggs are kept free of fungal hyphae by the grazing of adults. The juvenile instars are similar to the adult in general appearance, the only difference being smaller size, lack of genital apparatus, and a reduced arrangement of chaetae (chaetotaxy). Complete adult chaetotaxy and colour gradually develop during the pre-adult instars. There can be from 3 to 13 stages before maturity but four or five stages are common. The complete life-cycle from egg to adult can take (on average) from one to three months but this varies in different species and at different temperatures. Individuals of some species may live for five years, yet in other species, males mate with pre-adult females and then die, living only a few days. Adults continue moulting throughout life, and may undergo up to 60 moults in some species, although usually no further increase in size occurs after about moult 15.

Entomobrya aniwaniwaensis (Arthropleona, Entomobryidae).
From Salmon 1941

Ecology

Springtails are virtually ubiquitous, being found in all biomes including mountain tops, polar regions, and deserts, the only exception being the ocean. Terrestrial springtails are found in a wide variety of usually moist habitats, predominantly in soil and leaf litter and other decomposing materials such as logs and dung. Many species inhabit caves. Others are found on grasses, in flowers, under the bark of trees or in tree canopies. Some aquatic species live exclusively on water surfaces but others live more intimately with the aquatic environment

Number of introduced (naturalised-alien) species per family

Family	New Zealand no. of species	% of total	No. of alien species	% exotic
Brachystomellidae	3	0.8	1	33
Entomobryidae	98	25.8	8	8
Paronellidae	17	4.5	0	0
Hypogastruridae	22	5.8	7	32
Isotomidae	87	22.9	10	11
Neanuridae	41	10.8	2	5
Odontellidae	6	1.6	0	0
Onychiuridae	10	2.6	5	50
Tomoceridae	14	3.7	1	7
Tullbergiidae	7	1.8	1	14
Arrhopalitidae	1	0.3	1	100
Bourlettiellidae	5	1.3	4	80
Dicytomidae	4	1.1	1	25
Katiannidae	51	13.4	1	2
Sminthuridae	6	1.6	1	17
Sminthurididae	4	1.1	3	75
Spinothecidae	1	0.3	0	0
Neelidae	3	0.8	1	33
Onopoduridae	0	0	0	0
Totals	**380**	**100**	**47**	**12**

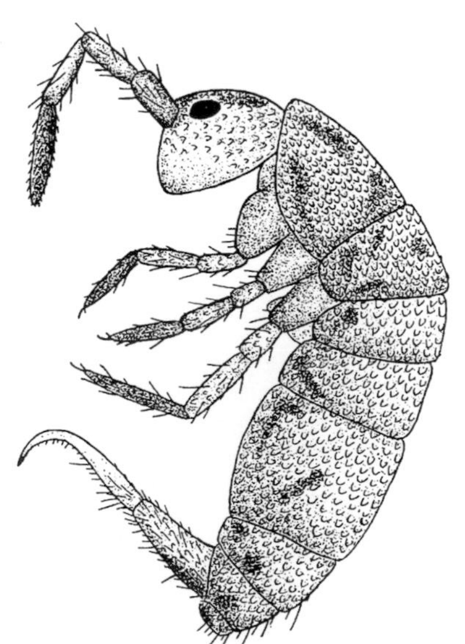

Novacerus spinosus (Arthropleona, Tomoceridae).
From Salmon, 1941

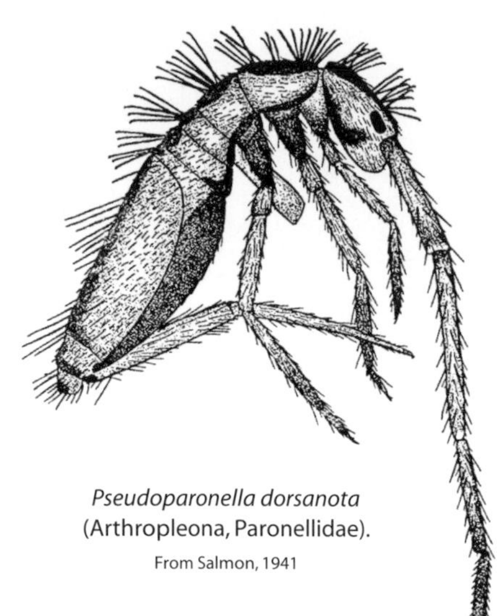

Pseudoparonella dorsanota (Arthropleona, Paronellidae).
From Salmon, 1941

interstitially in sand and under submerged stones and rocks in streams and on the seashore. As the cuticle of springtails is water repellent, being composed of triangular granules that may be fused into larger raised tubercles, rendering it hydrophobic, the animals are able to float on water. Some species seem to be distributed in surface run-off; others disperse on the wind or actively migrate in aggregations.

Most species feed on micro-organisms associated either with the soil around roots (rhizosphere), decomposing organic matter, the water surface, and fungal fruiting bodies, or to a lesser extent with the aerial surfaces of plants. They do not feed readily on sterile leaves. A few species are predatory and feed on small organisms such as rotifers and nematodes; others are saprophytic and pollen is also sometimes a food source. Springtails act as catalysts in the breakdown of organic matter and in the cycling of plant nutrients by grazing on and distributing propagules of micro-organisms and, through their feeding and other activities and the deposition of faecal material, they can alter the physical properties and structure of soils.

Average densities in soils are usually between 10,000 and 30,000 per square metre but can be as high as a million in rare situations. Some factors influencing the distribution and abundance of springtails are the location of food, moisture, and soil pore space. Springtails are also sensitive indicators of disturbances such as chemical pollution, fire, vegetation clearance, tillage, etc., and therefore are of value in environmental assessment.

Many arthropods prey on springtails as do coral-reef fish, birds, small reptiles, and frogs. The predatory arthropods, some of which have evolved elaborate catching devices, include carabid and staphylinid beetles, dacetine ants, Hemiptera, empidid and dolichopodid flies, spiders, harvestmen, pseudoscorpions, centipedes, and prostigmatid and mesostigmatid mites; for example, springtails comprise the main items of diet for bdellid, cunaxid, and anystid mites. Some springtail families, however, notably the Onychiuridae and Neanuridae, seem immune from predators, probably because they contain and sometimes exude a toxic or distasteful chemical. Apart from leaping and cryptic coloration, springtails have a number of defence mechanisms against predation. These include immobility, mimicry, and a spiny dorsum. Viruses, bacteria, fungi, and protozoans as well as nematodes have been found to be parasites of springtails.

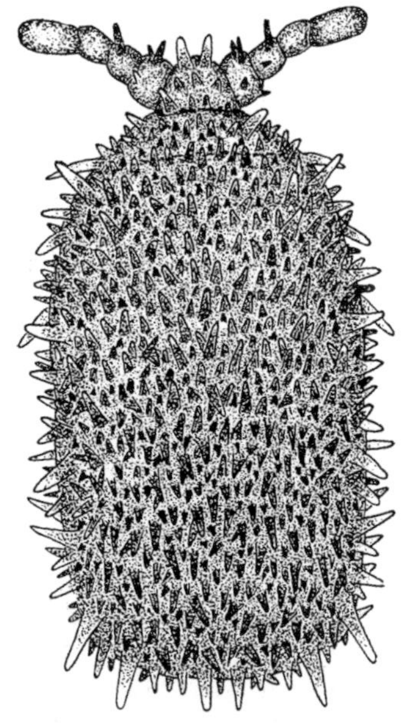

Holacanthella spinosa (Athropleona, Neanuridae).
From Salmon 1941

Characteristics of the New Zealand fauna

Nearly 80% (291) of the species listed here are currently considered endemic to New Zealand and 20% (105) of the genera. This level of endemism is very high and similar to that of Tasmania at the species level. The most highly endemic families are the Neanuridae, Paronellidae, and Katiannidae followed by the Entomobryidae and Isotomidae. Most taxa in the first three families are found in humid undisturbed forest leaf litter and include a large element with affinities with other southern regions such as southern South America and southeastern Australia. It has been shown for Tasmania that the highest level of endemism is found in temperate *Nothofagus* (southern beech) rainforests, on mountaintops, and in caves (Greenslade 1987) and it is to be expected that New Zealand would show similar patterns. However, much of the country has been cleared for agriculture or forestry, both in the North Island and in the eastern plains of the South Island, leaving only very little of the original vegetation. It is likely that suites of locally endemic species that were present before human settlement are now extinct so that a reliable comparison of endemism between vegetation types and regions cannot now be made.

The number of New Zealand species expressed as a percentage of the total world species is 4.8%. Coincidentally, this is the exact same percentage of the world fauna recorded for Australia, however the proportion that some families contribute to the total differs between the northern and southern hemispheres.

The Onychiuridae and Arrhopalitidae contribute a higher proportion of species to the total world fauna than they do in either Australia or New Zealand, three and four times respectively. The latter family is well developed in caves in the northern hemisphere but does not seem to occur in caves in New Zealand or Australia. Conversely the diversity of Katiannidae is greater in Australia (3 times) and New Zealand (10 times) than in the rest of the world. The two southern regions also differ. In Australia, the Brachystomellidae and Bourletiellidae contribute a higher percentage of species to the total than do these two families to the New Zealand fauna, five and seven times more, respectively. Species of both families are tolerant or prefer warmer conditions and possess strategies to survive in low humidities.

There are 47 naturalised-alien species recorded from New Zealand. The proportions of exotic species vary with family. Considering only those families with five or more species, exotics comprise a high proportion of the total in Hypogastruridae (32%), Onychiuridae (50%), and Bourletiellidae (80%). When the fauna of these families is revised, it may be that all Onychiuridae and Bourletiellidae species found in New Zealand are exotic. Alternatively, the Katiannidae, although a species-rich family, has only 2% exotic species recorded.

A genus of considerable interest from a conservation viewpoint is *Holacanthella,* comprising five species (Stevens et al. 2007a,b). All species are large and conspicuous with a dark bluish-black background and profusely covered with small brightly coloured orange, yellow, or red digitations. Specimens are rarely encountered and are confined in habitat to well-rotted logs in old-growth native forest. Phylogenetically they are close to the Australian genera *Acanthanura, Megalanura,* and *Womersleymeria,* also found only in old-growth vegetation, usually forest.

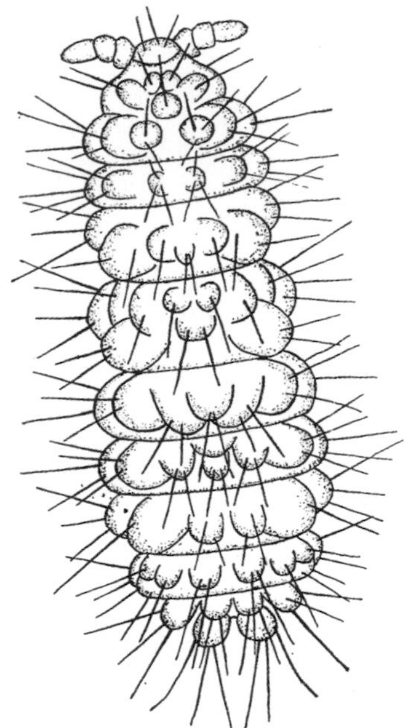

Crossodonthina radiata (Arthropleona, Neanuridae).
Salmon 1941

The publications of the late John Salmon, from 1937 until the mid-1970s, resulted in a very large increase in new genera and species in the New Zealand fauna. As there has been hardly any revision of Salmon's taxa since he first described them, it is likely that a number of his new genera and species are synonyms and that some of his species are incorrectly assigned to genus. For instance, based on the structure of the dens and tibiotarsus, it is likely that *Schoettella subcorta* belongs to the genus *Xenylla*. The genus *Metakatianna* was erected on the basis of an immature specimen of Katianninae and *M. nigraoculata* is now recognised as belonging to *Pseudokatianna*. Based on the figures and original description the genus *Isotomedia* may be synonymous with *Folsomotoma*. Additionally, species he described as endemic to New Zealand in some families, notably the Onychiuridae and Hypogastruridae, are now proving to be well-known cosmopolitan species. As not all of these synonyms and new combinations have been published, the level of endemism shown in the list published in this work may be less than stated. Notwithstanding, as many new species and some new genera remain to be discovered, the difference between the current levels and actual levels may eventually prove to be similar.

Class Diplura

Diplura means 'two tails', referring to the forked sensory structures (cerci) at the end of the body, which in one group are modified as pincers. Hence some Diplura resemble small wingless earwigs while others resemble primitive rove-beetle larvae (called campodeiform because of their resemblance to the dipluran genus *Campodea*). Two-pronged bristletails is another name for the group. Diplura have no compound eyes, only simple ocelli like springtails, but the antennae have many segments, like beads, and are up to three times as long as the head. The mouthparts are enclosed in the head capsule and the body is mainly white. The abdomen is long and slender with 10 segments and most species are 5–10 millimetres long, but some reach 40 millimetres. Diplura (and some stick insects) are the only terrestrial arthropods known to be able to regenerate lost body parts.

Heterojapyx novaezeelandiae.
Alastair Robertson and Maria Minor, Massey University

Legs, antennae, and cerci can be regenerated over the course of several moults.

Diplura inhabit damp sheltered places, mainly in soil and leaf litter, but also under stones and logs, where they feed on other soil dwellers (Collembola, mites, Symphyla, insect larvae, and even other Diplura). Some survive on plant litter and fungal mycelia but most seem to prefer animal prey. Some of the few New Zealand records are from forests (Moeed & Meads 1987a). There are 10 New Zealand species (cf. about 800 worldwide), distributed in three families in the single order Diplura (Tillyard 1924b; Hilton 1939; Pagés 1952; Townsend 1970; Wise 1970a). This diversity compares to only one family in Britain, but species diversity per square kilometre is similar in both countries. The four species of Campodeidae in New Zealand are likely to be herbivores, while the other species are likely to be predatory. An undetermined *Burmjapyx*, which is apparently an undescribed species, has been found in caves in Nelson (McGuiness 2001). The largest New Zealand dipluran is endemic *Heterojapyx novaezeelandiae,* which grows to 36 millimetres long. Townsend (1970) provided a translation of the original description of *H. novaezeelandiae* and commented on variation in the antennae. This species lays eggs in clusters in small cavities in the soil under stones or logs in the Nelson, Marlborough, and Westland districts from sea level (Greymouth) to 900 metres. An undetermined dipluran has been collected from the Poor Knights Islands (Watt 1982).

Class Insecta

What has made the insects so successful? There are more kinds of insects than any other group of multicellular organisms. Exact numbers are uncertain, but insects (around 909,000 species) account for about half of all named species of life. Moreover, among the insects some groups are much more diverse than the rest, including the ants, bees, and wasps (Hymenoptera) with ca. 110,000 species, flies (Diptera) ca. 120,000 species, moths and butterflies (Lepidoptera) ca. 170 000 species, and beetles (Coleoptera) ca. 360,000 species. Among insects, these groups have in common wings that flex and life cycles with complete metamorphosis, and these features have been implicated in their diversification along with parallel evolution with flowering plants. Famously, but possibly apocryphally, the evolutionary biologist J. B. S. Haldane was moved to quip that the Creator had an 'inordinate fondness for beetles', but it would be more accurate to include other winged insects in this perception – while beetles may have diversified faster than some sister lineages from their time of origin, they do not stand out relative to other closely related branches of the evolutionary tree (Mayhew 2002).

Following the classification used in the *Nomina Insecta Nearctica* for the insects of North America, and based on modern morphological and molecular studies, the Insecta is divided into two subclasses, the Archaeognatha, with a single order of the same name, and the Dicondylia, comprising all the remaining orders. The Archaeognatha ('ancient jaws') have several primitive features, one of which is that their mandibles are monocondylic, that is, with only one condyle articulating with the head capsule. All other insects, including the Thysanura (silverfish) with which they were previously included, have two condyles (dicondylic). Alternative classifications include the Archaeognatha and Thysanura in the subclass Pterygota (the primitively wingless insects), a group that used also to include the Protura, Collembola, and Diplura.

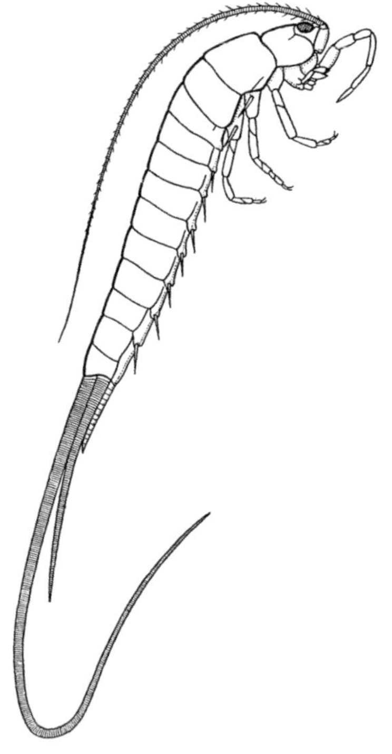

Nesomachilus maorica.
From Tillyard 1924

Subclass Archaeognatha

Order Archaeognatha: Bristletails

This small but ancient group of insects, sometimes called Microcoryphia ('small heads'), comprises about 350 described species worldwide, but there are only two in New Zealand. Based on Museum of New Zealand (Te Papa) collection records, *Nesomachilis maorica* (Tillyard 1924a) is found on Cuvier Island (Wygodzinsky 1948; Sturm 1980), the Poor Knights Islands (Watt 1982), and indeed most of the islands off the Northland east coast that have been checked for soil insects (Moeed & Meads 1987b). A second species, *N. novaezelandiae*, was described seven decades later (Mendes et al. 1994)

Bristletails resemble thysanurans with eyes, except for the simple, small, paired appendages known as styli under each abdominal segment. In addition, at the tip of the abdomen, the three cerci are parallel, unlike the splayed cerci of Thysanura. In *N. maorica*, all three cerci are directed backwards, with the inner one distinctly longer than outer pair. The compound eyes are prominent and meet near the middle – another bristletail distinctive. The antennae are directed backwards and are long and thin with many segments. The legs have three tarsal segments. Bristletails are omnivores and nocturnal, feeding on algae, lichens, vegetable debris from rock crevices, litter, and bark. When disturbed, they can jump, thanks to specialised muscles in the abdomen that snap it against the ground. It is this capacity that gives these creatures an alternate common name – jumping bristletails. A good source of information on the order is that of Sturm and Bach (1993), who revised generic concepts and gave an overview of ecological variation.

Subclass Dicondylia

Order Thysanura: Silverfish

Silverfish are well known from the introduced house pest *Lepisma saccharina* that feeds on books and clothing. The order includes about 200 species worldwide but should be split into two separate orders, Zygentoma and Monura, according to Larink (1997). These wingless insects are easily recognisable, with three cerci at the end of the abdomen that are long, thin and near-equal in length. The outer pair of cerci is splayed out, sometimes almost horizontally. Compared to bristletails, silverfish eyes are small or absent. The antennae are long and thin, with many segments, and pointed forward. The elongate body (up to 15 millimetres long) may have scales with short appendages under the last three abdominal segments. The legs have 2–4 tarsal segments. Silverfish are agile runners. Generally, they are scavengers, living under bark, in litter, among undisturbed paper, and in cupboards with dried food.

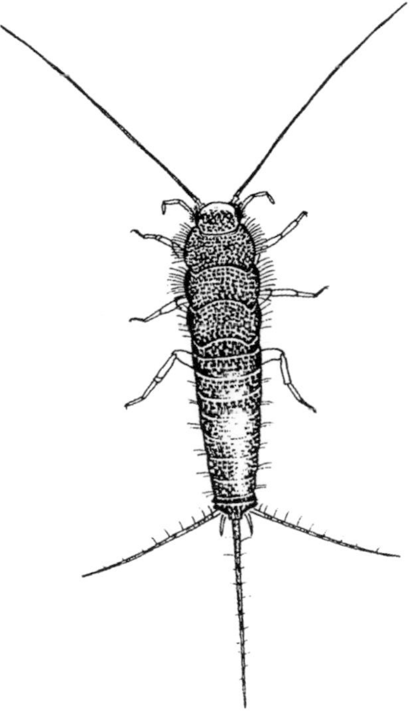

Lepisma saccharina.
From Grant 1999

The New Zealand fauna includes 10 or 11 species, not all named. Members of the family Lepismatidae have eyes, the scales can be dark, and the legs have only 3–4 tarsal segments. This family includes both of the accidentally introduced species, *Lepisma saccharina* from Europe and *Ctenolepisma longicauda* from South Africa. According to Scott (1984), *L. saccharina* is mainly found either in undisturbed piles of paper and tins with food in kitchens. These silverfish begin to lay eggs at the ninth or tenth instar (a stage between moults) and can live for over 2.5 years and undergo 30 moults. Some of the later instars undergo arrested development (diapause) at 25° C and below but there is no diapause at 28° C (Nishizaka et al. 1998).

Tillyard (1924b) described a native lepismatid species, *Heterolepisma zelandica*. There are apparently seven or eight undescribed species in several genera in the families Ateluridae and Tricholepidiidae (as Lepidotrichidae) (Wise 1970a). Species in these families lack eyes and scales, are white or yellow, and the tarsi have five segments, so they can be readily distinguished from the described species. An undetermined silverfish has been collected from the Poor Knights Islands (Watt 1982). The status of all these forms needs proper determination.

Order Ephemeroptera: Mayflies

The mayflies seen today in running waters throughout New Zealand are representatives of the most primitive winged insects still in existence. Their ancestors

Nesameletus ornatus adult male (imago).
W. J. Crawford

first appeared as fossils in the Carboniferous over 300 million years ago (Hubbard 1987). Mayflies comprise a relatively small portion of the global fauna of aquatic insects but are almost cosmopolitan, with species on all continents except Antarctica and many islands except those that are truly oceanic.

Naturalists have long been interested in mayflies because of their brief adult life – hence the ordinal name 'ephemeral wings'. The immature stages of mayflies are fully aquatic nymphs, usually inhabiting relatively unpolluted standing and running waters where they may remain for up to three years. Nymphs of most species are herbivores but a few are filter-feeders or predators (Peters & Campbell 1991). Once mayflies enter the terrestrial environment, however, they stop feeding; the two terrestrial stages, the subimago and imago, are essentially for reproduction and dispersal. For most species, the terrestrial stages last little more than 48 hours, but for some it is much shorter. In some North American species the female never reaches the adult state at all, moulting from nymph to subimago, mating, laying its eggs, and dying within the space of a few hours (Edmunds et al. 1976).

Because mayflies usually inhabit unpolluted waters, they have frequently been used as indicators of water quality (Williams 1980). In some circumstances, mayflies respond positively to nutrient enrichment of aquatic systems. In the USA, mayflies such as *Hexagenia* became so abundant following enrichment of the Mississippi River, the emerging adults blocked bridges and had to be removed using snowplows (Edmunds et al. 1976). This sensitivity to nutrient levels, dissolved-oxygen concentrations, and other chemical and physical attributes of fresh water, has led to the development of biotic indices, where combinations of species are used to rank water quality (Williams 1980). Such indices are now applied in many countries, including New Zealand (e.g. Stark 1993, 1998).

The value and use of mayflies to assess changes in aquatic systems is a relatively recent development. There is a longer history of interest in the role of mayflies as food for sports fish (especially salmonids), and a huge scientific and popular literature developed around the use of artificial baits to imitate aquatic insects such as mayflies following the first documented account in 1496: 'The Treatyse of Fysshynge wyth an Angle' by Dame Juliana Berners (see Williams 1980). This interest has produced a distinctive vocabulary based around stages of the life cycle of mayflies and the lures used to imitate them. The aquatic stages of nymph or larva and emerging subimago are thus imitated by the 'wet fly', and the terrestrial stages of subimago and imago are imitated by 'dry flies', referred to respectively as the dun and spinner. Local accounts are now available that recommend the construction and use of flies specifically to imitate New Zealand species of mayflies (e.g. Draper 1997).

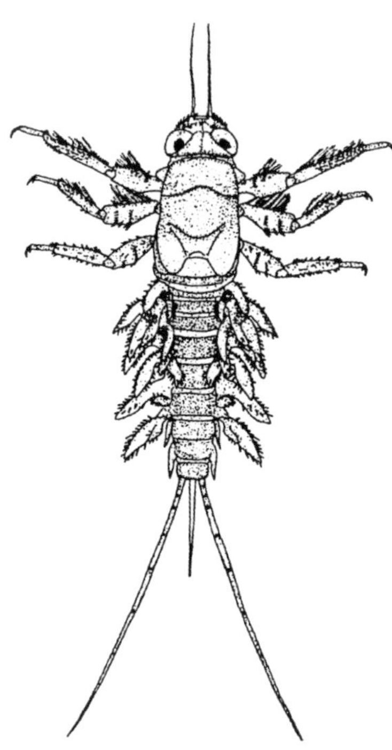

Coloburiscus humeralis larva.
Craig Dolphin, from Winterbourn et al. 2006

Historical overview of studies on mayfly diversity

The history of descriptions of selected groups of mayflies in New Zealand was outlined by Towns and Peters (1996) and Peters (2001). The first mayflies were described from New Zealand by Walker (1853) and are now known as *Coloburiscus humeralis* and *Neozephlebia scita*. With these two species there began a series of descriptions based largely on specimens collected, dried, and pinned in New Zealand and sent to recognised authorities in Britain. By the end of the 19th century, at least 11 species had been described in this way (Eaton 1899). But the use of pinned adult specimens, the distortions of male genitalia during drying, the limited series of specimens, and communication difficulties all contributed to subsequent identification problems, sometimes resulting in descriptions of the same species under different names. Some of these problems have yet to be resolved (see below). The first comprehensive account of the fauna added to this confusion when two of the species, *Deleatidium lillii* and *Neozephlebia scita*, were misidentified (Lillie 1898).

At the beginning of the 20th century, naturalists resident in New Zealand began collecting and describing new species of mayflies. The known fauna grad-

Families and species in mayfly faunas of New Zealand, Australia, and New Caledonia

Family	New Zealand	Australia[1]	New Caledonia[2]
Ameletopsidae	1	3	0
Baetidae	0	13	1?
Caenidae	0	5	0
Coloburiscidae	1	3	0
Ephemeridae	2	0	0
Ephemerellidae	0	1	0
Leptophlebiidae	35	54	46
Nesameletidae	6	1	0
Oniscigastridae	3	3	0
Prosopistomatidae	0	1	0
Rallidentidae	2	0	0
Siphlaenigmatidae[3]	1	0	0
Total endemic spp.	**51**	**84**	**46**

1. Peters and Campbell (1991)
2. Peters (2001)
3. Claims of Siphlaenigmatidae in Australia by Lugo-Ortiz and McCafferty (1998) have since been traced to curatorial errors (McCafferty 1999)

ually increased following work by Hudson (1904), Tillyard (1923c), and Phillips (1930b). The usefulness of this work was compromised, however, because neither Hudson nor Phillips identified specific type localities or nominated type series in recognised institutions. Nevertheless, for several decades the revision provided by Phillips (1930a, b) was the most comprehensive account of the mayfly fauna, recognising 22 species in what were then three families. By the middle of last century, the first ecological studies on New Zealand streams were under way (e.g. Allen 1951), especially when it was discovered that immature mayflies were amongst the most abundant organisms in some running waters. Biologists were hampered, however, by a lack of reference material – types were either held in Britain or had never been deposited. The confusion over identity prevailed within some genera until fairly recently, leading Winterbourn (1977) to describe the situation as a 'taxonomists' nightmare'.

The situation improved with the descriptions of new families, genera, and species by Penniket (1961, 1962a, b, 1966) and the description of *Atalophlebioides aucklandensis* by Peters (1971). The latter was the first comprehensive description of nymphs and adults to be accompanied by accurate, high-quality diagnostic illustrations for any species in the New Zealand Leptophlebiidae. A subsequent review of New Zealand aquatic insects by Wise (1973a) listed a mayfly fauna of 27 species in 10 genera and four families, representing a 23% increase in the number of species since 1930. The end-chapter checklist of Epheremoptera provided here is the first comprehensive update since Wise's list and represents an 89% increase in the number of species since 1973.

Most modern taxonomists have included New Zealand collections as repositories for part of the type series of species descriptions. The most comprehensive of these collections is the New Zealand Arthropod Collection held by Landcare Research in Auckland. Earlier workers were more likely to submit collections to the provincial museums and there are useful collections in the Auckland Institute and Museum, National Museum of New Zealand (Te Papa), and Canterbury Museum. Although Phillips collected mostly around Wellington, only a few specimens attributable to him are in the National Museum collections (Towns & Peters 1978); most of his material was found by T. Hitchings in the Canterbury Museum (Towns & Peters 1996) and some appears to have reached the Natural History Museum, London (Towns 1983a). This museum

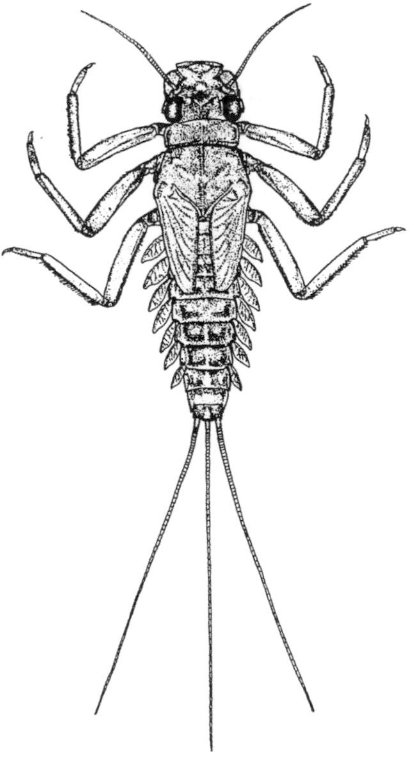

Deleatidium sp. larva.
Craig Dolphin, from Winterbourn et al. 2006

remains the repository of much of the material described in the late 19th century. Some material has also been deposited at Bernice P. Bishop Museum, Honolulu, Hawaii. Probably the largest overseas collection is held in the Entomology Department of Florida Agricultural & Mechanical University, Tallahassee, as part of a comparative collection of Ephemeroptera of the world. This collection also includes some reference material (especially Leptophlebiidae) previously deposited at the University of Utah, Salt Lake City. Reference collections of Siphlaenigmatidae, previously held at the University of Utah, are now held at the Entomology Department of Purdue University, West Lafayette, Indiana.

Species diversity, endemism, and biogeography of mayflies

A revision of the family-level classification of Ephemeroptera (Kluge et al. 1995) and revisions of the New Zealand Leptophlebiidae (Towns & Peters 1978, 1979a, b, 1996; Towns 1983a) reveal a fauna of at least 52 species in 20 genera and eight families. This fauna is distinctive in two ways – high levels of endemism and patchy levels of radiation within genera and families. All genera are endemic to New Zealand, but there are also two endemic families, Rallidentidae and Siphlaenigmatidae. The latter family is of great scientific interest as it shares features with both an ancient group like the Siphlonuridae and the more modern Baetidae (Penniket 1962b). Although the Baetidae are almost cosmopolitan in distribution, they did not reach New Zealand. Despite high distinctiveness within families, few of them have radiated into complexes of genera or species. Rather, six families comprise only a single genus with one to three species. In a few cases (e.g. *Coloburiscus tonnoiri* and *Oniscigaster intermedius*) species names may be synonymised. In contrast, two families have shown moderate to high levels of radiation, with more species present than had been realised. The Nesameletidae, currently under revision, has at least six species in one genus (Hitchings & Staniczek 2003) and the Leptophlebiidae has at least 35 species in 13 genera, the latter thus comprising almost 70% of the species and 65% of the genera in the fauna.

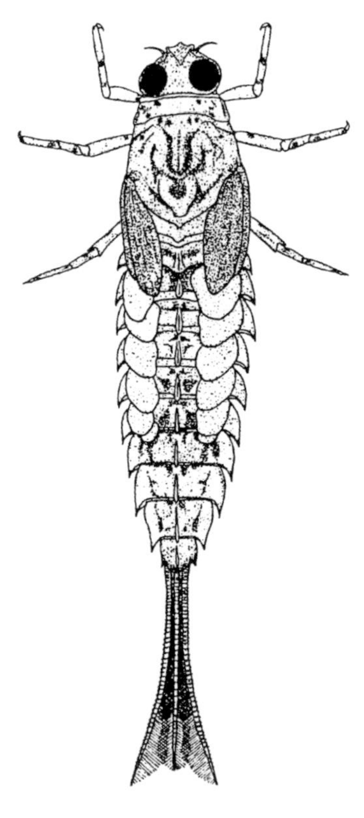

Oniscigaster distans larva.
Craig Dolphin, from Winterbourn et al. 2006

New Zealand and Australia have a similar number of mayfly families, but there are qualitative differences between them. New Zealand lacks modern families such as Baetidae, Caenidae, and Ephemerellidae and the tropical Prosopistomatidae. On the other hand, Australia lacks the apparently ancient families Rallidentidae and Siphlaenigmatidae. A particularly puzzling absence from the Australian fauna is the Ephemeridae, a family present in New Zealand, Madagascar, and all of the major continents (Edmunds et al. 1976). Like New Zealand, Australia shows low levels of species radiation within most mayfly families along with contrasting high levels in a few families. Most (64%) of Australian species are also leptophlebiids.

The most extreme predominance of Leptophlebiidae is found in New Caledonia, where the entire mayfly fauna, wholly endemic, belongs to this one family. Of the 19 genera known, 18 have close evolutionary relationships with New Zealand, the remaining one being most closely related to a genus in Madagascar (Peters 2001). In New Zealand, six evolutionary lineages of genera have been recognised (Towns & Peters 1996), four with close links to New Caledonia, and all six with links throughout much of the Southern Hemisphere. They are: *Deleatidium* plus *Atalophlebioides*, with other representatives in South America, Australia, Celebes, Sri Lanka, southern India and Madagascar; *Austroclima* and *Mauiulus*, with other representatives in South America; *Arachnocolus*, *Zephlebia*, and *Austronella*, with other representatives in South America and New Caledonia; an enigmatic group comprising *Isothraulus* and *Tepakia*, having affinities with *Arachnocolus* and its relatives as well as a tropical genus in Australia and genera in Madagascar and the Seychelles; *Neozephlebia*, represented also by genera in New Caledonia; and *Acanthophlebia*, as part of a lineage with representatives in Africa, South America, Australia and New Caledonia.

Sister-group relationships that link Chile and Australia with New Zealand were also demonstrated by Edmunds (1975) for the families Coloburiscidae, Oniscigastridae, Ameletopsidae, and Nesameletidae using nymphal morphology. These analyses indicate that the New Zealand fauna occupies a unique position when compared with the mayflies of Australia, South America, and New Caledonia. On the one hand, New Zealand has retained ancient families not represented elsewhere; on the other, there are families and genera that indicate very ancient linkages to Australia and Chile. In addition, some Leptophlebiidae show close taxonomic affinities with New Caledonia, indicating relatively recent past direct links to the north. These relationships between the families and genera of New Zealand mayflies and those elsewhere in the Southern Hemisphere loosely correlate with proposed sequences of the fragmentation of Gondwana (cf. Edmunds 1975). Some authors have also suggested that many New Zealand forms may have evolved more slowly than related groups elsewhere (Edmunds 1975; Lugo-Ortiz & McCafferty 1998). Consequently, whereas the fauna of New Caledonia has lost all but one family – and that of Australia has gained more recent and tropical elements – we conclude that the mayflies of New Zealand provide the most intact representation of a Cretaceous Gondwanan aquatic biota.

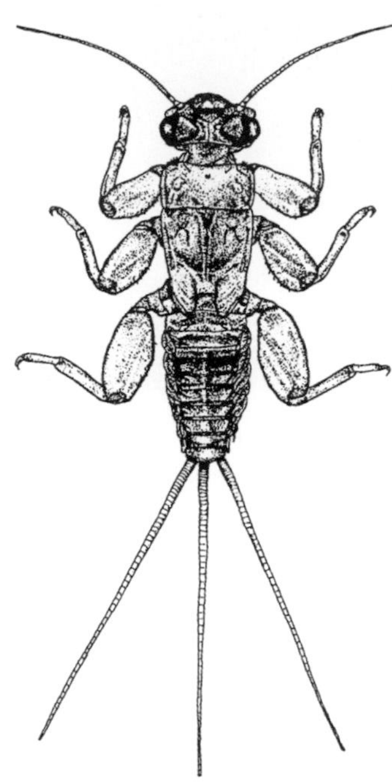

Austroclima sepia larva.
Craig Dolphin, from Winterbourn et al. 2006

Distribution and ecology

New Zealand mayflies show particularly close ties to running water; no species are primarily inhabitants of ponds, lakes, or slow-flowing rivers. Even where families or genera have strong representation in standing waters overseas, in New Zealand they inhabit streams, e.g. *Acanthophlebia cruentata* (Leptophlebiidae) and *Ichthybotus* spp. (Ephemeridae). The few species present in New Zealand lakes are found either near the mouths of streams or on lake margins where there is frequent wave action, which, presumably, maintains high oxygen concentration. Perhaps related to the predilection for running water, mayfly faunas on offshore islands are greatly reduced where permanent flowing water is limited (Towns 1987). Even where it is present, the fauna is usually small on all but the largest islands. For example, Riddell (1981) found only three species (all Leptophlebiidae) in a permanent stream on Cuvier Island (170 ha) but Watson (1972) was unable to find any in apparently permanent streams on Red Mercury Island (225 ha). Further afield, no mayflies have reached (or survived on) the Chatham Islands, and a single unique genus and species of Leptophlebiidae, *Cryophlebia aucklandensis,* inhabits streams on the subantarctic Auckland Islands.

Most New Zealand mayflies are thus confined to the two main (North and South), islands. Only 10 species in four families have been recorded on Stewart Island – Ameletopsidae, Coloburiscidae, Leptophlebiidae, and Nesameletidae, (Hitchings 2001). This low total may reflect the level of collecting effort, but Towns (1987) found a similarly low family diversity (five families) on Great Barrier Island, even after extensive sampling. Within the two main islands, several genera and species have enigmatic, patchy, or restricted distributions and a few genera have separate North Island and South Island representatives. Examples include a burrowing mayfly with one species, *Ichthybotus hudsoni,* in the North Island, and a second, *I. bicolor,* in the South Island. Several genera of leptophlebiids are confined to the North Island. For example, the distinctive orange mayfly *Acanthophlebia cruentata* is widespread through lowland North Island, is present in several streams on Great Barrier Island and southern North Island, but has not been found in the South Island (Towns 1987). *Deleatidium* also has separate North Island and South Island species. The patchy distribution of several other species in the North Island may reflect availability of specific habitats or variable collecting effort; if the latter, then some may yet be found to be more widely distributed.

A distinctive feature of the mayfly faunas of the North Island is the particularly high species diversity in some forested streams. The highest diversity so far recorded (28 species in the Waitakere River catchment) is more than twice

Subadult of *Zephlebia* sp..
Stephen Moore, Landcare Research

the number of species recorded in the entire South Island (Towns 1987 and references therein). Furthermore, northern North Island sites are dominated by species of Leptophlebiidae (19–21 species), which is an extraordinarily high diversity anywhere (Towns 1987). Although many mayflies appear associated with well-oxygenated waters (Collier 1994), the most species-rich faunas in northern New Zealand are not necessarily where water velocity (and hence dissolved oxygen levels) are highest. Instead, small, forested headwater streams of third order or less often support the most species-rich mayfly faunas (Towns 1987).

Where families are shared with southern South America and Australia (Peters & Campbell 1991; Edmunds 1975), the New Zealand species include highly mobile predators (Ameletopsidae), filter-feeders (Coloburiscidae), surface-feeders capable of burrowing (Oniscigastridae), free-swimming species that graze substratum surfaces (Nesameletidae), and sprawling species that graze the surfaces of rock, wood, leaves, and plants growing in water (Leptophlebiidae). Unlike Australia, New Zealand has Ephemeridae that burrow in stream sediments.

The diversity of leptophlebiid genera is a consequence of adaptive radiation into specific habitats, reflected, for example, in gill morphology. For example, the fringed highly tracheated gills of *Isothraulus abditus* may enable nymphs to inhabit pools in small headwater streams, whereas the single greatly expanded gills of some species of *Deleatidium* allow them to inhabit torrential waters in steep mountain streams. In some genera, body shape and gill morphology can vary greatly. This is particularly well demonstrated in *Deleatidium* – *D. myzobranchia*, which inhabits very fast-flowing waters, has compressed limbs, hairs on the ventral abdomen, and expanded gills to aid in adhesion, whereas *D. cerinum* is less compressed, has leaf-shaped gills, and occupies pools and slack water (Collier 1994; Towns & Peters 1996).

One of the most remarkable aspects of New Zealand leptophlebiids is the number of species in the same genus that can coexist in the same stream. For example, five species of *Deleatidium* and at least seven species of *Zephlebia* live together in streams in northern New Zealand, a degree of congeneric overlap that is rarely matched elsewhere in the world. How this overlap is maintained remains unclear but previously unsuspected ecological divergence into specific flow regimes and habitats may provide part of the explanation (Towns 1987). Perhaps this was a consequence of the poorly synchronised life-cycles shown by many New Zealand species. In northern hemisphere mayflies, closely related species often have staggered hatching and development periods, presumably in response to limited resources that are used sequentially. In contrast, leptophlebiids in one northern New Zealand stream have life-cycles ranging from univoltine (one generation per year) to bivoltine (two per year), as well as overlapping generations and cohorts and a high degree of habitat and life-history overlap (Towns 1983b). The role of competition in determining life-history strategies needs re-examining, and the weakly seasonal and poorly synchronised life histories found in some New Zealand mayflies also suggest that non-biological effects may play a larger role in influencing life-histories than was assumed in the past (Towns 1983b).

Deleatidium myzobranchia larva.
Stephen Moore, Landcare Research

Biological interactions, conservation, and management

Mayflies are significant components of freshwater food webs, forming part of the diet of a range of invertebrates, including stoneflies, caddisflies, freshwater crayfish, some fly larvae, and even a predatory mayfly (*Ameletopsis perscitus*) (McIntosh 2000). They are also a major part of the diet of native vertebrates; for example *Deleatidium* species are important food for bullies, torrentfish, and eels (McIntosh 2000) and species of *Deleatidium*, *Nesameletus*, and *Coloburiscus* dominate the diet of some whitebait (*Galaxias*) populations (Cadwallader 1975). The distinctive birds that inhabit braided rivers in the South Island – endemic black-fronted tern (*Sterna albostriata*), wrybill (*Anarhynchus frontalis*), and black stilt (*Himantopus novaezelandiae*) – all feed on mayflies, especially nymphs of *Deleatidium* (Robertson et al. 1983; Pierce 1986) and mayflies form part of the diet of an endemic torrent duck, the blue duck *Hymenolaimus malacorhynchos* (Veltman et al. 1995). Studies of primary production in streams in the South Island have demonstrated that grazing by invertebrates (including *Nesameletus* and *Deleatidium* mayflies) in streams lacking introduced trout is sufficient to suppress the standing crop of algae on stone surfaces. Once trout are introduced, the density of *Nesameletus* and *Deleatidium* nymphs foraging on stone surfaces is reduced and algae increase (McIntosh & Townsend 1996). Interestingly, this control on algae was not observed in streams containing native galaxiid fish (Huryn 1998).

Observations like these have conservation implications, not only for mayflies but also for their natural predators. Three species of *Galaxias* are listed amongst the rarest New Zealand vertebrates, as are black-fronted tern, wrybill, and black stilt (Molloy & Davis 1994). At least six species of mayflies are found in no more than three Ecological Regions, and are therefore regarded as highly restricted in distribution (Collier et al. 2000). There are thus three levels of concern. The first includes subtle ecosystem effects, such as trophic shifts resulting from the presence of trout, which may cause reductions in the diversity and density of mayflies in streams. The second level of concern includes reductions of species diversity caused through predation by introduced organisms. For both examples, streams may show little external evidence of modification, but ecological processes may be sufficiently altered to cause detrimental effects throughout a food web. The third level of concern includes extreme modification of stream channels through removal of riparian vegetation, channelisation, water abstraction, siltation, or pollution by substances that are either directly or indirectly toxic (Collier et al. 2000).

Extreme modification of the riparian vegetation of stream catchments is of particular concern, probably because it is so widespread, especially in lowland areas. Catchment modification can affect water temperature and this in turn may influence the distribution of particular species. Two species of leptophlebiid mayfly were the most temperature-sensitive of a range of aquatic species tested by Quinn et al. (1994), who suggested that elevated stream temperatures will very likely limit such species in the North Island, where many species – even in unmodified streams – are naturally near the upper limits of their thermal tolerance. Other studies have shown that removal of shade by deforestation results in reduced abundances of a range of mayfly species, apparently because of changes (some with combined effects) in temperature, suspended solids, suspended inorganic sediment, woody debris, periphyton (attached-algal) biomass and productivity, and the availability of leaf litter (Quinn et al. 1997). Furthermore, since the most species-rich faunas are often in headwater streams, removal of forest cover from headwaters could conceivably reduce mayfly diversity through entire catchments, even where native forest cover is retained downstream of the headwaters. Consequently, changes to stream systems, either subtly or through extreme modification, may have reinforced the isolation of several species of mayflies that now appear to exist in localised relict populations. Examples include *Austronella planulata* and *Tepakia caligata*, in scattered localities from Northland to Wellington, and *Deleatidium magnum*, in a few locations around the central North Island (Towns & Peters 1996). Collier (1993) has advocated the protection of rare and representative aquatic habitats as a means of maintaining species and genetic diversity of aquatic invertebrates. Perhaps it is necessary to go beyond this, and develop methods of restoring stream systems in order to re-establish connectivity between isolated catchments (e.g. Collier et al. 2000). Whether isolated populations of rare mayflies are capable of reinvading restored stream systems, however, remains unclear.

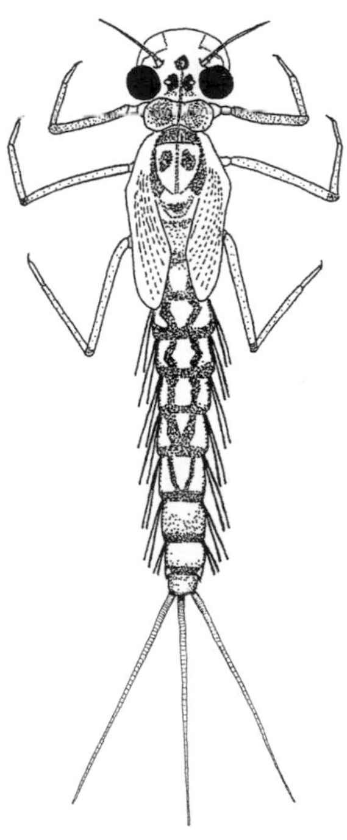

Aupouriella pohei.
From Winterbourn 2009

The effects of introduced salmonid fish on the species richness of mayfly assemblages in New Zealand streams are still unclear (McIntosh 2000). Early last century, Tillyard (1920) was of the opinion that in the 'hot springs region' (the lakes and streams around Rotorua) 'the largest May-flies, which form the very finest possible food for trout, have been practically exterminated, while the smaller forms have been reduced, at a moderate estimate, by over 50%.' This contention has yet to be tested. Approaches similar to those applied in terrestrial systems where introduced predators have been controlled locally (Saunders & Norton 2001), may well prove instructive in aquatic systems. There may also be subtle effects on mayflies in stream systems modified by species other than trout. For example, insectivorous birds feed on emerging mayflies (T. Hitchings pers. obs.), but whether the birds' effects are modified by changes in the extent of riparian forest cover remains unknown.

Order Odonata: Dragonflies and damselflies

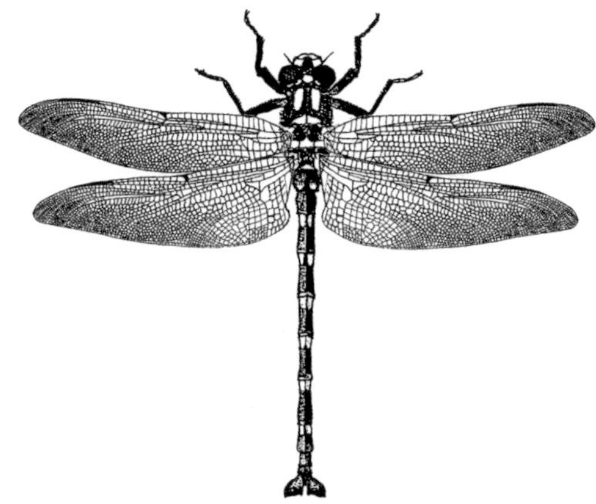

Male giant dragonfly *Uropetala carovei*

Modified after Grant 1999

Dragonflies are among the largest and most agile insect fliers in New Zealand. They are instantly recognisable from their long abdomen and two pairs of wings held at right angles to the body. The head has large eyes but the antennae are short, ending in a fine bristle. Other features include short appendages (cerci) at the end of the abdomen, a stigma (darkened area) on the front outer margin of each wing, and spiny legs (Rowe 1987). In New Zealand the order is represented by two suborders. The generally smaller damselflies (suborder Zygoptera) have widely separated eyes and wings of similar size with a narrow base, while dragonflies (suborder Anisoptera) have larger hind wings with a wide base and the eyes usually touch at the top (except for *Uropetala*). The bodies of seven of the smaller species (damselflies and the smaller dragonflies) are brightly coloured with red, blue, or green. In contrast, the larger dragonflies (41–86 millimetres long) have a contrasting pattern with yellow and dark-black marks or bands.

The diversity of species of Odonata in New Zealand (15, plus two non-breeding migrants) is low compared with Tasmania and Britain, with 26 species, i.e. seven times more species per square kilometre (Watson et al. 1991), and 45 species, i.e. two and a half times more species per square kilometre (Kloet & Hinks 1964), respectively. Nevertheless, what New Zealand lacks in diversity it makes up for in distinctiveness – three genera and 10 species are endemic. Rowe (1987) mapped known species distribution, which showed a lack of identified specimens from northeastern North Island and most of Marlborough. Rowe (1987) recorded *Uropetala chiltoni, Xanthocnemis sinclairi and X. zealandica, Austrolestes colensonis,* and *Procordulia smithii* and *P. grayi* from higher inland locations between altitudes of 600 and 900 metres (e.g. Lake Waikaremoana, the Hermitage) and noted that *X. zealandica* uses alpine tarns. Macfarlane (unpubl.) has noted both *U. chiltoni* up to 800–900 metres and *X. zealandica* above the bushline in subalpine snow tussock based in the Omarama saddle and Mt Dun areas. *Xanthocnemis zealandica* females were observed ovipositing in springs above 1550 metres and flying at the top of the Whether/Dunstan Range (1750 metres). *Xanthocnemis sinclairi* is known only from the upper Rakaia River at around 1250 metres.

Procordulia grayi

Stephen Moore, Landcare Research

The slender red, greenish (22–41 millimetres long), and blue (40–47 millimetres long) damselflies are represented by two families (Coenagrionidae, Lestidae), three genera, and six species. *Xanthocnemis sinclairi* is confined to a few known upland sites in snow tussock in Canterbury, *X. sobrina* to the Auckland province, and *X. tuanuii* to the Chatham islands. The self-introduced and smallest (22–33 millimetres long) gossamer damselfly *Ischnura aurora* is still confined to the North Island, so damselfly species diversity is limited over much of New Zealand. Conversely, *X. zealandica, Austrolestes colensonis,* and the dragonfly *Procordulia smithii* inhabit Stewart Island (Rowe 1987).

The stouter and mainly longer (41–86 millimetres length) dragonflies are represented by nine resident species in seven genera and four families. In addition, *Pantala flavescens* and *Tramea transmarina* are migratory dragonfly species

(Libellulidae), which are periodically recorded in the North Island and the west coast of the South Island (Rowe 1987). The other eight resident dragonfly species have black with yellow patterns (*Uropetala, Procordulia, Aeshna brevistyla*), greenish markings *Hemianax papuensis*, or dark with few markings *Antipodochlora braueri*. Their larger size allows them quickly to be distinguished from damselflies even in flight. Only *Diplacodes bipunctata* (smaller, at 29–34 millimetres length) is mainly red among the resident mainland dragonfly species. It is apparently confined to western New Zealand, where it is commoner in warmer areas. *Uropetala chiltoni* is the only dragonfly confined to the South Island on the east coast. Only *Procordulia smithii* extends to Stewart Island among the eight South Island species. A species of *Uropetala*, possibly *U. chiltoni*, has been reported from Banks Peninsula (Ward et al. 1999), which extends the known distribution of *Uropetela* (Rowe 1987). Generally, it appears these large dragonflies are unable to disperse and freely colonise above distances of 60 kilometres. *Antipodochlora braueri* and the recent immigrant *Hemianax papuensis* are confined to the North Island.

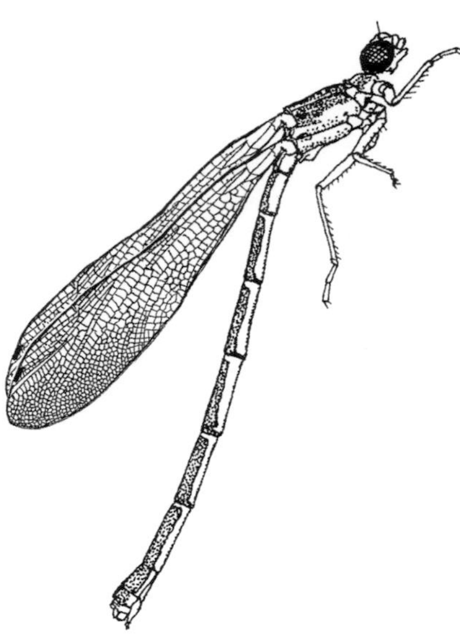

Blue damselfly *Austrolestes colensonis*.
Modified after Grant 1999

On offshore islands, *Tramea transmarina* resides in the Kermadec Islands and is only a migrant to the North Island. Wise (1983b) summarised northern records while Wilkinson and Wilkinson (1952) recorded four species, with records from Kapiti Island, while Moeed and Meads (1987b) reported two species (one undetermined) from Red Mercury Island. Thus *Ischnura aurora* has been seen at the Poor Knights, Cuvier, Mercury, and Kapiti Islands, *X. zealandica* and *A. colensonis* at Little Barrier, Great Barrier, and Kapiti Islands, and *X. zealandica* at Red Mercury and Mayor Islands. *Uropetala carovei* is present on Kapiti Island (Wilkinson & Wilkinson 1952). It is not always clear if these records represent breeding populations, so more critical observations are needed on the distribution of Odonata on offshore islands to understand the dispersal and colonisation ability of each species. The lack of apparent resident species on the Alderman Islands (Early 1995) and perhaps also the Poor Knights (Watt 1982) could be because seasonal stability and quantity of freshwater resources on these smaller islands is insufficient to allow for colonisation. At least *X. zealandica* has colonised the more substantial area of lakes and swamp on Mayor Island (Bayly et al. 1956).

Habitats and biological associations

Both nymphs and adults prey on insects and other small animals, and Rowe (1987) has compiled details on the limited records of species preyed on by the commoner species. His monograph also gives details about habitat, emergence, and other aspects of ecology as well as behavioural habits such as mating, oviposition, and flight. The nymphs of most species live in still to slowly running water (Rowe 1987; Winterbourn et al. 2000) with some apparently subtle differences in use of different vegetation. Among resident New Zealand species, only *Procordulia smithii* and the commonest damselfly, *Xanthocnemis zealandica*, may be found in slow-flowing streams among vegetation or in muddy sediments, but *P. smithii* mainly uses raupo (*Typha*) areas and ponds, while *X. zealandica* uses rush-sedge bases for shelter. The two *Uropetala* dragonfly species are associated with damp ground near bogs and seepages, either shaded by forest in the case of *U. carovei* or in more open sites for *U. chiltoni* (Rowe 1987), including snow-tussock grassland. Conversely, only the North Island dusk dragonfly *Antipodochlora braueri* inhabits stony streambeds. The limited occupation of Odonata in running fresh water contributes to their lack of importance in the diet of freshwater fish compared with other aquatic insects (Collier & Winterbourn 2000). All species require fresh water of at least moderate quality.

Damselfly enemies include large orb-web and *Tetragnatha* spiders, the larger robber flies (adults of *Neoitamus*), and dragonflies amongst invertebrate predators (Rowe 1987). Cats include dragonflies as a minor item of their prey (King 1990; Ryan 1994). Even the largest dragonfly species are not immune from predation

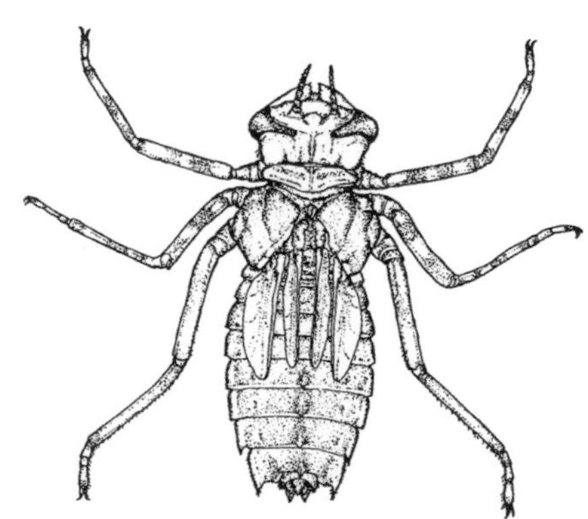

Larva of *Antipodochlora braueri*.
From Winterbourn et al. 2000

by the smaller German wasps, apart from being eaten by rats, kingfishers, harrier hawks, tuis, and sparrows (Rowe 1987). Of these predators, only the kingfisher is regularly associated with feeding in the freshwater environment. Harrier hawks include insects as a minor component (less than 10%) of their diet (Higgins & Marchant 1993). Other birds observed feeding on odonates include moreporks, shining cuckoos (Rowe 1987; Higgins et al. 1999), starlings, and mynas, which feed damselflies to nestlings (Moeed 1975).

Order Plecoptera: Stoneflies

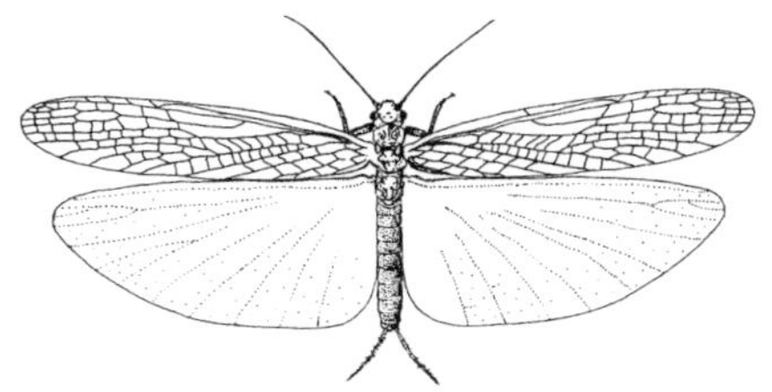

Large green stonefly *Stenoperla prasina*.
From Grant 1999

The Plecoptera ('folded wings') is another ancient order of insects, with a fossil record dating back about 260 million years ago to the Late Paleozoic era. A widespread group, with about 2000 species, they are found throughout the world except in polar regions. They are soft-bodied with clearly separated thoracic segments, and usually with four wings, which are folded straight back and closely applied to the abdomen. Superficially they resemble and share some primitive characters with orthopterans. Strictly, however, plecopterans do not show close relationships to any other neopteran ('new wing') insects, and may be an evolutionary sister group to all neopterans. A number of species do not have full wings and may be brachypterous (shortwinged) or apterous (without wings). Immature stoneflies are typically found under stones in streams or stream margins, hence the common name for the group.

Abdominal cerci may be long (e.g. Gripopterygidae) or reduced to one segment (e.g. Notonemouridae). Nymphs are normally aquatic and may have filamentous gills – in the form of an anal rosette in Gripopterygidae or in Austroperlidae as a few beaded filaments on the appendages of the abdominal apex. The Notonemouridae have no gills. Nymphs usually live in cool, running fresh water, but in New Zealand and southern South America some wingless species have terrestrial nymphs living in cool humid microclimates beneath stones or vegetation in alpine or subantarctic situations. At low altitudes there are winged species with semiterrestrial nymphs that, early in their existence, move out of the water to spend the winter under stones of stream flood-plains. The nymphs of stoneflies, being dependent on cool, well-oxygenated conditions, are very susceptible to human abuse of their environment. Even quite minor pollution sources such as farm drainage can eliminate stoneflies from nearby streams. Impoundment of water, which can raise water temperature, is also unfavourable to their existence. Plecopterans, then, are indicators of healthy streams and rivers.

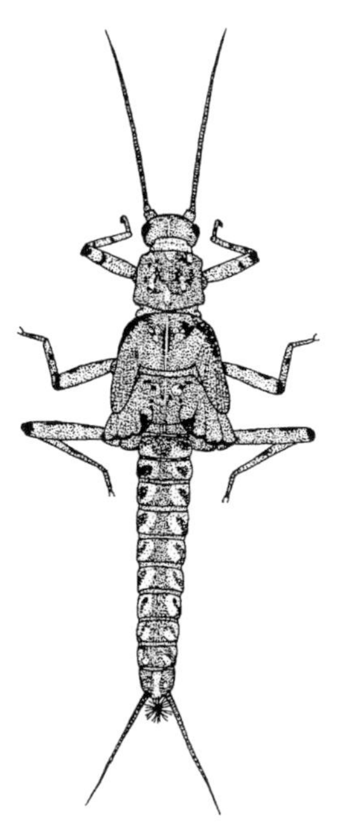

Larva of *Nesoperla fulvescens*.
From Winterbourn et al. 2000

New Zealand's stoneflies belong to four families that have an austral distribution only, on lands derived from Gondwana fragments, i.e. Australia, New Zealand, South America, Falkland Islands, South Africa, and Madagascar.

Zwick (1973) divided the order Plecoptera into two suborders – Arctoperlaria and Antarctoperlaria. The Antarctoperlaria belong exclusively to the Southern Hemisphere, whereas Arctoperlaria are those of Northern Hemisphere origin. Two families of Arctoperlaria are found into the Southern Hemisphere, however. The Perlidae, widespread in the Northern Hemisphere, extends to South Africa and South America, and the Notonemouridae is actually restricted to the south. McLellan (1991) revised New Zealand representatives of Notonemouridae.

The Antarctoperlaria is divided into two superfamilies. The Eusthenioidea comprises the families Diamphipnoidae (South America, five species) and Eustheniidae (Australia, 14 species; New Zealand, four species; South America, one species). The Leptoperloidea presents a less clear-cut picture – Zwick (1973) recognised the families Austroperlidae and Gripopterygidae but separated the subfamily Antarctoperlinae from the Gripopterygidae and the genus *Crypturoperla* from the Austroperlidae. Hence, the Gripopterygidae sensu Zwick (1973) is found in Australia, New Zealand, South America, and the Falkland Islands, with the subfamily Antarctoperlinae in New Zealand, South America, and the

Falkland Islands. The Austroperlidae occurs in New Zealand, South America, andn Australia. *Crypturoperla* is endemic to Australia.

The New Zealand fauna comprises 99 described species but there may be 21 more undescribed species. All genera and species are endemic to New Zealand apart from the genus *Notonemoura*, which is shared with Australia although the species are endemic to each country. Some New Zealand genera are locally restricted. *Rakiuraperla* is endemic to Stewart Island and *Aucklandobius* and *Rungaperla* are found only on subantarctic Auckland and Campbell Islands, respectively.

Order Blattodea: Cockroaches

Most New Zealanders might be surprised to learn that they share their country with 44 species of cockroach, only 13 of which are introduced. They are among the most cosmopolitan of insect pests and are associated with human dwellings throughout the world. Their flattened body is well suited to hiding in cracks and crevices. Although they do not sting or bite, some species are usually associated with unsanitary conditions and may carry a variety of human pathogens in some parts of the world. But cockroaches are truly remarkable creatures in many ways and have proved useful as research tools in the study of insect physiology and behaviour (one species can actually run on two legs when disturbed) and toxicology. Cockroaches are an ancient group, dating from the Carboniferous period 317 million years ago. Today there are about 4000 described species. The ordinal name Blattodea is derived from *blatta*, the Greek word for cockroach.

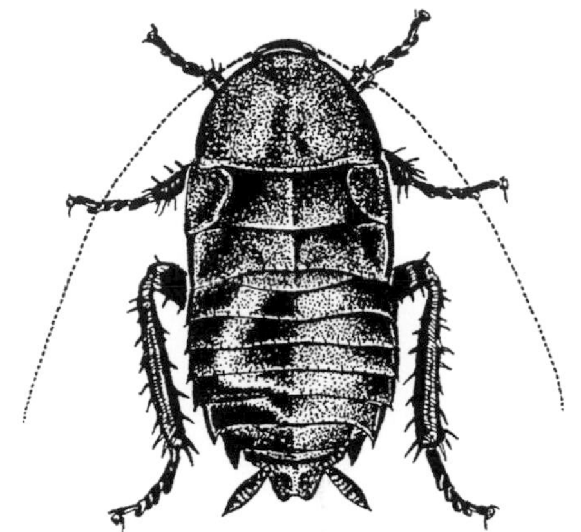

Native black cockroach *Maoriblatta novaeseelandiae*.
From Grant 1999

Taxonomic knowledge of the New Zealand fauna has changed little since the review of Johns (1966), but new knowledge has accrued concerning their distribution, through extensive collections now held in various museums, available from Peter M. Johns. Zervos (1983, 1984, 1987) has described bivoltine life-cycles (two generations per year) and the parasites of two lowland species. All species appear active throughout the year with only slight seasonal changes in age structure.

All native species are probably endemic. The largest genus is *Celatoblatta*, in which Princis (1974) included *Austrostylopyga*, thereby including eight Australian (mainly Queensland), one New Guinean, and three New Caledonian species. He also included two large native black cockroaches in the Australasian subgenus *Platyzosteria* (*Melanozosteria*), but these are now placed in *Maoriblatta* along with two New Caledonian species. As with several other native groups, flightlessness is a feature, and some species have strong local distributions, being confined to areas like Banks Peninsula and the Chatham Islands. Such geographically restricted species are not necessarily restricted to one habitat, however. One new Canterbury species is found in conditions ranging from subalpine shrub-tussock vegetation, rocky screes, and stony riverbed terraces, to salty coastal cliffs and stony beach ridges. Several species are well adapted to low temperatures. *Celatoblatta quinquemaculata*, *C. anisoptera*, *C. montana*, and two new species in the Otago mountains live above the winter snowline. None shows any degree of arrested development (diapause) – all between-moult instars may be found immediately after winter. *Celatoblatta anisoptera* tolerates freezing conditions by increasing blood glycerols, and presumably other species in similar habitats have a comparable cold-tolerance mechanism. *Celatoblatta qinquemaculata* is tolerant of freezing up to 74% of its water volume. It may go through up to 23 freeze/thaw cycles per month (Sinclair 1997a). Two widespread, short-winged, southern *Parellipsidion* species occur in many types of open shrubland, forest, tussock, and fellfield habitats from two to 2000 metres in the South Island. A related winged (but rarely flying) species is confined to lowland forests of the North Island and coastal Marlborough Sounds.

Celatoblatta undulivitta
Alastair Robertson and Maria Minor, Massey University

Most North Island species are widespread within the island, although those in North Auckland and especially on the Three Kings Islands need further taxonomic analysis – some may represent new species. *Parellipsidion pachycercum* is the only cockroach species to be found in the subantarctic islands and then only in a small area centred on the site of flax (*Phormium tenax*) introduced to the Auckland Islands in the middle of the 19th century. Six new species of *Celatoblatta* and two others, perhaps belonging to new genera, await description. All of these species are restricted in distribution. Five are known from a few specimens from single sites. It is thought that basic speciation patterns are related to the Early Tertiary archipelagic nature of New Zealand and that present ecological tolerances and distributional patterns are associated with Late Tertiary and Pleistocene events. This hypothesis is now being tested through DNA analyses of Canterbury mountain species. At higher taxon levels, the New Caledonian connection is of particular interest and examples of all genera need to be compared closely with species from Queensland and New Caledonia. Especially pertinent is the finding of a relatively large tryonicine roach (Blattidae) in Northland. This group is otherwise known only from eastern Australia (mainland Queensland) and New Caledonia.

Adventive cockroach species

There are at least 13 introduced species. Only two are at all widespread – *Blattella germanica* and *Drymaplaneta semivitta*. *Blattella germanica* probably became established many years ago, as it occurs throughout the country. It has been seen in houses, university and hospital restaurants, cafes, and old and new hotels everywhere. *Drymaplaneta semivitta* gradually spread through much of the North Island during the 1980s and 1990s and in 1999 was found established in Nelson. A new adventive is *Drymaplaneta heydeniana*, from Australia. Its colour pattern is very similar to the native species of *Celatoblatta*. It is spreading quite quickly out from Auckland. *Drymaplaneta semivitta*, *Paratemnopteryx couloniana*, and *Neotemnopteryx fulva* come from Western Australia, perhaps indicative of the huge numbers of telephone poles, wharf piles and planking, and railway sleepers imported from that state. *Paratemnopteryx couloniana* and *Blatta orientalis* are yet to get far (in terms of only hundreds of metres) beyond their original established sites in Auckland and Dunedin (Harris 1988), respectively. In late 1999, the Australian (NSW–Queensland) *Panesthia cribata* was found in Kerikeri. It is also in Auckland, as is another Australian species, *Tryonicus parvus*. *Periplaneta americana* is established in Auckland. For many years it has been reared by schools and universities for teaching. Cultures have to be carefully tended as the animals do not readily maintain natural populations in New Zealand. It is known from factories where heat processes are used regularly in production (e.g. tyre factories). Juveniles of what is thought to be *Periplaneta fuliginosa* have been collected from a garden in Mairangi Bay, Auckland. A single live *Periplaneta australasiae* was caught on a house site in Rangiora to which bricks had been imported from Rockhampton, Queensland. It has also been captured at Taumarunui, Masterton, and Dunedin, but as yet it has not become established. It was perhaps more because of good luck than good management that the pest *Periplaneta brunnea* did not become established in Timaru, where it was found in imported second-hand cars from Japan. It is, however, present at Raoul Island (Kermadecs). The worldwide household pest *Supella longipalpa* is now known from Wellington. *Shelfordina orchidae*, a species fairly well known in Australia and Japan associated with orchids, is now established in the orchid hothouse of the Auckland City Domain (Winter Gardens). Other tropical species still arrive occasionally in imported goods (fruit from the Americas and Pacific Islands, and, reputedly, oil-refinery pipes from India) but are yet to be confirmed as established. *Pelmatosilpha vagabunda* Princis, 1954 has as its type locality a ship in Auckland Harbour (ex banana cargo from Colombia), but it is not found in New Zealand! Cockroaches have been, and always will be, a biosecurity threat in both inward and outward goods.

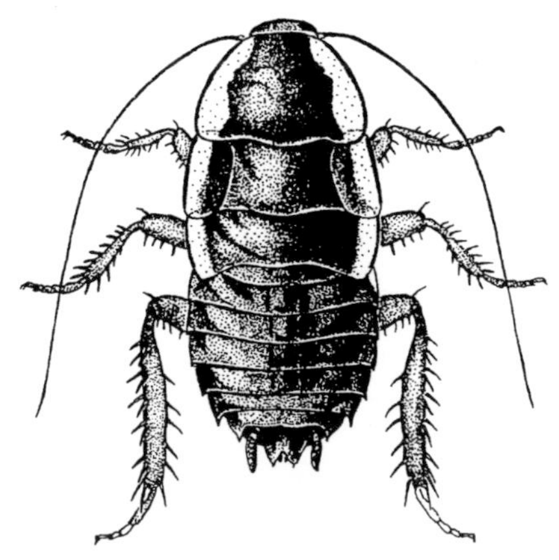

Introduced Gisborne cockroach
Drymaplaneta semivitta
From Grant 1999

Order Isoptera: Termites

In the minds of most New Zealanders, termites are generally associated with Australia. The New Zealand fauna is quite small, comprising only eight or nine named species including three endemics. In Australia there are ca. 350 species, but fewer than 270 have validated available names (Watson & Abbey, 1993). These social, wood-inhabiting insects with pale bodies are quite small (seldom exceeding 10 millimetres in length), and hence are sometimes referred to as white ants. The termite body is soft, the abdomen is tubular, and the wings, when present, are finely veined, a character that aids in family identification (Kelsey 1944; CSIRO 1991). The many-segmented, filamentous antennae are often longer than the head, and only the sexual forms are initially winged. The wings are longer than the body and clear and flat when at rest. The wingless workers have small heads, unlike the soldiers, which have enlarged heads and prominent mandibles. Keys, descriptions, and illustrations of the New Zealand species can be found in Hill (1942), Kelsey (1944), Gay (1969, 1976) and Bain and Jenkin (1983). Kelsey (1944) gave line drawings and measurements of adults and their cerci for four families and 14 species that had been reported from New Zealand.

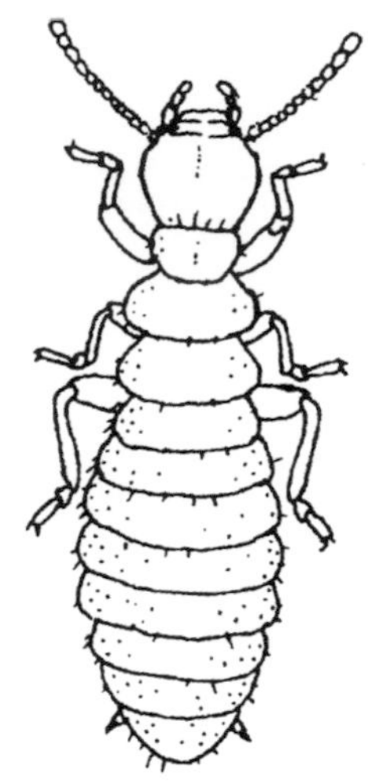

Kalotermes brouni worker.

From Grant 1999

Only seven or eight species of termites still exist on the main islands of New Zealand (Wise 1977; Bain & Jenkin 1983). Apart from the three known endemic species, the rest originated from Australia (Hill 1942; Wise 1977). At least one *Kalotermes* species has colonised and remained on White island despite periodic destruction of vegetation by the volcano (Hutcheson 1992) and at least one undetermined species exists on the Poor Knights Islands (Watt 1982). These very limited records from the offshore islands suggest that the native species may be present on many of the wooded islands, because the extent of forest on White island is quite limited and these are among two of the more remote northern offshore islands.

Symbiotic flagellates in the gut of termites (Nurse 1945) allow them to feed on dry or damp wood above or below the soil surface. The two *Stolotermes* species (Termopsidae) are damp-wood-inhabiting termites that colonise logs, poles, and posts, but not buildings. *Stolotermes ruficeps* inhabits wood of both native and introduced tree species and a royal pair can consume 6–16 cubic centimetres of wood per six months (Morgan 1959). This is more than 10 times the amount of wood consumed by a dry-wood-inhabiting species. Morgan (1959) also gave details on the habits, communication and defensive behaviour, colony structure, and growth of *S. ruficeps* as well as other insects associated with them in rotting wood. This species swarms in autumn (Morgan 1959). Typically, *Stolotermes* colonies have several hundred termites but can reach up to 6000 per colony (Milligan 1984a; Thorne & Lenz 2001). Malaise (Moeed & Meads 1984; Macfarlane unpubl.) and intercept traps (McWilliam & Death 1998) capture only a few of these termites in summer and autumn. *Stolotermes inopus* is mainly confined to the North Island. *Kalotermes brouni* (Kalotermitidae) is the commonest and most uniquitous species in New Zealand, extending to the Chatham Islands. It inhabits dry decaying wood of semi-hollow native trees, untreated power poles, fence posts, and buildings (Milligan 1984b). *Kalotermes cognitus* is confined to the Kermadec, Norfolk and Lord Howe Islands (Gay 1976).

Subterranean colonies are limited to two species of *Coptotermes* (Rhinotermitidae), which were serious pests in Auckland but are now apparently confined to four sites. *Glyptotermes brevicornis* has been recorded from imported poles, wharf timbers, and decaying logs and *Kalotermes banksiae* has been found in logs, driftwood, and dead trees in various North Island and Nelson locations (Bain & Jenkin 1983). None of these introduced species has been found in forests. *Neotermes insularis* is one of several species now considered to no longer exist (Bain & Jenkin 1983) in any of the North Island sites where they were originally detected (Kelsey 1944).

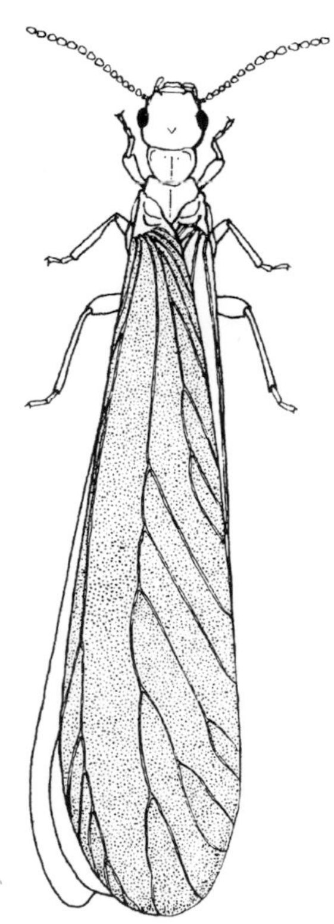

Winged reproductive form (alate) of New Zealand dry-wood termite *Kalotermes brouni*.

From Grant 1999

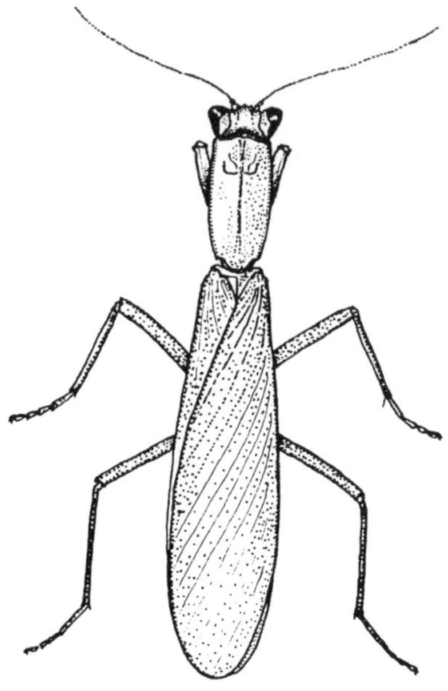

Praying mantis *Orthodera novaezelandiae*.
From Grant 1999

Order Mantodea: Mantids

There are two species of mantids in New Zealand, one endemic, the other introduced. For many years the endemic New Zealand species (*Orthodera novaezealandiae*) was considered to be conspecific with the Australian *O. ministralis*, but Ramsay (1984, 1990) has clarified their status, giving illustrations and noting differences in body features. A South African praying mantid, *Miomantis caffra*, was first detected in Auckland in 1978 (Ramsay 1984). *Orthodera novaezealandiae* is always green, and is found on both main islands of New Zealand and the nearer offshore islands (Moeed & Meads 1987a; Ramsay 1990) that remained connected to the main islands during the last ice age. *Miomantis caffra*, which is often green and only rarely brown, had by 1990 spread to the Waikato and near to the Bay of Plenty (Ramsay 1990). It has also been collected from Blenheim (LUNZ collection). For excellent general information and a bibliography of New Zealand studies on mantids the reader should consult Ramsay (1990).

Praying mantids are relatively large insects (32–52 millimetres long). Their front legs are enlarged, with spines to aid in grasping prey. Further, the basal segment (coxa) of the leg is large and mobile and the front thorax segment is much longer than combined middle and hind segments. The wings cover much of the abdomen and the front wing is narrower than the hind wing. After mating, praying mantids deposit leathery eggs cases (oothecae) containing, on average, 34 eggs (Suckling 1984). These predators either walk slowly, or quite fast when disturbed, and although can males fly the gravid female flies poorly (Castle 1988). They inhabit gardens, shrublands (Ramsay 1990), and native forest (Moeed & Meads 1992), and may also be found in bracken fern (Winterbourn 1987). Unlike the omnivorous European earwig (Macfarlane 2002), mantids have not been recorded in surveys of grassland, lucerne (alfalfa – *Medicago sativa*), fodder crops, or cereals, where small potential fly prey is common (Macfarlane & Andrew 2001). Perhaps they favour firm vegetation to perch on so they can see their prey properly and then attack it. Only insects that move are preyed on (Castle 1988). Prey can include blow flies, house flies, *Drosophila* fruit flies (Suckling 1984), the forest-inhabiting looper moth *Pseudocoremia suavis* (Valentine 1967a), honey bees (Ramsay 1990) and even *Vespula* wasps on ivy flowers (Andrew pers. comm.). Little is known of mantid prey under field conditions or their natural feeding capacity.

In New Zealand, mantid predators include sparrows, mynahs, bats, and cats. Other bird species and skinks probably prey on adults, too. Evidence of insect predation on mantid eggs is sketchy in New Zealand (Ramsay 1990). Hymenopteran parasites of three species affect only the eggs and the main species almost certainly are *Eupelmus antipoda* (Eupelmidae), and *Pachytomoides ?frater* and *Podagrion* sp. (Torymidae). Seasonal activity of these parasites remains unstudied. The limited information on the level of parasitism of the eggs suggests that this complex of parasites can cause considerable mortality (Ramsay 1990).

European earwig *Forficula auricularia*.
Alastair Robertson and Maria Minor, Massey University

Order Dermaptera: Earwigs

The Dermaptera is one of the most easily recognised insect orders, owing to the distinctive forceps or pincers at the end of the abdomen. Other characteristics include the leathery front wings that give the order its technical name (literally 'skinwings') – these cover only the thorax – and the semicircular hind wings that fold under the front wings when at rest. 'Earwig' is possibly a misinterpretation of 'earwing', a reference to the small, hardened forewings. If true, it may well be partially responsible for the mistaken belief that earwigs will crawl into a sleeping person's ear and then burrow into their brain, thus killing them. There are around 1200 species worldwide, of which rare, if not extinct, *Labidura herculeana* from St Helena is the largest at around eight centimetres long.

There are 22 earwig species in New Zealand, but the establishment of the two chelisochids and *Carcinophora occidentalis* needs to be rechecked. Of these 22 species, 13 (59%) are endemic and the rest are adventive. The adventive species contribute three families – viz *Forficula auricularia* (Forficulidae) from Europe, two species of Chelisochidae, three species of Spongiphoridae, and *Labidura truncata* (Labiduridae) (Hudson 1973; Cassis 1998). Hudson (1973) provided information on the distribution of each species based on the four main collections in New Zealand and provided an illustrated key. Since then, three genera have been redefined and Cassis (1998) also recorded *Hamaxas feae* from New Zealand. A review by CSIRO (1991) explains specialist descriptive terms (used also by Hudson) and has a key to families. It also illustrates three species found in New Zealand and complements Hudson's study. Brindle (1987) gave a key to the nymphs of most of the introduced genera of earwigs (not *Paraspania*). The habitat in which a specimen is found can aid species identification.

Biology and habits of endemic species

The biology and ecology of the 13 native earwigs are best known for the large coastal species *Anisolabis littorea*, thanks to the study by Giles (1953), who also studied copulation, egg-laying, and parental care. The body and legs of this earwig are uniformly yellow-brown, unlike the adventive European species *Euborellia annulipes* that also inhabits the seacoast. At 23–31 millimetres long, *A. littorea* is among the largest native species. It inhabits damp beach drifts of seaweed or wood, but its range may extend to gardens and wasteland provided that sheltering stones and logs do not rest on soil. Individuals shelter under stones and logs among damp debris, and the drier beaches are important for maintaining their warmth in winter. Adults feed on slaters and millipedes, cutting their prey with their forceps. They can be raised in captivity on a diet of flies, slaters, and raw liver. Introduced species have been reared on cat food (Shepard et al. 1975).

Despite being wingless, *Anisolabis littorea* is known from nine northern offshore islands as far as the Three Kings, Mokohinau, the recent volcano of White Island (Giles 1958; Wise 1970b; Hudson 1973), and Stephens Island (Wall 1981) but apparently not Kapiti Island (Moeed & Meads 1987b). The species is also present on Chatham and Pitt Islands and Martins Bay in the South Island (Giles 1953) through to Dunedin (Otago Museum records). On the Poor Knights Islands, which were separated from Northland during the last glacial period, a possibly endemic species of *Anisolabis* or a variety of *A. kaspar* is present (Watt 1982). *Anisolabis kaspar* inhabits bush, unlike *A. littorea* (Hudson 1973).

Parisolabis novaezeelandiae is the only species in this genus found in both North and South islands (Hudson 1973). Collection records, some of which include only the locality, indicate that *Parisolabis* species inhabit litter and moss and that some may live in both forest and grassland habitats. Species that occupy both habitats definitely include *P. novaezeelandiae* (Hudson 1973; Moeed & Meads 1987a) and *P. forsteri* from Otago (Hudson 1973; Barratt & Patrick 1987). In the Orongorongo Valley, Wellington, only pitfall trapping collected a few earwigs from the surface of the forest floor, and none was collected from the litter or on tree trunks. This might mean that forest species prefer to shelter and feed in decaying logs and stumps. *Parisolabis tapanuiensis* and an undetermined species have been collected from snow-tussock grassland (Barratt & Patrick 1987). Locality records indicate that both *P. nelsoni* and *P. boulderensis* from northwest Nelson (Hudson 1973) inhabit grassland. Conversely, *P. johnsi* inhabits broadleaf forest, and *P. setosa* from Fiordland is associated with shrubland and scree. The habitat association of *P. iti* is unclear from collection records. Perhaps the widespread species are more flexible in their use of habitat, but very little has been recorded about their ecology. The grasslands and forests of the former island of Banks Peninsula lack any native earwigs.

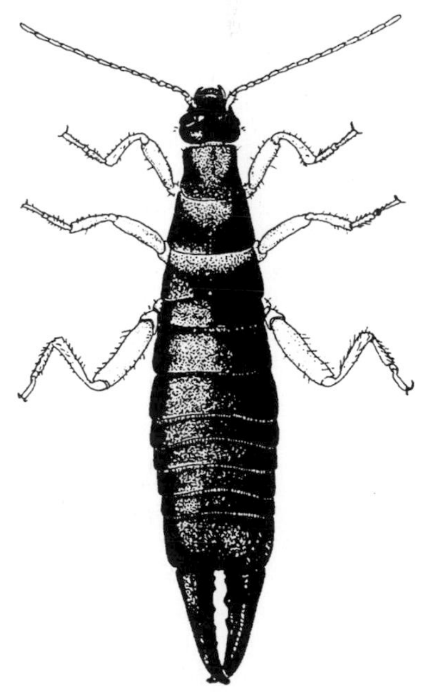

Female seashore earwig *Anisolabis littorea*.
From Grant 1999

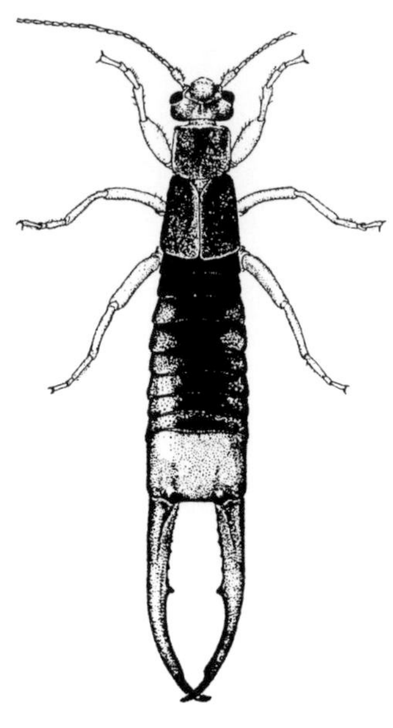

Male Australia earwig *Labidura truncata.*

From Grant 1999

Studies on the ecology, biology, behaviour, and biochemistry of the introduced species provide guidance on aspects of the life of the endemic species that could reveal valuable comparisons.

Biology of naturalised alien earwigs

Females of at least three introduced species (*Euborellia annulipes, Forficula auricularia*, and *Labidura riparia*) protect their eggs and first-instar nymphs and brood their young (Lamb 1976; Radl & Linsenmair 1991; Rankin et al. 1996). This subsocial behaviour is the first basic step towards social organisation. In the male the forceps are used for display and a tactile stimulus for the female (Walker & Fell 2001).

The European earwig *F. auricularia* is omnivorous. The damage it caused to ripening stone- and sometimes pipfruit as well as soft berry fruits (e.g. black currant, strawberry) drew attention to it as a horticultural pest in New Zealand. In addition, various flowers (e.g. dahlias) and vegetables (lettuce, celery, potatoes, seedling beans, beet, and sweetcorn flower stalks) can be fed on. As well, these earwigs are attracted to both the pollen resources gathered by bees and the sheltered habitat the bees inhabit. If they become numerous in shelters they can displace nesting lucerne leaf-cutter bees (*Megachile rotundata*) (Barthell et al. 1998). Aggregations in hives set out for occupation by bumble bees (*Bombus* spp.) appeared to discourage nesting and could also have caused the queens to desert their founding brood. How important pollen is in the diet of earwigs and whether it originates from flowers or bee-collected sources has not been investigated. Nocturnal activity ensures that the role of *F. auricularia* as a predator is much under-appreciated. Night sampling with light traps took only the European earwig from hill-country pasture (McGregor et al. 1987) as did night sweeping in lucerne (Macfarlane 1970; Leathwick & Winterbourn 1984). Day-time sweeping of pasture (Cumber 1958), fodder (Eyles 1960), and cereal crops (Cumber 1962; Bejakovich et al. 1998), or vacuuming in lucerne (Rohitha et al. 1985) failed to collect any earwigs. Pitfall traps were effective in pasture (Moeed 1976), while soil, pitfall, and sweep sampling in short dry pasture collected only one specimen each of *F. auricularia* and *L. riparia* (Martin 1983). In Christchurch insect-community studies, Macfarlane (unpubl.) found *F. auricularia* only among rotting trunks of kowhai trees in dry danthonia grassland or in leaf litter among the rushes and long grass in wetlands.

Solomon et al. (2000) found the European earwig to be the commonest predator in apple and pear orchards in Europe. In New Zealand apple orchards, Suckling et al. (2006) tested shelters and baiting to monitor these earwigs because of their potential importance in devouring leaf roller egg batches. Perhaps continental climates similar to that in central Otago favour this species. Its role as a predator of the currant clearwing (*Synanthedon tipuliformis*) has been examined in New Zealand (Scott 1979). Hill et al. (2005) considered earwigs to be the cause of heavy mortality of armoured scale (Diaspididae) on kiwifruit. In England there have been detailed studies of its role as an aphid predator in grasslands and barley (Carillo 1985). An analysis of its stomach contents has shown that *F. auricularia* can be one of the most important aphid predators during aphid buildup in cereal crops (Sunderland &Vickerman 1980). Since 1985, world literature on the European earwig has considered its predatory role on the black scale *Sissetia oleae,* red-legged earth mite *Halotydeus destructor,* psyllids, and moth caterpillars on grape vines. These earwigs are more effective predators of the woolly apple aphid *Eriosoma lanigerum* than ladybird species (Solomon et al. 2000). Unspecified earwigs, the commonest probably being *F. auricularia,* can also be opportunistic predators in ganging together to prey on the much larger puriri moths temporarily stunned by lights (Green 1983).

The European earwig has been caught in baited traps in open areas and not in forests (Kocarek 1998), although more are found where there is deciduous leaf litter (Thomas et al. 1992) such as from poplars (Macfarlane unpubl.). It is a better coloniser of offshore islands than *Labidura riparia*, which may inhabit salt marshes (Taglianti 1995). Earwigs aggregate together in response to a pheromone released in frass (Sauphanor & Sureau 1993) and the same pheromone is secreted from the tibia (Walker et al. 1993). Conversely, *L. riparia* does not aggregate, but earwigs are attracted to shelters previously occupied by their own species (Sauphanor & Sureau 1993), which suggests that a marker compound may be active.

Frost kills European earwigs in winter. Even the hardier females and eggs can withstand only –2° to –3° C, so females must find more-protected sites within soil or in cracks along house foundations (Gringas & Tournier 2001). Even so, the species is found in Central Otago and the Mackenzie Basin. In spring, a threshold of 6° C is needed to begin development in nymphs and an accumulation of 600–750 day degrees to reach the fourth and last nymphal instar (Helson et al. 1998). The number of segments on the antenna increases from the nymphal instar stage to the adult (Hincks 1949a). Scott (1984) and Thomas (1989a) summarised the pest status of European earwigs and the main aspects of their biology as well as control options.

Among the other accidentally introduced species, both *Labidura riparia* and *Euborellia annulipes* favour beaches. Ring-legged *E. annulipes*, at only 10–13 millimetres long, is distinctly smaller than both *L. riparia* and *Anisolabis littorea*. Both of these introduced species have spread around the North Island as well as to Nelson in the South Island (Hudson 1973). *Labidura riparia* has also spread as far offshore as Great Barrier Island. It prefers sandy habitats, but is a predator in orchards and among brassica, soybean, and maize crops. Prey includes codling moth caterpillars (CSIRO 1991), diamond-backed moths (Strandberry 1981), armyworm, aphid, scale insects, and mites (Schlinger et al. 1959; Shepard et al. 1973). One of these earwigs can consume two looper pupae and 5–38 caterpillars per day (Price & Shepard 1978). *Euborellia annulipes* can be a storage pest of potatoes, flour and corn, and roots in glasshouses as well as in gardens, nurseries, and even meat-packing plants, but it needs moist conditions to survive (Bharadwaj 1966). But it also preys on pests (Klostermeyer 1942) and may be associated with human corpses (Goff 1991). There is apparently no evidence that *Chelisoches morio* breeds in New Zealand (Hudson 1973) even though it was collected from a garden in Christchurch and at Titirangi, Auckland (Hincks 1949b). The collections of Lincoln and Canterbury Universities, Canterbury and Otago Museums, and Ministry of Agriculture & Forestry, Lincoln, have no specimens of any *Chelisoches* species. Unlike other species in New Zealand, *C. morio* is active during the day. Throughout its range in the Pacific and in Australasia, it shelters among debris, coconut shells, and decaying banana stalks and lays it eggs in leaf sheaths of grasses. It feeds on leafhoppers, scale insects, beetle larvae, and caterpillars (Brindle 1972, 1987). *Paraspania brunneri*, which originated from Australia, is mainly black centrally and red-brown at the body extremes and is about half the size of most European earwigs. It has been collected from logs and stumps in forests of pine and other introduced conifers in Canterbury (Hincks 1949b). *Labia minor* has been in New Zealand since before 1893, originating from Europe. The species can be associated with compost heaps (Ramsay 1997), straw, and dry dung (Allen 1985) and may prey on house fly eggs and larvae (Mourier 1986). Both *L. minor* and the doubtfully endemic *Paralabia kermadecensis* are small, with little colour difference in their brown to dark brown bodies (Hinks 1949a). The latter was initially found around Auckland in 1967 and identified as *Labia curvicauda* (Hudson 1973), but Hudson (1976) correctly identified this species.

Natural enemies of earwigs

The natural enemies of the European earwig have been studied extensively in the Northern Hemisphere (Crumb et al. 1941; Barthell & Stone 1995; Kuhlmann 1995). One enemy, the tachinid fly *Triarthia setipennis*, was released in New Zealand in 1928–29. It apparently failed to establish (Thomas 1989a) but discriminating it from other tachinids in New Zealand has not been easy (Macfarlane & Andrew 2001) – a recent redescription of *Triarthia* and its four species (O'Hara 1996) should help with recognition of this species. Any New Zealand tachinids introduced against the European earwig could be free of both main parasitic wasps (*Dibrachys cavus* and *Phygadeuon vexator*), but taxonomic work on *Dibrachys* and the large ichneumonid subfamily Phylacteophaginae has been minimal. By contrast, tachinid establishment was achieved in North America (Barthell & Stone 1995) where such wasp parasitism varies from at least 1% to 18% compared with up to 49% in Europe (Kuhlman 1995).

Cool moist habitats are thought to favour earwig parasites, and the first generation of earwigs may be more severely affected, as for example by the nematode *Mermis subnigrescens*, which is common in Canterbury (Thomas 1989a; Macfarlane unpubl.). This internal parasite sterilises the adult hosts. The effect on earwigs of the many other possible pathogens (fungi, gregarines) present in New Zealand has not been determined. Nor has the likely level of mortality caused by ground beetles (Miller & Walker 1984).

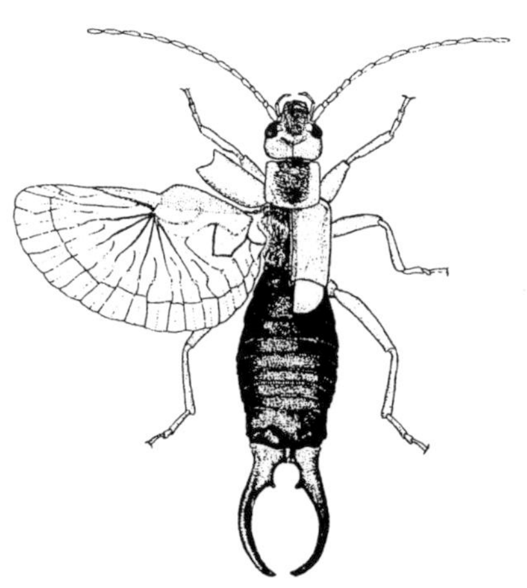

European earwig *Forficula auricularia*.
From Grant 1999

A decline in European earwig populations in New Zealand coincided with the spread of hedgehogs (Thomas 1989a) so predation may be important in keeping their populations in check. The hedgehog is commonest in open, warmer, coastal pastureland, dunes, and settled areas (King 1990). Earwigs (*F. auricularia, A. littorea*) are one of the most important dietary components of the hedgehog (Brockie 1959; Campbell 1973), with ca. 5–10% of hedgehog food in pasture, suburban areas, and dune habitats being earwigs, although fewer are eaten in orchards. Ship rats and mice consume earwigs (species not specified) as a minor part of their diet (Miller & Miller 1995). The tuatara includes *A. littorea* as a minor item of its diet (Wall 1981). The common skink *Oligosoma nigriplantare* was reported to feed on earwigs (possibly *F. auricularia* judging from the habitat) as a minor component (Berwick 1959).

Various species of bird capture earwigs, especially those that are active at night or which feed on the ground, unlike most of the common bush birds. The European earwig is a minor item of the food (1–10% frequency) gathered by ground- and day-feeding magpies (McIlroy 1968; Moeed 1976), mynahs (Moeed 1975), and starlings (Moeed 1975, 1976, 1980; Coleman 1977) with only a modest increase in seasonal incidence in the diet compared to the peak in numbers revealed by pitfall trapping (Moeed 1976). Sparrows infrequently feed nestlings on *F. auricularia* (MacMillan 1981; MacMillan & Pollock 1983), but if adult sparrows feed on them they do so less often (MacMillan 1981). This earwig is also a minor item in the diet of the little owl (Soper 1963). Giles (1953) examined *A. littorea* extensively for parasites without finding any. Introduced *Vespula* wasps capture earwigs in native forest very occasionally (Harris 1991) but apparently not elsewhere.

Order Orthoptera: Weta, grasshoppers, crickets, and kin

Orthopterans ('straight wing') have a generally cylindrical body, with hind legs elongated for jumping. They have mandibulate mouthparts and large compound eyes, and may or may not have ocelli, depending on the species. They have two pairs of wings, which are held overlapping the abdomen at rest. New Zealand has a very mixed bag of orthopteran insects. There are many endemics in a few families yet the commonest species seen by most people are introduced. The literature is large and scattered but early references can be obtained through the *BUGS* database (Ramsay & Crosby 1992).

Weta: A New Zealand icon

There are many weta species in New Zealand, classified in two families – Anostostomatidae (tree and ground weta) and Rhaphidophoridae (cave or jumping

Summary of species numbers of Orthoptera

Family	Described	Known undescribed	Adventive
Prophalangopsidae*	1	0	–
Anostostomatidae	29	>33	0
Rhaphidophoridae	54	>30	0
Tettigoniidae	5	0	3?
Gryllidae	8	1	4?
Gryllotalpidae	1	0	0
Acrididae	19	3	0
Totals	**117**	**>67**	**7?**

* Extinct

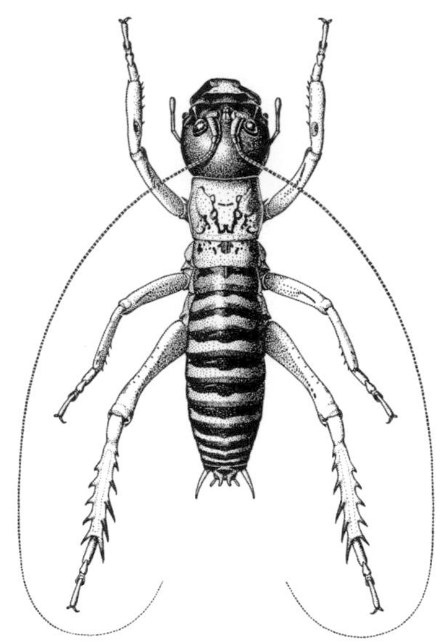

Male Canterbury weta *Hemideina femorata*.

Des Helmore

weta). All genera and species are endemic. A fossil prophalangopsid orthopteran indicates that the New Zealand fauna may have evolved during a long period of isolation, well before New Zealand's separation from the Gondwana supercontinent some 120 million years ago.

Anostostomatidae: Tree, giant, and ground weta

Everyone knows about weta, often referring only to *the* weta and implying that all the details of the two that are common can be applied to others. Adults often find weta repulsive, yet their supposed ugliness fascinates children. Most details apply to the well-known Auckland and Wellington tree weta, *Hemideina thoracica* and *H. crassidens*, respectively. It is the enlarged heads of the males and spiny legs that people find so abhorrent. Ironically, giant weta (*Deinacrida* species), among the world's heaviest insects, seem much less fearsome. Their conservation status is currently much discussed (Gibbs 1998). Discoveries of two species of horned weta on Red Mercury Island (Johns 1997) and the Raukumara Range of East Cape region (Gibbs 2002), four giant weta in South Island mountains (Gibbs 1999), and 35 species of ground weta throughout New Zealand (P. M. Johns, pers. obs.) have more than doubled the known fauna, indicating the real paucity of our understanding of these remarkable creatures. Furthermore, many gene sequences have now been prepared, pointing to additional 'hidden' species such as *Hemideina trewicki* (Morgan-Richards 1995). Population genetics has also shown that there are many separate populations that have developed through prehistoric restriction of gene flow by rising sea level or land sinking below the sea, leaving isolated islands and peninsulas (Morgan-Richards 1997). Similarly, populations of alpine species have diverged to a considerable degree, and studies have been made on their behaviour and behavioural evolution (McVean & Field 1996; Gwynne 1995) and genetic divergence (Trewick et al. 2000).

Rhaphidophoridae: Cave or jumping weta

Cave weta need revising. All species are endemic and there are many awaiting description or perhaps even discovery (Johns 1991, and subsequent observations). A name that was long forgotten (130 years) can now be applied to one recently collected form. A number have needed slight changes to their names, as indicated in the end-chapter checklist, to follow the ICZN rules of nomenclature concerning gender agreement. *Talitropsis* has recently been placed in its own tribe (Gorochov 1995) and is said to be primitive, and other genera may well be allied to it. But the entire family is in need of revision at higher taxonomic levels inasmuch as one nominally key character is contravened – normally, stridulatory pegs on the abdomen, similar to those present in the ground weta (Anostostomatidae), should be lacking but they are present and often well developed in many New Zealand species. There is still much confusion at the generic level, which the work of Ward (1997) partly remedied. But it is the dozen

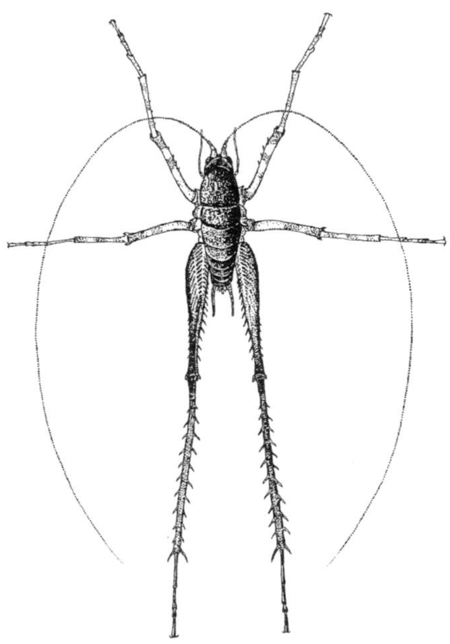

Cave weta *Gymnoplectron edwardsii*.

From Grant 1999

or so new species in *Isoplectron* and *Neonetus* that will pose a great problem with their merging variation in presently fixed generic characters. *Talitropsis* is already known to vary in 'key' characters. New Zealand's subantarctic islands (Antipodes, Auckland, Bounty, Campbell, and Snares) each have a monospecific genus (except Antipodes – a species has been seen there but not caught). These, too, may represent the ends of variation in one (especially *Pleioplectron*) or more mainland New Zealand genera. The species of *Gymnoplectron* should be combined with the older name *Pachyrhamma*. The situation concerning cave species in northwest Nelson and northern Westland (Johns 1981) is now even more complex as some, previously thought to be in *Gymnoplectron*, are certainly members of the poorly known genus *Macropathus*.

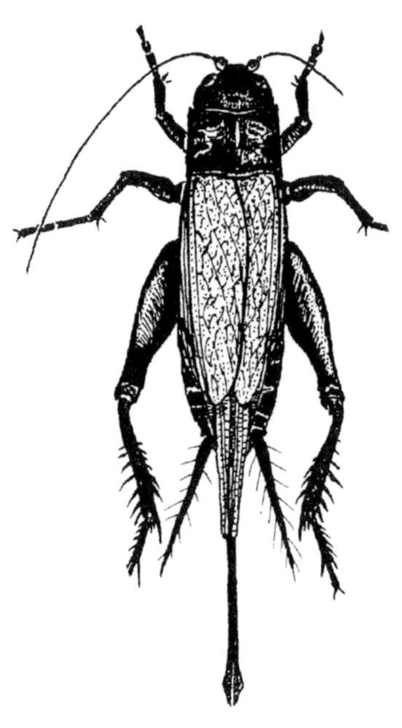

Female black field cricket
Teleogryllus commodus.
From Grant 1999

Crickets, mole crickets, and grasshoppers

Gryllidae: Crickets

There are relatively few black crickets in New Zealand – only nine species, comprising four shared with Australia, one introduced from Europe, and three small ones that are very similar to their close Australian relatives (Swan 1972; MacIntyre 1977a; Ramsay 1991). One very small species, known for at least 40 years and quite exceptional, is still undescribed. It has no wings, and thus neither sings nor has any organs for hearing cricket song. Crickets normally live in open, grassland/shrubland associations, but this unusual native species lives in thick moss in high-rainfall forests. It represents a new genus.

The large black cricket *Teleogryllus commodus* is widespread in Australia and the Pacific Islands. In New Zealand it is a well-known inhabitant of suburban gardens and lawns and a pest of pastures in the northern part of the North Island, but its range extends south to Kaikoura. The Kaikoura population is different, however (Chen et al. 1967), and the differences pose a problem concerning the time of its origination in Kaikoura and subsequent genetic differentiation. North Island and Nelson populations are genetically very close to Australian populations. The Kaikoura population is separate, yet if the species *was* introduced, even the 200 years since European colonisation would seem insufficient for genetic divergence. Perhaps it is part of a wider Polynesian introduction and the North Island populations have been genetically infused with later introductions from Australia.

Rarely seen little brown crickets are placed in the genus *Metioche*. Whether this placement is correct or merely convenient is not known; much study is needed. They appear to be throughout the country and, although only one species has been described, many species may be present. Thus the status of *Gryllodes maorius* Saussure, 1877, *Metioche maorica* (Walker, 1869), and *Ornebius* 'Paihia' of Ramsay (1993) need urgent revision.

Gryllotalpidae: Mole crickets

There is only the one burrowing cricket, *Trimescaptor aotea*, and it too is songless, unlike its very noisy Australian relatives.

Tettigoniidae: Katydids and long-horned grasshoppers

There are three long-horned grasshoppers in New Zealand, all within the worldwide genus *Conocephalus*, and one katydid, *Caedicia simplex*. One *Conocephalus* species has had a name change (Pitkin 1980) – *C. modestus* to *C. albescens*. The species are either shared with Australia or have close relatives there. A Kermadec Island record of *Solomona solida* needs confirmation.

Acrididae: Grasshoppers

There are 22 species of grasshopper in the fauna. The locust *Locusta migratoria* is widespread around the world, whereas the other 21 species are endemic. *Phaulacridium marginale* and those species presently in *Sigaus* have close relatives in Australia. There is a real need to re-examine *Sigaus* and Key (1991) has already suggested that it needs dividing into three genera. Three new species are thought to be present, two in Fiordland (Morris 2003).

All groups have been intensively studied biologically (MacIntyre 1977b; Richards 1954a, b, 1962, 1965a, b, 1973; White 1975, 1978; White & Sedcole 1991; Westerman & Ritchie 1984) and many examples have been successfully reared (Barrett 1991). Devolving from their icon status, intensive efforts are now being made to assess the rarity of the species (Gibbs 1998).

Grasshopper *Phaulacridium marginale*.
From Grant 1999

Order Phasmatodea: Stick insects

Phasmatodeans are stick and leaf insects but only the former occur in New Zealand. Masters of camouflage, their greatest diversity is in the tropics, where veritable giants have been recorded. The largest species, found on the island of Borneo, exceeds half a metre in length including the legs (Hennemann & Conle 2008). Phasmatodeans are related to other insect orders such as Embioptera and Orthoptera. Altogether, there are around 3000 species worldwide.

In New Zealand there are 22 described and several undescribed species (Jewell & Brock 2002; Buckley et al. 2009a). All of them are flightless slow-moving, night-feeding herbivores. They have a long body (40–150 millimetres) with widely separated tubular legs. Salmon (1991) provided keys to adults and eggs and provided coloured figures. Jewell and Brock (2002) published keys to adults (both sexes) and eggs of the genera, including two new genera, and gave notes on the taxonomic status of particular species, including synonyms and a bibliography for stick insects. Trewick et al. (2005) and Buckley et al. (2009a,b) proposed synonyms within *Argosarchus* and *Clitarchus* based on DNA sequence analysis. The most recent taxonomic addition is that of Buckley and Bradler (2010) for an endemic genus and species from Northland. Males are most easily identified by the shape of the claspers and the number, size, and arrangement of teeth on the claspers. Females are best identified by the terminalia including size and shape of the cerci, and internal valves. The distribution of spines on the body can be useful for identifying genera but are often too variable to be of use for species-level identification. Each species has diagnostic characters associated with the eggs (Salmon 1991; Jewell & Brock 2002).

All New Zealand species are endemic and, with the exception of *Clitarchus*, all genera are endemic. However, once a taxonomic revision of *Clitarchus* is completed this genus is likely to be endemic also (Buckley & Bradler pers. comm.). Following the taxonomic classification of Günther (1953) the genera *Acanthoxyla*, *Argosarchus*, *Clitarchus*, and *Pseudoclitarchus* and are placed in the Phasmatinae and the remaining genera – *Asteliaphasma*, *Micrarchus*, *Niveaphasma*, *Tectarchus*, and *Spinotectarchus* – are placed in the Pachymorphinae. However, it is known that neither of these groups is monophyletic (Günther 1953; Bradler 2001, 2009; Whiting et al. 2003; Trewick et al. 2008; Buckley et al. 2009c, 2010). Analysis of morphology and DNA sequences (Buckley et al. 2009c, 2010) shows that New Zealand taxa are all members of the clade Lanceocercata (Bradler 2001). This is supported by key apomorphies including absence of a vomer, the male clasping the female at the base of sternite X, claspers orientated medially, and cerci leaf like and flattened. The molecular phylogenetic studies of Buckley et al. (2009c, 2010) showed that the New Zealand fauna is not monophyletic and in fact forms two clades both nested within a larger and morphologically diverse New Caledonian radiation. The first clade contains only *Spinotectarchus acornutus* and the second clade contains all of the remaining genera. Many of the New Caledonian species are winged or have wing pads, suggesting that the ancestor of the New Zealand lineages may have been winged. The loss of flight in the New Zealand taxa is consistent with the evolution of island insects in general.

Tectarchus salebrosus.
Thomas Buckley

Most New Zealand stick-insect species are found only or mainly at lower altitudes below 900 metres, which reflects the tropical origin of these insects.

A notable exception is *Niveaphasma annulata* with several populations that are found in alpine areas up to 1500 metres in the central to southern South island (Jewell & Brock 2002; O'Neill et al. 2009). There is also at least one undescribed species of *Micrarchus* that is found above the tree line in northern Westland and Nelson (Salmon 1991). *Tectarchus salebrosus* has also been found in the Seaward Kaikoura Range well above the tree line (Jewell pers. obs.).

The genus *Asteliaphasma* contains two nominal species, *A. naomi* and *A. jucundum*, but it is not clear if these are taxonomically distinct. They are relatively common in forest in the upper North Island and usually associated with *Meterosideros* vines and sometimes *Leptospermum scoparium*. *Asteliaphasma* are relatively gracile phasmatodeans but some individuals are adorned with lobes and crests and vary in colour (Salmon 1991). This variation does not appear to be of taxonomic significance, with both body morphs being observed in the same population.

The genus *Tectarchus* contains four species and these species range from just south of Auckland to the central South Island. *Tectarchus huttoni* and *T. ovobessus* are found in both North and South Islands and *T. semilobatus* is restricted to the South Island. *Tectarchus salebrosus* is commonest in the South Island, but Salmon (1991) reported it from the lower North Island, although this observation has not been replicated (Buckley unpubl.). *Tectarchus huttoni* is found on Kapiti Island (Wilkinson & Wilkinson 1952; Moeed & Meads 1987) and islands in the Marlborough Sounds. *Tectarchus* species are found on a broad range of plant species including *Leptospermum scoparium* and *Meterosideros* spp. (Myrtaceae), *Rubus* spp. (Rosaceae), *Weinmannia racemosa* (Cunoniaceae), *Astelia* spp. (Asteliaceae), *Raukaua anomalus* (Araliaceae), and *Gahnia* spp. (Cyperaceae). Species of *Tectarchus* are most easily diagnosed by their smooth body and ridge-shaped tergites. Most species come in two colour morphs, green and brown of various shades.

Niveaphasma annulata is confined to the southern two thirds of the South Island where it is found from sea level to at least 1500 metres. It is associated with *Muehlenbeckia axillaris*, *M. complexa*, and more rarely *M. australis* (Polygonaceae). It is also commonly found on *Rubus* spp., especially *R. schmidelioides*, and also on *Coriaria angustissima* and *C. plumosa* (Coriariaceae), *Leptospermum scoparium*, and *Weinmannia racemosa* (Salmon 1991; Jewell and Brock 1991; O'Neill et al. 2009).

Micrarchus is another genus of small spiny stick insects and currently contains one recognised species and several undescribed species in the South Island. *Micrarchus hystriculeus* is found in lowland habitats and feeds on *Leptospermum scoparium*, *Rubus* sp., *Acaena* spp. (Rosaceae), *Muehlenbeckia* spp., and *Meterosideros perforata*. This species is found in the upper South Island and lower North Island. As mentioned above an undescribed species is present about the tree line in the upper South Island and can be found on tussock, sedges, *Astelia* spp., *Muehlenbeckia axillaris* and *Leptospermum scoparium* (Salmon 1991; Buckley & Jewell, unpubl.).

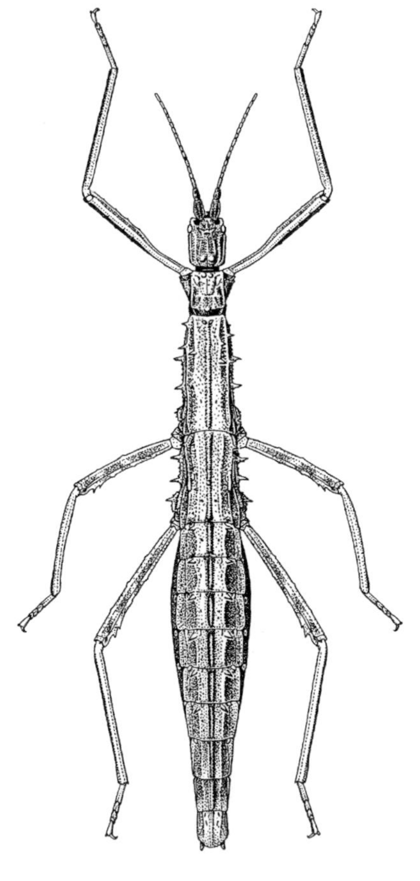

Micrarchus n. sp.

Des Helmore

Argosarchus horridus is the longest New Zealand stick insect, reaching slightly over 150 mm in the female. It is found most commonly on *Rubus* spp., *Plagianthus regius* (Malvaceae), *Lophomyrtus bullata*, and *Muehlenbeckia australis*. It is widespread throughout the North Island, with the exception of Northland, and at mainly coastal and low-elevation regions of the South Island (Salmon 1991; Buckley et al. 2009a). It is also recorded from the Chatham Islands (Brunner 1907), which is notable given that the species is flightless (Trewick et al. 2005; Buckley et al. 2009a). Within the Chathams group *A. horridus* is found on at least two of the islands (Buckley et al. 2009a). *Argosarchus horridus* is notable as being a geographic parthenogen (Trewick et al., 2005; Buckley et al., 2009a) with males extremely rare in the South Island and common in the North Island (Buckley et al., 2009a).

Pseudoclitarchus sentus is restricted to the Three Kings Islands where it is found on Great Island and South West Island. Salmon's (1991) record of this species from Great Barrier Island is dubious, having not been recorded there since, while other species are common. This species is commonly found on *Kunzea ericoides* (Myrtaceae) and Salmon (1948) also recorded it from *Litsea calicaris* (Lauraceae) and *Streblus smithii* (Moraceae).

Clitarchus hookeri is perhaps the commonest species on the North Island and is known from the northern offshore locations of Little Barrier, Mercury Islands, Cuvier, and Kapiti, and islands in the Marlborough Sounds (Johannesson 1972, Hicks et al. 1975; Moeed & Meads 1987; Salmon 1991; Buckley et al. 2009b). There is an undescribed species on the Poor Knights islands (Watt 1982; Buckley et al. 2009b). *Clitarchus hookeri* is most commonly found on *Leptospermum scoparium* but also feeds on *Rubus* spp. and *Meterosideros* spp. Populations in the upper North Island are predominantly bisexual whereas males are completely unknown from the South Island (Buckley et al. 2009b), a pattern similar to *A. horridus*.

The genus *Acanthoxyla* is remarkable for its complete lack of males (Salmon 1991), extremely unusual for an animal genus. There are nine recognised species (Jewell & Brock 2002) but these are difficult to identify and it is not clear how many species there actually are (Morgan-Richards & Trewick 2005; Buckley et al. 2008). Most individuals are green with black-tipped spines, but some individuals are brown and *A. inermis* lacks black-tipped spines. *Acanthoxyla* is found from Northland southwards to Stewart Island and in the South Island it is mainly restricted to lowland areas. *Acanthoxyla* species can be found on podocarps, particularly *Podocarpus totara, Dacrydium cupressinum*, and less commonly *Phyllocladus* spp. It is also common on *Metrosideros* vines, *Rubus* spp., *Leptospermum scoparium, Weinmannia racemosa*, and *Muehlenbeckia* spp. Despite the lack of males it appears that lineages of *Acanthoxyla* have repeatedly hybridised with *C. hookeri* (Morgan-Richards & Trewick 2005; Buckley et al. 2008) and it also appears that the *Acanthoxyla* lineages are themselves the result of hybrid events between sexual species of *Acanthoxyla*, which may now be extinct (Buckley et al. 2008).

The newly described genus *Tepakiphasma* contains one species, *T. ngatikuri*, and is known only from two specimens collected from Te Paki at the far north of the North Island (Buckley & Bradler 2010). Among other characters, this relatively gracile genus is diagnosed by having a perforated capitulum, which is unique amongst New Zealand phasmatodeans. The specimens were collected from *Metrosideros perforata* but reared on *M. excelsa* and *Lophomyrtus bullata*. It is likely to be restricted to the Te Paki region, which is well known as a centre of endemism for many taxa. Owing to the restricted known distribution and small number of specimens collected it is clear that *T. ngatikuri* requires urgent conservation attention.

Spinotectarchus acornutus has a very similar distribution to *Asteliaphasma* and is found in similar habitats. In addition to mainland localities, *Spinotectarchus acornutus* also occurs on Little Barrier Island (Salmon 1991), Great Barrier Island (Hutton 1898) and the Hen and Chickens Islands (Buckley, unpubl.). It is commonly found on *Meterosideros* vines as well as *Leptospermum scoparium, Cyathodes* spp., *Astelia* spp., *Gahnia* spp., *Weinmannia racemosa*, and *Podocarpus totara*. *Spinotectarchus acornutus* has highly distinctive male claspers with the teeth clustered at the apex of the hemitergites (Buckley et al. 2010). The structure of the claspers is similar to its New Caledonian sister taxon and reflects its phylogenetic isolation from the remaining New Zealand phasmatodean genera (Buckley et al. 2010).

Published details of the distribution of New Zealand species were based mainly on records from the Museum of New Zealand (Salmon 1948, 1954, 1991) and Canterbury Museum (Wise 1977) until Jewell and Brock (2002) re-evaluated the species and types. The Canterbury Museum has many unpublished locality records. The largest New Zealand phasmatodean collection is housed in the New

Pseudoclitarchus sentus mating pair.
Thomas Buckley

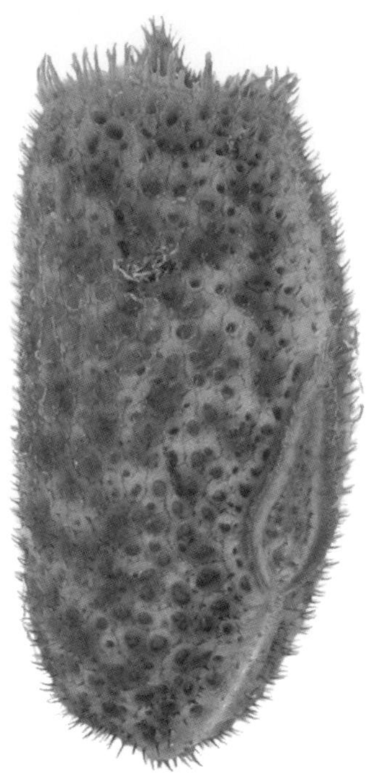

Egg of *Spinotectarchus acornutus*
Thomas Buckley

Zealand Arthropod Collection, which contains over 1500 specimens including several undescribed species and genera. Databasing and mapping of all specimens is underway and will reveal how many species are of conservation concern.

Stick insects can be found in almost all types of vegetation from just above high-tide mark to the high alpine zone. Eggs are dropped haphazardly to the ground and take several months to hatch (Salmon 1991). Identification of nymphs of most species is difficult as most are green on hatching, but *Argosarchus horridus* is brown with distinctive white banding on the legs (Salmon 1991). Young nymphs then walk up tree trunks or other plants in search of the host plant (Moeed & Meads 1983). Salmon (1991) remarked that phasmatodean requirements of host plants to achieve development through to adult are strict. However, more recent rearing studies reveal that many species can be successfully reared on hosts rarely supporting them in the wild. More critical study of food sources that favour particular species are warranted, as these would improve the ability to map and monitor their presence and obtain specimens for rearing in captivity. The number of moults from nymph to adult has not been determined for all species but Stringer (1970) reported that *C. hookeri* took six moults and Salmon (1991) claimed that other species had fewer moults.

The nocturnal feeding habits of phasmatodeans lower their risk of predation. This behaviour and their camouflage notwithstanding, several forest birds feed on stick insects. Some are captured by the mainly nectar-feeding tui (*Prosthemadera novaeseelandiae*) (Merton 1970; Higgins et al. 2000a) as well as by silvereye (*Zosterops lateralis*) (Gill 1979) and even fantail (*Rhipidura fuliginosa*) (Moeed & Fitzgerald 1982), which feeds mainly on flying insects (Gravatt 1971; Gill 1980). Robins (*Petroica* spp.) catch stick insects mainly in late summer–autumn but are occasionally deterred from capturing them when the spiny legs of the stick insect are used in defence (Powlesland 1981). Bellbirds (*Anthornis melanura*) possibly catch them too, because some nectar-rich flowers (rata, pohutukawa) used by bellbirds and tui (Higgins et al. 2000b) are also favoured by stick insects. During the night, moreporks (*Ninox novaeseelandiae*) (Clark 1992; Haw et al. 2001) and little owls (*Athene noctua*) (Soper 1963) take stick insects as a small part of their diet.

Possums are significant predators (Cowan & Moeed 1987) and among the smaller mammals in New Zealand, only ship rats (*Rattus rattus*) are known to feed on stick insects (Daniel 1973; Innes 1979). Mice (*Mus musculus*), Norway rats (*R. norvegicus*), and kiore rats (*R. exulans*) keep to the ground more than ship rats (King 1990), so they are less likely to feed on stick insects often but may take a toll on eggs. Cats and introduced birds also catch and eat stick insects. Ferrets (*Mustela furo*), which can feed in trees on birds, could possibly consume stick insects to some extent. Tuatara (*Sphenodon* spp.), now confined to offshore islands, occasionally consume stick insects (Wall 1981). Native lizards readily capture and eat stick insects and in some habitats could be a significant natural predator; some of the host plants such as *M. complexa*, *Leptospermum scoparium* and *Kunzea ericoides*, are also strongly favoured by certain lizard species. Both vespulid and other wasps are major predators and have devastated stick-insect communities in forests where they have reached plague proportions. Parasites of New Zealand stick insects have not been well studied, but mites are often observed on them and Yeates and Buckley (2009) recorded unknown species of mermithid nematodes emerging from *Clitarchus* spp., *Asteliaphasma naomi*, and *Acanthoxyla* spp.

Order Hemiptera: Bugs, cicadas, scale insects, aphids etc.

The Hemiptera is by far the largest and most successful of the hemimetabolic insects, i.e. those having young that look like wingless adults and an incomplete metamorphosis that does not involve a pupa. There are about 82,000 named species, and probably many more, and the Hemiptera has the dubious distinction of including probably more destructive and costly pest species than any other insect order.

These insects used to be grouped into the orders Heteroptera and Homoptera (the latter with two suborders, Sternorrhyncha and Auchenorrhyncha), based basically on wing structure. A number of recent studies, integrating molecular and morphological data, divide the order into four suborders – Sternorrhyncha, Auchenorrhyncha, Coleorrhyncha, and Heteroptera, with the Auchenorrhyncha believed to be more closely related to the Heteroptera than to the Sternorrhyncha. The Sternorrhyncha, especially members of the superfamily Psylloidea (jumping plant-lice), are closer to the basal hemipteran body-plan than the other groups.

Hemiptera means 'half wing' and refers to the fact that the basal part of the forewings is tough and hard while the outer part and the hindwings are membranous. The differentiation of the forewings is also implied in the name Heteroptera ('different wings'), in contrast to the Homoptera ('same wings'). All hemipterans are characterised by piercing and sucking mouthparts. Some suck plant juices and are plant pests, while others can bite painfully. The order dates back to the Permian period and is fairly well represented by fossils.

Suborder Sternorrhyncha

Included in this suborder are the psyllids or jumping plant-lice, whiteflies, aphids, coccids, mealybugs, and scale insects. All are plant suckers, mostly of the phloem (the food-conducting tissue) of flowering plants, but some also on conifers, on which they probably evolved.

Superfamily Psylloidea: Jumping plant-lice

These small plant-suckers attack plants as nymphs, feeding on the phloem via stomatal pores in leaves. Nymphs tend to be confined to one host species or to a group of closely related species. They are flattened and non-jumping. Adults, which are winged, are less discriminating and sometimes feed and oviposit on plants that do not support their nymphs. Leaf pit-galls and closed woody galls are a common feature of species of Triozidae and some Psyllidae also form galls.

The New Zealand psylloid fauna has 19 described genera, of which 12 are represented only by adventive species, mostly from Australia. Others are shared with Australia, with endemic species on both sides of the Tasman (e.g. *Acizzia, Anomalopsylla, Ctenarytaina*). Notable in the present classification is the absence of endemic genera, although a small number of endemic species are likely to fill some of this gap in the future. In all, there are more than 90 described and undescribed species. Of the six families of Psylloidea worldwide, only the Phacopteronidae and Carsidaridae are not found in New Zealand, but two other families – Calophyidae and Homotomidae – are each represented by only a single adventive species. The Psyllidae has 40 species and one subspecies, of which 23 (60%) are adventive. In contrast, the Triozidae has 54 species, only three of which are adventive (though not all yet determined to species); all the rest are endemic, including 19 undescribed species. The number of adventive species has grown over recent decades in parallel with the increase in aircraft traffic and the availability of exotic host plants such as plantation eucalypts. About a third of the New Zealand psyllid species are adventive.

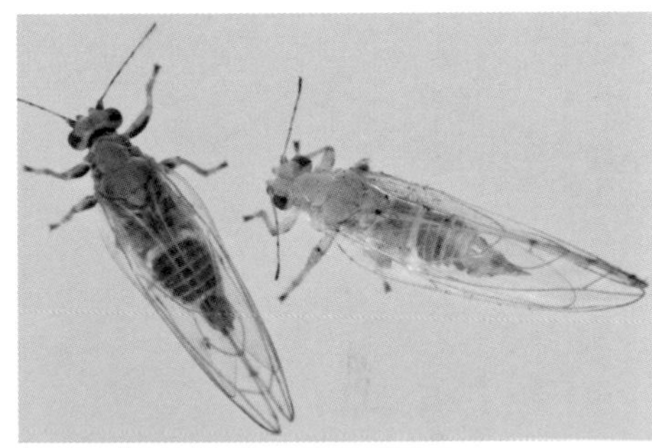

Trioza vitreoradiata – male (left) and female (right) (upper photo) and male in profile (lower photo).

Nicholas Martin

Psyllids are usually fairly narrowly host-specific, with related species feeding on related plant groups. In New Zealand, the genus *Trioza* has radiated on host-plants in five genera of Asteraceae, particularly *Olearia* – 10 of 15 described species of *Trioza* are on that genus. This loyalty to particular plant groups does not prevent psyllids from using more unusual host groups when circumstances allow or demand, e.g. the endemic spondyliaspidine *Ctenarytaina fuchsiae* on *Fuchsia* (Onagraceae). Although most psyllids worldwide use host plants among dicotyledonous angiosperms, in New Zealand two endemic triozine species complete their development on gymnosperm species (*Halocarpus*, Podocarpaceae).

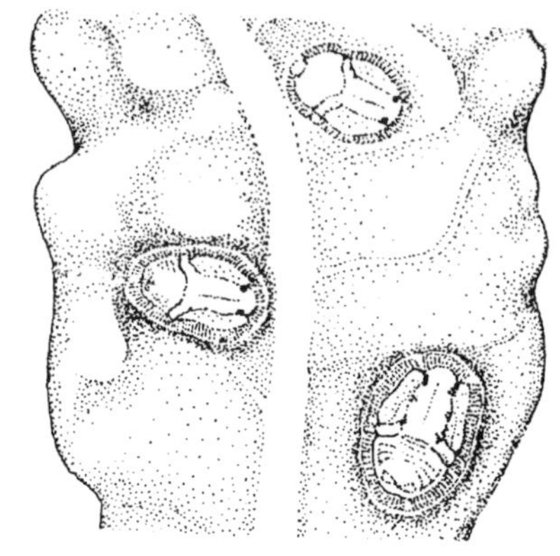

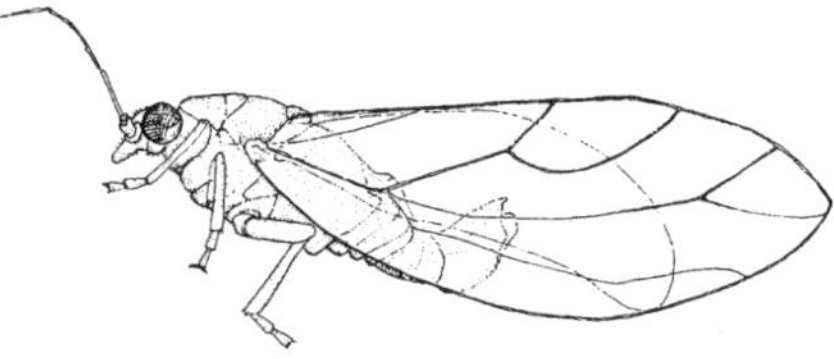

Trioza vitreoradiata adult (lower) and nymphs on leaf of lemonwood (*Pittosporum eugenioides*) (upper).
From Grant 1999

Parasitoids and predators recorded attacking psyllids in New Zealand

Order/Family	Parasitoid/predator species	Psyllid host/prey
Hymenoptera: Encyrtidae (parasitoid)	*Adelencyrtoides variabilis*	*Trioza irregularis*
	Coccidoctonus gemitus	*Cardiaspina fiscella**
	Psyllaephagus acaciae	*Acizzia acaciaebaileyanae*
	Psyllaephagus gemitus	*Cardiaspina fiscella*
	Psyllaephagus pilosus	*Ctenarytaina eucalypti*
	Psyllaephagus richardhenryi	*Anoeconeossa communis*
Coleoptera: Coccinellidae (predatory)	*Adalia bipunctata*	*Acizzia acaciae*
		Acizzia uncatoides
	Cleobora mellyi	*Acizzia acaciae*
		Acizzia acaciaebaileyanae
		Acizzia uncatoides
	Halmus chalybeus	*Trioza vitreoradiata*
		Acizzia acaciae
		Acizzia uncatoides
	Harmonia conformis	*Acizzia acaciae*
		Acizzia uncatoides
Hemiptera: Miridae (predatory)	*Idatiella albisignata*	*Psyllopsis fraxini*
		Psyllopsis fraxinicola
Neuroptera: Hemerobiidae (predatory)	*Boriomyia maorica*	*Trioza vitreoradiata*
	Drepanacra binocular	*Trioza vitreoradiata*
		Acizzia acaciae
		Acizzia albizziae
		Acizzia uncatoides
	Micromus tasmaniae	*Trioza vitreoradiata*

* via *Psyllaephagus gemitus*

Psylloids affect many native plant genera, including *Alseuosmia, Carmichaelia, Dacrydium, Discaria, Dodonaea, Fuchsia, Pseudopanax,* and *Schefflera*. Some psylloids severely modify their host, like the psyllid on *Pittosporum*, pitting and causing yellow streaks to appear on distorted leaves.

Superfamilies Aleyrodoidea and Coccoidea: Whiteflies, scale insects, and mealybugs

Scale insects (Coccoidea) are soft-bodied sap-suckers that produce honeydew through their anal apparatus and are sexually dimorphic, that is, the adult males are winged insects and the adult females are neotenic (larvaeform). Worldwide there are about 7600 species in 21 families. Individuals may be found feeding on all parts of a host plant, from the roots, stems, and leaves to the fruit. One New Zealand species, *Newsteadia myersi* (Ortheziidae), lives exclusively in litter and mosses and is thought to feed on roots. Soil-inhabiting, root-feeding genera of mealybugs such as *Rhizoecus* are pests in the garden nursery industry where plants are grown in pots. The vast majority of scale insects feed on leaves and young stems or through the bark of larger branches and tree trunks. Those that feed on fruit bear the risk of a short time-span before the fruit falls or is eaten by other animals, and are of economic importance where the fruit is a horticultural crop. Aleyrodoidea (whiteflies), with about 1450 species worldwide, have a similar life-style to scales, except that both male and female adults are fully winged.

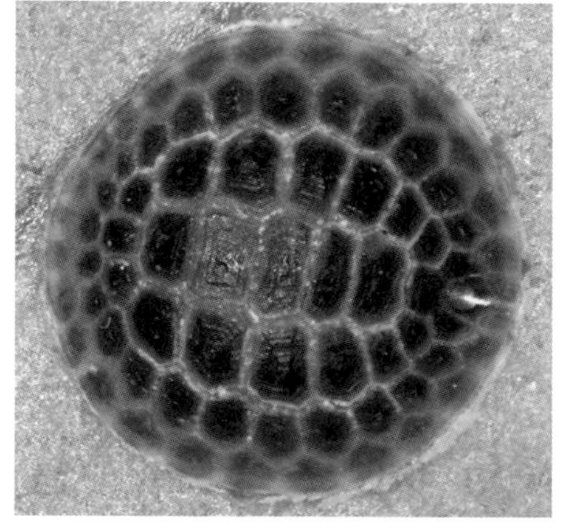

Endemic scale insect *Epelidochiton piperis* (Coccidae).
Rosa Henderson

Aspects of life-cycles

Male and female coccoids go through very different post-embryonic development. Males develop similarly to holometabolous insects, i.e. through prepupal and pupal stages to winged adults quite different from the females. They have no

Diversity of Coccoidea in New Zealand and globally

Superfamily/Family	No. of species in New Zealand Total	Endemic	Approx. no. world species	References*
Aleyrodoidea				
Aleyrodidae (whiteflies)	14	8	1450	Dumbleton 1957; Mound & Halsey 1978; Jesudasan & David 1991
Coccoidea				
Asterolecaniidae (pit scales)	3	2	232	Russell 1941; Stumpf & Lambdin 2006
Cerococcidae (false pit scales)	3	3	72	Lambdin & Kosztarab 1977; Lambdin 1998
Coccidae (soft scales)	59	45	1151	Hodgson & Henderson 1998, 2000; Henderson & Hodgson 2005
Diaspididae (armoured scales)	~90	~62	2413	Green 1914, 1929; Brittin 1915a,b, 1916, 1937; McKenzie 1960; Borchenius & Williams 1963; Morrison & Morrison 1966; Ben-Dov 1976; De Boer & Valentine 1977; Takagi 1985; Danzig 1993
Eriococcidae (felted scales)	102	96	554	Hoy 1961, 1962; Hodgson 1994; Hodgson & Henderson 1996; Henderson 2006, 2007a,b
Halimococcidae (halimococcids)	1	1	21	Deitz 1979a,b
Margarodidae (margarodids)	11	10	445	Morales 1991
Ortheziidae (ensign scales)	3	1	196	Green 1929; Kozár & Konczné Benedicty 2000
Phenacoleachiidae (phenacoleachiids)	2	2	2	Maskell 1891; Beardsley 1964
Pseudococcidae (mealy bugs)	116	96	2224	Cox 1987; Williams & Henderson 2005

* General reference: ScaleNet [www.selbarc.usda.gov/scalenet/scalenet.html]

mouthparts and are short-lived. The 1st-instars, known as crawlers, are usually indistinguishable as males or females, and are the main dispersal stage. They reach new sites by walking, either on their natal host plant or on suitable plants close by, or they may be blown by wind to more-distant sites. They soon settle, put their feeding stylets into the plant tissue and usually become sessile for varying periods. Nymphs in the families Asterolecaniidae, Cerococcidae, Diaspididae, Halimococcidae, and species of *Cryptococcus* (Eriococcidae) lose their legs at the first moult and their antennae reduce to several short setae on a sensory base, so individuals are quite sessile thereafter. The legs of whiteflies are reduced when nymphs and they remain sessile until becoming adult.

In New Zealand, the coccoid life-cycle shows greatest variation in endemic genera of the family Margarodidae. All crawlers have normal legs and antennae and all settled 1st- and 2nd-instars have reduced legs and antennae and inhabit waxy cysts. Females of *Coelostomidia* and *Platycoelostoma* species proceed through the 3rd-instar phase in this state, then redevelop normal legs and antennae and become mobile adult females; females of *Ultracoelostoma* species, on the other hand, retain reduced legs and antennae and continue to inhabit cysts. The male stages in *Coelostomidia* and *Platycoelostoma* revert to normal legs and antennae after the 2nd-instar, developing through prepupal and pupal stages to become fully winged adults (no males are known for *Platycoelostoma*). Species of Coccidae and Eriococcidae retain functional legs throughout their lives in most genera but females become more sessile when reaching adulthood. The mealybugs, phenacoleachiids, and ortheziids are mobile throughout their lives.

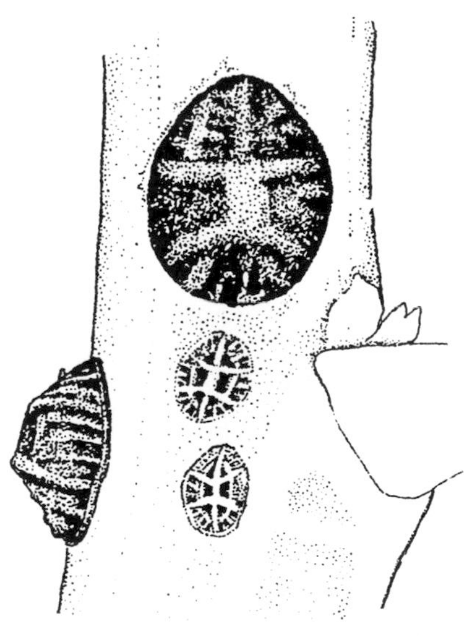

Introduced black scale *Saissetia oleae* (Coccidae).

From Grant 1999

Protective coverings

A distinctive feature of scale insects is their different types of protective waxy coverings. These can broadly identify each family or closely related groups of families. Of the more mobile families, the mealybugs have bodies coated with powdery wax like a dusting of flour, hence the name 'mealy'. Phenacoleachiids tend to live in clusters surrounded by fluffy cocoons. Ortheziids are also known as ensign scales, for the large white plates of wax extending posteriorly from their bodies. Margarodids inhabit stout waxy cysts, often hidden in crevices in the bark of host trees, once they are completely sessile with reduced legs. Their long anal tube extends from the cyst so that honeydew droplets fall away from the insect, incidentally providing a sweet, dripping 'tap' for birds and other invertebrates to feed from. Pit- and gall-formers in Halimococcidae and Eriococcidae need little covering other than some waxy strands near the opening of their dwelling. The common name for eriococcids – felted scales – derives from the adult covering (both female and the penultimate male instars) of a woven waxy sac on the majority of species. Members of the Aleyrodidae, Asterolecaniidae, Cerococcidae, and Coccidae in New Zealand all have covers of glassy wax called tests, in variable form, although females of the species in the coccid genus *Lecanochiton* lose this wax after the larval stage and develop a strongly sclerotised derm instead. Lastly, the so-called armoured scales (Diaspididae) construct a cap from the moulted skin of the previous instar, incorporating added waxy secretions whereby each succeeding instar can be seen as a ring on the cap.

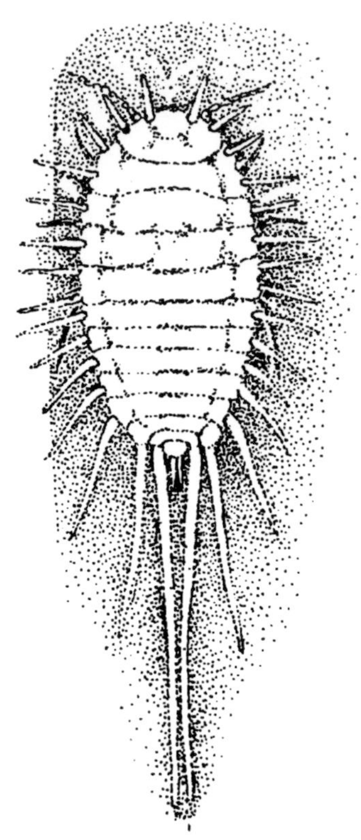

Introduced long-tailed mealybug *Pseudococcus longispinus* (Pseudococcidae).
From Grant 1999

Historical perspective

New Zealand scale insects were first studied by W. M. Maskell between 1879 and 1898 (e.g. Maskell 1887). He described about 300 species of Sternorrhyncha worldwide, including 94 species from New Zealand (Deitz & Tocker 1980). Since Maskell, there have been a number of studies on the New Zealand Coccoidea and Aleyrodidae. The Eriococcidae were completely revised by Hoy (1962), as were the Margarodidae by Morales (1991) and the Pseudococcidae by Cox (1987). A full revision of the Coccidae was published by Hodgson and Henderson (2000), with descriptions of 25 new species and seven new genera. Species in the other families have been described by various authors over many years.

Economic importance of whiteflies and scale insects

By the 1880s, the adventive species already present were causing economic damage to the horticultural enterprises of early settlers. Maskell (1887) summarised advice on how to control scale-insect pests in an extensive chapter in his book on the nuisance New Zealand species. In a reversal on controlling scales, they were subsequently used as a means of biological control. Between 1948 and 1952, the self-introduced felted scale *Eriococcus orariensis* (manuka blight) was deliberately spread by farmers wishing to control the growth of manuka scrub, *Leptospermum scoparium* (Hoy 1961). Eventually, populations of adventive scale species stabilised, including the manuka blight scale which has been overtaken by another less noxious scale, *Eriococcus leptospermi*, as the dominant eriococcid on manuka. Hoy (1961) attributed the decline of *E. orariensis* to the sudden appearance of the entomogenous (parasitic) fungus *Angatia thwaitesii* (generally known as *Myriangium thwaitesii*). It is interesting to note that manuka trees may still be covered in sooty mould growing on the honeydew produced by these scales and margarodids, leading the casual observer to assume that it is still 'manuka blight'.

Worldwide, whiteflies and scale insects (Aleyrodidae, Coccidae, Diaspididae, Margarodidae, Pseudococcidae, and to a lesser extent Eriococcidae) are the cause of millions of dollars of economic loss yearly to growers of agricultural and horticultural crops. By sucking the plant sap, these insects deplete their host plant's resources and vigour and may distort plant tissues. Further, by producing honeydew, they allow the usual growth of sooty moulds, with consequent spoilage of fruit and reduction in photosynthesis. Citrus crops in New Zealand suffer damage by adventive species of Coccidae and Diaspididae; the greenhouse whitefly *Trialeurodes vaporariorum* damages soft fruit, such as indoor and outdoor tomatoes. Scale insects may also transmit or cause plant diseases. For example, a strain of the sweet potato whitefly *Bemisia tabaci* causes leaf-silvering in cucurbits. Quarantine regulations between exporting and importing nations may require zero tolerance for certain listed pest species of scale insects, adding

Introduced rose scale *Aulacaspis rosae* (Diaspididae).
From Grant 1999

to compliance costs for fruit producers, and this is so for New Zealand's exports of pip-fruit and kiwifruit.

Endemic scale insects are not crop pests as they are restricted to native host plants and only rarely are found on exotic host plants – with some exceptions among grass-feeding mealybugs (Cox 1987). On the other hand, most adventive species prefer exotic host plants, although eight mealybug species, five soft-scale species, three armoured-scale species, and the single adventive margarodid species *Icerya purchasi* have migrated into natural ecosystems and also feed on native plants.

Endemic *Leucaspis podocarpi* (Diaspididae).
Rosa Henderson

Diversity
New Zealand has about 390 species of Coccoidea in 10 families and another 14 species in the Aleyrodidae (whiteflies). In comparison with the rest of the world, New Zealand is depauperate in the three largest families – the Pseudococcidae (mealybugs) have 116 species out of a worldwide total of 2224; Diaspididae (armoured scales) have ca. 90 species out of a total of 2413; and Coccidae (soft scales) have 59 known species out of a worldwide total of 1151. New Zealand is relatively species-rich, however, in the Eriococcidae, with 102 known species out of a total of 554 species worldwide. Endemism is high overall, at about 80%, but varies among families, e.g. 94% in the felt scales (Eriococcidae), 68% in the armoured scales, and 57% in the whiteflies. One family, the Phenacoleachiidae, is wholly endemic. Hodgson and Henderson (2000) considered all of the coccid genera to be endemic; exotic species ascribed to two of the genera, *Ctenochiton* and *Inglisia*, will need to be reassigned to other genera.

Inglisia patella (Coccidae).
Rosa Henderson

Special features
All of the indigenous soft scales (Coccidae), apart from the two *Pounamococcus* species, are probably unique in the way the wax test (protective covering) is constructed of rows of hexagonal wax plates. The males in this group have tests with a distinctive flexibly hinged plate that allows egress of the newly emerged adult male, itself more fragile than the protective waxy covering.

The large amount of honeydew produced by margarodids in beech forests is considered a very important food source for other invertebrates including honeybees and wasps, and for geckos and native birds. The availability of this food to the native fauna has recently been greatly reduced by the invasion of vespid wasps into South Island beech forests. The survival of New Zealand's endemic scale-insect species is dependent on the survival of their native host plants in their natural forest habitat. Owing to their limited dispersal capacity, few endemic scale species are able to recolonise patches of newly planted native forest unless they are very close to natural habitats. Hence the indigenous fauna will probably not benefit from new plantings in urban restoration schemes, but rather from enrichment of existing forest remnants.

Superfamily Aphidoidea: Aphids and kin

The great majority of Aphidoidea are in the family Aphididae (aphids) with the characteristics of polymorphism, complex life histories, the ability to reproduce both sexually and asexually, giving birth to live young, telescoping of generations, and high fecundity (Blackman & Eastop 1984; Minks & Harrewijn 1987; Dixon 1998). Aphids are major pests of temperate agricultural and horticultural crops and forest trees, causing damage either directly by feeding or indirectly by transmitting plant virus diseases (Minks & Harrewijn 1989). Two other families are included in the superfamily – the Adelgidae and Phylloxeridae. Adelgids and phylloxerids, although closely related to aphids, are quite distinct from them and retain a number of primitive features, including the absence of viviparity and the absence of siphunculi (the backward-pointing erect tubes found on the dorsal side of the last segment of the body of most aphid species).

The superfamily classification used here follows that of Carver et al. (1991),

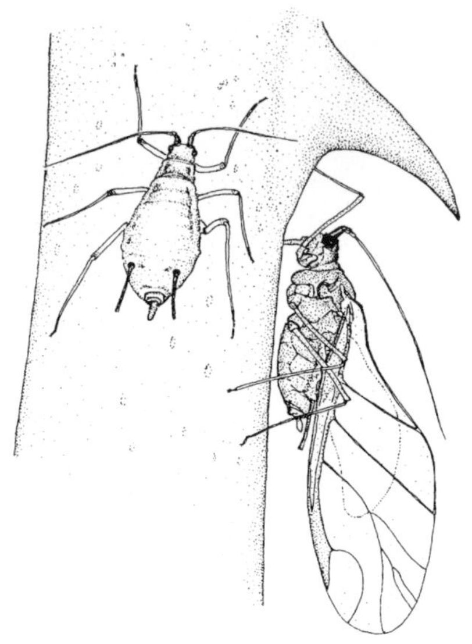

Rose aphids *Macrosiphium rosae*.
From Grant 1999

Blackman and Eastop (1994), and Remaudière and Remaudière (1997) in recognising three families. Classification of aphids follows Remaudière and Remaudière (1997) and Nieto Nafría et al. (1997). Key characteristics of the Aphidoidea are found in Heie (1980), Blackman and Eastop (1984), and Foottit and Richards (1992).

Aphids were first recorded in New Zealand in 1861 (probably *Eriosoma lanigerum*) (Thompson 1922) and 1862 (probably *Brevicoryne brassicae*) (Travers 1864). The first adelgid (*Pineus* sp.) and phylloxerid (*Viteus vitifoliae*) were found in about 1880 (Maskell 1885) and 1865 (Thompson 1922), respectively. Most of the early records of New Zealand aphidoids were of introduced species on economically important plants. Hutton (1904) listed nine species of Aphidoidea (including six aphids), Myers (1922) 15 species (12 aphids), and Tillyard (1926) 18 species (16 aphids). Cottier's (1953) seminal work *Aphids of New Zealand* significantly increased the number of aphid species to 59, with the most recently published estimate being 80 species (Lowe 1973). Adelgids and phylloxerids were not considered by Cottier and Lowe. Recent records of new species are found in Sunde (1973, 1984, 1988), Cox and Dale (1976), and Blackman and Eastop (1984, 1994), among others.

Major repositories of New Zealand specimens

The Natural History Museum, London, has the most comprehensive and best-organised collection of New Zealand aphidoid species. The New Zealand Arthropod Collection, Auckland, is extensive but has been neglected in recent years. Most type specimens of endemic species are found in these two collections. The two Ministry of Agriculture and Forestry National Plant Pest Reference Laboratory collections, in Tamaki (Auckland) and Lincoln (Canterbury), are comprehensive but comprise mostly pest species. A collection held by Plant & Food Research, Lincoln, contains most of the recent collections of indigenous aphid species. Some specimens, including some types of endemic species, are also found in the Australian National Insect Collection, Canberra.

Current known natural diversity

The Aphidoidea is predominantly a northern temperate group, richest in species in North America, Europe, and Central and East Asia (Blackman & Eastop 1984) with about 4700 species in 599 genera in the Aphididae (Remaudière & Remaudière 1997), 50 species in the Adelgidae, and fewer than 50 species in the Phylloxeridae (Foottit & Richards 1992).

Aphid genera and species are considered to be under-represented in the tropics and Southern Hemisphere (including New Zealand) compared to the Northern Hemisphere, possibly as a result of the tropics acting as a barrier to movement of the species-rich subfamilies Aphidinae and Lachninae, which underwent adaptive radiation in the Northern Hemisphere in the Late Tertiary (Heie 1994). Only seven (1%) of the world genera are from subtropical and temperate regions of the Southern Hemisphere (Heie 1994).

The present number of recognised aphidoid species in New Zealand is three adelgids, three phylloxerids, and at least 121 aphid species. The majority of aphidoid species in New Zealand (90%) are aliens (Teulon & Stufkens 2002). There are no endemic phylloxerids or adelgids and the relatively few endemic aphids (at least 12 species) are mostly rare. Only introduced aphids have been reported from subantarctic islands (Cottier 1964; Palmer 1974; Marris 2000; Horning unpubl.). Only seven of the endemic species have been described. *Thripsaphis foxtonensis*, which was first recorded in New Zealand and at one time considered to be endemic (Cottier 1953), is probably of North American origin.

In the Neophyllaphidinae, *Neophyllaphis totarae* lives on more than one species of *Podocarpus* and an undescribed *Neophyllaphis* is thought to occur on snow totara, *P. nivalis* (M. Carver pers. comm.). Worldwide, about 12 *Neophyllaphis* species are found on Podocarpaceae and Araucariaceae, with a distribution that includes the southern hemisphere and mountains of the tropics, extending northwards into China and Japan (Blackman & Eastop 1994). *Neophyllaphis* species exhibit a number of primitive characters and are considered to resemble the hypothesised ancestral aphid (Heie 1987; Carver et al. 1991).

Sensoriaphis nothofagi, subfamily Taiwanaphidinae, lives on several *Nothofagus* species in New Zealand. Three further species of *Sensoriaphis* are found on *Nothofagus* in Australia and the genus is closely related to *Neuquenaphis* (~10 species) on *Nothofagus* in South America and *Taiwanaphis* (about nine species) in southeastern Asia.

Within the Aphidinae, four *Aphis* species (*A. coprosmae, A. cottieri, A. healyi,* and *A. nelsonensis*) and two *Paradoxaphis* species (*P. aristoteliae, P. plagianthi*) have been described from New Zealand. Several other species have also been recognised. *Paradoxaphis* appears to be an endemic genus. At least four species of the *Aphis/Paradoxaphis* group appear to form a genetically distinct lineage within the subtribe Aphidina (von Dohlen & Teulon 2003).

At least two undescribed endemic species belonging to *Euschizaphis* have been recorded from *Dracophyllum* and *Aciphylla,* respectively. An undescribed *Casimira* species has been recorded from *Ozothamnus*.

Wingless *Rhopalosiphum maidis*, a pest of cereals.
Robert Lamberts, Plant and Food Research

Adventive aphid species

As already noted, about 90% of the New Zealand aphids are not endemic (Teulon & Stufkens 2002). A similar proportion of non-endemic species has been recorded from Australia (Carver et al. 1991). All adelgids and phylloxerids in New Zealand are aliens. In terms of the ratio between introduced and indigenous species (111/14), the aphids probably represent one of the most invasive insect groups in New Zealand. Furthermore, the introduced aphids also constitute a significant proportion of the 2600 exotic insect species estimated to be in New Zealand by Emberson (2000). Since the 1950s, an average of one new alien aphid species per year has been found in New Zealand.

Pest species

Aphids are *the* major pests of temperate agriculture, causing damage either directly by feeding or by transmitting plant virus diseases (Minks & Harrewijn 1987, 1989). In New Zealand, many of the introduced aphids, as well as the adelgids and phylloxerids, are also important pests of agricultural and horticultural crops and forest trees (see Lowe 1973; Scott 1984). Endemic species are not considered to be pests although *Neophyllaphis totarae* causes some damage to the growing tips of totara.

Taxonomic novelty

New Zealand endemic aphids constitute a distinctive taxonomic component of the New Zealand insect fauna and of the world aphid fauna. Specific characteristics of aphids and parasitic wasp associates include the following:

Neophyllaphis and *Sensoriaphis* are primitive genera with Gondwanan distributions (Carver et al. 1991). *Neophyllaphis* is considered the closest living relative of the ancestral aphid form.

Paradoxaphis appears to be endemic to New Zealand (Sunde 1988; Remaudière & Remaudière 1997).

Casimira might be an endemic austral genus. The type species, *C. canberrae,* is native to Australia. The only other nominal species, *C. bhutanensis,* was described from India, but the validity of its generic assignment is questionable.

The recent characterisation of the two New Zealand *Euschizaphis* species now means that half the known species of this genus are found in New Zealand.

Accepted dogma is that the austral Aphidinae (i.e. *Aphis, Casimira, Euschizaphis, Paradoxaphis*) are descendants of recent chance trans-tropical immigrants from the Northern Hemisphere (Eastop 2001; von Dohlen & Teulon 2002). However, recent molecular work has found that a group of four New Zealand

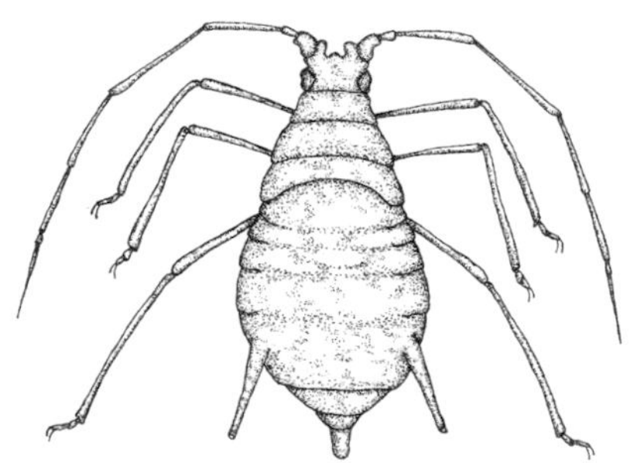

Wingless green peach aphid *Myzus persicae*.
Lincoln University

endemic aphids belonging to the genera *Aphis* and *Paradoxaphis* form a highly supported lineage (possibly basal in the tribe Aphidini) estimated to be ca. 15–30 million years old. These results place this New Zealand group as central to the evolution of the species-rich Aphidinae, which contains many agricultural pests (von Dohlen & Teulon 2003).

The braconid wasp subfamily Aphidiinae (all parasitoids of aphids), although now much more numerous in the Northern Hemisphere, probably originated in the Southern Hemisphere (Belshaw et al. 2000).

The 'mummy' of an external native parasitoid underneath the native aphid *Neophyllaphis totarae.*
Robert Lamberts, Plant and Food Research

Rarity of endemic aphids

Of the endemic taxa, only three – *Neophyllaphis totarae, Sensoriaphis nothofagi,* and the undescribed aphids on *Dracophyllum* – can be considered relatively common. Despite some effort in recent years seeking populations of the remaining species (see Teulon & Stufkens 1998) they have been difficult to find and should be considered rare. For example, *A. nelsonensis* has not been found for over 30 years. Only three sites (non-current) for *A. coprosmae* populations have ever been observed (Teulon & Stufkens 1998; Stufkens pers. obs.). Other species are only slightly commoner (see Teulon & Stufkens 1998).

Threats to endemic aphids

New Zealand's indigenous aphids face a number of threats to their continued survival, although the relative importance of these threats has yet to be determined. Threats include:

Habitat destruction, with consequent major disruptions in distribution owing to the complete removal of host plants as well as less obvious destruction in the form of animals browsing the young growing shoots of host plants, e.g. *Aphis healyi* on the native broom *Carmichaelia*.

The high ratio of alien to indigenous aphid species, which is increasingly steadily because of continued introductions (see above). Some indigenous aphid species may be threatened as a result of displacement from their host plants by introduced species. For example, *Aphis nelsonensis,* not recorded for over 30 years, may have been displaced on *Epilobium* by *Aphis* nr. *epilobii*.

Indigenous aphids may be threatened by attack from alien parasitoids and predators including vespid wasps. At least one introduced aphid predator, the ladybird *Coccinella unidecimpunctata,* is reported to have displaced its indigenous counterpart, *C. leonina,* in many areas of New Zealand (Watts 1986) and probably includes indigenous aphids among its prey. A number of introduced parasitoids have been found to attack and kill several indigenous aphid species in the laboratory (Stufkens & Farrell 1994; Teulon & Stufkens unpubl.) and an introduced parasitoid, *Aphidius ervi,* appears to attack *Aphis cottieri* in the field (Carver 2000).

Climate change represents a significant threat to global biodiversity and ecosystem integrity, including New Zealand indigenous aphid species and their host plants. For example, a species of *Paradoxaphis* that lives on *Plagianthus* may be susceptible to rising environmental temperature. A constant level of 25° C in the laboratory will kill it (J. Kean pers. comm.).

Life-histories

Phylloxerids and adelgids produce only eggs whereas aphids usually produce live young, with eggs produced only at certain times of the year. Aphids are peculiar in that they reproduce without mating during most of the growing season of their plant hosts. Complex life histories involving both winged and wingless generations are characteristic of many species of aphids (Blackman & Eastop 1994). In New Zealand, introduced aphid species tend to follow the life-histories found in their area of origin but with some tendency for the overwintering-egg diapause stage to be lost and replaced by continuous parthenogenesis throughout the year (e.g. *Rhopalosiphum padi*). Although very little is known about the biology of the indigenous aphids, sexual females, males, and/or eggs, indicating sexual reproduction, have been found in many of the Aphidini. *Neophyllaphis totarae* produces sexuales (sexual forms) and eggs in spring and early summer like some Australian species (Carver et al. 1991).

Ecological associations

All aphids, adelgids, and phylloxerids are phloem feeders and many are host specific. Most aphids are autoecious, living on one or a few closely related species of plants. Only about 10% are heteroecious, spending autumn, winter, and spring on a primary host plant and summer on a secondary host plant (sometimes more than one species) that is rarely closely related to the primary host plant (Dixon 1987). In general, introduced aphidoid species in New Zealand have similar associations with introduced plants as in their area of origin (Cottier 1953). Some introduced aphid species are also found on closely related indigenous hosts (e.g. *Cavariella aegopodii* on *Aciphylla*) and some polyphagous (having multiple host plants) aphid species are found on a number of indigenous plants (e.g. *Aulacorthum solani* on various species) (Cottier 1953). Indigenous aphids appear to be autoecious and mostly restricted to a single indigenous shrub or tree genus. Little is known about aphid–ant and aphid–microorganism associations in New Zealand.

Non-endemic and probably endemic species are prey to a number of generalist predators including ground beetles (Carabidae), ladybird beetles (Coccinellidae), lacewings, midges, nabid bugs, syrphid flies, and harvestmen and spiders (e.g. Valentine 1967a; Leathwick & Winterbourne 1984; Thomas 1989b). A large number of these predators, including the probable first purposeful attempt at classical biological control in New Zealand (*Coccinella undecimpunctata*), were introduced for control of aphids (Thomas 1989b). Braconid and aphelinid wasp parasitoids are known to attack a number of non-endemic species, which in turn are attacked by several hymenopteran hyperparasitoids (Valentine & Walker 1991). Some of these parasitoids were introduced as control agents for pest aphid species (Cameron & Walker 1989; Farrell & Stufkens 1990; Stufkens & Farrell 1994). A number of fungal pathogens from the Zygomycetes (Entomophthorales) have been recorded on aphids in New Zealand but no pathogenic viruses, bacteria, rickettsiae, protozoans, or nematodes (Glare et al. 1993).

It appears that most of the indigenous aphid species are hosts to hymenopteran parasitoids and hyperparasitoids but the taxonomic status of these has not yet been determined (D. A. J. Teulon unpubl.). A parasitoid very similar to the introduced biological control agent *Aphidius ervi* has been found to attack an indigenous aphid species (Carver 2000).

Gaps in knowledge

The total number of aphidoid species in New Zealand is about 130 (three adelgids, three phylloxerids, 124 aphids). There are at least 14 endemic aphid species of which at least six are undescribed. Only two endemic species have been described in the last 10 years (Carver 2000; Eastop 2001) but work is under way to describe up to three more in the near future. There is some uncertainty as to the exact identity of a number of introduced species (see below). Currently there are no New Zealand taxonomists working on this taxon.

Cottier's (1953) *Aphids of New Zealand* provides the most comprehensive information on New Zealand aphids but is now out of date. Teulon (1999) and Teulon et al. (1999) have developed illustrated multiple-entry keys for winged and wingless adult aphids in New Zealand for use by non-specialists. These focus on pest species and give no information about the phylogeny, biology, and ecology of New Zealand species.

One of the main areas of concern regarding this group in New Zealand is the state of collections. The major one in the New Zealand Arthropod Collection has

Wingless specimens of an undescribed native species of *Aphis* on *Hebe*.

Robert Lamberts, Plant and Food Research

been neglected in recent years. Specimen names have not been updated and new material has not been incorporated. This, along with the absence of an aphidoid taxonomist in New Zealand, makes it difficult for non-specialist workers in this field, who have to resort to international experts for accurate identifications and other advice. This is unfortunate considering the importance of aphids, phylloxerids, and adelgids as pests in New Zealand. A database specifying the location of specimens in the numerous collections in New Zealand and overseas is near completion (http://www.landcareresearch.co.nz/research/biosystematics/invertebrates/nzac/tfbis/index.asp) and will be very useful for aphid workers.

For indigenous aphids, the most important areas for future work include naming of the remaining species, locating and identifying new species, and determining their evolutionary relationships with faunas elsewhere. Very little is known about the biology of any indigenous species (Kean 2002), including explanations for their rarity. For the introduced aphids, the main issues relate to the exact identity of several species (e.g. *Akkaia ?taiwana*, *Aphis* nr *epilobii*, *Micromyzus* nr *katoi*) and the makeup of several potential species complexes (e.g. *Aphis gossypii* group, *Rhopalosiphum insertum* group, *Therioaphis* sp.).

Studies on introduced pest aphidoids would aid in managing their populations in New Zealand and elsewhere. Notwithstanding, it is indigenous aphids that are of most interest. Understanding their evolution would have importance for reconstructing the phylogenetic relationships of the major Aphidoidea lineages worldwide. New Zealand's native species are a key to resolving the debate over why aphids are common in the Northern Hemisphere compared with the tropics and the Southern Hemisphere (Dixon et al. 1987; Heie 1994; von Dohlen & Teulon 2002). Indigenous aphids also provide excellent models for research on rarity and the impact of introduced faunas on indigenous faunas. A New Zealand Marsden-funded project – The Population Dynamics of Rarity: Why are rare animals rare? – using indigenous aphid species as model organisms, was completed early in the new millennium (see Kean 2002; Kean & Barlow 2004; Kean & Stufkens 2005). Some other research funded by the New Zealand Foundation for Research, Science & Technology is being carried out on the impact of introduced aphid parasitoids on indigenous aphids but this could be expanded to look at displacement of indigenous aphids by introduced aphids and the impacts of other introduced natural enemies such as predators and pathogens.

Suborder Auchenorrhyncha: Cicadas, spittlebugs, leafhoppers, and planthoppers

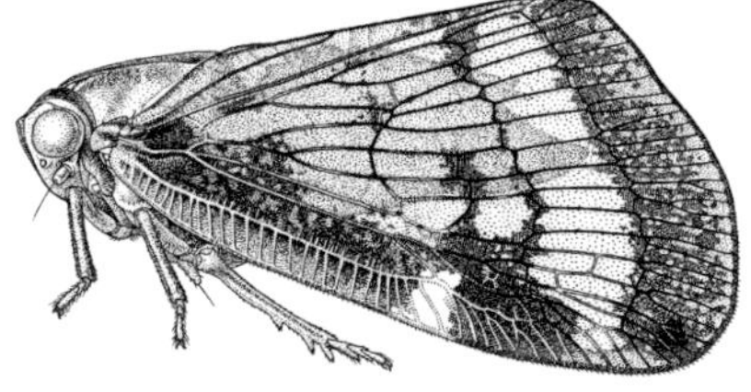

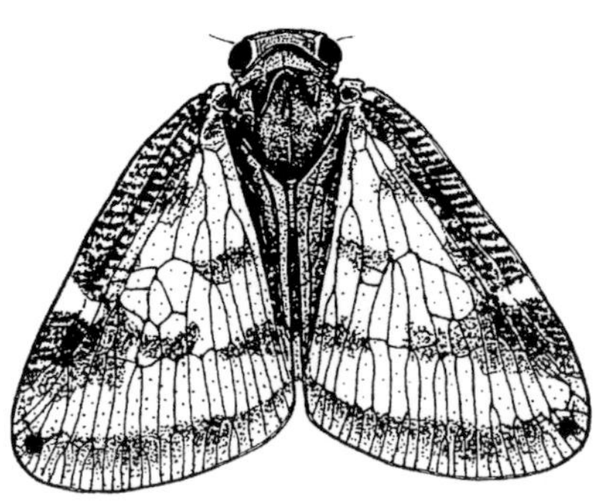

Alien passionvine hopper
Scolypopa australis (Ricaniidae).
Des Helmore (upper) and Grant 1999 (lower)

This overview of New Zealand Auchenorrhyncha has been extracted from Larivière (2005) and Larivière et al. (2006–10), with slight modifications.

The Auchenorrhyncha is a highly diverse suborder of Hemiptera that includes the cicadas, spittlebugs, leafhoppers, treehoppers (infraorder Cicadomorpha), and planthoppers (infraorder Fulgoromorpha). As a group they account for a major component of the plant-feeding insect fauna in most terrestrial ecosystems in New Zealand and around the world. They have adopted a variety of life habits on nearly all continents and islands (except Antarctica). Auchenorhhyncha have piercing-sucking mouthparts and most species feed on plant sap (phloem or xylem) or the content of plant cells (cell ruptures or parenchyma) although a number of species feed on mosses and fungi. This is an economically important group of insects that includes several plant pests and vectors of plant diseases (e.g. pathogens, including phytoplasmas, bacteria, and viruses). The world fauna is estimated to include around 42,000 described species distributed in 30 to 40 families.

The first modern checklist of New Zealand Auchenorrhyncha was provided by Wise (1977) and it included 64 genera, 160 species, and 11 families. Auchenorrhyncha have been collected extensively since the 1970s and became well repre-

Families, genera, and species of NZ Auchenorrhyncha

Australian and world figures are from Carver et al. (1991), Fletcher (1999), and other sources as indicated. Numbers of endemic taxa are bracketed.
(Prepared by M.-C. Larivière and M. J. Fletcher.)

Family	New Zealand genera	New Zealand species	Australian genera	Australian species	World species
Achilidae	2 (1)	2 (1)	15	21	350
Aphrophoridae*	4 (1)	15 (12)	21[1*]	34[1*]	2,400
Cicadellidae	29 (6)	78 (54)	191[2]	603[2]	20,000
Cicadidae	5 (3)	34 (34)	38[3]	202[3]	2,000
Cixiidae	11 (9)	26 (26)	17	49	>1,000
Delphacidae	10 (5)	18 (14)	40	71	300
Derbidae	1 (0)	1 (1)	18	48	800
Dictyopharidae	1 (0)	1 (1)	4	12	540
Flatidae	2 (0)	2 (0)	22	84	1,000
Membracidae	1 (0)	1 (0)	29[2]	74[2]	2,400
Myerslopiidae	2 (2)	16 (16)	0	0	>20
Ricaniidae	1 (0)	1 (0)	11	29	360
Totals	**69 (27)**	**195 (159)**	**406**	**1227**	**>31,170**

Sources: 1. Evans (1966); 2. Day & Fletcher (1994); 3. Moulds (1990)
* Includes Machaerotidae, absent from New Zealand

sented in New Zealand's entomological collections, leading to the publication of several taxonomic treatments since 1975. Larivière (2005) and Larivière et al. (2010, in press) provide the most up-to-date catalogue of this fauna, which now totals 68 genera and 196 species in 12 families. Once fully described, it is estimated that the fauna may comprise as many as 300–350 species.

Fabricius (1775) described the first native Auchenorrhyncha from New Zealand, the cicadas (Cicadidae) *Amphipsalta cingulata* and *Rhodopsalta cruentata*. Subsequently, until about the 1930s, several taxa were added to the fauna by European researchers such as Walker (1850–58) and White (1879a,b), and by two New Zealand workers, Hudson (1891) and Myers (1921–26). Little taxonomic discovery occurred during the 1940s and 1950s, although Evans (1941, 1942, 1947) and Hudson (1950) described a few leafhoppers (Cicadellidae) and cicadas (Cicadidae) respectively. The years from 1965 to 1984 were more prolific, resulting in several new taxa and important taxonomic revisions, mainly due to the efforts of Fennah (1965; Delphacidae), Evans (1966; Cicadellidae), Knight (1973–76; Cicadellidae), Fleming (1969, 1973, 1984; Cicadidae), Dugdale (1972; Cicadidae genera), and Dugdale and Fleming (1969, 1978; Cicadidae). The most recent period of active taxonomic research has occurred since 1992, as demonstrated by the following works: Hamilton and Morales (1992; Aphrophoridae), Larivière (1997a, 1999; Cixiidae), Larivière and Hoch (1998; Cixiidae), Hamilton (1999; Myerslopiidae genera), Emeljanov (2000; Cixiidae genera), Larivière and Fletcher (2004; identification of leafhopper genera and species), Szwedo (2004a; Myerslopiidae), Larivière et al. (2006–10), Larivière and Fletcher (2008; *Zeoliarus*, Cixiidae), and Fletcher and Larivière (2009; *Anzygina*, Cicadellidae).

Some groups or part of groups previously worked on are in need of further taxonomic research. A key to Aphrophoridae genera is urgently needed. The cicadellid genera *Arahura, Arawa, Horouta, Limotettix, Matatua, Novothymbris, Paradorydium, Scaphetus,* and *Zelopsis* need additional revisionary work. Knight's revisions of leafhoppers and Fennah's review of the Delphacidae need re-evaluation in view of large amount of new unidentified material accumulated in collections since the end of the 1970s. Four of five Cicadidae genera have never

Male chorus cicada *Amphipsalta zelandica* (Cicadidae).
From Grant 1999

High-alpine cicada *Maoricicada nigra nigra* (Cicadidae).
Des Helmore

been revised taxonomically. The available literature on Myerslopiidae and Ulopinae (Cicadellidae) is insufficient to provide a good understanding of these groups.

Compared to larger continental faunas, the New Zealand fauna may appear depauperate but New Zealand can still be regarded as a biodiversity 'hot spot' for Auchenorrhyncha with >80% of species and >40% of genera recognised as endemic.

Auchenorrhyncha are characterised by piercing-sucking mouthparts in the form of a beak extending from the back of the head – hence the name of this hemipteran suborder, which literally means 'neck-beaks', rather short and bristle-like antennae, and forewings of uniform texture, resting rooflike over the body.

These insects are generally active during the day and live from lowland to subalpine environments in a wide range of open or forested habitats. Native species usually live within the confines of natural habitats but some species also live in modified ecosystems. Depending on families and genera, species can be planticolous (occurring on low plants), arboreal (occurring on shrubs and trees), or sometimes epigean (living at the surface of the ground). Host plants are known for less than one-fifth of the fauna. The biology and morphology of immature stages are unknown for most species. There is anecdodal evidence of parasitic wasps, birds, predatory beetles, spiders, and mites being major natural enemies of New Zealand Auchenorrhyncha.

The described New Zealand fauna is about 13% the size of the known Australian fauna (about 1500 species), with 15 families present in Australia not represented in New Zealand. Twenty-four species (or 12% of the fauna) are currently recognised as introduced (adventive) in New Zealand. No family is endemic to this country but all ground-dwelling leafhoppers (family Myerslopiidae), or 70% of the world fauna, are endemic. The three most diverse families of Auchenorrhyncha in New Zealand are the leafhoppers (Cicadellidae), cicadas (Cicadidae), and cixiid planthoppers (Cixiidae). These families are also well represented in Australia.

The majority of species shared with Australia and elsewhere are cosmopolitan and probably introduced in New Zealand. Greatest faunal affinities are with eastern continental Australia and in the taxonomically diverse worldwide families Cicadellidae (leafhoppers) and Delphacidae (delphacid planthoppers). Faunal affinity is less between New Zealand and Tasmania or Norfolk Island, and even less so between New Zealand and Lord Howe Island or New Caledonia. These faunal relationships may be indicative of an old Gondwanan origin. Forty percent of native Auchenorrhyncha genera and 5% of native species are shared with Australia.

In terms of species distribution, a greater number of species (133) occur in the South Island, with 64 native species restricted to this island. A slightly lower number of species (119) occur in the North Island, with 44 native species restricted to this island. As many as 65 species are shared between North and South Islands. Offshore-island groups harbour a limited number of species: Chatham Islands (12) Kermadec Islands (10), and Three Kings Islands (21). Auchenorrhyncha are not known to occur on New Zealand's subantarctic islands.

On New Zealand's main islands, the North Island regions of Northland, Auckland, and Wellington and the South Island regions of Northwest Nelson and Mid Canterbury, show the highest overall species diversity but these regions contain many introduced species. The regions known to harbour the greatest number of local endemics – species only found in a single area and nowhere else in the world – are more interesting to the biologist. These regions are Northland and Wellington (North Island), northwest Nelson, Mid-Canterbury, Fiordland, and Southland (South Island). Fiordland is a largely unexplored and unspoilt area that may prove to be an even greater reservoir of endemic taxa than currently estimated.

The main trading ports or agricultural areas of New Zealand (Auckland, Hawke's Bay, Nelson, Christchurch) account for the greatest number of adventive (introduced) species, many of which have fully developed wings, a tendency to be attracted to lights, and an ability to adapt to life in partly or highly modified environments, hence they generally have good dispersal abilities.

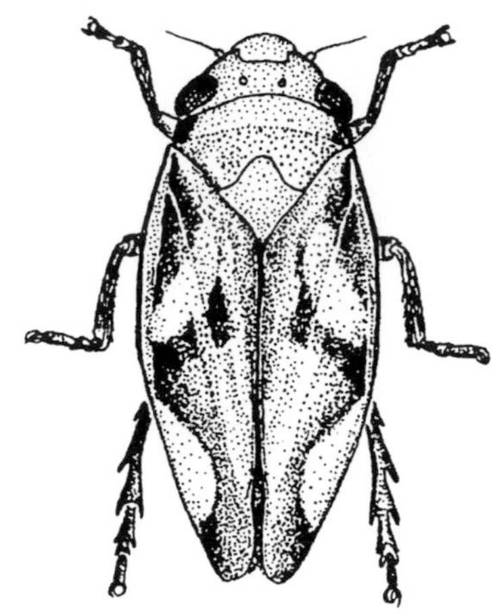

Spittlebug *Carystoterpa* sp. (Aphrophoridae).
From Grant 1999

Infraorder Cicadomorpha

Spittlebugs (family Aphrophoridae) are not abundant in New Zealand. They are xylem-feeders. In many cases nymphs are visible as frothy masses or 'cuckoo-spit' on the stems of small trees or shrubs. This froth is used by nymphs either to reduce the risk of dehydration or to deter enemies such as parasites. *Carystoterpa* is a native genus with species occurring mostly in the North Island. *Carystoterpa* species are usually found on native trees and shrubs such as species of *Coprosma* (Rubiaceae). *Pseudaphronella jactator* is a North Island species and the largest New Zealand spittlebug. It usually occurs on trees and shrubs in montane and subalpine environments. Both genera are characterised by wide-ranging species with fully winged adults.

Cicadas (family Cicadidae) are probably among the most familiar New Zealand insects owing to their loud song at the height of summer. The fauna comprises 34 endemic species distributed among five genera. New Zealand cicadas occur in a wide range of habitats from lowland coastal areas to subalpine and alpine zones (e.g. coastal sand dunes, riverbeds, grasslands, scrublands, shrublands and native forests, exotic tree plantations, and garden and orchard hedges). Most forest species live in the North Island while the South Island is mostly characterised by cicadas of rocky open spaces. Clapping cicadas (*Amphipsalta* species) can often be heard singing in urban environments on garden trees, buildings, fences, and even lamp posts. The clay-bank cicada *Notopsalta sericea* can also be an urban dweller in the North Island, often signing from any sun-warmed flat concrete surface. The genus *Kikihia* is native to New Zealand and comprises 13 endemic species traditionally placed in three groups mainly based on habitat preferences – the shade singers, the green-foliage cicadas, and the grass and scrub cicadas. 'Kikihi', the stem-base of the name *Kikihia*, is the word generally used by Maori, who have a considerable body of nature lore concerning insects, to refer to cicadas in general. The black cicadas (*Maoricicada* species), with 14 endemic species, favour open habitats from montane to subalpine environments where they occupy a wide range of ecological niches.

The closest relatives of New Zealand cicadas are Australian and New Caledonian. The evolution of the New Zealand fauna appears to have originated from multiple dispersal events (at least two) across the Tasman Sea from Australia and possibly New Caledonia within the last 12 million years. More recently, approximately within the last five million years, speciation events led to the highly diverse genera *Kikihia* and *Maoricicada*, most likely through adaptive radiation in new habitats created by the rise of the Southern Alps and the last glaciations.

Cicadas mostly spend their life in the nymphal stage, underground, feeding on the roots of plants. Limited knowledge is available about life span but some species are known to spend three to five years as nymphs, and two to four months as adults. Parasitic wasps, predatory beetles, fungal diseases, kiwis and various other birds, as well as spiders are among the main natural enemies of cicadas and other Auchenorrhyncha.

The economic importance of cicadas is low, but when they occur in large numbers damage may be caused by the female creating open cuts in plant tissue where eggs are laid, thus providing suitable entry points for pathogens and boring insects. Mass emergences of cicadas can become an annoyance to workers in horticulture and forestry owing to work disruption by loud song and repeated contact with flying individuals.

Leafhoppers (family Cicadellidae) are the most diverse group of Auchenorrhyncha in New Zealand. They occur in almost every type of vegetation. For example, the genera *Arahura, Arawa, Horouta,* and *Limotettix* live predominantly on low plants, the genera *Scaphetus* and *Novothymbris* occur on trees and shrubs, and the genus *Paradorydium* is found on or very close to the ground surface. In general, leafhoppers living close to the ground surface are more frequently short-winged. Most New Zealand leafhopper genera feed on phloem sap, but introduced and probably also native members of the subfamily Typhlocybinae feed on plant tissue.

A handful of leafhoppers have an economic impact on crops in New Zealand – *Edwardsiana froggatti* (adventive) on pipfruits, *Ribautiana tenerrima* (adventive) on commercial berries, *Batracomorphus angustatus* (possibly adventive) on potato and tomato, *Anzygina dumbletoni* (possibly adventive) on strawberries and cane fruit, *Anzygina zealandica* (native) on grass in orchards, and *Eupteryx melissae* (adventive) on commercial aromatic herbs. *Nesoclutha phryne* (native) and *Orosius argentatus* (adventive) are recognised plant-disease vectors in Australia but not in New Zealand.

Ground-dwelling leafhoppers (family Myerslopiidae) are an ancient group of small southern-hemisphere leafhoppers living as adults and nymphs in leaf litter and other ground debris of forests with high organic content. Species are characterised by large heads, spined hindlegs, compact almost barrel-shaped bodies with extensions on head and thorax, and prominences and punctures on their hardened forewings and other parts of the body (usually encrusted with soil and litter particles offering nearly perfect camouflage with their surroundings). All species lack functional hindwings and active dispersal by flight is excluded for this family. Myerslopiidae are thought to feed on fungi from the decomposing leaf and soil debris in which they live. Two endemic genera are known from New Zealand, viz *Myerslopia* (eight species-group taxa) and *Pemmation* (12 species-group taxa), together accounting for 70% of the world fauna.

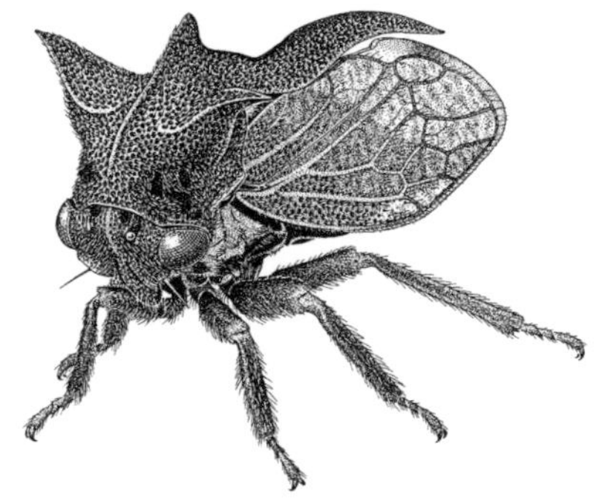
Alien treehopper *Acanthucus trispinifer* (Membracidae).
Des Helmore

Horned treehoppers (family Membracidae) are represented in New Zealand by a single species, *Acanthuchus trispinifer*, an Australian introduction. Membracids are usually sedentary but they are capable of jumping strongly if disturbed. Adults and nymphs are often gregarious (occurring in groups) and nymphs may be attended by ants.

Infraorder Fulgoromorpha

Achilid planthopper (family Achilidae) nymphs may generally be found in logs, under loose bark, or in leaf litter. They are believed to feed on fungi. Adults feed on phloem. Females generally lay their eggs by attaching them to woody particles in the leaf litter or to soil debris. Little is known about the biology of *Agandecca annectens*, New Zealand's only native and endemic species. Adults have fully developed wings, which may assist in their dispersal.

Cixiid planthopper (family Cixiidae) nymphs are thought to live primarily underground and feed on plant roots. Adults are phloem-feeders. Females usually lay their eggs in the soil and surround them with a waxy secretion. The majority of New Zealand species appear to favour woody plants, a lesser number are associated with ferns, and very few species feed on gymnosperms. New Zealand cixiids inhabit forested or bush environments such as scrublands and shrublands, and range from coastal lowlands to the subalpine zone although most genera are found in lowland to lower mountain mixed podocarp-broadleaf habitats. The genus *Aka* possibly represents an older lineage with an evolutionary history closely associated with *Nothofagus* forests. *Semo* is strictly a subalpine genus with highly similar species displaying entirely separate distribution ranges, which may indicate relatively recent speciation. *Confuga persephone* is the only cave-dwelling species known from New Zealand. In this country, cixiid planthoppers are mostly characterised by fully winged forms, with a tendency towards brachyptery (short-winged forms) more strongly demonstrated in the genera *Aka* and *Chathamaka*. The main economic importance of Cixiidae is as vectors of plant diseases (e.g. *Zeoliarus atkinsoni*, on flax).

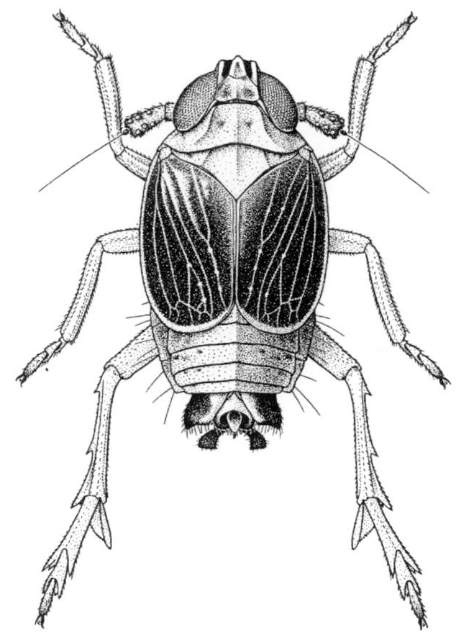

Planthopper *Sulix tasmani* (Delphacidae).
Des Helmore

Delphacid planthoppers (family Delphacidae) may be the most economically important planthopper family in the world. Delphacid species feed on or transmit virus diseases to cereals, an important food source for humans. Adult delphacids feed on the phloem of grassy plants. Nymphs roam freely as do adults. *Toya dryope* is the only species currently recorded as adventive in New Zealand. The native biostatus of *Opiconsiva dilpa* is uncertain. However, neither of these species is recognised as a plant pest or plant-disease vector in New Zealand. *Nilaparvata lugens* is a vector of virus disease of rice in South-East Asia but there is no evidence of this from its New Zealand relative *Nilaparvata myersi*. Other species of Delphacidae are all endemic to New Zealand. Multiple wing forms can be displayed by single species, but most New Zealand endemics are brachypterous (short winged) or, in a few cases (e.g. *Sulix*), species may have well-developed forewings and vestigial hindwings. Consequently, dispersal power by flight is thought to be low for New Zealand delphacids.

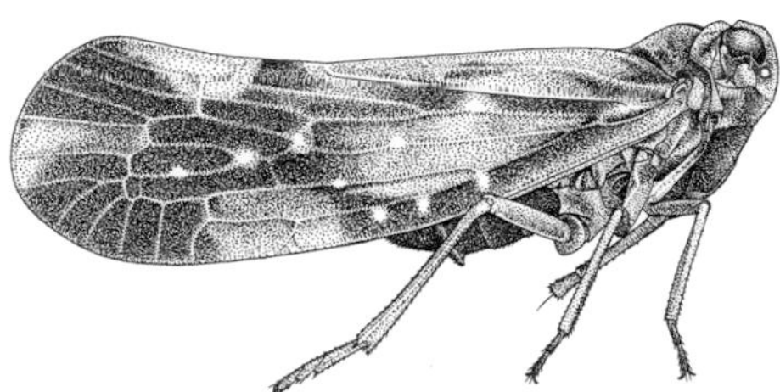

Treefern hopper *Eocenchrea maorica* (Derbidae).
Des Helmore

Derbid planthopper (family Derbidae) nymphs sometimes feed on fungi but most adults take their food from vascular-plant phloem. *Eocenchrea maorica* is the only native and endemic derbid so far known from New Zealand. Very little is known about the biology of this North Island lowland-montane forest species. Adults including newly emerged individuals (tenerals) have been found on *Astelia banksii* (Asteliaceae), which may serve as a food plant. Adults of *Eocenchrea maorica* have fully developed wings, which may assist dispersal.

Dictyopharid planthoppers (family Dictyopharidae) are poorly represented in the Australasian region and, in New Zealand, by a single endemic species *Thanatodictya tillyardi*. Species of this genus, also occurring in Australia, have the head extending considerably in front of the eyes into a long process. All life stages feed on grass.

Flatid planthoppers (family Flatidae) occurring in New Zealand are Australian introductions. Nymphs produce abundant wax filaments and cement themselves onto hostplants. Adults have long wings, feed on the phloem of a great variety of vascular plant families, and are wide-ranging in New Zealand. *Anzora unicolor* is a vector of fireblight on apple and pear.

Ricaniid planthoppers (family Ricaniidae) are represented by a single species in New Zealand, *Scolypopa australis* (the passionvine hopper), an Australian introduction that occurs on a wide range of hostplants. It is a pest of vine crops (e.g. kiwifruit). In addition, *S. australis* sometimes feeds on poisonous plants (e.g. tutu, *Coriaria arborea*) ands secrete honeydew that, in times of low nectar supply, may be gathered by honey bees and incorporated into honey that then becomes poisonous to humans. *Scolypopa australis* is a fully winged species and disperses easily. It occurs in large numbers during the summer months and it is not unusual to find individual plants (native or exotic) covered by hundreds of individuals.

The Auchenorrhyncha catalogue of Larivière et al. (2006–10) provides extensive additional information on nomenclature (including colour photos of types), geographic distribution, natural history, wing condition (including dispersal power), and a bibliography of over 500 references.

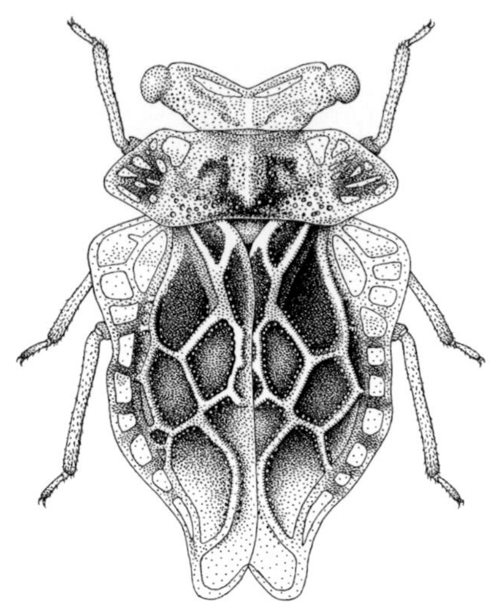

The moss bug *Oiophysa distincta* (Peloridiidae).
Des Helmorre

Suborder Coleorrhyncha: Moss bugs

All living Coleorrhyncha belong to the family Peloridiidae or moss bugs, with 17 genera and 32 species known from New Caledonia, New Zealand, southeastern Australia, and southern South America (Burckhardt 2009; Larivière et al. 2010). They are generally viewed as a relict Gondwanan group that was probably represented by a much richer world fauna in the Upper Permian and Upper Cretaceous. The phylogenetic position of moss bugs within the Hemiptera has been controversial for a long time although, since the 1990s, a sister-group relationship with Heteroptera has been supported by a number of studies based on morphological and molecular evidence.

Peloridiids are minute cryptically coloured bugs of 2–5 mm body length and have been said to be probably the rarest and most remarkable of all Hemiptera (Helmsing & China 1937). They are characterised by a flattened body, areolate dorsal surface (head, thorax and forewings), and a transverse opisthognathous head (with receding 'jaws' or suctorial mouthparts positioned posteroventrally). Apart from one South American species displaying wing dimorphism (winged and wingless forms) all known peloridiids are wingless and consequently incapable of flight. This makes them ideal test-organisms to study Gondwanan dispersal and vicariance hypotheses.

Moss bugs are found in permanently moist habitats among water-saturated mosses, on which they feed, and liverworts, often in association with southern beech (*Nothofagus*). Practically nothing is known about specific relationships between peloridiid bugs and mosses, and possibly also liverworts. Other known biological and behavioural attributes of peloridiids include five larval instars (Estévez and de Remes Lenícov 1990), spring mating and overwintering eggs (Cassis & Gross 1995), and, as in other sap-sucking Hemiptera, endosymbiotic micro-organisms present in special mycetomes (Müller 1951; Pendergrast 1962; Schlee 1969) as well as, in *Hacheriella veitchi* (Australia), vibrational communication (Hoch et al. 2006) and jumping ability (Burrows et al. 2007).

The New Zealand fauna is 100% endemic, with three genera and nine species described, representing about 28% of the known world species. Very little is known of the distribution and biology of individual New Zealand species and there are a few additional taxa remaining to be described.

Evans (1981) published a general world review of this very special group of bugs, based on a limited number of specimens (approximately 600). Burckhardt (2009) has revised the taxonomy and phylogeny of the world fauna based on substantially more material collected by Evans or accumulated in museums (mostly outside New Zealand) since the 1980s. Burckhardt (2009) described two new species for New Zealand (*Oiophysa paradoxa* and *Xenophysella greensladeae*) and synonymised three others (*Oiophysa fusca* = *O. ablusa*; *Xenophysella dugdalei* and *X. pegasusensis* = *X. stewartensis*). He also provided keys, descriptions, and figures to all genera and species as well as basic information on geographic distribution and habitat.

The distribution of the New Zealand taxa is known only in very general terms. Three species (33% of the fauna) occur in both North and South Islands (*Oiophysa cumberi, O. pendergrasti, Xenophyes cascus*). Of these, *Xenophyes cascus*, also found on Stewart Island, is by far the commonest and most wide-ranging New Zealand species. *Oiophysa cumberi* and *O. pendergrasti* are so far known from fewer than 10 populations in the North Island (only one population in the case of *O. pendergrasti*). These two species – the only ones shared between both islands – have also been recorded from a few populations (only one in *O. cumberi*) around Takaka Hill near Nelson in the South Island. No species is so far known to be endemic to the North Island.

Published records suggest that three species (33% of the fauna), *Oiophysa ablusa, Xenophyes kinlochensis,* and *Xenophysella greensladeae,* are endemic to the South Island. *Xenophyes kinlochensis* and *Xenophysella greensladeae* have so far been collected in a handful of locations around the greater Milford Sound area (Hollyford Valley, Key Summit, Routeburn Valley, Tutoko River Valley) and Secretary Island in Fiordland National Park (Burckhardt 2009; Larivière et al. 2010). *Oiophysa distincta* is shared between the southwestern areas of the South Island and Stewart Island. Two species (*Oiophysa paradoxa* and *Xenophysella stewartensis*) are endemic to Stewart Island. Peloridiidae have not been recorded from the Chatham Islands or from any of New Zealand's subantarctic islands.

The scope of Burckhardt's (2009) world revision did not allow him the opportunity to analyse critically the rather large amount of predominantly unidentified material deposited in New Zealand entomological collections and museums. This material is described in the most recent volume in the *Fauna of New Zealand* series (Larivière et al. 2010) and provides a comprehensive taxonomic treatment of the New Zealand fauna, including detailed information on distribution and biology.

Suborder Heteroptera: Bugs

The following overview of New Zealand Heteroptera and the end-chapter species list have been extracted from Larivière and Larochelle (2004) and Larivière (2005), with slight modifications.

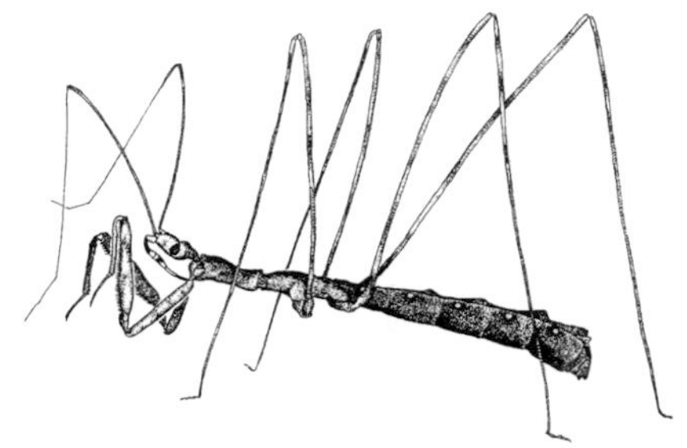

Antipodean assassin bug *Ploiaria antipodum* (Reduviidae).

From Grant 1999

Australasian and global diversity of Heteroptera

Australian and world figures are from Cassis and Gross (1995, 2002). Numbers of endemic taxa are bracketed (modified from Larivière & Larochelle (2004).

Family	New Zealand		Australia		World	
	Species	Genera	Species	Genera	Species	Genera
Acanthosomatidae	2 (1)	4 (4)	17 (12)	45 (43)	47	180
Aenictopecheidae	3 (2)	4 (4)	2 (1)	2 (2)	10	20
Anthocoridae	6 (1)	8 (4)	16 (5)	29 (19)	81	523
Aradidae	19 (12)	39 (38)	39 (18)	143 (127)	230	1,909
Artheneidae	1 (1)	1 (1)	1 (1)	2 (2)	8	20
Berytidae	1 (0)	1 (1)	6 (0)	7 (6)	36	172
Ceratocombidae	1 (0)	2 (2)	1 (0)	1 (1)	7	35
Cimicidae	1 (0)	1 (0)	1 (0)	1 (0)	23	108
Coreidae	1 (0)	1 (0)	43 (26)	83 (59)	252	1,802
Corixidae	2 (0)	6 (6)	5 (0)	31 (25)	36	556
Cydnidae	4 (1)	4 (1)	21 (9)	83 (76)	120	751
Cymidae	1 (0)	1 (0)	4 (0)	10 (6)	9	54
Enicocephalidae	3 (2)	4 (4)	3 (1)	5 (5)	50	180
Gerridae	1 (0)	1 (0)	10 (3)	29 (17)	69	586
Heterogastridae	1 (0)	1 (0)	3 (1)	5 (4)	23	97
Hydrometridae	1 (0)	1 (0)	1 (0)	6 (4)	7	119
Lygaeidae	4 (2)	33 (32)	22 (7)	81 (70)	101	972
Mesoveliidae	2 (1)	2 (1)	2 (0)	5 (3)	11	41
Miridae	39 (28)	115 (98)	91 (40)	186 (148)	1,300	9,800
Nabidae	2 (0)	4 (2)	7 (0)	22 (16)	31	380
Notonectidae	1 (0)	2 (2)	6 (2)	39 (25)	11	350
Pentatomidae	8 (1)	8 (1)	134 (94)	363 (333)	642	4,110
Reduviidae	3 (0)	7 (4)	100 (62)	226 (198)	961	6,601
Rhyparochromidae	22 (10)	42 (34)	75 (32)	185 (142)	368	1,824
Saldidae	1 (0)	7 (7)	3 (0)	10 (9)	28	274
Schizopteridae	1 (0)	1 (1)	13 (9)	61 (16)	42	221
Tingidae *sensu lato* (incl. Cantacaderidae)	4 (1)	1 (1)	56 (25)	147 (133)	250	2,025
Veliidae	1 (0)	1 (1)	4(0)	17(14)	46	673
Totals	**136 (55)**	**305 (249)**	**794 (390)**	**2,093 (1,734)**	**5,470**	**39,308**

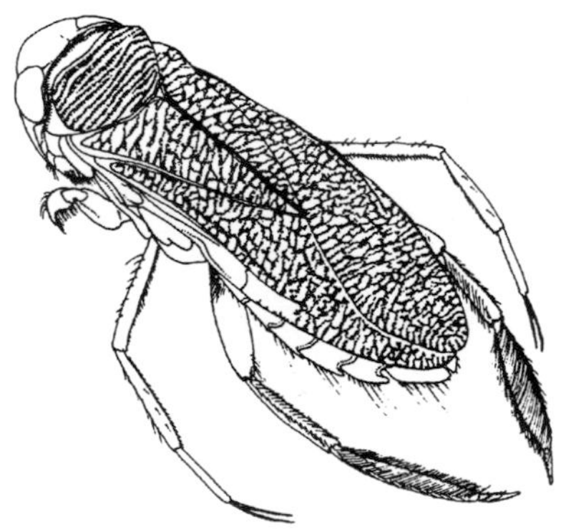

Common water boatman *Sigara arguta* (Corixidae).

From Grant 1999

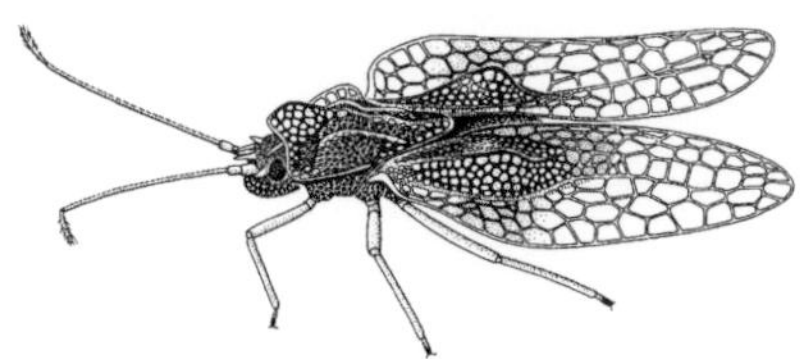

Lacebug *Tanybyrsa cumberi* (Tingidae).
Des Helmore

Around 37,000 species of true bugs are described worldwide, with possibly another 25,000 species remaining to be described (Schaefer & Panizzi 2000). The world fauna is divided into approximately 75 families. Better-known continental faunas such as those of North America, Europe, or Australia include thousands of species. By comparison, the New Zealand fauna – currently comprising 29 families, 138 genera, and 309 species – may appear relatively small, but what it lacks in size it makes up for in its uniqueness, with >80% of known species and 40% of known genera currently recognised as endemic. From this point of view New Zealand can be regarded as a biodiversity 'hot spot' for true bugs. Once fully described, the New Zealand fauna is likely to comprise 400 to 500 species.

The largest heteropteran families in New Zealand are the plant bugs (Miridae; 120 species or 39% of the fauna), rhyparochromid seed bugs (Rhyparochromidae; 42 species or 14%), flat bugs (Aradidae; 39 species or 13%), and lygaeid seed bugs (Lygaeidae; 33 species or 11%). In Australia, the four largest families are the stink bugs (Pentatomidae; 360 species or 18%), assassin bugs (Reduviidae; 226 species or 11%), flat bugs (Aradidae; 207 species or 10%), and plant bugs (Miridae; 186 species or 9%) (ABRS, 2009), but these numbers will change because large portions of the Australian fauna are still unrevised. The rhyparochromid seed bugs (Rhyparochromidae; 170 species) and lace bugs (Tingidae including 'Cantacaderidae'; 147 species) are also well represented in Australia. The largest heteropteran genus in New Zealand is *Chinamiris* (Miridae, 31 species). More than 30 unrevised heteropteran genera are currently represented in New Zealand by a single species.

Faunal affinities are greatest with southeastern continental Australia. A number of native taxa are also shared with Tasmania, Norfolk Island, Lord Howe Island, or southern Chile, suggesting a Gondwanan origin. The New Zealand fauna does not appear closely related to that of New Caledonia, with only a few generic and subgeneric level affinities supporting this relationship.

The New Zealand fauna is about 15% the size of the Australian fauna. More than 35 families present in Australia are not represented in New Zealand. New Zealand shares about 10% of its native true bug genera with Australia and only 5% of its native species. No family of Heteroptera is endemic to New Zealand.

Nearly all species of Heteroptera possess the following three diagnostic features – piercing-sucking mouthparts in the form of a segmented beak extending from the front of the head and running backward along its underside; slightly overlapping forewings lying almost flat over the abdomen; and each forewing base being much thicker than the tip (hence the name Heteroptera, derived from the Greek words *heteros* (different or other) and *pteron* (wing), referring to the non-uniform texture of the forewings).

The Heteroptera are the largest and most diverse group of insects with incomplete metamorphosis (hemimetabolous insects). Their life cycle involves an egg stage, a series of nymphs (usually 5), or growing stages that progressively look similar to the adult, and finally an adult stage.

The true bugs comprise a highly adaptable group that has managed to occupy most terrestrial as well as many aquatic and semi-aquatic habitats and to evolve remarkably diverse life habits on nearly all continents and most islands, suggesting a long evolutionary history for the group.

True bugs have been well collected in New Zealand and are well represented in entomological collections and museums. The first heteropteran species described from this country was the acanthosomatid stink bug *Oncacontias vittatus* (Fabricius, 1781). Subsequently, until the 1930s, most taxa were described by European workers, in particular White (1876–1879) and Bergroth (1927).

Renewed taxonomic activity from 1950 to 1970 yielded several new taxa and important revisions, mainly due to the efforts of Woodward (especially 1950, 1953, 1954, 1956, 1961) and Usinger and Matsuda (1959). These workers described more than 20 genera and 45 species in several families and provided

Seed bug *Arocatus rusticus* (Lygaeidae).
From Grant 1999

identification keys and detailed taxonomic descriptions. In addition, a number of other researchers described individual taxa from a range of families, which meant that by the end of the 1960s there were twice as many taxa known from New Zealand as there were 30–40 years earlier.

Much of the taxonomic effort between 1970 and 1977 was devoted to the family Lygaeidae *sensu lato* (Artheneidae, Cymidae, Heterogastridae, Lygaeidae, Rhyparochromidae, in this book). The solid contributions of Malipatil (especially 1976–79), particularly on the tribe Targaremini, are noteworthy.

The most active period of taxonomic description and revision, however, was still to come. Over the last 30 years or so the highly prolific work of one New Zealander, A. C. Eyles, especially on the families Lygaeidae and Miridae (e.g. Eyles 1990–2008) yielded more than 100 new species and several new genera. Other key publications are by Kirman (1985–1989) and Heiss (1990, 1998) who revised parts of the family Aradidae, Larivière (1995) who revised the Acanthosomatidae, Cydnidae, and Pentatomidae, Larivière and Larochelle (2004, 2006) who described the New Zealand Ceratocombidae and reviewed the genera of Aradidae, and Buckley and Young (2008) who updated the taxonomy of Corixidae.

As far as comprehensive taxonomic revisions are concerned, they currently cover approximately 175 species, or about 55% of the described New Zealand fauna. Consequently, apart from the Miridae, Lygaeidae, Pentatomoidea, and part of the Rhyparochromidae, all other families (>24) are in great need of modern revisionary treatment. Furthermore, so much new material has been collected and deposited in New Zealand collections in the last 30 years that numerous new taxa remain to be described, even in groups that have already been worked on.

Taxonomic works published until now generally deal with the adult stage. Less than 15% of described New Zealand Heteroptera have immature stages described. Only the last instar nymphs of Acanthosomatidae, Cydnidae, and Pentatomidae have been better documented, together with a few species of the superfamily Enicocephaloidea and the families Lygaeidae, Miridae, Rhyparochromidae, and Veliidae.

Identification keys are scarce. The most up-to-date keys to identify New Zealand Heteroptera at the family level are provided by *The Insects of Australia, Hemiptera* (Carver et al. 1991) and *A key to the bugs of Australia* (Elliott & Cassis 2001; LUCID key – http://www.faunanet.gov.au/).

Below the family level, identification is problematic and for most groups one has to rely on often inadequate original descriptions. With the exception of some recent works that include keys to taxa of Lygaeidae, Miridae, Pentatomoidea (Acanthosomatidae, Cydnidae, Pentatomidae), and Rhyparochromidae, the diagnostics literature is scattered. Eyles (2000b), however, provided a very useful overview of introduced Mirinae (Miridae).

Keys to Heteroptera so far recorded from New Zealand are being prepared by Marie-Claude Larivière (Landcare Research, Auckland) and electronic versions will be made available on the internet (The New Zealand Hemiptera website – http://hemiptera.landcareresearch.co.nz).

The majority of Heteroptera families occurring in New Zealand are terrestrial. Less than 7% of the fauna is semi-aquatic (living on or near water) or aquatic (living in water). The only species of Gerridae occurring in New Zealand is the sea skater *Halobates sericeus*, a true bug living on the surface of the ocean. Terrestrial species can be either epigean (living on the ground), planticolous (living on low-growing plants), or arboreal (living on trees and shrubs).

The two native terrestrial habitats harbouring the greatest number of species are forests and shrublands (in the lowlands and on mountains). Tussock grasslands and open subalpine environments also harbour their own suites of unique species. In general, native species tend to live within the confines of native habitats, but many species also survive in modified environments. Introduced (adventive) species seem to be able to invade natural habitats, but only to a slight degree.

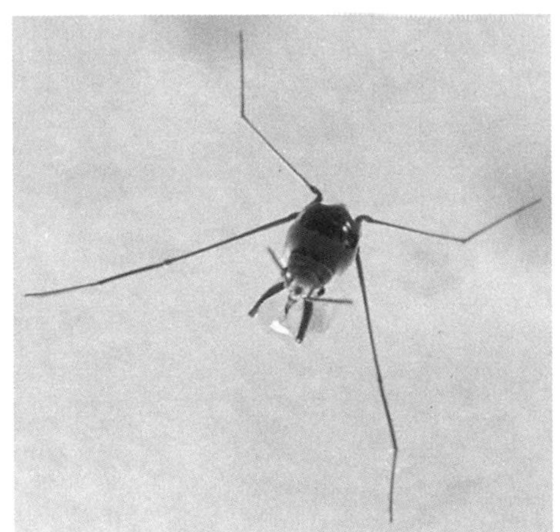

Sea skater *Halobates sericeus*.
Lanna Cheng, Scripps Institution of Oceanography

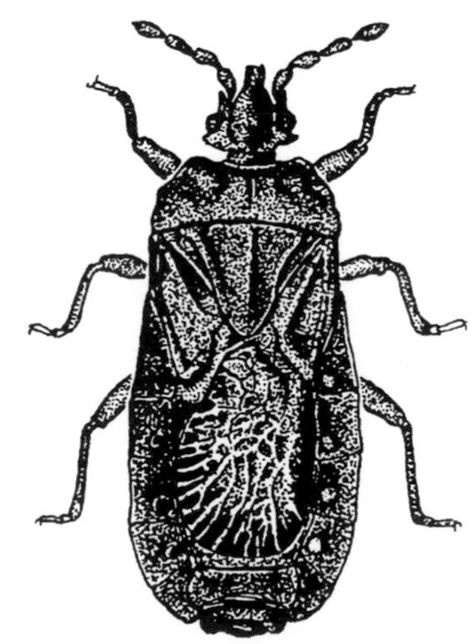

Bark bug *Ctenoneurus hochstetteri* (Aradidae).
From Grant 1999

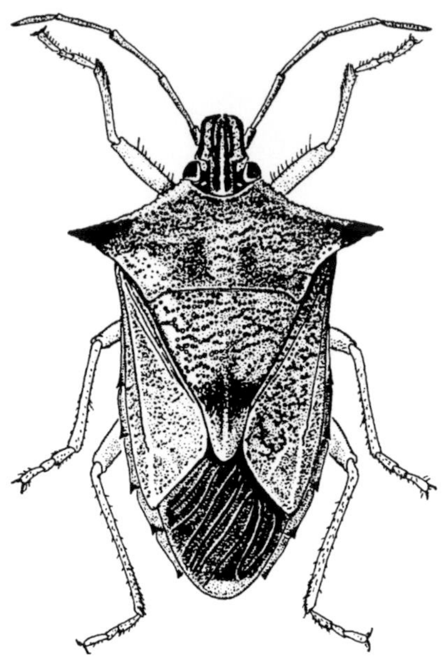

Schellenberg's soldier bug *Oechalia schellenburgii* (Pentatomidae).
From Grant 1999

Very few native species live exclusively in coastal lowlands. On the other hand, most coastal sand dunes, estuarine habitats, and coastal wetlands are typically inhabited by adventive species. Some adventive species are synanthropic (living around human dwellings).

Very little is known about the life history of native true bugs. Host plants (the plants on which true bugs breed, develop, and feed) have been confirmed for less than one-fourth of known species, and mainly for seed bugs (Lygaeidae, Rhyparochromidae) and plant bugs (Miridae). Adults of Heteroptera are probably diurnal in most families. Although adult true bugs of most species are active for much of the year, their peak of activity is between November and March, i.e. the end of spring (September–November), summer (December–February), and early autumn (March–May). The seasonal activity of immature stages (nymphs) as well as the time of the year when adults mate and reproduce, are mostly undocumented. Population biology and means of dispersal remain virtually unknown.

The majority of Heteroptera found in New Zealand are phytophagous (plant-feeding), extracting sap directly from the plant vascular system (in most families) or feeding on seeds, developing fruits or flowers, or sometimes pollen. The majority of species of the flat bugs (Aradidae) feed on the mycelia or fruiting bodies of various wood-rotting fungi.

Almost all families of Heteroptera also include species that prey on insects and other arthropods. There are also entire families that are predominantly predatory (e.g. Anthocoridae). Only the introduced bed bug *Cimex lectularius* (Cimicidae) is haematophagous (feeding on the blood of vertebrates, including humans); there does not appear to be any evidence of disease transmission.

Little is known about the natural enemies of New Zealand Heteroptera. Hymenopteran egg-parasites, some birds (e.g. pipits, rooks, starlings), spiders, damsel bugs, ground beetles, and mites have been observed as enemies of some true bugs in New Zealand, but published observations are rare. Spiders could be the most important predators, especially in open habitats such as alpine environments and tussock grasslands.

In terms of economic importance, direct damage to crops or disease transmission by a single species may be lower in Heteroptera than in other major insect groups including the hemipteran suborder Auchenorrhyncha. Nevertheless, some native and adventive species have limited economic impact in New Zealand, e.g. the green vegetable bug *Nezara viridula* (adventive Pentatomidae) on vegetable crops, *Engytatus nicotinianae* (adventive Miridae) on tobacco, *Closterotomus norwegicus* (adventive Miridae) on various seed and vegetable crops, and *Nysius huttoni* (native Lygaeidae) on crucifers and wheat. However, species with pest status in other parts of the world, including neighbouring island countries and other parts of Australasia, represent potential biosecurity risks for countries like New Zealand that rely heavily on primary industry for their economy. For example, chinch bugs (*Nysius* species) and other species in the seed bug family Lygaeidae have historically been among the most destructive plant-feeding pests in several countries of the world. Consequently there is a constant need to update the inventory of New Zealand's and neighbouring faunas, through sustained fieldwork and taxonomic reassessments.

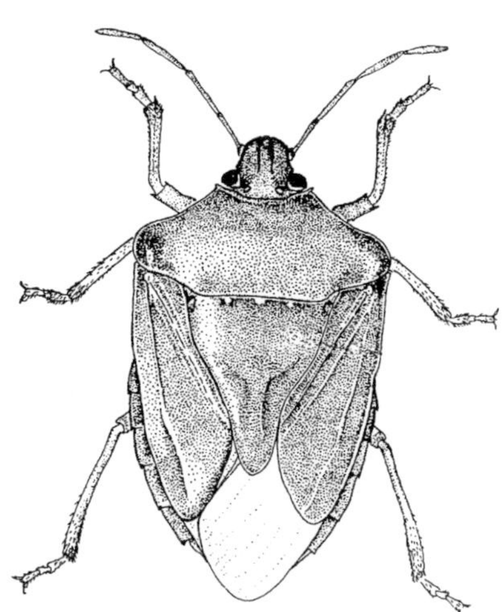

Green vegetable bug *Nezara viridula* (Pentatomidae).
From Grant 1999

As a group, Heteroptera can also serve humans and the environment in positive ways, especially those predatory species that can be useful biological control agents (e.g. in integrated pest-management programmes). As a general rule, predatory and zoophytophagous species native to New Zealand have not been investigated for use as biocontrol agents although true bugs belonging to the same families have been used overseas to control thrips, mites, moth eggs and caterpillars, leafhoppers, mosquitoes, and planthoppers.

Finally, seemingly economically unimportant groups of true bugs may be important to humans or to nature conservation. Aquatic and some semi-aquatic Heteroptera, for example, may prove important as foodstuffs for fish and as indicators of water quality. From an insect-conservation point of view, at least 130 endemic Heteroptera (42% of the total fauna) are known from 10 populations or fewer – many from the type locality only – and many of these species also live in habitats that are at risk or being lost or highly modified.

As for geographic distribution, about 73% of species occur in the South Island, although only around 25% of all native species are restricted to this island. A slightly lower number of species, about 68%, occur in the North Island, with about 20% of all native species restricted to this island. At least 45% of native species are shared between North and South Islands. Northland, Auckland, Wellington, northwest Nelson, and Mid-Canterbury are the regions currently showing the highest overall species diversity, taking into account several adventive species.

The areas of the country so far known to have the greatest number of local endemics are Northland, Wellington, northwest Nelson, and Fiordland. The warmer areas of New Zealand and its main trading ports and agricultural regions (Northland, Auckland, Gisborne, Bay of Plenty, Northwest Nelson, Mid Canterbury) include the largest number of adventive species. No true bugs have been recorded from the Antipodes Islands, Bounty Islands, Campbell Island, or Snares Islands.

Overall, about 25% of the fauna is flightless, but in flat bugs (Aradidae) and rhyparochromid seed bugs (Rhyparochromidae) flightlessness reaches 65–70%. Thus a large proportion of New Zealand species is limited in its dispersal abilities.

The Heteroptera catalogue of Larivière and Larochelle (2004) provides extensive additional information on nomenclature (including colour photos of types), geographic distribution, natural history, wing condition (including dispersal power), and a bibliography exceeding 1000 references.

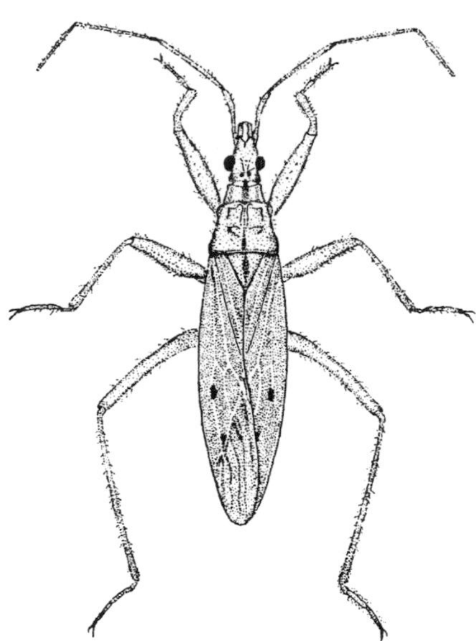

Tussock damsel bug *Nabis maoricus* (Nabidae).

From Grant 1999

Order Thysanoptera: Thrips

Thrips are small slender insects with a tubular abdomen and mouthparts that punch and suck. Worldwide, there are almost 5000 species. They range in length from half a millimetre to 15 millimetres, but most are in the range of 1.5–2.5 millimetres. The short antenna has 6–9 segments and a pointed tip. The two pairs of narrow wings, when present, are quite distinctive. They are fringed with long hairs (Thysanoptera means 'fringed wing') and have only 1–3 longitudinal veins and a pointed tip. In only one family (Aeolothripidae) are there species with cross-veined wings. Wing size (full, small, wingless) quite often varies within a species, some of which have wingless males or females. Development is peculiar in that there are nymph-like early instars (stages between moults) but a pre-adult pupa stage. A large number of species feed on sap in flowers, but some are predacious and feed on smaller insects and mites.

There are two suborders, most easily distinguished by differences in the wings. In the Terebrantia, the wings have three parallel veins, the fringe hairs are socketed, and the wings lie parallel when at rest. In the Tubulifera, the wings lack veins, the hairs are not socketed, and the wings overlap when at rest.

The New Zealand fauna consists of 121 species, of which about 65 are endemic. The systematic work of Mound and Masumoto (2005) and Rugman-Jones et al. (2006) builds on the earlier revisions of Mound and Walker (1982, 1986). Most species in New Zealand are small, but the adventive *Idolothrips spectrum* reaches 10.5 millimetres long. Among the 65 native species, 17.2% are wingless in both sexes.

Endemism is relatively high in both suborders – 32% in Terebrantia and 69% in Tubilifera. These percentages may decrease, however, when the thrips of Australia (CSIRO 1991) and New Caledonia are more adequately investigated. *Desmidothrips* is an example of a genus shared between New Zealand and New Caledonia. The genus *Physemothrips*, with two species, is endemic to Stewart Island and the subantarctic islands. *Adelphithrips dolus* is endemic to the Snares

Australian bottlebrush thrips *Teuchothrips disjunctus*.

From Grant 1999

Islands. None of the 11 species recorded from the Chatham Islands or the eight species known from the Three Kings Islands is endemic. No thrips have been identified from the Kermadec or Poor Knights Islands, where other orders have endemic species.

Some 47 species of herbivorous Thripidae and only three species of Phlaeothripidae feed on plants (see end-chapter checklist). They feed mainly on leaves but some species are abundant among flowers. Foliage and flower damage can become apparent as paler spots with dark speckles. Many of the species of Phlaeothripidae and the two species of Merothripidae dwell among leaf-litter or are found on dead twigs and branches. Analysis of gut contents shows that some genera feed on fungi. The three species of Aeolothripidae are apparently omnivorous, living on other immature thrips in flowers as well as feeding on the plants. *Haplothrips kurdjumovi* preys on moth eggs and mites in orchards. One of the alien species, the foliage thrips *Sericothrips staphylinus*, was deliberately introduced for release against gorse in 1990 and has become established (Hill et al. 2001; Hayes 2007).

Some native species, such as *Adelphithrips nothofagi*, *Thrips coprosmae*, and *Thrips phormiicola* are host-specific. Conversely, the New Zealand flower thrips, *Thrips obscuratus*, is found on at least 225 plant species (Teulon & Penman 1990) and is the main native pest species. Martin and Mound (2005) and Mound and Masumoto (2005) provided new well-defined host records for seven of the endemic species. The only introduced species found frequently among native vegetation is *Thrips tabaci*. Among four of the main pest species of the suborder Terebrantia, white backgrounds are preferred by *T. obscuratus* and yellow by *T. tabaci* and *Ceratothrips frici*, while foliage-feeding *Limothrips cerealium* shows no preference for different coloured traps (Teulon & Penman 1992).

Onion thrips *Thrips tabaci*.
Lincoln University

Several endemic species have been shown to have quite extensive distributions, but other species are recorded from only the type locality or a few places (Mound & Walker 1982, 1986). Beech forest has a greater diversity of Phlaeothripidae than broadleaf–podocarp forest (Mound & Walker 1986) and Evans et al. (2003) reported that species in this family favour soil under litter more than that under rotting logs in red beech (*Nothofagus fusca*) forests. The species of thrips and their diversity in some habitats remain unstudied. These include thrips of the grey shrubland (key taxa *Discaria toumatou* and *Olearia* spp.) of the South Island

Native *Thrips obscuratus* is a major pest of introduced plants, especially horticultural crops (Teulon & Penman 1990). Much damage is done to the surface of stonefruit, where white stippling and loss of colour can lead to rejection of fruit. There is zero tolerance for this pest in export fruit (McLaren & Fraser 2000). Populations gradually build up to an autumn peak on unsprayed trees (Teulon & Penman 1996; McLaren & Fraser 2001), but adults cause most damage during flowering while not affecting fruit set (McLaren & Fraser 2000). Reflective mulch has helped reduce populations in stonefruit orchards and carbaryl is the most effective spray (MacLaren & Fraser 2001). Thrips can also carry spores of the brownrot fungus (Ellis et al. 1988). This thrips is the main species among kiwifruit flowers, too, but on fruit and foliage the greenhouse thrips is the main pest (Tomkins et al. 1992). Post-harvest disinfestation may be needed to avoid rejection of cut flowers (Carpenter 1987) and other exported horticultural produce. *Thrips obscuratus* is apparently one of the main thrips species in lucerne (alfalfa) flowers (Macfarlane 1970, Teulon & Penman 1990), along with *Apterothrips secticornis* (Somerfield & Burnett 1976), but its role in disrupting pollination by consuming pollen remains unstudied.

Conversely, *T. obscuratus* is among the thrips species of broom flowers (Teulon & Penman 1990), where feeding on pollen may reduce pollen viability to some extent. Flowers of seed crops of red clover and white clover can be adversely affected by *Haplothrips niger* (Yates 1952), while the main thrips species in lotus seed crops remain unstudied. Timothy thrips (*Chirothrips manicatus*) may be found on timothy and other grass seedheads, where it can cause considerable (but inadequately documented) damage to developing seeds (Doull 1956; Scott 1984). *Limothrips cerealium* is the main thrips of wheat, barley, oats, and ryegrass, but it is apparently less common in brome grass and ryecorn crops (Bejakovich et al. 1998). Grass thrips *Aptinothrips rufus* and the American grass thrips *Anaphothrips obscurus* can be common among grass or cereal crops (Cumber 1959a, Mound & Walker 1982), where they feed mainly on the foliage. Flower thrips *Frankliniella occidentalis* and the onion thrips *Thrips tabaci* can transfer tomato spotted wild virus (Scott 1984; Cameron et al. 1992). Both introduced species have a wide host range, but the onion thrips is of main importance as a vegetable pest (Scott 1984); it is of secondary importance among nectarines and peaches in the spring (McLaren 1992; Teulon & Penman 1994). Reducing litter and weeds in the crop helps control populations of several pest thrips of asparagus (Townsend & Watson 1984). The greenhouse thrips *Heliothrips haemorrhoidalis* is of most importance as a pest in citrus fruit (Scott 1984). Other ornamental plants adversely affected by thrips include bottlebrushes by *Teuchothrips* species and gladiolus by *Thrips simplex* (Mound & Walker 1982; Scott 1984). Thrips populations can particularly build up in greenhouses because of warm conditions and few natural pests, producing up to 12 generations in a season. Among greenhouse capsicums, flower thrips *Frankliniella occidentalis* has developed resistance to tau-flavinate spray (Martin & Workman 1994), so it is important to develop biological controls where possible. *Frankliniella intonsa* has been found in the Auckland area since 2002 on strawberries, dahlias, capsicum, and sunflowers (Teulon & Nielson 2005; O'Donnell pers. comm.). Of the two species, *F. occidentalis* has a much wider distribution, as evidence by monitoring of capsicum flowers (Teulon & Nielson 2005). The main impact of the gorse thrips, *Sericothrips staphylinus*, in the field is on seedlings and young plants (Hill et al. 2001).

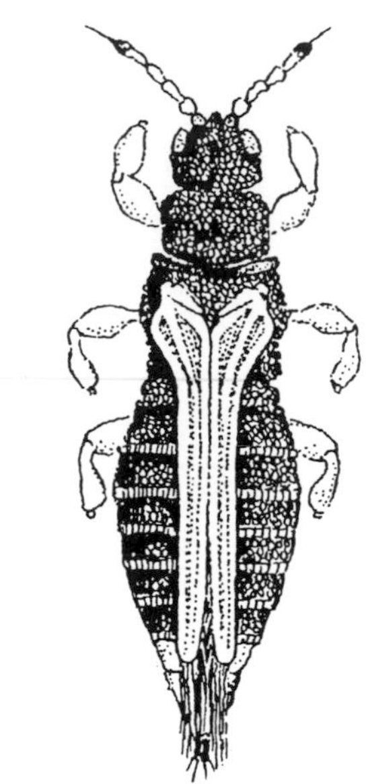

Greenhouse thrips *Heliothrips haemorrhoidalis*.

From Grant 1999

The natural enemies of thrips in New Zealand include four native species of *Spilonema* wasps (Sphecidae) (Harris 1994a), which nest in holes in twigs. Hymenopterans are important in control – thrips nymphs may be parasitised by the introduced eulophid *Ceranisus* (Valentine & Walker 1991), thrip eggs by *Megaphragma* (Trichogrammatidae) (Mound & Walker 1982), and *Thripobius semiluteus* (Eulophidae) has been imported for control of greenhouse thrips (Froud et al. 1996). So far, two pathogens are known to affect the New Zealand flower thrips – the fungus *Entomophthora* and the nematode *Thripinema* sp. (Allantonematodidae) (Teulon et al. 1997).

Order Psocoptera: Psocids, booklice, barklice

There are about 5500 species of Psocoptera, which probably represents at most 50% of existing species. Only 69 species are so far recorded from what must be a much greater New Zealand fauna. On present inadequate data, the fauna appears to be a reduced Australian fauna with a New Caledonian connection in some groups. Psocoptera are now known to be important components of ecosystems, especially through their role in freeing resources locked in the microflora. Detailed information on biology and ecology of most species other than some European and North American forms is meagre.

Surprisingly few entomologists, whether professional or not, have more than a nodding acquaintance with the Psocoptera. This is probably because most general textbooks include somewhere in their introductory paragraphs some such statement as 'a little-known order of insects' and an 'order of small insects, from 1–10 millimetres in length' or some such equally discouraging statement. They are then usually referred to as booklice or barklice and the member illustrated is often a member of the very atypical genus *Liposcelis,* relegated to the status of 'a pest in houses and granaries' or some similar emotively inferred undesirable status. Even the author of the words that you are reading has been unwittingly

Liposcelis corrodens.

Landcare Research

guilty of this type of denigration when preparing general texts. The truth of the matter is that, like most 'minor' orders of insects, there are many more species than most entomologists realise. Few species have pest status, they are often common in the field in a wide variety of habitats, not difficult to recognise or collect, and their biology poses many fascinating questions. Populations are at times extremely large, and many species are ecologically very important. Many are aesthetically very pleasing, a not-unimportant factor when many of the working hours of one's life are spent looking at them.

The 5500 described species of Psocoptera worldwide are currently grouped into more than 340 genera in 41 families (Lienhard & Smithers 2002). Estimates of the number of undescribed species cannot realistically be made on any logical basis at present, but intuitively there must be very many. There are already considerable numbers of undescribed species in some major museums and it is obvious that many smaller collections have undescribed species, too, which collectively could add many more species to the tally. Nearly 250 species are known from the Euro-Mediterranean region (Lienhard 1998), yet even in this otherwise well-worked part of the world many undescribed species have been found in recent years. There are several European countries for which the species list is very short. Considering that some parts of the world, such as South America, Asia, and much of Africa have very rich faunas in other insect orders, the relatively few species of Psocoptera that have been recorded indicates that there remains an enormous number of undescribed species to be collected and described. Until this is done, many of the most interesting problems associated with the order will remain unsolved.

There are 69 species listed here for New Zealand and its adjacent islands (including seven species probably not yet named). For comparison, in one study area of only just over 200 hectares in temperate eastern Australia, 78 species have been recorded. There is little doubt that there are additional species to come from the New Zealand region. This indicates the measure of neglect that has been accorded the Psocoptera in New Zealand and the ease with which interesting additions to the fauna would be found in a country that has as wide a range of altitudes, climatic regimes, and distinct ecological communities as New Zealand. We are fortunate to have an excellent general introductory account of the order that covers all the topics needed as background for their study, including collecting and study methods (Lienhard 1998). Perusal of this work will also soon reveal where the major gaps in knowledge lie.

Classification

The families of Psocoptera are currently grouped into three suborders, the Trogiomorpha, the Troctomorpha, and the Psocomorpha. This classification has been arrived at by modification of an outline sketch of a classification provided by Pearman (1936) that was far superior to any earlier scheme. In his outline, Pearman established a number of families without giving detailed definitions but with mention of a typical genus for each family. He did not erect formal superfamilies, but united families into a series of informal groups that he considered to indicate likely relationships. One of these, Homilopsocidea, contained families that he could not definitely place in his other groups. Roesler (1944) provided a very useful key to the genera, at the same time rearranging some genera in Pearman's suprafamilial groups and dispersing the elements of the Homilopsocidea. Unfortunately, the rearrangement obscured some relationships, but the work was a strong stimulus to studies of what had become a difficult group to approach. Badonnel (1951) used a classification that was a combination of those of Pearman and Roesler, with modifications suggested by other authors. Smithers (1972) proposed a more phylogenetically orientated reclassification but suggested at the time that, because so many psocopteran genera were still poorly known, it would be preferable to retain the earlier arrangement for practical descriptive purposes while the new arrangement could

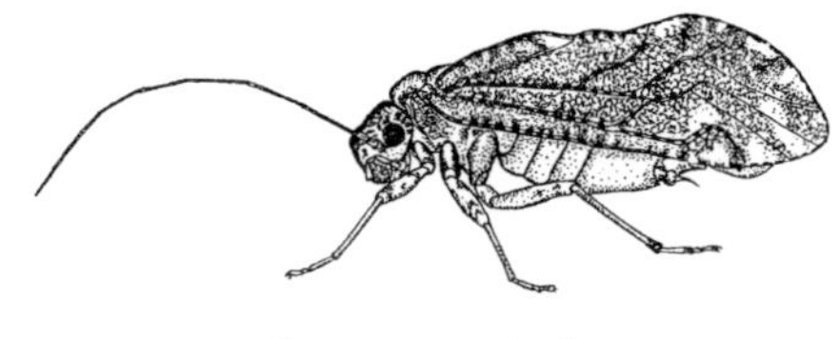

Myopsocus australis.
From Grant 1999

be tested against additional information as it became available. Smithers (1972) also tabulated these classifications in relation to one another. Mockford and Garcia Aldrete (1976) made the first attempt at formal definition of superfamilies by erecting two of them (Asiopsocoidea and Caecilioidea) to replace one of Pearman's informal groups, the Caecilietae, thus setting a welcome trend towards formalising the classification of the Psocoptera in line with terminology of other insect orders. Continuation of progress in this process will depend on more information becoming available on the insects themselves.

History of studies on the New Zealand psocopteran fauna

Kolbe (1883) described the first psocopteran from New Zealand, the large *Myopsocus novaezealandiae,* which had already been described from Australia as *Psocus australis* by Brauer (1865). Hudson (1892) described *Myopsocus zelandicus,* which McLachlan (1894) considered to be the same as *M. novaezealandiae*. Hutton (1899) redescribed this species and Enderlein (1903) repeated Kolbe's description. Not much progress was made until Tillyard (1923a) added 15 new species from New Zealand and gave a general account (Tillyard 1926) of the fauna. Cumber (1958, 1959b, 1962) recorded species found in pastures and cereal crops in the North Island, and Cumber and Eyles (1961a) recorded psocids from fodder crops from the same area. Thornton (1962) discussed the generic position of two of Tillyard's species. Three species were recorded from Campbell Island by Smithers (1964). Smithers (1969) described 15 new species, added further new records for New Zealand, and provided keys to all the species from the New Zealand subregion, bringing the number of species to 43. Gressitt and Wise (1971) included Psocoptera in their discussion of the insects of Auckland Island, Smithers (1973) added a species from the Kermadecs, and Smithers (1974) gave a summary of records from subantarctic islands, including those relevant to New Zealand. Wise (1977) provided an up-to-date list of species from the subregion. Smithers (1999) described a remarkable new genus and species from the Bounty Islands and recorded the Trichopsocidae for the first time in New Zealand (Smithers 2002). Progress in species recognition in the fauna can be followed in the literature cited in the references. The list deliberately includes papers not specifically mentioned in the text so that it effectively forms a bibliography of the New Zealand Psocoptera which should facilitate entry into the literature by anyone wishing to study the group.

With such a small proportion of species known, it is obvious that almost any conclusions regarding the fauna as a whole must be considered very preliminary because the proportion of species still to be recorded is so high that new discoveries will inevitably have considerable influence on conclusions.

Distribution of New Zealand psocopteran species

New Zealand species can be placed in four groups based on their distribution. These are not rigid categories, because of overlap, but are convenient for discussing the fauna. New Zealand's offshore territorial islands – Chathams, Snares, Auckland, Campbell, Bounty, and Kermadecs are included in the discussion.

Species associated with humans (six species)

There are many species of Psocoptera that have become habitual inhabitants of man-made environments, such as granaries and domestic and commercial buildings. They are, of course, found elsewhere in the wild. They tend to be distributed worldwide and have been at least partly spread through human activities. In New Zealand these include *Cerobasis guestfalica, Lepinotus inquilinus, L. patruelis, Liposcelis corrodens,* and *Trogium pulsatorium*. There is no doubt that many more species of *Liposcelis,* a cosmopolitan and economically significant genus, are present in domestic situations and in stored products in New Zealand. They simply await collection and identification, some of them having undoubtedly been listed in the past as '*L. divinatorius*', a species that cannot be recognised from its description and of which the types cannot be found. A few species in this first group (e.g. *Psyllipsocus ramburii*) appear to have spread naturally to human habitations from habitats such as caves. This species is widespread in caves and has probably been transported by bats.

Naturally occurring widespread species (six species)
Some naturally occurring widespread habitats, such as dead leaves or leaf litter, are rapidly and regularly colonised by some species of Psocoptera without human assistance. In New Zealand, *Echmepteryx madagascariensis* (pantropical, especially associated with dead banana leaves), *Ectopsocus briggsi*, *E. californicus*, *Propsocus pulchripennis* (widespread in tropical and temperate countries), and *Pteroxanium kelloggi* (sometimes domestic), probably fall into this group. Some species of *Ectopsocus* have, however, been found in packaging materials, which suggests that they could also be spread with human assistance. In the case of *E. briggsi*, *E. californicus*, and *P. pulchripennis*, they occur so widely in nature that it seems unlikely that human intervention has necessarily been a major factor in their spread. *Peripsocus milleri* appears to be widely self-spread, living on the bark of twigs and stems of woody plants. *Cerobasis guestfalica* can also be included in this category on the basis of its very frequent occurrence in natural habitats.

Species common to New Zealand and Australia (17 species)
Species of limited distribution that appear to be native to the Australasian Region, and common to New Zealand and Australia (including Tasmania) and/or Norfolk Island, include *Aaroniella rawlingsi*, *Austropsocus hyalinus*, *Blaste tillyardi*, *Ectopsocus petersi*, *Haplophallus maculatus*, *Chorocaecilius brunellus*, *Philotarsopsis guttatus*, *Lepinotus tasmaniensis*, *Maoripsocus semifuscatus*, *Paedomorpha gayi*, *Paracaecilius zelandicus*, *Pentacladus eucalypti*, *Peripsocus maoricus*, and *P. morulops*. One of these (*E. petersi*) has also been found on Lord Howe Island. Of the remaining three species in this group, *Myopsocus australis* has a somewhat more extensive range that includes Norfolk Island, the Kermadecs, and the Solomon Islands. *Trogium evansorum* occurs in New Zealand and Norfolk Island but has not yet been found in Australia. These species appear to have evolved in the Australasian Region and to have undergone some limited dispersal within it, somewhat extended in the case of *M. australis*. It is not possible to decide which of these have dispersed naturally and which, if any, may have been spread by importation on commercial material such as timber or horticultural produce. This group can be considered as essentially Australasian in origin. The distribution of *Ectopsocus axillaris* cannot be easily categorized. It is known from New Zealand (on introduced pines), Tasmania, and Ireland, which strongly suggests human influence, but in which direction this might have taken place is impossible to determine. On present records, it appears to be fairly localised in all three countries (Smithers 1978).

New Zealand endemic species (31 species)
Apparent New Zealand region endemics are listed in the end-chapter checklist. They belong to the following 14 genera: *Austropsocus*, *Bryopsocus*, *Echmepteryx*, *Ectopsocus*, *Latrobiella*, *Maoripsocus*, *Mepleres*, *Pteroxanium*, *Rhyopsocus*, *Sabulopsocus*, *Sandrapsocus*, *Spilopsocus*, *Valenzuela*, and *Zelandopsocus*.

Affinities of the New Zealand psocopteran fauna
The first two species groupings (above) can be expected to throw little reliable light on the likely affinities of the New Zealand fauna. The first group comprises widespread species, easily dispersed by human action and closely associated with domestic, suburban, and industrial habitats. The second comprises species

that habitually inhabit widely distributed ecosystems, which they can enter after natural long- or short-distance dispersal. They have probably not evolved in New Zealand or adjacent areas.

The third group comprises species that are not found beyond New Zealand or adjacent areas and have probably evolved within the Australasian region. The fourth group comprises probable endemics. The last two groupings can be expected to be the best indicators of faunal affinities. In reality, the fact that the New Zealand fauna is very poorly known makes any consideration of faunal affinities tentative at present. The faunas of Lord Howe Island and New Caledonia have also not yet been well documented for all families. The different psocopteran families are not equally represented in different parts of Australasia, making comparisons very unreliable because of the lack of equivalence of knowledge from group to group.

The two endemic species of Lepidopsocidae belong to a very large, mainly tropical, widespread genus, *Echmepteryx*. *Pteroxanium marrisi* from the Chatham Islands is the only species of the genus endemic to New Zealand – three others are endemic to Norfolk Island, another to Chile, and two to Mexico, in addition to the one worldwide, sometimes domestic, species. The only psoquillid in New Zealand (*Rhyopsocus conformis*) is an endemic member of an otherwise mainly North American family. It may well be a species that, while not yet recorded from its original country, was introduced to New Zealand and first recorded here, which is the case for a number of species in Europe and Kerguelen Island.

There is only one family endemic to New Zealand, the Bryopsocidae, which includes two species. It is clearly an offshoot of the Pseudocaeciliidae–Zealandopsocinae, discussed below. The Caeciliusidae (previously Caeciliidae) is a very large worldwide family. The relationships of the New Zealand species to those of other Australasian and Pacific areas are for the present unknown.

The Ectopsocidae is a family comprising species that mainly inhabit dried leaves. There are many species, with every zoogeographical region having its own complement of endemics. The New Caledonian fauna of this family has, unfortunately, not yet been fully studied. The New Zealand species of Elipsocidae are interesting – a large part of the family (subfamily Elipsocinae) is non-Australasian except for *Sabulopsocus*, a monotypic New Zealand endemic, and *Drymopsocus*, with three species in Chile and one in Australia. In the Propsocinae, two genera (*Kilauella* and *Palistreptus*) have undergone considerable speciation in the Hawaiian Islands. *Kilauella* may be closely related to *Drymopsocus* but the Hawaiian species of the group need closer study. *Pentacladus* is restricted to Australasia, with one species endemic to Lord Howe Island. *Sandrapsocus* is monotypic and apparently endemic to the Bounty Islands, the only known population being restricted to the few hectares of rocky, windswept Proclamation Island. *Spilopsocus* has species in New Zealand, Lord Howe Island, and Australia but, so far as is known, all are endemic to their own areas.

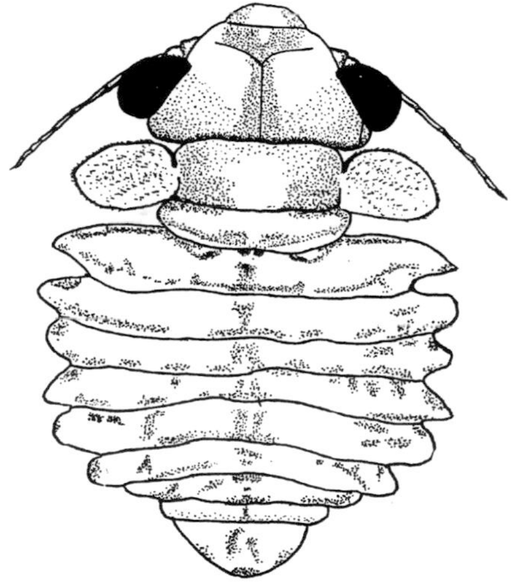

Endemic *Sandrapsocus clarki* from Bounty Island.
From Smithers 1999

There are no New Zealand endemics of Peripsocidae, a very widespread family. Neither this nor the equally widespread large family Philotarsidae, with one endemic in New Zealand, throw much light on faunal affinities. It should be noted that the Philotarsidae as presently constituted differs from that discussed by Thornton (1985) – the important genera *Austropsocus* and *Zelandopsocus* are now included in the Pseudocaeciliidae and his discussion of the Philotarsidae should be read with this in mind. The Pseudocaeciliidae–Zelandopsocinae includes three species of *Zelandopsocus* endemic to New Zealand, 27 endemic to New Caledonia, and one known only from Queensland. *Austropsocus,* the other genus in the subfamily, has eight species endemic to New Zealand, seven to New Caledonia, and 13 to Australia, with one found in both Australia and New Zealand. The Pseudocaeciliidae is otherwise a very large family with very strong Asian and Pacific Island representation. It does, however, indicate links between New Caledonia and New Zealand. The

Unidentified psocopteran.
Landcare Research

single New Zealand species of the extremely large worldwide family Psocidae also occurs also in Australia. The family can, for practical purposes, almost be considered absent from New Zealand, as is the case of the Myopsocidae, which also has a single New Zealand species. There is one unidentified species of Trichopsocidae known from New Zealand. The family is a small one, essentially western Palaearctic, with one other species in Chile and another in Australia and New Zealand.

Need for future work on Psocoptera

Considering the distributions given above, it would be unwise at present to make sweeping statements regarding the faunal relationships of the New Zealand Psocoptera. Thornton (1985) suggested that it is a 'reduced Australian fauna'. To this, perhaps, should be added: 'with the possibility of a New Caledonian connection'. There is clearly a need for a concerted effort to carry out the remaining descriptive taxonomy of the Psocoptera as a reliable basis for all other studies. In simple terms, it is necessary to find out how many and which psocopterans live in New Zealand and where they are found. The recent enormous interest in ecosystems and their sustainable management has resulted in an unprecedented number of faunal surveys. These are producing previously unheard-of quantities of material for taxonomic study. There is no reason to suppose that New Zealand is any less well endowed than any other part of the world so far as the Psocoptera are concerned.

It is known that many Psocoptera occur in very large populations. As a result, they are extremely important ecologically as grazers of microflora. The sheer bulk of this resource is not often appreciated, mainly because of the small size of its components. Psocoptera are prime intermediaries in the release of microbially derived nutrients into food webs. Psocopteran biology in general is very poorly known, so their ecological roles cannot be effectively quantified in any whole-ecosystem studies. Almost everything known (which is relatively little) about psocopteran biology and behaviour is based on work carried out in Europe and North America. There is remarkably little known about even elementary aspects of the biology of most of the species, so that what is known from the few case studies is applied uncritically to other parts of the world. In fact, there are many peculiar forms outside the Palaearctic Region, such as the morphologically unusual *Sandrapsocus clarki* from New Zealand's Bounty Islands and the wood-boring *Psilopsocus mimulus* from Australia.

There is a need for individual life-history studies to establish habitat and food preferences and behavioural interactions with other species. Closely related species of Psocoptera are sometimes found in very narrow habitats. These should prove fruitful areas for study of interspecific competition. There is a need for detailed studies in comparative morphology, only one species (European) having been thoroughly examined from this point of view. There are also interesting physiological phenomena displayed by psocopterans, such as the remarkable process of atmospheric moisture uptake involving their lingual sclerites (Rudolph 1982). Finally, phylogenetic studies and faunal comparisons are needed to understand the origins of the New Zealand taxa. Given that New Zealand has been an archipelago for much of its history makes it a potentially fruitful region to test hypotheses of island colonisation and the origins, importance, and functions of sexual polymorphism in adults.

All of these studies, however, rest squarely on a basis of sound taxonomy, which must, therefore, assume highest priority in planning research programmes.

Order Phthiraptera: Lice

Lice are wingless, highly modified, flat-bodied insects that live as obligate ectoparasites on birds and mammals, i.e. they spend their complete life-cycle on the host, being totally dependent on the heat, humidity, and secretions produced by the host. Both adults and nymphs have a similar diet that, depending on the species, may be blood, feathers, skin debris, mucus, or serum (Marshall 1981). Their geographical distribution is, with some exceptions, that of their hosts.

The name Phthiraptera is derived from the Greek *phtheir*, louse; the suffix *aptera* denotes winglessness. This highly specialised order probably evolved from psocopteroid ancestors. It is divided into four very different suborders – the Amblycera and Ischnocera (chewing or biting lice, previously grouped together as the Mallophaga), Rhynchophthirina (restricted to elephants and African hogs and also previously included in the Mallophaga), and Anoplura (sucking lice). More than 5000 species are known worldwide and there is a huge diversity of body form (Durden & Musser 1994; Price et al. 2003). Adults range in size from half a millimetre to 10 millimetres and males are usually smaller than females (Calaby & Murray 1991).

Because lice live as obligate parasites, their ecology, behaviour, and evolution are intimately linked to the attributes of the host; lice have developed morphological, behavioural, and physiological adaptations to survive on the host. Similarly, because lice are detrimental to host health and fitness, hosts have developed adaptations to control their lice (Murray 1990; Clayton 1991). This reciprocal natural-selection pressure has led to the co-evolution of hosts and lice (Clayton & Moore 1997). Thus the phylogenetic relationships of lice often parallel those of their hosts and may help both to elucidate the relationships of the latter and to distinguish closely related host taxa that are otherwise poorly defined.

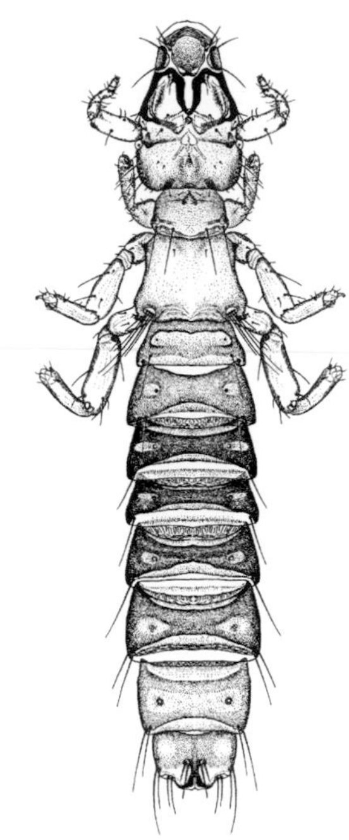

Male *Naubates thieli* from Providence petrel *Pterodroma solandri*, a vagrant species in New Zealand.

Ricardo Palma

A total of 347 identified species and subspecies of lice, belonging to 100 genera and subgenera in 13 families, have been recorded from birds and mammals in the New Zealand region and the Ross Sea area of Antarctica as defined in the Ornithological Society of New Zealand (1990) checklist (see Tenquist & Charleston 1981; Pilgrim & Palma 1982; Palma 1999; Palma & Price 2000, 2004, 2005). [Phthiraptera that may first have been recorded on birds in the Ross Sea area are included in the New Zealand checklist if their hosts range to the New Zealand EEZ.] This total represents about 7% of species worldwide. Only 29 species (8.42%) are endemic to the region (including one now extinct, in the *Rallicola* subgenus *Huiacola*, also extinct), while at higher taxonomic levels the degree of endemism is even lower (3%), with one endemic living genus and two endemic subgenera. Ninety species and subspecies (25.9%) have been introduced into the region, together with their hosts, by human agency.

As expected in a land where there are no native terrestrial mammals other than two bat species – neither of which is parasitised by lice – most of the New Zealand louse fauna is parasitic on birds. As many as 253 species (72.9%) live on native birds and 54 species (15.6%) on birds introduced by humans. In contrast, only four louse species (1.1%) parasitise the few native pinnipeds, with the remaining 36 species (10.4%) living on introduced mammals including humans. About 10 more louse species have been collected from native birds but have not been described yet. A much greater number of species is expected, however, especially from the considerable number of breeding and vagrant bird species that have not yet been sufficiently searched for lice.

Both chewing and sucking lice can have significant economic and health impacts. Heavy infestations on sheep, cattle, horses, goats, red deer, and poultry can cause anaemia and other conditions, resulting in production losses. Livestock have to be controlled by dipping or spraying with insecticides. Humans are not exempt from sucking lice. The head louse (*Pediculus humanus capitis*), a subspecies of the body louse, used to be found only in small, localised parts of New Zealand, but has been spreading in recent decades (Pilgrim 1975). Notwithstanding, archaeological evidence shows that, whatever recent introductions may have occurred, the presence of this louse in New Zealand predates European contact. Nits, the eggs of head lice, have been found attached

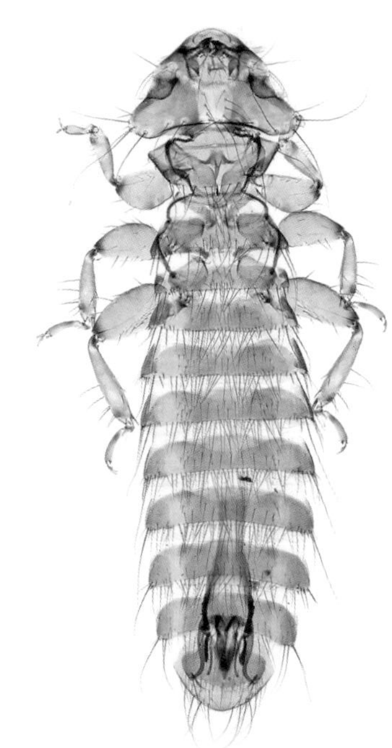

Male of *Apterygon okarito* from Okarito brown kiwi *Apteryx rowi*.

Jean-Claude Stahl, Museum of New Zealand Te Papa Tongarewa

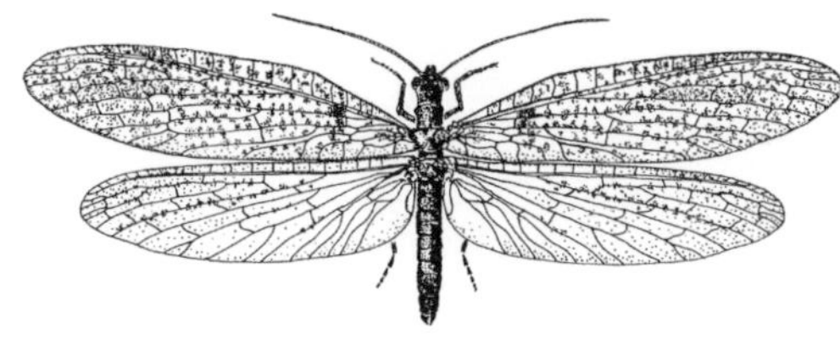

Dobsonfly *Archichauliodes diversus*.
From Grant 1999

to preserved Maori heads and bundles of hair, the latter from habitations dated at between 400 and 800 years ago (Savill 1990). There is a strong possibility that body and pubic lice were also present in pre-European populations. The latter species (*Phthirus pubis*), known as the crab louse from its stout, claw-like legs, is transferred during sexual intercourse. It is being found with increasing frequency in New Zealand, as in the rest of the world. Fortunately, it is not a disease carrier, though the body louse (*Pediculus humanus humanus*) has been associated with transmission of epidemic typhus (Calaby & Murray 1991).

Order Megaloptera: Dobsonflies

The Megaloptera ('great wing') are regarded as the most primitive of the holometabolous insects, i.e. those having a complete metamorphosis and three distinct life stages – larva, pupa, and adult. They used to be classified as a suborder of lacewings. This small order (ca. 300 species worldwide) is represented in New Zealand by only one species, *Archichauliodes diversus*. The flying adults are up to 40 millimetres long and the larvae, known as toebiters from their large mandibles, are of good size too, startling the discoverer when encountering them in streams. Dobsonfly adults are grey, unlike the four dark species of larger lacewings with broader wings (Osmylidae), and the prothorax is longer than mid- or hind thoracic segments. The four wings are flat and the veins delimit many (50 or so) small sectors (cells) but do not branch towards the end as in lacewings. The hind wings are broader at the base with an enlarged anal area that folds fanwise.

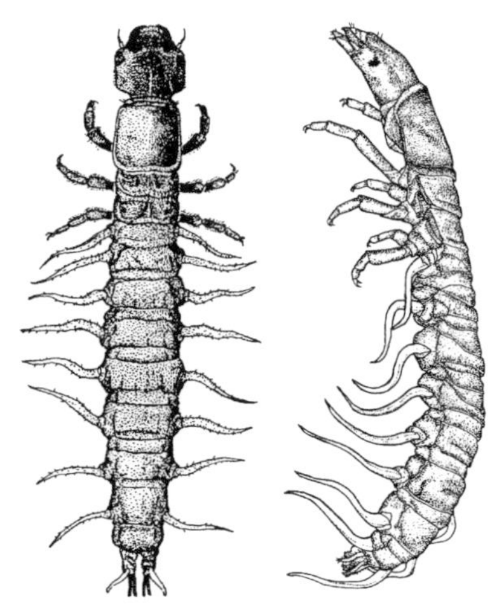

Dorsal and lateral views of the larva ('toe biter') of *Archichauliodes diversus*.
Left from Grant 1999, right from Winterbourn et al. 2000

Larvae are common in stony to gravelly streams, especially in shallower water, at temperatures that range from around 4.5 to 21.5° C (Hamilton 1940; Winterbourn et al. 2000). The sides of the larvae have finger-like gills. Adults begin flight at dusk and are noctural (Hamilton 1940). They are found in both partly wooded farmland (McGregor et al. 1987) and native hardwood bush, but are readily attracted to lights. The dobsonfly is widespread in North and South Islands and is found on those offshore islands that have dependable streams (Little and Great Barrier, Cuvier) (Riddell 1981; Wise 1983b, 1992a); none has been detected on the smaller Alderman and Poor Knights Islands (Watt 1982; Early 1995).

Larvae mainly consume *Deleatidium* mayflies or midges, but also capture various caseless caddisfly species (Devonport & Winterbourn 1976; Edwards 1986). Elmid beetles, their own larvae, and stoneflies are a very minor part of their diet. Hamilton (1940) and Edwards (1986) have provided further information on their biology. In turn, fish consume dobsonfly larvae. They constitute about 3.2 and 1.1% of the diet of Canterbury *Galaxias* and brown trout, respectively (Collier & Winterbourn 2000). Other galaxiids and some of the bully species apparently consume the larvae even less often. Notwithstanding, toebiters are the largest insects among the running-water species, so their contribution by volume to the diet of the smaller freshwater fish is more important than data on proportionate numerical consumption of stream insects would indicate.

Order Neuroptera: Lacewings

Although the Neuroptera ('nerve wing') is a relatively large order, with more than 4000 species worldwide, there are only 14 species in six families in New Zealand. Lacewings on the main islands have brown, dark brown, or grey wings of nearly equal size. The main veins define many (50–80) cells and have end branches. The antennae are thin and medium in size with many segments, while the chewing mouthparts are small. The larvae have prominent pincer mandibles (New 1988; Winterbourn et al. 2000) and consume mostly soft-bodied prey such as aphids.

Only eight of New Zealand's lacewing species are endemic, and new species are very unlikely to be discovered. The level of diversity of non-introduced

Species distributions and sources of information and figures of New Zealand lacewings

Species	Family illustrations	Distribution	References	Species
Cryptoscenea australiensis	Coniopteridae	N, S	8	3
Drepanacra binocula	Hemerobiidae	K, N, S, C	9	1, 8
Euosmylus stellae	Osmylidae	N, S, uplands	14	2
Heteroconis ornata	Coniopteridae	N	13	13
Kempynus citrinus	Osmylidae	N	14	2
Kempynus incisus	Osmylidae	N, S	14	2
Kempynus latiusculus	Osmylidae	N, S, uplands	14	2
Micromus bifasciatus	Hemerobiidae	N, S, St	12	1, 16
Micromus tasmaniae	Hemerobiidae	N, S, St, offshore islands	2, 3, 5, 6, 12	1, 9, 16
Protobiella zelandica	Berothidae	N, S	19	1, 15
Psectra nakaharai	Hemerobiidae	N, northern S	19	9, 10
Sisyra rufistigma	Sisyridae	N	19	17
Weeleus acutus	Myrmeleontidae	N, S	19	8, 18
Wesmaelius subnebulosus	Hemerobiidae	N, S, mainly lowland	19	1

C, Chatham Island; K, Kermadec Islands; N, North Island; S, South Island; St, Stewart Island.
Sources: 1, Tillyard 1923b; 2, Kimmins 1940; 3, Cumber 1959a; 4, Cumber & Eyles 1961b; 5, Kimmins & Wise 1962; 6, Hilson 1964; 7, Macfarlane 1970; 8, 9, New 1983, 1988; 10–17, Wise 1973b, 1983a, b, 1988, 1992a, b, 1993, 1998a; 18, Grant 1999; 19, Macfarlane unpubl.

species is only a fifth of the equivalent diversity per square kilometre of Great Britain (Kloet & Hinks 1964) and is also low compared with Australia (CSIRO 1991). Although Wise (1992a) assessed other species in common with Australia as indigenous, he later inferred they were adventive (Wise 1995). This conclusion is supported by the co-occurrence in New Zealand of *Anacharis zealandica*, a parasitoid of the brown lacewings *Micromus tasmaniae* and *Drepanacra binocula* in both countries (New 1982; Valentine & Walker 1991). This parasitoid is the only member of its family (Figitidae) in New Zealand and was probably introduced, with its host, on plant material.

Wise (1992a) mapped the general distribution of the species in New Zealand and there are further location records and illustrations for the relatively uncommon *Protobiella zelandica* (Wise 1992b), the two *Micromus* species (Wise 1993), and *Sisyra rufistigma* (Wise 1998a). Four indigenous or adventive species inhabit the more remote offshore islands (Kermadecs: four species; Chathams: two species; subantarctic islands: two species).

Since its introduction prior to 1869, *Micromus tasmaniae* has become widespread within New Zealand and is certainly the commonest species of offshore islands, even those as small as the Alderman Islands (Early 1995). In the south, *M. tasmaniae* extends to the seldom-visited Antipodes Islands as well as the more regularly visited Auckland Islands. The possible lack of indigenous lacewings on the Three Kings and Poor Knights Islands could mean North Island offshore-island distribution resembles dispersal achieved by European bumble bees (Macfarlane & Gurr 1995). Both of these island groups were isolated from the main islands by moderate distances only in the last glacial period about 12,000 years ago, but this isolation is enough for there to be modest levels of endemic flies (Macfarlane & Andrew 2001) and beetles (Watt 1982). Introduced insect species are the main prey, but *M. tasmaniae* is associated with native aphids on *Olearia* shrubland (Derraik et al. 2001; Teulon unpubl.). Typical prey species include the main aphids of grassland and cereals (Valentine 1967b; Farrell & Stufkens 1990) and aphids in lucerne (alfalfa) before (Macfarlane 1970) and after the accidental establishment of three new species (Thomas 1977; Henderson 1980; Bates & Miln 1982; Leathwick & Winterbourn 1984; Rohitha et al. 1985).

Clovers are also associated with their distribution (Scott 1984; Cameron et

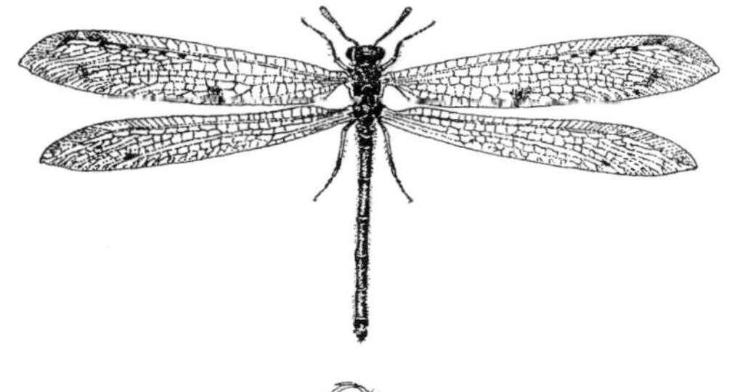

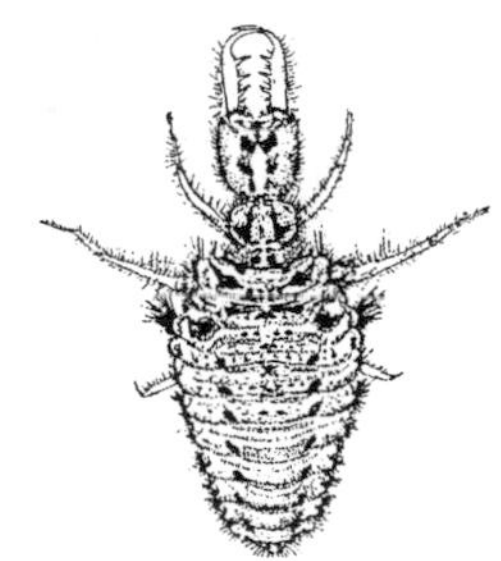

Adult and ant-lion larva of lacewing *Weeleus acutus*.
From Grant 1999

Tasmanian lacewing *Micromus tasmaniae*.
From Grant 1999

Kempynus incisus.
From Grant 1999

al. 1989). *Micromus tasmaniae* also occurs on fodder beet along with *Aulacorthum solani* and *Myzus persicae* (Pearson & Goldson 1980). In general, *M. tasmaniae* has been assessed as the most important enemy of these aphids, but the species also preys on mealy bugs, woolly apple aphid in orchards and vineyards, and aphids and the pine adelgid on conifers. The only recorded endemic prey other than aphids is the pittosporum psyllid on lemonwood (*Pittosporum eugenioides*). *Drepanacra binocula* is associated with endemic aphids on *Podocarpus totara*, the pittosporum psyllid on lemonwood, and accidentally introduced species (woolly apple aphid, spruce aphid, and pine adelgid) on apple trees and conifers. It can be associated with *Olearia* shrubland (Derraik et al. 2001), but is much less common in pastures and lucerne.

Three more Australian species have reached New Zealand in the last 30 years. *Sisyra rufistigma* was first found near Auckland in 1997 (Wise 1998a). The larva feeds on freshwater sponges (Spongillidae). *Psectra nakahari* was detected in Auckland in 1971 (recorded initially as *Sympherobius* (Wise 1988)) and is associated with wattles (Wise 1995). It reached the Manawatu (Canterbury Museum collection) and Wellington regions (Wise 1988) by 1986–87 and by 1996 it had spread at least as far south as Conway Flats (Canterbury Museum collection). *Heteroconis ornata* was collected from Auckland in 1988 (Wise 1988) and is assumed to have established (Wise 1992a). *Cryptoscenea australiensis* was first collected from Hastings in 1921 and has spread through the North Island and to Nelson (Wise 1963) and Christchurch (Canterbury Museum collection). It preys on mealy bugs on orchard trees and rushes and among undergrowth (Kimmins & Wise 1962).

These accidental introductions contrast with considerable efforts to establish green lacewings (Chrysopidae) in New Zealand, primarily against aphids (Cameron et al. 1989; Wise 1995). Between 1891 and 1928, importations of probably four *Chrysopa* species were made from the USA/Canada (five times) and Europe (once), as well as *Plesiochrysopa ramburi* from Australia (once) (Wise 1995). Six later importations of lacewings should have benefited from more rapid air transport that would have allowed them to arrive in better condition. *Chrysopa plorabunda* and *C. oculata* were imported from North America, and two releases were achieved against the oak aphid and aphids generally. Then *C. plorabunda* and *C. nigricornis* were imported again over three seasons between 1968 and 1976 for pasture and crop aphids (Cameron et al. 1989). Over 9000 Holarctic *Chrysoperla carnea* (= *downesi*) (Garland 1985) were sent from central British Columbia in 1927 for release against the pine adelgid or pine woolly aphid. In Europe, *C. carnea* inhabits decidous woodlands (Fraser 1959). Known prey that now exist in New Zealand include orchard pests (codling and oriental moth eggs, European red mite, and psyllid nymphs), cutworm eggs, and small caterpillars. It develops up to 20% faster on beetle eggs than on the aphid *Rhopalosiphum maidis* and greasy cutworm caterpillers (Obrycki et al. 1989), feeds on the early instars of grape mealy bug under bark (Grimes & Cone 1985), and is affected by its diet (Boszik 1992; Hodek 1993; Osman & Salmon 1996). Adults are nectar feeders, so release into evergreen, largely nectarless, pine forests may have been a major cause of this failure.

The type of vegetation and habitat is important for lacewings (Hodek 1993; Boszik 1994; Clark & Messina 1998). A lengthy shipment period and the strongly continental origin of *C. carnea* probably did not help either. *Chrysoperla carnea* of European origin was imported into New Zealand (Cameron et al. 1989; Wise 1995) for general aphid control, but *C. carnea* inhabits decidous woodland (Fraser 1959) so it is uncertain how well targeted the release sites were. Recent literature on electronic databases refers to *C. oculata* as predatory on aphids in sugarbeet and on apple trees. On the basis of suitable host availability, it is difficult to understand why green lacewing releases failed. Reviews of their role in biological control (Clausen 1978) and information on their biology (Carnard et al. 1984; Boszik 2000) may provide some alternative explanations. The extensive

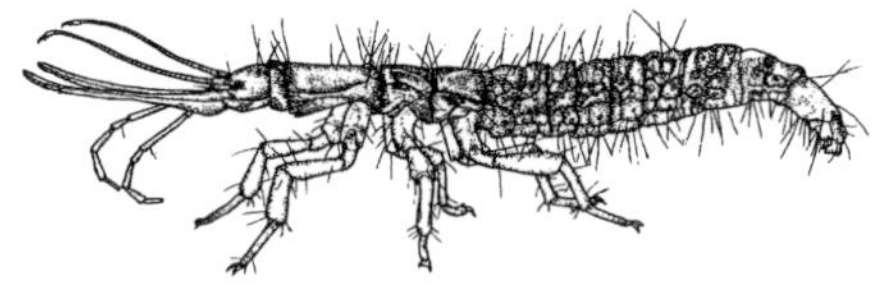

Aquatic larva of *Kempynus* sp.
From Winterbourn et al. 2000

information on *C. carnea* also provides guidance on how to conduct ecological studies of this species, which may be useful if applied in the New Zealand context. A 1904 European importation of lacewings was possibly the source of the brown lacewing *Wesmaelius subnebulosus*. This species was not captured until 1920 (Wise 1995), but only when Tillyard (1923b) provided the means for species identification was it possible to distinguish the species. In Australia, *Plesiochrysopa ramburi* is among the commonest and widespread chysopodids, but there are few records of it from the colder uplands and Victoria (New 1980).

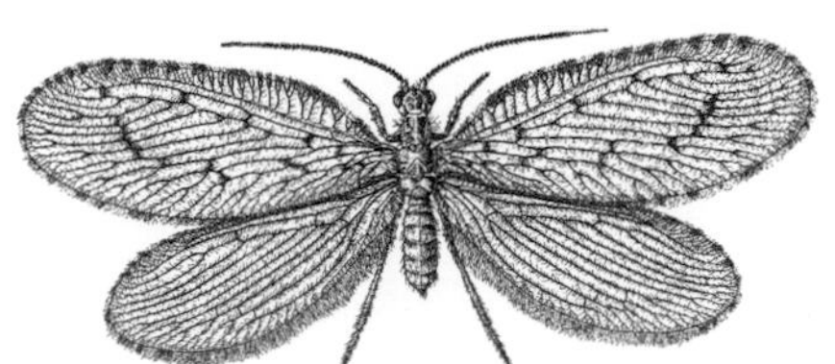

Protobiella zelandica.

J. Lyddiard, Auckland Museum

Since these introductions, extensive research and commercial rearing of *C. carnea* has been achieved (McEwan et al. 1999; Tauber et al. 2000; Vinkatensen et al. 2000). With this technical background, is *M. tasmaniae* sufficiently flexible in prey consumption of glasshouse and garden pests like rose aphids to merit research on its rearing? Also, is augmentative release of lacewings early in the season an option for control of certain crop pests, and can lessons from trials with *C. carnea* (Daane & Vokota 1997; Ehler et al. 1997) be applied or will the parasitoid *Anacharis zealandica* limit this possibility unduly?

Two species of *Mallada*, which are also green lacewings, inhabit the Kermadec Islands (Wise 1972a) and *Mallada basalis* has been recorded from Whale Island in the Bay of Plenty (Wise 1983b) while not yet being recorded from the mainland (Wise 1993). Also, a species of *Chrysopa* has been intercepted alive in Auckland (Wise 1988a).

Among endemic species, larvae of the four species of Osmylidae inhabit margins of streams, including the spray zone of waterfalls, where they shelter under stones (Winterbourn et al. 2000). In Britain, other osmylid species feed mainly on chironomids and other fly larvae (Fraser 1959). Larvae of the large antlion *Weeleus acutus* make pitfall shelters (Miller & Walker 1984) in sandy banks exposed to the sun (Hamilton 1921). Larvae feed on ants, flies, woodlice, spiders, and wetas. The adult is distinct from other Neuroptera because of the clubbed antenna, and the wingspan is up to 700 millimetres. *Protobiella zelandica* is associated with dead cabbage tree (*Cordyline australis*) and tree-fern foliage, but its prey is unknown (Wise 1992a). *Micromus bifasciatus* is associated with podocarp forest and mealy bugs (Wise 1993).

Lacewings are themselves prey of vertebrates. Most insectivorous bushbirds, for example the fantail *Rhipidura fuliginosa* (Moeed & Fitzgerald 1982), consume some. In more open country, known consumers of *Micromus tasmaniae* include rock wrens (*Xenicus gilviventris*) (Higgins et al. 2001).

Identification of adults of the larger lacewing species between 25 and 35 millimetres long can be quite readily achieved by consulting Kimmins (1940) and Grant (1999). However, identification of the smaller brown species, which are 3–8 millimetres long, is not easy for several reasons – details of wing venation patterns are scattered in the literature and the Australian family key (CSIRO 1991) is unduly complicated and not very user-friendly for beginners. New (1988) provided keys and illustrations of the Hemerobiidae of Australia and hence a means of distinguishing three of the genera in this family that also occur in New Zealand.

Order Coleoptera: Beetles

Beetles make up roughly one-fifth to one-quarter of the total insect fauna of the world and comprise one of the oft-cited cases of adaptive radiation of organisms. World estimates of the number of beetles vary, though there is general agreement that Coleoptera represent the highest number of described organisms. For example, Nielsen and Mound (1999) estimated 300,000 to 450,000 species worldwide. Calculating the total number of species is like counting stars, and indeed Grove and Stork (2000) emphasised that the question about the number of species overshadows more important questions about the taxonomy and biology of the species.

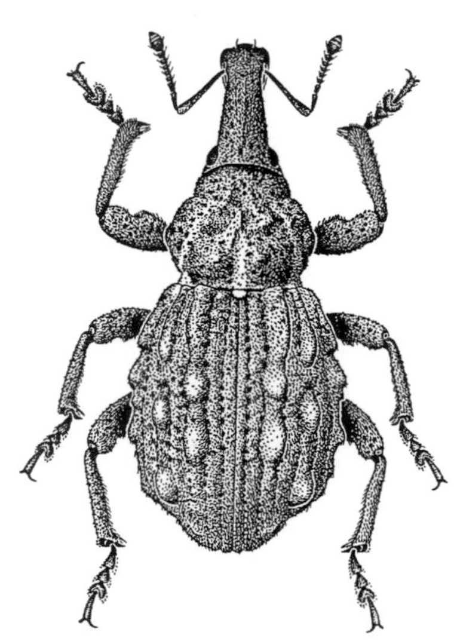

Canterbury knobbled weevil *Hadramphus tuberculatus* (Curculionidae), near extinction owing to habitat loss and predation by rodents.

From Young et al. 2008

Eleven-spotted ladybird *Coccinella undecimpunctata* (Coccinellidae).
Lincoln University

The order is well defined by having several diagnostic characters, of which the most distinctive is the hardened forewings, called elytra. It is from these that beetles get their formal name (*koleos*, sheath; *pteron*, wing). The elytra protect the more delicate hind wings and abdomen. During flight the forewings are opened enough to allow the hind wings to unfold and function. Other characters of beetles (see Lawrence & Britton 1994) include: holometabolous development (complete metamorphosis); adult antennae with 11 segments (sometimes fewer, or more in some Rhipiceridae and Cerambycidae); chewing mandibles that move in a horizontal plane; palp-bearing maxillae; well-developed prothorax that with the head forms a distinct tagma; body more or less dorsoventrally compressed so that leg bases (coxae) and the pleural region where they join the body lie ventrally; meta-sternum (hardened ventral part of hind thoracic segment) well developed with invaginated endosternite (cryptosterny); abdominal sternites typically more sclerotised (hardened) than the dorsal tergites and basal one or two sternites invaginated; and terminalia (genital and pregenital structures) usually enclosed within the apical segments of the abdomen. In larvae, the head capsule is complete and sclerotised with antennae and mandibulate mouthparts and the maxillae have well-developed palps; also the lip-like labium below the mouth lacks a silk gland and abdominal prolegs are usually absent. In the pupa, functional mouthparts are lacking (a condition known as adecticous) and the legs and wings are generally free from the body (exarate).

The oldest beetle fossil dates from 296 million years ago in the Mazon Creek (Illinois) deposit (Bethoux 2009) of upper Carboniferous (Pennsylvanian) age and since then the group has diversified globally into many different forms ranging in size from tiny featherwing beetles (Ptiliidae) as small as 0.3 millimetres long to giant Goliath and Hercules beetles (Scarabaeidae) well over 15 centimetres. The order is considered monophyletic (but see Whiting 2002a), although there are different opinions regarding the relationships of the Strepsiptera as a sister taxon to Coleoptera (cf. Kukalova-Peck & Lawrence 1993; Whiting et al. 1997). There are four suborders – Archostemata, Myxophaga, Adephaga, and Polyphaga. The latter two suborders comprise 99% of all beetles and the Archostemata are the most primitive group. No Archostemata or Myxophaga occur in New Zealand. Evolutionary relationships among the suborders are contentious (Beutel & Haas 2000; Caterino et al. 2002) and a full analysis using all available characters is warranted. There are at least 167 extant families and over 450 subfamilies contained in Coleoptera worldwide (Lawrence & Newton 1995), but new higher taxa are recognised annually, including at the family level.

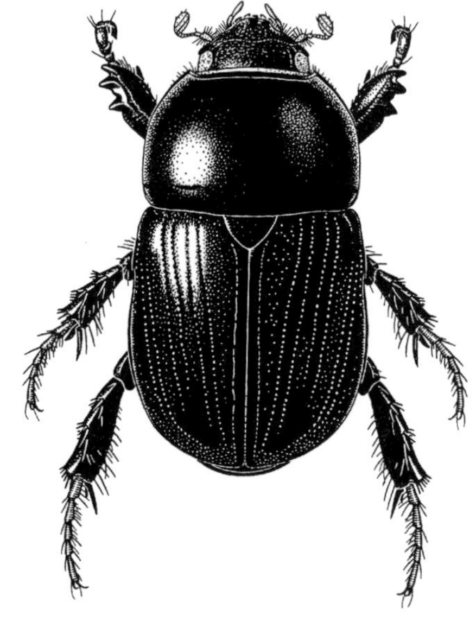

Black beetle *Heteronychus arator* (Scarabaeidae).
Des Helmore

The New Zealand beetle fauna

Most of the New Zealand beetle taxa were described between 1880 and 1923 by Thomas Broun, who named a total of 4323 species. His descriptions were based to a large extent on single specimens collected in the North Island lowlands, while the considerably more varied South Island fauna, in particular the rich but then still largely unknown subalpine and alpine component, had only scanty treatment. Hundreds of native and foreign species have since been added, though many groups require detailed taxonomic study, especially since there are many undescribed species and some that are incorrectly assigned to northern-hemisphere Holarctic genera.

In the absence of a full compilation of beetle species (such as that given in the end-chapter checklist), attempts were made in recent decades to estimate the total number of described and undescribed species in New Zealand. Watt (1983) estimated 4300 species, while Klimaszewski and Watt (1997) estimated over 5223 species and Emberson (1998) 6740 species. Based on the number of beetle species and potential host plants recorded in the Lynfield Survey in suburban Auckland, Kuschel (1990) estimated that 10,000 to 10,500 species are likely to occur in the fauna. Leschen et al. (2003) published a checklist of the known coleopteran genera of New Zealand, arriving at a total of 1091 in 82 families and 180 subfamilies. The species tally currently stands at 5479. The most diverse families in New Zealand are Staphylinidae (1232 species), Curculionidae (1225 species), Carabidae (557 species), and Zopheridae (191 species). These groups are also well represented in other parts of the world. The least diverse families, with one endemic species each, are Eucinetidae, Heteroceridae, Phycosecidae, Cucujidae, Prostomidae, Chelonariidae, Bostrichidae, and Monotomidae. The latter three families contain monotypic genera that may be primitive members of their group.

The fauna of New Zealand is disharmonic, consisting of ancient lineages that have been present long before the break-up of Gondwana and more-derived lineages and species that have arrived more recently from elsewhere (mainly Australia, the Pacific, Asia, and Indonesia). Very old amphitropical or bipolar groups are found in New Zealand and include broscine Carabidae, Derodontidae, and Byrrhidae, to name a few (Crowson 1980). Only one endemic family is present in New Zealand (Cyclaxyridae), but there are other groups representing more widespread Gondwanan elements. For example, the family Chaetosomatidae occurs only in New Zealand and Madagascar, and there are many examples of other family-group taxa found in New Zealand, southern South America, Australia, New Caledonia, and South Africa (e.g. migadopine Carabidae, camiarine Leiodidae, Cavognathidae, priasilphine Phloeostichidae, and Chalcodryidae).

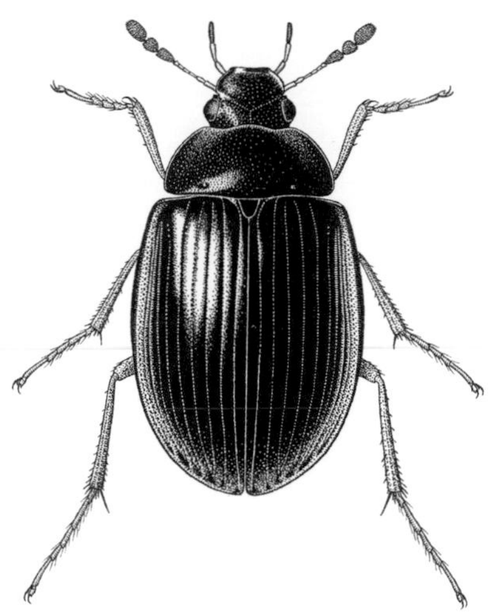

A terrestrial 'water' beetle *Rygmodus tibialis* (Hydrophilidae).

Des Helmore

Areas of endemism have not yet been established for New Zealand Coleoptera, but it is quite evident that certain groups are regionalised (Campbell Plateau, northwest Nelson area of the South Island, northern North Island, offshore islands, etc.) or restricted to certain communities (sooty moulds, *Nothofagus* forests, caves, and tussock grasslands).

With so many beetle species in New Zealand, it is clear that they will have significant ecological roles, economic impacts, and interesting stories to tell, and such is the case. Take, for instance, ladybirds (Coccinellidae), among the most commonly encountered and appreciated beetles. Several species are used in biological control, as they feed on aphids, mites, or scale insects. Probably the most successful is the cardinal ladybird *Rodolia cardinalis*, which reduced cottony cushion scale, a serious pest of citrus, to very low levels.

The tough forewings, the elytra, are probably one of the major contributing factors in the successful diversification and radiation of beetles into habitats that other insects with fragile, exposed wings cannot occupy. Beetles are also less vulnerable than other insects to predators and to weather in exposed situations. The elytra are also important in reducing water loss in dry conditions. New Zealand beetles are found in a wide range of situations, from beach sands to alpine heights, and, while the fauna of lakes, ponds, and streams is relatively small, a variety of species can be found in wet leaf axils of plants, saturated moss, and moist soil. Almost all native beetles are endemic, having evolved in isolation in New Zealand or are the sole survivors of groups that have died out overseas. Native species have suffered from habitat loss and very few have adapted themselves to modified landscapes, pastures, crops, and gardens. Most beetles in these situations have been introduced, usually accidentally, from overseas.

The longest New Zealand beetle is the slender giraffe weevil (*Lasiorhynchus barbicornis*), up to 80 millimetres long, but the heaviest is probably the huhu (*Prionoplus reticularis*). The smallest, around half a millimetre in length, are feather-winged beetles. The largest group of predatory beetles is the ground beetles (Carabidae). Many are large, powerful, and flightless, with small or vestigial hind wings and sometimes fused elytra (e.g. *Ctenognathus, Holcaspis, Mecodema, Megadromus*). Walking is their only means of dispersal so their geographic distribution tends to be limited. The presence of some flightless beetles on both the mainland and offshore islands is helpful in formulating hypotheses on the history and isolation of the islands. Some species of *Megadromus* and also *Plocamostethus planiusculus* appear to be unique among ground beetles in showing some degree of parental care (Watt 1974).

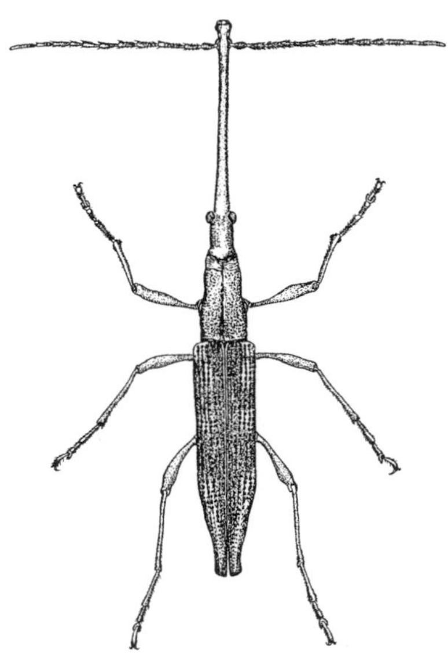

Male giraffe weevil *Lasiorhynchus barbicornis* (Brentidae).

From Grant 1999

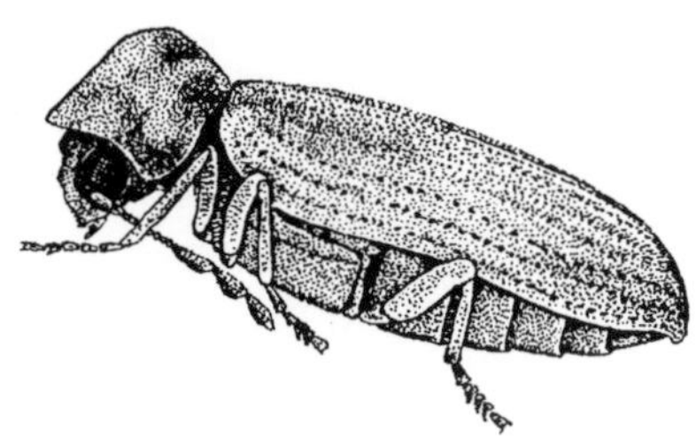

House borer *Anobium punctatum* (Anobiidae).
From Grant 1999

Rove beetles (Staphylinidae) are another important family of predatory beetles, as are tiger beetles (Cicindellidae), the larvae of which (known as penny doctors) inhabit tunnels in clay banks or sand waiting for unsuspecting prey, typically other small arthropods. Wireworms, the larvae of Elateridae (click beetles) are frequently predatory but can also feed on roots. Some beetles are parasites, like the larvae of *Bothrideres* (Colydiidae), which are external parasites on lavae of wood-boring beetles.

Mention wood-borers, and the tiny, brown introduced borer beetle *Anobium punctatum* comes to mind. Its galleries in timber and furniture cause great damage and the holes from which the adult emerges are unsightly. Yet, in its native habitat it serves an important role in breaking down dead wood too dry to support wood-rotting fungi. Other introduced borers include pine-bark beetles (*Ernobius mollis, Hylastes ater*), the powder-post beetle *Lyctus brunneus*, keyhole ambrosia beetle *Xyleborus saxeseni*, and pinestump longhorn beetle *Arhopalus ferus*, all from Europe. There is also an introduced Australian group of species, associated with eucalypts and acacias, including five species of longhorns, two ambrosia beetles, a bark beetle, and a weevil. There is also a Californian bark beetle (*Phloeosinus cupressi*) that infects macrocarpa and other cypresses (Watt 1974). It should not be assumed that all New Zealand's anobiids and other borers are exotic, however. Three native species – *Anobium magnum, Capnodes griseipilus*, and *Dorcatoma oblonga* – may occur in buildings, though normally they break down wood in decaying stumps of native trees. Other native borers include include two species of Buprestidae (jewel beetles). Larvae of *Nascioides enysi* mine galleries in the bark and sapwood of recently dead or felled *Nothofagus* species, and those of *Neocuris eremita* in *Nothofagus* and *Pittosporum* (Milligan 1975).

The main beetle families of aquatic habitats in New Zealand are the Dytiscidae, Hydrophilidae, Elmidae, and Hydraenidae. Dytiscids (diving water beetles) are aquatic in adult and larval stages. They are carnivorous and closely related to the Carabidae from which they are derived. The hind legs are flattened and paddle-like, acting like a pair of oars. Most dytiscids are able to fly and if they cannot find a suitable water body immediately they can survive without it. The New Zealand species include large-bodied forms that are related to Australian species. For example, New Zealand's largest dytiscid, 26mm-long *Homoeodytes hookeri* is closely related to two species in Australia. The common black water beetle *Rhantus pulverosus*, 10 millimetres long, is cosmopolitan. There are also small subterranean species, found in groundwater, that are among the world's most primitive water beetles. The name Hydrophiliidae ('water-lovers') notwithstanding, many members of this family are terrestrial. The aquatic species feed on decaying vegetation, are somewhat sluggish, and do not have modified legs for swimming. Elmids are small, elongate, crawling beetles of stream and river bottoms but are also very active fliers. Hydraenids are minute aquatic beetles living among dead leaves and algae, though one genus, *Meropathus*, is wholly terrestrial. The New Zealand fauna has not been well studied (Ordish 1975).

Shelf and bracket fungi are home to a diverse range of beetles. Most of these feed on the fungus itself, its spores, or other moulds growing on it. Some beetles are specific to a single species but others occupy a range of fungi of the same form or consistency. In fact, a succession of beetle species occupies a bracket fungus at different times as it ages. Almost all bracket fungi are attacked by *Cis* and its relatives (Ciidae), comprising beetles that resemble woodborers. Another group is the Zopheridae, including *Brouniphylax*. Mushrooms and toadstools are less vulnerable to beetle grazing owing to their life-history and fragility. *Triphyllus* species (Mycetophagidae), however, have a short life and may occur in large toadstools (Watt 1974).

A variety of beetles live under loose bark of native trees. Some merely shelter there, but others feed on moulds or the breakdown of plant tissues, including various Colyidiidae and larvae of stag beetles (Lucanidae). Many are flattened, even wafer-thin, including members of the families Cucujidae, Inopeplidae, and

Whirlygig beetle *Gyrinus convexiusculus* (Dytiscidae).
Des Helmore

Family-level diversity of New Zealand Coleoptera

Family	Subfamily	Described species and subspecies Endemic only	Adventive	Indigenous only	TOTAL	Undescribed species E = endemic A = adventive
Rhysodidae		6	0	0	6	
Carabidae	Cicindelinae	12	0	0	12	
	Carabinae	1	0	0	1	1E
	Migadopinae	7	0	0	7	
	Scaritinae	0	4	0	4	
	Trechinae	106+10	0	1	107+10	
	Psydrinae	109+5	5?	0	114+5	25E
	Harpalinae	207+2	25?	2	234+2	52E
Dytiscidae		11	0	8	19	
Gyrinidae		0	0	1	1	
Hydrophilidae		40	8	3	51	5E
Histeridae		21	6	0	27	3E
Hydraenidae		34	1	0	35	
Ptiliidae		44	11	1	56	5E
Agyrtidae		2	0	0	2	
Leiodidae		69	0	0	69	8E
Staphylinidae	Aleocharinae	139?	29?	0	168	45E,13A
	Euasthetinae	13	1	0	14	4E
	Habrocerinae	0	1	0	1	
	Microsilphinae	1	0	0	1	15E
	Omaliinae	55	4	1	60	2E
	Osoriinae	43	0	0	43	1E
	Oxytelinae	13	14?	2	29	2A
	Paederinae	19?	11	0	30	6E,1A?
	Phloeocharinae	1	0	1	2	
	Piestinae	1	0	0	1	
	Proteininae	5	0	0	5	3E
	Pselaphinae	374	2	1	377	21E,2A
	Pseudopsinae	1	0	0	1	1E
	Scaphidiinae	21	2	0	23	
	Scydmaeninae	185	1	0	186	10E
	Staphylininae	106	24	2	132	2E,1A
	Tachyporinae	23	3?	0	26	3E,1A
Lucanidae		35	2	0	37	
Trogidae		0	1	0	1	
Scarabaeidae		135	18	0	153	12E
Scirtidae		124	0	0	124	10E
Eucinetidae		2	0	0	2	
Clambidae		12+3	1	0	13+3	1E
Buprestidae		2	0	0	2	
Byrrhidae		49	0	0	49	2E
Dryopidae		4	0	0	4	
Elmidae		6	0	0	6	16E
Limnichidae		8	0	0	8	
Heteroceridae		1	0	0	1	
Ptilodactylidae		1	0	0	1	5–8E
Chelonariidae		1	0	0	1	
Eucnemidae		19	2	0	21	
Elateridae		117	4	0	121	
Lycidae		0	1	0	1	
Cantharidae		43+1	1	0	44+1	
Derodontidae		2	0	0	2	
Jacobsoniidae		1	0	0	1	4E
Nosodendridae		2	0	0	2	
Dermestidae		11	16	0	27	
Bostrichidae		1	5	0	6	2A
Anobiidae		34+1	6	0	40+1	1E
Ptinidae		6	9	0	15	
Trogossitidae		29	2	0	31	7E
Chaetosomatidae		3	0	0	3	1E
Cleridae		30	4	0	34	
Metaxinidae		1	0	0	1	

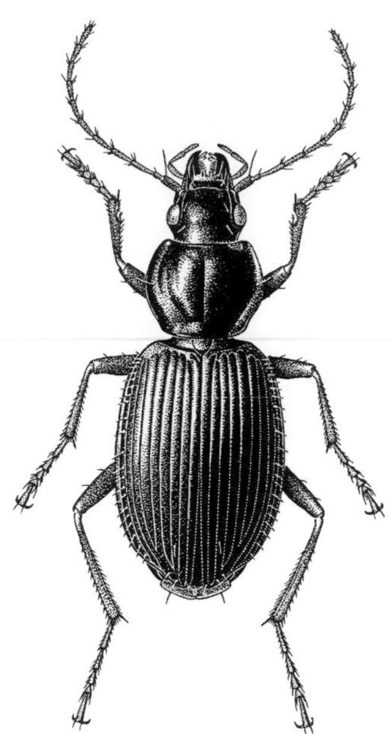

Coastal ground beetle *Ctenognathus novaezealandiae* (Carabidae).

Des Helmore

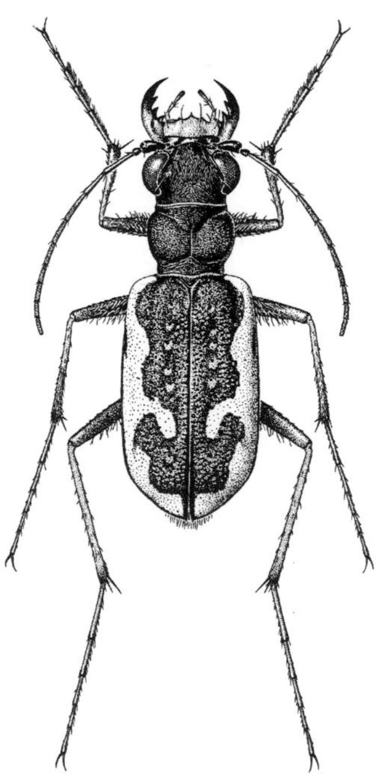

Common tiger beetle *Neocicindela tuberculata* (Carabidae).

Des Helmore

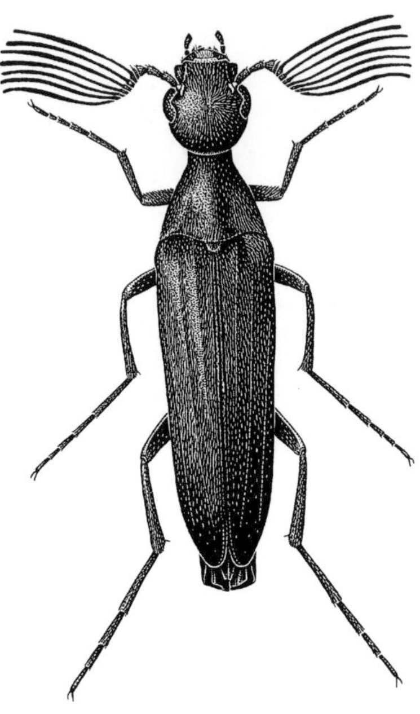
Antlered beetle *Rhipistena lugubris* (Ripiphoridae).
Des Helmore

Family	Subfamily	Described species and subspecies Endemic only	Adventive	Indigenous only	TOTAL	Undescribed species E = endemic A = adventive
Phycosecidae		1	0	0	1	
Melyridae		32	0	0	32	2E
Nitidulidae		17	17	0	34	
Monotomidae		1	5	0	6	
Agapythidae		1	0	0	1	
Priasilphidae		7	0	0	7	
Silvanidae		9	7	0	16	1E
Cucujidae		1	0	0	1	
Laemophloeidae		1	4	0	5	1E,1A
Phalacridae		0	1	0	1	1A
Cyclaxyridae		2	0	0	2	
Cavognathidae		4	0	0	4	
Cryptophagidae		19	4	0	23	1E
Erotylidae		16	1	0	17	
Bothrideridae		6	1	0	7	
Cerylonidae		5	0	0	5	
Endomychidae		2	1	0	3	4E
Coccinellidae		25	19	0	44	5E
Corylophidae		14	6	0	20	7E,2A
Latridiidae		39	15	0	54	8E,1A
Mycetophagidae		16	3	0	19	3E
Archeocrypticidae		0	1	0	1	
Ciidae		24	3	0	27	7E
Melandryidae		35	0	0	35	3E
Mordellidae		3	2	1	6	
Ripiphoridae		5	0	0	5	
Zopheridae		190	0	0	190	1A
Ulodidae		21	0	0	21	2E
Chalcodryidae		5	0	0	5	
Tenebrionidae		136	12	0	148	1A
Prostomidae		1	0	0	1	
Oedemeridae		18	3	0	21	
Pyochroidae		6	0	0	6	3E
Salpingidae		26	0	0	26	2E
Anthicidae		17	10?	0	27	2A
Aderidae		9	0	0	9	4E
Scraptiidae		4	0	0	4	2E,1A
Tenebrionoideaincertaesedis		2	0	0	2	
Cerambycidae		171	9	0	180	3E
Chrysomelidae		141+4	22	0	163+4	1E
Nemonychidae		4	0	0	4	
Anthribidae		58	3	0	61	2A
Belidae		5	0	0	5	
Oxycorinidae		1	0	0	1	
Brentidae		1	0	0	1	
Apionidae		5	1	0	6	
Dryophthoridae		0	4	0	4	1A
Erirhinidae		42	0	0	42	
Raymondionymidae		1	0	0	1	
Curculionidae	Baridinae	0	1	0	1	
	Ceutorhynchinae	0	3	0	3	
	Cossoninae	143	5	0	148	4E
	Cryptorhynchinae	259	1	0	260	19E
	Curculioninae	235	6	0	241	6E,1A
	Cyclominae	80+3	7	0	87+3	4E
	Entiminae	231	12	0	243	6E+1E,1A
	Lixinae	0	2	0	2	
	Molytinae	164	0	0	164	1E
	Platypodinae	4	0	0	4	
	Scolytinae	7	13	0	20	10E
Totals		**4620+29**	**418**	**24**	**5062+29**	**382–385+1E, 35A**

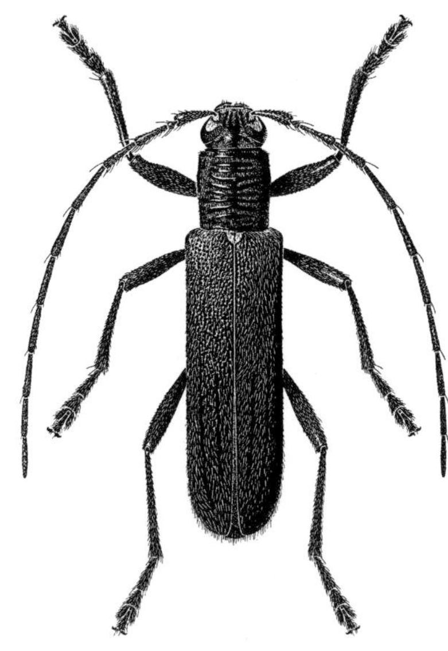
Lemon tree borer *Oemona hirta* (Cerambycidae).
Des Helmore

Prostomidae. Rotten wood is inhabited by larvae and adults of darkling beetles (Tenebrionidae), larvae of pintail beetles (Mordellidae), and others (Watt 1974).

Forest leaf litter is generally a favourable habitat for beetles, including larvae of scarabs (Scarabaeidae), which also inhabit grassland soils. One example is an economic pest, grass-grub (larvae of *Costelytra zealandica*), one of the few native beetles that have successfully adapted to improved pastures and can occur in staggering numbers if not checked. Other scarabs can be found at high altitudes in subalpine soils and litter. Some of the large scarabs of Otago and Southland (*Prodontria, Scythrodes*) are wingless, including rare *P. lewisi*, modified for life in sand under native grasses near Cromwell. Several beetles, in fact, are specialised to live in sand, having hard body parts, stouter limbs, and thick bristles. Tenebrionids *Choerodes* and *Actizeta*, the scarabs *Pericoptus*, and the sand weevil *Cecyropa* (Curculionidae) are examples (Watt 1974). Maori were familiar with grass-grub beetles, their larvae, and relatives, having several names for them. They used to eat one species – *kekerewai*, the manuka beetle – which was collected in large numbers, crushed, and baked with the pollen of bullrush (*Typha*) to make a kind of scone (Miller 1971).

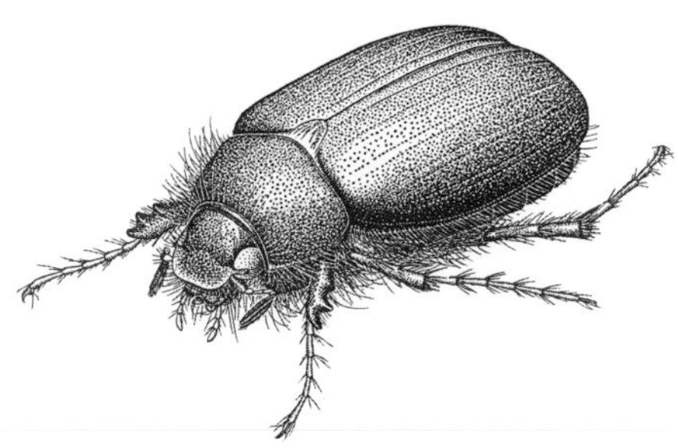

Grass grub beetle *Costelytra zealandica* (Scarabaeidae).
Des Helmore

Future work

Refining the end-chapter Coleoptera checklist will be an ongoing task; few genera and species have been revised since the major work done by Thomas Broun. Nevertheless, even some undescribed species have conservation status (McGuiness 2001), which illustrates the need for continuing taxonomic work. There is an ever-present time lag between the discovery of a new species and the availability of a published name, sometimes spanning decades. For example, a species of *Platisus* from the Three Kings that is the only known member of the genus in New Zealand was first collected by E. S. Gourlay in the 1960s and was not described until almost 40 years later by Watt et al. (2001). This process underpins the problems that systematists are faced with regarding modern taxonomy in general, which involves deep investigation in understanding character variation, evolutionary relationships, and classification, but all too often with a paucity of study material. Most systematists would certainly argue for more workers and students to describe New Zealand's Coleoptera fauna, but modern times are much different from those in the 19th and 20th centuries when taxonomic names were mass produced, and, in some ways, the taxonomic impediment (Heywood 1995) did not exist. Increased funding for training and research would go along way towards remedying the paucity of coleopterist expertise in New Zealand. Apart from obvious applications to conservation, biological control, and biosecurity and border control, future research should include surveys in New Zealand's inaccessible mountaintops, valleys, and offshore islands; descriptions of species and revising groups based on sound classification; and producing catalogues and databases to contain the vast amount of taxonomic information.

Order Mecoptera: Scorpionflies

Scorpionflies are an archaic group of insects dating from the Early Permian, some 260 million years ago. Today, there are only about 590 species worldwide, known variously as scorpionflies (from their upturned abdomen tip), hanging-flies, and snowfleas. (Mecoptera means long-winged.) There is a single representative in New Zealand, *Nannochorista philpotti*, widely distributed around small, stable streams in the South Island.

The Nannochoristidae to which *N. philpotti* belongs is a small family of just eight species, having a circum-Antarctic distribution – three species occur in Argentina and Chile, four in Australia, and one in New Zealand. It originated in the Late Permian, and, according to Willmann (1987), represents the earliest evolutionary branch among living scorpionfly families. No fossils are reported

Nannochorista philpotti.
Birgit Rhode, Landcare Research

from New Zealand, but nannochoristid wings have been found in the Late Permian of New South Wales. The family is distinct enough to have been separated as a separate order, Nonnomecoptera [sic], by Hinton (1981, vol. 2, p. 722, but as Nannomecoptera in Hinton 1981, vol. 1, p. 10). In Willmann's (1987) classification of the Mecoptera this group is regarded as a suborder. Whiting (2002b) concluded from gene-sequence data that the Mecoptera is paraphyletic with two major lineages – one comprising Nannochoristidae, the mecopteran family Boreidae, and the flea order Siphonaptera, and the other all other Mecoptera. On the other hand, detailed studies of head structures in adults and larvae give evidence that the Nannomecoptera may be segregated as an order in a clade that also includes Siphonaptera and Diptera (Beutel & Baum 2008; Beutel et al. 2009). The question remains open.

Tillyard (1917) erected the genus *Choristella* for the New Zealand species, separating it from his Australian genus *Nannochorista*, but the name *Choristella* was preoccupied by a gastropod mollusc so Byers (1974) proposed *Microchorista* as a replacement. Kristensen (1989) was of the opinion that differences in the New Zealand form were not significant above the species level and located *philpotti* in *Nannochorista*, but Ferrington (2008) has accepted *Microchorista* for the New Zealand species and an undescribed form from Tasmania.

Ecology

Known immature stages of other families of Mecoptera involve a relatively short-lived terrestrial larva, living in soil and feeding on plant roots and other vegetation, or emerging from burrows to eat dead insects and other animal matter. Pupation occurs in the burrows, and immature adults prey on other insects.

Immature stages of Nannochoristidae, however, are aquatic, a finding that was first documented by University of Canterbury's Bob Pilgrim (1972) in *Nannochorista philpotti*. He found pupae in the Hawdon River valley, near Cass, mid-Canterbury, alongside a small stream in *Nothofagus* forest. When first discovered they could not be placed in a known insect order, but were reared to adults, thus revealing their identity. Many larvae were subsequently reared through several instars to pupae and thence to adults. The larval habitat is fresh, slow-moving but well-aerated water of streams that are not subject to flooding and scouring. The fine silt accumulating in pockets and meanders contains an abundance of other insect larvae, particularly of small chironomid flies, especially species of *Paucispinigera* and *Polypedilum* (Chironominae) and *Macropelopia* (Tanypodinae). These appear to be a major food source for *N. philpotti* larvae – a large number of dissected larvae revealed predominantly chironomid larvae in the gut and smaller numbers of caddisflies, mayflies, and other insects, mites, and aquatic Oligochaeta.

First-instar larvae have been captured in November and the last prepupal forms of the previous generation may be present as late as December. The duration of larval life is thus very much longer than that reported for the terrestrial larvae of other families. Pupation occurs in wet moss and other vegetation on rocks and fallen tree trunks, just above the water line. The flying phase is shorter than in other families. Adults have been captured mostly from November to January but some have been collected in October and up to March in various localities; nevertheless, the species appears to have just one generation per year. Adults frequent streamside vegetation from which they may be collected by sweep-netting.

No observations on the food and feeding habits of the adult are available. It may be expected that, in common with most Mecoptera, they feed on other insects captured in flight. Eggs have not been found, but Australian species oviposit in wet litter near the water's edge (Lambkin 1996).

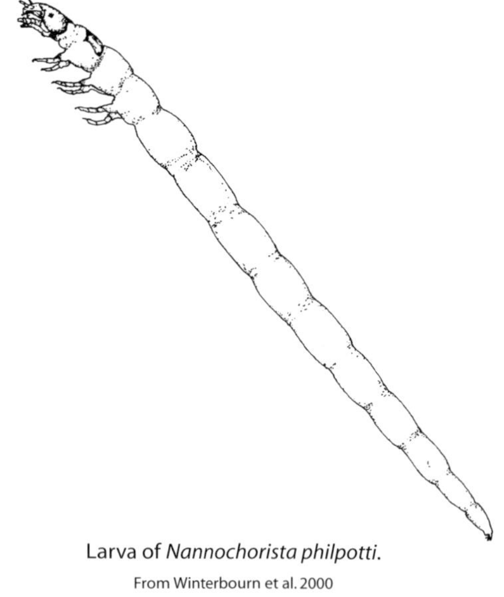

Larva of *Nannochorista philpotti*.
From Winterbourn et al. 2000

Larval morphology and behaviour

Larvae are very active and wriggle vigorously through the silt in streams. There are well-developed thoracic legs but no abdominal prolegs; the last segment bears a pair of hook-like structures only. The mandibles have several sharply pointed teeth and are well suited to impaling live prey. An unusual feature is the presence of a movable 'lacinia mobilis'-type of lobe on the inner face of the mandible. This lobe, itself well-toothed, presumably acts as a supplementary device for securing active prey. It is present from the first instar through to the early phase of the fourth, but is shed during that instar as the larva enters the prepupal phase. The mandible shows a scar where the lobe is lost and at this time the colour and general appearance of the larva undergo marked changes (Pilgrim 1972). Such conspicuous differentiation of phases within the fourth larval stadium are not reported elsewhere in the Mecoptera. Another feature unique to the larva is the presence of two papillae arising from the end of the hindgut; each contains the terminal portion of one of the six excretory Malpighian tubules. Pilgrim (1972) suggested that the papillae might function in relation to Malpighian tubule physiology by acting as a region for ion/metabolite exchange, but no experiments have been published on this possibility; they contain only minute tracheal branches, making it unlikely that they have a significant respiratory function.

Following the discovery of the aquatic stages of the New Zealand species, the same was made for Australian forms (see Riek in Lambkin 1996). Professor Pilgrim collected larvae of several species from many streams in Tasmania and Australian Capital Territory and also examined larvae (unidentified as to species) from streams in Chile. All of these forms are extremely similar, but no detailed comparison has been published. It is clear, however, that the overall morphology and habit of the larvae are characteristic of the family.

Distribution of Nannochoristidae

The family may be considered an outstanding example of Gondwanan distribution. Within New Zealand, adults and/or larvae have been collected from Stewart Island and Big South Cape Island, numerous localities in Southland (including Longwoods, the type locality), Otago, Westland, Nelson, and Marlborough. In Canterbury, it is present at lower altitudes along the main alpine range and in the foothill ranges; there is a gap over much of the Canterbury Plains, but the insect is widespread on Banks Peninsula.

It is typically found associated with the smaller streams in native forest or second-growth bush, but collections have been made in open country, e.g. in a peat-bog creek in the Umbrella Range region (Otago) and in the Glentui River (Canterbury). These occurrences are perhaps the result of downstream drift from small tributaries. Its altitudinal range is from near sea level (Stewart Island, Macandrew Bay (Dunedin), and Abel Tasman National Park) to about 1000 metres above sea level.

The only known specimens from the North Island are two pinned adults in the 'C. E. Clarke Collection' in the Auckland Museum, comprising a male and a male abdomen labelled 'Mamaku 28.12.[19]20'. No further specimens were found in recent visits to the Mamuku area, but the Mamaku forests have been severely reduced and the area now appears unlikely to harbour the insect. It is significant that no specimens were collected or reported in the Wellington area by G.V. Hudson, who certainly knew the insect (together with his daughter Stella he collected it in the Queenstown area and the specimens are in the Museum of New Zealand), yet there are habitats that appear suitable even in Karori, where Hudson lived for many years. Present evidence is that the insect is confined south of Cook Strait, but further collecting in the southeast of the North Island, including Kapiti Island, should be carried out. Although the insect is a weak flier, it could have been wind-assisted across Cook Strait (if indeed it had not been present in the area before the strait itself was established).

Future work

Investigations are needed to find sites where eggs are laid. More information is desirable on the physical parameters of the larval and pupal environment, as well as on detailed associations of the larval instars with their ambient fauna. The function of the anal papillae of the larva is worthy of experimentation. There is no published information on feeding, mating, or oviposition in the adult stage. Kristensen (1989, p. 113) has emphasised the phylogenetic interest of this archaic family of Mecoptera and called for '... penetrating comparisons of adults and immatures alike to reach a trustworthy phylogeny'. Additional molecular-phylogenetic work, currently being undertaken in laboratories in Germany and the USA, should clarify the status of the Nannomecoptera.

Order Siphonaptera: Fleas

Fleas are obligate blood-sucking parasites of warm-blooded vertebrates. About 95% of flea species worldwide parasitise mammals; the rest are found on birds. All are wingless (Siphonaptera means 'sucking-wingless') and their life-history includes complete metamorphosis, having larval and pupal stages. The larvae are with very rare exceptions non-parasitic 'maggot-like' grubs living in the nest debris of their adult hosts. Their food is organic detritus, including excess blood excreted by their adults and, in part, probably other tiny creatures inhabiting the nest. Fleas are famous for their prodigious ability to spring 150 times their body length – a feat achieved by a combination of muscle power and energy stored in a compressed springy pad of protein called resilin.

The fossil history of the order is poorly known. The very rare specimens recovered from Baltic amber (Eocene) and from Dominican amber (Miocene) belong to existing genera. No fossil fleas are reported from New Zealand. There are more than 2000 species worldwide. The largest is North American *Hystricopsylla schefferi*, a giant at 12 millimetres long.

New Zealand's flea fauna

The most recent account of fleas in New Zealand and its offshore islands is that of Smit (1979), who listed 34 species and subspecies, reviewed earlier literature, and described several new taxa. Later, Smit (1984) described a subspecies from the remote Bounty Islands. This is the latest described flea from the region. Collecting has been sporadic but widespread throughout New Zealand and the offshore islands, especially over the past several decades and it appears unlikely that many new taxa will be forthcoming, although many bird hosts have yet to be thoroughly investigated. It is probable, therefore, that we now have a complete taxonomic account of the adult flea fauna of the New Zealand region. This is unusual for most countries, but understandable in view of the size of New Zealand and its fauna, and of the efforts on the part of numerous collectors.

Eleven of the 35 species/subspecies are accidental introductions (10 on mammals, one on birds), six are naturally indigenous (all on birds), 17 are endemic (one on a bat, 16 on birds), and there is one deliberate introduction (on rabbit). New Zealand's flea fauna has many peculiarities:

1. There is a complete absence of members of the large family Ctenophthalmidae and the Hystrichopsyllidae, which together comprise ca. 30% of the world Siphonaptera.
2. The families Leptopsyllidae and Pulicidae are represented in New Zealand by introduced species only, but several have become well established.
3. The very large family Ceratophyllidae, c. 20% of the world fauna, typically Holarctic in character, is represented by three introduced taxa and one aberrant monospecific genus confined to Antarctica.
4. Of the world flea species, the vast majority normally parasitise mammals. In New Zealand, in contrast, the native/endemic flea taxa include only one on mammals and 22 on birds. These strange proportions are clearly related to the peculiarities of the New Zealand mammal and bird faunas – an extreme paucity of suitable land mammals but a large number (disproportionately so in

global terms) of seabirds, particularly of procellariiforms, penguins, and shags.
5. The affinities of the major families of endemic/native fleas are largely within the Southern Hemisphere.

Individual flea families

Ischnopsyllidae occur worldwide wherever bats are found. *Porribius pacificus* is host-specific to *Chalinolobus tuberculatus*; the record from *Mystacina tuberculata* (Smit 1979) has not been repeated and is probably an error owing to accidental laboratory contamination. Two other species of *Porribius* are found in Australia and one in New Guinea.

Pygiopsyllidae are well represented in Australia–Papua New Guinea, with some genera also in South America and southern Asia and a few reaching Japan, the Philippines, and India (see Mardon 1981). All six taxa of the genus *Notiopsylla* are found in the New Zealand region, four of them endemic. *Notiopsylla corynetes* is the only species from the mainland, where it is host-specific on *Puffinus huttoni*, nesting at high altitudes in the South Island; the remaining endemic and native taxa are associated with seabirds nesting near the shores of offshore and subantarctic islands. *Hoogstraalia imberbis* was described from a single female of unknown host association in the Auckland Islands. Extra-limital species of the genus are hosted by land birds in Tasmania, New Guinea, and the Philippines. Closely related *Pagipsylla galliralli* was described from a weka but has since been found more commonly on other ground-nesting birds and some introduced hosts. The two species of *Pygiopsylla* are associated with rats, especially *Rattus exulans*, presumably its primary host from which it has invaded the more recently introduced *R. norvegicus* and *R. rattus*; both species of the flea are native to Australia, where other species also occur, and there is one species in Papua New Guinea.

Rhopalopsyllidae are chiefly Neotropical in distribution where the hosts are mostly mammals, but the many species of *Parapsyllus* are bird fleas, distributed around coastal Australasia, South America, South Africa, and islands in the circumpolar seas south of about 60° S (see Smit 1987). Most species are found on seabirds nesting at or near the shore, but *P. lynnae alynnae*, like *Notiopsylla corynetes* (above), is host-specific on *Puffinus huttoni* in coastal mountains of the South Island. *Parapsyllus nestoris nestoris*, too, is unusual in specifically parasitising a parrot, *Nestor notabilis*, that also nests in alpine South Island, while *P. n. antichthones* is associated with parakeets in subantarctic islands. The remaining endemic species/subspecies are confined to the subantarctic islands.

Some introduced fleas

Pulicidae

Pulex irritans was formerly widespread on humans and pigs; it has become less abundant with improvements in human hygiene. *Ctenocephalides felis felis*, very common on domestic cats, is the flea most likely responsible for human infestation by the tapeworm *Dipylidium caninum*; it also occurs on dogs. *Ctenocephalides canis*, found on dogs, is less common than the previous species. *Xenopsylla cheopis* is the famous plague flea. It is probably not established in New Zealand but was recorded in the early 1900s at the time of an outbreak of bubonic plague in Auckland, which caused nine deaths (MacLean 1955). Specimens are still occasionally intercepted in shipping from overseas. As the rodent vectors (rats) are present in abundance here, the accidental introduction of plague-infected fleas could potentially lead to repeated outbreaks of the disease. *Xenopsylla vexabilis* is widespread in many parts of the Pacific and was presumably introduced to New Zealand along with the kiore *Rattus exulans*, its normal host. It is now known only in association with this rat, and confined to offshore islands.

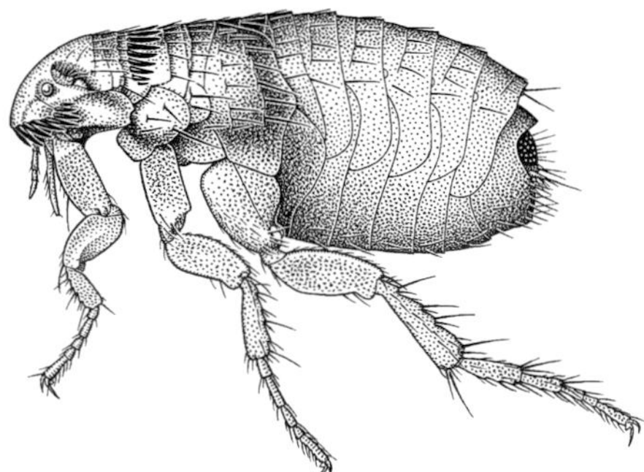

Cat flea *Ctenocephalides felis felis*.
Des Helmore

Leptopsyllidae

Leptopsylla segnis is abundant on mice, worldwide.

Ceratophyllidae
Ceratophyllus gallinae is abundant on introduced passerines (starling, house sparrow, blackbird). Although associated with domestic fowl in Europe, it is not widespread on this or other galliform hosts in Nw Zealand; it was perhaps introduced with chickens released on one of Captain Cook's voyages. It has been recorded from fernbird (*Bowdleria punctata caudata*) on the Snares Islands and has also been taken from *Petroica traversi* and *Gerygone albofrontata* nests in the Chatham Islands. Interestingly, *Parapsyllus struthophilus*, the typical flea of *Petroica traversi* and other endemic birds, has become a parasite of introduced birds (blackbird and hedge sparrow) in the Chatham Islands. *Nosopsyllus fasciatus* is abundant on brown and Norway rats. *Nosopsyllus londiniensis londiniensis* is taken mostly from mice within about 100 km of ports, suggesting that it is a relatively recent, but perhaps repeated, introduction, escaping from shipping.

Some noteworthy absences
Although the hedgehog *Erinaceus europaeus* is well established throughout much of mainland New Zealand, the flea *Archaeopsylla erinacei erinacei* has not been found here yet it is abundant on that host in the United Kingdom. Hedgehog mites, for example *Caparinia tripilis*, are, however, very abundant in New Zealand.

The opossum *Trichosurus vulpecula* lacks the several species of fleas normally found on it in Australia, but it does carry mites, e.g. *Trichosurolaelaps crassipes*.

In most of the New Zealand region, rabbits (*Oryctolagus cuniculus*) and hares (*Lepus europaeus*) are without the flea *Spilopsyllus cuniculi*, which is normally associated with them in Europe, even though they carry their normal complement of sucking lice (Anoplura) and mites (Acari). This flea has, however, been introduced to Macquarie Island to act as a vector in controlling rabbit populations there; it is the only deliberate introduction to the wider region.

Among introduced birds, the rock pigeon *Columba livia* does not carry *Ceratophyllus columbae*, but several species of feather lice are abundant (Pilgrim 1976) as are mites. The house sparrow *Passer domesticus* and hedge sparrow *Prunella modularis* lack their normal *Dasypsyllus* and *Ceratophyllus* spp. Domestic chickens are fortunately without *Echidnophaga gallinacea*, the 'stick-tight flea', which otherwise would be a great nuisance in the commercial poultry industry.

It might be speculated that the absence of the typical fleas from hosts that have been deliberately introduced is related to the disposal of cage litter during long voyages from the northern hemisphere to this country. This would have broken the life-cycles of the fleas since the non-parasitic larval phase occurs in nest debris; however, birds were presumably collected for introduction to New Zealand outside the breeding season, when they would be unlikely to carry fleas. Bird fleas are most abundant in nests, from which they make feeding forays onto the host, but the larvae are dependent on nest debris for their survival. Mammal fleas, on the other hand, are commonly found in the pelage of their hosts at all seasons; nevertheless, their larvae are also dependent on nest debris. If this were discarded *en route* to New Zealand, the life-cycle would likewise be broken.

The arrival of mites and lice on these same hosts, however, is not surprising as they are permanent obligate ectoparasites in all phases of their life-history. Information on the occurrence of the Mallophaga (chewing/ feather lice) and Acari (mites and ticks) of birds is given by Bishop and Heath (1998) and lists of bird Mallophaga are given by Pilgrim and Palma (1982, updated by Palma 1999). Mallophaga (chewing lice), Anoplura (sucking lice), and Acari of mammals are included in lists compiled by Tenquist and Charleston (1981).

Future work
Examination of a wider range of birds, especially indigenous and endemic species, is needed to seek possible additional taxa. Investigations are highly

desirable on the larval phase, especially in bird fleas. The families Pygiopsyllidae and Rhopalopsyllidae are found almost entirely in the Southern Hemisphere and their larvae possess many morphological features quite unknown among the families of the Northern Hemisphere, where most of the published work on larvae has been done. Some of these features are presented by Pilgrim (1998) for the six taxa of the genus *Notiopsylla* (Pygiopsyllidae).

Order Diptera: Flies

Many flying insects are referred to as flies, but only Diptera are true flies. The ordinal name name means two-winged, alluding to the two functional flight wings. The second pair of wings is reduced to a pair of tiny knob-ended halteres that probably function as stabilisers or air-speed detectors. These are easily seen in crane flies (so-called flying daddy-longlegs) but a flap (calypter) behind the wing base in the nine families of calyptrate flies obscures the haltere from above for the casual observer. Beginners may confuse members of the Hymenoptera (ants, bees, and kin) with Diptera, but winged hymenopterans always have two pairs of wings. The pattern of veins in Diptera also differs from that in Hymenoptera. Diptera have only sucking mouthparts, sometimes modified for piercing, but Hymenoptera also have mandibles for chewing. Flies of the division Cyclorrhapha and some of the Orthorrhapha have an arista (a plain or plumose bristle) on the last antennal segment an often a pad-ended sucking tongue and palps. This special tongue helps distinguish the more than 900 species of these flies in New Zealand from other insects. True flies are also generally characterised by large compound eyes. Their antennae may be long (Nematocera) or short (Brachycera).

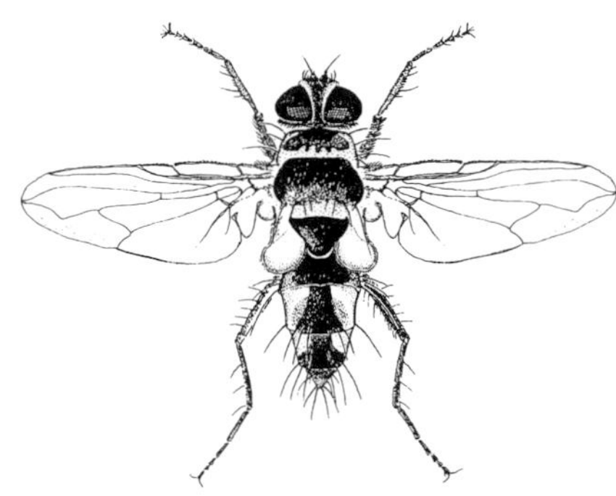

Male Australian leafroller parasite *Trigonospila brevifacies* (Tachinidae).

After Grant 1999

Dipterans include house flies, blow flies, mosquitoes, midges, sand flies, crane flies and a host of other forms. About 120,000 species have been described globally out of an estimated 153,000 species (Pape et al. 2009). They typically have sucking mouthparts but these may be modified as piercing structures.

Flies are among the largest and most important insect groups as pests of livestock, fruit, crops, and pasture and also include nuisance blood feeders. But many species are valued biological control agents of pests and weeds. A few introduced species help in the biological control of pest insects and some help to reduce weed vigour. Tachinid flies are especially important as natural enemies of caterpillars. Predatory and parasitic flies help keep populations of grass grub, stag and longhorn beetles, and sedge bugs in check. Flies can be common pollinators of some native plants. Flies can accelerate decomposition of carrion, dung, and leaf litter and they are important food sources for other invertebrates and for native birds, reptiles, and fish.

Certainly there is much more to flies in New Zealand than pest species that get up close and personal with humans and livestock. Fortunately, flies are much less of a health and veterinary menace in New Zealand than they are in some tropical countries. Nevertheless, New Zealanders and visitors need to assist quarantine services by not bringing into the country fruit and other produce that can establish further serious fly pests. At present, pest species account for only 0.5% of the 3225 known species of New Zealand Diptera.

Native flies live in a very wide range of habitats, from the tidal zone and salt pools to forests, grasslands, fresh waters (thermal to glacial), subalpine vegetation, and soil. Most species feed on various decaying materials, fungi, or algae in water. Herbivores include gall-midges, fruit flies, leaf miners and others that contribute part of the distinctive insect fauna dependent on native plants in New Zealand. For most species, much still needs to be recorded on habitat preferences, food use, season of activity, and egg-laying.

The comprehensive account that follows reviews what is known about the ecological and economic importance of Diptera and their diversity in New Zealand.

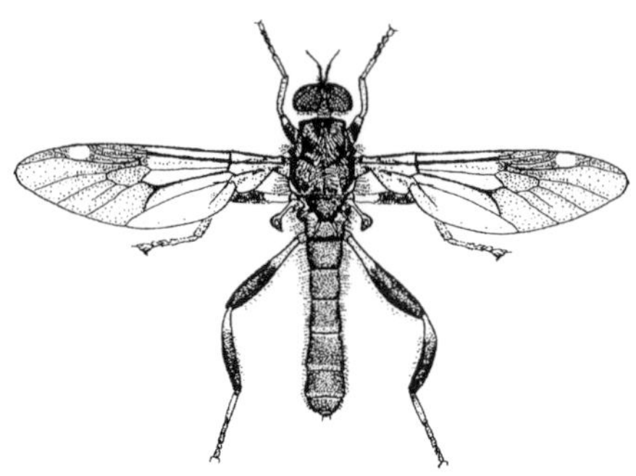

Garden soldier fly *Exaireta spinigera* (Stratiomyidae).

After Grant 1999

Beneficial roles of flies

Pollinators

Adult flies visit flowers to enhance their own survival and boost reproduction, and so flies contribute part of the pollinating service for plants. Flies visit pollen-rich kiwifruit quite readily (Macfarlane & Ferguson 1984) as well as spring-flowering fruit trees with nectar, where they make a minor contribution to pollination. Carrot and onion flowers are rather unattractive to honey bees and bumble bees (Macfarlane 1995a, unpubl.), so flies are likely to be relatively important in their pollination; introduced social bees (family Apidae) usually pollinate other crops and flowers in gardens adequately. Flies are more prominent on smaller and more open-flowered native species in: shrubland and bush, e.g. matagouri (*Discaria toumatou*), kanuka/manuka (*Kunzea*, *Leptospermum*), cabbage trees (*Cordyline* spp.), *Olearia* spp., tauhinu (*Cassinia* spp.) and ratas (*Meterosideros* spp.); in tussock and cushion-fields, e.g. spear grasses (*Aciphylla* spp.), *Pimelea*, *Raoulia*, *Dracophyllum*, *Phyllachne*; and in springs and bogs, e.g. *Eupatorium* (Heine 1937; Spencer 1977; Primack 1978; Heath 1982; Harrison 1990; Macfarlane unpubl.). At least on *Raoulia* they remain active in cooler and windier conditions than those that favour native bee flight.

During the day, the main flies that pollinate native and garden flowers are the hairier Syrphidae, Tabanidae, Bibionidae, and Stratiomyidae (*Odontomyia*) and the bristly blow flies, tachinid, and muscid flies. The former hairier flies carry much more pollen than the bristly flies (Macfarlane & Ferguson 1984). Survival of the pollinating flies, in contrast to bees, is independent of flowers, because the larvae do not require nectar. In gardens or wayside areas, energy-saving platform-like flower clusters (e.g. carrots, yarrow, thistles) attract fly visitors in summer and autumn when there is a narrower range of nectar and pollen sources for flies. Fly pollinators are attracted to a limited range of flower species, e.g. arum lilies (*Zantedeschia*), ivy (*Hedera helix*), *Fatsia japonica*) that have an aroma that attracts them. The role of midges such as *Forcipomyia* and Chironomidae during the evening or at night in New Zealand remains to be investigated, but overseas studies show they may visit flowers freely (McAlpine et al. 1981, 1987; Macfarlane 1995a).

Zygomyia sp. (Mycetophilidae).

Rod Macfarlane

Flies are likely to be considerably less efficient than bees as general pollinators for several reasons (Macfarlane 1995b). The major factor that limits the effectiveness of flies as pollinators is the inability of even the hairier flies to retain 10% of the pollen carried by bees, which have densely plumose-haired bodies. In addition, flies do not move so readily from plant to plant and so effect cross-pollination. Smaller fly species may often not contact the stigma during flower visits. In an extensive survey of flower visits to more than 50 native and 300 introduced plant species by bumble bees and native bees, flies were noted to visit more flowers per minute than beetles.

Small flies are specialist pollinators of certain ground orchids. Species of Mycetophilidae visit *Pterostylis* and *Corybas* in Australia (Bernhardt 1995) and both of these genera occur in New Zealand. The mycetophilid *Zygomyia* sp. has been observed bearing pollinia in *Pterostylis* flowers in New Zealand (C. Lehnebach, pers. comm.). Epiphytic orchids such as *Earina* spp. are pollinated by a wider range of insects.

Biocontrol agents of noxious weeds

Some flies, including species of Tephritidae, Anthomyiidae, and Agromyzidae, have been specifically introduced for controlling noxious weeds. Scott (1984) and Cameron et al. 1989) provided summaries for fly species (Tephritidae, two species; Anthomyziidae, *Botanophila seneciella*; Cecidomyiidae, *Zeuxidiplosis giardi*) that affect ragwort (*Senecio jacobaea*), mexican devil weed (*Ageratina adenophora*), Californian thistle (*Cirsium arvense*), and St John's wort (*Hypericum perforatum*). The website landcare research.co.nz/research/biocons/weeds/book has data for several species of biocontrol Diptera, viz Tephritidae (five species), Agromyzidae (two species), Cecidomyidae (two species), and Syrphidae (two European species of *Cheilosia*).

For each species a colour photo is provided, often with the overall wing pattern, release date and status, type of damage, and provisional effectiveness against thistle, *Hieraceum* and other Asteraceae, and old man's beard (*Clematis vitalba*).

Natural enemies of pests and other invertebrates

Some Diptera are important natural enemies of moths and soil pests. Most of the entomophagous (insect-eating) flies are members of the family Tachinidae, with 188 known species in New Zealand and records of hosts for 29 species on caterpillars, grassgrub, and longhorn larvae (Dugdale 1962; Valentine 1967b; Cantrell 1986; Cameron et al. 1989; Munro 1998, Schnitzler et al. 2004). Merton (1982) made the only critical investigation of the link between the biology of the host and a tachinid parasite for the main grassgrub pest, *Costelytra zealandica*. Other New Zealand examples include Australian-derived *Cryptochetum iceryae* (Cryptochetidae), which parasitises cottony-cushion scale (*Icerya purchasi*) on citrus (Valentine 1967b; Scott 1984), and South American *Leucopis tapiae* (Chamaemyiidae), which usually keeps pine-twig chermes (*Pineus laevis*) populations in check (Cameron et al. 1989).

In soil and sand, two important families with nearly 100 species between them (Therevidae and Asilidae) have larvae that are known or believed to be usually predatory. The mainly fine-haired grey stilletto flies (Therevidae) inhabit the driest sand (Sutherland 1966; Lyneborg 1992) and soil. Up to 20 species of stiletto flies are likely to prey on scarabs, weevils, and other insects in dune sands. In gravel, *Anabarhynchus* larvae prey on larvae of the solitary wasps *Podagritus* (Harris 1990). Adult collection data suggest that robber flies (Asilidae) prefer uncultivated grassland, but they are not found in wetland, where some crane-fly larvae may be predatory. Robber flies perch on elevated sites in vegetation among grassland or towards the edge of tree twigs. Females may prey on their own males in New Zealand, and prey selection is just one of several interesting aspects of behaviour and ecology that can be investigated and compared with overseas studies (Lavigne et al. 1978). Adults of Empididae can be active predators. Ian McLellan has seen *Thinempis takaka* adults attack and feed on beach amphipods and W. J. Crawford (pers. comm.) noticed adults of an undescribed species of *Hilara* preying on the mayfly *Coloburiscus humeralis*. An undescribed *Spilogona* species (Muscidae) has been observed to catch mayfly subadults as they emerged from a stream onto stones (Harris 1990). Most endemic muscids are probably predatory, at least as adults. John Early saw adults of apparent Hemerodromiinae with *Stylaclista* (Diapriidae) prey between the front legs at Waipoua forest. Adults of many Dolichopodidae may well be predatory on small soft-bodied insects such as aphids.

Larvae in the families Empididae, Dolichopodidae, and Muscidae are not often reported in freshwater surveys (Winterbourn et al. 2000) but there are some records of adults concentrated near running and boggy water. As well, there are a few records of concentrations of undetermined larvae associated with even more abundant midge larvae in lagoons (Knox & Bolton 1978). This evidence indicates a possible predatory role for larvae of such species on midge larvae and the smaller larvae of other characteristic flies active in the biological zone, e.g. shore flies *Scatella* and *Ephydrella*. The invertebrate fauna of the wet soil zone below and at the sides of fresh and semi-saline water bodies remains almost unstudied and difficult to assess (Collier & Winterbourn 2000).

Among hover flies of the subfamily Syrphinae (24 described species), large *Melangyna novaezealandiae* and smaller *Melanostoma fasciatum* have adapted best to modified grassland and cropping environments. Hover-fly larvae can be the major predators of white-butterfly caterpillars on cabbages (Ashby 1974; Ashby & Pottinger 1974). Adult hover flies are conspicuous in pastoral habitats from late spring when aphid populations increase (Cumber & Harrison 1959b; McLean et al. unpublished). There are at least three predatory genera of gall midges in New Zealand orchards. *Arthrocnodax* preys on pest mite species (Cottier 1934; Collyer 1964), *Diadiplosis koebelei* consumes long-tailed mealy bug (Charles 1981), and an

Bladder fly *Ogcodes brunneus* (Acroceridae).
After Grant 1999

unknown gall midge preys on the margarodid scale insect *Coelostomidia wairoensis* (Morales 1991). The latter is also found among Lucerne (alfalfa) infested with blue-green aphids (Thomas 1977). The hosts are unknown for the predatory gall midge genera *Trisopsis* and *Lestodiplosis* that occur in New Zealand.

More than 260 fly species are parasitic on other invertebrates. Valentine (1967b) reviewed the literature on flies as insect parasitoids up till 1966. In North America, Phasiinae species may be parasites of Heteroptera (McAlpine et al. 1987), but apart from the common and widespread *Huttonobesseria verecunda* there are only six rare phasiine species in New Zealand. Larvae of *H. verecunda* larvae live within the heteropteran sedge bugs *Rhopalimorpha.* Members of a few other fly families are also parasitic on invertebrates. Two species of cluster flies (*Pollenia rudis, P. pseudorudis*) became established in New Zealand by the early 1980s (Dear 1985; Macfarlane & Andrew 2001), and they parasitise *Aporrectodea* (= *Eisenia*) *rosea* (Rognes 1987), which forms a small component of the introduced earthworm species of cultivated grasslands (Fraser et al. 1995). The spider *Matachia ramulicola* (Psechridae) is a host for *Ogcodes brunneus* (Dumbleton 1941). Harris (unpub.) has reared a species resembling *Ogcodes nitens* from various spiders of the family Clubionidae. The hosts for 24 other species of Acroceridae, Pipunculidae, and Bombyliidae remain unknown in New Zealand. Among 26 species of Sciomyzidae, *Eulimnia* species feed on freshwater fingernail clams (Sphaeriidae) (Barnes 1980a), whereas *Neolimnia* species feed on gastropod snails, with aquatic snail hosts for the subgenus *Pseudolimnia* and terrestrial hosts for the subgenus *Neolimnia* (Homewood 1981).

Vegetation recycling

Flies are probably most beneficial as decomposers of litter and carrion. In forests and gorse shrubland, fungus gnats (Mycetophilidae, Keroplatidae, Ditomyiidae), together with some Phoridae and Cecidomyiidae, feed on fungi in leaf litter and in logs. Members of these five families, plus crane flies (Tipulidae) and other flies feeding in logs, must consume a high proportion of the fungi eaten by invertebrates (Skerman 1953; Somerfield 1974; Anderson 1982). This conclusion is based on their biomass compared to other fungus-consuming insects taken by malaise traps (Ward et al. 1999; Harris et al. 2004) or impact traps (Didham 1992).

Removers of carrion and dung

Blow flies and certain other flies (some Muscidae, Sphaeroceridae) consume much livestock, bird, rodent, and fish carrion biomass before it reaches the greatly reduced dry stage. Larger vertebrate carcasses (sheep, pigs) become heated during decomposition, when flies are most active (Dymock & Forgie 1993). Temperatures and features of the site affect the rate of decomposition and the availability of certain blow flies within medium-sized (hare, opposum) carcasses (Boswell 1967; Appleton 1993). Medium-sized mammal (opossum) or larger bird (magpie) carrion provide a higher percentage (2–2.5 times) of flystrike species than a blackbird (Heath & Appleton 2000). Mice carrion has been studied in New Zealand pastoral farmland, but published information states that no flystrike species were involved and that shade favours blow-fly colonisation (Heath & Appleton 2000). In North and South America, where *Lucilia sericata* and the secondary sheep strike *Calliphora vicina* are also present, these were the main species reared from rodent, fish and bird-gut carrion (Tomberlin & Butler 1998; Figueroa-Roa & Linhares 2002). The composition of blow-fly species is affected by the availability of kanuka (*Kunzea ericoides*) scrub and native bush (Robinsons Bay, Banks Peninsula) compared to areas with mainly nectarless conifer shelter belts (Lincoln University, Cheviot) or, on Chatham island, with sheep (based on a collection of 38,890 blow flies with baited western Australian traps in 1991–1992 (Macfarlane unpubl.)). *Lucilia sericata* averaged 91% of the blow flies caught at Lincoln and Cheviot, 74% at Robinsons Bay, but only 4% on Chatham Island. *Calliphora stygia* (primary blow-fly species) made up 22% at

Bladder fly *Ogcodes brunneus* (Acroceridae).
After Grant 1999

Robinsons Bay, 13% on the Chathams, and 5% and 3% at Lincoln and Cheviot, respectively. Similarly, *Calliphora quadrimaculata* made up 2.3% at Robinsons Bay, but only 0.1% elsewhere in Canterbury. *Calliphora vicina* averaged 5% (range 2–8%) at the Canterbury sites. *Xenocalliphora hortona* averaged 0.4% (range 0.2–0.6%) in Canterbury, but 81.7% of the catch on Chatham Island. *Xenocalliphora hortona* and *Lucilia sericata* are among the commoner species at beaches, where fish, mollusc, crab and seabird carrion is available. Thus rodent and small-bird carrion could well also suit these blow-fly species in lowland pastures in New Zealand. Some Sphaeroceridae, Phoridae, and Chloropidae feed on larger invertebrate carrion.

Flesh flies, for example *Oxysarcodexia varia* (Sarcophagidae), and also *Lasionemopoda hirsuta* (Sepsidae) and some Sphaeroceridae, complement earthworms in the consumption of livestock dung in summer.

Aid in forensic science, water evaluation, and reserve quality

As a corollary to their role in decomposition of carrion, the stages of fly development and species can be helpful in forensic science (Heath 1982), i.e. in determining how long a corpse has been around.

In the aquatic environment, fly species composition varies considerably with changes in water bodies (Quinn & Hickey 1990a,b; Stark 1993) so that some species can be useful as indicators of water quality. This role will become more apparent when Empididae, Muscidae, Chironomidae, and Ceratopogonidae have been adequately studied and mapped at the species level.

Surveys of urban reserves have shown that fly species can include regionally uncommon or characteristic habitat species (Macfarlane et al. 1998, 1999, 2005, 2007), and this can help to define the biological value of public reserves. Tenure surveys of South Island high-country farms are beginning to demonstrate the importance of some fly species as indicators of naturalness of relatively intact insect communities and for conservation.

Food for other animals

An additional major role of flies is as food for other creatures, mainly birds, spiders, and other invertebrates. The larger and less mobile forest flies provide valuable protein and at times fat-rich sources for bush birds. Fantails forage on flying insects and other birds capture inactive insects among twigs and on branches. Other flies provide prey for various fly, beetle, and spider species along beaches. These include kelp flies, which breed in stranded kelp on rocky shores and nearby beaches. About 80 species of flies, including many *Paralimnophora* species (Muscidae), make up the distinctive fly faunas of sandy beaches and rocky shores (Macfarlane & Andrew 2001; Harrison & Macfarlane unpubl.). On estuary foreshores, midges are abundant primary scavengers, while dolichopodid larvae and predatory saldid bugs are part of the estuary insect community (Knox et al. 1978; Knox & Bolton 1978). Most species of aquatic flies live in fresh water (Macfarlane & Andrew 2001) where they can be locally significant insects for biomass (Quinn & Hickey 1990a,b), especially in cool mountain waters or more degraded lowland waters. Immature flies, especially midges, are an important part of the invertebrate food of fish (McDowall 1990; Collier & Winterbourn 2000).

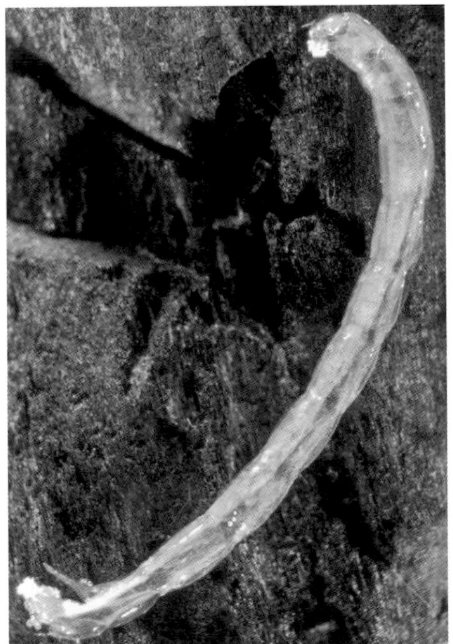

Larva of midge *Chironomus* sp. (Chironomidae).
Stephen Moore, Landcare Research

Introduced pest and other species

In 1950 only 45 introduced species of flies were listed among the 1731 species then known in New Zealand (Miller 1950). Now, with a sounder taxonomic base, almost four times more introduced species are known, which represents 7% of the described fly species in New Zealand. Most of the introduced species were accidentally derived from Europe (71–72%) or Australia (21%) (Evenhuis 1989). Many European and American species probably arrived well before 1945 (Macfarlane & Andrew 2001).

Flies are a familiar group of insects to the general public, although their

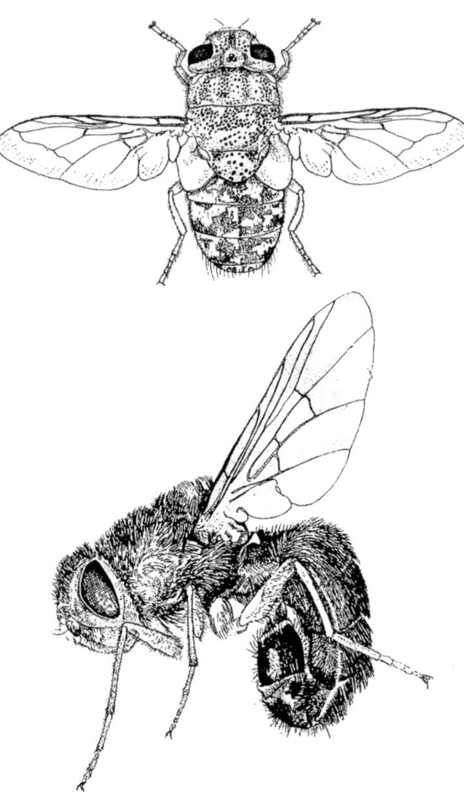

Two livestock pests – sheep nasal bot fly *Oestrus ovis* (upper) and *Gasterophilus intestinalis* of horses (lower) (Oestridae).
After Grant 1999

negative impact as pests to livestock and as disease carriers to humans is more widely understood and researched than their positive values. Only 0.5–1% of the species in New Zealand are pests and all the main fly pest species were accidentally introduced. Useful older works on pest species include those of Cottier (1956), who illustrated damage by many fly species, Helson (1974), who illustrated life-cycles of seven pest species including the threatening Mediterranean fruit fly, and Scott (1984), who provided an extensive outline of pest recognition, life history, economic significance, and control. The range of plant and animal species affected adversely by flies is moderate compared with the other main herbivorous and wood-feeding insect orders (Coleoptera, Lepidoptera, Hemiptera). In New Zealand, these three orders probably have more primary plant-consuming species than Diptera.

Flies in human and animal health

Some of the best-known flies affect the health of animals and humans, notably blow flies, bot flies, and blood-sucking flies such as mosquitoes and sand flies. On pastoral farms, pest blow-fly species can kill sheep or lower their wool quality (mainly the two *Lucilia* species and *Calliphora stygia*, which cause the initial blow-fly strike). Other blow-fly species such as *Calliphora vicina, C. quadrimaculata,* and *Xenocalliphora hortona* may then add to blow-fly strike as secondary species. Control measures and losses to farmers rank these blow flies among the major insect pests in New Zealand (Heath 1980; Dear 1985). With the arrival of *Lucilia cuprina,* strike activity has been extended from December until late autumn (Heath et al. 1983; Dymock et al. 1990) and it reached Cheviot by January 1992 (Macfarlane unpubl.). Nasal bot flies (*Oestrus ovis*) are common in sheep nostrils (Kettle 1973; Miller & Walker 1984) and they cause severe irritation and loss of condition in sheep (Scott 1984). The larvae of two bot-fly (*Gasterophilus*) species live in horse intestines, affecting horse health (Miller & Walker 1984). Both the adults and the larvae distress horses (Scott 1984).

Other pests, such as mosquitoes, are discussed below in the accounts of fly families.

Pests of pastures, crops, and gardens

There are no recent comprehensive reviews of plant hosts for flies. The latest is based on literature up to 1960 in Dale and Maddison (1982), who listed insect species and references concerning host-plant species. Only a minority of the plant-feeding flies can be considered pests, while some are actually beneficial.

In pastures and maize crops, the root-feeding Australian soldier fly *Inopus rubriceps* is a major pest in northern North Island (Helson 1974; Robertson & Blank 1982; Scott 1984; Macfarlane & Andrew 2001). In addition, several soil-dwelling *Leptotarsus* species may feed on both roots and dead organic matter, mostly in uncultivated pastures (Macfarlane & Andrew 2001). A number of root-gnat species (Sciaridae) can cause extensive damage to a wide range of greenhouse ornamental plants (Scott 1984). There are also sporadic reports of European-derived *Bradysia ?brunnipes* ('*Sciara annulata*') damaging cucumber and other cucurbit seedlings in New Zealand (Stacey 1969; Dennis 1978). Control of *Bradysia difformis* (= *pauparea*) has been investigated (Martin & Workman 1999) because this species can be a threat to root development in pots. Naturalised populations of daffodils or hyacinths may become depleted when narcissus bulb fly *Merodon equestris* and the lesser bulb- or onion flies (*Eumerus* species; Cottier 1956) feed in their bulbs. Commercial crops are vulnerable too (Scott 1984).

On cropping farms, other small flies can damage cereal and brassica crops. The hessian fly *Mayetiola destructor* feeds in stems of wheat, maize, other cereals, and prairie grass and can be common (Withers et al. 1995; Bejakovich et al. 1998). Resistant cereal varieties and burning of stubble mainly keep this pest in check (Cameron et al. 1989). *Cerodontha australis* (Agromyzidae), *Hydrellia tritici* (Ephydridae), and *Lonchoptera furcata* (Lonchopteridae) are all widespread and very common in grasslands (Macfarlane & Andrew 2001), but these small flies have been assessed to cause little or no economic damage (Barker et al. 1984; Cameron et al. 1989). *Cerodontha australis* is often found in cereal crops, but damage is slight (Bejakovich et al. 1998). A drosophilid leaf-miner, *Scaptomyza flava*, disfigures brassica and pea crops but causes economic damage only in exceptional circumstances (Seraj & Fenemore 1993). The beet leaf-miner *Liriomyza chenopodii* can cause extensive leaf damage to beet crops (Scott 1984). It is common in central Canterbury crops (Pearson & Goldson 1980). No control measures have been devised for it.

Several groups of flies, particularly the gall midges and leaf miners, damage aerial parts of flowers. In orchards and gardens, apple foliage can become badly distorted by the gall midge *Dasineura mali*, and *D. pyri* can cause similar damage to pears. At times, the leaf and flower galls of *Rhopalomyia chrysanthemi* may affect chrysanthemums in gardens (Wise 1957). This gall midge is a threat to the cut-flower industry. Leaves of cineraria (*Pericallis* × *hybrida*, often as *Senecio cruentus*) in gardens and chrysanthemum leaves in glasshouses can be attacked by the leaf-mining *Chromatomyia syngenesiae* (Cottier 1956; Scott 1984). Further breeding sites for it are common composite weeds like sowthistle (*Sonchus asper*), fireweed (*Senecio bipinnatisectus*), and other *Senecio* species (Spencer 1976; Macfarlane et al. 1999).

Flies affect native garden shrubs adversely. The rosette-making gall of *Oligotrophus oleariae* can disfigure at least three *Olearia* species, including *O. paniculata* (golden akeake), a useful hedge shrub. Kakabeak (*Clianthus* spp.) can be threatened by insect, mite, and mollusc pests, including the host-specific leaf miner *Liriomyza clianthi* (Watt 1923). Other gall midges affect seed crops of meadow foxtail (*Alopecurus pratensis*) and cocksfoot (*Dactylis glomerata*) (Jacks & Cottier 1948; Barnes 1940; Macfarlane & Andrew 2001).

Mushroom spoilers

Flies can be a threat to mushroom producers and often cause premature decay of mushrooms gathered in the field. The sciarid *Lycoriella castanescens* is the main pest species in New Zealand, but the sphaerocerid *Pullimosina heteroneura* can transmit the nematodes that attack commercial mushrooms (McAlpine et al. 1987) and *Phthitia emarginata* affects rotting mushrooms (Marshall et al. 2009). The phorid *Megaselia halterata* is often found in mushrooms in Europe and North America, so it will feed among mushrooms in New Zealand too. Most species of Mycetophilidae feed on fungi, but are probably of no economic significance. There is no summary of fly use of fungal species in New Zealand. Simon Hodge (unpubl.) recently investigated a range of habitats with mushrooms as bait in Canterbury and Wilding et al. (1989) summarised overseas findings.

Fruit flies

Fruit flies of the family Tephritidae are major potential pests, discussed below in the accounts of fly families. Frampton (1990) explained the extensive measures undertaken to monitor for any accidental establishment of fruit flies. White and Elson-Harris (1992) provided a modern summary of the pest species of fruit flies as well as species affecting weeds.

Natural enemies of flies

Insect numbers are to a large extent controlled by predators, pathogens, parasites, and parasitoids (parasitic insects that normally kill the host before it has matured). Birds, spiders, and many predatory insect species play a major role in controlling the populations of many fly species, as do insect parasitoids that develop on or inside a fly host. Nematode and protozoan parasites, fungi, and bacterial pathogens also profoundly affect fly populations. Larval mites (Arrenuridae and Microtrombidiidae) parasitise adult mosquitoes and crane flies (Snell & Heath 2006).

Predators

The role of birds and spiders is obvious. Small flies such as psychodids and sciarids are often caught in large numbers in the webs of the common orbweb spider (Laing 1988; Macfarlane et al. 1999). Apart from the ecological investigations of predation on pest species (Boswell 1967; Valentine 1967b; Allsopp & Robertson 1988; Appleton 1993), little has been recorded about predation of other invertebrates on fly larvae and pupae. Ground beetles, rove beetles, and click beetles can cause important mortalities in outbreaks of Australian soldier fly (Allsopp & Robertson 1988). Flies are also recorded as prey of predatory bugs (Hemiptera) (Valentine 1967b) and species of the sphecid wasp genera *Podagritus* and *Rhopalum* (Harris 1994a). In water, chironomid larvae prey on *Austrothaumalea* (Thaumaleidae) larvae (McLellan 1983).

Parasitoids

The most prominent parasitoids of Diptera and other arthropods belong to the order Hymenoptera. Thirty-eight species of parasitic Hymenoptera from 10 families have been reared from at least 39 New Zealand fly species in 14 families (Valentine 1967b; Valentine & Walker 1991). Gall midges, kelp flies, and the glow-worm (a fungus gnat), are known to host 10 diapriid species, three braconid species, two *Fusiterga* species (Pteromalidae), one species of Encyrtidae, and one of Torymidae.

The parasitoids of introduced and pest species have been more extensively investigated, with records from 14 introduced fly species. The remaining 130–143 species of introduced flies may have few parasitoids affecting them apart from the more adaptable native species that have extended their host range to include these species.

Striped dung fly *Oxysarcodexia varia* (Sarcophagidae).
After Grant 1999

Parasitoids of 26 species have been introduced deliberately or arrived with hessian flies, blow flies, pest muscid flies, and agromyzid leaf miners. Some have extended their host range to dung-inhabiting *Oxysarcodexia varia* and to kelp flies (Cameron et al. 1989; Valentine & Walker 1991). At least four species of beneficial hover flies are parasitised by the accidentally introduced *Diplazon laetatorius* (Ichneumonidae). The incidence of parasitism per host fly species varies with habititat, the origin of the species, and which of the three immature stages of the fly are parasitised. Also, species-rich fly genera are unlikely to support a high average of parasitoid species, especially if these genera are affected by the generally less host-specific external parasitoid species.

Hymenopterans very seldom parasitise immature stages of flies that live in water (Smith 1989; Stehr 1991; Goulet & Huber 1993; Ashe et al. 1998). Also, soil-inhabiting Therevidae and Asilidae larvae have no records of parasitic insects reared from them elsewhere in the world (Knutson 1972; Lavigne et al. 1978; Dennis & Knutson 1988). It seems that immature flies in soil and wetlands are more difficult for winged parasites to reach and detect, so parasitism per species is expected to be low.

Different parasitoid species may affect eggs, larvae, and pupae, and internal parasitoids tend to be more host-specific and allow the host to develop (Gauld & Bolton 1988; Quicke 1997). The undescribed species of Hymenoptera include a relatively diverse fauna of egg parasites (Noyes & Valentine 1989a,b; Berry, this chapter). Variation in the incidence of egg parasitism between fly taxa is unknown in New Zealand. The more-exposed eggs of Syrphidae on plants and of some Acroceridae on twigs (Miller & Walker 1984) may well host parasitoids more consistently than eggs protected within particular substrata (e.g. the Tipulidae, Sepsidae, and some Asilidae of soil or rotting logs, and the Anthomyiidae, Agromyzidae and some Cecidomyiidae of plants) (Watt 1923; Hinton 1981).

Most parasitic species of Hymenoptera in New Zealand remain undescribed, including about 200 species of Diapriidae, dipteran parasitoids. Many more may await discovery. Clearly, much more research on the immature stages of flies and their parasitoids is needed before a reasonably confident prediction can be made about their importance as hosts of New Zealand Hymenoptera.

Diversity and endemicity of New Zealand Diptera and offshore-island distributions

Fly family	New Zealand genera No. genera	New Zealand genera % endemic	New Zealand species† No. known	New Zealand species† % undet.	Described species % endemic	Described species % offshore	Families found on offshore islands
NEMATOCERA	260	30	1,638–1,654	19–20	95	4.1	
Bibionidae	1	0	8	0	100	0	Ch
Blephariceridae	3	100	13	38	100	0	No, Ch, **Ca**
Cecidomyiidae	31	19	229	59	85	4	All, **Sub**
Ceratopogonidae	10	0	39	13	100	9	All, **Sub**
Chironomidae	67	39	146	23	94	17	No, **Ch,** Sn, **Sub**
Culicidae	6	33	15	0	69	6	No, **Ch,** Sn, Su
Ditomyiidae	2	0	25	7	100	0	No
Dixidae	3	33	9	0	100	0	Ch, Sub
Keroplatidae	12	0	50	2	98	0	All, **Sub**
Mycetophilidae	33	45	266	27	99	2	All, Sn, Sub
Psychodidae	9	11	62–78	21–38	87	13	Ch, Sub
Rangomaramidae	4	**75**	12	0	92	8	
Scatopsidae*	5	0	10	50	50	0	Sub
Sciaridae	13	**8**	71	3	87	3	All, **Ch,** Ca
Simuliidae*	1	0	15	0	100	15	**Sub**
Thaumaleidae	2	0	11	0	100	0	
Tipulidae	48	33	619	8	100	2.5	**All**
Trichoceridae	4	25	19	5	94	0	**Sub**
Five other families	6	33	19	16	94	0	**Sub**
ORTHORRHAPHA	106	40	630	28	99	2.9	
Acroceridae	4	50	14	29	100	0	
Asilidae	6	33	30	43	100	6	**No,Ch**
Dolichopodidae	31	45	147	10	98	6	**All**
Empididae	41	42	283	37	100	1*	All, **Ak, Ca**
Stratiomyidae	13	46	59	46	88	3	No, **Ch,** Sn
Tabanidae*	3	0	20	15	100	0	
Therevidae	3	33	71	4	100	0	**TK, PK**
Four other families	5	40	6	57	100	0	
ASCHIZA	38	21	151	41	58	2.3	
Phoridae	18	44	44	14	74	0	All
Pipunculidae	3	0	14	64	100	0	
Syrphidae	13	**8**	90	51	86	4.5*	All, **Ch,** Sn, **Ca**
Three other families	4	0	3	33	33	0	Ch, Sn
SCHIZOPHORA Acalyptratae	115	22	386	12	81	7.3	
Agromyzidae	7	0	41	2	76	2.5	All, **Sn, An**
Canacidae	4	25	12	0	100	**27**	Sn, **Sub**
Chloropidae	12	**8**	45	2	86	5	?No, **Ch,** Sn
Drosophilidae*	3	0	19	21	27	0	All, **Ch,** Sn
Ephydridae	19	**5**	66	61	79	8	All, **No, Ch**
Heleomyzidae	8	63	32	3	90	6.5	**Bo,** Sn
Helosciomyzidae	6	**83**	11	0	100	27	No, **Ch**
Huttoninidae	2	100	9	0	100	0	Ch, **Sn, Ak**
Lauxaniidae	4	25	17	0	100	0	
Pallopteridae	1	100	10	10	100	0	
Sciomyzidae	2	100	18	0	100	0	
Sphaeroceridae	18	11	43	19	40	6	No, Sub
Tephritidae	5	0	20	5	80	0	Ch
Sixteen other families	24	17	43	17	80	19	All, **Ch, Sub, Sn, LB?**
SCHIZOPHORA Calyptratae	102	63	420	35	86	7.6	
Calliphoridae	8	25	58	5	82	15	All, **Sn, Sub**
Fanniidae	2	0	8	50	0	0	No, **Ch,** Sn
Muscidae	17	41	145	59	88	21	All, **Ch,** Sn, **Sub**
Tachinidae	61	90	190	30	98	0	No, **Ch**
Four other families	14	0	19	16	0	0	Ch, Sub
Total DIPTERA	**621**	**33**	**3,225–3,241**	**23**	**91**	**4.6**	

† not including subspecies

* All genera shared with Australia.

Ak, Auckland Is; All, all islands except subantarctic; An, Antipodes Is; Bo, Bounty Is; Cp, Campbell I.; Ch, Chatham I.; LB, Little Barrier; No, all northern islands; PK, Poor Knights Is; St, Stewart I.; Sub, subantarctic islands; TK, Three Kings Is. Island groups: northern = Northland–Bay of Plenty; subantarctic = Snares, Antipodes, Bounty, Auckland, Campbell.

Pathogens

Seven of the 10 specified records of fungi (eight species), nematodes (three species), and protozoans (three species) that affect Diptera in New Zealand pertain to aquatic species (Glare et al. 1993). International interest in the control of mosquitoes and black flies has centred on the possible use of bacteria (Lacey 1997). Sixteen papers from New Zealand on fly pathogens (Glare et al. 1993) provided records of the protozoan *Herpetomonas muscorum* from the native blow fly *Calliphora quadrimaculata* and the fungi *Tolypocladium extinguens* from the glow-worm and *Beauveria bassiana* from the hover fly *Melangyna novaezealandiae*. In the field, *Metarhizium anisopliae* is the main fungus known to affect larvae and pupae of the Australian soldier fly under a range of soil moistures. In Australia, efforts have been made to develop effective strains for control of this fly (Allsopp & Robertson 1988). More recently, strains of the bacterium *Serratia* have been investigated for their possible impact on one of the pest blow-fly species (O'Callaghan et al. 1996). There is not enough information on how many pathogens affect a particular fly host to gain any appreciation of how many protozoans, nematodes, and fungal species the flies of New Zealand may host, but many of the better-known soil-fungal pathogens affect a wide range of insect species (Steinhaus 1963; Tanada & Kaya 1993).

New Zealand fly diversity and scientific importance

Diversity and endemism

The Diptera is the second-largest order for family- and species-level diversity in New Zealand. This position will almost certainly still apply when the diversity of all New Zealand insects is eventually determined, even although Emberson (1998) predicted that species numbers of Hymenoptera may be similar to Diptera. Evenhuis (1989) catalogued New Zealand species among Australasian and Oceanian flies and his checklist has been updated to the present in the present review. The known biology of each family has been summarised in Evenhuis (1989) and in the Australian (CSIRO 1991), North American (McAlpine et al. 1981, 1987), and British reviews (Smith 1989). Macfarlane and Andrew (2001) summarised habitat use mainly at the family level for the New Zealand Diptera. They summarised what has been recorded on species mapping and the distribution of most dipteran pests within New Zealand. Prominent or readily identified species from the more restricted habitats of carrion, dung, caves, seashores, parasites, and aquatic habitats were also summarised, and known details for species are listed in the end-chapter Diptera checklist. The compilation of habitat use and feeding sources by species in the checklist demonstrates that much remains to be learnt about the basic ecology of flies in New Zealand. The checklist also compiles records for 739–758 undescribed species in 25 families, but the practical experience of the authors is insufficient to provide estimates for several major families (e.g. Cecidomyiidae, Chironomidae, Ceratopogonidae, Lauxaniidae and much of the Dolichopodidae). The checklist has records for subantarctic and Snares islands species; these are not listed in Evenhuis's (1989) catalogue of Diptera from Australasia and Oceania.

The tabulation below and the end-chapter checklist of Diptera cover 78 known families. The only certain endemic fly family in New Zealand is the Huttoninidae, collected from forest and subalpine sites (Harrison 1959, CMNZ, LUNZ), but the family status of *Starkomyia inexpecta* remains unresolved among Mycetophiloformia. The sole endemic subfamily Mystacinobiinae (Calliphoridae) has a single species, *Mystacinobia zelandica*. This fly survives on the guano and debris (Holloway 1976) of the New Zealand short tailed bat *Mystacina tuberculata*, which roosts in trees or caves.

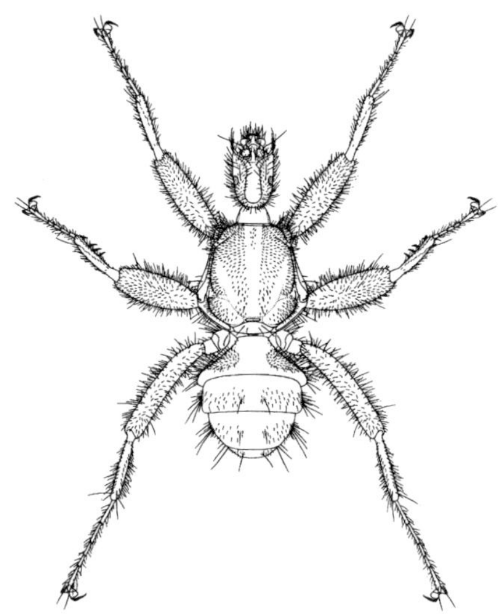

Bat fly *Mystacinobia zelandica* (Calliphoridae).
Des Helmore

The checklist includes 622 genera (17 undescribed), including six (*Ceonosia*, *Empis*, *Limnia*, *Platura*, *Sciara*, *Tachina*) that are almost certainly not valid for New Zealand and six others (*Anatopynia*, *Cerdistus*, *Dipsomyia*, *Neoitamus*, *Spilogona*, *Tanypus*) that are either endemic or need critical generic re-evaluation. Accidental introductions of flies, supplemented by biological-control introductions, have added 75 genera to the New Zealand fauna. In all, 211 of the genera in New Zealand are present in the British Isles (Chandler 1998) and 236 are shared with North America (McAlpine et al. 1981, 1987). New Zealand and Australia share 247 genera. About 33% of the genera existing in New Zealand prior to European settlement are endemic, a figure that is 4.7 times the level recorded for New Caledonia and Tasmania, which are our nearest relatively old or large neighbouring islands.

The apparent level of endemism is somewhat elevated owing to the existence of many monotypic genera, most notably in the Chironomidae. Overall, the percentage of all New Zealand Diptera that belong to endemic genera is quite high at about 22% (23% in pre-European times). The family that surpasses all others in numbers of endemic genera and species is the Tachinidae, with 90% of its 61 genera endemic and 83% of its 190 species in the endemic genera. For the rest of the calyptrate flies, the level of endemism is about average at 24%, but is low (2.9%) for Orthorrhapha. Five other smaller families, with at least eight species each, have high levels of endemism (60–100%). Conversely, three fully freshwater fly families (Simuliidae, Pelecorhynchidae, Tabanidae) and the mainly aquatic Ceratopogonidae each have around 12 genera, none of which is endemic. This may mean that these families became established in New Zealand later than Chironomidae, Blephariceridae and Tanyderidae, which have about average or higher levels of endemic genera for Nematocera flies. The level of generic endemism is quite variable among Acalyptrate flies.

There are no endemic genera of herbivorous Agromyzidae and Drosophilidae (fruit flies) and only one endemic genus in each of the families Ephydridae, Chloropidae, and Sphaeroceridae. The important herbivorous fly family Cecidomyiidae (gall midges) has some endemic genera, though six of the 30 genera were accidentally introduced and two were deliberately introduced for biocontrol of weeds. Further endemic genera are likely to be described from undescribed gall-making species associated with shrubs, forest trees, tree ferns, creepers, perennials, and grasses (Martin unpubl.).

The New Zealand dipteran fauna comprises 2483 described species, 736 known undescribed species, and a further 49 indeterminate species that might represent accidental introductions. Three methods were used by Macfarlane and Andrew (2001) to estimate a likely range of 999–2510 undescribed species for the Diptera. Despite a 160-year history of mainly accidental introductions, which have allowed for 171–190 species to become established, the species list still has a high level of endemicity (95%). Most of the estimated 2000 or so undescribed species of Diptera in New Zealand are probably mainly in forests and the mountains (Macfarlane & Andrew 2001).

The 15 groups of offshore islands have 121 known fly species confined to them (Evenhuis 1989; Bickel 1991; Lyneborg 1992; Marshall & Rohácek 2000; Macfarlane & Andrew 2001; checklist). The subantarctic islands more consistently appear to have their own endemic species. However, when the Diptera of the Three Kings, Poor Knights, and Chatham Islands have been thoroughly studied then perhaps the offshore islands could have 150–200 endemic species. Either way, these figures are indicative of the high conservation value of offshore-island habitats for fly species. The figures also suggest that ecological maintenance of offshore reserves is likely to be important to avoid placing further fly species under threat. Macfarlane and Andrew (2001) assessed the status of published information on fly diversity of offshore islands relative to the isolation, age, and botanical diversity of the islands. Speciation in several family-rank taxa that inhabit seashores is highest in Australasia and/or on subantarctic islands is highest on subantarctic islands, e.g. Canacidae (Zaleinae), Helcomyzidae, and Coelopidae (Macfarlane & Andrew 2001; McAlpine 2007). The Snares islands are the southern limit of distribution for Stratiomyzidae, Drosophilidae, Heleomyzidae, and Chloropidae (Harrison 1976; Macfarlane & Andrew 2001).

Miller (1950) listed the locations of the types of the 1731 species of New Zealand Diptera known to 1950. Since then most of the types have been lodged in the New Zealand Arthropod Collection with a minority kept in overseas collections (Macfarlane & Andrew 2001). Macfarlane & Andrew (2001) tabulated the main literature sources for the identification of each fly family and compiled isolated records of species additional to those listed in Evenhuis (1989) and other revisions. The larger fly faunas of Australia (CSIRO 1991) and North America (McAlpine et al. 1981, 1987) share all of the genera in New Zealand for 24 and 20 fly families, respectively, including eight large families. Consequently, for the mainly smaller families, overseas generic keys provide a guide for identification where there is no key to New Zealand genera.

Ecology, biology, and status of families

There are two suborders of Diptera – Nematocera and Brachycera. The latter is divided into two divisions, the Brachycera Orthorrhapha (referred to as Brachycera or Orthorrhapha) and the Cyclorrhapha (or Muscomorpha). The Cyclorrhapha include two sections – Aschiza (including hover flies and scuttle flies) and Schizophora. The latter are the 'acalyptrate' and 'calyptrate' flies, discussed under separate headings below and in the species list.

There has been no review of the immature stages of Diptera in New Zealand apart from those in fresh water (Winterbourn et al. 2000) and no recent review of the biology of New Zealand species, hence some new information is given here, particularly from the Manawatu (I. Andrew, pers. obs.) and Canterbury (R. Macfarlane, pers. obs.). Miller (1956) listed references to publications on immature stages and life-histories of New Zealand flies. An adapted family-level key to the larvae is needed to remove the difficulty of larval identification in New Zealand. It is not easy to investigate fly biology and glean information on habitat preferences if immature specimens cannot be distinguished on the basis of their morphology.

Suborder Nematocera

The 23 families of Nematocera form a major component of the litter-dwelling and fungus-feeding flies and of freshwater species.

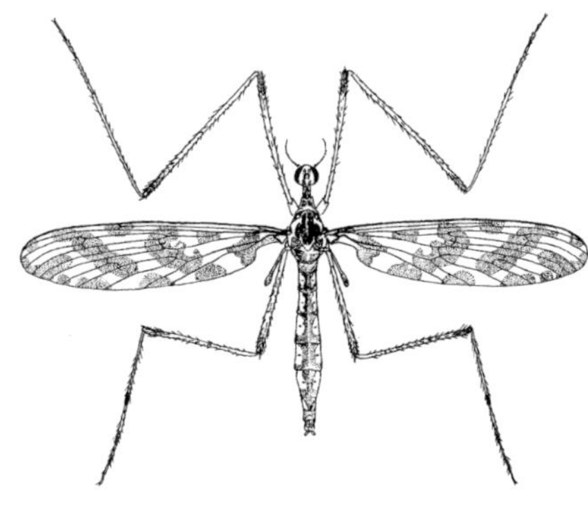

Mischoderus sp. (Tanyderidae).
From Grant 1999

Tipulidae

Crane flies have more species diversity than either the 29 acalyptrate families or the nine calyptrate families. Endemism of species is high at 99.6%. The range of habitat use in generally wetter environments is quite considerable, with major use of forest soil, litter, and rotting logs, as well as freshwater and even semi-saline water, and even reasonably drought-prone soils (Macfarlane & Andrew 2001). Habitats also range from the coastal foreshore to alpine fellfields (Edwards 1923a; Macfarlane & Andrew 2001. This family is treated in more detail below.

Trichoceridae, Tanyderidae. Winter crane flies (Trichoceridae) and the aquatic Tanyderidae resemble true crane flies quite closely (Edwards 1923a; McAlpine et al. 1981; Grant 1999) with their long legs and wings with several closed cells and quite complex venation. Most of the species are found in the Southern Hemisphere (McAlpine et al. 1981). Two overseas dipterists are investigating the taxonomy of both of these families. Trichoceridae are 3–9 millimetres long, the bodies range from brown to white and they have a strongly bent inner hind anal vein. Krzeminska (2001, 2003, 2005) has revised the 15 paracladurine species and allocated them to two new genera. These species are associated with native forest and are generally represented by limited specimens in the collections. Tanyderidae are 10–17 millimetres long and include species with circular spots in the wings (Edwards 1923a).

Ditomyiidae, Keroplatidae, Mycetophilidae. Among these, mycetophilidae is the main family of fungus gnats. The three families (including Keroplatidae and Ditomyiidae) are most readily distinguished from other Nematocera by their coxae being of a similar length to the femora and by their wings, with a moderate amount of venation and cells largely not closed. Tonnoir and Edwards (1927) illustrated the variation in wing-vein patterns and male genitalia in these three families. Zaitzev (2001, 2002a, b) has revised the genera *Brevicornu, Zygomyia,* and *Aneura*, describing a further 14 new species, but his *Zygomyia submarginata* is a preoccupied name. Jaschhof and Jaschhof (2009) and Jaschhof and Kallweit (2009) described three new genera and seven new species. Toft and Chandler (2005) provided the means to recognise three species that originated from Europe and Chile. Most species are expected to be fungus-feeders, but fungal associations have not been investigated to any extent. Some common species will be generalist feeders, but the high degree of sympatric speciation in some species (e.g. *Tetragoneura, Zygomyia*) suggests that some specialist associations will occur (R. J. Toft pers. comm.).

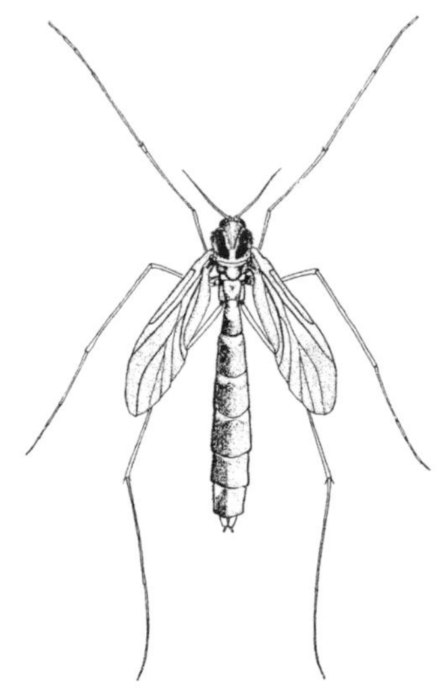

Adult of New Zealand glow-worm *Arachnocampa luminosa* (Keroplatidae)

From Grant 1999

The species are concentrated in forests (Didham 1992; Ward et al. 1999; Toft et al. 2001). Habitats with large quantities of dead wood, such as exotic gorse (*Ulex europaeus*) and kanuka (*Kunzea ericoides*), support 102 fungus-gnat species (Harris et al. 2004). Very few species (2–4) have been collected in extensive surveys of pasture grassland (Cumber & Harrison 1959b; Martin 1983), two from lucerne (Macfarlane 1970), and none in grass seed or cereal crops (Bejakovich et al. 1998). *Anomalomyia guttata* is apparently the most tolerant species, extending its habitat from forests to wet grasslands (Macfarlane 1970; Martin 1983; Macfarlane et al. 1998). There is still a great amount of taxonomic and faunistic work required before the species can be reliably identified and their distribution mapped in New Zealand.

Despite recent revisions, there are still more than 55 species of undescribed Mycetophilidae in New Zealand and even on a generic level our knowledge is far from complete. Phylogenetic and biogeographic relationships to fungus gnats in other parts of the world are poorly studied and understood, but relationships with other faunas of Gondwanan origin (Neotropics, Australia, New Caledonia) are obvious, as also with Indo-Malayan and Holarctic faunas (Freeman 1951; Matile 1990, 1993) Extensive malaise trapping of Mycetophilidae in and around regenerating forest at Hinewai Reserve on Banks Peninsula (Ward et al. 1999) produced late spring–early summer and autumn peaks in catches with some flight activity into winter. These flies form a considerable part of the biomass of forest insects and a few species account for large numbers of specimens in native forest (Didham 1996) or regenerating kanuka forest (Harris et al. 2004). These flies are available for insectivorous birds such as fantails (*Rhipidura fuliginosa*) to feed on.

The Keroplatidae include a number of predator species, including the famous New Zealand glow-worm *Arachnocampa luminosa*, which is the most thoroughly studied. The glow-worm is one of the few species of flies in the world whose larvae use light to attract insect prey (Richards 1960; Pugsley 1984). New Zealand glow-worms are a feature of 23 commercial caves, but the 103 New Zealand cave systems include none in the drier eastern areas of both islands (Worthy 1989). New Zealand glow-worms exist under overhanging banks in forests outside of caves and they are apparently absent from at least part of the drier non-forested east coast regions (Macfarlane & Andrew 2001). Among Keroplatidae, *Isoneuromyia* species can be common among gorse (Harris et al. 2004) and low numbers are also found among flax (*Phormium tenax*). The New Zealand Ditomyiidae are primarily associated with dead wood (R. J. Toft pers. comm.), and are therefore most commonly encountered in mature forests (Macfarlane et al. 1998, 1999; Ward et al. 1999) where *Nervijuncta* species may often be collected on mossy tree trunks.

'Fishing lines' of the New Zealand glow-worm.

Alastair Robertson and Maria Minor, Massey University

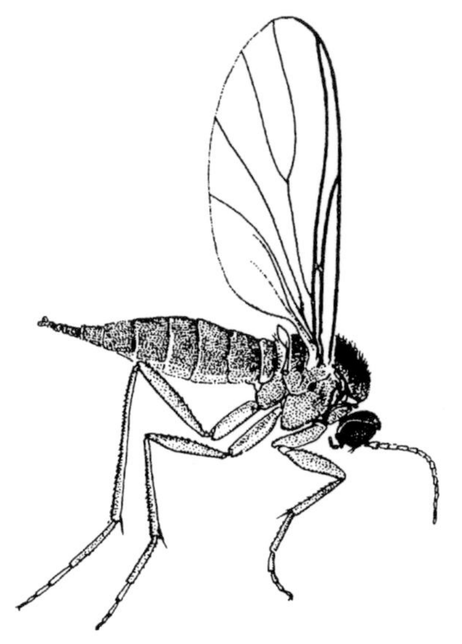

A species of Sciaridae.
After Grant 1999

Sciaridae. The number of New Zealand species of Sciaridae has more than doubled (Mohrig & Jaschhof 1999) since the last list was compiled (Evenhuis 1989). More species, including a flightless *Epidapus* species, await description. Species of *Bradysia* and *Lycoriella* still need verification from a competent authority so that endemic and introduced species can be reliably distinguished (see Checklist). Sciaridae are among the most abundant small flies in beech forest (Harrison & White 1969), broadleaf forest (Davies 1988; Didham 1992; Macfarlane et al. 1998), pine forest (Somerfield 1974; Macfarlane et al. 1999), and wet pasture (Cumber & Harrison 1959a,b; Macfarlane et al. 1998), but they extend to native grassland including braided riverbeds. The accidentally introduced species of root gnats (i.e. Sciaridae) in the genera *Bradysia* and *Lycoriella* are sporadic pests of horticultural crops, especially in glasshouses where these common flies can colonise potted plants before their natural enemies can slow population growth. '*Sciara annulata*' (probably *Bradysia brunnipes*) and *Scythropochroa nitida* have been reared from rotting willow logs (Wisely 1959) and the lack of native species recorded in pasture surveys indicates that most of the native species may favour forests and wetland habitats.

Rangomaramidae and Mycetophiloformia
The study of New Zealand Scairoidea fauna in the last decade led to the recognition of the new family Rangomaramidae, with three of four subfamilies from the Southern Hemisphere (Amorim & Rindal 2007). Placing these and other related genera has stimulated efforts to reinterpret the phylogeny of these flies (Hippa & Vilkamaa 2006) and inclusion of six infraorders within Nematocera (Amorim & Yates 2006). *Rangomarama* (Rangomaraminae) inhabits native forest, especially beech forest, from Auckland to Fiordland (Jaschhof & Didham 2002) and the other genera also inhabit podocarp or beech forest, but the larval habitat is unknown (Jaschhof 2004a,b, 2004, Jaschhof & Hippa 2003). *Starkomyia inexpecta*, which might be a new family or subfamily, is known only from a central North island forest.

Rangomarama edwardsi dorsal and lateral views (Rangomaramidae).
Uwe Kallweit, Senckenberg Naturhistorische Sammlungen, Dresden

Cecidomyiidae
Gall-midges are an abundant part of the New Zealand fauna, but very poorly known. A conservative estimate of the number of unnamed species lies between 200 and 300 and some will belong to Australian genera or unnamed endemic genera (Jaschhof, pers. comm.). Martin (unpubl.) has recorded a further 130 or so undescribed gall-making species. These small flies (0.7–3.0 millimetres long) have simple wing venation and no tibial spurs, which helps distinguish them from most Sciaridae. Most of the major collections have few if any quality slides or pinned Cecidomyiidae. There is a need to collect, rear, mount, and examine the species in New Zealand as well as a chronic need for an illustrated introduction to the family. Larvae of Lestremiinae and Porricondylinae are considered to feed on fungi or decaying vegetation while Cecidomyiinae are mainly herbivores. The fauna of free-living gall midges of the subfamilies Lestremiinae and Porricondylinae are relatively diverse compared to the herbivores or gall makers, and Porricondylinae could be more speciose than Lestremiinae (M. Jaschhof pers. comm.).

A monograph of the subfamily Lestremiinae (J&J 2003a,b, 2004) revealed 63 species in 11 genera, with *Peromyia* as the main genus. These flies dwell mainly in native forests and shrubland but a few species extend their range into tussock grassland, other open habitats, and exotic forests. Among indigenous forests, beech forests appear to support a more diverse lestremiine community than any kind of podocarp forest. Four predatory genera of Cecidomyiinae have been recorded – *Arthrocnodax* (Cottier 1934), *Trisopsis* (Crosby 1986), *Diadiplosis* (Evenhuis 1989), and *Lestodiplosis* (Macfarlane & Andrew 2001), but *Arthrocnodax* needs verification (Barnes 1936). Adults of the three subfamilies are quite readily distinguishable to subfamily.

However, only 22 of the 30 genera recorded from New Zealand are included in a North American key (McAlpine et al. 1981). A key to the Australian Cecidomyini (Kolesik et al. 2002) includes a considerable number of new genera, so at best only provisional generic identifications can be obtained using these summaries as guides. The generic affiliation of most New Zealand gall-midges, including Porricondylinae and Cecidomyinae, is in urgent need of revision, which will probably add further new genera (Jaschhof pers. comm.). Species recognition is impossible for about half the named New Zealand species (25) because the types named by P. Marshall have been lost (Miller 1950), the hosts are unknown, and Marshall's (1896) descriptions are inadequate for identification. Six further endemic species have been described, and their identification is facilitated by better descriptions, usually illustrated, and by association with plant galls. These are *Dryomyia shawiae* from *Olearia* leaves (Anderson 1935), *Kieffieria coprosmae* from coprosma stems (Barnes & Lamb 1954), the beech-leaf gall *Stephodiplosis nothofagi* (Barnes 1936), the beech-root gall *Protodiplosis radicis* (Wyatt 1963), *Dasineura hebefolia* from leaf galls on *Hebe salicifolia* (Lamb 1951; Miller & Walker 1984), and *Oligotrophus oleariae* from rosette twig galls on *Olearia paniculata* (Maskell 1889; Miller & Walker 1984). Lamb (1960) listed several undescribed gall-midges and others have been recorded since. These include stem galls on native broom (*Carmichaelia* spp.), bush lawyer (*Rubus* spp.), *Hebe* (Smith 1961; Lintott 1981; Miller & Walker 1984), and ngaio (*Myoporum laetum*) (Lintott 1974), bud, leaf and flower galls on *Coprosma* (Martin unpubl.), bud galls on *Raoulia*, *Helichrysum* (Smith 1961; Lintott 1974, 1981), and *Clematis foetida* (Macfarlane unpubl.), a leaf, stem, and fruit gall on putaputaweta (*Carpodetus serratus*) (Smith 1961; Hunt 1992), galls affecting *Parahebe decora* (Smith 1961), buds on a mountain heath (*Dracophyllum*) (Martin 2003), and an apparent gall midge on *Hebe elliptica* (Valentine & Walker 1991). The snow tussock midge *Eucalyptodiplosis chionochloae* feeds on the growing seed of *Chionochloa* species and does not make a gall (Kolesik et al. 2007), and danthonia seed is affected too (Cone 1995), but perhaps not by this species. Additional hosts or types of gall to those previously mentioned include trees of podocarps, broadleaf and beech forests, kowhais (*Sophora* spp.), cabbages trees (*Cordyline* spp.), and *Muehlenbeckia* creepers.

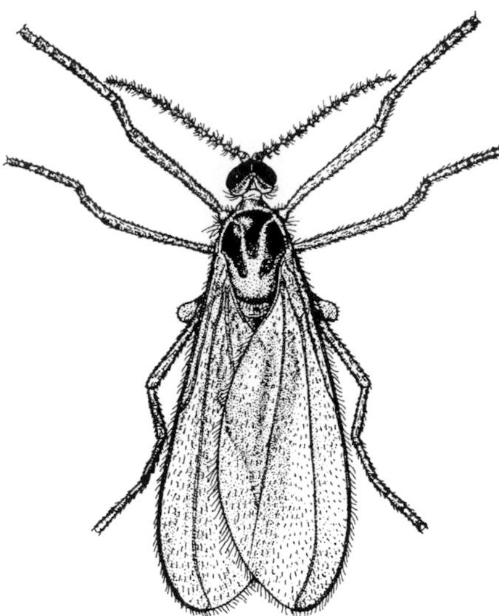

Snow tussock midge *Eucalyptodiplosis chionochloae* (Cecidomyiidae).
Tim Galloway, Lincoln University

Chironomidae

Midges are well known to New Zealanders, especially to Aucklanders who used to suffer plagues of them from the former sewage oxidation ponds in Manukau Harbour. The bright-red larvae of many species are known as blood-worms (not to be confused with tubificids and other annelid worms) and are an important food for aquarium animals in many parts of the world. Classification of midges in New Zealand has followed the European trend of splitting genera into fairly small units and an appreciable proportion (42%) of the genera are apparently endemic, with many monotypic. Freeman (1959) provided sparsely illustrated keys for 22 of the 62 known genera and 41 of 106 described species. Brundin (1966) dealt with four podonomine genera and 17 other species. Identification in field surveys remains a problem. Dubious generic names, probably derived from American keys, such as *Clunio* (Tortell 1981) and *Psectrotanypus* (Knox & Bolton 1978) have been applied to species among the tidal zone. Further species and genera have been described in the last decade (Martin 1998; Macfarlane & Andrew 2001) including some of the immature stages, so their generic placement is now soundly based (Boothroyd 1999, 2002, 2004; Boothroyd & Cranston 1999; Cranston 2007, 2009).

Midges occupy a wide range of freshwater habitats from the lower reaches of the west coast glaciers (Boothroyd & Cranston 1999) to geothermal waters and the seashore (Quinn & Hickey 1990a; Stark 1993; Winterbourn et al. 2000) and a few survive in ephemeral waterways. There are a few terrestrial species, including *Smittia verna* (Martin 1983), the adventive *Camptocladius stercorarius*,

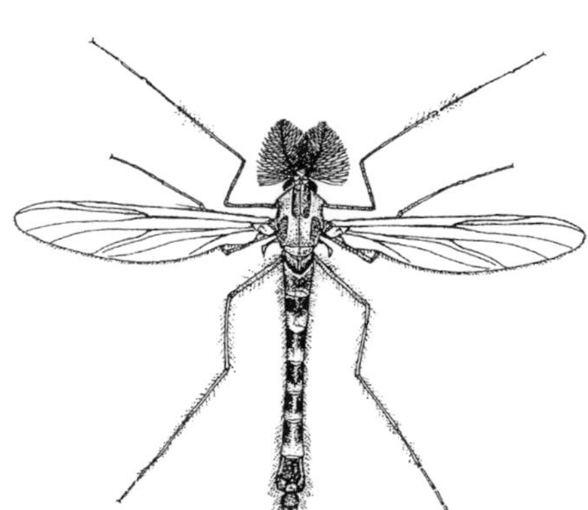

Common midge *Chironomus zealandicus* (Chironomidae).
After Grant 1999

and one or more undescribed endemic species. Midge larvae may be collector-gatherers, collector-filterers, scrapers, shredders, engulfers, or piercers. These differences, along with different uses of food and shelter, permit several species to coexist in the same area (Armitage et al. 1995). At least one species shelters in wood (Anderson 1982) and in North America small carrion in water may suit some species (Tomberlein & Adler 1998). Another New Zealand species is a predator (McLellan 1983). An undescribed *Eukiefferiella* species is a commensal on *Deleatidium* mayfly nymphs (Winterbourn 2004); *Tonnoircladius commensalis* is commmensal or phoretic on immature netwinged flies *Neocurupira* (Cranston 2007), and another shelters in wood (Anderson 1982). A full range of feeding strategies is likely to exist in New Zealand because the midge fauna is so diverse. Elsewhere in the world the larvae of most species are collector-gatherers for at least some larval instars and there is often flexibility in the use of food sources (algae, detritus and associated microorganisms, macrophytes, woody debris, invertebrates). Analysis of gut contents and behavioural investigations in the field have helped sort out differences in food use (Armitage et al. 1995).

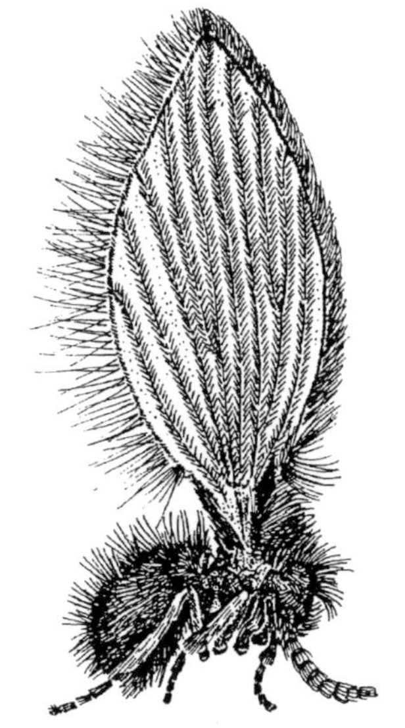

Moth fly *Psychoda* sp. (Psychodidae).
From Grant 1999

Psychodidae

Six of the 47 species of moth flies (Psychodidae) described from New Zealand are confined to the subantarctic islands (Quate 1964; Duckhouse 1971). Satchell (1950) described 27 species and included illustrations of wings, antennae, and genitalia for 16 species. He considered there were 40 or so undescribed species, prior to describing eight more (Satchell 1954) and keying 40 species from the main islands. Duckhouse (1980) described a further *Trichomyia* species and discussed limitations of zoogeographic interpretations owing to inadequate collections from a range of habitats. Duckhouse (1990) reassigned the 10 *Pericoma* species to *Didicrum*, *Satchellomyia*, and *Ancyroaspis*, and redescribed the species from New Zealand. Adults are found mainly beside freshwater streams or pools (Duckhouse 1990, 1995), among forests (Evenhuis 1989), and gardens around compost, and a few extend into cropland (Cumber & Eyles 1961a) but usually not grassland, especially during the day (Cumber & Harrison 1959b; Macfarlane 1970; Martin 1983; Bejakovich et al. 1998). Larvae are common in damp soil, rotting vegetation, sewage filter beds (Goldson 1977; Martin 1983; Miller & Walker 1984), dung (Duckhouse 1966; Goldson 1977; Martin 1983), and semiaquatic sites (Goldson 1977; Duckhouse 1980). There is likely to be a large, distributed, wet (alcohol) collection in New Zealand, resulting from malaise- and light-trapping and general collecting, that will need taxonomic analysis before any gaps in habitat, region, or method of collecting different species become apparent.

Ceratopogonidae

These are also poorly known in New Zealand. The last revision by Macfie (1932) listed 28 species from a total of 29 then known, based on about 170 specimens. Numerous specimens are now in collections, and sometimes more than 100 can be collected in a day without difficulty. Nevertheless, specimens on pins or slides are relatively uncommon and a systematic collection effort is desirable so the full diversity can become known. There is limited knowledge of phenology, distribution (Igram & Macfie 1931; Macfie 1932; Dumbleton 1971), and habitat use (Dumbleton 1971; Winterbourn et al. 2000; Andrew & Winterbourn unpubl.) of New Zealand species. However, knowledge of ceratopogonid ecology and biology in England (Smith 1989), Australia (Lee 1948; Debenham 1979; Elson-Harris & Kettle 1985), and North America (Saunders 1956, 1964; McAlpine et al. 1981) provides consistent indicators for genera and subgenera shared with New Zealand. On the basis of this literature, all *Atrichopogon* and most *Forcipomyia* species (Forcipomyiinae) are expected to live in wet terrestrial forest and wetland habitats, except for the aquatic *Forcipomyia* subgenus *Trichohelea*. On the other hand, *Dasyhelea* species (Dasyheleinae) live in small and temporary still waters, which might even be up in trees. In the Manawatu, *Forcipomyia* individuals have been reared from native forest litter, and an undetermined, possibly introduced, species of *Dasyhelea* from water in a garden bucket. Most species of the subfamily Ceratopogoninae are expected to live in still water and seepages. The main larvae found in running fresh water in New Zealand are of Ceratopogoninae, as illustrated in Winterbourn et al. (2000), but *Forcipomyia*-type larvae are also present (Winterbourn unpubl.). At least one species of blood-feeding midge, *Leptoconops myersi*, is present in tidal water and coastal seepages from Northland to Nelson (Dumbleton 1971). Although several species are often encountered in forests, only one species was collected in surveys of North Island pastures and fodder crops (Cumber & Harrison 1959a).

Simuliidae, Culicidae

There are only two prominent families of Nematocera flies with blood-feeding adults in New Zealand. Most New Zealanders will be bitten more often by the small, stout black *Austrosimulium* species than by the slender brown to black mosquitoes. Among the black flies/sand flies (Simuliidae), *A. australense* and *A. ungulatum* are the most likely to feed on humans (Winterbourn et al. 2000). Other species seldom bite humans, but they feed at least on birds. Cursed by trampers, black flies are most abundant in national parks and higher rainfall areas, which contain the premier hiking trails and wildlife reserves in New Zealand. In lowland areas, recreation spots in the vicinity of running water with shade trees are favoured by *A. ungulatum* (Crosby 1989; Glover & Sagar 1994, Winterbourn et al. 2000), while *A. australense* occurs throughout New Zealand.

West coast blackfly *Austrosimulium ungulatum* (Simuliidae).

From Grant 1999

There are 16 species of mosquitoes (Culicidae) (Miller & Phillips 1952; Snell 2005a), which breed in a range of standing and slow-flowing fresh to salt water and hot geothermal water in the case of *Culex rotoruae* (Belkin 1968; Holder et al. 1999; Winterbourn et al. 2000). Only the Australian striped mosquito *Aedes notoscriptus* and the native *Culex asteliae* are known to use epiphyte-axil water sources (Derriak 2005a,b), so this is a relatively vacant niche for colonisation by further accidentally transported exotic mosquitoes. Extensive use of discarded tyres on silage has contributed to the spread of *A. notoscriptus* southwards from Auckland (Laird 1995). Introduced *Aedes australis* is now the second mosquito species known on the Snares islands (Macfarlane & Andrew 2001, Snell 2005b).

All four introduced and five native species may bite humans, but *A. subalbirostris, A. antipodeus,* and *A. australis* are not known to bite livestock (Holder et al. 1999). The main period of biting or feeding activity is known for all but two species, with five species being nocturnal and four crespuscular to nocturnal (Derraik et al. 2005). *Aedes notoscriptus* and *Opifex fuscus* bite during the day, the others during the evening and night. Of these, the native vigilant mosquito *Culex pervigilans* is the commonest and most widespread species (Belkin 1968; Laird 1995). The winter mosquito (*Aedes antipodeus*) remains active into cooler weather compared to the other species (Miller & Walker 1984). Currently, none of the mosquito species transmits human disease, but the introduced species have potential to transmit viral and bacterial diseases including dengue fever.

New Zealand is also vulnerable to future disease transmission consequent upon the accidental establishment of further species and the potential for the temperate Australian *Anopheles annulipes* to establish here (Boyd & Weinstein 1996). *Anopheles* species are notorious for their ability to transmit malaria, and global warming would make more of New Zealand vulnerable to mosquito species that have already become temporarily established (e.g. *Aedes camptorhynchus*, the southern saltmarsh mosquito). Japanese *Aedes albopictus* is among four species intercepted arriving in New Zealand (Laird et al. 1994). *Culex quinquefasciatus* might transmit avian malaria; other species known to bite birds include *Opifex fuscus, Culiseta tonnoiri, Culex pervigilans,* and perhaps *A. notoscriptus* and these species can or might transmit viruses or avian flu (Holder et al. 1999). In native

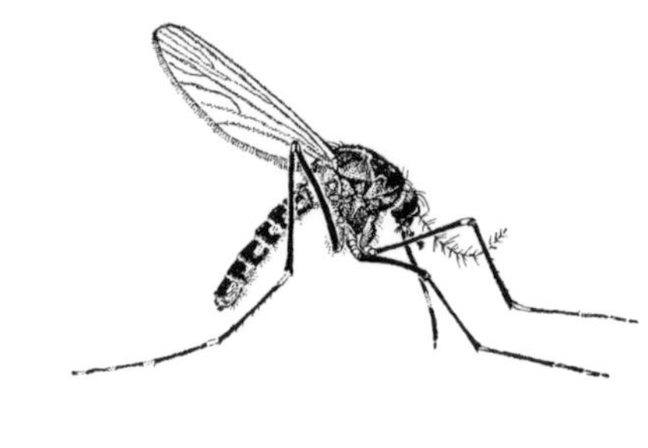

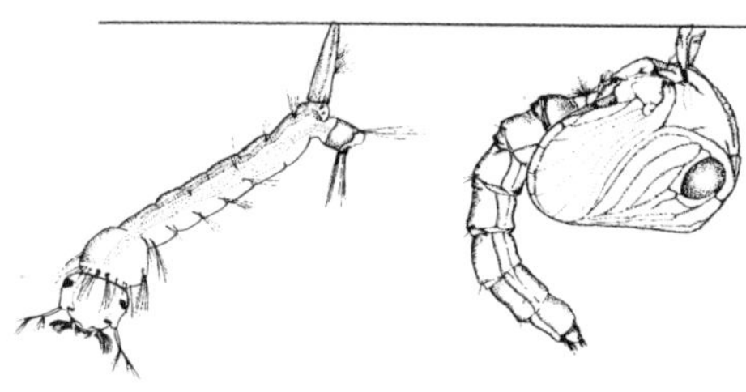

Adult female, larva, and pupa of the striped mosquito *Aedes notoscriptus* (Culicidae).
From Grant 1999

forest some species such as *C. asteliae* and *Maorigoeldia argyropa* are not attracted to carbon dioxide, unlike *C. pervigilans* and *A. notoscriptus*. *Maorigoeldia argyropa* does not bite humans at least (Snell et al. 2005) and the adult feeding habits and hosts also remain unkown for *Culex rotoruae*, *Culiseta tenuipalpis*, and *Aedes chathamensis*.

Blephariceridae

Net-winged midges are slender, long-legged and often black. Their wings (5–11 millimetres long) have numerous delicate cracks and folds that form a network, hence the common name. Adults are usually found close to turbulent bouldery streams where they can be seen either flying just above the rough water of rapids or resting or ovipositing in the splash zone of boulders. Females often have well-developed mouthparts and many species feed on smaller insects, which they catch with their specialised hind tarsi. Male adults generally do not feed but some groups (e.g. Apistomyiini) consume flower secretions. Larvae and pupae live in strong currents on smooth hard surfaces in streams (Dumbleton 1963; Craig 1969; Zwick & McLellan 1999). Larvae have broad bodies with little depth and appear to have only 6 or 7 segments, each separated by deep constrictions and bearing a mid-ventral sucker. Winterbourn et al. (2000) provided a key to larvae. Final-instar (4th) larvae are about 4–14 millimetres long. Pupae are oval, with gill lamellae projecting upwards from the anterior; the flat, soft, ventral face is cemented to the substratum. Larvae are scrapers, grazing on microscopic growths.

Within New Zealand, *Peritheates harrisi* is the only widespread species in the North Island where it has been recorded from Wellington to Te Aroha. *Peritheates turrifer*, a South Island species, has also been recorded in the Wellington region. No blephariceridss have been recorded north of Te Aroha despite the concentrated efforts of Towns (1978) and various collecting expeditions (Entomology Division DSIR, AMNZ). Three species with very restricted distribution are: *Neocurupira chiltoni* only on Banks Peninsula, *Nothohoraia micrognatha* in Fuchsia Creek and nearby streams in the Lower Buller Gorge, and *Neocurupira rotalapisculus* near Dunedin. There are 4–6 undescribed species known from the South Island. The other species are more widespread in the South Island from sea level to 1500 metres and *Neocurupira tonnoiri* extends to Stewart Island.

Thaumaleidae

These are among the less commonly collected aquatic fly species, because the stocky shiny yellow to brown adults are tiny (body 1.75–2.25 millimetres long). Immature stages live in a rather specialised way – larvae resemble midge larvae and they move in a typical sideways fashion with the body forming a U shape as one extremity moves while the other anchors (McLellan 1988). They are found in shaded areas on vertical rocks, clay, and other surfaces where there is a cool film of water thin enough not to completely submerge them. Reared species graze the surface of leaves and other substrata, ingesting diatoms or associated fungal spores and hyphae associated with decaying plant material. Pupae are secured by abdominal hooks to the substratum and take 10–20 days to emerge. It is likely that more undescribed species exist in New Zealand.

Dixidae

Dixids resemble mosquitoes, but no wing veins are distinctly hairy and there is a sharply bent radial vein near the wing tip (Belkin 1968). The wings, which may be smaller than the bodies, are 2.5–4.3 millimetres long in the New Zealand species. Adults are black to yellowish-brown. The larvae of *Nothodixa* inhabit very small, steep streams and quiet pools with moderate to strong currents (Belkin 1968; Winterbourn et al. 2000). Larvae of the four *Paradixa* species inhabit pools, seepages, lakes, swamps, and streams and two species prefer either shade or slower-flowing water.

Scatopsidae
Scatopsids can be readily distinguished from other fly families by their wings (prominent darkened front veins contrast with weak hind veins) and their short stout antennae. The two introduced species of Scatopsidae – *Scatopse notata* and to a lesser extent *Coboldia fuscipennis* – are commoner in collections than the two native species of *Anapausis* (Freeman 1989). The previously described *Scatopse carbonaria* is synonymised here with *Coboldia fuscipennis*. In the genera *Colobostemus* and *Rhegmoclemina* there are at least five undescribed species in collections, some of which are found in drier salt meadows and braided riverbeds. It remains to be seen how adequately alcohol collections represent New Zealand forest habitats.

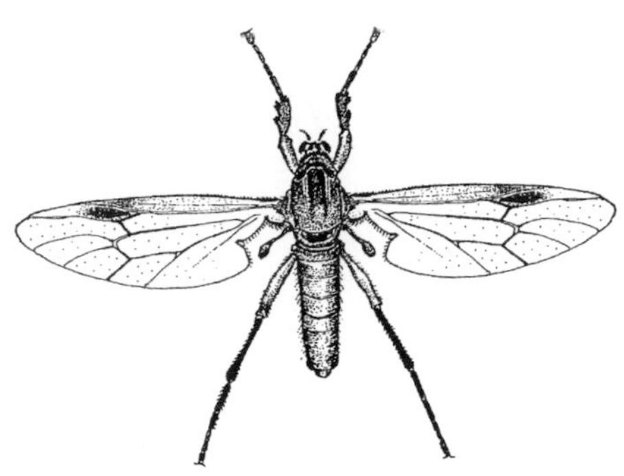

Female bicoloured swamp fly *Dilophus nigrostigma* (Bibionidae).
After Grant 1999

Bibionidae
These flies vary from 2 to 16 millimetres long. Harrison (1990) has described all eight native species, providing distribution maps and listing the flower species visited by the flies. Stigmata on the wings and their stout antennae, shorter than the head, distinguish these flies from other nematocerans. They show pronounced sexual dimorphism – males are shiny black with very large eyes while females are usually brown with small eyes placed forwards on an elongate head. The rather spiky larvae of *Dilophus nigrostigma* and *D. segnis* are readily found among fallen leaves (Harris 1983) or under rotting logs before they emerge in spring (Harrison 1990). Flight swarms of adults are often encountered during summer months.

Anisopodidae, Canthyloscelididae
The larger wood gnats or outhouse flies (*Sylvicola* spp.) have patterned wings and mostly live among rotting vegetation or in water in fallen plant material. *Sylvicola neozelandicus* commonly breeds in wet manure. It is the outhouse fly associated with long-drop toilets in forests and has also been recorded from Australia (Fuller 1935). There are illustrations of the wing pattern for three of the four described species in New Zealand (Edwards 1923a, 1923b). The less common *Canthyloscelis* (Canthyloscelididae) are apparently associated with forest areas, but nothing is known of their biology.

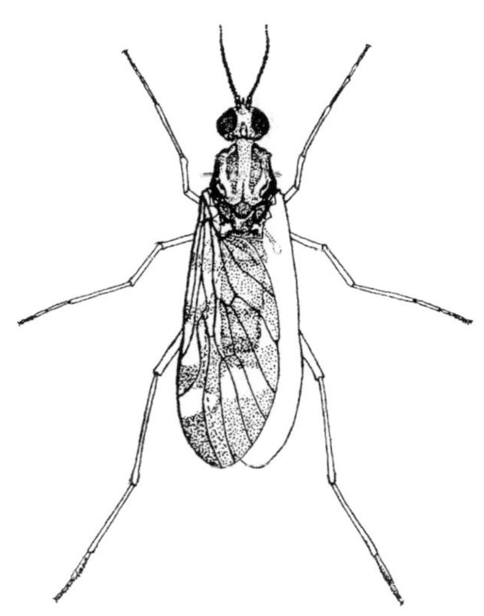

Outhouse fly *Sylvicola* cf. *neozelandicus* (Anisopodidae).
From Grant 1999

Corethrellidae
Corethrella novaezealandiae, which Belkin (1968) illustrated fully, frequents lake and pond margins along the South Island west coast. Corethrellids typically feed on frogs, and share with other biting Nematocera serrated mandibles and untoothed maxillae, unlike mammal- and bird-feeding mosquitoes. The few known corethrellids are vectors of trypanosome infections in frogs and the call of the frogs attracts female flies. The native species is assumed also to feed on frogs. *Corethrella* larvae prey on other aquatic insects (CSIRO 1991).

Suborder Brachycera: Division Orthorrhapha

There are 10 families of Orthorrhapha. They often favour moist habitats or live in seepages and freshwater, but this suborder includes soil predators of grassland (Asilidae and Therevidae) and beaches (Therevidae) and a few parasites of spiders (Acroceridae). There are some illustrations of immature stages of Brachycera and often sketchy information on the biology of species or genera for nine families. Lyneborg (1992) described techniques for rearing therevid larvae on tenebrionid beetle larvae and illustrated the larva of *Megathereva atritibia*.

Empididae. The great diversity of form in this family led to proposals to segregate it into four families (Chvála 1983). Sinclair (1995), among others, disputed this division, preferring to retain only the one family Empididae, which has seven subfamilies in New Zealand (Evenhuis 1989; Sinclair 1995). The monograph of Collin (1928) covered only 26 of the 41 genera and 87 of the 287 species

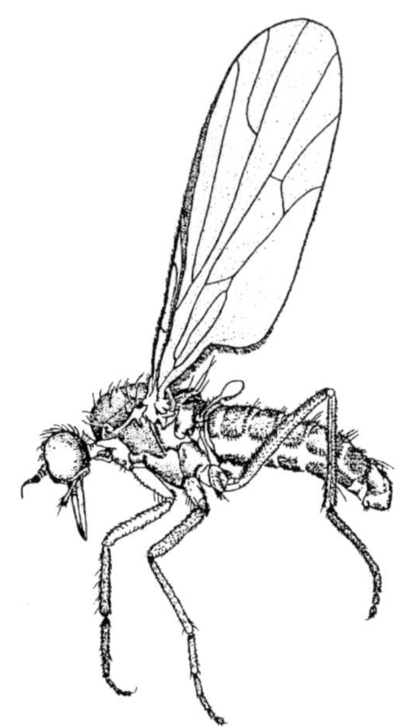
Unidentified dance fly (Empididae).
From Grant 1999

Tetrachaetus bipunctatus (Dolichopodidae).
Rod Macfarlane

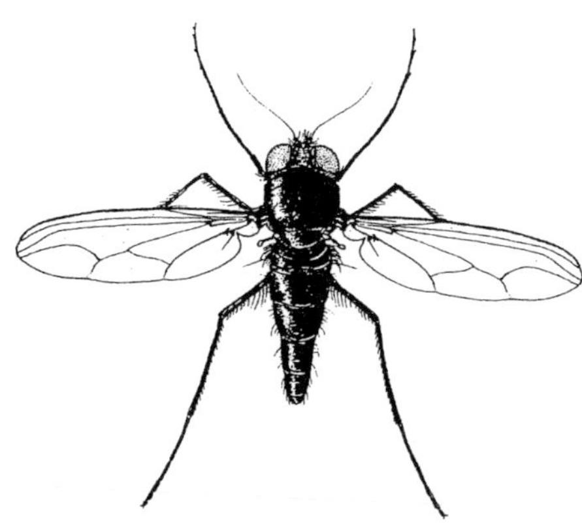
Green long-legged fly *Parentia* sp. (Dolichopodidae).
After Grant 1999

now known to occur in New Zealand, which limits its value for identification. Descriptions of new genera and revised keys to the genera of Clinocerinae (Sinclair 1995), Ocydromiinae (Plant 1989), and Ceratomerinae (Sinclair 1997b), the tribes of Empidinae (Bickel 1996) and the presence of Microphorinae (Shamshev & Grootaert 2002) update the higher classification. Further keys to species and descriptions or redescriptions apply to eight genera (Malloch 1931, 1932; Rogers 1982; Plant 1991, 1993, 1997, 2007; Shamshev & Grootaert 2002; Sinclair & McLellan 2004). Two more genera and at least 104 undescribed species are known to occur in New Zealand (see checklist). A consolidated summary of the family and much more systematic research is needed before the fauna and its affiliations are well known. Both adults and larvae prey on other insects, mainly dipterans. The elongated front legs (in eight genera of Hemerodromiinae), middle legs (*Platypalpus*, Tachydromiinae), or hind legs (*Pseudoscelolabes*, Ocydromiinae) with ventral spines are the most clearly adapted for holding captured prey. Adults may sometimes feed on nectar. Some genera, especially of the Empidinae, have an elaborate courtship display in which the prey is used as an essential stimulus initiating copulation. Some species form dancing swarms over vegetation on land (*Empis*, ?*Clinorhampha*) or over water (*Hilara*); such swarms are usually connected with mating behaviour. *Hilara* will take prey trapped in the surface film of water, and some *Hilarempis* species are found on boulders in streams or riparian vegetation. The three *Thinempis* species are confined to beaches from Stewart Island to Northland (Bickel 1996, LUNZ collection), but *Chersodromia* is so far known only from sandy beaches in Otago and at New Brighton (Macfarlane 2005) and *Chimerothalassius ismayi* from the stony beach at Birdlings flat, Canterbury (Shamshev & Grootaert 2002). The recent records from open sandy and gravelly habitats point to the need to collect more flies from these sites. Little has been published on the life history, biology, or behaviour of Empididae from New Zealand. Dumbleton (1966) described the aquatic larvae of *Chelifera tantula* and ?*Clinocera gressitti* from New Zealand. Elsewhere, larvae are known to be predaceous and occur in soil, leaf litter, rotting wood, and dung; others (Clinocerinae, Hemerodromiinae) are aquatic.

Dolichopodidae

The long-legged flies are likely to be among the top five to eight most diverse fly families in New Zealand. The revision of Parent (1933) is well illustrated but not readily available. Bickel (1991) published an undated key to the genera and descriptions of 30 species in the subfamilies Sciapodinae (*Parentia*) and Medeterinae and one new *Paraclius* is known from New Zealand (Bickel 2008). The metallic green *Parentia* species can be common in grassland (McLean et al. unpubl.) and may feed on insects attracted to honeydew secreted on manuka. Other members of the family are quite common in wetlands (Macfarlane et al. 1998) and forests (Didham 1997; Macfarlane & Andrew unpubl.). An undescribed *Scorporius* species exists in estuaries and lagoons in both main islands and several new species of *Abetatia* can be found along rocky shorelines. An examination of more than 1100 specimens of pinned Dolichopodidae has revealed four obvious undescribed species (Checklist), but the extent of undescribed species cannot be properly gauged until national specimens elsewhere and in alcohol have been adequately examined. There is also the possibility that specialist collecting near waterways and from alpine and northern regions may yield a considerable number of undescribed species. New Zealand collections may be partially deficient because of low use of yellow pan traps, which have been useful in revealing species in Australia (Bickel 1999).

Therevidae

Stiletto fly diversity and classification in New Zealand is relatively well documented (Lyneborg 1992) even though most species of these flies are not particularly common in most habitats apart from some beach and riverbed/

lakeside areas, where at least seven species can be conspicuous. Adults are 5–18 millimetres long. They probably ingest only water or honeydew because they do not have suitable piercing mouthparts for predation and are uncommon on flowers. In appearance they range from evenly dark with obvious white hairs (many *Anabarhynchus* species) to brownish *Ectinorhynchus* species and blackish *Megathereva* with whitish stripes on the thorax. Adult *Ectinorhynchus* species were collected from established broadleaf-podocarp forest margins in coastal Canterbury (Ward et al. 1999, Macfarlane 2007) and the Manawatu. However, at least one species of *Anabarhynchus* has been collected from bush margins in the Manawatu and Banks Peninsula. Therevid larvae have been found in soil under litter in red beech (*Nothofagus fusca*) forests (Evans et al. 2003). The closest biogeographic link is with Tasmanian *Anabarhynchus* and the genus also occurs in Chile. *Ectinorhynchus* is confined to Australia and New Zealand. The mobile larvae (Lyneborg 1992) are quite thin and streamlined (McAlpine et al. 1981; Smith 1989) and prey on other soil invertebrates. *Anabarhynchus harrisi* larvae will feed on paralysed spiders in nests of the spider-hunting *Priocnemus* (Lyneborg 1992; Harris pers. comm.), making such flies apex invertebrate predators.

Stiletto fly *Megathereva bilineata* (Therevidae).
After Grant 1999

Asilidae

If you have ever watched a seemingly large wasp as it quickly flies out to catch another insect in mid-air, you have probably seen a robber fly. Robber flies are often mistaken for bees or wasps but only robber flies hunt in this manner. In the lowlands, the robber-fly life-cycle may take more than one year, according to disparate overseas investigations (Musso 1978), which show that delays in development can occur in both eggs and the first of the five to seven instars. Macfarlane (unpubl.) analysed asiline seasonal occurrences based on New Zealand collection labels and the results showed that adults are mainly active between October and February, with most adult activity at a particular location being confined to two to three months. Most species were found below 920 metres elevation except for four species that extend into the subalpine zone, and most species including *Saropogon* (Dasypogoninae) were collected from uncultivated grassland, although *Neoitamus bulbus* has been reared from kowhai logs (FRNZ collection). *Zosteria novaezealandica* is known only from a single male and is unusual for New Zealand in having been collected from the upper reaches of beech forest near Nelson (Macfarlane et al. 1997). There are NZAC collection records of both *Neoitamus* auctt. and *Saropogon* adults reared from grassgrub infestations, apart from the literature records (Valentine 1967b). Overseas records show a similar prey relationship with grassgrubs for larval Asilini and Dasypogoninae (Knutson 1972; Musso 1978; Dennis & Knutson 1988). Information about identification and distribution of the New Zealand species is unfortunately inadequate. The only existing key (Hutton 1901), to 14 species, lacked illustrations and the descriptions were short. Attempts to classify two problematic smaller asiline genera have resulted in different generic placements in each catalogue (Miller 1950; Hull 1962; Evenhuis 1989) because the Australasian genera have been inadequately studied. Three subfamilies (Apocleinae, Asilinae, and Dasypogoninae) and two tribes of Asilinae (Lehr 1996) are represented in New Zealand compared to five subfamilies and 11 tribes in Australia (Evenhuis 1989; Lehr 1996).

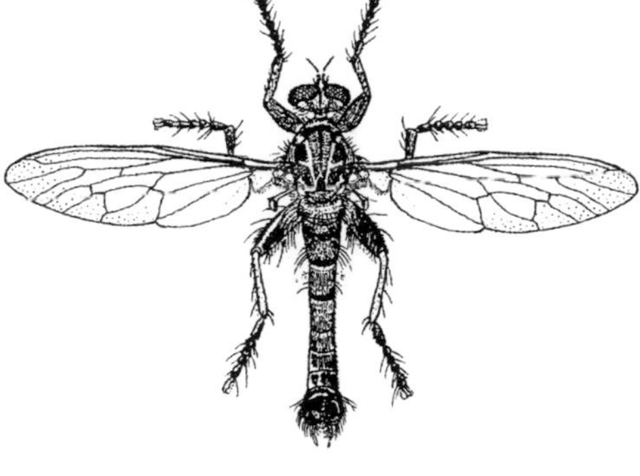

A robber fly, *Neoitamus* sp. (Asilidae).
After Grant 1999

Stratiomyidae

Soldier flies are rather large and stocky flies that resemble hover flies or wasps. They are often encountered 'standing sentry' on flowers. Most soldier flies breed in soil and litter, feeding on decaying vegetation, but Australian *Inopus rubriceps* is a pasture pest of much of the North Island (Macfarlane & Andrew 2001). Hutton (1901) keyed and briefly described 20 species in three subfamilies. His key to the eight described *Odontomyia* species (Stratiomyinae) is still the only published reference for these distinctive green or black flies. Miller (1917) described and

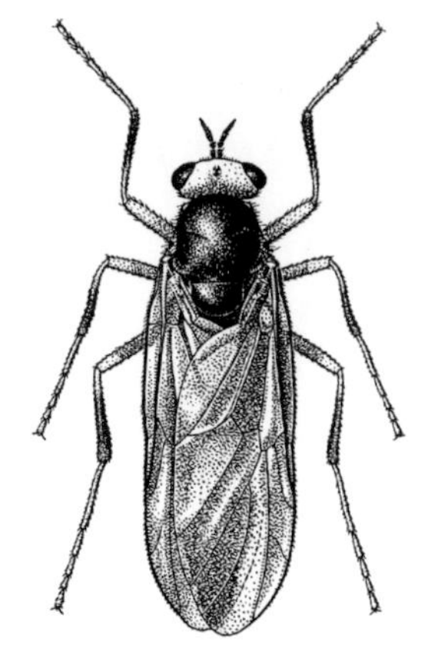

Australian soldier fly *Inopus rubriceps* (Stratiomyidae).
Des Helmore

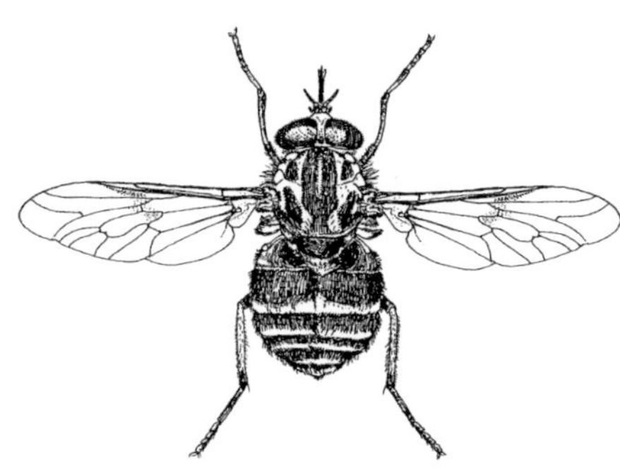

Scaptia sp. (Tabanidae).
After Grant 1999

illustrated 15 species of Beridinae and included a key, but six *Benhamyia* species were reduced to synonyms (Evenhuis 1989; Woodley 1995). All six genera of Beridinae when they were redefined were found to be endemic (Woodley 1995). Between 1939 and 1958 two more introduced species became established in the North Island (Macfarlane & Andrew 2001) and four new native species were described (Lindner 1958).

Tabanidae

These large, heavy-bodied flies with a strong beak-like proboscis are generally referred to as horse flies. Mackerras (1957) reviewed the New Zealand fauna, describing 17 species and illustrating various features of the head. Adults are often found at flowers such as manuka and cabbage trees, or in riverbeds. *Scaptia* species often hover for prolonged periods, patrolling invisible territories in open sunshine near shrubs or on forest edges. *Dasybasis* species can occasionally be a nuisance by biting swimmers in bush swimming pools. Larvae of some unidentified species are found sporadically among gravel in streams (Winterbourn et al. 2000), and these larvae are relatively tolerant of poor-quality water (Stark 1993). Successful rearing of two new *Dasybasis* species from moss on wet rocks by waterfalls and river banks from Coromandel and Southland and the record of a further new *Scaptia* species among mosses on fallen logs near Lake Brunner indicate that larvae of New Zealand species may occupy both better-quality freshwater sites and consistently moist sites away from flowing water. In Europe, some Tabanidae are even found in drier soils (Chvala et al. 1972). The 20 known New Zealand species of Tabanidae represent two subfamilies – Pangoniinae (*Ectenopsis*, *Scaptia*,) and Tabaninae (*Dasybasis*).

Acroceridae

Small-headed flies may sometimes be locally common but they are among the less common flies in insect collections. They can be swept from shrubs and grasses or seen on fences (Harris & Macfarlane unpubl.). Three genera occur in New Zealand, of which *Ogcodes* is most often encountered. The seven species of *Ogcodes* have straight thickset bodies (Schlinger 1960; CSIRO 1991). Schlinger (1960) commented that early instars of *Ogcodes* can be obtained by rearing immature spiders. The other two genera are very distinctive. *Helle*, with two species, is small, dark and almost V-shaped. *Apsona muscaria*, a metallic blue and turquoise fly, is known from Arthurs Pass National Park, Mt Hutt (CMNZ), and Motueka (OMNZ). Adult acrocerids were not readily collected in malaise traps during summer in forest on Banks Peninsula (Ward et al. 1999) and were not collected in a wetland survey (Macfarlane *et al.* 1998). The eggs of at least some *Ogcodes* species are unusual because they are black (Miller & Walker 1984; Ferrar 1987) and laid in clusters on twigs.

Bombyliidae, Rhagionidae

Bombyliidae (bee flies) and Rhagionidae (snipe flies) are each represented in New Zealand by one described species, and they are seldom seen. Tonnoir (1927) described both *Tillyardomyia gracilis* (Bombyliidae) and *Chrysopilus nitidiventris* (Rhagionidae). The slender *T. gracilis* with a blackened area on the wing is quite distinct and is apparently confined to the South Island. *Chrysopilus nitidiventris* is known from Westland forest (Tonnoir 1927, LUNZ), and one undescribed species occurs in the Paparoa ranges (LUNZ) and and another in Westland also (CMNZ). Overseas, the larvae of other *Chrysopilus* species are predators, preferring wet sites in litter or rotting logs (Oldroyd 1964; McAlpine et al. 1981).

Pelecorhynchidae

Larvae of the sole *Pelecorhynchus* species have been found in a stream near Dunedin (Winterbourn et al. 2000), but the species needs adults to be reared and identified to confirm the genus identification and determine if it is an endemic species.

Aspilocephalidae

This small family from North America, Tasmania, and New Zealand has just three genera and Kaurimyia from New Zealand is endemic (Winterton & Irwin 2008). The two known specimens have been collected from Northland kauri forest and from Dunedin.

Suborder Brachycera: Division Cyclorrhapha: Section Aschiza

The six families of Aschiza are dominated in species diversity and economic importance by hover flies (Syrphidae) and humpbacked or scuttle flies (Phoridae). Reviews of larvae (Smith 1989; Stehr 1991), Phoridae (Disney 1994), and of New Zealand insects generally (Miller & Walker 1984) provide illustrations of the larvae of *Dohrniphora cornuta, Megaselia rufipes, M. scalaris,* and *Spiniphora bergenstammi* (Phoridae), and of *Merodon equestris, Eristalis tenax, Eumerus strigatus,* and *E. tuberculus* (Syrphidae). For eggs, Ferrar (1987) has general family features including the range in size.

Syrphidae

Hover flies or flower flies are among the larger and more strikingly patterned flies in New Zealand. Endemic hover flies do not include mimics of bees and wasps such as exist in overseas countries but the introduced European drone fly *Eristalis tenax* and the narcissus bulb fly *Merodon equestris* mimic the honeybee and European species of bumble bees, respectively. Remarkably, the only overall review is that of Miller (1921), who also provided an illustrated key to the species. Miller and Walker (1984) and Grant (1999) respectively illustrated four common species in colour and black and white. Flies in the two syrphid subfamilies have different food sources and distinct wing venation. The five genera and 24 species of Syrphinae prey as larvae on other slow-moving, soft-bodied insects, while members of subfamily Eristalinae, with six genera and 18 species in New Zealand, feed mainly on decomposing material in wet sites and three European pest species feed on bulbs. Prey items have been recorded for only three common hover fly species (Valentine 1967b; Miller & Walker 1984). The small hover fly *Melanostoma fasciatum* is dominant in lucerne fields with blue-green aphid (Thomas 1977) and pea aphid (Bates & Miln 1982) among cabbages with aphid and caterpillar prey (Ashby 1974; White et al. 1994; 1995), and among grapevines with mealy bugs (Charles 1981). *Melanostoma fasciatum* may be better adapted than *Melangyna novaezealandiae* to relatively unstable pastoral and crop habitats dominated by introduced aphids. Buckwheat (*Fagopyrum esculentum*), coriander (*Coriandrum sativum*), and lacy phacelia (*Phacelia tanacetifolia*) can provide pollen food for hover flies and *M. novaezealandiae* can be dominant on both the latter (Lovei et al. 1993) and pine trees with the woolly pine aphid *Pineus laevis* (Macfarlane & Denholm unpubl.). Known hosts for *Allograpta ropala* are small caterpillars and scale insects among flax and other foliage while unidentified hover-fly larvae feed on cabbage-tree moth caterpillars (Miller & Walker 1984). *Allograpta ventralis* preys on the cabbage-tree mealy bug (Bowie 2001). The immature stages of predatory hover flies, including endemic *A. ropala* (Miller & Watt 1915), have an indistinctly tapering body compared to the 'rat-tailed' larvae of Eristalinae that live in fluids. The long breathing siphon ('rat tail') of drone fly (*Eristalis tenax*) larvae contrasts with the short siphon of the flax-inhabiting *Psilota* (= '*Lepidomyia*') *decessa* (Miller & Walker 1984). Eggs of the predatory syrphine species tend to be laid among vegetation or grass seedheads and singly (Miller & Walker 1984).

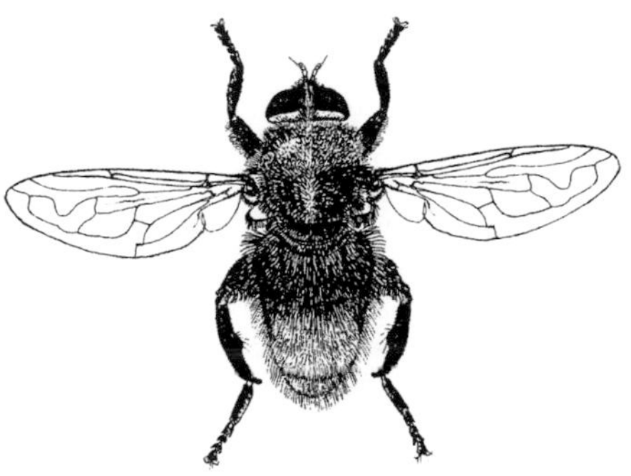

Narcissus bulb fly *Merodon equestris* (Syrphidae).
After Grant 1999

Phoridae

These are small flies, often black, and recognised by their distinctive dark and pale venation and head bristle pattern. The commonly encountered species mostly belong to the huge genus *Megaselia*, plus *Aphiura* and *Metopina*. They may be recognised as phorids by their jerky movements and humpbacked appearance. Identification of most genera is based on head bristle patterns, leg bristles, and wing venation (McAlpine et al. 1987). New Zealand has seven endemic genera (Disney, 1983; Brown & Oliver 2008), often with distinctive features including the head (Schmitz 1939). Phorids can be numerically abundant as scavengers or mushroom-feeders in open country (Macfarlane & Andrew 2001), and some introduced species may affect humans or they can be found in sand dunes (Brown & Oliver 2007). Eight species in New Zealand are introduced from Europe and Australia. Most native species live in forests, but part of the fauna is quite common in denser snow-tussock grassland. Didham (1992), for instance, recorded 20 species from one forest site. Different species may be found by sweep netting, malaise trapping, or by using baited traps or pitfall traps, but the family has not been investigated thoroughly enough to determine the extent of undescribed species.

Pipunculidae
Tonnoir (1925) described the blackish bigheaded fly species from sites with tussock grassland (three species) and bush (*Dasydorylas deansi*). At least two of the *Dasydorylas* species can be locally common (Macfarlane & Andrew 2001; LUNZ), but *Tomosvaryella novaezealandiae* is considerably less common (Macfarlane 2002; Skevington unpubl.). De Meyer (1991) described *Cephalops libidinosus* from New Zealand, which is also relatively well represented in collections. Only genitalia of the species are illustrated (Tonnoir 1925), but the reviews of Australian and North American species illustrate differences in the wing venation of the cosmopolitan *Dasydorylas* (then a subgenus of *Pipunculus*) and *Tomosvaryella*.

Lonchoptera furcata (Lonchopteridae).
Amber Sinton

Lonchopteridae, Sciadoceridae, Platypezidae
The remaining three families each have only one described species. Only the cosmopolitan small yellowish *Lonchoptera furcata* (Lonchopteridae) is often found in general collecting. Colloquially known as the spearwinged fly, it has a pointed wing tip with veins converging toward the tip (McAlpine et al. 1987; CSIRO 1991) and distinctive flanged larvae (Smith 1989). *Sciadocera rufomaculata* (Sciadoceridae) also occurs in Australia (CSIRO 1991). It is reddish with a dark patch on the wing tip (Oldroyd 1964; Brown 1992). It is somewhat reminiscent of Phoridae and feeds on carrion. Smoke flies (Platypezidae) are represented in New Zealand by *Microsania* and an undetermined species of Callomyiinae (Chandler 1994). Platypezids may need specialist collecting among fungi or smoke, and are expected to dwell in wetland or forests on the basis of overseas studies (Oldroyd 1964).

Suborder Brachycera: Division Cyclorrhapha: Section Schizophora

Schizophora can be divided into acalyptrate and calyptrate families. The calyptrate families (in which the haltere is typically concealed by a flap or calyptra) include a fairly compact group, while the acalyptrate groups (lacking the flap, so the halteres are clearly visible) are more disparate, being divided into several superfamilies, some of which show close affinities to calyptrate flies. Calyptrate flies are distinguished by a combination of features including antennal structure, wing venation, a suture across the top of the thorax, and the presence of upper and lower calypters (squamae) beneath the wing base (see Harrison 1959), although some of these features are sometimes lacking, especially in wingless forms. The most comprehensive guides, with illustrations of pupae, larvae, and eggs, are those of Ferrar (1987) and Smith (1989).

Acalyptrate flies
This division was last fully revised by Harrison (1959), who provided an excellent

account of all the known species. Some name changes have occurred since then and the work of Harrison (1976) and later reviews of particular families have added many new species. For recognition purposes, the acalyptrate flies can be artificially divided into the generally smaller flies that are often 2–6 millimetres long (e.g. *Drosophila* spp., *Cerodontha australis, Hydrellia tritici*) and a minority of larger species 8–16 millimetres long (e.g. some Coelopidae, Sciomyzidae) that are as large or somewhat larger than the house fly. The most extensive illustrations of whole flies of nearly all of the 29 acalyptrate families in New Zealand are in CSIRO (1991) and McAlpine et al. (1987).

Agromyzidae
These are small species (1.25–3.0 millimetres long) and the larvae are generally leaf miners. Some have a very narrow host-specificity, for example *Phytomyza clematadi* was released in 1996 for biocontrol of *Clematis vitalba* (old man's beard) (Fowler et al. 2006), and although it is now common and widespread its effectiveness seems to be adversely affected by parasites (Anon. 2007). The species most often encountered is *Cerodontha australis*, abundant everywhere, feeding on grasses and wheat. *Cerodontha angustipennis* can be abundant where hosts such as Yorkshire fog and tall oat grass are common (Martin 2007). Other abundant species include *Chromatomyia syngenesiae* (on *Sonchus*, cineraria, and other herbaceous Asteraceae), *Liriomyza brassicae* (on *Brassica* crops), *L. chenopodii* (on beet and spinach), and several *Phytomyza* species on *Hebe*. Two *Phytoliriomyza* species have fern hosts. Twenty-one species in various genera mine the leaves of angiosperm trees and shrubs including *Melicytus ramiflorus* (mahoe) and *Urtica ferox* (tree nettle) and species of *Olearia, Coprosma, Carmichaelia, Clianthus*, and *Clematis* or perennial herbs (species of *Ranunculus, Wahlenbergia, Cerastium, Lepidium, Gentianella, Plantago*, and *Juncus*) (Spencer 1976). Hosts are unknown for 13 of the 42 species (see revision by Spencer 1976; Macfarlane & Andrew 2001). Watt (1923) illustrated the pupae of nine agromyzid species.

Chloropidae
These are also small flies (1.25–5.0 millimetres long). Most species are endemic and there is wing reduction in a number of species of *Tricimba*, including some that are undescribed (see revision by Spencer 1977; Macfarlane unpubl.). Some subantarctic species have no wings. The genus *Gaurax* includes three common species (*G. neozealandicus, G. flavoapicalis, G. mesopleuralis*), often found in gardens. *Hippelates insignificans* often occurs in great numbers in semi-open country. *Apotropina tonnoiri* is one of the characteristic flies of beaches. Little is known of the larval habits of most New Zealand species. The introduced European species *Dicraeus tibialis* feeds on grass seed (Ismay 1991), endemic *Diplotoxa similis* feeds on seeds of six species of the snow tussock *Chionochloa* (Kelly et al. 1992; Cone 1995), and *D. moorei* feeds on hard tussock *Festuca novae-zelandiae* (Kelly et al. 1992). The grey *Conioscinella badia* can be common among grassland and beach areas, readily sucking on perspiration (Martin 1983) or weeping sores. Some of the other species have been associated with flowers, wetlands, rotting material, and subalpine areas (Spencer 1977; Macfarlane & Andrew 2001).

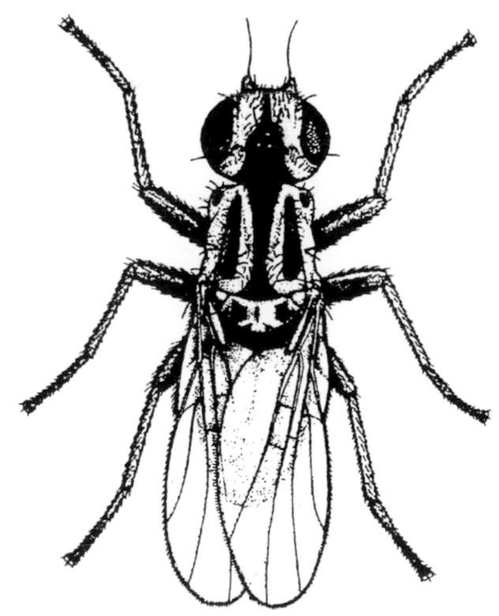

Diplotoxa similis (Chloropidae)
Jon Sullivan, Lincoln University

Ephydridae
Shore flies comprise a well-known family. An extensive revision has begun on the New Zealand fauna, with five new species described (Mathis et al. 2004; Edmiston & Mathis 2007). A distinctive character of the subfamily Ephydrinae is the strongly swollen lower face, while for most other genera in the other subfamilies the arista (a bristle on the last antennal segment) has long branches on the dorsal side only. Most of the common species of *Scatella* have a distinctly grey wing with a pattern formed by a few small white spots between the outer veins. Most species are found close to water, with larvae either aquatic (e.g.

Ephydrella) or semi-aquatic (e.g. *Scatella*). The genus *Hydrellia* is terrestrial and the larvae mine native and introduced grasses and rushes (Martin 2007). Australian *Hydrellia tritici*, with five proven host plants, is one of the commonest flies in New Zealand.

Drosophilidae

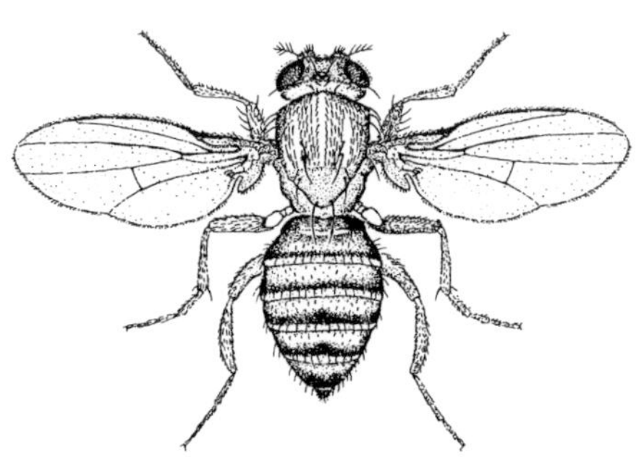

Fruit fly *Drosophila melanogaster* (Drosophilidae).
After Grant 1999

Pomace flies, vinegar flies, or fruit flies (not to be confused with tephritid fruit flies, below) are represented in New Zealand by two well-known genera and one or two that are largely unknown. As with the Sphaeroceridae (below), the number of drosophilid species in New Zealand is low, with combined totals for both families about 55 compared with 1130 for the entire Australasian/Oceanian region (Evenhuis 1989) and 189 for Britain (Chandler 1998). Members of the genus *Drosophila* are often known as fruit flies, although the actual association for many is with yeasts of ripe and rotting fruit. Colour illustrations of the introduced species in New Zealand can be found in Peterson's (1943) survey of species in southwestern USA. *Drosophila funebris*, *D. buskii*, and to a lesser extent *D. melanogaster*, *D. funebris*, and *D. hydei* are strongly associated with human settlement but in Texas they were much less common on dairy farms. Species such as *D. buskii* are associated more with rotting vegetables (including tomatoes) and grass clippings. Indigenous species are associated more with forest fungi. Traps with banana bait can be used to detect *Drosophila* populations. *Scaptomyza* includes six species, two of which are leaf miners (one of them, underscribed, in *Pratia angulata* (Campanulaceae)), and including adventive *S. flava*, which mines leaves of *Brassica* crops, *Tropaeolum*, and other plants (Martin 2004). *Paramycodrosophila* is represented by at least one unidentified species in New Zealand, sometimes found around exudates from holes of puriri moth caterpillars (*Aenetus virescens*). An unidentified species of *Mycodrosophila* was mentioned by Holloway (1976).

Heleomyzidae

The New Zealand species are placed in four tribes, one (Heleomyzini) comprising Palearctic species only, of which *Tephrochlamys rufiventris*, an elongate greyish fly with a reddish abdomen, is quite common, breeding in poultry manure. The South American *Prosopantrum flavifrons* is the sole representative of the Cnemospathidini and is found readily in open garden and pasture sites. The tribes Allophylopsini and Fenwickiini, with many endemic species, are commonly encountered in forests. Most familiar are the larger *Allophylopsis* species and the smaller *Fenwickia* species (Harrison 1959). *Xeneura* and *Aneuria* species have boldly patterned wings and some of the *Fenwickia* species have distinctive shading on the apical veins.

Lauxaniidae (Sapromyzidae)

Superficially resembling heleomyzids, these flies are readily distinguished by the lack of a vibrissa (a bristle on the lower front corner of the face) and details of the wing veins (Harrison 1959). There are four genera in New Zealand, including several common species of *Sapromyza*, which resemble oversized *Drosophila*. The genus *Trypetisoma* is distinctive for its patterned wings, similar to some of the Heleomyzidae and the tephritid *Trupanea*.

A species of subfamily Limosininae (Sphaeroceridae).
Rod Macfarlane

Sphaeroceridae

Lesser dung flies are small (1.5–3.5 millimetres long), dull black or brown species distinguished most readily by the short, robust hind metatarsus. Richards (1973) reviewed Australian species and provided a key covering many New Zealand species. The subfamilies Sphaerocerinae (*Sphaerocera*, *Ischiolepta*) and Copromyzinae (*Borborillus*) comprise only alien species, often encountered on dung (Macfarlane & Andrew 2001), although *Ischiolepta* in particular is often found in rotting vegetable matter such as grass clippings. Limosininae include both adventive and cosmopolitan species and few of them have any relationship with dung. They are easily distinguished from all other acalyptrates by their venation, with the veins from the discal cell not extending to the wing margin. Some species of Limosininae are extremely abundant, e.g. *Coproica hirtula* and *Spelobia bifrons*, both breeding in grass clippings and other decaying vegetation, and cosmopolitan *Opacifrons maculifrons* along freshwater shorelines (Marshall & Langstaff 1998). Steve Marshall is revising the endemic species, the most abundant of which belong to the genus *Phthitia* (Marshall & Smith 1992; Marshall et al. 2009). Some *Phthitia* species occur in open country but *P. plesioceris* is found along waterway shorelines and *P. emarginata* in rotting mushrooms. *Thoracochaeta* species are characteristic beach flies (Marshall & Rohácek 2000), found on seaweed and carrion. The two species of the endemic genus *Howickia* are wingless, and can be found actively jumping and scuttling about in disturbed forest floor litter.

Sciomyzidae

Marsh flies are relatively large flies (5–9 millimetres long), usually with many dark spots on the wings. The larvae feed on snails. *Neolimnia* species feed either on aquatic or terrestrial forest gastropods (Homewood 1981) while *Eulimnia* species feed on bivalves (Barnes 1980a).

Huttoninidae

Huttonina, formerly included in the Sciomyzidae, is now assigned to its own endemic family (CSIRO 1991) along with *Prosochaeta prima*. The species of these genera have a similar elongate shape to sciomyzids but *Huttonina* individuals are smaller, with two of eight species having many dark spots and three others with paler patterns (Harrison 1959). The wing of *Prosochaeta prima* is distinctive with a dark outer U band and a single inner band. Adults seem to be associated with shrubland or forest rather than open uncultivated grassland. The larvae of huttoninids are unknown.

Huttonina elegans (Huttoninidae).
Rod Macfarlane

Tephritidae

These picture-winged flies are small (3–5 millimetres long) but conspicuous creatures represented in New Zealand by three indigenous genera (*Austrotephritis, Sphenella, Trupanea* in subfamily Tephritinae) (Hancock & Drew 2003). Three *Urophora* and two *Procecidochares* have been introduced for biocontrol of thistle species, mist weed (*Ageratina riparia*) and Mexican devil weed (*Ageratina adenophora*) (Harman et al. 1996; Macfarlane & Andrew 2001; Hayes 2007). All five species introduced for the biological control of weeds have established (e.g. Cameron et al. 1989, Hayes 2007), but *U. cardui* is still uncommon and rather localized, partly because sheep graze the galls. White (1988) has an identification key for *Urophora* species in Great Britain, which includes illustrations of the wing pattern. *Urophora* and *Procecidochares* have two complete to three clear bands across the wings or the wing is largely clear with just a narrow dark band (*U. stylosa*). The native *Sphenella fascigera* almost has two bands with little or no spotting. By contrast *Austrotephritis* and *Trupanea* wings have at most one irregular band across the outer wing and many (20 plus) small to larger clear spots.

The name of fruit fly is applied generally to the family, although only a few genera feed on fruit. Some of these, including several *Bactrocera* species and *Ceratitis capitata* (Med fly), both in the subfamily Dacinae, are very serious pests in Australia, the Pacific islands and many other areas, causing fruit to disintegrate prematurely. A number of species are closely monitored in New Zealand and at least one accidental introduction has been successfully eradicated (Macfarlane & Andrew 2001). Some of these pest species are larger than native species (5–6 millimetres) and have distinctive wing patterns. All tephritids can be recognised by the disjointed wing, with the costa broken about a third of the way along the wing and the subcosta bent up toward this break at a right angle, together with the acutely pointed lower corner of the anal cell of the wing.

The New Zealand species can further be distinguished from similar-looking flies such as *Trypetisoma* (Lauxaniidae) and some Heleomyzidae (*Aneuria*, *Xeneura*) by their specific wing patterns, illustrated by Harrison (1959). Similar-sized *Huttonina* species (Huttoninidae) with spotted wings and *Zealandortalis* (Platystomatidae) with banded wings are also distinguished by wing pattern (Harrison 1959). Sciomyzids also have heavily spotted wings but they are larger than the endemic tephritids and the wing is unbroken. Endemic tephritids feed mainly on flowers or seedheads. Molloy (1975) found an average of 2.6–6.4 *Trupanea centralis* larvae per seed head in *Celmisia spectabilis*, with 64–96% of the seed consumed by these flies. He reared *Trupanea alboapicata* from *Raoulia mamillaris* and noted that Tephritidae larvae also affect *Senecio bellidioides*. *Sphenella fascigera* has a quite wide host range that includes native *Brachglottis kirki*, six species of introduced *Senecio* (Harrison 1959) and *Euryops pectinatus* (Keall pers. comm.). *Austrotephritis cassiniae* has been reared from *Cassinia leptophylla* (Harrison 1959), but the Asteraceae hosts for other endemic species remain unknown.

Heloscioinyzidae

This family comprises a southern hemisphere group of mostly forest-dwelling flies, centred in Australia and New Zealand and most recently revised by Barnes (1981). They are moderately large, resembling sciomyzids, but usually have unpatterned wings with prominent bristles along the costa. *Polytocus* is a subantarctic genus of two large species, of which the larvae of the Snares Islands species has been described (Barnes 1980c). Barnes (1980b) recorded *Helosciomyza subalpina* from ants' nests and described the larvae.

Scordalus femoratus (Helosciomyzidae).
Rod Macfarlane

Australimyzidae, Canacidae, Coelopidae, Helcomyzidae

Flies in these small families (28 species total) are confined to seashores or at least are more commonly found there. Kelp fly (Coelopidae) species breed in rotting seaweed and the beach stroller can be almost overwhelmed by their swarms. Whereas some species are widely distributed around the coasts others are confined to southern beaches. Adult kelp flies are dull brown with characteristic flattened bodies and bristly legs and a body length ranging from 3.5–13 millimetres (Harrison 1959; Grant 1999). The family Helcomyzidae (sometimes reduced to a subfamily of Dryomyzidae but here regarded as distinct following D. K. McAlpine, pers. comm.) includes the large endemic southern beach fly *Maorimyia bipunctata*. It is dark reddish-brown, about 10 millimetres long, and has two darker marks on the wing veins. It favours damp coastal sites (Harrison 1959). Canacidae are beach flies, which now include Tethininae. Eight *Zalea* species (Zaleinae, surge flies) have now been described from North and South Island rocky coasts and *Z. horningi* occurs on the South and Snares islands (McAlpine 2007). *Tethinosoma fulvifrons* is also a mainland species; it is a small (3.5 millimetres) grey fly with a deep cheek, widespread on sandy beaches where it is commonly on stranded seaweed with *Thoracochaeta* species (Sphaeroceridae). *Isocanace crosbyi* is so far known only from the Nelson area (Mathis 1999). Mathis and Munari (1996) ascribed the two subantarctic *Macrocanace* beach fly species to Tethinidae as well. Australimyzidae are small flies (1.5–2.5 millimetres long), dark brown, and suspected to be miners of seashore plants (Harrison 1959). They have a generally southern Australian distribution.

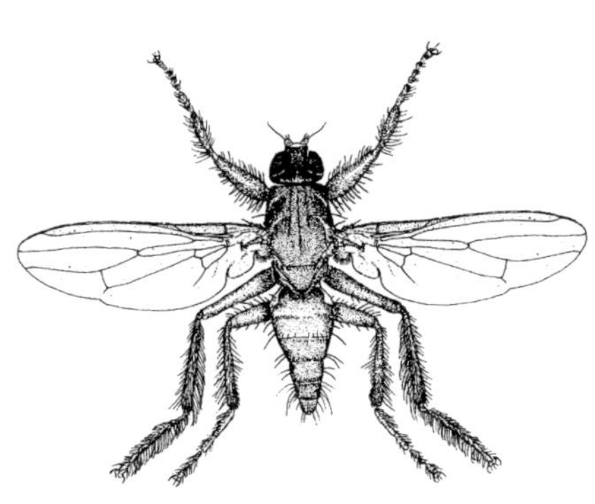

Kelp fly *Chaetocoelopa littoralis* (Coelopidae).
After Grant 1999

Pallopteridae, Platystomatidae, Asteiidae, Pseudopomyzidae, Milichiidae, Periscelididae, Teratomyzidae, Fergusoninidae, Anthomyzidae

These acalyptrate families together contain about 28 species, of which only about 10 have been described, several of them based on exciting recent discoveries. Seven of the nine species of Pallopteridae (*Maorina*) have a dark patch at the wing tip, which distinguishes them from all small acalyptrate flies in New Zealand except for the ephydrid *Scatella nubeculosa* (Harrison 1959). *Maorina* species are mostly found in forests, and occasionally in gardens, but their biology is unknown. Platystomatidae are picture-winged flies that sometimes associate with dung as adults. Asteiidae are easily recognised by their venation, having a short radial vein and no posterior crossvein. *Asteia* species are shiny brown flies only 23 millimetres long. Malaise trapping can gather modest numbers in broadleaf forest in coastal areas. Members of the small southern hemisphere family Teratomyzidae (fern flies) feed on ferns (McAlpine & Keyzer 1994) but the larvae of the New Zealand species are unknown. They have two short crossveins close to the wing base (Harrison 1959), which readily distinguishes them from other acalyptrates. *Teratomyza neozelandica* is an elongate yellowish-brown fly about 2.5 millimetres long, common in native forest. All the described New Zealand species of Pseudopomyzidae are in the genus *Pseudopomyza*. They are tentatively associated with bird dung but are also attracted to rotten fruit (Harrison 1959). They are shiny dark brown to black and 2.0–2.5 millimetres long. Milichiidae and Periscelididae have only recently been recognised in New Zealand, and the species have not yet been described. The family Fergusoninidae has likewise only recently been found in New Zealand (Taylor et al. 2007). *Fergusonina metrosiderosi* forms galls on pohutukawa (*Metrosideros excelsa*) in an obligate mutualistic association with the nematode *Fergusobia pohutukawa*. Described in the same year was *Zealanthus thorpei*, a monotypic endemic genus representing the first species of Anthomyzidae from New Zealand (Rohácek 2007). These records considerably extend the known distribution of these two families in Australasia.

Pohutukawa gall fly *Fergusonina metrosiderosi* (Fergusoninidae).
Birgit Rhode, Landcare Research

Sepsidae, Psilidae, Piophilidae, Chamaemyiidae, Cryptochetidae

Each of these families has only one or two introduced species. Dark, metallic Australian *Lasionemopoda hirsuta* (Sepsidae) is one of the larger acalyptrate species (body about 6 millimetres long) of New Zealand pastures, where it is attracted to the dung of livestock (Evenhuis 1989). It is found throughout most of the North Island and has recently reached Canterbury. It is recognised by the ant-like constriction at the base of the abdomen and the round head with no palps. Psilidae is quite a large family of herbivores (Evenhuis 1989), but the European carrot fly *Chamaepsila rosae*, illustrated in Somerfield (1982) and Scott (1984), is the only species in New Zealand. Piophilidae have larvae known as cheese or bacon skippers. Adults are shiny blackish and small (body 3–4 millimetres long) (Harrison 1959). Three species of Chamaemyiidae have been introduced to New Zealand for control of scale insects among fruit crops (*Leucopis* spp., Europe, South America) and gum trees (*Pseudoleucopis benefica*, Australia) (Cameron et al. 1989; Evenhuis 1989). Two of these species have established but were not mentioned in Harrison (1959). A further South American species, *Chamaemyia pilipes*, illustrated by McAlpine et al. (1987), is found at least in Canterbury (Bejackovich et al. 1998). These small (1–4 millimetres long) stout flies, silvery grey to brown, resemble a small lauxaniid. Tiny (1.5–1.75 millimetres), stout and shiny black *Cryptochetum iceryae* (Cryptochetidae) feeds on woolly aphids. These Australian flies have the distinctive features of a tiny arista on the enlarged third antennal segment (Harrison 1959), short abdomen, and large scutellum (Miller & Walker 1984).

European carrot fly *Chamaepsila rosae* (Psilidae).
Lincoln University

Calyptrate flies

This category includes house flies, blow flies, and related groups, including some vertebrate ectoparasites, as well as species whose larvae are endoparasitoids in insects or other invertebrates. Prior to the arrival of Europeans there were only five families of calyptrate flies in New Zealand – Muscidae, Fanniidae, Calliphoridae, Tachinidae, and the ectoparasitic Hippoboscidae. Investigations into the undescribed diversity of Muscidae (R. A. Harrison, R. P. Macfarlane) and Tachinidae (J. S. Dugdale), together with a report on Fanniidae (Holloway 1985) and a review of Calliphoridae (Dear 1985) demonstrate that species diversity of these calyptrate families averages 79% of the species diversity of these families in Britain. Despite clear evidence of limited colonisation of calyptrate flies in New Zealand compared to such a recently isolated island as Britain, the families that reached New Zealand have broadly comparable fly species diversity. This check is important in assessing the significance of area comparisons of species diversity to predict what diversity might exist in New Zealand (Macfarlane & Andrew 2001).

Muscidae

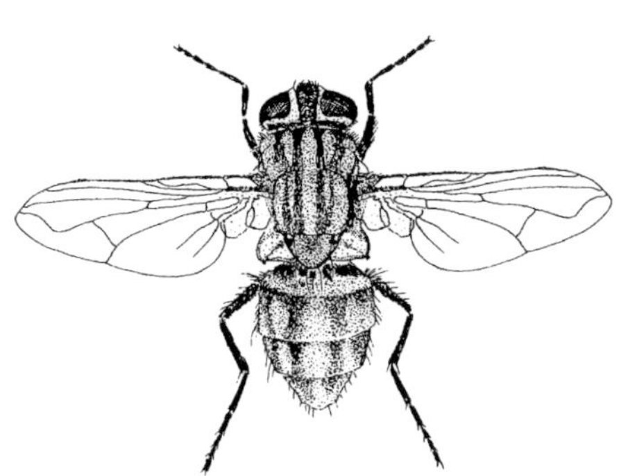

House fly *Musca domestica* (Muscidae).
From Grant 1999

These include four common introduced species and perhaps 134 endemic species. The introduced house fly *Musca domestica*, stable fly *Stomoxys calcitrans*, and false stable fly *Muscina stabulans* are all greyish, while the similar-sized *Hydrotaea rostrata* is shiny black (see Miller 1939; Miller & Walker 1984). Only *Musca domestica* has a medial vein that bends forward sharply to almost close the central cell, as in most blow flies and tachinid flies. However, the medial vein bends only moderately towards the wing tip in these other three introduced species (Miller 1939) and in endemic *Calliphoroides antennatis*, which resembles a blue blow fly or some *Pollenia* species. The orange antenna of *C. antennatis*, a species that breeds in rotting plant material, distinguishes it from these other partly metallic blue species. This vein is straight in the other New Zealand species. Many, if not all, ceonosiine adults are predaceous (Evenhuis 1989), based on observed habits (e.g. Harris 1990) and the presence of teeth at the outer edge of the tongue (Carvalho 2002). Females of the *Millerina* group of '*Spilogona*' and the *Exsul* species have 4–12 distinct black bristles on the ovipositor tip, unlike the rest of the endemic Muscidae. These more rigid tips may enhance penetration during egg-laying into gravelly sand and rotting logs. Coenosiine larvae may well prey on other insects such as shore fly larvae in moist, fine, gravelly to muddy waterway margins and invertebrates in rotting wood, judging by the concentrations of some species in forests and others among seashore driftwood. Species occupy various niches in estuaries, wetlands, and streamsides, extending up to the alpine zone. Many adults seem to need non-shaded waterway margins. The genus *Limnohelina*, whose somewhat elongate, smallish grey species often have patterned wings, are often associated with open, silty, freshwater shorelines. The diversity of forest-inhabiting species remains obscure at present. Most other endemic species are currently undescribed or included in '*Spilogona*'.

Tachinidae

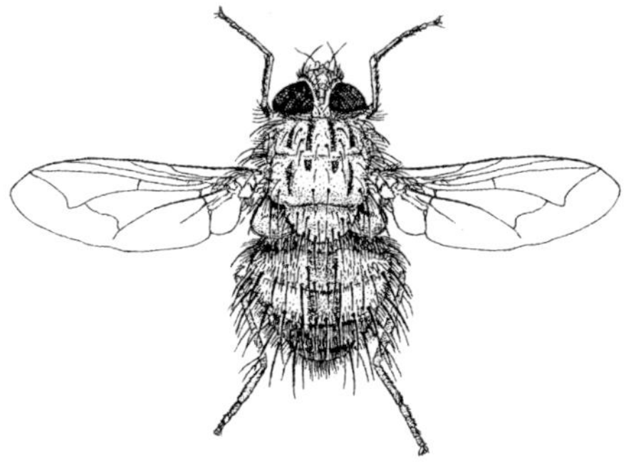

Porina parasite *Protohystricia alcis* (Tachinidae).
After Grant 1999

Tachinids parasitise other insects and the extent of information on the hosts of tachinids has already been summarised above in the section on beneficial insects. The Australian leafroller parasitoid *Trigonospila brevifacies* has become a successful natural enemy of pest species in the northern half of New Zealand but the establishment of introduced biological-control agents against grass-grubs has yet to succeed (Cameron et al. 1989; Munro 1998). The accidentally introduced Australian species *Chaetophthalmus bicolor* adds to the New Zealand arsenal against noctuid pests. Tachinids generally have bristly abdomens and similar wing venation to Sarcophagidae and Calliphoridae. Several tachinids, including species of *Pales*, *Occisor*, and *Perissina*, have a semimetallic blue abdomen like blue blow flies. Tachinids are distinguished from these families by a pouchlike bulge under and behind the scutellum. The bristly abdomen and the simple unfeathered arista will also often distinguish tachinids from the other families. There is a high level of endemism amongst the New Zealand genera, although some, such as *Pales*, with many New Zealand species, are widespread in the Australian and Oriental regions. About 180 species are recorded here and the family is particularly well represented in alpine regions. A review by Malloch (1938) is fairly extensive and quite well illustrated, but considerable effort is needed to describe the remaining undescribed genera and species.

Calliphoridae
Blow flies have been extensively investigated in New Zealand (Ramsay & Crosby 1992) because of their economic significance to pastoral farming. They range from well-known *Calliphora* and *Lucilia* species of farms to poorly known *Pollenia* species, which Dear (1985) revised. The three most prominent pest species, including recently arrived *Lucilia cuprina*, are all accidental introductions from Australia or Europe. Blow flies were one of the better-known groups of insects among Maori (Miller 1939). Among calliphorid blow flies, *Pollenia* is prominent, with 33 native New Zealand species (Dear 1985) compared to only eight in Britain (Chandler 1998) and three immigrant species in North America (McAlpine et al. 1987; Rognes 1987). Native species inhabit lagoonal to subalpine grassland habitats, but whether native species parasitise ground-inhabiting worms has yet to be resolved. The family Mystacinobiidae, erected for the New Zealand bat fly, which feeds on bat dung, is now included as subfamily Mystacinobiinae of Calliphoridae.

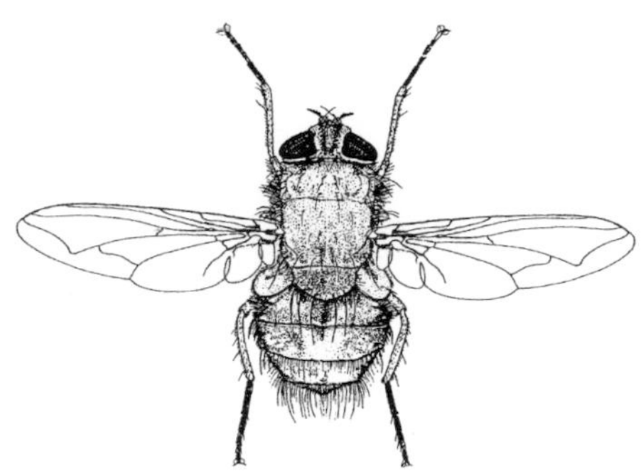

Brown blowfly *Calliphora stygia* (Calliphoridae).
After Grant 1999

Sarcophagidae
Dung flies are represented by the ubiquitous open-country *Oxysarcodexia varia* (previously known as *Sarcophaga milleri* or *Hybopygia varia*) from South America (Macfarlane & Andrew 2001). The distribution of the other four predominantly coastal species is inadequately known, and at least the larger *Sarcophaga crassipalpis* does not seem to have spread to Canterbury. There is also an unconfirmed report of an endemic species of the subfamily Miltogramminae, based on one specimen in an overseas collection.

Fanniidae
These include the lesser house fly *Fannia canicularis* and other less well-known species. Their distinctive pronged pupae (Miller & Walker 1984) have been investigated and a key produced for the species on the main islands (Holloway 1985). These flies tend to be associated with birds and bumble bee nests (Miller & Walker 1984; Holloway 1985; Macfarlane unpubl.) where they probably scavenge on carrion, broken eggs, and other detritus.

Anthomyiidae
Species include *Botanophila jacobeae*, introduced to control ragwort (Holloway 1983), the cornroot maggot *Delia platura*, a pest of gardens, and *Anthomyia punctipennis*, a common species distinguished by its dark wing-spot pattern (CSIRO 1991). In addition, one or two adventive species of the genus *Fucellia* are associated with decomposing kelp on beaches (Oldroyd 1964). Another species of *Delia* has also been recorded. All species are small (about 3–6 millimetres), and distinguished from other calyptrate flies by an inner hind vein (anal vein) that reaches the wing margin.

Louse fly *Ornithomyia* sp. (Hippoboscidae).
From Grant 1999

Oestridae, Hippoboscidae
These are the ectoparasitic bot flies and louse flies, respectively. The bibliographies of Miller (1956) and Ramsay and Crosby (1992) list the popular summaries on livestock fly pests in New Zealand. Macfarlane and Andrew (2001) reviewed what is known of the four species of louse flies that affect birds.

The Tipulidae or crane flies

Known as crane flies and daddy-long-legs, tipulids are commonly encountered by New Zealanders. Only a few species are strikingly colourful, mainly in the genera *Aurotipula* and *Chlorotipula*, but the patterned wings of many are delicate filigrees of golden-brown that would delight any miniaturist.

Some species can swarm and be a nuisance, but they do not bite like mosquitoes and have no appreciable effect on humans. The larvae of most species are very important in the breakdown of vegetable litter in forests, swamps, bogs, and streams. A few species (especially of *Gynoplistia*) are predators of worms and larvae. There is one record of a *Leptotarsus* species attacking potatoes but it has certainly not reached the pest status of other tipulines (*Tipula* spp.), as in Europe and North America. There is a great need for life-history studies to elucidate energy requirements relative to emergence, flight, egg production, and the phenomenon of wing reduction. The New Zealand fauna provides a wealth of exemplars.

Taxonomic history

In the late 1800s, collectors sent some specimens to European workers and some species were described at that time. In New Zealand, G. V. Hudson and F. W. Hutton also described a few species at the end of that century (4% of the present total). Hudson sent specimens to F. W. Edwards (British Museum) who provided the first major work on the group. His two publications (1923, 1924) account for 16% of the native species, and his keys and figures and the species are still useable today. His material is in the Natural History Museum, London.

Charles P. Alexander, of Amherst, Massachusetts, described most (74%) of the New Zealand species. He received specimens from a number of amateurs, especially two railway workers, two council water-supply employees, a dentist, and several professional entomologists. He engaged in considerable correspondence with them and always asked of their family life and for the biography, education, and of course collecting details of the flies. He carefully conserved those notes (about 3000 pages), now deposited in the U. S. National Museum. These could now give a deep insight into many aspects of New Zealand life and the professional politics of the 1920s to 1950s. Alexander gave short descriptions of each species, concentrating on aspects of colour, wing form, antennae, and male terminalia. These descriptions frustrate today's researchers by their inadequacy, yet one can only admire his conceptual ability as he rarely had to redescribe a species. Most of his types are in the American Museum of Natural History, Washington. Although there has been no detailed discussion since Oosterbroek (1989) the list given here does not follow his in its entirety. The extremely large genus *Limonia* (> 2200 species) is usually subdivided into subgenera (~ 63 for global coverage). Many of these are often informally used as genera, and such is the case here for the distinctive subgenera that occur in New Zealand – *Discobola, Idioglochina, Metalibnotes, Nealexandraria,* and *Zelandoglochina*. The others remain in *Limonia* (*Dicranomyia*), but there are several species, both named and new, that could be better placed in other subgenera or genera and one possibly belongs to a new genus. Likewise treated as genera are the *Leptotarsus* subgenera *Aurotipula, Brevicera, Chlorotipula,* and *Maoritipula,* and the *Trimicra* subgenus *Erioptera*.

Strictly, all these changes in species combinations would be new, or in some cases reverting to combinations used in the 1920s. All the subspecies listed in Oosterbroek (1989) are treated as full species, and eight species names are relegated to junior synonyms based on recent re-examination of type specimens (senior synonym first) – *Amphineurus campbelli/nox, Atarba connexa/confluenta, Discobola ampla/milleri, Chlorotipula elongata/angustior, Leptotarsus atridorsum/harrisi, Limonia monilicornis/gracilis, Zelandoglochina atrovittata/laterospina,* and *Z. cubitalis/huttoni*. Although synonymised by Hemmingsen and Johns (1975), the species *Zelandoglochina canterburiana* and *Z. allani* were listed separately by Oosterbroek (1989). The synonymy of the latter is also reconfirmed through examination of the types (P. Johns pers. obs.). It is unlikely that there will be many further synonymies (probably fewer than five). Several species have been the subject of recent reviews (Vane-Wright 1967; de Jong 1987; Johns & Jenner 2005, 2006) and *Trimicra* may soon be covered (Andrew 2000).

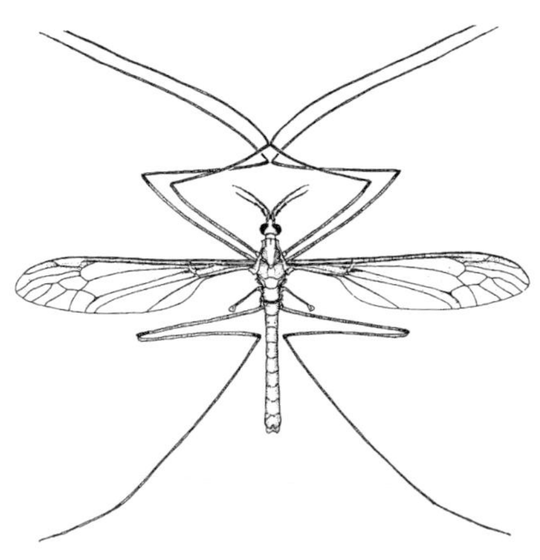

Male crane fly Chlorotipula albistigma (Tipulidae).
After Grant 1999

Some 46 new crane fly species from New Zealand are indicated in the checklist. All are so distinctive (mainly by their short wings) that they cannot be confused with any known species. The number of new species that have less noticeable characters cannot even be estimated from the meagre collections that have been made in recent years. The present state of taxonomy is still in 'catch-up' phase and even some described species, especially of the very large genera *Gynoplistia*, *Leptotarsus*, *Limonia*, and *Molophilus* are yet to be recognised in these modern collections.

Biology

Adult mouthparts are very varied. Most are simple, with the labiae or lips used for dipping into plant juices that are then sucked in. Three species have a greatly elongated tube-like proboscis that may well be used to get into nectaries of narrow, deep flowers such as those of *Dracophyllum*. *Zelandoglochina* species also have elongate mouthparts and are able to cut into the bases of flowers to sup the nectaries (of *Dracophyllum*, *Olearia*, and *Hoheria*). Larval mouthparts are also very varied. Those of the Tipulinae are strong and well muscled and easily chew through soils and litter, taking in much inorganic matter to which various unicellular algae, or fungi and bacteria are attached. *Zelandotipula* larvae browse on the last remnants of decayed vegetable matter in the margins of streams and seepages and are able to digest this material through having a very high alkaline pH in their gut, in striking contrast to the low-pH acidic stomachs of most other animals (Trought 1982).

Behaviours are poorly known. One species, *Rhamphophila sinistra*, flies in a vertically undulating path for several metres through the forest after being disturbed (Johns & Jenner 2005). It then folds its wings and suddenly drops to the forest floor where its brownish colours merge so well with the leafy litter. Two species of *Dicranomyia* vibrate quickly up and down on their legs – one while hanging on spiders' webs (the stickiness has no effect on the hairy legs), the other on the walls of seashore overhangs and caves. Wingless *Gynoplistia pedestris* sits on swamp grasses and sedges where it is a perfect mimic of tetragnathid spiders, with two legs forward and four backwards. Many species are crepuscular or night fliers, the most noticeable being species of *Discobola*, *Zelandoglochina*, and *Zelandotipula*.

Several species lists have been published for specific sites or presented as internal reports to forestry and conservation organisations. A classic is that of Alexander (1929) for Tongariro National Park and the most recent is by Ward et al. (1999). Up to 50 species may be found in any one forest site and it is strongly suspected that their larvae are sequentially partitioning the habitat as the adults merge in sequence over a relatively short period during spring through autumn. There are even a few winter-emergent species.

Distributions within New Zealand. Very high endemism usually points to very restricted distributions, speciation having taken place under the influence of a wide variety of environmental parameters. Such is the case in most New Zealand insects, but not so for the Tipulidae. All but the reduced-wing and alpine species have fairly broad distributions. Indeed, it is very difficult to envisage the parameters that may have stimulated divergence in the many widespread forest and tussockland species of *Gynoplistia*, *Leptotarsus*, *Limonia* and *Molophilus*.

There is a wealth of freshwater species. *Limonia nigrescens* inhabits sodden wood at or below the water surface (Anderson 1982) and *Limonia hudsoni* is a moss eater in montane streams (Suren & Winterburn 1991). The latter is widespread and often abundant, with the adults seen near fast-flowing well-oxygenated streams, usually in bright conditions. Such streams range from the former Kaikoura town water-supply overflow that cascaded over the limestone cliff adjacent to the Edward Percival Field Station at sea level, to alpine streamlets at 1000 metres or more in Arthur's Pass National Park and adjacent mountain ranges. The larvae live in crevices under or amongst the mosses and filamentous green algae near and in riffles and waterfalls. Within this habitat, they construct gelatinous tubes, binding the algal strands and detritus with a mucoid layer. There seems always to be a 'retreat', a crack in the rock, to which the larva goes when disturbed. Prior to pupation, larvae leave the water for a dryer, sheltered position and there secrete a mucoid lump, a mass of jelly; often several (1–6) larvae may occupy this lump.

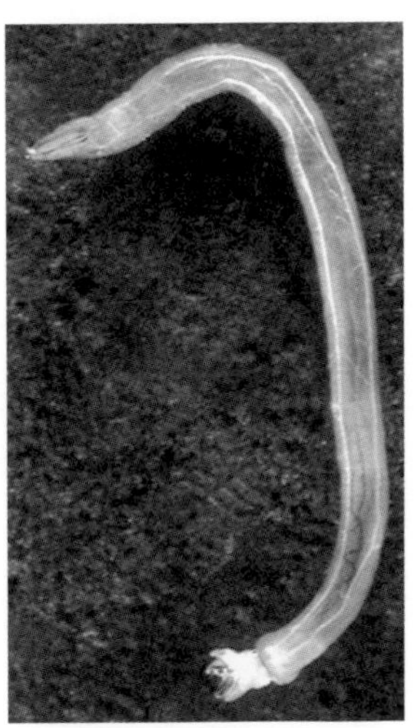

Aquatic larva of *Molophilus* sp. (Tipulidae).
Stephen Moore, Landcare Research

Aquatic pupa of *Aphrophila* sp. (Tipulidae).
Stephen Moore, Landcare Research

Pupation takes place within it and it is possible that other larvae may enter the lump after it has been formed. One example contained larvae and pupae at very different stages of development that must have been days or weeks apart. One pupa was kept for a week before the adult emerged. The adult sits flat on any nearby sheltered surface, apparently preferring the rough, broken surfaces of the rock banks or stony cliffs. When undisturbed the legs are almost completely spread and the body almost touches the ground. When disturbed the legs are shifted to raise the body and the animal then 'rocks and bounces'. Mating takes place 'tail to tail' on the same surfaces. Egg-laying has not been witnessed. *Limonia hudsoni* and *L. sperata* often occur together at higher altitudes.

Winterbourn (1996) described the larva and habitat of *Aphrophila neozelandica*. At least two other species of *Aphrophila* have been taken with it in light-traps and in all likelihood all species of *Aphrophila* are in flowing fresh waters. Winged and brachypterous species of *Rhabdomastix* occur in the sandy flats of many rivers. Adults of reduced-wing species in the Wanganui River at Harihari are able to withstand rapid rises of this flood-prone river. They hide under stones and wood and may remain submerged in up to a metre of water for hours if not days. As the flood recedes they emerge to run over the new sand banks and mate. The larvae are presumed to be in the margins of the normal river and nearby temporary ponds. *Paralimnophila skusei* is a widespread species, the larvae of which have regularly been taken amongst roots and bases of stream vegetation. Their mouthparts suggest that the larvae are predatory. Similarly, larvae thought to belong to *Gynoplistia* species have been taken in alpine seepages. One was seen feeding on an earthworm.

Endemic species

The one crane fly that is virtually worldwide, *Trimicra pilipes*, has had many forms described from various places around the world. These have been recognised at various taxonomic levels – species, subspecies, or varieties – and present usage favours the last (Alexander 1962). Only in New Zealand are there distinctive related species with different body form or habitats occupying brackish water/soil, highly organic sites (e.g. cowshed drains), and wet guano of bird colonies. The bird colonies of four subantarctic islands have their respective species (Alexander 1962, and two undescribed species).

The fauna in general is related to that of Australia, New Caledonia, and Chile. *Sigmatomera rufa* has a close relative in eastern Australia, and should a few species of *Macromastix* be transferred to *Leptotarsus* (*Habromastix*) then there would be a stronger connection with Australia. Several *Limonia* species are very similar to those in New Caledonia and Chile. *Zelandotipula* is rather odd in that all its close relatives are in the northern parts of South America. However, most others seem to be sufficiently different from known species at subgenus and species levels, suggesting that the New Zealand fauna has long been isolated. *Amphineurus*, *Gynoplistia*, *Limonia*, and *Molophilus* have seemingly explosively speciated in New Zealand, occupying unusual habitats, and many are flightless.

Although it is assumed that all species are native, many are tolerant of human environmental changes. *Limonia vicarians* has apparently emerged from household potplants and it is certainly common in household gardens. *Leptotarsus dichroithorax*, *L. obscuripennis*, *L. tapleyi*, and *L. zeylandiae* are all present in grassland-shrubland or pine plantations (Johns et al. 1980). *Chlorotipula albistigma* lies in rotting *Pinus radiata*, willow, and other exotic trees. Recently, *Leptotarsus* larvae have been studied to determine the effects of heavy-metal pollution on soils after distribution of treated sewage in pine plantations (Denholm pers. comm.).

Adventive species

There are no recorded adventive species in New Zealand, although one, *Trimicra pilipes*, is distributed almost worldwide and it favours wet to swampy, much modified, pastoral habitats.

Speciose genera and flightless species of New Zealand Tipulidae

Genus	Known species	New species	Flightless species
Amphineurus	44	1	0
Austrolimnophila	20	3	2
Gynoplistia	101	1	13
Leptotarsus	53	1	10
Limonia	62	12	9
Metalimnophila	18	2	2
Molophilus	92	1	1
Rhabdomastix/Limnophila	27	4	4
Subtotal	390	21	37
Other genera (37)	178	23	14
Unrecorded genera (2)	0	2	0
Totals (48)	**568**	**50**	**55**

Special features

Without doubt, the two features of New Zealand crane flies that separate them from the rest of the world are the inordinate number of flightless species and the richness of the marine intertidal fauna. Flightlessness is the result of a degeneration sequence. Firstly there is a reduction in the diameter of the muscle fibres, then loss of muscles and wing stenoptery (narrow aspect ratio), and brachyptery (short, flabby wing). Finally there is the almost total loss of wings (aptery) and the halteres may even be reduced to small non-functional stubs. In the female, eggs may occupy the thoracic space that holds the wing muscles of flighted species.

Brachyptery occurs in all habitats – marine intertidal, swamps and seepages, open stony and sandy flats of braided rivers, brush and tussock vegetation, dense forests, subalpine tussock, and alpine fellfield. With the progression to aptery, there is a concomitant increase in leg length and the animal has a long-legged, spider-like (or more like an opilione harvestman) form. It may occur in one sex (female) or both. Muscle atrophy is seen in a new species of *Leptotarsus* from the Snares Islands in which muscle diameter is reduced to approx 70% of similar members of the genus. The males are just able to fly, but the females are unable to do so until they have shed most of their eggs. *Leptotarsus zeylandiae* females have no muscles and are brachypterous, those of the males being fully formed. Females emerge from the soil late in an autumn afternoon, and, within a few hours, mate, lay their eggs, then die, even before they become fully sclerotised. The males swarm, especially if there is mist or light rain, and live for another day. The flight period is about four weeks in northern populations but shorter in southern populations. The condition of stenoptery and moderate brachyptery in both sexes occurs in about 10 *Leptotarsus* species, several *Gynoplistia*, *Austrolimnophila stewartiae*, *Trimicra campbellensis*, and in a magnificent 30 millimetre-long new species of *Zelandotipula*. Near-total wing loss is seen in several species of *Gynoplistia*, at least two *Metalimnophila*, *Molophilus*, *Zaluscodes*, *Tonnoiraptera*, two *Austrolimnophila*, three *Trimicra*, several *Rhabdomastix* and *Limnophila*, and at least nine species of *Limonia* (*Dicranomyia*). An estimated 10% of the New Zealand crane fly fauna is flightless.

The halophiles (found in saline, maritime areas) present interesting distributions. There are at least 10 species that are well adapted to the marine environment (Wise 1965; Hemmingsen & Johns 1975). Several (*Idioglochina fumipennis*, *I. kronei*, *Limonia nebulifera*, *Zelandoglochina canterburiana*, *Rhabdomastix* n. sp., and possibly *Geranomyia* n. sp.) live within the intertidal zone and inhabit various types of thick encrusting algae (*Corallina* sp., *Bostrychia* sp.). *Limonia kermadecensis* (also present on the mainland, near Auckland at least) makes a short burrow or occupies a small depression into soft rock in

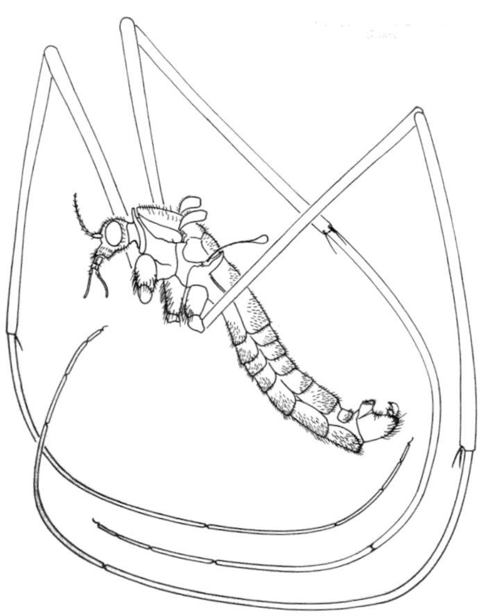

Leptotarsus sp., a wingless cranefly.
John Dugdale

the high intertidal zone. It encloses itself by secreting a silky cover to which stick sand grains. Larvae emerge at night to feed on fine algae by ingesting sand grains. It can achieve numbers of 100 per square metre and thus could be a minor degrader of soft mudstone shorelines in the north of New Zealand. A close relative, *Limonia fulviceps*, is known from saline marshes at Nelson and the head of Pelorus Sound near Havelock. *Limonia subviridis* has been reared from the intertidal red alga *Gelidium pusillum* (Wise 1965). Three undescribed flightless species are known respectively from Fiordland, Bluff, and Stewart Island, from the Snares, and from the wind- and spray-swept Bounty Islands, and a new *Rhabdomastix* occurs at the Antipodes Islands. *Limonia nebulifera* belongs to a very distinct group of halophiles. Two species are found on southern and eastern Australian coasts and another, its closest relative, in Chile. *Idioglochina* has representatives in Australia, several Pacific Islands, and Japan.

Dipteran zoogeography and speciation

The break-up of the southern landmass of Gondwana and the northern movement of New Zealand are considered to have occurred in the Cretaceous, around 110 to 85 million years ago during which the Tasman sea was formed (Stevens 1985). Stevens et al. (1995) summarised the changes in vegetation and climate that are understood to have occurred when flies began to occupy New Zealand, but the relevance of these changes to the present-day fauna is moot given the uncertainty regarding the possible complete submergence of the Zealandia continental mass at the end of the Oligocene, 23 million years ago (Landis et al. 2008). In a discussion of zoogeography and taxonomic methods, Hennig (1966) and Brundin (1967) advanced the theory of checking sister-group relationships between widely separated continental masses using New Zealand Diptera as an example. Hennig (1966) recorded 23 shared genera confined to Australia–New Zealand–South America and 15 genera shared between New Zealand and Australia. In the division Cyclorrhapha, Hennig (1966), Vockeroth (1969), and McAlpine (1991) reported a complete lack of transantarctic generic/subgeneric sister groups between New Zealand–South America and Australia–New Zealand–South America. However, *Ephydrella* is catalogued with an Australian, New Zealand, and Brazilian distribution (Mathis & Zatwarnicki 1995) and Amorim and Silva (2002) commented that there could be other, unrecognised, examples.

Hennig (1966) tabulated 14 genera with species diversity confined to New Zealand and Chile/South America and 138 endemic New Zealand genera. The 12 families of Nematocera that have been adequately studied show the following examples of genera confined to New Zealand and South America – *Aphrophila, Campbellomyia, Limnophilella, Zelandoglochina, Zelandotipula* (Tipulidae), *Nervijuncta* (Ditomyiidae), *Aneura, Parvicellula* (Mycetophilidae), *Isoneuromyia* (Keroplatidae), *Canthyloscelis* (Canthyloscelididae), *Corethrella* (Corethrellidae), and *Oterere* (Thaumaleidae). Jaschhof and Didham (2002) described the Rangomaramidae as a new family for the world. However, studies of South American species have enlarged this family to include three new subfamilies (Amorim & Rindal 2007), which allows most of the previously unplaced genera from New Zealand to be included in this family which has a relatively high southern-hemisphere diversity.

Among the larger Brachycera families, the only New Zealand–South American links are for *Filatopus* (Dolichopodidae) and *Oropezella* (Empididae). For the four smaller brachyceran families, there are closer affinities with Australia for Tabanidae and Bombyliidae (Hennig 1966; Greathead 1988), but for Acroceridae the more apparent affinities are with American genera (Paramonov 1955; Hennig 1966) or species groups (Schlinger 1960). Paramonov (1955) noted that New Zealand Acroceridae have their closest links with South America and not Australia and he discussed absences of fly families in New Zealand that cannot tolerate colder climates. In addition, information on the Diptera from the offshore islands has begun to demonstrate further families that are absent or depleted

Approximate numbers of described and recognised species of Coleoptera, Diptera, Hymenoptera, and Lepidoptera in New Zealand

Order	Number of described species	Number of recognised species
Coleoptera	5,235[M]	6,735[M]
Diptera	2,540[Mc]	3,213–3,230[Mc]
Hymenoptera	721[B]	1,535–1,649[B]
Lepidoptera	1,684	1,698

Sources: Maddison (this chapter), Macfarlane et al. (this chapter), Berry (this chapter), Dugdale (1988, with the addition of 22 adventive species recognised since, R. Hoare pers. comm.).

in diversity compared to the three main islands of New Zealand (Macfarlane & Andrew 2001).

Since 1966, revisionary studies have clarified affinities in four larger Brachycera families. For the Empididae, there are some closer links with Chile than for Australia, apart from links between the southern lands and subantarctic islands in the subfamilies Clinocerinae and Trichopezinae (Sinclair 1995). The genus *Heterophlebus* is one such example (Sinclair 1995), along with the 80 or so world species in the subfamily Ceratomerinae centred in New Zealand (Sinclair 1997b, unpubl.). Seashore-inhabiting *Thinempis* (Empidinae, Hilarini) species have affinities with Australia and an undescribed species from New Caledonia (Sinclair unpubl.). Species diversity in New Zealand is centred on the tribe Hilarini rather than Empidini in the Northern Hemisphere (McAlpine et al. 1981; Bickel 1996; Chandler 1998). New Zealand should also have a relatively diverse lot of *Chelipoda* (Plant 2007) even after description of new species from South America, New Caledonia and possibly Australia is achieved. An extensive examination of allied genera of robber flies in New Caledonia, Australia (Daniels 1987; Macfarlane unpubl.), and Chile (Artigas 1970) has revealed closer links with both Australia and New Caledonia for Asilinae genera, but there is very limited generic diversity compared to Australia. Two New Zealand species provisionally assigned to *Rhabdotoitamus* are clearly congeneric with Australian, New Caledonian, and Norfolk Island species. A second, much more distinct Australasian genus (undescribed) has only one certain described species, viz '*Cerdistus*' *lascus* from the North Island, plus a very similar undescribed species from New Caledonia. *Saropogon* (subfamily Dasypogoninae) is somewhat unusual for the Southern Hemisphere because this cosmopolitan genus (Hull 1962) has more species in New Zealand than the Americas or Australia (Artigas 1970; Evenhuis 1989). Malloch (1928), however, noted that New Zealand *Saropogon* species differ from the American ones in hair and bristle patterns on the head.

Of 23 genera with 15 or more species, nine are cosmopolitan, in which the number of New Zealand species is about the expected diversity given the size of New Zealand compared to North America, Australia, or Great Britain (Macfarlane & Andrew 2001). These are: *Molophilus* 93 species, *Limonia* 66 species, *Leptotarsus* 54 species (Tipulidae), *Mycetophila* 52 species, *Tetragoneura* 34 species (Mycetophilidae), *Hilara* 65 species (Empididae), *Corynoptera* 34 species (Sciaridae), *Peromyia* 33 species (Cecidomyiidae), and *Platycheirus* 25 species (Syrphidae). Seven of these genera are more diverse than in South America or Australia. Exceptional species diversity exists in most of the remaining 16 genera, which more than triples the British list (Chandler 1998) or is about equal to that of the much larger areas of Australia (Evenhuis 1989) or North America (McAlpine et al. 1981, 1987). These genera are: *Gynoplistia* 102 species, *Amphineurus* 45 species (Tipulidae, Limoniinae, Tipulinae), '*Spilogona*' auctt. 81 species (Muscidae, Ceonosiinae), *Anabarhynchus* 61 species (Therevidae), *Ceratomerus* 41 species, *Chelipoda* 37 species, *Hilarempis* 23 species (Empididae, Ceratomerinae, Empidinae, Hemerodrominae), *Zygomyia* 54

Mycetophila fagi (Mycetophilidae).
Rod Macfarlane

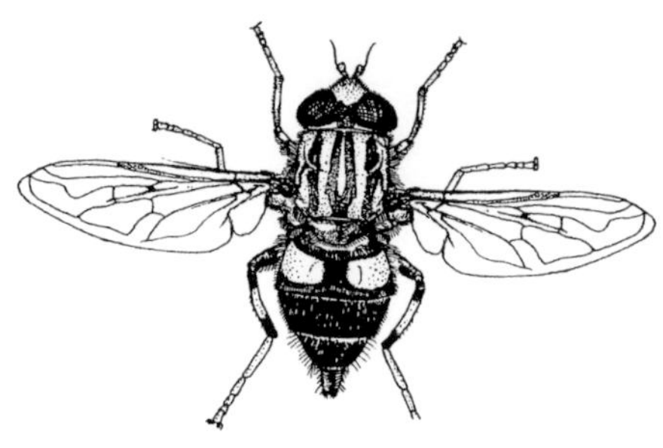

Rat-tail hover fly *Helophilus trilineatus* (Syrphidae).
After Grant 1999

species (Mycetophilidae, Mycetophilinae), *Pollenia* 34 endemic species (Calliphoridae, Polleniini), *Allograpta* 32 species, *Helophilus* 20 species (Syrphidae, Syrphinae, Eristalinae), *Parentia* 27 species, *Sympycnus* 21 species, *Micropygus* 15 species (Dolichopodidae, Sciapodinae, Sympycninae), *Nervijuncta* 20 species (Ditomyiidae), and *Cycloneura* 15 species (Mycetophilidae).

Conservation of vulnerable and endangered species

Currently, 27 species of flies have been placed on the indeterminate category of the endangered species list (Malloy & Davis 1994; McGuiness 2001). These include three aquatic fly species (*Nothohoraia micrognatha*, Blephariceridae; *Mischoderus*, two spp., Tanyderidae), the subalpine batwing muscid *Exsul singularis*, the New Zealand bat fly *Mystacinobia zelandica*, five robber flies including two undescribed offshore-island species, an undescribed flightless *Austrolimnophila* sp., the cave-dwelling crane fly *Gynoplistia troglophila*, and 12 species of stiletto flies. Some surveys have been conducted for only three of these species.

It is possible to specify five groups of flies that may be vulnerable to becoming endangered and at least five types of areas where species are susceptible to becoming rare. The first group is of larger predator flies (Therevidae, Asilidae) with limited mobility for almost all their life because they live as immatures in the soil during that time. Asilid adults feed on other smaller insects from perches, which may limit the time they can disperse compared to the non-predatory adults of Therevidae. A second group comprises any of the parasitic flies (Acroceridae, Bombyliidae, the calliphorid *Pollenia*, Pipunculidae, and Tachinidae) with restricted distributions and host ranges. All these families have reduced diversity on offshore islands in New Zealand, if they are even present there, which supports the theory they are potentially among the more vulnerable fly species. A third possible group is the smaller leaf-mining or gall-midge flies that may rely on rare plant species or those confined to offshore islands. These small flies, which can be encased in plant stems or buds, are vulnerable to extensive and repeated burning as well as spraying of grey shrubland, which is often dominated by matagouri (*Discaria toumatou*). Such rapid and often extensive and prolonged host loss probably decimates fly populations, especially when non-dominant shrub species develop small and fragmented populations. A fourth possible group comprises the more sensitive aquatic insect species such as some of the Empididae with restricted distributions or perhaps insects in thermal waters, e.g. *Tanytarsus* n. sp. (Winterbourn et al. 2000) if they become polluted. There are at least 84 flightless fly species known from New Zealand (Macfarlane & Andrew 2001), some of which occur on small offshore islands or in vulnerable habitats.

Seven types of areas – forests, grassland, offshore islands, wetlands, urban and alpine areas, dunes – can be susceptible to habitat destruction or degradation. Weed invasion, grazing pests, or farming and revegetation may change grasslands or forests. Small isolated offshore islands may be vulnerable, especially if they are more than 20 kilometres from other islands (beyond the flight range of most species), as also may subalpine or upland species on mountains and similar isolated areas, notably Mount Taranaki or Banks Peninsula. Wetlands adjacent to farmland or urban areas are often subject to draining and plant losses, while the loss of specialist native dune plants and removal of seaweed for garden fertiliser and driftwood for fires reduces habitat for coastal insect species that include flies. Extensive areas, especially of distinctive short tussock, have often been moderately to severely degraded, initially by rabbit grazing and burrowing then oversowing with pasture species, and more recently by invasion of four *Hieracium* species. Only a very low number of primary, generalist, foliage-, root-, and flower-feeding species exist in numbers where these weeds are dense (Syrett & Smith 1994; Macfarlane 2002). Snow-tussock grassland continues to decline in naturalness, including host-plant diversity, owing to oversowing, invasion by *Hieracium*, and firing. Grey shrubland in inland and dry regions, especially in the South Island, has already experienced a considerable retraction in area and in the availability of the less common host-plant species, before fly diversity has been adequately studied in these habitats. On the other hand, most fly species are probably less vulnerable than large flightless insects such as the wetas, many ground beetles, some weevils, and other beetles likely to be favoured as rodent food.

Order Strepsiptera: Twisted-wing parasites

There are only about 535 described species of Strepsiptera worldwide, and just two species in New Zealand. Both are endemic with mainly golden-brown or blackish bodies. Strepsiptera are parasites and all are small (0.5–4 mm long). Female Strepsiptera live within the host and only the males have wings. The hind wings are fan-shaped with few veins and the front wings are club-shaped halteres (in contradistinction to the hind-wing halteres of Diptera). In the New Zealand species *Elenchus maorianus* each antenna has a two-pronged acute-angled fork (Gourlay 1953) whereas *Coriophagus casui* has five close-set projections (Cowley 1984). The compound eyes of the males are distinctive too, looking somewhat like raspberries.

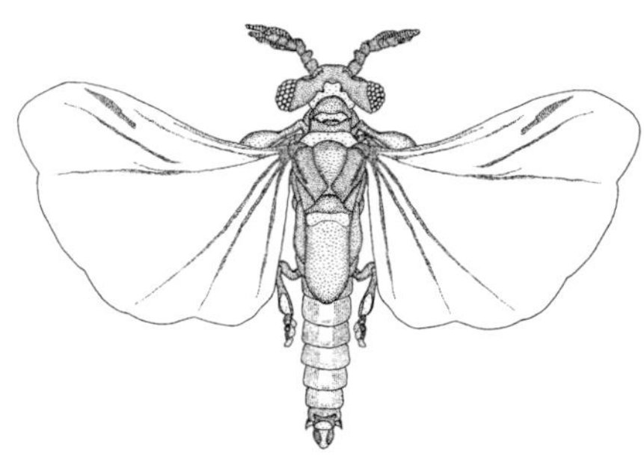

Coriophagus casui.
After Cowley 1984

Overseas, strepsipterans are mainly parasites of leafhoppers, planthoppers, and spittlebugs (suborder Auchenorrhyncha; Hemiptera), and a considerable range of bees and non parasitic wasps (Hymenoptera), and the host range can be quite broad (Kathirithamby 1989ab, 1992, Kathirithamby & Taylor 2005). Silverfish, cockroaches, mantids, and some orthopterans and flies are also hosts. In New Zealand, the recorded hosts are from the leafhoppers *Arawa variegata* (Cicadellidae) (Gourlay 1954, 1964; Prestige 1989) and *Novothymbris vagans* (Cowley 1984) as well as the plant hopper *Notogryps ithoma* (Delphacidae) (Cowley 1984). Male strepsipteran pupae in *Arawa* project on the upper side of the abdomen, while the larva-like adult females are on the underside with an obscure small projecting tip. Most of each strepsipteran pupa is within the host, whereas parasitic Dryinidae larvae (Hymenoptera) are fully exposed. Females produce a free living planidium, which seeks out and parasitises the next host.

Finding strepsipterans in New Zealand is achieved by knowing where to find the host species. *Novothymbris vagans* has been collected only from the west and southwest of the South island where it is associated with beech forest and shrubs of *Coprosma* and *Hebe* (Knight 1974b). A 2010 sampling of a salt-marsh meadow found 3.7% of the adults of the plant hopper *Anchodelphax ?olenus* (Delphacidae) to be affected by Strepsiptera; a male *Elenchus maorianus* was collected at the same site. Arawa species from adjacent shore ribbonwood (*Plagianthus divaricatus*) is known to be affected by an undetermined strepsipteran species. *Arawa variegata*, which is found in pasture and tussock grassland (Knight 1975), reportedly hosts *E. maorianus* (Gourlay 1964) but this could be based on an association and not from rearing. At least 12 other *Elenchus* species usually rely on delphacid planthopper hosts (Hassan 1939; Kathirithamby 1989). Cowley (1984) described *Coriophagus casui* from a male and so the host is unknown. Known hosts of *Coriophagus* (Pentatomidae, Coreidae) are based on three of the eight species in the world (Kathirithamby 1992).

The known level of parasitism in New Zealand (4.2%, Prestige 1989) is generally low (i.e. below 5%), but parasitism of *Arawa* was noted to be as high as 61.5% in February and 20.5% in October/November at one shore ribbonwood site and below 1.1% at a second site in spring. Several factors may contribute to males being rare in New Zealand collections, including the apparently overall low levels of parasitism, short male longevity, inadequate collection methods, and unsuitability of the habitat sampled.

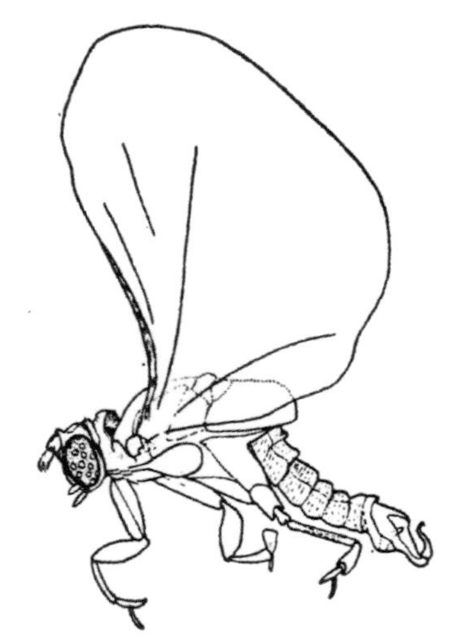

Elenchus maorianus.
From Gourlay 1953

Parasitised Hemiptera and/or Strepsitera males have been collected from pasture and subalpine grassland, salt-marsh meadow, mangove shrubland, and shore ribbonwood. The Auckland collection site for *C. casui* was close to kanuka (*Kunzea ericoides*) and broadleaf/podocarp restored bush (Cowley 1984), and specimens in the Lincoln University collection may be from within or adjacent to broadleaf forest. Adult males have been collected in malaise traps, pan traps, and even light traps. The Lincoln University collection possibly has undescribed species of Halictophagidae. Identification of delphacids and cicadellids hosting Strepsiptera pupae could soon improve knowledge of host range and seasonal activity. Rearing males from hosts, collecting more males in spring and summer, and descriptions of females will aid species recognition and clarify the hosts in New Zealand.

Order Hymenoptera: Sawflies, wasps, ants, and bees

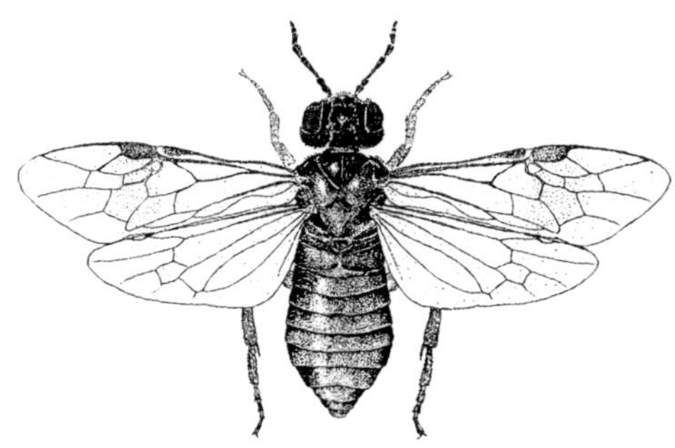

Cherryslug sawfly *Caliroa cerasi* (Tenthredinidae).
From Grant 1999

The order Hymenoptera ('membrane-winged') is one of the most morphologically and biologically diverse of all the insect orders. This is reflected in the order having no common name in English, unlike the three other megadiverse insect orders Diptera (flies), Coleoptera (beetles), and Lepidoptera (moths and butterflies). Hymenopterans include famous examples of social insects, like bees and ants, with regimented social systems in which members are divided into castes. They are also important pollinators of flowering plants, some species of which have flowers specially adapted to attract and receive the attention of specific hymenopteran visitors.

Like other holometabolous insects, hymenopterans undergo complete metamorphosis. Adults have mandibulate (biting) mouthparts, two pairs of membranous wings (in alates) with the fore and hind wings linked by hooks called hamuli, a lepismatid form of ovipositor (formed from modified vestiges of appendages on the eighth and ninth abdominal segments in females), and prominent antennae (usually with nine or more segments, but fusion or reduction may occur). Gauld and Bolton (1988) have provided an excellent chapter on adult morphology and Goulet and Huber (1993) published a list of diagnostic characters for adults and larvae. Keys to families can be found in the latter work and in Naumann (1991).

There are two hymenopteran suborders – the Symphyta (generally accepted as a paraphyletic assemblage, i.e. including ancestors of a number of evolutionary lineages) and the Apocrita (probably monophyletic). The symphytans (sawflies and woodwasps) are the most primitive members of the order. Most symphytan larvae are caterpillar-like and feed externally on the leaves of plants, but a few, such as siricids, are borers in wood and some species of the Orussidae are carnivorous. Where they impact economically on humans, they tend to be pest species, for example *Sirex* woodwasps, which can be important pests of pine trees.

The suborder Apocrita is traditionally divided into the Parasitica and the Aculeata, approximating to the parasitic Hymenoptera and the stinging Hymenoptera. While the Aculeata is demonstrably a monophyletic grouping, the Parasitica is paraphyletic (Gauld & Bolton 1988). These authors and Quicke (1997) have discussed the problems with higher-level classification in the order. The Parasitica represents the most species-rich group of Hymenoptera. Many of its members are physically tiny wasps – possibly the smallest insect known is a mymarid, *Dicopomorpha echmepterygis*, at about 0.13 of a millimetre in length (Mockford 1997). For these reasons (rich species diversity and often small physical size), as well as the high level of morphological variation and the presence of cryptic species (Austin 1999), the taxonomy and biology of the Parasitica is still extremely poorly known in almost all parts of the world. LaSalle and Gauld (1993) estimated that at least 75% of parasitic hymenopteran species (Parasitica and parasitic aculeates) are yet to be described. Some taxa within the Parasitica are secondarily plant feeders or predatory but the majority are insect-eating parasitoids. Possibly no other group of insects is as beneficial to humans as the parasitic wasps, through their role as regulators of other insect populations in natural and managed systems. Most species of natural enemies used in biological control programmes against pest insects are members of the Parasitica.

A leafhopper bearing a parasitic larva of a species of Dryinidae (Chrysidoidea).
Rod Macfarlane

The Aculeata includes the spider-hunting wasps, vespid wasps, ants, and bees. The ovipositor (egg-laying tube) and associated structures have been modified into a sting, and some species are health hazards for humans. Others are important pollinators. The development of social organisation has reached a peak for invertebrates in the aculeates. Ants, in particular, impact on ecosystems in a way unequaled by any other organism (LaSalle & Gauld 1993), and rank alongside soils and climate as probably one of the most important factors controlling vegetation. In *Science* magazine, Harvard biologist E. O. Wilson (1985) cited studies estimating that about a third of the entire animal biomass of the Amazonian rainforest could be composed of ants and termites.

Striated ant *Huberia striata* (Formicidae).
Des Helmore

Current known diversity

There are more than 115,000 described species of Hymenoptera worldwide (LaSalle & Gauld 1993), establishing it as one of the four great insect orders – Coleoptera, Diptera, Hymenoptera and Lepidoptera – each with over 100,000 described species (Goulet & Huber 1993). Despite the well-known quip about God's 'inordinate fondness for beetles', there has been debate in the literature as to which of these orders is the most species-rich (LaSalle & Gauld 1993; Grissell 1999). Stork (1991) found the Hymenoptera to be the most speciose order in tree canopies in Borneo and a number of surveys in tropical forests have supported this finding. Masner (1990) estimated that the Hymenoptera is the largest order in North America, and Gauld and Bolton (1988) likewise showed it to be the most speciose order in the much better known British fauna. Estimates of the true diversity of the Hymenoptera vary widely in the literature, from 250,000 to 2.5 million species (Grissell 1999). Gauld and Bolton (1988) estimated a minimum worldwide species tally of 250,000 in 83 families and Goulet and Huber (1993) more than 300,000 species in 99 families. LaSalle and Gauld (1993) also recognised about 80 families. These authors and Gauld and Bolton (1988) divided the superfamily Apoidea into just two families (Apidae and Sphecidae) whereas Goulet and Huber (1993) recognised 20 apoid families. These numbers are certain to change with increased knowledge and further research into phylogenetic relationships. For convenience, the treatment here generally follows that of LaSalle and Gauld (1993) with some recent changes and some exceptions that allow the classification in the end-chapter checklist to be consistent with New Zealand revisions.

New Zealand hymenopteran diversity

The New Zealand hymenopteran fauna is particularly poorly known. Valentine (1970) provided a historical survey of the order in New Zealand, from the first named hymenopterans (three ichneumonid wasps and a pompilid, described by Fabricius in 1775) until 1970, when 180 species were described. At this time, more than half the named species were described by just four authors – Ashmead, Cameron, Cockerell, and Smith. Valentine and Walker (1991) produced the first comprehensive catalogue, listing 549 species of Hymenoptera recorded in the literature as occurring in New Zealand up to 1988. The present review records about 200 more, for a total nearing 750 described species. Additionally, 775 to 892 undescribed or undetermined species are recorded here, providing a minimum species tally of around 1500 hymenopterans from New Zealand. More than 500 genera are recorded here, compared with 291 in Valentine and Walker (1991). A total of 47 families are recorded, nine of which are represented only by introduced species (Tenthredinidae, Pergidae, Siricidae, Agaonidae, Scoliidae, Scolebythidae, Mutillidae, Vespidae, and Megachilidae). The table given here compares family-level taxon diversity in New Zealand and globally.

Bumble bee *Bombus terrestris* (Apidae).
From Grant 1999

Repositories

The major repository of New Zealand hymenopteran specimens is the New Zealand Arthropod Collection (NZAC), which holds more than 150,000

Common wasp *Vespula vulgaris* (Vespidae).
From Grant 1999

mounted specimens, a small slide collection, and a large collection of specimens in ethanol. The NZAC also houses the largest collection of New Zealand hymenopteran types in the world (Berry 1991) and also an important collection of voucher specimens of arthropods (mostly hymenopterans) introduced into New Zealand for the biological control of weeds and pests. Examples of the value of retaining vouchers of biocontrol introductions have been demonstrated (Berry 1998, 2003). Significant collections of New Zealand Hymenoptera are also held by Lincoln University, Canterbury Museum, Canterbury University, Otago Museum, Forest Research Institute and Auckland Museum.

Endemism

A taxon is endemic if it has only ever been recorded from New Zealand and/or its offshore islands. However, since knowledge of the hymenopteran fauna from most parts of the world is at best incomplete, any statements regarding endemism must remain open to question. For example, the encyrtids *Alamella mira* and *Psyllaephagus acaciae* and the pteromalid *Nambouria xanthops* (all superfamily Chalcidoidea) are presently known only from New Zealand but, based on phylogenetic and/or host relationships, all three are almost certainly Australian (Berry & Withers 2002).

These examples illustrate the point that poor knowledge of both the New Zealand fauna and that of surrounding regions makes discussion of endemism at the species level in a quantitative fashion very difficult. Nonetheless, estimates of endemism for those groups where revisions have been undertaken recently, and where faunas of other areas are well known, are likely to have some significance. The table below compares percentage endemism at the species and genus levels for some such groups. It is evident that endemism at the species level is often reportedly high, whereas at the genus level it is

Family-level diversity of Hymenoptera

Showing the number of families worldwide vs New Zealand. Classification and numbers are based on Goulet and Huber (1993), except Xiphydridae, Proctotrupoidea, and Spheciformes.

Taxon	No. of known families worldwide	No. of known families in New Zealand
Tenthredinoidea	6	2
Siricoidea	2	1
Xiphydrioidea	1	1
Orussoidea	1	1
Xyeloidea	1	0
Evanioidea	3	1
Ceraphronoidea	2	2
Proctotrupoidea	8	1
Diaprioidea sensu Sharkey (2007)	3	2
Platygastroidea	1	1
Cynipoidea	5	3
Chalcidoidea	19	14
Parasitica	48	30
Mymarommatoidea	1	1
Ichneumonoidea	2	2
Chrysidoidea	7	4
Vespoidea	11	5
Apoidea	10	6
Total families	**82**	**47**

Endemism at genus and species levels for selected groups of New Zealand Hymenoptera

Taxon		Percentage endemism genus level	Percentage endemism species level
Symphyta[B]		25	33
Parasitica			
Chalcidoidea	Encyrtidae[No]	17–20	51
Chalcidoidea	Moranilini[B95]	0	65
Chalcidoidea	Mymaridae[N&V]	48	†
Proctotrupoidea	Ambositrinae[N]	70	100
Aculeata			
Vespoidea	Pompilidae[H87]	0	91
Vespoidea	Sphecidae[H94]	0	82
Apoidea	Colletidae	0	77

Sources: Berry (unpublished), Berry (1995), Donovan (pers. comm.), Harris (1987), Harris (1994a), Naumann (1988), Noyes (1988), and Noyes & Valentine (1989b)
† high number of undetermined species prevents calculation of this figure

extremely variable. At the family level, the chalcidoid family Rotoitidae was thought to be endemic until a second undescribed genus was recognised from Chile (Noyes & Valentine 1989b) and described by Gibson and Huber (2000). The recently described proctotrupoid family Maamingidae is known only from New Zealand (Early et al. 2001).

The groups Symphyta and Aculeata are small and relatively well known in New Zealand. At the species level, endemism is low amongst the Symphyta and high amongst the three aculeate groups that have been revised recently. Within the Parasitica, the Chalcidoidea is the only superfamily well known at the genus level, mainly owing to the works of Boucek (1988), Noyes (1988), and Noyes and Valentine (1989a, b). Boucek (1988) considered the New Zealand chalcidoid fauna to be much poorer than in other parts of Australasia, and he cited only 14 New Zealand chalcidoid genera as being endemic. Conversely, Noyes and Valentine (1989b) considered that the New Zealand chalcidoid fauna shows a high level of endemism, listing 43 genera endemic to New Zealand, which they estimated to be 25% of the total chalcidoid fauna. They estimated endemism at the species level to be up to 50%. Part of the reason for this discrepancy is that Boucek (1988) did not cover the families Encyrtidae and Mymaridae, which both show high generic levels of endemism.

Among other groups of Hymenoptera, Naumann (1988) found that five out of seven (71%) of the genera and all of the species of subfamily Ambositrinae (Proctotrupoidea) occurring in New Zealand are endemic. The superfamily Proctotrupoidea is an older group than the chalcidoids, which may explain the very high levels of endemism found in New Zealand at the genus level. There are several ancestral elements in the New Zealand hymenopteran fauna, for example species of Rotoitidae and the pteromalid genera *Errolia*, *Fusiterga*, and *Zeala*, but the remaining 10 endemic chalcidoid genera listed by Boucek (1988) are closely related to Australian forms. Boucek (1988) considered the latter genera to be probably descendants of forms that arrived at various times from Australia, carried by the wind.

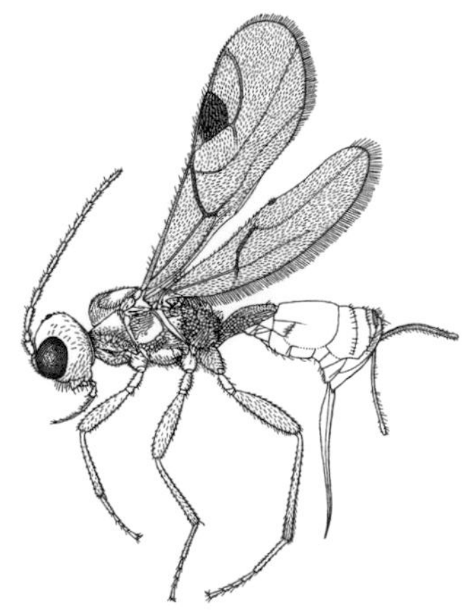

Cryptoxilos thorpei (Braconidae).
Des Helmore

Special features of the New Zealand hymenopteran fauna

Faunal composition

The most striking features of the New Zealand hymenopteran fauna are the paucity of stinging (aculeate) and sawfly/woodwasp (symphytan) species and

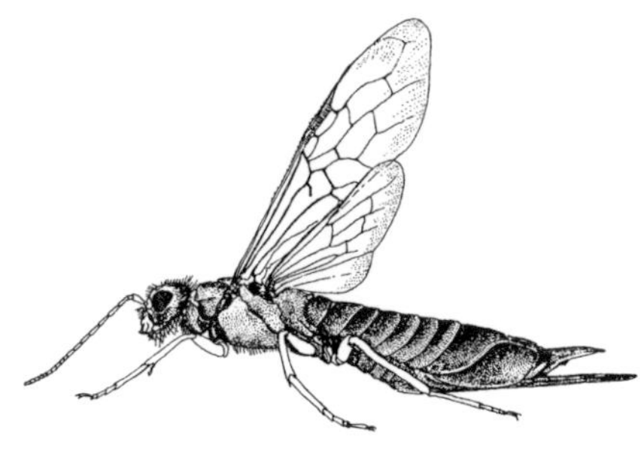

Female sirex wood wasp *Sirex noctilio* (Siricidae).
From Grant 1999

the radiation of certain parasitic groups. The Symphyta is represented by five of a total of 12 families, suggestive of a more substantial fauna than actually exists in New Zealand. In fact only nine species are known (of a total of around 4300 worldwide), six of which are accidentally introduced pest species and a mere three endemic. The Australian symphytan fauna is also depauperate, with the family Pergidae predominating (the one pergid known from New Zealand is Australian). Likewise, the Aculeata is poorly represented as a group in New Zealand – only about 130 species are known of a total of 49,000 worldwide. As with the symphytans, a large majority of the species of several aculeate families are introduced, for example the ants and the vespid wasps. New Zealand's ant fauna is strikingly depauperate, in sharp contrast to that of Australia, which is rich in both genera and species – 1275 described native species and probably around the same number undescribed (Shattuck 1999). For New Zealand, the ant fauna is around 40 species, including only 11 endemic species (Don 2007). The family Vespidae is represented solely by five introduced species, but two of these have had significant effects on some native ecosystems (see Alien Species). Three aculeate families are known only from one introduced species each (Mutillidae, Scolebythidae, and Scoliidae) (Berry et al. 2001). Within this sparse aculeate fauna the bees, or Apoidea, are better represented, with more than 40 species (from a world total of more than 20,000). The family Colletidae is dominant, with more than 20 endemic species (Donovan 2007). The colletids, probably basal in the apoid lineage, are abundant in Australia. Donovan (1983) proposed a dispersal model to account for the New Zealand apoid fauna, with species here derived mainly from very few Australian founder species. As evidence for this theory, he cited the similarity and relationships of groups of bees to the Australian bees, and the lack of distinctive characters (apomorphies) peculiar to the New Zealand fauna.

Hairy native bee *Leioproctus fulvescens* (Colletidae).
From Grant 1999

The Parasitica is much more strongly represented in New Zealand, but while there have been several radiations, there are also curious absences. Unlike any other zoogeographic region, the New Zealand hymenopteran fauna is apparently dominated in abundance of individuals by 'proctotrupoids' (in the broad sense), particularly diapriids, and the Chalcidoidea is represented disproportionately by the family Mymaridae (Austin 1988). Noyes and Valentine (1989a) estimated more than 160 New Zealand species of Mymaridae in 42 genera, twice the number known from Britain and around 10% of the described world species. Further, they consider that the study of additional material will certainly yield many more species belonging to the new genera they described. The chalcidoid families Agaonidae, Chalcididae, Eurytomidae, and Signiphoridae are poorly represented and probably have few endemic species (Noyes & Valentine 1989b). Other special features of the chalcidoid fauna noted by these authors are the large number of species of *Pteroptrix* (Aphelinidae) and elachertine eulophids and the moderately speciose endemic encyrtid genus *Adelencyrtoides*. The superfamily Cynipoidea (gall wasps and kin) is very poorly represented – from a total of over 3000 worldwide, fewer than a dozen species are known from New Zealand.

The superfamily Ichneumonoidea is poorly known (see Gaps in Knowledge), but the apparent absence of the cosmopolitan braconid genus *Chelonus* (Austin 1988) is unlikely to be an artefact. *Ascogaster* is the only genus known from the subfamily Cheloninae in New Zealand, and in no other area of the world does *Ascogaster* occur without *Chelonus* (Walker & Huddleston 1987). Among the Ichneumonidae, the campoplegine genus *Diadegma* is represented by around 50 endemic species, and the endemic ichneumonine genera *Aucklandella* and *Degithina* by a total of around 75 species. *Aucklandella* and *Degithina* are likely to be radiations of the large cosmopolitan genus *Cratichneumon* (D. B. Wahl, I. D. Gauld pers. comm.). A single specimen of the ichneumonine genus *Poecilocryptus* has been identified from Fiordland by I. D. Gauld. *Poecilocryptus*, the only known plant-feeding ichneumonid, was previously thought to be restricted to Australia where several species have been reared from galls in *Eucalyptus* and *Acacia*. The New Zealand species is endemic (I. D. Gauld pers. comm.) but its biology is unknown.

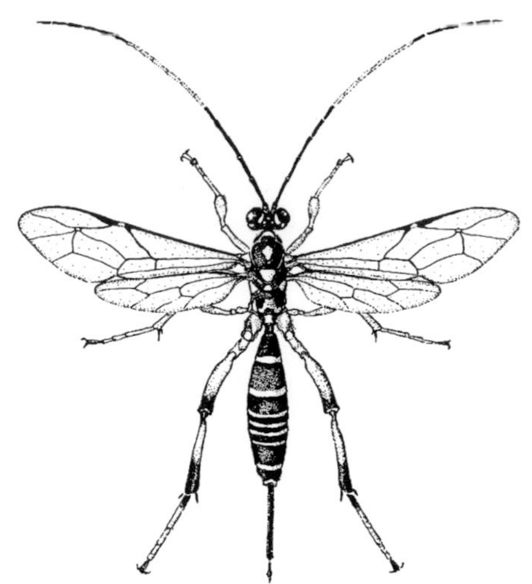

Xanthocryptus novozealandicus (Ichneumonidae).
From Grant 1999

Wing reduction and flightlessness

The New Zealand hymenopteran fauna appears to contain a high proportion of flightless species. Noyes and Valentine (1989a) noted that amongst New Zealand Mymaridae, 40% of the genera include species with abbreviated wings. Naumann (1988) noted wing reduction in 89% of species of New Zealand ambositrine diapriids, as compared with 66% of Australian species. In moraniline pteromalids, 36% of endemic New Zealand species exhibit wing reduction, as opposed to 7% of endemic Australian species (Berry 1995). In a number of species, females exhibit brachyptery (short wings) or aptery (no wings), while males are fully winged (Berry 1999). Wing reduction is often associated with wet forests, seasonally damp habitats, or alpine environments. Noyes and Valentine (1989a) speculated that in dense habitats, such as leaf litter and alpine tussock grasses, flight is not advantageous and wings may be an encumbrance when searching for hosts. Recolonisation of the lowlands by taxa isolated in mountain refugia and subject to such selective pressures may explain the high percentage of aptery and brachyptery in the New Zealand fauna.

Variation within species

A number of authors have commented on the high levels of intraspecific variation exhibited by insects in New Zealand, including ichneumonids and proctotrupoids (broad sense). Noyes (1988) reported generally high levels of variation amongst New Zealand Encyrtidae and further noted that the degree of variation shown by *Tetracnemoidea bicolor* in New Zealand is possibly greater than that recognised in any other species of encyrtid in the world. LaSalle and Boler (1994) discussed the extreme variation found in a new endemic species of eulophid, and Naumann (1988) reported pronounced sexual dimorphism (a form of intraspecific variation) in New Zealand Ambositrinae, particularly amongst wing-reduced forms.

Alien species

Exotic hymenopterans are a significant component of the New Zealand fauna, in terms of numbers of species and the dominance of certain species in natural and managed ecosystems. Although some have been introduced deliberately as pollinators or biological-control agents, many of the exotic hymenopterans in New Zealand have arrived accidentally. Some have very clearly become pests, and some are beneficial, but the impact of many species is less easily definable.

Accidental introductions

Wasps and ants (eusocial aculeates) have perhaps the most readily observable impact on New Zealand's existing biodiversity, affecting natural and human environments dramatically. Two European species of vespid wasp are now considered major pests in both habitats. The German wasp *Vespula germanica* has been present in New Zealand since 1945 and the common wasp *V. vulgaris* probably first established in the late 1970s (Donovan 1984). New Zealand's mild climate, abundance of food, and absence of natural enemies have contributed to the establishment of large populations of both species. In South Island beech (*Nothofagus*) forests, there is documented evidence that *Vespula* species compete for nectar and honeydew, which are important food sources for native bird species (Beggs 2001). Social wasps also prey on native and introduced arthropods (including pest species), and at high densities they may restructure insect communities in native forests, causing flow-on effects throughout the food web, including insectivorous birds (Beggs 2001).

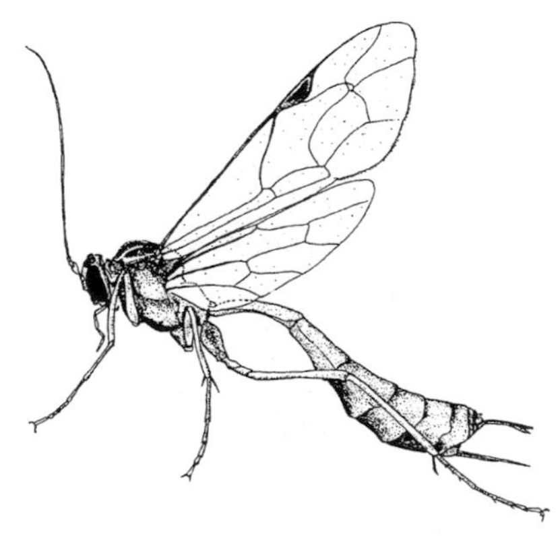

Orange ichneumon *Netelia producta* (Ichneumonidae).
From Grant 1999

The Asian paper wasp *Polistes chinensis* is a significant urban nuisance. Apart

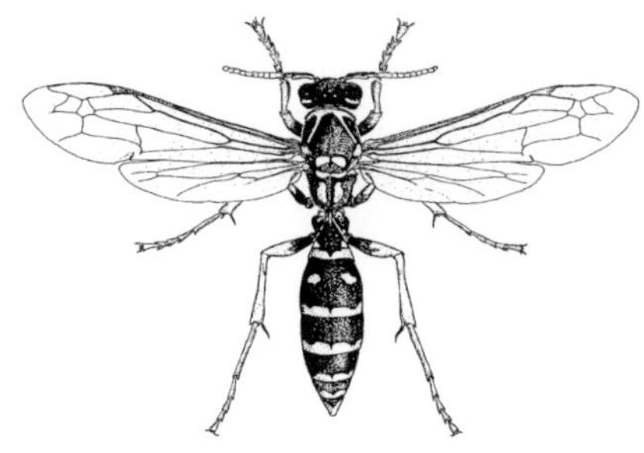

Asian paper wasp *Polistes chinensis* (Vespidae).
From Grant 1999

from its defensive sting, it has an unpopular habit of preying on monarch butterfly caterpillars, though it also includes garden pests in its prey.

Among the ants, several tramp species that have established in New Zealand are known to be highly invasive and to have caused severe ecological disturbances in other places, e.g. *Pheidole megacephala* (big-headed ant) and *Linepithema humile* (Argentine ant). The big-headed ant is believed to be native to Africa, but now rates as one of the most widespread and economically important of all pest ants in the world. It invades houses, stores, and factories, is known to tend homopteran bugs injurious to garden and horticultural crops, and is generally held by naturalists to be one of the most destructive agents of the lowland native Hawaiian invertebrate fauna (Hölldobler & Wilson 1990). In New Zealand the big–headed ant has been collected only from the Auckland region and the Kermadec Islands. Since it is pantropical in distribution, Berry et al. (1997) considered that climatic factors will probably restrict its spread, so that it is unlikely to become an important threat to the endemic invertebrate fauna throughout New Zealand. The Argentine ant has been recorded from as far south as Christchurch (R. Harris pers. comm.), and recently on Tiritiri Matangi, a bird reserve in the Hauraki Gulf. Like another tramp species, *Technomyrmex albipes* (white-footed house ant), the Argentine ant is a significant domestic and commercial nuisance, but it has also had negative impacts on native fauna elsewhere (Human & Gordon 1997; Cole et al. 1992).

Other exotic hymenopteran pests include accidentally introduced phytophagous species. The gall-forming eulophid *Ophelimus eucalypti* is a serious pest of commercially-grown *Eucalyptus* species (Withers et al. 2000), as was the pergid *Phylacteophaga froggatti* prior to the introduction of a successful biocontrol agent (Faulds 1991).

There is little doubt that the impact of these exotic species – in economic terms, in nuisance value, and/or on New Zealand's existing biodiversity – is overwhelmingly negative. The effects of other accidentally (and deliberately) introduced species are more open to debate, for various reasons. The recently introduced willow sawfly (*Nematus oligospilus*) is host-specific to willows (*Salix* spp.) and can cause severe defoliation (Berry 1997b). The pest status of this sawfly is debatable since on the one hand willows are an integral component of soil-conservation and erosion-control programmes, but on the other, feral crack willows (*Salix fragilis*) threaten conservation wetlands. Another example, the European eumenid wasp *Ancistrocerus gazella*, is a solitary predaceous species, first recorded in 1988 (Berry 1989), that provisions its nest with caterpillars. Recorded hosts include pest species, both introduced and endemic (Harris 1994b), as well as endemic non-pest species (Berry 1997a). *Ancistrocerus gazella* is now common throughout the Auckland area, and has also been recorded from Whangarei, the Waikato, Hawke's Bay, Wellington, and Central Otago (Green et al. 1994), as well as Canterbury (M. Bowie pers. comm.).

The impact of accidental introductions of parasitic species is more complex. Some records suggest that accidentally introduced natural enemies play an important role in managing pest species in modified habitats, forming natural enemy guilds. Charles (1998) listed 82 hymenopteran natural enemies of fruit pests from the New Zealand literature. Of these, around 90% are exotic and 10% endemic. The majority of exotic species (around 80%) have apparently established in New Zealand accidentally, with only about 20% being deliberate introductions as part of classical biological-control programmes. This research also suggested that, based on records from fruit crops and more specifically for mealybug hosts, exotic parasitoids are largely restricted to exotic hosts and native parasitoids to native hosts. This may also be true in other systems, but very few data are available. One exotic parasitoid known to attack endemic species is the highly polyphagous braconid *Meteorus pulchricornis*, which was first recorded from New Zealand in 1996 (Berry 1997a). It has an extremely wide host range, including hosts from 11 lepidopteran families worldwide, some of

Ancistrocerus gazella (Vespidae).
Jocelyn Berry

which are economically important pests (Berry & Walker 2004). Most of the hosts recorded from New Zealand are exotic pest species, but native hosts have also been reported, albeit from modified habitats (suburban and orchard). Collection details from native vegetation suggest that *M. pulchricornis* may be moving onto native hosts in native habitats.

Anecdotal evidence suggests that exotic parasitoid species may often enter New Zealand along with their hosts. New records of a number of recently introduced pest species have been followed closely by or even recorded at the same time as their respective parasitoids. Some examples are:

- the ash whitefly *Siphoninus phillyreae* and its parasitoid, the encyrtid *Encarsia inaron*; in fact, when *E. inaron* was first recorded from the field in New Zealand, the species had already been deliberately imported and was being reared in quarantine for possible release as a biocontrol agent of the ash whitefly;
- the psyllid *Glycaspis granulata* and its parasitoid *Psyllaephagus quadricyclus* (Berry 2007b);
- the brown lace lerp *Cardiaspina fiscella* (Psyllidae) and the encyrtid *Psyllaephagus gemitus* (Berry 2006);
- and the psyllid *Anoeconeossa communis*, first recorded in Mangere in 2002, and its parasitoid *Psyllaephagus richardhenryi*, which was reared and described from the first New Zealand collections of the psyllid (Berry 2007b).

These examples suggest that founder members of many pest species may enter the country complete with their own parasitoid complement.

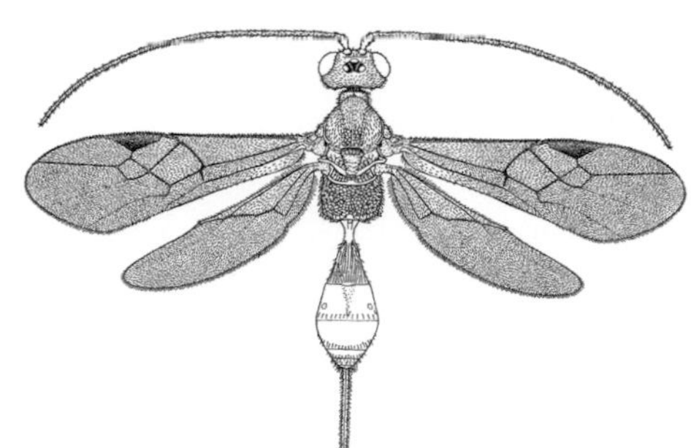

Meteorus pulchricornis (Braconidae).
Des Helmore

Deliberate introductions

These are made for the biological control of weeds and other arthropod pests and for the pollination of various crops. Four species of bumble bee were introduced into New Zealand from England between 1876 and 1906 for the pollination of clover. All four species established successfully and are now a familiar part of the landscape. There is no doubt that these bees are of direct economic benefit to New Zealand agriculture as pollinators of red clover and of lucerne (Donovan 1974). The ubiquitous honey bee (*Apis mellifera*) is also an extremely important introduced pollinator.

Classical biological control, using exotic predators, parasitoids, and pathogens against pests, has a long and in some cases undistinguished history in New Zealand. Increased consumer demand and higher chemical costs, along with ecological concerns and resistance to various pesticides, gave impetus to such projects in the last half of the 20th century. Cameron et al. (1989) comprehensively reviewed the period 1874 to 1987, with evaluations of the impact of the imported organisms. Estimates of the percentage of established agents that have had some impact on the specified target species in New Zealand are around 24% (Cameron et al. 1993) to 30% (Barratt 1996). Some programmes have resulted in very successful control of the target pest (for example, control of *Sirex noctilio* and *Phylacteophaga froggatti* by *Megarhyssa nortoni* and *Bracon phylacteophaga*, respectively). Others have provided significant control of major pest species, for example the control of *Mythimna separata* (cosmopolitan armyworm) by *Cotesia ruficrus*. Estimates of the monetary savings by this biological control programme alone range from NZ$4.5–10 million per year (Hill in Cameron et al. 1989).

These programmes provide undeniable economic benefits, but it is inevitable that the importation of control agents will change the environment in some way, although this change may be small measured against the impact of accidental introductions and the effects of habitat destruction. The environmental implications of biocontrol programmes are imperfectly understood, and their safety in terms of non-target effects, especially on the native fauna, has been questioned by a number of authors (see Barratt et al. 1997; Barratt et al. 2000). Although few biocontrol agents have been shown to have significant non-target effects, in New Zealand or elsewhere, this is more probably because of a lack of

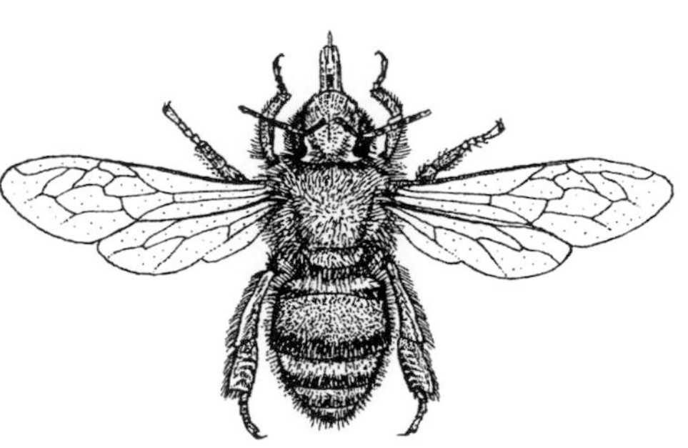

Honey bee *Apis mellifera* (Apidae).
From Grant 1999

research than the absence of such effects. Currently the only well-documented New Zealand case is that of *Microctonus aethiopoides*, a braconid wasp released against sitona weevil from 1982–1985. *Microctonus aethiopoides* has since been shown to attack 10 native and three other non-target species of weevils, including the weed biocontrol agent *Rhinocyllus conicus* (Barratt et al. 1997). Other examples of biocontrol agents that are suggested to have had significant impacts on non-target species in New Zealand are *Xanthopimpla rhopaloceros* (Ichneumonidae) and *Trigonospila brevifacies* (Tachinidae), both introduced against pest leaf-roller species, and *Pteromalus puparum* (Pteromalidae), introduced against cabbage white butterfly (Barratt 1996). *Xanthopimpla rhopaloceros* and *T. brevifacies* were reported to dominate the parasitoid complex attacking the native moth *Hierodoris atchychioides* (Berry 1990b).

It must be said that most criticism of biocontrol agents affecting non-target organisms relates to releases made 20 years or more ago. Increasingly extensive criteria for approval to release agents have evolved over this time. Since July 1998, the regulatory body ERMA (Environmental Risk Management Authority) has had the authority to make decisions on the importations of new organisms. ERMA is required to weigh up the risks, costs, and benefits in each case. There is no question that progressive legislative development has provided increased consideration of environmental effects (Moeed 2000), but the disadvantages, particularly the increased cost both in time and funding, has led to speculation that biological-control projects may be unsustainable for applicants. One certain impediment to the development of sustainable pest-management programmes is the lack of taxonomic knowledge, emphasising the need for work to be carried out on the New Zealand insect fauna.

Gaps in knowledge

Despite the economic and environmental importance of the group, the taxonomic diversity and ecological roles of New Zealand Hymenoptera are very poorly known indeed. The number of described species of Hymenoptera is a fraction of that of Coleoptera, Diptera, and Lepidoptera. This is unlikely to be a reflection of the real size of the fauna, but instead of a lack of taxonomic resources.

More than 1500 species of Hymenoptera are recognised from New Zealand in the following list. This includes about 700 described and about 900 undescribed or undetermined species. But the actual number of species will be considerably higher. Many species will not yet have been discovered or recognised. Some may already be in collections as unexamined material or they may not have been collected. Estimating the diversity of this portion of the New Zealand hymenopteran fauna is highly problematic because for a number of taxa, in particular the families Braconidae and Platygastridae (see below), our knowledge is too limited to make any meaningful estimate of their diversity. As Packer and Taylor (1997) commented, this is literally 'an exercise in estimating the magnitude of our ignorance'.

Of the three major hymenopteran groups, the symphytans and aculeates are relatively well known because their numbers are few and they are generally large and conspicuous insects. Conversely, owing to their generally small size and high diversity, the Parasitica is the least well-known group. This is an important gap for a number of reasons. The parasitic Hymenoptera, in their role as density-dependent regulators of host populations, play an important part in maintaining the diversity of natural communities (Quicke 1997). They are of direct economic importance in the biological and integrated control of horticultural, agricultural and forestry pests. While other orders of arthropods are used as control agents, parasitic hymenopterans are the most commonly imported agents, particularly ichneumonids and braconids.

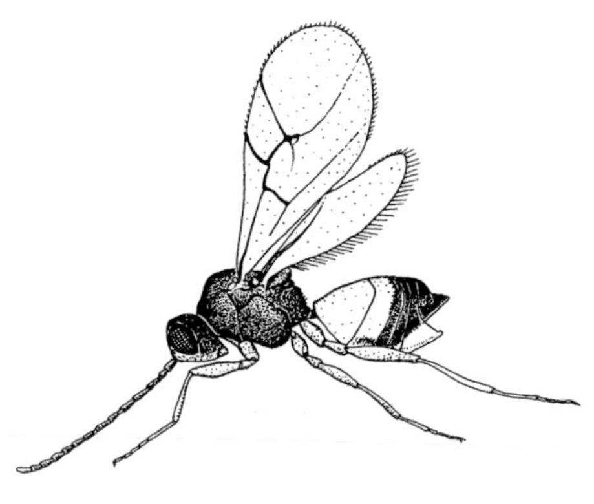

Gall wasp *Phanacis hypochoeridis* (Cynipidae).
From Grant 1999

The Chalcidoidea is the only superfamily in the Parasitica that is more or less well known at the generic level, largely thanks to the works of Zdenek

Boucek (1988), John Noyes, and Errol Valentine (Noyes 1988; Noyes & Valentine 1989a,b). The remaining groups – Ceraphronoidea, Cynipoidea, Evanioidea, Ichneumonoidea, and the proctotrupomorphs – are all poorly documented. Taxa within these groups have generally been described sporadically, as parts of offshore-island faunas for example, or as they have been associated with pest species. The superfamilies Ceraphronoidea, Cynipoidea, and Evanoidea are unlikely to be highly speciose in New Zealand. In contrast, the Ichneumonoidea and proctotrupomorphs contain families or lower-rank taxa that have radiated spectacularly in New Zealand, but no generic framework is available for any of these groups. The most important gaps in our knowledge, and thus priority groups for taxonomic revision, are discussed below.

Superfamily Ichneumonoidea (Ichneumonidae + Braconidae)

There are an estimated 60,000 species in the family Ichneumonidae worldwide (Gauld & Bolton 1988). About 2000 species are known from Britain and around 1200 recognised from the much less well-known Australian fauna (Gauld 1984). Valentine and Walker (1991) recorded just 77 described species from New Zealand in 35 genera. More recent estimates, based on the work of I. D. Gauld, suggest more than 300 species in about 70 genera. Of these, 12 are new endemic genera and 14 are described non-endemic genera newly recorded from New Zealand. The only subfamily-level revisions are for the small subfamily Metopiinae, koinobiont endoparasitoids (those that allow the host to continue its development) of lepidopteran larvae (Berry 1990a), and a planned revision of the Tersilochinae, parasitoids of wood-boring Coleoptera, with an estimated fauna of more than 40 species in five genera.

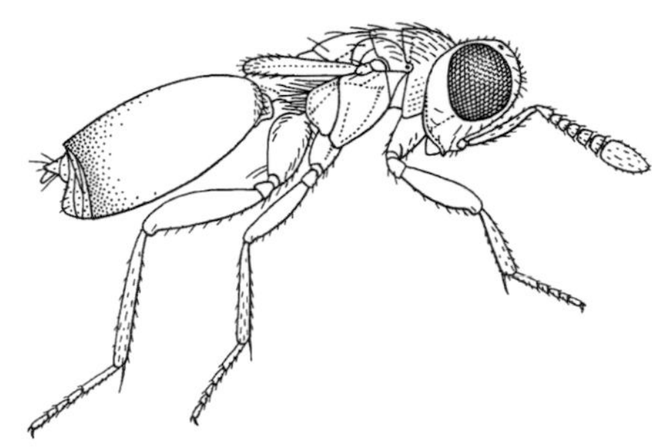

Ophelosia mcglashani (Pteromalidae).
Des Helmore

The sister-group of the Ichneumonidae, the Braconidae, is likewise a large family, with an estimated 40,000 species worldwide (Goulet & Huber 1993). Braconid wasps are important parasitoids of many economically significant pest species, yet New Zealand's fauna is exceedingly poorly understood. Around 100 described species are recorded in this list, but there are only 20 or so further undescribed or undetermined species recognised from New Zealand. This is almost certainly a fraction of the true number of braconid species. Subfamily-level revisions have never been carried out on the New Zealand fauna, with the exception of the Alysiinae or 'jaw wasps', important in the biological control of dipteran pests (Berry 2007a). Valentine and Walker (1991) recorded five species of alysiines in four genera in New Zealand; the subfamily revision recorded 21 species in nine genera.

Platygastroidea (Platygastridae and Scelionidae)

Although New Zealand has a small landmass, its platygastrid diversity is apparently high, and it is likely that the centre of platgastrine diversity is in the Southern Hemisphere (Goulet & Huber 1993). However, the extant generic framework was developed in the Northern Hemisphere, making it extremely difficult to apply generic concepts to austral faunas without unrealistically altering the limits of the northern genera. It is almost impossible to work on the New Zealand platygastrid fauna without extensive input from northern-hemisphere collections and taxonomists and no systematic revisions have been carried out.

From 1775, when the first hymenopterans were described from New Zealand, until 1988, when the fauna was first completely catalogued (Valentine & Walker 1991), the average rate of description was around 2.5 species per year. From 1988 to 1999 about 7.5 species were described on average per year. Even at the latter rate, it is clear that centuries will be required even to approach a completely described New Zealand hymenopteran fauna.

Inostemma boscii (Platygastridae).
Birgit Rhode, Landcare Research

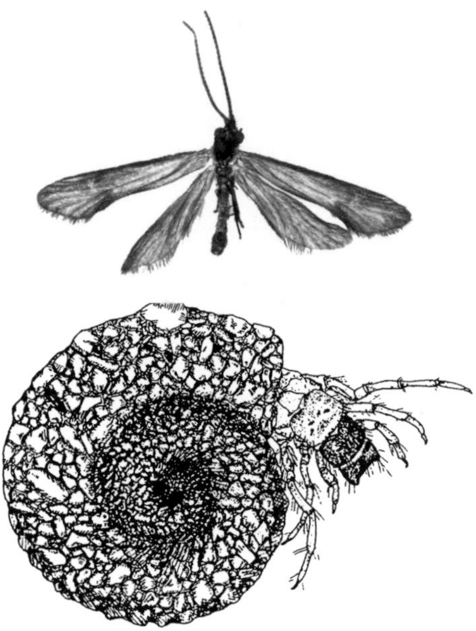

Helicopsyche zealandica adult (upper)
Simon Pollard, Canterbury Museum

Helicopsyche sp. larva (lower)
Craig Dolphin, from Winterbourn et al. 2006

Order Trichoptera: Caddis or caddisflies

The Trichoptera ('hairy wings') comprises more than 12,500 named species inhabiting all continents except Antarctica. Caddis look rather like moths (Lepidoptera), and are generally accepted as their evolutionary sister group, although caddis wings are covered with hairs and not flattened scales. Adults have two pairs of subequal wings, which in New Zealand species range in length from less than two millimetres to about 35 millimetres. The wings are generally dull brown, grey, or black without much colour pattern, although many species can be recognised immediately by wing colour alone. Most caddis, when at rest, hold their wings flat against the sides of the abdomen (some species hold them over the abdomen) with the antennae extended forwards. Caddis adults are unable to chew solid food, but need liquid food to survive and have been seen visiting flowers. Mostly they hide up and rest during daytime, become active at night, and come freely to light. Their aquatic larvae have four feeding regimes – predation, feeding on decomposing vegetable matter or filamentous algae, scraping of microflora of stone surfaces (periphyton), and filtering of food from water currents. Winterbourn (2000) has thoroughly reviewed his own and others' work on larval ecology, diet and lifestyles.

Caddis adults and larvae are a significant component of freshwater biodiversity and of the food chain. They are preyed on by fish (introduced and native), birds (such as New Zealand's endemic blue duck, an endangered species), and many invertebrates. By feeding on coarse organic matter (leaves, twigs) they contribute to the breakdown of this material and release it in a more finely divided form that smaller organisms can utilise. Caddis larvae (in conjunction with other aquatic macroinvertebrates) are useful indicators of water purity in streams because of their sensitivity to pollution. A numerical analysis incorporating them can be performed rapidly (Boothroyd & Stark 2000). The results can reflect periodic sources of pollution not revealed by conventional testing of the water, say for oxygen content and organic matter.

All stream sizes contain caddis larvae, as do trickles and seepages. Medium-sized streams in untouched forest are usually the most speciose, because most human enterprises reduce biodiversity. However, even modified streams in farmland or built-up areas can be quite species-rich. A few caddis species are found in lakes, but New Zealand lacks the distinctive fauna that is specialised to this habitat, such as is found in the northern hemisphere, perhaps because there are no ancient lakes here.

There is a distinct alpine caddis fauna known to extend up to at least 1900 metres, first explored by A. G. McFarlane in the 1930s around Cass and Arthur's Pass, by John Child in the Central Otago uplands in the 1970s, and more recently by Brian Patrick of the Otago Museum. About 17 caddis species live predominantly or solely above 600 metres. Their larvae and pupae pass the winter under snow in mountain streams. Adults emerge early in spring and can be found in snow caves (Patrick 1992), under stones, or on stream banks. They are usually day-active. At higher altitudes, adults tend to be short-winged and are often unable to fly. Alpine species can be found at lower altitudes at the limits of their ranges. For example, *Tiphobiosis childi*, a typical alpine species, also occurs on Banks Peninsula where it is found almost to sea level. This may be the result of its isolation on this gradually eroding 'island'.

The New Zealand caddis fauna

The New Zealand fauna of about 244 known species is highly endemic, and the 49 New Zealand caddis genera are 73% endemic. Only one species is shared with Australia, the marine caddis *Philanisus plebeius* (but New Zealand *Triplectides cephalotes* is very close to Australian *Triplectides australis*, with which species it was formerly confused). It is surprising that more species are not shared, but caddis are apparently not long-distance dispersers (they are often good short-distance dispersers). There are also interesting examples of local endemism, such as on Banks Peninsula (five endemic species).

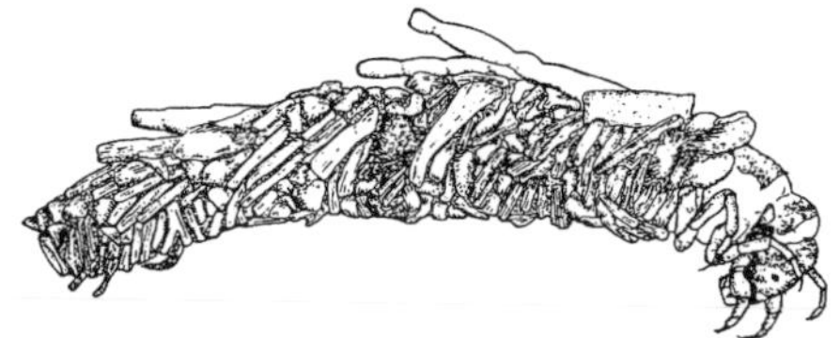

Larval marine caddis *Philanisus plebeius* (Chathamiidae).
Craig Dolphin, from Winterbourn et al. 2006

Caddis larvae are nearly all aquatic (a few species live in damp terrestrial situations but this has not yet been demonstrated in the New Zealand fauna). They produce silk from modified salivary glands and this is the foundation of their various lifestyles as larvae – building shelters (moveable or otherwise), spinning capture-nets, and constructing cocoons before pupation. The popular concept of caddis larvae living inside portable tubular cases is true for less than half of the New Zealand species, many of whose larvae rove freely on stream beds, or construct fixed nets or shelters. Classification into three suborders based on adult and larval morphological and molecular data (Morse 1997) also reflects these lifestyles, as follows. (The Spicipalpia may not be a natural group, however (see Kjer et al. 2002.)

Suborder Annulipalpia ('net spinners and fixed retreat-makers')
Family Ecnomidae – the sole New Zealand species, *Ecnomina zealandica* (forewing length 4 millimetres), is one of the most rarely captured New Zealand caddis. Only one male specimen has ever been caught and the larva is unknown. The family occurs worldwide.

Family Hydropsychidae (12 New Zealand species in three genera, forewing length 7–17 millimetres) – adults of *Diplectrona* do not appear at light traps whereas other species come freely. Larvae build tunnel-like retreats in streams on submerged wood and rocks. Coarse-meshed nets, close to the anterior opening of the retreat, filter food materials from the current. Hydropsychid larvae pupate in a cocoon of loose silk inside an enclosure of small stones. The family is distributed worldwide.

Family Philopotamidae (10 New Zealand species in 4 genera, forewing length 5–10 millimetres) – all adults come to light and at least one species flies commonly during the day. Larvae are filter-feeders and build stocking-shaped fine-meshed capture-nets. Larvae pupate inside cocoons in closed stony shelters. Family distribution is worldwide.

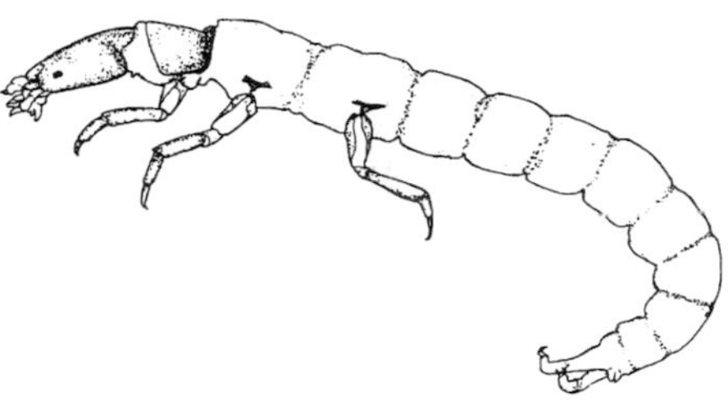

Larva of *Hydrobiosella stenocerca* (Philopotamidae).
Craig Dolphin, from Winterbourn et al. 2006

Families Polycentropodidae (6 New Zealand species in two genera, forewing length 5–14 millimetes) and *Psychomyiidae* (1 species, *Zelandoptila moselyi*, forewing length 3.5–5 millimetres) – adults in both families come to light freely. Larvae spin silk nets on the surface of rocks in streams. Polycentropodids are found worldwide. Psychomyiidae occur in all faunal regions, but in the Neotropical Region are present only in Mexico.

Suborder Spicipalpia ('cocoon-makers')
Family Hydrobiosidae (105 New Zealand species in 11 genera, forewing length 3.5–25 millimetres) – this largely southern-hemisphere family comprises 43% of the known New Zealand caddis fauna, considerably more than the 7–8% found in the other Gondwanan regions (Ward 1998). The carnivorous larvae are prognathous, with foreleg tibiae, tarsi, and tarsal claws modified into pincer-like structures. Most larvae live in streams, although a few are restricted to seepages (some *Tiphobiosis* and *Edpercivalia* species). The larva of one stream species, *Hydrobiosis parumbripennis*, has been given the common name 'green free-living caddis'. At high altitudes, tarns can also be colonised. To pupate, the larva builds a closed semi-ellipsoidal shelter of small stones attached to the underside of a larger stone, and inside this spins a cigar-shaped cocoon of fine dark-brown silk. All species come freely to light. Mature females swim underwater and lay flat egg masses, enclosed in a film of jelly, on the underside of stones.

Family Hydroptilidae ('micro-caddis', 'purse-case caddis') (21 New Zealand species in three genera, forewing length 2–5 millimetres) – these are tiny caddis less than five millimetres long, with pointed wings fringed with long hairs. Adults come freely to light and on mass-emergence nights can almost cover lighted windows near streams and lakes, running madly around on the glass. Early instar larvae are free-living with long hairs and legs. In the final instar they construct flattened cases of silk and their bodies become swollen (this change is called hypermetamorphosis). They feed on algae by piercing individual cells and pupate inside their cases, which are attached to stones, often in large groups. The family is found worldwide.

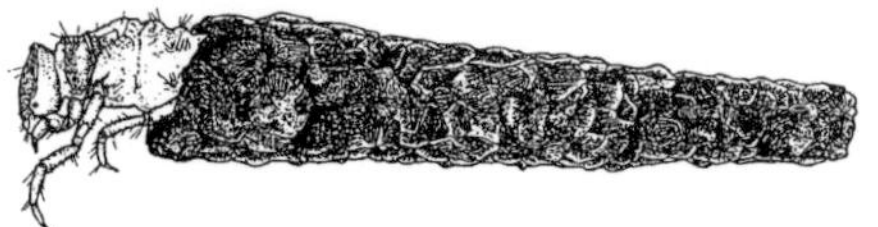

Larva of *Oeconesus maori* (Oeconesidae).
Craig Dolphin, from Winterbourn et al. 2006

Suborder Integripalpia ('portable case-makers'): Infraorder Plenitentoria
Family Oeconesidae (23 New Zealand species in five genera, forewing length 9–30 millimetres) – New Zealand is the centre of distribution of this family with one species in Tasmania and none elsewhere. They are medium- to large-sized caddis with broad wings, the larvae of which build straight, slightly tapering cases made of gravel and/or plant fragments, and are primarily vegetarian. All species come to light but many more males than females are usually captured in this way. The larvae inhabit bogs, seepages, and small-to-medium-sized streams. *Pseudoeconesus* seems to be confined to boggy sites. *Zelandopsyche ingens* is New Zealand's largest caddis species. In beech-forest streams it feeds on, and constructs its case of, beech-leaf (*Nothofagus*) fragments. If food is scarce, the larva will feed on its case (Winterbourn 1976).

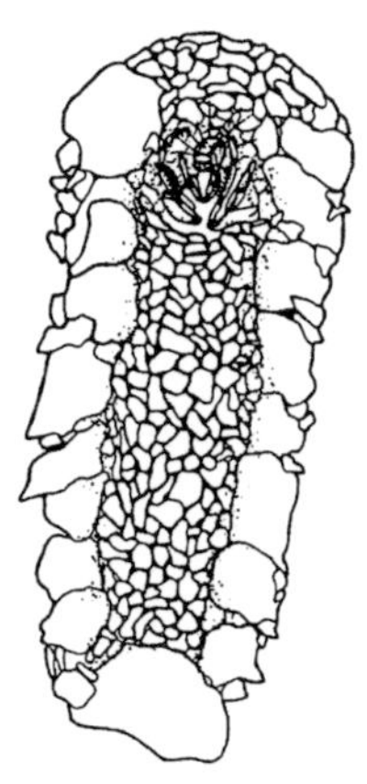

Larva of *Kokiria miharo* (Kokiriidae).
Craig Dolphin, from Winterbourn et al. 2006

Family Kokiriidae (one New Zealand species, *Kokiria miharo*, forewing length 7–8 millimetres) – this family is also Gondwanan, with three genera and five species in Australia, a few species in New Caledonia, and one known species in South America. Adults have elongated mouthparts, which may be longer than the rest of the head in some species. *Kokiria* larvae inhabit slow streams with a sandy bottom. The distinctive larval case is dorsoventrally flattened with an overhang in front and lateral flanges. Both fore- and midlegs of the larvae have fused tibiae and tarsi, forming raptorial limbs.

Suborder Integripalpia ('portable case-makers): Infraorder Brevitentoria
Family Philorheithridae (seven New Zealand species, all in *Philorheithrus*, forewing length 11–16 millimetres) – this is another Gondwanan family, most speciose in Australia but also occurring in South America and Madagascar. Adults have broad, squarish forewings with a sclerotised projection basally on the rear margin. Larvae are carnivorous, with the midleg tibia and tarsus fused. Their tapered and slightly curved cases are constructed of large sand particles held together with dark silk.

Family Leptoceridae ('longhorned caddis') (13 New Zealand species in four genera, forewing length 6–19 millimetres) – the common name is a literal translation of the family name (Greek *leptos*, long; *ceros*, horn), alluding to the antennae that are usually much longer than the wings. The larvae construct a variety of cases of various materials depending on the species. In *Triplectides*, small twigs and pieces of leaf are used and the popular name is 'stick caddis'. Frequently a single twig is used, with a hole bored through it. *Oecetis* and *Hudsonema amabile* make a sandgrain case; *Hudsonema alienum* uses short lengths of vegetable matter arranged longitudinally in a distinctive spiral pattern. Distribution is worldwide.

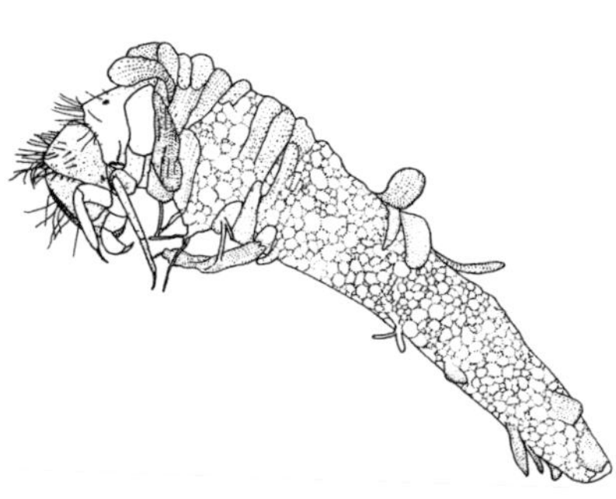

Larva of *Alloecentrella magnicornis* (Helicophidae).
Craig Dolphin, from Winterbourn et al. 2006

Family Chathamiidae ('marine caddis') (five species in two genera, forewing length 5–11 millimetres, except for *Chathamia brevipennis* which has reduced wings) – larvae live in the intertidal zone and sublittoral fringe to at least one metre depth (Riek 1976). Their cases are composed of sand and seaweed. Female adults have a strong ovipositor; those of *Philanisus plebeius* have shown the bizarre behaviour of ovipositing inside the body cavity of a starfish (Anderson & Lawson-Kerr 1977; Winterbourn & Anderson 1980). The family is confined to New Zealand, the Chatham Islands, and the Kermadec Islands except for *P. plebeius*, which also occurs in Eastern Australia.

Families Calocidae (one New Zealand species, in *Pycnocentrella*, forewing length 8–11 millimetres, with an Australian and New Zealand distribution), *Conoesucidae* (24 New Zealand species in six genera, forewing length 5–12 millimetres, Australia and New Zealand only), and *Helicophidae* (two New

Zealand species in *Zelolessica*, forewing length 2.5–6.5 millimetres, and four in *Alloecentrella*, the family restricted to Australia, New Zealand, and southern South America) – larvae of these three families make curved, tubular, tapered cases of plant material or small stones depending on the species – mosses and liverworts (*Alloecentrella*) or small sand grains (*Pycnocentrella, Zelolessica*). The following species of Conoesucidae have acquired popular names – *Olinga feredayi*, the 'horny-case caddis', with a red-brown case of silk only; *Pycnocentria evecta*, the 'sandy-case caddis', with a case of uniformly sized sand grains arranged circumferentially; and *Pycnocentrodes* species, 'stony-case caddis', with a case of sand with much larger grains placed laterally.

Family Helicopsychidae ('snail-shell case caddis') (eight New Zealand species in two genera, forewing length 4–7 millimetres) – larvae have very distinct spiralled sandgrain cases resembling a small snail. Larvae graze on the periphyton and pupate communally. The family occurs worldwide.

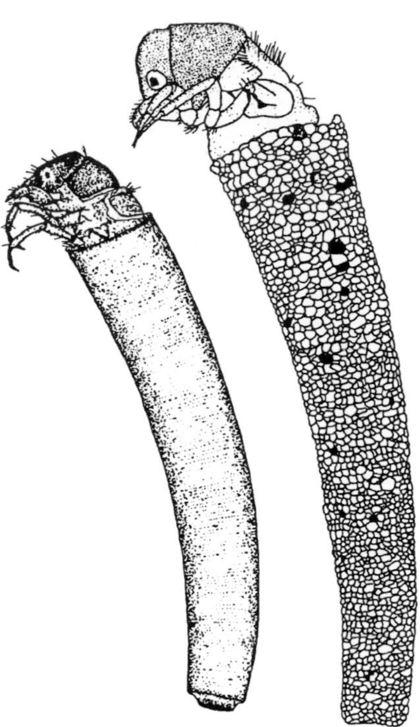

Larvae of *Olinga feredayi* (left) and *Pycnocentria evecta* (right) (Conoesucidae).

Craig Dolphin, from Winterbourn et al. 2006

Neboiss' (1986) *Atlas of Trichoptera of the SW Pacific–Australian Region* is a good source of illustrations of wings and genitalia for identification of caddis adults, brought up to date for the New Zealand species by Ward (2002). Recent papers describing new species include those by Ward and Henderson (2004), Ward et al. (2004), Ward (2005), Henderson and Ward (2006, 2007), Henderson (2008), and Smith (2008). Larvae of only about one-third of the New Zealand species have been described, however (Cowley 1976, 1978; Harding 1991; Smith 1998, 2000, 2001, 2002; Winterbourn et al. 2000; Stark 2000 and references therein). There is a key to the larvae by genus by McFarlane (1990), and a more recent one including coloured images is on the NIWA website at http://niwa.co.nz/ncabb/tools. A complete world checklist of Trichoptera taxa is maintained by J. C. Morse on the Clemson University website at http://entweb.clemson.edu/database/trichopt.

Regarding the deposition of New Zealand material – early workers favoured the British Museum (Natural History) (now the Natural History Museum), London, and the primary types of 35 species are held there. Thirty-six are in the New Zealand Arthropod Collection, 14 in Auckland Museum, five in Te Papa, and two in Oxford Museum, UK. The largest collection, comprising the primary types of more than 100 species, is in Canterbury Museum, Christchurch, New Zealand, and this museum also has the largest holding of the New Zealand Trichoptera, at about 50,000 identified specimens. The second largest holding of about 25,000 identified specimens is in the personal collection of Dr Ian Henderson, Massey University, Palmerston North.

Order Lepidoptera: Moths and butterflies

Butterflies and moths have traditionally been among the best loved and most studied of all insects. Their most characteristic feature, and the one that gives the order Lepidoptera its name, is the presence of scales, which are modified setae, on the wings and body. The scales are often highly coloured or refractive, creating the beautiful patterns that have made the order so popular with entomologists and the public alike. Most Lepidoptera also have distinctive mouthparts, with a long, usually coilable proboscis formed from the modified maxillae; the proboscis is absent in only two primitive families (Micropterigidae and Agathiphagidae) and in some more derived lineages with short-lived, non-feeding adults (e.g. Psychidae, Cossidae). Larvae of almost all butterflies and moths feed on flowering plants, most species being host-specific, and therefore lepidopteran diversity throughout the world tends to reflect floristic diversity, with the greatest number of species occurring in the tropics. However, a number of groups, especially amongst the more primitive families of moths, are associated as larvae with lower plants (mosses, liverworts, algae), with fungi and lichens, or with detritus (dead wood and leaf-litter); these groups (Micropterigidae, Psychidae, Tineidae, Oecophoridae) are well represented in New Zealand. The

Hierodoris tygris (Oecophoridae).

Des Helmore

total diversity of the Lepidoptera worldwide is unknown, but there are estimated to be more than 165,000 named species (Robinson et al. 1994).

Overall, the New Zealand lepidopteran fauna (1,703 named species and subspecies in the end-chapter checklist) is not diverse by world standards, but shows a number of features that make it of great interest and importance – for example the high level of endemism, the number of species and groups showing wing reduction and flightlessness, and the representation of primitive groups, including one endemic superfamily, that enhance scientific understanding of lepidopteran evolution. As with so many other animal groups in New Zealand and on other oceanic islands, the fauna is disharmonious, i.e. there are no native species of some otherwise ubiquitous families (e.g. Lymantriidae, Notodontidae, Hesperiidae, Pieridae, Papilionidae), whilst other groups have undergone considerable endemic radiations (e.g. Crambidae: Scopariinae, Oecophoridae: Oecophorinae).

New Zealand's largest lepidopteran, the puriri moth *Aenetus virescens* (Hepialidae).

After Grant 1999

Ecological and economic importance of Lepidoptera in New Zealand

The ecology of adult Lepidoptera has received little systematic study in New Zealand. Even straightforward data on abundance and flight times have rarely been gathered in a quantitative fashion, with the notable exception of the long-term light-trapping studies carried out in the South Island by White (1991, 2002). Adult butterflies and moths undoubtedly have an important role in native ecosystems as pollinators of flowering plants; however, it is generally accepted that most New Zealand flowers are 'generalists' that do not rely on pollination by particular species or even particular orders of insects (Lloyd 1985). There is no known New Zealand equivalent to the long-spurred Madagascan orchid *Angraecum* that relies for pollination on a particular subspecies of hawkmoth (*Xanthopan morgani praedicta*) with an equally long tongue (Pinhey 1975), hence the overall importance of butterflies and moths in comparison to other insect pollinators is very hard to assess in New Zealand.

Likewise, adult Lepidoptera hold significant but largely unquantified importance in the food chain, particularly as food for native (and introduced) insectivorous birds. The New Zealand lepidopterist will be familiar with the experience of disturbing a moth from the forest undergrowth, only to see it caught and consumed by a hungry fantail (*Rhipidura fuliginosa*). The morepork (*Ninox novaeseelandiae*) is a well-known predator of New Zealand's largest lepidopteran, the puriri moth (*Aenetus virescens*) in North Island forests. The two native species of bats must also rely heavily on Lepidoptera for sustenance. The most significant invertebrate predators of adult moths are probably spiders (Arachnida) – New Zealand has an especially diverse fauna, including one genus of orbweb spiders (*Celaenia*) whose species mimic the female pheromones of moths and thus attract the males into their grasp (Forster & Forster 1970).

Knowledge of the ecology of immature Lepidoptera is much better though still incomplete. Larvae in particular form an abundant and important part of the plant-feeding and litter-feeding guilds of invertebrates, as well as providing food for birds and for the larvae of insect parasitoids, especially Hymenoptera and Diptera (Tachinidae). A number of endemic species with polyphagous larvae are considered pests – this is especially true of some leaf-rollers (Tortricidae) in the genera *Ctenopseustis* and *Planotortrix*, which commonly feed on orchard trees (e.g. Dugdale 1990). Late-instar larvae of some endemic *Wiseana* species (Hepialidae) compete with stock for spring pasture growth and have in the past been considered serious pests, although modern pasture management has alleviated the problem (Dugdale 1994). Of greater concern are the exotic pest species that have occasionally established themselves in New Zealand. The most notable examples are the white-spotted tussock moth *Orgyia thyellina* and the painted apple moth *Teia anartoides* (both Lymantriidae), which were accidentally introduced to Auckland in 1996 and 1999 respectively (Hoare 2001a). Such species are believed to represent a serious threat to New Zealand's economy as potential defoliators of orchard and commercial forestry trees. Being highly polyphagous, they may also threaten native forests should they be allowed to establish. The white-spotted tussock moth was successfully eradicated in 1997 by aerial spraying with *Bacillus thuringiensis* var. *kurstaki* (a pathogen of lepidopteran larvae); eradication of the painted apple moth involved a more prolonged spray campaign, but in 2006 was also achieved.

Australia is the main source of foreign Lepidoptera that become established in New Zealand (Hoare 2001a). Much of the influx is natural; moths migrate to New Zealand or are transported by the prevailing westerly winds, especially when a large anticyclone lies over the Tasman Sea (Fox 1978). However, establishment is much more a matter of chance, and has certainly been aided since European settlement of New Zealand by the planting of large numbers of favoured Australian host-plants such as *Eucalyptus* and *Acacia* species. The warmer, drier microclimates created by the destruction of New Zealand's original forest cover also more closely mimic Australian conditions. Thus we should not be surprised that the rate of establishment seems to have increased (Hoare 2001a). Recent arrivals from Australia include the banksia leaf-miner *Stegommata sulfuratella* (Lyonetiidae), the eucalyptus leaf-miner *Acrocercops laciniella* (Gracillariidae), and the gum-leaf skeletonizer *Uraba lugens* (Nolidae).

As noted above, New Zealand is especially rich in Lepidoptera with detritivorous larvae, i.e. those that feed on dead wood, fungi, or leaf-litter (Dugdale 1998). In the Oecophorinae, for example, more than 250 New Zealand species are currently recognized – this compares with a figure of approximately 85 for the whole of the former USSR (Lvovsky 2003), a land area more than 80 times as great (Hoare 2005). Undoubtedly the diversity and abundance of these detritivores render them important contributors to the breakdown of organic material and the recycling of nutrients, especially in New Zealand's forest ecosystems. There may be other less benign interactions – for example, in seasons of abundance, high availability of oecophorid larvae as prey may boost numbers of introduced social wasps (*Vespula* spp.) and/or of rodents in certain habitats. The remains of leaf-litter-feeding larvae have been recovered from the stomachs of mice (*Mus musculus*) in beech forest in the Orongorongo Valley, near Wellington (Dugdale 1996). Such interactions are much in need of investigation. Likewise, the impact on New Zealand native detritivores of the increasing number of Australian species that have become established (see Hoare 2001a) is unknown; possibly the introduced species favour different microhabitats than the natives, but there may still be examples of competition. At least one suspected detritivore (*'Schiffermuelleria' orthophanes* – Oecophoridae) appears to have become much scarcer in recent years (cf. Hudson 1928).

One endemic species of Tineidae (*Archyala opulenta*) has been reared only from larvae feeding on the guano of the short-tailed bat *Mystacina tuberculata* (Patrick & Dugdale 2000). The moth is rarely caught as an adult, but its presence in a locality could indicate the presence of the bat, itself regarded as a threatened species. This is the only known example amongst the endemic New Zealand Lepidoptera of a close association with a specific vertebrate 'host'. Amongst herbivorous taxa, and especially leaf-miners, there are many examples of host specificity; a recently discovered example is *Houdinia flexilissima*, the only member of a newly described genus (Hoare et al. 2006). The exceedingly thin larva mines and pupates inside the living stems of the large endemic restiad *Sporadanthus ferrugineus*. The habitat of the rush is threatened, which means that this unique lepidopteran is automatically considered a species of high conservation status.

Several species of Lepidoptera have been introduced to New Zealand as biocontrol agents for weeds. Their impact is seldom spectacular, but they are undoubtedly beneficial in keeping these plants in check. Gorse (*Ulex europaeus*) is attacked by the European *Cydia succedana* (Tortricidae), *Pempelia genistella* (Pyralidae), and *Agonopterix umbellana* (Depressariidae), all deliberately intro-

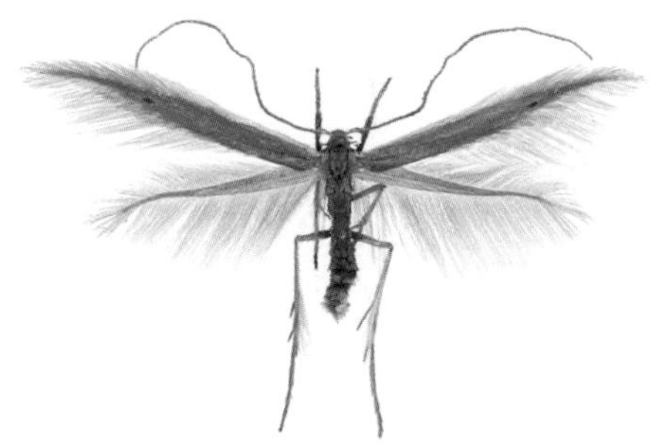

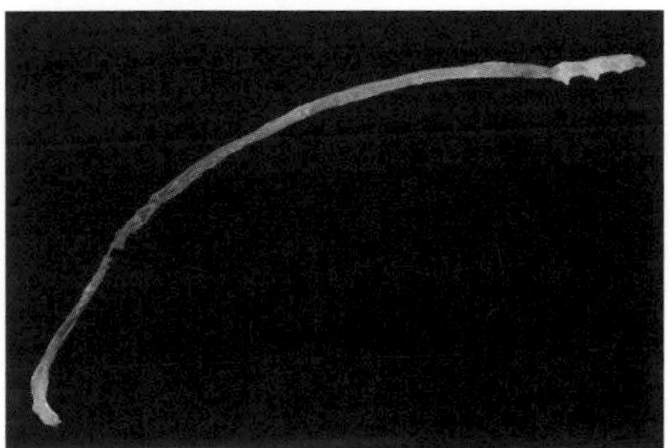

Houdinia flexilissima and its exceedingly slender larva (Batrachedridae).

Birgit Rhode, Landcare Research

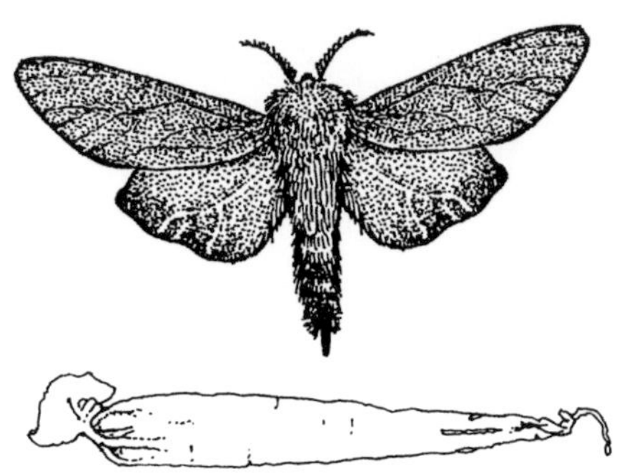

Adult and larval/pupal case of the bag moth *Liothula omnivora* (Tortricidae).

After Grant 1999

duced. The twigs of Scotch broom (*Cytisus scoparius*) are mined by the tiny larvae of *Leucoptera spartifoliella* (Lyonetiidae), another northern-hemisphere species that has become well established in New Zealand, apparently without being deliberately released. The European cinnabar moth (*Tyria jacobaeae* – Arctiidae) is now found throughout New Zealand following an initial release in the late 1920s. At first it became firmly established only in the Wairarapa, but further releases from this stock in the late 1980s have seen it take hold from Auckland to Southland (C. Winks pers. comm.). Its larvae feed on ragwort (*Senecio jacobaea*), a pasture weed that is poisonous to stock. The moth appears to be a less effective control agent than the introduced flea beetle *Longitarsus jacobaeae*.

The current New Zealand species list

Dugdale (1988) provided an annotated catalogue of New Zealand Lepidoptera and discussed the composition of the fauna in some detail. The current listing is based very firmly on this foundation, and departs from it only in some details of nomenclature and in adopting some subsequent changes in family classification (Nielsen et al. 1996; Kristensen 1999). Changes in the generic assignment of Pterophoridae follow Gielis (2003). Additions to the list are the adventives treated by Hoare (2001a), together with the native Nepticulidae described by Donner and Wilkinson (1989), the Psychidae described by Hättenschwiler (1989), the Tortricidae, Hepialidae, and Plutellidae described by Dugdale (1990, 1994, 1996 respectively), the Geometridae described by Stephens and Gibbs (2003), Wientraub and Scoble (2004), and Stephens et al. (2007), the Oecophoridae described by Hoare (2005), and monotypic genera described by Matthews and Patrick (1998), Dugdale (1995), Hoare and Dugdale (2003), and Hoare et al. (2006). The nomenclatural changes of White (2002) have also been adopted. No new synonymies are introduced here, but *Izatha griseata* is omitted from the list as it will shortly be synonymised with an adventive Australian taxon listed elsewhere. Known undescribed species are not treated, except where they are given separate listing by Dugdale (1988) or Hoare (2001a). The exclusion here of known undescribed species accounts for the differences in numbers between the present list and Table 1 of Dugdale (1988). *Mnesarchaea loxoscia* has been removed from synonymy with *M. fusilella* on the advice of G. W. Gibbs (pers. comm.).

Some incorrect subsequent spellings in Dugdale (1988) have been emended to their original form in accordance with Article 32.5 of the ICZN; these are (with incorrect spelling in parentheses): *Batrachedra tristicta* (*tristictica*), *Asaphodes periphaea* (*peripheraea*), *Orthoclydon praefectata* (*praefactata*), *Tatosoma topea* (*topia*), *Xanthorhoe orophyla* (*orophylla*), *Acrocercops panacivermiforma* (*panacivermiformis*), *Aletia* (s.l.) *sollennis* (*sollenis*), *Euxoa ceropachoides* (*cerapachoides*), *Phalaenoides glycinae* (*Phalaenodes*), *Tinearupa sorenseni aucklandica* (*aucklandiae*), *Rhathamictis perspersa* (*perspera*), *Tephrosara cimmeria* (*Tephrosaria*). Three species named from the Snares (in the genera *Elachista, Stigmella* and *Apoctena*) had the species name changed from *laquaeorum* to *laqueorum* by Dugdale (1988); since the error was from incorrect latinization, these corrections are deemed unjustified emendations by the ICZN (2000), and the original spelling is restored.

The following adventive taxa not recorded by Hoare (2001a) are added to the New Zealand list based on material held in NZAC; further details of these finds will be published elsewhere: *Musotima ochropteralis, Uresiphita polygonalis maorialis, Ephestiopsis oenobarella, Pantydia sparsa, Proteuxoa sanguinipuncta, Oinophila v-flava, Isotenes miserana*. All these species are Australian, with the exception of *O. v-flava*, a cosmopolitan species not recorded from Australia. The three adventive Psychidae listed as unnamed species by Dugdale (1988, p. 69) have been determined as *Lepidoscia heliochares, L. protorna*, and *L.* cf. *lainodes*; all are Australian in origin. Finally, one native phycitine (Pyralidae), shared with Australia, has been determined as *Ptyomaxia trigonogramma* by M. Horak.

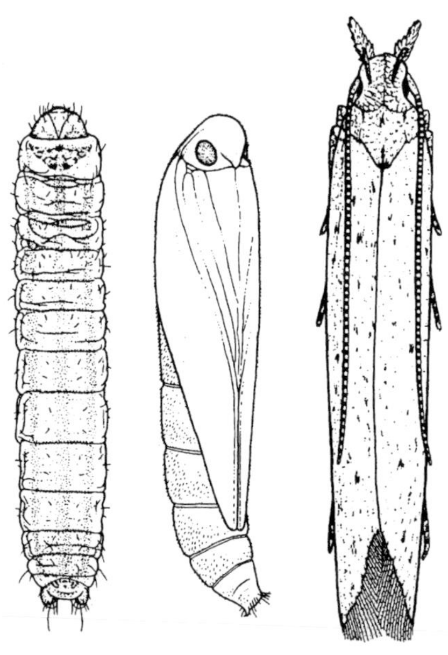

Larva, pupa, and adult of *Megacraspedus calamogonus* (Gelechiidae).

Jon Sullivan, Lincoln University

The family classification follows Kristensen (1999), with the exception of the Gelechioidea, where the more conservative classification of Nielsen et al. (1996) is retained. Families are listed alphabetically rather than in systematic order, and genera and species are listed alphabetically within their respective subfamilies. No subfamily divisions have been adopted for Tineidae because of confusion over subfamily definitions (cf. Dugdale 1988; Robinson & Nielsen 1993) and because the New Zealand species cannot be assigned to subfamily without detailed revision. Likewise, the subfamily classification of world Noctuidae has been substantially modified in recent years (e.g. Kitching & Rawlins 1999; Lafontaine & Fibiger 2006) and is still in a state of flux; a subfamily division of the New Zealand fauna has not been attempted here.

Gilled aquatic larva of *Hygraula nitens* (Crambidae).

Stephen Moore, Landcare Research

Endemism

Of the 1703 species and subspecies listed here, 1490 can be regarded as native and 174 as adventive or vagrant (including species that have subsequently become extinct in New Zealand) (see Dugdale 1988; Hoare 2001a). A further 25 species are known in the New Zealand region only from the Kermadecs, and 13 species represent biological control agents, deliberately or accidentally introduced (including species that are not known to have established). The status of two species (*Cateristis eustyla* and *Hydriomena* (s.l.) *iolanthe*) is considered uncertain. Approximately 33 species are naturally shared with Australia, leaving an endemic fauna of 1457 species, or 85.5%. This is a very high level of endemism, but not unexpected for a country so long isolated from other landmasses. Since most additions to the New Zealand list will be from undescribed endemic species, this percentage is certainly an underestimate. It is difficult to judge endemism at the generic level because many New Zealand Lepidoptera are currently placed in inappropriate (usually Australian or Palaearctic) genera to which they probably or certainly do not belong. At the family level, one taxon, the Mnesarchaeidae (with 14 species, most undescribed), is currently recognised as endemic, and indeed these tiny moths are so distinctive that they also merit their own superfamily, the Mnesarchaeoidea. They represent the sister-group to the diverse and worldwide Hepialoidea (swift or ghost moths), and as such constitute one of the most significant and remarkable elements of the New Zealand fauna. Adult mnesarchaeids are often common in damp bush where their larvae feed in webs amongst mats of liverworts and mosses, but tend to be overlooked by all but the most careful observer. A revision of the family by G. W. Gibbs and N. P. Kristensen is in progress.

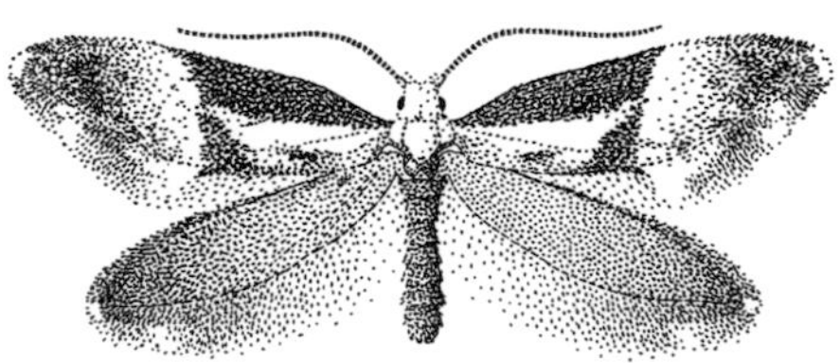

Mnesarchaea fusilella (Mnesarchaeidae).

Des Helmore

There are other enigmatic species in New Zealand, for which new taxa of suprageneric level may eventually need to be described. *Titanomis sisyrota* is a large, very rarely collected moth that has some characters in common with the Cossidae or goat moths, a group otherwise absent from New Zealand except as importations. However, further detailed study is required before its true affinities can be ascertained (Dugdale 1988; Hoare 2001b).

Undescribed species

It is difficult to estimate the number of undescribed lepidopteran species currently in New Zealand collections. Certainly, the majority of the fauna has been described, but only a few groups have been recently revised, and in these a surprisingly large number of new species have been added. Thus, of 27 hepialid species treated by Dugdale (1994), 10 (37%) were described as new, and of 28 species of Nepticulidae in Donner and Wilkinson's (1989) revision, 14 (50%) were new. Even more surprisingly, at least 12 new species of Nepticulidae have been recognised or collected since Donner and Wilkinson's work. Some genera of New Zealand Lepidoptera seem to be in an active state of radiation, creating a bewildering array of colour forms or biological races of uncertain taxonomic status. This is the case, for example, with the copper butterflies (Lycaenidae), leading some workers to propose the existence of a number of undescribed species. Much more work on these problems, preferably involving 'reciprocal illumination' by parallel molecular and morphological studies, is much to be desired. Only then can we hope to know the true diversity of New Zealand's Lepidoptera.

Fossil insects

Fossil fragments representing 121 species of insects have been found from the Triassic up to and including the Pleistocene. One of them is a contender for a role in the next Jurassic Park movie. It is based on a wing impression found in Upper Jurassic rock near Port Waikato (Grant-Mackie et al. 1996). The animal to which it belonged may have been ancestral to the living weta families.

The only fossil fly recorded from New Zealand is based on an impression of a larval bibionid (*Dilophus campbelli*) in Early to Middle Eocene coal near Livingston, Central Otago (Harris 1983).

Fragmentary lepidopteran fossils have been found in New Zealand. Evans (1931) described a wing scale from Late Eocene coal measures in the Waikato District. Recently, a fragment found in sediment at Rakaia Gorge, South Island, has been interpreted as a possible sclerite from the genitalia of a geometrid (Harris & Raine 2002). This is only the second Cretaceous insect fossil known from New Zealand.

The oldest beetle fossils in New Zealand are Triassic and Late Cretaceous fragments of elytra. Predictably, beetle exoskeletal remains are fairly well represented in the Quaternary (Pleistocene and Holocene), including a nearly complete ulodid, representing a new genus, from a cave in the Waitomo region (Leschen & Rhode 2002). The distribution and diversity of Coleoptera in the Holocene allows direct comparison with modern faunas from the same geographical areas and from other parts of New Zealand. Kuschel and Worthy (1996) noted the drastic reduction in the present-day range of large weevils, and one extinction, compared to the Holocene, attributing the cause to extensive clearing of native vegetation for pastures and to the introduction of rats and mice into the country. Beetle remains have proven particularly useful in paleoclimate interpretation, allowing reconstruction of vegetational history and paleotemperatures in interglacial periods at locations in both North and South Islands (Marra 2003; Marra & Leschen 2004; Marra et al. 2004, 2006).

Gaps in knowledge and potential for further research on Hexapoda

Major gaps in our knowledge of the New Zealand hexapod fauna can be divided into five topic areas. They include taxonomic knowledge of the size and distribution of the fauna, the conservation status and security of rare species, the effects of introduced organisms, life-histories and feeding biology, and the capacity for restoration of depleted habitats.

The most pressing need is to obtain a clearer understanding of the size, status, and distribution of the fauna. More specialist monographs, translated into user-friendly interactive electronic keys, would enable new workers to enter the field and encourage non-taxonomists to attempt initial identifications, thus releasing experienced taxonomists to engage in work requiring more specialised experience.

Based on existing collections and what is known of the fauna and what might be expected, it is clear that only a small proportion of the collembolan fauna has been described. Insect groups known to include at least 20 new undescribed species include Orthoptera, Hemiptera, Coleoptera, Diptera, Hymenoptera, and Trichoptera. But most orders will have some undiscovered species. At the present rate of new species description it will take centuries to complete the task of describing the fauna.

Among the endemic whiteflies (Hemiptera), there appears to be a complex of fern-feeding species, currently known as *Trialeurodes asplenii* (Maskell), and another complex of undescribed species on beech (*Nothofagus* spp.). It is likely that there is more than the one named species of Ortheziidae in New Zealand. The list of Diaspididae in the end-chapter checklist is based on out-of-date taxonomic knowledge and the earlier-named species are likely to harbour some undetected synonymies. A thorough revision of the armoured scales is needed, both for taxonomy of the indigenous fauna and to provide keys to both indigenous and exotic species. This is particularly important for biosecurity, as the potential for the introduction of invasive species is high in this family. The tribe Leucaspidini may need molecular studies to sort out the indigenous species within it. Associated hymenopteran parasitoids are poorly understood and require concerted efforts in collection and rearing-out from scale-insect hosts. Nothing is known about the potential of adventive parasitoid species to cross over into the endemic fauna.

The Coleoptera and Diptera are important groups that are still significantly understudied, with many gaps in knowledge and numerous undescribed species. In the Diptera, families known to include at least 20 undescribed species include Cecidomyiidae, Chironomidae, Dolichopodidae, Empididae, Ephydridae, Muscidae, Mycetophilidae, Stratiomyiidae, Syrphidae, Tachinidae, Tipulidae, and probably also Ceratopogonidae, Lauxaniidae, Phoridae, and Psychodidae. As taxonomic gaps are filled there will still be plenty of interesting behavioural, sex-attractant, and other biochemical studies to flesh out our knowledge of flies, whose species currently outnumber the scientists and hobbyists investigating them by a ratio exceeding 200 species per person. For those with a penchant for natural history, flies offer a rich resource for serious discovery.

The identification of flies would benefit greatly from the publication of a consolidated series similar to the North American summary of flies (McAlpine et al. 1981, 1987) and of English immature flies (Smith 1989). There are few quality line drawings of fly species in New Zealand, but there are extensive illustrations of fly wings, heads, and legs that help to identify genera and species. However, these illustrations are scattered in a considerable number of publications that are often over 40 years old. For the families Phoridae and Dolichopodidae, illustrations and descriptions are neither in English nor in journals that are readily available in New Zealand.

Much more taxonomic work is needed to determine what species are vulnerable or endangered so that a realistic list of insects in the largest insect orders can be developed for the various Department of Conservation conservancies in New Zealand. For Diptera alone, at least two full-time taxonomists dealing with flies should be employed in New Zealand. In fact, 4–5 would be better, in order to described formally the likely 2000 or so remaining species. One impediment to the development of workable keys is lack of information on variation within species. Many species are known from only a single specimen, reflecting patchy collection effort if not actual species and population losses. Although distributional information is patchy on the two main islands, it tends to be poorer for Stewart Island and some of the larger islands with permanent flowing water in the case of aquatic insects.

The number of described species of Hymenoptera that have been recorded from the New Zealand biogeographic region is a fraction of that in the three other mega-diverse insect orders (Coleoptera, Diptera, and Lepidoptera). One of the major impediments to futher ecological research involving the native hymenopteran fauna is scant knowledge of the group's alpha taxonomy. Within the order, the symphytans and aculeates are relatively well known because their numbers are few and they are generally large, conspicuous insects. Conversely, owing to the generally small size and high diversity of its members, the Parasitica is the least known group. Of the seven superfamilies of Parasitica represented in New Zealand, six are poorly documented. Three of these are unlikely to be highly speciose here, but the remaining three contain families or lower-rank taxa that have radiated spectacularly in New Zealand. The most important gaps in knowledge of the order are within these three superfamilies – Ichneumonoidea, Proctotrupoidea, and Platygastroidea. The priority groups for taxonomic revision include the families Braconidae and Platygastridae.

Coupled with limited knowledge of morphological variation is the lack of knowledge of life-history stages for most hexapod species. Further, studies of life-cycles of the same species in a range of temperature regimes, particularly in aquatic environments, could prove to be useful in determining the role of abiotic versus biotic influences on life-history patterns. Such studies rely on available expertise to accurately identify species and funding to develop and sustain this expertise continues to decline precipitously in New Zealand.

Acknowledgements

Numerous colleagues helped in the preparation of this chapter, supplying information and answering questions. Partial support for the Hexapoda studies in Landcare Research was made possible by FRST (Contract C09X002 to Landcare Research).

Kevin Collier (Environment Waikato, Hamilton), Chris Green (Department of Conservation, Auckland), and Jaap Jasperse (SPREP, Apia) provided helpful comments on the Ephemeroptera review.

Chris Simon (University of Connecticut, Storrs) and J. S. Dugdale (Landcare Research, Auckland) made helpful comments on the cicada checklist. Carol von Dohlen (Utah State University) is thanked for her ideas on aphid phylogenetic relationships. Unpublished correspondence between Don Horning and various workers is held at Crop & Food Research, Lincoln, New Zealand.

Richard Leschen, John Lawrence, Guillermo Kuschel, and Qiao Wang contributed to the initial review and/or compilation of Coleoptera. Rowan Emberson, John Marris, John Nunn, and Margaret Thayer are thanked for helpful comments on the generic list of Coleoptera. Tony Jewell and Steve Trewick gave advice on Phasmatodea.

The advice of the various scientists who co-authored the Diptera checklist and supplied records of undescribed Diptera species have allowed more accurate perspective on the status of dipteran taxonomy in New Zealand. Thomas Pape kindly supplied information on the presence of Miltogramminae in New Zealand. Access to 14 New Zealand collections has been invaluable in reviewing the fauna and compiling the checklist and is gratefully acknowledged. Rod Macfarlane studied Australian genera of robber flies from the CSIRO (Canberra), and Natural History Museum (London) collections and New Caledonian species from the Bishop Museum (Honolulu) and the IRD collection (Nouméa). Access to these and Chilean collections greatly assisted in clarification of generic relationships relevant to New Zealand. Antony Harris (Otago Museum) and Matthias Jaschhof (Greifswald University, Germany) supplied comments and unpublished information on the Diptera and discussions with Dan Bickel (Australian Museum) assisted in the preparation of the review. Trevor Crosby thoroughly checked the Diptera list.

The following people are thanked for information on the Hymenoptera – Eric Grissell (USDA), *Megastigmus* (Torymidae); Tamas Megyaszai (Hungarian Natural History Museum), *Stylaclista* (Diapriidae); John Early (Auckland Museum), all other Diapriidae; Richard Harris (Landcare Research, Auckland), Warwick Don (University of Otago, Dunedin), Olwyn Green and John Keall (MAF), Formicidae; Barry Donovan (Donovan Scientific Insect Research, Christchurch), Apoidea; Serguei Triapitsyn (University of California, Riverside), *Anagrus* (Mymaridae); John LaSalle (CSIRO Division of Entomology), Eulophidae; Celso Azevedo (Universidade Federal do Espirito Santo, Brazil), Bethylidae; Till Osten (Staatliches Museum für Naturkunde, Stuttgart), Scoliidae; and Ian Gauld (NHM, London), Ichneumonidae. Graham Walker (Crop & Food, Christchurch), John Charles (HortResearch, Auckland), Malcolm Kay (FRI, Rotorua), and Richard Harris contributed to the discussion on alien species. Leonie Clunie and M. Anne Austin (Landcare Research, Auckland) assisted with gathering data and helpfully commenting on the text, respectively. Andrew Austin (University of Adelaide) and Ian Gauld commentd on the manuscript generally and Jan Klimaszewski (Canadian Forest Service) discussed aspects of cataloguing.

Authors

Dr Ian G. Andrew Institute of Molecular Sciences, Massey University, Private Bag 11222, Palmerston North, New Zealand [iands.andrew@xtra.co.nz] Diptera

Dr Jocelyn A. Berry Policy and Risk Directorate, MAF Biosecurity New Zealand, P.O. Box 2526, Wellington 6140, New Zealand [jo.berry@maf.govt.nz] Hymenoptera

Dr Thomas R. Buckley Landcare Research, Private Bag 92170, Auckland, New Zealand [buckleyt@landcareresearch.co.nz] Phasmatodea

Dr Daniel Burckhardt Naturhistorisches Museum, Augustinergasse 2, CH-4001 Basel, Switzerland [daniel.burckhardt@unibas.ch] Coleorrhyncha

Dr Pamela J. Dale 401 Hillsborough Road, Mt Roskill, Auckland 1004 [pdale@ihug.co.nz] Psylloidea

Dr Victor F. Eastop Department of Entomology, The Natural History Museum, Cromwell Rd, London SW7 5BD, UK [v.eastop@btinternet.com] Aphidoidea

Dr Murray J. Fletcher Industry & Investment NSW, Orange Agricultural Institute, Orange, NSW 2800, Australia [murray.fletcher@industry.nsw.gov.au] Auchenorrhyncha

Ms Penelope Greenslade Centre for Environmental Management, University of Ballarat, GPO Box 663, Ballarat, VIC 3353 (formerly Department of Biology, Australian National University, ACT 0200), Australia [p.greenslade@ballarat.edu.au] Collembola

Mrs Rosa Henderson Landcare Research, Private Bag 92170, Auckland, New Zealand [hendersonr@landcareresearch.co.nz] Sternorrhyncha

Mr Terry R. Hitchings Canterbury Museum, Rolleston Avenue, Christchurch 8001, New Zealand [thitchings@internet.co.nz] Ephemeroptera

Dr Robert J. B. Hoare Landcare Research, Private Bag 92-170, Auckland, New Zealand [hoarer@landcareresearch.co.nz] Lepidoptera

Mr Peter M. Johns Canterbury Museum, Rolleston Avenue, Christchurch 8001, New Zealand [majohns@xtra.co.nz] Blattodea, Orthoptera, Diptera (Tipulidae)

Dr Marie-Claude Larivière Landcare Research, Private Bag 92-170, Auckland, New Zealand [larivierem@landcareresearch.co.nz] Hemiptera epidoptera

Dr Rod P. Macfarlane 5 McAllister Place, Kaiapoi 7630, New Zealand [rodpam@clear.net.nz] Protura, Diplura, Archaeognatha, Thysanura, Odonata, Isoptera, Mantodea, Dermaptera, Thysanoptera, Phasmatodea, Megaloptera, Neuroptera, Diptera, Strepsiptera

Dr Peter Maddison Research Associate, Landcare Research, Private Bag 92-170, Auckland, New Zealand [maddisonpa@yahoo.com.au] Coleoptera checklist

Dr Nicholas Martin Plant and Food Research, Private Bag 92169, Auckland Mail Centre, Auckland 1142, New Zealand [nicholas.martin@plantandfood.co.nz] Diptera: Cecidomyiidae, Fergusoninidae, plant hosts of Tephritidae

Dr Ian D. McLellan Deceased 2010. Plecoptera, Diptera (Blephariceridae, Thaumaleidae)

Ricardo L. Palma Museum of New Zealand Te Papa Tongarewa, P.O. Box 467, Wellington, New Zealand [ricardop@tepapa.govt.nz] Phthiraptera

Professor Robert L. C. Pilgrim Deceased 2010. Mecoptera, Siphonaptera

Dr David A. J. Teulon Plant and Food Research, Private Bag 4704, Lincoln, New Zealand [david.teulon@plantandfood.co.nz] Aphidoidea

Dr David R. Towns Department of Conservation, Private Bag 68-908, Newton, Auckland, New Zealand [dtowns@doc.govt.nz] Ephemeroptera

Dr Steven A. Trewick Institute of National Resources, Massey University, Private Bag 11-222, Palmerston North 4222, New Zealand [s.trewick@massey.ac.nz] Phasmatodea

Dr Courtenay Smithers Research Associate, Entomology Department, Australian Museum, 6 College Street, Sydney, NSW 2010, Australia [smithers@sydney.net] Psocoptera

Dr Marlon A. W. Stufkens Plant and Food Research, Private Bag 4704, Lincoln, New Zealand [stufkens@clear.net.nz] Aphidoidea

Dr John B. Ward Canterbury Museum, Rolleston Avenue, Christchurch 8001, New Zealand [jward@cantmus.govt.nz] Trichoptera

References

ABRS 2009: *Australian Faunal Directory*. Australian Biological Resources Study, Canberra. [http://www.environment. gov.au/biodiversity/abrs/online-resources/fauna/afd/index.html]

ACHTERBERG, C. van 1990: *Pronkia*, a new genus of the Meteorideinae (Hymenoptera: Braconidae) from New Zealand. *Zoologische Mededelingen Leiden 64* : 169–175.

ACHTERBERG, C. van 1995: Generic revision of the subfamily Betylobraconinae (Hymenoptera: Braconidae) and other groups with modified fore tarsus. *Zoologische Verhandelingen 298*: 1–242.

ACHTERBERG, C. van 2004: New Indo-Australian subgenera and species of the genera *Xynobius* Foerster and *Ademoneuron* Fischer (Hymenoptera: Braconidae: Rogadinae). *Zoologische Mededelingen Leiden 78*: 313–329.

ACHTERBERG, C. van; BERRY, J. 2004: *Neptihormius* gen. nov. (Hymenoptera: Braconidae: Hormiinae), a parasitoid of Nepticulidae (Lepidoptera) from New Zealand. *Zoologische Mededelingen Leiden 78*: 291–299.

ACHTERBERG, C. van; BERNDT, L.; BROCKERHOFF, E.; BERRY, J. 2005: A new species of the genus *Aleiodes* from New Zealand (Hymenoptera: Braconidae: Rogadinae). *Zoologische Mededelingen Leiden 78*: 301–311.

ALEXANDER, C. P. 1929: Tipuloidea of the Tongariro National Park and Ohakune District, New Zealand (Diptera). *Philippine Jouirnal of Science 38*: 157–199.

ALEXANDER, C. P. 1953: New or little-known (Diptera): 95 Oriental–Australasian species. *Annals and Magazine of Natural History, ser. 12, 6*: 739–757.

ALEXANDER, C. P. 1955: The craneflies of the subantarctic islands of New Zealand (Diptera). *Records of the Dominion Museum 2*: 233–239.

ALEXANDER, C. P. 1962: Insects of Macquarie Iasland. Diptera, Tipulidae. *Pacific Insects 4*: 939–944.

ALLEN, A. A. 1985: *Labia minor* L. (Dermaptera) in east London. *Entomologists Record and Journal of Variation 97*: 66.

ALLEN, K. R. 1951: The Horokiwi Stream. A study of a trout population. *New Zealand Marine Department Fisheries Bulletin 10*: 1–231.

ALLSOPP, P. G.; ROBERTSON, L. N. 1988: Biology, ecology and control of soldier flies *Inopus* spp. (Diptera: Stratiomyidae): a review. *Australian Journal of Zoology 36*: 627–648.

AMORIN, D. S.; RINDAL, E. 2007: Phylogeny of the Mycetophiliformia, with proposal of the subfamilies Hetertrichinae, Ohakuneinae, and Chiletrichinae for the Rangomaramidae (Diptera, Bibionomorpha). *Zootaxa 1535:* 1–92.

AMORIM, D. S.; SILVA, V. C. 2002: How far advanced was Diptera evolution in the Pangaea? *Annales de la Société Entomologique de France, n.s., 38*: 177–200.

AMORIN, D. S.; YEATES, D. K. 2006: Pesky gnats: getting rid of Nematocera in Diptera classification. *Studia Dipterologia 13*: 3–9.

ANDERSON, J. A. T. 1935: The description, bionomics, morphology and anatomy of a new *Dryomyia*. (Cecidomyiidae, Diptera). *Proceedings of the Zoological Society of London 1935*: 421–430.

ANDERSON, N. H. 1982: A survey of aquatic insects associated with wood debris in New Zealand streams. *Mauri Ora 10*: 21–33.

ANDERSON, D. T.; LAWSON-KERR, C. 1977: The embryonic development of the marine caddis fly, *Philanisus plebeius* Walker (Trichoptera: Chathamiidae). *Biological Bulletin 153*: 98–105.

ANDREW, I. G. 2000: Species diversity in the *Trimicra pilipes* complex (Diptera: Tipulidae). *New Zealand Entomologist 23*: 3–8.

APPLETON, C. 1993: Habitat and seasonal effects on blowfly ecology in possum carcasses in Manawatu. MSc thesis, Massey University, Palmerston North. 86 p.

ARENSBURGER, P.; BUCKLEY, T. R.; SIMON, C.; MOULDS, M.; HOLSINGER, K. 2004a: Biogeography and phylogeny of the New Zealand cicada genera (Hemiptera: Cicadidae) based on nuclear and mitochondrial DNA data. *Journal of Biogeography 31*: 557–569.

ARENSBURGER, P.; SIMON, C.; HOLSINGER, K. 2004b: Evolution and phylogeny of the New Zealand cicada genus *Kikihia* Dugdale (Homoptera: Auchenorrhyncha: Cicadidae) with special reference to the origin of the Kermadec and Norfolk Islands species. *Journal of Biogeography 31i*: 1769–1783.

ARMITAGE, P. D.; CRANSTON, P. S.; PINDER, L. C.V. (Eds) 1995: *The Chironomidae. Biology and Ecology of Non–biting Midges*. Chapman & Hall, London. 572 p.

ARTIGAS, J. N. 1970: The Asilidae of Chile. *Gayana, Zoología 17*: 1–472. [In Spanish.]

ASHBY, J. W. 1974: A study of arthropod predation of *Pieris rapae* L. using serological and exclusion techniques. *Journal of Applied ecology 11*: 419–426.

ASHBY, J. W.; POTTINGER, R. P. 1974: Natural regulation of *Pieris rapae* Linnaeus (Lepidoptera: Pieridae) in Canterbury, New Zealand. *New Zealand Journal of Agricultural Research 17*: 229–239.

ASHE, F.; O'CONNOR, J. P.; MURRAY D. A. 1998: A check list of Irish aquatic insects. *Irish Biogeographical Society Occasional Publication 3*: 1–80.

AUSTIN, A. D. 1988: A new genus of baeine wasp (Hymenoptera: Scelionidae) from New Zealand associated with moss. *New Zealand Journal of Zoology 15*: 173–183.

AUSTIN, A. D. 1999: The importance of'species' in biodiversity studies: lessons from a mega-diverse group–the parasitic Hymenoptera. Pp. 159–165 *in*: Ponder, W.; Lunney, D. (Eds), *The Other 99%. The Conservation and Biodiversity of Invertebrates*. The Royal Zoological Society of New South Wales, Mosman. vii + 454 pp.

AZIDAH, A. A.; FITTON, M. G.; QUICKE, D. L. J. 2000: Identification of the *Diadegma* species (Hymenoptera: Ichneumonidae, Campopleginae) attacking the Diamondback Moth, *Plutella xylostella* (Lepidoptera: Plutellidae). *Bulletin of Entomological Research 90*: 375–389.

BADONNEL, A. 1951: Psocoptères. *In*: Grassé, P.-P. *Traité de Zoologie 10(2)*: 1301–1340.

BAIN, J.; JENKIN, M. J. 1983: *Kalotermes banksiae, Glyptotermes brevicornis*, and other termites (Isoptera) in New Zealand. *New Zealand Journal of Entomology 7*: 365–371.

BANKS, J. C.; PALMA, R. L. 2003: A new species and new host records of *Austrogoniodes* (Insecta: Phthiraptera: Philopteridae) from penguins (Aves: Sphenisciformes). *New Zealand Journal of Zoology 30*: 69–75.

BARKER, G. M.; POTTINGER, R. P.; ADDISON, P. J.; OLIVER, E. H. A. 1984: Pest status of *Cerodontha* species and other shoot flies in Waikato pastures. *Proceedings of the Weed and Pest Control Conference 37*: 96–100.

BARNES, H. F. 1936: Notes on Cecidomyidae. II. *Annals and Magazine of Natural History, ser. 10, 17*: 272–279.

BARNES, H. F. 1940: The gall midges attacking the seed heads of cocksfoot, *Dactylis glomerata* L. *Bulletin Entomological Research 31*: 111–119.

BARNES, H. F.; LAMB, K. P. 1954: Gall midges (Cecidomyidae) on *Coprosma* (Rubiaceae) in New Zealand. *Transactions of the Royal Society of New Zealand 82*: 813–816.

BARNES, J. K. 1979: Revision of the New Zealand genus *Neolimnia* (Diptera, Sciomyzidae). *New Zealand Journal of Zoology 6*: 241–265.

BARNES, J. K. 1980a: Taxonomy of the New Zealand genus *Eulimnia*, and biology and immature stages of *E. philpotti* (Diptera: Sciomyzidae). *New Zealand Journal of Zoology 7*: 91–103.

BARNES, J. K. 1980b: Biology and immature stages of *Helosciomyza subalpina* (Diptera: Helosciomyzidae), an ant killing fly from New Zealand. *New Zealand Journal of Zoology 7*: 221–229.

BARNES, J. K. 1980c: Immature stages of *Polytocus costatus* (Diptera: Helosciomyzidae) from the Snares Islands, New Zealand. *New Zealand Journal of Zoology 7*: 231–233.

BARNES, J. K. 1981: Revision of the Helosciomyzidae (Diptera). *Journal of the Royal Society of New Zealand 11*: 45–72.

BARRATT, B. I. P. 1996: Biological control: is it environmentally safe? *Forest & Bird*, November: 36–41.

BARRATT, B. I. P.; EVANS, A. A.; FERGUSON, C. M.; BARKER, G. M.; McNEILL, M. R.; PHILLIPS, C. B. 1997: Laboratory nontarget host range of the introduced parasitoids *Microctonus aethiopoides* and *M. hyperode* (Hymenoptera: Braconidae) compared with field parasitism in New Zealand. *Environmental Entomology 26*: 694–702.

BARRATT, B. I. P.; FERGUSON, C. M.; GOLDSON, S. L.; PHILLIPS, C. M.; HANNAH, D. J. 2000: Predicting the risk from biological control agent introductions: a New Zealand approach. Pp. 167–194 *in*: Follett, P.A.; Duan, J.J. (eds), *Nontarget Effects of Biological Control*. Kluwer Academic Publishing, Boston. 316 pp.

BARRATT, B. I. P.; PATRICK, B. H. 1987: Insects of snow tussock grassland on the east Otago plateau. *New Zealand Entomologist 10*: 69–98.

BARRETT, P. 1991: *Keeping Wetas in Captivity: A Series of Nine Articles for Schools and Nature Lovers*. [Ramsay, G. W. (ed.)] Wellington Zoological Gardens, Wellington. 60 p.

BARTHELL, J. F.; FRANKIE, G. W.; THORP, R. W. 1998: Invader effects on a community of cavity nesting megachilid bees (Hymenoptera: Megachilidae). *Environmental Entomology 27*: 240–247.

BARTHELL, J. F.; STONE, R. 1995: Recovery of the parasite *Triarthria spinipennis* (Meigen) (Diptera: Tachinidae) from an inland Californian population of the introduced European earwig. *Pan–Pacific Entomologist 71*: 137–141.

BATES, L. H.; MILN, A. J. 1982: Parasites and predators of lucerne aphids at Flock House, Bulls. *Proceedings of the New Zealand Weed and Pest Control Conference 35*: 123–126.

BAYLY, I. A. E.; EDWARDS, J. S.; CHAMBERS, T. C. 1956: The crater lakes of Mayor Island. *Tane,*

Journal of the Auckland University Field Club 7: 36–46.

BEARDSLEY, J. W. 1964: Insects of Campbell Island. Homoptera: Coccoidea. *Pacific Insects Monograph 7*: 238–252.

BEGGS, J. 2001: The ecological consequences of social wasps (*Vespula* spp.) invading an ecosystem that has an abundant carbohydrate resource. *Biological Conservation 99*: 17–28.

BEJAKOVICH, D.; PEARSON, W. D.; O'DONNELL, M. R. 1998: Nationwide survey of pests and diseases of cereals and grass seed crops in New Zealand. 1. Arthropods and molluscs. *Proceedings of the New Zealand Plant Protection Conference 51*: 38–50.

BELKIN, J. N. 1968: Mosquito studies (Diptera: Culicidae) VII. The Culicidae of New Zealand. *American Entomological Institute Contribution 3*: 1–182.

BELLINGER, P. F.; CHRISTIANSEN, K. A.; JANSSENS, F. 1996–2005: Checklist of the Collembola of the World. http://www.collembola.org http://www.geocities.com/CapeCanaveral/Lab/1300/index.html.

BELOKOBYLSKIJ, S. A.; IQBAL, M.; AUSTIN, A. D. 2004: Systematics, distribution and diversity of the Australasian doryctine wasps (Hymenoptera, Braconidae, Doryctinae). *Records of the South Australian Museum Monograph Series 7*: 1–150.

BELSHAW, R.; DOWTON, M.; QUICKE, D. L. J.; AUSTIN, A. D. 2000: Estimating ancestral distribution: a Gondwanan origin for aphid parasitoids? *Proceedings of the Royal Society of London, ser. B, 267*: 491–496.

BEN-DOV, Y. 1976: Redescription of *Natalaspis leptocarpi* n. comb. (Homoptera: Diaspididae). *New Zealand Journal of Zoology 3*: 27–29.

BEN-DOV, Y. 1993: *A Systematic Catalogue of the Soft Scale Insects of the World (Homoptera: Coccoidea: Coccidae).* [*Flora and Fauna Handbook 9.*] Sandhill Crane Press, Gainsville. 536 p.

BEN-DOV, Y.; MILLER, D. R.; GIBSON, G. A. P. 2007: *ScaleNet: A database of the Scale Insects of the World.* http://www.sel.barc.usda. gov/scalenet/scalenet.htm. (Accessed 20 October 2007.)

BERGROTH, E. 1927: Hemiptera Heteroptera from New Zealand. *Transactions and Proceedings of the New Zealand Institute 57*: 671–684.

BERNHARD, T. P. 1995: Notes on the anthecology of *Pterostylis curta* (Orchidaceae). *Cunninghamia 4*: 1–8.

BERRY, J. A. 1989: *Ancistrocerus gazella* (Vespoidea: Eumenidae): a first record for New Zealand. *New Zealand Entomologist 12:* 63–65.

BERRY, J. A. 1990a: The New Zealand species of the subfamily Metopiinae (Hymenoptera: Ichneumonidae). *New Zealand Journal of Zoology 17*: 607–614.

BERRY, J. A. 1990b: Two parasitoid complexes: *Hierodoris atychioides* (Butler) (Lepidoptera: Oecophoridae) and *Icerya purchasi* Maskell (Homoptera: Margarodidae). *New Zealand Entomologist 13*: 60–62.

BERRY, J. A. 1991: Hymenoptera primary types in the New Zealand Arthropod Collection. *New Zealand Journal of Zoology 18*: 323–341.

BERRY, J. A. 1995: Moranilini (Insecta: Hymenoptera). *Fauna of New Zealand 33*: 1–82.

BERRY, J. A. 1997a: *Meteorus pulchricornis* Wesmael (Hymenoptera: Braconidae: Euphorinae), a new record for New Zealand. *New Zealand Entomologist 20*: 45–48.

BERRY, J. A. 1997b: *Nematus oligospilus* (Hymenoptera: Tenthredinidae), a recently introduced sawfly defoliating willows in New Zealand. *New Zealand Entomologist 20*: 51–54.

BERRY, J. A. 1998: The bethyline species (Hymenoptera: Bethylidae: Bethylinae) imported into New Zealand for biological control of pest leafrollers. *New Zealand Journal of Zoology 25*: 329–333.

BERRY, J. A. 1999: Revision of the New Zealand endemic eulophid genus *Zealachertus* Boucek (Hymenoptera: Chalcidoidea). *Invertebrate Taxonomy 13*: 883–915.

BERRY, J. A. 2003: *Neopolycystus insectifurax* Girault (Hymenoptera: Pteromalidae) is established in New Zealand, but how did it get here? *New Zealand Entomologist 26*: 113–114.

BERRY, J. A. 2006: Brown lace lerp hyperparasitoid found in New Zealand. *Biosecurity Magazine 68*: 18–19.

BERRY, J. A. 2007a: Alysiinae (Insecta: Hymenoptera: Braconidae). *Fauna of New Zealand 58*: 1–95.

BERRY, J. A. 2007b: A key to the New Zealand species of *Psyllaephagus* Ashmead (Hymenoptera: Encyrtidae) with descriptions of three new species and a new record of the psyllid hyperparasitoid *Coccidoctonus psyllae* Riek (Hymenoptera: Encyrtidae). *Australian Journal of Entomology 46*: 294–299.

BERRY, J. A.; GREEN, O. R.; SHATTUCK, S. O. 1997: Species of *Pheidole* Westwood (Hymenoptera: Formicidae) established in New Zealand. *New Zealand Journal of Zoology 24*: 25–33.

BERRY, J. A.; OSTEN, T.; EMBERSON, R. M. 2001: *Radumeris tasmaniensis* (Saussure, 1855), the first record of a scoliid wasp from New Zealand (Hymenoptera, Scoliidae, Campsomerini). *Entomofauna 22*: 41–48.

BERRY, J. A.; WALKER, G. P. 2004: *Meteorus pulchricornis* (Wesmael) (Hymenoptera: Braconidae: Euphorinae): an exotic polyphagous parasitoid in New Zealand. *New Zealand Journal of Zoology 31*: 33–44.

BERRY, J. A.; WITHERS, T. M. 2002. New, gall-forming species of ormocerine pteromalid (Hymenoptera: Pteromalidae: Ormocerinae) described from New Zealand. *Australian Journal of Entomology 41*: 18–22.

BERWICK, R. E. 1959: The life history of the common New Zealand skink *Leiolopisma zealandica* (Gray 1843). *Transactions of the Royal Society of New Zealand 86*: 331–380.

BETHOUX, O. 2009: The earliest beetle identified. *Journal of Paleontology 83*: 931–937.

BEUTEL, R. G.; BAUM, E. 2008: A longstanding entomological problem finally solved? Head morphology of *Nannochorista* (Mecoptera, Insecta) and possible phylogenetic implications. *Journal of Zoological Systematics and Evolutionary Research 46*: 346–367.

BEUTEL, R. G.; HAAS, F. 2000: Phylogenetic relationships of the suborders of Coleoptera (Insecta). *Cladistics 16*: 103–141.

BEUTEL, R. G.; KRISTENSEN, N. P.; POHL, H. 2009: Resolving insect phylogeny: the significance of cephalic structures of Nannomecoptera in understanding endopterygote relationships. *Arthropod Structure and Development 38*: 427–460.

BHARADWAJ, R. D. S. 1966: Observations on the bionomics of *Euboriella annulipes* (Dermaptera: Labiduriidae). *Annals of Entomological Society of America 59*: 441–450.

BICKEL, D. J. 1991: Sciapodinae, Medeterinae (Insecta: Diptera) with a generic review of the Dolichopodidae. *Fauna of New Zealand 23*: 1–74.

BICKEL, D. J. 1996: *Thinempis* a new genus from Australia and New Zealand (Diptera: Empididae) with notes on the tribal classification on the Empidinae. *Systematic Entomologist 21*: 115–128.

BICKEL, D. J. 1999: Australian Sympycininae II. *Syntormon* Leow and *Nothorhapium* gen. nov., with a treatment of the West Pacific fauna, and notes on the subfamily Rhaphiinae and *Dactylonotus* Parent (Diptera: Dolichopodidae). *Invertebrate Taxonomy 13*: 179–206.

BICKEL, D. J. 2008: The Dolichopodinae (Diptera: Dolichopodidae) of New Caledonia, with descriptions and records from Australia, New Zealand and Melanesia. *In*: Grandcolas, P. (ed.), *Zoologia Neocaledonica 6. Biodiversity Studies in New Caledonia. Mémoires du Muséum national d'Histoire naturelle 196*: 13–47.

BIGELOW, R. S. 1967: *The Grasshoppers of New Zealand: Their Taxonomy and Distribution.* University of Canterbury, Christchurch. 112 p.

BISHOP, D. M.; HEATH, A. C. G. 1998: Checklist of ectoparasites of birds in New Zealand. *Surveillance 25, Special Issue*: 13–31.

BITSCH, C.; BITSCH, J. 1998: Internal anatomy and phylogenetic relationships among apterygote insect clades (Hexapoda). *Annales de la Société Entomologique de France 34*: 339–363.

BITSCH, C.; BITSCH, J. 2000: The phylogenetic interrelationships of the higher taxa of apterygote hexapods. *Zoologica Scripta 29*: 131–156.

BLACKMAN, R. L.; EASTOP, V. F. 1984: *Aphids on the World's Crops: An Identification Guide.* John Wiley & Sons, Chichester.

BLACKMAN, R. L.; EASTOP, V. F. 1994: *Aphids on the World's Trees. An Identification and Information Guide.* CAB International, Wallingford, Oxford.

BOOTHROYD, I. K. G. 1999: Description of *Kaniwhaniwhanus* genus n. (Diptera: Chironomidae: Othocladinae) from New Zealand. *New Zealand Journal of Marine and Freshwater Research 33*: 341–349.

BOOTHROYD, I. K. G. 2002: *Crictopus* and *Paratrichocladius* (Chironomidae: Insecta) in New Zealand, with description of *C. hollyfordensis* n. sp. and redescription of adult and immature stages of *C. zealandicus* and *P. pleuriserialis*. *New Zealand Journal of Marine and Freshwater Research 36*: 725–788.

BOOTHROYD, I. K. G.; CRANSTON, P. 1999: The 'iceworm'–the immature stages, phylogeny and biology of the glacier midge *Zelandochlus* (Diptera: Chironomidae). *Aquatic Insects 21*: 303–316.

BOOTHROYD, I.; STARK, J. 2000: Use of invertebrates in monitoring. Pp. 344–373 *in*: Collier, K.J.; Winterbourn, M.J. (eds), *New Zealand Stream Invertebrates: Ecology and Implications for Management.* Caxton Press, Christchurch. 415 p.

BORSCHENIUS, N. S.; WILLIAMS, D. J. 1963: A study of the types of some little-known genera of Diaspididae with descriptions of new genera (Hemiptera: Coccoidea). *Bulletin of the British Museum (Natural History) Entomology 13*: 1–394.

BOSWELL, C. C. 1967: An ecological study of the decomposition of the carrion of the European hare *Lepus europaeus* Pallas with special reference to the insect fauna. MAgrSci Thesis, Lincoln University, New Zealand. 127 p.

BOSZIK, A. 1992: Natural adult food of some important *Chrysopa* species (Plannipennia: Chrysopidae). *Acta Phytopathologia et Entomologica Hungarica 27*: 91–146.

BOSZIK, A. 1994: Impact of vegetative diversity and structure parameters of chrysopid assemblages. *Redia 77*: 69–77.

BOSZIK, A. 2000: Analysis of the food of some adult central European Chrysopidae (Neuroptera: Chrysopidae). *Beitrage zur Entomologie 50*: 237–246.

BOUCEK, Z. 1988: *Australasian Chalcidoidea (Hymenoptera). A biosystematic revision of genera of fourteen families, with a reclassification of species*. CAB International, Wallingford. 832 p.

BOWIE, M. H. 2001: Ecology and morphology of *Allograpta ventralis* (Diptera: Syrphidae) a predator on the cabbage tree mealy bug *Balanococcus cordylinidis* (Hemiptera: Pseudococcidae). *New Zealand Natural Sciences 26*: 1–11.

BOWIE, M. H.; MARRIS, W. M.; EMBERSON, R. M.; ANDREW, I. G.; BERRY, J. A.; VINK, C. J.; WHITE, E. G.; STUFKINS, M. A. W.; OLIVER, E. H. A.; EARLY, J. W.; KLIMASZEWSKI, J.; JOHNS, P. M.; WRATTEN, S. D.; MAHLFIELD, K.; BROWN, B.; EYLES, A. C.; PAWSON, S. M. & MACFARLANE, R. P. 2004: A terrestrial invertebrate inventory of Quail Island (Otamahua): towards the restoration of the invertebrate community. *New Zealand Natural Sciences, 28*: 1–29.

BOYD, A. M.; WEINSTEIN, P. 1996. *Anopheles annulipes*: an under-rated temperate climate malaria vector. *New Zealand Entomologist 19*: 35–41.

BRAGG, P. E. 1993: Parasites of Phasmida. *Entomologist 112*: 37–42.

BRADLER, S. 2001: The Australian stick insects, a monophyletic group within the Phasmatodea? *Zoology 104*, Suppl. 3: 69.

BRADLER, S. 2009: Phylogeny of the stick and leaf insects (Insecta: Phasmatodea). *Species, Phylogeny and Evolution 2*: 3–139.

, I.; MATHIS, W. N. 2006: Revision of *Australimyza* Harrison (Diptera; Australimyzidae). *Systematic Entomology 32*: 252–275.

BRAUER, F. 1865: Vierter Bericht über die auf der Weltfahrt der kais. Fregatte Novara gesammelten Neuropteren. *Verhandlungen des zoologische–botanischen Gesellschaft in Wien 15*: 903–908.

BRETFELD, G. 1999: Symphypleona. *In*: Dunger, W. (ed.), *Synopses of Palaearctic Collembola, Volume 2. Abhandlungen und Berichte des Naturhistorischen Museums zu Görlitz 71*: 1–318.

BRINDLE, A. 1972: Insects of Micronesia: Dermaptera. *Insects of Micronesia 5*: 97–171.

BRINDLE, A. 1987: Order Dermaptera. Pp. 171–178 *in*: Stehr, F. W. (ed.), *Immature Insects*. Vol. 1. Kendal Hunt, Dubuque, Iowa.

BRITTIN, G. 1915a: Some new Coccidae. *Transactions and Proceedings of the New Zealand Institute 47*: 149–156.

BRITTIN, G. 1915b: New Coccidae. *Transactions and Proceedings of the New Zealand Institute 47*: 156–160.

BRITTIN, G. 1916: Notes on some Coccidae in the Canterbury Museum, together with a description of a new species. *Transactions and Proceedings of the New Zealand Institute 48*: 423–426.

BRITTIN, G.; 1937: Notes on the genus *Leucaspis*, with descriptions of thirteen New Zealand species and re-description of eight foreign species. *Transactions and Proceedings of the Royal Society of New Zealand 67*: 281–302.

BROCKIE, R. E. 1959: Observations of the food of the hedgehog in New Zealand. *New Zealand Journal of Science 2*: 121–136.

BROWN, B.V. 1992: Generic revision of Phoridae of the neartic region and phylogenetic classification of Phoridae, Sciadoceridae and Ironyiidae (Diptera: Phoridae). *Memoirs of the Entomological Society of Canada 164*: 3–144.

BROWN, B.V.; OLIVER, E. H. A. 2007: First records of *Megaselia scalaris* (Loew) and *Megaselia spiracularis* Schmitz (Diptera: Phoridae) from New Zealand with additional information on other worldwide species. *New Zealand Entomologist 30*: 85–87.

BROWN, B.V.; OLIVER, H. 2008: Two new genera of Phoridae (Diptera) from New Zealand. *Zootaxa 1933*: 1–11.

BROWN, W. L. 1958: A review of the ants of New Zealand. *Acta Hymenopterologica 1*: 1–50.

BRUNDIN, L. 1966: Transantarctic relationships and their significance as evidenced by chironomid midges. *Kungliga Svenska Vetenskapsakademiens Handlinger*, ser. *4, 11*: 1–472.

BRUNDIN, L. 1967: Insects and the problem of austral disjunctive distribution. *Annual Review of Entomology 12*: 149–168.

BRUNNER VON WATTENWYL, K. 1907: Die Insektenfamilie der Phasmiden. II. Phasmidae Anareolatae (Clitumnini, Lonchodini, Bacunculini). Wilhelm Engelmann, Leipzig. Pp. 181–340, pls 7–15.

BUCKLEY, T. R.; ARENSBURGER, P.; SIMON, C.; CHAMBERS, G. K. 2002: Combined data; Bayesian phylogenetics, and the origin of the New Zealand cicada genera. *Systematic Biology 51*: 4–18.

BUCKLEY, T. R.; ATTANAYAKE, D; BRADLER S. 2009c: Extreme convergence in stick insect evolution: phylogenetic placement of the Lord Howe Island tree lobster. Proceedings of the Royal Society B, 276: 1055–1062.

BUCKLEY, T. R.; ATTANAYAKE, D.; NYLANDER, J. A. A.; BRADLER, S. 2010: The phylogenetic placement and biogeographical origins of the New Zealand stick insects (Phasmatodea). Systematic Entomology 35: 207–225.

BUCKLEY, T. R.; ATTANAYAKE, D.; PARK, D.-C.; RAVINDRAN, S; JEWELL, T. R.; NORMARK, B. B. 2008: Investigating hybridization in the parthenogenetic New Zealand stick insect Acanthoxyla (Phasmatodea) using single-copy nuclear loci. Molecular Phylogenetics and Evolution 48: 335–349.

BUCKLEY, T. R.; BRADLER, S. 2010: Tepakiphasma ngatikuri, a new genus and species of stick insect (Phasmatodea) from the Far North of New Zealand. New Zealand Entomologist 33: 118–126.

BUCKLEY, T. R.; CORDEIRO, M.; MARSHALL, D. C.; SIMON, C. 2006: Differentiating between hypotheses of lineage sorting and introgression in New Zealand alpine cicadas (*Maoricicada* Dugdale). *Systematic Biology 51*: 4–18.

BUCKLEY, T. R.; MARSKE, K.; ATTANAYAKE, D. 2009A: Phylogeography and ecological niche modelling of the New Zealand stick insect *Clitarchus hookeri* (White) support survival in multiple coastal refugia. *Journal of Biogeography 37*: 682–695.

BUCKLEY, T. R.; MARSKE, K.; ATTANAYAKE, D. 2009B: Identifying glacial refugia in a geographic parthenogen using palaeoclimate modeling and phylogeography: the New Zealand stick insect *Argosarchus horridus* (White). *Molecular Ecology 18*: 4650–4663.

BUCKLEY, T. R; SIMON, C. 2007: Evolutionary radiation of the cicada genus *Maoricidada* Dugdale (Hemiptera: Cicadoidea) and the origins of the New Zealand alpine biota. *Biological Journal of the Linnean Society 91*: 419–435.

BUCKLEY, T. R.; SIMON, C.; CHAMBERS, G. K. 2001: Phylogeography of the New Zealand cicada *Maoricicada campbelli* based on mitochondrial DNA sequences: ancient clades associated with Cenozoic environmental change. *Evolution 55*: 1395–1407.

BUCKLEY, T. R.; YOUNG, E. C. 2008: A revision of the taxonomic status of *Sigara potamius* and *S. limnochares* (Hemiptera: Corixidae), water boatmen of braided rivers in New Zealand. *New Zealand Entomologist 31*: 47–57.

BURCKHARDT, D. 2009: Taxonomy and phylogeny of the Gondwanan moss bugs or Peloridiidae (Hemiptera, Coleorrhyncha). *Deutsche Entomologische Zeitschrift 56*: 173–235.

BURMEISTER, T. 2001: Molecular evolution of the arthropod hemocyanin superfamily. *Molecular Biology and Evolution 18*: 184–195.

BURROWS, M.; HARTUNG, V.; HOCH, H. 2007: Jumping behaviour in a Gondwanan relict insect (Hemiptera: Coleoptera: Peloridiidae). *Journal of Experimental Biology 210*: 3311–3318.

BYERS, G. W. 1974: New generic names for Mecoptera of Australia and New Zealand. *Journal of the Australian Entomological Society 13*: 165–167.

CADWALLADER, P. L. 1975: Feeding habits of two fish species in relation to invertebrate drift in a New Zealand river. *New Zealand Journal of Marine and Freshwater Research 9*: 11–26.

CALABY, J. H.; MURRAY, M. D. 1991: Phthiraptera. Pp. 421–428 *in*: CSIRO (ed.), *The Insects of Australia. A Textbook for Students and Research Workers.* 2nd edn. Melbourne University Press, Carlton. Vol. 1, 542 p., Vol. 2, 600 p.

CAMERON, P. J.; HERMAN, T. J. B.; FLETCHER, J. D. 1992: Incidence of thrips and aphids as potential virus vectors in field tomatoes. *New Zealand Weed and Pest Control Conference 45*: 38–47.

CAMERON, P. J.; HILL, R. L.; BAIN, J.; THOMAS, W. P. (Eds) 1989: *A Review of Biological Control of Invertebrate Pests and Weeds in New Zealand 1874 to 1987. [CAB Inter-national Institute of Biological Control Technical Communication 10.]* CAB International, Wallingford. 423 p.

CAMERON, P. J.; HILL, R. L.; BAIN, J.; THOMAS, W. P. 1993: Analysis of importations for biological control of insect pests and weeds in New Zealand. *Biocontrol Science and Technology 3*: 387–404.

CAMERON, P. J.; WALKER, G. P. 1989: *Acyrthosiphom kondoi* Shinji, bluegreen lucerne aphid and *Acyrthosiphon pisum* (Harris), pea aphid (Homoptera: Aphididae). Pp. 3–8 *in*: Cameron, P. J.; Hill, R. L.; Bain, J.; Thomas, W. P. (eds), *A Review of Biological Control of Invertebrate Pests and Weeds in New Zealand 1874 to 1987. CAB International Technical Communication 10*: 1–423.

CAMERON, P. J.; WALKER, G. P.; KELLER, M. A; POPAY, A. J. 1995: Introduction of *Cotesia rubecula*, a parasitoid of white butterfly. *Proceedings of the New Zealand Plant Protection Conference 48*: 345–347.

CAMPBELL, J. W. 1928: Notes on egg-laying and mating habits of *Myopsocus novae-zelandiae* Kolbe. *Bulletin of the Brooklyn Entomological Society 23*: 124–128.

CAMPBELL, P. A. 1973: Feeding behaviour of the hedgehog (*Erinaceus europaeus* L.) in pasture land in New Zealand. *Proceedings of the New Zealand Ecological Society 20*: 35–40.

CANTRELL, B. K. 1986: An updated host catalogue for the Australian Tachinidae (Diptera). *Journal of the Australian Entomological Society 25*: 255–265.

CARAPELLI, A.; LIÒ, P.; NARDI, F.; van der WATH, E.; FRATI, F. 2007: Phylogenetic analysis of mitochondrial protein genes confirms the reciprocal paraphyly of Hexapoda and Crustacea. *BMC Evolutionary Biology 7(Suppl. 2):S8*: [1–13]. doi:10.1186/1471-2148-7-S2-S8.

CARILLO, R. 1985: Ecology of and aphid predation by the European earwig; *Forficula auricularia* L. in grassland and barley. PhD Thesis, Sutherland University, UK. 265 pp. (*Dissertation Abstracts 48*: 673–674).

CARNARD, M.; SEMERIN, V.; NEW, T. R. 1984: *Biology of Chrysopidae*. Junk, Boston.

CARPENTER, A. 1987: Insecticidal control of thrips on nerines. *New Zealand Weed and Pest Control Conference 40*: 44–46.

CARVALHO, C. J. B. (Ed.) 2002: *Muscidae (Diptera) of the Neotropical Region: Taxonomy*. CUFPR, Curitiba. 287 p.

CARVER, M. 2000: A new, indigenous species of *Aphis* Linnaeus (Hemiptera: Aphididae) on *Muehlenbeckia* (Polygonaceae) in New Zealand. *New Zealand Entomologist 22*: 3–7.

CARVER, M.; GROSS, G. F.; WOODWARD, T. E. 1991: Hemiptera (Bugs, leafhoppers, cicadas, aphids, scale insects etc.). Pp. 429–509 *in*: CSIRO, (eds), *The Insects of Australia. A Textbook for Students and Research Workers*. 2nd edn, Volume I. Melbourne University Press, Melbourne. 542 p.

CASSIS, G. 1998: Dermaptera. Pp. 279–345 *in*: Houston, W. W. K.; Welles, A. (eds), Zoological Catalogue of Australia. Vol. 23: Archaeognatha, Zygentoma, Blattodea, Isoptera, Mantodea, Dermaptera, Phasmatodea, Embioptera, Zoraptera. CSIRO Publishing, Melbourne.

CASSIS, G.; GROSS, G.F. 1995: Hemiptera: Heteroptera (Coleorrhyncha to Cimicomorpha). *Zoological Catalogue of Australia, 27.3A*: 1–506.

CASSIS, G.; GROSS, G. F. 2002: Hemiptera: Heteroptera (Penta- tomomorpha). *In*: Houston, W. W. K.; Maynard, G.V. (eds), *Zoological Catalogue of Australia Vol. 27.3B*. CSIRO, Melbourne. xiv + 737 p.

CASTLE, J. F. 1988: Behaviour of the mantid *Orthodera ministralis* Fabr. *The Weta* 11(2): 25–29.

CATERINO, M. S.; SHULL, V. L.; HAMMOND, P. M.; VOGLER, A. P. 2002: Basal relationships of Coleoptera inferred from 18S rDNA sequences. *Zoologica Scripta 31*: 41–49.

CAUGHLEY, G. 1994: Directions in conservation biology. *Journal of Animal Ecology 63*: 215–244.

CHANDLER, P. J. 1994: The oriental and Australasian species of Platypezidae (Diptera). *Invertebrate Taxonomy 8*: 351–434.

CHANDLER, P. J. 1998: Checklists of the British Isles. Part 1 Diptera. *Handbooks for the Identification of British Insects, n.s., 12*: 1–234.

CHANDLER, P. J. 2002: *Heterotricha* Leow and allied genera (Diptera: Sciaroidea): offshoots of the stem group of Mycetophilidae and/or Sciaridae? *Annales de la Société Entomologique de France, n.s., 38*: 101–144.

CHARLES, J. G. 1981: Distribution and life history of the long tongued mealy bug *Pseudococcus longispinus* (Homoptera: Pseudococcidae), in Auckland vineyards. *New Zealand Journal of Zoology 8*: 285–293.

CHARLES, J. G. 1998: The settlement of fruit crop arthropod pests and their natural enemies in New Zealand: an historical guide to the future. *Biocontrol News and Information 19(2)*: 47N–58N.

CHEN, G.-T.; VICKERY, V. R.; KEVAN, D. K. M. 1967: A morphological; comparison of Antipodean *Teleogryllus* species (Orthoptera: Gryllidae). *Canadian Journal of Zoology 45*: 1215–1224.

CHRISTIANSEN, K.; BELLINGER, P. 2000: Redescriptions of some of Salmon's isotomid types. *Contributions from the Biology Laboratory of Kyoto University 29*: 103–115.

CHRISTIANSEN, K.; BELLINGER, P. F. 2001: Checklist of the Collembola: Generic names of the Entomobryidae. http://www.geocities.com/Cape Canaveral/Lab/1300/ index.html

CHUI, V. W. D.; THORNTON, I. W. B. 1972: A numerical taxonomic study of the endemic *Ptycta* species of the Hawaiian Islands (Psocoptera: Psocidae). *Systematic Zoology 21*: 7–22.

CHVÁLA, M. 1983: The Empidoidea (Diptera) of Fennoscandia and Denmark. II. *Fauna Entomologica Scandinavica 12:* 1–279.

CHVÁLA, M.; LYNEBORG, L.; MOUCHA, J. 1972: *The Horse Flies of Europe (Diptera, Tabanidae).* Entomological Society of Copenhagen, Copenhagen. 499 p.

CLARK, J. M. 1992: Food of the morepork in Taranaki. *Notornis 39*: 94.

CLARK, T. L.; MESSINA, F. J. 1998: Foraging behaviour of lacewing larvae (Neuroptera: Chrysopidae) on plants with divergent architecture. *Journal of Insect Behaviour 11*: 303–317.

CLAUSEN C. P. (Ed.) 1978: Introduced parasites and predators of arthropod pests and weeds: a world review. *U.S. Department of Agriculture Handbook 480*: 1–545.

CLAYTON, D. H. 1991: Coevolution of avian grooming and ectoparasite avoidance. Pp. 258–289 *in*: Loye, J. E.; Zuk, M. (eds), *Bird–parasite Interactions: Ecology, Evolution and Behaviour.* Oxford University Press, New York. xvi + 406 p.

CLAYTON, D. H.; MOORE, J. (Eds) 1997: *Host–Parasite Evolution. General Principles and Avian Models.* Oxford University Press Inc., New York. xiii + 473 p.

COLE, F. R.; MEDEIROS, A. C.; LOOPE, L. L.; ZUEHLKE, W. W. 1992: Effects of the Argentine ant on arthropod fauna of Hawaiian high-elevation shrubland. *Ecology 73*: 1313–1322.

COLEMAN, J. D. 1977: The food and feeding of starlings in Canterbury. *Proceedings of the New Zealand Ecological Society 24*: 94–105.

COLLIER, K. J. 1993: Review of the status, distribution, and conservation of freshwater invertebrates in New Zealand. *New Zealand Journal of Marine and Freshwater Research 27*: 219–226.

COLLIER, K. J. 1994: Influence of nymphal size, sex and morphotype on microdistribution of *Deleatidium* (Ephemeroptera: Leptophlebiidae) in a New Zealand river. *Freshwater Biology 31*: 35–42.

COLLIER, K. J.; WINTERBOURN, M. J. (Eds) 2000: *New Zealand Stream Invertebrates: Ecology and Implications for Management.* New Zealand Limnological Society & NIWA, Christchurch. 415 p.

COLLIER, K. J.; FOWLES, C.; HOGG. D. 2000: Management, education and conservation. Pp. 374–400 *in* Collier, K. J.; Winterbourn, M. J. (eds), *New Zealand Stream Invertebrates: Ecology and Implications for Management.* New Zealand Limnological Society & NIWA, Christchurch. 415 p.

COLLIN, J. E. 1928: *New Zealand Empididae: Based on Material in the British Museum (Natural History).* Trustees of the British Museum, London. 110 p.

COLLYER, E. 1964: Phytophagous mites and their predators in New Zealand orchards. *New Zealand Journal of Agricultural Research 7*: 551–568.

CONE, A. 1995: Mast seeding and the biologies of *Chionochloa* pre-dispersal seed products. M.Sc. thesis, University of Canterbury, Christchurch. 251 p.

COOK, C. E.; SMITH, M. L.; TELFORD, M. E.; BASTIONELLO, A.; AKAM, M. 2001: *Hox* genes and the phylogeny of the arthropods. *Current Biology 11*: 759–763.

COTTIER, W. 1934: The natural enemies of the European red-mite in New Zealand. (*Paratetranychus pilosus* Can. and Fanz.) *New Zealand Journal of Science and Technology 16*: 54–55.

COTTIER, W. 1953: Aphids of New Zealand. *New Zealand Department of Scientific and Industrial Research Bulletin 106*: 1–382.

COTTIER, W. 1956: Insect pests. Pp. 209–481 *in:* Atkinson, J. D.; Brien, R. M.; Chamberlain, E. E.; Cottier, W.; Dingley, J. M.; Jacks, H.; Reid, W. D.; Taylor, G. G. (eds), *Plant Protection in New Zealand.* Government Printer, Wellington. 699 p.

COTTIER, W. 1964: Insects of Campbell Island. Hemiptera. Homoptera: Aphididae. *Pacific Insects Monograph 7*: 236–237.

COWAN, P. E.; MOEED, A. 1987: Invertebrates in the diet of brushtail possums, *Trichosurus vulpecula*, in lowland podocarp/broadleaf forest, Orongorongo Valley, Wellington, New Zealand. *New Zealand Journal of Zoology 14*: 163–177.

COWLEY, D. R. 1976a: Additions and amendments to the New Zealand Trichoptera. *New Zealand Journal of Zoology 3*: 21–26.

COWLEY, D. R. 1976b: Family characteristics of the pupae of New Zealand Trichoptera. *New Zealand Journal of Zoology 3*: 99–109.

COWLEY, D. R. 1978: Studies on the larvae of New Zealand Trichoptera. *New Zealand Journal of Zoology 5*: 639–750.

COWLEY, D. R. 1984 [1985]: *Coriophagus casui* sp. n. (Halictophagidae: Coriophaginae): a male strepsipteran from Auckland, New Zealand. *New Zealand Journal of Zoology 11*: 351–353.

COX, J. M. 1987: Pseudococcidae (Insecta: Hemiptera: Pseudococcidae). *Fauna of New Zealand 11*: 1–230.

COX, J.; DALE, P. S. 1976: New records of plant pests in New Zealand. II. *New Zealand Journal of Agricultural Research 20*: 109–111.

CRAIG, D. A. 1969: A taxonomic revision of New Zealand Blepharoceridae and the origin and evolution of the Australasian Blepharoceridae (Diptera: Nematocera). *Transactions of the Royal Society of New Zealand Biological Sciences 11:* 101–151.

CRANSTON, P. S. 2007: The identity of *Dactylocladius commensalis* (Diptera: Chironomidae) revealed. *Aquatic Insects 29*: 103–114.

CRANSTON, P. S. 2009: A new genus of trans-Tasman midge: *Anzacladius* gen. n. (Diptera: Chironomidae : Orthocladiinae). *Australian Journal of Entomology 48*: 130–139.

CROSBY, T. K. 1986: The genus *Trisopsis* in New Zealand (Diptera: Cecidomyiidae). *Weta* 9: 30–32.

CROWSON, R. A. 1980: On amphipolar distribution patterns in some cool climate groups of Coleoptera. *Entomologia Generalis 6*: 281–292.

CRUMB, S. E.; EIDE, P. M.; BONN, A. E. 1941: The European earwig. *Technical Bulletin USDA 766*: 1–76.

CSIRO (Ed.) 1991: *The Insects of Australia. A Textbook for Students and Research Workers.* 2nd edn. Melbourne University Press, Carlton. Vol. 1, 542 p., Vol. 2, 600 p.

CUMBER, R. A. 1958: The insect complex of sown pastures in the North Island. I. The general picture revealed by summer sweep sampling. *New Zealand Journal of Agricultural Research 1*: 719–749.

CUMBER, R. A. 1959a: The insect complex of sown pastures in the North Island. VI. The Psocoptera and Neuroptera as revealed by summer sweep-sampling. *New Zealand Journal of Agricultural Research 2*: 898–902.

CUMBER, R. A. 1959b: The insect complex of sown pastures in the North Island VII. The Thysanoptera as revealed by summer sweep sampling. *New Zealand Journal of Agricultural Research 2*: 1123–1130.

CUMBER, R. A. 1962: Insects associated with wheat, barley, and oat crops in the Rangitikei, Manawatu, Southern Hawkes Bay and Waipara Districts during the 1960–61 season. *New Zealand Journal of Agricultural Research 5*:163–178.

CUMBER, R. A. 1966: Factors influencing population levels of *Scolypopa australis* Walker (Hemiptera–Homoptera: Ricaniidae) in New Zealand. *New Zealand Journal of Science 9*: 336–356.

CUMBER, R. A.; EYLES, A. C. 1961a: Insects associated with the major fodder crops of the North Island VI. Odonata, Orthoptera, Isoptera, Psocoptera, Thysanoptera, Neuroptera, Lepidoptera. *New Zealand Journal of Agricultural Research 4*: 426–440.

CUMBER, R. A.; EYLES, A. C. 1961b: Insects associated with the major fodder crops in the North Island. V. Diptera. *New Zealand Journal of Agricultural Research 4*: 409–425.

CUMBER, R. A.; HARRISON, R. A 1959a: Preliminary flight records of Diptera taken with a light trap operated at Piaka. *New Zealand Journal of Science 2*: 237–239.

CUMBER, R. A.; HARRISON, R. A 1959b: The insect complex of sown pastures in the North Island. 2 The Diptera as revealed by summer sweep sampling. *New Zealand Journal of Agricultural Research 2*: 741–762.

DAANE, K. M.; VOKOTA, G. Y. 1997: Release strategies affect survival and distribution of green lacewings (Neuroptera: Chrysopidae) in augumentative programs. *Environmental Entomology 26*: 455–464.

Da GAMA ASSALINO, M. M.; GREENSLADE, P. 1981: Relationships between the distribution and phylogeny of *Xenylla* (Collembola, Hypogastruridae) species in Australia and New Zealand. *Revue d'Écologie et Biologie du Sol 18*: 269–284.

DALE, P. S.; MADDISON, P. A. 1982: A catalogue (1860–1960) of New Zealand insects and their host plants (Revision). *New Zealand Department of Scientific and Industrial Research Bulletin 231*: 1–260.

DANIEL, M. J. 1973: Seasonal diet of the ship rat (*Rattus r. rattus*) in lowland forest in New Zealand. *Proceedings of the New Zealand Ecological Society 20*: 21–30.

DANIELS, G. 1987: A revision of *Neoaratus* Ricardo, with the description of six allied new genera from the Australian region (Diptera: Asilidae: Asilini). *Invertebrate Taxonomy 1*: 473–592.

DANZIG, E. M. 1993: Rhyncota. Scale insects in families Phoenicoccidae and Diaspididae. *Fauna of Russia, n.s., 144, 10*: 1–452.

DAVIDSON, M. M.; CILGI, T.; PETERSEN, M. K.; WRATTEN, S. D.; FRAMPTON, F. 1997: Resilience of springtail (Collembola) populations in farmland following exposure to insecticides. *Australasian Journal of Ecotoxicology 3*: 99–108.

DAVIES, T. H. 1988: List of Mycetophilidae and Sciaridae (Diptera) collected in Hawkes Bay. *New Zealand Entomologist 11*: 12–14.

DAY, M. F.; FLETCHER, M. J. 1994: An annotated catalogue of the Australian Cicadelloidea (Hemiptera: Auchenorrhyncha). *Invertebrate Taxonomy 8*: 117–1288.

DEAR, J. P. 1985: Calliphoridae (Insecta: Diptera). *Fauna of New Zealand 8*: 1–88.

De Barro, P. J.; Driver, F.; Naumann, I. D.; Schmidt, S.; Clarke, G. M.; Curran, J. 2000: Descriptions of three species of *Eretmocerus* Haldeman (Hymenoptera: Aphelinidae) parasitising *Bemisia tabaci* (Gennadius) (Hemiptera: Aleyrodidae) and *Trialeurodes vaporariorum* (Westwood) (Hemiptera: Aleyrodidae) in Australia based on morphological and molecular data. *Australian Journal of Entomology 39*: 259–269

DEBENHAM, M. L. 1979: An annotated checklist and bibliography of the Australasian region Ceratopogonidae (Diptera: Nematocera). *School of Public health in Tropical medicine, University of Sydney, Commonwealth Entomological Monograph 1*: 1–671.

De BOER, J. A.; VALENTINE, E. W. 1977: The identity of *Leucaspis gigas* (Homoptera: Diaspididae) with descriptions of four similar species in New Zealand. *New Zealand Journal of Zoology 4*: 153–164.

DEHARVENG, L.; WISE, K. A. J. 1987: A new genus of Collembola (Neanuridae: Neanurinae) from southern New Zealand. *Records of the Auckland Institute and Museum 24*: 143–146

DEITZ, L. L. 1979a: Two new species of *Colobopyga* (Homoptera: Halimococcidae) from the Australian region. *New Zealand Journal of Zoology 6*: 453–457.

DEITZ, L. L. 1979b: Selected references for identifying New Zealand Hemiptera (Homoptera and Heteroptera), with some notes on nomenclature. *New Zealand Entomologist 7*: 20–29.

DEITZ, L. L.; TOCKER, M. F. 1980: W. M. Maskell's Homoptera: species–group names and type material. *New Zealand Department of Scientific and Industrial Research Information Series 146*: 1–76.

De JONG, H. 1987: Some remarks on the New Zealand crane-fly taxon *Austrotipula* Alexander, 1920 (Diptera, Tipulidae). *New Zealand Journal of Zoology 14*: 95–98.

DELL'AMPIO, E.; SZUCSICH, U.; CARAPELLI, A.; FRATI, F.; STEINER, G.; STEINACHER, A.; PASS, G. 2008: Testing for misleading effects in the phylogenetic reconstruction of ancient lineages of hexapods: influence of character dependence and character choice in analyses of 28S rRNA sequences. *Zoologica Scripta 38*: 155–170.

De MEYER, M. 1991: A new *Cephalops* Fallen, 1810, species from New Zealand (Dipt.,Pipunculidae). *Entomologist's Monthly magazine 1528–1531*: 215–218.

DENNIS, D. J. 1978: Observations of fungus gnat damage to glasshouse curcubits. *New Zealand Journal of Experimental Agriculture 6*: 83–84.

DENNIS, D. S.; KNUTSON, L. 1988: Descriptions of pupae of South American robber flies (Diptera: Asilidae). *Annals of the Entomological Society of America 81*: 851–864.

DEPARTMENT OF CONSERVATION 1996: *The Chatham Islands*. Canterbury University Press, Christchurch. 136 p.

DERRAIK, J. S. B. 2004: A survey of the mosquito (Diptera) fauna of the Auckland Zoological Park. *New Zealand Entomologist 27*: 51–53.

DERRAIK, J. S. B. 2005a: Mosquitoes breeding in phytotelmata in native forests in the Wellington area. *New Zealand Journal of Ecology 29*: 185–191.

DERRAIK, J. S. B. 2005b: Presence of *Culex asteliae* larvae and *Ochlerotatus notoscriptus* (Diptera: Culicidae) in a native tree canopy in the Auckland region. *Weta 29*: 9–11.

DERRAIK, J. S. B.; BARRATT, B. I. P.; SIRVID, P.; MACFARLANE, R. P.; PATRICK, B. H.; EARLY, J. W., EYLES, A. C.; JOHNS, P. M.; FRASER, P. M.; BARKER, G. M.; HENDERSON, R.; DALE, P. M.; HARVEY, M. S.; FENWICK, G.; McLELLAN, I. D.; DICKINSON, K. J. M.; CLOSS, G. R. 2001: Invertebrate survey of a modified native shrubland, Brookvale Covenant, Rock and Pillar Range, Otago, New Zealand. *New Zealand Journal of Zoology 28*: 273–290.

DERRAIK, J. S. B.; HEATH, A. C. G. 2005: Immature Diptera (excluding Culicidae) inhabiting phytotelmata in Auckland and Wellington regions. *New Zealand Journal of Marine and Freshwater Research 39*: 981–987.

DERRAIK, J. S. B.; SNELL A.; SLANEY D. 2005: An investigation into the circadian response of adult mosquiotoes (Diptera: Culicidae) to host cues in west Auckland. *New Zealand Entomologist 28*: 89–94.

DESSART, P. 1987: Révision des Lagynodinae (Hymenoptera: Ceraphronidae: Megaspilidae). *Bulletin de l'Institut Royale des Sciences Naturelles de Belgique, Entomologie 57*: 5–30.

DESSART, P. 1997: Les Megaspilinae ni européens, ni américains. 1. Le genre *Conostigmus* Dahlbom, 1858 (Hym. Ceraphronidae: Megaspilidae). *Mémoires de la Société Royale Belge d'Entomologie 37*: 3–144.

DEVONPORT, B. F.; WINTERBOURN, M. J. 1976: The feeding relationships of two invertebrate predators in a New Zealand river. *Freshwater Biology 6*: 167–176.

DIDHAM, R. 1992: Faunal composition, diversity and spatial heterogeneity of the arthropods in rainforest canopy, New Zealand. MSc thesis, University of Canterbury, Christchurch. 355 p.

DIDHAM, R. K. 1997: Dipteran tree-crown assemblages in a diverse southern temperate rainforest. Pp. 321–343 *in*: Stork, N. E.; Adis, J.; Didham, R. K. (eds), *Canopy Arthropods*. Chapman and Hall, London.

DISNEY, R. H. L. 1994: *Scuttle flies: The Phoridae*. Chapman Hall, London. 467 p.

DIXON, A. F. G. 1987: The way of life of aphids: host specificity, speciation and distribution. Pp. 197–207 *in*: Minks, A. K.; Harrewijn, P. (eds), *Aphids: Their Biology, Natural Enemies and Control Volume A*. Elsevier, Amsterdam.

DIXON, A. F. G. 1998: *Aphid ecology. An Optimization Approach*. Chapman & Hall, London.

DIXON, A. F. G.; KINDLMANN, P.; LEPS, J.; HOLMAN, J. 1987: Why are there so few species of aphids, especially in the tropics? *American Naturalist 129*: 580–592.

DOHLEN, C. D. von; TEULON, D. A. J. 2003: Phylogeny and historical biogeography of New Zealand indigenous Aphidini aphids (Hemiptera: Aphididae): an hypothesis. *Annals of the Entomological Society of America 96*: 107–116.

DOLLING, W. R. 1991: *The Hemiptera*. Oxford University Press, Oxford. 273 p.

DON, W. 2007: *Ants of New Zealand*. Otago University Press, Dunedin. 240 p.

DONNER, J. H.; WILKINSON, C. 1989: Nepticulidae (Insecta: Lepidoptera). *Fauna of New Zealand 16*: 1–88.

DONOVAN, B. J. 1974: Bees of the world: Part III: Bumble bees in New Zealand. *The New Zealand Beekeeper 36(3)*: 49–55.

DONOVAN, B. J. 1983: Comparative biogeography of native Apoidea of New Zealand and New Caledonia. *GeoJournal 7*: 511–516.

DONOVAN, B. J. 1984: Occurrence of the common wasp, *Vespula vulgaris* (L.) (Hymenoptera: Vespidae) in New Zealand. *New Zealand Journal of Zoology 11*: 417–427.

DONOVAN, B. J. 2007: Apoidea (Insecta: Hymenoptera). *Fauna of New Zealand 57*: 1–295.

DOULL, K. M. 1956: Thrips infesting cocksfoot in New Zealand. II. The biology and economic importance of the cocksfoot thrip *Chirothrips manicatus* Halliday. *New Zealand Journal of Science and Technology 38A*: 56–65.

DOWTON, M.; AUSTIN, A. D.; DILLON, N.; BARTOWSKY, E. 1997: Molecular phylogeny of the apocritan wasps: the Proctotrupomorpha and Evaniomorpha. *Systematic Entomology 22*: 245–255.

DRAPER, K. 1997: *Choose the Right Fly! A Streamside Guide for New Zealand Anglers.* Shoal Bay Press, Christchurch. 80 pp.

DUCKHOUSE, D. A. 1966: Psychodidae (Diptera: Nematocera) of southern Australia: subfamily Psychodinae. *Transactions of the Royal Entomological Society, London 118*: 153–220.

DUCKHOUSE, D. A. 1971: Entomology of the Aucklands and other islands south of New Zealand: Diptera Psychodidae). *Pacific Insects Monograph 27*: 317–325.

DUCKHOUSE, D. A. 1980: *Trichomyia* species (Diptera: Psychodidae) from southern Africa and New Zealand, with a discussion of their affinities and the concept of monophyly in southern hemisphere biogeography. *Annals of the Natal Museum 24*: 177–191.

DUCKHOUSE, D.A. 1990: The Australasian genera of pericomoid Psychodidae (Diptera) and the status of related Enderlein genera in the neotropics. *Invertebrate Taxonomy 3*: 721–746.

DUCKHOUSE, D. A. 1995: The final stage larvae of *Brunettia* Diptera: Psychodinae) and their evolutionary significane. *Invertebrate Taxonomy 9*: 83–105

DUGDALE, J. S. 1962: Description of *Perrisinoides cerambycivorae* gen and sp. nov. (Diptera: Tachinidae). *Transactions of the Royal Society of New Zealand (Zoology) 1*: 241–248.

DUGDALE, J. S. [1971] 1972: Genera of New Zealand Cicadidae (Homoptera). *New Zealand Journal of Science 14*: 856–882.

DUGDALE, J. S. 1988: Lepidoptera: annotated catalogue, and keys to family-group taxa. *Fauna of New Zealand 14*: 1–262.

DUGDALE, J. S. 1990: Reassessment of *Ctenopseustis* Meyrick and *Planotortrix* Dugdale with descriptions of two new genera (Lepidoptera: Tortricidae). *New Zealand Journal of Zoology 17*: 437–465.

DUGDALE, J. S. 1994: Hepialidae (Insecta: Lepidoptera). *Fauna of New Zealand 30*: 1–164.

DUGDALE, J. S. 1996: Natural history and identification of litter–feeding Lepidoptera larvae (Insecta) in beech forests, Orongorongo Valley, New Zealand, with especial reference to the diet of mice (*Mus musculus*). *Journal of the Royal Society of New Zealand 26*: 251–274.

DUGDALE, J. S. 1998: A preliminary overview of Lepidoptera herbivore–host plant associations in New Zealand. *In*: *Ecosystems, Entomology and Plants. Royal Society of New Zealand Miscellaneous Series 48*: 25–29.

DUGDALE, J. S.; FLEMING, C. A. 1969: Two New Zealand cicadas collected on Cook's Endeavour voyage, with description of a new genus. *New Zealand Journal of Science 12*: 929–957.

DUGDALE, J. S.; FLEMING, C. A. 1978: New Zealand cicadas of the genus *Maoricicada* (Homoptera: Tibicinidae). *New Zealand Journal of Zoology 5*: 295–340.

DUMBLETON, L. J. 1941: *Oncodes brunneus* Hutton: A dipterous spider parasite. *New Zealand Journal of Science and Technology 22*: 97–102.

DUMBLETON, L. J. 1957: The New Zealand Aleyrodidae (Hemiptera: Homoptera). *Pacific Science 11*:141–160.

DUMBLETON, L. J. 1963: New Zealand Blepharaceridae (Diptera: Nematocera) *New Zealand Journal of Science 6*: 234–258.

DUMBLETON, L. J. 1966: Immature stages of two aquatic Empididae (Dipt.). *New Zealand Journal of Science 9*: 565–568.

DUMBLETON, L. J. 1971: The biting midge *Styloconops myseri* (Tonnoir) (Diptera: Ceratopogonidae) description of male and redescription of female. *New Zealand Journal of Science 14*: 270–275.

DURDEN, L. A.; MUSSER, G. G. 1994: The sucking lice (Insecta: Anoplura) of the world: a taxonomic checklist with records of mammalian hosts and geographical distributions. *Bulletin of the American Museum of Natural History 218*: 1–90.

DYMOCK, J. J.; FORGIE, S. A. 1993: Habitat preferences and carcass colonization by sheep blow flies in the Northern North Island of New Zealand. *Medical and Veterinary Entomology 7*: 155–160.

DYMOCK, J. J.; PETERS, M. O. E.; HERMAN, T. J. B.; FROUDE, K. J. 1990: The relative importance of the Australian green blowfly *Lucilia cuprina* (Wiedmann), as a flystrike blowfly at a Waikato sheep station. *Proceedings of the Weed and Pest Control Conference 43*: 356–358.

EARLY, J. W. 1995: Insects of the Alderman Islands. *Tane, Journal of the Auckland University Field Club 35*: 1–14.

EARLY, J. W.; DUGDALE, J. S. 1994: *Fustiserphus* (Hymenoptera: Proctotrupidae) parasitises Lepidoptera in leaf litter in New Zealand. *New Zealand Journal of Zoology 21*: 249–252.

EARLY, J. W.; MASNER, L.; JOHNSON, N. F. 2007: Revision of *Archaeoteleia* Masner (Hymenoptera: Platygastroidea, Scelionidae). *Zootaxa 1655*: 1–48.

EARLY, J. W.; MASNER, L; NAUMANN, I. D; AUSTIN, A. D. 2001: Maamingidae, a new family of proctotrupoid wasp (Insecta: Hymenoptera) from New Zealand. *Invertebrate Taxonomy 15*: 341–352.

EASTOP, V. F. 2001: A new native *Paradoxaphis* Sunde (Hemiptera: Aphididae) from New Zealand. *New Zealand Entomologist 24*: 11–13.

EATON, A. E. 1899: An annotated list of the Ephemeridae of New Zealand. *Transactions of the Entomological Society of London, 1899*: 285–293.

EDMISTON, J. F.; MATHIS, W. N. 2007: New Zealand species of the shore-fly genus *Nostima* Coquillett (Diptera : Ephydridae). *Zootaxa 1661*: 1–16.

EDMUNDS, G. F. Jr 1975: Phylogenetic biogeography of mayflies. *Annals of the Missouri Botanical Garden 62*: 251–263.

EDMUNDS, G. F. Jr; JENSEN, S. L.; BERNER, L. 1976: *The Mayflies of North and Central America.* University of Minnesota Press, Minneapolis. x + 330 p.

EDWARDS, F. W. 1923a: A preliminary revision of the crane flies of New Zealand (Anisopodidae, Tanyderidae, Tipulidae). *Transactions and Proceedings of the New Zealand Institute 54*: 265–352, pls 27–56.

EDWARDS, F. W. 1923b: Notes on the dipterous family Anisopodidae. *Annals and Magazine of Natural History, ser. 9, 12*: 475–493.

EDWARDS, F. W. 1924: New species of craneflies collected by Mr G.V. Hudson in New Zealand, part II. *Annals and Magazine of Natural History, ser. 9, 13*: 159–163.

EDWARDS, R. J. 1986: The biology of *Archichauliodes diversus* (Megaloptera) in some Waihi Basin streams. PhD thesis, University of Waikato, Hamilton.

EHLER, L. E.; LONG, F. R.; KINSEY, M. G.; KELLEY, S. K. 1997: Potential for augmentative biological control of black bean aphid in California sugar beet. *Entomophaga 42*: 241–256.

ELLIOTT, M.; CASSIS, G. 2001: *The Heteroptera of Australia* [LUCID key to the families]. http://faunanet.gov.au/ faunakeys]

ELLIS, E. C.; PENMAN, D. R.; GAUNT, R. E. 1988: Thrips as potential vectors of brown rot of stonefruit in New Zealand. *New Zealand Weed and Pest Control Conference 41*: 286–28.

ELSON-HARRIS, M. M.; KETTLE, D. S. 1985: A new species of *Paradasyhelea* Macfie (Diptera: Ceratopogonidae), with descriptions of the immature stages of Australian *Paradasyhelea. Journal of Australian Entomological Society 24*: 233–240.

EMBERSON, R. M. 1998: The size and shape of the New Zealand insect fauna. *In*: *Ecosystems, Entomology and Plants. The Royal Society of New Zealand Miscellaneous Series 48*: 31–37.

EMBERSON, R. M. 2000: Endemic biodiversity, natural enemies, and the future of biological control. Pp. 875–880 *in*: Spencer, N. R. (ed.), *Proceedings of the 10th International Symposium on Biological Control of Weeds.* Montana State University, Bozeman, Montana.

EMELJANOV, A. F. 2000: New genera of the family Cixiidae (Homoptera, Fulgoroidea) from Australia and neighbouring territories. *Entomological Review 80*: 251–270. [Transl. From *Entomologicheskoe Obozrenie 79*: 12–34.]

ENDERLEIN, G. 1903: Die Copeognathen des indo-australischen Faunengebietes. *Annales Historico-naturales Musei Nationalis Hungarici 1*: 179–344, pls 3, 4.

ENDERLEIN, G. 1909: Die Insekten des Antarktischen Gebiets. *Deutsche Südpolar-Expedition, 1901–1903, 10, Zoologie 2* (4): 361–528, pls 40–63.

ENGEL, M. S.; GRIMALDI, D. A. 2004: New light shed on the oldest insect. *Nature 427*: 627–630.

ESTÉVEZ, A. L.; REMES LENICOV, A. M. M. de 1990: Peloridiidae (Hompteга) sudamericanos. I. Sobre la bionomía de *Peloridium hammoniorum* Breddin 1897 en Tierra del Fuego, Argentina. *Animalia, Catania 17*: 111–122.

EVANS, A. M.; CLINTON, P. W., ALLEN, R. R., FRAMPTON, C. M. 2003: The influence of logs on the spacial litter dwelling invertebrates and forest floor processes in New Zealand forests. *Forest Ecology and Management 194*: 251–262.

EVANS, J. W. 1941: A new genus of New Zealand leaf-hoppers (Jassoidea, Homoptera). *Transactions of the Royal Society of New Zealand 71*: 162–163, pl. 28.

EVANS, J. W. 1942a: New leafhoppers (Homoptera, Jassoidea) from Western Australia. *Journal of the Royal Society of Western Australia 27*: 143–163.

EVANS, J. W. 1942b: Some new leafhoppers from

Australia and Fiji. *Proceedings of the Royal Society of Queensland 54*: 49–51.

EVANS, J. W. 1947a: A natural classification of leafhoppers (Jassoidea, Homoptera). Part 3. Jassidae. *Transactions of the Royal Entomological Society of London 98*: 105–271.

EVANS, J. W. 1947b: Some new Ulopinae (Homoptera, Jassidae). *Annals and Magazine of Natural History, ser. 11, 14*: 140–150.

EVANS, J. W. 1947c: A new leafhopper from Victoria. (Homoptera: Jassidae). *Memoirs of the National Museum, Melbourne 15*: 126–127.

EVANS, J. W. 1966: The leafhoppers and froghoppers of Australia and New Zealand (Homoptera: Cicadelloidea and Cercopoidea). *Australian Museum Memoir 12*: 1–347.

EVANS, J. W. 1981: A review of the present knowledge of the Family Peloridiidae and new genera and species from New Zealand and New Caledonia (Hemiptera: Insecta). *Records of the Australian Museum 34*: 381–406.

EVANS, W. P. 1931: Traces of a lepidopterous insect from the Middle Waikato Coal Measures. *Transactions and Proceedings of the New Zealand Institute 62*: 99–101.

EVENHUIS, N. L. 1989: *Catalog of the Diptera of the Australasian and Oceanian Regions*. Bishop Museum Press, Honolulu. 1155 p.

EVENHUIS, N. L. 1994: *Catalogue of the Fossil Flies of the World (Insecta: Diptera)*. Backhuys Publications, Leiden. 600 p.

EYLES, A. C. 1960: Insects associated with the major fodder crops in the North Island. I. The general picture. *New Zealand Journal of Agricultural Research 3*: 779–791.

EYLES, A. C. 1990: A review and revision of the genus *Rhypodes* Stål (Hemiptera: Lygaeidae). *New Zealand Journal of Zoology 17*: 347–418.

EYLES, A. C. 1996: *Josemiris*, a new genus of Orthotylinae (Hemiptera: Miridae) from New Zealand. *New Zealand Journal of Zoology 23*: 211–214.

EYLES, A. C. 1998: The identity of *Romna marginicollis* (Reuter), a new name for *marginicollis* sensu Eyles & Carvalho, and notes on two other mirids (Hemiptera). *New Zealand Journal of Zoology 25*: 43–46.

EYLES, A. C. 1999a: New genera and species of the *Lygus*-complex (Hemiptera: Miridae) in the New Zealand subregion compared with subgenera (now genera) studied by Leston (1952) and *Niastama* Reuter. *New Zealand Journal of Zoology 26*: 303–354.

EYLES, A. C. 1999b: Introduced Mirinae of New Zealand (Hemiptera: Miridae). *New Zealand Journal of Zoology 26*: 355–372.

EYLES, A. C. [1999] 2000a: New genera and species of the *Lygus*-complex (Hemiptera: Miridae) in the New Zealand subregion compared with subgenera (now genera) studied by Leston (1952) and *Niastama* Reuter. *New Zealand Journal of Zoology 26*: 303–354.

EYLES, A. C. 2000b: Introduced Mirinae of New Zealand (Hemiptera: Miridae). *New Zealand Journal of Zoology (1999), 26*: 355–372.

EYLES, A.C. 2000c: *Tinginotum* Kirkaldy in New Zealand and Australia: a shared new species, and a new species of *Tinginotopsis* Poppius from Norfolk Island (Hemiptera: Miridae). *New Zealand Journal of Zoology 27*: 111–119.

EYLES, A. C. 2001: Key to the genera of Mirinae (Hemiptera: Miridae) in New Zealand and descriptions of new taxa. *New Zealand Journal of Zoology 28*: 197–221.

EYLES, A. C.; CARVALHO, J. C. M. 1991: Revision of the genus Chinamiris Woodward (Hemiptera: Miridae). *New Zealand Journal of Zoology 18*: 267–321.

EYLES, A. C.; CARVALHO, J. C. M. 1995: Further endemic new genera and species of Mirinae (Hemiptera: Miridae) from New Zealand. *New Zealand Journal of Zoology 22*: 49–90.

EYLES, A. C.; SCHUH, R. T. 2003: Revision of New Zealand Bryocorinae and Phylinae (Insecta: Hemiptera: Miridae). *New Zealand Journal of Zoology 30*: 263–325.

FABRICIUS, J. C. 1775: *Systema Entomologiae, sistens insectorum classes, ordines, genera, species, adiectis synonymis, locis, descriptionibus, observationibus*. Officina Libraria Kortii, Flensburgi et Lipsiae. xxxii + 832 p.

FABRICIUS, J. C. 1781: *Species Insectorum exhibentes eorum differentias specificas, synonyma auctorum, loca natalia, metamorphosin adiectis observationibus, descriptionibus*. Impensis Carol Ernest Bohnii, Hamburgi et Kolnii. Vol. 2, 517 p.

FARRELL, J. A.; STUFKENS, M. W. 1990: The impact of *Aphidius rhopalosiphi* (Hymenoptera: Aphidiidae) on populations of the rose grain aphid (*Metopolophium dirhodum*) (Hemiptera: Aphididae) on cereals in Canterbury New Zealand. *Bulletin of Entomological Reseach 80*: 377–383.

FAULDS, W. 1991: Spread of *Bracon phylacteophagus*, a biocontrol agent of *Phylacteophaga froggatti*, and impact on host. *New Zealand Journal of Forestry Science 21*: 185–193.

FENNAH, R. G. 1965: Delphacidae from Australia and New Zealand. Homoptera: Fulgoroidea. *Bulletin of the British Museum (Natural History), Entomology 17*: 1–59.

FERRAR, P. 1987: A guide to the breeding habits and immature stages of Diptera Cyclorrhapha. *Entomograph 8*: 1–907.

FERRINGTON, L. C. Jr 2008: Global diversity of scorpionflies and hangingflies (Mecoptera) in freshwater. *Hydrobiologia 595*: 443–445.

FERRIS, G. F.; KLYVER, F. D. 1932: Report upon a collection of Chermidae (Homoptera) from New Zealand. *Transactions and Proceedings of the New Zealand Institute 63*: 34–173.

FIGUEROA-ROA, L.; LINHARES, A.X. 2002: Synathropy of the Calliphoridae (Diptera) from Valdivia, Chile. *Neotropical Entomology 31*: 233-239.

FLEMING, C. A. 1969: [Genus *Amphipsalta* Fleming, nov.]. Pp. 932–934. *In*: Dugdale, J. S.; Fleming, C. A. 1969: Two New Zealand Cicadas collected on Cook's Endeavour voyage, with description of a new genus. *New Zealand Journal of Science 12*: 929–957.

FLEMING, C. A. 1973: The Kermadec Islands cicada and its relatives (Hemiptera: Homoptera). *New Zealand Journal of Science 16*: 315–332.

FLEMING, C. A. 1975a: Cicadas (1). *New Zealand's Nature Heritage 4(56)*: 1568–1572.

FLEMING, C. A. 1975b: Cicadas (2). *New Zealand's Nature Heritage 4(57)*: 1591–1595.

FLEMING, C. A. 1984: The cicada genus *Kikihia* Dugdale (Hemiptera, Homoptera). Part 1. The New Zealand green foliage cicadas. *Records of the National Museum of New Zealand 2*: 191–206.

FLETCHER, M. J. 1999: *Identification Key and Checklists for the Australian Planthoppers (Superfamily Fulgoroidea)*. http://www.agric.nsw.gov.au/hort/ascu/fulgor/fulg0.htm

FOOTTIT, R. G.; RICHARDS, W. R. 1992: The genera of the aphids in Canada. Homoptera: Aphidoidea and Phylloxeridae. *The Insects and Arachnids of Canada 22*: 1–766.

FÖRSTER, A. 1854: Neue Blattwespen. *Verhandlungen des Naturhistorischen Vereins der Preussischen Rheinlande 11*: 265–350.

FORSTER, R. R.; FORSTER, L. M. 1970: *Small Land Animals of New Zealand*. John McIndoe, Dunedin. 175 p.

FOX, K. J. 1978: The transoceanic migration of Lepidoptera to New Zealand: a history and a hypothesis on colonisation. *New Zealand Entomologist 6*: 368–380.

FRAMPTON, E. R. 1990: Surveillance and monitoring for exotic pests. *Bulletin of the Entomological Society of New Zealand 10*: 101–107.

FRASER, F. C. 1959: Mecoptera, Megaloptera, Neuroptera. *Handbooks for the Identification of British Insects 1(12/13)*: 1–40.

FRASER, P. M.; WILLIAMS, P. H.; HAYNES, R. J. 1995: Earthworm species, population size and biomass under different cropping systems across the Canterbury Plains, New Zealand. *Applied Soil Ecology 3*: 49–57.

FREEMAN, P. 1951: *Diptera of Patagonia and South Chile. Part III – Mycetophilidae*. The British Museum (Natural History), London. 138 p.

FREEMAN, P. 1959: A study of the New Zealand Chironomidae (Diptera, Nematocera). *Bulletin of the British Museum (Natural History), B. Entomology 7*: 395–437.

FREEMAN, P 1989: Some non British species of *Anapausis* (Diptera, Scatopsidae). *Entomologists Monthly Magazine 125*: 37–43.

FROUD, K. J.; STEVENS, P. S.; COWLEY, D. R. 1996: A potential biocontrol agent for greenhouse thrips. *New Zealand Weed and Pest Control Conference 49*: 17–20.

FULLER, M. E. 1935: Notes on Australian Anisopodidae (Diptera). *Proceedings of the Linnean Society of New South Wales 60*: 291–302.

GALLOWAY, I. D.; AUSTIN, A. D. 1984: Revision of the Scelioninae (Hymenoptera: Scelionidae) in Australia. *Australian Journal of Zoology, Supplementary Series 99*:1–138.

GARLAND, J. A. 1985: Identification of Chrysopidae in Canada, with bionomic notes (Neuroptera). *Canadian Entomologist 117*: 737–762.

GAULD, I. D. 1980: Notes on the New Zealand Anomaloninae (Hymenoptera : Ichneumonidae) with a description of a new species of *Aphanistes* Förster of possible economic importance in forestry. *New Zealand Entomologist 7*: 130–134.

GAULD, I. D. 1984: *An Introduction to the Ichneumonidae of Australia*. Trustees of the British Museum (Natural History), London. 413 p.

GAULD, I. D.; BOLTON, B. (Eds) 1988: *The Hymenoptera*. Trustees of the British Museum (Natural History) & Oxford University Press. xi + 332 p.

GAUTHIER, N.; LA SALLE, J.; QUICKE, D. L. J.; GODFRAY, H. C. J. 2000: Phylogeny of Eulophidae (Hymenoptera: Chalcidoidea), with a reclassification of Eulophinae and the recognition that Elasmidae are derived eulophids. *Systematic Entomology 25 (4)*: 521–539.

GAY, F. J. 1969: A new species of *Stolotermes* (Isoptera: Termopsidae, Stolotermitinae) from New Zealand. *New Zealand Journal of Science 12*: 748–753.

GAY, F. J. 1976: Isoptera of Kermadec islands. *New Zealand Entomologist 6*: 149–153.

GIBBS, G. W. 1998: Why are some weta (Orthoptera: Stenopelmatidae) vulnerable yet others are common? *Journal of Insect Conservation 2*: 161–166.

GIBBS, G. W. 1999: Four new species of giant weta, *Deinacrida* (Orthoptera: Anostostomatidae: Deinacridinae) from New Zealand. *Journal of the*

Royal Society of New Zealand 29: 307–324.

GIBBS, G. W. 2002: A new species of tusked weta from the Raukumara Range, North Island, New Zealand (Orthoptera: Anostomatidae: *Motuweta*). *New Zealand Journal of Zoology 29*: 293–301.

GIBSON, G. A. P.; HUBER, J. T. 2000: Review of the family Rotoitidae (Hymenoptera: Chalcidoidea), with description of a new genus and species from Chile. *Journal of Natural History 34*: 2293–2314.

GIBSON, G. A. P; READ, J; HUBER, J. T. 2007: Diversity, classification and higher relationships of Mymarommatoidea (Hymenoptera). *Journal of Hymenoptera Research 16*: 51–146.

GILES, E. T. 1953: The biology of *Anisolabis littorea* (White) (Dermaptera: Labiduridae). *Transactions and Proceedings of the Royal Society of New Zealand 80*: 383–398.

GILES, E. T. 1958: Dermaptera from the Three Kings Islands, with the description of a new species of *Brachylabis* (Dohrn) (Labiduridae). *Records of the Auckland Institute and Museum 5*: 43–48.

GILL, B. J. 1980: Abundance, feeding, and morphology of passerine birds at Kowhai Bush, Kaikoura, New Zealand. *New Zealand Journal of Zoology 7*: 235–246.

GIRIBET, G.; EDGECOMBE, G. D.; CARPENTER, J. M.; D'HAESE, C. A.; WHEELER, W. C. 2004: Is Ellipura monophyletic? A combined analysis of basal hexapod relationships with emphasis on the origin of insects. *Organisms, Diversity & Evolution 4*: 319–340.

GLARE, T. R.; O'CALLAGHAN, M. 1993: Checklist of naturally occurring entomopathogenic microbes and nematodes in New Zealand. *New Zealand Journal of Zoology 20*: 95–120.

GLARE, T. R.; O'CALLAGHAN, M.; WIGLEY, P. J. 1993: Check list of naturally occurring entomopathogenic microbes and nematodes in New Zealand. *New Zealand Journal of Zoology 20*: 95–120.

GLOVER, G. J.; SAGAR, P. M. 1994: Comparison of fish and microinvertebrate standing stocks in relation to riparian vegetation (*Salix* species) in three New Zealand streams. *New Zealand Journal of Marine and Freshwater Research 28*: 255–266.

GOFF, M. L. 1991: Comparison of insect species associated with decomposing remains recovered inside dwellings and outdoors in the Island of Oahu, Hawaii. *Journal of Forensic Sciences 36*: 748–753.

GOLDSON, S. L. 1977: Larvae of four species of *Psychoda* (Diptera: Psychodidae). *New Zealand Entomologist 6*: 279–284.

GORCZYCA, J.; EYLES, A. C. 1997: A new species of *Peritropis* Uhler, the first record of Cylapinae (Hemiptera: Miridae) from New Zealand. *New Zealand Journal of Zoology 24*: 225–230.

GORDH, G.; HARRIS, A. 1996: New records and a new species of Eupsenella (Hymenoptera: Bethylidae) in New Zealand. *Journal of the Royal Society of New Zealand 26*: 529–536.

GOROCHOV, A. V. 1995: System and evolution of the suborder Ensifera (Orthoptera). Parts 1 & 2. *Proceedings of the Zoological Institute, St Petersburg 260*: 1– 224, 1–214. [In Russian.]

GOULET, H.; HUBER, J. T. 1993: Hymenoptera of the world: An identification guide to families. *Agriculture Canada Publication 1894/E*: vii, 1–668.

GOULSON, D.; STOUT, J. C.; KELLS, A. R. 2002: Do exotic bumble bees and honeybees compete with native flower-visiting insects in Tasmania? *Journal of Insect Conservation 6*: 179–189.

GOURLAY, E. S. 1953: The Strepsiptera, an insect order new to New Zealand. *New Zealand Entomologist 1(3)*: 3–8.

GOURLAY, E. S. 1954: The Dryinidae: a family of Hymenoptera new to New Zealand. *New Zealand Entomologist 1(4)*: 3–5.

GOURLAY, E. S. 1964: Notes of New Zealand insects and records of introduced species. *New Zealand Entomologist 3(3)*: 45–51.

GRAHAM, M. W. R. de V. 1991: A reclassification of the European Tetrastichinae (Hymenoptera: Eulophidae): revision of the remaining genera. *Memoirs of the American Entomological Institute 49*: 1–322.

GRANT, E. A. 1999: *An Illustrated Guide to Some New Zealand Insect Families*. Manaaki Whenua Press, Lincoln. 196 p.

GRANT-MACKIE, J. A.; BUCKERIDGE, J. S.; JOHNS, P. M. 1996: Two new Upper Jurassic arthropods from New Zealand. *Alcheringa 20*: 31–39.

GRAVATT, D. J. 1971: Aspects of habitat use by New Zealand honey eaters, with special reference to other forest species. *Emu 71*: 65–72.

GREATHEAD, D. J. 1988: The relationship of *Tillyardomyia* Tonnoir with a redefinition of the subfamily Ecliminae (Diptera: Bombyliidae). *New Zealand Entomologist 11*: 21–24.

GREEN, E. E. 1914: Some remarks on the coccid genus *Leucaspis*, with descriptions of two new species. *Transactions of the Entomological Society of London, ser. 3, 4*: 459–467.

GREEN, E. E. 1929: Coccidae collected by J. G. Myers in New Zealand. *Bulletin of Entomological Research 21*: 377–389.

GREEN, O. 1983: Puriri moth predation by earwigs. *The Weta 6*: 36.

GREEN, O. R.; GREEN, C. J.; HERMAN, T. J. B.; TAYLOR, M. J.; OLIVER, H. A. 1994: *Ancistrocerus gazella* (Vespoidea: Eumenidae): new locality and date records in New Zealand. *The Weta 17*: 20–23.

GREEN, O. R.; PONT, A. C. 2003: A new record, *Pygophora apicalis* Schiner (Diptera: Muscidae: Coenosiini), established in Auckland, New Zealand. *New Zealand Entomologist 26*: 101–104.

GREEN, O. R.; RAMSAY, G. W. 2003: A winged weta, *Pterapotrechus* (Orthoptera: Gryllacrididae) established in New Zealand. *New Zealand Entomologist 26*: 75–77.

GREENSLADE, P. 1982: Revision of the Spinothecinae (Collembola: Sminthuridae) including a new Australian genus. *Journal of the Australian Entomological Society 21*: 81–95.

GREENSLADE, P. 1984: The identity of *Orchezelandia rubra* (Collembola: Entomobryidae). *Transactions of the Royal Society of South Australia 108*: 129–130.

GREENSLADE, P. 1986: Identity and synonymy of *Isotoma* (*Folsomotoma*) Bagnall (Isotomidae). Pp. 53–59 *in*: Dallai, R. (ed.), *Proceedings of the 2nd International Seminar on Apterygota, Siena 1986*. Università di Siena, Siena. 334 p.

GREENSLADE, P. 1987: Generic biogeography of Tasmanian Collembola. Pp. 653–660 *in*: Striganova, B.R. (ed.), *Soil Fauna and Soil Fertility* [*Proceedings of the 9th International Colloquium on Soil Zoology, Moscow, 1985*]. Nauka, Moscow. 774 p.

GREENSLADE, P. 1989: Genera of Isotomidae with spined dens from southern regions. Pp. 107–118 *in*: Dallai, R. (ed.), *Proceedings of the 3rd International Seminar on Apterygota, Siena 1989*. Università di Siena, Siena. 489 p.

GREENSLADE, P. 1994: Collembola. Pp. 19–138 *in*: Houston, W.W.K. (ed.), *Zoological Catalogue of Australia. Volume 22. Protura, Collembola, Diplura*. CSIRO, Melbourne. 188 p.

GRESSITT, J. L.; WEBER, N. A. 1959: Bibliographic introduction to antarctic–subantarctic entomology. *Pacific Insects 1*: 441–480.

GRESSITT, J. L.; WISE, K. A. J. 1971: Entomology of the Aucklands and other islands south of New Zealand: Introduction. *Pacific Insects Monographs 27*: 1–45.

GRIMES, E. W.; CONE, W. W. 1985: Life history, sex attraction, mating and natural enemies of grape mealy bug *Pseudococcus maritimus* (Homoptera: *Pseudococcus*). *Annals of the Entomological Society America 78*: 554–558.

GRINGAS, J.; TOUNIER, J. C. 2001: Timing of adult mortality: oviposition, and hatching during the underground phase of *Forficula auricularia* (Dermaptera: Forficulidae). *Canadian Entomologist 133*: 269–278.

GRISSELL, E. E. 1999: Hymenopteran biodiversity: some alien notions. *American Entomologist 45*: 235–244.

GROSS, G. F. CASSIS, G. 1995: Hemiptera: Heteroptera (Coleorrhyncha to Cimicomorpha). *In*: Houston, W. W. K.; Maynard, G. V. (eds), *Zoological Catalogue of Australia 27.3A*: 1–506.

GROVE, S. J; STORK, N. E. 2000: An inordinate fondness for beetles. *Invertebrate Taxonomy 14*: 733–739.

GUERRIERI, E.; NOYES, J. S. 2000: Revision of European species of genus *Metaphycus* Mercet (Hymenoptera: Chalcidoidea: Encyrtidae), parasitoids of scale insects. *Systematic Entomology 25*: 147–222.

GWYNNE, D. T. 1995: Phylogeny of the Ensifera (Orthoptera): A hypothesis supporting multiple origins of acoustical signalling, complex spermatophores and maternal care in crickets, katydids and weta. *Journal of Orthopteran Research 4*: 203–218.

HAMILTON A. 1940: The New Zealand dobson-fly (*Archichauliodes diversus* Walk.): life history and bionomics. *New Zealand Journal of Science and Technology A 22*: 44–55.

HAMILTON, H. 1921: Natural history notes on the larval habits of the ant-lion *Mymelion acutus* Walker. *New Zealand Journal of Science and Technology 4*: 203.

HAMILTON, K. G. A. 1999: The ground-dwelling leafhoppers Myerslopiidae, new family, and Sagmatiini, new tribe (Homoptera: Membracoidea). *Invertebrate Taxonomy 13*: 207–235.

HAMILTON, K. G. A.; MORALES, C. F. 1992: Cercopidae (Insecta: Homoptera). *Fauna of New Zealand 25*: 1–37.

HARDING, J. S. 1991: The larva of *Neurochorema forsteri* McFarlane (Trichoptera: Hydrobiosidae). *New Zealand Natural Sciences 18*: 51–54.

HARDY, N. B.; GULLAN, P. J.; HENDERSON, R. C.; COOK, L. G. 2008: Relationships among felt scale insects (Hemiptera: Coccoidea: Eriococcidae) of southern beech, *Nothofagus* (Nothofagaceae), with the first descriptions of Australian species of the *Nothofagus*-feeding genus *Madarococcus* Hoy. *Invertebrate Systematics 22*: 365–405.

HARMAN, H. M.; SYRETT, P.; HILL, R. L.; JESSEP, C. T. 1996: Arthropod introductions for biological control of weeds in New Zealand. 1929–1995. *New Zealand Entomologist 19*: 71–80.

HARRIS, A. C. 1983: An Eocene larval insect fossil (Diptera: Bibionidae) from North Otago, New Zealand. *Journal of the Royal Society of New Zealand 13*: 93–105.

HARRIS, A. C. 1987: Pompilidae (Insecta: Hymenoptera). *Fauna of New Zealand 12*: 1–154.

HARRIS, A. C. 1988: *Blatta orientalis* (Blattodea: Blattidae) established in Dunedin, New Zealand. *The Weta 11*: 46.

HARRIS, A. C. 1990: *Podagrites cora* and *P. albipes* (F.Smith) (Hymenoptera: Sphecidae: Craboninae) preying on Ephemeroptera and Trichoptera. *Pan-Pacific Entomologist 66*: 55–61.

HARRIS, A. C. 1994a: Sphecidae (Insecta: Hymenoptera). *Fauna of New Zealand 32*: 1–111.

HARRIS, A. C. 1994b: Biology of *Ancistrocerus gazella* (Hymenoptera: Vespoidea: Eumenidae) in New Zealand. *New Zealand Entomologist 17*: 29–36.

HARRIS, A. C.; RAINE, J. I. 2002: A sclerite from a Late Cretaceous moth (Insecta: Lepidoptera) from Rakaia Gorge, Canterbury, New Zealand. *Journal of the Royal Society of New Zealand 32*: 457–462.

HARRIS, R. J. 1991: Diet of wasps *Vespula vulgaris* and *vespula germanica* in honey dew beech forest of the South Island of New Zealand. *New Zealand Journal of Zoology 18*: 159–169.

HARRIS, R. J.; BERRY, J. A. 2001: Confirmation of the establishment of three adventive ants (Hymenoptera: Formicidae) in New Zealand: *Cardiocondyla minutior* Forel, *Ponera leae* Forel, *Mayriella abstinens* Forel. *New Zealand Entomologist 24*: 53–56.

HARRIS, R. J.; TOFT, R. J.; DUGDALE, J. S.; WILLIAMS, P. A.; REES, J. S. 2004: Insect assemblages in a native (*Kunzea ericoides*) and an invasive (gorse *Ulex europeaus*) shrubland. *New Zealand Journal of Ecology 28*: 35–47.

HARRISON, R. A. 1953: The Diptera of the Antipodes and the Bounty Islands. *Transactions of the Royal Society of New Zealand 81*: 269–282.

HARRISON, R. A. 1959: Acalypterate Diptera of New Zealand. *Department of Scientific and Industrial Research Bulletin 128*: 1–382.

HARRISON, R. A. 1976: The arthropods of the Southern Islands of New Zealand (9): Diptera. *Journal Royal Society of New Zealand 6*: 107–152.

HARRISON, R. A. 1990: Bibionidae (Insecta: Diptera). *Fauna of New Zealand 20*: 1–25.

HARRISON, R. A.; WHITE, E. G. 1969: Grassland invertebrates. Pp. 379–390 *in:* Knox, G. A. (ed.), *The Natural History of Canterbury*. A. H. & A. W. Reed, Wellington. 620 p.

HASSAN A. I. 1939: The biology and hosts of some British Delphacidae (Homoptera) and their parasites with special reference to Strepsiptera. *Transactions of the Royal Entomological Society, London 89*: 345–384.

HÄTTENSCHWILER, P. 1989: Genus *Scoriodyta* Meyrick, 1888, a new subfamily and description of a new species and forms (Lepidoptera: Psychidae). *New Zealand Journal of Zoology 16*: 51–63.

HAUSRER, M.; IRWIN, M. E. 2005: Fossil Therevidae (Insecta: Diptera) from Florissant, Colorado (Upper Eocene). *Journal of Systematic Palaeontology 3*: 393–401.

HAW, J. M.; CLOUT, M. W.; POWLESLAND, R. G. 2001: Diet of moreporks (*Ninox novaeseelandiae*) in Pureora Forest determined by prey remains in regurgitated pellets. *New Zealand Journal of Ecology 25*: 61–67.

HAYES, I. (Comp.) 2007. *The biological control of weeds book: a New Zealand guide.* [http://www.landcareresearch. co.nz/research/biocons/weeds/book.asp]

HEATH, A. C. G. 1980: Ticks and blow flies: economic importance and control. *Proceedings of the Ruakura Farmers Conference 32*: 87–95.

HEATH, A. C. G. 1982: Beneficial aspects of blowflies (Diptera: Calliphoridae). *New Zealand Entomologist 7*: 343–348.

HEATH, A. C. G., APPLETON, C. 2000: Small vertebrate carrion and its use by blowflies (Calliphoridae) causing ovine myiasis (flystrike) in New Zealand. *New Zealand Entomologist 22:* 81–87.

HEATH, A. C. G.; TENQUIST, J. D.; NEILSON, F. J. A.; WALLACE, G.V. 1983: Fly-strike, biology, importance and control, parts I and II. *New Zealand Farmer 104*: 114–116, 120–125.

HEIE, O. E. 1980: The Aphidoidea (Hemiptera) of Fennoscandia and Denmark. I. *Fauna Entomologica Scandinavica 9*: 1–236.

HEIE, O. E. 1987: Palaeontology and phylogeny. Pp. 367–391 *in*: Minks, A. K.; Harrewijn, P. (eds), *Aphids: Their Biology, Natural Enemies and Control.* Volume A. Elsevier, Amsterdam.

HEIE, O. E. 1994: Why are there so few aphid species in the temperate areas of the Southern Hemisphere? *European Journal of Entomology 91*: 127–133.

HEINE, E. M. 1937: Observations on the pollination of New Zealand flowering plants. *Transactions and Proceedings of the Royal Society of New Zealand 67*: 133–148.

HEISS, E. 1990: New apterous Carventinae from New Zealand (Heteroptera: Aradidae). *Journal of the New York Entomological Society 98*: 393–401.

HEISS, E. 1998: Review of the genus *Aneurus* from New Zealand with description of three new species (Heteroptera, Aradidae). *New Zealand Journal of Zoology 25*: 29–42.

HELMSING, I. W.; CHINA, W. E. 1937: On the biology and ecology of *Hemiodoecus veitchi* Hacker. *Annals and Magazine of Natural History, ser. 10, 19*: 473–489.

HELSON, G. A. H. 1974: Insect pests. *New Zealand Minisrty of Agriculture and Fisheries Bulletin 413*: 1–196.

HELSON, H.; VAAL, F.; BLOOMER, L. 1998: Phenology of the common earwig *Forficula auricularia* L. (Dermaptera: Forficulidae) in an apple orchard. *Insect Journal of Pest Management 44*: 75–79.

HEMMINGSON, A. M.; JOHNS, P. M. 1975: The dark colouration of marine cranefly eggs (Diptera: Tipulidae: Limoniinae). *Videnskabelige Meddelelser fra Dansk Naturhistorisk Forening 138*: 127–136.

HENDERSON, I. M. 1983: A contribution to the systematics of New Zealand Philopotamidae (Trichoptera). *New Zealand Journal of Zoology 10*: 163–176.

HENDERSON, I. M. 2008: A new species of *Traillochorema* (Trichoptera: Hydrobiosidae) from North Island, New Zealand, and a revised diagnosis of the genus. *Records of the Canterbury Museum 22*: 23–28.

HENDERSON, I. M.; WARD, J. B. 2006: Four new species of the caddis genus *Philorheithrus* (Trichoptera: Philorheithridae) from New Zealand. *Records of the Canterbury Museum 20*: 21–33.

HENDERSON, I. M.; WARD, J. B. 2007: Three new species in the endemic New Zealand genus *Alloecentrella* (Trichoptera), and a re-evaluation of its family placement. *Aquatic Insects* 29: 79–96.

HENDERSON, N. C. 1980: Studies on the predators, a parasite an entomophagous fungus affecting *Acyrthosiphon kondoi* Shinji (Homoptera: Aphididae) in the Manawatu. *Proceedings of the Australasian Conference of Grassland Invertebrate Ecology 2*: 278–282.

HENDERSON, R. C. 2006: Four new species and a new monotypic genus *Hoheriococcus* (Hemiptera: Coccoidea: Eriococcidae) associated with plant galls in New Zealand. *New Zealand Entomologist 29*: 37–57.

HENDERSON, R. C. 2007a: A new genus and species of felt scale (Hemiptera: Coccoidea: Eriococcidae) from epiphyte communities of northern rata (*Metrosideros robusta* Cunn.: Myrtaceae) canopy in New Zealand. *New Zealand Entomologist 30*: 25–33.

HENDERSON, R. C. 2007b: Three new genera and six new species of felt scales (Hemiptera: Coccoidea: Eriococcidae) from mountain habitats in New Zealand. *Zootaxa 1449*: 1–29.

HENDERSON, R. C.; HODGSON, C. J. 2005: Two new specis of *Umbonichiton* (Hemiptera: Sternorrhyncha: Coccoidea: Coccidae) from New Zealand. *Zootaxa 854*: 1–11.

HENNEMANN, F. H.; CONLE, O.V. 2008: Revision of Oriental Phasmatodea: the tribe Pharnaciini Günther, 1953, including the description of the world's longest insect, and a survey of the family Phasmatidae Gray, 1835 with keys to the subfamilies and tribes (Phasmatodea: 'Anareolatae': Phasmatidae). *Zootaxa 1906*: 1–316.

HENNIG, W. 1966: The Diptera fauna of New Zealand as a problem in systematics and zoogeography. *Pacific Insects Monograph 9*: 1–81.

HENRY, T. J. 1997a: Cladistic analysis and revision of the stilt bug genera of the world (Heteroptera: Berytidae). *Contributions of the American Entomological Institute 30*: 3–100.

HENRY, T. J. 1997b: Phylogenetic analysis of family groups withing the infraorder Pentatomomorpha (Hemiptera: Heteroptera), with emphasis on the Lygaeoidea. *Annals of the Entomological Society of America 90*: 275–301.

HEYWOOD, V. H. (Ed.). 1995. *Global Biodiversity Assessment*. Cambridge University Press for UNEP, Cambridge. 1140 p.

HICKS, G .R. F.; McCOLL, H. P.; MEADS, M. J.; HARDY, G. S.; ROSER, R. J. 1975: An ecological reconnaissance of Korapuki Island, Mercury Islands. *Notornis 2*: 195–220.

HIGGINS, P. J. (Ed.) 1999: *Handbook of Australian, New Zealand and Antarctic Birds. Volume 4. Parrots to Dollar Bird.* Oxford University Press, Melbourne. 1248 p.

HIGGINS, P. J.; MARCHANT, S. (Eds) 1993: *Handbook of Australian, New Zealand and Antarctic Birds. Volume 2. Raptors to Lapwings.* Oxford University Press, Melbourne. 984 p.

HIGGINS, P. J.; PETER, J. M.; STEELE, W. K. (Eds) 2001: *Handbook of Australian, New Zealand and Antarctic Birds. Volume 5. Tyrant-flycatchers to Chats.* Oxford University Press, Melbourne. 1268 p.

HILL, G. F. 1942: *Termites (Isoptera) from the Australian Region*. CSIRO, Melbourne. 479 p.

HILL, K. B. R.; MARSHALL, D. C.; COOLEY, J. R. 2005: Crossing Cook Strait: Possible human transportation and establishment of two New Zealand cicadas from North Island to South Island (*Kikihia scutellaris* and *K. ochrina*, Hemiptera: Cicadidae). *New Zealand Entomologist 28*: 71–80.

HILL, K. B. R.; SIMON, C.; MARSHALL, D. C.; CHAMBERS, G. K. 2009: Surviving glacial ages within the Biotic Gap: phylogeography of the New Zealand cicada *Maoricicada campbelli*. *Journal of Biogeography 36*: 675–692.

HILL, R. L.; MARKIN, G. P.; GOURLAY, A. H.;

FOWLER, S.V.; YOSHIOKA, E. 2001: Host range, release and establishment of *Sericothrips staphylinus* Haliday (Thysanoptera: Thripidae) as a biological control agent for gorse, *Ulex europeus* L. (Fabaceae), in New Zealand and Hawaii. *Biological Control 21*: 63–74.
HILL, R. L.; WITTENBERG, R.; GOURLAY, A. H. 2001: Biology and host range of *Phytomyza vitalbae* and its establishment for the biological control of *Clematis vitalba* in New Zealand. *Biocontrol Science and Technology 11*: 459–473.
HILSON, R. J. D. 1964: The ecology of *Micromus tasmaniae* (Walker). Unpublished MSc (Hons) thesis, University of Canterbury, Christchurch. 100 p.
HILTON, A. W. 1939: *Campodea zelandica* n.sp. *Journal of Entomology and Zoology 3*: 6–7.
HINCKS, W. D. 1949a: Dermaptera, Orthoptera. *Handbooks for the Identification of British Insects 1*: 1–20.
HINCKS, W. D. 1949b: Some earwigs (Dermaptera) from New Zealand. *Proceedings of the Royal Entomological Society of London, ser. B, 18*: 201–206.
HINTON, H. E. 1981: *Biology of Insect Eggs*. Volumes 1–3. Pergamon Press, Oxford.
HIPPA, H.; VILKAMAA, P. 2006: Phylogeny of the Sciaroidea (Diptera): the implications of additional taxa and character data. *Zootaxa 1132*: 63–68.
HITCHINGS, T. R. 2001: The Canterbury Museum Ephemeroptera (mayfly) collection and database (Insecta: Ephemeroptera). *Records of the Canterbury Museum 15*: 11–33.
HITCHINGS, T. R.; STANISZEC, A. H. 2003: Nesameletidae (Insecta: Ephemeroptera). *Fauna of New Zealand 46*: 1–72.
HOARE, R. J. B. 2001a: Adventive species of Lepidoptera recorded for the first time in New Zealand since 1988. *New Zealand Entomologist 24*: 23–47.
HOARE, R. J. B. 2001b: New Zealand's most enigmatic moth–what we know about *Titanomis sisyrota*. *DOC Science Internal Series 5*: 1–17.
HOARE, R.; DUGDALE, J.; WATTS, C. 2006: The world's thinnest caterpillar? A new genus and species of Batrachedridae (Lepidoptera) from *Sporodanthus ferrugineus* (Restionaceae), a threatened New Zealand plant. *Invertebrate Systematics 20*: 571–583.
HOCH, H.; DECKERT, J.; WESSEL, A. 2006: Vibrational signalling in a Gondwanan relict insect (Hemiptera: Coleorrhyncha: Peloridiidae). *Biology Letters 2*: 222–224.
HODEK, I. 1993: Habitat and food specificity in aphidophagous predators. *Biocontrol Science and Technology 3*: 91–100.
HODGSON, C. J. 1994: *Eriochiton* and a new genus of the scale insect family Eriococcidae (Homoptera: Coccoidea). *Journal of the Royal Society of New Zealand 24*: 171–208.
HODGSON, C. J.; HENDERSON, R. C. 1996: A review of the *Eriochiton spinosus* (Maskell) species-complex (Eriococcidae: Coccoidea), including a phylogenetic analysis of its relationships. *Journal of the Royal Society of New Zealand 26*: 143–204.
HODGSON, C. J.; HENDERSON, R. C. 1998: A new genus with two new species of soft scale insect (Hemiptera: Coccoidea: Coccidae) from New Zealand. *Journal of the Royal Society of New Zealand 28*: 605–639.
HODGSON, C. J.; HENDERSON, R. C. 2000: Coccidae (Insecta: Hemiptera: Coccoidea). *Fauna of New Zealand 41*: 1–264.
HOLDER, P.; BULLIANS, M.; BROWN, G. 1999: The mosquitoes of New Zealand and their animal disease significance. *Surveillance 26(4)*: 12-15.
HÖLLDOBLER, B.; WILSON, E. O. 1990: *The Ants*. The Belknap Press of Harvard University Press, Cambridge. 732 p
HOLLOWAY, B. A. 1976: A new bat fly family from New Zealand (Diptera: Mystacinobiidae). *New Zealand Journal of Zoology 3*: 279–301.
HOLLOWAY, B. A. 1983: Species of ragwort seedflies imported into New Zealand (Diptera: Anthomyiidae). *New Zealand Journal of Agricultural Research 26*: 245–249.
HOLLOWAY, B. A. 1985: Larvae of New Zealand Fanniidae (Diptera: Calypteratae). *New Zealand Journal of Zoology 11*: 239–257.
HOMEWOOD, B. S. D. 1981: Predator-prey relationships and reproductive biology of the marsh fly *Neolimnia sigma* (Diptera: Sciomyzidae). BScHons thesis, Zoology Department, University of Canterbury, Christchurch. 59 p.
HOPKIN, S. P. 1997: *Biology of the Springtails (Insecta: Collembola)*. Oxford University Press, Oxford. 330 p.
HOY, J. M. 1961: *Eriococcus orariensis* Hoy and other Coccoidea (Homoptera) associated with *Leptospermum* Forst. species in New Zealand. *New Zealand Department of Scientific and Industrial Research Bulletin 141*: 1–70.
HOY, J. M. 1962: Eriococcidae of New Zealand. *New Zealand Department of Scientific and Industrial Research Bulletin 146*: 1–219.
HUBBARD, M.D. 1987: Ephemeroptera. Pp. iii + 1–94 *in*: Westphal, F. (ed.), *Fossilium Catalogus 1: Animalia*. Kugler Publications, Amsterdam.
HUDSON, G. V. 1891: On the New Zealand Cicadidae. *Transactions and Proceedings of the New Zealand Institute 23*: 49–55.
HUDSON, G. V. 1892: *An Elementary Manual of New Zealand Entomology*. London. 122 p., frontis., 20 pls.
HUDSON, G. V. 1904: *New Zealand Neuroptera; a popular introduction to the life histories and habits of mayflies, dragonflies and allied insects inhabiting New Zealand, including notes on their relation to angling*. West, Newman & Co., London. ix + 102 p.
HUDSON, G. V. 1928. *The Butterflies and Moths of New Zealand*. Ferguson and Osborn, Wellington. 386 p., 52 pl.
HUDSON, G. V. 1950: The New Zealand Cicadas. Pp. 123–151 *in*: *Fragments of New Zealand Entomology. A popular account of all New Zealand cicadas. The natural history of the New Zealand glow-worm. A second supplement to the butterflies and moths of New Zealand and notes on many other native insects*. Ferguson & Osborn, Wellington. 188 p., 18 pls.
HUDSON, L. 1973: A systematic revision of the New Zealand Dermaptera. *Journal of the Royal Society of New Zealand 3*: 219–254.
HUDSON, L. 1976: A note on the Dermaptera from the Kermadec islands. *New Zealand Entomologist* 6(2): 154.
HULL, F. M. 1962. Robber flies of the world: the genera of the family Asilidae. *Bulletin of the U.S. National Museum 24*: 1–907. [2 vols.]
HUMAN, K. G.; GORDON, D. M. 1997: Effects of argentine ants on invertebrate biodiversity in Northern California. *Conservation Biology 11*: 1242–1248.
HUNT, G. R. 1992: Life cycles of a gall-forming midge (Diptera: Cecidomyiidae) and associated parasitoids, on putaputaweta. *New Zealand Entomologist 15*: 14–21.
HURYN, A. D. 1998: Ecosystem-level evidence for top-down and bottom-up control of production in a grassland stream ecosystem. *Oecologia 115*: 173–183.
HUTCHESON, J. 1992: Observations on the effects of volcanic activity on insects and their habitat on White Island. *New Zealand Entomologist 15*: 72–76.
HUTTON, F. W. 1898: Revision of the New Zealand Phasmidae. *Transactions of the New Zealand Institute 30*: 50–59.
HUTTON, F. W. 1899: The Neuroptera of New Zealand. *Transactions and Proceedings of the New Zealand Institute 31*: 208–249.
HUTTON, F. W. 1901: Synopsis of the Diptera Brachycera of New Zealand. *Transactions and Proceedings of the New Zealand Institute 33*: 1–95.
HUTTON, F. W. 1904: *Index Faunæ Novæ Zealandiæ*. Dulau and Co., London. viii + 372 p.
INGRAM, A.; MACFIE, J. W. S. 1931: New Zealand Ceratopogonidae. *Annals of the Tropical Medicine and Parasitology 25*: 195–209.
INNES, J. G. 1979: Diet and reproduction of ship rats in the northern Tararuas. *New Zealand Journal of Ecology 2*: 85–86.
IKEDA, E. 1999: A revision of the world species of *Quadrastichodella* Girault, with descriptions of four new species (Hymenoptera, Eulophidae). *Insecta Matsumurana 55*: 13–35.
IQBAL, M.; AUSTIN A. D. 2000: Systematics of *Ceratobaeus* Ashmead (Hymenoptera: Scelionidae) from Australasia. *Records of the South Australian Museum Monographic Series 6*: 1–164.
ISMAY, J. W. 1991: *Dicraceus tibialis* (Macquart: Chloropidae) new to New Zealand. *New Zealand Entomologist 14*: 21–24.
JACKS, H.; COTTIER, W. 1948: Cecidomyid midges on meadow foxtail and cocksfoot in New Zealand. *New Zealand Journal of Science and Technology, A, 30*: 9–12.
JASCHHOF, M. 1998: Revision der'Lestremiinae' (Diptera: Cecidomyiidae) der Holarktis. *Studia Dipterologica, Suppl. 6*: 1–552.
JASCHHOF, M. 2004a: *Starkomyia* gen nov. from New Zealand and its implications for the phylogeny of the Sciaroidea (Diptera: Bibiomorpha). *Studia Dipterologia* 11: 63–74.
JASCHHOF, M. 2004b: The *Heterotricha* group in New Zealand (Diptera: Sciaroidea). *Beitrage zur Entomologie 54*: 3–30.
JASCHHOF, M.; DIDHAM, R. K. 2002: Rangomaramidae fam. n. from New Zealand and implications for the phylogeny of the Sciaroidea (Diptera). *Studia Dipterologia, Supplement 11*: 1–60.
JASCHHOF, M.; HIPPA, H. 2003: Sciaroid but not sciarid: a review of the genus *Ohakunea* Tonnoir & Edwards, with the description of two new species (Insecta: Diptera: Bibionomorpha). *Entomologische Abhandlungen 60*: 23–44.
JASCHHOF, M.; JASCHHOF, C. 2003a: Wood midges of New Zealand (Cecidomyiidae, Lestremiinae). Part I: Introductory notes and tribes Lestremiini, Stroblielini, Campylomyzini and Pteridomyiini Jaschhof trib. nov. *Studia Dipterologica 10*: 97–132.
JASCHHOF, M.; JASCHHOF, C. 2003b: Wood midges of New Zealand (Cecidomyiidae, Lestremiinae). Part II. Tribes Micromyini and Aprioini. *Studia Dipterologica 10*: 423–440.
JASCHHOF, M.; JASCHHOF, C. 2004: Wood midges of New Zealand (Cecidomyiidae, Lestremiinae). Part III. Tribe peromyiini and

remarks on the composition, origin and relationship of the fauna as a whole. *Studia Dipterologica 11*: 75–127.

JASCHHOF, M.; JASCHHOF, C. 2009: *Cowanomyia hillaryi* gen. and sp. n., a remarkable new gnoristine (Diptera: Mycetophilidae: Sciophilinae) for New Zealand. *Zootaxa 2117*: 43–48.

JASCHHOF, M.; KALLWEIT, U. 2009: The *Cycloneura* Marshall group of genera in New Zealand (Diptera: Mycetophilidae: Leiini). *Zootaxa 2090*: 1–39.

JAMIESON, C. D. 1999: A new species of *Sigaus* from Alexandra, New Zealand. *New Zealand Journal of Zoology 26*: 43–48.

JESUDASAN, R. W. A.; DAVID, B.V. 1991: Taxonomic studies on Indian Aleyrodiae (Insecta: Homoptera) *Oriental Insects 25*: 231–434.

JEWELL, A. 2007: Two new species of *Hemiandrus* (Orthoptera: Anostostomatidae) from Fiordland National Park, New Zealand. *Zootaxa 1542*: 49–57.

JEWELL, A.; BROCK, P. D. 2002: A review of the New Zealand stick insects: New genera and synonymy, keys and a catalogue. *Journal of Orthoptera Research 11*: 189–197.

JOHANNESSON, J. M. 1972: Insects of Red Mercury Island. *Tane, Journal of the Auckland University Field Club 18*: 81–86.

JOHANSON, K. A. 1999: Revision of the New Zealand *Helicopsyche* (Trichoptera: Helicopsychidae). *Entomologica Scandinavica 30*: 263–280.

JOHNS, P. M. 1966: The cockroaches of New Zealand. *Records of the Canterbury Museum 8*: 93–136.

JOHNS, P. M. 1972: Notes on aquatic tipulids. *Limnological Society Newsletter 8*: 24–26.

JOHNS, P. M. 1991: Distribution of cave species of Northwest Nelson, Westland and Canterbury. *The Weta 14*: 11–21.

JOHNS, P. M. 1997: The Gondwanaland weta: family Anostostomatidae (formerly in Stenopelmatidae, Henicidae or Mimnermidae): nomenclatural problems, world checklist, new genera and species. *Journal of Orthopteran Research 6*: 125–138.

JOHNS, P. M.; DEACON, K. J.; HERRON, S.; HOMEWOOD, B.; LITTLE, G.; NOTMAN, P.; RAPLEY, M.; WILSON, K. 1980: *Hanmer State Forest Arthropod Survey. Report to the Hanmer State Forest Park Advisory Committee.* New Zealand Forest Service, Christchurch. 90 p.

JOHNS, P. M.; JENNER, L. 2005: The nomenclatural state of species of *Rhamphophila* Edwards, 1923 (Tipulidae: Limoniinae: Hexatomini). *Records of the Canterbury Museum 19*: 23–29.

JOHNS, P. M.; JENNER, L. 2006: The crane-fly genus *Discobola* (Diptera: Tipulidae: Limoniinae) in New Zealand. *Records of the Canterbury Museum 20*: 35–54.

JOHNSON, N. D. 1991: Revision of Australasian *Trissolcus* Species (Hymenoptera: Scelionidae). *Invertebrate Taxonomy 5*: 211–239.

KATHIRITHAMBY, J. 1989a: Review of the order Strepsiptera. *Systematic Entomology 14*: 41–92.

KATHIRITHAMBY, J. 1989b: Descriptions and biological notes of the Australian Elenchidae (Strepsiptera). *Invertebrate Taxonomy 3*: 175–195.

KATHIRITHAMBY, J. 1992: Descriptions and biological hosts of Halictophagidae (Strepsiptera) from Australia with a check list of world genera and species. *Invertebrate Taxonomy 6*: 159–196.

KATHIRITHAMBY, J.; TAYLOR, J. S. 2005: New species of *Halictophagus* (Insecta: Strepsiptera: Halictophagidae) from Texas and a checklist of Strepsiptera from the United States and Canada. *Zootaxa 1056*: 1–18.

KEAN, J. 2002: Population patterns of *Paradoxaphis plagianthi*, a rare New Zealand aphid. *New Zealand Journal of Ecology 26*: 95–99.

KEAN, J.; BARLOW, N. 2004: Exploring rarity using a general model for distribution and abundance. *The American Naturalist 163*: 407–416.

KEAN, J.; STUFKENS, M. A. W. 2005: Phenology, population ecology, and rarity of the New Zealand ribbonwood aphid, *Paradoxaphis plagianthi*. *New Zealand Journal of Zoology 32*: 143–153.

KELLY, D.; McKONE, M. J.; BATCHELOR, K. J.; SPENCE, J. R. 1992: Mast seeding of the *Chionochloa* (Poaceae) and predispersal predation by a seed specialist fly (*Diplotoxa*: Diptera, Chloropidae). *New Zealand Journal of Botany 30*: 125–133.

KELSEY, J. M. 1944: The identification of termites in New Zealand. *New Zealand Journal of Science and Technology, ser. B, 27*: 446–457.

KELSEY, J. M. 1946: Insects attacking milled timber, poles and posts in New Zealand. *New Zealand Journal of Science and Technology, ser. B, 28*: 65–100.

KETTLE, P. R. 1973: A study of the sheep botfly, *Oestrus ovis* (Diptera: Oestridae) in New Zealand. *New Zealand Entomologist 5*: 185–191.

KEY, K. H. L. 1991: On four endemic genera of Tasmanian Acrididae (Orthoptera). *Invertebrate Taxonomy 5*: 241–288.

KHALAIM, A. I. 2004: New tersilochines from Australia and New Zealand (Hymenoptera: Ichneumonidae, Tersilochinae). *Zoosystematica Rossica 13*: 43–45.

KIMMINS, D. E. 1940: A revision of the osmylid subfamilies Stenosmylinae and Kalosmylinae. *Novitates Zoologicae 42*: 165–202.

KIMMINS, D. E.; WISE, K. A. J. 1962: A record of *Cryptoscenea australiensis* (Enderlein) (Neuroptera: Coniopterygidae) in New Zealand. *Transactions of the Royal Society of New Zealand 2*: 35–39.

KING, C. M. 1990: *Handbook of New Zealand Mammals.* Oxford University Press, Auckland. 600 p.

KIRMAN, M. 1985a: *Clavaptera ornata* n. gen. et sp., a new genus and species of Carventinae (Hemiptera: Heteroptera: Aradidae) from Northland, New Zealand. *New Zealand Journal of Zoology 12*: 125–129.

KIRMAN, M. 1985b: A new genus and species of Prosympiestinae (Hexapoda: Hemiptera: Aradidae) from Northland, New Zealand. *Records of the Auckland Institute and Museum 22*: 77–83.

KIRMAN, M. 1989a: A new genus and species of Carventinae (Hexapoda: Hemiptera: Aradidae) from Northland, New Zealand. *Records of the Auckland Institute and Museum 26*: 25–32.

KIRMAN, M. 1989b: A redescription of the genus *Neocarventus* (Hexapoda: Hemiptera: Aradidae) and a description of a new species from Northland, New Zealand. *Records of the Auckland Institute & Museum 26*: 33–38.

KITCHING, I. J.; RAWLINS, J. E. 1999: The Noctuoidea. *In*: Kristensen, N. P. (ed.), *Handbuch der Zoologie 4(35)*: 355–401.

KJER, K. M.; BLAHNIK, R. J.; HOLZENTHAL, R. W. 2002: Phylogeny of caddisflies (Insecta: Trichoptera). *Zoologica Scripta 31*: 83–91.

KLIMASZEWSKI, J. 1997: Biodiversity of New Zealand beetles (Insecta: Coleoptera). *Memoirs of the Museum of Victoria 56*: 659–666.

KLIMASZEWSKI, J; WATT, J. C. 1997: Coleoptera: family-group review and keys to identification. *Fauna of New Zealand 37*: 1–199.

KLOET, G. S.; HINCKS, W. D. 1964: Checklist of British insects. Part 1. Small orders. *Handbook for the Identification of British Insects 11*: 1–119.

KLOSTERMEYER, E. C. 1942: Life history and habits of *Euboriella annulipes* (Lucas) (Dermaptera). *Journal of the Kansas Entomological Society 15*: 13–18.

KLUGE, N. J.; STUDEMANN, D; LANDOLT, P.; GONSER, T. 1995: A reclassification of Siphlonuroidea (Ephemeroptera). *Mitteilungen der Schweizerischen Entomologischen Gesellschaft 68*: 103–132.

KNIGHT, W. J. 1973a: Hecalinae of New Zealand (Homoptera: Cicadellidae). *New Zealand Journal of Science 16*: 957–969.

KNIGHT, W. J. 1973b: Ulopinae of New Zealand. *New Zealand Journal of Science 16*: 971–1007.

KNIGHT, W. J. 1974a: Leafhoppers of New Zealand: subfamilies Aphrodinae, Jassinae, Xestocephalinae, Idiocerinae and Macropsinae. *New Zealand Journal of Zoology 1*: 475–493

KNIGHT, W. J. 1974b: Revision of the New Zealand genus *Novothymbris* (Homoptera: Cicadellidae). *New Zealand Journal of Zoology 1*: 453–473.

KNIGHT, W. J. 1975: Deltocephalinae of New Zealand (Homoptera: Cicadellidae). *New Zealand Journal of Zoology 2*: 169–208

KNIGHT, W. J. 1976a: Typhlocybinae of New Zealand (Homoptera: Cicadellidae). *New Zealand Journal of Zoology 3*: 71–87.

KNIGHT, W. J. 1976b: The leafhoppers of Lord Howe, Norfolk, Kermadec, and Chatham Islands and their relationship to the fauna of New Zealand (Homoptera: Cicadellidae). *New Zealand Journal of Zoology 3*: 89–98.

KNILL, J. S. A. 1999: Investigations into the feeding habits of kampods (Diplura: Campodidae). *Entomologist's Record and Journal of Variation 111*: 96–100.

KNOX, G. A.; BOLTON, L. A. 1978: The ecology and benthic macrofauna and fauna of Brooklyn lagoon, Waimakariri River estuary. *Canterbury University Zoology Department Estuarine Unit Report 16*: 1–128.

KNOX, G. A.; BOLTON, L. A.; SAGAR, P. 1978: The ecology of the westshore lagoon, Ahuriri estuary, Hawkes Bay. *Canterbury University Zoology Department Estuarine Unit Report 15*: 1–89.

KNUTSON, L.V. 1972: Pupa of *Neomochtherus angustipennis* (Hine), with notes on feeding habits of robber flies and a review of publications on morphology of immature stages (Diptera: Asilidae). *Proceedings of the Biological Society of Washington 85*: 163–178.

KOCAREK, P. 1998: Life cycles and habitat associations of three earwig (Dermaptera) species in lowland forest and its surroundings. *Biologia Bratislavia 53*: 205–211.

KOCH, M. 1997: Monophyly and phylogenetic position of the Diplura (Hexapoda). *Pedobiologia* 41: 9–12.

KOLBE, H. J. 1883: Ueber das Genus *Myopsocus* und dessen species. *Entomologische Nachrichten, Berlin 9*: 141–146.

KOLESIK, P.; TAYLOR, G. S.; KENT, D. S. 2002: New genus and two new species of gall midge

(Diptera: Cecidomyiidae) damaging buds on *Eucalyptus* in Australia. *Australian Journal of Entomology 41*: 23–29.

KOLESIK, P.; SARFATI, M. S., BROCKERHOFF, E. G., KELLY, D. 2007: Description of *Eucalypterosis chionchloae* sp. nov., a cecidomyiid feeding on inflorescences of *Chionochloa* (Poaceae) in New Zealand. *New Zealand Journal of Zoology 34:* 107–115.

KORMILEV, N. A.; FROESCHNER, R. C. 1987: *Flat Bugs of the World: A Synonymic Checklist (Heteroptera: Aradidae)*. Entomography Publications, Sacramento. 246 p.

KOZÁR, F.; KONCZNÉ BENEDICTY, Z. 2000: Revision of *Newsteadia* of the Australian and Pacific regions, with description of eleven new species (Homoptera: Coccoidea, Ortheziidae). *Acta Zoologica Academiae Scientiarum Hungaricae 46*: 197–229.

KRISTENSEN, N. P. 1989: The New Zealand scorpionfly (*Nannochorista philpotti* comb.n.): wing morphology and its phylogenetic significance. *Zeitschrift für zoologische Systematik und Evolutionsforschung 27*: 106–114.

KRISTENSEN, N. P. (Ed.) 1999: *Lepidoptera, Moths and Butterflies, Volume 1: Evolution, Systematics and Biogeography*. [*Handbuch der Zoologie Volume IV, Part 35*.] Walter de Gruyter, Berlin, New York. 491 p.

KRZEMINSKA, E. 2001: Genus *Paracladura* Brunetti of the Australian region. 1. Characteristics of the *antipoda* group of species; a new species described (Diptera: Trichoceridae). *New Zealand Journal of Zoology 28*: 373–385.

KRZEMINSKA, E. 2003a: Genus *Paracladura* Brunetti of the Australian region. II. Characteristics of the *lobifera* and *curtisi* group of species; two new species described (Diptera: Trichoceridae). *New Zealand Journal of Zoology 30*: 127–139.

KRZEMINSKA, E. 2003: subfamily Paracladurinae. III Phylogenetic biogeography: two new genera and three species described (Diptera: Trichoceridae). *New Zealand Journal of Zoology 32*: 317-352.

KUHLMANN, U. 1995: Biology of *Triarthria setipennis* (Fallen) (Diptera: Tachinidae), a native parasitoid of the European earwig, *Forficula auricularia* L. (Dermaptera: Forficulidae) in Europe. *Canadian Entomologist 127*: 507–517.

KUKALOVÁ-PECK, J.; LAWRENCE, J.F. 1993: Evolution of the hind wing in Coleoptera. *Canadian Entomologist 125*: 181–258.

KUSCHEL, G. 1990: Beetles in a suburban environment: a New Zealand case study. The identity and status of Coleoptera in the natural and modified habitats of Lynfield, Auckland (1974–1989). *DSIR Plant Protection Report 3*: 1–119.

KUSCHEL, G.; WORTHY, T. H,. 1996: Past distribution of large weevils (Coleoptera: Curculionidae) in the South Island, New Zealand, based on Holocene fossil remains. *New Zealand Entomologist 19*: 15–22.

LACEY, L. A. (Ed.) 1997: *Manual of Techniques in Insect Pathology*. Academic Press, London. 409 p.

LAING, D. J. 1988: A comparison of the prey of tree common web building spiders in open country, bush fringe and urban areas. *Tuatara 30*: 23–25.

LAIRD, M. 1995: Background and findings of the 1993–1994, New Zealand mosquito survey. *New Zealand Entomologist* 18: 77–90.

LAIRD, M.; CALDER, L.; THORNTON, R. C. 1994: Japanese *Aedes albopictus* among four mosquito species reaching New Zealand in used tyres. *Journal of the American Mosquito Control Association 10*: 14–23.

LAMB, K. P. 1951: A new species of gall midge (Cecidomyiidae) from *Hebe salicifolia* Forst. leaf galls. *Transactions of the Royal Society of New Zealand 79*: 210–212.

LAMB, K. P. 1960: A check list of New Zealand plant galls (Zoocecida). *Transactions of the Royal Society of New Zealand 88*: 121–139.

LAMB, R. J. 1976: Parental behaviour in the Dermaptera, with special reference to *Forficula auricularia* (Dermaptera: Forficulidae). *Canadian Entomologist 108*: 609–619.

LAMB, R. J.; WELLINGTON, W. G. 1974: Techniques for studying the behaviour and ecology of the European earwig *Forficula auricularia* (Dermaptera: Forficulidae). *Canadian Entomologist 106*: 881–888.

LAMBDIN, P. L. 1998: *Cerococcus michaeli* (Hemiptera: Cerococcidae): a new species of false pit scale from New Zealand. *Entomological News 109*: 297–300.

LAMBDIN, P. L.; KOSZTARAB, M. 1977: Morphology and systematics of scale insects No. 10. Morphology and systematics of the adult females of the genus *Cerococcus* (Homoptera: Coccoidea: Cerococcidae). *Research Division Bulletin, Virginia Polytechnic Institute and State University 128*: 1–251.

LAMBKIN, K. J. 1996: Mecoptera. Pp. 123–135, 184 (App. 3), 220–221 (Index) *in* Wells, A. (ed.), *Zoological Catalogue of Australia. Vol. 28. Neuroptera, Strepsiptera, Mecoptera, Siphonaptera*. CSIRO Publishing, Melbourne.

LANDIS, C. A.; CAMPBELL, H. J.; BEGG, J. C.; MILDENHALL, D. C.; PATERSON, A. M.; TREWICK, S. A. 2008: The Waipounamu Erosion Surface: questioning the antiquity of the New Zealand land surface and terrestrial fauna and flora. *Geological Magazine 145*: 173–197.

LARINK, O. 1997: Apomorphic and plesiomorphic characterization of Archeognatha, Monura, and Zygentoma. *Pedobiologia 41*: 3–8.

LARIVIÈRE, M.-C. 1995: Cydnidae, Acanthosomatidae, and Pentatomidae (Insecta: Heteroptera); systematics, geographical distribution, and bioecology. *Fauna of New Zealand 35*: 1–112.

LARIVIÈRE, M.-C. 1997a: Taxonomic review of *Koroana* Myers (Hemiptera: Cixiidae), with description of a new species. *New Zealand Journal of Zoology 24*: 213–223.

LARIVIÈRE, M.-C. 1997b: Composition and affinities of the New Zealand heteropteran fauna (including Coleorrhyncha). *New Zealand Entomologist 20*: 37–44.

LARIVIÈRE, M.-C. 1999: Cixiidae (Insecta: Hemiptera: Auchenorrhyncha). *Fauna of New Zealand 40*: 1–93.

LARIVIÈRE, M.-C. 2000: Primary types of Heteroptera in the New Zealand Arthropod Collection: Enicocephalomorpha, Dipsocoromorpha and Cimicomorpha described from New Zealand before the year 2000. *New Zealand Entomologist 23*: 37–45.

LARIVIÈRE, M.-C. 2003: Primary types of Heteroptera in the New Zealand Arthropod Collection: Pentatomomorpha described from New Zealand before the year 2000. *New Zealand Entomologist 25*: 79–85.

LARIVIÈRE, M.-C. (ed.) 2005 (and updates): Checklist of New Zealand Hemiptera (excluding: Sternorrhyncha). *The New Zealand Hemiptera Website, NZHW 04.* [http://hemiptera. landcareresearch.co.nz/]

LARIVIÈRE, M.-C.; BURCKHARDT, D.; LAROCHELLE, A. 2010: Peloridiidae (Insecta: Hemiptera: Coleorrhyncha). *Fauna of New Zealand.*

LARIVIÈRE, M.-C.; FLETCHER, M. J. 2004 (and updates): The New Zealand leafhoppers and treehoppers (Hemiptera: Auchenorrhyncha): web–based identification keys and checklists. *The New Zealand Hemiptera Website, NZHW 02.* [http://hemiptera. landcareresearch.co.nz/]

LARIVIÈRE, M.-C.; FLETCHER, M. J.; JACOB, H. 2004: English–Maori identification key to leafhopper and treehopper genera occurring in New Zealand. *The New Zealand Hemiptera Website, NZHW 03.*

LARIVIÈRE, M.-C.; FLETCHER, M. J. 2008: A new genus, *Zeoliarus*, for the endemic New Zealand species *Oliarus atkinsoni* Myers and *O. oppositus* (Walker) (Hemiptera: Fulgoromorpha: Cixiidae: Cixiinae: Pentastirini). *Zootaxa 1891*: 66–68.

LARIVIÈRE, M.-C.; FLETCHER, M. J.; LAROCHELLE, A. 2010. Auchenorrhyncha (Insecta: Hemiptera): Catalogue. *Fauna of New Zealand 63*: 1–232.

LARIVIÈRE, M.-C.; FROESCHNER, R. C. 1994: *Chilocoris neozelandicus*, a new species of Burrowing Bug from New Zealand (Heteroptera: Cydnidae). *New Zealand Journal of Zoology 21*: 245–248.

LARIVIÈRE, M.-C.; HOCH, H. 1998: The New Zealand planthopper genus *Semo* White (Hemiptera: Cixiidae): taxonomic review, geographical distribution, and biology. *New Zealand Journal of Zoology 25*: 429–442.

LARIVIÈRE, M.-C.; LAROCHELLE, A. 2004: Heteroptera (Insecta: Hemiptera): Catalogue. *Fauna of New Zealand 50*: 1–330.

LARIVIÈRE, M.-C; LAROCHELLE, A. 2006: An overview of flat bug genera (Hemiptera, Aradidae) from New Zealand, with considerations on faunal diversification and affinities. *Denisia 19*: 181–214.

LARIVIÈRE, M.-C.; RHODE, B. E. 2002 (and updates): Virtual collection of primary types of New Zealand Hemiptera (excluding Sternorrhyncha). *The New Zealand Hemiptera Website, NZHW 01*. [http://hemiptera. landcareresearch.co.nz]

LARIVIÈRE, M.-C.; RHODE, B. E. 2006 (and updates): New Zealand Cicadas (Hemiptera: Cicadidae): A virtual identification guide. *The New Zealand Hemiptera Website, NZHW 05.* [http:// hemiptera.landcareresearch.co.nz]

LARIVIÈRE, M.-C.; WEARING, C. H. 1994: *Orius vicinus* Ribaut (Heteroptera: Anthocoridae), a predator of orchard pests new to New Zealand. *New Zealand Entomologist 17*: 17–21.

LAROCHELLE, A.; LARIVIÈRE, M.-C. 2001: Carabidae (Insecta: Coleoptera): Catalogue. *Fauna of New Zealand 43*: 1–285.

LAROCHELLE, A.; LARIVIÈRE, M.-C. 2005: Harpalini (Insecta: Coleoptera: Carabidae: Harpalinae). *Fauna of New Zealand 53*: 1–160.

LAROCHELLE, A.; LARIVIÈRE, M.-C. 2007: Carabidae (Insecta: Coleoptera). *Fauna of New Zealand 60*: 1–188.

LAROCHELLE, A.; LARIVIÈRE, M.-C. 2007 (and updates): New Zealand ground-beetles (Coleoptera: Carabidae): A virtual identification guide to genera. *The New Zealand Carabidae Website, NZCW 02.* [http:// carabidae. landcareresearch.co.nz/]

LAROCHELLE, A.; LARIVIÈRE, M.-C.; RHODE, B.E. 2004 (and updates): Checklist of New Zealand ground-beetles (Coleoptera: Carabidae). *The New Zealand Carabidae Website, NZCW 01.* [http://carabidae. landcareresearch.

co.nz/]

LaSALLE, J.; BOLER, I. 1994: *Hadranellus anomalus* n.gen. *et* n.sp. (Hymenoptera: Eulophidae): an example of extreme intraspecific variation in an endemic New Zealand insect. *New Zealand Entomologist 17*: 37–46.

LaSALLE, J.; GAULD, I. D. (Eds). 1993: *Hymenoptera: their Diversity, and their Impact on the Diversity of Other Organisms*. CAB International, Wallingford. xi +348 p.

LAVIGNE, R. J.; DENNIS, S. D.; GOWAN, J. A. 1978: Asilid literature update 1956–1976 including a brief review of fly biology. *University of Wyoming Science Monograph 36*: 1–134.

LAWRENCE, J. F.; BRITTON, E. B. 1994: *Australian Beetles*. Melbourne University Press, Melbourne. 192 p.

LAWRENCE, J. F.; HASTINGS, A.; DALLWITZ, M. J.; PAINE, T. A.; ZURCHER, E. J. 1999: *Beetles of the World: A Key and Information System for Families and Subfamilies*. CD–ROM, Version 1.0 for MS–DOS. CSIRO Publishing, East Melbourne.

LAWRENCE, J. F.; NEWTON, A. F. 1995: Families and subfamilies of Coleoptera (with selected genera, notes, references and data on family–group names). Pp. 779–1006 *in*: Pakaluk, J.; Slipinski, S.A. (eds), *Biology, Phylogeny and Classification of Coleoptera: Papers Celebrating the 80th Birthday of Roy A. Crowson*. Museum i Instytut Zoologii PAN, Warszawa.

LEADER, J. P. 1972: The New Zealand Hydroptilidae (Trichoptera). *Journal of Entomology, ser. B, 41*: 191–200.

LEATHWICK, D. M., WINTERBOURN, M. J. 1984: Arthropod predation on aphids in a lucerne crop. *New Zealand Entomologist 8*: 75–80.

LEE, D. J. 1948: Australasian Ceratopogonidae (Diptera: Nematocera). Part 1 Relation to disease, biology, general characters and generic classification of the family, with a note on the genus *Ceratopogon*. *Proceedings of the Linnean Society of New South Wales* 72: 313–331.

LEE, S. S.; THORNTON, I. W. B. 1967: The family Pseudocaeciliidae (Psocoptera): a reappraisal based on the discovery of new Oriental and Pacific species. *Pacific Insects Monographs 16*: 1–116.

LEHR, P. A. 1996: Palaearctic Robberflies of Subfamily Asilinae (Diptera, Asilidae): Ecological and Morphological Analysis, Taxonomy and Evolution. Russian Academy of Science, Far Eastern Branch, Vladivostok. 194 p. [In Russian.]

LESCHEN, R. A. B.; LAWRENCE, J. F.; KUSCHEL, G.; THORPE, S.; WANG, Q. 2003: *Coleoptera* genera of New Zealand. *New Zealand Entomologist 26*: 15–28.

LESCHEN, R. A. B.; RHODE, B. E. 2002: A new genus and species of large extinct Ulodidae (Coleoptera) from New Zealand. *New Zealand Entomologist 25*: 57–64.

LIENHARD, C. 1998: Psocoptères euro-mediterranéens. *Faune de France* 83: xx, 1–517.

LIENHARD, C.; SMITHERS, C. N. 2002: Psocoptera (Insecta). World catalogue and bibliography. *Instrumenta Biodiversitatis 5*: xli, 1–745.

LILLIE, C. O. 1898: On New Zealand Ephemeridae: two species. *Transactions of the New Zealand Institute 31*: 164–169.

LINDNER, E. 1958: Ueber einige neuseelandische Stratiomyiiden Osten–Sackens im Deutschen Entomologischen Institut in Berlin (Diptera). *Beitrage zur Entomologia 8*: 431–437.

LINDSAY, C. J.; ORDISH, R.G. 1964: Food of the morepork. *Notornis 11*: 154–158.

LINTOTT, W. H. 1974: Plant galls of the Canterbury region. *Canterbury Botanical Society Journal 7*: 3–9.

LINTOTT, W. H. 1981: Midge galls of New Zealand Inuleae (Compositae). *Cecidologia Internationale 2*: 25–34.

LIS, J. A. 1995: A synonymic cheklist of burrower bugs of the Australian Region (Heteroptera: Cydnidae). *Genus 6*: 137–149.

LIS, J. A. 1996: A review of burrower bugs of the Australian Region, with a discussion on the distribution of the genera (Hemiptera: Heteroptera: Cydnidae). *Genus 7*: 177–238.

LLOYD, D. G. 1985: Progress in understanding the natural history of New Zealand plants. *New Zealand Journal of Botany 23*: 707–722.

LOAN, C. C.; LLOYD, D. G. 1974: Description and biology of *Microctonus hyperodae* Loan, n.sp. (Hymenoptera: Braconidae: Euphorinae) a parasite of *Hyperodes bonariensis* in South America (Coleoptera: Curculionidae). *Entomophaga 19(1)*: 7–12.

LOVEI, G. L.; HICKMAN, J. M.; McDOUGALL, D.; WRATTEN, S. D. 1993: Field penetration of beneficial insects from habitat islands: hoverfly dispersal from flowering crop strips. *Proceedings of the New Zealand Plant Protection Conference 46*: 325–328.

LOWE, A. D. 1973: Aphid biology in New Zealand. *Bulletin of the Entomological Society of New Zealand 2*: 7–9.

LU, W.; WANG, Q. 2005: Systematics of the New Zealand longicorn beetle genus *Oemona* Newman with discussion of the taxonomic position of the Australian species, *O. simplex* White (Coleoptera: Cerambycidae: Cerambycinae). *Zootaxa 971*: 1–31.

LU, W.; Wang, Q. 2005: Systematics of the longicorn beetle genus *Ophryops* (Coleoptera: Cerambycidae). *Invertebrate Systematics 19*: 169–188.

LUGO-ORTIZ, C. R.; McCAFFERTY, W. P. 1998: First report of the genus *Siphlaenigma* Penniket and the family Siphlaenigmatidae (Ephemeroptera) from Australia. *Proceedings of the Entomological Society of Washington 100*: 209–213.

LVOVSKY, A.L. 2003: Check-list of the broad-winged moths (Oecophoridae *s.l.*) of Russia and adjacent countries. *Nota Lepidopterologica 25*: 213–220.

LYNEBORG, L. 1992: Therevidae (Insecta: Diptera). *Fauna of New Zealand 24*: 1–40.

LYSAGHT, A. M. 1925: Orthoptera and Dermaptera from the Chatham islands. *Records of the Canterbury Museum 2*: 301–310.

MAA, T.C. 1986: A taxonomic revision of the Hippoboscidae (Diptera) of Tasmania, Australia. *Journal of the Tasmanian Museum 39*: 83–92.

MACFARLANE, R. P. 1970: A preliminary study of the fauna associated with lucerne *Medicago sativa* L. variety Wairau in New Zealand, with special reference to pests, sampling methods, and lucerne seed production. Unpublished M. Agric. Sci. thesis, Lincoln University, Canterbury. 681 + 71 p. (app.).

MACFARLANE, R. P. 1995a: Applied pollination in temperate areas. *In*: Roubik, D.W. (ed.), *Pollination of Cultivated Plants in the Tropics. FAO Agricultural Services Bulletin 118*: 20–39.

MACFARLANE, R. P. 1995b: Evaluating pollinators. *In*: Roubik, D. W. (ed.), *Pollination of Cultivated Plants in the Tropics. FAO Agricultural Services Bulletin 118*: 101–108.

MACFARLANE, R. P. 2002: Irishman Creek invertebrate survey. Department of Conservation Tenure Report, Christchurch. 27 p.

MACFARLANE, R. P. 2005: New Brighton sand dune invertebrates. Christchurch City Council, Greenspace Unit Report. 47 p

MACFARLANE, R. P. 2007: Styx Mill conservation reserve invertebrate assessment: implications for management. Christchurch City Council, Greenspace Unit Report 134 p.

MACFARLANE, R. P.; ANDREW, I. G. 2001: New Zealand Diptera identification, diversity and biogeography: summary. *Records of the Canterbury Museum 15*: 33–72.

MACFARLANE, R. P.; FERGUSON, A. M. 1984: Kiwifruit pollination: a survey of the pollinators in New Zealand. *International Symposium of Pollination Proceedings 6*: 367–373.

MACFARLANE, R. P.; GURR, L. 1995: Distribution of bumble bees in New Zealand. *New Zealand Entomologist 18*: 29–36.

MACFARLANE, R. P.; MILLAR, I.; PATRICK, B. H. 1997: Nelson uplands insects: conservation status of a predatory robber fly and ground beetle and associated beech forest insect ecology. Department of Conservation Report, Nelson–Marlborough Conservancy, Nelson. 21 p.

MACFARLANE, R. P.; PATRICK, B. H.; JOHNS, P. M.; VINK, C. J. 1998: Travis Marsh: Invertebrate inventory and analysis. Christchurch City Council Report, Parks and Recreation Section, Christchurch. 66 p.

MACFARLANE, R. P.; PATRICK, B. H.; VINK, C. J. 1999: McLeans Island: Invertebrate inventory and analysis. Christchurch City Council Report, Parks and Recreation Section, Christchurch. 44 p.

MACFIE, J. W. S. 1932: New Zealand biting midges (Diptera, Ceratopogonidae). *Annals of Tropical Medicine and Parasitology 26*: 23–53.

MacINTYRE, M. E. 1977a: Chromosome counts and colour pattern variation in the New Zealand *Pteronemobius* species (Orthoptera: Gryllidae). *New Zealand Entomologist 6*: 319–323.

MacINTYRE, M. E. 1977b: Acoustical communication in the field crickets *Pteronemobius nigrovus* and *P. bigelowi* (Orthoptera: Gryllidae). *New Zealand Journal of Zoology 4*: 63–72.

MACKERRAS, F. M. 1957: Tabanidae (Diptera) of New Zealand. *Transactions and Proceedings of the Royal Society of New Zealand 84*: 581–610.

MacLEAN, F. S. 1955: The history of plague in New Zealand. *New Zealand Medical Journal 54*: 131–143.

MacMILLAN, B. W. H. 1981: Food of house sparrows and greenfinches in a mixed district, Hawkes Bay, New Zealand. *New Zealand Journal of Zoology 8*: 93–104.

MacMILLAN, B. W. H.; POLLOCK, B. J. 1983: Food of nestling house sparrows (*Passer domesticus*) in mixed farmland of Hawkes Bay, New Zealand. *New Zealand Journal of Zoology 12*: 307–317.

MALDONADO CAPRILES, J. 1990: Systematic catalogue of the Reduviidae of the World (Insecta: Heteroptera). *Caribbean Journal of Science, Special Edition*: 1–694.

MALIPATIL, M. B. 1976: *Metagerra* White (Heteroptera: Lygaeidae); a review. *New Zealand Journal of Zoology 3*: 303–312.

MALIPATIL, M. B. 1977a: Additions to the Drymini of New Zealand (Heteroptera: Lygaeidae). *New Zealand Journal of Zoology 4*: 177–182.

MALIPATIL, M. B. 1977b: On *Nothochromus maoricus* Slater, Woodward & Sweet (Heteroptera: Lygaeidae). *New Zealand Journal of Zoology 4*: 217–219.

MALIPATIL, M. B. 1977c: The Targaremini of New Zealand (Hemiptera: Lygaeidae) a revision. *New Zealand Journal of Zoology 4*: 333–367.

MALIPATIL, M. B. 1978: Revision of the Myodochini (Hemiptera: Lygaeidae: Rhyparochrominae) of the Australian Region. *Australian Journal of Zoology, Supplementary Series 56*: 1–178.

MALIPATIL, M. B. 1979: An unusual orsilline genus from New Zealand (Heteroptera: Lygaeidae). *New Zealand Journal of Zoology 6*: 237–239.

MALIPATIL, M. B. 1980: Revision of Australian *Microvelia* Westwood (Hemiptera: Veliidae) with a description of two new species from Eastern Australia. *Australian Journal of Marine and Freshwater Research 31*: 85–108.

MALLOCH, J. R. 1928: Notes on Australian Diptera. No. XIV. *Proceedings of the Linnean Society of New South Wales 53*: 295–309.

MALLOCH, J. R. 1931: Notes on New Zealand Empididae (Diptera). *Records of the Canterbury Museum 3*: 423–429.

MALLOCH, J. R. 1932: Notes on New Zealand Empididae (Diptera) 2. *Records of the Canterbury Museum 3*: 457–458.

MALLOCH, J. R. 1938: The calypterate Diptera of New Zealand: Parts 8 and 9. *Transactions of the Royal Society of New Zealand 68*: 161–258.

MARDON, D. K. 1981: *An Illustrated Catalogue of the Rothschild Collection of Fleas (Siphonaptera) in the British Museum (Natural History). Volume VI: Pygiopsyllidae.* Trustees of the British Museum (Natural History), London. 298 p.

MARRA, M. J. 2003: Last interglacial beetle fauna from New Zealand. *Quaternary Research 59*: 122–131.

MARRA, M.; LESCHEN, R. A. B. 2004: Late Quaternary paleoecology from fossil beetle communities in the Awatere Valley, South Island, New Zealand. *Journal of Biogeography 31*: 571–586.

MARRA, M. J.; SHULMEISTER, J.; SMITH, E. G. C. 2006: Reconstructing temperature during the Last Glacial Maximum from Lyndon Stream, South Island, New Zealand using fossil beetles and maximum likelihood envelopes. *Quaternary Science Reviews 25*: 1841–1849.

MARRA, M. J.; SMITH, E. G. C.; SHULMEISTER, J.; LESCHEN, R. 2004: Late Quaternary climate change in the Awatere Valley, South Island, New Zealand using a sine model with a maximum likelihood envelope on fossil beetle data. *Quaternary Science Reviews 23*: 1637–1650.

MARRIS, J. M. 2000: The beetle (Coleoptera) fauna of the Antipodes Islands, with comments on the impact of mice; and an annotated checklist of the insect and arachnid fauna. *Journal of the Royal Society of New Zealand 30*: 169–195.

MARSHALL, A. G. 1981: *The Ecology of Ectoparasitic Insects.* Academic Press, London. xvi + 459 p.

MARSHALL, D. C.; SLON, K.; COOLEY, J. R.; HILL, K. B. R.; SIMON, C. 2008: Steady Plio-Pleistocene diversification and a 2-million-year sympatry threshold in a New Zealand cicada radiation. *Molecular Phylogenetics and Evolution 48*: 1054–1066.

MARSHALL, P. 1896: New Zealand Diptera. No. 1. *Transactions and Proceedings of the New Zealand Institute 28*: 216–250; 3 pls.

MARSHALL, S. A.; HALL, S. C. B.; HODGE, S. 2009: A review of the genus *Phthitia* Enderlein (Diptera; Sphaeroceridae; Limosininae) in New Zealand with a description of two new species. *New Zealand Entomologist 32*: 48–54.

MARSHALL, S. A., LANGSTAFF, R. 1998: Revision of the new world species of *Opacifrons* Duda (Diptera: Sphaeroceridae: Limosininae). *Contributions in Science 474*: 1–27.

MARSHALL, S. A.; ROHÁCEK, J. 2000: World revision of *Thoracochaeta* (Diptera: Sphaeroceridae, Limosininae). *Studia Dipterologica 7*: 259–311.

MARSHALL, S. A.; SMITH, I. P. 1992: A revision of the new world and Pacific *Phthitia* Enderlein (Diptera: Sphaeroceridae, Limosininae) including *Kimosina* Rohácek new synonymn and *Aubertinia* Richards new synonym. *Memoirs of the Entomological Society of Canada 161*: 1–83.

MARTIN, J. 1998: *Chironomus forsythi* n.sp. from New Zealand, a member of the *Chironomus zealandicus* group with *salinarius* type larvae. *Journal of Kansas Entomological Society 71*: 243–255.

MARTIN, N. A. 1983: Miscellaneous observations on a pasture fauna: an annotated species list. *New Zealand Department of Scientific and Industrial Research Entomology Division Report 3*: 1–98.

MARTIN, N. A. 2003: Are *Dracophyllum* bud galls caused by *Eriophyes dracophylli* (Acari: Eriophyidae) or Cecidomyiidae (Diptera)? *Weta 26*: 20–22.

MARTIN, N. A. 2004: History of an invader, *Scaptomyza flava* (Fallen, 1823) (Diptera: Drosophilidae). *New Zealand Journal of Zoology* 31: 27–32.

MARTIN, N. A. 2007: Ecological observations on grass leaf mining flies in New Zealand. *New Zealand Entomologist 30:* 35-39

MARTIN, N. A.; MOUND, L. A. 2005: Host plants for some New Zealand thrips (Thysanoptera: Terebrantia). *New Zealand Entomologist 27*: 119–123.

MARTIN, N. A.; WORKMAN, P. J. 1994: Confirmation of a pesticide resistant strain of western flowers thrips in New Zealand. *Proceedings of the New Zealand Plant Protection Conference 47*: 144–148.

MARTIN, N. A.; WORKMAN, P. J. 1999: A comparison between preventative and curative treatments for control of gnats. *Proceedings of the New Zealand Plant Protection Conference 52*: 50–55.

MARY, N.; MARMONIER, P. 2000: First survey of interstital fauna of New Caledonia rivers, influence of geological and geomorphological characteristics. *Hydrobiologia 418*: 197–208.

MASKELL, W. M. 1885: Note on an aphidian insect infesting pine trees, with observations on the name 'Chermes' or 'Kermes'. *Transactions and Proceedings of the New Zealand Institute 17*: 13–19.

MASKELL, W. M. 1887: *An Account of the Insects Noxious to Agriculture and Plants in New Zealand. The Scale Insects (Coccidae).* Government Printer, State Forests & Agricultural Department, Wellington. 116 p.

MASKELL, W. M. 1889: On some gall–producing insects in New Zealand. *Transactions and Proceedings of the New Zealand Institute 21*: 253–258; 2 pls.

MASKELL, W. M. 1890: On some species of Psyllidae in New Zealand. *Transactions and Proceedings of the New Zealand Institute 22*: 157–170.

MASKELL, W. M. 1891: Further coccid notes: with descriptions of new species from New Zealand, Australia and Fiji. *Transactions and Proceedings of the New Zealand Institute 23*: 1–36.

MASKELL, W. M. 1894: On a new species of *Psylla*. *Entomologist's Monthly Magazine 30*: 171–173.

MASNER, L. 1976: Revisionary notes and keys to world genera of Scelionidae (Hymenoptera: Proctotrupoidea). *Memoirs of the Entomological Society of Canada 97*: 1–87.

MASNER, L. 1990: Status report on taxonomy of Hymenoptera in North America. Pp. 231–240 *in*: Kosztarab, M.; Schaefer, C. W. (eds), *Systematics of the North American Insects and Arachnids: Status and Needs.* [*Virginia Agricultural Experiment Station Information Series 90–1.*] Virginia Polytechnic Institute and State University, Blacksburg.

MASNER, L.; HUGGERT, L. 1989: World review and keys to genera of the subfamily Inostemmatinae with reassignment of the taxa to the Platygastrinae and Sceliotrachelinae (Hymenoptera: Platygastridae). *Memoirs of the Entomological Society of Canada 147*: 1–214.

MATHIS, W. N. 1999: A review of the beach-fly genus *Isocanace* Mathis (Diptera: Canacidae). *Proceedings of the Entomological Society of Washington 101*: 347–358.

MATHIS, W. N.; MUNARI, L. 1996: A catalog of the Tethinidae (Diptera). *Smithsonian Contributions in Zoology 584*: 1–27.

MATHIS, W. N.; ZATWARNICKI, T. 1995: World catalogue of shore flies (Diptera: Ephydridae). *Memoirs on Entomology International 4*: 1–423.

MATHIS, W. M; ZATWARNICKI, T.; MARRIS, J. W. T. 2004: Review of unreported shore-fly genera of the tribe Scatellini from the New Zealand subregion (Diptera: Ephydridae) with descriptions of new species. *Zootaxa 622*: 1–27.

MATILE, L. 1990: Recherches sur la systématique et l'évolution des Keratoplatidae (Diptera: Mycetophiloidea). *Mémoires du Muséum national d'Histoire naturelle, A, 148*: 1–682.

MATILE, L. 1993: Diptères Mycetophiloidea de Nouvelle-Calédonie. 5. Mycetophilidae Leiinae et Manotinae. *In*: Matile, L.; Najit, J.; Tillier, S. (eds), *Zoologica Neocaledonica, Volume 3. Mémoires du Muséum national d'Histoire naturelle 157*: 165–211.

MAYHEW, P. J. 2002: Shifts in hexapod diversification and what Haldane could have said. *Proceedings of the Royal Society of London, ser. B, 269*: 969–974.

McALPINE, D. K. 1991: Review of the Australian kelp flies (Diptera: Coelopidae). *Systematic Entomology 16*: 29–84.

McALPINE, D. K. 2007: The surge flies (Diptera: Canacidae: Zaleinae) of Australasia and notes on tethinid-Canacid morphology and relationships. *Records of the Australian Museum 59:* 27-64.

McALPINE, D. K.; DE KEYZER, P. G. 1994: Generic classification of the fern flies (Diptera: Teratomyzidae) with a larval description. *Systematic Entomology 19*: 305–326.

McALPINE, J. F.; PETERSON, B.V.; SHEWELL, G. E.; TESKEY, H. J.; VOCKEROTH, J. R.; WOOD, D. M. 1981: Manual of Nearctic Diptera. 1. *Monograph, Agriculture Research Branch, Agriculture Canada 27*: 1–674.

McALPINE, J. F.; PETERSON, B.V.; SHEWELL, G. E.; TESKEY, H. J.; VOCKEROTH, J. R.; WOOD, D. M. 1987: Manual of Nearctic Diptera. 2. Monograph, *Agriculture Research Branch, Agriculture Canada 28*: 675–1332.

McCAFFERTY, W. P. 1999: Distribution of Siphlaenigmatidae (Ephemeroptera). *Entomological News 110*: 191.

McDOWALL, R. M. 1990: *New Zealand Freshwater Fishes:A Natural History Guide.* Heinemann

Reed, Auckland. 553 p.
McEWAN, P. K.; KIDD, N. A. C.; BAILEY, E.; ECCLESTON, L. 1999: Small scale production of the common green lacewing *Chrysoperla carnea* (Stephens) (Neuropt., Chrysopidae): minimizing costs and maximizing output. *Journal of Applied Entomology 123*: 303–305.
McFARLANE, A. G. 1939: Additions to New Zealand Rhyacophilidae. Part 1. *Transactions and Proceedings of the Royal Society of New Zealand 69*: 330–340.
McFARLANE, A. G. 1951a: A note on the genus *Neurochorema* Till: and the addition of a species thereto. *Records of the Canterbury Museum 5*: 253–254.
McFARLANE, A. G. 1951b: Additions to the New Zealand Rhyacophilidae Part 2. *Records of the Canterbury Museum 5*: 255–265.
McFARLANE, A. G. 1956: Additions to the New Zealand Trichoptera (part 3). *Records of the Canterbury Museum 7*: 29–41.
McFARLANE, A. G. 1960: Additions to the New Zealand Trichoptera (part 4). *Records of the Canterbury Museum 7*: 203–218.
McFARLANE, A. G. 1964: A new endemic subfamily, and other additions and emendations to the Trichoptera of New Zealand (part 5). *Records of the Canterbury Museum 8*: 55–79.
McFARLANE, A. G. 1966: New Zealand Trichoptera (part 6). *Records of the Canterbury Museum 8*: 137–161.
McFARLANE, A. G. 1973: Five new species of Trichoptera from New Zealand. *Journal of the Royal Society of New Zealand 3*: 23–34.
McFARLANE, A. G. 1990: A generic key to late instar larvae of the New Zealand Trichoptera (caddis flies). *Records of the Canterbury Museum 10*: 25–38.
McFARLANE, A. G. *in* McFARLANE, A. G.; COWIE, B. 1981: Descriptions of new species and notes on some genera of New Zealand Trichoptera. *Records of the Canterbury Museum 9*: 353–385.
McFARLANE, A. G.; WARD, J. B. 1990: *Triplectidina moselyi* n.sp., a previously misidentified New Zealand caddis-fly (Trichoptera: Leptoceridae). *New Zealand Entomologist 13*: 55–59.
McGREGOR, P. G.; WATTS, P. J.; ESSON, M. J. 1987: Light trap records from southern North Island hill country. *New Zealand Entomologist 10*: 104–121.
McGUINESS, C. A. 2001: The conservation requirements of New Zealand's nationally threatened invertebrates. *New Zealand Department of Conservation Threatened Species Occasional Publication 20*: 1–657.
McILROY, J. C. 1968: The biology of magpies (*Gymnorhina* spp.) in New Zealand. Master of Agricultural Science thesis, Lincoln College, Canterbury. 344 p.
McINTOSH, A. R. 2000: Aquatic predator-prey interactions. Pp. 125–156 *in*: Collier, K. J.; Winterbourn, M. J. (eds), *New Zealand Stream Invertebrates: Ecology and Implications for Management.* New Zealand Limnological Society and NIWA, Hamilton. 415 p.
McINTOSH, A. R.; TOWNSEND, C. R. 1996: Interactions between fish, grazing invertebrates and algae in a New Zealand stream: a trophic cascade mediated by fish-induced changes to grazer behaviour? *Oecologia 107*: 174–181.
McKENZIE, H. L. 1960: Taxonomic position of *Parlatoria virescens* Maskell, and descriptions of related species (Homoptera; Coccoidea; Diaspididae). Scale studies–Part 15. *California Department of Agriculture Bulletin 49*: 204–211.
McLACHLAN, R. 1862: Characters of new species of exotic Trichoptera; also of one new species inhabiting Britain. *Transactions of the Entomological Society of London, ser. 3, 1*: 301–311.
McLACHLAN, R. 1866: Descriptions of new or little-known genera and species of exotic Trichoptera; with observations on certain species described by Mr. F. Walker. *Transactions of the Entomological Society of London, ser. 3, 5*: 247–278, 3 pls.
McLACHLAN, R. 1868: On some new forms of trichopterous insects from New Zealand; with a list of the species known to inhabit those colonies. *Journal of the Linnean Society of London, Zoology 10*: 196–214, 1 pl.
McLACHLAN, R. 1871: On new forms etc., of extra-european trichopterous insects. *Journal of the Linnean Society of London, Zoology 11*: 98–141, 3 pls.
McLACHLAN, R. 1894: Some additions to the neuropterous fauna of New Zealand, with notes on certain described species. *Entomologist's Monthly Magazine 30(365)*: 238–243.
McLAREN, G. E. 1992: Thrips on nectarines in the spring. *New Zealand Weed and Pest Control Conference 45*: 111–115.
McLAREN, G. E.; FRASER, J. A. 2000: Development of thresholds for insecticidal control of New Zealand flower thrips on nectarines in spring. *Proceedings of the New Zealand Plant Protection Conference 53*: 194–199
McLAREN, G. E.; FRASER, J. A. 2001: Alternative strategies to control New Zealand flower thrip on nectarines. *Proceedings of the New Zealand Plant Protection Conference 54*: 10–14.
McLEAN, J. A.; KILVERT, S.; JONES, D.; MACFARLANE, R. P.; ECROYD, C. 1998: Insects as ecological indicators of natural and modified landscapes in the Whangamata area. Forest Research Institute Project Report. 35 p.
McLELLAN, I. D. 1967: New gripopterygids (Plecoptera) of New Zealand. *Transactions of the Royal Society of New Zealand 9*: 1–15.
McLELLAN, I. D. 1977: New alpine and southern Plecoptera from New Zealand. *New Zealand Journal of Zoology* 4: 119–147.
McLELLAN, I. D. 1983: New diagnosis for genus *Austrothaumalea*, and redescription of *A. neozealandica* (Diptera: Thaumaleidae). *New Zealand Journal of Zoology 10*: 267–270.
McLELLAN, I. D. 1988: A revision of New Zealand Thaumaleidae (Diptera: Nematocera) with descriptions of new species and a new genus. *New Zealand Journal of Zoology 15*: 563–575.
McLELLAN, I. D. 1991: Notonemouridae (Insecta: Plecoptera). *Fauna of New Zealand 22*: 1–64.
McLELLAN, I. D. 1993: Antarctoperlinae (Insecta: Plecoptera). *Fauna of New Zealand 27*: 1–70.
McLELLAN, I. D. 1997: *Austroperla cyrene*–an adaptable and unpalatable New Zealand stonefly. Pp. 117–118 *in*: Landholt, P.; Sartori, M. (eds), *Ephemeroptera and Plecoptera: Biology–Ecology–Systematics.* Mauron, Tinguely, & Lachat, Fribourg. 569 p.
McLELLAN, I. D. 2003: Six new species and a new genus of stoneflies (Plecoptera) from New Zealand. *New Zealand Journal of Zoology 30*: 101–113.
McMILLAN, B. W. H. 1980: Food of house sparrows and greenfinches in a mixed district, Hawkes Bay, New Zealand. *New Zealand Journal of Zoology 8*: 93–104.
McMILLAN, B. W. H.; POLLOCK, B. J. 1983: Food of nestling house sparrows (*Passer domesticus*) in mixed farmland of Hawkes Bay, New Zealand. *New Zealand Journal of Zoology 12*: 307–317.
McVEAN, A.; FIELD, L. H. 1996: Communication by substratum vibration in the New Zealand tree weta, *Hemideina femorata* (Stenopelmatidae: Orthoptera). *Journal of Zoology 239*: 101–122.
McWILLIAM, H. A.; DEATH, R. G. 1998: Arboreal arthropod communities of remnant podocarp–hardwood rain forests in North Island, New Zealand. *New Zealand Journal of Zoology 25*: 157–169.
MENDES, L. F.; ROCA, C. B. de; GAJU-RICART, M. 1994: Descricão de uma nova especie de *Nesomachilis* Tillyard: *N. novaezelandiae* sp. n. (Microcoryphia; Meinertellidae). *Garcia De Orta Serie de Zoologia 20*: 141–148.
MERTON, D. V. 1970: Kermadec Island expedition reports: a general account of birdlife. *Notornis 17*: 147–199.
MERTON, J. M. 1982: Interactions of the tachinid, *Procissio cana* with its host the New Zealand grass grub, *Costelytra zelandica. Proceedings of the Australasian Conference of Grassland Invertebrate Ecology 3*: 161–168.
MEY, E. 2004: Zur Taxonomie, Verbreitung und parasitophyletischer Evidenz des *Philopterus*-Komplexes (Insecta, Phthiraptera, Ischnocera). *Ornithologischer Anzeiger 43*: 149–203.
MILLER, C. J.; MILLER, T. K. 1995: Population dynamics and diet of rodents on Rangitoto Island, New Zealand including the effect of a 1080 poison operation. *New Zealand Journal of Ecology 19*: 19–27.
MILLER, D. 1913: New species of New Zealand Empididae (order Diptera). *Transactions of the New Zealand Institute 45*: 198–206.
MILLER, D. 1917: Contributions to the Diptera fauna of New Zealand: Part I *Transactions of the New Zealand Institute 49*: 172–194.
MILLER, D. 1921: Material for a monograph on the Diptera fauna of New Zealand: Part II Family Empididae *Transactions of the New Zealand Institute 53*: 289–333.
MILLER, D. 1923: Material for a monograph on the Diptera fauna of New Zealand: Part III Family Syrphidae *Transactions of the New Zealand Institute 54*: 437–464.
MILLER, D. 1939: Blow flies (Calliphoridae) and their associates in New Zealand. *Cawthron Institute Monograph 2*: 1–75.
MILLER, D. 1950: Catalogue of New Zealand Diptera of the New Zealand subregion. *New Zealand Department of Scientific and Industrial Research Bulletin 100*: 1–194.
MILLER, D. 1956: Bibliography of New Zealand entomology 1775–1952. *New Zealand Department of Scientific and Industrial Research Bulletin 120*: 1–492.
MILLER, D. 1971: *Common Insects in New Zealand.* A.H. & A.W. Reed, Wellington. xix + 178 p.
MILLER, D.; PHILLIPS, W. J. 1952: *Identification of New Zealand Mosquitoes.* Cawthron Institute, Nelson. 28 p.
MILLER, D.; WALKER A. K. 1984: *Common Insects in New Zealand.* A.H. & A.W. Reed, Wellington. 178 p.
MILLER, D.; WATT, M. N. 1915: Contributions to the study of New Zealand entomology, from an economical and biological standpoint: No 6: *Syrphus ropalus* Walk. *Transactions and Proceedings of the New Zealand Institute 47*: 274–284.
MILLIGAN, R. H. 1974: Wood-borers (1). *New Zealand's Nature Heritage 4(58)*: 1611–1616.
MILLIGAN, R. H. 1984a: *Stolotermes rufipes* Braeur,

Stolotermes inopus Gay (Isoptera: Termopsidae) New Zealand wetwood termites. *Forest and Timber insects in New Zealand, FRI, New Zealand Forest Service Bulletin* 60: 1–4.

MILLIGAN, R. H. 1984b: *Kalotermes brouni* Froggat (Isoptera: Kalotermitidae) New Zealand drywood termites. *Forest and Timber insects in New Zealand, FRI, New Zealand Forest Service Bulletin* 59: 1–4.

MINKS, A. K.; HARREWIJN, P. (Eds) 1987: *Aphids: Their Biology, Natural Enemies and Control.* Volume A. Elsevier, Amsterdam.

MINKS, A.K.; HARREWIJN, P. (Eds) 1988: *Aphids: Their Biology, Natural Enemies and Control.* Volume B. Elsevier, Amsterdam.

MINKS, A.K.; HARREWIJN, P. (Eds) 1989: *Aphids: Their Biology, Natural Enemies and Control.* Volume C. Elsevier, Amsterdam.

MINOR, M. A. 2008: Protura in native and exotic forests in the North Island of New Zealand. *New Zealand Journal of Zoology 25*: 271–279.

MOCKFORD, E. L. 1997: A new species of *Dicopomorpha* (Hymenoptera: Mymaridae) with diminutive apterous males. *Annals of the Entomological Society of America 90*: 115–120.

MOCKFORD, E. L.; GARCIA ALDRETE, A. N. 1976: A new species and notes on the taxonomic position of *Asiopsocus* Günther (Psocoptera). *Southwestern Naturalist 21*: 335–346.

MOEED, A. 1975: Diets of nestling starlings and mynas at Havelock North, Hawkes Bay. *Notornis 22*: 291–294.

MOEED, A. 1976: Birds and their food resources in at Christchurch International Airport, New Zealand. *New Zealand Journal of Zoology 3*: 373–390.

MOEED, A. 1980: Diet of adult and nestling starlings (*Sturnus vulgaris*) in Hawkes Bay, New Zealand. *New Zealand Journal of Zoology 7*: 247–256.

MOEED, A. 2000: Current context of biocontrol in New Zealand. *ERMA New Zealand Evaluation and Review Report: NOR99001 Import for Release or Release from Containment any New Organism.* Environmental Risk Management Authority, Wellington.

MOEED, A.; FITZGERALD, B. M. 1982: Foods of insectivorous birds in forest of the Orongorongo valley, Wellington New Zealand. *New Zealand Journal of Zoology 9*: 391–403.

MOEED, A., MEADS, M. J. 1983: Invertebrate fauna of four tree species in Orongorongo valley, as revealed by tree trunk traps. *New Zealand Journal of Ecology* 6: 39–53.

MOEED, A.; MEADS, M. J. 1984: Vertical and seasonal distribution of airborne invertebrates in Orongorongo Valley, Wellington, New Zealand. *New Zealand Journal of Zoology 11*: 49–58.

MOEED, A., MEADS, M. J. 1985: Seasonality of pitfall trapped invertebrates in three types of native forest, Orongorongo valley, New Zealand. *New Zealand Journal of Zoology* 12: 17–53.

MOEED, A.; MEADS, M. J. 1987a: Seasonality and density of litter and humus inhabiting invertebrates in two forest floors in Orongorongo valley, New Zealand. *New Zealand Journal of Zoology 14*: 51–63.

MOEED, A.; MEADS, M. J. 1987b: Invertebrate survey of offshore islands in relation to potential food sources for the little spotted kiwi, *Apteryx oweni* (Aves: Apterygidae). *New Zealand Entomologist 10*: 50–64.

MOEED, A.; MEADS, M. J. 1992: A survey of invertebrates in scrublands and forest, Hawke's Bay, New Zealand. *New Zealand Entomologist 15*: 63–71.

MOHRIG, W.; JASCHHOF, M. 1999: Sciarid flies (Diptera, Sciaridae) of New Zealand. *Studia Dipterologica Suppl. 7*: 1–101.

MOLLOY, B. M. 1975: Insects and seed production in *Celmisia*. *Canterbury Botanical Society Journal 8*: 1–6.

MOLLOY, J.; DAVIS, A. 1994: *Setting Priorities for the Conservation of New Zealand's Threatened Plants and Animals.* 2nd edn. Department of Conservation, Wellington. 64 p.

MORALES, C. F. 1991: Margarodidae (Insecta: Hemiptera). *Fauna of New Zealand 21*: 1–123.

MORGAN, F. D. 1959: The ecology and external morphology of *Stolotermes ruficeps* Brauer (Isoptera: Hoeltermitidae). *Transactions of the Royal Society of New Zealand 86*: 155–195.

MORGAN-RICHARDS, M. 1995: A new species of tree weta from the North Island of New Zealand (*Hemideina;* Stenopelmatidae: Orthoptera). *New Zealand Entomologist 18*: 15–23.

MORGAN-RICHARDS, M. 1997: Intraspecific karyotype variation is not concordant with allozyme variation in the Auckland tree weta of New Zealand, *Hemideina thoracica* (Orthoptera: Stenopelmatidae). *Biological Journal of the Linnean Society 60*: 423–442.

MORRIS, S. J. 2003: Two new species of *Sigaus* from Fiordland, New Zealand (Orthoptera: Acrididae). *New Zealand Entomologist 26*: 65–74.

MORRISON, H.; MORRISON, E. R. 1966: An annotated list of generic names of the scale insects (Homoptera: Coccoidea). *Miscellaneous Publications. US Department of Agriculture 1015*: 1–206.

MORSE, J. C. 1977: Phylogeny of Trichoptera. *Annual Review of Entomology 42*: 427–450.

MOSELY, M. E. 1924: New Zealand Hydroptilidae (order Trichoptera). *Transactions and Proceedings of the New Zealand Institute 55*: 670–673.

MOSELY, M. E. *in* MOSELY, M. E.; KIMMINS, D. E. 1953: *The Trichoptera (Caddis-Flies) of Australia and New Zealand.* Trustees of the British Museum (Natural History), London. 550 p.

MOULDS, M. S. 1990: *Australian Cicadas.* New South Wales University Press, Kensington. 217 p.

MOUND, L. A.; HALSEY, S. H. 1978: *Whitefly of the World. A Systematic Catalogue of the Aleyrodidae (Hemiptera: Homoptera) with Host Plant and Natural Enemy Data.* Trustees of the British Museum (Natural History) & John Wiley & Sons, Chichester. 340 p.

MOUND, L. A.; MASUMOTO M. 2005: The genus *Thrips* (Thysanopetra, Thripidae) in Australia, New Caledonia and New Zealand. *Zootaxa 1020*: 3–64.

MOUND, L. A.; WALKER, A. K. 1982: Terebrantia (Insecta: Thysanoptera). *Fauna of New Zealand 1*: 1–113.

MOUND, L. A.; WALKER, A. K. 1986: Tubulifera (Insecta: Thysanoptera). *Fauna of New Zealand 8*: 1–140.

MOURIER, H. 1986: Notes on the life history of *Labia minor* (L.) (Dermaptera), a potential predator of housefly eggs and larvae (Diptera: *Musca domestica*). *Entomologiske Meddelelser 53*: 143–148.

MÜLLER, H. J. 1951: Über die intrazellulare Symbiose der Peloridiidae *Hemiodoecus fidelis* Evans und ihre Stellung unter den Homopterensymbiosen. *Zoologischer Anzeiger 146*: 150–167.

MUNRO, V. M. 1998: A record of the releases and recoveries of the Australian parasitoids *Xanthopimpla rhopalocerus* Kreiger (Hymenoptera: Ichneumonidae) and *Trigonospila brevifacies* Hardy (Diptera: Tachinidae) introduced to New Zealand for leafroller control. *New Zealand Entomologist 21*: 81–91.

MURPHY, N. P.; CAREY, D.; CASTRO, L.; DOWTON, M.; AUSTIN, A. D. 2007: Phylogeny of the platygastroid wasps (Hymenoptera) based on sequences from the 18S rRNA, 28S rRNA and CO1 genes: implications for classification and the evolution of host relationships. *Biological Journal of the Linnean Society 91*: 653–669.

MURRAY, M. D. 1990: Influence of host behaviour on some ectoparasites of birds and mammals. Pp. 290–315 *in*: Barnard, C. J.; Behnke, J. M. (eds), *Parasitism and Host Behaviour.* Taylor & Francis Ltd, London. xii + 332 p.

MUSSO, J.-J. 1978: Research on the development, nutrition and ecology of Asilidae (Diptera–Brachycera). Ph.D. thesis, Université de Droit, France. 312 p.

MYERS, J. G. 1921a: A revision of the New Zealand Cicadidae (Homoptera) with descriptions of new species. *Transactions and Proceedings of the New Zealand Institute 53*: 238–250.

MYERS, J. G. 1921b: Notes on the Hemiptera of the Kermadec Islands, with an addition to the Hemiptera fauna of the New Zealand subregion. *Transactions and Proceedings of the New Zealand Institute 53*: 256–257.

MYERS, J. G. 1922a: The order Hemiptera in New Zealand. *New Zealand Journal of Science and Technology 5*: 1–12.

MYERS, J. G. 1922b: Life-history of *Siphanta acuta* (Walk.), the large green plant-hopper. *New Zealand Journal of Science and Technology 5*: 256–263.

MYERS, J. G. 1923a: A contribution to the study of the New Zealand leaf-hoppers and plant-hoppers (Cicadellidae and Fulgoroidea). *Transactions and Proceedings of the New Zealand Institute 54*: 407–429.

MYERS, J. G. 1923b: New species of New Zealand Cicadidae. *Transactions and Proceedings of the New Zealand Institute 54*: 430–431.

MYERS, J. G. 1924a: The Hemiptera of the Chatham Islands. *Records of the Canterbury Museum 2*: 171–183.

MYERS, J. G. 1924b: The New Zealand plant-hoppers of the family Cixiidae (Homoptera). *Transactions and Proceedings of the New Zealand Institute 55*: 315–326, pls 20–24.

MYERS, J. G. 1926: New or little-known Australasian cicadas of the genus *Melampsalta* with notes on songs by Iris Myers. *Psyche 33*: 61–76.

MYERS, J. G. 1927: On the nomenclature of New Zealand Homoptera. Cicadidae, Jassoidea, Cixiidae and Coccidae. *Transactions and Proceedings of the New Zealand Institute 57*: 685–690.

NARDI, F.; SPINSANTI, G.; BOORE, J. L.; CARAPELLI, A.; DALLAI, R.; FRATI, F. 2003: Hexapod origins: monophyletic or paraphyletic? *Science 299*: 1887–1889.

NARTSHUK, P. 1993: On New Zealand Chloropidae of New Zealand (Diptera). *Zoosystematica Rossica 2*: 185–188.

NAUMANN, I. D. 1988: Ambositrinae (Insecta: Hymenoptera: Diapriidae). *Fauna of New Zealand 15*: 1–165.

NAUMANN, I. D. 1991: Hymenoptera (wasps, bees, ants, sawflies). Pp. 916–1000 *in*: CSIRO (ed.), *The Insects of Australia. A Textbook for Students and Research Workers.* Melbourne University Press, Carlton. Vol. 1, 542 pp.; Vol. 2, 543–1137.

NEBOISS, A. 1986: *Atlas of Trichoptera of the SW Pacific–Australian Region*. Dr W. Junk Publishers. Dordrecht. 286 p.

NEES von ESENBECK, C. G. 1834: *Hymenopterorum Ichneumonibus affinium Monographiae, Genera Europaea et Species Illustrantes*. Cotta, Tübingen, Stuttgart.

NEW, T. R. 1975: Aerial dispersal of some Victorian Psocoptera as indicated by suction trap catches. *Journal of the Australian Entomological Society 14*: 179–184.

NEW, T. R. 1980: Revision of the Australian Chrysopidae (Insecta: Neuroptera). *Australian Journal of Zoology Supplement 77*: 1–143.

NEW, T. R. 1982: A new synonomy in *Anacharis* Dahlman (Hymenoptera: Figitidae). *New Zealand Entomologist 7*: 320–321.

NEW, T. R. 1983: Notes on the New Zealand antlion *Weeleus acutus* (Neuroptera). *New Zealand Journal of Zoology 10*: 281–284.

NEW, T. R. 1988: A revision of the Australian Hemerobiidae (Insecta: Neuroptera). *Invertebrate Taxonomy 2*: 339–411.

NIELSEN, E. S.; EDWARDS, E. D.; RANGSI, V. 1996: Checklist of the Lepidoptera of Australia. *Monographs on Australian Lepidoptera 4*: xiv, 1–529.

NIELSEN, E. S.; MOUND, L. A. 1999: Global diversity of insects: the problems of estimating numbers. Pp. 213–22 *in*: Raven, P. H.; Williams, T. (eds), *Nature and Human Society: The Quest for a Sustainable World*. National Academic Press, Washington, DC.

NIETO NAFRÍA, J. M.; MIER DURANTE, M. P.; REMAUDIÈRE, G. 1997: Les noms des taxa du group-famille chez les Aphididae. *Revue française d'Entomologie, n.s., 19*: 77–92.

NISHIZAKA, M.; AZUMA, A.; MASAKI, S. 1998: Diapause response to photoperiod and temperature of *Lepisma saccharina* Linnaeus (Thysanura : Lepismatidae). *Entomological Science 1*: 7–14.

NOYES, J. S. 1988: Encyrtidae (Insecta: Hymenoptera). *Fauna of New Zealand 13*: 1–188.

NOYES, J. S. 1998: *Catalogue of the Chalcidoidea of the World*. [Biodiversity catalogue database and image library CD-ROM Series.] ETI, Amsterdam & The Natural History Museum, London.

NOYES, J. S.; VALENTINE, E. W. 1989a: Mymaridae (Insecta: Hymenoptera): introduction, and review of genera. *Fauna of New Zealand 17*: 1–95.

NOYES, J. S.; VALENTINE, E. W. 1989b: Chalcidoidea (Insecta: Hymenoptera): introduction, and review of genera in smaller families. *Fauna of New Zealand 18*: 1–91.

NURSE, F. R. 1945: Protozoa from New Zealand termites. *Transactions of the Royal Society of New Zealand 74*: 306–314.

OBRYCKI, J. J.; HAMID, M. N.; SAJAP, A. S.; LEWIS, L. C. 1989: Suitability of corn pests for development and survival of *Chrysoperla carnea* and *Chrysoperla oculata* (Neuroptera: Chrysopidae). *Environmental Entomology 18*: 1126–1130.

O'CALLAGHAN, M.; GARNHAM, M. L.; NELSON, T. L.; BAIRD, D.; JACKSON, D. A. 1996: The pathogenicity of *Serratia* strains to *Lucilia sericata* (Diptera: Calliphoridae). *Journal of Invertebrate Pathology 68*: 22–27.

O'HARA, J. E. 1996: Earwig parasitoids of the genus *Triarthria* Stephens (Diptera: Tachinidae) in the New World. *Canadian Entomologist 128*: 15–26.

OLDROYD, H. 1964: *The Natural History of Flies*. Weidenfeld & Nicholson, London. 324 p.

OLMI, M. 1996: A revision of the world Embolemidae (Hymenoptera Chrysidoidea). *Frustula Entomologica 18(31)*: 85–146.

OLMI, M. 2007: New Zealand Dryinidae and Embolemidae (Hymenoptera: Chrysidoidea): new records and description of *Bocchus thorpei* new species. *Records of the Auckland Museum 44*: 5–16.

O'NEILL, S. B.; BUCKLEY, T. R.; JEWELL, T. R.; RITCHIE, P. A. 2009: Phylogeographic history of the New Zealand stick insect *Niveaphasma annulata* (Phasmatodea) estimated from mitochondrial and nuclear loci. *Molecular Phylogenetics and Evolution 53*: 523–536

OOSTERBROEK, P. 1989: Tipulidae. Pp. 53–116 *in*: Evenhuis, N. L. (ed.), *Catalog of the Diptera of the Australasian and Oceanian Regions*. Bishop Museum Press, Honolulu. 1155 p.

ORDISH, R.G. 1975: Aquatic insects (3). *New Zealand's Nature Heritage 4(50)*: 1391–1395.

ORNITHOLOGICAL SOCIETY OF NEW ZEALAND 1990: *Checklist of the Birds of New Zealand and the Ross Dependency, Antarctica*. The Checklist Committee (E. G. Turbott, Convener). Third Edition. Ornithological Society of New Zealand & Random Century New Zealand Ltd, Auckland. xvi + 247 p.

OSMAN, M. Z.; SALMON, B. J. 1996: Effect of larval diet on the performance of the predator *Chrysoperla carnea* Stephen (Neuroptera: Chrysopidae). *Journal of Applied Entomology 120*: 115–117.

OTTE, D.; ALEXANDER, R. D.; CADE, W. 1987: The Crickets of New Caledonia. *Proceedings of the Academy of Natural Sciences of Philadelphia 139*: 375–475.

PACKER, L.; TAYLOR, J. S. 1997: How many hidden species are there? An application of the phylogenetic species concept to genetic data for some comparatively well-known bee 'species'. *Canadian Entomologist 129*: 587–594.

PAGÉS, J. 1952: Diploures japygides de Nouvelle-Zélande. *Records of the Canterbury Museum 6*: 149–162.

PALMA, R. L. 1999: Amendments and additions to the 1982 list of chewing lice (Insecta: Phthiraptera) from birds in New Zealand. *Notornis 46*: 373–387.

PALMA, R. L. 2000: The species of *Saemundssonia* (Insecta: Phthiraptera: Philopteridae) from skuas (Aves: Stercorariidae). *New Zealand Journal of Zoology 27*: 121–128.

PRICE, R. D.; HELLENTHAL, R. A.; PALMA, R. L. 2003: World checklist of chewing lice with host associations and keys to families and genera. Pp. 1–448 *in*: Price, R. D.; Hellenthal, R. A.; Palma, R. L.; Johnson, K. P.; Clayton, D. H. (eds), *The Chewing Lice: World Checklist and Biological Overview. Illinois Natural History Survey Special Publication 24*: x + 501.

PALMA, R. L.; LOVIS, P. M.; TITHER, C. 1989: An annotated list of primary types of the phyla Arthropoda (except Crustacea) and Tardigrada held in the National Museum of New Zealand. *National Museum of New Zealand Miscellaneous Series 20*: 1–49.

PALMA, R. L.; PILGRIM, R. L. C. 2002: A revision of the genus *Naubates* (Insecta: Phthiraptera: Philopteridae). *Journal of the Royal Society of New Zealand 32*: 7–60.

PALMA, R. L.; PRICE, R. D. 2000: *Philopterus novaezealandiae*, a new species of chewing louse (Phthiraptera: Philopteridae) from the kokako (Passeriformes: Callaeidae). *Journal of the Royal Society of New Zealand 30*: 293–297.

PALMA, R. L.; PRICE, R. D. 2004: *Apterygon okarito*, a new species of chewing louse (Insecta: Phthiraptera: Menoponidae) from the Okarito brown kiwi (Aves: Apterygiformes: Apterygidae). *New Zealand Journal of Zoology 31*: 67–73.

PALMA, R. L.; PRICE, R. D. 2005: *Menacanthus rhipidurae*, a new species of chewing louse (Insecta: Phthiraptera: Menoponidae). From South Island fantails, *Rhipidura fuliginosa fuliginosa* (Aves: Passeriformes: Dicruridae). *New Zealand Journal of Zoology 32*: 111–115.

PALMER, J. M. 1974: Arthropoda of the subantarctic islands of New Zealand (2). Hemiptera: Aphididae. *Journal of the Royal Society of New Zealand 4*: 303–306.

PAPE, T.; BICKEL, D.; MEIER, R. 2009: *Diptera Diversity: Status, Challenges and Tools*. E. J. Brill, Leiden. xix + 459 p.

PAPP, J. 1977: *Phaenocarpa (Asobara) persimilis* sp. n. (Hymenoptera, Braconidae, Alysiinae) from Australia. *Opuscula Zoologica, Budapest 13*: 73–77.

PARAMONOV, S. J. 1955: The New Zealand Cyrtidae (Diptera) the problem of the Pacific Island fauna. *Pacific Science 9*: 16–25.

PARENT, O. 1933: Étude monographique sur les Diptères dolichopodides de Nouvelle-Zélande. *Annales de la Société Scientifique de Bruxelles, sér. B, 53*: 325–441.

PATRICK, B. H. 1992: Snow caving for caddisflies. *The Weta 15*: 10–13.

PATRICK, B. H.; DUGDALE, J. S. 2000: Conservation status of the New Zealand Lepidoptera. *Science for Conservation 136*: 1–33.

PEARMAN, J.V. 1936: The Taxonomy of the Psocoptera: Preliminary sketch. *Proceedings of the Royal Entomological Society of London, ser. B, 5*: 58–62.

PEARSON, J. F.; GOLDSON, S. L. 1980: A preliminary examination of pests in fodder beet in Canterbury. *Proceedings of the New Zealand Weed and Pest Control Conference 33*: 211–214.

PENDERGRAST, J. G. 1962: The internal anatomy of the Peloridiidae. *Transactions of the Royal Entomological Society of London 114*: 49–65.

PENNIKET, J. G. 1961: Notes on New Zealand Ephemeroptera. I. The affinities with Chile and Australia, and remarks on *Atalophlebia* Eaton (Leptophlebiidae). *New Zealand Entomologist 2*: 1–11.

PENNIKET, J. G. 1962a: Notes on New Zealand Ephemeroptera. II. A preliminary account of *Oniscigaster wakefieldi* McLachlan, recently rediscovered (Siphlonuridae). *Records of the Canterbury Museum 7*: 375–388.

PENNIKET, J. G. 1962b: Notes on New Zealand Ephemeroptera. III. New family, genus and species. *Records of the Canterbury Museum 7*: 389–398.

PENNIKET, J. G. 1966: Notes on New Zealand Ephemeroptera. IV. A new siphlonurid subfamily: Rallidentinae. *Records of the Canterbury Museum 8*: 163–175.

PETERS, W. L. 1971: Entomology of the Aucklands and other islands south of New Zealand: Ephemeroptera: Leptophlebiidae. *Pacific Insects Monographs 27*: 47–51.

PETERS, W. L. 2001: The Ephemeroptera of New Zealand and New Caledonia. Pp. 43–45 *in*: Dominguez, E. (ed.), *Trends in Research in Ephemeroptera and Plecoptera*. Kluwer Academic, New York.

PETERS, W. L.; CAMPBELL, I. C. 1991: Ephemeroptera (Mayflies). Pp. 279–293 *in*: Naumann, I. D.; Carne, P. B.; Lawrence, J.

F.; Neilson, E. S.; Spradbery, J. P.; Taylor, R. W.; Whitten, M. J.; Littlejohn, M. J. (eds), *The Insects of Australia*. 2nd edn. CSIRO/Melbourne University Press, Melbourne.

PETERSON, J. T. 1943: Studies on the genetics of *Drosophila* III. The Drosophilidae of the South West. *University of Texas Publication 4113:* 1–327.

PHILLIPS, J. S. 1930a: A revision of New Zealand Ephemeroptera. Part 1. *Transactions and Proceedings of the New Zealand Institute 61*: 271–334.

PHILLIPS, J. S. 1930b: A revision of New Zealand Ephemeroptera. Part 2. *Transactions and Proceedings of the New Zealand Institute 61*: 335–390.

PICKETT, J. A.; WADHAMS, L. J.; WOODCOCK, C. M.; HARDIE, J. 1992: The chemical ecology of aphids. *Annual Review of Entomology 37*: 67–90.

PIERCE, R. J. 1986: Foraging responses of stilts (*Himantopus* spp.: Aves) to changes in behaviour and abundance of their riverbed prey. *New Zealand Journal of Marine and Freshwater Research 20*: 17–28.

PILGRIM, R. L. C. 1972: The aquatic larva and the pupa of *Choristella philpotti* Tillyard, 1917 (Mecoptera: Nannochoristidae). *Pacific Insects 14*: 151–168.

PILGRIM, R. A. 1975: Lice and fleas (1). *New Zealand's Nature Heritage 4(37)*: 1030–1036.

PILGRIM, R. L. C. 1976: Mallophaga on the rock pigeon (*Columba livia*) in New Zealand, with a key to their identification. *New Zealand Entomologist 6*: 160–164.

PILGRIM, R. L. C. 1998: Larvae of the genus *Notiopsylla* (Siphonaptera: Pygiopsyllidae) with a key to their identification. *Journal of Medical Entomology 35*: 362–376.

PILGRIM, R. L. C.; PALMA, R. L. 1982: A list of the chewing lice (Insecta: Mallophaga) from birds in New Zealand. *National Museum of New Zealand Miscellaneous Series 6*: 1–32. [Also published as *Notornis 29* (suppl.).]

PINHEY, E. C. G. 1975: *Moths of Southern Africa*. Tafelberg Publishers, Cape Town. 273 p., 63 pls.

PINTO, J. D.; OATMAN, E. R. 1996: Description of three new *Trichogramma* (Hymenoptera: Trichogrammatidae) from New Zealand and their relationship to New World species. *Proceedings of the Entomological Society of Washington 98*: 396–406.

PITKIN, L. M. 1980: A revision of the Pacific species of *Conocephalus* Thunberg (Orthoptera: Tettigoniidae). *Bulletin of the British Museum (Natural History), Entomology Series 41*: 315–355.

PLANT, A. R. 1989: A revision of the Ocydromiinae (Diptera: Empidoidea: Hybotidae) of New Zealand with descriptions of new species. *New Zealand Journal of Zoology 16*: 231–241.

PLANT, A. R. 1991: A revision of the genus *Ceratomerus* (Diptera: Empididae: Ceratomerinae) of New Zealand. *Journal of Natural History 25*: 1313–1330.

PLANT, A. R. 1993: Sexual dimorphism in the genus *Monodromia* Collin (Diptera: Empidioidea: Hemerodromiinae). *New Zealand Journal of Zoology 20*: 207–210.

PLANT, A. R. 1997: *Atodrapetis*, a new genus of empidoid fly (Diptera, Empidoidea, Hybotidae, Tachydromiinae) from New Zealand. *Studia Dipterologica 4*: 435–440.

PLANT A. R. 2005: The Hemedromiinae (Diptera: Empididae) of New Zealand. I *Phyllodromia* Zetterstedt. *Studia Dipterologia 12:* 119-128.

PLANT A. R. 2007: The Hemedromiinae (Diptera: Empididae) of New Zealand II *Chelipoda* Macquart. *Zootaxa 1537*: 1–88.

POPHAM, E. J. 1959: The anatomy in relation to feeding habits of *Forficula auricularia* L. and other Dermaptera. *Proceedings of the Zoological Society of London 133*: 251–300.

POTAPOV, M. 2001: Isotomidae. *In*: Dunger, W. (ed.), *Synopses of Palaearctic Collembola, Volume 3. Abhandlungen der Berliner Naturkundemuseum Görlitz. 73(2)*: 1–603.

POWLESLAND, R. G. 1981: The foraging behaviour of South Island robin. *Notornis 28*: 89–102.

PRESTIGE, R. A. 1989: Preliminary observations on the grassland leafhopper fauna of the central North Island volcanic plateau. *New Zealand Entomologist* 12: 54–57

PRICE, J. F.; SHEPARD, M. 1978: *Colosoma sayi* and *Labidura riparia* predation on noctuid prey in soya beans and locomotor activity. *Environmental Entomology 7*: 653–656.

PRIMACK, R. B. 1978: Variability in New Zealand montane and alpine pollinator assemblages. *New Zealand Journal of Ecology 1*: 66–73.

PRINCIS, K. 1974: Ergebniss der Osterreichischen NeukaledonianExpedition 1965. Blattaria – Schaben. *Annalen Naturhistorisches Museum, Wien 78*: 513–521.

PUGSLEY, C. W. 1984: Ecology of the New Zealand glowworm, *Arachnocampa luminosa* (Diptera: Keroplatidae), in the Glowworm Cave, Waitomo. *Journal of the Royal Society of New Zealand 14*: 387–407.

QUAIL, A. 1901: Hymenopterous parasites of ovum of *Vanessa gonerilla*. *Transactions of the New Zealand Institute 33*: 153–154.

QUATE, L. W. 1964: Insects of Campbell Island. Diptera: Psychodidae. *Pacific Insects Monograph 7*: 280–288.

QUICKE, D. L. J. 1997: *Parasitic Wasps*. Chapman & Hall, London. 470 p.

QUILTER, C. G. 1971: Insects of Whale Island. *Tane, Journal of the Auckland University Field Club 17*: 75–76.

QUINN, J. M.; COOPER, A. B.; DAVIES–COLLEY, R. J.; RUTHERFORD, J. C.; WILLIAMSON, R. B. 1997: Land use effects on habitat, water quality, periphyton, and benthic invertebrates in Waikato, New Zealand hill-country streams. *New Zealand Journal of Marine and Freshwater Research 31*: 579–597.

QUINN, J. M.; HICKEY, C. W. 1990a: Characterisation and classification of benthic invertebrate communities in 88 New Zealand rivers in relation to environmental factors. *New Zealand Journal of Marine and Freshwater Research 24*: 387–409.

QUINN, J. M.; HICKEY, C. W. 1990b: Magnitude of effects of substrate particle size, recent flooding and catchment development on benthic invertebrates in 88 New Zealand rivers. *New Zealand Journal of Marine and Freshwater Research 24*: 387–409.

QUINN, J. M.; STEELE, G. L.; HICKEY, C. W.; VICKERS, M. L. 1994: Upper thermal tolerances of twelve New Zealand invertebrate species. *New Zealand Journal of Marine and Freshwater Research 28*: 391–397.

RADL, R. C.; LINSENMAIR, K. E. 1991: Maternal behaviour and nest recognition in the subsocial earwig *Labidura riparia* Pallas (Dermaptera: Labiduridae). *Ethology 89*: 287–296.

RAMLOV, H.; WESTH, P. 1993: Ice formation in the freeze tolerant alpine weta *Hemideina maori* Hutton (Orthoptera: Stenopelmatidae). *Cryo–Letters 14*: 169–176.

RAMSAY, A. 1997: Some interesting invertebrate captures for a dung heap in West Lothian with reference to recent status of *Labia minor* (L.) (Dermaptera: Labiidae) in Scotland, and a new habitat for this species. *Entomologist Record and Journal of Variation 109*: 262.

RAMSAY, G. W. 1984: *Miomantis caffra*, a new mantid record (Mantodea: Mantidae) for New Zealand. *New Zealand Entomologist* 8: 102–104.

RAMSAY G. W. 1990: Mantodea (Insecta), with a review of functional aspects of morphology and biology. *Fauna of New Zealand 19*: 1–96.

RAMSAY, G. W. 1991: The cricket *Ornebius aperta* (Orthoptera: Gryllidae) established in New Zealand. *New Zealand Entomologist 14*: 9–14.

RAMSAY, G. W.; CROSBY, T. K. 1992: Bibliography of New Zealand terrestrial invertebrates 1775–1985, and a guide to the associated information retrieval database BUGS. *Bulletin of the Entomological Society of New Zealand 11*: 1–440. [Diptera pp. 67–97.]

RANKIN, S. M.; PALMER, J. O.; LAROQUE, L.; RISSER, A. L. 1995: Life history characteristics of ring legged earwig (Dermaptera: Labiduridae): emphasis on ovarian development. *Annals of Entomological Society of America 88*: 887–893.

RANKIN, S. M.; STORM, S. K.; PIETO, D. L.; RISSER, A. L. 1996: Maternal behaviour and clutch manipulation in the ring legged earwig (Dermaptera: Carcinophoridae). *Journal of Insect Behaviour 9*: 85–103.

RAPP, G. 1995: Eggs of the stick insect *Graeffea crouanii* Le Guillou (Orthoptera, Pharmidae). Mortality after exposure to natural enemies and high temperature. *Journal of Appled Entomology 119(2)*: 89–91.

REGIER, J. C.; SCHULTZ, J. W.; KAMBIC, R. E. 2005: Pancrustacean phylogeny: hexapods are terrestrial crustaceans and maxillopods are not monophyletic. *Proceedings of the Royal Society, B, 272*: 395–401.

REMAUDIÈRE, G.; REMAUDIÈRE, M. 1997: *Catalogue des Aphididae du Monde*. INRA, Paris.

RICHARDS, A. M. 1954a: Notes on food and cannibalism in *Macropathus filifer* Walker, 1869 (Rhaphidophoridae, Orthoptera). *Transactions of the Royal Society of New Zealand 82*: 733–737.

RICHARDS, A. M. 1954b: Notes on behaviour and parasitism in *Macropathus filifer* Walker, 1869. *Transactions of the Royal Society of New Zealand 82*: 821–822.

RICHARDS, A. M. 1960: Observations on the New Zealand glow-worm *Arachnocampa luminosa* (Skuse) 1890. *Transactions of the Royal Society of New Zealand 88*: 559–574.

RICHARDS, A. M. 1962: Feeding behaviour and enemies of Rhaphidophoridae (Orthoptera) from Waitomo caves, New Zealand. *Transactions of the Royal Society of New Zealand, Zoology 2*: 121–129.

RICHARDS, A. M. 1965a: The effect of weather on Rhaphidophoridae (Orthoptera) in New Zealand and Australia. *Annals of Speleology 20*: 391–400.

RICHARDS, A. M. 1965b: Movements of Rhaphidophoridae (Orthoptera) in caves at Waitomo, New Zealand. *Helictite 3*: 65–78.

RICHARDS, A. M. 1973: A comparative study of the biology of the giant wetas *Deinacrida heteracantha* and *D. fallai* (Orthoptera: Henicidae) from New Zealand. *Journal of Zoology 169*: 195–236.

RICHARDS, O. W. 1973: The Sphaeroceridae (= Boboridae or Cypselidae; Diptera, Cyclorrhapha) of the Australian region. *Australian Journal of Zoology, Suppl. Ser., 22*: 297–401.

RIDDELL, D. J. 1981: Notes on the stream macrofauna of Cuvier Island. *Tane, Journal of the Auckland University Field Club 27*: 33–36.

RIEK, E. F. 1962: The Australian species of *Psyllaephagus* (Hymenoptera: Encyrtidae), parasites of psyllids (Homoptera). *Australian Journal of Zoology 10*: 684–757.

RIEK, F. 1976: The marine caddisfly family Chathamiidae (Trichoptera). *Journal of the Australian Entomological Society 15*: 405–419.

ROBERTSON, C. J. R.; O'DONNELL, C. F. J.; OVERMARS, F. B. 1983: Habitat requirements of wetland birds in the Ahuriri catchment, New Zealand. *New Zealand Wildlife Service Occasional Publication No. 3*: 1–455.

ROBERTSON, L. N.; BLANK, R. H. 1982: The survey and monitoring of recently established Australian soldier fly populations in the North Island, New Zealand (Diptera: Stratiomyidae). *New Zealand Entomologist 7*: 256–262.

ROBINSON, G. S.; NIELSEN, E. S. 1993: Tineid Genera of Australia. *Monographs on Australian Lepidoptera 2*: 1–344.

ROBINSON, G. S.; TUCK, K. R., SHAFFER, M. 1994: *A Field Guide to the Smaller Moths of South-East Asia*. Natural History Museum, London. 309 p., 32 pls.

RODGERS, D.; GREENSLADE, P. 1996: A new diagnosis for *Dinaphorura* (Collembola: Onychiuridae: Tullbergiinae) and description of new species from Australia. *Journal of Natural History 30*: 1367–1376.

ROESLER, R. 1944: Die Gattungen der Copeognathen. *Stettiner entomologische Zeitung 195*: 117–166.

ROGERS, E. 1982: *Chersodromia* Walker discovered in New Zealand (Diptera: Empididae), and the description of a new species. *New Zealand Entomologist 7*: 340–343.

ROGNES, K. 1987. The taxonomy of the *Pollenia rudis* species group in the Holarctic region (Diptera: Calliphoridae). *Systematic Entomology 12*: 475–502.

ROHÁCEK, J. 2007: *Zealantha thorpei* gen. et sp. nov. (Diptera: Anthomyzidae) – first family report from New Zealand. *Zootaxa 1576*: 1–13.

ROHITHA, B. H.; POTTINGER, R. P.; FIRTH, H.C. 1985. Population monitoring studies of lucerne aphids and their predators in the Waikato. *Proceedings of the New Zealand Weed and Pest Control Conference 38*: 3134.

RONQUIST, F. 1999: Phylogeny, classification and evolution of the Cynipoidea. *Zoologica Scripta 28*: 139–164.

ROSENZWEIG, V.Ye. 1997: Revised classification of the *Calocoris* complex and related genera (Heteroptera: Miridae). *Zoosystematica Rossica* 6: 139–169.

ROWE, R. J. 1987: *The Dragonflies of New Zealand*. Auckland University Press, Auckland. 260 p.

RUDOLPH, D. 1982: Site, process and mechanism of active uptake of water vapour from the atmosphere in the Psocoptera. *Journal of Insect Physiology 28*: 205–212.

RUGMAN-JONES, P. F; HODDLE M. S.; MOUND, L. A.; STOUTHAMMER, R. 2006: Molecular identification key for pest species of *Scirtothrips* (Thysanoptera: Thripidae). *Journal of Economic Entomology 99*: 1813–1819.

RUSSELL, L. M. 1941: A classification of the scale insect genus *Asterolecanium*. *Miscellaneous Publications. U.S. Department of Agriculture 424*: 1–324.

RYAN, A. N. J. 1994: Diet of Feral Cats and Energy Received from Invertebrate Prey. BSc(Hons) thesis, University of Canterbury, Christchurch.

SAEED, A.; AUSTIN, A. D.; DANGERFIELD, P. C. 1999: Systematics and host relationships of Australasian Diolcogaster (Hymenoptera: Braconidae: Microgastrinae). *Invertebrate Taxonomy 13* 117–178.

SALMON, J. T. 1948: New genera, species and records of Orthoptera from the Three Kings Islands, New Zealand. *Records of the Auckland Institute and Museum* 3: 301–307.

SALMON, J. T. 1954: A new genus and species of Phasmidae from New Zealand. *Transactions of the Royal Society of New Zealand 82*: 161–168.

SALMON, J. T. 1964: An index to the Collembola. *Bulletin of the Royal Society of New Zealand 7*: 1–644.

SALMON, J. T. 1991: *The Stick Insects of New Zealand*. Reed Books, Auckland. 124 p.

SATCHELL, G. H. 1950: The New Zealand Psychodidae: a study based upon the collection and manuscript notes of the late Dr A. L. Tonnoir. *Transactions of the Royal Entomological Society, London 101*: 147–178.

SATCHELL, G. H. 1954: Keys to the described species of New Zealand Psychodidae with descriptions of eight new species. *Transactions of the Royal Entomological Society, London 105*: 475–491.

SAUNDERS, A. J.; NORTON, D. A. 2001: Ecological restoration at mainland islands in New Zealand. *Biological Conservation 99*: 109–119.

SAUNDERS, L. G. 1956: Revision of the genus *Forcipomyia* based on characters of all stages (Diptera: Ceratopogonidae). *Canadian Journal of Zoology 34*: 657–705

SAUNDERS, L. G. 1964: New species of the genus *Forcipomyia* in the *Lasiohelea* complex described in all stages (Diptera: Ceratopogonidae). *Canadian Journal of Zoology 42*: 463–482.

SAUPHANOR, B.; SUREAU, F. 1993: Aggregation behaviour and interspecific relationships in Dermaptera. *Oecologia 96*: 360–364.

SAVILL, R. A. 1990: Early records of the human head louse *Pediculus humanus capitis* (Phthirapthera: Pediculidae) in New Zealand. *Records of the Canterbury Museum 10*: 69–72.

SCHAEFER, C. W.; PANIZZI, A. R. (Eds.) 2000: *Heteroptera of Economic Importance*. CRC Press, Boca Raton. 828 p.

SCHLEE, D. 1969: Morphologie und Symbiose: Ihre Beweiskraft für die Verwandtschafts-beziehungen der Coleorrhyncha. *Stuttgarter Beiträge zur Naturkunde 210*: 1–27.

SCHLINGER, E. I. 1960: A revision of the genus *Ogcodes* Latreille with particular reference to species of the western Hemisphere. *Proceedings of the US National Museum 111(3429)*: 227–336.

SCHLINGER, E. I.; BOSCH, R. van den; DIETRACK, E. J. 1959: Biological notes on the predaceous earwig *Labidura riparia* (Pallas) a recent immigrant to California (Dermaptera: Labiduriidae). *Journal of Economic Entomology 52*: 247–249.

SCHMITZ, H. 1939: Neuseeländische Phoriden. *Natuurhistorich Maandblad 28*: 34–37, 55–56, 67–68, 75–76, 86–89, 98–101, 110–116, 124–129.

SCHNITZLER, D. R.; WANG, Q. 2005: Revision of *Zorion* Pascoe (Coleoptera: Cerambycidae: Cerambycinae), an endemic genus of New Zealand. *Zootaxa. 1066*: 1–42.

SCHNITZLER, F.-R.; SARTY, M., LESTER, P. J. 2004: Larval parasitoids reared from *Cleora scrptaria* (Geometiridae, Ennominae). *Weta 28*: 13–18.

SCHUH, R. T. 1995: *Plant Bugs of the World (Insecta: Heteroptera: Miridae): Systematic Catalog, Distributions, Host List, and Bibliography*. The New York Entomological Society, New York. 1329 p.

SCHUH, R. T.; SLATER, J. A. 1995: *True Bugs of the World (Hemiptera: Heteroptera): Classification and Natural History*. Cornell University Press, Ithaca. 336 p.

SCHWARTZ, M. D.; EYLES, A. C. 1999: Identity of *Lygus buchanani* Poppius (Heteroptera: Miridae: Mirini): a deletion from the New Zealand fauna. *New Zealand Journal of Zoology 26*: 221–227.

SCOTT, R. R. 1979: The biology and life history of the current clear wing *Synanthedon tipuliformis* (Lepidoptera: Sessidae) in Canterbury, New Zealand. *New Zealand Journal of Zoology 6*: 145–163.

SCOTT, R. R. (Ed.) 1984: *Pest and Beneficial Insects in New Zealand*. Lincoln Agricultural College, Canterbury. 373 p.

SERAJ, A. A.; FENEMORE, P. G. 1993: Leaf miner damage assessment experiments in laboratory and field with *Scaptomyza flava*. *Proceedings of the New Zealand Plant Protection Conference 46*: 45–48.

SHAMSHEV, I.V.; GROOTAERT, P. 2003: A new genus of Microphorinae (Diptera: Empidoidea) from New Zealand. *Belgium Journal of Entomology 4*: 129–144.

SHARKEY, M. J. 2007: Phylogeny and Classification of Hymenoptera. *Zootaxa 1668*: 521–548.

SHATTUCK, S. 1999: Australian ants: their biology and identification. *Monographs on Invertebrate Taxonomy 3*: xi, 1–226.

SHAW, S. R. 1993: Three new *Microctonus* species indigenous to New Zealand (Hymenoptera: Braconidae). *New Zealand Entomologist 16*: 29–39.

SHAW, S. R; BERRY, J. A. 2005: Two new *Cryptoxilos* species (Hymenoptera: Braconidae: Euphorinae) from New Zealand and Fiji parasitizing adult Scolytidae (Coleoptera). *Invertebrate Systematics 19*: 371–381.

SHEPARD, M.; WADDILL, V.; KLOFT, W. 1973: Biology of the predaceous earwig *Labidura riparia* (Dermaptera: Labiduridae). *Annals of Entomological Society of America 66*: 837–841.

SIMON, C. 2009: Using New Zealand examples to teach Darwin's 'Origin of Species': lessons from molecular phylogenetic studies of cicadas. *New Zealand Science Review 66*: 102–112.

SINCLAIR, B.J. 1995: Generic review of the Clinocerinae (Empididae) and description and phylogenetic relationships of the Trichopezinae, new status (Diptera: Empidoidea). *Canadian Entomologist 127*: 665–752.

SINCLAIR, B. 1997a: Freeze tolerance of alpine cockroaches. *Ecological Entomology 22*: 462–467.

SINCLAIR, B. J. 1997b: *Icasma* Collin and an allied new genus *Glyphidopeza* from New Zealand (Diptera: Empidoidea; Ceratomerinae). *Records of the Australian Museum 49*: 195–211.

SINCLAIR, B. J. 1999: Review of the genera *Dipsomyia* Bezzi, *Zanclotus* Wilder, and an allied new Gondwanan genus (Diptera: Empidoidea, *Ragas*-group). *Entomological Science 2*: 131–145.

SINCLAIR, B. J.; McLELLAN, I. D. 2004: Revision of the New Zealand species of *Hydropeza* Sinclair (Diptera: Empidoidea: *Ragas*-group). *Invertebrate Systematics 18*: 627-647.

SKERMAN, T.M. 1953: The structure, morphology and adaptive features of the head in larval Tipulidae. MSc thesis, University of Otago, Dunedin. 114 p.

SLATER, J. A.; O'DONNELL, J. E. 1995: *A Catalogue of the Lydaeidae of the World (1960–1994)*. The New York Entomological Society, New York. 410 p.

SMIT, F. G. A. M. 1979: The fleas of New Zealand (*Siphonaptera*). *Journal of the Royal Society of New Zealand 9*: 143–232.

SMIT, F. G. A. M. 1984: *Parapsyllus magellanicus largificus*, a new flea from the Bounty Islands. *New Zealand Journal of Zoology 11*: 13–16.

SMIT, F. G. A. M. 1987: *An Illustrated Catalogue of the Rothschild Collection of Fleas (Siphonaptera) in the British Museum (Natural History). Volume VII: Malacopsylloidea (Malacopsyllidae and Rhopalopsyllidae).* Oxford University Press; British Museum (Natural History), Oxford & London. 380 p., 5 pls.

SMITH, B. J. 1998: The larva of *Hydrobiosis gollanis* Mosely (Trichoptera: Hydrobiosidae). *New Zealand Journal of Zoology 25*: 421–428.

SMITH, B. J. 2000: The larva of *Hydrobiosis torrentis* Ward (Trichoptera: Hydrobiosidae). *New Zealand Journal of Zoology 27*: 15–20.

SMITH, B. J. 2001: The larva of *Traillochorema rakiura* McFarlane (Trichoptera: Hydrobiosidae), a caddisfly endemic to Stewart Island, New Zealand. *New Zealand Entomologist 24*: 71–74.

SMITH, B. J. 2002: The larvae of *Costachorema* McFarlane (Trichoptera: Hydrobiosidae) from New Zealand. *Aquatic Insects 24*: 2–35.

SMITH, B. J. 2008: Two new species of caddisflies (Trichoptera) from New Zealand. *Aquatic Insects* 30: 43–50.

SMITH, K. G.V. 1989: An introduction to the immature stages of British flies. *Handbook for the Identification of British Insects 10(14)*: 1–280.

SMITH, P. H. 1961: The anatomy and morphology of various New Zealand plant galls with especial reference to the bud rosette gall of *Olearia paniculata* (*Shawia paniculata* Forst.). MSc thesis, University of Canterbury, Christchurch. 138 p.

SMITHERS, C. N. 1963: The generic position of two species of Philotarsidae. (Psocoptera). *Journal of the Entomological Society of Queensland 2*: 60.

SMITHERS, C. N. 1964: Insects of Campbell Island. Psocoptera. *Pacific Insects Monographs 7*: 226–229.

SMITHERS, C. N. 1969: The Psocoptera of New Zealand. *Records of the Canterbury Museum 8*: 259–344.

SMITHERS, C. N. 1970: Knowledge of New Zealand Psocoptera. *New Zealand Entomologist 4*: 71.

SMITHERS, C. N. 1970a: Some thoughts on Trans-Tasman relationships in the Psocoptera. *New Zealand Entomologist 4*: 79–85.

SMITHERS, C. N. 1972: The classification and phylogeny of the Psocoptera. *Australian Museum Memoirs 14*: 1–349.

SMITHERS, C. N. 1973: A new species and new records of Psocoptera from the Kermadec Islands. *New Zealand Entomologist 5*: 147–150.

SMITHERS, C. N. 1974: Arthropoda of the subantarctic islands of New Zealand. 4. Psocoptera. *Journal of the Royal Society of New Zealand 4*: 315–318.

SMITHERS, C. N. 1975: The names of Australian and New Zealand Myopsocidae (Psocoptera). *Australian Entomological Magazine 2*: 76–78.

SMITHERS, C. N. 1978: A new species and new records of Psocoptera from Ireland. *Irish Naturalists' Journal 19*: 141–148.

SMITHERS, C. N. 1999: *Sandrapsocus clarki*, an unusual new genus and species of Elipsocidae (Insecta: Psocoptera) from the Bounty Islands. *Journal of the Royal Society of New Zealand 29*: 159–164.

SMITHERS, C. N. 2000: First records of *Pteroxanium marrisi* sp. n. and *Haplophallus maculatus* (Tillyard) for the Chatham Islands and list of Psocoptera from the subantarctic islands of New Zealand. *New Zealand Entomologist 22*: 9–13.

SMITHERS, C. N. 2002: First record of the family Trichopsocidae (Psocoptera) from New Zealand. *Entomologists' Monthly Magazine 138*: 155.

SMITHERS, C. N.; O'CONNOR, J. P. 1991: New records of Psocoptera (Insecta) (booklice, barklice) from Ireland, including a species previously known from New Zealand. *Irish Naturalists' Journal 23*: 477–486.

SMITHERS, C. N.; THORNTON, I. W. B. 1982: The role of New Guinea in the evolution and biogeography of some families of psocopteran insects. *Monographiae Biologicae 42*: 621–638.

SMITHERS, C. N.; THORNTON, I. W. B. 1990: Systematics and distribution of the Melanesian Psocidae (Psocoptera). *Invertebrate Taxonomy 3*: 431–468.

SNELL, A. 2005a: Identification keys to larval and adult female mosquitoes (Diptera: Culicidae) of New Zealand. *New Zealand Journal of Zoology 32:* 99-110.

SNELL, A. 2005b: The discovery of the exotic mosquito *Ochlerotatus australis* and endemic *Opifex fuscus* (Diptera: Culicidae) on North East island, Snares islands. *The Weta 30:* 10-13.

SNELL, A.; DERRAIK, J. G. A.; McINTYRE, M. 2005: *Maorigoeldia argropus* Walker (Diptera: Culicidae): is this another threatened endemic species? *New Zealand Entomologist 28:* 95-99.

SNELL, A. E.; HEATH, A. C. G. 2006: Parasitism of mosquitoes (Diptera: Culicidae) by larvae of Arrenuridae and Microtrombidiidae (Acari: Parasitengona) in the Wellington region, New Zealand. *New Zealand Journal of Zoology 33*: 9–15.

SOLOMON, M. G.; CROSS, J. W.; FITZGERALD, J. D.; CAMPBELL, C. A. M.; JOLLY, R. L.; OLSZAK, R. W.; NIEMCYZK, E.; VOGT, H. 2000: Biocontrol of pests of apples and pears in northern and central Europe: 3. Predators. *Biocontrol Science and Technology 10*: 91–128.

SOMERFIELD, K. G. 1974: Ecological studies on the fauna associated with decaying logs and leaf litter in New Zealand *Pinus* forests with particular reference to Coleoptera. PhD thesis, University of Auckland. 312 p.

SOMERFIELD, K. G. 1982: Carrot rust fly in Canterbury (Diptera: Psilidae). *New Zealand Entomologist 7*: 338–340.

SOMERFIELD, K. G.; BURNETT, P. A. 1976: Lucerne insect survey. *Proceedings of the New Zealand Weed and Pest Control Conference 29*: 14–18.

SOMERFIELD, K. G.; MASON, D. C. M.; DALE, P. S. 1980: Insects and mites associated with dried milk product storage areas in New Zealand. *New Zealand Journal of Experimental Agriculture 8*: 83–85.

SONG, D. P.; WANG, Q. 2001: Taxonomy and phylogeny of the New Zealand longicorn beetle genus *Calliprason* (Coleoptera: Cerambycidae). *Invertebrate Taxonomy 15*: 53–71.

SONG, D. P.; WANG, Q. 2003: Systematics of the longicorn beetle genus *Coptomma* (Coleoptera: Cerambycidae). *Invertebrate Systematics 17*: 429–447

SOPER, M. F. 1963: *New Zealand Bird Portraits*. Whitcomb and Tombs, Christchurch. 103 p.

SPAIN, A.V. 1967: A study of the arthropods associated with *Olearia colensoi* Hook f. M.Agr. Sc. thesis, Lincoln College, University of Canterbury, Christchurch. 288 p.

SPENCER, K. A. 1976: The Agromyzidae of New Zealand (Insecta: Diptera). *Journal of the Royal Society of New Zealand 6*: 153–211.

SPENCER, K. A. 1977: A revision of the New Zealand Chloropidae. *Journal of the Royal Society New Zealand 7*: 433–472.

STACEY, W. 1969: Fly pests can affect some Northland crops. *New Zealand Journal of Agriculture 118*: 74–75.

STARK, J. D. 1993: Performance of the Macroinvertebrate Community Index: effects of sampling method, sample replication, water depth, current velocity, and substratum index values. *New Zealand Journal of Marine and Freshwater Research 27*: 463–478.

STARK, J. D. 1998: SQMCI: a biotic index for freshwater macroinvertebrate code-abundance data. *New Zealand Journal of Marine and Freshwater Research 32*: 55–66.

STARK, J. D. 2000: Hydroptilidae. P. 32 *in*: Winterbourn, M. J.; Gregson, K. L. D.; Dolphin, C. H. (eds), *Guide to the Aquatic Insects of New Zealand. Bulletin of the Entomological Society of New Zealand 13*: 1–102.

STEHR, F. W. 1991. *Immature Insects*. Kendall Hunt, Dubuque, Iowa.

STEINHAUS, E. A. 1963: *Insect Pathology*. Academic press, New York. Vol. 1, 686 p.

STEVENS, G. R. 1985: Lands in collision: Discovering New Zealand's past geography. *Department of Scientific and Industrial Research Information Series 161*: 1–129.

STEVENS, G. R.; McGLONE, H.; McCULLOCH, B. 1995: *Prehistoric New Zealand*. Reed, Auckland. 128 p.

STEVENS, M.; McCARTNEY, J.; STRINGER, I. 2007a: New Zealand's forgotten biodiversity: different techniques reveal new records for 'giant' springtails. *New Zealand Entomologist 30*: 79–84.

STEVENS, M. I.; WINTER, D. J.; MORRIS, R.; McCARTNEY, J.; GREENSLADE, P. 2007b: New Zealand's giant Collembola: new information on distribution and morphology for *Holacanthella* Börner, 1906 (Neanuridae: Uchidanurinae). *New Zealand Journal of Zoology 34*: 63–78.

STEVENS, N. B.; AUSTIN, A. D. 2007: Systematics, distribution and biology of the Australian 'micro-flea' wasps, *Baeus* spp. (Hymenoptera: Scelionidae): parasitoids of spider eggs. *Zootaxa 1499*: 1–45.

STORK, N. E. 1991: The composition of the arthropod fauna of Bornean lowland rain forest trees. *Journal of Tropical Ecology 7*: 161–180.

STRANDBERRY, J. D. O. 1981: Activity and abundance of the earwig *Labidura riparia,* in a winter cabbage production ecosystem. *Environmental Entomology 10*: 701–704.

STRINGER, I. A. N. 1970: The nymphal and imaginal stages of the bisexual stick insect *Clitarchus hookeri* (Phasmida: Phasminae). *New Zealand Entomologist 4*: 85–95.

STUFKENS, M. A. W.; FARRELL, J. A. 1994: Quarantine host range tests on two exotic parasitoids imported for aphid control. *Proceedings of the New Zealand Plant Protection Conference 47*: 149–154.

STUMPF, C. F.; LAMBDIN, P. L. 2006: *Pit scales (Sternorrhyncha: Coccoidea) of North and South America*. University of Tennessee Institute of Agriculture, Knoxville. 231 p.

STURM, H. 1980: Redescription of *Nesomachilis* (Archeognatha: Meinertellidae) with descriptions of new species from the Australian region. *New Zealand Journal of Zoology 7*: 533–550.

STURM, H.; BACH, R.C. de 1993: On the systematics of the Archaeognatha (Insecta). *Entomologia Generalis 19*: 55–90.

STUSAK, J. M.1989: Two new genera and one new subgenus of Berytinae, with nomenclatorial changes (Heteroptera, Berytidae). *Acta Entomologica Bohemoslovaca 86*: 286–294.

STYLES, J. H. 1967: Decomposition of *Pinus radiata* litter on the forest floor. 2. Changes in microfauna population. *New Zealand Journal of Science 10*: 1045–1060.

SUCKLING, D. M. 1984: Laboratory studies on the praying mantis *Orthodera ministralis* (Mantodea: Mantidae). *New Zealand Entomologist 8*: 96–101.

SUCKLING, D. M.; BURNIP, G. M.; HARKETT, J.; DALY, J. C. 2006: Frass sampling and baiting indicate European earwig (*Forficlua auricularia*) foraging in orchards. *Journal of Applied Entomology 130*: 263–267.

SUNDE, R.G. 1973: New records of Aphidoidea (Homoptera) in New Zealand. *New Zealand Entomologist 5*: 127–130.

SUNDE, R. G. 1984: New records of plant pests in New Zealand 4. 7 aphid species (Homoptera: Aphidoidea). *New Zealand Journal of Agricultural Research 27*: 575–579.

SUNDE, R. G. 1988: A new indigenous aphid from New Zealand (Homoptera: Aphididae). *New Zealand Journal of Zoology 14*: 587–592.

SUNDERLAND, K. D.; VICKERMAN, G. P. 1980: Aphid feeding by some polyphagous predators in relation to aphid density in cereal fields. *Journal of Applied Ecology 17*: 389–396.

SUREN, A. M.; WINTERBOURN, M. J. 1991: Consumption of aquatic bryophytes by alpine stream invertebrates in New Zealand. *New Zealand Journal of Marine and Freshwater Research 25*: 331–343.

SWAN, D. I. 1973: Evaluation of biological control of the oak leaf-miner *Phyllonorycter messaniella* (Zell.) (Lep., Gracillariidae) in New Zealand. *Bulletin of Entomological Research 63*: 49–55.

SWENEY, W. J. 1980: Insects of Mount Cook National Park. M. Agric. Sci. thesis, Lincoln University, Canterbury. 328 p.

SYRETT, P.; SMITH, L. A. 1994: The insect fauna of four weedy *Hieraceum* (Asteraceae) species in New Zealand. *New Zealand Journal of Zoology* 25: 73–83.

SZWEDO, J. 2004: A new genus and six new species of ground-dwelling leafhoppers from Chile and New Zealand (Hemiptera: Cicado-morpha: Myerslopiidae). *Zootaxa 424*: 1–20.

TAGLIANTI, A.V. 1995: Ricerche zoologiche della nave oceanografica 'Minerva' (C.N.R.) sulle isole circumsarde. 22. I dermatteri delle isole circumsarde (1). (Insecta, Dermaptera). *Annali del Museo civico di storia naturale Giacomo Doria 90*: 529–552.

TAKAGI, S. 1985: The scale insect genus *Chionaspis* (Homoptera: Coccoidea: Diaspididae). *Insecta Matsumurana, n.s., 33*: 1–77.

TANADA, Y.; KAYA, H. K. 1993: *Insect Pathology*. Academic Press, New York. 666 p.

TAUBER, M. J.; TAUBER, C. A.; DAANE, K. M.; HAGEN, K. S. 2000: Commercialization of predators : recent lessons from green lacewings (Neuroptera: Chrysopidae: *Chrysoperla*). *American Entomologist 46*: 26–38.

TAYLOR, G.; DAVIES, K.; MARTIN, N.; CROSBY, T. 2007: First record of *Fergusonina* (Diptera: Fergusoninidae) and associated *Fergusobia* (Tylenchida: Neotylenchidae) forming galls on *Metrosideros* (Myrtaceae) from New Zealand. *Systematic Entomology 32*: 548–557.

TAYLOR, R. C.; EWERS, R.M. 2003: The invertebrate fauna inhabiting tree holes in a red beech (*Nothofagus fusca*) tree. *The Weta 25*: 24–28.

TAYLOR, R. W. 1971: The ants (Hymenoptera: Formicidae) of the Kermadec Islands. *New Zealand Entomologist 5*: 81–82.

TENQUIST, J. D.; CHARLESTON, W. A. G. 1981: An annotated checklist of ectoparasites of terrestrial mammals in New Zealand. *Journal of the Royal Society of New Zealand 11*: 257–285.

TEULON, D. A. J. 1999: Illustrated multiple-entry key for winged aphids in New Zealand. *CropInfo Confidential Report 614*: 1–41.

TEULON, D. A. J.; EASTOP, V. F.; STUFKENS, M. A. W.; HARCOURT, S. J. 1999: Illustrated multiple-entry key for apterous aphids of economic importance in New Zealand. *CropInfo Confidential Report 612*: 1–140.

TEULON, D. A. J.; NEILSON, M.-C. 2005: Distribution of western (Glasshouse strain) and intonsa flower thrips in New Zealand. *New Zealand Plant Protection 58*: 208–212.

TEULON, D. A. J.; PENMAN, D. R. 1990: Host records for the New Zealand flower thips (*Thrips obscuratus* (Crawford) (Thysanoptera: Thripidae) on nectarine and peach flowers. *New Zealand Entomologist 13*: 46–51.

TEULON, D. A. J.; PENMAN, D. R. 1992: Colour preferences of New Zealand thrips (Terebrantia: Thysanoptera). *New Zealand Entomologist 15*: 8–13.

TEULON, D. A. J.; PENMAN, D. R. 1994: Phenology of the New Zealand flower thrips, *Thrips obscuratus* (Crawford) (Thysanoptera: Thripidae). *New Zealand Entomologist 17*: 70–77.

TEULON, D. A. J.; PENMAN, D. R. 1996: Thrips (Thysanoptera) seasonal flight activity and infestation of ripe fruit in Canterbury, New Zealand. *Journal of Economic Entomology 89*: 722–734.

TEULON, D. A. J.; WOUTS, W.; PENMAN, D. R. 1997: A nematode parasitoid of the New Zealand flower thrips (Thysanoptera: Thripidae). *New Zealand Entomologist 20*: 67–69.

TEULON, D. A. J.; STUFKENS, M. A. W. 1998: Current status of New Zealand indigenous aphids. *Conservation Advisory Science Notes 216*: 1–23.

TEULON, D. A. J.; STUFKENS, M. A. W. 2002: Biosecurity and aphids in New Zealand. *New Zealand Plant Protection 55*: 12–17.

THOMAS, W. P. 1977: Biological control of the blue-green lucerne aphid: the Canterbury situation. *Proceedings of the New Zealand Weed and Pest Control Conference 30*: 182–187.

THOMAS, M. B.; SOTHERTON, N. W.; COOMBES, D. S.; WRATTEN, S. D. 1992: Habitat factors influencing the distribution of polyphagous predatory insects between field boundaries. *Annals of Applied Biology 120*: 197–202.

THOMAS, W. P. 1989a: *Forficula auricularia* L. European earwig (Dermaptera: Forficulidae). Pp. 201–205 *in*: Cameron, P. J.; Hill, R. L.; Bain, J.; Thomas, W. P. (eds), *A Review of Biological Control of Invertebrate Pests and Weeds in New Zealand 1874 to 1987. CAB International Technical Communication 10*: 1–423.

THOMAS, W. P. 1989b: Aphididae, aphids (Homoptera). Pp. 55–66 *in*: Cameron, P. J.; Hill, R. L.; Bain, J.; Thomas, W. P. (eds), *A Review of Biological Control of Invertebrate Pests and Weeds in New Zealand 1874 to 1987. CAB International Technical Communication 10*: 1–423.

THOMPSON, F. C. 2008: A conspectus of New Zealand flower flies (Diptera: Syrphidae). *Zootaxa 1716*: 1–20.

THOMPSON, G.M. 1922: *The Naturalisation of Animals and Plants in New Zealand*. Cambridge University Press, London.

THOMPSON, W. R. 1954: Part 3. Hosts of the Hymenoptera (Calliceratid to Evaniid).Pp. 191–332 *in*: *A Catalogue of the Parasites and Predators of Insect Pests. Section 2. Host Parasite Catalogue*. Commonwealth Agricultural Bureaux, Commonwealth Institute of Biological Control, Ottawa.

THOMSON, M. S. 1934: An account of the systematics and bionomics of *Austroperla cyrene* Newman. Unpublished MSc(Hons) thesis, Canterbury University College, Christchurch.

THORNE, B. L.; LENZ, M. 2001: Population and colony structure of *Stolotermes inopus* and *Stolotermes ruficeps* (Isoptera: Stolotermitinae) in New Zealand. *New Zealand Entomologist* 24: 63–70.

THORNTON, I. W. B. 1962: Note on the genitalia of two New Zealand Philotarsids (Insecta: Psocoptera). *New Zealand Journal of Science 5*: 241–245.

THORNTON, I. W. B. 1985: The geographical and ecological distribution of arboreal Psocoptera. *Annual Review of Entomology 30*: 175–196.

THORNTON, I. W. B.; LEE, S. S.; CHUI, V. W. D. 1972: Insects of Micronesia: Psocoptera. *Insects of Micronesia 8*: 45–144.

THORNTON, I. W. B.; NEW, T. R. 1977: The Philotarsidae (Insecta: Psocoptera) of Australia. *Australian Journal of Zoology, Supplementary Series 54*: 1–62.

THORNTON, I. W. B.; WONG, S. K. 1968: The peripsocid fauna (Psocoptera) of the Oriental Region and the Pacific. *Pacific Insects Monographs 19*: 1–158.

THORNTON, I. W. B.; WONG, S. K.; SMITHERS, C. N. 1977: The Philotarsidae (Psocoptera) of New Zealand and the islands of the New Zealand Plateau. *Pacific Insects 17*: 197–228.

TILLYARD, R. J. 1917: Studies in Australian *Mecoptera*. No. 1. The new family Nanno-choristidae, with descriptions of a new genus and four new species: and an appendix descriptive of a new genus and species from New Zealand. *Proceedings of the Linnean Society of New South Wales 42*: 284–301.

TILLYARD, R. J. 1920: Neuropteroid insects of the hot springs region, New Zealand, in relation to the problem of trout food. *Proceedings of the Linnean Society of New South Wales 45*: 205–213.

TILLYARD, R. J. 1921: Studies of New Zealand Trichoptera, or caddis-flies. No. 1. Descriptions of a new genus and species belonging to the family Sericostomatidae. *Transactions and Proceedings of the New Zealand Institute 53*: 346–350.

TILLYARD, R. J. 1923a: Two new species of may-flies (order Plectoptera) from New Zealand. *Transactions and Proceedings of the New Zealand Institute 54*: 226–230.

TILLYARD, R. J. 1923b: A monograph of the Psocoptera, or Copeognatha, of New Zealand. *Transactions of the New Zealand Institute 54*: 170–196, pl. 18.

TILLYARD, R. J. 1923c: Descriptions of new species and varieties of lacewings (order Neuroptera Planipennia) from New Zealand belonging to the families Berothinidae and Hemerobiidae. *Transactions and Proceedings of the New Zealand Institute 54*: 217–225.

TILLYARD, R. J. 1924a: Studies of New Zealand Trichoptera, or caddis-flies. No. 2. Descriptions of new genera and species. *Transactions and Proceedings of the New Zealand Institute 55*: 285–314, 1 pl.

TILLYARD, R. J. 1924b: Primitive wingless insects. Part 1: The silverfish, bristletails, and their allies (order Thysanura). *New Zealand Journal of Science and Technology 7*: 232–242.

TILLYARD, R. J. 1925: Caddis-flies (order Trichoptera) from the Chatham Islands. *Records of the Canterbury Museum 2*: 277–284.

TILLYARD, R. J. 1926: *The Insects of Australia and New Zealand*. Angus & Robertson, Sydney. xiv + 560 pp., 44 pls.

TIMMERMANS, M. J. T. N.; ROELOFS, D.; MARIËN, J.; van STRAALEN, N. M. 2008: Revealing pancrustacean relationships: phylogenetic analysis of ribosomal protein genes places Collembola (springtails) in a monophyletic Hexapoda and reinforces the discrepancy between mitochondrial and nuclear DNA markers. *BMC Evolutionary Biology 8:83*: [1–10]. doi:10.1186/1471-2148-8-83

TOFT, R. J.; CHANDLER P. J. 2004: Three introduced species of Mycetophilidae (Diptera: Sciaroidea) established in New Zealand. *New Zealand Entomologist* 27: 43-49.

TOFT, R. J.; HARRIS, R. J.; WILLIAMS, P. A. 2001: Impacts of the weed *Tradescantia fluminensis* on insect communities in fragmented forests in New Zealand. *Biological Conservation 102*: 31–46.

TOMBERLIN, J.K.; ADLER, P.H. 1998: Seasonal colonization and decomposition of rat carrion in water and land in an open field in South Carolina. *Journal of Medical Enotomology 35*: 704–709.

TOMKINS, A. R.; NILSON, D. J.; THOMSON, C. 1992: Thrip control with fluvalinate on kiwifruit. *New Zealand Weed and Pest Control Conference 45*. 162–166.

TONNOIR, A. L. 1925: New Zealand Pipunculidae (Dipt.). *Records of the Canterbury Museum 2*: 313–316.

TONNOIR, A. L. 1927: Descriptions of new and remarkable New Zealand Diptera. *Records of the Canterbury Museum 3*: 101–112.

TONNOIR, A. L.; EDWARDS, F. W. 1927: New Zealand fungus gnats (Diptera: Mycetophilidae). *Transactions and Proceedings of the New Zealand Institute 57*: 747–878.

TORTELL, P. 1981: *Atlas of Coastal Resources*. Government Printer, Wellington. 59 p.

TOWNS, D. R. 1978: Some little-known benthic insect taxa from a northern New Zealand river and its tributaries. *New Zealand Entomologist 6*: 409–419.

TOWNS, D. R. 1983a: A revision of the genus *Zephlebia* (Ephemeroptera: Leptophlebiidae). *New Zealand Journal of Zoology 10*: 1–52.

TOWNS, D. R. 1983b: Life history patterns of six sympatric species of Leptophlebiidae (Ephemeroptera) in a New Zealand stream and the role of interspecific competition in their evolution. *Hydrobiologia 99*: 37–50.

TOWNS, D. R. 1987: The mayflies (Ephemeroptera) of Great Barrier Island, New Zealand: macro- and micro-distributional comparisons. *Journal of the Royal Society of New Zealand 17*: 349–361.

TOWNS, D. R.; PETERS, W. L. 1978: A revision of genus *Atalophlebioides* (Ephemeroptera: Leptophlebiidae). *New Zealand Journal of Zoology 5*: 607–614.

TOWNS, D. R.; PETERS, W. L. 1979a: Three new genera of Leptophlibiidae (Ephemeroptera) from New Zealand. *New Zealand Journal of Zoology 6*: 213–235.

TOWNS, D. R.; PETERS, W. L. 1979b: New genera and species of Leptophlebiidae (Ephemeroptera) from New Zealand. *New Zealand Journal of Zoology 6*: 439–452.

TOWNS, D. R.; PETERS, W. L. 1996: Leptophlebiidae (Insecta: Ephemeroptera). *Fauna of New Zealand 36*: 1–141.

TOWNSEND, J. I. 1970: Some notes on *Heterojapyx novaezeelandiae* (Verhoeff) (Diplura: Japygidae). *New Zealand Entomologist 4*: 100–102.

TOWNSEND, R. J.; WATSON, R. N. 1981: Stand management of asparagus for control of thrips. *New Zealand Weed and Pest Control Conference 37*: 151–155.

TRAVERS, W. T. L. 1864: Additional observations on the diffusion of European weeds, and their replacement of the indigenous vegetation, in New Zealand. *Natural History Review 16*: 617–619.

TREWICK, S. A. 1999: A new weta from the Chatham Islands (Orthoptera: Rhaphidophoridae). *Journal of the Royal Society of New Zealand 29*: 165–173.

TREWICK, S. A.; GOLDBERG, J.; MORGAN-RICHARDS, M. 2005: Fewer species of *Argosarchus* and *Clitarchus* stick insects (Phasmida, Phasmatinae): evidence from nuclear and mitochondrial DNA sequence data. *Zoologica Scripta 34*: 483–491.

TREWICK, S. A.; MORGAN-RICHARDS, M.; COLLINS, L. 2008: Are you my mother? Phylogenetic analysis reveals orphan hybrid stick insect genus is part of a monophyletic New Zealand clade. *Molecular Phylogenetics and Evolution 48*: 799–808.

TREWICK, S. A.; WALLIS, G. P.; MORGAN-RICHARDS, M. 2000: Phylogenetic pattern correlates with Pliocene mountain-building in the alpine scree weta (Orthoptera: Anostostomatidae). *Molecular Entomology 9*: 657–666.

TRIAPITSYN, S.V.; BEREZOVSKIY, V.V. 2001. Review of the Mymaridae (Hymenoptera, Chalcidoidea) of Primorskii krai: genus *Mymar* Curtis. *Far Eastern Entomologist 100*: 1–20.

TRIAPITSYN, S.V.; BEREZOVSKIY, V.V. 2007: Review of the Oriental and Australasian species of *Acmopolynema*, with taxonomic notes on *Palaeoneura* and *Xenopolynema stat. rev.* and description of a new genus (Hymenoptera: Mymaridae). *Zootaxa 1455*: 1–68.

TROUGHT, K. F. 1982: Aspects of the gut structure and function affecting digestion in several species of aquatic insect larvae. BScHons Dissertation, Department of Zoology, University of Canterbury, Christchurch. 41 p.

TUTHILL, L. D. 1952: On the Psyllidae of New Zealand (Homoptera). *Pacific Science* 6: 83–125.

TUXEN, S. L. 1985: The Protura (Insecta). *Fauna of New Zealand* 9: 1–50.

USINGER, R. L.; MATSUDA, R. 1959: *Classification of the Aradidae*. Trustees of the British Museum (Natural History), London. 410 pp.

VALENTINE, E.W. 1967a: Biological control of aphids. *Proceedings of the New Zealand Weed and Pest Control Conference 20*: 204–207.

VALENTINE, E. W. 1967b: A list of the hosts of entomophagous insects in New Zealand. *New Zealand Journal of Science 10*: 1100–1210.

VALENTINE, E. W. 1970: Hymenoptera. *New Zealand Entomologist* 4: 47–50.

VALENTINE, E. W.; WALKER, A. W. 1991: Annotated catalogue of New Zealand Hymenoptera. *Department of Scientific and Industrial Research Plant Protection Report 4*: 1–84.

VANE-WRIGHT, R. I. 1967: A re-assessment of the genera *Holorusia* Loew (= *Ctenacroscelis* Enderlein), *Ischnotoma* Skuse and *Zelandotipula* Alexander (Diptera: Tipulidae), with notes on their phylogeny and zoogeography. *Journal of Natural History 1*: 511–547.

VELTMAN, C. J.; COLLIER, K. J.; HENDERSON, I. M.; NEWTON, L. 1995: Foraging ecology of blue ducks *Hymenolaimus malacorhynchos* on a New Zealand river: implications for conservation. *Biological Conservation 74*: 187–194.

VINKATENSEN, T.; SINGH, S. P.; JULALI, S. K. 2000: Rearing *Chrysoperla carnea* (Stephens) (Neuroptera: Chrysoperlidae) on semi-synthetic diet and its predatory effect against cotton pests. *Entomophaga 25*: 81–89.

VOCKEROTH, J. R. 1969: A revision of the genera of the Syrphini (Diptera: Syphidae). *Memoirs of the Entomological Society of Canada 62*: 1–176.

WAHL, D. B. 1991: A new species of *Dusona* from New Zealand, and the application of *Dusona* vs. *Delopia* (Hymenoptera: Ichneumonidae, Campopleginae). *Proceedings of the Entomological Society of Washington 93*: 946–950.

WALKER, A. K. 1996: A new species of *Choeras* (Braconidae: Microgastrinae) widespread in New Zealand. *New Zealand Entomologist 19*: 43–48.

WALKER, A. K.; HUDDLESTON, T. 1987: New Zealand chelonine braconid wasps (Hymenoptera). *Journal of Natural History 21*: 339–361.

WALKER, F. 1850: *List of the Specimens of Homopterous Insects in the Collection of the British Museum. Part 1*. Trustees of the British Museum (Natural History), London. 260 p.

WALKER, F. 1851a: *List of the Specimens of Homopterous Insects in the Collection of the British Museum. Part 2*. Trustees of the British Museum (Natural History), London. Pp. 261–636.

WALKER, F. 1851b: *List of the Specimens of Homopterous Insects in the Collection of the British Museum. Part 3*. Trustees of the British Museum (Natural History), London. 637–907.

WALKER, F. 1852: *Catalogue of the Specimens of Neuropterous Insects in the Collection of the British Museum*. Trustees of the British Museum (Natural History), London. 192 p.

WALKER, F. 1853: Ephemeridae. Pp. 533–585 *in*: *List of the Specimens of Neuropterous Insects in the Collection of the British Museum, Volume 3*. Trustees of the British Museum (Natural History), London.

WALKER, F. 1857: Catalogue of the Homopterous insects collected at Singapore and Malacca by Mr. A.R. Wallace, with descriptions of new species. *Journal of the Proceedings of the Linnean Society of London, Zoology* 1: 82–100.

WALKER, F. 1858a: *List of the Specimens of Homopterous Insects in the Collection of the British Museum. Supplement*. Trustees of the British Museum (Natural History), London. Pp. 1–307.

WALKER, F. 1858b: *List of the Specimens of Homopterous Insects in the Collection of the British Museum. Addenda*. Trustees of the British Museum (Natural History), London. Pp. 308–369.

WALKER, F. 1858c: *Homoptera. Insecta Saundersiana: or Characters of Undescribed Insects in the Collection of William Wilson Saunders Esq*. Trustees of the British Museum (Natural History), London. 117 p.

WALKER, F. A.; FELL, R. D. 2001: Courtship roles of male and female European earwigs *Forficula auricularia* L. (Dermaptera : Forficulidae), and sexual use of forceps. *Journal of Insect Behaviour 14*: 1–17.

WALKER, K. A.; JONES, T. H.; FELL, R. D. 1993:

Pheromonal basis of aggregation in European earwig *Forficula auricularia* L. (Dermaptera : Forficulidae). *Journal of Chemical Ecology 19*: 2029–2038.

WALL, G.Y. 1981: Feeding ecology of the tuatara (*Sphenodon punctatus*) on Stephens Island, Cook Strait. *New Zealand Journal of Ecology 4*: 89–97.

WANG, Q.; LESCHEN, R. 2003: Identification and distribution of *Arhopalus* species (Coleoptera: Cerambycidae: Aseminae) in Australia and New Zealand. *New Zealand Entomologist 26*: 53–59.

WANG, Q.; LU, W. 2004: A systematic revision of the New Zealand longicorn beetle genus *Drototelus* Broun (Coleoptera: Cerambycidae: Cerambycinae). *Invertebrate Systematics 18*: 649–659.

WARD, D. F. 1997: A new generic key to the New Zealand cave weta genera. *New Zealand Natural Sciences 23*: 13–17.

WARD, J. B. 1991: Two new species of New Zealand caddisflies (Trichoptera: Hydrobiosidae). *New Zealand Entomologist 14*: 15–21.

WARD, J. B. 1995: Nine new species of New Zealand caddis (Trichoptera). *New Zealand Journal of Zoology 22*: 91–103.

WARD, J. B. 1997: Twelve new species in the New Zealand caddis (Trichoptera) fauna, corrected type localities and new synonyms. *New Zealand Journal of Zoology 24*: 173–191.

WARD, J. B. 1998: Five new species of New Zealand Hydrobiosidae (Insecta: Trichoptera). *Records of the Canterbury Museum 12*: 1–16.

WARD, J. B. 2002: Additions and Amendments to New Zealand caddis species in the 'Atlas of Trichoptera ...'. *Records of the Canterbury Museum 16*: 32–59.

WARD, J. B. 2005: Four new species of caddis in the endemic New Zealand genus Edpercivalia (Insecta: Trichoptera: Hydrobiosidae). *Records of the Canterbury Museum 19*: 61–70.

WARD, J. B.; HENDERSON, I. M.; 2004: Eleven new species of micro-caddis (Trichoptera: Hydroptilidae) from New Zealand. *Records of the Canterbury Museum 18*: 9–22.

WARD, J. B.; LESCHEN, R. A. B.; SMITH, B. J.; DEAN, J. C. 2004: Phylogeny of the caddisfly (Trichoptera) family Hydrobiosidae using larval and adult morphology, with the description of a new genus and species from Fiordland, New Zealand. *Records of the Canterbury Museum 18*: 23–43.

WARD, J. B.; MACFARLANE, R. P.; QUINN, P. J.; MORRIS, S. J.; HITCHINGS, T. R.; GREEN, E. H.; EARLY, J. W.; EMBERSON, R. W.; JOHNS, P. M.; FENWICK, G. D.; HENDERSON, I. M.; HENDERSON, R.; LARIVIERE, M.-C.; MARRIS, J. M. M.; MATILE, L.; McLELLAN, I. D.; PATRICK, B. H.; SMITHERS, C.; STUFKENS, M.; VINK, C.; WILSON, H. D. 1999. Insects and other arthropods of Hinewai Reserve, Banks Peninsula, New Zealand. *Records of the Canterbury Museum 19*: 97–121.

WARD, J. B.; McKENZIE, J.C. 1998: Synopsis of the genus *Olinga* (Trichoptera: Conoesucidae) with a comparative SEM study of the male forewing androconia and the description of a new species. *New Zealand Natural Sciences 23*: 1–11.

WATSON, G.W. 1972: The lotic fauna of Red Mercury Island (Whakau). *Tane: Journal of the Auckland University Field Club 18*: 67–79.

WATSON, J. A. L.; ABBEY, H. M. 1993: *Atlas of Australian Termites*. CSIRO Entomology Division, Canberra. 155 p.

WATSON, J. A. L.; THEISCHINGER, G.; ABBEY, H. M. 1991: *The Australian Dragonflies: A Guide to the Identification, Distributions and Habitat of Australian Odonata*. CSIRO, Canberra. 278 p.

WATT, J. C. 1974: Beetles. *New Zealand's Nature Heritage 2(28)*: 761–768.

WATT, J. C. 1982: Terrestrial arthropods from the Poor Knights Islands, New Zealand. *Journal of the Royal Society of New Zealand 12*: 283–320.

WATT, J. C. 1983: Hexapoda, Myriapoda and Arachnida. *In*: Brownsey, P.J.; Baker, A.N. (eds). *The New Zealand Biota–What Do We Know after 200 Years? National Museum of New Zealand Miscellaneous Series 7*: 62–67

WATT, J. C. 1986: Beetles (Coleoptera) of the offshore islands of northern New Zealand. *In*: Wright, A. E.; Beever, R. E. (eds), *The Offshore Islands of Northern New Zealand. Department of Lands and Survey Inofromation Series 16*: 221–228.

WATT, J. C.; MARRIS, J. W. M.; KLIMASZEWSKI, J. 2001: A new species of *Platisus* (Coleoptera: Cucujidae) from New Zealand, described from the adult and larva. *Journal of the Royal Society of New Zealand 31*: 327–339.

WATT, M. N. 1923: The leaf–mining insects of New Zealand: Part 3–Species belonging to the genera *Agromyza* (Fallen) and *Phytomyza* (Fallen) (Diptera). *Transactions and Proceedings of the New Zealand Institute 54*: 465–489.

WESTERMAN, M.; RITCHIE, J. M. 1984: The taxonomy, distribution and origins of two species of *Phaulacridium* (Orthoptera: Acrididae) in the South Island of New Zealand. *Biological Journal of the Linnean Society 21*: 283–298.

WHITE, A. J.; WRATTEN, S. D.; BERRY, N. A.; WEIGMANN, U. 1995: Habitat manipulation to enhance biological control of *Brassica* pests by hover flies (Diptera: Syrphidae). *Journal of Economic Entomology 88*: 1171–76

WHITE, E. G. 1964: A survey and investigation of the insect fauna associated with some tussock grasslands. M.Hort.Sc. thesis, Lincoln College, University of Canterbury, Christchurch. 279 p.

WHITE, E. G. 1975: A survey and assessment of grasshoppers as herbivores in the South Island alpine tussock grasslands of New Zealand. *New Zealand Journal of Agricultural Research 18*: 73–85.

WHITE, E. G. 1978: Energetics and consumption rates of alpine grasshoppers (Orthoptera: Acrididae) in New Zealand. *Oecologia 33*: 17–44.

WHITE, E. G. 1991: The changing abundance of moths in a tussock grassland, 1962–1989, and 50- to 70-year trends. *New Zealand Journal of Ecology 15*: 5–22.

WHITE, E. G. 2002: *New Zealand Tussock Grassland Moths*. Manaaki Whenua Press, Lincoln. 362 p.

WHITE, E. G.; SEDCOLE, J. R. 1991: A 20-year record of grasshopper abundance, with interpretations for climate change. *New Zealand Journal of Ecology 15*: 139–152.

WHITE, F. B. 1876: Descriptions of three new species of Hemiptera–Heteroptera from New Zealand. *Entomologist's Monthly Magazine 13*: 105–106.

WHITE, F. B. 1878–1879a: List of the Hemiptera of New Zealand. *Entomologist's Monthly Magazine 14 (1878)*: 274–277; *15*: 31–34, 73–76, 130–133, 159–161; *(1879)* 213–220.

WHITE, F. B. 1879a: List of Hemiptera of New Zealand. *Entomologist's Monthly Magazine 15*: 213–220.

WHITE, F. B. 1879b: Descriptions of new Anthocoridae. *Entomologist's Monthly Magazine 16*: 142–148.

WHITE, I. M. 1988: Tephritid flies (Diptera :Tephritidae. *Handbooks for Identification of British Insects 10(5a)*: 1–134.

WHITE, I. M.; ELSON–HARRIS, M. M. 1992: *Fruit Flies of Economic Significance: Their Identification and Bionomics*. Commonwealth Agricultural Bureau International, Wallingford. 601 p.

WHITING, M. F. 2002a: Phylogeny of the holometabolous insect orders: molecular evidence. *Zoologica Scripta 31*: 3–15.

WHITING, M. F. 2002b: Mecoptera is paraphyletic: multiple genes and phylogeny of Mecoptera and Siphonaptera. *Zoologica Scripta 31*: 93–104.

WHITING, M. F., CARPENTER, J. C., WHEELER, Q. D.; WHEELER, W. C. 1997: The Strepsiptera problem: Phylogeny of the holometabolous insect orders inferred from 18S and 28S ribosomal DNA sequences and morphology. *Systematic Biology 46*: 1–68.

WILDING, N.; COLLINS, N. M.; HAMMOND, P. M.; WEBBER, J. F. (Eds) 1989: *Insect-Fungus Interactions*. Academic Press, London. 344 p.

WILKINSON, A. S.; WILKINSON, A. 1952: *Kapiti Bird Sanctuary*. Masterton Printing, Masterton. 190 p.

WILLIAMS, D. D. 1980: Applied aspects of mayfly biology. Pp. 1–17 *in*: Flannagan, J. F.; Marshall, K. E. (eds), *Advances in Ephemeroptera Biology*. Plenum Press, London.

WILLIAMS, D. J.; HENDERSON, R. C. 2005: A new species of the mealybug genus *Rastrococcus* Ferris (Hemiptera: Coccoidea: Pseudococcidae) from New Zealand. *Zootaxa 1085*: 47-60.

WILLMANN, R. 1987: The phylogenetic system of the Mecoptera. *Systematic Entomology 12*: 519–524.

WILSON, E. O. 1985: *The Diversity of Life*. Norton, New York. 424 p.

WINTERBOURN, M. J. 1973: Ecology of the Copland river warm springs, South Island, New Zealand. *Proceedings of the New Zealand Ecological Society 20*: 72–78.

WINTERBOURN, M. J. 1977: Biology aspects of the stream fauna. Pp. 279–290 *in:* Burrows, C. J. (ed.), *Cass*. University of Canterbury, Christchurch.

WINTERBOURN, M. J. 1982: Food utilization by a stream detritivore *Zelandopsyche ingens* (Trichoptera: Oeconesidae). *Internationale Revue der gesamten Hydrobiologie 67*: 209–222.

WINTERBOURN, M. J. 1987: The arthropod fauna of bracken (*Pteridium aquilinum*) on the Port Hills, South Island, New Zealand. *New Zealand Entomologist 10*: 99–104.

WINTERBOURN, M. J. 1996: Life history, production and food of *Aphrophila neozelandica* (Diptera: Tipulidae) in a New Zealand stream. *Aquatic Insects 18*: 45–53.

WINTERBOURN, M. J. 2000: Feeding ecology. Pp. 100–124 *in*: Collier, K. J.; Winterbourn, M. J. (eds), *New Zealand Stream invertebrates: Ecology and Implications for Management*. Caxton Press, Christchurch. 415 p.

WINTERBOURN, M. J. 2004: Association between a commensal chironomid and its mayfly host in rivers of north Westland. *New Zealand Journal of Natural Sciences 29:* 21-31.

WINTERBOURN, M. J.; ANDERSON, N. H. 1980: The life history of *Philanisus plebeius* Walker (Trichoptera: Chathamiidae), a caddisfly whose eggs were found in a starfish. *Ecological Entomology 5*: 293–303.

WINTERBOURN, M. J.; DAVIS, S. F. 1976: Ecological role of *Zelandopsyche ingens* (Trichoptera: Oeconesidae) in a beech forest stream ecosystem. *Australian Journal of Marine and Freshwater Research 27*: 192–215.

WINTERBOURN, M. J.; GREGSON, K. L. D.;

DOLPHIN, C. H. 2000: Guide to the aquatic insects of New Zealand. 3rd edn. *Bulletin of the Entomological Society of New Zealand 13*: 1–102.
WINTERBOURN, M. J.; GREGSON, K. L. D.; DOLPHIN, C. H. 2006: Guide to the aquatic insects of New Zealand. 4th edn. *Bulletin of the Entomological Society of New Zealand 14*: 1–108.
WINTERTON, S. L.; IRWIN, M. E. 2008: *Kaurimyia* gen. nov.: discovery of Aspilocephalidae (Diptera: therevoid clade) in New Zealand. *Zootaxa 1779*: 38–44.
WISE, K. A. J. 1957: Trials for the control of chrysanthemum gall midge (*Diarthonomyia chrysanthemi* Ahjlberg). *New Zealand Journal of Science and Technology, A, 38*: 728–234.
WISE, K. A. J. 1958: Trichoptera of New Zealand. 1. A catalogue of the Auckland Museum collection with descriptions of new genera and new species. *Records of the Auckland Institute and Museum 5*: 49–63.
WISE, K. A. J. 1959: Insects on citrus. *New Zealand Entomologist 2*: 22–24.
WISE, K. A. J. 1962: A new genus and three new species of Trichoptera. *Records of the Auckland Institute and Museum 5*: 247–250.
WISE, K. A. J. 1963: A list of the Neuroptera of New Zealand. *Pacific Insects 5*: 53–58
WISE, K. A. J. 1965: An annotated list of the aquatic and semi-aquatic insects of New Zealand. *Pacific Insects 7*: 191–216.
WISE, K. A. J. 1970a: Apterygota of New Zealand. *New Zealand Entomologist 4*: 62–65.
WISE, K. A. J. 1970b: On the terrestrial invertebrate fauna of White Island, New Zealand. *Records of the Auckland Institute and Museum 10*: 217–252.
WISE, K. A. J. 1972a: Trichoptera of the Auckland Islands. *Records of the Auckland Institute and Museum 9*: 253–267.
WISE, K. A. J. 1972b: Neuroptera of the Kermadec Islands. *Records of the Auckland Institute and Museum 9*: 269–272.
WISE, K. A. J. 1973a: A list and bibliography of the aquatic and water-associated insects of New Zealand. *Records of the Auckland Institute and Museum 10*: 143–187.
WISE, K. A. J. 1973b: New records in the New Zealand Neuroptera: Hemerobiidae. *New Zealand Entomologist 5*: 181–185.
WISE, K. A. J. 1974: Lacewings. *New Zealand's Nature Heritage 4(46)*: 1273–1275.
WISE, K. A. J. 1977: A synonymic checklist of the Hexapoda of the New Zealand subregion. The smaller orders. *Bulletin of the Auckland Institute and Museum 11*: 1–176.
WISE, K. A. J. 1982: Two new species of Trichoptera from the Murchison Mountains, South Island, New Zealand. *Records of the Auckland Institute and Museum 19*: 149–151.
WISE, K. A. J. 1983a: Lacewings and aquatic insects of New Zealand. 2. Fauna of the northern offshore islands. *Records of the Auckland Institute and Museum 20*: 259–271.
WISE, K. A. J. 1983b: Lacewings and aquatic insects of New Zealand. 1. Three new northern distribution records. *Records of the Auckland Institute and Museum 20*: 255–257.
WISE, K. A. J. 1988: Lacewings and aquatic insects of New Zealand. 4. New records and further distributions for Neuroptera. *Records of the Auckland Institute and Museum 25*: 181–184.
WISE, K. A. J. 1992a: Distribution and zoogeography of New Zealand Megaloptera and Neuroptera (Insecta). *Proceedings of the International Symposium on Neuropterology 1991*: 393–395.
WISE, K. A. J. 1992b: The New Zealand species of Berothidae. *Records of the Auckland Institute and Museum 30*: 93–117.
WISE, K. A. J. 1993: Species of *Micromus* (Neuroptera: Hemorobiidae) in New Zealand. *Records of the Auckland Institute and Museum 30*: 93–117.
WISE, K. A. J. 1995: Records concerning biological control of insect pests by Neuroptera (Insecta) in New Zealand. *Records of the Auckland Institute and Museum 32*: 101–117.
WISE, K. A. J. 1998a: A species of the family Sisyridae (Insecta: Neuroptera) in New Zealand. *New Zealand Entomologist 21*: 11–16.
WISE, K. A. J. 1998b: Two new species of *Oxyethira* (Trichoptera: Hydroptilidae) in New Zealand. *New Zealand Entomologist 21*: 17–23.
WISELY, H. B. 1959: A contribution to the life histories of two fungus gnats, *Scythropochroa nitida* Edw. and *Sciara annulata* Mg. (Diptera, Mycetophilidae, Sciarinae). *Transactions of the Royal Society of New Zealand 86*: 59–64.
WITHERS, T. M.; HARRIS, M. O.; DAVIS, L. K. 1995: The incidence of hessian fly and other pests in New Zealand wheat crops. *Proceedings of the New Zealand Plant Protection Conference 48*: 165–169.
WITHERS, T. M.; RAMAN, A.; BERRY, J. A. 2000: Host range and biology of *Ophelimus eucalypti* (Gahan) (Hym.: Eulophidae), a pest of New Zealand eucalypts. *Proceedings of the New Zealand Plant Protection Conference 53*: 339–344.
WOMERSLEY, H.; SOUTHCOTT, R.V. 1941: Notes on the Smarididae (Acarina) of Australia and New Zealand. *Transactions of the Royal Society of South Australia 65*: 61–78.
WOODLEY, N. E. 1995: The genera of Berinae (Diptera: Stratiomyidae). *Memoirs of the Entomological Society of Washington 16*: 1–231.
WOODWARD, T. E. 1950a: New records of Miridae (Heteroptera) from New Zealand, with descriptions of a new genus and four new species. *Records of the Auckland Institute and Museum 4*: 9–23.
WOODWARD, T. E. 1950b: A new species of *Cermatulus* from the Three Kings Islands, New Zealand (Heteroptera: Pentatomidae). *Records of the Auckland Institute and Museum 4*: 24–30.
WOODWARD, T. E. 1953a: The Heteroptera of New Zealand. Part I – Introduction; Cydnidae Pentatomidae. *Transactions of the Royal Society of New Zealand 80*: 299–321.
WOODWARD, T. E. 1953b: New genera and species of Rhyparochrominae from New Zealand (Heteroptera: Lygaeidae). *Records of the Canterbury Museum 6*: 191–218.
WOODWARD, T. E. 1954: New records and descriptions of Hemiptera–Heteroptera from the Three Kings Islands. *Records of the Auckland Institute and Museum 4(4)*: 215–233.
WOODWARD, T. E. 1956: The Heteroptera of New Zealand. Part 2. –The Enicocephalidae with a supplement to part 1 (Cydnidae and Pentatomidae). *Transactions of the Royal Society of New Zealand 84*: 391–430.
WOODWARD, T. E. 1961: The Heteroptera of New Zealand. Part 3 – Coreidae, Berytidae, Tingidae, Cimicidae. *Transactions of the Royal Society of New Zealand, Zoology 1(11)*: 145–158.
WORTHY, T. H. 1989: Inventory of New Zealand caves and karst of international, national or regional importance. *Geological Society of New Zealand Miscellaneous Publication 45*: 1–41.
WYATT, I. J. 1963: A root-infecting gall midge (Diptera: Cecidomyiidae) from New Zealand. *Proceedings of the Royal Entomological Society, London, series B, 32*: 103–107.
WYGODZINSKY, P. 1948: Redescription of *Nesomachilis maoricus* Tillyard 1924, with notes on the family Machilidae (Thysanura). *Dominion Museum Records of Entomology, Wellington 1*: 69–78.
YATES, J. 1952: *Haplothrips niger* Osb., the red clover thrips. *New Zealand Journal of Science and Technology 34B*: 166–172.
YEATES, G. W.; BUCKLEY, T. R. 2009: First records of mermithid nematodes (Nematoda: Mermithidae) parasitising stick insects (Insecta: Phasmatodea). *New Zealand Journal of Zoology 36*: 35–39.
ZAITZEV, A. J. 2001: The Sciaroidea (Diptera) excluding Sciaridae of New Zealand. I. Genus *Aneura* Marshall. *International Journal of Dipterological Research 12*: 33–42.
ZAITZEV, A. J. 2002a: The Sciaroidea (Diptera) excluding Sciaridae of New Zealand. II. Genus *Brevicornu* Marshall. *International Journal of Dipterological Research 13*: 9–14.
ZAITZEV, A. J. 2002b: The Sciaroidea (Diptera) excluding Sciaridae of New Zealand. III. New and little known species of the genus *Zygomyia* Winn.. *International Journal of Dipterological Research 13*: 109–119.
ZERVOS, S. 1983: *Blatticola monandros* n. sp. (Nematoda: Thelastomatidae) from the blatellid cockroach *Parellipsidion pachycercum*. *New Zealand Journal of Zoology 10*: 329–334.
ZERVOS, S. 1984: Seasonality in a field population of two New Zealand cockroaches (Blattodea). *New Zealand Journal of Zoology 11*: 307–312.
ZERVOS, S. 1987: Notes on the size distribution of a New Zealand cockroach, *Celatoblatta vulgaris* *New Zealand Journal of Zoology 14*: 295–297.
ZWICK, P. 1973: Insecta Plecoptera. Phylogenetisches System und Katalog. *Tierreich 94*: i–xxxii, 1–465.
ZWICK, P.; McLELLAN, I. D. 1999: The first instar larva of *Nothohoraia* (Diptera: Blephariceridae). *Aquatic Insects 21*: 317–320.

Checklist of New Zealand Hexapoda

Checklists of New Zealand arthropods from Landcare Research's *Fauna of New Zealand* are given at: www.landcareresearch.co.nz/research/biosystematics/invertebrates/faunaofnz/.

Taxon status and origin: ? before genus = genus uncertain; ? before species = species uncertain; ? after species = doubt about establishment or presence *in* New Zealand. SI, species inquirenda; *, new record. Single letter codes indicate if a species is E, endemic, or A, adventive (naturalised alien). A? indicates uncertainty as to whether a species is indigenous or introduced. If a putative adventive species has not established with certainty, this is stated. A BC, Biological Control Agent; Do, species of doubtful status; V, vagrant/visitor/migrant (non-breeding). All other species are automatically considered indigenous. Endemic families are indicated by E, wholly adventive families by A. Endemic genera are underlined (first entry only). Offshore island endemics: Ch, Chatham Is.; K, Kermadec Islands; Sn, The Snares; Su, subantarctic islands; TK, Three Kings Islands.

Habitat codes, where given, include: Al, alpine; Be, beach; Ca, carrion; Cr, crops, cereals, vegetables; Cv, cave; D, dung; Fo, forest; F, freshwater (Fr, running freshwater; Fs, still or static pond or lakewater); Ga, garden; Gl, glacier, meltwater; Gr, grassland and tussock; Hs, hot spring; Lc, lichen; Li, litter, compost, or decaying vegetation; Mo, moss; Nb, nest of bird; Nm, nest of mammal; Ns, of social insect – bee, wasp, ant; Or, orchard; Sa, subalpine; Se, sewerage and dairy effluent; Sh, shrubland; So, soil; St, stored products; SW (fully saline to brackish); We, wetland-rush, sedge, raupo, flax, marsh, swamp to bare mud/silt.

Carnivore/predator/parasite codes: BF, blood feeder (mammals, birds); BP, bird parasite or commensal; Ci/IPr, insect predator; Cm, mite predator; EP, earthworm parasitoid; IP, insect parasitoid; LP, livestock parasite; MoP, snail or molluscan predator/parasite; SP, spider parasitoid.

Herbivore codes: Ff, fungal feeder; H, herbivore (generalised); Hb, bulb feeder; Hf, flower specialist; HG, galls; HLG, leaf galls; HL, leaves, including seedlings; HLM, leaf miner; HR, roots; HS, seeds, fruits; HSG seed galls; Sc, scavenger; Tw, tree or shrub wood to twigs.

SUBPHYLUM HEXAPODA
Class PROTURA
[Compiled by R. P. Macfarlane]
Order EOSENTOMOIDEA
EOSENTOMIDAE
Eosentomon australicum (Womersley, 1939)
Eosentomon dawsoni Conde, 1952 E
Eosentomon gracile Tuxen, 1985 E
Eosentomon macronyx Tuxen, 1986 E
Eosentomon maximum Tuxen, 1986 E
Eosentomon wygodzinskyi Bonet, 1950
Eosentomon zelandicum Tuxen, 1986 E
PROTENTOMIDAE A
Proturentomon minimum (Berlese, 1908) A

Order ACERENTOMOIDEA
ACERENTOMIDAE
Acerentulus kermadecensis Ramsay & Tuxen, 1978 E
Amphientulus zelandicus Tuxen, 1986 E
Australentulus tillyardi (Womersley, 1932)
Australentulus sp. E
Berberentulus capensis (Womersley, 1931)
Berberentulus nelsoni Tuxen, 1986
Gracilentulus gracilis (Berlese, 1908) A
Kenyentulus kenyanus Conde, 1948
Tasmanentulus intermedius Tuxen, 1986 E
Yinentulus paedocephalus Tuxen, 1986 E

Class COLLEMBOLA
[Wise 1973, updated by P. Greenslade]
Order ARTHROPLEONA
BRACHYSTOMELLIDAE
Brachystomella parvula (Schäffer, 1896) A
Brachystomella terrafolia Salmon, 1944
Setanodosa tetrabrachta Salmon, 1942 E

ENTOMOBRYIDAE
ENTOMOBRYINAE
Coecobrya caeca (Schött, 1896) A
Drepanura aurifera Salmon, 1941 E
Entomobrya aniwaniwaensis Salmon, 1941 E
Entomobrya atrocincta Schött, 1896 A
Entomobrya auricorpa Salmon, 1941 E
Entomobrya divafusca Salmon, 1941 E
Entomobrya duofascia duofascia Salmon, 1941 E
Entomobrya d. maxima Salmon, 1941 E
Entomobrya d. variabilia Salmon, 1941 E
Entomobrya egmontia Salmon, 1941 E
Entomobrya ephippiaterga Salmon, 1941 E
Entomobrya exfoliata Salmon, 1943 E
Entomobrya exoricarva Salmon, 1941 E
Entomobrya fusca (Salmon, 1943) E
Entomobrya hurunuiensis Salmon, 1941 E
Entomobrya lamingtonensis Schött, 1917
Entomobrya livida Salmon, 1941 E
Entomobrya multifasciata Tullberg, 1871 A
Entomobrya nigranota nigranota Salmon, 1941 E
Entomobrya n. sinfascia Salmon, 1941 E
Entomobrya nigraoculata Salmon, 1944 E
Entomobrya nivalis (Linnaeus, 1758) A
Entomobrya n. immaculata Schäffer, 1896
Entomobrya obscuroculata Salmon, 1941 E
Entomobrya opotikiensis Salmon, 1941 E
Entomobrya penicillata Salmon, 1941 E
Entomobrya rubra (Salmon, 1937) E
Entomobrya salta Salmon, 1941 E
Entomobrya saxatila Salmon, 1941 E
Entomobrya totapunctata Salmon, 1941 E
Entomobrya varia Schött, 1917
Entomobrya (Mesentotoma) exalga Salmon, 1942 E
Entomobrya glaciata glaciata (Salmon, 1941) E
Entomobrya g. nigralata (Salmon, 1941 E
Entomobrya intercolorata (Salmon, 1943) E
Entomobrya interfilixa (Salmon, 1941) E
Entomobrya miniparva (Salmon, 1941) E
Entomobrya proceraseta (Salmon, 1941) E
Entomobrya processa (Salmon, 1941) E
Lepidobrya mawsoni (Tillyard, 1920) (= *L. aurantiaca* Salmon, 1949) E
Lepidobrya thalassarchia Salmon, 1949 E
Lepidobrya violacea Salmon, 1949 E
Lepidocyrtus assymetrica Salmon, 1937 E
Lepidocyrtus caeruleacrura (Salmon, 1941) E
Lepidocyrtus cyaneus cinereus Folsom, 1924 A
Lepidocyrtus c. cyaneus Tullberg, 1871 A
Lepidocyrtus elongata (Salmon, 1944) E
Lepidocyrtus fimbriatus Salmon, 1944 E
Lepidocyrtus kauriensis Salmon, 1941 E
Lepidocyrtus lindensis Salmon, 1941 E
Lepidocyrtus moorei Salmon, 1941 E
Lepidocyrtus nigrofasciatus Womersley, 1934
Lepidocyrtus rataensis Salmon, 1941 E
Lepidocyrtus submontanus Salmon, 1941 E
Lepidocyrtus unafascius Salmon, 1941 E
Lepidosira anomala Salmon, 1944 E
Lepidosira arborea arborea Salmon, 1944 E
Lepidosira a. pigmenta Salmon, 1944 E
Lepidosira bidentata Salmon, 1938 E
Lepidosira bifasciata (Salmon, 1944) E
Lepidosira bisecta (Salmon, 1944) E
Lepidosira flava dorsalis (Salmon, 1941) E
Lepidosira f. flava (Salmon, 1938) E
Lepidosira fuchsiata (Salmon, 1938) E
Lepidosira fuscata Womersley, 1930 E
Lepidosira glebosa Salmon, 1941 E
Lepidosira ianthina (Salmon, 1941) E

Lepidosira inconstans (Salmon, 1938) E
Lepidosira indistincta Salmon, 1938 E
Lepidosira magna lichenata (Salmon, 1938) E
Lepidosira m. magna Salmon, 1937 E
Lepidosira m. violacea (Salmon, 1938) E
Lepidosira minima Salmon, 1938 E
Lepidosira minuta Salmon, 1938 E
Lepidosira obscura (Salmon, 1944) E
Lepidosira okarita Salmon, 1938 E
Lepidosira omniofusca Salmon, 1941 E
Lepidosira parva (Salmon, 1941) E
Lepidosira purpurea purpurea (Salmon, 1938) E
Lepidosira p. reducta (Salmon, 1938) E
Lepidosira quadradentata (Salmon, 1941) E
Lepidosira rotorua Salmon, 1938 E
Lepidosira sagmaria (Schött, 1917)
Lepidosira sexmacula Salmon, 1938 E
Lepidosira splendida (Salmon, 1941) E
Lepidosira terraereginae Ellis & Bellinger, 1973
Pseudosinella eudyptidus (Salmon, 1949) E
Pseudosinella alba (Packard, 1873) A
Pseudosinella dispadentata Salmon, 1949 E
Pseudosinella fasciata Womersley, 1934
Pseudosinella insoloculata Salmon, 1941 E
Pseudosinella nonoculata Salmon, 1941 E
Pseudosinella spelunca Salmon, 1958 E
Seira setapartita (Salmon, 1944) E
Sinella castanea (Salmon, 1949) E
Sinella pulverafusca Salmon, 1941 E
Sinella termitum Schött, 1917
ORCHESELLINAE A
Heteromurus nitidus (Templeton, 1935) A
PARONELLIDAE
Glacialoca caerulea (Salmon, 1941) E
Micronellides oliveri Salmon, 1941 E
Parachaetoceras pritchardi (Womersley, 1936) E
Paronana bidenticulata (Carpenter, 1925)
Paronana karoriensis (Salmon, 1937) E
Paronana maculosa (Salmon, 1937) E
Paronana pigmenta Salmon, 1941 E
Paronana pilosa (Salmon, 1941) E
Paronana tasmasecta boldensis (Salmon, 1944) E
Paronana t. tasmasecta (Salmon, 1941) E
Paronellides novaezealandiae novaezealandiae Salmon, 1941 E
Paronellides n. purpurea Salmon, 1941 E
Pseudoparonella dorsanota dorsanota (Salmon, 1941) E
Pseudoparonella d. intermedia Salmon, 1944 E
Pseudoparonella d. sufflava (Salmon, 1941) E
Pseudoparonellides badius Salmon, 1941 E
Pseudoparonellides cryptodontus Salmon, 1944 E
HYPOGASTRURIDAE
Ceratophysella armata (Nicolet, 1842) A
Ceratophysella guthriei (Folsom, 1916) A
Ceratophysella longispina (Tullberg, 1876) A
Hypogastrura campbelli Womersley, 1930 E
Hypogastrura manubrialis (Tullberg, 1869) A
Hypogastrura morbillata (Salmon, 1941) E
Hypogastrura obliqua (Salmon, 1949) E
Hypogastrura omnigra (Salmon, 1941) E
Hypogastrura purpurescens (Lubbock, 1868) A
Hypogastrura rossi (Salmon, 1941) E
Hypogastrura viatica (Tullberg, 1872) A
Schoettella subcorta Salmon, 1941 E
Triacanthella alba Carpenter, 1909 E
Triacanthella enderbyensis Salmon, 1949 E
Triacanthella purpurea Salmon, 1943 E
Triacanthella rosea Wahlgren, 1906
Triacanthella rubra Salmon, 1941 E
Triacanthella setacea Salmon, 1941 E
Triacanthella sorenseni Salmon, 1949 E
Triacanthella terrasilvatica Salmon, 1943 E
Xenylla atrata (Salmon, 1944) E
Xenylla maritima Tullberg, 1869 A
ISOTOMIDAE
Acanthomurus alpinus alpinus Salmon, 1941 E
Acanthomurus a. obscuratus Salmon, 1943 E
Acanthomurus rivalis Wise, 1964 E
Acanthomurus setosus setosus Salmon, 1941 E
Acanthomurus s. violaceus Salmon, 1941 E
Acanthomurus womersleyi Salmon, 1941 E
Archisotoma brucei (Carpenter, 1907)
Ballistura aqualata (Salmon, 1941) E
Cryptopygus caecus Wahlgren, 1906
Cryptopygus campbellensis Wise, 1964 E
Cryptopygus decemoculatus (Salmon, 1949) E
Cryptopygus granulatus Salmon, 1943
Cryptopygus lamellatus Salmon, 1941 E
Cryptopygus loftyensis Womersley, 1934
Cryptopygus minimus Salmon, 1941 E
Cryptopygus novaezealandiae (Salmon, 1943) E
Cryptopygus novazealandia (Salmon, 1941) E
Cryptopygus parasiticus (Salmon, 1943) E
Cryptopygus subalpina (Salmon, 1944) E
Desoria fasciata (Salmon, 1941) E
Folsomia candida Willem, 1902 A
Folsomia diplophthalma (Axelson, 1902) A
Folsomia fimetarioides (Axelson, 1903) A
Folsomia miradentata Salmon, 1943 E
Folsomia pusilla Salmon, 1944 E
Folsomia quadrioculata (Tullberg, 1871) A
Folsomia salmoni Stach, 1947 E
Folsomia sedecimoculata (Salmon, 1943) E
Folsomides neozealandia Salmon, 1948 E
Folsomina onychiurina Denis, 1931
Folsomotoma anomala (Salmon, 1949) E
Folsomotoma minuta (Salmon, 1949) E
Folsomotoma octooculata (Willem, 1901) rec. dub.
Folsomotoma ovata (Salmon, 1949) E
Folsomotoma subflava (Salmon, 1949) E
Halisotoma maritima (Tullberg, 1871)
Halisotoma pritchardi (Womersley, 1936)
Halisotoma sindentata (Salmon, 1943) E
Hemisotoma thermophilus (Axelson, 1900) A
Isotoma exiguadentata Salmon, 1941 E
Isotoma pallida fasciata Salmon, 1941 E
Isotoma raffi Womersley, 1934
Isotoma turbotti (Salmon, 1949)
Isotomedia triseta Salmon, 1944 E
Isotomiella minor (Schäffer, 1896) A
Isotomodes productus (Axelson, 1906) A
Isotomurus chiltoni (Carpenter, 1925)
Isotomurus lineatus lineatus (Salmon, 1941) E
Isotomurus l. violaceus (Salmon, 1941) E SI?
Isotomurus novaezealandiae (Salmon, 1941) E
Isotomurus palustris (O.F. Müller, 1776) A
Isotomurus papillatus (Womersley, 1934)
Papillomurus magnificus Salmon, 1949 E SI
Papillomurus parvus (Salmon, 1937) E SI? (Stach placed this species *in Tomocerura*)
Parisotoma confusoculata Salmon, 1944 E
Parisotoma dividua Salmon, 1944 E
Parisotoma notabilis (Schäffer, 1896) A
Parisotoma picea Salmon, 1949 E
Parisotoma postantennala Salmon, 1949 E
Parisotoma quinquedentata Salmon, 1943 E
Procerura dissimilis (Salmon, 1944) E
Procerura fasciata Salmon, 1941
Procerura fusca fusca (Salmon, 1941) E
Procerura f. pallida (Salmon, 1941) E
Procerura montana Salmon, 1941 E
Procerura ochracea (Salmon, 1949) E
Procerura purpurea Salmon, 1941 E
Procerura serrata Salmon, 1941 E
Procerura violacea aequaoculata Salmon, 1941 E
Procerura v. violacea Salmon, 1941 E
Proisotoma atrata (Salmon, 1941) E
Proisotoma haweaensis (Salmon, 1941) E
Proisotoma minuta (Tullberg, 1871) A
Proisotoma niger (Carpenter, 1925) E
Proisotoma octojuga Salmon, 1949 E
Proisotoma okukensis (Salmon, 1941) E
Proisotoma terrigenus (Salmon, 1943) E
Proisotoma xanthella Salmon, 1949 E
Proisotomurus fuscus Salmon, 1944 E SI
Proisotomurus lineatus violaceus Salmon, 1944 E SI
Proisotomurus lapidosus Salmon, 1949 E SI
Setocerura maruiensis (Salmon, 1941) E
Setocerura rubenota (Salmon, 1941) E
Spinocerura capillata Salmon, 1941 E
Tibiolatra latronigra Salmon, 1941 E
Tomocerura colonavia Salmon, 1949 E
Womersleyella niveata Salmon, 1944 E
NEANURIDAE
ANURIDINAE
Anurida granaria (Nicolet, 1847) A
Delamarellina ubiquata (Salmon, 1944) E
Forsteramea megacephala (Salmon, 1954) E
Platanurida lata Carpenter, 1925 E
Platanurida marplesi (Salmon, 1941) E
Platanurida marplesioides Massoud, 1967 E
Pseudachorudina brunnea (Carpenter, 1925) E
Pseudachorudina osextara (Salmon, 1941) E
Pseudachorudina pacifica (Womersley, 1936) E
Quatacanthella proprieta (Salmon, 1941) E
FRIESIINAE
Friesea flava (Salmon, 1949) E
Friesea litoralis (Wise, 1964) E
Friesea mirabilis (Tullberg, 1871) A
Friesea parva (Womersley, 1936) E
Friesea salmoni Massoud, 1967 E
NEANURINAE
Australonura meridionalis (Stach, 1951)
Crossodonthina radiata (Salmon, 1941) E
Gnatholonche angularis (Salmon, 1944) E
Gnatholonche sensilla Salmon, 1948 E
Hemilobella newmani(Womersley, 1933)
Neanura muscorum (Templeton, 1835) A
Neanura rosacea (Schött, 1917) SI
Paleonura guadalcanarae (Yosii, 1960) var. *novaezelandiae* Salmon, 1941 E SI
Zelandanura bituberculata Deharveng & Wise, 1987 E
Zealandmeria harrisi (Salmon, 1944) E
Zealandmeria novaezealandiae (Womersley, 1936) E
PSEUDACHORUTINAE
Ceratrimeria aurea Salmon, 1944 E
Ceratrimeria harrisi Salmon, 1942 E
Ceratrimeria novaezealandiae (Womersley, 1936) E
Pseudachorutes algidensis Carpenter, 1925 E
Pseudachorutes conspicuatus conspicuatus Salmon, 1944 E
Pseudachorutes c. flavus Salmon, 1944 E
Pseudachorutes c. lineatus Salmon, 1944 E
Pseudachorutes c. maximus Salmon, 1944 E
Pseudachorutes decussus (Salmon 1941) E?
Pseudachorutes puniceus Salmon, 1944 E
UCHIDANURINAE
Holacanthella brevispinosa (Salmon, 1942) E
Holacanthella duospinosa (Salmon, 1942) E
Holacanthella laterospinosa (Salmon, 1944) E
Holacanthella paucispinosa (Salmon, 1941) E
Holacanthella spinosa (Lubbock, 1899) E
ODONTELLIDAE
Odontella emineodentata Salmon, 1944 E
Odontella forsteri (Salmon, 1942) E
Odontella anomala Salmon, 1944 E
Odontella caerulumbrosa Salmon, 1944 E
Odontella conspicuata Salmon, 1944 E
Odontella minutissima Salmon, 1941 E
ONYCHIURIDAE
Clavaphorura septemseta Salmon, 1943 E
Deuteraphorura acicindelis (Salmon, 1958) E
Onychiurus ambulans (Linnaeus, 1758) A
Onychiurus a. inermis Agren, 1903 A
Onychiurus fimetarius (Linnaeus, 1758) SI A

Orthonychiurus novaezealandiae (Salmon, 1942) E
Orthonychiurus subantarcticus Salmon, 1949 E
Protaphorura armata (Tullberg, 1869) A
Protaphorura a. inermis (Axelson, 1950) A
Thalassaphorura petallata Salmon, 1958 E
TOMOCERIDAE
Antennacyrtus insolitus Salmon, 1941 E
Lepidophorella australis australis Carpenter, 1925
Lepidophorella a. fusca Salmon, 1941 E
Lepidophorella brachycephala (Moniez, 1894)
Lepidophorella communis Salmon, 1937
Lepidophorella nigra Salmon, 1943
Lepidophorella rubicunda Salmon, 1941 E
Lepidophorella spadica Salmon, 1944 E
Lepidophorella unadentata Salmon, 1941 E
Novacerus insoliatus (Salmon, 1941) E
Novacerus spinosus (Salmon, 1941) E
Pseudolepidophorella longiterga (Salmon, 1937) E
Tomocerus minor (Lubbock, 1862) A
Tomocerus setoserratus Salmon, 1941
TULLBERGIIDAE
Dinaphorura laterospina Salmon, 1941 E
Dinaphorura novaezealandeae Womersley, 1935 E
Mesaphorura krausbaueri Börner, 1901 A ?misid.
Mesaphorura minutissima Salmon, 1944 E
Tullbergia gambiense Womersley, 1935
Tullbergia mixta Wahlgren, 1906
Tullbergia bisetosa Börner, 1902

Order SYMPHYPLEONA
ARRHOPALITIDAE
Arrhopalites caecus (Tullberg, 1871) *sensu* Stach 1956, Bretfeld 1999
BOURLETIELLIDAE
Bourletiella arvalis (Fitch, 1863) *sensu* Stach 1956 A
Bourletiella a. dorsobscura Salmon, 1941 E
Bourletiella hortensis (Fitch, 1963)
Deuterosminthurus pallipes (Bourlet, 1842) A?
Deuterosminthurus sulphureus sulphureus Koch, 1840
DICYRTOMIDAE
Calvatomina superba (Salmon, 1943) E
Dicyrtomina minuta (O. Fabricius, 1783) *sensu* Stach, 1957 A
Dicyrtomina novazealandica Salmon, 1941 E
Dicyrtomina turbotti Salmon, 1948 E
KATIANNIDAE
Katianna antennapartita Salmon, 1941 E
Katianna australis australis Womersley, 1932
Katianna a. tillyardi Womersley, 1932
Katianna gloriosa Salmon, 1946 E
Katianna perplexa Salmon, 1944 E
Katianna purpuravirida Salmon, 1941 E
Katianna ruberoculata reducta Salmon, 1944 E
Katianna r. ruberoculata Salmon, 1944 E
Parakatianna albirubrafrons albirubrafrons Salmon, 1943 E
Parakatianna a. niveanota Salmon, 1943 E
Parakatianna cortica Salmon, 1943 E
Parakatianna diversitata diversitata Salmon, 1943 E
Parakatianna d. viridis Salmon, 1943 E
Parakatianna hexagona Salmon, 1941 E
Parakatianna homerica (Salmon, 1946) E
Parakatianna prospina (Salmon, 1946) E
Parakatianna salmoni (Wise, 1964) E
Parakatianna superba (Salmon, 1946) E
Polykatianna cremea Salmon, 1949 E
Polykatianna davidi (Tillyard, 1920)
Polykatianna flammea Salmon, 1946 E
Pseudokatianna fasciata (Salmon, 1944) E
Pseudokatianna campbellensis Salmon, 1949 E
Pseudokatianna fagophila Salmon, 1946 E
Pseudokatianna livida (Salmon, 1943) E
Pseudokatianna lutea Salmon, 1946 E
Pseudokatianna minuta Salmon, 1946 E
Pseudokatianna nigra Salmon, 1946 E
Pseudokatianna nigraoculata Salmon, 1948 E SI
Pseudokatianna nigretalba aurea Salmon, 1944 E
Pseudokatianna n. nigretalba Salmon, 1944 E
Pseudokatianna niveovata nigra Salmon, 1946 E
Pseudokatianna n. niveovata Salmon, 1946 E
Pseudokatianna triclavata Salmon, 1949 E
Pseudokatianna triverrucata Salmon, 1944 E
Pseudokatianna umbrosalata Salmon, 1946 E
Pseudokatianna zebra Salmon, 1946 E
Sminthurinus aureus (Lubbock, 1862) A
Sminthurinus discordipes Salmon, 1949 E
Sminthurinus duplicatus duplicatus Salmon, 1941 E
Sminthurinus d. obscurus Salmon, 1944 E
Sminthurinus glaucus Salmon, 1943 E
Sminthurinus granulatus Salmon, 1946 E
Sminthurinus kerguelensis Salmon, 1964
Sminthurinus lichenatus Salmon, 1943 E
Sminthurinus muscophilus Salmon, 1946 E
Sminthurinus nigrafuscus Salmon, 1941 E
Sminthurinus oculatus Schött, 1917
Sminthurinus procerasetus Salmon, 1946 E
Sminthurinus mime Börner, 1907 A
Sminthurinus tunicatus Salmon, 1954 E
SMINTHURIDAE
Temeritas denisii (Womerseley, 1934)
Novokatianna cummyxa Salmon, 1944 E
Novokatianna radiata Salmon, 1946 E
Novokatianna venusta (Salmon, 1943) E
Sminthurus multidentatus Salmon, 1943 E
Sminthurus viridis (Linnaeus, 1758)
SMINTHURIDIDAE
Jeannenotia stachi (Jeannenot, 1955) A
Sphaeridia pumilis (Krausbauer, 1898) A
Sphaeridia serrata (Folsom & Mills, 1938)
Sphaeridia sphaera (Salmon, 1946) E
SPINOTHECIDAE
Spinotheca magnasetacea (Salmon, 1941) E

Order NEELIPLEONA
NEELIDAE
Megalothorax incertus Börner, 1903 A
Megalothorax rubidus Salmon, 1946 E
Zelandothorax novaezealandiae (Salmon, 1944) E

Class DIPLURA
[Compiled by R. P. Macfarlane]
Order DIPLURA
CAMPODEIDAE
Campodea fragilis Meinert, 1865 A?
Campodea zelanda Hilton, 1939 E
Campodea sp. indet. Silvestri 1931 ?A
Tricampa philpotti (Tillyard, 1924) E
HETEROJAPYGIDAE
Heterojapyx novaezeelandiae (Verhoeff, 1903) E
JAPYGIDAE
Burmjapyx forsteri archeyi (Pagés, 1952) E
Burmjapyx f. forsteri (Pagés, 1952) E
Burmjapyx michaelseni (Silvestri, 1930) A?
Burmjapyx punamuensis (Pagés, 1952) E
Burmjapyx n. sp. E
Notojapyx tillyardi (Silvestri, 1930) A?

Class INSECTA
Subclass ARCHAEOGNATHA
[Compiled by R. P. Macfarlane]
Order ARCHAEOGNATHA
MEINERTELLIDAE
Nesomachilis maorica Tillyard 1924 E Fo
Nesomachilis novaezelandiae Mendes, Bach de Rocha & Gaju-Ricart, 1994 E

Subclass DICONDYLIA
Infraclass THYSANURA
[Compiled by R. P. Macfarlane]
Order THYSANURA
ATELURIDAE [P. M. Johns]
Atopatelura sp.
Gen. nov. et n. spp. (3) Wise 1970 3E
LEPIDOTRICHIDAE
Gen. nov. et n. spp. (4) Wise 1970 4E
LEPISMATIDAE
Ctenolepisma longicaudata Escherich, 1905 A
Heterolepisma zelandica (Tillyard, 1924) E
Lepisma saccharina Linnaeus, 1758 A St

Infraclass PTERYGOTA
Superorder EPHEMEROPTERA
[Compiled by D. R. Towns & T. R. Hitchings]
Order EPHEMEROPTERA
AMELETOPSIDAE
Ameletopsis perscitus (Eaton, 1899) E
COLOBURISCIDAE
Coloburiscus humeralis (Walker, 1853) E
EPHEMERIDAE
Ichthybotus bicolor Tillyard, 1923 E
Ichthybotus hudsoni (McLachlan, 1894) E
LEPTOPHLEBIIDAE
Acanthophlebia cruentata (Hudson, 1904) E
Arachnocolus phillipsi Towns & Peters, 1979 E
Atalophlebioides cromwelli (Phillips, 1930) E
Aupouriella pohei Winterbourn, 2009 E
Austroclima jollyae Towns & Peters, 1979 E
Austroclima sepia (Phillips, 1930) E
Austronella planulata (Towns, 1983) E
Cryophlebia aucklandensis (Peters, 1971) E
Deleatidium (Deleatidium) angustum Towns & Peters, 1996 E
Deleatidium (*D.*) *atricolor* Hitchings, 2009 E
Deleatidium (*D.*) *autumnale* Phillips, 1930 E
Deleatidium (*D.*) *branchiola* Hitchings, 2009 E
Deleatidium (*D.*) *cerinum* Phillips, 1930 E
Deleatidium (*D.*) *fumosum* Phillips, 1930 E
Deleatidium (*D.*) *lillii* Eaton, 1899 E
Deleatidium (*D.*) *magnum* Towns & Peters, 1996 E
Deleatidium (*D.*) *myzobranchia* Phillips, 1930 E
Deleatidium (*D.*) *townsi* Hitchings, 2009 E
Deleatidium (*D.*) *vernale* Phillips, 1930 E
Deleatidium (*D.*) n. sp. E
Deleatidium (*Penniketellum*) *insolitum* (Towns & Peters, 1979b) E
Deleatidium (*P.*) *cornutum* Towns & Peters, 1996 E
Isothraulus abditus Towns & Peters, 1979 E
Mauiulus aquilus Towns & Peters, 1996 E
Mauiulus luma Towns & Peters, 1979 E
Neozephlebia scita (Walker, 1853) E
Tepakia caligata Towns & Peters, 1996 E
Zephlebia borealis (Phillips, 1930) E
Zephlebia dentata (Eaton, 1871) E
Zephlebia inconspicua Towns, 1983 E
Zephlebia nebulosa Towns & Peters, 1996 E
Zephlebia pirongia Towns & Peters, 1996 E
Zephlebia spectabilis Towns, 1983 E
Zephlebia tuberculata Towns & Peters, 1996 E
Zephlebia versicolor (Eaton, 1899) E
NESAMELETIDAE
Nesameletus austrinus Hitchings & Staniczek, 2003 E
Nesameletus flavitinctus (Tillyard, 1923) E
Nesameletus murihiku Hitchings & Staniczek, 2003 E
Nesameletus ornatus (Eaton, 1882) E
Nesameletus vulcanus Hitchings & Staniczek, 2003 E
Nesameletus n. sp. (1) E
ONISCIGASTRIDAE
Oniscigaster distans Eaton, 1899 E
Oniscigaster intermedius Eaton, 1899 E
Oniscigaster wakefieldi McLachlan, 1873 E
RALLIDENTIDAE E
Rallidens mcfarlanei Penniket, 1966 E
Rallidens n. sp. E
SIPHLAENIGMATIDAE E
Siphlaenigma janae Penniket, 1962 E

Superorder ODONATA
[Compiled by R. P. Macfarlane]
Order ODONATA
Suborder ZYGOPTERA
COENAGRIONIDAE
Ischnura aurora (Brauer, 1865) A Fs Ci
Xanthocnemis sobrina (McLachlan, 1873) E Fs Ci
Xanthocnemis sinclairi Rowe, 1987 E Fs Ci
Xanthocnemis tuanuii Rowe, 1981 E Fs Ci
Xanthocnemis zealandica (McLachlan, 1873) E Fsr Ci
LESTIDAE
Austrolestes colensonis (White, 1846) E Fs Ci

Suborder ANISOPTERA
AESHNIDAE
Aeshna brevistyla (Rambur, 1842) Fsr Ci
Hemianax papuensis (Burmeister, 1839) A Fs Ci
CORDULIIDAE
Antipodochlora braueri (Selys, 1871) E Fr Ci
Hemicordulia australiae Rambur, 1842 A Fsr Ci
Procordulia grayi (Selys, 1871) E Fs Ci
Procordulia smithii (White, 1846) E Fsr Ci
LIBELLULIDAE
Diplacodes bipunctata (Brauer, 1865) Fs Ci
Pantala flavescens (Fabricius, 1798) V Fs Ci
Tramea transmarina Braeur, 1867 V Fs Ci
PETALURIDAE
Uropetala carovei (White, 1843) E We Ci
Uropetala chiltoni Tillyard, 1921 E We Ci

Superorder NEOPTERA
Order PLECOPTERA
[Compiled by I. D. McLellan]
Suborder ANTARCTOPERLARIA
AUSTROPERLIDAE
Austroperla cyrene (Newman, 1845) E
EUSTHENIIDAE
Stenoperla helsoni McLellan, 1996 E
Stenoperla hendersoni McLellan, 1996 E
Stenoperla maclellani Zwick, 1979 E
Stenoperla prasina (Newman, 1845) E
GRIPOPTERYGIDAE
Acroperla flavescens (Kimmins, 1938) E
Acroperla christinae McLellan 1998 E
Acroperla spiniger (Tillyard, 1923) E
Acroperla samueli McLellan, 1977 E
Acroperla trivacuata (Tillyard, 1923) E
Apteryoperla monticola Wisely, 1953 E
Apteryoperla illiesi McLellan, 1977 E
Apteryoperla nancyae McLellan, 1977 E
Apteryoperla ramsayi McLellan, 1977 E
Apteryoperla tillyardi McLellan, 1977 E
Apteryoperla n. spp. (2) 2E
Aucklandobius complementarius Enderlein, 1909 E
Aucklandobius gressitti Illies, 1963 E
Aucklandobius kuscheli (Illies, 1974) E
Aucklandobius turbotti (Illies, 1963) E
Holcoperla angularis (Wisely, 1953) E
Holcoperla jacksoni McLellan, 1977 E
Holcoperla magna McLellan, 1983 E
Megaleptoperla diminuta Kimmins, 1938 E
Megaleptoperla grandis (Hudson, 1913) E
Rungaperla campbelli (Illies, 1963) E
Rungaperla longicauda (Illies, 1963) E
Nesoperla fulvescens Tillyard, 1923 E
Nesoperla johnsi McLellan, 1977 E Sn
Nesoperla n. sp. 1 E
Rakiuraperla nudipes McLellan, 1977 E
Taraperla ancilis (Harding & Chadderton, 1995) E
Taraperla howesi (Tillyard, 1923) E
Taraperla pseudocyrene McLellan, 1998 E
Taraperla n. sp. 1 E
Vesicaperla dugdalei McLellan, 1977 E
Vesicaperla eylesi McLellan, 1977 E
Vesicaperla kuscheli McLellan, 1977 E
Vesicaperla substirpes McLellan, 1967 E
Vesicaperla townsendi McLellan, 1977 E
Vesicaperla n. spp. (2) 2E
Zelandobius alatus McLellan, 1993 E
Zelandobius albofasciatus McLellan, 1993 E
Zelandobius auratus McLellan, 1993 E
Zelandobius brevicauda McLellan, 1977 E
Zelandobius childi McLellan, 1993 E
Zelandobius confusus Tillyard, 1923 E
Zelandobius cordatus McLellan, 1993 E
Zelandobius dugdalei McLellan, 1993 E
Zelandobius edensis Gray, 2009 E
Zelandobius foxi McLellan, 1993 E
Zelandobius furcillatus Tillyard, 1923 E
Zelandobius gibbsi McLellan, 1993 E
Zelandobius illiesi McLellan 1969 E
Zelandobius inversus McLellan, 1993 E
Zelandobius jacksoni McLellan, 1993 E
Zelandobius kuscheli McLellan, 1993 E
Zelandobius macburneyi McLellan, 1993 E
Zelandobius mariae McLellan, 1993 E
Zelandobius montanus McLellan, 1993 E
Zelandobius ngaire McLellan, 1993 E
Zelandobius patricki McLellan, 1993 E
Zelandobius peglegensis McLellan, 1993 E
Zelandobius pilosus Death, 1990 E
Zelandobius takahe McLellan, 1993 E
Zelandobius truncus McLellan, 1993 E
Zelandobius unicolor Tillyard, 1923 E
Zelandobius uniramus McLellan, 1993 E
Zelandobius wardi McLellan, 1993 E
Zelandobius n. spp. (11) 11E
Zelandoperla agnetis McLellan, 1967 E
Zelandoperla decorata Tillyard, 1923 E
Zelandoperla denticulata McLellan, 1967 E
Zelandoperla fenestrata Tillyard, 1923 E
Zelandoperla pennulata McLellan, 1967 E
Zelandoperla tillyardi McLellan, 1999 E
Zelandoperla n. sp. E

Suborder ARCTOPERLARIA
NOTONEMOURIDAE
Cristaperla eylesi McLellan, 1991 E
Cristaperla fimbria (Winterbourn, 1965) E
Cristaperla waharoa McLellan, 1991 E
Halticoperla gibbsi McLellan, 1991 E
Halticoperla tara McLellan, 1991 E
Halticoperla viridans McLellan & Winterbourn, 1968 E
Notonemoura alisteri McLellan, 1968 E
Notonemoura hendersoni McLellan, 2000 E
Notonemoura latipennis Tillyard, 1923 E
Notonemoura spinosa McLellan, 1991 E
Notonemoura winstanleyi McLellan, 1991 E
Omanuperla bruningi McLellan, 1972 E
Omanuperla hollowayae McLellan, 1991 E
Spaniocerca acuta McLellan, 1991 E
Spaniocerca bicornuta McLellan, 1987 E
Spaniocerca hamishi McLellan, 2000 E
Spaniocerca longicauda McLellan, 1977 E
Spaniocerca minor Kimmins, 1938 E
Spaniocerca zelandica Tillyard, 1923 E
Spaniocerca zwicki McLellan, 1991 E
Spaniocerca n. spp. (2) 2E
Spaniocercoides cowleyi (Winterbourn, 1965) E
Spaniocercoides foxi McLellan, 1984 E
Spaniocercoides howesi McLellan, 1984 E
Spaniocercoides hudsoni Kimmins, 1938 E
Spaniocercoides jacksoni McLellan, 1991 E
Spaniocercoides philpotti Winterbourn, 1965 E
Spaniocercoides townsendi McLellan, 1984 E
Spaniocercoides watti McLellan, 1984 E
Spaniocercoides n. sp. E

Order BLATTODEA
[Compiled by P. M. Johns]
BLABERIDAE A
Panesthia cribrata Saussure, 1874* A
BLATTELLIDAE
Blatella germanica Linnaeus, 1767 A
Neotemnopteryx fulva (Saussure, 1863) A
Ornatiblatta maori (Rehn, 1904) E
Parellipsidion conjunctum (Walker, 1868) E
Parellipsidion inaculeatum Johns, 1966 E
Parellipsidion pachycercum Johns, 1966 E
Paratemnopteryx couloniana (Saussure, 1863) A
Shelfordina orchidae (Asahina, 1985) A
Supella longipalpa (Fabricius, 1798) A
Gen. nov. et n. spp. (2)* 2E
BLATTIDAE
Blatta orientalis Linnaeus, 1758 A
Celatoblatta anisoptera Johns, 1966 E
Celatoblatta brunni (Alfken, 1901) E
Celatoblatta fuscipes Johns, 1966 E
Celatoblatta hesperia Johns, 1966 E
Celatoblatta laevispinata Johns, 1966 E
Celatoblatta montana Johns, 1966 E
Celatoblatta notialis Johns, 1966 E
Celatoblatta pallidicauda Johns, 1966 E
Celatoblatta peninsularis Johns, 1966 E
Celatoblatta quinquemaculata Johns, 1966 E
Celatoblatta sedilloti (Bolivar, 1882) E
Celatoblatta subcorticaria Johns, 1966 E
Celatoblatta undulivitta (Walker, 1868) E
Celatoblatta vulgaris Johns, 1966 E
Celatoblatta n. spp. (6)* 6E
Drymaplaneta semivitta (Walker, 1868) A
Drymaplaneta heydeniana (Saussure, 1864) A
Maoriblatta novaeseelandiae (Brunner von Wattenwyl, 1865) E
Maoriblatta rufoterminata (Brunner von Wattenwyl, 1865) E
Periplaneta americana (Linnaeus, 1758) A
Periplaneta brunnea Burmeister, 1868 A
Periplaneta fuliginosa (Serville, 1839) A
Tryonicus parvus (Tepper, 1895) A
Gen et. sp. indet. Tryonicini* E
CHORISONEURIDAE
Celeriblattina major Johns, 1966 E
Celeriblattina minor Johns, 1966 E

Order ISOPTERA
[Compiled by R. P. Macfarlane]
KALOTERMITIDAE
Kalotermes banksiae Hill, 1942 A
Kalotermes brouni Froggatt, 1897 E Fo
Kalotermes cognatus Gay, 1976
RHINOTERMITIDAE A
Coptotermes acinaciformis (Froggatt, 1898) A
Coptotermes frenchi Hill, 1932 A
TERMITIDAE A
Glyptotermes brevicornis Froggatt, 1896 A
TERMOPSIDAE
Stolotermes ruficeps Brauer, 1865 E Fo
Stolotermes inopus Gay, 1967 E Fo
Porotermes adamsoni (Froggatt, 1897) A (doubtful)

Order MANTODEA
[Compiled by R. P. Macfarlane]
MANTIDAE
Miomantis caffra (Saussure, 1871) A Ga
Orthodera novaezealandiae (Colenso, 1882) E Ga Fo Ci

Order DERMAPTERA
[Compiled by R. P. Macfarlane & P. M. Johns]
ANISOLABIDIDAE
Anisolabis kaspar Hudson, 1975 E
Anisolabis littorea (White, 1846) E Be
Brachylabis manawatawhi Giles, 1958 E
Carcinophora occidentalis Kirby, 1896 A?
Euborellia annulipes (Lucas, 1847) A

Parisolabis boulderensis Hudson, 1975 E
Parisolabis forsteri Hudson, 1975 E
Parisolabis iti Hudson, 1975 E
Parisolabis johnsi Hudson, 1975 E
Parisolabis nelsonensis Hudson, 1975 E
Parisolabis novaezeelandiae Verhoeff, 1904 E Fo
Parisolabis setosa Hudson, 1975 E
Parisolabis tapanuiensis Hudson, 1975 E
Parisolabis n. sp. E
CHELISOCHIDAE A?
Chelisoches morio (Fabricius, 1775) A?
Hamaxas feae (Bormans, 1894) A?
FORFICULIDAE A
Forficula auricularia (Linnaeus, 1758) A Ga Gr Or
LABIDURIDAE A
Labidura truncata Kirby, 1903 A
SPONGIPHORIDAE
Labia minor (Linnaeus, 1758) A
Nesogaster halli (Hincks, 1949) A Ga
Paralabella kermadecensis (Giles, 1973) ?E
Paraspania brunneri (Bormans, 1883) A

Order ORTHOPTERA
[Compiled by P. M. Johns]
Suborder ENSIFERA
ANOSTOSTOMATIDAE
Anisoura nicobarica Ander, 1938 E
Deinacrida carinata Salmon, 1950 E
Deinacrida connectens (Ander, 1939) E
Deinacrida elegans Gibbs, 1999 E
Deinacrida fallai Salmon, 1950 E
Deinacrida heteracantha White, 1842 E
Deinacrida mahoenui Gibbs, 1999 E
Deinacrida pluvialis Gibbs, 1999 E
Deinacrida rugosa Buller, 1871 E
Deinacrida talpa Gibbs, 1999 E
Deinacrida tibiospina (Salmon, 1950) E
Hemiandrus bilobatus Ander, 1938 E
Hemiandrus fiordensis (Salmon, 1950) E
Hemiandrus focalis (Hutton, 1897) E
Hemiandrus lanceolatus (Walker, 1869) E
Hemiandrus maculifrons (Walker, 1869) E
Hemiandrus nitaweta Jewell, 2007 E
Hemiandrus pallitarsis (Walker, 1869) E
Hemiandrus subantarcticus (Salmon, 1950) E
Hemiandrus superbus Jewell, 2007 E
Hemiandrus n. spp. (>33) >33E
Hemideina broughi (Buller, 1896) E
Hemideina crassidens (Blanchard, 1851) E
Hemideina femorata Hutton, 1898 E
Hemideina maori (Pictet & Saussure, 1891) E
Hemideina ricta Hutton, 1898 E
Hemideina thoracica (White, 1842) E
Hemideina trewicki Morgan Richards, 1995 E
Motuweta isolata Johns, 1997 E
Motuweta riparia Gibbs, 2002 E
GRYLLIDAE
Bobilla bigelowi (Swan, 1972) E
Bobilla bivitatta (Walker, 1869) A
Bobilla nigrova (Swan, 1972) E
Metioche maorica (Walker, 1869) E
Modicogryllus lepidus (Walker, 1869) A
Ornebius aperta Otte & Alexander, 1983 A
Ornebius novarae (Saussure, 1877)
Teleogryllus commodus (Walker, 1869) A?
Gen. nov. et. n. sp. E
GRYLLOTALPIDAE
Trimescaptor aotea Tindale, 1928 E
RHAPHIDOPHORIDAE
Dendroplectron aucklandense Richards, 1964 E
Insulanoplectron spinosum Richards, 1970 E
Ischyroplectron isolatum Hutton, 1897 E
Isoplectron aciculatum Karny, 1937 E
Isoplectron armatum Hutton, 1897 E
Isoplectron calcaratum Hutton, 1897 E
Isoplectron cochleatum Karny, 1935 E
Isoplectron n. spp. (3) 3E
Macropathus filifer Walker, 1869 E
Macropathus huttoni Kirby, 1906 E
Neonetus huttoni Chopard, 1923 E
Neonetus pilosus (Hutton, 1904) E
Neonetus poduroides (Walker, 1869) E
Neonetus variegatus Brunner, 1888 E
Neonetus n. spp. (9) 9E
Novoplectron serratum Hutton, 1897 E
Notoplectron campbellense Richards, 1964 E
Pachyrhamma acanthocerum (Milligan, 1926) E
Pachrhamma altum (Walker, 1869) E
Pachyrhamma delli (Richards, 1954) E
Pachyrhamma edwardsii (Scudder, 1869) E
Pachyrhamma fuscum (Richards, 1959) E
Pachyrhamma giganteum Richards, 1962 E
Pachyrhamma longicaudum (Richards, 1959) E
Pachyrhamma longipes (Colenso, 1887) E
Pachyrhamma ngongotahaense Richards, 1961 E
Pachyrhamma spinosum Richards, 1961 E
Pachyrhamma tuarti Richards, 1961 E
Pachyrhamma uncatum (Richards, 1959) E
Pachyrhamma waipuense (Richards, 1960) E
Pachyrhamma waitomoense (Richards, 1958) E
Pachyrhamma n. spp. (>11) 11E
Pharmacus brewsterensis Richards, 1972 E
Pharmacus chapmanae Richards, 1972 E
Pharmacus dumbletoni Richards, 1972 E
Pharmacus montanus Pictet & Saussure, 1891 E
Pharmacus? n. spp. (3) 3E
Pallidoplectron peniculosum Richards, 1960 E
Pallidoplectron subterraneum Richards, 1965 E
Pallidoplectron turneri Richards, 1958 E
Pleioplectron diversum Hutton, 1897 E
Pleioplectron hudsoni Hutton, 1897 E
Pleioplectron simplex Hutton, 1897 E
Pleioplectron n. spp. (3) 3E
Paraneonetus multispinus Salmon, 1948 E
Petrotettix cupolensis Richards, 1972 E
Petrotettix nigripes Richards, 1972 E
Petrotettix serratus Richards, 1972 E
Petrotettix spinosus Richards, 1972 E
Setascutum ohauense Richards, 1972 E
Setascutum pallidum Richards, 1972 E
Turbottoplectron cavernae (Hutton, 1897) E
Turbottoplectron unicolor Salmon, 1948 E
Talitropsis crassicruris Hutton, 1897 E
Talitropsis irregularis Hutton, 1897 E
Talitropsis sedilloti Bolivar, 1883 E
Talitropsis megatibia Trewick, 1999 E
Talitropsis n. sp. E
Weta thomsoni Chopard, 1923 E
'Weta' chopardi Karny, 1937 E
TETTIGONIIDAE
Caedicia simplex (Walker, 1869) E
Conocephalus albescens (Walker, 1869) A
Conocephalus bilineatus (Erichson, 1842) A?
Conocephalus semivitatus (Walker, 1869) E
Salomona solida (Walker, 1869) A?

Suborder CAELIFERA
ACRIDIDAE
Alpinacris crassicauda Bigelow, 1967 E
Alpinacris tumidicauda Bigelow, 1967 E
Brachaspis collinus (Hutton, 1897) E
Brachaspis nivalis (Hutton, 1897) E
Brachaspis robustus Bigelow, 1967 E
Locusta migratoria (Linnaeus, 1758)
Paprides dugdali Bigelow, 1967 E
Paprides nitidus Hutton, 1898 E
Phaulacridium marginale (Walker, 1870) E
Phaulacridium otagense Ritchie & Westerman, 1984 E
Phaulacridium n. spp. (3) 3E
Sigaus australis (Hutton, 1898) E
Sigaus campestris (Hutton, 1898) E
Sigaus childi Jamieson, 1999 E
Sigaus homerensis Morris, 2003 E
Sigaus minutus Bigelow, 1967 E
Sigaus obelisci Bigelow, 1967 E
Sigaus piliferus Hutton, 1898 E
Sigaus takahe Morris, 2003 E
Sigaus villosus (Salmon, 1950) E

Order PHASMATODEA
[Compiled by R. P. Macfarlane and T. R. Buckley]
PHASMIDAE
Acanthoxyla fasciata (Hutton, 1899) E
Acanthoxyla geisovii (Kaup, 1866) E
Acanthoxyla huttoni Salmon, 1955, E
Acanthoxyla inermis Salmon, 1955 E
Acanthoxyla intermedia Salmon, 1955 E
Acanthoxyla prasina (Westwood, 1859) E Fo
Acanthoxyla speciosa Salmon, 1955 E
Acanthoxyla suteri (Hutton, 1899) E
Argosarchus horridus (White, 1846) E Fo
Asteliaphasma jucundum (Salmon, 1991) E Fo
Asteliaphasma naomi (Salmon, 1991) E Fo
Asteliaphasma n. sp. E
Clitarchus hookeri (White, 1846) E
Clitarchus tuberculatus Salmon, 1991 E
Clitarchus n. spp. (2) 2E
Micrarchus hystriculeus Westwood, 1859 E Sh
Micrarchus n. spp. (3) 3E
Niveaphasma annulatum (Hutton, 1898) E
Pseudoclitarchus sentus (Salmon, 1948) E
Spinotectarchus acornutus (Hutton, 1899) E
Tectarchus huttoni (Brunner, 1907) E
Tectarchus ovobessus Salmon, 1954 E
Tectarchus salebrosus (Hutton, 1899) E Fo/Sh
Tectarchus semilobatus Salmon, 1954 E
Tectarchus n. sp. E
Tepakiphasma ngatikuri Buckley & Bradler, 2010 E

Order HEMIPTERA
Suborder STERNORRHYNCHA
[Compiled by R. Henderson, D. A. J. Teulon, P. J. Dale, V. F. Eastop & M. A. W. Stufkens]
The host genus/genera are listed after each entry for Psylloidea. † Belongs to a new genus; genus name to be changed.
PSYLLOIDEA
PSYLLIDAE
Acizzia acaciae (Maskell, 1894) A Acacia
Acizzia acaciaebaileyanae (Froggatt, 1901) A Acacia
Acizzia albizziae (Ferris & Klyver, 1932) A Acacia
Acizzia conspicua (Tuthill, 1952) A Acacia
Acizzia dodonaeae (Tuthill, 1952) E Dodonaea
Acizzia exquisita (Tuthill, 1952) A Acacia
Acizzia hakeae (Tuthill, 1952) A Hakea, Grevillea
Acizzia jucunda (Tuthill, 1952) A Acacia
Acizzia uncatoides (Ferris & Klyver, 1932) A Acacia, Albizzia, Paraserianthes
Acizzia n. sp. A Acacia
Anoecoconeossa communis Taylor, 1987 A Eucalyptus
Anomalopsylla insignita Tuthill, 1952 E Olearia
Anomalopsylla n. spp. (2) 2E Olearia
Arytainilla spartiophila (Foerster, 1848) A Cytisus
Atmetocranium myersi (Ferris & Klyver, 1932) E Weinmannia
Baeopelma foersteri (Flor, 1861) A Alnus
Blastopsylla occidentalis Taylor, 1985 A Eucalyptus
Cardiaspina fiscella Taylor, 1962 A Eucalyptus
Creiis liturata Froggatt, 1990 A Eucalyptus
Cryptoneossa triangula Taylor, 1990 A Eucalyptus
Ctenarytaina clavata Ferris & Klyver, 1932 E Kunzea, Leptospermum
Ctenarytaina eucalypti (Maskell, 1890) A Eucalyptus
Ctenarytaina fuchsiae (Maskell, 1890) E Fuchsia
Ctenarytaina longicauda Taylor, 1987 A Lophostemon

Ctenarytaina pollicaris Ferris & Klyver, 1932 E *Kunzea*
Ctenarytaina spatulata Taylor, 1997 A *Eucalyptus*
Ctenarytaina thysanura Ferris & Klyver, 1932 A *Boronia*
Ctenarytaina n. sp. A *Acmena*
Ctenarytaina n. spp. (2) 2E *Leptospermum*
Eucalyptolyma maideni Froggatt, 1901 A *Eucalyptus*
Glycaspis granulata (Froggatt, 1901) A *Eucalyptus*
†*Gyropsylla zealandica* (Ferris & Klyver, 1932) E host unknown
†*Psylla apicalis* Ferris & Klyver, 1932 E *Sophora*
†*Psylla carmichaeliae carmichaeliae* Tuthill, 1952 E *Carmichaelia*
†*Psylla c. indistincta* Tuthill, 1952 E *Carmichaelia*
Psyllopsis fraxini (Linnaeus, 1758) A *Fraxinus*
Psyllopsis fraxinicola (Foerster, 1848) A *Fraxinus*
CALOPHYIDAE A
Calophya schini Tuthill, 1959 A *Schinus*
HOMOTOMIDAE A
Mycopsylla fici (Tryon, 1895) A *Ficus*
TRIOZIDAE
Bactericera cockerelli (Sulc, 1909) A *Capsicum*, *Solanum* (incl. *Lycopersicon*)
Trioza acuta (Ferris & Klyver, 1932) E *Ozothamnus*
Trioza adventicia Tuthill, 1952 A *Syzygium*
Trioza alseuosmiae Tuthill, 1952 E *Alseuosmia*
Trioza australis Tuthill, 1952 E *Brachyglottis*
Trioza bifida (Ferris & Klyver, 1932) E *Olearia*
Trioza colorata (Ferris & Klyver, 1932) E *Halocarpus*
Trioza compressa Tuthill, 1952 E *Olearia*
Trioza crinita Tuthill, 1952 E *Olearia*
Trioza curta (Ferris & Klyver, 1932) E *Metrosideros*
Trioza dacrydii Tuthill, 1952 E *Halocarpus*
Trioza decurvata Tuthill, 1952 E *Dracophyllum*
Trioza dentiforceps Dumbleton, 1967 E *Olearia*
Trioza discariae Tuthill, 1952 E *Discaria*
Trioza doryphora (Maskell, 1880) E *Olearia*
Trioza emarginata (Ferris & Klyver, 1932) E *Coprosma*
Trioza equalis (Ferris & Klyver, 1932) E *Aristotelia*
Trioza falcata (Ferris & Klyver, 1932) E *Aristotelia*
Trioza fasciata (Ferris & Klyver, 1932) E *Muehlenbeckia*
Trioza flavida Tuthill, 1952 E *Olearia*
Trioza gourlayi Tuthill, 1952 E *Olearia*
Trioza hebicola Tuthill, 1952 E *Hebe*
Trioza irregularis (Ferris & Klyver, 1932) E *Pseudopanax*
Trioza latiforceps Tuthill, 1952 E *Olearia*
Trioza obfusca (Ferris & Klyver, 1932) E *Hebe*
Trioza obscura Tuthill, 1952 E *Hebe*
Trioza panacis Maskell, 1890 E *Pseudopanax*
Trioza parvipennis Tuthill, 1952 E *Brachyglottis*
Trioza schefflericola Tuthill, 1952 E *Schefflera*
Trioza scobina Tuthill, 1952 E *Olearia*
Trioza styligera (Ferris & Klyver, 1932) E *Brachyglottis*
Trioza subacuta (Ferris & Klyver, 1932) E *Brachglottis*
Trioza subvexa Tuthill, 1952 E *Olearia*
Trioza vitreoradiata (Maskell, 1879) E *Pittosporum*
Trioza n. spp. (18) 18E
Gen. et spp. indet. (2) 2A *Casuarina*
ALEYRODOIDEA
ALEYRODIDAE
Bemisia sp. N. Martin A?
Dumbletoniella eucalypti (Dumbleton, 1956) A
Aleyrodes fodiens (Maskell, 1896) E
Aleyrodes proletella (Linnaeus, 1758) A
Aleyrodes winterae Takahashi, 1937 E
Asterochiton aureus Maskell, 1879 E
Asterochiton cerata (Maskell 1896) E
Asterochiton fagi (Maskell, 1890) E
Asterochiton pittospori Dumbleton, 1956 E
Asterochiton simplex (Maskell, 1890) E
Bemisia tabaci (Gennadius, 1889) A
Orchamoplatus citri (Takahashi, 1940) A
Pealius azaleae (Baker & Moles, 1920) A
Siphoninus phillyreae (Haliday, 1835) A
Trialeurodes asplenii (Maskell, 1890) E
Trialeurodes vaporariorum (Westwood, 1856) A
Trialeurodes sp. N. Martin E
APHIDOIDEA
ADELGIDAE A
Adelges nordmannianae (Eckstein, 1890) A
Pineus boerneri Annand, 1928 A
Pineus pini (Macquart, 1819) A
APHIDIDAE
Aphidinae
Macrosiphini A
Acyrthosiphon kondoi Shinji, 1938 A
Acyrthosiphon malvae (Mosley, 1841) A
Acyrthosiphon pisum (Harris, 1776) A
Acyrthosiphon primulae (Theobald, 1913) A
Akkaia ?taiwana Takahashi, 1933 A
Amphorophora rubi (Kaltenbach, 1843) A
Aulacorthum solani (Kaltenbach, 1843) A
Brachycaudus helichrysi (Kaltenbach, 1843) A
Brachycaudus persicae (Passerini, 1860) A
Brachycaudus rumexicolens (Patch, 1917) A
Brevicoryne brassicae (Linnaeus, 1758) A
Capitophorus elaeagni (Del Guercio, 1894) A
Capitophorus hippophaes javanicus (Hille Ris Lambers, 1953) A
Cavariella aegopodii (Scopoli, 1763) A
Chaetosiphon fragaefolii (T.D.A. Cockerell, 1901) A
Chaetosiphon sp. nr *fragaefolii* (T.D.A. Cockerell, 1901) A
Chaetosiphon tetrarhodum (Walker, 1849) A
Coloradoa rufomaculata (Wilson, 1908) A
Dysaphis apiifolia (Theobald, 1923) A
Dysaphis aucupariae (Buckton, 1879) A
Dysaphis foeniculus (Theobald, 1923) A
Dysaphis tulipae (Boyer de Fonscolombe, 1841) A
Elatobium abietinum (Walker, 1849) A
Hyadaphis passerinii (Del Guercio, 1911) A
Hyperomyzus lactucae (Linnaeus, 1758) A
Idiopterus nephrelepidis Davis, 1909 A
Illinoia azaleae (Mason, 1925) A
Jacksonia papillata Theobald, 1923 A
Liosomaphis berberidis (Kaltenbach, 1843) A
Lipaphis pseudobrassicae (Davis, 1914) A
Macrosiphoniella sanborni (Gillette, 1908) A
Macrosiphum euphorbiae (Thomas, 1878) A
Macrosiphum hellebori (Theobald & Walton, 1923) A
Macrosiphum rosae (Linnaeus, 1758) A
Macrosiphum stellariae Theobald, 1913 A
Metopolophium dirhodum (Walker, 1849) A
Metopolophium festucae (Theobald, 1917) A
Micromyzus nr *katoi* (Takahashi, 1925) A
Myzaphis rosarum (Kaltenbach, 1843) A
Myzus ascalonicus Doncaster, 1946 A
Myzus cerasi (Fabricius, 1775) A
Myzus cymbalariae Stroyan, 1954 A
Myzus hemerocallis Takahashi, 1921 A
Myzus ornatus Laing, 1932 A
Myzus persicae (Sulzer, 1776) A
Nasanovia ribisnigri (Mosley, 1841) A
Neomyzus circumflexus (Buckton, 1876) A
Neotoxoptera formosana Takahashi, 1965 A
Neotoxoptera oliveri (Essig, 1935) A
Neotoxoptera violae (Pergande, 1900) A
Ovatus crataegarius (Walker, 1850) A
Pseudacaudella rubida (Börner, 1939) A
Rhopalosiphoninus latysiphon (Davidson, 1912) A
Rhopalosiphoninus staphyleae (Koch, 1854) A
Sitobion nr *fragariae* (Walker, 1848) A
Sitobion miscanthi (Takahashi, 1921) A
Uroleucon sonchi (Linnaeus, 1767) A
Aphidini
Aphis coprosmae Laing *in* Tillyard, 1926 E
Aphis cottieri Carver, 1999 E
Aphis craccivora Koch, 1854 A
Aphis nr *epilobii* Kaltenbach, 1843 A
Aphis gossypii Glover, 1877 A
Aphis healyi Cottier, 1953 E
Aphis hederae Kaltenbach, 1843 A
Aphis idaei Van der Groot, 1912 A
Aphis nelsonensis Cottier, 1953 E
Aphis nerii Boyer de Fonscolombe, 1841 A
Aphis sedi Kaltenbach, 1843 A
Aphis spiraecola Patch, 1914 A
Aphis sp. (ex *Olearia*) E
Casimira n. sp. [ex *Ozothamnus*] E?
Euschizaphis n. spp. (1-2?) [ex *Dracophyllum*] 2E
Euschizaphis n. sp. (ex *Aciphylla*) E
Paradoxaphis aristoteliae Sunde, 1988 E
Paradoxaphis plagianthi Eastop, 2001 E
Rhopalosiphum insertum (Walker, 1849) A
Rhopalosiphum maidis (Fitch, 1856) A
Rhopalosiphum nymphaeae (Linnaeus, 1761) A
Rhopalosiphum padi (Linnaeus, 1758) A
Rhopalosiphum rufiabdominale (Sasaki, 1899) A
Rhopalosiphum sp. Bulman A?
Rhopalosiphum sp. (ex *Cordyline*) A?
Toxoptera aurantii (Boyer de Fonscolombe, 1841) A
Toxoptera citricidus (Kirkaldy, 1907) A
Calaphidinae (Myzocallidinae) A
Betulaphis brevipilosa Börner, 1940 A
Calaphis flava Mordvilko, 1928 A
Eucallipterus tiliae (Linnaeus, 1758) A
Euceraphis betulae (Koch, 1855) A
Myzocallis boerneri Stroyan, 1957 A
Myzocallis carpini (Koch, 1855) A
Myzocallis castanicola Baker, 1917 A
Myzocallis coryli (Goeze, 1778) A
Pterocallis alni (de Geer, 1773) A
Takecallis arundinariae (Essig, 1917) A
Takecallis taiwanus (Takahashi, 1926) A
Therioaphis trifolii (Monell, 1882) A
Tuberculatus annulatus (Hartig, 1841) A
Chaitophorinae A
Periphyllus californiensis (Shinji, 1917) A
Periphyllus testudinaceus (Fernie, 1852) A
Drepanosiphini A
Drepanosiphum platanoidis (Schrank, 1801) A
Eriosomatinae (Pemphiginae) A
Aploneura lentisci (Passerini, 1856) A
Colophina clematicola (Shinji, 1922) A
Eriosoma lanigerum (Hausmann, 1802) A
Eriosoma pyricola Baker & Davidson, 1916 A
Geoica lucifuga (Zehntner, 1879) A
Melaphis rhois (Fitch, 1866) A
Pemphigus bursarius (Linnaeus, 1758) A
Pemphigus populitransversus Riley *in* Riley & Monell, 1879 A
Smynthurodes betae Westwood, 1849 A
Tetraneura nigriabdominalis (Sasaki, 1899) A
Hormaphidinae A
Cerataphis orchidearum (Westwood, 1879) A
Pseudoregma panicola (Takashi, 1921) A
Lachninae A
Cinara fresai Blanchard, 1939 A
Cinara juniperi (de Geer, 1773) A
Cinara louisianensis Boudreaux, 1949 A
Cinara pilicornis (Hartig, 1841) A
Cinara tujafilina (Del Guercio, 1909) A
Essigella californica (Essig, 1909) A
Eulachnus brevipilosus Börner, 1940 A
Neophyllaphidinae
Neophyllaphis totarae Cottier, 1953 E
Neophyllaphis n. sp. [ex *Podocarpus nivalis*] E
Phyllaphidinae A
Phyllaphis fagi (Linnaeus, 1767) A
Saltusaphidinae A

Thripsaphis foxtonensis Cottier, 1953 A
TAIWANAPHIDINAE
Sensoriaphis nothofagi Cottier, 1953 E
PHYLLOXERIDAE A
Viteus vitifoliae (Fitch, 1855) A
Moritziella corticalis (Kaltenbach, 1867) A
Phylloxera glabra (von Heyden, 1837) A
COCCOIDEA
ASTEROLECANIIDAE
Asterodiaspis variolosa (Ratzeburg, 1870) A
Asterolecanium vitreum Russell, 1941 E
Asterolecanium n. sp. E
CEROCOCCIDAE
Cerococcus corokiae Maskell, 1890 E
Cerococcus michaeli Lambdin, 1998 E
Solenophora fagi Maskell, 1890 E
COCCIDAE
Aphenochiton chionochloae Henderson & Hodgson, 2000 E
Aphenochiton dierama Henderson & Hodgson, 2000 E
Aphenochiton grammicus Henderson & Hodgson, 2000 E
Aphenochiton inconspicuus (Maskell, 1892) E
Aphenochiton kamahi Henderson & Hodgson, 2000 E
Aphenochiton matai Henderson & Hodgson, 2000 E
Aphenochiton pronus Henderson & Hodgson, 2000 E
Aphenochiton pubens Henderson & Hodgson, 2000 E
Aphenochiton subtilis Henderson & Hodgson, 2000 E
Ceroplastes ceriferus (Fabricius, 1798) A
Ceroplastes destructor Newstead, 1917 A
Ceroplastes sinensis Del Guercio, 1900 A
Coccus hesperidum Linnaeus, 1758 A
Coccus longulus (Douglas, 1887) A
Crystallotesta fagi (Maskell, 1891) E
Crystallotesta fusca (Maskell, 1884) E
Crystallotesta leptospermi (Maskell, 1882) E
Crystallotesta neofagi Henderson & Hodgson, 2000 E
Crystallotesta ornata (Maskell, 1885) E
Crystallotesta ornatella Henderson & Hodgson, 2000 E
Ctenochiton chelyon Henderson & Hodgson, 2000 E
Ctenochiton paraviridis Henderson & Hodgson, 2000 E
Ctenochiton toru Henderson & Hodgson, 2000 E
Ctenochiton viridis Maskell, 1879 E
Epelidochiton piperis (Maskell, 1882) E
Inglisia patella Maskell, 1879 E
Kalasiris depressa (Maskell, 1884) E
Kalasiris paradepressa Henderson & Hodgson, 2000 E
Kalasiris perforata (Maskell, 1879) E
Lecanochiton actites Henderson & Hodgson, 2000 E
Lecanochiton metrosideri Maskell, 1882 E
Lecanochiton minor Maskell, 1891 E
Lecanochiton scutellaris Henderson & Hodgson, 2000 E
Parasaissetia nigra (Nietner, 1861) A
Parthenolecanium corni (Bouché, 1844) A
Parthenolecanium persicae (Fabricius, 1776) A
Plumichiton diadema Henderson & Hodgson, 2000 E
Plumichiton elaeocarpi (Maskell, 1885) E
Plumichiton flavus (Maskell, 1884) E
Plumichiton nikau Henderson & Hodgson, 2000 E
Plumichiton pollicinus Henderson & Hodgson, 2000 E
Plumichiton punctatus Henderson & Hodgson, 2000 E
Poropeza cologabata Henderson & Hodgson, 2000 E
Poropeza dacrydii (Maskell, 1892) E
Pounamococcus cuneatus Henderson & Hodgson, 2000 E
Pounamococcus tubulus Henderson & Hodgson, 2000 E
Pulvinaria floccifera (Westwood, 1870) A
Pulvinaria hydrangeae Steinweden, 1946 A
Pulvinaria mesembryanthemi (Vallot, 1829) A
Pulvinaria vitis (Linnaeus, 1758) A
Saissetia coffeae (Walker, 1852) A
Saissetia oleae (Olivier, 1791) A
Umbonichiton adelus Henderson & Hodgson, 2000 E
Umbonichiton bispinatus Henderson & Hodgson, 2005 E
Umbonichiton bullatus Henderson & Hodgson, 2000 E
Umbonichiton hymenantherae (Maskell, 1885) E
Umbonichiton jubatus Henderson & Hodgson, 2000 E
Umbonichiton pellaspis Henderson & Hodgson, 2000 E
Umbonichiton rimu Henderson & Hodgson, 2005 E
DIASPIDIDAE
Abgrallaspis cyanophylli (Signoret, 1869) A
Anoplaspis maskelli Morrison & Morrison, 1922 E
Anoplaspis metrosideri (Maskell, 1880) E
Aonidiella aurantii (Maskell, 1879) A
Aspidioides corokiae (Maskell, 1891) E
Aspidiotus nerii Bouché, 1833 A
Aulacaspis rosae (Bouché, 1833) A
Aulacaspis rosarum Borchsenius, 1958 A
Carulaspis juniperi (Bouché, 1851) A
Carulaspis minima (Signoret, 1869) A
Chionaspis angusta Green, 1904
Diaspidiotus ostreaeformis (Curtis, 1843) A
Diaspidiotus perniciosus (Comstock, 1881) A
Diaspis boisduvali Signoret, 1869 A
Eulepidosaphes pyriformis (Maskell, 1879) E
Furchadaspis zamiae (Morgan, 1890) A
Hemiberlesia lataniae (Signoret, 1869) A
Hemiberlesia rapax (Comstock, 1881) A
Kuwanaspis pseudoleucaspis (Kuwana, 1902) A
Labidaspis myersi (Green, 1929) E
Lepidosaphes beckii (Newman, 1869) A
Lepidosaphes lactea (Maskell, 1895) E
Lepidosaphes multipora (Leonardi, 1904) A
Lepidosaphes pallida (Maskell, 1895) A
Lepidosaphes pinnaeformis (Bouché, 1851) A
Lepidosaphes ulmi (Linnaeus, 1758) A
Leucaspis brittini Green, 1929 E
Leucaspis carpodeti Brittin, 1937 E
Leucaspis cordylinidis Maskell, 1893 E
Leucaspis elaeocarpi Brittin, 1937 E
Leucaspis gigas (Maskell, 1879) E
Leucaspis greeni Brittin, 1937 E
Leucaspis hoheriae Brittin, 1937 E
Leucaspis maskelli (Brittin, 1915) E
Leucaspis melicytidis Brittin, 1937 E
Leucaspis mixta de Boer, 1977 E
Leucaspis morrisi (Brittin 1915) E
Leucaspis ohakunensis Brittin, 1937 E
Leucaspis pittospori Brittin, 1937 E
Leucaspis podocarpi Green, 1929 E
Leucaspis portaeaureae Ferris, 1942 A/E?
Leucaspis senilobata Green, 1929 E
Leucaspis stricta (Maskell, 1884) E
Leucaspis n. spp. (30) 30E
Lindingaspis rossi (Maskell, 1891) A
Parlatoria desolator McKenzie, 1960 A
Parlatoria fulleri Morrison, 1939 A
Parlatoria pittospori Maskell, 1891 A
Pinnaspis aspidistrae (Signoret, 1869) A
Pinnaspis dysoxyli (Maskell, 1885) E
Poliaspis argentosis Brittin, 1915 E
Poliaspis media Maskell, 1880 E
Poliaspis n. sp. E
Poliaspoides leptocarpi (Brittin, 1916) E
Pseudaulacaspis brimblecombei Williams, 1973 A
Pseudaulacaspis cordylinidis (Maskell, 1879) E
Pseudaulacaspis epiphytidis (Maskell, 1885) E
Pseudaulacaspis eugeniae (Maskell, 1892) A
Pseudaulacaspis phymatodidis (Maskell, 1880) E
Pseudoparlatoria parlatorioides (Comstock, 1883) A
Scrupulaspis intermedia (Maskell, 1891) E
Symeria leptospermi (Maskell, 1882) E
Trullifiorinia acaciae (Maskell, 1892) A
Gen. nov. et n. sp. E
ERIOCOCCIDAE
Affeldococcus kathrinae Henderson, 2007 E
Alpinococcus elongatus Henderson, 2007 E
Bryococcus hippodamus Henderson, 2007 E
Capulinia orbiculata Hoy, 1958 E
Cryptococcus nudatus Brittin, 1915 E
Eriochiton armatus Brittin, 1915 E
Eriochiton brittini Hodgson & Henderson, 1996 E
Eriochiton deboerae Hodgson & Henderson, 1996 E
Eriochiton dracophylli Hodgson & Henderson, 1996 E
Eriochiton dugdalei Hodgson & Henderson, 1996 E
Eriochiton hispidus Maskell, 1887 E
Eriochiton hoheriae Hodgson, 1994 E
Eriochiton propespinosus Hodgson, 1994 E
Eriochiton pseudohispidus Hodgson & Henderson, 1996 E
Eriochiton spinosus (Maskell, 1879) E
Eriococcus abditus Hoy, 1962 E
Eriococcus aconeae Henderson, 2007 E
Eriococcus acutispinatus Hoy, 1962 E
Eriococcus albatus Hoy, 1962 E
Eriococcus araucariae Maskell, 1879 A
Eriococcus arcanus Hoy, 1962 E
Eriococcus argentifagi Hoy, 1962 E
Eriococcus asteliae Hoy, 1962, E
Eriococcus beilschmiediae Hoy, 1962 E
Eriococcus brittini Hoy, 1962 E
Eriococcus campbelli Hoy, 1959 A
Eriococcus cavelli (Maskell, 1890) E
Eriococcus celmisiae (Maskell, 1884) E
Eriococcus chathamensis Hoy, 1962 E
Eriococcus coccineus Cockerell, 1894 A
Eriococcus coprosmae Hoy, 1962 E
Eriococcus coriaceus Maskell, 1893 A
Eriococcus crenilobatus Hoy, 1962 E
Eriococcus dacrydii Hoy, 1962 E
Eriococcus danthoniae Maskell, 1891 E
Eriococcus detectus Hoy, 1962 E
Eriococcus elaeocarpi Hoy, 1962 E
Eriococcus elytranthae Hoy, 1962 E
Eriococcus fagicorticis Maskell, 1892 E
Eriococcus fossor (Maskell, 1884) E
Eriococcus fuligitectus Hoy, 1962 E
Eriococcus gaultheriae Hoy, 1962 E
Eriococcus hebes Hoy, 1962 E
Eriococcus hispidus Hoy, 1962 E
Eriococcus humatus Hoy, 1962 E
Eriococcus kamahi Hoy, 1958 E
Eriococcus kowhai Hoy, 1962 E
Eriococcus latilobatus Hoy, 1962 E
Eriococcus leptospermi Maskell, 1891
Eriococcus maskelli Hoy, 1962 E
Eriococcus matai Hoy, 1962 E
Eriococcus meridianus Hoy, 1962 E
Eriococcus mimus Hoy, 1962 E
Eriococcus montanus Hoy, 1962 E
Eriococcus montifagi Hoy, 1962 E
Eriococcus multispinus (Maskell, 1879) E
Eriococcus myrsinae Hoy, 1962 E
Eriococcus nelsonensis Hoy, 1962 E
Eriococcus neomyrti Hoy, 1962 E
Eriococcus nitidulus Hoy, 1962 E
Eriococcus nothofagi Hoy, 1962 E
Eriococcus orariensis Hoy, 1954 A
Eriococcus pallidus Maskell, 1885 E
Eriococcus parabilis Hoy, 1962 E
Eriococcus parsonsiae Henderson, 2006 E
Eriococcus parvulus Hoy, 1962 E
Eriococcus phyllocladi Maskell, 1892 E

Eriococcus pimeliae Hoy, 1962 E
Eriococcus podocarpi Hoy, 1962 E
Eriococcus pohutukawa Hoy, 1958 E
Eriococcus raithbyi Maskell, 1890 E
Eriococcus rata Hoy, 1958 E
Eriococcus rotundus Hoy, 1962 E
Eriococcus rubrifagi Hoy, 1962 E
Eriococcus setulosus Hoy, 1962 E
Eriococcus sophorae Green, 1929 E
Eriococcus n. spp. (4) 4E
Hoheriococcus fionae Henderson, 2006 E
Kuwanina kiwiana Henderson, 2007 E
Madarococcus cruriamplus Hoy, 1962 E
Madarococcus cunicularius Hoy, 1962 E
Madarococcus maculatus (Maskell, 1980) E
Madarococcus pulchellus (Maskell, 1890) E
Madarococcus totarae (Maskell, 1890) E
Madarococcus viridulus Hoy, 1962 E
Montanococcus graemei Henderson, 2007 E
Montanococcus petrobius Henderson, 2007 E
Montanococcus thriaticus Henderson, 2007 E
Neoeriochiton clareae Hodgson, 1994 E
Noteococcus hoheriae (Maskell, 1880) E
Phloeococcus cordylinidis Hoy, 1962 E
Phloeococcus loriceus Hoy, 1962 E
Scutare fimbriata Brittin, 1915 E
Scutare lanuginosa Hoy, 1962 E
Scutare pittospori Hoy, 1962 E
Sisyrococcus intermedius (Maskell, 1891) E
Sisyrococcus papillosus Hoy, 1962 E
Stegococcus flagellatus Henderson, 2006 E
Stegococcus oleariae Hoy, 1962 E
Tolypecoccus latebrosus Hoy, 1962 E
HALIMOCOCCIDAE
Colobopyga hedyscapes Deitz, 1979 E
MARGARODIDAE
Coelostomidia deboerae Morales, 1991 E
Coelostomidia jenniferae Morales, 1991 E
Coelostomidia montana (Green, 1929) E
Coelostomidia pilosa (Maskell, 1891) E
Coelostomidia wairoensis (Maskell, 1884) E
Coelostomidia zealandica (Maskell, 1880) E
Ultracoelostoma assimile (Maskell, 1890) E
Ultracoelostoma brittini Morales, 1991 E
Ultracoelostoma dracophylli Morales, 1991 E
Platycoelostoma compressa (Maskell, 1892) E
Icerya purchasi Maskell, 1879 A
ORTHEZIIDAE
Newsteadia caledoniensis Kozár & Konczné Benedicty, 2000
Newsteadia gullanae Kozár & Konczné Benedicty, 2000
Newsteadia myersi Green, 1929 E
PHENACOLEACHIIDAE
Phenacoleachia australis Beardsley, 1964 E
Phenacoleachia zealandica (Maskell, 1891) E
PSEUDOCOCCIDAE
Acrochordonus chionochloae Cox, 1987 E
Acrochordonus curtatus Cox, 1987 E
Agastococcus zelandiensis Cox, 1987 E
Antonina socialis Newstead, 1901 A
Asaphococcus agninus Cox, 1987 E
Asaphococcus amissus Cox, 1987 E
Asaphococcus montanus (Brittin, 1938) E
Asteliacoccus zelandigena Cox, 1987 E
Balanococcus aberrans Cox, 1987 E
Balanococcus acerbus Cox, 1987 E
Balanococcus agnostus Cox, 1987 E
Balanococcus alpigenus Cox, 1987 E
Balanococcus botulus Cox, 1987 E
Balanococcus celmisiae Cox, 1987 E
Balanococcus cockaynei (Brittin, 1915) E
Balanococcus conglobatus Cox, 1987 E
Balanococcus contextus Cox, 1987 E
Balanococcus cordylinidis (Brittin, 1938) E
Balanococcus cortaderiae Cox, 1987 E
Balanococcus danthoniae (Morrison, 1925) E
Balanococcus diminutus (Leonardi, 1918) E
Balanococcus dracophylli Cox, 1987 E
Balanococcus gahniicola Cox, 1987 E
Balanococcus mayae Cox, 1987 E
Balanococcus nelsonensis Cox, 1987 E
Balanococcus notodanthoniae Cox, 1987 E
Balanococcus poae (Maskell, 1879) E
Balanococcus sexaspinus (Brittin, 1915) E
Balanococcus tunakinensis Cox, 1987 E
Balanococcus turriseta Cox, 1987 E
Balanococcus wisei (Williams & de Boer, 1973) E
Chorizococcus oreophilus Williams, 1985 A
Chryseococcus arecae (Maskell, 1890) A
Chryseococcus longispinus (Beardsley, 1964) E
Crisicoccus australis Cox, 1987 E
Crisicoccus comatus Cox, 1987 E
Crisicoccus indigenus Cox, 1987 E
Crisicoccus tokaanuensis Cox, 1987 E
Crocydococcus cottieri (Brittin, 1938) E
Cyphonococcus alpinus (Maskell, 1884) E
Cyphonococcus furvus Cox, 1987 E
Cyphonococcus iceryoides (Maskell, 1892) E
Dysmicoccus ambiguus (Morrison, 1925) E
Dysmicoccus arcanus Cox, 1987 E
Dysmicoccus celmisicola (Cox, 1987) E
Dysmicoccus delitescens Cox, 1987 E
Dysmicoccus formicicola (Maskell, 1892) E
Dysmicoccus ornatus Cox, 1987 E
Dysmicoccus rupestris Cox, 1987 E
Dysmicoccus viticis (Green, 1929) E
Eurycoccus antiscius Williams, 1985 A
Laminicoccus asteliae Cox, 1987 E
Laminicoccus eastopi Cox, 1987 E
Laminicoccus flandersi Williams, 1985 A
Maskellococcus nothofagi Cox, 1987 E
Maskellococcus obtectus (Maskell, 1890) E
Nipaecoccus aurilanatus (Maskell, 1890) A
Paracoccus abnormalis Cox, 1987 E
Paracoccus acaenae Cox, 1987 E
Paracoccus albatus Cox, 1987 E
Paracoccus aspratilis Cox, 1987 E
Paracoccus butcherae Cox, 1987 E
Paracoccus canalis (Brittin, 1938) E
Paracoccus cavaticus Cox, 1987 E
Paracoccus coriariae (Brittin, 1938) E
Paracoccus cryptus Cox, 1987 E
Paracoccus deboerae Cox, 1987 E
Paracoccus deceptus Cox, 1987 E
Paracoccus definitus Cox, 1987 E
Paracoccus drimydis (Brittin, 1938) E
Paracoccus glaucus (Maskell, 1879) E
Paracoccus hebes Cox, 1987 E
Paracoccus insolitus (Brittin, 1938) E
Paracoccus leptospermi Cox, 1987 E
Paracoccus longicauda Cox, 1987 E
Paracoccus miro (de Boer, 1967) E
Paracoccus multiductus Cox, 1987 E
Paracoccus nothofagicola Cox, 1987 E
Paracoccus parvicirculus Cox, 1987 E
Paracoccus podocarpi Cox, 1987 E
Paracoccus redactus Cox, 1987 E
Paracoccus zealandicus (Ezzat & McConnell, 1956) E
Paraferrisia podocarpi (Brittin, 1938) E
Phenacoccus graminicola Leonardi, 1908 A
Planococcus dubius Cox, 1987 E
Planococcus mali Ezzat & McConnell, 1956 A
Pseudococcus calceolariae (Maskell, 1879) A
Pseudococcus hypergaeus Williams, 1985 A
Pseudococcus longispinus (Targioni Tozzetti, 1867) A
Pseudococcus viburni (Signoret, 1875) A
Pseudococcus zelandicus Cox, 1987 E
Rastrococcus asteliae (Maskell, 1884) E
Rastrococcus namartini Williams & Henderson, 2005 E
Renicaula chionochloae (de Boer, 1968) E
Renicaula junci (de Boer, 1968) E
Renicaula pauca Cox, 1987 E
Renicaula raouliae (de Boer, 1968) E
Rhizoecus cacticans (Hambleton, 1946) A
Rhizoecus californicus Ferris, 1953 A
Rhizoecus deboerae Hambleton, 1974 E
Rhizoecus dianthi Green, 1926 A
Rhizoecus falcifer Künckel d'Herculais, 1878 A
Rhizoecus graminis Hambleton, 1946 A
Rhizoecus oliveri Cox, 1978 E
Rhizoecus puhiensis Hambleton, 1974 E
Rhizoecus rumicis Maskell, 1892 A
Sarococcus comis Cox, 1987 E
Sarococcus deplanatus Cox, 1987 E
Sarococcus fagi (Maskell, 1891) E
Sarococcus undatus Cox, 1987 E
Spilococcus geoffreyi Cox, 1987 E
Spilococcus mamillariae (Bouché, 1844) A
Ventrispina crebrispina Cox, 1987 E
Ventrispina dugdalei Cox, 1987 E
Ventrispina otagoensis (Brittin, 1938) E
Vryburgia amaryllidis (Bouché, 1837) A

Suborder AUCHENORRHYNCHA
[Compiled by M.-C. Larivière & M. J. Fletcher]
Infraorder CICADOMORPHA
CICADOIDEA
CICADIDAE
Amphipsalta cingulata (Fabricius, 1775) E
Amphipsalta strepitans (Kirkaldy, 1909) E
Amphipsalta zelandica (Boisduval, 1835) E
Kikihia angusta (Walker, 1850) E
Kikihia cauta (Myers, 1921) E
Kikihia cutora cumberi Fleming, 1973 E
Kikihia c. cutora (Walker, 1850) E
Kikihia c. exulis (Hudson, 1950) E K
Kikihia dugdalei Fleming, 1984 E
Kikihia horologium Fleming, 1984 E
Kikihia laneorum Fleming, 1984 E
Kikihia longula (Hudson, 1950) E
Kikihia muta muta (Fabricius, 1775) E
Kikihia m. pallida (Hudson, 1950) E
Kikihia ochrina (Walker, 1858) E
Kikihia paxillulae Fleming, 1984 E
Kikihia rosea (Walker, 1850) E
Kikihia scutellaris (Walker, 1850) E
Kikihia subalpina (Hudson, 1891) E
Maoricicada alticola Dugdale & Fleming, 1978 E
Maoricicada campbelli (Myers, 1923) E
Maoricicada cassiope (Hudson, 1891) E
Maoricicada clamitans Dugdale & Fleming, 1978 E
Maoricicada hamiltoni (Myers, 1926) E
Maoricicada iolanthe (Hudson, 1891) E
Maoricicada lindsayi (Myers, 1923) E
Maoricicada mangu celer Dugdale & Fleming, 1978 E
Maoricicada m. gourlayi Dugdale & Fleming, 1978 E
Maoricicada m. mangu (White, 1879) E
Maoricicada m. multicostata Dugdale & Fleming, 1978 E
Maoricicada myersi (Fleming, 1971) E
Maoricicada nigra frigida Dugdale & Fleming, 1978 E
Maoricicada n. nigra (Myers, 1921) E
Maoricicada oromelaena (Myers, 1926) E
Maoricicada otagoensis maceweni Dugdale & Fleming, 1978 E
Maoricicada o. otagoensis Dugdale & Fleming, 1978 E
Maoricicada phaeoptera Dugdale & Fleming, 1978 E
Maoricicada tenuis Dugdale & Fleming, 1978 E
Notopsalta sericea (Walker, 1850) E
Rhodopsalta cruentata (Fabricius, 1775) E
Rhodopsalta leptomera (Myers, 1921) E
Rhodopsalta microdora (Hudson, 1936) E
CERCOPOIDEA

APHROPHORIDAE
Basilioterpa bullata Hamilton & Morales, 1992
Bathyllus albicinctus (Erichson, 1842) A
Carystoterpa aurata Hamilton & Morales, 1992 E
Carystoterpa chelyon Hamilton & Morales, 1992 E
Carystoterpa fingens (Walker, 1851) E
Carystoterpa ikana Hamilton & Morales, 1992 E
Carystoterpa maori Hamilton & Morales, 1992 E
Carystoterpa minima Hamilton & Morales, 1992 E
Carystoterpa minor Hamilton & Morales, 1992 E
Carystoterpa subtacta (Walker, 1858) K
Carystoterpa subvirescens (Butler, 1874) E
Carystoterpa trimaculata (Butler, 1874) E
Carystoterpa tristis (Alfken, 1904) E
Carystoterpa vagans Hamilton & Morales, 1992 E
Philaenus spumarius (Linnaeus, 1758) A
Pseudaphronella jactator (White, 1879) E
MEMBRACOIDEA
CICADELLIDAE
Anzygina agni (Knight, 1976) E
Anzygina barrattae Fletcher & Larivière, 2009 E
Anzygina dumbletoni (Ghauri, 1963) E
Anzygina ramsayi (Knight, 1976) E
Anzygina toetoe (Cumber, 1952) E
Anzygina zealandica (Myers, 1923)
Arahura dentata Knight, 1975 E
Arahura gourlayi Knight, 1975 E
Arahura reticulata Knight, 1975 E
Arawa dugdalei Knight, 1975 E
Arawa negata (White, 1879) E
Arawa novella (Metcalf, 1968)
Arawa pulchra Knight, 1975
Arawa variegata Knight, 1975 E
Balclutha incisa (Matsumura, 1902) A?
Balclutha lucida (Butler, 1877) A? K
Balclutha viridinervis Matsumura, 1914 A? K
Batracomorphus adventitiosus Evans, 1966
Batracomorphus angustatus (Osborn, 1934)
Batracomorphus punctatus Evans, 1940
Edwardsiana froggatti (Baker, 1925) A
Edwardsiana lethierryi (Edwards, 1881) A
Euacanthella palustris Evans, 1938 A
Eupteryx melissae Curtis, 1837 A
Exitianus plebeius (Kirkaldy, 1906) K
Horouta inconstans Knight, 1975 E
Idiocerus distinguendus Kirschbaum, 1868 A
Kybos lindbergi (Linnavuori, 1951) A
Kybos smaragdula (Fallén, 1806) A
Limotettix awae (Myers, 1924) E
Limotettix harrisi Knight, 1975 E
Limotettix pallidus Knight, 1975 E
Limotettix pullatus (Evans, 1942)
Macrosteles fieberi (Edwards, 1889) A
Maiestas knighti Webb & Viraktamath, 2009
Maiestas samuelsoni (Knight, 1976) K
Maiestas vetus (Knight, 1975)
Matatua maorica (Myers, 1923) E
Matatua montivaga Knight, 1976 E
Nesoclutha phryne (Kirkaldy, 1907)
Novolopa falcata Knight, 1973 E
Novolopa infula Knight, 1973 E
Novolopa kuscheli Knight, 1973 E
Novolopa maculata Knight, 1973 E
Novolopa montivaga Knight, 1973 E
Novolopa townsendi Evans, 1966 E
Novothymbris cassiniae (Myers, 1923) E
Novothymbris castor Knight, 1974 E
Novothymbris cithara Knight, 1974 E
Novothymbris extremitatis Knight, 1974 E
Novothymbris eylesi Knight, 1974 E
Novothymbris hinemoa (Myers, 1923) E
Novothymbris maorica (Myers, 1923) E
Novothymbris notata Knight, 1974 E
Novothymbris notialis Knight, 1974 E
Novothymbris peregrina Knight, 1974 E
Novothymbris pollux Knight, 1974 E
Novothymbris punctata Knight, 1974 E
Novothymbris solitaria Knight, 1974 E
Novothymbris tararua (Myers, 1923) E
Novothymbris vagans Knight, 1974 E
Novothymbris zealandica (Myers, 1923) E
Orosius argentatus (Evans, 1938) A K
Paracephaleus curtus Knight, 1973 E
Paracephaleus hudsoni (Myers, 1923) E
Paradorydium aculeatum Knight, 1973 E
Paradorydium cuspis Knight, 1973 E
Paradorydium gourlayi Evans, 1966 E
Paradorydium insulare Evans, 1966 E
Paradorydium philpotti Myers, 1923 E
Paradorydium sertum Knight, 1973 E
Paradorydium stewartensis Evans, 1966
Paradorydium watti Knight, 1973 E
Paradorydium westwoodi (White, 1879) E
Rhytidodus decimaquartus (Schrank, 1776) A
Ribautiana tenerrima (Herrich-Schäffer, 1834) A
Scaphetus brunneus Evans, 1966 E
Scaphetus simus Knight, 1975 E
Xestocephalus ovalis Evans, 1966 E
Zelopsis nothofagi Evans, 1966 E
MEMBRACIDAE A
Acanthucus trispinifer (Fairmaire, 1846) A
MYERSLOPIIDAE
Myerslopia magna amplificata Knight, 1973 E
Myerslopia m. magna Evans, 1947 E
Myerslopia m. scabrata Knight, 1973 E
Myerslopia rakiuraensis Szwedo, 2004 E
Myerslopia tawhai Szwedo, 2004 E
Myerslopia tearohai Szwedo, 2004 E
Myerslopia triregia Knight, 1973 E
Myerslopia whakatipuensis Szwedo, 2004 E
Pemmation asperum asperum (Knight, 1973) E
Pemmation a. cognatum (Knight, 1973) E
Pemmation bifurca (Knight, 1973) E
Pemmation insulare (Knight, 1973) E
Pemmation montis (Knight, 1973) E
Pemmation parvum (Evans, 1947) E
Pemmation simile (Knight, 1973) E
Pemmation terrestre (Knight, 1973) E
Pemmation townsendi (Knight, 1973) E
Pemmation variabile austrinum (Knight, 1973) E
Pemmation v. variabile (Knight, 1973) E
Pemmation verrucosum (Knight, 1973) E

Infraorder FULGOROMORPHA
ACHILIDAE
Achilus flammeus Kirby, 1818 A
Agandecca annectens White, 1879 E
CIXIIDAE
Aka dunedinensis Larivière, 1999 E
Aka duniana (Myers, 1924) E
Aka finitima (Walker, 1858) E
Aka rhodeae Larivère, 1999 E
Aka westlandica Larivière, 1999 E
Cermada aspilus Walker, 1858
Cermada inexspectata (Larivière, 1999) E
Cermada kermadecensis (Myers, 1924) E K
Cermada punctimargo (Walker, 1858) E
Cermada triregia (Larivière, 1999) E
Chathamaka andrei Larivière, 1999 E
Confuga persephone Fennah, 1975 E
Huttia nigrifrons Myers, 1924 E
Huttia northlandica Larivière, 1999 E
Koroana arthuria Myers, 1924 E
Koroana helena Myers, 1924
Koroana lanceloti Larivière, 1997 E
Koroana rufifrons (Walker, 1858) E
Malpha cockcrofti Myers, 1924 E
Malpha iris Myers, 1924
Malpha muiri Myers, 1924 E
Parasemo hutchesoni Larivière, 1999 E
Semo clypeatus White, 1879 E
Semo harrisi (Myers, 1924) E
Semo southlandiae Larivière & Hoch, 1998 E
Semo transinsularis Larivière & Hoch, 1998 E
Tiriteana clarkei Myers, 1924 E
Zeoliarus atkinsoni (Myers, 1924) E
Zeoliarus oppositus (Walker, 1850) E
DELPHACIDAE
Anchodelphax hagnon Fennah, 1965 E
Anchodelphax olenus Fennah, 1965 E
Eorissa cicatrifrons Fennah, 1965 E
Nilaparvata myersi Muir, 1923 E
Notogryps ithoma Fennah, 1965 E
Notogryps melanthus Fennah, 1965 E
Notohyus erosus Fennah, 1965 E
Opiconsiva dilpa (Kirkaldy, 1907) native?
Sardia rostrata pluto (Kirkaldy, 1906) K
Sulix insecutor Fennah, 1965 E
Sulix miridianalis (Muir, 1917) E
Sulix tasmani (Muir, 1923) E
Sulix vetranio Fennah, 1965 E
Toya dryope (Kirkaldy, 1907) A
Ugyops (Paracona) pelorus Fennah, 1965 E
Ugyops (P.) raouli (Muir, 1923) E K
Ugyops (Ugyops) caelatus (White, 1879) E
Ugyops (U.) rhadamanthus Fennah, 1965 E
DERBIDAE
Eocenchrea maorica (Kirkaldy, 1909) E
DICTYOPHARIDAE
Thanatodictya (Niculda) tillyardi Myers, 1923 E
FLATIDAE A
Anzora unicolor (Walker, 1862) A
Siphanta acuta (Walker, 1851) A
RICANIIDAE A
Scolypopa australis (Walker, 1851) A

Suborder COLEORRHYNCHA
[Compiled by M.-C. Larivière & D. Burckhardt]
PELORIDIIDAE
Oiophysa ablusa Drake & Salmon, 1950 E
Oiophysa cumberi Woodward, 1958 E
Oiophysa distincta Woodward, 1952 E
Oiophysa paradoxa Burckhardt, 2009 E
Oiophysa pendergrasti Woodward, 1956 E
Xenophyes cascus Bergroth, 1924 E
Xenophyes kinlochensis Evans, 1981 E
Xenophysella greensladeae Burckhardt, 2009 E
Xenophysella stewartensis (Woodward, 1952) E

Suborder HETEROPTERA
[Compiled by M.-C. Larivière]
Infraorder ENICOCEPHALOMORPHA
AENICTOPECHEIDAE
Aenictocoris powelli Woodward, 1956 E
Maoristolus parvulus Woodward, 1956 E
Maoristolus tonnoiri (Bergroth, 1927) E
Nymphocoris maoricus Woodward, 1956 E
ENICOCEPHALIDAE
Gourlayocoris mirabilis (Gourlay, 1952) E
Phthirostenus magnus (Woodward, 1956) E
Systelloderes maclachlani (Kirkaldy, 1901) E
Systelloderes notialis Woodward, 1956 E

Infraorder DIPSOCOROMORPHA
CERATOCOMBIDAE
Ceratocombus aotearoae Larivière & Larochelle, 2004 E
Ceratocombus novaezelandiae Larivière & Larochelle, 2004 E
SCHIZOPTERIDAE
Hypselosoma acantheen Hill, 1991 E

Infraorder GERROMORPHA
GERRIDAE
Halobates sericeus Eschscholtz, 1822 SW

HYDROMETRIDAE
Hydrometra strigosa (Skuse, 1893) F
MESOVELIIDAE
Mesovelia hackeri Harris & Drake, 1941 A F
Mniovelia kuscheli Andersen & Polhemus, 1980 E
VELIIDAE
Microvelia macgregori (Kirkaldy, 1899) E F

Infraorder NEPOMORPHA
CORIXIDAE
Diaprepocoris zealandiae Hale, 1924 E F
Sigara (Tropocorixa) arguta (White, 1878) E F
Sigara (T.) infrequens Young, 1962 E F
Sigara (T.) potamius Young, 1962 E F
Sigara (T.) uruana Young, 1962 E F
NOTONECTIDAE
Anisops assimilis White, 1878 E F
Anisops wakefieldi White, 1878 E F

Infraorder LEPTOPODOMORPHA
SALDIDAE
Saldula australis (White, 1876) E
Saldula butleri (White, 1878) E
Saldula laelaps (White, 1878) E
Saldula maculipennis Cobben, 1961 E
Saldula parvula Cobben, 1961 E
Saldula stoneri Drake & Hoberlandt, 1950 E
Saldula trivialis Cobben, 1961 E

Infraorder CIMICOMORPHA
ANTHOCORIDAE
Buchananiella whitei (Reuter, 1884) E
Cardiastethus brounianus White, 1878 E
Cardiastethus consors White, 1879 E
Cardiastethus poweri White, 1879 E
Lyctocoris (Lyctocoris) campestris (Fabricius, 1794) A
Maoricoris benefactor China, 1933 E
Orius (Heterorius) vicinus (Ribaut, 1923) A
Xylocoris (Proxylocoris) galactinus (Fieber, 1836) A
CANTACADERIDAE
Carldrakeana socia (Drake & Ruhoff, 1961)
Cyperobia carectorum Bergroth, 1927 E
CIMICIDAE A
Cimex lectularius Linnaeus, 1758 A
MIRIDAE
Anexochus crassicornis Eyles, 2001 E
Basileobius gilviceps Eyles & Schuh, 2003 E
Bipuncticoris cassinianus Eyles & Carvalho, 1995 E
Bipuncticoris chlorus Eyles & Carvalho, 1995 E
Bipuncticoris convexus Eyles & Carvalho, 1995 E
Bipuncticoris gurri Eyles & Carvalho, 1995 E
Bipuncticoris irroratus Eyles & Carvalho, 1995 E
Bipuncticoris lineatus Eyles & Carvalho, 1995 E
Bipuncticoris longicerus Eyles & Carvalho, 1995 E
Bipuncticoris minor Eyles & Carvalho, 1995 E
Bipuncticoris olearinus Eyles & Carvalho, 1995 E
Bipuncticoris planus Eyles & Carvalho, 1995 E
Bipuncticoris robustus Eyles & Carvalho, 1995 E
Bipuncticoris triplex Eyles & Carvalho, 1995 E
Bipuncticoris vescus Eyles & Carvalho, 1995 E
Bipuncticoris xestus Eyles & Carvalho, 1995 E
Campylomma novocaledonica Schuh, 1984 A
Chaetedus longiceps Eyles, 1975
Chaetedus plumalis Eyles, 1975
Chaetedus reuterianus (White, 1878) E
Chinamiris acutospinosus Eyles & Carvalho, 1991 E
Chinamiris aurantiacus Eyles & Carvalho, 1991 E
Chinamiris brachycerus Eyles & Carvalho, 1991 E
Chinamiris citrinus Eyles & Carvalho, 1991 E
Chinamiris cumberi Eyles & Carvalho, 1991 E
Chinamiris daviesi Eyles & Carvalho, 1991 E
Chinamiris dracophylloides Eyles & Carvalho, 1991 E
Chinamiris elongatus Eyles & Carvalho, 1991 E
Chinamiris fascinans Eyles & Carvalho, 1991 E
Chinamiris guttatus Eyles & Carvalho, 1991 E
Chinamiris hamus Eyles & Carvalho, 1991 E
Chinamiris indeclivis Eyles & Carvalho, 1991 E
Chinamiris juvans Eyles & Carvalho, 1991 E
Chinamiris laticinctus (Walker, 1873) E
Chinamiris marmoratus Eyles & Carvalho, 1991 E
Chinamiris minutus Eyles & Carvalho, 1991 E
Chinamiris muehlenbeckiae Woodward, 1950 E
Chinamiris niculatus Eyles & Carvalho, 1991 E
Chinamiris nigrifrons Eyles & Carvalho, 1991 E
Chinamiris opacus Eyles & Carvalho, 1991 E
Chinamiris ovatus Eyles & Carvalho, 1991 E
Chinamiris punctatus Eyles & Carvalho, 1991 E
Chinamiris quadratus Eyles & Carvalho, 1991 E
Chinamiris rufescens Eyles & Carvalho, 1991 E
Chinamiris secundus Eyles & Carvalho, 1991 E
Chinamiris testaceus Eyles & Carvalho, 1991 E
Chinamiris unicolor Eyles & Carvalho, 1991 E
Chinamiris virescens Eyles & Carvalho, 1991 E
Chinamiris viridicans Eyles & Carvalho, 1991 E
Chinamiris whakapapae Eyles & Carvalho, 1991 E
Chinamiris zygotus Eyles & Carvalho, 1991 E
Closterotomus norwegicus (Gmelin, 1790) A
Coridromius chenopoderis Tatarnic & Cassis, 2008 A
Cyrtodiridius aurantiacus Eyles & Schih, 2003 E
Cyrtorhinus cumberi Woodward, 1950 E
Deraeocoris maoricus Woodward, 1950 E
Diomocoris fasciatus Eyles, 2000 E
Diomocoris granosus Eyles, 2000 E
Diomocoris maoricus (Walker, 1873) E
Diomocoris ostiolum Eyles, 2000 E
Diomocoris punctatus Eyles, 2000 E
Diomocoris raoulensis Eyles, 2000 E K
Diomocoris russatus Eyles, 2000 E
Diomocoris sexcoloratus Eyles, 2000 E
Diomocoris woodwardi Eyles, 2000 E
Engytatus nicotianae (Koningsberger, 1903) A
Felisacus elegantulus (Reuter, 1904)
Halormus velifer Eyles & Schuh, 2003 E
Halticus minutus Reuter, 1885 A
Josemiris carvalhoi Eyles, 1996 E
Kiwimiris bipunctatus Eyles & Carvalho, 1995 E
Kiwimiris coloratus Eyles & Carvalho, 1995 E
Kiwimiris concavus Eyles & Carvalho, 1995 E
Kiwimiris melanocerus Eyles & Carvalho, 1995 E
Kiwimiris niger Eyles & Carvalho, 1995 E
Lincolnia lucernina Eyles & Carvalho, 1988 E
Lopus decolor (Fallén, 1807) A
Macrolophus pygmaeus (Rambur, 1839) A
Maoriphylina dimorpha Cassis & Eyles, 2006 E
Mecenopa albiapex Eyles & Schuh, 2003 E
Megaloceroea recticornis (Geoffroy, 1785) A
Monopharsus annulatus Eyles & Carvalho, 1995 E
Monospatha distincta Eyles & Schuh, 2003 E
Peritropis aotearoae Gorczyca & Eyles, 1997 E
Pimeleocoris luteus Eyles & Schuh, 2003 E
Pimeleocoris roseus Eyles & Schuh, 2003 E
Pimeleocoris viridis Eyles & Schuh, 2003 E
Poecilomiris longirostris Eyles, 2006 E
Poecilomiris planus Eyles, 2006 E
Polyozus galbanus Eyles & Schuh, 2003 E
Reuda mayri White, 1878 E
Romna albata Eyles & Carvalho, 1988 E
Romna bicolor Eyles & Carvalho, 1988 E
Romna capsoides (White, 1878) E
Romna cuneata Eyles & Carvalho, 1988 E
Romna nigrovenosa Eyles & Carvalho, 1988 E
Romna oculata Eyles & Carvalho, 1988 E
Romna ornata Eyles & Carvalho, 1988 E
Romna pallescens Eyles, 2006 E
Romna pallida Eyles & Carvalho, 1988 E
Romna rubisura Eyles, 2006 E
Romna scotti (White, 1878) E
Romna tenera Eyles, 1998 E
Romna uniformis Eyles & Carvalho, 1988 E
Romna variegata Eyles & Carvalho, 1988 E
Sejanus albisignatus (Knight, 1938)
Sidnia kinbergi (Stål, 1859) A
Stenotus binotatus (Fabricius, 1794) A
Sthenarus myersi Woodward, 1950 E
Taylorilygus apicalis (Fieber, 1861) A
Tinginotum minutum Eyles, 2000
Trigonotylus tenuis Reuter, 1893 A
Tuicoris excelsus Eyles & Carvalho, 1995 E
Tuicoris lipurus Eyles, 2001 E
Tytthus chinensis (Stål, 1859) A
Wekamiris auropilosus Eyles & Carvalho, 1995 E
Xiphoides badius Eyles & Schuh, 2003 E
Xiphoides luteolus Eyles & Schuh, 2003 E
Xiphoides multicolor Eyles & Schuh, 2003 E
Xiphoides myersi (Woodward, 1950) E
Xiphoides regis Eyles & Schuh, 2003 E
Xiphoides vacans Eyles & Schuh, 2003 E
NABIDAE
Alloeorhynchus (Alloeorhynchus) myersi Bergroth, 1927 E
Nabis (Australonabis) biformis (Bergroth, 1927)
Nabis (Tropiconabis) kinbergii Reuter, 1872 A
Nabis (T.) maoricus Walker, 1873 E
REDUVIIDAE
Empicoris aculeatus (Bergroth, 1927) E
Empicoris angulipennis (Bergroth, 1926) E
Empicoris rubromaculatus (Blackburn, 1888)
Empicoris seorsus (Bergroth, 1927) E
Ploiaria antipodum (Bergroth, 1927) E
Ploiaria chilensis (Philippi, 1862)
Stenolemus fraterculus Wygodzinsky, 1956 A
TINGIDAE
Stephanitis rhododendri Horváth, 1905 A
Tanybyrsa cumberi Drake, 1959 E

Infraorder PENTATOMOMORPHA
ACANTHOSOMATIDAE
Oncacontias vittatus (Fabricius, 1781) E
Rhopalimorpha (Lentimorpha) alpina Woodward, 1953 E
Rhopalimorpha (Rhopalimorpha) lineolaris Pendergrast, 1950 E
Rhopalimorpha obscura Dallas, 1851 E
ARADIDAE
Acaraptera myersi Usinger & Matsuda, 1959 E
Acaraptera waipouensis Heiss, 1990 E
Adenocoris brachypterus Usinger & Matsuda, 1959 E
Adenocoris spiniventris Usinger & Matsuda, 1959 E
Aneuraptera cimiciformis Usinger & Matsuda, 1959 E
Aneurus (Aneurodellus) brevipennis Heiss, 1998 E
Aneurus (A.) brouni White, 1876 E
Aneurus (A.) maoricus Heiss, 1998 E
Aneurus (A.) prominens Pendergrast, 1965 E
Aneurus (A.) salmoni Pendergrast, 1965 E
Aneurus (A.) zealandensis Heiss, 1998 E
Aradus australis Erichson, 1842
Calisius zealandicus Pendergast, 1968 E
Carventaptera spinifera Usinger & Matsuda, 1959 E
Chinamyersia cinerea (Myers & China, 1928) E
Chinamyersia viridis (Myers & China, 1928) E
Clavaptera ornata Kirman, 1985 E
Ctenoneurus hochstetteri (Mayr, 1866) E
Ctenoneurus myersi Kormilev, 1953 E
Ctenoneurus pendergrasti Kormilev, 1971 E
Ctenoneurus setosus Lee & Pendergrast, 1977 E
Isodermus crassicornis Usinger & Matsuda, 1959 E
Isodermus maculosus Pendergrast, 1965 E
Isodermus tenuicornis Usinger & Matsuda, 1959 E
Leuraptera yakasi Heiss, 1990 E
Leuraptera zealandica Usinger & Matsuda, 1959 E
Lissaptera completa (Usinger & Matsuda, 1959) E
Mesadenocoris robustus Kirman, 1985 E
Modicarventus wisei Kirman, 1989 E
Neadenocoris abdominalis Usinger & Matsuda, 1959 E
Neadenocoris acutus Usinger & Matsuda, 1959 E
Neadenocoris glaber Usinger & Matsuda, 1959 E

Neadenocoris ovatus Usinger & Matsuda, 1959 E
Neadenocoris reflexus Usinger & Matsuda, 1959 E
Neadenocoris spinicornis Usinger & Matsuda, 1959 E
Neocarventus angulatus Usinger & Matsuda, 1959 E
Neocarventus uncus Kirman, 1989 E
Tretocoris grandis Usinger & Matsuda, 1959 E
Woodwardiessa quadrata Usinger & Matsuda, 1959 E
ARTHENEIDAE
Nothochromus maoricus Slater, Woodward & Sweet, 1962 E
BERYTIDAE
Bezu wakefieldi (White, 1878) E
COREIDAE A
Acantholybas brunneus (Breddin, 1900) A
CYDNIDAE
Chilocoris neozealandicus Larivière & Froeschner, 1994 E
Cydnochoerus nigrosignatus (White, 1878) E
Macroscytus australis (Erichson, 1842)
Microporus thoreyi (Signoret, 1882) A
CYMIDAE
Cymus novaezelandiae Woodward, 1954
HETEROGASTRIDAE A
Heterogaster urticae (Fabricius, 1775) A
LYGAEIDAE
Arocatus rusticus (Stål, 1867) A
Lepiorsillus tekapoensis Malipatil, 1979 E
Nysius convexus (Usinger, 1942) E
Nysius huttoni White, 1878
Nysius liliputanus Eyles & Ashlock, 1969 E
Rhypodes anceps (White, 1878) E
Rhypodes argenteus Eyles, 1990 E
Rhypodes atricornis Eyles, 1990 E
Rhypodes brachypterus Eyles, 1990 E
Rhypodes brevifissas Eyles, 1990 E
Rhypodes brevispilis Eyles, 1990 E
Rhypodes bucculentus Eyles, 1990 E
Rhypodes celmisiae Eyles, 1990 E
Rhypodes chinai Usinger, 1942 E
Rhypodes clavicornis (Fabricius, 1794) E
Rhypodes cognatus Eyles, 1990 E
Rhypodes crinitus Eyles, 1990 E
Rhypodes depilis Eyles, 1990 E
Rhypodes eminens Eyles, 1990 E
Rhypodes gracilis Eyles, 1990 E
Rhypodes hirsutus Eyles, 1990 E
Rhypodes jugatus Eyles, 1990 E
Rhypodes koebelei Eyles, 1990 E
Rhypodes longiceps Eyles, 1990 E
Rhypodes longirostris Eyles, 1990 E
Rhypodes myersi Usinger, 1942 E
Rhypodes rupestris Eyles, 1990 E
Rhypodes russatus Eyles, 1990 E
Rhypodes sericatus Usinger, 1942 E
Rhypodes spadix Eyles, 1990 E
Rhypodes stewartensis Usinger, 1942 E
Rhypodes townsendi Eyles, 1990 E
Rhypodes triangulus Eyles, 1990 E
ORSILLIDAE
Nysius caledoniae Distant, 1920 A
PENTATOMIDAE
Cermatulus nasalis hudsoni Woodward, 1953 E
Cermatulus n. nasalis (Westwood, 1837)
Cermatulus n. turbotti Woodward, 1950 E
Cuspicona simplex Walker, 1867 A
Dictyotus caenosus (Westwood, 1837)
Glaucias amyoti (Dallas, 1851)
Hypsithocus hudsonae Bergroth, 1927 E
Monteithiella humeralis (Walker, 1868) A
Nezara viridula (Linnaeus, 1758) A
Oechalia schellenbergii (Guérin, 1831)
RHYPAROCHROMIDAE
Brentiscerus putoni (White, 1878) E
Dieuches notatus (Dallas, 1852) A
Forsterocoris bisinuatus Woodward, 1953 E
Forsterocoris salmoni (Woodward, 1953) E
Forsterocoris sinuatus Woodward, 1953 E
Forsterocoris stewartensis Malipatil, 1977 E
Geratarma eylesi Malipatil, 1977 E
Geratarma manapourensis Malipatil, 1977 E
Grossander major (Gross, 1965) A
Horridipamera robusta Malipatil, 1978 A
Margareta dominica White, 1878 E
Metagerra angusta Eyles, 1967 E
Metagerra helmsi (Reuter, 1890) E
Metagerra kaikourica Eyles, 1967 E
Metagerra obscura White, 1878 E
Metagerra truncata Malipatil, 1976 E
Millerocoris conus (Eyles, 1967) E
Millerocoris ductus Eyles, 1967 E
Paradrymus exilirostris Bergroth, 1916 A
Paramyocara iridescens Woodward & Malipatil, 1977
Paratruncala insularis (Woodward, 1953) E
Plinthisus (Locutius) woodwardi Slater & Sweet, 1977 A
Regatarma forsteri Woodward, 1953 E
Remaudiereana inornata (Walker, 1872)
Remaudiereana nigriceps (Dallas, 1852)
Stizocephalus brevirostris Eyles, 1970
Targarema electa White, 1878 E
Targarema stali White, 1878 E
Tomocoris ornatus (Woodward, 1953) E
Tomocoris truncatus Woodward, 1953 E
Truncala hirsuta Woodward, 1953 E
Truncala hirta Woodward, 1953 E
Truncala insularis Malipatil, 1977 E
Truncala sulcata Woodward, 1953 E
Trypetocoris aucklandensis Woodward, 1953 E
Trypetocoris rudis Woodward, 1953 E
Trypetocoris separatus Woodward, 1953 E
Udeocoris levis Eyles, 1971 E
Woodwardiana evagorata (Woodward, 1953) E
Woodwardiana nelsonensis (Woodward, 1953) E
Woodwardiana notialis (Woodward, 1953) E
Woodwardiana paparia Malipatil, 1977 E

Order THYSANOPTERA
[Compiled by R. P. Macfarlane]
Suborder TEREBRANTIA
AEOLOTHRIPIDAE
Aeolothrips fasciatus (Linnaeus, 1758) A Cr Ga Gr O Ci
Aeolothrips melaleucus Haliday, 1852 A Fo Cm
Desmidothrips walkerae Mound, 1977 E Sh Ci
MEROTHRIPIDAE A
Merothrips brunneus Hood, 1912 A Fo Nb Li Tw
Merothrips floridensis Watson, 1927 A Fo Nb Lc Li Tw
THRIPIDAE
Adelphithrips cassinae Mound & Palmer, 1980 E Sh Hf
Adelphithrips dolus Mound & Walker, 1982 E Su Sh Hf
Adelphithrips nothofagi Mound & Palmer, 1980 E Fo Hf
Adelphithrips sp. E Fo Hf
Anaphothrips dubius (Girault, 1926) A Ga Hl
Anaphothrips obscurus (Müller, 1776) A Cr Hl
Anaphothrips varii Moulton, 1935 A Gr H
Anaphothrips woodi Pitkin, 1978 A Gr Al
Anaphothrips zealandicus Mound, 1978 E Gr Al Be We
Anaphrygymothrips otagensis Mound & Walker, 1982 E Sa
Apterothrips apteris (Daniel, 1904) A Gr We Sh Ga Hl
Aptinothrips rufus (Haliday, 1836) A Gr Hl
Aptinothrips stylifer Trybom, 1894 A Gr Hl
Ceratothrips ericae (Haliday, 1836) A Sh H
Chirothrips manicatus (Haliday, 1836) A Gr Hf
Dichromothrips spiranthidis (Bagnall, 1926) E Gr H
Dikrothrips diphyes Mound & Walker, 1982 E Fo
Frankliniella intonsa (Trybom, 1895) A Cr Ga H
Frankliniella occidentalis (Fergande, 1895) A Or Sh Hf
Heliothrips haemorrhoidalis (Bouche, 1833) A Cr Sh Hl
Hercinothrips bicinctus (Bagnall, 1919) A Cr Hl
Karphothrips dugdalei Mound & Walker, 1982 E Gr Hl
Limothrips cerealium (Haliday, 1836) A Cr Hl
Lomatothrips paryphis Mound & Walker, 1982 E Fo Mo Li
Microcephalothrips abdominalis (Crawford, 1910) A Hf
Neohydatothrips samayunkur (Kudo, 1996) A Ga
Parthenothrips draconae (Heeger, 1854) A Or Hl
Pezothrips kellyanus (Bagnall, 1926) A Or Ga Hl
Physemothrips chrysodermus Stannard, 1962 E Su Gr
Physemothrips hadrus Mound, 1978 E Su
Pseudanaphothrips achetus (Bagnall, 1916) A Gr Hf
Pseudanaphothrips annettae Mound & Palmer, 1980 E Sa
Scirtothrips inermis Priesner, 1933 A Or Hl
Scirtothrips pan Mound & Walker, 1982 E Fo Hl
Sericothrips staphylinus Haliday, 1836 A Sh Hl
Sigmothrips aotearoana Ward, 1970 E Fo Hl
Tenothrips frici (Uzel, 1895) A Gr Cr Or Hf
Thrips austellus Mound, 1978 E Fo Hf
Thrips australis (Bagnall, 1915) A Or Fo Hf
Thrips coprosmae Mound, 1978 E Sh H
Thrips ?hawaiensis (Morgan, 1913) A Or H
Thrips imaginis Bagnall, 1926 A O Hf
Thrips martini Mound & Masumoto, 2005 E Sh HL
Thrips nigropilosus Uzel, 1895 A Gr Hf
Thrips obscuratus (Crawford, 1941) E Cr Or Hf
Thrips phormicola Mound, 1978 E We Hl
Thrips physapus Linnaeus, 1758 A Gr H
Thrips simplex (Morrison, 1930) A Ga Cr H
Thrips tabaci Lindeman, 1888 A Ga Cr Gr Hf
Thrips vulgatissimus Haliday, 1836 A Ga Hf

Suborder TUBULIFERA
PHLAEOTHRIPIDAE
Anaglyptothrips dugdalei Mound & Palmer, 1983 E Gr H
Apterygothrips australis Pitkin, 1973 A We Gr Li
Apterygothrips collyerae Mound & Walker, 1986 E Fo Sh Nb
Apterygothrips kohai Mound & Walker, 1986 E Gr
Apterygothrips sparsus Mound & Walker, 1986 E Gr Nb
Apterygothrips viretum Mound & Walker, 1986 E We Gr
Azaleothrips neatus Mound & Walker, 1986 E Al
Baeonothrips moundi (Stannard, 1970) A Gr
Carientothrips badius (Hood, 1918) A Gr
Carientothrips loisthus Mound, 1974 A Gr We Li
Cartomothrips manukae Stannard, 1962 E Sh
Cartomothrips neboissi Mound & Walker, 1982 E Sh
Cleistothrips idolothripoides Bagnall, 1932 E Fo
Cryptothrips okiwiensis Mound & Walker, 1986 E Fo
Deplorothrips bassus Mound & Walker, 1986 E Fo Li Tw
Emprosthiothrips bogong Mound, 1969 A Gr
Gynaikothrips ficorum (Marchal, 1908) A Ga HLG
Haplothrips kurdjumovi Karny, 1913 A O Ci Cm
Haplothrips niger (Osborn, 1883) A Cr Hf
Haplothrips salicorniae Mound & Walker, 1986 E We Be
Heptathrips cottieri Mound & Walker, 1986 E Fo Tw
Heptathrips cumberi Mound & Walker, 1986 E Sh Tw
Heptathrips kuscheli Mound & Walker, 1986 E Fo Li
Heptathrips tillyardi Mound & Walker, 1986 E Fo Tw
Heptathrips tonnoiri Moulton, 1942 E Fo Tw
Hoplandothrips bidens (Bagnall, 1910) A Tw
Hoplandothrips choritus Mound & Walker, 1986 E

Fo ?Ff
Hoplandothrips ingenuus Mound & Walker, 1986 E Fo Tw
Hoplandothrips vernus Mound & Walker, 1986 E Fo Tw
Hoplothrips anobii Mound & Walker, 1986 E Sh Tw
Hoplothrips corticus (De Geer, 1773) A Or Fo
Hoplothrips kea Mound & Walker, 1986 E Fo
Hoplothrips orientalis (Anathakristan, 1969) A Tw
Hoplothrips oudeus Mound & Walker, 1986 ?E Fo Mo Li
Hoplothrips poultoni (Bagnell & Kelly, 1929) A Or
Hoplothrips semicaceus (Uzel, 1895) A Su
Idolothrips? spectrum Haliday, 1852 A Fo Ff
Klambothrips annulosus (Priesner, 1928) A Sh Sa
Liothrips vaneeckei Priesner, 1920 A Ga Hb
Lissothrips dentatus Mound & Walker, 1986 E Fo Lc
Lissothrips dugdalei Mound & Walker, 1986 E Fo Mo
Lissothrips gersoni Mound & Walker, 1986 E Fo Mo
Macrophthalmothrips argus (Karny, 1920) A Fo Tw
Nesothrips alexandrae Mound & Walker, 1986 E Or Fo Tw
Nesothrips doulli (Mound, 1974) E Fo
Nesothrips eastopi (Mound, 1974) E Gr We
Nesothrips pintadus Mound & Walker, 1986 E Fo Tw
Nesothrips propinquus (Bagnall, 1916) A Gr Nb Lc Li
Nesothrips rangi Mound & Palmer, 1983 E We
Nesothrips zondagi (Mound, 1974) E Fo Tw
Ozothrips eurytis Mound & Palmer, 1983 E Fo Nb
Ozothrips janus Mound & Palmer, 1983 E We
Ozothrips priscus Mound & Palmer, 1983 E Fo Tw
Ozothrips tubulatus Mound & Walker, 1986 E We Ga
Ozothrips vagus Mound & Walker, 1986 E Tw
Plectrothrips orientalis Okajima, 1981 A
Podothrips orarius Mound & Walker, 1986 E Be Gr
Podothrips turangi Mound & Walker, 1986 E Gr Sa
Poecilothrips albopictus Uzel, 1895 A Or Tw
Priesneriella gnomus Mound & Palmer, 1983 ?E Fo Tw
Psalidothrips moeone Mound & Walker, 1986 E Fo Li
Psalidothrips tane Mound & Walker, 1986 E Fo Tw
Psalidothrips taylori Mound & Walker, 1986 E Fo Mo Li
Sophiothrips alceurodisci Mound & Walker, 1982 E Fo ?Ff
Sophiothrips duvali Mound & Walker, 1982 E Fo
Sophiothrips greensladei Mound & Walker, 1982 A
Strepterothrips tuberculatus (Girault, 1929) E Fo Sh Tw Lc Li
Teuchothrips disjunctus (Hood, 1918) A Sr Hl
Yarnkothrips kolourus Mound & Walker, 1986 E Be Lc

Order PSOCOPTERA
[Compiled by C. Smithers]
Suborder TROGIOMORPHA
LEPIDOPSOCIDAE
Echmepteryx (Oxypsocus) hamiltoni (Tillyard, 1923) E
Echmepteryx (Thylacomorpha) stylesi Smithers, 1969 E
Echmepteryx (Thylacopsis) madagascariensis (Kolbe, 1885)
Pteroxanium kelloggi (Ribaga, 1905)
Pteroxanium marrisi Smithers, 2000 E
PSOQUILLIDAE
Rhyopsocus conformis Smithers, 1969 E
PSYLLIPSOCIDAE
Psyllipsocus ramburii Selys-Longchamps, 1872
TROGIIDAE
Cerobasis guestfalica (Kolbe, 1880) A
Lepinotus inquilinus Heyden, 1850 A
Lepinotus patruelis Pearman, 1931 A
Lepinotus tasmaniensis Hickman, 1934
Trogium evansorum Smithers, 1994
Trogium pulsatorium (Linnaeus, 1758) A

Suborder TROCTOMORPHA
LIPOSCELIDIDAE A
Liposcelis corrodens (Heymons, 1909) A [*Liposcelis divinatorius* Mueller, 1776]

Suborder PSOCOMORPHA
BRYOPSOCIDAE E
Bryopsocus angulatus (Smithers, 1969) E
Bryopsocus townsendi (Smithers, 1969) E
CAECILIUSIDAE
Caecilius sp. 1
Caecilius sp. 2
Maoripsocus fastigatus (Smithers, 1969) E
Maoripsocus semifasciatus Tillyard, 1923
Paracaecilius sp.
Paracaecilius zelandicus (Tillyard, 1923)
Valenzuela flavistigma (Tillyard, 1923) E
Valenzuela flavus (Smithers, 1969) E
ECTOPSOCIDAE
Ectopsocus axillaris (Smithers, 1969)
Ectopsocus briggsi McLachlan, 1899
Ectopsocus californicus (Banks, 1903)
Ectopsocus coronatus Smithers, 1969 E
Ectopsocus dialeptus Thornton & Wong, 1968 E
Ectopsocus gracilis Thornton & Wong, 1968 E
Ectopsocus petersi Smithers, 1978
Ectopsocus sp. 1
Ectopsocus sp. 2
ELIPSOCIDAE
Paedomorpha gayi Smithers, 1963
Pentacladus eucalypti Enderlein, 1906
Propsocus pulchripennis (Perkins, 1899)
Sabulopsocus tractuosus Smithers, 1969 E
Sandrapsocus clarki Smithers, 1999 E
Spilopsocus annulatus Smithers, 1969 E
Spilopsocus avius Smithers, 1964 E
Spilopsocus stigmaticus (Tillyard, 1923) E
MYOPSOCIDAE
Myopsocus australis (Brauer, 1865)
PERIPSOCIDAE
Peripsocus maoricus (Tillyard, 1923)
Peripsocus milleri (Tillyard, 1923)
Peripsocus morulops (Tillyard, 1923)
Peripsocus sp.
PHILOTARSIDAE
Aaroniella rawlingsi Smithers, 1969
Haplophallus maculatus (Tillyard, 1923)
Philotarsopsis basipunctata (Thornton, Wong & Smithers, 1977) E
Philotarsopsis guttatus (Tillyard, 1923)
Philotarsopsis parda (Thornton, Wong & Smithers, 1977) E
PSEUDOCAECILIIDAE
Austropsocus apicipunctatus (Tillyard, 1923) E
Austropsocus australis Thornton, Wong & Smithers, 1977 E
Austropsocus chathamensis Thornton, Wong & Smithers, 1977 E
Austropsocus delli Smithers, 1969 E
Austropsocus fasciatus Thornton, Wong & Smithers, 1977 E
Austropsocus hyalinus Thornton, Wong & Smithers, 1977
Austropsocus insularis Smithers, 1962 E
Austropsocus nimbosus Thornton, Wong & Smithers, 1977 E
Austropsocus ramsayi Thornton, Wong & Smithers, 1977 E
Austropsocus salmoni Smithers, 1969 E
Austropsocus sp.
Chorocaecilius brunellus (Tillyard, 1923)
Mepleres watti (Smithers, 1973) E
Zelandopsocus formosellus Tillyard, 1923 E
Zelandopsocus kuscheli Thornton, Wong & Smithers, 1977 E
Zelandopsocus tectus Thornton, Wing & Smithers, 1977 E
PSOCIDAE
Blaste tillyardi Smithers, 1969
TRICHOPSOCIDAE
Trichopsocus clarus (Banks, 1908)

Order PHTHIRAPTERA
[Compiled by R. Palma]
Suborder AMBLYCERA
BOOPIIDAE A
Boopia notafusca Le Souëf, 1902 A
Heterodoxus ampullatus Kéler, 1971 A
GYROPIDAE A
Gyropus ovalis Burmeister, 1838 A
Gliricola porcelli (Schrank, 1781) A
LAEMOBOTHRIIDAE
Laemobothrion tinnunculi (Linnaeus, 1758)
MENOPONIDAE
Actornithophilus bicolor (Piaget, 1880) M
Actornithophilus ceruleus (Timmermann, 1954) M
Actornithophilus grandiceps (Piaget, 1880) M
Actornithophilus hoplopteri (Mjöberg, 1910)
Actornithophilus limosae (Kellogg, 1908) M
Actornithophilus ochraceus (Nitzsch, 1818) M
Actornithophilus pediculoides (Mjöberg, 1910) M
Actornithophilus piceus (Packard, 1870)
Actornithophilus spinulosus (Piaget, 1880) M
Actornithophilus umbrinus (Burmeister, 1838) M
Amyrsidea (Argimenopon) minuta Emerson, 1961 A
Amyrsidea (A.) perdicis (Denny, 1842) A
Ancistrona vagelli (J.C. Fabricius, 1787) M
Apterygon dumosum Tandan, 1972 E
Apterygon hintoni Clay, 1966 E
Apterygon mirum Clay, 1961 E
Apterygon okarito Palma & Price, 2004 E
Austromenopon aegialitidis (Durrant, 1906)
Austromenopon affine (Piaget, 1890) M
Austromenopon atrofulvum (Piaget, 1880) M
Austromenopon beckii (Kellogg, 1906) M
Austromenopon brevifimbriatum (Piaget, 1880) M
Austromenopon bulweriae Timmermann, 1963 M
Austromenopon elliotti Timmermann, 1954 M
Austromenopon enigki Timmermann, 1963 M
Austromenopon fuscofasciatum (Piaget, 1880) M
Austromenopon haematopi Timmermann, 1954 M
Austromenopon himantopi Timmermann, 1954 M
Austromenopon limosae Timmermann, 1954 M
Austromenopon lutescens (Burmeister, 1838) M
Austromenopon meyeri (Giebel, 1874) M
Austromenopon navigans (Kellogg, 1896) M
Austromenopon ossifragae (Eichler, 1949) M
Austromenopon paululum (Kellogg & Chapman, 1899) M
Austromenopon phaeopodis (Schrank, 1802) M
Austromenopon pinguis (Kellogg, 1896) M
Austromenopon popellus (Piaget, 1890) M
Austromenopon stammeri Timmermann, 1963
Austromenopon transversum (Denny, 1842)
Austromenopon spp. (2) M
Bonomiella columbae Emerson, 1957 A
Ciconiphilus decimfasciatus (Boisduval & Lacordaire, 1835) F
Ciconiphilus pectiniventris (Harrison, 1916) F A
Colpocephalum eucarenum Burmeister, 1838 M
Colpocephalum fregili Denny, 1842 A
Colpocephalum leptopygos Nitzsch *in* Giebel, 1874 F
Colpocephalum pilgrimi Price, 1967
Colpocephalum tausi (Ansari, 1951) A
Colpocephalum turbinatum Denny, 1842
Colpocephalum subzerafae Tendeiro, 1988
Eidmanniella albescens (Piaget, 1880) F
Eidmanniella pellucida (Rudow, 1869) F
Eidmanniella pustulosa (Nitzsch, 1866) F
Eidmanniella subrotunda (Piaget, 1880) F

Eucolpocephalum femorale (Piaget, 1880) F
Franciscoloa (Franciscoloa) pallida (Piaget, 1880) A
Heteromenopon (Keamenopon) kea (Kellogg, 1907) E
Hohorstiella lata (Piaget, 1880) A
Hohorstiella sp.
Holomenopon leucoxanthum (Burmeister, 1838) F
Holomenopon tadornae (Gervais, 1844) F
Holomenopon sp. F
Kurodaia cryptostigmatia (Nitzsch, 1861)
Longimenopon galeatum Timmermann, 1957 M
Menacanthus eurysternus (Burmeister, 1838)
Menacanthus pallidulus Neumann, 1912 A
Menacanthus rhipidurae Palma & Price, 2005 E
Menacanthus stramineus (Nitzsch, 1818) A
Menopon gallinae (Linnaeus, 1758) A
Myrsidea serini (Séguy, 1944) A
Myrsidea thoracica (Giebel, 1874) A
Myrsidea sp. F
Nosopon lucidum (Rudow, 1869)
Plegadiphilus plegadis (Dubinin, 1938) F
Plegadiphilus threskiornis Bedford, 1939 F
Pseudomenopon concretum (Piaget, 1880)
Pseudomenopon pilgrimi Price, 1974 E
Pseudomenopon pilosum (Scopoli, 1763) F
Pseudomenopon scopulacorne (Denny, 1842)
Trinoton nigrum Le Souëf, 1902 F A
Trinoton querquedulae (Linnaeus, 1758) F

Suborder ISCHNOCERA
PHILOPTERIDAE
Acidoproctus gottwaldhirschi Eichler, 1958 F E
Alcedoecus alatoclypeatus (Piaget, 1885) F
Anaticola anseris (Linnaeus, 1758) F A
Anaticola crassicornis crassicornis (Scopoli, 1763) F
Anaticola magnificus Ansari, 1955 F
Anatoecus dentatus dentatus (Scopoli, 1763) F
Anatoecus d. magnicornutus Zlotorzycka, 1970 F A
Anatoecus icterodes icterodes (Nitzsch, 1818) F
Anatoecus i. oloris Zlotorzycka, 1970 F A
Aquanirmus australis Kettle, 1974 F E
Aquanirmus sp. F
Ardeicola expallidus Blagoveshtchensky, 1940 F
Ardeicola neopallidus Price, Hellenthal & Palma, 2003 M
Ardeicola pilgrimi Tandan, 1972 F
Ardeicola plataleae (Linnaeus, 1758) F
Ardeicola rhaphidius (Nitzsch *in* Giebel, 1866) F
Ardeicola stellaris (Denny, 1842) F
Austrogoniodes antarcticus Harrison, 1937 M
Austrogoniodes concii (Kéler, 1952) M
Austrogoniodes cristati Kéler, 1952 M
Austrogoniodes hamiltoni Harrison, 1937 M
Austrogoniodes macquariensis Harrison, 1937 M
Austrogoniodes mawsoni Harrison, 1937 M
Austrogoniodes strutheus Harrison, 1915 M
Austrogoniodes vanalphenae Banks & Palma, 2003 M E
Austrogoniodes waterstoni (Cummings, 1914) M
Bedfordiella unica Thompson, 1937 M
Brueelia amsel (Eichler, 1951) A
Brueelia cyclothorax (Burmeister, 1838) A
Brueelia delicata (Nitzsch *in* Giebel, 1866) A
Brueelia merulensis (Denny, 1842) A
Brueelia nebulosa (Burmeister, 1838) A
Brueelia semiannulata (Piaget, 1883) A
Brueelia turdinulae Ansari, 1956 A
Brueelia sp.
Campanulotes bidentatus compar (Burmeister, 1838) A
Carduiceps cingulatus (Denny, 1842) M
Carduiceps zonarius (Nitzsch *in* Giebel, 1866) M
Chelopistes meleagridis (Linnaeus, 1758) A
Coloceras novaeseelandiae (Tendeiro, 1972) E
Coloceras harrisoni (Tendeiro, 1972) E
Columbicola columbae columbae (Linnaeus, 1758) A
Cuclotogaster heterographus (Nitzsch *in* Giebel, 1866) A
Cuclotogaster synoicus (Clay, 1938) A
Cuculicola kui Kettle, 1980
Cuculicola latirostris (Burmeister, 1838)
Cuculiphilus (Cuculiphilus) fasciativentris Carriker, 1955
Cuculiphilus (C.) *platygaster* (Giebel, 1874)
Degeeriella fusca (Denny, 1842)
Degeeriella rufa rufa (Burmeister, 1838)
Docophoroides brevis (Dufour, 1835) M
Docophoroides harrisoni Waterston, 1917 M
Docophoroides murphyi (Kellogg, 1914) M
Docophoroides simplex (Waterston, 1914) M
Episbates pederiformis (Dufour, 1835) M
Forficuloecus meinertzhageni Guimarães, 1974 E
Forficuloecus pilgrimi Guimarães, 1985 E
Fulicoffula lurida (Nitzsch, 1818) F
Goniocotes chrysocephalus Giebel, 1874 A
Goniocotes gallinae (De Geer, 1778) A
Goniocotes pusillus (Nitzsch (in Giebel), 1866) A
Goniodes colchici Denny, 1842 A
Goniodes dispar Burmeister, 1838 A
Goniodes dissimilis Denny, 1842 A
Goniodes ortygis Denny, 1842 A
Goniodes pavonis (Linnaeus, 1758) A
Goniodes retractus Le Souëf, 1902 A
Goniodes stefani Clay & Hopkins, 1955 A
Haffneria grandis (Piaget, 1880) M
Halipeurus (Halipeurus) bulweriae Timmermann, 1960 M
Halipeurus (H.) *consimilis* Timmermann, 1960 M
Halipeurus (H.) *diversus* (Kellogg, 1896) M
Halipeurus (H.) *falsus pacificus* Edwards, 1961 M
Halipeurus (H.) *gravis priapulus* Timmermann, 1961 M
Halipeurus (H.) *kermadecensis* (Johnston & Harrison, 1912) M
Halipeurus (H.) *leucophryna* Timmermann, 1960 M
Halipeurus (H.) *marquesanus* (Ferris, 1932) M
Halipeurus (H.) *mirabilis* Thompson, 1940 M
Halipeurus (H.) *mundae* Edwards, 1961 M
Halipeurus (H.) *noctivagus* Timmermann, 1960 M
Halipeurus (H.) *placodus* Edwards, 1961 M
Halipeurus (H.) *procellariae* (J.C. Fabricius, 1775) M
Halipeurus (H.) *spadix* Timmermann, 1961 M
Halipeurus (H.) *theresae* Timmermann, 1969 M
Halipeurus (H.) *thompsoni* Edwards, 1961 M
Halipeurus (H.) *turtur* Edwards, 1961 M
Halipeurus sp. M
Halipeurus (Synnautes) pelagicus (Denny, 1842) M
Harrisoniella ferox (Giebel, 1867) M
Harrisoniella hopkinsi Eichler, 1952 M
Ibidoecus bisignatus (Nitzsch *in* Giebel, 1866) F
Ibidoecus dianae Tandan, 1958 F
Ibidoecus plataleae (Denny, 1842) F
Incidifrons fulicae (Linnaeus, 1758) F
Lagopoecus docophoroides (Piaget, 1880) A
Lipeurus caponis (Linnaeus, 1758) A
Lipeurus maculosus maculosus Clay, 1938 A
Lunaceps actophilus (Kellogg & Chapman, 1899) M
Lunaceps drosti Timmermann, 1954 M
Lunaceps incoenis (Kellogg & Chapman, 1899) M
Lunaceps limosella Timmermann, 1954 M
Lunaceps numenii numenii (Denny, 1842) M
Lunaceps n. oliveri (Johnston & Harrison, 1912) M
Lunaceps sp. M
Naubates (Guenterion) clypeatus (Giebel, 1874) M
Naubates (G.) *damma* Timmermann, 1961 M
Naubates (G.) *heteroproctus* Harrison, 1937 M
Naubates (G.) *lessonii* Palma & Pilgrim, 2002 M
Naubates (G.) *prioni* (Enderlein, 1908) M
Naubates (G.) *pterodromi* Bedford, 1930 M
Naubates (Naubates) fuliginosus (Taschenberg, 1882) M
Naubates (N.) *harrisoni* Bedford, 1930 M
Naubates (N.) *thieli* Timmermann, 1965 M
Neopsittaconirmus albus (Le Souëf & Bullen, 1902) A
Neopsittaconirmus kea (Kellogg, 1907) E
Nesiotinus demersus Kellogg, 1903 M
Ornithobius bucephalus (Giebel, 1874) F A
Ornithobius fuscus Le Souëf, 1902 F A
Ornithobius goniopleurus Denny, 1842 F A
Oxylipeurus clavatus (McGregor, 1917) A
Oxylipeurus ellipticus (Kéler, 1958) A
Oxylipeurus mesopelius colchicus Clay, 1938 A
Oxylipeurus polytrapezius polytrapezius (Burmeister, 1838) A
Paraclisis diomedeae (J.C. Fabricius, 1775) M
Paraclisis hyalina (Neumann, 1911) M
Paraclisis miriceps (Kellogg & Kuwana, 1902) M
Paraclisis obscura (Rudow, 1869) M
Pectinopygus annulatus (Piaget, 1880) M
Pectinopygus australis Thompson, 1948 M
Pectinopygus bassani (O. Fabricius, 1780) M
Pectinopygus carunculatus Timmermann, 1964 F E
Pectinopygus dispar (Piaget, 1880) F
Pectinopygus garbei (Pessôa & Guimarães, 1935) F
Pectinopygus gyricornis (Denny, 1842) F
Pectinopygus punctatus Timmermann, 1964 F E
Pectinopygus setosus (Piaget, 1880) F
Pectinopygus turbinatus (Piaget, 1890) F
Pectinopygus varius Timmermann, 1964 F
Pelmatocerandra setosa (Giebel, 1876) M
Perineus circumfasciatus Kéler, 1957 M
Perineus concinnoides Kéler, 1957 M
Perineus macronecti Palma & Pilgrim, 1988 M
Perineus nigrolimbatus (Giebel, 1874) M
Philoceanus fasciatus (Carriker, 1958) M
Philoceanus garrodiae (Clay, 1940) M
Philoceanus robertsi (Clay, 1940) M
Philopteroides novaezelandiae Mey, 2004 E
Philopteroides xenicus Mey, 2004 E
Philopterus novaezealandiae Palma & Price, 2000 E
Philopterus turdi (Denny, 1842) A
Pseudonirmus charcoti (Neumann, 1907) M
Pseudonirmus gurlti (Taschenberg, 1882) M
Pseudonirmus lugubris (Taschenberg, 1882) M
Psittoecus vanzolini Guimarães, 1974 A
Quadraceps assimilis (Piaget, 1890) M
Quadraceps auratus (Haan, 1829) M
Quadraceps birostris (Giebel, 1874) M
Quadraceps caspius (Giebel, 1874) M
Quadraceps cedemajori Timmermann, 1969 M
Quadraceps charadrii charadrii (Linnaeus, 1758) M
Quadraceps coenocoryphae Timmermann, 1955 M E
Quadraceps dominella Timmermann, 1953 M E
Quadraceps ellipticus (Nitzsch *in* Giebel, 1866) M
Quadraceps hemichrous (Nitzsch *in* Giebel, 1866) M
Quadraceps hopkinsi apophoretus Timmermann, 1969 M
Quadraceps h. hopkinsi Timmermann, 1952 M
Quadraceps houri Hopkins, 1949 M
Quadraceps normifer alpha (Kellogg, 1914) M
Quadraceps n. normifer (Grube, 1851) M
Quadraceps n. parvopallidus (Eichler, 1951) M
Quadraceps novaeseelandiae Timmermann, 1953 M
Quadraceps nychthemerus (Burmeister, 1838) M
Quadraceps ornatus fuscolaminulatus (Enderlein, 1908) M
Quadraceps punctatus (Burmeister, 1838)
Quadraceps renschi Timmermann, 1954
Quadraceps ridgwayi (Kellogg, 1906) M
Quadraceps sellatus (Burmeister, 1838) M
Quadraceps semifissus (Nitzsch *in* Giebel), 1866) M
Quadraceps separatus (Kellogg & Kuwana, 1902) M
Quadraceps strepsilaris (Denny, 1842) M
Rallicola (Rallicola) fulicae (Denny, 1842) F
Rallicola (R.) *harrisoni* Emerson, 1955
Rallicola (R.) *lugens* (Giebel, 1874)
Rallicola (R.) *ortygometrae* (Schrank, 1781)

Rallicola (*R.*) *tabuensis* Emerson, 1966
Rallicola (*R.*) *takahe* Holloway, 1956 E
Rallicola (*Aptericola*) *gadowi* Harrison, 1915 E
Rallicola (*A.*) *gracilentus* Clay, 1953 E
Rallicola (*A.*) *pilgrimi* Clay, 1972 E
Rallicola (*A.*) *rodericki* Palma, 1991 E
Rallicola (*Huiacola*) *extinctus* (Mey, 1990) E
Saemundssonia sp. M
Saemundssonia (*Puffinoecus*) *enderleini* (Eichler, 1949) M
Saemundssonia (*P.*) *puellula* Timmermann, 1965 M
Saemundssonia (*P.*) *valida* (Kellogg & Chapman, 1899) M
Saemundssonia (*Saemundssonia*) *albemarlensis* (Kellogg & Kuwana, 1902) M
Saemundssonia (*S.*) *antarctica* (Wood, 1937) M
Saemundssonia (*S.*) *bicolor* (Rudow, 1870) M
Saemundssonia (*S.*) *cephalus* (Denny, 1842) M
Saemundssonia (*S.*) *chathamensis* Timmermann, 1977 M E
Saemundssonia (*S.*) *conica conica* (Denny, 1842) M
Saemundssonia (*S.*) *desolata* Timmermann, 1959 M
Saemundssonia (*S.*) *euryrhyncha* (Giebel, 1874) M
Saemundssonia (*S.*) *gaini* (Neumann, 1913) M
Saemundssonia (*S.*) *haematopi* (Linnaeus, 1758) M
Saemundssonia (*S.*) *hexagona* (Giebel, 1874) M
Saemundssonia (*S.*) *incisa* Timmermann, 1950 M
Saemundssonia (*S.*) *inexspectata* Timmermann, 1951 M
Saemundssonia (*S.*) *lari* (O. Fabricius, 1780)
Saemundssonia (*S.*) *limosae* (Denny, 1842) M
Saemundssonia (*S.*) *lobaticeps remota* Timmermann, 1951 M
Saemundssonia (*S.*) *lockleyi* Clay, 1949 M
Saemundssonia (*S.*) *marina* Timmermann, 1956 M
Saemundssonia (*S.*) *melanocephalus* (Burmeister, 1838) M
Saemundssonia (*S.*) *nereis* Timmermann, 1956 M
Saemundssonia (*S.*) *platygaster* (Denny, 1842) s.l. M
Saemundssonia (*S.*) *p. balati* Timmermann, 1969 M
Saemundssonia (*S.*) *pterodromae* Timmermann, 1959 M
Saemundssonia (*S.*) *scolopacisphaeopodis scolopacisphaeopodis* (Schrank, 1803) M
Saemundssonia (*S.*) *stammeri* Timmermann, 1959 M
Saemundssonia (*S.*) *sternae* (Linnaeus, 1758) M
Saemundssonia (*S.*) *thompsoni* Timmermann, 1951 M
Saemundssonia (*S.*) *tringae* (O. Fabricius, 1780) M
Saemundssonia (*S.*) *uppalensis* (Rudow, 1870) M
Strigiphilus aitkeni Clay, 1966
Strigiphilus cursitans (Nitzsch, 1861) A
Strigiphilus vapidus Clay, 1977
Sturnidoecus sturni (Schrank, 1776) A
Trabeculus aviator (Evans, 1912) M
Trabeculus flemingi Timmermann, 1959 M
Trabeculus fuscoclypeatus (Johnston & Harrison, 1912) M
Trabeculus hexakon (Waterston, 1914) M
Trabeculus mirabilis (Kellogg, 1896) M
Trabeculus schillingi Rudow, 1866 M
TRICHODECTIDAE A
Bovicola (*Bovicola*) *bovis* (Linnaeus, 1758) A
Bovicola (*B.*) *caprae* (Gurlt, 1843) A
Bovicola (*B.*) *limbatus* (Gervais, 1844) A
Bovicola (*B.*) *longicornis* (Nitzsch, 1818) A
Bovicola (*B.*) *ovis* (Schrank, 1781) A
Bovicola (*Spinibovicola*) *hemitragi* (Cummings, 1916) A
Tricholipeurus lipeuroides (Megnin, 1884) A
Tricholipeurus parallelus (Osborn, 1896) A
Felicola (*Felicola*) *subrostratus* (Burmeister, 1838) A
Trichodectes (*Trichodectes*) *canis* (De Geer, 1778) A
Trichodectes (*Stachiella*) *ermineae* (Hopkins 1941) A
Trichodectes (*S.*) *mustelae* (Schrank 1803) A
Werneckiella equi (Denny, 1842) A
Werneckiella ocellata (Piaget, 1880) A

Suborder ANOPLURA
ECHINOPHTHIRIIDAE
Antarctophthirus lobodontis Enderlein, 1909 M
Antarctophthirus ogmorhini Enderlein, 1906 M
Antarctophthirus microchir (Trouessart & Neumann, 1888) M
Lepidophthirus macrorhini Enderlein, 1904 M
HAEMATOPINIDAE A
Haematopinus asini (Linnaeus, 1758) A
Haematopinus eurysternus (Nitzsch, 1818) A
Haematopinus suis (Linnaeus, 1758) A
HOPLOPLEURIDAE A
Hoplopleura pacifica Ewing, 1924 A
LINOGNATHIDAE A
Linognathus ovillus (Neumann, 1907) A
Linognathus pedalis (Osborn, 1896) A
Linognathus setosus (von Olfers, 1816) A
Linognathus stenopsis (Burmeister, 1838) A
Linognathus vituli (Linnaeus, 1758) A
Solenopotes burmeisteri (Fahrenholz, 1919) A
Solenopotes capillatus Enderlein, 1904 A
PEDICULIDAE A
Pediculus humanus capitis (De Geer, 1778) A
Pediculus h. humanus (Linnaeus, 1758) A
POLYPLACIDAE A
Haemodipsus lyriocephalus (Burmeister, 1839) A
Haemodipsus ventricosus (Denny, 1842) A
Polyplax serrata (Burmeister, 1839) A
Polyplax spinulosa (Burmeister, 1839) A
PTHIRIDAE A
Pthirus pubis (Linnaeus, 1758) A

Order MEGALOPTERA
[Compiled by R. P. Macfarlane]
CORYDALIDAE
Archichauliodes diversus (Walker, 1853) E Fr Ci

Order NEUROPTERA
[Compiled by R. P. Macfarlane]
BEROTHIDAE
Protobiella zelandica Tillyard, 1926 E Fo IPr
CONIOPTERYGIDAE A
Cryptoscenea australiensis (Enderlein, 1906) A Or We IPr
Heteroconis ornata Enderlein, 1905 A
MYRMELEONTIDAE
Weeleus acutus (Walker, 1853) E IPr
HEMEROBIIDAE
Drepanacra binocula (Newman, 1838) A Fo Or IPr
Micromus bifasciatus (Walker, 1860) E IPr
Micromus tasmaniae (Walker, 1860) A Gr Cr Ga IPr
Psectra nakaharai New, 1988 A IPr
Wesmaelius subnebulosus (Stephens, 1836) A IPr
OSMYLIDAE
Euosmylus stellae (McLachlan, 1899) E We IPr
Kempynus citrinus (McLachlan, 1873) E We IPr
Kempynus incisus (McLachlan, 1863) E We IPr
Kempynus latiusculus (McLachlan, 1894) E We IPr
SISYRIDAE A
Sisyra rufistigma Tillyard, 1916 A Fr IPr

Order COLEOPTERA
[Compiled by P. Maddison]
Suborder ADEPHAGA
CARABOIDEA
RHYSODIDAE
CLINIDIINI
Rhyzodiastes (*Rhyzoarca*) *proprius* (Broun, 1880) E
RHYSODINI
Kupeus arcuatus (Chevrolat, 1873) E
Kaveinga (*Ingevaka*) *bellorum* Emberson, 1995 E
Kaveinga (*Ingevaka*) *orbitosa* (Broun, 1880) E
Kaveinga (*Vakeinga*) *lusca* (Chevrolat, 1875) E
Tangarona pensus (Broun, 1880) E
CARABIDAE

NEBRIIFORMES
CICINDELLINAE
CICINDELINI
S. CICINDELINA
Cicindela (*Neocicindela*) *austromontana* Bates, 1878 E
Cicindela (*N.*) *brevilunata* Horn, 1926 E
Cicindela (*N.*) *dunedensis* Laporte de Castelnau, 1867 E
Cicindela (*N.*) *feredayi* Bates, 1867 E
Cicindela (*N.*) *hamiltoni* Broun, 1921 E
Cicindela (*N.*) *helmsi* Sharp, 1886 E
Cicindela (*N.*) *latecincta* White, 1846 E
Cicindela (*N.*) *parryi* White, 1846 E
Cicindela (*N.*) *perhispida campbelli* Broun, 1886 E
Cicindela (*N.*) *p. giveni* (Brouerius van Nidek, 1965) E
Cicindela (*N.*) *p. perhispida* Broun, 1880 E
Cicindela (*N.*) *spilleri* (Brouerius van Nidek, 1965) E
Cicindela (*N.*) *tuberculata* Fabricius, 1775 E
Cicindela (*N.*) *waiouraensis* Broun, 1914 E
CARABINAE
PAMBORINI
Maoripamborus fairburni Brookes, 1944 E
Maoripamborus n. sp. E
LOXOMERIFORMES
MIGADOPINAE
AMAROTYPINI
Amarotypus edwardsii Bates, 1872 E
MIGADOPINI
Calathosoma rubromarginatum (Blanchard, 1843) E Su
Loxomerus (*Loxomerus*) *nebrioides* (Guérin-Méneville, 1841) E Su
Loxomerus (*Pristancylus*) *brevis* (Blanchard, 1843) E Su
Loxomerus (*P.*) *capito* Jeannel, 1938 E
Loxomerus (*P.*) *huttoni* (Broun, 1902) E Su
Loxomerus (*P.*) *philpotti* (Broun, 1914) E
SCARITINAE A
CLIVININI
S. CLIVININA
Clivina australasiae Boheman, 1858 A Do
Clivina basalis Chaudoir, 1843 A
Clivina heterogena Putzeys, 1866 A
Clivina vagans Putzeys, 1866 A
MELAENIFORMES
TRECHINAE
BROSCINI
S. CREOBIINA
Bountya insularis Townsend, 1971 E Su
S. NOTHOBROSCINA
Brullea antarctica Laporte de Castelnau, 1867 E
Diglymma castigatum Broun, 1909 E Su
Diglymma clivinoides (Laporte de Castelnau, 1867) E
Diglymma marginale Broun, 1914 E
Diglymma obtusum (Broun, 1886) E
Diglymma seclusum (Johns, 2007) E
Mecodema allani Fairburn, 1945 E
Mecodema a. alternans Laporte de Castelnau, 1867 E
Mecodema a. hudsoni Broun, 1909 E Su
Mecodema angustulum Broun, 1914 E
Mecodema atrox Britton, 1949 E
Mecodema brittoni Townsend, 1965 E
Mecodema bullatum Lewis, 1902 E
Mecodema chiltoni Broun, 1917 E
Mecodema costellum costellum Broun, 1903 E
Mecodema c. gordonense Broun, 1917 E
Mecodema c. lewisi Broun, 1908 E
Mecodema c. obesum Townsend, 1965 E
Mecodema costipenne Broun, 1914 E
Mecodema crenaticolle Redtenbacher, 1868 E
Mecodema crenicolle Laporte de Castelnau, 1867 E
Mecodema curvidens (Broun, 1915) E
Mecodema ducale Sharp, 1886 E
Mecodema dunense Townsend, 1965 E
Mecodema dux Britton, 1949 E

Mecodema elongatum Laporte de Castelnau, 1867 E
Mecodema femorale Broun, 1921 E
Mecodema florae Britton, 1949 E
Mecodema fulgidum Broun, 1881 E
Mecodema gourlayi Britton, 1949 E
Mecodema hector Britton, 1949 E
Mecodema howittii Laporte de Castelnau, 1867 E
Mecodema huttense Broun, 1915 E
Mecodema impressum Laporte de Castelnau, 1867 E
Mecodema infimate Lewis, 1902 E
Mecodema integratum Townsend, 1965 E
Mecodema laeviceps Broun, 1904 E
Mecodema laterale Broun, 1917 E
Mecodema litoreum Broun, 1886 E
Mecodema longicolle Broun, 1923 E
Mecodema lucidum Laporte de Castelnau, 1867 E
Mecodema metallicum Sharp, 1886 E
Mecodema minax Britton, 1949 E
Mecodema morio (Laporte de Castelnau, 1867) E
Mecodema nitidum Broun, 1903 E
Mecodema oblongum (Broun, 1882) E
Mecodema occiputale Broun, 1923 E
Mecodema oconnori Broun, 1912 E
Mecodema oregoides (Broun, 1894) E
Mecodema pavidum Townsend, 1965 E
Mecodema persculptum Broun, 1915 E
Mecodema pluto Britton, 1949 E
Mecodema politanum Broun, 1917 E
Mecodema proximum Britton, 1949 E
Mecodema puiakium Johns & Ewers, 2007 E
Mecodema pulchellum Townsend, 1965 E
Mecodema punctatum (Laporte de Castelnau, 1867) E
Mecodema punctellum Broun, 1921 E
Mecodema quoinense Broun, 1912 E
Mecodema rectolineatum Laporte de Castelnau, 1867 E
Mecodema regulus Britton, 1964 E TK
Mecodema rex Britton, 1949 E
Mecodema rugiceps anomalum Townsend, 1965 E
Mecodema r. rugiceps Sharp, 1886 E
Mecodema sculpturatum puncticolle Broun, 1914 E
Mecodema s. sculpturatum Blanchard, 1843 E
Mecodema simplex Laporte de Castelnau, 1867 E
Mecodema spiniferum Broun, 1880 E
Mecodema striatum Broun, 1904 E
Mecodema strictum Britton, 1949 E
Mecodema sulcatum (Sharp, 1886) E
Mecodema validum Broun, 1923 E
Metaglymma aberrans Putzeys, 1868 E
Metaglymma moniliferum Bates, 1867 E
Metaglymma tibiale (Laporte de Castelnau, 1867) E
Oregus aereus (White, 1846) E
Oregus crypticus Pawson, 2003 E
Oregus inaequalis (Laporte de Castelnau, 1867) E
Oregus septentrionalis Pawson, 2003 E
TRECHINI
S. AEPINA
Kenodactylus audouini (Guérin-Méneville, 1830)
Kiwitrechus karenscottae Larochelle & Larivière, 2007 E
Maoritrechus nunni Townsend, 2010 E
Maoritrechus rangitotoensis Brookes, 1932 E
Maoritrechus stewartensis Townsend, 2010 E
Neanops caecus (Britton, 1960) E
Neanops pritchardi Valentine, 1987 E
Oarotrechus gracilentus Townsend, 2010 E
Waiputrechus cavernicola Townsend, 2010 E
S. HOMALODERINA
Erebotrechus infernus Britton, 1964 E
Kupetrechus gracilis Townsend, 2010 E
Kupetrechus lamberti (Britton, 1960) E
Kupetrechus larsonae Townsend, 2010 E
S. TRECHINA
Duvaliomimus (Duvaliomimus) australis Townsend, 2010 E
Duvaliomimus (D.) chrystallae Townsend, 2010 E
Duvaliomimus (D.) crypticus Townsend, 2010 E
Duvaliomimus (D.) maori (Jeannel, 1928) E
Duvaliomimus (D.) megawattus Townsend, 2010 E
Duvaliomimus (D.) obscurus Townsend, 2010 E
Duvaliomimus (D.) orientalis Giachino, 2005 E
Duvaliomimus (D.) pseudostyx Townsend, 2010 E
Duvaliomimus (D.) styx Britton, 1959 E
Duvaliomimus (D.) taieriensis Townsend, 2010 E
Duvaliomimus (D.) walkeri brittoni Jeannel, 1938 E
Duvaliomimus (D.) w. walkeri (Broun, 1903) E
Duvaliomimus (D.) watti Britton, 1958 E
Duvaliomimus (Mayotrechus) mayae mayae Britton, 1958 E
Duvaliomimus (M.) m. mayorum Townsend, 2010 E
Kettlotrechus edridgei Townsend, 2010 E
Kettlotrechus marchanti Townsend, 2010 E
Kettlotrechus millari Townsend, 2010 E
Kettlotrechus orpheus (Britton, 1962) E
Kettlotrechus pluto (Britton, 1964) E
Scototrechus hardingi hardingi Townsend, 2010 E
Scototrechus h. worthyi Townsend, 2010 E
Scototrechus morti Townsend, 2010 E
Scototrechus orcinus Britton, 1962 E
PSYDRIFORMES
PSYDRINAE
MECYCLOTHORACINI
Mecyclothorax ambiguus (Erichson, 1842) A
Mecyclothorax oopteroides Liebherr & Marris, 2009 E
Mecyclothorax otagoensis Liebherr & Marris, 2009 E
Mecyclothorax rotundicollis (White, 1846) E
MEONINI
Meonochilus amplipennis amplipennis (Broun, 1912) E
Meonochilus a. labralis (Broun, 1912) E
Meonochilus eplicatus (Broun, 1923) E
Meonochilus placens (Broun, 1880) E
Selenochilus fallax (Broun, 1893) E
Selenochilus frontalis (Broun, 1917) E
Selenochilus oculator (Broun, 1893) E
Selenochilus piceus (Blanchard, 1843) E
Selenochilus ruficornis (Broun, 1882) E
Selenochilus syntheticus (Sharp, 1886) E
Selenochilus n. spp. (2) 2E
TROPOPTERINI
Molopsida alpinalis (Broun, 1893) E
Molopsida antarctica (Laporte de Castelnau, 1867) E
Molopsida carbonaria (Broun, 1908) E
Molopsida cincta (Broun, 1893) E
Molopsida convexa (Broun, 1917) E
Molopsida cordipennis (Broun, 1912) E
Molopsida debilis (Sharp, 1886) E
Molopsida diversa (Broun, 1917) E
Molopsida dubia (Broun, 1894) E
Molopsida fovealis (Broun, 1917) E
Molopsida fuscipes (Broun, 1923) E
Molopsida halli (Broun, 1917) E
Molopsida longula (Broun, 1917) E
Molopsida marginalis (Broun, 1882) E
Molopsida optata (Broun, 1917) E
Molopsida oxygona (Broun, 1886) E
Molopsida phyllocharis (Broun, 1912) E
Molopsida polita White, 1846 E
Molopsida pretiosa (Broun, 1910) E
Molopsida propinqua (Broun, 1917) E
Molopsida puncticollis (Sharp, 1883) E
Molopsida robusta (Broun, 1921) E
Molopsida seriatoporus (Bates, 1874) E
Molopsida simplex (Broun, 1903) E
Molopsida simulans (Broun, 1894) E
Molopsida southlandica (Broun, 1908) E
Molopsida strenua (Broun, 1894) E
Molopsida sulcicollis (Bates, 1874) E
ZOLINI
S. ZOLINA
Oopterus atratus (Broun, 1893) E
Oopterus basalis Broun, 1915 E
Oopterus carinatus Broun, 1882 E
Oopterus clivinoides Guérin-Méneville, 1841 E Su
Oopterus collaris Broun, 1893 E
Oopterus femoralis (Broun, 1894) E
Oopterus frontalis Broun, 1908 E
Oopterus fulvipes Broun, 1886 E
Oopterus helmsi (Sharp, 1886) E
Oopterus labralis (Broun, 1921) E
Oopterus laevicollis Bates, 1871 E
Oopterus laevigatus Broun, 1912 E
Oopterus laeviventris (Sharp, 1883) E
Oopterus latifossus Broun, 1917 E
Oopterus latipennis Broun, 1903 E
Oopterus lewisi (Broun, 1912) E
Oopterus marrineri Broun, 1909 E Su
Oopterus minor Broun, 1917 E
Oopterus nigritulus Broun, 1908 E
Oopterus ocularius (Broun, 1917) E
Oopterus pallidipes Broun, 1893 E
Oopterus parvulus Broun, 1903 E
Oopterus patulus (Broun, 1881) E
Oopterus plicaticollis Blanchard, 1843 E Su
Oopterus probus Broun, 1903 E
Oopterus puncticeps Broun, 1893 E
Oopterus pygmeatus Broun, 1907 E
Oopterus sculpturatus ovinotatus Broun, 1908 E
Oopterus s. sculpturatus Broun, 1908 E
Oopterus sobrinus Broun, 1886 E
Oopterus strenuus Johns, 1974 E Su
Oopterus suavis Broun, 1917 E
Oopterus subopacus (Broun, 1915) E
Oopterus spp. n. (21) 21E
Synteratus ovalis Broun, 1909 E Su
BEMBIDIINI
S. BEMBIDIINA
Bembidion (Ananotaphus) rotundicolle eustictum Bates, 1878 E
Bembidion (A.) r. rotundicolle Bates, 1874 E
Bembidion (Notaphus) brullei Gemminger & Harold, 1868 A
Bembidion (Zeactedium) musae Broun, 1882 E
Bembidion (Z.) orbiferum giachinoi Toledano, 2005 E
Bembidion (Z.) o. orbiferum Bates, 1878 E
Bembidion (Zecillenus) alacre (Broun, 1921) E
Bembidion (Z.) albescens (Bates, 1878) E
Bembidion (Z.) chalmeri (Broun, 1886) E
Bembidion (Z.) embersoni (Lindroth, 1980) E
Bembidion (Z.) tillyardi (Brookes, 1927) E
Bembidion (Zemetallina) anchonoderus Bates, 1878 E
Bembidion (Z.) chalceipes Bates, 1878 E
Bembidion (Z.) hokitikense Bates, 1878 E
Bembidion (Z.) parviceps Bates, 1878 E
Bembidion (Z.) solitarium Lindroth, 1976 E
Bembidion (Z.) stewartense Lindroth, 1976 E
Bembidion (Z.) tekapoense Broun, 1886 E
Bembidion (Z.) urewerense Lindroth, 1976 E
Bembidion (Z.) wanakense Lindroth, 1976 E
Bembidion (Zeperyphodes) callipeplum Bates, 1878 E
Bembidion (Zeperyphus) actuarium Broun, 1903 E
Bembidion (Zeplataphus) charile Bates, 1867 E
Bembidion (Z.) dehiscens Broun, 1893 E
Bembidion (Z.) granuliferum Lindroth, 1976 E
Bembidion (Z.) maorinum levatum Lindroth, 1976 E
Bembidion (Z.) m. maorinum Bates, 1867 E
Bembidion (Z.) tairuense Bates, 1878 E
Bembidion (Z.) townsendi Lindroth, 1976 E
S. TACHYINA
Kiwitachys antarcticus (Bates, 1874) E
Kiwitachys latipennis (Sharp, 1886) E
Paratachys crypticola (Britton, 1960) A?
Pericompsus (Upocompsus) australis (Schaum, 1863) A
Polyderis captus Blackburn, 1888 A
S. ANILLINA
Hygranillus kuscheli Moore, 1980 E

Nesamblyops oreobius (Broun, 1893) E
Nesamblyops subcaecus (Sharp, 1886) E
Pelodiaetodes prominens Moore, 1980 E
Pelodiaetodes n. sp. E
Pelodiaetus lewisi Jeannel, 1937 E
Pelodiaetus sulcatipennis Jeannel, 1937 E
Pelodiaetus n. sp. E
Zeanillus pallidus (Broun, 1884) E
Zeanillus phyllobius (Broun, 1893) E
Zeanillus punctiger (Broun, 1914) E
HARPALINAE
PTEROSTICHINI
S. PTEROSTICHINA
Aulacopodus brouni (Csiki, 1930) E
Aulacopodus calathoides (Broun, 1886) E
Aulacopodus maorinus (Bates, 1874) E
Aulacopodus sharpianus (Broun, 1893) E
Aulacopodus n. spp. (2) 2E
Gourlayia regia Britton, 1964 E TK
Holcaspis abdita Johns, 2003 E
Holcaspis algida Britton, 1940 E
Holcaspis angustula (Chaudoir, 1865) E
Holcaspis bathana Butcher, 1984 E
Holcaspis bessatica Johns, 2003 E
Holcaspis bidentella Johns, 2003 E
Holcaspis brevicula Butcher, 1984 E
Holcaspis brouniana (Sharp, 1886) E
Holcaspis catenulata Broun, 1882 E
Holcaspis delator (Broun, 1893) E
Holcaspis dentifera (Broun, 1880) E
Holcaspis egregialis (Broun, 1917) E
Holcaspis elongella (White, 1846) E
Holcaspis falcis Butcher, 1984 E
Holcaspis hispida (Broun, 1877) E
Holcaspis hudsoni Britton, 1940 E
Holcaspis impigra Broun, 1886 E
Holcaspis implica Butcher, 1984 E
Holcaspis intermittens (Chaudoir, 1865) E
Holcaspis mordax Broun, 1886 E
Holcaspis mucronata Broun, 1886 E
Holcaspis obvelata Johns, 2003 E
Holcaspis odontella (Broun, 1908) E
Holcaspis oedicnema Bates, 1874 E
Holcaspis ohauensis Butcher, 1984 E
Holcaspis ovatella (Chaudoir, 1865) E
Holcaspis placida Broun, 1881 E
Holcaspis sinuiventris (Broun, 1908) E
Holcaspis sternalis Broun, 1881 E
Holcaspis stewartensis Butcher, 1984 E
Holcaspis subaenea (Guérin-Méneville, 1841) E
Holcaspis suteri (Broun, 1893) E
Holcaspis tripunctata Butcher, 1984 E
Holcaspis vagepunctata (White, 1846) E
Holcaspis vexata (Broun, 1908) E
Megadromus (Megadromus) alternus (Broun, 1886) E
Megadromus (M.) antarcticus (Chaudoir, 1865) E
Megadromus (M.) asperatus (Broun, 1886) E
Megadromus (M.) bucolicus (Broun, 1903) E
Megadromus (M.) bullatus (Broun, 1915) E
Megadromus (M.) capito (White, 1846) E
Megadromus (M.) compressus (Sharp, 1886) E
Megadromus (M.) curtulus (Broun, 1884) E
Megadromus (M.) enysi (Broun, 1882) E
Megadromus (M.) fultoni (Broun, 1882) E
Megadromus (M.) guerinii (Chaudoir, 1865) E
Megadromus (M.) haplopus (Broun, 1893) E
Megadromus (M.) lobipes (Bates, 1878) E
Megadromus (M.) memes (Broun, 1903) E
Megadromus (M.) meritus (Broun, 1884) E
Megadromus (M.) omaramae Johns, 2007 E
Megadromus (M.) rectalis (Broun, 1881) E
Megadromus (M.) rectangulus (Chaudoir, 1865) E
Megadromus (M.) sandageri (Broun, 1893) E
Megadromus (M.) speciosus Johns, 2007 E
Megadromus (M.) temukensis (Bates, 1878) E
Megadromus (M.) turgidiceps (Broun, 1908) E
Megadromus (M.) vigil (White, 1846) E
Megadromus (M.) virens (Broun, 1886) E
Megadromus (M.) walkeri (Broun, 1903) E
Megadromus (M.) wallacei (Broun, 1912) E
Megadromus (M.) n. spp. (10) 10E
Neoferonia ardua (Broun, 1893) E
Neoferonia edax (Chaudoir, 1878) E
Neoferonia fossalis (Broun, 1914) E
Neoferonia integrata (Bates, 1878) E
Neoferonia prasignis (Broun, 1903) E
Neoferonia procerula (Broun, 1886) E
Neoferonia prolixa (Broun, 1880) E
Neoferonia straneoi Britton, 1940 E
Neoferonia truncatula (Broun, 1923) E
Neoferonia n. sp. (20) 20E
Onawea pantomelas (Blanchard, 1843) E
Plocamostethus planiusculus (White, 1846) E
Plocamostethus scribae Johns, 2007 E
Prosopogmus oodiformis (Macleay, 1871) A
Psegmatopterus politissimus (White, 1846) E
Rhytisternus liopleurus (Chaudoir, 1865) A
Rhytisternus miser (Chaudoir, 1865) A
Zeopoecilus calcaratus (Sharp, 1886) E
Zeopoecilus caperatus Johns, 2007 E
Zeopoecilus putus (Broun, 1882) E
LICININI
S. DICROCHILINA
Dicrochile anchomenoides Guérin-Méneville, 1846 E
Dicrochile anthracina Broun, 1893 E
Dicrochile aterrima Bates, 1874 E
Dicrochile cephalotes Broun, 1894 E
Dicrochile cordicollis Broun, 1903 E
Dicrochile fabrii Guérin-Méneville, 1846 E
Dicrochile flavipes Broun, 1917 E
Dicrochile insignis Broun, 1917 E
Dicrochile maura Broun, 1880 E
Dicrochile nitida Broun, 1882 E
Dicrochile novaezelandiae (Laporte de Castelnau, 1867) E
Dicrochile rugicollis Broun, 1917 E
Dicrochile subopaca Bates, 1874 E
Dicrochile thoracica Broun, 1908 E
Dicrochile whitei (Csiki, 1931) E
S. LICININA
Physolaesthus insularis Bates, 1878
Physolaesthus limbatus (Broun, 1880) E
HARPALINI
S. ANISODACTYLINA
Allocinopus angustulus Broun, 1912 E
Allocinopus belli Larochelle & Larivière, 2005 E
Allocinopus bousqueti Larochelle & Larivière, 2005 E
Allocinopus latitarsis Broun, 1911 E Ch
Allocinopus sculpticollis Broun, 1903 E
Allocinopus smithi Broun, 1912 E
Allocinopus wardi Larochelle & Larivière, 2005 E
Anisodactylus (Anisodactylus) binotatus (Fabricius, 1787) A
Gaioxenus pilipalpis Broun, 1910 E
Gnathaphanus melbournensis (Laporte de Castelnau, 1867) A
Hypharpax antarcticus (Laporte de Castelnau, 1867) E
Hypharpax australis (Dejean, 1829) A
Maoriharpalus sutherlandi Larochelle & Larivière, 2005 E TK
Notiobia (Anisotarsus) quadricollis (Chaudoir, 1878) A?
Parabaris atratus Broun, 1881 E
Parabaris hoarei Larochelle & Larivière, 2005 E
Parabaris lesagei Larochelle & Larivière, 2005 E
Triplosarus novaezelandiae (Laporte de Castelnau, 1867) E
Tuiharpalus clunieae Larochelle & Larivière, 2005 E
Tuiharpalus crosbyi Larochelle & Larivière, 2005 E TK
Tuiharpalus gourlayi (Britton, 1964) E
Tuiharpalus hallae Larochelle & Larivière, 2005 E
Tuiharpalus moorei Larochelle & Larivière, 2005 E
S. HARPALINA
Harpalus (Harpalus) affinis (Schrank, 1781) A
Harpalus (Harpalus) tardus (Panzer, 1797) A
Harpalus australasiae Dejean, 1829 A
S. PELMATELLINA
Hakaharpalus cavelli (Broun, 1893) E
Hakaharpalus davidsoni Larochelle & Larivière, 2005 E
Hakaharpalus maddisoni Larochelle & Larivière, 2005 E
Hakaharpalus patricki Larochelle & Larivière, 2005 E
Hakaharpalus rhodeae Larochelle & Larivière, 2005 E
Kupeharpalus barrattae Larochelle & Larivière, 2005 E
Kupeharpalus embersoni Larochelle & Larivière, 2005 E
Kupeharpalus johnsi Larochelle & Larivière, 2005 E
Lecanomerus atriceps (Macleay, 1871) A
Lecanomerus insignitus Broun, 1880 E
Lecanomerus obesulus Bates, 1878 E
Lecanomerus latimanus Bates, 1874 E
Lecanomerus marrisi Larochelle & Larivière, 2005 E
Lecanomerus sharpi (Csiki, 1932) E
Lecanomerus verticalis (Erichson, 1842) A
Lecanomerus vestigialis (Erichson, 1842) A
Syllectus anomalus Bates, 1878 E
Syllectus gouleti Larochelle & Larivière, 2005 E
Syllectus magnus Britton, 1964 E
S. STENOLOPHINA
Egadroma picea (Guérin-Méneville, 1830) A
Euthenarus bicolor Moore, 1985 A
Euthenarus brevicollis Bates, 1874 E
Euthenarus promptus (Erichson, 1842) A
Euthenarus puncticollis Bates, 1874 E
Haplanister crypticus Moore, 1996 A
Kiwiharpalus townsendi Larochelle & Larivière, 2005 E TK
Pholeodytes cerberus Britton, 1964 E
Pholeodytes helmorei Larochelle & Larivière, 2005 E
Pholeodytes nunni Larochelle & Larivière, 2005 E
Pholeodytes palmai Larochelle & Larivière, 2005 E
Pholeodytes townsendi Britton, 1962 E
PLATYNINI
S. SPHODRINA
Laemostenus (Laemostenus) complanatus (Dejean, 1828) A
S. PLATYNINA
Cerabilia aphela (Broun, 1912) E
Cerabilia major (Broun, 1912) E
Cerabilia maori Laporte de Castelnau, 1867 E
Cerabilia oblonga (Broun, 1910) E
Cerabilia rufipes (Broun, 1893) E
Cerabilia striatula (Broun, 1893) E
Cerabilia n. spp. (15) 15E
Ctenognathus actochares Broun, 1894 E
Ctenognathus adamsi (Broun, 1886) E
Ctenognathus arnaudensis (Broun, 1921) E
Ctenognathus bidens (Chaudoir, 1878) E
Ctenognathus cardiophorus (Chaudoir, 1878) E
Ctenognathus cheesemani (Broun, 1880) E
Ctenognathus colensonis (White, 1846) E
Ctenognathus crenatus (Chaudoir, 1878) E
Ctenognathus deformipes (Broun, 1880) E
Ctenognathus edwardsii (Bates, 1874) E
Ctenognathus helmsi (Sharp, 1881) E
Ctenognathus integratus (Broun, 1908) E
Ctenognathus intermedius (Broun, 1908) E
Ctenognathus libitus (Broun, 1914) E
Ctenognathus littorellus Broun, 1908 E
Ctenognathus lucifugus (Broun, 1886) E
Ctenognathus macrocoelis (Broun, 1908) E
Ctenognathus montivagus (Broun, 1880) E

Ctenognathus munroi Broun, 1893 E
Ctenognathus neozelandicus (Chaudoir, 1878) E
Ctenognathus novaezelandiae (Fairmaire, 1843) E
Ctenognathus oreobius (Broun, 1886) E
Ctenognathus otagoensis (Bates, 1878) E
Ctenognathus parabilis (Broun, 1880) E
Ctenognathus perrugithorax (Broun, 1880) E
Ctenognathus pictonensis Sharp, 1886 E
Ctenognathus politulus (Broun, 1880) E
Ctenognathus punctulatus (Broun, 1877) E
Ctenognathus sandageri (Broun, 1882) E
Ctenognathus simmondsi Broun, 1912 E
Ctenognathus sophronitis (Broun, 1908) E
Ctenognathus suborbithorax (Broun, 1880) E
Ctenognathus sulcitarsis (Broun, 1880) E
Ctenognathus xanthomelas (Broun, 1908) E
Notagonum chathamense (Broun, 1909) E Ch
Notagonum feredayi (Bates, 1874) E
Notagonum lawsoni (Bates, 1874) E
Notagonum submetallicum (White, 1846)
Notagonum n. spp. (2) 2E
Platynus macropterus (Chaudoir, 1879) E
Prosphodrus occultus Britton, 1960 E
Prosphodrus waltoni Britton, 1959 E
PERIGONINI
Perigona (*Trechicus*) *nigriceps* (Dejean, 1831) A
PENTAGONICINI
Pentagonica vittipennis Chaudoir, 1877
Scopodes basalis Broun, 1893 E
Scopodes bryophilus Broun, 1886 E
Scopodes cognatus Broun, 1886 E
Scopodes edwardsii Bates, 1878 E
Scopodes fossulatus (Blanchard, 1843) E
Scopodes laevigatus Bates, 1878 E
Scopodes levistriatus Broun, 1886 E
Scopodes multipunctatus Bates, 1878 E
Scopodes prasinus Bates, 1878 E
Scopodes pustulatus Broun, 1882 E
Scopodes versicolor Bates, 1878 E
LEBIINI
S. PERICALINA
Agonocheila antipodum (Bates, 1867)
Philophlaeus luculentus (Newman, 1842) A
S. ACTENONYCINA
Actenonyx bembidioides White, 1846 E
Actenonyx n. spp. (2) 2E
S. CALLEIDINA
Anomotarus (*Anomotarus*) *illawarrae* (Macleay, 1873) A
Anomotarus (*A.*) *variegatus* Moore, 1967 A
Demetrida (*Demetrida*) *dieffenbachii* (White, 1843) E
Demetrida (*D.*) *lateralis* Broun, 1910 E
Demetrida (*D.*) *lineella* White, 1846 E
Demetrida (*D.*) *longula* Sharp 'n. sp.' [*nomen nudum*] SI
Demetrida (*D.*) *moesta atra* Broun, 1880 E
Demetrida (*D.*) *m. moesta* Sharp, 1878 E
Demetrida (*D.*) *nasuta* White, 1846 E
Demetrida (*D.*) *sinuata maculata* Britton, 1941 E
Demetrida (*D.*) *s. sinuata* Broun, 1917 E
Trigonothops (*Trigonothops*) *pacifica* (Erichson, 1842) A
S. DROMIINA
Dromius (*Dromius*) *meridionalis* Dejean, 1825 A
PSEUDOMORPHINAE A
PSEUDOMORPHINI
Adelotopus macilentus Baehr, 1997 A?
DYTISCIDAE
COLYMBETINAE
COLYMBETINI
Rhantus plantaris Sharp, 1882 E SI F
Rhantus schauinslandi Ordish, 1989 E Ch F
Rhantus suturalis (Macleay, 1825) F
COPELATINAE
COPELATINI
Exocelina australis (Clark, 1863) F-M
DYTISCINAE
CYBISTRINI
Onychohydrus hookeri (White, 1846) E F
Onychohydrus scutellaris (Germar, 1848) V F
DYTISCINI
Dytiscus semisulcatus Müller, 1776 V F
HYDATICINI
Hydaticus consanguineus Aubé, 1838 K F
HYDROPORINAE
BIDESSINI
Allodessus oliveri (Ordish, 1966) [?= *A. bistrigatus* (Clark, 1862)] E K F
Huxelhydrus syntheticus Sharp, 1882 E F
Kuschelydrus phreaticus Ordish, 1976 E F
Liodessus deflectus Ordish, 1966 E F
Liodessus plicatus (Sharp, 1882) E F
Phreatodessus hades Ordish, 1976 E F
Phreatodessus pluto Ordish, 1991 E F
HYDROPORINI
Antiporus femoralis (Boheman, 1858) F
Antiporus uncifer Sharp, 1882 E F
HYPHYDRINI
Hyphydrus (*Apriophorus*) *elegans* (Montrouzier, 1860) V F
LANCETINAE
LANCETINI
Lancetes lanceolatus (Clark, 1863) F
GYRINIDAE
GYRININAE
Gyrinus convexiusculus Macleay, 1871 V F

Suborder POLYPHAGA
STAPHYLINIFORMIA
HYDROPHILOIDEA
HYDROPHILIDAE
HORELOPHINAE
Horelophus walkeri d'Orchymont, 1913 E
HYDROPHILINAE
SPERCHOPSINI
Cylomissus glabratus Broun, 1903 E
BEROSINI
Berosus (*Phelerosus*) *pallidipennis* (Sharp, 1884) E F
Berosus n. spp. (2) 2E F
ANACAENINI
Paracymus pygmaeus (Macleay, 1871) F
Paracymus n. sp. E F
LACCOBIINI
Laccobius (*Platylaccobius*) *arrowi* d'Orchymont, 1925 E F
HYDROPHILINI
Enochrus (*Lumetus*) *abditus* (Sharp, 1884) E SI F
Enochrus (*Lumetus*) *tritus* (Broun, 1880) F
Limnoxenus zealandicus (Broun, 1880) F
Sternolophus (*Neosternolophus*) *marginicollis* (Hope, 1841) A Do F
SPHAERIDIINAE
COELOSTOMATINI
Adolopus altulus (Broun, 1880) E
Adolopus badius (Broun, 1880) E
Adolopus convexus Broun, 1893 E
Adolopus helmsi Sharp, 1884 E
Adolopus rugipennis Broun, 1886 E
Cyloma flemingi (Ordish, 1974) E Su
Cyloma guttulatus Sharp, 1884 E
Cyloma lawsonus Sharp, 1872 E
Cyloma lineatus (Broun, 1893) E
Cyloma nigratus (Broun, 1915) E
Cyloma pictus (Kirsch, 1877) E Su
Cyloma stewarti Broun, 1894 E
Cyloma thomsonus Sharp, 1884 E
Dactylosternum abdominale (Fabricius, 1792) A
Dactylosternum marginale (Sharp, 1876) A
MEGASTERNINI
Cercyodes laevigatus Broun, 1886 E
Cercyon (*Cercyon*) *depressus* Stephens, 1829 A
Cercyon (*C.*) *haemorrhoidalis* (Fabricius, 1775) A
Cercyon (*C.*) *nigriceps* (Marsham, 1802) A
Cercyon (*Paracercyon*) *analis* (Paykull, 1798) A
RYGMODINI
Rygmodus alienus Broun, 1893 E
Rygmodus antennatus (Sharp, 1884) E
Rygmodus cyaneus Broun, 1881 E
Rygmodus femoratus Sharp, 1884 E
Rygmodus incertus Broun, 1880 E
Rygmodus longulus (Sharp, 1884) E
Rygmodus modestus White, 1846 E
Rygmodus oblongus Broun, 1880 E
Rygmodus opimus Broun, 1880 E
Rygmodus pedinoides White, 1846 E
Rygmodus tibialis Broun, 1893 E
Rygmodus n. sp. E
Saphydrus monticola Broun, 1893 E
Saphydrus obesus Sharp, 1884 E
Saphydrus suffusus Sharp, 1884 E
SPHAERIDIINI
Sphaeridium lunatum Fabricius, 1792 A
TORMISSINI
Tormus femoralis (Broun, 1910) E
Tormus helmsi Sharp, 1884 E
Tormus nitidus (Broun, 1893) E
Exydrus gibbosus (Broun, 1880) E
Hydrostygnus frontalis (Broun, 1880) E
Tormissus guanicola (Broun, 1904) E Su
Tormissus linsi (Sharp, 1884) E
Tormissus sp. n. E TK
HISTERIDAE
ABRAEINAE A
ACRITINI
Acritus nigricornis (Hoffman, 1803) A
SAPRININAE
Gnathoncus communis (Marseul, 1862) A
Gnathoncus rotundatus (Kugelann, 1792) A
Neopachylopus lepidulus (Broun, 1881) E
Reichardtia pedatrix (Sharp, 1876) E
Saprinus detritus (Fabricius, 1775) E
Saprinus n. spp. (2) 2E Ch
Tomogenius australis Dahlgren, 1976 E
Tomogenius kuscheli Dahlgren, 1976 E
Tomogenius latipes (Broun, 1881) E
Tomogenius n. sp. E Ch
DENDROPHILINAE
'Abraeus' brunneus Broun, 1881 E
'Abraeus' vividulus Broun, 1880 E
PAROMALINI
Carcinops (*Carcinops*) *pumilio* (Erichson, 1834) A
HISTERINAE
PLATYSOMATINI
Aulacosternus zelandicus Marseul, 1853 E
Eblisia bakewelli (Marseul, 1864) A
Eblisia carolinum (Paykull, 1811) A Do SI
ONTHOPHILINAE
Parepierus abrogatus (Broun, 1886) E
Parepierus crenulatus (Broun, 1886) E
Parepierus planiceps (Broun, 1886) E
Parepierus punctulipennis (Broun, 1880) E
Parepierus purus (Broun, 1880) E
Parepierus rufescens (Reitter, 1880) E
Parepierus rusticus (Broun, 1886) E
Parepierus simplex (Broun, 1886) E
Parepierus spinellus (Broun, 1921) E
Parepierus sylvanus (Lewis, 1879) E
Tribalus brouni (Broun, 1880) E
Tribalus phyllobius (Broun, 1914) E
STAPHYLINOIDEA
HYDRAENIDAE
ORCHYMONTIINAE
Podaena dentipalpis Ordish, 1984 E F
Podaena glabriventris Ordish, 1984 E F
Podaena kuscheli Ordish, 1984 E F

Podaena latipalpis Ordish, 1984 E F
Podaena maclellani (Zwick, 1975) E F
Podaena obscura Ordish, 1984 E F
Podaena trochanteralis Ordish, 1984 E F
Homalaena acuta Ordish, 1984 E F
Homalaena carinata Ordish, 1984 E F
Homalaena dilatata Ordish, 1984 E F
Homalaena dispersa Ordish, 1984 E F
Homalaena nelsonensis Ordish, 1984 E F
Homalaena setosa Ordish, 1984 E F
Homalaena spatulata Ordish, 1984 E F
Orchymontia banksiana Ordish, 1984 E F
Orchymontia bidentata Ordish, 1984 E F
Orchymontia calcarata Ordish, 1984 E F
Orchymontia ciliata Ordish, 1984 E F
Orchymontia crassifemur Ordish, 1984 E F
Orchymontia curvipes Ordish, 1984 E F
Orchymontia dilatata Ordish, 1984 E F
Orchymontia dugdalei Ordish, 1984 E F
Orchymontia laminifera Ordish, 1984 E F
Orchymontia latispina Ordish, 1984 E F
Orchymontia nunni Delgado & Palma 2000 E F
Orchymontia otagensis Ordish, 1984 E F
Orchymontia spinipennis Broun, 1919 E F
Orchymontia vulgaris Ordish, 1984 E F
HYDRAENINAE
HYDRAENINI
Hydraena (Hydraena) ambiflagellata Zwick, 1977 A F
Hydraena (H.) ordishi Delgado & Palma, 1997 E F
Hydraena (H.) zealandica Ordish, 1984 E F
OCTHEBIINAE
OCHTHEBIINI
S. MEROPATHINA
Meropathus aucklandicus Ordish, 1971 E Su
Meropathus campbellensis Brookes, 1951 E Su
Meropathus johnsi Ordish, 1971 E Su
Meropathus zelandicus Ordish, 1984 E
PTILIIDAE
PTILIINAE
NANOSELLINI
Mikado sp. nr *parvicornis* (Deane, 1932) E
Nellosana elegantula Johnson, 1982 E
Nellosana grandis Johnson, 1982 E
Nellosana intermedia Johnson, 1982 E
Nellosana minima Johnson, 1982 E
DISCHERAMOCEPHALINI
Cissidium crowsoni Johnson, 1982 E
Cissidium foveolatum Johnson, 1982 E
PTILIINI
Actidium angulicolle Johnson, 1982 E
Actidium delicatulum Johnson, 1982 E
Actidium lineare Matthews, 1874 E
Dipentium zelandicum Johnson, 1982 E
Kuschelidium maori Johnson, 1982 E
Notoptenidium apterum Johnson, 1982 E TK
Notoptenidium aubrooki Johnson, 1982 E
Notoptenidium crassum Johnson, 1982 E
Notoptenidium kuscheli Johnson, 1982 E
Notoptenidium lawsoni (Matthews, 1873) E
Notoptenidium oblongum Johnson, 1982 E
Notoptenidium parvum Johnson, 1982 E
Notoptenidium similatum Johnson, 1982 E
Notoptenidium sparsum Johnson, 1982 E
Notoptenidium subitum Johnson, 1982 E
Oligella foveolata (Allibert, 1844) A
Ptenidium (Ptenidium) laevigatum Erichson, 1845 A
Ptenidium (P.) punctatum (Gyllenhal, 1827) A
Ptenidium (P.) pusillum (Gyllenhal, 1808) A
Ptenidium n. sp. E Ch
Ptenidotonium longicorne Johnson, 1982 E
PTINELLINI
Ptinella acaciae Johnson, 1982 E
Ptinella atrata Johnson, 1975 E Su
Ptinella bitumida Johnson, 1982 E Ch
Ptinella brunnescens Johnson, 1982 E Ch
Ptinella cavelli (Broun, 1893) E
Ptinella chathamensis Johnson, 1982 E Ch
Ptinella confusa Johnson, 1982 E
Ptinella errabunda Johnson, 1975 A Do
Ptinella fallax Johnson, 1975 E K
Ptinella ferruginea Johnson, 1982 E
Ptinella kermadecensis Johnson, 1975 E K
Ptinella lucida Johnson, 1982 E
Ptinella octopunctata Johnson, 1975 E Su
Ptinella propria (Broun, 1893) E
Ptinella pustulata Johnson, 1982 E
Ptinella simsoni (Matthews, 1878)
Ptinella snarensis Johnson, 1975 E Su
Ptinella taylorae Johnson, 1977 E
Ptinella watti Johnson, 1982 E
Ptinella n. spp. (2) E Ch
Ptinella n. sp. E K
ACROTRICHINAE A
ACROTRICHINI
Acrotrichis (Acrotrichis) fascicularis (Herbst, 1793) A Do
Acrotrichis (A.) inconspicua (Matthews, 1874) A
Acrotrichis (A.) insularis (Mäklin, 1852) A
Acrotrichis (A.) josephi (Matthews, 1872) A
Acrotrichis (Ctenopteryx) montandoni (Allibert, 1844) A
NEPHANINI
Nephanes titan (Newman, 1834) A
Ptiliodes amplicollis Johnson, 1982 E
Ptiliodes austerus Johnson, 1982 E
Ptiliodes curtus Johnson, 1982 E
Ptiliodes naufragus Johnson, 1982 E TK
Ptiliodes posticalis (Broun, 1893) E
AGYRTIDAE
NICROPHILINAE
Zeanecrophilus prolongatus (Sharp, 1881) E
Zeanecrophilus thayerae Newton, 1997 E
LEIODIDAE
CAMIARINAE
AGYRTODINI
Agyrtodes bicolor (Broun, 1880) E
Agyrtodes hunuensis (Broun, 1893) E
Agyrtodes labralis (Broun, 1921) E
Agyrtodes lescheni Seago, 2009 E
Agyrtodes monticola (Broun, 1893) E
Agyrtodes nebulosus (Broun, 1880) E
Agyrtodes nemoralis (Broun, 1909) E
Chelagyrtodes crowsoni Szymczakowski, 1973 E
Chelagyrtodes '*davidi* Seago, 2008' [*nomen nudum*] E SI
Chelagyrtodes '*glacicola* Seago, 2008' [*nomen nudum*] E SI
Chelagyrtodes '*haasti* Seago, 2008' [*nomen nudum*] E SI
Chelagyrtodes '*newtoni* Seago, 2008' [*nomen nudum*] E SI
Chelagyrtodes '*nunni* Seago, 2008' [*nomen nudum*] E SI
Chelagyrtodes '*rotundus* Seago, 2008' [*nomen nudum*] E SI
Zeagyrtes antennalis (Broun, 1880) E
Zeagyrtes vitticollis Broun, 1917 E
Zeagyrtoma separanda Szymczakowski, 1966 E
Zeagyrtoma undulata (Broun, 1880) E
Zearagytodes brouni Jeannel, 1936 E
Zearagytodes concinnus (Broun, 1880) E
Zearagytodes maculifer (Broun, 1880) E
CAMIARINI
Baeosilpha rufescens Broun, 1895 E
Camiarites convexus (Sharp, 1876) E
Camiarites indiscretus (Broun, 1880) E
Camiarus estriatus Broun, 1912 E
Camiarus thoracicus (Sharp, 1876) E
Inocatops compactus Broun, 1893 E
Inocatops concinnus (Broun, 1880) E
Inocatops elongellus Broun, 1917 E
Inocatops flectipes Broun, 1893 E
Inocatops granipennis Broun, 1917 E
Inocatops impressus Broun, 1921 E
Inocatops nigrescens Broun, 1893 E
Inocatops separatus Broun, 1917 E
Inocatops spinifer Broun, 1917 E
Zenocolon laevicollis Broun, 1917 E
Camiarini gen. nov. (2) et n. spp. 2E
NEOPELATOPINI
Catopsolius laevicollis Sharp, 1886 E
Catopsolius nitidus (Broun, 1893) E
CHOLEVINAE
ANEMADINI
S. EUNEMADINA
Pseudonemadus (Pseudonemadus) lituratus (Broun, 1880) E
S. PARACATOPINA
Mesocolon caecum (Broun, 1912) E
Mesocolon castaneum (Broun, 1912) E
Mesocolon clathratum Broun, 1880 E
Mesocolon crassipes Szymczakowski, 1973 E
Mesocolon crenatellum (Broun, 1921) E
Mesocolon microps Jeannel, 1936 E
Mesocolon nesobium Jeannel, 1936 E
Mesocolon puncticeps Broun, 1880 E
Mesocolon retractum Szymczakowski, 1963 E
Paracatops acantharius Szymczakowski, 1973 E
Paracatops alacris (Broun, 1880) E
Paracatops antipoda (Kirsch, 1877) E Su
Paracatops brounianus Jeannel, 1936 E
Paracatops brunneipes (Broun, 1911) E Ch
Paracatops campbellicus (Brookes, 1951) E Su
Paracatops dickensis Jeannel, 1936 E
Paracatops fulvitarsis (Broun, 1886) E
Paracatops granifer (Broun, 1886) E
Paracatops lugubris (Sharp, 1882) E
Paracatops phyllobius (Broun, 1893) E
Paracatops pogonomerus Szymczakowski, 1973 E
Paracatops relatus (Broun, 1893) E
Paracatops suturalis (Broun, 1895) E
Paracatops triangulus Jeannel, 1936 E
COLONINAE
Colon (Mesagyrtes) hirtale (Broun, 1880) E
LEIODINAE
LEIODINI
Zeadolopus maoricus Daffner, 1985 E
Zeadolopus spinipes Broun, 1903 E
Zeadolopus validipes Daffner, 1985 E
PSEUDOLIODINI
Colenisia zelandica Leschen 2000 E
Zelodes kuscheli Leschen, 2000 E
Zelodes minutus Leschen, 2000 E
SOGDINI
Isocolon frontale Broun, 1921 E
Isocolon hilare Broun, 1893 E
Isocolon modestum Broun, 1921 E
Isocolon oruruense Broun, 1923 E
Isocolon ovale (Broun, 1893) E
STAPHYLINIDAE
ALEOCHARINAE
ALEOCHARINI
S. ALEOCHARINA
Aleochara (Aleochara) aucklandica Klimaszewski, 1997 E
Aleochara (A.) complexa Klimaszewski, 1997 E K
Aleochara (A.) hammondi Klimaszewski, 1997 E
Aleochara (A.) subaenea Fauvel, 1878 E
Aleochara (A.) watti Klimaszewski, 1997 E TK
Aleochara (Xenochara) puberula Klug, 1834 A
ATHETINI
S. ATHETINA
Acronota aterrima (Gravenhorst, 1802) A? Do
Acronota lugens (Motschulsky, 1858) A
Aloconota (Aloconota) planifrons (Waterhouse, 1863) A

Aloconota plicata (Cameron, 1945) E
Aloconota sulcifrons (Stephens, 1832) A
Amischa analis (Gravenhorst, 1802) A
Amischa decipiens (Sharp, 1869) A
Amischa nigrofusca (Stephens, 1832) A
Amriathaea antipodum (Cameron, 1947) E
Amriathaea microps Cameron, 1948 E
Amriathaea n. spp. (14) 14E
Atheta amicula (Stephens, 1832) A
Atheta brouni Bernhauer & Scheerpeltz, 1926 E
Atheta cottieri Cameron, 1945 E
Atheta kingorum Klimaszewski & Marris, 2003 E TK
Atheta luridipennis (Mannerheim, 1830) A
Atheta maruiana Cameron, 1950 E
Atheta muggeridgei Bernhauer, 1943 E
Atheta nigra (Kraatz, 1856) A
Atheta splendidicollis Bernhauer, 1943 E
Atheta trinotata (Kraatz, 1856) A
Atheta zealandica Cameron, 1945 A
Atheta n. sp. E
Atheta n. spp. (4) 4A
Brundinia semipallidula (Bernhauer, 1943) E
Dalotia coriaria (Kraatz, 1856) A
'Geostiba' n. sp. E? SI
Geostibasoma antipodum (Bernhauer, 1943) E
Halobrecta algophila (Fenyes, 1909) A
Halobrecta flavipes Thomson, 1861 A?
Leptostiba neozelandensis Pace, 2003 E
Leptostiba politula (Fauvel, 1878) A
Leptostiba pseudopolitula Pace, 2003 E
Liogluta n. sp. A?
Mocyta fungi (Gravenhorst, 1806) A
Nehemitropia lividipennis (Mannerheim, 1830) A
Tramiathea cornigera (Broun, 1880) E
S. THAMIARAEINA
Thamiaraea aucklandica Cameron, 1950 E
Thamiaraea fuscicornis (Broun, 1880) E
DIGLOTTINI
Paradiglotta nunni Ashe & Ahn, 2005 E
DIGRAMMINI
Digrammus miricollis Fauvel, 1900 E
FALAGRIINI
Cordalia obscura (Gravenhorst, 1802) A
Dasytricheta funesta (Broun, 1912) E
Dasytricheta haastiana Paśnik, 2007 E
Dasytricheta hookeriana Paśnik, 2007 E
Dasytricheta intermedia Paśnik, 2007 E
Dasytricheta kapuniana Paśnik, 2007 E
Dasytricheta mahitahiana Paśnik, 2007 E
Dasytricheta periana Paśnik, 2007 E
Dasytricheta shotoveriana Paśnik, 2007 E
Dasytricheta spectabilis Bernhauer, 1943 E
Dasytricheta testacea Paśnik, 2007 E
Dasytricheta waihoana Paśnik, 2007 E
Ecomorypora densepunctata Paśnik, 2007 E
Ecomorypora granulata (Broun, 1912) E
Ecomorypora longelytrata Paśnik, 2007 E
Ecomorypora pseudogranulata Paśnik, 2007 E
Ecomorypora n. spp. (3) 3E
Falagria subopaca Broun, 1893 E
Falagria n. spp. (2) E, A
Galafria rufa Cameron, 1945 A
Myrmecocephalus concinnus (Erichson, 1839) A
Myrmecocephalus micans (Broun, 1880) E
Myrmecopora paradoxa Bernhauer, 1943 E
Plesiosipalia arrowi Bernhauer, 1943 E
Plesiosipalia n. sp. E
GYMNUSINI
Stylogymnusa subantarctica Hammond, 1975 E Su
HOMALOTINI
S. BOLITOCHARINA
Austrasilida zealandica (Cameron, 1948) E
Euryusa aliena Cameron, 1945 E
Phymatura neozelandensis Pace 2003 E
S. GYROPHAENINA
?Adelarthra n. sp. A?
'Brachida' minuta Bernhauer, 1941 E
Encephalus latulus Broun, 1894 E
Encephalus zealandicus Cameron, 1945 E
Gyrophaena brookesi Cameron, 1947 E
Gyrophaena densicornis Broun, 1880 E
Gyrophaena glabricollis Bernhauer, 1941 E
Gyrophaena nugax Broun, 1880 E
Gyrophaena oligotina Cameron, 1945 E
Gyrophaena punctata Broun, 1880 E
Gyrophaena n. spp. (6) 6E
Notiomerinx zealandica Ashe, 2003 E
Pseudoligota n. sp. E?
Sternotropa versicolor Broun, 1880 E
S. HOMALOTINA
Coenonica puncticollis Kraatz, 1857 A
Homalota n. spp. (2) 2E
Leptusa (Halmaeusa) antarctica (Kiesenwetter, 1877) E Su
Pseudopisalia turbotti Cameron, 1950 E TK
Pseudopisalia n. sp. E
Silusa parallela Bernhauer, 1943 E
Silusa puber (Broun, 1880) E
Silusa n. sp. (2) 2E
Stenomastax dentata Cameron, 1945 E
Stenomastax sulcicollis Cameron, 1945 E
HYPOCYPHTINI
Oligota carinulata (Broun, 1914) E
Oligota excavata Williams, 1976 E
Oligota ferruginea Williams, 1976 E
Oligota fungicola Williams, 1976 E
Oligota grandis Williams, 1976 E
Oligota hudsoni Williams, 1976 E
Oligota inconspicua Williams, 1976 E
Oligota inflata (Mannerheim, 1930) A
Oligota longula Cameron, 1945 E
Oligota masculina (Cameron, 1947) E
Oligota parva Kraatz, 1862 A
Oligota pumilio Kiesenwetter, 1858 A
Oligota punctum Williams, 1976 E
Oligota setigera Williams, 1976 E
Oligota speculicollis (Cameron, 1945) E
Oligota transversalis Williams, 1976 E
Oligota watti Williams, 1976 E
Oligota wendyi Williams, 1976 E
Oligota zealandica Bernhauer, 1941 E
LIPAROCEPHALINI
Baeostethus chiltoni Broun, 1909 E Su
Ianmoorea zealandica (Ahn, 2004) E
MESOPORINI
Paraconosoma naviculare Bernhauer, 1941 E
Paraconosoma polita Steel, 1960 E
Paraconosoma n. sp. E
MYLLAEINI
Myllaena intermedia Erichson, 1837 A
Myllaena magnicollis Cameron, 1947 E
Myllaena neozelandensis Pace, 2008 E
OXYPODINI
Anocalea n. sp. A?
Aphytopus gracilis Sharp, 1886 E
Aphytopus granifer Broun, 1912 E
Aphytopus guinnessi Broun, 1912 E
Aphytopus pictulus Broun, 1914 E
Aphytopus porosus Broun, 1912 E
Austrocalea brookesi Cameron, 1950 E?
Austrocalea lewisi Cameron, 1948 E
Brouniana lucida (Cameron, 1945) E
Brouniana vulcanica (Cameron, 1945) E
Brouniana n. sp. (2) 2E
Calodera diversa Broun, 1894 E
Calodera fultoni Broun, 1912 E
Calodera glabra Bernhauer, 1943 E
Calodera grandipennis Bernhauer, 1943 E
Calodera minima Bernhauer, 1943 E
Calodera rhopalicornis Bernhauer, 1943 E
Calodera sericophora Broun, 1894 E
Calodera strandi Bernhauer, 1943 E
Calodera thoracica (Broun, 1880) E
Calodera tumidella Broun, 1894 E
Calodera vestita Broun, 1894 E
Calodera wallacei Broun, 1912 E
Colle campbellensis Steel, 1964 E Su
Crataraea suturalis (Mannerheim, 1830) A
Gastrolamprusa helmsi (Bernhauer, 1941) E
'Gyronotus' rufipennis (Broun, 1880) E
Ischnoglossa bituberculata (Broun, 1894) E
Ischnoglossa parciventris Cameron, 1945 E
Ischnoglossa pectinata Cameron, 1945 E
Ischnoglossa rufa Cameron, 1945 E
Makara hudsoni (Cameron, 1945) E
Neodoxa giachinoi Pace 2008 E
Neodoxa secreta (Cameron, 1950) E TK
Neozelandusa giachinoi Pace 2008 E
Ocalea abdominalis (Cameron, 1945) E
Ocalea brouni Cameron, 1945 E
Ocalea fungicola (Broun, 1894) E
Ocalea hudsoni Cameron, 1945 E
Ocalea rufa (Cameron, 1945) E
Ocalea socialis (Broun, 1880) E
Ocalea suturalis Cameron, 1945 E
Ocalea zelandica Klimaszewski and Marris, 2003 E TK
Ocyusa brouni Bernhauer, 1943 E
Ocyusa brouniana Bernhauer, 1943 E
Ocyusa n. sp. E
Oxypoda haemorrhoa (Mannerheim, 1830) A
Oxypoda zelandica Bernhauer, 1943 E
Paraphytopus brookesi Cameron, 1948 E
Paraphytopus minutus Cameron, 1948 E
Paraphytopus n. sp. E?
Polylobus sternalis (Broun, 1880) E
Sytus aerarius (Broun, 1880) E
Sytus bifossuta (Cameron, 1945) E
Sytus curiosus (Bernhauer, 1941) E
Sytus flavescens (Broun, 1880) E
Sytus fulgens (Broun, 1880) E
Sytus granifer (Broun, 1894) E
Sytus optabilis (Broun, 1880) E
Sytus n. spp. (2) 2E
PHYTOSINI
Arena fultoni Cameron, 1945 E
Aleocharinae n. spp. (10) 5E, 5A
EUAESTHETINAE
AUSTROESTHETINI
Kiwiaesthetus biimpressus Puthz, 2008 E
Kiwiaesthetus carltoni Puthz, 2008 E
Kiwiaesthetus kieneri Puthz, 2008 E
Kiwiaesthetus kuscheli Puthz, 2008 E
Kiwiaesthetus lescheni Puthz, 2008 E
Kiwiaesthetus ramsayi Puthz, 2008 E
Kiwiaesthetus whitehorni (Broun, 1912) E
Kiwiaesthetus n. sp. E
EUASTHETINI
Edaphus beszedesi Reitter, 1914 A
Protopristus minutus Broun, 1909 E
Protopristus n. spp. (2) 2E
STENAESTHETINI
Agnosthaetus bisulciceps (Broun, 1917) E
Agnosthaetus brouni (Broun, 1910) E
Agnosthaetus brouni Bernhauer, 1939, non *Dimerus brouni* (Broun, 1910) [homonym] E
Agnosthaetus cariniceps Bernhauer, 1939 E
Agnosthaetus stilbus (Broun, 1910) E
Agnosthaetus vicinus (Broun, 1921) E
HABROCERINAE A
Habrocerus capillaricornis (Gravenhorst, 1806) A
MICROSILPHINAE
Microsilpha litorea Broun, 1886 E
Microsilpha n. spp. (15) 15E
OMALIINAE
CORNEOLABIINI

<u>Corneolabium</u> *mandibulare* Steel, 1950 E
Metacorneolabium convexum Thayer, 1985 E
Metacorneolabium gigas Thayer, 1985 E
Metacorneolabium hokitika Thayer, 1985 E
Metacorneolabium minus Steel, 1950 E
Metacorneolabium pyriforme Thayer, 1985 E
Metacorneolabium rangipo Thayer, 1985 E
Metacorneolabium zanotium Thayer, 1985 E
Metacorneolabium n. spp. (2) 2E
<u>Paracorneolabium</u> *brouni* Steel, 1950 E
OMALIINI
Allodrepa decipiens Steel, 1964 E Su
Allodrepa subcylindrica (Kiesenwetter, 1877) E Su
Austrolophrum cribriceps (Fauvel, 1878) A
Brouniellum australe (Broun, 1894) E
Brouniellum parcum Bernhauer, 1939 E
Brouniellum sagoloide (Sharp, 1886) E
Brouniellum zealandicum Cameron, 1947 E
Crymus kronii (Kiesenwetter, 1877) E
Ischnoderus cognatus Broun, 1910 E
Ischnoderus curtipennis Broun, 1915 E
Ischnoderus fultoni Broun, 1893 E
Ischnoderus genalis (Broun, 1880) E
Ischnoderus morosus Broun, 1893 E
Ischnoderus opaciceps Cameron, 1947 E
Ischnoderus politulus (Broun, 1880) E
Ischnoderus tectus (Broun, 1880) E
<u>Macralymma</u> *punctiventre* Cameron, 1945 E
<u>Nesomalium</u> *campbellense* Steel, 1964 E Su
Nesomalium imitator Steel, 1964 E Su
Nesomalium insulare (Kiesenwetter, 1877) E Su
Nesomalium pacificum (Kiesenwetter, 1877) E Su
<u>Omaliomimus</u> *actobius* (Broun, 1893) E
Omaliomimus albipennis (Kiesenwetter, 1877) E Su
Omaliomimus carinigerus (Broun, 1893) E
Omaliomimus chalmeri (Broun, 1893) E
Omaliomimus conicum (Fauvel, 1878) E
Omaliomimus laetipennis (Broun, 1910) E
Omaliomimus litoreus (Broun, 1886) E
Omaliomimus robustus (Broun, 1911) E Ch
Omaliomimus setipes (Broun, 1909) E
Omaliomimus venator (Broun, 1909) E
Omalium allardii Fairmaire & Brisout de Barneville, 1859 A
Omalium hebes Broun, 1880 E
Paraphloeostiba gayndahensis (MacLeay, 1871) A
<u>Selonomus</u> *linearis* Steel, 1964 E Su
Stenomalium antipodum (Broun, 1893) E
Stenomalium debile (Broun, 1893) E
Stenomalium micrarthrum (Broun, 1893) E
Stenomalium moniliferum (Broun, 1893) E
Stenomalium parkeri (Bernhauer, 1939) E
Stenomalium philpotti (Broun, 1894) E
Stenomalium tenellum (Broun, 1893) E
Xylodromus (*Omalissus*) *concinnus* (Marsham, 1802) A
<u>Zeolymma</u> *brachypterum* Steel, 1950 E
Omaliinae Genus E ('*Omalium*') *helmsi* (Cameron, 1945)
Omaliinae Genus F ('*Omalium*') *cognatum* (Broun, 1893) E
Omaliinae Genus F ('*Omalium*') *cottieri* (Bernhauer, 1939) E
Omaliinae Genus F ('*Omalium*') *sulcithorax* (Broun, 1880) E
Omaliinae Genus F? ('*Omalium*') *agreste* (Broun, 1880) E
Omaliinae Genus F? ('*Omalium*') *brookesi* (Cameron, 1947) E
Omaliinae Genus F? ('*Omalium*') *spadix* (Broun, 1880) E
OSORIINAE
ELEUSININI
Zeoleusis virgula (Fauvel, 1889) E
Zeoleusis n. sp. E
OSORIINI
<u>Nototorchus</u> *ferrugineus* (Broun, 1893) E
Nototorchus montanus (Broun, 1910) E
<u>Paratorchus</u> *aculeatus* (McColl, 1982) E
Paratorchus alifer (McColl, 1982) E
Paratorchus angulatus (McColl, 1982) E
Paratorchus angustus (McColl, 1982) E
Paratorchus anophthalmus (Fauvel, 1900) E
Paratorchus arrowi (Bernhauer, 1939) E
Paratorchus bifurcatus (McColl, 1982) E
Paratorchus brevipennis (Broun, 1893) E
Paratorchus brevisetis (McColl, 1982) E
Paratorchus bucinifer (McColl, 1982) E
Paratorchus caecus (Broun, 1910) E
Paratorchus curvisetis (McColl, 1982) E
Paratorchus decipiens (McColl, 1982) E
Paratorchus falcifer (McColl, 1984) E
Paratorchus fiordensis (McColl, 1982) E
Paratorchus flexuosus (McColl, 1982) E
Paratorchus foveatus (McColl, 1982) E
Paratorchus grandis (McColl, 1984) E
Paratorchus hamatus (McColl, 1982) E
Paratorchus helmsi (Fauvel, 1900) E
Paratorchus hermes (McColl, 1982) E
Paratorchus homerensis (McColl, 1982) E
Paratorchus humilis (McColl, 1982) E
Paratorchus insuetus (McColl, 1982) E
Paratorchus maritimus (McColl, 1984) E
Paratorchus microphthalmus (Fauvel, 1900) E
Paratorchus minutus (McColl, 1984) E
Paratorchus monstrosus (Bernhauer, 1939) E
Paratorchus parvulus (McColl, 1982) E
Paratorchus pelorensis (McColl, 1982) E
Paratorchus phaseolinus (McColl, 1982) E
Paratorchus pubescens (McColl, 1982) E
Paratorchus relictus (McColl, 1984) E
Paratorchus retroflexus (McColl, 1982) E
Paratorchus scapulifer (McColl, 1982) E
Paratorchus tardus (McColl, 1982) E
Paratorchus tricarinatus (McColl, 1982) E
Paratorchus trivialis (McColl, 1982) E
Paratorchus tubifer (McColl, 1982) E
Paratorchus vagepunctus (Fauvel, 1900) E
OXYTELINAE
COPROPHILINI
<u>Coprostygnus</u> *curvipes* Broun, 1921 E
Coprostygnus optandus Broun, 1893 E
Coprostygnus picipennis Broun, 1921 E
Coprostygnus sculptipennis Sharp, 1886 E
OXYTELINI
Anotylus brunneipennis (MacLeay, 1873) A
Anotylus complanatus (Erichson, 1839) A
Anotylus cribriceps (Fauvel, 1878) A
Anotylus pusillimus (Kraatz, 1859) A
Anotylus rugosus (Fabricius, 1775) A
Anotylus semirufus (Fauvel, 1877) A
Anotylus sparsus (Fauvel, 1877) A? Do
Anotylus varius (Fauvel, 1877) A
Anotylus vinsoni (Cameron, 1936) A
Anotylus wattsensis (Blackburn, 1902) A
Anotylus n. sp. A?
Oxytelus sculptus Gravenhorst, 1806 A
Oxytelus n. sp. A?
THINOBIINI
Blediotrogus cordicollis (Broun, 1907) E
Blediotrogus cribricollis Fauvel, 1900 E
Blediotrogus guttiger Sharp, 1900 E
Bledius amplicollis Fauvel, 1900
Bledius bidentifrons Broun, 1912 E
Bledius salinus Cameron, 1947 E
Carpelimus bilineatus (Stephens, 1834) A
Carpelimus corticinus (Gravenhorst, 1806) A
Carpelimus persimilis (Cameron, 1947) E
Carpelimus pusillus (Gravenhorst, 1802) A
Carpelimus zealandicus (Sharp, 1900) E
Teropalpus coloratus (Sharp, 1900) E
Teropalpus maritimus (Broun, 1903) E
Teropalpus unicolor (Sharp, 1900)
PAEDERINAE
PAEDERINI
S. ASTENINA
Astenus guttulus (Fauvel, 1877) A
Astenus zealandicus Cameron, 1945 E?
S. CRYPTOBIINA
Hyperomma discrepans Broun, 1921 E
Hyperomma dispersum Broun, 1893 E
Hyperomma duplicatum Broun, 1893 E
Hyperomma flavipes Broun, 1923 E
Hyperomma lobatum Broun, 1921 E
Hyperomma mandibulare Broun, 1893 E
Hyperomma picipenne Broun, 1921 E
Hyperomma sanguineum Broun, 1894 E
Hyperomma subcaecum Broun, 1921 E
Hyperomma tenellum Broun, 1909 E
Hyperomma n. spp. (4) 4E
S. LATHROBIINA
Lathrobium bipartitum Fauvel, 1878 A
Lathrobium '*longipenne* (Broun, 1912)' [preoccupied name] E
<u>Phanophilus</u> *comptus* (Broun, 1880) E
Scymbalium laetum Blackburn, 1888 A
S. MEDONINA
Lithocharis nigriceps Kraatz, 1859 A
Lithocharis ochracea (Gravenhorst, 1802) A
Lithocharis vilis Kraatz, 1859 A
Lithocharis n. sp. A?
Medon coecus (Broun, 1894) E
Medon granipennis (Broun, 1910) E
Medon mandibularis (Broun, 1880) E
Medon microps Cameron, 1947 E
Medon ventralis (Broun, 1880) E
Medon zeelandicus (Redtenbacher, 1867) E
Pseudomedon obscurellus (Erichson, 1840) A
Pseudomedon obsoletus (Nordmann, 1837) A
Sunius debilicornis (Wollaston, 1857) A
Sunius propinquus (Brisout de Barneville, 1867) A
S. SCOPAEIINA
Scopaeus apterus Cameron, 1950 E TK
Scopaeus n. sp. E?
S. STILICINA
Rugilus orbiculatus (Paykull, 1789) A
PHLOEOCARINAE
<u>Phloeognathus</u> *monticola* Steel, 1953 E
Pseudophloeocharis australis (Fauvel, 1903)
PIESTINAE
<u>Parasiagonum</u> *hudsoni* (Cameron, 1927) E
PROTEININAE
ANEPIINI
<u>Eupsorus</u> *costatus* Broun, 1904 E
Eupsorus n. spp. (3) 3E
NESONEINI
<u>Nesoneus</u> *acuticeps* Bernhauer, 1939 E
<u>Paranesoneus</u> *sparsior* (Cameron, 1945) E
SILPHOTELINI
Silphotelus nitidus Broun, 1895 E
Silphotelus obliquus Broun, 1912 E
PSELAPHINAE
EUPLECTITAE
EUPLECTINI
Euplectus auripilus Broun, 1886 E
Euplectus caviceps Broun, 1904 E
Euplectus '*cephalotes* Reitter, 1880', non Motschulsky, 1845 [preoccupied name] E
Euplectus frontalis Broun, 1880 E
Euplectus incomptus Broun, 1884 E
Euplectus lepiphorus Broun, 1893 E
Euplectus longulus Broun, 1880 E
Euplectus opacus Sharp, 1874 E
Euplectus ovicollis Broun, 1880 E
Euplectus personatus Broun, 1893 E

Euplectus scruposus Broun, 1893 E
Euplectus sculpturatus Broun, 1880 E
Euplectus semiopacus Broun, 1895 E
Euplectus sulciceps Broun, 1904 E
Euplectus tuberigerus Broun, 1882 E
Euplectus unicus Broun, 1893 E
Euplectus vacuus Broun, 1884 E
Euplectus 'verticalis Broun, 1893', non Reitter, 1884 [preoccupied name] E
Euplectus n. spp. (7) 7E
Leptoplectus n. spp. (2) 2E
TRICHONYCHINI
S. PANAPHANTINA
Adalmus velutinus Reitter, 1885 E
Dalma gigantea Broun, 1914 E
Dalma graniceps Broun, 1921 E
Dalma pubescens Sharp, 1874 E
Dalma tuberculata Broun, 1880 E
Dalmisus batrisodes Sharp, 1886 E
Eleusomatus 'acuminatus (Broun, 1893)', non (Schaufuss 1882) [preoccupied name] E
Eleusomatus allocephalus (Broun, 1893) E
Eleusomatus caudatus (Broun, 1893) E
Eleusomatus oculatus Broun, 1921 E
Eleusomatus ovicollis Broun, 1915 E
Eleusomatus subcaecus Broun, 1921 E
Eleusomatus vidamoides Broun, 1921 E
Euglyptus abnormis Broun, 1921 E
Euglyptus costifer Broun, 1893 E
Euglyptus elegans Broun, 1893 E
Euglyptus foveicollis Broun, 1912 E
Euglyptus iracundus (Broun, 1893) E
Euglyptus longiceps Broun, 1921 E
Euglyptus longicornis Broun, 1912 E
Euglyptus punctatus (Broun, 1893) E
Euglyptus sublaevis Broun, 1921 E
Euplectopsis antennalis Broun, 1912 E
Euplectopsis antiqua (Broun, 1893) E
Euplectopsis biimpressa Broun, 1912 E
Euplectopsis blandiata Broun, 1915 E
Euplectopsis brevicollis (Reitter, 1880) E
Euplectopsis bryocharis Broun, 1915 E
Euplectopsis carinata Broun, 1912 E
Euplectopsis clavatula Broun, 1913 E
Euplectopsis crassipes (Broun, 1884) E
Euplectopsis crassula Broun, 1921 E
Euplectopsis cuneiceps Broun, 1915 E
Euplectopsis curvipennis Broun, 1914 E
Euplectopsis dorsalis Broun, 1915 E
Euplectopsis duplex Broun, 1915 E
Euplectopsis duplicata Broun, 1913 E
Euplectopsis elongella Broun, 1915 E
Euplectopsis eminens (Broun, 1886) E
Euplectopsis eruensis Broun, 1912 E
Euplectopsis fastigiata Broun, 1911 E
Euplectopsis femoralis Broun, 1914 E
Euplectopsis granulata Broun, 1911 E
Euplectopsis heterarthra Broun, 1912 E
Euplectopsis impressa Broun, 1915 E
Euplectopsis inscita (Broun, 1893) E
Euplectopsis longicollis (Reitter, 1880) E
Euplectopsis microcephala (Reitter, 1880) E
Euplectopsis mirifica (Broun, 1884) E
Euplectopsis modesta (Broun, 1895) E
Euplectopsis 'monticola (Broun, 1884)', non (Wollaston, 1864) [preoccupied name] E
Euplectopsis mucronella Broun, 1911 E
Euplectopsis nitipennis Broun, 1915 E
Euplectopsis obnisa (Broun, 1884) E
Euplectopsis ovithorax (Broun, 1884) E
Euplectopsis parvula (Broun, 1895) E
Euplectopsis patruelis (Broun, 1884) E
Euplectopsis perpunctata Broun, 1915 E
Euplectopsis 'pusilla (Broun, 1895)' non (Denny, 1825) [preoccupied name] E
Euplectopsis rotundicollis (Reitter, 1880) E
Euplectopsis sanguinea Broun, 1913 E
Euplectopsis schizocnemis Broun, 1912 E
Euplectopsis terrestris Broun, 1914 E
Euplectopsis tibialis Broun, 1914 E
Euplectopsis trichoniformis (Reitter, 1880) E
Euplectopsis tumida Broun, 1911 E
Euplectopsis tumipes (Broun, 1895) E
Kenocoelus dimorphus Broun, 1911 E
Paraplectus n. sp. E
Patreus lewisi Broun, 1904 E
Philiopsis n. sp. A?
Placodium zenarthrum Broun, 1893 E
Plectomorphus anguliferus Broun, 1921 E
Plectomorphus brevicornis Broun, 1913 E
Plectomorphus collinus Broun, 1921 E
Plectomorphus egenus Broun, 1913 E
Plectomorphus insignis Broun, 1921 E
Plectomorphus laminifer Broun, 1915 E
Plectomorphus longiceps Broun, 1913 E
Plectomorphus longipes Broun, 1912 E
Plectomorphus munroi (Broun, 1893) E
Plectomorphus optandus Broun, 1912 E
Plectomorphus rugiceps Broun, 1921 E
Plectomorphus scitiventris Broun, 1921 E
Plectomorphus 'spinifer (Broun, 1893)' non Casey, 1884 [preoccupied name] E
Plectomorphus trisulcicollis (Broun, 1880) E
Sagolonus arohaensis (Broun, 1895) E
Sagolonus impressus Broun, 1910 E
Sagolonus patronus (Broun, 1893) E
Vidamodes furvus Broun, 1921 E
Vidamus armiferus Broun, 1911 E
Vidamus brevitarsis (Broun, 1880) E
Vidamus bryophilus Broun, 1914 E
Vidamus calcaratus Broun, 1912 E
Vidamus cereus (Broun, 1884) E
Vidamus clavipes Broun, 1915 E
Vidamus congruus Broun, 1915 E
Vidamus convexus (Sharp, 1874) E
Vidamus fossalis Broun, 1921 E
Vidamus gracilipes Broun, 1917 E
Vidamus incertus (Reitter, 1880) E
Vidamus modestus Broun, 1913 E
Vidamus muscicola Broun, 1921 E
Vidamus nitidus Broun, 1921 E
Vidamus ovicollis Broun, 1921 E
Vidamus punctulatus Broun, 1915 E
Vidamus simplex Broun, 1921 E
Vidamus sternalis Broun, 1913 E
Vidamus trochanteralis Broun, 1911 E
Vidamus u-impressus (Broun, 1884) E
Vidamus validus (Broun, 1893) E
Whitea laevifrons (Broun, 1893) E
Zelandius asper (Broun, 1880) E
Zelandius basalis (Broun, 1914) E
Zelandius brookesi (Broun, 1915) E
Zelandius clevedonensis (Broun, 1893) E
Zelandius coxalis (Broun, 1893) E
Zelandius fovealis Broun, 1913 E
Zelandius foveiceps (Broun, 1895) E
Zelandius fulgens (Broun, 1911) E
Zelandius illustris (Broun, 1911) E
Zelandius moerens (Broun, 1893) E
Zelandius obscurus (Broun, 1893) E
Zelandius raffrayi (Broun, 1911) E
Zelandius sandageri (Broun, 1893) E
Zelandius spinifer (Broun, 1914) E
Zelandius tuberalis (Broun, 1915) E
Zelandius usitatus (Broun, 1910) E
S. TRICHONYCHINA
Macroplectus bifoveata (Broun, 1921) E
Macroplectus parallela (Broun, 1921) E
Macroplectus spinipes (Broun, 1910) E
S. TRIMIINA
Allopectus claviger (Broun, 1893) E
Allopectus picipennis Broun, 1911 E
Allopectus subcaecus Broun, 1911 E
TROGASTRINI
S. TROGASTRINA
Neosampa granulata Broun, 1921 E
Platomesus n. sp. E
FARONITAE
FARONINI
Exeirarthra angustula Broun, 1917 E
Exeirarthra enigma Broun, 1893 E
Exeirarthra longiceps Broun, 1917 E
Exeirarthra pallida Broun, 1893 E
Exeirarthra parviceps Broun, 1921 E
Logasa sp. n. A?
Sagola acuminata Broun, 1921 E
Sagola aemula Broun, 1921 E
Sagola affinis Broun, 1921 E
Sagola angulifera Broun, 1911 E
Sagola anisarthra Broun, 1893 E
Sagola arboricola Broun, 1921 E
Sagola auripila Broun, 1911 E
Sagola basalis Broun, 1911 E
Sagola bifida Broun, 1915 E
Sagola bifoveiceps Broun, 1912 E
Sagola biimpressa Broun, 1912 E
Sagola bilobata Broun, 1921 E
Sagola bipunctata Broun, 1886 E
Sagola bipuncticeps Broun, 1921 E
Sagola bituberata Broun, 1914 E
Sagola brevicornis Raffray, 1893 E
Sagola brevifossa Broun, 1921 E
Sagola brevisternis Broun, 1915 E
Sagola brevitarsis Broun, 1886 E
Sagola carinata Broun, 1912 E
Sagola castanea Broun, 1886 E
Sagola cilipes Broun, 1921 E
Sagola citima Broun, 1893 E
Sagola clavatella Broun, 1912 E
Sagola cognata Broun, 1911 E
Sagola colorata Broun, 1914 E
Sagola concolorata Broun, 1915 E
Sagola confusa Broun, 1915 E
Sagola convexa Broun, 1886 E
Sagola cordiceps Broun, 1921 E
Sagola crassulipes Broun, 1915 E
Sagola deformipes Broun, 1880 E
Sagola denticollis Broun, 1880 E
Sagola dickensis Broun, 1917 E
Sagola dilucida Broun, 1914 E
Sagola disparata Broun, 1914 E
Sagola dissonans Broun, 1921 E
Sagola distorta Broun, 1921 E
Sagola diversa Broun, 1911 E
Sagola duplicata Broun, 1886 E
Sagola electa Broun, 1914 E
Sagola elevata Broun, 1886 E
Sagola elongata Broun, 1893 E
Sagola eminens Broun, 1895 E
Sagola excavata Broun, 1886 E
Sagola fagicola Broun, 1921 E
Sagola fasciculata Broun, 1921 E
Sagola flavipes Broun, 1893 E
Sagola fovealis Broun, 1886 E
Sagola foveiventris Broun, 1921 E
Sagola frontalis Raffray, 1893 E
Sagola fulva Broun, 1893 E
Sagola fulvipennis Broun, 1915 E
Sagola furcata Broun, 1921 E
Sagola fuscipalpis Broun, 1914 E
Sagola genalis Broun, 1881 E
Sagola grata Broun, 1912 E
Sagola guinnessi Broun, 1911 E
Sagola halli Broun, 1914 E
Sagola hectorii Broun, 1917 E

Sagola hirtalis Broun, 1893 E
Sagola ignota Broun, 1921 E
Sagola immota Broun, 1893 E
Sagola indiscreta Broun, 1915 E
Sagola insignis Broun, 1893 E
Sagola insolens Broun, 1893 E
Sagola insueta Broun, 1914 E
Sagola laetula Broun, 1915 E
Sagola laminata Broun, 1893 E
Sagola laticeps Broun, 1911 E
Sagola latistriata Broun, 1911 E
Sagola latula Broun, 1912 E
Sagola lawsoni Broun, 1912 E
Sagola lineata Broun, 1893 E
Sagola lineiceps Broun, 1921 E
Sagola longicollis Broun, 1911 E
Sagola longipennis Broun, 1911 E
Sagola longipes Broun, 1915 E
Sagola longula Broun, 1912 E
Sagola macronyx Broun, 1893 E
Sagola major Sharp, 1874 E
Sagola mimica Broun, 1893 E
Sagola minuscula Broun, 1921 E
Sagola misella Sharp, 1874 E
Sagola monstrosa Reitter, 1880 E
Sagola monticola Broun, 1912 E
Sagola nitida Broun, 1911 E
Sagola notabilis Broun, 1880 E
Sagola occipitalis Broun, 1912 E
Sagola opercularis Broun, 1915 E
Sagola osculans Broun, 1886 E
Sagola pallidula Broun, 1912 E
Sagola parallela Broun, 1893 E
Sagola parva Sharp, 1874 E
Sagola pertinax Broun, 1893 E
Sagola planicula Broun, 1921 E
Sagola planipennis Broun, 1921 E
Sagola posticalis Broun, 1915 E
Sagola prisca Sharp, 1874 E
Sagola pulchra Broun, 1880 E
Sagola punctata Broun, 1893 E
Sagola puncticeps Broun, 1911 E
Sagola puncticollis Broun, 1911 E
Sagola punctulata Raffray, 1893 E
Sagola rectipennis Broun, 1921 E
Sagola rectipes Broun, 1893 E
Sagola remixta Broun, 1921 E
Sagola robusta Broun, 1893 E
Sagola robustula Broun, 1917 E
Sagola rotundiceps Broun, 1915 E
Sagola rufescens Broun, 1921 E
Sagola ruficeps Broun, 1893 E
Sagola rugifrons Broun, 1895 E
Sagola rustica Broun, 1915 E
Sagola setiventris Broun, 1915 E
Sagola sharpi Raffray, 1893 E
Sagola sobrina Broun, 1893 E
Sagola socia Broun, 1915 E
Sagola spinifer Broun, 1895 E
Sagola spiniventris Broun, 1912 E
Sagola strialis Broun, 1921 E
Sagola striatifrons Broun, 1921 E
Sagola subcuneata Broun, 1921 E
Sagola sulcator Broun, 1886 E
Sagola suturalis Broun, 1914 E
Sagola tenebrica Broun, 1921 E
Sagola tenuis Broun, 1886 E
Sagola terricola Broun, 1886 E
Sagola unicalis Broun, 1917 E
Sagola valida Broun, 1921 E
Sagola ventralis Broun, 1912 E
<u>*Stenosagola*</u> *connata* (Broun, 1911) E
Stenosagola crassicornis (Broun, 1911) E
Stenosagola gracilis (Broun, 1893) E
Stenosagola griseipila Broun, 1921 E
Stenosagola oblongiceps Broun, 1921 E
Stenosagola planiocula Broun, 1921 E
Stenosagola n. spp. (2) 2E
GONIACERITAE
BRACHYGLUTINI
S. BRACHYGLUTINA
Anabaxis electrica (King, 1863) A Do
Anabaxis foveolata (Broun, 1880) E
Eupines (*Eupines*) *calcarata* (Broun, 1886) E
Eupines (*E.*) *grata* (Sharp, 1874) E
Eupines (*E.*) *micans* (Sharp, 1874) E
Eupines (*E.*) *nasuta* (Broun, 1880) E
Eupines (*E.*) *nesobia* Broun, 1914 E
Eupines (*E.*) *piciceps* (Broun, 1880) E
Eupines (*E.*) *platynota* (Broun, 1893) E
Eupines (*E.*) *simplex* Broun, 1913 E
Eupines (*E.*) *sternalis* (Broun, 1893) E
Eupines (*Byraxis*) *acceptus* (Broun, 1923) E
Eupines (*B.*) *allocera* (Broun, 1893) E
Eupines (*B.*) *anisarthra* (Broun, 1914) E
Eupines (*B.*) *bisulcifrons* (Broun, 1914) E
Eupines (*B.*) *clemens* Broun, 1921 E
Eupines (*B.*) *conspicua* (Broun, 1893) E
Eupines (*B.*) *costata* (Broun, 1893) E
Eupines (*B.*) *crassicornis* (Broun, 1880), non *Bryaxis crassicornis* Motschulsky, 1835, 1851 [preoccupied name] E
Eupines (*B.*) *decens* (Broun, 1893) E
Eupines (*B.*) *deformis* (Sharp, 1874) E
Eupines (*B.*) *dispar* (Sharp, 1874) E
Eupines (*B. diversa* (Broun, 1893), non *Reichenbachia diversa* Raffray, 1887, nec *R. diversa* Sharp, 1887 [preoccupied name] E
Eupines (*B.*) *forficulida* (Broun, 1890) E
Eupines (*B.*) *foveatissima* (Broun, 1890) E
Eupines (*B.*) *fraudulenta* (Broun, 1886) E
Eupines (*B.*) *glabrata* (Broun, 1886) E
Eupines (*B.*) *halli* (Broun, 1921) E
Eupines (*B.*) *hectori* (Broun, 1895) E
Eupines (*B.*) *ignotus* (Broun, 1881) E
Eupines (*B.*) *illustris* (Broun, 1914) E
Eupines (*B.*) *impar* (Sharp, 1874) E
Eupines (*B.*) *impressifrons* (Broun, 1880) E
Eupines (*B.*) *lewisi* Broun, 1910 E
Eupines (*B.*) *longiceps* Raffray, 1904 E
Eupines (*B.*) *monstrosa* (Reitter, 1880) E
Eupines (*B.*) *mundula* (Schaufuss, 1888) E
Eupines (*B.*) *munroi* (Broun, 1890) E
Eupines (*B.*) *nemoralis* (Broun, 1886) E
Eupines (*B.*) *paganus* (Broun, 1881) E
Eupines (*B.*) *platyarthra* (Broun, 1893) E
Eupines (*B.*) *rhyssarthra* (Broun, 1912) E
Eupines (*B.*) *rudicornis* Broun, 1882 E
Eupines (*B.*) *sanguinea* (Broun, 1880) E
Eupines (*B.*) *setifer* (Broun, 1893) E
Eupines (*B.*) *sylvicola* (Broun, 1884) E
<u>*Eupinogitus*</u> *sulcipennis* Broun, 1921 E
Eupinolus altulus (Broun, 1880) E
Eupinolus punctatus (Broun, 1886) E
<u>*Gastrobothus*</u> *abdominalis* (Broun, 1880), non *Brachygluta abdominalis* Aubé, 1833 E
Gastrobothus sharpi (Broun, 1880) E
<u>*Physobryaxis*</u> *inflata* (Sharp, 1874) E
<u>*Simkinion*</u> *bimanum* Park & Pearce, 1962 E
Simkinion prelaticum Park & Pearce, 1962 E
Startes foveata Broun, 1893 E
Startes sculpturata Broun, 1886 E
PSELAPHITAE
PSELAPHINI
Pselaphogenius citimus (Broun, 1893) E
Pselaphogenius delicatus (Broun, 1886) E
Pselaphogenius ventralis (Broun, 1895) E
Pselaphophus atriventris (Westwood, 1856)
<u>*Pselaphotheseus*</u> *hippolytae* Park, 1964 E Su
Pselaphotheseus ihupuku Carlton & Leschen, 2001 E Su
Pselaphus caecus Broun, 1886 E
Pselaphus cavelli Broun, 1893 E
Pselaphus cavidorsis Broun, 1923 E
Pselaphus dulcis Broun, 1881 E
Pselaphus fuscopilus Broun, 1886 E
Pselaphus meliusculus Broun, 1893 E
Pselaphus oviceps Broun, 1917 E
Pselaphus pauper Sharp, 1874 E
Pselaphus pilifrons Broun, 1914 E
Pselaphus pilistriatus Broun, 1880 E
Pselaphus sulcicollis Broun, 1893, non *Trichonyx sulcicollis* (Reichenbach, 1816) E
Pselaphus trifoveatus Broun, 1914 E
Pselaphus urquharti Broun, 1917 E
TYRINI
S. TYRINA
Agatyrus fulvihirtus Broun, 1917 E
Gerallus punctipennis Schaufuss, 1880 A
Hamotulus angulipes Broun, 1914 E
Hamotulus armatus (Broun, 1893) E
Hamotulus cornutus Broun, 1915 E
Hamotulus curvipes (Broun, 1893) E
Hamotulus frontalis Broun, 1914 E
Hamotulus fuscipalpis Broun, 1915 E
Hamotulus mutandus (Sharp, 1874) E
Hamotulus robustus Broun, 1915 E
Hamotulus spinipes (Broun, 1893) E
Hamotulus sternalis (Broun, 1893) E
<u>*Phormiobius*</u> *halli* Broun, 1917 E
<u>*Plesiotyrus*</u> *crassipes* (Broun, 1893) E
Tyrogetus optandus Broun, 1893 E
Tyrogetus palpalis Broun, 1910 E
<u>*Zeatyrus*</u> *lawsoni* Sharp, 1881 E
PSEUDOPSINAE
Pseudopsis arrowi Bernhauer, 1939 E
Pseudopsis n. sp. E
SCAPHIDIINAE
CYPARIINI
Cyparium earlyi Löbl & Leschen, 2003 E
Cyparium thorpei Löbl & Leschen, 2003 E
SCAPHISOMATINI
Baeocera abrupta Löbl & Leschen, 2003 E
Baeocera actuosa (Broun, 1881) E
Baeocera benolivia Löbl & Leschen, 2003 E
Baeocera elenae Löbl & Leschen, 2003 E
Baeocera epipleuralis Löbl & Leschen, 2003 E
Baeocera hillaryi Löbl & Leschen, 2003 E
Baeocera karamea Löbl & Leschen, 2003 E
Baeocera punctatissima Löbl & Leschen, 2003 E
Baeocera sternalis Broun, 1914 E
Baeocera tekootii Löbl & Leschen, 2003 E
Baeocera tensingi Löbl & Leschen, 2003 E
Baeocera tenuis Löbl & Leschen, 2003 E
Brachynopus latus Broun, 1881 E
Brachynopus scutellaris (Redtenbacher, 1867) E
<u>*Notonewtonia*</u> *thayerae* Löbl & Leschen, 2003 E
Notonewtonia watti Löbl & Leschen, 2003 E
Scaphisoma corcyricum Löbl, 1964 A Do
Scaphisoma funereum Löbl, 1977 A
Scaphisoma hanseni Löbl & Leschen, 2003 E
Spinoscapha rufa (Broun, 1881) E
Vickibella apicella (Broun, 1880) E
SCYDMAENINAE
SCYDMAENITAE
CYRTOSCYDMINI
<u>*Chathamaenus*</u> *chathamensis* Franz, 1980 E Ch
Euconnus (*Tetramelus*) *castawayensis* Franz, 1975 E TK
Euconnus (*T.*) *curvicrus* Franz, 1975 E TK
Euconnus (*T.*) *northlandensis* Franz, 1975 E
Euconnus (*T.*) *ohenae* Franz, 1980 E
Euconnus (*T.*) *picicollis* (Broun, 1880) E
Euconnus (*T.*) *pseudoramsayi* Franz, 1975 E TK
Euconnus (*T.*) *ramsayi* Franz, 1975 E TK
Euconnus (*T.*) *stewartensis* Franz, 1980 E

Euconnus (*T.*) *threekingensis* Franz, 1975 E TK
Euconnus (*T.*) *wairauensis* Franz, 1986 E
Euconnus arthuris Franz, 1975 E
Euconnus brouni Franz, 1975 E
Euconnus calvus (Broun, 1880) E
Euconnus clarkei Franz, 1975 E
Euconnus horridus Franz, 1975 E
Euconnus impressipennis Franz, 1975 E
Euconnus kaimanawae Franz, 1986 E
Euconnus microcilipes Franz, 1986 E
Euconnus paracilipes Franz, 1986 E
Euconnus russatus (Broun, 1893) E
Euconnus setosus (Sharp, 1874) E
Euconnus n. sp. E
Maorinus alacer (Broun, 1915) E
Maorinus angulatus (Broun, 1893) E
Maorinus '*australis* (Franz, 1975)', non (Franz, 1867) [preoccupied name] E
Maorinus brookesi (Franz, 1975) E
Maorinus bullockensis (Franz, 1986) E
Maorinus cilipes (Broun, 1893) E
Maorinus codfishensis (Franz, 1975) E
Maorinus divaricatus (Franz, 1980) E
Maorinus dunensis (Franz, 1975) E
Maorinus dunsdalensis (Franz, 1975) E
Maorinus egmontiensis (Franz, 1975) E
Maorinus eruensis (Franz, 1975) E
Maorinus erythronotus (Broun, 1893) E
Maorinus fabiani (Franz, 1975) E
Maorinus fugax (Franz, 1980) E
Maorinus hawkesi (Franz, 1975) E
Maorinus helmsi (Franz, 1975) E
Maorinus hollyfordensis (Franz, 1975) E
Maorinus hunuae (Franz, 1980) E
Maorinus hunuaeformis (Franz, 1986) E
Maorinus inangahuae (Franz, 1975) E
Maorinus kuscheli (Franz, 1975) E
Maorinus maketuensis (Franz, 1980) E
Maorinus marionensis (Franz, 1975) E
Maorinus milfordensis (Franz, 1975) E
Maorinus monilifer (Broun, 1893) E
Maorinus pandorae (Franz, 1975) E
Maorinus pelorianus (Franz, 1975) E
Maorinus pelorii (Franz, 1975) E
Maorinus pseudoalacer (Franz, 1975) E
Maorinus pseudoangulatus (Franz, 1975) E
Maorinus sanguineus (Broun, 1893) E
Maorinus tangihuae (Franz, 1980) E
Maorinus toronouii (Franz, 1986) E
Maorinus tunakinoi (Franz, 1975) E
Maorinus turangii (Franz, 1986) E
Maorinus turrethi (Franz, 1975) E
Maorinus wakamarinae (Franz, 1975) E
Maorinus walkerianus (Franz, 1975) E
Maorinus wellingtonensis (Franz, 1975) E
Microscydmus (*Microscydmus*) *lynfieldi* Franz, 1977 E
Microscydmus (*M.*) *perpusillus* Franz, 1980 E
Microscydmus (*M.*) n. spp. (3) 3E
Phaganophana (?) *palpalis* Broun, 1915 E
Sciacharis (*Magellanoconnus*) *brevivestis* (Franz, 1980) E
Sciacharis (*M.*) *catharactae* Franz, 1980 E
Sciacharis (*M.*) *galerus* (Broun, 1885) E
Sciacharis (*M.*) *kuscheli* (Franz, 1977) E
Sciacharis (*M.*) *maruiensis* (Franz, 1975) [corrected spelling] E
Sciacharis (*M.*) *remissus* (Franz, 1980) E
Sciacharis (*M.*) *sericeus* (Franz, 1980) E
Sciacharis (*M.*) n. spp. (3) 3E
Sciacharis (*Sciacharis*) *alackensis* (Franz, 1975) E
Sciacharis (*S.*) *ambigua* (Broun, 1880) E
Sciacharis (*S.*) *angustata* (Broun, 1885) E
Sciacharis (*S.*) *antennalis* (Broun, 1893) E
Sciacharis (*S.*) *arohana* (Franz, 1975) E
Sciacharis (*S.*) *arthurensis* Franz, 1986 E
Sciacharis (*S.*) *aurifascis* Franz, 1980 E
Sciacharis (*S.*) *bluffensis* (Franz, 1975) E
Sciacharis (*S.*) *brachycera* (Broun, 1893) E
Sciacharis (*S.*) *calcaritibia* (Franz, 1975) E TK
Sciacharis (*S.*) *cedia* (Broun, 1893) E
Sciacharis (*S.*) *collega* Franz, 1980 E
Sciacharis (*S.*) *dobsoni* (Franz, 1975) E
Sciacharis (*S.*) *dublinensis* (Franz, 1975) E
Sciacharis (*S.*) *dugdalei* Franz, 1980 E
Sciacharis (*S.*) *dunedinensis* Franz, 1980 E
Sciacharis (*S.*) *dunicola* (Franz, 1975) E
Sciacharis (*S.*) *durvillei* (Franz, 1975) E
Sciacharis (*S.*) *eruana* (Franz, 1975) E
Sciacharis (*S.*) *fiordlandensis* (Franz, 1975) E
Sciacharis (*S.*) *fletcheri* Franz, 1980 E
Sciacharis (*S.*) *florana* Franz, 1986 E
Sciacharis (*S.*) *fragilis* (Broun, 1915) E
Sciacharis (*S.*) *fulva* Broun, 1893 E
Sciacharis (*S.*) *glacialis* Franz, 1980 E
Sciacharis (*S.*) *gunnensis* (Franz, 1975) E
Sciacharis (*S.*) *hakatarameana* (Franz, 1975) E
Sciacharis (*S.*) *halli* (Broun, 1915) E
Sciacharis (*S.*) *haurokana* (Franz, 1975) E
Sciacharis (*S.*) *heterartha* (Broun, 1893) E
Sciacharis (*S.*) *hokianoae* (Franz, 1975) E
Sciacharis (*S.*) *hopeana* (Franz, 1975) E
Sciacharis (*S.*) *humpensis* Franz, 1980 E
Sciacharis (*S.*) *hutti* (Franz, 1975) E
Sciacharis (*S.*) *kaimanawana* Franz, 1986 E
Sciacharis (*S.*) *kuscheliana* (Franz, 1975) E
Sciacharis (*S.*) *labiata* (Franz, 1975) E
Sciacharis (*S.*) *lanosa* (Broun, 1885) E
Sciacharis (*S.*) *lanosiformis* Franz, 1986 E
Sciacharis (*S.*) *lewisi* Franz, 1986 E
Sciacharis (*S.*) *macburneyi* Franz, 1980 E
Sciacharis (*S.*) *marlboroughensis* (Franz, 1975) E
Sciacharis (*S.*) *maruiae* (Franz, 1975) E
Sciacharis (*S.*) *moerakensis* Franz, 1986 E
Sciacharis (*S.*) *montana* Franz, 1980 E
Sciacharis (*S.*) *ohakunei* (Franz, 1975) E
Sciacharis (*S.*) *oreas* (Broun, 1885) E
Sciacharis (*S.*) *orrhillensis* (Franz, 1975) E
Sciacharis (*S.*) *ovipennis* (Broun, 1893) stat. rev. E
Sciacharis (*S.*) *oweni* (Franz, 1975) E
Sciacharis (*S.*) *paralatuliceps* (Franz, 1975) E
Sciacharis (*S.*) *paraoveni* Franz, 1986 E
Sciacharis (*S.*) *parawhangamoana* (Franz, 1975) E TK
Sciacharis (*S.*) *peckiana* Franz, 1986 E
Sciacharis (*S.*) *perspicax* Franz, 1980 E
Sciacharis (*S.*) *pictonensis* (Franz, 1975) E
Sciacharis (*S.*) *pontis* Franz, 1986 E
Sciacharis (*S.*) *portae* Franz, 1980 E
Sciacharis (*S.*) *pseudowhangamoana* (Franz, 1975) E
Sciacharis (*S.*) *puncticollis* (Broun, 1880) E
Sciacharis (*S.*) *relata* (Broun, 1893) E
Sciacharis (*S.*) *restinga* (Franz, 1975) E
Sciacharis (*S.*) *sannio* (Franz, 1977) E
Sciacharis (*S.*) *sinuata* (Broun, 1915) E
Sciacharis (*S.*) *stenocera* (Broun, 1893) E
Sciacharis (*S.*) *sulcifera* (Broun, 1915) E
Sciacharis (*S.*) *takakae* (Franz, 1975) E
Sciacharis (*S.*) *tapapana* Franz, 1986 E
Sciacharis (*S.*) *taranakii* Franz, 1980 E
Sciacharis (*S.*) *taupoensis* Franz, 1986 E
Sciacharis (*S.*) *tautukuensis* Franz, 1986 E
Sciacharis (*S.*) *tennysoni* Franz, 1986 E
Sciacharis (*S.*) *tennysoniana* Franz, 1986 E
Sciacharis (*S.*) *townsendiana* (Franz, 1975) E
Sciacharis (*S.*) *waikawensis* (Franz, 1975) E
Sciacharis (*S.*) *waiporiensis* (Franz, 1975) E
Sciacharis (*S.*) *waipouana* (Franz, 1975) E
Sciacharis (*S.*) *waipouensis* (Franz, 1975) E
Sciacharis (*S.*) *wairauensis* Franz, 1980 E
Sciacharis (*S.*) *whakatanensis* (Franz, 1975) E
Sciacharis (*S.*) *whangamoana* (Franz, 1975) E
Sciacharis (*S.*) *wilmotensis* Franz, 1980 E
Sciacharis (*S.*) *xanthopa* (Broun, 1893) E
Sciacharis (*S.*) *yakasensis* Franz, 1986 E
Stenichnaphes newtoni Franz, 1986 E
Stenichnaphes urbanus Franz, 1980 E
Stenichnaphes n. sp. E
Stenichnus (*Austrostenichnus*) *insignis* (Broun, 1893) E
Stenichnus (*A.*) *kuschelianus* Franz, 1977 E
Stenichnus (*A.*) n. sp. E
EUTHEIINI
Eutheia schaumi Kiesenwetter, 1858 A Do
SCYDMAENINI
Adrastia angulifrons (Broun, 1915) E
Adrastia angustissima (Franz, 1975) E
Adrastia anophthalma (Franz, 1977) E
Adrastia anthicoides (Franz, 1980) E
Adrastia brookesi (Franz, 1975) E
Adrastia brouni (Franz, 1975) E
Adrastia clarkei (Franz, 1975) E
Adrastia clavatella Broun, 1915 E
Adrastia curticornis (Franz, 1975) E
Adrastia decipiens (Franz, 1975) E
Adrastia dentipes (Franz, 1975) E
Adrastia edwardsi (Sharp, 1874) E
Adrastia gourlayi (Franz, 1975) E
Adrastia greymouthi (Franz, 1975) E
Adrastia haastensis (Franz, 1980) E
Adrastia laetans Broun, 1881 E
Adrastia nelsoni (Franz, 1975) E
Adrastia novaezeelandiae (Franz, 1975) E
Adrastia paravillosipennis (Franz, 1975) E
Adrastia peckorum (Franz, 1986) E
Adrastia stokesi (Franz, 1975) E
Adrastia subinermis (Franz, 1980) E
Adrastia townsendi (Franz, 1977) E
Adrastia vilosipennis (Franz, 1975) E
Adrastia wilmotensis (Franz, 1977) E
Scydmaenus elongellus Broun, 1893 E SI
STAPHYLININAE
MAOROTHIINI
Maorothius adustus (Broun, 1880) E
Maorothius brevispinosus Assing, 2000 E
Maorothius brookesi (Cameron, 1952) E
Maorothius brouni (Steel, 1949) E
Maorothius coalitus Assing, 2000 E
Maorothius dispar Assing, 2000 E
Maorothius effeminatus Assing, 2000 E
Maorothius hamifer Assing, 2000 E
Maorothius hammondi Assing, 2000 E
Maorothius insulanus Assing, 2000 E
Maorothius longispinosus Assing, 2000 E
Maorothius pectinatus Assing, 2000 E
Maorothius pubescens Assing, 2000 E
Maorothius puncticeps (Broun, 1894) E
Maorothius setiger Assing, 2000 E
Maorothius solus Assing, 2000 E
Maorothius tonsor Assing, 2000 E
Maorothius torquatus Assing, 2000 E
Maorothius tridens Assing, 2000 E
Maorothius volans Assing, 2000 E
STAPHYLININI
S. PHILONTHINA
Bisnius parcus (Sharp, 1874) A
Bisnius sordidus (Gravenhorst, 1802) A
Cafius algophilus Broun, 1894 E
Cafius litoreus (Broun, 1880) E
Cafius maritimus (Broun, 1880) E
Cafius puncticeps (White, 1846) E
Cafius quadriimpressus (White, 1846) E
Cafius zealandicus Cameron, 1947 E
Gabrius nigritulus (Gravenhorst, 1802) A
Gabronthus sulcifrons (Sharp, 1889) A
Neobisnius n. sp. A?
Philonthus burrowsi (Broun, 1915) E
Philonthus discoideus (Gravenhorst, 1802) A

Philonthus hepaticus Erichson, 1840 A
Philonthus insularis Bernhauer & Schubert, 1916 E
Philonthus longicornis Stephens, 1832 A
Philonthus novaezeelandiae Duvivier, 1883 E
Philonthus ohianensis (Broun, 1923) E
Philonthus politus (Linnaeus, 1758) A
Philonthus pyropterus Kraatz, 1859 A
Philonthus rectangulus Sharp, 1874 A
Philonthus umbratilis (Gravenhorst, 1802) A
Philonthus ventralis (Gravenhorst, 1802) A
Thinocafius insularis Steel, 1949 E Ch
S. QUEDIINA
Cafioquedius gularis Sharp, 1886 E
Cafioquedius n. sp. E
Heterothops minutus Wollaston, 1860 A
Quediocafius insolitus (Sharp, 1886) E
Quediocafius taieriensis (Broun, 1894) E
Quediomimus brookesi Cameron, 1948 E
Quediomimus hybridus (Erichson, 1840) A Do
Quedius aeneiventris Broun, 1910 E
Quedius aeneoceps (Broun, 1880) E
Quedius agathis Broun, 1893 E
Quedius aliiceps Cameron, 1948 E Su
Quedius ambiguus Broun, 1894 E
Quedius antarcticus Bernhauer & Schubert, 1916 E
Quedius antipodum Sharp, 1886 E
Quedius arctifrons (Broun, 1880) E
Quedius arrowi Bernhauer, 1941 E
Quedius aucklandicus Cameron, 1947 E
Quedius badius Broun, 1923 E
Quedius brookesi Cameron, 1947 E
Quedius brouni Cameron, 1947 E
Quedius brunneorufus Bernhauer, 1941 E
Quedius bryocharis Broun, 1923 E
Quedius cavelli Broun, 1893 E
Quedius collinus Broun, 1893 E
Quedius conspicuellus Broun, 1894 E
Quedius discrepans Broun, 1893 E
Quedius diversicollis Cameron, 1945 E
Quedius edwardsi Sharp, 1886 E
Quedius enodis (Broun, 1880) E
Quedius eruensis Broun, 1912 E
Quedius fulgidus (Fabricius, 1792), non (Fabricius, 1787) A
Quedius fultoni Cameron, 1945 E
Quedius fuscatus Broun, 1893 E
Quedius hallianus Broun, 1917 E
Quedius hilaris Broun, 1909 E
Quedius hirtipennis Broun, 1915, non Stephens, 1832 E
Quedius iridescens Broun, 1921 E
Quedius latifrons Sharp, 1886 E
Quedius latimanus Broun, 1893 E
Quedius longiceps Broun, 1910 E
Quedius mannaiaensis Cameron, 1945 E
Quedius maorinus Broun, 1923 E
Quedius megophthalmus Broun, 1917 E
Quedius mesomelinus (Marsham, 1802) A
Quedius ophthalmicus Cameron, 1945 E
Quedius quadripunctus Bernhauer, 1941 E
Quedius recticeps Broun, 1917 E
Quedius sciticollis Broun, 1894 E
Quedius secretus Cameron, 1948 E Su
Quedius sericeicollis Bernhauer, 1941 E
Quedius slipsensis Broun, 1923 E
Quedius subapterus Cameron, 1950 E TK
Quedius tinctellus Broun, 1910 E
Quedius tripunctatus Bernhauer, 1941 E
Quedius urbanus Broun, 1921 E
Quedius variegatus Bernhauer, 1941 E
Quedius veteratorius (Broun, 1880) E
Quedius vividus (Broun, 1880) E
Quedius wakefieldi Bernhauer, 1941 E
Quedius xenophaenus Broun, 1912 E
Sphingoquedius strandi Bernhauer, 1941 E
S. STAPHYLININA
Creophilus huttoni (Broun, 1880) E
Creophilus oculatus (Fabricius, 1775)
Hadrotes wakefieldi Cameron, 1945 E
Tasgius ater (Gravenhorst, 1802) A
XANTHOLININI
Gyrohypnus fracticornis (Müller, 1776) A
Leptacinus pusillus (Stephens, 1833) A
Neohypnus andinus (Fauvel, 1866) A
Neoxantholinus brouni (Sharp, 1876) E
Neoxantholinus chathamicus Bordoni, 2005 E Ch
Neoxantholinus pseudorufulus Bordoni, 2005 E
Neoxantholinus rufulus (Broun, 1880)
Notolinus socius (Fauvel, 1877)
Otagonia chathamensis Bordoni, 2005 E
Otagonia labralis (Broun, 1880) E
Otagonia nunni Bordoni, 2010 E
Paracorynus arecae (Broun, 1880) E
Pseudocorynus archaicus Bordoni, 2005 E
Pseudocorynus cultus (Broun, 1880) E
Pseudocorynus mediocris (Broun, 1880) E
Pseudocorynus nelsonianus Bordoni, 2005 E
Pseudocorynus neozelandicus Bordoni, 2005 E
Pseudoxantholinus sharpi (Broun, 1880) E
Thyreocephalus chalcopterus (Erichson, 1839) A?
Thyreocephalus chloropterus (Erichson, 1839) A
Thyreocephalus orthodoxus (Olliff, 1887) A
Thyreocephalus taitensis (Boheman, 1858) A
Waitatia bellicosa Bordoni, 2005 E
Waitatia maoriana Bordoni, 2005 E
Whangareiella fulvipes (Broun, 1880) E
Xantholinus linearis (Olivier, 1795) A
TACHYPORINAE
TACHYPORINI
'Coproporus' n. sp. A?
Sepedophilus acerbus (Broun, 1880) E
Sepedophilus antennalis (Broun, 1923) E
Sepedophilus asperellus (Broun, 1914) E
Sepedophilus atricapillus (Broun, 1880) E
Sepedophilus auricomus (Broun, 1880) E
Sepedophilus austerus (Broun, 1880) E
Sepedophilus badius (Broun, 1880) E
Sepedophilus basipennis (Bernhauer, 1941) E
Sepedophilus brevicornis (Broun, 1893) E
Sepedophilus convexus (Bernhauer, 1941) E
Sepedophilus flavithorax (Broun, 1880) E
Sepedophilus helmsi (Bernhauer, 1941) E
Sepedophilus hudsoni (Cameron, 1945) E
Sepedophilus laetulus (Broun, 1914) E
Sepedophilus largulus (Broun, 1880) E
Sepedophilus maculosus (Broun, 1880) E
Sepedophilus maorinus (Broun, 1893) E
Sepedophilus morosus (Broun, 1923) E
Sepedophilus niticollis (Broun, 1893) E
Sepedophilus nubilus (Broun, 1880) E
Sepedophilus phoxus (Olliff, 1886) A
Sepedophilus seminudus (Broun, 1923) E
Sepedophilus suavis (Fauvel, 1895) A?
Sepedophilus subruber (Broun, 1880) E
Sepedophilus zealandicus (Bernhauer, 1941) E
Sepedophilus n. spp. (3) 3E
Tachyporus nitidulus (Fabricius, 1781) A
SCARABAEOIDEA
LUCANIDAE
AESALINAE
Holloceratognathus cylindricus (Broun, 1895) E
Holloceratognathus helotoides (Thomson, 1862) E
Holloceratognathus passaliformis (Holloway, 1962) E
Mitophyllus alboguttatus (Bates, 1867) E
Mitophyllus angusticeps Broun, 1895 E
Mitophyllus arcuatus Holloway, 2007 E
Mitophyllus dispar (Sharp, 1882) E
Mitophyllus falcatus Holloway, 2007 E
Mitophyllus foveolatus (Broun, 1880) E
Mitophyllus fusculus (Broun, 1886) E
Mitophyllus gibbosus (Broun, 1885) E
Mitophyllus insignis Broun, 1923 E
Mitophyllus irroratus Parry, 1843 E
Mitophyllus macrocerus (Broun, 1886) E
Mitophyllus parrianus Westwood, 1863 E
Mitophyllus reflexus Broun, 1909 E Ch
Mitophyllus solox Holloway, 2007 E
LAMPRIMINAE
Dendroblax earlii White, 1846 E
SYNDESINAE A
Syndesus cornutus (Fabricius, 1801) A
LUCANINAE
Geodorcus alsobius Holloway, 2007 E
Geodorcus auriculatus (Broun, 1903) E
Geodorcus capito (Deyrolle, 1873) E Ch
Geodorcus helmsi (Sharp, 1881) E
Geodorcus ithaginis (Broun, 1893) E
Geodorcus montivagus Holloway, 2007 E
Geodorcus novaezealandiae (Hope, 1845) E
Geodorcus philpotti (Broun, 1914) E
Geodorcus servandus Holloway, 2007 E
Geodorcus sororum Holloway, 2007 E Ch
Paralissotes mangonuiensis (Brookes, 1927) E
Paralissotes oconnori (Holloway, 1961) E
Paralissotes planus (Broun, 1880) E
Paralissotes reticulatus (Westwood, 1844) E
Paralissotes rufipes (Sharp, 1886) E
Paralissotes stewarti (Broun, 1881) E
Paralissotes triregius (Holloway, 1963) E TK
Ryssonotus nebulosus (Kirby, 1818) A
TROGIDAE A
Trox (Trox) scaber (Linnaeus, 1767) A
SCARABAEIDAE
MELOLONTHINAE
LIPARETRINI
Costelytra austrobrunnea Given, 1952 E
Costelytra brookesi Given, 1966 E
Costelytra brunnea (Broun, 1880) E
Costelytra distincta Given, 1966 E
Costelytra diurna Given, 1960 E
Costelytra gregoryi Given, 1966 E
Costelytra macrobrunnea Given, 1952 E
Costelytra piceobrunnea Given, 1952 E
Costelytra pseudobrunnea Given, 1952 E
Costelytra symmetrica Given, 1966 E
Costelytra zealandica (White, 1846) E
Odontria albonotata Broun, 1893 E
Odontria aurantia Given, 1952 E
Odontria aureopilosa Given, 1952 E
Odontria australis Given, 1952 E
Odontria autumnalis Given, 1952 E
Odontria borealis Given, 1960 E
Odontria carinata Given, 1954 E
Odontria cassiniae Given, 1952 E
Odontria cinnamomea White, 1846 E
Odontria communis Given, 1952 E
Odontria convexa Given, 1952 E
Odontria decepta Given, 1952 E
Odontria fusca Broun, 1893 E
Odontria giveni Watt, 1984 E
Odontria halli Broun, 1921 E
Odontria inconspicua Given, 1952 E
Odontria macrothoracica Given, 1952 E
Odontria magna Given, 1952 E
Odontria marmorata Broun, 1893 E
Odontria monticola Broun, 1912 E
Odontria nesobia Broun, 1921 E
Odontria nitidula Broun, 1912 E
Odontria obscura Broun, 1895 E
Odontria obsoleta Broun, 1917 E
Odontria occiputale Broun, 1893 E
Odontria regalis Given, 1952 E
Odontria rufescens Given, 1952 E
Odontria sandageri Broun, 1881 E
Odontria smithii Broun, 1893 E

Odontria striata White, 1846 E
Odontria suavis Broun, 1880 E
Odontria subnitida Given, 1952 E
Odontria sylvatica Broun, 1880 E
Odontria varicolorata Given, 1952 E
Odontria variegata Given, 1952 E
Odontria velutina Given, 1952 E
Odontria xanthosticta White, 1846 E
Odontria n. spp. (12) 12 E
Prodontria capito (Broun, 1909) E
Prodontria grandis Given, 1964 E
Prodontria jenniferae Emerson, 1997 E
Prodontria lewisi Broun, 1904 E
Prodontria longitarsis (Broun, 1909) E Su
Prodontria matagouriae Emerson, 1997 E
Prodontria minuta Emerson, 1997 E
Prodontria modesta (Broun, 1909) E
Prodontria montis Emerson, 1997 E
Prodontria patricki Emerson, 1997 E
Prodontria pinguis Given, 1952 E
Prodontria praelatella (Broun, 1909) E
Prodontria rakiurensis Emerson, 1997 E
Prodontria regalis Emerson, 1997 E
Prodontria setosa Given, 1952 E
Prodontria truncata Given, 1960 E
Scythrodes squalidus Broun, 1886 E
Sericospilus advena Sharp, 1882 E
Sericospilus aenealis (Broun, 1909) E
Sericospilus brevis Given, 1952 E
Sericospilus costellus (Broun, 1880) E
Sericospilus cumberi Given, 1953 E
Sericospilus eximius (Broun, 1917) E
Sericospilus glabratus (Broun, 1893) E
Sericospilus intermediatus Given, 1952 E
Sericospilus minor Given, 1952 E
Sericospilus obscurus Given, 1952 E
Sericospilus ornatus Given, 1960 E
Sericospilus piliventris (Broun, 1921) E
Sericospilus rossii (White, 1846) E
Sericospilus truncatus Given, 1952 E
Sericospilus watti Given, 1960 E
SCITALINI
Gnaphalopoda brookesi (Broun, 1921) E
Gnaphalopoda picea (Broun, 1886) E
XYLONYCHINI
Mycernus elegans Broun, 1904 E
Mycernus intermediatus Given, 1952 E
Psilodontria viridescens Broun, 1895 E
Pyronota edwardsi Sharp, 1876 E
Pyronota festiva (Fabricius, 1775) E
Pyronota inconstans Brookes, 1926 E
Pyronota lugubris Sharp, 1886 E
Pyronota minor Given, 1952 E
Pyronota munda Sharp, 1876 E
Pyronota pallida Broun, 1893 E
Pyronota punctata Given, 1952 E
Pyronota rubra Given, 1952 E
Pyronota setosa Given, 1952 E
Pyronota sobrina Sharp, 1876 E
Pyronota splendens Given, 1952 E
Stethaspis convexa (Given, 1952) E
Stethaspis discoidea (Broun, 1893) E
Stethaspis intermediata (Given, 1952) E
Stethaspis lineata (Arrow, 1924) E
Stethaspis longicornis (Arrow, 1924) E
Stethaspis prasina (Broun, 1893) E
Stethaspis pulchra (Broun, 1895) E
Stethaspis simmondsi (Broun, 1893) E
Stethaspis suturalis (Fabricius, 1775) E
SERICINI
S. PHYLLOTOCINA
Phyllotocus macleayi macleayi Fischer, 1823 A Do
SCARABAEINAE
COPRINI
Copris incertus Say, 1835 A BC
CANTHONINI
Epirinus aeneus (Wiedemann, 1823) A BC
Saphobiamorpha maoriana Brookes, 1944 E
Saphobius brouni Paulian, 1935 E
Saphobius curvipes Broun, 1893 E
Saphobius edwardsi Sharp, 1873 E
Saphobius fulvipes Broun, 1893 E
Saphobius fuscus Broun, 1893 E
Saphobius inflatipes Broun, 1893 E
Saphobius laticollis Broun, 1914 E
Saphobius lepidus Broun, 1912 E
Saphobius lesnei Paulian, 1935 E
Saphobius nitidulus Broun, 1880 E
Saphobius setosus Sharp, 1886 E
Saphobius squamulosus Broun, 1886 E
Saphobius tibialis Broun, 1895 E
Saphobius wakefieldi Sharp, 1877 E
ONTHOPHAGINI
Onthophagus granulatus Boheman, 1858 A BC
Onthophagus posticus Erichson, 1842 A BC
DYNASTINAE
PENTODONTINI
Adoryphorus couloni (Burmeister, 1847) A
Dasygnathus dejeani MacLeay, 1819 A
Heteronychus arator (Fabricius, 1775) A
Pericoptus frontalis Broun, 1904 E
Pericoptus nitidulus Broun, 1880 E
Pericoptus punctatus (White, 1846) E
Pericoptus stupidus Sharp, 1876 E
Pericoptus truncatus Fabricius, 1775 E
APHODIINAE
APHODIINI
Aphodius (Calamosternus) granarius (Linnaeus, 1767) A
Aphodius (Nialus) lividus (Olivier, 1789) A
Acrossidius tasmaniae (Hope, 1847) A
Phycocus graniceps Broun, 1883 E
Tesarius sulcipennis (Lea, 1904) A Ch
PROCTOPHANINI
Proctophanes sculptus (Hope, 1846) A
Australaphodius frenchi (Blackburn, 1892) A
EUPARIINI
Parataenius simulator (Harold, 1868) A
Ataenius brouni (Sharp, 1876) A
Ataenius picinus Harold, 1867 A
Saprosites communis (Broun, 1880) E
Saprosites distans (Sharp, 1876) E
Saprosites exsculptus (White, 1846) E
Saprosites fortipes (Broun, 1881) E
Saprosites kaimai Stebnicka, 2005 E
Saprosites kingsensis Stebnicka, 2001 E TK
Saprosites mendax (Blackburn, 1892) A?
Saprosites raoulensis (Broun, 1910) E K
Saprosites sulcatissimus (Broun, 1911) E Ch
Saprosites watti Stebnicka, 2001 E
ELATERIFORMIA
SCIRTOIDEA
SCIRTIDAE
SCIRTINAE
Amplectopus fuscus Broun, 1893 E
Amplectopus latulus Broun, 1893 E
Amplectopus ovalis Sharp, 1886 E
Amplectopus pallicornis Broun, 1893 E
Atopida basalis Broun, 1912 E
Atopida brouni Sharp, 1878 E
Atopida castanea White, 1846 E
Atopida dorsale Broun, 1893 E
Atopida grahami Broun, 1910 E
Atopida hirta Broun, 1880 E
Atopida impressa Broun, 1914 E
Atopida lawsoni Sharp, 1878 E
Atopida montana Broun, 1921 E
Atopida pallidula Broun, 1921 E
Atopida proba Sharp, 1878 E
Atopida sinuata Broun, 1893 E
Atopida suffusa Broun, 1910 E
Atopida suturalis White, 1846 E
Atopida testacea Broun, 1880 E
Atopida villosa Broun, 1921 E
Brounicyphon sericeum Pic, 1947 E
Byrrhopsis gravidus (Sharp, 1878) E
Cyphanodes vestitus Broun, 1893 E
Cyphanus capax Broun, 1880 E
Cyphanus debilis Sharp, 1878 E
Cyphanus dubius Broun, 1893 E
Cyphanus granulatus Broun, 1880 E
Cyphanus granulosus Broun, 1910 E
Cyphanus laticeps Sharp, 1878 E
Cyphanus maculifer Broun, 1910 E
Cyphanus mandibularis Broun, 1883 E
Cyphanus medius Broun, 1880 E
Cyphanus mollis Sharp, 1878 E
Cyphanus ocularius Broun, 1910 E
Cyphanus punctatus Sharp, 1878 E
Cyphanus scaber Broun, 1893 E
Cyphon acerbus Broun, 1883 E
Cyphon aequalis Sharp, 1878 E
Cyphon aethiops Broun, 1886 E
Cyphon albidosparsus Nyholm, 2000 E
Cyphon arduus Sharp, 1878 E
Cyphon brouni Pic, 1913 E
Cyphon burrowsi Broun, 1915 E
Cyphon decussatus Nyholm, 2000 E
Cyphon deterius Broun, 1914 E
Cyphon dilutus Broun, 1883 E
Cyphon discedens Broun, 1893 E
Cyphon genalis Sharp, 1878 E
Cyphon graniger Sharp, 1878 E
Cyphon huttoni Sharp, 1878 E
Cyphon ignoratus Nyholm, 2000 E
Cyphon lateralis Broun, 1883 E
Cyphon laticeps Sharp, 1878 E
Cyphon laticollis Broun, 1883 E
Cyphon nigritulus Broun, 1893 E
Cyphon nigropictus Broun, 1883 E
Cyphon ornatus Broun, 1893 E
Cyphon oscillans Sharp, 1878 E
Cyphon parviceps Sharp, 1878 E
Cyphon pauper Broun, 1893 E
Cyphon pictulus Sharp, 1878 E
Cyphon poecilopterus Nyholm, 2000 E
Cyphon princeps Nyholm, 2000 E
Cyphon pumilio Sharp, 1878 E
Cyphon pusillus Nyholm, 2000 E
Cyphon rectalis Broun, 1886 E
Cyphon rectangulus Broun, 1883 E
Cyphon remotus Broun, 1883 E
Cyphon subvariegatus Nyholm, 2000 E
Cyphon suffusus Sharp, 1878 E
Cyphon thomasi Pic, 1914 E
Cyphon trivialis Broun, 1886 E
Cyphon umbricolor Nyholm, 2000 E
Cyphon variegatus Sharp, 1882 E
Cyphon viridipennis Broun, 1880 E
Cyphon waikatoensis Broun, 1886 E
Cyphon zealandicus Sharp, 1878 E
Gen. nr *Cyphon* Paykull, 1799, sensu Nyholm, 2000, *amplus* Broun, 1880 E
Gen. nr *Cyphon cincticollis* Broun, 1915 E
Gen. nr *Cyphon fuscifrons* Broun, 1893 E
Gen. nr *Cyphon granulicollis* Broun, 1915 E
Gen. nr *Cyphon mackerrowi* Broun, 1895 E
Gen. nr *Cyphon pachymerus* Broun, 1912 E
Gen. nr *Cyphon plagiatus* Broun, 1883 E
Gen. nr *Cyphon plumatellus* Broun, 1915 E
Gen. nr *Cyphon propinquus* Broun, 1883 E
Gen. nr *Cyphon* n. spp. (8) 8E
Cyphotelus angustifrons Sharp, 1878 E
Cyprobius nitidus Sharp, 1878 E
Cyprobius terrenus Broun, 1895 E

Cyprobius undulatus Broun, 1883 E
Mesocyphon bifoveatus Broun, 1914 E
Mesocyphon capito Broun, 1914 E
Mesocyphon divergens Sharp, 1878 E
Mesocyphon granulatus Broun, 1886 E
Mesocyphon lateralis Broun, 1914 E
Mesocyphon laticeps Broun, 1893 E
Mesocyphon longicornis Broun, 1914 E
Mesocyphon mandibularis Broun, 1912 E
Mesocyphon marmoratus Sharp, 1878 E
Mesocyphon monticola Broun, 1886 E
Mesocyphon pallidus Broun, 1893 E
Mesocyphon setiger Sharp, 1878 E
Mesocyphon tristis Broun, 1910 E
Mesocyphon vestitus Broun, 1914 E
Mesocyphon wakefieldi Sharp, 1878 E
Veronatus amplus Broun, 1895 E
Veronatus antennalis Broun, 1883 E
Veronatus apterus Broun, 1921 E
Veronatus brevicollis Broun, 1921 E
Veronatus capito Broun, 1880 E
Veronatus frontalis Broun, 1880 E
Veronatus fulgidulus Broun, 1915 E
Veronatus granicollis Broun, 1910 E
Veronatus longicornis Sharp, 1878 E
Veronatus nubilus Broun, 1893 E
Veronatus punctipennis Broun, 1914 E
Veronatus reversus Broun, 1921 E
Veronatus scabiosus Broun, 1880 E
Veronatus sharpi Broun, 1880 E
Veronatus sternalis Broun, 1921 E
Veronatus tarsalis Broun, 1915 E
Veronatus tricostellus (White, 1846) E
Veronatus versicolor Broun, 1921 E
Veronatus vestitus Broun, 1921 E
Gen. et sp(p.) indet. (2) 2E
EUCINETIDAE
Eucinetus stewarti (Broun, 1881) E
Noteucinetus nunni Bullians & Leschen, 2005 E
CLAMBIDAE
CLAMBINAE
Clambus bulla bulla Endrödy-Younga, 1974 E
Clambus b. elongatus Endrödy-Younga, 1990 E
Clambus b. marmoratus Endrödy-Younga, 1990 E
Clambus domesticus Broun, 1886 E
Clambus jupiter Endrödy-Younga, 1999 E
Clambus neptunus Endrödy-Younga, 1990 E
Clambus pluto Endrödy-Younga, 1990 E
Clambus saturnus annulus Endrödy-Younga, 1990 E
Clambus s. saturnus Endrödy-Younga, 1990 E
Clambus uranus Endrödy-Younga, 1999 E
Clambus venus Endrödy-Younga, 1990 E
Clambus n. sp. E
Sphaerothorax brevisternalis Endrödy-Younga, 1990 E
Sphaerothorax kuscheli Endrödy-Younga, 1990 E
Sphaerothorax suffusus (Broun, 1886) E
Sphaerothorax tierensis (Blackburn, 1902) A
Sphaerothorax zealandicus Endrödy-Younga, 1990 E
BUPRESTOIDEA
BUPRESTIDAE
BUPRESTINAE
Maoraxia eremita (White, 1846) E
Nascioides enysi (Sharp, 1877) E
BYRRHOIDEA
BYRRHIDAE
BYRRHINAE
Chlorobyrrhulus coruscus (Pascoe, 1875) E
Curimus squamifer Broun, 1893 E
Curimus striatus Broun, 1880 E
Curimus vestitus Broun, 1904 E
Curimus zeelandicus Redtenbacher, 1868 E
Cytilissus claviger Broun, 1893 E
Epichorius aucklandiae Kiesenwetter & Kirsch, 1877 E Su
Epichorius longulus (Broun, 1909) E Su
Epichorius sorenseni (Brookes, 1951) E Su
Epichorius tumidellus (Broun, 1909) E Su
Liochoria huttoni Pascoe, 1875 E
Liochoria insueta (Broun, 1886) E
Liochoria mixta (Broun, 1886) E
Liochoria nigralis (Broun, 1893) E
Liochoria nigricans (Broun, 1881) E
Liochoria orbicularis (Broun, 1880) E
Liochoria sternalis Broun, 1893 E
Pedilophorus aemulator Broun, 1914 E
Pedilophorus bryobius Broun, 1910 E
Pedilophorus cognatus Broun, 1910 E
Pedilophorus creperus Broun, 1893 E
Pedilophorus foveigerus Broun, 1910 E
Pedilophorus gemmeus (Broun, 1883) E
Pedilophorus helmsi (Reitter, 1880) E
Pedilophorus humeralis Broun, 1914 E
Pedilophorus laetus (Broun, 1893) E
Pedilophorus laevipennis Broun, 1893 E
Pedilophorus lewisi Broun, 1907 E
Pedilophorus nigrescens Broun, 1915 E
Pedilophorus opaculus Broun, 1912 E
Pedilophorus ornatus Broun, 1914 E
Pedilophorus picipes Broun, 1893 E
Pedilophorus probus Broun, 1893 E
Pedilophorus pulcherrimus Broun, 1909 E
Pedilophorus puncticeps Broun, 1893 E
Pedilophorus sculpturatus Broun, 1910 E
Pedilophorus tibialis Broun, 1893 E
Synorthus anomalus (Broun, 1880) E
Synorthus granulatus (Broun, 1893) E
Synorthus insularis Watt, 1971 E Su
Synorthus laevigatus Broun, 1910 E
Synorthus mandibularis Broun, 1910 E
Synorthus pygmaeus Broun, 1910 E
Synorthus rectifrons Broun, 1915 E
Synorthus rotundus (Broun, 1881) E
Synorthus setarius (Broun, 1880) E
Synorthus substriatus Broun, 1914 E
Synorthus versipilus Broun, 1914 E
Synorthus villosus (Broun, 1886) E
Synorthus n. sp. E Su
SYNCALYPTINAE
Microchaetes n. sp. E Ch
DRYOPIDAE
Parnida agrestis Broun, 1880 E
Parnida longulus (Sharp, 1886) E
Parnida scutellaris (Broun, 1910) E
Parnida vestitus (Sharp, 1883) E
ELMIDAE
LARAINAE
Hydora angusticollis (Pascoe, 1877) E F
Hydora lanigera Broun, 1914 E F
Hydora nitida Broun, 1885 E F
Hydora obsoleta Broun, 1885 E F
Hydora picea (Broun, 1881) E F
Hydora subaenea Broun, 1914 E F
Hydora n. sp. E F
Elmidae gen. et spp. indet. (15) 15E F
LIMNICHIDAE
HYPHALINAE
Hyphalus kuscheli Britton, 1977 E M
Hyphalus prolixus Britton, 1977 E M
Hyphalus ultimus Britton, 1977 E M
Hyphalus wisei Britton, 1973 E M
LIMNICHINAE
LIMNICHINI
Limnichus decorus Broun, 1880 E
Limnichus nigripes Broun, 1893 E
Limnichus picinus Broun, 1881 E
Limnichus simplex Broun, 1910 E
HETEROCERIDAE
HETEROCERINAE
HETEROCERINI
Heterocerus novaeselandiae Charpentier, 1968 E F
PTILODACTYLIDAE
ANCHYTARSINAE
Byrrocryptus urquharti (Broun, 1893) E
Byrrocryptus n. spp. (4–7) 4–7 E
Ptilodactylidae gen. et. sp. indet. E
CHELONARIIDAE
CHELONARIINAE
Brounia thoracica Sharp, 1878 E
ELATEROIDEA
EUCNEMIDAE
MELASINAE
XYLOBIINI
Agalba cylindrata (Broun, 1883) E
Agalba nigrescens Broun, 1893 E
Agalba ruficornis Broun, 1893 E
NEOCHARINI
Neocharis concolor Sharp, 1877 E
Neocharis lobitarsis Broun, 1910 E
Neocharis osculans Broun, 1881 E
Neocharis pubescens Sharp, 1877 E
Neocharis simplex Sharp, 1877 E
Neocharis varia Sharp, 1877 E
DIRHAGINI
Balistica foveata (Broun, 1881) E
Talerax capax Broun, 1881 E
Talerax distans Sharp, 1877 E
Talerax dorsalis Broun, 1912 E
Talerax micans Broun, 1893 E
Talerax niger Broun, 1881 E
Talerax rusticus Broun, 1881 E
Talerax spinitarsis Broun, 1910 E
Talerax tenuis Broun, 1883 E
MACRAULACINAE
MACRAULACINI
Dromaeolus australasiae Bonvouloir, 1871 A Do
Dromaeolus nigellus (White, 1846) E
Nematodes major Bonvouloir, 1872 A
ELATERIDAE
AGRYPNINAE A
AGRYPNINI
Agrypnus variabilis (Candèze, 1857) A
CONODERINI
Conoderus exsul (Sharp, 1877) E
Conoderus maritimus (Broun, 1893) E
Conoderus planatus Schwarz, 1906 E
Conoderus posticus (Eschscholtz, 1822) A
Conoderus submarmoratus Macleay, 1872 A
Conoderus subrufus (Broun, 1880) E
Silene brunnea Broun, 1893 E SI
HEMIRHIPINI
Thoramus angustus Broun, 1881 E
Thoramus cervinus Broun, 1881 E
Thoramus feredayi Sharp, 1877 E
Thoramus foveolatus Broun, 1880 E
Thoramus huttoni Sharp, 1886 E
Thoramus laevithorax (White, 1846) E Ch
Thoramus parryi (Candèze, 1863) E
Thoramus parvulus Broun, 1881E
Thoramus perblandus Broun, 1880 E
Thoramus rugipennis Broun, 1880 E
Thoramus wakefieldi Sharp, 1877 E
CARDIOPHORINAE
CARDIOPHORINI
Exoeolus brouni Douglas, 2005 E
Exoeolus rufescens Broun, 1893 E
PROSTERNINAE
PITYOBIINI
Metablax acutipennis (White, 1846) E
Metablax approximans (White, 1846) E
Metablax brouni Sharp, 1877 E
Metablax cinctiger (White, 1846) E
Metablax gourlayi Calder, 1976 E TK
DENTICOLLINAE
Acritelater barbatus (Candèze, 1865) E
Acritelater elongatus (Sharp, 1877) E

Acritelater reversus (Sharp, 1877) E
Acritelater setiger (Broun, 1881) E
Amphiplatys lawsoni Sharp, 1877 E
Amychus candezei Pascoe, 1876 E
Amychus granulatus (Broun, 1883) E
Amychus manawatawhi Marris & Johnson, 2010 E TK
Asymphus insidiosus Sharp, 1886 E
Elatichrosis aeneola (Candèze, 1865) E
Elatichrosis castanea (Broun, 1881) E
Elatichrosis certa (Broun, 1881) E
Elatichrosis fulvipes (Broun, 1881) E
Elatichrosis polita (Sharp, 1877) E
Elatichrosis violacea (Sharp, 1877) E
Hapatesus (*Hapatesus*) *electus* Neboiss, 1957 A
Oxylasma basale Broun, 1886 E
Oxylasma carinale Broun, 1893 E
Oxylasma pannosum Broun, 1881 E
Oxylasma tectum Broun, 1881 E
Oxylasma vittiger Broun, 1883 E
Parinus villosus Sharp, 1877 E
Poemnites (*Poemnites*) *agriotoides* (Sharp, 1877) E
Poemnites (*P.*) *antipodus* (Candèze, 1863) E
Poemnites (*P.*) *approximans* (Broun, 1912) E
Poemnites (*P.*) *canaliculatus* (Broun, 1893) E
Poemnites (*P.*) *dubius* (Sharp, 1877) E
Poemnites (*P.*) *fulvescens* (Broun, 1912) E
Poemnites (*P.*) *irregularis* (Sharp, 1886) E
Poemnites (*P.*) *megops* (White, 1846) E
Poemnites (*P.*) *mundus* (Sharp, 1886) E
Poemnites (*P.*) *munroi* (Broun, 1893) E
Poemnites (*P.*) *olivascens* (White, 1846) E
Poemnites (*P.*) *sternalis* (Broun, 1912) E
Poemnites (*P.*) *strangulatus* (White, 1846) E
Poemnites (*P.*) *vitticollis* (Broun, 1912) E
Zeaglophus pilicornis Broun, 1895 E
HYPNOIDINAE
HYPNOIDINI
Hypnoidus meinertzhageni (Broun, 1881) E
PRISAHYPNINI
Australeeus humilis (Sharp, 1877) E
Australeeus powelli (Sharp, 1877) E
Insulahypnus kuscheli Stibick, 1981 E
Insulahypnus lancea Stibick, 1981 E
Insulahypnus longicornis (Sharp, 1877) E
Insulahypnus mayae Stibick, 1981 E
Insulahypnus wisei Stibick, 1981 E
Prisahypnus attenuatus (Broun, 1893) E
Prisahypnus frontalis (Sharp, 1877) E
LISSOMINAE
PROTELATERINI
Protelater atriceps Broun, 1886 E
Protelater costiceps Broun, 1893 E
Protelater diversus Broun, 1912 E
Protelater elongatus Sharp, 1877 E
Protelater guttatus Sharp, 1877 E
Protelater huttoni Sharp, 1877 E
Protelater nigricans Sharp, 1881 E
Protelater opacus Sharp, 1877 E
Protelater picticornis Sharp, 1877 E
Protelater pubescens Broun, 1893 E
Protelater urquharti Broun, 1893 E
Protelater vitticollis Broun, 1886 E
Sphaenelater collaris (Pascoe, 1877) E
Sphaenelater lineicollis (White, 1846) E
Sphaenelater nitidofuscus (Blanchard, 1853) E Su
ELATERINAE
AMPEDINI
Aglophus modestus Sharp, 1877 E
Lomemus collaris Sharp, 1877 E
Lomemus elegans Sharp, 1877 E
Lomemus flavipes Sharp, 1877 E
Lomemus frontalis Broun, 1893 E
Lomemus fulvipennis Broun, 1893 E
Lomemus fuscicornis Broun, 1893 E
Lomemus fuscipes Broun, 1893 E
Lomemus maurus Broun, 1893 E
Lomemus obscuripes Sharp, 1877 E
Lomemus pictus Sharp, 1877 E
Lomemus pilicornis Sharp, 1877 E
Lomemus puncticollis Broun, 1895 E
Lomemus rectus Broun, 1883 E
Lomemus sculpturatus Broun, 1893 E
Lomemus similis Sharp, 1877 E
Lomemus suffusus Sharp, 1877 E
Lomemus vittatus Broun, 1883 E
Lomemus vittipennis Broun, 1910 E
Mecastrus convexus Sharp, 1877 E
Mecastrus discedens Sharp, 1877 E
Mecastrus intermedius Broun, 1893 E
Mecastrus lateristrigatus (White, 1846) E
Mecastrus vicinus Sharp, 1877 E
Ochosternus zealandicus (White, 1864) E K
MEGAPENTHINI
Megapenthes kieneri Schimmel, 1997 E
POMACHILINI
Betarmonides flavipilus (Broun, 1893) E
Betarmonides frontalis (Sharp, 1877) E
Betarmonides gracilipes (Sharp, 1877) E
Betarmonides laetus (Sharp, 1877) E
Betarmonides obscurus (Sharp, 1877) E
Betarmonides sharpi (Candèze, 1882) E
Panspoeus guttatus Sharp, 1877 E
LYCIDAE
METRIORRHYNCHINAE A
Porrostoma (*Porrostoma*) *rufipennis* (Fabricius, 1801) A
CANTHARIDAE
DYSMORPHOCERINAE
Asilis (*Asilis*) *alticola* Wittmer, 1979 E
Asilis (*A.*) *annulicornis* Wittmer, 1979 E
Asilis (*A.*) *apicalis* Broun, 1909 E
Asilis (*A.*) *arcuata* Wittmer, 1979 E
Asilis (*A.*) *brevicornis* Broun, 1910 E
Asilis (*A.*) *calleryensis* Wittmer, 1979 E
Asilis (*A.*) *collaris* Broun, 1910 E
Asilis (*A.*) *cornuta* Wittmer, 1979 E
Asilis (*A.*) *dentata* Wittmer, 1979 E
Asilis (*A.*) *dugdalei* Wittmer, 1979 E
Asilis (*A.*) *dunensis* Wittmer, 1979 E
Asilis (*A.*) *fiordensis* Wittmer, 1979 E
Asilis (*A.*) *forcipifera* Wittmer, 1979 E
Asilis (*A.*) *fulvithorax* (Broun, 1880) E
Asilis (*A.*) *grossepunctata* Wittmer, 1979 E
Asilis (*A.*) *homerica* Wittmer, 1979 E
Asilis (*A.*) *kuscheli* Wittmer, 1979 E
Asilis (*A.*) *laevigata* (Broun, 1886) E
Asilis (*A.*) *laeviuscula* Wittmer, 1979 E
Asilis (*A.*) *lyriformis* Wittmer, 1979 E
Asilis (*A.*) *maori* Wittmer, 1979 E
Asilis (*A.*) *nelsonensis* Wittmer, 1979 E
Asilis (*A.*) *paralella* Broun, 1910 E
Asilis (*A.*) *pilicornis* Broun, 1909 E
Asilis (*A.*) *piliventer* (Broun, 1881) E
Asilis (*A.*) *planata* Wittmer, 1979 E
Asilis (*A.*) *platygona* Wittmer, 1979 E
Asilis (*A.*) *pugiunculus* Wittmer, 1979 E
Asilis (*A.*) *ramosa* Wittmer, 1979 E
Asilis (*A.*) *reflexa* Wittmer, 1979 E
Asilis (*A.*) *reflexodentata* Wittmer, 1979 E
Asilis (*A.*) *sinuella* Broun, 1909 E
Asilis (*A.*) *tenuicula* (Broun, 1880) E
Asilis (*A.*) *waipouana* Wittmer, 1979 E
Asilis (*Heterasilis*) *flavipennis* Broun, 1914 E
Asilis (*H.*) *intermixta* Wittmer, 1979 E
Asilis (*H.*) *nigricans* (Broun, 1880) E
Asilis (*H.*) *subnuda* (Broun, 1880) E
Asilis (*H.*) *tumida aptera* Wittmer, 1979 E
Asilis (*H.*) *t. tumida* (Broun, 1881) E
Neoontelus bifurcatus Wittmer, 1979 E
Neoontelus elongatus Wittmer, 1979 E
Neoontelus punctipennis (Broun, 1910) E
Neoontelus striatus (Broun, 1880) E
MALTHININAE A
MALTHODINI
Malthodes pumilus (Brébisson, 1835) A
BOSTRICHIFORMIA
DERODONTOIDEA
DERODONTIDAE
LARICOBIINAE
Nothoderodontus gourlayi Crowson, 1959 E
Nothoderodontus watti Lawrence, 1985 E
BOSTRICHOIDEA
JACOBSONIIDAE
Derolathrus n. spp. (3) 3E
Saphophagus minutus Sharp, 1886 E
Saphophagus n. sp. E
NOSODENDRIDAE
Nosodendron (*Nosodendron*) *ovatum* Broun, 1880 E
Nosodendron (*N.*) *zealandicum* Sharp, 1882 E
DERMESTIDAE
DERMESTINAE A
DERMESTINI
Dermestes (*Dermestes*) *ater* De Geer, 1774 A
Dermestes (*D.*) *haemorrhoidalis* Küster, 1852 A
Dermestes (*D.*) *lardarius* Linnaeus, 1758 A
Dermestes (*D.*) *peruvianus* Laporte de Castelnau, 1840 A
Dermestes (*Dermestinus*) *carnivorus* Fabricius, 1775 A Do
Dermestes (*D.*) *maculatus* De Geer, 1774 A
TRINODINAE
THYLODRINI
Hexanodes vulgata (Broun, 1880) E
ATTAGENINAE A
ATTAGENINI
Attagenus pellio (Linnaeus, 1758) A Do
MEGATOMINAE
MEGATOMINI
Anthrenocerus australis (Hope, 1843) A
Reesa vespulae (Milliron, 1939) A
Trogoderma antennale Broun, 1893 E
Trogoderma carteri Armstrong, 1942 A Do
Trogoderma granulatum Broun, 1886 E
Trogoderma grassmani Beal, 1954 A Do
Trogoderma inclusum LeConte, 1854 A Do
Trogoderma maestum Broun, 1880 E
Trogoderma ornatum Say, 1825 A Do
Trogoderma pictulum Broun, 1911 E Ch
Trogoderma punctatum Broun, 1886 E
Trogoderma puncticolle Broun, 1914 E
Trogoderma quadrifasciatum Broun, 1893 E
Trogoderma serrigerum Sharp, 1877 E
Trogoderma signatum Sharp, 1877 E
Trogoderma suffusum Broun, 1886 E
Trogoderma variabile Ballion, 1878 A
ANTHRENINI
Anthrenus (*Florilinus*) *museorum* (Linnaeus, 1761) A
Anthrenus (*Nathrenus*) *verbasci* (Linnaeus, 1767) A
BOSTRICHIDAE
DINODERINAE A
Dinoderus minutus (Fabricius, 1775) A
Rhyzopertha dominica (Fabricius, 1792) A
Dinoderinae sp. indet. A?
LYCTINAE A
LYCTINI
Lyctus brunneus (Stephens, 1830) A
Lyctus linearis (Goeze, 1777) A Do
Lyctus planicollis LeConte, 1858 A Do
Lyctinae sp. indet. A?
EUDERINAE
Euderia squamosa Broun, 1880 E
ANOBIIDAE
ANOBIINAE
Anobium punctatum De Geer, 1774 A
Australanobium inaequale inaequale (Broun, 1912) E
Australanobium i. trapezicolle Español, 1976 E

Hadrobregmus (Megabregmus) australiensis Pic, 1901 E
Hadrobregmus (M.) crowsoni Español, 1976 E
Hadrobregmus (M.) magnus (Dumbleton, 1941) E
Macranobium truncatum Broun, 1886 E
Stegobium paniceum (Linnaeus, 1758) A
Xenocera ambigua Broun, 1881 E
Xenocera furca Broun, 1881 E
Xenocera granulata (Broun, 1880) E
Xenocera notata (Broun, 1880) E
Xenocera obscura (Sharp, 1886) E
Xenocera plagiata Broun, 1881 E
Xenocera pulla Broun, 1881 E
Xenocera sericea (Broun, 1880) E
Xenocera versuta Broun, 1881 E
DORCATOMINAE
CRYPTORAMORPHINI
Dorcatomiella ornata Español, 1977 E K
Mirosternomorphus crowsoni (Español, 1970) E
Mirosternomorphus oblongus (Broun, 1880) E
Mirosternomorphus n. sp. E TK
Serianotus (Serianotus) punctilatera kermadecensis Español, 1979 E K
DORCATOMINI
Cyphanobium illustre (Broun, 1880) E
Dorcatoma lauta Broun, 1881 E
Dorcatoma pilosella Hinton, 1941 E
PROTHECINI
Methemus griseipilus (Broun, 1881) E
DRYOPHILINAE
Sphinditeles atriventris Broun, 1881 E
Sphinditeles debilis (Sharp, 1882) E
Sphinditeles dorsalis Broun, 1893 E
Sphinditeles nigricornis Broun, 1893 E
Sphinditeles niticollis (Broun, 1912) E
Sphinditeles rufescens Broun, 1893 E
Sphinditeles ruficornis Broun, 1893 E
ERNOBIINAE A
Ernobius mollis (Linnaeus, 1758) A
MESOCOELOPODINAE
TRICORYNI
Tricorynus herbarium (Gorham, 1883) A
XYLETININAE
Deroptilinus granicollis Lea, 1924 A
Lasioderma serricorne (Fabricius, 1792) A
Leanobium flavomaculatum Español, 1979 E K
Leanobium marmoratum (Lea, 1924) E K
Leanobium undulatum (Broun, 1881) E
Xyletobius (Holcobius) watti (Español, 1982) E
Xyletobius (Xyletobius) kuscheli Español, 1982 E
PTINIDAE
GIBBIINAE
Mezium affine Boieldieu, 1856 A
Mezium americanum (Laporte de Castelnau, 1840) A
PTININAE
Niptus hololeucus (Faldermann, 1836) A
Ptinus clavipes Panzer, 1792 A
Ptinus fur (Linnaeus, 1758) A Do
Ptinus littoralis Broun, 1893 E
Ptinus maorianus Brookes, 1926 E
Ptinus murinus White, 1846 E
Ptinus plagiatus Broun, 1914 E
Ptinus sexpunctatus Panzer, 1792 A Do
Ptinus speciosus Broun, 1880 E
Ptinus suturalis White, 1846 E
Ptinus tectus Boieldieu, 1856 A
Sphaericus (Sphaericus) gibboides (Boieldieu, 1854) A
Trigonogenius globulus Solier, 1849 A
CUCUJIFORMIA
CLEROIDEA
TROGOSSITIDAE
LOPHOCATERINAE
ANCYRONINI
Grynoma albosparsa Broun, 1909 E
Grynoma clavalis Broun, 1917 E
Grynoma diluta Sharp, 1877 E
Grynoma fusca Sharp, 1877 E
Grynoma pallidula Broun, 1917 E
Grynoma proxima Broun, 1917 E
Grynoma regularis Sharp, 1882 E
Grynoma rugosa Broun, 1893 E
Grynoma setigera Broun, 1917 E
Grynoma varians Broun, 1893 E
Neaspis variegata (MacLeay, 1871) A
LOPHOCATERINI
Promanus auripilus Broun, 1893 E
Promanus depressus Sharp, 1877 E
Promanus subcostatus Broun, 1909 E
PELTINAE
THYMALINI
Australiodes vestitus (Broun, 1883) E
Parentonium magnum (Crowson, 1966) E
Protopeltis pulchella (Broun, 1915) E SI
Protopeltis viridescens (Broun, 1886) E
Rentonellum apterum Crowson, 1966 E
Rentonellum n. sp. (3) 3E
Rentonidium costiventris Crowson, 1966 E
Rentonidium daldiniae Crowson, 1966 E
Rentonium n. sp. (2) 2E
'Peltidae' indet. spp. (2) 2E
TROGOSSITINAE
GYMNOCHILINI
Lepidopteryx ambigua (Broun, 1880) E
Lepidopteryx brounii (Pascoe, 1876) E
Lepidopteryx farinosa (Sharp, 1877) E
Lepidopteryx interrupta Brookes, 1932 E
Lepidopteryx nigrosparsa (White, 1846) E
Lepidopteryx shandi (Broun, 1910) E Ch
Lepidopteryx sobrina (White, 1846) E
Lepidopteryx wakefieldi (Sharp, 1877) E
Tenebroides affinis (White, 1846) E
Tenebroides mauritanicus (Linnaeus, 1758) A
CHAETOSOMATIDAE sensu Crowson 1952
Chaetosoma colossa Opitz, 2010
Chaetosoma scaritides Westwood, 1851 E
Chaetosoma n. sp. E
Chaetosomodes halli Broun, 1921 E
CLERIDAE
CLERINAE
Balcus signatus Broun, 1880 E
Balcus violacea (Fabricius, 1787) E
Thanasimus formicarius (Linnaeus, 1758) A
ENOPLIINAE
Phymatophaea aeraria (Pascoe, 1876) E
Phymatophaea aquila Opitz, 2009 E
Phymatophaea atrata Broun, 1881 E
Phymatophaea auripila Opitz, 2009 E
Phymatophaea breviclava Broun, 1914 E
Phymatophaea deirolinea Opitz, 2009 E TK
Phymatophaea earlyi Opitz, 2009 E Ch
Phymatophaea enodis Opitz, 2009 E TK
Phymatophaea fuscitarsis Broun, 1914 E
Phymatophaea guttigera (Waterhouse, 1877) E
Phymatophaea hudsoni Broun, 1923 E
Phymatophaea insula Opitz, 2009 E
Phymatophaea longula Sharp, 1877 E
Phymatophaea lugubris Broun, 1909 E
Phymatophaea maorias Opitz, 2009 E
Phymatophaea oconnori Broun, 1914 E
Phymatophaea opacula Broun, 1893 E
Phymatophaea opiloides (Pascoe, 1876) E
Phymatophaea pantomelas (Boisduval, 1835) E
Phymatophaea pustulifera (Westwood, 1852) E
Phymatophaea testacea Broun, 1881 E
Phymatophaea trachelogloba Opitz, 2009 E
Phymatophaea watti Opitz, 2009 E
KORYNETINAE A
Necrobia ruficollis (Fabricius, 1775) A
Necrobia rufipes (De Geer, 1775) A
HYDNOCERINAE
Lemidia aptera (Sharp, 1877) E
Lemidia debilis (Sharp, 1877) E
Lemidia longipes (Sharp, 1877) E
Lemidia rugosa (Broun, 1893) E
Lemidia violacea (Broun, 1912) E
TARSOSTENINAE A
Paratillus carus (Newman, 1840) A
METAXINIDAE E
Metaxinida ornata Broun, 1909 E
PHYCOSECIDAE
Phycosecis limbatus (Fabricius, 1781) E
MELYRIDAE
DASYTINAE
'Arthracanthus' atriceps Broun, 1914 E
'Arthracanthus' fossicollis Broun, 1914 E
'Arthracanthus' foveicollis Broun, 1912 E
'Arthracanthus' fulvipes Broun, 1914 E
Dasytes aethiops Broun, 1893 E
Dasytes anacharis Broun, 1909 E
Dasytes aurisetifer Broun, 1909 E
Dasytes cheesemani Broun, 1886 E
Dasytes cinereohirtus Broun, 1880 E
Dasytes clavatus Broun, 1921 E
Dasytes constrictus Broun, 1883 E
Dasytes fuscitarsis Broun, 1914 E
Dasytes helmsi Sharp, 1882 E
Dasytes laevulifrons Broun, 1914 E
Dasytes laticeps Broun, 1880 E
Dasytes littoralis Broun, 1893 E
Dasytes minutus (Fabricius, 1781) E
Dasytes obscuricollis Broun, 1880 E
Dasytes occiputalis Broun, 1883 E
Dasytes opaculus Broun, 1886 E
Dasytes oreocharis Broun, 1893 E
Dasytes philpotti Broun, 1915 E
Dasytes pittensis Broun, 1911 E Ch
Dasytes planifrons (Broun, 1883) E
Dasytes stewarti Broun, 1881 E
Dasytes veronicae Broun, 1910 E
Dasytes violascens Broun, 1921 E
Dasytes wakefieldi Sharp, 1877 E
Dasytes n. spp. (2) 2E
Halyles brevicornis Broun 1883 E
Halyles nigrescens Broun, 1883 E
Halyles semidilutus Broun, 1883 E
MALACHIINAE
CARPHURINI
Carphurus venustus Kirsch, 1877 E Su
CUCUJOIDEA
NITIDULIDAE
EPURAEINAE
EPURAEINI
Epuraea antarctica (White, 1846) E
Epuraea imperialis Reitter, 1877 A
Epuraea scutellaris (Broun, 1880), non *scutellaris* Kraatz, 1895 E
Epuraea signata Broun, 1880 E
Epuraea zelandica Sharp, 1878 E
CARPOPHILINAE A
Carpophilus davidsoni Dobson, 1952 A
Carpophilus dimidiatus (Fabricius, 1792) A
Carpophilus gaveni Dobson, 1964 A
Carpophilus hemipterus (Linnaeus, 1758) A
Carpophilus ligneus Murray, 1864 A
Carpophilus maculatus Murray, 1864 A Do
Carpophilus marginellus Motschulsky, 1858 A
Carpophilus mutilatus Erichson, 1843 A Do [not established]
Carpophilus oculatus gilloglyi Dobson, 1993 A Do
Carpophilus o. oculatus Murray, 1864 A
Urophorus humeralis (Fabricius, 1798) A
NITIDULINAE
NITIDULINI
Nitidula carnaria (Schaller, 1783) A
Omosita (Saprobia) colon (Linnaeus, 1758) A

Omosita (Saprobia) discoidea (Fabricius, 1775) A
Omosita (Saprobia) spinipes Broun, 1880 E
Soronia asperella (Broun, 1893) E
Soronia micans Broun, 1893 E
Soronia morosa Broun, 1893 E
Soronia oculata Reitter, 1880 E
Soronia optata Sharp, 1878 E
Hisparonia hystrix (Sharp, 1876) E
Neopocadius (Brounthina) aequalis (Kirejtschuk, 1997) E
Aethina (Idaethina) concolor (MacLeay, 1872) A
Thalycrodes australis (Germar, 1848) A
CYLLODINI
Cerylollodes dacnoides Kirejtschuk, 2006 E
CILLAEINAE
Brachypeplus brevicornis Sharp, 1878 E
CRYPTARCHINAE
Cryptarcha nitidisssima Reitter, 1873 A
Crytptarcha optanda (Broun, 1881) E
Homepuraea amoena (Broun, 1880) E
Homepuraea halli (Broun, 1921) E
MONOTOMIDAE
LENACINAE
LENACINI
Lenax mirandus Sharp, 1877 E
MONOTOMINAE A
Monotoma (Monotoma) bicolor Villa & Villa, 1835 A
Monotoma (M.) longicollis (Gyllenhal, 1827) A
Monotoma (M.) picipes Herbst, 1793 A
Monotoma (M.) spinicollis Aubé, 1838 A
Monotoma (M.) testacea Motschulsky, 1845 A
AGAPYTHIDAE E
Agapytho foveicollis Broun, 1921 E
PRIASILPHIDAE
Priasilpha angulata Leschen, Lawrence & Ślipiński, 2005 E
Priasilpha aucklandica Leschen, Lawrence & Ślipiński, 2005 E
Priasilpha bufonia Leschen, Lawrence & Ślipiński, 2005 E
Priasilpha carinata Leschen, Lawrence & Ślipiński, 2005 E
Priasilpha earlyi Leschen, Lawrence & Ślipiński, 2005 E
Priasilpha embersoni Leschen, Lawrence & Ślipiński, 2005 E
Priasilpha obscura Broun, 1893 E
SILVANIDAE
BRONTINAE
BRONTINI
Brontopriscus pleuralis (Sharp, 1877) E
Brontopriscus sinuatus Sharp, 1886 E
Dendrophagella capito (Pascoe, 1876) E
Protodendrophagus antipodes Thomas, 2004 E
SILVANINAE
CRYPTAMORPHINI
Cryptamorpha brevicornis (White, 1846) E
Cryptamorpha curvipes Broun, 1880 E
Cryptamorpha desjardinsi (Guérin-Méneville, 1829) A
Cryptamorpha picturata (Reitter, 1880) E
Cryptamorpha rugicollis (Broun, 1910) E
Cryptamorpha n. sp. E TK
SILVANINI
Ahasverus advena (Waltl, 1832) A
Nausibius clavicornis (Kugelann in Schneider, 1794) A
Oryzaephilus mercator (Fauvel, 1889) A
Oryzaephilus surinamensis (Linnaeus, 1758) A
Silvanus bidentatus (Fabricius, 1792) A Do
Silvanus lateritius (Broun, 1880) E
Silvanus unidentatus (Olivier, 1790) A Do
CUCUJIDAE
Platisus zelandicus Marris & Klimaszewski, 2001 E TK
LAEMOPHLOEIDAE
Cryptolestes capensis (Waltl, 1834) A
Cryptolestes ferrugineus (Stephens, 1831) A Do
Cryptolestes pusilloides (Steel & Howe, 1952) A
Cryptolestes pusillus (Schönherr, 1817) A Do
Microbrontes lineatus (Broun, 1893) E
Microbrontes n. sp. E
Notolaemus n. sp. A
PHALACRIDAE A
PHALACRINAE
Phalacrus uniformis frigoricola Thompson & Marshall, 1980 A
Phalacrus sp. indet. A
CYCLAXYRIDAE E
Cyclaxyra jelineki Gimmel, Leschen & Ślipiński, 2009 E
Cyclaxyra politula (Broun, 1881) E
CAVOGNATHIDAE
Taphropiestes chathamensis (Watt, 1980) E Ch
Taphropiestes dumbletoni (Crowson, 1973) E
Taphropiestes electa (Broun, 1921) E
Taphropiestes watti Ślipiński & Tomaszewska, 2010 E
CRYPTOPHAGIDAE
CRYPTOPHAGINAE
CRYPTOPHAGINI
Cryptophagus pilosus Gyllenhal, 1828 A
Henoticus californicus (Mannerheim, 1843) A Do
CRYPTOSOMATULINI
Antarcticotectus aucklandicus Brookes, 1951 E Su
Brounina distincta (Broun, 1893) E
Micrambina amoena (Broun, 1912) E
Micrambina angulifera (Broun, 1880) E
Micrambina australis (Redtenbacher, 1867) E
Micrambina discoidea (Broun, 1893) E
Micrambina helmsi Reitter, 1880 E
Micrambina hispidula (Broun, 1880) E
Micrambina insignis Reitter, 1880 E
Micrambina obscura (Broun, 1893) E
Micrambina rufescens (Blanchard, 1853) E
Micrambina rutila (Broun, 1880) E
'Micrambina' silvana (Broun, 1880) E
'Micrambina' tumida (Broun, 1893) E
Ostreacryptus clarkae Leschen, 2001 E
Picrotus thoracicus Sharp, 1886 E
Thortus ovalis Broun 1893 E
ATOMARIINAE
ATOMARIINI
Atomaria lewisi Reitter, 1877 A
Ephistemus globulus (Paykull, 1798) A
Paratomaria crowsoni Leschen, 1996 E
Salltius ruficeps (Broun, 1880) E
Gen. et sp. indet. E
EROTYLIDAE
CRYPTOPHILINAE
CRYPTOPHILINI
Cathartocryptus maculosus (Broun, 1881) E
Cryptophilus integer (Heer, 1841) A
EROTYLINAE A
DACNINI
Cryptodacne brounii (Pascoe, 1876) E
Cryptodacne ferrugata Reitter, 1880 E
Cryptodacne lenis Broun, 1880 E
Cryptodacne nui Skelley & Leschen, 2007 E
Cryptodacne pubescens Broun, 1893 E
Cryptodacne rangiauria Skelley & Leschen, 2007 E
Cryptodacne synthetica Sharp, 1878 E
Kuschelengis politus (White, 1846) E
LANGURIINAE
HAPALIPINI
Hapalips prolixus (Sharp, 1876) E
LOBERINAE
Loberus anthracinus (Broun, 1893) E
Loberus borealis Leschen, 2003 E
Loberus depressus (Sharp, 1876) E
Loberus nitens (Sharp, 1876) E
Loberus watti Leschen, 2003 E TK
XENOSCELINAE
Loberonotha olivascens (Broun, 1893) E
BOTHRIDERIDAE
ANOMMATINAE A
Anommatus duodecimstriatus (Müller, 1821) A
BOTHRIDERINAE
Ascetoderes cognatus (Sharp, 1886) E
Ascetoderes diversus (Broun, 1912) E
Ascetoderes moestus (Sharp, 1877) E
Ascetoderes obsoletus (Broun, 1895) E
Ascetoderes paynteri (Broun, 1911) E Ch
Ascetoderes picipes (Broun, 1903) E
CERYLONIDAE
EUXESTINAE
Hypodacnella rubripes (Reitter, 1880) E
CERYLONINAE
Philothermus bicavus Reitter, 1880 E
Philothermus nitidus Sharp, 1876 E
Philothermus notabilis Broun, 1880 E
Philothermus sanguineus Broun, 1880 E
ENDOMYCHIDAE
MYCETAEINAE A
Mycetaea subterranea (Fabricius, 1801) A
MEROPHYSIINAE
Holoparamecus castaneus Broun, 1893 E
Holoparamecus tenuis Reitter, 1880 E
Holoparamecus n. sp. E
Gen. et spp. indet. (3) 3E
COCCINELLIDAE
COCCINELLINAE
CHILOCORINI
Halmus chalybeus (Boisduval, 1835) A BC
COCCIDULINI
Adoxellus flavihirtus (Broun, 1880) E
Adoxellus picinus (Broun, 1880) E
Apolinus lividigaster (Mulsant, 1853) A
Cryptolaemus montrouzieri Mulsant, 1853 A
Hoangus venustus (Pascoe, 1875) E
Rhyzobius acceptus (Broun, 1880) E
Rhyzobius consors (Broun, 1880) E
Rhyzobius eximius (Broun, 1880) E
Rhyzobius fagus (Broun, 1880) E
Rhyzobius forestieri (Mulsant, 1853) E
Rhyzobius minutulus (Broun, 1880) E
Rhyzobius nigritulus (Broun, 1914) E
Rhyzobius pallidiceps (Broun, 1880) E
Rhyzobius prolongatus (Broun, 1914) E
Rhyzobius rarus (Broun, 1880) E
Rhyzobius sedatus (Broun, 1886) E
Rhyzobius suffusus (Broun, 1880) E
Rhyzobius terrenus (Broun, 1880) E
Rhyzobius tristis (Broun, 1880) E
Rhyzobius ventralis (Erichson, 1843) A
Rhyzobius villosus (Broun, 1886) E
Rhyzobius n. spp. (3) 3E
Rodolia cardinalis (Mulsant, 1850) A BC
Rodolia koebelei Coquillett, 1893 A
Scymnus loewii Mulsant, 1850 A
Scymnus pygmaeus (Blackburn, 1892), non *pygmaeus* (Fourcroy, 1785), nec *pygmaea* Brullé, 1832 A
Stethorus bifidus Kapur, 1948 A
Stethorus griseus Chazeau, 1979 E
Stethorus (Parastethorus) histrio Chazeau 1974 E
Veronicobius aucklandiae Kirsch, 1877 E Su
Veronicobius hirtus Broun, 1893 E
Veronicobius macrostictus (Broun, 1911) E
Veronicobius n. spp. (2) E Ch
DIOMINI
Diomus notescens (Blackburn, 1889) A
Diomus subclarus (Blackburn, 1895) A
COCCINELLINI
Adalia (Adalia) bipunctata bipunctata (Linnaeus, 1758) A BC
Cleobora mellyi (Mulsant, 1850) A BC
Coccinella leonina Fabricius, 1775 E

Coccinella undecimpunctata Linnaeus, 1758 A BC
Coelophora inaequalis (Fabricius, 1775) A BC
Harmonia antipoda (Mulsant, 1848) E
Harmonia conformis (Boisduval, 1835) A BC
Illeis galbula (Mulsant, 1850) A
EPILACHNINI
Epilachna vigintioctopunctata (Fabricius, 1775) A
MICROWEISEINAE A
SERANGIINI
Serangium maculigerum Blackburn, 1892 A
CORYLOPHIDAE
CORYLOPHINAE
AENIGMATICINI
Stanus bowesteadi Ślipiński, Tomaszewska & Lawrence, 2009 E
CORYLOPHINI
Corylophus n. sp. E?
PELTINODINI
Holopsis lawsoni Broun, 1883 E
Holopsis nigella Broun, 1883 E
Holopsis oblonga Endrödy-Younga, 1964 E
Holopsis pallida Broun, 1883 E
Holopsis pictula Broun, 1893 E
Holopsis rotundata Broun, 1893 E
Holopsis n. spp. (4) 4E
SERICODERINI
Sericoderus apicalis Lea, 1895 A
Sericoderus ater (Matthews, 1886) E
Sericoderus brevicornis Matthews, 1890 A
Sericoderus brouni Csiki, 1910 E
Sericoderus seelandicus Csiki, 1910 E
Sericoderus sharpi (Matthews, 1886) E
Sericoderus thoracicus (Erichson, 1842), non Stephens, 1829 A
Sericoderus sp. indet. A
ORTHOPERINAE A
Orthoperus aequalis Sharp, 1885 A
Orthoperus atomarius (Heer, 1841) A
Orthoperus sp. indet. A
PARMULINAE
Arthrolips curtula (Broun, 1914) E
Arthrolips laetans (Broun, 1914) E
Arthrolips oblonga Broun, 1893 E
Clypastraea pulchella (Lea, 1895) A
Gen. et spp. indet. (2) E?
LATRIDIIDAE
LATRIDIINAE
Adistemia watsoni (Wollaston, 1871) A Do
Cartodere (Aridius) bifasciatus (Reitter, 1877) A
Cartodere (A.) costatus (Erichson, 1842) A
Cartodere (A.) nodifer (Westwood, 1839) A
Cartodere (Cartodere) constricta (Gyllenhal, 1827) A
Dienerella (Cartoderema) ruficollis (Marsham, 1802) A
Dienerella (Dienerella) filiformis (Gyllenhal, 1827) A
Dienerella (D.) filum (Aubé, 1850) A
Enicmus bifoveatus (Broun, 1886) E
Enicmus caviceps (Broun, 1893) E
Enicmus floridus (Broun, 1880) E
Enicmus foveatus Belon, 1884 E
Enicmus priopterus (Broun, 1886) E
Enicmus puncticeps (Broun, 1886) E
Enicmus rufifrons (Broun, 1914) E
Enicmus sharpi Belon, 1884 E
Latridius minutus (Linnaeus, 1767) A
Lithostygnus serripennis Broun, 1914 E
Lithostygnus sinuosus (Belon, 1884) E
Metophthalmus minor (Broun, 1893) E
Metophthalmus n. spp. (3) 3E
CORTICARIINAE
Bicava alacris (Broun, 1880) E
Bicava angusticollis (Broun, 1880) E
Bicava castanea (Broun, 1914) E
Bicava discoidea (Broun, 1880) E
Bicava diversicollis (Belon, 1884) E
Bicava erythrocephala (Broun, 1886) E
Bicava fuscicollis (Broun, 1912) E
Bicava gilvipes (Broun, 1886) E
Bicava globipennis (Reitter, 1881) E Su
Bicava illustris (Reitter, 1880) E
Bicava obesa (Broun, 1880) E
Bicava platyptera (Broun, 1886) E
Bicava pubera (Broun, 1880) E
Bicava pudibunda (Broun, 1880 E
Bicava semirufa (Broun, 1886) E
Bicava unicolor (Broun, 1914) E
Bicava variegata (Broun, 1880) E
Bicava zelandica (Belon, 1884) E
Corticaria clavatula Broun, 1914 E
Corticaria elongata (Gyllenhal, 1827) A
Corticaria ferruginea Marsham, 1802 A
Corticaria latulipennis Broun, 1914 E
Corticaria picicornis Broun, 1914 E
Corticaria pubescens (Gyllenhal, 1827) A
Corticaria serrata (Paykull, 1798) A
Corticarina clarula Broun, 1895 E
Corticarina pacata (Broun, 1886) E
Cortinicara hirtalis (Broun, 1880) A
Cortinicara meridiana Johnson, 1975 A
Cortinicara vagepunctata (Broun, 1914) E
Diarthrocera formicaephila Broun, 1893 E
'Melanophthalma' n. spp. (5) 5E
Rethusus fulvescens Broun, 1921 E
Rethusus lachrymosus Broun, 1886 E
Rethusus pustulosus (Belon, 1884) E
Gen. et sp. indet. (1) A
TENEBRIONOIDEA
HETEROMERA
MYCETOPHAGIDAE
MYCETOPHAGINAE
MYCETOPHAGINI
Litargus (Alitargus) balteatus LeConte, 1856 A
Litargus (Litargus) vestitus Sharp, 1879 A
Triphyllus aciculatus (Broun, 1880) E
Triphyllus adspersus (Broun, 1880) E
Triphyllus concolor Sharp, 1886 E
Triphyllus confertus Sharp, 1886 E
Triphyllus constans Broun, 1914 E
Triphyllus fuliginosus (Broun, 1880) E
Triphyllus hispidellus (Broun, 1880) E
Triphyllus huttoni Sharp, 1886 E
Triphyllus integritus Broun, 1893 E
Triphyllus maculosus Sharp, 1886 E
Triphyllus pubescens Broun, 1909 E
Triphyllus punctulatus (Broun, 1880) E
Triphyllus rubicundus Sharp, 1886 E
Triphyllus serratus (Broun, 1880) E
Triphyllus substriatus (Broun, 1880) E
Triphyllus zealandicus Sharp, 1886 E
Triphyllus n. spp. (2) 2E Ch
TYPHAEINI
Typhaea stercorea (Linnaeus, 1758) A
Mycetophagidae gen. et sp. indet. E
ARCHEOCRYPTICIDAE A
Archeocrypticus topali Kaszab, 1964 A
CIIDAE
CIINAE
CIINI
Cis anthracinus Broun, 1880 E
Cis asperrimus Broun, 1880 E
Cis bilamellatus Wood, 1884 A
Cis boettgeri (Reitter, 1880) E
Cis cornutičeps Broun, 1880 E
Cis flavitarsis Broun, 1880 E
Cis fulgens Broun, 1895 E
Cis fultoni Broun, 1886 E
Cis fuscipes Mellié, 1848 A
Cis illustris Broun, 1880 E
Cis lineicollis Broun, 1880 E
Cis lobipes Broun, 1895 E
Cis minutus Bayford, 1931 E
Cis obesulus Broun, 1886 E
Cis obsolctus (Reitter, 1880) E
Cis perpinguis Broun, 1880 E
Cis piciceps Broun, 1886 E
Cis picicollis Broun, 1883 E
Cis picturatus Broun, 1886 E
Cis recurvatus Broun, 1883 E
Cis rufulus Broun, 1880 E
Cis viridiflorus Broun, 1883 E
Cis zeelandicus Reitter, 1880 E
Cis n. spp. (5) 5E
Ennearthron n. sp. E
Orthocis assimilis (Broun, 1880) E
Orthocis undulatus (Broun, 1880) E
OROPHILINI
Octotemnus dilutipes (Blackburn, 1891) A
XYLOGRAPHELLINI
XYLOGRAPHELLINA
Scolytocis novaezelandiae Lopes-Andrade, 2008 E
Scolytocis n. sp. 1 E
MELANDRYIDAE
MELANDRYINAE
HYPULINI
Doxozilora punctata Broun, 1909 E
ORCHESIINI
Allorchesia guinnessi (Broun, 1912) E
Allorchesia validipes Broun, 1914 E
Hylobia acuminata Broun, 1915 E
Hylobia arboricola Broun, 1915 E
Hylobia bifasciata Broun, 1880 E
Hylobia calida Broun, 1880 E
Hylobia cylindrata Broun, 1880 E
Hylobia minor Broun, 1880 E
Hylobia nigricornis Broun, 1880 E
Hylobia nubeculosa Broun, 1880 E
Hylobia plagiata Broun, 1912 E
Hylobia pulla Broun, 1880 E
Hylobia sexnotata Broun, 1915 E
Hylobia undulata Broun, 1880 E
Hylobia velox Broun, 1880 E
Hylobia n. sp. E
Lyperocharis agilis Broun, 1914 E
Lyperocharis n. sp. E
Neorchesia divergens Broun, 1914 E
Neorchesia terricola Broun, 1914 E
Orchesia rennelli Gressitt & Samuelson, 1964 E Su
SERROPALPINI
Allopterus cavelli Broun, 1893 E
Allopterus instabilis Broun, 1886 E
Allopterus ornatus (Broun, 1880) E
Allopterus reticulatus Broun, 1883 E
Allopterus simulans Broun, 1910 E
Axylita sericophora Broun, 1914 E
Ctenoplectron coloratum Broun, 1886 E
Ctenoplectron costatum Broun, 1881 E
Ctenoplectron dignum Broun, 1886 E
Ctenoplectron fasciatum Redtenbacher, 1867 E
Ctenoplectron fuliginosum Broun, 1880 E
Ctenoplectron maculatum Broun, 1881 E
Ctenoplectron vittatum Broun, 1886 E
Mecorchesia brevicornis Broun, 1914 E
Mecorchesia spectabilis Broun, 1914 E
Melandryidae gen. et sp. indet. E
MORDELLIDAE
MORDELLINAE
MORDELLINI
Hoshihananomia antarctica (White, 1846)
Mordella jucunda (Broun, 1880) E
Mordella promiscua Erichson, 1842 A
Stenomordellaria neglecta (Broun, 1880) E
Zeamordella monacha Broun, 1886 E
MORDELLISTENINI
Tolidopalpus nitidocoma (Lea, 1929) A
RIPIPHORIDAE
PELECATOMINAE

Allocinops *brookesi* Broun, 1921 E
Rhipistena *cryptarthra* Broun, 1904 E
Rhipistena lugubris Sharp, 1878 E
Rhipistena sulciceps Broun, 1904 E
Sharpides *hirtella* (Broun, 1880) E
ZOPHERIDAE
COLYDIINAE
SYNCHITINI
Ablabus *brevis* Broun, 1886 E
Ablabus crassulus (Broun, 1914) E
Ablabus crassus Broun, 1881 E
Ablabus demissus (Broun, 1912) E
Ablabus discors (Broun, 1921) E
Ablabus facetus (Broun, 1893) E
Ablabus fervidulus Broun, 1880 E
Ablabus libentus (Broun, 1886) E
Ablabus lobiferus (Broun, 1909) E
Ablabus longipes (Broun, 1914) E
Ablabus nodosus Broun, 1886 E
Ablabus ornatus Broun, 1880 E
Ablabus pallidipictus Broun, 1880 E
Ablabus punctipennis Broun, 1880 E
Ablabus scabrus Broun, 1880 E
Ablabus sellatus (Sharp, 1886) E
Ablabus sparsus (Broun, 1886) E
Ablabus truncatus (Broun, 1914) E
Ablabus varicornis (Broun, 1910) E
Allobitoma *halli* Broun, 1921 E
Bitoma auriculata Sharp, 1886 E
Bitoma brouni Hetschko, 1928 E
Bitoma costicollis Reitter, 1880 E
Bitoma discoidea Broun, 1880 E
Bitoma distans Sharp, 1876 E
Bitoma distincta Broun, 1880 E
Bitoma guttata Broun, 1886 E
Bitoma insularis White, 1846 E
Bitoma lobata Broun, 1886 E
Bitoma morosa Broun, 1921 E
Bitoma mundula Sharp, 1886 E
Bitoma nana Sharp, 1876 E
Bitoma novella Hetschko, 1929 E
Bitoma picicornis Broun, 1909 E
Bitoma rugosa Sharp, 1876 E
Bitoma scita Broun, 1886 E
Bitoma serraticula Sharp, 1886 E
Bitoma vicina Sharp, 1876 E
Chorasus *costatus* (Broun, 1893) E
Chorasus costicollis (Broun, 1893) E
Chorasus incertus (Broun, 1895) E
Chorasus lateralis (Broun, 1921) E
Chorasus posticalis (Broun, 1921) E
Chorasus purus (Broun, 1921) E
Chorasus setarius (Broun, 1921) E
Chorasus subcaecus Sharp 1882 E
Chorasus suturalis (Broun, 1921) E
Ciconissus granifer Broun, 1893 E
Coxelus bicavus Broun, 1909 E
Coxelus chalmeri Broun, 1886 E
Coxelus clarus Broun, 1882 E
Coxelus dubius Sharp, 1876 E
Coxelus elongatus Broun, 1909 E
Coxelus graniceps Broun, 1893 E
Coxelus grossanus Broun, 1886 E
Coxelus instabilis Broun, 1914 E
Coxelus longulus Broun, 1893 E
Coxelus mucronatus Broun, 1911 E
Coxelus oculator Broun, 1893 E
Coxelus ovicollis Broun, 1893 E
Coxelus picicornis Broun, 1893 E
Coxelus posticalis Broun, 1893 E
Coxelus punctatus Broun, 1910 E K
Coxelus regularis Broun, 1893 E
Coxelus robustus Broun, 1880 E
Coxelus rufus Broun, 1893 E
Coxelus similis Sharp, 1876 E
Coxelus thoracicus Broun, 1895 E
Coxelus variegatus Broun, 1909 E
Coxelus xanthonyx Broun, 1910 E K
Epistranus *fulvus* Reitter, 1880 E
Epistranus hirtalis Broun, 1893 E
Epistranus humeralis Broun, 1880 E
Epistranus lawsoni (Sharp, 1876) E
Epistranus optabilis Broun, 1893 E
Epistranus parvus Broun, 1886 E
Epistranus sharpi Reitter, 1880 E
Epistranus valens Broun, 1881 E
Glenentela *costata* Broun, 1921 E
Glenentela serrata Broun, 1893 E
Heterargus *angulifer* (Broun, 1914) E
Heterargus decorus (Broun, 1914) E
Heterargus fuscus (Broun, 1923) E
Heterargus indentatus (Broun, 1893) E
Heterargus interruptus (Broun, 1923) E
Heterargus nodosus (Broun, 1893) E
Heterargus obliquecostatus (Broun, 1909) E
Heterargus pallens (Broun, 1914) E
Heterargus parallelus Broun, 1914 E
Heterargus posticalis (Broun, 1909) E
Heterargus rudis Sharp, 1886 E
Heterargus ruficornis (Broun, 1893) E
Heterargus serricollis Broun, 1893 E
Heterargus subaequus Broun, 1914 E
Heterargus tricavus (Broun, 1909) E
Lasconotus gracilis (Sharp, 1876) E
Namunaria sp. indet. A
Norix *crassus* Broun, 1893 E
Notocoxelus *helmsi* (Reitter, 1880) E
Pristoderus *aberrans* (Broun, 1880) E
Pristoderus acuminatus (Broun, 1880) E
Pristoderus aemulus (Broun, 1923) E
Pristoderus affinis (Broun, 1923) E
Pristoderus antarcticus (White, 1846) E
Pristoderus asper (Sharp, 1876) E
Pristoderus atratus (Broun, 1880) E
Pristoderus bakewelli (Pascoe, 1866) E
Pristoderus brouni (Sharp, 1876) E
Pristoderus carus (Broun, 1886) E
Pristoderus cinereus (Broun, 1886) E
Pristoderus contractifrons (Broun, 1880) E
Pristoderus cucullatus (Sharp, 1886) E
Pristoderus discalis (Broun, 1921) E
Pristoderus discedens (Sharp, 1877) E
Pristoderus dissimilis (Sharp, 1886) E
Pristoderus dorsalis (Broun, 1882) E
Pristoderus exiguus (Broun, 1882) E
Pristoderus fulvus (Broun, 1893) E
Pristoderus fuscatus (Broun, 1886) E
Pristoderus insignis (Broun, 1880) E
Pristoderus integratus (Broun, 1886) E
Pristoderus isostictus (Broun, 1886) E
Pristoderus lawsoni (Wollaston, 1873) E
Pristoderus philpotti (Broun, 1914) E
Pristoderus plagiatus (Broun, 1911) E
Pristoderus planiceps (Broun, 1915) E
Pristoderus probus (Broun, 1893) E
Pristoderus proprius (Broun, 1914) E
Pristoderus punctatus (Broun, 1886) E
Pristoderus reitteri (Sharp, 1882) E
Pristoderus rudis (Sharp, 1877) E
Pristoderus rufescens (Broun, 1886) E
Pristoderus salebrosus (Broun, 1880) E
Pristoderus scaber (Fabricius, 1775) E
Pristoderus tuberculatus (Broun, 1880) E
Pristoderus undosus (Broun, 1882) E
Pristoderus uropterus (Broun, 1912) E
Pristoderus viridipictus (Wollaston, 1873) E
Pristoderus wakefieldi (Sharp, 1877) E
Pristoderus wallacei (Broun, 1912) E
Rytinotus *squamulosus* Broun, 1880 E
Syncalus *explanatus* Broun, 1912 E
Syncalus granulatus (Broun, 1880) E
Syncalus hystrix Sharp, 1876 E
Syncalus munroi Broun, 1893 E
Syncalus oblongus (Broun, 1880) E
Syncalus optatus Sharp, 1876 E
Syncalus piciceps Broun, 1893 E
Syncalus politus Broun, 1880 E
Syncalus solidus Broun, 1923 E
Tarphiomimus *indentatus* Wollaston, 1873 E
Tarphiomimus tuberculatus Broun, 1912 E
Tarphiomimus wollastoni Sharp, 1882 E
PYCNOMERINAE
Pycnomerodes peregrinus Broun, 1886 E
Pycnomerus acutangulus Reitter, 1878 E
Pycnomerus aequicollis Reitter, 1878 E
Pycnomerus angulatus Broun, 1893 E
Pycnomerus arboreus Broun, 1886 E
Pycnomerus arcuatus Broun, 1914 E
Pycnomerus basalis Broun, 1882 E
Pycnomerus caecus Broun, 1886 E
Pycnomerus candidus Broun, 1912 E
Pycnomerus carinellus Broun, 1886 E
Pycnomerus cognatus Broun, 1886 E
Pycnomerus depressiusculus (White, 1846) E
Pycnomerus ellipticus Broun, 1880 E
Pycnomerus elongellus Broun, 1893 E
Pycnomerus frontalis Broun, 1893 E
Pycnomerus helmsi Sharp, 1886 E
Pycnomerus hirtus Broun, 1886 E
Pycnomerus impressus Broun, 1893 E
Pycnomerus lateralis Broun, 1886 E
Pycnomerus latitans Sharp, 1886 E
Pycnomerus longipes Broun, 1893 E
Pycnomerus longulus Sharp, 1886 E
Pycnomerus marginalis Broun, 1893 E
Pycnomerus mediocris Broun, 1911 E
Pycnomerus minor Sharp, 1876 E
Pycnomerus nitidiventris Broun, 1903 E
Pycnomerus ocularius Broun, 1914 E
Pycnomerus parvulus Broun, 1921 E
Pycnomerus reversus Broun, 1912 E
Pycnomerus rufescens Broun, 1882 E
Pycnomerus ruficollis Broun, 1909 E
Pycnomerus simplex Broun, 1880 E
Pycnomerus simulans Sharp, 1876 E
Pycnomerus sinuatus Broun, 1893 E
Pycnomerus sophorae Sharp, 1876 E
Pycnomerus sulcatissimus Reitter, 1880 E
Pycnomerus suteri Broun, 1909 E
Pycnomerus tenuiculus Broun, 1914 E
ULODIDAE
Arthopus *brouni* Sharp, 1876 E
Brouniphylax *binodosus* (Broun, 1895) E
Brouniphylax exiguus (Broun, 1914) E
Brouniphylax squamiger (Broun, 1880) E
Brouniphylax sternalis (Broun, 1904) E
Brouniphylax varius (Broun, 1880) E
Exohadrus *volutithorax* (Broun, 1880) E
Syrphetodes bullatus Sharp, 1886 E
Syrphetodes cordipennis Broun, 1893 E
Syrphetodes crenatus Broun, 1880 E
Syrphetodes decoratus Broun, 1880 E
Syrphetodes dorsalis Broun, 1893 E
Syrphetodes marginatus Pascoe, 1875 E
Syrphetodes nodosalis Broun, 1904 E
Syrphetodes pensus Broun, 1921 E
Syrphetodes punctatus Broun, 1893 E
Syrphetodes simplex Broun, 1903 E
Syrphetodes sylvius Broun, 1893 E
Syrphetodes thoracicus Broun, 1921 E
Syrphetodes truncatus Broun, 1912 E
Syrphetodes variegatus Broun, 1917 E
Syrphetodes n. spp. (2) 2E
CHALCODRYIDAE E
Chalcodrya *hilaris* Watt, 1974 E

Chalcodrya variegata Redtenbacher, 1868 E
Onysius anomalus Broun, 1886 E
Philpottia levinotis Watt, 1974 E
Philpottia mollis (Broun, 1886) E
TENEBRIONIDAE
ALLECULINAE
Omedes nitidus Broun, 1893 E
Omedes substriatus (Broun, 1880) E
Tanychilus metallicus White, 1846 E
Tanychilus sophorae Broun, 1880 E
Xylochus dentipes Broun, 1883 E
Xylochus spinifer Broun, 1893 E
Xylochus tibialis Broun, 1880 E
Xylochus triregius Watt, 1992 E TK
Zomedes borealis Watt, 1992 E TK
COELOMETOPINAE
COELOMETOPINI
Chrysopeplus expolitus (Broun, 1880) E
Chrysopeplus triregius Watt, 1992 E TK
DIAPERINAE A
DIAPERINI
Gnatocerus cornutus (Fabricius, 1798) A
GNATHIDIINI
Menimus batesi Sharp, 1876 E
Menimus borealis Watt, 1992 E
Menimus brouni Watt, 1992 E
Menimus caecus Sharp, 1876 E
Menimus crassus Sharp, 1876 E
Menimus crinalis Broun, 1880 E
Menimus crosbyi Watt, 1993 E
Menimus curtulus Broun, 1883 E
Menimus dubius Broun, 1880 E
Menimus elongatus Watt, 1992 E
Menimus helmorei Watt, 1992 E
Menimus laevicollis Broun, 1895 E
Menimus moehauensis Watt, 1992 E
Menimus oblongus Broun, 1880 E
Menimus obscurus Broun, 1880 E
Menimus pubiceps Broun, 1921 E
Menimus puncticeps Broun, 1880 E
Menimus sinuatus Broun, 1886 E
Menimus thoracicus Broun, 1880 E
PLATYDEMINI
Platydema sp. indet. A
LAGRIINAE
ADELIINI
Adelium brevicolle Blessig, 1861 A
Exadelium rufilabrum (Broun, 1886) E
Kaszabadelium aucklandicum (Broun, 1880) E
Mesopatrum granulosum Broun, 1893 E
Mitua triangularis Watt, 1992 E
Mitua tuberculicostata (White, 1846) E
Periatrum carinatum Watt, 1992 E
Periatrum edentatum Watt, 1992 E
Periatrum helmsi Sharp, 1886 E
Periatrum manapouricum Watt, 1992 E
Periatrum rotundatum Watt, 1992 E
Periatrum tumipes Broun, 1893 E
Pheloneis amaroides (Lacordaire, 1859) E
Pheloneis simulans (Redtenbacher, 1868) E
Pheloneis triregius Watt, 1992 E TK
Stenadelium striatum Watt, 1992 E
Wattadelium alienum (Broun, 1880) E
Wattadelium curtulum (Watt, 1992) E
Wattadelium pleurale Broun, 1893) E
Zeadelium aeratum (Broun, 1880) E
Zeadelium arthurense Watt, 1992 E
Zeadelium australe Watt, 1992 E
Zeadelium bullatum (Pascoe, 1876) E
Zeadelium chalmeri (Broun, 1883) E
Zeadelium complicatum (Broun, 1912) E
Zeadelium femorale (Broun, 1910) E
Zeadelium gratiosum (Broun, 1893) E
Zeadelium hanseni (Broun, 1885) E
Zeadelium hudsoni (Broun, 1908) E
Zeadelium indigator (Broun, 1886) E
Zeadelium intermedium (Sharp, 1886) E
Zeadelium intricatum (Broun, 1880) E
Zeadelium lentum (Broun, 1880) E
Zeadelium nigritulum (Broun, 1885) E
Zeadelium parvum Watt, 1992 E
Zeadelium senile Watt, 1992 E
Zeadelium simplex (Sharp, 1886) E
Zeadelium thoracicum (Broun, 1880) E
Zeadelium zelandicum (Bates, 1874) E
CHAERODINI
Chaerodes laetus Broun, 1880 E
Chaerodes trachyscelides White, 1846 E
LUPROPINI
Lorelus crassicornis Broun, 1880 E
Lorelus kaszabi Watt, 1992 E
Lorelus laticornis Watt, 1992 E
Lorelus latulus Broun, 1910 E
Lorelus marginalis Broun, 1910 E
Lorelus obtusus Watt, 1992 E
Lorelus opacus Watt, 1992 E
Lorelus politus Watt, 1992 E
Lorelus priscus Sharp, 1876 E
Lorelus pubescens Broun,1880 E
Lorelus punctatus Watt, 1992 E
Lorelus quadricollis Broun, 1883 E
Lorelus tarsalis Broun, 1910 E
PHRENAPATINAE
Archaeoglenes costipennis Broun, 1893 E
PIMELIINAE
CNEMEPLATIINI
Actizeta albata Pascoe, 1875 E
Actizeta fusca Watt, 1992 E
STENOCHINAE A
CNODALONINI
Hypaulax crenata (Boisduval, 1835) A Do
TENEBRIONINAE
ALPHITOBIINI
Alphitobius diaperinus (Panzer, 1797) A
Alphitobius laevigatus (Fabricius, 1781) A
AMARYGMINI
Amarygmus watti Bremer, 2005 A
HELEINI
Mimopeus buchanani (Broun, 1880) E
Mimopeus clarkei Watt, 1988 E
Mimopeus convexus Watt, 1988 E
Mimopeus costellus (Broun, 1905) E
Mimopeus elongatus (Brême, 1842) E
Mimopeus granulosus (Brême, 1842) E
Mimopeus humeralis (Bates, 1873) E
Mimopeus impressifrons (Bates, 1873) E
Mimopeus insularis Watt, 1988 E
Mimopeus johnsi Watt, 1988 E
Mimopeus lateralis (Broun, 1908) E
Mimopeus lewisianus (Sharp, 1903) E
Mimopeus neglectus Watt, 1988 E
Mimopeus opaculus (Bates, 1873) E
Mimopeus parallelus Watt, 1988 E
Mimopeus parvus Watt, 1988 E
Mimopeus pascoei (Bates, 1873) E Ch
Mimopeus rugosus (Bates, 1873) E
Mimopeus subcostatus (Sharp, 1903) E Ch
Mimopeus thoracicus (Bates, 1873) E
Mimopeus tibialis (Bates, 1873) E
Mimopeus turbotti Watt, 1988 E TK
Mimopeus vallis Watt, 1988 E
OPATRINI
Gonocephalum elderi (Blackburn, 1892) A
TENEBRIONINI
Tenebrio molitor Linnaeus, 1758 A
Tenebrio obscurus Fabricius, 1792 A
TITAENINI
Artystona erichsoni (White, 1846) E
Artystona lata Watt, 1992 E
Artystona obscura Sharp, 1886 E
Artystona richmondiana Watt, 1992 E
Artystona rugiceps Bates, 1874 E
Artystona wakefieldi Bates, 1874 E
Cerodolus arthurensis Watt, 1992 E
Cerodolus chrysomeloides Sharp, 1886 E
Cerodolus curvellus Broun, 1912 E
Cerodolus genialis Broun, 1893 E
Cerodolus manepouricus Watt, 1992 E
Cerodolus sinuatus Watt, 1992 E
Cerodolus tuberculatus Broun, 1917 E
Partystona metallica Watt, 1992 E TK
Pseudhelops antipodensis Watt, 1971 E Su
Pseudhelops capitalis (Broun, 1917) E
Pseudhelops chathamensis Watt, 1992 E Ch
Pseudhelops clandestinus Watt, 1971 E Su
Pseudhelops liberalis Watt, 1971 E Su
Pseudhelops posticalis Broun, 1909 E Su
Pseudhelops quadricollis Broun, 1909 E Su
Pseudhelops tuberculatus Guérin-Méneville, 1841 E Su
TRIBOLIINI
Tribolium castaneum (Herbst, 1797) A
Tribolium confusum Jacquelin du Val, 1868 A
ULOMINI
Aphtora rufipes Bates, 1872 E
Uloma sanguinipes (Fabricius, 1775) A Do
Uloma tenebrionoides (White, 1846) E
Ulomotypus laevigatus Broun, 1886 E
INCERTAE SEDIS (Tenebrioninae)
Demtrius carinulatus Broun, 1895 E
ZOLODININAE
Zolodinus zelandicus Blanchard, 1853 E
PROSTOMIDAE
Dryocora howitti Pascoe, 1868 E
OEDEMERIDAE
NACERDINAE
NACERDINI
Nacerdes melanura (Linnaeus, 1758) A
OEDEMERINAE
Baculipalpus clarencensis Hudson, 1975 E
Baculipalpus mollis (Broun, 1886) E
Baculipalpus oconnori Hudson, 1975 E
Baculipalpus prolatus Hudson, 1975 E
Baculipalpus rarus Broun, 1880 E
Baculipalpus strigipennis (White, 1846) E
Parisopalpus macleayi (Champion, 1895) A
Parisopalpus nigronotatus (Boheman, 1858) A
Parisopalpus thoracicus (Broun, 1893) E
Koniaphassa obscura (Broun, 1880) E
Selenopalpus aciphyllae Broun, 1886 E
Selenopalpus cyaneus (Fabricius, 1775) E
Selenopalpus rectipes Broun, 1909 E
Thelyphassa brouni Hudson, 1975 E
Thelyphassa chrysophana Hudson, 1975 E
Thelyphassa diaphana Pascoe, 1876 E
Thelyphassa latiuscula (Broun, 1880) E
Thelyphassa lineata (Fabricius, 1775) E
Thelyphassa nemoralis Broun, 1886 E
Thelyphassa pauperata Pascoe, 1876 E
PYROCHROIDAE
PILIPALPINAE
Exocalopus pectinatus Broun, 1893 E
Techmessa concolor Bates, 1874 E
Techmessa longicollis Broun, 1903 E
Techmessa telephoroides Bates, 1874 E
Techmessodes picticornis (Broun, 1880) E
Techmessodes versicolor Broun, 1893 E
Pyrochroidae gen. et spp. indet. (3) 3E
SALPINGIDAE
AEGIALITINAE
Antarcticodomus fallai Brookes, 1951 E Su
INOPEPLINAE
Diagrypnodes wakefieldi Waterhouse, 1876 E
SALPINGINAE
Salpingus angusticollis Broun, 1880 E

Salpingus aterrimus Broun, 1921 E
Salpingus atrellus Broun, 1914 E
Salpingus bilunatus Pascoe, 1876 E
Salpingus cognatus Broun, 1910 E
Salpingus denticollis Broun, 1914 E
Salpingus fossulatus Broun, 1893 E
Salpingus hirtus Broun, 1883 E
Salpingus hudsoni Blair, 1925 E
Salpingus laticollis Blair, 1925 E
Salpingus lautus Broun, 1880 E
Salpingus lepidulus Broun, 1910 E
Salpingus nigricans Broun, 1921 E
Salpingus ornatus Broun, 1895 E
Salpingus pallidipes Blair, 1925 E
Salpingus perpunctatus Broun, 1880 E
Salpingus quisquilius Broun, 1883 E
Salpingus rugulosus Broun, 1910 E
Salpingus semilaevis Broun, 1914 E
Salpingus simplex Broun, 1883 E
Salpingus swalei Blair, 1925 E
Salpingus tarsalis Broun, 1910 E
Salpingus testaceus Blair, 1925 E
Salpingus unguiculus Broun, 1880 E
Salpingus sp. n. E
Vincenzellus (= *Trichoocolposinus*) sp. indet. E?
ANTHICIDAE
ANTHICINAE
ANTHICINI
Anthicus (*Anthicus*) *hesperi* King, 1869 A
Anthicus (*A.*) *kreusleri* King, 1869 A
Floydwernerius gushi (Werner & Chandler, 1995) A
Floydwernerius troilus (Hinton, 1945) A
Omonadus floralis (Linnaeus, 1758) A
Omonadus formicarius (Goeze, 1777) A
Pseudocyclodinus glaber (King, 1869) A
Pseudocyclodinus minor (Broun, 1886) E
Pseudocyclodinus otagensis (Werner & Chandler, 1995) E
Sapintus argenteofasciatus Telnov, 2003 A
Sapintus aucklandensis Werner & Chandler, 1995 E
Sapintus deitzi Werner & Chandler, 1995 E
Sapintus obscuricornis (Broun, 1880) E
Sapintus pellucidipes (Broun, 1880) E
Stricticollis tobias (Marseul, 1879) A
LEMODINAE
Cotes bullata Broun, 1923 E
Cotes crispi (Broun, 1880) E
Cotes gourlayi Werner & Chandler, 1995 E
Cotes halliana Broun, 1923 E
Cotes optima Broun, 1893 E
Cotes proba Broun, 1881 E
Cotes rufa Broun, 1893 E
Cotes vestita Sharp, 1877 E
Trichananca fulgida Werner & Chandler, 1995 A?
Trichananca spp. indet. (2) 2A
Zealanthicus sulcatus Werner & Chandler, 1995 E
MACRATRIINAE
Macratria aotearoa Werner & Chandler, 1995 E
Macratria exilis Pascoe, 1877 E
ADERIDAE
Scraptogetus anthracinus Broun, 1893 E
Scraptogetus arboreus (Broun, 1914) E
Xylophilus antennalis Broun, 1893 E
Xylophilus brouni (Pic, 1901) E
Xylophilus coloratus Broun, 1893 E
Xylophilus luniger Champion, 1916 E
Xylophilus nitidus Broun, 1893 E
Xylophilus pictipes Broun, 1893 E
Xylophilus xenarthrus Broun, 1910 E
Xylophilus n. spp. (4) 4E
SCRAPTIIDAE
SCRAPTIINAE
SCRAPTIINI
Nothotelus nigellus (Broun, 1880) E
Nothotelus ocularius Broun, 1914 E
Nothotelus usitatus (Broun, 1880) E
Nothotelus n. sp. E
Phytilea propera Broun, 1893 E
Phytilea n. sp. E
Scraptia sp. indet. A
TENEBRIONOIDEA
LAGRIOIDINAE (INCERTAE SEDIS)
Lagrioida brouni Pascoe, 1876 E
INCERTAE SEDIS
Rhizonium antiquum Sharp, 1876 E
CHRYSOMELOIDEA
CERAMBYCIDAE
CERAMBYCINAE
APHNEOPINI
<u>*Gnomodes*</u> *piceus* Broun, 1893 E
<u>*Zorion*</u> *angustifasciatum* Schnitzler, 2005 E
Zorion australe Schnitzler, 2005 E
Zorion batesi Sharp, 1875 E
Zorion dugdalei Schnitzler, 2005 E
Zorion guttigerum (Westwood, 1845) E
Zorion kaikouraiensis Schnitzler, 2005 E
Zorion minutum (Fabricius, 1775) E
Zorion nonmaculatum Schnitzler, 2005 E
Zorion opacum Sharp, 1903 E
Zorion taranakiensis Schnitzler, 2005 E
CALLIDIOPINI
Bethelium signiferum (Newman, 1840) A
Callidiopis scutellaris (Fabricius, 1801) A
Didymocantha flavopicta Mckeown, 1948 E
Didymocantha obliqua Newman, 1840 A
Didymocantha quadriguttata Sharp, 1886 E
Didymocantha sublineata White, 1846 E
<u>*Oemona*</u> *hirta* (Fabricius, 1775) E
Oemona plicicollis Sharp, 1886 E
Oemona separata (Broun, 1921) E
Oemona simplicicollis (Broun, 1880) E
COPTOMMATINI
Coptomma lineatum (Fabricius, 1775) E
Coptomma marrisi Song & Wang, 2003 E
Coptomma sticticum (Broun, 1893) E
Coptomma sulcatum (Fabricius, 1775) E
Coptomma variegatum (Fabricius, 1775) E
ELAPHIDIINI
Coptocercus rubripes (Boisduval, 1835) A
Epithora dorsalis (MacLeay, 1827) A
<u>*Liogramma*</u> *zelandica* (Blanchard, 1853) E
HESPEROPHANINI
<u>*Xuthodes*</u> *batesi* Sharp, 1877 E
Xuthodes punctipennis Pascoe, 1875 E
HETEROPSINI
Aridaeus thoracicus (Donovan, 1805) A
MOLORCHINI
<u>*Anencyrus*</u> *discedens* Sharp, 1886 E
<u>*Gastrosarus*</u> *lautus* Broun, 1893 E
Gastrosarus nigricollis Bates, 1874 E
Gastrosarus picticornis Broun, 1893 E
Gastrosarus urbanus Broun, 1893 E
Gastrosarus n. sp. E
PHLYCTAENODINI
<u>*Agapanthida*</u> *morosa* (Sharp, 1886) E
Agapanthida pulchella White, 1846 E
<u>*Ambeodontus*</u> *tristis* (Fabricius, 1775) E
<u>*Astetholea*</u> *aubreyi* Broun, 1880 E
Astetholea lepturoides Bates, 1876 E
Astetholea pauper Bates, 1874 E
<u>*Astetholida*</u> *lucida* Broun, 1880 E
<u>*Ophryops*</u> *aegrotus* (Bates, 1876) E
Ophryops dispar Sharp, 1886 E
Ophryops fuscicollis (Broun, 1913) E
Ophryops medius (Broun, 1913) E
Ophryops pallidus White, 1846 E
Ophryops pseudofuscicollis Lu & Wang, 2005 E
<u>*Pseudosemnus*</u> *retifer* (Lacordaire, 1869) E
<u>*Votum*</u> *mundum* Broun, 1880 E
PHORACANTHINI
Phoracantha semipunctata (Fabricius, 1775) A
PYTHEINI
<u>*Brounopsis*</u> *hudsoni* Blair, 1937 E
STENODERINI
Cacodrotus bifasciatus Broun, 1893 E
<u>*Calliprason*</u> *costifer* (Broun, 1886) E
Calliprason elegans (Sharp, 1877) E
Calliprason marginatum White, 1846 E
Calliprason pallidus (Pascoe, 1875) E
Calliprason sinclairii White, 1843 E
<u>*Drototelus*</u> *elegans* (Brookes, 1927) E
Drototelus politus Broun, 1903 E
Drototelus rarus Wang and Lu, 2004 E
<u>*Eburilla*</u> *sericea* (White, 1855) E
TESSAROMMATINI
Tessaromma undatum Newman, 1840 A
TRIBUS INCERTAE SEDIS
<u>*Leptachrous*</u> *strigipennis* (Westwood, 1845) E
Nesoptychias simpliceps (Broun, 1880) E
<u>*Ochrocydus*</u> *huttoni* Pascoe, 1876 E
LAMIINAE
ACANTHOCINI
<u>*Mesolamia*</u> *aerata* Broun, 1893 E
Mesolamia marmorata Sharp, 1882 E
<u>*Metalamia*</u> *cuprea* Breuning, 1940 E
Metalamia obtusipennis (Bates, 1876) E
<u>*Microlamia*</u> *elongata* Breuning, 1940 E
Microlamia pygmaea Bates, 1874 E
<u>*Polyacanthia*</u> *flavipes* (White, 1846) E
<u>*Psilocnaeia*</u> *aegrota* (Bates, 1874) E
Psilocnaeia asteliae Kuschel, 1990 E
Psilocnaeia brouni Bates, 1876 E
Psilocnaeia bullata (Bates, 1876) E
Psilocnaeia linearis Bates, 1874 E
Psilocnaeia nana (Bates, 1874) E
Psilocnaeia parvula (White, 1846) E
<u>*Spilotrogia*</u> *elongata* (Broun, 1883) E
Spilotrogia fragilis (Bates, 1874) E
Spilotrogia hilarula Broun, 1880 E
Spilotrogia maculata Bates, 1874 E
Spilotrogia pictula (Bates, 1876) E
Spilotrogia pulchella (Bates, 1874) E
<u>*Stenellipsis*</u> *bimaculata* (White, 1846) E
Stenellipsis cuneata Sharp, 1886 E
Stenellipsis gracilis (White, 1846) E
Stenellipsis grata (Broun, 1880) E
Stenellipsis latipennis Bates, 1874 E
Stenellipsis longula Breuning, 1940 E
Stenellipsis sculpturata (Broun, 1915) E
<u>*Tetrorea*</u> *cilipes* White, 1846 E
Tetrorea discedens Sharp, 1882 E
Tetrorea longipennis Sharp, 1886 E
Tetrorea sellata Sharp, 1882 E
Tetrorea variegata (Broun, 1880) E
PARMENINI
<u>*Adriopea*</u> *pallidata* Broun, 1910 E
<u>*Hexatricha*</u> *pulverulenta* (Westwood, 1845) E
<u>*Nodulosoma*</u> *angustum* (Broun, 1880) E
Nodulosoma flavidorsis (Broun, 1914) E
Nodulosoma halli (Broun, 1914) E
Nodulosoma helmsi (Sharp, 1882) E
Nodulosoma laevinotatum (Broun, 1914) E
Nodulosoma laevior (Broun, 1893) E
Nodulosoma laevithorax (Breuning, 1940) E
Nodulosoma maculatum (Broun, 1921) E
Nodulosoma picticornis (Broun, 1895) E
Nodulosoma pinguis (Broun, 1913) E
Nodulosoma posticalis (Broun, 1913) E
Nodulosoma rufescens (Breuning, 1940) E
Nodulosoma spectabilis (Broun, 1914) E
Nodulosoma suffusum (Broun, 1914) E
Nodulosoma testaceum (Broun, 1909) E
<u>*Ptinosoma*</u> *ampliatum* (Breuning, 1940) E
Ptinosoma convexum (Broun, 1893) E
Ptinosoma fulvipes (Broun, 1923) E

Ptinosoma lineiferum (Broun, 1909) E
Ptinosoma ptinoides (Bates, 1874) E
Ptinosoma spinicollis (Broun, 1893) E
Ptinosoma waitei (Broun, 1911) E
Ptinosoma n. sp. E
Somatidia antarctica (White, 1846) E
Somatidia grandis Broun, 1893 E
Somatidia longipes Sharp, 1878 E
Somatidia simplex Broun, 1893 E
Tenebrosoma albicoma (Broun, 1893) E
Tenebrosoma corticola (Broun, 1913) E
Tenebrosoma crassipes (Broun, 1883) E
Tenebrosoma diversa (Broun, 1880) E
Tenebrosoma nitida (Broun, 1880) E
Tenebrosoma parvula (Broun, 1914) E
Tenebrosoma pictipes (Broun, 1880) E
Tenebrosoma tenebrica (Broun, 1893) E
Tenebrosoma terrestre (Broun, 1880) E
Tenebrosoma testudo (Broun, 1904) E
Xylotoles (*Trichoxylotoles*) *apicalis* (Broun, 1923) E
Xylotoles (*T.*) *phormiobius* Broun, 1893 E
Xylotoles (*Xylotoles*) *costatus* Pascoe, 1875 E
Xylotoles (*X.*) *costipennis* (Breuning, 1982) E
Xylotoles (*X.*) *griseus* (Fabricius, 1775) E
Xylotoles (*X.*) *humeratus* Bates, 1874 E
Xylotoles (*X.*) *inornatus* Broun, 1880 E
Xylotoles (*X.*) *laetus* White, 1846 E
Xylotoles (*X.*) *lynceus* (Fabricius, 1775) E
Xylotoles (*X.*) *nanus* Bates, 1874 E
Xylotoles (*X.*) *nudus* Bates, 1874 E
Xylotoles (*X.*) *rugicollis* Bates, 1874 E
Xylotoles (*X.*) *sandageri* Broun, 1886 E
Xylotoles (*X.*) *scissicauda* Bates, 1874 E
Xylotoles (*X.*) *traversii* Pascoe, 1876 E Ch
Xylotoloides huttoni (Sharp, 1882) E
POGONOCHERINI
Hybolasiopsis trigonellaris (Hutton, 1898) E
Hybolasius castaneus Broun, 1893 E
Hybolasius cristatellus Bates, 1876 E
Hybolasius cristus (Fabricius, 1775) E
Hybolasius dubius Broun, 1893 E
Hybolasius femoralis Broun, 1893 E
Hybolasius lanipes Sharp, 1877 E
Hybolasius modestus Broun, 1880 E
Hybolasius optatus Broun, 1893 E
Hybolasius parvus Broun, 1880 E
Hybolasius pedator Bates, 1876 E
Hybolasius picitarsis Broun, 1883 E
Hybolasius postfusciatus Breuning, 1940 E
Hybolasius promissus Broun, 1880 E
Hybolasius pumilus Pascoe, 1876 E
Hybolasius vegetus Broun, 1881 E
Hybolasius viridescens Bates, 1874 E
Hybolasius wakefieldi Bates, 1876 E
Hybolasius n. sp. E Ch
Poecilippe femoralis Sharp, 1886 E
Poecilippe medialis Sharp, 1886 E
Poecilippe simplex Bates, 1874 E
Poecilippe stictica Bates, 1874 E
Sphinohybolasius spinicollis Breuning, 1959 E
LEPTURINAE
LEPTURINI
Blosyropus spinosus Redtenbacher, 1868 E
PRIONINAE
ANACOLINI
Prionoplus reticularis White, 1843 E
SPHONDYLIDINAE
ASEMINI
Arhopalus ferus (Mulsant, 1839) A
CHRYSOMELIDAE
ALTICINAE
Agasicles hygrophila Selman & Vogt, 1971 A BC
Alema paradoxa Sharp, 1876 E
Alema spatiosa Broun, 1880 E
Altica carduorum (Guérin-Méneville, 1858) A BC
Chaetocnema (*Chaetocnema*) *paspalae* (Broun, 1923) E
Chaetocnema (*Tlanoma*) *aotearoa* Samuelson, 1973 E
Chaetocnema (*T.*) *graminicola* (Broun, 1893) E
Chaetocnema (*T.*) *littoralis* (Broun, 1893) E
Chaetocnema (*T.*) *moriori* Samuelson, 1973 E Ch
Chaetocnema (*T.*) *nitida* (Broun, 1880) E
Longitarsus fuliginosus (Broun, 1880) E
Longitarsus jacobaeae (Waterhouse, 1858) A BC
Phyllotreta undulata (Kutschera, 1860) A
Pleuraltica cyanea (Broun, 1880) E
Psylliodes brettinghami Baly, 1862 A
Trachytetra robusta Broun, 1923 E
Trachytetra rugulosa (Broun, 1880) E
BRUCHINAE A
ACANTHOSCELIDINI
Acanthoscelides obtectus (Say, 1931) A
BRUCHIDINI
Bruchidius villosus (Fabricius, 1792) A BC
Callosobruchus maculatus (Fabricius, 1775) A Do
BRUCHINI
Bruchus pisorum (Linnaeus, 1758) A Do
Bruchus rufimanus Boheman, 1833 A
CRIOCERINAE A
Lema cyanella (Linnaeus, 1758) A BC
GALERUCINAE
LUPERINI
S. LUPERINA
Adoxia aenea Broun, 1880 E
Adoxia aenescens (Sharp, 1886) E
Adoxia angularia (Broun, 1909) E
Adoxia anthracina (Broun, 1914) E
Adoxia asperella (Broun, 1909) E
Adoxia atripennis (Broun, 1913) E
Adoxia attenuata Broun, 1880 E
Adoxia aurella (Broun, 1914) E
Adoxia axyrocharis (Broun, 1909) E
Adoxia brevicollis (Broun, 1893) E
Adoxia bullata (Broun, 1914) E
Adoxia calcarata (Broun, 1893) E
Adoxia cheesemani (Broun, 1910) E
Adoxia cyanescens (Broun, 1917) E
Adoxia dilatata (Broun, 1914) E
Adoxia dilucida (Broun, 1917) E
Adoxia dilutipes (Broun, 1915) E
Adoxia discrepans (Broun, 1914) E
Adoxia diversa (Broun, 1910) E
Adoxia foveigera (Broun, 1913) E
Adoxia fuscata (Broun, 1893) E
Adoxia fuscifrons (Broun, 1910) E
Adoxia gracilipes (Broun, 1917) E
Adoxia halli (Broun, 1917) E
Adoxia insolita (Broun, 1914) E
Adoxia iridescens (Broun, 1914) E
Adoxia lewisi (Broun, 1909) E
Adoxia mediocris (Broun, 1917) E
Adoxia minor (Broun, 1917) E
Adoxia mollis (Broun, 1893) E
Adoxia monticola (Broun, 1893) E
Adoxia nigricans Broun, 1880 E
Adoxia nigricornis (Sharp, 1886) E
Adoxia nitidicollis Broun, 1880 E
Adoxia nodicollis (Broun, 1915) E
Adoxia obscura (Broun, 1910) E
Adoxia oconnori (Broun, 1913) E
Adoxia oleareae (Broun, 1893) E
Adoxia palialis (Broun, 1909) E
Adoxia perplexa (Broun, 1917) E
Adoxia princeps (Broun, 1893) E
Adoxia proletaria (Weise, 1924) E
Adoxia pubicollis (Broun, 1915) E
Adoxia puncticollis (Sharp, 1886) E
Adoxia quadricollis (Broun, 1917) E
Adoxia rectipes (Broun, 1893) E
Adoxia rugicollis (Broun, 1893) E
Adoxia scutellaris (Broun, 1909) E
Adoxia simmondsi (Broun, 1913) E
Adoxia sordidula (Weise, 1924) E
Adoxia sulcifera (Broun, 1893) E
Adoxia truncata (Broun, 1893) E
Adoxia vestitus (Weise, 1924) E
Adoxia vilis (Weise, 1924) E
Adoxia viridis Broun, 1880 E
Adoxia vulgaris Broun, 1880 E
Adoxia xenoscelis (Broun, 1917) E
Allastena eminens Broun, 1917 E
Allastena nitida Broun, 1893 E
Allastena piliventris Broun, 1915 E
Allastena quadrata Broun, 1893 E
Bryobates aeratus Broun, 1914 E
Bryobates coniformis Broun, 1886 E
Bryobates nigricans Broun, 1914 E
Bryobates rugidorsis Broun, 1917 E
Lochmaea suturalis Thomson, 1866 A BC
CRYPTOCEPHALINAE
CRYPTOCEPHALINI
Aporocera melanocephala Saunders, 1843 A
Diachus auratus (Fabricius, 1801) A
Ditropidus compactus (Sharp, 1881) E?
STYLOSOMINI
Arnomus brouni Sharp, 1876 E
Arnomus curtipes Broun, 1893 E
Arnomus fulvus Broun, 1915 E
Arnomus marginalis Broun, 1893 E
Arnomus signatus Broun, 1909 E
Arnomus vicinus Broun, 1915 E
Arnomus viridicollis Broun, 1909 E
EUMOLPINAE
Atrichatus aeneicollis Broun, 1895 E
Atrichatus ochraceus (Broun, 1880) E
Eucolaspis antennata Shaw, 1957 E
Eucolaspis brunnea (Fabricius, 1792) E
Eucolaspis colorata Broun, 1893 E
Eucolaspis hudsoni Shaw, 1957 E
Eucolaspis jucunda (Broun, 1880) E
Eucolaspis pallidipennis (White, 1846) E
Eucolaspis picticornis Broun, 1893 E
Peniticus antiquus Sharp, 1876 E
Peniticus plicatus Broun, 1921 E
Peniticus robustus Broun, 1880 E
Peniticus suffusus Sharp, 1876 E
Peniticus wallacei Broun, 1910 E
Pilacolaspis angulatus Broun, 1913 E
Pilacolaspis huttoni (Broun, 1880) E
Pilacolaspis latipennis Broun, 1913 E
Pilacolaspis rugiventris Broun, 1914 E
Pilacolaspis wakefieldi Sharp, 1886 E
CHRYSOMELINAE
CHRYSOMELINI
S. CHRYSOLININA
Chrysolina hyperici (Förster, 1771) A BC
Chrysolina quadrigemina (Suffrian, 1851) A BC
S. DICRANOSTERNINA
Dicranosterna semipunctata (Chapuis, 1877) A
S. GONIOCTENINA
Gonioctena (*Spartophila*) *olivacea* (Förster, 1771) A BC
S. PAROPSINA
Paropsis charybdis Stål, 1860 A
Peltoschema suturalis Germar, 1848 A Do [eradicated]
Trachymela catenata (Chapuis, 1877) A
Trachymela sloanei (Blackburn, 1896) A
S. PHYLLOCHARINA
Allocharis fuscipes Broun, 1917 E
Allocharis limbata Broun, 1893 E
Allocharis marginata Sharp, 1882 E
Allocharis media Broun, 1917 E
Allocharis morosa Broun, 1893 E
Allocharis nigricollis Broun, 1917 E
Allocharis picticornis Broun, 1917 E
Allocharis praestans Broun, 1917 E

Allocharis robusta Broun, 1917 E
Allocharis subsulcata Broun, 1917 E
Aphilon convexum Broun, 1893 E
Aphilon enigma Sharp, 1876 E
Aphilon impressum Broun, 1914 E
Aphilon laticolle Broun, 1893 E
Aphilon latulum Broun, 1893 E
Aphilon minutum Broun, 1880 E
Aphilon monstrosum Broun,1886 E
Aphilon praestans Broun, 1893 E
Aphilon pretiosum Broun, 1880 E
Aphilon punctatum Broun, 1880 E
Aphilon scutellare Broun, 1893 E
Aphilon sobrinum Broun,1886 E
Aphilon sternale Broun, 1921 E
Aphilon n. sp. E
Caccomolpus amplus Broun, 1921 E
Caccomolpus cinctiger Broun, 1921 E
Caccomolpus flectipes Broun, 1914 E
Caccomolpus fuscicornis Broun, 1917 E
Caccomolpus globosus Sharp, 1886 E
Caccomolpus hallianus Broun, 1917 E
Caccomolpus maculatus Broun, 1893 E
Caccomolpus montanus Broun, 1921 E
Caccomolpus nigristernis Broun, 1917 E
Caccomolpus ornatus Broun, 1910 E
Caccomolpus plagiatus Sharp, 1886 E
Caccomolpus pullatus Broun, 1893 E
Caccomolpus subcupreus Broun, 1921 E
Caccomolpus substriatus Broun, 1917 E
Caccomolpus tibialis Broun, 1917 E
Caccomolpus viridescens Broun, 1917 E
Chalcolampra (*Eualema*) *speculifera* Sharp, 1882 E
Cyrtonogetus crassus Broun, 1915 E
CASSIDINAE
CASSIDINI
S. CASSIDINA
Cassida rubiginosa Müller, 1776 A BC
CURCULIONOIDEA
RHYNCHOPHORA
NEMONYCHIDAE
RHINORHYNCHINAE
Rhinorhynchus halli Kuschel, 2003 E
Rhinorhynchus halocarpi Kuschel, 2003 E
Rhinorhynchus phyllocladi Kuschel, 2003 E
Rhinorhynchus rufulus (Broun, 1880) E
ANTHRIBIDAE
ANTHRIBINAE
CORRHECERINI
Cacephatus aucklandicus (Brookes, 1951) E Su
Cacephatus huttoni (Sharp, 1876) E
Cacephatus incertus (White, 1846) E
Cacephatus inornatus (Sharp, 1886) E
Cacephatus propinquus (Broun, 1911) E Ch
Cacephatus vates (Sharp, 1876) E
Etnalis obtusus (Sharp, 1886) E
Etnalis spinicollis Sharp, 1873 E
PLATYSTOMINI
Arecopais spectabilis (Broun, 1880) E
Euciodes suturalis Pascoe, 1866 A
Lawsonia variabilis Sharp, 1873 E
STENOCERINI
Helmoreus sharpi (Broun, 1880) E
TRIBUS INCERTAE SEDIS
Androporus discedens (Sharp, 1876) E
Caliobius littoralis Holloway, 1982 E
Cerius otagensis Holloway, 1982 E
Cerius triregius Holloway, 1982 E TK
Dasyanthribus purpureus (Broun, 1880) E
Eugonissus conulus (Broun, 1880) E
Garyus altus (Sharp, 1876) E
Gynarchaeus ornatus (Sharp, 1876) E
Hoherius meinertzhageni (Broun, 1880) E
Hoplorhaphus nodifer Holloway, 1982 E
Hoplorhaphus spinifer (Sharp, 1876) E
Isanthribus dracophylli Holloway, 1982 E
Isanthribus phormii Holloway, 1982 E
Isanthribus proximus (Broun, 1880) E
Lichenobius littoralis Holloway, 1970 E
Lichenobius maritimus Holloway, 1982 E Su
Lichenobius silvicola Holloway, 1970 E Ch
Lophus cristatellus (Broun, 1911) E Ch
Lophus lewisi (Broun, 1909) E
Lophus rudis (Sharp, 1876) E
Phymatus cucullatus (Sharp, 1886) E
Phymatus hetaera (Sharp, 1876) E
Phymatus phymatodes (Redtenbacher, 1868) E
Pleosporius bullatus (Sharp, 1876) E
Sharpius brouni (Sharp, 1876) E
Sharpius chathamensis Holloway, 1982 E Ch
Sharpius imitarius (Broun, 1914) E
Sharpius sandageri (Broun, 1893) E
Sharpius venustus (Broun, 1914) E
Tribasileus noctivagus Holloway, 1982 E
Xenanthribus hirsutus Broun 1893 E
CHORAGINAE A
ARAECERINI
Araecerus fasciculatus (De Geer, 1775) A
Araecerus palmaris (Pascoe, 1882) A
Araeocerodes sp. A
Xanthoderopygus sp. A
TRIBUS INCERTAE SEDIS
Dysnocryptus balthasar Holloway, 1982 E TK
Dysnocryptus dignus (Broun, 1880) E
Dysnocryptus gaspar Holloway, 1982 E TK
Dysnocryptus inflatus (Sharp, 1876) E
Dysnocryptus maculifer Broun, 1893 E
Dysnocryptus melchior Holloway, 1982 E TK
Dysnocryptus pallidus Broun, 1893 E
Dysnocryptus pilicornis (Broun, 1911) E Ch
Dysnocryptus rugosus (Sharp, 1876) E
Liromus pardalis (Pascoe, 1876) E
Micranthribus atomus (Sharp, 1876) E
Notochoragus chathamensis Holloway, 1982 E Ch
Notochoragus crassus (Sharp, 1876) E
Notochoragus fungicola (Broun, 1893) E
Notochoragus nanus (Sharp, 1876) E
Notochoragus thoracicus (Broun, 1893) E
BELIDAE
PACHYURINAE
AGNESIOTIDINI
Agathinus tridens (Fabricius, 1787) E
PACHYURINI
Pachyurinus sticticus (Broun, 1893) E
Rhicnobelus aenescens (Broun, 1915) E
Rhicnobelus metallicus (Pascoe, 1877) E
Rhicnobelus rubicundus (Broun, 1880) E
OXYCORYNIDAE
AGLYCYDERINAE
Aralius wollastoni (Sharp, 1876) E
BRENTIDAE
TRACHELIZINAE
ITHYSTENINI
Lasiorhynchus barbicornis (Fabricius, 1775) E
APIONIDAE
APIONINAE
EXAPIINI
Exapion (*Ulapion*) *ulicis* (Forster, 1771) A BC
RHADINOCYBINI
Cecidophyus nothofagi Kuschel, 2003 E
Neocyba metrosideros (Broun, 1880) E
Neocyba regalis Kuschel, 2003 E TK
Strobilobius libocedri Kuschel, 2003 E
Zelapterus terricola (Broun, 1923) E
DRYOPHTHORIDAE A
DRYOPHTHORINAE A
Dryophthorus sp. indet. A
RHYNCHOPHORINAE A
LITOSOMINI
Sitophilus granarius (Linnaeus, 1758) A
Sitophilus oryzae (Linnaeus, 1763) A
Sitophilus zeamais Motschulsky, 1855 A
SPHENOPHORINI
Sphenophorus brunnipennis (Germar, 1824) A
ERIRHINIDAE
ERIRHININAE
ERIRHININI
Colabotelus dealbatus Broun, 1914 E
Praolepra albopicta Broun, 1881 E
Praolepra asperirostris Broun, 1881 E
Praolepra fultoni Broun, 1886 E
Praolepra infusca Broun, 1880 E
Praolepra pallida Broun, 1881 E
Praolepra rufescens Broun, 1881 E
Praolepra squamosa Broun, 1880 E
Praolepra uniformis Marshall, 1938 E
Praolepra varia Broun, 1881 E
STENOPELMINI
Athor arcifera Broun, 1909 E
Baeosomus alternans (Broun, 1914) E
Baeosomus amplus (Broun, 1915) E
Baeosomus angustus (Broun, 1923) E
Baeosomus burrowsi (Broun, 1915) E
Baeosomus crassipes (Broun, 1923) E
Baeosomus crassirostris (Broun, 1921) E
Baeosomus diversus (Broun, 1923) E
Baeosomus elegans (Broun, 1921) E
Baeosomus fordi (Broun, 1923) E
Baeosomus humeratus (Broun, 1921) E
Baeosomus iridescens (Broun, 1921) E
Baeosomus jugosus (Broun, 1914) E
Baeosomus lugubris (Broun, 1921) E
Baeosomus nigrirostris (Broun, 1914) E
Baeosomus niticollis (Broun, 1921) E
Baeosomus nodicollis (Broun, 1914) E
Baeosomus ovipennis (Broun, 1923) E
Baeosomus plicatus (Broun, 1923) E
Baeosomus polytrichi (Kuschel, 1990) E
Baeosomus quadricollis (Broun, 1921) E
Baeosomus rubidus (Broun, 1921) E
Baeosomus rugosus (Broun, 1921) E
Baeosomus scapularis (Marshall, 1937) E
Baeosomus serripes Kuschel, 1964 E Su
Baeosomus tacitus Broun, 1904 E
Baeosomus thoracicus (Broun, 1923) E
TRIBUS INCERTAE SEDIS
Aganeuma rufula Broun, 1893 E
Euprocas scitulus Broun, 1893 E
Philacta maculifera Broun, 1904 E
Philacta testacea Broun, 1880 E
Stilbopsis polita Broun, 1893 E
RAYMONDIONYMIDAE
MYRTONYMINAE
MYRTONYMINI
Myrtonymus zelandicus Kuschel, 1990 E
CURCULIONIDAE
BARIDINAE A
MADOPTERINI
S. ZYGOBARIDINA
Linogeraeus urbanus (Boheman, 1859) A
CEUTORHYNCHINAE A
CEUTORHYNCHINI
Trichosirocalus mortadelo Alonso-Zarazaga & Sánchez-Ruiz, 2002 A
HYPURINI
Hypurus bertrandi (Perris, 1852) A BC
PHYTOBIINI
Rhinoncus australis Oke, 1931 A BC
COSSONINAE
ARAUCARIINI
Inosomus rufopiceus (Broun, 1881) E
Xenocnema spinipes Wollaston, 1873 E
COSSONINI
Exomesites optimus Broun, 1886 E
Exomesites n. sp. E

Mesites pallidipennis Boheman, 1838 A
Phloeophagosoma (Amorphorhynchus) brouni Kuschel, 1982 E
Phloeophagosoma (A.) corvinum Wollaston, 1873 E
Phloeophagosoma (Phloeophagosoma) abdominale Broun, 1881 E
Phloeophagosoma (P.) dilutum Wollaston, 1874 E
Phloeophagosoma (P.) pedatum Wollaston, 1874 E
Phloeophagosoma (P.) rugipenne Broun, 1881 E
Phloeophagosoma (P.) thoracicum Wollaston, 1874 E
Stenotrupis debilis (Sharp, 1878) E
Stenotrupis wollastonianum (Sharp, 1878) E
DRYOTRIBINI
Allaorus carinifer Broun, 1921 E
Allaorus impressus Broun, 1917 E
Allaorus ovatus Broun, 1893 E
Allaorus pedatus Broun, 1893 E
Allaorus piciclavus Broun, 1909 E
Allaorus pyriformis (Broun, 1893) E
Allaorus rugosus (Broun, 1893) E
Allaorus scutellaris Broun, 1914 E
Allaorus sternalis Broun, 1893 E
Allaorus urquharti Broun, 1893 E
Allaorus versutus (Broun, 1880) E
Catolethrobius silvestris (Kolbe, 1910) A
Eiratus costatus Broun, 1886 E
Eiratus nitirostris Broun, 1910 E
Eiratus ornatus Broun, 1886 E
Eiratus parvulus Pascoe, 1877 E
Eiratus setulifer (Marshall, 1953) E
Eiratus suavis Broun, 1886 E
Eiratus tetricus Broun, 1880 E
Microtribus brouni (Wollaston, 1874) E
Microtribus huttoni Wollaston, 1873 E
Microtribus sculpturatus Broun, 1910 E
Microtribus sp. n. E
Stilbocara constricticollis (Broun, 1880) E
Stilbocara nitida Broun, 1893 E
Stilbocara serena Broun, 1893 E
Stilboderma impressipennis Broun, 1909 E
ONYCHOLIPINI
Pselactus ferrugineus Broun, 1909 E
Pselactus spadix (Herbst, 1795) A
Stenoscelis hylastoides Wollaston, 1861 A
PENTARTHRINI
Adel crenatum (Broun, 1883) E
Agastegnus aeneopiceus (Broun, 1880) E
Agastegnus coloratus Broun, 1886 E
Agastegnus concinnus Broun, 1914 E
Agastegnus distinctus Broun, 1893 E
Agastegnus femoralis Broun, 1886 E
Agastegnus ornatus Broun, 1911 E
Agastegnus rarus (Broun, 1886) E
Agastegnus rufescens Broun, 1907 E
Agastegnus rugipennis Broun, 1914 E
Agastegnus simulans (Sharp, 1878) E
Agastegnus thoracicus Broun, 1914 E
Agrilochilus prolixus Broun, 1880 E
Arecocryptus bellus (Broun, 1880) E
Camptoscapus planiusculus (Broun, 1880) E
Entium aberrans Sharp, 1878 E
Eucossonus antennalis Broun, 1910 E
Eucossonus comptus Broun, 1886 E
Eucossonus constrictus Broun, 1921 E
Eucossonus discalis Broun, 1910 E
Eucossonus disparilis Broun, 1921 E
Eucossonus elegans Broun, 1893 E
Eucossonus gracilis Broun, 1893 E
Eucossonus nasalis Broun, 1921 E
Eucossonus orneobius Broun, 1921 E
Eucossonus rostralis Broun, 1909 E
Eucossonus setiger (Sharp, 1878) E
Eucossonus sulcicollis Broun, 1921 E
Euophryum confine (Broun, 1881) E
Euophryum rufum (Broun, 1880) E
Macroscytalus cheesemani (Broun, 1893) E
Macroscytalus confertus (Sharp, 1886) E
Macroscytalus constrictus (Sharp, 1886) E
Macroscytalus elongatus (Broun, 1909) E
Macroscytalus fusiformis (Broun, 1914) E
Macroscytalus glabrus (Broun, 1881) E
Macroscytalus gracilis (Broun, 1909) E
Macroscytalus halli (Broun, 1914) E
Macroscytalus lewisi (Broun, 1909) E
Macroscytalus parvicornis (Sharp, 1878) E
Macroscytalus remotus (Sharp, 1878) E
Macroscytalus sagax Broun, 1886 E
Morronella latirostris (Marshall, 1926) E
Morronella lawsoni (Wollaston, 1873) E
Morronella n. sp. 1 E
Novitas dispar Broun, 1893 E
Novitas nigrans Broun, 1880 E
Novitas rufus Broun, 1880 E
Pentarthrum amicum Broun, 1893 E
Pentarthrum assimilatum Broun, 1880 E
Pentarthrum auricomus Broun, 1881 E
Pentarthrum auripilum Broun, 1911 E Ch
Pentarthrum brevicorne Broun, 1915 E
Pentarthrum brunneum Broun, 1880 E
Pentarthrum carmichaeli Waterhouse, 1884 A
Pentarthrum castum Broun, 1881 E
Pentarthrum dissimile Broun, 1911 E Ch
Pentarthrum ferrugineum Broun, 1883 E
Pentarthrum fultoni Broun, 1893 E
Pentarthrum gracilicorne Broun, 1910 E
Pentarthrum impressum Broun, 1913 E
Pentarthrum melanosternum Broun, 1886 E
Pentarthrum nubilum Broun, 1893 E
Pentarthrum philpotti Broun, 1895 E
Pentarthrum planicolle Broun, 1909 E
Pentarthrum proximum Broun, 1886 E
Pentarthrum punctirostre Broun, 1881 E
Pentarthrum reductum Broun, 1881 E
Pentarthrum ruficorne Broun, 1881 E
Pentarthrum rugirostre Broun, 1881 E
Pentarthrum subsericatum Wollaston, 1873 E
Pentarthrum tenebrosum Broun, 1913 E
Pentarthrum triste Broun, 1909 E
Pentarthrum vestitum Broun, 1880 E
Pentarthrum zealandicum Wollaston, 1873 E
Proconus asperirostris (Broun, 1880) E
Proconus crassipes Broun, 1884 E
Sericotrogus ovicollis Broun, 1909 E
Sericotrogus plexus Broun, 1913 E
Sericotrogus stramineus Broun, 1909 E
Sericotrogus subaenescens Wollaston, 1873 E
Stenotoura exilis (Broun, 1893) E
Stenotoura lateritia (Broun, 1880) E
Tanysoma aciphyllae Broun, 1913 E
Tanysoma angustum (Broun, 1886) E
Tanysoma comata (Broun, 1886) E
Tanysoma fuscicollis Broun, 1909 E
Tanysoma impressella Broun, 1913 E
Torostoma apicale Broun, 1880 E
Toura fulva (Broun, 1893) E
Toura helmsianum (Sharp, 1882) E
Toura longirostre (Wollaston, 1873) E
Toura morosa (Broun, 1886) E
Toura sharpiana (Wollaston, 1874) E
Touropsis brevirostris (Sharp, 1878) E
Touropsis n. sp. E
Unas conirostris Marshall, 1953 E
Unas piceus (Broun, 1880) E
Unas pictonensis (Sharp, 1886) E
Zenoteratus cephalotes (Sharp, 1886) E
Zenoteratus diversus Broun, 1909 E
Zenoteratus macrocephalus (Broun, 1886) E
Zenoteratus servulus (Broun, 1886) E
PROECINI
Conarthrus cylindricus (Broun, 1893) E
Conarthrus parvulus (Broun, 1893) E
Eutornopsis picea Broun, 1910 E
RHYNCOLINI
S. RHYNCOLINA
Macrorhyncolus littoralis (Broun, 1880) E
Pachyops dubius (Wollaston, 1873) E
CRYPTORHYNCHINAE
CRYPTORHYNCHINI
S. CRYPTORHYCHINA
Allanalcis altostethus (Broun, 1893) E
Allanalcis aulacus (Broun, 1893) E
Allanalcis eruensis (Broun. 1913) E
Allanalcis laticollis Broun, 1914 E
Allanalcis melastictus Broun, 1921 E
Allanalcis n. sp. E
Baeorhynchodes cristatus Broun, 1909 E
Indecentia nubila Broun, 1880 E
Mitrastethus baridioides Redtenbacher, 1868 E
Rhynchodes ursus White, 1846 E
S. MECISTOSTYLINA
Mecistostylus douei Lacordaire, 1866 E
S. TYLODINA
Adstantes arctus (Broun, 1881) E
Adstantes rudis (Broun, 1881) E
Adstantes n. sp. E Ch
Agacalles comptus (Broun, 1893) E
Agacalles formosus Broun, 1886 E
Agacalles gracilis (Broun, 1913) E
Agacalles integer (Broun, 1893) E
Agacalles tortipes (Broun, 1880) E
Ampagia rudis (Pascoe, 1877) E
Anaballus amplicollis (Fairmaire, 1849) A Do
Andracalles canescens (Broun, 1881) E
Andracalles diversus (Broun, 1883) E
Andracalles horridus (Broun, 1881) E
Andracalles pani Lyal, 1993 E TK
Andracalles spurcus (Broun, 1881) E
Andracalles vividus (Broun, 1880) E
Andracalles n. spp. (2) 2E
Clypeolus binodes (Broun, 1921) E
Clypeolus brookesi (Broun, 1923) E
Clypeolus cilicollis (Broun, 1921) E
Clypeolus cineraceus Broun, 1909 E
Clypeolus complexus (Broun, 1921) E
Clypeolus dux (Broun, 1893) E
Clypeolus fuscidorsis (Broun, 1909) E
Clypeolus lachrymosus (Broun, 1881) E
Clypeolus maritimus (Broun, 1893) E
Clypeolus notoporhinus (Broun, 1914) E
Clypeolus pascoei (Broun, 1880) E
Clypeolus robustus (Broun, 1909) E
Clypeolus signatus (Broun, 1880) E
Clypeolus simulans (Broun, 1921) E
Clypeolus squamosus (Broun, 1914) E
Clypeolus sympedioides (Broun, 1893) E
Clypeolus terricola (Broun, 1921) E
Clypeolus veratrus (Broun. 1893) E
Crisius anceps (Broun, 1921) E
Crisius baccatellus (Broun, 1917) E
Crisius bicinctus (Broun, 1915) E
Crisius bicristaticeps (Broun, 1914) E
Crisius binotatus Pascoe, 1876 E
Crisius brouni Lyal, 1993 E
Crisius cinereus (Broun, 1883) E
Crisius confusus (Broun, 1914) E
Crisius contiguus Broun, 1921 E
Crisius curtus (Broun, 1881) E
Crisius decorus Broun, 1913 E
Crisius dives Broun, 1921 E
Crisius dorsalis Broun, 1904 E
Crisius eucoelius (Broun, 1921) E
Crisius eximius Broun, 1921 E
Crisius fasciatus (Broun, 1914) E
Crisius fasciculatus Broun, 1893 E
Crisius flavisetosus (Broun, 1909) E

Crisius fulvicornis (Broun, 1914) E
Crisius fuscatus (Broun, 1907) E
Crisius grisealis (Broun, 1921) E
Crisius griseicollis (Broun, 1909) E
Crisius hopensis (Broun, 1921) E
Crisius humeralis Broun, 1913 E
Crisius humeratus (Broun, 1893) E
Crisius latirostris Broun, 1914 E
Crisius lineirostris (Broun, 1911) E
Crisius longulus Broun, 1921 E
Crisius lunalis (Broun, 1917) E
Crisius minor (Broun, 1893) E
Crisius nodigerus (Broun, 1917) E
Crisius obesulus Sharp, 1886 E
Crisius oblongus (Broun, 1914) E
Crisius obscurus (Broun, 1921) E
Crisius ornatus Broun, 1893 E
Crisius picicollis Broun, 1893 E
Crisius posticalis (Broun, 1914) E
Crisius postpuncta Lyal, 1993 E
Crisius rostralis (Broun, 1893) E
Crisius scutellaris Broun, 1880 E
Crisius semifuscus Broun, 1913 E
Crisius signatus Broun, 1893 E
Crisius sparsus (Broun, 1914) E
Crisius sternalis (Broun, 1917) E
Crisius subcarinatus (Broun, 1911) E
Crisius variegatus Broun, 1880 E
Crisius variellus (Broun, 1914) E
Crisius ventralis (Broun, 1885) E
Crisius zenomorphus (Broun, 1917) E
Crooktacalles abruptus (Marshall, 1937) E
Crooktacalles certus (Broun, 1880) E
Dermothrius asaphus (Broun, 1921) E
Dermothrius brevipennis (Broun, 1921) E
Dermothrius farinosus (Broun, 1893) E
Dermothrius porcatus (Broun, 1893) E
Dermothrius puncticollis (Broun, 1893) E
Dermothrius ruficollis (Broun, 1893) E
Dermothrius sanguineus (Broun, 1881) E
Didymus bicostatus (Broun, 1921) E
Didymus erroneus (Pascoe, 1876) E
Didymus impexus (Pascoe, 1877) E
Didymus intutus (Pascoe, 1876) E
Didymus metrosideri (Broun, 1910) E K
Didymus n. spp. (2) 2E TK
Ectopsis ferrugalis Broun, 1881 E
Ectopsis foveigera Broun, 1917 E
Ectopsis simplex Broun, 1893 E
Hadracalles fuliginosus Broun, 1893 E
Hiiracalles dolosus (Broun, 1893) E
Hiiracalles scitus (Broun, 1880) E
Maneneacalles concinnus (Broun, 1893) E
Metacalles aspersus Broun, 1893 E
Metacalles aterrimus (Broun, 1909) E
Metacalles cordipennis (Broun, 1881) E
Metacalles crinitulus Hustache, 1936 E
Metacalles crinitus (Broun, 1881) E
Metacalles exiguus (Broun, 1881) E
Metacalles irregularis (Broun, 1913) E
Metacalles lanosus Broun, 1913 E
Metacalles latisulcatus (Broun, 1909) E
Metacalles latus (Broun, 1881) E
Metacalles ornatus (Broun, 1909) E
Metacalles picatus Broun, 1914 E
Metacalles rugicollis Broun, 1893 E
Metacalles sentus (Broun, 1883) E
Metacalles sticticus (Broun, 1921) E
Microcryptorhynchus (*Microcryptorhynchus*) *albistrigalis* (Broun, 1909) E
Microcryptorhynchus (*M.*) *contractus* (Broun, 1913) E
Microcryptorhynchus (*M.*) *ferrugo* (Kuschel, 1971) E
Microcryptorhynchus (*M.*) *kronei* (Kirsch, 1877) E
Microcryptorhynchus (*M.*) *latitarsis* (Kuschel, 1964) E
Microcryptorhynchus (*M.*) *linagri* Kuschel, 1997 E
Microcryptorhynchus (*M.*) *mayae* Kuschel, 1997 E
Microcryptorhynchus (*M.*) *multisetosus* (Broun, 1907) E
Microcryptorhynchus (*M.*) *perpusillus* (Pascoe, 1877) E
Microcryptorhynchus (*M.*) *praesetosus* (Broun, 1909) E
Microcryptorhynchus (*M.*) *quietus* (Broun, 1893) E
Microcryptorhynchus (*M.*) *setifer* (Broun, 1886) E
Microcryptorhynchus (*M.*) *suillus* (Kuschel, 1964) E
Microcryptorhynchus (*M.*) *vafer* (Broun, 1881) E
Microcryptorhynchus (*M.*) n. spp. (4) 4E
Microcryptorhynchus (*Notacalles*) *floricola* (Broun, 1886) E
Microcryptorhynchus (*N.*) *leviculus* (Broun, 1881) E
Microcryptorhynchus (*N.*) *piciventris* (Broun, 1909) E
Microcryptorhynchus (*N.*) *planidorsis* (Kirsch, 1877) E
Microcryptorhynchus (*N.*) n. spp. (4) 4E
Omoeacalles crisioides (Broun, 1880) E
Omoeacalles ovatellus (Broun, 1881) E
Omoeacalles perspicuus Broun, 1909 E
Pachyderris nigricans (Broun, 1917) E
Pachyderris nodifer (Broun, 1914) E
Pachyderris punctiventris Broun, 1909 E
Pachyderris squamiventris (Broun, 1911) E
Pachyderris triangulatus (Broun, 1883) E
Pachyderris n. sp. TK
Paromalia nigricollis (Broun, 1895) E
Paromalia setiger Broun, 1880 E
Paromalia vestita Broun, 1880 E
Patellitergum rectirostre Lyal, 1993 E Ch
Postacalles rangirua Lyal, 1993 E
Rainacalles volens (Broun, 1881) E
Sceledolichus altulus Broun, 1886 E
Sceledolichus celsus (Broun, 1880) E
Sceledolichus decorus Broun, 1923 E
Sceledolichus denotans (Broun, 1881) E
Sceledolichus flectipes Broun, 1914 E
Sceledolichus hilaris Broun, 1893 E
Sceledolichus juncobius Broun, 1893 E
Sceledolichus lineithorax (Broun, 1880) E
Sceledolichus politus Broun, 1895 E
Sceledolichus pyriformis Broun, 1923 E
Sceledolichus setosus (Broun, 1881) E
Sceledolichus squamosus Broun, 1895 E
Sceledolichus villosus (Broun, 1881) E
Sympedius bufo (Sharp, 1883) E
Sympedius costatus (Broun, 1913) E
Sympedius densus (Broun, 1880) E
Sympedius ferrrugatus (Pascoe, 1876) E
Synacalles hystriculus (Pascoe, 1876) E
Sympedius lepidus Broun, 1885 E
Sympedius minor Broun, 1921 E
Sympedius rectirostris Broun, 1909 E
Sympedius testudo Pascoe, 1876 E
Synacalles cingulatus (Broun, 1883) E
Synacalles dorsalis (Broun, 1881) E
Synacalles mundus (Broun, 1881) E
Synacalles peelensis (Broun, 1913) E
Synacalles posticalis (Broun, 1886) E
Synacalles trinotatus (Broun, 1880) E
Synacalles n. sp. 1 E
Trinodicalles adamsi (Broun, 1893) E
Trinodicalles altus (Broun, 1909) E
Trinodicalles conicollis (Broun, 1913) E
Trinodicalles cristatus (Broun, 1881) E
Trinodicalles decemcristatus (Broun, 1883) E
Trinodicalles latirostris (Broun, 1883) E
Trinodicalles lepirhinus (Broun, 1893) E
Trinodicalles mimus (Broun, 1893) E
Trinodicalles terricola (Broun, 1885) E
Tychanopais dealbatus Broun, 1921 E
Tychanopais flavisparsus Broun, 1913 E
Tychanopais fougeri (Hutton, 1898) E
Tychanopais hudsoni (Marshall, 1926) E
Tychanopais pictulus Broun, 1893 E
Tychanopais tuberosus (Broun, 1923) E
Tychanus gibbus Pascoe, 1876 E
Tychanus verrucosus Pascoe, 1876 E
Tychanus vexatus (Pascoe, 1876) E
Whitiacalles ignotus (Broun, 1914) E
Zeacalles aeratus Broun, 1921 E
Zeacalles albipictus (Broun, 1921) E
Zeacalles alpestris (Broun, 1893) E
Zeacalles binodosus Broun, 1910 E
Zeacalles bisulcatus Broun, 1921 E
Zeacalles blanditus (Broun, 1921) E
Zeacalles brookesi Marshall, 1937 E
Zeacalles carinellus Broun, 1914 E
Zeacalles coarctalis Broun, 1921 E
Zeacalles cordipennis Broun, 1921 E
Zeacalles dilatatus (Broun, 1913) E
Zeacalles estriatus Broun, 1914 E
Zeacalles femoralis Broun, 1913 E
Zeacalles finitimus Broun, 1921 E
Zeacalles flavescens Broun, 1893 E
Zeacalles formosus (Broun, 1893) E
Zeacalles ignealis (Broun, 1913) E
Zeacalles igneus (Broun, 1909) E
Zeacalles incultus (Broun, 1893) E
Zeacalles inornatus Broun, 1921 E
Zeacalles latulus Broun, 1921 E
Zeacalles lepidulus Broun, 1909 E
Zeacalles oculatus (Broun, 1913) E
Zeacalles parvus Broun, 1921 E
Zeacalles picatus (Broun, 1893) E
Zeacalles pictus Broun, 1913 E
Zeacalles scaber Broun, 1915 E
Zeacalles scruposus Broun, 1921 E
Zeacalles seticollis (Broun, 1921) E
Zeacalles sparsus Broun, 1915 E
Zeacalles speciosus Broun, 1917 E
Zeacalles variatus (Broun, 1921) E
Zeacalles varius Broun, 1893 E
GASTEROCERCINI
Eutyrhinus squamiger White, 1846 E
PSEPHOLACINI
Homoreda flavisetosa (Broun, 1911) E
Homoreda murina (Broun, 1880) E
Mesoreda brevis (Pascoe, 1876) E
Mesoreda orthorhina (Broun, 1886) E
Mesoreda sulcifrons Broun, 1909 E
Nothaldonus peacei (Broun, 1880) E
Oreda notata White, 1846 E
Psepholax acanthomerus Broun, 1913 E
Psepholax coronatus White, 1846 E
Psepholax crassicornis Broun, 1895 E
Psepholax femoratus Broun, 1880 E
Psepholax macleayi (Schönherr, 1847) E
Psepholax mediocris Broun, 1886 E
Psepholax mystacinus Broun, 1886 E
Psepholax simplex Pascoe, 1876 E
Psepholax sulcatus White, 1843 E
Psepholax tibialis (Broun, 1880) E
Psepholax n. spp. (3) 3E
Strongylopterus chathamensis (Sharp, 1903) E Ch
Strongylopterus hylobioides (White, 1846) E
CURCULIONINAE
CIONINI
Cleopus japonicus Wingelmüller, 1914 A BC
EUGNOMINI
S. EUGNOMINA
Amylopterus pilosus (Broun, 1880) E
Amylopterus prasinus (Broun, 1883) E
Ancistropterus brouni Sharp, 1876 E
Ancistropterus helmsi Sharp, 1886 E
Ancistropterus mundus Sharp, 1876 E
Ancistropterus quadrispinosus White, 1846 E
Eugnomus aenescens Broun, 1893 E
Eugnomus albisetosus Broun, 1921 E
Eugnomus alternans Broun, 1917 E
Eugnomus antennalis Broun, 1909 E

Eugnomus argutus Sharp, 1883 E
Eugnomus aspersus Broun, 1893 E
Eugnomus atratus Broun, 1921 E
Eugnomus calvulus Broun, 1913 E
Eugnomus carbonarius (Broun, 1921) E
Eugnomus dennanensis Broun, 1913 E
Eugnomus dispar (Broun, 1886) E
Eugnomus durvillei Schönherr, 1847 E
Eugnomus elegans Pascoe, 1876 E
Eugnomus fasciatus Broun, 1881 E
Eugnomus femoralis Broun, 1909 E
Eugnomus fervidus Pascoe, 1876 E
Eugnomus flavipilus (Broun, 1883) E
Eugnomus fucosus Pascoe, 1877 E
Eugnomus interstitialis Broun, 1880 E
Eugnomus lituratus Broun, 1893 E
Eugnomus luctuosus Broun, 1886 E
Eugnomus maculosus Broun, 1881 E
Eugnomus maurus Broun, 1893 E
Eugnomus monachus Broun, 1886 E
Eugnomus nobilis Broun, 1893 E
Eugnomus nubilans Broun, 1881 E
Eugnomus picipennis Pascoe, 1876 E
Eugnomus robustus Marshall, 1926 E
Eugnomus squamifer Broun, 1893 E
Eugnomus tristis Broun, 1917 E
Eugnomus wakefieldi Pascoe, 1877 E
Goneumus bryobius (Broun, 1917) E
Gonoropterus spinicollis Broun, 1904 E
Hoplocneme cyanea Broun, 1893 E
Hoplocneme forcipata Marshall, 1938 E
Hoplocneme hookeri White, 1846 E
Hoplocneme inaequale Broun, 1893 E
Hoplocneme propinqua Broun, 1914 E
Hoplocneme punctatissima Pascoe, 1876 E
Hoplocneme squamosa Broun, 1880 E
Hoplocneme vicina Broun, 1913 E
Icmalius abnormis (Broun, 1886) E
Nyxetes bidens (Fabricius, 1792) E
Oreocalus albosparsa (Broun, 1913) E
Oreocalus carinulata (Broun, 1914) E
Oreocalus castanea (Broun, 1913) E
Oreocalus cinnamomea (White, 1846) E
Oreocalus congruens (Broun, 1917) E
Oreocalus dealbata (Broun, 1893) E
Oreocalus fasciata (Broun, 1917) E
Oreocalus hebe (Marshall, 1938) E
Oreocalus latipennis (Broun, 1914) E
Oreocalus lineirostris (Broun, 1914) E
Oreocalus nigrescens (Broun, 1886) E
Oreocalus nigriceps (Broun, 1886) E
Oreocalus picigularis (Broun, 1886) E
Oreocalus pleuralis (Broun, 1915) E
Oreocalus pullata (Broun, 1904) E
Oreocalus uniformis (Broun, 1913) E
Oreocalus veronicae (Broun, 1913) E
Oreocalus vittata (Broun, 1893) E
Oropterus coniger White, 1846 E
Pactola fuscicornis Broun, 1913 E
Pactola demissa Pascoe, 1876 E
Pactola fairburni Marshall, 1938 E
Pactola hudsoni Marshall, 1938 E
Pactola nigra (Hudson, 1950) E
Pactola posticalis Marshall, 1938 E
Pactola variabilis Pascoe, 1876 E
Pactolotypus depressirostris (Kirsch, 1877) E Su
Pactolotypus humeralis (Broun, 1895) E
Pactolotypus prolixus (Broun, 1914) E
Pactolotypus striatus (Broun, 1909) E
Pactolotypus subantarcticus Kuschel, 1964) E Su
Pactolotypus n. sp. E Ch
Rhopalomerus tenuirostris Blanchard, 1851 E
Scolopterus aequus Broun, 1880 E
Scolopterus penicillatus White, 1846 E
Scolopterus tetracanthus White, 1846 E
Stephanorhynchus aper Sharp, 1886 E
Stephanorhynchus attelaboides (Fabricius, 1775) E
Stephanorhynchus brevipennis Pascoe, 1876 E
Stephanorhynchus costifer Broun, 1893 E
Stephanorhynchus crassus Broun, 1880 E
Stephanorhynchus curvipes White, 1846 E
Stephanorhynchus griseipictus Broun, 1886 E
Stephanorhynchus halli Broun, 1914 E
Stephanorhynchus insolitus Broun, 1893 E
Stephanorhynchus lawsoni Sharp, 1876 E
Stephanorhynchus nigrosparsus Broun, 1893 E
Stephanorhynchus purus Pascoe, 1876 E
Stephanorhynchus pygmaeus Broun, 1903 E
Stephanorhynchus tuberosus Broun, 1881 E
Tysius bicornis (Fabricius, 1781) E
S. MERIPHINA
Geochus apicalis Broun, 1921 E
Geochus certus Broun, 1921 E
Geochus convexus Broun, 1921 E
Geochus distinguens Broun, 1914 E
Geochus frontalis Broun, 1893 E
Geochus inaequalis (Broun, 1880) E
Geochus lateralis Broun, 1914 E
Geochus marginatus Broun, 1893 E
Geochus morosus Broun, 1914 E
Geochus nigripes Broun, 1893 E
Geochus nodosus Broun, 1893 E
Geochus pictulus Broun, 1921 E
Geochus plagiatus Broun, 1893 E
Geochus politus Broun, 1881 E
Geochus posticalis Broun, 1913 E
Geochus puncticollis Broun, 1893 E
Geochus pyriformis Broun, 1914 E
Geochus rufipictus Broun, 1923 E
Geochus rugulosus Broun, 1885 E
Geochus setiger Broun, 1893 E
Geochus similis Broun, 1893 E
Geochus squamosus Broun, 1893 E
Geochus suffusus Broun, 1914 E
Geochus sulcatus Broun, 1914 E
Geochus tibialis Broun, 1893 E
Geochus variegatus Broun, 1914 E
Geochus n. sp. E
GONIPTERINI
Gonipterus scutellatus Gyllenhal, 1833 A
MECININI
Mecinus pascuorum (Gyllenhal, 1813) A
RHADINOSOMINI
Rhadinosomus acuminatus (Fabricius, 1775) E?
STOREINI
Abantiadinus gratulus (Broun, 1917) E
Abantiadinus nodipennis (Broun, 1914) E
Abantiadinus pusillus (Broun, 1914) E
Aneuma compta Broun, 1885 E
Aneuma conspersa Broun, 1921 E
Aneuma erubescens Broun, 1910 E
Aneuma fasciata (Broun, 1880) E
Aneuma ferruginea Broun, 1886 E
Aneuma fulvipes Pascoe, 1876 E
Aneuma oblonga Broun, 1921 E
Aneuma rostralis Broun, 1921 E
Aneuma rubricale (Broun, 1880) E
Aneuma rufa Broun, 1921 E
Aneuma spinifera Broun, 1913 E
Aneuma stramineipes (Broun, 1886) E
Emplesis bifoveata Lea, 1927 A
Gerynassa sp. A
Phorostichus linearis (Broun, 1881) E
Storeus albosignatus Blackburn, 1890 A
TYCHIINI
S. TYCHIINA
Tychius schneideri (Herbst, 1795) A
S. INCERTAE SEDIS
Notinus aucklandicus Kuschel, 1964 E Su
Peristoreus acalyptoides (Pascoe, 1876) E
Peristoreus acceptus (Broun, 1881) E
Peristoreus aciphyllae (Broun, 1886) E
Peristoreus aericomus (Broun, 1883) E
Peristoreus albisetosus (Broun, 1914) E
Peristoreus altivagans (Broun, 1921) E
Peristoreus anchoralis (Broun, 1881) E
Peristoreus anxius (Broun, 1893) E
Peristoreus australis (Broun, 1921) E
Peristoreus bicavus (Broun, 1886) E
Peristoreus castigatus (Broun, 1909) E
Peristoreus celmisiae (Broun, 1917) E
Peristoreus cheesemani (Broun, 1886) E
Peristoreus confusus (Broun, 1886) E
Peristoreus consonus (Broun, 1913) E
Peristoreus cordipennis (Broun, 1915) E
Peristoreus crucigerus (Broun, 1881) E
Peristoreus decussatus (Marshall, 1926) E
Peristoreus difformipes (Broun, 1886) E
Peristoreus dilucidus (Broun, 1921) E
Peristoreus discoideus (Broun, 1880) E
Peristoreus dolosus (Broun, 1881) E
Peristoreus durus (Broun, 1886) E
Peristoreus elegans (Sharp, 1883) E
Peristoreus eustictus (Broun, 1886) E
Peristoreus exilis (Broun, 1913) E
Peristoreus fascialis (Broun, 1881) E
Peristoreus femoralis (Broun, 1881) E
Peristoreus flavitarsis (Broun, 1880) E
Peristoreus floricola (Broun, 1914) E
Peristoreus fulvescens (Broun, 1914) E
Peristoreus fulvus (Broun, 1886) E
Peristoreus fuscipes (Broun, 1893) E
Peristoreus fusconotatus (Broun, 1880) E
Peristoreus fuscoventris (Broun, 1886) E
Peristoreus glottis (Pascoe, 1877) E
Peristoreus gracilirostris (Broun, 1881) E
Peristoreus grossus (Broun, 1893) E
Peristoreus innocens Kirsch, 1877 E Su
Peristoreus insignis (Broun, 1909) E
Peristoreus insolitus (Broun, 1909) E
Peristoreus lateralis (Broun, 1881) E
Peristoreus leucocomus (Broun, 1921) E
Peristoreus limbatus (Pascoe, 1877) E
Peristoreus maorinus (Broun, 1913) E
Peristoreus melastictus (Broun, 1914) E
Peristoreus melastomus (Broun, 1886) E
Peristoreus methvenensis (Broun, 1915) E
Peristoreus nesobius (Broun, 1886) E
Peristoreus nocens (Broun, 1881) E
Peristoreus obscurus (Broun, 1921) E
Peristoreus ochraceus (Broun, 1881) E
Peristoreus oleariae (Broun, 1913) E
Peristoreus pardalis (Marshall, 1926) E
Peristoreus pectoralis (Broun, 1914) E
Peristoreus poecilus (Broun, 1921) E
Peristoreus rufirostris (Broun, 1880) E
Peristoreus sexmaculatus (Broun, 1881) E
Peristoreus spadiceus (Broun, 1909) E
Peristoreus stramineus (Broun, 1881) E
Peristoreus subconicollis (Broun, 1923) E
Peristoreus sudus (Broun, 1881) E
Peristoreus sylvaticus (Broun, 1914) E
Peristoreus terrestris (Broun, 1914) E
Peristoreus thomsoni (Broun, 1886) E
Peristoreus titahensis (Broun, 1913) E
Peristoreus trilobus (Pascoe, 1877) E
Peristoreus veronicae (Broun, 1886) E
Peristoreus viridipennis (Broun, 1880) E
Peristoreus vittatus (Broun, 1921) E
Peristoreus xenorhinus (Broun, 1886) E
Peristoreus n. spp. (3) 3E
CURCULIONINAE INCERTAE SEDIS
Alloprocas muticus Broun, 1914 E
Alloprocas niger Broun, 1893 E
Alloprocas rufus Broun, 1893 E

Celetotelus fulvus Broun, 1893 E
Hypotagea castanea (Broun, 1881) E
Hypotagea concolor (Broun, 1881) E
Hypotagea creperus (Broun, 1881) E
Hypotagea dissona Broun, 1886 E
Hypotagea lewisi Broun, 1913 E
Hypotagea rubida Pascoe, 1876 E
Hypotagea simulans (Broun, 1881) E
Hypotagea testaceipennis Broun, 1880 E
Hypotagea tibialis Broun, 1893, non Broun, 1921 E
Hypotagea 'tibialis Broun, 1921', non Broun, 1893 E
Hypotagea variegata Broun, 1880 E
Hypotagea vestita (Broun, 1881) E
Neomycta pulicaris Pascoe, 1877 E
Neomycta rubida Broun, 1880 E
Neomycta seticeps Broun, 1914 E
Simachus cuneipennis Broun, 1914 E
Simachus montanus Broun, 1886 E
Simachus placens Broun, 1921 E
CYCLOMINAE
ATERPINI
S. ATERPINA
Anagotus aterrimus (Broun, 1913) E
Anagotus carinirostris Marshall, 1953 E
Anagotus costipennis (Broun, 1915) E
Anagotus fairburni (Brookes, 1932) E
Anagotus gourlayi (Brookes, 1932) E
Anagotus graniger (Broun, 1886) E
Anagotus halli (Broun, 1915) E
Anagotus hamiltoni (Broun, 1913) E
Anagotus helmsi Sharp, 1882 E
Anagotus laevicostatatus(Broun, 1914) E
Anagotus latirostris (Broun, 1904) E
Anagotus lewisi (Broun, 1904) E
Anagotus oconnori (Broun, 1910) E
Anagotus pascoi (Broun, 1917) E
Anagotus peelensis (Marshall, 1937) E
Anagotus rugosus (Broun, 1883) E
Anagotus stephenensis Kuschel, 1982 E
Anagotus turbotti (Spiller, 1942) E
Anagotus n. spp. (2) 2E
Heterotyles argentatus Broun, 1883 E
Lyperopais alternans Broun, 1914 E
Lyperopais mirus Broun, 1893 E
Lyperopais stellae Marshall, 1937 E
LISTRODERINI
Gromilus anthracinus (Broun, 1921) E
Gromilus aucklandicus Kuschel, 1971 E Su
Gromilus bicarinatus (Broun, 1921) E
Gromilus bifoveatus (Broun, 1923) E
Gromilus brevicornis (Broun, 1893) E
Gromilus calvulus (Broun, 1913) E
Gromilus caudatus (Broun, 1913) E
Gromilus clarulus (Broun, 1917) E
Gromilus cockaynei (Broun, 1905) E Su
Gromilus cordipennis (Broun, 1893) E
Gromilus cristatus (Broun, 1893) E
Gromilus dorsalis (Broun, 1921) E
Gromilus exiguus (Brookes, 1951) E Su
Gromilus fallai (Brookes, 1951) E Su
Gromilus foveirostris (Broun, 1913) E
Gromilus furvus (Broun, 1921) E
Gromilus gracilipes (Sharp, 1883) E
Gromilus granissimus (Broun, 1917) E
Gromilus halli (Broun, 1917) E
Gromilus impressus (Broun, 1893) E
Gromilus inophloeoides (Broun, 1904) E
Gromilus insularis antipodarum Kuschel, 1964 E Su
Gromilus i. insularis Blanchard, 1853 E Su
Gromilus i. robustus (Brookes, 1951) E Su
Gromilus laqueorum Kuschel, 1964 E Su
Gromilus majusculus (Broun, 1915) E
Gromilus merus (Broun, 1917) E
Gromilus narinosus Kuschel, 1971 E Su
Gromilus nitidellus (Broun, 1917) E
Gromilus nitidulus (Broun, 1915) E
Gromilus nodiceps (Broun, 1914) E
Gromilus philpotti (Broun, 1917) E
Gromilus setosus (Broun, 1893), non (Broun, 1917) E
Gromilus 'setosus (Broun, 1917)', non (Broun, 1893) E
Gromilus sparsus (Broun, 1921) E
Gromilus striatus (Broun, 1915), non (Broun, 1921) E
Gromilus 'striatus (Broun, 1921)', non (Broun, 1915) E
Gromilus sulcicollis (Broun, 1913) E
Gromilus sulcipennis (Broun, 1917) E
Gromilus tenuiculus (Broun, 1921) E
Gromilus thoracicus (Broun, 1893) E
Gromilus variegatus (Broun, 1893) E
Gromilus veneris setarius (Broun, 1909) E Su
Gromilus v. veneris (Kirsch, 1877) E Su
Liparogetus sulcatissimus Broun, 1915 E
Listroderes delaiguei Germain 1895 A
Listroderes difficilis Germain, 1895 A
Listroderes foveatus (Lea, 1928) A
Listronotus bonariensis (Kuschel, 1955) A
Nestrius bifurcus Kuschel, 1964 E Su
Nestrius cilipes Broun, 1909 E
Nestrius crassicornis Broun, 1915 E
Nestrius foveatus (Broun, 1893) E
Nestrius hudsoni Marshall, 1953 E
Nestrius irregularis (Broun, 1910) E
Nestrius laqueorum Kuschel, 1964 E Su
Nestrius ovithorax (Broun, 1893) E
Nestrius prolixus Broun, 1917 E
Nestrius rubidus (Broun, 1904) E
Nestrius sculpturatus (Broun, 1909) E
Nestrius serripes Broun, 1893 E
Nestrius simmondsi Broun, 1921 E
Nestrius sulcirostris Broun, 1917 E
Nestrius zenoscelis Broun, 1921 E
Steriphus ascitus (Pascoe, 1876) E
Steriphus diversipes lineatus (Pascoe, 1873) A
Steriphus pullus (Broun, 1910) E
Steriphus variabilis (Broun, 1885) E
NOTIOMIMETINI
Aphela algarum Pascoe, 1870 A
Neosyagrius cordipennis Lea, 1904 A Do
ENTIMINAE
CELEUTHETINI
Platysimus planidorsis (Broun, 1910) E K
GEONEMINI
Lyperobates ardens Broun, 1910 E
Lyperobates asper Broun, 1893 E
Lyperobates carinifer Broun, 1910 E
Lyperobates elegantulus Broun, 1913 E
Lyperobates guinnessi Broun, 1913 E
Lyperobates punctatus Broun, 1913 E
Lyperobates rostralis Broun, 1913 E
Lyperobates virilis Broun, 1910 E
Lyperobates waterworthi Broun, 1910 E
NAUPACTINI
Asynonychus cervinus (Boheman, 1840 A
Atrichonotus sordidus (Hustache, 1939) A
Atrichonotus taeniatulus (Berg, 1881) A
Naupactus leucoloma Boheman, 1840 A
OOSOMINI
Phlyctinus callosus Boheman, 1834 A
OTIORHYNCHINI
Epitimetes bicolor Broun, 1921 E
Epitimetes cupreus Broun, 1917 E
Epitimetes densus Broun, 1921 E
Epitimetes foveiger Broun, 1913 E
Epitimetes grisealis Broun, 1913 E
Epitimetes lutosus Pascoe 1877 E
Epitimetes wakefieldi Sharp, 1886 E
Homodus cuprealis Broun, 1923 E
Homodus fumeus Broun, 1881 E
Homodus longicornis Broun, 1923 E
Homodus posticalis Broun, 1923 E
Hygrochus cordipennis Broun, 1910 E
Hygrochus granifer Broun, 1909 E
Hygrochus illepidus Broun, 1893 E
Hygrochus monilifer Broun, 1921 E
Hygrochus oculatus Broun, 1893 E
Hygrochus oscitans Broun, 1881 E
Hygrochus scutellaris Broun, 1914 E
Hygrochus verrucosus Broun, 1893 E
Otiorhynchus (Cryphiphorus) sulcatus (Fabricius, 1775) A
Otiorhynchus (Pendragon) ovatus (Linnaeus, 1758) A
Otiorhynchus rugostriatus Goeze, 1877 A
Phaeocharis cuprealis Broun, 1913 E
Phaeocharis punctatus Broun, 1913 E
OTTISTIRINI
Maleuterpes spinipes Blackburn, 1894 A
PHYLLOBIINI
Nonnotus albatus (Broun, 1881) E
Nonnotus albicans (Broun, 1880) E
Nonnotus eclectus Broun, 1893 E
Nonnotus griseolus Sharp, 1886 E
Nonnotus nigricans Broun, 1913 E
Nonnotus pallescens Broun, 1893 E
SCAIAPHILINI
Barypeithes pellucidus Jacquelin du Val, 1854 A
SITONINI
Sitona (Sitona) discoidea Gyllenhal, 1834 A
Sitona (Sitona) lepidus Gyllenhal, 1834 A
TRACHYPHLOEINI
Trachyphloeus sp. indet. A? SI
TROPIPHORININI
Agatholobus waterhousei Broun, 1913 E
Brachyolus albescens Broun, 1903 E
Brachyolus asperatus Broun, 1914 E
Brachyolus bagooides Sharp, 1886 E
Brachyolus bicostatus Broun, 1917 E
Brachyolus breviusculus (Broun, 1880) E
Brachyolus cervalis Broun, 1903 E
Brachyolus elegans Broun, 1893 E
Brachyolus fuscipictus Broun, 1914 E
Brachyolus huttoni Sharp, 1886 E
Brachyolus inaequalis Sharp, 1886 E
Brachyolus labeculatus Broun, 1913 E
Brachyolus longicollis Sharp, 1886 E
Brachyolus nodirostris Broun, 1921 E
Brachyolus obscurus Broun, 1921 E
Brachyolus posticalis Broun, 1893 E
Brachyolus punctatus White, 1846 E
Brachyolus sylvaticus Broun, 1910 E
Brachyolus terricola Broun, 1917 E
Brachyolus varius Broun, 1913 E
Brachyolus viridescens Broun, 1893 E
Brachyolus n. sp. E
Catodryobiolus antipodus Brookes, 1951 E Su
Catoptes acuminatus (Broun, 1913) E
Catoptes aequus (Broun, 1886) E
Catoptes agrestis (Marshall, 1938) E
Catoptes amotus (Broun, 1886) E
Catoptes amplus (Broun, 1921) E
Catoptes apicalis Broun, 1923 E
Catoptes argentalis Broun, 1914 E
Catoptes asperellus Broun, 1893 E
Catoptes asteliae (Broun, 1917) E
Catoptes attenuatus Broun, 1886 E
Catoptes aulicus (Broun, 1893) E
Catoptes bicostatus (Broun, 1886) E
Catoptes binodis (White, 1846) E
Catoptes binodulus (Sharp, 1886) E
Catoptes brevicornis Sharp, 1886, non (Broun, 1904) E
Catoptes 'b.' australis (Kuschel, 1964) E
Catoptes 'b. brevicornis' (Broun, 1904) E
Catoptes caliginosus Broun, 1893 E
Catoptes cavelli (Broun, 1893) E
Catoptes censorius (Pascoe, 1876) E
Catoptes chalmeri Broun, 1893 E
Catoptes cheesemani Broun, 1893 E

Catoptes citimus Broun, 1917 E
Catoptes constrictus Broun, 1910 E
Catoptes coronatus (Sharp, 1886) E
Catoptes cuspidatus Broun, 1881 E
Catoptes dehiscens Broun, 1917 E
Catoptes dispar (Broun, 1904) E
Catoptes dorsalis (Broun, 1917) E
Catoptes duplex Broun, 1904 E
Catoptes enysi (Broun, 1886) E
Catoptes flaviventris Broun, 1917 E
Catoptes fraudator Marshall, 1931 E
Catoptes fumosus Broun, 1914 E
Catoptes funestus (Broun, 1921) E
Catoptes furvus Broun, 1893 E
Catoptes hamiltoni (Broun, 1913) E
Catoptes humeralis Broun, 1893, non (Broun, 1910) E
Catoptes 'humeralis (Broun, 1910)', non Broun, 1893 E
Catoptes instabilis Marshall, 1931 E
Catoptes interruptus (Fabricius, 1781) E
Catoptes latipennis Broun, 1893, non (Broun, 1917) E
Catoptes 'latipennis (Broun, 1917)', non Broun, 1893 E
Catoptes limbatus Broun, 1909 E
Catoptes lobatus Broun, 1921 E
Catoptes longulus Sharp, 1886 E
Catoptes murinus (Broun, 1917) E
Catoptes nasalis (Broun, 1917) E
Catoptes nigricans Broun, 1917 E
Catoptes pallidipes Broun, 1917 E
Catoptes planus (Broun, 1881) E
Catoptes postrectus Marshall, 1931 E
Catoptes robustus (Sharp, 1886), non Broun, 1917 E
Catoptes 'robustus Broun, 1917', non (Sharp, 1886) E
Catoptes scutellaris Sharp, 1886, non (Broun, 1893) E
Catoptes 'scutellaris (Broun, 1893)', non Sharp, 1886 E
Catoptes simulator (Sharp, 1886) E
Catoptes spectabilis Broun, 1914 E
Catoptes spermophilus Broun, 1895 E
Catoptes subnitidus Broun, 1914 E
Catoptes subplicatus Broun, 1917 E
Catoptes tenebricus Broun, 1893 E
Catoptes vastator Broun, 1893 E
Catoptes versicolor (Broun, 1911) E
Catoptes vexator Broun, 1904 E
Cecyropa discors Broun, 1881 E
Cecyropa fumosa Broun, 1893 E
Cecyropa litorea Broun, 1921 E
Cecyropa modesta (Fabricius, 1781) E
Cecyropa setigera Broun, 1886 E
Cecyropa striata Broun, 1903 E
Cecyropa sulcifrons Broun, 1917 E
Cecyropa tychioides Pascoe, 1875 E
Echinopeplus dilatatus Broun, 1886 E
Echinopeplus dorsalis Broun, 1914 E
Echinopeplus insolitus (Sharp, 1886) E
Echinopeplus verrucatus Broun, 1914 E
Eurynotia enysi (Broun, 1886) E
Eurynotia hochstetteri (Redtenbacher, 1868) E
Haplolobus aethiops Broun, 1893 E
Haplolobus frontalis Broun, 1914 E
Haplolobus granulatus Broun, 1914 E
Haplolobus gregalis Broun, 1893 E
Haplolobus saevus Broun, 1893 E
Heterexis sculptipennis (Brookes, 1951) E Su
Heterexis seticostatus (Brookes, 1951) E
Inophloeus alacer Broun, 1893 E
Inophloeus albonotatus Broun, 1893 E
Inophloeus aplorhinus Broun, 1915 E
Inophloeus collinus Broun, 1917 E
Inophloeus costifer Broun, 1886 E
Inophloeus cuprellus Broun, 1921 E
Inophloeus discrepans Broun, 1904 E
Inophloeus egregius Broun, 1886 E
Inophloeus festucae Broun, 1921 E
Inophloeus fuscatus Broun, 1917 E
Inophloeus inuus Pascoe, 1875 E
Inophloeus laetificus Broun, 1909 E
Inophloeus longicornis Broun, 1904 E
Inophloeus medius Broun, 1893 E
Inophloeus nigellus Broun, 1881 E
Inophloeus nodifer Broun, 1893 E
Inophloeus obsoletus Broun, 1921 E
Inophloeus pensus Broun, 1914 E
Inophloeus praelatus Broun, 1886 E
Inophloeus punctipennis Sharp, 1886 E
Inophloeus quadricollis Broun, 1909 E
Inophloeus quadrinodosus Brookes, 1934 E
Inophloeus rhesus Pascoe, 1875 E
Inophloeus rubidus Broun, 1881 E
Inophloeus sexnodosus Broun, 1921 E
Inophloeus sternalis Broun, 1904 E
Inophloeus sulcicollis Broun, 1914 E
Inophloeus sulcifer Broun, 1886 E
Inophloeus suturalis Broun, 1893 E
Inophloeus traversii Pascoe, 1875 E
Inophloeus tricostatus Broun, 1915 E
Inophloeus turricolus Marshall, 1926 E
Inophloeus vestitus Broun, 1893 E
Inophloeus villaris Pascoe, 1875 E
Inophloeus vitiosus Pascoe, 1875 E
Irenimus aemulator (Broun, 1893) E
Irenimus aequalis (Broun, 1895) E
Irenimus albosparsus (Broun, 1917) E
Irenimus compressus (Broun, 1880) E
Irenimus curvus Barratt and Kuschel, 1996 E
Irenimus dugdalei Barratt and Kuschel, 1996 E
Irenimus egens (Broun, 1904) E
Irenimus parilis Pascoe, 1876 E
Irenimus patricki Barratt and Kuschel, 1996 E
Irenimus pilosellus Broun, 1886 E
Irenimus posticalis (Broun, 1893) E
Irenimus similis Barratt and Kuschel, 1996 E
Irenimus stolidus Broun, 1886 E
Irenimus tibialis Broun, 1886 E
Leptopius robustus Olivier, 1807 A
Mandalotus albosparsus (Broun, 1914) E
Mandalotus cecyropioides (Broun, 1886) E
Mandalotus hariolus (Broun, 1886) E
Mandalotus irritus (Pascoe, 1877) E
Mandalotus miricollis (Broun, 1917) E
Mandalotus pallidus (Broun, 1893) E
Mandalotus scapalis (Broun, 1921) E
Mandalotus n. sp.1 E
Neoevas celmisiae Broun, 1921 E
Nicaeana catoptoides Broun, 1914 E
Nicaeana cervina Broun, 1893 E
Nicaeana cinerea Broun, 1885 E
Nicaeana concinna Broun, 1886 E
Nicaeana cordipennis Broun, 1921 E
Nicaeana crassifrons Broun, 1917 E
Nicaeana gracilicornis Broun, 1914 E
Nicaeana infuscata Broun, 1909 E
Nicaeana modesta Pascoe, 1877 E
Nicaeana nesophila Broun, 1913 E
Nicaeana placida Broun, 1914 E
Nicaeana tarsalis Broun, 1893 E
Oclandius cinereus Blanchard, 1853 E Su
Oclandius laeviusculus (Broun, 1902) E Su
Oclandius vestitus (Broun, 1909) E Su
Paelocharis clarus (Broun, 1880) E
Paelocharis corpulentus (Broun, 1880) E
Paelocharis inflatus Broun, 1893 E
Paelocharis setiferus (Broun, 1893) E
Paelocharis sternalis (Broun, 1893) E
Paelocharis terricola (Broun, 1913) E
Paelocharis vestita Broun, 1893 E
Protolobus granicollis Broun, 1914 E
Protolobus nodosus Broun, 1917 E
Protolobus obscurus Sharp, 1886 E
Protolobus porculus (Pascoe, 1876) E
Sargon carinatus Broun, 1903 E
Sargon hudsoni Broun, 1909 E
Thesius inophloeoides Broun, 1909 E
Thotmus halli Broun, 1911 E Ch
Zenagraphus albinotatus Broun, 1921 E
Zenagraphus garviensis Marshall, 1938 E
Zenagraphus metallescens Broun, 1915 E
LIXINAE A
RHINOCYLLINI
Rhinocyllus conicus (Frölich, 1792) A BC
MESOPTILIINAE
LAEMOSACCINI
Neolaemosaccus narinus (Pascoe, 1872) A
MOLYTINAE
MOLYTINI
S. LEIOSOMATINA
Lyperobius australis Craw, 1999 E
Lyperobius barbarae Craw, 1999 E
Lyperobius carinatus Broun, 1881 E
Lyperobius clarkei Craw, 1999 E
Lyperobius coxalis Kuschel, 1987 E
Lyperobius cupiendus Broun, 1886 E
Lyperobius eylesi Craw, 1999 E
Lyperobius fallax Broun, 1917 E
Lyperobius glacialis Craw, 1999 E
Lyperobius hudsoni Broun, 1914 E
Lyperobius huttoni Pascoe, 1876 E
Lyperobius montanus Craw, 1999 E
Lyperobius nesidiotes Kuschel, 1987 E Su
Lyperobius patricki Craw, 1999 E
Lyperobius spedenii Broun, 1917 E
Lyperobius townsendii Craw, 1999 E
S. MOLYTINA
Hadramphus spinipennis Broun, 1911 E Ch
Hadramphus stilbocarpae Kuschel, 1971 E
Karocolens pittospori Kuschel, 1987 E
Karocolens tuberculatus (Pascoe, 1877) E
PHRYNIXINI
Abrotheus placitus Broun, 1917 E
Allaorops carinatus Broun, 1917 E
Allostyphlus jugosus Broun, 1921 E
Amphiskirra umbricola Broun, 1909 E
Araeoscapus ardens Broun, 1909 E
Araeoscapus brevicollis Broun, 1910 E
Araeoscapus estriatus Broun, 1909 E
Araeoscapus flavipes (Broun, 1893) E
Araeoscapus ocularius Broun, 1914 E
Araeoscapus ovipennis Broun, 1893 E
Araeoscapus punctipennis Broun, 1910 E
Araeoscapus subcostatus Broun, 1921 E
Astyplus brevicornis Broun, 1921 E
Astyplus conicus Broun, 1893 E
Bradypatae armiger Broun, 1893 E
Bradypatae capitalis (Broun, 1886) E
Bradypatae dilaticollis Broun, 1909 E
Bradypatae impressum Broun, 1921 E
Bradypatae interstitialis Broun, 1909 E
Bradypatae minor Broun, 1913 E
Bradypatae subnodifer Broun, 1921 E
Chamaepsephis aurisetifer Broun, 1893 E
Cuneopterus conicus Sharp, 1886 E
Dermotrichus elegantalis Broun, 1917 E
Dermotrichus multicristatus Broun, 1917 E
Dermotrichus mundulus Sharp, 1886 E
Dermotrichus vicinus Broun, 1921 E
Dolioceuthus dumetosus Broun, 1893 E
Dolioceuthus granulatus (Broun, 1880) E
Dolioceuthus vestitus Broun, 1893 E
Erymneus castaneus Broun, 1880 E
Erymneus celatus (Broun, 1880) E
Erymneus coenosus Broun, 1886 E
Erymneus crassipes Broun, 1893 E
Erymneus ferrugatus Broun, 1893 E
Erymneus firmus Broun, 1893 E
Erymneus irregularis Broun, 1893 E

Erymneus longulus Broun, 1886 E
Erymneus probus Broun, 1893 E
Erymneus scabiosus Broun, 1880 E
Erymneus sharpi Pascoe, 1877 E
Erymneus terrestris Broun, 1921 E
Halliellara antennalis Broun, 1917 E
Halliellara cuneata Broun, 1921 E
Halliellara longicollis Broun, 1917 E
Halliellara squamipes Broun, 1917 E
Lithocia acuminata Broun, 1913 E
Lithocia angustula Broun, 1914 E
Lithocia basalis Broun, 1917 E
Lithocia ciligera Broun, 1917 E
Lithocia fimbriata Broun, 1893 E
Lithocia nigricrista Broun, 1917 E
Lithocia rectisetosa Broun, 1917 E
Lithocia setirostris Broun, 1917 E
Lithocia stictica Broun, 1923 E
Megacolabus bifurcatus May, 1973 E
Megacolabus decipiens Marshall, 1938 E
Megacolabus garviensis May, 1963 E
Megacolabus harrisi (Brookes, 1926) E
Megacolabus obesus May, 1963 E
Megacolabus pteridosus May, 1963 E
Megacolabus reductus Marshall, 1938 E
Megacolabus sculpturatus Broun, 1893 E
Notonesius aucklandicus Kuschel, 1964 E Su
Pachyprypnus longiusculus (Broun, 1880) E
Pachyprypnus modicus Broun, 1904 E
Pachyprypnus pyriformis Broun, 1883 E
Phemus constrictus Broun 1913 E
Phemus curvipes Broun 1913 E
Phemus rufipes Broun, 1893 E
Phemus scabralis Broun, 1893 E
Phrynixodes scruposus Broun, 1921 E
Phrynixus amoenus Broun, 1921 E
Phrynixus asper Broun, 1911 E
Phrynixus astutus Pascoe, 1876 E
Phrynixus bicarinellus Broun, 1909 E
Phrynixus binodosus Broun, 1913 E
Phrynixus blandus Broun, 1921 E
Phrynixus brevipennis Broun, 1893 E
Phrynixus cedius Broun, 1893 E
Phrynixus conspicuus Broun, 1921 E
Phrynixus costirostris Broun, 1893 E
Phrynixus differens Broun, 1886 E
Phrynixus facetus Broun, 1881 E
Phrynixus humeralis Broun, 1893 E
Phrynixus humilis Broun, 1921 E
Phrynixus intricatus Broun, 1886 E
Phrynixus laqueorum Kuschel, 1964 E Su
Phrynixus longulus Broun, 1910 E
Phrynixus modicus Broun, 1880 E
Phrynixus rufipes Broun, 1886 E
Phrynixus rufiventris Broun, 1914 E
Phrynixus setipes Broun, 1913 E
Phrynixus simplex Broun, 1893 E
Phrynixus squamalis Broun, 1921 E
Phrynixus terreus Pascoe, 1875 E
Phrynixus thoracicus (Broun, 1893) E
Phrynixus tuberculatus Broun, 1886 E
Phrynixus ventralis Broun, 1909 E
Phrynixus n. sp. 1 E
Rachidiscodes altipennis Broun, 1917 E
Rachidiscodes glaber Broun, 1921 E
Rachidiscus granicollis Broun, 1893 E
Rachidiscus multinodosus Broun, 1913 E
Reyesiella caecus (Broun, 1893) E
Rystheus fulvosetosus Marshall, 1926 E
Rystheus hudsoni Marshall, 1926 E
Rystheus notabilis Broun, 1917 E
Rystheus ocularius Broun, 1893 E
Styphlotelus fascicularis Broun, 1893 E
Styphlotelus foveatus Broun, 1893 E
Tymbopiptus valeas Kuschel, 1987 E (subfossil)

TRYPETIDINI
Arecophaga varia Broun, 1880 E
Etheophanus pinguis Broun, 1893 E
Exeiratus laqueorum Kuschel, 1984 E Su
Exeiratus setarius Broun, 1914 E
Exeiratus turbotti Brookes, 1951 E Su
Inososgenes acerbus Broun, 1921 E
Inososgenes longiventris Broun, 1917 E
Paedaretus hispidus Pascoe, 1876 E
Paedaretus rufulus Broun, 1880 E
Phronira aspera (Broun, 1880) E
Phronira costosa (Broun, 1880) E
Phronira nodosa Broun, 1893 E
Phronira osculans (Broun, 1880) E
Phronira simplex (Broun, 1880) E
Phronira striata (Broun, 1880) E
Phronira sulcirostris (Broun, 1880) E
Pogonorhinus opacus (Broun, 1880) E
Pogonorhinus punctithorax (Broun, 1886) E
TRIBUS INCERTAE SEDIS
Bantiades cupiendus Broun, 1914 E
Bantiades cylindricus Broun, 1917 E
Bantiades fuscatus Broun, 1893 E
Bantiades morosus Broun, 1917 E
Bantiades nodosus Broun, 1914 E
Bantiades notatus Broun, 1917 E
Bantiades rectalis Broun, 1921 E
Bantiades suturalis Broun, 1914 E
Bantiades trifoveatus Broun, 1921 E
Bantiades valgus Broun, 1893 E
Memes rufirostris Broun, 1903 E
Sosgenes carinatus Broun, 1893 E
Sosgenes discalis Broun, 1917 E
Sosgenes longicollis Broun, 1914 E
Sosgenes planirostris Broun, 1913 E
PLATYPODINAE
PLATYPODINI
Crossotarsus externedentatus Fairmaire, 1850 E K
Platypus apicalis White, 1846 E
Platypus gracilis Broun, 1893 E
Treptoplatypus caviceps (Broun, 1880) E
SCOLYTINAE
HYLESININI
S. HYLASTINA
Hylastes ater (Paykull, 1800) A
S. PHLOEOSININA
Phloeosinus cupressi Hopkins, 1903 A
S. TOMICINA
Chaetoptelius mundulus (Broun, 1881) E
Chaetoptelius versicolor Wood, 1988 A
Dendrotrupes costiceps Broun, 1881 E
Dendrotrupes vestitus Broun, 1881 E
Dendrotrupes zealandicus Wood, 1992 E
Dendrotrupes n. sp. E
Hylurgus ligniperda (Fabricius, 1787) A
Pachycotes peregrinus (Chapuis, 1869) A
SCOLYTINI
S. CRYPHALINA
Cryphalus wapleri Eichhoff, 1871 A
Hypocryphalus asper (Broun, 1881) E
Hypocryphalus longipennis (Browne, 1970) E
Hypocryphalus n. spp. (7) 7E?
Hypothenemus sp. indet. E K
S. DRYOCOETINA
Coccotrypes dactyliperda (Fabricius, 1801) A
S. SCOLYTINA
Scolytus multistriatus (Marsham, 1802) A
S. XYLEBORINA
Amasa truncata (Erichson, 1842) A
Ambrosiodmus compressus (Lea, 1894) A
Cnestus pseudosolidus (Schedl, 1936) A
Microperus eucalypticus (Schedl, 1938) A
Xyleborinus saxesenii (Ratzeburg, 1837) A
Xyleborus inurbanus (Broun, 1880) E
Scolytinae gen. et sp. indet. E

Order MECOPTERA
[Compiled by R. L. C. Pilgrim]
Suborder NANNOMECOPTERA
NANNOCHORISTIDAE
Nannochorista philpotti (Tillyard, 1917) E F

Order SIPHONAPTERA
[Compiled by R. L. C. Pilgrim]
Major hosts are shown *in* brackets after each entry; 'sea birds' includes a wide variety, but especially penguins, procellariiforms, shags, and gulls. Further details – Smit (1979, 1984).

CERATOPHYLLIDAE A
Ceratophyllus (*Ceratophyllus*) *gallinae* (Schrank, 1803) A poultry, passerines
Nosopsyllus (*Nosopsyllus*) *fasciatus* (Bosc, 1800) A rat
Nosopsyllus (*N.*) *londiniensis londiniensis* (Rothschild, 1903) A black rat, mouse
Glaciopsyllus antarcticus Smit & Dunnet, 1962 petrels
ISCHNOPSYLLIDAE
Porribius pacificus Jordan, 1946 E long-tailed bat
LEPTOPSYLLIDAE A
Leptopsylla (*Leptopsylla*) *segnis* (Schönherr, 1811) A mouse
PULICIDAE A
Ctenocephalides canis (Curtis, 1826) A dog
Ctenocephalides felis felis (Bouché, 1835) A cat, dog
Pulex (*Pulex*) *irritans* Linnaeus, 1758 A human, pig
Xenopsylla cheopis (Rothschild, 1903) A human, rats
Xenopsylla vexabilis Jordan, 1925 A kiore (Polynesian rat)
PYGIOPSYLLIDAE
Notiopsylla corynetes Smit, 1979 E Hutton's shearwater
Notiopsylla enciari enciari Smit, 1957 seabirds, parakeets
Notiopsylla e. regula Smit, 1979 E seabirds
Notiopsylla kerguelensis kerguelensis (Taschenberg, 1800) seabirds
Notiopsylla k. tenuata Smit, 1979 E seabirds, parakeets
Notiopsylla peregrinus Smit, 1979 E seabirds
Hoogstraalia imberbis Smit, 1979 E ?landbirds
Pagipsylla gallinalli (Smit, 1965) E passerines
Pygiopsylla hoplia Jordan & Rothschild, 1922 A rats, especially kiore
Pygiopsylla phiola Smit, 1979 A rats, especially kiore
RHOPALOPSYLLIDAE
Parapsyllus longicornis (Enderlein, 1901) seabirds
Parapsyllus magellanicus largificus Smit, 1984 E albatross
Parapsyllus m. magellanicus Jordan, 1938 seabirds
Parapsyllus mangarensis Smit, 1979 E seabirds
Parapsyllus struthophilus Smit, 1979 E parakeets, passerines
Parapsyllus valedictus Smit, 1979 E weka
Parapsyllus cardinis Dunnet, 1961 seabirds
Parapsyllus lynnae alynnae Smit, 1979 E Hutton's shearwater
Parapsyllus l. lynnae Smit, 1965 E seabirds
Parapsyllus l. mariae Smit, 1979 E seabirds
Parapsyllus jacksoni Smit, 1965 E seabirds
Parapsyllus nestoris antichthones Smit, 1979 E parakeets
Parapsyllus nestoris nestoris Smit, 1965 E kea

Order DIPTERA
[Compiled by R. P. Macfarlane and T. K. Crosby]
Suborder NEMATOCERA
ANISOPODIDAE [Toft]
Mycetobia sp. indet. Toft* ?E
Sylvicola festivus (Edwards, 1928) E Fo
Sylvicola neozelandicus (Schiner, 1868) A? Fo-D

Sylvicola notatus (Hutton, 1902) E Fo Ga Li
Sylvicola undulatus (Lamb, 1909) E Fo
BIBIONIDAE
Dilophus alpinus Harrison, 1990 E
Dilophus crinitus (Hardy, 1951) E
Dilophus fumipennis Harrison, 1990 E
Dilophus harrisoni (Hardy, 1953) E
Dilophus neoinsolitus Harrison, 1990 E
Dilophus segnis Hutton, 1902 E
Dilophus nigrostigma (Walker, 1848) E We
Dilophus tuthilli (Hardy, 1953) E
BLEPHARICERIDAE [McLellan]
Neocurupira campbelli Dumbleton, 1963 E Fr
Neocurupira chiltoni (Campbell, 1921) E Fr
Neocurupira hudsoni Lamb, 1913 E Fr
Neocurupira rotalapiscula Craig, 1969 E Fr
Neocurupira tonnoiri Dumbleton, 1963 E Fr
Neocurupira n. spp. (>4)* McLellan 5E
Nothohoraia micrognathia Craig, 1969 I[C] E Fr
Peritheates harrisi (Campbell, 1921) E Sa Fr
Peritheates turrifer Lamb, 1913 E Fr
CANTHYLOSCELIDIDAE
Canthyloscelis antennata Edwards, 1922 E Fo
Canthyloscelis balaena Hutson, 1977 E Fo
Canthyloscelis brevicornis Nagatomi, 1983 E Fo
Canthyloscelis claripennis Edwards, 1922 E Fo
Canthyloscelis nigricosta Edwards, 1922 E Fo
CECIDOMYIIDAE gall midges
[Martin]
Amediella involuta Jaschhof *in* Jaschhof & Jaschhof, 2003 E Ff
Aprionus bullerensis Jaschhof *in* Jaschhof & Jaschhof, 2003, E Ff
Aprionus mycophiloides Jaschhof *in* Jaschhof & Jaschhof, 2003, E Ff
Aprionus remotus Jaschhof *in* Jaschhof & Jaschhof, 2003, E Ff
Arthrocnodax sp. indet. A? Cm
Campylomyza flavipes Meigen, 1818* A Ff
'Cecidomyia' dubiella Gagne, 1989 E
'Cecidomyia' flavella Kieffer, 1913 E
'Cecidomyia' fragilina Gagne, 1989 E
'Cecidomyia' hirta Marshall, 1896 E
'Cecidomyia' marshalli Gagne, 1989 E
'Cecidomyia' melana Marshall, 1896 E
'Cecidomyia' minuscula Gagne, 1989 E
'Cecidomyia' scoparia Marshall, 1896 E
'Cecidomyia' wanganuiensis Marshall, 1896 E
Dasineura? alopecuri (Reuter, 1895) A Gr HS
Dasineura hebefolia Lamb, 1951 E Fo HLG
Dasineura mali (Kieffer, 1904) A Ga-Or HL
Dasineura pyri (Bouche, 1847) A Ga-Or HL
Diadiplosis koebelei (Koebele, 1893) A IPr
Dryomyia shawiae Anderson, 1935 E Fo HSG
Eucalyptodiplosis chionochloae Kolesik *in* Kolesik, Sarfati, Brockerhoff & Kelly, 2007 E Gr HS
Holoneurus aliculatus Yukawa, 1964 E Su Gr Ff
Kiefferia coprosmae Barnes & Lamb, 1954 E Fo H
Lestodiplosis sp. indet.* A? Cr IPr
Lestremia cinerea Macquart, 1826* A? Ff
Lestremia leucophaea (Meigen, 1818) A? Ff
Lestremia novaezealandiae Marshall, 1896 E
Mayetiola destructor (Say, 1817) A Gr HS
Miastor agricolae Marshall, 1896 E Ff
Miastor difficilis Marshall, 1896 E Ff
Monardia dividua Jaschhof *in* Jaschhof & Jaschhof, 2003, E Ff
Monardia fumea Jaschhof *in* Jaschhof & Jaschhof, 2003, E Ff
Monardia furcillata Jaschhof *in* Jaschhof & Jaschhof, 2003, 2003 E Ff
Monardia modica Jaschhof *in* Jaschhof & Jaschhof, 2003, E Ff
Monardia stirpium Kieffer, 1895* A Ff
Monardia sp. E Ff Fo
Mycophila fungicola Felt, 1911* A? Ff
Oligotrophus coprosmae Barnes & Lamb, 1954 E Fo
Oligotrophus oleariae (Maskell, 1889) E Fo
Peromyia carinata Jaschhof, 2001 E Ff Fo
Peromyia clandestina Jaschhof *in* Jaschhof & Jaschhof, 2004 E Ff Fo
Peromyia culta Jaschhof *in* Jaschhof & Jaschhof, 2004 E Ff Fo
Peromyia derupta Jaschhof *in* Jaschhof & Jaschhof, 2004 E Ff Fo
Peromyia didhami Jaschhof *in* Jaschhof & Jaschhof, 2004 E Ff Fo
Peromyia dissona Jaschhof *in* Jaschhof & Jaschhof, 2004 E Ff Fo
Peromyia doci Jaschhof *in* Jaschhof & Jaschhof, 2004 E Ff Fo
Peromyia insueta Jaschhof *in* Jaschhof & Jaschhof, 2004 E Ff Fo
Peromyia intecta Jaschhof *in* Jaschhof & Jaschhof, 2004 E Ff Fo
Peromyia intonsa Jaschhof *in* Jaschhof & Jaschhof, 2004 E Ff Fo
Peromyia katieae Jaschhof *in* Jaschhof & Jaschhof, 2004 E Ff Fo
Peromyia latebrosa Jaschhof *in* Jaschhof & Jaschhof, 2004 E Ff Fo
Peromyia memoranda Jaschhof *in* Jaschhof & Jaschhof, 2004 E Ff Fo
Peromyia mountalbertiensis Jaschhof *in* Jaschhof & Jaschhof, 2004 E Ff Fo
Peromyia multifurcata Jaschhof *in* Jaschhof & Jaschhof, 2004 E Ff Fo
Peromyia muscorum (Kieffer, 1895) A
Peromyia novaezealandiae Jaschhof *in* Jaschhof & Jaschhof, 2004 E Ff Fo
Peromyia obunca Jaschhof *in* Jaschhof & Jaschhof, 2004 E Ff Fo
Peromyia palustris (Kieffer, 1895) A
Peromyia perardua Jaschhof *in* Jaschhof & Jaschhof, 2004 E Ff Fo
Peromyia pertrita Jaschhof *in* Jaschhof & Jaschhof, 2004 E Ff Fo
Peromyia plena Jaschhof *in* Jaschhof & Jaschhof, 2004 E Ff Fo
Peromyia praeclara Jaschhof *in* Jaschhof & Jaschhof, 2004 E Ff Fo
Peromyia rara Jaschhof *in* Jaschhof & Jaschhof, 2004 E Ff Fo
Peromyia rotoitiensis Jaschhof *in* Jaschhof & Jaschhof, 2004 E Ff Fo
Peromyia sera Jaschhof *in* Jaschhof & Jaschhof, 2004 E Ff Fo
Peromyia serrata Jaschhof *in* Jaschhof & Jaschhof, 2004 E Ff Fo
Peromyia setosa Jaschhof *in* Jaschhof & Jaschhof, 2004 E Ff Fo
Peromyia sinuosa Jaschhof *in* Jaschhof & Jaschhof, 2004 E Ff Fo
Peromyia spinigera Jaschhof *in* Jaschhof & Jaschhof, 2004 E Ff Fo
Peromyia squamigera Jaschhof *in* Jaschhof & Jaschhof, 2004 E Ff Fo
Peromyia tecta Jaschhof *in* Jaschhof & Jaschhof, 2004 E Ff Fo
Peromyia tumida Jaschhof *in* Jaschhof & Jaschhof, 2004 E Ff Fo
Polyardis illustris Jaschhof *in* Jaschhof & Jaschhof, 2003 E Ff
Polyardis triangula Jaschhof *in* Jaschhof & Jaschhof, 2003 E Ff
Porricondyla agricolae (Marshall, 1896) E Ff
Porricondyla aurea (Marshall, 1896) E Ff
Porricondyla magna (Marshall, 1896) E Ff
Porricondyla ordinaria (Marshall, 1896) E Ff
Proterodiplosis radicis Wyatt, 1963 E FoHR
Pseudomonardia australis Jaschhof *in* Jaschhof & Jaschhof, 2003 E Ff
Pseudomonardia communis Jaschhof *in* Jaschhof & Jaschhof, 2003 E Ff
Pseudomonardia elongata Jaschhof *in* Jaschhof & Jaschhof, 2003 E Ff
Pseudomonardia glacialis Jaschhof *in* Jaschhof & Jaschhof, 2003 E Ff
Pseudomonardia hutchesoni Jaschhof *in* Jaschhof & Jaschhof, 2003 E Ff
Pseudomonardia invisitata Jaschhof *in* Jaschhof & Jaschhof, 2003 E Ff
Pseudomonardia neurolygoides Jaschhof *in* Jaschhof & Jaschhof, 2003 E Ff
Pseudomonardia pallida Jaschhof *in* Jaschhof & Jaschhof, 2003 E Ff
Pseudomonardia parva Jaschhof *in* Jaschhof & Jaschhof, 2003 E Ff
Pseudomonardia parvalobata Jaschhof *in* Jaschhof & Jaschhof, 2003 E Ff
Pseudomonardia vicina Jaschhof *in* Jaschhof & Jaschhof, 2003 E Ff
Pteridomyia bilobata Jaschhof *in* Jaschhof & Jaschhof, 2003 E Ff
Pteridomyia gressitti[NC] (Yukawa, 1964) E Su Ff
Rhopalomyia chrysanthemi Ahlberg, 1939 A Ga HG
Stenodiplosis geniculati (Reuter, 1895) A Gr HS
Stephodiplosis nothofagi Barnes, 1936 E Fo H
Trisopsis sp. indet. Crosby 1986 E or A IPr
Zeuxidiplosis giardi (Kieffer, 1896) A BC Gr HLG
Gen. indet. et n. sp. E Gr HS
Gen. indet. (gall makers) et n. spp. (130) 130E Fo HG
CERATOPOGONIDAE
Atrichopogon fitzroyi Macfie, 1932 E
Atrichopogon greyi Macfie, 1932 E
Atrichopogon hobsoni Macfie, 1932 E
Atrichopogon shortlandi Macfie, 1932 E
Atrichopogon 'vestitipennis' Kieffer, 1917 ?E Sa
Austrohelea antipodalis (Ingram & Macfie, 1931) E
Austrohelea campbellensis (Tokunaga, 1964) E Su
Austrohelea ferruginea (Macfie, 1932) E
Austrohelea tonnoiri (Macfie, 1932) E
Dasyhelea aucklandensis Sublette & Wirth, 1980 E Su ?F
Dasyhelea jucunda Macfie, 1932 E ?F
Dasyhelea oribates Macfie, 1932 E ?F Sa
Dasyhelea sp. indet. Andrew A? Fs
Dasyhelea sp. indet. Macfie 1932 E?
Forcipomyia antipodum (Hudson, 1892) E
Forcipomyia austrina Macfie, 1932 E F
Forcipomyia belkini de Meillon & Wirth, 1979 E
Forcipomyia cooki Macfie, 1932 E
Forcipomyia desurvillei Macfie, 1932 E Sa
Forcipomyia kuscheli Sublette & Wirth, 1980 E Su F
Forcipomyia parvicellula Ingram & Macfie, 1931 E Sa
Forcipomyia tapleyi Ingram & Macfie, 1931 E
Forcipomyia tasmani Macfie, 1932 E
Leptoconops myersi (Tonnoir, 1924) E Be F
Monohelea clavipes Macfie, 1932 E Sa
Monohelea nubeculosa Macfie, 1932 E
Monohelea n. spp. (2) Green, Holder, Macfarlane 2E?Gr/We
?*Neurohelea* sp. indet. Andrew
Palpomyia cantuaris Ingram & Macfie, 1931 E ?F
Palpomyia nelsoni Macfie, 1932 E ?F
Palpomyia rastellifera Macfie, 1932 E ?F
Palpomyia urpicifemoris Macfie, 1932 E ?F
Paradasyhelea egregia (Macfie, 1932) E
Paradasyhelea harrisoni Wirth, 1981 E
Stilobezzia antipodalis Ingram & Macfie, 1931 E ?WeF
Stilobezzia badia Macfie, 1932 E ?WeF
Stilobezzia ohakunei Ingram & Macfie, 1931 E ? WeF
Stilobezzia tonnoiri Macfie, 1932 E ?WeF

CHIRONOMIDAE [Boothroyd, Forsyth]
Ablabesmyia mala (Hutton, 1902) E Fr&Fs
'Anatopynia' boninensis Tokunaga, 1964 E F
'Anatopynia' elongata Tokunaga, 1964 E F
'Anatopynia' pennipes Freeman, 1961 E F
Anzacladius kiwi Cranston, 2009 E F
?*Apsectrotanypus cana* (Freeman, 1959) E F
?*Apsectrotanypus quadricincta* (Freeman, 1959) E F
Austrocladius harrisi (Freeman, 1959) E F
Austrocladius n. sp. E F
Camptocladius stercorarius (De Geer, 1776) A So D
Chironomus antipodensis Sublette & Wirth, 1980 E Su F
Chironomus analis Freeman, 1959 E F
Chironomus forsythi Martin, 1999 E F
Chironomus subantarcticus Sublette & Wirth, 1980 E Su F
Chironomus zealandicus Hudson, 1892 E Fs
Chironomus n. sp. Winterbourn & Gregson, 1989 E F/SW
Chironomus n. spp. (3) 3E F Forsyth
Chironomus n. sp. E Ch F Forsyth
Cladopelma curtivalva (Kieffer, 1917) Fs
?*Clunio* sp. Tortell 1981 E SW
Corynocera n. sp. Boothroyd E F
?*Corynoneura scutellata* Winnertz, 1846 A Gr Fs
Cricotopus aucklandicus Sublette & Wirth, 1980 E Fr
Cricotopus cingulatus (Hutton, 1902) E Fr
Cricotopus hollyfordensis Boothroyd, 2002 E Fr
Cricotopus planus Boothroyd, 1990 E Fr
Cricotopus vincenti Boothroyd, 1990 E Fr
Cricotopus zealandicus Freeman, 1959 E Fr
Cryptochironomus n. sp. Winterbourn & Gregson 1989 E Fr,Fs
Eukiefferiella brundini Boothroyd & Cranston, 1995 E Fr
?*Eukiefferiella commensalis* (Tonnoir, 1923) E F
Eukiefferiella heveli Sublette & Wirth, 1980 E Su Fr
Eukiefferiella n. spp. (3) 3E F
<u>*Gressittius*</u> *antarcticus* (Hudson, 1892) E F
Gressittius umbrosus (Freeman, 1959) E F
Gymnometriocnemus lobifer (Freeman, 1959) E F
<u>*Gynnidocladius*</u> *pilulus* Sublette & Wirth, 1980 E Su F
Harrisius pallidus Freeman, 1959 E Fo Fr
<u>*Hevelius*</u> *carinatus* Sublette & Wirth, 1980 E Su F
<u>*Kaniwhaniwhanus*</u> *chapmani* Boothroyd 1999 E Fr
Kiefferulus opalensis Forsyth, 1975 E Fs
<u>*Kuschelius*</u> *dentifer* Sublette & Wirth, 1980 E Su F
Larsia n. spp. (2) Boothroyd 2E F
Limnophyes vestitus (Skuse, 1889) A Fr
<u>*Lobodiamesa*</u> *campbelli* Pagast, 1947 E Sa Fr
Lobodiamesa n. sp. Boothroyd E Sa F
Macropelopia apicincta (Freeman, 1959) E F
Macropelopia apicinella (Freeman, 1959) E F
Macropelopia flavipes (Freeman, 1959) E F
Macropelopia quinquepunctata (Freeman, 1959) E F
<u>*Maoridiamesa*</u> *glacialis* Brundin, 1966 E Sa Fr
Maoridiamesa harrisi Pagast, 1947 E Sa Fr
Maoridiamesa insularis Brundin, 1966 E Su Fr
Maoridiamesa intermedia Brundin, 1966 E Sa Fr
Maoridiamesa stouti Brundin, 1966 E Sa Fr
<u>*Maryella*</u> *reducta* Sublette & Wirth, 1980 E Su F
<u>*Mecaorus*</u> *elongatus* Sublette & Wirth, 1980 E Su F
?*Microtendipes* n. sp. Winterbourn & Gregson 1989 E Fs
<u>*Nakataia*</u> *cisdentifer* Sublette & Wirth, 1980 E Su F
<u>*Naonella*</u> *forsythi* Boothroyd, 1994 E F
Naonella kimihia Boothroyd, 2005 E F
<u>*Nesiocladius*</u> *gressitti* Sublette & Wirth, 1980 E Su F
<u>*Ophryophorus*</u> *ramiferus* Freeman, 1959 E F
Orthocladius pictipennis Freeman, 1959 E F
'Orthocladius' publicus Hutton, 1902 E F
Parachironomus cylindricus (Freeman, 1959) E Fs
Parakiefferiella? sp. indet. Macfarlane
Paratanytarsus grimmii (Schneider, 1885) A Fr,Fs
Paratrichocladius pluriserialis (Freeman, 1959) Fr,Fs
Parochlus aotearoae Brundin, 1966 E Sa Fr
Parochlus brevis Sublette & Wirth, 1980 E Su F
Parochlus carinatus Brundin, 1966 E Sa Fr
Parochlus conjugens Brundin, 1966 E Sa Fr
Parochlus glacialis Brundin, 1966 E Sa Fr
Parochlus gressitti Sublette & Wirth, 1980 E Su F
Parochlus longicornis Brundin, 1966 E Sa Fr
Parochlus maorii Brundin, 1966 E Sa Fr
Parochlus novaezelandiae Brundin, 1966 E Sa Fr
Parochlus ohakunensis (Freeman, 1959) E Sa Fr
Parochlus pauperatus Brundin, 1966 E Sa Fr
Parochlus reductus Sublette & Wirth, 1980 E Su F
Parochlus rennelli Sublette & Wirth, 1980 E Su F
Parochlus spinosus Brundin, 1966 E Sa Fr
<u>*Paucispinigera*</u> *approximata* Freeman, 1959 E Fo Fr
Paucispinigera n. sp. Stark E F
<u>*Pirara*</u> *matakiri* Boothroyd & Cranston, 1995 E Fr
Podochlus cockaynei Brundin, 1966 E Sa Fr
Podochlus grandis Brundin, 1966 E Sa Fr
Podochlus knoxi Brundin, 1966 E Sa Fr
Podochlus stouti Brundin, 1966 E Sa Fr
Podonomus parochloides Brundin, 1966 E Sa Fr
Podonomus pygmaeus Brundin, 1966 E Sa Fr
Podonomus waikukupae Brundin, 1966 E Sa Fr
Podonomus n. spp. (4) Brundin 1966 4E Sa Fr
Polypedilum alternans Forsyth, 1971 E F
Polypedilum canum Freeman, 1959 E Fr
Polypedilum cumberi Freeman, 1959 E F
Polypedilum digitulum Freeman, 1959 E F
Polypedilum harrisi Freeman, 1959 E Fr
Polypedilum lentum (Hutton, 1902) E F
Polypedilum longicrus Kieffer, 1921 A F
Polypedilum luteum Forsyth, 1971 E Fr
Polypedilum opimum (Hutton, 1902) E Fr
Polypedilum pavidum (Hutton, 1902) E Fs
Polypedilum n. sp. Boothroyd E F
<u>*Pterosis*</u> *wisei* Sublette & Wirth, 1980 E Su F
Riethia zeylandica Freeman, 1959 F
Semiocladius kuscheli Sublette & Wirth, 1980 E Su
Semiocladius reinga (Leader, 1975) E Be SW ?MoP
Semiocladius whangaroa (Leader, 1975) E SW ?MoP
Semiocladius n. sp. Boothroyd E
Smittia verna (Hutton, 1902) E Gr ?S or Li
Stempellina n. sp. Boothroyd E F
Stictocladius lacuniferus (Freeman, 1959) E F
Stictocladius pictus (Freeman, 1959) E F
'Tanypus' debilis (Hutton, 1902) E F
'Tanypus' languidus Hutton, 1902 E F
Tanytarsus albanyensis Forsyth, 1971 E F
Tanytarsus funebris Freeman, 1959 E Fr,Fs
Tanytarsus vespertinus Hutton, 1902 E Fr,Fs
Tanytarsus n. sp. Winterbourn et al. 2000 E Hs F
Tanytarsus n. sp. Boothroyd E F
Telmatogeton antipodensis Sublette & Wirth, 1980 E Su Be SW
Telmatogeton mortoni Leader, 1975 E Be ?F
Thienimanniella n. sp. Boothroyd E
<u>*Tonnoirocladius*</u> *commensalis* (Tonnoir, 1923) E F
Xenochironomus canterburyensis Freeman, 1959 E Fs MoP
Zavrelimyia harrisi (Freeman, 1959) E Fr&Fs
<u>*Zelandochlus*</u> *latipalpis* Brundin, 1966 E Sa Fr Gl
<u>Gen. nov. (8)</u> et n. spp. (9) Boothroyd 9E F
CORETHRELLIDAE
Corethrella novaezealandiae Tonnoir, 1927 E F IPr ?BF
CULICIDAE [Holder]
Aedes antipodeus (Edwards, 1920) Fs
Aedes australis (Erichson, 1842) A SW
Aedes chathamicus Dumbleton, 1962 E SW
Aedes notoscriptus (Skuse, 1889) A Fo Fs BF
Aedes subalbirostris Klein & Marks, 1960 E Fs
Coquillettidia iracunda (Walker, 1848) E F LP
Coquillettidia tenuipalpis (Edwards, 1924) E F
Culex asteliae Belkin, 1968 E Fs
Culex pervigilans Bergroth, 1889 E FoGr Fs BF
Culex quinquefasciatus Say, 1823 A Fs
Culex rotoruae Belkin, 1968 E F
Culiseta novaezealandiae Pillai, 1966 E Fs
Culiseta tonnoiri (Edwards, 1925) E Fo Fs BF
<u>*Maorigoeldia*</u> *argyropa* (Walker, 1848) E Fs
<u>*Opifex*</u> *fuscus* Hutton, 1902 E RsSW BF
DITOMYIIDAE [Toft]
Australosymmerus basalis (Tonnoir *in* Tonnoir & Edwards, 1927) E Fo
Australosymmerus fumipennis (Tonnoir *in* Tonnoir & Edwards, 1927) E Fo
Australosymmerus nitidus (Tonnoir *in* Tonnoir & Edwards, 1927) E Fo
Australosymmerus tillyardi (Tonnoir *in* Tonnoir & Edwards, 1927) E Fo
Australosymmerus trivittatus Edwards *in* Tonnoir & Edwards, 1927 E Fo
Nervijuncta bicolor Edwards *in* Tonnoir & Edwards, 1927 E Fo
Nervijuncta flavoscutellata Tonnoir *in* Tonnoir & Edwards, 1927 E Fo
Nervijuncta harrisi Edwards *in* Tonnoir & Edwards, 1927 E Fo
Nervijuncta hexachaeta Edwards *in* Tonnoir & Edwards, 1927 E Fo
Nervijuncta hudsoni (Marshall, 1896) E Fo
Nervijuncta longicauda Edwards *in* Tonnoir & Edwards, 1927 E Fo
Nervijuncta marshalli Edwards *in* Tonnoir & Edwards, 1927 E Fo
Nervijuncta nigrescens Marshall, 1896 E Fo
Nervijuncta nigricornis Tonnoir *in* Tonnoir & Edwards, 1927 E Fo
Nervijuncta nigricoxa Edwards *in* Tonnoir & Edwards, 1927 E Fo
Nervijuncta ostensackeni Tonnoir *in* Tonnoir & Edwards, 1927 E Fo
Nervijuncta parvicauda Edwards *in* Tonnoir & Edwards, 1927 E Fo
Nervijuncta pilicornis Edwards *in* Tonnoir & Edwards, 1927 E Fo
Nervijuncta pulchella Edwards *in* Tonnoir & Edwards, 1927 E Fo
Nervijuncta punctata Tonnoir *in* Tonnoir & Edwards, 1927 E Fo
Nervijuncta ruficeps Edwards *in* Tonnoir & Edwards, 1927 E Fo
Nervijuncta tridens (Hutton, 1881) E Fo
Nervijuncta wakefieldi (Edwards, 1921) E Fo
Nervijuncta n. spp. (2) Toft 2E
DIXIDAE
Dixella fuscinervis (Tonnoir, 1924) E Sa Fs
Dixella harrisi (Tonnoir, 1925) E Fo Fs
Dixella neozelandica (Tonnoir, 1924) E Fs
Dixella tonnoiri (Belkin, 1968) E Fo Fs
<u>*Neodixa*</u> *minuta* (Tonnoir, 1924) E F
Nothodixa campbelli (Alexander, 1922) E Fs
Nothodixa otagensis (Alexander, 1922) E F
Nothodixa philpotti (Tonnoir, 1924) E F
Nothodixa septentrionalis (Tonnoir, 1924) E F
KEROPLATIDAE mostly fungus feeders Ff [Toft]
Arachnocampa luminosa (Skuse, 1891) E Cv-Fo
Cerotelion bimaculatum Tonnoir *in* Tonnoir & Edwards, 1927 E
Cerotelion dendyi (Marshall, 1896) E
Cerotelion hudsoni (Marshall, 1896) E
Cerotelion leucoceras (Marshall, 1896) E
Cerotelion nigrum Tonnoir *in* Tonnoir & Edwards, 1927 E
Cerotelion tapleyi Edwards *in* Tonnoir & Edwards, 1927 E
Cerotelion vitripenne Tonnoir *in* Tonnoir & Edwards,

1927 E
Chiasmoneura fenestrata (Edwards *in* Tonnoir & Edwards, 1927) E Fo
Chiasmoneura milligani (Tonnoir *in* Tonnoir & Edwards, 1927) E Fo
Isoneuromyia harrisi (Tonnoir *in* Tonnoir & Edwards, 1927) E Fo
Isoneuromyia novaezelandiae (Tonnoir *in* Tonnoir & Edwards, 1927) E Fo
Macrocera annulata Tonnoir *in* Tonnoir & Edwards, 1927 E Fo
Macrocera antennatis Marshall, 1896 E Fo
Macrocera campbelli Edwards *in* Tonnoir & Edwards, 1927 E Fo
Macrocera fusca Tonnoir *in* Tonnoir & Edwards, 1927 E Fo
Macrocera glabrata Tonnoir *in* Tonnoir & Edwards, 1927 E Fo
Macrocera gourlayi Tonnoir *in* Tonnoir & Edwards, 1927 E Fo
Macrocera howletti Marshall, 1896 E Fo
Macrocera hudsoni Tonnoir *in* Tonnoir & Edwards, 1927 E Fo
Macrocera ngaireae Edwards *in* Tonnoir & Edwards, 1927 E Fo
Macrocera obsoleta Edwards *in* Tonnoir & Edwards, 1927 E Fo
Macrocera pulchra Tonnoir *in* Tonnoir & Edwards, 1927 E Fo
Macrocera ruficollis Edwards *in* Tonnoir & Edwards, 1927 E Fo
Macrocera scoparia Marshall, 1896 E Fo
Macrocera tonnoiri Matile, 1989 E Fo
Macrocera unipunctata Tonnoir *in* Tonnoir & Edwards, 1927 E Fo
Neoplatyura brookesi (Edwards *in* Tonnoir & Edwards, 1927) E
Neoplatyura lamellata (Tonnoir *in* Tonnoir & Edwards, 1927) E
Neoplatyura marshalli (Tonnoir *in* Tonnoir & Edwards, 1927) E
Neoplatyura proxima (Tonnoir *in* Tonnoir & Edwards, 1927) E
Orfelia nemoralis (Meigen, 1818) A Gr
Paramacrocera brevicornis Edwards *in* Tonnoir & Edwards, 1927 E
'Platyura' albovittata (Tonnoir *in* Tonnoir & Edwards, 1927) E
Pseudoplatyura truncata Tonnoir *in* Tonnoir & Edwards, 1927 E
Pyrtaula agricolae (Marshall, 1896) E
Pyrtaula campbelli (Tonnoir *in* Tonnoir & Edwards, 1927) E
Pyrtaula carbonaria (Tonnoir *in* Tonnoir & Edwards, 1927) E
Pyrtaula chiltoni (Tonnoir *in* Tonnoir & Edwards, 1927) E
Pyrtaula curtisi (Edwards *in* Tonnoir & Edwards, 1927) E
Pyrtaula maculipennis (Tonnoir *in* Tonnoir & Edwards, 1927) E
Pyrtaula ohakunensis (Edwards *in* Tonnoir & Edwards, 1927) E Fo
Pyrtaula philpotti (Tonnoir *in* Tonnoir & Edwards, 1927) E
Pyrtaula punctifusa (Edwards *in* Tonnoir & Edwards, 1927) E
Pyrtaula ruficauda (Tonnoir *in* Tonnoir & Edwards, 1927) E
Pyrtaula rufipectus (Tonnoir *in* Tonnoir & Edwards, 1927) E Fo
Pyrtaula rutila (Edwards *in* Tonnoir & Edwards, 1927) E
Pyrtaula n. sp. Toft E
Rypatula brevis (Tonnoir *in* Tonnoir & Edwards, 1927) E
Rypatula subbrevis (Tonnoir *in* Tonnoir & Edwards, 1927) E Fo

MYCETOPHILIDAE ~ fungus feeders Ff
[Toft, Didham, Matile, Jaschhof]
Allocotocera anaclinoides (Marshall, 1896) E
Allocotocera cephasi Edwards *in* Tonnoir & Edwards, 1927 E
Allocotocera crassipalpis Tonnoir *in* Tonnoir & Edwards, 1927 E
Allocotocera dilatata Tonnoir *in* Tonnoir & Edwards, 1927 E
Aneura appendiculata Tonnoir *in* Tonnoir & Edwards, 1927 E
Aneura bispinosa Edwards *in* Tonnoir & Edwards, 1927 E
Aneura boletinoides Marshall, 1896 E
Aneura defecta Edwards *in* Tonnoir & Edwards, 1927 E
Aneura fagi (Marshall, 1896) E
Aneura filiformis Tonnoir *in* Tonnoir & Edwards, 1927 E
Aneura fusca Tonnoir *in* Tonnoir & Edwards, 1927 E
Aneura jaschhofi Zaitzev, 2001 E
Aneura longicauda Tonnoir *in* Tonnoir & Edwards, 1927 E
Aneura longipalpis Tonnoir *in* Tonnoir & Edwards, 1927 E
Aneura nitida Tonnoir *in* Tonnoir & Edwards, 1927 E
Aneura pallida Edwards *in* Tonnoir & Edwards, 1927 E
Aneura tonnoiri Zaitzev, 2001 E
Anomalomyia affinis Tonnoir *in* Tonnoir & Edwards, 1927 E
Anomalomyia basalis Tonnoir *in* Tonnoir & Edwards, 1927 E
Anomalomyia flavicauda Edwards *in* Tonnoir & Edwards, 1927 E
Anomalomyia guttata (Hutton, 1901) E Fo-Gr
Anomalomyia immaculata Edwards *in* Tonnoir & Edwards, 1927 E
Anomalomyia minor (Marshall, 1896) E
Anomalomyia obscura Tonnoir *in* Tonnoir & Edwards, 1927 E
Anomalomyia subobscura Tonnoir *in* Tonnoir & Edwards, 1927 E
Anomalomyia thompsoni Tonnoir *in* Tonnoir & Edwards, 1927 E
Anomalomyia viatoris Edwards *in* Tonnoir & Edwards, 1927 E
Austrosynapha apicalis (Tonnoir *in* Tonnoir & Edwards, 1927) E
Austrosynapha cawthroni (Tonnoir *in* Tonnoir & Edwards, 1927) E
Austrosynapha claripennis (Tonnoir *in* Tonnoir & Edwards, 1927) E
Austrosynapha gracilis (Tonnoir *in* Tonnoir & Edwards, 1927) E
Austrosynapha parva (Edwards *in* Tonnoir & Edwards, 1927) E
Austrosynapha pulchella (Tonnoir *in* Tonnoir & Edwards, 1927) E
Austrosynapha similis (Tonnoir *in* Tonnoir & Edwards, 1927) E
Brevicornu antennata (Harrison, 1964) E Su
Brevicornu brunnea (Harrison, 1964) E Su
Brevicornu flavum Marshall, 1896 E
Brevicornu fragile Marshall, 1896 E
Brevicornu maculatum (Tonnoir *in* Tonnoir & Edwards, 1927) E
Brevicornu marshalli Zaitzev, 2002 E
Brevicornu matilei Zaitzev, 2002 E
Brevicornu quadriseta (Edwards *in* Tonnoir & Edwards, 1927) E
Brevicornu rufithorax (Tonnoir *in* Tonnoir & Edwards, 1927) E
Brevicornu subrufithorax Zaitzev, 2002 E
Brevicornu tongariro Zaitzev, 2002 E
<u>*Cawthronia*</u> *nigra* Tonnoir *in* Tonnoir & Edwards, 1927 E
<u>*Cowanomyia*</u> *hillaryi* Jaschhof & Jaschhof, 2009 E
<u>*Cycloneura*</u> *flava* Marshall, 1896 E Fo
Cycloneura triangulata Tonnoir *in* Tonnoir & Edwards, 1927 E Fo
Cycloneura n. spp. (6) 6E
Exechia biseta Edwards *in* Tonnoir & Edwards, 1927 E
Exechia filata Edwards *in* Tonnoir & Edwards, 1927 E
Exechia hiemalis (Marshall, 1896) E
Exechia howesi Edwards *in* Tonnoir & Edwards, 1927 E
Exechia novaezelandiae Tonnoir *in* Tonnoir & Edwards, 1927 E
Exechia thomsoni Miller, 1918 E
Exechia n. sp. Matile E
Leia arsona Hutson, 1978 A
Manota birgitae Jaschhof & Jaschhof, 2010 E Fo
Manota granvillensis Jaschhof & Jaschhof, 2010 E Fo
Manota maorica Edwards *in* Tonnoir & Edwards, 1927 E
Manota purakaunui Jaschhof & Jaschhof, 2010 E Fo
Manota regineae Jaschhof & Jaschhof, 2010 E Fo
<u>*Morganiella*</u> *fusca* Tonnoir *in* Tonnoir & Edwards, 1927 E
Morganiella n. sp. E
Mycetophila campbellensis Harrison, 1964 E Su
Mycetophila clara Tonnoir *in* Tonnoir & Edwards, 1927 E Fo
Mycetophila colorata Tonnoir *in* Tonnoir & Edwards, 1927 E Fo
Mycetophila conica Tonnoir *in* Tonnoir & Edwards, 1927 E Fo
Mycetophila consobrina Tonnoir *in* Tonnoir & Edwards, 1927 E Fo
Mycetophila crassitarsis Edwards *in* Tonnoir & Edwards, 1927 E Fo
Mycetophila curtisi Edwards *in* Tonnoir & Edwards, 1927 E Fo
Mycetophila diffusa Tonnoir *in* Tonnoir & Edwards, 1927 E Fo
Mycetophila dilatata Tonnoir *in* Tonnoir & Edwards, 1927 E Fo
Mycetophila elegans Tonnoir *in* Tonnoir & Edwards, 1927 E Fo
Mycetophila elongata Tonnoir *in* Tonnoir & Edwards, 1927 E Fo
Mycetophila fagi Marshall, 1896 E Fo
Mycetophila filicornis Tonnoir *in* Tonnoir & Edwards, 1927 E Fo
Mycetophila fumosa Tonnoir *in* Tonnoir & Edwards, 1927 E Fo
Mycetophila furtiva Tonnoir *in* Tonnoir & Edwards, 1927 E Fo
Mycetophila grandis Tonnoir *in* Tonnoir & Edwards, 1927 E Fo
Mycetophila griseofusca griseofusca Tonnoir *in* Tonnoir & Edwards, 1927 E Fo
Mycetophila g. nigriclava Edwards *in* Tonnoir & Edwards, 1927 E Fo
Mycetophila grisescens Edwards *in* Tonnoir & Edwards, 1927 E Fo
Mycetophila harrisi Edwards *in* Tonnoir & Edwards, 1927 E Fo
Mycetophila howletti Marshall, 1896 E
Mycetophila impunctata Edwards *in* Tonnoir & Edwards, 1927 E Fo
Mycetophila integra Tonnoir *in* Tonnoir & Edwards, 1927 E Fo
Mycetophila intermedia Edwards *in* Tonnoir & Edwards, 1927 E Fo

Mycetophila latifascia Edwards *in* Tonnoir & Edwards, 1927 E Fo
Mycetophila lomondensis Edwards *in* Tonnoir & Edwards, 1927 E Fo
Mycetophila luteolateralis Edwards *in* Tonnoir & Edwards, 1927 E Fo
Mycetophila marginepunctata marginepunctata Tonnoir *in* Tonnoir & Edwards, 1927 E Fo
Mycetophila m. ruapehensis Tonnoir *in* Tonnoir & Edwards, 1927 E Fo
Mycetophila m. rotundipennis Tonnoir *in* Tonnoir & Edwards, 1927 E Fo
Mycetophila marshalli Enderlein, 1910 E Fo
Mycetophila media Tonnoir *in* Tonnoir & Edwards, 1927 E Fo
Mycetophila minima Edwards *in* Tonnoir & Edwards, 1927 E Fo
Mycetophila nigricans Tonnoir *in* Tonnoir & Edwards, 1927 E Fo
Mycetophila nigripalpis Edwards *in* Tonnoir & Edwards, 1927 E Fo
Mycetophila nitens Tonnoir *in* Tonnoir & Edwards, 1927 E Fo
Mycetophila nitidula Edwards *in* Tonnoir & Edwards, 1927 E Fo
Mycetophila ornatissima Tonnoir *in* Tonnoir & Edwards, 1927 E Fo
Mycetophila phyllura Edwards *in* Tonnoir & Edwards, 1927 E Fo
Mycetophila pollicata Edwards *in* Tonnoir & Edwards, 1927 E
Mycetophila pseudomarshalli Tonnoir *in* Tonnoir & Edwards, 1927 E Fo
Mycetophila solitaria Tonnoir *in* Tonnoir & Edwards, 1927 E Fo
Mycetophila spinigera Tonnoir *in* Tonnoir & Edwards, 1927 E Fo
Mycetophila submarshalli Tonnoir *in* Tonnoir & Edwards, 1927 E Fo
Mycetophila subnitens Edwards *in* Tonnoir & Edwards, 1927 E Fo
Mycetophila subspinigera Tonnoir *in* Tonnoir & Edwards, 1927 E Fo
Mycetophila subtenebrosa Tonnoir *in* Tonnoir & Edwards, 1927 E Fo
Mycetophila subtillis Tonnoir *in* Tonnoir & Edwards, 1927 E Fo
Mycetophila sylvatica Marshall, 1896 E Fo
Mycetophila tapleyi Edwards *in* Tonnoir & Edwards, 1927 E Fo
Mycetophila tenebrosa Edwards *in* Tonnoir & Edwards, 1927 E Fo
Mycetophila tonnoiri Matile, 1989 E Fo
Mycetophila trispinosa Tonnoir *in* Tonnoir & Edwards, 1927 E Fo
Mycetophila unispinosa Tonnoir *in* Tonnoir & Edwards, 1927 E Fo
Mycetophila virgata Tonnoir *in* Tonnoir & Edwards, 1927 E Fo
Mycetophila viridis Edwards *in* Tonnoir & Edwards, 1927 E Fo
Mycetophila vulgaris Tonnoir *in* Tonnoir & Edwards, 1927 E Fo
Mycetophila n. spp. (4) Matile, Toft 4E
Mycomya flavilatera Tonnoir *in* Tonnoir & Edwards, 1927 E Fo
Mycomya furcata Edwards *in* Tonnoir & Edwards, 1927 E Fo
Mycomya plagiata Tonnoir *in* Tonnoir & Edwards, 1927 E Fo
Mycomya n. sp. Matile E
Neoaphelomera elongata Tonnoir *in* Tonnoir & Edwards, 1927 E
Neoaphelomera forcipata Edwards *in* Tonnoir & Edwards, 1927 E
Neoaphelomera longicauda Edwards *in* Tonnoir & Edwards, 1927 E
Neoaphelomera majuscula Edwards *in* Tonnoir & Edwards, 1927 E
Neoaphelomera marshalli Edwards *in* Tonnoir & Edwards, 1927 E
Neoaphelomera opaca Tonnoir *in* Tonnoir & Edwards, 1927 E
Neoaphelomera skusei Marshall, 1896 E Sn Su
<u>*Neotrizygia*</u> *obscura* Tonnoir *in* Tonnoir & Edwards, 1927 E
Neotrizygia n. spp. (6) 6E
<u>*Paracycloneura*</u> *apicalis* Tonnoir *in* Tonnoir & Edwards, 1927 E Fo
Paracycloneura inopinata Jaschhof & Kallweit, 2009 E Fo
Paracycloneura sp. Jaschhof & Kallweit 2009
<u>*Paradoxa*</u> *fusca* Marshall, 1896 E
Parvicellula apicalis Tonnoir *in* Tonnoir & Edwards, 1927 E
Parvicellula fuscipennis Edwards *in* Tonnoir & Edwards, 1927 E
Parvicellula gracilis Tonnoir *in* Tonnoir & Edwards, 1927 E
Parvicellula hamata Edwards *in* Tonnoir & Edwards, 1927 E
Parvicellula nigricoxa Tonnoir *in* Tonnoir & Edwards, 1927 E
Parvicellula obscura Tonnoir *in* Tonnoir & Edwards, 1927 E
Parvicellula ruficoxa Tonnoir *in* Tonnoir & Edwards, 1927 E
Parvicellula subhamata Tonnoir *in* Tonnoir & Edwards, 1927 E
Parvicellula triangula Marshall, 1896 E
Phthinia longiventris Tonnoir *in* Tonnoir & Edwards, 1927 E
Phthinia n. sp. Toft E
Platurocypta dilatata Tonnoir 1927 E Fo
Platurocypta immaculata Tonnoir *in* Tonnoir & Edwards, 1927 E Fo
Sciophila parviareolata Santos Abreu, 1920 A
Sciophila ocreata Phillipi, 1865 A
<u>*Sigmoleia*</u> *melanoxantha* Edwards *in* Tonnoir & Edwards, 1927 E
Sigmoleia peterjohnsi Jaschhof & Kallweit, 2009 E
Sigmoleia separata Jaschhof & Kallweit, 2009 E
Sigmoleia similis Jaschhof & Kallweit, 2009 E
<u>*Taxicnemis*</u> *flava* Edwards *in* Tonnoir & Edwards, 1927 E
Taxicnemis marshalli Matile, 1989 E
Taxicnemis n. spp. (2) 2E
Tetragoneura exigua Matile, 1989 E
Tetragoneura flexa Edwards *in* Tonnoir & Edwards, 1927 E
Tetragoneura fusca Tonnoir *in* Tonnoir & Edwards, 1927 E
Tetragoneura minima Tonnoir *in* Tonnoir & Edwards, 1927 E
Tetragoneura nigra Marshall, 1896 E
Tetragoneura obliqua Edwards *in* Tonnoir & Edwards, 1927 E Fo
Tetragoneura obscura Tonnoir *in* Tonnoir & Edwards, 1927 E
Tetragoneura opaca Tonnoir *in* Tonnoir & Edwards, 1927 E
Tetragoneura proxima Tonnoir *in* Tonnoir & Edwards, 1927 E
Tetragoneura rufipes Tonnoir *in* Tonnoir & Edwards, 1927 E
Tetragoneura spinipes Edwards *in* Tonnoir & Edwards, 1927 E Fo
Tetragoneura tonnoiri Matile, 1989 E
Tetragoneura ultima Tonnoir *in* Tonnoir & Edwards, 1927 E Fo
Tetragoneura venusta Tonnoir *in* Tonnoir & Edwards, 1927 E
Tetragoneura n. spp. (20) 20E
<u>*Tonnwardsia*</u> *aberrans* (Tonnoir *in* Tonnoir & Edwards, 1927) E Fo
Tonnwardsia n. spp. (6) 6E
<u>*Trichoterga*</u> *monticola monticola* Tonnoir *in* Tonnoir & Edwards, 1927 E Fo
Trichoterga m. incisurata Edwards *in* Tonnoir & Edwards, 1927 E Fo
<u>*Waipapamyia*</u> *dentata* Jaschhof & Kallweit, 2009 E
Waipapamyia elongata Jaschhof & Kallweit, 2009 E
Waipapamyia truncata Jaschhof & Kallweit, 2009 E
Zygomyia acuta Tonnoir *in* Tonnoir & Edwards, 1927 E Fo
Zygomyia albinotata Tonnoir *in* Tonnoir & Edwards, 1927 E Fo
Zygomyia apicalis Tonnoir *in* Tonnoir & Edwards, 1927 E Fo
Zygomyia bifasciola Matile, 1989 E Fo
Zygomyia bivittata Tonnoir *in* Tonnoir & Edwards, 1927 E Fo
Zygomyia brunnea Tonnoir *in* Tonnoir & Edwards, 1927 E Fo
Zygomyia costata Tonnoir *in* Tonnoir & Edwards, 1927 E Fo
Zygomyia crassicauda Tonnoir *in* Tonnoir & Edwards, 1927 E Fo
Zygomyia crassipyga Tonnoir *in* Tonnoir & Edwards, 1927 E Fo
Zygomyia diffusa Edwards *in* Tonnoir & Edwards, 1927 E
Zygomyia distincta Tonnoir *in* Tonnoir & Edwards, 1927 E Fo
Zygomyia egmontensis Zaitzev, 2002 E Fo
Zygomyia eluta Edwards *in* Tonnoir & Edwards, 1927 E Fo
Zygomyia filigera Edwards *in* Tonnoir & Edwards, 1927 E Fo
Zygomyia flavicoxa Marshall, 1896 E Fo
Zygomyia fusca Marshall, 1896 E Fo
Zygomyia grisescens Tonnoir *in* Tonnoir & Edwards, 1927 E Fo
Zygomyia guttata Tonnoir *in* Tonnoir & Edwards, 1927 E Fo
Zygomyia humeralis Tonnoir *in* Tonnoir & Edwards, 1927 E Fo
Zygomyia immaculata Tonnoir *in* Tonnoir & Edwards, 1927 E Fo
Zygomyia longicauda Tonnoir *in* Tonnoir & Edwards, 1927 E Fo
Zygomyia marginata Tonnoir *in* Tonnoir & Edwards, 1927 E Fo
Zygomyia multiseta Zaitzev, 2002 E Fo
Zygomyia nigrita Tonnoir *in* Tonnoir & Edwards, 1927 E Fo
Zygomyia nigriventris Tonnoir *in* Tonnoir & Edwards, 1927 E Fo
Zygomyia nigrohalterata Tonnoir *in* Tonnoir & Edwards, 1927 E Fo
Zygomyia obsoleta Tonnoir *in* Tonnoir & Edwards, 1927 E Fo
Zygomyia ovata Zaitzev, 2002 E Fo
Zygomyia penicillata Edwards *in* Tonnoir & Edwards, 1927 E Fo
Zygomyia ruficollis Tonnoir *in* Tonnoir & Edwards, 1927 E Fo
Zygomyia rufithorax Tonnoir *in* Tonnoir & Edwards, 1927 E Fo
Zygomyia similis Tonnoir *in* Tonnoir & Edwards, 1927 E Fo
Zygomyia submarginata Harrison, 1955 E Su Fo
Zygomyia submarginata Zaitzev, 2002 E [preoccupied name]
Zygomyia taranakiensis Zaitzev, 2002 E Fo

Zygomyia trifasciata Tonnoir in Tonnoir & Edwards, 1927 E Fo
Zygomyia trispinosa Zaitzev, 2002 E Fo
Zygomyia truncata Tonnoir in Tonnoir & Edwards, 1927 E Fo
Zygomyia unispinosa Tonnoir in Tonnoir & Edwards, 1927 E Fo
Zygomyia varipes Edwards in Tonnoir & Edwards, 1927 E Fo
Gen. nov. 1 et n. spp. (3) 3E
Gen. nov. 2 et n. spp. (2) 2E
Gen. nov. 3 et n. sp. E
Gen. nov. 4 et n. sp. E
MYCETOPHILIFORMIA INCERTAE SEDIS
Starkomyia inexpecta Jaschhof, 2004 E Fo
PSYCHODIDAE
Ancyroaspis funebris (Hutton, 1902) E We
Ancyroaspis multimaculata (Satchell, 1950) E
Brunettia novaezelandiae Satchell, 1950 E Fo
Didicrum clarkei (Satchell, 1950) E
Didicrum claviatum (Satchell, 1950) E
Didicrum drepanatum (Satchell, 1950) E
Didicrum maurum (Satchell, 1950) E
Didicrum solitarium (Satchell, 1954) E
Didicrum triuncinatum (Satchell, 1950) E
Logima surcoufi (Tonnoir, 1922) A Ga Gr Li D
Nemapalpus zelandiae Alexander, 1921 E
Psychoda acutipennis Tonnoir, 1920 E Su
Psychoda alternata Say, 1824 A Se
Psychoda parthenogenetica Tonnoir, 1940 A
Psychoda pseudoalternata Williams, 1946 A
Psychoda solivaga Duckhouse, 1971 E Su
Psychodocha brachyptera (Quate, 1964) E Su
Psychodocha campbellica (Quate, 1964) E Su
Psychodocha cinerea Banks, 1894 A
Psychodocha eremita (Quate, 1964) E Su
Psychodocha formosa (Satchell, 1954) E
Psychodocha inaequalis (Satchell, 1950) E
Psychodocha lloydi (Satchell, 1950) E
Psychodocha novaezealandica (Satchell, 1950) E
Psychodocha penicillata (Satchell, 1950) A
Psychodocha pulchrima (Satchell, 1954) E
Psychodocha setistyla (Satchell, 1950) E
Psychodocha simplex (Satchell, 1954) E
Psychodocha squamulata (Satchell, 1950) E
Psychodocha triaciculata (Satchell, 1950) E
Psychodocha tridens (Satchell, 1954) E
Psychodocha zonata (Satchell, 1950) E Li
Psychodula harrisi (Satchell, 1950) A D
Satchellomyia barbata (Satchell, 1954) E
Satchellomyia bifalcata (Satchell, 1950) E
Satchellomyia bilobata (Satchell, 1954) E
Satchellomyia diffusa (Satchell, 1954) E
Satchellomyia gourlayi (Satchell, 1950) E
Satchellomyia lobisterna (Satchell, 1950) E
Satchellomyia serratipenis (Satchell, 1950) E
Satchellomyia spiralifera (Satchell, 1950) E
Sycorax cryptella Satchell, 1950 E
Sycorax dispar Satchell, 1950 E
Sycorax impatiens Satchell, 1950 E
Sycorax milleri Satchell, 1950 E
Threticus philpotti (Satchell, 1950) E
Threticus tortuosus Duckhouse, 1971 E Su
Trichomyia fusca Satchell, 1950 E Fo
Trichomyia capsulata Duckhouse, 1980 E Fo
Trichomyia n. sp. E Fo
Gen. indet. et n. spp. (12–29) ?12–29 E Satchell 1950
RANGOMARAMIDAE
Anisotricha novaezelandiae (Tonnoir in Tonnoir & Edwards, 1927) E Fo
Anisotricha similis Jaschhof, 2004 E Fo
Insulatricha catrinae Jaschhof, 2004 E Fo
Insulatricha chandleri Jaschhof, 2004 E Fo
Insulatricha hippai Jaschhof, 2004 E Fo
Ohakunea australiensis Colless, 1963*
Ohakunea bicolor Edwards in Tonnoir & Edwards, 1927 E
Rangomarama edwardsi Jaschhof & Didham, 2002 E Fo
Rangomarama humboldti Jaschhof & Didham, 2002 E Fo
Rangomarama leopoldinae Jaschhof & Didham, 2002 E Fo
Rangomarama matilei Jaschhof & Didham, 2002 E Fo
Rangomarama tonnoiri Jaschhof & Didham, 2002 E Fo
SCATOPSIDAE [Andrew, Macfarlane]
Anapausis stapedifortmis Freeman, 1989 E Fo
Anapausis zealandica Freeman, 1989 E Fo
Coboldia fuscipes (Meigen, 1830) A Ga Li
Colobostemus n. spp. (3) Andrew & Macfarlane 3E Fo
Rhegmoclemina n. spp. (2) 2E Andrew
Scatopse notata (Linnaeus, 1758) A Ga Li
SCIARIDAE
Bradysia ?amoena (Winnertz, 1867) A
Bradysia ?brunnipes (Meigen, 1804)* A Gr Fo Hr
Bradysia campbellensis Steffan, 1964 E Gr
Bradysia difformis Frey, 1948 A GrGa HR
Bradysia rubra (Harrison, 1955) E Su Gr
Bradysia sp. indet.* GrFo (MAFNZ)
Corynoptera ancylospina Mohrig in Mohrig & Jaschhof, 1999 E
Corynoptera basisetosa Mohrig in Mohrig & Jaschhof, 1999 E
Corynoptera coronospina Mohrig in Mohrig & Jaschhof, 1999 E
Corynoptera cowanorum Mohrig in Mohrig & Jaschhof, 1999 E
Corynoptera densisetosa Mohrig in Mohrig & Jaschhof, 1999 E
Corynoptera densospica Mohrig in Mohrig & Jaschhof, 1999 E
Corynoptera didymistyla Mohrig in Mohrig & Jaschhof, 1999 E
Corynoptera dividospica Mohrig in Mohrig & Jaschhof, 1999 E
Corynoptera expressospina Mohrig in Mohrig & Jaschhof, 1999 E
Corynoptera facticia Mohrig in Mohrig & Jaschhof, 1999 E
Corynoptera filisetosa Mohrig in Mohrig & Jaschhof, 1999 E
Corynoptera filispica Mohrig in Mohrig & Jaschhof, 1999 E
Corynoptera fuscispica Mohrig in Mohrig & Jaschhof, 1999 E
Corynoptera harrisi (Edwards in Tonnoir & Edwards, 1927) E
Corynoptera hemisetosa Mohrig in Mohrig & Jaschhof, 1999 E
Corynoptera microsetosa Mohrig in Mohrig & Jaschhof, 1999 E
Corynoptera nigrospina Mohrig in Mohrig & Jaschhof, 1999 E
Corynoptera nigrotegminis Mohrig in Mohrig & Jaschhof, 1999 E
Corynoptera oririclausa Mohrig in Mohrig & Jaschhof, 1999 E
Corynoptera parasetosa Mohrig in Mohrig & Jaschhof, 1999 E
Corynoptera? philpotti (Tonnoir in Tonnoir & Edwards, 1927) E
Corynoptera pentaspina Mohrig in Mohrig & Jaschhof, 1999 E
Corynoptera plasiosetosa Mohrig in Mohrig & Jaschhof, 1999 E
Corynoptera prinospina Mohrig in Mohrig & Jaschhof, 1999 E
Corynoptera priscospina Mohrig in Mohrig & Jaschhof, 1999 E
Corynoptera pronospica Mohrig in Mohrig & Jaschhof, 1999 E
Corynoptera propriospina Mohrig in Mohrig & Jaschhof, 1999 E
Corynoptera prosospina Mohrig in Mohrig & Jaschhof, 1999 E
Corynoptera psilospina Mohrig in Mohrig & Jaschhof, 1999 E
Corynoptera quasisetosa Mohrig in Mohrig & Jaschhof, 1999 E
Corynoptera semiaggregata Mohrig in Mohrig & Jaschhof, 1999 E
Corynoptera subantarctica Steffan, 1964 E Gr
Corynoptera tapleyi (Edwards in Tonnoir & Edwards, 1927) E Li
Corynoptera variospina Mohrig in Mohrig & Jaschhof, 1999 E
Cratyna zealandica Mohrig in Mohrig & Jaschhof, 1999 E
Ctenosciara constrictans (Edwards in Tonnoir & Edwards, 1927) E
Ctenosciara griseinervis (Edwards in Tonnoir & Edwards, 1927) E
Ctenosciara hyalipennis? (Meigen, 1804) A
Ctenosciara nigrostyla Mohrig in Mohrig & Jaschhof, 1999 E
Ctenosciara nudopterix Mohrig in Mohrig & Jaschhof, 1999 E
Ctenosciara ovalis (Edwards in Tonnoir & Edwards, 1927) E
Ctenosciara rufulenta (Edwards in Tonnoir & Edwards, 1927) E Li
Ctenosciara xanthonota (Edwards in Tonnoir & Edwards, 1927) E
Epidapus chaetovenosus Mohrig in Mohrig & Jaschhof, 1999 E
Epidapus ctenosciaroides Mohrig in Mohrig & Jaschhof, 1999 E
Epidapus espinosalus Mohrig in Mohrig & Jaschhof, 1999 E
Epidapus parvus Mohrig in Mohrig & Jaschhof, 1999 E
Epidapus n. sp. E
Lycoriella castanescens (Lengersdorf, 1940)* A Gr-Fo Ff Ca
Lycoriella ?ingenua (Dufour, 1839)* A Gr HR
Neophnyxia nelsonia Tonnoir in Tonnoir & Edwards, 1927 ?E
Pseudolycoriella bispina Mohrig in Mohrig & Jaschhof, 1999 E
Pseudolycoriella breviseta Mohrig in Mohrig & Jaschhof, 1999 E
Pseudolycoriella jejuna (Edwards in Tonnoir & Edwards, 1927) E
Pseudolycoriella macrotegmenta Mohrig in Mohrig & Jaschhof, 1999 E
Pseudolycoriella zealandica (Edwards in Tonnoir & Edwards, 1927) E
Scatopsciara unicalcarata (Edwards in Tonnoir & Edwards, 1927) E
'Sciara' marcilla (Hutton, 1902) A?
'Sciara' neorufescens Miller, 1950 E
'Sciara' zealandica Kieffer, 1910 E
Scythropochroa nitida Edwards in Tonnoir & Edwards, 1927 E Fo
Xylosciara brevipes Steffan, 1964 E Su
Zygoneura contractans (Edwards in Tonnoir & Edwards, 1927) E
SIMULIIDAE
Austrosimulium albovelatum Dumbleton, 1973 E Fr
Austrosimulium australense (Schiner, 1868) E Fr
Austrosimulium bicorne Dumbleton, 1973 E Fr
Austrosimulium campbellense Dumbleton, 1973 E Su Fr

Austrosimulium dumbletoni Crosby, 1976 E Fr
Austrosimulium laticorne Tonnoir, 1925 E Fr
Austrosimulium laticorne alveolatum Dumbleton, 1973 E Fr
Austrosimulium longicorne Tonnoir, 1925 E Fr
Austrosimulium multicorne Tonnoir, 1925 E Fr
Austrosimulium multicorne stewartense Dumbleton, 1973 E Fr
Austrosimulium stewartense Dumbleton, 1973 E Fr
Austrosimulium tillyardianum Dumbleton, 1973 E Fr
Austrosimulium ungulatum Tonnoir, 1925 E Fr
Austrosimulium unicorne Dumbleton, 1973 E Fr
Austrosimulium vexans (Mik, 1881) E Su Fr
TANYDERIDAE [Judd]
Mischoderus annuliferus (Hutton, 1901) E F
Mischoderus forcipatus (Osten Sacken, 1880) E F
Mischoderus marginatus (Edwards, 1923) I[C] E F
Mischoderus neptunus (Edwards, 1923) I[C] E F
Mischoderus varipes (Edwards, 1923) E F
Mischoderus n. spp. (2) 2E Judd
THAUMALEIDAE
Austrothaumalea appendiculata Tonnoir, 1927 E F
Austrothaumalea crosbyi McLellan, 1988 E F
Austrothaumalea gibbsi McLellan, 1988 E F
Austrothaumalea macfarlanei McLellan, 1988 E F
Austrothaumalea maxwelli McLellan, 1988 E F
Austrothaumalea neozelandica Tonnoir, 1927 E F
Austrothaumalea ngaire McLellan, 1988 E F
Austrothaumalea pala McLellan, 1988 E F
Austrothaumalea walkerae McLellan, 1988 E F
Austrothaumalea zwicki McLellan, 1988 E F
Oterere oliveri McLellan, 1988 E F
TIPULIDAE
Acantholimnophila bispina (Alexander, 1922) E
Acantholimnophila maorica (Alexander, 1922) E
Amphineurus bicinctus Edwards, 1923 E
Amphineurus bicorniger Alexander, 1924 E
Amphineurus blackballensis Alexander, 1953 E
Amphineurus breviclavus Alexander, 1924 E
Amphineurus cacoxenus Alexander, 1925 E
Amphineurus campbelli Alexander, 1922 E
Amphineurus cyathetanus Alexander, 1952 E
Amphineurus edentulus Alexander, 1939 E
Amphineurus fatuus (Hutton, 1902) E
Amphineurus fimbriatulus Alexander, 1925 E
Amphineurus flexuosus Alexander, 1923 E
Amphineurus gracilisentis Alexander, 1922 E
Amphineurus harrisi Alexander, 1922 E
Amphineurus hastatus Alexander, 1925 E
Amphineurus horni Edwards, 1923 E
Amphineurus hudsoni Edwards, 1923 E
Amphineurus insulsus (Hutton, 1902) E
Amphineurus kingi Alexander, 1950 E
Amphineurus longi Alexander, 1950 E
Amphineurus lyriformis Alexander, 1923 E
Amphineurus meridionalis Alexander, 1924 E
Amphineurus minor Alexander, 1923 E
Amphineurus molophilinus Alexander, 1922 E
Amphineurus niveinervis Edwards, 1923 E
Amphineurus nothofagi Alexander, 1925 E
Amphineurus ochroplacus Alexander, 1925 E
Amphineurus operculatus Alexander, 1924 E
Amphineurus otagensis Alexander, 1922 E
Amphineurus patruelis Alexander, 1925 E
Amphineurus perarmatus Alexander, 1924 E
Amphineurus perdecorus Edwards, 1923 E
Amphineurus pressus Alexander, 1922 E
Amphineurus pulchripes Alexander, 1925 E
Amphineurus recurvans Alexander, 1922 E
Amphineurus senex Alexander, 1922 E
Amphineurus spinulistylus Alexander, 1925 E
Amphineurus stewartiae Alexander, 1924 E
Amphineurus subdecorus Edwards, 1924 E
Amphineurus subfatuus Alexander, 1922 E
Amphineurus subglaber Edwards, 1923 E
Amphineurus submolophilinus Alexander, 1923 E
Amphineurus tenuipollex Alexander, 1952 E
Amphineurus tortuosus Alexander, 1923 E
Amphineurus tumidus Alexander, 1923 E
Amphineurus n. sp. E
Aphrophila flavopygialis (Alexander, 1922) E
Aphrophila luteipes Alexander, 1926 E
Aphrophila monacantha Alexander, 1926 E
Aphrophila neozelandica (Edwards, 1923) E F
Aphrophila tridentata Alexander, 1926 E
Aphrophila trifida Alexander, 1926 E
Aphrophila triton (Alexander, 1922) E
Aphrophila vittipennis Alexander, 1925 E
Aphrophila n. spp. (3) 3E
Atarba connexa (Alexander, 1923) E
Atarba eluta (Edwards, 1923) E
Atarba filicornis Alexander, 1922 E
Atarba viridicolor Alexander, 1922 E
Aurotipula aperta (Edwards, 1923) E
Aurotipula atroflava (Alexander, 1922) E
Aurotipula auroatra (Edwards, 1923) E
Aurotipula bivittata (Edwards, 1923) E
Aurotipula brevitarsis (Edwards, 1923) E
Aurotipula clara (Kirby, 1884) E
Aurotipula dux (Kirby, 1884) E
Aurotipula ferruginosa (Edwards, 1923) E
Aurotipula flavidipennis (Alexander, 1923) E
Aurotipula flavoscapa (Alexander, 1922) E
Aurotipula occlusa (Edwards, 1924) E
Aurotipula orion (Hudson, 1895) E
Aurotipula ruapehuensis (Alexander, 1923) E
Aurotipula n. sp. E
Austrolimnophila agathicola Alexander, 1952 E
Austrolimnophila argus (Hutton, 1900) E
Austrolimnophila atripes (Alexander, 1922) E
Austrolimnophila chrysorrhoea (Edwards, 1923) E
Austrolimnophila crassipes (Hutton, 1900) E
Austrolimnophila cyatheti (Edwards, 1923) E
Austrolimnophila geographica (Hutton, 1900) E
Austrolimnophila lambi (Edwards, 1923) E
Austrolimnophila leucomelas (Edwards, 1923) E
Austrolimnophila marshalli (Hutton, 1900) E
Austrolimnophila nigrocincta (Edwards, 1923) E
Austrolimnophila obliquata (Alexander, 1922) E
Austrolimnophila oculata (Edwards, 1923) E
Austrolimnophila oriunda (Alexander, 1952) E
Austrolimnophila proximata (Alexander, 1926) E
Austrolimnophila stemma (Alexander, 1922) E
Austrolimnophila stewartiae (Alexander, 1924) E
Austrolimnophila strigimacula (Edwards, 1923) E
Austrolimnophila subinterventa (Edwards, 1923) E
Austrolimnophila wilfredlongi Alexander, 1952 E
Austrolimnophila n. spp. (3) 3E
Austrotipula hudsoni Hutton, 1900 E
Brevicera aenigmatica (Alexander, 1926) E
Brevicera heterogama (Hudson, 1913) E
Brevicera mesocera (Alexander, 1922) E
Brevicera waitakerensis (Alexander, 1952) E
Brevicera n. sp. E
Cerozodia hemiptera Alexander, 1922 E
Cerozodia hudsoni Edwards, 1923 E
Cerozodia laticosta (Alexander, 1930) E
Cerozodia paradisea Edwards, 1923 E
Cerozodia plumosa Osten Sacken, 1888 E
Cerozodia pulverulenta Edwards, 1923 E
Cerozodia striata Edwards, 1923 E
Cheilotrichia hamiltoni (Alexander, 1939) E
Chlorotipula albistigma (Edwards, 1923) E
Chlorotipula elongata (Edwards, 1923) E
Chlorotipula holochlora (Nowicki, 1875) E
Chlorotipula virescens (Edwards, 1923) E
Chlorotipula viridis (Walker, 1856) E
Ctenolimnophila alpina (Alexander, 1922) E
Ctenolimnophila brevitarsis (Alexander, 1926) E
Ctenolimnophila fulvipleura (Alexander, 1923) E
Ctenolimnophila fumipennis (Alexander, 1923) E
Ctenolimnophila harrisiana (Alexander, 1924) E
Ctenolimnophila pallipes (Alexander, 1926) E
Ctenolimnophila venustipennis (Alexander, 1925) E
Dicranota n. sp.* E
Discobola ampla (Hutton, 1900) E
Discobola chathamica Alexander, 1924 E
Discobola dicycla Edwards, 1923 E
Discobola dohrni (Osten-Sacken, 1894) E
Discobola gibberina (Alexander, 1948) E
Discobola haetara Johns & Jenner, 2006 E
Discobola striata chathamica Alexander, 1924 E
Discobola s. striata Edwards, 1923 E
Discobola tessellata (Osten-Sacken, 1894) E
Discobola venustula Alexander, 1929 E
Dolichopeza atropos (Hudson, 1895) E
Dolichopeza fenwicki Alexander, 1923 E
Dolichopeza howesi Alexander, 1922 E
Dolichopeza parvicauda Edwards, 1923 E
Elephantomyia ruapehuensis Alexander, 1923 E
Elephantomyia zealandica Edwards, 1923 E
Geranomyia n. sp.* E
Gonomyia banksiana Alexander, 1924 E
Gonomyia bispina Alexander, 1924 E
Gonomyia circumcincta Alexander, 1924 E
Gonomyia longispina Alexander, 1922 E
Gonomyia ludibunda Alexander, 1926 E
Gonomyia nigrohalterata Edwards, 1923 E
Gonomyia oliveri Alexander, 1924 E
Gonomyia tenuistyla Alexander, 1926 E
Gonomyia n. spp. (3) 3E
Gynoplistia aculeata Alexander, 1924 E
Gynoplistia albicincta Edwards, 1923 E
Gynoplistia ambulator Alexander, 1924 E
Gynoplistia angustipennis Edwards, 1924 E
Gynoplistia anthracina Alexander, 1920 E
Gynoplistia arthuriana Edwards, 1923 E
Gynoplistia auriantopyga Alexander, 1922 E
Gynoplistia bicornis Alexander, 1924 E
Gynoplistia bidentata Alexander, 1922 E
Gynoplistia bilobata Alexander, 1923 E
Gynoplistia bituberculata Alexander, 1923 E
Gynoplistia bona Alexander, 1920 E
Gynoplistia bucera Alexander, 1923 E
Gynoplistia campbelli Alexander, 1922 E
Gynoplistia canterburiana Edwards, 1923 E
Gynoplistia chathamica Alexander, 1924 E
Gynoplistia cladophora Alexander, 1922 E
Gynoplistia clarkeana Alexander, 1951 E
Gynoplistia clavipes Edwards, 1923 E
Gynoplistia concava Alexander, 1922 E
Gynoplistia conjuncta Edwards, 1923 E
Gynoplistia cuprea Hutton, 1900 E
Gynoplistia dactylophora Alexander, 1926 E
Gynoplistia digitifera Alexander, 1953 E
Gynoplistia dilatata Alexander, 1924 E
Gynoplistia dimidiata Alexander, 1922 E
Gynoplistia dispila Alexander, 1923 E
Gynoplistia dispiloides Alexander, 1926 E
Gynoplistia eluta Alexander, 1923 E
Gynoplistia fimbriata Alexander, 1920 E
Gynoplistia flavohalterata Alexander, 1926 E
Gynoplistia formosa Hutton, 1900 E
Gynoplistia fulgens Hutton, 1900 E
Gynoplistia fuscoplumbea Edwards, 1923 E
Gynoplistia generosa Alexander, 1926 E
Gynoplistia glauca Edwards, 1923 E
Gynoplistia hamiltoni Alexander, 1924 E
Gynoplistia harrisi Alexander, 1922 E
Gynoplistia heighwayi Alexander, 1930 E
Gynoplistia hiemalis (Alexander, 1923) E
Gynoplistia hirsuticauda Alexander, 1923 E
Gynoplistia hirtamera Alexander, 1922 E
Gynoplistia hyalinata Alexander, 1923 E
Gynoplistia incisa Edwards, 1923 E
Gynoplistia inconjuncta Alexander, 1926 E
Gynoplistia inflata Alexander, 1926 E
Gynoplistia lobulifera Alexander, 1923 E

Gynoplistia luteibasis Alexander, 1922 E
Gynoplistia luteicincta Alexander, 1924 E
Gynoplistia lyrifera Alexander, 1922 E
Gynoplistia magnifica Edwards, 1923 E
Gynoplistia moanae Alexander, 1951 E
Gynoplistia myersae Alexander, 1924 E
Gynoplistia nebulipennis Alexander, 1922 E
Gynoplistia nebulosa Edwards, 1923 E
Gynoplistia nematomera Alexander, 1926 E
Gynoplistia neonebulosa Alexander, 1923 E
Gynoplistia nigrobimbo Alexander, 1923 E
Gynoplistia nigronitida Edwards, 1923 E
Gynoplistia niveicincta Alexander, 1922 E
Gynoplistia notabilis Alexander, 1926 E
Gynoplistia notata Edwards, 1923 E
Gynoplistia ocellifera Alexander, 1923 E
Gynoplistia orophila Alexander, 1923 E
Gynoplistia otagana Alexander, 1930 E
Gynoplistia pallidistigma Alexander, 1923 E
Gynoplistia pedestris Edwards, 1923 E
Gynoplistia percara Alexander, 1926 E
Gynoplistia persimilis Alexander, 1926 E
Gynoplistia philpotti Alexander, 1939 E
Gynoplistia pleuralis Alexander, 1923 E
Gynoplistia plutonis Alexander, 1926 E
Gynoplistia polita Edwards, 1923 E
Gynoplistia princeps Alexander, 1923 E
Gynoplistia purpurea Alexander, 1922 E
Gynoplistia pygmaea Alexander, 1923 E
Gynoplistia recurvata Alexander, 1923 E
Gynoplistia resecta Edwards, 1924 E
Gynoplistia romae Alexander, 1930 E
Gynoplistia sackeni Alexander, 1920 E
Gynoplistia serrulata Alexander, 1926 E
Gynoplistia speciosa Edwards, 1923 E
Gynoplistia speighti Edwards, 1923 E
Gynoplistia spinicalcar Alexander, 1922 E
Gynoplistia spinigera Alexander, 1922 E
Gynoplistia splendens Alexander, 1922 E
Gynoplistia subclavipes Alexander, 1924 E
Gynoplistia subfasciata Walker, 1848 E
Gynoplistia subformosa Alexander, 1924 E
Gynoplistia subobsoleta Alexander, 1923 E
Gynoplistia tridactyla Edwards, 1923 E
Gynoplistia trifasciata Edwards, 1923 E
Gynoplistia trispinosa Alexander, 1922 E
Gynoplistia troglophila Alexander, 1962 E
Gynoplistia tuberculata Edwards, 1923 E
Gynoplistia unimaculata Alexander, 1922 E
Gynoplistia vexator Alexander, 1952 E
Gynoplistia violacea Edwards, 1923 E
Gynoplistia vittinervis Alexander, 1924 E
Gynoplistia waitakerensis Alexander, 1952 E
Gynoplistia wakefieldi Westwood, 1881 E
Gynoplistia n. sp. E
Harrisomyia bicuspidata Alexander, 1923 E
Harrisomyia terebrella Alexander, 1932 E
Helius harrisi (Alexander, 1923) E
Heterolimnophila subtruncata Alexander, 1923 E
Heterolimnophila truncata Alexander, 1922 E
Idioglochina fumipennis (Butler, 1875) E
Idioglochina kronei (Mik, 1881) E Su
Leptotarsus albiplagia (Alexander, 1923) E
Leptotarsus alexanderi (Edwards, 1923) E
Leptotarsus amissionis (Alexander, 1952) E
Leptotarsus angusticosta (Alexander, 1923) E
Leptotarsus atridorsum (Alexander, 1922) E
Leptotarsus binotatus (Hutton, 1900) E
Leptotarsus campbelli (Alexander, 1923)
Leptotarsus cinereus (Edwards, 1923) E
Leptotarsus cubitalis (Edwards, 1923) E
Leptotarsus decoratus (Edwards, 1923) E
Leptotarsus dichroithorax (Alexander, 1920) E
Leptotarsus fumibasis (Edwards, 1923) E
Leptotarsus fucatus (Hutton, 1900) E
Leptotarsus fuscolateratus (Alexander, 1922) E
Leptotarsus glaucocapillus (Alexander, 1952) E
Leptotarsus greyanus (Alexander, 1922) E
Leptotarsus halteratus (Alexander, 1923) E
Leptotarsus hudsonianus (Alexander, 1922) E
Leptotarsus huttoni (Edwards, 1923) E
Leptotarsus incertus (Edwards, 1923) E
Leptotarsus intermedius (Alexander, 1922) E
Leptotarsus longioricornis (Alexander, 1923) E
Leptotarsus lunatus (Hutton, 1900) E
Leptotarsus mesocerus (Alexander, 1922) E
Leptotarsus minor (Edwards, 1923) E
Leptotarsus minutissimus (Alexander, 1923) E
Leptotarsus monstratus (Alexander, 1924) E
Leptotarsus montanus (Hutton, 1900) E
Leptotarsus neali (Oosterbroek, 1989) E
Leptotarsus obliquus (Edwards, 1923) E
Leptotarsus obliteratus (Alexander, 1923) E
Leptotarsus obscuripennis (Kirby, 1884) E
Leptotarsus ohakunensis (Alexander, 1923) E
Leptotarsus pallidistigmus (Alexander, 1922) E
Leptotarsus pallidus (Hutton, 1900) E
Leptotarsus pedestris (Alexander, 1939) E
Leptotarsus rufibasis (Alexander, 1922) E
Leptotarsus rufiventris (Edwards, 1923) E
Leptotarsus sessilis (Alexander, 1924) E
Leptotarsus simillimus (Alexander, 1924) E
Leptotarsus sinclairi (Edwards, 1923) E
Leptotarsus submancus (Alexander, 1923) E
Leptotarsus submontanus (Edwards, 1923) E
Leptotarsus subobsoletus (Alexander, 1926) E
Leptotarsus subtener (Alexander, 1922) E
Leptotarsus subvittatus (Alexander, 1939) E
Leptotarsus tapleyi (Alexander, 1923) E
Leptotarsus tenuifrons (Alexander, 1926) E
Leptotarsus variegatus (Edwards, 1923) E
Leptotarsus vittatus (Edwards, 1923) E
Leptotarsus vulpinus (Hutton, 1881) E
Leptotarsus zeylandiae (Alexander, 1920) E
Leptotarsus n. sp. E Su
Limnophila bryobia Mik, 1881 E Su
Limnophila campbelliana Alexander, 1932 E
Limnophila latistyla Alexander, 1923 E
Limnophila luteicauda Alexander, 1924 E
Limnophila mira Alexander, 1926 E
Limnophila miroides Alexander, 1932 E
Limnophila nebulifera Alexander, 1923 E
Limnophila oliveri Alexander, 1923 E
Limnophila perscita Alexander, 1926 E
Limnophila platyna Alexander, 1952 E
Limnophila quaesita Alexander, 1923 E
Limnophila scitula Alexander, 1926 E
Limnophila spissigrada Alexander, 1926 E
Limnophila tonnoiri Alexander, 1926 E
Limnophila n. sp. E
Limnophilella delicatula (Hutton, 1900) E
Limnophilella serotina (Alexander, 1922) E
Limonia acanthophallus (Alexander, 1924) E
Limonia aegrotans (Edwards, 1923) E
Limonia annulifera (Alexander, 1922) E
Limonia archeyi (Alexander, 1924) E
Limonia arthuriana (Alexander, 1924) E
Limonia brookesi (Edwards, 1923) E
Limonia chlorophylloides (Alexander, 1925) E
Limonia cinerella (Alexander, 1923) E
Limonia conulifera (Edwards, 1923) E
Limonia crassispina (Alexander, 1923) E
Limonia cuneipennis (Alexander, 1923) E
Limonia diversispina (Alexander, 1923) E
Limonia fasciata (Hutton, 1900) E
Limonia fulviceps (Alexander, 1925) E
Limonia fulvinota (Alexander, 1923) E
Limonia funesta (Alexander, 1922) E
Limonia gubernatoria (Alexander, 1924) E
Limonia hemimelas (Alexander, 1922) E
Limonia heteracantha (Alexander, 1923) E
Limonia hudsoni (Edwards, 1923) E
Limonia incompta (Alexander, 1922) E
Limonia insularis (Mik, 1881) E Su
Limonia kermadecensis Alexander, 1973 E
Limonia lindsayi (Alexander, 1924) E
Limonia luteipes (Alexander, 1923) E
Limonia luteonitens (Edwards, 1923) E
Limonia maoriensis (Alexander, 1923) E
Limonia megastigmosa (Alexander, 1922) E
Limonia melina (Alexander, 1924) E
Limonia moesta (Alexander, 1923) E
Limonia monilicornis (Hutton, 1900) E
Limonia multispina (Alexander, 1922) E
Limonia nebulifera (Alexander, 1922) E
Limonia nelsoniana (Alexander, 1925) E
Limonia nephelodes (Alexander, 1922) E
Limonia nigrescens (Hutton, 1900) E
Limonia otagensis (Alexander, 1924) E
Limonia pendulifera (Alexander, 1923) E
Limonia pictithorax (Alexander, 1923) E
Limonia plurispina (Alexander, 1925) E
Limonia primaeva Alexander, 1929 E
Limonia repanda (Edwards, 1923) E
Limonia reversalis (Alexander, 1922) E
Limonia seducta (Alexander, 1923) E
Limonia semicuneata (Alexander, 1924) E
Limonia sperata (Alexander, 1922) E
Limonia sponsa (Alexander, 1922) E
Limonia subfasciata (Alexander, 1924) E
Limonia subviridis (Alexander, 1922) E
Limonia sulphuralis (Edwards, 1923) E
Limonia tapleyi (Alexander, 1924) E
Limonia tarsalba (Alexander, 1922) E
Limonia tenebrosa (Edwards, 1923) E
Limonia torrens (Alexander, 1923) E
Limonia tricuspis (Alexander, 1923) E
Limonia tristigmata (Alexander, 1925) E
Limonia veenmani Oosterbroek, 1986 E
Limonia vicarians (Schiner, 1868) E Su
Limonia waitakeriae Alexander, 1952 E
Limonia weschei (Edwards, 1923) E
Limonia wilfredi Alexander, 1952 E
Limonia wiseana Alexander, 1955 E
Limonia n. spp. (3) 3E Su
Limonia n. spp. (9) 9E
Maoritipula hudsoni (Alexander, 1924) E
Maoritipula maori (Alexander, 1920) E
Metalibnotes perhyalina (Alexander, 1973) E K
Metalibnotes watti (Alexander, 1973) E K
Metalimnophila alpina Alexander, 1926 E
Metalimnophila apicispina (Alexander, 1923) E
Metalimnophila banksiana (Alexander, 1923) E
Metalimnophila greyana Alexander, 1926 E
Metalimnophila howesi (Alexander, 1922) E
Metalimnophila integra Alexander, 1926 E
Metalimnophila longi Alexander, 1952 E
Metalimnophila mirifica (Alexander, 1922) E
Metalimnophila montivaga Alexander, 1926 E
Metalimnophila nemocera (Alexander, 1923) E
Metalimnophila nigroapicata (Alexander, 1922) E
Metalimnophila palmata Alexander, 1932 E
Metalimnophila penicillata (Alexander, 1922) E
Metalimnophila productella Alexander, 1926 E
Metalimnophila protea Alexander, 1923 E
Metalimnophila simplicis (Alexander, 1922) E
Metalimnophila unipuncta (Alexander, 1922) E
Metalimnophila yorkensis Alexander, 1926 E
Metalimnophila n. spp. (2) 2E
Molophilus abruptus Alexander, 1923 E
Molophilus acanthus Alexander, 1923 E
Molophilus aenigmaticus Alexander, 1925 E
Molophilus analis Alexander, 1923 E
Molophilus aucklandicus Alexander, 1923 E
Molophilus banksianus Alexander, 1922 E
Molophilus basispina Alexander, 1923 E
Molophilus bidens Alexander, 1923 E
Molophilus bifalcatus Alexander, 1925 E

Molophilus brevinervis Alexander, 1923 E
Molophilus campbellianus Alexander, 1924 E
Molophilus coloratus Alexander, 1923 E
Molophilus coronarius Alexander, 1952 E
Molophilus crassistylus Alexander, 1952 E
Molophilus cristiferus Alexander, 1950 E
Molophilus cruciferus Alexander, 1922 E
Molophilus curtivena Alexander, 1925 E
Molophilus curvistylus Alexander, 1925 E
Molophilus cyatheticolus Alexander, 1950 E
Molophilus denticulatus Alexander, 1923 E
Molophilus evanidus Alexander, 1923 E
Molophilus flagellifer Alexander, 1922 E
Molophilus flavidulus Alexander, 1923 E
Molophilus flavomarginalis Alexander, 1923 E
Molophilus gladiator Alexander, 1939 E
Molophilus greyensis Alexander, 1925 E
Molophilus harrisianus Alexander, 1925 E
Molophilus heteracanthus Alexander, 1925 E
Molophilus hexacanthus Alexander, 1924 E
Molophilus hilaris Alexander, 1923 E
Molophilus howesi Alexander, 1923 E
Molophilus imberbis Alexander, 1923 E
Molophilus improcerus Alexander, 1939 E
Molophilus infantulus Edwards, 1923 E
Molophilus inornatus Edwards, 1923 E
Molophilus irregularis Alexander, 1923 E
Molophilus jenseni Alexander, 1924 E
Molophilus latipennis Alexander, 1923 E
Molophilus lindsayi Alexander, 1922 E
Molophilus longiclavus Alexander, 1924 E
Molophilus luteipennis Alexander, 1923 E
Molophilus luteipygus Alexander, 1922 E
Molophilus macrocerus Alexander, 1922 E
Molophilus macrophallus Alexander, 1925 E
Molophilus morosus Alexander, 1923 E
Molophilus multicinctus Edwards, 1923 E
Molophilus multispinosus Alexander, 1923 E
Molophilus myersi Alexander, 1925 E
Molophilus niveicinctus Alexander, 1922 E
Molophilus ohakunensis Alexander, 1923 E
Molophilus oliveri Alexander, 1922 E
Molophilus oppositus Alexander, 1923 E
Molophilus pallidulus Alexander, 1925 E
Molophilus parvulus Alexander, 1922 E
Molophilus pediformis Alexander, 1925 E
Molophilus perlucidus Alexander, 1950 E
Molophilus phalacanthus Alexander, 1950 E
Molophilus philpotti Alexander, 1922 E
Molophilus pictipleura Alexander, 1922 E
Molophilus picturatus Alexander, 1923 E
Molophilus pilosulus Edwards, 1924 E
Molophilus plagiatus Alexander, 1922 E
Molophilus porrectus Alexander, 1925 E
Molophilus pugnax Alexander, 1925 E
Molophilus pulcherrimus Edwards, 1923 E
Molophilus pullatus Alexander, 1924 E
Molophilus quadrifidus Alexander, 1922 E
Molophilus quinquespinosus Alexander, 1952 E
Molophilus recisus Alexander, 1924 E
Molophilus reduncus Alexander, 1925 E
Molophilus remotus Alexander, 1923 E
Molophilus repandus Alexander, 1923 E
Molophilus satyr Alexander, 1925 E
Molophilus secundus Alexander, 1923 E
Molophilus semiermis Alexander, 1926 E
Molophilus sepositus Alexander, 1923 E
Molophilus speighti Alexander, 1939 E
Molophilus stewartensis Alexander, 1924 E
Molophilus sublateralis Alexander, 1922 E
Molophilus submorosus Alexander, 1924 E
Molophilus subscaber Alexander, 1952 E
Molophilus subuliferus Alexander, 1925 E
Molophilus sylvicolus Alexander, 1924 E
Molophilus tanypus Alexander, 1922 E
Molophilus tenuissimus Alexander, 1923 E
Molophilus tenuistylus Alexander, 1923 E
Molophilus terminans Alexander, 1922 E
Molophilus tillyardi Alexander, 1922 E
Molophilus tonnoiri Alexander, 1925 E
Molophilus uniplagiatus Alexander, 1923 E
Molophilus variegatus Edwards, 1923 E
Molophilus verecundus Alexander, 1924 E
Molophilus n. sp. E Su
Nealexandriaria conveniens (Walker, 1848) E
Notholimnophila exclusa (Alexander, 1922) E
Nothophila fuscana Alexander, 1922 E
Nothophila nebulosa Alexander, 1922 E
Paralimnophila kumarensis (Alexander, 1939) E
Paralimnophila skusei (Hutton, 1902) E
Pedicia arthuriana (Alexander, 1924) E
Pedicia furcata (Alexander, 1926) E
Pedicia novaezelandiae (Alexander, 1922) E
Pedicia n. sp. E
Rhabdomastix brunneipennis Alexander, 1926 E
Rhabdomastix callosa Alexander, 1923 E
Rhabdomastix monilicornis Alexander, 1926 E
Rhabdomastix neozelandiae Alexander, 1922 E
Rhabdomastix optata Alexander, 1923 E
Rhabdomastix otagana Alexander, 1922 E
Rhabdomastix sagana Alexander, 1925 E
Rhabdomastix tonnoirana Alexander, 1934 E
Rhabdomastix trichiata Alexander, 1923 E
Rhabdomastix trilineata Alexander, 1939 E
Rhabdomastix unilineata Alexander, 1939 E
Rhabdomastix vittithorax Alexander, 1923 E
Rhamphophila lyrifera Edwards, 1923 E
Rhamphophila sinistra (Hutton, 1900) E
Sigmatomera rufa (Hudson, 1895) E
Symplecta antipodarum (Alexander, 1953) E Su
Symplecta brachyptera (Alexander, 1955) E Su
Symplecta campbellicola (Alexander, 1964) E SU
Symplecta confluens (Alexander, 1922) E
Symplecta inconstans (Alexander, 1922)
Symplecta pilipes (Fabricius, 1787)
Symplecta n. spp. (4) 4E
Tasiocera aproducta Alexander, 1952 E
Tasiocera bituberculata Alexander, 1924 E
Tasiocera cervicula Alexander, 1925 E
Tasiocera diaphana Alexander, 1932 E
Tasiocera divaricata Alexander, 1932 E
Tasiocera gourlayi (Alexander, 1922) E
Tasiocera longiana Alexander, 1952 E
Tasiocera paulula (Alexander, 1923) E
Tasiocera semiermis Alexander, 1932 E
Tasiocera tonnoirana Alexander, 1932 E
Tasiocera tridentata Alexander, 1922 E
Tasiocera triton Alexander, 1925 E
Tinemyia margaritifera Hutton, 1900 E
Tonnoiraptera neozelandica (Tonnoir, 1926) E
Toxorhina levis (Hutton, 1900) E
Toxorhina ochraceum (Edwards, 1923) E
Zaluscodes aucklandicus Lamb, 1909 E
Zelandoglochina atrovittata (Alexander, 1922) E
Zelandoglochina canterburiana (Alexander, 1923) E
Zelandoglochina circularis (Alexander, 1924) E
Zelandoglochina circumcincta (Alexander, 1924) E
Zelandoglochina crassipes (Edwards, 1923) E
Zelandoglochina cubitalis (Edwards, 1923) E
Zelandoglochina decincta (Edwards, 1923) E
Zelandoglochina flavidipennis (Edwards, 1923) E
Zelandoglochina harrisi (Alexander, 1923) E
Zelandoglochina melanogramma (Edwards, 1923) E
Zelandoglochina myersi (Alexander, 1924) E
Zelandoglochina octava (Edwards, 1923) E
Zelandoglochina paradisea (Alexander, 1923) E
Zelandoglochina sublacteata (Edwards, 1923) E
Zelandoglochina unicornis (Alexander, 1923) E
Zelandoglochina unijuga (Alexander, 1923) E
Zelandoglochina n. spp. (4-5) 4-5E Johns
Zelandomyia angusta (Alexander, 1923) E
Zelandomyia atridorsum Alexander, 1932 E
Zelandomyia cinereipleura (Alexander, 1922) E
Zelandomyia deviata (Alexander, 1922) E
Zelandomyia otagensis (Alexander, 1923) E
Zelandomyia pallidula Alexander, 1924 E
Zelandomyia penthoptera Alexander, 1924 E
Zelandomyia pygmaea Alexander, 1923 E
Zelandomyia ruapehuensis (Alexander, 1922) E
Zelandomyia tantula Alexander, 1926 E
Zelandomyia watti (Alexander, 1922) E
Zelandotipula fulva (Hutton, 1900) E
Zelandotipula novarae (Schiner, 1868) E
Zelandotipula otagana (Alexander, 1922) E
Zelandotipula n. spp. (8) 8E
TRICHOCERIDAE
Asdura decussata (Alexander, 1924) E
Asdura lyrifera (Alexander, 1923) E
Asdura howesi (Alexander, 1923) E
Asdura obtusicornis (Alexander, 1923) E
Nothotrichocera antarctica (Edwards, 1923) E
Nothotrichocera aucklandica Johns, 1975 E
Nothotrichocera johnsi Krzeminska, 2006 E
Trichocera annulata Meigen, 1818 A Gr
Zedura antipodum (Mik, 1881) E
Zedura aperta (Alexander, 1922) E
Zedura complicata (Alexander, 1924) E
Zedura curtisi (Alexander, 1924) E
Zedura dololabella Krzeminska, 2005 E
Zedura harrisi (Alexander, 1924) E
Zedura lobifera (Alexander, 1922) E
Zedura macrotrichiata (Alexander, 1922) E
Zedura maori (Alexander, 1921) E
Zedura oparara (Krzeminska, 2001) E
Zedura tautuku (Krzeminska, 2003) E

Suborder BRACHYCERA
Division ORTHORRHAPHA
ACROCERIDAE [Schlinger]
Apsona muscaria Westwood, 1876 E Sa SP
Apsona n. sp. Schlinger E SP
Helle longirostris (Hudson, 1892) E SP
Helle rufescens Brunetti, 1926 E SP
Ogcodes argigaster Schlinger, 1960 E SP
Ogcodes brunneus (Hutton, 1881) E SP
Ogcodes consimilis Brunetti, 1926 E SP
Ogcodes leptisoma Schlinger, 1960 E SP
Ogcodes nitens (Hutton, 1901) E SP
Ogcodes paramonovi Schlinger, 1960 E SP
Ogcodes similis Schlinger, 1960 E SP
Ogcodes n. spp. (2) 2E SP (Schlinger)
Pterodontia n. sp. E SP (Schlinger)
ASILIDAE (Macfarlane)
'Cerdistus' lascus (Walker, 1849) E Gr IPr
Cerdistus meridionalis (Hutton, 1901) Gr Ipr
Cerdistus? n. sp. E Sa IPr
'Neoitamus' bulbus (Walker, 1849) E Gr-?Fo IPr
'Neoitamus' melanopogon (Schiner, 1868) E Gr IPr
'Neoitamus' smithii (Hutton, 1901) I^C E Gr IPr
'Neoitamus' walkeri Daniels, 1989 E Gr IPr
'Neoitamus' n. spp. (3) 3E Gr Ipr
'Neoitamus' n. spp. 2E TK IPr
Saropogon antipodus Schiner, 1868 E Gr IPr
Saropogon chathamensis Hutton, 1901 I^C E Ch Gr IPr
Sarapogon clarkii Hutton, 1901 E
Saropogon discus (Walker, 1849) E Gr IPr
Saropogon extenuatus Hutton, 1901 E Gr IPr
Saropogon fascipes Hutton, 1902 E Gr IPr
Saropogon fugiens Hutton, 1901 E Gr IPr
Saropogon hudsoni Hutton, 1901 E Gr IPr
Saropogon proximus Hutton, 1901 E Gr IPr
Saropogon viduus (Walker, 1849) E Gr IPr
Saropogon n. spp. (6) 6E Gr IPr
Zosteria novaezealandica Daniels, 1987 I^C E Fo IPr
Gen. nov. 1 Asilini et n. sp. I^C E Gr IPr
APSILOCEPHALIDAE
Kaurimyia thorpei Winterton & Irwin, 2008 E
BOMBYLIIDAE [Schnitzler]

Geron sp. E
Tillyardomyia gracilis Tonnoir, 1927 E Gr IP
DOLICHOPODIDAE [Bickel, Andrew, Macfarlane]
Abatetia robusta (Parent, 1933) E Be
Abatetia n. spp. (4) Andrew, Bickel 4E Be
Aphrosylopsis lineata Lamb, 1909 E Su
Apterachalcus borboroides (Oldroyd, 1955) E Su
Australachalcus chaetifemoratus Parent, 1933 E
Australachalcus luteipes Parent, 1933 E
Australachalcus medius Parent, 1933 E
Australachalcus minor Parent, 1933 E
Australachalcus miniscululus Parent, 1933 E
Australachalcus minutus Parent, 1933 E
Australachalcus nigroscutatus Parent, 1933 E
Australachalcus relictus Parent, 1933 E
Australachalcus separatus Parent, 1933 E
Austrosciapus proximus (Parent, 1928) A Ga
Brevimyia pulverea (Parent, 1933) E
Chrysotimus bilineatus Parent, 1933 E
Chrysotimus lunulatus Parent, 1933 E
Chrysotimus nigrichaetus (Parent, 1933) E
Chrysotimus scutatus Parent, 1933 E
Chrysotimus n. spp. (2) 2E LCNZ
Chrysotus albisignatus Becker, 1924 E
Chrysotus bellax Parent, 1933 E
Chrysotus chaetipalpus Parent, 1933 E
Chrysotus chaetoproctus Parent, 1933 E
Chrysotus diversus Parent, 1933 E
Chrysotus neoselandensis Parent, 1933 E Ch
Chrysotus uniseriatus Parent, 1933 E
Chrysotus vicinus Parent, 1933 E
Chrysotus n. sp. W SW
Dactylonotus formosus (Parent, 1933) E
Diaphorus infumatus Parent, 1933 E
Diaphorus obscurus Parent, 1933 E
Diaphorus parapraestans Dyte, 1980 E
Diaphorus stylifer Parent, 1933 E
Diaphorus tetrachaetus Parent, 1933 E
Filatopus ciliatus (Parent, 1933) E
Filatopus mirabilis (Parent, 1933) E
Filatopus ornatus (Parent, 1933) E
Halteriphorus mirabilis Parent, 1933 E
Helichochaetus discifer Parent, 1933 E
Hercostomus argentifacies Parent, 1933 E
Hercostomus aurifacies Parent, 1933 E
Hercostomus philpotti Parent, 1933 E Ch
Hercostomus pollinifrons Parent, 1933 E
Hercostomus n. sp. Macfarlane 2007 E
Hydrophorus praecox (Lehmann, 1822) A Fs
Ischiochaetus lenis Parent, 1933 E
Ischiochaetus ornatipes Parent, 1933 E
Ischiochaetus rotundicornis Parent, 1933 E
Ischiochaetus spinosus Parent, 1933 E
?*Liancalus* n. sp. E
Micromorphus albipes (Zetterstedt, 1843) A
Micropygus bifenestratus Parent, 1933 E
Micropygus bipunctatus Parent, 1933 E
Micropygus brevicornis Parent, 1933 E
Micropygus brevithorax Parent, 1933 E
Micropygus divergens Parent, 1933 E
Micropygus inornatus Parent, 1933 E
Micropygus lacustris Parent, 1933 E
Micropygus nigripes Parent, 1933 E
Micropygus puerulus Parent, 1933 E
Micropygus pulchellus Parent, 1933 E
Micropygus ripicola Parent, 1933 E
Micropygus serratus Parent, 1933 E
Micropygus striatus Parent, 1933 E
Micropygus tarsatus Parent, 1933 E
Micropygus transiens Parent, 1933 E
Micropygus vagans Parent, 1933 E
Micropygus n. sp. E LCNZ
Naufraga hexachaeta (Parent, 1933) E
Ostenia robusta Hutton, 1901 E
Paraclius aeotearoa Bickel, 2008 E
Parentia anomalicosta Bickel, 1992 E
Parentia aotearoa Bickel, 1992 E
Parentia argentifrons Bickel, 1992 E TK
Parentia calignosa Bickel, 1992 E
Parentia chathamensis Bickel, 1992 E Ch
Parentia cilifoliata (Parent, 1933) E
Parentia defecta Bickel, 1992 E
Parentia fuscata (Hutton, 1901) E
Parentia gemmata (Walker, 1849) E
Parentia griseicollis (Becker, 1924) E
Parentia insularis Bickel, 1992 E TK
Parentia johnsi Bickel, 1992 E
Parentia lyra Bickel, 1992 E
Parentia magniseta Bickel, 1992 E
Parentia malitiosa (Hutton, 1901) E
Parentia milleri (Parent, 1933) E
Parentia mobile (Hutton, 1901) E
Parentia modesta (Parent, 1933) E
Parentia nova (Parent, 1933) E
Parentia pukakiensis Bickel, 1992 E
Parentia recticosta (Parent, 1933) E
Parentia restricta (Hutton, 1901) E
Parentia schlingeri Bickel, 1992 E
Parentia titirangi Bickel, 1992 E
Parentia tonnoiri (Parent, 1933) E
Parentia varifemorata Bickel, 1992 E
Parentia whirinaki Bickel, 1992 E
Parentia n. sp. E
Scelloides armatus Parent, 1933 E
Scelloides brunneifrons Parent, 1933 E
Scelloides conspicuus Parent, 1933 E
Scelloides fulvifrons Parent, 1933 E
Scelloides maculatus Parent, 1933 E
Scelloides ornatipes Parent, 1933 E
Scelloides parcespinosus Parent, 1933 E
Scelloides parvus Parent, 1933 E
Scelloides pollinosus Parent, 1933 E
Scelloides raptorius Parent, 1933 E
Scelloides spinosus Parent, 1933 E
Scelloides vicinus Parent, 1933 E
?'*Schoenophilus*' *campbellensis* Harrison, 1964 E Su
Scorpiurus aenescens Parent, 1932 E
Scorpiurus n. sp. E SW
Sympycnus albinotatus Parent, 1933 E
Sympycnus alchymicus[NC] (Parent, 1933) E
Sympycnus amplitarsus Parent, 1933 E
Sympycnus brevicornis Parent, 1933 E
Sympycnus campbelli Parent, 1933 E
Sympycnus contemptus Parent, 1933 E
Sympycnus distinctus Parent, 1933 E
Sympycnus edwardsi Parent, 1933 E
Sympycnus gracilipes Parent, 1933 E
Sympycnus harrisi Parent, 1933 E
Sympycnus humilis Parent, 1933 E
Sympycnus ignavus Parent, 1933 E
Sympycnus longicornis Parent, 1933 E
Sympycnus longipilus Parent, 1933 E
Sympycnus luteinotatus Parent, 1933 E
Sympycnus modestus Parent, 1933 E
Sympycnus moestus Parent, 1933 E
Sympycnus normalis Parent, 1933 E
Sympycnus ornatipes Parent, 1933 E
Sympycnus ornatus Parent, 1933 E
Sympycnus tenueciliatus Parent, 1933 E
Syntormon aotearoa Bickel, 1999 E We
Tetrachaetus bipunctatus Parent, 1933 E We
Tetrachaetus simplex Parent, 1933 E
Thinophilus milleri Parent, 1933 E
Thinophilus n. spp. (2–3) 2–3E
Thrypticus arahakiensis Bickel, 1992 E We H
EMPIDIDAE [Sinclair, Plant, Andrew, Macfarlane]
Abocciputa pilosa Plant, 1989 E Fo/F
Adipsomyia mutabilis Collin, 1928 E
Adipsomyia stigmosa (Smith, 1964) E Su
Adipsomyia n. spp. (2) 2E Sinclair
Apalocnemis fumosa (Hutton, 1901) E
Apalocnemis simulans Collin, 1928 E
Asymphyloptera n. spp. (4–5) Sinclair 1995 4–5E
Atodrapetis infrapratula Plant, 1997 E
Atrichopleura compitalis Collin, 1928 E
Atrichopleura conjuncta Malloch, 1931 E
Atrichopleura n. spp. (2) 2E
Austropeza insolita (Collin, 1928) E
Ceratomerus biseriatus Plant, 1991 E Fo
Ceratomerus brevifurcatus Plant, 1991 E
Ceratomerus crassinervis Malloch, 1931 E
Ceratomerus dorsatus Collin, 1928 E ?F
Ceratomerus earlyi Plant, 1991 E
Ceratomerus exiguus Collin, 1928 E Fo
Ceratomerus flavus Plant, 1991 E
Ceratomerus longifurcatus Collin, 1931 E
Ceratomerus melaneus Plant, 1991 E Sa
Ceratomerus prodigiosus Collin, 1928 E
Ceratomerus tarsalis Plant, 1991 E Fo
Ceratomerus virgatus Collin, 1928 E Fo
Ceratomerus vittatus Plant, 1991 E Sa
Ceratomerus n. spp. (29) Sinclair 29E
Chelifera apicata Collin, 1928 E
Chelifera fontinalis (Miller, 1923) E ?Fr
Chelifera tacita Collin, 1928 E
Chelifera tantula Collin, 1928 E
Chelipoda abdita Collin, 1928 E
Chelipoda abjecta Collin, 1928 E
Chelipoda aritarita Plant, 2007 E
Chelipoda atrocitax Plant, 2007 E
Chelipoda australpina Plant, 2007 E
Chelipoda brevipennis Plant, 2007 E
Chelipoda consignata Collin, 1928 E
Chelipoda cornigera Plant, 2007 E
Chelipoda cycloseta Plant, 2007 E
Chelipoda delecta Collin, 1928 E
Chelipoda didhami Plant, 2007 E
Chelipoda digressa Collin, 1928 E
Chelipoda dominatrix Plant, 2007 E
Chelipoda ferocitrix Plant, 2007 E
Chelipoda fuscoptera Plant, 2007 E
Chelipoda gracilis Plant, 2007 E
Chelipoda inconspicua Collin, 1928 E
Chelipoda interposita Collin, 1928 E
Chelipoda lateralis Plant, 2007 E
Chelipoda longicornis Collin, 1928 E
Chelipoda macrostigma Plant, 2007 E
Chelipoda mediolinea Plant, 2007 E
Chelipoda mirabilis Collin, 1928 E
Chelipoda moderata Collin, 1928 E
Chelipoda modica Collin, 1928 E
Chelipoda monorhabdos Plant, 2007 E
Chelipoda oblata Collin, 1928 E
Chelipoda oblinata Collin, 1928 E
Chelipoda otiraensis (Miller, 1923) E
Chelipoda puhihuroa Plant, 2007 E
Chelipoda rakiuraensis Plant, 2007 E
Chelipoda rangopango Plant, 2007 E
Chelipoda secreta Collin, 1928 E
Chelipoda tainuia Plant, 2007 E
Chelipoda tangerina Plant, 2007 E
Chelipoda trepida Collin, 1928 E
Chelipoda ultraferox Plant, 2007 E
Chelipoda venatrix Plant, 2007 E
Chersodromia zelandica Rogers, 1982 E
Chimerothalassius ismayi Shamshev & Grootaert, 2002 E Be
Cladodromia futilis Collin, 1928 E
Cladodromia insignita Collin, 1928 E
Cladodromia inturbida (Collin, 1928) E
Cladodromia negata Collin, 1928 E
Cladodromia soleata Collin, 1928 E
Cladodromia n. sp. Andrew E ?Fr
Clinocera gressitti Smith, 1964 E Su ?F
?*Clinorhampha politella* (Malloch, 1931) E
?*Clinorhampha* n. sp. Andrew E
Doliodromia avita Collin, 1928 E
Empidadelpha propria Collin, 1928 E

Empidadelpha torrentalis (Miller, 1923) E
Empidadelpha n. sp. CMNZ E Su
'Empis' probata Collin, 1928 E
Glyphidopeza fluviatilis Sinclair, 1997 E F
Glyphidopeza longicornis Sinclair, 1997 E F
Gynatoma atra Malloch, 1931 E
Gynatoma continens Collin, 1928 E
Gynatoma evanescens Collin, 1928 E
Gynatoma pygmaea Collin, 1928 E
Gynatoma quadrilineata Collin, 1928 E
Gynatoma subfulva Collin, 1928 E
Gynatoma n. sp. Andrew E
Hemerodromia radialis Collin, 1928 E
Heterophlebus maculipennis (Collin, 1928) E F [as genus *Oreogeton* in A/O Catalogue]
Heterophlebus rostratus (Collin, 1928) E F
Heterophlebus undulatus (Collin, 1928) E F
Heterophlebus n. spp. (3) 3E Macfarlane
Hilara anisonychia Collin, 1928 E
Hilara consanguinea Collin, 1928 E
Hilara dracophylli Miller, 1923 E
Hilara flavinceris Miller, 1923 E
Hilara fossalis Miller, 1923 E
Hilara hudsoni (Hutton, 1901) E
Hilara intuta Collin, 1928 E
Hilara littoralis Miller, 1923 E
Hilara macrura Collin, 1928 E
Hilara philpotti Miller, 1923 E
Hilara retecta Collin, 1928 E
Hilara spinulenta Collin, 1928 E
Hilara urophora Collin, 1928 E
Hilara urophylla Collin, 1928 E
Hilara vector Miller, 1923 E
Hilara n. spp. (50) Andrew 50E
Hilarempis argentella Collin, 1928 E
Hilarempis benhami (Miller, 1913) E
Hilarempis brevistyla Collin, 1928 E
Hilarempis cineracea Collin, 1928 E
Hilarempis dichropleura Collin, 1928 E
Hilarempis diversimana Collin, 1928 E
Hilarempis huttoni Bezzi, 1904 E
Hilarempis immota Collin, 1928 E
Hilarempis kaiteriensis (Miller, 1913) E
Hilarempis longistyla Collin, 1928 E
Hilarempis minthaphila Collin, 1928 E
Hilarempis nigra Miller, 1923 E
Hilarempis ochrozona Collin, 1928 E Fo Li
Hilarempis simillima Collin, 1928 E
Hilarempis smithii (Hutton, 1901) E
Hilarempis subdita[1] Collin, 1928 E
Hilarempis trichopleura Collin, 1928 E
Hilarempis uniseta Collin, 1928 E
Hilarempis n. spp. (5) Andrew 5E
Homalocnemis adelensis (Miller, 1913) E
Homalocnemis inexpleta Collin, 1928 E
Homalocnemis maculipennis Malloch, 1932 E
Homalocnemis perspicua (Hutton, 1901) E
Hybomyia oliveri Plant, 1995 E
Hydropeza agnetis Sinclair & McLellan, 2004 E F
Hydropeza akatarawa Sinclair & McLellan, 2004 E F
Hydropeza clarae Sinclair & McLellan, 2004 E F
Hydropeza daviesi Sinclair & McLellan, 2004 E F
Hydropeza longipennae (Miller, 1923) E F
Hydropeza milleri Sinclair & McLellan, 2004 E F
Hydropeza paniculata Sinclair & McLellan, 2004 E F
Hydropeza tutoko Sinclair & McLellan, 2004 E F
Hydropeza vockerothi Sinclair & McLellan, 2004 E F
Hydropeza wardi Sinclair & McLellan, 2004 E F
Icasma aequabilis Plant, 1990 E
Icasma fascipennis Sinclair, 1997 E
Icasma longicauda Sinclair, 1997 E
Icasma masneri Sinclair, 1997 E
Icasma setosa Sinclair, 1997 E
Icasma singularis Collin, 1928 E
Icasma tararua Sinclair, 1997 E
Isodrapetis excava Plant, 1999 E
Isodrapetis hyalina Plant, 1999 E
Isodrapetis nitidula Collin, 1928 E
Isodrapetis nitidiuscula Plant, 1999 E
Isodrapetis rauparaha Plant, 1999 E
Isodrapetis spinositibia Plant, 1999 E
Isodrapetis subpollinosa Collin, 1928 E
Isodrapetis suda Collin, 1928 E
Monodromia fragilis Collin, 1928 E
Neoplasta n. sp. Plant E
Ngaheremyia fuscipennis Plant & Didham, 2006 E
Oropezella antennata Collin, 1928 E
Oropezella bifurcata Collin, 1928 E Fo
Oropezella diminuloruma Plant, 1989 E Fo
Oropezella loripes Plant, 1989 E Fo
Oropezella trucispicata Plant, 1989 E Fo
Oropezella nigra (Miller, 1923) E
Oropezella tanycera Collin, 1928 E
Phyllodromia falcata Plant, 2005 E
Phyllodromia flexura Plant, 2005 E
Phyllodromia floridula Plant, 2005 E
Phyllodromia nigricoxa Plant, 2005 E
Phyllodromia proiecta Plant, 2005 E
Phyllodromia scopulifera Collin, 1928 E
Phyllodromia striata Collin, 1928 E
Platypalpus ementitus (Collin, 1928) E
Platypalpus scambus (Collin, 1928) E
Platypalpus spatiosus (Collin, 1928) E
Pseudoscelolabes fulvescens (Miller, 1923) E Fo Li
Pseudoscelolabes n. sp. Macfarlane E
Sematopoda elata Collin, 1928 E
Thinempis brouni NC (Hutton, 1901) E Be Andrew
Thinempis otakouensis (Miller, 1910) E ?SW/Be
Thinempis takaka Bickel, 1996 E Be
Gen. nov. 1 Hemerodromiinae n. sp. E
Gen. nov. 2 Hilarini nr *Hilarempis* et n. spp. (2) 2E
Gen. nov. 3 nr Oreogetoninae et n. sp. E
PELECORHYNCHIDAE
Pelecorhynchus sp. indet.* Winterbourn F
RHAGIONIDAE
Chrysopilus nitidiventris Tonnoir, 1927 E Fo
Chrysopilus n. spp. (2) LCNZ, CMNZ 2E Gr
STRATIOMYIDAE [Woodley, Andrew]
Australoberis amoena Lindner, 1958 E
Australoberis refugians (Miller, 1917) E
Australoberis n. spp. (2) Woodley 2E
Benhamyia alpina (Hutton, 1901) E
Benhamyia apicalis (Walker, 1849) E IPr
Benhamyia hoheria (Miller, 1917) E
Benhamyia smaragdina Lindner, 1958 E
Benhamyia straznitzkii (Nowicki, 1875) E Fo
Benhamyia n. spp. (4) Woodley 4E
Berisina caliginosa (Miller, 1917) E
Berisina maculipennis Malloch, 1928 E
Berisina saltusans (Miller, 1917) E
Berisina n. spp. (2) Woodley 2E
Boreoides tasmaniensis Bezzi, 1922 A Ga
Dysbiota parvula Lindner, 1958 E
Dysbiota peregrina (Hutton, 1901) E Fo
Exaireta spinigera (Wiedemann, 1830) A Ga Li
Hermetia illucens (Linnaeus, 1758) A Ga-We
Inopus rubriceps (Macquart, 1847) A Gr
Neactina opposita (Walker, 1854) E Fo
Neactina ostensackeni (Lindner, 1958) E Fo
Neactina simmondsii (Miller, 1917) E Fo
Neactina n. spp. (3) Woodley 2E
Odontomyia angusta Walker, 1854 E F
Odontomyia australensis Schiner, 1868 E F
Odontomyia atrovirens Bigot, 1879 E F
Odontomyia chathamensis Hutton, 1901 E Ch F
Odontomyia chloris (Walker, 1854) E F
Odontomyia collina Hutton, 1901 E F
Odontomyia fulviceps (Walker, 1854) E F
Odontomyia neodorsalis (Miller, 1950) E F
Odontomyia n. spp. (10) Andrew 10E F
Tytthoberis cuprea (Hutton, 1901) E
Zealandoberis lacuans (Miller, 1917) E
Zealandoberis micans (Hutton, 1901) E
Zealandoberis substituta (Walker, 1854) E
Zealandoberis violacea (Hutton, 1901) E
Zealandoberis n. spp. (3) 3E Woodley
Gen. indet. Chiromyzinae et n. spp. (3)* 3E Andrew
TABANIDAE [Macfarlane]
Dasybasis bratrankii (Nowicki, 1875) E ?F
Dasybasis difficilis (Krober, 1931) E ?F
Dasybasis loewi (Enderlein, 1925) E ?F
Dasybasis nigripes (Krober, 1931) E ?F
Dasybasis opla (Walker, 1850) E ?F
Dasybasis sarpa (Walker, 1850) E ?F
Dasybasis thereviformis Mackerras, 1957 E ?F
Dasybasis transversa (Walker 1854) E ?F
Dasybasis truncata (Walker, 1850) E ?F
Dasybasis viridis (Hudson, 1892) E ?F
Dasybasis n. spp. (2) Hayakawa 2E ?F
Ectenopsis lutulenta (Hutton, 1901) E ?F
Ectenopsis n. sp. Hayakawa E ?F
Scaptia adrel (Walker, 1850) E ?F
Scaptia brevipalpis Krober, 1931 E ?F
Scaptia lerda (Walker, 1850) E ?F
Scaptia milleri Mackerras, 1957 E ?F
Scaptia montana (Hutton, 1901) E ?F
Scaptia ricardoae (Hutton, 1901) E ?F
THEREVIDAE
Anabarhynchus acuminatus Lyneborg, 1992 E So IPr
Anabarhynchus albipennis Lyneborg, 1992 I^C E So IPr
Anabarhynchus arenarius Lyneborg, 1992 E Be So IPr
Anabarhynchus atratus Lyneborg, 1992 I^C E So IPr
Anabarhynchus atripes Lyneborg, 1992 E So IPr
Anabarhynchus aureosericeus Krober, 1932 E So IPr
Anabarhynchus brevicornis Lyneborg, 1992 E So IPr
Anabarhynchus brunninervis Krober, 1932 E So IPr
Anabarhynchus caesius Krober, 1912 E So IPr
Anabarhynchus completus Lyneborg, 1992 E Be So IPr
Anabarhynchus curvistylus Lyneborg, 1992 E Be So IPr
Anabarhynchus diversicolor Lyneborg, 1992 E So IPr
Anabarhynchus dugdalei Lyneborg, 1992 E So IPr
Anabarhynchus dysmachiiformis Krober, 1932 E So IPr
Anabarhynchus embersoni Lyneborg, 1992 I^C E So IPr
Anabarhynchus exiguus Hutton, 1901 E So IPr
Anabarhynchus farinosus Lyneborg, 1992 E Be So IPr
Anabarhynchus femoralis Krober, 1932 E So IPr
Anabarhynchus fenwicki Lyneborg, 1992 E So IPr
Anabarhynchus flaviventris Lyneborg, 1992 I^C E So IPr
Anabarhynchus fluviatilis Lyneborg, 1992 I^C E So IPr
Anabarhynchus fuscofemoratus Lyneborg, 1992 I^C E So IPr
Anabarhynchus gibbsi Lyneborg, 1992 E Be So IPr
Anabarhynchus grossus Lyneborg, 1992 E So IPr
Anabarhynchus harrisi Lyneborg, 1992 E So IPr
Anabarhynchus hayakawai Lyneborg, 1992 E So IPr
Anabarhynchus hudsoni Lyneborg, 1992 I^C E So IPr
Anabarhynchus huttoni Lyneborg, 1992 E Be So IPr
Anabarhynchus indistinctus Lyneborg, 1992 I^C E So IPr
Anabarhynchus innotatus (Walker, 1856) E Be-We So IPr
Anabarhynchus lacustris Lyneborg, 1992 E So IPr
Anabarhynchus lateripilosus Lyneborg, 1992 E Be So IPr
Anabarhynchus latus Lyneborg, 1992 E So IPr
Anabarhynchus limbatinervis Krober, 1932 E So IPr
Anabarhynchus longepilosus Lyneborg, 1992 E So IPr
Anabarhynchus longipennis Krober, 1932 E So IPr
Anabarhynchus macfarlanei Lyneborg, 1992 E So IPr
Anabarhynchus major Lyneborg, 1992 E So IPr
Anabarhynchus maori Hutton, 1901 E So IPr
Anabarhynchus megalopyge Lyneborg, 1992 E So IPr

Anabarhynchus microphallus Lyneborg, 1992 E Be/So IPr
Anabarhynchus monticola Lyneborg, 1992 E Sa So IPr
Anabarhynchus nebulosus Hutton, 1901 E Be So IPr
Anabarhynchus neglectus Krober, 1932 E Sa So IPr
Anabarhynchus nigrofemoratus Krober, 1932 E Be-Gr So IPr
Anabarhynchus olivaceus Lyneborg, 1992 I[C] E Gr So IPr
Anabarhynchus ostentatus Lyneborg, 1992 E Sa So IPr
Anabarhynchus postocularis Lyneborg, 1992 E Gr So IPr
Anabarhynchus robustus Lyneborg, 1992 E Gr So IPr
Anabarhynchus ruficoxa Lyneborg, 1992 E Be-Gr So IPr
Anabarhynchus rufobasalis Lyneborg, 1992 E Sa So IPr
Anabarhynchus schlingeri Lyneborg, 1992 E Be So IPr
Anabarhynchus similis Lyneborg, 1992 E Sa So IPr
Anabarhynchus simplex Lyneborg, 1992 E Gr So IPr
Anabarhynchus spiniger Lyneborg, 1992 E Gr So IPr
Anabarhynchus spitzeri Lyneborg, 1992 E Gr-Sa So IPr
Anabarhynchus triangularis Lyneborg, 1992 E Gr So IPr
Anabarhynchus tricoloratus Lyneborg, 1992 E Be So IPr
Anabarhynchus waitarerensis Lyneborg, 1992 I[C] E Be So IPr
Anabarhynchus westlandensis Lyneborg, 1992 E Be So IPr
Anabarhynchus wisei Lyneborg, 1992 I[C] E So IPr
Anabarhynchus n. spp. (3) 3E
Ectinorhynchus castaneus (Hutton, 1901) E Fo-Sa
Ectinorhynchus cupreus (Hutton, 1901) E Ga-Gr
Ectinorhynchus furcatus Lyneborg, 1992 I[C] E Sa
Ectinorhynchus micans (Hutton, 1901) E
Megathereva albopilosa Lyneborg, 1992 E Be
Megathereva atritibia Lyneborg, 1992 E Be So IPr
Megathereva bilineata (Fabricius, 1775) E Be

Division CYCLORRHAPHA
Section ASCHIZA
LONCHOPTERIDAE A
Lonchoptera bifurcata (Fallen, 1810) A Gr
PHORIDAE [Oliver]
Antipodiphora austrophila (Schmitz, 1939) E
Antipodiphora brevicornis (Schmitz, 1939) E
Antipodiphora nana (Schmitz, 1939) E
Antipodiphora similicornis (Schmitz, 1939) E
Antipodiphora subarcuata (Schmitz, 1939) E
Antipodiphora tonnoiri (Schmitz, 1939) E
Aphiura breviceps Schmitz, 1939 E D
Beckerina polysticha Schmitz, 1939 E
Bothroprosopa mirifica Schmitz, 1939 E
Ceratoplatus fullerae Schmitz, 1939 E
Diplonevra caudata Schmitz, 1939 E
Diplonevra n. spp (2) Oliver 2E
Distichophora crassimana Schmitz, 1939 E
Dohrniphora cornuta (Bigot, 1857) A Gr-Ga ScCa IPr
Kierania grata Schmitz, 1939 E
Kierania n. sp. Oliver E
Macroselia longiseta Schmitz, 1939 E
Megaselia castanea Bridarolli, 1937 E
Megaselia comparabilis Schmitz, 1929 E
Megaselia curtineura (Brues, 1909) A Ca
Megaselia dolichoptera Bridarolli, 1937 E
Megaselia dupliciseta Bridarolli, 1937 E
Megaselia halterata (Wood, 1910) A GrGa Ff
Megaselia impariseta Bridarolli, 1937 A?
Megaselia longinqua Bridarolli, 1937 E
Megaselia lucida Bridarolli, 1937 E
Megaselia rufipes (Meigen, 1804) A Gr-Ga Sc-IPr
Megaselia scalaris (Loew, 1866) A Gr-Ga Sc-IPr
Megaselia spiracularis Schmitz, 1938 A
Megaselia n. spp. (2) Oliver 2E
Metopina australiana Borgmeier, 1963 Ga I/A
Metopina climieorum Disney, 1994 Ga
Metopina n. sp. Oliver E
Minicosta mollyae Brown & Oliver, 2008 E
Palpocrates obscurior Schmitz, 1939 E
Palpocrates rufipalpis Schmitz, 1939 E
Spiniphora bergistammi (Mik, 1864) A Ca
Tarsocrates niger Schmitz, 1939 E
'Triphleba' atripalpis Schmitz, 1939 E
'Triphleba' fuscithorax Schmitz, 1939 E
'Triphleba' rufithorax Schmitz, 1939 E
Wharia willcocksorum Brown & Oliver, 2008 E
PIPUNCULIDAE [Skevington]
Cephalops libidinosus De Meyer, 1991 E IP
Dasydorylas arthurianus[NC] (Tonnoir, 1925) E Gr IP
Dasydorylas deansi[NC] (Tonnoir, 1925) E Fo IP
Dasydorylas harrisi[NC] (Tonnoir, 1925) E Gr IP
Dasydorylas n. spp. (9) 9E IP
Tomosvaryella novaezealandiae (Tonnoir, 1925) E Gr IP
PLATYPEZIDAE
Microsania tonnoiri Collart, 1934 E
Callomyiinae n. sp. Chandler 1994 E
SCIADOCERIDAE A
Sciadocera rufomaculata White, 1916 A Fo
SYRPHIDAE (Thompson)
Allograpta atkinsoni (Miller, 1921) E IPr
Allograpta dorsalis (Miller, 1924) E IPr
Allograpta flavofaciens (Miller, 1921) E IPr
Allograpta hirsutifera (Hull, 1949) E IPr
Allograpta hudsoni (Miller, 1921) E IPr
Allograpta pseudoropala (Miller, 1921) E IPr
Allograpta ropala (Walker, 1849) E IPr
Allograpta ventralis (Miller, 1921) E IPr
Allograpta n. spp. (25) Thompson 25E
Anu una Thompson, 2008 E
Eristalinus aeneus (Scopali, 1763)* A
Eristalis tenax (Linnaeus, 1758) A Ga-We-Se
Eumerus strigatus (Fallen, 1817) A GaHb
Eumerus tuberculatus Rondani, 1857 A Ga Hb
Helophilus antipodus Schiner, 1868 E
Helophilus campbelli (Miller, 1921) E
Helophilus campbellicus Hutton, 1902 E
Helophilus cargilli Miller, 1911 E
Helophilus chathamensis Hutton, 1901 E Ch
Helophilus cingulatus (Fabricius, 1775) E
Helophilus hectori Miller, 1924 E
Helophilus hochstetteri Nowicki, 1875 E F
Helophilus ineptus Walker, 1849 E
Helophilus montanus (Miller, 1921) E
Helophilus seelandicus (Gmelin, 1790) E
Helophilus taruensis (Miller, 1924) E
Helophilus n. spp. (8) Thompson 8E
Melangyna novaezealandiae (Macquart, 1855) E Gr Ga IPr
Melanostoma fasciatum (Macquart, 1850) E Cr Gr Ga IPr
Merodon equestris (Fabricius, 1794) A Ga Hb
Orthoprosopa bilineata (Walker, 1849) E
Platycheirus antipodus (Hull, 1949) E IPr
Platycheirus captalis (Miller, 1924) E IPr
Platycheirus clarkei Miller 1921 E IPr
Platycheirus cunninghami (Miller, 1921) E IPr
Platycheirus fulvipes (Miller, 1924) E IPr
Platycheirus harrisi (Miller, 1921) E IPr
Platycheirus howesii (Miller, 1921) E IPr
Platycheirus huttoni Thompson, 1989 E IPr Ns
Platycheirus leptospermi (Miller, 1921) E IPr
Platycheirus lignudus Miller, 1921 E IPr
Platycheirus myersii (Miller, 1924) E IPr
Platycheirus notatus (Bigot, 1884) E IPr
Platycheirus ronanus (Miller, 1921) E IPr
Platycheirus n. spp. (13) Thompson 13E IPr
Psilota decessa (Hutton, 1901) E
Simosyrphus grandicornis (Macquart, 1842) A K IPr

Section SCHIZOPHORA
ACALYPTRATAE
AGROMYZIDAE
Cerodontha angustipennis Harrison, 1959 E Gr HLM
Cerodontha australis Malloch, 1925 A Gr HLM
Cerodontha sylvesterensis Spencer, 1976 E Sa H
Cerodontha triplicata (Spencer, 1963) A
Chromatomyia syngenesiae Hardy, 1849 A Ga-Gr HLM
Hexomyza coprosmae Spencer, 1976 E Sh HG
Liriomyza antipoda Harrison, 1976 E Su Cr-Gr HLM
Liriomyza brassicae (Riley, 1885) A Ga HLM
Liriomyza chenopodii (Watt, 1924) A Gr HLM
Liriomyza citreifemorata (Watt, 1923) E Fo HLM
Liriomyza clianthi (Watt, 1923) E Ga HLM
Liriomyza craspediae Spencer, 1976 E H
Liriomyza flavocentralis (Watt, 1923) E HLM
Liriomyza flavolateralis (Watt, 1923) E H
Liriomyza hebae Spencer, 1976 E Sh HLM
Liriomyza homeri Spencer, 1976 E Sa H
Liriomyza lepidii Harrison, 1976 E Gr HLM
Liriomyza oleariae Spencer, 1976 E Fo HLM
Liriomyza penita Spencer, 1976 E Sa H
Liriomyza plantaginella Spencer, 1976 E Gr HLM
Liriomyza umbrina (Watt, 1923) E FoSh H
Liriomyza umbrinella (Watt, 1923) E FoSh HLM
Liriomyza umbrosa (Watt, 1923) E FoSh H
Liriomyza urticae (Watt, 1924) E Fo HLM
Liriomyza vicina Spencer, 1976 E Gr H
Liriomyza wahlenbergiae Spencer, 1976 E Gr HLM
Liriomyza watti Spencer, 1976 E HLM
Melanagromyza senecionella Spencer, 1976 E Gr HS
Phytoliriomyza bicolorata Spencer, 1976 E Sa H
Phytoliriomyza convoluta Spencer, 1976 E Fo H
Phytoliriomyza cyatheae Spencer, 1976 E Fo HLM
Phytoliriomyza flavopleura (Watt, 1923) E Fo HLM
Phytoliriomyza huttensis Spencer, 1976 E Sa H
Phytoliriomyza tearohensis Spencer, 1976 E Fo H
Phytoliriomyza sp. indet. Spencer 1976 A
Phytomyza clematadi Watt, 1923 E BC Fo HLM
Phytomyza costata Harrison, 1959 E Sa HLS
Phytomyza improvisa Spencer, 1976 E Gr H
Phytomyza lyalli Spencer, 1976 E Sa H
Phytomyza plantaginis Goreau, 1851 A Gr-Sa H
Phytomyza vitalbae Kaltenberg, 1872 A Fo HG
ANTHOMYZIDAE
Zealanthus thorpei Rohácek, 2007 E H
ASTEIIDAE
Asteia crassinervis Malloch, 1930 E Fo Ff
Asteia levis Hutton, 1902 E Fo Ff
Asteia tonnoiri Malloch, 1930 E Fo Ff
AUSTRALIMYZIDAE
Australimyza anisotomae Harrison, 1953 E Su Gr
Australimyza australensis (Mik, 1881) E Su
Australimyza kaikoura Brake & Mathis, 2007 E
Australimyza longiseta Harrison, 1959 E
Australimyza salicorniae Harrison, 1959 E Be
Australimyza setigera Harrison, 1959 E Ca
CANACIDAE
Apetaenus australis (Hutton, 1902) E Su
Apetaenus littoreus (Hutton, 1902) E Su
Isocanace crosbyi Mathis, 1999 E Be
Tethinosoma fulvifrons (Hutton, 1901) E
Zalea horningi (Harrison, 1976) E Su
Zalea earlyi McAlpine, 2007 E Rs
Zalea johnsi McAlpine, 2007 E Rs
Zalea lithax McAlpine, 2007 E Rs
Zalea mathisi McAlpine, 2007 E Rs
Zalea ohauorae McAlpine, 2007 E Rs
Zalea uda McAlpine, 2007 E Rs
Zalea wisei McAlpine, 2007 E Rs
Zalea n. spp. (2) 2E
CHAMAEMYIIDAE A

Chamaemyia polystigma (Meigen, 1830) A Gr IPr
Leucopis tapiae Blanchard, 1964 A BC Fo IPr
Pseudoleucopis benefica? Malloch, 1930 A IPr
CHLOROPIDAE
Aphanotrigonum huttoni (Malloch, 1931) E Fo-Sa
Apotropina quadriseta (Harrison, 1959) E
Apotropina shewelliana (Spencer, 1977) E
Apotropina sulae (Spencer, 1977) E Nb Sc
Apotropina tonnoiri (Sabrosky, 1955) E Be
Apotropina wisei (Harrison, 1959) E
Chlorops multisulcatus Malloch, 1931 E Fo
Chlorops occipitalis Malloch, 1931 E Fo
Conioscinella apterina Spencer, 1977 E Ch
Conioscinella badia (Hutton, 1901) E Fo-Gr
Conioscinella chathamensis Spencer, 1977 E Ch Fo
Conioscinella fulvithorax Spencer, 1977 E Fo
Conioscinella grandis Spencer, 1977 E Fo
Conioscinella speighti (Malloch, 1931) E
Conioscinella spenceri Nartshuk 1993 E
Dicraeus tibialis (Macquart, 1835) A Gr HS
Diplotoxa anorbitalis Malloch, 1931 E
Diplotoxa gemina Spencer, 1977 E We
Diplotoxa harrisoni Spencer, 1977 E
Diplotoxa knighti Spencer, 1977 E Fo
Diplotoxa lineata Malloch, 1931 E We Sa
Diplotoxa moorei (Salmon, 1939) E Gr HS
Diplotoxa neozelandica Harrison, 1959 E
Diplotoxa orbitalis Malloch, 1931 E
Diplotoxa similis Spencer, 1977 E Gr Sa HS
Diplotoxa stepheni Spencer, 1977 E
Gaurax flavoapicalis (Malloch, 1931) Ga-Gr Ca
Gaurax mesopleuralis (Becker, 1911) A Gr-Ga
Gaurax neozealandicus (Malloch, 1931) E Fo
Gaurax solidus Becker, 1910 A We
Hippelates insignificans (Malloch, 1931) E GrD
Lieparella zentae Spencer, 1986 A
Melanum neozelandicum Malloch, 1931 E Gr
Siphunculina breviseta Malloch, 1924 Ga
Siphunculina montana Spencer, 1977 E Sa
Tricimba anglemensis Spencer, 1977 E Sa
Tricimba deansi (Malloch, 1931) E Fo-S
Tricimba dugdalei Spencer, 1977 E
Tricimba flaviseta Malloch, 1931 E Fo
Tricimba fuscipes Malloch, 1931 E Fo
Tricimba kuscheli (Spencer, 1977) E
Tricimba tinctipennis (Malloch, 1931) E Fo
Tricimba walkerae (Spencer, 1977) E
Tricimba watti Spencer, 1977 E Fo
Tricimba n. sp. Macfarlane E Fo
COELOPIDAE
Baeopterus philpotti (Malloch, 1933) E Rs/Be
Baeopterus robustus Lamb, 1909 E Su Be
Chaetocoelopa littoralis (Hutton, 1881) E Be
Chaetocoelopa huttoni Harrison, 1959 E
Coelopella curvipes (Hutton, 1902) E Rs/Be
Icaridion debile (Lamb, 1909) E Rs/Be
Icaridion nasutum Lamb, 1909 E Su Rs/Be
Icaridion nigrifrons (Lamb, 1909) E Rs/Be
CRYPTOCHETIDAE A
Cryptochetum iceryae (Williston, 1888) A BC Ga-Or IP
DROSOPHILIDAE [Hodge, Martin]
Drosophila busckii Coquillett, 1901 A Ga Ff
Drosophila funebris (Fabricius, 1787) A Ga Li Ff D
Drosophila hydei Sturtevant, 1921 A Ga Li
Drosophila immigrans Sturtevant, 1921 A Ga Li Ff
Drosophila melanogaster Meigen, 1830 A Ga Li
Drosophila pseudoobscura Frolova *in* Frolova & Astaurov, 1929 A Ga
Drosophila repleta Woolaston, 1858 A
Drosophila simulans Sturtevant, 1919 A Ga Li
Drosophila n. sp. E
Scaptodrosophila enigma Malloch, 1927 A
Scaptodrosophila kirki (Harrison, 1959) E
Scaptodrosophila neozelandica (Harrison, 1959) E Ff?
Scaptodrosophila n. sp. McEvery E
Scaptomyza elmoi Takada, 1970 A
Scaptomyza flava Fallen, 1823 A Gr HL
Scaptomyza flavella Harrison, 1959 E
Scaptomyza fuscitarsis Harrison, 1959 E Gr
Scaptomyza n. spp. (2) Martin 2E
EPHYDRIDAE [Mathis]
Atissa suturalis Cresson, 1929 A
Brachydeutera sydneyensis Malloch, 1924 A Fs
Ditrichophora flavitarsis (Tonnoir & Malloch, 1926) A
Ditrichophora n. spp. (2) 2E
Eleleides chloris Cresson 1948* A We
Ephydrella aquaria (Hutton, 1901) E F
Ephydrella assimilis (Tonnoir & Malloch, 1926) E F
Ephydrella novaezealandiae (Tonnoir & Malloch, 1926) E F
Ephydrella spathulata Cresson, 1935 E F
Ephydrella thermara Dumbleton, 1969 E F Hs
Ephydrella n. sp. E Ch F
Haloscatella balioptera Mathis, Zatwarnicki & Marris, 2004 E Ch
Haloscatella karekare Mathis, Zatwarnicki & Marris, 2004 E Su
Haloscatella harrisoni Mathis, Zatwarnicki & Marris, 2004 E
Hecamede granifera Thomson, 1869 A Be
Hecamedoides affinis (Tonnoir & Malloch, 1926) E
Hecamedoides n. sp. E
Hyadina irrorata Tonnoir & Malloch, 1926 E We
Hyadina obscurifrons Tonnoir & Malloch, 1926 E
Hyadina n. sp. E
Hydrellia acutipennis Harrison, 1959 E
Hydrellia enderbii (Hutton, 1902) A? Gr
Hydrellia mareeba Bock 1991 A
Hydrellia novaezealandiae Harrison, 1959 E
Hydrellia tritici Coquillett, 1903 A Gr H
Hydrellia velutinifrons Tonnoir & Malloch, 1926 E
Hydrellia williamsi Cresson, 1936 A
Hydrellia n. spp. (3) 3E
Hydrellia n. sp. E Ch
Leptopsilopa n. sp. E
Limnellia abbreviata (Harrison, 1976) E Sn
Limnellia maculipennis Malloch, 1925 A
Nostima duoseta Cresson, 1943
Nostima kiwistriata Edmiston & Mathis, 2007 E
Nostima negramaculata Edmiston & Mathis, 2007 E
Parahyadina lacustris Tonnoir & Malloch, 1926 E
Parahyadina n. spp. (7) 7E Ch
Parydra neozelandica Tonnoir & Malloch, 1926 E
Psilopa metallica (Hutton, 1901) E Gr Li Ff
Scatella acutipennis Harrison, 1964 E Su
Scatella brevis Harrison, 1964 E Su
Scatella nelsoni Tonnoir & Malloch, 1926 E
Scatella nitidithorax Malloch, 1925 A?
Scatella nubeculosa Tonnoir & Malloch, 1926 E We
Scatella subvittata Tonnoir & Malloch, 1926 E We
Scatella tonnoiri Hendel, 1931 E
Scatella unguiculata Tonnoir & Malloch, 1926 E
Scatella vittithorax Malloch, 1925 A? Fs
Scatella n. spp. (5) 5E
Scatella n. spp. (2) 2E Ch
Subpelignus n. sp. E
Zeros invenatus (Lamb, 1912) A
FERGUSONINIDAE [Martin]
Fergusonina metrosiderosi Taylor *in* Taylor, Davies, Martin & Crosby, 2007 E HG
HELCOMYZIDAE
Maorimyia bipunctata (Hutton, 1901) E Be ?Li
HELEOMYZIDAE
Allophylina albitarsis Tonnoir & Malloch, 1927 E
Allophylopsis bivittata Harrison, 1959 E Li
Allophylopsis chathamensis Tonnoir & Malloch, 1927 E Ch
Allophylopsis distincta Tonnoir & Malloch, 1927 E
Allophylopsis fulva (Hutton, 1901) E
Allophylopsis fuscipennis Tonnoir & Malloch, 1927 E
Allophylopsis hudsoni (Hutton, 1901) E
Allophylopsis inconspicua Tonnoir & Malloch, 1927 E
Allophylopsis laquei (Hutton, 1902) E Su
Allophylopsis lineata Tonnoir & Malloch, 1927 E
Allophylopsis minuta Tonnoir & Malloch, 1927 E
Allophylopsis obscura Tonnoir & Malloch, 1927 E
Allophylopsis philpotti Tonnoir & Malloch, 1927 E
Allophylopsis rufithorax Tonnoir & Malloch, 1927 E
Allophylopsis scutellata (Hutton, 1901) E
Allophylopsis subscutellata Tonnoir & Malloch, 1927 E
Allophylopsis sp. indet.* Sa Andrew
Aneuria angusta Harrison, 1959 E
Aneuria bipunctata Malloch, 1930 E
Aneuria imitatrix Malloch, 1930 E
Aneuria sexpunctata Malloch, 1930 E
Aneuria tripunctata Malloch, 1930 E
Fenwickia affinis Harrison, 1959 E
Fenwickia caudata Harrison, 1959 E
Fenwickia claripennis Malloch, 1930 E
Fenwickia hirsuta Malloch, 1930 E
Fenwickia nuda Malloch, 1930 E
Fenwickia similis Malloch, 1930 E
Oecothea fenestralis (Fallen, 1820) A Ga
Prosopantrum flavifrons (Tonnoir & Malloch, 1927) A Gr
Tephrochlamys rufiventris (Meigen, 1830) A Ga
Xeneura picata (Hutton, 1902) E
HELOSCIOMYZIDAE
Dasysciomyza pseudosetuligera (Tonnoir & Malloch, 1928) E
Dasysciomyza setuligera (Malloch, 1922) E
Helosciomyza subalpina Tonnoir & Malloch, 1928 E IP
Napaeosciomyza rara (Hutton, 1901) E
Napaeosciomyza spinicosta (Malloch, 1922) E
Napaeosciomyza subspinicosta (Tonnoir & Malloch, 1928) E
Polytocus costatus Harrison, 1976 E Su
Polytocus spinicosta Lamb, 1909 E Su
Scordalus femoratus (Tonnoir & Malloch, 1928) E
Xenosciomyza prima Tonnoir & Malloch, 1928 E
Xenosciomyza turbotti Harrison, 1955 E Su
HUTTONINIDAE E
Huttonina (Huttonina) abrupta Tonnoir & Malloch, 1928 E
Huttonina (*H.*) *brevis* Malloch, 1930 E
Huttonina (*H.*) *elegans* Tonnoir & Malloch, 1928 E
Huttonina (*H.*) *furcata* Tonnoir & Malloch, 1928 E
Huttonina (*H.*) *glabra* Tonnoir & Malloch, 1928 E
Huttonina (*H.*) *scutellaris* Tonnoir & Malloch, 1928 E
Huttonina (Huttoninella) angustipennis Tonnoir & Malloch, 1928 E
Huttonina (*H.*) *claripennis* Harrison, 1959 E
Prosochaeta prima Malloch, 1935 E
LAUXANIIDAE
Poecilohetaerella antennata Harrison, 1959 E
Poecilohetaerella bilineata (Hutton, 1901) E Gr
Poecilohetaerella dubiosa Tonnoir & Malloch, 1926 E
Poecilohetaerella minuta (Tonnoir & Malloch, 1926) E
Poecilohetaerella punctatifrons (Tonnoir & Malloch, 1926) E
Poecilohetaerella punctatifrons obscura (Tonnoir & Malloch, 1926) E
Poecilohetaerella scutellata Harrison, 1959 E
Poecilohetaerella watti Tonnoir & Malloch, 1926 E
Poecilohetaerus punctatifacies Tonnoir & Malloch, 1926 E
Sapromyza arenaria Tonnoir & Malloch, 1926 E
Sapromyza dichromata Walker, 1849 E
Sapromyza neozelandica Tonnoir & Malloch, 1926 E Li
Sapromyza persimillima Harrison, 1959 E
Sapromyza simillima Tonnoir & Malloch, 1926 E
Trypetisoma guttatum (Tonnoir & Malloch, 1926) E

Trypetisoma costatum (Harrison, 1959) E
Trypetisoma tenuipenne (Malloch, 1930) E
MILICHIIDAE [Andrew, Thorpe]
Desmonetopa sp. indet. Thorpe
?*Milichillia* sp. indet. Thorpe
Paramyia sp. indet. Andrew
Stomosis sp. indet. Andrew
PALLOPTERIDAE
Maorina apicalis (Walker, 1849) E Fo
Maorina aristata Malloch, 1930 E
Maorina bimacula Malloch, 1930 E
Maorina gourlayi (Harrison, 1959) E
Maorina lamellata (Harrison, 1959) E
Maorina macronycha Malloch, 1930 E
Maorina palpalis Malloch, 1930 E
Maorina pseudoapicalis (Harrison, 1959) E
Maorina scutellata Malloch, 1930 E
Maorina n. sp. Macfarlane, CMNZ E Ch
PERISCELIDIDAE [Andrew]
Cyamops sp. indet.* Khoo 1985 ?E
PIOPHILIDAE A
Piophila australis (Harrison, 1959) A
Piophila casei (Linnaeus, 1758) A St
PLATYSTOMATIDAE
Zealandortalis interrupta Malloch, 1930 E
Zealandortalis philpotti Harrison, 1959 E
PSEUDOPOMYZIDAE
Pseudopomyza antipoda (Harrison, 1955) E Su Gr
Pseudopomyza aristata (Harrison, 1959) E Fo
Pseudopomyza brevicaudata (Harrison, 1964) E Su
Pseudopomyza brevis (Harrison, 1976) E Su
Pseudopomyza flavitarsis (Harrison, 1959) E Fo
Pseudopomyza neozelandica (Malloch, 1933) E
PSILIDAE A
Chamaepsila rosae (Fabricius, 1794) A Ga-Cr H
SCIOMYZIDAE
Eulimnia milleri Tonnoir & Malloch, 1928 E
Eulimnia philpotti Tonnoir & Malloch, 1928 E MoP F
'Limnia' transmarina? Schiner, 1868 E
Neolimnia castanea (Hutton, 1904) E
Neolimnia diversa Tonnoir & Malloch, 1928 E
Neolimnia irrorata Tonnoir & Malloch, 1928 E
Neolimnia minuta Tonnoir & Malloch, 1928 E
Neolimnia nitidiventris Tonnoir & Malloch, 1928 E
Neolimnia obscura (Hutton, 1901) E
Neolimnia pepikeiti Barnes, 1979 E
Neolimnia raiti Barnes, 1979 E
Neolimnia repo Barnes, 1979 E
Neolimnia sigma (Walker, 1849) E MoP F
Neolimnia striata (Hutton, 1904) E
Neolimnia striata brunneifrons Tonnoir & Malloch, 1928 E
Neolimnia tranquilla (Hutton, 1901) E
Neolimnia ura Barnes, 1979 E
Neolimnia vittata Harrison, 1959 E
SEPSIDAE A
Lasionemopoda hirsuta (de Meijere, 1906) A Gr D
SPHAEROCERIDAE (Marshall)
Biroina myersi (Richards, 1973) E
Biroina n. spp. (3) Marshall 3E
Coproica ferruginata (Stenhammar, 1855) A
Coproica hirtula (Rondani, 1880) A Li
Coproica hirticula Collin 1956* A
Howickia trilineata (Hutton, 1901) E Li
Howickia n. sp. Marshall E
Ischiolepta pusilla (Fallen, 1820) A D-Ca Li
Leptocera caenosa (Rondani, 1880) A
Leptocera n. sp. E Sa
Minilimosina knightae (Harrison, 1959) E
Minilimosina sp. indet. E Marshall
Minilimosina n. sp. E Marshall
Norrbomia sordida (Zetterstedt, 1847)* A D
Opacifrons maculifrons (Becker, 1907) A We
Opalimosina mirabilis (Collin, 1902) A Li
Phthitia emarginata Marshall in Marshall, Hall & Hodge, 2009 E Ff
Phthitia empirica (Hutton, 1901) A Ca-Nb
Phthitia lobocercus Marshall *in* Marshall & Smith, 1992 E
Phthitia notthomasi Marshall *in* Marshall & Smith, 1992 E
Phthitia plesiocerus Marshall in Marshall, Hall & Hodge, 2009 E Fs
Phthitia rennelli (Harrison, 1964) E Su Ca
Phthitia thomasi (Harrison, 1959) A Ca
Pullimosina heteroneura (Haliday, 1836) A
Pullimosina pullula (Zetterstedt, 1847) * A
Rachispoda fuscipennis (Haliday, 1833) A
Rachispoda sp. indet.* Marshall
Spelobia bifrons (Stenhammar, 1855) A Li
Spelobia luteilabris (Rondani, 1880) A
Spelobia pseudosetaria (Duda, 1918) * A
Sphaerocera curvipes Latreille, 1805 A D
Telomerina flavipes (Meigen, 1830) A
Thoracochaeta alia Marshall & Rohácek, 2000 E Be Li
Thoracochaeta ancudensis (Richards, 1931) Be Li
Thoracochaeta conglobata Marshall & Rohacek, 2000 E Be Li
Thoracochaeta harrisoni Marshall & Rohacek, 2000 E Be Li Ch
Thoracochaeta imitatrix Marshall & Rohacek, 2000 E Be Li
Thoracochaeta mucronata Marshall & Rohacek, 2000 E Be Li
Thoracochaeta zealandica (Harrison, 1959) E Be Li
Trachyopella lineafrons Spuler, 1925* A Li Marshall
Gen. nov. *mediospinosa* (Duda, 1925) A Li
TEPHRITIDAE [Macfarlane]
Austrotephritis cassiniae (Malloch, 1931) E
Austrotephritis marginata (Malloch, 1931) E
Austrotephritis plebeia (Malloch, 1931) E
Austrotephritis thoracica (Malloch, 1931) E
Procecidochares alani Steyskal, 1974 A BC HG
Procecidochares utilis Stone, 1947 A BC HG
Sphenella fascigera (Malloch, 1931) E
Trupanea alboapicata Malloch, 1931 E
Trupanea centralis Malloch, 1931 E
Trupanea completa Malloch, 1931 E
Trupanea dubia Malloch, 1931 E
Trupanea extensa Malloch, 1931 E
Trupanea fenwicki Malloch, 1931 E
Trupanea vitiosa Foote, 1989 E
Trupanea vittigera Malloch, 1931 E
Trupanea watti Malloch, 1931 E
Trupanea n. sp. E
Urophora cardui? (Linnaeus, 1758) A BC Gr HSG
Urophora solstitialis (Linnaeus, 1758) A BC Gr H
Urophora stylata (Fabricius, 1775) A BC Gr HG
TERATOMYZIDAE
Teratomyza neozelandica Malloch, 1933 E
Teratomyza n. sp. McAlpine E

CALYPTRATAE
ANTHOMYIIDAE A? [Andrew]
Anthomyia punctipennis Wiedemann, 1830 A Gr
Botanophila jacobaeae (Hardy, 1872) A BC Gr HS
Delia platura (Meigen, 1826) A Gr HL
Delia ?urbana (Malloch, 1924)* A Andrew
Fucellia spp. indet. (2)* 2A? Be Andrew
CALLIPHORIDAE
Calliphora hilli Patton, 1925 A Fo-Gr Ca
Calliphora h. kermadecensis Kurahashi, 1971 E K Fo-Gr Ca
Calliphora quadrimaculata (Swederus, 1787) E Fo-Gr Ca
Calliphora stygia (Fabricius, 1782) A Gr Ca-LP
Calliphora vicina Robineau-Desvoidy, 1830 A Gr Ca
Chrysomya megacephala (Fabricius, 1794) A Ca
Chrysomya rufifacies (Macquart, 1843) A Ca
Hemipyrellia ligurriens? (Wiedemann, 1830) A Ca-Li
Lucilia cuprina (Wiedemann, 1830) A Gr Ca-LP
Lucilia sericata (Meigen, 1826) A Gr-Be Ca-LP
Mystacinobia zelandica Holloway, 1976 I[C] E Fo D
Pollenia advena Dear, 1986 E Fo-Ga
Pollenia aerosa Dear, 1986 E
Pollenia antipodea Dear, 1986 E Gr
Pollenia astrictifrons Dear, 1986 E Sa
Pollenia atricoma Dear, 1986 E ?Gr-Sa
Pollenia atrifemur? Malloch, 1930 E
Pollenia commensurata Dear, 1986 E
Pollenia consanguinea Dear, 1986 E GrSa
Pollenia consectata Dear, 1986 E Ga
Pollenia cuprea? Malloch, 1930 E
Pollenia demissa (Hutton, 1901) E
Pollenia dysaethria Dear, 1986 E ?Gr
Pollenia dyscheres Dear, 1986 E GrSa
Pollenia enetera Dear, 1986 E
Pollenia eurybregma Dear, 1986 E GrSa
Pollenia fulviantenna Dear, 1986 E Gr
Pollenia fumosa (Hutton, 1901) E GrSa
Pollenia hispida Dear, 1986 E GrSa
Pollenia immanis Dear, 1986 E GrSa
Pollenia insularis Dear, 1986 E
Pollenia lativertex Dear, 1986 E GrSa
Pollenia limpida Dear, 1986 E GrSa
Pollenia nigripalpis Dear, 1986 E TK Fo
Pollenia nigripes Malloch, 1930 E
Pollenia nigrisquama Malloch, 1930 E Fo-Gr
Pollenia notialis Dear, 1986 E GrSa
Pollenia opalina Dear, 1986 E ?GrSa
Pollenia oreia Dear, 1986 E GrSa
Pollenia pernix (Hutton, 1901) E Gr-Sa
Pollenia primaeva Dear, 1986 E
Pollenia pseudorudis Rognes, 1985 A Gr EP
Pollenia pulverea Dear, 1986 E FoWe
Pollenia sandaraca Dear, 1986 E
Pollenia scalena Dear, 1986 E Su Gr
Pollenia uniseta Dear, 1986 E GrSa
Ptilonesia auronotata (Macquart, 1855) A? Be-Gr Ca
Xenocalliphora antipodea (Hutton, 1902) E Su Ca
Xenocalliphora clara Dear, 1986 E Ca
Xenocalliphora divaricata Dear, 1986 E Ca
Xenocalliphora eudypti (Hutton, 1902) E Su Ca
Xenocalliphora flavipes (Lamb, 1909) E Su Ca
Xenocalliphora hortona (Walker, 1849) E Gr-Be Ca
Xenocalliphora neohortona (Miller, 1939) E Ca
Xenocalliphora neozealandica (Murray, 1954) E Ca
Xenocalliphora solitaria Dear, 1986 E Ch Ca
Xenocalliphora vetusta Dear, 1986 E TK Ca
Xenocalliphora viridiventris Malloch, 1930 E Su
FANNIIDAE
Euryomma perigrinum (Meigen, 1826) A
Fannia albitarsis Stein, 1911 A
Fannia canicularis (Linnaeus, 1761) A Ga Li
?*Fannia scalaris* (Fabricius, 1794) A
Fannia n. sp. (larva described) A
Fannia n. spp. (2) (larva described) 2E Be Nb
Fannia n. sp. (larva described) E Ch Be Nb
HIPPOBOSCIDAE [Sinclair]
Melophagus ovinus (Linnaeus, 1758) A Gr LP
Ornithoica ?exilis (Walker, 1861) A? Fo BP
Ornithoica stipituri (Schiner, 1868) A Gr-Ga BP
Ornithomya nigricornis Erichson, 1842 Gr BP
Ornithomya variegata Bigot, 1885 A? FoGa BP
MUSCIDAE [Harrison]
Calliphoroides antennatis (Hutton, 1881) E
'Coenosia' algivora Hutton, 1901 E
'Coenosia' rubriceps Hutton, 1901 E
Exsul singularis Hutton, 1901 I[C] E Sa IPr
Exsul tenuis Malloch, 1923 E
Exsul n. spp. (2) 2E
Helina sexmaculata (Preyssler, 1791) A
Hydrotaea rostrata (Robineau-Desvoidy, 1830) A

Gr Ca
Idiohelina nubeculosa Malloch, 1921 E
Idiohelina nelsoni Malloch 1929 }
Idiohelina setifemur Malloch, 1929
Limnohelina bivittata Malloch, 1930 E F
Limnohelina debilis (Hutton, 1901) E
Limnohelina dorsovittata Malloch, 1930 E
Limnohelina grisea Malloch, 1930 E F
Limnohelina huttoni Malloch, 1930 E
Limnohelina nelsoni Malloch, 1930 E
Limnohelina nigripes Malloch, 1930 E
Limnohelina smithii (Hutton, 1901) E F
Limnohelina spinipes (Walker, 1849) E F
Limnohelina uniformis Malloch, 1930 E
Macrorchis meditata (Fallen, 1825) A?
Millerina hudsoni Malloch, 1925 E
Millerina nigrifemur Malloch, 1925 E
Millerina pennata Malloch, 1925 E
Millerina rapax (Hutton, 1901) E
Muscina stabulans (Fallen, 1817) A Li Ca D
Musca domestica Linnaeus, 1758 A Ga Li-D
Paracoenosia tonnoiri Malloch, 1938 E
Paralimnophora depressa Lamb, 1909 E Su
Paralimnophora filipennis [NC] (Lamb, 1909) E Su
Paralimnophora fumipennis [NC] (Lamb, 1909) E Su
Paralimnophora purgatoria[NC] (Hutton, 1901) E
Paralimnophora n. spp. (7) Harrison 7E
Pygophora apicalis Schiner, 1868 A
'Spilogona' albifrons Malloch, 1931 E
'Spilogona' argentifrons Malloch, 1931 E
'Spilogona' aucklandica (Hutton, 1902) E Su We
'Spilogona' aureifacies Malloch, 1931 E
'Spilogona' badia (Hutton, 1901) E
'Spilogona' brunneinota (Harrison, 1955) E Su
'Spilogona' brunneivittata (Harrison, 1955) E Su
'Spilogona' carbonaria (Hutton, 1901) E
'Spilogona' curvipes (Lamb, 1909) E Su
'Spilogona' dolosa (Hutton, 1901) E
'Spilogona' flaviventris Malloch, 1931 E
'Spilogona' fuliginosa (Hutton, 1901) E Su
'Spilogona' fulvescens (Hutton, 1901) E
'Spilogona' fumicosta Malloch, 1931 E
'Spilogona' insularis (Lamb, 1909) E Su
'Spilogona' lasiophthalma (Lamb, 1909) E Su
'Spilogona' latimana Malloch, 1931 E
'Spilogona' maculipennis (Hutton, 1901) E
'Spilogona' melas (Schiner 1868) E
'Spilogona' minuta (Harrison, 1955) E Su
'Spilogona' ordinata (Hutton, 1901) E
'Spilogona' sorenseni (Harrison, 1955) E Su
'Spilogona' tenuicornis (Malloch, 1923) E
'Spilogona' villosa (Hutton, 1902) E
'Spilogona' n. spp. (57) 57E
Stomoxys calcitrans (Linnaeus, 1758) A D-Li BF
Gen. nov. 1 *limpida* (Hutton, 1901) E
Gen. nov. 1 n. spp. (20) 20E
OESTRIDAE A
Gasterophilus haemorrhoidalis (Linnaeus, 1758) A Gr LP
Gasterophilus nasalis (Linnaeus, 1758) A Gr LP
Hypoderma bovis (Linnaeus, 1758) A Gr LP
Oestrus ovis Linnaeus, 1758 A Gr LP
SARCOPHAGIDAE [Andrew]
Oxysarcodexia varia (Walker, 1836) A Gr D
Sarcophaga bifrons Walker, 1853 A
Sarcophaga crassipalpis Macquart, 1939 A Ga Li
Sarcophaga peregrina (Robineau-Desvoidy, 1830) A
Tricharaea brevicornis Wiedemann, 1830* A?
Miltogramminae gen. et sp. indet.* Pape
TACHINIDAE [Dugdale]
Altaia geniculata Malloch, 1938 E IP
Asetulia nigropolita Malloch, 1938 E IP
Asetulia n. sp. NZAC E IP
Austromacquartia claripennis (Malloch, 1932) E IP
Avibrissia longirostris Malloch, 1932 E IP
Avibrissia n. sp. NZAC E IP
Avibrissina brevipalpis Malloch, 1932 E Gr So IP
Avibrissina laticornis Malloch, 1938 E IP
Avibrissina n. spp. (2) NZAC 2E IP
Bothrophora lupina (Swederus, 1787) E IP
Calcager apertum Hutton, 1901 E IP
Calcager dubius Malloch, 1938 E IP
Calcager n. sp. NZAC E IP
Calcageria incidens Curran, 1927 E IP
Calcageria varians Malloch, 1938 E IP
Calosia binigra (Malloch, 1938) E IP
Calosia n. spp. (3) NZAC 3E IP
Calotachina tricolor Malloch, 1938 E IP
Campbellia campbelli Miller, 1928 E IP
Campbellia cockaynei Miller, 1928 E IP
Campbellia lancifer (Malloch, 1930) E IP
Campylia nudara Malloch, 1938 E IP
Campylia temerarium (Hutton, 1901) E IP
Campylia n. sp. NZAC E IP
Chaetophthalmus bicolor (Macquart, 1848) A GrGa IP
Chaetopletha centralis Malloch, 1938 E IP
Erythronychia aliena Malloch, 1932 E IP
Erythronychia aperta Malloch, 1932 E IP
Erythronychia australiensis (Schiner, 1868) E IP
Erythronychia defecta Malloch, 1932 E IP
Erythronychia grisea Malloch, 1932 E IP
Erythronychia hirticeps Malloch, 1932 E IP
Erythronychia minor Malloch, 1932 E IP
Erythronychia princeps (Curran, 1927) E IP
Erythronychia velutina Malloch, 1932 E IP
'Evibrissa' huttoni Malloch, 1931 E IP
Genotrichia minor Malloch, 1938 E IP
Genotrichia tonnoiri Malloch, 1938 E IP
Gracilicera monticola (Malloch, 1938) E IP
Gracilicera pallipes (Malloch, 1938) E IP
Gracilicera politiventris (Malloch, 1938) E IP
Graphotachina sinuata Malloch, 1938 E IP
Heteria appendiculata Malloch, 1930 E IP
Heteria atripes Malloch, 1930 E IP
Heteria extensa Malloch, 1930 E IP
Heteria flavibasis Malloch, 1930 E IP
Heteria plebeia Malloch, 1930 E IP
Heteria punctigera Malloch, 1930 E IP
Heteria n. spp. (2) NZAC 2E IP
Huttonobesseria verecunda (Hutton, 1901) E We IP
Mallochomacquartia flavohirta (Malloch, 1938) E IP
Mallochomacquartia nigrihirta (Malloch, 1938) E IP
Mallochomacquartia vexata (Hutton, 1901) E IP
Medinella albifrons Malloch, 1938 E IP
Medinella flavofemorata Malloch, 1938 E IP
Medinella nigrifemorata Malloch, 1938 E IP
Medinella varipes Malloch, 1938 E IP
Medinella n. spp. (2) NZAC 2E IP
Microhystricia gourlayi Malloch, 1938 E IP
Montanarturia dimorpha (Malloch, 1938) E IP
Neoerythronychia hirta Malloch, 1932 E IP
Neotachina angusticornis Malloch, 1938 E IP
Neotachina depressa Malloch, 1938 E Gr-Sh So IP
Neotachina laticornis Malloch, 1938 E IP
Neotachina obtusa Malloch, 1938 E IP
Neotachina n. spp. (2) NZAC 2E IP
Neotryphera atra Malloch, 1938 E IP
Occisor atratus Malloch, 1938 E IP
Occisor inscitus Hutton, 1901 E IP
Occisor versutus Hutton, 1901 E Gr IP
Occisor n. spp. (2) NZAC 2E IP
Pales atrox (Hutton, 1901) E IP
Pales aurea (Hutton, 1901) E IP
Pales brouni (Hutton, 1901) E IP
Pales casta (Hutton, 1904) E IP
Pales clathrata[1] (Nowicki, 1875) E IP
Pales efferata (Hutton, 1901) E GaGr IP
Pales exitiosa (Hutton, 1901) E IP
Pales feredayi (Hutton, 1901) E GaCr IP
Pales funesta (Hutton, 1901) E GaOr Fo IP
Pales inconspicua (Hutton, 1901) E IP
Pales marginata (Hutton, 1901) E IP
Pales nefaria (Hutton, 1901) E IP
Pales nyctemeriana (Hudson, 1883) E GrCr IP
Pales orasus[1] (Walker, 1849) E IP
Pales perniciosa (Hutton, 1901) E IP
Pales tecta (Hutton, 1901) E IP
Pales usitata[1] (Hutton, 1901) E Gr IP
Pales n. spp. (10) NZAC 10E IP
Peremptor egmonti (Hutton, 1901) E IP
Peremptor kumaraensis (Miller, 1913) E IP
Peremptor modicus Hutton, 1901 E IP
Peremptor n. sp. NZAC E IP
Perrissina albiceps Malloch, 1938 E IP
Perrissina brunniceps Malloch, 1938 E IP
Perrissina crocea Malloch, 1938 E Fo IP
Perrissina variceps Malloch, 1938 E IP
Perrissina xanthopyga Malloch, 1938 E IP
Perrissinoides cerambycivorae Dugdale, 1962 E Fo IP
Phaoniella bifida Malloch, 1938 E IP
Plagiomyia achaeta Malloch, 1938 E IP
Plagiomyia alticeps Malloch, 1938 E IP
Plagiomyia longicornis Malloch, 1938 E GaGr IP
Plagiomyia longipes Malloch, 1938 E IP
Plagiomyia turbida (Hutton, 1901) E IP
Platytachina angustifrons Malloch, 1938 E IP
Platytachina atricornis Malloch, 1938 E IP
Platytachina difficilis Malloch, 1938 E IP
Platytachina latifrons Malloch, 1938 E IP
Platytachina major Malloch, 1938 E IP
Plethochaetigera fenwicki Malloch, 1938 E IP
Plethochaetigera isolata Malloch, 1938 E IP
Plethochaetigera setiventris Malloch, 1938 E IP
Plethochaetigera n. spp. (4) NZAC 4E IP
Procissio albiceps Malloch, 1938 E IP
Procissio cana Hutton, 1901 E Gr So IP
Procissio lateralis Malloch, 1938 E IP
Procissio milleri Malloch, 1938 E Gr So IP
Procissio montana Hutton, 1901 E IP
Procissio n. sp.'clear wing' NZAC E IP
Prosenosoma greyi Malloch, 1938 E IP
Protohystricia alcis (Walker, 1849) E Gr IP
Protohystricia gourlayi (Tonnoir, 1935) E IP
Protohystricia huttoni Malloch, 1930 E IP
Protohystricia orientalis (Schiner, 1868) E IP
Pygocalcager humeratum (Hutton, 1901) E IP
Pygocalcager n. spp. (2) NZAC 2E IP
Senostoma rubricarinatum? [7] (Macquart, 1846) IP
'Tachina' mestor?[22] Walker, 1849 E IP
'Tachina' sosilus?[22] Walker, 1849 E IP
Tachineo clarkii Hutton, 1901 E IP
Triarthria setipennis? (Fallen, 1810) A Gr-Ga IP
Trigonospila brevifacies (Hardy, 1934) A BC GaOr Fo IP
Truphia grisea Malloch, 1930 E IP
Truphia n. sp. NZAC E IP
Trypherina grisea Malloch, 1938 E IP
Trypherina n. spp. (2) NZAC 2E IP
Uclesiella irregularis Malloch, 1938 E IP
Uclesiella n. spp. (2) 2E IP
Veluta albicincta Malloch, 1938 E IP
Wattia ferruginea Malloch, 1938 E IP
Wattia petiolata Malloch, 1938 E IP
Wattia sessilis Malloch, 1938 E IP
Xenorhynchia peeli Malloch, 1938 E IP
Xenorhynchia n. spp. (3) NZAC 3E IP
Zealandotachina infuscata Malloch, 1938 E IP
Zealandotachina lamellata Malloch, 1938 E IP
Zealandotachina latifrons Malloch, 1938 E IP
Zealandotachina nigrifemorata Malloch, 1938 E IP
Zealandotachina quadriseta Malloch 1938 E IP
Zealandotachina quadrivittata Malloch, 1938 E IP
Zealandotachina setigera Malloch, 1938 E IP
Zealandotachina subtilis (Hutton, 1901) E IP
Zealandotachina tenuis Malloch, 1938 E IP
Zealandotachina varipes Malloch, 1938 E IP
Gen. nov. 1 et n. spp. (2) NZAC 2E IP
Gen. nov. 2 et n. spp. (2) NZAC 2E IP

Gen. nov. 3 et n. sp. NZAC E IP
Gen. nov. 4 et n. sp. NZAC E IP
Gen. nov. 5 et n. sp. NZAC E IP
Gen. nov. 6 et n. sp. NZAC E IP
Gen. nov. 7 et n. sp. NZAC E IP
Gen. nov. 8 Voriinae et n. sp. NZAC E IP

Order STREPSIPTERA
[Compiled by R. P. Macfarlane]
ELENCHIDAE
Elenchus maoricus Gourlay, 1953 E Gr IP
HALICTOPHAGIDAE
Coriophagus casui Cowley 1984 E Fo IP
Coriophaginae n. spp. (2) 2E IP

Order HYMENOPTERA
[Compiled by J. A Berry]
Suprafamilial classification generally follows that used by LaSalle and Gauld (1993), with some updates, which are noted. Families and genera are arranged alphabetically. (?) indicates the record of the species *in* New Zealand is doubtful. 'n.sp.' indicates known undescribed species; 'sp.' indicates indeterminate status.

For generic records where no species are reported from New Zealand, the origin of the generic record is indicated in square brackets [...]. In some cases this is a publication, in others the record is based on expert opinion. Authorities thus cited are: ADA – Andrew D. Austin; AH – Allen Heath; SB – Sergey Belokobylskij; JAB – Jocelyn A. Berry; ZB – Z. Boucek; PD – Paul Dessart; BJD – Barry J. Donovan; JWE – John W. Early; NDMF – N.D.M. Fergusson; IDG – Ian D. Gauld; GAPG – Gary Gibson; EEG – Eric E. Grissell; JH – John Heraty; JL – John LaSalle; JBM – J. B. Munro; MM – Manfred Mackauer; TM – T. Megyaszai; IDN – Ian D. Naumann; JSN – John S. Noyes; SRS – Scott R. Shaw; SET – Stephen Thorpe; ST – Serguei Triapitsyn; DBW – David B. Wahl.

Suborder SYMPHYTA
TENTHREDINOIDEA
PERGIDAE A
Phylacteophaga froggatti Riek, 1955 A
TENTHREDINIDAE A
Caliroa cerasi (Linnaeus, 1758) A
Nematus oligospilus Förster, 1854 A
Pontania proxima (Lepeletier, 1823) A
Priophorus morio (Lepeletier, 1823) A
SIRICOIDEA A
SIRICIDAE A
Sirex noctilio Fabricius, 1793 A
XIPHYDRIOIDEA
XIPHYDRIIDAE
Moaxiphia deceptus (Smith, 1876) E
Moaxiphia duniana (Gourlay, 1927) E
ORUSSOIDEA
ORUSSIDAE
Guiglia schauinslandi (Ashmead, 1903) E

Suborder APOCRITA
APOCRITA-PARASITICA
EVANIOIDEA
GASTERUPTIIDAE
Gasteruption expectatum Pasteels, 1957 E
Gasteruption flavicuspis Kieffer, 1911 E
Gasteruption scintillans Pasteels, 1957 E
Pseudofoenus crassipes (Smith, 1876) E
Pseudofoenus nocticolor Kieffer, 1911 E
Pseudofoenus pedunculatus (Schletterer, 1889) E
Pseudofoenus unguiculatus (Westwood, 1841) E
CERAPHRONOIDEA
CERAPHRONIDAE
Aphanogmus sp. [PD]
Ceraphron sp. [PD]
MEGASPILIDAE
Conostigmus variipilosus Dessart, 1997 E
Dendrocerus aphidum (Rondani, 1877) A
Dendrocerus carpenteri (Curtis, 1829) A
Dendrocerus laticeps (Hedicke, 1929) A
Lagynodes coxivillosus coxivillosus Dessart, 1987 E
Lagynodes gastroleius Dessart, 1987 E
Lagynodes hecaterapterus Dessart, 1981 E
Lagynodes velutinus Dessart & Masner *in* Dessart, 1977 E
Trichosteresis glaber (Boheman, 1832) A
PROCTOTRUPOIDEA
PROCTOTRUPIDAE
Exallonyx trifoveatus Kieffer, 1908 A
Fustiserphus intrudens (Smith, 1878) E
Fustiserphus longiceps Townes *in* Townes & Townes, 1981 E
Fustiserphus spp. (2-3) [Early & Dugdale 1994]
Oxyserphus baini Townes *in* Townes & Townes, 1981 E
Oxyserphus maculipennis (Cameron, 1888) E
Oxyserphus pediculatus Townes *in* Townes & Townes, 1981 E
DIAPRIOIDEA *sensu* Sharkey (2007)
DIAPRIIDAE
AMBOSITRINAE
Archaeopria eriodes Naumann, 1988 E
Archaeopria pelor Naumann, 1988 E
Archaeopria pristina Naumann, 1988 E
Betyla auriger Naumann, 1988 E
Betyla eupepla Naumann, 1988 E
Betyla fulva Cameron, 1889 E
Betyla karamea Naumann, 1988 E
Betyla midas Naumann, 1988 E
Betyla paparoa Naumann, 1988 E
Betyla prosedera Naumann, 1988 E
Betyla rangatira Naumann, 1988 E
Betyla thegalea Naumann, 1988 E
Betyla tuatara Naumann, 1988 E
Betyla wahine Naumann, 1988 E
Diphoropria sinuosa Naumann, 1988 E
Diphoropria kuscheli Naumann, 1988 E
Maoripria annettae Naumann, 1988 E
Maoripria earlyi Naumann, 1988 E
Maoripria masneri Naumann, 1988 E
Maoripria verticillata Naumann, 1988 E
Pantolytomyia flocculosa Naumann, 1988 E
Pantolytomyia insularis Naumann, 1988 E
Pantolytomyia polita Naumann, 1988 E
Pantolytomyia takere Naumann, 1988 E
Pantolytomyia taurangi Naumann, 1988 E
Pantolytomyia tungane Naumann, 1988 E
Pantolytomyia wairua Naumann, 1988 E
Parabetyla nauhea Naumann, 1988 E
Parabetyla ngarara Naumann, 1988 E
Parabetyla pipira Naumann, 1988 E
Parabetyla pokorua Naumann, 1988 E
Parabetyla spinosa Brues, 1922 E
Parabetyla tahi Naumann, 1988 E
Parabetyla tika Naumann, 1988 E
Zealaptera chambersi Naumann, 1988 E
BELYTINAE
Gladicauda aucklandica Early, 1980 E
Gladicauda spp. (~10) [JWE] E
Stylaclista quasimodo Early, 1980 E
Stylaclista spp. (~9) [TM]
Synacra sp. [JWE]
Gen. nov. et n. spp. (~10) [JWE] E
DIAPRIINAE
Antarctopria campbellana Yoshimoto, 1964 E
Antarctopria coelopae Early, 1978 E
Antarctopria diomedeae Early, 1978 E
Antarctopria latigaster Brues, 1920 E
Antarctopria rekohua Early, 1978 E
Basalys spp. (> 3) [JWE]
Cardiopria sp. [JWE]
Diapria sp. [JWE]
Entomacis subaptera Early, 1980 E
Entomacis spp. (3-19) [JWE]
Hemilexomyia spinosa Early, 1980 E
Hemilexomyia sp. [JWE]
Idiotypa spp. (7-35) [JWE]
Malvina helosciomyzae Early & Horning, 1978 E
Malvina insulae Early, 1980 E
Malvina punctata Cameron, 1889 E
Malvina quadriceps (Smith, 1878) E
Malvina spp. (~8) [JWE] E
Neurogalesus spp. (~2) [JWE]
Paramesius spp. (6-30) [JWE]
Pentapria sp. [JWE]
Probetyla subaptera Brues, 1922 E
Spilomicrus barnesi Early & Horning, 1978 E
Spilomicrus carolae Early, 1980 E
Spilomicrus pilgrimi Early, 1978 E
Spilomicrus spp. (54-100) [JWE]
Trichopria spp. (5) [JWE]
Gen. nov. et n. spp. (5) [JWE] E
Gen. nov. et n. sp. [JWE] E
Gen. nov. et n. spp. (2) [JWE.] E
Gen. nov. et n. sp. [JWE]
MAAMINGIDAE E
Maaminga marrisi Early, Masner, Naumann & Austin, 2001 E
Maaminga rangi Early, Masner, Naumann & Austin, 2001 E
PLATYGASTROIDEA *sensu* Sharkey (2007)
PLATYGASTRIDAE [Includes former Scelionidae (Sharkey 2007)]
PLATYGASTRINAE
Allostemma sp. [Masner & Huggert 1989]
Annettella gracilis Masner & Huggert, 1989 E
Inostemma boscii (Jurinc, 1807) A BC
Iphitrachelus sp. [Masner & Huggert 1989]
Orseta sp. [Masner & Huggert 1989]
Platygaster hiemalis Forbes, 1888 A BC
Platygaster demades Walker, 1835 A BC
Zelostemma oleariae (Maskell, 1888) E
Zelostemma chionochloae Buhl *in* Buhl, Sarfati, Brockerhoff & Kelly, 2008 E
Zelostemma spp. (~4) [Masner & Huggert 1989] E
SCELIONINAE
Archaeoteleia chambersi Early, 2007 E
Archaeoteleia gilbertae Early, 2007 E
Archaeoteleia karere Early, 2007 E
Archaeoteleia novaezealandiae Masner, 1968 E
Archaeoteleia onamata Early, 2007 E
Archaeoteleia waipoua Early, 2007 E
Baeus seminulum Haliday, 1883 (?)
Baeus leai Dodd, 1914
Baeus saliens (Hickman, 1967)
Baeus n. spp. (3) [Stevens & Austin 2007] E
Calliscelio teleogrylli Hill, 1983 E
Ceratobaeus turneri Dodd, 1920
Ceratobaeus mussiae Iqbal & Austin, 2000 E
Cremastobaeus spp. (1-2) [Masner 1976]
Duta spp. (1-2) [ADA]
Gryon spp. (1-2) [ADA]
Hickmanella n. sp. [Austin 1988]
Idris n. spp. (4) [Austin 1988] E
Mirobaeus n. spp. (2) [Austin 1988] E
Mirotelenomus spp. (1-2) [ADA]
Neobaeus novazealandensis Austin, 1988 E
Odontacolus n. spp. (2) [Austin 1988] E
Opisthacantha spp. (1-2) [Masner 1976]
Probaryconus dubius (Nixon, 1931) A
Scelio spp. (1-2) [ADA]
Triteleia sp(p). [ADA]
SCELIOTRACHELINAE
Aphanomerus pusillus Perkins, 1905 A
Errolium piceum Masner & Huggert, 1989 E
Errolium spp. (~6) [Masner & Huggert 1989] E
Fidiobia sp. [Masner & Huggert 1989]

Fidiobia ?citri (Nixon, 1969) A
Zelamerus amicorum Masner & Huggert, 1989 E
Zelamerus spp. (~4) [Masner & Huggert 1989] E
Zelandonota kiwi Masner & Huggert, 1989 E
Zelandonota spp. (~12) [Masner & Huggert 1989] E
TELEASINAE
Trimorus castaneus (Brues, 1922) E
Trimorus spp. (~10) [SET]
Trimorus novaezealandiae Brues, 1922 E
TELENOMINAE
Eumicrosoma spp. (1-2) [ADA]
Telenomus spp. (1-2) [ADA]
Telenomus crinisacri Quail, 1901 E
Trissolcus basalis (Wollaston, 1858) A BC
Trissolcus maori Johnson, 1991 E
Trissolcus oenone (Dodd, 1913)
Trissolcus sp. [JAB]
CYNIPOIDEA
CYNIPIDAE A
Aulacidea subterminalis Niblett, 1946 A BC
Phanacis hypochoeridis (Kieffer, 1887) A
FIGITIDAE [Includes former Eucoilidae and Charapidae (Ronquist 1999)]
Alloxysta victrix Westwood, 1833 A
Alloxysta sp. [Valentine 1975]
Anacharis zealandica Ashmead, 1900
Hexacola sp. [JAB]
Kleidotoma (Pentakleidota) subantarcticana Yoshimoto, 1964 E
Leptopilina heterotoma (Thomson, 1862) A
Phaenoglyphis villosa (Hartig, 1841) A
?*Trybliographa* sp. [NDMF]
IBALIIDAE A
Ibalia leucospoides (Hockenwarth, 1785) A BC
CHALCIDOIDEA
AGAONIDAE A
Herodotia subatriventris (Girault, 1923) A
Odontofroggatia galili Wiebes, 1980 A
Pleistodontes froggatti Mayr, 1906 A
Pleistodontes imperialis Saunders, 1883 A
Pseudidarnes minerva Girault, 1927 A
Sycoscapter australis (Froggatt, 1900) A
APHELINIDAE
Ablerus spp. (2) [JBM]
Aphelinus abdominalis (Dalman, 1820) A
Aphelinus asychis Walker, 1839 A
Aphelinus gossypii Timberlake, 1924 A
Aphelinus humilis Mercet, 1928 A
Aphelinus mali (Haldeman, 1851) A BC
Aphelinus subflavescens (Westwood, 1837) A BC
Aphelinus sp. [Noyes & Valentine 1989b]
Aphytis chilensis Howard, 1900 A
Aphytis chrysomphali (Mercet, 1912) A
Aphytis diaspidis (Howard, 1881) A BC
Aphytis ignotus Compere, 1955
Aphytis mytilaspidis (LeBaron, 1870) A
Aphytis spp. [Noyes & Valentine 1989b]
Cales n. sp. [JH]
Centrodora scolypopae Valentine, 1966
Centrodora xiphidii (Perkins, 1906) A
Centrodora spp. (2) [Noyes & Valentine 1989b]
Coccophagoides sp. [Noyes & Valentine 1989b]
Coccophagus gurneyi Compere, 1929 A BC
Coccophagus ochraceus Howard, 1895 A
Coccophagus philippiae (Silvestri, 1915) A
Coccophagus scutellaris (Dalman, 1825) A
Coccophagus spp. (~7) [Noyes & Valentine 1989b]
Encarsia citrina (Craw, 1891) A
Encarsia formosa Gahan, 1924 A BC
Encarsia inaron (Walker, 1839) A
Encarsia koebelei (Howard, 1908) A
Encarsia pergandiella Howard, 1907 A
Encarsia perniciosi (Tower, 1913) A
Encarsia spp. (~5) [Noyes & Valentine 1989b]
Eretmocerus eremicus Rose & Zolnerowich, 1997
Euryischia sp. [Berry 1990b]
Eutrichosomella sp. [Noyes & Valentine 1989b]
Pteroptrix spp. (~25) [Noyes & Valentine 1989b] (most E)
CHALCIDIDAE
Antrocephalus sp. [Noyes & Valentine 1989b]
Brachymeria phya (Walker, 1838) (?) A BC
Brachymeria rubrifemur (Girault, 1913)
Brachymeria teuta (Walker, 1841) (?) A BC
Brachymeria sp. (1-2) [Noyes & Valentine 1989b]
Proconura sp. [Noyes & Valentine 1989b]
ENCYRTIDAE
Adelencyrtoides acutus Noyes, 1988 E
Adelencyrtoides blastothrichus Noyes, 1988 E
Adelencyrtoides inconstans Noyes, 1988 E
Adelencyrtoides mucro Noyes, 1988 E
Adelencyrtoides novaezealandiae Tachikawa & Valentine, 1969 E
Adelencyrtoides otago Noyes, 1988 E
Adelencyrtoides palustris Noyes, 1988 E
Adelencyrtoides pilosus Noyes, 1988 E
Adelencyrtoides proximus Noyes, 1988 E
Adelencyrtoides similis Noyes, 1988 E
Adelencyrtoides suavis Noyes, 1988 E
Adelencyrtoides tridens Noyes, 1988 E
Adelencyrtoides unicolor Noyes, 1988 E
Adelencyrtoides variabilis Noyes, 1988 E
Adelencyrtoides spp. (~3) [Noyes 1988]
Adelencyrtus aulacaspidis (Brethes, 1914) A
Alamella mira Noyes, 1988
Anagyrus costalis (Noyes, 1988) E
Anagyrus cyrenis (Noyes, 1988) E
Anagyrus fusciventris (Girault, 1915) A
Anagyrus regis (Noyes, 1988) E
Arrhenophagoidea coloripes Girault, 1915
Arrhenophagus chionaspidis Aurivillius, 1888 A
Austrochoreia antipodis Noyes, 1988 E
Baeoanusia sp. [JAB]
Cheiloneurus antipodis Noyes, 1988
Cheiloneurus gonatopodis Perkins, 1906 A
Coccidoctonus dubius (Girault, 1915) A BC
Coccidoctonus psyllae (Riek, 1962)
Coelopencyrtus australis Noyes, 1988 E
Coelopencyrtus maori Noyes, 1988 E
Copidosoma exvallis Noyes, 1988 E
Copidosoma floridanum (Ashmead, 1900) A BC
Cryptanusia sp. [JSN] A
Encyrtus aurantii (Geoffroy, 1785) A
Encyrtus infelix (Embleton, 1902) A
Epiblatticida minutissimus (Girault, 1923)
Epitetracnemus intersectus (Fonscolombe, 1832) A
Eusemion cornigerum (Walker, 1838) A
Gyranusoidea advena Beardsley, 1969 A
Habrolepis dalmanni (Westwood, 1837) A BC
Ixodiphagus n. sp. [AH] E
Lamennaisia ambigua (Nees, 1834) A
Metanotalia maderensis (Walker, 1872) A
Metaphycus anneckei Guerrieri & Noyes, 2000 A
Metaphycus aurantiacus Annecke & Mynhardt, 1981
Metaphycus claviger (Timberlake, 1916)
Metaphycus lounsburyi (Howard, 1898) A BC
Metaphycus maculipennis (Timberlake, 1916) A
Metaphycus reductor Noyes, 1988 E
Microterys nietneri (Motschulsky, 1859) A BC
Notodusmetia coroneti Noyes, 1988 E
Odiaglyptus biformis Noyes, 1988 E
Parectromoides varipes (Girault, 1915)
Protyndarichoides cinctiventris (Girault, 1934)
Pseudaphycus maculipennis Mercet, 1923 A BC
Pseudococcobius annulipes Noyes, 1988 E
Psyllaephagus acaciae Noyes, 1988 A
Psyllaephagus bliteus Riek, 1962 A
Psyllaephagus breviramus Berry, 2007 A
Psyllaephagus cornwallensis Berry, 2007 A
Psyllaephagus gemitus Riek, 1962 A
Psyllaephagus pilosus Noyes, 1988 A
Psyllaephagus richardhenryi Berry, 2007 A
Rhopus anceps Noyes, 1988 E
Rhopus garibaldia (Girault, 1933) (?) A
Rhopus spp. (1-2) [Noyes 1988]
Subprionomitus ferus (Girault, 1922)
Syrphophagus aphidivorus (Mayr, 1876)
Tachinaephagus australiensis (Girault, 1914) (?)
Tachinaephagus zealandicus Ashmead, 1904 A
Tetracnemoidea bicolor (Girault, 1915)
Tetracnemoidea brevicornis (Girault, 1915) A BC
Tetracnemoidea brounii (Timberlake, 1929) E
Tetracnemoidea peregrina (Compere, 1939) A
Tetracnemoidea sydneyensis (Timberlake, 1929) A
Tetracnemoidea zelandica Noyes, 1988 E
Zaomma lambinus (Walker, 1838) A
Zelaphycus aspidioti (Tachikawa & Valentine, 1969) E
Zelencyrtus latifrons Noyes, 1988 E
Gen. nov. et n. sp. [Noyes 1988] ?E
EULOPHIDAE [Includes former Elasmidae (Gauthier et al. 2000)]
Achrysocharoides latreillii (Curtis, 1826) A
Achrysocharoides sp. [Boucek 1988]
Apleurotropis sp. [JAB]
Aprostocetus zosimus (Walker, 1839) A
Aprostocetus (Ootetrastichus) sp. [ZB]
Arachnoobius austini Boucek, 1988
Asecodes spp. (2) [Boucek 1988 as *Teleopterus*]
Astichus spp. (2) [Boucek 1988]
Australsecodes sp. [Boucek 1988]
Baryscapus bruchophagi (Gahan, 1913) A
Baryscapus galactopus (Ratzeburg, 1844) A
Ceranisus menes (Walker, 1839) A
Ceranisus sp. [Boucek 1988]
Chrysocharis gemma (Walker, 1839) A
Chrysocharis pubicornis (Zetterstedt, 1838) A
Chrysonotomyia spp. (5) [Boucek 1988]
Cirrospilus variegatus (Masi, 1907)
Cirrospilus vittatus Walker, 1838
Cirrospilus sp. [Boucek 1988]
Closterocerus cruy (Girault, 1918) A
Closterocerus formosus Westwood, 1833 A BC
Deutereulophus n. sp. [Boucek 1988 as *Entedonomorpha*]
Diaulomorpha spp. (2) [Boucek 1988]
Diglyphus isaea (Walker, 1838) A BC
Elachertus sp. (?) [Thompson 1954]
Elasmus spp. (2) [Noyes & Valentine 1989b]
Entedon methion Walker, 1839 A
Entedonastichus dei (Girault, 1922)
Euderus spp. (2) [Boucek 1988]
Eulophus spp. (?) (1-2) [Thompson 1954]
Euplectrus agaristae Crawford, 1911
Euplectrus flavipes (Fonscolombe, 1832)
Eupronotius scaposus Boucek, 1988 E
Eupronotius n. sp. [Boucek 1988]
Hadranellus anomalus LaSalle & Boler, 1994 E
Hemiptarsenus varicornis (Girault, 1913) A BC
Hemiptarsenus spp. [Boucek 1988]
Hyssopus spp. (~2) [Boucek 1988]
Makarora obesa Boucek, 1988 E
Melittobia acasta (Walker, 1839) A
Melittobia australica Girault, 1912 A
Melittobia hawaiiensis Perkins, 1907 A
Neotrichoporoides viridimaculatus (Fullaway, 1955) A
Nesympiesis venosa Boucek, 1988 E
Noyesius metallicus Boucek, 1988 E
Noyesius testaceus Boucek, 1988 E
Noyesius n. sp. [Boucek 1988]
Omphale spp. (2) [Boucek 1988]
Oomyzus scaposus (Thompson, 1878) (?) A
Ophelimus eucalypti (Gahan, 1922) A
Ophelimus maskelli (Ashmead, 1900) A
Parasecodella n. sp. [Boucek 1988]
Pediobius bruchicida (Rondani, 1872) A
Pediobius epigonus (Walker, 1839) A BC
Pediobius metallicus (Nees, 1834) A
Pnigalio pectinicornis (Linnaeus, 1758) (?) A

Pnigalio soemius (Walker, 1839) A
Proacrias sp. [JAB]
Quadrastichodella aena Girault, 1913 (?)
Quadrastichodella pilosa Ikeda, 1999
Stenomesius n. spp. (~2) [Boucek 1988] E
Sympiesis campbellensis (Kerrich & Yoshimoto, 1964) E
Sympiesis sericeicornis (Nees, 1834)
Sympiesis spp. (~2) [JAB]
Tamarixia sp. [JL]
Tetrastichus spp. (2) [Boucek 1988]
Thripobius javae (Girault, 1917) A BC
Trielacher forticornis Boucek, 1988 E
Trielacher n. spp. (2) [Boucek 1988] E
Zasympiesis pilosa Boucek, 1988 E
Zasympiesis n. spp. (3) [Boucek 1988] E
Zealachertus abbreviatus Berry, 1999 E
Zealachertus aspirensis Berry, 1999 E
Zealachertus bildiri Berry, 1999 E
Zealachertus binarius Berry, 1999 E
Zealachertus conjunctus Berry, 1999 E
Zealachertus holderi Berry, 1999 E
Zealachertus longus Berry, 1999 E
Zealachertus nephelion Berry, 1999 E
Zealachertus nothofagi Boucek, 1978 E
Zealachertus pilifer Berry, 1999 E
Zealachertus planus Berry, 1999 E
Zealachertus tortriciphaga Berry, 1999 E
Zeastichus asper Boucek, 1988 E
Zeastichus n. sp. [Boucek 1988] E
EUPELMIDAE
Eupelmus antipoda Ashmead, 1900
Eupelmus cyaneus (Gourlay, 1928) E
Eupelmus vesicularis (Retzius, 1783) A
Eusandalum barteli (Gourlay, 1928) E
Tineobius sp. [Noyes & Valentine 1989b]
EURYTOMIDAE
Axanthosoma ?io Girault, 1915
Bruchophagus acaciae (Cameron, 1910)
Bruchophagus gibbus (Boheman, 1836) A
Bruchophagus roddi (Gussakovsky, 1933) A
Systole foeniculi Otten, 1941 A
Tetramesa linearis (Walker, 1832) A
Tetramesa spp. (~2) [Noyes & Valentine 1989b]
MYMARIDAE
Acmotemnus luteiclava Noyes & Valentine, 1989a E
Alaptus spp. (~5) [Noyes & Valentine 1989a]
Allanagrus magniclava Noyes & Valentine, 1989 E
Allarescon ochroceras Noyes & Valentine, 1989 E
Allarescon sp. [Noyes & Valentine 1989a]
Anagroidea n. spp. (~4) [Noyes & Valentine 1989a] E
Anagrus atomus (Linnaeus, 1767) A
Anagrus avalae Soyka, 1955 A
Anagrus frequens Perkins, 1905 A
Anagrus incarnatus Haliday, 1833 A
Anagrus optabilis (Perkins, 1905) A
Anagrus ustulatus Haliday, 1833 A
Anaphes nitens (Girault, 1928) A BC
Anaphes spp. (~17) [Noyes & Valentine 1989a]
Apoxypteron grandiscapus Noyes & Valetine, 1989 E
Arescon spp. (2) [Noyes & Valentine 1989a]
Australomymar n. spp. (18) [Noyes & Valentine 1989a] E
Boccacciomymar (Prosto) pobeda Triapitsyn & Berezovskiy, 2007 E
Boccacciomymar (Prosto) tak Triapitsyn & Berezovskiy, 2007 E
Camptoptera spp. (~15) [Noyes & Valentine 1989a]
Camptopteroides verrucosa (Noyes & Valentine, 1989) E
Ceratanaphes mandibularis (Noyes & Valentine, 1989) E
Ceratanaphes monticola Noyes & Valentine, 1989 E
Cleruchus n. spp. (~11) [Noyes & Valentine 1989a] E
Cybomymar fasciifrons Noyes & Valentine, 1989 E
Dicopomorpha n. sp. [Noyes & Valentine 1989] E
Dicopus spp. (2) [Noyes & Valentine 1989a]
Dorya pilosa Noyes & Valentine, 1989 E
Dorya n. sp. [Noyes & Valentine 1989a] E
Gonatocerus spp. [Noyes & Valentine 1989a]
Ischiodasys occulta Noyes & Valentine, 1989 E
Ischiodasys n. spp. (~14) [Noyes & Valentine 1989] E
Mimalaptus obscurus Noyes & Valentine, 1989 E
Mimalaptus n. spp. (1-3) [Noyes & Valentine 1989] E
Mymar schwanni Girault, 1912 A
Mymar taprobanicum Ward, 1875 A
Neserythmelus zelandicus Noyes & Valentine, 1989 E
Nesomymar magniclave Valentine, 1971 E
Nesomymar n. sp. [Noyes & Valentine 1989] E
Nesopatasson flavidus Valentine, 1971 E
Ooctonus vulgatus Haliday, 1833 [S.Triapitsyn]
Paracmotemnus potanus Noyes & Valentine, 1989 E
Paracmotemnus spp. (~5) [Noyes & Valentine 1989a]
Paranaphoidea (Idiocentrus) mirus (Gahan, 1927) E
Paranaphoidea sp. [Noyes & Valentine 1989a]
Polynema spp. (11) [Noyes & Valentine 1989a]
Prionaphes depressus Hincks, 1961 E
Prionaphes n. spp. (2) [Noyes & Valentine 1989] E
Pseudanaphes hirtus Noyes &Valentine, 1989 E
Richteria lamennaisi Girault, 1920
Scleromymar breve Noyes & Valentine, 1989 E
Scleromymar n. spp. (~4) [Noyes & Valentine 1989] E
Steganogaster silvicola Noyes & Valentine, 1989 E
Steganogaster n. spp. (~4) [Noyes & Valentine 1989a] E
Stephanodes reduvioli (Perkins, 1905) A
Stethynium spp. (~2) [Noyes & Valentine 1989a]
Zelanaphes lamprogonius Noyes & Valentine, 1989 E
Gen. nov. et n. sp. [Noyes & Valentine 1989a] E
Gen. nov. et n. sp. [Noyes & Valentine 1989a] E
Gen. nov. et n. sp. [Noyes & Valentine 1989a] E
PERILAMPIDAE
Austrotoxeuma kuscheli Boucek, 1988 E
Austrotoxeuma n. spp. (2) [Noyes & Valentine 1989b] E
PTEROMALIDAE
Acoelocyba n. sp. [Boucek 1988]
Acroclisoides sp. [IDN] A
Amerostenus sp. [Boucek 1988]
Anisopteromalus calandrae (Howard, 1881) A
Aphobetus cultratus Berry, 1995 E
Aphobetus cyanea (Boucek, 1988) E
Aphobetus erroli Berry, 1995 E
Aphobetus maskelli Howard, 1896 E
Aphobetus nana (Boucek, 1988) E
Aphobetus paucisetosus Berry, 1995 E
Asaphes vulgaris Walker, 1834 A
Callitula viridicoxa (Girault, 1913)
Cleonymus sp. [Boucek 1988]
Dibrachys sp. [JAB]
Dipareta spp. (~3) [Boucek 1988]
Enoggera nassaui Girault, 1915 A BC
Errolia cyanea Boucek, 1988 E
Errolia n. spp. (3) [Boucek 1988] E
Epanogmus sp. [Boucek 1988]
Fusiterga gallarum Boucek, 1988 E
Fusiterga lativentris Boucek, 1988 E
Gastrancistrus spp. (1-2) [Boucek 1988]
Homoporus nypsius Walker, 1839 A
Inkaka quadridentata Girault, 1939
Lariophagus distinguendus (Förster, 1840) A
Macromesus sp. [Boucek 1988]
Maorita reticulata Boucek, 1988 E
Mesopolobus incultus (Walker, 1834) A
Mesopolobus nobilis (Walker, 1834) A
Mesopolobus sp. [Boucek 1988]
Moranila aotearoae Berry, 1995 E
Moranila californica (Howard, 1881) A
Moranila comperei (Ashmead, 1904)
Moranila strigaster Berry, 1995 E
Muscidifurax raptor Girault & Sanders, 1910 A BC
Nambouria xanthops Berry & Withers, 2002 A
Nasonia vitripennis (Walker, 1836) A BC
Neocalosoter n. spp. (2) [Boucek 1988] E
Neopolycystus insectifurax Girault, 1915 A
Notoglyptus scutellaris (Dodd & Girault, 1915) A
Ophelosia australis Berry, 1995 E
Ophelosia bifasciata Girault, 1916
Ophelosia charlesi Berry, 1995
Ophelosia crawfordi Riley, 1890
Ophelosia keatsi Girault, 1927
Ophelosia mcglashani Berry, 1995 E
Ophelosia stenopteryx Berry, 1995 E
Omphalodipara sp. [Boucek 1988]
Pachyneuron aphidis (Bouché, 1834) A
Proshizonotus resplendens (Gourlay, 1928) E
Proshizonotus spp. (3-4) [GAPG]
Pseudanogmus silanus (Walker, 1843)
Pseudoceraphron sp. [IDN]
Pteromalus puparum (Linnaeus, 1758) A BC
Pteromalus semotus (Walker, 1834) (?) A BC
Pteromalus sequester Walker, 1835 A
Pteromalus sp. [JAB]
Rhaphitelus maculatus Walker, 1834 A
Spalangia cameroni Perkins, 1910 A
Spalangia endius Walker, 1839 A
Spalangia nigra Latreille, 1805 (?) A
Spalangia nigroaenea Curtis, 1839
Spalangia spp. (2) [Boucek 1988]
Stinoplus etearchus (Walker, 1848) A
Systasis lelex (Walker, 1839) E
Theocolax formiciformis Westwood, 1832 A
Theocolax sp. [Boucek 1988]
Trichomalopsis hemiptera Walker, 1935 A
Trichomalopsis iambe (Walker, 1839) E
Trichomalopsis spp. (~8) [Boucek 1988]
Trigonogastrella spp. (1-2) [Boucek 1988]
Zeala walkerae Boucek, 1988 E
ROTOITIDAE
Rotoita basalis Boucek & Noyes, 1987 E
Rotoita n. spp. (2) [Noyes & Valentine 1989b] E
SIGNIPHORIDAE
Chartocerus sp. [Noyes & Valentine 1989b]
Signiphora flavella Girault, 1913 A
Signiphora flavopalliata Ashmead, 1880 (?) A
Signiphora merceti Malenotti, 1916 A
TORYMIDAE
Idiomacromerus terebrator Masi, 1916 A
Megastigmus aculeatus (Swederus, 1795) A
Megastigmus spermotrophus Wachtl, 1893 A
Megastigmus sp. [ex *Procecidochares*, Noyes & Valentine 1989b]
Megastigmus sp. [ex *Ficus*, Early 2000]
Megastigmus sp. [ex *Ophelimus*, EEG]
Palmon sp. [Noyes & Valentine 1989b]
Podagrion sp. [Noyes & Valentine 1989b]
Torymoides antipoda (Kirby, 1883) E
Torymoides spp. (~4) [Noyes & Valentine 1989b]
Torymus varians (Walker, 1833) A
TRICHOGRAMMATIDAE
Aphelinoidea spp. (5) [Noyes & Valentine 1989b]
Brachyia n. sp. [Noyes & Valentine 1989b]
Lathromeris spp. (2) [Noyes & Valentine 1989b]
Megaphragma n. spp. (2) [Noyes & Valentine 1989b]
Oligosita spp. (3) [Noyes & Valentine 1989b]
Pseudogrammina n. spp. (3) [Noyes & Valentine 1989b]
Trichogramma falx Pinto & Oatman, 1996 E
Trichogramma (Trichogrammanza) funiculatum Carver, 1978
Trichogramma maori Pinto & Oatman, 1996 E
Trichogramma minutum Riley, 1871 (?)
Trichogramma valentinei Pinto & Oatman, 1996 E
Trichogramma spp. (~4) [Pinto & Oatman 1996]
Trichogrammatoidea bactrae Nagaraja, 1978 A
Trichogrammatoidea sp. [Noyes & Valentine 1989b]
Trichogrammatomyia sp.[Noyes & Valentine 1989b]
Ufens spp. (2) [Noyes & Valentine 1989b]

Zelogramma maculatum Noyes & Valentine, 1989 E
MYMAROMMATOIDEA
MYMAROMMATIDAE
Zealaromma insulare (Valentine, 1971) E
Zealaromma valentinei Gibson, Read & Huber 2007 E
Mymaromma sp. [Gibson, Read & Huber 2007]
ICHNEUMONOIDEA
BRACONIDAE
AGATHADINAE
Gen. et sp. indet. [AKW]
ALYSIINAE
Alysia manducator (Panzer, 1799) A BC
Aphaereta aotea Hughes & Woolcock, 1976 E
Aphaereta pallipes (Say, 1829) A
Asobara ajbelli Berry, 2007 E
Asobara albiclava Berry, 2007 E
Asobara antipoda Ashmead, 1900 E
Asobara persimilis (Papp, 1977)
Asobara tabida (Nees von Esenbeck, 1834) A
Aspilota andyaustini Wharton, 2002
Aspilota albertica Berry, 2007 E
Aspilota angusta Berry, 2007 E
Aspilota parecur Berry, 2007 E
Aspilota villosa Berry, 2007 E
Chaenusa helmorei Berry, 2007 E
Chorebus paranigricapitis Berry, 2007 E
Chorebus rodericki Berry, 2007 E
Chorebus thorpei Berry, 2007 E
Dacnusa areolaris (Nees, 1812) A
Dinotrema barrattae Berry, 2007 E
Dinotrema longworthi Berry, 2007 E
Dinotrema philipi Berry, 2007 E
APHIDIINAE
Aphidius colemani Viereck, 1912 A
Aphidius eadyi Stary, Gonzales & Hall, 1980 A BC
Aphidius ervi Haliday, 1834 A BC
Aphidius pelargonii Stary & Carver, 1979
Aphidius rhopalosiphi De Stefani Perez, 1902 A BC
Aphidius salicis Haliday, 1834 A
Aphidius similis Stary & Carver, 1979
Aphidius sonchi Marshall, 1896 A
Diaeretiella rapae (M'Intosh, 1855) A
Ephedrus plagiator (Nees ab Esenbeck, 1811) A BC
Lysiphlebus testaceipes (Cresson, 1880) A
Trioxys complanatus Quilis, 1931 A BC
Gen. nov. et n. sp. [MM] E
BETYOBRACONINAE
Gen. et sp. indet. [SB]
BLACINAE
Gen. et sp. indet. [JAB]
BRACONINAE A
Bracon phylacteophagus Austin, 1989 A BC
Bracon variegator Spinola, 1808 A BC
Habrobracon hebetor Say, 1836 A
CHELONINAE
Ascogaster bicolorata Walker & Huddleston, 1987 E
Ascogaster crenulata Cameron, 1898 E
Ascogaster elongata Lyle, 1923 E
Ascogaster erroli Walker & Huddleston, 1987 E
Ascogaster gourlayi Walker & Huddleston, 1987 E
Ascogaster iti Walker & Huddleston, 1987 E
Ascogaster mayae Walker & Huddleston, 1987 E
Ascogaster parrotti Walker & Huddleston, 1987 E
Ascogaster quadridentata Wesmael, 1835 A BC
Ascogaster strigosa Walker & Huddleston, 1987 E
Ascogaster tekapoense Walker & Huddleston, 1987 E
Ascogaster vexator Walker & Huddleston, 1987 E
DORYCTINAE
Caenophanes spp. (~2) [Belokobylskij et al. 2004]
Doryctopsis neozealandicus (Belokobylskij et al., 2004) E
Monolexis fuscicornis Förster, 1862 A
Ontsira antica (Wollaston, 1858) A
Parallorhogas pallidiceps (Perkins, 1910)
Pseudosyngaster pallidus (Gourlay, 1928) E
Spathius exarator (Linnaeus, 1758)
EUPHORINAE
Aridelus sp. [SET, SRS]
Cryptoxilos thorpei Shaw & Berry, 2005 E
Dinocampus coccinellae (Schrank, 1802) A
Leiophron sp. [SRS]
Meteorus annettae Huddleston, 1986 E
Meteorus cespitator (Thunberg, 1822) A
Meteorus cinctellus (Spinola, 1808) A
Meteorus cobbus Huddleston, 1986 E
Meteorus luteus (Cameron, 1911)
Meteorus novazealandicus Cameron, 1898 E
Meteorus pulchricornis (Wesmael, 1835) A
Meteorus quinlani Huddleston, 1986 E
Meteorus n. sp. [brachypterous female; SRS, JAB] E
Microctonus aethiopoides Loan, 1975 A BC
Microctonus alpinus Shaw, 1993 E
Microctonus falcatus Shaw, 1993 E
Microctonus hyperodae Loan *in* Loan & Lloyd, 1974 A BC
Microctonus zealandicus Shaw, 1993 E
Syntretus sp. [SRS]
HELCONINAE
Aspicolpus hudsoni Turner, 1922 E
Aspicolpus penetrator (Smith, 1878) E
Diospilus antipodum Turner, 1922 E
Schauinslandia alfkenii Ashmead, 1900 E
Schauinslandia femorata Ashmead 1900 E
Schauinslandia pallidipes Ashmead, 1900 E
HORMIINAE
Austrohormius punctatus van Achterberg, 1995 E
Chremylus elaphus Haliday, 1833 A
Neptihormius stigmellae van Achterberg & Berry, 2004 E
Neptihormius sp. (ex cecidomyid; CvA, JAB) E
MACROCENTRINAE
Macrocentrus rubromaculatus (Cameron, 1901)
METEORIDEINAE
Pronkia antefurcalis van Achterberg, 1990 E
MICROGASTRINAE
Apanteles carpatus (Say, 1836) A
Apanteles galleriae Wilkinson, 1932 A
Apanteles subandinus Blanchard, 1947 A BC
Choeras helespas Walker, 1996 E
Cotesia glomerata (Linnaeus, 1758) A BC
Cotesia kazak (Telenga, 1949) A BC
Cotesia rubecula (Marshall, 1885) A BC
Cotesia ruficrus (Haliday, 1834) A BC
Diolcogaster perniciosus (Wilkinson, 1929)
Dolichogenidea carposinae (Wilkinson, 1938) E
Dolichogenidea tasmanica (Cameron, 1912) A BC
Dolichogenidea spp. (~4) [JAB]
Glyptapanteles aucklandensis (Cameron, 1909) E
Glyptapanteles demeter (Wilkinson, 1934) E
Microplitis croceipes (Cresson, 1872) A BC
Pholetesor arisba Nixon, 1973 A
Pholetesor bicolor (Nees, 1834) A
Sathon spp. (1-2) [JAB]
OPIINAE
Opius carpocapsae (Ashmead, 1900) E
Opius cinerariae Fischer, 1963
Opius spp. (~5) [JAB]
Xynobius (Paraxynobius) albobasalis van Achterberg, 2004 E
Xynobius (Paraxynobius) granulatus van Achterberg, 2004 E
RHYSSALINAE
Doryctomorpha antipoda Ashmead, 1900 E
Metaspathius apterus Brues, 1922 E
Rhyssaloides ambeodonti (Muesebeck, 1941) E
Rhyssaloides antipoda Belokobylskij, 1999 E
ROGADINAE
Aleiodes declanae van Achterberg *in* van Achterberg et al., 2005 E
Aleiodes gressitti (Muesebeck, 1964) E
Rhinoprotoma masneri van Achterberg, 1995 E
ICHNEUMONIDAE
ANOMALONINAE
Aphanistes kayi Gauld, 1980 E
Habronyx n. spp. (3) [IDG]
Spolas spp. (~4) [Gauld 1980]
Gen. nov. et n. sp. [IDG]
BANCHIINAE
Lissonota albopicta Smith, 1878 E
Lissonota aspera Bain, 1970 E
Lissonota atra Bain, 1970 E
Lissonota comparata Cameron 1898
Lissonota flavopicta Smith, 1878 E
Lissonota fulva Bain, 1970 E
Lissonota multicolor Colenso, 1885 E
CAMPOPLEGINAE
Campoletis obstructor (Smith, 1878) E
Campoletis sp. [IDG].
Campoplex disjunctus Townes, 1964 E
Campoplex hudsoni (Cameron, 1901) E
Campoplex spp. (>20) [IDG]
Casinaria spp. (~3) [IDG]
Diadegma agens Townes, 1964 E
Diadegma muelleri (White, 1874) E
Diadegma novaezealandiae Azidah, Fitton & Quicke, 2000 E
Diadegma semiclausum (Hellén, 1949) A BC
Diadegma n. spp. (~50) [IDG] E
Dusona destructor Wahl, 1991
Dusona stramineipes Cameron, 1901 E
Venturia canescens (Gravenhorst, 1829) A
Venturia intrudens (Smith, 1878) E
CREMASTINAE
Temelucha sp. [IDG]
Trathala (Trathala) agnina (Kerrich, 1959) E
CRYPTINAE
Aclastus spp. (5) [IDG]
Aclosmation spp. (~30) [IDG]
Amblyaclastus sp. [IDG]
Anacis sp. [IDG]
'Gelis' campbellensis Townes, 1964 E
'Gelis' cinctus (Linnaeus, 1758)
'Gelis' philpottii (Brues, 1922) E
Gelis tenellus (Say, 1836) A
Glabridorsum stokesii (Cameron, 1912) A BC
Sphecophaga vesparum vesparum (Curtis, 1828) A BC
Xanthocryptus novozealandicus (Dalla Torre, 1902)
Xenolytus bitinctus (Gmelin, 1790) A
Gen. nov. et n. sp. [IDG] E
Gen. nov. et n. sp. [IDG] E
Gen. nov. et n. sp. [IDG] E
Gen. nov. et n. sp. [IDG] E
CTENOPELMATINAE A
Lathrolestes luteolator (Gravenhorst, 1829) A BC
DIPLAZONTINAE
Diplazon laetatorius (Fabricius, 1781) A
Woldstedtius spp. (3) [IDG]
EUCEROTINAE
Euceros coxalis Barron, 1978 E
Euceros sp. [IDG]
ICHNEUMONINAE
Aucklandella conspirata (Smith, 1876) E
Aucklandella flavomaculata Cameron, 1909 E
Aucklandella geiri (Dalla Torre, 1902) E
Aucklandella hudsoni (Cameron, 1901) E
Aucklandella machimia (Cameron, 1898) E
Aucklandella minuta (Ashmead, 1890) E
Aucklandella novazealandica (Cameron, 1898) E
Aucklandella pyrastis (Cameron, 1901) E
Aucklandella thyellma (Cameron, 1898) E
Aucklandella ursula (Cameron, 1898) E
Aucklandella utetes (Cameron, 1898) E
Aucklandella wellingtoni (Cameron, 1901) E
Aucklandella n. spp. (~50) [IDG] E
Ctenochares bicolorus (Linnaeus, 1767) A
Degithina actista (Cameron, 1898) E
Degithina apicalis (Ashmead, 1890) E

Degithina davidi Cameron, 1901 E
Degithina decepta (Smith, 1876) E
Degithina exhilarata (Smith, 1876) E
Degithina hersilia (Cameron, 1898) E
Degithina huttonii (Kirby, 1881) E
Degithina melanopus (Cameron, 1901) E
Degithina sollicitoria (Fabricius, 1775) E
Degithina n. spp. (~24) [IDG] E
Diadromus collaris (Gravenhorst, 1829) A BC
Eutanyacra licitatoria (Erichson, 1842)
Ichneumon lotatorius Fabricius, 1775 E
Ichneumon promissorius Erichson, 1842
Levansa decoratoria (Fabricius, 1775) E
Levansa leodacus (Cameron, 1898) E
Levansa spp. (2) [IDG]
Lusius sp. [IDG] E
?*Pterocormus cinctus* (Ashmead, 1890) E
LABENINAE
Certonotus fractinervis (Vollenhoven, 1873) E
Poecilocryptus n. sp. [IDG] E
MESOCHORINAE
Mesochorus sp. [IDG]
METOPIINAE
Carria fortipes (Cameron, 1898) E
Hypsicera femoralis (Geoffroy, 1785) A
Hypsicera nelsonensis Berry, 1990a E
Sciron enolae Berry, 1990a E
Sciron glaber Berry, 1990a E
OPHIONINAE
Enicospilus insularis (Kirby, 1881)
Enicospilus skeltonii (Kirby, 1881)
Ophion inutilis Smith, 1876 E
Ophion oculatus Parrott, 1954 E
Ophion peregrinus Smith, 1876a E
Ophion punctatus Cameron, 1898 E
ORTHOCENTRINAE
Helictes n. spp. (3) [IDG] E
Megastylus n. spp. (~10) [IDG] E
Stenomacrus sp. [IDG] (?)
Gen. nov. et n. spp. (2) [IDG] E
PIMPLINAE
Camptotypus sp. [IDG]
Echthromorpha intricatoria (Fabricius, 1804)
Liotryphon caudatus (Ratzeburg, 1848) A BC
Lissopimpla excelsa (Costa, 1864)
Megarhyssa nortoni (Cresson, 1864) A BC
Rhyssa lineolata (Kirby, 1837) A
Rhyssa persuasoria (Linnaeus, 1758) A BC
Xanthopimpla rhopaloceros Krieger, 1914 A BC
TERSILOCHINAE
Allophroides sp. [IDG]
Diaparsis sp. [IDG]
Zealochus gauldi Khalaim, 2006
Zealochus postfurcalis Khalaim, 2006
Zealochus supergranulatus Khalaim, 2004
Gen. nov. et n. spp. (~8) [NZAC-IDG] E
Gen. nov. et n. spp. (~6) [NZAC-IDG] E
Gen. nov. et n. spp. (2) [NZAC-IDG] E
Gen. nov. et n. spp. (~4) [NZAC-IDG] E
Gen. nov. et n. spp. (~10) [NZAC-IDG] E
TRYPHONINAE
Netelia (Netelia) ephippiata (Smith, 1876) E
Netelia (Netelia) producta (Brullé, 1846)
Phytodietus (Euctenopus) zealandicus (Ashmead, 1900) E
INCERTAE SEDIS
Gen. et sp. indet. [IDG]

APOCRITA-ACULEATA
CHRYSIDOIDEA
BETHYLIDAE
Chilepyris platythelys Sorg & Walker, 1988
Eupsenella insulana Gordh & Harris, 1996 E
Goniozus jacintae Farrugia, 1971 A BC
Goniozus sp. [Berry 1998]
Rhabdepyris sp. [Berry 1998]
Sierola spp. (2) [Berry 1998]
DRYINIDAE
Anteon bribianum Olmi, 1987
Anteon caledonianum Olmi, 1984
Bocchus thorpei Olmi, 2007
Dryinus koebelei (Perkins, 1905)
Gonatopus alpinus (Gourlay, 1953) E
Gonatopus zealandicus Olmi, 1984 E
EMBOLEMIDAE
Embolemus zealandicus Olmi, 1996 E
SCOLEBYTHIDAE
Ycaploca sp. [SET]
VESPOIDEA
FORMICIDAE
Amblyopone australis Erichson, 1842 A
Amblyopone saundersi Forel, 1892 E
Cardiocondyla minutior Forel, 1899 A
Discothyrea antarctica Emery, 1895 E
Doleromyrma darwiniana (Forel, 1907) A
Heteroponera brouni (Forel, 1892) E
Huberia brounii Forel, 1895 E
Huberia striata (Smith, 1876) E
Hypoponera confinis (Roger 1860) A
Hypoponera eduardi (Forel, 1894) A
Hypoponera punctatissima (Roger, 1859) A
Iridomyrmex spp. (not *anceps*) A
Linepithema humile (Mayr, 1868) A
Mayriella abstinens Forel, 1902 A
Monomorium antarcticum (F. Smith, 1858) E
Monomorium fieldi Forel, 1910 SI
Monomorium pharaonis (Linnaeus, 1758) A
Monomorium smithii Forel, 1892 E
Monomorium sydneyense Forel, 1902 A
Nylanderia tasmaniensis (Forel, 1913) A
Nylanderia sp(p). A
Ochetellus glaber (Mayr, 1862) A
Orectognathus antennatus Smith, 1853
Pachycondyla castanea (Mayr, 1865) E
Pachycondyla castaneicolor (Dalla Torre, 1893) ?E
Pheidole megacephala (Fabricius, 1793) A
Pheidole proxima Mayr, 1876
Pheidole rugosula Forel, 1902 A
Pheidole vigilans (Smith, 1858) A
Ponera leae Forel, 1913 A
Prolasius advenus (Smith, 1862) E
Rhytidoponera chalybaea Emery, 1901 A
Rhytidoponera metallica (Smith, 1858) A
Solenopsis sp. (cryptic Australian species) A
Strumigenys perplexa (Smith, 1876)
Strumigenys xenos Brown, 1955
Technomyrmex jocosus Forel 1910 A
Tetramorium bicarinatum (Nylander, 1846) A
Tetramorium grassii Emery, 1895 A
MUTILLIDAE A
Ephutomorpha bivulnerata (André, 1901) A
POMPILIDAE
Epipompilus insularis Kohl, 1884 E
Cryptocheilus australis (Guérin, 1830) A
Priocnemis (Trichocurgus) carbonarius (Smith, 1855) E
Priocnemis (Trichocurgus) conformis Smith, 1876 E
Priocnemis (Trichocurgus) crawi Harris, 1987 E
Priocnemis (Trichocurgus) monachus (Smith, 1855) E
Priocnemis (Trichocurgus) nitidiventris Smith, 878 E
Priocnemis (Trichocurgus) ordishi Harris, 1987 E
Sphictostethus calvus Harris, 1987 E
Sphictostethus fugax (Fabricius, 1775) E
Sphictostethus nitidus (Fabricius, 1775) E
SCOLIIDAE A
Radumeris tasmaniensis (Saussure, 1855) A
VESPIDAE A
Ancistrocerus gazella (Panzer, 1798) A
Polistes (Polistes) chinensis (Fabricius, 1793) A
Polistes (Polistela) humilis (Fabricius, 1781) A
Vespula (Vespula) germanica (Fabricius, 1793) A
Vespula (V.) vulgaris (Linnaeus, 1758) A
APOIDEA
APIDAE A
Apis mellifera Linnaeus, 1758
Bombus (Bombus) terrestris (Linnaeus, 1758)
Bombus (Megabombus) hortorum (Linnaeus, 1761)
Bombus (M.) ruderatus (Fabricius, 1775)
Bombus (Subterraneobombus) subterraneus (Linnaeus, 1758)
COLLETIDAE
Euryglossina (Euryglossina) hypochroma Cockerell, 1916 (?) A
Euryglossina (E.) proctotrypoides Cockerell, 1913 A
Hylaeus (Prosopisteron) agilis (Smith, 1876) E
Hylaeus (P.) asperithorax (Rayment, 1927) A
Hylaeus (P.) capitosus Smith, 1876 E
Hylaeus (P.) kermadecensis Donovan, 2007 E
Hylaeus (P.) matamoko Donovan, 2007 E
Hylaeus (P.) murihiku Donovan, 2007 E
Hylaeus (P.) perhumilis (Cockerell, 1914) A
Hylaeus (P.) relegatus (Smith, 1876) E
Hyleoides concinna (Fabricius, 1775) A
Leioproctus (Leioproctus) boltoni Cockerell, 1904 E
Leioproctus (L.) huakiwi Donovan, 2007 E
Leioproctus (L.) imitatus Smith, 1853 E
Leioproctus (L.) kanapuu Donovan, 2007 E
Leioproctus (L.) keehua Donovan, 2007 E
Leioproctus (L.) launcestonensis (Cockerell, 1914) [?not established; BJD]
Leioproctus (L.) metallicus (Smith, 1853) E
Leioproctus (L.) pango Donovan, 2007 E
Leioproctus (L.) purpureus (Smith, 1853) E
Leioproctus (L.) vestitus (Smith, 1876) E
Leioproctus (L.) waipounamu Donovan, 2007 E
Leioproctus (Nesocolletes) fulvescens (Smith, 1876) E
Leioproctus (N.) hudsoni (Cockerell, 1925) E
Leioproctus (N.) maritimus (Cockerell, 1936) E
Leioproctus (N.) monticola (Cockerell, 1925) E
Leioproctus (N.) nunui Donovan, 2007 E
Leioproctus (N.) paahaumaa Donovan, 2007 E
Leioproctus (N.) pekanui Donovan, 2007 E
CRABRONIDAE
Argogorytes (Argogorytes) carbonarius (Smith, 1856) E
Pison morosum Smith, 1856 E
Pison spinolae Shuckard, 1837
Pison ?ruficorne Smith, 1856 [JAB]
Podagritus (Parechuca) albipes (Smith, 1878) E
Podagritus (P.) carbonicolor (Dalla Torre, 1897) E
Podagritus (P.) chambersi Harris, 1994a E
Podagritus (P.) cora (Cameron, 1888) E
Podagritus (P.) digyalos Harris, 1994a E
Podagritus (P.) parrotti (Leclercq, 1955) E
Rhopalum (Zelorhopalum) aucklandi Leclercq, 1955 E
Rhopalum (Aporhopalum) perforator Smith, 1876 E
Rhopalum (Zelorhopalum) zelandum Leclerq, 1955 E
Spilomena earlyi Harris, 1994a E
Spilomena elegantula Turner, 1916
Spilomena emarginata Vardy, 1987 E
Spilomena nozela Vardy, 1987
Tachysphex nigerrimus (Smith, 1856) ?E
HALICTIDAE
Lasioglossum (Chilalictus) cognatum (Smith, 1853) A
Lasioglossum (Austrevylaeus) mataroa Donovan, 2007 E
Lasioglossum (A.) maunga Donovan, 2007 E
Lasioglossum (A.) sordidum (Smith, 1853) E
Nomia (Acunomia) melanderi Cockerell, 1906 P
MEGACHILIDAE A
Anthidium (Anthidium) manicatum (Linnaeus, 1758) A
Megachile (Eutricharaea) rotundata (Fabricius, 1787) P
Osmia (Helicosmia) coerulescens (Latreille, 1758) P
SPHECIDAE
Podalonia tydei suspiciosa (Smith, 1856) E

Order TRICHOPTERA
[Compiled by J. B. Ward]

Suborder ANNULIPALPIA
ECNOMIDAE
Ecnomina zealandica Wise, 1958 E
HYDROPSYCHIDAE
Aoteapsyche catherinae (McFarlane, 1960) E
Aoteapsyche colonica (McLachlan, 1871) E
Aoteapsyche philpotti (Tillyard, 1924) E
Aoteapsyche raruraru (McFarlane, 1973) E
Aoteapsyche tepoka (Mosely, 1953) E
Aoteapsyche tipua (McFarlane, 1964) E
Aoteapsyche winterbourni Smith, 2008 E
Aoteapsyche n. sp. E
Diplectrona bulla Wise, 1958 E
Diplectrona zealandensis Mosely, 1953 E
Hydropsyche auricoma Hare, 1910 E SI
Hydropsyche occulta (Hare, 1910) E SI
Orthopsyche fimbriata (McLachlan, 1862) E
Orthopsyche thomasi (Wise, 1962) E
PHILOPOTAMIDAE
Cryptobiosella furcata Henderson, 1983 E
Cryptobiosella hastata Henderson, 1983 E
Cryptobiosella spinosa Henderson, 1983 E
Cryptobiosella tridens Henderson, 1983 E
Hydrobiosella aorere Henderson, 1983 E
Hydrobiosella mixta (Cowley, 1976) E
Hydrobiosella stenocerca Tillyard, 1924 E
Hydrobiosella tonela (Mosely, 1953) E
Neobiosella irrorata Wise, 1958 E
Xenobiosella motueka Henderson, 1983 E
POLYCENTROPODIDAE
Plectrocnemia maclachlani Mosely, 1953 E
Plectrocnemia tuhuae Ward, 1995 E
Polyplectropus altera McFarlane, 1981 E
Polyplectropus aurifusca McFarlane, 1956 E
Polyplectropus impluvii Wise, 1962 E
Polyplectropus puerilis (McLachlan, 1868) E
PSYCHOMYIIDAE
Zelandoptila moselyi Tillyard, 1924 E

Suborder SPICIPALPIA
HYDROBIOSIDAE
Atrachorema macfarlanei Ward, 1991 E
Atrachorema mangu McFarlane, 1964 E
Atrachorema tuarua McFarlane, 1966 E
Costachorema brachypterum McFarlane, 1939 E
Costachorema callistum McFarlane, 1939 E
Costachorema hebdomon McFarlane, 1981 E
Costachorema hecton McFarlane, 1981 E
Costachorema notopterum Wise, 1972 E
Costachorema peninsulae Ward, 1995 E
Costachorema psaropterum McFarlane, 1939 E
Costachorema xanthopterum McFarlane, 1939 E
Edpercivalia banksiensis (McFarlane, 1939) E
Edpercivalia borealis (McFarlane, 1951b) E
Edpercivalia cassicola (McFarlane, 1939) E
Edpercivalia dugdalei Ward, 1998 E
Edpercivalia flintorum Ward, 2005 E
Edpercivalia fusca (McFarlane, 1939) E
Edpercivalia harrisoni Wise, 1982 E
Edpercivalia maxima (McFarlane, 1939) E
Edpercivalia morrisi Ward, 1998 E
Edpercivalia oriens Ward, 1997 E
Edpercivalia shandi (McFarlane, 1951b) E
Edpercivalia schistaria Ward, 2005 E
Edpercivalia smithi Ward, 2005 E
Edpercivalia spaini McFarlane, 1973 E
Edpercivalia tahatika Ward, 2005 E
Edpercivalia thomasoni (McFarlane, 1960) E
Erichorema basale Ward, Leschen, Smith & Dean, 2004 E
Hydrobiosis budgei McFarlane, 1960 E
Hydrobiosis centralis Ward, 1997 E
Hydrobiosis chalcodes McFarlane, 1981 E
Hydrobiosis charadraea McFarlane, 1951b E
Hydrobiosis clavigera McFarlane, 1951b E
Hydrobiosis copis McFarlane, 1960 E
Hydrobiosis falcis Wise, 1958 E
Hydrobiosis frater McLachlan, 1868 E
Hydrobiosis gollanis Mosely, 1953 E
Hydrobiosis harpidiosa McFarlane, 1951b E
Hydrobiosis ingenua Hare, 1910 E SI
Hydrobiosis johnsi McFarlane, 1981 E
Hydrobiosis kiddi McFarlane, 1951b E
Hydrobiosis lindsayi Tillyard, 1925 E
Hydrobiosis neadelphus Ward, 1997 E
Hydrobiosis parumbripennis McFarlane, 1951b E
Hydrobiosis sherleyi Ward, 1998 E
Hydrobiosis silvicola McFarlane, 1951b E
Hydrobiosis soror Mosely, 1953 E
Hydrobiosis spatulata McFarlane, 1951b E
Hydrobiosis styracine McFarlane, 1960 E
Hydrobiosis styx McFarlane, 1951b E
Hydrobiosis taumata Ward, 1997 E
Hydrobiosis torrentis Ward, 1995 E
Hydrobiosis umbripennis McLachlan, 1868 E
Hydrobiosis n. sp. E
Hydrochorema crassicaudatum Tillyard, 1924 E
Hydrochorema lyfordi Ward, 1997 E
Hydrochorema tenuicaudatum Tillyard, 1924 E
Hydrochorema n. spp. (3) 3E
Neurochorema armstrongi McFarlane, 1951a E
Neurochorema confusum (McLachlan, 1868) E
Neurochorema forsteri McFarlane, 1964 E
Neurochorema pilosum McFarlane, 1964 E
Neurochorema n. sp. E
Psilochorema acheir McFarlane, 1981 E
Psilochorema bidens McFarlane, 1951b E
Psilochorema cheirodes McFarlane, 1981 E
Psilochorema donaldsoni McFarlane, 1960 E
Psilochorema embersoni Wise, 1982 E
Psilochorema folioharpax McFarlane, 1956 E
Psilochorema leptoharpax McFarlane, 1951b E
Psilochorema macroharpax McFarlane, 1951b E
Psilochorema mataura McFarlane, 1956 E
Psilochorema mimicum McLachlan, 1866 E
Psilochorema nemorale McFarlane, 1951b E
Psilochorema spiniharpax Ward, 1995 E
Psilochorema tautoru McFarlane, 1964 E
Psilochorema vomerharpax McFarlane, 1964 E
Synchorema tillyardi McFarlane, 1964 E
Synchorema zygoneurum Tillyard, 1924 E
Tiphobiosis cataractae Ward, 1995 E
Tiphobiosis childella Ward, 1995 E
Tiphobiosis childi McFarlane, 1981 E
Tiphobiosis cowiei Ward, 1991 E
Tiphobiosis fulva Tillyard, 1924 E
Tiphobiosis hinewai Ward, 1995 E
Tiphobiosis intermedia Mosely, 1953 E
Tiphobiosis kleinpastei Ward, 1998 E
Tiphobiosis kuscheli Wise, 1972 E
Tiphobiosis montana Tillyard, 1924 E
Tiphobiosis plicosta McFarlane, 1960 E
Tiphobiosis quadrifurca Ward, 1997 E
Tiphobiosis salmoni McFarlane, 1981 E
Tiphobiosis schmidi Ward, 1998 E
Tiphobiosis trifurca McFarlane, 1981 E
Tiphobiosis veniflex McFarlane, 1960 E
Tiphobiosis n. spp. (7) 7E
Traillochorema rakiura McFarlane, 1981 E
Traillochorema wardorum Henderson, 2008 E
HYDROPTILIDAE
Oxyethira (*Trichoglene*) *ahipara* Wise, 1998 E
Oxyethira (*Trichoglene*) *albiceps* (McLachlan, 1862) E
Oxyethira (*Trichoglene*) *kirikiriroa* Smith, 2008 E
Oxyethira (*Trichoglene*) *waipoua* Wise, 1998 E
Paroxyethira auldorum Ward & Henderson, 2004 E
Paroxyethira dundensis Ward & Henderson, 2004 E
Paroxyethira eatoni Mosely, 1924 E
Paroxyethira hendersoni Mosely, 1924 E
Paroxyethira hintoni Leader, 1972 E
Paroxyethira hughwilsoni Ward & Henderson, 2004 E
Paroxyethira kimminsi Leader, 1972 E
Paroxyethira manapouri Ward & Henderson, 2004 E
Paroxyethira pounamu Ward & Henderson, 2004 E
Paroxyethira ramifera Ward & Henderson, 2004 E
Paroxyethira sarae Ward & Henderson, 2004 E
Paroxyethira takitimu Ward & Henderson, 2004 E
Paroxyethira teika Ward & Henderson, 2004 E
Paroxyethira tillyardi Mosely, 1924 E
Paroxyethira zoae Ward & Henderson, 2004 E
Paroxyethira n. sp. E
Xuthotrichia aotea Ward & Henderson, 2004 E

Suborder INTEGRIPALPIA
Infraorder PLENITENTORIA
KOKIRIIDAE
Kokiria miharo McFarlane, 1964 E
OECONESIDAE
Oeconesus angustus Ward, 1997 E
Oeconesus incisus Mosely, 1953 E
Oeconesus maori McLachlan, 1862 E
Oeconesus similis Mosely, 1953 E
Pseudoeconesus bistirpis Wise, 1958 E
Pseudoeconesus geraldinae Ward, 1997 E
Pseudoeconesus haasti Ward, 1997 E
Pseudoeconesus hendersoni Ward, 1997 E
Pseudoeconesus hudsoni Mosely, 1953 E
Pseudoeconesus mimus McLachlan, 1894 E
Pseudoeconesus paludis Ward, 1997 E
Pseudoeconesus squamosus Mosely, 1953 E
Pseudoeconesus stramineus McLachlan, 1894 E
Pseudoeconesus n. spp. (6) 6E
Tarapsyche olis McFarlane, 1960 E
Zelandopsyche ingens Tillyard, 1921 E
Zelandopsyche maclellani McFarlane, 1981 E
Zepsyche acinaces McFarlane, 1960 E

Infraorder BREVITENTORIA
CALOCIDAE
Pycnocentrella eruensis Mosely, 1953 E
CHATHAMIIDAE
Chathamia brevipennis Tillyard, 1925 E
Chathamia integripennis Riek, 1976 E
Philanisus fasciatus Riek, 1976 E
Philanisus mataua Ward, 1995 E
Philanisus plebeius Walker, 1852 SW
CONOESUCIDAE
Beraeoptera roria Mosely, 1953 E
Confluens hamiltoni (Tillyard, 1924) E
Confluens olingoides (Tillyard, 1924) E
Confluens n. sp. E
Olinga christinae Ward & McKenzie, 1998 E
Olinga feredayi (McLachlan, 1868) E
Olinga fumosa Wise, 1958 E
Olinga jeanae McFarlane, 1966 E
Periwinkla childi McFarlane, 1973 E
Pycnocentria evecta McLachlan, 1868 E
Pycnocentria forcipata Mosely, 1953 E
Pycnocentria funerea McLachlan, 1866 E
Pycnocentria gunni (McFarlane, 1956) E
Pycnocentria hawdonia McFarlane, 1956 E
Pycnocentria mordax Ward, 1997 E
Pycnocentria patricki Ward, 1995 E
Pycnocentria sylvestris McFarlane, 1973 E
Pycnocentria n. spp. (4) 4E
Pycnocentrodes aeris Wise, 1958 E
Pycnocentrodes aureolus (McLachlan, 1868) E
Pycnocentrodes modestus Cowley, 1976 E
HELICOPHIDAE
Alloecentrella cirratus Henderson & Ward, 2007 E
Alloecentrella incisus Henderson & Ward, 2007 E
Alloecentrella linearis Henderson & Ward, 207 E
Alloecentrella magnicornis Wise, 1958 E
Zelolessica cheira McFarlane, 1956 E
Zelolessica meizon McFarlane, 1981 E
HELICOPSYCHIDAE
Helicopsyche (*Saetotricha*) *albescens* Tillyard, 1924 E
Helicopsyche (*S.*) *cuvieri* Johanson, 1999 E

Helicopsyche (S.) haurapango Johanson, 1999 E
Helicopsyche (S.) howesi Tillyard, 1924 E
Helicopsyche (S.) poutini McFarlane, 1964 E
Helicopsyche (S.) torino Johanson, 1999 E
Helicopsyche (S.) zealandica Hudson, 1904 E
Rakiura vernale McFarlane, 1973 E
LEPTOCERIDAE
Hudsonema alienum (McLachlan, 1868) E
Hudsonema amabile (McLachlan, 1868) E
Hudsonema n. sp. E
Oecetis chathamensis Tillyard, 1925 E
Oecetis iti McFarlane, 1964 E
Oecetis unicolor (McLachlan, 1868) E
Triplectides cephalotes (Walker, 1852) E
Triplectides dolichos McFarlane, 1981 E
Triplectides obsoletus (McLachlan, 1862) E
Triplectidina moselyi McFarlane & Ward, 1990 E
Triplectidina oreolimnetes (Tillyard, 1924) E
PHILORHEITHRIDAE
Philorheithrus agilis (Hudson, 1904) E
Philorheithrus aliciae Henderson & Ward, 2006 E
Philorheithrus harunae Henderson & Ward, 2006 E
Philorheithrus lacustris Tillyard, 1924 E
Philorheithrus latentis Henderson & Ward, 2006 E
Philorheithrus litoralis Henderson & Ward, 2006 E

Order LEPIDOPTERA
[Compiled by R. J. B. Hoare]
ARCTIIDAE
Metacrias erichrysa Meyrick, 1886e E
Metacrias huttoni (Butler, 1879a) E
Metacrias strategica (Hudson, 1889) E
Nyctemera amica (White, 1841) V
Nyctemera annulata (Boisduval, 1832) E
Tyria jacobaeae (Linnaeus, 1758) A BC
Utetheisa lotrix (Cramer, 1777) V
Utetheisa pulchelloides Hampson, 1907 V
AUTOSTICHIDAE A
Oegoconia caradjai Popescu-Gorj & Capuse, 1965 A
BATRACHEDRIDAE
Batrachedra agaura Meyrick, 1901 E
Batrachedra arenosella (Walker, 1864b)
Batrachedra astricta Philpott, 1930b E
Batrachedra eucola Meyrick, 1889b E
Batrachedra filicicola Meyrick, 1917a E
Batrachedra litterata Philpott, 1928a E
Batrachedra psithyra Meyrick, 1889b E
Batrachedra tristicta Meyrick, 1901 E
Houdinia flexilissima Hoare, Dugdale & Watts, 2006 E
BEDELLIIDAE
Bedellia psamminella Meyrick, 1889b
Bedellia somnulentella (Zeller, 1847)
BLASTOBASIDAE A
Blastobasis sp. nr *tarda* Meyrick, 1902d A
BLASTODACNIDAE
Circoxena ditrocha Meyrick, 1916b E
Microcolona limodes Meyrick, 1897a E
BOMBYCIDAE A
Bombyx mori (Linnaeus, 1758) A
CARPOSINIDAE
Campbellana attenuata Salmon & Bradley, 1956 E Su
Coscinoptycha improbana Meyrick, 1881 A
Ctenarchis cramboides Dugdale, 1995 E
Glaphyrarcha euthrepta Meyrick, 1938 E
Heterocrossa adreptella (Walker, 1864a) E
Heterocrossa canescens (Philpott, 1930c) E
Heterocrossa contactella (Walker, 1866b) E
Heterocrossa cryodana Meyrick, 1885b E
Heterocrossa epomiana Meyrick, 1885b E
Heterocrossa eriphylla Meyrick, 1888d E
Heterocrossa exochana Meyrick, 1888d E
Heterocrossa gonosemana Meyrick, 1882b E
Heterocrossa hudsoni Dugdale, 1988 E
Heterocrossa ignobilis (Philpott, 1930c) E
Heterocrossa iophaea Meyrick, 1907c E
Heterocrossa literata (Philpott, 1930b) E
Heterocrossa maculosa (Philpott, 1927a) E
Heterocrossa morbida (Meyrick, 1912c) E
Heterocrossa philpotti hudsoni Dugdale, 1988 E
Heterocrossa p. philpotti (Dugdale, 1971) E Su
Heterocrossa rubophaga Dugdale, 1988 E
Heterocrossa sanctimonea (Clarke, 1926) E
Heterocrossa sarcanthes (Meyrick, 1918) E
Paramorpha marginata (Philpott, 1931) E
CECIDOSIDAE
Xanadoses nielseni Hoare & Dugdale, 2003 E
CHOREUTIDAE
Asterivora albifasciata (Philpott, 1924a) E
Asterivora analoga (Meyrick, 1912c) E
Asterivora antigrapha (Meyrick, 1911b) E
Asterivora barbigera (Meyrick, 1915a) E
Asterivora chatuidea (Clarke, 1926) E
Asterivora colpota (Meyrick, 1911b) E
Asterivora combinatana (Walker, 1863c) E
Asterivora exocha (Meyrick, 1907c) E
Asterivora fasciata (Philpott, 1930b) E
Asterivora inspoliata (Philpott, 1930b) E
Asterivora iochondra (Meyrick, 1911b) E
Asterivora marmarea (Meyrick, 1888e) E
Asterivora microlitha (Meyrick, 1888e) E
Asterivora ministra (Meyrick, 1912c) E
Asterivora nivescens (Philpott, 1926a) E
Asterivora oleariae Dugdale, 1979 E Sn
Asterivora symbolaea (Meyrick, 1888e) E
Asterivora tillyardi (Philpott, 1924c) E
Asterivora tristis (Philpott, 1930b) E
Asterivora urbana (Clarke, 1926) E
Tebenna micalis (Mann, 1857)
COLEOPHORIDAE A
Coleophora alcyonipennella (Kollar, 1832) A
Coleophora mayrella (Huebner, [1813]) A
Coleophora striatipennella Tengstrom, 1848 A
Coleophora versurella Zeller, 1849 A
COPROMORPHIDAE
Isonomeutis amauropa Meyrick, 1888d E
Isonomeutis restincta Meyrick, 1923 E
Phycomorpha metachrysa Meyrick, 1914a E
COSMOPTERIGIDAE
Cosmopterix attenuatella (Walker, 1864)
Labdia anarithma (Meyrick, 1889b)
Limnaecia phragmitella Stainton, 1851
Pyroderces aellotricha (Meyrick, 1889b)
Pyroderces apparitella (Walker, 1864b) E
Pyroderces deamatella (Walker, 1864a) E
COSSIDAE A
Endoxyla cinereus (Tepper, 1890) A
CRAMBIDAE
CRAMBINAE
Bleszynskia malacelloides (Bleszynski, 1955) A
Culladia cuneiferellus (Walker, 1863) A
Gadira acerella Walker, 1866b E
Gadira leucophthalma (Meyrick, 1882a) E
Gadira petraula (Meyrick, 1882a) E
Glaucocharis auriscriptella (Walker, 1864b) E
Glaucocharis bipunctella (Walker, 1866b) E
Glaucocharis chrysochyta (Meyrick, 1882a) E
Glaucocharis elaina (Meyrick, 1882a) E
Glaucocharis epiphaea (Meyrick, 1885a) E
Glaucocharis harmonica (Meyrick, 1888c) E
Glaucocharis helioctypa (Meyrick, 1882a) E
Glaucocharis holanthes (Meyrick, 1885a) E
Glaucocharis interrupta (Felder & Rogenhofer, 1875) E
Glaucocharis lepidella (Walker, 1866b) E
Glaucocharis leucoxantha (Meyrick, 1882a) E
Glaucocharis metallifera (Butler, 1877) E
Glaucocharis microdora (Meyrick, 1905) E
Glaucocharis parorma (Meyrick, 1924a) E
Glaucocharis planetopa (Meyrick, 1923) E
Glaucocharis pyrsophanes (Meyrick, 1882a) E
Glaucocharis selenaea (Meyrick, 1885a) E
Glaucocharis stella (Meyrick, 1938) E
Kupea electilis Philpott, 1930a E
Maoricrambus oncobolus (Meyrick, 1885a) E
Orocrambus abditus (Philpott, 1924a) E
Orocrambus aethonellus (Meyrick, 1882a) E
Orocrambus angustipennis (Zeller, 1877) E
Orocrambus apicellus (Zeller, 1863) E
Orocrambus callirrhous (Meyrick, 1882a) E
Orocrambus catacaustus (Meyrick, 1885a) E
Orocrambus clarkei Philpott, 1930b E
Orocrambus corruptus (Butler, 1877) E
Orocrambus crenaeus (Meyrick, 1885a) E
Orocrambus cultus Philpott, 1917b E
Orocrambus cyclopicus (Meyrick, 1882a) E
Orocrambus dicrenellus (Meyrick, 1883b) E
Orocrambus enchophorus (Meyrick, 1885a) E
Orocrambus ephorus (Meyrick, 1885a) E
Orocrambus eximia (Salmon, 1946) E
Orocrambus flexuosellus (Doubleday *in* White & Doubleday, 1843) E
Orocrambus fugitivellus (Hudson, 1950) E
Orocrambus geminus Patrick, 1991 E
Orocrambus haplotomus (Meyrick, 1882a) E
Orocrambus harpophorus (Meyrick, 1882a) E
Orocrambus heliotes (Meyrick, 1888c) E
Orocrambus heteraulus (Meyrick, 1905) E
Orocrambus horistes (Meyrick, 1902c) E
Orocrambus isochytus (Meyrick, 1888c) E
Orocrambus jansoni Gaskin, 1975 E
Orocrambus lectus (Philpott, 1929a) E
Orocrambus lewisi Gaskin, 1975 E
Orocrambus machaeristes Meyrick, 1905 E
Orocrambus melampetrus Purdie, 1884 E
Orocrambus melitastes (Meyrick, 1909a) E
Orocrambus mylites Meyrick, 1888c E
Orocrambus oppositus (Philpott, 1915) E
Orocrambus ordishi Gaskin, 1975 E
Orocrambus ornatus (Philpott, 1927d) E
Orocrambus paraxenus (Meyrick, 1885a) E
Orocrambus philpotti Gaskin, 1975 E
Orocrambus punctellus (Hudson, 1950) E
Orocrambus ramosellus (Doubleday *in* White & Doubleday, 1843) E
Orocrambus scoparioides Philpott, 1914 E
Orocrambus scutatus (Philpott, 1917b) E
Orocrambus simplex (Butler, 1877) E
Orocrambus siriellus (Meyrick, 1882a) E
Orocrambus sophistes Meyrick, 1905 E
Orocrambus sophronellus (Meyrick, 1885a) E
Orocrambus thymiastes Meyrick, 1901 E
Orocrambus tritonellus (Meyrick, 1885a) E
Orocrambus tuhualis (Felder & Rogenhofer, 1875) E
Orocrambus ventosus Meyrick, 1920a E
Orocrambus vittellus (Doubleday *in* White & Doubleday, 1843) E
Orocrambus vulgaris (Butler, 1877) E
Orocrambus xanthogrammus (Meyrick, 1882a) E
Tauroscopa gorgopis Meyrick, 1888c E
Tauroscopa notabilis Philpott, 1923 E
Tauroscopa trapezitis Meyrick, 1905 E
Tawhitia glaucophanes (Meyrick, 1907c) E
Tawhitia pentadactyla (Zeller, 1863)
CYBALOMIINAE A
Trichophysetis sp. 1 A
GLAPHYRIINAE
Hellula hydralis Guenée, 1854 V
MUSOTIMINAE
Musotima aduncalis (Felder & Rogenhofer, 1875) E
Musotima nitidalis (Walker, 1866a)
Musotima ochropteralis (Guenée, 1854) * A
NYMPHULINAE
Dracaenura aegialitis Meyrick, 1910b K
Eranistis pandora Meyrick, 1910b K
Hygraula nitens (Butler, 1880) F
PYRAUSTINAE
Achyra affinitalis (Lederer, 1863) A

Botys (s.l.) sp. Dugdale 1988 K
Deana hybreasalis (Walker, 1859b) E
Diasemia grammalis Doubleday *in* White & Doubleday, 1843
Diasemiopsis ramburialis (Duponchel, [1834]) K
Glyphodes onychinalis (Guenée, 1854) A
Herpetogramma licarsisalis (Walker, 1859) A
Proternia philocapna Meyrick, 1884e E
Proteroeca comastis Meyrick, 1884e E
Sceliodes cordalis (Doubleday *in* White & Doubleday, 1843)
Spoladea recurvalis (Fabricius, 1775) A
Udea adversa (Philpott, 1917b) E
Udea antipodea (Salmon *in* Salmon & Bradley, 1956) E Su
Udea daiclesalis (Walker, 1859c) E
Udea flavidalis (Doubleday *in* White & Doubleday, 1843) E
Udea marmarina Meyrick, 1884e E
Udea notata (Butler, 1879a) E
Udea pantheropa (Meyrick, 1902c) E Ch
Uresiphita polygonalis maorialis (Felder & Rogenhofer, 1875) E
Uresiphita p. ornithopteralis (Guenée, 1854) A*
SCOPARIINAE
Antiscopa acompa (Meyrick, 1884d) E
Antiscopa elaphra (Meyrick, 1884d) E
Antiscopa epicomia (Meyrick, 1884d) E
Eudonia alopecias (Meyrick, 1901) E
Eudonia aspidota (Meyrick, 1884d) E
Eudonia asterisca (Meyrick, 1884d) E
Eudonia atmogramma (Meyrick, 1915a) E
Eudonia axena (Meyrick, 1884d) E
Eudonia bisinualis (Hudson, 1928) E
Eudonia campbellensis (Munroe, 1964) E Su
Eudonia cataxesta (Meyrick, 1884d) E
Eudonia chalara (Meyrick, 1901) E
Eudonia characta (Meyrick, 1884d) E
Eudonia chlamydota (Meyrick, 1884d) E
Eudonia choristis (Meyrick, 1907c) E
Eudonia colpota (Meyrick, 1888c) E
Eudonia critica (Meyrick, 1884d) E
Eudonia crypsinoa (Meyrick, 1884d) E
Eudonia cymatias (Meyrick, 1884d) E
Eudonia cyptastis (Meyrick, 1909a) E
Eudonia deltophora (Meyrick, 1884d) E
Eudonia dinodes (Meyrick, 1884d) E
Eudonia dochmia (Meyrick, 1905) E
Eudonia epicremna (Meyrick, 1884d) E
Eudonia feredayi (Knaggs, 1867) E
Eudonia gressitti (Munroe, 1964) E Su
Eudonia gyrotoma (Meyrick, 1909a) E
Eudonia hemicycla (Meyrick, 1884d) E
Eudonia hemiplaca (Meyrick, 1889b) E
Eudonia legnota (Meyrick, 1884d) E
Eudonia leptalaea (Meyrick, 1884d) E
Eudonia leucogramma (Meyrick, 1884d) E
Eudonia linealis (Walker, 1866a) E
Eudonia locularis (Meyrick, 1912c) E
Eudonia luminatrix (Meyrick, 1909a) E
Eudonia manganeutis (Meyrick, 1884d) E
Eudonia melanaegis (Meyrick, 1884d) E
Eudonia meliturga (Meyrick, 1905) E
Eudonia microphthalma (Meyrick, 1884d) E
Eudonia minualis (Walker, 1866a) E
Eudonia octophora (Meyrick, 1884d) E
Eudonia oculata (Philpott, 1927d) E
Eudonia oreas (Meyrick, 1884d) E
Eudonia organaea (Meyrick, 1901) E
Eudonia pachyerga (Meyrick, 1927a) E
Eudonia paltomacha (Meyrick, 1884d) E
Eudonia periphanes (Meyrick, 1884d) E
Eudonia philerga (Meyrick, 1884d) E
Eudonia philetaera (Meyrick, 1884d) E
Eudonia pongalis (Felder & Rogenhofer, 1875) E
Eudonia psammitis (Meyrick, 1884d) E
Eudonia quaestoria (Meyrick, 1929) E
Eudonia rakaiaensis (Knaggs, 1867) E
Eudonia sabulosella (Walker, 1863b) E
Eudonia steropaea (Meyrick, 1884d) E
Eudonia subditella (Walker, 1866b) E
Eudonia submarginalis (Walker, 1863b) E
Eudonia thyridias (Meyrick, 1905) E
Eudonia torodes (Meyrick, 1901) E
Eudonia triclera (Meyrick, 1905) E
Eudonia trivirgata (Felder & Rogenhofer, 1875) E
Eudonia ustiramis (Meyrick, 1931a) E
Eudonia xysmatias (Meyrick, 1907c) E
Eudonia zophochlaena (Meyrick, 1923) E
Exsilirarcha graminea Salmon & Bradley, 1956 E Su
Heliothela atra (Butler, 1877) E
Protyparcha scaphodes Meyrick, 1909b E Su
Scoparia acharis Meyrick, 1884d E
Scoparia apheles (Meyrick, 1884d) E
Scoparia augastis Meyrick, 1907c E
Scoparia autochroa Meyrick, 1907c E
Scoparia chalicodes Meyrick, 1884d E
Scoparia cyameuta (Meyrick, 1884d) E
Scoparia dryphactis Meyrick, 1911b E
Scoparia ejuncida Knaggs, 1867 E
Scoparia encapna Meyrick, 1888c E
Scoparia fragosa Meyrick, 1910b E K
Scoparia halopis Meyrick, 1909b E
Scoparia harpalaea (Meyrick, 1884d) E
Scoparia nomeutis (Meyrick, 1884d) E
Scoparia parachalca Meyrick, 1901 E
Scoparia parmifera Meyrick, 1909b E
Scoparia petrina (Meyrick, 1884d) E
Scoparia rotuella (Felder & Rogenhofer, 1875) E
Scoparia sideraspis Meyrick, 1905 E
Scoparia triscelis Meyrick, 1909b E
Scoparia ustimacula Felder & Rogenhofer, 1875 E
Scoparia (s.l.) *albafascicula* Salmon *in* Salmon & Bradley, 1956 E Su
Scoparia (s.l.) *animosa* Meyrick, 1914a E
Scoparia (s.l.) *asaleuta* Meyrick, 1907c E
Scoparia (s.l.) *astragalota* (Meyrick, 1884d) E
Scoparia (s.l.) *autumna* Philpott, 1927d E
Scoparia (s.l.) *caesia* (Philpott, 1926a) E
Scoparia (s.l.) *caliginosa* Philpott, 1918 E
Scoparia (s.l.) *cinefacta* Philpott, 1926a E
Scoparia (s.l.) *claranota* Howes, 1946 E
Scoparia (s.l.) *clavata* Philpott, 1912 E
Scoparia (s.l.) *contexta* Philpott, 1931 E
Scoparia (s.l.) *crepuscula* Salmon, 1946 E
Scoparia (s.l.) *declivis* Philpott, 1918 E
Scoparia (s.l.) *diphtheralis* Walker, 1866a E
Scoparia (s.l.) *ergatis* Meyrick, 1884d E
Scoparia (s.l.) *exilis* Knaggs, 1867 E
Scoparia (s.l.) *falsa* Philpott, 1924a E
Scoparia (s.l.) *famularis* Philpott, 1930b E
Scoparia (s.l.) *fimbriata* Philpott, 1917b E
Scoparia (s.l.) *fumata* Philpott, 1915 E
Scoparia (s.l.) *gracilis* Philpott, 1924a E
Scoparia (s.l.) *humilialis* Hudson, 1950 E
Scoparia (s.l.) *illota* Philpott, 1919 E
Scoparia (s.l.) *indistinctalis* (Walker, 1863b) E
Scoparia (s.l.) *limatula* Philpott, 1930c E
Scoparia (s.l.) *lychnophanes* Meyrick, 1927a E
Scoparia (s.l.) *minusculalis* Walker, 1866a E
Scoparia (s.l.) *molifera* Meyrick, 1926b E
Scoparia (s.l.) *monochroma* Salmon, 1946 E
Scoparia (s.l.) *niphospora* (Meyrick, 1884d) E
Scoparia (s.l.) *pallidula* Philpott, 1928a E
Scoparia (s.l.) *panopla* Meyrick, 1884d E
Scoparia (s.l.) *parca* Philpott, 1928a E
Scoparia (s.l.) *pascoella* Philpott, 1920 E
Scoparia (s.l.) *phalerias* Meyrick, 1905 E
Scoparia (s.l.) *pura* Philpott, 1924a E
Scoparia (s.l.) *scripta* Philpott, 1918 E
Scoparia (s.l.) *sinuata* Philpott, 1930c E
Scoparia (s.l.) *subita* (Philpott, 1912) E
Scoparia (s.l.) *sylvestris* Clarke, 1926 E
Scoparia (s.l.) *tetracycla* Meyrick, 1884d E
Scoparia (s.l.) *trapezophora* Meyrick, 1884d E
Scoparia (s.l.) *tuicana* Clarke, 1926 E
Scoparia (s.l.) *turneri* Philpott, 1928a E
Scoparia (s.l.) *valenternota* Howes, 1946 E
Scoparia (s.l.) *vulpecula* Meyrick, 1927a E
INCERTAE SEDIS
Argyria s.l. *strophaea* Meyrick, 1905 E
Clepsicosma iridia Meyrick, 1888c E
CTENUCHIDAE A
Antichloris viridis Druce, 1884 A
ELACHISTIDAE s.l.
Agonopterix alstromeriana (Clerck, 1759) A
Agonopterix assimilella (Treitschke, 1832) * A BC
Agonopterix umbellana (Fabricius, 1794) A BC
Agriophara colligatella (Walker, 1864a) E
Cryptolechia (s.l.) *rhodobapta* Meyrick, 1923 E
Cryptolechia (s.l.) *semnodes* Meyrick, 1911b E
Donacostola notabilis (Philpott, 1928a) E
Eutorna caryochroa Meyrick, 1889b E
Eutorna inornata Philpott, 1927d E
Eutorna phaulocosma Meyrick, 1906 A
Eutorna symmorpha Meyrick, 1889b E
Heliostibes (s.l.) *vibratrix* Meyrick, 1927a E
Nymphostola galactina (Felder & Rogenhofer, 1875) E
Proteodes carnifex (Butler, 1877) E
Proteodes clarkei Philpott, 1926a E
Proteodes melographa Meyrick, 1927a E
Proteodes profunda Meyrick, 1905 E
Proteodes smithi Howes, 1946 E
ELACHISTIDAE s.s.
Elachista antipodensis (Dugdale, 1971b) E Su
Elachista archaeonoma (Meyrick, 1889b) E
Elachista eurychora (Meyrick, 1919) E
Elachista exaula (Meyrick, 1889b) E
Elachista galatheae (Viette, 1954) E Su
Elachista gerasmia Meyrick, 1889b E
Elachista helonoma (Meyrick, 1889b) E
Elachista hookeri (Dugdale, 1971b) E Su
Elachista laquaeorum (Dugdale, 1971b) E Sn
Elachista melanura Meyrick, 1889b E
Elachista napaea Philpott, 1930c E
Elachista ochroleuca (Meyrick, 1923) E
Elachista ombrodoca (Meyrick, 1889b) E
Elachista plagiaula (Meyrick, 1938) E
Elachista pumila (Dugdale, 1971b) E Su
Elachista sagittifera Philpott, 1927d E
Elachista thallophora Meyrick, 1889b E
Elachista watti (Philpott, 1924a) E
EPERMENIIDAE
Thambotricha vates Meyrick, 1922b E
GELECHIIDAE
Anarsia dryinopa Lower, 1897 A
Anisoplaca achyrota (Meyrick, 1885) E
Anisoplaca acrodactyla (Meyrick, 1907c) E
Anisoplaca cosmia Bradley, 1956b A
Anisoplaca fraxinea Philpott, 1928a E
Anisoplaca ptyoptera Meyrick, 1885 E
Aristotelia paradesma (Meyrick, 1885) E
Athrips (s.l.) *zophochalca* (Meyrick, 1918) E
Bilobata subsecivella (Zeller, 1852) A
Chrysoesthia drurella (Fabricius, 1775) A
Epiphthora melanombra Meyrick, 1888e E
Epiphthora nivea (Philpott, 1930b) E
Helcystogramma sp. nr *phryganitis* (Meyrick, 1911e) A
Hierodoris (s.l.) *insignis* Philpott, 1926a E
Kiwaia aerobatis (Meyrick, 1924a) E
Kiwaia brontophora (Meyrick, 1885i) E
Kiwaia caerulaea (Hudson, 1925) E
Kiwaia calaspidea (Clarke, 1934) E
Kiwaia cheradias (Meyrick, 1909a) E
Kiwaia contraria (Philpott, 1930b) E
Kiwaia dividua (Philpott, 1921) E
Kiwaia eurybathra (Meyrick, 1931b) E

Kiwaia glaucoterma (Meyrick, 1911b) E
Kiwaia heterospora (Meyrick, 1924a) E
Kiwaia hippeis (Meyrick, 1901) E
Kiwaia jeanae Philpott, 1930a E
Kiwaia lapillosa (Meyrick, 1924a) E
Kiwaia lenis (Philpott, 1929a) E
Kiwaia lithodes (Meyrick, 1885i) E
Kiwaia matermea (Povolny, 1974) E
Kiwaia monophragma (Meyrick, 1885i) E
Kiwaia neglecta (Philpott, 1924c) E
Kiwaia parapleura (Meyrick, 1885i) E
Kiwaia parvula (Philpott, 1930b) E
Kiwaia pharetria (Meyrick, 1885i) E
Kiwaia plemochoa (Meyrick, 1916b) E
Kiwaia pumila (Philpott, 1928c) E
Kiwaia schematica (Meyrick, 1885i) E
Kiwaia thyraula (Meyrick, 1885i) E
Megacraspedus calamogonus Meyrick, 1885i E
Monochroa leptocrossa (Meyrick, 1926) A
Phthorimaea operculella (Zeller, 1873) A
Scrobipalpa obsoletella (Fischer von Roeslerstamm, 1841) A
Sitotroga cerealella (Olivier, 1789) A
Symmetrischema tangolias (Gyen, 1913) A
Thiotricha lindsayi Philpott, 1927d E
Thiotricha oleariae Hudson, 1928 E
Thiotricha tetraphala (Meyrick, 1885i) E
Thiotricha thorybodes (Meyrick, 1885i) E
GEOMETRIDAE
ENNOMINAE
Chalastra aristarcha (Meyrick, 1892) E
Chalastra ochrea (Howes, 1911) E
Chalastra pellurgata Walker, 1862b E
Cleora scriptaria (Walker, 1860b) E
Declana atronivea (Walker, 1865a) E
Declana egregia (Felder & Rogenhofer, 1875) E
Declana feredayi Butler, 1877 E
Declana floccosa Walker, 1858c E
Declana glacialis Hudson, 1903 E
Declana griseata Hudson, 1898 E
Declana hermione Hudson, 1898 E
Declana junctilinea (Walker, 1865a) E
Declana leptomera (Walker, 1858c) E
Declana niveata Butler, 1879a E
Declana toreuta Meyrick, 1929 E
Gellonia dejectaria (Walker, 1860b) E
Gellonia pannularia (Guenée, 1868) E
Ischalis dugdalei Weintraub & Scoble, 2004 E
Ischalis fortinata (Guenée, 1868) E
Ischalis gallaria (Walker, 1860a) E
Ischalis nelsonaria (Felder & Rogenhofer, 1875) E
Ischalis variabilis (Warren, 1895) E
Pseudocoremia albafasciata (Philpott, 1915) E
Pseudocoremia amaculata Stephens & Gibbs, 2003 E
Pseudocoremia berylia (Howes, 1943) E
Pseudocoremia campbelli (Philpott, 1927a) E
Pseudocoremia colpogramma (Meyrick, 1936) E
Pseudocoremia dugdalei Stephens & Gibbs, 2003 E
Pseudocoremia fascialata (Philpott, 1903) E
Pseudocoremia fenerata (Felder & Rogenhofer, 1875) E
Pseudocoremia flava Warren, 1896b E
Pseudocoremia fluminea (Philpott, 1926a) E
Pseudocoremia foxi Stephens, Gibbs & Patrick, 2007 E
Pseudocoremia hollyae Stephens, Gibbs & Patrick, 2007 E
Pseudocoremia hudsoni Stephens, Gibbs & Patrick, 2007 E
Pseudocoremia indistincta Butler, 1877 E
Pseudocoremia insignita (Philpott, 1930b) E
Pseudocoremia lactiflua (Meyrick, 1912c) E
Pseudocoremia leucelaea (Meyrick, 1909a) E
Pseudocoremia lupinata (Felder & Rogenhofer, 1875) E
Pseudocoremia lutea (Philpott, 1914) E
Pseudocoremia melinata (Felder & Rogenhofer, 1875) E
Pseudocoremia modica (Philpott, 1921) E
Pseudocoremia monacha (Hudson, 1903) E
Pseudocoremia ombrodes (Meyrick, 1902c) E Ch
Pseudocoremia pergrata (Philpott, 1930b) E
Pseudocoremia productata (Walker, 1862a) E
Pseudocoremia rudisata ampla (Hudson, 1923b) E
Pseudocoremia r. rudisata (Walker, 1862b) E
Pseudocoremia suavis Butler, 1879a E
Pseudocoremia terrena (Philpott, 1915) E
Pseudocoremia (s.l.) *cineracia* (Howes, 1942) E
Sarisa muriferata (Walker, 1863a) E
Sestra flexata (Walker, 1862b) E
Sestra humeraria (Walker, 1861) E
Zermizinga indocilisaria Walker, 1863a[a] A?
LARENTIINAE
Anachloris subochraria (Doubleday *in* White & Doubleday, 1843)
Aponotoreas anthracias (Meyrick, 1883d) E
Aponotoreas dissimilis (Philpott, 1914) E
Aponotoreas incompta Philpott, 1918 E
Aponotoreas insignis (Butler, 1877) E
Aponotoreas orphnaea (Meyrick, 1883d) E
Aponotoreas synclinalis Hudson, 1903 E
Aponotoreas villosa Philpott, 1917b E
Arctesthes catapyrrha (Butler, 1877) E
Arctesthes siris (Hudson, 1908) E
Asaphodes abrogata (Walker, 1862a) E
Asaphodes adonis (Hudson, 1898) E
Asaphodes aegrota (Butler, 1879a) E
Asaphodes albalineata (Philpott, 1915) E
Asaphodes aphelias (Prout, 1939) E
Asaphodes beata (Butler, 1877) E
Asaphodes camelias (Meyrick, 1888b) E
Asaphodes campbellensis (Dugdale, 1964) E Su
Asaphodes cataphracta (Meyrick, 1883d) E
Asaphodes chionogramma (Meyrick, 1883d) E
Asaphodes chlamydota (Meyrick, 1883d) E
Asaphodes chlorocapna (Meyrick, 1925a) E Ch
Asaphodes cinnabari (Howes, 1912) E
Asaphodes citroena (Clarke, 1934) E
Asaphodes clarata (Walker, 1862a) E
Asaphodes cosmodora (Meyrick, 1888b) E
Asaphodes declarata (Prout, 1914) E
Asaphodes dionysias (Meyrick, 1907c) E
Asaphodes exoriens (Prout, 1912) E
Asaphodes frivola (Meyrick, 1913a) E
Asaphodes glaciata (Hudson, 1925) E
Asaphodes helias (Meyrick, 1883d) E
Asaphodes ida (Clarke, 1926) E
Asaphodes imperfecta (Philpott, 1905) E
Asaphodes limonodes (Meyrick, 1888b) E
Asaphodes mnesichola (Meyrick, 1888b) E
Asaphodes nephelias (Meyrick, 1883d) E
Asaphodes obarata (Felder & Rogenhofer, 1875) E
Asaphodes omichlias (Meyrick, 1883d) E
Asaphodes oraria (Philpott, 1903) E
Asaphodes oxyptera (Hudson, 1909) E Su
Asaphodes periphaea (Meyrick, 1905) E
Asaphodes philpotti (Prout, 1927) E
Asaphodes prasinias (Meyrick, 1883d) E
Asaphodes prymnaea (Meyrick, 1911b) E
Asaphodes recta (Philpott, 1905) E
Asaphodes sericodes (Meyrick, 1915a) E
Asaphodes stephanitis Meyrick, 1907c E
Asaphodes stinaria (Guenée, 1868) E
Austrocidaria anguligera (Butler, 1879a) E
Austrocidaria arenosa (Howes, 1911) E
Austrocidaria bipartita (Prout, 1958) E
Austrocidaria callichlora (Butler, 1879a) E
Austrocidaria cedrinodes (Meyrick, 1911b) E
Austrocidaria gobiata (Felder & Rogenhofer, 1875) E
Austrocidaria haemophaea (Meyrick, 1925a) E
Austrocidaria lithurga (Meyrick, 1911b) E
Austrocidaria parora (Meyrick, 1884c) E
Austrocidaria praerupta (Philpott, 1918) E
Austrocidaria prionota (Meyrick, 1883d) E
Austrocidaria similata (Walker, 1862b) E
Austrocidaria stricta (Philpott, 1915) E
Austrocidaria umbrosa (Philpott, 1917b) E
Austrocidaria venustatis (Salmon, 1946) E
Cephalissa siria Meyrick, 1883d E
Chloroclystis (s.l.) *filata* (Guenée, 1857b) A
Chloroclystis (s.l.) *impudicis* Dugdale, 1964 E Su
Chloroclystis (s.l.) *inductata* (Walker, 1862b) E
Chloroclystis (s.l.) *lichenodes* (Purdie, 1887) E
Chloroclystis (s.l.) *nereis* (Meyrick, 1888b) E
Chloroclystis (s.l.) *sphragitis* (Meyrick, 1888b) E
Chrysolarentia subrectaria (Guenée, 1858)
Dasyuris anceps (Butler, 1877) E
Dasyuris austrina Philpott, 1928a E
Dasyuris callicrena (Meyrick, 1883d) E
Dasyuris catadees Prout, 1939 E
Dasyuris enysii (Butler, 1877) E
Dasyuris fulminea Philpott, 1915 E
Dasyuris grisescens Prout, 1939 E
Dasyuris hectori (Butler, 1877) E
Dasyuris leucobathra Meyrick, 1911b E
Dasyuris micropolis Meyrick, 1929 E
Dasyuris octans Hudson, 1923c E
Dasyuris partheniata Guenée, 1868 E
Dasyuris pluviata Hudson, 1928 E
Dasyuris strategica (Meyrick, 1883d) E
Dasyuris transaurea Howes, 1912 E
Elvia glaucata Walker, 1862b E
Epicyme rubropunctaria (Doubleday *in* White & Doubleday, 1843)
Epiphryne charidema autocharis (Meyrick, 1924a) E
Epiphryne c. charidema (Meyrick, 1909b) E
Epiphryne undosata (Felder & Rogenhofer, 1875) E
Epiphryne verriculata (Felder & Rogenhofer, 1875) E
Epiphryne xanthaspis (Meyrick, 1883d) E
Epyaxa lucidata (Walker, 1862a) E
Epyaxa rosearia (Doubleday *in* White & Doubleday, 1843) E
Epyaxa venipunctata (Walker, 1863a) E
Gingidiobora nebulosa (Philpott, 1917b) E
Gingidiobora subobscurata (Walker, 1862b) E
Helastia alba Craw, 1987 E
Helastia angusta Craw, 1987 E
Helastia christinae Craw, 1987 E
Helastia cinerearia (Doubleday *in* White & Doubleday, 1843) E
Helastia clandestina (Philpott, 1921) E
Helastia corcularia (Guenée, 1868) E
Helastia cryptica Craw, 1987 E
Helastia cymozeucta (Meyrick, 1913a) E
Helastia expolita (Philpott, 1917b) E
Helastia farinata (Warren, 1896b) E
Helastia mutabilis Craw, 1987 E
Helastia ohauensis Craw, 1987 E
Helastia plumbea (Philpott, 1915) E
Helastia salmoni Craw, 1987 E
Helastia scissa Craw, 1987 E
Helastia semisignata (Walker, 1862a) E
Helastia siris (Hawthorne, 1897) E
Helastia triphragma (Meyrick, 1883d) E
Homodotis amblyterma (Meyrick, 1931a) E
Homodotis falcata (Butler, 1879a) E
Homodotis megaspilata (Walker, 1862a) E
Horisme (s.l.) *suppressaria* (Walker, 1863a) E
Hydriomena (s.l.) *arida* (Butler, 1879a) E
Hydriomena (s.l.) *canescens* Philpott, 1918 E
Hydriomena (s.l.) *clarkei* (Howes, 1917) E
Hydriomena (s.l.) *deltoidata* (Walker, 1862b) E
Hydriomena (s.l.) *hemizona* Meyrick, 1897b E
Hydriomena (s.l.) *purpurifera* (Fereday, 1884) E
Hydriomena (s.l.) *rixata* (Felder & Rogenhofer, 1875) E
Microdes epicryptis Meyrick, 1897b E
Microdes quadristrigata Walker, 1862a E
Notoreas arcuata Philpott, 1921 E

Notoreas atmogramma Meyrick, 1911b E
Notoreas blax Prout, 1939 E
Notoreas chioneres Prout, 1939 E
Notoreas chrysopeda (Meyrick, 1888b) E
Notoreas galaxias Hudson, 1928 E
Notoreas hexaleuca (Meyrick, 1914a) E
Notoreas ischnocyma Meyrick, 1905 E
Notoreas isoleuca Meyrick, 1897b E
Notoreas isomoera Prout, 1939 E
Notoreas mechanitis (Meyrick, 1883d) E
Notoreas niphocrena (Meyrick, 1883d) E
Notoreas ortholeuca Hudson, 1923b E
Notoreas paradelpha (Meyrick, 1883d) E
Notoreas perornata (Walker, 1863a) E
Notoreas regilla (Philpott, 1928a) E
Notoreas simplex Hudson, 1898 E
Orthoclydon chlorias (Meyrick, 1883d) E
Orthoclydon praefectata (Walker, 1861) E
Orthoclydon pseudostinaria (Hudson, 1918) E
Paradetis porphyrias (Meyrick, 1883d) E
Paranotoreas brephosata (Walker, 1862a) E
Paranotoreas ferox (Butler, 1877) E
Paranotoreas fulva (Hudson, 1905a) E
Paranotoreas opipara Philpott, 1915 E
Paranotoreas zopyra (Meyrick, 1883d) E
Pasiphila acompsa (Prout, 1927) E
Pasiphila aristias (Meyrick, 1897b) E
Pasiphila bilineolata (Walker, 1862a) E
Pasiphila charybdis (Butler, 1879a) E
Pasiphila cotinaea (Meyrick, 1913a) E
Pasiphila dryas Meyrick, 1891 E
Pasiphila erratica (Philpott, 1916) E
Pasiphila fumipalpata (Felder & Rogenhofer, 1875) E
Pasiphila furva (Philpott, 1917b) E
Pasiphila halianthes (Meyrick, 1907c) E
Pasiphila heighwayi (Philpott, 1927a) E
Pasiphila humilis (Philpott, 1917b) E
Pasiphila lunata (Philpott, 1912) E
Pasiphila magnimaculata (Philpott, 1915) E
Pasiphila malachita (Meyrick, 1913a) E
Pasiphila melochlora (Meyrick, 1911b) E
Pasiphila muscosata (Walker, 1862a) E
Pasiphila nebulosa Dugdale, 1971b E Su
Pasiphila plinthina Meyrick, 1888b E
Pasiphila punicea (Philpott, 1923) E
Pasiphila rivalis (Philpott, 1916) E
Pasiphila rubella (Philpott, 1915) E
Pasiphila sandycias (Meyrick, 1905) E
Pasiphila semochlora (Meyrick, 1919) E
Pasiphila suffusa (Hudson, 1928) E
Pasiphila testulata (Guenée, 1857b)
Pasiphila urticae (Hudson, 1939) E
Pasiphila vieta (Hudson, 1950) E
Phrissogonus laticostatus (Walker, 1862a) A
Poecilasthena pulchraria (Doubleday *in* White & Doubleday, 1843)
Poecilasthena schistaria (Walker, 1861) E
Poecilasthena subpurpureata (Walker, 1863a) E
Tatosoma agrionata (Walker, 1862b) E
Tatosoma alta Philpott, 1913 E
Tatosoma apicipallida Prout, 1914 E
Tatosoma fasciata Philpott, 1914 E
Tatosoma lestevata (Walker, 1862b) E
Tatosoma monoviridisata Clarke, 1920 E
Tatosoma tipulata (Walker, 1862b) E
Tatosoma topea Philpott, 1903 E
Tatosoma transitaria (Walker, 1862b) E
Xanthorhoe bulbulata (Guenée, 1868) E
Xanthorhoe frigida Howes, 1946 E
Xanthorhoe lophogramma Meyrick, 1897b E
Xanthorhoe occulta Philpott, 1903 E
Xanthorhoe orophyla (Meyrick, 1883d) E
Xanthorhoe orophylloides Hudson, 1909 E Su
Xanthorhoe semifissata (Walker, 1862b) E
OENOCHROMINAE s.l.
Adeixis griseata (Hudson, 1903) E
Dichromodes cynica Meyrick, 1911b E
Dichromodes gypsotis Meyrick, 1888b E
Dichromodes ida Hudson, 1905a E
Dichromodes niger (Butler, 1877) E
Dichromodes simulans Hudson, 1905a E
Dichromodes sphaeriata (Felder & Rogenhofer, 1875) E
Samana acutata Butler, 1877 E
Samana falcatella Walker, 1863b E
Theoxena scissaria (Guenée, 1868) E
Xyridacma alectoraria (Walker, 1860a) E
Xyridacma ustaria (Walker, 1863a) E
Xyridacma veronicae Prout, 1934 E
STERRHINAE
Scopula rubraria (Doubleday *in* White & Doubleday, 1843)
NOMEN DUBIUM
Hydriomena (s.l.) *iolanthe* Hudson, 1939 [Based on a single specimen, now lost. Hudson's figure has not been convincingly matched with any known species, endemic or adventive. Omitted from Dugdale (1988).]
GLYPHIPTERIGIDAE
Glyphipterix achlyoessa (Meyrick, 1880b) E
Glyphipterix acronoma Meyrick, 1888e E
Glyphipterix acrothecta Meyrick, 1880b E
Glyphipterix aenea Philpott, 1917b E
Glyphipterix aerifera Meyrick, 1912d E
Glyphipterix astrapaea Meyrick, 1880b E
Glyphipterix ataracta (Meyrick, 1888e) E
Glyphipterix aulogramma Meyrick, 1907c E
Glyphipterix bactrias Meyrick, 1911b E
Glyphipterix barbata Philpott, 1918 E
Glyphipterix brachydelta Meyrick, 1916b E
Glyphipterix calliactis Meyrick, 1914a E
Glyphipterix cionophora (Meyrick, 1888e) E
Glyphipterix codonias Meyrick, 1909a E
Glyphipterix dichorda Meyrick, 1911b E
Glyphipterix erastis Meyrick, 1911b E
Glyphipterix euastera Meyrick, 1880b E
Glyphipterix iocheaera Meyrick, 1880b E
Glyphipterix leptosema Meyrick, 1888e E
Glyphipterix metasticta Meyrick, 1907c E
Glyphipterix morangella Felder & Rogenhofer, 1875 E
Glyphipterix necopina Philpott, 1927d E
Glyphipterix nephoptera Meyrick, 1888e E
Glyphipterix octonaria Philpott, 1924a E
Glyphipterix oxymachaera (Meyrick, 1880b) E
Glyphipterix rugata Meyrick, 1915a E
Glyphipterix scintelella Walker, 1864b E
Glyphipterix scintilla Clarke, 1926 E
Glyphipterix scolias Meyrick, 1910b KE
Glyphipterix similis Philpott, 1928a E
Glyphipterix triselena Meyrick, 1880b E
Glyphipterix tungella Felder and Rogenhofer, 1875 E
Glyphipterix xestobela (Meyrick, 1888e) E
Glyphipterix zelota Meyrick, 1888e E
Pantosperma holochalca Meyrick, 1888e E
GRACILLARIIDAE
GRACILLARIINAE
Acrocercops (s.l.) *alysidota* (Meyrick, 1880a) A
Acrocercops (s.l.) *laciniella* (Meyrick, 1880) A
Acrocercops (s.l.) *leucocyma* (Meyrick, 1889b) E
Caloptilia azaleella (Brants, 1913) A
Caloptilia chalcodelta (Meyrick, 1889b) E
Caloptilia chrysitis (Felder & Rogenhofer, 1875) E
Caloptilia elaeas (Meyrick, 1911b) E
Caloptilia linearis (Butler, 1877) E
Caloptilia octopunctata (Turner, 1894) K
Caloptilia selenitis (Meyrick, 1909a) E
Conopomorpha cyanospila Meyrick, 1885i E
Dialectica scalariella (Zeller, 1850) A
Macarostola miniella (Felder & Rogenhofer, 1875) E
Polysoma eumetalla (Meyrick, 1880a) A
LITHOCOLLETINAE A
Phyllonorycter hardenbergiella (Wise, 1957) A
Phyllonorycter messaniella (Zeller, 1846) A
PHYLLOCNISTINAE
Acrocercops (s.l.) *aellomacha* (Meyrick, 1880a) E
Acrocercops (s.l.) *aethalota* (Meyrick, 1880a) E
Acrocercops (s.l.) *panacicorticis* (Watt, 1920) E
Acrocercops (s.l.) *panacifinens* (Watt, 1920) E
Acrocercops (s.l.) *panacitorsens* (Watt, 1920) E
Acrocercops (s.l.) *panacivagans* (Watt, 1920) E
Acrocercops (s.l.) *panacivermiforma* (Watt, 1920) E
Acrocercops (s.l.) *zorionella* (Hudson, 1918) E
HELIOZELIDAE A
Heliozela cf. *catoptrias* Meyrick, 1897 A
HEPIALIDAE
Aenetus virescens (Doubleday *in* White & Doubleday, 1843) E
Aoraia aspina Dugdale, 1994 E
Aoraia aurimaculata (Philpott, 1914) E
Aoraia dinodes (Meyrick, 1890) E
Aoraia enysii (Butler, 1877) E
Aoraia flavida Dugdale, 1994 E
Aoraia hespera Dugdale, 1994 E
Aoraia insularis Dugdale, 1994 E
Aoraia lenis Dugdale, 1994 E
Aoraia macropis Dugdale, 1994 E
Aoraia oreobolae Dugdale, 1994 E
Aoraia orientalis Dugdale, 1994 E
Aoraia rufivena Dugdale, 1994 E
Aoraia senex (Hudson, 1908) E
Cladoxycanus minos (Hudson, 1905a) E
Dioxycanus fuscus (Philpott, 1914) E
Dioxycanus oreas (Hudson, 1920) E
Dumbletonius characterifer (Walker, 1865a) E
Dumbletonius unimaculatus (Salmon, 1948) E
Heloxycanus patricki Dugdale, 1994 E
Wiseana cervinata (Walker, 1865a) E
Wiseana copularis (Meyrick, 1912c) E
Wiseana fuliginea (Butler, 1879a) E
Wiseana jocosa (Meyrick, 1912c) E
Wiseana mimica (Philpott, 1923) E
Wiseana signata (Walker, 1856a) E
Wiseana umbraculata (Guenée, 1868) E
incertae sedis
Gen. indet. *mairi* (Buller, 1873) E [*Porina mairi* was based on a single specimen, now lost. *Porina* is preoccupied (Bryozoa); Lepidoptera formerly assigned to *Porina* have been reassigned to other genera. It is not known to which *mairi* belongs.]
LECITHOCERIDAE
Compsistis bifaciella (Walker, 1864a) E
Lecithocera micromela (Lower, 1897) A
Sarisophora leucoscia Turner, 1919 A
LYCAENIDAE
Lampides boeticus (Linnaeus, 1767) A
Lycaena boldenarum boldenarum (White, 1862) E
Lycaena b. caerulaea (Salmon, 1946) E
Lycaena b. ianthina (Salmon, 1946) E
Lycaena feredayi (Bates, 1867) E
Lycaena rauparaha (Fereday, 1877b) E
Lycaena salustius (Fabricius, 1793) E
Zizina otis labradus (Godart, 1824) A?
Zizina oxleyi (Felder & Felder, 1865) E
LYMANTRIIDAE A
Orgyia antiqua (Linnaeus, 1758) A extinct
Orgyia thyellina Butler, 1881c A eradicated
Teia anartoides Walker, 1855 A eradicated
LYONETIIDAE
Cateristis eustyla Meyrick, 1889b Do
Leucoptera spartifoliella (Huebner, [1813]) A
Stegommata leptomitella Meyrick, 1880a A
Stegommata sulfuratella Meyrick, 1880 A
MICROPTERIGIDAE
Micropardalis aurella (Hudson, 1918) E
Micropardalis doroxena (Meyrick, 1888e) E
Sabatinca aemula Philpott, 1924c E
Sabatinca aenea Hudson, 1923c E
Sabatinca aurantiaca Philpott, 1924c E

Sabatinca barbarica Philpott, 1918 E
Sabatinca calliarcha Meyrick, 1912c E
Sabatinca caustica Meyrick, 1912c E
Sabatinca chalcophanes (Meyrick, 1885i) E
Sabatinca chrysargyra (Meyrick, 1885i) E
Sabatinca demissa Philpott, 1923 E
Sabatinca heighwayi Philpott, 1927d E
Sabatinca ianthina Philpott, 1921 E
Sabatinca incongruella Walker, 1863c E
Sabatinca lucilia Clarke, 1920 E
Sabatinca passalota Meyrick, 1923 E
Sabatinca quadrijuga Meyrick, 1912c E
Sabatinca rosicoma Meyrick, 1814a E
Sabatinca zonodoxa Meyrick, 1888e E
MNESARCHAEIDAE E
Mnesarchaea acuta Philpott, 1929a E
Mnesarchaea fallax Philpott, 1927a E
Mnesarchaea fusca Philpott, 1922 E
Mnesarchaea fusilella (Walker, 1864b) E
Mnesarchaea hamadelpha Meyrick, 1888e E
Mnesarchaea loxoscia Meyrick, 1888e E
Mnesarchaea paracosma Meyrick, 1885i E
MOMPHIDAE
Zapyrastra calliphana Meyrick, 1889b E
Zapyrastra stellata (Philpott, 1931) E
NEPTICULIDAE
Stigmella aigialeia Donner & Wilkinson, 1989 E
Stigmella aliena Donner & Wilkinson, 1989 E
Stigmella atrata Donner & Wilkinson, 1989 E
Stigmella cassiniae Donner & Wilkinson, 1989 E
Stigmella childi Donner & Wilkinson, 1989 E
Stigmella cypracma (Meyrick, 1916b) E
Stigmella erysibodea Donner & Wilkinson, 1989 E
Stigmella fulva (Watt, 1921a) E
Stigmella hakekeae Donner & Wilkinson, 1989 E
Stigmella hamishella Donner & Wilkinson, 1989 E
Stigmella hoheriae Donner & Wilkinson, 1989 E
Stigmella ilsea Donner & Wilkinson, 1989 E
Stigmella insignis (Philpott, 1927d) E
Stigmella kaimanua Donner & Wilkinson, 1989 E
Stigmella laquaeorum (Dugdale, 1971b) E Su
Stigmella lucida (Philpott, 1919) E
Stigmella maoriella (Walker, 1864b) E
Stigmella microtheriella (Stainton, 1854) A
Stigmella ogygia (Meyrick, 1889b) E
Stigmella oriastra (Meyrick, 1917a) E
Stigmella palaga Donner & Wilkinson, 1989 E
Stigmella platina Donner & Wilkinson, 1989 E
Stigmella progama (Meyrick, 1924b) E
Stigmella progonopis (Meyrick, 1921) E
Stigmella propalaea (Meyrick, 1889b) E
Stigmella sophorae (Hudson, 1939) E
Stigmella tricentra (Meyrick, 1889b) E
Stigmella watti Donner & Wilkinson, 1989 E
NOCTUIDAE
Achaea janata (Linnaeus, 1758) V
Agrotis infusa (Boisduval, 1832) V
Agrotis innominata Hudson, 1898 E
Agrotis ipsilon (Hufnagel, 1766) E
Agrotis munda Walker, 1857a V
Aletia (s.l.) *argentaria* Howes, 1945 E
Aletia (s.l.) *cucullina* (Guenée, 1868) E
Aletia (s.l.) *cuneata* Philpott, 1916 E
Aletia (s.l.) *cyanopetra* (Meyrick, 1927b) E
Aletia (s.l.) *dentata* Philpott, 1923 E
Aletia (s.l.) *falsidica fasidica* (Meyrick, 1911b) E
Aletia (s.l.) *f. hamiltoni* Hampson, 1913b E
Aletia (s.l.) *fibriata* Meyrick, 1913a E
Aletia (s.l.) *inconstans* (Butler, 1880) E
Aletia (s.l.) *lacustris* Meyrick, 1934 E
Aletia (s.l.) *longstaffi* (Howes, 1911) E
Aletia (s.l.) *mitis* (Butler, 1877) E
Aletia (s.l.) *moderata* (Walker, 1865a) E
Aletia (s.l.) *nobilia* Howes, 1946 E
Aletia (s.l.) *obsecrata* Meyrick, 1914a E
Aletia (s.l.) *panda* Philpott, 1920 E
Aletia (s.l.) *parmata* Philpott, 1926a E
Aletia (s.l.) *probenota* Howes, 1945 E
Aletia (s.l.) *sistens* (Guenée, 1868) E
Aletia (s.l.) *sollennis* Meyrick, 1914a E
Aletia (s.l.) *temperata* (Walker, 1858c) E
Aletia (s.l.) *virescens* (Butler, 1879a) E
Andesia pessota (Meyrick, 1887) E
Anomis flava (Fabricius, 1775) V
Anomis involuta (Walker, 1858b) V
Anticarsia irrorata (Fabricius, 1781) K
Arcte coerula (Guenée, 1852c) V
Artigisa melanephele Hampson, 1914 A
Athetis tenuis (Butler, 1886) V
Australothis volatilis Matthews & Patrick, 1998 E
Austramathes purpurea (Butler, 1879a) E
Bityla defigurata (Walker, 1865b) E
Bityla sericea Butler, 1877 E
Callopistria maillardi (Guenée, 1862) K
Chasmina sp. of Hoare, 2001 V
Chrysodeixis argentifera (Guenée, 1852b) V
Chrysodeixis eriosoma (Doubleday *in* White & Doubleday, 1843)
Condica illecta (Walker, 1865a) V
Cosmodes elegans (Donovan, 1805) A
Ctenoplusia albostriata (Bremer & Grey, 1853) A
Dasypodia cymatodes Guenée, 1852c A
Dasypodia selenophora Guenée, 1852c A
Diarsia intermixta (Guenée, 1852a)
Dipaustica epiastra (Meyrick, 1911b) E
Ectopatria aspera (Walker, 1857b)
Eudocima fullonia (Clerck, 1764) V
Eudocima materna (Linnaeus, 1767) V
Euxoa admirationis (Guenée, 1868) E
Euxoa ceropachoides (Guenée, 1868) E
Feredayia graminosa (Walker, 1857b) E
Grammodes pulcherrima T.P. Lucas, 1892 V
Graphania agorastis (Meyrick, 1887) E
Graphania averilla (Hudson, 1921) E
Graphania beata (Howes, 1906) E
Graphania bromias (Meyrick, 1902c) E
Graphania brunneosa (Fox, 1970) E
Graphania chlorodonta (Hampson, 1911) E
Graphania chryserythra (Hampson, 1905) E
Graphania disjungens (Walker, 1858c) E
Graphania erebia (Hudson, 1909) E Su
Graphania fenwicki (Philpott, 1921) E
Graphania homoscia (Meyrick, 1887) E
Graphania infensa (Walker, 1857b) E
Graphania insignis (Walker, 1865b) E
Graphania lignana (Walker, 1857b) E
Graphania lindsayi Dugdale, 1988 E
Graphania lithias (Meyrick, 1887) E
Graphania maya (Hudson, 1898) E
Graphania mollis (Howes, 1908) E
Graphania morosa (Butler, 1880) E
Graphania mutans (Walker, 1857b) E
Graphania nullifera (Walker, 1857b) E
Graphania olivea (Watt, 1916) E
Graphania oliveri (Hampson, 1911) E
Graphania omicron (Hudson, 1898) E
Graphania omoplaca (Meyrick, 1887) E
Graphania pagaia (Hudson, 1909) E Sn
Graphania paracausta (Meyrick, 1887) E
Graphania pelanodes (Meyrick, 1931a) E
Graphania petrograpta (Meyrick, 1929) E
Graphania phricias (Meyrick, 1888a) E
Graphania plena (Walker, 1865b) E
Graphania prionistis (Meyrick, 1887) E
Graphania rubescens (Butler, 1879a) E
Graphania scutata (Meyrick, 1929) E
Graphania sequens (Howes, 1912) E
Graphania sericata (Howes, 1945) E
Graphania tetrachroa (Meyrick, 1931a) E
Graphania ustistriga (Walker, 1857b) E
Helicoverpa armigera conferta (Walker, 1857b) A
Helicoverpa punctigera (Wallengren, 1860) A
Homohadena (s.l.) *fortis* (Butler, 1880) E
Hydrillodes surata Meyrick, 1910b K
Hypenodes gonospilalis (Walker, [1866]) K
Hypocala deflorata (Fabricius, 1792) V
Ichneutica cana Howes, 1914 E
Ichneutica ceraunias Meyrick, 1887 E
Ichneutica dione Hudson, 1898 E
Ichneutica empyrea (Hudson, 1918) E
Ichneutica homerica Howes, 1943 E
Ichneutica lindsayi Philpott, 1926a E
Ichneutica marmorata (Hudson, 1924) E
Ichneutica nervosa Hudson, 1922 E
Ichneutica notata Salmon, 1946 E
Leucania stenographa Lower, 1900 A
Leucocosmia nonagrica (Walker, 1864) K
Meterana alcyone (Hudson, 1898) E
Meterana asterope (Hudson, 1898) E
Meterana badia (Philpott, 1927d) E
Meterana coctilis (Meyrick, 1931a) E
Meterana coeleno (Hudson, 1898) E
Meterana decorata (Philpott, 1905) E
Meterana diatmeta (Hudson, 1898) E
Meterana dotata (Walker, 1857b) E
Meterana exquisita (Philpott, 1903) E
Meterana grandiosa (Philpott, 1903) E
Meterana inchoata (Philpott, 1920) E
Meterana levis (Philpott, 1905) E
Meterana merope (Hudson, 1898) E
Meterana meyricci (Hampson, 1911) E
Meterana ochthistis (Meyrick, 1887) E
Meterana octans (Hudson, 1898) E
Meterana pansicolor (Howes, 1912) E
Meterana pascoi (Howes, 1912) E
Meterana pauca (Philpott, 1910) E
Meterana pictula (White *in* Taylor, 1855) E
Meterana praesignis (Howes, 1911) E
Meterana stipata (Walker, 1865b) E
Meterana tartarea (Butler, 1877) E
Meterana vitiosa (Butler, 1877) E
Mocis alterna (Walker, 1858c) V
Mocis frugalis (Fabricius, 1775) K
Mocis trifasciata (Stephens, 1830) K
Mythimna separata (Walker, 1865a) A
Pantydia sparsa Guenée, 1852 * A
Persectania aversa (Walker, 1856c) E
Phalaenoides glycinae Lewin, 1805 A
Physetica caerulea (Guenée, 1868) E
Proteuxoa comma (Walker, 1856c) E
Proteuxoa sanguinipuncta (Guenée, 1852) * A
Rhapsa scotosialis Walker, 1866a E
Schrankia costaestrigalis (Stephens, 1834)
Speiredonia spectans (Guenée, 1852c) V
Spodoptera exempta (Walker, 1857a) V
Spodoptera litura (Fabricius, 1775) A
Spodoptera mauritia (Boisduval, 1833) V
Tathorhynchus exsiccata fallax Swinhoe, 1902 V
Thysanoplusia orichalcea (Fabricius, 1775) A
Tiracola plagiata (Walker, 1857) K
Tmetolophota acontistis (Meyrick, 1887) E
Tmetolophota alopa (Meyrick, 1887) E
Tmetolophota arotis (Meyrick, 1887) E
Tmetolophota atristriga (Walker, 1865b) E
Tmetolophota blenheimensis (Fereday, 1883c) E
Tmetolophota hartii (Howes, 1914) E
Tmetolophota lissoxyla (Meyrick, 1911b) E
Tmetolophota micrastra (Meyrick, 1897b) E
Tmetolophota paraxysta (Meyrick, 1929) E
Tmetolophota phaula (Meyrick, 1887) E
Tmetolophota propria (Walker, 1856c) E
Tmetolophota purdii (Fereday, 1883c) E
Tmetolophota semivittata (Walker, 1865a) E
Tmetolophota similis (Philpott, 1924a) E
Tmetolophota steropastis (Meyrick, 1887) E
Tmetolophota stulta (Philpott, 1905) E
Tmetolophota sulcana (Fereday, 1880) E
Tmetolophota toroneura (Meyrick, 1901) E

Tmetolophota unica (Walker, 1856c) E
Trigonistis anticlina (Meyrick, 1901) E
NOLIDAE
Nola parvitis (Howes, 1917) E
Pseudoips fagana (Fabricius, 1781) A [once only]
Uraba lugens Walker, 1863 A
NYMPHALIDAE
Danainae
Danaus petilia (Stoll, 1790) V
Danaus plexippus (Linnaeus, 1758) A
Tirumala hamata (Macleay, 1826) V
Morphinae A
Opsiphanes cassina Felder & Felder, 1862 A
Opsiphanes tamarindi Felder & Felder, 1861 A
Nymphalinae
Hypolimnas bolina (Linnaeus, 1758) V
Junonia villida (Fabricius, 1787) V
Vanessa gonerilla gonerilla (Fabricius, 1775) E
Vanessa g. ida (Alfken, 1899) E
Vanessa itea (Fabricius, 1775)
Vanessa kershawi (McCoy, 1868) V
Satyrinae
Argyrophenga antipodum Doubleday, 1845 E
Argyrophenga harrisi Craw, 1978 E
Argyrophenga janitae Craw, 1978 E
Dodonidia helmsii Butler, 1884a E
Erebiola butleri Fereday, 1879 E
Melanitis leda (Linnaeus, 1758) V
Oreixenica lathoniella (Westwood, 1851) V
Percnodaimon merula (Hewitson, 1875) E
OECOPHORIDAE
Barea group
Atalopsis n. sp. of Hoare 2001 A
Atomotricha chloronota Meyrick, 1914a E
Atomotricha exsomnis Meyrick, 1913a E
Atomotricha isogama Meyrick, 1909a E
Atomotricha lewisi Philpott, 1927d E
Atomotricha oeconoma Meyrick, 1914a E
Atomotricha ommatias Meyrick, 1883c E
Atomotricha prospiciens Meyrick, 1924b E
Atomotricha sordida (Butler, 1877) E
Atomotricha versuta Meyrick, 1914a E
Barea codrella (Felder & Rogenhofer, 1875) A
Barea confusella (Walker, 1864a) A
Barea confusella sensu Philpott 1927f A
Barea consignatella Walker, 1864a A
Barea exarcha (Meyrick, 1883e) A
Borkhausenia (s.l.) *morella* Hudson, 1939 A
Chersadaula ochrogastra Meyrick, 1923 E
Corocosma memorabilis Meyrick, 1927a E
Euchersadaula lathriopa (Meyrick, 1905) E
Euchersadaula tristis Philpott, 1926a E
Eulechria (s.l.) *zophoessa* Meyrick, 1883a E
Euthictis (s.l.) *chloratma* (Meyrick, 1916b) E
Gymnobathra (s.l.) *rufopunctella* Hudson, 1950 E
Leptocroca sanguinolenta Meyrick, 1886a A
Leptocroca (s.l.) *amenena* (Meyrick, 1888e) E
Leptocroca (s.l.) *aquilonaris* Philpott, 1931 E
Leptocroca (s.l.) *asphaltis* (Meyrick, 1911b) E
Leptocroca (s.l.) *lenita* Philpott, 1931 E
Leptocroca (s.l.) *lindsayi* Philpott, 1930a E
Leptocroca (s.l.) *porophora* (Meyrick, 1929) E
Leptocroca (s.l.) *scholaea* (Meyrick, 1883c) E
Leptocroca (s.l.) *variabilis* Philpott, 1926a E
Leptocroca (s.l.) *vinaria* (Meyrick, 1914a) E
Leptocroca (s.l.) *xyrias* Meyrick, 1931b E
Locheutis (s.l.) *fusca* Philpott, 1930b E
Locheutis (s.l.) *pulla* Philpott, 1928g E
Locheutis (s.l.) *vagata* Meyrick, 1916b E
Mermeristis ocneropis Meyrick, 1936 E
Mermeristis spodiaea Meyrick, 1915c E
Sphyrelata amotella Walker, 1864 A
Tachystola acroxantha (Meyrick, 1885c) A
Tingena actinias (Meyrick, 1901) E
Tingena affinis (Philpott, 1926a) E
Tingena afflicta (Philpott, 1926b) E
Tingena aletis (Meyrick, 1905) E
Tingena amiculata (Philpott, 1926b) E
Tingena anaema (Meyrick, 1883c) E
Tingena ancogramma (Meyrick, 1919) E
Tingena apanthes (Meyrick, 1883c) E
Tingena apertella (Walker, 1864a) E
Tingena aphrontis (Meyrick, 1883c) E
Tingena armigerella (Walker, 1864a) E
Tingena aurata (Philpott, 1931) E
Tingena basella (Walker, 1863c) E
Tingena berenice (Meyrick, 1929) E
Tingena brachyacma (Meyrick, 1909a) E
Tingena chloradelpha (Meyrick, 1905) E
Tingena chloritis (Meyrick, 1883c) E
Tingena chrysogramma (Meyrick, 1883c) E
Tingena clarkei (Philpott, 1928a) E
Tingena collitella (Walker, 1864a) E
Tingena compsogramma (Meyrick, 1920a) E
Tingena contextella (Walker, 1864a) E
Tingena crotala (Meyrick, 1915b) E
Tingena decora (Philpott, 1928a) E
Tingena enodis (Philpott, 1927d) E
Tingena epichalca (Meyrick, 1886a) E
Tingena epimylia (Meyrick, 1883c) E
Tingena eriphaea (Meyrick, 1914a) E
Tingena eumenopa (Meyrick, 1926b) E
Tingena falsiloqua (Meyrick, 1932) E
Tingena fenestrata (Philpott, 1926b) E
Tingena grata (Philpott, 1927d) E
Tingena griseata (Butler, 1877) E
Tingena hastata (Philpott, 1916) E
Tingena hemimochla (Meyrick, 1883c) E
Tingena homodoxa (Meyrick, 1883c) E
Tingena honesta (Philpott, 1929a) E
Tingena honorata (Philpott, 1918) E
Tingena hoplodesma (Meyrick, 1883c) E
Tingena horaea (Meyrick, 1883c) E
Tingena idiogama (Meyrick, 1924b) E
Tingena innotella (Walker, 1864a) E
Tingena lassa (Philpott, 1930b) E
Tingena laudata (Philpott, 1930b) E
Tingena letharga (Meyrick, 1883c) E
Tingena levicula (Philpott, 1930b) E
Tingena loxotis (Meyrick, 1905) E
Tingena macarella (Meyrick, 1883c) E
Tingena maranta (Meyrick, 1886a) E
Tingena marcida (Philpott, 1927a) E
Tingena melanamma (Meyrick, 1905) E
Tingena melinella (Felder & Rogenhofer, 1875) E
Tingena monodonta (Meyrick, 1911b) E
Tingena morosa (Philpott, 1926b) E
Tingena nycteris (Meyrick, 1890) E
Tingena ombrodella (Hudson, 1950) E
Tingena opaca (Philpott, 1926b) E
Tingena ophiodryas (Meyrick, 1936) E
Tingena oporaea (Meyrick, 1883c) E
Tingena oxyina (Meyrick, 1883c) E
Tingena pallidula (Philpott, 1924a) E
Tingena paratrimma (Meyrick, 1910a) E
Tingena paula (Philpott, 1927a) E
Tingena penthalea (Meyrick, 1905) E
Tingena perichlora (Meyrick, 1907c) E
Tingena pharmactis (Meyrick, 1905) E
Tingena phegophylla (Meyrick, 1883c) E
Tingena plagiatella (Walker, 1863c) E
Tingena pronephela (Meyrick, 1907c) E
Tingena robiginosa (Philpott, 1915) E
Tingena seclusa (Philpott, 1921) E
Tingena serena (Philpott, 1926b) E
Tingena siderodeta (Meyrick, 1883c) E
Tingena siderota (Meyrick, 1888e) E
Tingena sinuosa (Philpott, 1928g) E
Tingena tephrophanes (Meyrick, 1929) E
Tingena terrena (Philpott, 1926a) E
Tingena thalerodes (Meyrick, 1916b) E
Tingena vestita (Philpott, 1926a) E
Tingena xanthodesma (Philpott, 1923) E
Tingena xanthomicta (Meyrick, 1916b) E
Trachypepla amphileuca Meyrick, 1914a E
Trachypepla anastrella Meyrick, 1883c E
Trachypepla angularis (Philpott, 1929a) E
Trachypepla aspidephora Meyrick, 1883c E
Trachypepla conspicuella (Walker, 1864a) E
Trachypepla contritella (Walker, 1864a) E
Trachypepla cyphonias Meyrick, 1927b E
Trachypepla euryleucota Meyrick, 1883c E
Trachypepla festiva Philpott, 1930b E
Trachypepla galaxias Meyrick, 1883c E
Trachypepla hieropis Meyrick, 1892 E
Trachypepla importuna Meyrick, 1914a E
Trachypepla ingenua Meyrick, 1911b E
Trachypepla leucoplanetis Meyrick, 1883c E
Trachypepla lichenodes Meyrick, 1883c E
Trachypepla minuta Philpott, 1931 E
Trachypepla nimbosa Philpott, 1930b E
Trachypepla photinella (Meyrick, 1883a) E
Trachypepla protochlora Meyrick, 1883c E
Trachypepla roseata Philpott, 1923 E
Trachypepla semilauta Philpott, 1918 E
Trachypepla spartodeta Meyrick, 1883c E
Chelaza group
Endrosis sarcitrella (Linnaeus, 1758) A
Hofmannophila pseudospretella (Stainton, 1849) A
Prepalla austrina (Meyrick, 1914) E
Hierodoris group
Gymnobathra ambigua (Philpott, 1926a) E
Gymnobathra bryaula Meyrick, 1905 E
Gymnobathra caliginosa Philpott, 1927a E
Gymnobathra calliploca Meyrick, 1883c E
Gymnobathra callixyla (Meyrick, 1888e) E
Gymnobathra cenchrias (Meyrick, 1909a) E
Gymnobathra dinocosma (Meyrick, 1883c) E
Gymnobathra flavidella (Walker, 1864a) E
Gymnobathra hamatella (Walker, 1864a) E
Gymnobathra hyetodes Meyrick, 1883c E
Gymnobathra inaequata Philpott, 1928a E
Gymnobathra jubata (Philpott, 1918) E
Gymnobathra levigata Philpott, 1928a E
Gymnobathra omphalota Meyrick, 1888e E
Gymnobathra parca (Butler, 1877) E
Gymnobathra philadelpha Meyrick, 1883c E
Gymnobathra primaria Philpott, 1928a E
Gymnobathra sarcoxantha Meyrick, 1883c E
Gymnobathra tholodella Meyrick, 1883c E
Hierodoris atychioides (Butler, 1877) E
Hierodoris bilineata (Salmon, 1948) E
Hierodoris callispora (Meyrick, 1912d) E
Hierodoris electrica (Meyrick, 1889b) E
Hierodoris eremita Philpott, 1930c E
Hierodoris frigida Philpott, 1923 E
Hierodoris gerontion Hoare, 2005 E
Hierodoris huia Hoare, 2005 E
Hierodoris illita (Felder & Rogenhofer, 1875) E
Hierodoris iophanes Meyrick, 1912d E
Hierodoris pachystegiae Hoare, 2005 E
Hierodoris polita Hoare, 2005 E
Hierodoris s-fractum Hoare, 2005 E
Hierodoris squamea (Philpott, 1915) E
Hierodoris stella (Meyrick, 1914a) E
Hierodoris torrida Hoare, 2005 E
Hierodoris tygris Hoare, 2005 E
Izatha amorbas (Meyrick, 1910a) E
Izatha apodoxa (Meyrick, 1888e) E
Izatha attactella Walker, 1864a E
Izatha austera (Meyrick, 1883c) E
Izatha balanophora (Meyrick, 1897b) E
Izatha caustopa (Meyrick, 1892) E
Izatha churtoni Dugdale, 1988 E
Izatha convulsella (Walker, 1864a) E
Izatha copiosella (Walker, 1864b) E
Izatha epiphanes (Meyrick, 1883c) E
Izatha florida Philpott, 1927d E

Izatha heroica Philpott, 1926a E
Izatha hudsoni Dugdale, 1988 E
Izatha manubriata Meyrick, 1923 E
Izatha mesoschista Meyrick, 1931a E
Izatha metadelta (Meyrick, 1905) E
Izatha mira Philpott, 1913 E
Izatha oleariae Dugdale, 1971b E Sn
Izatha peroneanella (Walker, 1864a) E
Izatha phaeoptila (Meyrick, 1905) E
Izatha picarella (Walker, 1864a) E
Izatha plumbosa Philpott, 1927d E
Izatha prasophyta (Meyrick, 1883c) E
Izatha psychra (Meyrick, 1883c) E
Izatha rigescens Meyrick, 1929 E
Lathicrossa leucocentra Meyrick, 1883c E
Lathicrossa prophetica Meyrick, 1927a E
Phaeosaces apocrypta Meyrick, 1885i E
Phaeosaces coarctatella (Walker, 1864a) E
Phaeosaces compsotypa Meyrick, 1885i E
Phaeosaces lindsayae (Philpott, 1928c) E
Schiffermuelleria (s.l.) *orthophanes* (Meyrick, 1905) E
Scieropepla typhicola Meyrick, 1885i
Thamnosara sublitella (Walker, 1864a) E
Tinearupa sorenseni aucklandica Dugdale, 1971b E Su
Tinearupa s. sorenseni Salmon & Bradley, 1956 E
Wingia group A
Heteroteucha dichroella (Zeller, 1877) A
Trachypepla (s.l.) *indolescens* Meyrick, 1927 A
PAPILIONIDAE A
Papilio xuthus Linnaeus, 1760 A
PIERIDAE
Catopsilia pomona (Fabricius, 1775) V
Pieris rapae (Linnaeus, 1758) A
PLUTELLIDAE
Charixena iridoxa (Meyrick, 1916b) E
Chrysorthenches argentea Dugdale, 1996 E
Chrysorthenches drosochalca (Meyrick, 1905) E
Chrysorthenches glypharcha (Meyrick, 1919) E
Chrysorthenches halocarpi Dugdale, 1996 E
Chrysorthenches phyllocladi Dugdale, 1996 E
Chrysorthenches polita (Philpott, 1918) E
Chrysorthenches porphyritis (Meyrick, 1885i) E
Chrysorthenches virgata (Philpott, 1920) E
Doxophyrtis hydrocosma Meyrick, 1914a E
Hierodoris (s.l.) *stellata* Philpott, 1918 E
Leuroperna sera (Meyrick, 1885i) E
Orthenches chartularia Meyrick, 1924a E
Orthenches chlorocoma Meyrick, 1885i E
Orthenches dictyarcha Meyrick, 1927b E
Orthenches disparilis Philpott, 1931 E
Orthenches homerica (Salmon, 1956) E
Orthenches prasinodes Meyrick, 1885i E
Orthenches saleuta Meyrick, 1913a E
Orthenches semifasciata Philpott, 1915 E
Orthenches septentrionalis Philpott, 1930b E
Orthenches similis Philpott, 1924a E
Orthenches vinitincta Philpott, 1917b E
Phylacodes cauta Meyrick, 1905 E
Plutella antiphona Meyrick, 1901 E
Plutella xylostella (Linnaeus, 1758) A
Plutella(s.l.) *psammochroa* Meyrick, 1885i
Prays nephelomima Meyrick, 1907b A
Prays sp. 1 K
Prays sp. 2 K
Proditrix chionochloae Dugdale, 1987a E
Proditrix gahniae Dugdale, 1987a E
Proditrix megalynta (Meyrick, 1915a) E
Proditrix tetragona (Hudson, 1918) E
Protosynaema eratopis Meyrick, 1885i E
Protosynaema hymenopis Meyrick, 1935 E
Protosynaema matutina Philpott, 1928g E
Protosynaema quaestuosa Meyrick, 1924a E
Protosynaema steropucha Meyrick, 1885i E
PSYCHIDAE
Cebysa leucotelus Walker, 1854 A
Grypotheca horningae Dugdale, 1987b E Sn
Grypotheca pertinax Dugdale, 1987b E
Grypotheca triangularis (Philpott, 1930b) E
Lepidoscia heliochares (Meyrick, 1893) A*
Lepidoscia cf. *lainodes* Meyrick, 1921 A*
Lepidoscia protorna (Meyrick, 1893) A*
Liothula omnivora Fereday, 1878b E
Mallobathra abyssina (Clarke, 1934) E
Mallobathra angusta Philpott, 1928g E
Mallobathra aphrosticha Meyrick, 1912c E
Mallobathra campbellica Dugdale, 1971b E Su
Mallobathra cana Philpott, 1927d E
Mallobathra cataclysma Clarke, 1934 E
Mallobathra crataea Meyrick, 1888e E
Mallobathra fenwicki Philpott, 1924a E
Mallobathra homalopa Meyrick, 1891 E
Mallobathra lapidosa Meyrick, 1914a E
Mallobathra memotuina Clarke, 1934 E
Mallobathra metrosema Meyrick, 1888e E
Mallobathra obscura Philpott, 1928a E
Mallobathra perisseuta Meyrick, 1920a E
Mallobathra petrodoxa (Meyrick, 1923) E
Mallobathra scoriota Meyrick, 1909a E
Mallobathra strigulata Philpott, 1924a E
Mallobathra subalpina Philpott, 1930a E
Mallobathra tonnoiri Philpott, 1927d E
Orophora unicolor (Butler, 1877) E
Reductoderces araneosa (Meyrick, 1914a) E
Reductoderces aucklandica Dugdale, 1971b E Su
Reductoderces cawthronella (Philpott, 1921) E
Reductoderces fuscoflava Salmon & Bradley, 1956 E
Reductoderces illustris (Philpott, 1917b) E
Reductoderces microphanes (Meyrick, 1888e) E
Rhathamictis nocturna (Clarke, 1926) E
Rhathamictis perspersa Meyrick, 1924b E
Scoriodyta conisalia Meyrick, 1888e E
Scoriodyta dugdalei Hättenschwiler, 1989 E
Scoriodyta patricki Hättenschwiler, 1989 E
Scoriodyta rakautarensis Hättenschwiler, 1989 E
Scoriodyta sereinae Hättenschwiler, 1989 E
Scoriodyta suttonensis Hättenschwiler, 1989 E
Scoriodyta virginella Hättenschwiler, 1989 E
PTEROPHORIDAE
Amblyptilia aeolodes (Meyrick, 1902c) E
Amblyptilia deprivatalis (Walker, 1864b) E
Amblyptilia epotis (Meyrick, 1905) E
Amblyptilia falcatalis (Walker, 1864b) E
Amblyptilia heliastis (Meyrick, 1885a) E
Amblyptilia lithoxesta (Meyrick, 1885a) E
Amblyptilia repletalis Walker, 1864b E
Lantanophaga pusillidactyla (Walker, 1864b) A BC
Oxyptilus pilosellae (Zeller, 1852) A BC
Platyptilia campsiptera Meyrick, 1907c E
Platyptilia carduidactyla (Riley, 1869) A
Platyptilia celidota (Meyrick, 1885a) E
Platyptilia charadrias (Meyrick, 1885a) E
Platyptilia hokowhitalis Hudson, 1939 E
Platyptilia isodactyla (Zeller, 1852) A BC
Platyptilia isoterma Meyrick, 1909a E
Platyptilia pulverulenta Philpott, 1923 E
Pterophorus furcatalis (Walker, 1864b) E
Pterophorus innotatalis Walker, 1864b E
Pterophorus monospilalis (Walker, 1864b) E
Sphenarches anisodactylus (Walker, 1864) K
Stenoptilia orites (Meyrick, 1885a) E
Stenoptilia zophodactyla (Duponchel, [1840])
PYRALIDAE
Galleriinae A
Achroia grisella (Fabricius, 1794) A
Galleria mellonella (Linnaeus, 1758) A
Phycitinae
Arcola malloi (Pastrana, 1961) A BC
Cadra cautella (Walker, 1863b) A
Crocydopora cinigerella (Walker, 1866b)
Cryptoblabes sp. K
Delogenes limodoxa Meyrick, 1918 E
Ephestia elutella (Huebner, 1796) A
Ephestia kuehniella Zeller, 1879 A
Ephestiopsis oenobarella (Meyrick, 1879) * A
Etiella behrii (Zeller, 1848) V
Homoeosoma anaspila Meyrick, 1901 E
Homoeosoma ischnomorpha Meyrick, 1931a E
Morosaphycita oculiferella (Meyrick, 1879b) A
Patagoniodes farinaria (Turner, 1904)
Pempelia genistella (Duponchel, 1836) A BC
Plodia interpunctella (Huebner, [1813]) A
Ptyomaxia trigonogramma (Turner, 1947) *
Sporophyla oenospora (Meyrick, 1897b) E
Vinicia sp. of Horak 1997 * E
Pyralinae
Aglossa caprealis (Huebner, [1809]) A
Diplopseustis perieresalis (Walker, 1859c)
Endotricha mesenterialis of Meyrick, 1910 K
Endotricha pyrosalis Guenée, 1854 V
Gauna aegusalis (Walker, 1859c) A
Pyralis farinalis (Linnaeus, 1758) A
ROESLERSTAMMIIDAE
Dolichernis chloroleuca Meyrick, 1891 E
Roeslerstammiidae s.l.
Vanicela disjunctella Walker, 1864b E
SATURNIIDAE A
Actias selene (Huebner, [1819]) A extinct
Antheraea pernyi (Guerin-Meneville, 1855) A extinct
Caligula simla (Westwood, 1847) A extinct
Hyalophora cecropia (Linnaeus, 1758) A extinct
Opodiphthera eucalypti (Scott, 1864) A
Samia cynthia (Drury, 1773) A extinct
SCYTHRIDIDAE
Scythris epistrota (Meyrick, 1889b) E
Scythris grandipennis (Haworth, 1828) A BC
Scythris lacustris (Philpott, 1930a) E
Scythris nigra Philpott, 1931 E
Scythris niphozela Meyrick, 1931b E
Scythris triatma Meyrick, 1935 E
SESIIDAE A
Synanthedon tipuliformis (Clerck, 1759) A
SPHINGIDAE
Agrius convolvuli (Linnaeus, 1758)
Cizara ardeniae (Lewin, 1805) V
Daphnis placida (Walker, 1856b) V
Hippotion celerio (Linnaeus, 1758) V
STATHMOPODIDAE
Calicotis crucifera Meyrick, 1889b
Pachyrhabda antinoma Meyrick, 1910b KE
Stathmopoda albimaculata Philpott, 1931 E
Stathmopoda aposema Meyrick, 1901 E
Stathmopoda aristodoxa Meyrick, 1926b E
Stathmopoda caminora Meyrick, 1890 E
Stathmopoda campylocha Meyrick, 1889b E
Stathmopoda cephalaea Meyrick, 1897 A
Stathmopoda coracodes Meyrick, 1923 E
Stathmopoda distincta Philpott, 1923 E
Stathmopoda endotherma Meyrick, 1931b E
Stathmopoda holochra Meyrick, 1889b E
Stathmopoda horticola Dugdale, 1988 E
Stathmopoda melanochra Meyrick, 1897a A BC
Stathmopoda mysteriastis Meyrick, 1901 E
Stathmopoda plumbiflua Meyrick, 1911b E
Stathmopoda skelloni (Butler, 1880) E
Stathmopoda trimolybdias Meyrick, 1926b E
Thylacosceles acridomima Meyrick, 1889b E
Thylacosceles radians Philpott, 1918 E
THYRIDIDAE
Morova subfasciata Walker, 1865a E
THYATIRIDAE A
Thyatira batis Linnaeus, 1758 A BC extinct
TINEIDAE
Amphixystis hapsimacha Meyrick, 1901 E
Archyala culta Philpott, 1931 E
Archyala lindsayi (Philpott, 1927a) E
Archyala opulenta Philpott, 1926a E
Archyala paraglypta Meyrick, 1889b E

Archyala pentazyga Meyrick, 1915a E
Archyala terranea (Butler, 1879a) E
Astrogenes chrysograpta Meyrick, 1921 E
Astrogenes insignita Philpott, 1930b E
Bascantis sirenica Meyrick, 1914a E
Crypsitricha agriopa (Meyrick, 1888e) E
Crypsitricha generosa Philpott, 1926a E
Crypsitricha mesotypa (Meyrick, 1888e) E
Crypsitricha pharotoma (Meyrick, 1888e) E
Crypsitricha roseata (Meyrick, 1913a) E
Crypsitricha stereota (Meyrick, 1914a) E
Dryadaula castanea Philpott, 1915 E
Dryadaula myrrhina Meyrick, 1905 E
Dryadaula pactolia Meyrick, 1901 E
Dryadaula terpsichorella (Busck, 1910) K
Endophthora omogramma Meyrick, 1888e E
Endophthora pallacopis Meyrick, 1918 E
Endophthora rubiginella Hudson, 1939 E
Endophthora tylogramma Meyrick, 1924a E
Erechthias acrodina (Meyrick, 1912c) E
Erechthias charadrota Meyrick, 1880b E
Erechthias chasmatias Meyrick, 1880b E
Erechthias chionodira Meyrick, 1880b E
Erechthias crypsimima (Meyrick, 1920a) E
Erechthias decoranda (Meyrick, 1925a) E
Erechthias exospila (Meyrick, 1901) E
Erechthias externella (Walker, 1864b) E
Erechthias flavistriata (Walsingham, 1907) K
Erechthias fulguritella (Walker, 1863c) E
Erechthias hemiclistra (Meyrick, 1911b) E
Erechthias indicans Meyrick, 1923 E
Erechthias lychnopa Meyrick, 1927a E
Erechthias macrozyga Meyrick, 1916b E
Erechthias stilbella (Doubleday *in* White & Doubleday, 1843) E
Erechthias terminella (Walker, 1863c) E
Eschatotypa derogatella (Walker, 1863c) E
Eschatotypa halosparta (Meyrick, 1919) E
Eschatotypa melichrysa Meyrick, 1880b E
Eugennaea laquearia (Meyrick, 1914a) E
Habrophila compseuta Meyrick, 1889b E
Lindera tessellatella Blanchard, 1852 A
Lysiphragma epixyla Meyrick, 1888e E
Lysiphragma howesii Quail, 1901 E
Lysiphragma mixochlora Meyrick, 1888e E
Monopis argillacea (Meyrick, 1893) A
Monopis crocicapitella (Clemens, 1859) A
Monopis dimorphella Dugdale, 1971b E
Monopis ethelella (Newman, 1856) A
Monopis ornithias (Meyrick, 1888e) E
Monopis typhlopa Meyrick, 1925a E
Nemapogon granella (Linnaeus, 1758) A
Niditinea fuscella (Linnaeus, 1758) A
Oinophila v-flava (Haworth, 1828) * A
Opogona aurisquamosa (Butler, 1881) K
Opogona comptella (Walker, 1864b) A
Opogona omoscopa (Meyrick, 1893) A
Petasactis technica (Meyrick, 1888e) E
Proterodesma byrsopola Meyrick, 1909b E
Proterodesma chathamica Dugdale, 1971a E Ch
Proterodesma turbotti (Salmon & Bradley, 1956) E Su
Prothinodes grammocosma (Meyrick, 1888e) E
Prothinodes lutata Meyrick, 1914a E
Sagephora exsanguis Philpott, 1918 E
Sagephora felix Meyrick, 1914a E
Sagephora jocularis Philpott, 1926a E
Sagephora phortegella Meyrick, 1888e E
Sagephora steropastis Meyrick, 1891 E
Sagephora subcarinata Meyrick, 1931a E
Tephrosara cimmeria (Meyrick, 1914a) E
Thallostoma eurygrapha Meyrick, 1913a E
Tinea dubiella Stainton, 1859 A
Tinea pallescentella Stainton, 1851 A
Tinea pellionella (Linnaeus, 1758) A
Tinea (s.l.) *accusatrix* Meyrick, 1916b E
Tinea (s.l.) *aetherea* Clarke, 1926 E
Tinea (s.l.) *argodelta* Meyrick, 1915a E
Tinea (s.l.) *astraea* Meyrick, 1911b E
Tinea (s.l.) *atmogramma* Meyrick, 1927b E
Tinea (s.l.) *belonota* Meyrick, 1888e E
Tinea (s.l.) *conferta* Meyrick, 1914a E
Tinea (s.l.) *conspecta* Philpott, 1931 E
Tinea (s.l.) *dicharacta* sensu Meyrick, 1911b nec Meyrick, 1893 E
Tinea (s.l.) *dividua* Philpott, 1928a E
Tinea (s.l.) *fagicola* Meyrick, 1921 E
Tinea (s.l.) *furcillata* Philpott, 1930b E
Tinea (s.l.) *margaritis* Meyrick, 1914a E
Tinea (s.l.) *mochlota* Meyrick, 1888e E
Tinea (s.l.) *munita* Meyrick, 1932 E
Tinea (s.l.) *sphenocosma* Meyrick, 1919 E
Tinea (s.l.) *texta* Meyrick, 1931a E
Tineola bisselliella (Hummel, 1823) A
Trichophaga tapetzella (Linnaeus, 1758) A
Trithamnora certella (Walker, 1863c) E

TORTRICIDAE

CHLIDANOTINAE

Lopharcha insolita (Dugdale, 1966b) E

OLETHREUTINAE

Acroclita discariana Philpott, 1930a E
Argyroploce (s.l.) *chlorosaris* Meyrick, 1914a E
Bactra noteraula Walsingham, 1907 E
Bactra optanias Meyrick, 1911c E
Crocidosema plebejana Zeller, 1847
Cryptaspasma querula (Meyrick, 1912c) E
Cydia pomonella (Linnaeus, 1758) A
Cydia succedana (Denis & Schiffermueller, 1776) A BC
Grapholita molesta (Busck *in* Quaintance & Wood, 1916) A
Hendecasticha aethaliana Meyrick, 1881b E
Holocola charopa Meyrick, 1888d E
Holocola dolopaea Meyrick, 1905 E
Holocola emplasta Meyrick, 1901 E
Holocola parthenia Meyrick, 1888d E
Holocola n. sp. Hoare 2001 A
Holocola zopherana Meyrick, 1881b
Parienia mochlophorana (Meyrick, 1882c) E
Polychrosis meliscia Meyrick, 1910b K
Protithona fugitivana Meyrick, 1882c E
Protithona potamias (Meyrick, 1909a) E
Strepsicrates ejectana (Walker, 1863c)
Strepsicrates infensa (Meyrick, 1911d) A
Strepsicrates macropetana (Meyrick, 1881b) A
Strepsicrates melanotreta (Meyrick, 1910b) K
Strepsicrates (s.l.) *sideritis* (Meyrick, 1905) E
Zomariana doxasticana (Meyrick, 1881) A

TORTRICINAE

Acleris comariana Lienig &d Zeller, 1846 A
Apoctena clarkei (Philpott, 1930b) E
Apoctena conditana (Walker, 1863c) E
Apoctena laquaeorum Dugdale, 1971b E Sn
Apoctena orthocopa (Meyrick, 1924b) E
Apoctena persecta (Meyrick, 1914a) E
Apoctena spatiosa (Philpott, 1923) E
Apoctena syntona (Meyrick, 1909b) E
Apoctena taipana (Felder & Rogenhofer, 1875) E
Apoctena tigris (Philpott, 1914) E
Apoctena (s.l.) *fastigata* (Philpott, 1916) E
Apoctena (s.l.) *flavescens* (Butler, 1877) E
Apoctena (s.l.) *orthropis* (Meyrick, 1901) E
Apoctena (s.l.) *pictoriana* (Felder & Rogenhofer, 1875) E
Ascerodes prochlora Meyrick, 1905 E
Capua intractana (Walker, 1869) A
Capua (s.l.) *semiferana* (Walker, 1863c) E
Catamacta alopecana (Meyrick, 1885g) E
Catamacta gavisana (Walker, 1863c) E
Catamacta lotinana (Meyrick, 1882c) E
Catamacta rureana (Felder & Rogenhofer, 1875) E
Cnephasia (s.l.) *holorphna* Meyrick, 1911b E
Cnephasia (s.l.) *incessana* (Walker, 1863c) E
Cnephasia (s.l.) *jactatana* (Walker, 1863c) E
Cnephasia (s.l.) *latomana* (Meyrick, 1885b) E
Cnephasia (s.l.) *melanophaea* Meyrick, 1927a E
Cnephasia (s.l.) *microbathra* Meyrick, 1911b E
Cnephasia (s.l.) *ochnosema* Meyrick, 1936 E
Cnephasia (s.l.) *paterna* Philpott, 1926a E
Cochylis atricapitata (Stephens, 1852) A BC
Ctenopseustis filicis Dugdale, 1990 E
Ctenopseustis fraterna Philpott, 1930b E
Ctenopseustis herana (Felder & Rogenhofer, 1875) E
Ctenopseustis obliquana (Walker, 1863c) E
Ctenopseustis servana (Walker, 1863c) E
Dipterina imbriferana Meyrick, 1881a E
Ecclitica hemiclista (Meyrick, 1905) E
Ecclitica philpotti Dugdale, 1978 E
Ecclitica torogramma (Meyrick, 1897b) E
Ecclitica triorthota (Meyrick, 1927a) E
Epalxiphora axenana Meyrick, 1881b E
Epichorista aspistana (Meyrick, 1882c) E
Epichorista hemionana (Meyrick, 1882c) E
Epichorista (s.l.) *abdita* Philpott, 1924c E
Epichorista (s.l.) *allogama* (Meyrick, 1914a) E
Epichorista (s.l.) *crypsidora* (Meyrick, 1909a) E
Epichorista (s.l.) *elephantina* (Meyrick, 1885b) E
Epichorista (s.l.) *emphanes* (Meyrick, 1901) E
Epichorista (s.l.) *eribola* (Meyrick, 1889b) E
Epichorista (s.l.) *fraudulenta* (Philpott, 1928a) E
Epichorista (s.l.) *lindsayi* Philpott, 1928c E
Epichorista (s.l.) *mimica* Philpott, 1930b E
Epichorista (s.l.) *siriana* (Meyrick, 1881a) E
Epichorista (s.l.) *tenebrosa* Philpott, 1917b E
Epichorista (s.l.) *zatrophana* (Meyrick, 1882c) E
Epiphyas postvittana (Walker, 1863c) A
Ericodesma aerodana (Meyrick, 1881a) E
Ericodesma argentosa (Philpott, 1924a) E
Ericodesma cuneata (Clarke, 1926) E
Ericodesma melanosperma (Meyrick, 1916b) E
Ericodesma scruposa (Philpott, 1924a) E
Eurythecta robusta (Butler, 1877) E
Eurythecta zelaea Meyrick, 1905 E
Eurythecta (s.l.) *curva* Philpott, 1918 E
Eurythecta (s.l.) *eremana* (Meyrick, 1885b) E
Eurythecta (s.l.) *leucothrinca* (Meyrick, 1931b) E
Eurythecta (s.l.) *loxias* (Meyrick, 1888d) E
Eurythecta (s.l.) *phaeoxyla* Meyrick, 1938 E
Gelophaula aenea (Butler, 1877) E
Gelophaula aridella Clarke, 1934 E
Gelophaula lychnophanes (Meyrick, 1916b) E
Gelophaula palliata (Philpott, 1914) E
Gelophaula praecipitalis Meyrick, 1934 E
Gelophaula siraea (Meyrick, 1885b) E
Gelophaula tributaria (Philpott, 1913) E
Gelophaula trisulca (Meyrick, 1916b) E
Gelophaula vana Philpott, 1928g E
Harmologa amplexana (Zeller, 1875) E
Harmologa columella Meyrick, 1927a E
Harmologa festiva Philpott, 1915 E
Harmologa oblongana (Walker, 1863c) E
Harmologa petrias Meyrick, 1901 E
Harmologa pontifica Meyrick, 1911b E
Harmologa reticularis Philpott, 1915 E
Harmologa sanguinea Philpott, 1915 E
Harmologa scoliastis (Meyrick, 1907c) E
Harmologa sisyrana Meyrick, 1882c E
Harmologa speciosa (Philpott, 1927d) E
Harmologa (s.l.) *toroterma* Hudson, 1925 E
Isotenes miserana (Walker, 1863) A*
Leucotenes coprosmae (Dugdale, 1988) E
Maoritenes cyclobathra (Meyrick, 1907c) E
Maoritenes modesta (Philpott, 1930b) E
Merophyas divulsana (Walker, 1863c) A
Merophyas leucaniana (Walker, 1863c) E
Merophyas paraloxa (Meyrick, 1907c) E
Ochetarcha miraculosa (Meyrick, 1917a) E
Philocryptica polypodii (Watt, 1921b) E
Planotortrix avicenniae Dugdale, 1990 E

Planotortrix excessana (Walker, 1863c) E
Planotortrix flammea (Salmon, 1956) E
Planotortrix notophaea (Turner, 1926) E
Planotortrix octo Dugdale, 1990 E
Planotortrix octoides Dugdale, 1990 E
Planotortrix puffini Dugdale, 1990 E
Prothelymna antiquana (Walker, 1863c) E
Prothelymna niphostrota (Meyrick, 1907c) E
Pyrgotis arcuata (Philpott, 1915) E
Pyrgotis calligypsa (Meyrick, 1926b) E
Pyrgotis chrysomela (Meyrick, 1914a) E
Pyrgotis consentiens Philpott, 1916 E
Pyrgotis eudorana Meyrick, 1885b E
Pyrgotis humilis Philpott, 1930b E
Pyrgotis plagiatana (Walker, 1863c) E
Pyrgotis plinthoglypta Meyrick, 1892 E
Pyrgotis pyramidias Meyrick, 1901 E
Pyrgotis transfixa (Meyrick, 1924a) E
Pyrgotis zygiana Meyrick, 1882c E
Sorensenata agilitata Salmon & Bradley, 1956 E Su
Tortrix (s.l.) *antichroa* Meyrick, 1919 E
Tortrix (s.l.) *demiana* Meyrick, 1882c E
Tortrix (s.l.) *fervida* (Meyrick, 1901) E
Tortrix (s.l.) *incendiaria* (Meyrick, 1923) E
Tortrix (s.l.) *molybditis* Meyrick, 1907c E
Tortrix (s.l.) *sphenias* (Meyrick, 1909a) E
Tortrix (s.l.) *zestodes* Meyrick, 1924a E
Tortrix (s.l.) sp. Boneseed leafroller A BC
YPONOMEUTIDAE
Kessleria copidota (Meyrick, 1889b) E
Zelleria (s.l.) *maculata* Philpott, 1930b E
Zelleria (s.l.) *porphyraula* Meyrick, 1927b E
Zelleria (s.l.) *rorida* Philpott, 1918 E
Zelleria (s.l.) *sphenota* (Meyrick, 1889b) E
ZYGAENIDAE A
Artona martini Efetov, 1997 A
INCERTAE SEDIS
Cadmogenes literata Meyrick, 1923 E
Gymnobathra (s.l.) *origenes* Meyrick, 1936 E
Lysiphragma (s.l.) *argentaria* Salmon, 1948 E
Tanaoctena dubia Philpott, 1931 E
Thectophila acmotypa Meyrick, 1927a E
Titanomis sisyrota Meyrick, 1888 E

Checklist of New Zealand fossil and subfossil Hexapoda

All species listed are taken to be endemic. Stratigraphic occurrence: Tri, Triassic; LJur, Late Jurassic; LCre, Late Cretaceous; Eoc, Eocene; Ple, Pleistocene; Hol, Holocene (last 10,000 years or so). Quaternary records compiled from Kuschel & Worthy (1996), Leschen & Rhode (2002), Marra (2003), Marra & Leschen (2004), and Marra et al. (2006).

PHYLUM ARTHROPODA
SUBPHYLUM HEXAPODA
Class INSECTA
Subclass DICONDYLIA
Superorder NEOPTERA
Order ORTHOPTERA
Suborder ENSIFERA
PROPHALANGOPSIDAE
Notohagla mauii Johns *in* Grant-Mackie et al., 1996 LJur

Order HEMIPTERA
Suborder HETEROPTERA
Infraorder LEPTOPODOMORPHA
SALDIDAE
Saldula sp. Marra et al. 2006 Ple

Infraorder PENTATOMOMORPHA
LYGAEIDAE
Nysius huttoni White, 1878 Ple

Order COLEOPTERA
Suborder ADEPHAGA
CARABIDAE
Actenonyx bemibidioides White, 1846 Ple
Bembidion hokitikense Bates, 1878 Ple
Bembidion rotundicolle Bates, 1874 Hol
Bembidion sp. 1 Marra & Leschen 2004 Ple
Bembidion sp. 2 Marra & Leschen 2004 Hol
Bembidion sp. 3 Marra & Leschen 2004 Hol
Cicindela feredayi Bates, 1867 Ple
Euthenaris sp. 1 Marra & Leschen 2004 Hol
Haplanister crypticus Moore, 1996 Hol
Megadromus sp. 1 Marra & Leschen 2004 Hol
Notagonum feredayi (Bates, 1874) Ple
Notagonum lawsoni (Bates, 1874) Hol
Pelodiaetodes prominens Moore, 1980 Hol
Carabidae sp. 1 Marra 2003 Ple
Pterostichini sp. 1 Marra & Leschen 2004 Hol
DYTISCIDAE
Antiporus strigulosus Broun, 1880 Ple
Huxelhydrus syntheticus Sharp, 1882 Hol
Lancetes lanceolatus (Clark, 1863) Hol
Liodessus deflectus Ordish, 1966 Ple-Hol
Liodessus plicatus (Sharp, 1882) Ple
Liodessus sp. 2 Marra et al. 2006 Ple
Rhantus pulverosus (Stephens, 1828) Hol
Gen. et sp. indet. 1 Marra & Leschen 2004 Hol

Suborder POLYPHAGA
ANOBIIDAE
Hadrobregmus crowsoni (Español, 1976) Hol
Leanobium flavomaculatum Español, 1979 Ple
Ptinus littoralis Broun, 1893 Hol
Ptinus maorianus Brookes, 1926 Hol
Sphinditeles sp. indet. Marra 2003 Ple
Xenogonus sp. 1 Marra & Leschen 2004 Hol
Zenocera sp. 1 Marra 2003 Ple
ANTHICIDAE
Anthicus minor Broun, 1886 Ple
Anthicus sp. 1 Marra & Leschen 2004 Hol
Sapintus obscuricornis (Broun, 1880) Ple
ANTHRIBIDAE
Etnalis spinicollis Sharp, 1873 Ple
Isanthribis phormii Holloway, 1982 Hol
BELIDAE
Rhicnobelus metallicus (Pascoe, 1877) Ple
BYRRHIDAE
Curimus sp. Marra et al. 2006 Ple
Epichorius sp. indet. Marra 2003 Ple
CANTHARIDAE
Gen. et sp. indet. 1 Marra & Leschen 2004 Hol
CERAMBYCIDAE
Eburida quadriguttata (Broun, 1893) Hol
Hybolasius cristatellus Bates, 1876 Hol
Hybolasius lanipes Sharp, 1877 Hol
Psilocnaeia nana (Bates, 1874) Ple
Ptinosoma sp. 1 Marra & Leschen 2004 Hol
Somatidia antarctica (White, 1846) Ple
Lamiinae sp. Marra et al. 2006 Ple
CHAETOSOMATIDAE
Chaetosomodes halli Broun, 1921 Hol
CHRYSOMELIDAE
Adoxia dilutipes Broun, 1915 Ple
Eucolaspis hudsoni Shaw, 1957 Hol
Eucolaspis pallidipennis (White, 1846) Hol
Trachytetra rugulosa (Broun, 1880) Hol
Chrysomelidae sp. 1 Marra & Leschen 2004 Hol
Chrysomelidae sp. 2 Marra & Leschen 2004 Hol
Phyllocharitini n. gen. et sp. Marra et al. 2006 Ple
CHRYSOMELIDAE?
Gen. et sp. indet. (elytron) Craw & Watt 1987 LCre
CURCULIONIDAE
Anagotus rugosus (Broun, 1883) Hol
Anagotus stephenensis Kuschel, 1982 Hol
Baeosomus sp. 1 Marra 2003, Marra & Leschen 2004, Marra et al. 2006 Ple
Baeosomus sp. 2 Marra & Leschen 2004 Hol
Catoptes sp. 1 Marra Ple
Cecyropa modesta (Fabricius, 1781) Ple
Chaetoptelius mundulus Wood & Bright, 1992 Hol
Crisius sp. 1 Marra 2003, Marra & Leschen 2004 Ple-Hol
Ectopsis ferrugalis Broun, 1881 Hol
Erymneus probus Broun, 1893 Hol
Eucossonus setiger (Broun, 1909) Ple
Euophryum ?rufum Broun, 1880 Ple-Hol
Euthyrhinus squamiger White, 1846 Ple
Hadramphus tuberculatus (Pascoe, 1877) Hol
Hypotagea lewisi Broun, 1913 Ple
Irenimus compressus (Broun, 1880) Ple
Irenimus sp. 1 Marra & Leschen 2004 Ple
Irenimus sp. 2 Marra et al. 2006 Ple
Lyperopais sp. 1 Marra & Leschen 2004 Ple
Macroscytalus parvicornis (Sharp, 1878) Ple
Mandalotus ?irritus (Pascoe, 1877) Ple
Mandalotus sp. 1 Marra & Leschen 2004 Hol
Mandalotus sp. Marra et al. 2006 Ple
Mesoreda orthorhina Houstache, 1936 Hol
Microcryptorhynchus perpusillus (Pascoe, 1877) Ple
Nestrius sp. Marra 2003 Ple
Nicaeana cinerea Broun, 1885 Ple
Nicaeana sp. 1 Marra & Leschen 2004 Ple
Nicaeana sp. 2 Marra & Leschen 2004 Hol
Novitas nigrans Broun, 1880 Ple
Pentarthrum reductum Broun, 1881 Hol
Pentarthrum zealandicum (Wollaston, 1873) Ple
Peristoreus fusconotatus (Broun, 1880) Ple
Phloephagosoma pedatum Wollaston, 1874 Ple-Hol
Phrynixus terreus Pascoe, 1875 Hol
Platypus apicalis White, 1846 Hol
Psepholax coronatus White, 1846 Hol
Psepholax crassicornis Broun, 1910 Hol
Rhopalomerus tenuirostris Blanchard, 1851 Ple
Scelodolichus sp. indet. Marra 2003 Ple
Scolopterus penicillatus White, 1846 Hol
Steriphus ascitus (Pascoe, 1876) Hol
Tymbopiptus valeas Kuschel, 1987 E Hol
Toura longirostris (Wollaston, 1873) Hol
Zeacalles binodosus Broun, 1910 Hol
Zeacalles sp. Marra & Leschen 2004 Hol
Cryptorhynchini sp. 1 Marra & Leschen 2004 Hol
Cryptorhynchini sp. 2 Marra & Leschen 2004 Hol
Cryptorhynchini sp. 3 Marra & Leschen 2004 Hol

Cryptorhynchini sp. 4 Marra & Leschen 2004 Hol
Cryptorhynchini sp. 5 Marra & Leschen 2004 Hol
Gen. et sp. indet. 1 Marra 2003 Ple
Gen. et sp. indet. 2 Marra 2003 Ple
Gen. et sp. indet. 3 Marra 2003 Ple
Gen. et sp. indet. 4 Marra 2003 Ple
CYCLAXYRIDAE
Cyclaxyra politula (Broun, 1881) Hol
CYRTOPHAGIDAE
Paratomaria crowsoni Leschen, 1996 Hol
DERMESTIDAE
Trogoderma serrigerum Sharp, 1877 Hol
Trogoderma sp. 1 Marra & Leschen 2004 Hol
ELATERIDAE
Acritelater reversus (Sharp, 1877) Ple
ELMIDAE
Hydora sp. Marra et al. 2006 Ple
EROTYLIDAE
Loberus depressus (Sharp, 1876) Hol
EUCINETIDAE
Eucinetus stewarti (Broun, 1881) Hol
HISTERIDAE
Parepierus purus (Broun, 1880) Hol
HYDROPHILIDAE
Adolopus helmsi Sharp, 1884 Ple
Adolopus sp. 1 Marra & Leschen 2004 Hol
Limnoxenus zealandicus (Broun, 1880) Hol
Paracymus pygmaeus (Macleay, 1871) Ple-Hol
Rygmodus sp. 1 Marra & Leschen 2004 Hol
Gen. et sp. indet. 1 Marra & Leschen 2004 Hol
LAEMOPHLOEIDAE
Microbrontes lineatus (Broun, 1893) Ple
LATRIDIIDAE
Bicava sp. 1 Marra 2003 Ple
cf. *Enicmus foveatus* Belon, 1884 Ple
Melanophthalma sp. 1 Marra 2003 Ple
Melanophthalma sp. 2 Marra 2003 Ple
Melanophthalma sp. 3 Marra 2003 Ple
Rethusus pustulosus (Belon, 1884) Ple
LEIODIDAE
Inocatops sp. 1 Marra & Leschen 2004 Hol
Mesocolon sp. 1 Mara & Leschen 2004 Hol
LIMNICHIDAE
Limnichus decorus Broun, 1880 Ple
Limnichus simplex Broun, 1910 Ple
Gen. et sp. 1 Marra 2003, Mara & Leschen 2004 Ple-Hol
Gen. et sp. 2 Marra 2003 Ple
LUCANIDAE
Mitophyllus parrianus Westwood, 1863 Hol
MELYRIDAE
'Dasytes' laticeps Broun, 1880 Ple
MORDELLIDAE
cf. *Stenomordellaria neglecta* (Broun, 1880) Hol
NITIDULIDAE
Hisparonia hystrix (Sharp, 1876) Hol
Soronia sp. 1 Marra & Leschen 2004 Hol
PTILIIDAE
Notoptenidium sp. 1 Marra & Leschen 2004 Ple
SALPINGIDAE
Gen. et sp. indet. 1 Marra & Leschen 2004 Ple
SCARABIDAE
Saprosites communis Broun, 1880 Hol
Saprosites sp. 1 Marra & Leschen 2004 Hol
SCIRTIDAE
Cyphon sp. 1 Marra 2003 Ple
Cyphon sp. 2 Marra 2003 Ple
Cyphon sp. 3 Marra 2003 Ple
Cyphon sp. 4 Marra 2003 Ple
Scirtidae sp. 1 Marra & Leschen 2004 Hol
Scirtidae sp. 2 Marra & Leschen 2004 Hol
Scirtidae sp. 3 Marra & Leschen 2004 Hol
Scirtidae sp. 4 Marra & Leschen 2004 Hol
Scirtidae sp. 5 Marra & Leschen 2004 Hol
Scirtidae sp. 6 Marra & Leschen 2004 Hol
Scirtidae sp. 7 Marra & Leschen 2004 Ple
Scirtidae sp. 8 Marra & Leschen 2004 Ple
Cyprobius nitidius Sharp, 1878 Ple
SCYDMAENIDAE
Scydmaenidae sp. 1 Marra 2003 Ple
Scydmaenidae sp. 2 Marra 2003 Ple
STAPHYLINIDAE
Aleochara hammondi Klimszewski *in* Klimaszewski & Crosby, 1997 Ple
Bledius sp. 1 Marra 2003, Marra & Leschen 2004 Ple
Bledius sp. 2 Marra 2003, Marra & Leschen 2004 Ple-Hol
Bledius sp. 3 Marra 2003 Ple
Carpelimus zealandicus (Sharp, 1900) Ple-Hol
Carpelimus sp. 1 Marra & Leschen 2004 Ple
Carpelimus sp. 2 Marra & Leschen 2004 Ple
Carpelimus sp. 3 Marra & Leschen 2004 Hol
Carpelimus sp. 4 Marra & Leschen 2004 Hol
Eupines sp. Marra 2003 Ple
Euplectus sp. Marra 2003 Ple
cf. *Digrammus miracollis* Fauvel, 1900 Ple
Ischnoderus curtpennis Broun, 1915 Ple
Ischnoderus tectus (Broun, 1880) Ple
Ischnoderus sp. 1 Marra & Leschen 2004 Hol
Ischnoderus sp. 2 Marra & Leschen 2004 Hol
Metacorneolabium n. sp. Marra & Leschen 2004 Ple
Microsilpha sp. 1 Marra & Leschen 2004 Hol
Philonthus sp. Marra 2003 Ple
Pselaphus sp. Marra 2003 Ple
Oxytelus sp. 1 Marra & Leschen 2004 Ple
Quedius sp. 1 Marra & Leschen 2004 Hol
Quedius sp. 2 Marra & Leschen 2004 Hol
Sagola sp. Marra 2003 Ple
Stenomalium antipodum (Broun, 1893) Ple
Stenomalium planimarginatum (Broun, 1909) Ple
'Stenomalium' sulcithorax (Broun, 1880) Ple
Stenosagola sp. Marra 2003 Ple
Zelandius sp. 1 Marra 2003 Ple
Zelandius sp. 2 Marra 2003 Ple
Aleocharinae sp. 1 Marra 2003, Marra & Leschen 2004 Ple
Aleocharinae sp. 2 Marra 2003, Marra & Leschen 2004 Ple-Hol
Aleocharinae sp. 3 Marra 2003, Marra & Leschen 2004 Ple
Pselaphinae sp. 1 Marra & Leschen 2004 Ple
Pselaphinae sp. 2 Marra & Leschen 2004 Ple
Pselaphinae sp. 3 Marra & Leschen 2004 Ple
Pselaphinae sp. 4 Marra & Leschen 2004 Hol
Pselaphinae sp. 5 Marra & Leschen 2004 Hol
Pselaphinae sp. 6 Marra & Leschen 2004 Hol
Pselaphinae sp. 7 Marra & Leschen 2004 Hol
Pselaphinae sp. 8 Marra & Leschen 2004 Hol
Gen. et sp. indet. 1 Marra & Leschen 2004 Ple
TENEBRIONIDAE
Artystona rugiceps Bates, 1874 Hol
Lorelus crassicornis Broun, 1880 Ple
Mimopeus sp. 1 Marra & Leschen 2004 Hol
Zolodinus zelandicus Blanchard, 1853 Hol
Tenebrionidae sp. 1 Marra & Leschen 2004 Hol
TROGOSSITIDAE
Grynoma sp. 1 Marra & Leschen 2004 Hol
Lepidopteryx wakefieldi (Sharp, 1877) Hol
Promanus depressus Sharp, 1877 Hol
ULODIDAE
Waitomophylax worthyi (Leschen & Rhode, 2002) Hol
ZOPHERIDAE
Notocoxelus helmsi (Reitter, 1880) Hol
Notocoxelus similis (Sharp, 1876) Hol
Pycnomerus sp. 1 Marra & Leschen 2004 Hol
Pycnomerus sp. 2 Marra & Leschen 2004 Hol
Tarphiomimus wollastoni Sharp, 1882 Hol
INCERTAE SEDIS
Gen. et sp. indet. (?beetle elytron) Grant-Mackie 1958 Tri

Order DIPTERA
Suborder NEMATOCERA
BIBIONIDAE
Dilophus campbelli Harris, 1983 Eoc
CHIRONOMIDAE
Chironomus zealandicus Hudson, 1892 Ple

Order TRICHOPTERA
Suborder SPICIPALPIA
HYDROBIOSIDAE
Hydrobiosidae sp. Marra et al. 2006 Ple

Order LEPIDOPTERA
GEOMETRIDAE?
Helastia? sp. Harris & Raine 2002 LCre
HEPIALIDAE?
Gen. et sp. indet. (wing scale) Evans 1931 LEoc

Taxonomic and establishment status of selected species of Diptera (uncertain status or validity)

1. *Ctenosciara hyalipennis* Meigen, 1804 (Sciaridae) was recorded from New Zealand by Evenhuis (1989) as a senior synonym of *Sciara annulata* sensu Tonnoir & Edwards, 1927, but the latter is a species of *Bradysia*, probably *B. brunnipes* (Mohrig & Jaschof, 1999).
2. *Hemipyrellia ligurriens* Wiedemann, 1830 (Calliphoridae) may be intercepted occasionally in New Zealand, but there is no certain record of its establishment (Dear 1985).
3. *Macrorchis meditata* Rondani, 1877 (Muscidae) was mentioned in Smith (1989) as occurring in New Zealand, but has not been listed in the catalogues of New Zealand species (Miller 1950; Evenhuis 1989).
4. *Fannia scalaris* Fabricius, 1794 (Fanniidae) reportedly present in New Zealand (Miller 1984), but on limited New Zealand material from a minor collection Pont (1977) stated it was absent.
5. *Pollenia cuprea* Malloch, 1930 and *P. atrifemur* Malloch, 1930 (Calliphoridae) have types that are badly damaged and only the latter species has been tentatively retained as valid (Dear 1985).
6. *Pseudoleucopis benefica* Malloch, 1930 (Chamaemyiidae) establishment is doubtful (Cameron et al. (1989).
7. *Senostoma rubricarinatum* Malloch, 1930 (Tachinidae) is a parasite of earwigs. With no identified specimens from the field, it is assumed that no establishment has been achieved. However, insufficient investigations have been made to be sure.

Taxonomic notes (synonyms, nomia dubia, new combinations, misidentifications)

1. *Atrichopogon 'vestitipennis'* Kieffer, 1917 (Ceratopogonidae) is a probable misidentification. Macfie (1932) described New Zealand material as this species, but it probably requires a new name. This name is preoccupied by a species from Papua New Guinea (Evenhuis 1989).
2. *'Cecidomyia'* (Cecidomyiidae) – Marshall (1896) described nine species, but their generic affinity within Cecidomyiinae cannot be determined from his descriptions and the types have been lost (Miller 1950). Many of these species are probably among the undescribed species known from galls of certain plants in Canterbury (Lintott 1974; Johns 1977; Hunt 1992; Cone 1995) and considerably more from a wide range of plants (Martin unpubl.). These descriptions are *nomia dubia* especially because they can not be linked to hosts. These names might still be used when future revisionary studies designate appropriate types.
3. *'Ceonosia'/Paralimnophora* (Muscidae) – all the described species from New Zealand except for *C. algivora* and *C. rubriceps* Hutton were investigated by R. A. Harrison (now deceased) in a partly completed review of New Zealand Muscidae, so these species names could displace current names in these genera. Three *'Ceonosia'* species have been reassigned to *Paralimnophora* and so are new combinations. *Paralimnophora antipoda* Harrison, 1955 and *Paralimnophora indistincta* Lamb, 1909 are new synonyms for *P. depressa* Lamb, 1909.
4. *Clunio* (Chironomidae) is a probable misidentification. The genus is reputed to exist at seashores in New Zealand (Tortell 1981), but may be just a misidentified *Telmatogeton*.
5. ?*Dasineura alopecuri* (Reuter, 1895) (Cecidomyiidae) – identification doubtful (see note on *Stenodiplosis geniculati*).
6. *'Dipsomyia'* (Empididae) – an undescribed endemic genus as noted by Sinclair (1995).
7. *Drosophila* n. spp. (Drosophilidae). Hodge and MAF staff have noted an undescribed species, but it is uncertain whether one or two species are involved.
8. *Exechia 'thomasi'* Miller, 1918, from Mt Taranaki – according to Tonnoir and Edwards (1927) it could not be distinguished from *Exechia hiemalis* on the basis of the incomplete type specimen. Eventually, an appropriate revisionary study may apply a new adequate type.
9. *Hilarempis subdita* Collin, 1928 (Empididae) appears to be a possible synonym of *Hilarempis huttoni* Bezzi, 1904.
10. *Ischiodon scutellaris* (Fabricius, 1805) was collected from Hamilton, New Zealand, in 1890 (Thompson pers. comm.), but we are unaware of any more recent specimens or its presence here, so establishment seems doubtful.
11. *'Limnia' transmarina* Schiner, 1868 (Sciomyzidae) – both Miller (1950) and Harrison (1959) overlooked this species and some of Schiner's types were lost in the Second World War.
12. *Lycoriella ingenua* (Dufour, 1839) is the probable correct name for the species previously known as *Sciara agraria* Felt, 1898 (Mohrig & Jaschhof 1999).
13. ?*Neurohelea* (Ceratopogonidae) – specimens that appear to belong to this genus await verification from an overseas specialist.
14. *Ornithoica exilis* Walker, 1861 (Hippoboscidae) was used by Watt (1972) for Kermadec species but listed as *O. stipituri* in Evenhuis (1989).
15. *Ornithomya avicularia* (Linnaeus, 1758) (Hippoboscidae) as used by Bishop and Heath (1988) is invalid (Evenhuis 1989. Maa (1986) considered *O. nigricornis* Erichson as valid for Australasian temperate regions rather than *O. avicularia*. They are both very similar morphologically, but *O. nigricornis* does not exist in Britain (Chandler 1998) where the introduced birds came from. Also the two species have widely disjunct distributions. We accept *O. nigricornis* as the valid name for this species.
16. *Pales orasus* (Walker, 1849) (Tachinidae) is apparently the senior synonym of both *Pales clathrata* (Nowicki, 1875), which was lost in the war, and *P. usitata* (Hutton, 1901) based on non-type material at the Canterbury Museum identified by Hutton and checked by Malloch and Macfarlane.
17. ?*Rhabdotoitamus* (Asilidae) – this genus has at least eight species in Australia (e.g. *Cerdistus caliginosus, C. neoclaripes, C. rusticanus*) and one in New Caledonia, apart from New Zealand, based on wing venation, head features, thorax hairs patterns and male genitalia. Until the holotype of the type species *R. vittipes* has been checked we are not sure that the New Zealand species are congeneric or if they belong to an undescribed genus. We have not seen the holotype of the type species of *Rhabdotoitamus*.
18. *Schoenophilus campbellensis* Harrison, 1964 (Dolichopodidae) awaits correct generic alignment (Bickel 1991).
19. *Stenodiplosis geniculati dactylidis* Barnes, 1940 (Cecidomyiidae) is included under the species *S. geniculati*. It is unclear to which genus this taxon belongs, because four gall-midge species affect cocksfoot in Great Britain (Chandler 1998). In a South American review, Gagne (1994) apparently reallocated the grass-seed-infesting *Dasineura* to *Stenodiplosis* (Chandler 1998).
20. *'Sciara'* (Sciaridae) species are not *Sciara*, but their descriptions are too imprecise to place them. *'Sciara' marcilla* Hutton, 1902 (Sciaridae) is possibly a synonym of *Bradysia amoena* (Winnertz, 1867) (see Mohrig & Jaschhof 1999).
21. *Scythropochroa antarctica* (Hudson, 1892) (Sciaridae) – *nomia dubia* (Evenhuis 1989). The description relies mostly on the illustration and has insufficient detail.
22. *'Tachina' mestor* and *'T.' sosilus* Walker, 1849 (Tachinidae) – Walker's descriptions are inadequate for generic attribution. When, Malloch (1938) described *Platytachina latifrons*, he found that Hutton had misidentified it as *T. mestor* (p. 211) and similarly *T. sosilus* was actually *Trypherina grisea* (p. 219).
23. *Thinempis brouni* (Hutton, 1901) Andrew n. comb. (Empididae) – previously *Empis*.
24. *'Tonnoirina'* (Phoridae) – Schmitz used this as a subgenus of *Triphleba*, but Brown (1962) noted that the genus needs revision.
25. *Zygomyia submarginata* Zaitzev 2002 is an invalid name, because it is preoccupied by *Zygomyia submarginata* Harrison 1955, so a new name is needed.

Excluded species (including a new synonym)

1. *Aedes nocturnus* (Theobold, 1903) (Culicidae), catalogued in both Miller (1950) and Evenhuis (1989), is based on a 1939 misidentification of larvae, probably of *A. notoscriptus* (Belkin 1968). It has not been found in any of the more recent mosquito surveys (Laird 1995; Holder et al. 1999).
2. *Asphondylia ulicis* Traill, 1873 (Cecidomyiidae), listed as possibly being in New Zealand (Evenhuis 1989), is treated as a failed introduction (Macfarlane & Andrew 2001).
3. *Austromaquartia isolita* (Tachinidae) – excluded from the checklist because it is almost certainly not established in New Zealand.
4. *Calliphora nothocalliphoralis* Miller, 1939 (Calliphoridae) – a very dubious species record. We do not know this species and Dear (1985) stated he did not consider it represented any existing New Zealand calliphorid.
5. ?*Coccomyza brittini* Del Guercio, 1918 (Cecidomyiidae) – a probable invalid record for New Zealand (Miller 1950).
6. *Drosophila marmoria* Hutton, 1901 (Drosophilidae) – a probable synonym of *D. hydei*. The type specimen is in too poor a condition to describe properly and is similar in many respects to *D. hydei* (Harrison, 1959) so we have allocated it as a probable synonym.
7. *Mallochomacquartia setiventris* (Tachinidae) has not been captured in New Zealand.
8. *Pseudoleria placata* (Hutton, 1901) (Heleomyzidae) – not known in New Zealand collections and should be excluded from a New Zealand distribution (McAlpine).
9. *Scatopse carbonaria* Hutton, 1902 Andrew n. syn. and *Scatopse nicarbonaria* Miller, 1950 Andrew n. syn. (Scatopsidae) – new junior synonyms of *Coboldia fuscipes* (Meigen, 1830).
10. *Sciomyza nigricornis* Macquart, 1851 (Lauxaniidae), listed from New Zealand and Tasmania (Evenhuis 1989), is the endemic Australian species *Tapeigaster nigricornis* (Heleomyzidae) (McAlpine & Kent 1982).
11. *Pseudonapomyza* sp. (Agromyzidae), recorded by Bowie et al. (2004) is based on a misidentification.
12. *Pollenia rudis* (Fabricius, 1794) – not shown to be established. The only record is the original interception (two specimens in NZAC). All other specimens similar to this species are *Pollenia pseudorudis*, which is well established.

TEN

Phylum KINORHYNCHA

mud dragons

BIRGER NEUHAUS, ROBERT P. HIGGINS, BRIAN L. PAAVO

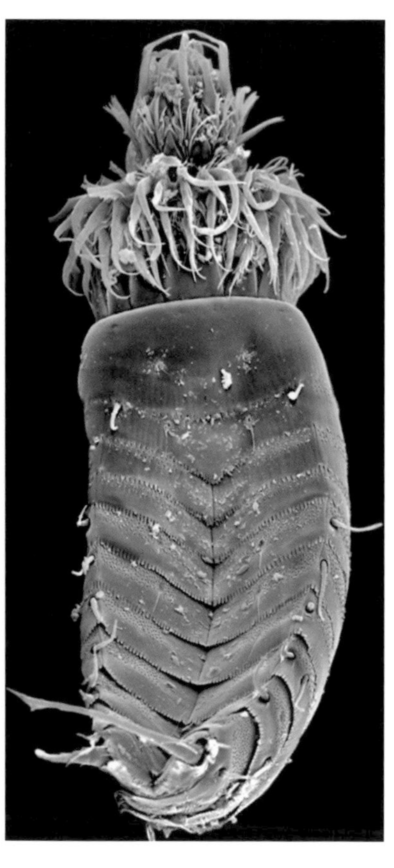

An unidentified mud dragon from 15 m depth in Blueskin Bay, Otago.

Brian Paavo

Mud dragons or kinorhynchs are minute marine creatures, mostly less than a millimetre long (the largest barely exceeds one millimetre), with a jointed body. Kinorhynch means 'moveable snout' – the head, which carries distinctive circlets of spines called scalids, can be completely retracted. Mud dragons are part of the meiofauna that inhabits the spaces between sediment grains and organic particles. They can be found in the intertidal zone, in the surface layer of mud or several centimetres deeper in sand, in shallow coastal waters, and into the deep sea as far as 5300 metres depth. They are sometimes associated with algae, sponges, or other invertebrates that are also found in sediment. They are completely devoid of cilia and are unable to swim, moving in somewhat worm-like fashion by extending the snout-like head, grasping substratum particles with the scalids, and drawing up the trunk. The head is then withdrawn before being thrust further forward. Some 141 species have been described worldwide based on adult stages, with another 38 species based on juvenile stages.

The kinorhynch body is colourless or yellowish-brown, with three regions. The near-spherical head has a mouth at the end of a protrusible cone armed with spines that project forward, whereas the much longer scalids project backwards. Behind the head is a short neck covered with large plates (placids); their arrangement differs in the two main groups of kinorhynchs. In some forms, the placids form a kind of closing apparatus when the head is retracted. In others, both the head and neck are withdrawn into the third body section, the trunk, which has 11 segments. These are not so obvious internally, which has led to some debate as to whether they are truly segments, but there is serial repetition of some muscles, the nervous system, glands, and sensory structures (Neuhaus & Higgins 2002). Externally, the trunk bears cuticular plates, setae, and spines, and structures or openings associated with reproduction.

Internal organs include a straight digestive tract from mouth to anus, with a muscular pharynx, longitudinal nerve cords (the two ventral ones with segmental ganglia), a pair of excretory structures opening on the seventh segment, and musculature (Higgins 1982). Food consists of bacteria, diatoms, and other unicells, and diets vary according to the fine anatomy of the mouth parts and the microenvironment in which particular species live. Kinorhynchs are themselves food for some organisms. Milward (1982) reported them in the diet of mudskippers (up to 20% by volume) from Queensland mudflats fronting mangroves, and one of the authors (RPH) discovered four species of kinorhynchs (*Condyloderes storchi, Kinorhynchus anomalus, Pycnophyes argentinensis,* and *P. neuhausi*) in the gut of an Argentinian shrimp (Martorelli & Higgins 2004). Several species of shrimp are likely to include kinorhynchs in their diet – as the shrimps forage, they kick up bottom sediment, eating the meiofaunal organisms that cannot rapidly swim to safety.

Summary of New Zealand kinorhynch diversity

Taxon	Described species	Known undes./undet. species	Estimated unknown species	Endemic species	Endemic genera
Kinorhyncha	6	39	30	6	0

Male and female kinorhynchs are externally distinguishable in only a few species. Some spines on the posterior trunk segments of species of *Echinoderes*, *Kinorhynchus*, and *Pycnophyes* are modified as penial spines that may assist in keeping worms together and genital apertures open during mating. This has been observed only once. Posterior ends are held together while heads face in opposite directions and a mucous mass containing sperm packages is produced (Neuhaus 1999). Very little is known about the development of the embryo. Hatching of the late embryo has been observed once (Kozloff 2007) and juveniles have only nine of the 11 trunk segments well defined. Juveniles moult via a series of six stages to the adult, but only adults of *Antygomonas oreas* and *Zelinkaderes floridensis* have been observed to moult to a second adult stage. This postembryonic developmental mode confirms inclusion of the Kinorhyncha among the moulting animals (Ecdysozoa).

Of the two main groups of Kinorhyncha, the largest is the Cyclorhagida. Species in this group tend to have a pharynx that is round in cross-section, 14–16 plates in the neck region, and a trunk that is triangular to round in cross-section. Lateral, dorsal, and tail spines are acicular or cuspidate. In the Homalorhagida, which includes the largest species, there are only 6–8 neck plates, the trunk and pharynx are triangular in cross-section, and jointed trunk spines are few or absent. Higgins (1982) and Sørensen and Pardos (2008) gave a pictorial guide to most of the kinorhynch genera. Adrianov and Malakhov (1999) provided descriptions and illustrations of almost all species as well as a guide to identification.

Although tiny, kinorhynchs are so distinctive it may be asked what their closest relatives are. Candidates include penis worms (Priapulida) and the tiny loricate animals known as Loricifera (Neuhaus & Higgins 2002). Among the three groups, kinorhynchs uniquely possess segmentation, a mouth cone with distinctive structural anatomy, and seven dilatator muscles of the hind gut. All three groups have an eversible snout (introvert) with scalid spines and inner and outer retractor muscles, a similar excretory filter (protonephridium), and similar sense organs, so there is some justification for uniting them in a single group, the Scalidophora (Lemburg 1995). To these groups Adrianov and Malakhov (1995) added the Nematomorpha (horsehair worms), in which the larva has a spinose extrovert, in a phylum Cephalorhyncha. Here, we prefer to treat each group as a separate phylum. Scalidophora, which has priority over Cephalorhyncha, may be thought of as a superphylum.

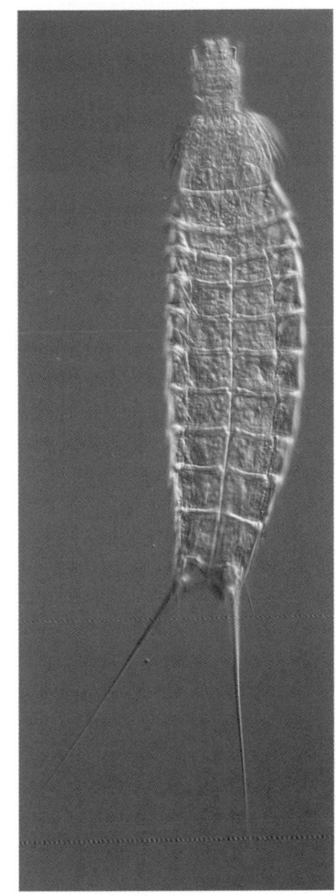

Echinoderes sp.
Birger Neuhaus

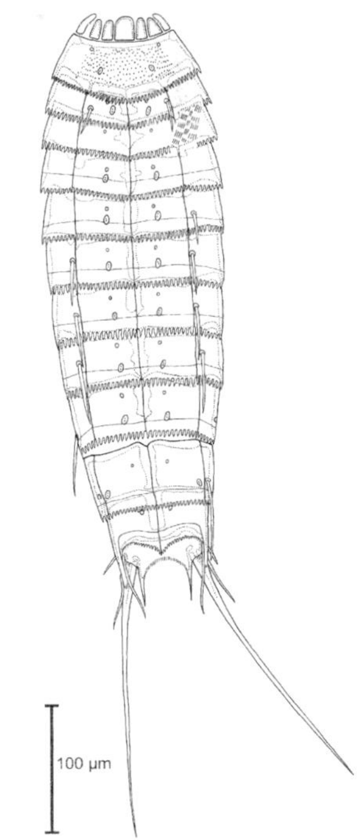

Ventral view of *Fissuroderes higginsi*.
Birger Neuhaus

The New Zealand fauna

Few New Zealanders, including professional zoologists, have ever seen a mud dragon even though they are apparently very common. Schminke and Noodt (1968) observed: 'Until now, nothing has been reported on marine interstitial animals in New Zealand. Preliminary random samples on intertidal beaches, however, have revealed a rich and varied fauna of nematodes, gastrotrichs, polychaetes, archiannelids, oligochaetes, kinorhynchs, mites, copepods, ostracods, amphipods and isopods. Thus the cosmopolitan psammon community of marine sandy beaches proves to be fully represented.' Kinorhynchs are very tiny and thus easily overlooked. The first named record from New Zealand was that of Coull and Wells (1981), who found *Echinoderes* cf. *coulli* (actually representing a new genus

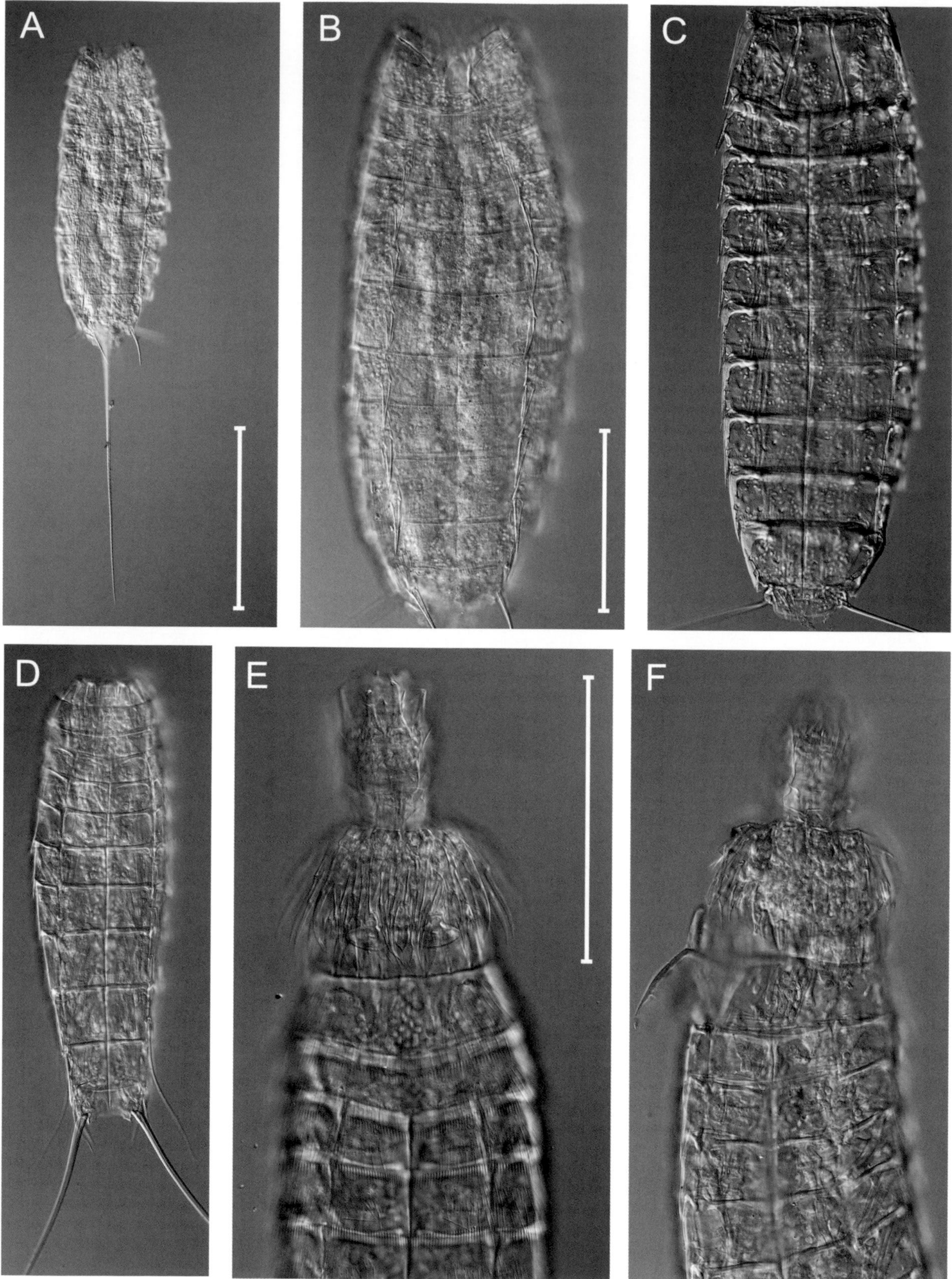

A. *Sphenoderes* sp., ventral side, with its characteristic long midterminal spine. B. Ventral view of the body of *Sphenoderes* with a deeply indented first trunk segment. C. Ventral view of the body of male *Pycnophyes* sp. with a more box-like form. The terminal segment has two pairs of penile spines. D. Ventral view of *Echinoderes* sp. with large sperm inside. E, F. Anterior end of *Echinoderes* sp. (E) and *Fissuroderes papai* (F) with the head protruded. The second trunk segment is ring-like in *Echinoderes* but split into two ventral plates in *Fissuroderes*. Scale bars: A, 200 µm (C at same scale); B, 100 µm (D at same scale); E, 100 µm (F at same scale).

Birger Neuhaus

and species) at low tide at Ration Point, Pauatahanui Inlet, Porirua. Remarkably, they found densities of this species to be as high as 80 per 10 square centimetres of surface mud, the second-highest kinorhynch abundance ever recorded.

Other kinorhynchs were discovered during a Russian expedition to study sublittoral hydrothermal vents in the western Pacific, including New Zealand. Specimens were collected from 4–12-metre-deep sediments off Whale Island in the Bay of Plenty at densities of c. 330 per 1000 square centimetres. More were found nearby at 60 metres depth (Kamenev *et al.* 1993). These species – *Echinoderes malakhovi* and *Pycnophyes newzealandiensis* – were the first to be formally described from New Zealand (Adrianov *in* Adrianov & Malakhov 1999). Other New Zealand species in overseas collections will be described in the near future. The most recent deliberate collection was made by B. Neuhaus in 2002 and 2004, from which four new species of the new genus *Fissuroderes* were described (Neuhaus & Blasche 2006). It appears that another 25 species can be added to the list of known undescribed species below (for a total of 39), and the genera *Campyloderes* and *Sphenoderes* can also be added to the fauna.

Authors

Dr Birger Neuhaus Museum für Naturkunde Berlin, Invalidenstrasse 43, D-10115 Berlin, Germany [birger.neuhaus@mfn-berlin.de]

Dr Robert P. Higgins 122 Strawbridge Court , Asheville, North Carolina 28803-1745, USA [rphigginsphd@yahoo.com]

Dr Brian L. Paavo Benthic Science Ltd, 1 Porterfield Street, Macandrew Bay, Dunedin 9014, New Zealand [paavo@benthicscience.com]

References

ADRIANOV, A.V.; MALAKHOV, V.V. 1995: The phylogeny and classification of the phylum Cephalorhyncha. *Zoosystematica Rossica 3*: 181–201.

ADRIANOV, A.V.; MALAKHOV, V.V. 1999: *Cephaloryhncha of the World Ocean*. KMK Scientific Press Ltd, Moscow. 328 p. [In Russian with English summary and diagnoses.]

COULL, B. C.; WELLS, J. B. J. 1981: Density of mud-dwelling meiobenthos from three sites in the Wellington region. *New Zealand Journal of Marine and Freshwater Research 15*: 411–415.

HIGGINS, R. P. 1982: Kinorhyncha. Pp. 873–877 *in*: Parker, S.P. (ed.), *Synopsis and Classification of Living Organisms*. 2 vols. McGraw-Hill. 1232 p.

KAMENEV, G. M.; FADEEV, V. I.; SELIN, N. I.; TARASOV, V. G.; MALAKHOV, V.V. 1993: Composition and distribution of macro- and meiobenthos around sublittoral hydrothermal vents in the Bay of Plenty, New Zealand. *New Zealand Journal of Marine and Freshwater Research 27*: 407–418.

KOZLOFF, E. N. 2007: Stages of development, from first cleavage to hatching, of an *Echinoderes* (Phylum Kinorhyncha: Class Cyclorhagida). *Cahiers de Biologie Marine 48*: 199–206.

LEMBURG, C. 1995: Ultrastructure of sense organs and receptor cells of the neck and lorica of the *Halicryptus spinulosus* larva (Priapulida). *Microfauna Marina 10*: 7–30.

MARTORELLI, S.; HIGGINS, R. P. 2004: Kinorhyncha from the stomach of the shrimp *Pleoticus muelleri* (Bate, 1888) from Comodo Rivadavia, Argentina. *Zoologischer Anzeiger 243*: 85–98.

MILWARD, N. E. 1982: Mangrove-dependent biota. Pp. 121-139 *in*: Clough, B.F. (ed.), *Mangrove Ecosystems in Australia. Structure, Function and Management*. Australian Institute of Marine Science, Townsville. xvi + 302 p.

NEUHAUS, B. 1999: Kinorhyncha. Pp. 933–937 *in*: Knobil, E.; Neill, J.D. (eds), *Encyclopedia of Reproduction*. Vol. 2. Academic Press, New York.

NEUHAUS, B.; HIGGINS, R. P. 2002: Ultrastructure, biology, and phylogenetic relationships of Kinorhyncha. *Integrative and Comparative Biology 42*: 619–632.

NEUHAUS, B.; BLASCHE, T. 2006: *Fissuroderes*, a new genus of Kinorhyncha (Cyclorhagida) from the deep sea and continental shelf of New Zealand and from the continental shelf of Costa Rica. *Zoologischer Anzeiger 245*: 19–52.

SØRENSEN, M.V; PARDOS, F. 2008: Kinorhynch systematics and biology – an introduction to the study of kinorhynchs, inclusive identification keys to the genera. *Meiofauna Marina 16*: 21–73.

SCHMINKE, H. K.; NOODT, W. 1968: Discovery of Bathynellacea, Stygocaridacea and other interstitial Crustacea in New Zealand. *Zeitschrift die Naturwissenschaften 54*: 184–185.

Checklist of New Zealand Kinorhyncha

PHYLUM KINORHYNCHA
Order CYCLORHAGIDA
ANTYGOMONIDAE
Antygomonas sp. indet. B. Paavo unpubl.
CENTRODERIDAE
Centroderes sp. indet. R.P. Higgins unpubl.
Condyloderes? sp. indet. R.P. Higgins unpubl.
Condyloderes sp. indet. B. Neuhaus unpubl.
Campyloderes sp. indet. B. Neuhaus unpubl.
ECHINODERIDAE
Echinoderes malakhovi Adrianov *in* Adrianov & Malakhov, 1999 **E**
Echinoderes [?] cf. *E. coulii* Higgins, 1977 Coull & Wells 1981
Echinoderes n. sp. R.P. Higgins unpubl.
Echinoderes spp. indet. (6) R.P. Higgins unpubl.
Echinoderes spp. indet. (14) B. Neuhaus unpubl.
Fissuroderes higginsi Neuhaus *in* Neuhaus & Blasche, 2006 E
Fissuroderes novaezealandia Neuhaus *in* Neuhaus & Blasche, 2006 E
Fissuroderes papai Neuhaus *in* Neuhaus & Blasche, 2006 E
Fissuroderes rangi Neuhaus *in* Neuhaus & Blasche, 2006 E
SEMNODERIDAE
Sphenoderes spp. indet. (7) B. Neuhaus unpubl.
Order HOMALORHAGIDA
PYCNOPHYIDAE
Kinorhynchus sp. indet. R.P. Higgins unpubl.
Pycnophyes newzealandiensis Adrianov *in* Adrianov & Malakhov, 1999 **E**
Pycnophyes spp. indet. (3) R.P. Higgins unpubl.
Pycnophyes spp. indet. (2) B. Neuhaus unpubl.

ELEVEN

Phylum LORICIFERA

loriciferans, corset animals

IBEN HEINER, BIRGER NEUHAUS

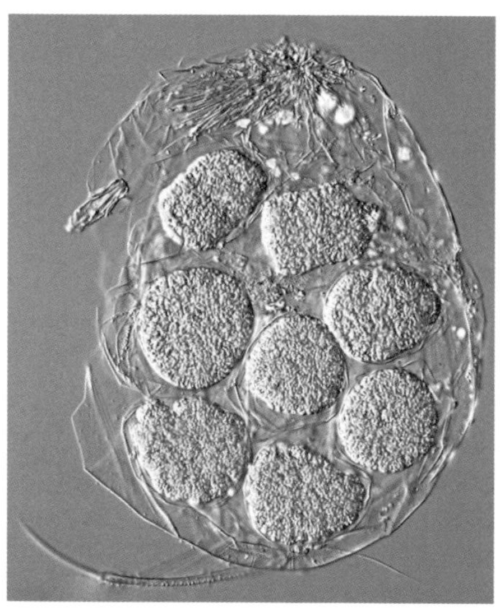

Higgins larva of *Pliciloricus* n. sp. This asexual form has several daughter larvae inside and has ruptured on the left-hand side. The structure at upper left is the displaced internal armature.

Iben Heimer

Loriciferans were first observed in the 1970s but it was not until Danish scientist Reinhardt M. Kristensen discovered all the different life stages from the coast of France that the phylum was described (Kristensen 1983). Individuals are tiny, with adults and larvae ranging from a tenth to less than half a millimetre in length. Adults are somewhat pear-shaped, with a swollen abdomen covered in a plated or sculptured cuticular exoskeleton called a lorica (Latin, *loricus*, corset, girdle). There is a concertina-like neck region and a spine-bearing head, also called an introvert, with a tubular mouth cone. Loriciferans can retract their introvert into the abdomen.

The introvert of the adults bears nine circles of articulated spine-like appendages called scalids, which consist of two or more elements. They can fold together like the ribs of an umbrella. The scalids are extraordinarily varied, depending on the species – they are flattened, or club-shaped, or like pea-pods, or spiny, hairy, and jointed. Some may be so thin that they may function like long cilia, whereas in more posterior rows they are cusp-like.

Internally, the body contains a cuticularized and eversable buccal tube, a pharynx, and a digestive system. The muscular sucking pharynx is three-sided in cross section. The food of loriciferans is still unknown but they are presumed to feed on bacteria and algae, in the latter case by piercing and sucking out the contents. The excretory system is very complex and varies greatly between species. In general, it consists of a pair of protonephridia, with a compound filtration area constructed by at least two terminal cells that open into a urogenital system (Kristensen 1991a; Neuhaus & Kristensen 2007). Fluid in the body cavity probably serves for both circulatory and respiratory functions. The brain, a collar-shaped arrangement around the gut in the anterior part of the introvert, is relatively large. Nerves run to each scalid, and these can have a sensory function.

Larval loriciferans, the so-called Higgins larvae (named for U.S. meiofaunal biologist Robert Higgins), have body regions similar to those of the adults but with a simpler mouth cone, a smaller number of shorter, more hook-like scalids, a thin lorica, and a collar between the introvert and neck, the latter being proportionately longer. The lorica is equipped on the ventral side with 2–3 pairs of appendages (setae) and, at the posterior end, paddle-like toes as well as 2–3 pairs of setae. Detailed accounts of anatomy of both the adults and the Higgins larvae were given by Kristensen (1991a,b).

Loriciferans are generally separate-sexed, but hermaphrodites have been found in two undescribed species. The differences between males and females in the family Nanaloricidae, apart from their gonads, is in the most anterior row of scalids, which are branched in males. Males also have modified appendages that are presumed to be used during copulation. Ovaries are paired, but only a single large egg develops at a time. Loriciferans have a very complex life history. In the

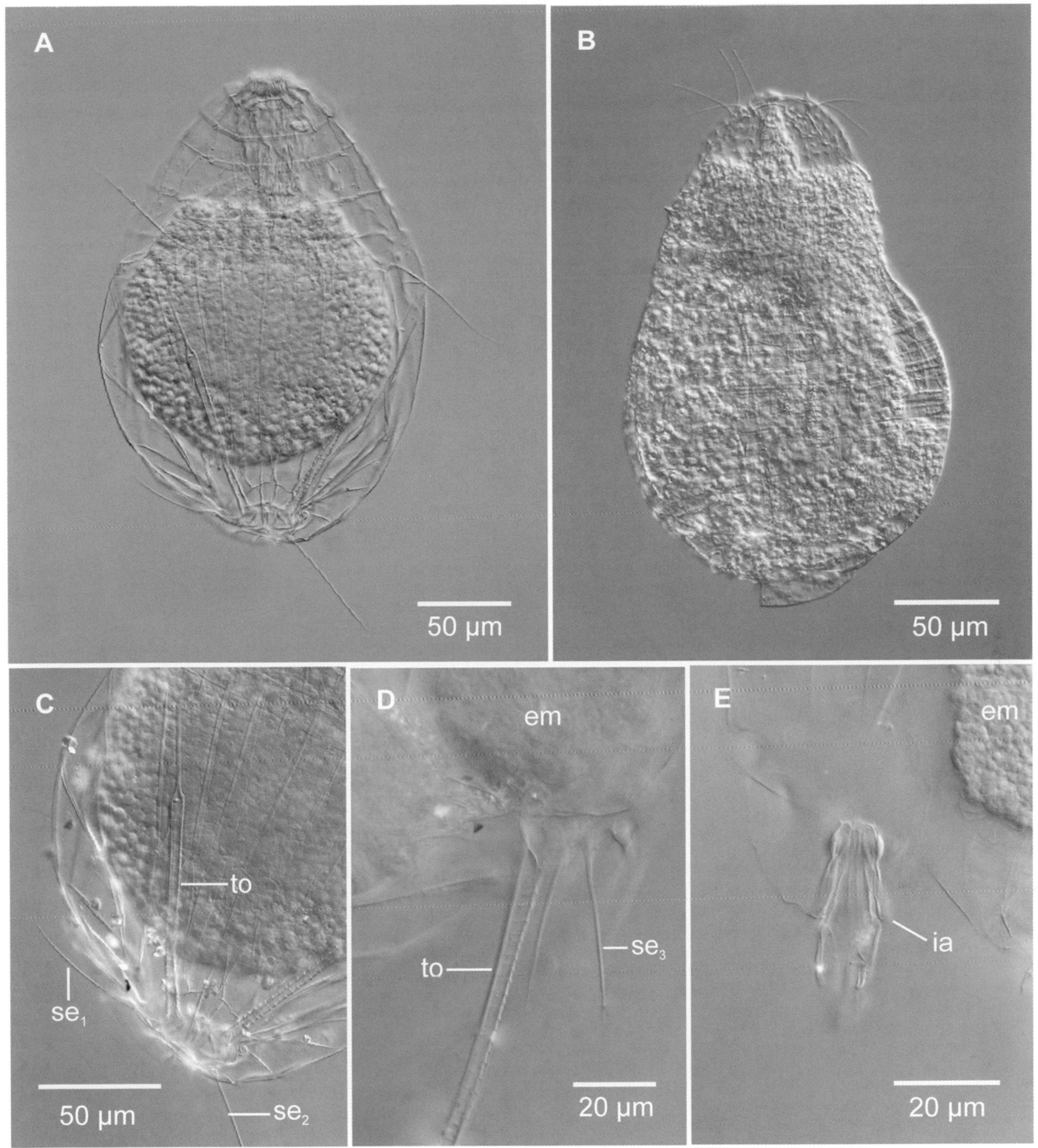

Loriciferans from New Zealand. A, C-E, *Pliciloricus* n. sp.; B, *Rugiloricus* n. sp. A, Higgins larva with adult inside; B, postlarva; C, ventral view of toes; D, lateral view of toes; E, internal armature. Abbreviations: em = embryo, ia = internal armature, se1–3 = posterior setae 1–3, to = toes.

family Nanaloricidae, only the sexual life-cycle occurs and here there are several larval stages, punctuated by moults. Afterwards, the larva may moult directly to the adult stage (*Rugiloricus carolinensis* and *Pliciloricus* species) or via a postlarval stage (*R. cauliculus* and Nanaloricidae species) (Kristensen 2002; Kristensen & Brooke 2002). Some species, especially in the family Pliciloricidae, have a complicated life history with both sexual and asexual reproduction. In *Rugiloricus bacatus* (see Heiner 2008) and *Titaniloricus inexpectatovus* (see Gad 2005), reduced larvae, called ghost larvae, reproduce paedogenetically, hence they develop eggs that mature inside either the ghost larva itself or in the penultimate Higgins

Summary of New Zealand loriciferan diversity

Taxon	Described species	Known undes./undet. species	Estimated unknown species	Endemic species	Endemic genera
Loricifera	0	4	10	2	0

larva. The eggs hatch as typical Higgins larvae (for further information concerning life cycles see Heiner 2008).

Loriciferans are closely related to mud dragons (Kinorhyncha), penis worms (Priapulida), nematodes (Nematoda), and horsehair worms (Nematomorpha). Members of these five phyla moult their cuticles and are regarded as part of the same evolutionary branch (clade). The loriciferans, kinorhynchs, and priapulids are particularly closely related and are collectively referred to as Scalidophora (Lemburg 1995). Cephalorhyncha is sometimes also used for this grouping plus Nematomorpha (Adrianov & Malakhof 1995). Recent findings of chitin in the body cuticle of Loricifera (Neuhaus et al. 1997a), Kinorhyncha (Neuhaus & Higgins 2002), larval Nematomorpha (Neuhaus et al. 1996), and pharyngeal cuticle of a nematode (Neuhaus et al. 1997b) may ally these taxa closer to the other Ecdysozoa.

Although an estimated 200 or so species are known from around the world, only 28 species destributed in nine genera and two families have been described since 1983 (see Heiner et al. 2009). They are thus a little-known meiofaunal group. Nevertheless, they may turn out to be a significant cosmopolitan component of the interstitial milieu. In general, they inhabit the supposedly well-oxygenated surface layers of the seafloor and are very widespread in the ocean, ranging from 7 metres depth in coarse shelly-sand sediments to 8300 metres in fine mud (Kristensen & Shirayama 1998).

Only a few species have been described from the South Pacific – *Phoeniciloricus simplidigitatus* from volcanic sediments at 1813 metres in a deep-sea basin north of Papua New Guinea (Gad 2004), plus a newly described species of *Pliciloricus, P. cavernicola,* and a new genus of Nanaloricidae, *Nanaloricus oculatus,* from marine caves in New South Wales, Australia (Heiner et al. 2009).

No loriciferan has yet been formally described from the New Zealand region but the 'Zealandia' research cruise SO 168 of the German vessel *Sonne,* conducted in 2002–03, captured specimens from 526–5918 metres deep east of Cook Strait in sediments of the Hikurangi Plateau, the north slope of Chatham Rise, and the abyssal plain east of the EEZ towards the Louisville Ridge seamount chain. Inasmuch as this chapter constitutes the first formal report of the phylum from New Zealand waters, station data are included here. Collection devices included a geological dredge (gDR) and a grab with a video TV camera (TVG):

SO 168 gDR 05: 42°42.687′S, 179°57.049′W – 42°42.830′S, 179°57.182′W, 1025–1185 m
SO 168 gDR 35: 36°02.05′S, 178°16.55′W – 36°01.45′S, 178°17.22′W, 5167–5918 m
SO 168 gDR 54: 40°38.53′S, 169°44.34′W – 40°38.05′S, 169°44.96′W, 3030–3750 m
SO 168 gDR 56: 40°45.89′S, 169°50.57′W – 40°45.39′S, 169°50.70′W, 2752–3548 m
SO 168 gDR 64: 43°40.64′S, 168°12.79′W – 43°41.15′S, 168°13.95′W, 3411–3778 m
SO 168 TVG 16: 40°20.713′S, 179°23.617′W, 3014 m
SO 168 TVG 92: 43°3.659′S, 178°39.058′W, 526 m

Pliciloricus n. sp. was collected from stations 54 and 56, *Rugiloricus* n. sp. 1 from station 5, and R. n. sp. 2 from station 64. Specimens were observed in sediments collected from the other stations but were not later recoverable from the sediment tubes back in the laboratory. Strictly, only *Rugiloricus* sp. 1 and the unrecoverable, unidentified loriciferans pertain to the EEZ. *Pliciloricus* n. sp. is included here, however, as it is likely to be found within the EEZ boundary and it certainly belongs to the broader New Zealand region.

Despite the paucity of material (two specimens only), the new New Zealand species of *Pliciloricus* has some interesting attributes; one specimen is a Higgins larva that contains an adult and the other is a parthenogenetic larva with several embryos within. No ghost larva can be observed inside the parthenogenetic larva, hence this is the reproductive last-stage Higgins larva. The new species resembles *Pliciloricus gracilis* from the U.S., especially in the sculpturing of the lorica, the serrated margins of the toes, and the ventral setae with several hairs (Higgins & Kristensen 1986).

Authors

Dr Iben Heiner* Zoological Museum, Natural History Museum of Denmark, Universitetsparken 15, DK-2100 Copenhagen Ø, Denmark [iheiner@snm.ku.dk]
*now **Dr Iben Heiner Bang-Berthelsen**, DTU, National Food Institute, Mørkhøj Bygade 19, Bldg. H, DK-2860 Søborg, Denmark [iber@food.dtu.dk]

Dr Birger Neuhaus Museum für Naturkunde, Invalidenstrasse 43, D-10115 Berlin, Germany [birger.neuhaus@mfn-berlin.de]

References

ADRIANOV, A.V.; MALAKHOV, V.V. 1995: The phylogeny and classification of the phylum Cephalorhyncha. *Zoosystematica Rossica 3*: 181–201.

GAD, G. 2004: A new genus of Nanaloricidae (Loricifera) from deep-sea sediments of volcanic origin in the Kininailau Trench north of Papua New Guinea. *Helgoland Marine Research 58*: 40–53.

GAD, G. 2005: Giant Higgins-larva with paedogenetic reproduction from the deep sea of the Angola Basin – evidence for a new life cycle and for abyssal gigantism in Loricifera? *Organisms Diversity & Evolution* 5: 59–75.

HEINER, I. 2008: *Rugiloricus bacatus* nov. sp. (Loricifera – Pliciloricidae) and a ghostlarva with paedogenetic reproduction. *Systematics & Biodiversity 6*: 225–247.

HEINER, I; BOESGARD, T.M.; KRISTENSEN, R.M. 2009: First time discovery of Loricifera from Australian waters and marine caves. *Marine Biology Research 5*: 529–546.

HIGGINS, R. P.; KRISTENSEN, R. M. 1986: New Loricifera from Southeastern United States Coastal Waters. *Smithsonian Contributions to Zoology* 438: 1–70.

KRISTENSEN, R.M. 1983: Loricifera, a new phylum with Aschelminthes characters from the meiobenthos. *Zeitschrift für zoologische Systematik und Evolutionsforschung 21*: 163–180.

KRISTENSEN, R. M. 1991a: Loricifera. Pp. 351–375 *in*: Harrison, F. W. (ed.), *Micrscopic Anatomy of Invertebrates. Volume 4. Aschelminthes*. Wiley-Liss, New York.

KRISTENSEN, R. M. 1991b: Loricifera – a general biological and phylogenetic overview. *Verhandlungen der Deutschen zoologischen Gesellschaft 84*: 231–246.

KRISTENSEN, R. M. 2002: An introduction to Loricifera, Cycliophora, and Micrognathozoa. *Integrative and Comparative Biology 42*: 641–651.

KRISTENSEN, R. M.; BROOKE, S. 2002: Phylum Loricifera. Pp. 179–187 *in*: Young, C. M.; Sewell, M. A.; Rice, M. E. (eds), *Atlas of Marine Invertebrate Larvae*. Academic Press, London.

KRISTENSEN, R.M.; SHIRAYAMA, Y. 1988: *Pliciloricus hadalis* (Pliciloricidae), a new loriciferan species collected from the Izu-Ogasawara Trench. *Zoological Science 5*: 875–881.

LEMBURG, C. 1995: Ultrastructure of sense organs and receptor cells of the neck and lorica of the *Halicryptus spinulosus* larva (Priapulida). *Microfauna Marina 10*: 7–30.

NEUHAUS, B.; BRESCIANI, J.; PETERS, W. 1997b: Ultrastructure of the pharyngeal cuticle and lectin labelling with wheat germ agglutinin-gold conjugate indicating chitin in the pharyngeal cuticle of *Oesophagostomum dentatum* (Strongylida, Nematoda). *Acta Zoologica* 78: 205–213.

NEUHAUS, B.; HIGGINS, R. P. 2002: Ultrastructure, biology, and phylogenetic relationships of Kinorhyncha. *Integrative and Comparative Biology 42*: 619–632.

NEUHAUS, B.; KRISTENSEN, R. M. 2007: Ultrastructure of the protonephridia of larval *Rugiloricus* cf. *cauliculus*, male *Armorloricus elegans*, and female *Nanaloricus mysticus* (Loricifera). *Journal of Morphology* 268: 357–370.

NEUHAUS, B.; KRISTENSEN, R. M.; LEMBURG, C. 1996. Ultrastructure of the cuticle of the Nemathelminthes and electron microscopical localization of chitin. *Verhandlungen der Deutschen Zoologischen Gesellschaft* 89: 221.

NEUHAUS, B; KRISTENSEN, R. M.; PETERS, W. 1997a: Ultrastructure of the cuticle of Loricifera and demonstration of chitin using wheat germ agglutinin. *Acta Zoologica* 78: 215–225.

Checklist of New Zealand Loricifera

PHYLUM LORICIFERA
PLICILORICIDAE
Rugiloricus nov. sp. 1
Rugiloricus nov. sp. 2
Pliciloricus nov. sp.
INCERTAE SEDIS
Gen. et sp. indet. Heiner & Neuhaus

TWELVE

Phylum

PRIAPULIDA

penis worms

JACOB VAN DER LAND

Priapulopsis australis from 41 m depth at Omaha Bay, Northland

Richard Taylor

The Priapulida (from *priapulus*, little penis) is a very small phylum of marine worms. There are only about 18 living species known, but the group is very ancient, and a variety of species have been found in fossiliferous shales dating from the Middle Cambrian about 530 million years ago. These dwelt in muddy seafloor sediments, much like their living counterparts, the largest of which range from a few centimetres in length to almost 40 centimetres in the case of an Alaskan species of *Halicryptus*. About half the living species are tiny meiobenthic species (less than half a millimetre in length) that live between sand grains feeding on bacteria.

Most of the larger species are thought to lie with their mouths at the seafloor with their sausage-like bodies buried beneath the sediment in which they burrow. They are carnivores, feeding on polychaetes and other worms, which they capture with the strong curved teeth on the eversible proboscis (or introvert). During its retraction, the teeth point inwards, securing the captured prey. The priapulid body is relatively simple. The muscular body wall encloses a non-segmented body cavity with blood cells that contain the oxygen-carrying pigment haemerythrin. Macrobenthic priapulids inhabit oxygen-poor and anoxic sediments and their physiology has been studied to understand the conditions in which Cambrian forms may have lived. There is a straight gut terminating in an anus. Species of *Priapulus*, *Priapulopsis*, and *Acanthopriapulus* have one or two tail-like processes of uncertain function; they may be involved in respiration. The excretory system consists of a pair of protonephridia (waste-collecting ciliated tubules) that share ducts with the gonads. Sexes are separate but males have not yet been found in the tube-dwelling species of *Maccabeus*.

Van der Land (1970) summarised all that was known about priapulid biology, taxonomy, and distribution in a comprehensive review to that time. Until 1968, only cold-water macrobenthic species were known. Since 1968, the year in which the first meiobenthic species was discovered, 11 additional species have been described. The seven living genera are distributed among macrobenthic *Acanthopriapulus*, *Halicryptus*, *Priapulopsis*, and *Priapulus*, and meiobenthic *Maccabeus*, *Meiopriapulus*, and *Tubiluchus*. All but one (*Meiopriapulus*) of these genera have a larval stage in which the trunk (neck and abdomen) is encased in armoured plates of cuticle called the lorica.

A number of zoologists have recognised that the structure and disposition of proboscis spines in priapulids is similar to that found in kinorhynchs and loriciferans and, following their suggestions, the three groups were united by Lemburg (1995) under the collective name Scalidophora. Based on the anterior spines in horsehair larvae, Adrianov and Malakhov (1995) added the Nematomorpha to this assemblage, which they called phylum Cephalorhyncha. Neuhaus and Higgins (2002) have noted that conflicting evidence exists for every one of the possible sister-group relationships among the four phyla, preferring to keep them separate. Scalidophora, which has priority, may be thought of as a superphylum, including Kinorhyncha, Priapulida, and Loricifera.

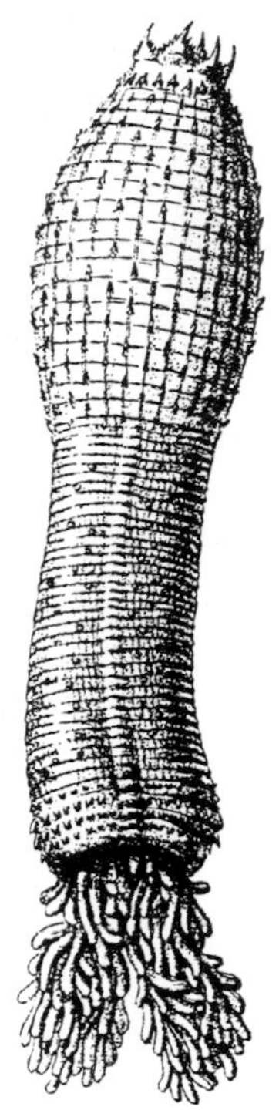

Priapulopsis australis, length about 40 mm.
After de Guerne (1888).

The New Zealand fauna

Three named genera and species of priapulids have been reported from the New Zealand region. The most comprehensive collection of New Zealand priapulids is that held at the National Institute of Water & Atmospheric Research (NIWA) in Wellington, comprising just one of the species, *Priapulopsis australis*. The other two species were reported by Russian expeditions in New Zealand waters. These records are commented on below.

Priapulopsis australis is a rather large species – contracted preserved specimens may be more than 60 millimetres long, so live specimens may be up to about 100 millimetres. Generally, the genus *Priapulopsis* is characterised by the possession of two tails. In the NIWA material, however, the paired tail is not always in evidence or may be easily overlooked. This is either because the papillae of the two tails are entangled so as to form only one mass (Hurley (1962) or one of the tails is lost early in development (Adrianov & Malakhov 1996). Consequently, the paired tail is an unreliable character to separate *Priapulopsis* from *Priapulus* and one should always look at other characters, notably the scalids, which means that a high-power dissecting microscope is needed for reliable identification. Van der Land (1970) illustrated the characteristics of the scalids in *P. australis*. Those of a series are close together and their pubescent basal parts are often fused. There are large gaps between the series of a row. Another characteristic is the near absence of spines on the tail vesicles. Mostly there is only a distal group of very small spinulae on the posteriormost vesicles. The larva of this species was described by Adrianov and Malakhov (1996), probably from New Zealand.

Priapulopsis australis has a circumantarctic distribution but it has not been collected very often. There are records from South America (Chile, Argentina) and South Africa as well as New Zealand. The NIWA material represents the only large collection in the world. The species was first recognised in New Zealand waters by Hurley (1962). It is almost certain, however, that two earlier records attributed to another species (*Priapulus tuberculatospinosus*, common in Antarctic waters) also refer to *Priapulopsis australis*, especially as no specimens of *P. tuberculatospinosus* occur in any New Zealand collections. One of the records is that of Benham (1932) from 15 metres depth at Islington Bay, Waitemata Harbour. The other is of Dell (1955) of material from 185 metres depth off Hicks Bay, East Cape, and 73 metres depth off Cape Campbell, Cook Strait. More recent records are those by Estcourt (1967) from a number of locations in the Marlborough Sounds and Storch et al. (1995) from 2–3 metres depth in Pauatahanui Inlet, Porirua. The latter authors described the introvert in some detail. Unpublished NIWA records based on about 80 samples, plotted here, indicate a depth range of 2–1190 metres for the species. One other unpublished New Zealand record is of a specimen from 1195 metres from c. 38° S, 178° W (Dr Thomas J. Trott, Woods Hole Oceanographic Institution, pers. comm.).

Summary of New Zealand priapulid diversity

Taxon	Described species	Known undes./undet. species	Estimated unknown species	Endemic species	Endemic genera
Priapulida	3	1	4	0	0

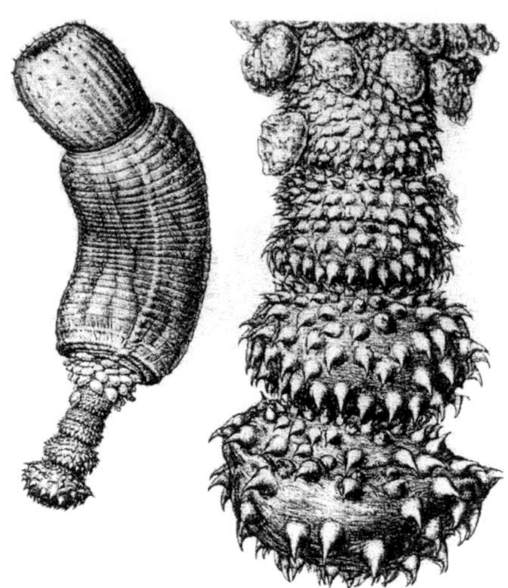

Acanthopriapulus horridus, type specimen, South Atlantic, off Uruguay (length 6 mm) and detail of tail.
After Théel (1911).

Acanthopriapulus horridus can easily be recognised by its spiny tail. A small species (c. 6 millimetres long), it has been collected only a few times. Théel (1911) described a single specimen (the type specimen) from the South Atlantic Ocean near Uruguay from a depth of 80 metres. This specimen was re-examined at the Stockholm Museum and additional observations made by van der Land (1970). Amor (1975) reported one specimen of this species in the intestine of a fish off Argentina, taken from a depth of 60 to 80 metres. Adrianov and Malakhov (1996) mentioned the occurrence of this species in New Zealand waters, but no details were given. Subsequent correspondence with Professor Malakhov ascertained that a single specimen was captured 20 April 1990 in the Bay of Plenty, halfway between Whale Island and White Island, at a depth of about 120 metres.

Priapulus abyssorum is a deep-sea species with a wide distribution in all oceans, but the number of specimens in collections is still very small (less than about 20). Most specimens are small, less than 50 millimetres long. There is one record of a single specimen from the New Zealand region. It was reported as *Priapulus tuberculatospinosus abyssorum* by Murina and Starobogatov (1961). The details given were: *Vitjaz* Stn 3838, Hikurangi Trench east of Cape Palliser, 41°29′7 S, 177°38′5 E, in grey clayey-silty mud, depth 3013 metres, taken 11 January 1958. This specimen was re-examined by van der Land (1972) during a comparative study of all known deep-sea priapulids. The specimen proved to be of taxonomic value because it was a very small postlarva showing no species-specific characters, and, at the time, it was not possible to reach a conclusion about the specific identity of deep-sea *Priapulus*. Since then, following a study of larvae, it became apparent that *P. abyssorum* is a good species (van der Land 1985). Larvae have four long hairs on their lorica instead of the usual four tubuli. At least for the time being, all deep-sea specimens may be regarded as belonging to a single species, *Priapulus abyssorum*, with a worldwide distribution at abyssal depths.

The ecology of New Zealand priapulids has not been studied, although the habitat in which *Priapulopsis australis* occurs at Pauatahanui Inlet has been well described (Healy 1980). Until very recently, no microfaunal priapulid had been discovered in New Zealand waters, but an interstitial species is now known to live in sediments on the eastern part of the Chatham Rise. Its discoverer reports that it is probably new to science and looks 'quite different from what I have seen in the literature' (B. Neuhaus pers. comm.). No doubt further sampling of offshore sedimentary environments will also yield more microfaunal species. Representatives of the genera *Maccabeus*, *Meiopriapulus*, and *Tubiluchus* are known from the South Pacific and Indian Oceans.

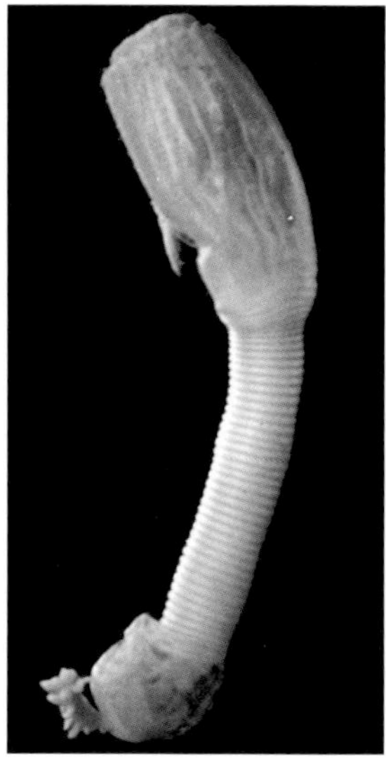

Priapulus abyssorum, Pacific Ocean, off California (length 25 mm).
After van der Land (1972).

Acknowledgements

The author wishes to thank Dr Steve O'Shea (formerly curator at NIWA, Wellington, now at Auckland University of Technology) for making available a considerable number of priapulid samples for examination. These were collected in the New Zealand region from the 1950s to the 1980s.

Author

Dr Jacob van der Land National Museum of Natural History, P.O. Box 9517, 2300 RA Leiden, The Netherlands [land@naturalis.nnm.nl]

References

ADRIANOV, A.V.; MALAKHOV, V.V. 1995: The phylogeny and classification of of the phylum Cephalorhyncha. *Zoosystematica Rossica 3*: 181–201.

ADRIANOV, A.V.; MALAKHOV, V.V. 1996: Priapulida: Structure, development, phylogeny, and classification: 1–266. KMK Scientific Press, Moscow. [In Russian with English summary]

AMOR, A. 1975: Algunas estructuras cuticulares desaconocidas de *Acanthopriapulus horridus* (Théel, 1911) (Priapulida, Priapulidae). *Physis (A) 34*: 441–444.

BENHAM, W. B. 1932: *Priapulus caudatus* in New Zealand waters. *Nature 130*: 890.

DELL, R. K. 1955: The occurrence of *Priapulus* in New Zealand waters. *Transactions of the Royal Society of New Zealand 82*: 1129–1133.

ESTCOURT, I. N. 1967: Distributions and associations of benthic invertebrates in a sheltered water soft-bottom environment (Marlborough Sounds, New Zealand). *New Zealand Journal of Marine and Freshwater Research 1*: 352–370.

GUERNE, J. de 1886: Sur les Géphyriens de la famille des Priapulides receuillis par la Mission du cap Horn. *Comptes Rendus de l'Académie des Sciences, Paris 103*: 760–762.

GUERNE, J. de 1888: Priapulides. *Mission Scientifique du Cap Horn. 1882–1883, 6*: 3–20.

HEALY, W. B. (Coord.) 1980: Pauatahanui Inlet – an environmental study. *New Zealand Department of Scientific and Industrial Research Information Series 141*: 1–198.

HURLEY, D. E. 1962: A second species of *Priapulus* from New Zealand waters. *New Zealand Journal of Science 5*: 13–16.

LAND, J. van der 1970: Systematics, geography, and ecology of the Priapulida. *Zoologische Verhandelingen, Leiden 112*: 1–118.

LAND, J. van der 1972: *Priapulus* from the deep sea. *Zoologische Mededelingen, Leiden 47*: 358–368.

LAND, J. van der 1985: Abyssal *Priapulus* (Vermes, Priapulida). Pp. 379–863 *in*: Laubier, L.; Monniot, C. (eds), *Peuplements profonds du Golfe de Gascogne*. IFREMER, Paris.

LEMBURG, C. 1995: Ultrastructure of sense organs and receptor cells of the neck and lorica of the *Halicryptus spinulosa larva Microfauna Marina 10*: 7–30.

MURINA, V.V.; STAROBOGATOV, J. I. 1961: [Classification and zoogeography of Priapuloidea]. *Trudy Instituta Okeanologii 46*: 178–200. [In Russian]

NEUHAUS, B.; HIGGINS, R. P. 2002: Ultrastructure, biology, and phylogenetic relationships of Kinorhyncha. *Integrative and Comparative Biology 42*: 619–632.

STORCH, V.; HIGGINS, R. P.; ANDERSON, P.; SVAVARSSON, J. 1995: Scanning and transmission electron microscopic analyses of the introvert of *Priapulopsis australis* and *Priapulopsis bicaudatus* (Priapulida). *Invertebrate Biology 114*: 64–72.

THÉEL, H. 1911: Priapulids and sipunculids dredged by the Swedish Antarctic Expedition 1901–1903 and the phenomenon of bipolarity. *Kungliga svenska Vetenskaps-Akademiens Handlingar 47*: 3–36.

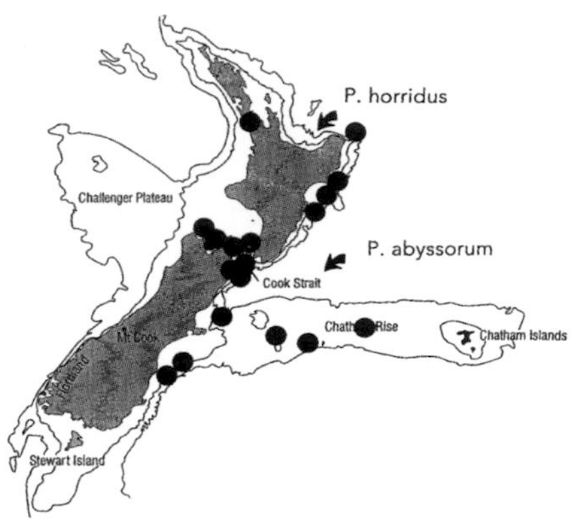

Distribution of the commonest priapulid (*Priapulopsis australis*) in New Zealand waters and indication of records of *Acanthopriapulus horridus* and *Priapulus abyssorum.*

Checklist of New Zealand Priapulida

PHYLUM PRIAPULIDA
Order PRIAPULOMORPHA
PRIAPULIDAE
Acanthopriapulus horridus (Théel, 1911)
Priapulopsis australis (de Guerne, 1886)
Priapulus abyssorum Menzies, 1959
INCERTAE SEDIS
Gen. et sp. indet.

THIRTEEN

Phylum NEMATODA

roundworms, eelworms

GREGOR W. YEATES

Pselionema sp., a free-living marine nematode from the Chatham Rise.

Daniel Leduc

Nematodes are numerous, diverse, and all-pervasive, occupying most conceivable moist habitats that can physically support them. Cobb (1915) famously wrote: '[If] all the matter in the universe except the nematodes were swept away, our world would still be dimly recognisable, and if, as disembodied spirits, we could then investigate it, we should find its mountains, hills, vales, rivers, lakes, and oceans represented by a film of nematodes. The location of towns would be decipherable, since for every massing of human beings there would be a corresponding massing of certain nematodes. Trees would stand in ghostly rows representing our streets and highways. The location of various plants and animals would still be decipherable, and, had we sufficient knowledge, in many cases even their species could be determined by an examination of their erstwhile nematode parasites.'

Nematodes – roundworms of vertebrates, eelworms of plants – are the most numerous multicellular animals on earth. On a global scale, some 25,000 nematode species have been described, and estimates of the total number of species range from a hundred thousand to a million (Heywood & Watson 1995, p. 118). Numbers for New Zealand have not previously been determined, but the preliminary checklist below gives a fauna of some 760 species.

While they are best known for their adverse effects on animal health and plant growth, variations in feeding structures and adaptation in the four juvenile stages that precede the adults enable nematodes to occupy a wide range of niches, hence they are one of the most successful groups in modern (Quaternary) environments. Although nematodes are generally unseen, their diversity and abundance contribute to so many fundamental ecological processes that, unless the nematode fauna of the region is described and understood, significant factors controlling the 'visible world', often taken to represent the 'natural environment', may be overlooked.

Despite a very simple body plan, nematodes achieve high levels of diversity. They are essentially constructed of a tube (the cuticular body wall) containing a second tube (the intestinal tract) with reproductive organs lying in the body cavity between. The head structure of nematodes is very diverse and it is largely on head anatomy (through which they obtain their food requirements) that classification of genera and species is based. Reproductive/copulatory structures may also be distinctive. The four juvenile stages present in most nematode life cycles provide opportunities for species to use a range of resources. Stages can differ in size, with thinner early stages being able to traverse narrower pore spaces; migratory and sedentary stages may have different substratum specificity and they can be dormant or resistant, free-living or parasitic, and inhabit primary or secondary hosts. Given free water and suitable temperatures (c. 0–40°C), nematodes can use available food resources; when these abilities are combined with the range of head structures and the flexibility that four juvenile stages give to life-cycle development, the potential for species diversity is enormous.

In suitable habitats nematodes are very numerous. Herbage ingested by cattle may have more than 2000 parasitic nematode juveniles per kilogram and the abomasal wall may contain 106 arrested juveniles (Bisset 1994). A parasitised wax moth larva may release 800,000 infective *Heterorhabditis* (Poinar 1983) and heavily infested ryegrass roots in the Waikato district have been reported to contain 330 root-lesion nematodes (*Pratylenchus*) per gram of root. Even in the sea, as Cobb (1914) observed, 'the seabottoms of the farthest south swarm with these little beings. Hundreds of them, male, female and young, were taken from a mere thimbleful of the dredgings.'

Scanning electron micrograph images of the head and tail regions of *Blandicephalanema serratum*, a soil nematode with a highly ornamented cuticle, found in southern beech (*Nothofagus*) forest.

Piet Look, Wageningen

Origins and relations of nematodes

The abundance of nematodes testifies to their ecological and reproductive success in modern environments. Owing to their small soft bodies, however, they have no fossil record apart from those found associated with insects, and the earliest record is from 120–135 million-year-old Cretaceous Lebanese amber (Poinar et al. 1994). Although they are probably ancient organisms, we cannot give an exposition on nematodes comparable to that of Gould (1989) on the macrofauna of the 500 million-year-old Burgess Shale. While the geographic range of nematodes may reduce the chance of extinction in mass-extinction events, the presence of traits of no apparent previous significance may equally lead to survival. In nematodes, life-history strategy and the range of food resources utilised may be key traits for success. Nematodes have survived, indeed radiated, to fill many niches, presumably through a series of cycles of diversification and decimation (Gould 1989). Using molecular techniques, evolutionary relationships among the various groups of nematodes are becoming clearer (Blaxter et al. 1998) as alternative phylogenetic scenarios (e.g. Maggenti 1983; Siddiqi 1983) are evaluated and multiple transitions between terrestrial and marine habitats become apparent (Holterman et al. 2008).

There has been speculation on how plant-parasitic nematodes and their hosts have co-evolved during about 400 million years since the Devonian (Siddiqi 1983, 2000), and the divergence of groups of nematodes parasitising vertebrates has often been linked to the divergence of their hosts. However, nematodes of the orders Rhabditida, Tylenchida, and Aphelenchida are solely terrestrial [exceptions being littoral forms like *Rhabditis marina* (Rhabditida), *Halenchus fucicola* (Tylenchida), and *Aphelenchoides gynotylurus* (Aphelenchida)] and, according to Maggenti (1981), all nematode parasitism had its developmental origin terrestrially, so the traditional listing of certain parasites matched to their hosts in an evolutionary sequence actually serves to show that host and parasite evolution need not follow.

The study of ribosomal DNA by Blaxter et al. (1998) has not only shown the Rhabditida, Chromadorida, and Enoplida to be paraphyletic but also questions the validity of the two traditional classes within the phylum, Adenophorea and Secernentea (or Aphasmidea and Phasmidea). To date, these findings have not been combined with traditional morphological information to produce a unified classification. On a more positive note, experiments in several groups have suggested that traditional morphological species are reproductively isolated (e.g. *Heterodera carotae*, *H. cruciferae*, and *H. goettingiana* among the plant-parasitic cyst nematodes (Tylenchida); *Rhabditis mariannae* and *R. synpapillata* in the free-living bacterial-feeding nematodes (Rhabditida)) (Yeates 1970; Rehfeld & Sudhaus 1985). Supplementing the 'typological species concept' with a study of gene flow between *Haemonchus contortus* (Strongylida: Trichostrongylidae), which tends to parasitise sheep, and *H. placei*, which tends to parasitise cattle, confirmed that host specificity was not absolute, and mixed infections produced hybrids (Le Jambre 1981); rapid separation between these forms was apparently occurring. A

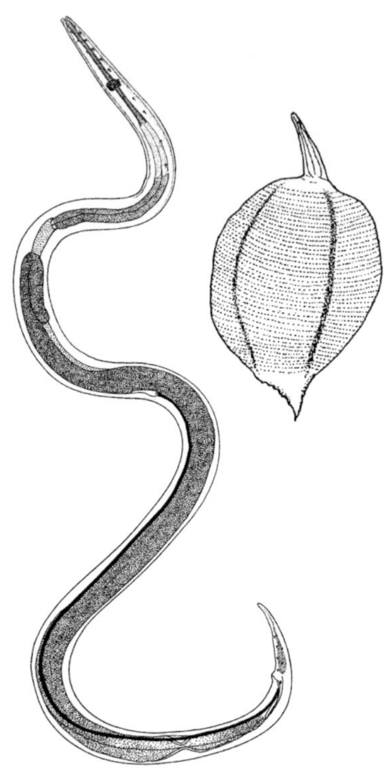

Male (left) and female of *Tetrameres tarapungae* from the proventriculus of the red-billed gull *Larus novaehollandiae*.

From Clark 1978

recent study using ribosomal DNA has indicated that three morphological forms of trichostrongylid parasites of livestock (*Teladorsagia* (= *Ostertagia*) *circumcincta*, *T. davtiana*, and *T. trifurcata*) (Strongylida: Trichostrongylidae) represent a single species (Stevenson et al. 1996). In plant-feeding taxa, similar patterns are shown by 'races' and 'pathotypes' among *Heterodera*, *Globodera* (Tylenchida: Heteroderidae), *Pratylenchus*, and *Radopholus* (Tylenchida: Pratylenchidae). Use of anthelmintic drenches to control gastrointestinal nematodes of grazing animals has selected 'strains' resistant to one or more of those drenches. There is also a reciprocal effect, with the genetics of plant and animal hosts impacting on their parasites.

Twentieth-century biology has made nematodes unique in that one of them, *Caenorhabditis elegans* (Rhabditida), was the first multicellular organism in which the entire genome was mapped (CSC 1998). Nematode species show, apart from the reproductive cells, cell constancy and highly determined cleavage of the embryo. These characters permitted experiments in which cells were ablated using laser beams to study cell lineages (Sulston et al. 1983). It now seems that the 'pseudocoelomic body cavity' is confined to Nematoda and Nematomorpha – the traditional grouping of Aschelminthes not being a monophyletic group (Nielsen 1996; de Rosa et al. 1999).

The New Zealand nematode fauna

Early natural-history expeditions recorded nematodes from New Zealand (e.g. Chatin 1884). The first records of nematodes of economic importance in New Zealand appeared late in the 19th century (Kirk 1899; Thomson 1922), freshwater species were first described in 1904 (Cobb 1904), and marine nematodes described in 1921 (Ditlevsen 1921). However, systematic research has been sporadic. Apart from the nematodes of plants and introduced mammals that have been studied for their perceived economic and public health impacts, the New Zealand nematode fauna is as poorly known as that of most southern hemisphere countries; there are tantalising links but too few data.

Given our present limited knowledge, it is not possible to discuss the

Summary of New Zealand nematode diversity

Paucity of records (from both the New Zealand region and the Southern Hemisphere) significantly limits the ability to identify taxa as endemic or alien. For those groups where an allocation is made it must be regarded are tentative.

Habitat code	Described species+ subspecies	Known undes./undet. species	Estimated undiscov. species	Alien species	Endemic species	Endemic genera
Invertebrate hosts	48	0	500	1	25	2
Invertebrate/Soil[1]	2	0	50	?	?	0
Soil (free-living)	151[2]	0	400	?	?	0
Plants (parasites)	157 + 1	0	150	18[3]	47[4]	0
Marine substrata	129[5]	0	500	?	9	0
Vertebrate hosts	221	52[6]	200	133[7]	9	2
Totals	**708 + 1**	**52**	**1800**	**152**	**90**	**4**

1. Species with both invertebrate host and soil generations.
2. Includes five species from freshwater sediments.
3. Eight species of cyst nematode, one root-knot nematode, seven criconematids, *Tylenchus* and two longidorids are placed here.
4. Wouts (2006) indicated 47 criconematids to be endemic but no other taxa included here.
5. Includes 32 taxa of uncertain status marked (?) in lists.
6. Comprises 52 undescribed fish-parasite species listed by Brunsdon (1956) (not in checklist).
7. Includes 11 species that may be able to maintain populations in the region but cannot be regarded as established (naturalised).

biogeographic relationships of the nematode fauna of the New Zealand region. Allgén (1932) listed 94 species of marine nematodes from Campbell Island. One of these has been synonymised and, according to Gerlach and Riemann (1973, 1974), at least 19 of them are of dubious status. Endemic earthworms, wetas, and similar terrestrial invertebrates may each have morphologically distinct nematode parasites (Dale 1967; Clark 1978; Yeates & Spiridonov 1996) but nothing is known of the relative divergence of nematodes and their hosts, nor of how the nematodes relate to those of other Gondwanan landmasses. Human-mediated dispersal of parasites of introduced mammals is expected, as it is for the parasites of introduced plants – e.g. *Globodera rostochiensis, Heterodera trifolii,* and *Anguina tritici* (Tylenchida). In listing helminth parasites of introduced mammals, McKenna (2009) regarded most of them as endemic; this is in the parasitological sense of 'established' (naturalised) rather than the biogeographic sense of 'native'.

The following examples illustrate the diversity of nematodes in the region and the present knowledge of species distribution and endemicity.

Parasites of kiwi

The first record of nematodes from kiwi is that of Chatin (1884). Valid endemic species are *Heterakis gracilicauda* (Ascaridida: Heterakidae) and *Cyrena* (*Cyrena*) *apterycis* (Spirurida: Spiruridae). Inglis and Harris (1990) proposed transferring *H. gracilicauda,* as *Kiwinema gracilicauda,* to a new family, Kiwinematidae, which would also contain *Hatterianema hollandei* from the tuatara. They considered that the morphology of the two species, both from indigenous New Zealand hosts, reinforces the possibility that the superfamily Heterakoidea arose in Gondwana. Kiwis have acquired infections of *Porrocaecum ensicaudatum* (Ascaridida: Anisakidae) from introduced blackbirds and thrushes and *Toxocara cati* (Ascaridida: Ascarididae) from cats (Clark 1982).

Parasites of possum

New Zealand populations of the introduced brush-tailed opossum *Trichosurus vulpecula* contain not only *Parastrongyloides trichosuri* (Rhabditida: Strongyloididae) and *Paraustrostrongylus trichosuri* (Strongylida: Trichostrongyloidea: Herpetostrongylidae) (which were first described from the possum in Australia) but also *Trichostrongylus colubriformis, T. retortaeformis,* and *T. vitrinus* (Strongylida: Trichostrongyloidea: Trichostrongylidae) (Stankiewicz et al. 1998). The three trichostrongylids are presumably derived from introduced grazing mammals and it is not known whether the populations in possums can be sustained in the absence of these eutherian mammals.

Parasites of domestic ruminants

Advertisements for drenches represent the greatest public exposure nematodes receive in New Zealand. However, such drenches lack specificity and their worldwide use in recent decades has led to selection of strains of several species of gastrointestinal nematodes that are resistant to one or more of the three main families of drenches (benzimidazoles, levamizoles, and macrocyclic lactones). McKenna (1997) listed 29 nematode parasites of sheep from New Zealand and 27 species from cattle. Such parasites occur in various parts of the animal – Vlassoff and McKenna (1994) recorded three species from the lungs, seven from the abomasum, 15 from the small intestine, and four from the large intestine of sheep. As land managers move towards integrated control of gastrointestinal nematodes, the practical ability to identify the species present will be critical in integrating rotational, biological, and chemical control tools.

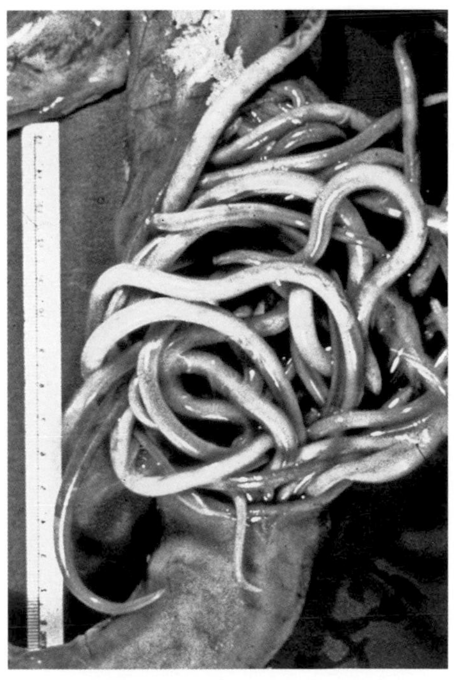

Ascaridid nematodes in the intestine of a farm animal.
Bill Pomroy

Biological control of insects

One of the success stories of biological control is that of the use of the nematode *Deladenus siricidicola* (Tylenchida) to control the woodwasp (*Sirex noctilio*)

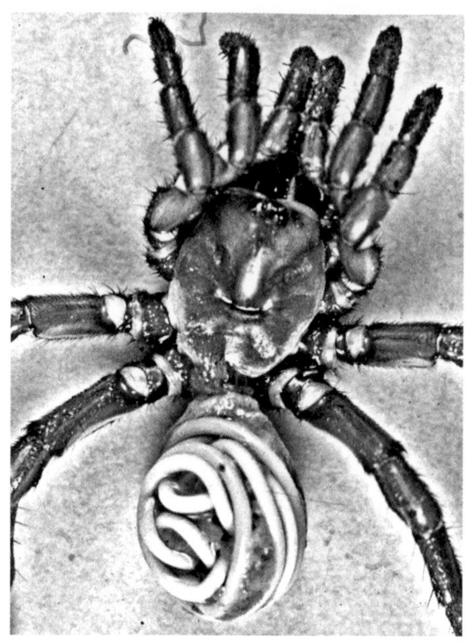

Aranimermis giganteus, the largest-known mermithid nematode, achieving 32 cm length, coiled in the opisthosoma of the native mygalomorph spider *Cantuaria borealis*, Banks Peninsula.

David Hollander

outbreaks that can cause serious damage to *Pinus radiata* in both New Zealand and Australia. This work resulted from the 1962 discovery of the nematode parasite in New Zealand (Zondag 1969). *Deladenus siricidicola* is one of a group of nematodes that has two life cycles, one free-living and fungal-feeding and the other parasitic (Bedding 1973). Presumably both the nematode and the host were introduced into New Zealand. The description of *Neoaplectana leucaniae* Hoy, 1954 (= *Steinernema feltiae*) represented a significant early New Zealand contribution to the study of entomopathogenic nematodes. Historically, the identification of *Steinernema* isolates worldwide has been uncertain, with differing schemes allocating a given isolate from or used in New Zealand to either of the two species (*S. feltiae, S. carpocapsae*) listed in the present species list.

Nematode parasites presumably exert natural control in populations of native arthropods as evidenced by the recording of mermithid parasites in a range of groups including stick insects, mygalomorph spiders (Poinar & Early 1990), diptera, grass grubs, and intertidal amphipods. Mermithids have been associated with mortaility of animals brought into laboratories and their wider population effects are worthy of investigation.

Parasites of humans

All thirteen species of nematodes that have been recorded from humans in New Zealand are, by definition, introduced. While six of them are clearly established in the human population, others are not, having been recorded only from individuals who have arrived from overseas (Spratt et al. 1999; McKenna 2009). With increasing movement of people, changing climate, and the potential establishment of intermediate hosts, it is important to maintain an adequate science base to characterise and identify such nematodes. Medical practitioners should be aware of the possibility of new and unfamiliar parasitological infections.

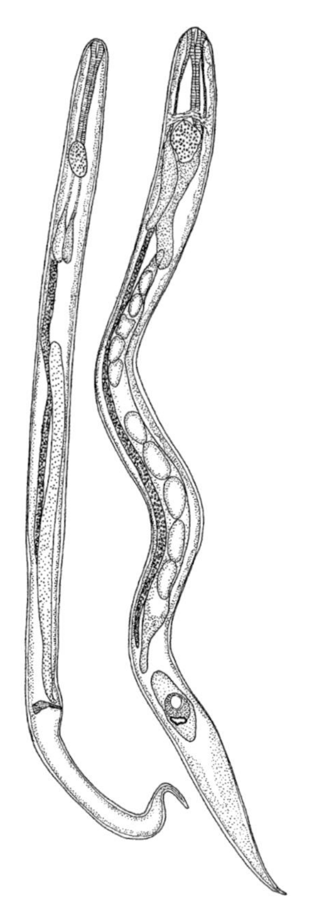

Female (left) and male of *Haycocknema perplexum*, an intramuscular parasite of humans.

From Spratt et al. 1999

Oxyuroid parasites of millipedes

The work of Bowie (1986) on the nematode parasites of New Zealand millipedes suggests that there are at least nine nematode species in the guts of five millipede species. Some estimates put the number of New Zealand millipedes at 200 – hence there are clearly many nematode taxa awaiting discovery.

Drilonematid parasites of earthworms

An initial study of this group of nematodes, which occurs in the coelomic cavity of earthworms, revealed seven species from four species of native Megascolecidae (Yeates & Spiridonov 1996). If all 178 native megascolecids are similarly endowed with such parasites, are there more than 300 species awaiting discovery in this group alone? What are their relationships to the drilonematids found in native Australian earthworms?

Cyst and root-knot nematodes of agriculture

That 'The Potato Cyst Nematode Regulations 1974' were gazetted shows the economic and quarantine significance of *Globodera rostochiensis* and *G. pallida* (Tylenchida). Work is continuing to develop resistant crop cultivars as a management tool, but the situation is complicated by the presence of 'pathotypes' of the two species; these pathotypes are presumed to reflect ancestral host variants or geographic variants in the Andes and to have reached New Zealand via Europe (e.g. Mercer 1994). Root-knot nematodes (*Meloidogyne* spp.) (Tylenchida) are widespread, and probably have a significant effect on the health of many plant species, both native and introduced. The 1997 differentiation of *M. trifoliophila*, parasitising white clover in both New Zealand and Tennessee, was critical to future plant-breeding programmes (Mercer & Watson 2007). It also raised questions about the geographic range and origin of the populations.

Summary of New Zealand Recent nematode diversity by major environment

For parasitic nematodes the environment is that of the primary host.

Taxon	Marine	Freshwater	Terrestrial
Nematoda	211*	9*	540

* includes 52 fish-parasite species listed by Brunsdon (1956) (not in checklist)

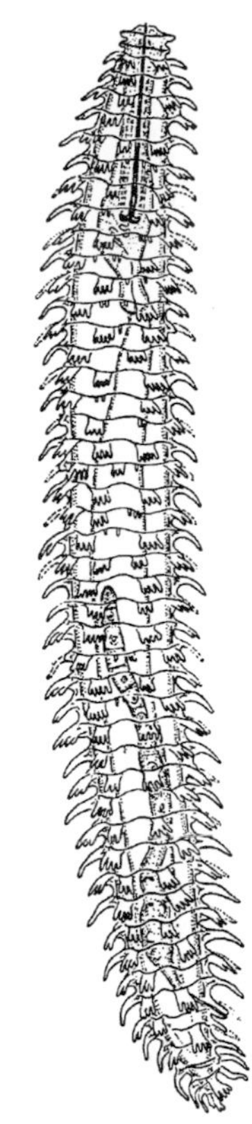

Female *Ogma palmatum*, a root-feeding nematode of soils.

From Siddiqi & Southey 1962

Criconematid nematodes of native vegetation

Many Criconematidae (Tylenchida) are small nematodes with robust stylets that feed on woody plants; some also feed on grasses. Recent publications have shown the criconematid fauna of New Zealand to include 30 species that so far appear to have local (20 species), global (7), Pacific (2), and Australian (1) distributional patterns. Those with global patterns may have spread with agricultural materials (Wouts et al. 1999). Their recorded distribution raises several questions. For example:

1. *Hemicycliophora chathami major* Yeates, 1978 (Tylenchida) was described from the Chatham Islands and the only other published record is from the Falkland Islands (Orton Williams 1986). What is its actual global distribution?
2. The four species of *Blandicephalanema* (Tylenchida: Criconematidae) known from New Zealand are all associated with *Nothofagus* forest and related *B. bossi* was described from *Nothofagus* forest in New South Wales. Other species are described from Tonga and Western Samoa.
3. In the North and South Islands, several related species of *Criconema* are found in either tussock grasslands or forests. Are these valid species or does the microclimate, soil pH, or host vegetation significantly influence their morphology?
4. A recent monographic study using light and scanning electron microscopy has recorded 68 species of Criconematina from New Zealand, 47 of which were regarded as endemic by Wouts (2006) and five as introduced.

Nematode vectors of plant viruses

The distribution in New Zealand of the potential virus vector *Longidorus elongatus* (Dorylaimida) could be simply related to climatic factors (Boag et al. 1997). Even so, it is not clear whether the species is native to New Zealand or whether it was so widely dispersed following its introduction by humans that it successfully colonised the appropriate areas of the region. *Longidorus elongatus* populations have been found to increase 3–4-fold in pastures exposed to elevated atmospheric carbon dioxide (Yeates & Newton 2009). The identity of New Zealand collections of *L. elongatus* and *Xiphinema brevicollum* have been confirmed using molecular techniques. A climate-based model suggests that the geographic distribution of *Paratrichodorus minor* (a stubby root nematode) (Triplonchida) in Australia could significantly increase in southeastern Australia following a 2C° increase in mean annual temperature. The study by Yeates et al. (1998) indicates it may also become more widespread in New Zealand; it is known only from Ruakura and Wairakei at present. The ability to identify nematodes such as *P. minor* may become critical for horticulture.

A rhizosphere nematode

The genus *Falcihasta* Clark, 1964 (Dorylaimida: Belondiridae), with a distinctive asymmetric cephalic region, was erected for *F. palustris* Clark, 1964 on the basis of material from Canterbury, the Auckland Islands, and Campbell Island. It has since been found at a range of higher-rainfall sites throughout New Zealand, and appears to be associated with the rhizosphere. By 1995 three further species had been described from New Caledonia and New Guinea (Andrássy 1995) with unpublished records from Queensland. Does this distribution indicate a genus of some antiquity, and what is its actual modern range?

Predatory nematodes

The predatory nematode *Anatonchus killicki* Clark, 1963 (Mononchida) was first described from soil around grape vines at Te Kauwhata. The next published record was from grape vines in Spain (Peña Santiago et al. 1992). *Anatonchus killicki* is the only species of the genus recorded from New Zealand but several species occur in Spain. If this predatory nematode has been dispersed by trade it makes any attempt to reconstruct the native nematode fauna of New Zealand, indeed of any region, difficult. Many other species of Mononchida are apparently endemic to New Zealand (Clark 1963; Yeates 1967).

The genus *Nygolaimus* (Dorylaimida) preys on enchytraeid earthworms. *Nygolaimus directus* Heyns, 1968 has been reported from coastal sand dunes in New Zealand (Patea, Castlecliff, Himatangi, Sumner, Taylors Mistake). It has also been found under grass and onions in South Australia, citrus in Queensland, sugarbeet in England, and in sand near a beach in Eastern Cape Province, South Africa (Heyns 1968).

Maritime environments

Rhabditis marina (Rhabditida) has been found under debris on the New Zealand seashore and from similar situations in numerous other countries (Sudhaus 1974). It is one of the few species of Rhabditida to live in maritime environments. Its wide distribution demonstrates the ability of individual nematode species to colonise suitable habitats, provided that propagules arrive and can establish. A more recent record of *R. mediterranea* has been confirmed by cross-breeding in cultures (Leduc & Gwyther 2008).

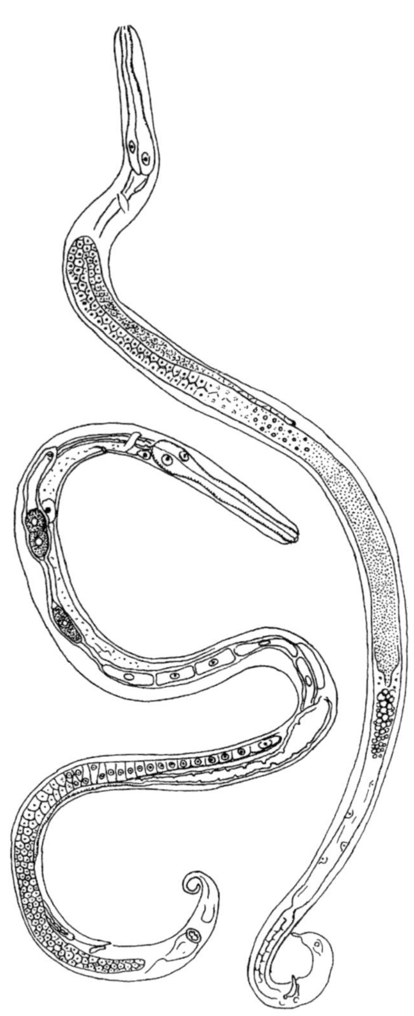

Female (left) and male of *Iponema australe*, an obligate parasite of earthworm body cavities.
From Yeates & Spiridonov 1996

Nematodes of marine fishes

There are many records of nematodes found in fish in the New Zealand region. When the fish is the final host there is often an intermediate host, while on the other hand the fish may be the intermediate host for nematodes maturing in marine mammals and birds (Anderson 2000). Hine et al. (2000) listed 77 nominal species of nematodes, of which only 25 are fully identified and included in the end-chapter checklist. Unpublished studies such as those of Brunsdon (1956) contain diagnostic information of value for future unravelling of life-cycles and the economic and medical importance of these nematodes.

The New Zealand species list

The higher classification used in the checklist of New Zealand species is based on that of Maggenti (1981), supplemented by specialised compilations such as those of the Tylenchida (Maggenti et al. 1988), Dorylaimida (Jairajpuri & Ahmad 1992), and parasites of vertebrates (Anderson 2000). In the past 30 years there has been a tendency for taxonomic inflation in the phylum, with various groups whose affinities were regarded as 'difficult' being given ordinal status, species-rich genera being split, and subfamilies being elevated. In the groupings below, the superfamilies are grouped into higher units that are somewhat arbitrary and not necessary monophyletic.

The New Zealand checklist represents the first attempt to bring together regional taxonomic information from the disciplines of plant nematology, animal helminthology, marine biology, soil zoology, and invertebrate pathology. Because of our incomplete knowledge of the nematode fauna, it is very much a preliminary, partial assessment. Parasites of marine mammals are particularly poorly represented. Key literature used includes Hewitt and Hine (1972), Bowie (1984), Glare et al. (1993), Knight et al. (1997), McKenna (1998, 2009) and Hine et al. (2000). The status of marine nematode species given by Gerlach and Rieman

(1973, 1974) and Leduc and Gwyther (2008) is generally followed, an initial search of primary literature having been supplemented by those sources. Species have been given a single habitat code based on recorded adult habitat, with no attempt to differentiate feeding habits, span life cycles, or include alternation of generations. Given the as-yet-undefined biogeographic relationships of the New Zealand nematode fauna, only species parasitic in endemic animals, or marine species not recorded elsewhere by Gerlach and Rieman (1973, 1974) and Leduc and Gwyther (2008), are regarded as truly endemic. Host-specific parasites of introduced plants and animals are regarded as alien, even if naturalised.

Gaps in knowledge and scope for future research

Native nematodes of the New Zealand region have scarcely been studied. A major coordinated effort is required to give a basic picture of the nematodes of marine sediments, plants, soils, and terrestrial invertebrates. A few coastal collections of marine nematodes have been made while the great expanse of the seabed remains unsampled. While some nematodes have been recorded from native vegetation, no integrated picture of their diversity, relationships, distribution, and possible impacts on the flora will be possible until representative samples have been collected from each environmental domain. Because of changed land use and vegetation, it is no longer possible to sample some natural combinations of soil, climate, and vegetation. Nematode parasites of native invertebrates are just as much part of our fauna as their hosts; extinction of an endemic invertebrate probably implies the extinction of an endemic nematode parasite. While nematode parasites of introduced vertebrates are well-known, more work is required to characterise undescribed species in native vertebrates and to clarify life-cycles so as better to manage those populations that impact on economic production or on the survival of native vertebrates.

In addition to the need for border protection, there is a clear need to maintain an ability to identify nematodes of direct economic impact. The interaction of nematode genetics with plant breeding, anthelminthic use, and other management techniques is a reality. The need to establish the impacts of climate change on species distribution is likely.

Acknowledgements

This review attempts to consolidate the contributions of many workers over a century. In particular, it draws on compilations initially made by Pat Dale and Phil McKenna. Phil gave advice on authorities for parasites of vertebrates. The author is also grateful for advice received from Karen Knight, Alex Vlassoff, Wim Wouts, George Poinar, and Bill Pomroy. Others have contributed in various ways.

Author

Dr Gregor W. Yeates 406 Albert Street, Palmerston North, New Zealand [gregor.yeates@gmail.com]

References

ALLGÉN, C. 1932: Weitere Beiträge zur Kenntis der mariner Nematodenfauna der Campbell-insel. *Nyt Magazin for Naturvidenskaberne 70*: 97–198.

ANDERSON, R. C. 2000: *Nematode Parasites of Vertebrates: their development and transmission.* 2nd edn. CAB International, Wallingford. 650 p.

ANDRÁSSY, I. 1995: Tropical nematodes of rare genera (Dorylaimida). *Opuscula Zoologica Budapest 27-28*: 5–24.

BEDDING, R. A. 1973: Biology of *Deladenus siricidicola* (Neotylenchidae), an entomophagous-mycetophagous nematode parasitic in siricid woodwasps. *Nematologica 18*: 482–493.

BISSET, S. A. 1994: Helminth parasites of economic importance in cattle in New Zealand. *New Zealand Journal of Zoology 21*: 9–22.

BLAXTER, M. L.; DE LEY, P.; GAREY, J. R.; LIU, L. X.; SCHELDEMAN, P.; VIERSTRAETE, A.; VANFLETEREN, J. R.; MACKEY, L.Y.; DORRIS, M.; FRISSE, L. M.; VIDA, J. T.; THOMAS, W. K. 1998: A molecular evolutionary framework for the phylum Nematoda. *Nature, London 392*: 71–75.

BOAG, B.; EVANS, K. A.; YEATES, G. W.; BROWN, D. J. F.; NEILSEN, R. 1997: Global potential distribution of European longidorid virus-vector nematodes. *Nematologica 43*: 99–106.

BOWIE, J.Y. 1984: Parasites from an Atlantic bottle-nose dolphin (*Tursiops truncatus*), and a revised checklist. *New Zealand Journal of Zoology 11*: 395–398.

BOWIE, J.Y. 1986: New species of rhigonematid and thelastomatid nematodes from indigenous New Zealand millipedes. *New Zealand Journal of Zoology 12*: 485–503.

BRUNSDON, R.V. 1956: Studies on nematode parasites of New Zealand fishes. 2 vols. Unpublished PhD thesis, Victoria University of Wellington. 356 p.

CHATIN, M. J. 1884: Parasites de l'Apterix. *Compte Rendu des Séances de la Société de Biologie, sér. 8, 1*: 770–771

CLARK, W. C. 1963: Notes on the Mononchidae (Nematoda) of the New Zealand region with descriptions of new species. *New Zealand Journal of Science 6*: 612–632.

CLARK, W. C. 1978: New species of rhigonematid and thelastomatid nematodes from the pill millipede *Procyliosoma tuberculata* (Diplopoda: Oniscomorpha). *New Zealand Journal of Zoology 5*: 1–6.

CLARK, W. C. 1982: Nematodes of kiwis. *New Zealand Journal of Zoology 10*: 129. [Abstract]

COBB, N. A. 1904: Free-living fresh-water New Zealand nematodes. *Proceedings of the Cambridge Philosophical Society 12*: 363–374.

COBB, N. A. 1914: Antarctic marine free-living nematodes of the Shackleton Expedition. *Contributions to a Science of Nematology* 1: 1–33.

COBB, N. A. 1915: Nematodes and their relationships. *Yearbook of the United States Department of Agriculture, 1914*: 457–490.

CSC [The *C. elegans* Sequencing Consortium] 1998: Genome sequence of the nematode *C. elegans*: a platform for investigating biology. *Science 282*: 2012–2018.

DALE, P. S. 1967: *Wetanema hula* n. gen. et sp., a nematode from the weta *Hemideina thoracica*. *New Zealand Journal of Science 10*: 402–406.

DE ROSA, R.; GRENIER, J. K.; ANDREEVA, T.; COOK, C. E.; ADOUTTE, A.; AKAMI, M.; CARROLL, S. B.; BALAVOINE, G. 1999: Hox genes in brachiopods and priapulids and protostome evolution. *Nature, London 399*: 772–776.

DITLEVSEN, H. 1921: Papers from Dr. Th. Mortensen's Pacific Expedition 1914–10. III. Marine free-living nematodes from the Auckland and Campbell Islands. *Videnskabelige Meddelelser fra Dansk Naturhistorisk Forening, København 73*: 1–32.

GERLACH, S. A.; RIEMANN, F. 1973, 1974: The Bremerhaven checklist of aquatic nematodes. *Veröffentlichungen des Instituts für Meeres-forschung in Bremerhaven, Supplement 4(1, 2)*: 1–404, 405–736.

GLARE, T. R.; O'CALLAGHAN, M.; WIGLEY, P. J. 1993: Checklist of naturally occurring entomopathogenic microbes and nematodes in New Zealand. *New Zealand Journal of Zoology 20*: 95–120.

GOULD, S. J. 1989: *Wonderful Life: The Burgess Shale and the Nature of History*. Hutchinson Radius, London. 347 p.

HEWITT, G. C.; HINE, P. M. 1972: Checklist of parasites of New Zealand fishes and of their hosts. *New Zealand Journal of Marine and Freshwater Research 6*: 69–114.

HEYNS, J. 1968: A monographic study of the nematode families Nygolaimidae and Nygolaimellidae. *Department of Agricultural Technical Services, Entomology Memoirs No 19*: 1–144.

HEYWOOD, V. H.; WATSON, R. T. (Eds) 1995: *Global Biodiversity Assessment*. Cambridge University Press for UNEP, Cambridge. 1140 p

HINE, P. M.; JONES, J. B.; DIGGLES, B. K. 2000: A checklist of the parasites of New Zealand fishes, including previously unpublished records. *NIWA Technical Report 75*: 1–95.

HOLTERMAN, M.; HOLOVACHOV, O.; VAN DEN ELSEN, S.; VAN MEGEN, H.; BONGERS, T.; BAKKER, J.; HELDER, J. 2008: Small subunit ribosomal DNA-based phylogeny of basal Chromadoria (Nematoda) suggests that transitions from marine to terrestrial habitats (and vice versa) require relatively simple adaptations. *Molecular Phylogenetics and Evolution 48*: 758–763.

INGLIS, W. G.; HARRIS, E. A. 1990: Kiwinematidae n. fam. (Nematoda) for *Kiwinema* n. g. and *Hatterianema* Chabaud & Dolfus, 1966: heterakoids of native New Zealand vertebrates. *Systematic Parasitology 15*: 75–79.

JAIRAJPURI, M. S.; AHMAD, W. 1992: *Dorylaimida: Free-living, Predaceous and Plant-parasitic Nematodes*. E.J. Brill, Leiden. 458 p.

KIRK, T. W. 1899: Ear-cockle, peppercorns, purplers in wheat (*Tylenchus scandens = T. tritici*). *New Zealand Department of Agriculture Leaflets for Farmers No. 48*: 1–2.

KNIGHT, K. W. L.; BARBER, C. J.; PAGE, G. D. 1997: Plant-parasitic nematodes of New Zealand recorded by host association. *Journal of Nematology 29*: 640–656.

LEDUC, D.; GWYTHER, J. 2008: Description of new species of *Setosabatieria* and *Desmolaimus* (Nematoda: Monhysterida) and a checklist of New Zealand free-living marine nematode species. *New Zealand Journal of Marine and Freshwater Research 42*: 339–362.

LE JAMBRE, L. F. 1981: Hybridization of Australian *Haemonchus placei* (Place, 1893), *Haemonchus contortus cayugensis* (Das & Whitlock, 1960) and *Haemonchus contortus* (Rudolphi, 1803) from Louisiana. *International Journal for Parasitology 11*: 323–330.

MAGGENTI, A. 1981: *General Nematology*. Springer-Verlag, New York. 372 p.

MAGGENTI, A. R. 1983: Nematode higher classification as influenced by species and family concepts. Pp. 25–40 *in* Stone, A. R.; Platt, H. M.; Khalil, L. F. (eds) *Concepts in Nematode Systematics*. Academic Press, London. 388 p.

MAGGENTI, A. R.; LUC, M.; RASKI, D. J.; FORTUNER, R.; GERAERT, E. 1988: A reappraisal of Tylenchina (Nemata). 11. List of generic and supra-generic taxa, with their junior synonyms. *Revue de Nématologie 11*: 177–188.

McKENNA, P. B. 1997: Checklist of helminth parasites of terrestrial mammals in New Zealand. *New Zealand Journal of Zoology 24*: 277–290.

McKENNA, P. B. 1998: Checklist of helminth and protozoan parasites of birds in New Zealand. *Surveillance 25*: 3–31.

McKENNA, P. B. 2009: An updated checklist of helminth and protozoan parasites of terrestrial mammals in New Zealand. *New Zealand Journal of Zoology 36*: 89–113.

MERCER, C. F. 1994: Plant-parasitic nematodes in New Zealand. *New Zealand Journal of Zoology 21*: 57–65.

MERCER, C. F.; WATSON, R. N. 2007: Effects of nematicides and plant resistance on white clover performance and seasonal populations of nematodes parasitizing white clover in grazed pasture. *Journal of Nematology 39*: 298–304.

NIELSEN, C. 1996: *Animal Evolution: Inter-relationships of the Living Phyla*. Oxford University Press, Oxford. 467 p.

ORTON WILLIAMS, K. J. 1986: Descriptions of *Scutellonema southeyi* n. sp. and a population of *Hemicycliophora chathami* Yeates, 1978 (Tylenchida: Nematoda) from the Falkland Islands. *Systematic Parasitology 8*: 207–214.

PEÑA SANTIAGO, R.; JIMÉNEZ GUIRADO, D.; QUIJANO, R.; PERALTA, M.; LIEBANAS, G. 1992: Observations on the postembryonic development in species of the genus *Anatonchus* (Cobb, 1916) de Coninck, 1939 (Nematoda: Anatonchidae). *Nematologica 38*: 428. [Abstract]

POINAR, G. O. 1983: *The Natural History of Nematodes*. Prentice-Hall, Englewood Cliffs. 323 p.

POINAR, G. O.; ACRA, A.; ACRA, F. 1994: Earliest fossil nematode (Mermithidae) in Cretaceous Lebanese amber. *Fundamental and Applied Nematology 17*: 475–477.

POINAR, G. O. Jr; EARLY, J. 1990: *Aranimermis giganteus* n. sp. (Mermithidae: Nematoda), a parasite of New Zealand mygalomorph spiders (Araneae: Arachnidae). *Revue de Nematologie 13*: 395–402.

REHFELD, K.; SUDHAUS, W. 1985: Vergleichende Untersuchung des Sexualverhaltens eines Zwillingsarten-Paares von *Rhabditis* (Nematoda). *Zoologischer Jahrbucher Systematik 112*: 435–454.

SIDDIQI, M. R. 1983: Evolution of plant parasitism in nematodes. Pp. 113–129 *in* Stone, A. R.; Platt, H. M.; Khalil, L. F. (eds) *Concepts in Nematode Systematics*. Academic Press, London. 388 p.

SIDDIQI, M. R. 2000: *Tylenchida: Parasites of*

Plants and Insects, 2nd edn. CABI Publishing, Wallingford. 833 p.

SPRATT, D. M.; BEVERIDGE, I.; ANDREWS, J. R. H.; DENNETT, X. 1999: *Haycocknema perplexum* n. g., n. sp. (Nematoda : Robertdollfusidae): an intramyofibre parasite in man. *Systematic Parasitology 43*: 123–131.

STANKIEWICZ, M.; COWAN, P. C.; HEATH, D. D. 1998: Endoparasites of possums from selected areas of North Island, New Zealand. *New Zealand Journal of Zoology 25*: 91–97.

STEVENSON, L. A.; GASSER, R. B.; CHILTON, N. B. 1996: The ITS-2DNA *of Teladorsagia circumcincta, T. trifurcata* and *T. davtiana* (Nematoda, Trichstrongylidae) indicates that these taxa are one species. *International Journal for Parasitology 26*: 1123–1126.

SUDHAUS, W. 1974: Regarding the systematics, distribution, ecology and biology of new and lesser-known Rhabditidae (Nematoda). Part II. *Zoologische Jahrbücher, Systematik 101*: 417–465.

SULSTON, J. E.; SCHIERENBERG, E.; WHITE, J. G.; THOMSON, J. N. 1983: The embryonic cell lineage of the nematode *Caenorhabditis elegans*. *Developmental Biology 100*: 64–119.

THOMSON, G. M. 1922: *The Naturalisation of Animals and Plants in New Zealand*. Cambridge University Press, Cambridge. 607 p.

VLASSOFF, A.; McKENNA, P. B. 1994: Nematode parasites of economic importance in sheep in New Zealand. *New Zealand Journal of Zoology 21*: 1–8.

WOUTS, W. M. 2006: Criconematina (Nematoda: Tylenchida). *Fauna of New Zealand 55*: 1–232.

WOUTS, W. M.; YEATES, G. W.; LOOF, P. A. A. 1999: Criconematidae (Nematoda : Tylenchida) from the New Zealand region: genera *Ogma* Southern, 1914 and *Blandicephalanema* Mehta & Raski, 1971. *Nematology 1*: 561–590.

YEATES, G. W. 1967: Studies on nematodes from dune sands. 3. Oncholaimidae, Ironidae, Alaimidae and Mononchidae. *New Zealand Journal of Science 10*: 299–321.

YEATES, G. W. 1970: Failure of *Heterodera carotae, H. cruciferae* and *H. goettingiana* to interbreed in vitro. *Nematologica 16*: 153–154.

YEATES, G. W.; BOAG, B.; EVANS, K. A.; NEILSEN, R. 1998: Impact of climatic changes on the distribution of *Paratrichodorus minor* (Nematoda: Trichodoridae) as estimated using 'CLIMEX'. *Nematologica 44*: 293–301.

YEATES, G W.; NEWTON, P. C. D. 2009: Long-term changes in topsoil nematode populations in grazed pasture under elevated atmospheric carbon dioxide. *Biology and Fertility of Soils 45*: 799–808.

YEATES, G. W.; SPIRIDONOV, S. E. 1996: New nematodes of the families Drilonematidae, Ungellidae and Mesidionematidae from New Zealand megascolecid earthworms. *New Zealand Journal of Zoology 23*: 381–399.

ZONDAG, R. 1969: A nematode infection of *Sirex noctilio* (F.) in New Zealand. *New Zealand Journal of Science 12*: 732–747.

Checklist of New Zealand Nematoda

Environment/habitat/host information codes: A, adventive; E, endemic; I, invertebrate host; M, marine; P, plant host; S, soil; V, vertebrate host. Endemic genera are underlined (first entry only). Species of 'marine' genera described from stabilised sand dunes are regarded as soil nematodes while Rhabditidae from decaying seaweed are regarded as marine.

† is used to indicated parasites of vertebrates that have been recorded from the region but for which there is insufficient evidence to regard the population as being able to maintain itself.

(?) signifies species of doubtful status or placement and ? signifies those taxa possibly arbitrarily ascribed to species from other countries.

PHYLUM NEMATODA
Class ADENOPHOREA
Subclass ENOPLIA
Order ENOPLIDA
Suborder ENOPLINA
ENOPLOIDEA
ANOPLOSTOMATIDAE
Anoplostoma campbelli Allgén, 1932 M
ANTICOMIDAE
Anticoma acuminata (Eberth, 1863) M
Anticoma campbelli Allgén, 1932 M
Anticoma pellucida Bastian, 1865 M
ENCHELIDIIDAE
Eurystomina eurylaima (Ditlevsen, 1930) M
Eurystomina stenolaima (Ditlevsen, 1930) M
Eurystomina terricola de Man, 1907 (?) M
Eurystomina tenuicaudata Allgén, 1932 M
Eurystomina whangae Yeates, 1967 S
Polygastrophora hexabulba (Filipjev, 1918) M
Symplocostoma tenuicolle (Eberth, 1863) M
ENOPLIDAE
Enoplus benhami Ditlevsen, 1930 M
Enoplus parabrevis Allgén, 1928 (?) M
Mesacanthion infantile (Ditlevsen, 1930) M
Mesacanthion paradentatum (Allgén, 1931) (?) M
Mesacanthion virilie (Ditlevsen, 1930) M
Oxyonchus australis (de Man, 1904) M
Oxyonchus dentatus (Ditevsen, 1918) M
Paramesacanthion microsetosum (Allgén, 1931) M
Ruamowhitia orae Yeates, 1967 S
PHANODERMIDAE
Crenopharynx crassa (Ditlevsen, 1930) M E
Phanoderma cocksi Bastian, 1865 M
Phanoderma campbelli Allgén, 1928 M
Phanoderma serratum Ditlevsen, 1930 M
Phanoderma tuberculatum campbelloides (Allgen, 1927) (?) M
ONCHOLAIMOIDEA
ONCHOLAIMIDAE
Mononcholaimus glaberoides Allgén, 1931 (?) M
Oncholaimus aegypticus Steiner, 1921 M
Oncholaimus brachycercus de Man, 1889 M
Oncholaimus chiltoni Ditlevsen, 1930 (?) M
Oncholaimus conicaudatus Allgén, 1928 (?) M
Oncholaimus cylindricaudatus Allgén, 1932 (?) M
Oncholaimus dujardini de Man, 1876 M
Oncholaimus incurvatus Ditlevsen, 1930 M E
Oncholaimus moanae Leduc, 2008 M E
Oncholaimus tenuicaudatus Allgén, 1932 (?) M
Oncholaimus viridis Bastian, 1865 M
Pelagonema obtusicauda Filipjev, 1918 M
Phaenoncholaimus monodon (Ditlevsen, 1930) M E
Viscosia carnleyensis (Ditlevsen, 1921) M
OXYSTOMINOIDEA
ALAIMIDAE
Alaimus himatangiensis Yeates, 1967 S
Alaimus primitivus de Man, 1880 S
Alaimus arcuatus Thorne, 1939 S
Alaimus ?uniformis Thorne, 1939 S
Alaimus tasmaniensis (Allgén, 1929) S
Amphidelus papuanus Andrássy, 1973 S
OXYSTOMINIDAE
Halalaimus (Halalaimus) ciliocaudatus Allgén, 1932 M
Nemanema campbelli (Allgén, 1932) M
Thalassolaimus septentrionales Filipjev, 1927 M

Suborder TRIPYLINA
TRIPYLOIDEA
IRONIDAE
Dolicholaimus marioni de Man, 1888 M
Ironus ignavus Bastian, 1865 S
Syringolaimus stratocaudatus de Man, 1888 M
Trissonchulus littoralis Yeates, 1967 S
Trissonchulus quinquepapillatus Yeates, 1967 S
LEPTOSOMATIDAE
Dentostoma aucklandiae (Ditlevsen, 1921) M
Thoracostoma bruuni Wieser, 1956 M
Thoracostoma campbelli Ditlevsen, 1921 M
Thoracostoma galatheae Wieser, 1956 M
Thoracostoma papillosum Ditlevsen, 1921 M
Thoracostoma vallini Allgén, 1928 M
TRIPYLIDAE
Tripyla affinis de Man, 1880 S
Tripyla bioblitz Zhao, 2009 S
Tripyla filicaudata de Man, 1880 S
Tripylina kaikoura Zhao, 2009 S
Tripylina manurewa Zhao, 2009 S
Tripylina tamaki Zhao, 2009 S
Tripylina tearoha Zhao, 2009 S
Tripylina yeatesi Zhao, 2009 S
Trischistoma stramenti Yeates, 1972 S
Trischistoma monohystera (de Man, 1880) S
TRIPYLOIDIDAE
Arenasoma terricola Yeates, 1967 S
Bathylaimus australis Cobb, 1894 M

Order MONONCHIDA
MONONCHOIDEA
Anatonchus killicki Clark, 1963 S
Clarkus campbelli (Allgén, 1929) S
Clarkus propapillatus (Clark, 1960) S
Cobbonchus australis Clark, 1963 S
Cobbonchus chauloidus Clark, 1961 S
Cobbonchus longicaudatus Jairajpuri, Ahmad & Wouts, 1998 S

Cobbonchus pounamua Clark, 1961 S
Coomansus composticola (Clark, 1960) S
Coomansus meridionalis Jiménez Guirado, Wouts & Bello, 1998 S
Coomansus mesadenus (Clark, 1960) S
Iotonchus basidontus Clark, 1961 S
Iotonchus maragnus Clark, 1961 S
Iotonchus paraschokkei (Allgén, 1929) S
Iotonchus percivali Clark, 1961 S
Iotonchus ophiocercus Clark, 1961 S
Iotonchus stockdilli Yeates, 1988 S
Judonchulus monhystera Jairajpuri, Ahmad & Wouts, 1998 S
Miconchus kirikiri Yeates, 1967 S
Miconchus reflexus Yeates, 1967 S
Miconchus rex (Cobb, 1904) S
Mylonchulus arenicolus Clark, 1961 S
Mylonchus psammophilus Yeates, 1967 S
Mylonchus sigmaturus (Cobb, 1917) S
Mylonchus striatus (Thorne, 1924) S
Mylonchus ubis Clark, 1961 S
Prionchulus muscorum (Dujardin, 1845) S

Order DORYLAIMIDA
ACTINOLAIMOIDEA
ACTINOLAIMIDAE
Neoactinolaimus zealandicus (Clark, 1963) S
BELONDIROIDEA
BELONDIRIDAE
Axonchium (Axonchium) sabulum (Yeates, 1967) S
Axonchium (Epaxonchium) valvatum Nair & Coomans, 1974 S
Axonchium (Dactyluraxonchium) watti Yeates, 1979 S
Axonchium (Discaxonchium) coxi Yeates, 1979 S
Dorylaimellus ?basiri Jairajpuri, 1965 S
Dorylaimellus egmonti Yeates & Ferris, 1984 S
Dorylaimellus monticolous Clark, 1963 S
Dorylaimellus pastura Yeates, 1979 S
FALCIHASTIDAE
Falcihasta palustris Clark, 1964 S
OXYDIRIDAE
Oxydirus visseri Yeates, 1979 S
DORYLAIMOIDEA
APORCELAIMIDAE
Aporcelaimellus maitai Yeates, 1967 S
Aporcelaimellus taylori Yeates, 1967 S
Idiodorylaimus novaezealandiae (Cobb, 1904) S
Mesodorylaimus profundis (Cobb, 1904) S
Makatinus silvaticus Ahmad, Sturhan & Wouts, 2003 S
Sectonema sica Clark, 1964 S
Takamangai waenga Yeates, 1967 S
Torumanawa wahapuensis Yeates, 1967 S
DORYLAIMIDAE
Dorylaimus stagnalis Dujardin, 1845 S
Eudorylaimus parvus (de Man, 1880) S
Eudorylaimus ?intermedius (de Man, 1880) S
Eudorylaimus ?eremitus (Thorne, 1939) S
Mesodorylaimus exilis (Cobb, 1913) S
LEPTONCHIDAE
Capilonchus capitatus Ahmad, Sturhan & Wouts, 2003 S
Dorylaimoides chathami Yeates, 1979 S
Doryllium uniforme Cobb, 1920 S
Proleptonchus attenuatus Ahmad, Sturhan & Wouts, 2003 S
Proleptonchus saccatus (Clark, 1962) S
LONGIDORIDAE
Longidorus elongatus (de Man, 1876) P
Longidorus orongorongensis Yeates, van Etteger & Hooper, 1992 P
Longdorus taniwha Clark, 1963 P
Longidorus waikouaitii Yeates, Boag & Brown, 1997 P
Xiphinema americanum Cobb, 1913 P
Xiphinema brevicollum Lordello & Da Costa, 1961 P
Xiphinema diversicauadatum (Micoletzky, 1927) P
Xiphinema krugi Lordello, 1955 P
Xiphinema radicicola T. Goodey, 1936 P
Xiphinema waimungui Yeates, Boag & Brown, 1997 P
QUDSIANEMATIDAE
Crassolabium australe Yeates, 1967 S
Discolaimus arenicolus Yeates, 1967 S
Hulqus zelandicus Ahmad, Sturhan & Wouts, 2003 S
Labronema rikia Yeates, 1967 S
Pungentus maorium Clark, 1963 S
TYLENCHOLAIMIDAE
Meylis dicephalus (Yeates, 1967) S
Tylencholaimus australis Yeates, 1979 S
Tylencholaimus chathami Yeates, 1979 S
Tylencholaimus micronanus Yeates, 1979 S
Tylencholaimus nanus Thorne, 1939 S
Tylencholaimus tahatikus (Yeates, 1967) S
Tylencholaimus rossi Yeates, 1979 S
Tylencholaimus stecki Steiner, 1914 S
Tylencholaimellus montanus Thorne, 1939 S
NYGOLAIMOIDEA
NYGOLAIMIDAE
Nygolaimus (Nygolaimus) directus Heyns, 1968 S

Order TRIPLONCHIDA
DIPHTHEROPHOROIDEA
DIPHTHEROPHORIDAE
Longibulbophora ammophilae Yeates, 1967 S
TRICHODOROIDEA
TRICHODORIDAE
Paratrichodorus allius (Jensen, 1963) P
Paratrichodorus lobatus (Colbran, 1965) P
Paratrichodorus minor (Colbran, 1956) P
Paratrichodorus pachydermus (Seinhorst, 1954) P
Paratrichodorus porosus (Allen, 1957) P
Trichodorus cotteri Clark, 1963 P
Trichodorus primitivus (de Man, 1880) P

Order TRICHOCEPHALIDA
MUSPICEOIDEA
ROBERTDOLLFUSIDAE
Haycocknema perplexum Spratt, Beveridge, Andrews & Dennett, 1999 V†
TRICHINELLOIDEA
TRICHINELLIDAE
Capillaria aerophila (Creplin, 1839) V
Capillaria bovis (Schnyder, 1906) V
Capillaria anatis (Schrank, 1790) V
Capillaria annulata (Molin, 1858) V
Capillaria cadovulvata Madsen, 1945 V
Capillaria caudinflata (Molin, 1858) V
Capillaria contorta (Creplin, 1839) V
Capillaria emberizae Yamaguti, 1961 V
Capillaria erinacei Rudolphi, 1819 V
Capillaria hepatica (Bancroft, 1893) V
Capillaria obsignata Madsen, 1945 V
Capillaria plagiaticia Freitas & Mendonça, 1959 V
Capillaria strigis Johnston & Mawson, 1941 V
Capillaria tasmanica Johnston & Mawson, 1945 V
Capillaria tenuissima (Rudolphi, 1803) V
Trichinella pseudospiralis (Garkavi, 1972) V A
Trichinella spiralis (Owen, 1835) V
Trichosomoides crassicauda (Bellingham, 1840) V
Trichuris discolor (von Linstow, 1906) V
Trichurus globulosa (von Linstow, 1901) V
Trichurus ovis (Abildgaard, 1795) V
Trichurus parvispiculum Ortlepp, 1937 V
Trichuris skrjabini (Baskakov, 1924) V A
Trichurus suis (Schrank, 1788) V
Trichuris tenuis Chandler, 1930 V A
Trichurus vulpis (Froelich, 1789) V
Trichurus trichiura (Linnaeus, 1771) V

Order MERMITHIDA
MERMITHOIDEA
MERMITHIDAE
Aranimermis giganteus Poinar & Early, 1990 I E
Austromermis namis Poinar, 1990 I E
Bispiculum inaequale Zervos, 1980 I E
Blepharomermis craigi Poinar, 1990 I E
Mermis novaezealandiae Cobb, 1904 I E
Psammomermis canterburiensis Poinar & Jackson, 1992 I E
Thaumamermis zealandica Poinar, Latham & Poulin, 2002 I E

Subclass CHROMADORIA
Order CHROMADORIDA
Suborder CHROMADORINA
CHROMADOROIDEA
CHROMADORIDAE
Atrochromadora parva (de Man, 1893) M
Chromadora macrolaimoides Steiner, 1915 M
Chromadora nudicapitata Bastian, 1865 M
Chromadorina cylindricauda (Allgen, 1927) (?) M
Chromadorita brachypharynx (Allgén, 1932) M
Chromadorita heterophya (Steiner, 1916) M
Chromadorita minor (Allgén, 1927) M
Dichromadora cephalata (Steiner, 1916) M
Graphonema amokurae (Ditlevsen, 1921) M
Innocuonema paraheterophya (Allgén, 1932) (?) M
Innocuonema spectabile (Allgén, 1932) (?) M
Neochromadora craspedota (Steiner, 1916) M
Prochromadorella affinis (Allgén, 1930) (?) M
Prochromadorella paramucrodonta (Allgén, 1929) M
Prochromadorella ungulidentata (Allgén, 1932) (?) M
Rhips ornata Cobb, 1920 M
Spiliphera dolichura de Man, 1893 M
Spilophera amokuroides (Allgén, 1928) (?) M
Spilophorella campbelli Allgén, 1928 M
Spilophorella paradoxa (de Man, 1888) M
CYATHOLAIMIDAE
Halichoanolaimus ovalis Ditlevsen, 1921 M
Halichoanolaimus robustus (Bastian, 1865) M
Paracanthonchus caecus (Bastian, 1865) M
Paracanthonchus conicaudatus (Allgén, 1928) (?) M
SELACHINEMATIDAE
Synonchium pacificum Yeates, 1967 S
DESMODOROIDEA
DESMODORIDAE
Acanthopharynx similis (Allgén, 1932) (?) M
Croconema stateni (Allgén, 1928) M
Desmodora aucklandiae Ditlevsen, 1921 (?) M
Desmodora campbelli Allgén, 1932 (?) M
Desmodorella tenuispiculum (Allgén, 1928) M
Molgolaimus tenuilaimus (Allgén, 1932) (?) M
Molgolaimus tenuispiculum Ditlevsen, 1921 M
Paradesmodora campbelli (Allgén, 1932) M
Pselionema annulatum (Filipjev, 1922) M
DRACONEMATIDAE
Desmoscolex cristatus (Allgén, 1932) (?) M
Draconactussuillum (Allgén, 1932) M
Draconema cephalatum Cobb, 1913 M
Onyx ferox (Ditlevsen, 1921) (?) M
Prochaetosoma campbelli (Allgén, 1932) M
Prochaetosoma longicapitatum (Allgén, 1932) M
MICROLAIMOIDEA
MICROLAIMIDAE
Aponema subtile Leduc & Wharton, 2008 M E
Microlaimus falciferus Leduc & Wharton, 2008 M E
Microlaimus problematicus Allgén, 1932 (?) M
Microlaimus tenuilaimus Allgén, 1932 M
MONOPOSTHIIDAE
Nudora campbelli (Schulz, 1935) M

Suborder LEPTOLAIMINA
AULOLAIMIDAE
Aulolaimus mowhitius (Yeates, 1967) S
CHRONOGASTERIDAE

Chronogaster brasiliensis Meyl, 1957 S
Chronogaster typica (de Man, 1921) S
HALIPLECTIDAE
Haliplectus onepui Yeates, 1967 S
LEPTOLAIMIDAE
Acontiolaimus zostericola Filipjev, 1918 M
Aphanolaimus solitudinus Andrássy, 1968 S
Camacolaimus tardus de Man, 1889 M
Cricolaimus coronatus Ditlevsen, 1930 M E
Ionema cobbi (Steiner, 1916) M
PLECTIDAE
Anaplectus granulosus (Bastian, 1865) S
Anaplectus porosus Allen & Noffsinger, 1968 S
Neotylocephalus inflatus (Yeates, 1967) S
Pakira orae Yeates, 1967 S
Plectus annulatus Magenti, 1961 S
Plectus acuminatus Bastian, 1865 S
Plectus armatus Bütschli, 1873 S
Plectus assimilis Bütschli, 1873 S
Plectus cirratus Bastian, 1865 S
Plectus longicaudatus Bütschli, 1873 S
Plectus parietinus Bastian, 1865 S
Plectus parvus Bastian, 1865 S
Tylocephalus auriculatus (Bütschli, 1873) S
Wilsonema otophorum (de Man, 1880) S
PRISMATOLAIMIDAE
Prismatolaimus waipukea (Yeates, 1967) S
RHABDOLAIMIDAE
Rhabdolaimus terrestris de Man, 1880 S
TERATOCEPHALIDAE
Euteratocephalus crassidens (de Man, 1880) S
Teratocephalus decarinus Anderson, 1969 S

Order MONHYSTERIDA
AXONOLAIMOIDEA
COMESOSOMATIDAE
Comesoma tenuispiculum (Ditlevsen, 1921) M
Sabatieria annulata Leduc & Wharton, 2008 M
Sabatiera punctata (Kreis, 1924) M
Setosabatieria australis Leduc & Gwyther, 2008 M E
MONHYSTEROIDEA
DIPLOPELTIDAE
Araeolaimus elegans de Man, 1888 M
Diplopeltis cirrhatus (Eberth, 1863) M
Diplopeltula cylindricauda (Allgén, 1932) M
Southerniella simplex Allgén, 1932 M
LINHOMOEIDAE
Desmolaimus courti Leduc & Gwyther, 2008 M E
Linhomoeus elongatus Bastian, 1865 M
Paralinhomoeus litoralis Allgén, 1932 (?) M
MONHYSTERIDAE
Daptonema cuspidospiculum (Allgén, 1932) (?) M
Daptonema filispiculum (Allgén, 1932) (?) M
Halomonhystera disjuncta (Bastian, 1865) M
Monhystera elegans Allgén, 1928 (?) M
Monhystera paraambiguoides Allgén, 1932 (?) M
Monhystera praevulvata Allgén, 1932 (?) M
Monhystera tasmaniensis Allgén, 1927 (?) M
Monhystera vulgaris de Man, 1880 S
Terschellingia longicaudata de Man, 1907 M
Theristus acer Bastian, 1865 M
Theristus chitinolaimus (Allgén, 1932) (?) M
Theristus heterospiculum (Allgén, 1932) M
Theristus problematica (Allgén, 1928) M
Theristus oistospiculum (Allgén, 1932) M
Theristus velox Bastian, 1865 M
SIPHONOLAIMIDAE
Siphonolaimus pellucidus Allgén, 1932 (?) M
SPHAEROLAIMIDAE
Sphaerolaimus campbelli Allgén, 1928 M

Class SECERNENTEA
Order RHABDITIDA
CEPHALOBOIDEA
CEPHALOBIDAE
Acrobeles kotingotingus Yeates, 1967 S
Acrobeles maeneeneus Yeates, 1967 S
Acrobeles taraus Yeates, 1967 S
Acrobeloides syrtisus Yeates, 1967 S
Acrobeloides ellesmerensis Yeates, 1967 S
Heterocephalobus multicinctus (Cobb, 1893) S
Penjatinema novaezeelandiae Holovachov, Bostrom, de Lay, Nadler & de Ley, 2009 S
Stegelleta iketaia Yeates, 1967 S
Stegelleta tuarua Yeates, 1967 S
Zeldia punua Yeates, 1967 S
PANAGROLAIMOIDEA
PANAGROLAIMIDAE
Anguilluloides zondagi Dale, 1967 I
Panagrolaimus australis (Cobb, 1893) S
RHABDITOIDEA
HETERORHABDITIDAE
Heterorhabditis heliothidis (Khan, Brooks & Hirschmann, 1976) I
Heterorhabditis zealandica (Wouts, 1979) I
RHABDIASIDAE
Angiostoma schizoglossae Morand & Barker, 1994 I
RHABDITIDAE
Parasitorhabditis ateri (Fuchs, 1937) I
Protorhabditis wirthi Sudhaus, 1974 M
Rhabditis (Cephaloboides) valida Sudhaus, 1974 M
Rhabditis (Mesorhabditis) monhystera Bütschli, 1873 S
Rhabditis (M.) spiculigera Steiner, 1936 S
Rhabditis (Pellioditis) marina (Bastian, 1865) M
Rhabditis (P.) mediterranea (Sudhaus, 1974) M
Rhabditis (Rhabditoides) intermediformis Sudhaus, 1974 M
STEINERNEMATIDAE
Steinernema carpocapsae (Weiser, 1955) I
Steinernema feltiae (Filipjev, 1934) I
STRONGYLOIDIDAE
Parastrongyloides trichosuri Mackerras, 1959 V A
Strongyloides papillosus (Wedl, 1856) V A
Strongyloides stercoralis (Bavay, 1876) V A†
Strongyloides westeri Ihle, 1917 V A

Order STRONGYLIDA
ANCYLOSTOMATOIDEA
ANCYLOSTOMATIDAE
Ancylostoma brasiliense de Faria, 1910 V A†
Ancylostoma caninum (Ercolani, 1859) Hall, 1913 V A
Ancylostoma tubaeforme (Zeder, 1800) V A†
Bunostomum phlebotomum (Railliet, 1900) V A
Bunostomum trigonocephalum (Rudolphi, 1808) V A
Globocephaloides trifidospicularis Kung, 1948 V A
Globocephalus urosubulatus (Alessandrini, 1909) V A
Uncinaria stenocephala (Railliet, 1884) V A
METASTRONGYLOIDEA
ANGIOSTRONGYLIDAE
Aelurostrongylus abstrusus (Railliet, 1898) V A
CRENOSOMATIDAE
Crenosoma striatum (Zeder, 1800) V A
FILAROIDIDAE
Filaroides martis (Werner, 1782) V A
Oslerus osleri (Cobbold, 1876) V A
METASTRONGYLIDAE
Metastrongylus apri (Gmelin, 1790) V A
Metastrongylus pudendodectus Wostokow, 1905 V A
PROTOSTRONGYLIDAE
Apteragia odocoilei (Hobmaier & Hobmaier, 1934) V A
Apteragia quadrispiculata Jansen, 1958 V A
Elaphostrongylus cervi Cameron, 1931 V A
Muellerius capillaris (Müller, 1889) V A
Protostrongylus rufescens (Leuckart, 1865) V A
Varestrongylus sagittatus (Müller, 1890) V A
PSEUDALIIDAE
Skrjabinalius cryptocephalus Delyamure, 1942 V A
Stenurus ovatus (von Linstow, 1910) V A
SKRJABINGYLIDAE
Skrjabingylus nasicola (Leuckart, 1842) V A
STRONGYLOIDEA
CHABERTIIDAE
Chabertia ovina (Fabricus, 1788) V A
CLOACINIDAE
Labiostrongylus communis (Yorke & Maplestone, 1926) V A
HERPETOSTRONGYLIDAE
Paraustrostrongylus trichosuri Mawson, 1973 V A
PHARYNGOSTRONGYLIDAE
Pararugopharynx protemnodontis Mayzoub, 1964 V A
Rugopharynx australis (Mönnig, 1926) V A
Rugopharynx longibursaris (Kung, 1948) V A
Rugopharynx longispicularis Beveridge, 1999 V A
Rugopharynx omega (Monnig, 1927) V A
STRONGYLIDAE
Coronocyclus coronatus (Looss, 1900) V A
Coronocyclus labiatus (Looss, 1902) V A
Cyathostomum catinatum Looss, 1900 V A
Cyathostomum coronatum Looss, 1900 V A
Cylicocyclus calciatus (Looss, 1900) V A
Cylicocyclus insigne (Boulenger, 1917) V A
Cylicocyclus leptostomum (Kotlán, 1920) V A
Cylicocyclus nassatus (Looss, 1900) V A
Cylicocyclus radiatus (Looss, 1900) V A
Cylicostephanus calicatus (Looss, 1900) V A
Cylicostephanus goldi (Boulenger, 1917) V A
Cylicostephanus longibursatus (Yorke & Macfie, 1918) V A
Cylicostephanus minutus (Yorke & Macfie, 1918) V A
Gyalocephalus capitatus Looss, 1900 V A
Oesophagodontus robustus (Giles, 1892) V A
Oesophagostomum columbianum Curtice, 1890 V A†
Oesophagostomum dentatum (Rudolphi, 1803) V A
Oesophagostomum radiatum (Rudolphi, 1803) V A
Oesophagostomum venulosum (Rudolphi, 1809) V A
Petrovinema poculatum Looss, 1900 V A
Strongylus edentatus (Looss, 1900) V A
Strongylus equinus Müller, 1780 V A
Strongylus vulgaris (Looss, 1900) V A
Triodontophorus brevicauda Boulenger, 1916 V A
Triodontophorus minor (Looss, 1900) V A
Triodontophorus serratus Looss, 1902 V A
Triodontophorus tenuicollis Boulenger, 1916 V A
SYNGAMIDAE
Cyathostoma cacatua (Blanchard, 1849) V
Stephanurus dentatus Diesing, 1839 V A
Syngamus trachea (Montagu, 1811) V A
TRICHOSTRONGYLOIDEA
AMIDOSTOMATIDAE
Amidostomum acutum Seurat, 1918 V A
DICTYOCAULIDAE
Dictyocaulus arnfieldi (Cobbold, 1884) V A
Dictyocaulus eckerti Skrjabin, 1931 V A
Dictyocaulus filaria (Rudolphi, 1809) V A
Dictyocaulus viviparus (Bloch, 1782) V A
HELIGOSOMIDAE
Nippostrongylus brasiliensis (Travassos, 1914) V A
MOLINEIDAE
Lamanema ?chavezi Becklund, 1963 V A
Nematodirus abnormalis May, 1920 V A
Nematodirus battus Crofton & Thomas, 1951 V A†
Nematodirus filicollis (Rudolphi, 1802) V A
Nematodirus helvetianus May, 1920 V A
Nematodirus spathiger (Railliet, 1896) V A
Ollulanus tricuspis Leuckart, 1865 V A
ORNITHOSTRONGYLIDAE
Ornithostrongylus quadriradiatus (Stevenson, 1904) V A
TRICHOSTRONGYLIDAE
Camelostrongylus mentulatus (Railliet & Henry, 1908) V A
Cooperia curticei (Giles, 1892) V A

Cooperia oncophora (Railliet, 1898) V A
Cooperia pectinata Ransom, 1907 V A
Cooperia punctata (von Linstow, 1907) V A
Cooperia surnabada Antipin, 1931 V A
Graphidium strigosum (Dujardin, 1845) V A
Haemonchus contortus (Rudolphi, 1803) V A
Heligosomoides polygyrus (Dujardin, 1845) V A
Hyostrongylus rubidus (Hassall & Stiles, 1892) V A
Libyostrongylus douglasi (Cobbold, 1882) V A
Ostertagia circumcincta (Stadelmann, 1894) V A
Ostertagia kolchida Popova, 1937 V A
Ostertagia leptospicularis Asadov, 1953 V A
Ostertagia lyrata Sjöberg, 1926 V A
Ostertagia ostertagi (Stiles, 1892) V A
Ostertagia pinnata Daubney, 1933 V A
Ostertagia trifurcata Ransom, 1907 V A
Rinadia mathevossiana (Ruchliadev, 1948) V A
Spiculopteragia asymmetrica (Ware, 1925) V A
Spiculopteragia spiculoptera (Guschanskaya, 1931) V A
Trichostrongylus askivali Dunn, 1964 V A
Trichostrongylus axei (Cobbold, 1879) V A
Trichostrongylus capricola Ransom, 1907 V A
Trichostrongylus colubriformis (Giles, 1892) V A
Trichostrongylus longispicularis Gordon, 1933 V A
Trichostrongylus retortaeformis (Zeder, 1800) V A
Trichostrongylus tenuis (Mehlin, 1846) V A
Trichostrongylus vitrinus Looss, 1905 V A

Order OXYURIDA
OXYUROIDEA
OXYURIDAE
Enterobius vermicularis (Linnaeus, 1758) V A
Oxyuris equi (Schrank, 1788) V A
Passalurus ambiguus (Rudolphi, 1819) V A
Skrjabinema ovis (Skrjabin, 1915) V A
Syphacia muris Yamaguti, 1941 V A
Syphacia obvelata (Rudolphi, 1802) V A
PHARYNGODONIDAE
Skrjabinodon poicilandri Ainsworth, 1990 V E
Skrjabinodon trimorphi Ainsworth, 1990 V E
RHIGONEMATIDAE
Dudekemia alpinensis Bowie, 1986 I E
Dudekemia hirsutus Bowie, 1986 I E
Dudekemia kaorinus Bowie, 1986 I E
Dudekemia prolifica Bowie, 1986 I E
Dudekemia zealandica (Clark, 1978) I E
THELASTOMATIDAE
Blatticola barryi Zervos, 1987 I E
Blatticola monandros Zervos, 1983 I E
Blatticola tuapakae Dale, 1966 I E
Cephalobellus costelytrae Dale, 1964 I E
Cephalobellus fluxi Dale, 1966 I E
Heth hamatus Bowie, 1986 I E
Protrellina gurri Dale, 1966 I E
Protrellus dalei Zervos, 1987 I E
Protrellus dixoni Zervos, 1986 I E
Suifunema mackenziei Zervos, 1987 I E
Tetleyus clarki Dale, 1965 I E
Tetleyus lissotetus Dale, 1966 I E
Tetleyus miersi Dale, 1965 I E
Tetleyus pericopti Dale, 1964 I E
Thelastoma meadsi Clark, 1978 I E
Thelastoma moko Bowie, 1986 I E
Thelastoma rigo Bowie, 1986 I E
Wetanema hula Dale, 1967 I E
Wetanema ripariae Yeates & McCartney, 2006 I E

Order ASCARIDIDA
ASCARIDOIDEA
ANISAKIDAE
Anisakis simplex (Rudolphi, 1809) V
ASCARIDIDAE
Acanthocheilus bicuspis (Wedl, 1855) V
Acanthocheilus quadridentatus Molin, 1858 V
Ascaris lumbricoides Linnaeus, 1758 V A
Ascaris suum Goeze, 1782 V A
Contracaecum diomedeae (von Linstow, 1888) V
Contracaecum eudyptes Johnston & Mawson, 1953 V
Contracaecum magnicollare Johnston & Mawson, 1941 V
Contracaecum microcephalum (Rudolphi, 1809) V
Contracaecum rudolphii Hartwich, 1964 V
Contracaecum scotti (Leiper & Atkinson, 1914) V
Contracaecum spiculigerum (Rudolphi, 1809) V
Contracaecum (Thynnascaris) aduncum (Rudolphi, 1802) V
Hysterothylacium aduncum (Rudolphi, 1802) V
Hysterothylacium marinum (Linnaeus, 1767) V
Hysterothylacium seriolae (Yamaguti, 1941) V
Hysterothylacium zenopsis (Yamaguti, 1941) V
Paranisakiopsis australiensis Johnston & Mawson, 1945 V
Parascaris equorum (Goeze, 1782) V A
Porrocaecum crassum (Deslongchamps, 1824) V
Porrocaecum decipiens (Krabbe, 1878) V
Porrocaecum ensicaudatum (Zeder, 1800) V
Terranova antarctica Leiper & Atkinson, 1914 V
Toxascaris leonina (Linstow, 1902) V A
Toxocara canis (Werner, 1782) V A
Toxocara cati (Schrank, 1788) V A
COSMOCERCOIDEA
COSMOCERCIDAE
Aplectana novaezelandiae Baker & Green, 1988 V E
Cosmocerca archeyi Baker & Green, 1988 V E
Probstmayria vivipara (Probstmayr, 1865) V A
HETERAKOIDEA
HETERAKIDAE
Ascaridia galli (Schrank, 1788) V
Ascaridia platycerci Hartwich & Tscherner, 1979 V
Ascaridia zelandica Clark, 1981 V
<u>*Hatterianema*</u> *hollandei* Chabaud & Dolfus, 1966 V E
Heterakis gallinarum (Schrank, 1788) V
Heterakis isolonche Linstow, 1906 V
Heterakis spumosa Schneider, 1866 V A
Heterakis vesicularis (Froelich, 1791) V
<u>*Kiwinema*</u> *gracilicauda* (Harris, 1975) V E
SEURATOIDEA
CUCULLANIDAE
Cucullanellus cnidoglanis Johnston & Mawson, 1945 V
Cucullanellus pleuronectidis Yamaguti, 1935 V
Cucullanellus sheardi Johnston & Mawson, 1945 V
Cucullanus antipodeus Baylis, 1932 V
Cucullanus robustus Yamaguti, 1935 V

Order SPIRURIDA
Suborder SPIRURINA
ACUARIOIDEA
ACUARIIDAE
Acuaria skrjabini Ozerska, 1926 V
Cosmocephalus jaenschi Johnston & Mawson, 1941 V
Cosmocephalus tanakai Rodrigues & Vincente, 1963 V
Echinuria australis Clark, 1979 V
Echinuria uncinata (Rudolphi, 1819) V
Sciadiocara tarapunga Clark, 1978 V
Streptocara cirrohamata (von Linstow, 1888) V
Viktorocara torea Clark, 1978 V
DRACUNCULOIDEA
ANGUILLICOLIDAE
Anguillicola australiensis Johnston & Mawson, 1940 V
Anguillicola novaezelandiae Moravec & Tarashewski, 1988 V
PHILOMETRIDAE
Philometra lateolabricus Yamaguti, 1935 V
Phlyctainophora lamnae Steiner, 1921 V
DRILONEMATOIDEA
DRILONEMATIDAE
Iponema australe Yeates & Spiridonov, 1996 I E
<u>*Nemanoke*</u> *aotearoa* Yeates & Spiridonov, 1996 I E
Nemanoke stomiculata Yeates & Spiridonov, 1996 I E
Siconema neozealandicum Yeates & Spiridonov, 1996 I E
FILARIOIDEA
ONCHOCERCIDAE
Dipetalonema reconditum (Grassi, 1889) V A
Dirofilaria immitis (Leidy, 1856) V A†
Dirofilaria repens Railliett & Henry, 1911 V A†
Loa loa (Cobbold, 1864) V A†
Wuchereria bancrofti (Cobbold, 1877) V A†
HABRONEMATOIDEA
CYSTIDICOLIDAE
Ascarophis cooperi Johnston & Mawson, 1945 V
Ctenascarophis lesteri Crites, Overstreet & Muang, 1993 V
Prospinitectus exiguus Crites, Overstreet & Muang, 1993 V
HABRONEMATIDAE
Cyrnea (Cyrnea) aptercyis Harris, 1975 V E
Draschia megastoma (Rudolphi, 1819) V
Hedruris minuta Andrews, 1974 V E
Hedruris spinigera Baylis, 1931 V
Microtetrameres nestoris Clark, Black & Rutherford, 1979 V
Procyrnea kea Clark, 1978 V
Tetrameres tarapungae Clark, 1978 V
PHYSALOPTEROIDEA
PHYSALOPTERIDAE
Physaloptera getula Seurat, 1917 V
SPIRUROIDEA
GONGYLONEMATIDAE
Gongylonema pulchrum Molin, 1857 V A
SPIROCERCIDAE
Spirocerca lupi (Rudolphi, 1809) V A†
SPIRURIDAE
Cylicospirura advena Clark, 1981 V A†
Mastophorus muris (Gmelin, 1790) V A

Suborder CAMALLANINA
CAMALLANIDAE
Camallanus aotea Slankis & Krottaeva, 1974 V

Order DIPLOGASTERIDA
DIPLOGASTEROIDEA
DIPLOGASTERIDAE
<u>*Hugotdiplogaster*</u> *neozelandica* Morand & Barker, 1994 I E
NEODIPLOGASTERIDAE
Mononchoides colobocercus (Andrássy, 1964) S
Micoletzkya thalenhorsti (Rühm, 1956) I

Order APHELENCHIDA
APHELENCHOIDEA
APHELENCHIDAE
Aphelenchus avenae Bastian, 1965 S
APHELENCHOIDOIDEA
APHELENCHOIDIDAE
Aphelenchoides bicaudatus (Imamura, 1931) S
Aphelenchoides blastophorus Franklin, 1952 S
Aphelenchoides composticola Franklin, 1957 S
Aphelenchoides fragariae Ritzema Bos, 1890 P
Aphelenchoides ritzemabosi (Schwartz, 1911) P
Aphelenchoides subtenuis (Cobb, 1926) S
Bursaphelenchus eggersi (Rühm, 1956) I
Bursaphelenchus fungivorus Franklin & Hooper, 1962 S
Seinura demani (T. Goodey, 1928) S

Order TYLENCHIDA
CRICONEMATOIDEA
CRICONEMATIDAE
Blandicephalanema pilatum Mehta & Raski, 1971 P
Blandicephalanema serratum Mehta & Raski, 1971 P
Blandicephalanema inserratum Wouts, 2006 P

Blandicephalanema nothofagi Wouts, 2006 P
Criconema (Criconema) aucklandicum Loof, Wouts & Yeates, 1997 P
Criconema (C.) cristulatum Loof, Wouts & Yeates, 1997 P
Criconema (C.) mackenziei Wouts, 2006 P
Criconema (C.) makahuense Wouts, 2006 P
Criconema (C.) nelsonense Wouts, 2006 P
Criconema (C.) spinicaudatum (Raski & Pinochet, 1976) P
Criconema (Nothocriconema) annuliferum (de Man, 1921) P A
Criconema (N.) grandisoni Wouts, 2006 P
Criconema (N.) lineatum Loof, Wouts & Yeates, 1997 P
Criconema (N.) magnum Loof, Wouts & Yeates, 1997 P
Criconema (N.) undulatum Loof, Wouts & Yeates, 1997 P
Criconema (Nothocriconemella) acuticaudatum Loof, Wouts & Yeates, 1997 P
Criconema (N.) alpinum Loof, Wouts & Yeates, 1997 P
Criconema (N.) californicum Diab & Jenkins, 1966 P
Criconema (N.) crosbyi Wouts, 2000 P
Criconema (N.) dugdalei Wouts 2000 P
Criconema (N.) farrelli Wouts, 2000 P
Criconema (N.) graminicola Loof, Wouts & Yeates, 1997 P
Criconema (N.) macilentum (Raski & Pinochet, 1976) P
Criconema (N.) mutabile (Taylor, 1936) P A
Criconema (N.) pasticum (Raski & Pinochet, 1976) P
Crioconema (N.) ramsayi Wouts, 2000 P
Criconema (N.) sphagni Micoletzky, 1925 P
Criconemoides (Criconemoides) informis (Micoletzky, 1922) P A
Criconemoides (Criconemella) parvus Raski, 1952 P
Hemicriconemoides cocophilus (Loos, 1949) P
Macroposthonia campbelli Wouts, 2006 P
Macroposthonia rustica (Micoletzky, 1915) P A
Macroposthonia xenoplax (Raski, 1952) P A
Ogma alternum Wouts, 2006 P
Ogma campbelli Wouts, Yeates & Loof, 1999 P
Ogma capitatum Wouts, 2006 P
Ogma catherinae Wouts, 2006 P
Ogma civellae (Steiner, 1914) P A
Ogma crenulatum Wouts, Yeates & Loof, 1999 P
Ogma inaequale Wouts, 2006 P
Ogma latens (Mehta & Raski, 1971) P
Ogma mucronatum Wouts, 2006 P
Ogma niagarae Wouts, 2006 P
Ogma palmatum (Siddiqi & Southey, 1962) P A
Ogma paucispinatum Wouts, Yeates & Loof, 1999 P
Ogma polyandra Wouts, Yeates & Loof, 1999 P
Ogma semicrenatum Wouts, Yeates & Loof, 1999 P
Ogma sexcostatum Wouts, Yeates & Loof, 1999 P
Ogma sturhani Wouts, 2006 P
Ogma subantarcticum Wouts, Yeates & Loof, 1999 P
Patercephalanema imbricatum (Colbran, 1956) P
Syro glabellus Wouts, 2006 P
Syro tribulosus Wouts, 2006 P
HEMICYCLIOPHORIDAE
Hemicycliophora chathami chathami Yeates, 1978 P
Hemicyciophora c. major Yeates, 1978 P
Hemicycliophora halophila Yeates, 1967 P
TYLENCHULIDAE
Paratylenchus (Gracilacus) aonli Misra & Edward, 1971 P
Paratylenchus (G.) straeleni (de Coninck, 1931) P
Paratylenchus (Paratylenchus) elachistus Steiner, 1949 P
Paratylenchus (P.) halophilus Wouts, 1966 P
Paratylenchus (P.) tateae Wu & Townshend, 1973 P
Paratylenchus (P.) minutus Linford *in* Linford, Oliveria & Ishi, 1949 P
Paratylenchus (P.) nainianus Edward & Misra, 1963 P
Paratylenchus (P.) nanus Cobb, 1923 P
Paratylenchus (P.) neoamblycephalus Geraert, 1965 P
Paratylenchus (P.) projectus Jenkins, 1956 P
Sphaeronema californicum Raski & Sher, 1952 P
Tylenchulus semipenetrans Cobb, 1913 P A
Trophotylenchulus okamotoi (Mingawa, 1983) P
NEOTYLENCHOIDEA
NEOTYLENCHIDAE
Deladenus durus (Cobb, 1922) Thorne, 1941 I/S
Deladenus siricidicola Bedding, 1968 I/S
Fergusobia pohutukawa Davies *in* Taylor, Davies, Martin & Crosby, 2007 I
Parasitylenchus hylastis (Wülker, 1923) I
SPHAERULARIOIDEA
SPHAERULARIIDAE
Sphaerularia bombi Dofour, 1837 I
TYLENCHOIDEA
ANGUINIDAE
Anguina agrostis (Steinbuch, 1799) P
Anguina tritici (Steinbuch, 1799) P
Ditylenchus destructor Thorne, 1945 P
Ditylenchus dipsaci (Kuhn, 1857) P
Ditylenchus drepanocercus T. Goodey, 1953 P
Ditylenchus myceliophagus J.B. Goodey, 1958 S
Subanguina radicicola (Greeff, 1872) P
BELONOLAIMIDAE
Geocenamus nanus (Allen, 1955) P
Morulaimus geniculatus Sauer, 1966 P
Quinisculcius capitatus (Allen, 1955) P
Tylenchorhynchus dubius (Bütschli, 1873) P
Tylenchorhynchus maximus Allen, 1955 P
DOLICHODORIDAE
Neodolichodorus arenarius (Clark, 1963) P
HETERODERIDAE
Cryphodera coxi (Wouts, 1973) P
Cryphodera nothofagi (Wouts, 1973) P
Cryphodera podocarpi (Wouts, 1973) P
Globodera pallida (Stone, 1973) P A
Globodera rostochiensis (Wollenweber, 1923) P A
Globodera zealandica Wouts, 1984 P
Heterodera aucklandica Wouts & Sturhan, 1995 P
Heterodera avenae Wollenweber, 1924 P A
Heterodera cacti Filipjev & Schuurmans Stekhoven, 1941 P A
Heterodera fici Kir'yanova, 1954 P A
Heterodera humuli Filipjev, 1934 P A
Heterodera schachtii Schmidt, 1871 P A
Heterodera trifolii Goffart, 1932 P A
Meloidogyne arenaria (Neal, 1889) P
Meloidogyne fallax Karssen, 1996 P
Meloidogyne hapla Chitwood, 1949 P
Meloidogyne incognita (Kofoid & White, 1919) P
Meloidogyne incognita acrita Chitwood, 1949 P
Meloidogyne javanica (Treub, 1885) P
Meloidogyne naasi Franklin, 1965 P
Meloidogyne trifoliophila Bernard & Eisenbach, 1997 P A
Paradolichodera tenuissima Sturhan, Wouts & Subbotin, 2007 P
HOPLOLAIMIDAE
Antarctylus humus Sher, 1973 P
Helicotylenchus canadensis Waseem, 1961 P
Helicotylenchus depressus Yeates, 1967 P
Helicotylenchus digonicus Perry, 1959 P
Helicotylenchus dihystera (Cobb, 1893) P
Helicotylenchus erythrinae (Zimmermann, 1904) P
Helicotylenchus exallus Sher, 1966 P
Helicotylenchus labiatus Roman, 1965 P
Helicotylenchus leiocephalus Sher, 1966 P
Helicotylenchus lissocaudatus Fernández, Razjivin, Ortega & Quincosa, 1980 P
Helicotylenchus minzi Sher, 1966 P
Helicotylenchus paxilli Yuen, 1964 P
Helicotylenchus paraplatyurus Siddiqi, 1972 P
Helicotylenchus pseudorobustus (Steiner, 1914) P
Helicotylenchus serenus Siddiqi, 1963 P
Helicotylenchus spitsbergensis Loof, 1971 P
Helicotylenchus varicaudatus Yuen, 1964 P
Helicotylenchus vulgaris Yuen, 1964 P
Rotylenchus buxophilus Golden, 1956 P
Rotylenchus labiodiscus Wouts & Sturhan, 1999 P
Rotylenchus nelsonensis Ryss & Wouts, 1997 P
Rotylenchus robustus (de Man, 1876) P
Rotylenchus uniformis (Thorne, 1949) P
Rotylenchus vacuus Colbran, 1971 P
Rotylenchus kahikateae Ryss & Wouts, 1997 P
Scutellonema brachyurus (Steiner, 1949) P
PRATYLENCHIDAE
Pratylenchus coffeae (Zimmermann, 1898) P
Pratylenchus crenatus Loof, 1960 P
Pratylenchus neglectus (Rensch, 1924) P
Pratylenchus penetrans (Cobb, 1917) P
Pratylenchus pratensis (de Man, 1880) P
Pratylenchus thornei Sher & Allen, 1953 P
Pratylenchus vulnus Allen & Jensen, 1951 P
Radopholus cavenessi Egunjobi, 1968 P
Radopholus nativus Sher, 1968 P
Zygotylenchus guevarai (Tobar Jiménez, 1963) P
TYLENCHIDAE
Boleodorus spiralis Egunjobi, 1968 S
Campbellenchus filicauda Wouts, 1977 S
Campbellenchus poae Wouts, 1977 S
Cephalenchus emarginatus (Cobb, 1893) S
Cephalenchus hexalineatus (Geraert, 1962) S
Cephalenchus tahus Wood, 1973 S
Coslenchus areolatus (Egunjobi, 1968) S
Coslenchus costatus (de Man, 1921) S
Filenchus clarki (Egunjobi, 1968) S
Filenchus ruatus (Egunjobi, 1967) S

INCERTAE SEDIS
Ascaris apterycis Chatin, 1884 V E
Nuadella primitiva Allgén, 1928 M
Rhinonema paradoxum Allgén, 1928 M

FOURTEEN

Phylum NEMATOMORPHA

horsehair worms, gordian worms

GEORGE POINAR JR

Two specimens of *Gordionus diblastus* emerging from the ground weta *Hemiandrus pallitarsis*, Banks Peninsula.
From Poinar 1991

As the common name implies, horsehair worms (or hairworms) are long, very thin, and generally brown in colour. Although often classified with nematodes, priapulids, and kinorhynchs as 'Aschelminthes', they are an enigmatic group of parasitic worms that have no close relationships with any other living phylum of invertebrates (Poinar 1999a, 2000). Although New Zealand has only four mainland and one marine species, some of them are exceptionally abundant, especially in the South Island. In fact, New Zealand is one of the best countries in the world to study hairworms, being one of the few remaining places where large numbers can be obtained throughout the year. This unusual abundance of material is ideal for conducting studies on the behaviour, life-history, host-parasite associations, and intraspecific variation of these interesting invertebrates. It is quite likely that additional species and host-parasite relations may be found in New Zealand. If so, it is likely that they may include relictual species of groups that are near extinction in other parts of the world.

The basic body plan of adult and preparasitic larval hairworms is unique, supporting the contention that hairworms arose from a basic lineage of the Aschelminthes quite early and evolved independently thereafter. The origin of hairworms is obscure. The only irrefutable fossils are two specimens of *Paleochordodes protus* (Poinar 1999b), associated with a cockroach in 15–40 million-year-old Dominican amber, and *Cretachordodes burmitis* (Poinar & Buckley, 2006) in 100 million-year-old Burmese amber. A partial fossil of a questionable hairworm has been described from the Eocene brown coals (Voigt 1938). The resemblance between Paleozoic paleoscolecid worms and Nematomorpha (Xianguang & Bergstrom 1994) is superficial and the former cannot be considered as hairworms at this time (Poinar 1999a). There are two classes in the phylum Nematomorpha — the hairworms proper, or Gordioidea, and their marine kin the Nectonematoidea.

The free-living stages of New Zealand gordian hairworms (eggs, preparasitic larvae and adults) can occur in any type of freshwater supply, including lakes, rivers, creeks, ponds, tarns, ditches, caves, watering troughs, and even domestic water supplies (reservoirs). They can also be found in bays and are able to survive for some time in brackish water. They occur from sea level to mountain-tops, the highest recorded elevation for them in New Zealand being 1646 metres (Poinar 1991), but they probably occur at even higher elevations.

The four species of New Zealand gordian worms belong to three worldwide genera – *Euchordodes, Gordionus,* and *Gordius*. Three of the New Zealand species are considered endemic (Poinar 1991) while the fourth also occurs in South America (Schmidt-Rhaesa et al. 2000). Both sexual and age dimorphism occurs in relation to body size, colouration, cuticular spotting, and size and colour of the areoles (raised areas on the cuticle that may be associated with bristles and warts in the adult). The major diagnostic characters for discriminating species of hairworms are the presence and structure of the areoles, the nature of the calotte (head region of the adult), and features of the male tail (whether the end is simple or forked, presence/absence of bristle fields around the cloacal opening, etc).

Scanning electron microscopy is a useful tool for observing the fine details of morphological structures of hairworms and has been used in conjunction with intraspecific variation studies of *E. nigromaculatus* from the South Island (Schmidt-Rhaesa et al. 1998).

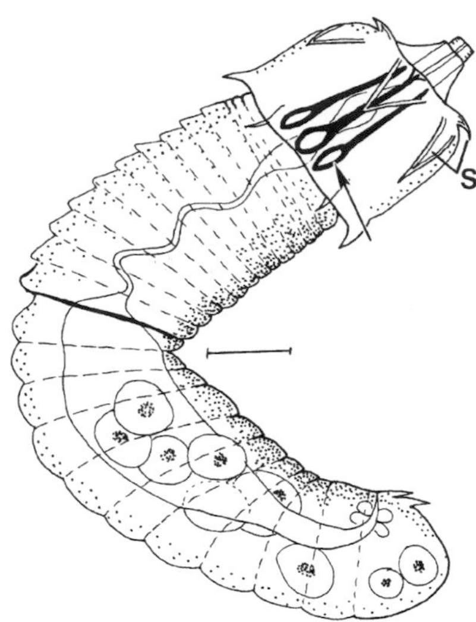

Preparasitic larva removed from its cyst in the caddisfly *Olinga feredayi* (S= spines; arrow = stylet). Scale 4 µm.

From Poinar 1991

Class Gordioidea

Gordian hairworms are arthropod parasites as juveniles, emerging from their hosts into aquatic habitats for a brief reproductive period as adults. In fact, the adult digestive tract may be non-functional in some species. Adult males tend to be shorter than females and may be coiled, curved, or forked at the posterior end for mating. As far as we know, the New Zealand gordian hairworms have an indirect life-cycle, involving an aquatic paratenic host and a terrestrial developmental host. The paratenic host (one that carries a parasite internally but does not support its growth or reproduction) ingests hairworm eggs or preparasitic larvae, normally along with debris in the water habitat. The preparasitic larva enters the body cavity of the paratenic host by burrowing through the intestinal wall. Larvae are well equipped for this task since the presoma (anterior portion of the body, also called the preseptum) contains an extensible proboscis bearing circlets of backward-pointing spines as well as three stylets that are used to pierce the tissues of the host. The preparasitic stage burrows into the body cavity of the host and then encysts in various tissues. The location chosen for encystment can vary from host to host. In the dobsonfly *Archichauliodes diversus* most cysts occur in the fat body whereas the intestinal wall and connective tissues are preferred locations in other hosts. However, encystment can occur in various tissues as the range of paratenic hosts is quite extensive. In New Zealand, fish and their trematode parasites can also serve as paratenic hosts (Blair 1983).

If a paratenic host is attacked by another host not suitable for hairworm development, then it is possible for the preparasite larva to re-encyst in this

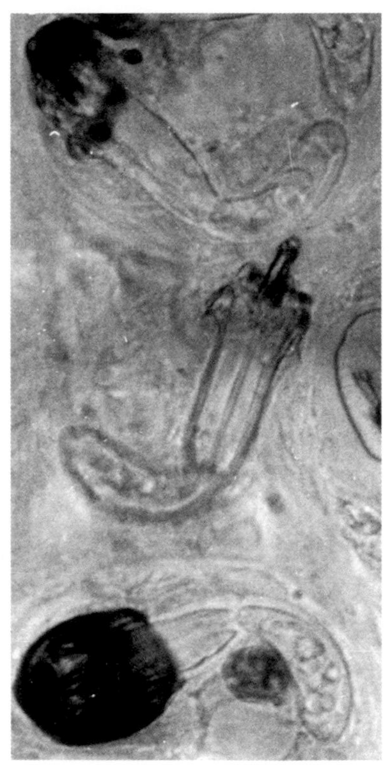

Stages of host melanisation (darkening) reactions against preparasitic hairworms (some dead) in mature larvae of the stonefly *Stenoperla prasina*.

George Poinar Jr

Paratenic hosts of New Zealand hairworms

Host species	Hairworm species	Reference
TREMATODA (flukes)		Blair 1983
Coitocaecum sp.	?	
DIPTERA (flies)		Poinar 1991
Chironomus zealandicus	prob. *Gordius dimorphus*	
EPHEMEROPTERA (mayflies)		Poinar 1991
Deleatidium sp.	prob. *Gordius dimorphus*	
MEGALOPTERA (dobsonflies)		Poinar 1991
Archichauliodes diversus	prob.*Gordius dimorphus*	
PLECOPTERA (stoneflies)		Poinar 1991
Stenoperla prasina	*Gordius dimorphus* and *Euchordodes nigromaculatus*	
TRICHOPTERA (caddisflies)		Poinar 1991
Hydrobiosis parumbripennis and *Olinga feredayi*	*Gordius dimorphus* and *Euchordodes nigromaculatus*	
OSTEICHTHYES (fish)		Blair 1983
Galaxias vulgaris	?	
Gobiomorphus breviceps	?	

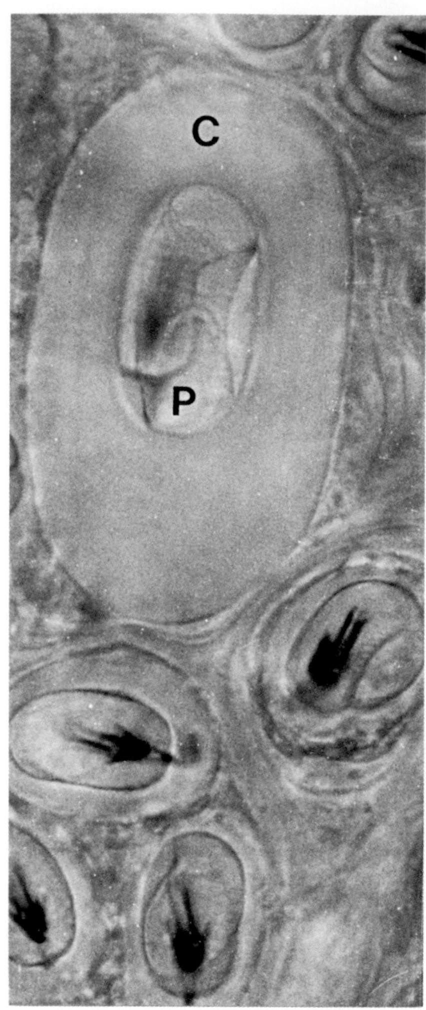

Small and large preparasitic hairworms (some dead) encysted in the gut wall of a young stonefly larva, *Stenoperla prasina*. (P = preparasitic larva; C = cyst).

George Poinar Jr

Developmental hosts of New Zealand gordian hairworms

Host species	Hairworm species	Range	Reference
ACRIDIDAE			
Sigaus australis	*Euchordodes nigromaculatus*	S. Island	Poulin 1995
Sigaus obelisci	*Euchordodes nigromaculatus*	S. Island	Poulin 1995
BLATTELLIDAE			
Parellipsidion pachycercum	?	S. Island	Zervos 1989
BLATTIDAE			
Celatoblatta peninsularis	?	S. Island	Zervos 1989
CARABIDAE			
Oregus aereus	?	?	Townsend 1970
Undescribed broscine	?	?	Townsend 1970
RHAPHIDOPHORIDAE			
Macropathus filifer	?	?	Richards 1954
Pleioplectron simplex	*Gordius paranensis*	?	Schmidt-Rhaesa 2000
Pleioplectron spp.	*Gordius dimorphus*	N/S Islands	Poulin 1996
ANOSTOSTOMATIDAE			
Deinacrida connectens	*Euchordodes nigromaculatus*	S. Island	Poinar 1991
Deinacrida rugosa	*Euchordodes nigromaculatus*	S. Island	Poinar 1991
Hemideina crassidens	?	?	?
Hemideina pallitarsis	*Gordionus diblastus*	N/S Islands	Poinar 1991
Hemideina thoracica	*Gordionus diblastus*	N/S Islands	Poinar 1991
Hemiandrus maculifrons	*Gordius dimorphus*	N/S Islands	Poinar 1991
Hemiandrus sp.	*Gordius dimorphus*	N/S Islands	Poinar 1991

secondary paratenic host. The cyst is composed of a clear material that is of parasite origin, apparently being secreted by glands in the posterior portion of the body. The secretions swell and form a clear protective cover around the larva. If the preparasitic larva remains for a long period in the body cavity of a paratenic host, then a host melanisation reaction (deposition of melanin) can destroy the hairworm larvae. Presumably there is a chemical cue that signals the preparasitic larva whether a host is suitable for development or for encystment. Whether encystment in a paratenic host is necessary for completion of the hairworm's life-cycle is also not known. Experimental studies on these aspects of hairworm biology should be possible with the abundance and easy access to these animals in New Zealand.

Growth of the hairworm preparasite occurs only after the paratenic host is consumed by a developmental host (one in which the parasite can complete its development – also called a definitive host). This is why developmental hosts are either predators or omnivores. After ingestion of a paratenic host by a developmental host, the preparasitic larva leaves its cyst and burrows through the gut into the body cavity, slowly growing to maturity there. Surprisingly, many of the New Zealand aquatic insects that serve as paratenic hosts of hairworms also serve as paratenic hosts to those mermithid nematodes that possess a similar indirect life-cycle involving both aquatic and terrestrial hosts. Larvae of the large green stonefly *Stenoperla prasina* can have their intestinal walls packed with hairworm cysts as well as containing preparasitic stages of mermithids in their body cavities. Both parasite groups have also been found in larval mayflies (*Deleatidium* species) and caddis flies (*Hydrobiosis parumbripennis, Olinga feredayi*) (Poinar 1990, 1991, pers. obs.).

The parasitic development of hairworms has been little studied. At least one moult probably occurs, since the basic body plan of the preparasitic larva, including both stylets and spines, completely disappears and the adult hairworm has completely different features. Food is probably taken up through the body wall since the mouth and pharynx are degenerate in the developing stages.

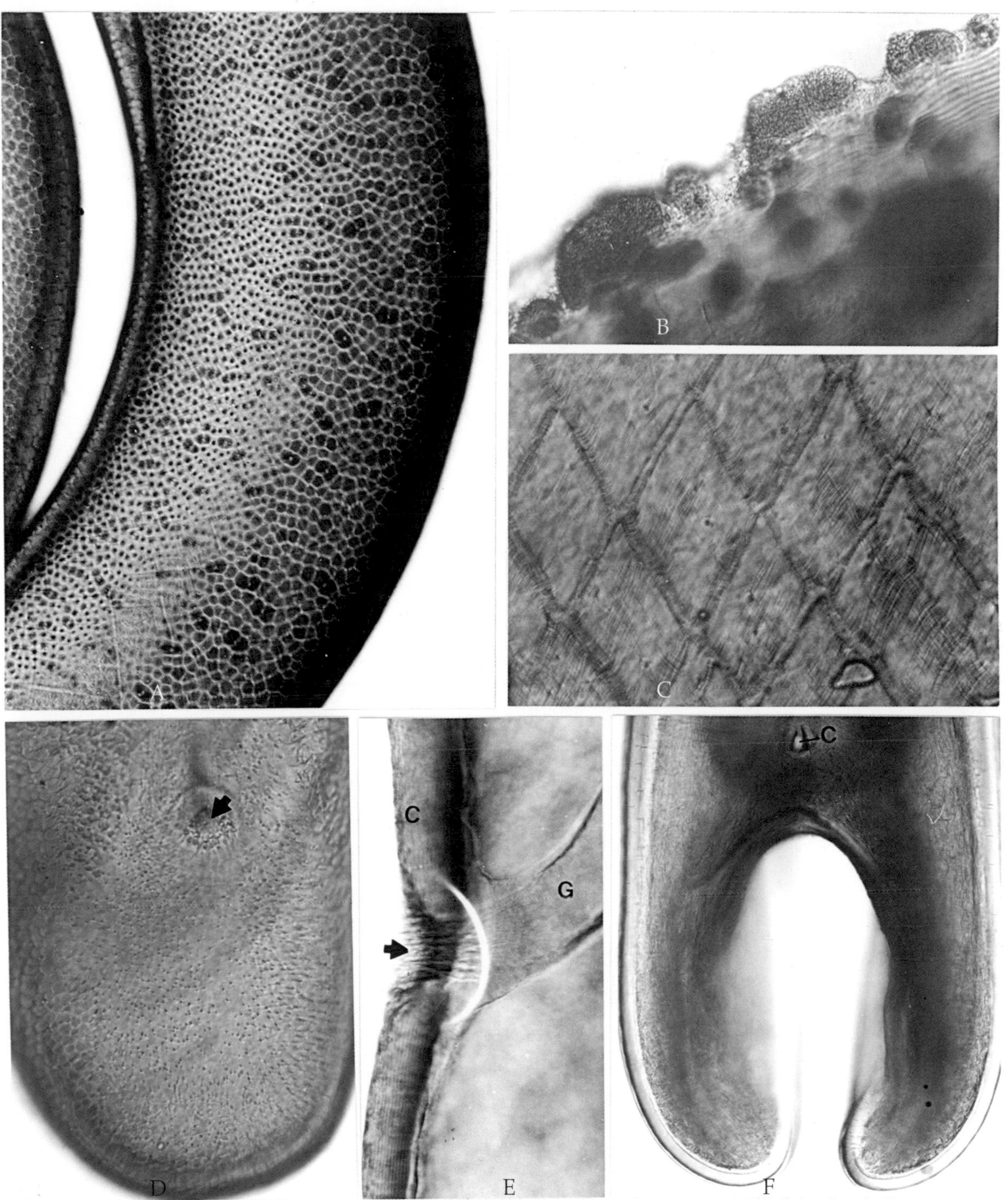

Surface morphological features of New Zealand horsehair worms.
A: Body surface of *Euchordodes nigromaculatus* showing several types of areoles, including the double black ones that give a pepper-like appearance to the worm. B: Raised, irregular areoles on body surface of *Gordionus diblastus*. C: Crossing rhomboidal lines and fine criss-cross fibres on the body wall of *Gordius dimorphus*. D: Ventral view of rounded simple tail of *Euchordodes nigromaculatus* (arrow = cloacal opening). E: Longitudinal section through cloacal area of *E. nigromaculatus* (C = cuticle; G = cloaca; arrow shows bristles surrounding cloacal opening. F: Lobed tail of *Gordius dimorphus* (C = cloacal opening).
George Poinar Jr

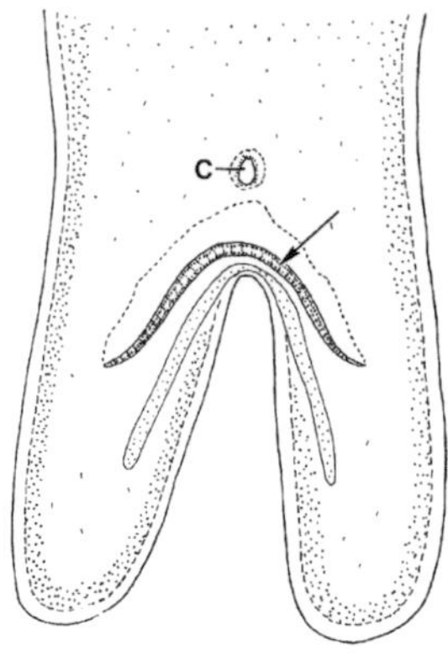

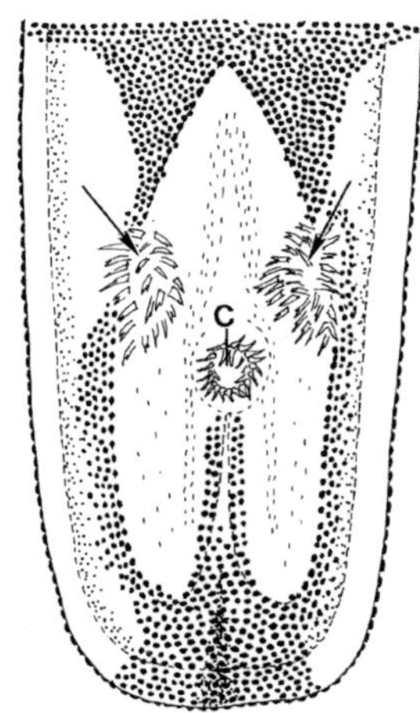

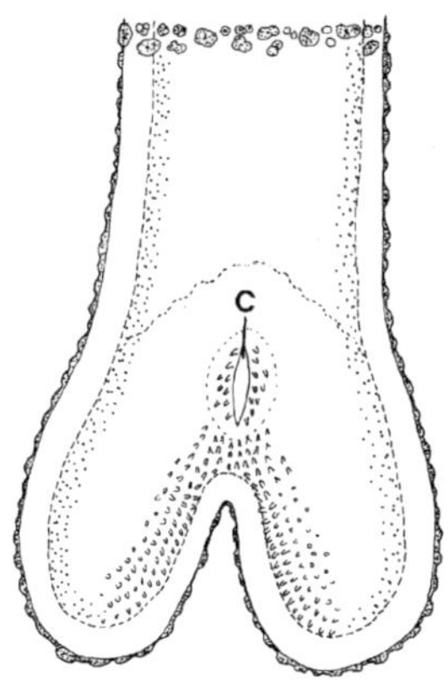

Ventral view of male tails of *Gordius dimorphus* (upper), *Euchordodes nigromaculatus* (middle), and *Gordionus diblastus* (lower). (C = cloacal opening; upper arrow = post-cloacal fold; middle arrows = patches of forked bristles).

From Poinar 1991

Since hairworms breed in water, the manner in which the pre-emergent forms manoeuvre their hosts to water sources is a striking survival adaptation. How this is accomplished is not known, but possibly at the final stages the parasite absorbs a considerable amount of water from its host, thus driving the latter to a water source. At any rate, the hairworm senses when the host is drinking water and then makes its exit, normally killing the host at the time of exit or shortly after. However, weta have been kept alive for several days after their hairworms emerged. Yet it is not uncommon to find the bodies of weta in the same water sources as their hairworm parasites.

The abundance of New Zealand hairworms offers a unique opportunity to study aspects of sexual attraction and mating behaviour. Observations on the adult stages of *G. dimorphus* have indicated that the hairworms form mating groups more or less at random (Poulin 1996). Field studies also suggested that the immediate microhabitat of male *E. nigromaculatus* can determine their probability of mating (Thomas et al. 1999).

Whereas most accounts of egg-laying by hairworms state that the eggs are deposited in strings, this is not always the case as hairworms have been seen depositing eggs in masses on the bottom of an aquarium. In such cases the eggs were attached to the surface by a sticky secretion. A unique egg-laying behaviour occurs in at least one New Zealand hairworm (probably *E. nigromaculatus*) in the mountainous streams of the South Island. The eggs are deposited as a flat sheet on the undersurface of rocks, being attached with a very durable waterproof deposit that makes them almost impossible to remove (Poinar 2000 pers. obs.). Such an adaptation would serve to keep the eggs from being washed downstream.

Insects are the normal developmental hosts of developing gordian worms. Thus far, all confirmed cases of developmental hosts for New Zealand hairworms are members of five orthopteroid families – the Anostostomatidae (ground and tree weta) (Poinar 1991), Rhaphidophoridae (cave weta) (Poulin 1996), Acrididae (grasshoppers) (Poulin 1996), and Blattidae and Blatellidae (cockroaches) (Zervos 1989). There are earlier New Zealand reports of hairworms from Rhaphidophoridae (Richards 1954) and carabid (tiger) beetles (Townsend 1970), but the specimens need to be re-examined to be certain of their identity. Mermithid nematodes are commonly confused with hairworms since they also are elongate worms that parasitise insects and other invertebrates. In fact, New Zealand is home to some endemic aquatic mermithids that occur in the same habitats as adult hairworms and parasitise sandflies (Simuliidae), grasshoppers, mantids, netwing midges (Blepharoceridae), and spiders (Poinar 1990, 1991; Poinar & Early 1990). A close examination of the body wall and anterior end of

Key to New Zealand adult gordian hairworms

1. Body surface smooth, lacking areoles; male tail bilobed 2

   Body surface not smooth, containing areoles (raised protuberances) that stand out against the lighter background (some magnification is helpful to see the areoles); tip of head (calotte) dark, or if white, then colour is a result of a gradual lightening of body colour and is not distinctly separated; male tail simple or lobed 3

2. Body colour uniformly brown (females) or dark with diffuse white spots (males); tip of head (calotte) with a white cap distinctly separated from the rest of the body; male with one post-cloacal fold and one post-cloacal cuticular strap .... *Gordius dimorphus*

   Body olive-brown with yellowish female tail; tip of head yellow-white; male with one post- and one precloacal fold .... *Gordius paranensis*

3. Body colour varies from brown to black with small flattened areoles; dark areoles cause a pepper-like spotting on the body (more noticeable in males); male tip simple or slightly knobbed with a short ventral groove .... *Euchordodes nigromaculatus*

   Body colour of various hues of brown, with larger, raised areoles ranging from yellow to brown, producing a rough speckled appearance; male tail distinctly bifid ... *Gordionus diblastus*

the adult worm can distinguish mermithids from hairworms. The unidentified hairworms from two endemic New Zealand cockroaches lacked conspicuous areoles but males had bilobed posterior ends (Zervos 1989).

Class Nectonematoidea

Marine hairworms of the genus *Nectonema* are an enigmatic group of worldwide occurrence but low diversity. Until recently, only four species were known, distributed from Norway, through Western European coasts to the Mediterranean and Black Sea, Eastern North America, and Indonesia. Unlike gordian hairworms, *Nectonema* adults have natatory bristles to aid in swimming in the pelagic coastal environment. They also have ventral and dorsal nerve cords, whereas gordian worms have a ventral cord only. Hosts for developmental stages include crabs and shrimps.

An endemic species, *Nectonema zealandica*, parasitises the purple rock crab *Hemigrapsus sexdentatus*. The discovery of this marine hairworm at a rocky intertidal beach in Canterbury constituted the first record from the South Pacific Ocean and the southernmost locality for the genus (Poinar & Brockerhoff 2001). Living juveniles and preadults of *N. zealandica* are white to cream in colour and soft-textured. Preadults attain 30 centimetres in length. Crabs collected between 1998 and 2000 showed that monthly parasitism rates vary from zero to 31.6%, with females having a generally higher rate of infection than adults. Encapsulated dead *Nectonema* worms in a number of crabs suggest that *Hemigrapsus sexdentatus* may not be the preferred host of the worm.

Nectonema zealandica coiled in the hepatopancreas of the rock crab *Hemigrapsus sexdentatus*.
Annette Brockerhoff

Conclusions

New Zealand serves as one of the few refuges for large populations of hairworms and offers a unique opportunity to study these interesting creatures. Maori recognised hairworms from weta as 'ngaio' (restless), which may reflect the continuous movements of adult hairworms in water (Williams 1975). While hairworms are probably the most important parasites of the now-protected weta, these parasitic associations have continued for millions of years. Unfortunately, human activities such as habitat destruction and the introduction of animals that prey on weta have reduced the populations of both host and parasite. The role that hairworms play in the environment is poorly understood. They do serve as food for fish in New Zealand, including trout (Stokell 1936; McLennan & MacMillan 1984), and are eaten by birds in other parts of the world (Poinar 2000). Their eggs may serve as food for a number of aquatic microarthropods. Although there is understandable concern for the fate of New Zealand's unique weta, the future of hairworms should be of equal concern and they deserve the same protection. Humans have a tendency to preserve those creatures that have attributes such as attractiveness, usefulness, or uniqueness. This doesn't usually include parasites. But creatures such as hairworms also deserve protection so that their roles in ecosystems can be fully elucidated, especially in the South Island where populations are still high. Already, horsehair abundance in the more populated North Island has appeared to decline, based on museum records (Poinar 1991). It is hoped that weta and hairworm populations can be maintained in forested areas of the South Island so both groups do not suffer the fate of other New Zealand creatures.

Author

Professor George O. Poinar Jr Department of Zoology, Oregon State University, Corvallis, Oregon 97331, USA [poinarg@science.oregonstate.edu]

References

BLAIR, D. 1983: Larval horsehairworms (Nematomorpha) from the tissues of native freshwater fish in New Zealand. *New Zealand journal of Zoology 10*: 341–344.

BULLER, W. L. 1867: Notes on the genus *Deinacridida* in New Zealand. *Zoologist 2*: 849–850.

McLENNAN, J. A.; MacMILLAN, B. W. H. 1984: The food of rainbow and brown trout in the Mohaka and other rivers of Hawke's Bay, New Zealand. *New Zealand Journal of Marine and Freshwater Research 18*: 143–158.

ORLEY, L. 1881: On hair-worms in the collection of the British Museum. *Annal and Magazine of Natural History, ser. 5*, 8: 325–332.

POINAR, G. O. Jr 1990: *Austromermis* n. gen. and *Blepharomermis* n. gen. (Mermithidae: Nematoda) from New Zealand Simuliidae and Blepharoceridae (Diptera). *Revue de Nematologie 13*: 395–402.

POINAR, G .O. Jr 1991: Hairworm (Nematomorpha: Gordioidea) parasites of New Zealand wetas (Orthoptera: Stenopelmatidae). *Canadian Journal of Zoology 69*: 1592–1599.

POINAR, G. O. Jr 1999a: Nematomorpha. *In: Nature Encyclopedia of Life Sciences*. Nature Publishing Group, London. [www.els.net]

POINAR, G. O. Jr 1999b: *Paleochordodes protus* n. g., n. sp. (Nematomorpha, Chordodidae), parasites of a fossil cockroach, with a critical examination of other fossil hairworms and helminths of extant cockroaches (Insecta: Blattaria). *Invertebrate Biology 118*: 109–115.

POINAR, G. O. Jr 2000: Nematoda and Nematomorpha. Pp. 255–280 *in*: Thorp, J. H.; Covich, A.P. (eds), *Ecology and Systematics of North American Fresh Water Invertebrates*. 2nd edn. Academic Press, NewYork.

POINAR, G. O. Jr; BROCKERHOFF, A. M. 2001: *Nectonema zealandica* n. sp. (Nematomorpha: Nectonematoidea) parasitising the purple rock crab *Hemigrapsus edwardsi* (Brachyura: Decapoda) in New Zealand, with notes on the prevalence of infection and host defence reactions. *Systematic Parasitology 50*: 149–157.

POINAR, G. O. Jr; EARLY, J. W. 1990: *Aranimermis giganteus* n. sp. (Mermithidae: Nematoda), a parasite of New Zealand mygalomorph spiders (Araneae: Arachnida). *Revue de Nematologie 13*: 403–410.

POULIN, R. 1995: Hairworms (Nematomorpha: Gordioidea) infecting New Zealand short-horned grasshoppers (Orthoptera: Acrididae). *Journal of Parasitology 81*: 121–122.

POULIN, R. 1996: Observations on the free-living adult stage of *Gordius dimorphus* (Nematomorpha: Gordioidea). *Journal of Parasitology 82*: 845–846.

RICHARDS, A. M. 1954: Notes on behaviour and parasitism in *Macropathus filifer* Walker, 1869. *Transactions of the Royal Society of New Zealand 82*: 821–822.

SCHMIDT-RHAESA, A.; THOMAS, F.; POULIN, R. 1998: Scanning electron microscopy and intraspecific variation in *Euchordodes nigromaculatus* from New Zealand. *Journal of Helminthology 72*: 65–70.

SCHMIDT-RHAESA, A.; THOMAS, F.; POULIN, R. 2000: Redescription of *Gordius paranensis* Camerano, 1892 (Nematomorpha), a species new for New Zealand. *Journal of Natural History 34*: 333–340.

STOKELL, G. 1936: The nematode parasites of Lake Ellesmere trout. *Transactions of the Royal Society of New Zealand 66*: 80–96.

TOWNSEND, J.I. 1970: Records of Gordian worms (Nematomorpha) from New Zealand Carabidae. *The New Zealand Entomologist 4*: 98–99.

THOMAS, F.; SCHMIDT-RHAESA, A.; POULIN, R. 1999: Microhabitat characteristics and reproductive status of male *Euchordodes nigromaculatus* (Nematomorpha). *Journal of Helminthology 73*: 91–93.

VOIGT, E. 1938: Ein fossiler Saitenwurm (*Gordius tenuifibrosus* n. sp.) aus der eozanen Braunkohle des Geiseltales. *Nova Acta Leopoldina 5*: 351–360.

WILLIAMS, H. W. 1975: *A Dictionary of the Maori Language*. 7th edn. Government Printer, Wellington.

XIANGUANG, H.; BERGSTROM, J. 1994: Paleoscolecid worms may be nematomorphs rather than annelids. *Lethaia 27*: 11–17.

ZERVOS, S. 1989: Stadial and seasonal occurrence of gregarines and nematomorphs in two New Zealand cockroaches. *New Zealand Journal of Zoology 16*: 143–146.

Checklist of New Zealand Nematomorpha

Habitat and status codes: E, endemic; F, freshwater; M, marine.

PHYLUM NEMATOMORPHA
Class GORDIOIDEA
Order CHORDODEA
CHORDODIDAE
Euchordodes nigromaculatus Poinar, 1991 F E
Gordionus diblastus (Orley, 1881) F E
(= *G. pachydermus* Orley, 1881)

Order GORDEA
GORDIIDAE
Gordius dimorphus Poinar, 1991 F E
Gordius paranensis Camerano, 1892 F

Class NECTONEMATOIDEA
Order NECTONEMATIDA
NECTONEMATIDAE
Nectonema zealandica Poinar & Brockerhoff, 2001 M E

Summary of New Zealand nematomorph diversity

Taxon	Described species	Known undes./undet. species	Estimated undisc. species	Endemic species	Endemic genera
Gordioidea	4	0	2	3	0
Nectonematoidea	1	0	0	1	0
Totals	5	0	2	4	0

FIFTEEN

ICHNOFOSSILS

(TRACE FOSSILS)

whispers of ancient life

MURRAY R. GREGORY

Trace fossils (ichnofossils), include a diversity of surface trails, tracks, and trackways as well as burrows, of both vertebrate and invertebrate origin in unconsolidated sedimentary substrata, together with borings made in hard substrata such as wood, shell, and rock, and also etchings and scrapings or other evidence of bioerosion. Trace fossils are sometimes known as 'biohieroglyphs' (e.g. van der Lingen 1969). Their classification and paleobiological significance lie outside the ambit of most systematists. Many workers place coprolites and faecal pellets and regurgitations in the trace-fossil category and there is also increasing acceptance of plant-root and other penetration structures (rhizoliths) while some would also include algal stromatolites (see Frey 1975; Ekdale et al. 1984; Bromley 1990). For many workers, but not all, 'lebensspuren' is often restricted to modern traces. Pseudofossils, structures of inorganic origin, are often and mistakenly identified as trace fossils to which they may bear superficial resemblances. Naturally, there has to be a formal definition – 'trace fossils are structures resulting from life activities of organisms', as in the proposed emendation for the fourth edition of the International Code of Zoological Nomenclature (ICZN) (see Bertling 1999). The published Code (ICZN 1999) states that a zoological ichnotaxon – the formal name attached to an animal trace fossil –'is based on the fossilized work of an organism, including fossilized trails, tracks or burrow (trace fossils) made by an animal.' The work of an animal is defined as 'the result of the activity of an animal (e.g. burrows, borings, galls, nests, worm tubes, cocoons, tracks).'

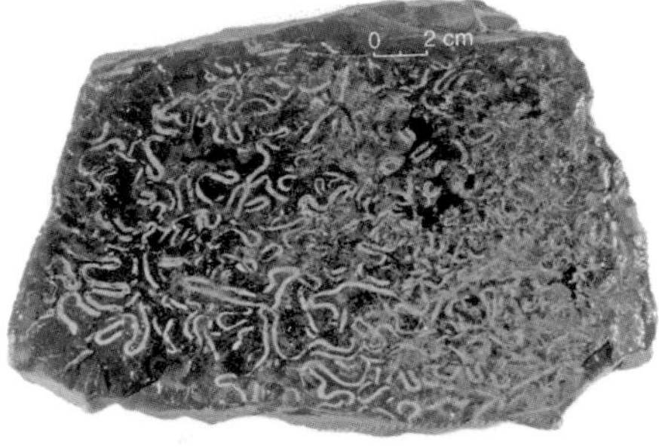

The trace fossil *Cosmorhaphe* in slatey argillite from the Tasman Glacier.

Murray Gregory

Over the past two centuries the history of trace-fossil study (ichnology) has been labyrinthine (e.g. Osgood 1975; Rindsberg 1990), with a specialist journal (*Ichnos*) devoted to the subdiscipline making an appearance in 1990. This historical development can be summarised and simplified as follows (after Osgood 1975):

- the age of fucoids – *Fucoides*, and use of the suffix *-phycus* (e.g. Brongniart 1823; Saporta 1884);
- the age of controversy – *Cruziana* and *Rusophycus* (trilobite trails) (e.g. Dawson 1864; Nathorst 1873; James 1894);
- the age of development – the German School and 'actuopalaeontology' (e.g. Richter 1927, 1941; Abel 1935);
- the modern age – contemporary thought (e.g. Häntzschel 1962, 1975; Seilacher 1964; Schafer 1972; Frey 1975);
- a golden age – expanding horizons and applications (e.g. Ekdale *et al*. 1984; *Ichnos* volume 1, 1990; Pickerill 1994; Pemberton & MacEachern 1995; Hasiotis 2002).

In the early 1800s, many trace fossils were considered to be faithful impressions of seaweeds, and at one time some 100 species of '*Fucoides*' had been recorded

(James 1894). Many of the ichnotaxa recognised today continue to carry the stigma of an 'algal' origin (e.g. *Algacites*, *Chondrites*, *Phycodes*, and use of the suffix *-phycus*). Osgood (1975) has recorded how the 'age of fucoides' gave way to an extended 'period of controversy' in the late 1800s, when several authors demonstrated animal-activity origins for many traces from crossings within some meandering trails and cutting of sedimentary structures such as laminations and cross-bedding. Through the first half of the 20th century, and with but few exceptions, ichnological developments lay with a German School that placed emphasis on 'actuopalaeontology' (or more correctly neoichnology) (e.g. Richter 1927, 1941; Abel 1935) and confirmed animal progenitors for most trace fossils. The fruits of these pioneering works later became available to English speakers through Häntzschel's (1962, 1975) *Treatise on Invertebrate Paleontology Part W* compilations, as well as Schafer (1972) and the innovative syntheses of Seilacher (e.g. 1967, 1992). Fucoid and fucoidal markings or impressions, generally with algal implications, had early recognition in New Zealand geology (e.g. Hochstetter 1864; Haast 1867) and continued wide local usage thereafter (e.g. Thomson 1920; Bartrum 1948; Eade 1966; Mossman & Force 1969) even though fossil 'worm trails, tracks and tubes' had also been noted in early reports (e.g. Haast 1867; Davis 1871; McKay 1888). Hochstetter's (1864) report of vermiform markings in the 'Maitai Slates', Nelson Province, is probably the first published and illustrated account of a trace fossil (?*Scalarituba*) in New Zealand, although early Maori also had an appreciation for some structures of this kind (Gregory & Lockley 2004). It should also be mentioned that Stevens (1968), in referring to the 'Amuri fucoid' (= *Zoophycos*), followed Plicka (1965) in suggesting (somewhat hesitantly) that it was the impression of a sabellid worm's tentacle fan – an interpretation quickly refuted by Webby (1969) and Lewis (1970), who both recognised that infaunal mining activities were involved. The origin of *Zoophycos* and identification of possible progenitors is a continuing challenge, and one yet to be fully resolved (e.g. Bradley 1973; Wetzel 1992).

By the late 20th century, ichnology was a recognised subdiscipline lying at the boundaries between sedimentology and paleontology. In marine and other sedimentary rocks, trace fossils provide important and additional insights for paleoenevironmental and paleoecological reconstructions because they are largely representative of a soft-bodied invertebrate biota that otherwise has little prospect of being preserved in the fossil record, and furthermore, post-mortem transport is highly improbable.

Ichnotaxonomy

For the purposes of comparison and communication, some kind of formalised hierarchical classification system for trace fossils is required. Unnamed or informally named trace fossils, e.g. Ballance's (1964) and Jones's (1969) use of the designation Types A, B, C, etc. for material in the Miocene Waitemata Group of the Auckland isthmus, are prone to escape notice in later publications even when attention is drawn to formally described and named ichnotaxa (Gregory 1969). For this reason, and because trace fossils are the preserved evidence of, or responses to, organisms' behavioural patterns, rather than skeletal remains or body impressions, a phylogenetic approach that follows traditional principles of evolution and common descent as used in the International Codes of Botanical and Zoological Nomenclature (ICBN and ICZN) is not a realistic option. Informal groups based on non-interpretative morphological criteria (e.g. circular, simple, branched, rosetted, spreiten, winding, spiral, meandering, winding and meandering with branches, networks – see Ksiazkiewicz 1977, p. 13) may seem simple and be easy to apply but have some disadvantages (Uchman 1995). This approach has not met with wide acceptance. From similar groupings, Lindholm (1987) developed a key that when used with the *Treatise of Invertebrate Paleontology* (Häntzschel 1975) could separate and identify over 130 ichnogenera. Although a system using Linnaean binominal nomenclature has been developed, it is independent of both ICBN and ICZN, even if now included in the latter (ICZN 1999). Bromley (1990, p. 143) recognised a 'Dark Age' in ichnotaxonomy that persisted from the 1930s to the 1990s, during which time nomenclatural order and austerity were maintained, with practitioners 'abiding by ICZN rules even though they were not bound by them.' While the ICZN rule of priority is generally followed, difficulties arise when 'old' names are resurrected from seldom cited and/or historically obscure publications, to replace those of long standing. Thus, Uchman's (1995) emendation of *Echinospira* for New Zealand examples of *Zoophycos* is unlikely to find acceptance.

Häntzschel (1975) presented a comprehensive historical account of approaches to trace-fossil classification and nomenclature. Approaches to, and difficulties encountered with, developing ichnotaxobases and in constructing an ichnotaxonomic code have been reviewed by Ekdale et al. (1984), Magwood (1992), and, in more detail, by Pickerill (1994), and it now has formal recognition and acceptance in the ICZN (see Rindsberg 1990).

Trace-fossil classification is developed around several descriptive (i.e. objective) and interpretive (i.e. subjective) factors. These are (summarised from Ekdale et al. 1984 and Magwood 1992):

1. formational – position of emplacement relative to stratal boundaries
2. preservational – toponomy, distinguish between full and partial relief
3. ethological – interpretation of behaviour that generated the trace
4. morphology – distinctive and/or characteristic features (shape, form, etc.)
5. environmental – depth, substrate and setting
6. taxonomy – knowledge of the presumed trace-maker.

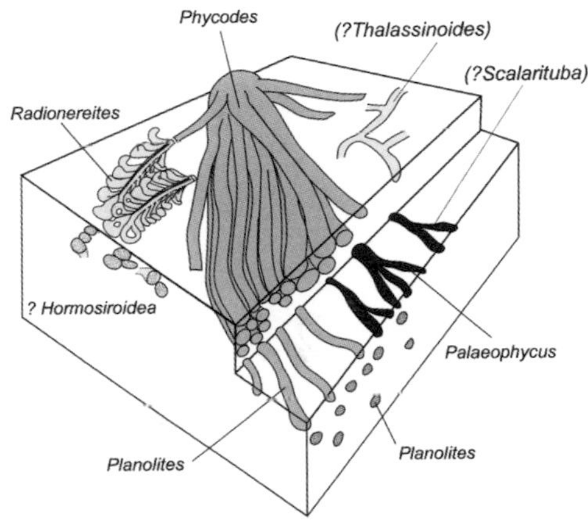

Schematic representation of spatial relationships within the *Radionereites* x *Phycodes* x *Planolites* x (?*Palaeophycus* x ?*Hormosiroidea*) complex.

After Gregory 1969 (and later unpublished field observations) and Bradley 1980, 1981

The basic unit in ichnotaxonomy is the ichnogenus (ichnogen. or igen.). Ichnospecies (ichnosp. or isp.) also has formal recognition. Ichnosubgenera and ichnosubspecies or varieties have limited acceptance. Higher categories (e.g. ichnofamily) have yet to be formalised, although at least 12 are considered valid under ICZN rules (Rindsberg pers. comm.). While it is accepted practice to erect an ichnogenus with a designated type species, this was not a formal requirement of the ICZN third edition (ICZN 1985; Bromley 1990) but became so in the fourth edition (ICZN 1999) for those erected after 1 January 2000. In many sedimentological and paleontological publications, identification of trace fossils has been at the ichnogenus level only, and no doubt this practice will continue. Curated holotypes are desirable, but soft crumbly sediments and large size often make this an impossibility – in these circumstances, recourse of necessity has to be made to iconotypes (photographs, line drawings, or field sketches) as is acceptable under ICBN (Silva 1993). It is also conventional to use the suffix *-ichnus* when erecting a new ichnogenus.

Seilacher (1964) established five ethological (or behavioural) categories – Domichnia (dwelling structures), Repichnia (crawling or locomotion traces), Cubichnia (resting traces and impressions), Pascichnia (meandering grazing trails), and Fodinichnia (feeding burrow systems). Added subsequently are Agrichnia (complex 'gardening' traces and traps), Fugichnia (escape structures), Praedichnia (predation traces), and Equilibrichnia (equilibrium traces). Some authorities have incorrectly given these categories the status of (ichno-)order or family. For instance, Fleming (1973) in his description of 'fossil cufflinks' (*Diplocraterion morgani*) from Miocene mudstones of the Blue Bottom Formation in Westland, used Domichnia in the sense of an order and Rhizocorallidae as an (ichno)family – the latter usage may ultimately be considered valid.

It is important to note that changes in an individual organism's behaviour, and/or the character of the substratum in which it is active, can give rise to distinctly different trace-fossil structures. On the other hand, biologically unrelated organisms behaving in the same way may produce identical structures. Biological identification of trace makers (progenitors) is fraught with uncertainty (see Bromley 1990) and it led Bradley (1981), who worked with New Zealand ichnofossils, to suggest that the trace itself could be a neutral and unbiased indicator of its maker (e.g. *Chondrites* – chondritor, *Phycodes* – phycodor, and *Radionereites* – radionereitor). This approach has not found favour with ichnologists. Presence of a body fossil, or a part (such as a crustacean chela) in a once-open burrow system is not necessarily evidence of an original inhabitant or any causal relationship.

Other difficulties in trace-fossil classification arise with complex or compound structures in which there are sharp-to-intergradational changes between two or more otherwise separately identifiable and discrete ichnotaxa. They can be considered integral parts of a single system (Pickerill 1994). New Zealand examples include variation of structures produced by irregular echinoids and the intimate (and causal) association of *Cardioichnus* (the 'bow tie' trace), *Imbrichnus*, and *Scolicia*, illustrated by Gregory (1985a), and those between *Scolicia* and *Phycodes* and between *Radionereites*, *Phycodes*, and *Chondrites* figured by Bradley (1980, 1981). In circumstances such as these it may be as inappropriate to name a complex structure after an identifiable or discrete part that has formal priority, but which is a seldom encountered or an otherwise minor component, as it is to use a later-established ichnotaxon that dominates (cf. Pickerill 1994). A practical approach to this dilemma could be to use the nomenclature presented in the captions to the adjacent figures. In some instances it may be more appropriate to unite a trace-fossil complex under a single ichnotaxon – the recently identified *Hillichnus lobosensis* produced by tellinacean bivalves, and in which at least nine discrete ichnotaxa elements have been recognised, is just such an example (Bromley et al. 2003).

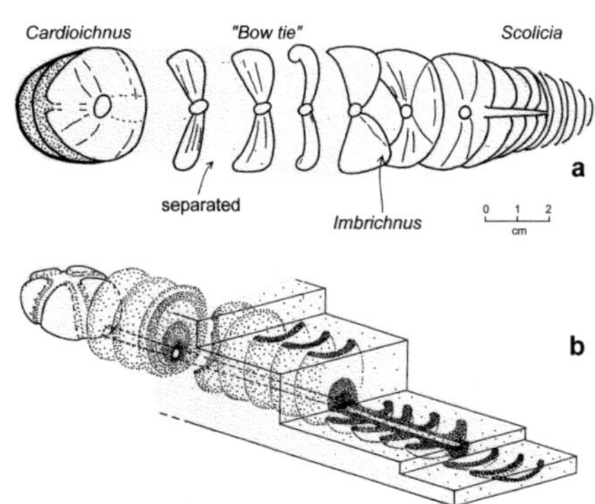

Schematic reconstructions of the compound trace fossil *Cardioichnus* 'bow tie' x *Imbrichnus* x *Scolicia*. The entire structure reflects movement of an irregular echinoid (sea urchin) through soft sediment and the identified ichnogenus is an expression of the level to which it has been exposed or etched by erosional and weathering processes.

After Gregory 1985a

In ichnotaxonomy, fossil borings create some special problems. With a few exceptions they are generally identified by comparison with equivalent modern structures and their makers. Thus clionaid (i.e. sponge) and pholadid borings are often recorded in fossil lists. These borings should be referred to the appropriate ichnotaxa (e.g. *Entobia* and *Gastrochaenolites*, respectively).

The status of root and other plant-produced traces is the subject of debate. Some workers consider they are adequately covered in the ICZN while others (including paleobotanists) are firmly of the opinion that they are covered by the ICBN. These nomenclatural difficulties would not have arisen if a separate ichnological code, independent of both the ICZN and ICBN had been adopted (see Sarjeant & Kennedy 1973; Basan 1979).

Using trace fossils

While the common trace fossils typically have long time ranges and are hence of limited use in stratigraphic correlation, they have become important tools in paleoecological and paleoenvironmental reconstructions. Seilacher (1964, 1967) introduced the concept of five universally recurring, soft-ground trace-fossil associations (ichnofacies) whose spatial and temporal distributions were considered to be depth-controlled. Since that time, further archetypal ichnofacies have been recognised, although it is now appreciated that substratum character is of more importance than depth, and the use of these features as relative paleobathometers needs to be tempered with caution (Ekdale et al. 1984; Pemberton et al. 1994). Other well-established attributes of trace fossils include: presence in otherwise unfossiliferous strata; improbability of secondary displacement; and evidence for a soft-bodied biota that otherwise escapes preservation. Trace-fossil studies have become integral to the recognition of omission surfaces and gaps in the stratigraphic record, and, through composite ichnofabrics, help identify multiple lithification episodes (Lewis 1992; Lewis & Ekdale 1992), event beds such as tempestites (e.g. Pemberton et al. 1992b; Pemberton & MacEachern 1997), and developments in sequence stratigraphy (e.g. Pemberton et al. 1994). The utility of trace fossils in reconstructing redox-

Woodground	Hardground	Firmground	Softground		
			Scoyenia	Freshwater	
Teredolites	*Trypanites*	*Glossifungites*	*Skolithos*	High energy	
		(*Psilonichnus*)			
			Cruziana	Medium energy	Marine
?	?	?	*Zoophycos*		
			Nereites	Low energy	

The currently recognised and commonly occurring ichnofacies. Softground ichnofacies are differentiated on environmental factors such as depth and hydrodynamic energy levels. Hard-, firm-, and woodground ichnofacies are distinguished on substratum type and await separation into characteristic trace-fossil assemblages.

After Ekdale et al. 1984 and Pemberton 1992a

related benthic events and paleo-oxygenation histories is now widely appreciated (e.g. Savrda et al. 1991; Savrda 1992). Trace fossils, through ichnofabric indices, also lend themselves to semiquantitative investigations in stratigraphy (e.g. Droser & Bottjer 1986).

The most commonly recorded trace fossils are the burrows, tracks, and trails of marine, softground, epi- and infaunal invertebrate benthos. While the *Scoyenia* ichnofacies is considered indicative of a marginal marine realm, there is growing appreciation of what continental (terrestrial, fluvial, lacustrine, and eolian settings) ichnology has to offer (e.g. Hasiotis & Bown 1992; Hasiotis 2002). While trackways of large vertebrates (e.g. dinosaurs) will continue to attract popular acclaim, the important advances will come only when synthesis of a suite of terrestrial ichnofacies, comparable in stature to those presently recognised in marine substrata, has been accomplished. Two examples illustrating the possibilities in paleoenvironmental studies come from Holocene floodplain deposits (Ratcliffe & Fagerstrom 1980) and Triassic paleosols (Retallack 1984).

The New Zealand ichnofossil record

Only 100 or so of an estimated >970 described and named ichnogenera globally, of which 300–400 may possibly be valid (Rindsberg pers. comm.), can be considered common. Of the >100 ichnotaxa that have been recorded from New Zealand strata (see end-chapter Checklist), fewer than 20 ichnogenera are widely represented.

Cullen's (1967) assertion of a paucity of trace fossils in New Zealand strata was premature. As noted previously, several earlier studies recorded 'fucoides' (e.g. Thomson 1920; Bartrum 1948) and animal traces or 'worm' burrows (e.g. Haast 1867; Davies 1871; McKay 1888). Later studies have revealed a relatively rich and diverse record, particularly in Miocene and other weak 'papa' lithologies (e.g. Glennie 1959; Ghent & Henderson 1966; Gregory 1969; Nodder et al. 1990; Manley & Lewis 1998) as well as in poorly consolidated Holocene sediments (e.g. Ekdale & Lewis 1991). Ichnogenera commonly encountered in New Zealand Tertiary sediments include *Anconichnus, Chondrites* (in several size categories), *Ophiomorpha, Palaeophycus, Paleodictyon, Planolites, Phycodes, Scolicia, Thalassinoides, Tigillites,* and *Zoophycos.* Cullen (1967) reported an enteropneust (acorn worm) faecal cast from mid-Tertiary rock near Castlepoint, North Island, noting its close similarity to trails made by living, unnamed deep-sea worms. The originators of these living trails were recently identified as a new family of deep-sea enteropneust – Torquaratoridae (Holland et al. 2005). The ichnogenera

The trace fossil *Paleodictyon* from a shore platform on the Mataikona coastline, southern North Island (Early Miocene).

Marianna Terezow, GNS

Spirodesmos and *Spiroraphe* may represent fossil torquaratorid traces. Deep-water flysch-like sequences of alternating sandstones and mudstones are commonly characterised by the *Nereites* ichnofacies, and *Zoophycos* dominates many marls and deposits of slope and outer-shelf environments (e.g. Lewis 1970). A *Scolicia* ichnofabric is well developed in many of New Zealand's Oligocene cool-temperate shelf limestones. Trace fossils may be sharply colour-defined in ash-rich sediments such as the 'hieroglyph' (*Teichichnus*-bearing) units of the late Miocene and early Pliocene shelf-to-slope deposits near East Cape (see Ballance et al. 1984). Application of the ICD/BI (ichnofabric constituent diagram/ bioturbation index) approach (see Taylor & Goldring 1993; Taylor et al. 2003) has expanded understanding of paleoenvironmental changes and furthered identification of complex event stratigraphy in this section (Tonkin 2002; Tonkin et al. 2004).

Non-diagnostic burrows and/or traces are not uncommon in local lower Paleozoic strata. The earliest trace fossils in the New Zealand record are reasonably abundant *Skolithos* and rare *Rusophycus* in the Murray Creek Formation (Devonian) at Reefton (Bradshaw 1999). In rocks of similar age near Baton River, escape burrows (fugichnia) and *Spirophyton* have been noted (Bradshaw pers. comm.), while at nearby Upper Takaka Hickey (1986) has recognised *Planolites, Rhizocorallium,* and *Zoophycos.* Metasediments of the Greenhills Group (Permian) near Bluff have yielded a diverse and well-preserved soup- or softground ichnobiota dominated by *Scalarituba* and *Chondrites* – Mossman and Force (1969) considered the latter to be fucoidal (i.e. an algal imprint). The locality is important because of the presence of *Lophoctenium* and also for the excellent evidence of tiering (to depths exceeding 25 centimetres) revealed by the ichnofauna (Gregory & Campbell 1988). An abundance of 'worm tubes' and 'animal trails' (as well as lebensspuren) have been recorded and illustrated from the Tramway Sandstone (Permian) of Nelson (Waterhouse 1964).

Knowledge of ichnotaxa in local Mesozoic rocks is limited, although attention is often drawn to (probable) bioturbation phenomena (e.g. Hudson 1999). These effects may be particularly evident in silty and fine sandy lithologies, where intense burrow mottling with few recognisable ichnotaxa may pass laterally and vertically into completely homogeneous packages of sediment. When identifiable, smaller traces can generally be ascribed to *Chondrites* (e.g. Hudson 1999) and perhaps *Anconichnus* and *Rorschachichnus* (unpublished) and larger ones to *Phycodes, Planolites,* and *Tigillites.* Isolated examples of *Zoophycos* are known from the Torlesse Supergroup of North Canterbury (Fordyce 1976) and Murihiku Supergroup of South Otago (Cave 1982) and South Auckland (unpublished), but, as noted previously, this ichnotaxon is more characteristic of Tertiary marls. Paired burrows have been recorded on several occasions. This led Spörli and Grant-Mackie (1976) to identify *Rhizocorallium* in the Waipapa Supergroup of North Auckland and suggest a shallow-water marine depositional setting for these strata. In the absence of spreiten and any U-shaped connection between the tubes, Ballance (1976) questioned both this ichnogenus designation and the environmental interpretation. However, the unidentified fossil burrow illustrated by Marwick (1946, plate 16, fig. 1) from Murihiku argillites at Kiritehere, South Auckland, is an excellent example of *Rhizocorallium* – the U-shaped tube with spreiten between is clearly evident. Slatey argillite found as float in moraines of both the Tasman and Fox Glaciers often have surfaces extensively covered with several representatives of the deep-water *Nereites* ichnofacies, including spectacular *Helminthoida, Helminthopsis,* and *Spirorhaphe* as well as *Chondrites* and ?*Halimedides*.

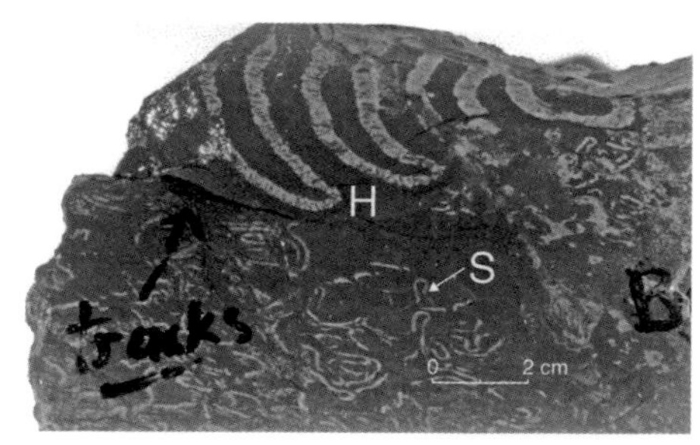

The trace fossils *Helminthoida* (H) and *Helminthopsis* (S) in slatey argillite from Franz Josef Glacier.
Murray Gregory

Non-diagnostic faecal pellets, small in size and generally considered to be of invertebrate origin, often receive passing mention, or are recorded in faunal lists, e.g. Hudson's (1999) account of Murihiku Supergroup (Middle Jurassic) strata from both North and South Islands. Examples of more substance are sharply defined concentrations and felted masses of *Inoceramus* prisms in the Cretaceous of Raukumara Peninsula that Speden (1971) considered to be coprolites (and/or regurgitations) indicative of predation by marine reptiles and fish. There are also several accounts of Quaternary bird-regurgitated pellets as well as gastrolith accumulations (e.g. Worthy & Holdaway 1994; Twigg 2001).

Borings in shelly fossils and other hard or firm substrata are widely reported from throughout the local stratigraphic record, but rarely are they given any formal ichnotaxonomic designation. An exception was Bradshaw's (1980) redefinition of the ichnogenus *Teredolites* for a Cenozoic bivalve boring better attributed to *Gastrochaenolites.* Identification of the *Trypanites* ichnofacies assisted Lewis (1992) in the recognition of hardground omission surfaces in the Amuri Limestone (mid-Oligocene) of North Canterbury. Borings are often identified through comparisons with equivalent structures produced by modern bioeroders – see examples of contemporary sponge, fungal, algal and prokaryote attack on Hauraki Gulf materials described and illustrated by Smith (1992). Exceptions to use of 'clionid' (i.e. clionaid) and 'pholadid' borings or 'bryozoan' etchings, etc. are Bradshaw's (1999) recognition of *Clionolithes priscus* for sponge borings in the Reefton Group (Devonian), and illustrations by Taylor et al. (1999) of the new ichnogenus *Leptichnus* for bryozoan etchings in carbonate substrates from the Pliocene of Chatham Island and Wanganui, as well as Early Oligocene, Middle Miocene, and Pleistocene strata elsewhere. It is strongly recommended that, in future, a formal ichnotaxonomic designation is used for borings, rather than identifications based on extant taxa – thus *Rogerella* for barnacle etchings, *Entobia* for clionaid sponge borings, etc. (see Checklist). Attachment scars, e.g. *Capulus* on *Pecten* from the Castlecliffian Te Piki Bed (Grant-Mackie & Chapman-Smith 1971), are probably commoner than generally credited. It is also appropriate to mention the significant role that bioerosion by rock browsers (e.g. *Chiton pelliserpentis* and *Lunella smaragda*) and rock borers (e.g. *Anchomasa similis, Evechinus chloroticus, Sphaeroma quoyanum,* and *Cliona celata*) have to play in the sculpturing of weak (papa) rocky-shore platforms and calcareous substrata (e.g. Healy 1968; Wilkie 1998).

Trace fossils from terrestrial environments are poorly represented in New Zealand. The ichnogenus *Cochlichnus,* thought to have been made by a nematode, has been identified in the freshwater Ohika Formation (Early Cretaceous) in the Buller Gorge (Fordyce 1980). A structure (?cf. *Imbrichnus*) considered to reflect activities of the freshwater bivalve *Hyridella* is known from the Gore Lignite Measures (Miocene) (Lindqvist 1987) and similar traces are associated with Holocene tephra in a lacustrine setting in Poverty Bay. There are a number of illustrated accounts of moa footprints and trackways (e.g. Gillies 1872; Williams 1872; Owen 1879; Voy 1880; Wilson & Benham 1913). Following Hitchcock (1839), Owen (1879) called moa footprints *Ornithichnites,* an unfortunate name, as later it was shown that the trails described by Hitchcock were made by dinosaurs. Today it is more appropriate to use the term avian ichnite. Most moa ichnites appear to be of Holocene or sub-Recent age (Hill 1914). Many specimens held in museum collections appear to be preserved in tephric deposits (unpublished). In the absence of photographic evidence or a specimen, Collen and Vella's (1984) record of a moa footprint in middle to late Pleistocene sediments of the Wairarapa must be considered suspect. Their description is strongly suggestive of the ichnogenus *Asterosoma,* which has a probable crustacean origin (Gregory 1985b). Horn (1989) has described and figured well-beaten, broad dish-shaped pathways on farmland near Poukawa, Hawke's Bay, that are crossed by contour-hugging, modern sheep and cattle tracks. These are considered to be relict, having been made by moa and predating the forest clearances that accompanied early pakeha settlement. Similar pathways at Paremoremo near Auckland, and also rising obliquely towards ridge crests, were known to local Maori as *ara-moa* (Graham 1919). Hutton's (1899) mention of a kiwi-like footprint in a Tertiary sandstone from the Pelorus Sound area requires confirmation. Fossil trackways or solitary footprints ('*Rhynchocephalichnus*') of the tuatara await discovery. Human footprints, including a trackway of at least seven impressions, and dog prints with evidence that it was drinking at a puddle, are known from c. 600-year-old Rangitoto ash on nearby Motutapu Island at the entrance to the Waitemata Harbour (Nichol 1981, 1982).

Plant stems (often preserved as rhizoliths) and the molds and casts of root systems are known from some Quaternary deposits of northern New Zealand and elsewhere (Gregory & Campbell 2003; Gregory et al. 2004). They are typically associated with aeolian dune systems and paleosols and in some instances give rise to a phytoturbate texture with vaguely meandroid patterned traces resembling *Thalassinoides* (and *Planolites*?), as well as a vertical fabric suggestive of *Skolithos,* and one that could easily be confused with a marine soft-ground ichnofacies (Gregory & Campbell 2000). As an example of the interpretational dilemmas one may encounter, Gregory and Campbell (2003), in identifying the rooting system of a nikau palm as a '*Phoebichnus* look-alike', were noting its close pattern similarity to the outer-shelf marine ichnogenus *Phoebichnus.* Non-descript small burrows with structureless fill typically paler in colour than the hosting brownish paleosols are not uncommon in these deposits. Similar structures have been noted in paleosols buried by tephra. Large borings in fossil driftwood from Eocene and Holocene estuarine and inner-shelf sediments are similar to those made by the modern shipworm *Teredo* (e.g. Lindqvist 1986) and are often mistakenly referred to that taxon – they should be placed in the ichnogenus *Teredolites.* Other recorded woodground (or xylic substratum) borings include complex termite galleries in silicified Miocene mangrove logs (Sutherland 1985, 2003). Silicified, hexagonal faecal pellets are stuffed in some of these galleries.

Because of their 'worm-like' character, the enigmatic but well known tubular and agglutinated local fossils *Terebellina* (= *Torlessia*) (see Begg et al. 1983) and the larger *Titahia* (Webby 1958) have often been considered annelids (e.g. Webby 1967; Fleming 1971). While it is now widely accepted that many of the former should be referred to the large agglutinated foraminiferid genus *Bathysiphon* (see Moore 1987; Miller 1995), confusion in the designation persists. Some materials, including New Zealand specimens, are more suitably accommodated in the ichnogenera *Schaubcylindrichnus* and/or possibly *Palaeophycus*.

While modern structures (lebensspuren) are not generally considered true trace fossils, Morton and Miller's (1973) illustrations of in- and epifaunal activities in local tidal flats can be the basis for any student interested in the neoichnology of these environments (e.g. Zuraida 2002 Whangateau estuary). However, there are several local examples that warrant inclusion in this review. Ekdale and Lewis (1993) described modern sabellariid reefs in Ruby Bay, near Motueka, and noted that they were close analogues to fossil *Skolithos* 'pipe rock'. Similarly Gregory (1991) noted the nearly identical characteristics of modern ray-feeding depressions and a large Miocene structure now known as *Piscichnus waitemata.* Horse hoof-prints in modern West Coast beach sands have given some insight into the development of heavy-mineral-rich laminae (van der Lingen & Andrews 1969). Unusual paddling trails made by the black-backed gull (*Larus dominicanus*) in high-intertidal, estuarine and beach, well-sorted fine sands from Northland and described by Gregory (1999) have preservational potential, perhaps a little less likely than that of ray-feeding depressions (cf. Gregory et al. 1979; Gregory 1991). Gregory (1994) also saw some similarities between graphoglyptid trace fossils and the tailings (*Aureffodioichnus exsculptum*) of modern placer gold dredges.

Scope for future work

It is unlikely that the coming years will see great numbers of ichnotaxa newly erected or identified from New Zealand material. However, the potential value of trace fossils will find expression in broadening the bases upon which paleoecological and paleoenvironmental reconstructions are made. They will continue to be an important factor in those sedimentary sequences with an otherwise poor or non-existent fossil record. The study of trace fossils in local terrestrial settings may well be rewarding.

In many contemporary New Zealand stratigraphic and regional geological studies, attention is drawn to varying degrees of bioturbation, without or with minimal comment on identifiable ichnotaxa (e.g. King et al. 1993). Nevertheless trace fossils, and in particular recognition of the *Glossifungites* ichnofacies, have been important factors in deciphering sequence boundaries of Neogene Wanganui Basin strata (e.g. Abbott & Carter 1997; Abbott 1998) and allostratigraphy of the Great South Basin (McMillan & Wilson 1997). Variations in bioturbation intensity (Jin & Lin 1991) and semiquantitave investigations of stratigraphy using the ichnofabric indices of Droser and Bottjer (1986, 1993), as well as the tiering concept to establish paleo-oxygenation histories (e.g. Manley & Lewis 1998), seem to have found limited acceptance with local hydrocarbon explorers. These are all aspects of ichnology with important practical applications and ones in which future developments and advances can be predicted with some confidence. Examples of the possibilities are to be found in McIlroy (2004).

Author

Dr Murray R. Gregory Geology Department, University of Auckland, Private Bag 92-019, Auckland, New Zealand [m.gregory@auckland.ac.nz]

References

ABBOTT, S. T. 1998: Transgressive system tracts and onlap shellbeds from Mid-Pleistocene sequences, Wanganui Basin, New Zealand. *Journal of Sedimentary Research, B, 68*: 253–268.

ABBOTT, S. T.; CARTER, R. M. 1997: Macrofossil associations from Mid-Pleistocene cyclothems, Castlecliff Section, New Zealand: implications for sequence stratigraphy. *Palaios 12*: 88–210.

ABEL, O. 1935: *Vorzeitliche Lebensspuren*. Gustav Fischer, Jena. 644 p.

ADAMS, A. G. 1985: Late Cretaceous fauna and sediments of southern Hawkes Bay – New Zealand. Unpublished PhD thesis, University of Auckland. 214 p.

BALLANCE, P. F. 1964: The sedimentology of the Waitemata Group in the Takapuna section, New Zealand. *New Zealand Journal of Geology and Geophysics 7*: 466–499.

BALLANCE, P. F. 1976: Tawharanui fossils and depth of deposition of Torlesse sediments (comment). *New Zealand Journal of Geology and Geophysics 19*: 949–953.

BALLANCE, P. F.; GREGORY, M. R.; GIBSON, G. W.; CHAPRONIÈRE, G. C. H.; KADAR, A. P.; SAMAESHIMA, T. 1984: A late Miocene and early Pliocene upper slope-to-shelf sequence of calcareous fine sediment from the Pacific margin of New Zealand. Pp. 331–342 *in*: Stow, D. A. V.; Piper, D. J. W. (eds), *Fine-grained Sediments – Deep water Processes and Facies*. The Geological Society, London.

BARTRUM, J. A. 1948: Two undetermined New Zealand Tertiary fossils. *Journal of Paleontology 22*: 448–489.

BASAN, P. B. 1979: Trace fossil nomenclature: the developing picture. *Palaeogeography, Palaeoclimatology, Palaeoecology 28*: 143–167.

BEGG, J. G.; CAVE, M. P.; CAMPBELL, J. D. 1983: *Terebellina mackayi* Bather in Oretian Murihiku rocks, Wairaki Hills, Southland (note). *New Zealand Journal of Geology and Geophysics 26*: 121–122.

BERTLING, M. 1999: What's hot in Ichnofossils! *Priscum 9*: 9–10.

BLOM, W. M. 1982: Sedimentology of the Tokomaru Formation, Waipapu Subdivision, Raukumara Peninsula. Unpublished MSc thesis, University of Auckland. 157 p.

BRADLEY, J. 1973: *Zoophycos* and *Umbellula* (Pennatulacea): their synthesis and identity. *Palaeogeography, Palaeoclimatology, Palaeoecology 13*: 103–108.

BRADLEY, J. 1980: *Scolicia* and *Phycodes*, trace fossils of *Renilla* (Pennatulacea). *Pacific Geology 14*: 73–86.

BRADLEY, J. 1981: *Radionereites*, *Chondrites* and *Phycodes*: trace fossils of anthoptiloid sea pens. *Pacific Geology 15*: 1–16.

BRADSHAW, M. A. 1980: Boring bivalves in the New Zealand Cenozoic with a redefinition of *Teredolites*. *Records of the Canterbury Museum 9*: 289–293.

BRADSHAW, M. A. 1999: Lower Devonian bivalves from the Reefton Group, New Zealand. *Association of Australasian Palaeontologists Memoir 20*: 1–171.

BRONGNIART, A. T. 1823: Observations sur les Fucoïdes. *Mémoire de la Société d'Histoire Naturelle 1*: 301–320.

BROMLEY, R.G. 1990: *Trace Fossils: Biology and Taphonomy*. Unwin Hyman, London. 280 p.

BROMLEY, R. G.; UCHMAN, A.; GREGORY, M. R.; MARTIN, A. J. 2003: *Hillichnus lobosensis* igen. et isp nov., a complex trace fossil produced by tellinacean bivalves, Paleocene, Monterey, California, USA. *Palaeogeography, Palaeoclimatology, Palaeoecology 192*: 157–186.

CAMPBELL, K. A.; GRANT-MACKIE, J. A.; BUCKERIDGE, J. S.; HUDSON, N.; ALFARO, A. C.; HOVERD, J.; MORGAN, S.; HORNE, N.; BANFIELD, A. 2004: Paleoecology of an early Miocene, rapidly submerged rocky shore, Motuketekete Island, Hauraki Gulf, New Zealand. *New Zealand Journal of Geology and Geophysics 47*: 731–748.

CARTER, R. M.; LINDQVIST, J. K. 1977: Balleny Group, Chalky Island, southern New Zealand: an inferred Oligocene submarine canyon and fan complex. *Pacific Geology 12*: 1–46.

CAVE, M. P. 1982: Occurrence of *Zoophycos* in Oretian rocks of the Murihuku Supergroup, South Otago, New Zealand. *New Zealand Journal of Geology and Geophysics 25*: 367–369.

COLLEN, J. D.; VELLA, P. 1984: Hautotara, Te Muna and Ahiaruhe Formations, middle to late Pleistocene, Wairarapa, New Zealand. *Journal of the Royal Society of New Zealand 14*: 297–317.

CULLEN, D. J. 1967: Ecological implications of possible enteropneust faecal casts in Tertiary deposits near Castlepoint, New Zealand. *New Zealand Journal of Marine and Freshwater Research 1*: 283–290.

DAVIS, E. H. 1871: On the geology of certain districts of the Nelson Province. *Reports of Geological Exploration during 1870–1*: 103–135.

DAWSON, J. W. 1864: On the fossils of the genus *Rusophycus*. *Canadian Naturalist and Geologist 1*: 363–367.

DOUGLAS, B. J.; LINDQVIST, J. K. 1987: Late Cretaceous–Paleocene fluvial and shallow marine deposits, Kaitangata Coalfield: Taratu and Wangaloa Formations. *In*: One day trips. *Geological Society of New Zealand, Miscellaneous Publication 37B*: 29–51.

DROSER, M. L.; BOTTJER, D. J. 1986: A semi-=quantitative field classification of ichnofabric. *Journal of Sedimentary Petrology 56*: 558–559.

DROSER, M. L.; BOTTJER, D. J. 1993: Trends and patterns of Phanerozoic ichnofabrics. *Annual Review of Earth and Planetary Sciences 21*: 205–225.

EADE, J.V. 1966: Stratigraphy and structure of the Mount Adams area, eastern Wairarapa. *Transactions of the Royal Society of New Zealand, Geology 4*: 103–117.

EKDALE, A. A.; BROMLEY, R. G.; PEMBERTON, S. G. 1984: *Ichnology: the Use of Trace Fossils in Sedimentology and Stratigraphy. [SEPM Short Course No. 15.]* Society of Economic Paleontologists and Mineralogists. Tulsa. 317 p.

EKDALE, A. A.; LEWIS, D. W. 1991: Trace fossils and paleoenvironmental control of ichnofacies in a late Quaternary gravel and loess fan delta complex, New Zealand. *Palaeogeography, Palaeoclimatology, Palaeoecology 81*: 253–279.

EKDALE, A. A.; LEWIS, D. W. 1993: Sabellarid reefs in Ruby Bay, New Zealand: a modern analogue of *Skolithos* 'pipe rock' that is *not* produced by burrowing activity. *Palaios 8*: 614–620.

FELDMAN, R. M.; MAXWELL, P. A. 1990: Late Eocene decapod Crustacea from north Westland, South Island, New Zealand. *Journal of Paleontology 64*: 779–797.

FLEMING, C. A. 1971: A preliminary list of New Zealand fossil polychaetes. *New Zealand Journal of Geology and Geophysics 14*: 742–750.

FLEMING, C. A. 1973: 'Fossil cuff links': a new Miocene trace fossil of the genus *Diplocriterion* from New Zealand. *Tohoku University, Science Reports, 2nd ser. (Geology), Special Volume no. 6 (Hatai Memorial Volume)*: 415–418.

FLORES, R. M.; SYKES, R. 1996: Depositional controls on coal distribution and quality in the Eocene Brunner Coal Measures, Buller Coalfield, South Island, New Zealand. *International Journal of Coal Geology 29*: 291–336.

FORDYCE, R. E. 1976: *Zoophycos* from the Torlesse Supergroup, North Canterbury, New Zealand. *New Zealand Journal of Geology and Geophysics 19*: 289–291.

FORDYCE, R. E. 1980: Trace fossils from Ohika Formation (Pororari Group, Lower Cretaceous), lower Buller Gorge, Buller, New Zealand. *New Zealand Journal of Geology and Geophysics 23*: 121–124.

FREY, R.W. (Ed.) 1975: *The Study of Trace Fossils*. Springer-Verlag, New York. 562 p.

GHENT, E. D.; HENDERSON, R. A. 1966: Petrology, sedimentation, and paleontology of Middle Miocene graded sandstones and mudstones, Kaiti Beach, Gisborne. *Transactions of the Royal Society of New Zealand, Geology 4*: 147–169.

GILLIES, R. 1877: Notes on some changes in the fauna of Otago. *Transactions and Proceedings of the New Zealand Institute 10*: 306–324.

GLENNIE, K. W. 1959: The graded sediments of the Mahoenui Formation (King Country, North Island). *New Zealand Journal of Geology and Geophysics 2*: 613–621.

GRAHAM, G. 1919: Rangi-hua-moa. A legend of the Moa in Waitemata District, Auckland. *Journal of the Polynesian Society 28*: 107–110.

GRANT-MACKIE, J. A.; CHAPMAN-SMITH, M. 1971: Paleontological notes on the Castlecliffian Te Piki Bed, with descriptions of new molluscan taxa. *New Zealand Journal of Geology and Geophysics 14*: 655–704.

GREGORY, M. R. 1969: Trace fossils from the turbidite facies of the Waitemata Group, Whangaparaoa Peninsula, Auckland. *Transactions of the Royal Society of New Zealand, Earth Sciences 7*: 1–20.

GREGORY, M. R. 1985a: The 'bow tie' trace fossil from East Cape, North Island, New Zealand. *New Zealand Geological Survey Record 9 (Hornibrook Symposium)*: 56–58.

GREGORY, M. R. 1985b: Taniwha footprints or fossilised starfish impressions? A re-interpretation: the fodinichnial trace fossil *Asterosoma. Geological Society of New Zealand Newsletter 70*: 61–64.

GREGORY, M. R. 1991: New trace fossils from the Miocene of Northland, New Zealand: *Rorschachichnus amoeba* and *Piscichnus waitemata. Ichnos 1*: 195–205.

GREGORY, M. R. 1994: *Aureffodioichnus exsculptum* ichnogen. et ichnosp. nov. A very large sub-Recent trace fossil.*Geological Society of New Zealand Newsletter 103*: 80–85.

GREGORY, M. R. 1999: Unusual 'paddling' trails left by gulls and 'aviturbation' in Rarawa and Mangawhai Estuaries, Northland, New Zealand. *New Zealand Natural Sciences 24*: 27–34.

GREGORY, M. R. 2001: *Protovirgularia* – local record of a seldom-recognised ichnotaxon. *Geological Society of New Zealand Miscellaneous Publication 110A*: 49. Programme and Abstracts, Annual Conference, 'Advances in Geosciences', 2001, Hamilton, New Zealand.

GREGORY, M. R.; BALLANCE, P. F.; GIBSON, G. W.; AYLING, A. M. 1979: On how some rays (Elasmobranchia) excavate feeding depressions by jetting water. *Journal of Sedimentary Petrology 49*: 1125–1130.

GREGORY, M. R.; CAMPBELL, J. H. 1988: Permian sea floor, near Bluff. *Geological Society of New Zealand Miscellaneous Publication 41A*: 73. [Programme and Abstracts, Annual Conference, 28 Nov–1 Dec 1988, Hamilton, New Zealand.]

GREGORY, M. R.; CAMPBELL, K. A. 2000: Towards a phytoturbation index. *Geological Society of New Zealand Miscellaneous Publication 108A*: 57. [Geological Society of New Zealand and New Zealand Geophysical Society, Joint Annual Conference, Wellington, 2000.]

GREGORY, M. R.; CAMPBELL, K. A. 2003: A '*Phoebichnus* look-alike': a fossilised root system from Quaternary coastal dune sediments, New Zealand. *Palaeogeography, Palaeoclimatology, Palaeoecology 192*: 247–258.

GREGORY, M. R.; LOCKLEY, M. G. 2004: Voy's missing moa footprints. *Geological Society of New Zealand, Historical Studies Group Newsletter, no 28*: 16–23.

GREGORY, M. R.; MARTIN, A. J.; CAMPBELL, K. A. 2004: Compound trace fossils formed by plant and animal interactions: Quaternary of northern New Zealand and Sapelo Island, Georgia (USA). *Fossils and Strata 51*: 88–105.

GRENFELL, H. R.; HAYWARD, B. W. 1995: Fossilised casts of shrimp burrows at Pollen Island, Waitemata Harbour, Auckland. *Tane (Journal of the Auckland University Field Club) 35*: 149–159.

HAAST, J. 1867: Notes on the geology of the Province of Canterbury, New Zealand, principally in reference to the deposits of the Glacial Epochs at the western base of the Southern Alps. *Quarterly Journal of the Geological Society, London, 23*: 342–352.

HANNA, M. J.; CAMPBELL, M. J. 1996: *Torlessia mackayi* and other foraminifers from the Torlesse Terrane, New Zealand. *New Zealand Journal of Geology and Geophysics 39*: 75–82.

HÄNTZSCHEL, W. 1962: Trace fossils and problematica. Pp. W177–W245 *in*: Moore, R. C. (ed.), *Treatise on Invertebrate Paleontology, Part W, Miscellanea*. Geological Sociey of America and University of Kansas, Boulder & Lawrence.

HÄNTZSCHEL, W. 1975: Trace fossils and problematica (2nd edition). *In* Teichert, C. (ed), *Treatise on Invertebrate Paleontology, Part W, Miscellanea, Supplement 1*. Geological Sociey of America and University of Kansas, Boulder & Lawrence. 269 p.

HASIOTIS, S. T. 2002: Continental trace fossils. SEPM Short Course Notes No. 51: SEPM (Society for Sedimentary Research), Tulsa, Oklahoma. 132 p.

HASIOTIS, S. T.; BOWN, T. M. 1992: Invertebrate trace fossils: the backbone of continental ichnology: *In*: Maples, C. G.; West, R. R. (eds). *Trace fossils. Paleontological Society Short Courses in Paleontology 5*: 64–104.

HAYWARD, B. W. 1976: Lower Miocene bathyal and submarine canyon ichnocoenoses from Northland, New Zealand. *Lethaia 9*: 149–162.

HEALY, T. R. 1968: Bioerosion on shore platforms developed in the Waitemata Formation, Auckland. *Earth Science Journal 2*: 26–37.

HICKEY, K. A. 1986: Geology of the Paleozoic and Tertiary rocks between Upper Takaka and the Waingaro River, North-West Nelson. Unpublished MSc thesis, University of Auckland. 73 p.

HILL, H. 1914: The moa – legendary, historical, and geological: why and when the moa disappeared. *Transactions of the New Zealand Institute 46*: 330–351.

HITCHCOCK, E. 1839: *Ornithichnology* – description of the foot marks of birds, (*Ornithichnites*) on new Red Sandstone in Massachussetes. *American Journal of Science 29*: 307–340.

HOCHSTTER, F. von. 1864: *Reise der Österreichischen Fregatte Novara um die erde in den jahren 1857, 1858, 1859 unter den behlen des Commodore B.von Wüllerstorf-Urbair. Geologischer Theil, Erster Band: Beiträge zur Geologie der Provinzen Auckland und Nelson*. Aus der Kaiserlich-Königlichen Hof- und Staatsdruckerei, in Commission bei Karl Gerold's Sohn, Wien. 273 p.

HOLLAND, N. D.; CLAGUE, D. A.; GORDON, D. P.; GEBRUK, A.; PAWSON, D. L.; VECCHIONE, M. 2005: 'Lophenteropneust' hypothesis refuted by collection and photos of new deep-sea hemichordates. *Nature 434*: 374–376.

HORN, P. L. 1989: Moa tracks: an unrecognised legacy from an extinct bird? *New Zealand Journal of Ecology 12 (Supplement)*: 45–50.

HUDSON, N. 1999: The Middle Jurassic of New Zealand: A study of the lithostratigraphy and biostratigraphy of the Ururoan, Temaikan and Lower Heterian Stages (?Pliensbachian to ?Kimmeridgian). Unpublished PhD thesis, University of Auckland. 316 p.

HUNT, A. P. 1992: Late Pennsylvanian coprolites from the Kinney Brick Quarry, central New Mexico with notes on the classification and utility of coprolites. *New Mexico Bureau of Mines and Mineral Resources Bulletin 138*: 221–229.

HUTTON, F. W. 1899: On the footprints of a kiwi-like bird from Manaroa. *Transactions of the New Zealand Institute 31*: 486.

ICZN 1985: *International Code of Zoological*

Nomenclature, Third Edition. International Trust for Zoological Nomenclature, London. xx + 338 p.

ICZN 1999: *International Code of Zoological Nomenclature, Fourth Edition*. International Trust for Zoological Nomenclature, London. 306 p.

JAMES, J. F. 1894: Studies in problematic organisms. No 2, the genus *Fucoides*. *Cincinnati Society of Natural History Journal 16*: 62–81.

JIN, H.; LIN, H. 1991: Biogenic sedimentary structure and its environmental significance in Pakawau and Kapuni Groups, Taranaki Basin, New Zealand. *Acta Sedimentologica Sinica 9*: 40–49.

JONES, B. G. 1969: Sedimentology of the Waitemata Group in the Stanley Point–Devonport area, Auckland, New Zealand. *New Zealand Journal of Geology and Geophysics 12*: 215–247.

KING, P. R.; SCOTT, G. H.; ROBINSON, P. H. 1993: Description, correlation and depositional history of Miocene sediments outcropping along North Taranaki coast. *Institute of Geological and Nuclear Sciences Monograph 5*: 1–199.

KSIAZKIEWICZ, M. 1977: Trace fossils in the flysch of the Polish Carpathians. *Palaeontologia Polonica 36*: 1–208, 29 pl.

LEWIS, D. W. 1970: The New Zealand *Zoophycos*. *New Zealand Journal of Geology and Geophysics 13*: 295–315.

LEWIS, D. W. 1992: Anatomy of an unconformity on mid-Oligocene Amuri Limestone. *New Zealand Journal of Geology and Geophysics 35*: 253–256.

LEWIS, D. W.; EKDALE, A. A. 1992: Composite ichnofabric of a mid-Tertiary unconformity on a pelagic limestone. *Palaios 7*: 222–235.

LINDHOLM, R. 1987: *A Practical Approach to Sedimentology*. Allen & Unwin, London. 276 p.

LINDQVIST, J. 1986: Teredinid-bored Araucariaceae logs preserved in shoreface sediments, Wangaloa Formation (Paleocene), Otago, New Zealand. *New Zealand Journal of Geology and Geophysics 29*: 253–256.

LINDQVIST, J. K. 1987: Trace fossils associated with the freshwater bivalve *Hydridella* (*sic*) sp.: Gore Lignite Measures (Miocene) Southland. *Geological Society of New Zealand Miscellaneous Publication 37A*: –. [unpaginated abstract.]

LINDQVIST, J. 1998: Nonmarine, estuarine and shallow marine ichno-sedimentary facies associations, Taratu and Wangaloa Formations, Late Cretaceous – Paleocene Otago. New Zealand *Geological Society, Miscellaneous Publication 101A*: 148. [Geological Society of New Zealand and New Zealand Geophysical Society, Joint Annual Meeting, Programme and Abstracts. Christchurch, 30 Dec–3 Nov 1998.]

MAGWOOD, J. P. A. 1992: Ichnotaxonomy: a burrow by another name ...? *In*: Maples, C. G.; West, R. R. (eds), *Trace Fossils. Paleontological Society Short Courses in Paleontology 5*: 15–33.

MANLEY, R.; LEWIS, D. W. 1998: Ichnocoenoses of the Mount Messenger Formation, a Miocene submarine fan system, Taranaki, New Zealand. *New Zealand Journal of Geology and Geophysics 41*: 15–33.

MARWICK, J. 1946: The geology of the Te Kuiti Subdivision. *New Zealand Geological Survey Bulletin 41*: 1–89.

MAZENGARB, C.; FRANCIS, D. 1985: The occurrence of Paramoudra concretions in the Gisborne District. *Geological Society of New Zealand Newsletter 68*: 52–53.

McILROY, D. 2004: The application of ichnology to palaeoenvironmental and stratigraphic analysis. Geological Society, London. [Special Publication 228.] 490 p.

McKAY, A. 1888: On the Tauherenikau and Waiohine Valleys, Tararua Range. *Reports of Geological exploration during 1887–88*: 58–67.

McMILLAN, S. G.; WILSON, G. J. 1997: Allostratigraphy of coastal south and east Otago: a stratigraphic framework for interpretation of the Great South Basin, New Zealand. *New Zealand Journal of Geology and Geophysics 40*: 91–107.

MILLER, M. 1995: '*Terebellina*' (= *Schaubcylindrichnus freyi* ichnosp. nov.) in Pleistocene outer-shelf mudrocks of northern California. *Ichnos 4*: 141–149.

MOORE, P. R. 1987: *Terebellina* – sponge or foraminiferid? A comparison with *Makiyama* and *Bathysiphon*. *New Zealand Geological Survey Record 20*: 43–50.

MORTON, J.; MILLER, M. 1973: *The New Zealand Sea Shore*. Collins, Auckland. 653 p.

MOSSMAN, D. J.; FORCE, L. M. 1969: Permian fossils from the Greenhills Group, Bluff, Southland, New Zealand. *New Zealand Journal of Geology and Geophysics 12*: 659–672.

NATHORST, A. G. 1873: Om nagra formodade vaxfossilier: *Oversigt af Kongliga Vetenskaps-Akademiens Förhandlingar 9*: 25–32.

NEEF, G. 1978: *Ophiomorpha* ichnofossils from the Late Miocene sandstone near Little Wanganui settlement, Buller, South Island, New Zealand. Note. *New Zealand Journal of Geology and Geophysics 21*: 419–421.

NICHOL, R. K. 1981: Preliminary report on excavation at the Sunde site, N38/24. *New Zealand Archaeological Association Newsletter 24*: 237–256.

NICHOL, R. 1982: Fossilised human footprints in Rangitoto ash on Motutapu Island. *Geological Society of New Zealand Newsletter 55*: 11–13.

NODDER, S.D.; NELSON, C.S.; KAMP, P.J.J. 1990: Mass-emplaced siliciclastic-volcaniclastic-carbonate sediments in Middle Miocene shelf-to-slope environments at Waikawau, northern Taranaki, and some implications for Taranaki Basin development. *New Zealand Journal of Geology of and Geophysics 33*: 599–615.

OSGOOD, R. G. 1975: The history of invertebrate ichnology. Pp. 3–12 *in*: Frey, R. W. (ed), *The Study of Trace Fossils*. Springer-Verlag, New York.

OWEN, R. 1879: Memoir on the Ornithichnites, or footprints of species of *Dinornis*. Pp. 451–453 (Vol. 1), pl. 116 (Vol. 2): *In*: *Memoirs on the Extinct Wingless birds of New Zealand; with an appendix on those of England, Australia, Newfoundland , Mauritius, and Rodrigues*. John Van Voorst, London.

PEMBERTON, S. G.; MacEACHERN, J. A.; FREY, R. W. 1994: Trace fossil facies models: environmental and allostratigraphic significance. Pp. 47–71 *in* Walker, R. G.; James, N. P. (eds), *Facies Models: Response to Sea Level Change*. Geological Association of Canada, Stittsville, Ontario.

PEMBERTON, S. G.; FREY, R. W.; RANGER, M. J.; MacEACHERN, J. A. 1992a: The conceptual framework of ichnology. *In* Pemberton, S. G. (ed.), *Applications of Ichnology to Petroleum Exploration. SEPM Core Workshop 17*: 1–32.

PEMBERTON, S. G.; MacEACHERN, J.A.; RANGER, M. J. 1992b: Ichnology and event stratigraphy: the use of trace fossils in recognising tempestites. *In*: Pemberton S.G. (ed.), *Applications of Ichnology to Petroleum Exploration. SEPM Core Workshop 17*: 85–117.

PEMBERTON, S. G.; MacEACHERN, J. A. 1995: The sequence stratigraphic significance of trace fossils: examples from the Cretaceous foreland basin of Alberta, Canada. *In*: Van Wagoner, J.C.; Bertram, G.T. (eds), *Sequence Stratigraphy of Foreland Basin Deposits. American Association of Petroleum Geology Memoir 64*: 429–475.

PEMBERTON, S. G.; MacEACHERN, J. A. 1997: The ichnological signature of storm deposits. Pp. 73–109 *in* Brett, C. E.; Baird, G. C. (eds), *Paleontological Events. Stratigraphic, Ecological and Evolutionary Implications*. Columbia University Press, New York.

PICKERILL, R. K. 1994: Nomenclature and taxonomy of invertebrate trace fossils. Pp. 3–42 *in* Donovan, S.K. (ed.), *The Palaeobiology of Trace Fossils*. The Johns Hopkins University Press, Baltimore.

PLICKA, M. 1965: Origin of the fossil '*Zoophycos*'. *Nature 208*: 579.

RATCLIFFE, B. C.; FAGERSTROM, J. A. 1980: Invertebrate lebensspuren of Holocene floodplains: their morphology, origin, and paleoecological significance. *Journal of Paleontology 54*: 614–630.

RETALLACK, G. J. 1980: Middle Triassic megafossil plants and trace fossils from Tank Gully, Canterbury, New Zealand. *Journal of the Royal Society of New Zealand 10*: 31–63.

RETALLACK, G. J. 1984: Trace fossils of burrowing beetles and bees in an Oligocene paleosol, badlands National Park, South Dakota. *Journal of Paleontology 58*: 571–592.

RICHTER, R. 1927: Die fossilien Fahrten und Bauten der Wurmer, ein Uberblick über ihre biologischen Grundformen und deren geologische Bedeutung. *Palaontologische Zeitschrchrift 9*: 193–240.

RICHTER, R. 1941: Marken und Spuren im Hunsruckschiefer. 3. Faharten als Zeugnisse des Lebens auf dem Meeresgrunde. *Senckenbergiana 23*: 218–260.

RINDSBERG, A. K. 1990: Ichnological consequences of the 1985 International Code of Zoological Nomenclature. *Ichnos 1*: 59–63.

SAPORTA, G. de 1884: *Les Organismes Problématiques des Anciennes Mers*. Masson, Paris. 100 p.

SARJEANT, W. A. S.; KENNEDY, W. J. 1973: Proposal of a code for the nomenclature of trace-fossils. *Canadian Journal of Earth Sciences 10*: 460–475.

SAVRDA, C. E. 1992: Trace fossils and benthic oxygenation. *In*: Maples, C. G.; West, R. R. (eds), *Trace fossils. Paleontological Society Short Courses in Paleontology 5*: 172–196.

SAVRDA, C. E.; BOTTJER, D. J.; SEILACHER, A. 1991: Redox-related benthic events. Pp. 524–541 *in*: Einsele, G.; Ricken, W.; Seilacher, A. (eds), *Cycles and Events in Stratigraphy*. Springer-Verlag, Berlin.

SCHAFER, W. 1972: *Ecology and Palaeoecology of Marine Environments*. University of Chicago Press, Chicago. 568 p.

SEILACHER, A. 1964: Biogenic sedimentary structures. Pp. 269–316 *in*: Imbrie, J.; Newell, N.D. (eds), *Approaches to Paleoecology*. Wiley, New York. SEILACHER, A. 1967: Bathymetry of trace fossils. *Marine Geology 5*: 413–428.

SEILACHER, A. 1992: Quo Vadis Ichnology. *In*: Maples, C.G.; West, R.R. (eds), *Trace Fossils. Paleontological Society Short Courses in Paleontology 5*: 224–238.

SILVA, P. C. 1993: Proposal to amend the code. (262–263) Two proposals to incorporate the term

'iconotype' in the Code. *Taxon 42*: 165–166.
SMITH, A. M. 1992: Bioerosion of bivalve shells in Hauraki Gulf, North Island, New Zealand. Pp. 175–181 *in* Battershill, C. N.; Schiel, R. R.; Jones, G. P.; Creese, R. G.; MacDiarmid, A. B. (eds), *Proceedings of the Second International Temperate Reef Symposium*. NIWA, Wellington. 251 p.
SPEDEN, I. G. 1971: Notes on New Zealand fossil Mollusca – 2. Predation on New Zealand Cretaceous species of *Inoceramus* (Bivalvia). *New Zealand Journal of Geology and Geophysics 14*: 56–70.
SPÖRLI, K. B.; GRANT-MACKIE, J. A. 1976: Upper Jurassic fossils from the Waipapa Group of Tawharanui Peninsula, North Auckland, New Zealand. *New Zealand Journal of Geology and Geophysics 29*: 21–34.
STEVENS, G. 1968: The Amuri fucoid. *New Zealand Journal of Geology and Geophysics 11*: 253–261.
STEVENS, G. R. 1972: Paleontology of the Torlesse Supergroup. *New Zealand Geological Survey Report 54*: 1–18.
SUTHERLAND, J. I. 1985: Miocene wood from Kaipara Harbour, New Zealand. Unpublished MSc thesis, University of Auckland. 148 p.
SUTHERLAND, J. I. 2003: Miocene petrified wood and associated borings and termite faecal pellets from Hukatere Peninsula, Kaipara Harbour, North Auckland, New Zealand. *Journal of the Royal Society of New Zealand 33*: 395–414.
TAYLOR, A. M.; GOLDRING, R. 1993: Description and analysis of bioturbation and ichnofabric. *Journal of the Geological Society, London 150*: 141–148.
TAYLOR, A. M.; GOLDRING, R.; GOWLAND, S. 2003: Analysis and application of ichnofabrics. *Earth-Science Reviews 60*: 227–259.
TAYLOR, P. D.; WILSON, M. A.; BROMLEY, R. G. 1999: A new ichnogenus for etchings made by cheilostome bryozoans with calcareous substrates. *Palaeontology 42*: 595–604.
THOMSON, J. A. 1920: The Notocene Geology of the Middle Wairarapa and Weka Pass district, Canterbury, New Zealand. *Transactions of the New Zealand Institute 52*: 322–415.
TONKIN, N. 2002: Application of ichnofabric analysis and event stratigraphy to paleoenvironmental reconstruction in Late Miocene strata, East Cape – Te Araroa, New Zealand. Unpublished MSc thesis, University of Auckland. 106 p.
TONKIN, N.; GREGORY, M. R.; CAMPBELL, K. A. 2004: Icnofabrics, event beds and paleoenvironmental interpretations: Upper Miocene–Lower Pliocene, Te Araroa to East Cape, North Island, New Zealand: P. 79 *in*: Buatois, L. A.; Mángano, M. G. (eds), *Ichnia 2004, First International Conference on Ichnology, Trelew Argentina*. Abstract Book.
TWIGG, J. S. 2001: Differentiating moa gizzard stones. Unpublished MSc thesis, University of Auckland. 118 p.
UCHMAN, A. 1995: Taxonomy and palaeoecology of flysch fossils: the Marnoso-arenacea Formation and associated facies (Miocene, Northern Apennines, Italy. *Beringeria 15*: 1–115.
VAN DER LINGEN, G. J. 1969: The turbidite problem. *New Zealand Journal of Geology and Geophysics 12*: 7–50.
VAN DER LINGEN, G. J.; ANDREWS, P. B. 1969: Hoof-print structures in beach sand. *Journal of Sedimentary Petrology 39*: 350–357.
VOY, C. D. 1880: On the occurrence of footprints of *Dinornis* at Poverty Bay, New Zealand. *American Naturalist 14*: 682–684.
WARD, D. M.; LEWIS, D. W. 1975: Paleoenvironmental implications of storm-scoured ichnofossiliferous Mid-Tertiary limestones, Waihao District, South Canterbury, New Zealand. *New Zealand Journal of Geology and Geophysics 18*: 881–908.
WATERHOUSE, J. B. 1964: Permian stratigraphy and faunas of New Zealand. *New Zealand Geological Survey Bulletin 72*: 1–101.
WEBBY, B. D. 1958: A Lower Mesozoic annelid from Rock Point, South-western Wellington, New Zealand. *New Zealand Journal of Geology and Geophysics 1*: 509–513.
WEBBY, B. D. 1967: Tube fossils from the Triassic of southwest Wellington. *Transactions of the Royal Society of New Zealand, Geology 5*: 181–191.
WEBBY, B. D. 1969: Trace fossils *Zoophycos* and *Chondrites* from the Tertiary of New Zealand. *New Zealand Journal of Geology and Geophysics 12*: 208–214.
WETZEL, A. 1992: The New Zealand *Zoophycos* revisited: morphology, ethology, and paleoecology – some notes for clarification. *Ichnos 2*: 91–92.
WILKIE, M. 1998: The grazing impact of *Chiton pelliserpentis*, with emphasis on bioerosion. Unpublished MSc thesis, University of Auckland. 123 p.
WILLIAMS, W. L. 1872: On the occurrence of a large bird, found at Turanganui, Poverty Bay. *Transactions of the New Zealand Institute 4*: 124–127.
WILSON, K.; BENHAM, W.B. 1913: Footprints of the moa. *Transactions of the New Zealand Institute 4*: 124–127.
WORTHY, T. H.; HOLDAWAY, R. N. 1994: Quaternary fossil faunas from caves in Takaka Valley and on Takaka Hill, northwest Nelson, South Island, New Zealand. *Journal of the Royal Society of New Zealand 24*: 297–392.
ZURAIDA, R. 2002: Neoichnology and sedimentology of an extensive tidal flat, Whangateau Harbour, New Zealand. Unpublished MSc thesis, University of Auckland. 87 p.

Checklist of New Zealand ichnotaxa

These have been subdivided into groups as follows and in each they are listed in alphabetical order:

A. Invertebrate trails and burrows – these are further separated into sections dealing with:

1. ichnotaxa formally erected and described from New Zealand material with appropriate reference citation;
2. well-identified ichnotaxa, generally accompanied by illustrations and/or adequate descriptions;
3. records, often in discussion or passing conversation and unpublished, and/or of lesser veracity.

In most instances for 2 and 3 the cited authority is the first New Zealand reference to that ichnotaxon if it is accompanied by any decriptive notes or illustration. Where a later reference is used it reflects a fuller or more detailed description. No authorities are given with 3 and this list must be considered indicative only. Because ichnofossils have long time ranges and are of little use in stratigraphic correlation, age ranges have been omitted.

B. Coprolites and regurgitations.

C. Invertebrate borings.

D. Vertebrate tracks.

Invertebrate trails and burrows

Described from New Zealand material

Arborichnus sparsus Ekdale & Lewis, 1991
Auroffodioichnus exsculptum Gregory, 1994
'Bow tie' structure (Gregory 1985a) (?= *Imbrichnus*)
Cycloichnus waitemataensis Gregory, 1969
Diplocraterion asymmetrium Ekdale & Lewis, 1991
Diplocraterion morgani Fleming, 1973
Diplocraterion parallelum var. *arcum* Ekdale & Lewis, 1991
Diplocraterion parallelum var. *lingum* Ekdale & Lewis, 1991
Diplocraterion parallelum var. *quadrum* Ekdale & Lewis, 1991
Laminites kaitiensis Ghent & Henderson, 1966 (= *Scolicia*)
Macanopsis erowhonensis Retallack, 1980
Piscichnus waitemata Gregory, 1991
Radionereites ballancei Gregory, 1969
Rorschachichnus amoeba Gregory, 1991

Reported

Anconichnus horizontalis (Ekdale & Lewis 1991)
Anconichnus ichspp. *a, b, c* (Manly & Lewis 1998)
Anemonichnus (Carter & Lindqvist 1977)
Arenicolites (Ekdale & Lewis 1991)
Asterichnites (Adams 1985)
Asterosoma (Gregory 1985b)
Bathichnus paramoudrae (as 'Paramoudra', *in* Mazengarb & Francis 1985)
Cardioichnus (Gregory 1985a)
Chondrites (Gregory 1969) (large and small varieties; Webby 1969)
Cochlichnus (Fordyce 1980)
Corophioides (Douglas & Lindqvist 1987 noted its absence)
Cosmorhaphe (Gregory pers. obs.)
Cylindrichnus concentricus (Ekdale & Lewis 1991)
Cylindrites (Jones, 1969) (invalid, Häntzschel 1975)
Echinospira pauciradiata (Uchman 1995) [= *Zoophycos* (Lewis 1970)]
Glockeria (Nodder et al. 1990)
Gordia (Ekdale & Lewis 1991)
Granularia (Nodder et al. 1990)
Gyrolithes (Gregory 1985b) (= *Xenohelix*)
Gyrophyllites (Gregory 1969)
Helminthoida (Gregory 1969)
Helminthopsis (Gregory pers. obs.)
Himanthalites (Jones, 1969) (= large *Chondrites*, Häntzschel 1975)
Hormosiroidea (Gregory pers. obs.)
Hydrancylus (Manley & Lewis 1998)
Imbrichnus (Ward & Lewis 1975)
Lophoctenium (Gregory & Campbell 1988)
Macaronichnus (Lindqvist 1998)
Margaritichnus (Feldmann & Maxwell 1990)
Neonereites (Gregory pers. obs.)
Nereites (Gregory 1969)
Ophiomorpha (Ward & Lewis 1975; Neef 1978)
Ophiomorpha nodosa (Manley & Lewis 1998)
Paleodictyon cf. *menhegenii* (Gregory 1969)
Paleodictyon cf. *regulare* (Gregory 1969)
Palaeophycus (Manley & Lewis 1998)
Phycodes (Gregory 1969)
Phycodes circinnatum (Bradley 1981)
Phycosiphon (?= *Anconichnus*) (Gregory pers. obs.)
Piscichnus ichsp. (Ekdale & Lewis 1991)
Planolites ichspp. *a, b* (Gregory 1969)
Planolites montanus (Lindqvist 1997)
Protovirgularia (Gregory 2001)
Psilonichnus (Linqvist pers. comm.) (Zuraida 2002, lebensspuren)
Ramidictyon (Blom 1982)
Rhizocorallium (Hayward 1976)
Rosselia (Gregory pers. obs.)
Rusophycus (Bradshaw 1999)
Scalarituba (Mossman & Force 1969)
Schaubcylindrichnus (Gregory pers. obs.) (in part *'Terebellina'*)
Scolicia (Gregory 1969)
Scolicia ichspp. *a, b* (Manley & Lewis 1998)
Skolithos (Bradshaw 1999)
Skolithos linearis (Ekdale & Lewis 1991)
Spirophyton (Bradshaw pers. comm.)
Spirorhaphe (Gregory pers. obs)
Spirophycus (Manley & Lewis 1998)
Spongeliomorphia (Grenfell & Hayward 1995)
Subphyllochorda (Manley & Lewis 1998)
Taenidium (marine, Jones 1969)
Taenidium (terrestrial, Gregory et al. 2004)
Teichichnus (Ballance *et al.* 1984)
Thalassinoides (Hayward 1976)
Thalassinoides suevicus (Flores & Sykes 1996)
Tigillites (Gregory 1969)
Tigillites ichspp. *a, b, c, d* (Hayward 1976)
Urohelminthoida (Stevens 1972)
Zoophycos (Lewis 1970)
Zoophycos plicatus (= *Pinna plicata* Hutton) (see Fleming, 1971)

Status uncertain

Ancorichnus (?)
Arthrophycus
Bergaueria
Bifasciculus
Dictydora
Gyrochorte
Gyrophyllites
Halimedides
Helicodromites
Kulindrichnus
Lorenzina
'Mycellia'
Pelycepodichnus
Polykampton
'Spongia'
Stellascolites
Taphrhelminthopis
Trichichnus

Coprolites and regurgitations

Hunt (1992) erected 'Bromalite' as a major category that includes 'Regurgitalite' (from oral cavity), 'Cololite' (intestinal contents), and 'Coprolites' (fossilised faeces).

Vertebrate

Reptilian and/or fish coprolites/ regurgitations (Speden 1971)
Avian regurgitation pellets (Worthy & Holdaway 1994)
Avian gastroliths (Twigg 2001)

Invertebrate

Termite faecal pellets (Sutherland 1988)
Faecal pellets (indet. marine biota)

Borings (and etchings)

Ichnotaxa, with examples of modern progenitors

Centrichnus (bivalve mollusca and *Verruca*)
Clionolithes priscus (Bradshaw, 1999)
Entobia (often as clionaid or sponge borings)
Entobia cretacea (Campbell *et al.* 2004)
Gastrochaenolites (bivalve mollusca, pholads, etc.)
Gastrochaenolites ornatus (Campbell et al. 2004)
Gnathichnus (regular echinoid)
Leptichnus peristroma (Taylor et al. 1999)
Leptichnus dromeus (Taylor et al. 1999)
Oichnus (cephalopod, naticid, and nematode predation)
Podichnus (brachiopod)
Polydora (polychaeta)
Radulichnus (chiton)
Rogerella (cirripedia)
Teredolites (teredinid borings; Lindqvist 1986)
Teredolites clavatus (Bradshaw, 1980) (= *Gastrochaenolites*)=
Tremnichnus (gastropoda and foraminiferan pits)
Trypanites (sipunculan/polychaete borings; Lewis 1992)
Zapfella (Campbell et al. 2004: Cirripedia)

Vertebrate tracks

'Rhynchocephalichnus' (tuatara tracks; yet to be recorded)
Avian ichnites (bird footprints, moa Gillies 1872, Williams 1872, Hill 1914) (= *Ornithichnites*, Owen 1879), (kiwi and other birds) (trails and pathways, *ara-moa*, Graham 1919; Horn 1989)
Undichna (?) (sinuous fish trails)
Human and dog footprints (sub-Recent) (Nichol 1982)

CREDITS AND ACKNOWLEDGEMENTS

Captions and credits for thumbnail images on page 16

Top to bottom, left to right:

Chaetognatha: *Pterosagitta draco* (Cheryl Clarke, Institute of Marine Science, University of Alaska, Fairbanks, USA)

Tardigrada: *Echiniscus elaeinae* (Diane Nelson, East Tennessee State University, Johnson City, Tennessee, USA)

Onychophora: *Ooperipatellus viridimaculatus* (Hilke Ruhberg and Hubert Bosch, Hamburg University, Germany)

Arthropoda Introduction: *Lepidurus apus viridis* (Stephen Moore, Landcare Research, Auckland, New Zealand)

Arthropoda Trilobitomorpha: *Koptura* sp. (Marianna Terezow, GNS Science, Lower Hutt, New Zealand)

Arthropoda Chelicerata: *Dolomedes minor* (Stephen Moore, Landcare Research, Auckland, New Zealand)

Arthropoda Myriapoda: *Procyliosoma striolatum* (Alastair Robertson and Maria Minor, Massey University, Palmerston North, New Zealand)

Arthropoda Crustacea: *Stenopus hispidus* (Roger V. Grace, Leigh, New Zealand)

Arthropoda Hexapoda: *Hemideina maori* (Alastair Robertson and Maria Minor, Massey University, Palmerston North, New Zealand)

Kinorhyncha: *Echinoderes* sp. (Birger Neuhaus, Museum für Naturkunde, Berlin, Germany)

Loricifera: *Pliciloricus* sp. (Iben Heiner, National Food Institute, Søborg, Denmark)

Priapulida: *Priapulopsis australis* (Richard Taylor, Leigh Laboratory, University of Auckland, New Zealand)

Nematoda: *Pselionema* sp. (Daniel Leduc, National Institute of Water & Atmospheric Research, Wellington, New Zealand)

Nematomorpha: *Gordius dimorphus* (Stephen Moore, Landcare Research, Auckland, New Zealand)

Ichnofossils: *Paleodictyon* sp. (Marianna Terezow, GNS Science, Lower Hutt, New Zealand)

Captions and credits for other figures

All illustrations are acknowledged as to source, either below or in the chapter references. In many cases, authors or authors' colleagues freely supplied photographs and drawings. Illustrations requiring permission to be reproduced or adapted from published literature are credited below. Attempts were made to obtain permissions for all illustrations as appropriate; if any have been missed, please alert the publisher of this volume.

Line figures on pages 22–25, 104, 117, 120, 125–127, 146, 148 (top), 150, 154 (top), 156 (bottom), 160–163, 168, 181, 183, 184 (top), 189 reproduced from *New Zealand Oceanographic Institute Memoirs*. By permission of NIWA.

Line figures on pages 28, 30, 31, 57 (top), 61, 151, 249, 299 (bottom), 349 (top), 482, 486 reproduced from *New Zealand Journal of Zoology*. By permission of the Royal Society of New Zealand.

Line figures on pages 52, 53, 55, 63, 64 reproduced from *Spiders of New Zealand*. By permission of Doug Forster.

Line figures on pages 99, 103, 107, 124, 132, 147 (bottom), 152 (bottom), 153 (top), 155 (top), 293, 503 reproduced from *Journal of the Royal Society of New Zealand*. By permission of the Royal Society of New Zealand.

Line figures on pages 105, 110, 162, 153 (bottom), 156 (top), 164, 180 reproduced from *Zootaxa*. By permission of the editor.

Figure on page 113 reproduced from *New Zealand Geological Survey Paleontological Bulletin 50*. By permission of GNS Science.

Line figures on pages 119, 138, 149 (middle), 190 reproduced from *New Zealand Journal of Marine and Freshwater Research*. By permission of the Royal Society of New Zealand.

Line figures on page 122, 123 reproduced from *Contributions to Zoology*. By permission of the Netherlands Centre for Biodiversity Naturalis, Leiden.

Line figure on page 128 (bottom) reproduced from *Korean Journal of Systematic Zoology*. By permission of author.

Line figure on page 137 reproduced from *NIWA Science and Technology Series*. By permission of NIWA.

Line figure on pages 147 (top), 149 (bottom) reproduced from *Transactions of the Royal Society of New Zealand*. By permission of the Royal Society of New Zealand.

Line figure on page 152 (top) reproduced from *Records of the Dominion Museum*. By permission of Te Papa Tongarewa.

Line figure on page 154 (bottom) reproduced from *Journal of Natural History*. By permission of Taylor & Francis.

Line figure on page 155 (top) reproduced from *Crustaceana*. By Permission of E. J. Brill.

Line figures on pages 155 (bottom), 157 (bottom) reproduced from *United States National Museum Bulletin*. Open source via Biodiversity Heritage Library.

Line figures on page 157, 187 reproduced from *Tuhinga*. By permission of Te Papa Tongarewa.

Line figure on page 159 reproduced from *Memoirs of Museum Victoria*. By permission of Museum Victoria, Melbourne.

Line figure L on page 168 reproduced from the *Proceedings of the Biological Society of Washington*, volume 115, page 413, fig. 1. By permission of the Biological Society of Washington.

Line figures on pages 234, 485 reproduced from *Fauna of New Zealand*. By permission of Landcare Research.

Line figures on pages 237–241 (top), 242 reproduced from the *New Zealand Journal of Science and Technology*. By permission of the Royal Society of New Zealand.

Line drawings on pages 243, 250, 251 (top), 252 (top), 253–258, 260, 261 (bottom), 263, 268–279, 283, 284 (bottom), 286, 287, 289, 290, 296–298 (top), 301 (bottom), 302 (top), 311, 312, 314, 316, 318, 322 333, 336, 338, 340–342, 348, 350–352, 354–358, 364, 366 (top) reproduced from *An Illustrated Guide to Some New Zealand Insect Families*. By permission of Elizabeth A. Grant.

Line drawings on pages 244–247, 251 (bottom), 296 (middle right), 306, 360–363 reproduced from *Guide to the Aquatic Insects of New Zealand*. By permission of Dr Michael Winterbourn.

Line drawing on page 349 (bottom) reproduced from *New Zealand Entomologist*. By permission of the Entomological Society of New Zealand.

Line figure on page 484 reproduced from *Systematic Parasitology*. By permission of the publisher.

Line figures on pages 495, 498 reproduced from *Canadian Journal of Zoology*. By permission of NRC Research Press.

Dennis Gordon acknowledges support for this project from the New Zealand Foundation for Research, Science & Technology (Contracts C01421, C01X0219, C01X0026, C01X0502).

SPECIES INDEX

To keep this index to a manageable size, it has been applied to the main text and figures only, and not to the checklists or summary tables. Full binominals are given where they appear in the text, e.g. *Spelungula cavernicola*; otherwise, the generic name only is given without the qualifier 'sp.', e.g. *Phoroncidia*. Scientific names of Orders and Families are not indexed, e.g. Decapoda, but the common names for them are, e.g. decapods, shrimps, lobsters, crabs. Where such groups appear in the index, the page reference is to the first that deals substantively with that group, e.g. isopods 158 signifies that the main description of isopods begins on page 158. Common names for individual species have not normally been indexed.

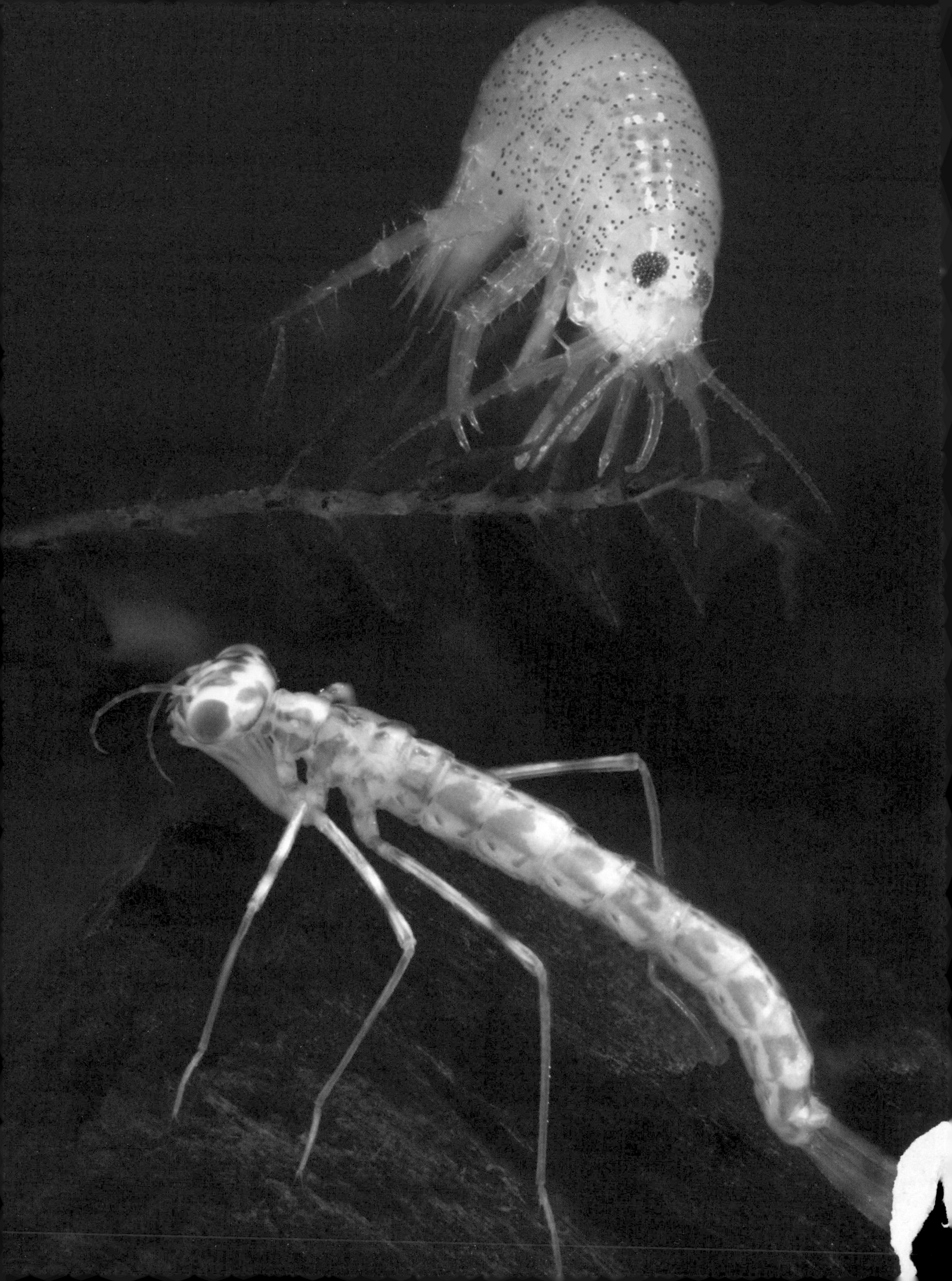